CALCULUS SET FREE

Calculus Set Free: Infinitesimals to the Rescue

Charles Bryan Dawson

University Professor of Mathematics, Union University, USA

OXFORD
UNIVERSITY PRESS

Great Clarendon Street, Oxford, OX2 6DP,
United Kingdom

Oxford University Press is a department of the University of Oxford. It furthers the University's objective of excellence in research, scholarship, and education by publishing worldwide. Oxford is a registered trade mark of Oxford University Press in the UK and in certain other countries

Impression: 1

Published in the United States of America by Oxford University Press
198 Madison Avenue, New York, NY 10016, United States of America

British Library Cataloguing in Publication Data
Data available

Library of Congress Control Number: 2021937201

ISBN 978-0-19-289559-2 (hbk.)

ISBN 978-0-19-289560-8 (pbk.)

DOI: 10.1093/oso/9780192895592.001.0001

Printed and bound by
CPI Group (UK) Ltd, Croydon, CR0 4YY

Contents

Preface for the Student

For many, the study of calculus is seen as a rite of passage—to conquer calculus is to pass through the gateway to the sciences, engineering, mathematics, business, economics, technology, and many other fields. For some, the study of calculus is indicative of achievement, a hallmark of a quality education. A few can't wait to study calculus, their curiosity overflowing with enthusiasm. Yet others see calculus as an annoyance, something to tolerate in pursuit of more important or more interesting subjects. This book is for all of you.

Whatever your reason for studying calculus, it is my hope that this text facilitates not just the mastering of technical skills and the understanding of mathematical concepts, but also training in thinking in a patient, systematic, disciplined, and logical manner. Although technical skills can be useful for some students in their careers, and the understanding of mathematical concepts can be of use to even more, the habits of mind created by careful thinking can be of use to everyone, at any time, in any place.

Preparation for success

If you have learned to drive a car, then you may recall how driving took much conscious thought at first, but later, with practice, driving became much more of a background task. The same is true of addition and multiplication facts; the task $2+4$ takes very little mental energy. This is the hallmark of deep learning: when a task has been learned thoroughly, then it can be performed accurately with little effort.

Success in calculus is much easier if basic algebraic and trigonometric skills have been learned this deeply. If the quadratic formula and laws of exponents can be applied accurately upon demand, then the mind is free to concentrate on the concepts at hand. If not, then instead of juggling three new concepts consciously, a dozen or more distracting items that must be relearned compete with the new concepts for mental energy, hampering one's learning of the new material.

Mental skill-building is much the same as physical skill-building; it takes time and consistent effort on the part of the learner. Lifting weights several times per week for a month is a much more effective strategy than waiting until the night before the skills test to try to cram the entire month's reps into one evening's workout. Body-building simply does not work that way, and neither does learning mathematics.

One final bit of advice: learn from failure. Everyone makes errors, even textbook authors with decades of experience in the subject. No one is perfect. But when you make an error, make sure you understand why it was an error, why a different approach must be used, and how to avoid making the same error in the future. There is often something to be learned from your errors. Mistakes are not to be feared, but to be used to your advantage!

Features of this textbook

What makes this textbook different? The most obvious answer is that it uses infinitesimals, which are infinitely small numbers that you might not have encountered in your previous courses.

Although infinitesimals were an essential part of the development of calculus, they have been absent from nearly all calculus textbooks for more than a century. The largest factor in the switch away from infinitesimals was the fact that, at the time, no one had been able to develop rigorously the required number system. This state of affairs changed during the 1960s, and now the use of infinitesimals is once again seen as mathematically legitimate. Using notation and procedures that I have developed and published, the study of certain portions of calculus in this text is both more intuitive and simpler algebraically than in other calculus textbooks.

Additional features include:

- **A readable and student-friendly narrative.** The narrative is written to help you think through the development of concepts and think through solutions to examples. Following the thinking process helps you create meaning and retain ideas more easily.
- **Reading exercises.** Reading exercises are meant to be worked when encountered during reading. The solution is placed in the margin one to three pages later.
- **Hundreds of diagrams.** Consistency of color use throughout the text's diagrams helps with interpretation.
- **Margin notes.** Margin notes are used to add explanations, tips, cautions against making common errors, and historical notes.
- **Examples.** Hundreds of examples with complete solutions are included. Some solutions are written compactly, demonstrating the level of detail expected of student work. Others include more details of how to think through the solution.
- **Thousands of exercises.** Exercises range from the routine to the challenging. Many sections include "rapid-response" exercises meant to help you distinguish between objects or algebraic forms. Some exercises are very similar to examples in the narrative. Other exercises require you to think creatively or explore the ideas more deeply. Sometimes exercises from much older textbooks are included, such as those labeled "(GSL)" from the classic early-20th-century text of Granville, Smith, and Longley.
- **Answers to odd-numbered exercises.** The answers to odd-numbered exercises sometimes include hints, brief explanations of why some attempted answers are incorrect, alternate forms of answers, or both simplified and non simplified answers to help you determine the source of an error.
- **An extensive index.**

It is my prayer that this textbook is a blessing to you, that it helps you understand the concepts and develop the skills of calculus as you continue your educational journey. Enjoy!

Bryan Dawson
University Professor of Mathematics
Union University

July 2021

Preface for the Instructor

This textbook covers single-variable calculus through sequences and series, and corresponds to the first two semesters of college-level calculus at most universities in the United States. The organization is similar to that of other popular calculus texts.

What makes this textbook different is its use of infinitesimals (and hyperreal numbers in general) for all limiting processes, including the definitions of derivative and integral. The notation and procedures used with hyperreal numbers in this textbook (which differ from those used in other nonstandard analysis sources) were developed by me[1] and were introduced in articles in *The American Mathematical Monthly* (February 2018) and *The College Mathematics Journal* (November 2019), with additional articles planned. The utility of the hyperreal numbers reaches into other areas of the calculus as well, such as comparing rates of growth of functions and a related procedure for testing series.

In addition to making calculus concepts more intuitive, the use of infinitesimals corresponds more closely to the way our colleagues in other disciplines teach students to analyze their ideas. The procedures used in this textbook for limits are also connected more directly to the definitions, are algebraically simpler, and are met with a much greater degree of student success.

Additional differences from other textbooks include the splitting of some material into two sections to facilitate more easily the multiple-day coverage typical of those topics, as well as some minor reorganization of topics compared to other books.

Features of this textbook

Features of this textbook include:

- **A readable and student-friendly narrative.** The narrative is written to help students think through the development of concepts and think through solutions to examples.
- **Reading exercises.** Reading exercises are meant to be worked when encountered during reading. The solution is placed in the margin one to three pages later.
- **Hundreds of diagrams.** Consistency of color use throughout the text's diagrams helps with interpretation. For instance, graphs of functions are blue whereas tangent lines are orange.
- **Margin notes.** Margin notes are used to add explanations, tips, cautions against making common errors, and links to biographies on the MacTutor History of Mathematics website.
- **Examples.** Hundreds of examples with complete solutions are included. Some solutions are written compactly, demonstrating the level of detail expected of student work. Others include more details of how to think through the solution.
- **Thousands of exercises.** Exercises range from the simple and the routine to the challenging. Many sections include "rapid-response" exercises meant to help students distinguish between objects or

[1] See the Acknowledgments section for one exception.

algebraic forms; these exercises could be considered for in-classroom use. Some exercises are very similar to examples in the narrative. Other exercises require students to think creatively or explore the ideas more deeply. Sometimes exercises from much older textbooks are included (or adapted for inclusion), such as those labeled "(GSL)" from the classic early-20th-century text of Granville, Smith, and Longley.

Enough exercises are included to allow you to have choices of which odd-numbered exercises to include to craft an appropriate homework set. Even-numbered exercises correspond roughly to odd-numbered exercises for additional student practice. Although it is now common to find solutions to textbook exercises on the internet, I still follow the custom of only providing the answers to odd-numbered exercises in the textbook.

- **Answers to odd-numbered exercises.** The answers to odd-numbered exercises sometimes include hints, brief explanations of why some attempted answers are incorrect, alternate forms of answers, or both simplified and non simplified answers to help students determine the source of an error.
- **A review section on avoiding common errors.** Each subsection of section 0.6 focuses on one particular type of error so that you can refer students to help as needed. For instance, students who are prone to cancellation errors can be referred to section 0.6 "Cancellation." The exercises in each subsection are designed to help students recognize whether such an error has been made, in the hopes of helping them avoid such errors in the future.
- **An extensive index.**

Another, perhaps unusual, feature of this textbook is that it does not contain fictitious names in word problems and it does not contain any gender-specific words after the prefaces.

Teaching infinitesimals

Sections 1.1–1.3 contain basic ideas, notation, concepts, and procedures for manipulating hyperreal numbers. Just as we allow calculus students to use real numbers without first subjecting them to Dedekind cuts, students should be allowed to use hyperreal numbers without reference to ultrafilters. As with a student's introduction to any other type of number, these sections help a student learn what infinitesimals and other hyperreals are, how to manipulate them algebraically, and where they fit on number lines. Note that the symbols ε and δ are not used for infinitesimals in this book, because these symbols may be used for real numbers in later courses.

Sections 1.1–1.3 are fundamental for working with hyperreals and therefore should be covered thoroughly and mastered by the student. I cover one section per 50-minute class period, spending three class days total on this material.

Section 1.3 gives students the opportunity to practice the calculations involved in finding limits; then, in section 1.4, students can concentrate on the concept of limits having already learned the manipulations. This separation also allows the flexibility to skip limits and cover sections 1.8 and 2.1 on the derivative immediately after section 1.3, or cover sections 4.2–4.4 on sums and the definite integral immediately after section 1.3. However, it is assumed that at some point the students learn the material in sections 1.4–1.7.

What is sometimes known as the *direct substitution property* of limits, which is called *evaluating limits using continuity* in this textbook, is not covered until section 1.7. This allows the verification of continuity in section 1.6 to be both natural and meaningful. What other textbooks call *limit laws* are not necessary in this curriculum; the equivalent is implicit in the manipulations of sections 1.1–1.3.

More about section dependencies

Chapter 0 is entirely optional. I usually cover sections 0.4 and 0.5 prior to beginning chapter 1. Section 1.0 is also optional and may be skipped; much of this material is repeated in section 1.8.

Section 3.6, limits at infinity, can be covered immediately after section 1.5. Although sections 3.5 and 3.7 on curve sketching are not strictly necessary for later material, it is my opinion that most students learn properties of graphs best if they sketch a few by hand.

Section 4.1 can be delayed until just before section 4.5, on the fundamental theorem. Sections 4.8 and 4.9 can be delayed as long as you desire.

The order of chapters 5 and 6 can be swapped, and the order of sections in chapter 6 can be varied. Sections 5.13, 7.7, and 7.9 are easily omitted. Although the use of l'Hospital's rule in sections 5.11 and 5.12 is traditional (and I still cover it), many of the limits of these sections can be calculated in other ways, such as using techniques from section 5.10 and from series in section 10.10.

Sections 9.2 and 9.4 depend on portions of chapter 8, but other sections in chapter 9 can be covered after chapter 7 or, with a judicious selection of exercises, along with chapter 6. Sections 8.1–8.4 can be covered as early as immediately following chapter 5, if desired.

Finally, as is usual in calculus textbooks, numerous sections contain subsections upon which no later material depends. This allows you to cover favorite (sub)topics. However, coverage of every problem type demonstrated in the narrative may not always be practical.

in section 1.6 to be both [illegible] and meaningful. What other textbooks call [illegible] are not necessary [illegible] in the manipulation of equations in [illegible].

More about section dependencies

Chapter 0 is [illegible] sections 0.4 and 0.5 prior to beginning chapter 1. Section 1.6 is also optional and may be skipped (much of the material is covered in section [illegible]).

Section 2.4, [illegible] arbitrary [illegible], can be covered immediately after section 1.5, although sections 2.5 and 3.7 on [illegible] are not strictly necessary [illegible] properties of [illegible].

Section 4.1 can be delayed until just before section [illegible] the fundamental theorem. Sections [illegible] can be delayed as long as [illegible].

The order of chapters 5 and 6 can be reversed, and the order of sections in chapter 6 can be varied. Sections [illegible] are easily omitted. Although the use of [illegible] sections [illegible] and [illegible] traditional [illegible] of the [illegible], such [illegible] using techniques from section [illegible].

Sections 9.2 and 9.4 depend on portions of chapter [illegible] other sections [illegible] chapter 7 [illegible] section of exercises [illegible] with chapter [illegible] can be covered as early as immediately following chapter [illegible] if desired.

Finally, [illegible] optional [illegible] sections [illegible] upon which no later material depends. This allows [illegible] favorite topics. However [illegible] by the [illegible] may not always be practical.

Acknowledgments

I thank my wife for her patience with me while embarking on the long journey of writing this book. Without her support, this project would not exist.

The first colleague to join me in teaching calculus using these infinitesimal methods was Troy Riggs (Union University). Troy suggested the use of the symbol $\doteq$ for rendering a real result (see section 1.3) and was the first to use the symbol in the classroom. He and I spent more hours than either of us wish to admit discussing the use of infinitesimal ideas, and these discussions were essential to this project's success. Others who classroom-tested a preliminary version of this text are George Moss (Union University), Nicholas Zoller (Southern Nazarene University), and Mo Niazi (Southern Nazarene University).

Thanks also go to the hundreds of students who provided encouragement and feedback, both explicitly and implicitly. Knowing how students interact with the material has shaped many aspects of this book.

Additional thanks to colleagues include editorial boards and referees of articles and book proposals; administrators who granted a research leave and other release time to write; and many others who participated in hallway discussions, attended my workshops, or provided encouragement.

The seed for investigating infinitesimal methods in calculus was planted by a student in a calculus class in 2004, Robert Michael, who asked many questions about infinitesimals to which I did not have adequate answers. That seed sat dormant for several years, but when it sprouted, it grew larger than I could have imagined.

Some journeys take generations. My paternal grandfather's formal education ended after the eighth grade (as it did for all but one of my grandparents). He was a sharecropper, leasing the same land year by year for nearly four decades. Someone once asked him why he never purchased land of his own. His reply was that if he purchased land, he could leave a legacy for one of his children; if he instead spent that money for college educations, he could leave a legacy for all five of his children. My father majored in mathematics and then enriched my mathematical education enough as a child to instill a curiosity that led to an academic career. In addition to passing that legacy on to my own children, it is my privilege as a professor to help other families build their legacies. What a blessing!

Last, I thank my Creator, not only for giving me life, but also for giving me the insights key to the development of the infinitesimal methods and for not letting me quit when I wearied of the journey.

> I will bless the Lord who has given me counsel;
> My heart also instructs me in the night seasons.
>
> —*Psalm 16:7 NKJV*

I have often prayed that this textbook will be a blessing to both students and instructors. May it always be so.

Bryan Dawson
University Professor of Mathematics
Union University

July 2021

Chapter 0

Review

Algebra Review, Part I

0.1

Some of you have already learned algebra and trigonometry at a deep level and are ready to jump in to chapter 1 with confidence. Others need to spend time in this chapter, perhaps much time. The more accurately and readily one can perform algebra, the more easily one can learn calculus. If it is needed, an investment in time and effort now will pay great dividends later.

Because the material of chapter 0 is a review, the narrative is relatively sparse. Motivation and development of formulas are not always presented.

The word *algebra* derives from the name of the first book about algebra, *Hisab al-jabr w'al-muqabala*, written by Abu Ja'far Muhammad ibn Musa Al-Khwarizmi.

Abu Ja'far Muhammad ibn Musa Al-Khwarizmi, 780–850 (approximately) http://www-history.mcs.st-andrews.ac.uk/Biographies/Al-Khwarizmi.html. The Mac Tutor History of Mathematics archive, hosted by the University of St. Andrews in Scotland, is one of the most trusted sources for the history of mathematics. The website has been in operation since before the turn of the century. Links to biographies of mathematicians in this book are to the MacTutor archive.

Real number line

The first numbers a child learns are the counting numbers, $1, 2, 3, \ldots$, and so on, also known as the *natural numbers* **N**. These numbers are pictured on a horizontal line, equally spaced, with larger numbers to the right and smaller numbers to the left (figure 1).

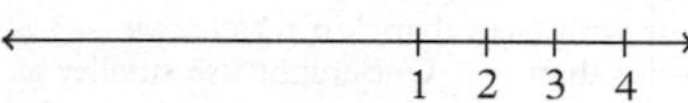

Figure 1 *Counting numbers on the number line*

Next one learns the *integers* **Z**, which include the natural numbers, their negatives, and zero. These are also placed on the number line (figure 2).

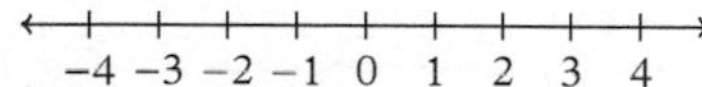

Figure 2 *Integers on the number line*

Then came numbers of the form $\frac{a}{b}$, where a and b are integers. These are the *rational numbers* **Q**. Rational numbers are also placed proportionally on the number line, as always with larger numbers to the right and smaller numbers to the left. It is common to represent numbers as points on a line, as shown in figure 3. Rational numbers have decimal expansions that either terminate or repeat, such as

The term *rational* is derived from the word *ratio*. Rational numbers are ratios of integers.

$$\frac{1}{2} = 0.5$$

or

$$\frac{4}{3} = 1.\overline{3} = 1.33333\ldots.$$

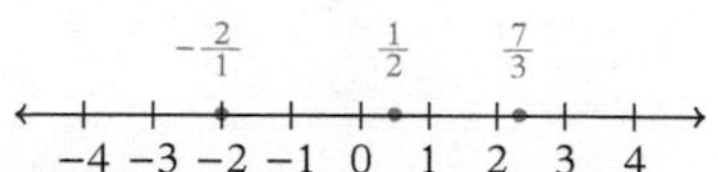

Figure 3 *Rational numbers (in orange) as points on the number line*

Irrational numbers have decimal expansions that neither terminate nor repeat, such as $\sqrt{2} = 1.4142\ldots$ and $\pi = 3.14159\ldots$. They also find their place on the number line. The collection of all of these numbers is called the *real numbers* **R** (figure 4). The real numbers fill out the number line; they can be placed in one-to-one correspondence with the points on the line. Even so, stay tuned for more numbers in chapter 1!

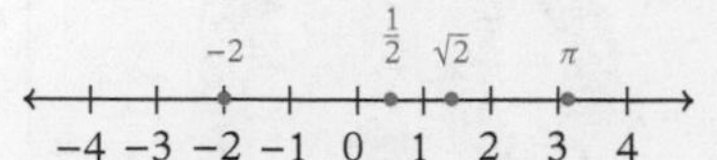

Figure 4 *Real numbers (in orange) as points on the number line*

Inequalities

The statement $a < b$ means the number a is less than the number b, and a is to the left of b on the number line. The statement $a > b$ means the number a is greater than the number b, and a is to the right of b on the number line.

Any inequality can be written in two different ways: $4 < 9$ means the same thing as $9 > 4$.

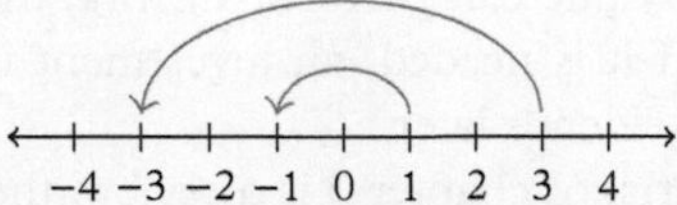

Figure 5 *A number line. Because 1 is to the left of 3, $1 < 3$. Multiply both numbers by -1 and the order reverses: $-1 > -3$*

Notice in figure 5 that although 1 is to the left of 3 on the number line, -1 is to the right of -3. Thus, $1 < 3$, but $-1 > -3$.When multiplying (or dividing) an inequality by a negative number, the direction of the inequality must be reversed.

CAUTION: The word *smaller* can be ambiguous. One might use smaller as synonymous with "less than," in which case -3 is smaller than -1. One might use smaller as synonymous with "closer to zero" (smaller in magnitude), in which case -1 is smaller than -3.

INEQUALITIES: MULTIPLYING OR DIVIDING BY NEGATIVES

When multiplying or dividing both sides of an inequality by a negative number, the direction of the inequality must be reversed.

The same is true when taking reciprocals. Although $2 < 7$, notice in figure 6 that $\frac{1}{2} > \frac{1}{7}$. The larger the denominator, the smaller the fraction.

The *reciprocal* of a number is 1 divided by that number; the reciprocal of 35 is $\frac{1}{35}$.

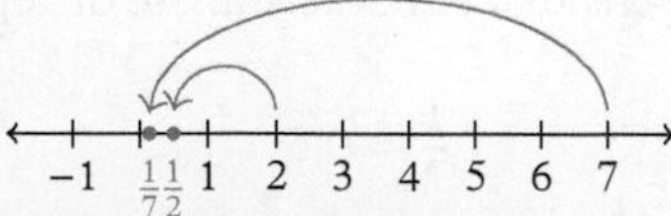

Figure 6 *A number line. Because 2 is to the left of 7, $2 < 7$. Take reciprocals of both numbers and the order reverses: $\frac{1}{2} > \frac{1}{7}$*

> **INEQUALITIES: RECIPROCALS**
>
> When taking reciprocals of both sides of an inequality, the direction of the inequality must be reversed.

Solving linear inequalities is similar to solving linear equations. Care must be taken, however, to change the direction of the inequality under the circumstances just described.

Example 1 *Solve the inequality* $4x + 2 < x - 7$.

Solution First we subtract x from both sides:

The goal is to isolate the variable x on one side of the equation by itself.

$$3x + 2 < -7.$$

Next we subtract 2 from both sides:

$$3x < -9.$$

Finally, we divide both sides by 3:

Because we are dividing by a positive number, the direction of the inequality does not change.

$$x < -3.$$

The solution to the inequality is $x < -3$. ■

Solutions to inequalities may also be presented in graphical form. The idea is to indicate which numbers on the number line satisfy the inequality. For $x < -3$, the variable x satisfies the inequality as long as x is to the left of -3 (figure 7):

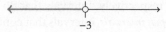

−3

Figure 7 *The inequality $x < -3$, shaded in green. The open circle indicates that -3 is not included*

Drawing an open circle indicates the point is not included. An alternate version that is sometimes used is to draw a parenthesis instead (figure 8).

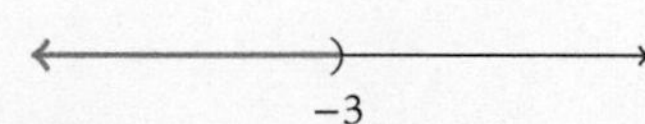

Figure 8 *The inequality $x < -3$, shaded in green. The parenthesis indicates that -3 is not included*

Recall that $a \le b$ means that either $a < b$ or $a = b$.

Example 2 *Solve the inequality* $-7x + 1 \le 4$.

Solution First we subtract 1 from both sides:

$$-7x \le 3.$$

Next we divide both sides by -7, which requires switching the direction of the inequality:

$$x \ge \frac{3}{-7}.$$

The solution to the inequality is $x \ge -\frac{3}{7}$. ■

Because the value $x = -\frac{3}{7}$ is included in the solution to example 2, when presenting the solution in graphical form, the circle is filled in (figure 9):

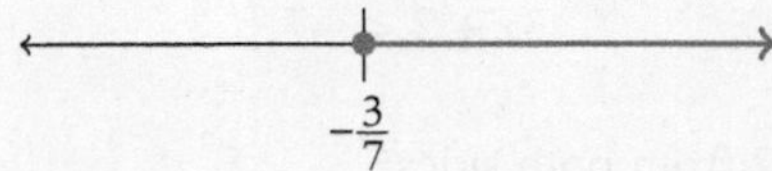

Figure 9 *The inequality $x \ge -\frac{3}{7}$, shaded in green. The filled circle indicates that $-\frac{3}{7}$ is included*

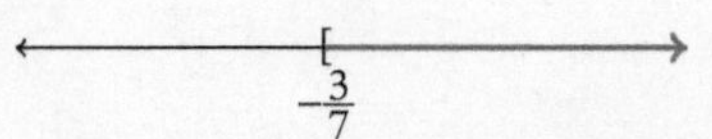

Figure 10 *The inequality $x \ge -\frac{3}{7}$, shaded in green. The bracket indicates that $-\frac{3}{7}$ is included*

An alternate version of the diagram is to use a bracket instead of a filled circle (figure 10).

Intervals

Interval notation, and the various types of intervals, are summarized in table 1. Parentheses indicate the endpoint is not included; brackets indicate the endpoint is included. The symbols ∞ and $-\infty$ are not numbers, but merely indicators that the interval has no endpoint on the right or the left, respectively. Intervals that do not contain any of their endpoints are *open intervals*; intervals that contain all of their endpoints are *closed intervals*. Because neither ∞ nor $-\infty$ are endpoints, then $[a, \infty)$ is a closed interval. Bounded intervals have two endpoints; unbounded intervals range to ∞ or $-\infty$. The two numbers or symbols in the interval are always written with the smaller value or $-\infty$ on the left and the larger value or ∞ on the right. In table 1, the green shading on the graphs indicates the solution set.

Example 3 *For the inequality $3 < x \le 5$, (a) write the inequality in interval notation and (b) classify the interval (open/closed, bounded/unbounded).*

Table 1 *Notation, visualization, and classification of intervals*

inequality	interval	graph	open/closed	bounded?
$a < x < b$	(a, b)	a b	open	bounded
$a \le x \le b$	$[a, b]$	a b	closed	bounded
$a < x \le b$	$(a, b]$	a b	neither	bounded
$a \le x < b$	$[a, b)$	a b	neither	bounded
$a < x$	(a, ∞)	a	open	unbounded
$a \le x$	$[a, \infty)$	a	closed	unbounded
$x < b$	$(-\infty, b)$	b	open	unbounded
$x \le b$	$(-\infty, b]$	b	closed	unbounded
all $x \in \mathbf{R}$	$(-\infty, \infty)$		both	unbounded

Solution (a) Because 3 is not included, we use a parenthesis for that endpoint, and because 5 is included, we use a bracket for that endpoint. The interval is $(3, 5]$.

(b) The interval is neither open nor closed because it contains one, but not the other, endpoint. The interval is bounded because it does not range to ∞ or to $-\infty$. ∎

Example 4 *Graph the interval: (a)* $[4, \infty)$*, (b)* $(-5, -3)$*.*

Solution (a) The interval $[4, \infty)$ includes 4, so we draw a filled circle at 4 to indicate its inclusion. The shading (green) has no bound on the right.

4

(b) For the interval $(-5, -3)$ we do not include either endpoint, so we draw open circles at -5 and at -3. The shading (green) is between those two numbers.

−5 −3

∎

Bounded open intervals such as $(-5, -3)$ share a notation with points in the xy-plane. Context is nearly always enough to determine which is meant.

Intervals are characterized by the property that all numbers between two of the interval's numbers are also in the interval. If there is any "gap" in the set, then it is not an interval. The set $(-4, 0) \cup (3, 5)$ is not an interval; see figure 11.

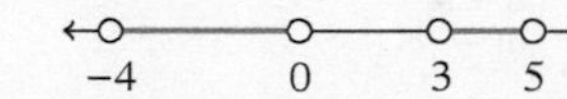

Figure 11 *The set* $(-4, 0) \cup (3, 5)$ *is not an interval because there is a "gap" from 0 to 3*

Absolute value

The idea of absolute value can be visualized as the distance from the number to zero on the number line (figure 12).

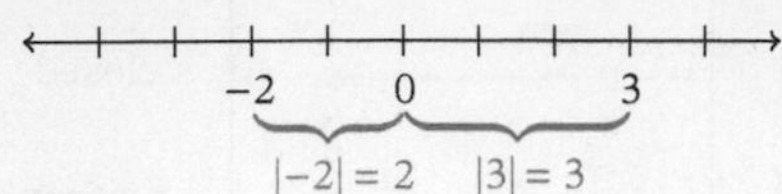

Figure 12 *Absolute value as the distance on the number line between the number and zero*

Because the number 3 is located 3 units away from zero on the number line, $|3| = 3$. Because the number -2 is located 2 units away from zero on the number line, $|-2| = 2$.

Alternately, we can think of folding the number line at the number 0, folding the left side of the number line onto the right side. The absolute value of a number is where it is located after the folding. The formal definition of absolute value says that we leave positive numbers alone but "reflect" or "fold" the negative numbers by taking their negatives.

CAUTION: MISTAKE TO AVOID
The heuristic "absolute value changes − to +" is the source of many algebraic errors and should be avoided. For instance, if $x = -5$, then

$$|-x| = |-(-5)| = |5| = 5 = -x,$$

so for this value of x,

$$|-x| \neq x.$$

"Removing the negative" does not work for absolute values of variable expressions, because $-x$ is not necessarily a negative number.

Definition 1 ABSOLUTE VALUE *For any real number* a,

$$|a| = \begin{cases} a, & \textit{if } a \geq 0 \\ -a, & \textit{if } a < 0. \end{cases}$$

Because $-2 < 0$, the definition (using $a = -2$) says that $|-2| = -(-2) = 2$.

PROPERTIES OF ABSOLUTE VALUE

For any real numbers a and b,

$$|a \cdot b| = |a| \cdot |b| \text{ and}$$

$$\left|\frac{a}{b}\right| = \frac{|a|}{|b|} \text{ (assuming } b \neq 0\text{).}$$

CAUTION: MISTAKE TO AVOID
$|a + b| \neq |a| + |b|$
$|a - b| \neq |a| - |b|$

The properties of absolute value give us rules we can use to simplify expressions. For instance,

$$|-5x| = |-5| \cdot |x| = 5|x|.$$

How can we simplify $\sqrt{x^2}$? Many assume the answer is x, but this is not correct, because the square root operation always gives us the nonnegative number with the square that is inside. For instance, $\sqrt{16} = 4$ and not -4 and not ± 4. Therefore,

The solutions to $x^2 = 16$ are $x = \pm\sqrt{16} = \pm 4$, but the symbol $\sqrt{16}$ means only the positive value, 4.

$$\sqrt{4^2} = \sqrt{16} = 4,$$

whereas

CAUTION: MISTAKE TO AVOID

$$\sqrt{x^2} \neq x$$

$$\sqrt{(-4)^2} = \sqrt{16} = 4 \text{ (not } -4).$$

This illustrates the following fact:

$$\sqrt{x^2} = |x|.$$

SQUARE ROOTS OF SQUARES

$$\sqrt{x^2} = |x|$$

Absolute value equations and inequalities

Which numbers satisfy $|x| = 3$? The two numbers $x = 3$ and $x = -3$ are the only solutions to this equation. See figure 13.

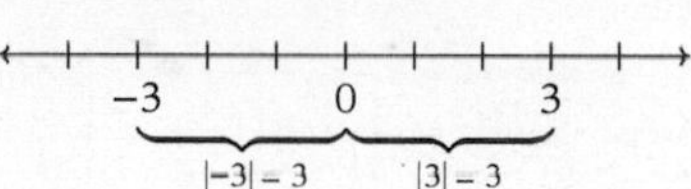

Figure 13 *The two numbers located 3 units from zero*

ABSOLUTE VALUE EQUATIONS

If $a \geq 0$, then

$$|x| = a \text{ if and only if } x = \pm a.$$

The phrase "$x = \pm a$" has the same meaning as "$x = a$ or $x = -a$."

Example 5 *Solve* $x^2 = 16$.

Solution We begin by taking the square root of both sides of the equation and simplifying:

$$\sqrt{x^2} = \sqrt{16}$$
$$|x| = 4.$$

Remember that $\sqrt{x^2} = |x|$.

Next we solve the absolute value equation:

$$x = \pm 4.$$

The two solutions to the equation are $x = 4$ and $x = -4$. ■

It is customary to skip the middle two steps when solving the equation of example 5, writing

$$x^2 = 16,$$
$$x = \pm 4.$$

CAUTION: MISTAKE TO AVOID
Writing

$$x^2 = 16$$
$$x = 4$$

misses one of the solutions!

ABSOLUTE VALUE INEQUALITIES, <

If $a > 0$, then

$$|x| < a \text{ if and only if } -a < x < a.$$

Fold the number line at zero, folding the left side onto the right side. Then $|x| < 7$ represents those numbers to the left of 7 when folded. Shade those values. Unfold the line, and what is shaded? The numbers between -7 and 7:

−7 0 7

Example 6 *Solve* $|4x - 1| < 3$.

Solution First we rewrite the equation as a compound inequality:

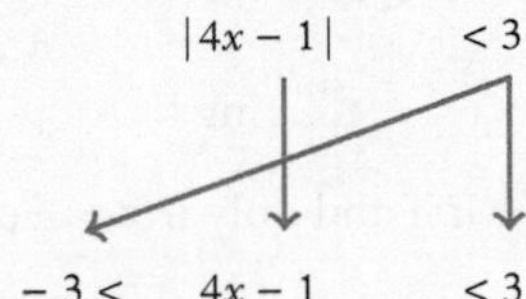

A *compound inequality* is two inequalities that are satisfied simultaneously: $-2 < x < 5$ means that both $-2 < x$ and $x < 5$ at the same time.

Then we proceed to isolate x in the middle of the inequality by adding 1 to all three parts of the compound inequality and dividing all parts by 4:

$$-3 < 4x - 1 < 3$$
$$-2 < 4x < 4$$
$$-\frac{1}{2} < x < 1.$$

The solution is $-\frac{1}{2} < x < 1$. ■

The solution can be written in interval notation as $\left(-\frac{1}{2}, 1\right)$. The solution can also be illustrated graphically:

$-\frac{1}{2}$ 1

If the variable is both inside and outside the absolute values, then the situation can quickly become more complicated. The solution method of example 6 needs to be modified and expanded to solve an inequality such as $|2x - 3| < 5x$.

Fold the number line at zero, folding the left side onto the right side. Then $|x| > 7$ represents those numbers to the right of 7 when folded. Shade those values. Unfold the line, and what is shaded? The numbers to the left of -7 and the numbers to the right of 7:

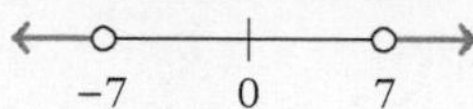

ABSOLUTE VALUE INEQUALITIES, >

If $a > 0$, then

$$|x| > a \text{ if and only if } x < -a \text{ or } x > a.$$

Example 7 *Solve* $|9 - x| > 1$.

Solution First we rewrite the equation as two separate inequalities:

CAUTION: MISTAKE TO AVOID
Writing $-1 > 9 - x > 1$ is incorrect because this says that $-1 > 1$, which is false. A compound inequality has two inequalities satisfied simultaneously; "and" is implied. Here, "or" is needed instead.

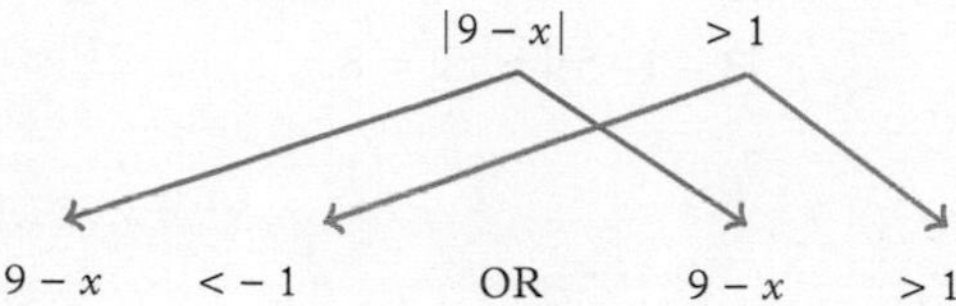

9 − x < − 1 OR 9 − x > 1

Each inequality is solved separately by subtracting 9 from both sides and dividing both sides by -1, changing the direction of the inequality:

An alternative is to add one to both sides (left equation) or subtract one from both sides (right equation) and then add x to both sides.

$$9 - x < -1 \quad \text{OR} \quad 9 - x > 1,$$
$$-x < -10 \quad \text{OR} \quad -x > -8,$$
$$x > 10 \quad \text{OR} \quad x < 8.$$

The solution is $x < 8$ or $x > 10$. ■

Although the solution set is not an interval, it may still be written using interval notation as

$$(-\infty, 8) \cup (10, \infty).$$

The solution can also be illustrated graphically:

8 10

Distance on the number line

Consider the distance on the number line between the numbers 3 and 7 (figure 14). Subtracting the two numbers may or may not give that distance:

$$7 - 3 = 4,$$
$$3 - 7 = -4.$$

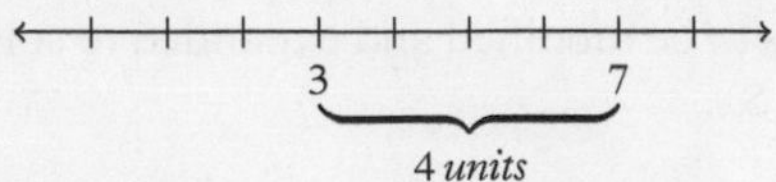

Figure 14 *The distance between the numbers* 3 *and* 7

However, taking the absolute value of the difference between the numbers does the trick:

$$|7 - 3| = |4| = 4,$$
$$|3 - 7| = |-4| = 4.$$

The absolute value of the difference still works even if one number is positive and the other is negative (see figure 15):

$$|(-5) - 3| = |-8| = 8$$
$$|3 - (-5)| = |8| = 8.$$

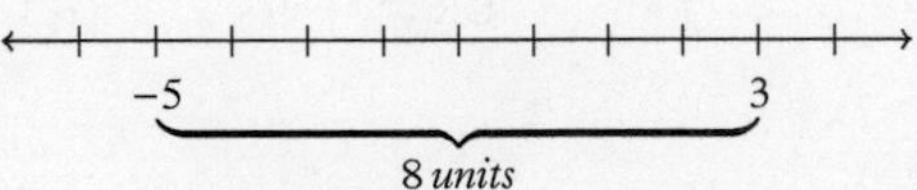

Figure 15 *The distance between the numbers* 3 *and* −5

The calculation works equally well when both numbers are negative (try it).

Often in calculus, formulas for intuitive ideas such as length, area, volume, average value, and many others can be developed in a manner similar to the development of the formula for distance on the number line. Rather than offer a "proof" of such formulas, which would require us to have already defined (or described axiomatically) the term in question, we offer the formula as the definition of the term instead.

Definition 2 DISTANCE ON THE NUMBER LINE

The distance *between two real numbers* a *and* b *is* $|a - b|$.

Example 8 *Determine the distance between the numbers* −2 *and* −8.

Solution Using the definition, the distance between −2 and −8 is

$$|-2 - (-8)| = |-2 + 8| = |6| = 6. \quad ■$$

An alternate solution is $|(-8)-(-2)| = |-8+2| = |-6| = 6$. The order in which the two numbers are subtracted does not matter because we take the absolute value of that difference.

Example 9 *Write an expression for the distance between z and π.*

Solution Using the definition, the distance between z and π can be written as either

$$|z-\pi|$$

or

$$|\pi - z|.$$

■

Exercises 0.1

Rapid-response exercises are to be answered very quickly, and often involve object recognition. These types of exercises usually do not appear on exams. Rapid-response exercises do not appear in every section.

1–10. Rapid response: has correct interval notation been used? Answer "correct" or "wrong."

1. $(4, 7]$	**6.** $[5, \infty]$
2. $)4, 7]$	**7.** $(-\infty, \frac{4}{7}]$
3. $(7, 4)$	**8.** $(3, -15)$
4. $(5, \infty)$	**9.** $[5, 9[$
5. $(\infty, 5)$	**10.** $(4 < x \leq 5]$

11–16. Rapid response: when rewriting the absolute value inequality, is a compound inequality appropriate? If so, answer "compound inequality"; if not, answer "separate inequalities."

11. $|4x-7| < 9$	**14.** $2 > |9x+5|$
12. $|6x| > 5$	**15.** $|27x-12| < 4$
13. $4 < |2x+3|$	**16.** $|7x-3| < 11$

17–24. (a) Solve the inequality. Express the answer in interval notation. (b) Graph the solution on a number line.

17. $9x - 5 < 8$	**19.** $4 - 3x \geq x - 2$
18. $6x - 11 < 7$	**20.** $7 - x \leq 9x + 5$

21. $3x < 7 - 4x < 3x + 12$

22. $-4x - 9 > 11$

23. $-2x + 1 \leq -8$

24. $14 < 8x + 2 \leq 22$

25–26. (a) Solve the inequality. Express the answer in interval notation. (b) Graph the solution on a number line.

25. $5(x - 2) - (x + 1) > 11x$

26. $4(2x + 1) - (x - 3) \geq 4 - x$

27–30. Simplify the expression.

27. $x - \sqrt{x^2}$

28. $|\sqrt{16}x|$

29. $|-12x|$

30. $\sqrt{9x^2}$

31–40. Solve the inequality.

31. $|4x - 7| < 9$

32. $|6x| > 5$

33. $4 < |2x + 3|$

34. $2 \geq |9x + 5|$

35. $|27x - 12| \leq 4$

36. $|7x - 3| < 11$

37. $|3x + 5| < -2$

38. $|8 - x| \geq 6$

39. $|3 - 4x| \geq 5$

40. $|12x + 56| > -35$

41–46. Find the distance between the two given numbers on the number line.

41. -5 and 7

42. -5 and -7

43. -6 and -9

44. 6 and 9

45. 12 and 7

46. -12 and 7

47–50. Write an expression for the distance between the two quantities.

47. x and y

48. $-x$ and $-y$

49. $x + 4$ and $y - 3$

50. $3x$ and 7

Algebra Review, Part II

0.2

A line has one dimension. Lengths or distances are one-dimensional units. We describe the length of a board or the length of a race using units such as feet (ft) or meters (m). A measuring tape used to find lengths or distances is a physical number line. Locations on roads are marked with mile or kilometer markers that essentially turn the road into a physical number line.

Both English units and metric units are used freely in this book.

A plane has two dimensions. Areas are two-dimensional units. One dimension—a length or distance—is inadequate to describe the size of a field. A rectangular field that is 1 km long and 100 m wide is not as large as a rectangular field that is 1 km long and 1 km wide; saying that a field is 1 km long is not a complete description. Areas are given in square units, such as square feet (ft^2) or square kilometers (km^2). Locations in a plane (or other two-dimensional surface, such as the surface of the earth) are given using two values. We may say "go 1 mile east and 2 miles north" or give a location's longitude and latitude. Instead of a number line, we need a coordinate plane.

Specialized units of area such as acres and hectares are also used, but they are still two-dimensional units.

Space has three dimensions. Volumes are three-dimensional units. The same truckload of dirt could be spread thinly over an entire acre or dumped in a pile; neither length nor area gives an adequate description of its size. To say that a rectangular storage bin is 2 m wide and 1 m long does not indicate its capacity, for such a bin could be 20 cm high or it could be 4 m high. Volumes are measured in cubic units, such as cubic inches (in^3) or cubic meters (m^3). Locations in space require three values. An airplane's location is described not just by latitude and longitude, but by altitude as well. Instead of a number line or a coordinate plane, we need a three-dimensional coordinate system.

In the previous section we reviewed one-dimensional algebra using number lines. In this section we review two-dimensional algebra using coordinate planes. Three-dimensional coordinate systems are discussed later in calculus.

Coordinate plane

One way to describe the location of a point in a plane, which has two dimensions, is to use two number lines placed perpendicularly. We call the horizontal number line the *x-axis* (positive coordinates to the right, negative to the left) and the vertical number line the *y-axis* (positive

We study a separate coordinate system for two dimensions, polar coordinates, in chapter 8.

coordinates up, negative coordinates down), with the axes intersecting at their zero points. We call that intersection the *origin*. See figure 1.

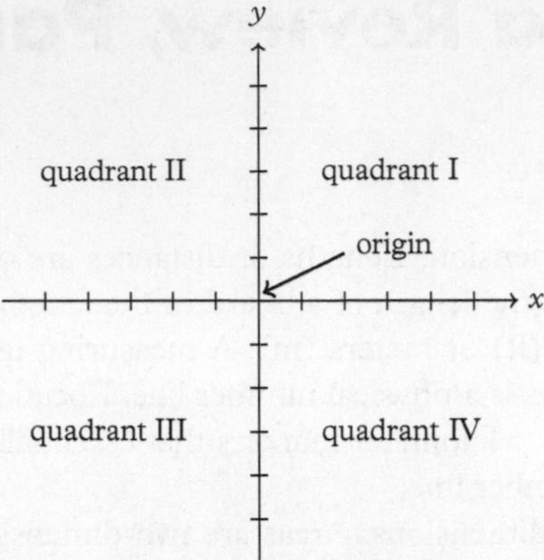

Figure 1 *The coordinate plane, also called the xy-plane or the Cartesian plane*

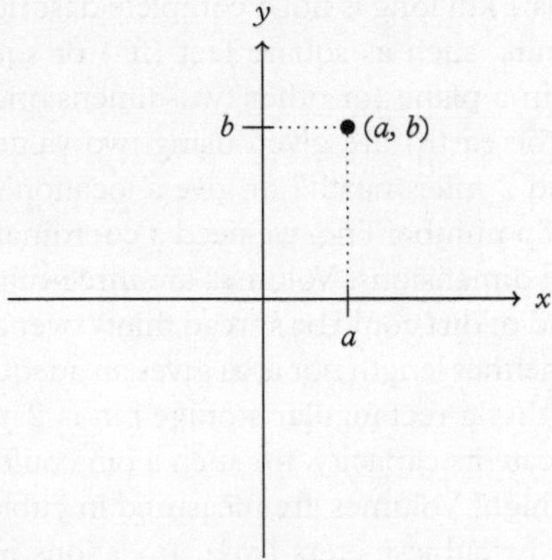

Figure 2 *The location of a point described by coordinates in the xy-plane*

The axes partition the plane into four regions, called *quadrants*, numbered as in figure 1. The location of a point is given by the x- and y-coordinates. The x-coordinate of a point is the number on the x-axis for which the point aligns vertically; the y-coordinate of a point is the number on the y-axis for which the point aligns horizontally. See figure 2. The coordinates are denoted by a pair of numbers in parentheses, separated by a comma, with the first number being the x-coordinate and the second number, the y-coordinate. This plane is called the *xy-plane*, the *coordinate plane*, or the *Cartesian plane* (in honor of René Descartes). These coordinates are sometimes called *rectangular coordinates*.

René Descartes, 1596–1650
http://www-history.mcs.st-andrews.ac.uk/Biographies/Descartes.html

To *plot* a point means to identify and mark its location in the xy-plane.

Example 1 *Plot the points (a)* $(2,1)$*, (b)* $(-1,-3)$*, and (c)* $(3,-\frac{5}{2})$.

Solution (a) To plot the point $(2,1)$, we locate 2 on the x-axis and move 1 unit up, or we locate 1 on the y-axis and move 2 units right. See figure 3.

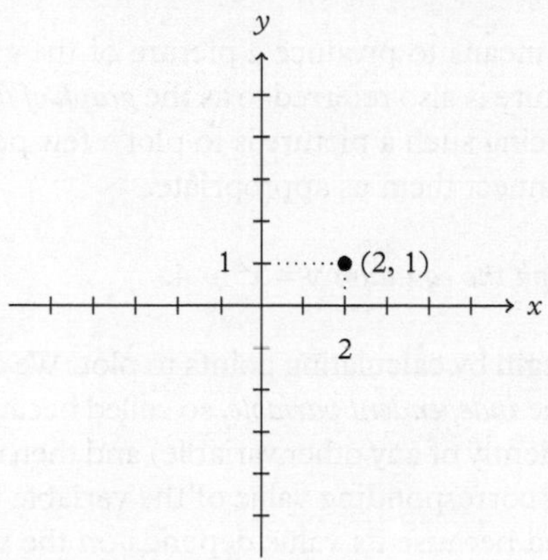

Figure 3 *Plotting the point* $(2,1)$ *in the xy-plane*

(b) To plot $(-1,-3)$, we locate -1 on the x-axis and move 3 units down or we locate -3 on the y-axis and move 1 unit left. (c) Likewise, to plot $(3,-\frac{5}{2})$ we locate 3 on the x-axis and move 2.5 units down, or we locate $-\frac{5}{2}$ on the y-axis and move 3 units right. See figure 4.

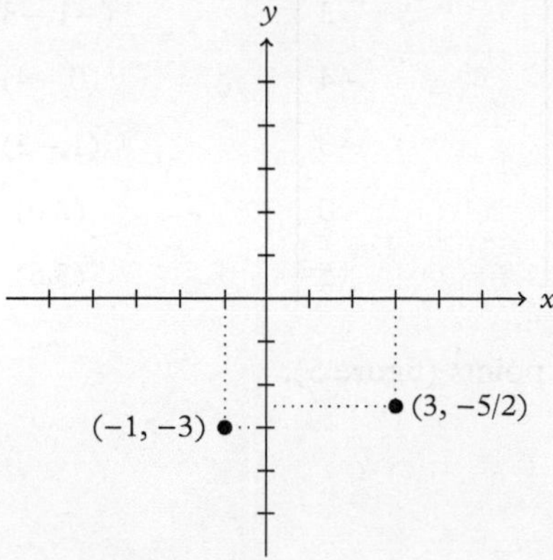

Figure 4 *Plotting the points* $(-1,-3)$ *and* $(3,-\frac{5}{2})$ *in the xy-plane* ■

Graphs of equations

Set-builder notation describes a set by naming the type of object on the left and conditions satisfied by the object on the right:

$$\{\text{object} \mid \text{condition}\}.$$

The set

$$\{(x, y) \mid y = x^2 - 4\}$$

gives a point in the xy-plane as the object, with the condition that the y-coordinate is the x-coordinate squared, minus 4. All points that satisfy the condition are in the set. The symbol | can be read as "such that."

The word *graph* is used as both a noun and a verb. As a noun, it represents a set of points. For instance, the graph of the equation $y = x^2 - 4$ is the set of points

$$\{(x, y) \mid y = x^2 - 4\}.$$

As a verb, *graph* means to produce a picture of the graph (the set of points). This picture is also referred to as the *graph of the equation.* One method of producing such a picture is to plot a few points that are on the graph and connect them as appropriate.

Example 2 *Graph the equation* $y = x^2 - 4$.

For some types of equations, there are alternative methods to graph the equation quickly or to visualize the graph quickly.

Choosing values of x can be seen as an art, influenced by the form of the equation and based on skill gained with practice.

Solution We begin by calculating points to plot. We choose values of the variable x (the *independent variable,* so called because its values are chosen independently of any other variable) and then use the equation to determine the corresponding value of the variable y (the *dependent variable,* so called because its value depends on the value of x). The values of x and y are the x- and y-coordinates of the points to be plotted. We choose the values $x = -3, \ldots, 3$, as shown in Table 1.

Table 1 *Points on the graph of* $y = x^2 - 4$

x	y	point
−3	5	(−3,5)
−2	0	(−2,0)
−1	−3	(−1,−3)
0	−4	(0,−4)
1	−3	(1,−3)
2	0	(2,0)
3	5	(3,5)

Next we plot the points (figure 5).

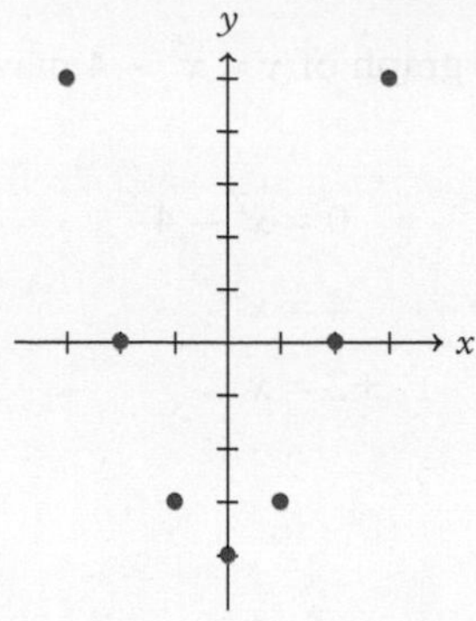

Figure 5 *A few plotted points on the graph of* $y = x^2 - 4$

Finally, we connect the dots—not by using straight line segmentsbetween points, but by making a smooth, connected curve (figure 6). ■

The connect-the-dot game is justified in chapter 1. Drawing the graph in a smooth manner (without corners) is justified in chapter 2.

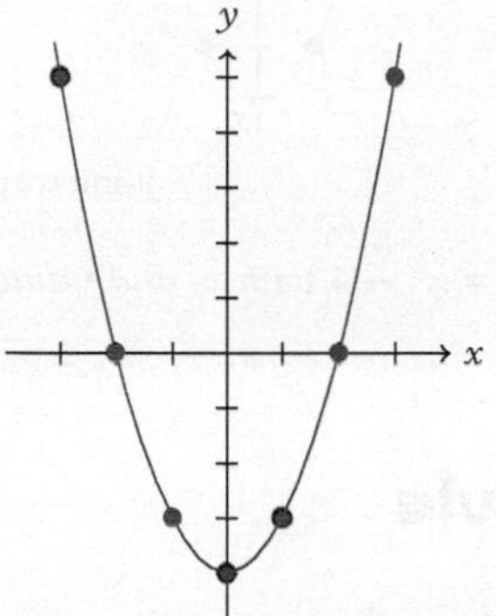

Figure 6 *The graph of* $y = x^2 - 4$. *The graph is the set of points colored blue*

The points where the graph of an equation intersects the x-axis are called *x-intercepts*; points where the graph of an equation intersects the y-axis are called *y-intercepts*. See figure 7.

The y-intercept on the graph of $y = x^2 - 4$ may be found by setting $x = 0$:

$$y = 0^2 - 4 = -4.$$

The x-intercepts on the graph of $y = x^2 - 4$ may be found by setting $y = 0$:

$$0 = x^2 - 4$$
$$4 = x^2$$
$$\pm 2 = x.$$

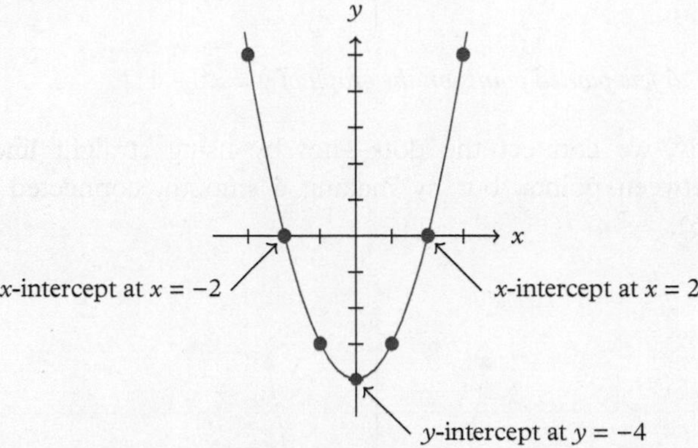

Figure 7 *The graph of $y = x^2 - 4$ with x- and y-intercepts as indicated*

Distance formula

How do we find the distance between two points in the plane, P_1 and P_2? If their coordinates are given as (x_1, y_1) and (x_2, y_2), respectively, then consider the third point (x_2, y_1) that aligns vertically with P_2 and horizontally with P_1, as pictured in figure 8.

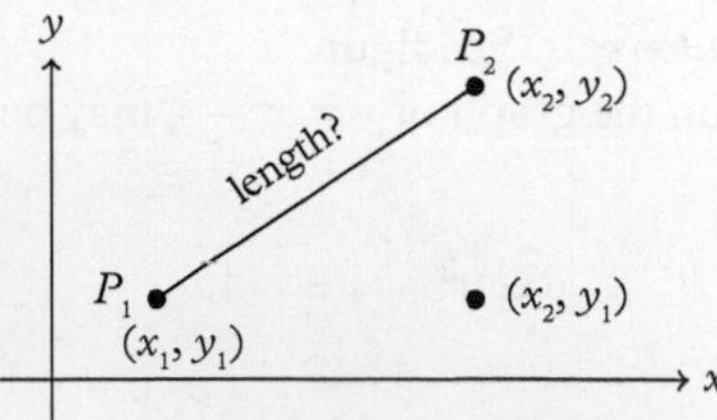

Figure 8 *Determining the distance between points P_1 and P_2*

Drawing the vertical and horizontal segments forms a right triangle, so the Pythagorean theorem applies. The length of the horizontal segment is the difference in the x-coordinates: $|x_2 - x_1|$. The length of the vertical segment is the difference in the y-coordinates: $|y_2 - y_1|$. See figure 9. Using the Pythagorean theorem, the distance between P_1 and P_2, written $d(P_1, P_2)$, is

$$d(P_1, P_2) = \sqrt{|x_2 - x_1|^2 + |y_2 - y_1|^2}$$
$$= \sqrt{(x_2 - x_1)^2 + (y_2 - y_1)^2}.$$

In the Pythagorean theorem, $a^2 + b^2 = c^2$, with leg lengths $a, b \geq 0$ and hypotenuse $c \geq 0$. Taking square roots of both sides gives

$$c = \sqrt{a^2 + b^2}.$$

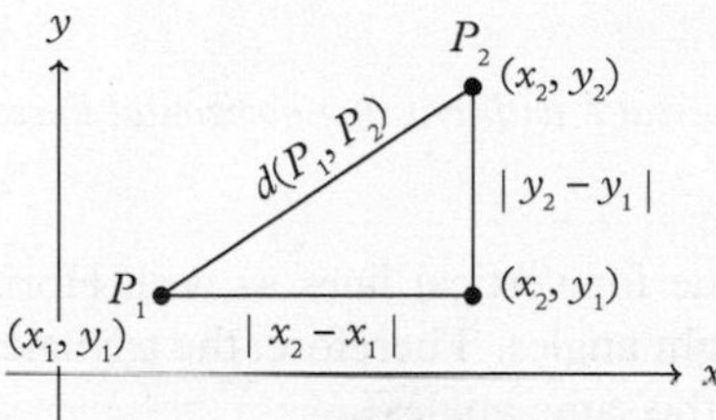

Figure 9 *Determining the distance between points P_1 and P_2*

Alternate definitions of distance in the plane exist. One, called the *taxicab metric*, is motivated by considering a city with blocks laid out in a square grid. The route from point (x_1, y_1) to point (x_2, y_2) by taxi has to go along the streets, not through all the buildings. The distance traveled by the taxi is the sum of the east–west (map horizontal) and north–south (map vertical) distances, or along the legs of the right triangle of figure 9. The distance is defined as $d(P_1, P_2) = |x_2 - x_1| + |y_2 - y_1|$.

This is called the *distance formula* and it serves as our definition of the distance between two points in the plane.

Definition 3 DISTANCE FORMULA *The distance between two points $P_1 = (x_1, y_1)$ and $P_2 = (x_2, y_2)$ is given by*

$$d(P_1, P_2) = \sqrt{(x_2 - x_1)^2 + (y_2 - y_1)^2}.$$

Example 3 *Find the distance between the points* $(2, 6)$ *and* $(1, -4)$.

Solution Setting $(x_1, y_1) = (2, 6)$ and $(x_2, y_2) = (1, -4)$, the distance formula gives

$$d((2,6),(1,-4)) = \sqrt{(1-2)^2 + (-4-6)^2}$$
$$= \sqrt{(-1)^2 + (-10)^2} = \sqrt{1 + 100}$$
$$= \sqrt{101}.$$

CAUTION: MISTAKE TO AVOID

$\sqrt{1^2 + 10^2} \neq 1 + 10$

The distance between the two points is $\sqrt{101}$. If desired, a decimal approximation can be given. ■

Slopes of lines

Consider a line ℓ that is neither vertical nor horizontal (figure 10). Draw two horizontal lines that cross line ℓ. Because horizontal lines are parallel to one another, the corresponding angles at which the horizontal lines meet ℓ must be congruent (by the corresponding angles theorem from geometry).

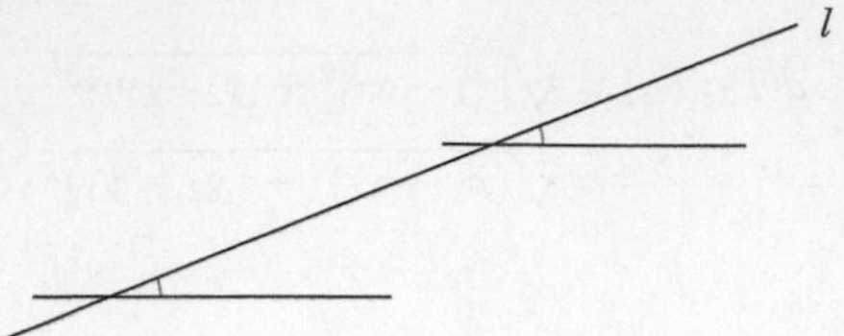

Figure 10 *A transversal ℓ with crossing horizontal lines; angles (marked in green) are congruent*

The same is true for vertical lines as well. Horizontal and vertical lines meet at right angles. Therefore, the triangles in figure 11 are similar (they have the same angles).

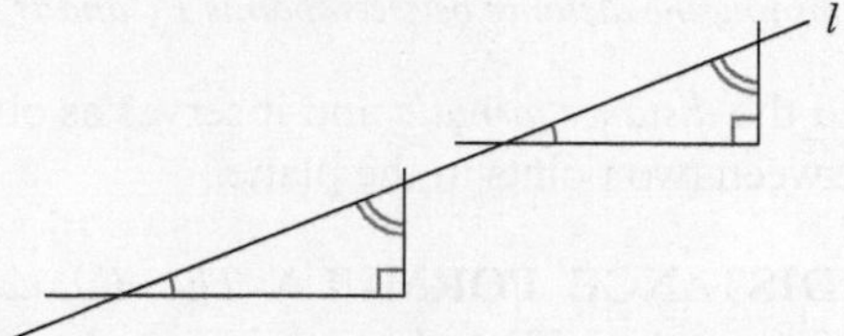

Figure 11 *A transversal ℓ with crossing horizontal and vertical lines; the triangles are similar*

One important fact from geometry about similar triangles is that the ratios of corresponding sides are equal in the two triangles. Labeling sides a, b, c, and d as in figure 12, a pair of equal ratios is

$$\frac{a}{b} = \frac{c}{d}.$$

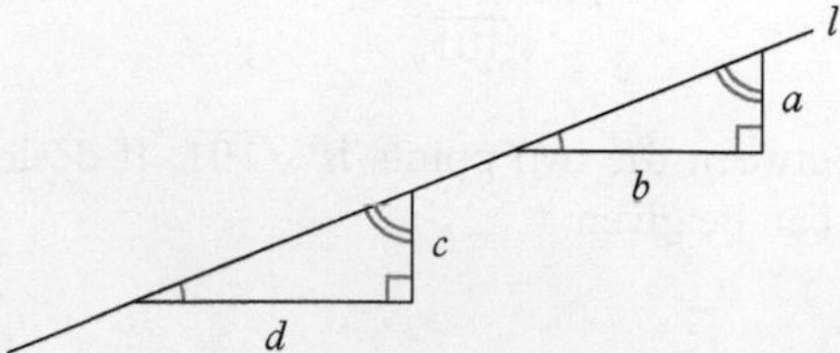

Figure 12 *A transversal ℓ with crossing horizontal and vertical lines; $\frac{rise}{run}$ is the same wherever the horizontal and vertical lines are drawn*

The "rise" of the line as it moves along the hypotenuse of one of the right triangles is a or c, whereas the "run" of the line is b or d. This quantity of $\frac{\text{rise}}{\text{run}}$ is therefore the same for any such triangle we draw; it is a property of the line. We call this property the *slope* of the line.

How can the value of the slope be calculated? If we know the coordinates of two points on the line, (x_1, y_1) and (x_2, y_2), then the rise is $y_2 - y_1$ and the run is $x_2 - x_1$ (see figure 13), and the slope is

$$\text{slope of line} = \frac{\text{rise}}{\text{run}} = \frac{y_2 - y_1}{x_2 - x_1}.$$

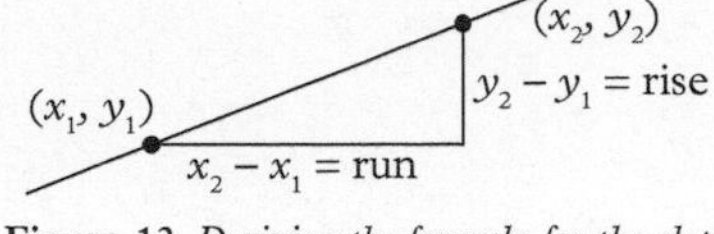

Figure 13 *Deriving the formula for the slope of a line*

Definition 4 SLOPE FORMULA

The slope of a nonvertical line through the points (x_1, y_1) and (x_2, y_2) is

$$slope = \frac{y_2 - y_1}{x_2 - x_1}.$$

Example 4 *Find the slope of the line through the points* $(1, 7)$ *and* $(2, 4)$.

In the construction industry, the slope of a roof is called its *pitch* and is expressed in terms of rise to run. A "6-in-12 pitch" is a rise of 6 inches for a run of 12 inches, or what the slope formula says is a slope of $\frac{1}{2}$. In northern areas of Europe and North America where heavy snowfall occurs, it is common for building codes to specify a minimum pitch for roofs.

Solution Using $(x_1, y_1) = (1, 7)$ and $(x_2, y_2) = (2, 4)$ in the slope formula yields

$$\text{slope} = \frac{y_2 - y_1}{x_2 - x_1} = \frac{4 - 7}{2 - 1} = \frac{-3}{1} = -3.$$

The slope of the line is -3. ■

Choosing $(x_1, y_1) = (2, 4)$ and $(x_2, y_2) = (1, 7)$ gives the same result:

$$\text{slope} = \frac{y_2 - y_1}{x_2 - x_1} = \frac{7 - 4}{1 - 2} = \frac{3}{-1} = -3.$$

It does not matter which point is chosen for (x_1, y_1) and which is chosen for (x_2, y_2).

As we move to the right, the run is positive. As we move upward, the rise is positive, and as we move downward, the rise is negative. Therefore, a line that slopes upward to the right has a positive slope, whereas a line that slopes downward to the right has a negative slope. See figure 14.

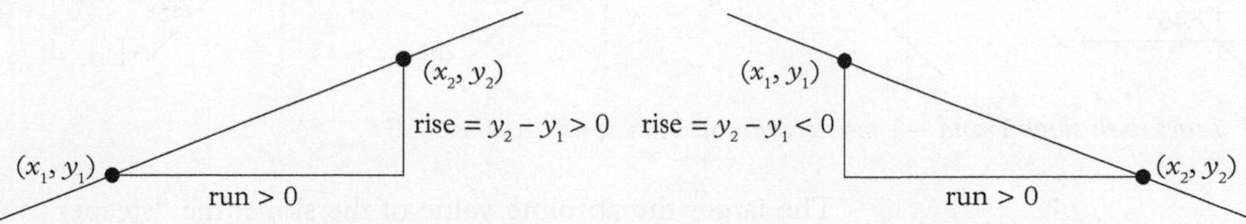

Figure 14 *(left) Lines sloping upward to the right have a positive slope; (right) lines sloping downward to the right have a negative slope*

A horizontal line (figure 15) has the same y-coordinate for every point. Therefore, its rise is always 0 and its slope is 0. Horizontal lines and their slopes of 0 play a special role in calculus and are used quite often.

Figure 15 *The horizontal line* $y = c$ *has slope* 0*; by the slope formula, slope =* $\frac{c - c}{x_2 - x_1} = 0$

A vertical line (figure 16) has the same x-coordinate for every point. Therefore, its run is always 0, and the slope formula results in division by zero, which is undefined. It is for this reason that we define slope for nonvertical lines only. Vertical lines have meaning in calculus, although they are not encountered as often as horizontal lines.

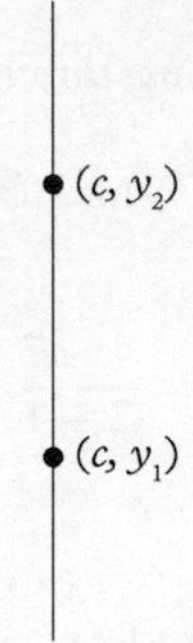

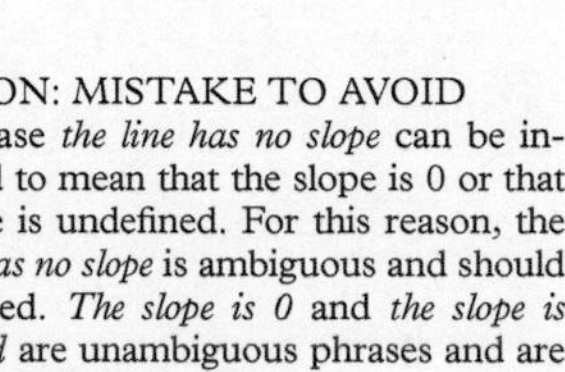
CAUTION: MISTAKE TO AVOID
The phrase *the line has no slope* can be interpreted to mean that the slope is 0 or that the slope is undefined. For this reason, the phrase *has no slope* is ambiguous and should be avoided. *The slope is 0* and *the slope is undefined* are unambiguous phrases and are therefore much better to use.

Figure 16 *The slope of the vertical line* $x = c$ *is undefined; trying the slope formula results in slope =* $\frac{y_2 - y_1}{c - c} = \frac{y_2 - y_1}{0}$*, which is undefined*

Lines with slope 1 and -1 meet the coordinate axes at angles of 45°; see figure 17.

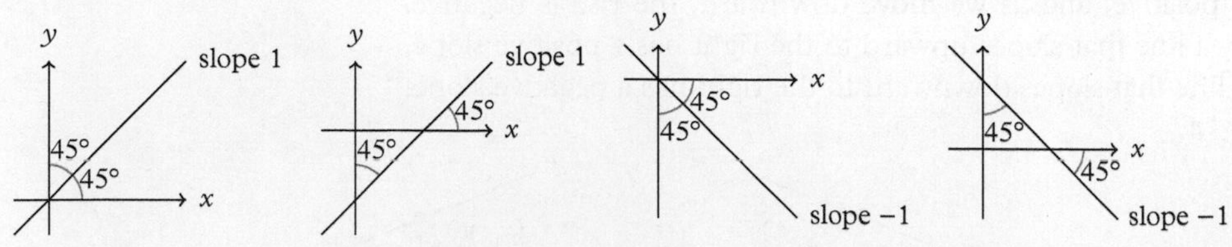

Figure 17 *Lines with slope* 1 *and* -1 *meet coordinate axes at* 45° *angles*

The larger the absolute value of the slope, the "steeper" the line; see figure 18.

The slope concept is generalized in calculus, where the slopes of curves, and not just lines, are studied.

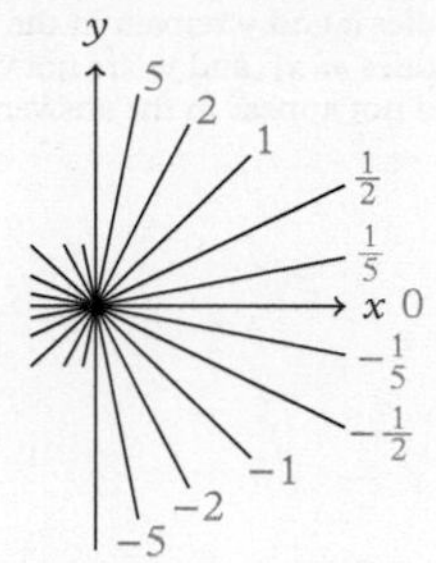

Figure 18 *The larger the |slope|, the "steeper" the line; slopes of lines are labeled in orange*

Point-slope form of the equation of a line

Suppose we wish to know the equation of a line with slope m through the point (x_1, y_1) (figure 19). The equation of the line gives the relationship between the x- and y-coordinates of points on the line. If the point (x, y) is on the line, then the slope formula applied to the points (x, y) and (x_1, y_1) must yield the number m, which is given as the slope of the line:

$$\text{slope} = m = \frac{y - y_1}{x - x_1}.$$

Clearing the fraction by multiplying both sides of the equation by the denominator results in

$$y - y_1 = m(x - x_1).$$

We call this the *point-slope form of the equation of a line.*

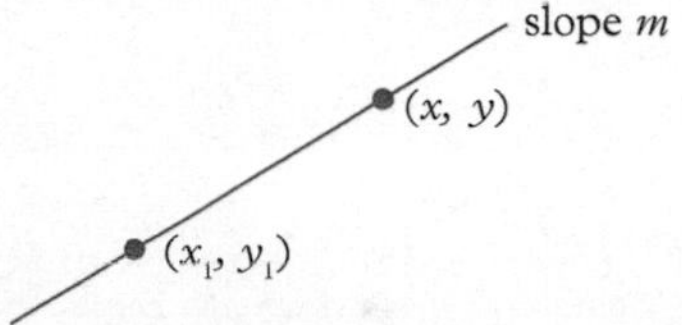

Figure 19 *A line with slope m through the point (x_1, y_1). Any point (x, y) on the line must result in a slope of m when the slope formula is applied*

> **POINT-SLOPE FORM OF THE EQUATION OF A LINE**
>
> The equation of the line with slope m through the point (x_1, y_1) is
>
> $$y - y_1 = m(x - x_1).$$

In the formula, x and y are variables whereas x_1, y_1, and m represent specific numbers.

Example 5 *Find the equation of the line with slope* −2 *through the point* $(1, 4)$.

Solution We are given the slope of the line, $m = -2$, and a point on the line, $(x_1, y_1) = (1, 4)$. Using $m = -2$, $x_1 = 1$, and $y_1 = 4$ in the point-slope form of the equation of a line yields

$$y - y_1 = m(x - x_1)$$
$$y - 4 = -2(x - 1).$$

The variables x and y remain in the answer. The quantities m, x_1, and y_1 are not variables and should not appear in the answer.

The equation of the line is $y - 4 = -2(x - 1)$. However, the answer is traditionally expressed in a different form, the form $y = mx + b$. To place the equation in the traditional form, we first distribute the slope -2 through the parentheses:

$$y - 4 = -2x + 2.$$

We finish by adding 4 to both sides of the equation:

$$y = -2x + 6.$$

The equation of the line is $y = -2x + 6$. ■

The reason for expressing the equation in the form $y = mx + b$ is that the form lends itself readily to graphing and interpretation, as we shall see shortly.

In geometry you learned that two points determine a line. Therefore, given two points we ought to be able to determine the equation of the line through those points.

Example 6 *Find the equation of the line through the points* $(1, 2)$ *and* $(5, -1)$.

Solution If we are to use the point-slope form of the equation of a line, we need to know the line's slope. This information is not given, but the slope formula tells us the slope of the line through two points. Using $(x_1, y_1) = (1, 2)$ and $(x_2, y_2) = (5, -1)$, we have

$$\text{slope} = \frac{y_2 - y_1}{x_2 - x_1} = \frac{-1 - 2}{5 - 1} = \frac{-3}{4} = m.$$

Now we have the information we need to use the point-slope form of the equation of a line. Using $m = -\frac{3}{4}$, $x_1 = 1$, and $y_1 = 2$ yields

$$\begin{aligned} y - y_1 &= m(x - x_1) \\ y - 2 &= -\frac{3}{4}(x - 1) \\ y - 2 &= -\frac{3}{4}x + \frac{3}{4} \\ y &= -\frac{3}{4}x + \frac{11}{4}. \end{aligned}$$

The equation of the line is $y = -\frac{3}{4}x + \frac{11}{4}$. ■

We can use the other point as (x_1, y_1) instead and reach the same solution. With $x_1 = 5$ and $y_1 = -1$,

$$y - y_1 = m(x - x_1)$$
$$y - (-1) = -\frac{3}{4}(x - 5)$$
$$y + 1 = -\frac{3}{4}x + \frac{15}{4}$$
$$y = -\frac{3}{4}x + \frac{11}{4}.$$

After the first step, the equations look different; but, after placing them in the traditional form, they are seen to be the same.

Slope-intercept form of the equation of a line

Suppose we know that a line has slope m and y-intercept b. The y-intercept gives the value of y at which the line crosses the y-axis, so the point $(0, b)$ is on the graph of the line (figure 20).

Using the point-slope form of the equation of a line with slope m and point $(x_1, y_1) = (0, b)$ gives

$$y - y_1 = m(x - x_1)$$
$$y - b = m(x - 0)$$
$$y - b = mx$$
$$y = mx + b.$$

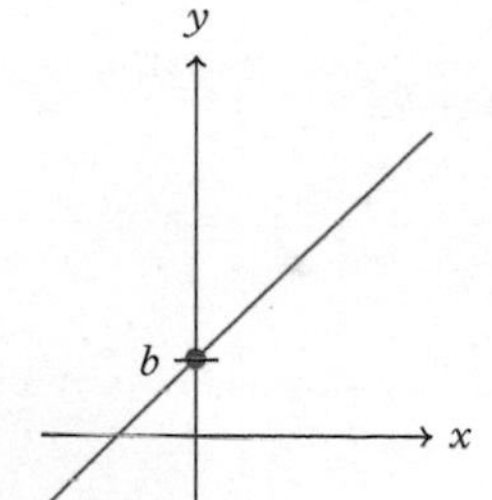

Figure 20 *A line with y-intercept b goes through the point* $(0, b)$

SLOPE-INTERCEPT FORM OF THE EQUATION OF A LINE

The equation of a line with slope m and y-intercept b is

$$y = mx + b.$$

The traditional form of the answer as used in examples 5 and 6 is the slope-intercept form. That makes it easy to recognize the y-intercept;

Because the y-intercept of an equation is determined by letting $x = 0$, the y-intercept of the equation $y = mx + b$ is at $y = b$.

for the equation $y = -2x + 6$, the y-intercept is 6 and the slope is -2. For the equation $y = -\frac{3}{4}x + \frac{11}{4}$, the y-intercept is $\frac{11}{4}$ and the slope is $-\frac{3}{4}$. This combination of information is all we need to graph the line relatively quickly.

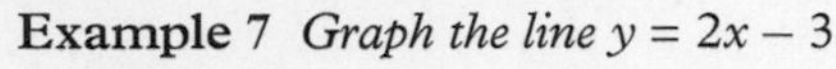

Example 7 *Graph the line $y = 2x - 3$.*

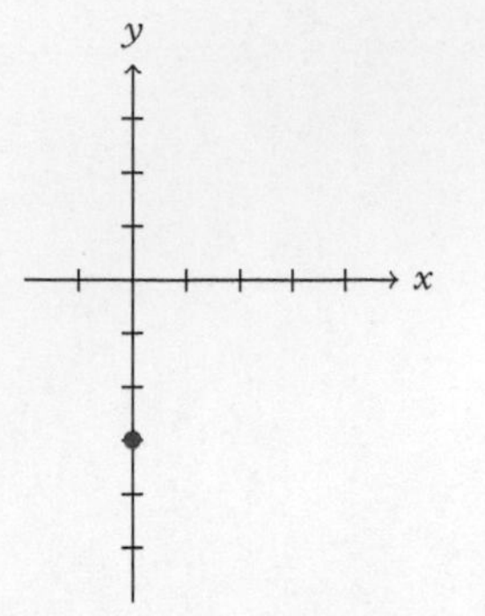

Figure 21 *The y-intercept of the line $y = 2x - 3$, marked by a blue dot*

Solution Instead of plotting points as in example 2, we use the slope and y-intercept of the line to mark points quickly on the graph. The line is in slope-intercept form, with slope $m = 2$ and y-intercept -3. We begin by placing the y-intercept on the graph (figure 21).

Next we interpret the slope as $\frac{\text{rise}}{\text{run}}$:

$$\text{slope} = \frac{\text{rise}}{\text{run}} = \frac{2}{1},$$

making the rise 2 when the run is 1. Starting at the y-intercept, we move to the right 1 unit and up 2 units to find another point on the graph, and repeat as often as desired (figure 22).

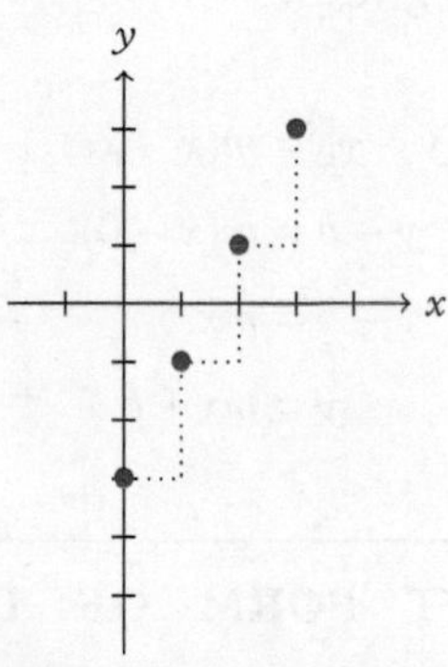

Figure 22 *Points on the line $y = 2x - 3$. The dotted lines represent a run of 1 and a rise of 2 between points*

Because $\frac{-2}{-1} = 2$ as well, we may also use a run of -1 and a rise of -2. In other words, beginning at the y-intercept, we may also move 1 unit left and 2 units down to find more points (figure 23).

Two points determine a line, thus five points is overkill. But, increasing the number of points plotted also increases accuracy when drawing graphs by hand.

We finish by drawing the line through the points (figure 24). With practice, this process can be performed relatively quickly. ■

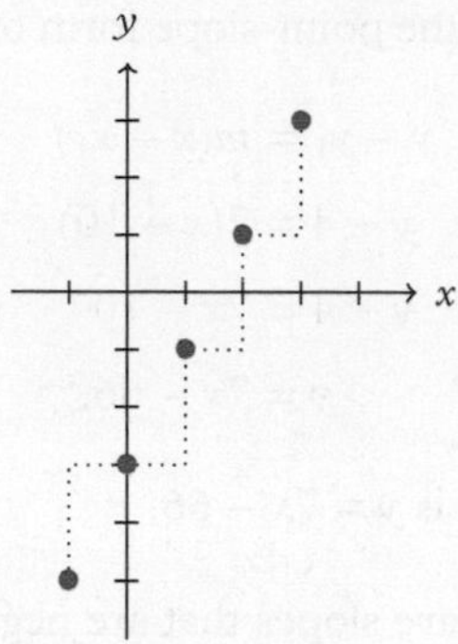

Figure 23 *Points on the line $y = 2x - 3$. The dotted lines represent a run of 1 and a rise of 2 between points (equivalent to a run of -1 and a rise of -2 between points)*

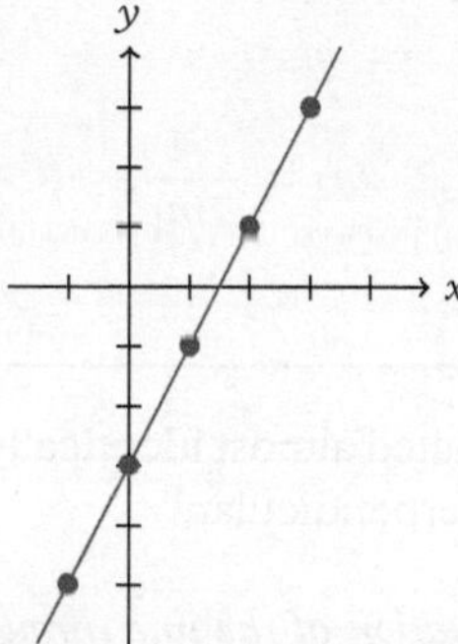

Figure 24 *The graph of the line $y = 2x - 3$*

Parallel and perpendicular lines

Parallel lines have the same slope. Since $y = -4x - 29$ has slope -4, any line parallel to $y = -4x - 29$ has slope -4.

The line $y = -4x - 29$ is written in slope-intercept form $y = mx + b$. Its slope is $m = -4$ and its y-intercept is $b = -29$.

Example 8 *Find the equation of the line through* $(10, 4)$ *parallel to* $y = 7x + 1$.

Solution Whichever form of the equation of a line is used, the point-slope form or the slope-intercept form, the slope is required information. Our line must be parallel to $y = 7x + 1$, which has slope $m = 7$; therefore, the slope of our line is $m = 7$. We have a point and the slope, so we can use the point-slope form of the equation of a line:

$$
\begin{aligned}
y - y_1 &= m(x - x_1)\\
y - 4 &= 7(x - 10)\\
y - 4 &= 7x - 70\\
y &= 7x - 66.
\end{aligned}
$$

The equation of the line is $y = 7x - 66$. ■

Perpendicular lines have slopes that are negative reciprocals. If the slopes of the lines are m_1 and m_2, then $m_2 = -\frac{1}{m_1}$. See figure 25.

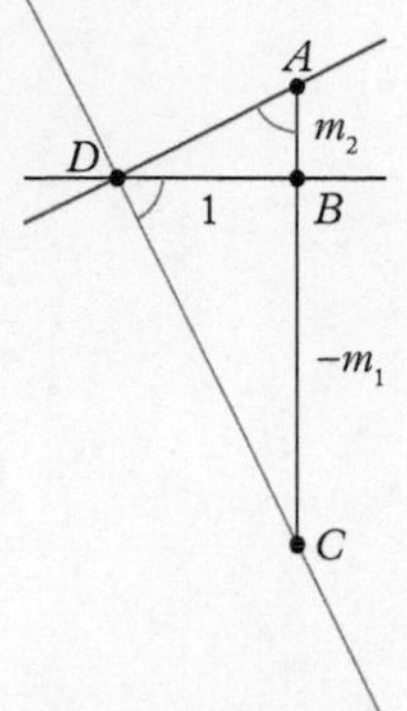

Figure 25 *Perpendicular lines (blue and brown), the horizontal line through their intersection, and a vertical line 1 unit to its right. With a run of DB for the two lines, their slopes are* $\frac{rise}{run} = \frac{rise}{1}$*, so the rise AB is the slope* m_2 *of the blue line, whereas the rise BC of the brown line, a negative number, is the slope* m_1*. Using positive numbers to label segment lengths, we label BC with the positive number* $-m_1$*. Because the lines are perpendicular,* $m(\angle ADB) + m(\angle CDB) = 90°$*. Because horizontal and vertical meet at a right angle,* $m(\angle ADB) + m(\angle BAD) + 90° = 180°$*, and hence* $m(\angle ADB) + m(\angle BAD) = 90°$*. Thus,* $\angle CDB \cong \angle BAD$*. Then the triangles* $\triangle BAD$ *and* $\triangle BDC$ *are similar. Ratios of corresponding sides are equal, hence* $\frac{BA}{BD} = \frac{BD}{BC}$*, or* $\frac{m_2}{1} = \frac{1}{-m_1}$

PERPENDICULAR LINES

If the line $y = m_2x + b_2$ is perpendicular to the line $y = m_1x + b_1$, then

$$m_2 = -\frac{1}{m_1}.$$

The next example is stated almost identically to example 8, but with "parallel" changed to "perpendicular."

Example 9 *Find the equation of the line through* $(10, 4)$ *perpendicular to* $y = 7x + 1$.

Solution Whichever form of the equation of a line is used, the point-slope form or the slope-intercept form, the slope is required information. Our line must be perpendicular to $y = 7x + 1$, which has slope $m = 7$; therefore, the slope of our line is $m = -\frac{1}{7}$. We have a point and the slope, so we may use the point-slope form of the equation of a line:

$$
\begin{aligned}
y - y_1 &= m(x - x_1)\\
y - 4 &= -\frac{1}{7}(x - 10)\\
y - 4 &= -\frac{1}{7}x + \frac{10}{7}\\
y &= -\frac{1}{7}x + \frac{38}{7}.
\end{aligned}
$$

The equation of the line is $y = -\frac{1}{7}x + \frac{38}{7}$. ■

Vertical lines

Vertical lines consist of points with the same x-coordinate, so their equations have the form $x = c$ for a real number c (see figure 16).

Example 10 *Find the equation of the vertical line through the point* $(5, 3)$.

Solution All points on a vertical line have the same x-coordinate. The given point on the line has x-coordinate 5. Therefore, all points on the line have x-coordinate 5. The equation of the line is $x = 5$. ■

Exercises 0.2

1–12. Rapid response: state the slope and y-intercept of the line.

1. $y = 4x - 7$

2. $y = 2x + 5$

3. $y = -91x + 225$

4. $y = -3x - 11$

5. $y = 17x$

6. $y = 17$

7. $y + 1 = 5x$

8. $y = \frac{1}{4}x - 15$

9. $y = \dfrac{2x}{3} + 12$

10. $y - 3 = -2x$

11. $y = 2(x + 3)$

12. $y = -(x + 4)$

13. Plot the points $(4, 7)$, $(-1, \frac{3}{2})$, and $(0, 5)$.

14. Plot the points $(-2, 0)$, $(6, -1)$, and $(2, \frac{1}{2})$.

15–24. Graph the equation.

15. $y = x^2$

16. $y = x^2 - 5$

17. $y = 7 - x^2$

18. $y = \sqrt{x}$

19. $y = \sqrt{x + 2}$

20. $y = x^3$

21. $y = \dfrac{x^2}{3}$

22. $y = 3x^2 - 11$

23. $y = \dfrac{x + 1}{2}$

24. $y = 1 + \sqrt{x - 2}$

25–32. Use the method of example 7 to graph the line.

25. $y = 3x - 5$
26. $y = x + 1$
27. $y = -x + 3$
28. $y = -2x - 1$
29. $y = \frac{1}{2}x$
30. $y = \dfrac{x}{2} + 1$
31. $y = \dfrac{x}{4} + 5$
32. $y = 5 + 6x$

33–38. Find *x*- and *y*-intercepts on the graph of the equation.

33. $y = x^2 - 9$
34. $y = 8 - 2x$
35. $y = 5x + 7$
36. $y = 8 - 2x^2$
37. $y = \sqrt{x + 2}$
38. $y = x^2 - 5x$

39–48. Find the distance between the points.

39. $(4, 1)$ and $(2, 3)$
40. $(-3, 4)$ and $(1, -9)$
41. $(0, 4)$ and $(5, 0)$
42. $(6, 11)$ and $(6, 4)$
43. $(2, 9)$ and $(2, 4)$
44. $(-2, -2)$ and $(5, 5)$
45. $(-1, 0)$ and $(c, 2)$
46. $(4, c)$ and $(7, c)$
47. $(7, 3)$ and $(15, -12)$
48. $(2, 3)$ and $(-3, -9)$

49–58. Find the slope of the line through the points.

49. $(1, 4)$ and $(2, 3)$
50. $(-3, 4)$ and $(1, -9)$
51. $(0, 4)$ and $(5, 0)$
52. $(6, 11)$ and $(6, 4)$
53. $(2, 9)$ and $(2, 4)$
54. $(-2, -2)$ and $(5, 5)$
55. $(-1, 0)$ and $(5, 0)$
56. $(4, 9)$ and $(7, 9)$
57. $(7, 3)$ and $(15, 12)$
58. $(2, 3)$ and $(-3, -9)$

Answers should be expressed in slope-intercept form.

59–74. Find the equation of the line with the given information.

59. slope 5, through point $(1, 9)$
60. slope $\frac{1}{2}$, *y*-intercept 11
61. slope -2, *y*-intercept 7
62. through points $(-1, 0)$ and $(11, 9)$
63. through points $(2, 1)$ and $(4, -2)$
64. slope $-\frac{1}{2}$, through point $(4, 4)$
65. horizontal line through $(1, -3)$

66. vertical line through $(-291, 4\pi)$

67. vertical line through $(8, 12)$

68. horizontal line through $(9, 0)$

69. through $(1, 5)$ parallel to $y = -2x + 1$

70. through $(2, 3)$ parallel to $y = 4x + \sqrt{2}$

71. through $(8, 4)$ perpendicular to $y = -\frac{1}{4}x + 27$

72. through $(3, -2)$, negative slope, makes 45° angle with coordinate axes

73. through $(2, 0)$, positive slope, makes 45° angle with coordinate axes

74. through $(-5, 8)$, perpendicular to $y = -3x$

Trigonometry Review

0.3

Certain portions of the study of trigonometry are used repeatedly in calculus. This review focuses on these facts and skills.

Angles

One way of defining an *angle* is to say that it is the union of two rays, as pictured in figure 1. It is common to consider angles in *standard position*, with the *vertex* at the origin and the *initial side* along the positive x-axis (figure 2). The other ray is called the *terminal side* of the angle. Initial and terminal conjure the idea of movement, and we can think of the ray rotating around the vertex from the initial side to the terminal side. If the rotation is counterclockwise, then the angle has a positive measure; if the rotation is clockwise, then the angle has a negative measure. Angles are sometimes labeled near the vertex between the initial and terminal sides.

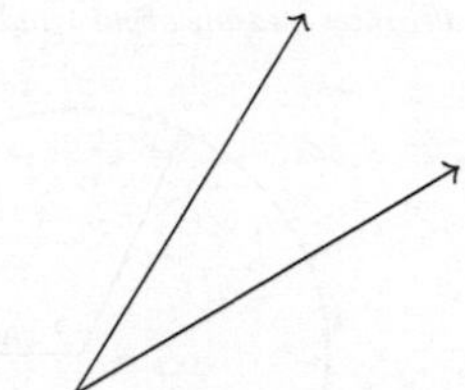

Figure 1 *An angle as the union of two rays*

It is common to use a lowercase Greek letter such as θ (theta) to label an angle.

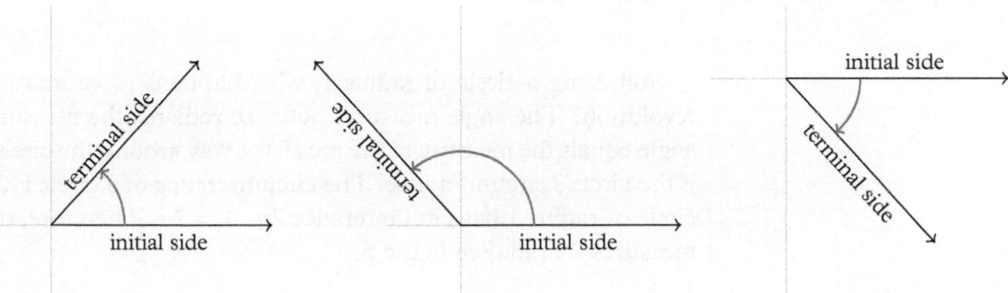

Figure 2 *Angles in standard position. Counterclockwise rotation is positive (left, middle) and clockwise rotation is negative (right)*

Radians

The two common units of measure used to describe the size of an angle are degrees and radians. Although most students of trigonometry find degrees easier to use than radians, many calculations in calculus

A third unit of measure, the *gradian*, is similar to degrees but with a right angle measuring 100 gradians. Gradians see occasional use in certain settings, especially in a few European countries.

The superiority of radians compared to degrees for use in calculus is explained in chapter 2.

are much easier if radians are used instead of degrees for measuring angles.

One *radian* is the size of an angle that subtends an arc of the same length as the radius of the circle. For instance, in a circle of radius 1, an angle of 1 radian (rad) captures an arc of length 1, as in figure 3. The sector pictured in figure 3 looks almost like an equilateral triangle, but with one side curved. Angles in an equilateral triangle measure 60°, but adding a slight curve to the side opposite the angle reduces the size of the angle to a little under 60° (to approximately 57.3°).

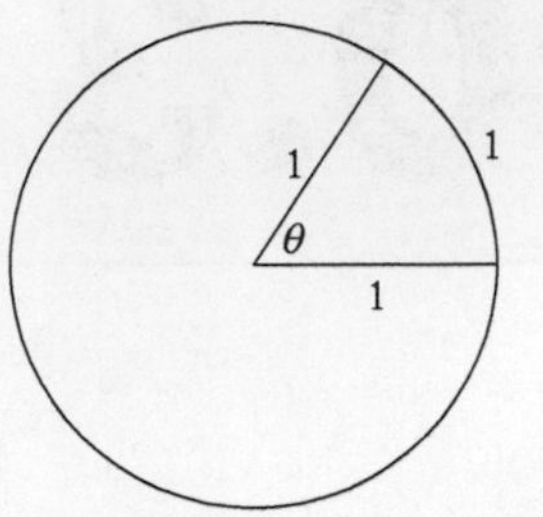

Figure 3 *An angle θ measuring* 1 *rad. If a pizza is cut into six pieces with nearly equal size, then the smaller slices measure about* 1 *rad*

Double the angle, and the arc it subtends is doubled (figure 4, left). Halve the angle, and the arc it subtends is halved (figure 4, middle). In a circle of radius 1, an angle of θ rad subtends an arc of length θ (figure 4, right).

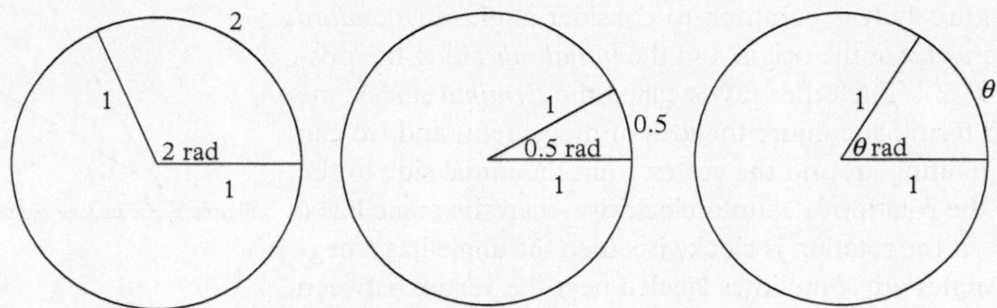

Figure 4 *Angles measuring* 2 *rad (left),* 0.5 *rad (middle), and θ rad (right) in a circle of radius* 1. *The arc subtended by the angle is equal to the measure of the angle in radians*

Still using a circle of radius 1, what happens if we make one full revolution? The angle measures 360°. In radians, the measure of the angle equals the measure of the arc all the way around the circle, which is the circle's circumference. The circumference of a circle is $2\pi r$, so a circle of radius 1 has circumference $2\pi \cdot 1 = 2\pi$. Therefore, the angle measures 2π rad. See figure 5.

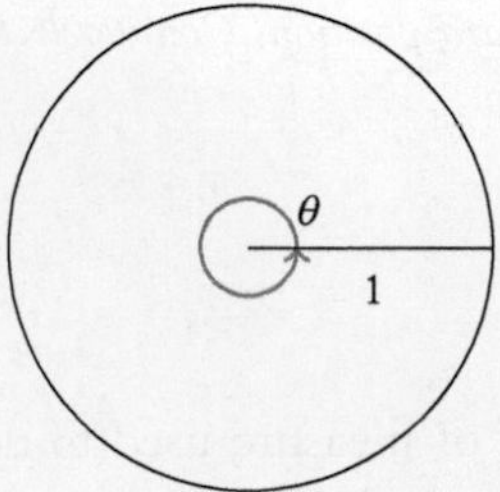

Figure 5 *An angle θ measuring* 360°. *The circumference of the circle is* 2π, *and therefore the angle measures* 2π *rad*

Rearranging the equation $2\pi\,\text{rad} = 360°$ to solve for 1 rad or 1° gives the conversion formulas.

The abbreviation *rad* stands for radians and should be read (pronounced) "radians."

CONVERTING BETWEEN DEGREES AND RADIANS

$$1\,\text{rad} = \left(\frac{180}{\pi}\right)^{\circ}$$

$$1^{\circ} = \frac{\pi}{180}\,\text{rad}$$

It is common to remember the conversions by one formula: $\pi\,\text{rad} = 180°$.

Example 1 *Convert the measurement 135° to radians.*

Solution Using the conversion factor $1° = \frac{\pi}{180}$ rad, we have

$$135 \cdot 1^{\circ} = 135 \cdot \frac{\pi}{180}\,\text{rad}$$

$$135^{\circ} = \frac{135}{180}\pi\,\text{rad} = \frac{3\pi}{4}\,\text{rad.}$$

The measurement 135° is equal to $\frac{3\pi}{4}$ rad. ■

It should be obvious that not having instant recall of arithmetic facts such as $2 + 4 = 6$ is a huge impediment to learning algebra. Imagine simplifying $2x + 4x$ while having to count on one's fingers to reach the result $6x$, counting on fingers *every time* arithmetic is needed. Working algebra problems would be interminably slow! The same is true of the need for instant recall of certain facts in trigonometry. Not having instant recall of facts such as $30° = \frac{\pi}{6}$ rad or $\frac{3\pi}{2}$ rad $= 270°$ can easily turn a 1-hour calculus assignment into a 3-hour assignment. It can also make the difference between finishing an exam with ample time to review work and not having time to attempt all the problems, leaving some blank.

Degrees and radians are considered "dimensionless" units. It is not necessary to keep track of degrees or radians as part of a unit calculation, as long as only degrees or only radians is used.

Among the facts needing instant recall are the following comparisons between degrees and radians, and the resulting angle in standard position (table 1).

Table 1 *Measures of angles in degrees and radians, with angles pictured in standard position. Instant recall of these relationships should be attained.*

degrees	radians	angle in standard position
0°	0	
30°	$\frac{\pi}{6}$	
45°	$\frac{\pi}{4}$	
60°	$\frac{\pi}{3}$	
90°	$\frac{\pi}{2}$	
180°	π	
270°	$\frac{3\pi}{2}$	
360°	2π	

If an angle measure is given without any units listed, it is implied that the angle is measured in radians.

Angles can have measures greater than 2π. If the rotation from the initial side to the terminal side is three full revolutions (three times around the circle), then the angle measures $3 \cdot 2\pi = 6\pi$. Various angles with their rotations marked are illustrated in figure 6.

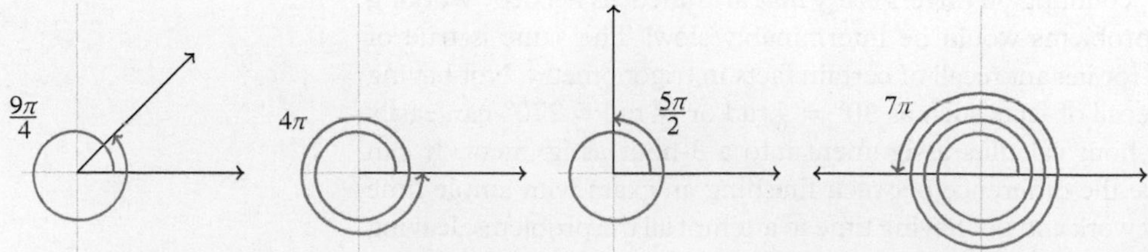

Figure 6 *Various angles with positive measure and more than one revolution. (left) The angle features one complete revolution (2π) and then rotates an additional $\frac{\pi}{4}$, for a total of $2\pi + \frac{1}{4}\pi = (2 + \frac{1}{4})\pi = \frac{9}{4}\pi$. (right) The angle features three complete revolutions ($3 \cdot 2\pi$) plus another half revolution (π) for a total of 7π*

Negative angles feature clockwise rotation. Instant recall of many negative angles is also important. They need not be memorized separately if their relationship to the corresponding positive angle is understood. See figure 7.

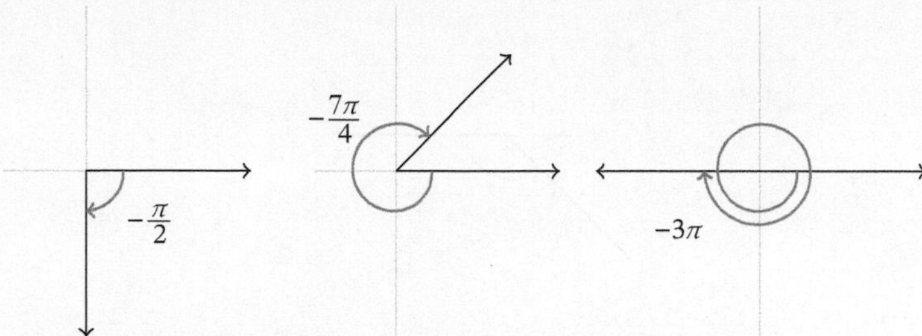

Figure 7 *Various angles with negative measures*

Example 2 *Draw the angles in standard position: (a) $\frac{7\pi}{2}$ rad and (b) $-\frac{3}{4}\pi$ rad.*

Solution (a) Notice that π rad is a half revolution. Since $\frac{7}{2} = 3.5$, we need 3.5 half-revolutions. The angle is positive, so the rotation is counterclockwise. Counting the half-revolutions is illustrated in figure 8.

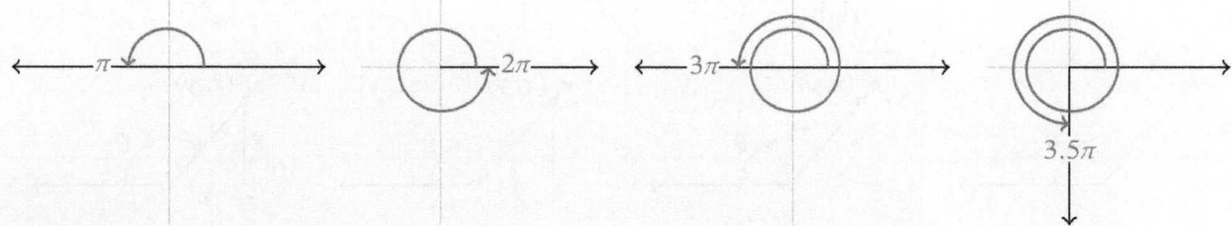

Figure 8 *Counting 3.5 half-revolutions*

The solution is in figure 9.

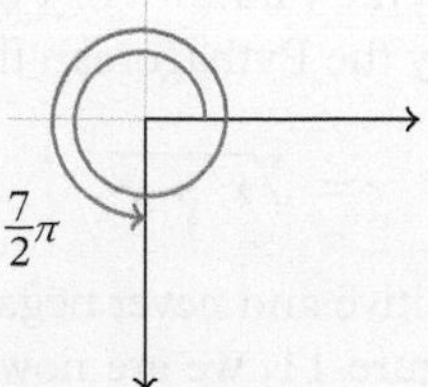

Figure 9 *The angle $\frac{7}{2}\pi$ rad of example 2(a)*

(b) Notice that $\frac{3}{4} = \frac{1}{2} + \frac{1}{4}$, so we rotate 90° plus another 45°. Because the angle is negative, the rotation is clockwise. The result is in figure 10.

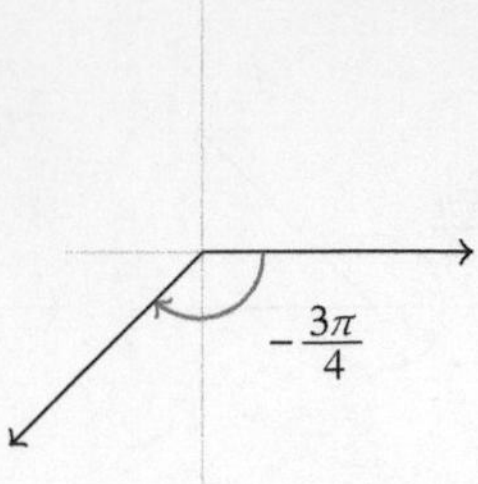

Figure 10 *The angle $-\frac{3}{4}\pi$ rad of example 2(b)* ■

Trigonometric functions and their values

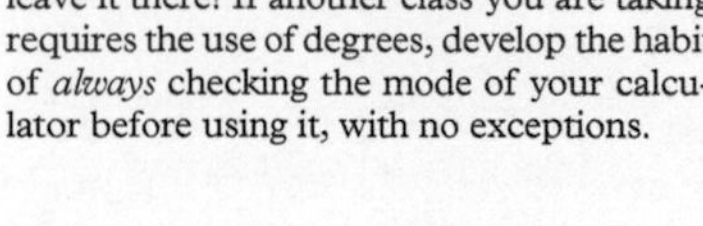

Tip: put your calculator in radian mode and leave it there! If another class you are taking requires the use of degrees, develop the habit of *always* checking the mode of your calculator before using it, with no exceptions.

Among the methods of defining the trigonometric functions is the following. Begin with an angle θ in standard position, measured in radians. Choose a point (x, y) on the terminal side of the angle and drop a perpendicular from the point (x, y) to the x-axis, forming a right triangle. Label the horizontal leg of the triangle x, the vertical leg y, and the hypotenuse r. These steps are illustrated in figure 11, from left to right.

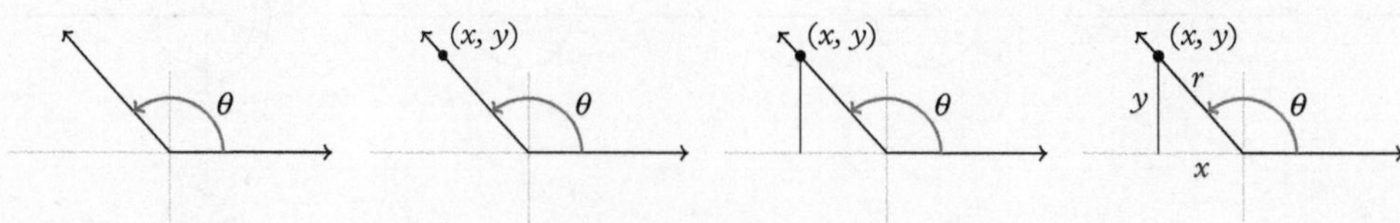

Figure 11 *Steps in the development of the trig functions. As pictured, x is negative and y is positive*

Notice that by labeling the legs of the triangle x and y, the numbers are allowed to be negative. In fact, if the terminal side of the triangle lies along a coordinate axis, the value of x or y can also be zero. Because we have a right triangle, by the Pythagorean theorem,

If the value of either x or y is zero, there is no triangle, but we may still use the same formulas for r and for the trigonometric functions.

$$r = \sqrt{x^2 + y^2}.$$

The abbreviation *trig* is often used instead of the word *trigonometric*.

Notice that r is always positive and never negative. Using the diagram on the far right side of figure 11, we are now ready to define the six trigonometric functions, which represent the six possible ratios of sides of the triangle.

Definition 5 TRIGONOMETRIC FUNCTIONS

Let (x, y) be a point on the terminal side of an angle θ (measured in radians) in standard position and let $r = \sqrt{x^2 + y^2}$. Then,

$$\sin\theta = \frac{y}{r} \qquad \csc\theta = \frac{r}{y}$$

$$\cos\theta = \frac{x}{r} \qquad \sec\theta = \frac{r}{x}$$

$$\tan\theta = \frac{y}{x} \qquad \cot\theta = \frac{x}{y}.$$

The names of the trigonometric functions are sine, cosine, tangent, cotangent, secant, and cosecant, with emphasis always on the first syllable. The abbreviations are pronounced by their function names. For instance, "sin θ" is pronounced "sine theta" and "csc θ" is pronounced "cosecant theta."

As an aid to memorizing the six trig ratios, notice that each row in definition 5 contains one "co-" function (top row, cosecant; second row, cosine; third row, cotangent). The two ratios in each row are reciprocals. Whether the ratios are memorized by their letters (x, y, r) or by their descriptions (adjacent, opposite, hypotenuse; see figure 12) is a matter of personal preference.

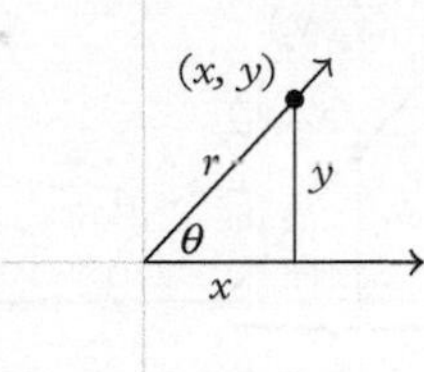

Figure 12 *A first-quadrant angle θ. The side x is "adjacent" to θ and the side y is "opposite" θ. The trig functions are therefore sometimes remembered by the mnemonic SOHCAHTOA (soh·cah· to'·a), which stands for Sin is Opposite over Hypotenuse, Cosine is Adjacent over Hypotenuse, Tangent is Opposite over Adjacent*

Example 3 *A point on the terminal side of the angle θ (in standard position) is $(5, -12)$. Find the values of the six trigonometric functions.*

Solution We use $x = 5$, $y = -12$, and

$$\begin{aligned} r &= \sqrt{x^2 + y^2} \\ &= \sqrt{5^2 + (-12)^2} \\ &= \sqrt{25 + 144} \\ &= \sqrt{169} = 13. \end{aligned}$$

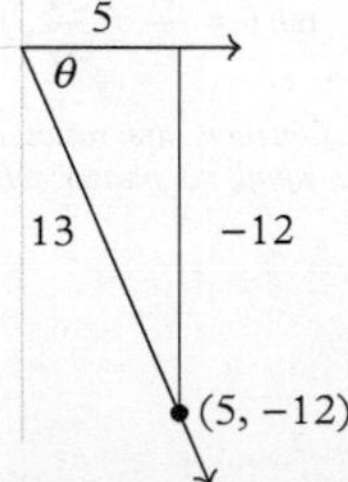

Figure 13 *The angle θ of example 3. Even though we use the phrase "drop a perpendicular to the x-axis", when the point is below the x-axis the segment extends vertically upward to the x-axis*

It is helpful to draw the triangle and label its sides, so that a visual connection is made between the trig functions and their formulas and values (figure 13).

The values of the six trig functions are

$$\sin\theta = \frac{-12}{13} \qquad \csc\theta = \frac{13}{-12}$$

$$\cos\theta = \frac{5}{13} \qquad \sec\theta = \frac{13}{5}$$

$$\tan\theta = \frac{-12}{5} \qquad \cot\theta = \frac{5}{-12}.$$

■

Developing the habit of writing the trig function values in three rows and two columns as presented in example 3 is helpful. When the ratios in the first column are applied, the values in the second column are written quickly as the reciprocals of those in the first column.

No matter which point on the terminal side of the angle is used (figure 14), the values of the trig functions are the same. This is important, for we would not want two different values for $\sin\theta$, as that would make it fail to be a function. Because the results are the same no matter which point is chosen, we say that the trig functions are *well-defined.*

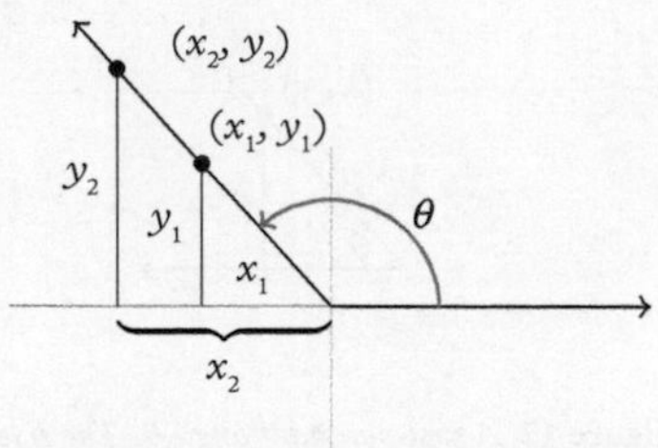

Figure 14 *If (x_1, y_1) and (x_2, y_2) are two different points on the terminal side of the angle θ and we drop perpendiculars to the x-axis, then similar triangles are formed (same angle at vertex, each triangle has a right angle, third angle matches by subtraction from 180°). Therefore, the ratios of sides are equal. For instance,*

$$\tan\theta = \frac{y_1}{x_1} = \frac{y_2}{x_2}.$$

All the trig functions are ratios of sides and therefore are equal no matter which point is chosen

Two "special triangles" from geometry are helpful for quickly determining the values of trig functions: the 45°–45°–90° triangle and the 30°–60°–90° triangle. With angles labeled in radians, these triangles are pictured in figure 15. To use these triangles effectively, they must be on instant recall.

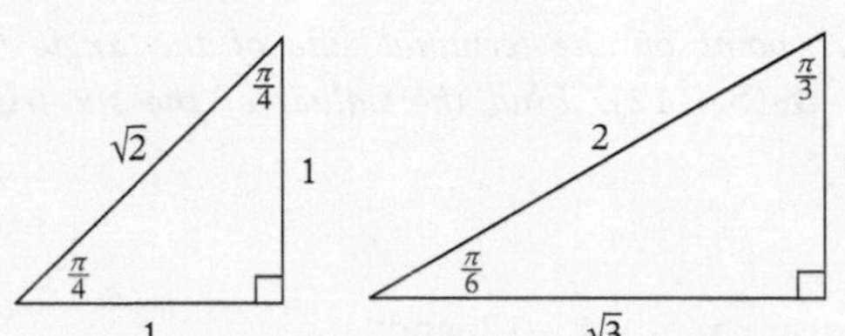

Figure 15 *Two special triangles, with angles labeled in radians. Which side goes opposite which angle can be remembered by recalling that the shortest side is opposite the smallest angle and the longest side is opposite the largest angle. The triangle on the left is sometimes referred to as an isosceles right triangle*

Example 4 *Determine the values of (a)* $\sin\frac{\pi}{4}$, *(b)* $\tan\frac{\pi}{4}$, *and (c)* $\sec\frac{\pi}{4}$.

Solution Referring to the isosceles right triangle on the left side of figure 15 (redrawing the triangle as a visual aid), (a) sine is opposite over hypotenuse:

$$\sin\frac{\pi}{4} = \frac{1}{\sqrt{2}}.$$

If desired, the denominator may be rationalized by multiplying the numerator and the denominator by $\sqrt{2}$:

$$\sin\frac{\pi}{4} = \frac{1}{\sqrt{2}} \cdot \frac{\sqrt{2}}{\sqrt{2}} = \frac{\sqrt{2}}{2}.$$

Definition 5 defines the sine of an angle. When a number such as $\frac{\pi}{4}$ is given as the *argument* to sine, we mean the sine of an angle in standard position with a measure that is $\frac{\pi}{4}$ (the given number).

(b) Tangent is opposite over adjacent:

$$\tan\frac{\pi}{4} = \frac{1}{1} = 1.$$

(c) Secant is hypotenuse over adjacent:

$$\sec\frac{\pi}{4} = \frac{\sqrt{2}}{1} = \sqrt{2}.$$

■

By definition, the secant is the reciprocal of cosine. Cosine is adjacent over hypotenuse, so secant is its reciprocal, hypotenuse over adjacent.

Instant recall of facts such as $\tan\frac{\pi}{4} = 1$ is even quicker than consulting the special triangle; but, if the special triangles are on instant recall, it is still easy to come up with these values relatively quickly. Relying on a calculator to determine common values of trig functions is slower; misses exact values such as $\frac{\sqrt{2}}{2}$, which leads to simplifications in some problems; and sometimes reduces the making of connections that help with the learning process.

Example 5 *Determine the values of (a)* $\cos\frac{\pi}{6}$ *and (b)* $\tan\frac{\pi}{3}$.

Solution The angles $\frac{\pi}{6}$ and $\frac{\pi}{3}$ are in one of the special triangles (figure 15, right). Redrawing the triangle as a visual aid (every time!) helps; see figure 16.

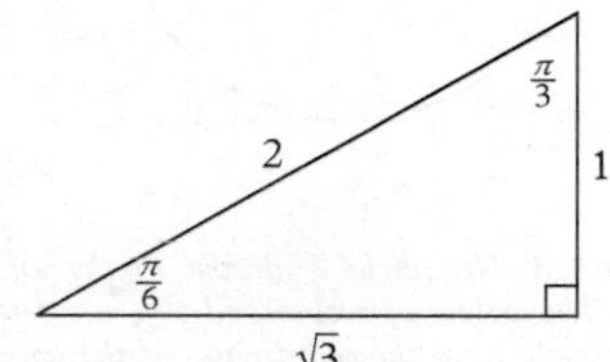

Figure 16 *Special triangle, redrawn*

(a) Cosine is adjacent over hypotenuse. Referring to the angle $\frac{\pi}{6}$, the adjacent side has length $\sqrt{3}$ whereas the hypotenuse has length 2:

$$\cos\frac{\pi}{6} = \frac{\sqrt{3}}{2}.$$

(b) Tangent is opposite over adjacent. Referring to the angle $\frac{\pi}{3}$, the opposite side has length $\sqrt{3}$ and the adjacent side has length 1:

$$\tan\frac{\pi}{3} = \frac{\sqrt{3}}{1} = \sqrt{3}.$$

■

The *unit circle* is the circle of radius 1 centered at the origin. If the point (x, y) on the terminal side of the angle θ lies on the unit circle, then $r = 1$. Then, $\sin\theta = \frac{y}{r} = \frac{y}{1} = y$ and $\cos\theta = \frac{x}{r} = \frac{x}{1} = x$. Therefore, $(x, y) = (\cos\theta, \sin\theta)$ (figure 17).This fact can be used to help find

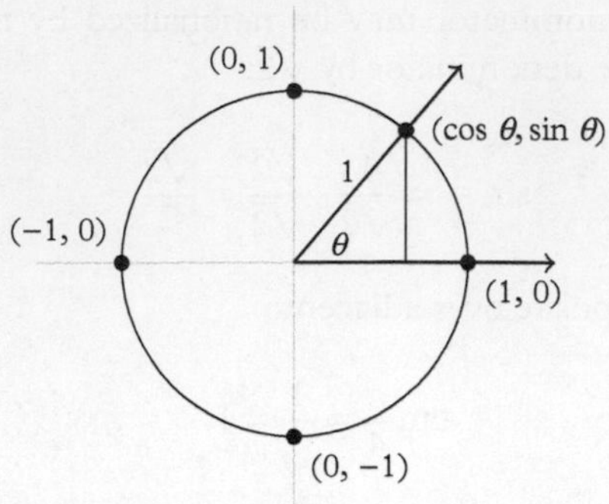

Figure 17 *The coordinates of a point on the unit circle are* $(\cos\theta, \sin\theta)$

values of trig functions, especially when the terminal side of the angle lies on a coordinate axis.

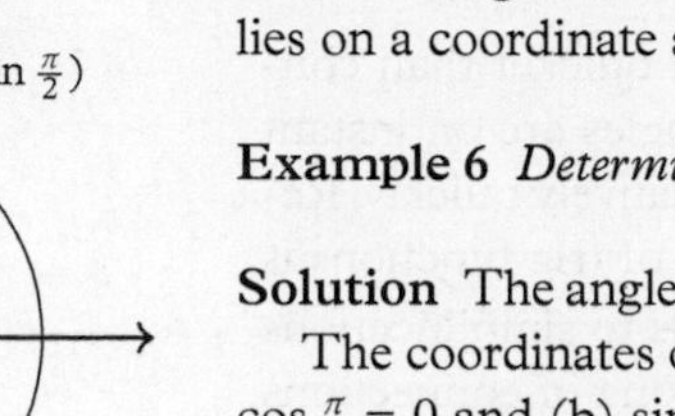

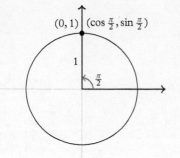

Figure 18 *The angle* $\frac{\pi}{2}$ *drawn in the unit circle. The values of cosine and sine are seen immediately from the coordinates of the point where the terminal side of the angle meets the unit circle*

Example 6 *Determine the values of (a)* $\cos\frac{\pi}{2}$*, (b)* $\sin\frac{\pi}{2}$*, and (c)* $\tan\frac{\pi}{2}$.

Solution The angle is $\frac{\pi}{2}$. Draw that angle in the unit circle (figure 18).

The coordinates of the point $(\cos\frac{\pi}{2}, \sin\frac{\pi}{2})$ are $(0, 1)$. Therefore (a) $\cos\frac{\pi}{2} = 0$ and (b) $\sin\frac{\pi}{2} = 1$.

(c) Reverting back to the original definition,

$$\tan\frac{\pi}{2} = \frac{y}{x} = \frac{1}{0}, \text{ undefined.}$$ ■

If the angle $\frac{\pi}{2}$ is on instant recall, then so should the point $(0, 1)$ on the unit circle, and nearly immediately we have the values of $\cos\frac{\pi}{2}$ and $\sin\frac{\pi}{2}$. The values of other trig functions are a little less immediate.

Example 7 *Determine the values of (a)* $\sin\frac{3\pi}{2}$*, (b)* $\cos\frac{3\pi}{2}$*, (c)* $\cos\pi$*, and (d)* $\sin\pi$.

Solution For (a) and (b) the angle is $\frac{3\pi}{2}$. Draw the angle in the unit circle (figure 19, left). Then, (a) using the y-coordinate, $\sin\frac{3\pi}{2} = -1$ and (b) using the x-coordinate, $\cos\frac{3\pi}{2} = 0$.

For (c) and (d), the angle is π. Draw the angle in the unit circle (figure 19, right). Then, (c) using the x-coordinate, $\cos\pi = -1$ and (d) using the y-coordinate, $\sin\pi = 0$. ■

When instant recall of table 1 is achieved, the solution to example 7 can be completed in mere seconds.

Example 8 *Determine the values of (a)* $\sin\frac{2\pi}{3}$ *and (b)* $\sec\frac{2\pi}{3}$.

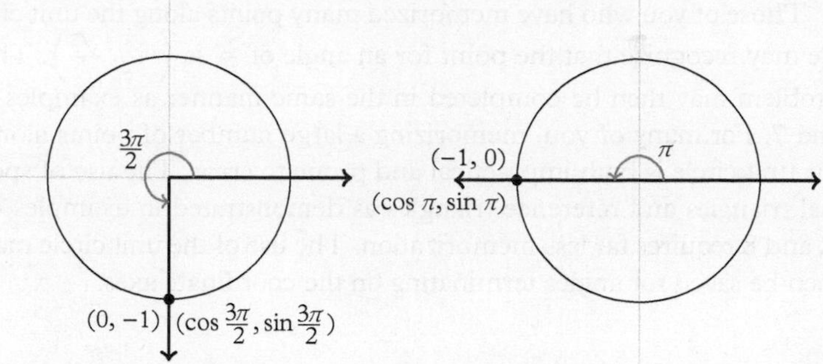

Figure 19 *The angles $\frac{3\pi}{2}$ (left) and π (right) drawn in the unit circle*

Solution We begin by drawing the angle $\frac{2\pi}{3}$. Because $\frac{2}{3}$ is between $\frac{1}{2}$ and 1, the angle $\frac{2\pi}{3}$ is between the angles $\frac{\pi}{2}$ (up) and π (left) (figure 20, left). Choose a point on the terminal side of the angle and drop a perpendicular to the x-axis, forming a right triangle, which is called the *reference triangle.* The angle at the origin inside the triangle is not $\frac{2\pi}{3}$, but rather $\pi - \frac{2\pi}{3} = \frac{\pi}{3}$. We label this angle in the diagram (figure 20, middle). At this point we recognize a special triangle and label the sides of the triangle accordingly (figure 20, right).

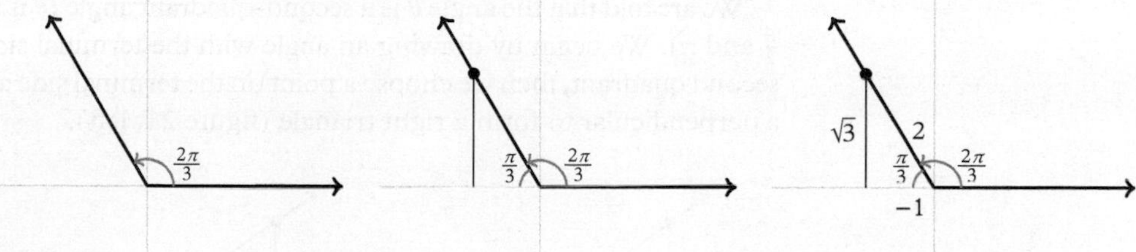

Figure 20 *The angle $\frac{2\pi}{3}$ (left), with its reference triangle (middle), labeled using values from a special triangle (right)*

The sides are to be labeled with the x- and y-coordinates of a point on the terminal side of the angle. Because our point is in the second quadrant, the value of x must be negative. That side of the triangle is labeled −1 (instead of 1).

When the triangle has been labeled, we are ready to determine the values of the trig functions. (a) Using $y = \sqrt{3}$ and $r = 2$, $\sin \frac{2\pi}{3} = \frac{\sqrt{3}}{2}$. (b) Using $x = -1$ and $r = 2$, $\sec \frac{2\pi}{3} = \frac{2}{-1} = -2$. ■

Those of you who have memorized many points along the unit circle may recognize that the point for an angle of $\frac{2\pi}{3}$ is $\left(-\frac{1}{2}, \frac{\sqrt{3}}{2}\right)$. The problem may then be completed in the same manner as examples 6 and 7. For many of you, memorizing a large number of points along the unit circle is both impractical and prone to error. The use of special triangles and reference triangles as demonstrated in examples 4, 5, and 8 requires far less memorization. The use of the unit circle may then be saved for angles terminating on the coordinate axes.

Finding additional trig values

Given the value of one trig function of an angle and the quadrant in which the angle lies, the values of the other trig functions can be determined as well.

Based on where the values of x and y are positive, we can determine in which quadrants the values of the trig functions are positive:

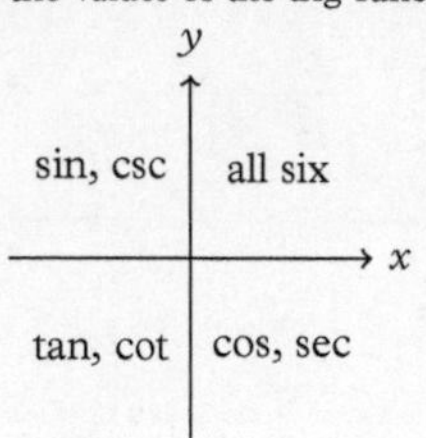

Example 9 *Given* $\sin\theta = \frac{4}{7}$ *and* $\frac{\pi}{2} < \theta < \pi$*, find the other five trig functions of* θ*.*

Solution There are always two quadrants in which a given trig function has positive values. Knowing that $\sin\theta$ is a positive number is not enough to know in which quadrant the terminal side of θ lies.

We are told that the angle θ is a second-quadrant angle (θ is between $\frac{\pi}{2}$ and π). We begin by drawing an angle with the terminal side in the second quadrant, then we choose a point on the terminal side and drop a perpendicular to form a right triangle (figure 21, left).

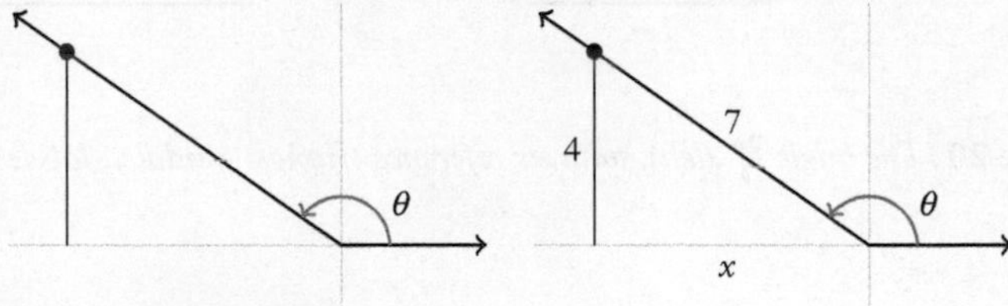

Figure 21 *Steps in drawing the diagram for example 9*

Since $\sin\theta = \frac{y}{r} = \frac{4}{7}$, we label the vertical leg of the triangle 4 and the hypotenuse 7 (figure 21, right). The label for the horizontal leg, x, should be negative because we are in the second quadrant.

The value of x may be found using the Pythagorean theorem:

$$\begin{aligned} x^2 + 4^2 &= 7^2 \\ x^2 + 16 &= 49 \\ x^2 &= 33 \\ x &= -\sqrt{33}. \end{aligned}$$

Pythagoras of Samos, 569–475 BC http://www-history.mcs.st-andrews.ac.uk/Biographies/Pythagoras.html

The solutions to the equation $x^2 = 33$ are $x = \pm\sqrt{33}$. We know that we need the negative solution, so we discard the positive solution.

We finish the diagram by labeling the horizontal leg (figure 22).

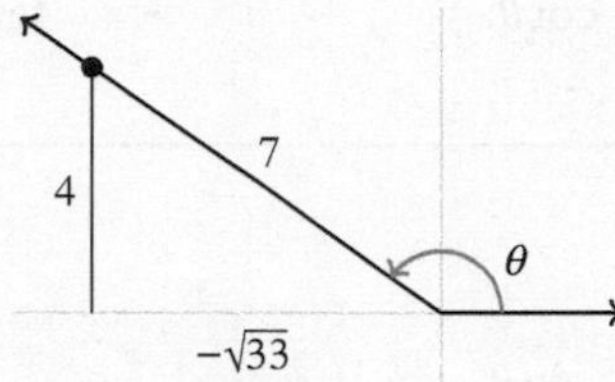

Figure 22 *The completed figure for example 9*

We are now ready to determine the values of the other five trig functions using the definition:

$$\csc\theta = \frac{7}{4}$$

$$\cos\theta = \frac{-\sqrt{33}}{7} \qquad \sec\theta = \frac{7}{-\sqrt{33}}$$

$$\tan\theta = \frac{4}{-\sqrt{33}} \qquad \cot\theta = \frac{-\sqrt{33}}{4}.$$

■

Useful trig identities

There are a very large number of *trigonometric identities*, equations that relate trig functions to one another and are true for all values for which the functions are defined. Some identities are much more commonly used than others, and the ability to recognize the applicability of an identity readily can be crucial to completing some calculus exercises. The identities described next are the ones that all calculus students should know.

The first set of identities includes the *reciprocal identities*. Since $\sin\theta = \frac{y}{r}$ and $\csc\theta = \frac{r}{y}$, the two trig functions are reciprocals of one another.

Most calculators have buttons for sine, cosine, and tangent only. To find csc 3.79, we ask the calculator for $\frac{1}{\sin 3.79}$.

RECIPROCAL IDENTITIES

For any angle θ for which the functions are defined,

$$\sin\theta = \frac{1}{\csc\theta} \qquad \csc\theta = \frac{1}{\sin\theta}$$
$$\cos\theta = \frac{1}{\sec\theta} \qquad \sec\theta = \frac{1}{\cos\theta}$$
$$\tan\theta = \frac{1}{\cot\theta} \qquad \cot\theta = \frac{1}{\tan\theta}.$$

CAUTION: MISTAKE TO AVOID
Do not use a negative exponent to express the reciprocal of a trig function, because what appears to be an exponent of -1 is reserved for a different type of function: an *inverse trigonometric* function. In other words,

$$\sin^{-1} x \neq \frac{1}{\sin x} = \csc x.$$

Notice that

$$\frac{\sin\theta}{\cos\theta} = \frac{\frac{y}{r}}{\frac{x}{r}} = \frac{y}{x} = \tan\theta.$$

This is called a *ratio identity.*

RATIO IDENTITIES

For any angle θ for which the functions are defined,

$$\tan\theta = \frac{\sin\theta}{\cos\theta} \qquad \cot\theta = \frac{\cos\theta}{\sin\theta}.$$

The reciprocal identities for secant and cosecant along with the ratio identities for tangent and cotangent give us the ability to write every trigonometric function in terms of sines and cosines. This ability can be very helpful in certain calculus exercises.

Example 10 *Rewrite the expression* $\tan\theta\csc\theta + 2\sec\theta$ *in terms of sines and cosines.*

Solution Using the reciprocal and ratio identities,

$$\begin{aligned}\tan\theta\csc\theta + 2\sec\theta &= \frac{\sin\theta}{\cos\theta}\cdot\frac{1}{\sin\theta} + 2\cdot\frac{1}{\cos\theta}\\ &= \frac{1}{\cos\theta} + \frac{2}{\cos\theta}\\ &= \frac{3}{\cos\theta}.\end{aligned}$$

■

Another often-used identity is proved as follows, remembering that $r = \sqrt{x^2 + y^2}$ and therefore $r^2 = x^2 + y^2$:

The expression $\sin^2\theta$ is a shortcut for writing $(\sin\theta)^2$.

$$\sin^2\theta + \cos^2\theta = \frac{y^2}{r^2} + \frac{x^2}{r^2} = \frac{y^2 + x^2}{r^2} = \frac{r^2}{r^2} = 1.$$

Because the Pythagorean theorem is inherent in $x^2 + y^2 = r^2$ (consult figure 11), it is called a *Pythagorean identity.*

PYTHAGOREAN IDENTITIES

For any angle θ for which the functions are defined,

$$\sin^2\theta + \cos^2\theta = 1,$$

$$1 + \cot^2\theta = \csc^2\theta, \text{ and}$$

$$\tan^2\theta + 1 = \sec^2\theta.$$

The second and third Pythagorean identities can be derived easily from the first by division and use of the reciprocal and ratio identities. For instance, divide both sides by $\cos^2\theta$:

$$\frac{\sin^2\theta + \cos^2\theta}{\cos^2\theta} = \frac{1}{\cos^2\theta}$$

$$\frac{\sin^2\theta}{\cos^2\theta} + \frac{\cos^2\theta}{\cos^2\theta} = \left(\frac{1}{\cos\theta}\right)^2$$

$$\tan^2\theta + 1 = \sec^2\theta.$$

An *even function* is a function with a graph that is symmetric about the x-axis, such as the graph of $y = x^2$ or the graph of $y = x^4$. An *odd function* is a function with a graph that is symmetric about the origin, such as the graph of $y = x^3$ or $y = x^5$. As we shall see shortly, the graph of $y = \cos x$ is symmetric about the x-axis because $\cos(-\theta) = \cos\theta$, and the graph of $y = \sin x$ is symmetric about the origin because $\sin(-\theta) = -\sin\theta$. These are called *odd-even identities* (see figure 23).

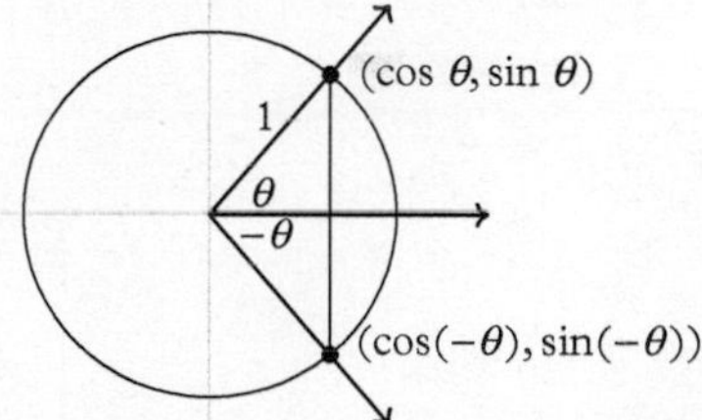

Figure 23 *The odd-even identities illustrated. The two triangles are congruent. Therefore, the x-coordinates of the two points are the same and we conclude* $\cos(-\theta) = \cos\theta$. *The y-coordinates of the two points are negatives of one another, hence* $\sin(-\theta) = -\sin\theta$

ODD-EVEN IDENTITIES

For any angle θ,

$$\sin(-\theta) = -\sin\theta \text{ and}$$

$$\cos(-\theta) = \cos\theta.$$

More than one angle can have the same terminal side (figure 24). Because the values of the trigonometric functions are determined from the terminal side of the angle, angles with the same terminal side have the same trig values. Because adding 2π to the measure of an angle adds one full revolution to the angle and brings the terminal side to the same location, the trig values repeat every 2π.

We say that $y = \sin\theta$ and $y = \cos\theta$ have *period* 2π.

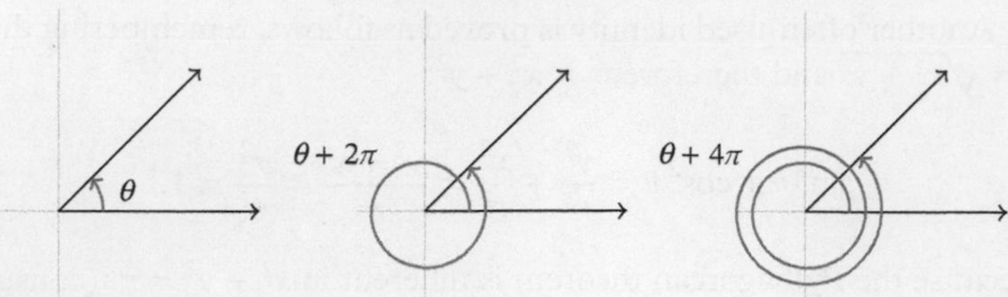

Figure 24 *Three angles with the same terminal side and therefore the same trig values*

PERIODICITY IDENTITIES

For any angle θ,

$$\sin(\theta + 2\pi) = \sin\theta \text{ and}$$
$$\cos(\theta + 2\pi) = \cos\theta.$$

Other identities that eventually need to be on instant recall, although not in the first few chapters, include a double-angle formula and two half-angle formulas. Their proofs are omitted.

DOUBLE-ANGLE IDENTITY

For any angle θ,

$$\sin 2\theta = 2\sin\theta\cos\theta.$$

HALF-ANGLE IDENTITIES

For any angle θ,

$$\cos^2\theta = \frac{1}{2}(1 + \cos 2\theta) \text{ and}$$
$$\sin^2\theta = \frac{1}{2}(1 - \cos 2\theta).$$

Other trigonometric identities may see occasional or one time use in calculus and can be brought up when needed.

Basic trig graphs

Points on the unit circle have coordinates $(\cos\theta, \sin\theta)$, as in figure 25 (right). Points on the graphs of $y = \sin\theta$ and $y = \cos\theta$ have the form $(\theta, \sin\theta)$ and $(\theta, \cos\theta)$, respectively (figure 25, left).

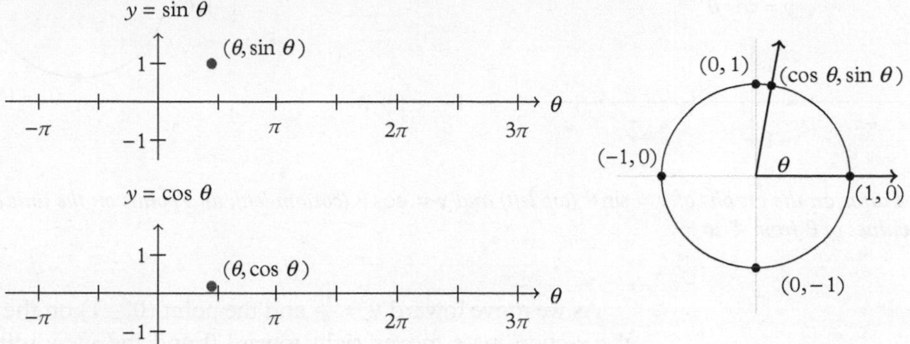

Figure 25 *Form of points on the graphs of* $y = \sin\theta$ *(top left) and* $y = \cos\theta$ *(bottom left), and points on the unit circle (right)*

Tracing the points on the unit circle and the graphs as we allow θ to vary helps us see the relationship between the unit circle, values of $\sin\theta$ and $\cos\theta$, and their graphs. Start with $\theta = 0$. The coordinates on the unit circle are $(1, 0)$; therefore, $\cos 0 = 1$ and $\sin 0 = 0$. As θ gets larger, the point on the unit circle moves left and upward, meaning that the x-coordinate ($\cos\theta$) decreases and the y-coordinate ($\sin\theta$) increases. Eventually, as θ reaches $\frac{\pi}{2}$, the point on the unit circle reaches $(0, 1)$, meaning $\cos\frac{\pi}{2} = 0$ and $\sin\frac{\pi}{2} = 1$. This is illustrated in figure 26.

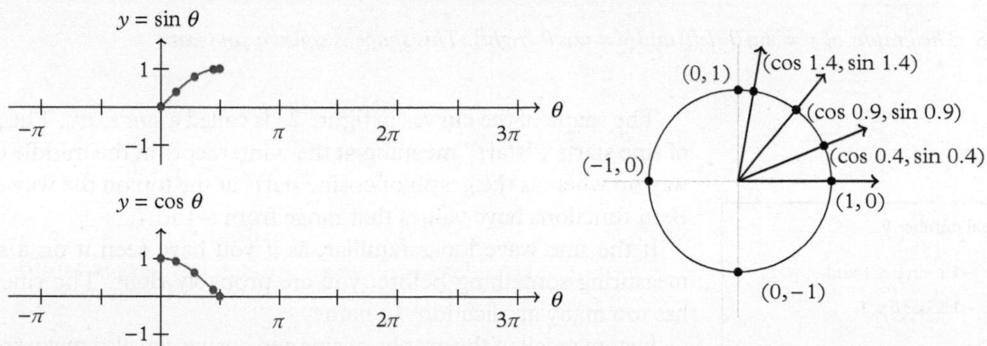

Figure 26 *Points on the graphs of* $y = \sin\theta$ *(top left) and* $y = \cos\theta$ *(bottom left), and points on the unit circle (right), for various values of* θ *from* 0 *to* $\frac{\pi}{2}$

Continuing toward $\theta = \pi$ moves the point on the unit circle left and downward; therefore, $\cos\theta$ is becoming more negative and $\sin\theta$ is decreasing, eventually reaching $(-1, 0)$, meaning that $\cos\pi = -1$ and $\sin\pi = 0$. See figure 27.

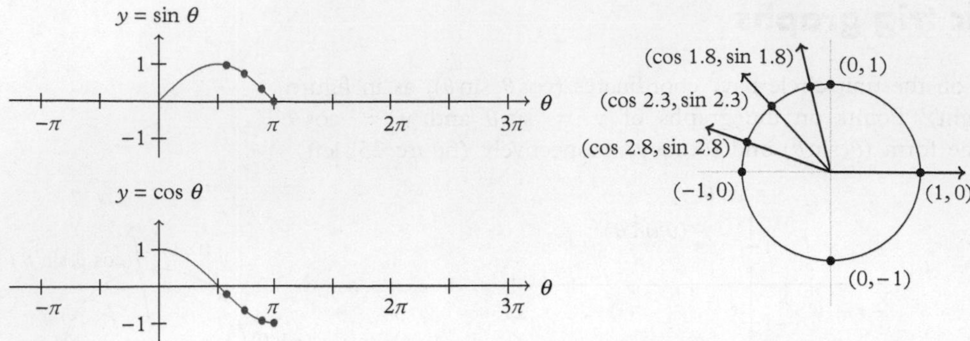

Figure 27 *Points on the graphs of $y = \sin\theta$ (top left) and $y = \cos\theta$ (bottom left), and points on the unit circle (right), for various values of θ from $\frac{\pi}{2}$ to π*

As we move toward $\theta = \frac{3\pi}{2}$ and the point $(0, -1)$ on the unit circle, the x-coordinate moves right toward 0 and the y-coordinate moves more negative toward -1. Then, as θ moves toward 2π and the point $(1, 0)$ on the unit circle, the x-coordinate continues moving to the right toward 1 and the y-coordinate starts moving upward toward 0. Then the entire process repeats, over and over, ad infinitum, every length of 2π. We may move clockwise into negative values of θ, with the pattern still repeating. The final results are in figure 28.

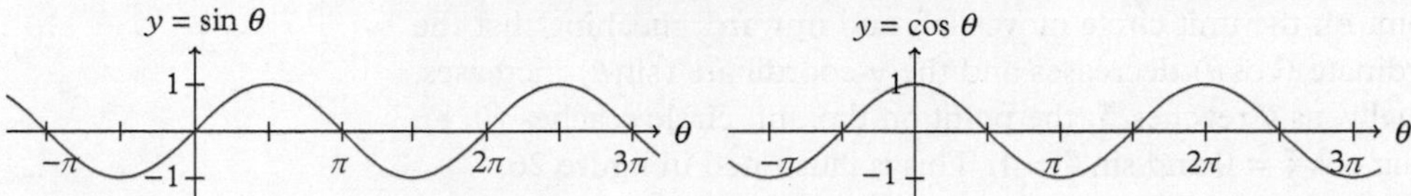

Figure 28 *The graphs of $y = \sin\theta$ (left) and $y = \cos 0$ (right). This shape is called a sine wave*

The shape of the curves in figure 28 is called a *sine wave*. The graph of sine starts ("start" meaning at the y-intercept) in the middle on the way up whereas the graph of cosine starts at the top on the way down. Both functions have values that range from -1 to 1.

For any real number θ,

$-1 \leq \sin\theta \leq 1$ and

$-1 \leq \cos\theta \leq 1$.

If the sine wave looks familiar, as if you have seen it on a screen measuring something before, you are probably right. The sine wave has too many applications to name.

Instant recall of the graphs of sine and cosine can also make recall of the values of sine and cosine very quick. If you need the value of $\sin\pi$, just read the value of the y-coordinate, 0, from the picture generated in your mind.

Example 11 *What are the x-intercepts on the graph of* $y = \sin x$*?*

Solution Checking the graph of the sine function (figure 28, left), we see that the x-intercepts are at $x = 0$, $x = \pi$, $x = 2\pi$, …, as well as $x = -\pi$, $x = -2\pi$, …. This can be written compactly as

$$x = k\pi \text{ for any integer } k. \qquad \blacksquare$$

An even more compact form of the solution is

$$x = k\pi \text{ for all } k \in \mathbf{Z}.$$

Example 12 *Solve the equation* $\cos x = 0$.

Solution This is the same as asking for the x-intercepts of $y = \cos x$. Checking the graph of the cosine function (figure 28, right), we see that the solutions are

$$x = \frac{\pi}{2} + k\pi \text{ for any integer } k. \qquad \blacksquare$$

The graphs of the other four trig functions can be generated from the reciprocal and ratio identities. For instance, $y = \csc\theta$ is the same as $y = \frac{1}{\sin\theta}$, so the y-coordinates on the graph of $y = \csc\theta$ are the reciprocals of the y-coordinates on the graph of $y = \sin\theta$. Where $\sin\theta = 1$, $\csc\theta = \frac{1}{1} = 1$. Where $\sin\theta = 0$, $\csc\theta = \frac{1}{0}$, which is undefined. There are vertical asymptotes on the graph of $y = \csc\theta$ where there are x-intercepts on the graph of $y = \sin\theta$. See figure 29, where the variable is named x instead of θ. The period of $y = \csc\theta$ is 2π.

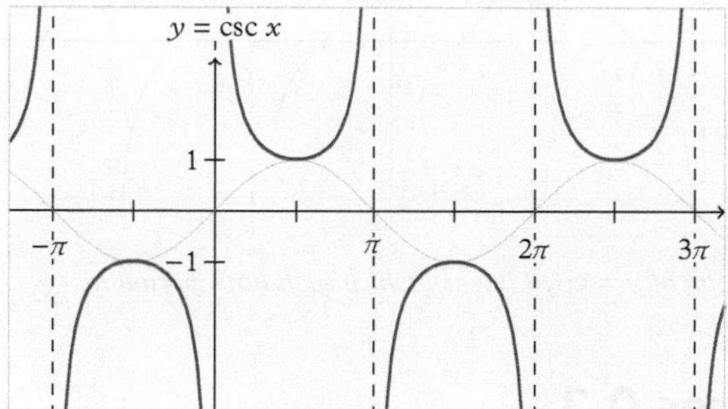

Figure 29 *The graph of* $y = \csc x$ *(blue) features vertical asymptotes where the graph of* $y = \sin x$ *(light gray) has x-intercepts. The pieces of the cosecant curve are not parabolas*

Just as the graph of $y = \csc\theta$ can be remembered from its relationship to the graph of $y = \sin\theta$, the exact same relationship holds comparing the graph of $y = \sec\theta$ to that of $y = \cos\theta$ (figure 30). Its period is also 2π.

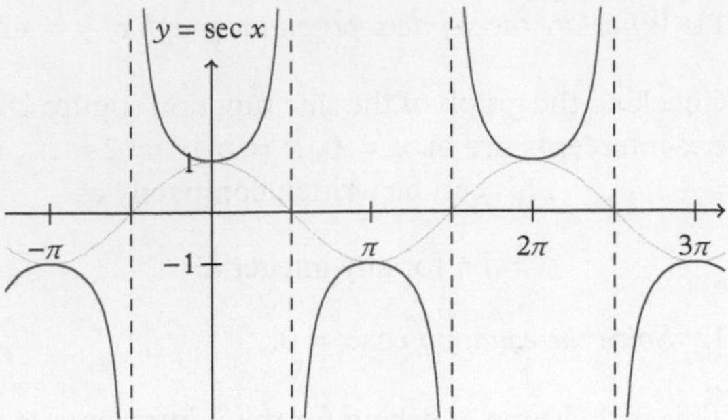

Figure 30 *The graph of* $y = \sec x$ *(blue) features vertical asymptotes where the graph of* $y = \cos x$ *(light gray) has x-intercepts. The pieces of the secant curve are not parabolas*

Because $\tan\theta = \frac{\sin\theta}{\cos\theta}$ and $\sec\theta = \frac{1}{\cos\theta}$, the two functions have the same zero denominators and therefore the same vertical asymptotes. The same is true of cotangent and cosecant. The graphs of tangent and cotangent are in figure 31, from which it is apparent that the values of $y = \tan\theta$ and $y = \cot\theta$ repeat with a period of π.

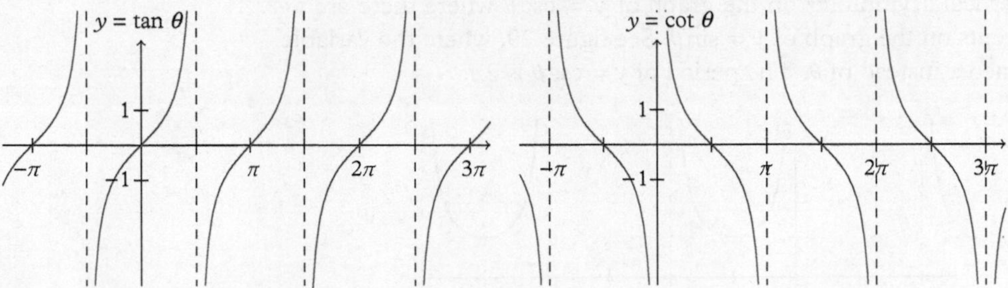

Figure 31 *The graph of* $y = \tan\theta$ *(left) and the graph of* $y = \cot\theta$ *(right), which each have period* π

Exercises 0.3

1–10. Rapid response: give the measure of the angle in radians.

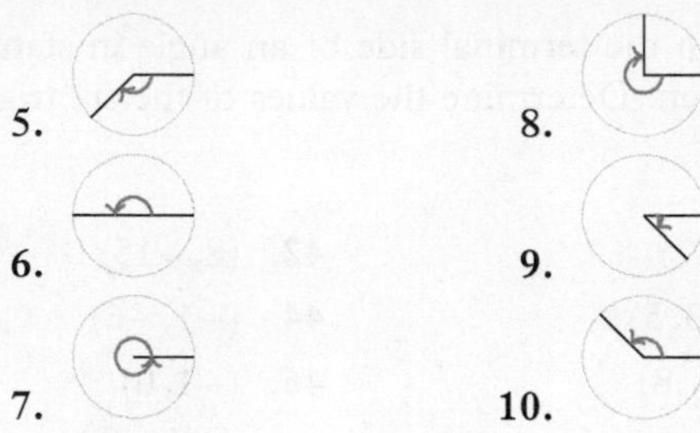

11–20. Rapid response: state the value of the trig function.

11. $\sin \pi$ **12.** $\cos(-\frac{\pi}{2})$
13. $\sin 2\pi$ **14.** $\cos 6\pi$
15. $\sin \frac{3\pi}{2}$ **16.** $\cos 0$
17. $\sin \frac{\pi}{2}$ **18.** $\cos 3\pi$
19. $\sin(-\pi)$ **20.** $\cos \frac{3\pi}{2}$

21–26. Convert to radians.

21. 135° **22.** 720°
23. −120° **24.** −240°
25. 210° **26.** 150°

27–32. An angle measurement is given in radians. Convert to degrees.

27. $\frac{7\pi}{6}$ **28.** 7π
29. $\frac{11\pi}{4}$ **30.** $-\frac{5\pi}{2}$
31. $-\frac{5\pi}{3}$ **32.** $\frac{11\pi}{6}$

33–40. Draw the angle in standard position.

33. $\frac{7\pi}{6}$ **34.** 7π
35. $\frac{11\pi}{4}$ **36.** $-\frac{5\pi}{2}$
37. $-\frac{5\pi}{3}$ **38.** $\frac{11\pi}{6}$
39. -3π **40.** 5π

41–50. A point on the terminal side of an angle in standard position is given. Determine the values of the six trigonometric functions.

41. $(3,7)$
42. $(8,-15)$
43. $(-3,5)$
44. $(-1,-6)$
45. $(.6,.8)$
46. $(-1,0)$
47. $(0,1)$
48. $\left(\frac{1}{-3},\frac{\sqrt{8}}{3}\right)$
49. $\left(\frac{\sqrt{24}}{7},-\frac{5}{7}\right)$
50. $(9,40)$

51–60. Determine the exact value of the trig function.

51. $\sin\frac{\pi}{3}$
52. $\tan\frac{\pi}{4}$
53. $\tan\frac{5\pi}{6}$
54. $\cos\frac{-\pi}{3}$
55. $\csc\frac{3\pi}{4}$
56. $\sin\frac{5\pi}{3}$
57. $\cot\frac{-\pi}{4}$
58. $\sec\frac{7\pi}{6}$
59. $\sec\frac{2\pi}{3}$
60. $\cot\left(-\frac{7\pi}{4}\right)$

61–66. The value of a trig function is given, as well as additional information regarding the quadrant in which the terminal side of the angle lies. Determine the values of the other five trig functions.

61. $\cos\theta = \frac{5}{8}$ and $0 < \theta < \frac{\pi}{2}$
62. $\tan\theta = 3$ and θ is a first-quadrant angle
63. $\sin\theta = -0.2$ and θ is a fourth-quadrant angle
64. $\sec\theta = \frac{9}{7}$ and $-\frac{\pi}{2} < \theta < 0$
65. $\cot\theta = \frac{2}{5}$ and $\pi < \theta < \frac{3\pi}{2}$
66. $\cos\theta = -\frac{2}{7}$ and $\sin\theta < 0$

67–74. Rewrite the expression in terms of sines and cosines, simplifying as appropriate.

67. $\dfrac{\tan\theta}{\sin\theta}$
68. $\sec^2\theta\cot\theta$
69. $\sin\theta\sec\theta - \tan\theta$
70. $\dfrac{\cot\theta}{\tan\theta}$
71. $\csc x - \cot^2 x \sin x$
72. $\sec^3 t \tan^2 t$

73. $\sec\theta(\cot\theta + \cos\theta)$

74. $(\sin\theta + \cos\theta)(\sec\theta + \csc\theta)$

75. Graph the function $y = \sin x$ over the intervals (a) $(0, 2\pi)$, (b) $(-2\pi, 2\pi)$, (c) $(0, 6\pi)$, and (d) $(-5\pi, \pi)$.

76. Graph the function $y = \cos x$ over the intervals (a) $(0, 2\pi)$, (b) $(-\pi, 3\pi)$, (c) $(4\pi, 6\pi)$, and (d) $(-2\pi, 0)$.

77. Graph the function $y = \tan x$ over the intervals (a) $(0, 2\pi)$, (b) $(-\pi, \pi)$, (c) $\left(-\frac{\pi}{2}, \frac{\pi}{2}\right)$, and (d) $\left(-\frac{3\pi}{2}, 0\right)$.

78. Graph the function $y = \cot x$ over the intervals (a) $(0, 2\pi)$, (b) $(2\pi, 3\pi)$, (c) $\left(-\frac{\pi}{2}, \frac{\pi}{2}\right)$, and (d) $(-\pi, 3\pi)$.

79. Graph the function $y = \csc x$ over the intervals (a) $(0, 2\pi)$, (b) $(-2\pi, 2\pi)$, (c) $(0, 6\pi)$, and (d) $(-5\pi, \pi)$.

80. Graph the function $y = \sec x$ over the intervals (a) $(0, 2\pi)$, (b) $(-\pi, 3\pi)$, (c) $(4\pi, 6\pi)$, and (d) $(-2\pi, 0)$.

81. What are the x-intercepts on the graph of $y = \tan x$?

82. Solve the equation $\cot x = 0$.

83. Solve the equation $\sec x = 0$.

Functions Review, Part I

0.4

The world around us is full of processes that take in inputs and produce outputs. An apple tree takes in sunshine, water, and nutrients and produces apples. Bees process pollen and make honey. A baker processes several ingredients and makes bread. If mathematics is to help describe the world, then it needs objects that process inputs and produce outputs. We call these objects *functions*.

Machine description of a function

The idea of a function is sometimes represented as a machine on a production line. A real-number input rolls along a conveyor belt into the function machine, the function does its job, and a real-number output comes out the other side. Imagine "The Squarer," a function machine that squares all inputs. If the number 2 is the input, then the function spits out the number 4 on the other side, as pictured in figure 1.

Functions can operate on other types of inputs and outputs; real numbers are not the only possibilities.

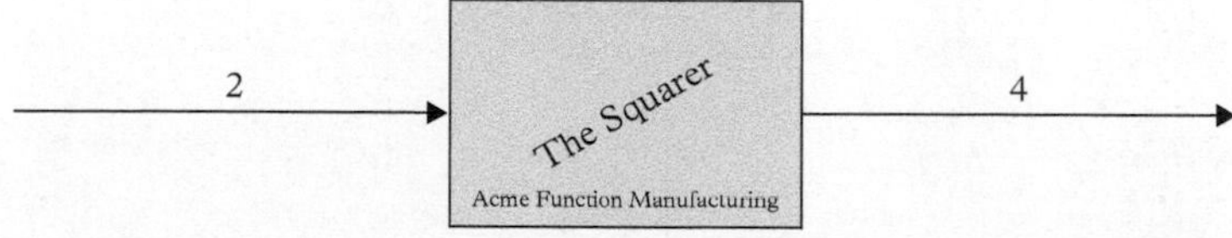

Figure 1 *Function as a machine*

Use the input -1 and the machine outputs 1. Use the input 7 and the machine outputs 49. In general, if the input is x, then the output is x^2 (figure 2).

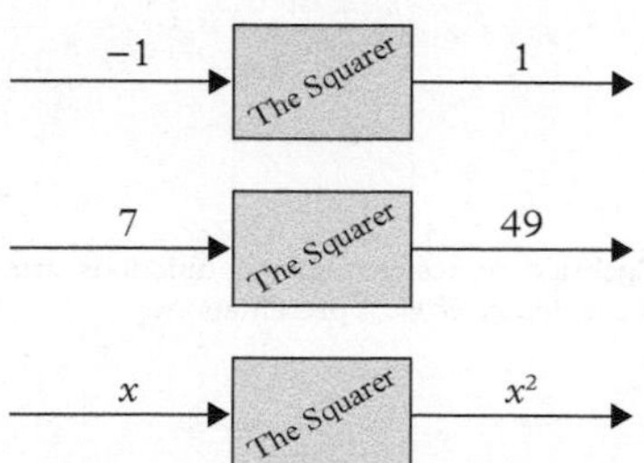

Figure 2 *Inputs processed by The Squarer. The input x with output x^2 best describes the action of the machine*

The *domain* of a function is the set of all inputs. Unless otherwise stated, we use the *natural domain* of a function, which is the set of all inputs that the function can actually handle without running into problems, such as division by zero or square roots of negative numbers. Because any real number can be squared, the domain of The Squarer is the set of all real numbers, written in interval notation as $(-\infty, \infty)$.

The *range* of a function is the set of all outputs as the input varies throughout the domain. Squares of numbers cannot be negative, so the range of The Squarer is the interval $[0, \infty)$.

To be called a function, the machine is only allowed to have one output for a given input. When the input to a function has been chosen, its output is determined.

Representations of functions

Functions can be represented in various ways. Before symbols for algebra were developed, words were used. A function may still be represented *verbally*. For instance, The Squarer can be described by the sentence "the output of function f is the square of its input."

A function can be represented *numerically* by giving a list of inputs and outputs. A numerical representation of The Squarer is in table 1. Notice that a numerical description is somewhat limited as it does not describe what happens to every input; inferences must be made if one wants to apply a numerical representation to other inputs. Numerical descriptions often arise in scientific and engineering settings from observing a process and measuring inputs and outputs. Values from an experiment might be the starting point for the exploration of a function.

Table 1 *A numerical representation of The Squarer*

input	output
−3	9
−2	4
−1	1
0	0
1	1
2	4

Algebraic representations of functions are also called *symbolic* representations.

An *algebraic* representation of a function is a complete description of the function using the power of symbolic notation. The Squarer can be represented algebraically in function notation as

$$f(x) = x^2$$

or as the equation

$$y = x^2.$$

The form of function notation is name(input) = output:

$$\underbrace{f}_{\text{name}}(\underbrace{x}_{\substack{\text{input,}\\ \text{independent}\\ \text{variable}}}) = \underbrace{x^2}_{\text{output}}.$$

In the equation $y = x^2$, x is the independent variable and y is the dependent variable; the inputs are represented by x and the outputs, by y.

Functions may also be represented *graphically*. Using the horizontal axis for the inputs and the vertical axis for the outputs, points on the graph represent input–output pairs (figure 3).

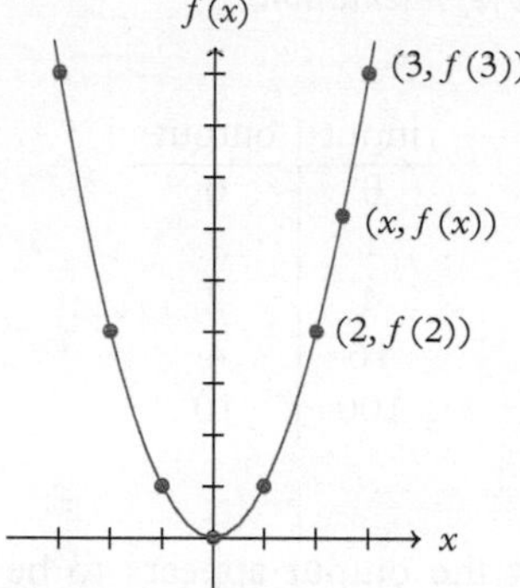

Figure 3 *The graph of The Squarer (blue). Points on the graph have the form* $(x, f(x))$

The ability to change from one type of representation of a function to another contributes to a complete understanding of the function. For instance, upon seeing the algebraic representation $f(x) = x^2$, we should be able to give the verbal representation that the output of function f is the square of its input. If we are making observations of a process and our data are given in table 1 (numerical representation), then we should recognize that the function just might be $f(x) = x^2$ (algebraic representation). The algebraic representation $f(x) = x^2$ should render immediate recognition of its graph in figure 3, and vice versa. Success in calculus requires that these connections be made.

Example 1 *Give an algebraic representation of the function with verbal representation "the output is three more than the input."*

Solution If we call the input x, then the output is three more, or $x+3$. An algebraic representation is therefore

$$f(x) = x + 3,$$

and we are done. ■

There were choices to be made in example 1. We could call the function by a different name if we want; $g(x) = x + 3$ is the same function ("a rose by any other name ..."). We can represent the input by a different variable if we want. If we call the input t, then the output is three more, or $t + 3$, and $h(t) = t + 3$ is still the same function.

Example 2 *Conjecture an algebraic representation for a function given the following numerical representation:*

The given information is not enough to ensure that the relationship holds for every possible input; there are other functions that pass through all these points. But, recognizing relationships is an important part of the scientific process, and without making such conjectures, progress is very difficult to come by. Sometimes one needs to try various possibilities, fitting different types of curves to imperfect data; this task is part of what is called *mathematical modeling.*

input	output
0	0
1	1
4	2
16	4
100	10

Solution Notice that the output appears to be the square root of the input (verbal representation). Therefore, we conjecture that an algebraic representation of the function is $f(x) = \sqrt{x}$. ■

The phrase "the function $f(x) = \sqrt{x}$" seems to identify the function with its algebraic representation. Although this phrase is used quite frequently, the function itself may be represented in various ways.

Example 3 *Give a graphical representation of the function $f(x) = \sqrt{x}$.*

See section 0.2 for more details on plotting points and graphing equations.

Solution The instructions to this example may be rephrased as "Graph the function $f(x) = \sqrt{x}$." This is the exact same as saying "graph the equation $y = \sqrt{x}$," although we label the vertical axis $f(x)$ instead of y. After plotting a few points, we connect the dots in a nice, smooth manner (figure 4). ■

Vertical line test

Not all equations have graphs that represent functions. To be a function, there must be only one output for any given input. Because inputs

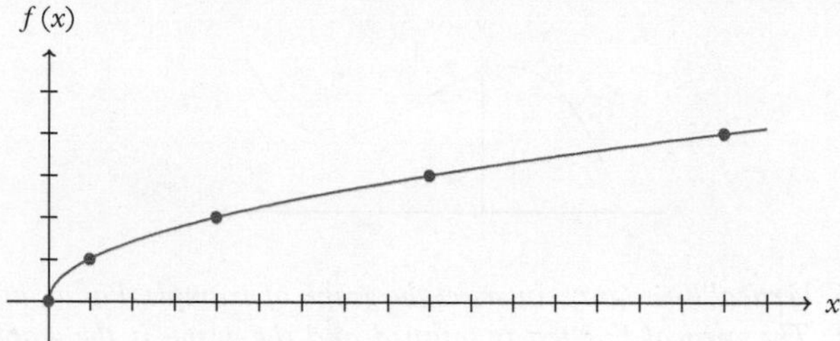

Figure 4 *The graph of* $f(x) = \sqrt{x}$

are horizontal coordinates and outputs are vertical coordinates, for any horizontal coordinate (input), there can be only one vertical coordinate (output). In other words, no vertical line should intersect the graph at more than one point; this is called the *vertical line test.*

> **VERTICAL LINE TEST**
>
> Given a graph, if there is a vertical line that intersects the graph at more than one point, then the graph is not the graph of a function. If there is no such line, then the graph is the graph of a function.

Example 4 *For each graph in figure 5, is the graph the graph of a function?*

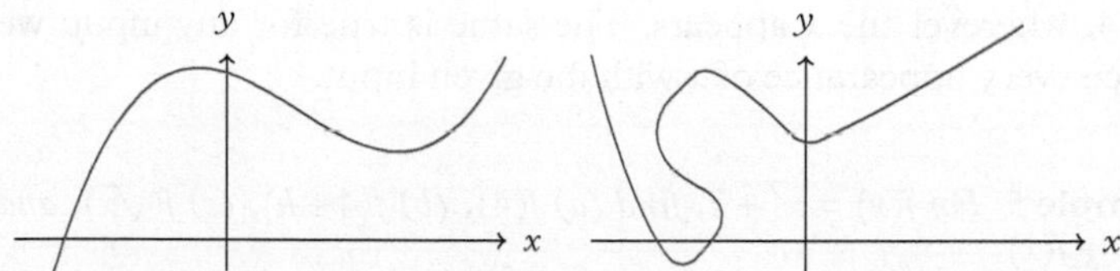

Figure 5 *The graphs for example 4*

Solution (a) Any vertical line (in gray in figure 6) intersects the graph at, at most, one point. The graph passes the vertical line test, so the graph does represent the graph of a function.

(b) There is a vertical line that intersects the graph at more than one point (see figure 7), so the graph fails the vertical line test. The graph does not represent the graph of a function.

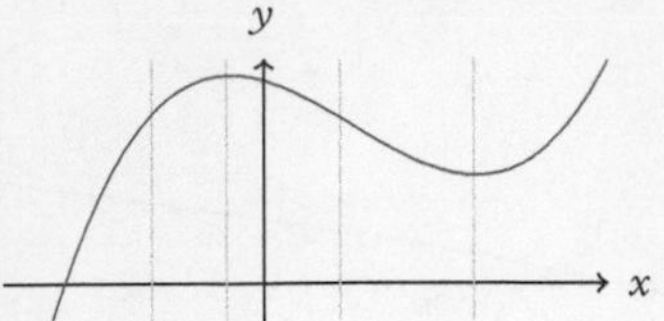

Figure 6 *Vertical lines (gray) intersect the graph of example 4(a) at, at most, one point. The vertical line test is satisfied and the graph is the graph of a function*

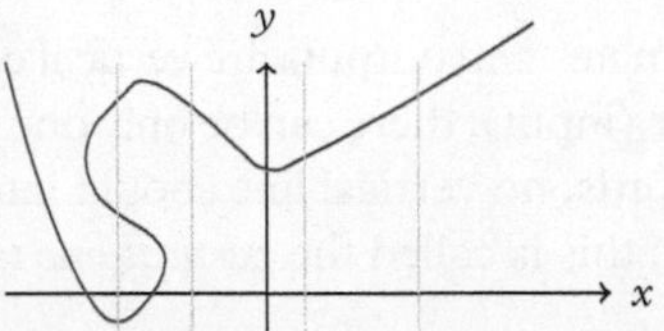

Figure 7 *The left-most of the vertical lines (gray) intersects the graph of example 4(b) at more than one point. The graph fails the vertical line test. (To pass the vertical line test, all, not just some, of the vertical lines must intersect the graph at, at most, one point)* ■

Function values

To evaluate $f(x)$ at $x = 4$, we simply replace x in the output expression with 4, wherever the x appears. The same is true for any input; we replace every appearance of x with the given input.

Example 5 *For* $f(x) = x^2 + 2$, *find (a)* $f(4)$, *(b)* $f(4+h)$, *(c)* $f(\sqrt{x})$, *and (d)* $\dfrac{7 + f(x)}{f(3)}$.

Solution (a) To find $f(4)$, we replace x with 4 (emphasized using bold):

$$f(x) = x^2 + 2$$
$$f(\mathbf{4}) = \mathbf{4}^2 + 2 = 18.$$

(b) To find $f(4 + h)$, we replace x with $4 + h$. Because of the manner in which we write expressions, if the input expression has more than one term, it should be surrounded by parentheses:

$$f(x) = x^2 + 2$$
$$f(\mathbf{4+h}) = (\mathbf{4+h})^2 + 2$$
$$= 16 + 8h + h^2 + 2$$
$$= 18 + 8h + h^2.$$

CAUTION: MISTAKE TO AVOID
This is not the same as $f(4) + h$:

$$f(4+h) \neq f(4) + h.$$

In other words,

$$f(4+h) \neq 4^2 + 2 + h.$$

(c) Again, the replacement is emphasized using bold:

$$f(x) = x^2 + 2$$
$$f(\sqrt{\mathbf{x}}) = (\sqrt{\mathbf{x}})^2 + 2$$
$$= x + 2.$$

Although $\sqrt{x^2} = |x|$, because x must be positive to take its square root, $(\sqrt{x})^2 = x$.

(d) Function values can sit anywhere in an expression. They are evaluated and placed in the expression where written:

$$\frac{7 + f(x)}{f(3)} = \frac{7 + (x^2 + 2)}{3^2 + 2}$$
$$= \frac{x^2 + 9}{11}.$$

■

If $f(x)$ contains more than one occurrence of the variable x, we replace each x with the input expression.

Example 6 *For* $f(x) = \dfrac{x^2 - 5x + 2}{x - 7}$, *find* $f(3)$.

Solution We replace each occurrence of x with 3, emphasized in bold:

$$f(3) = \frac{\mathbf{3}^2 - 5 \cdot \mathbf{3} + 2}{\mathbf{3} - 7}$$
$$= \frac{-4}{-4}$$
$$= 1.$$

■

Finding domains and ranges

Unless otherwise specified, when we are asked for the domain of a function, this means the natural domain, the set of all possible acceptable inputs. What must be avoided are operations that are not defined (at least for real numbers), such as division by zero or square roots of negative numbers. As our list of functions grows, our list of operations that must be avoided may also grow (for instance, logarithms are defined only for positive numbers); but, for now, our list of what to look for is short.

Finding the range of a function is often not nearly as straightforward algebraically as finding the domain, and, as a result, we often resort to checking the graph of the function to determine the function's range.

Example 7 *Find the domain and range of $f(x) = x^2 + 7$.*

Solution The expression $x^2 + 7$ is defined for every real number. Any number can be squared; we can add 7 to any number. There is no division by zero, no square root of a negative number, nothing that must be avoided. The domain of f is the set of all real numbers, which may be expressed as **R** or as $(-\infty, \infty)$.

Because boldface fonts are difficult to write by hand, it is customary when writing **R** by hand to write $\mathbb{R}$ instead (the left side of the letter has a double line).

Checking the graph (figure 8), we see that the y-coordinates are 7 or larger. Therefore the range is $[7, \infty)$, which may also be written in inequality form as $y \geq 7$.

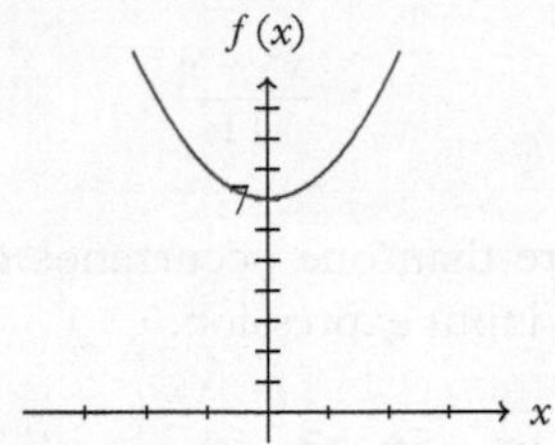

Figure 8 *The graph of $f(x) = x^2 + 7$*

The range can sometimes be determined by algebraic reasoning without resorting to looking at the graph. In example 7, because we know that $x^2 \geq 0$, adding 7 to both sides gives the range: $x^2 + 7 \geq 7$.

We write the final answer as

$$\text{domain: } (-\infty, \infty),$$
$$\text{range: } [7, \infty). \qquad \blacksquare$$

Example 8 *Find the domain and range of $f(x) = \sqrt{x - 3}$.*

Solution There is no division and therefore no division by zero, but there is a square root. We can not take square roots of negative numbers. Therefore, we set what is inside the square root to be zero or larger:

In calculus we do not use complex numbers.

$$x - 3 \geq 0$$
$$x \geq 3.$$

The domain of f is $[3, \infty)$.

For the range, we graph the function where $x \geq 3$ (figure 9) and see that the y-coordinates are zero or larger. The range of f is $[0, \infty)$. ■

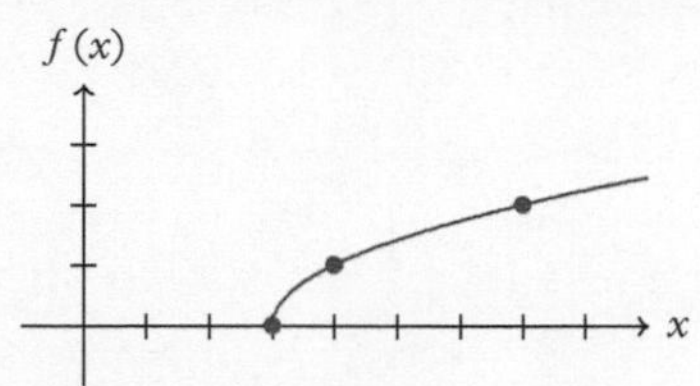

Figure 9 *The graph of* $f(x) = \sqrt{x-3}$

Example 9 *Find the domain of* $g(x) = \sqrt[3]{x^2 - 2}$.

Solution Before setting $x^2 - 2 \geq 0$ and solving, notice that we have a cube root, not a square root. Odd roots of negative numbers are defined (for instance, $\sqrt[3]{-8} = -2$, since $(-2)(-2)(-2) = -8$). Therefore, there is no disallowed operation and the domain of g is $(-\infty, \infty)$. ■

We can not take even roots of negative numbers, but odd roots are allowed.

Example 10 *Find the domain of* $f(x) = \dfrac{x^2 + 7}{x - 3}$.

Solution There are no square roots to worry about, but there is division. The denominator must not be zero:

$$x - 3 \neq 0$$
$$x \neq 3.$$

The domain of f can be expressed in a variety of ways. One is $x \neq 3$. Another is using interval notation (even though the solution set is not an interval): $(-\infty, 3) \cup (3, \infty)$. Yet another way is to use setminus notation, $\mathbf{R} \setminus \{3\}$. ■

The set $(-\infty, 3) \cup (3, \infty)$ does not include the number 3, but it includes all numbers less than 3 and all numbers greater than 3.

The *setminus notation* $\mathbf{R} \setminus \{3\}$ means to take all real numbers except for the number 3.

Example 11 *Find the domain of* $f(x) = \sqrt{x^2 + 2}$.

Solution There is no division, but there is a square root. We set what is under the square root to be zero or larger:

$$x^2 + 2 \geq 0$$
$$x^2 \geq -2.$$

How do we solve $x^2 \geq -2$? By remembering that $x^2 \geq 0$, so that x^2 is always larger than -2. The solution is the set of all real numbers, $(-\infty, \infty)$. ■

What if the inequality is $x^2 \geq 2$ instead? We proceed by taking square roots of both sides of the inequality:

$$\begin{aligned} x^2 &\geq 2, \\ \sqrt{x^2} &\geq \sqrt{2}, \\ |x| &\geq \sqrt{2}, \\ x \leq -\sqrt{2} \text{ or } x &\geq \sqrt{2}. \end{aligned}$$

Recall that $\sqrt{x^2} = |x|$ and that the absolute value inequality $|x| > a$ can be rewritten as $x < -a$ or $x > a$.

Piecewise-defined functions

A piecewise-defined function is a function in which separate expressions are given for various pieces of the domain.

Example 12 *For the function* $f(x) = \begin{cases} x^2, & x > 0 \\ x+2, & x \leq 0, \end{cases}$ *find (a)* $f(3)$, *(b)* $f(-2)$, *and (c)* $f(0)$.

Solution (a) For $f(3)$, the input $x = 3$ is greater than zero. Therefore, we use the rule for $x > 0$, and

$$\begin{aligned} f(x) &= x^2, \ x > 0 \\ f(3) &= 3^2 = 9. \end{aligned}$$

(b) For $f(-2)$, the input $x = -2$ is less than zero. Therefore, we use the rule for $x \leq 0$:

$$\begin{aligned} f(x) &= x + 2, \ x \leq 0 \\ f(-2) &= (-2) + 2 = 0. \end{aligned}$$

(c) For $f(0)$, the input is $x = 0$. Therefore, we use the rule for $x \leq 0$:

$$\begin{aligned} f(x) &= x + 2, \ x \leq 0 \\ f(0) &= 0 + 2 = 2. \end{aligned}$$

■

Graphing a piecewise-defined function is illustrated in the next example.

Example 13 *Graph* $f(x) = \begin{cases} x^2, & x \leq 1 \\ x-1, & x > 1. \end{cases}$

Solution The graph of $f(x) = x^2$ (same as $y = x^2$) should be familiar. The graph of $f(x) = x - 1$ (same as $y = x - 1$) is a line with slope 1 and y-intercept -1. See figure 10.

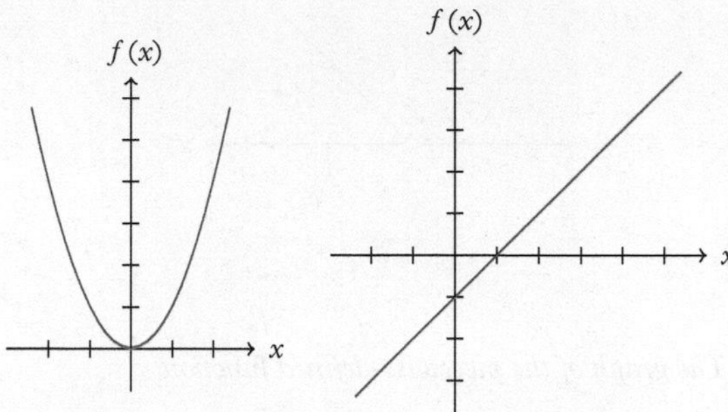

Figure 10 *The graphs of $y = x^2$ (left) and $y = x - 1$ (right)*

But, notice that $f(x) = x^2$ only where $x \leq 1$ and that $f(x) = x - 1$ only where $x > 1$. In figure 11, the portion of the picture for $f(x) = x^2$ with x-coordinates less than or equal to 1 is shaded tan, whereas the portion of the picture for $f(x) = x - 1$ with x-coordinates larger than 1 is shaded pink.

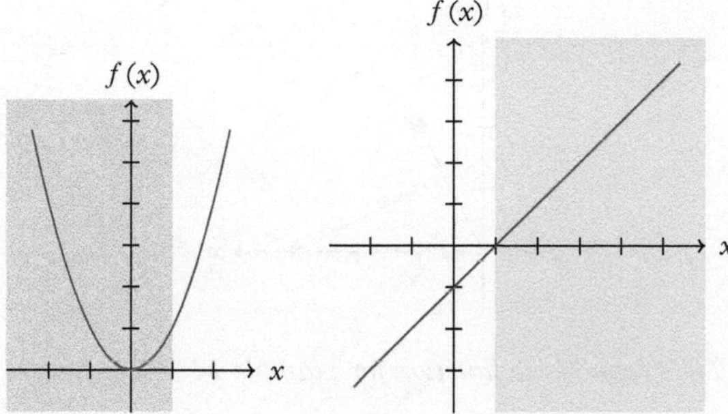

Figure 11 *The graphs of $y = x^2$ (left) and $y = x - 1$ (right). The portions of the pictures applicable to the piecewise-defined function are shaded*

The function f consists of the two shaded regions put together; the result is in figure 12. To show that the rule $f(x) = x^2$ applies when $x = 1$, we place a filled circle at the end of its piece; to show that the rule $f(x) = x - 1$ does not apply when $x = 1$, we place an open circle at the end of its piece. ■

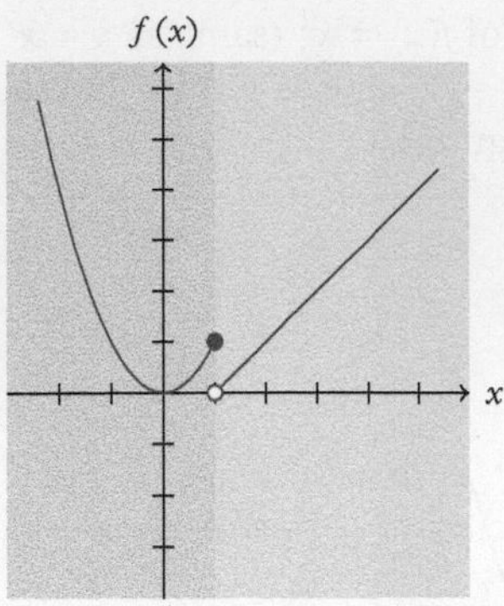

Figure 12 *The graph of the piecewise-defined function*

$$f(x) = \begin{cases} x^2, & x \leq 1 \\ x - 1, & x > 1. \end{cases}$$

The shading is for illustrating in which region each piece applies; shading is not a required part of the graph

Example 14 *Find an algebraic representation for the function graphed in figure 13.*

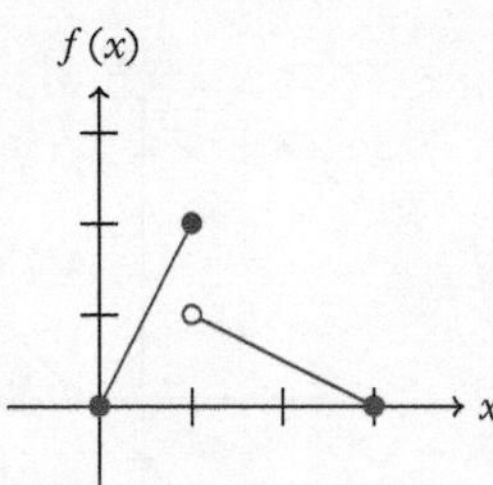

Figure 13 *The graph of the function for example 14*

Solution The function appears to consist of two line segments. The segment on the left is part of a line that has y-intercept 0 and slope 2 (the line goes right 1 unit and up 2 units, thus $\frac{\text{rise}}{\text{run}} = \frac{2}{1} = 2$). Using the slope-intercept form of the equation of a line, its equation is $y = 2x+0$, or $y = 2x$. So far we have

$$f(x) = \begin{cases} 2x, & \\ & \end{cases}.$$

Consulting the graph, the values of x for which the line $y = 2x$ applies are from $x = 0$ to $x = 1$. Because filled dots are used at each endpoint, both $x = 0$ and $x = 1$ are included. Therefore, we use $0 \leq x \leq 1$, and so far we have

$$f(x) = \begin{cases} 2x, & 0 \le x \le 1 \\ & \end{cases}.$$

To find the equation of the line segment on the right, notice that we have two points on the line: $(1, 1)$ and $(3, 0)$. We can find the slope of the line using the slope formula:

Although the point $(1, 1)$ is not included in the graph of the function, the line of which the segment is a part includes the point $(1, 1)$.

$$\begin{aligned} \text{slope} &= \frac{y_2 - y_1}{x_2 - x_1} \\ &= \frac{0 - 1}{3 - 1} \\ &= \frac{-1}{2}. \end{aligned}$$

Then, using the point-slope form of the equation of a line gives

$$\begin{aligned} y - y_1 &= m(x - x_1) \\ y - 1 &= -\frac{1}{2}(x - 1) \\ y - 1 &= -\frac{1}{2}x + \frac{1}{2} \\ y &= -\frac{1}{2}x + \frac{3}{2}. \end{aligned}$$

Now we have

$$f(x) = \begin{cases} 2x, & 0 \le x \le 1 \\ -\frac{1}{2}x + \frac{3}{2}, & \end{cases}.$$

Consulting the graph again, the piece on which the segment on the right applies is for $x = 1$ to $x = 3$, with $x = 1$ not included. We therefore use $1 < x \le 3$ and our final answer is

The order in which the portions of the expression for f are determined is not important.

$$f(x) = \begin{cases} 2x, & 0 \le x \le 1 \\ -\frac{1}{2}x + \frac{3}{2}, & 1 < x \le 3. \end{cases}$$

■

Many useful functions are defined piecewise. An example is absolute value. Notice that in definition 1, absolute value is given as a piecewise-defined function.

Catalog of essential functions

Some functions are used so often that their algebraic representations and their typical graph shapes should be on instant recall. For instance,

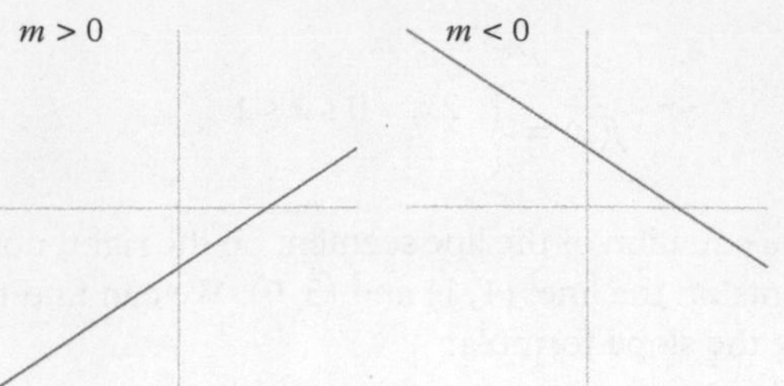

Figure 14 *Typical graph shapes for $y = mx + b$, lines*

upon reading $f(x) = x^2$, a vision of its graph should leap instantly to your mind, and you should know that its graph shape is called a *parabola.* Such knowledge is extremely helpful in visualizing and understanding various ideas in calculus.

In what follows, typical graph shapes are graphed in green and specific functions with graphs that should be known are graphed in blue.

Polynomial functions. Polynomial functions have an algebraic representation of the form

$$f(x) = a_n x^n + a_{n-1}x^{n-1} + \cdots + a_2x^2 + a_1x + a_0,$$

where n is the *degree* of the polynomial, x is the variable, and the a's are real numbers called the *coefficients* of the polynomial. An example is

$$f(x) = 4x^7 - 2x^3 + \frac{4}{13}x - 1,$$

a polynomial with degree 7 and coefficients $a_7 = 4$, $a_3 = -2$, $a_1 = \frac{4}{13}$, and $a_0 = -1$. To be classified as a polynomial, the powers of x must be nonnegative integers.

A function with a graph that is a line is called a *linear function.*

Lines: degree 1 polynomials. A function of the form

$$f(x) = mx + b$$

should be recognized instantly as a line. Typical graph shapes for $m > 0$ and $m < 0$ are in figure 14. In addition to recognizing the slope as positive or negative, the y-intercept should also be immediately recognized and pictured.

Two specific lines with graphs that should be known are $y = x$ and $y = -x$ (figure 15).

Parabolas: degree 2 polynomials. A function of the form

$$f(x) = ax^2 + bx + c$$

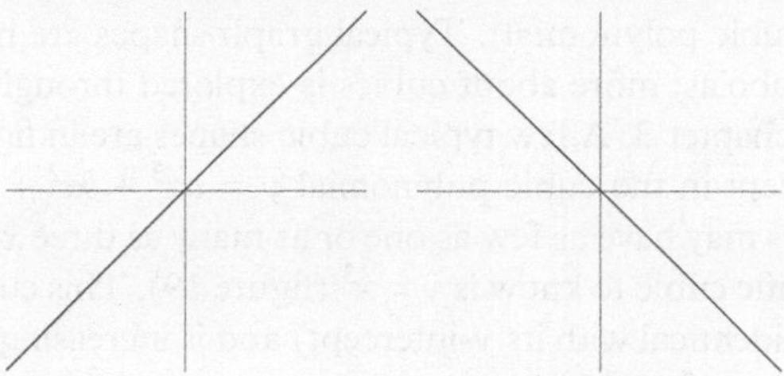

Figure 15 *The lines $y = x$ (left) and $y = -x$ (right), which meet the coordinate axes at 45° angles*

is a parabola. Typical graph shapes for $a > 0$ (opens up) and $a < 0$ (opens down) should be on instant recall, although the exact position of the parabola may not be (figure 16). The y-intercept of the parabola $y = ax^2 + bx + c$ is at $y = c$. Parabolas may or may not have x-intercepts; the quadratic formula may be used to find them.

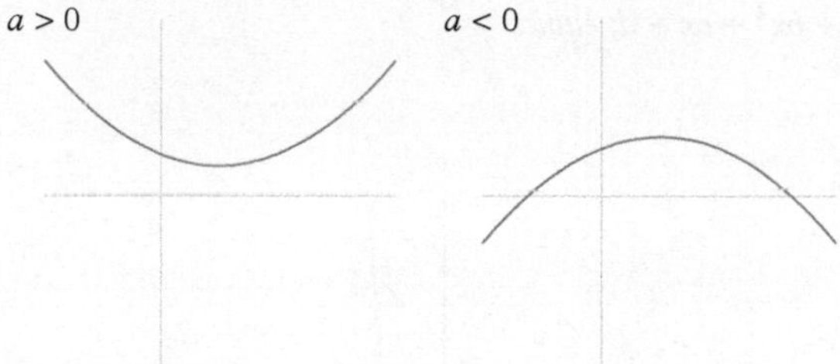

Figure 16 *Typical graph shapes for $y = ax^2 + bx + c$, parabolas*

Two specific parabolas with graphs that should be known are $y = x^2$ and $y = -x^2$ (figure 17).

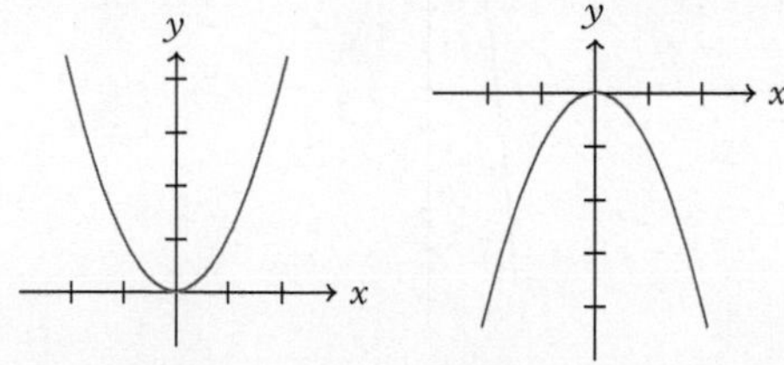

Figure 17 *The graphs of $y = x^2$ (left) and $y = -x^2$ (right)*

Cubics: degree 3 polynomials. A function of the form

$$f(x) = ax^3 + bx^2 + cx + d$$

is a cubic (cubic polynomial). Typical graph shapes are more varied than for parabolas; more about cubics is explored through the use of calculus, in chapter 3. A few typical cubic shapes are in figure 18. The y-intercept in the cubic polynomial $y = ax^3 + bx^2 + cx + d$ is at $y = d$. Cubics may have as few as one or as many as three x-intercepts.

One specific cubic to know is $y = x^3$ (figure 19). This cubic has one x-intercept (identical with its y-intercept) and is increasing (the graph goes up as we move to the right).

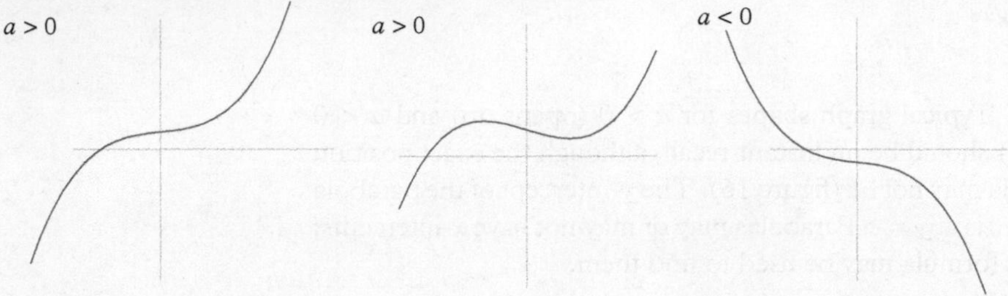

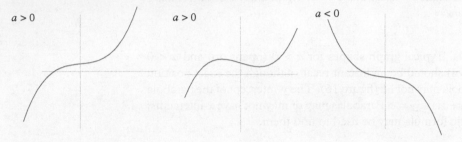

Figure 18 *A few typical graph shapes for $y = ax^3 + bx^2 + cx + d$, cubics*

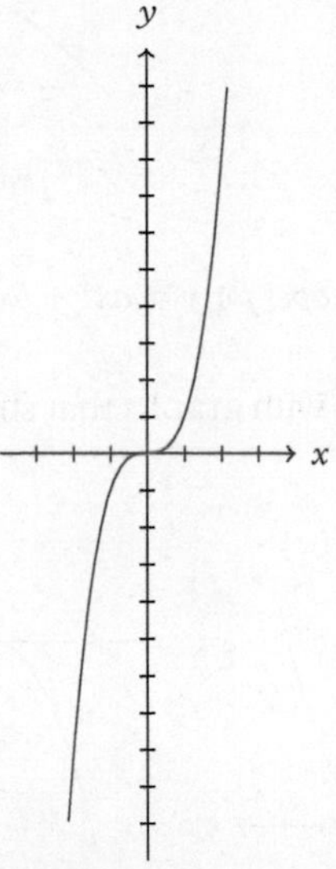

Figure 19 *The graph of $y = x^3$*

Polynomials of larger degrees are also explored in chapter 3, although their typical shapes are less important to recognize immediately. It should be known that even powers of x (such as $y = x^4$ and $y = x^{10}$) have shapes similar to that of $y = x^2$, but with "broader shoulders," whereas the shapes of odd powers of x are similar to that of $y = x^3$.

Power functions. Power functions are functions of the form $f(x) = x^n$. Some power functions are also polynomials, such as $y = x^2$, $y = x^3$, and $y = x^{10}$. If the exponent is negative or a fraction or anything other than a positive integer, then the power function is not a polynomial.

Root functions. Root functions are power functions with exponent $\frac{1}{k}$ for an integer $k \geq 2$. Specific root functions with graphs that should be known are the square root $y = \sqrt{x}$ and the cube root $y = \sqrt[3]{x}$ (figure 20). In general, graphs of even root functions have shapes like that of the square root, and graphs of odd root functions have shapes like that of the cube root.

Roots are fractional powers: $\sqrt{x} = x^{1/2}$ and $\sqrt[3]{x} = x^{1/3}$.

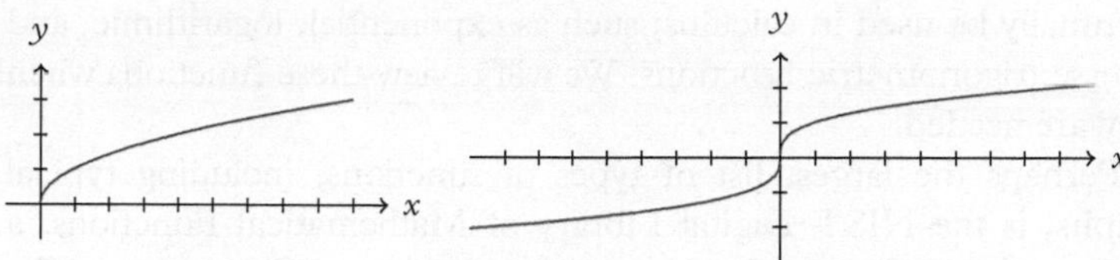

Figure 20 *The graphs of* $y = \sqrt{x}$ *(left) and* $y = \sqrt[3]{x}$ *(right)*

Reciprocal functions. Power functions with negative exponents are called *reciprocal functions.* Two specific reciprocal functions to know are $f(x) = \frac{1}{x}$ and $f(x) = \frac{1}{x^2}$. The graphs of these functions have vertical asymptotes at $x = 0$ and horizontal asymptotes at $y = 0$. They do not have x- or y-intercepts. See figure 21.

Recall that $\frac{1}{x^k} = x^{-k}$.

One of the most commonly mispronounced and misspelled words in mathematics is the word *asymptote*, usually by misplacing the letters "p" and "t." If you are prone to this mistake, take the time to practice the correct pronunciation several times, and repeat daily until the incorrect pronunciation is overcome.

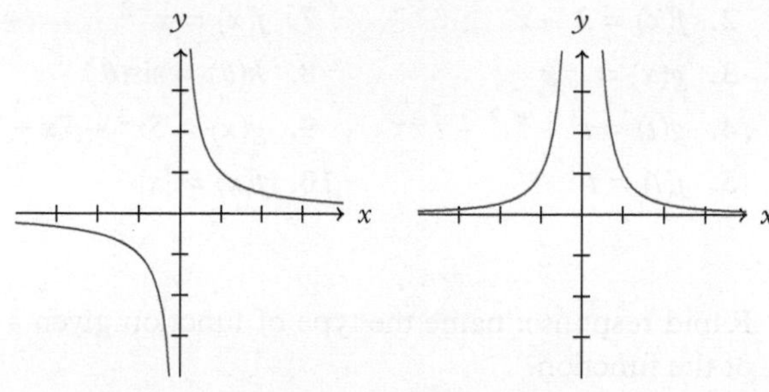

Figure 21 *The graphs of* $y = \dfrac{1}{x}$ *(left) and* $y = \dfrac{1}{x^2}$ *(right)*

Absolute value. The graph of $y = |x|$ should be known. The graph consists of the line $y = x$ to the right of $x = 0$ and $y = -x$ to the left of $x = 0$, with a corner at $x = 0$ (figure 22).

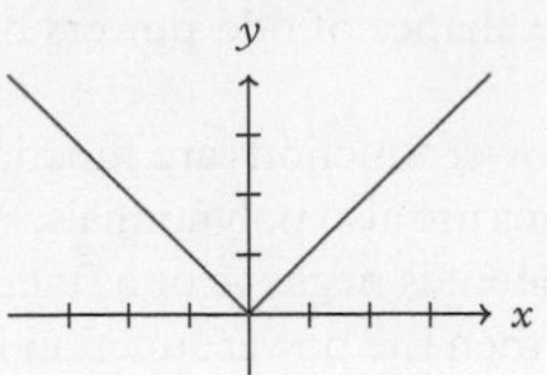

Figure 22 *The graph of $y = |x|$*

Trig functions. The six basic trigonometric functions and their graphs were discussed in the previous section. Their graphs should be memorized as well.

Other functions. You may have studied other functions that will eventually be used in calculus, such as exponential, logarithmic, and inverse trigonometric functions. We will review these functions when they are needed.

Perhaps the largest list of types of functions, including typical graphs, is the NIST Digital Library of Mathematical Functions, a project of the National Institute of Standards and Technology. The URL is http://dlmf.nist.gov.

Exercises 0.4

1–10. Rapid response: name the type of function given its algebraic representation.

1. $f(x) = 5x - 7$
2. $f(x) = 2 - x$
3. $g(x) = \sqrt[4]{x}$
4. $g(t) = t^3 + 5t^2 - t$
5. $f(t) = t^7$
6. $f(x) = x^{1/3}$
7. $f(x) = x^{-2}$
8. $h(\theta) = \sin\theta$
9. $g(x) = 5x^2 - 7x + 3$
10. $f(x) = |x|$

All these functions are graphed in blue so that no color clue is given.

11–18. Rapid response: name the type of function given a graph of the function.

11.

12.

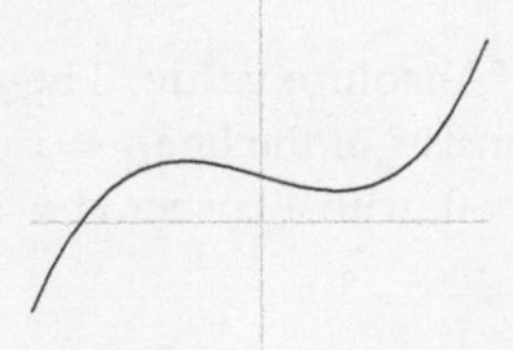

13. **16.**

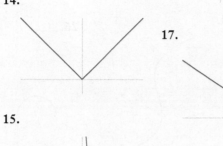

14. **15.** **17.** **18.**

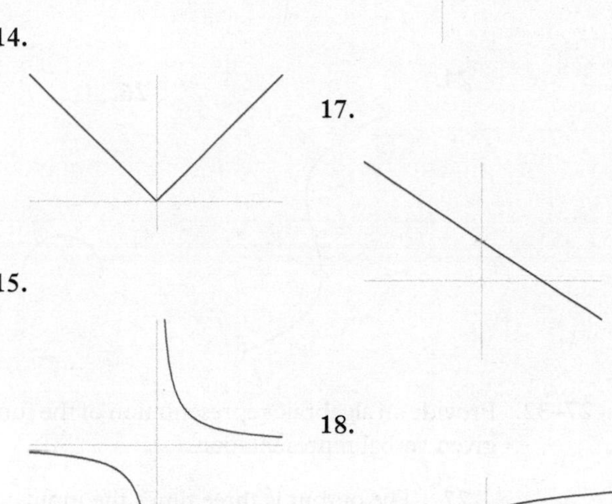

19–26. Rapid response: does the graph represent the graph of a function?

19. **21.**

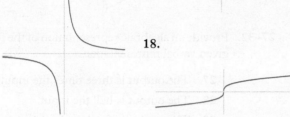

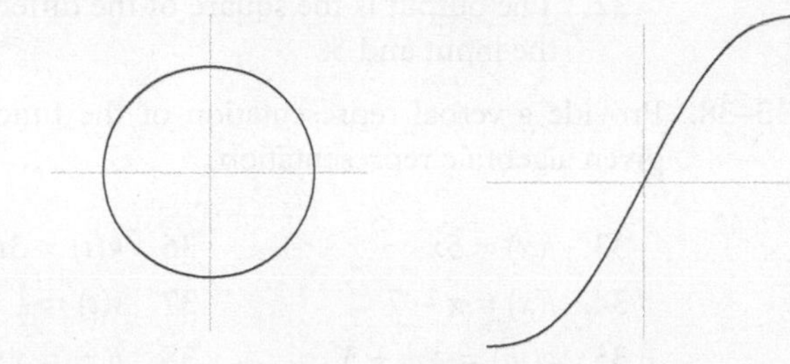

20.

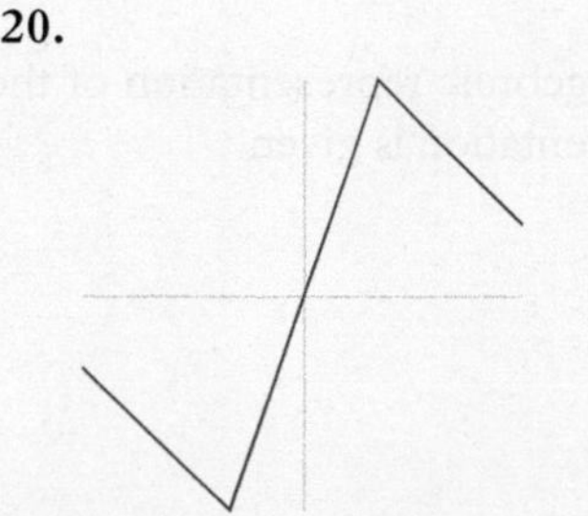

22.

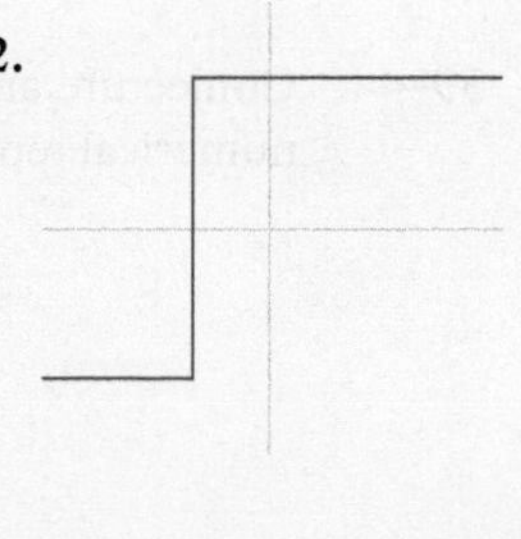

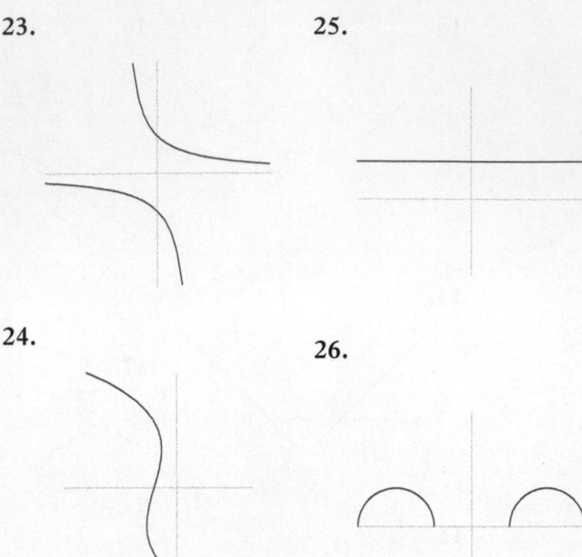

27–32. Provide an algebraic representation of the function with the given verbal representation.

27. The output is three times the input.
28. The output is half the input.
29. The output is five less than the input.
30. The output is the negative of the square of the input.
31. The output is the sum of the input and the input cubed.
32. The output is the square of the difference between the input and 5.

33–38. Provide a verbal representation of the function with the given algebraic representation.

33. $f(x) = 5x$
34. $f(x) = x - 7$
35. $g(m) = \frac{1}{2}m + 3$
36. $k(t) = 3t^2$
37. $s(t) = \frac{1}{t}$
38. $f(x) = x + x^2 + x^3$

39–44. Conjecture an algebraic representation of the function. A numerical representation is given.

39.

input	output
0	0
1	2
2	4
3	6
4	8

40.

input	output
0	1
1	3
2	5
3	7
4	9

41.

input	output
0	0
1	1
2	8
3	27
4	64

42.

input	output
0	0
5	1
10	2
15	3
20	4

43.

input	output
−3	18
−1	2
0	0
1	2
3	18

44.

input	output
−6	106
1	99
22	78
40	60
75	25

45–52. Prove a graphical representation of the function with the given algebraic representation.

45. $f(x) = (x-1)^2$

46. $f(x) = x^2 - 3$

47. $f(x) = 2\sqrt{x}$

48. $f(x) = \frac{1}{4}x^3$

49. $f(x) = 1 + \sin x$

50. $f(x) = -\sqrt{x}$

51. $f(x) = 3 - x$

52. $f(x) = 6$

53. For $f(x) = 4 - 5x$, find (a) $f(2)$, (b) $f(3-\theta)$, (c) $f(4x)$, and (d) $2 + f(x)$.

54. For $f(x) = \sqrt{x} + 1$, find (a) $f(0)$, (b) $f(x^2)$, (c) $f(x^2+25)$, and (d) $f(x^2) + 25$.

55. For $f(x) = 17$, find (a) $f(4)$, (b) $f(3-h)$, (c) $f(x^2)$, and (d) $f(x-7)+3$.

56. For $f(x) = 4x + 1$, find (a) $f(9)$, (b) $f(2+h)$, (c) $f(2) + f(h)$, and (d) $f(t)$.

57. For $f(x) = x + x^2$, find (a) $f(5)$, (b) $f(y)$, (c) $f(3+2h)$, and (d) $f(4x) - f(3x)$.

58. For $f(x) = \dfrac{x+5}{x-2}$, find (a) $f(3)$, (b) $f(2+h)$, (c) $f(\frac{1}{x})$, and (d) $\frac{1}{f(x)}$.

59. For $f(x) = \begin{cases} 2x-1, & x < 3 \\ -x^2, & x \geq 3, \end{cases}$ find (a) $f(0)$, (b) $f(5)$, and (c) $f(3)$.

60. For $f(x) = \begin{cases} \sqrt{x-1}, & x > 1 \\ x, & x \le 1, \end{cases}$ find (a) $f(0)$, (b) $f(5)$, and (c) $f(1)$.

61. For $f(x) = \begin{cases} \sin(\frac{1}{x}), & x \ne 0 \\ 0, & x = 0, \end{cases}$ find (a) $f(1/\pi)$, (b) $f(0)$, and (c) $f(\frac{2}{3\pi})$.

62. For $f(x) = \begin{cases} x^2, & x < -3 \\ x, & -3 \le x \le 1 \\ 5, & x > 1, \end{cases}$ find (a) $f(-4)$, (b) $f(0)$, (c) $f(1)$, and (d) $f(4)$.

Answers in the back of the book use interval notation.

63–68. Find (a) the domain and (b) the range of the given function.

63. $f(x) = \dfrac{2}{x-1}$

64. $f(x) = \sqrt{x-12}$

65. $g(x) = 2 - x^2$

66. $s(t) = |t-1|$

67. $f(t) = \dfrac{4t-1}{5}$

68. $f(x) = x^3 - 1$

69-76. Find the domain of the function.

69. $f(x) = \sqrt{4-x}$

70. $f(x) = \dfrac{5}{2x+6}$

71. $g(x) = \dfrac{5+x}{x^2-6x+8}$

72. $h(x) = \sqrt[3]{8+x}$

73. $f(x) = \dfrac{\sqrt{x+5}}{x-2}$

74. $f(x) = \dfrac{3}{x\sqrt{x+1}}$

75. $s(t) = \sqrt{9-t^2}$

76. $g(t) = \sqrt{4-x} + \sqrt{x+2}$

77–84. Graph the piecewise-defined function.

77. $f(x) = \begin{cases} 2x-1, & x < 3 \\ -x^2, & x \ge 3 \end{cases}$

78. $f(x) = \begin{cases} 2-x, & x < 0 \\ x+1, & x \ge 0 \end{cases}$

79. $f(x) = \begin{cases} 4, & x < -1 \\ 2, & x > 1 \end{cases}$

80. $f(x) = \begin{cases} x^2, & x < -3 \\ x, & -3 \le x \le 1 \\ 5, & x > 1 \end{cases}$

81. $f(x) = \begin{cases} x, & x \ne 2 \\ 5, & x = 2 \end{cases}$

82. $f(x) = \begin{cases} x+2, & x < -1 \\ -x, & -1 < x \leq 1 \\ x-2, & x > 1 \end{cases}$

83. $\text{sgn}(x) = \begin{cases} -1, & x < 0 \\ 0, & x = 0 \\ 1, & x > 0 \end{cases}$

The function sgn(x) is called the *signum function* (Latin for "sign") because it gives the sign (positive, negative, zero) of the number.

84. $f(x) = \begin{cases} x^2, & x \neq 0 \\ 1, & x = 0 \end{cases}$

85. Find an algebraic representation for the function with the following graph.

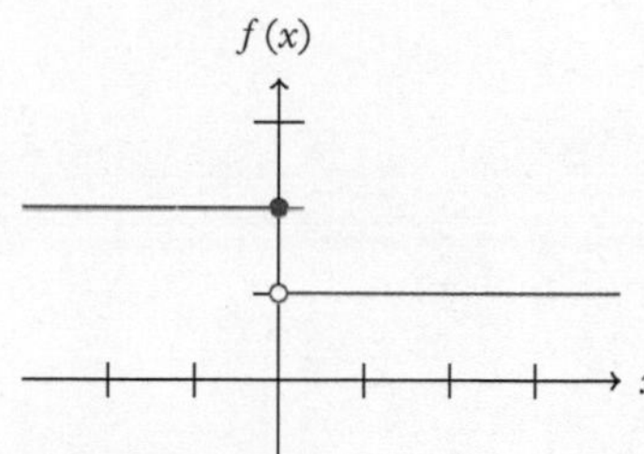

86. Find an algebraic representation for the function with the following graph.

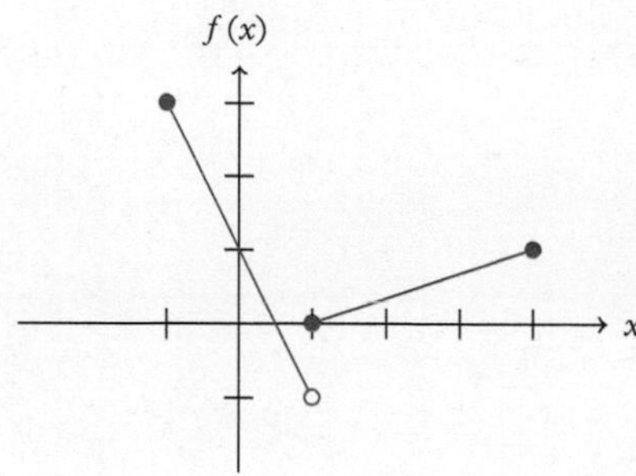

Functions Review, Part II

0.5

Our work with functions continues with transformations of graphs and combinations of functions.

Vertical shifts

Start with the graph of the function $y = f(x)$. Consider the graph of $y = f(x) + k$:

$$y = f(x)$$
$$y = \underbrace{f(x) + k}_{\text{output}}.$$

Here, k has been added to the output. The outputs are represented by vertical coordinates on a graph, so the graph of $y = f(x) + k$ shifts k units vertically from the graph of $y = f(x)$. If $k > 0$, the y-coordinates are larger so the shift is upward; if $k < 0$, the y-coordinates are smaller so the shift is downward. See figure 1.

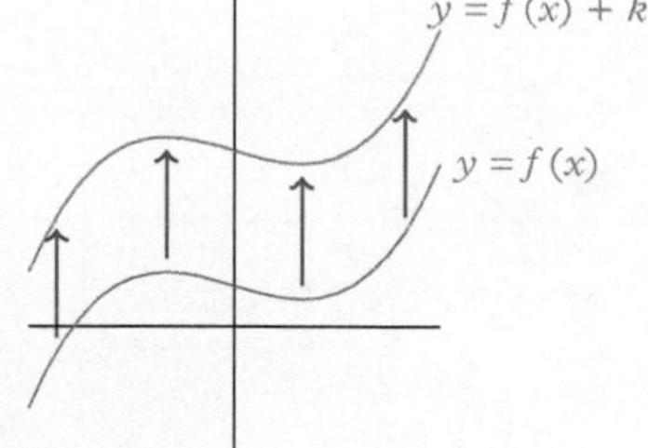

Figure 1 *The graph of a function* $y = f(x)$ *(green) and the vertically shifted function* $y = f(x) + k$ *(brown). The red arrows indicate the shift. Here,* $k > 0$*. The two graphs may look closer together toward the left and the right than in the middle, but this is because we are judging the distance visually and nonvertically. The vertical distance between the two graphs is constant*

VERTICAL SHIFTS

Compared to the graph of $y = f(x)$, the graph of $y = f(x) + k$ is shifted vertically by k units:

$k > 0$, shift upward k units;

$k < 0$, shift downward k units.

If a vertical shift applies to a graph that is on instant recall, then the new graph can be sketched very quickly.

Example 1 *Quickly sketch the graph of* $y = x^2 + 1$.

Solution The graph of $f(x) = x^2$ should be on instant recall. What we want adds 1 to the output:

$$y = x^2 \qquad \text{(instant recall)}$$

$$y = x^2 + 1. \qquad \text{(shift up 1)}$$

Here, $k = 1$. Because k is positive, the shift is upward. To get the graph of $y = x^2 + 1$, we shift the graph of $y = x^2$ up 1. The result is in figure 2.

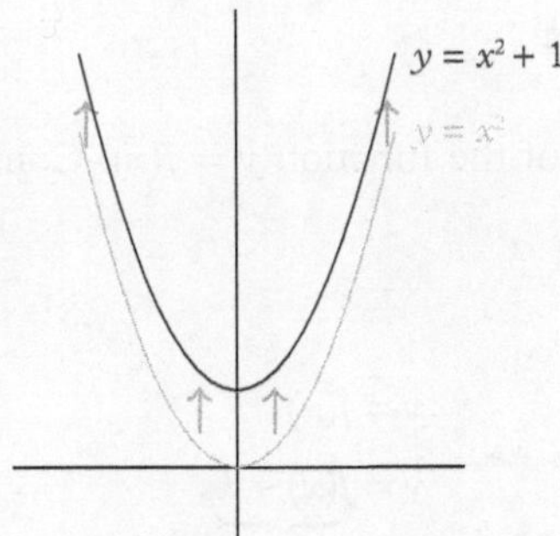

Figure 2 *The graph of $y = x^2 + 1$ (blue), shifted upward from the instantly recalled graph of $y = x^2$ (gray). Look at the left-hand tips of the two graphs; the vertical distance between the blue and gray tips is the same as the vertical distance between the graphs at their bottoms*

■

If $k < 0$, then the shift is downward.

Example 2 *Quickly sketch the graph of $y = x^2 - 2$.*

Solution The graph of $f(x) = x^2$ should be on instant recall. What we want subtracts 2 from the output:

$$y = x^2 \qquad \text{(instant recall)}$$

$$y = x^2 - 2. \qquad \text{(shift down 2)}$$

Here, $k = -2$. Because k is negative, the shift is downward. To get the graph of $y = x^2 - 2$, we shift the graph of $y = x^2$ down 2. The result is in figure 3.

■

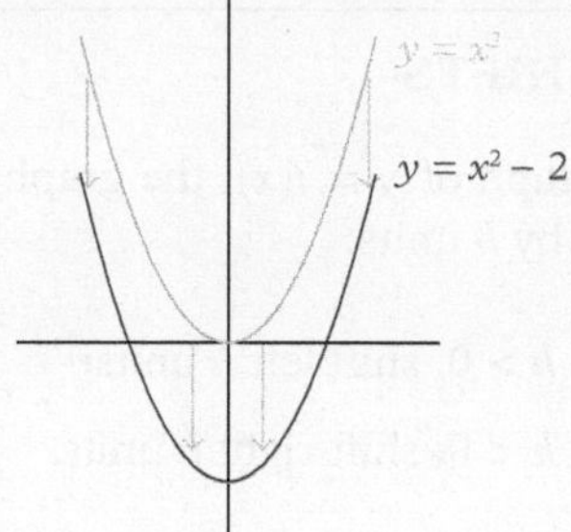

Figure 3 *The graph of* $y = x^2 - 2$ *(blue), shifted downward from the instantly recalled graph of* $y = x^2$ *(gray)*

Horizontal shifts

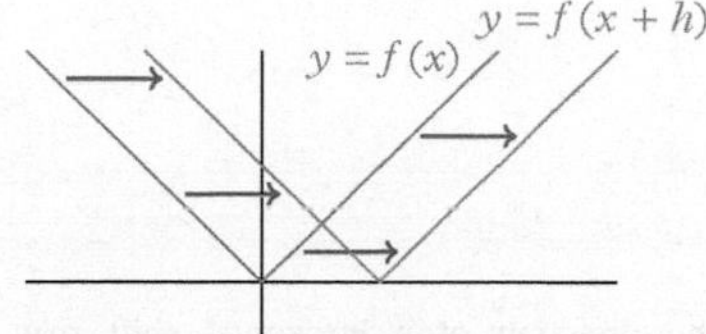

Figure 4 *The graph of a function* $y = f(x)$ *(green) and the horizontally shifted function* $y = f(x + h)$ *(brown). The red arrows indicate the shift. Here,* $h < 0$

Instead of adding k to the output, what if we added h to the input? Start with the graph of the function $y = f(x)$. Consider the graph (figure 4) of $y = f(x + h)$:

$$y = f(x),$$

$$y = f(\underbrace{x + h}_{\text{input}}).$$

Here, h has been added to the input. The inputs are represented by horizontal coordinates on a graph, so the graph of $y = f(x + h)$ shifts h units horizontally from the graph of $y = f(x)$. But in which direction? A numerical comparison of $y = x^2$ to $y = (x - 1)^2$ is helpful in this regard (table 1).

If $f(x) = x^2$, then $f(x - 1) = (x - 1)^2$.

Table 1 *Comparing y-coordinates of* $y = x^2$ *and* $y = (x - 1)^2$*. The graph of* $y = (x - 1)^2$ *is a shift to the right of 1 unit from the graph of* $y = x^2$*.*

x	$y = x^2$	$y = (x - 1)^2$
−2	4	9
−1	1	4
0	0	1
1	1	0
2	4	1
3	9	4

HORIZONTAL SHIFTS

Compared to the graph of $y = f(x)$, the graph of $y = f(x + h)$ is shifted horizontally by h units:

$$h > 0, \text{ shift left } h \text{ units;}$$

$$h < 0, \text{ shift right } h \text{ units.}$$

In table 1, where $h = -1$, when moving from the middle column to the right-hand column, the shift is in the direction of larger x-coordinates. Larger x-coordinates are to the right on a graph. Therefore, when $h < 0$, we shift to the right. Similarly, if $h > 0$, we shift to the left.

Just as with vertical shifts, horizontal shifts can help us gain a quick mental image of a function's graph.

The direction of a horizontal shift may seem counterintuitive. It may help to realize that a horizontal shift changes the variable's value **before** evaluating the original function f, whereas in a vertical shift, the value is changed **after** the original function f is evaluated.

Example 3 *Quickly sketch the graph of* $y = \sqrt{x - 4}$.

Solution The graph of $y = \sqrt{x}$ should be on instant recall. Our function replaces x with $x - 4$:

$$y = \sqrt{x}, \qquad \text{(instant recall)}$$

$$y = \sqrt{x - 4}. \qquad \text{(shift right 4)}$$

Here, $h = -4$. Because h is negative, the shift is to the right 4 units (figure 5).

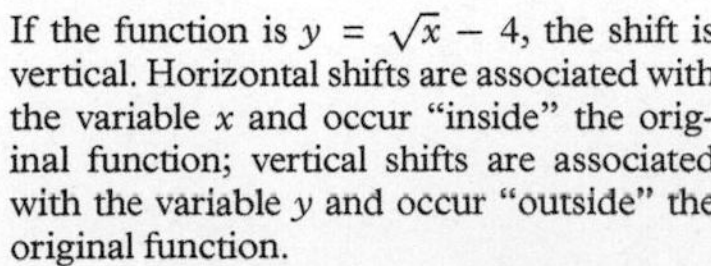
If the function is $y = \sqrt{x} - 4$, the shift is vertical. Horizontal shifts are associated with the variable x and occur "inside" the original function; vertical shifts are associated with the variable y and occur "outside" the original function.

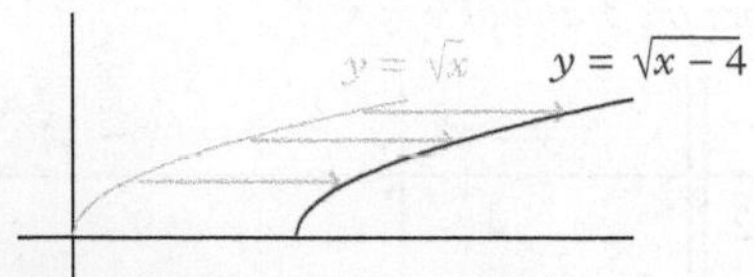

Figure 5 *The graph of* $y = \sqrt{x - 4}$ *(blue), shifted right from the instantly recalled graph of* $y = \sqrt{x}$ *(gray)*

■

Reflections

Start with the graph of $y = f(x)$ and consider the graph of $y = -f(x)$:

$$y = f(x),$$
$$y = \underbrace{-f(x)}_{\text{output}}.$$

Again, the output is changed, this time by taking its negative. Instead of a y-coordinate of 2, we have -2; instead of -5, we have 5. Look at the scale on the y-axis and you can see that points above the x-axis go below it by the same amount, whereas points below the x-axis come back above it by the same amount (figure 6). The effect is to reflect the image across the x-axis (vertical reflection).

If instead we start with $y = f(x)$ and consider the graph of $y = f(-x)$, the negative is associated with the input instead of the output, and the reflection is across the y-axis (horizontal reflection). The reflections are illustrated in figure 6.

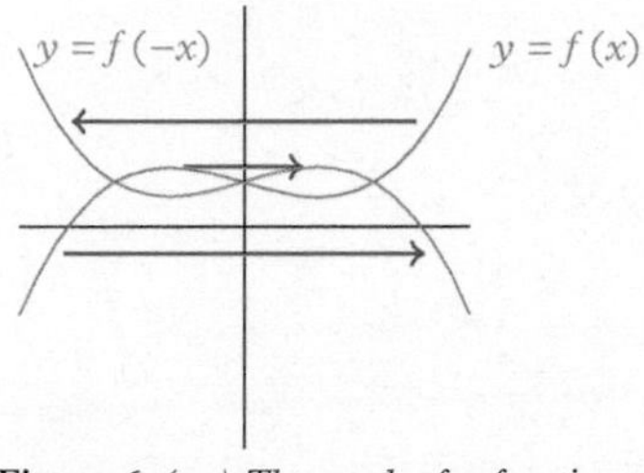

Figure 6 *(top) The graph of a function $y = f(x)$ (green) and the vertically reflected function $y = -f(x)$ (brown). The red arrows indicate the reflection. (bottom) The graph of a function $y = f(x)$ (green) and the horizontally reflected function $y = f(-x)$ (brown). The red arrows indicate the reflection*

> **REFLECTIONS**
>
> Compared to the graph of $y = f(x)$, the graph of $y = -f(x)$ is reflected vertically across the x-axis.
>
> Compared to the graph of $y = f(x)$, the graph of $y = f(-x)$ is reflected horizontally across the y-axis.

Example 4 *Quickly sketch the graph of $y = \sqrt{-x}$.*

Solution The graph of $y = \sqrt{x}$ should be on instant recall. Our function replaces x with $-x$, which causes a horizontal reflection:

$$y = \sqrt{x}, \qquad \text{(instant recall)}$$
$$y = \sqrt{-x}. \qquad \text{(horizontal reflection)}$$

The result is in figure 7.

■

Notice that in example 4, the domain of the function $y = \sqrt{-x}$ is $(-\infty, 0]$.

To find the domain of $y = \sqrt{-x}$, set what's inside the square root to be zero or larger and solve:

$$-x \geq 0$$
$$x \leq 0.$$

Example 5 *Quickly sketch the graph of $y = -x^2 + 4$.*

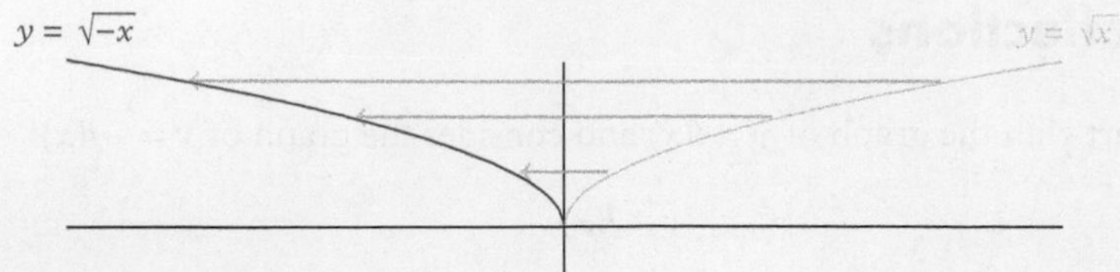

Figure 7 *The graph of $y = \sqrt{-x}$ (blue), reflected horizontally from the instantly recalled graph of $y = \sqrt{x}$ (gray)*

Solution The graph of $y = x^2$ should be on instant recall. Our function features $-x^2$, which causes a vertical reflection. The +4 causes a vertical shift upward. The order of the graph effects (reflection then shift) is determined by the order of operations:

$$y = x^2, \qquad \text{(instant recall)}$$
$$y = -x^2, \qquad \text{(vertical reflection)}$$
$$y = -x^2 + 4. \qquad \text{(vertical shift up 4)}$$

Had the function been $y = -(x^2 + 4)$, we shift first and reflect second:

$$y = x^2, \qquad \text{(instant recall)}$$
$$y = x^2 + 4, \qquad \text{(vertical shift up 4)}$$
$$y = -(x^2 + 4). \qquad \text{(vertical reflection)}$$

First we reflect vertically (figure 8, left) and then we shift vertically up 4 (figure 8, right).

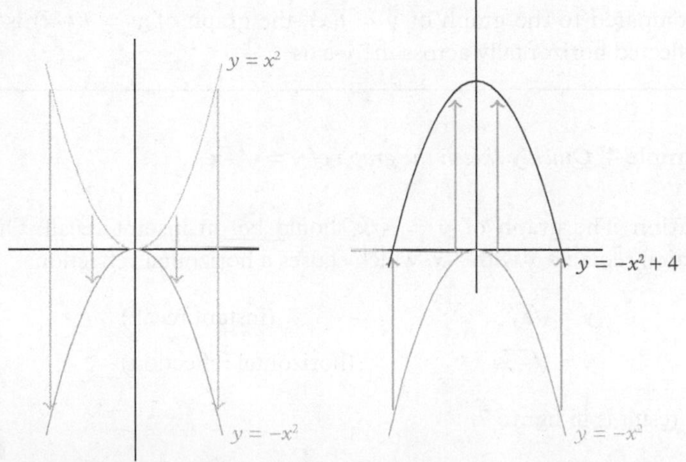

Figure 8 *(left) The graph of $y = -x^2$ (pink), reflected vertically from the instantly recalled graph of $y = x^2$ (gray); (right) the graph of $y = -x^2 + 4$ (blue), shifted upward from the graph of $y = -x^2$ (pink)*

■

With practice, these exercises can be accomplished with mental images in a few seconds.

Stretching

A trick that many children of my generation tried was to press Silly Putty on the Sunday comics and then stretch the putty, distorting the faces from the comics. The same effect can be gained on the graph of a function by multiplying the input or output of a function by a positive stretching factor c.

Start with the graph of $y = f(x)$ and multiply the output by a factor of c to arrive at $y = c \cdot f(x)$. The effect is to stretch the graph vertically. If $c > 1$, the outputs are larger, so the graph expands; if $0 < c < 1$, the outputs are smaller, so the graph shrinks (expand and shrink as a visual effect can sometimes be in the eye of the beholder; expand means that y-coordinates get farther from the x-axis, whereas shrink means that y-coordinates get closer to the x-axis). See figure 9.

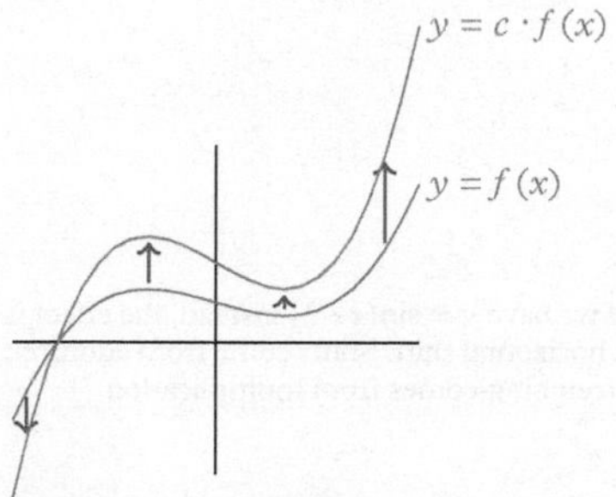

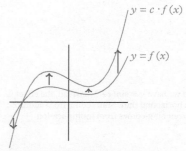

Figure 9 *The graph of $y = f(x)$ (green) and the vertically stretched $y = c \cdot f(x)$ (brown). The red arrows indicate the direction of stretching. Here, $c > 1$. Notice that the x-intercepts do not change*

The same visual effect can be accomplished by changing the *aspect ratio* on a graph. (The aspect ratio is the ratio of the length of one unit as it appears on the vertical scale to one unit as it appears on the horizontal scale.) Some computer algebra systems change the aspect ratio on graphs rather freely, reducing or increasing the "steepness" of a graph so that it can be seen and interpreted more readily. As a result, some "stretched" graphs may not appear stretched at all.

> **VERTICAL STRETCHING**
>
> Compared to the graph of $y = f(x)$, the graph of $y = c \cdot f(x)$ is stretched vertically by a factor of c:
>
> $c > 1$, graph expands;
>
> $0 < c < 1$, graph shrinks.

Multiplying the input by a factor of c to arrive at $y = f(cx)$ affects the inputs and stretches the graph horizontally. Just as with shifting, the effect is opposite of intuition: $c > 1$ shrinks (the x-coordinates get closer to the y-axis) and $c < 1$ expands (the x-coordinates get farther from the y-axis). See figure 10.

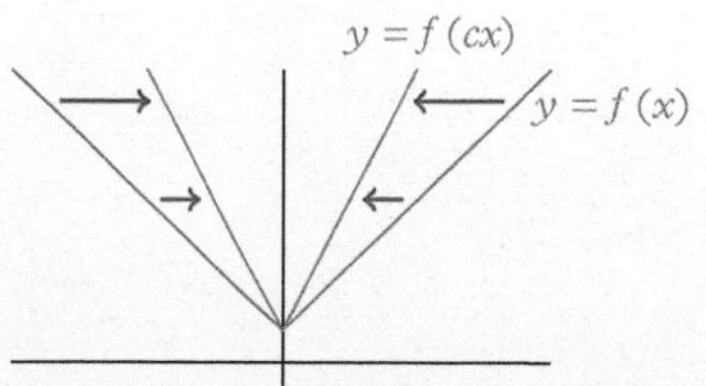

Figure 10 *The graph of $y = f(x)$ (green) and the horizontally stretched $y = f(cx)$ (brown). The red arrows indicate the direction of stretching. Here, $c > 1$. Notice that the y-intercept does not change*

HORIZONTAL STRETCHING

Compared to the graph of $y = f(x)$, the graph of $y = f(cx)$ is stretched horizontally by a factor of c:

$$c > 1\text{, graph shrinks;}$$

$$0 < c < 1\text{, graph expands.}$$

Example 6 *Quickly sketch the graph of* $y = \sin 3x$.

If we have $y = \sin(x+3)$ instead, the effect is a horizontal shift. Shifts come from addition; stretching comes from multiplication.

Solution The graph of $y = \sin x$ should be on instant recall. In our function, we replace x with $3x$, and the result is horizontal stretching. Since $c = 3 > 1$, the x-coordinates move closer to the y-axis by a factor of 3 (one-third as far from the axis). The graph is in figure 11.

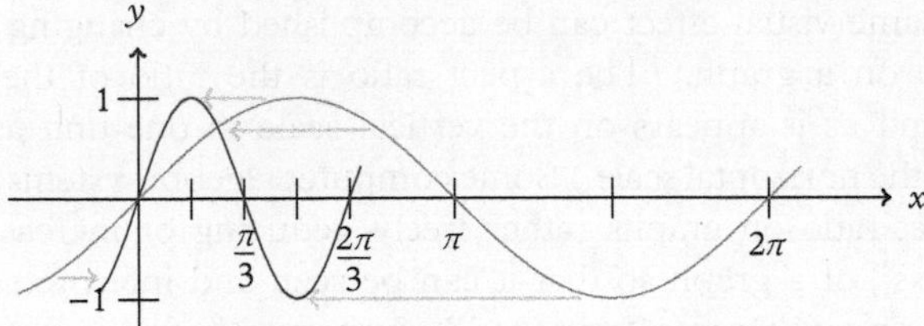

Figure 11 *The graph of* $y = \sin 3x$ *(blue), stretched horizontally from the instantly recalled graph of* $y = \sin x$ *(gray), with light-gray arrows indicating the direction of stretching*

■

Notice from figure 11 that the period of $y = \sin 3x$ is not 2π but rather $\frac{2\pi}{3}$. Horizontal stretching changes the period of trig functions (and other periodic functions as well).

Combining several types of graph transformations—shifting, stretching, reflecting—no longer makes the mental image as quick to come by, but it can still be useful to think in these terms.

Example 7 *Sketch the graph of* $y = -2(x - 3)^2 + 4$.

Solution Graphs of this type are quite often sketched by plotting points, but shifting, reflecting, and stretching can be used instead. Start with the instantly recalled graph of $y = x^2$ and go from there:

The order of graph effects comes from the order of operations. We start inside with the x and move outward.

$$y = x^2, \qquad \text{(instant recall)}$$

$$y = (x-3)^2, \qquad \text{(shift right 3)}$$

$$y = 2(x-3)^2, \qquad \text{(stretch vertically)}$$

$$y = -2(x-3)^2, \qquad \text{(reflect vertically)}$$

$$y = -2(x-3)^2 + 4. \qquad \text{(shift up 4)}$$

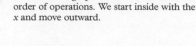

The sequence of graphs, without arrows indicating the changes, is in figure 12.

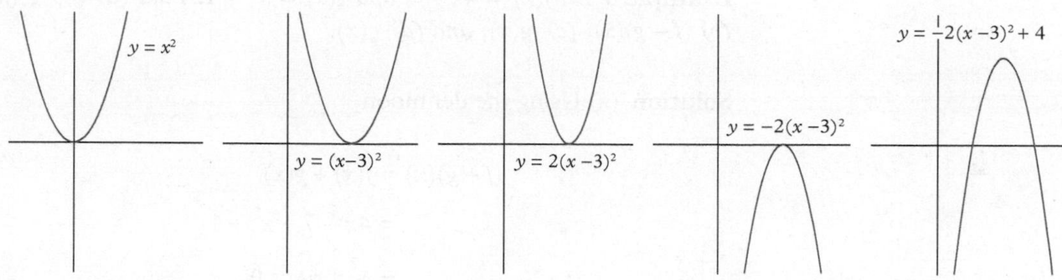

Figure 12 *Stages in determining the graph of $y = -2(x-3)^2 + 4$ using shifting, reflection, and stretching. All graphs are in the same graphing "window" for the sake of comparison*

■

In figure 11, it is clear that the stretching of example 6 is horizontal and not vertical. Yet, there are some functions for which it seems difficult visually to tell which type of stretching is occurring, because the stretching could be interpreted as being either vertical or horizontal. Consider, for example, $g(x) = \sqrt{4x}$. This is the instantly recalled $y = \sqrt{x}$, with x replaced by $4x$, so the graph is stretched horizontally (toward the y-axis) by a factor of 4. But, extracting the square root of 4 gives us $g(x) = 2\sqrt{x}$, which is $y = \sqrt{x}$, stretched vertically by a factor of 2 (away from the x-axis). The two interpretations of the stretching are shown in figure 13.

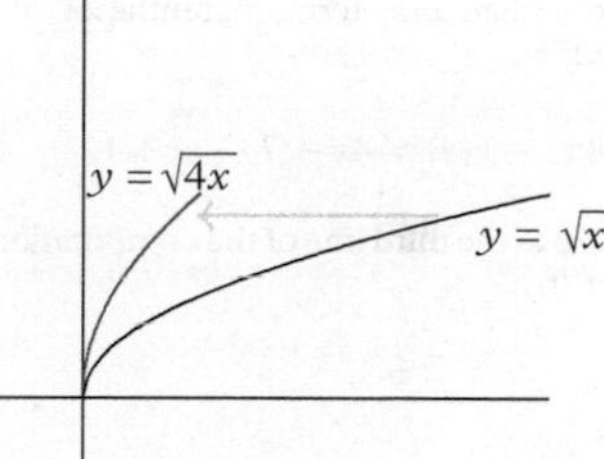

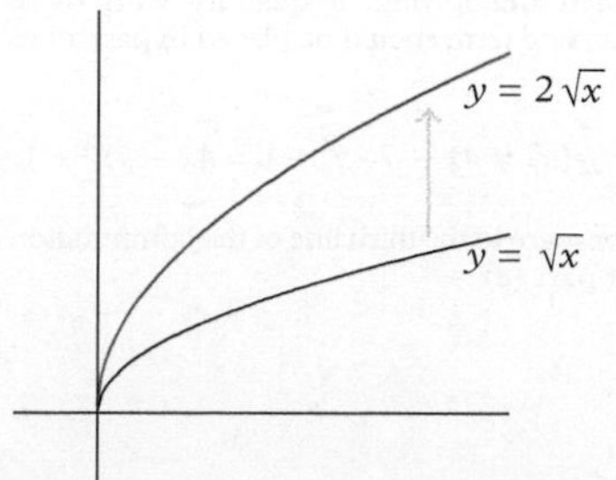

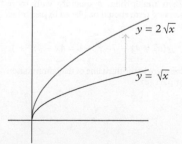

Figure 13 *The graph of $g(x) = \sqrt{4x} = 2\sqrt{x}$ (blue) stretched from the instantly recalled graph of $y = \sqrt{x}$ (gray), pictured as horizontal stretching (top) and as vertical stretching (bottom)*

Combining functions

Given two functions named f and g, we call their sum by the name $f + g$, their difference $f - g$, their product fg and their quotient $\frac{f}{g}$. These are names of functions, but they represent operations on their outputs.

Definition 6 COMBINATIONS OF FUNCTIONS *Given two functions f and g, for values of x in the domain of both functions we define*

$$(f+g)(x) = f(x) + g(x),$$
$$(f-g)(x) = f(x) - g(x),$$
$$fg(x) = f(x)g(x), \text{ and}$$
$$\frac{f}{g}(x) = \frac{f(x)}{g(x)} \text{ (where } g(x) \neq 0).$$

Example 8 *Let* $f(x) = 4x - 7$ *and* $g(x) = x^2 + 1$. *Find (a)* $(f+g)(x)$, *(b)* $(f-g)(x)$, *(c)* $fg(x)$, *and (d)* $\frac{f}{g}(x)$.

Solution (a) Using the definition,

$$\begin{aligned}(f+g)(x) &= f(x) + g(x)\\ &= 4x - 7 + x^2 + 1\\ &= x^2 + 4x - 6.\end{aligned}$$

CAUTION: MISTAKE TO AVOID
When subtracting a quantity $g(x)$ containing more than one term, parentheses are required:

$$f(x) - g(x) \neq 4x - 7 - x^2 + 1.$$

Compare to the third line of the computation for part (b).

(b) Using the definition,

$$\begin{aligned}(f-g)(x) &= f(x) - g(x)\\ &= (4x-7) - (x^2+1)\\ &= 4x - 7 - x^2 - 1\\ &= -x^2 + 4x - 8.\end{aligned}$$

CAUTION: MISTAKE TO AVOID
When multiplying, a quantity with more than one term should be placed in parentheses:

$$f(x)g(x) \neq 4x - 7 \cdot x^2 + 1 = 4x - 7x^2 + 1.$$

Compare to the third line of the computation for part (c).

(c) Using the definition,

$$\begin{aligned}fg(x) &= f(x)g(x)\\ &= (4x-7)(x^2+1)\\ &= 4x^3 + 4x - 7x^2 - 7.\end{aligned}$$

(d) Using the definition,

$$\begin{aligned}\frac{f}{g}(x) &= \frac{f(x)}{g(x)}\\ &= \frac{4x-7}{x^2+1}.\end{aligned}$$ ■

Composite functions

Often a manufacturing process is not accomplished all in one step; more than one machine is required, with the output from one machine comprising the input to another. The same can be true when we picture functions as machines—one function's output can be another function's input, as in figure 14.

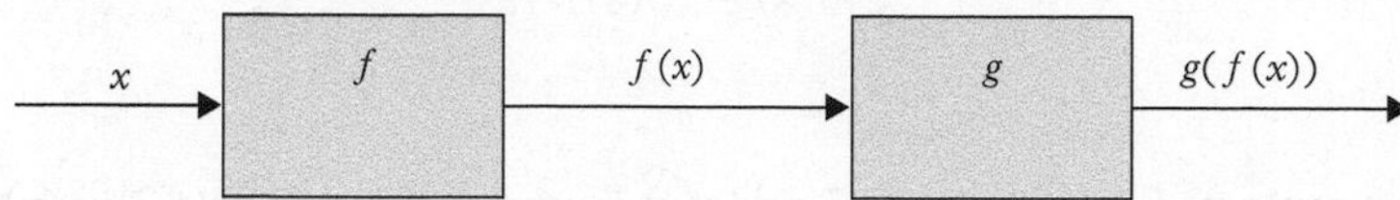

Figure 14 *Composition of functions*

Sometimes what appears to be a single machine actually consists of several machines inside, but with one shell placed around the entire process. We can place a box around the two functions of figure 14 to make it look like one function, as in figure 15:

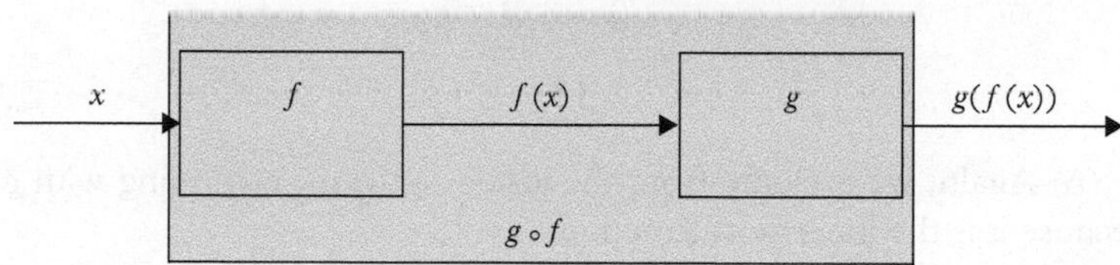

Figure 15 *Composition of functions, pictured as a single function*

To the outside observer unaware of the inner mechanism, the outer shell makes it appear to be a single function, as in figure 16. We call this single function the *composition* of the functions f and g, denoted by $g \circ f$.

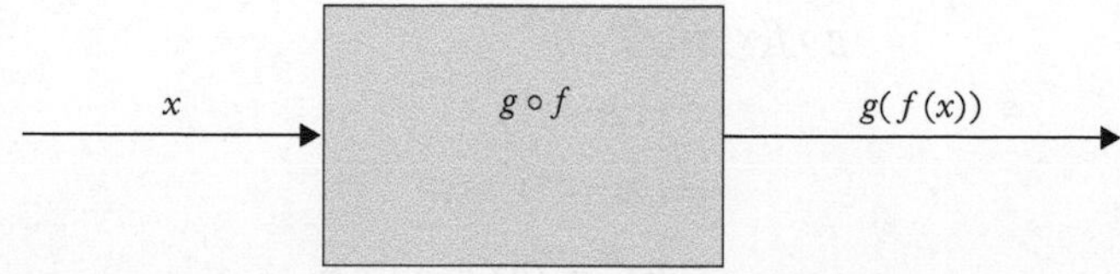

Figure 16 *Composition of functions, pictured as a single function, with inner functions hidden*

The notation $g \circ f(x)$ means $g(f(x))$ and is the reason for the order (g on the left, f on the right), even though f is executed first and g,

The expression $g \circ f(x)$ can be read "g of f of x" or "g composed with f of x." The name of the composite function is $g \circ f$.

second. When evaluating expressions, we move from inside the parentheses outward, and the composition notation is consistent with that rule.

Definition 7 COMPOSITION OF FUNCTIONS *Where x is in the domain of f and f(x) is in the domain of g, we define the composite function $g \circ f$ by*

$$g \circ f(x) = g(f(x)).$$

Example 9 *Let $f(x) = 2x - 7$ and $g(x) = x^2 + 4$. Find (a) $g(f(2))$, (b) $f(g(x))$, and (c) $g \circ f(x)$.*

Color-coding can help. Let's use bold to color the computation of $f(2)$, which takes place "inside" g:

$$g(f(2)) = g(2(2) - 7)$$
$$= g(-3)$$
$$= (-3)^2 + 4 = 13.$$

When $g(-3)$ is reached, the computation proceeds as expected.

Solution (a) We evaluate from the inside outward. First we evaluate f at 2 to get -3 and then we evaluate g at -3 to get 13:

$$\begin{aligned} g(f(2)) &= g(2(2) - 7) \\ &= g(-3) \\ &= (-3)^2 + 4 \\ &= 13. \end{aligned}$$

(b) Again, we evaluate from the inside outward, beginning with g because it is the innermost function:

The first step replaces $g(x)$ by its value $x^2 + 4$.

$$\begin{aligned} f(g(x)) &= f(x^2 + 4) \\ &= 2(x^2 + 4) - 7 \\ &= 2x^2 + 8 - 7 \\ &= 2x^2 + 1. \end{aligned}$$

(c) Here we use the definition of composition:

In line 2, $f(x)$ is replaced by its value $2x - 7$.

$$\begin{aligned} g \circ f(x) &= g(f(x)) \\ &= g(2x - 7) \\ &= (2x - 7)^2 + 4 \\ &= 4x^2 - 28x + 49 + 4 \\ &= 4x^2 - 28x + 53. \end{aligned}$$ ■

Parts (b) and (c) demonstrate that, in general, $f(g(x)) \neq g(f(x))$; that is, $f \circ g \neq g \circ f$. The order matters in a composition.

Putting on socks and shoes is a composition of two processes. The order in which these operations are performed (socks first, shoes second or shoes first, socks second) definitely affects the outcome!

Example 10 *Write* $y = \sqrt{(2x+5)^3}$ *as a composition of functions, using* $f(x) = \sqrt{x}$, $g(x) = 2x + 5$, *and* $h(x) = x^3$.

Solution We evaluate an expression from the inside out, and that gives the order of the composition. We take $2x + 5$ first, then its cube, then its square root:

Recognizing the pieces of a composition is an important skill for success in calculus, even though the compositions are not usually written as explicitly as in the solution to example 10.

$$\begin{aligned} y &= \sqrt{(2x+5)^3} \\ &= f(h(g(x))) \\ &= f \circ h \circ g(x). \end{aligned}$$ ■

Exercises 0.5

1–10. Rapid response: which techniques are used—shifting, reflecting, and/or stretching—to obtain the graph from that of $y = x^2$?

1. $y = (x-3)^2$
2. $y = 4x^2$
3. $y = x^2 - 3$
4. $y = (x+4)^2$
5. $y = -3x^2$
6. $y = x^2 + 4$
7. $y = \dfrac{x^2}{-3}$
8. $y = \dfrac{x^2}{4}$
9. $y = (3x)^2$
10. $y = 4(x+4)^2$

11–20. Rapid response: which instantly recognized graph is a good starting point for visualizing the graph of the given function?

11. $y = \sqrt{x} - 7$
12. $y = x^3 + 5$
13. $y = |x-9|$
14. $y = \frac{1}{x-5}$
15. $y = 5 \sin 3x$
16. $y = 4 + \sqrt[3]{x}$
17. $y = 4(x-3)^2 + 9$
18. $y = \cos(4x + \pi)$
19. $y = 3 - \frac{1}{x^2}$
20. $y = 9 - 4|x-1|$

21–30. Rapid response: is the shift up, down, left, or right?

21. $y = \sqrt{x-2}$
22. $y = \sin(x+\pi)$
23. $y = \cos x + \pi$
24. $y = 2 + \sqrt[3]{x}$
25. $y = x^2 - 5$
26. $y = -2 + x^2$
27. $y = |x+9|$
28. $y = (x+1)^3$
29. $y = 5 + \frac{1}{x}$
30. $y = |x-2|$

31–46. Quickly sketch the graph of the given function.

31. $f(x) = x^2 - 5$
32. $f(x) = \sqrt{x+2}$
33. $y = |x+1|$
34. $h(x) = 1 + \cos x$
35. $g(x) = \dfrac{1}{x-6}$
36. $f(t) = 4 - t^2$
37. $s(t) = 2 - \sqrt{t}$
38. $y = \dfrac{1}{x} - 6$
39. $f(x) = 3\sin x + 2$
40. $f(x) = -x^3 + 7$
41. $f(x) = \frac{1}{2}\sqrt{x+4}$
42. $f(x) = 3 + 2|x-1|$
43. $f(x) = \sqrt[3]{x+1} - 1$
44. $f(x) = 2(x+3)^2 - 1$
45. $f(x) = -\dfrac{x^2}{3}$
46. $f(x) = 5\sqrt{x} + 1$

47–54. For the given functions, find (a) $(f+g)(x)$, (b) $(f-g)(x)$, (c) $fg(x)$, and (d) $\frac{f}{g}(x)$.

47. $f(x) = 3x + 5$, $g(x) = x - 2$
48. $f(x) = x^2$, $g(x) = 3\sqrt{x}$
49. $f(x) = \dfrac{1}{x+1}$, $g(x) = x^2 + x$
50. $f(x) = x + 8$, $g(x) = 6 - x^2$
51. $f(x) = x + \sin x$, $g(x) = x - \sin x$
52. $f(x) = |x+2|$, $g(x) = |x+3|$
53. $f(x) = \sqrt{x^2 - 9}$, $g(x) = x - 3$
54. $f(x) = 8x^3$, $g(x) = 2x^2$

55–64. Find (a) $g(f(0))$, (b) $f \circ g(x)$, and (c) $g \circ f(x)$.

55. $f(x) = 8x^3$, $g(x) = 2x^2$
56. $f(x) = x^2$, $g(x) = 3\sqrt{x}$
57. $f(x) = \dfrac{1}{x}$, $g(x) = x^2$

58. $f(x) = x + 8$, $g(x) = 6 - x^2$
59. $f(x) = 5x + \sqrt{x}$, $g(x) = x - 7$
60. $f(x) = \dfrac{x+5}{x-2}$, $g(x) = \dfrac{1}{x}$
61. $f(x) = |x + 12|$, $g(x) = x^2 + 1$
62. $f(x) = \sin x$, $g(x) = 3x + 7$
63. $f(x) = \dfrac{3}{x^2}$, $g(x) = \dfrac{x^2}{3}$
64. $f(x) = 5x^2 - 1$, $g(x) = x^2 + 1$

65. Find (a) $f(f(x))$, (b) $g \circ g(x)$, and (c) $g \circ f \circ g(x)$ for $f(x) = 3x+1$, $g(x) = x^2$.

66. Find (a) $f(f(x))$ and $f \circ f \circ f(x)$ for $f(x) = 5x + 7$.

67–72. Write the given function as a composition of functions. Use $f(x) = x^2$, $g(x) = x - 17$, and $h(x) = \sqrt{x}$.

67. $y = (x - 17)^2$
68. $y = \sqrt{x - 17}$
69. $y = (\sqrt{x} - 17)^2$
70. $y = \sqrt{x^2 - 17}$
71. $y = (x - 17)^2 - 17$
72. $y = \sqrt{\sqrt{x - 17} - 17}$

73–78. Write the given function as a composition of functions.

73. $y = (11x + 5)^4$
74. $y = \tan(5x - \pi)$
75. $y = \sin^2(x - 3)$
76. $y = \cos(x + 5)^2$
77. $y = \dfrac{1}{\sqrt{19x + 1}}$
78. $y = \sqrt{4 + \cos(x + x^2)}$

Avoiding Common Errors

0.6

Learning to drive a car consists not only of what to do, such as using the pedals, mirrors, and steering wheel, but also of how to avoid wrecks (defensive driving), often by learning what not to do—what mistakes to avoid. Unfortunately, many students who do a reasonably good job of learning what to do in calculus still fail as a result of having never learned what *not* to do in algebra.

If you are hungry while walking through the woods, is it advisable to pick just any mushroom or berry you find and pop it in your mouth? Likewise, just trying a random algebraic manipulation is fraught with danger if you do not know how to recognize what's toxic and what isn't.

One approach to making good use of this section is to work (or rework) the section corresponding to each mistake made, each time the mistake is made, until the error has been eliminated from one's habits.

Each subsection consists of a common error made by calculus students, an explanation of why it is incorrect, and exercises designed to help learn how to recognize and avoid making the error. The exercises are presented at the end of every subsection to facilitate the use of this material as a reference to consult each time a particular type of mistake is made.

Squaring a binomial

CAUTION: MISTAKE TO AVOID

$$(a+b)^2 \neq a^2 + b^2$$

Try it with $a = 2$, $b = 5$:

$$(2+5)^2 \stackrel{?}{=} 2^2 + 5^2$$

$$7^2 \stackrel{?}{=} 4 + 25$$

$$49 \neq 29.$$

Although one example is not enough to prove a theorem, one counterexample is enough to disprove a theorem.

The correct formula is $(a + b)^2 = a^2 + 2ab + b^2$. Remembering the formula when prompted is not the issue; remembering to use the formula when it applies is the issue.

SQUARING A BINOMIAL

$$(a+b)^2 = a^2 + 2ab + b^2$$

The formula is derived as follows:

$$\begin{aligned}(a+b)^2 &= (a+b)(a+b)\\ &= a^2 + ab + ba + b^2\\ &= a^2 + 2ab + b^2.\end{aligned}$$

Example 1 *Which of the following calculations is correct and which is incorrect? Answer "correct" or "wrong."*

(a) $(x+3)^2 = x^2 + 9$
(b) $(y-1)^2 = y^2 - 2y + 1$
(c) $(5+2t)^2 = 25 + 4t^2$
(d) $(x-y)^2 = x^2 + y^2$
(e) $(x-y)^2 = x^2 - y^2$
(f) $(x-y)^2 = x^2 - 2xy + y^2$

The "middle term" refers to the "$2ab$" in the formula for squaring a binomial.

Solution (a) Wrong, because the middle term is missing.
(b) Correct!
(c) Wrong, because the middle term is missing.
(d) Wrong, because the middle term is missing.
(e) Wrong, because the middle term is missing.
(f) Correct! ■

Example 2 *Expand* $(8+3x)^2$.

Notice the use of parentheses with $(3x)^2$. Writing $3x^2$ instead is incorrect:

$$(3x)^2 \neq 3x^2.$$

Solution We use the formula for squaring a binomial with $a = 8$, $b = 3x$:

$$(\underbrace{8}_{a} + \underbrace{3x}_{b})^2 = \underbrace{8^2}_{a^2} + \underbrace{2\cdot 8\cdot 3x}_{2\cdot a\cdot b} + \underbrace{(3x)^2}_{b^2}$$
$$= 64 + 48x + 9x^2.$$ ■

Example 3 *Expand* $(1-2x)^2$.

Subtraction is the same as adding the negative:

$$1 - 2x = 1 + (-2x).$$

Solution We use the formula for squaring a binomial with $a = 1$, $b = -2x$:

$$(\underbrace{1}_{a} + \underbrace{(-2x)}_{b})^2 = \underbrace{1^2}_{a^2} + \underbrace{2\cdot 1\cdot(-2x)}_{2\cdot a\cdot b} + \underbrace{(-2x)^2}_{b^2}$$
$$= 1 - 4x + 4x^2.$$ ■

Exercises for squaring a binomial

1–30. Which of the following calculations is correct and which is incorrect? Answer "correct" or "wrong."

1. $(x-4)^2 = x^2 - 16$
2. $(x+10)^2 = x^2 + 100$
3. $(x+1)^2 = x^2 + 2x + 1$
4. $(t-7)^2 = t^2 + 49$
5. $(x+0)^2 = x^2$
6. $(3x+2)^2 = 9x^2 + 4$
7. $(3+\sqrt{x})^2 = 9 + x$
8. $(6x+1)^2 = 36x^2 + 12x + 1$
9. $(2r-12)^2 = 4r^2 - 48r + 144$
10. $(7t-9)^2 = 49t^2 - 81$
11. $(3m+k)^2 = 9m^2 + k^2$
12. $(x+h)^2 = x^2 + 2xh + h^2$
13. $(\sin\theta + \cos\theta)^2 = \sin^2\theta + \cos^2\theta = 1$
14. $(x+\sin x)^2 = x^2 + 2x\sin x + \sin^2 x$
15. $\left(\frac{1}{x}+1\right)^2 = \frac{1}{x^2} + 1$
16. $(\pi+\sqrt{3})^2 = \pi^2 + 2\pi\sqrt{3} + 3$
17. $(x+\omega)^2 = x^2 + \omega^2$
18. $(x+\omega)^2 = x^2 + 2\omega + \omega^2$
19. $(11x-15)^2 = 121x^2 + 330x + 225$
20. $(y-7)^2 = y^2 - 49$
21. $(2xy-k)^2 = 4x^2y^2 + k^2$
22. $(3+x+2y)^2 = 9 + x^2 + 4y^2$
23. $(2x-3)^3 = 8x^3 - 27$
24. $(\sqrt{x}+\sqrt{y})^2 = x + 2\sqrt{xy} + y$
25. $(x+\sqrt{x})^2 = x^2 + 2x^{3/2} + x$
26. $(3x+2)^4 = 81x^4 + 16$
27. $(x^2+4)^2 = x^4 + 16$
28. $(1+\tan x)^2 = 1 + \tan^2 x = \sec^2 x$
29. $(x^2+x^3)^2 = x^4 + 2x^5 + x^6$
30. $\left(x+\frac{1}{x}\right)^2 = x^2 + \frac{1}{x^2}$

ω is the lowercase Greek letter omega.

The formula for cubing a binomial is

$$(a+b)^3 = a^3 + 3a^2b + 3ab^2 + b^3.$$

The formulas for the fourth power, fifth power, and so on, can be remembered most easily through the use of *Pascal's triangle.* (see https://mathworld.wolfram.com/PascalsTriangle.html for more information).

31–42. Expand the expression.

31. $(x+2)^2$	**37.** $(3+\sqrt{x})^2$
32. $(x-5)^2$	**38.** $(2xy-k)^2$
33. $(5x+3)^2$	**39.** $(\sin\theta+\cos\theta)^2$
34. $(2y-4x)^2$	**40.** $(x^2+4)^2$
35. $\left(x+\frac{1}{x}\right)^2$	**41.** $(2x-3)^3$
36. $(2-x)^2$	**42.** $\left(\frac{1}{x}+1\right)^2$

Square root of a sum

CAUTION: MISTAKE TO AVOID

$$\sqrt{a^2+b^2} \neq a+b$$

Also,

$$\sqrt[3]{a^3+b^3} \neq a+b,$$

and the same for other roots as well.

Try it with $a = 3$, $b = 4$:

$$\sqrt{3^2+4^2} \stackrel{?}{=} 3+4$$
$$\sqrt{9+16} \stackrel{?}{=} 7$$
$$\sqrt{25} \stackrel{?}{=} 7$$
$$5 \neq 7.$$

The expression $\sqrt{a^2+b^2}$ cannot be simplified, as tempting as it is.

Since $\sqrt{x^2} = |x|$, the expression $\sqrt{(a+b)^2}$ can be simplified to $|a+b|$.

NO SIMPLIFICATION

The expression $\sqrt{a^2+b^2}$ cannot be simplified.

Recall that the square root is the number that, if squared, gives the expression inside the square root. But,

$$(3+x)^2 \neq 9+x^2.$$

The square-root-of-a-sum error and the squaring-a-binomial error (previous subsection) are actually the same error.

Example 4 *Which of the following calculations is correct and which is incorrect? Answer "correct" or "wrong."*

(a) $\sqrt{9+x^2} = 3+x$
(b) $\sqrt{25-x^2} = 5-x$
(c) $\sqrt{25x^2} = 5|x|$
(d) $\sqrt{x^2+3} = x+\sqrt{3}$

Solution (a) Wrong, because the square root of a sum cannot be simplified.

(b) Wrong, because the square root of a difference cannot be simplified either.

(c) Correct! This is not the square root of a sum, but rather the square root of a product.

(d) Wrong, because the square root of a sum cannot be simplified. ■

Example 5 *Simplify the expression, if possible:*

(a) $\sqrt{x^2+9}$
(b) $\sqrt{4+t^2}$
(c) $\sqrt{4t^2}$
(d) $\sqrt{4x^2+9y^2}$

Solution (a) No simplification is possible.

(b) No simplification is possible.

(c) This expression is not of the same form as parts (a) and (b); instead of a sum inside the square root, there is a product. We simplify by taking the product of the square roots:

SQUARE ROOT OF A PRODUCT

$$\sqrt{ab} = \sqrt{a}\sqrt{b}$$

$$\begin{aligned}\sqrt{4t^2} &= \sqrt{4}\cdot\sqrt{t^2}\\ &= 2|t|.\end{aligned}$$

(d) No simplification is possible. ■

Exercises for square root of a sum

1–24. Which of the following calculations is correct and which is incorrect? Answer "correct" or "wrong."

1. $\sqrt{x^2+4} = x+2$
2. $\sqrt{9x^2+16} = 3x+4$
3. $\sqrt{9x^2} = 3|x|$
4. $\sqrt{x+25} = \sqrt{x}+5$
5. $\sqrt{x+y} = \sqrt{x}+\sqrt{y}$
6. $\sqrt{xy} = \sqrt{x}\sqrt{y}$
7. $\sqrt{y+4} = \sqrt{y}+2$
8. $\sqrt{\sin^2\theta+\cos^2\theta} = \sin\theta+\cos\theta$
9. $\sqrt{16x^4y^2} = 4x^2|y|$

10. $\sqrt{(x^2+3)^2} = x^2 + 3$
11. $\sqrt{16 + \sec^2 x} = 4 + \sec x$
12. $\sqrt{1 + x^2 + y^2} = 1 + |x| + |y|$
13. $\sqrt{(25 - 3y)^2} = |25 - 3y|$
14. $\sqrt[3]{8 + x^3} = 2 + x$
15. $\sqrt{a^2 - b^2} = a - b$
16. $\sqrt{1 + \dfrac{4}{x^2}} = 1 + \dfrac{2}{x}$
17. $\sqrt{x^2 + \alpha^2} = x + \alpha$

α is the lowercase Greek letter alpha.

18. $\sqrt{x^2 + 2 + \dfrac{1}{x^2}} = \sqrt{(x + \dfrac{1}{x})^2} = \left|x + \dfrac{1}{x}\right|$
19. $\sqrt[4]{81x^4 + 1} = 3x + 1$
20. $\sqrt{1 + 10^4} = 1 + 10^2 = 101$
21. $\sqrt{x^2 - 10x + 25} = \sqrt{(x-5)^2} = |x - 5|$
22. $\sqrt{t^2 + 9} = t + 3$
23. $\sqrt{4x - 9} = 2\sqrt{x} - 3$
24. $\sqrt{(x+5)^2 + 49} = x + 5 + 7 = x + 12$

25–36. Simplify the expression, if possible.

25. $\sqrt{36 + x^2}$
26. $\sqrt{144 + t}$
27. $\sqrt{x + 12}$
28. $\sqrt{25 + x^2 + y^2}$
29. $\sqrt{(x + 18)^2}$
30. $\sqrt{9x^4}$
31. $\sqrt{100x^2y^2 + \pi}$
32. $\sqrt{9x^2 - 25}$
33. $\sqrt{\sin^2\theta + \cos^2\theta}$
34. $\sqrt{x^2 + \sin^2 x + \cos^2 x}$
35. $\sqrt{16x^3}$
36. $\sqrt{1 + \tan^2\theta}$

Cancellation

Cancellation (more specifically, multiplicative cancellation) is derived from the fact that $1 \cdot x = x$ for any number x. This is what allows us to reduce fractions:

$$\frac{24}{18} = \frac{6 \cdot 4}{6 \cdot 3} = \frac{6}{6} \cdot \frac{4}{3} = 1 \cdot \frac{4}{3} = \frac{4}{3}.$$

We often write this calculation using *cancellation*:

$$\frac{24}{18} = \frac{\not{6} \cdot 4}{\not{6} \cdot 3} = \frac{4}{3}.$$

The point is that cancellation requires what's being canceled (the 6s, in this case) to be something that can be factored out of both the numerator and the denominator.

CANCELLATION

For any numbers a, b, and c,

$$\frac{ab}{ac} = \frac{b}{c},$$

which can be written as

$$\frac{ab}{ac} = \frac{\not{a} \cdot b}{\not{a} \cdot c} = \frac{b}{c}.$$

Notice the multiplication in the numerator and the denominator. Here, a is a *factor* of the numerator and of the denominator.

Just because a number appears in both the numerator and denominator of a fraction doesn't mean that it can be canceled:

$$\frac{6+8}{6+1} = \frac{14}{7} = 2;$$

but,

$$\frac{\not{6}+8}{\not{6}+1} = \frac{8}{1} = 8.$$

Therefore, canceling the 6s is incorrect!

CAUTION: MISTAKE TO AVOID

Canceling numbers that are not factors of the entire numerator and denominator gives incorrect results:

$$\frac{a+b}{a+c} \neq \frac{b}{c}.$$

Notice the addition in the numerator and the denominator. Here, a is a *term* of the numerator and of the denominator.

Example 6 *Which of the following calculations is correct and which is incorrect? Answer "correct" or "wrong."*

(a) $\frac{4\cancel{x}}{5\cancel{x}} = \frac{4}{5}$

(b) $\frac{\cancel{x}+4}{\cancel{x}} = 4$

(c) $\frac{x\cancel{^2} + 4\cancel{x} + 7}{\cancel{x} + 3} = \frac{x+4+7}{3} = \frac{x+11}{3}$

(d) $\frac{x\cancel{^2} + 5\cancel{x}}{\cancel{x}} = x + 5$

(e) $\frac{\cancel{x}(x+3)}{\cancel{x}(x^2+5)} = \frac{x+3}{x^2+5}$

(f) $\frac{\cancel{x}}{\cancel{x}(x+4)} = x + 4$

Solution (a) Correct! The x's are factors of the numerator and the denominator.

(b) Wrong, because the x in the numerator is a term, not a factor. The x is added to the 4, not multiplied by the 4.

(c) Wrong, because cancellation in some terms (x^2, $4x$, x), but not others (7, 3), is incorrect.

(c) Put another way, the x cannot be factored from the entire numerator x^2+4x+7 because it is not present in the last term, nor can x be factored from the entire denominator $x+3$ because it is not present in the last term. Therefore, the x's cannot be canceled.

(d) Correct! Although this calculation is correct, it is written in a form that often leads one to make cancellation errors of the type in part (c). A better way to write the calculation is

$$\frac{x^2+5x}{x} = \frac{\cancel{x}(x+5)}{\cancel{x}} = x + 5.$$

(d) Although addition is present in the numerator, the addition is inside parentheses. When canceling, there is still just one term with two factors, x and $(x + 5)$, in the numerator.

Notice that by factoring the x out of the numerator, the result is written in a form in which x is a factor of both the numerator and the denominator.

(e) Correct! The x is a factor of both the numerator and the denominator.

(f) Wrong, because although the x's in the numerator and the denominator cancel, the $x + 4$ does not move to the numerator. The correct calculation is

$$\frac{\cancel{x}}{\cancel{x}(x+4)} = \frac{1}{x+4}.$$

■

Perhaps different details may help with part (f):

$$\frac{x}{x(x+4)} = \frac{x}{x} \cdot \frac{1}{x+4} = 1 \cdot \frac{1}{x+4} = \frac{1}{x+4}.$$

Example 7 *Simplify the expression, if possible:*

(a) $\dfrac{6x^2}{7x}$

(b) $\dfrac{12x^5}{x^2(x-3)}$

(c) $\dfrac{4x+3}{x-7}$

(d) $\dfrac{x^3+5x}{2x^2+7x}$

(e) $\dfrac{x^3+5x+4}{2x^2+7x}$.

Solution (a) Version 1, written using factoring:

$$\frac{6x^2}{7x} = \frac{\cancel{x}(6x)}{\cancel{x}(7)} = \frac{6x}{7}.$$

Version 2, written as cancellation in the exponent (cancellation of one of the x's in the numerator and the denominator):

$$\frac{6x^{\cancel{2}}}{7\cancel{x}} = \frac{6x}{7}.$$

Version 3, written using the laws of exponents:

Recall that $\dfrac{x^a}{x^b} = x^{a-b}$.

$$\frac{6x^2}{7x} = \frac{6}{7}\cdot\frac{x^2}{x^1} = \frac{6}{7}x^{2-1} = \frac{6}{7}x^1 = \frac{6}{7}x.$$

(b) Version 1, written using factoring:

$$\frac{12x^5}{x^2(x-3)} = \frac{\cancel{x^2}(12x^3)}{\cancel{x^2}(x-3)} = \frac{12x^3}{x-3}.$$

Version 2, written as cancellation in the exponent:

$$\frac{12x^{\cancel{5}\,3}}{\cancel{x^2}(x-3)} = \frac{12x^3}{x-3}.$$

Version 3, written using the laws of exponents:

$$\frac{12x^5}{x^2(x-3)} = \frac{12}{x-3}x^{5-2} = \frac{12}{x-3}x^3.$$

(c) Because there is no common factor in the terms of the numerator or the denominator, no simplification (no cancellation) is possible. The x's do not cancel.

(d) There is a common factor of x in each term of the numerator and in each term in the denominator. Therefore factoring the x from the numerator and the denominator allows us to use cancellation:

$$\frac{x^3+5x}{2x^2+7x} = \frac{\cancel{x}(x^2+5)}{\cancel{x}(2x+7)} = \frac{x^2+5}{2x+7}.$$

(e) There is a term in the numerator, x^3+5x+4, that does not have an x in it. Therefore, we cannot factor x out of the entire numerator, and consequently cancellation cannot be used. ■

(e) There is an alternate algebraic manipulation that results in $\frac{x^2+5}{2x+7} + \frac{4}{2x^2+7x}$, in which cancellation is used in obtaining the first fraction, but whether this is considered simpler than the original form of the expression can be debated.

For many of you, the safest procedure is always factor before canceling anything, as in example 7 (a) version 1, (b) version 1, and (d).

Exercises for cancellation

1–30. Which of the following calculations is correct and which is incorrect? Answer "correct" or "wrong."

1. $\frac{\cancel{x}(x^2-6)}{\cancel{x}(x+11)} = \frac{x^2-6}{x+11}$
2. $\frac{3\cancel{x}+4}{\cancel{x}} = 3+4 = 7$
3. $\frac{x^3+\cancel{4x^2}+5}{\cancel{4x^2}+6x-8} = \frac{x^3+5}{6x-8}$
4. $\frac{\cancel{5x^2}(3x-7)}{\cancel{5x^2}(8)} = \frac{3x-7}{8}$
5. $\frac{(x-5)\cancel{(x+2)}}{(x^2+4)\cancel{(x+2)}} = \frac{x-5}{x^2+4}$
6. $\frac{\cancel{x}-5}{4\cancel{x}+2} = \frac{-5}{6}$
7. $\frac{4\cancel{x}}{\cancel{x}-7} = \frac{4}{-7}$
8. $\cancel{x}(x+5)\frac{1}{\cancel{x}} = x+5$
9. $\frac{2\cancel{x+5}}{7\cancel{x+5}} = \frac{2}{7}$
10. $\frac{\cancel{x^2}(x+11)}{3\cancel{x^2}} = \frac{x+11}{3}$
11. $\frac{\cancel{x-11}}{7x\cancel{(x-11)}} = 7x$

12. $\dfrac{\cancel{2x}+4}{x^2+\cancel{2x}} = \dfrac{4}{x^2}$

13. $\dfrac{x\cancel{(x+4)}}{3\cancel{(x+4)}} = \dfrac{x}{3}$

14. $(\cancel{x}+3)\dfrac{1}{\cancel{x}} = 3$

15. $\dfrac{5\cancel{(x-8)}}{\cancel{x-8}} = 5$

16. $\dfrac{x^{\cancel{4}\,3}+3\cancel{x}}{4x^{\cancel{3}\,2}+3x^{\cancel{2}}} = \dfrac{x^3+3}{4x^2+3x}$

In exercise 16, the numbers in gray are not part of the original expression. They have been added as part of the cancellation process, as in example 7(b) version 2.

17. $\dfrac{\cancel{x^2}(2\cancel{x}-7)}{\cancel{x}(4\cancel{x^2}+6)} = \dfrac{-5}{10} = -\dfrac{1}{2}$

18. $\dfrac{\cancel{x^2}(2x-7)}{\cancel{x}(4x+1)} = \dfrac{2x-7}{4x+1}$

19. $\dfrac{\cancel{3}x+\cancel{9}}{\cancel{9}x+\cancel{3}} = \dfrac{x}{x} = 1$

20. $\dfrac{x^3+2x}{7x^2+15x} = \dfrac{\cancel{x}(x^2+2)}{\cancel{x}(7x+15)} = \dfrac{x^2+2}{7x+15}$

21. $\dfrac{\frac{3}{\cancel{2x}}}{\cancel{2x}} = 3$

In a compound fraction the longer fraction bar indicates the order of division. For exercise 21, the original expression is to be interpreted as $(3/2x)/2x$.

22. $\dfrac{x^{\cancel{2}}+4\cancel{x}+\frac{1}{\cancel{x}}}{7\cancel{x}} = \dfrac{x+4+1}{7} = \dfrac{x+5}{7}$

23. $\dfrac{\sqrt{\cancel{x}(x-3)}}{4\cancel{x}} = \dfrac{\sqrt{x-3}}{4}$

24. $\dfrac{\sin(\cancel{x^2}(x+2))}{\cancel{x^2}(4x+9)} = \dfrac{\sin(x+2)}{4x+9}$

25. $\dfrac{{}^{2}\cancel{4}x^2+\cancel{2}x}{{}_{3}\cancel{6}x+1} = \dfrac{2x^2+x}{3x+1}$

In exercise 25, the numbers in gray are not part of the original expression. They have been added as part of the cancellation process.

26. $\dfrac{x\cancel{(x+1)}}{x^2+\cancel{x+1}} = \dfrac{\cancel{x}}{x^{\cancel{2}}} = \dfrac{1}{x}$

27. $\dfrac{\sqrt{x^{\cancel{2}}+7\cancel{x}}}{\sqrt{\cancel{x}}(x-4)} = \dfrac{\sqrt{x+7}}{x-4}$

28. $\dfrac{\cancel{(x+3)}(x-1)}{(2\cancel{x+3})(x+5)} = \dfrac{x-1}{2x+10}$

29. $\dfrac{x(x+1)}{x^2+x+1} = \dfrac{\cancel{x^2}+\cancel{x}}{\cancel{x^2}+\cancel{x}+1} = 1$

30. $\dfrac{x^4+\cancel{3}x}{4x^3+\cancel{3}x^2} = \dfrac{x^4+x}{4x^3+x^2} = \dfrac{\cancel{x}(x^3+1)}{\cancel{x}(4x^2+x)} = \dfrac{x^3+1}{4x^2+x}$

31–46. Simplify the expression, if possible.

31. $\dfrac{3x^2 + 4x}{x^2 - 6x}$

32. $\dfrac{x^4 - 9x^3}{x^5 + x^2}$

33. $\dfrac{3x + 9}{9x + 3}$

34. $\dfrac{7x + 3}{x}$

35. $\dfrac{5x^2 + 10x + 15}{4x^2 + 8x}$

36. $\dfrac{x^4 + 3x}{4x^3 + 3x^2}$

37. $\dfrac{\sqrt{x + 7}}{(x - 3)(x + 7)}$

38. $\dfrac{\sqrt{x + 7}}{(x - 3)\sqrt{x + 7}}$

39. $\dfrac{x \sin x}{x^2 + x \cos x}$

40. $\dfrac{x \sin x}{\sin x \cos x}$

41. $\dfrac{4x + 7}{2x}$

42. $\dfrac{x \sin x}{\sin x + \cos x}$

43. $\dfrac{(x + 4)(2x + 5)}{x^2(x + 4)}$

44. $\dfrac{(2x + 5)(x - 7)}{(2x + 5) + (3x + 1)}$

45. $\dfrac{(x + 4)(2x + 5)}{x^2 + x + 4}$

46. $\dfrac{(5x + 2) - (3x - 1)}{(2x + 3)(3x - 1)}$

Polynomial equations

When solving the polynomial equation $x^2 + 5x + 6 = 0$, we begin by factoring:

$$x^2 + 5x + 6 = 0$$

$$(x + 2)(x + 3) = 0.$$

Next we set each factor equal to zero and solve:

$$x + 2 = 0 \qquad\qquad x + 3 = 0$$

$$x = -2 \qquad\qquad x = -3.$$

The solutions to the equation are $x = -2$ and $x = -3$.

Why is this our procedure? It's because of the zero product property.

> ZERO PRODUCT PROPERTY
>
> If $a \cdot b = 0$, then either $a = 0$ or $b = 0$.

It is not difficult to be convinced that the zero product property is true. We know that $0 \cdot c = 0$ for any number c, and if we multiply any two nonzero numbers, such as $4 \cdot 5 = 20$, we cannot arrive at an answer of zero.

Now return to the equation $(x + 2)(x + 3) = 0$. This is a product of two numbers, $(x + 2)$ and $(x + 3)$, and the product is zero. Therefore, one of the two numbers must be zero; either $x + 2 = 0$ or $x + 3 = 0$. This gives us the justification for setting each factor equal to zero.

Example 8 *Solve the equation* $x^2 - 9x + 14 = 0$.

Solution We factor and then use the zero product property:

$$x^2 - 9x + 14 = 0$$
$$(x - 2)(x - 7) = 0$$
$$x - 2 = 0 \text{ or } x - 7 = 0$$
$$x = 2 \text{ or } x = 7.$$

The solutions are $x = 2$ and $x = 7$. ■

Next, consider the equation $x^2 + 3x + 7 = 0$. Factoring is not helpful; the polynomial cannot be factored using only integers. Some of you then make the mistake of trying the following:

CAUTION: MISTAKE TO AVOID
The steps at the left are not incorrect, but they are terribly misleading and serve as a precursor to errors.

$$x^2 + 3x + 7 = 0$$
$$x^2 + 3x = -7$$
$$x(x + 3) = -7.$$

The previous steps are not incorrect, but it is never helpful and nearly always is performed to facilitate making the following mistake: *setting each factor equal to −7 is incorrect.*

Let's try it and we'll see that the resulting "solutions" are incorrect:

An alternate explanation is as follows. If $x(x + 3) = -7$ and we set $x = -7$, then $x(x + 3) = (-7)(-7 + 3) = (-7)(-4) = 28 \neq -7$. If instead we set $x + 3 = -7$, then $x(x + 3) = (-10)(-7) = 70 \neq -7$.

$$x = -7 \qquad\qquad x + 3 = -7$$
$$x = -10.$$

Evaluating the polynomial at either −7 or at −10 does not give the desired value of zero:

$$(-7)^2 + 3(-7) + 7 \stackrel{?}{=} 0 \qquad\qquad (-10)^2 + 3(-10) + 7 \stackrel{?}{=} 0$$
$$49 - 21 + 7 \stackrel{?}{=} 0 \qquad\qquad 100 - 30 + 7 \stackrel{?}{=} 0$$
$$35 \neq 0 \qquad\qquad 77 \neq 0$$

Knowing the zero product property and why the usual solution method works is the key to avoiding this error.

Therefore, $x = -7$ and $x = -10$ are not solutions to the equation.

Why does example 8 work but the previous calculation fail? It's because there is a zero product property but there is not a negative-seven product property. If the product of two numbers is zero, then one of the numbers must be zero. But this is not true of -7. Many different pairs of numbers can multiply to give -7, such as $-3.5 \cdot 2 = -7$ or $-0.7 \cdot 10 = -7$. Just because $x(x+3) = -7$ does not mean that one of x or $x+3$ must equal -7.

CAUTION: MISTAKE TO AVOID

If $(x+a)(x+b) = c$ where c is not zero, then

$$x + a \neq c, \qquad x + b \neq c.$$

What should we do instead? Any time that a quadratic equation cannot be factored easily, use the quadratic formula.

SOLVING QUADRATIC EQUATIONS

When solving a quadratic equation, try factoring first. If the equation cannot be factored quickly, use the quadratic formula.

The algebraic method of completing the square can be used instead of the quadratic formula, and this method can be used to derive the quadratic formula (see sidebar).

To derive the quadratic formula, we begin with the equation

$$ax^2 + bx + c = 0.$$

First we move the constant to the right-hand side:

$$ax^2 + bx = -c.$$

Next we factor a from the left-hand side:

$$a\left(x^2 + \frac{b}{a}x\right) = -c.$$

To complete the square inside the parentheses, we need to add

$$\left(\frac{1}{2} \cdot \frac{b}{a}\right)^2 = \frac{b^2}{4a^2}.$$

To keep the equation balanced, we need to add

$$a \cdot \frac{b^2}{4a^2} = \frac{b^2}{4a}$$

to the right-hand side. We now have

$$a\left(x^2 + \frac{b}{a}x + \frac{b^2}{4a^2}\right) = -c + \frac{b^2}{4a}.$$

Factoring the perfect square gives

$$a\left(x + \frac{b}{2a}\right)^2 = -c + \frac{b^2}{4a}$$

$$\left(x + \frac{b}{2a}\right)^2 = -\frac{c}{a} + \frac{b^2}{4a^2},$$

then we take the square root of both sides and solve for x:

$$x + \frac{b}{2a} = \pm\sqrt{-\frac{c}{a} + \frac{b^2}{4a^2}}$$

$$x + \frac{b}{2a} = \pm\sqrt{\frac{-4ac}{4a^2} + \frac{b^2}{4a^2}}$$

$$x + \frac{b}{2a} = \pm\sqrt{\frac{b^2 - 4ac}{4a^2}}$$

$$x + \frac{b}{2a} = \pm\frac{\sqrt{b^2 - 4ac}}{2a}$$

$$x = -\frac{b}{2a} \pm \frac{\sqrt{b^2 - 4ac}}{2a}$$

$$x = \frac{-b \pm \sqrt{b^2 - 4ac}}{2a}.$$

QUADRATIC FORMULA

Given constants a, b, and c, the equation $ax^2 + bx + c = 0$ has solutions

$$x = \frac{-b \pm \sqrt{b^2 - 4ac}}{2a}.$$

Example 9 *Solve* $x^2 + 3x + 7 = 0$.

Solution Trying to factor $x^2 + 3x + 7 = 0$ quickly is unsuccessful, so we try the quadratic formula with $a = 1$, $b = 3$, and $c = 7$. (The a is the coefficient on x^2, the b is the coefficient on x, and the c is the constant term, regardless of the order in which the terms appear. The only requirement is that everything be on one side of the equation with zero on the other side.) We have

$$\begin{aligned} x &= \frac{-b \pm \sqrt{b^2 - 4ac}}{2a} \\ &= \frac{-3 \pm \sqrt{3^2 - 4(1)(7)}}{2(1)} \\ &= \frac{-3 \pm \sqrt{9 - 28}}{2} \\ &= \frac{-3 \pm \sqrt{-19}}{2}, \end{aligned}$$

which are not real numbers. There are no real-number solutions to the equation. ■

The value under the square root, $b^2 - 4ac$, is called the *discriminant*. If the discriminant is negative, there are no real solutions; if the discriminant is zero, then there is exactly one real solution; and if the discriminant is positive, there are two real solutions.

Using the quadratic formula for an exercise in which factoring works is also acceptable; the quadratic formula always works, so factoring (as in example 8) should be used only when factoring is quick and easy.

Example 10 *Instead of factoring, use the quadratic formula to solve* $x^2 - 9x + 14 = 0$.

This is the same equation as in example 8.

Solution We use the quadratic formula with $a = 1$, $b = -9$, and $c = 14$:

If the equation is given as $x^2 + 14 = 9x$, then we rearrange to have everything on one side with zero on the other,

$$x^2 - 9x + 14 = 0,$$

before assigning values to a, b, and c.

$$\begin{aligned} x &= \frac{-b \pm \sqrt{b^2 - 4ac}}{2a} \\ &= \frac{-(-9) \pm \sqrt{(-9)^2 - 4(1)(14)}}{2(1)} \\ &= \frac{9 \pm \sqrt{81 - 56}}{2} \\ &= \frac{9 \pm \sqrt{25}}{2} \\ &= \frac{9 \pm 5}{2} \\ &= \frac{14}{2} = 7 \text{ or } \frac{4}{2} = 2. \end{aligned}$$

The solutions to the equation are $x = 2$ and $x = 7$, just as we obtained in example 8. ■

If the equation involves only x^2 and constants (no x term), then it can be solved by taking square roots.

Example 11 *Solve* $5x^2 - 20 = 0$.

Solution Although this equation can be solved by factoring and also by the quadratic formula, the quickest method is to take square roots. First we solve for x^2:

$$5x^2 - 20 = 0$$
$$5x^2 = 20$$
$$x^2 = 4.$$

Then we take square roots:

$$\sqrt{x^2} = \sqrt{4}$$
$$|x| = 2$$
$$x = \pm 2.$$

(It is traditional to go straight from $x^2 = 4$ to $x = \pm 2$.) The solutions are $x = 2$ and $x = -2$. ■

Solving by factoring requires the formula for factoring a difference of squares:

$$5x^2 - 20 = 0$$
$$5(x^2 - 4) = 0$$
$$5(x - 2)(x + 2) = 0$$
$$x - 2 = 0 \text{ or } x + 2 = 0$$
$$x = 2 \text{ or } x = -2.$$

Solving by the quadratic formula requires recognizing that $b = 0$ (with $a = 5$, $c = -20$):

$$x = \frac{-b \pm \sqrt{b^2 - 4ac}}{2a} = \frac{-0 \pm \sqrt{0^2 - 4(5)(-20)}}{2(5)} = \frac{\pm\sqrt{400}}{10} = \frac{\pm 20}{10} = \pm 2.$$

The history of solving polynomial equations of degree 3 and higher is quite interesting (see http://www-history.mcs.st-andrews.ac.uk/HistTopics/Quadratic_etc_equations.html). There are formulas for solving any cubic or quartic equation, but they are so time-consuming to use that virtually no one uses these formulas by hand. Many such equations can be solved through factoring; other methods useful for some equations, such as synthetic division and the rational roots theorem, are no longer widely used by hand. Factoring, the quadratic formula, and simple extraction of roots have become the lone expectations for solving polynomial equations by hand in calculus.

Example 12 *Solve the equation* $x^3 - x^2 - 12x = 0$.

Solution Notice that x appears in every term of the equation. It can therefore be factored out:

$$x^3 - x^2 - 12x = 0$$
$$x(x^2 - x - 12) = 0.$$

What remains in parentheses is quadratic and can also be factored:

$$x(x - 4)(x + 3) = 0.$$

All that remains is to apply the zero product property and solve those equations:

$$x = 0 \qquad x - 4 = 0 \qquad x + 3 = 0$$
$$x = 4 \qquad x = -3.$$

The solutions to the equation are $x = 0$, $x = 4$, and $x = -3$. ■

CAUTION: MISTAKE TO AVOID
A common error when solving equations in factored form such as

$$x^2(x - 5)(x + 7) = 0$$

is to forget to set the power of x equal to zero. We would have not only $x-5 = 0$ and $x+7 = 0$, but $x^2 = 0$ as well.

Exercises for polynomial equations

1–22. Solve the quadratic equation.

Exact solutions are required; approximate solutions from a calculator are not acceptable.

1. $x^2 + 7x + 10 = 0$
2. $4x^2 = 3x$
3. $x^2 - 3x = 4$
4. $x^2 + 2x - 15 = 0$
5. $2x^2 = 14$
6. $2x - 5 = -3x^2$
7. $x^2 + 4x + 4 = 0$
8. $35 + 12x + x^2 = 0$
9. $x^2 + 5x + 9 = 0$
10. $x^2 - 16 = 0$
11. $28 - 11x + x^2 = 0$
12. $x^2 - 3x = -11$
13. $2x^2 = 5x + 3$
14. $21x^2 - 5x - 4 = 0$
15. $7x^2 + 8x = 0$
16. $3x = x^2 - 40$
17. $4 + x^2 = 7x$

18. $x^2 + 6x = -9$
19. $8x^2 + 26x + 15 = 0$
20. $3x^2 + 12x - 8 = 0$
21. $9x - 1 + 5x^2 = 0$
22. $2x^2 = 5 - x$

23–34. Solve the polynomial equation.

23. $x^3 = 8$
24. $x^4 - 81 = 0$
25. $x^5 = 4x^3$
26. $x^4 + x^2 = 0$
27. $x^3 + 10x^2 + 25x = 0$
28. $x^3 - x^2 = 6x$
29. $x^4 - 5x^2 + 6 = 0$
30. $x^4 - 4x^2 + 3 = 0$
31. $12x^3 + x^5 - 7x^4 = 0$
32. $2x^4 + 3x^3 = x^2$
33. $3x^2 + 5x^3 = x$
34. $77x^3 + 18x^4 + x^5 = 0$

Inside vs. outside: being careful with function arguments

An old maxim in real estate is that the three most important things are location, location, and location. A similar truth is present in mathematics: where an expression is located is very important. We cannot simply change the location of an item at will without changing its value. Numbers inside parentheses, inside square roots, or inside the argument of a trig function cannot come outside unscathed.

This type of error is called a *function argument error.*

CAUTION: MISTAKE TO AVOID
In general, given a number c,

$$\sqrt{c \cdot x} \neq c\sqrt{x},$$
$$(c \cdot x)^2 \neq c \cdot x^2,$$
$$\sin(c \cdot x) \neq c \cdot \sin x.$$

The same problem applies to nearly every function argument.

Sometimes there are algebraic or trigonometric identities (or other types of identities) that help us take numbers from inside a function to outside. For instance, we know that $\sqrt{16x} = 4\sqrt{x}$ because we may extract the square root of 16. The double-angle identity for the sine function says that $\sin 2x = 2\sin x\cos x$. The general principle is that whenever we want to bring a number from inside a function argument to outside the function, an identity is required.

> INSIDE-OUTSIDE PRINCIPLE
>
> Whenever a number is moved from inside a function to outside, an identity is required, such as
>
> $$\sqrt{cx} = \sqrt{c}\sqrt{x},$$
> $$(cx)^2 = c^2x^2, \text{ or}$$
> $$\sin 2x = 2\sin x\cos x.$$

Example 13 *Which of the following calculations is correct and which is incorrect? Answer "correct" or "wrong."*

(a) $\sqrt{5x} = 5\sqrt{x}$

(b) $(5x^2)^2 = 25x^4$

(c) $(4x^2)^3 = 4x^6$

(d) $\dfrac{\cos 2x}{\cos x} = \dfrac{2\cos x}{\cos x} = 2$

(e) $\dfrac{\sin 2x}{\sin x} = \dfrac{2\sin x\cos x}{\sin x} = 2\cos x$

Solution (a) Wrong, because bringing the 5 from under the square root to outside it without taking its square root is incorrect. A correct manipulation is $\sqrt{5x} = \sqrt{5}\sqrt{x}$.

(b) Correct! More detail: $(5x^2)^2 = 5^2(x^2)^2 = 25x^4$.

(c) Wrong, because bringing the 4 outside the parentheses without cubing it is incorrect. A correct simplification is $(4x^2)^3 = 4^3(x^2)^3 = 64x^6$.

(d) Wrong, because bringing the 2 outside the cosine function without using an appropriate identity is incorrect.

(e) Correct! An appropriate identity is used when bringing the 2 outside the sine function. ■

The same principle applies in reverse; bringing quantities from outside to inside a function argument is the same mistake. Canceling an inside quantity with an outside quantity is also the same mistake.

Example 14 *Which of the following calculations is correct and which is incorrect? Answer "correct" or "wrong."*

$$\text{(a)}\quad x\sin x = \sin x^2$$

$$\text{(b)}\quad 6\sqrt{x} = \sqrt{36x}$$

$$\text{(c)}\quad \cancel{3}\tan\frac{x}{\cancel{3}} = \tan x$$

$$\text{(d)}\quad \frac{\sqrt{4x\cancel{(x+1)}}}{(x+3)\cancel{(x+1)}} = \frac{\sqrt{4x}}{x+3}$$

$$\text{(e)}\quad \frac{(\cancel{2x}\sin x)^2}{\cancel{2x}} = (\sin x)^2$$

$$\text{(f)}\quad \sin\frac{x^{\cancel{2}}}{\cancel{x}} = \sin x$$

(a) In this case, such an identity does not exist.

Solution (a) Wrong, because bringing the x from outside the sine function to inside it without using an appropriate identity is incorrect.

(b) Correct! Instead of just bringing the 6 inside the square root, it is squared when bringing it inside.

(c) Wrong, because cancellation of a 3 outside the tangent function with a 3 inside the tangent function's argument is incorrect.

(d) Wrong, because cancellation of $x + 1$ inside (under) the square root with an $x + 1$ outside the square root is incorrect.

(e) Wrong, because cancellation of $2x$ inside the quantity being squared with a $2x$ outside the quantity being squared is incorrect.

(f) Correct! The cancellation takes place entirely inside the argument of the sine function, so we are not violating the inside-outside principle. ■

Simplifying an expression is often very helpful, but simplifying expressions incorrectly is one of the main sources of errors in calculus. If you are stuck on a problem and unsure of what to try next, just trying something, as long as it is a correct algebraic manipulation, may lead to a solution. But knowing what not to do—what algebraic manipulations are incorrect—can be the key to knowing when to stop and just deal with the expression you have instead of forging ahead with an incorrect manipulation and losing all chances of arriving at a correct answer. Mathematical discretion is highly underrated.

Example 15 *Simplify the expression, if possible:*

(a) $\dfrac{\sin x^2}{x}$

(b) $\dfrac{3x+1}{\sqrt{3x+1}}$

(c) $x\cos\dfrac{1}{x}$

(d) $\dfrac{\sin 6x}{\cos 3x}$

Solution (a) As tempting as it is, we cannot cancel an x from the numerator and the denominator. The x^2 is inside the trig function and the x is outside the trig function, so cancellation is incorrect:

$$\frac{\sin x^2}{x} \neq \sin x.$$

No simplification is possible.

(b) As tempting as it is, we cannot cancel the $3x + 1$ from the numerator and the denominator. The $3x+1$ in the denominator is inside (under) the square root whereas the $3x+1$ in the numerator is outside the square root. Cancellation is incorrect:

$$\frac{3x+1}{\sqrt{3x+1}} \neq \frac{1}{\sqrt{1}} = 1.$$

This is an example of how knowing what not to do can eventually lead to trying something else that is correct.

We can, however, rewrite the square root as the $\frac{1}{2}$ power and remember that the numerator is to the power 1, and use the laws of exponents to simplify the expression:

$$\frac{3x+1}{\sqrt{3x+1}} = \frac{(3x+1)^1}{(3x+1)^{\frac{1}{2}}} = (3x+1)^{1-\frac{1}{2}} = (3x+1)^{\frac{1}{2}} = \sqrt{3x+1}.$$

(c) As tempting as it is, we cannot cancel the x outside the cosine function with the x in the denominator inside the cosine function:

$$x\cos\frac{1}{x} \neq \cos 1.$$

No simplification is possible.

(d) As tempting as it is, we cannot cancel the $6x$ inside the sine function with the $3x$ inside the cosine function and have 2 left:

$$\frac{\sin 6x}{\cos 3x} \neq \tan 2.$$

$\sin 6x = \sin(2 \cdot 3x) = 2 \sin 3x \cos 3x$

However, observing that $6x = 2 \cdot 3x$, perhaps the double-angle identity for sine could help:

$$\frac{\sin 6x}{\cos 3x} = \frac{2 \sin 3x \cos 3x}{\cos 3x} = 2 \sin 3x.$$ ■

Exercises for inside vs. outside

1–18. Which of the following calculations is correct and which is incorrect? Answer "correct" or "wrong."

1. $\sqrt{100x} = 10\sqrt{x}$
2. $(7x^2)^{3/2} = 7x^3$
3. $\sqrt{10x^3} = 10x^{3/2}$
4. $\sin 4x = 4 \sin x$
5. $(7x)^{-1} = \dfrac{7}{x}$
6. $(5x)^3 = 5x^3$
7. $(8x)^{-1} = \dfrac{1}{8x}$
8. $(3x)^2 = 9x^2$
9. $|7x| = 7|x|$
10. $|-6x| = -6|x|$
11. $\dfrac{\tan 5\cancel{x}}{\tan 3\cancel{x}} = \dfrac{\tan 5}{\tan 3}$
12. $\dfrac{\cancel{4}\sqrt{3\cancel{x}}}{\cancel{4x}} = \sqrt{3}$
13. $\sin \cancel{4x} \cos \dfrac{1}{\cancel{4x}} = \sin 1 \cos 1$
14. $x^2 \sec x = \sec x^3$
15. $\dfrac{\sqrt{\cancel{4x}}}{\sin \cancel{4x}} = \dfrac{\sqrt{1}}{\sin 1}$
16. $\sqrt{\dfrac{\cancel{x} \sin x}{\cancel{x} \cos x}} = \sqrt{\tan x}$
17. $\tan \dfrac{5\cancel{x}}{3\cancel{x}} = \tan \dfrac{5}{3}$

18. $\dfrac{|\cancel{x}\sin x|}{x^{\cancel{2}}} = \dfrac{|\sin x|}{x}$

19–20. Which of the following calculations is correct and which is incorrect? Answer "correct" or "wrong."

19. $\dfrac{(x+1)\cos\cancel{(x+2)}}{\cancel{(x+2)}\cos(x+3)} = \dfrac{x+1}{\cos(x+3)}$

20. $\cos\left(\dfrac{(x+1)\cancel{(x+2)}}{\cancel{(x+2)}(x+3)}\right) = \cos\left(\dfrac{x+1}{x+3}\right)$

21–32. Simplify the expression, if possible.

21. $\sin\dfrac{x^2}{3x}$

22. $\dfrac{(2x\sin x)^2}{2x}$

23. $\dfrac{\sin x^2}{3x}$

24. $4\cos\dfrac{x}{4}$

25. $\dfrac{1}{x}\sqrt{x}$

26. $\dfrac{\sin 5x}{\cos 5x}$

27. $x^2\cos x^3$

28. $\dfrac{\sin 5x}{\cos 4x}$

29. $|-6x|$

30. $\dfrac{\cos x}{\sin 2x}$

31. $2\sin x$

32. $\dfrac{\sqrt{16x}}{8x}$

Trig notation

The name of a function is just that—a name. It is *not* a number. Names such as *tan* or *sin* are not numbers, and are therefore incomplete without their *argument*, the input into the function.

> CAUTION: MISTAKE TO AVOID
>
> The name of a trig function must be followed by a number or variable expression. For example, the following attempts at expressions are incomplete:
>
> $5 + \sin \qquad x\tan,$
>
> whereas the following expressions are complete:
>
> $5 + \sin x \qquad x\tan\pi.$

Example 16 *Which of the following expressions or equations is written correctly and which is incomplete? Answer "correct" or "wrong."*

$$(a)\quad \sin^2 + \cos^2 = 1$$
$$(b)\quad \sin^2 5x + \cos^2 5x = 1$$
$$(c)\quad \frac{\sin}{\cos} = \tan$$
$$(d)\quad \frac{\sin 5x}{\cos 5x} = \tan 5x$$
$$(e)\quad \frac{\sin \not{x}}{\not{x}} = \sin$$

Although one might say verbally that "sine squared plus cosine squared equals one," it must be understood that an argument to the trig functions is necessary.

Although one might say "sine over cosine equals tangent," it must be understood that an argument to the trig functions is necessary.

Solution (a) Wrong, because the arguments of the sine and cosine functions are missing. The left-hand side is a name, not a number.

(b) Correct! Each trig function has an argument.

(c) Wrong, because the arguments of the trig functions are missing. Both sides of the equation are names, not numbers.

(d) Correct! Each trig function has an argument.

(e) Wrong, because the sine function on the right-hand side is missing an argument; the left-hand side is a number whereas the right-hand side is a name. The cancellation also violates the inside-outside principle (see the previous subsection "inside vs. outside"). ■

Example 17 *For the right triangle in figure 1, which of the following expressions are written correctly and which are written incorrectly? Answer "correct" or "wrong."*

$$(a)\quad \sin = \frac{5}{13}$$
$$(b)\quad \tan = \frac{5}{12}$$
$$(c)\quad \cos\theta = \frac{12}{13}$$
$$(d)\quad \csc = \frac{1}{\sin\theta} = \frac{13}{5}$$

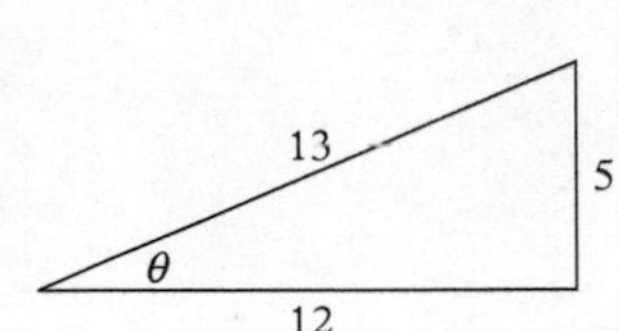

Figure 1 The figure for example 17

Solution (a) Wrong, because the argument of the sine function is missing; the correct notation is $\sin\theta = \frac{5}{13}$.

(b) Wrong, because the argument of the tangent function is missing; the correct notation is $\tan\theta = \frac{5}{12}$.

(c) Correct!

(d) Wrong, because the argument of the cosecant function is missing; the correct notation is $\csc\theta = \frac{1}{\sin\theta} = \frac{13}{5}$. ■

Some cancellation errors are propagated by thinking of names as algebraic quantities. Names cannot be canceled.

Example 18 *Which of the following calculations is correct and which is incorrect? Answer "correct" or "wrong."*

(a) $\dfrac{\cancel{\sin}\,7x}{\cancel{\sin}\,2x} = \dfrac{7\cancel{x}}{2\cancel{x}} = \dfrac{7}{2}$

(b) $\dfrac{4\cancel{\sin x}}{3\cancel{\sin x}} = \dfrac{4}{3}$

Solution (a) Wrong, because sin is a name and not a number; it is not a quantity that can be canceled.

(b) Correct! Because $\sin x$ is a number, it is a quantity that can be canceled. ■

This example demonstrates that proper notation is not just about the instructor being picky; learning to use proper notation is essential to avoiding errors.

Exercises for trig notation

1–14. Which of the following expressions or equations is written correctly and which is incomplete? Answer "correct" or "wrong."

1. $3 - \cos$
2. $4x \tan$
3. $-\cos 3$
4. $\tan 4x$
5. $-1 \le \sin \le 1$
6. $\cot = \dfrac{\cos}{\sin}$
7. $\sin 2x = 2 \sin \cos x$
8. $1 + \tan^2 \theta = \sec^2 \theta$
9. $3 \sin -5 \cos x$
10. $7x \cos \sec$
11. $(8 \cos)(9 \sin 5x)$
12. $\cos(\pi + x)$
13. $\dfrac{\sin}{5x}$
14. $(\pi + x) \cos$

A student wrote $\frac{\sin x}{n} = 6$. The professor asked how the student got that answer. The student replied, "I canceled the n's and read it: $\frac{\operatorname{si}\cancel{n}\,x}{\cancel{n}} = 6$."

15–22. Which of the following calculations is correct and which is incorrect? Answer "correct" or "wrong."

15. $\sin 1 = \sin$
16. $\dfrac{1 + \sin \cancel{3x}}{\cancel{3x}} = 1 + \sin$
17. $\dfrac{3\cancel{\tan}\,5x}{2\cancel{\tan}\,4x} = \dfrac{15x}{8x} = \dfrac{15}{8}$
18. $\dfrac{3\cancel{\tan 5x}}{2\cancel{\tan 5x}} = \dfrac{3}{2}$
19. $\dfrac{\cos 3x}{\cos 3} = x$
20. $x \sec \dfrac{1}{x} = \sec$
21. $\dfrac{3 \cos x}{\cos x} = 3$
22. $x \dfrac{\sec 1}{x} = \sec$

Parentheses

There are a wide variety of errors involving parentheses or the lack thereof. Some of these errors come from operations on a quantity with more than one term. For instance, suppose $f(x) = x^3 - 5$. What is $-f(x)$? It is tempting to put the negative sign in front of the expression for $f(x)$ to get $-x^3 - 5$. But, by the order of operations, this places the negative only on the first term, not the entire quantity:

$$-x^3 - 5 \neq -(x^3 - 5) = -f(x).$$

What is $(f(x))^2$? It is tempting to put the square at the end of the expression for $f(x)$ to get $x^3 - 5^2$, but this only squares the second term, not the entire quantity:

$$x^3 - 5^2 \neq (x^3 - 5)^2 = (f(x))^2.$$

The first known use of parentheses as grouping symbols occurred in the sixteenth century in a work by Tartaglia.

Nicolo Tartaglia, 1500–1557
http://www-history.mcs.st-andrews.ac.uk/Biographies/Tartaglia.html

> CAUTION: MISTAKE TO AVOID
>
> An operation that is to be done to a quantity with more than one term requires parentheses:
>
> $$-a + b \neq -(a + b), \qquad a + b^2 \neq (a + b)^2,$$
>
> $$1/a + b \neq 1/(a + b) = \frac{1}{a+b}, \qquad ca + b \neq c(a + b).$$

Example 19 *Let* $f(x) = 4x - 7x^2$. *Which of the following calculations is correct and which is incorrect? Answer "correct" or "wrong."*

(a) $-f(x) = -4x - 7x^2$
(b) $3f(x) = 3 \cdot 4x - 7x^2$
(c) $(f(x))^5 = (4x - 7x^2)^5$
(d) $1/f(x) = 1/4x - 7x^2$
(e) $4 - f(x) = 4 - 4x - 7x^2$

Solution (a) Wrong, because the negative is applied to the first term only. The correct calculation is

$$-f(x) = -(4x - 7x^2) = -4x + 7x^2.$$

(b) Wrong, because multiplying by 3 is applied to the first term only. By the order of operations, $3 \cdot 4x - 7x^2 = 12x - 7x^2$. The correct calculation is

$$3f(x) = 3(4x - 7x^2) = 12x - 21x^2.$$

(c) Correct! The entire quantity is taken to the fifth power.

(d) Wrong, because the first term only was placed in the denominator. By the order of operations, $1/4x - 7x^2 = \frac{1}{4x} - 7x^2$. The correct calculation is

$$1/f(x) = 1/(4x - 7x^2) = \frac{1}{4x - 7x^2}.$$

(e) Wrong, because the first term only was subtracted. The correct calculation is

$$4 - f(x) = 4 - (4x - 7x^2) = 4 - 4x + 7x^2.$$

Notice that in each correct calculation, parentheses are placed around the quantity that is $f(x)$; we use $(4x - 7x^2)$ every time. In each of the incorrect calculations, the first term was still treated correctly, but the second term was not. ■

Example 20 *Let* $f(x) = 11 - 5x^2$. *Evaluate the following expressions: (a)* $4f(x)$, *(b)* $1/f(x)$, *and (c)* $6 - 2f(x)$.

Solution The expression for $f(x)$ has two terms, so we use parentheses around $(11 - 5x^2)$ in each calculation.

(a) We multiply by 4, distributing through the parentheses:

$$4f(x) = 4(11 - 5x^2) = 44 - 20x^2.$$

(b) We take the reciprocal:

(b) Although parentheses are not present in the expression $\frac{1}{11-5x^2}$, they are implied by the fraction bar.

$$1/f(x) = 1/(11 - 5x^2) = \frac{1}{11 - 5x^2}.$$

Either of the expressions for $1/f(x)$ answers the question.

(c) We replace $f(x)$ with $(11 - 5x^2)$:

$$6 - 2f(x) = 6 - 2(11 - 5x^2) = 6 - 22 + 10x^2 = -16 + 10x^2.$$

Notice that the -2 is distributed through the parentheses. ■

Another type of parentheses error turns a one-term expression into a two-term expression. Suppose $g(x) = 5x$. What is $g(-\sin x)$? Although

$$g(-\sin x) = 5 \cdot -\sin x$$

is correct,

$$g(-\sin x) \neq 5 - \sin x.$$

Because those two expressions are so easily conflated, the better method of writing the calculation is

$$g(-\sin x) = 5(-\sin x) = -5\sin x.$$

CAUTION: MISTAKE TO AVOID

When multiplying a negative, use parentheses or move the negative to the front:

$a(-c) = -ac$

$a(-c) = a \cdot -c$ (correct but misinterpreted easily)

$a(-c) \neq a - c.$

Example 21 *Let $f(x) = 6x$. Which of the following calculations is correct and which is incorrect? Answer "correct" or "wrong."*

(a) $f(-\cos x) = 6 - \cos x$
(b) $f(-5x) = 6 - 5x$
(c) $f(-3x^2) = 6(-3x^2) = -18x^2$
(d) $f(-\tan x) = 6(-\tan x) = -6\tan x$

Solution (a) Wrong, because placing the 6 in front of $-\cos x$ turns the multiplication into subtraction. The correct calculation is

$$f(-\cos x) = 6(-\cos x) = -6\cos x.$$

(b) Wrong, because placing the 6 in front of the $-5x$ turns the multiplication into subtraction. The correct calculation is

$$f(-5x) = 6(-5x) = -30x.$$

(c) Correct! Parentheses are used and the multiplication is carried out.

(d) Correct! Parentheses are used and the multiplication is carried out. ■

Example 22 *Let $f(x) = 3x$. Evaluate the following expressions: (a) $f(-5x)$, (b) $f(-\sin^2 x)$, and (c) $5 - f(-7x^2 + 4)$.*

Solution (a) We use parentheses, replacing the x in the function rule with $(-5x)$:

$$f(-5x) = 3(-5x) = -15x.$$

(b) We use parentheses, replacing the x in the function rule with $(-\sin^2 x)$:

$$f(-\sin^2 x) = 3(-\sin^2 x) = -3\sin^2 x.$$

(c) We use parentheses, replacing the x in the function rule with $(-7x^2 + 4)$:

$$5 - f(-7x^2 + 4) = 5 - 3(-7x^2 + 4) = 5 + 21x^2 - 12 = -7 + 21x^2.$$

■

If the argument of a trig function contains more than one term, parentheses are required. If $h(x) = \sin x$ and we want to find $h(x + 7)$, the correct calculation is

$$h(x + 7) = \sin(x + 7),$$

where the $+7$ is inside the argument of the trig function. If parentheses are not used, then the $+7$ is not inside the argument of the trig function and we have erred:

$$h(x + 7) \neq \sin x + 7 = (\sin x) + 7.$$

> CAUTION: MISTAKE TO AVOID
>
> Multiple-term function arguments require parentheses: $\sin(a + b) \neq \sin a + b = (\sin a) + b$.

This caution applies to all trig functions and all other functions with notation of the form name(argument).

Example 23 *Let* $f(x) = x + \cos x$. *Find (a)* $f(2x - 8)$, *(b)* $f(4 - x^2)$, *and (c)* $f(5 - x + x^3)$.

Solution (a) Because $2x - 8$ has two terms, parentheses are needed for the trig function argument:

$$f(2x - 8) = 2x - 8 + \cos(2x - 8).$$

(a) By the order of operations,

$$2x - 8 + \cos 2x - 8 = 2x + \cos 2x,$$

which is not the same as $f(2x - 8)$.

(b) Because $4 - x^2$ has two terms, parentheses are needed for the trig function argument:

$$f(4 - x^2) = 4 - x^2 + \cos(4 - x^2).$$

(b) Without parentheses, the expression is incorrect:

$$\cos 4 - x^2 \neq \cos(4 - x^2).$$

(c) Because $5 - x + x^3$ has three terms, parentheses are needed for the trig function argument:

$$f(5 - x + x^3) = 5 - x + x^3 + \cos(5 - x + x^3).$$ ■

Exercises for parentheses

1–12. Let $f(x) = 2x + 5$. Which of the following calculations is correct and which is incorrect? Answer "correct" or "wrong." For incorrect calculations, give the correct calculation.

1. $5f(x) = 5(2x + 5) = 10x + 25$
2. $-f(x) = -2x + 5$
3. $(f(x))^3 = 2x + 5^3$
4. $1/f(x) = 1/2x + 5$
5. $5x - f(x) = 5x - 2x + 5 = 3x + 5$
6. $\sqrt{f(x)} = \sqrt{2x + 5}$
7. $\frac{2}{3}f(x) = 2/3 \cdot 2x + 5$
8. $f(3 - x^2) = 2 \cdot 3 - x^2 + 5$
9. $f(-7x) = 2 - 7x + 5 = 7 - 7x$
10. $f(-3 + \sin x) = 2(-3 + \sin x) + 5$
11. $f(\tan x) = 2\tan(x + 5)$
12. $f(-\tan x) = 2 - \tan x + 5$

13–18. Let $f(x) = 4x^2 - 12$. Evaluate the given expression.

13. $1/f(x)$
14. $(f(x))^2$
15. $6f(x)$
16. $-f(x)$
17. $2 - f(x)$
18. $\dfrac{5}{f(x) + 12}$

19–24. Let $f(x) = 1 + 2x + x^2 + \sin x$. Evaluate the given expression.

19. $f(x+1)$

20. $f(2x+3)$

21. $f(x^2+5)$

22. $f(-6x)$

23. $f(3x)$

24. $1-f(x)$

25–32. Let $f(x) = 9x$. Evaluate the given expression.

25. $f(-4x+2)$

26. $f(-1-x)$

27. $f(-x^2)$

28. $f(-\sin x)$

29. $f(-\tan x)$

30. $f(-9x)$

31. $f(2x-7)$

32. $f(x+2)$

When not to distribute

Parentheses often signal the need for the distributive property, but the distributive property only applies with numbers, not with names.

> CAUTION: MISTAKE TO AVOID
>
> The distributive property
>
> $$a(b+c) = ab + ac$$
>
> is valid for numbers a, b, and c, not for names. For instance,
>
> $$\sin(b+c) \neq \sin b + \sin c.$$

The correct identity for sine of a sum is

$$\sin(b+c) = \sin b \cos c + \cos b \sin c.$$

Example 24 *Which of the following calculations is correct and which is incorrect? Answer "correct" or "wrong."*

(a) $\sin(x+\pi) = \sin x + \sin \pi$
(b) $\tan(2x+3) = \tan 2x + \tan 3$
(c) $\cos(x-2\pi) = \cos x - \cos 2\pi$

Solution (a) Wrong, because function names such as sin are not numbers and do not distribute. The correct calculation, using the

The previous subsections "Squaring a binomial" and "Square root of a sum" discuss similar errors when squaring or taking the square root of a sum.

identity for sine of a sum, is

$$\sin(b+c) = \sin b \cos c + \cos b \sin c$$

$$\begin{aligned}\sin(x+\pi) &= \sin x \cos \pi + \cos x \sin \pi \\ &= (\sin x)(-1) + (\cos x)(0) \\ &= -\sin x.\end{aligned}$$

(b) Wrong, because the function name tan does not distribute. The correct calculation needs to use the identity for the tangent of a sum.

(c) Wrong, because the function name cos does not distribute. The correct calculation needs to use the identity for the cosine of a sum. ■

Which operations are present dictates whether the distributive property applies.

More precisely, the distributive property says that multiplication distributes over addition and subtraction. Multiplication does not distribute over multiplication or division.

> CAUTION: MISTAKE TO AVOID
>
> The distributive property applies to addition and subtraction, but not to multiplication or to division:
>
> $$a(b+c) = ab + ac,$$
> $$a(b-c) = ab - ac,$$
> $$a(b \cdot c) \neq ab \cdot ac,$$
> $$a\left(\frac{b}{c}\right) \neq \frac{ab}{ac}.$$

Example 25 *Which of the following calculations is correct and which is incorrect? Answer "correct" or "wrong."*

(a) $5(x + \sin^2 x) = 5x + 5\sin^2 x$

(b) $5(x \sin^2 x) = 5x \cdot 5 \sin^2 x = 25x \sin^2 x$

(c) $4(2x - 5) = 8x - 20$

(d) $4\left(\frac{2x}{5}\right) = \frac{4 \cdot 2x}{4 \cdot 5} = \frac{8x}{20}$

Solution (a) Correct! The distributive property applies to addition.

(b) Wrong, because the distributive property does not apply. There is no addition or subtraction. The correct calculation is

$$5(x\sin^2 x) = 5x\sin^2 x.$$

(c) Correct! The distributive property applies to subtraction.

(c) Subtraction is addition of the negative: $2x - 5 = 2x + (-5)$. The distributive property applies to addition, so it must also apply to subtraction.

(d) Wrong, because the distributive property does not apply. There is no addition or subtraction. The correct calculation is

$$4\left(\frac{2x}{5}\right) = \frac{4 \cdot 2x}{5} = \frac{8x}{5}.$$

■

Exercises for when not to distribute

1–20. Which of the following calculations is correct and which is incorrect? Answer "correct" or "wrong."

1. $3(7x + 1) = 21x + 3$
2. $8(2x - 3) = 16x - 24$
3. $10(2x - 1) = 20x - 1$
4. $8\left(\frac{2x}{3}\right) = \frac{16x}{3}$
5. $5x(2\sin x) = 10x\sin x$
6. $2(x - 5) = 2x - 5$
7. $5x(2 + \sin x) = 10x + 5x\sin x$
8. $\sin(5x + 7) = \sin 5x\cos 7 + \cos 5x\sin 7$
9. $\sin(5x + 2) = \sin 5x + \sin 2$
10. $\cos(5x + 7) = \cos 5x + \cos 7$
11. $2(\sin 5x) = 2\sin 10x$
12. $5(\cos(x + 7)) = 5\cos(5x + 35)$
13. $(5 - x)\sin 2x = 5\sin 2x - x\sin 2x$
14. $5(\cos x + 7) = 5\cos x + 35$
15. $\csc(2(x + 1)) = \csc 2\csc(x + 1)$
16. $5(x - 7) = 5x - 35$
17. $\csc(2(x + 1)) = \csc(2x + 1)$
18. $5(x \cdot 7) = 5x \cdot 35$
19. $\sec(5x + 12) = \sec 5\sec x + \sec 12$
20. $\cot(9 - x + x^2) = \cot 9 - \cot x + \cot x^2$

21–32. Simplify the expression, if possible.

21. $3(5x + 1)$

22. $2(6x - 9)$

23. $7(x \sin x)$

24. $3(5x\sqrt{x+1})$

25. $7(x - \sin x)$

26. $3(5x + \sqrt{x+1})$

27. $\sin(7 + x)$

28. $\sin(3 + 5x)$

29. $3\,((x+1)(x+2))$

30. $3(5/x)$

31. $4(x/5)$

32. $3(5 + x)$

Chapter I

Hyperreals, Limits, and Continuity

Motivation

1.0

Author's note: Much of the discussion in this section is repeated in section 1.8, and this content can be saved until then. The purpose of this section is to motivate the use of a new type of number.

The concept of a tangent line to a curve has many uses in the study of calculus. What is a tangent line?

A tangent line to a circle is a line that intersects the circle at exactly one point (figure 1). Although this description of a tangent line works well for some curves that are not circles, for others it does not. Consider the curves in figure 2. A line that just touches the curve at one point might or might not pass through the curve at some other point. It might even cut through the curve altogether. We need a different description of the term *tangent*. One way to think of a tangent line is the direction of momentum when traveling along the curve. If you are on the edge of a spinning merry-go-round and release a ball from your hand, it will not keep going in a circle along with you. Ignoring the fact that the ball drops toward the ground, it travels in a straight line in the direction it was heading at the time of release; this is the tangent line to the curve (figure 3). The same is true for any curve; the tangent line is the direction of momentum.

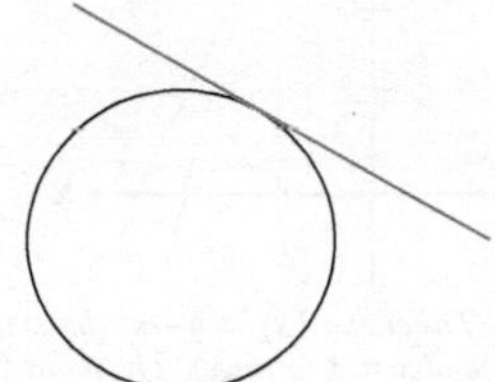

Figure 1 *A line tangent to a circle*

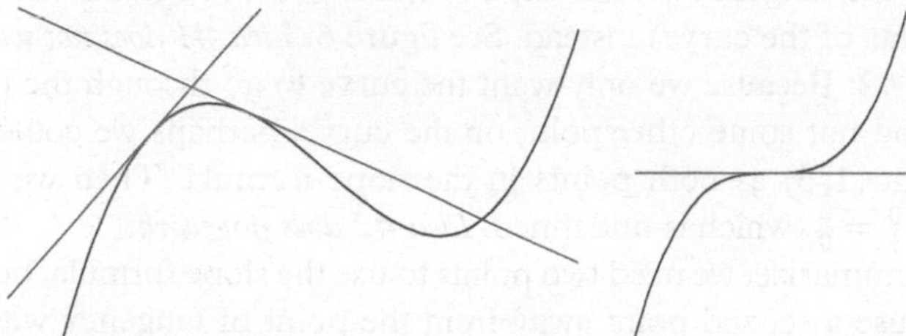

Figure 2 *Tangents to curves*

Another way to think of a tangent line is the direction of travel if you are traveling along a curve in a vehicle (figure 4). The headlight beam of the vehicle does not go uphill and downhill following the curve; it goes straight ahead. This is the direction of the tangent line.

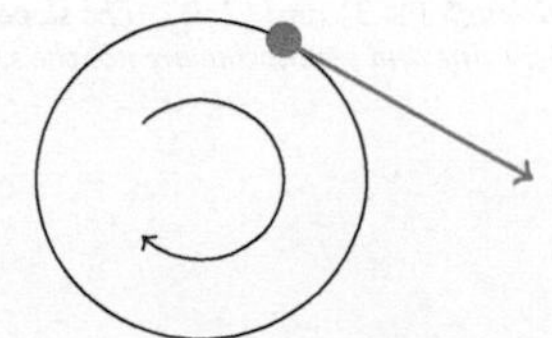

Figure 3 *A ball released at the edge of a merry-go-round*

Figure 4 *A vehicle traveling on a curve. The headlights shine in the direction of the tangent*

Tangent line to a curve

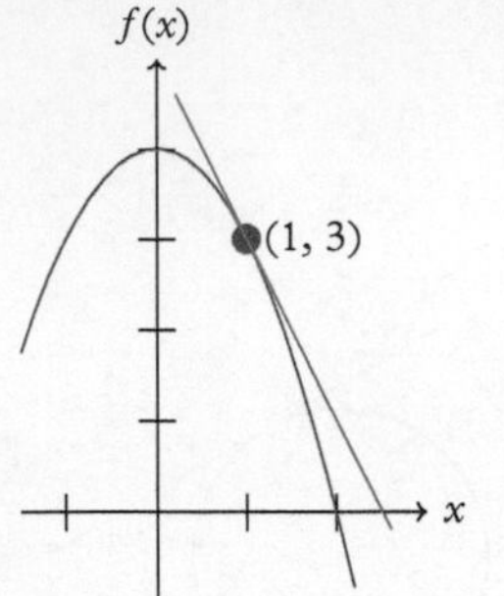

Figure 5 *The curve $f(x) = 4-x^2$ (blue) with a tangent line at $x = 1$ (orange). The point $(1,3)$ is called the* point of tangency

Suppose we wish to find the equation of the tangent line to the curve $f(x) = 4 - x^2$ at $x = 1$. By "at $x = 1$," we mean the point with x-coordinate 1. The y-coordinate of this point is $f(1) = 4 - 1^2 = 3$, so the point at which we want the tangent line is $(1, f(1)) = (1, 3)$. The tangent is pictured in figure 5.

One formula used for the equation of a line is the point-slope equation of a line: $y - y_1 = m(x - x_1)$. All we need is a point (which we have) and the slope of the line (which we need to find). There is also a formula for slope:

$$m = \frac{y_2 - y_1}{x_2 - x_1}.$$

The slope formula needs two points on the line, (x_1, y_1) and (x_2, y_2). We only have one point, $(x_1, y_1) = (1, 3)$. What should we use for the second point?

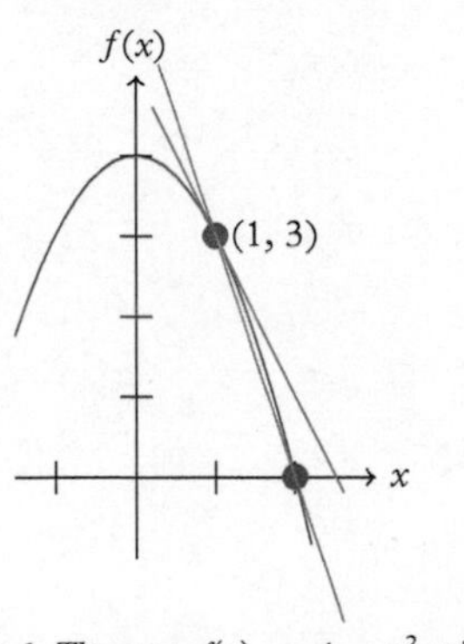

Figure 6 *The curve $f(x) = 4 - x^2$ with a tangent line (orange) at $x = 1$ and a secant line (green) through $(1,3)$ and $(2,0)$. The slopes of the tangent line and secant line are not the same*

Idea #1: The first idea we might try is to use a second point from the curve. To illustrate, try using the point with $x = 2$ as the second point. Since $f(2) = 4 - 2^2 = 0$, the point is $(x_2, y_2) = (2, 0)$. We can find the slope using the slope formula, $m = \frac{0-3}{2-1} = -3$, but it is not the slope of the tangent line; it is the slope of a *secant line* (a line through two points of the curve) instead. See figure 6. *Idea #1 does not work.*

Idea #2: Because we only want the curve to go through the point $(1, 3)$ and not some other point on the curve, perhaps we could use the point $(1, 3)$ as both points in the slope formula. Then we have $m = \frac{3-3}{1-1} = \frac{0}{0}$, which is undefined. *Idea #2 does not work.*

To summarize, we need two points to use the slope formula, but we cannot use a second point away from the point of tangency without changing the slope.

One way to solve this dilemma is to introduce a new type of number. Our concept of "number" has expanded before—for instance, when we were introduced to fractions or to negative numbers. This time we look at a type of number that helps us overcome the problem of idea #2: division by zero. No, we do not actually divide by zero, but we will learn to do the next best thing: divide by a number that is infinitely close to zero. This also overcomes the problem of idea #1, by allowing the second point to be infinitely close to the point of tangency, making the slope of the line between the two points infinitely close to the slope of the tangent line.

It turns out that all the major concepts of calculus can be understood with the use of these new, infinitely small numbers. Bring on the *infinitesimals!*

Authors' note: by "new," we mean new to the reader who has not used them before.

Infinitesimals

1.1

This section begins our study of material that is traditionally called *calculus*. Material that you learned in a course called *precalculus* falls in the categories of algebra and trigonometry.

The word *infinitesimal* has entered general use in the English language as an adjective, meaning something very small. The word began life as a noun in mathematics centuries ago and was an object that played a leading role in the development of calculus. In this section, we learn about infinitesimals and their larger context: the hyperreal numbers.

It can't be done ... or can it?

At various times in our mathematical past, we have been told there are certain operations that can be done and operations that cannot be done. But which operations can be done changed as we learned more mathematics.

For instance, think about when we first learned subtraction. We worked problems such as $11 - 4 = 7$. However, some of us may have been told that if we try to subtract a larger number from a smaller number, such as $4 - 7$, it cannot be done. The problem was that when our universe of numbers was restricted to the *whole numbers* 0, 1, 2, 3, and so on, then there was no number that could represent the result of $4 - 7$. Subtracting a larger number from a smaller number could not be done.

There was a solution to this problem, which was to learn about a new kind of number: a *negative number*. Then we were allowed to subtract the smaller number from the larger number, and $4 - 7 = -3$.

The same thing happened with division. We first learned facts such as $12 \div 3 = 4$, but then some of us might have been told we cannot divide 5 by 3 because 3 does not go into 5 evenly. The problem was that if we are using only the *integers*, the numbers $\ldots, -2, -1, 0, 1, 2, \ldots$, then there is no number to represent $5 \div 3$; the operation cannot be done. The solution was, once again, to learn about another type of number, to enlarge our system to the *rational numbers*, which are fractions of integers. We wanted something to represent $5 \div 3$, so we called it the number $\frac{5}{3}$. It may have seemed like we were just making things up; but, when we learned how to work with the rational numbers, how to add fractions, multiply fractions, divide fractions, and so on, then

The set of integers, which is comprised of whole numbers and their negatives, is commonly represented by the symbol **Z**. Why the letter z? It comes from the German word *zahlen*, meaning "numbers."

The set of rational numbers is commonly represented by the symbol **Q**.

the mystery dissipated and we eventually grew comfortable with an expanded number set.

Next came square roots. We learned that $\sqrt{25} = 5$ and $\sqrt{36} = 6$, but what about $\sqrt{2}$? Perhaps some of us were told we cannot take the square root of 2 because there is no number that, when multiplied by itself, gives 2. This is true if we are restricted to the rational numbers; we have an operation that cannot be done. Once again, we enlarged our view of number to contain *irrational numbers*, such as $\sqrt{2}$. This was one reason to learn how to write numbers in decimal form; as long as we could use a calculator to approximate $\sqrt{2}$ as 1.4142, we felt comfortable with the number, even knowing that it's not exactly equal to 1.4142 or any other terminating or repeating decimal. The set of *real numbers* may have been described as the set of all numbers with any decimal representation, not just the terminating or repeating decimal representations of the rational numbers.

An example of a terminating decimal is $\frac{1}{8} = 0.125$ and an example of a repeating decimal is $\frac{5}{11} = 0.\overline{45} = 0.454545\ldots.$

The set of rational numbers can be described both as the set of all fractions of integers and as the set of all numbers with terminating or repeating decimal expansions.

The set of real numbers is represented by the symbol **R**.

The real numbers didn't take care of all problems with square roots. Because any number multiplied by itself is nonnegative (zero or positive), then there are no square roots of negative numbers and we cannot do $\sqrt{-1}$. But again, by fiat, we were told something like, "Let there be *i*," and an *imaginary number* was available. By creating one such number and then using arithmetic and algebra to figure out what else could be created, we arrived at the *complex numbers*.

The phrase "*i* is imaginary" is not only correct grammatically, but also it is not a sign of insanity.

The symbol for the set of complex numbers is **C**.

At every point along the way, we were expanding our number system; the previous numbers were included in the new set of numbers. It turns out that we should take one step back to the real numbers before proceeding, because calculus does not concern itself with the complex numbers.

Calculus using complex numbers is called *complex analysis* and is usually taught in an upper-level course. Physics and electrical engineering are among the fields of study that sometimes use complex analysis.

Can you think of another "it can't be done?" There is perhaps no answer to this question more well known than division by zero. Why is it that we cannot divide by zero? The answer lies in multiplication. To divide 12 by 3, we want to find a number that, when multiplied by 3, gives 12. Because $4 \cdot 3 = 12$, $12 \div 3 = 4$. Because $\frac{5}{3} \cdot 3 = 5$, $5 \div 3 = \frac{5}{3}$ (okay, that last one seems like cheating, but it really is the same thing). To find $7 \div 0$, we need a number that, when multiplied by 0, gives 7. But of course there is no such number. Because $0 \cdot r = 0$ for any real number r, we cannot divide by zero. So this time the solution to our problem is to ...?! Well, sorry. We still won't be able to divide by zero. But we will be able to do the next best thing! We can divide by a number that is so close to zero that, for all practical purposes, it is zero, even though it is not actually equal to zero. The reason we care is that much of calculus arises from dealing with situations in which we seemingly need to divide by 0.

So what is meant by a number that is so close to zero that, for all practical purposes, it is zero? This question can be turned into yet another "it can't be done" situation: is there a number that is less than every positive real number but is still positive? If so, the number has to be less than .01258, so its decimal expansion has to start with 0.0. It has to be less than 0.00592, so it must have zero in the second decimal place as well. And it has to be less than 0.0000001, and so on, and therefore its decimal expansion must begin with infinitely many zeros. The number must be zero, and we have shown that it can't be done.

Can't subtract 4 from 7? Call the answer a negative number: -3. Can't divide 5 by 3? Give it the name $\frac{5}{3}$ and introduce the rational numbers. Can't take the square root of -1? Call it i and expand our view of numbers to the complex numbers. Can't find a number that is positive yet smaller than every positive real number? Give it a name; call it an *infinitesimal*—a number that can be described as "infinitely small," and write it using the symbol ω. If we assume that arithmetic and algebra work as usual with such a number, what other numbers must exist? What does the resulting system of numbers, called the *hyperreal numbers*, look like? If we can get familiar with the numbers, learn how to add and multiply them and otherwise manipulate them, then we can get as comfortable with the hyperreals as with any other type of number. This is our task for the remainder of this section.

ω is the lowercase Greek letter omega. It is not the letter "w" in the English alphabet and should not be pronounced "double u."

Among the symbols used for the hyperreals are **H** and ***R**.

Arithmetic with infinitesimals

Arithmetic with infinitesimals is accomplished by treating ω the same as any other algebraic quantity (more on this idea later in this section).

Example 1 *Simplify* $7 + \omega - (4 + 3\omega)$.

Solution We distribute the negative through the parentheses and then collect like terms, just as we do if ω is a variable (think "x") instead:

$$7 + \omega - (4 + 3\omega) = 7 + \omega - 4 - 3\omega = 3 - 2\omega. \qquad \blacksquare$$

In other words, we already know how to do quite a bit of arithmetic involving infinitesimals. The next example should not be challenging either.

Example 2 *Expand* $(3 + \omega)^2$.

Don't forget to use $(a+b)^2 = a^2 + 2ab + b^2$.

Solution

$$(3+\omega)^2 = 9 + 6\omega + \omega^2. \qquad \blacksquare$$

Reading exercises are to be worked when encountered during reading. Answers to reading exercises appear in the margin, usually one to four pages after the exercise.

Reading Exercise 1 *Perform the arithmetic on infinitesimals:* $5(7+4\omega)$.

Example 2 brings up an interesting question: what is 6ω? How big is it? Knowing that an infinitesimal such as ω must have a decimal expansion starting with infinitely many zeros helps us answer the question. Computing 6ω, we have $6(0.00000\ldots) = 0.00000\ldots$, which also has to start with infinitely many zeros. It must therefore also be an infinitesimal, for it is infinitely small. The same is true for $38\,576.53\omega$, or even for -14ω, which should be negative because ω is positive, but still starts $-0.00000\ldots$. It seems that any nonzero real number times an infinitesimal is infinitesimal. But before writing and proving this theorem, we need a definition of just what an infinitesimal is.

Definition 1 INFINITESIMALS *A number h is* infinitesimal *if either (1)* $h > 0$ *and* $h < r$ *for every positive real number r or (2)* $h < 0$ *and* $h > r$ *for every negative real number r.*

The definition asserts that any number that is smaller than every positive real number but still positive is an infinitesimal, and that any number that is larger than every negative real number but still negative is also an infinitesimal. Notice that zero is not an infinitesimal.

Some sources include the number 0 as an infinitesimal, so care must be taken in interpreting statements found elsewhere.

Theorem 1 PRODUCT OF A REAL NUMBER AND AN INFINITESIMAL *A nonzero real number times an infinitesimal is infinitesimal.*

Proof. We want to show that for any nonzero real number k and any infinitesimal α that the number $k\alpha$ is infinitesimal. Let's show this for the case when both k and α are positive. Then, $k\alpha$ is also positive, and we need to show that $k\alpha < r$ for every positive real number r. Since $\frac{r}{k}$ is a positive real number and α is infinitesimal, by the definition of infinitesimal, $\alpha < \frac{r}{k}$. Multiply by k and we have $k\alpha < r$, as desired. The proof for the other cases is similar. $\blacksquare$

This theorem is a fact to be remembered, similar to knowing that a negative times a negative is a positive. Just as knowing such rules helps us work with negative numbers, knowing facts such as theorem 1 helps us work with the hyperreal numbers.

Over the years, as we learned different types of numbers, we learned how to place them on a number line, such as the one in figure 1. So where should the infinitesimals be placed? They cannot go to the right

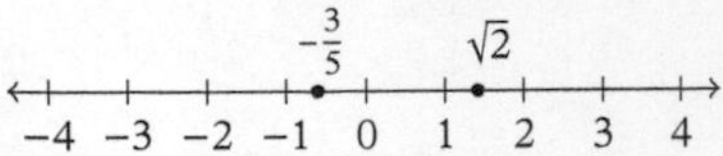

Figure 1 *Points on a number line*

of zero. Because they are less than every positive real number, they must go to the left of all the positive numbers. Similarly, they must go to the right of all the negative numbers because they are larger than all the negative numbers. The only place left to put them on the real number line is at zero, but there are a lot of numbers to place there, such as ω, 12ω, and -14ω. The solution is to place them on their own number line, the ω number line.

Ans. to reading exercise 1
$35 + 20\omega$

In many video games there is a portal leading to another level. Going through that portal brings you to another world, another level of the game. Think of zero as a portal that takes you to a lower level of numbers, the ω level. It is pictured in figure 2. The entire number line at the ω level, when infinitely shrunk, fits in the real number line at zero. And these are not the only two levels of hyperreal numbers.

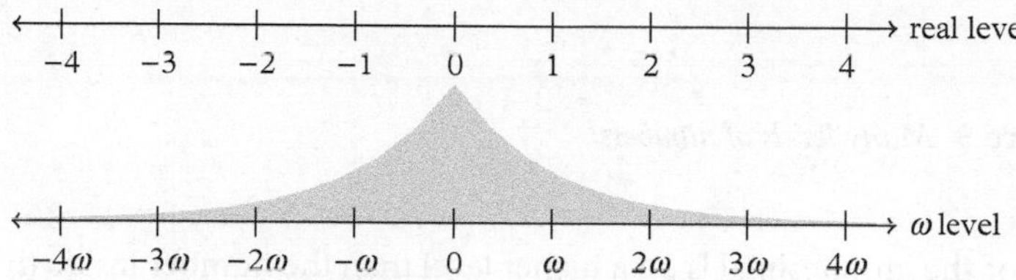

Figure 2 *Two levels of numbers*

More levels of infinitesimals

We now have a number-line picture of numbers of the form $k\omega$. But these are not the only infinitesimals. For instance, what about ω^2, a number we ran across in example 2? It is not pictured in figure 2. Where does it lie?

We know that the infinitesimal ω is positive but smaller than any positive real number; we have $\omega < 0.003$ and $\omega < 0.000072$. Multiply each of these inequalities by the positive number ω and we have $\omega^2 < 0.003\omega$ and $\omega^2 < 0.000072\omega$. In fact, we must have $\omega^2 < 0.0000\ldots\omega$, so that ω^2 is infinitely smaller than ω. It is infinitesimal by comparison; it must be on a yet lower level. Just as zero on the real number line is a portal to the ω level, zero on the ω level is a portal to a yet lower level, the ω^2 level. And why stop there? The number ω^3 is on a yet lower level by the same reasoning, and we can continue ad infinitum. It's like many video games where you keep going to another level with seemingly no end in sight. A partial picture is in figure 3.

And what about $\sqrt{\omega}$? We know that $\sqrt{\omega^2} = \omega$ is on a higher level than ω^2, and that $\sqrt{\omega^4} = \omega^2$ is on a higher level than ω^4. The square

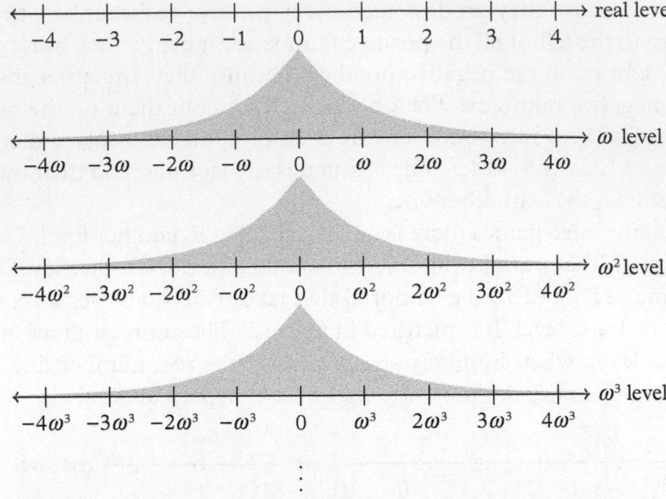

Figure 3 *Many levels of numbers*

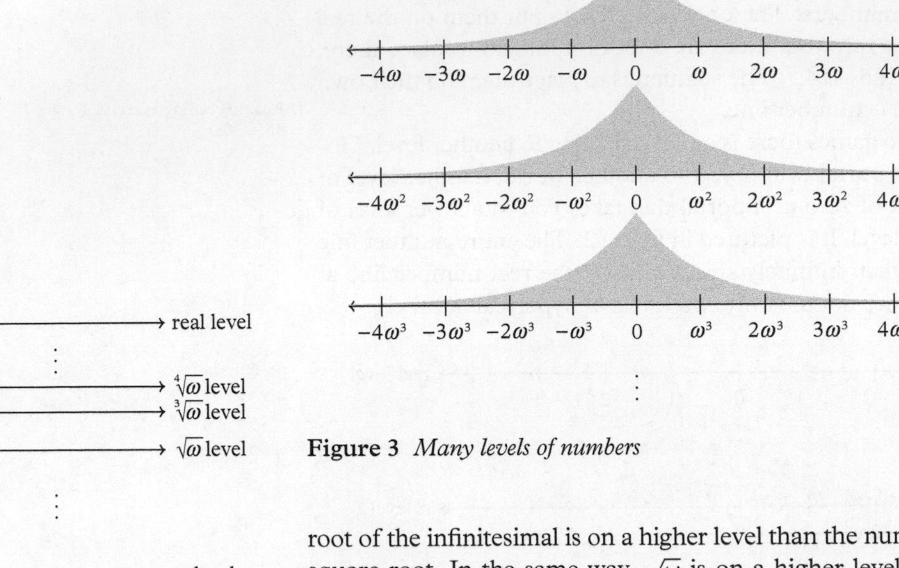

Figure 4 *Even more levels of numbers. The scales on the various number lines and the shading have been removed in this figure. The spacing between the levels is based on the value of the exponent. There are more levels between the ones pictured*

root of the infinitesimal is on a higher level than the number inside the square root. In the same way, $\sqrt{\omega}$ is on a higher level than ω, but it is still infinitesimal. There are levels in between the levels pictured in figure 3; the higher the exponent on ω, the lower the level, and there are infinitely many exponents, so there are infinitely many levels—in fact, infinitely many levels between any two levels! For instance, the number $\omega^{5/2}$ is on a level in between ω^2 and ω^3. The number $\sqrt[4]{\omega}$ is on a higher level than $\sqrt[3]{\omega}$, which is on a higher level than $\sqrt{\omega}$, but all lie below the real level. A revised, but still incomplete, picture is in figure 4.

Reading Exercise 2 *Which number is on a lower level than the other? 3ω or $47\omega^3$?*

Infinite numbers

You may recall that reciprocals of small numbers are large numbers. For instance,

$$\frac{1}{0.1} = 10, \quad \frac{1}{0.01} = 100, \text{ and } \frac{1}{0.000\,000\,1} = 1\,000\,000.$$

The smaller the denominator, the larger the resulting number. Then what about

$$\frac{1}{\omega}?$$

The denominator is infinitely small, so the result must be infinitely large! This can be proved in the following manner. If we can show that $\frac{1}{\omega}$ is larger than any positive real number r, then it must be infinite. To this end, let r be any positive real number. Then, $\frac{1}{r}$ is also a positive real number, and because ω is infinitesimal, we know that $\omega < \frac{1}{r}$. Taking reciprocals of numbers reverses the direction of the inequality (such as $\frac{1}{2} > \frac{1}{3}$); hence,

$$\frac{1}{\omega} > r,$$

as desired. The opposite is also true; the reciprocal of an infinite number is infinitesimal.

Theorem 2 INFINITESIMAL RECIPROCALS *The reciprocal of an infinitesimal is infinite. The reciprocal of an infinite number is infinitesimal.*

Because we will work with these infinite numbers often, it is convenient to write Ω in place of $\frac{1}{\omega}$. We do so throughout this text, and follow the convention that infinitesimals are represented by lowercase Greek letters whereas infinite numbers are represented by uppercase Greek letters—"little" letters for little numbers, "big" letters for big numbers.

Ω is the uppercase Greek letter omega. When necessary, to distinguish Ω from ω when speaking, it is customary to refer to "capital omega" and "little omega."

$$\frac{1}{\omega} = \Omega$$

$$\frac{1}{\Omega} = \omega$$

Boxes such as this one contain formulas or rules that should be learned.

Arithmetic with the infinite number Ω works just like arithmetic with ω.

Example 3 *Simplify* $\Omega + 4 - (5\Omega)$.

Solution We treat Ω like any other algebraic quantity and collect like terms:

$$\Omega + 4 - 5\Omega = 4 - 4\Omega. \quad \blacksquare$$

When working with both Ω and ω it is helpful to remember the reciprocal relationships.

Ans. to reading exercise 2
$47\omega^3$

Example 4 *Perform the arithmetic:* $4\Omega^2 \cdot \omega$.

Recall that $\frac{x^2}{x^1} = x^{2-1} = x^1 = x$ by the laws of exponents.

Solution Remembering that $\omega = \frac{1}{\Omega}$, we write

$$4\Omega^2 \cdot \omega = 4\Omega^2 \cdot \frac{1}{\Omega} = 4\Omega. \quad \blacksquare$$

It is also sometimes helpful to think of cancellation of ω and Ω. Just as we would write $\cancel{7} \cdot \frac{1}{\cancel{7}} = 1$, we can write

$$\cancel{\Omega} \cdot \frac{1}{\cancel{\Omega}} = 1$$

$\cancel{\Omega \cdot \omega} = 1$

or

$$\cancel{\Omega \cdot \omega} = 1.$$

Reading Exercise 3 *Simplify* $6\Omega^5\omega^2$.

Now let's go back to the picture of different levels of numbers. Where does Ω fit? It certainly isn't infinitesimal, so it does not go on a level below the real numbers. And it is larger than any real number, so it cannot fit on the real number line, either. But it must go somewhere!

Recall that we think of ω as infinitely smaller than a real number. This is the same as saying that a real number is infinitely larger than an infinitesimal, and because the reals are on a higher level than ω, it seems that if one number is infinitely larger than another, then it goes on a higher level. In a similar way, infinite numbers, being infinitely larger than real numbers, belong on higher levels than the real numbers. The Ω level is above the real level, and the real numbers fit in the Ω level at 0Ω.

Now recall your playground days when someone said something like, "I'm infinity better than you," followed by the inevitable "Oh yeah? Well, I'm infinity times infinity better than you," or maybe even "I'm infinity to the infinity power better than you!" Even the intuition of children says that multiplying an infinite number by an infinite number gives a much larger number. This reasoning is reflected in the reality of higher levels of infinite numbers, such as the ones pictured in figure 5. The larger the exponent on Ω, the higher the level. Just as with the levels of infinitesimals, there are infinitely many levels between each of the levels pictured, such as the $\sqrt{\Omega}$ level between the real level and the Ω level. The Ω^Ω level is above all the real-number power of Ω levels, and there are even levels above that! It is a true video gamer's paradise of unending varieties of levels to explore. The collection of all these numbers is called the *hyperreal numbers*. The hyperreals contain both infinitesimal and infinite numbers in addition to real numbers.

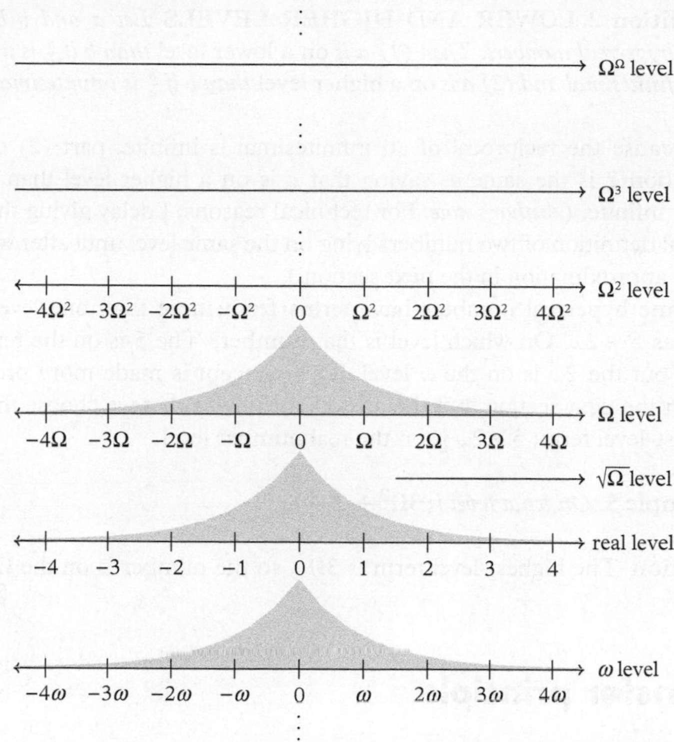

Figure 5 *A few of the infinitesimal and infinite levels of hyperreal numbers*

It seems as though the real number line is lost in this vast wonderland of numbers; the diagram may make it seem as though there is nothing special about the real numbers, but the real-number level is the only one where you can multiply two numbers and stay on the same level; the real numbers are *closed under multiplication.* By contrast, if you multiply two infinitesimals, the result is much smaller; it is on a lower level than either of the numbers being multiplied. If you multiply two infinite numbers, the result is on an even higher level. The real numbers are special after all!

Ans. to reading exercise 3
$6\Omega^3$

- Multiply by a real number and the level stays the same.
- Multiply by an infinitesimal and the result is on a lower level.
- Multiply by an infinite number and the result is on a higher level.

Reading Exercise 4 *Which number is larger,* 5Ω *or* $2\Omega^3$*?*

We have seen what it means for two numbers to be on different levels, but we have not yet written the formal definition. To be on a lower level means to be infinitely smaller, or infinitesimal by comparison.

Definition 2 LOWER AND HIGHER LEVELS *Let a and b be hyperreal numbers. Then (1) a is* on a lower level *than b if $\frac{a}{b}$ is infinitesimal and (2) a is* on a higher level *than b if $\frac{b}{a}$ is infinitesimal.*

Because the reciprocal of an infinitesimal is infinite, part (2) of definition 2 is the same as saying that a is on a higher level than b if $\frac{a}{b}$ is infinite. (*Author's note:* For technical reasons, I delay giving the formal definition of two numbers lying on the same level until after we study approximation in the next section.)

Some hyperreal numbers have terms from more than one level, such as $5 + 2\omega$. On which level is that number? The 5 is on the real level, but the 2ω is on the ω level. This concept is made more precise in the next section, but the answer for now is that we choose the highest-level term; $5 + 2\omega$ is on the real number level.

Example 5 *On what level is* $3\Omega^2 + \Omega - 8$*?*

Solution The highest-level term is $3\Omega^2$, so the number is on the Ω^2 level. ■

Transfer principle

We asserted earlier that algebra in the hyperreal numbers works the same as it does for real numbers. This is part of what is called the *transfer principle.*

> **TRANSFER PRINCIPLE (CALCULUS VERSION)**
> Algebraic formulas are true in the real numbers if and only if they are true in the hyperreal numbers.

The full version of the transfer principle is more complicated, and its technical details are beyond calculus. As long as a statement can be written using only certain types of symbols and quantifiers, then it is true in the reals if and only if it is true in the hyperreals. Although there are types of statements that cannot be written in a form for use with the transfer principle, these types of statements are easily avoided in calculus. Furthermore, all the algebraic statements transfer, so they can be used worry-free.

Applications of the transfer principle go as follows. First we start with an algebraic statement that we know is true for the real numbers:

$$-1 \le \sin x \le 1 \text{ for every real number } x.$$

Notice that the statement, as written, is true for every real number. Then, by the transfer principle, the statement is also true for every hyperreal number:

$$-1 \leq \sin x \leq 1 \text{ for every hyperreal number } x.$$

Therefore, we can conclude, for instance, that

$$-1 \leq \sin(47\Omega - 500 + 2\omega^2) \leq 1.$$

Ans. to reading exercise 4
$2\Omega^3$

We can put any hyperreal number we want inside sin and the result is still between -1 and 1.

The transfer principle is almost like a magic wand that we can wave over algebraic statements and say, "Be true in the hyperreals!" It is perhaps the most powerful tool used by *nonstandard analysts*, mathematicians who study the hyperreal numbers and their applications to calculus and beyond.

Reading Exercise 5 *True or false:* $-1 \leq \cos\omega \leq 1$.

The transfer principle is part of a larger theorem known as Łoś' theorem, named after its discoverer Jerzy Łoś, a Polish mathematician. This discovery led to the work of Abraham Robinson, who gave the first rigorous proof, in the 1960s, of the existence of a number system that includes infinitesimals and can be used to develop calculus.

The work of Robinson is beyond the scope of a course in calculus. Fortunately, everything we need for calculus can be developed from the assumptions of this section: that one infinitesimal exists and that the transfer principle applies.

Abraham Robinson, 1918–1974
http://www-history.mcs.st-and.ac.uk/Biographies/Robin son.html

The MacTutor History of Mathematics archive, hosted by the University of St. Andrews in Scotland, is one of the most trusted sources for the history of mathematics. The website has been in operation since before the turn of the century. Links to biographies of mathematicians in this book are to the MacTutor archive.

Exercises 1.1

1–12. Rapid response: is the given hyperreal number infinitesimal, infinite, or neither?

Rapid-response exercises are to be answered very quickly, and often involve object recognition. Students should not expect such exercises to appear on exams. Rapid-response exercises do not appear in every section.

1. ω
2. Ω
3. 7ω
4. ω^4
5. 0
6. 5
7. $5 + \omega$
8. $\Omega^2 + 7$
9. $\dfrac{1}{\omega}$
10. $\dfrac{1}{\Omega}$
11. $\Omega^2\omega$
12. $\Omega\omega^2$

13–32. Perform the arithmetic.

13. $2(7 - 4\omega)$
14. $3\omega - 5(\omega^2 - \omega)$
15. $\frac{1}{2}\omega - \frac{1}{4}(3\omega - \omega^2)$
16. $4\Omega(5 + 6\Omega)$
17. $(\Omega + 7) + 8(6 + \Omega)$
18. $(\omega + 3\omega^2) - \frac{1}{2}(4\omega - \omega^3)$
19. $(4 - \omega)^2$
20. $(2 - 3\omega)^2$
21. $(2 + \omega)^3$
22. $(\Omega + 1)^3$
23. $\dfrac{3 - \omega}{\omega}$
24. $(\omega + \Omega)^2$
25. $4\Omega(1 - \omega)$
26. $\dfrac{4\Omega}{\Omega + \Omega^2}$
27. $\dfrac{\Omega^3}{\omega}$
28. $\dfrac{\omega}{3\Omega}$
29. $\omega^3\Omega^5$
30. $\sqrt{\Omega^4}$
31. $(\Omega - \sqrt{2}\omega)^2$
32. $\omega(\Omega + 1)^2$

33–38. Which of the two given numbers is smaller?

33. .000000000000000000000000000000000000000000254 or $289\,475\,638\,239\,457\,856\sqrt{\omega}$
34. $\dfrac{\Omega}{10\,000\,000\,000\,000}$ or the number of dollars representing the national debt of the United States
35. $\Omega(3\omega^2)$ or ω
36. $12\omega^3$ or $3\omega^{12}$
37. $12\Omega^3$ or $3\Omega^{12}$
38. $5\omega^7$ or $7\Omega^5$

39–46. On which level is the number?

39. $\dfrac{5\omega^2}{2\sqrt{\omega}}$

40. $\Omega^7\omega^4$

41. $\Omega^{12}\omega^3$

42. $\dfrac{3\Omega}{\sqrt{\omega}}$

43. $(\Omega + 5)^2 - \Omega^2$

44. $(\Omega - 2)^2 - 2\Omega^2$

45. $(\omega + 5)^2 - \omega^2$

46. $(4 + \omega) - 2(\omega + 2)$

47. True or false: $-1 \leq \cos(6\Omega - 127) \leq 1$

48. True or false: $\sin^2 \omega + \cos^2 \omega = 1$

49. A rod has length 45 cm. It is divided into Ω pieces of equal length. What is the length of each piece?

50. A container of height 6 m is sliced horizontally into Ω pieces of equal height. What is the height of each piece?

51. A rod lies along the x-axis from $x = 3$ to $x = 7$. It is divided into Ω pieces of equal length. The x-axis is measured in centimeters. What is the length of each piece?

52. A container is positioned so that its bottom lies at $y = 1$ and its top lies at $y = 8$. It is sliced horizontally into Ω pieces of equal height. The y-axis is measured in meters. What is the height of each piece?

Ans. to reading exercise 5
true

Although it is not physically possible to divide an actual object into infinitely many pieces, this type of abstract thinking is useful in calculus.

Approximation

1.2

You are familiar with the idea of approximating numbers using a calculator:

Example 1 *Approximate $\sqrt{2}$ to four decimal places.*

Solution Using a calculator, $\sqrt{2} \approx 1.4142$. ■

You may be used to writing $\sqrt{2} = 1.4142$. However, technically speaking, this statement is false. The numbers $\sqrt{2}$ and 1.4142 are not exactly the same. But, for most practical purposes, they are the same, which is one reason why we often calculate with 1.4142 instead of $\sqrt{2}$.

We use decimal approximations because of their convenience. As long as we keep enough digits during intermediate calculations, our final answers are close enough. If we need a rod of length $10\sqrt{2}$ cm, we write the answer as 14.142 cm because we know how to measure the decimal expression much more easily than the square root expression, and if we are off in the fourth decimal place, it's just thousandths of a millimeter and, for most purposes, is totally inconsequential.

Remember approximations of π? You may have used $\pi = \frac{22}{7}$ or $\pi = 3.14$. Again, these statements as written are actually false, so we often write $\pi \approx \frac{22}{7}$ and $\pi \approx 3.14$ instead; but, for many applications, these approximations do just fine.

The difference between an approximation and the actual value it is meant to approximate is called the *error in the approximation.* For instance, the error in the approximation $\sqrt{2} \approx 1.4142$ can be calculated as

The error in the approximation is defined as approximation minus actual.

$$\begin{array}{rcl} 1.4142 & = & 1.4142 \\ -\quad \sqrt{2} & = & 1.414213562\ldots \\ \hline & & -0.000013562\ldots, \end{array}$$

whereas the error in $\pi \approx \frac{22}{7}$ is

$$\begin{array}{rcl} \frac{22}{7} & = & 3.14285714\ldots \\ -\quad \pi & = & 3.14159265\ldots \\ \hline & & 0.00126449\ldots. \end{array}$$

Now consider approximating an expression such as $7+\omega$. The infinitesimal ω is so much smaller than the real number 7 that $7+\omega$ is, for all practical purposes, the same as 7. We therefore say that they are approximately equal.

Example 2 *Approximate* $7+\omega$.

Solution Because ω is infinitely smaller than 7, we write $7+\omega \approx 7$. ■

How close are we in this approximation? Let's calculate the error, recalling that the decimal expansion of an infinitesimal starts with infinitely many zeros:

$$\begin{array}{rcl} 7 & = & 7.0000000\ldots \\ -\quad(7+\omega) & = & 7.0000000\ldots \\ \hline & & -0.0000000\ldots. \end{array}$$

This approximation is much closer than the approximations for $\sqrt{2}$ and π given earlier! The error in the approximation is $-\omega$, which is infinitesimal, whereas the earlier approximations had real-number errors.

A better way to compare approximations is to compare their *relative errors*, or their *percent relative errors*. The percent relative error is given by

$$\%\text{ relative error} = 100 \cdot \frac{\text{error}}{\text{actual value}}.$$

Example 3 *Determine the percent relative error in the approximation* $\pi \approx \frac{22}{7}$.

Solution We have already calculated the error as $0.00126449\ldots$, so we have

$$\%\text{ relative error} = 100 \cdot \frac{0.00126449\ldots}{3.14159\ldots} = 0.04\%. \qquad ■$$

Did the use of = here bother you? We are so used to writing = instead of ≈ in such instances that we often don't even catch that the two numbers are not equal, and that, instead, the given percent relative error has also been approximated! It is more accurate to write $\approx 0.04\%$.

Example 4 *Determine the percent relative error in the approximation* $\sqrt{2} \approx 1.4142$.

Solution We have already calculated the error as $-0.000013562\ldots$, so we have

$$\% \text{ relative error} = 100 \cdot \frac{-0.000013562\ldots}{1.4142\ldots} = -0.00096\%. \quad \blacksquare$$

Not a bad approximation at all.

Example 5 *Determine the percent relative error in the approximation* $7 + \omega \approx 7$.

Solution Calculating using decimals,

$$\% \text{ relative error} = 100 \cdot \frac{-\omega}{7+\omega} = 100 \cdot \frac{0.00000\ldots}{7.00000\ldots} = 0.0000\ldots\%. \quad \blacksquare$$

The percent relative error is infinitesimal! The exact value of the percent relative error is $100 \cdot \frac{-\omega}{7+\omega}$. As long as what we "throw away" is infinitesimal by comparison to the original value, then the relative error is infinitesimal and the approximation is exceedingly good. Because terms on a lower level are infinitesimal compared to higher-level terms, we may throw away the lower-level terms.

Terms are separated by subtraction and addition. In the expression $39.6 + 2\omega - \omega^2$, there are three terms: 39.6, 2ω, and ω^2.

Example 6 *Approximate* $39.6 + 2\omega - \omega^2$.

Solution Because the terms 2ω and ω^2 are on lower levels than the term 39.6, we throw them away:

$$39.6 + 2\omega - \omega^2 \approx 39.6. \quad \blacksquare$$

A picture of approximation may help. The figures of section 1.1 show many levels of hyperreal numbers, such as those in figure 1. Although one can find where any of the numbers 39.6 or 2ω or $-\omega^2$ fit in the vast array of number lines given in these pictures, where is $39.6+2\omega-\omega^2$? Using reasoning similar to that of section 1.1, the number must fit in the real number line at 39.6, and this number acts as a portal to lower levels, containing all the numbers approximately equal to 39.6. In figure 2, the location of $39.6 + 2\omega - \omega^2$ is marked with a dot on each level. The lower of the three dots marks the exact location of the number, whereas the higher of the three dots marks the location where it fits in the scheme of figure 1. Determining the location of a point in the scheme of figure 1, where we look for the number's nonzero appearance in that diagram, is one picture of approximation. Rising as high as possible in a diagram of the style of figure 2 without rising to zero in a yet higher level is a second picture of approximation. When an expression is written in a form amenable to a figure 2-style diagram, approximation can be accomplished by throwing away the lower-level terms.

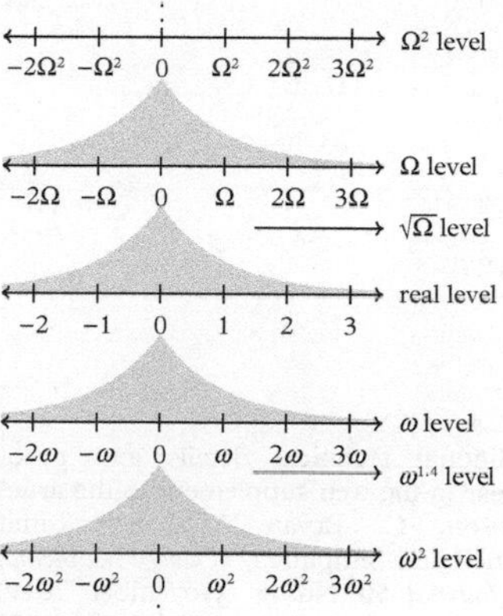

Figure 1 *A few levels of hyperreals, as given in section 1.1*

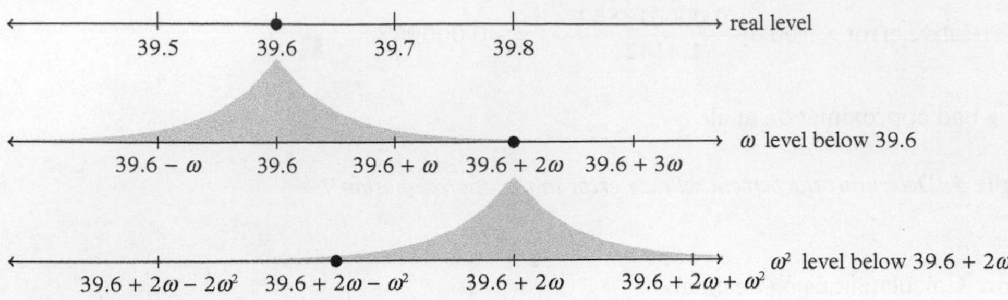

Figure 2 *Levels of hyperreals below 39.6, with the location of* $39.6 + 2\omega - \omega^2$ *on each level marked by a dot*

Example 7 *Approximate* $27\omega + 3\omega^2$.

Solution The highest-level term is 27ω, so we throw away the lower-level term $3\omega^2$. Our approximation is

$$27\omega + 3\omega^2 \approx 27\omega.$$

■

Notice that we don't throw away all the infinitesimals, just the ones on the lower levels. The highest-level term remains.

Reading Exercise 6 *Approximate* $35 - 8.4\omega$.

Hyperreal approximations: a few details

Additional technical details and proofs appear in the web supplement to the article Dawson, C. Bryan, "Calculus Limits Unified and Simplified," *College Mathematics Journal* 50 No. 5 (November 2019), 331–342, doi: https://doi.org/10.1080/07468342.2019.1662706 (accessed May 10, 2021).

Now that we have studied approximation, it is time to look at a few of the technical details before returning to more examples. Two hyperreal numbers are approximately equal if their relative difference is infinitesimal.

Definition 3 APPROXIMATION *Let x and y be hyperreal numbers. If* $x \neq 0$, $x \approx y$ *if* $\frac{y-x}{x}$ *is zero or infinitesimal. Also,* $0 \approx 0$.

The phrase "$x \approx y$" can be read "x is approximately equal to y" or "x is approximately y" or "x approximates y."

By this definition, numbers that are equal are also approximately equal. Also, the only number that is approximately equal to zero is zero itself. This definition does not give approximation as an operation, but rather as an "equivalence relation" (think "=" rather than "+"). But theorem 3 shows how we can often use approximation as if it is an operation, by throwing away lower-level terms.

Among other things, "approximation is an equivalence relation" means that if $x \approx y$, then $y \approx x$.

Theorem 3 SIMPLIFICATION THROUGH APPROXIMATION *If x is on a lower level than y, then $y+x \approx y$.*

Based on the previous marginal note, this is the same as saying $y \approx y + x$.

Proof. Notice $\frac{y+x-y}{y} = \frac{x}{y}$ is infinitesimal by definition of lower level. Therefore, $y + x \approx y$ by definition of approximation. ■

In section 1.1 we looked at lower and higher levels, which were defined formally, but not the same level. Approximation allows us to assign this definition now.

Definition 4 SAME LEVEL *Two hyperreal numbers x and y are* on the same level *if $\frac{x}{y} \approx r$ for some nonzero real number r.*

Therefore, $3\omega+\omega^2$ and 12ω are on the same level because $\frac{3\omega+\omega^2}{12\omega} \approx \frac{3\omega}{12\omega} = \frac{1}{4}$ is a nonzero real number.

Both $3\omega + \omega^2$ and 12ω are on the ω level.

Additional examples

When instructions are given to "approximate," use theorem 3 and throw away lower-level terms.

Example 8 *Approximate* $497 + 5\omega$.

Solution The number is on the real-number level, and by theorem 3, we can throw away the lower-level term 5ω:

$$497 + 5\omega \approx 497.$$ ■

Throwing away lower-level terms works even if the number is on an infinite level.

Example 9 *Approximate* $3\Omega + 49\,241$.

Solution The number is on the Ω level, and by theorem 3, we can throw away the lower-level term $49\,241$:

$$3\Omega + 49\,241 \approx 3\Omega.$$ ■

Answer to reading exercise 6
35

Notice that in example 9, we threw away a term that was not infinitesimal. As long as the term is on a lower level than the number as a whole, it can be thrown away. The term does not have to be infinitesimal; it just has to be infinitesimal by comparison.

This is not as foreign a concept as you may think. For instance, consider approximating the national debt of the United States. As of the time of this writing, the U.S. national debt is approximately \$18 trillion. It's not exactly \$18 trillion; in fact, that figure is off by hundreds of billions of dollars! But approximated to the nearest trillion, it's \$18 trillion. The error is very, very large, but the relative error is not that large. What is relatively small depends on context. We can even throw away an infinite amount when approximating, if the number is on an even higher level.

Example 10 *Approximate* $4\Omega^3 + 7\Omega$.

Solution The number is on the Ω^3 level, so we can throw away the lower-level term 7Ω:

$$4\Omega^3 + 7\Omega \approx 4\Omega^3. \qquad \blacksquare$$

Reading Exercise 7 *Approximate* $24 - 5\Omega$.

Approximation principle

Approximation requires us to know on which level the number is located. If we are wrong about the level of the number, we could make a mistake. For instance, suppose we wish to approximate the quantity

$$(\Omega + 7)^2 - \Omega^2.$$

If we say the number is on the Ω^2 level, throw away the lower-level terms Ω and 7, and approximate as

If you see the warning CAUTION: MISTAKE TO AVOID, the work shown is purposefully incorrect to demonstrate what not to do.

$$(\Omega + 7)^2 - \Omega^2 \approx -\Omega^2,$$

we are incorrect. The number is not on the Ω^2 level, but rather the Ω level, as a little arithmetic shows:

Example 11 *Approximate* $(\Omega + 7)^2 - \Omega^2$.

This is the correct version.

Solution Performing some arithmetic first to determine the level of the number,

$$(\Omega + 7)^2 - \Omega^2 = \cancel{\Omega^2} + 14\Omega + 49 - \cancel{\Omega^2} = 14\Omega + 49 \approx 14\Omega. \qquad \blacksquare$$

In contrast to examples 8–10, the preceding example contained an operation that could be performed to simplify the expression (squaring $\Omega + 7$) before determining the level of the number. This is what to look for. If there are such simplifications possible, then you may need to simplify before undertaking an approximation.

Example 12 *Approximate* $(7+\omega)-4-(3+\omega^2)$.

Solution We notice that some simplification can be performed, so we perform that arithmetic:

$$(7+\omega)-4-(3+\omega^2) = \not{7}+\omega-\not{4}-\not{3}-\omega^2 = \omega-\omega^2.$$

The number is then on the ω level and we may only throw away ω^2, arriving at the answer

$$\approx \omega.$$ ■

This is the correct version.

Suppose instead that we decided the number is on the real level and we throw away all the terms on lower levels than the reals. We then have

$$(7+\omega)-4-(3+\omega^2) \approx 7-4-3 = 0.$$

CAUTION: MISTAKE TO AVOID

This is incorrect for multiple reasons. One is that when we approximate, we want the relative error to be infinitesimally small. But, if we approximate a number as zero, we throw away all of its value, and the percent relative error is 100%! Remember, the only number approximately equal to zero is zero itself. If we have a nonzero number, approximate, and get zero, we have made an error. We call this the *approximation principle.*

Ans. to reading exercise 7
-5Ω

Actually, using the percent relative error formula gives a result of -100%, but it's the same idea; all the value of the number has been thrown away.

APPROXIMATION PRINCIPLE (CALCULUS VERSION)

If the result of an approximation is zero, then the approximation was performed incorrectly. Go back and perform exact arithmetic and possibly some other algebraic manipulation before determining the level of the number.

Example 13 *Approximate* $(2+\omega)^2-4$.

Solution If we decide the number is on the real level and throw away ω, we have $(2+\omega)^2-4 \approx 2^2-4 = 0$, which violates the approximation principle. Instead, let's try exact arithmetic first and see if anything cancels:

$$(2+\omega)^2-4 = \not{4}+4\omega+\omega^2-\not{4} = 4\omega+\omega^2 \approx 4\omega.$$

When we throw anything away and the result of our approximation is zero, we say that we have *violated the approximation principle.*

It turns out that our number is on the ω level, not the real level. ■

Reading Exercise 8 *Approximate* $(3+\omega)^2-9$.

Approximations with fractions

Approximations with fractions are as simple as approximating the numerator and denominator separately and then dividing. You may already use this procedure in the real-number setting; for instance,

$$\frac{\sqrt{2}+3}{5-\sqrt{7}} \approx \frac{4.4142}{2.3542} \approx 1.8750.$$

To enter the expression all at once in a calculator, be sure to use parentheses around numerator and denominator: (√(2)+3)/(5-√(7)).

Of course, we could enter the entire fraction into the calculator all at once, rather than calculate the numerator and denominator separately before dividing, in order to reduce roundoff error. This is good practice; but behind the scenes, the calculator is actually calculating the numerator and denominator separately before dividing, just to a greater precision than it shows on the screen. It is still an approximation of the type demonstrated.

Example 14 *Approximate* $\dfrac{2+\omega-\omega^2}{6+\sqrt{\omega}}$.

Solution We approximate the numerator and denominator separately. The numerator is on the real level, so we throw away the infinitesimals, giving $2+\omega-\omega^2 \approx 2$. The same is true for the denominator, and $6+\sqrt{\omega} \approx 6$. Therefore,

$$\frac{2+\omega-\omega^2}{6+\sqrt{\omega}} \approx \frac{2}{6} = \frac{1}{3}. \quad \blacksquare$$

The numerator and denominator do not need to be on the same level for the procedure to work; we still approximate the numerator and denominator separately.

Example 15 *Approximate* $\dfrac{\Omega^2+3}{2\omega+\omega^4}$.

Recall that the larger the exponent on Ω, the higher the level; the larger the exponent on ω, the lower the level.

Solution The numerator is on the Ω^2 level and the denominator is on the ω level. Throwing away the lower-level terms in each numerator and denominator yields

In step 2, the 2 remains in the denominator; $\frac{1}{2\omega} = \frac{1}{2}\Omega$, not 2Ω.

$$\frac{\Omega^2+3}{2\omega+\omega^4} \approx \frac{\Omega^2}{2\omega} = \frac{1}{2}\Omega^2 \cdot \Omega = \frac{1}{2}\Omega^3. \quad \blacksquare$$

The second step of the calculation used the fact that $\frac{1}{\omega} = \Omega$. Why do this? If we stop at $\frac{\Omega^2}{2\omega}$, it is not clear where the number lies in the scheme of figure 1, and we fall short of our goal. Approximating before using the reciprocal relationships is convenient, for if there is just one term in the numerator and denominator, the algebra of using the reciprocal relationships is much easier.

Ans. to reading exercise 8
6ω

Reading Exercise 9 *Approximate* $\dfrac{6+7\omega^2}{2+\sqrt{\omega}}$.

It is time to prove what we practiced.

Theorem 4 APPROXIMATION OF PRODUCTS AND FRACTIONS *Let* $x \approx y$ *and* $x_1 \approx y_1$. *Then,*

(a) $xx_1 \approx yy_1$ *and*

(b) $\frac{x}{x_1} \approx \frac{y}{y_1}$, *provided* $x_1, y_1 \neq 0$.

Proof. Suppose $x = 0$. Then $y = 0$ as well (because the only number approximately zero is zero), and $xx_1 = 0$, $yy_1 = 0$, and therefore $xx_1 \approx yy_1$. Similarly for part (b), $\frac{x}{x_1} = 0$, $\frac{y}{y_1} = 0$, and $\frac{x}{x_1} \approx \frac{y}{y_1}$.

Suppose now that none of x, x_1, y, y_1 are zero.

(a) Since $y \approx x$, y and x are on the same level and therefore $\frac{y}{x}$ is on the real level. Then,

$$\frac{yy_1 - xx_1}{xx_1} = \frac{yy_1 - yx_1 + yx_1 - xx_1}{xx_1} = \frac{y}{x} \cdot \frac{y_1 - x_1}{x_1} + \frac{y - x}{x},$$

which is real times (zero or infinitesimal) + (zero or infinitesimal), which then must be zero or infinitesimal. By definition of approximation, $xx_1 \approx yy_1$.

(b) First we wish to show that $\frac{1}{x_1} \approx \frac{1}{y_1}$. We have

$$\frac{\frac{1}{y_1} - \frac{1}{x_1}}{\frac{1}{x_1}} = x_1\left(\frac{1}{y_1} - \frac{1}{x_1}\right) = \frac{x_1}{y_1} - 1 = \frac{x_1 - y_1}{y_1},$$

which is zero or infinitesimal since $x_1 \approx y_1$ (by definition of approximation). Therefore, $\frac{1}{x_1} \approx \frac{1}{y_1}$ by the definition of approximation. Last, by part (a), $x \cdot \frac{1}{x_1} \approx y \cdot \frac{1}{y_1}$. ■

Part (a) shows that approximating products can be accomplished by approximating each factor. It is also true that sums can be approximated by approximating each term separately, provided we do not violate the approximation principle. This is more difficult to prove, so the proof is omitted. One last fact is useful (although the proof isn't provided until section 2.6): square roots can be approximated by approximating inside the square root.

Example 16 *Approximate* $\sqrt{9+4\omega} - \sqrt{4-\omega^2}$.

Solution Approximating inside the square root gives

$$\sqrt{9+4\omega} - \sqrt{4-\omega^2} \approx \sqrt{9} - \sqrt{4} = 1.$$

■

An example that requires a little more thought is this: approximate $\frac{\sqrt{9+\omega}-3}{\omega}$. If we throw out ω on top, then we get $\frac{\sqrt{9}-3}{\omega} = \frac{0}{\omega} = 0$. But, with the approximation in the numerator being zero, we have made a mistake; the approximation principle has been violated. We must find a way of combining the real numbers in the numerator, or otherwise manipulating the expression, before we can proceed. We can't simplify $\sqrt{9+\omega}$ algebraically by itself, so we must try something different. In an algebra class, you learned how to rationalize denominators involving square roots by multiplying the numerator and the denominator by the *conjugate* of the expression involving the square root. The same strategy works here. If we multiply the numerator and the denominator by $\sqrt{9+\omega}+3$, then we can rid ourselves of the square root in the numerator and proceed from there. The full calculation is shown in the solution to example 17.

Ans. to reading exercise 9
3
Recall that $\sqrt{a+b} \neq \sqrt{a}+\sqrt{b}$, and therefore $\sqrt{9+\omega}$ is not the same as $\sqrt{9}+\sqrt{\omega}$.

The conjugate of $\sqrt{9+\omega}-3$ is $\sqrt{9+\omega}+3$, and vice versa. The conjugate changes $\sqrt{a}-b$ to $\sqrt{a}+b$ or vice versa, but it does not change what is inside the square root.

Example 17 *Approximate* $\dfrac{\sqrt{9+\omega}-3}{\omega}$.

Solution If we approximate the numerator as $\sqrt{9+\omega}-3 \approx \sqrt{9}-3 = 0$, we are in violation of the approximation principle. Therefore, we multiply the numerator and the denominator by the conjugate of the square root expression:

$$\begin{aligned}\frac{\sqrt{9+\omega}-3}{\omega} &= \frac{\sqrt{9+\omega}-3}{\omega}\cdot\frac{\sqrt{9+\omega}+3}{\sqrt{9+\omega}+3}\\ &= \frac{9+\omega-9}{\omega(\sqrt{9+\omega}+3)}\\ &= \frac{\cancel{\omega}}{\cancel{\omega}(\sqrt{9+\omega}+3)}\\ &= \frac{1}{\sqrt{9+\omega}+3}.\end{aligned}$$

It does not matter whether the square root expression is in the numerator or the denominator. We use the conjugate of the square root expression that causes the problem.

Recall that $(a-b)(a+b) = a^2 - b^2$, so $(\sqrt{9+\omega}-3)(\sqrt{9+\omega}+3) = (\sqrt{9+\omega})^2 - 3^2 = 9+\omega-9$.

In the second step, it is convenient not to perform the multiplication in the denominator. With the denominator in factored form, it is easy to see when something from the numerator might cancel, leading to a simplified fraction.

Now the denominator may be approximated separately, and if we throw out ω, we do not get zero like we did before; we have a plus instead of a minus. We can continue without violating the approximation principle and get

$$= \frac{1}{\sqrt{9}+3} = \frac{1}{6}.$$

■

Our first thought when approximating a square root should be to approximate inside the square root. But, if this leads to a violation of the approximation principle, try multiplying the numerator and the denominator by the conjugate of the square root expression, even if the expression is not originally a fraction.

Absolute value approximations

Recall that the absolute value of a positive number is that number, whereas the absolute value of a negative number changes the sign of the number. Therefore, evaluating an absolute value requires us to know whether the number is positive or negative. Because ω is a positive infinitesimal,

Recall that $|x| = \begin{cases} x \text{ if } x\geq 0 \\ -x \text{ if } x<0. \end{cases}$

$$|\omega| = \omega.$$

Because $-\omega$ is negative,

$$|-\omega| = -(-\omega) = \omega.$$

The sign changes to make the result positive. But what about more complicated hyperreals?

Consider $|\omega - \omega^2|$. To evaluate the absolute value, we need to know whether the number inside is positive or negative, and a first glance shows that ω is positive whereas $-\omega^2$ is negative. So which is it? Because $\omega - \omega^2 \approx \omega$, the number must be positive (it is approximately equal to a positive number). Therefore, absolute value does not change the number, and $|\omega - \omega^2| = \omega - \omega^2$.

Recall that $|a - b| \neq a + b$, so we cannot just change the subtraction to addition.

If two nonzero numbers are approximately equal, either both are positive or both are negative.

Example 18 *Perform the arithmetic:* $|38 - 15\omega|$.

Solution Since $38 - 15\omega \approx 38$, the number inside is positive and

$$|38 - 15\omega| = 38 - 15\omega.$$

■

If the number inside is negative, its sign is changed by taking the negative of what's inside: $|-5| = -(-5) = 5$.

Example 19 *Perform the arithmetic:* $|38 - 15\Omega|$.

Solution Since $38 - 15\Omega \approx -15\Omega$, the number is negative. Therefore,

$$|38 - 15\Omega| = -(38 - 15\Omega) = -38 + 15\Omega.$$

■

Approximations of absolute value quantities can be made by first approximating the quantity inside the absolute values, as long as the approximation principle is not violated.

Example 20 *Approximate* $\dfrac{|\Omega - 8\Omega^2|}{4+\omega}$.

Solution Approximating inside the absolute values and approximating the numerator and denominator separately give

$$\frac{|\Omega - 8\Omega^2|}{4+\omega} \approx \frac{|-8\Omega^2|}{4} = \frac{8\Omega^2}{4} = 2\Omega^2. \quad \blacksquare$$

Just as any other time, we must be careful to avoid violating the approximation principle.

Example 21 *Approximate* $\dfrac{|\omega - 5| - 5}{3+\sqrt{\omega}}$.

Solution If we try to approximate inside the absolute values first,

$$\frac{|\omega - 5| - 5}{3+\sqrt{\omega}} \approx \frac{|-5| - 5}{3} = \frac{0}{3},$$

we are in violation of the approximation principle. We must therefore perform exact arithmetic in the numerator, as in examples 18 and 19. Since $\omega - 5 \approx -5$, the quantity inside the absolute values is negative, and $|\omega - 5| = -(\omega - 5) = -\omega + 5$. Therefore,

$$\frac{|\omega - 5| - 5}{3+\sqrt{\omega}} = \frac{-\omega + 5 - 5}{3+\sqrt{\omega}} \approx \frac{-\omega}{3}. \quad \blacksquare$$

When approximating absolute value expressions, our first thought should be to try to approximate inside the absolute values. If this results in a violation of the approximation principle, we need to go back and perform exact arithmetic, evaluating the absolute value.

Because their relative errors are infinitesimal, the hyperreal approximations of this section are far more accurate that any decimal approximation we have encountered in the real-number system. In fact, a hyperreal approximation is a closer approximation than being off by one atom when counting the atoms of the universe!

This amazing statement assumes that the current theories of a finite universe are correct.

Exercises 1.2

1–4. Calculate the percent relative error for the given approximation.

In calculus, trigonometric functions are always given in radians unless otherwise specified. Make sure that your calculator is in radian mode before working exercise 2.

1. $\sqrt{3} \approx 1.732$

2. $\sin(0.05) \approx 0.05$

3. $2^{20} \approx 1\,000\,000$

4. $1.001^{1000} \approx 2.72$

5–10. Perform the arithmetic.

5. $|4 + \omega|$

6. $|4 - \Omega|$

7. $|-3\Omega + 5|$

8. $|\Omega - 4\Omega^2|$

9. $|\sqrt{\omega} - \omega|$

10. $|-3 - \omega^2 + 5|$

11–32. Approximate the given quantity.

11. $2(7 - 4\omega)$

12. $5\omega^2 - \omega^3$

13. $(\Omega + 7) + 8(6 + \Omega)$

14. $\sqrt{\Omega} + 2\Omega - \pi$

15. $(\omega + 1)^2 + 4$

16. $\dfrac{3 - \omega}{\omega}$

17. $\dfrac{4 + 3\omega + 5\omega^2}{5 - \omega}$

18. $\dfrac{\Omega^2 - 7}{4\Omega + 3}$

19. $(\omega + \sqrt{\omega})^2 - \omega$

20. $4 + \frac{1}{5\omega}$

21. $(6 + \omega) - 2(\omega^2 + 3)$

22. $\dfrac{3 + \omega - \omega^7}{12 - \sqrt{\omega}}$

23. $\dfrac{(5 + \omega)^2 - 5^2}{\omega}$

24. $\dfrac{\sqrt{5 + \omega} - \sqrt{5}}{\omega}$

25. $\dfrac{|4 - 3\omega| - 4}{2\omega}$

26. $\dfrac{|\omega^2 - 5\omega|}{2 + \omega}$

27. $|3 + 2\omega| - |3 - 2\omega|$

28. $|2 - \omega| + |5\omega - 2|$

29. $\dfrac{\Omega^2 + \sqrt{\Omega} - 7\omega}{\sqrt{\Omega} + 284}$

30. $\dfrac{\Omega - \Omega^2}{\omega - \omega^2}$

31. $\dfrac{\sqrt{49 - \sqrt{\omega}} - 7}{2\omega}$

32. $\dfrac{\sqrt{9\Omega^2 - \Omega + 1}}{4\Omega}$

33. Draw a diagram in the style of figure 2, showing the location of the number $13 - \omega^2 + \omega^3$.

34. Draw a diagram in the style of figure 2, showing the location of the number $\Omega^2 - 3\Omega$.

Hyperreals and Functions

1.3

To say that functions are important in calculus is an understatement. In algebra, we learned to evaluate functions at real-number inputs. In this section, we learn how to evaluate functions at hyperreal number inputs.

Evaluating functions at hyperreals

Let's begin with a review example.

Example 1 *For* $f(x) = x^2 - 7x + 1$, *find* $f(1 + \sqrt{2})$.

Solution We start by replacing the x in the function's output expression with the quantity $1 + \sqrt{2}$:

$$f(1 + \sqrt{2}) = (1 + \sqrt{2})^2 - 7(1 + \sqrt{2}) + 1.$$

We often then wish to simplify the expression. We can square the quantity in parentheses, distribute the -7 through the next parentheses, and collect like terms:

$$= 1 + 2\sqrt{2} + 2 - 7 - 7\sqrt{2} + 1 = -3 - 5\sqrt{2}. \qquad \blacksquare$$

The idea of a function is sometimes represented as a machine on a production line. A real-number input rolls along a conveyor belt into the function machine, the function does its job, and a real-number output comes out the other side, as pictured in figure 1. One

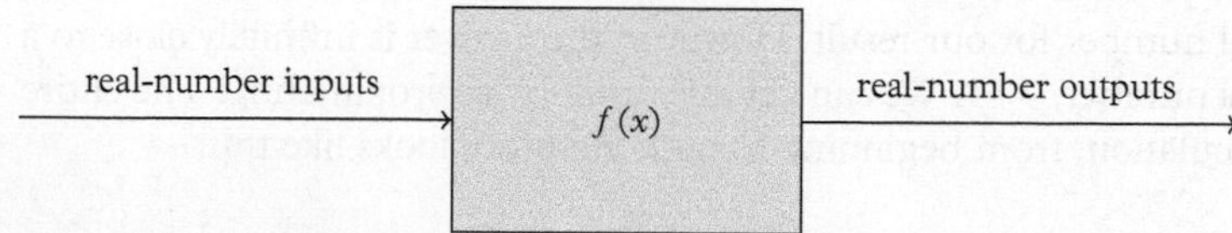

Figure 1 *Function as a machine*

of the main tricks in infinitesimal calculus is to extract information from a real-number function by using infinitesimals or, more generally, by using hyperreal inputs instead of real inputs. It may seem like cheating, but any function designed to handle real-number inputs can also handle hyperreal inputs (more on this momentarily). However, we will still want a real-number output; we want our function machine to work something like figure 2. Throwing hyperreal inputs into an

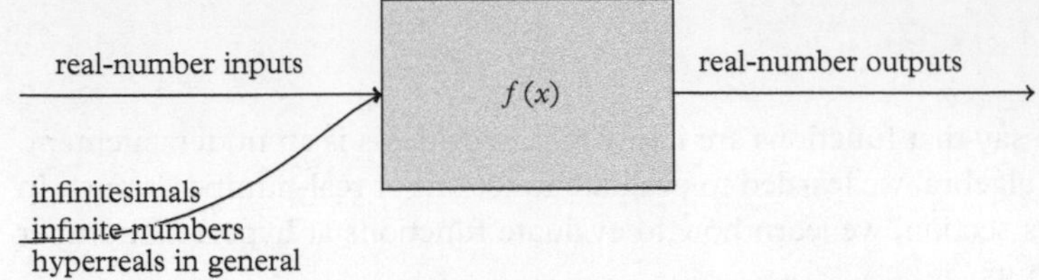

Figure 2 *Allowing hyperreal inputs in a real-number function*

algebraic function such as the one in example 1 is not difficult, because we know that algebra and arithmetic work the same with the hyperreals as it does with the reals. The only issue is to ensure a real-number output.

Example 2 *For* $f(x) = x^2 - 7x + 1$, *find* $f(3 + \omega)$.

Solution Just as in example 1, we evaluate the function at $3 + \omega$ by replacing the x in the function's output expression with the quantity $3 + \omega$:

$$f(3 + \omega) = (3 + \omega)^2 - 7(3 + \omega) + 1.$$

We can then simplify using the same algebraic steps as example 1, including collecting like terms:

$$= 9 + 6\omega + \omega^2 - 21 - 7\omega + 1 = -11 - \omega + \omega^2.$$

The result is not a real number, but a hyperreal number. We want a real number for our result. However, the answer is infinitely close to a real number, -11. We can get this result by approximating! The entire calculation, from beginning to end, therefore looks like this:

$$\begin{aligned} f(3 + \omega) &= (3 + \omega)^2 - 7(3 + \omega) + 1 \\ &= 9 + 6\omega + \omega^2 - 21 - 7\omega + 1 \\ &= -11 - \omega + \omega^2 \\ &\approx -11. \end{aligned}$$

■

This is not the whole story, but it is a good portion of the story. Our steps so far are to evaluate the expression, simplify as needed, and approximate.

Reading Exercise 10 *For $f(x) = x^2$, find $f(1 - \omega)$.*

Nonalgebraic functions such as the trigonometric functions can also be evaluated at hyperreal numbers. In fact, any function with a domain that includes an interval of real numbers can be evaluated at hyperreal numbers in that interval. For instance, if f is defined for real numbers x with $a \le x \le b$, then there is a hyperreal version of this function that is defined for hyperreals x with $a \le x \le b$. This is true for any type of interval. The proof is beyond the scope of this book.

Such intervals include the types $a < x < b$, $a < x \le b$, $a \le x < b$, $a \le x \le b$, $a < x$, $a \le x$, $x < b$, $x \le b$, and $x \in \mathbf{R}$.

The 3 Rs principle

Evaluating a function at a hyperreal and then approximating does not always result in a real number. For instance, given $f(x) = x^2 - 4$ and trying $f(2 - \omega)$ yields

$$\begin{aligned} f(2-\omega) &= (2-\omega)^2 - 4 \\ &= \not{4} - 4\omega + \omega^2 - \not{4} \\ &= -4\omega + \omega^2 \\ &\approx -4\omega. \end{aligned}$$

The result is an infinitesimal instead of a real number. As stated before, we wish to have a real-number output from the function. But, an infinitesimal is infinitely close to zero, so our result should be zero. We say that we *render the real result* zero. Nor is an infinitesimal the only possibility; approximation can also result in an infinite value. In this case, we render the real result ∞ or $-\infty$, depending on whether the infinite number is positive or negative.

Recall that ∞ and $-\infty$ are not numbers, but rather descriptive symbols. For instance, in the interval $(3, \infty)$ the ∞ represents lack of an upper boundary. Hyperreal numbers such as 4Ω or $-27\Omega^3$ are infinite numbers, but ∞ and $-\infty$ are not specific hyperreal numbers. Therefore, we do not say "render the real number," but rather "render the real result," so that we are careful not to call ∞ or $-\infty$ a number.

Definition 5 3 Rs PRINCIPLE: RENDERING A REAL RESULT *Let x be a hyperreal number.*

(1) *If x is zero or infinitesimal, we say that x* renders the real result *zero, denoted $x \doteq 0$.*

The phrase "$x \doteq 0$" is read "x renders zero." Similarly, "$x \doteq \infty$" is read "x renders infinity." The use of the symbol $\doteq$ for rendering was suggested by and first used in the classroom by my colleague Troy Riggs.

Part (2) of the definition is included for technical reasons that may become clear later in this chapter. Parts (1) and (3) are the portions of the definition that are applied during calculations.

(2) *If $x \approx r$ for a nonzero real number r, then we say that x* renders the real result *r, denoted $x \doteq r$.*

(3) *If $x \approx A$ for an infinite hyperreal A, then we say that x* renders the real result *∞ if A is positive or $-\infty$ if A is negative, denoted $x \doteq \infty$ or $x \doteq -\infty$, respectively.*

The previous calculation is then completed as follows:

Example 3 *For $f(x) = x^2 - 4$, find $f(2 - \omega)$.*

Solution We evaluate, simplify, approximate, and render a real result:

$$\begin{aligned} f(2-\omega) &= (2-\omega)^2 - 4 && \text{(evaluate)} \\ &= \not{4} - 4\omega + \omega^2 - \not{4} && \text{(simplify)} \\ &= -4\omega + \omega^2 \\ &\approx -4\omega && \text{(approximate)} \\ &\doteq 0. && \text{(render a real result)} \end{aligned}$$ ■

If the preliminary result after approximation is an infinitesimal, we finish the calculation by rendering the real result zero. This is part (1) of the 3 Rs principle.

This is called the *EAR procedure: evaluate* (and simplify), *approximate, render* a real result. The EAR procedure has many applications throughout calculus.

Ans. to reading exercise 10
1

In example 3, when we reached $-4\omega + \omega^2$, we could skip the approximation step and render zero because we have an infinitesimal. It may also be tempting to skip other steps. However, it is generally unsafe to skip steps, because incorrect answers may result. *Using the EAR procedure in the correct order is the best practice every time.*

Reading Exercise 11 *For $f(x) = x^2 - 1$, find $f(1 + \omega)$.*

Another twist: arbitrary infinitesimals

α is the lowercase Greek letter alpha. It should not be called by the same name as the letter "a" in the English language.

To say that α is an arbitrary infinitesimal is to say that α can be any infinitesimal. It can be ω. It can be -14.3ω. It can be $-\omega+3\omega^{3/2}$. It can be any other infinitesimal. An arbitrary infinitesimal α is a specific, but unidentified, infinitesimal. It can be positive; it can be negative. All we know is that it is an infinitesimal. Often, this is still enough to work a problem.

"Let α be an infinitesimal" is the same as saying, "Let α be an arbitrary infinitesimal."

Example 4 *Let $f(x) = \dfrac{x+2}{x^2-4}$ and let α be an infinitesimal. Find (a) $f(1)$ and (b) $f(1+\alpha)$.*

Solution Part (a) is algebra review: $f(1) = \frac{1+2}{1^2-4} = \frac{3}{-3} = -1$.

(b) Although we do not know much about α, all we need to know is that it is infinitesimal. Let's try the EAR procedure:

$$f(1+\alpha) = \frac{(1+\alpha)+2}{(1+\alpha)^2-4} = \frac{3+\alpha}{1+2\alpha+\alpha^2-4}$$

$$= \frac{3+\alpha}{-3+2\alpha+\alpha^2} \approx \frac{3}{-3} = -1.$$

If the preliminary result after approximation is a real number, the calculation is finished. According to the 3 Rs principle part (2), a real number renders itself, and nothing further needs to be done.

In the approximation step, knowing that α is infinitesimal means that $3+\alpha \approx 3$ and $-3+2\alpha+\alpha^2 \approx -3$. We do not need to know which specific infinitesimal is represented by α; this does not matter. All we need to know is that α is infinitesimal. ■

If we compare parts (a) and (b) in the solution, we should notice similarities in the calculations.

Let's work a very similar problem without the extra discussion.

Example 5 *Let* $f(x) = \dfrac{x+2}{x^2-4}$ *and let* α *be an infinitesimal. Find (a)* $f(0)$ *and (b)* $f(\alpha)$.

In this book, α is often used to represent an arbitrary infinitesimal.

Solution (a) $f(0) = \dfrac{2}{-4} = -\dfrac{1}{2}$.

(b) $f(\alpha) = \dfrac{\alpha+2}{\alpha^2-4} \approx \dfrac{2}{-4} = -\dfrac{1}{2}$. ■

In example 4, we first evaluated f at one and then at a hyperreal number infinitely close to one—namely, $1+\alpha$. The answers were identical. The same thing happened in example 5; we first evaluated f at zero and then at a hyperreal number infinitely close to zero—namely, $0+\alpha = \alpha$. Again, the results are identical. Does it always work this way? No. It does happen often, however, and we study this property later in this chapter in section 1.6.

Reading Exercise 12 *Let* $f(x) = \dfrac{x-3}{x^2-9}$ *and let* α *be an infinitesimal. Find* $f(\alpha)$.

Almost dividing by zero

Our next example continues to use the same function, but with a different outcome.

Ans. to reading exercise 11
0

Example 6 *Let* $f(x) = \dfrac{x+2}{x^2-4}$ *and let* α *be an infinitesimal. Find (a)* $f(-2)$ *and (b)* $f(-2+\alpha)$.

Solution (a) $f(-2) = \dfrac{-2+2}{(-2)^2-4} = \dfrac{0}{0}$, which is undefined. In other words, -2 is not in the domain of the function f because we are then dividing by zero. But, if we scoot over an infinitesimal amount and evaluate the function at $-2+\alpha$ instead (this is part (b); see figure 3), then perhaps we can avoid division by zero.

(b) Using the EAR procedure, we have

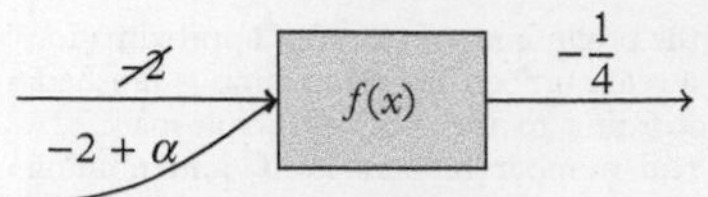

Figure 3 Machine function diagram for example 6. Although -2 is not in the domain of f, using $-2+\alpha$ instead gives a result

$$f(-2+\alpha) = \frac{(-2+\alpha)+2}{(-2+\alpha)^2-4} = \frac{\alpha}{\not{4} - 4\alpha + \alpha^2 - \not{4}}$$
$$\approx \frac{\alpha}{-4\alpha} = -\frac{1}{4}.$$ ■

We succeeded in evaluating the function at a number infinitely close to the hole in the domain; instead of dividing by zero in our calculation, we divided by a number that was not zero, but for all practical purposes is zero (the denominator in example 6(b) is an infinitesimal whereas the denominator in example 6(a) is zero). We may not have found a way to divide by zero, but we came awfully close!

Sometimes, instead of getting a real-number result when we almost divide by zero, we get an infinite result. The next example illustrates this. We need the notation $\frac{1}{\alpha} = A$, similar to $\frac{1}{\omega} = \Omega$. The difference is that although we know that Ω is positive (because ω is positive), we do not know the sign of A because α is an arbitrary infinitesimal and can be either positive or negative. If α is positive, so is A; but, if α is negative, A is negative as well.

The infinite hyperreal number A should be read "capital alpha." It may be the same symbol as the capital letter "A" used in the English language, but we are keeping with the heuristic of little Greek letters for infinitesimals and capital Greek letters for infinite numbers.

$\frac{1}{\alpha} = A$

$\frac{1}{A} = \alpha$

Example 7 *Let* $f(x) = \dfrac{x+2}{x^2-4}$ *and let* α *be an infinitesimal. Find (a)* $f(2)$*, (b)* $f(2+\omega)$*, (c)* $f(2-\omega)$*, and (d)* $f(2+\alpha)$.

Solution (a) $f(2) = \dfrac{2+2}{2^2-4} = \dfrac{4}{0}$, which is undefined because of division by zero. The number 2 is not in the domain of f.

(b) Using the EAR procedure, we have

According to the 3 Rs principle part (3), if the preliminary result after approximation is a positive infinite number, then we finish the calculation by rendering the real result ∞.

$$f(2+\omega) = \frac{(2+\omega)+2}{(2+\omega)^2-4} = \frac{4+\omega}{\not{4} + 4\omega + \omega^2 - \not{4}} \quad \text{(evaluate)}$$
$$\approx \frac{4}{4\omega} = \frac{1}{\omega} = \Omega \quad \text{(approximate)}$$
$$\doteq \infty. \quad \text{(render)}$$

Ans. to reading exercise 12
$\frac{1}{3}$

(c) Similarly,

$$f(2-\omega) = \frac{(2-\omega)+2}{(2-\omega)^2-4} = \frac{4-\omega}{\not{4}-4\omega+\omega^2-\not{4}} \quad \text{(evaluate)}$$
$$\approx \frac{4}{-4\omega} = -\frac{1}{\omega} = -\Omega \quad \text{(approximate)}$$
$$\doteq -\infty. \quad \text{(render)}$$

According to the 3 Rs principle part (3), if the preliminary result after approximation is a negative infinite number, then we finish the calculation by rendering the real result $-\infty$.

(d) Last,

$$f(2+\alpha) = \frac{(2+\alpha)+2}{(2+\alpha)^2-4} = \frac{4+\alpha}{\not{4}+4\alpha+\alpha^2-\not{4}} \quad \text{(evaluate)}$$
$$\approx \frac{4}{4\alpha} = \frac{1}{\alpha} = A, \quad \text{(approximate)}$$

An infinite result with a sign that is unknown cannot be rendered ∞ or $-\infty$. We may either stop the computation, leaving the infinite hyperreal as written in example 7 (d), or we may finish by writing the word *stuck* instead.

and we are stuck! We cannot render ∞ because A might be negative; we cannot render $-\infty$ because A might be positive. Remember that α is an arbitrary infinitesimal; it can be ω, as in part (b); it can be $-\omega$, as in part (c); or it can be something else. We know that the result is infinite; but, because we do not know its sign, we cannot go further. ■

Notice that division by zero, which is undefined, can produce multiple types of possible results nearby. In example 6, division by zero produced a real-number result nearby, whereas in example 7, division by zero produced an infinite result nearby. This is an important fact to remember: *division by zero does not always produce an infinite result.*

Division by zero does not always produce an infinite result.

It is also true that in example 7 (b) and (c), the inputs $2+\omega$ and $2-\omega$ are infinitely close (their difference is 2ω, an infinitesimal), but the outputs are not identical; in fact, they are infinitely far apart! Therefore, infinitely close inputs do not always give the same results. More on this phenomenon later in sections 1.4–1.6.

Reading Exercise 13 *Let* $f(x) = \frac{x-3}{x^2-9}$ *and let* α *be an infinitesimal. Find* $f(3+\alpha)$.

Evaluating a function at an infinite number

If a function is defined on an infinite interval such as (a, ∞), then it can be evaluated at infinite hyperreal numbers. The EAR procedure still applies.

Example 8 *Let* $f(x) = \dfrac{x+2}{x^2-4}$. *Find* $f(\Omega)$.

Solution

$$\begin{aligned} f(\Omega) &= \frac{\Omega+2}{\Omega^2-4} \quad \text{(evaluate)} \\ &\approx \frac{\Omega}{\Omega^2} = \frac{1}{\Omega} = \omega \quad \text{(approximate)} \\ &\doteq 0. \quad \text{(render)} \end{aligned}$$

■

Evaluating a function at an infinite number can have any variety of results, finite or infinite, just as when evaluating at finite values.

Reading Exercise 14 *Let* $f(x) = \dfrac{x^2-3}{x^2-9}$. *Find* $f(\Omega)$.

All the computations in this section have implications for further work. It is important to learn these procedures well, for they are used repeatedly. Master the EAR procedure; learn to evaluate, approximate, and render. A little investment of effort now will pay huge dividends down the road.

Exercises 1.3

Any time a function is to be evaluated at a hyperreal number, the EAR procedure is to be used. In other words, final answers should not be infinitesimals (they are rendered as zero), nor should they be positive infinite hyperreal numbers (rendered ∞) or negative infinite hyperreal numbers (rendered $-\infty$). Infinite numbers that cannot be rendered because the sign is unknown may be left as is or may be given the final answer *stuck*.

1–5. Use $f(x) = x^3 - 1$ and find the given quantity, assuming that α is an arbitrary infinitesimal.

1. $f(1)$
2. $f(1+\omega)$
3. $f(1-\omega)$
4. $f(1+\alpha)$
5. $f(1+\sqrt{\omega - 14\omega^2 + 3\omega^{14}})$, without calculations, based on the answer to exercise 4.

6–14. Use $g(x) = \dfrac{x-7}{x^2-9x+14}$ and find the given quantity, assuming that α is an arbitrary infinitesimal.

6. $g(2)$
7. $g(2+\omega)$
8. $g(2-\omega)$
9. $g(2+\alpha)$
10. $g(7)$
11. $g(7+\alpha)$
12. $g(\Omega)$
13. $g(\Omega^2)$
14. $g(-\Omega)$

15–22. Use $h(x) = \dfrac{x^2 - 3x}{x^2 + 2x - 15}$ and find the given quantity, assuming that α is an arbitrary infinitesimal.

15. $h(3)$
16. $h(3+\alpha)$
17. $h(-5)$
18. $h(-5+\alpha)$
19. $h(-5+\omega)$
20. $h(-5-\omega)$
21. $h(\Omega)$
22. $h(-\Omega)$

23. For $f(x) = x^2$, find $\dfrac{f(1+\alpha) - f(1)}{\alpha}$.

Ans. to reading exercise 13
$\frac{1}{6}$

24. For $f(x) = 3x$, find $\dfrac{f(2+\alpha) - f(2)}{\alpha}$.
25. For $f(x) = |x-3|$, find $f(3+\omega)$ and $f(3-\omega)$.
26. For $f(x) = \dfrac{|4-x|}{x-4}$, find $f(4+\omega)$ and $f(4-\omega)$.
27. For $f(x) = \dfrac{x^6 - x^3 + 4}{x^3 - 465\sqrt{2}}$, find $f(\Omega)$ and $f(-\Omega)$.
28. For $f(x) = \dfrac{x^2 - x}{x^3 + \sqrt{x}}$, find $f(\Omega)$ and $f(-\Omega)$.
29. For $f(x) = \dfrac{x+4}{\sqrt{x^2 - 7x + 1}}$, find $f(\Omega)$ and $f(-\Omega)$.
30. For $f(x) = \dfrac{\sqrt{x^2+6}}{2x+1}$, find $f(\Omega)$ and $f(-\Omega)$.
31. For $f(x) = \sqrt{x-7}$, find $f(11-\omega)$.

Ans. to reading exercise 14
1

32. For $f(x) = 2 + \sqrt{5x+1}$, find $f(3-8\omega)$.

Limits, Part I

Consider the function $f(x) = \dfrac{x^2 + 2x - 15}{x - 3}$. The function is not defined when $x = 3$, because $x = 3$ causes us to divide by zero. As a result, there is no point on the graph of $y = f(x)$ at $x = 3$. But when we look at a graph of f (see figure 1), it looks like there is a point with $x = 3$. It appears the point $(3, 8)$ is on the graph, even though we know it should not be there.

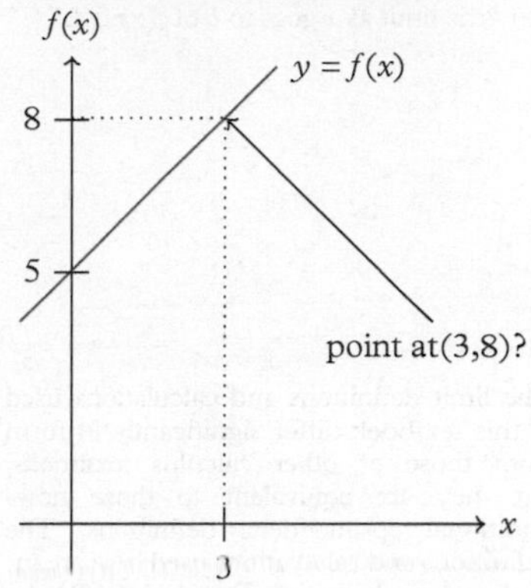

Figure 1 *A view of* $f(x) = \frac{x^2+2x-15}{x-3}$

Now "zoom in" toward $(3, 8)$ on the graph, viewing it at smaller and smaller scales, such as in figures 2 and 3. It still appears the point $(3, 8)$ is on the graph, even though we are absolutely sure it cannot be there. The graphs indicate that as x gets close to 3, $f(x)$ gets close to 8.

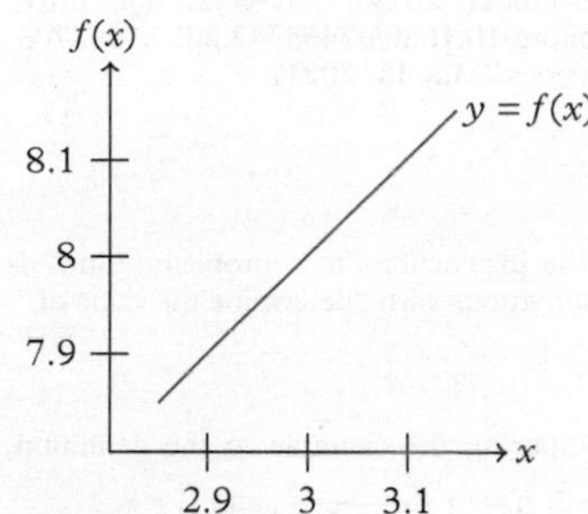

Figure 2 *A closer view of the function*

One way to explore the behavior of the function near $x = 3$ is to choose values of x that are close to 3, evaluate the function, and see if the values of $f(x)$ are close to 8. For instance, we could use $x = 3.00001$. This number is close to 3, but not infinitely close. So why not try $x = 3 + \alpha$ for an infinitesimal α? This is much, much closer than any number that is a nonzero real-number distance from 3.

If we use $x = 3 + \alpha$ as the input to our function, the division-by-zero problem goes away, and we can calculate an output:

$$\begin{aligned} f(3+\alpha) &= \frac{(3+\alpha)^2 + 2(3+\alpha) - 15}{(3+\alpha) - 3} \\ &= \frac{9 + 6\alpha + \alpha^2 + 6 + 2\alpha - 15}{\alpha} = \frac{8\alpha + \alpha^2}{\alpha} \\ &\approx \frac{8\alpha}{\alpha} = 8. \end{aligned}$$

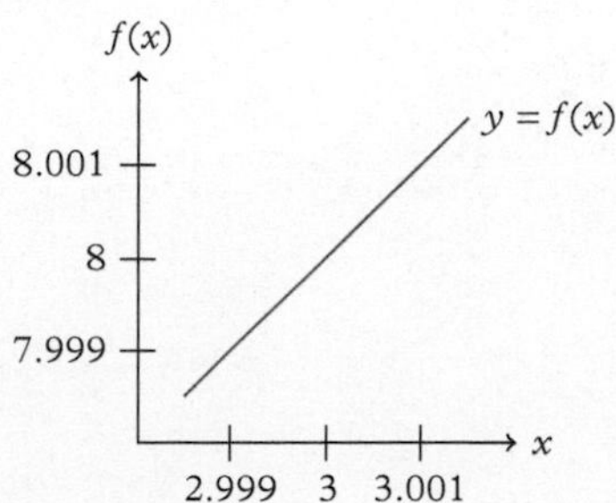

Figure 3 *An even closer view of the function*

The calculation shows that if the input to our function is infinitely close to 3, then the output from our function is infinitely close to 8. This explains why the graph seems to have a point at $(3, 8)$ even though it does not! What happens on the infinitesimal scale also happens on the real scale: as x gets close to 3, $f(x)$ gets close to 8.

The preceding discussion is the idea of a *limit*. The notation for the result is

$$\lim_{x \to 3} \frac{x^2 + 2x - 15}{x - 3} = 8.$$

If we want to know what is happening on the graph of a function near $x = b$ (in the previous discussion b is 3), we calculate $f(b + \alpha)$ for an arbitrary infinitesimal α.

Ways of reading $\lim_{x \to b} f(x) = L$ include "the limit of $f(x)$ as x approaches b is equal to L" and "the limit as x goes to b of $f(x)$ is L."

Definition 6 LIMIT *Let f be a function and let b be a real number. If $f(b + \alpha)$ is defined and renders the same real result L for every infinitesimal α, then we write*

$$\lim_{x \to b} f(x) = L.$$

When L is a real number (i.e., not ∞ or $-\infty$), then we say that the limit exists.

The limit definitions and calculations used in this textbook differ significantly in form from those of other calculus textbooks, but they are equivalent to those now-traditional epsilon–delta definitions. The definitions and calculations used here are introduced in the article Dawson, C. Bryan, "Calculus Limits Unified and Simplified," *College Mathematics Journal* 50 No. 5 (November 2019), 331–342, doi: https://doi.org/10.1080/07468342.2019.1662706 (accessed May 11, 2021)

For the purposes of calculation we write

$$\lim_{x \to b} f(x) = f(b + \alpha),$$

with the understanding that α is an arbitrary infinitesimal (and therefore may be either positive or negative) and that we must render the same real result for every such α or the limit does not exist.

The earlier calculation can be rewritten as shown in example 1.

In the instructions to a problem, "find" is synonymous with "determine the value of."

Example 1 *Find* $\lim_{x \to 3} \dfrac{x^2 + 2x - 15}{x - 3}$.

Comparing the example to the definition, $b = 3$, $f(x) = \frac{x^2 + 2x - 15}{x - 3}$, and $L = 8$.

Solution We evaluate the expression at $x = 3 + \alpha$ and use the EAR procedure:

$$\begin{aligned}\lim_{x \to 3} \frac{x^2 + 2x - 15}{x - 3} &= \frac{(3 + \alpha)^2 + 2(3 + \alpha) - 15}{(3 + \alpha) - 3} \\ &= \frac{9 + 6\alpha + \alpha^2 + 6 + 2\alpha - 15}{\alpha} = \frac{8\alpha + \alpha^2}{\alpha} \\ &\approx \frac{8\alpha}{\alpha} = 8.\end{aligned}$$

The limit exists because 8 is a real number. The value of the limit is 8. ■

The graphical meaning of limit is pictured in figure 4. If $\lim_{x \to b} f(x) = L$ exists, then the graph of f appears to contain the point (b, L). Wheather the point (b, L) is actually on the graph is another matter; it might or might not be there. But, it appears to be there, because when the values of x are infinitely close to b, the values of $f(x)$ are infinitely close to L.

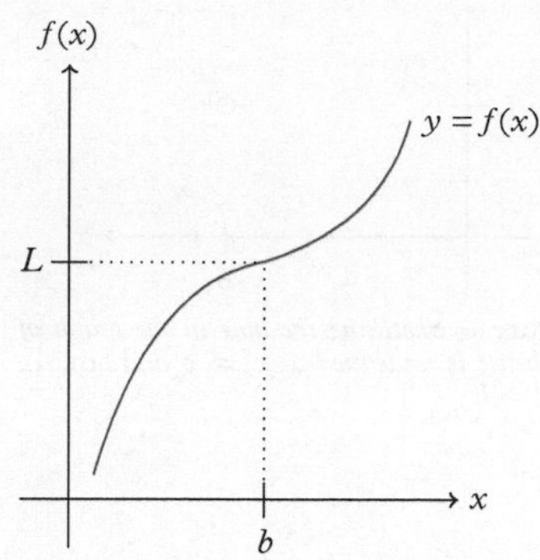

Figure 4 *A picture of* $\lim_{x \to b} f(x) = L$. *There may or may not actually be a point on the graph at* (b, L)

Reading Exercise 15 *Find* $\lim\limits_{x \to 1} \dfrac{x^2 - 1}{x - 1}$.

Limits: holes in a graph

Let's look at another example of calculating a limit and then interpret the answer graphically.

Example 2 *Find* $\lim\limits_{x \to 4} \dfrac{x - 4}{x^2 + x - 20}$.

Solution Evaluate the expression at $x = 4 + \alpha$:

$$\lim_{x \to 4} \frac{x - 4}{x^2 + x - 20} = \frac{4 + \alpha - 4}{(4 + \alpha)^2 + (4 + \alpha) - 20}$$
$$= \frac{\alpha}{16 + 8\alpha + \alpha^2 + 4 + \alpha - 20} = \frac{\alpha}{9\alpha + \alpha^2}$$
$$\approx \frac{\alpha}{9\alpha} = \frac{1}{9}.$$

When compared to the definition of limit, $f(x) = \frac{x - 4}{x^2 + x - 20}$ and $b = 4$. We calculate $f(b + \alpha) = f(4 + \alpha)$ for an arbitrary infinitesimal α and see if it renders the same real result every time. Because it renders $\frac{1}{9}$ for any such α, the limit exists.

The limit is $\frac{1}{9}$. ■

A graphical interpretation of the limit is that the point $(4, \frac{1}{9})$ appears to be on the graph of $f(x) = \frac{x - 4}{x^2 + x - 20}$. But is this point really on the graph? Let's try evaluating the function at $x = 4$:

The "hole" is called by a different name later in this chapter, in section 1.6.

$$f(4) = \frac{4 - 4}{4^2 + 4 - 20} = \frac{0}{0} \text{ (undefined)}.$$

There is no point on the graph with $x = 4$, because the function is undefined there as a result of division by zero; $x = 4$ is not in the domain of f.

Putting all this information together, we see that there must be a "hole" in the graph of f at $(4, \frac{1}{9})$. There is no point there; but, on a graph produced by a calculator or computer, it appears there is. Because points have no width, the hole is too small to show up in the graph, so we must make it show up. We denote the hole with an unfilled circle (an "open circle") in the graph (figure 5).

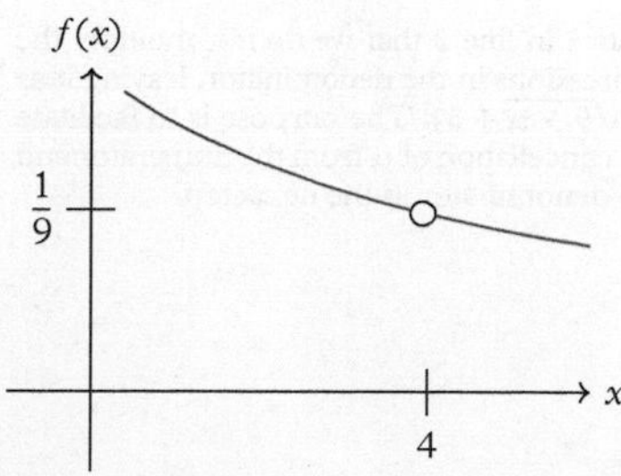

Figure 5 *The hole in the graph of* $f(x) = \frac{x - 4}{x^2 + x - 20}$ *at* $x = 4$

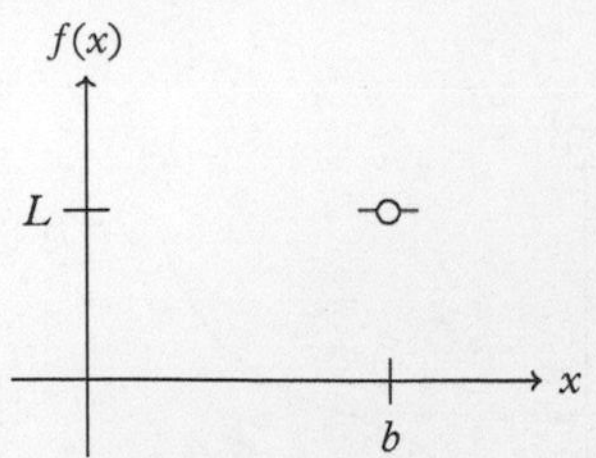

Figure 6 *Sketching the hole in the graph of f when f is undefined at $x = b$ and $\lim_{x\to b} f(x) = L$*

To summarize, a hole appears in a graph at (b, L) when $f(b)$ is undefined and $\lim_{x\to b} f(x) = L$. A rough sketch of the hole can be made even without knowing the rest of the function, by drawing a small circle at the point (b, L) and drawing a very short snippet of curve on both sides of the hole, as in figure 6. When more is known about the function, the general direction of the curve near the hole can be fixed.

Reading Exercise 16 *For $\lim\limits_{x\to 1} \dfrac{x^2-1}{x-1} = 2$, make a rough sketch of the "hole" in the graph corresponding to the limit.*

Limit example with a square root

The next example involves the limit of a square root expression.

Example 3 *Find $\lim\limits_{x\to 2} \dfrac{\sqrt{x+7}-3}{x-2}$.*

Solution We begin by evaluating the expression at $x = 2 + \alpha$:

$$\lim_{x\to 2} \frac{\sqrt{x+7}-3}{x-2} = \frac{\sqrt{2+\alpha+7}-3}{2+\alpha-2} = \frac{\sqrt{9+\alpha}-3}{\alpha}.$$

Ans. to reading exercise 15
2

Our first thought when approximating a square root expression is to approximate under the square root, so let's try that:

$$\approx \frac{\sqrt{9}-3}{\alpha} = \frac{0}{\alpha}.$$

A more detailed algebraic explanation of the steps shown here is found in section 1.2 example 17.

But, in doing so, we violate the approximation principle because the result of the approximation in the numerator is zero. We must therefore try something different, and as we learned before, multiplying the numerator and the denominator by the conjugate of the square root expression can help:

Notice in line 2 that we do not multiply the expressions in the denominator, leaving it as $\alpha(\sqrt{9+\alpha}+3)$. The purpose is to facilitate the cancellation of α from the numerator and the denominator in the next step.

$$\begin{aligned}\frac{\sqrt{9+\alpha}-3}{\alpha} &= \frac{\sqrt{9+\alpha}-3}{\alpha}\cdot\frac{\sqrt{9+\alpha}+3}{\sqrt{9+\alpha}+3}\\ &= \frac{9+\alpha-9}{\alpha(\sqrt{9+\alpha}+3)}\\ &= \frac{\alpha}{\alpha(\sqrt{9+\alpha}+3)} = \frac{1}{\sqrt{9+\alpha}+3}\\ &\approx \frac{1}{\sqrt{9}+3} = \frac{1}{6}.\end{aligned}$$

The limit is $\frac{1}{6}$. ■

Because the expression is undefined at $x = 2$ (division by zero), the graphical interpretation is that there is a hole in the graph of $f(x) = \frac{\sqrt{x+7}-3}{x-2}$ at the point $(2, \frac{1}{6})$. See figure 7.

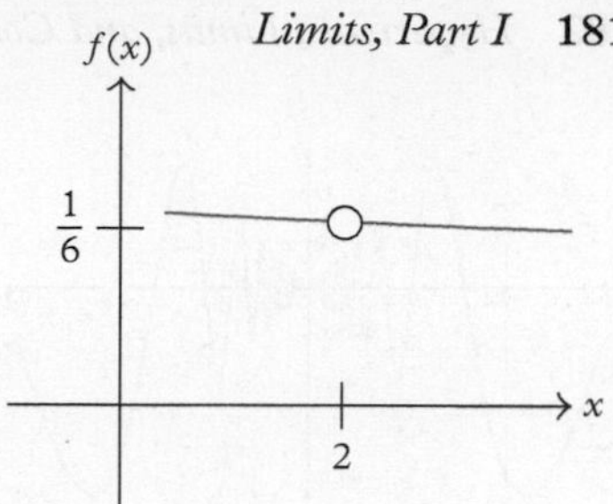

Figure 7 *The hole in the graph of* $f(x) = \frac{\sqrt{x+7}-3}{x-2}$ *at* $x = 2$

Limit example with trig

Example 4 *Find* $\lim_{x\to 0} x \sin \frac{1}{x}$.

Solution We evaluate the expression at $x = 0 + \alpha$, which is the same as $x = \alpha$:

$$\lim_{x\to 0} x \sin \frac{1}{x} = \alpha \sin \frac{1}{\alpha}.$$

Ans. to reading exercise 16

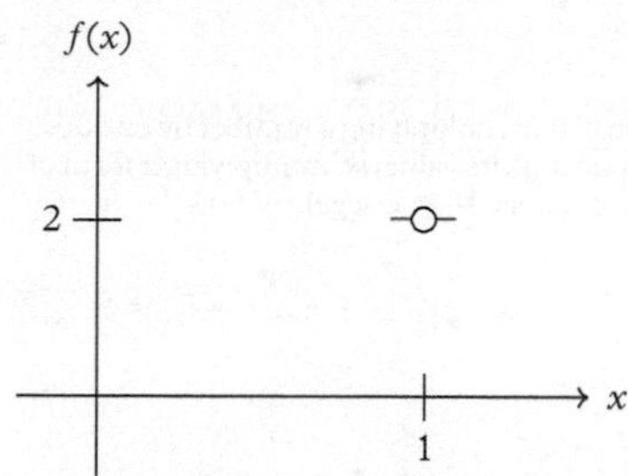

At this point, perhaps we are stuck, because we do not know how to evaluate trigonometric functions at hyperreal numbers. However, the transfer principle does tell us something about the trig functions. In section 1.1, we learned that since $-1 \le \sin x \le 1$ for any real number x, the same is true for any hyperreal number x. Therefore, $-1 \le \sin \frac{1}{\alpha} \le 1$. Then, multiplying through that inequality by α gives either

$$-\alpha \le \alpha \sin \frac{1}{\alpha} \le \alpha$$

or

$$-\alpha \ge \alpha \sin \frac{1}{\alpha} \ge \alpha,$$

Recall that multiplying through an inequality by a negative number changes the direction of the inequality.

depending on whether α is positive or negative. In either case, $\alpha \sin \frac{1}{\alpha}$ is between α and $-\alpha$, so the expression must be infinitesimal, and therefore renders the real result zero. The entire calculation can be written as follows:

Alternate reasoning: since $\sin \frac{1}{\alpha}$ is real, multiplying by the infinitesimal α results in an infinitesimal by theorem 1.

$$\lim_{x\to 0} x \sin \frac{1}{x} = \underbrace{\alpha \cdot \underbrace{\sin \frac{1}{\alpha}}_{\substack{\text{between} \\ 1\,\text{and}\,-1}}}_{\text{infinitesimal}} \doteq 0.$$

The limit is 0. ■

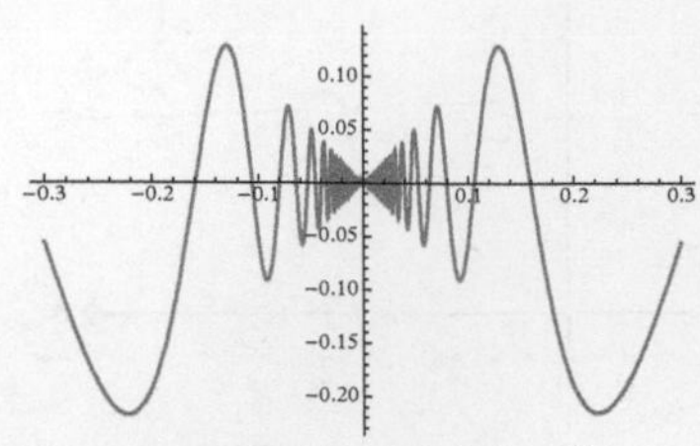

Figure 8 *The graph of $f(x) = x\sin\frac{1}{x}$, generated by Mathematica. The hole in the graph is not visible*

Because the expression is undefined at $x = 0$ (division by zero), the graphical interpretation is that there is a hole in the graph of $f(x) = x\sin\frac{1}{x}$ at the point $(0, 0)$. See figure 8.

Limit example with compound fraction

Example 5 *Evaluate* $\lim\limits_{x\to 1}\dfrac{\frac{1}{x}-1}{x-1}$.

Solution We evaluate the expression at $x = 1 + \alpha$:

$$\lim_{x\to 1}\frac{\frac{1}{x}-1}{x-1} = \frac{\frac{1}{1+\alpha}-1}{1+\alpha-1} = \frac{\frac{1}{1+\alpha}-1}{\alpha}.$$

Approximating at this point gives

$$\approx \frac{\frac{1}{1}-1}{\alpha} = \frac{0}{\alpha},$$

which violates the approximation principle. More algebra must be done before approximating.

The compound fraction (fraction within a fraction) can be made into a simple fraction (noncompound fraction) by multiplying the numerator and the denominator of the "big fraction" by the denominators appearing in the "little fractions." We therefore multiply by $\frac{1+\alpha}{1+\alpha}$ to simplify:

Recall that multiplying a number by one does not change its value, so multiplying a form of one such as $\frac{1+\alpha}{1+\alpha}$ is legal.

$$\begin{aligned}\frac{\frac{1}{1+\alpha}-1}{\alpha} &= \frac{\frac{1}{1+\alpha}-1}{\alpha}\cdot\frac{1+\alpha}{1+\alpha}\\ &= \frac{\frac{1}{\cancel{1+\alpha}}\cancel{(1+\alpha)}-1(1+\alpha)}{\alpha(1+\alpha)}\\ &= \frac{1-1-\alpha}{\alpha(1+\alpha)} = \frac{-\cancel{\alpha}}{\cancel{\alpha}(1+\alpha)} = \frac{-1}{1+\alpha}\\ &\approx \frac{-1}{1} = -1.\end{aligned}$$

The limit is -1. ■

Limit example with polynomial

There is nothing in the definition of limit that requires us to find limits only at places where we are dividing by zero. The procedure is the same, regardless of whether it is at a place where we are dividing by zero.

Example 6 *Find* $\lim\limits_{x\to 2}(x^2+7x-3)$.

Solution We evaluate the expression at $x = 2 + \alpha$:

$$\lim_{x\to 2}(x^2 + 7x - 3) = (2+\alpha)^2 + 7(2+\alpha) - 3$$
$$= 4 + 4\alpha + \alpha^2 + 14 + 7\alpha - 3 = 15 + 11\alpha + \alpha^2$$
$$\approx 15.$$

The limit is 15. ∎

In this example, the function is defined at $x = 2$. In fact, $f(2) = 2^2 + 7(2) - 3 = 15$, and the value of the function matches the limit. In this case there is no hole in the graph (figure 9); the point (2, 15) is on the graph. In the last several examples, the function is not defined at $x = b$, so there is no point at $x = b$; but, this is not the case with this example. This idea is discussed again later in this chapter, in section 1.6.

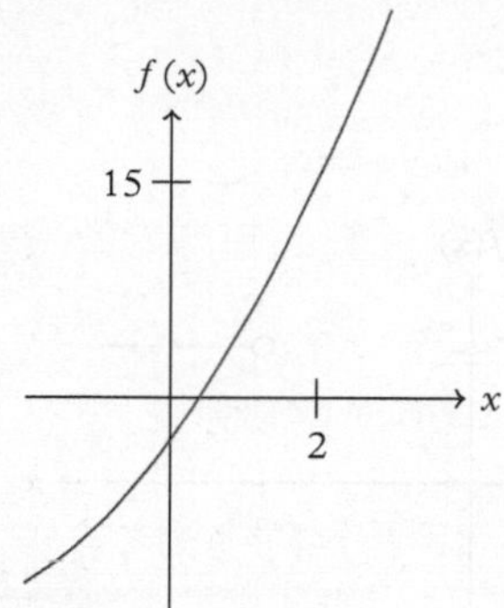

Figure 9 *The graph of $f(x) = x^2 + 7x - 3$, including the point with $x = 2$. There is no hole in the graph*

Reading Exercise 17 *Find* $\lim_{x\to 2}(x^2 + 1)$.

One-sided limits

The definition of limit requires that we render the same real result for every infinitesimal α. In our previous examples, this has happened, but sometimes it does not happen.

Example 7 *Find* $\lim_{x\to 3}\frac{|x-3|}{x-3}$.

Solution We evaluate the expression at $x = 3 + \alpha$:

$$\lim_{x\to 3}\frac{|x-3|}{x-3} = \frac{|3+\alpha-3|}{3+\alpha-3} = \frac{|\alpha|}{\alpha}.$$

At this point we are stuck! To evaluate $|\alpha|$, we need to know whether α is positive or negative, but because α is an arbitrary infinitesimal, it could be either. If $\alpha > 0$, we continue with

Recall the definition of $|x|$:
$$|x| = \begin{cases} x & \text{if } x \geq 0 \\ -x & \text{if } x < 0. \end{cases}$$

$$\frac{|\alpha|}{\alpha} = \frac{\alpha}{\alpha} = 1.$$

If $\alpha < 0$, we continue with

$$\frac{|\alpha|}{\alpha} = \frac{-\alpha}{\alpha} = -1.$$

We may write $\lim_{x\to3}\frac{|x-3|}{x-3}$ DNE or $\lim_{x\to3}\frac{|x-3|}{x-3}$ d.n.e. as abbreviated ways to express that the limit does not exist. Also, some instructors may prefer the phrase "does not exist as a real number."

For some values of α, the result is 1; for other values of α, the result is -1. Because we do not render the same real result for every infinitesimal α, the limit does not exist. ■

A peek at the graph (figure 10) reveals why the limit of example 7 does not exist. If x is close to 3 but larger than 3 (to the right of 3), then the values of the expression are close to 1. If x is close to 3 but smaller than 3 (to the left of 3), the values of the expression are close to -1.

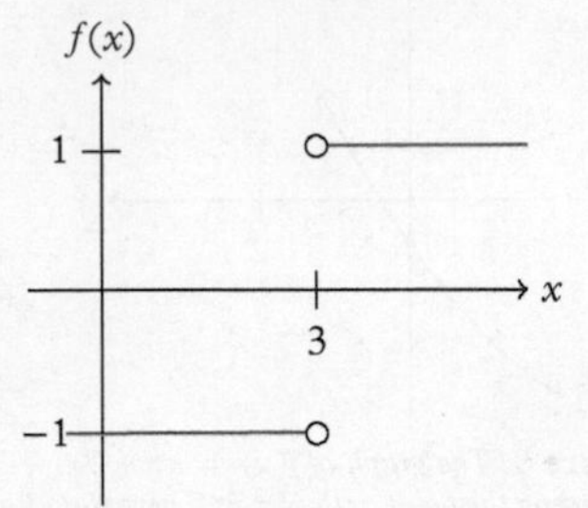

Figure 10 *The jump in the graph of* $f(x) = \frac{|x-3|}{x-3}$ *at* $x = 3$

Instead of a hole in the graph, we have a jump in the graph. It appears there are limits from each side, but they are different. We call them *one-sided limits*. To distinguish these one-sided limits from the regular (two-sided) limit, we denote the limit from the right by $\lim_{x\to b^+} f(x)$ and we denote the limit from the left by $\lim_{x\to b^-} f(x)$.

How can these one-sided limits be calculated? If we want to see what happens infinitely close to $x = 3$, but only to the right of $x = 3$, this means we are adding a positive infinitesimal amount to 3. Instead of calculating $f(3+\alpha)$ for an arbitrary infinitesimal, which checks both sides, we can calculate $f(3+\omega)$ for an arbitrary positive infinitesimal ω, which only checks to the right of 3. Similarly, if we check $f(3-\omega)$, still with ω positive, we are checking values just smaller than 3—that is, to the left of 3. The definition of a one-sided limit is like that of a limit, but with the modifications just discussed.

Ans. to reading exercise 17
5

Definition 7 ONE-SIDED LIMIT *Let f be a function and let b be a real number.*

(a) If $f(b+\omega)$ is defined and renders the same real result L for every positive infinitesimal ω, then we write

$$\lim_{x\to b^+} f(x) = L.$$

When L is a real number (i.e., not ∞ or $-\infty$), then we say that the limit exists.

(b) If $f(b-\omega)$ is defined and renders the same real result L for every positive infinitesimal ω, then we write

$$\lim_{x\to b^-} f(x) = L.$$

When L is a real number (i.e., not ∞ or $-\infty$), then we say that the limit exists.

The phrase $\lim_{x\to b^+} f(x) = L$ can be read "the limit of $f(x)$ as x goes to b from the right is L," whereas $\lim_{x\to b^-} f(x) = L$ can be read "the limit of $f(x)$ as x goes to b from the left is L."

For the purposes of calculation, we write

$$\lim_{x \to b^+} f(x) = f(b + \omega) \text{ and}$$

$$\lim_{x \to b^-} f(x) = f(b - \omega),$$

with the understanding that ω is an arbitrary positive infinitesimal and that we must render the same real result for every such ω or the limit does not exist.

Notice that the notation is suggestive of the formula for calculation: for $x \to b^+$, we use $x = b + \omega$ whereas for $x \to b^-$, we use $x = b - \omega$.

We can revisit the previous example to examine the calculation of one-sided limits.

Example 8 *Find* $\lim_{x \to 3^+} \frac{|x-3|}{x-3}$ *and* $\lim_{x \to 3^-} \frac{|x-3|}{x-3}$.

Solution For the limit from the right, we evaluate at $x = 3 + \omega$:

$$\lim_{x \to 3^+} \frac{|x-3|}{x-3} = \frac{|3+\omega-3|}{3+\omega-3} = \frac{|\omega|}{\omega} = \frac{\omega}{\omega} = 1.$$

The limit from the right is 1. For this calculation we are not stuck, because we know ω is positive.

For the limit from the left, we evaluate at $x = 3 - \omega$:

$$\lim_{x \to 3^-} \frac{|x-3|}{x-3} = \frac{|3-\omega-3|}{3-\omega-3} = \frac{|-\omega|}{-\omega} = \frac{\omega}{-\omega} = -1.$$

The limit from the left is -1. We are not stuck on this calculation either, because we know ω is positive. ■

Note that the only way for the two-sided limit to exist is for both one-sided limits to exist and have the same value.

EFFECT OF ONE-SIDED LIMITS ON TWO-SIDED LIMITS

- If either $\lim_{x \to b^+} f(x)$ or $\lim_{x \to b^-} f(x)$ does not exist, then $\lim_{x \to b} f(x)$ does not exist.
- If $\lim_{x \to b^+} f(x) = L$ and $\lim_{x \to b^-} f(x) = M$ with $L \neq M$, then $\lim_{x \to b} f(x)$ DNE.
- If $\lim_{x \to b^+} f(x) = L$ and $\lim_{x \to b^-} f(x) = L$, then $\lim_{x \to b} f(x) = L$.

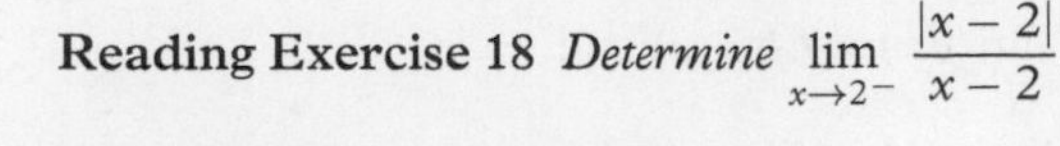

Reading Exercise 18 *Determine* $\lim_{x\to 2^-} \dfrac{|x-2|}{x-2}$.

To summarize, a jump appears in a graph at $x = b$ when $\lim_{x\to b^+} f(x) = L$ and $\lim_{x\to b^-} f(x) = M$ each exist but have different values. This is true regardless of whether $f(b)$ exists. A rough sketch of the jump can be given even without knowing the rest of the function by drawing small circles at the points (b, L) and (b, M), and drawing a very short snippet of curve to the right of (b, L) and to the left of (b, M), as in figure 11. When more is known about the function, the general direction of the curve near the jump can be fixed and the point $f(b)$ can be placed in the diagram if it exists.

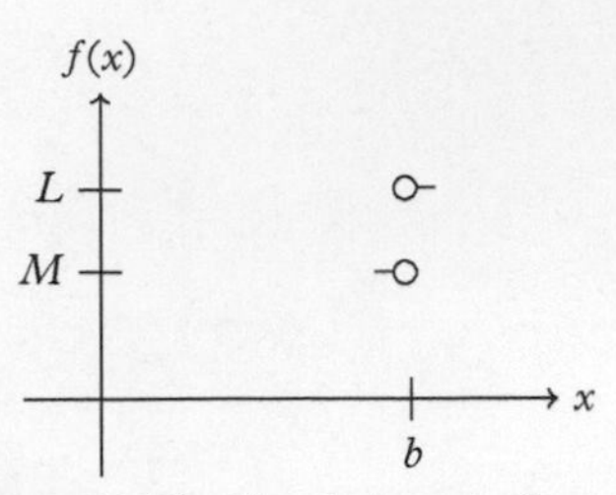

Figure 11 *Sketching the jump in the graph of f when $\lim_{x\to b^+} f(x) = L$ and $\lim_{x\to b^-} f(x) = M$. No point is shown at $x = b$, but such a point may be added if it exists*

Vertical asymptotes

One way for a limit not to exist is if it renders the real result ∞ or $-\infty$.

Example 9 *Determine* $\lim_{x\to 4} \dfrac{x+3}{(x-4)^2}$.

CAUTION: MISTAKE TO AVOID
Although the limit $\lim_{x\to 4} \dfrac{x+3}{(x-4)^2}$ does not exist, writing $\lim_{x\to 4} \dfrac{x+3}{(x-4)^2}$ DNE is insufficient because it does not convey all the available information. Writing $\lim_{x\to 4} \dfrac{x+3}{(x-4)^2} = \infty$ gives more information; it says in what manner the limit does not exist.

Solution This is not a one-sided limit (we have $x \to 4$, not $x \to 4^+$ or $x \to 4^-$). As mentioned, such limits are sometimes called *two-sided limits*. We therefore use $x = 4 + \alpha$:

$$\lim_{x\to 4} \frac{x+3}{(x-4)^2} = \frac{4+\alpha+3}{(4+\alpha-4)^2} = \frac{7+\alpha}{\alpha^2}$$
$$\approx \frac{7}{\alpha^2} = 7A^2$$
$$\doteq \infty.$$

We are not stuck, because squared values cannot be negative; we know that $7A^2$ is positive. Because the limit is infinite, the limit does not exist. ■

What is the graphical interpretation of this limit? Because the limit's value is ∞, we know that near $x = 4$, the values of the expression must become infinite. Therefore, the graph must rise above a y-coordinate of 100, above 1000, above 1 000 000, above any real number! The function's values must therefore rise above the top of our picture. We express this idea by saying that the function has a *vertical asymptote* at $x = 4$. We denote the vertical asymptote in a graph by a dotted line. The graph of this function is in figure 12.

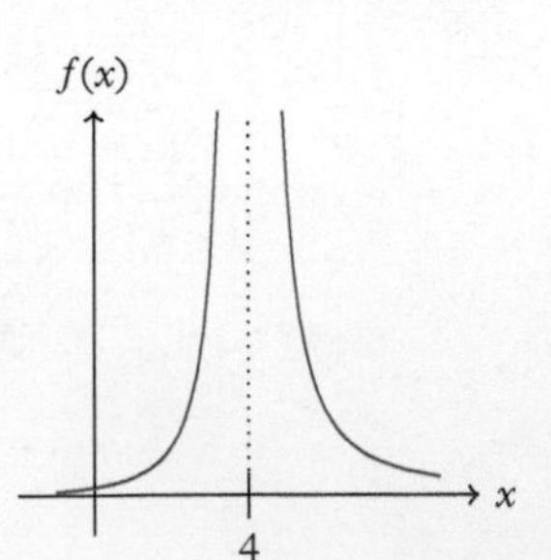

Figure 12 *The graph of $f(x) = \frac{x+3}{(x-4)^2}$, with a vertical asymptote at $x = 4$ marked by a dotted line*

In example 2, division by zero resulted in a hole in the graph. In example 9, division by zero resulted in a vertical asymptote. Merely knowing that we are dividing by zero does not tell us what is happening on the graph; we can only tell by determining the limit.

DIVISION BY ZERO, TWO-SIDED LIMIT

- If the two-sided limit exists and is a real number, then there is a hole in the graph.
- If the two-sided limit is ∞ or $-\infty$, then there is a vertical asymptote in the graph.
- If the two-sided limit cannot be rendered, check the one-sided limits to see what is happening.

Ans. to reading exercise 18
-1

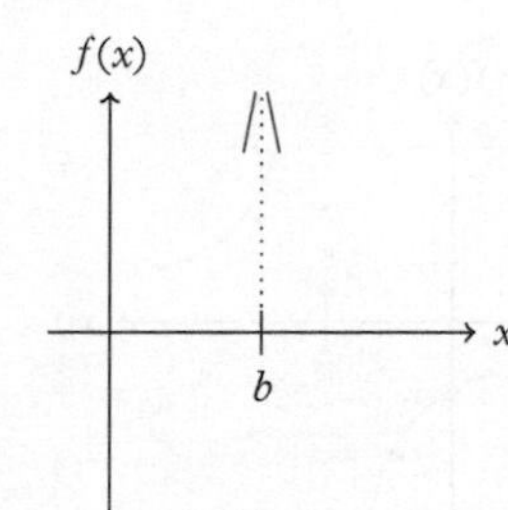

Figure 13 *Sketching the vertical asymptote in the graph of f when* $\lim_{x\to b} f(x) = \infty$

If $\lim_{x\to b} f(x) = \infty$, a rough sketch of the vertical asymptote can be made, even without knowing the rest of the function, by drawing a dotted line at $x = b$ and drawing very short snippets of curve at the top of the picture near $x = b$, as in figure 13. When more is known about the function, these snippets can be connected to the rest of the curve. A very similar sketch can be made if the limit is $-\infty$ instead; see figure 14.

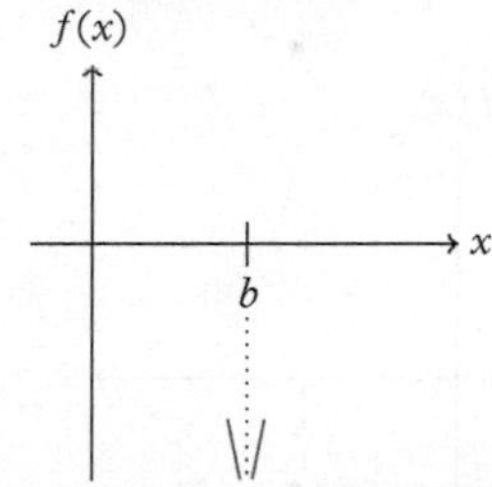

Figure 14 *Sketching the vertical asymptote in the graph of f when* $\lim_{x\to b} f(x) = -\infty$

Reading Exercise 19 *Determine* $\lim\limits_{x\to 1} \dfrac{-3}{(x-1)^2}$.

Remember, if we are stuck on a two-sided limit, we can check the one-sided limits.

Example 10 *Determine* $\lim\limits_{x\to 3} \dfrac{5}{x-3}$. *Check one-sided limits if necessary.*

Solution The limit is two-sided, so we begin by using $x = 3 + \alpha$:

$$\lim_{x\to 3} \frac{5}{x-3} = \frac{5}{3+\alpha-3} = \frac{5}{\alpha} = 5A, \text{ DNE.}$$

Although we are stuck at $5A$, we do know that the result is infinite, so we may safely conclude there is a vertical asymptote on the graph of the function at $x = 3$.

Because A can be either positive or negative, we are stuck. Because the two-sided limit can not be rendered, we check the one-sided limits instead. First we check the limit from the right using $x = 3 + \omega$:

We may check the limit from the left first if we wish. The order does not matter.

$$\lim_{x\to 3^+} \frac{5}{x-3} = \frac{5}{3+\omega-3} = \frac{5}{\omega} = 5\Omega \doteq \infty.$$

We are not stuck because we know that Ω is positive. Now let's try the limit from the left using $x = 3 - \omega$:

$$\lim_{x \to 3^-} \frac{5}{x-3} = \frac{5}{3-\omega-3} = \frac{5}{-\omega} = -5\Omega \doteq -\infty.$$

Although the two-sided limit is infinite, we must write $\lim_{x \to 3} \frac{5}{x-3}$ DNE because we cannot render either ∞ or $-\infty$. We can render ∞ and $-\infty$ for the right- and left-hand limits, respectively, so we do so and write the results accordingly.

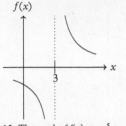

The one-sided limits exist and are different. This also means the two-sided limit does not exist. ■

Figure 15 *The graph of* $f(x) = \frac{5}{x-3}$*, with a vertical asymptote at* $x = 3$ *marked by a dotted line*

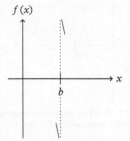

In this example the limits were infinite, but from the right the limit was positive infinite and from the left the limit was negative infinite. This means the graph exits the top of the picture just to the right of $x = 3$ and the graph exits the bottom of the picture just to the left of $x = 3$. See figure 15.

If $\lim_{x \to b^+} f(x) = \infty$ and $\lim_{x \to b^-} f(x) = -\infty$, a rough sketch of the vertical asymptote can be made even without knowing the rest of the function by drawing a dotted line at $x = b$, drawing a very short snippet of curve at the top of the picture just to the right of $x = b$, and drawing a very short snippet of curve at the bottom of the picture just to the left of $x = b$, as in figure 16. When more is known about the function, these snippets can be connected to the rest of the curve. A very similar sketch can be made if ∞ and $-\infty$ are swapped; see figure 17.

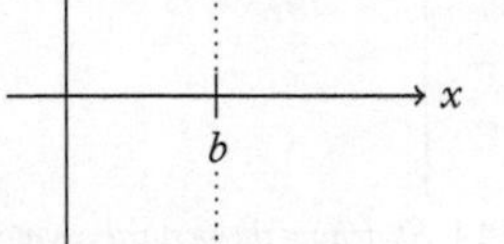

Figure 16 *Sketching the vertical asymptote in the graph of f when* $\lim_{x \to b^+} f(x) = \infty$ *and* $\lim_{x \to b^-} f(x) = -\infty$

In example 7, division by zero resulted in a jump in the graph. In example 10, division by zero resulted in a vertical asymptote. Even when the one-sided limits are different, there is more than one possible graphical outcome.

DIVISION BY ZERO, ONE-SIDED LIMITS

- If the one-sided limits exist and are different real numbers, then there is a jump in the graph.
- If one of the one-sided limits is ∞ and the other is $-\infty$, then there is a vertical asymptote on the graph.
- If the one-sided limits are identical, then the two-sided limit can be rendered; check the box "division by zero, two-sided limit" for graphical interpretations.

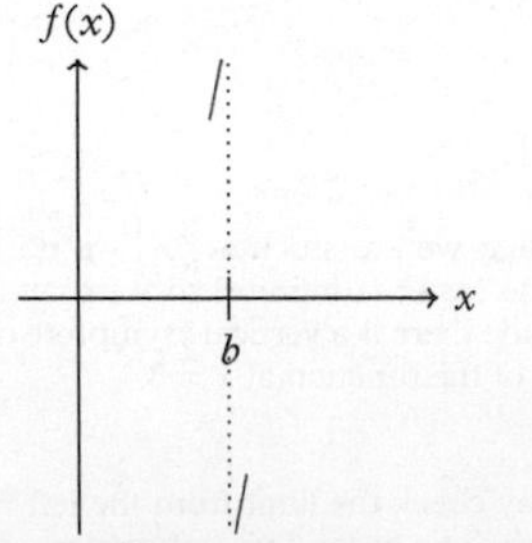

Figure 17 *Sketching the vertical asymptote in the graph of f when* $\lim_{x \to b^+} f(x) = -\infty$ *and* $\lim_{x \to b^-} f(x) = \infty$

The calculations in this section are much the same as in the previous section. What is new is the connection between the calculations and the interpretation on the graph of a function. This very important connection is provided by limits.

Exercises 1.4

1–4. Sketch the hole in the graph indicated by the limit.

Ans. to reading exercise 19
$-\infty$

1. $\lim_{x\to 3} f(x) = -2$
2. $\lim_{x\to -4} f(x) = 3$
3. $\lim_{x\to 5} f(x) = 7$
4. $\lim_{x\to -2} f(x) = 0$

5–8. Sketch the jump in the graph indicated by the limits.

The notation $\lim_{x\to 3} f(x) = -2$ and the notation $\lim\limits_{x\to 3} f(x) = -2$ mean the same thing.

5. $\lim_{x\to 1^+} f(x) = 2$, $\lim_{x\to 1^-} f(x) = 4$
6. $\lim_{x\to 0^+} f(x) = -3$, $\lim_{x\to 0^-} f(x) = \sqrt{2}$
7. $\lim_{x\to -1^+} f(x) = -1$, $\lim_{x\to -1^-} f(x) = 1$
8. $\lim_{x\to 3^+} f(x) = 2$, $\lim_{x\to 3^-} f(x) = 2$

9–14. Sketch the vertical asymptote in the graph indicated by the limit(s).

9. $\lim_{x\to 1} f(x) = \infty$
10. $\lim_{x\to -1} f(x) = -\infty$
11. $\lim_{x\to -1^+} f(x) = \infty$, $\lim_{x\to -1^-} f(x) = -\infty$
12. $\lim_{x\to -2^+} f(x) = -\infty$, $\lim_{x\to -2^-} f(x) = \infty$
13. $\lim_{x\to 0^+} f(x) = -\infty$, $\lim_{x\to 0^-} f(x) = -\infty$
14. $\lim_{x\to 4^+} f(x) = 2$, $\lim_{x\to 4^-} f(x) = \infty$

15–38. Calculate the limit.

"Calculate the limit" implies the use of the EAR procedure as demonstrated by the examples in this section.

15. $\lim\limits_{x\to 1} \dfrac{x^2 + 3x - 4}{2x - 2}$
16. $\lim\limits_{x\to 2} \dfrac{4x - 8}{x^2 - 5x + 6}$
17. $\lim\limits_{x\to 3^-} \dfrac{-4}{3 - x}$
18. $\lim\limits_{x\to 4^-} \dfrac{x + 1}{x - 4}$
19. $\lim\limits_{x\to -2} \dfrac{x + 2}{\sqrt{x + 6} - 2}$
20. $\lim\limits_{x\to 3} \dfrac{\sqrt{x - 2} - 1}{x - 3}$
21. $\lim\limits_{x\to 2^+} \dfrac{x + 1}{2 - x}$
22. $\lim\limits_{x\to 3} \dfrac{3x - 9}{x^2 - 2x - 3}$
23. $\lim\limits_{x\to -7} \dfrac{3x + 21}{x^2 + 12x + 35}$
24. $\lim\limits_{x\to 3^-} \dfrac{2 - x}{x^2 - 9}$
25. $\lim\limits_{x\to 5^-} \dfrac{\sqrt{x + 4} - 3}{x - 5}$
26. $\lim\limits_{x\to 4^+} \dfrac{(x - 4)^2}{\sqrt{2x + 1} - 3}$
27. $\lim\limits_{x\to 3^-} \dfrac{4|2x - 6|}{3 - x}$
28. $\lim\limits_{x\to 2^-} \dfrac{4|x - 2|}{6 - 3x}$

29. $\lim_{x\to 0}(4x^2 - 5x + 12)$

30. $\lim_{x\to 1}(4x^2 - 5x + 12)$

31. $\lim_{x\to 2}\dfrac{\frac{1}{x}-\frac{1}{2}}{x-2}$

32. $\lim_{x\to 0^+}\dfrac{\frac{1}{x}-\frac{1}{2}}{x-2}$

33. $\lim_{x\to 0} x\sin(2x^2+1)$

34. $\lim_{x\to 0} x^2\cos x$

35. $\lim_{x\to 0}\dfrac{1-x\sin x}{3+x}$

36. $\lim_{x\to 1^+}\dfrac{1-x\sin 3}{x-1}$

37. $\lim_{x\to 4^+}\dfrac{\sqrt{x-4}}{\sqrt{2x+1}-3}$

38. $\lim_{x\to 2^+}\dfrac{\sqrt{8x+9}-5}{\sqrt{x-2}}$

39–44. Determine the limit. If the two-sided limit cannot be rendered, check the one-sided limits.

39. $\lim_{x\to 1}\dfrac{|x+2|}{(x-1)^2}$

40. $\lim_{x\to 3}\dfrac{-2}{(x-3)^2}$

41. $\lim_{x\to 0}\dfrac{4}{x}$

42. $\lim_{x\to 3}\dfrac{-2}{(x-3)^3}$

43. $\lim_{x\to 3}\dfrac{x^2-9}{|x-3|}$

44. $\lim_{x\to 1}\dfrac{|x^2-1|}{x}$

If the limit is one-sided, just sketch the graph snippet on that side.

45–54. Sketch the graphical interpretation of the limit from the given exercise; do not sketch the function itself, just give the snippet(s) of curve indicated by the limit(s). Then, use a calculator or computer to check the graph of the appropriate function and compare to your sketch.

45. exercise 15

46. exercise 16

47. exercise 17

48. exercise 28

49. exercise 27

50. exercise 30

51. exercise 29

52. exercise 32

53. exercise 43

54. exercise 44

Limits, Part II

1.5

In this section we continue our study of limits.

Limit examples with piecewise-defined functions

Sometimes a piecewise-defined function can be evaluated as simply as any other limit.

Example 1 *Let* $f(x) = \begin{cases} 2x, & x \le 0 \\ 4x, & x > 0 \end{cases}$. *Investigate* $\lim_{x \to -3} f(x)$.

Solution Because the limit is two-sided, we use $x = -3 + \alpha$:

Because $-3 + \alpha \approx -3 < 0$, we use the rule for $x \le 0$, which is $f(x) = 2x$.

$$\lim_{x \to -3} f(x) = f(-3 + \alpha) = 2(-3 + \alpha) = -6 + 2\alpha \approx -6.$$

The limit is -6. ■

Recall that when evaluating a piecewise-defined function, the first task is to determine which rule applies to the function's input. This was not a difficult task in example 1. If, however, the limit is requested at a value where the function rules change, the situation can be a little more complicated.

Example 2 *Let* $f(x) = \begin{cases} x, & x \le 0 \\ x^2, & x > 0 \end{cases}$. *Investigate* $\lim_{x \to 0} f(x)$.

Solution Because the limit is two-sided, we use $x = 0 + \alpha = \alpha$:

$$\lim_{x \to 0} f(x) = f(\alpha),$$

and we are stuck because we do not know which rule to use to evaluate f. If $\alpha > 0$, then we use the rule $f(x) = x^2$; but, if $\alpha < 0$, we use the rule $f(x) = x$. Just as when we are stuck for other reasons, let's use one-sided limits to help. We evaluate the limit from the right by using $x = 0 + \omega = \omega$:

In section 1.4 example 7, we were stuck because we could not evaluate $|\alpha|$. In section 1.4 example 10, we were stuck because we could not render ∞ or $-\infty$ for $5A$. In both cases, one-sided limits helped.

$$\lim_{x \to 0^+} f(x) = f(\omega) = \omega^2 \doteq 0.$$

We are not stuck because ω is positive and we can use the rule for $x > 0$, which is $f(x) = x^2$. The limit from the right is 0.

We evaluate the limit from the left using $x = 0 - \omega = -\omega$:

$$\lim_{x \to 0^-} f(x) = f(-\omega) = -\omega \doteq 0.$$

We are not stuck because $-\omega$ is negative and we can use the rule for $x \leq 0$, which is $f(x) = x$. The limit from the left is 0.

Because both one-sided limits have the same value, 0, the two-sided limit exists:

$$\lim_{x \to 0} f(x) = 0.$$

The limit is 0. The matching one-sided limits can be seen visually in figure 1. ■

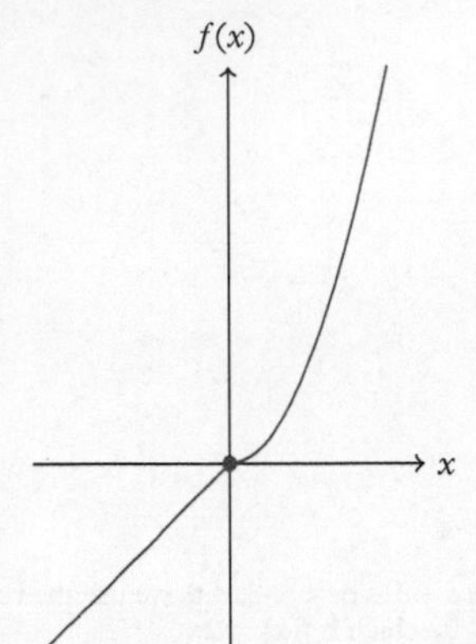

Figure 1 The graph of the function f from example 2 with a dot at $(0, 0)$ to emphasize the inclusion of that point. There is no hole in the graph

Although we were stuck initially when calculating the two-sided limit, the two-sided limit exists because the one-sided limits are equal. Just because we get stuck does not mean we can conclude the limit does not exist. More information is required.

Example 3 *Let* $f(x) = \begin{cases} x+1, & x < 2 \\ 11, & x = 2 \\ x^2, & x > 2 \end{cases}$. *Investigate* $\lim_{x \to 2} f(x)$.

Solution Because the limit is requested at the same place as the change in function rules ($x = 2$), we'll go straight to the one-sided limits. For the limit from the right, we use $x = 2 + \omega$:

$$\lim_{x \to 2^+} f(x) = f(2 + \omega) = (2 + \omega)^2 = 4 + 4\omega + \omega^2 \approx 4.$$

Because ω is positive, $2 + \omega > 2$ and we have to use the rule for $x > 2$, which is $f(x) = x^2$.

For the limit from the left, we use $x = 2 - \omega$:

$$\lim_{x \to 2^-} f(x) = f(2 - \omega) = (2 - \omega) + 1 = 3 - \omega \approx 3.$$

When finding $\lim_{x \to b} f(x)$, the value $f(b)$ is completely irrelevant, even if it is undefined.

Because ω is positive, $2 - \omega < 2$ and we have to use the rule for $x < 2$, which is $f(x) = x + 1$.

The limit from the right is 4 whereas the limit from the left is 3. Because the one-sided limits do not agree, the two-sided limit does not exist:

$$\lim_{x\to 2} f(x) \text{ DNE.} \qquad \blacksquare$$

The differing one-sided limits indicate a jump in the graph, as pictured in figure 2. Notice that the value of the function at $x = 2$ has absolutely no bearing on the calculation of the limit. Nowhere in example 3 did we use the value 11. We calculated $f(2+\omega)$ and $f(2-\omega)$, but we did not need $f(2)$. We could have made $f(2)$ anything we wanted, or we could have left $f(2)$ undefined, without affecting the limit at all.

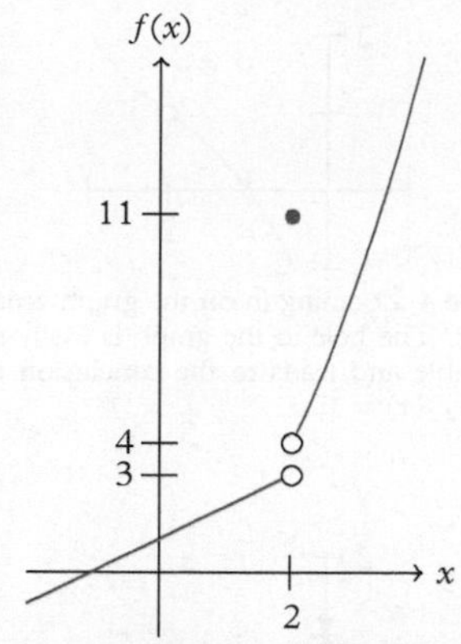

Figure 2 The graph of the function f from example 3 with the point at $x = 2$ denoted by a dot. The open circles emphasize there are not points at these locations

If the two endpoints of the pieces of a piecewise-defined curve match up, as in figure 1, then the one-sided limits are equal and the two-sided limit exists, as in example 2. If the two endpoints of the pieces of a piecewise-defined curve do not match up, as in figure 2, then the one-sided limits are not equal and the two-sided limit does not exist, as in example 3.

Reading Exercise 20 *Let* $f(x) = \begin{cases} x, & x > 1 \\ x+2, & x \leq 1 \end{cases}$. *Investigate* $\lim_{x\to 1} f(x)$.

Finding limits graphically

In section 1.4 we evaluated limits "algebraically" and then considered what the graph might look like. Given the limit, we sketched a snippet of graph. The reverse can also be done: from the graph, we should be able to determine the limit. We call this *determining a limit graphically.* Knowing which limits give rise to holes, jumps, and vertical asymptotes, when we see such features on a graph, we can identify the limit accordingly.

One of the limitations of determining a limit from a graph is the resolution of our picture. Is the limit really 5 or is the point a few atoms higher at 5.00000034? I will be kind and assume the limits are easily discernible to the naked eye and that they line up nicely on the grid.

Example 4 *Use figure 3 to determine the following limits:*

(a) $\lim_{x\to 2} f(x)$ *(b)* $\lim_{x\to 4} f(x)$ *(c)* $\lim_{x\to 6} f(x)$
(d) $\lim_{x\to 6^+} f(x)$ *(e)* $\lim_{x\to 6^-} f(x)$ *(f)* $\lim_{x\to -2} f(x)$
(g) $\lim_{x\to -1} f(x)$ *(h)* $\lim_{x\to -1^-} f(x)$ *(i)* $\lim_{x\to -1^+} f(x)$
(j) $\lim_{x\to 1} f(x)$

Solution (a) To find $\lim_{x\to 2} f(x)$, we look on the graph near $x = 2$; see figure 4. We see a hole in the graph at the point (2, 1). The value of f at exactly $x = 2$ is irrelevant, so we do not care whether the hole is filled in or whether there is a point for $x = 2$ at some other y-coordinate. The hole in the graph is what corresponds to a graphical interpretation of limit, so this is what matters. Compare the graph in figure 4 to the graph in section 1.4 figure 6. We see that the value of the limit at $x = 2$ is the y-coordinate of the hole, which is 1. We conclude that $\lim_{x\to 2} f(x) = 1$.

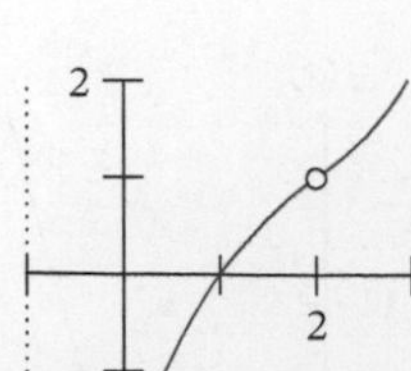

Figure 4 Zooming in on the graph around $x = 2$. The hole in the graph is easily recognizable and leads to the conclusion that $\lim_{x\to 2} f(x) = 1$

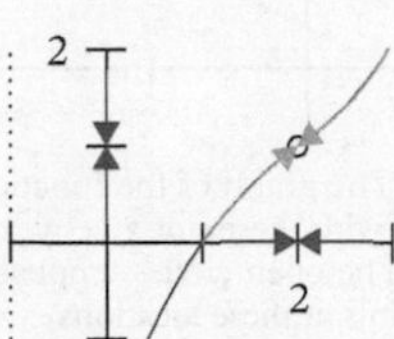

Figure 5 Zooming in on the graph around $x = 2$. Red arrows illustrate x nearing 2, green arrows illustrate the curve as x nears 2, and magenta arrows illustrate the effect on $f(x)$

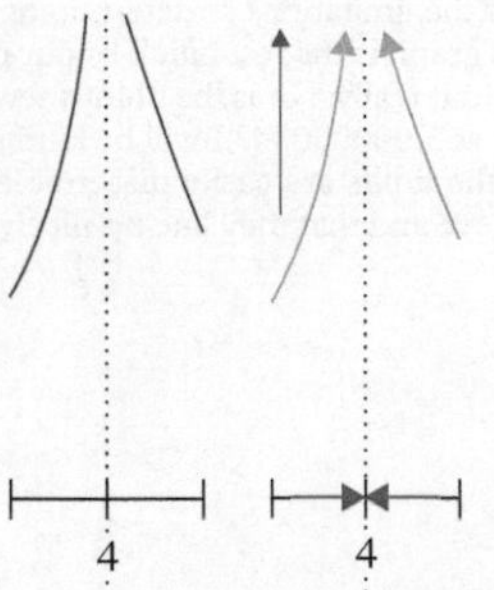

Figure 6 Zooming in on the graph around $x = 4$. The vertical asymptote is easily recognizable and leads to the conclusion that $\lim_{x\to 4} f(x) = \infty$

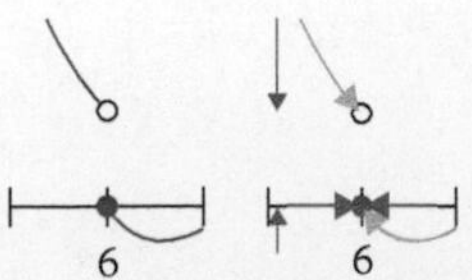

Figure 7 Zooming in on the graph around $x = 6$. The jump in the graph is easily recognizable and leads to the conclusion that the two-sided limit $\lim_{x\to 6} f(x)$ does not exist

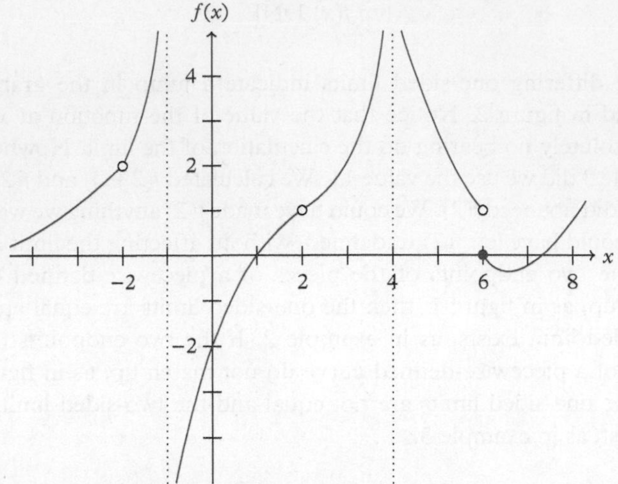

Figure 3 *The graph for example 4*

Another way of seeing this limit is to ask: As x nears 2, what happens to the y-coordinates on the graph? In figure 5, the red arrows on the horizontal axis point to $x = 2$. Follow the curve above these red arrows; the result is in green. Then, go from the green arrows to the vertical axis, where the resulting arrows are magenta. We conclude that as x nears 2, $f(x)$ nears 1. Therefore, $\lim_{x\to 2} f(x) = 1$.

(b) To determine $\lim_{x\to 4} f(x)$, we look at the graph near $x = 4$, as in figure 6. We see a vertical asymptote that exits the top of the picture on both sides of $x = 4$. This matches the picture in section 1.4 figure 13, and we conclude that $\lim_{x\to 4} f(x) = \infty$. The implication in the picture given by the dotted line representing the vertical asymptote is that near $x = 4$, the values of $f(x)$ become infinitely large. Using arrows instead, we see that the magenta arrow in figure 6 is pointing, not to a specific real number, but rather off the top of the diagram to an infinite height, so that the limit is ∞.

(c) For $\lim_{x\to 6} f(x)$, we look near $x = 6$, as in figure 7. We see a jump in the graph, which means the limits from the left and the right are not equal. We conclude that $\lim_{x\to 6} f(x)$ DNE. Compare to section 1.4 figure 11. Using arrows instead, note that the two magenta arrows in figure 7 do not point to the same number, so the two-sided limit does not exist.

(d) To determine a one-sided limit such as $\lim_{x\to 6^+} f(x)$, we only wish to look at one side, not both sides. Instead of looking at figure 7, we look at figure 8, which shows the graph only on the right-hand side of $x = 6$. As always, the value of the function at $x = 6$

is irrelevant, so it does not matter which circle is filled in and which is not. Instead, we pay attention to the curve and where it is headed, which is toward the y-coordinate 0. We conclude that $\lim_{x\to 6^+} f(x) = 0$.

(e) The limit from the left is just like the limit from the right, except that we look on the left side of $x = 6$ instead of the right side. See figure 9. Once again, the value of f at $x = 6$ is irrelevant, so we do not care which circle is open or closed. We only care about where the curve leads us, which is to the open circle. Looking at the original diagram (figure 3) to see at which y-coordinate this open circle lines up, we conclude that $\lim_{x\to 6^-} f(x) = 1$.

(f) For $\lim_{x\to -2} f(x)$, look at the original diagram, figure 3, and notice the hole at $x = -2$. This is the same scenario as in part (a). The y-coordinate of the hole is 2 and we conclude that $\lim_{x\to -2} f(x) = 2$.

(g) For $\lim_{x\to -1} f(x)$, we look near $x = -1$ and see a vertical asymptote (figure 10). This is similar but not identical to part (b), for here the graph exits the top of the picture to the left of $x = -1$, but exits the bottom of the picture to the right of $x = -1$. This matches section 1.4 figure 17, where the limit from the right is $-\infty$ and the limit from the left is ∞. We conclude that the two-sided limit does not exist: $\lim_{x\to -1} f(x)$ DNE. Notice also that the magenta arrows in figure 10 do not point in the same direction.

(h) Considering only what is to the left of $x = -1$ in figure 10, we see that $\lim_{x\to -1^-} f(x) = \infty$. The magenta arrow corresponding to the portion of the curve to the left of $x = -1$ points off the top of the picture.

(i) Considering only what is to the right of $x = -1$ in figure 10, we see that $\lim_{x\to -1^+} f(x) = -\infty$. The magenta arrow corresponding to the portion of the curve to the right of $x = -1$ points off the bottom of the picture.

(j) Looking at the graph close to $x = 1$ (figure 11), we cannot tell whether the point at $x = 1$ is included; a hole is not indicated. But, when using a calculator- or computer-generated graph, such holes are generally not indicated even when present. In any case, it does not matter! Recall that the value of the function at $x = 1$, or even whether it is defined at all at $x = 1$, is irrelevant. The curve appears to pass through the point $(1, 0)$, so we conclude the limit is 0: $\lim_{x\to 1} f(x) = 0$. ■

It is not necessary to follow the arrows as indicated in the solution to example 4. With a little practice, limits can be read directly off a graph very quickly.

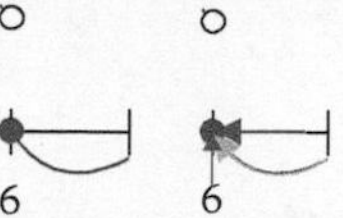

Figure 8 Zooming in on the graph around $x = 6$, but only to the right of $x = 6$. We conclude that $\lim_{x\to 6^+} f(x) = 0$

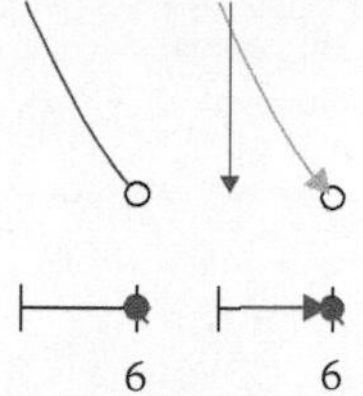

Figure 9 Zooming in on the graph around $x = 6$, but only on the left-hand side of $x = 6$. We conclude that $\lim_{x\to 6^-} f(x) = 1$

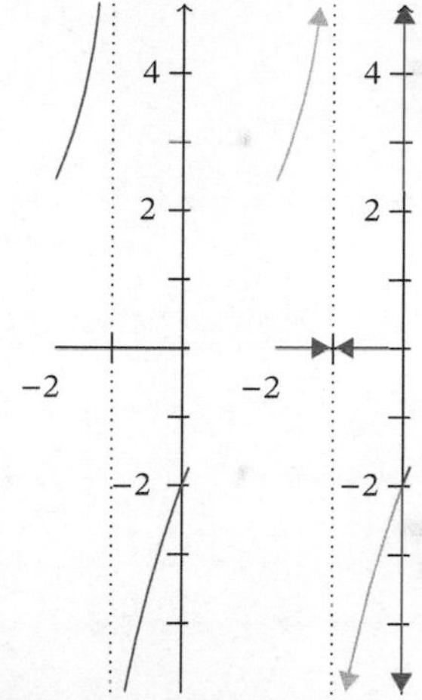

Figure 10 Zooming in on the graph around $x = -1$. The vertical asymptote is easily recognizable, but because the two sides do not point in the same direction, we conclude that $\lim_{x\to -1} f(x)$ DNE

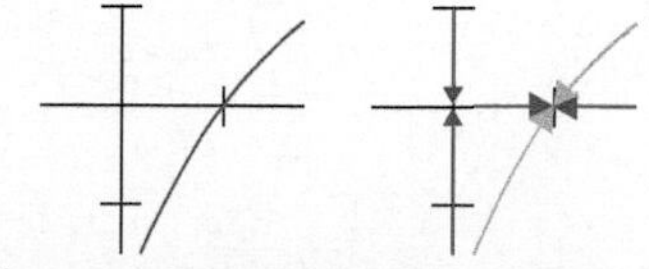

Figure 11 Zooming in on the graph around $x = 1$. We conclude that $\lim_{x\to 1} f(x) = 0$

Ans. to reading exercise 20
$\lim_{x\to 1^-} f(x) = 3$, $\lim_{x\to 1^+} f(x) = 1$, $\lim_{x\to 1} f(x)$ DNE

Reading Exercise 21 *Use figure 12 to determine the following limits:*

(a) $\lim_{x \to -3} f(x)$ (b) $\lim_{x \to 1} f(x)$ (c) $\lim_{x \to 1^+} f(x)$
(d) $\lim_{x \to 1^-} f(x)$ (e) $\lim_{x \to 3} f(x)$ (f) $\lim_{x \to 3^+} f(x)$
(g) $\lim_{x \to 3^-} f(x)$

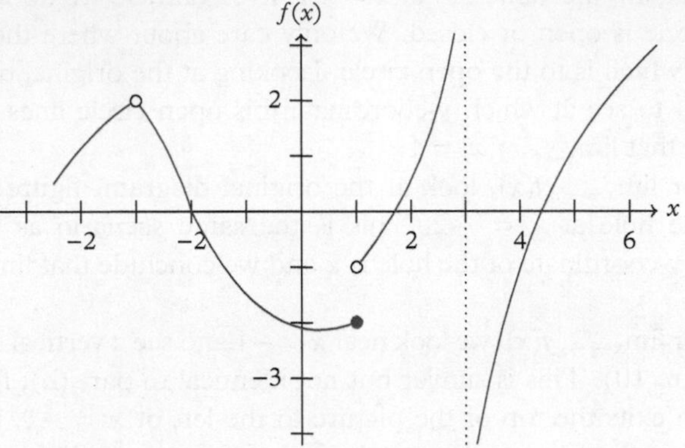

Figure 12 *The graph for reading exercise 21*

Sketching functions from limit information

We have previously sketched snippets of the graph of a function from information provided by limits. To practice the connection between graphs and limits even further, we can give several limits and sketch the graph of a function that meets all the given information. The general idea is to sketch the graph snippets as before, all on one graph, and then connect the snippets, yielding the graph of a function.

Example 5 *Sketch the graph of a function f satisfying* $\lim_{x \to 2^-} f(x) = 1$, $\lim_{x \to 2^+} f(x) = 3$, $\lim_{x \to -1} f(x) = -1$, *and* $f(2) = 0$.

It is best to draw open circles for each of these limits rather than filled circles. We can always go back and fill in a circle later if we run across information that requires it.

Solution We begin by drawing the graph snippets corresponding to each of the limits. For $\lim_{x \to 2^-} f(x) = 1$, we draw an open circle at the point $(2, 1)$ with a snippet of graph on the left side. For $\lim_{x \to 2^+} f(x) = 3$, we draw an open circle at the point $(2, 3)$ with a snippet of graph on the right side. For $\lim_{x \to -1} f(x) = -1$, we draw an open circle at the point $(-1, -1)$ with a snippet of graph on both sides. Last, we have one more piece of information—namely, $f(2) = 0$. Just as when plotting points in algebra, we place a point on the graph at the point $(2, 0)$. These steps are shown in figure 13.

Now all that is left is to connect the pieces of graph. This can be done in any manner one wishes, as long as the result is a function, and the required limits and function values are preserved. Two of the infinitely many possible answers are in figure 14. ■

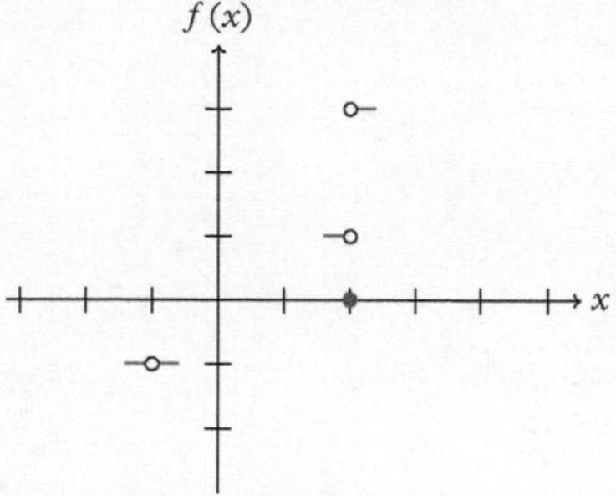

Figure 13 A partially completed graph for example 5. Graph snippets are short so they do not overlap

It is apparent that just knowing a few limits on the graph of a function is not enough to specify the function completely. Thankfully, there are additional tools of calculus that can produce information to shape a function's graph. Stay tuned!

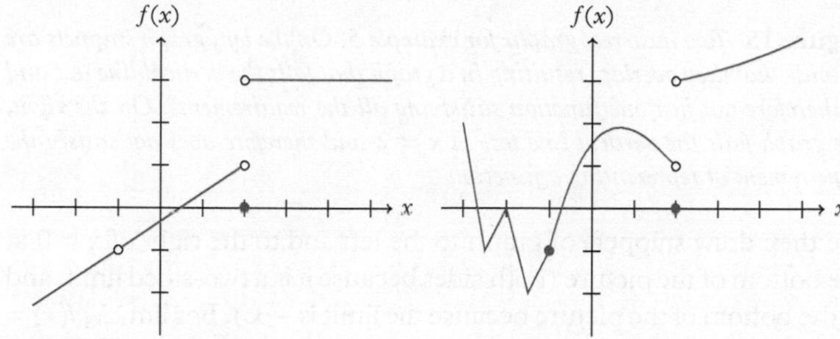

Figure 14 *Two possible solutions to example 5. Many other possible solutions exist*

Ans. to reading exercise 21
(a) 2, (b) DNE, (c) −1, (d) −2, (e) DNE, (f) $-\infty$, (g) ∞

Although there are many possible answers to example 5, it is also possible to make an incorrect graph, even if the graph snippets of each limit are still represented correctly. This occurs most often when drawing very wide snippets of graph around the limits instead of very short snippets of graph. For instance, the graph on the left in figure 15 is not a solution to example 5 because it is not the graph of a function. The graph of a function ("a" function meaning "one" function) is required. Recall that functions must pass the vertical line test; a function cannot have more than one output (y-coordinate) for any given input (x-coordinate). The graph on the left in figure 15 fails the vertical line test between $x = 0$ and $x = 1$. The moral of this story is to make your graph snippets short!

A more subtle way of making the same mistake is to fill in circles for limits instead of using open circles. Even if the vertical line test fails for just one value of x, it fails and the result is not a function. See the graph on the right in figure 15.

CAUTION: MISTAKE TO AVOID

Example 6 *Sketch the graph of a function f satisfying* $\lim_{x\to 0} f(x) = -\infty$, $\lim_{x\to 4} f(x) = 1$, *and* $f(4) = 2$.

Solution We begin by drawing graph snippets for the limits. For $\lim_{x\to 0} f(x) = -\infty$, we draw a dotted line at $x = 0$ to represent a vertical asymptote.

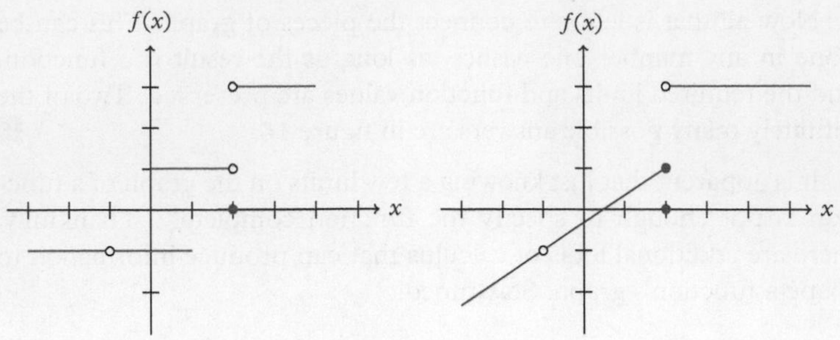

Figure 15 *Two incorrect graphs for example 5. On the left, graph snippets are so wide that they overlap, resulting in a graph that fails the vertical line test and is therefore not just one function satisfying all the requirements. On the right, the graph fails the vertical line test at $x = 2$ and therefore does not satisfy the requirement of representing a function*

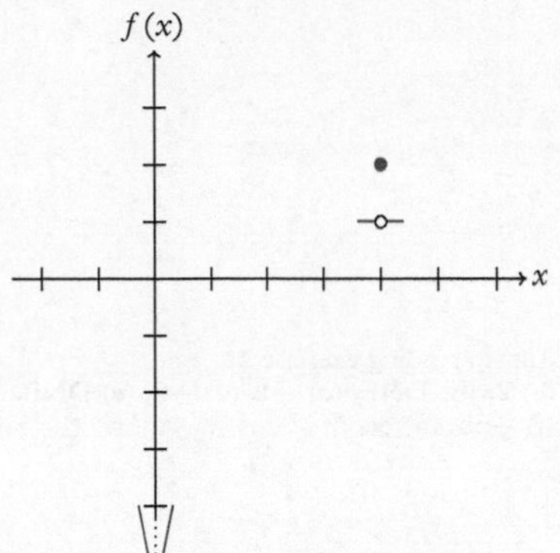

Figure 16 A partially completed graph for example 6. Graph snippets are short so they do not overlap

We then draw snippets of graph to the left and to the right of $x = 0$ at the bottom of the picture (both sides because it is a two-sided limit, and at the bottom of the picture because the limit is $-\infty$). For $\lim_{x\to 4} f(x) = 1$, we draw an open circle at the point (4, 1) and draw short snippets of graph on both sides. For $f(4) = 2$, we place a dot at the point (4, 2). This is done in figure 16.

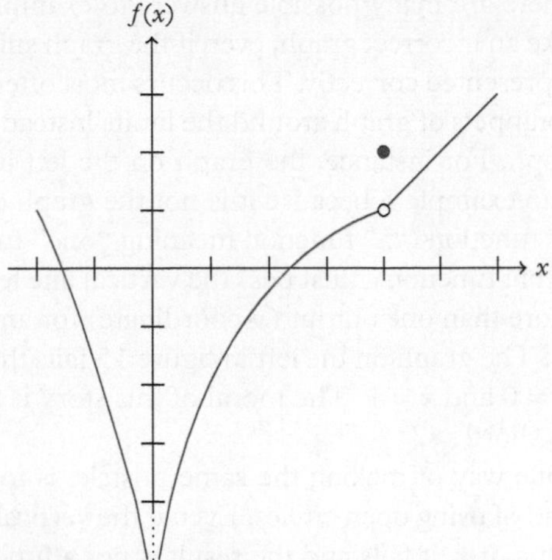

Figure 17 *A solution to example 6. Many other solutions exist*

■

All that remains is to connect the pieces. It is customary to continue the graph to the right side and the left side of the picture. A solution is shown in figure 17.

Numerical estimation of limits

Suppose we wish to find $\lim_{x\to 0} \frac{1-\cos x}{x}$. This is a two-sided limit, so we use $x = 0 + \alpha = \alpha$:

$$\lim_{x\to 0} \frac{1-\cos x}{x} = \frac{1-\cos \alpha}{\alpha} = ?$$

Although we will eventually learn the value of $\cos \alpha$, at this point we have not studied it. And even if we had, there will always be other functions for which an algebraic approach to finding limits hits some roadblock. It is not possible to determine algebraically the exact value of every limit we might want.

We have also studied the graphical approach. A graph of $f(x) = \frac{1-\cos x}{x}$ is presented in figure 18. Because the point $(0, 0)$ appears to be on the graph (even though it can't be there because of division by zero), it appears that the value of the limit is 0, but we have no assurance that it is exactly 0. We can zoom in repeatedly to smaller scales and look again, but at no point can we be exactly sure; we just have a very strong suspicion that the limit is 0.

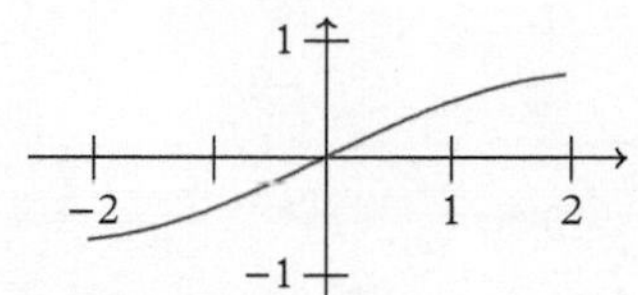

Figure 18 The graph of $f(x) = \frac{1-\cos x}{x}$. The hole at $x = 0$ is not marked

When generating the graph, a calculator or computer must calculate a large number of values of the function. An alternative is to calculate a few of these values ourselves. Then, instead of checking the graph to see which y-coordinate the curve seems to be approaching, we can check the numbers for the same thing. Because numbers exhibit greater precision much more easily than points in a picture, we have greater assurance that the limit has been calculated relatively precisely. We call this process *estimating a limit numerically*.

We can always zoom in to just as many decimal places as we calculate numerically, but the numerical procedure is much more efficient than graphing, especially when programmed on a computer.

Example 7 *Estimate* $\lim_{x\to 0} \frac{1-\cos x}{x}$ *numerically.*

Solution Our general strategy is to estimate the limit from the right and from the left by choosing values closer and closer to the target value of x. Choosing values that are 0.1, 0.01, 0.001, and so on, to the right and to the left of the target value is generally effective.

For $\lim_{x\to 2} f(x)$, the values to the right of the target value are 2.1, 2.01, 2.001, ..., and the values to the left of the target value are 1.9, 1.99, 1.999, ...

First we check the limit from the right by checking values larger than the target value of 0:

Tips for good results: (1) be consistent in the number of decimal places used for $f(x)$ (right-hand column of table); (2) round correctly (for instance, using six decimal places, $f(.01) = .004999|958\ldots$ rounds to 0.005000 because the seventh digit, in bold, is 5 or larger); (3) write the values of x in numerical order where the lower the line in the table, the closer x is to the target value; and (4) use more values of x if you are unsure of the limit.

x	$\frac{1-\cos x}{x}$
0.1	0.049958
0.01	0.005000
0.001	0.000500
0.0001	0.000050

It appears that the values of the expression (the values in the right-hand column) are approaching zero. We therefore estimate the value of the limit from the right as $\lim_{x\to 0^+} \frac{1-\cos x}{x} = 0$.

Next we check the limit from the left by checking values smaller than the target value of 0:

The values of the left- and right-hand limits are not always symmetric. They can be quite different.

x	$\frac{1-\cos x}{x}$
−0.1	−0.049958
−0.01	−0.005000
−0.001	−0.000500
−0.0001	−0.000050

The values appear to be approaching zero from the left as well; we estimate $\lim_{x\to 0^-} \frac{1-\cos x}{x} = 0$.

Because our estimates of the one-sided limits agree, the two-sided limit should be the same. We finish by concluding the estimate $\lim_{x\to 0} \frac{1-\cos x}{x} = 0$. ■

numerical	algebraic
estimated answer	exact answer
less assurance	complete assurance
use when necessary	preferred method

When estimating a limit numerically, we are checking values of x that are close, but not infinitely close, to the target value. When calculating a limit algebraically using infinitesimals, we are checking values infinitely close to the target value. By comparison, the values used for numerical estimation are infinitely farther away than the ones we use with infinitesimals! For this reason, although algebraic methods with infinitesimals give the exact answer to the limit with complete assurance, a numerical estimate only estimates the answer, and with less assurance. In practice, we should only use a numerical estimate when an exact answer cannot be found algebraically.

The mathematical subject *numerical analysis* is the study of numerical methods for estimating quantities that cannot be determined easily by exact algebraic methods. We encounter additional numerical methods throughout this book.

Limitations of graphical and numerical methods for estimating limits

Although numerical and graphical estimates of limits can be helpful, they are sometimes imperfect.

Example 8 *Estimate* $\lim_{x\to 0^+} \dfrac{\sqrt{x^2+16}-4}{x^2}$ *numerically.*

Solution The limit is one-sided and we only check values to the right of (larger than) zero:

x	$\frac{\sqrt{x^2+16}-4}{x^2}$
0.1	0.1249805
0.01	0.1249998
0.001	0.1250000

At this point, we conclude that the limit is 0.125, or $\frac{1}{8}$, writing our answer $\lim_{x\to 0^+} \frac{\sqrt{x^2+16}-4}{x^2} = 0.125$. This is the correct answer, as can be seen by calculating the limit using the methods of the previous section. ■

What happens, though, if we check more values of x? The following is an expanded table, as calculated with a TI-83 calculator:

x	$\frac{\sqrt{x^2+16}-4}{x^2}$
0.1	0.1249805
0.01	0.1249998
0.001	0.1250000
0.0001	0.1250000
0.00001	0.1250000
0.000001	0.0000000
0.0000001	0.0000000

The last two entries of the right-hand column are different from the others. If we conclude the limit is 0, we are wrong! We were right the first time with an answer of 0.125. So what is going on?

The problem is that when computing $\sqrt{x^2+4}$ for small values of x, the answers are of the form 4.00000000 . . . 1, and when the number of 0 digits after the 4 becomes larger than the number of digits carried by the calculator (or computer software) in the calculation, the number is equated with 4. Then, $\sqrt{x^2+4}-4$ becomes zero, and the entire quantity we are estimating is erroneously said to be zero.

Avoiding such computational difficulties is part of what makes the subject of numerical analysis both challenging and interesting. It also illustrates the need for caution in accepting answers naively calculated on a calculator or computer.

Because graphs are produced by computing values of functions, estimating the value of a limit graphically is fraught with the same difficulties. Let's use the same function to illustrate this point, but switch to the two-sided limit.

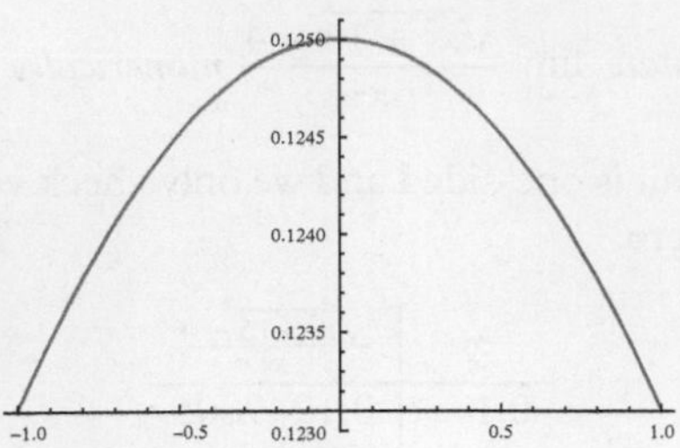

Figure 19 *The graph of* $f(x) = \frac{\sqrt{x^2+16}-4}{x^2}$*, generated by Mathematica. The hole in the graph at* $x = 0$ *is not visible*

Example 9 *Use graphs to estimate the value of* $\lim_{x\to 0} \frac{\sqrt{x^2+16}-4}{x^2}$.

Solution To estimate a limit graphically, we look at a graph of the function for values of x near the target point. Because our limit is as $x \to 0$, we first choose to graph the function between $x = -1$ and $x = 1$, as in figure 19. From this diagram, it appears that the y-coordinate on the graph corresponding to the x-coordinate 0 is 0.125, and we conclude that $\lim_{x\to 0} \frac{\sqrt{x^2+16}-4}{x^2} = 0.125$. This is the correct value of the limit. ■

What happens if we choose narrower intervals on which to view the graph? The three graphs in figure 20 use the intervals $[-0.1, 0.1]$, $[-0.01, 0.01]$, and $[-0.001, 0.001]$ with widely differing results! The source of the difficulties is the same as for the numerical estimate of the limit, as explained after example 8.

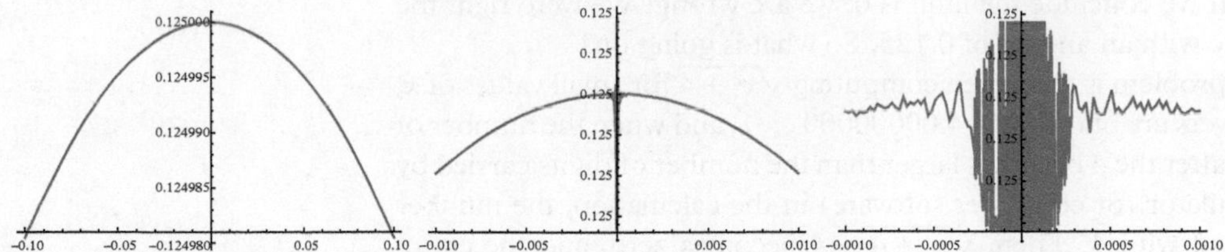

Figure 20 *Three graphs of* $f(x) = \frac{\sqrt{x^2+16}-4}{x^2}$*, generated by Mathematica. The graph on the left leads to the correct conclusion, the graph in the middle leaves doubt because of its fuzziness around* $x = 0$*, and the graph on the right leaves one thoroughly confused*

Don't believe everything you read or see from a calculator or computer!

Squeeze theorem

The squeeze theorem, sometimes called the *sandwich theorem*, is illustrated in figure 21. Two functions, f and h, have the same limit L as $x \to b$. Another function, g, is "squeezed" between f and h. Under

Although we are convinced by the picture, a picture does not a proof make.

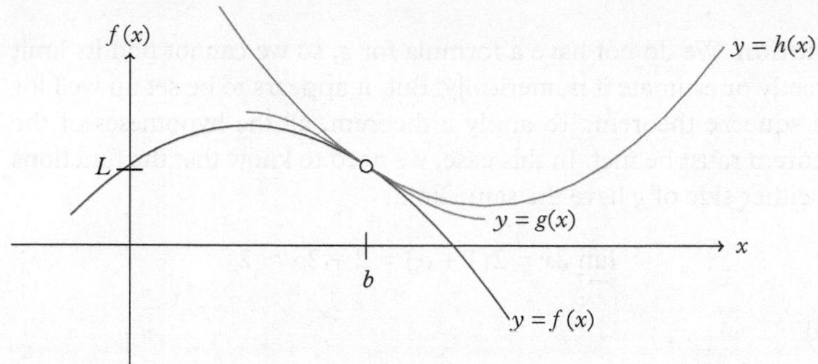

Figure 21 *The squeeze theorem, illustrated*

these constraints, it appears that g has no choice but to take the same limit at $x = b$.

Theorem 5 SQUEEZE THEOREM *If $f(x) \le g(x) \le h(x)$ for all values of x in an open interval containing b, except possibly at b, and $\lim_{x\to b} f(x) = L = \lim_{x\to b} h(x)$, then $\lim_{x\to b} g(x) = L$ also.*

Theorem 5 is sometimes useful when we do not have a formula for g.

Proof. Let's prove the case that L is a real number. Let α be an infinitesimal. Since $\lim_{x\to b} f(x) = L$, $f(b+\alpha)$ must render L. Therefore, $f(b+\alpha) = L+\gamma_1$ for some infinitesimal γ_1. Similarly, $h(b+\alpha) = L+\gamma_2$ for some infinitesimal γ_2. Since $f(x) \le g(x) \le h(x)$ on an open interval around $x = b$, then for any infinitesimal α,

γ is the lowercase Greek letter "gamma." Technically, γ_1 and γ_2 could also be zero. Such details are often left out of proofs in this text.

$$f(b+\alpha) \le g(b+\alpha) \le h(b+\alpha)$$

This step uses the transfer principle.

and thus

$$L+\gamma_1 \le g(b+\alpha) \le L+\gamma_2$$

and

$$\gamma_1 \le g(b+\alpha) - L \le \gamma_2.$$

Then, $g(b+\alpha) - L$ must be an infinitesimal because it is between two infinitesimals, so that $g(b+\alpha) = L +$ infinitesimal $\doteq L$. Since $g(b+\alpha)$ renders L for any infinitesimal α, then $\lim_{x \to b} g(x) = L$.

The cases $L = \infty$ and $L = -\infty$ are similar. ■

Example 10 *Suppose* $2x \le g(x) \le x^2 + 1$ *for all x. Find* $\lim_{x \to 1} g(x)$.

Solution We do not have a formula for g, so we cannot find its limit directly or estimate it numerically. But, it appears to be set up well for the squeeze theorem. To apply a theorem, all the hypotheses of the theorem must be met. In this case, we need to know that the functions on either side of g have the same limit:

$$\lim_{x \to 1} 2x = 2(1+\alpha) = 2 + 2\alpha \approx 2$$

and

$$\lim_{x \to 1} (x^2 + 1) = (1+\alpha)^2 + 1 = 1 + 2\alpha + \alpha^2 + 1 = 2 + 2\alpha + \alpha^2 \approx 2.$$

Because the functions on either side of g have the same limit, then g is squeezed to that same limit and we conclude $\lim_{x \to 1} g(x) = 2$. ■

Reading Exercise 22 *Suppose* $1 - x^2 \le g(x) \le x^2 + 1$ *for all x. Determine* $\lim_{x \to 0} g(x)$.

Exercises 1.5

1–14. Calculate the limit.

1. $\lim_{x \to 3^-} f(x)$ for $f(x) = \begin{cases} 5x + 2, & x < 3 \\ x + 7, & x > 3 \end{cases}$
2. $\lim_{x \to 0^+} f(x)$ for $f(x) = \begin{cases} \sqrt{x}, & x \ge 0 \\ x + 3, & x < 0 \end{cases}$
3. $\lim_{x \to 3^+} f(x)$ for $f(x) = \begin{cases} 5x + 2, & x < 3 \\ x + 7, & x > 3 \end{cases}$
4. $\lim_{x \to 0^-} f(x)$ for $f(x) = \begin{cases} \sqrt{x}, & x \ge 0 \\ x + 3, & x < 0 \end{cases}$
5. $\lim_{x \to 5} f(x)$ for $f(x) = \begin{cases} 5x + 2, & x < 3 \\ x + 7, & x > 3 \end{cases}$
6. $\lim_{x \to -1} f(x)$ for $f(x) = \begin{cases} \sqrt{x}, & x \ge 0 \\ x + 3, & x < 0 \end{cases}$

7. $\lim_{x\to 3} f(x)$ for $f(x) = \begin{cases} x^2, & x < 3 \\ 2x+1, & x > 3 \end{cases}$

8. $\lim_{x\to 4} f(x)$ for $f(x) = \begin{cases} \frac{x^2}{8}, & x > 4 \\ x-2, & x < 4 \end{cases}$

9. $\lim_{x\to 0} f(x)$ for $f(x) = \begin{cases} \frac{x-6}{x-2}, & x \geq 0 \\ x^2+3, & x < 0 \end{cases}$

10. $\lim_{x\to 1} f(x)$ for $f(x) = \begin{cases} 2x+3, & x \leq 1 \\ x-2, & x > 1 \end{cases}$

11. $\lim_{x\to 1} g(x)$ where $x \leq g(x) \leq |2x-2| + 1$

12. $\lim_{x\to 0} g(x)$ where $x^2 \leq g(x) \leq 2x^2$

13. $\lim_{x\to 4} g(x)$ where $\sqrt{x} \leq g(x) \leq \frac{1}{4}x + 1$

14. $\lim_{x\to 2} g(x)$ where $-\frac{1}{4}x + 1 \leq g(x) \leq \frac{1}{x}$

15–20. Use the following graph of the function f to find the limit.

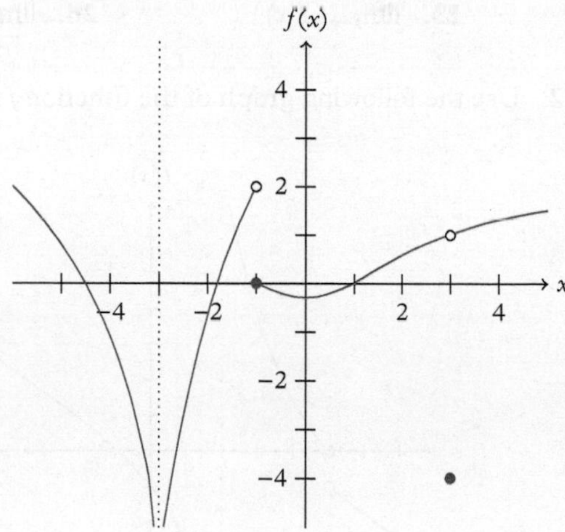

15. $\lim_{x\to -3} f(x)$
16. $\lim_{x\to 1} f(x)$
17. $\lim_{x\to -1} f(x)$
18. $\lim_{x\to -1^+} f(x)$
19. $\lim_{x\to 3} f(x)$
20. $\lim_{x\to -1^-} f(x)$

21–26. Use the following graph of the function f to find the limit.

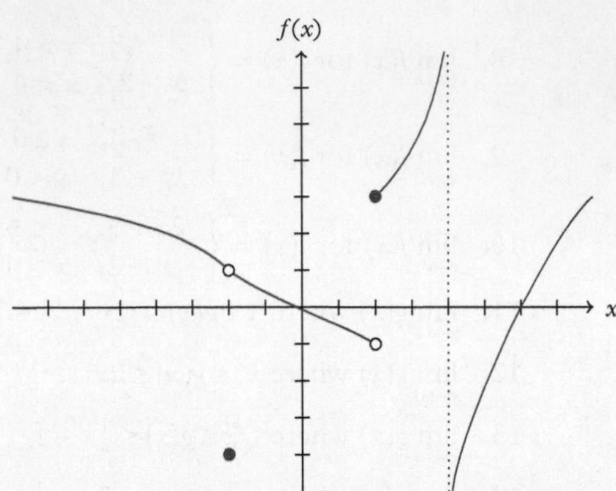

21. $\lim_{x\to -2} f(x)$

22. $\lim_{x\to 4^-} f(x)$

23. $\lim_{x\to 2} f(x)$

24. $\lim_{x\to 4^+} f(x)$

25. $\lim_{x\to 2^-} f(x)$

26. $\lim_{x\to 6} f(x)$

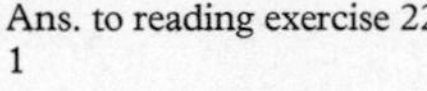
Ans. to reading exercise 22
1

27–32. Use the following graph of the function f to find the limit.

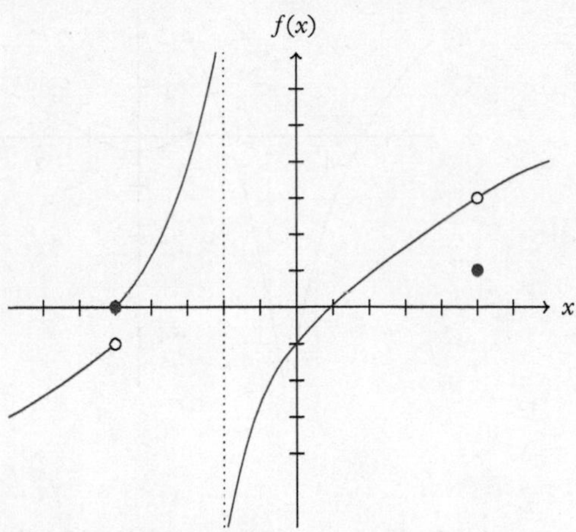

27. $\lim_{x\to 5} f(x)$

28. $\lim_{x\to 2} f(x)$

29. $\lim_{x\to -2^+} f(x)$

30. $\lim_{x\to -5^-} f(x)$

31. $\lim_{x\to -5} f(x)$

32. $\lim_{x\to -2} f(x)$

33–38. Use the following graph of the function f to find the limit.

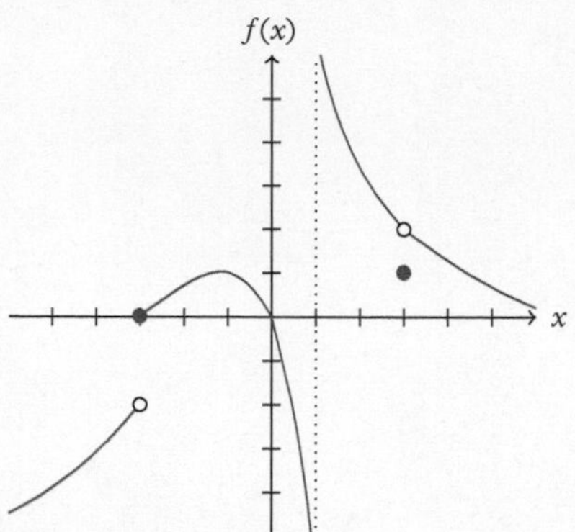

33. $\lim_{x\to -3^+} f(x)$

34. $\lim_{x\to 1^-} f(x)$

35. $\lim_{x\to 3} f(x)$

36. $\lim_{x\to 1} f(x)$

37. $\lim_{x\to -1} f(x)$

38. $\lim_{x\to -3} f(x)$

39–44. Sketch the graph of a function f that has the given properties.

39. $\lim_{x\to 3^+} f(x) = \infty$, $\lim_{x\to 3^-} f(x) = -\infty$, $\lim_{x\to 1^+} f(x) = 2$, $\lim_{x\to 1^-} f(x) = -1$

40. $\lim_{x\to -3} f(x) = 2$, $f(-3) = 0$, $\lim_{x\to 1} f(x) = \infty$

41. $\lim_{x\to 3^+} f(x) = 4$, $\lim_{x\to 3^-} f(x) = 1$, $\lim_{x\to 1^+} f(x) = -\infty$, $\lim_{x\to 1^-} f(x) = \infty$, $\lim_{x\to -3} f(x) = 2$, $f(-3) = 0$

42. $\lim_{x\to 1} f(x) = 3$, $f(1) = 2$, $\lim_{x\to 2^+} f(x) = -1$, $\lim_{x\to 2^-} f(x) = \infty$, $f(4) = 0$, $\lim_{x\to -1^+} f(x) = 1$, $\lim_{x\to -1^-} f(x) = 0$

43. $\lim_{x\to -2} f(x) = \infty$, $\lim_{x\to 1^+} f(x) = 0$, $\lim_{x\to 1^-} f(x) = 2$, $f(1) = 2$, $\lim_{x\to 4} f(x) = 1$

44. $\lim_{x\to 4} f(x) = \infty$, $\lim_{x\to 1^+} f(x) = 0$, $\lim_{x\to 1^-} f(x) = 2$, $f(1) = 2$, $\lim_{x\to -2} f(x) = 1$

45–50. Estimate the limit numerically.

45. $\lim\limits_{x\to 0} \dfrac{\sin(3x^2 - x)}{x}$

46. $\lim\limits_{x\to 0} \dfrac{\tan x}{x}$

47. $\lim\limits_{x\to 0} \dfrac{\sqrt{1 - \cos x}}{x}$

48. $\lim\limits_{x\to 0^+} x^x$

49. $\lim\limits_{x\to 2^+} \dfrac{\sin(x - 2)}{\sqrt{x^2 - 4}}$

50. $\lim\limits_{x\to -1} \dfrac{x + 1}{x^5 + 1}$

Continuity, Part I

1.6

When you first learned to graph functions, you likely were taught to evaluate the function at a few values of x, plot the resulting points, and then play the connect-the-dot game through those points in a nice, smooth manner. For instance, given $f(x) = x^2$, we can produce this table of values:

x	x^2	point
-2	4	$(-2, 4)$
-1	1	$(-1, 1)$
0	0	$(0, 0)$
1	1	$(1, 1)$
2	4	$(2, 4)$

plot the points (figure 1):

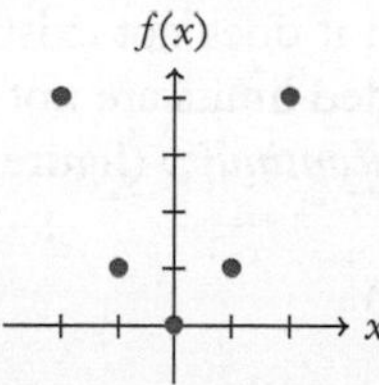

Figure 1 *A few plotted points on the graph of* $f(x) = x^2$

and then connect the dots, not by using straight line segments between points, but by making a smooth, connected curve (figure 2):

We examine "smoothness" in section 2.1.

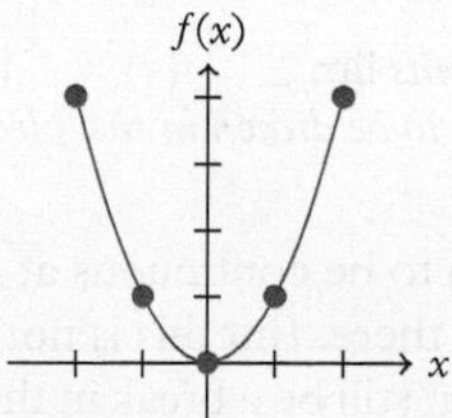

Figure 2 *A curve through plotted points on the graph of* $f(x) = x^2$

What justifies connecting the dots? The answer is *continuity*. To investigate how we might define this property, recall that there were times you were taught *not* to connect the dots, such as across a vertical asymptote (figure 3):

CAUTION: MISTAKE TO AVOID

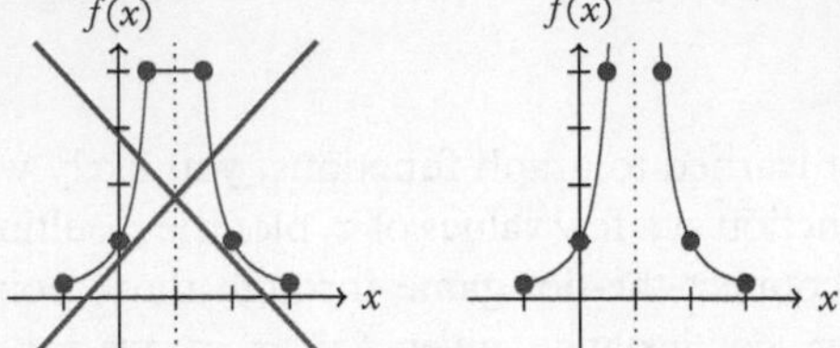

Figure 3 *Do not connect points across a vertical asymptote! The graph on the left is incorrect; the graph on the right is correct*

Instead of drawing the curve in one piece, we drew the curve in two pieces, one on each side of the vertical asymptote. A break in the graph that separates two pieces of the curve is called a *discontinuity*. How do we recognize a vertical asymptote on a curve? We have seen that if $\lim_{x\to b} f(x)$ is infinite, then there is a vertical asymptote on the graph of f at $x = b$. For this reason, we call this type of discontinuity an *infinite discontinuity*. Evidently, then, in order for a function to be continuous at $x = b$, we need the limit not to be infinite. What else do we need?

The effect on the graph is a vertical asymptote. The type of discontinuity is an infinite discontinuity. We use separate terminology for graph effects and discontinuities.

If a limit is infinite, then it does not exist. Another way for a limit not to exist is if the one-sided limits are not equal, causing a jump in the graph, called a *jump discontinuity* (figure 4):

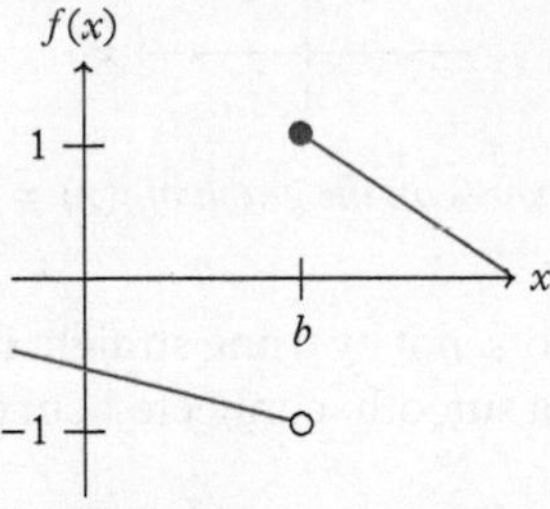

Figure 4 *The one-sided limits* $\lim_{x\to b^-} f(x) = -1$ *and* $\lim_{x\to b^+} f(x) = 1$ *are not equal, causing the graph to be drawn in two pieces*

In order for a function to be continuous at $x = b$, it appears that we need for the limit to exist there. But this is not all we need. A limit can exist at $x = b$ and there can still be a break in the graph there (figure 5).

This time the limit exists but the point is missing. Perhaps, then, we need to require that the point needs to exist—that is, that $f(b)$ exists

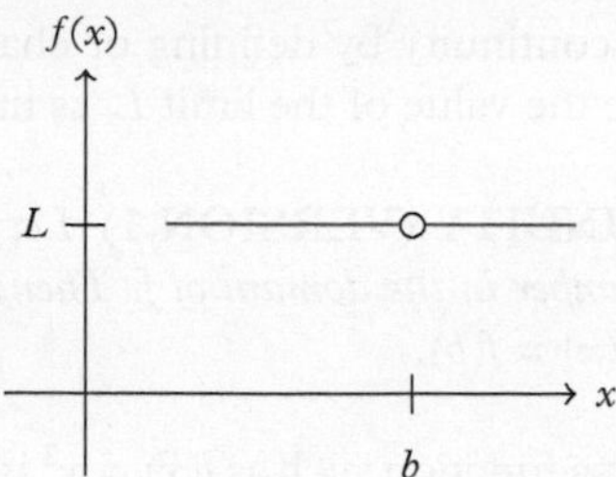

Figure 5 $\lim_{x\to b} f(x) = L$, *but the graph must still be drawn in two pieces: one to the left of the hole and the other to the right of the hole*

in order for the function to be continuous at $x = b$. But even this is not enough (figure 6):

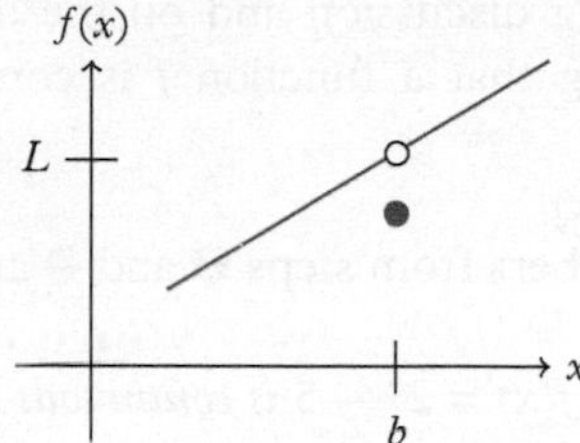

Figure 6 *The limit exists at* $x = b$, $f(b)$ *exists, but the graph must still be drawn in more than one piece*

We appear to need that the limit exists at $x = b$, $f(b)$ exists, and that these two values be equal. Is this enough? Yes, for it requires that the curve approach the point (b, L) and that (b, L) is on the graph, meaning there cannot be a break in the graph (figure 7):

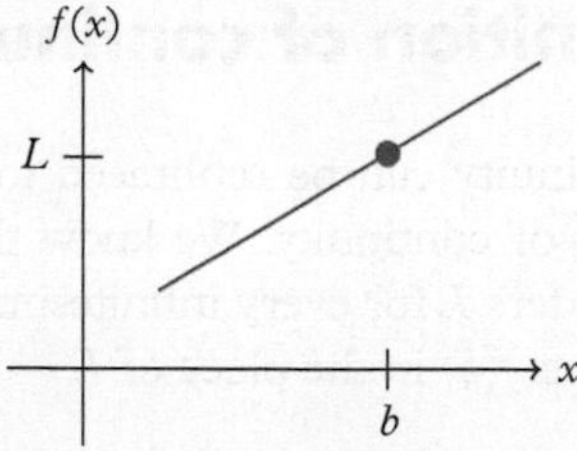

Figure 7 *The limit exists at* $x = b$, $f(b)$ *exists, and they are equal. There is no break in the graph at* $x = b$

Before stating the definition of continuity, we note that the discontinuity in figures 5 and 6 is called a *removable discontinuity*, because we

The effect on the graph is a hole. The type of discontinuity is a removable discontinuity.

can "remove" the discontinuity by defining or changing the value of the function at b to be the value of the limit L, as in figure 7.

Definition 8 CONTINUITY (VERSION 1) *Let f be a function and let b be a real number in the domain of f. Then, f is* continuous *at $x = b$ if $\lim_{x\to b} f(x) = f(b)$.*

If we can verify that a function such as $f(x) = x^2$ is continuous everywhere (that is, for every value of x), then when graphing the function, the connect-the-dot game is justified because there are no breaks in the graph. This goal is reached fully in the next section.

Verifying continuity

Based on the preceding discussion and on the definition, there are three steps to verifying that a function f is continuous at $x = b$:
❶ Calculate $f(b)$.
❷ Calculate $\lim_{x\to b} f(x)$.
❸ Verify that the numbers from steps ❶ and ❷ are equal.

Writing steps ❶ and ❷ only is insufficient because we need more than just the existence of the function and the limit at $x = b$. We need these two values to be equal.

The "answer" to an exercise asking for verification of continuity is not merely "continuous." The "answer" is the verification, which consists of all three steps.

Example 1 *Verify that $f(x) = 2x - 5$ is continuous at $x = -1$.*

Solution We complete the three steps:
❶ $f(-1) = 2(-1) - 5 = -7$
❷ $\lim_{x\to -1}(2x - 5) = 2(-1 + \alpha) - 5 = -7 + 2\alpha \approx -7$
❸ $\lim_{x\to -1} f(x) = -7 = f(-1)$, continuous
By definition, f is continuous at $x = -1$. ∎

Reading Exercise 23 *Verify that $f(x) = x + 3$ is continuous at $x = 1$.*

Alternate definition of continuity

The definition of continuity can be rephrased to give additional insight into the meaning of continuity. We know that $\lim_{x\to b} f(x) = L$ means that $f(b + \alpha)$ renders L for every infinitesimal α. The definition of continuity simply puts $f(b)$ in the place of L.

Except for the use of the rendering terminology and the use of the letter b, this is identical to Cauchy's definition of continuity.

Augustin Louis Cauchy, 1789–1857
http://www-history.mcs.st-andrews.ac.uk/Biographies/Cauchy.html

Definition 9 CONTINUITY (VERSION 2) *Let f be a function and let b be a real number in the domain of f. Then, f is* continuous *at $x = b$ if $f(b + \alpha)$ renders the real result $f(b)$ for every infinitesimal α.*

This means that a small change in the value of x (from $x = b$ to $x = b + \alpha$) produces only a small change in the value of $y = f(x)$

(because $f(b + \alpha)$ renders $f(b)$ and therefore must differ from $f(b)$ by, at most, an infinitesimal amount). In other words, small changes in the x-coordinate on the graph produce small changes in the y-coordinate on the graph. This is what keeps the points close together and prevents the graph from jumping or heading off to infinity. Small changes in x producing small changes in y, rather than the ability to draw a graph in one piece, is the traditional intuitive concept of continuity.

Identifying discontinuities graphically

Certain types of discontinuities in a function may be recognized easily from a graph. If we see what we have called a *hole* in the graph, then the function has a removable discontinuity; if we see a vertical asymptote, then the function has an infinite discontinuity; and if we see a jump in the graph, the function has a jump discontinuity. The locations of these discontinuities should be recorded using their x-coordinates, because the use of y-coordinates does not make sense for some types of discontinuities.

Example 2 *Identify by type the discontinuities in the function f graphed in figure 8.*

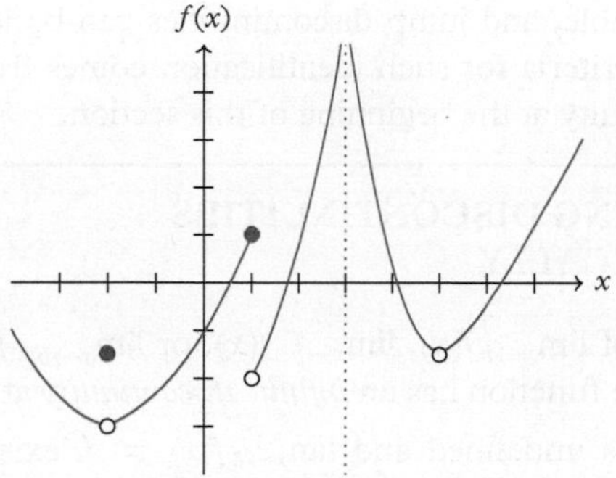

Figure 8 *The graph of the function f for example 2*

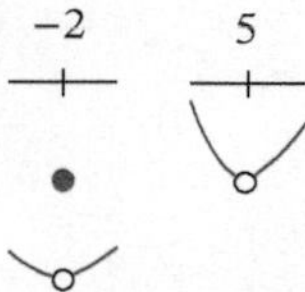

Figure 9 Removable discontinuities identified from the graph of the function f for example 2

Solution We recognize two "holes" in the graph, which represent removable discontinuities. The locations of these removable discontinuities are $x = -2$ and $x = 5$ (figure 9).

Next we recognize a jump in the graph, which represents a jump discontinuity, at $x = 1$ (figure 10).

Last, we recognize a vertical asymptote in the graph at $x = 3$, which represents an infinite discontinuity (figure 11).

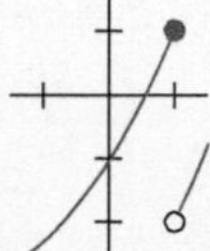

Figure 10 Jump discontinuity identified from the graph of the function f for example 2

Figure 11 Infinite discontinuity identified from the graph of the function f for example 2

The final answer can be written as follows:

removable discontinuities at $x = -2$ and $x = 5$
jump discontinuity at $x = 1$
infinite discontinuity at $x = 3$ ■

Reading Exercise 24 *Identify by type the discontinuities in the function f graphed in figure 12.*

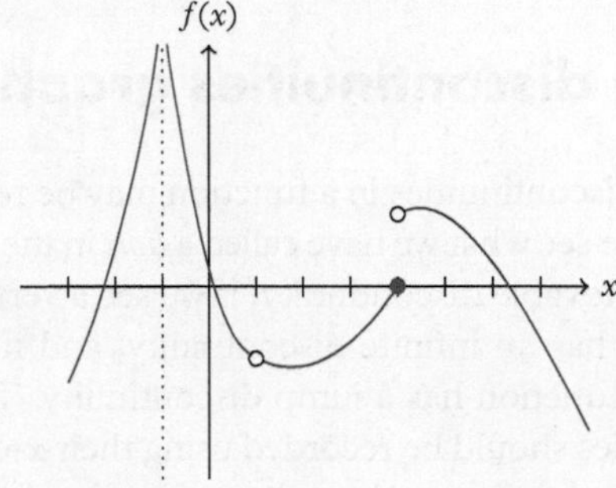

Figure 12 *The graph of the function f for reading exercise 24*

Ans. to reading exercise 23
❶ $f(1) = 1 + 3 = 4$

❷ $\lim_{x\to 1}(x+3) = (1+\alpha)+3 = 4+\alpha \approx 4$

❸ $\lim_{x\to 1} f(x) = f(1)$, continuous.

Recall that if $\lim_{x\to b} f(x) = f(b)$, then the function is continuous at $x = b$ and there is no discontinuity at $x = b$.

Identifying discontinuities algebraically

Infinite, removable, and jump discontinuities can be identified algebraically. The criteria for such identification comes from our exploration of continuity at the beginning of this section.

> **IDENTIFYING DISCONTINUITIES ALGEBRAICALLY**
>
> - If any of $\lim_{x\to b} f(x)$, $\lim_{x\to b^+} f(x)$, or $\lim_{x\to b^-} f(x)$ are infinite, the function has an *infinite discontinuity* at $x = b$.
> - If $f(b)$ is undefined and $\lim_{x\to b} f(x) = L$ exists, then the function has a *removable discontinuity* at $x = b$.
> - If $f(b)$ and $\lim_{x\to b} f(x) = L$ both exist but are unequal, then the function has a *removable discontinuity* at $x = b$.
> - If $\lim_{x\to b^+} f(x) = L$ and $\lim_{x\to b^-} f(x) = M$ both exist but are unequal, then the function has a *jump discontinuity* at $x = b$.

Example 3 *Is* $f(x) = \dfrac{x-2}{x^2-3x+2}$ *continuous at* $x = 2$*? If not, what type of discontinuity does it have?*

Solution To determine whether the function is continuous at $x = 2$ we check to see if $\lim_{x\to 2} f(x) = f(2)$. We start with determining $f(2)$:

$$f(2) = \frac{2-2}{2^2 - 3(2) + 2} = \frac{0}{0},$$

which is undefined. The function f is not continuous at $x = 2$.

This does not yet tell us what type of discontinuity we have. Therefore, we check the limit:

$$\lim_{x\to 2} \frac{x-2}{x^2-3x+2} = \frac{2+\alpha-2}{(2+\alpha)^2 - 3(2+\alpha)+2}$$
$$= \frac{\alpha}{4+4\alpha+\alpha^2-6-3\alpha+2} = \frac{\alpha}{\alpha+\alpha^2}$$
$$\approx \frac{\alpha}{\alpha} = 1.$$

Because the two-sided limit exists (the limit is a real number) but $f(2)$ is undefined, f has a removable discontinuity at $x = 2$. ■

When we checked $f(2)$, the form of the expression was $\frac{0}{0}$. This is sometimes called an *indeterminate form,* because the form itself does not say what is happening to the function or to the limit. Removable discontinuities, infinite discontinuities, jump discontinuities, and other types of discontinuities we have not yet studied can all occur with the form $\frac{0}{0}$. The limit must be checked to determine what is happening.

This is similar to "verify that $f(x) = \ldots$ is continuous at $x = b$," but the instructions are different. We may try the same three steps, and if the function is continuous our answer is "$f(x)$ is continuous at $x = b$." Otherwise, we seek to determine the type of discontinuity.

It is often convenient to check the two-sided limit first. We can always check the one-sided limits if the two-sided limit does not exist.

Don't forget to collect like terms before approximating.

Ans. to reading exercise 24
infinite discontinuity at $x = -1$, removable discontinuity at $x = 1$, jump discontinuity at $x = 4$

Example 4 *Is* $f(x) = \dfrac{x}{x+5}$ *continuous at* $x = -5$*? If not, what type of discontinuity does it have?*

Solution To determine whether the function is continuous at $x = -5$, we check to see if $\lim_{x\to -5} f(x) = f(-5)$. We start with determining $f(-5)$:

$$f(-5) = \frac{-5}{-5+5} = \frac{-5}{0},$$

which is undefined. The function f is not continuous at $x = -5$.

To determine the type of discontinuity, we check the limit:

$$\lim_{x\to -5} \frac{x}{x+5} = \frac{-5+\alpha}{-5+\alpha+5} = \frac{-5+\alpha}{\alpha} \approx \frac{-5}{\alpha} = -5A,$$

and although we are stuck on the limit, we do know that the limit is infinite. This is enough to know that f has an infinite discontinuity at $x = -5$. ■

Although calculating $\lim_{x\to -5^+} f(x) = -\infty$ and $\lim_{x\to -5^-} f(x) = \infty$ is not difficult, it is unnecessary for this question. We do not need to know exactly how the function approaches the vertical asymptote to determine there is a vertical asymptote.

Example 5 *Is* $f(x) = \begin{cases} 2-x, & x<1 \\ 1+x, & x>1 \end{cases}$ *continuous at* $x = 1$*? If not, what type of discontinuity does it have?*

Solution To determine whether the function is continuous at $x = 1$, we check to see if $\lim_{x\to 1} f(x) = f(1)$. We start by trying $f(1)$. Notice that f is literally not defined at $x = 1$; the rules only say what happens when $x > 1$ and when $x < 1$. The function f is not continuous at $x = 1$.

Functions are undefined when division by zero is encountered, but they can also be undefined in the literal sense, as in example 5. All the types of discontinuities we have studied are still possible and the limit must still be checked.

To determine the type of discontinuity, we check the limit. Because this is a piecewise-defined function and the rules change at the value of interest $x = 1$, we go straight to one-sided limits:

$$\lim_{x\to 1^-} f(x) = \lim_{x\to 1^-} (2-x) = 2-(1-\omega) = 1+\omega \approx 1$$

$$\lim_{x\to 1^+} f(x) = \lim_{x\to 1^+} (1+x) = 1+(1+\omega) = 2+\omega \approx 2.$$

Because the one-sided limits are different, f has a jump discontinuity at $x = 1$. ■

If the one-sided limits are equal, then the two-sided limit exists and the result is a removable discontinuity (all else the same).

Reading Exercise 25 *Is* $f(x) = \dfrac{x+1}{x-2}$ *continuous at* $x = 2$*? If not, what type of discontinuity does it have?*

Example 6 *Is* $f(x) = \dfrac{x}{x+5}$ *continuous at* $x = 2$*? If not, what type of discontinuity does it have?*

Solution To determine whether the function is continuous at $x = 2$, we check to see if $\lim_{x\to 2} f(x) = f(2)$. We start with determining $f(2)$:

This is the same function as in example 4, but we are checking a different value of x for continuity.

$$f(2) = \frac{2}{2+5} = \frac{2}{7}.$$

Next we check the value of the limit:

$$\lim_{x\to 2} \frac{x}{x+5} = \frac{2+\alpha}{2+\alpha+5} = \frac{2+\alpha}{7+\alpha} \approx \frac{2}{7}.$$

Because $\lim_{x\to 2} f(x) = f(2)$, f is continuous at $x = 2$. ■

Compare this solution to that of example 1. Here the "answer" is "f is continuous at $x = 2$." The difference in instructions gives rise to the different presentations of the solutions and the different compositions of the answers.

This function is continuous at $x = 2$ (example 6) but not at $x = -5$ (example 4). It should not be surprising that a function can be continuous at some values and not continuous at others.

Oscillatory discontinuities

There are other types of discontinuity besides those already mentioned. One such type is called an *oscillatory discontinuity.* Consider the function $f(x) = \sin \frac{1}{x}$, which is undefined at $x = 0$ and therefore not continuous at $x = 0$. The function is graphed in figure 13.

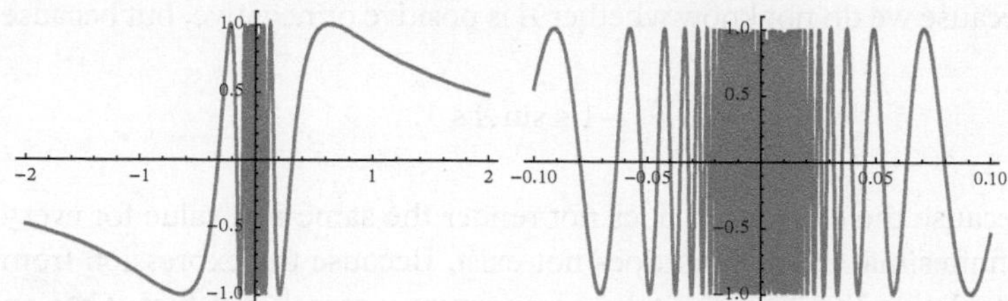

Figure 13 *Two graphs of $f(x) = \sin \frac{1}{x}$, generated by Mathematica. Unlike the graph on the right side of section 1.5 figure 20, the graphs in this figure give the correct impression and are not misleading, even though they have a similar look*

To see what is going on with this function, recall that $y = \sin x$ between $x = 2\pi$ and $x = 4\pi$ goes through one complete cycle of the graph (one *period* of the function). But, since $\frac{1}{1/(2\pi)} = 2\pi$ and $\frac{1}{1/(4\pi)} = 4\pi$, the function f goes through one complete cycle of the sin function between $1/(2\pi)$ and $1/(4\pi)$, as seen in figure 14. It goes through another cycle between $1/(4\pi)$ and $1/(6\pi)$, another between $1/(6\pi)$ and $1/(8\pi)$, and so on. This continues ad infinitum, and therefore the function oscillates infinitely many times as we near $x = 0$. It is for this reason that we call this discontinuity an *oscillatory discontinuity.*

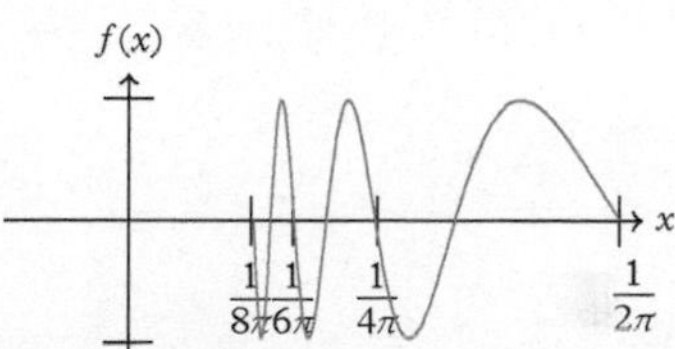

Figure 14 The graph of $f(x) = \sin \frac{1}{x}$ between $\frac{1}{8\pi}$ and $\frac{1}{2\pi}$

This type of discontinuity can be detected algebraically using the same procedures as demonstrated in examples 3–5.

Example 7 *Is $f(x) = \sin \frac{1}{x}$ continuous at $x = 0$? If not, what type of discontinuity does it have?*

Solution To determine whether the function is continuous at $x = 0$, we check to see if $\lim_{x\to 0} f(x) = f(0)$. We start with $f(0)$:

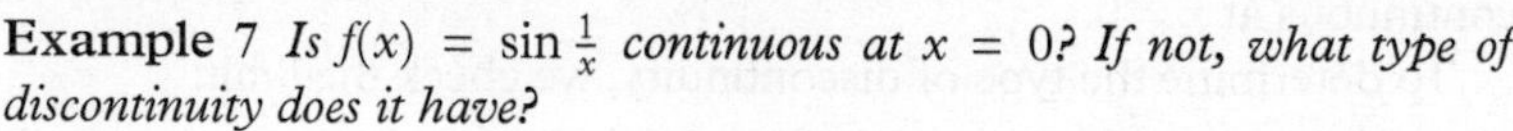

$$f(0) = \sin \frac{1}{0},$$

which is undefined because of division by zero. The function f is not continuous at $x = 0$.

Ans. to reading exercise 25
infinite discontinuity at $x = 2$

To determine the type of discontinuity, we check the limit:

$$\lim_{x\to 0} \sin\frac{1}{x} = \sin\frac{1}{\alpha} = \underbrace{\sin A}_{\substack{\text{between}\\ 1\text{and}-1}}.$$

In trying to render $\sin A$, we are stuck, not because it is infinite or because we do not know whether A is positive or negative, but because

$$-1 \le \sin A \le 1.$$

If $f(b + \alpha)$ can take on an entire interval of values, then f has an oscillatory discontinuity at $x = b$.

Because the expression does not render the same real value for every infinitesimal α, the limit does not exist. Because the expression from calculating the limit can take on an entire interval of values, f has an oscillatory discontinuity at $x = 0$. ■

We may recognize an oscillatory discontinuity graphically by its similarity to figure 13.

A function may oscillate without having a discontinuity, but the oscillations must become infinitely small.

Example 8 *Is $f(x) = x\sin\frac{1}{x}$ continuous at $x = 0$? If not, what type of discontinuity does it have?*

Solution To determine whether the function is continuous at $x = 0$, we check to see if $\lim_{x\to 0} f(x) = f(0)$. We start with $f(0)$:

$$f(0) = 0\sin\frac{1}{0},$$

which is undefined because of division by zero. The function f is not continuous at $x = 0$.

This limit was calculated previously in section 1.4 example 4.

To determine the type of discontinuity, we check the limit:

$$\lim_{x\to 0} x\sin\frac{1}{x} = \underbrace{\alpha\cdot\underbrace{\sin\frac{1}{\alpha}}_{\substack{\text{between}\\ 1\text{and}-1}}}_{\text{infinitesimal}} \doteq 0.$$

Because the two-sided limit exists (the limit is a real number) but $f(0)$ is undefined, f has a removable discontinuity at $x = 0$. ■

The graph in figure 15 shows oscillations, but these oscillations do not take on an entire interval of values infinitely close to $x = 0$, whereas those in figure 13 do.

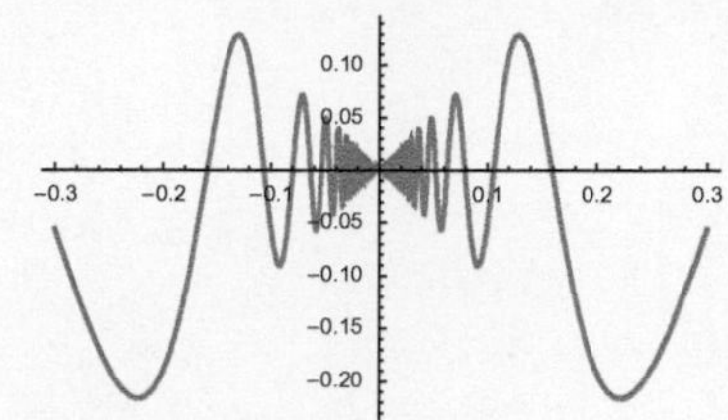

Figure 15 The graph of $f(x) = x \sin \frac{1}{x}$, generated by Mathematica. The oscillations become infinitely small. There is a removable discontinuity, not an oscillatory discontinuity, at $x = 0$

Exercises 1.6

1. Verify that $f(x) = x - 7$ is continuous at $x = 1$.
2. Verify that $f(x) = 3x + 1$ is continuous at $x = 2$.
3. Verify that $f(x) = x^3 - 1$ is continuous at $x = 0$.
4. Verify that $f(x) = x^4 + 6$ is continuous at $x = 0$.
5. Verify that $f(x) = x^2 + x$ is continuous at $x = 2$.
6. Verify that $f(x) = x^2 - 2\pi$ is continuous at $x = -1$.
7. Verify that $f(x) = \dfrac{x+2}{x-3}$ is continuous at $x = 1$.
8. Verify that $f(x) = \dfrac{1}{x}$ is continuous at $x = 2$.

9–12. Using the following graph, determine whether the function is continuous at the given value of x and, if it is not, indicate the type of discontinuity.

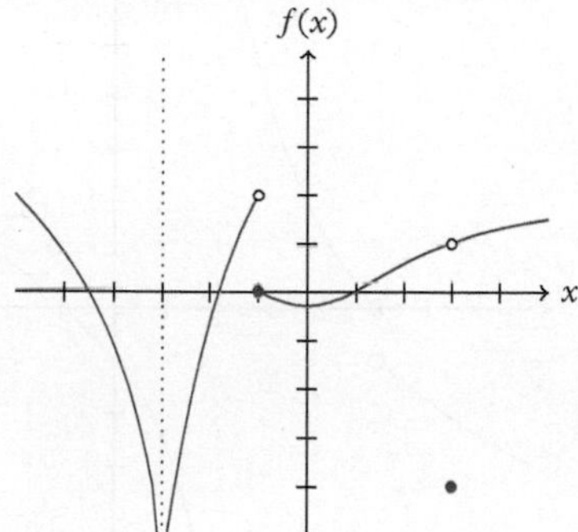

9. at $x = -3$
10. at $x = 1$
11. at $x = -1$
12. at $x = 3$

13–16. Using the following graph, determine whether the function is continuous at the given value of x and, if it is not, indicate the type of discontinuity.

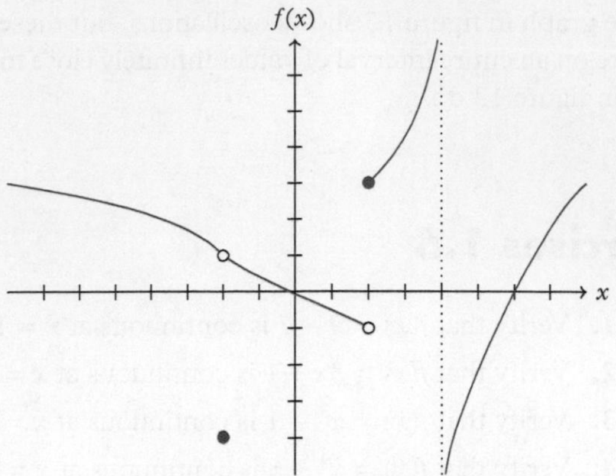

13. at $x = -2$

14. at $x = 2$

15. at $x = 0$

16. at $x = 4$

17–20. Using the following graph, determine whether the function is continuous at the given value of x and, if it is not, indicate the type of discontinuity.

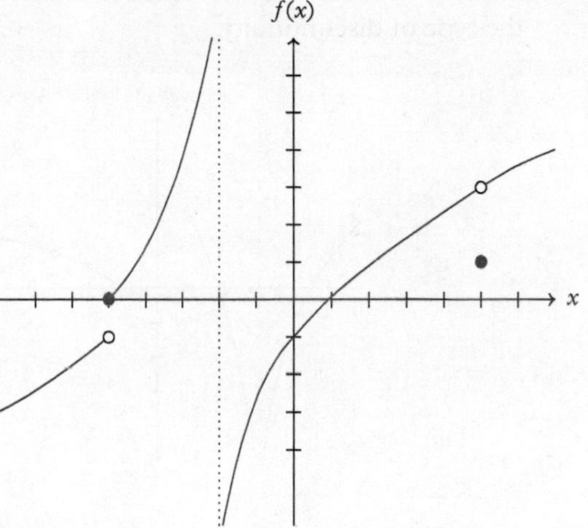

17. at $x = -5$

18. at $x = 2$

19. at $x = -2$

20. at $x = 5$

21–24. Using the following graph, determine whether the function is continuous at the given value of x and, if it is not, indicate the type of discontinuity.

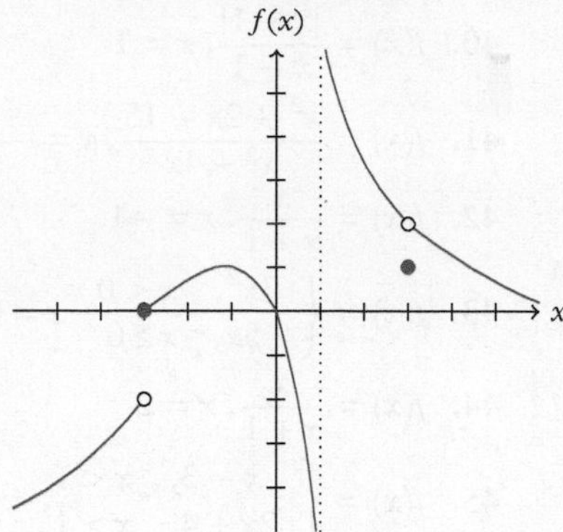

21. at $x = -3$

22. at $x = 1$

23. at $x = -1$

24. at $x = 3$

25–36. Use the information provided about the function f to determine whether f is continuous at the given value of x and, if not, to determine the type of discontinuity.

25. $\lim_{x \to 3} f(x) = 4, f(3) = 7$

26. $\lim_{x \to 0^+} f(x) = 6, \lim_{x \to 0^-} f(x) = 6, f(0) = 6$

27. $\lim_{x \to 3} f(x) = 4, f(3) = 4$

28. $\lim_{x \to 0^+} f(x) = 6, \lim_{x \to 0^-} f(x) = 6, f(0) = 5$

29. $\lim_{x \to 3} f(x) = 4$, $f(3)$ undefined

30. $\lim_{x \to 0^+} f(x) = 6, \lim_{x \to 0^-} f(x) = 5, f(0) = 5$

31. $\lim_{x \to 3^+} f(x) = 4, \lim_{x \to 3^-} f(x) = 7$

32. $\lim_{x \to 0} f(x) = \infty, f(0) = 5$

33. $\lim_{x \to 3} f(x) = \infty$

34. $\lim_{x \to 0} f(x) = \infty$, $f(0)$ undefined

35. $\lim_{x \to 3^+} f(x) = -\infty, \lim_{x \to 3^-} f(x) = \infty$

36. $\lim_{x \to 0} f(x) = 6$, $f(0)$ undefined

37–48. Determine whether the function is continuous at the given value of x and, if it is discontinuous, determine the type of discontinuity.

It is implied that these problems are to be worked algebraically, as in examples 3–8.

37. $f(x) = \dfrac{3}{x-4}, x = 4$

38. $f(x) = \dfrac{x^2 - x - 6}{x^2 + 3x + 2}, x = -2$

39. $f(x) = \dfrac{x^2 - x - 2}{x^2 + x}, x = -1$

40. $f(x) = \dfrac{|x-1|}{x-1}, x = 1$

41. $f(x) = \dfrac{x^2 - 2x - 15}{x^2 + 4x}, x = -4$

42. $f(x) = \dfrac{x}{x+1}, x = -1$

43. $f(x) = \begin{cases} 3x^2, & x < 0 \\ 5x, & x \geq 0 \end{cases}, x = 0$

44. $f(x) = \dfrac{x}{x+1}, x = 2$

45. $f(x) = \begin{cases} x - 3, & x < 1 \\ 2x + 4, & x > 1 \end{cases}, x = 1$

46. $f(x) = \begin{cases} x + 2, & x < 1 \\ \frac{1}{x}, & x \geq 1 \end{cases}, x = 1$

47. $f(x) = x^2 \cos \dfrac{1}{2x}, x = 0$

48. $f(x) = \cos \dfrac{1}{x}, x = 0$

49. Verify that $f(x) = 3x - 1$ is continuous at $x = k$, where k is any real number.

50. Verify that $f(x) = x^2$ is continuous at $x = k$, where k is any real number.

Continuity, Part II

1.7

It is finally time to justify the connect-the-dot game! Let's begin with linear functions.

Linear functions are continuous

A linear function is a function with a graph that is a line. Such functions are of the form $f(x) = mx + b$. There are no breaks in the graph of a line, so linear functions should be continuous everywhere.

Theorem 6 CONTINUITY OF LINEAR FUNCTIONS *If* $f(x) = mx + b$ *for real numbers* m *and* b, *then* f *is continuous at every real number* k.

Proof. We proceed as before, but use generic constants m, b, and k in place of specific real numbers.

Compare the procedure here to that of the solution to example 6 in section 1.6.

To determine whether f is continuous at $x = k$, we check to see if $\lim_{x\to k} f(x) = f(k)$. We start by determining $f(k)$:

$$f(k) = mk + b.$$

Next we check the limit:

In the approximation step, the terms on the real-number level are kept, but the term with an infinitesimal is discarded because it is on a lower level.

$$\lim_{x\to k}(mx + b) = m(k + \alpha) + b = mk + m\alpha + b \approx mk + b.$$

Because $\lim_{x\to k} f(x) = f(k)$, f is continuous at $x = k$. ■

Now transport yourself back to algebra and graph $f(x) = 2x - 1$ by plotting points. We use the following table of values:

x	$2x - 1$
-2	-5
-1	-3
0	-1
1	1
2	3

and plot the points (figure 1):

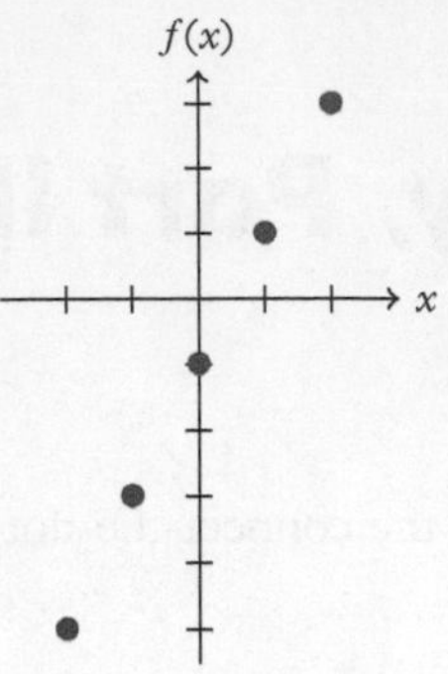

Figure 1 *A few plotted points on the graph of $f(x) = 2x - 1$*

and because we now know that every linear function is continuous, we know that we can safely connect the dots (figure 2). There are no holes, jumps, vertical asymptotes, oscillations, or any other discontinuities in the graph of the function!

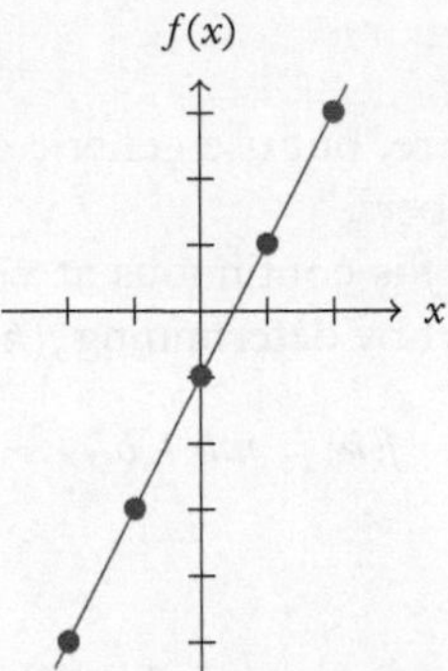

Figure 2 *The graph of the continuous function $f(x) = 2x - 1$*

There are, of course, some functions that have discontinuities. But, a quick perusal of examples 3–8 in section 1.6 reveals that whenever we had an algebraic formula for a function that was not piecewise-defined, the discontinuities of the function occurred only where the function was undefined. If we know that for all other values of x that the function is continuous, we can play the connect-the-dot game on each piece. This helps motivate the following definition:

Definition 10 CONTINUOUS FUNCTION *A function f is* continuous *if it is continuous at $x = k$ for every real number k in its domain.*

In other words, a function is called just plain *continuous* if it is continuous everywhere it is defined. We can restate theorem 6 using the new terminology:

Theorem 6 RESTATED *If f is a linear function, then f is continuous.*

Using this terminology, it turns out that all the commonly used functions are continuous. In other words, the only discontinuities of the functions occur where they are not defined (such as at division by zero). Otherwise, we can draw the graphs of the functions in connected pieces.

Continuity of polynomial and rational functions

The same type of proof that linear functions are continuous applies to quadratic functions as well.

Theorem 7 CONTINUITY OF QUADRATIC FUNCTIONS *If* $f(x) = bx^2 + cx + d$, *then f is continuous.*

This proof uses the same strategy as theorem 6.

Proof. We wish to prove that f is continuous at $x = k$ for every real number k. To do this, we need to show that $\lim_{x \to k} f(x) = f(k)$. We start by determining $f(k)$:

$$f(k) = bk^2 + ck + d.$$

Next we check the limit:

$$\begin{aligned} \lim_{x \to k} (bx^2 + cx + d) &= b(k + \alpha)^2 + c(k + \alpha) + d \\ &= b(k^2 + 2k\alpha + \alpha^2) + ck + c\alpha + d \\ &= bk^2 + 2bk\alpha + b\alpha^2 + ck + c\alpha + d \\ &\approx bk^2 + ck + d. \end{aligned}$$

In the approximation step, the terms on the real-number level are kept, but the terms with infinitesimals are discarded because they are on a lower level.

Because $\lim_{x \to k} f(x) = f(k)$, f is continuous at $x = k$. This is true for every real number k, so f is continuous. ■

Parabolas can therefore be drawn using connect-the-dots. In a very similar manner, so can all polynomial functions. One proof that a polynomial of arbitrary degree is continuous is very similar to theorems 6 and 7. A different proof is presented at the end of this section.

Recall that a rational function has the form polynomial divided by a polynomial.

Rational functions can get into the act too.

Theorem 8 CONTINUITY OF SOME RATIONAL FUNCTIONS *If* $f(x) = \dfrac{bx + c}{rx + s}$, *then f is continuous.*

The phrase "f is continuous" means that f is continuous where it is defined, which *does not* include where we would divide by zero. For instance, the function $f(x) = \frac{3}{x-7}$ is not continuous at $x = 7$, but it is continuous at all other values of x.

Proof. We wish to prove that f is continuous at $x = k$ for every real number k in the domain of f. To do this, we need to show that $\lim_{x\to k} f(x) = f(k)$. We start by determining $f(k)$:

$$f(k) = \frac{bk + c}{rk + s}.$$

Next we check the limit:

$$\begin{aligned}\lim_{x\to k} \frac{bx + c}{rx + s} &= \frac{b(k + \alpha) + c}{r(k + \alpha) + s} \\ &= \frac{bk + b\alpha + c}{rk + r\alpha + s} \\ &\approx \frac{bk + c}{rk + s}.\end{aligned}$$

In the approximation step, the terms on the real-number level are kept, but the terms with infinitesimals are discarded because they are on a lower level.

Because $\lim_{x\to k} f(x) = f(k)$, f is continuous at $x = k$. This is true for every k in the domain of f, so f is continuous. ■

Because they are continuous, graphs of rational functions of the type in theorem 8 may be drawn playing connect-the-dots. These functions have one discontinuity (where the denominator is zero), so their graphs must be drawn in two pieces, one on each side of the discontinuity. The discontinuity must not be "crossed" when drawing the graph (see figure 3). But, the connect-the-dot game is still justified!

Figure 3 *(left) A rational function continuous everywhere except at $x = 1$; the graph is drawn correctly using connect-the-dots, but in two separate pieces: one to the left of the discontinuity and one to the right. (right) The same rational function is graphed incorrectly if connect-the-dot extended across the discontinuity, for we put a point at $x = 1$, where the function is not defined*

It turns out that all rational functions are continuous, not just those of the form in theorem 8. A proof is at the end of this section. By now, it is tempting to conclude that all functions are continuous, with proofs similar to the ones shown earlier. In a sense, this sentiment is not far off, for all "elementary" functions—functions of the types studied in algebra, trigonometry, and calculus—are continuous. However, there are functions that are not continuous at points in their domains, such as some of the piecewise-defined functions we have encountered. In addition to the proofs already mentioned, a proof of the continuity of trig functions is also given at the end of this section. That said, let's forge ahead and use the following theorem, even though part of its proof is delayed.

Elementary functions include algebraic functions (which include polynomial, rational, and root functions), trigonometric, inverse trigonometric, hyperbolic, inverse hyperbolic, logarithmic, and exponential functions, along with any combination or composition thereof. Many of these functions are not studied until later in calculus.

Theorem 9 CONTINUITY OF VARIOUS FUNCTIONS *Polynomial, rational, root, and trigonometric functions, as well as combinations and compositions thereof, are continuous.*

The proof that root functions are continuous is delayed until chapter 2.

Because so many functions can be graphed by plotting points and connecting the dots, this is exactly what most graphing routines in calculators and computer software do. It may seem rather unsophisticated, but the effect of hundreds or even thousands of points calculated and plotted accurately is quite good. One effect that must be watched for, however, is that sometimes points are connected across a vertical asymptote unless care is taken to avoid doing so. The result can be to draw the vertical asymptote as if it is part of the curve. Care must therefore be taken when interpreting results of technology-generated graphs.

Where is *f* continuous?

Because elementary functions are continuous throughout their domains, asking where such a function is continuous is the exact same thing as asking where it is defined.

Example 1 *On what intervals is* $f(x) = \dfrac{2x-9}{3x-5}$ *continuous?*

A function is continuous on an interval if it is continuous at every real number in that interval.

Solution This is the exact same question as: What is the domain of $f(x) = \frac{2x-9}{3x-5}$? A rational function is defined everywhere except where we divide by zero, so we set the denominator equal to zero and solve:

$$3x - 5 = 0$$
$$3x = 5$$
$$x = \frac{5}{3}.$$

A picture of the domain is given in figure 4. The domain of the function is $\left(-\infty, \frac{5}{3}\right) \cup \left(\frac{5}{3}, \infty\right)$. Our solution can be written as "the intervals $\left(-\infty, \frac{5}{3}\right)$ and $\left(\frac{5}{3}, \infty\right)$." ■

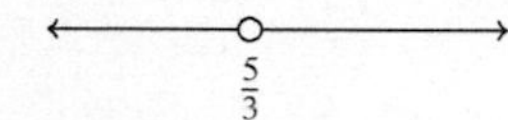

Figure 4 *The domain of* $f(x) = \frac{2x-9}{3x-5}$

A consequence of the answer to example 1 is that we can draw the graph of $f(x) = \frac{2x-9}{3x-5}$ in two pieces: one to the left of $\frac{5}{3}$ and one to the right of $\frac{5}{3}$. Figure 5 shows the graph of *f* as drawn by Mathematica.

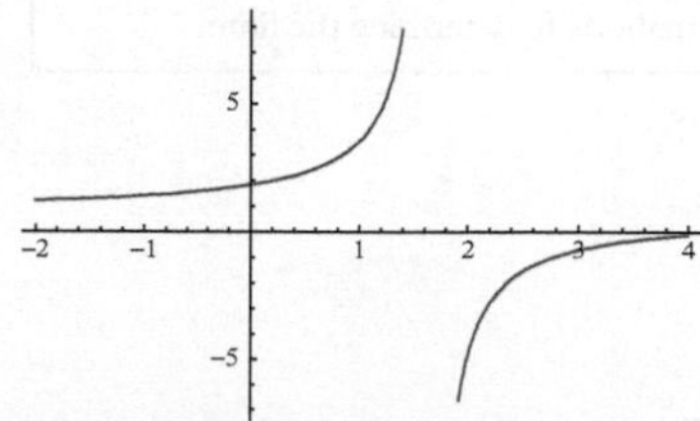

Figure 5 *The graph of the continuous function* $f(x) = \frac{2x-9}{3x-5}$*, generated by Mathematica. The graph is drawn in two pieces: one to the left of the discontinuity at* $x = \frac{5}{3}$ *and one to its right*

Example 2 *On what intervals is* $f(x) = \dfrac{\sqrt{x+4}}{x+1}$ *continuous?*

Solution We need to know (1) where we are dividing by zero and (2) where the square root is defined. For (1), set the denominator equal to zero and solve, yielding $x = -1$, which must be excluded from the domain. For (2), we set the expression under the square root to be at least zero and solve:

$$x + 4 \geq 0$$
$$x \geq -4.$$

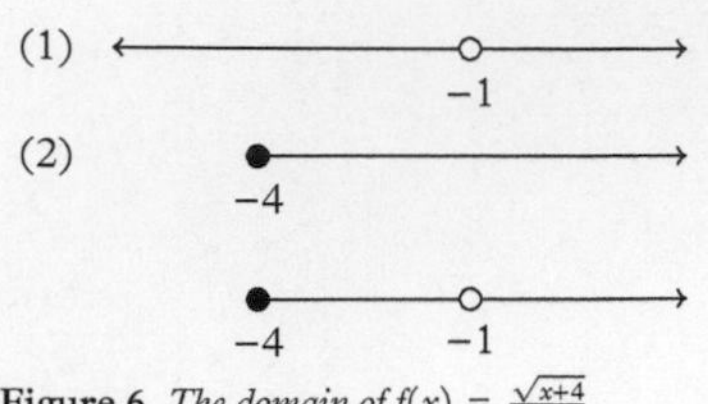

Figure 6 *The domain of* $f(x) = \frac{\sqrt{x+4}}{x+1}$

Figure 6 contains number lines picturing solutions to (1) and (2) and their intersection; f is continuous on the intervals $[-4, -1)$ and $(-1, \infty)$. ■

A consequence of the answer to example 2 is that we can draw the graph of $f(x) = \frac{\sqrt{x+4}}{x+1}$ in two pieces: one between -4 and -1 (including -4 but not -1), and one to the right of -1.

Reading Exercise 26 *On what intervals is* $f(x) = \dfrac{4x+5}{2x+1}$ *continuous?*

Using continuity to evaluate limits

Another very useful consequence of knowing the continuity of certain types of functions lies in the evaluation of limits. Suppose that f is a continuous function. If k is in the domain of f, then f is continuous at $x = k$, and—by definition of continuity—$\lim_{x\to k} f(x) = f(k)$. This gives us a simpler, alternate way of calculating limits:

> **EVALUATING LIMITS USING CONTINUITY**
>
> If f is a continuous function and k is a real number in the domain of f, then $\lim_{x\to k} f(x) = f(k)$; that is, we can evaluate $\lim_{x\to k} f(x)$ by finding $f(k)$.

> If k is not in the domain of f, then continuity is of no help and we must use other methods to determine the limit.

Let's take a look at this by comparing two methods for finding $\lim\limits_{x\to 3}(x^2 + 7)$:

using the definition of limit, we "plug in" $x = 3 + \alpha$:	using continuity, we "plug in" $x = 3$:
$\lim\limits_{x\to 3}(x^2+7) = (3+\alpha)^2 + 7$	$\lim\limits_{x\to 3}(x^2+7) = 3^2 + 7$
$= 9 + 6\alpha + \alpha^2 + 7$	$= 9 + 7$
≈ 16	$= 16$

Because $f(x) = x^2 + 7$ is continuous at $x = 3$, the two answers must be the same; the limit $\lim_{x\to 3} f(x)$ (calculated on the left) must equal the value of the function $f(3)$ (calculated on the right). This means that when finding limits of continuous functions, we can try to evaluate the function first. If the function is defined, then we have found the limit. Otherwise, we must try other methods.

Continuous functions include any function type studied in calculus with the possible exception of piecewise-defined functions, for which the evaluation rules change.

Example 3 *Find* $\lim\limits_{x\to 5} \dfrac{x^2 - 1}{x + 4}$.

Solution Rational functions are continuous, so we start by trying to evaluate using continuity:

$$\lim_{x\to 5} \frac{x^2 - 1}{x + 4} = \frac{5^2 - 1}{5 + 4} = \frac{24}{9}.$$

Because the expression is defined at $x = 5$ (the answer is a real number, $\frac{24}{9}$), evaluating using continuity is successful and the limit is $\frac{24}{9}$. ■

Because we evaluated the limit by substituting 5 for x, this method of calculating limits is sometimes called the *direct substitution property*.

Reading Exercise 27 *Use continuity to evaluate* $\lim\limits_{x\to 2} \dfrac{x + 1}{x^2 + 3}$.

Example 4 *Find* $\lim\limits_{x\to 2} \dfrac{x^2 - 4}{x - 2}$.

Solution We start by trying to evaluate using continuity:

$$\lim_{x\to 2} \frac{x^2 - 4}{x - 2} = \frac{2^2 - 4}{2 - 2} = \frac{0}{0},$$

which is undefined. Therefore, $x = 2$ is not in the domain of the expression and we cannot evaluate the limit using continuity; we must use other methods.

Any time we evaluate the function at the given value of x and the result is undefined (such as division by zero), we are not finished. We must use other methods to determine the limit.

Returning to our previous methods, we use $x = 2 + \alpha$:

$$\lim_{x\to 2} \frac{x^2 - 4}{x - 2} = \frac{(2 + \alpha)^2 - 4}{2 + \alpha - 2} = \frac{4 + 4\alpha + \alpha^2 - 4}{\alpha}$$
$$\approx \frac{4\alpha}{\alpha} = 4.$$

This is the type of situation for which infinitesimal methods are designed. The function is undefined, but using hyperreals helps us because we can divide by an infinitesimal instead of dividing by zero.

The value of the limit is 4. ■

From now on, our first thought when evaluating the limit of a continuous function should be to try to evaluate using continuity. When that fails, we use other methods.

Ans. to reading exercise 26
$(-\infty, -\frac{1}{2})$ and $(-\frac{1}{2}, \infty)$

Because trigonometric functions are continuous and so are combinations and compositions of continuous functions, we can use continuity to evaluate limits in a wide variety of situations. Again, our first step is to try evaluating using continuity.

Example 5 *Find* $\lim_{x\to 0} \dfrac{(x-7)\cos(x^2+3\sin x)}{x+1}$.

Solution Polynomial and trigonometric functions are continuous, so this function is a combination and composition of continuous functions. We therefore start by trying to evaluate using continuity:

$$\lim_{x\to 0} \frac{(x-7)\cos(x^2+3\sin x)}{x+1} = \frac{-7\cos(0+3\sin 0)}{1} = \frac{-7\cos 0}{1} = -7.$$

Because the value of the expression at $x = 0$ is a real number, evaluating by continuity succeeded. The value of the limit is -7. ■

Reading Exercise 28 *Find* $\lim_{x\to 0} \sin(x^2 - 1 + \cos x)$.

Intermediate value theorem

Many theorems of calculus can be understood visually, but only if enough attention is paid to learning the graphical interpretation or meaning of various terms. Among these theorems is the intermediate value theorem (IVT).

Suppose that f is a continuous function defined on the closed interval $[a, b]$ and that $f(a) \neq f(b)$, as in figure 7.

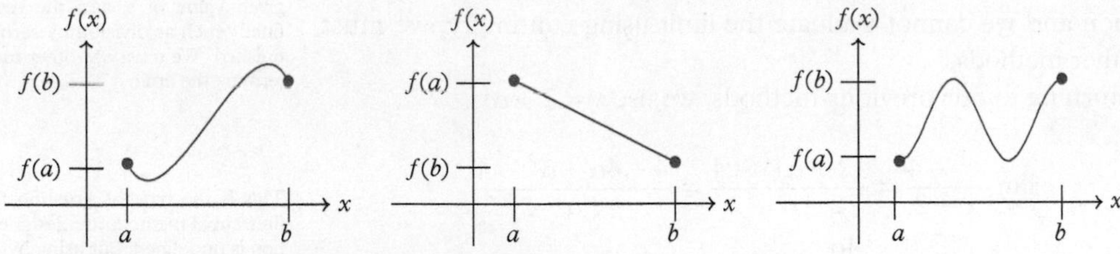

Figure 7 *Continuous functions defined on* $[a, b]$ *with* $f(a) \neq f(b)$

Ans. to reading exercise 27
$\lim_{x\to 2} \frac{x+1}{x^2+3} = \frac{2+1}{2^2+3} = \frac{3}{7}$

Each of the functions pictured in figure 7 passes through all the y-coordinates between $f(a)$ and $f(b)$ at least once. If N is a real number between $f(a)$ and $f(b)$, it appears that the function must take on that value at least once between a and b. That is, if the line $y = N$ is drawn, as long as f is continuous, it must cross that line at least once, as in figure 8.

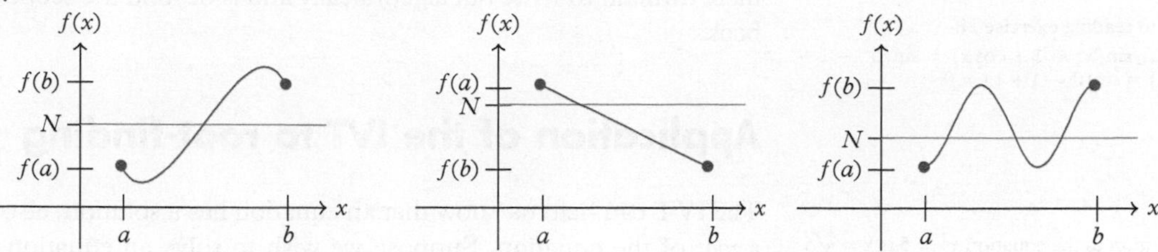

Figure 8 *Continuous functions defined on $[a, b]$ with $f(a) \neq f(b)$. The functions cross line $y = N$, where N is between $f(a)$ and $f(b)$*

For each point of intersection of the line $y = N$ with the curve, there is an x-coordinate of that point. If we label such an x-coordinate c, then its y-coordinate is $f(c)$, and therefore $f(c) = N$. See figure 9.

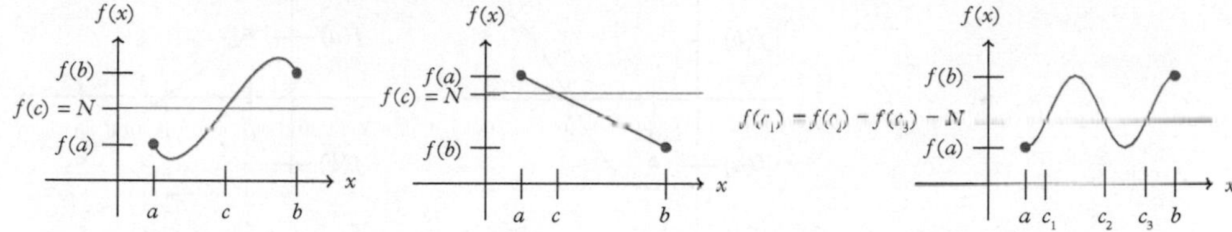

Figure 9 *Continuous functions defined on $[a, b]$ with $f(a) \neq f(b)$. The functions cross line $y = N$, where N is between $f(a)$ and $f(b)$*

We can summarize our conclusion by saying that for any real number N between $f(a)$ and $f(b)$, there is at least one real number c between a and b such that $f(c) = N$. As long as f is continuous, this must be the case, because one of the two points $(a, f(a))$ and $(b, f(b))$ is above the line $y = N$ and the other is below that line. If we must draw the graph in one piece between those two points, the graph must intersect the line. But, if the function is not continuous, then the conclusion might not hold, because there could be a jump in the graph so that the function's graph jumps the line and never intersects it, as in figure 10. Continuity is the key to the IVT.

Because N is a value of the function that is between the two function values $f(a)$ and $f(b)$, it can be called an *intermediate value* of the function, hence the name *intermediate value theorem.*

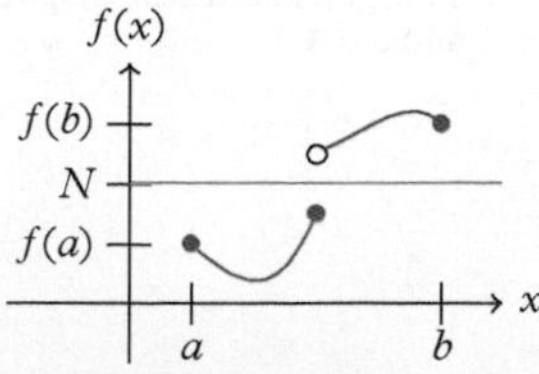

Figure 10 *Function with a discontinuity defined on $[a, b]$ with $f(a) \neq f(b)$ that does not intersect line $y = N$, where N is between $f(a)$ and $f(b)$. The conclusion of the IVT does not hold*

Theorem 10 INTERMEDIATE VALUE THEOREM *Suppose f is a continuous function defined on the closed interval $[a, b]$ and N is a real number between $f(a)$ and $f(b)$. Then, there is at least one real number c between a and b such that $f(c) = N$.*

Ans. to reading exercise 28
$\lim_{x\to 0} \sin(x^2 - 1 + \cos x) = \sin(0^2 - 1 + \cos 0) = \sin(0 - 1 + 1) = 0$

Although the pictures are convincing, a proof of the IVT is a little more difficult to write out algebraically and is beyond the scope of this book.

Application of the IVT to root-finding

A solution to the equation $x^2 = 5$ is $x = \sqrt{5}$. The terms *root of an equation* and *square root* have a common root (pun intended).

The equation $x^2 + 2x = 0$ is of the form $f(x) = 0$.

The IVT can help us know that an equation has a solution, also called a *root* of the equation. Suppose we wish to solve an equation of the form $f(x) = 0$. If f is continuous on $[a, b]$ and one of the two numbers $f(a)$ and $f(b)$ is positive whereas the other is negative, then the number $N = 0$ is between $f(a)$ and $f(b)$ and, by the IVT, there is a number c between a and b such that $f(c) = 0$. Then, $x = c$ is a solution to the equation $f(x) = 0$. This is illustrated in figure 11.

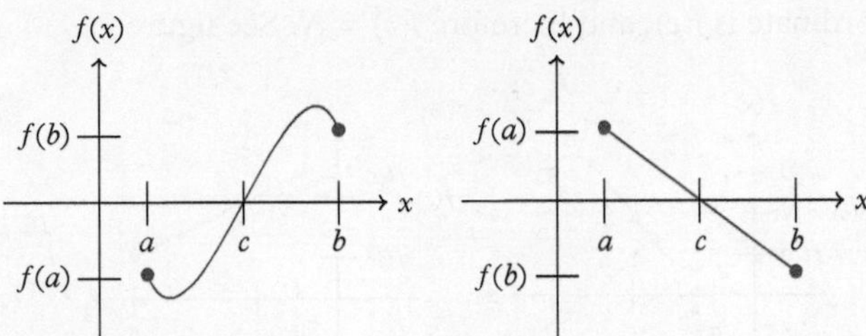

Figure 11 *One of $f(a)$ and $f(b)$ is positive; the other is negative. If f is continuous, then the equation $f(x) = 0$ must have a solution $x = c$ between $x = a$ and $x = b$*

Example 6 *Show that there is a solution to the equation $x^2 - 1 + \sin x = 0$ between $x = 0$ and $x = 1$.*

If the equation is not in the form $f(x) = 0$, then it must be manipulated to get it in that form to determine $f(x)$.

Solution The equation is in the form $f(x) = 0$, where $f(x) = x^2 - 1 + \sin x$. Next we calculate $f(0)$ and $f(1)$:

Compared to the previous discussion, we are using $a = 0$ and $b = 1$.

$$f(0) = 0^2 - 1 + \sin 0 = -1$$
$$f(1) = 1^2 - 1 + \sin 1 = \sin 1 = 0.8415$$

Because one of $f(0)$ and $f(1)$ is positive and the other is negative, there is a solution to the equation between $x = 0$ and $x = 1$. ■

Now that we have determined that the interval $(0, 1)$ contains a solution to the equation, we can use the same procedure repeatedly to home in on a solution, as illustrated in the next example.

Example 7 *Determine the first two digits of a solution to the equation* $x^2 - 1 + \sin x = 0$.

One might think that we should begin by solving the equation exactly. Good luck with that. Even computer algebra systems can't solve that one. Equations that involve both trig functions and polynomial functions are very difficult, if not impossible, to solve exactly, so we must try to approximate a solution using a numerical method. In chapter 3 section 3.9, we see another, more efficient method for solving equations numerically, but this method will do for now.

Solution As in example 6, with the equation in the form $f(x) = 0$, we have $f(x) = x^2 - 1 + \sin x$. We already know there is a solution between $x = 0$ and $x = 1$, since $f(0) = -1$ and $f(1) = 0.8415$. Plotting these two points and knowing that f is continuous, we can visualize the curve crossing the x-axis (the line $y = 0$) somewhere between $x = 0$ and $x = 1$ (figure 12). We wish to estimate more closely where the curve crosses the line.

Where $y = f(x)$ crosses the line $y = 0$, it must be the case that $f(x) = 0$ and that value of x solves the equation.

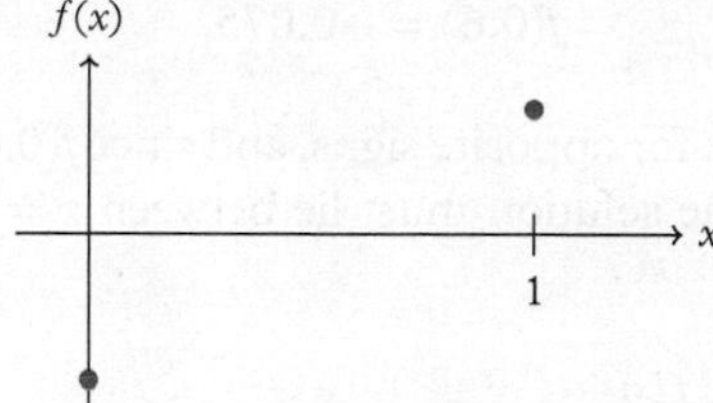

Figure 12 *Two points on the graph of the continuous function* $f(x) = x^2 - 1 + \sin x$. *We see that the function must cross the line* $y = 0$ *somewhere between* $x = 0$ *and* $x = 1$

Next we calculate the value of f somewhere in between $x = 0$ and $x = 1$. Let's use $x = 0.5$:

$$f(0.5) = -0.27.$$

To visualize the next step, we place this point on the graph as well (figure 13).

Must the curve cross the x-axis between $x = 0$ and $x = 0.5$? No. Because both $f(0)$ and $f(0.5)$ are negative, both points are below the x-axis and the curve can easily stay below the line when traveling from one point to the other. But between $x = 0.5$ and $x = 1$, the curve must cross the line. Since $f(0.5)$ is negative and $f(1)$ is positive, there must be a solution between $x = 0.5$ and $x = 1$. Our interval containing a solution is now only half as wide—it has been *bisected*, which gives rise

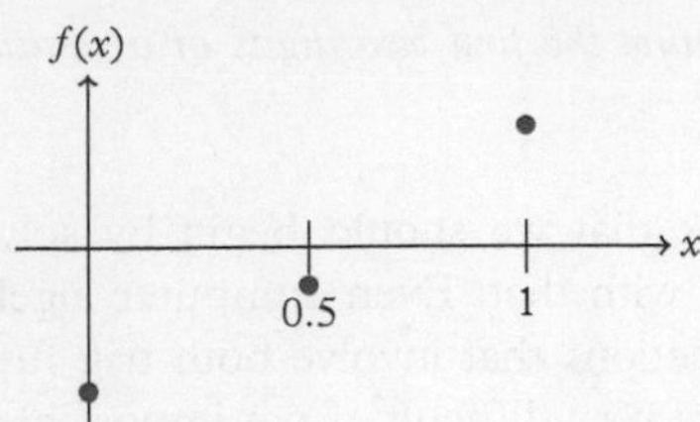

Figure 13 *Three points on the graph of the continuous function $f(x) = x^2 - 1 + \sin x$. We see that the function must cross the line $y = 0$ somewhere between $x = 0.5$ and $x = 1$*

to the name of this method: the *bisection method.* We can try it again; this time, let's choose a number between 0.5 and 1, such as 0.7:

The bisection method always cuts the length of the interval in half. This is conveniently programmed on a computer using binary, but it is much less convenient when performing the method by hand. The method demonstrated here is a *modified bisection method.*

$$f(0.7) = 0.134.$$

Because $f(0.5)$ is negative and $f(0.7)$ is positive, a solution must lie between $x = 0.5$ and $x = 0.7$. We split the difference and find that

$$f(0.6) = -0.075.$$

Once again, we look for opposite signs, and since $f(0.6)$ is negative and $f(0.7)$ is positive, the solution must lie between $x = 0.6$ and $x = 0.7$ (figure 14).

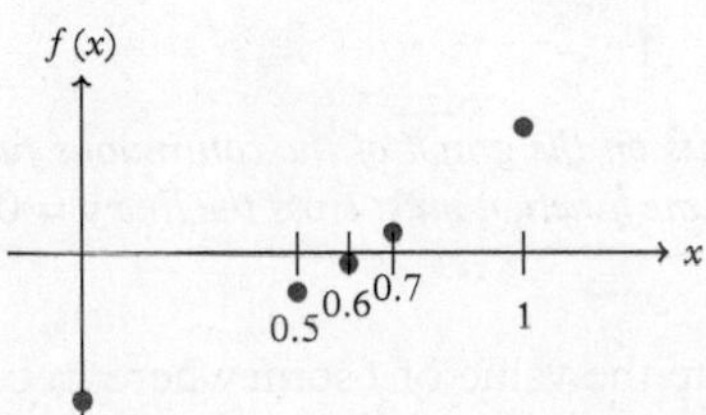

Figure 14 *Several points on the graph of the continuous function $f(x) = x^2 - 1 + \sin x$. We see that the function must cross the line $y = 0$ somewhere between $x = 0.6$ and $x = 0.7$*

At this point, we know the first digit in the solution. Because the solution lies between $x = 0.6$ and $x = 0.7$, it must start $x = 0.6\ldots$. To determine the next digit, we repeat the process again, first choosing $x = 0.65$ followed in the next two steps by $x = 0.63$ and then $x = 0.64$:

x	$f(x)$	interval containing solution
0.65	0.027	(0.6, 0.65)
0.63	−0.014	(0.63, 0.65)
0.64	0.0068	(0.63, 0.64)

Our solution begins $x = 0.63\ldots$ because it is between 0.63 and 0.64. This could continue to a third or fourth digit or beyond, but at this point our task is finished. ■

A solution beginning with $x = 0.63\ldots$ does not necessarily round to $x = 0.63$. In fact, since $f(0.635) = -0.0036$ is negative, the solution is between 0.635 and 0.64, and therefore rounds to $x = 0.64$.

If the bisection method looks vaguely familiar, it might be because it is essentially equivalent to using a graphing calculator to "zoom in" on a solution. After all, there is not much difference between a graph such as that in figure 14 and a graph of the function; the only difference is the number of points plotted: five vs. hundreds. Zooming in to see a small section of the graph to inspect it to see where it crosses the x-axis is effectively the same as the visual version of inspecting a table of values to see where the function changes from negative to positive. The comparison in figure 15 should make it clear that the two methods are two versions of the same idea.

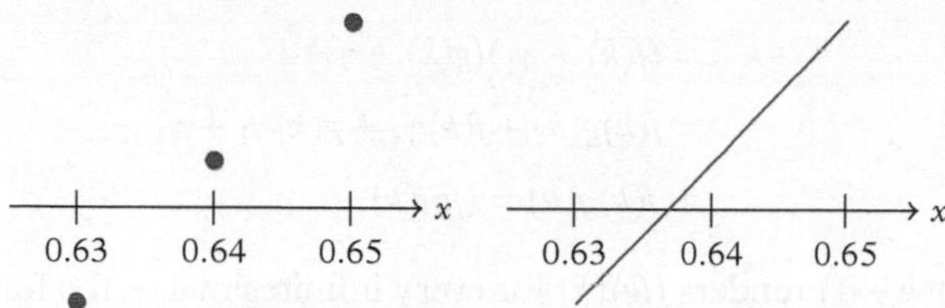

Figure 15 *(left) Three points on the graph of the continuous function* $f(x) = x^2 - 1 + \sin x$ *visualizing the bisection method; (right) the graph of that same function as zoomed in to the interval* $[0.63, 0.65]$

Reading Exercise 29 *Show there is a solution to the equation* $x^3 - \cos x = 0$ *between* $x = 0$ *and* $x = 2$.

Proof of continuity of polynomial and rational functions

Now that we have seen how continuity can be used, all that remains is to prove some of the remaining portions of theorem 9. We begin by proving that combinations of continuous functions are continuous.

Theorem 11 COMBINATIONS OF CONTINUOUS FUNCTIONS *If f and g are continuous functions and c is a real number, then the functions $f + g$, $f - g$, fg, f/g, and cf are also continuous.*

Proof. If k is a real number in the domain of both f and g, then for any infinitesimal α,

Here we are using version 2 of the definition of continuity.

$$f(k + \alpha) \doteq f(k)$$

and

$$g(k+\alpha) \doteq g(k).$$

This means that

$$f(k+\alpha) = f(k) + \gamma_1$$

and

$$g(k+\alpha) = g(k) + \gamma_2,$$

where γ_1 and γ_2 are infinitesimals (or zero). To prove that fg is continuous, we need $(fg)(k+\alpha)$ to render $(fg)(k)$. We have

It is more accurate to use $\doteq$ instead of $\approx$ in the last step. If the quantity $f(k)g(k)$ happens to be zero, then with approximation we are violating the approximation principle. But, what remains are infinitesimals, which renders $0 = f(k)g(k)$ anyway. Use of the symbol $\approx$ emphasizes how the calculation works in all other cases.

A similar note applies to the proofs of theorems 6–8, as well as the answer to section 1.6 exercise 49.

$$\begin{aligned}(fg)(k+\alpha) &= f(k+\alpha)\cdot g(k+\alpha)\\ &= (f(k)+\gamma_1)(g(k)+\gamma_2)\\ &= f(k)g(k) + f(k)\gamma_2 + g(k)\gamma_1 + \gamma_1\gamma_2\\ &\approx f(k)g(k) = (fg)(k).\end{aligned}$$

Since $(fg)(k+\alpha)$ renders $(fg)(k)$ for every infinitesimal α, the function fg is continuous at $x = k$. Because this is true for every real number k, fg is a continuous function.

Similarly, as long as $g(k) \neq 0$,

$$\begin{aligned}\left(\frac{f}{g}\right)(k+\alpha) &= \frac{f(k+\alpha)}{g(k+\alpha)} = \frac{f(k)+\gamma_1}{g(k)+\gamma_2}\\ &\approx \frac{f(k)}{g(k)} = \left(\frac{f}{g}\right)(k)\end{aligned}$$

If $f(k) = 0$, then the expression is infinitesimal and renders $0 = f(k)/g(k)$.

and

If $f(k) = 0$, then the expression is infinitesimal and renders $0 = c \cdot f(k)$.

$$\begin{aligned}(cf)(k+\alpha) &= c\cdot f(k+\alpha) = c(f(k)+\gamma_1) = c\cdot f(k) + c\gamma_1\\ &\approx c\cdot f(k) = (cf)(k),\end{aligned}$$

so both f/g and cf are also continuous functions. The continuities of $f+g$ and $f-g$ are left as exercises. ■

Recall that we have already proved that linear functions are continuous (theorem 6). That, along with theorem 11, is enough to establish the continuity of all polynomial and rational functions, because all such functions are combinations of linear functions. For instance, $x^3 = x{\cdot}x{\cdot}x$ is the product of continuous functions and hence $f(x) = x^3$ is continuous. Then, so is $4\cdot x^3$ by the continuity of cf, and $\frac{4x^3}{x+1}$ by the continuity of f/g. A formal proof uses the same reasoning.

Ans. to reading exercise 29

❶ $f(x) = x^3 - \cos x$

❷ $f(0) = -1$, $f(2) = 8.416$ (If you got 7.0006, your calculator is in degree mode. Switch to radian mode.)

❸ By the IVT, there is a solution between $x = 0$ and $x = 2$.

Proof of continuity of trigonometric functions

To prove that the trig functions are continuous, we need to recall four identities, which are true for hyperreals as well as reals by the transfer principle:

$$\sin(-x) = -\sin x,$$
$$\cos^2 x = 1 - \sin^2 x,$$
$$\sin(a+b) = \sin a \cos b + \cos a \sin b, \text{ and}$$
$$\cos(a+b) = \cos a \cos b - \sin a \sin b.$$

Another fact that is useful is that

$$0 < \sin x < x \text{ when } 0 < x < \frac{\pi}{2},$$

which is also true for hyperreals by the transfer principle. If this inequality is unfamiliar to you, it can be justified by figure 16.

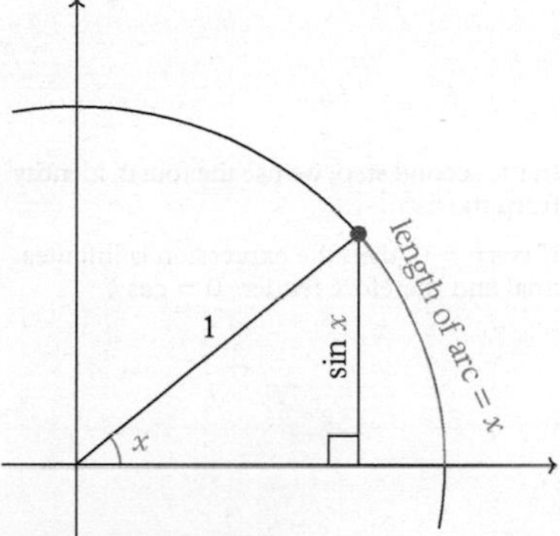

Figure 16 *A point on the unit circle and the angle x radians. The arc in orange is longer than the segment labeled* sin *x because the shortest distance from a point (in blue) to a line (the x-axis) is along a perpendicular*

Let ω be a positive infinitesimal. Then, by the previous inequality,

$$0 < \sin \omega < \omega,$$

and because $\sin \omega$ is between zero and an infinitesimal, it must also be infinitesimal. Multiplying through the inequality by -1 and changing the direction of the inequalities yields $0 > -\sin \omega > -\omega$, rewritten easily as

$$-\omega < -\sin \omega < 0.$$

From the first of the four identities listed earlier, $\sin(-\omega) = -\sin \omega$. Therefore,

$$-\omega < \sin(-\omega) < 0,$$

and $\sin(-\omega)$ is also infinitesimal. Therefore, the sine of an infinitesimal, either positive or negative, is infinitesimal.

> $\sin \alpha$ is infinitesimal for any infinitesimal α. (This fact is refined in section 2.4.)

Using the second of the four identities, for any infinitesimal α,

$$\cos^2 \alpha = 1 - \sin^2 \alpha \approx 1,$$

because $\sin \alpha$ is infinitesimal.

> $\cos^2 \alpha \approx 1$ for any infinitesimal α.

We are now ready to prove the continuity of the sine function.

Theorem 12 CONTINUITY OF SINE *The function* $f(x) = \sin x$ *is continuous.*

Proof. To prove that the sine function is continuous, we need to show that $\lim_{x\to k} \sin x = \sin k$ for every real number k. To this end,

In the second step, we use the third identity from the list.

$$\lim_{x\to k} \sin x = \sin(k+\alpha)$$
$$= \underbrace{\sin k}_{\text{real}}\underbrace{\cos\alpha}_{\approx 1} + \underbrace{\cos k}_{\text{real}}\ \underbrace{\sin\alpha}_{\text{infinitesimal}}$$
$$\approx \sin k \cdot 1 + 0 = \sin k.$$

If $\sin k = 0$, then the expression is infinitesimal and therefore renders $0 = \sin k$.

Since $\lim_{x\to k} \sin x = \sin k$, f is continuous at $x = k$ (for any real number k) and the sine function is continuous. ■

The proof for cosine is similar.

Theorem 13 CONTINUITY OF COSINE *The function* $f(x) = \cos x$ *is continuous.*

Proof. To prove that the cosine function is continuous, we need to show that $\lim_{x\to k} \cos x = \cos k$ for every real number k. To this end,

In the second step, we use the fourth identity from the list.

If $\cos k = 0$, then the expression is infinitesimal and therefore renders $0 = \cos k$.

$$\lim_{x\to k} \cos x = \cos(k+\alpha)$$
$$= \underbrace{\cos k}_{\text{real}}\underbrace{\cos\alpha}_{\approx 1} - \underbrace{\sin k}_{\text{real}}\ \underbrace{\sin\alpha}_{\text{infinitesimal}}$$
$$\approx \cos k \cdot 1 + 0 = \cos k.$$

Since $\lim_{x\to k} \cos x = \cos k$, f is continuous at $x = k$ (for any real number k) and the cosine function is continuous. ■

Because the other four basic trig functions are quotients or reciprocals of sine and cosine, they are also continuous by theorem 11.

Recall that continuous means "continuous where defined". These functions are not defined everywhere and their graphs have vertical asymptotes, but otherwise they are continuous.

Theorem 14 CONTINUITY OF MORE TRIGONOMETRIC FUNCTIONS *The tangent, cotangent, secant, and cosecant functions are continuous.*

Expressions like those in example 5 are more than just sums, products, and quotients of continuous functions. They also involve compositions of functions, the subject of our next theorem.

Theorem 15 COMPOSITION OF CONTINUOUS FUNCTIONS *If* f *and* g *are continuous functions, then so is* $f \circ g$.

Proof. Let k be a real number in the domain of g such that $g(k)$ is in the domain of f. Since g is continuous at $x = k$, for any infinitesimal α, $g(k+\alpha)$ renders $g(k)$, meaning $g(k+\alpha) = g(k)+\gamma$ where γ is infinitesimal (or zero). Then,

Here we use version 2 of the definition of continuity.

$$f \circ g(k + \alpha) = f(g(k + \alpha)) = f(g(k) + \gamma).$$

Since $g(k)$ is a real number in the domain of f, f is continuous at $g(k)$. Because γ is an infinitesimal, continuity requires that $f(g(k)+\gamma)$ render $f(g(k))$. Our calculation from the beginning is therefore

$$f \circ g(k + \alpha) = f(g(k + \alpha)) = f(g(k) + \gamma) \doteq f(g(k)) = f \circ g(k).$$

By version 2 of the definition of continuity, $f \circ g$ is continuous. ■

Approximating inside a continuous function

Consider again the second version of the definition of continuity. If f is a continuous function and b is a real number in the domain of f, then $f(b+\alpha) \doteq f(b)$ for every infinitesimal α. By the definition of rendering, if the real number $f(b)$ is not zero then we can replace $\doteq$ with $\approx$—that is,

$$f(b + \alpha) \approx f(b).$$

Therefore, as long as the result is not zero (which violates the approximation principle) and what is inside approximates to a real number, then we can safely approximate inside any continuous function.

Theorem 16 APPROXIMATING INSIDE A CONTINUOUS FUNCTION *If f is a continuous function, b is a real number in the domain of f, $k \approx b$, and $f(b) \neq 0$, then $f(k) \approx f(b)$.*

This theorem does not say anything about what happens when the quantity inside the continuous function is infinitesimal or infinite.

Example 8 *Find* $\lim_{x \to 2} \sin\left(\dfrac{x^2 - 4}{x - 2}\right)$.

Solution Evaluating using continuity fails because we are dividing by zero. Using $x = 2 + \alpha$,

$$\lim_{x\to 2} \sin\left(\frac{x^2-4}{x-2}\right) = \sin\left(\frac{(2+\alpha)^2-4}{2+\alpha-2}\right)$$
$$= \sin\left(\frac{4+4\alpha+\alpha^2-4}{\alpha}\right) = \sin\left(\frac{4\alpha+\alpha^2}{\alpha}\right)$$
$$\approx \sin\frac{4\alpha}{\alpha} = \sin 4.$$

Because the sine function is continuous, the approximation on the inside results in a real number (4), and the result isn't zero ($\sin 4 \neq 0$), the use of approximation is justified as written. ■

If the approximation principle is violated (that is, if $f(b) = 0$), if the quantity inside the continuous function is an infinitesimal (that is, if b is infinitesimal), or if the quantity inside the continuous function is infinite (that is, if b is infinite), then theorem 16 does not apply. For some continuous functions, we can only approximate "inside" under the circumstances of theorem 16. For others, we may always approximate inside. For instance, we may always approximate under a square root, even when the inside quantity is infinite or infinitesimal.

The proof that we may always approximate under a square root is given in section 2.6.

Knowing when we may or may not approximate inside a continuous function is important for maintaining correct notation, but it is sometimes the case that when the conditions of theorem 16 are not met, then we may resort to the use of $\doteq$ instead of $\approx$ to save the calculation. Examples are given in section 10.2.

Exercises 1.7

1–14. On what intervals is the given function continuous?

1. $f(x) = \dfrac{5}{x-7}$
2. $f(x) = \dfrac{3}{x^2-3x+2}$
3. $f(x) = \sqrt{3x-2}$
4. $f(x) = \dfrac{x^2+4x-7}{5}$
5. $f(x) = x^2 - 3x + 5$
6. $f(t) = \sqrt{5-3t}$
7. $g(x) = \dfrac{3x+4}{x^2-5x-6}$
8. $r(t) = \sqrt{t^2-9}$
9. $f(t) = \sqrt{4-t^2}$
10. $g(x) = \dfrac{4x+1}{\sqrt{x}-3}$
11. $h(x) = \dfrac{2-\sqrt{x+1}}{x^2+5x}$
12. $h(x) = \sqrt{\dfrac{x+2}{x^2+1}}$

13. $f(x) = \sqrt[3]{x-8}$ 14. $f(x) = \dfrac{x-6}{2-\sqrt{x-5}}$

15–26. Find the limit.

Graphical and numerical methods should not be used for these problems.

15. $\lim_{x\to 9} \dfrac{\sqrt{x}+2}{3x-20}$

16. $\lim_{x\to 1} \dfrac{x^2+5}{x+4}$

17. $\lim_{x\to 4} \dfrac{|3-x|}{x^2+1}$

18. $\lim_{x\to 0} \dfrac{x^3+6x-5}{3x+10}$

19. $\lim_{x\to 3} \dfrac{x+1}{(x-3)^2}$

20. $\lim_{x\to -2} \dfrac{x^2+7x+10}{x+2}$

21. $\lim_{x\to 1} \dfrac{x^2-7}{\sqrt{x+3}}$

22. $\lim_{x\to \pi/2} \dfrac{\sin x}{x}$

23. $\lim_{x\to 0} \sin\left(\dfrac{x+\pi}{x^2-2}\right)$

24. $\lim_{x\to 0} \sin^2\left(\dfrac{x+3}{\sqrt{9-x^2}} - \cos 5x\right)$

25. $\lim_{x\to \pi} \dfrac{\tan\left(\frac{1}{4}x\right)}{x^2}$

26. $\lim_{x\to 4} \dfrac{5}{x^2-6x+8}$

27. Find $\lim_{x\to 2} f(x)$, where $f(x) = \begin{cases} 2x-3, & x<2 \\ 3x^2-11, & x>2 \end{cases}$.

28. Find $\lim_{x\to 0} g(x)$, where $g(x) = \begin{cases} x-5, & x<0 \\ x^2+2, & x>0 \end{cases}$.

29. Show that there is a solution to the equation $\cos(x+\sin x) = 0$ between $x = 0$ and $x = 1$.

30. Show that there is a solution to the equation $\sqrt{x^3-5x}+12x+3 = 0$ between $x = -1$ and $x = 0$.

31. Show that there is a solution to the equation $\sqrt{x+1} = 3\sin x$ between $x = 2$ and $x = 3$.

32. Show that there is a solution to the equation $x^5 - 4x^3 + x^2 = 8$ between $x = 2$ and $x = 3$.

33. Determine the first two digits of a solution to the equation $\cos(x+\sin x) = 0$.

34. Determine the first two digits of a solution to the equation $\sqrt{x^3-5x}+12x+3 = 0$.

35. Determine the first three digits of a solution to the equation $\sqrt{x+1} = 3\sin x$.

36. Determine the first three digits of a solution to the equation $x^5 - 4x^3 + x^2 = 8$.

37. Determine $\lim_{x\to 0} \tan\left(\dfrac{\pi\sqrt{x+1}}{x^2-4}\right)$.

38. Find $\lim_{x\to\infty} \cos\left(\dfrac{\pi x}{\sqrt{x^2+7}}\right)$.

Answers to odd-numbered proof exercises are often not in the back of the book.

39. Use a proof similar to that of theorem 7 to show that cubic functions (form $f(x) = bx^3 + cx^2 + dx + r$) are continuous.

40. Use a proof similar to that of theorem 8 to show that rational functions of the form $\dfrac{bx^2 + cx + d}{rx + s}$ are continuous.

41. Prove that if f and g are continuous functions, then so is $f + g$.

42. Prove that if f and g are continuous functions, then so is $f - g$.

Slope, Velocity, and Rates of Change

1.8

A tangent line to a circle is a line that intersects the circle in exactly one point (figure 1). Although this description of a tangent line works well on some curves that are not circles, for others it does not. Consider the curves in figure 2. A line that just touches the curve at one point might

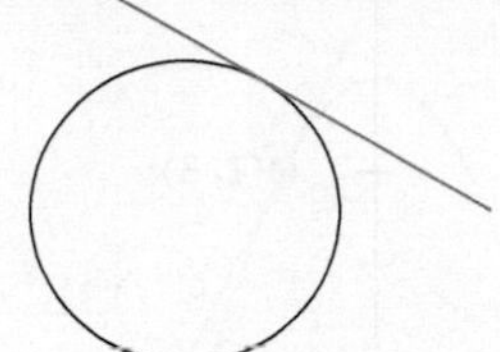

Figure 1 *A line tangent to a circle*

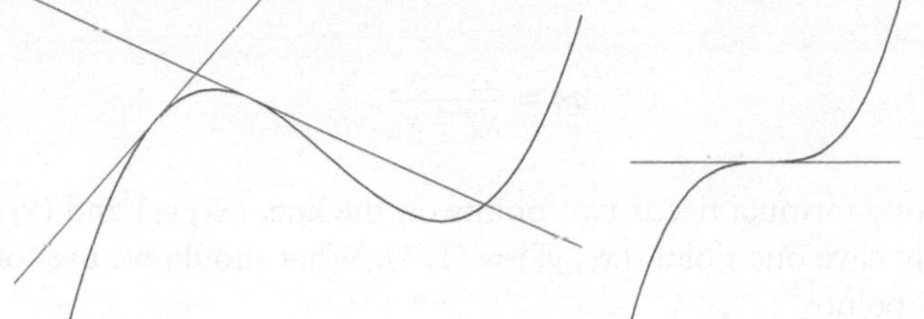

Figure 2 *Tangents to curves*

or might not pass through the curve at some other point. It might even cut through the curve altogether. We need a different description of the term *tangent.*

This section does not depend strictly on sections 1.4–1.7 and may be read immediately after section 1.3 if desired, although the practice of the EAR procedure gained in sections 1.4–1.7 is beneficial for this material.

One way to think of a tangent line is the direction of momentum when traveling along the curve. If you are on the edge of a spinning merry-go-round and release a ball from your hand, it does not keep going in a circle along with you. Ignoring the fact that the ball drops toward the ground, it travels in a straight line in the direction it was heading just at the time of release; this is the tangent line to the curve (figure 3). The same is true for any curve; the tangent line is the direction of momentum.

Another way to think of a tangent line is the direction of travel if you are traveling along the curve in a vehicle (figure 4). The head-

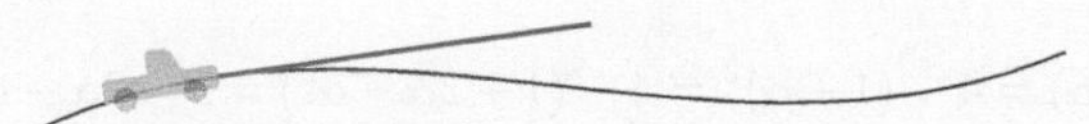

Figure 4 *A vehicle traveling on a curve. The headlights shine in the direction of the tangent*

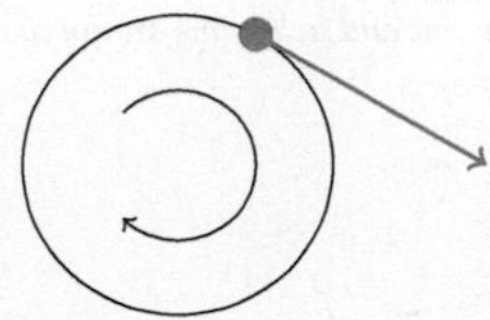

Figure 3 *A ball released at the edge of a merry-go-round*

light beam of the vehicle does not go uphill and downhill following the curve; it goes straight ahead. This is the direction of the tangent line.

Tangent line to a curve

Suppose we wish to find the equation of the tangent line to the curve $f(x) = 4 - x^2$ at $x = 1$. By "at $x = 1$," we mean the point with the x-coordinate 1. The y-coordinate of that point is $f(1) = 4 - 1^2 = 3$, so the point at which we want the tangent line is $(1, f(1)) = (1, 3)$. The tangent is pictured in figure 5.

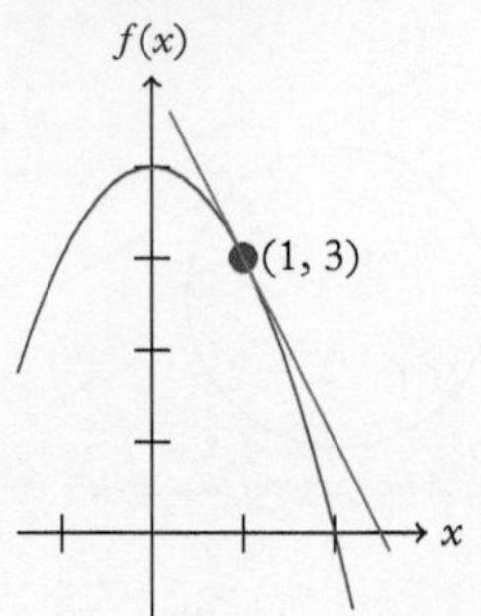

Figure 5 *The curve $f(x) = 4 - x^2$ with tangent line at $x = 1$. The point $(1, 3)$ is called the* point of tangency

One formula used for the equation of a line is the point-slope equation of a line: $y - y_1 = m(x - x_1)$. All we need is a point (which we have) and the slope of the line (which we need to find). There is also a formula for slope:

$$m = \frac{y_2 - y_1}{x_2 - x_1}.$$

The slope formula needs two points on the line, (x_1, y_1) and (x_2, y_2). We only have one point, $(x_1, y_1) = (1, 3)$. What should we use for the second point?

Idea #1: The first idea we might try is to use a second point from the curve. To illustrate, let's use the point with $x = 2$ as the second point. Since $f(2) = 4 - 2^2 = 0$, the point is $(x_2, y_2) = (2, 0)$. We can find the slope using the slope formula, $m = \frac{0-3}{2-1} = -3$, but it is not the slope of the tangent line; it is the slope of a *secant line* (a line through two points of the curve) instead. See figure 6. *Idea #1 does not work.*

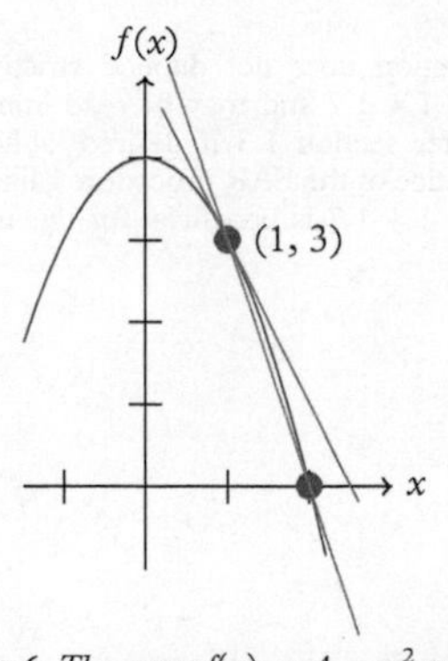

Figure 6 *The curve $f(x) = 4 - x^2$ with the tangent line (orange) at $x = 1$ and the secant line (green) through $(1, 3)$ and $(2, 0)$. The slopes of the tangent line and secant line are not the same*

Idea #2: Because we only want the curve to go through the point $(1, 3)$ and not some other point on the curve, perhaps we could use the point $(1, 3)$ as both points in the slope formula. Then we have $m = \frac{3-3}{1-1} = \frac{0}{0}$, which is undefined. *Idea #2 does not work.*

We have seen other instances of division by zero. Each time, using infinitesimals helped us avoid division by zero. Perhaps we could combine ideas #1 and #2 into a third idea, and use a second point on the curve that is, for all practical purposes, the same point as $(1, 3)$.

Idea #3: Instead of moving as far away from $x = 1$ as $x = 2$, we can try moving over just an infinitesimal amount to $x = 1 + \alpha$. Then the y-coordinate of the point is

$$f(1 + \alpha) = 4 - (1 + \alpha)^2 = 4 - (1 + 2\alpha + \alpha^2) = 3 - 2\alpha - \alpha^2.$$

Our second point on the curve is $(x_2, y_2) = (1 + \alpha, 3 - 2\alpha - \alpha^2)$. The slope formula gives

$$m = \frac{y_2 - y_1}{x_2 - x_1} = \frac{3 - 2\alpha - \alpha^2 - 3}{1 + \alpha - 1} = \frac{-2\alpha - \alpha^2}{\alpha}$$
$$\approx \frac{-2\alpha}{\alpha} = -2.$$

Instead of $\frac{3-3}{1-1} = \frac{0}{0}$, by moving over an infinitesimal amount we have $\frac{3 - 2\alpha - \alpha^2 - 3}{1 + \alpha - 1}$ and we have successfully avoided division by zero. We have an infinitesimal in the denominator instead.

This works because the two points are so close together they are practically the same point. We avoided the problem of idea #1, which gave us the wrong slope, and we avoided the problem of division by zero in idea #2 as well.

To finish, we use the point-slope equation of the line with point $(x_1, y_1) = (1, 3)$ and slope $m = -2$:

$$y - y_1 = m(x - x_1)$$
$$y - 3 = -2(x - 1)$$
$$y - 3 = -2x + 2$$
$$y = -2x + 5.$$

The equation $y - 3 = -2(x - 1)$ is the equation of the tangent line, and at this point our task has been fulfilled. It is customary, however, to finish by writing the equation of the line in slope-intercept form $y = mx + b$.

This is the correct equation of the tangent line, as pictured in figure 7.

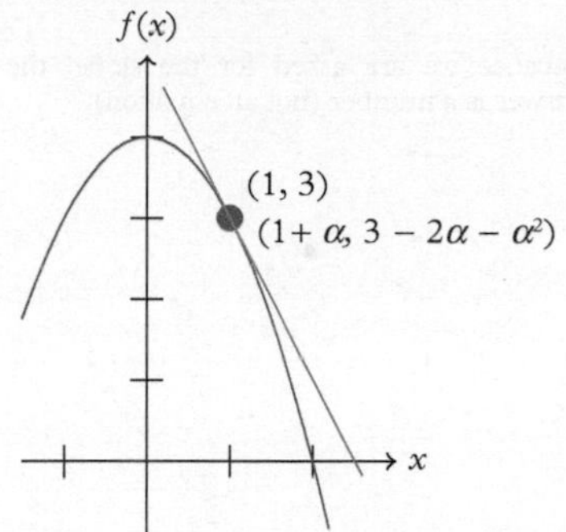

Figure 7 *The curve $f(x) = 4 - x^2$ with tangent line at $x = 1$. The curve passes through the points $(1, 3)$ and $(1 + \alpha, 3 - 2\alpha - \alpha^2)$, both of which are located at the blue dot*

If we run through the procedure more generally, then we can develop a formula for the slope of the tangent line. Suppose we want the slope of the tangent line to the curve $y = f(x)$ at $x = k$ (previously, k was 1). Then our point of tangency is $(x_1, y_1) = (k, f(k))$ (previously, it was $(1, 3)$). To get the second point, we scoot over an infinitesimal amount using $x = k + \alpha$. Then our y-coordinate is $f(k + \alpha)$, and $(x_2, y_2) = (k + \alpha, f(k + \alpha))$. See figure 8. Using the slope formula,

$$m = \frac{y_2 - y_1}{x_2 - x_1} = \frac{f(k + \alpha) - f(k)}{k + \alpha - k} = \frac{f(k + \alpha) - f(k)}{\alpha}.$$

For this to work, we need the formula to render the same real result for every infinitesimal α.

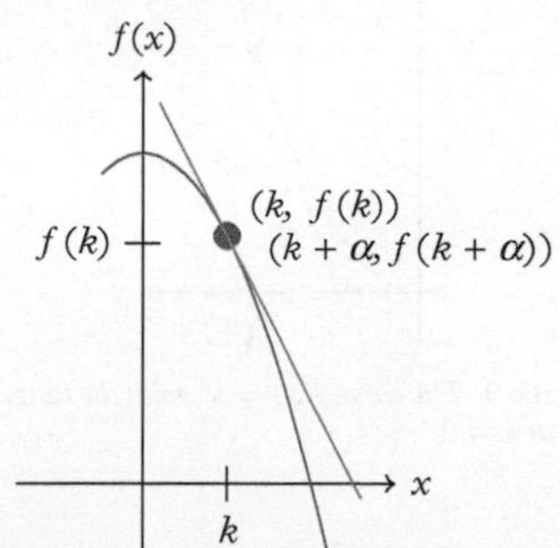

Figure 8 The curve $y = f(x)$ with the tangent line at $x = k$. The curve passes through the points $(k, f(k))$ and $(k+\alpha, f(k+\alpha))$, both of which are located at the blue dot

SLOPE OF THE TANGENT LINE

$$\text{slope of the tangent line} = \frac{f(k + \alpha) - f(k)}{\alpha},$$

provided this formula renders the same real result for every infinitesimal α.

The use of the formula is demonstrated in the next few examples.

Slope of the tangent line example with a parabola

Example 1 *Find the slope of the tangent line to the curve* $f(x) = x^2$ *at* $x = 3$.

The formula is for the slope of the tangent line at $x = k$. We want the slope at $x = 3$. Therefore, we use 3 in place of k in the formula.

Solution We use the formula for the slope of the tangent line, with $k = 3$:

$$\text{slope} = \frac{f(3+\alpha) - f(3)}{\alpha} = \frac{(3+\alpha)^2 - 3^2}{\alpha}$$
$$= \frac{9 + 6\alpha + \alpha^2 - 9}{\alpha} = \frac{6\alpha + \alpha^2}{\alpha}$$
$$\approx \frac{6\alpha}{\alpha} = 6.$$

Because we are asked for the slope, the answer is a number (not an equation).

The slope of the tangent line is 6. ■

Be sure to check the wording of the question carefully to note whether the slope or the equation of the tangent line is requested. In example 1, the slope is requested, and the final form of the answer is a number. In example 2, the equation is requested, and the final form of the answer is an equation.

Example 2 *Find the equation of the tangent line to the curve* $f(x) = x^2$ *at* $x = 3$.

Solution This is identical to example 1 with one exception: the word *slope* has been replaced by the word *equation*. To find the equation of a tangent line, we begin by finding the slope. This was completed in example 1; the slope is 6. We also need the point. The x-coordinate is $x = 3$. We find the y-coordinate by evaluating the function: $f(3) = 3^2 = 9$. The point is therefore $(3, 9)$. Using the point-slope equation, we have

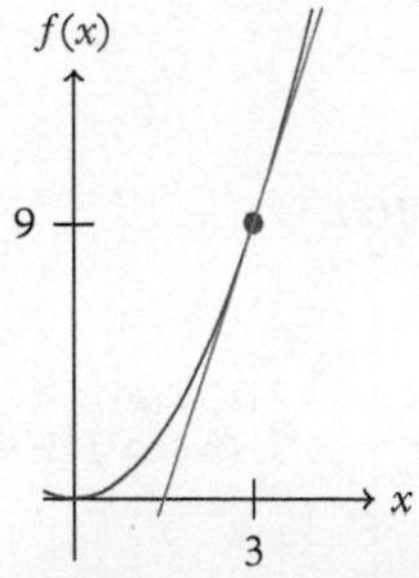

$$y - y_1 = m(x - x_1)$$
$$y - 9 = 6(x - 3)$$
$$y - 9 = 6x - 18$$
$$y = 6x - 9.$$

Figure 9 *The curve* $f(x) = x^2$ *with the tangent line at* $x = 3$

Because we are asked for the equation, the answer is an equation (not a number).

The equation of the tangent line is $y = 6x - 9$. The curve and the tangent line are graphed in figure 9. ■

Reading Exercise 30 *Find the slope of the tangent line to the curve $f(x) = x^2$ at $x = -1$.*

Slope of the tangent line example with a rational function

Example 3 *Find the slope of the tangent line to the curve $f(x) = \dfrac{x^2 - 7}{x + 2}$ at $x = 1$.*

Solution We use the formula for the slope of the tangent line, with $k = 1$:

$$\text{slope} = \frac{f(1+\alpha) - f(1)}{\alpha} = \frac{\frac{(1+\alpha)^2 - 7}{(1+\alpha)+2} - \frac{1^2 - 7}{1+2}}{\alpha}$$

$$= \frac{\frac{1 + 2\alpha + \alpha^2 - 7}{3+\alpha} - (-2)}{\alpha}.$$

CAUTION: MISTAKE TO AVOID
If we approximate now, we have

$$\approx \frac{\frac{-6}{3} - (-2)}{\alpha} = \frac{0}{\alpha},$$

which violates the approximation principle. More algebra must be done before approximating.

Compare the algebra of this example to that of section 1.4 example 5.

To rid ourselves of the compound fraction, we multiply the "big" fraction by $\frac{3+\alpha}{3+\alpha}$:

$$= \frac{\left(\frac{-6 + 2\alpha + \alpha^2}{3+\alpha} + 2\right)}{\alpha} \cdot \frac{(3+\alpha)}{(3+\alpha)}$$

$$= \frac{\frac{-6 + 2\alpha + \alpha^2}{\cancel{3+\alpha}} \cdot \cancel{(3+\alpha)} + 2(3+\alpha)}{\alpha \cdot (3+\alpha)}$$

$$= \frac{-6 + 2\alpha + \alpha^2 + 6 + 2\alpha}{3\alpha + \alpha^2} = \frac{4\alpha + \alpha^2}{3\alpha + \alpha^2}$$

$$\approx \frac{4\alpha}{3\alpha} = \frac{4}{3}.$$

The slope of the tangent line is $\frac{4}{3}$. ■

Because we are asked for the slope, the answer is a number (not an equation).

Slope of the tangent line example with a square root

Example 4 *Find the equation of the tangent line to the curve $f(x) = \sqrt{3x+1}$ at $x = 2$.*

Compare the algebra of this example to that of section 1.4 example 3.

Solution We begin by applying the formula for the slope of the tangent line, with $k = 2$:

$$\text{slope} = \frac{f(2+\alpha) - f(2)}{\alpha}$$

$$= \frac{\sqrt{3(2+\alpha)+1} - \sqrt{3(2)+1}}{\alpha} = \frac{\sqrt{7+3\alpha} - \sqrt{7}}{\alpha}.$$

CAUTION: MISTAKE TO AVOID
If we approximate now, we have

$$\approx \frac{\sqrt{7} - \sqrt{7}}{\alpha} = \frac{0}{\alpha},$$

which violates the approximation principle. More algebra must be done before approximating. This is true for every calculation for the slope of the tangent line. Approximating too early leads to a violation of the approximation principle.

If we were to approximate now, we violate the approximation principle. With square root expressions, we can avoid this error by multiplying the numerator and the denominator by the conjugate of the square root expression:

$$= \frac{\sqrt{7+3\alpha} - \sqrt{7}}{\alpha} \cdot \frac{\sqrt{7+3\alpha} + \sqrt{7}}{\sqrt{7+3\alpha} + \sqrt{7}}$$

$$= \frac{7 + 3\alpha - 7}{\alpha(\sqrt{7+3\alpha} + \sqrt{7})} = \frac{3\alpha}{\alpha(\sqrt{7+3\alpha} + \sqrt{7})} = \frac{3}{\sqrt{7+3\alpha} + \sqrt{7}}$$

$$\approx \frac{3}{\sqrt{7} + \sqrt{7}} = \frac{3}{2\sqrt{7}}.$$

Tip: do not multiply the α throughout the denominator. The α must eventually cancel, and this is easiest if the denominator remains in factored form.

The slope of the tangent line is $\frac{3}{2\sqrt{7}}$. Because we are asked for the equation of the tangent line, and not just the slope, we are not yet finished. We use the point-slope equation with slope $\frac{3}{2\sqrt{7}}$ and point $(2, f(2)) = (2, \sqrt{7})$,

Ans. to reading exercise 30
−2

$$y - \sqrt{7} = \frac{3}{2\sqrt{7}}(x - 2),$$

and we are finished. If the slope-intercept form is desired, we continue:

$$y - \sqrt{7} = \frac{3}{2\sqrt{7}}x - \frac{3}{\sqrt{7}}$$

$$y = \frac{3}{2\sqrt{7}}x - \frac{3}{\sqrt{7}} + \sqrt{7},$$

or we can also rationalize the denominator and collect like terms, ending with

$$y = \frac{3\sqrt{7}}{14}x + \frac{4\sqrt{7}}{7}.$$

■

Some computer algebra systems might factor a little and give the answer in the form $y = \frac{\sqrt{7}}{14}(3x + 8)$. There are many possible forms of the final answer that are equivalent algebraically, each with advantages and disadvantages. But that's algebra; the calculus was finished when the equation of the tangent line was found. Students should check with their instructor for the desired form of final answers.

Average velocity

Mile markers or kilometer markers on a highway are the same idea as numbers on a number line. Think of the beginning of the road (the physical beginning of the road or when mile markers start over at a state or national boundary) as the origin on the number line, and think of the mile markers as the numbers to the right of the origin. Some roads even use decimals, marking tenths. These markers can be used to calculate average speed.

In some countries "kilometer zero" is a reference point marked by a monument located in the capital city.

I confess to having used this method in an automobile with a broken speedometer.

Example 5 *At 2:11 p.m. a car enters a highway at mile marker 83. At 3:26 p.m. the car exits the highway at mile marker 172. What is the car's average speed?*

Solution Recalling the formula $d = r \cdot t$ for distance = rate × time, we wish to know the average rate of travel r. Then $r = \frac{d}{t}$, and

$$\begin{aligned}\text{average speed} &= \frac{\text{distance traveled}}{\text{time elapsed}}\\ &= \frac{172 - 83}{\text{3:26 p.m.} - \text{2:11 p.m.}} = \frac{89 \text{ miles}}{75 \text{ minutes}} = \frac{89 \text{ miles}}{\frac{75}{60} \text{ hours}}\\ &= 71.2 \text{mi/hr}.\end{aligned}$$

■

The units miles per hour are commonly abbreviated mph.
71.2 mph = 114.6 km/hr

What if the car in example 5 enters at mile marker 100 and exits at mile marker 11 instead? Using $\frac{11-100}{\text{3:26 p.m.}-\text{2:11 p.m.}}$ gives −71.2 mph. The negative does not mean the car's transmission is in reverse; the negative means the car is traveling west instead of east. The average speed is still 71.2 mph, but the average *velocity*, which indicates both speed and direction, is −71.2 mph (figure 10).

West and east are illustrative only. On east–west highways in the United States, the mile marker numbers increase as one travels east.

Taking the highway/number line analogy a little further, think of position on a number line as a function f of time t. For example 5, we have

$$f(\text{2:11 p.m.}) = \text{mile marker } 83,$$

$$f(\text{3:26 p.m.}) = \text{mile marker } 172.$$

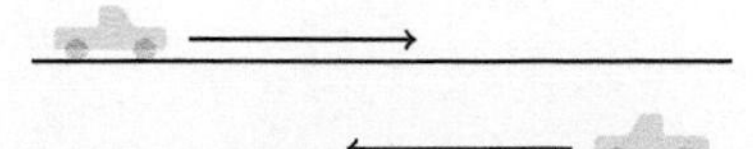

Figure 10 Positive velocity (top): the vehicle travels in the direction of increasing mile marker numbers, or to the right on the number line; negative velocity (bottom): the vehicle travels in the direction of decreasing mile marker numbers, or to the left on the number line

Then, with a = 2:11 p.m. and b = 3:26 p.m., the distance traveled is $f(b) - f(a)$, whereas the time elapsed is $b - a$. Our average velocity is $\frac{f(b) - f(a)}{b - a}$.

This is the formula for average velocity, not for instantaneous velocity.

AVERAGE VELOCITY

$$\begin{array}{c}\text{average velocity from} \\ t = a \text{ to } t = b\end{array} = \frac{f(b) - f(a)}{b - a}$$

The average velocity formula is really just the slope formula in disguise! It is still the difference in function values (y-coordinates) divided by the difference in inputs (x-coordinates). It is just $m = \frac{y_2 - y_1}{x_2 - x_1}$ rewritten.

Average velocity example

Instead of a vehicle moving on a highway, we can envision a particle moving on a number line. The two situations are identical; the same formulas are used.

The position function is also sometimes called the *equation of motion*.

Example 6 *A particle travels on a number line according to the position function $f(t) = t^2 - 2t$. Find the average velocity of the particle from $t = 1$ to $t = 3$. Units are meters and seconds.*

The alternative phrases "on the interval $[1, 3]$" or simply "on $[1, 3]$" have the same meaning as the phrase "from $t = 1$ to $t = 3$."

Solution With $a = 1$ and $b = 3$, we use the formula for average velocity:

$$\text{average velocity} = \frac{f(3) - f(1)}{3 - 1} = \frac{(3^2 - 2 \cdot 3) - (1^2 - 2 \cdot 1)}{2} = 2\text{m/s}. \blacksquare$$

Because average velocity is distance traveled divided by time elapsed, the units for velocity are units of distance divided by units of time.

Reading Exercise 31 *A particle travels on a number line according to the equation of motion $f(t) = 1 - t^2$. Units are millimeters and hours. Find the average velocity of the particle from $t = 1$ to $t = 3$.*

Instantaneous velocity

A vehicle's speedometer gives the speed at an instant, not the average speed over an interval of time. The two ideas are not identical. After all, saying that you have only averaged 50 mph on today's trip does not get you out of a ticket for going 80 mph when the radar gun measures your speed.

Example 7 *A particle travels on a number line. Its position at time t is given by $f(t) = t^2 - 2t$, with t in seconds and $f(t)$ in meters. What is the particle's velocity at time $t = 5$?*

The solution is given shortly, but first note that we are asked for velocity at an instant in time ("at time $t = 5$"), not for the average velocity over an interval of time. We face the same conundrum as when finding the slope of the tangent line. Just as the slope formula requires two points on the line, the average velocity formula requires two points in time. When faced with the same formula and the same problem, why not try the same solution? We scoot over an infinitesimal amount for our second point and we find the average velocity of the particle from $t = 5$ to $t = 5 + \alpha$.

The word *instantaneous* need not appear for the question to be about instantaneous velocity. A question about velocity at one instant in time is a question about instantaneous velocity.

Intuitively, this should work. The reason our average velocity over the course of an hour may be quite different from our velocity at one instant in time is that we have enough time for the instantaneous velocity to change. If we only have a time interval from $t = 5$ to $t = 5 + \alpha$, its length is infinitesimal and there is no time at all to change speed. Our instantaneous velocity should be essentially the same over the entire time interval, and therefore the average velocity should match our instantaneous velocity.

Our formula for velocity at time $t = k$ can therefore be found from the average velocity from $t = k$ to $t = k + \alpha$, or

Here we are using the average velocity formula with $a = k$ and $b = k + \alpha$.

$$\frac{f(k+\alpha) - f(k)}{k + \alpha - k} = \frac{f(k+\alpha) - f(k)}{\alpha},$$

which is identical to the formula for the slope of the tangent line!

INSTANTANEOUS VELOCITY

$$\text{(instantaneous) velocity at time } t=k = \frac{f(k+\alpha) - f(k)}{\alpha},$$

provided this formula renders the same real result for every infinitesimal α.

This is the formula for velocity at an instant in time, not the formula for average velocity.

We are now ready to return to the solution of the example.

Ans. to reading exercise 31
-4mm/hr

The calculation is the same as for finding the slope of the tangent line. See examples 3 and 4 for how to deal with rational functions and square roots, respectively.

Solution (example 7). Using the formula for instantaneous velocity, we have

$$\begin{aligned}\text{velocity} &= \frac{f(5+\alpha)-f(5)}{\alpha} = \frac{(5+\alpha)^2-2(5+\alpha)-(25-10)}{\alpha}\\ &= \frac{25+10\alpha+\alpha^2-10-2\alpha-15}{\alpha} = \frac{8\alpha+\alpha^2}{\alpha}\\ &\approx \frac{8\alpha}{\alpha} = 8\text{m/s}.\end{aligned}$$

Just as with average velocity, units are of the form unit of distance divided by unit of time.

The velocity of the particle at time $t = 5$ is 8m/s. ■

Reading Exercise 32 *A particle travels on a number line. Its position at time t is given by $f(t) = t^2$, with t in seconds and $f(t)$ in meters. What is the particle's velocity at time $t = -2$?*

Instantaneous rate of change

Instead of asking how fast an object is moving, we can ask how fast a quantity is changing.

Example 8 *The price of a gallon of ChemForm B this year is given by $f(t) = 0.005t^2 + 0.01t + 7.924$, where t is in days and $f(t)$ is in dollars. How fast is the price changing when $t = 60$?*

Before we can find the solution, we need to develop the solution method. To begin, what does it mean by "how fast is the price changing?" Think about the price of gasoline (example 8 uses ChemForm B instead of gasoline, but it's the same idea). If the price of gas goes up \$0.21 this week, this is a rate of change of \$0.03/day. "How fast the price is changing" means the rate at which the price changes over time. The units are units of price (dollars) divided by units of time (days).

How do we calculate the rate of change of the price of gasoline, \$0.03/day? We first find the change in price of \$0.21 for the week as

price at end of week – price at beginning of week.

To find the rate of change, we then divide by the amount of time. If the week in question was noon July 3 to noon July 10, we calculate the number of days by $10 - 3 = 7$, or

day at end of week – day at beginning of week.

Using a = noon July 3 and b = noon July 10, and by calling the price of gasoline $f(t)$, then $f(a)$ is the price of gasoline at noon on July 3 and

$f(b)$ is the price of gasoline at noon on July 10. The average rate of change of the price of gasoline over the time period is then

$$\frac{f(b)-f(a)}{b-a}.$$

It is the same formula as for average velocity. In fact, velocity can be considered the rate of change of position over time (how fast the position of the object is changing).

AVERAGE RATE OF CHANGE

$$\text{average rate of change from } t=a \text{ to } t=b = \frac{f(b)-f(a)}{b-a}$$

This is the formula for average rate of change, not for instantaneous rate of change.

Because the formula for average rate of change is the same as for the slope of the tangent line and the same as the average velocity, the exact same reasoning as before can be used to find the instantaneous rate of change. We scoot over an infinitesimal amount and use the formula for average rate of change over the time period from $t = k$ to $t = k + \alpha$. This gives an infinitesimal amount of time over which the instantaneous rate cannot vary meaningfully and should be the same as the average rate.

Ans. to reading exercise 32
-4m/s

INSTANTANEOUS RATE OF CHANGE

$$\text{(instantaneous) rate of change at time } t = k = \frac{f(k+\alpha)-f(k)}{\alpha},$$

provided this formula renders the same real result for every infinitesimal α.

This is the formula for rate of change at an instant in time, not the formula for average rate of change.

Now we are ready for the solution to example 8.

Solution (example 8). Using the formula for instantaneous rate of change, we have

$$\begin{aligned}\text{rate of change} &= \frac{f(60+\alpha)-f(60)}{\alpha}\\ &= \frac{0.005(60+\alpha)^2+0.01(60+\alpha)+7.924-(0.005(60)^2+0.01(60)+7.924)}{\alpha}\\ &= \cdots = \frac{0.6\alpha+0.005\alpha^2+0.01\alpha}{\alpha}\\ &\approx \frac{0.61\alpha}{\alpha} = \$0.61/\text{day}.\end{aligned}$$

An alternative expression of the answer is 61¢/day. Note that 61¢/day and 0.61¢/day are not the same thing, so only the former is correct.

The rate of change of price of ChemForm B at time $t = 60$ is \$0.61/day. ■

Because a very large number of applications involve quantities changing over the course of time, the idea of the rate of change is very important. For this reason we study this idea extensively in the next chapter.

Exercises 1.8

1–6. Rapid response: is the line tangent to the curve at $x = k$?

1.

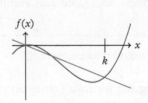

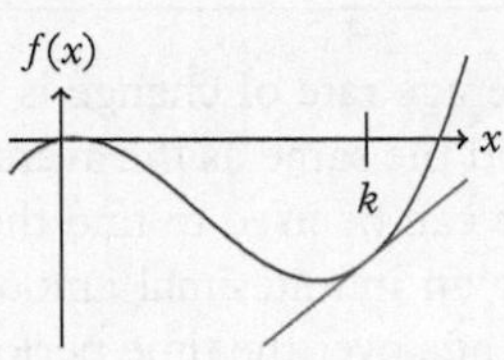

2.

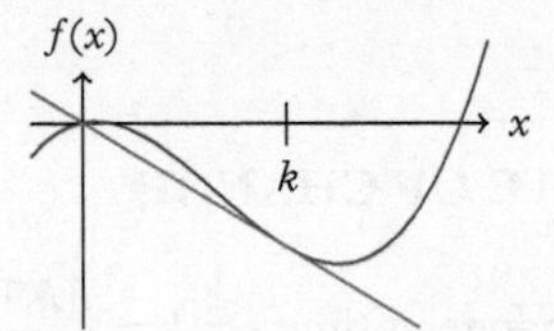

3.

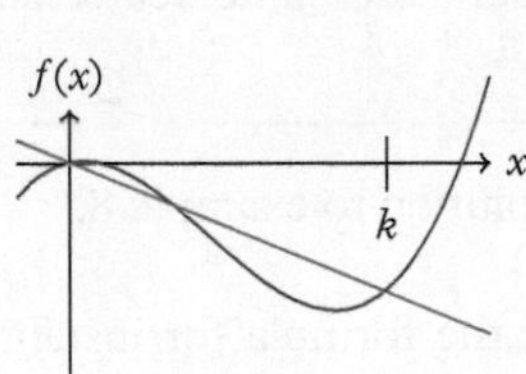

4.

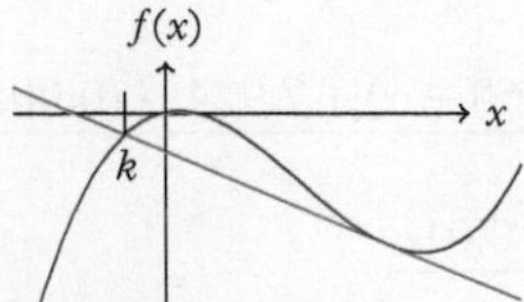

5.

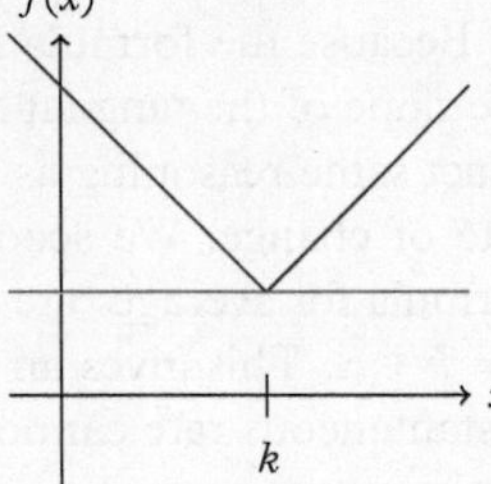

6.

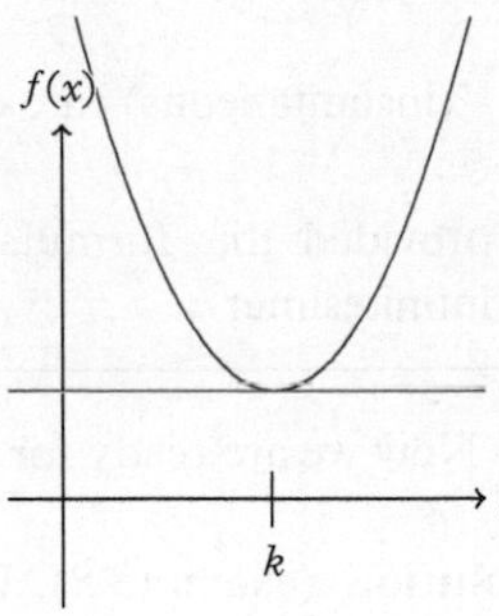

f(x)

x

k

7. Find the equation of the tangent line to the curve $f(x) = x^2 - 2$ at $x = 1$, then draw the curve and the tangent line together as in figure 5.

8. Find the equation of the tangent line to the curve $f(x) = 2x - \frac{1}{2}x^2$ at $x = 3$, then draw the curve and the tangent line together as in figure 5.

9. Find the slope of the tangent line to the curve $y = \dfrac{4}{x-1}$ at $x = 2$.

10. Find the slope of the tangent line to the curve $y = x^3 - 3x^2$ at $x = 1$.

11–20. Find the equation of the tangent line to the given curve at the given point.

11. $y = 2x^2$ at $(-1, 2)$
12. $y = x^2 + 1$ at $(2, 5)$
13. $y = \dfrac{3x-1}{2}$ at $(3, 4)$
14. $y = \sqrt{x}$ at $(4, 2)$
15. $y = \dfrac{6}{x}$ at $(3, 2)$
16. $y = \dfrac{7}{2x+1}$ at $(0, 7)$
17. $y = x^3 + 7$ at $(-1, 6)$
18. $y = \dfrac{1}{5-x}$ at $(4, 1)$
19. $y = \sqrt{x}$ at $(1, 1)$
20. $y = x^2 - 4x$ at $(-1, 5)$

21. Find the slope of the tangent line to the curve $f(x) = \dfrac{3x-5}{x+1}$ at $x = 0$.

22. Find the slope of the tangent line to the curve $y = \sqrt{5 - x^2}$ at $x = 1$.

23. Find the slope of the tangent line to the curve $y = \sqrt{2 - 7x}$ at $x = -1$.

24. Find the slope of the tangent line to the curve $f(x) = \dfrac{5x-1}{2x+4}$ at $(1, \frac{2}{3})$.

25. A particle moves on a number line according to the position function $f(t) = t^3$. Find its average velocity from $t = -2$ to $t = 3$. Units are meters and seconds.

26. A particle moves on a number line according to the position function $f(t) = t^3$. Find its average velocity from $t = -1$ to $t = 2$. Units are meters and seconds.

27. A particle moves on a number line according to the position function $f(t) = \sqrt{1+t}$. Find its average velocity from $t = 0$ to $t = 8$. Units are centimeters and hours.

28. A particle moves on a number line according to the position function $f(t) = 2\sqrt{t+4}$. Find its average velocity from $t = -3$ to $t = 5$. Units are centimeters and hours.

Some exercises in this section are taken from or slightly modified from the classic text *Elements of the Differential and Integral Calculus* by Granville, Smith, and Longley, from either the 1904 (Granville) or 1941 (Granville, Smith, Longley) edition. In later sections this is denoted as "from GSL" or "(GSL)."

For all section 1.8 exercises, it is required that you use the formulas given in this section.

Be sure to choose carefully between the formulas for average velocity and instantaneous velocity.

29. An arrow is shot into the air. Its height in feet t seconds after shooting is given by $y = 70t - 16t^2$. Find the average velocity of the arrow for the time period $[1.5, 2]$.

30. A model rocket is fired into the air. Its height in feet t seconds after launch is given by $y = 100t - 16t^2$. Find the average velocity of the rocket for the time period $[2, 2.5]$.

31. A particle moves on a number line according to the position function $f(t) = \sqrt{1 + t}$. Find its velocity at time $t = 2$. Units are centimeters and hours.

32. A particle moves on a number line according to the position function $f(t) = 2\sqrt{t + 4}$. Find its velocity at time $t = 0$. Units are centimeters and hours.

33. An arrow is shot into the air. Its height in feet t seconds after shooting is given by $y = 70t - 16t^2$. Find the velocity of the arrow 2 seconds after shooting.

34. A model rocket is fired into the air. Its height in feet t seconds after launch is given by $y = 100t - 16t^2$. Find the velocity of the rocket 1 second after launch.

35. Find the average rate of change of $f(x) = \sqrt{x}$ on the interval $[1, 4]$.

36. Find the average rate of change of $f(x) = \dfrac{1}{x^2}$ on the interval $[1, 4]$.

37. Find the average rate of change of $f(x) = \dfrac{1}{\sqrt{x + 3}}$ on the interval $[-2, 6]$.

38. Find the average rate of change of $f(x) = x^2 + 3$ on the interval $[1, 5]$.

39. P. K.'s weight on January 14 is 104 kg, but after a new diet and exercise regimen P. K.'s weight on January 30 is 98 kg. What is P. K.'s average weight loss?

40. After starting a vehicle's air conditioner, the temperature in the vehicle cools from 49 °C to 29 °C in a period of 10 minutes. What is the average rate of change of temperature in the vehicle?

41. The manufacturing cost of a gram of Compound C this year is given by $f(t) = 13.425 - 0.01t + 0.0002t^2$, where t is in days and $f(t)$ is in Mexican pesos (symbol \$). How fast is the price changing when $t = 100$?

42. The price of a gallon of ChemForm B this year is given by $f(t) = 0.005t^2 + 0.01t + 24.49$, where t is in days and $f(t)$ is in dollars. How fast is the price changing when $t = 20$?

43. The time it takes a worker to assemble a widget after t days of experience is $3 + \frac{20}{\sqrt{t}}$ minutes. At what rate is the assembly time changing after 4 days?

44. The time it takes a worker to assemble a widget after t days of experience is $3 + \frac{20}{\sqrt{t}}$ minutes. At what rate is the assembly time changing after 25 days?

45. Challenge: The time it takes a worker to assemble a widget after t days of experience is $12 + \frac{1}{\sqrt[3]{4x}}$ minutes. At what rate is the assembly time changing after 16 days? *Hint:* use the difference of cubes formula from algebra to help you modify the square root procedure to fit cube roots.

Chapter II

Derivatives

The Derivative

2.1

The formula

$$\frac{f(k+\alpha)-f(k)}{\alpha}$$

is used for the slope of the tangent line, instantaneous velocity, and instantaneous rate of change (see section 1.8). Because it is so versatile, the formula deserves its own name and deserves to be studied. It is called the *derivative* and we study it in this chapter.

The definition of derivative

The derivative is defined in a similar manner as limit and continuity.

Definition 1 THE DERIVATIVE *Let f be a function and let k be a real number in the domain of f. If*

$$\frac{f(k+\alpha)-f(k)}{\alpha}$$

is defined and renders the same real result L for every infinitesimal α, then we write $f'(k) = L$ and call $f'(k)$ the **derivative** *of f at k. When L is a real number (not ∞ or $-\infty$), then we say that f is* **differentiable** *at k.*

The apostrophe symbol in $f'(k)$ is read "prime." We read "$f'(k)$" as "f prime of k."

For the purpose of calculations we write

$$f'(k) = \frac{f(k+\alpha)-f(k)}{\alpha},$$

DERIVATIVE FORMULA

$$f'(k) = \frac{f(k+\alpha)-f(k)}{\alpha}$$

with the understanding that α is an arbitrary infinitesimal that can be either positive or negative and that we must render the same real result for every α.

Example 1 *Let $f(x) = x^2$. Find $f'(5)$.*

Solution We use the derivative formula with $k = 5$:

$$f'(5) = \frac{f(5+\alpha) - f(5)}{\alpha} = \frac{(5+\alpha)^2 - 5^2}{\alpha}$$
$$= \frac{25 + 10\alpha + \alpha^2 - 25}{\alpha} = \frac{10\alpha + \alpha^2}{\alpha}$$
$$\approx \frac{10\alpha}{\alpha} = 10.$$

This is the same type of calculation as in section 1.8. The only thing new is the notation for the derivative.

The value of the derivative is 10, and we write $f'(5) = 10$. ■

Reading Exercise 1 *Let $f(x) = 5x + 1$. Find $f'(3)$.*

INTERPRETATIONS OF DERIVATIVE

- The slope of the tangent line to the curve $y = f(x)$ at $x = k$ is $f'(k)$.
- Given position function $f(t)$, the velocity at time $t = k$ is $f'(k)$.
- The rate of change of f at $x = k$ is $f'(k)$.

Interpretations of the answer to example 1 include the following:

- The slope of the tangent line to the curve $f(x) = x^2$ at $x = 5$ is 10.
- If a particle's position on a number line at time x is $f(x) = x^2$, then its velocity at time $x = 5$ is 10.
- The rate of change of $f(x) = x^2$ at $x = 5$ is 10.

If we repeat example 1 to find the value of the derivative of $f(x) = x^2$ at several different values of x, we make the same type of calculation over and over again. Instead, we can calculate the value of the derivative at a generic value k and then replace k with whatever number is desired.

Example 2 *Let $f(x) = x^2$. Find $f'(k)$ (assume k is a real number). Then, determine the values of $f'(5)$, $f'(3)$, and $f'(-4)$.*

Solution We use the derivative formula:

$$f'(k) = \frac{f(k+\alpha) - f(k)}{\alpha} = \frac{(k+\alpha)^2 - k^2}{\alpha}$$
$$= \frac{k^2 + 2k\alpha + \alpha^2 - k^2}{\alpha} = \frac{2k\alpha + \alpha^2}{\alpha}$$
$$\approx \frac{2k\alpha}{\alpha} = 2k.$$

Because k is a real number, the number $2k\alpha$ is on the α level and we may drop the lower-level number α^2 when approximating the numerator of the expression.

If k is zero, this violates the approximation principle, but then the expression is $\frac{\alpha^2}{\alpha} = \alpha \doteq 0 = 2k$. We often ignore such violations of the approximation principle because they render the same real result anyway.

Since $f'(k) = 2k$, we have

$$f'(5) = 2(5) = 10$$
$$f'(3) = 2(3) = 6$$
$$f'(-4) = 2(-4) = -8.$$ ■

Writing $f'(k) = 2k$ is essentially the same as writing $f'(x) = 2x$. We can think of the derivative as being its own function! The derivative formula can be used with the variable x in place of k to emphasize this fact. We still keep the understanding that α is an arbitrary infinitesimal and that we must render the same real result for every such α.

DERIVATIVE FORMULA (FUNCTION VERSION)

$f'(x) = \frac{f(x+\alpha)-f(x)}{\alpha}$

Example 3 *Find $f'(x)$ for $f(x) = x^3$.*

Solution Using the derivative formula (function version),

$$f'(x) = \frac{f(x+\alpha)-f(x)}{\alpha} = \frac{(x+\alpha)^3 - x^3}{\alpha}$$
$$= \frac{x^3 + 3x^2\alpha + 3x\alpha^2 + \alpha^3 - x^3}{\alpha}$$
$$= \frac{3x^2\alpha + 3x\alpha^2 + \alpha^3}{\alpha}$$
$$\approx \frac{3x^2\alpha}{\alpha} = 3x^2.$$

■

The formula for cubing a binomial is $(a+b)^3 = a^3 + 3a^2b + 3ab^2 + b^3$.

The final answer should not contain any infinitesimals (α should not appear in the answer), but the variable x is nearly always present in the expression for $f'(x)$.

Reading Exercise 2 *Find $f'(x)$ for $f(x) = 4x^2$.*

The derivative function can be used to find slopes, velocities, and rates of change.

Ans to reading exercise 1

$f'(3) = \frac{f(3+\alpha) - f(3)}{\alpha} = \cdots = 5$

Example 4 *Find the slope of the tangent line to the curve $f(x) = x^3$ at (a) $x = -3$, (b) $x = 0$, and (c) $x = 15$.*

Solution We already know from example 3 that $f'(x) = 3x^2$. One interpretation of the derivative is that it represents the slope of the tangent line.

(a) The slope of the tangent line at $x = -3$ is $f'(-3) = 3(-3)^2 = 27$.

(b) The slope of the tangent line at $x = 0$ is $f'(0) = 3(0)^2 = 0$.

(c)The slope of the tangent line at $x = 15$ is $f'(15) = 3(15)^2 = 675$. ■

Differentiability implies continuity

One of the phrases defined in definition 1 is "differentiable at $x = k$," which means the derivative exists at $x = k$. This requires two things: first, that the derivative formula yields the same real result for every infinitesimal α; and second, that it does not render ∞ or $-\infty$, but renders a real number. This tells the story algebraically but not visually.

Definition and theorem numbers start over in each chapter. If no chapter number is given, the definition or theorem that is referenced is to be found in the current chapter.

Our task for most of the remainder of this section is to determine what differentiability means on the graph of a function.

If f is differentiable at $x = k$, then the function has a tangent line at $x = k$, as in figure 1. The function must then have a point at $x = k$, meaning that $f(k)$ exists. We have seen the requirement that $f(k)$ exists before—namely, with continuity. Perhaps there is a connection.

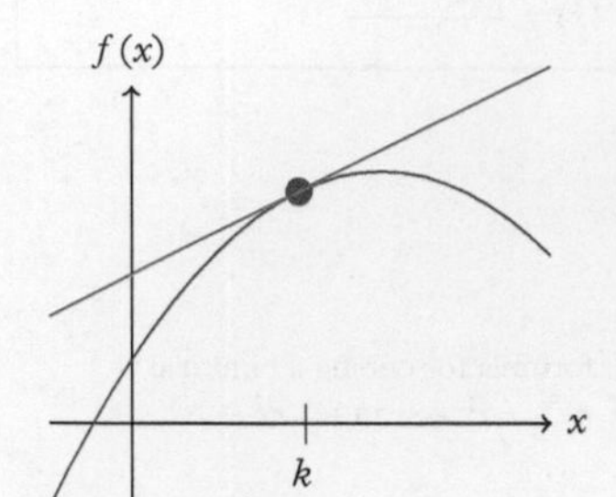

Figure 1 *A curve $y = f(x)$ with tangent line at $x = k$. To have a tangent line, there must be a point of tangency; $f(k)$ must exist*

Pictures of discontinuities can help us think about differentiability. In figure 2, the function is not defined at the discontinuities and therefore is not differentiable there.

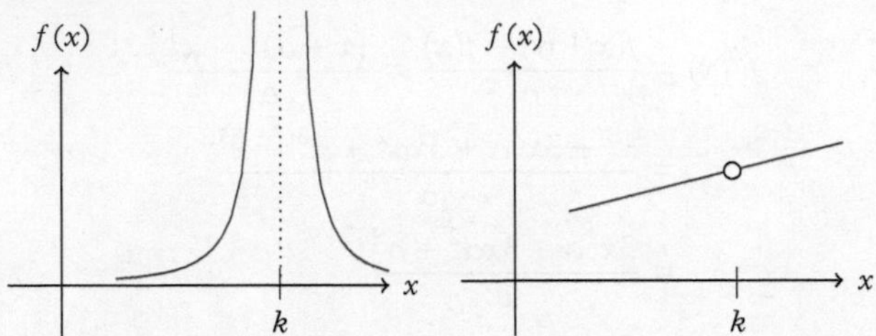

Figure 2 *Functions with discontinuities at $x = k$. If $f(k)$ does not exist, then there can not be a tangent line at $x = k$*

We studied the slope of the tangent line in section 1.8.

Ans to reading exercise 2
$f'(x) = 8x$

What if the value of the function exists at $x = k$, but the function is still discontinuous there? In this case, imagine what the tangent line must look like. The slope of the tangent line is found as the slope of the line through the two points $(k, f(k))$ and $(k + \alpha, f(k + \alpha))$ for an infinitesimal α. The second point is located on the curve at a point with the x-coordinate essentially the same as $x = k$. Such a line is drawn for various possibilities in figure 3.For an appropriately chosen value of α, the result in each graph is a vertical line, and the derivative

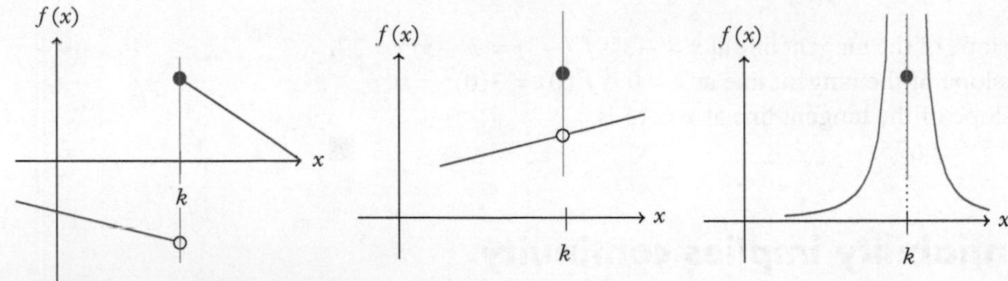

Figure 3 *Functions with discontinuities at $x = k$. If $f(k)$ exists, then trying to draw a tangent line at $x = k$ gives a vertical result for at least some values of α*

formula renders either ∞ or $-\infty$ for that value of α. The function is not differentiable at $x = k$. Although the graphs in figure 3 do not constitute a proof, it is reasonable to conjecture that if a function is not continuous at $x = k$, then it is not differentiable there either.

Returning to an algebraic perspective, as stated earlier, f differentiable at $x = k$ means that $\frac{f(k+\alpha)-f(k)}{\alpha}$ renders the real number $f'(k)$ for every infinitesimal α. As long as $f'(k) \neq 0$, this is the same as

$$\frac{f(k+\alpha)-f(k)}{\alpha} \approx f'(k).$$

Rearranging to solve for $f(k+\alpha)$ gives

$$f(k+\alpha) - f(k) \approx \alpha f'(k)$$
$$f(k+\alpha) \approx \underbrace{f(k)}_{\text{real}} + \underbrace{\alpha f'(k)}_{\text{infinitesimal}} \approx f(k).$$

If $f(k) = 0$, then the quantity is infinitesimal and renders zero, which is the same as $f(k)$.

Then, for every infinitesimal α, the quantity $f(k+\alpha)$ renders the real result $f(k)$, so f is continuous at $x = k$.

Theorem 1 DIFFERENTIABILITY IMPLIES CONTINUITY *If f is differentiable at $x = k$, then f is continuous at $x = k$.*

A more formal proof of theorem 1, with all the details, is given at the end of this section.

Another fact from the previous calculation that is very useful to us is that $f(k+\alpha) \approx f(k) + \alpha f'(k)$. This fact is called the *local linearity formula*; the reason for the name is made clear in section 2.3.

LOCAL LINEARITY FORMULA

$f(k+\alpha) \approx f(k) + \alpha f'(k)$

The local linearity formula is part of the local linearity theorem, which gives details regarding when the formula is valid. The theorem and its proof are at the end of this section.

The converse of theorem 1 is not true. A function can be continuous but not differentiable. An example of how this can happen, and what it means on the graph of a function, is next.

Differentiability, corners, and smoothness

Example 5 *Let $f(x) = |x|$. Find $f'(0)$.*

See section 1.4 example 7 for an expanded discussion of why we are stuck with the expression $|\alpha|$.

Solution We use the derivative formula with $k = 0$:

$$f'(0) = \frac{f(0+\alpha)-f(0)}{\alpha} = \frac{|\alpha|-|0|}{\alpha} = \frac{|\alpha|}{\alpha}$$

For the derivative to exist, the derivative formula must render the same real result for any infinitesimal we use, including ω, $-\omega$, $\omega^3 - \sqrt{\omega}$, or any other infinitesimal.

The quantities $\frac{f(k+\omega)-f(k)}{\omega}$ and $\frac{f(k-\omega)-f(k)}{-\omega}$ are one-sided derivatives, the derivatives from the right and left, respectively. We do not study one-sided derivatives in this book.

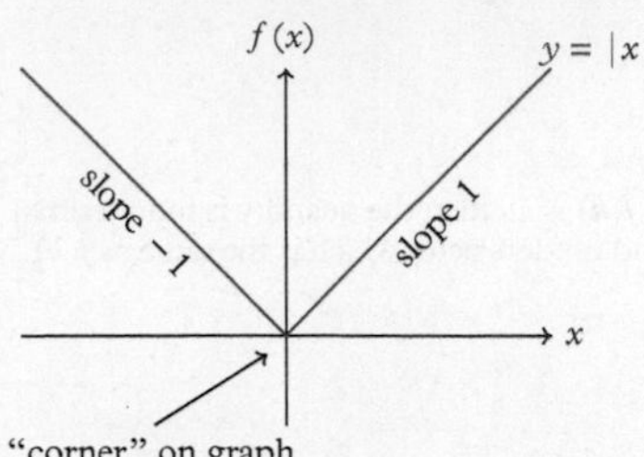

Figure 4 *The graph of $f(x) = |x|$, which has a corner at $x = 0$*
A function is continuous at a corner, but it is not differentiable there.

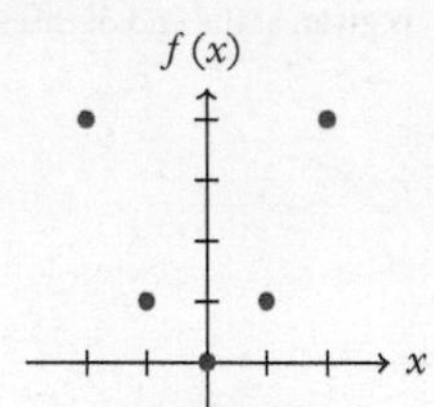

CAUTION: MISTAKE TO AVOID

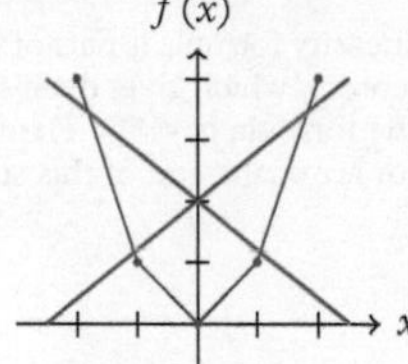

This is the correct version.

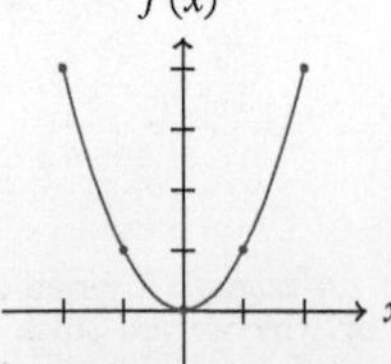

Figure 5 *(top) Plotting points for the graph of the function $f(x) = x^2$. (middle) Straight line segments connecting the plotted points result in corners, which cannot appear on the graph of a differentiable function; the graph is incorrect. (bottom) A smooth curve through plotted points on the graph of $f(x) = x^2$ is required*

and we are stuck. Because α may be either positive or negative, we do not know how to take its absolute value. When we encountered this same issue with limits, we tried using ω and $-\omega$ to help us. We can try it here as well.

Using the derivative formula with the positive infinitesimal ω in place of α gives

$$f'(0) = \frac{f(0+\omega) - f(0)}{\omega} = \frac{|\omega| - |0|}{\omega} = \frac{|\omega|}{\omega} = \frac{\omega}{\omega} = 1.$$

We are not stuck with $|\omega|$ because we know that ω is positive.

Using the derivative formula with the negative infinitesimal $-\omega$ in place of α gives

$$f'(0) = \frac{f(0+(-\omega)) - f(0)}{-\omega} = \frac{|-\omega| - |0|}{-\omega} = \frac{|-\omega|}{-\omega} = \frac{\omega}{-\omega} = -1.$$

We are not stuck with $|-\omega|$ because we know that ω is positive.

Remember that in order for the derivative to exist, we must render the same real result for every infinitesimal α, but here we rendered different results when we used the infinitesimals ω and $-\omega$. The derivative does not exist (DNE); $f'(0)$ DNE. The function f is not differentiable at $x = 0$. ■

The results of example 5 are easily explained from the graph of $y = |x|$ (figure 4). When x is positive, $y = |x|$ is the same as $y = x$, which is a line with slope 1. When x is negative, $y = |x|$ is the same as $y = -x$, which is a line with slope -1. This is exactly what we calculated for the slope near zero. Using a positive infinitesimal, the slope is 1; using a negative infinitesimal, the slope is -1. Where the two lines meet at $(0, 0)$ there is a "corner." At a corner in a graph, the slope on one side of the corner appears to be different from the slope on the other side of the corner, and the function is not differentiable there.

Because a corner in a graph indicates a place where the function is not differentiable, then the graph of a function that is differentiable everywhere must have no corners. We say that such a function is *smooth.* At the beginning of section 1.6 we examined the connect-the-dot game with the function $f(x) = x^2$, and saw that the instructions in an algebra course were to connect the points, not by straight line segments, but by making a smooth, connected curve. In example 2, we saw that if $f(x) = x^2$, then $f'(x) = 2x$, which is defined everywhere; since $f(x) = x^2$ is differentiable everywhere, then its graph must have no corners. Differentiability is the justification for the instructions you have followed since algebra to draw most functions in a smooth manner. See figure 5.

Vertical tangents

There is yet one more feature on a graph that can indicate nondifferentiability.

Example 6 *Let* $f(x) = x^{1/3}$. *Find* $f'(0)$.

Solution We use the derivative formula with $k = 0$:

$$f'(0) = \frac{f(0+\alpha) - f(0)}{\alpha} = \frac{\alpha^{1/3} - 0^{1/3}}{\alpha}$$
$$= \frac{\alpha^{1/3}}{\alpha} = \frac{1}{\alpha^{2/3}} = A^{2/3}$$
$$\doteq \infty.$$

We are not stuck at $A^{2/3}$ because of the even numerator in the exponent; $A^{2/3} = (A^2)^{1/3}$, and A^2 is positive. We can render ∞. Because we rendered ∞ rather than a real number, $f'(0)$ does not exist. ■

Regardless of whether we are stuck, the value is infinite and not a real number. We conclude that $f'(0)$ does not exist.

Since $f'(0) = \infty$, it seems the slope of the tangent line at $x = 0$ is positive infinite. This is true, as verified by the graph; see figure 6.

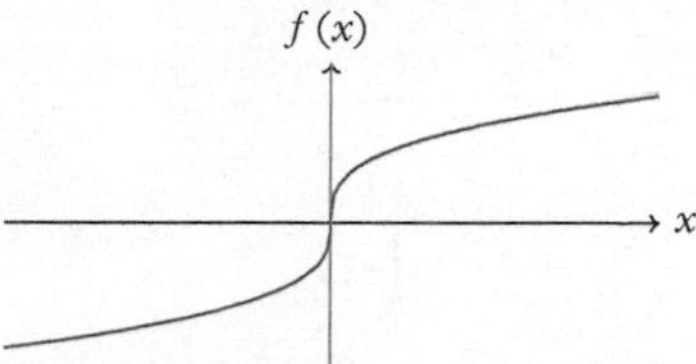

Figure 6 *The graph of* $y = x^{1/3}$ *(blue), with a vertical tangent line (orange) at* $x = 0$

The procedure for determining the equation of the tangent line to the curve $y = x^{1/3}$ at $x = 0$ is not the same as before. Because the equation of a vertical line is of the form $x = k$, we cannot use the point-slope equation of the line. Instead, we simply use the x-coordinate at which we are finding the derivative; the equation of the tangent line is $x = 0$.

If the derivative renders an infinite value, the tangent line is vertical. Figure 7 shows four of the possibilities.

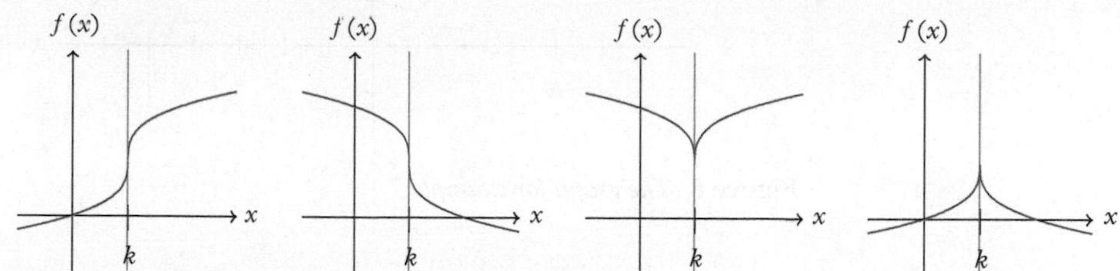

Figure 7 *Four scenarios yielding vertical tangent lines to* $y = f(x)$ *at* $x = k$: *(far left)* $f'(k) = \infty$, *(middle left)* $f'(k) = -\infty$, *(middle right)* $f'(k) = A$, *(far right)* $f'(k) = -A$

The two graphs on the right contain a *cusp*, where there is a corner with a vertical tangent line. The cusp is a point on the graph; the function is defined there. Because of the manner in which calculators and computer software calculate and display graphs of functions, it is sometimes difficult to distinguish cusps from vertical asymptotes using only a graph, but calculus can help you determine the difference!

There are other types of cusps in addition to the ones shown here; not all cusps involve vertical tangents.

Recognizing nondifferentiability visually

We have now seen three different situations in which a function is nondifferentiable at $x = k$.

RECOGNIZING NONDIFFERENTIABILITY

- If f is not continuous at $x = k$, then f is not differentiable at $x = k$.
- If f has a corner at $x = k$, then f is not differentiable at $x = k$.
- If f has a vertical tangent at $x = k$, then f is not differentiable at $x = k$.

Example 7 *Use the graph of $y = f(x)$ in figure 8 to determine where f is not differentiable and why.*

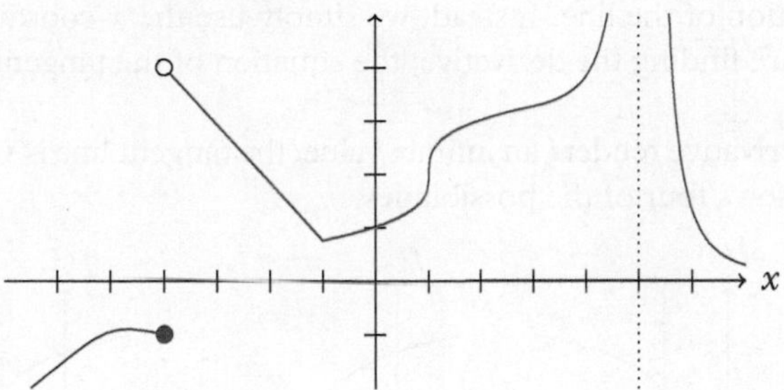

Figure 8 *The graph for example 7*

Solution We look for discontinuities, corners, and vertical tangents in the graph. Just as with discontinuities, we record their locations by their x-coordinates.

f is not differentiable at $x = -4$ (jump discontinuity)
f is not differentiable at $x = -1$ (corner)

f is not differentiable at $x = 1$ (vertical tangent)
f is not differentiable at $x = 5$ (infinite discontinuity) ■

It is sometimes difficult to be sure that a tangent line really is vertical. For exercises of this type, assume that if the graph looks very steep, the tangent line is vertical.

Reading Exercise 3 *Use the graph of $y = f(x)$ in figure 9 to determine where f is not differentiable and why.*

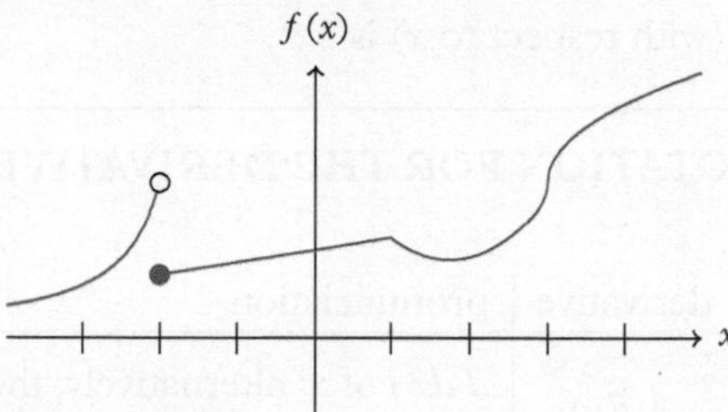

Figure 9 *The graph for reading exercise 3*

Notation for derivatives

Because the derivative has many applications in many different settings, various notations have arisen and remain in current use. The prime notation used earlier was introduced by Joseph Louis Lagrange. We know from example 3 that if $f(x) = x^3$, then $f'(x) = 3x^2$. This can also be expressed as "the derivative of $y = x^3$ is $y' = 3x^2$."

Joseph Louis Lagrange 1736–1813
http://www-history.mcs.st-andrews.ac.uk/Biographies/Lagrange.html

PRIME NOTATION FOR THE DERIVATIVE

function	derivative	pronunciation
$f(x)$	$f'(x)$	f prime of x
y	y'	y prime

Dot notation is similar and is often used in physics and engineering. We write "the derivative of $y = x^3$ is $\dot{y} = 3x^2$."

DOT NOTATION FOR THE DERIVATIVE

function	derivative	pronunciation
$f(x)$	$\dot{f}(x)$	f dot of x
y	$\dot{y}$	y dot

Leibniz is widely considered to be one of the cofounders of the subject of calculus.

Gottfried Wilhelm von Leibniz 1646–1716
http://www-history.mcs.st-andrews.ac.uk/Biographies/Leibniz.html

Leibniz notation is based upon the idea of slope. The slope formula is the change in y divided by the change in x, sometimes denoted $\frac{\Delta y}{\Delta x}$, where the uppercase Greek letter delta (Δ) represents change. Using d instead of Δ, the derivative is represented by $\frac{dy}{dx}$. Further interpretation of these symbols is given in section 2.3. We write "the derivative of $y = x^3$ is $\frac{dy}{dx} = 3x^2$." We can also write $\frac{d}{dx}x^3 = 3x^2$, which is read "the derivative of x^3 (with respect to x) is $3x^2$."

Leibniz notation is also called *differential notation*. An explanation of this term is provided in section 2.3.

LEIBNIZ NOTATION FOR THE DERIVATIVE

function	derivative	pronunciation
$f(x)$	$\frac{d}{dx}f(x)$	*d dx f* of *x*; alternatively, the derivative with respect to *x* of *f* of *x*
$f(x)$	$\frac{df}{dx}$	*df dx*; alternatively, the derivative of *f* with respect to *x*
y	$\frac{dy}{dx}$	*dy dx*; alternatively, the derivative of *y* with respect to *x*

To find the slope of the tangent line to $y = f(x)$ at $x = 2$, we find $f'(2)$ (prime notation) or $\dot{y}(2)$ (dot notation) or $\left.\frac{dy}{dx}\right|_{x=2}$ (Leibniz notation). The symbol $|$ is an *evaluation symbol*; $|_{x=2}$ is read "evaluated at x equals 2."

NOTATION FOR EVALUATING THE DERIVATIVE

The derivative of $y = f(x)$ at $x = a$ is written as:

prime notation	$f'(a)$
dot notation	$\dot{f}(a)$
Leibniz notation	$\left.\frac{dy}{dx}\right|_{x=a}$

Ans. to reading exercise 3
f is not differentiable at $x = -2$ (jump discontinuity), f is not differentiable at $x = 1$ (corner), f is not differentiable at $x = 3$ (vertical tangent)

In this textbook, prime and Leibniz notation are used primarily.

Proofs of theorems

The local linearity formula is part of the local linearity theorem.

Theorem 2 LOCAL LINEARITY THEOREM *Let f be differentiable at $x = k$ and let α be an infinitesimal.*

(1) If not both $f(k) = 0$ *and* $f'(k) = 0$*, then* $f(k+\alpha) \approx f(k) + \alpha f'(k)$.

(2) If $f(k) = 0 = f'(k)$*, then* $f(k+\alpha)$ *is zero or is on a lower level than* α.

Proof. (1) Case 1: $f'(k) \neq 0$.

Since f is differentiable at k, $\frac{f(k+\alpha)-f(k)}{\alpha} \doteq f'(k)$. Since $f'(k) \neq 0$ is real, $\frac{f(k+\alpha)-f(k)}{\alpha} \approx f'(k)$, and in fact $\frac{f(k+\alpha)-f(k)}{\alpha} = f'(k) + \beta$, where β is zero or infinitesimal. Then, $f(k+\alpha) - f(k) = \alpha f'(k) + \alpha\beta$, and $f(k+\alpha) = f(k) + \alpha f'(k) + \alpha\beta$. Because $\alpha\beta$ is either zero or is on a lower level than $\alpha f'(k)$, this term can be discarded and $f(k+\alpha) \approx f(k) + \alpha f'(k)$.

(1) Case 2: $f(k) \neq 0, f'(k) = 0$.

Then, since $\frac{f(k+\alpha)-f(k)}{\alpha} \doteq f'(k) = 0$, $\frac{f(k+\alpha)-f(k)}{\alpha} = \beta$, where β is zero or infinitesimal. Then, $f(k+\alpha) = f(k) + \alpha\beta \approx f(k) = f(k) + \alpha f'(k)$.

(2) Now suppose $f'(k) = f(k) = 0$. Then, $\frac{f(k+\alpha)-f(k)}{\alpha} = \frac{f(k+\alpha)}{\alpha} \doteq 0$, so $\frac{f(k+\alpha)}{\alpha}$ is zero or infinitesimal. Thus, either $f(k+\alpha) = 0$ or $f(k+\alpha)$ is on a lower level than α. ∎

It remains to furnish a complete proof of theorem 1. Let f be a function that is differentiable at $x = k$. We wish to prove that f is continuous at $x = k$.

Proof. (theorem 1) Case (1): $f(k) \neq 0$. By the local linearity theorem part (1), $f(k+\alpha) \approx f(k) + \alpha f'(k) \approx f(k)$. By definition, f is continuous at $x = k$.

Case (2): $f(k) = 0, f'(k) \neq 0$. By the local linearity theorem part (1), $f(k+\alpha) \approx f(k) + \alpha f'(k) = \alpha f'(k) \doteq 0 = f(k)$. By definition, f is continuous at $x = k$.

Case (3): $f(k) = 0, f'(k) = 0$. By the local linearity theorem part (2), $f(k+\alpha) \doteq 0 = f(k)$. By definition, f is continuous at $x = k$. ∎

Formal proofs that involve the local linearity formula $f(k+\alpha) \approx f(k) + \alpha f'(k)$ nearly always require the use of both parts of the local linearity theorem. In this text, results are usually derived using the local linearity formula only and the rest of the details are left unstated.

Exercises 2.1

1–6. Rapid response: does the given notation represent a number, a function, or neither?

1. $f'(2)$
2. $\left.\frac{dy}{dx}\right|_{x=2}$
3. $f'(x)$
4. $\frac{dy}{dx}$
5. $\frac{d}{dx}$
6. $\dot{y}(2)$

7. Use the following graph of $y = f(x)$ to determine where f is not differentiable and why.

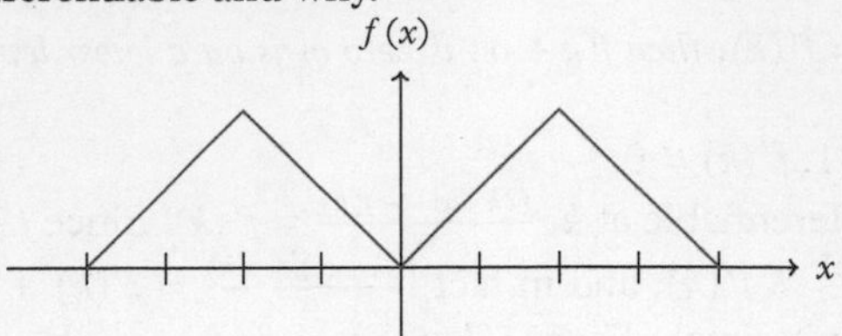

8. Use the following graph of $y = f(x)$ to determine where f is not differentiable and why.

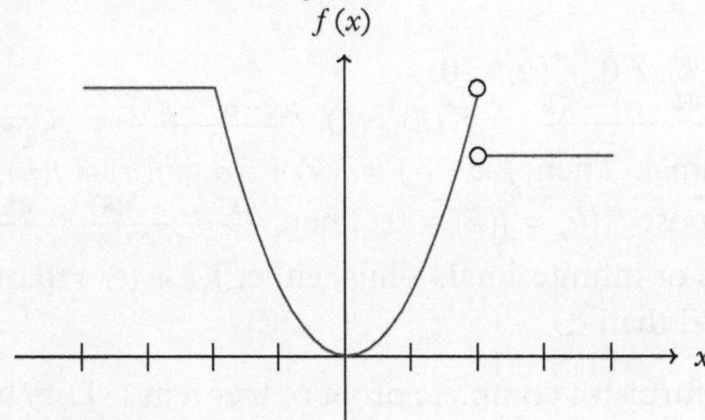

9. Use the following graph of $y = f(x)$ to determine where f is not differentiable and why.

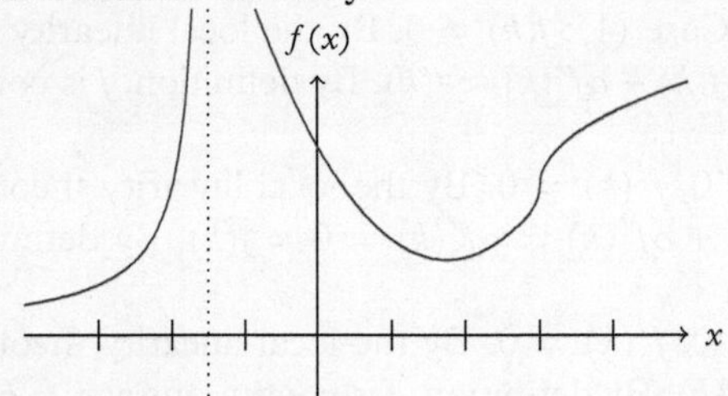

10. Use the following graph of $y = f(x)$ to determine where f is not differentiable and why.

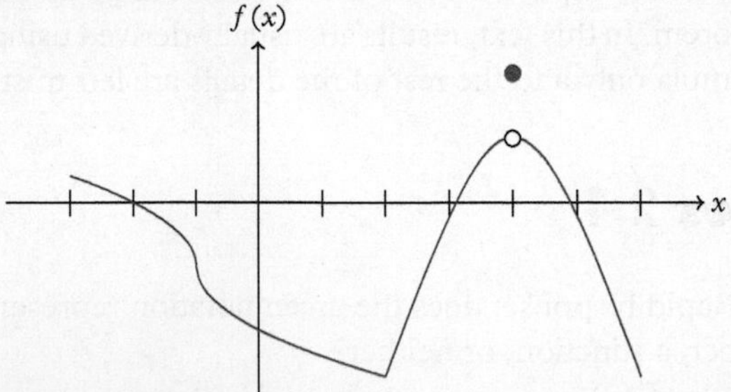

In this section's exercises, you must use the definition of derivative even if the exercise instructions do not say so explicity.

11. Let $f(x) = 2x - 7$. Find $f'(3)$.
12. Let $f(x) = 5x^2$. Find $f'(1)$.

13. Let $y = x + x^2$. Find $y'(-2)$.
14. Let $y = 5 - 8x$. Find $y'(4)$.
15. Let $y = x^4$. Find $\left.\frac{dy}{dx}\right|_{x=0}$.
16. Let $f(x) = x^{12}$. Find $\left.\frac{d}{dx}f(x)\right|_{x=0}$.
17. Find the slope of the tangent line to the curve $y = \frac{1}{x}$ at $x = -1$.
18. Find the slope of the tangent line to the curve $y = \sqrt{x}$ at $x = 9$.

19-24. Use the definition of derivative to find $f'(x)$.

19. $f(x) = x^2$
20. $f(x) = 4x$
21. $f(x) = 3x$
22. $f(x) = \sqrt{x}$
23. $f(x) = x^2 + 3x$
24. $f(x) = 4x + \sqrt{x}$

25. Based on exercises 19–24, make a conjecture regarding the derivative of the sum of two functions.
26. Based on exercises 20 and 21, make a conjecture regarding the derivative of $f(x) = mx$.
27. Let $g(x) = x^{1/5}$. Find $g'(0)$.
28. Let $s(t) = t^{2/3}$. Find $s'(0)$.
29. Let $f(x) = |x - 4|$. Find $f'(4)$.
30. Let $h(t) = |t - 4|$. Find $h'(1)$.
31. Let $h(y) = |y^3|$. Find $h'(0)$.
32. Find $\frac{d}{dx}cx^2$, where c is a real number.

33-38. Differentiate the function using the definition of derivative. From (GSL)

33. $y = \dfrac{3}{x^2 + 2}$
34. $s = \dfrac{t + 4}{t}$
35. $\rho = \dfrac{\theta}{\theta + 2}$
36. $y = \dfrac{1}{1 - 2x}$
37. $y = x^4$
38. $y = \dfrac{1}{x^2 + a^2}$

Derivative Rules

2.2

So far we have calculated derivatives by using a formula developed from the slope of a line, using infinitesimals so that our two points are infinitely close. Doing so has helped build our intuition, but if the function is very complicated, then using this formula becomes impractical. Our next task is to begin simplifying the process of finding derivatives by learning rules for derivatives that apply to a wide variety of situations.

Derivative of a constant function

The first rule is for the derivative of a constant function. Let's begin with an example.

Example 1 *Let* $f(x) = 7$. *Find* $f'(x)$.

Solution We use the derivative formula (function version) as before:

$$f'(x) = \frac{f(x+\alpha) - f(x)}{\alpha} = \frac{7-7}{\alpha} = \frac{0}{\alpha} = 0.$$

■

Note that $f(\text{anything}) = 7$, so $f(x) = 7$ and $f(x+\alpha) = 7$.

Look at the graph of $f(x) = 7$ (figure 1) and this result should make sense. The graph of a constant function is a horizontal line, the derivative gives the slope of the tangent line, and the slope of a horizontal line is zero.

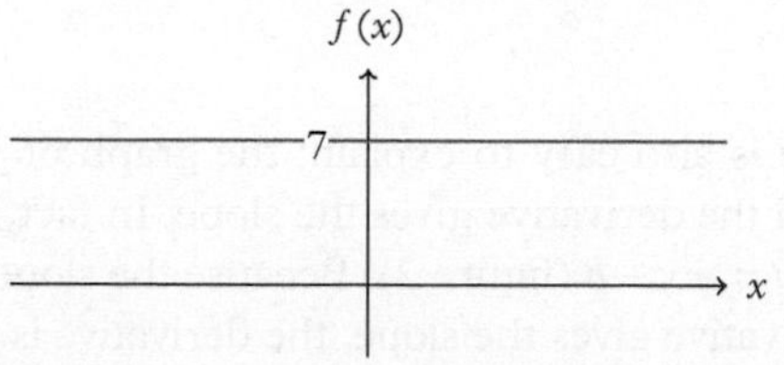

Figure 1 *The graph of the constant function* $f(x) = 7$

The number 7 is not special. Any constant function (a function of the form $f(x) = c$) has a derivative of zero.

Derivative formulas are written compactly using Leibniz notation. They may be remembered more easily in this form. For convenience, Leibniz-notation versions of formulas appear in the margin.

DERIVATIVE OF A CONSTANT FUNCTION

If $f(x) = c$, then $f'(x) = 0$.

$\frac{d}{dx}c = 0$

To prove a derivative rule, we use the definition of derivative by applying the derivative formula (function version). Here, we want the derivative of $f(x) = c$, where c is a real number:

This does not violate the approximation principle because we did not approximate.

$$f'(x) = \frac{f(x+\alpha) - f(x)}{\alpha} = \frac{c-c}{\alpha} = \frac{0}{\alpha} = 0.$$

Now that we know the derivative of any constant function is zero, we need not use the definition of derivative; we can apply the rule for the derivative of a constant function instead.

Example 2 *Let* $y = \frac{347.9\pi\sqrt{14}}{\sin\frac{\pi}{9}}$. *Find* y'.

Solution By the rule for the derivative of a constant function, $y' = 0$. ■

It is not necessary to write out which derivative rule is being applied; the solution is written just as easily as follows:

Solution $y' = 0$. ■

Reading Exercise 4 *Let* $f(x) = 48$. *Find* $f'(x)$.

Derivatives of linear functions

Example 3 *For* $f(x) = x$, *find* $f'(x)$.

Solution Using the derivative formula (function version), we have

$\frac{d}{dx}x = 1$

$$f'(x) = \frac{f(x+\alpha) - f(x)}{\alpha} = \frac{x+\alpha-x}{\alpha} = \frac{\alpha}{\alpha} = 1.$$ ■

This example is also easy to explain: the graph of $y = x$ is a line with slope 1, and the derivative gives the slope. In fact, this should be true of any line $y = mx + b$ (figure 2). Because the slope of $y = mx + b$ is m and the derivative gives the slope, the derivative is m.

DERIVATIVE OF A LINEAR FUNCTION

If $f(x) = mx + b$, then $f'(x) = m$.

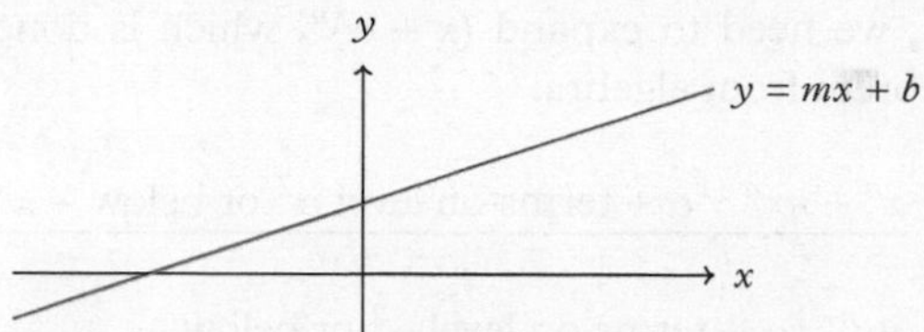

Figure 2 *The graph of the linear function $y = mx + b$, which has slope m*

Once again, a proof uses the derivative formula (function version):

$$f'(x) = \frac{f(x+\alpha) - f(x)}{\alpha} = \frac{m(x+\alpha) + b - (mx+b)}{\alpha}$$

$$= \frac{mx + m\alpha + b - mx - b}{\alpha} = \frac{m\alpha}{\alpha} = m.$$

$\frac{d}{dx}(mx + b) = m$

When the rule has been proved, it can be used instead of the definition.

Example 4 *Find* $\frac{d}{dx}(37x - 14)$.

Solution Using the rule for the derivative of a linear function, $\frac{d}{dx}(37x - 14) = 37$. ■

Reading Exercise 5 *Find* y' *for* $y = 75x + 1$.

Power rule (version 1)

Three derivative results collected from previous examples and exercises are:

$$\frac{d}{dx}x^2 = 2x,$$

$$\frac{d}{dx}x^3 = 3x^2, \text{ and}$$

$$\frac{d}{dx}x^4 = 4x^3.$$

Do you see the pattern? Let's try calculating the derivative of $f(x) = x^n$ for an arbitrary positive integer n to see if the pattern continues. We use the derivative formula (function version):

Ans. to reading exercise 4
$f'(x) = 0$

$$f'(x) = \frac{f(x+\alpha) - f(x)}{\alpha} = \frac{(x+\alpha)^n - x^n}{\alpha}.$$

$(a+b)^n = a^n + na^{n-1}b + \frac{n(n-1)}{2}a^{n-2}b^2 + \cdots + nab^{n-1} + b^n$
The powers of b in the formula increase by one each term. Here, $a = x$ and $b = \alpha$. Therefore, the terms have increasing powers of α.

To continue, we need to expand $(x+\alpha)^n$, which is done using the binomial theorem from algebra:

$$= \frac{x^n + nx^{n-1}\alpha + \text{terms on level } \alpha^2 \text{ or below} - x^n}{\alpha}$$

$$= \frac{nx^{n-1}\alpha + \text{terms on level } \alpha^2 \text{ or below}}{\alpha}.$$

Compare the algebra here to that of section 2.1 examples 2 and 3.

Next, we need to approximate. In the numerator, the term with α is on the higher level and the terms on level α^2 or below are lower-level terms, which are discarded when approximating:

$$\approx \frac{nx^{n-1}\alpha}{\alpha} = nx^{n-1}.$$

Using algebra only, the binomial theorem is proved for positive integer exponents only. This theorem was used in our proof, so at this point the power rule has only been proved for positive integer exponents.

$\frac{d}{dx}x^n = nx^{n-1}$

The pattern does continue!

POWER RULE (VERSION 1)

Let $f(x) = x^n$ for a positive integer n. Then, $f'(x) = nx^{n-1}$.

Example 5 *Find* $\frac{d}{dx}x^{49}$.

Solution Using the power rule,

$$\frac{d}{dx}x^{49} = 49x^{48}.$$ ■

It should now be apparent that the derivative rules save time and effort.

Reading Exercise 6 *Find* $\frac{d}{dx}x^{12}$.

Constant multiple rule

Look for a pattern in the following results of derivative calculations:

$$\frac{d}{dx}x^2 = 2x,$$

$$\frac{d}{dx}4x^2 = 8x = 4 \cdot 2x, \text{ and}$$

$$\frac{d}{dx}7x^2 = 14x = 7 \cdot 2x.$$

Let $f(x) = x^2$. Then, the previous second and third lines translate as

$$\frac{d}{dx}4 \cdot f(x) = 4 \cdot f'(x) \text{ and}$$
$$\frac{d}{dx}7 \cdot f(x) = 7 \cdot f'(x).$$

Multiplying the function by a constant appears to multiply the derivative by that same constant.

Ans. to reading exercise 5
$y' = 75$

CONSTANT MULTIPLE RULE

If f is differentiable at x and $g(x) = c \cdot f(x)$, then $g'(x) = c \cdot f'(x)$.

$\frac{d}{dx}c \cdot f(x) = c \cdot f'(x)$

As with the other rules, the constant multiple rule is proved using the derivative formula (function version):

$$g'(x) = \frac{g(x+\alpha) - g(x)}{\alpha} = \frac{cf(x+\alpha) - cf(x)}{\alpha}.$$

Now how do we calculate using $f(x+\alpha)$? One method is to use the local linearity formula, $f(x+\alpha) \approx f(x) + \alpha f'(x)$:

$$\approx \frac{c\left(f(x) + \alpha f'(x)\right) - cf(x)}{\alpha} = \frac{cf(x) + c\alpha f'(x) - cf(x)}{\alpha}$$
$$= \frac{c\alpha f'(x)}{\alpha} = cf'(x).$$

Because the local linearity formula is used, a formal proof includes a case for when the formula does not apply and the second part of the local linearity theorem must be used. This formality for the derivations of derivative rules is skipped here.

Example 6 *Find* $\frac{d}{dx}7x^3$.

Solution Using the constant multiple rule and the power rule,

$$\frac{d}{dx}7 \cdot x^3 = 7 \cdot 3x^2 = 21x^2. \quad \blacksquare$$

Reading Exercise 7 *Find* $\frac{d}{dx}4x^7$.

Sum and difference rules

In the exercise set for section 2.1, it was conjectured that the derivative of the sum of two functions is the sum of the derivatives of the two functions.

Ans. to reading exercise 6
$12x^{11}$

$\frac{d}{dx}(f(x)+g(x)) = f'(x) + g'(x)$

SUM RULE

If f and g are differentiable at x, then

$$(f+g)'(x) = f'(x) + g'(x).$$

The proof uses the local linearity formula for both f and g. Assuming $f'(x)$ and $g'(x)$ both exist,

$$\frac{d}{dx}(f(x)+g(x)) = \frac{f(x+\alpha)+g(x+\alpha)-(f(x)+g(x))}{\alpha}$$

$$\approx \frac{f(x)+\alpha f'(x)+g(x)+\alpha g'(x)-f(x)-g(x)}{\alpha}$$

(using the local linearity formula)

$$= \frac{\alpha f'(x)+\alpha g'(x)}{\alpha} = f'(x)+g'(x).$$

Example 7 *Let* $f(x) = 3x^4 + x^2 + 2x$. *Find* $f'(x)$.

Solution The derivatives of the three terms can be found separately using the power, constant multiple, and linear function rules:

$$\frac{d}{dx}3x^4 = 3 \cdot 4x^3 = 12x^3$$

$$\frac{d}{dx}x^2 = 2x$$

$$\frac{d}{dx}2x = 2.$$

The results can then be combined using the sum rule:

$$f'(x) = 12x^3 + 2x + 2.$$ ■

The difference rule is similar to the sum rule.

$\frac{d}{dx}(f(x)-g(x)) = f'(x) - g'(x)$

DIFFERENCE RULE

If f and g are differentiable at x, then

$$(f-g)'(x) = f'(x) - g'(x).$$

Example 8 *Find* $\frac{d}{dx}\left(x^3 - 4x^2\right)$.

Solution We find the derivative of x^3 and of $4x^2$ and subtract:

$$\frac{d}{dx}\left(x^3 - 4x^2\right) = 3x^2 - 4 \cdot 2x = 3x^2 - 8x. \qquad \blacksquare$$

Ans. to reading exercise 7
$28x^6$

Derivatives of polynomials

The derivative of any polynomial may be found using the derivative rules. Let's look at the next example, which illustrates the procedure step-by-step.

Example 9 *Determine* $\frac{d}{dx}\left(5x^7 + x^4 - 8x^2 + 7x - 3\right)$.

Solution First we use the sum and difference rules:

$$\frac{d}{dx}\left(5x^7 + x^4 - 8x^2 + 7x - 3\right)$$
$$= \frac{d}{dx}5x^7 + \frac{d}{dx}x^4 - \frac{d}{dx}8x^2 + \frac{d}{dx}7x - \frac{d}{dx}3.$$

Next we use the constant multiple rule:

$$= 5\frac{d}{dx}x^7 + \frac{d}{dx}x^4 - 8\frac{d}{dx}x^2 + \frac{d}{dx}7x - \frac{d}{dx}3.$$

Last we use the power, linear function, and constant rules:

$$= 5 \cdot 7x^6 + 4x^3 - 8 \cdot 2x + 7 - 0 = 35x^6 + 4x^3 - 16x + 7. \qquad \blacksquare$$

With practice, most of these rules can be applied rather quickly and the solution written down directly. It looks like this:

Example 10 *Find* y' *for* $y = 4x^3 - 2x^2 + 5$.

Solution $y' = 12x^2 - 4x.$ $\blacksquare$

Reading Exercise 8 *Find* $f'(x)$ *for* $f(x) = x^2 + 2x - 7$.

We see that the derivative of a polynomial function is also a polynomial function, which is defined everywhere. This means that polynomial functions are differentiable everywhere, and therefore their graphs are smooth and have no corners, as advertised previously.

Higher-order derivatives

Consider the function $f(x) = 3x^3 + 7x - 5$. It has a derivative:

$$f'(x) = 3 \cdot 3x^2 + 7 - 0 = 9x^2 + 7.$$

The derivative f' is a function in its own right. Therefore, it also has a derivative, $(f')'$, customarily written as f'':

$$f''(x) = 9 \cdot 2x + 0 = 18x.$$

$f'(x)$ is also referred to as the *first derivative*.

We call $f''(x)$ the *second derivative* of f and read it "f double-prime of x." Why stop there? Since f'' is also a function, let's take its derivative:

$$(f'')'(x) = f'''(x) = 18.$$

This is the *third derivative* of f and is read "f triple-prime of x." For the sake of readability, beginning with the fourth derivative it is customary to write the number of the derivative in parentheses:

$$(f''')'(x) = f^{(4)}(x) = 0.$$

Because the degree of a polynomial decreases by one with each derivative taken, the derivatives of a polynomial eventually reach zero. This is not the case with other types of functions, as we shall see in section 2.4 and chapter 5.

The peculiarities of Leibniz notation for higher derivatives become less mysterious when studying multivariable calculus.

Leibniz notation for the second derivative is $\frac{d^2y}{dx^2}$, which can be read "the second derivative of y with respect to x both times." Third, fourth, and higher derivatives are similar.

Reading Exercise 9 *Find* $f''(x)$ *for* $f(x) = x^4$.

Product rule

The derivative of a sum is the sum of the derivatives. The derivative of a difference is the difference of the derivatives. Is the same true for products? Consider $x^5 = x^2 \cdot x^3$. Is the derivative of x^5 the product of the derivatives of x^2 and x^3?

$$x^5 = x^2 \cdot x^3$$
$$\frac{d}{dx}x^5 \stackrel{?}{=} \frac{d}{dx}x^2 \cdot \frac{d}{dx}x^3$$
$$5x^4 \stackrel{?}{=} 2x \cdot 3x^2$$
$$5x^4 \neq 6x^3$$

Ans. to reading exercise 8
$f'(x) = 2x + 2$

The derivative of a product is *not* the product of the derivatives. So what is the derivative of a product?

Using the derivative formula (function version) and the local linearity formula, we can calculate the derivative of a product:

CAUTION: MISTAKE TO AVOID
$\frac{d}{dx}f(x)g(x) \neq f'(x)g'(x)$
Calculating the derivative of a product as the product of the derivatives is one of the most common mistakes in calculus. This mistake may have been made a billion times by humans. Its cumulative toll on grades and its contributions to changes of career plans—even its total economic impact—is scary to contemplate.

$$\begin{aligned}
\frac{d}{dx}(f(x) \cdot g(x)) &= \frac{f(x+\alpha)g(x+\alpha) - f(x)g(x)}{\alpha} \\
&\approx \frac{(f(x) + \alpha f'(x))(g(x) + \alpha g'(x)) - f(x)g(x)}{\alpha} \\
&\qquad \text{(using the local linearity formula)} \\
&= \frac{f(x)g(x) + \alpha f(x)g'(x) + \alpha f'(x)g(x) + \alpha^2 f'(x)g'(x) - f(x)g(x)}{\alpha} \\
&= \frac{\alpha f(x)g'(x) + \alpha f'(x)g(x) + \alpha^2 f'(x)g'(x)}{\alpha} \\
&\approx \frac{\alpha f(x)g'(x) + \alpha f'(x)g(x)}{\alpha} \\
&= f(x)g'(x) + f'(x)g(x).
\end{aligned}$$

PRODUCT RULE

If f and g are differentiable at x, then

$$\frac{d}{dx}(f(x) \cdot g(x)) = f(x)g'(x) + g(x)f'(x).$$

$\frac{d}{dx}(f(x) \cdot g(x)) = f(x)g'(x) + g(x)f'(x)$

The product rule is often remembered as

$$\text{first} \cdot \text{second}' + \text{second} \cdot \text{first}'.$$

Because multiplication and addition are both commutative, other orders for remembering the product rule are in common use as well, such as first · second′ + first′ · second.

Example 11 *Use the product rule to differentiate*

$$y = (3x^2 + 7)(x^4 + 12x - 9).$$

CAUTION: MISTAKE TO AVOID $y' \neq (6x)(4x^3 + 12)$. **Do not** calculate the derivative of a product by taking the product of the derivatives.

Solution The function is written as a product of two factors. Label the two factors "first" and "second":

$$\underset{\text{first}}{(3x^2 + 7)} \cdot \underset{\text{second}}{(x^4 + 12x - 9)}$$

It is not necessary to write out the formula template in the manner described here, but doing so is often helpful.

The product rule is first · second′ + second · first′:

$$y' = \underset{\text{first}}{(3x^2 + 7)} \cdot \underset{\text{second}'}{(4x^3 + 12)} + \underset{\text{second}}{(x^4 + 12x - 9)} \cdot \underset{\text{first}'}{(6x)}.$$ ■

Ans. to reading exercise 9 $f''(x) = 12x^2$

It is not necessary to multiply the factors and add them at the end of example 11. If we want to multiply the factors, then we do so at the beginning and avoid using the product rule. There are other types of functions for which the product rule cannot be avoided, and exercises such as example 11 provide practice and an opportunity to learn the rule.

Reading Exercise 10 *Use the product rule to find y' for*

$$y = (x^2 + 1)(x^3 - 5).$$

Quotient rule

CAUTION: MISTAKE TO AVOID $\frac{d}{dx}\frac{f(x)}{g(x)} \neq \frac{f'(x)}{g'(x)}$
Calculating the derivative of a quotient as the quotient of the derivatives is also among the most common mistakes in calculus.

A rule for finding derivatives of quotients is required as well. As always, we begin with the derivative formula (function version):

$$\frac{d}{dx}\frac{f(x)}{g(x)} = \frac{\frac{f(x+\alpha)}{g(x+\alpha)} - \frac{f(x)}{g(x)}}{\alpha}.$$

Next we eliminate the compound fraction:

$$= \frac{\frac{f(x+\alpha)}{g(x+\alpha)} - \frac{f(x)}{g(x)}}{\alpha} \cdot \frac{g(x+\alpha)g(x)}{g(x+\alpha)g(x)}$$

$$= \frac{\frac{f(x+\alpha)}{\cancel{g(x+\alpha)}} \cdot \cancel{g(x+\alpha)}g(x) - \frac{f(x)}{\cancel{g(x)}} \cdot g(x+\alpha)\cancel{g(x)}}{\alpha \cdot g(x+\alpha)g(x)}$$

$$= \frac{f(x+\alpha)g(x) - f(x)g(x+\alpha)}{\alpha g(x+\alpha)g(x)}.$$

Now we use the local linearity formula, simplify, and approximate:

$$\approx \frac{(f(x)+\alpha f'(x))g(x) - f(x)(g(x)+\alpha g'(x))}{\alpha(g(x)+\alpha g'(x))g(x)} \text{ (using local linearity)}$$

$$= \frac{f(x)g(x)+\alpha f'(x)g(x) - f(x)g(x) - \alpha f(x)g'(x)}{\alpha(g(x))^2+\alpha^2 g(x)g'(x)}$$

$$= \frac{\alpha f'(x)g(x) - \alpha f(x)g'(x)}{\alpha(g(x))^2+\alpha^2 g(x)g'(x)}$$

$$\approx \frac{\alpha f'(x)g(x) - \alpha f(x)g'(x)}{\alpha(g(x))^2}$$

$$= \frac{f'(x)g(x) - f(x)g'(x)}{(g(x))^2}.$$

QUOTIENT RULE

If f and g are differentiable at x and $g(x) \neq 0$, then

$$\frac{d}{dx}\left(\frac{f(x)}{g(x)}\right) = \frac{g(x)f'(x) - f(x)g'(x)}{(g(x))^2}.$$

$$\frac{d}{dx}\left(\frac{f(x)}{g(x)}\right) = \frac{g(x)f'(x) - f(x)g'(x)}{(g(x))^2}$$

The quotient rule is often remembered as

$$\left(\frac{\text{top}}{\text{bot}}\right)' = \frac{\text{bot}\cdot\text{top}' - \text{top}\cdot\text{bot}'}{(\text{bot})^2}.$$

Another common expression of the quotient rule is "lo d hi minus hi d lo over lo lo."

Example 12 *Find* $f'(x)$ *for* $f(x) = \dfrac{4x^2-7x+8}{x^5+x^3}$.

Solution The function is a quotient, so the quotient rule is required. Label the numerator "top" and the denominator "bot":

$$\underset{\text{bot}}{\overset{\text{top}}{\frac{4x^2-7x+8}{x^5+x^3}}}.$$

The quotient rule is $\dfrac{\text{bot}\cdot\text{top}'-\text{top}\cdot\text{bot}'}{(\text{bot})^2}$:

$$f'(x) = \frac{\overset{\text{bot}}{(x^5+x^3)}\cdot\overset{\text{top}'}{(8x-7)} \overset{-}{-} \overset{\text{top}}{(4x^2-7x+8)}\cdot\overset{\text{bot}'}{(5x^4+3x^2)}}{\underset{\text{bot}^2}{(x^5+x^3)^2}}.$$

■

CAUTION: MISTAKE TO AVOID
$f'(x) \neq \frac{8x-7}{5x^4+3x^2}$. **Do not** calculate the derivative of a quotient by taking the quotient of the derivatives.

CAUTION: MISTAKE TO AVOID
A common mistake in applying the quotient rule is to remember it in the wrong order. Because subtraction is not commutative ($4 - 3 \neq 3 - 4$), the terms in the numerator cannot be switched.

Ans. to reading exercise 10
$y' = (x^2+1)(3x^2) + (x^3-5)(2x)$

For derivatives of rational functions, it is often convenient to simplify the derivative by multiplying the factors in the numerator and subtracting, collecting like terms. For instance, if a second derivative is required, the form of the solution in example 12 is not convenient.

Because we can use the quotient rule to find the derivatives of all rational functions and the derivative is undefined precisely when the rational function is undefined, all rational functions are differentiable throughout their domains. Therefore, graphs of rational functions are smooth and have no corners.

Reading Exercise 11 *Find* $\frac{d}{dx}\left(\frac{x^2+1}{x^3-5}\right)$.

Power rule for negative integer exponents

The power rule was proved earlier for positive integer exponents only. We can use it for $\frac{d}{dx}x^6 = 6x^5$, but we do not yet know that it works for $\frac{d}{dx}x^{-3}$. Since $x^{-3} = \frac{1}{x^3}$, the quotient rule can help:

$$\frac{d}{dx}x^{-3} = \frac{d}{dx}\left(\frac{1}{x^3}\right) = \frac{x^3 \cdot 0 - 1 \cdot 3x^2}{(x^3)^2} = \frac{-3x^2}{x^6} = -3x^{-4}.$$

This still conforms to the pattern $\frac{d}{dx}x^{-3} = -3x^{-3-1}$. We can turn this calculation into a proof of the power rule for negative integers.

> **POWER RULE (VERSION 2)**
>
> Let $f(x) = x^n$ for a negative integer n. Then, $f'(x) = nx^{n-1}$.

For the proof, let n be a negative integer. Then $-n$ is a positive integer (if $n = -3$, then $-n = 3$), and the power rule can be applied to x^{-n}. Then,

$$\frac{d}{dx}x^n = \frac{d}{dx}\left(\frac{1}{x^{-n}}\right) = \frac{x^{-n}\cdot 0 - 1 \cdot -nx^{-n-1}}{(x^{-n})^2}$$

$$= \frac{nx^{-n-1}}{x^{-2n}} = nx^{-n-1-(-2n)} = nx^{n-1}.$$

Laws of exponents were used in the calculation; $(x^a)^b = x^{a\cdot b}$ was used for $(x^{-n})^2 = x^{-2n}$ and $\frac{x^a}{x^b} = x^{a-b}$ was used for $\frac{x^{-n-1}}{x^{-2n}} = x^{-n-1-(-2n)}$.

Example 13 *Find* $f''(x)$ *for* $f(x) = \frac{1}{4x}$.

Solution We could use the quotient rule to find the derivative, but we can also use the power rule if we rewrite the function using a negative exponent:

$$f(x) = \frac{1}{4}x^{-1}$$

$$f'(x) = \frac{1}{4}(-1 \cdot x^{-2}) = -\frac{1}{4}x^{-2} = -\frac{1}{4x^2}.$$

CAUTION: MISTAKE TO AVOID
The 4 does not rise to the numerator:

$$\frac{1}{4x} \neq 4x^{-1}.$$

To bring the 4 to the numerator requires that it also have a negative exponent:

$$\frac{1}{4x} = 4^{-1}x^{-1}.$$

The last step, expressing the derivative in terms of a fraction instead of a negative exponent, is not necessary but can be helpful for interpreting the result. On the other hand, thinking of $f'(x) = -\frac{1}{4}x^{-2}$ is helpful for taking the second derivative using the power rule:

$$f''(x) = -\frac{1}{4}(-2x^{-3}) = \frac{1}{2}x^{-3} = \frac{1}{2x^3}.$$ ■

Notice that as we take derivatives, the exponent in the denominator is getting larger, not smaller. A polynomial's derivatives eventually reach zero, but this function's derivatives never reach zero.

Ans. to reading exercise 11

$$\frac{(x^3-5)(2x)-(x^2+1)(3x^2)}{(x^3-5)^2}$$
$$= \frac{-x^4-3x^2-10x}{(x^3-5)^2}$$

Reading Exercise 12 *Find y' for $y = \frac{1}{x}$.*

Example 14 *Find $\frac{d}{dx}\left(\frac{x^4+2x^3-x+5}{x^2}\right)$.*

Solution The quotient rule may be used for this problem, but because the denominator has only one term, the expression can be simplified using negative exponents, and then the power rule can be used:

$$\frac{d}{dx}\left(\frac{x^4+2x^3-x+5}{x^2}\right) = \frac{d}{dx}\left(x^2+2x-x^{-1}+5x^{-2}\right)$$
$$= 2x+2-(-1\cdot x^{-2})+5(-2x^{-3})$$
$$= 2x+2+x^{-2}-10x^{-3}$$
$$= 2x+2+\frac{1}{x^2}-\frac{10}{x^3}.$$ ■

In Leibniz notation, $\frac{d}{dx}$ can be thought of as a command to take the derivative. In the first step, we used algebra to rewrite what was inside the parentheses. Because we did not take the derivative there, the command to take the derivative has not yet been completed and is still needed. In the second step, the derivative is taken, so the $\frac{d}{dx}$ is dropped.

General power rule

The power rule actually works for any real-number exponent, even though we have only proved it so far for integer exponents. The proofs come later (proof for rational exponents in section 2.6, proof for irrational exponents in section 5.5), but let's proceed and use the general power rule now.

GENERAL POWER RULE

Let $f(x) = x^n$ for a real number n. Then, $f'(x) = nx^{n-1}$.

Example 15 *Find $f'(x)$ for $f(x) = \sqrt{x} + \sqrt[3]{x}$.*

Solution Rewriting the roots as fractional exponents helps us use the power rule:

$$f(x) = x^{\frac{1}{2}} + x^{\frac{1}{3}}$$
$$f'(x) = \frac{1}{2}x^{\frac{1}{2}-1} + \frac{1}{3}x^{\frac{1}{3}-1}$$
$$= \frac{1}{2}x^{-\frac{1}{2}} + \frac{1}{3}x^{-\frac{2}{3}}.$$

If desired, the exponents can be rewritten in terms of roots:

Recall that $x^{\frac{2}{3}} = (x^2)^{\frac{1}{3}} = \sqrt[3]{x^2}$.

$$= \frac{1}{2x^{\frac{1}{2}}} + \frac{1}{3x^{\frac{2}{3}}} = \frac{1}{2\sqrt{x}} + \frac{1}{3\sqrt[3]{x^2}}.$$ ■

Exercises 2.2

Ans. to reading exercise 12
$y' = -x^{-2} = -\frac{1}{x^2}$

1–10. Rapid response: state the derivative rule(s) needed to find $\frac{dy}{dx}$. (Do not find the derivative.)

1. $y = x^4$
2. $y = 7x - 5$
3. $y = \frac{1}{x^3}$
4. $y = 9$
5. $y = 6x^2$
6. $y = \frac{x^2+7}{x^4-12x}$
7. $y = (x^2+5)(x^3-7x^2)$
8. $y = \sqrt[4]{x}$
9. $y = \sqrt{x} + x^3$
10. $y = 4(x^2 + 3x - 1)$

From now on, unless otherwise specified, you may use derivative rules and formulas to calculate derivatives.

11–22. Determine the derivative using the appropriate derivative rules and formulas.

11. $y = x^{27}$
12. $y = 4x^2$
13. $f(x) = 12x - 55$
14. $g(x) = 1987$
15. $y = \frac{1}{x^5}$
16. $y = \sqrt[3]{x}$
17. $s(t) = 3t + 5 - \sqrt[3]{t} + \frac{1}{t^4}$
18. $f(x) = 5x^3 - \sqrt{x} + \frac{1}{x^2}$

19. $h(x) = (x^3 + 2x + 8)(4x^5 - 2x^2 + 38)$

20. $f(x) = (6x^4 - x^2)(5x - 1)$

21. $f(x) = (12x^2 + 7x - 1)(x^3 - 9x - 5)$

22. $s(t) = (t^{2.7} - t^{1.2})(6t^5 - 5t^6)$

23–36. Differentiate the function.

Because there may be more than one way to work a problem, your answer might not look like the answer in the back of the book even if your answer is correct. Try using a calculator or computer algebra system to graph or simplify the quantity **your answer minus the book's answer** to see if the result is zero.

23. $g(x) = \dfrac{x^2 - 7}{4x + 9}$

24. $s(t) = \dfrac{t^2 - 4}{t + 17}$

25. $f(t) = t^5 - 7t^3 + 4t$

26. $y = 15x^3 - 10x^2 + 14x$

27. $y = \dfrac{3x - 9}{x^3 + 5x}$

28. $f(x) = \dfrac{\sqrt{x}}{x^2}$

29. $y = \dfrac{5x^2 - 7}{x}$

30. $g(x) = \dfrac{x^3 + 4x^2 - 7x + 12}{x^3}$

31. $f(x) = \sqrt{x} \cdot x^2$

32. $y = \dfrac{\sqrt{t}}{t + 1}$

33. $f(x) = \dfrac{\frac{1}{x} + 4}{5x^2 + 3}$

34. $y = x\left(\dfrac{1}{x + 2} + \dfrac{2}{x + 5}\right)$

35. $y = \dfrac{\frac{2}{3}t^3 - \frac{1}{4}t}{22t - 1}$

36. $f(x) = \dfrac{x^4 + x^2}{3 - \frac{1}{x^2}}$

37–46. Find the indicated derivative.

37. $y = -2x^2 - 5x + 1.3, y''$

38. $y = x^6 - 5x^4, y'''$

39. $f(x) = \sqrt{x}, f''(x)$

40. $f(x) = \dfrac{5}{x}, f''(x)$

41. $y = \dfrac{x^3}{4}, \frac{d^2y}{dx^2}$

42. $y = -x^{-4}, \frac{d^2y}{dx^2}$

43. $\frac{d^3}{dx^3}x(x^4 - 12)$

44. $\frac{d^2}{dx^2}\left(-x^{3/2}\right)$

45. $f(t) = t^7, f^{(4)}(t)$

46. $g(x) = x^6 - 7x^3, g^{(10)}(x)$

47. (GSL) Differentiate $y = \dfrac{3x^3}{\sqrt[5]{x^2}} - \dfrac{7x}{\sqrt[3]{x^4}} + 8\sqrt[7]{x^3}$.

48. (GSL) Differentiate $y = \dfrac{a - x}{a + x}$, where a is a real number.

49. Derive the difference rule.

50. Use a derivation like that of the product rule to derive a formula for $\frac{d}{dx}(f(x)g(x)h(x))$.

51. Let $x(f) = f^2 + 4f - 7$. Find $x'(f)$.

Tangent Lines Revisited

2.3

One of the interpretations of a derivative is that it represents the slope of the tangent line. Now that we have derivative rules to simplify the calculation of derivatives, it will be easier to calculate the equations of tangent lines. A broad exploration of tangent lines is the subject of this section.

Equations of tangent lines revisited

The procedure we used in section 1.8 to find the equation of the tangent line to the curve $y = f(x)$ at $x = a$ (figure 1) is summarized in three steps:

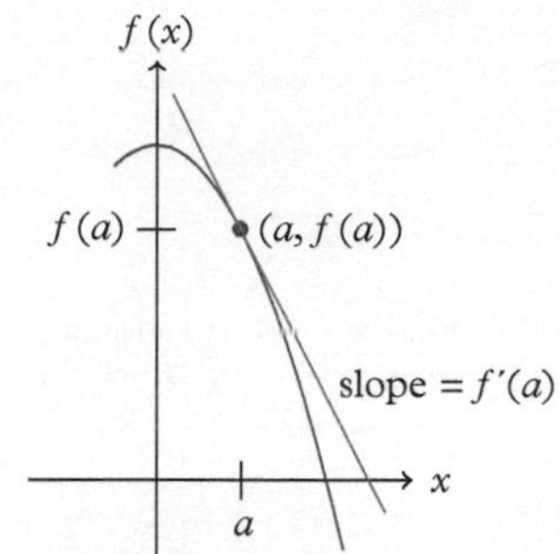

Figure 1 *The curve $y = f(x)$ (blue) with tangent line (orange) at $x = a$*

❶ Find the slope $f'(a)$.

❷ If necessary, calculate the y-coordinate $f(a)$ of the point of tangency.

❸ Use the point-slope form of the equation of a line with point $(a, f(a))$ and slope $f'(a)$.

Example 1 *Find the equation of the tangent line to the curve $y = 7x^3 - 4x^2 + 1$ at $x = 1$.*

Referring to the three-step procedure, $a = 1$.

Solution ❶ To find the slope, we find the derivative using the derivative rules:

$$y' = 21x^2 - 8x.$$

Next we evaluate the derivative at $x = 1$:

$$\text{slope} = y'(1) = 21(1)^2 - 8(1) = 13.$$

❷ We have not been given the y-coordinate of the point of tangency, so it must be calculated:

$$y(1) = 7(1)^3 - 4(1)^2 + 1 = 4.$$

Referring to the three-step procedure, this is the point $(a, f(a))$.

The point of tangency is $(1, 4)$.

❸ We finish by using the point-slope equation of the line with point $(1, 4)$ and slope 13:

$$y - y_1 = m(x - x_1)$$
$$y - 4 = 13(x - 1)$$
$$y - 4 = 13x - 13$$
$$y = 13x - 9$$

Recall that it is traditional to express the answer in slope-intercept form.

The equation of the tangent line is $y = 13x - 9$. ■

Reading Exercise 13 *Find the equation of the tangent line to the curve $y = x^2 + 5$ at $x = 3$.*

Detecting horizontal tangent lines

Suppose we wish to find the locations of any horizontal tangent lines on the graph of a function (figure 2). Recall that a horizontal line has

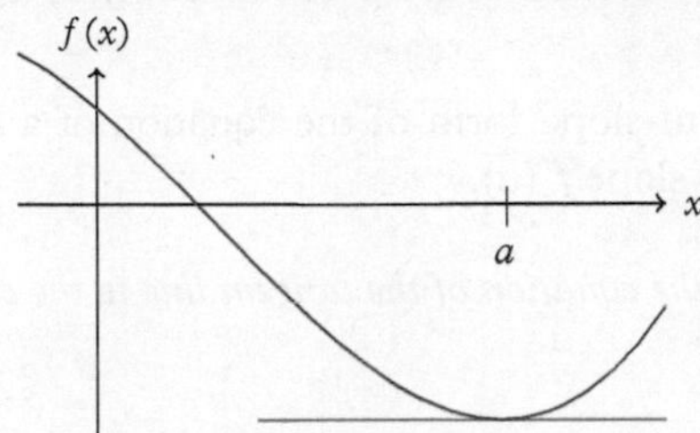

Figure 2 *The curve $y = f(x)$ (blue) with horizontal tangent line (orange) at $x = a$. This requires $f'(a) = 0$*

slope 0, and that the derivative represents the slope of the tangent line. Therefore, we want to know where the derivative (slope) is zero. We can find such places by solving the equation $f'(x) = 0$.

Horizontal tangents occur where $f'(x) = 0$.

The steps to finding horizontal tangent lines are as follows:

❶ Calculate the derivative $f'(x)$.

❷ Solve the equation $f'(x) = 0$ (slope $= 0$) for x. The solutions are the locations of the horizontal tangent lines.

Example 2 *Find the locations of all horizontal tangent lines on the graph of* $f(x) = \dfrac{x^3}{3} - \dfrac{3x^2}{2} - 10x + 14$ *(figure 3).*

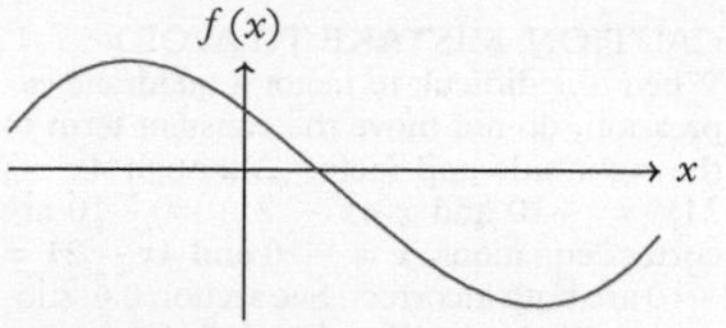

Figure 3 *The graph of* $f(x) = \frac{x^3}{3} - \frac{3x^2}{2} - 10x + 14$

Solution ❶ We begin by finding $f'(x)$ using the derivative rules:

$$f(x) = \frac{1}{3}x^3 - \frac{3}{2}x^2 - 10x + 14$$

$$f'(x) = \frac{1}{3}(3x^2) - \frac{3}{2}(2x) - 10$$
$$= x^2 - 3x - 10.$$

If a denominator is a constant (a real number), then the expression can be rewritten to avoid the use of the quotient rule.

❷ Next we set $f'(x) = 0$ and solve for x:

Various techniques for solving equations might be needed in this step, such as factoring or the quadratic equation.

$$x^2 - 3x - 10 = 0$$
$$(x - 5)(x + 2) = 0$$
$$x = 5,\ x = -2.$$

Horizontal tangent lines are located at $x = 5$ and at $x = -2$. ■

Reading Exercise 14 *Find the locations of all horizontal tangent lines on the graph of* $y = x^3 - 3x$.

Example 3 *Find the locations of all horizontal tangent lines on the graph of* $f(x) = x^4 - 7x^3 + 5x^2$.

Solution ❶ We begin by finding $f'(x)$:

Ans. to reading exercise 13
$y = 6x - 4$

$$f'(x) = 4x^3 - 7(3x^2) + 5(2x) = 4x^3 - 21x^2 + 10x.$$

❷ Next we set $f'(x) = 0$ and solve for x:

$$4x^3 - 21x^2 + 10x = 0$$
$$x(4x^2 - 21x + 10) = 0 \text{ (factor out common x)}$$
$$x = 0,\ 4x^2 - 21x + 10 = 0.$$

CAUTION: MISTAKE TO AVOID
When it is difficult to factor a quadratic expression, do not move the constant term to the right side and factor. Although $4x^2 - 21x = -10$ and $x(4x - 21) = -10$ are correct equations, $x = -10$ and $4x - 21 = -10$ are both incorrect! See section 0.6, subsection "Polynomial equations", for further explanation.

Whenever it seems difficult to factor a quadratic expression, the quadratic formula can be used:

$$\begin{aligned} x &= \frac{-(-21) \pm \sqrt{(-21)^2 - 4(4)(10)}}{2(4)} \\ &= \frac{21 \pm \sqrt{441 - 160}}{8} \\ &= \frac{21 \pm \sqrt{281}}{8}. \end{aligned}$$

Horizontal tangent lines are located at $x = 0$, $x = \frac{21+\sqrt{281}}{8}$, and $x = \frac{21-\sqrt{281}}{8}$, or at approximately $x = 0$, $x = 0.5296$, and $x = 4.7204$. ■

Comparing the graphs of f and f′

How do the graphs of a function f and its derivative f' compare? We can visualize the comparison by considering slopes of tangent lines. Figure 4 shows the graph of $f(x) = x^2 + 3$ along with the graph of its derivative $f'(x) = 2x$. Slopes on the graph of f are the values of the

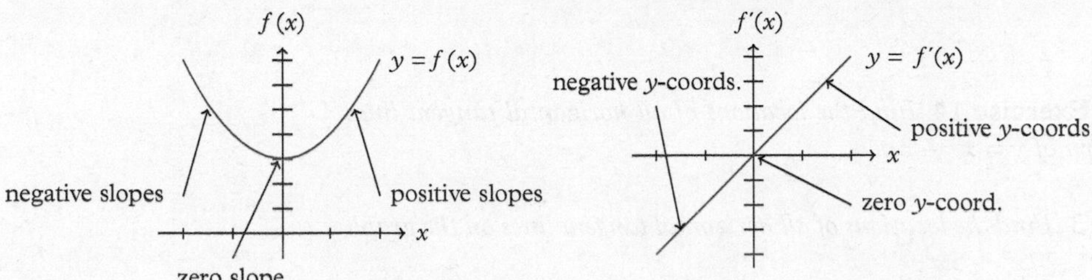

Figure 4 *The curves $f(x) = x^2 + 3$ (left) and $f'(x) = 2x$ (right). The slopes on f correspond to the y-coordinates on f'*

derivative f', which are the y-coordinates on the graph of f'. To the left of $x = 0$, slopes on the graph of f are negative; to the left of $x = 0$, the y-coordinates on the graph of f' are negative (below the x-axis). To the right of $x = 0$, slopes on the graph of f are positive; to the right of $x = 0$, the y-coordinates on the graph of f' are positive (above the x-axis).

Ans. to reading exercise 14
$x = 1, x = -1$

This correspondence between the graph of a function and the graph of its derivative can help us build the graph of one from the graph of the other.

Example 4 *The graph of a function $y = f(x)$ is in figure 5. Sketch the graph of f'.*

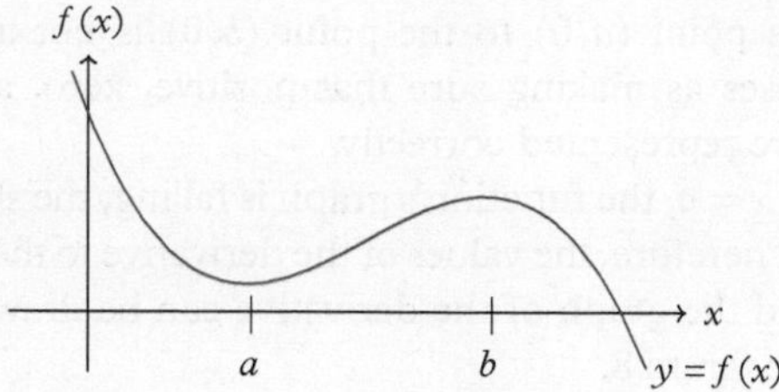

Figure 5 *The graph of $y = f(x)$ for example 4*

Solution First we should look for discontinuities, corners, and vertical tangents—places where the function is not differentiable. Seeing none, we next look for horizontal tangent lines. There appear to be two horizontal tangents, one at $x = a$ and the other at $x = b$. Because the slope of a horizontal line is zero, we must have $f'(a) = 0$ and $f'(b) = 0$, and the points $(a, 0)$ and $(b, 0)$ must then be on the graph of f', and we can begin the graph by plotting those two points. See figure 6.

Where the function is not differentiable, there should be no point on the graph of f'.

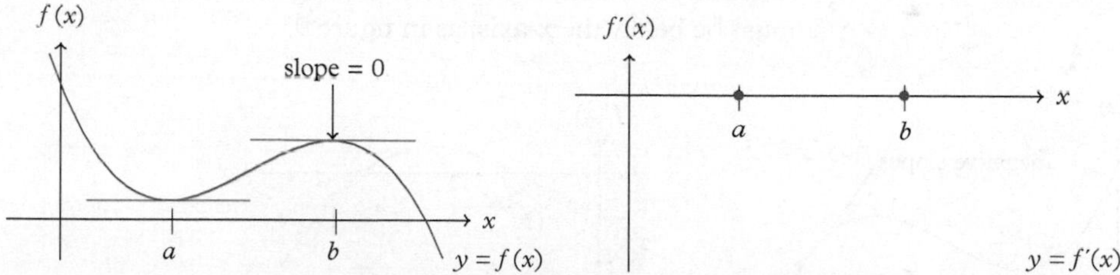

Figure 6 *The graph of $y = f(x)$ (left) along with the graph of $y = f'(x)$ (right), in progress*

On the graph of f between $x = a$ and $x = b$, the function's graph is rising; the slopes appear to be positive. Therefore, the values of the derivative are positive between $x = a$ and $x = b$—that is, the graph of f' must have positive y-coordinates. We can therefore draw the graph of f' between $x = a$ and $x = b$ above the x-axis (positive y-coordinates are above the x-axis); see figure 7. The exact manner in which the graph

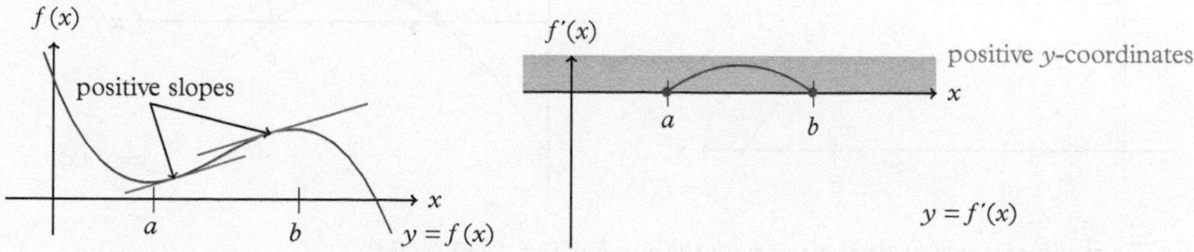

Figure 7 *The graph of $y = f(x)$ (left) along with the graph of $y = f'(x)$ (right), in progress*

travels from the point $(a, 0)$ to the point $(b, 0)$ is not as important for these exercises as making sure that positive, zero, and negative y-coordinates are represented correctly.

To the left of $x = a$, the function's graph is falling; the slopes appear to be negative. Therefore, the values of the derivative to the left of $x = a$ are negative, and the graph of the derivative can be drawn below the x-axis there. See figure 8.

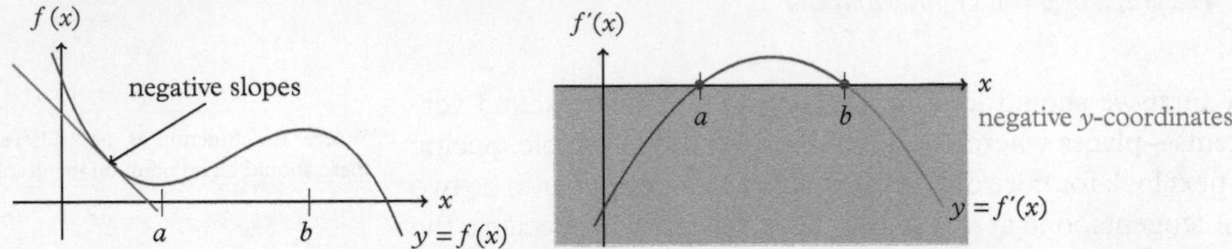

Figure 8 *The graph of $y = f(x)$ (left) along with the graph of $y = f'(x)$ (right), in progress*

We finish by noting that to the right of $x = b$, the graph of the function f is falling, so the slopes are negative—meaning, the derivative must be below the x-axis, as in figure 9.

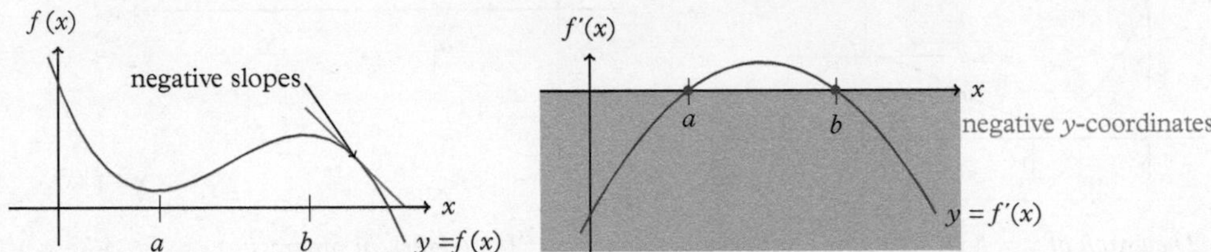

Figure 9 *The graph of $y = f(x)$ (left) along with the graph of $y = f'(x)$ (right), in progress*

The final solution, comparing the graphs of the function and its derivative, is in figure 10. ■

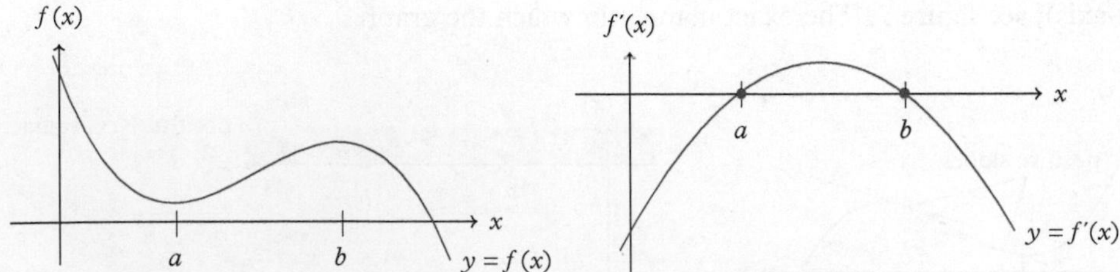

Figure 10 *The graph of $y = f(x)$ (left) along with the graph of $y = f'(x)$ (right)*

Notice that in example 4, the graph of the function looks like the graph of a cubic function (a polynomial of degree 3) whereas the graph of the derivative looks like a parabola (a polynomial of degree 2). This fits what we know about derivatives of polynomials.

Going the opposite direction, recreating the graph of a function f from the graph of its derivative f', has an additional issue or two that we are not yet ready to examine.

Reading Exercise 15 *The graph of a function $y = f(x)$ is in figure 11. Sketch the graph of f'.*

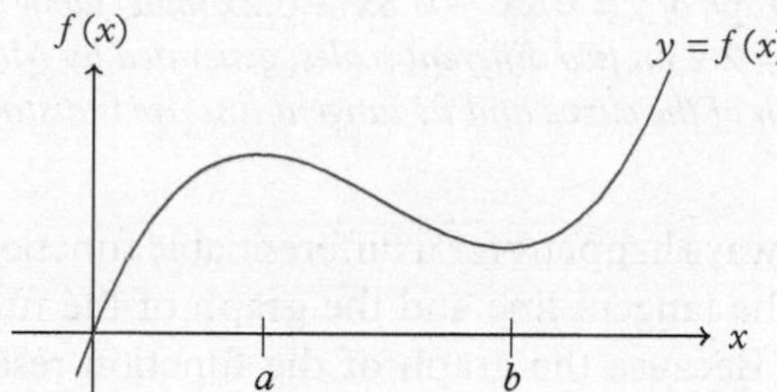

Figure 11 *The graph of $y = f(x)$ for reading exercise 15*

Local linearity

Figure 12 shows the graph of the function $y = 0.2x^3 - 0.8x^2 + 0.2x$ with its tangent line at $x = 2.2$. Viewed between $x = -2$ and $x = 4$, the graph of the function is very clearly curved and the tangent line appears to be quite different from the function. If we zoom in and view the graph only between $x = 2.1$ and $x = 2.3$, as in figure 13 (left), the graph of the function still seems slightly curved, but the tangent line and function

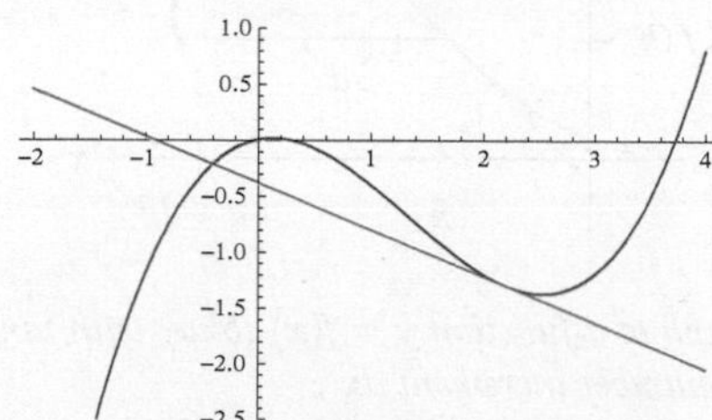

Figure 12 *The graph of $y = 0.2x^3 - 0.8x^2 + 0.2x$ (blue) along with its tangent line (orange) at $x = 2.2$, generated by Mathematica*

are much closer in appearance. If we zoom in even further and view the graph only between $x = 2.19$ and $x = 2.21$ as in figure 13 (right), the function and its tangent line are indistinguishable.

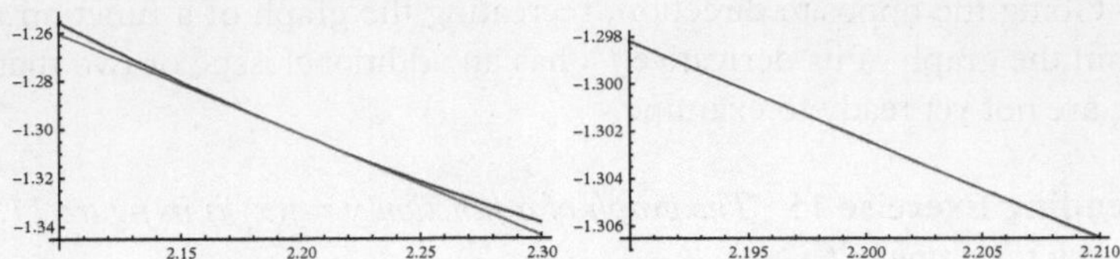

Figure 13 *The graph of* $y = 0.2x^3 - 0.8x^2 + 0.2x$ *(blue) along with its tangent line (orange) at* $x = 2.2$ *on two different scales, generated by Mathematica. On the right, the graph of the curve and its tangent line are indistinguishable*

This effect always happens for a differentiable function! If we zoom in far enough, the tangent line and the graph of the function are indistinguishable. Because the graph of the function resembles a line, we call this property *local linearity*. ("Local" refers to the small region in which the graph is viewed, as opposed to a "global" view of the function.)

Local linearity always shows up for a differentiable function on a sufficiently small scale as long as a constant aspect ratio is used as we zoom in. The aspect ratio is the ratio of the length of one unit as it appears on the vertical scale to the length of one unit as it appears on the horizontal scale. Some computer algebra systems change the aspect ratio rather freely, so the aspect ratio may sometimes need to be controlled to "see" the local linearity.

The traditional diagram that illustrates these relationships is presented in figure 14. The curve $y = f(x)$ and its tangent line at $x = k$ are

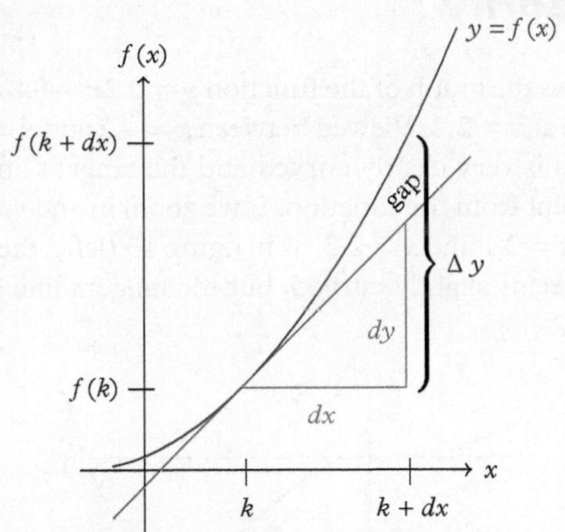

Figure 14 *The graph of a function* $y = f(x)$ *(blue) with tangent line (orange) at* $x = k$ *and a real-number increment* dx

shown. If we "increment" in the x-direction by a real number dx, moving to $k + dx$, then the y-coordinate of the function (blue) increases by Δy to $f(k + dx)$ whereas the y-coordinate of the tangent line (orange) increases by a different amount, dy. This leaves a "gap" between the function and the tangent line. The size of the gap is the difference between Δy and dy. When we zoom in to show local linearity, the relative size of the gap gets small enough that we cannot see the difference between the curve and the tangent line.

We need to notice two more facts from figure 14 before continuing. The first is that Δy is the difference in y-coordinates on the graph of the function at $x = k$ and $x = k + dx$:

$$\Delta y = f(k + dx) - f(k).$$

For the second fact, recall that the slope of the tangent line (orange) at $x = k$ is $f'(k)$ and that slope $= \frac{\text{rise}}{\text{run}}$. Then, from figure 14, $f'(k) = \frac{dy}{dx}$.

What happens when we zoom in infinitely far, so far that the increment dx is no longer a real number, but an infinitesimal? Because f is differentiable at $x = k$, the derivative formula must hold for all infinitesimals, including dx, so that

$f'(k) = \frac{dy}{dx}$ in figures 14 and 15 illustrates the reasoning behind Leibniz notation.

$$f'(k) = \frac{f(k + dx) - f(k)}{dx}.$$

Ans. to reading exercise 15

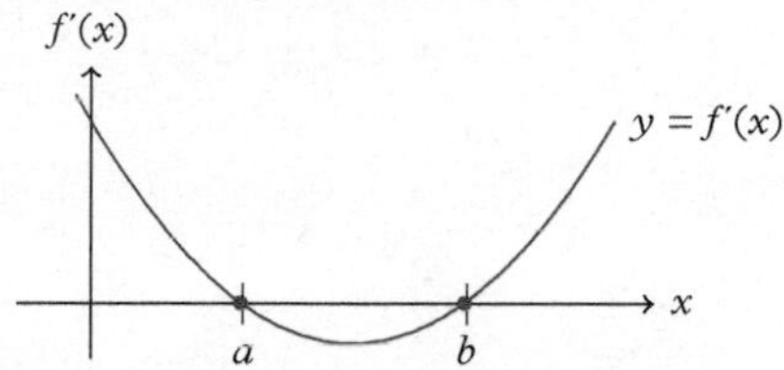

Using both our facts from the previous paragraph yields

$$\frac{dy}{dx} = f'(k) = \frac{f(k + dx) - f(k)}{dx} = \frac{\Delta y}{dx}.$$

Then, dy and Δy must be the same—at least approximately the same—because there is some approximation and rendering when calculating the derivative, even though we used the equal sign. Then, at the infinitesimal scale, the gap disappears! With no gap, the function and

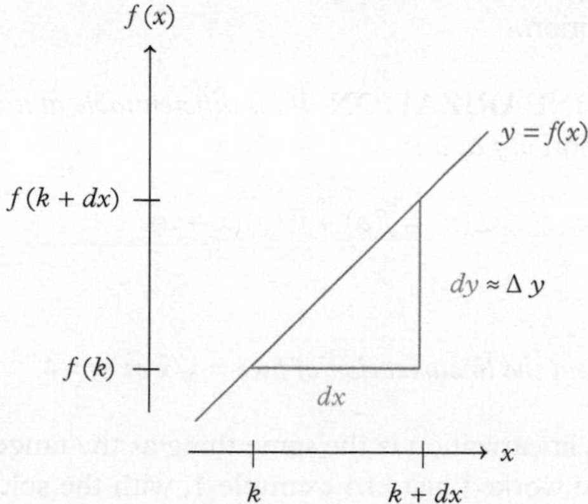

Figure 15 *The graph of a function $y = f(x)$ indistinguishable from its tangent line at $x = k$, where dx is an infinitesimal*

tangent line must be indistinguishable, and we have local linearity. This situation is pictured in figure 15.

The local linearity formula gets its name from this situation as well. The tangent line in figure 15 has equation $y = f'(k)(x - k) + f(k)$. At $x = k + dx$, the y-coordinate of the tangent line is approximately the same as the y-coordinate on the graph of the function, $f(k+dx)$. Using the equation of the tangent line and replacing x with $k+dx$ and y with $f(k+dx)$ gives

$$f(k+dx) \approx f'(k)(k+dx-k) + f(k) = f(k) + dx\,f'(k).$$

Rewriting using the infinitesimal α in place of the infinitesimal dx gives the local linearity formula

$$f(k+\alpha) \approx f(k) + \alpha f'(k).$$

Linearizations

Because the graph of a differentiable function is indistinguishable from its tangent line when we stick close enough to the point of tangency, the tangent line is sometimes called the *linearization* of the function. At the point $(a, f(a))$, with slope $f'(a)$ (see figure 1 in this section), the equation of the tangent line is

$$\begin{aligned} y - y_1 &= m(x - x_1) \\ y - f(a) &= f'(a)(x - a) \\ y &= f(a) + f'(a)(x - a). \end{aligned}$$

Writing the tangent line as a function L (L for linearization) yields the following definition.

Definition 2 LINEARIZATION *If f is differentiable at $x = a$, then the* linearization *of f at a is*

$$L(x) = f(a) + f'(a)(x - a).$$

Example 5 *Find the linearization of $f(x) = \sqrt{x}$ at $x = 4$.*

Because the linearization is the same thing as the tangent line, this example can be worked just like example 1, with the solution placed in the linearization format. A slightly rearranged version is given next.

Solution We want the linearization at $x = 4$, so a is 4. Next we calculate the derivative:

$$f(x) = x^{1/2}$$
$$f'(x) = \frac{1}{2}x^{\frac{1}{2}-1} = \frac{1}{2}x^{-\frac{1}{2}} = \frac{1}{2x^{\frac{1}{2}}} = \frac{1}{2\sqrt{x}}.$$

Then, we calculate $f(a)$ and $f'(a)$:

$$f(x) = \sqrt{x} \qquad f(a) = f(4) = \sqrt{4} = 2$$
$$f'(x) = \tfrac{1}{2\sqrt{x}} \qquad f'(a) = f'(4) = \tfrac{1}{2\sqrt{4}} = \tfrac{1}{4}.$$

Using the formula for linearization,

$$L(x) = f(a) + f'(a)(x - a)$$
$$L(x) = 2 + \frac{1}{4}(x - 4).$$

The format of the answer is important; we leave the final answer in the form $L(x) = 2 + \frac{1}{4}(x - 4)$. ■

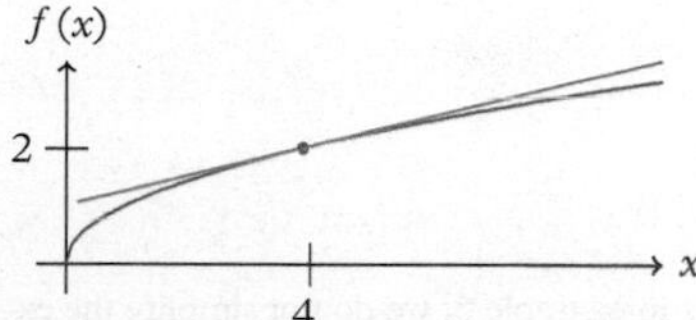

Figure 16 *The curve* $y = \sqrt{x}$ *(blue) with its linearization (orange) at* $x = 4$

Reading Exercise 16 *Find the linearization of* $f(x) = x^2$ *at* $x = 1$.

The linearization and the function of example 5 are close when x is close to 4. The farther we stray from $x = 4$, the farther apart the two are, as shown in figure 16. If we use the linearization, which is easy to calculate by hand, as an approximation to the function, the approximation is good close to the point of tangency, but not as good if we stray too far.

In addition to viewing figure 16, a table of values (table 1) can help us determine how close the linearization and the function are as we stray away from the point of tangency.

Table 1 shows that, depending upon the desired accuracy, using the line to estimate values of the square root is reasonable as long as we stay close enough to the point of tangency. In practice, this method of estimation is only rarely used, but it is related to other important methods of estimating functions presented in section 10.10. For this reason, we should look at an example.

Table 1 also illustrates a method of mental estimation of square roots (or other functions). We know $\sqrt{4} = 2$. To estimate $\sqrt{4.1}$ we follow the linearization, using $x = 4.1$: $L(x) = 2 + \frac{1}{4}(4.1 - 4) = 2 + 0.25(0.1) = 2 + 0.025 = 2.025$, which agrees with $\sqrt{4.1}$ to four significant digits. With practice, this can make an impressive party trick. An alternate version uses differentials.

Example 6 *Use the linearization of* $f(x) = \sqrt[3]{x}$ *at* $x = 8$ *to estimate* $\sqrt[3]{8.05}$.

Table 1 *Comparing the values of a function and its linearization*

x	$f(x) = \sqrt{x}$	$L(x) = 2 + \frac{1}{4}(x-4)$	difference
4	2	2	0
4.1	2.02485	2.025	0.00015
4.5	2.12132	2.125	0.00368
5	2.23607	2.25	0.01393
10	3.16228	3.5	0.33772
100	10	26	16

Solution First we find the linearization. We begin by calculating the derivative:

$$f'(x) = \frac{1}{3}x^{-2/3} = \frac{1}{3(\sqrt[3]{x})^2}.$$

Next we calculate $f(a)$ and $f'(a)$, using $a = 8$:

$$f(a) = f(8) = \sqrt[3]{8} = 2$$

$$f'(a) = f'(8) = \frac{1}{3(\sqrt[3]{8})^2} = \frac{1}{3 \cdot 2^2} = \frac{1}{12}.$$

As in example 5, we do not simplify the expression, but leave the linearization in the prescribed format.

Using the formula for linearization,

$$L(x) = f(a) + f'(a)(x-a) = 2 + \frac{1}{12}(x-8).$$

We finish estimating $\sqrt[3]{8.05}$ by finding $L(8.05)$:

$$L(8.05) = 2 + \frac{1}{12}(8.05-8) = 2 + \frac{1}{12}(0.05) = 2 + 0.004167 = 2.004167.$$

The result, which can be calculated without use of technology, is accurate to five significant digits. ■

Differentials

One idea gained from figure 15 (repeated in the margin as figure 17) is that $f'(k) = \frac{dy}{dx}$. As long as dx is an infinitesimal, so is dy, and the derivative $\frac{dy}{dx}$ can be thought of as a ratio of infinitesimals. Rearranging,

$$dy = f'(k)\,dx.$$

The quantities dy and dx are also called *differentials*. Until the turn of the 20th century, calculus was written in terms of differentials instead of derivatives. The differential notation is still used in certain contexts.

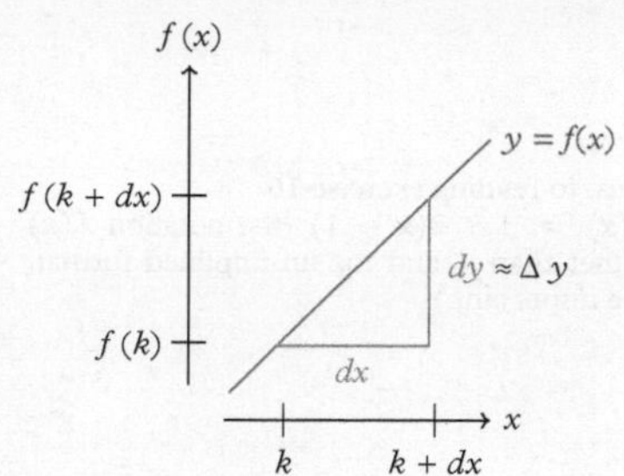

Figure 17 *The graph of a function $y = f(x)$ indistinguishable from its tangent line at $x = k$, where dx is an infinitesimal*

Definition 3 DIFFERENTIAL *Let $y = f(x)$ be a differentiable function. We let the* differential *dx be an independent variable. Then the* differential *dy is given by*

$$dy = f'(x)\,dx.$$

Example 7 *Find dy for $y = x^4 + 2x - 1$.*

Solution We begin by finding the derivative:

$$\frac{dy}{dx} = 4x^3 + 2.$$

Rearranging, we can think of multiplying both sides by the differential dx:

$$dy = (4x^3 + 2)\,dx. \qquad \blacksquare$$

In the 18th and 19th centuries, the differential dy was calculated directly. Instead of derivative rules there were differential rules, such as "If $y = x^n$, then $dy = nx^{n-1}\,dx$." The product rule was written

$$d(uv) = u\,dv + v\,du.$$

More than a mere curiosity, in the contexts in which differential notation is still used it can be very helpful for avoiding certain types of errors. An investment in learning the notation now pays dividends later.

Reading Exercise 17 *Find dy for $y = \sqrt{x}$.*

Exercises 2.3

1–8. Rapid response: state whether the given expression is a derivative or a differential.

1. $y' = 4x^2$
2. $f'(x) = 7x - 9$
3. $dy = (5x - 17)\,dx$
4. $\frac{dy}{dx} = \sqrt{x}$

Ans. to reading exercise 16
$L(x) = 1 + 2(x - 1)$ (the notation $L(x)$ rather than y, and the unsimplified format, are important.)

5. $g'(t) = t$
6. $dy = 4x^5\,dx$
7. $dy = 47t\,dt$
8. $du = (4t - 9)\,dt$

9. Find the equation of the tangent line to the curve $y = x^2 - 4x + 7$ at $x = 5$.
10. Find the equation of the tangent line to the curve $f(x) = x^5 + x$ at $x = 1$.
11. Find the linearization of the function $y = x^2 - 4x + 7$ at $x = 5$.
12. Find the linearization of the function $f(x) = x^5 + x$ at $x = 1$.

13–22. Find the equation of the tangent line to the curve at the given value of the variable.

13. $y = \dfrac{x+1}{x-4}$ at $x = 0$
14. $f(t) = \dfrac{1}{t^3}$ at $t = 2$
15. $g(z) = 4 - z^2$ at $z = -5$
16. $y = (x^{12} - 4)(5 - 3x)$ at $x = 0$
17. $s = -3t + 7$ at $t = 14.7$
18. $h(x) = x^3 - 4x^2 + 5$ at $x = -1$
19. $y = \sqrt[3]{x}$ at $x = 1$
20. $f(x) = \dfrac{x^2 + 6}{3x + 1}$ at $x = 2$
21. $f(x) = (5x - 1)(x^{-2})$ at $x = -1$
22. $y = x^3$ at $x = 0$

23–28. Calculate the linearization of the function at the given value of the variable.

23. $y = x^4 - 11x^2$ at $x = -2$
24. $f(x) = x(x + 1)(x - 3)$ at $x = 5$
25. $f(x) = \dfrac{1}{x + \frac{1}{x^2}}$ at $x = 2$
26. $y = \sqrt[5]{x^8} - \sqrt[7]{142}$ at $x = 0$
27. $g(z) = \dfrac{(z-2)(z-5)}{z+1}$ at $z = 1$
28. $f(t) = \dfrac{t^4 + 3t^2}{t^2}$ at $t = -3$

29–38. Calculate the differential dy.

29. $y = x^2$

30. $y = 5x^{20}$

31. $y = \dfrac{22}{t^2}$

32. $y = 6\theta - 87$

33. $y = x(4x - 1)$

34. $y = \sqrt[5]{x^2}$

35. $y = \dfrac{\sqrt{x} - 1}{\sqrt{x} + 1}$

36. $y = \dfrac{x^2 + x^3 + x^4}{x^5}$

37. $y = 3s + \dfrac{s}{s + 1}$

38. $y = \dfrac{t - 6}{5 - t}$

39–50. Find the locations of all horizontal tangent lines on the graph of the function.

39. $f(x) = \dfrac{x^3}{3} - \dfrac{x^2}{2} - 6x + 11$

40. $y = \dfrac{x^3}{3} + \dfrac{5x^2}{2} + 4x + 9$

41. $y = 15x^4 - 20x^3 + 7$

42. $f(x) = 5x^4 + 20x^3 + 12$

43. $y = 10x + \dfrac{3x^2}{2} - \dfrac{x^3}{3} + \sqrt{3}$

44. $f(x) = \dfrac{1}{1 + x^2}$

45. $f(x) = 2x^3 + 6x^2 + x$

46. $y = -x^3 + 2x^2 + 3x$

47. $f(t) = t^3 - 4t^2 + 9t - 6$

48. $f(x) = \dfrac{3x + 7}{x - 4}$

49. $y = \dfrac{x + 5}{x^2 - 4}$

50. $g(t) = \sqrt{t} - t$

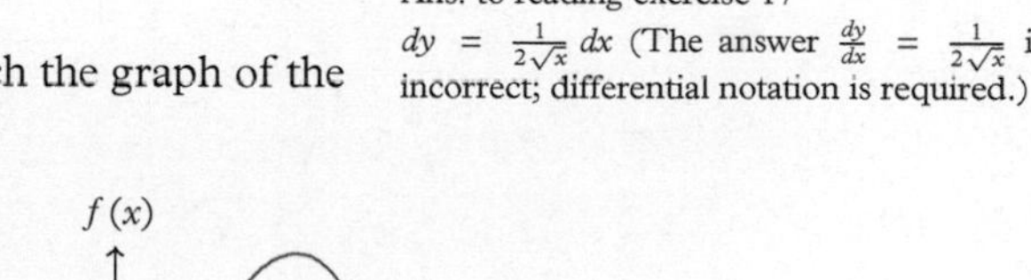

Ans. to reading exercise 17
$dy = \frac{1}{2\sqrt{x}}\,dx$ (The answer $\frac{dy}{dx} = \frac{1}{2\sqrt{x}}$ is incorrect; differential notation is required.)

51–56. The graph of a function f is given. Sketch the graph of the derivative f'.

51.

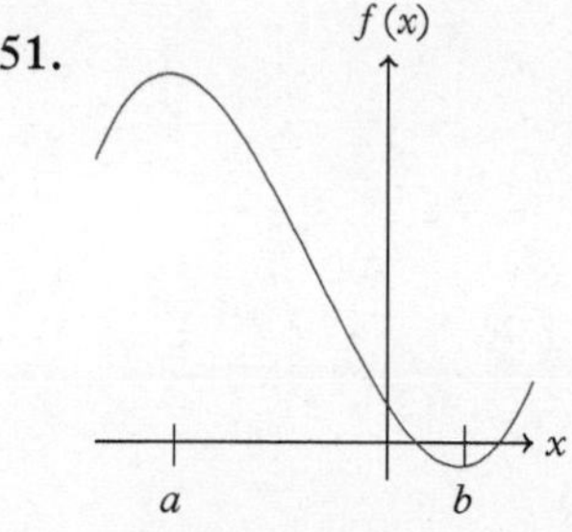

52.

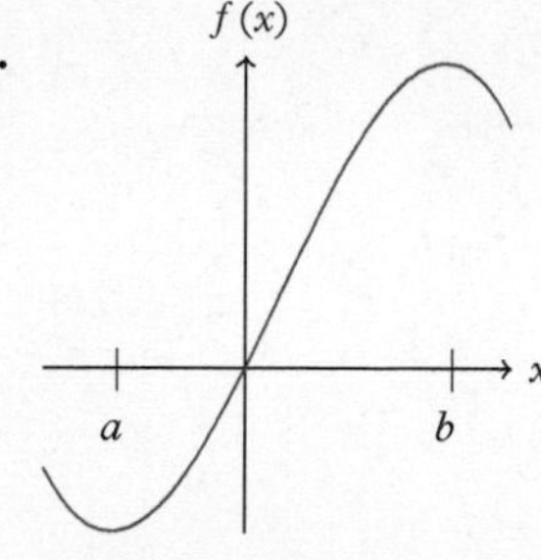

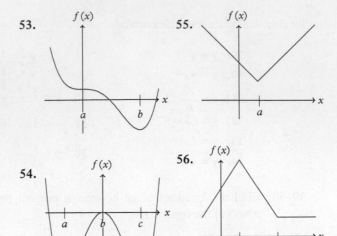

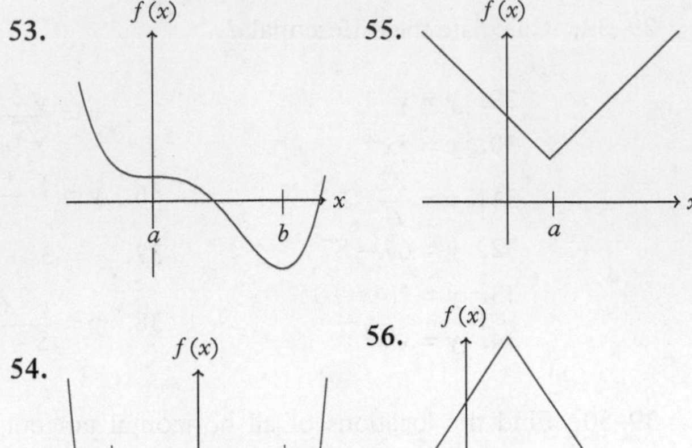

54.

56.

57. Use the linearization of $f(x) = \sqrt{x}$ at $x = 16$ to estimate $\sqrt{15.7}$.

58. Use the linearization of $f(x) = \sqrt[3]{x}$ at $x = -1$ to estimate $\sqrt[3]{-0.9}$.

59. (a) For $f(x) = \frac{1}{x}$, find the equation of the tangent line at $x = a$. (b) Find the x-intercept and y-intercept of the tangent line. (c) Conclude that the point of tangency is halfway between the tangent line's x- and y-intercepts.

Derivatives of Trigonometric Functions

2.4

Trig functions have derivatives too. First we need to know a little more about limits of trigonometric functions.

Two trig limits

Theorem 3 USEFUL LIMIT INVOLVING THE SINE FUNCTION

$$\lim_{x\to 0} \frac{\sin x}{x} = 1$$

Why do we not just calculate the limit? Because we would have $\lim_{x\to 0} \frac{\sin x}{x} = \frac{\sin \alpha}{\alpha}$ and we do not yet know how to handle $\sin \alpha$. We know that $\sin \alpha$ is infinitesimal, but we do not know which infinitesimal it is.

Proof. From section 1.7 figure 16, we learned that

$$0 < \sin x < x \text{ when } 0 < x < \frac{\pi}{2},$$

which rearranges to

$$\frac{\sin x}{x} < 1.$$

A similar argument (see figure 1) shows that

$$x < \tan x \text{ when } 0 < x < \frac{\pi}{2},$$

and therefore $x < \frac{\sin x}{\cos x}$, which rearranges to $\cos x < \frac{\sin x}{x}$. Combining inequalities we have

$$\cos x < \frac{\sin x}{x} < 1.$$

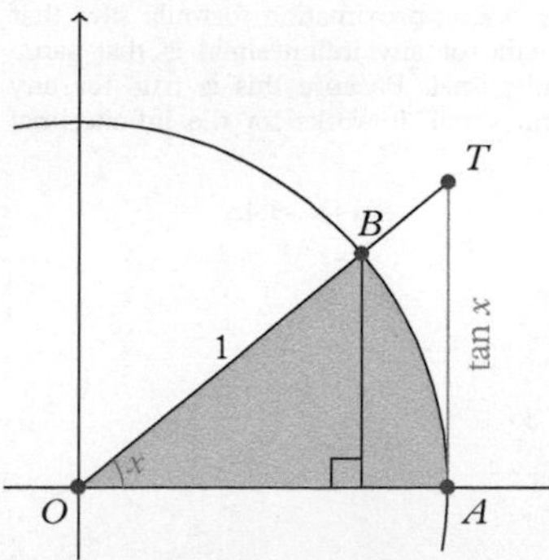

Figure 1 *Proof that $x < \tan x$ for $0 < x < \frac{\pi}{2}$. Notice area(sectorOAB) $= \frac{1}{2}r^2\theta = \frac{1}{2}1^2x = \frac{1}{2}x$ and area($\triangle OAT$) $= \frac{1}{2}bh = \frac{1}{2} \cdot 1 \cdot \tan x$, and the area of the sector must be smaller than the area of the triangle, yielding $\frac{1}{2}x < \frac{1}{2}\tan x$*

We are now set up to try the squeeze theorem. On the left, evaluating using continuity,

Because $\lim_{x\to 0} \cos x$ is calculated as $\lim_{x\to 0} \cos x = \cos \alpha$, it must be true that $\cos \alpha \approx 1$.

For any infinitesimal α, $\cos \alpha \approx 1$.

$$\lim_{x\to 0} \cos x = \cos 0 = 1.$$

For the right-hand side,

$$\lim_{x\to 0} 1 = 1.$$

Because the limits on both sides of the inequality are the same number, by the squeeze theorem, so is the limit of the middle:

$$\lim_{x\to 0^+} \frac{\sin x}{x} = 1.$$

Note that we used inequalities valid for $0 < x < \frac{\pi}{2}$, so our limit is only proven for values larger than zero, or to the right of zero, but a similar argument (omitted) works for the limit from the left, completing the proof. ■

Now we return to the "calculation" of the limit. We have $\lim_{x\to 0} \frac{\sin x}{x} = \frac{\sin \alpha}{\alpha}$, and that quantity must render one. Therefore, $\frac{\sin \alpha}{\alpha} \approx 1$, or

$$\sin \alpha \approx \alpha$$

for any infinitesimal α.

SINE APPROXIMATION FORMULA

For any infinitesimal α,

$$\sin \alpha \approx \alpha.$$

The sine approximation formula says that the sine of any infinitesimal is that same infinitesimal. Because this is true for any infinitesimal, it works for the infinitesimal 4α:

$$\sin 4\alpha \approx 4\alpha.$$

Example 1 *Find* $\lim_{x\to 0} \dfrac{\sin 3x}{7x}$.

Solution Evaluating using continuity does not work (division by zero), so we use $x = 0 + \alpha = \alpha$:

The next-to-last step uses the sine approximation formula for $\sin 3\alpha \approx 3\alpha$.

$$\lim_{x\to 0} \frac{\sin 3x}{7x} = \frac{\sin 3\alpha}{7\alpha} \approx \frac{3\alpha}{7\alpha} = \frac{3}{7}.$$

The limit is $\frac{3}{7}$. ■

Compare the sine approximation formula $\sin\alpha \approx \alpha$ to the formula $\cos\alpha \approx 1$. Both are used in the next example.

Since $\cos\alpha \approx 1$ for any infinitesimal α, and 3α is infinitesimal, $\cos 3\alpha \approx 1$.

Example 2 *Find* $\lim\limits_{\theta\to 0} \theta\cot 3\theta$.

Solution Evaluating using continuity does not work ($\cot 0$ is undefined), so we use $\theta = 0 + \alpha = \alpha$:

$$\lim_{\theta\to 0} \theta\cot 3\theta = \alpha\cot 3\alpha.$$

We have not looked at a cotangent approximation formula, but we can use an identity to rewrite cotangent in terms of sine and cosine:

Throughout calculus, it is occasionally helpful to rewrite tan, cot, sec, and csc in terms of sin and cos, especially if formulas or techniques for dealing with sin and cos are available.

$$= \alpha \cdot \frac{\cos 3\alpha}{\sin 3\alpha} \approx \alpha \cdot \frac{1}{3\alpha} = \frac{1}{3}.$$

The sine approximation formula and $\cos\alpha \approx 1$ are both used in example 2.

The limit is $\frac{1}{3}$. ■

Reading Exercise 18 *Find* $\lim\limits_{x\to 0} \dfrac{\sin 5x}{\sin 7x}$.

The sine approximation formula can be used to prove the next theorem.

Theorem 4 A USEFUL LIMIT INVOLVING THE COSINE FUNCTION

Compare theorem 4 to section 1.5 example 7.

$$\lim_{x\to 0} \frac{\cos x - 1}{x} = 0$$

Proof. We start with $x = 0 + \alpha = \alpha$, giving

$$\lim_{x\to 0} \frac{\cos x - 1}{x} = \frac{\cos\alpha - 1}{\alpha}.$$

Using $\cos\alpha \approx 1$ results in $\frac{1-1}{\alpha} = \frac{0}{\alpha}$, which violates the approximation principle. The trick is to multiply the numerator and the denominator by the conjugate expression $\cos\alpha + 1$:

$$= \frac{\cos\alpha - 1}{\alpha} \cdot \frac{\cos\alpha + 1}{\cos\alpha + 1}$$

$$= \frac{\cos^2\alpha - 1}{\alpha(\cos\alpha + 1)} = \frac{-\sin^2\alpha}{\alpha(\cos\alpha + 1)} = \frac{-(\sin\alpha)^2}{\alpha(\cos\alpha + 1)}.$$

This step uses the Pythagorean identity $\sin^2\theta + \cos^2\theta = 1$, rearranged to $\cos^2\theta - 1 = -\sin^2\theta$.

Now we can use the sine approximation formula and the fact that $\cos\alpha \approx 1$:

$$\approx \frac{-\alpha^2}{\alpha(1+1)} = \frac{-\alpha}{2} \doteq 0.$$

■

The proof of theorem 4 also demonstrates the following formula:

COSINE APPROXIMATION FORMULA

For any infinitesimal α,

$$\frac{\cos\alpha - 1}{\alpha} \approx \frac{-\alpha}{2},$$

which is infinitesimal.

Our primary use for the cosine approximation formula is in the derivation of derivatives of trig functions.

Derivatives of sine and cosine

To calculate the derivatives of sine and cosine, we use the sine and cosine approximation formulas along with the sum formulas from trigonometry:

$$\sin(a+b) = \sin a\cos b + \cos a\sin b,$$

$$\cos(a+b) = \cos a\cos b - \sin a\sin b.$$

Using the derivative formula (function version) and a sum formula,

$$\frac{d}{dx}\sin x = \frac{\sin(x+\alpha) - \sin x}{\alpha}$$

$$= \frac{\sin x\cos\alpha + \cos x\sin\alpha - \sin x}{\alpha}.$$

Using $\cos\alpha \approx 1$ gives, for the first and last terms in the numerator, $\sin x \cdot 1 - \sin x = 0$, violating the approximation principle for that portion of the numerator.

Next we factor the $\sin x$ from the first and last terms in the numerator, rearrange algebraically, and use the sine and cosine approximation formulas:

$$= \frac{\sin x(\cos\alpha - 1) + \cos x \sin\alpha}{\alpha}$$

$$= \underbrace{\sin x}_{\text{real}} \underbrace{\left(\frac{\cos\alpha - 1}{\alpha}\right)}_{\text{infinitesimal}} + \underbrace{\cos x}_{\text{real}} \underbrace{\left(\frac{\sin\alpha}{\alpha}\right)}_{\approx 1}$$

$$\approx 0 + \cos x \cdot 1 = \cos x.$$

The term on the left is infinitesimal and the term on the right is real, so the term on the left is discarded when approximating.

DERIVATIVE OF SINE

$$\frac{d}{dx}\sin x = \cos x$$

Ans. to reading exercise 18

$\frac{5}{7}$

The derivative of cosine is similar:

$$\frac{d}{dx}\cos x = \frac{\cos(x+\alpha) - \cos x}{\alpha}$$

$$= \frac{\cos x \cos\alpha - \sin x \sin\alpha - \cos x}{\alpha}$$

$$= \underbrace{\cos x}_{\text{real}} \underbrace{\left(\frac{\cos\alpha - 1}{\alpha}\right)}_{\text{infinitesimal}} - \underbrace{\sin x}_{\text{real}} \underbrace{\left(\frac{\sin\alpha}{\alpha}\right)}_{\approx 1}$$

$$\approx 0 - \sin x \cdot 1 = -\sin x.$$

The term on the left is infinitesimal and the term on the right is real, so the term on the left is discarded when approximating.

DERIVATIVE OF COSINE

$$\frac{d}{dx}\cos x = -\sin x$$

Derivatives of the remaining trig functions

The derivatives of tangent, cotangent, secant, and cosecant can all be derived from the derivatives of sine and cosine using the quotient rule and trig identities. For instance,

$$\frac{d}{dx}\tan x = \frac{d}{dx}\frac{\sin x}{\cos x} = \frac{\cos x \cos x - \sin x(-\sin x)}{\cos^2 x}$$
$$= \frac{\cos^2 x + \sin^2 x}{\cos^2 x} = \frac{1}{\cos^2 x}$$
$$= \sec^2 x$$

Although the derivative of tangent can be written as $\frac{d}{dx}\tan x = \frac{1}{\cos^2 x}$, it is traditional to write the derivative as $\sec^2 x$. Likewise, the derivatives of cotangent, secant, and cosecant are presented in their traditional forms.

and

$$\frac{d}{dx}\sec x = \frac{d}{dx}\frac{1}{\cos x} = \frac{\cos x \cdot 0 - 1(-\sin x)}{\cos^2 x}$$
$$= \frac{\sin x}{\cos^2 x} = \frac{\sin x}{\cos x}\cdot\frac{1}{\cos x}$$
$$= \tan x \sec x.$$

The derivatives of $y = \cot x$ and $y = \csc x$ are left as exercises.

TRIG DERIVATIVES

$$\frac{d}{dx}\sin x = \cos x \qquad \frac{d}{dx}\csc x = -\csc x \cot x$$
$$\frac{d}{dx}\cos x = -\sin x \qquad \frac{d}{dx}\sec x = \sec x \tan x$$
$$\frac{d}{dx}\tan x = \sec^2 x \qquad \frac{d}{dx}\cot x = -\csc^2 x$$

Tips for remembering the formulas: (1) the derivative formulas come in cofunction pairs; (2) the derivatives of functions beginning with "co" (cosine, cotangent, cosecant) all have a negative sign in the formula, whereas the others do not.

Now that the derivatives of the trig functions have been derived, there is no need to rederive them each time they are needed; we can simply use the formulas as presented in the trig derivatives box.

Derivative examples with trig functions

Derivative rules such as the product and quotient rules apply (where appropriate) to all functions, including trig functions.

Example 3 *Find $f'(x)$ for $f(x) = x \sin x$.*

When the variable (x) appears in two or more places in a term, it's an indication that the term is a product of two or more factors.

Solution Notice that $x \sin x$ is a product; it's $x \cdot \sin x$. We therefore use the product rule:

$$f'(x) = \underset{\text{first}}{x} \cdot \underset{\text{second}'}{\cos x} + \underset{\text{second}}{\sin x} \cdot \underset{\text{first}'}{1} = x\cos x + \sin x. \quad \blacksquare$$

Sometimes more than one derivative rule must be used.

Example 4 *Find y' given* $y = \dfrac{x^2 \tan x}{3x - 7}$.

Solution The function is a quotient, but there is also a product in the numerator. Which rule comes first? The key to deciding this is to determine which is the "outside" or "overall" structure; in this case, the function is a quotient with a product inside. We start by using the quotient rule:

We can rewrite the function as a product with a quotient inside:

$$y = \frac{x^2}{3x - 7} \cdot \tan x.$$

To find the derivative from this form, we start with the product rule.

$$y' = \frac{\overset{\text{bot}}{(3x-7)} \cdot \overset{\text{top}'}{\overbrace{\underset{\text{first}\,\cdot\,\text{second}'\;+\;\text{second}\,\cdot\,\text{first}'}{\left(x^2 \sec^2 x + (\tan x)(2x)\right)}}} \overset{-\text{top}}{-(x^2 \tan x)} \cdot \overset{\text{bot}'}{(3)}}{\underset{\text{bot}^2}{(3x-7)^2}}.$$

When finding top′, we are finding $\frac{d}{dx} x^2 \tan x$, which is the derivative of a product; therefore, the product rule is needed.

To avoid potential confusion, the polynomial factors of a term are traditionally written in front of the trig factors so that it is clear what is in the argument of the trig function and what is not:

$$= \frac{(3x - 7)(x^2 \sec^2 x + 2x \tan x) - 3x^2 \tan x}{(3x - 7)^2}.$$

If desired, the result can be simplified. ■

Writing the templates for the quotient and product rules can be very helpful for keeping track of all the pieces, especially the first few times an exercise requiring multiple rules is encountered.

Example 5 *Find* $\frac{d^2y}{dx^2}$ *for* $y = \sec x$.

Solution The first derivative is merely one of the trig derivative formulas:

$$y' = \sec x \tan x.$$

We need a second derivative. Because y' is a product, we use the product rule:

The variable x appears in both $\sec x$ and $\tan x$; the implication is that we have the product $\sec x \cdot \tan x$.

$$y'' = \overset{\text{first}}{\sec x} \cdot \overset{\text{second}'}{\sec^2 x} + \overset{\text{second}}{(\tan x)} \cdot \overset{\text{first}'}{(\sec x \tan x)} = \sec^3 x + \sec x \tan^2 x. \;\blacksquare$$

Reading Exercise 19 *Find* $\dfrac{d}{dx}\dfrac{\sin x}{x + \cos x}$.

Example 6 *Find the equation of the tangent line to the curve* $y = \sin x$ *at* $x = 0$.

Compare this example to section 2.3 example 1.

Solution Recall that there are three steps in finding the equation of a tangent line.

❶ To find the slope, we need the derivative:

$$y' = \cos x.$$

We then evaluate the derivative at $x = 0$:

$$\text{slope} = y'(0) = \cos 0 = 1.$$

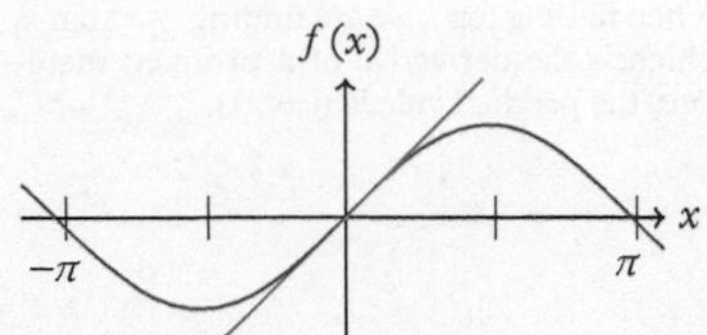

Figure 2 *The curve $y = \sin x$ (blue) with tangent line (orange) at $x = 0$*

❷ We have not been given the y-coordinate of the point of tangency, so we calculate it next:

$$y(0) = \sin 0 = 0.$$

The point of tangency is $(0, 0)$.

CAUTION: MISTAKE TO AVOID
Some students make the mistake of drawing $y = \sin x$ so that the curve appears to have slopes that are much too steep, even to the point of appearing to have vertical tangent lines. If you are prone to making this mistake, practice drawing the sine curve with appropriate slopes several times.

❸ We finish by using the point-slope equation of the line with point $(0, 0)$ and slope 1:

$$y - 0 = 1(x - 0)$$
$$y = x.$$

The equation of the tangent line is $y = x$. ■

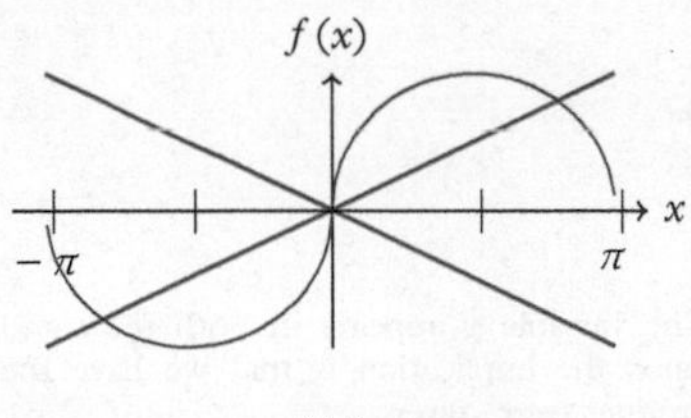

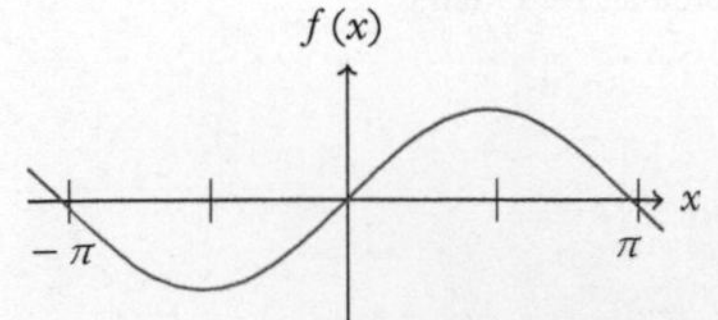

Figure 3 *(top) Incorrectly drawn graph of $y = \sin x$; (bottom) correctly drawn graph of $y = \sin x$*

The graph of $y = \sin x$ with its tangent line at $x = 0$ is in figure 2. Because the derivative of $y = \sin x$ is $y' = \cos x$, and the values of $\cos x$ are always between 1 and -1, the slopes on the sine curve are always between 1 and -1. Do not make the mistake of drawing the sine curve with slopes that are too steep. See figure 3.

Why study derivatives of trig functions?

We study the derivatives of trigonometric functions because of their applicability to physics, engineering, and other sciences. Among the untold numbers of such applications is *harmonic motion*. Consider a mass attached to the end of a spring, as in figure 4. On the left, the

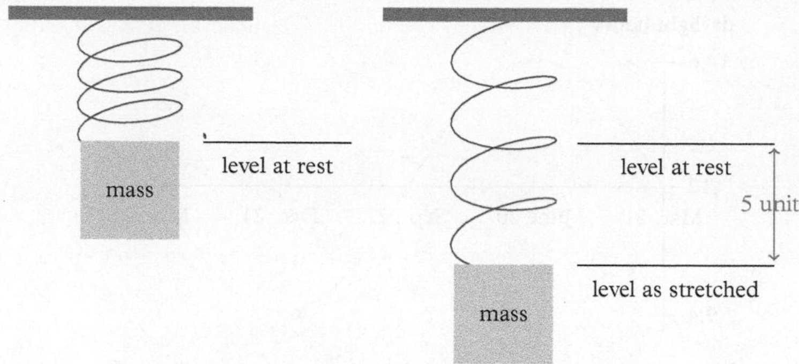

Figure 4 *Spring–mass system: (left) at rest; (right) pulled down 5 units*

spring is at rest in its natural position. On the right, someone has pulled the mass down 5 units and is ready to release it. When released, the mass does not just return to the resting position and stay; it moves above the resting position, then below it, and repeats, oscillating up and down. In fact, in the absence of any friction or other similar forces, the mass would continue forever with equation of motion $y = 5\cos t$. Then the derivative would describe the velocity of the object.

Ans. to reading exercise 19

$$\frac{(x+\cos x)\cos x - (\sin x)(1-\sin x)}{(x+\cos x)^2} = \frac{x\cos x - \sin x + 1}{(x+\cos x)^2}$$

Another application of trigonometric functions is to determine hours of daylight. For Tahlequah, OK, the number of hours of daylight is given by

$$12 + 2.6\sin\left(\frac{2\pi}{365.25}t\right),$$

Because the orbit of earth is not exactly circular, the actual function is a little more complicated, but the number of hours of daylight can be closely approximated by a sine curve.

where t is measured in days and $t = 0$ corresponds to March 21. With an appropriate value in place of 2.6 (higher values for higher latitudes), this equation is a good approximation for locations in the northern hemisphere above the tropic of cancer and below the arctic circle.

For the southern hemisphere, $t = 0$ corresponds to September 22.

The graph of the number of hours of daylight for Tahlequah is in figure 5. The curve is steepest near the spring and fall equinoxes. The number of hours of daylight is changing most rapidly then. The largest value of the derivative is at March 21 whereas the most negative value of the derivative is at September 22. The curve is flattest (in orange) near the summer and winter solstices; the number of hours of daylight is changing least rapidly then. The derivative is zero at June 20 and December 21.

The dates mentioned vary slightly from year to year.

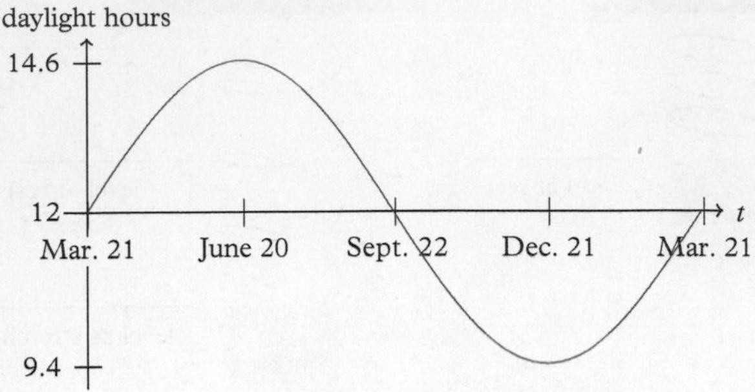

Figure 5 *Hours of daylight in Tahlequah, OK; the orange portions of the graph are where the number of hours of daylight is fairly consistent (the rate of change is small)*

Exercises 2.4

1–8. Rapid response: state the derivative rule(s), if any, from section 2.2 that are needed to find y'. (Do not find the derivative.)

1. $y = x \tan x$
2. $y = \dfrac{x}{\tan x}$
3. $y = \dfrac{1}{\tan x}$
4. $y = x + \tan x$
5. $y = \dfrac{x^2 - 1}{x \sin x}$
6. $y = 62 \csc x$
7. $y = \sin x \cos x$
8. $y = \dfrac{\sin x}{\cos x}$

9–20. Evaluate the limit.

9. $\lim\limits_{x \to 0} \dfrac{\sin 4x}{9x}$
10. $\lim\limits_{x \to 0} \dfrac{\sin 2x}{5x}$
11. $\lim\limits_{x \to 0} \dfrac{\sin 5x}{x + \sin 6x}$
12. $\lim\limits_{x \to 0} \dfrac{\sin x^2}{4x - 2x^2}$
13. $\lim\limits_{x \to 0^+} \dfrac{\sin 5x}{x^2}$
14. $\lim\limits_{x \to 0} (4 - \tan 3x)$
15. $\lim\limits_{x \to 0} \dfrac{\tan 7x}{2x}$
16. $\lim\limits_{x \to 0} \cos \left(\dfrac{\sin 4x^2}{5x} \right)$
17. $\lim\limits_{x \to 0} \dfrac{\sin 4x + \cos 2x}{6x - 1}$
18. $\lim\limits_{x \to 0} \dfrac{x^2 + \sin x}{x^4}$
19. $\lim\limits_{x \to 0} \left(\dfrac{\sin 5x}{x^2 - \sin 6x} \right)^2$
20. $\lim\limits_{x \to 0} \dfrac{\sin 3x}{6x - \sin 2x}$

21–34. Differentiate the given function.

21. $y = x^2 \sec x$
22. $f(x) = x^4 \cos x$
23. $f(x) = \dfrac{4x^3 - 7}{\sec x}$
24. $f(t) = \dfrac{\cos t}{3t^2}$
25. $f(x) = 5x^2 \sin x$
26. $y = (x^2 + 2x + 7) \sec x$
27. $g(y) = \dfrac{6y}{\csc y}$
28. $f(x) = \dfrac{4}{x^3 \sin x}$
29. $z = \dfrac{x \sec x}{x^3 + 9x}$
30. $h(x) = \sin x \cot x$
31. $g(t) = \dfrac{(t^3 - 4t^2) \cos t}{t \sin t}$
32. $s = \dfrac{t^2 \tan t}{t^2 - 7}$
33. $y = \sin x \tan x$
34. $y = 5\sqrt{x} + 4 \sec x + \tan x$

35–44. Find the required derivative.

35. $y = \csc x$, y''
36. $y = x^2 \sin x$, $\frac{d^2y}{dx^2}$
37. $f(x) = \sin x$, $f^{(4)}(x)$
38. $g(t) = \cos t$, $g^{(4)}(t)$
39. $f(x) = \sin x$, $f^{(4000)}(x)$
40. $g(t) = \cos t$, $g^{(401)}(t)$
41. $y = t \cos t$, $\frac{d^3y}{dt^3}$
42. $f(x) = \dfrac{\sin x}{x}$, $f''(x)$
43. $h(\theta) = \sin\theta \cos\theta$, $h''(\theta)$
44. $y = \sin^2\theta + \cos^2\theta$, y''

45–50. Find the equation of the tangent line to the curve $y = f(x)$ at the given value of x.

45. $y = \cos x$, $x = \frac{\pi}{2}$
46. $y = \sin x$, $x = -\frac{\pi}{2}$
47. $y = \sec x$, $x = 0$
48. $y = \tan x$, $x = 4\pi$
49. $y = x \sin x$, $x = \pi$
50. $y = 3x - 4 \cos x$, $x = 0$

51. Where does the curve $y = \sin x$ have horizontal tangent lines?
52. Where does the curve $y = \tan x$ have horizontal tangent lines?
53. The equation of motion for a spring–mass system is $y = 5 \cos t$. Find the velocity of the object at time $t = \frac{\pi}{2}$. Units are seconds and millimeters.
54. The equation of motion for a spring–mass system is $y = 2 \cos t$. Find the velocity of the object at time $t = \frac{\pi}{4}$. Units are seconds and millimeters.
55. Find the differential dy for $y = 4\theta \sin\theta$.

56. Find the differential dy for $y = \dfrac{x^2}{\cos x}$.

57. Find the linearization of $f(x) = x + \tan x$ at $x = 0$.

58. Find the linearization of $f(\theta) = \frac{1}{2} \sin \theta$ at $\theta = 0$.

59. Derive the formula $\frac{d}{dx} \csc x = -\csc x \cot x$.

60. Derive the formula $\frac{d}{dx} \cot x = -\csc^2 x$.

61. A century ago, some textbooks such as GSL included a seventh trig function, *versine*, defined as $\operatorname{vers} x = 1 - \cos x$. Find $\frac{d}{dx} \operatorname{vers} x$.

62. Evaluate $\lim\limits_{x \to 0} \dfrac{\sin(3 \sin(2 \sin x))}{x}$. (From *College Mathematics Journal*, November 2016, p. 365.)

Chain Rule

2.5

So far, we have derivative rules for sums, products, and quotients. We have derivative formulas for powers and for trig functions. But, we have no derivative rule for compositions of functions. Look over the exercises for calculating derivatives (not limits) in section 2.4. Every trig function has "just plain x" as its argument. We can differentiate $y = x^2 \sin x$ by the product rule, we can differentiate $y = x^2 + \sin x$ by the sum rule, and we can differentiate $y = \frac{x^2}{\sin x}$ by the quotient rule. But how do we differentiate $y = \sin(x^2)$?

Similarly, every root function in the exercises for sections 2.2 and 2.3 can be written as a power of the variable. We can differentiate $y = \sqrt{x}$ or $y = \sqrt[5]{x^3}$, but we can not yet differentiate $y = \sqrt{4 - 5x}$.

The function $y = \sin x^2$ is a composition of two functions. If we set $f(u) = \sin u$ and $g(x) = x^2$, then $f(g(x)) = f(x^2) = \sin x^2$. We need a rule for differentiating a composition of functions.

The chain rule: derivatives of compositions

We wish to find the derivative of a composition of functions, or

$$\frac{d}{dx} f(g(x)).$$

Just as with many other derivative rules, the local linearity formula

$$f(k + \alpha) \approx f(k) + \alpha f'(k)$$

is helpful. In the local linearity formula, notice that k is a real number and α is an infinitesimal, so

$$f(\textbf{real} + \text{infinitesimal}) \approx f(\textbf{real}) + \text{infinitesimal} \cdot f'(\textbf{real}).$$

As part of our derivative calculation, we need to find $f(g(x+\alpha))$, where α is infinitesimal. Using local linearity on the function g, we have

$$f(g(x + \alpha)) \approx f(g(x) + \alpha g'(x)).$$

This step requires g to be differentiable at x and f to be continuous at $g(x)$.

Notice that $g(x)$ is a real number and $\alpha g'(x)$ is infinitesimal. Color-coding the application of the local linearity formula for the function f gives

This step requires f to be differentiable at $g(x)$.

$$\begin{aligned} f(g(x+\alpha)) &\approx f(\mathbf{g}(\mathbf{x}) + \alpha g'(x)) \\ &\approx f(\mathbf{g}(\mathbf{x})) + \alpha g'(x) f'(\mathbf{g}(\mathbf{x})). \end{aligned}$$

Now we are ready to calculate the derivative of $f(g(x))$ using the derivative formula (function version):

The approximation step uses the previously calculated formula for $f(g(x+\alpha))$.

The last step rewrites the formula in the traditional order.

$$\begin{aligned} \frac{d}{dx} f(g(x)) &= \frac{f(g(x+\alpha)) - f(g(x))}{\alpha} \\ &\approx \frac{f(g(x)) + \alpha g'(x) f'(g(x)) - f(g(x))}{\alpha} \\ &= \frac{\alpha g'(x) f'(g(x))}{\alpha} = g'(x) f'(g(x)) \\ &= f'(g(x)) g'(x). \end{aligned}$$

If $y = f(u)$ and $u = g(x)$, then $g'(x) = \frac{du}{dx}$, $f(g(x)) = f(u)$, and $f'(g(x)) = f'(u)$, so the formula can be rewritten as

$$\frac{d}{dx} f(u) = f'(u) g'(x) = f'(u) \frac{du}{dx}.$$

The derivative of a composition is a product of derivatives.

CHAIN RULE

Suppose $y = f(u)$ and $u = g(x)$, and also suppose that g is differentiable at x and f is differentiable at $g(x)$. Then, $f \circ g(x)$ is differentiable at x and

$$\frac{d}{dx} f(g(x)) = f'(g(x)) g'(x),$$

which can also be written as

$$\frac{d}{dx} f(u) = f'(u) \frac{du}{dx}.$$

In pure Leibniz notation, the chain rule formula is

$$\frac{dy}{dx} = \frac{dy}{du}\frac{du}{dx}.$$

Applying the chain rule: trig functions

The trick to learning how to apply the chain rule is to use it to rewrite the derivative formulas we already know. For instance, we know $\frac{d}{dx}\sin x = \cos x$. Letting $y = f(u) = \sin u$, we have $f'(u) = \cos u$. Applying the chain rule formula

$$\frac{d}{dx}f(u) = f'(u)\frac{du}{dx}$$

gives

$$\frac{d}{dx}\sin u = \cos u \cdot \frac{du}{dx}.$$

This is called the *chain rule form* or the *general form* of the derivative formula for sine.

This means that the derivative of sine is cosine, but that we also must multiply by the derivative of what is "inside" the sine function.

Example 1 *Find* $\frac{d}{dx}\sin(3x-5)$.

Solution This is $\frac{d}{dx}\sin u$, where $u = 3x-5$. Then,

Here, we use the formula $\frac{d}{dx}\sin u = \cos u \cdot \frac{du}{dx}$.

$$\begin{aligned}\frac{d}{dx}\sin(3x-5) &= \cos(3x-5)\cdot\frac{d}{dx}(3x-5)\\ &= \cos(3x-5)\cdot 3\\ &= 3\cos(3x-5).\end{aligned}$$

■

Recall that we write polynomial parts (including constants) first and trig parts second for the sake of clarity.

The other trig functions are combined with the chain rule in the same manner. It is worth rewriting or even memorizing the trig derivatives in this more general form.

TRIG DERIVATIVES (GENERAL FORM)

$\frac{d}{dx}\sin u = \cos u \cdot \frac{du}{dx}$	$\frac{d}{dx}\csc u = -\csc u \cot u \cdot \frac{du}{dx}$
$\frac{d}{dx}\cos u = -\sin u \cdot \frac{du}{dx}$	$\frac{d}{dx}\sec u = \sec u \tan u \cdot \frac{du}{dx}$
$\frac{d}{dx}\tan u = \sec^2 u \cdot \frac{du}{dx}$	$\frac{d}{dx}\cot u = -\csc^2 u \cdot \frac{du}{dx}$

Whenever a trig function has "not just plain x" as its argument, the chain rule is needed and we must multiply by the derivative of what is "inside."

Example 2 *Differentiate* $f(x) = \sec(x^2 + 3x)$.

Here, $u = x^3 + 3x$.

Solution This is a trig function of "not just plain x," so the chain rule is needed. We use the general form of the derivative:

$$\begin{aligned} f'(x) &= \sec(x^2 + 3x)\tan(x^2 + 3x) \cdot \frac{d}{dx}(x^2 + 3x) \\ &= \sec(x^2 + 3x)\tan(x^2 + 3x) \cdot (2x + 3) \\ &= (2x + 3)\sec(x^2 + 3x)\tan(x^2 + 3x). \end{aligned}$$

■

Reading Exercise 20 *Find* $\frac{d}{dx}\cos(4x^2)$.

If the independent variable in the function is t instead of x, we just use the formula with t in place of x.

Example 3 *Find* $g'(t)$ *for* $g(t) = \tan 4t$.

Solution This is a trig function of "not just plain t," so the chain rule is needed:

$$\begin{aligned} g'(t) &= \sec^2 4t \cdot \frac{d}{dt}4t \\ &= \sec^2 4t \cdot 4 \\ &= 4\sec^2 4t. \end{aligned}$$

■

Here, $u = 4t$ in the formula $\frac{d}{dt}\tan u = \sec^2 u \cdot \frac{du}{dt}$. Because the variable is t, we use t instead of x.

Compositions of trig functions may look more complicated, but they are handled just the same as the previous examples.

Example 4 *Find* $\frac{d}{dx}\sin(\cos x)$.

Solution Although $\cos x$ is a trig function of "just plain x," the sine function is not; it is $\sin(\cos x)$, not $\sin x$. We must use the chain rule for the derivative of sine:

Compare to example 1. This is $\sin u$ with $u = \cos x$. We use the formula $\frac{d}{dx}\sin u = \cos u \cdot \frac{du}{dx}$.

$$\begin{aligned} \frac{d}{dx}\sin(\cos x) &= \cos(\cos x) \cdot \frac{d}{dx}\cos x \\ &= \cos(\cos x) \cdot (-\sin x). \end{aligned}$$

Writing the final answer as $(-\sin x)(\cos(\cos x))$ is perhaps more clear. ■

Note that $\cos(\cos x)$ is not the same as $\cos x \cos x$. That is, $\cos(\cos x) \neq \cos^2 x$.

Example 5 *Find* $\frac{d}{dx}\sin x \cos x$.

Solution This is not a composition of functions; it is a product, $\sin x \cdot \cos x$, and requires the product rule:

$$\frac{d}{dx}\sin x \cos x = \sin x \cdot (-\sin x) + \cos x \cdot \cos x$$
$$= -\sin^2 x + \cos^2 x. \quad \blacksquare$$

How can one tell the difference between the need for the chain rule in example 4 and the need for the product rule in example 5? The key is in recognizing a complete expression. The expression $\sin x$ is a complete trigonometric expression; sin is a name of a function and is not a complete expression. A complete trigonometric expression requires something inside the trig function, as in $\tan(x^2)$ or $\sec(3 + \tan x - x^4)$. The expression $\sin x \cos x$ has two complete trigonometric expressions, $\sin x$ and $\cos x$, which are multiplied. The expression $\sin \cos x$ cannot be the product of two expressions because sin by itself is not a complete expression; it needs something inside of it, so $\cos x$ (which is a complete expression) must be the "inside," or the argument, of sin. A summary is provided in table 1.

Those of you who have cultivated the bad habit of writing sin when you mean $\sin x$ will be hard-pressed to tell the difference between a product and a composition. Using proper notation is important!

Ans. to reading exercise 20
$-8x \sin(4x^2)$

Table 1 *Discriminating between product and composition*

algebraic form:	product	composition
rule required:	product rule	chain rule
example:	$\underbrace{\sin x}_{\text{complete}}\ \underbrace{\cos x}_{\text{complete}}$	$\underbrace{\underbrace{\sin}_{\text{incomplete}}\ \cos x}_{\text{complete}}$
visual recognition:	x in more than one spot separated by (implied) multiplication	x in only one spot, no multiplication of complete expressions

Applying the chain rule: algebraic functions

The power rule also has a chain rule form:

> **POWER RULE (CHAIN RULE FORM)**
>
> $$\frac{d}{dx}u^n = nu^{n-1} \cdot \frac{du}{dx}$$

We recognize the need for the chain rule form of the power rule when we have "not just plain x" to a power.

Example 6 *Differentiate* $y = (x^2 + 1)^5$.

We use the formula

$$\frac{d}{dx}u^n = nu^{n-1} \cdot \frac{du}{dx}.$$

Solution The function has the form u^n, where $u = x^2 + 1$ and $n = 5$. In other words, this is "not just plain x" to a power, so the chain rule is needed:

$$\begin{aligned} y' &= 5(x^2 + 1)^4 \cdot \frac{d}{dx}(x^2 + 1) \\ &= 5(x^2 + 1)^4(2x) \\ &= 10x(x^2 + 1)^4. \end{aligned}$$

■

With practice, the first step can be skipped and the solution can be written as

$$y' = 5(x^2 + 1)^4(2x) = 10x(x^2 + 1)^4.$$

Example 7 *Differentiate* $y = 4(x + 7)^3$.

Here, $u = x + 7$.

Solution This is "not just plain x" to a power, so the chain rule is needed. We also use the constant multiple rule:

$$\begin{aligned} y' &= 4 \cdot 3(x + 7)^2 \cdot \frac{d}{dx}(x + 7) \\ &= 12(x + 7)^2 \cdot 1 \\ &= 12(x + 7)^2. \end{aligned}$$

■

Here, the chain rule only has us multiply by one, which has no effect. But this is not always the case, and trying to skip the chain rule in some instances leaves us vulnerable to skipping it when it is needed.

Reading Exercise 21 *Differentiate* $y = (x^2 + 3)^7$.

An alternate, but much longer, method of working examples 6 and 7 is to expand the powers, obtaining polynomials of degree 10 and 3, respectively, that can be differentiated using the derivative rules of section 2.2.

Because roots can be rewritten as fractional exponents, they also are differentiated using the power rule.

Example 8 *Find* $\frac{d}{dx}\sqrt{x^3 + 4x + 7}$.

Solution This is the square root of "not just plain x," which is the same thing as "not just plain x" to a power. The chain rule is needed:

$$\begin{aligned}\frac{d}{dx}\sqrt{x^3+4x+7} &= \frac{d}{dx}(x^3+4x+7)^{\frac{1}{2}}\\ &= \frac{1}{2}(x^3+4x+7)^{-\frac{1}{2}}\cdot\frac{d}{dx}(x^3+4x+7)\\ &= \frac{1}{2}(x^3+4x+7)^{-\frac{1}{2}}(3x^2+4)\\ &= \frac{3x^2+4}{2\sqrt{x^3+4x+7}}.\end{aligned}$$

■

Here, $u = x^3+4x+7$. We still use the formula

$$\frac{d}{dx}u^n = nu^{n-1}\cdot\frac{du}{dx}.$$

Reading Exercise 22 *Find* $\frac{d}{dx}\sqrt{x^2+6x}$.

Chain rule with product rule

Just as the product and quotient rules are sometimes needed in the same exercise, the chain rule can be required along with other rules as well.

Example 9 *Find* y' *for* $y = 5x^2(11x-8)^4$.

Solution The expression is a product: $5x^2\cdot(11x-8)^4$. Therefore, the product rule is needed. Setting "first" as $5x^2$ and "second" as $(11x-8)^4$, we have

Because juxtaposition means multiplication, products are often harder to recognize visually than quotients, which present themselves as fractions. Looking intentionally for two expressions written next to one another can help.

$$y' = \underset{\text{first}\,\cdot}{5x^2}\cdot\underset{\text{second}'}{} + \underset{\text{second}}{(11x-8)^4}\cdot\underset{\text{first}'}{10x}.$$

When finding second′, we see that we must find $\frac{d}{dx}(11x-8)^4$, which is "not just plain x" to a power, and requires the chain rule. Filling in the blank, we have

$$\begin{aligned}y' &= \underset{\text{first}\,\cdot}{5x^2}\cdot\underset{\text{second}'}{\left(4(11x-8)^3\frac{d}{dx}(11x-8)\right)} + \underset{\text{second}}{(11x-8)^4}\cdot\underset{\text{first}'}{10x}\\ &= 5x^2(4(11x-8)^3\cdot 11) + (11x-8)^4\cdot 10x\\ &= 220x^2(11x-8)^3 + 10x(11x-8)^4.\end{aligned}$$

Further simplification of the answer is possible but is not necessarily advantageous. ■

Ans. to reading exercise 21
$14x(x^2+3)^6$

Reading Exercise 23 *Find* y' *for* $y = 3x^2(4x+1)^4$.

Multiple-link chains

The word *chain* in "chain rule" evokes a picture of a chain with many links. The chain rule may have to be applied repeatedly when differentiating one expression.

Example 10 *Find y' for $y = \sin^2(5x^2)$.*

Tip: when differentiating a power of a trig function, the form $\tan^3\theta$ is likely to conceal the correct order for use of the chain rule. The form $(\tan\theta)^3$ is helpful to indicate the correct order visually for use of the chain rule. Always rewrite in the helpful form.

Solution Recall that the expression $\sin^2\theta$ is the same as $(\sin\theta)^2$. Rewriting, we have

$$y = (\sin(5x^2))^2.$$

Looking at the outermost part of the expression first, we see that it is of the form "not just plain x" to a power. The chain rule is needed:

Here, $u = \sin(5x^2)$. We use the formula

$$\frac{d}{dx}u^n = nu^{n-1}\cdot\frac{du}{dx}.$$

$$y' = 2(\sin(5x^2))^1 \cdot \frac{d}{dx}\sin(5x^2).$$

When finding $\frac{d}{dx}\sin(5x^2)$, the chain rule is needed again; this is a trig function of "not just plain x":

Here, $u = 5x^2$. We use the formula

$$\frac{d}{dx}\sin u = \cos u \cdot \frac{du}{dx}.$$

$$\begin{aligned} &= 2\sin(5x^2)\cdot\cos(5x^2)\cdot\frac{d}{dx}5x^2 \\ &= 2\sin(5x^2)\cdot\cos(5x^2)\cdot 10x \\ &= 20x\sin(5x^2)\cos(5x^2). \end{aligned}$$

■

Ans. to reading exercise 22 $\frac{2x+6}{2\sqrt{x^2+6x}} = \frac{x+3}{\sqrt{x^2+6}}$

Color-coding can help visualize multiple-link chain rule exercises such as example 10. Color the outer power and its portion of the derivative bold, the sine function and its part of the derivative the usual black, and the inner polynomial and its part of the derivative gray:

$$\begin{aligned} &\frac{d}{dx}(\sin(5x^2))^2 \\ &= \mathbf{2(\sin(5x^2))^1}\cdot\cos(5x^2)\cdot 10x \\ &= 20x\sin(5x^2)\cos(5x^2). \end{aligned}$$

The derivative of each portion of the expression is taken, one at a time, outside-in. With practice, one can write the answer to a multiple-link chain all at once (pending simplification). However, if there is any question of how to proceed, the solution method of example 10 is the most straightforward and the easiest to get right.

Reading Exercise 24 *Find y' for $y = \cos^2(4x^3)$.*

Why radians instead of degrees?

Why are all trig functions in calculus in terms of radians instead of degrees? Is it because calculus instructors like making things hard and torturing their students? On the contrary. It's because we like making things easier.

Ans. to reading exercise 23
$48x^2(4x+1)^3 + 6x(4x+1)^4$

When we write $y = \sin x$, x is in terms of radians. Our formulas such as $\frac{d}{dx}\sin x = \cos x$ and $\sin\alpha \approx \alpha$ also require radians. What if, instead, we tried degrees?

Recall that $x^\circ = x \cdot \frac{\pi}{180}$ rad. Therefore, $y = \sin x^\circ$ is a trig function of "not just plain x," because it is the same as $\sin\left(\frac{\pi}{180}x\right)$ (in terms of radians), and the chain rule is required if we desire its derivative:

$$\begin{aligned}\frac{d}{dx}\sin x^\circ &= \frac{d}{dx}\sin\left(\frac{\pi}{180}x\right)\\ &= \cos\left(\frac{\pi}{180}x\right)\cdot\frac{d}{dx}\left(\frac{\pi}{180}x\right)\\ &= \cos\left(\frac{\pi}{180}x\right)\cdot\frac{\pi}{180}\\ &= \frac{\pi}{180}\cos x^\circ.\end{aligned}$$

Here, $u = \frac{\pi}{180}x$ and we use the formula

$$\frac{d}{dx}\sin u = \cos u \cdot \frac{du}{dx}.$$

When trig functions are given in terms of degrees, the derivative formula is not as simple. It is

$$\frac{d}{dx}\sin x^\circ = \frac{\pi}{180}\cos x^\circ.$$

Similarly, for any infinitesimal α,

$$\sin\alpha^\circ = \sin\left(\frac{\pi}{180}\alpha\right) \approx \frac{\pi}{180}\alpha,$$

which is also not as simple as $\sin\alpha \approx \alpha$.

All calculus formulas involving trig functions are much less simple if we use degrees. Be thankful for radians!

Differentials and the chain rule

Using local linearity diagrams can help us make intuitive sense of the chain rule. As before, suppose $y = f(u)$ and $u = g(x)$. Suppose also that g is differentiable at $x = k$ and that f is differentiable at $u = g(k)$. The local linearity diagrams for $y = f(u)$ and $u = g(x)$ are in figure 1. These diagrams are the same as section 2.3 figure 15, with variable names changed as appropriate.

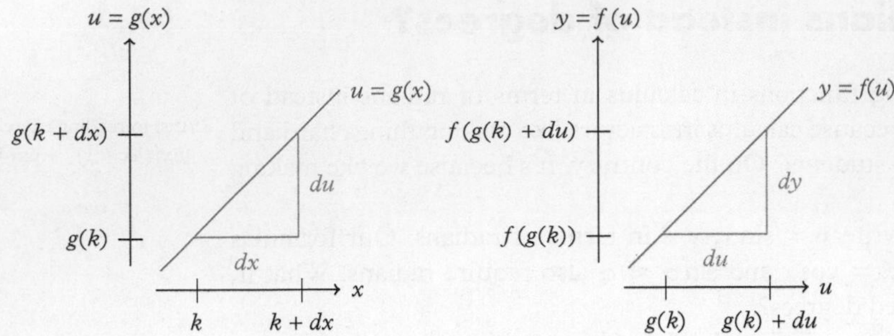

Figure 1 *(left) Local linearity diagram for $u = g(x)$, with dx infinitesimal; (right) local linearity diagram for $y = f(u)$, with du infinitesimal*

Ans. to reading exercise 24
$-24x^2 \cos(4x^3) \sin(4x^3)$

Suppose now that $g'(k) = 2$, meaning that in figure 1, $\frac{du}{dx} = 2$. Then $du = 2\,dx$, and we conclude that u is changing twice as fast as x. Suppose further than $f'(g(k)) = 5$, meaning that in figure 1 $\frac{dy}{du} = 5$. Then, $dy = 5\,du$, and we conclude that y is changing five times as fast as du.

The local linearity diagram for $u = g(x)$ works for any infinitesimal dx, and the local linearity diagram for $y = f(u)$ works for any infinitesimal du. Therefore, we can suppose that du represents the same infinitesimal in both diagrams. Then, $dy = 5\,du = 5 \cdot 2\,dx = 10\,dx$. In other words, because y is changing five times as fast as u and u is changing twice as fast as x, then y is changing 10 times as fast as x. We conclude that $\frac{dy}{dx} = 10$, and we have

$$\frac{dy}{dx} = \frac{dy}{du} \cdot \frac{du}{dx},$$

which is the Leibniz-notation version of the chain rule. The resulting diagram for $y = f(g(x))$ is in figure 2.

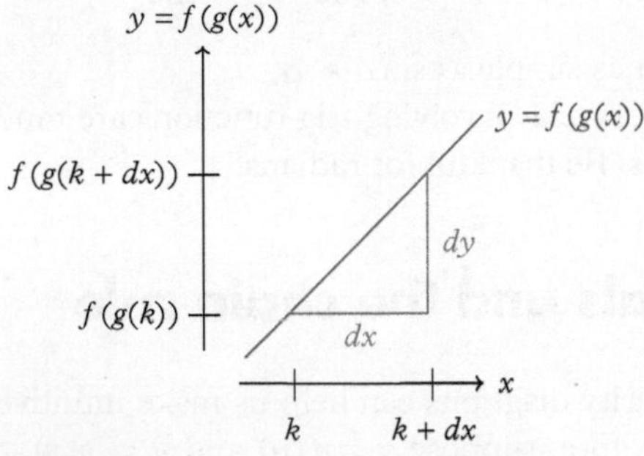

Figure 2 *Local linearity diagram for $y = f(g(x))$, with dx infinitesimal. Since $g(k) + du \approx g(k + dx)$ by local linearity of g (see figure 1), $f(g(k) + du) = f(g(k + dx))$*

Exercises 2.5

1–8. Rapid response: which is needed, the product rule or the chain rule?

1. $\frac{d}{dx}\sqrt{x}\sin x$
2. $\frac{d}{dx}\sqrt{\sin x}$
3. $\frac{d}{dx}\cos x^2$
4. $\frac{d}{dx}x^2\cos x$
5. $\frac{d}{dx}\tan\sec x$
6. $\frac{d}{dx}\tan x\sec x$
7. $\frac{d}{dx}\sin^2 x$
8. $\frac{d}{dx}4\sin^2 4$

9–18. Rapid response: is the chain rule needed?

9. $\frac{d}{dx}x^2\sin x$
10. $\frac{d}{dx}x^2\sin 4x$
11. $\frac{d}{dx}\sqrt{5-9x}$
12. $\frac{d}{dx}\sqrt{x^3}$
13. $\frac{d}{dx}(6x+1)^2$
14. $\frac{d}{dx}(6x+1)^{21}$
15. $\frac{d}{dx}(7x-5)\cos^3 x$
16. $\frac{d}{dx}\cos^3(7x-5)$
17. $\frac{d}{dx}\tan(4x-9)$
18. $\frac{d}{dx}\frac{3x}{\sqrt{x^3-x}}$

19–48. Differentiate the function.

19. $y=\cos(4x-3)$
20. $f(x)=\sqrt{5x-7}$
21. $f(x)=(7x+5)^4$
22. $y=\sin(x^2+1)$
23. $g(x)=\sqrt{\tan 5x}$
24. $g(y)=\frac{(y^3-7)^4}{2y+9}$
25. $y=\frac{(2x+1)^3}{x^3-5}$
26. $y=t^2\sin(2t+1)$
27. $f(t)=t^3\cos(6t+1)$
28. $f(\theta)=\cos(3\theta-4)^7$
29. $g(x)=\sec((x^2+5)^3)$
30. $g(x)=\tan((x^2+\pi)^5)$
31. $h(x)=\csc\sqrt{x^3+x}$
32. $f(x)=(3+\sin(x^2))^4$
33. $s=\theta\sin(2\theta-3)$
34. $f(t)=\sin^3(1-2t)$
35. $y=(\tan(x^2+1))^4$
36. $K(x)=\frac{5x-1}{(3x^2+4)^4}$
37. $f(x)=(5+\cos(x^2))^7$
38. $y=(x^2+2x+7)^4\sec x$
39. $y=\cos^2(6-3x)$
40. $f(x)=\sec\cot x$
41. $k(x)=\frac{(4x^2+1)^3}{2x+7}$
42. $k(s)=\sin^2\frac{s}{2}$
43. $y=\cos\tan\theta$
44. $y=\sin\cos t^3$
45. $f(x)=\sqrt{\cos(x^4+\frac{1}{x})}$
46. $f(x)=\sin^2((4x-1)^{\frac{1}{2}})$
47. $x=t^2\sec(5t-8)^4$
48. $y=\tan^3\sqrt{4-\frac{1+(x-6)^3}{\cos^2 5x^3}}$

49–52. Find the indicated derivative.

49. $y = \tan x$, y''

50. $y = \cot x$, $\frac{d^2y}{dx^2}$

51. $\frac{d^2}{dx^2} \sin 5x$

52. $y = \sin x^2$, y''

53. Find the equation of the tangent line to the curve $h(x) = \sqrt{9x}$ at the point $(1, 3)$.

54. Find the equation of the tangent line to the curve $f(x) = (2x + 1)^{14}$ at the point $(0, 1)$.

55. Find the equation of the tangent line to the curve $y = (4x - 1)^2$ at the point $(1, 9)$.

56. Find the equation of the tangent line to the curve $y = \sqrt{2x + 3}$ at the point $(11, 5)$.

57. Find the differential dy for $y = \sqrt{4x + 1}$.

58. Find the differential dy for $y = \sin 2x$.

59. Find the differential dy for $y = \cos 7x$.

60. Find the differential dy for $y = (23x - 52)^3$.

61. Find the linearization of $f(x) = \sin(x - \frac{\pi}{4})$ at $x = 0$.

62. Find the linearization of $f(x) = \cos(3x - \frac{\pi}{2})$ at $x = 0$.

63. Find the locations of all horizontal tangent lines on the graph of $y = \sqrt{x^2 - 4x + 10}$.

64. Find the locations of all horizontal tangent lines on the graph of $y = (x^2 + 5)^4$.

65. For Tahlequah, OK, the number of hours of daylight is approximated by

$$12 + 2.6 \sin\left(\frac{2\pi}{365.25} t\right),$$

where t is measured in days and $t = 0$ corresponds to March 21. Find the rate at which the number of hours of daylight is changing on April 1.

66. Rewrite $y = \frac{x^2 - 7}{x^3 - 4x}$ as a product (using a negative exponent), and find y' using the product and chain rules instead of the quotient rule.

Implicit Differentiation

2.6

Suppose we wish to know the slope of the tangent line to the curve $x^2 + y^2 = 1$ at the point $\left(\frac{1}{\sqrt{2}}, -\frac{1}{\sqrt{2}}\right)$. So far, we have only found equations of tangent lines to functions. Not only is our equation not in the form of a function, $y = f(x)$, the equation does not represent a function at all! Its graph is a circle (figure 1) and it fails the vertical line test. The purpose of this section is to learn how to handle such situations.

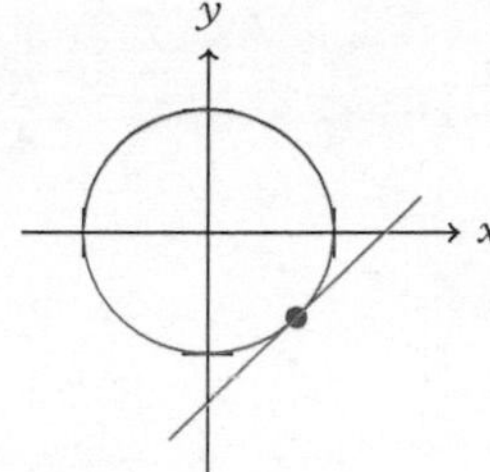

Figure 1 *The graph of $x^2 + y^2 = 1$ with the tangent line at point $\left(\frac{1}{\sqrt{2}}, -\frac{1}{\sqrt{2}}\right)$*

Differentiating implicitly defined functions

Although the curve $x^2 + y^2 = 1$ is not a function, it is a circle and therefore does have a tangent line at any given point. It even looks as though we should be able to find the tangent. If we zoom in near the point of tangency, the picture looks like the graph of a function with a tangent line (figure 2).

Figure 2 *The graph of $x^2 + y^2 = 1$ with the tangent line at point $\left(\frac{1}{\sqrt{2}}, -\frac{1}{\sqrt{2}}\right)$, near the point of tangency*

Figure 2 illustrates the key idea: keep only a portion of the curve so that what remains passes the vertical line test and therefore represents a function. This can be done in many different ways. For instance, we could keep the top half of the curve or the bottom half of the curve, as in figure 3. Either way, what remains passes the vertical line test and

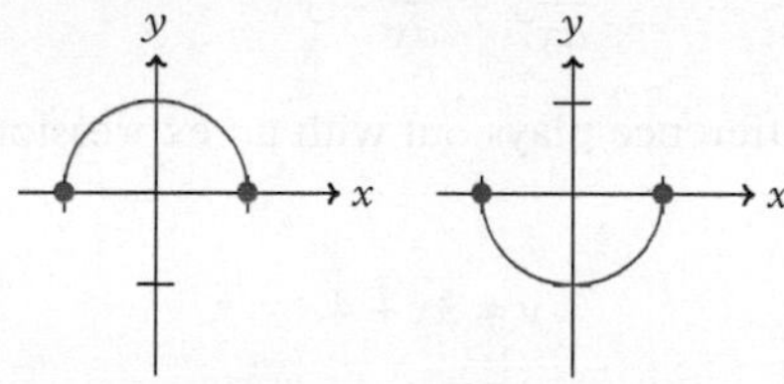

Figure 3 *The graphs of two functions defined implicitly by $x^2 + y^2 = 1$*

represents the graph of a function. These are not the only two possibilities; we could also keep the top left and bottom right quarters of the circle, or even pick and choose other pieces seemingly at random, as long as what remains passes the vertical line test. See figure 4. There is a sense in which the equation $x^2 + y^2 = 1$ defines all of these different functions. These functions are not written explicitly, but they are implied; therefore, we say that these functions are *defined implicitly* by the equation $x^2 + y^2 = 1$.

Implicitly defined functions are as opposed to explicitly defined functions, such as $f(x) = x^2 + 3$. The equation of the function is stated for an explicitly defined function, and the equation of the function is not stated in the form $f(x) = \ldots$ for an implicitly defined function.

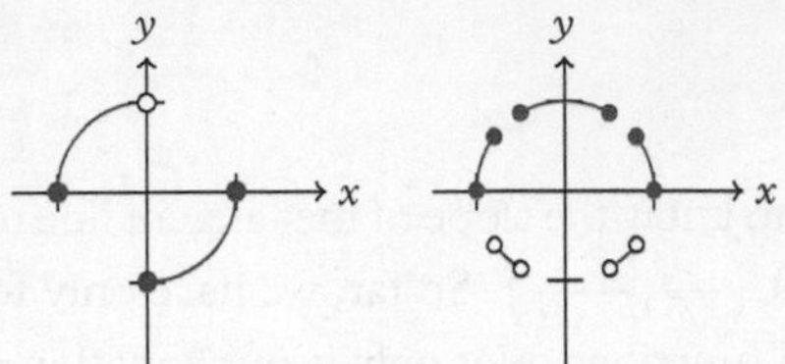

Figure 4 *The graphs of two functions defined implicitly by* $x^2 + y^2 = 1$

There are two ideas of how we might proceed to find the slope of the tangent line. One is to determine the equation of the bottom half of the circle and then use our previous methods (find the derivative and evaluate it) to find the slope. This is demonstrated later (example 6), but the problem with this method is that it is sometimes hard or even impossible to find a function representing the appropriate piece of the curve. We need a different method.

The second idea, called *implicit differentiation,* is to go ahead and take the derivative of the equation in its current form. We need rules for differentiating an equation such as $x^2+y^2 = 1$, instead of a function.

If we want the slope of the tangent line, we want to know $\frac{dy}{dx}$. That is, we want to find the derivative with respect to the variable x. The variables x and y are not treated the same:

$$\frac{d}{dx}x = 1,$$

whereas

$$\frac{d}{dx}y = \frac{dy}{dx} = y'.$$

To see how this difference plays out with an expression such as $\frac{d}{dx}y^2$, suppose

If y defines a function of x implicitly, then we may assume we are dealing with

$$y = \ldots,$$

and we just may not know exactly what the ellipsis represents.

$$y = 3x + 4,$$

for which

$$\frac{dy}{dx} = y' = 3.$$

Then,

$$y^2 = (3x + 4)^2$$

and

$$\frac{d}{dx}y^2 = \frac{d}{dx}(3x+4)^2 = 2(3x+4)^1 \cdot 3$$
$$= 2y \cdot y'.$$

Since $y = 3x + 4$ and $y' = 3$,
$$2(3x+4) \cdot 3 = 2y \cdot y'.$$

The chain rule applies to $(3x+4)^2$ because the expression is "not just plain x" to a power; therefore, the chain rule applies to y^2 because the expression is "not just plain x" to a power. In fact, applying the chain rule form of the power rule exactly as we did in section 2.5 examples 6–8 gives

The variable y is not the same as the variable x, so y^2 is not x to a power.

Here, $u = y$ and we use the formula
$$\frac{d}{dx}u^n = nu^{n-1} \cdot \frac{du}{dx}.$$

$$\frac{d}{dx}y^2 = 2y^1 \cdot \frac{d}{dx}(y) = 2y \cdot y'.$$

Our procedure, therefore, is to apply the chain rule every time we see y in an expression. This amounts to multiplying by y' (the same as $\frac{dy}{dx}$) whenever an expression involving y is differentiated. For instance,

$$\frac{d}{dx}y^4 = 4y^3 \cdot y'$$
$$\frac{d}{dx}\sin y = \cos y \cdot y' = y' \cos y,$$
$$\frac{d}{dx}(3y^2+7)^5 - 5(3y^2+7)^4 \cdot 6y \cdot y'.$$

We are now ready to return to our original example.

Example 1 *Find the slope of the tangent line to the curve* $x^2 + y^2 = 1$ *at the point* $\left(\frac{1}{\sqrt{2}}, -\frac{1}{\sqrt{2}}\right)$.

Solution Let's look at the solution in three steps.

❶ Differentiate both sides of the equation implicitly with respect to x:

Note that we are performing the same operation on both sides of the equation.

$$\frac{d}{dx}(x^2+y^2) = \frac{d}{dx}(1)$$
$$2x + 2y \cdot y' = 0.$$

The rules for differentiating expressions of x remain the same. For differentiating expressions in y, the chain rule requires that we multiply by $\frac{dy}{dx}$ (the same as y').

❷ Solve for y':

$$2yy' = -2x$$
$$y' = \frac{-2x}{2y} = -\frac{x}{y}.$$

When implicit differentiation is used, the expression for y' nearly always involves both x and y, just as the original equation does.

❸ Apply the derivative as required (when necessary). We want the slope of the tangent line, so we need to evaluate the derivative at our point. Note that this requires the use of values for both x and y:

The point at which we wish to know the slope is $\left(\frac{1}{\sqrt{2}}, -\frac{1}{\sqrt{2}}\right)$, so we use $x = \frac{1}{\sqrt{2}}$ and $y = -\frac{1}{\sqrt{2}}$.

$$y'\Big|_{\left(\frac{1}{\sqrt{2}}, -\frac{1}{\sqrt{2}}\right)} = -\frac{\frac{1}{\sqrt{2}}}{-\frac{1}{\sqrt{2}}} = 1.$$

The slope of the tangent line is 1. ■

A peek at figure 1 reveals why we need to evaluate the derivative using both the x- and y-coordinates of the point. There are two points with the same x-coordinate, and the slopes at these two points are not the same. Therefore, the derivative y' must involve both coordinates to give us the information we desire.

Implicit differentiation examples

Because the variables in the equation are x and y, it is implied that y' means $\frac{dy}{dx}$.

Example 2 *Find y' for $(y+3)^4 = x^7 - 3\sin x + y$.*

Solution ❶ We begin by differentiating both sides with respect to x:

Recall that $\frac{d}{dx}y = \frac{dy}{dx} = y'$.

$$\frac{d}{dx}\left((y+3)^4\right) = \frac{d}{dx}\left(x^7 - 3\sin x + y\right)$$
$$4(y+3)^3 \cdot 1 \cdot y' = 7x^6 - 3\cos x + y'.$$

The algebra of step ❷ of an implicit differentiation exercise may vary, but it nearly always involves the three substeps described here.

❷ Next we solve for y'. We start by moving terms involving y' to one side of the equation and terms not involving y' to the other side:

$$4(y+3)^3y' - y' = 7x^6 - 3\cos x.$$

If necessary (if y' is in more than one term), we can factor out y':

$$y'\left(4(y+3)^3 - 1\right) = 7x^6 - 3\cos x.$$

Finally, we divide to isolate y':

$$y' = \frac{7x^6 - 3\cos x}{4(y+3)^3 - 1}.$$

Because y' is all that was requested, there is no third step. ■

Reading Exercise 25 *Find y' for $(y+1)^2 = x^2 + 4$.*

Recall that we are thinking of y as defining a function of x implicitly. Just as both $x\sin x$ and $x\sqrt{x^3 - 4x + 7}$ are products and require the product rule to differentiate, so does $x \cdot y$.

Example 3 *Find y' for $3xy = 2x^2 + y^2$.*

Solution ❶ We begin by differentiating both sides implicitly with respect to x. On the left side we have a product $3x \cdot y$, which requires the product rule.

Here, first = $3x$ and second = y. Recall that $\frac{d}{dx}y = \frac{dy}{dx} = y'$.

$$\frac{d}{dx}(3xy) = \frac{d}{dx}\left(2x^2 + y^2\right)$$
$$3x \cdot y' + y \cdot 3 = 4x + 2y \cdot y'.$$

❷ Next we solve for y':

The same three algebraic substeps of example 2 are used here: move terms involving y' to one side of the equation and terms without y' to the other side, factor out the y', and divide.

$$3xy' - 2yy' = 4x - 3y$$
$$y'(3x - 2y) = 4x - 3y$$
$$y' = \frac{4x - 3y}{3x - 2y}.$$ ■

Reading Exercise 26 *Find y' for $2xy + y^2 = x^2$.*

The quotient rule can get in on the act too.

Example 4 *Find $\frac{dy}{dx}$ given $\dfrac{2x - 3}{7y^4 + 5x} = \sin y$.*

Solution ❶ We begin by differentiating both sides with respect to x:

An alternative to using the quotient rule is to rewrite the equation in the form

$$2x - 3 = (7y^4 + 5x)\sin y$$

and use the product rule instead.

$$\frac{d}{dx}\left(\frac{2x - 3}{7y^4 + 5x}\right) = \frac{d}{dx}(\sin y)$$
$$\frac{(7y^4 + 5x) \cdot 2 - (2x - 3)(28y^3 \cdot y' + 5)}{(7y^4 + 5x)^2} = \cos y \cdot y'.$$

❷ Next we solve for y', which begins by trying to move every term involving y' to one side and every term not involving y' to the other side. The fraction complicates matters only slightly; we first clear the fraction by multiplying each side by $(7y^4 + 5x)^2$:

$$2(7y^4 + 5x) - (2x - 3)(28y^3y' + 5) = y'(7y^4 + 5x)^2 \cos y$$
$$14y^4 + 10x = y'(7y^4 + 5x)^2 \cos y + (2x - 3)(28y^3y' + 5).$$

The next step is supposed to be to factor out y', but in its present form, y' is not a factor of the term on the far right. Instead, we must distribute the $(2x - 3)$ through the factor $(28y^3y' + 5)$ to prepare for factoring out y':

$$14y^4 + 10x = y'(7y^4 + 5x)^2 \cos y + (2x - 3)28y^3y' + (2x - 3)5$$

$$14y^4 + 10x - 5(2x - 3) = y'(7y^4 + 5x)^2 \cos y + (2x - 3)28y^3y'$$

$$14y^4 + 10x - 10x + 15 = y'\left((7y^4 + 5x)^2 \cos y + (56x - 84)y^3\right).$$

If the equation is rewritten for use with the product rule, the solution is in the form

$$y' = \frac{2 - 5\sin y}{28y^3 \sin y + (7y^4 + 5x)\cos y},$$

which appears to differ algebraically from the main solution presented. The original equation gives a relationship between the two variables that needs to be used to show that the two solutions are equivalent.

We finish by dividing:

$$y' = \frac{14y^4 + 15}{(7y^4 + 5x)^2 \cos y + (56x - 84)y^3}.$$ ■

A simpler solution to example 4 is possible using the tools of multivariable calculus.

Second derivatives implicitly

Ans. to reading exercise 25
$y' = \frac{x}{y+1}$

Example 5 *Find y'' for $y^3 = 4\sin x - y^2 + 7$.*

Solution The usual procedure for finding the second derivative is to start by finding the first derivative. We begin by differentiating each side with respect to x:

$$\frac{d}{dx}\left(y^3\right) = \frac{d}{dx}\left(4\sin x - y^2 + 7\right)$$

$$3y^2 \cdot y' = 4\cos x - 2y \cdot y'.$$

Next we solve for y':

The same three algebraic substeps of example 2 are used here: move terms involving y' to one side of the equation and terms without y' to the other side, factor out the y', and divide.

$$3y^2y' + 2yy' = 4\cos x$$

$$y'(3y^2 + 2y) = 4\cos x$$

$$y' = \frac{4\cos x}{3y^2 + 2y}.$$

$$\frac{d}{dx}y' = \frac{d}{dx}\frac{dy}{dx} = \frac{d^2y}{dx^2} = y''$$

We wish to find the second derivative. Differentiating both sides again with respect to x gives

$$\frac{d}{dx}y' = \frac{d}{dx}\left(\frac{4\cos x}{3y^2 + 2y}\right)$$

$$y'' = \frac{(3y^2 + 2y)\cdot 4(-\sin x) - (4\cos x)(6y\cdot y' + 2\cdot y')}{(3y^2 + 2y)^2}.$$

There is no need to solve for y''. However, what we have now is listed in terms of x, y, and y'. To have an expression for y'' that depends on x and y only, we can replace y' with its previous expression:

Because we have solved for y', differentiating again to find y'' always leaves just y'' on the left-hand side and not on the right-hand side.

$$y'' = \frac{(3y^2 + 2y)\cdot 4(-\sin x) - (4\cos x)(6y\cdot \frac{4\cos x}{3y^2 + 2y} + 2\cdot \frac{4\cos x}{3y^2 + 2y})}{(3y^2 + 2y)^2}.$$

The solution can be simplified if desired. ■

Ans. to reading exercise 26
$y' = \frac{2x - 2y}{2x + 2y} = \frac{x - y}{x + y}$

Reading Exercise 27 *Find y'' given that $y' = \dfrac{x^2}{3x + y^2}$.*

Making implicit explicit

Sometimes an equation in x and y can be solved explicitly for y (more generally, solved for the dependent variable in terms of the independent variable) to avoid implicit differentiation. Let's reprise example 1 to illustrate.

Example 6 *Find the slope of the tangent line to the curve $x^2 + y^2 = 1$ at the point $\left(\frac{1}{\sqrt{2}}, -\frac{1}{\sqrt{2}}\right)$.*

Solution We begin by solving the equation for y (see figure 5):

$$x^2 + y^2 = 1,$$

$$y^2 = 1 - x^2,$$

$$y = \pm\sqrt{1 - x^2}.$$

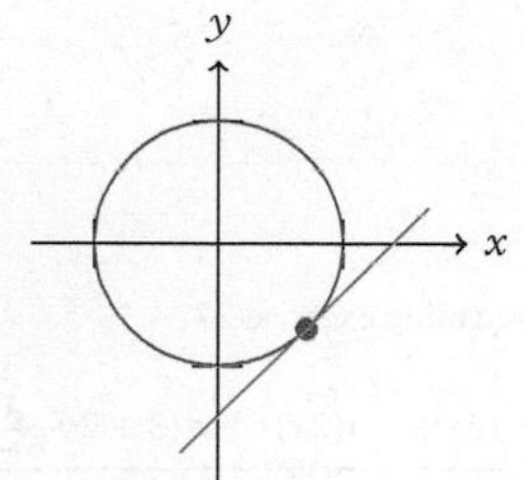

Figure 5 *The graph of $x^2 + y^2 = 1$ with the tangent line at point $\left(\frac{1}{\sqrt{2}}, -\frac{1}{\sqrt{2}}\right)$*

Notice that this gives not one, but two, functions (or more; see the earlier discussion). The top half of the curve has positive y-coordinates and is therefore represented by $y = \sqrt{1 - x^2}$, whereas the bottom half of the curve has negative y-coordinates and is represented by $y = -\sqrt{1 - x^2}$ (figure 6).

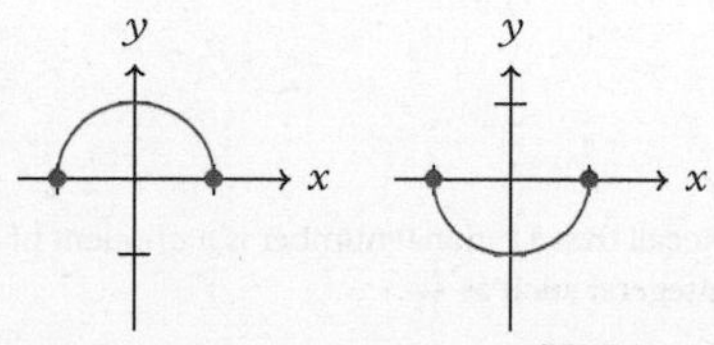

Figure 6 *(left) The graph of $y = \sqrt{1 - x^2}$ and (right) the graph of $y = -\sqrt{1 - x^2}$*

Consulting the graph of the equation (see figure 5), we see that the desired point of tangency lies on the bottom half of the curve, so we use $y = -\sqrt{1-x^2}$ and find the equation of the tangent line to that curve at $x = \frac{1}{\sqrt{2}}$. First we find y':

$$y' = -\frac{1}{2}(1-x^2)^{-\frac{1}{2}} \cdot (-2x) = \frac{x}{\sqrt{1-x^2}}.$$

The steps in this solution are (1) solve for y, (2) determine the portion of the curve to use, and (3) evaluate the derivative at the given value of x.

Then, we evaluate the derivative at $x = \frac{1}{\sqrt{2}}$:

$$y'\left(\frac{1}{\sqrt{2}}\right) = \frac{\frac{1}{\sqrt{2}}}{\sqrt{1-\frac{1}{2}}} = \frac{\frac{1}{\sqrt{2}}}{\frac{1}{\sqrt{2}}} = 1.$$

The slope of the tangent line is one, matching the solution to example 1. ■

The problem with using the method of example 6 is that the first step, solving for y, can be difficult or even impossible. Example 3 can be worked in this manner, but it is more difficult. Solving for y in example 5 is very difficult, so that the messy algebra of implicit differentiation is still preferred. Solving for y in example 2 is so difficult that it takes two printed pages just to present the solution, and then we would still need to take the derivative of that mess. And solving for y in example 4 cannot be done at all. Implicit differentiation is, by far, the preferred method.

Proof of the power rule for rational exponents

Ans. to reading exercise 27

$$y'' = \frac{(3x+y^2)(2x) - x^2(3 + 2y\frac{x^2}{3x+y^2})}{(3x+y^2)^2}$$

We have proved the power rule for integer exponents only. Implicit differentiation can be used to provide a proof of the power rule for rational exponents.

Theorem 5 POWER RULE FOR RATIONAL EXPONENTS
If $n = \frac{p}{q}$ is a rational number, then

$$\frac{d}{dx}x^n = nx^{n-1}.$$

Recall that a rational number is a quotient of integers, such as $\frac{52}{17}$.

Proof. Let $y = x^n = x^{p/q}$ with p, q integers. Then,

$$y^q = \left(x^{p/q}\right)^q = x^{(p/q)q} = x^p.$$

We have already proved the power rule for integer exponents, so it can be used. Differentiate both sides implicitly with respect to x:

Because p and q are integers, the power rule can be used.

$$\frac{d}{dx}(y^q) = \frac{d}{dx}(x^p)$$

$$qy^{q-1} \cdot y' = px^{p-1}.$$

Next we solve for y':

$$y' = \frac{px^{p-1}}{qy^{q-1}}.$$

This expression can be simplified if we use the fact that $y = x^{p/q}$ and use the laws of exponents:

$$\begin{aligned} y' &= \frac{p}{q} \cdot \frac{x^{p-1}}{(x^{p/q})^{q-1}} \\ &= \frac{p}{q} \cdot \frac{x^{p-1}}{x^{(p/q)q-(p/q)}} \\ &= \frac{p}{q} \cdot \frac{x^{p-1}}{x^{p-(p/q)}} \\ &= \frac{p}{q} \cdot x^{p-1-(p-(p/q))} \\ &= \frac{p}{q} \cdot x^{(p/q)-1}. \end{aligned}$$

Since $(x^a)^b = x^{ab}$, we multiply the exponents. The exponent is $\frac{p}{q}(q-1)$, and distributing yields an exponent of $(p/q)q - (p/q)$.

Since $\frac{x^a}{x^b} = x^{a-b}$, we subtract exponents.

But, $n = \frac{p}{q}$, and therefore

$$y' = nx^{n-1}. \qquad \blacksquare$$

Consider $y = \sqrt{x} = x^{\frac{1}{2}}$. We have stated that the square root function is continuous without giving a proof. Theorem 5 justifies the use of the power rule (because $\frac{1}{2}$ is a rational number) and therefore $y' = \frac{1}{2\sqrt{x}}$. The derivative is defined for $x > 0$, so $y = \sqrt{x}$ is differentiable for $x > 0$ and the graph of the square root curve is smooth. A function is continuous where it is differentiable (theorem 1), and therefore $y = \sqrt{x}$ is continuous for $x > 0$. Continuity, in turn, helps justify approximating under roots, as practiced in chapter 1.

The same reasoning applies to root functions in general. With the exception of $x = 0$, $y = \sqrt[n]{x}$ is differentiable and therefore continuous where defined. See section 2.1 example 6 and the following discussion for examples of what can happen to a root function at $x = 0$.

Theorem 6 APPROXIMATION INSIDE ROOT FUNCTIONS

If $x \approx y$ and $y \neq 0$, then $\sqrt[n]{x} \approx \sqrt[n]{y}$.

Proof. Suppose $x \approx y$ and $y \neq 0$. Then,

$$\frac{x}{y} \approx \frac{y}{y} = 1.$$

Therefore, $\frac{x}{y} = 1 + \gamma$, where γ is zero or infinitesimal. To apply the definition of approximation, we calculate

$$\frac{\sqrt[n]{x} - \sqrt[n]{y}}{\sqrt[n]{y}} = \sqrt[n]{\frac{x}{y}} - 1 = \sqrt[n]{1 + \gamma} - 1.$$

Since $f(x) = \sqrt[n]{x}$ is continuous at $x = 1$, $\sqrt[n]{1 + \gamma} \doteq \sqrt[n]{1} = 1$. Therefore,

$$\sqrt[n]{1 + \gamma} = 1 + \eta,$$

where η is zero or infinitesimal. Thus,

$$\frac{\sqrt[n]{x} - \sqrt[n]{y}}{\sqrt[n]{y}} = 1 + \eta - 1 = \eta,$$

which is zero or infinitesimal. By the definition of approximation,

$$\sqrt[n]{x} \approx \sqrt[n]{y}.$$ ■

It still remains to show that the power rule holds for irrational exponents. This proof is provided in section 5.5.

Exercises 2.6

1–10. Rapid response: is a function given explicitly or implicitly?

1. $y = x^2 + 3$
2. $f(x) = 6x^2 - 7\sqrt{x}$
3. $x + \sin y = 4y^2 - x^2$
4. $s = r^2 + 7r + r^3$
5. $x = y^4 - 3y$
6. $s = s^2 + 7r + r^3$
7. $m(k) = k^3 - 5$
8. $x^5 - 7\cos x = y$
9. $a^2 + b^2 = 49$
10. $z + 4c - 55c^2 = 11$

11–20. Find y'.

11. $7x^2 - 5y^3 = y + 53$
12. $7xy = y^3 + 22$
13. $4xy - y^3 = 7x$
14. $2x^2 + 3y^2 = 5x + 9y$
15. $\frac{7x}{y} = y^2 + 27$
16. $5x^3y^2 = 4x - 17y$
17. $4y^3 + 2x^2 + y = 1$
18. $x + y^2 - x^3 = y^4$
19. $y + 1 = 5x - xy^2$
20. $\frac{x - 1}{y^2} = 4 - x$

21–30. Find $\frac{dy}{dx}$.

21. $(2y-7)^3 = y - x^5$

22. $x^2y - 5x = y^3 + \sqrt{7}$

23. $3xy = y^2 - 4x + \sqrt[3]{5}$

24. $\tan y = y + \sec x$

25. $\cos y + x^2 = x$

26. $4 + y^{4/3} = 5x - y^3$

27. $11 + \sqrt{y-5} = x^2 + y^2$

28. $(6y-19)^4 + y^2 = 7x$

29. $x \sin y = y \sin x$

30. $(5x-2)(3y-8) = y^2$

31–38. Find y''.

31. $x^2 + y^2 = 1$

32. $x^3 - y^3 = 9$

33. $y = x^2 - \sin y$

34. $4\cos y^2 = 9x - x^2$

35. $4xy = 5y - 8x^3$

36. $\sqrt{9y+x} = 5x - 7$

37. $(4y-3)^2 = x - 7y$

38. $x^2y^{-1} = 5y + 3x$

39. Find $\frac{dx}{dt}$ for $x^2 + 3t = x + t^3$.

40. Find $\frac{ds}{dr}$ for $r - s = r^2 - s^2$.

41. Find the slope of the tangent line to the curve $xy^3 = x^2 - y$ at the point $(2, 1)$.

42. Find the slope of the tangent line to the curve $y^2 - \sqrt{x} = 4y - 4x$ at the point $(1, 3)$.

43. Find the equation of the tangent line to the curve $4xy - x^2 - y^2 = -13$ at the point $(-1, 2)$.

44. Find the equation of the tangent line to the curve $\sin x \cos y = y - \pi$ at the point $(0, \pi)$.

45. Find the linearization of $x^2 + y^2 = 16$ at the point $(3, \sqrt{7})$.

46. Find the linearization of $x^2 - y^2 = 5$ at the point $(3, 2)$.

47. Find dy for $x^3 - 5y^2 = x - 4y$.

48. Find dy for $y^2 - 5y = 12x$.

49. Prove the quotient rule by rewriting $y = \frac{f(x)}{g(x)}$ as $f(x) = y \cdot g(x)$ and by using implicit differentiation and the product rule.

Rates of Change: Motion and Marginals

2.7

In section 2.1, we looked at three interpretations of the derivative:

❶ The slope of the tangent line to the curve $y = f(x)$ at $x = k$ is $f'(k)$.

❷ Given position function $f(t)$, the velocity at time $t = k$ is $f'(k)$.

❸ The rate of change of f at $x = k$ is $f'(k)$.

Considerable time was spent on the first interpretation, the slope of the tangent line, in section 2.3. In this section, we revisit the other two interpretations.

Motion

Suppose a particle is traveling on a number line such that its position at time t is $s(t)$. We call s the *position function* or the *equation of motion* for the particle. Its velocity $v(t)$, which combines speed and direction, is then given by the derivative of the position function: $v(t) = s'(t)$. When the velocity is positive, the particle is moving to the right (moving in the positive direction); when the velocity is negative, the particle is moving to the left (moving in the negative direction).

The traditional notation for the position function is $s(t)$.

Unless specifically stated otherwise, the term *velocity*, if unmodified, refers to instantaneous velocity.

See section 1.8 for a more complete discussion of velocity.

Speed is the rate at which we are moving without regard to direction. It is found by removing direction from the velocity, which is done by taking its absolute value: speed $= |v(t)| = |s'(t)|$.

Velocity is the rate of change of position (with respect to time). What is the rate of change of velocity, $v'(t)$? If a car's speed is increasing, it is said to be accelerating. If a car's speed is decreasing, it is said to be decelerating. We use the term *acceleration* to mean the rate of change of velocity (with respect to time), whether velocity is increasing or decreasing: $a(t) = v'(t) = s''(t)$. Acceleration is the second derivative of the position function. If acceleration is positive, the velocity is increasing; if acceleration is negative, the velocity is decreasing.

Units for velocity are units of distance (position) over units of time.

Units of acceleration are units of velocity over units of time. If velocity is measured in centimeters per second, the acceleration is measured in centimeters per second per second, which is equal to centimeters per second squared.

What about the rate of change of acceleration? Think again about a car. The gas pedal and brake on a car help control the car's speed.

Pressing the gas pedal farther downward increases the car's speed (which is a component of velocity), so it affects the vehicle's acceleration. If a car is stopped at a light, the light turns green, and the driver stomps on the gas pedal to accelerate quickly, the passengers feel a jerk. The rate of change of acceleration is therefore called *jerk*: $j(t) = a'(t) = v''(t) = s'''(t)$. Jerk is the third derivative of position. A summary of derivatives applied to motion appears in table 1.

If the vehicle accelerates too quickly, a passenger may also refer to the driver as a jerk.

Table 1 *Motion*

quantity	function	derivative	typical units
position	$s(t)$		cm
velocity	$v(t) = s'(t)$	first derivative of position	cm/s
acceleration	$a(t) = v'(t) = s''(t)$	second derivative of position	cm/s^2
jerk	$j(t) = a'(t) = v''(t) = s'''(t)$	third derivative of position	cm/s^3
speed	$\lvert v(t)\rvert = \lvert s'(t)\rvert$	absolute value of velocity	cm/s

This type of motion is sometimes called *rectilinear motion*, especially in older sources.

Motion examples

The equation of motion is given, so that is the position function. The name of the function, *s*, is consistent with the notation used earlier, but it does not follow that every source uses the same notation, so be careful.

Example 1 *A particle moves along a number line according to the equation* $s(t) = \frac{t^3}{3} - t$*, starting at time* $t = 0$*. Units are centimeters and seconds. (a) In what direction is the particle moving at time* $t = 0$*? (b) What is the average velocity of the particle during the interval* $0 \le t \le 2$*? (c) Does the particle ever change directions? (d) What is the speed of the particle at time* $t = 0$*? (e) What is the acceleration of the particle at time* $t = 2$*?*

Solution (a) Direction is part of velocity (velocity is speed and direction), so we need to calculate the velocity at time $t = 0$:

$$v(t) = s'(t) = \frac{1}{3} \cdot 3t^2 - 1 = t^2 - 1$$

$$v(0) = 0^2 - 1 = -1\ cm/s.$$

Because the velocity at time $t = 0$ is negative, the particle is moving in the negative direction (left).

The answer to part (a) is not $-1cm/s$, because this is not a direction.

(b) Average velocity was discussed in section 1.8. The formula for average velocity over the time interval $[a, b]$ is

$$\frac{s(b) - s(a)}{b - a}.$$

Using $a = 0$ and $b = 2$ gives

$$\frac{s(2) - s(0)}{2 - 0} = \frac{\frac{8}{3} - 2 - (\frac{0}{3} - 0)}{2} = \frac{\frac{2}{3}}{2} = \frac{1}{3}.$$

The average velocity during the interval $0 \le t \le 2$ is $\frac{1}{3}$cm/s.

(c) When might a particle change directions? Consider a car on a narrow driveway. In order for the car to change directions (for instance, change from moving forward into the driveway to moving backward out of the driveway), the car must stop. Stopping means that the speed is zero (the same as velocity is zero). Therefore, we can determine where the particle might change directions by solving the equation $v(t) = 0$:

$$v(t) = t^2 - 1 = 0$$
$$t^2 = 1$$
$$t = \pm 1.$$

Notice, though, that we are told in the statement of the example that the equation of motion is valid "starting at $t = 0$." Therefore, negative values of t are invalid and our only possibility is $t = 1$.

In order for the particle to change directions, it must stop. But it does not follow that every time the particle stops it changes direction. Think again about a car; if you come to a stop sign and stop, if no one is coming you can start traveling again in the same direction. Stopping does not automatically mean the direction changes, so just because the velocity is zero at $t = 1$ does not mean that our particle actually changed directions. We must check the direction in which the particle is traveling before and after the stop.

One way to check the direction before and after stopping is to check right before and right after – that is, infinitesimally before and after the stop. We do this by checking the velocity at $t = 1 - \omega$ (right before the stop) and at $t = 1 + \omega$ (right after the stop):

$$v(1 - \omega) = (1 - \omega)^2 - 1 = 1 - 2\omega + \omega^2 - 1 = -2\omega + \omega^2 \approx -2\omega,$$
$$v(1 + \omega) = (1 + \omega)^2 - 1 = 1 + 2\omega + \omega^2 - 1 = 2\omega + \omega^2 \approx 2\omega.$$

The velocity $v(t)$ is the instantaneous velocity, not the average velocity.

The position does not matter; we can stop anywhere. Acceleration does not matter; we can press the brake hard to stop quickly or lightly to stop slowly.

When finding values of time t, always check the statement of the exercise to determine which values of t are valid and discard those that are not valid.

Before the stop, the velocity is negative (-2ω is a negative number), so the particle is moving in the negative direction (left). After the stop, the velocity is positive (2ω is a positive number), so the particle is moving in the positive direction (right). Therefore, the particle changes directions at time $t = 1$.

(d) The speed is the absolute value of velocity, so we want to know $|v(0)|$:

$$|v(0)| = |0^2 - 1| = 1.$$

The speed at time $t = 0$ is 1 cm/s.

(e) We wish to find $a(2)$:

$$a(t) = v'(t) = 2t$$
$$a(2) = 2(2) = 4.$$

The acceleration at time $t = 2$ is 4 cm/s^2. ■

Some portions of example 1 are relatively straightforward; we only needed to calculate the requested quantity. Part (c) required some reasoning. When reasoning is required, visualizing the motion (picturing it in your mind) is usually necessary.

Reading Exercise 28 *A particle moves on a number line according to the position function* $s(t) = t^3 - 7t$. *Find the velocity and acceleration functions.*

This is still motion on a straight line, even though the motion is vertical.

Example 2 *A package is dropped (with an initial velocity of zero) from a hovering helicopter that is* 1960 m *above the ground. Ignoring air resistançe, its height above the ground* t *seconds after the drop is* $s(t) = 1960 - 4.9t^2$, *with height measured in meters. (a) Determine the time at which the package hits the ground. (b) How fast is the package falling when it hits the ground? (c) What is the package's acceleration?*

Solution (a) The height of the package above the ground is $s(t)$. When the package hits the ground, its height above the ground is zero. Therefore, we need to know when $s(t) = 0$. We want the time t when that happens, so we set $s(t) = 0$ and solve for t:

$$s(t) = 1960 - 4.9t^2 = 0$$
$$1960 = 4.9t^2$$
$$t^2 = \frac{1960}{4.9} = 400$$
$$t = \pm\sqrt{400} = \pm 20.$$

The package is dropped at time $t = 0$, so $t = -20$ is 20 seconds before the package is dropped, when it is still sitting in the helicopter and can not have hit the ground. Therefore, our solution is that the package hits the ground at time $t = 20$, or 20 seconds after it is dropped.

We found a value of time t, so we need to check which times are valid and which are not.

(b) "How fast" does not need a direction, so we are asked for the speed, not the velocity. The package hits the ground at time $t = 20$, so we are asked for the speed at time $t = 20$:

$$v(t) = s'(t) = -9.8t$$
$$\text{speed} = |v(20)| = |-9.8(20)| = 196.$$

The speed at which the package is falling when it hits the ground is 196 m/s.

The speed 196 m/s is equivalent to 438 mph or 706 km/h. Unless the package is exceedingly aerodynamic, air resistance applies and the package takes longer than 20 seconds to reach the ground and is not moving quite as fast as we calculated. Still, this illustrates the need for parachutes.

(c) Acceleration is $a(t)$:

$$a(t) = v'(t) = -9.8.$$

The package's acceleration is $-9.8\,\text{m/s}^2$. ■

For objects near the surface of the earth, acceleration resulting from gravity is $9.8\,\text{m/s}^2$ (or $32\,\text{ft/s}^2$) toward the center of the earth. Working backward, the equation of motion can be determined from the acceleration, the initial velocity, and the initial position of the package. We work such problems in section 4.10.

Ans. to reading exercise 28
velocity $v(t) = 3t^2 - 7$, acceleration $a(t) = 6t$

Rates of change in economics

A business that sells goods or services receives money, called *revenue*, from its sales. The amount of revenue received depends on the amount of goods or services sold. The *revenue function* describes the relationship between revenue and sales; $R(x)$ represents the revenue from selling x units of goods or services. Likewise, the *cost function* is represented by $C(x)$ and gives the cost of procuring, manufacturing, or providing x units of goods and services. Profit is revenue minus cost, so the *profit function* $P(x) = R(x) - C(x)$ represents the profit from producing and selling x units.

$R(x)$	revenue from selling x units
$C(x)$	cost of producing x units
$P(x)$	profit from x units

The adjective *marginal* means the effect of "one more." Marginal revenue is the additional revenue from selling one more unit, marginal cost is the additional cost of producing one more unit, and marginal profit is the additional profit from producing and selling one more unit.

It turns out that under most circumstances, "marginal" can be considered a synonym for "derivative." To see why, let's consider the local linearity diagram in figure 1. The slope of the tangent line is $R'(k) = \frac{dy}{dx}$. Note that dx represents additional units sold whereas Δy represents additional revenue. If $dx = 1$ and the gap is small enough, then $R'(k) = \frac{dy}{1} = dy$ is close to Δy, which is the additional revenue from one additional unit sold (i.e., the marginal revenue). Marginal revenue is therefore usually equated with $R'(x)$, marginal cost with $C'(x)$, and marginal profit with $P'(x)$.

Another interpretation of $R'(x)$ is the "instantaneous marginal revenue."

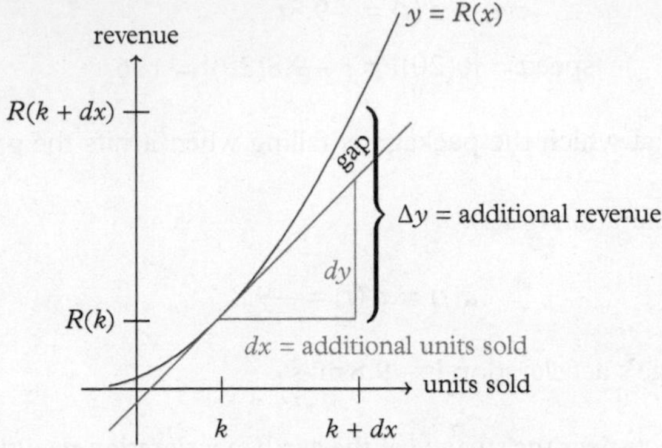

Figure 1 *The graph of the revenue function $y = R(x)$ (blue) with tangent line (orange) at $x = k$ and a real-number increment dx*

Estimated functions such as the one in example 3 are called *models*. Modeling can be highly situation specific, but tools for modeling can be studied in courses with names such as "Mathematical Modeling."

Example 3 *Suppose Big Telecom estimates revenue from selling x units of call time to be $R(x) = \$4.3\sqrt{x}$. What is the marginal revenue when 1 000 000 units have been sold? Interpret your answer.*

Solution The marginal revenue is $R'(x)$:

$$R'(x) = 4.3 \cdot \frac{1}{2}x^{-\frac{1}{2}} = \frac{2.15}{\sqrt{x}}.$$

We want the marginal revenue when $x = 1\,000\,000$:

$$R'(1\,000\,000) = \frac{2.15}{\sqrt{1\,000\,000}} = \frac{2.15}{1000} = 0.00215.$$

The marginal revenue when 1 000 000 units are sold is \$0.00215/unit sold.

Units for marginal revenue are units of revenue (dollars or other monetary unit) divided by units sold (however measured). The same units are used for marginal cost and marginal profit.

Interpretation: At this level of sales, Big Telecom's revenue increases by 0.215¢ for each additional call time unit sold. ■

This calculation is what is meant by the instructions to find the marginal revenue. As an alternative, we can use the meaning of marginal revenue as the additional revenue produced by one additional unit sold, which is

$$R(1\,000\,001) - R(1\,000\,000) = \$0.0021499996,$$

which is practically identical to the amount we calculated using the derivative. The point, however, is not that we are seeking an easier method of calculating the marginal revenue, but rather that we are seeking to understand the relationships involved. Much economic theory is developed using derivatives for "marginals." Understanding the connection is therefore very beneficial.

Reading Exercise 29 *AAA Fencing Co. estimates its cost of installing x feet of fencing for a customer as $C(x) = 500 + 2.7x$ dollars. Find the marginal cost function.*

Exercises 2.7

1–8. Given the following equations of motion, find the position, velocity, and acceleration at the instant indicated (GSL).

Units were not given in (GSL). They are left off the answers for these exercises.

Because these equations are equations of motion, each represents $s(t)$ in the notation of this section, regardless of whether the function or variable is named s.

1. $s(t) = 4t^2 - 6t,\ t = 2$

2. $s(t) = 6t^2 - 2t^3,\ t = 1$

3. $s = 120t - 16t^2,\ t = 4$

4. $y = 16t^2 - 20t + 4,$ $t = 2$

5. $x = 32t - 8t^2,\ t = 2$

6. $s = 100 - 4t - 8t^2,\ t = 3$

7. $s(t) = \dfrac{t}{t+1},\ t = 2$

8. $s(t) = \sqrt{5t} + \dfrac{10}{\sqrt{5t}},\ t = 5$

9. (GSL) Given the equation of motion $s = 16t^2 - 64t + 64$, find the position and acceleration when the particle first comes to rest (when the particle first stops).

10. (GSL) Given the equation of motion $s = 120t - 16t^2$, find the position and acceleration when the particle first comes to rest (when the particle first stops).

11. A particle moves on a number line according to the position function $s(t) = t^3 - 6t + 1$ for $t \geq 0$. Time t is given in seconds and $s(t)$ is given in meters. (a) In what direction is the particle moving at time $t = 1$? (b) What is the acceleration of the particle at time $t = 2$? (c) Does the particle ever stop? If so, when? (d) What is the speed of the particle at time $t = 0$?

12. A particle moves on a number line according to the position function $s(t) = 2t^3 - 5t + 7$ for $t \geq 0$. Time is given in seconds and $s(t)$ is given in meters. (a) In what direction is the particle moving at time $t = 1$? (b) What is the acceleration of the particle at time $t = 3$? (c) Does the particle ever stop? If so, when? (d) What is the speed of the particle at time $t = 2$?

13. A projectile is fired vertically from a point 5 feet above the ground so that its height after t seconds is $h = 5 + 100t - 16t^2$ ft. (a) Find the velocity of the projectile at time $t = 2$. (b) What is the acceleration of the projectile at time $t = 3$? (c) When does the projectile reach its maximum height? (d) What is its maximum height?

14. A projectile is fired vertically from a point 10 m above the ground so that its height after t seconds is $h = 10+98t-4.9t^2$ m. (a) Find the velocity of the projectile at time $t = 8$. (b) What is the acceleration of the projectile at time $t = 4$? (c) When does the projectile reach its maximum height? (d) What is its maximum height?

15. A particle moves along a number line according to the position function $s(t) = \frac{t^3}{3} + t^2 - 8t$, for $t \geq 0$. Units are meters and seconds. (a) Find the velocity of the particle at time $t = 3$. (b) Does the particle ever stop? If so, when? (c) What is the total distance traveled during the first 5 seconds?

16. A particle moves along a number line according to the position function $s(t) = \frac{t^3}{3} + \frac{t^2}{2} - 20t + 11$, for $t \geq 0$. Units are meters and seconds. (a) Find the velocity of the particle at time $t = 3$. (b) Does the particle ever stop? If so, when? (c) What is the total distance traveled during the first 5 seconds?

Ans. to reading exercise 29
$C'(x) = 2.7$. The units are dollars per foot.

17. A particle moves along a number line according to the position function $s(t) = \frac{t^3}{3} + \frac{t^2}{2} - 6t + 1$, for $t \geq -10$. Units are meters and seconds. (a) Find the velocity of the particle at time $t = 4$. (b) What is the acceleration of the particle at time $t = 5$? (c) What is the jerk at time $t = \sqrt[4]{\frac{12}{19}}$? (d) Does the particle ever stop? If so, when?

18. A particle moves along a number line according to the position function $s(t) = \frac{t^3}{3} + t^2 - 15t + 7$, for $t \geq -8$. Units are meters and seconds. (a) Find the velocity of the particle at time $t = 1$. (b) Find the acceleration of the particle at time $t = 5$. (c) What is the jerk at time $t = \frac{47}{19}$?

19. A book is dropped accidentally from a third-floor window in a library. Its height above the ground after t seconds is given by $s(t) = 38 - 16t^2$, measured in feet. (Do not actually try this because injury to persons below could result.) (a) What is the book's acceleration? (b) What is the book's velocity after 1 second? (c) When does the book hit the ground? (d) How fast is the book traveling when it hits the ground?

20. A sandwich is dropped accidentally from a fourth-floor balcony of a hotel. Its height above the ground after t seconds is $s(t) = 10 - 4.9t^2$, measured in meters. (Do not try this because injury to persons below could result.) (a) What is the sandwich's acceleration? (b) What is the sandwich's velocity after 1.2 seconds? (c) When does the sandwich hit the ground? (d) How fast is the sandwich traveling when it hits the ground?

21. A full water bottle is thrown upward from a high platform. Its height above the ground after t seconds is given by $s(t) = 48 + 32t - 16t^2$, measured in feet. (a) What is the bottle's acceleration when $t = 2.1$? (b) When does the bottle stop its ascent and begin to fall? (c) When does the bottle hit the ground? (d) What is the speed of the bottle when it hits the ground?

22. (GSL) A ball thrown directly upward moves according to the law $s(t) = 80t - 16t^2$ (units are feet and seconds). Find (a) its position and velocity after 2 seconds and after 3 seconds, (b) how high it will rise, and (c) how far it will move in the fourth second.

23. Hypercomplex Manufacturing finds that its total profit from manufacturing and selling x widgets in 1 month is $P(x) = 112x - 8200 - 0.003x^2$ dollars. (a) Find the marginal profit when 20 000 widgets have been manufactured and sold. (b) When 20 000 widgets have been manufactured and sold, is it desirable to increase production and sales?

24. Acme Manufacturing finds that its total profit from manufacturing and selling x roadrunner elimination kits in 1 month is $P(x) = 13x - 4600 - 0.001x^2$. (a) Find the marginal profit when 5000 kits have been manufactured and sold. (b) When 5000 kits have been manufactured and sold, is it desirable to increase production and sales?

25. Calculus R Us finds that its revenue from selling x units of consulting services in a month is $R(x) = 59x^{0.8}$, measured in dollars. What is the marginal revenue when 14 000 units have been sold? Interpret your answer.

26. Infinitesimals Rule Corporation finds that its revenue from selling x units of services in a month is $R(x) = 214x^{0.9}$, measured in dollars. What is the marginal revenue when 3000 units have been sold? Interpret your answer.

Related Rates: Pythagorean Relationships

2.8

Time to rejoice—word problems! One interpretation of the derivative is that it represents a rate of change. In this section we study two or more related items that are changing simultaneously, where the items are related through the Pythagorean theorem.

Sliding ladder

The method of related rates can be explained through an example.

The term *related rates* is short for "related rates of change."

Example 1 *A ladder* 20 *feet long leans against a vertical building. If the bottom of the ladder slides away from the building horizontally at a rate of* 2 *ft/s, how fast is the ladder sliding down the building when the top of the ladder is* 12 *feet above the ground?*

Solution As with any word problem, the first task is to draw a picture. We can begin by drawing a building with a 20-foot ladder leaning against it, as in figure 1.

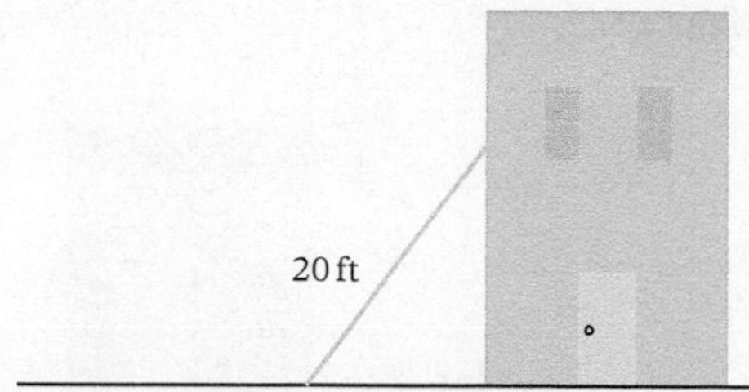

Figure 1 *A* 20-*foot ladder leaning against a building*

Next we notice that the ladder is not stationary, but moving. We are also given the rate at which the bottom of the ladder is moving, and we are asked to find the rate at which the top of the ladder is moving. These are the distinguishing features of a related rates problem: things are changing (moving, shrinking, growing, rising, rotating, etc.), we know the rates at which some things are changing, and we are asked to find the rate at which something else is changing.

The medium through which things change is time, so these rates are all rates of change with respect to time.

We began by drawing a diagram. Figure 1 labels values that are not changing. Do not place in the diagram numbers that represent a changing quantity, for they can easily mislead. Instead, quantities that are changing are variable quantities, so we label them as variables. The distance of the bottom of the ladder to the building is changing, so we label that quantity x. The distance from the ground to the top of the ladder is changing, so we label that quantity y. It is also helpful to include arrows in the diagram to mark the direction of change. See figure 2.

We are accustomed to naming horizontal quantities x and vertical quantities y, so they seem like good choices of variable names.

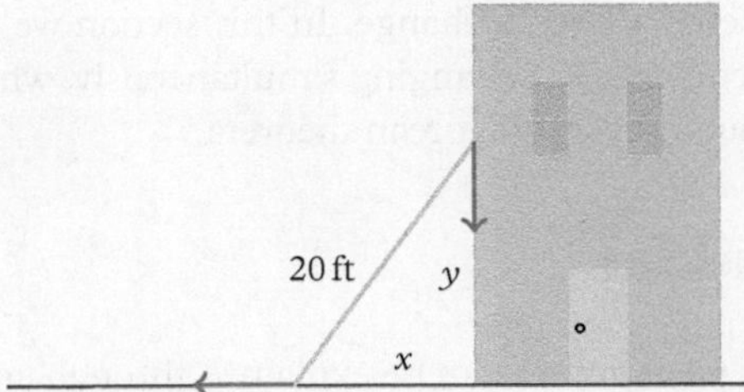

Figure 2 *A 20-foot ladder leaning against a building, with changing quantities marked by variables and arrows (orange) representing direction of change*

Our next step is to determine which value we are asked to find. Notice that we are not asked to find the value of the variable x or the variable y. Instead, we are asked to determine how fast the top of the ladder is sliding down the building, which is a rate of change—specifically, the rate of change of y with respect to time, $\frac{dy}{dt}$. This quantity can be placed in the diagram next to the arrow representing its change. See figure 3.

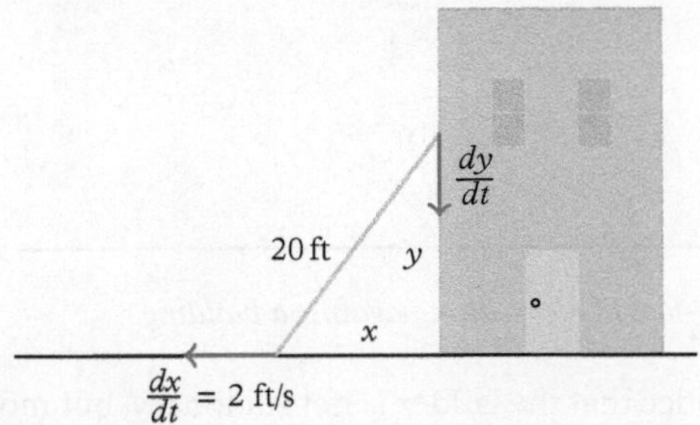

Figure 3 *A 20-foot ladder leaning against a building, with changing quantities marked by variables and arrows (orange) representing direction of change, labeled by their rates of change*

The other arrow in the diagram also represents change; it is the change in the variable x with respect to time, $\frac{dx}{dt}$. We are told the value of that rate: the bottom of the ladder is sliding away from the building at a rate of 2 ft/s. This can also be placed in the diagram next to its arrow. Our diagram, in figure 3, is finally complete. A diagram that indicates changing quantities, such as figure 3, is called a *dynamic diagram*. A diagram that does not indicate any changing quantity, such as in figure 4, is called a *static diagram*. Related rates exercises require dynamic diagrams.

CAUTION: MISTAKE TO AVOID

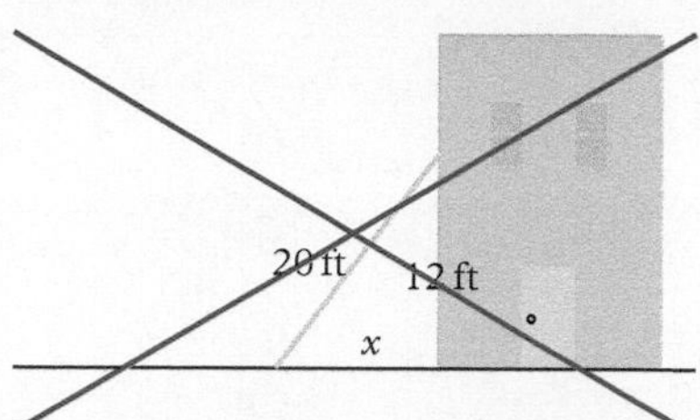

Figure 4 A 20-foot ladder leaning against a building, where the top of the ladder is 12 feet from the ground and not moving. A static diagram such as this is insufficient for illustrating changing quantities

Now that the diagram is drawn, how do we find the desired rate of change, $\frac{dy}{dt}$? The next step in the solution process is to determine a geometric relationship between the variables in the diagram. Because vertical and horizontal meet in a right angle, we recognize a right triangle in the diagram. The variables x and y are therefore related through the Pythagorean theorem:

$$x^2 + y^2 = 20^2.$$

However, this still does not tell us the relationship we really want to know, which is the relationship between the rates of change $\frac{dx}{dt}$ and $\frac{dy}{dt}$. This relationship can be found by differentiating both sides of the equation implicitly with respect to time t:

$$\frac{d}{dt}\left(x^2 + y^2\right) = \frac{d}{dt}\left(20^2\right)$$

$$2x \cdot \frac{dx}{dt} + 2y \cdot \frac{dy}{dt} = 0.$$

Rewriting the chain rule form of the power rule for differentiating with respect to t instead of x gives

$$\frac{d}{dt}u^n = nu^{n-1}\frac{du}{dt}.$$

Using $u = y$, $n = 2$ gives

$$\frac{d}{dt}y^2 = 2y\frac{dy}{dt},$$

and using $u = x$, $n = 2$ gives

$$\frac{d}{dt}x^2 = 2x\frac{dx}{dt}.$$

We know that $\frac{dx}{dt} = 2$; thus,

$$2x \cdot 2 + 2y \cdot \frac{dy}{dt} = 0.$$

If we know values of x and y, then we can easily solve for $\frac{dy}{dt}$. Rereading the exercise, we see that we want to know $\frac{dy}{dt}$ when the top of the ladder is 12 feet above the ground—that is, when $y = 12$. Now we need only find x. Drawing a second diagram for the instant in which $y = 12$ (figure 5), we recognize that the Pythagorean theorem can help us find x at that instant in time:

$$x^2 + 12^2 = 20^2$$

$$x^2 + 144 = 400$$

$$x^2 = 256$$

$$x = \pm 16.$$

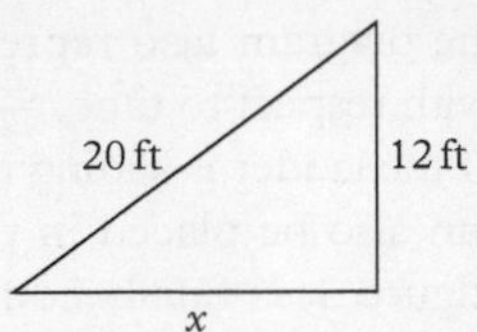

Figure 5 *The moment when $y = 12$. This static diagram should only be drawn after the implicit differentiation step; it is insufficient for representing changing quantities*

In figure 3, x is a positive quantity, therefore we use $x = 16$:

$$2(16) \cdot 2 + 2(12) \cdot \frac{dy}{dt} = 0$$

$$64 + 24\frac{dy}{dt} = 0$$

$$24\frac{dy}{dt} = -64$$

$$\frac{dy}{dt} = -\frac{64}{24} = -\frac{8}{3} \text{ ft/s.}$$

Checking figure 3, we see that the negative rate means that y is getting smaller, that the top of the ladder is sliding down the building as expected.

It is always good practice to reread the question before deciding upon the form of the final answer. "How fast" indicates speed, which does not contain direction. The top of the ladder is sliding down the building at a rate of $\frac{8}{3}$ ft/s. ■

Sliding "down" indicates direction; the speed at which it is moving is the absolute value of the rate, just as speed is the absolute value of velocity in problems involving motion.

The solution method used in example 1 is summarized here.

RELATED RATES SOLUTION METHOD

1. Draw a dynamic diagram.
   - a. Draw the objects and label quantities that are not changing (as in figure 1).
   - b. Use variable names to label quantities that are changing (as in figure 2).
   - c. Use arrows to indicate how the variables are changing (as in figure 2).

> ⓓ Label the arrows using the appropriate rates of change with respect to time *t*. Be sure that all rates with values that are given are included, as well as the rate that is requested (as in figure 3).
>
> ❷ Determine a geometric relationship between the variables of step ❶ ⓑ.
>
> ❸ Differentiate both sides of the geometric relationship with respect to time *t*.
>
> ❹ Input the values of known rates into the equation from step ❸.
>
> ❺ Input the values of the variables at the instant in which the rate's value is requested.
>
> ❻ Solve the equation for the requested rate.
>
> ❼ Reread the question to determine the final form of the answer.

Reading Exercise 30 *Differentiate the geometric relationship* $x^2 + 57.4^2 = z^2$ *with respect to time t.*

Submarine passing under radar station

We tackle the next example using the related rates solution method.

Example 2 *At 1600 hours a submarine in the arctic ocean passes directly under a radar station that sits on top of the ice. The submarine's depth is* 600 m. *Maintaining a constant depth, the submarine keeps a straight course with speed* 100 m/min. *At 1610 hours, how fast is the distance between the radar station and the submarine increasing?*

The time 1600 hours corresponds to 4:00 p.m.

Solution ❶ We begin by drawing a dynamic diagram. ⓐ First we draw the submarine and radar station and label the depth because it does not change (figure 6).

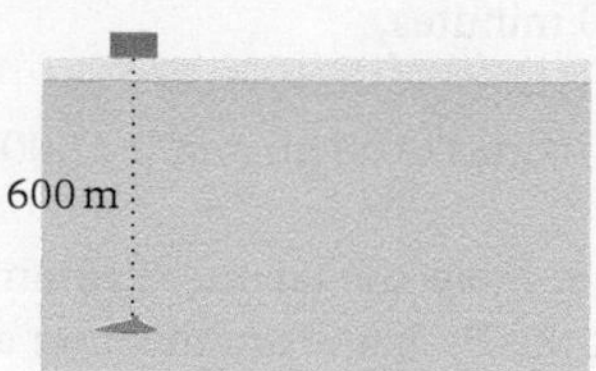

Figure 6 *A submarine* 600 *m directly beneath a radar station*

The sub is moving horizontally, so a good choice is the variable x. Using y for a nonvertical distance can be confusing, so we choose z for the distance from the radar station to the sub.

ⓑ Next we use variables to label quantities that are changing. The submarine is moving, so we may change the sub's position and let that distance be one of the variables (x). We are asked a question about the distance between the radar station and the submarine, so we also label that distance (z). ⓒ We also use an arrow to indicate the movement of the sub and ⓓ label that arrow $\frac{dx}{dt}$. See figure 7.

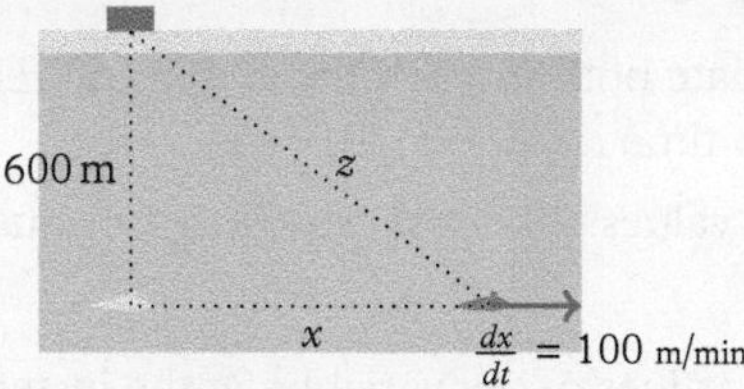

Figure 7 *A submarine after passing 600 m directly beneath a radar station. The movement of the sub changes both x and z*

❷ We need a geometric relationship between the variables x and z from step ❶ ⓑ. Because the submarine is moving horizontally (moving at a constant depth), there is a right triangle in our picture and the Pythagorean theorem applies:

$$x^2 + 600^2 = z^2.$$

Using Leibniz notation instead of prime notation helps emphasize the fact that we are differentiating with respect to time t, not with respect to x or some other variable.

❸ Differentiate with respect to time t:

$$\frac{d}{dt}\left(x^2 + 360\,000\right) = \frac{d}{dt}\left(z^2\right)$$

$$2x \cdot \frac{dx}{dt} + 0 = 2z \cdot \frac{dz}{dt}.$$

Ans. to reading exercise 30
$2x \cdot \frac{dx}{dt} + 0 = 2z \cdot \frac{dz}{dt}$

❹ We know $\frac{dx}{dt} = 100$ m/min:

$$2x \cdot 100 = 2z \cdot \frac{dz}{dt}.$$

❺ We wish to know the value of $\frac{dz}{dt}$ at 1610 hours, which is 10 minutes after the sub passes under the radar station. Because the sub is traveling at 100 m/min, after 10 minutes,

Recall that distance = rate · time.

$$x = 10\,\text{min} \cdot 100\,\text{m/min} = 1000\,\text{m}.$$

We also need to know z. Draw the (static) diagram for the instant 1610 hours and we see that the Pythagorean theorem applies (figure 8):

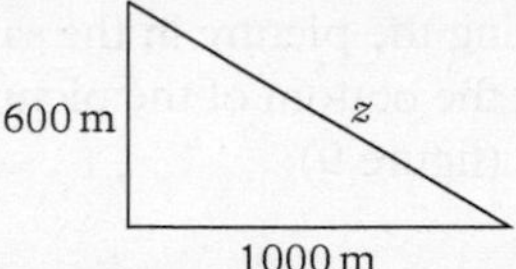

Figure 8 *The submarine at 1610 hours*

$$600^2 + 1000^2 = z^2$$
$$360\,000 + 1\,000\,000 = z^2$$
$$z = \pm\sqrt{1\,360\,000} = \pm 1166.$$

The distance z is positive. Using $x = 1000$ and $z = 1166$ gives

$$2 \cdot 1000 \cdot 100 = 2 \cdot 1166 \cdot \frac{dz}{dt}.$$

❻ Solving yields

$$200\,000 = 2332\frac{dz}{dt}$$
$$\frac{dz}{dt} = \frac{200\,000}{2332} = 85.76.$$

Notice that the submarine is moving at a rate of 100 m/min, but is only increasing its distance from the radar station at a rate of 85.76 m/min. The reason is that the sub is not moving directly away from the radar station. When the sub is very distant from the station, it is closer to moving directly away from the station and the rate at which the distance is changing is much closer to 100 m/min.

❼ At 1610 hours, the distance between the radar station and the submarine is increasing at a rate of 85.76 m/min. ■

Passing ships

Example 2 featured one stationary object (the radar station) and one object in motion (the submarine). The next example features two objects in motion.

Example 3 *At 1:00 p.m., ship A is 25 miles due south of ship B. If ship A is sailing west at a rate of* 16 *mph and ship B is sailing south at a rate of* 20 *mph, what is the rate at which the distance between the ships is changing at 1:30 p.m.?*

The abbreviation mph is short for "miles per hour," or mi/h.

Solution ❶ The diagram should contain two ships, *A* and *B*. Both ships are moving, and the movement on the water is horizontal. Instead of drawing a diagram from a side view, such as for the sliding ladder or the submarine under the radar station, a top view, as from a satellite, is more helpful. ⓐ Orienting the picture in the same manner as a map, at 1:00 p.m., ship A is at the bottom of the picture and ship B is at the top, aligned "vertically" (figure 9):

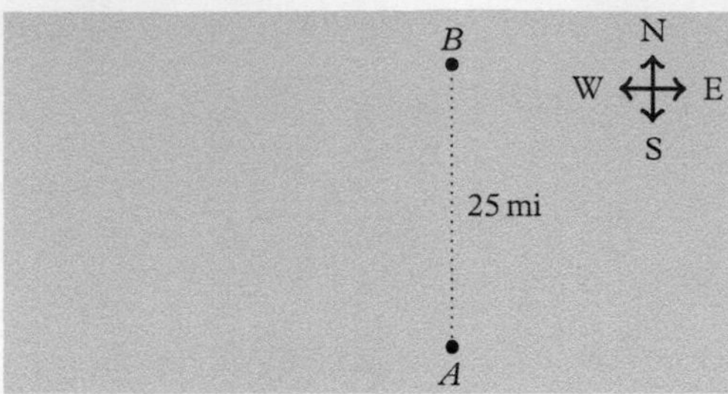

Figure 9 *Ship A* 25 miles *south of ship B at 1:00 p.m., view from above, orientation the same as a map*

To show the movement of the ships, we retain points marked at the ships' positions at 1:00 p.m. and move the ships to new positions as in figure 10.

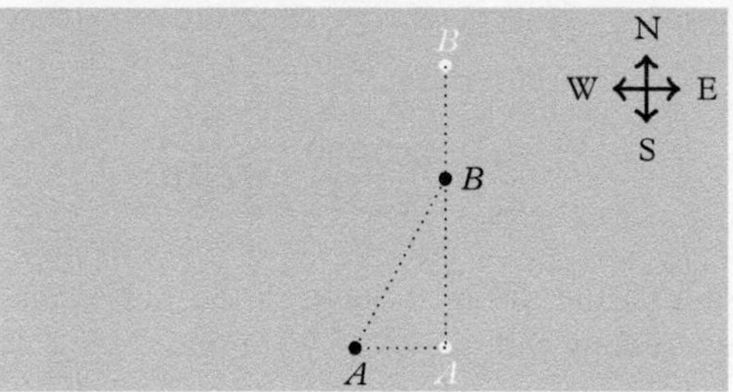

Figure 10 *Ships A and B after 1:00 p.m., view from above, orientation the same as a map*

ⓑ We see a right triangle in figure 10, and we notice that all three sides of the triangle change as the ships move. We label these distances as variables x (east–west), y (north–south), and s (diagonal). ⓒ We also need to place arrows in the diagram indicating the direction the ships are moving, and ⓓ we can place the given rates of change next to their arrows (figure 11). The rate of change $\frac{dy}{dt}$ is negative because the distance y is decreasing, whereas the rate of change $\frac{dx}{dt}$ is positive because the distance x is increasing.

❷ We need a relationship between the three variable quantities identified in ❶ⓑ. Once again, we have a right triangle and the Pythagorean theorem applies:

$$x^2 + y^2 = s^2.$$

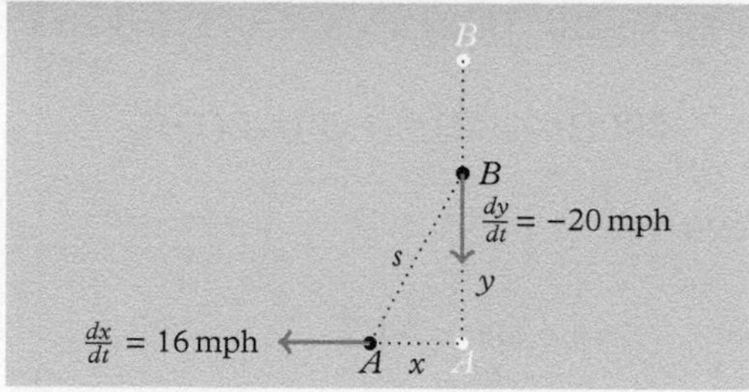

Figure 11 *Ships A and B after 1:00 p.m., with directions and speeds marked*

❸ Next we differentiate with respect to time t:

$$\frac{d}{dt}\left(x^2 + y^2\right) = \frac{d}{dt}\left(s^2\right)$$
$$2x \cdot \frac{dx}{dt} + 2y \cdot \frac{dy}{dt} = 2s \cdot \frac{ds}{dt}.$$

❹ We know the rates $\frac{dx}{dt} = 16$ mph and $\frac{dy}{dt} = -20$ mph:

$$2x(16) + 2y(-20) = 2s \cdot \frac{ds}{dt}.$$

❺ We wish to know the value of $\frac{ds}{dt}$ at 1:30 p.m., which is half an hour after 1:00 p.m. Because ship A is traveling west at 16 mph,

$$x = \frac{1}{2}\,\text{h} \cdot 16\,\text{mi/h} = 8\,\text{mi}.$$

Because ship B is traveling south at 20 mph, it has traveled

$$\frac{1}{2}\,\text{h} \cdot 20\,\text{mi/h} = 10\,\text{mi}$$

south from its original position. Therefore, $y = 25 - 10 = 15$ miles. We also need to know s. Draw the (static) diagram for the instant 1:30 p.m. and we see that the Pythagorean theorem applies (figure 12):

$$8^2 + 15^2 = s^2$$
$$289 = s^2$$
$$s = \pm\sqrt{289} = \pm 17.$$

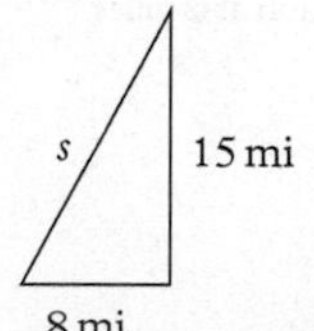

Figure 12 The ships at 1:30 p.m.

The distance s is positive. Using $x = 8$, $y = 15$, and $s = 17$ gives

$$2(8)(16) + 2(15)(-20) = 2(17)\frac{ds}{dt}.$$

❻ Solving yields

$$256 - 600 = 34\frac{ds}{dt},$$
$$\frac{ds}{dt} = -\frac{344}{34} \approx -10.12.$$

Because $\frac{ds}{dt}$ is negative, the ships are getting closer to one another—that is, the distance between them is decreasing.

❼ At 1:30 p.m., the rate at which the distance between the ships is changing is $-\frac{344}{34}$ mi/h ≈ -10.12 mph. ■

In this section, the geometric relationship of step ❷ has always been the Pythagorean theorem. The next section explores related rates exercises with many other geometric relationships.

Exercises 2.8

1. At 3:00 p.m., ship A is 8 miles directly north of ship B. Ship A is moving north at a rate of 12 mph whereas ship B is moving east at 20 mph. At what rate is the distance between the ships increasing at 4:00 p.m.?
2. At 1:00 a.m., ship A is 10 km directly east of ship B. Ship A is moving west at a rate of 7 km/h whereas ship B is moving north at a rate of 20 km/h. At what rate is the distance between the ships changing at 2:00 a.m.?
3. A plane flying horizontally at an altitude of 1 mile and a speed of 500 mi/h passes directly over a radar station. Find the rate at which the distance from the plane to the station is increasing when it is 2 miles away from the station.

The phrase "when it is 2 miles away from the station" gives the distance from the plane to the radar station as 2 miles.

4. A model airplane flies horizontally at a rate of 20 m/s and passes directly above its operator at a height of 100 m. At what rate is the distance between the operator and the airplane changing when the airplane is 200 m from its operator?
5. At 3:00 p.m., ship A is 20 miles due east of ship B. Ship A is sailing south at a rate of 12 mph whereas ship B is sailing east at a rate of 18 mph. At what rate is the distance between the ships changing at 3:30 p.m.?

Exercises marked "(Love)" are from *Differential and Integral Calculus* by Clyde E. Love, revised edition, 1926.

6. (Love) Two railroad tracks intersect at right angles. At 12:00 noon, there is a train on each track approaching the crossing

at 40 mph, one being 100 miles and the other being 200 miles distant. Find how fast they are approaching each other.

7. (Love) A ladder 10 feet long leans against a wall. If the lower end moves outward at the rate of 18 inches per second, find how fast the top is descending when the lower end is 6 feet from the wall.

12 inches = 1 foot

8. (GSL) One end of a ladder 50 feet long is leaning against a perpendicular wall standing on a horizontal plane. Suppose the foot of the ladder is to be pulled away from the wall at the rate of 3 ft/min. How fast is the top of the ladder descending when its foot is 14 feet from the wall?

9. (Love) The city of A is 10 miles north of B. A car starts east from A at 30 mph; at the same time, another car starts east from B at 60 mph. How fast are the cars separating after 10 minutes?

10. (GSL) One ship is sailing south at a rate of 6 mph; another is sailing east at a rate of 8 mph. At 4:00 p.m., the second ship crosses the track of the first where the first was 2 hours earlier. How is the distance between the ships changing at 3:00 p.m.?

11. (Love) A kite is 120 feet high. If the kite moves horizontally at 4 mph directly away from the child flying it, how fast is the cord being paid out when 130 feet of cord is out?

1 mile = 5280 feet

12. A child flies a kite at a height of 200 feet. The wind is carrying the kite horizontally away from the child at a rate of 16 ft/s. How fast must the string be let out when the kite is 350 feet away from the child?

13. (GSL) A barge with a deck that is 12 feet below the level of a dock is drawn up to it by means of a cable attached to a ring in the floor of the dock, with the cable being hauled in by a windlass on deck at the rate of 8 ft/min. How fast is the barge moving toward the dock when it is 16 feet away?

The phrase "moving toward the dock" is to be interpreted as the horizontal movement of the barge.

14. (Love) A person on a wharf 20 feet above the water pulls in a rope, to which a boat is attached, at a rate of 4 ft/s. At what rate is the boat approaching the shore when there is 25 feet of rope out?

15. A submarine is attempting to surface and is moving vertically upward at a rate of 1.5 m/s. A stationary ship is nearby, 800 m from the point at which the submarine will surface. At what rate is the distance between the ship and the submarine changing when the sub is at a depth of 100 m?

16. An antelope is happily springing along through a grassy field in a straight line at a rate of 2 m/s. A stalking bobcat is sneaking

through the grass on a straight line perpendicular to the antelope's path at a rate of $\frac{1}{2}$ m/s. Assuming the bobcat has timed its intercept perfectly and the animals maintain constant rates of travel, how fast is the distance between the animals decreasing 5 seconds before they meet?

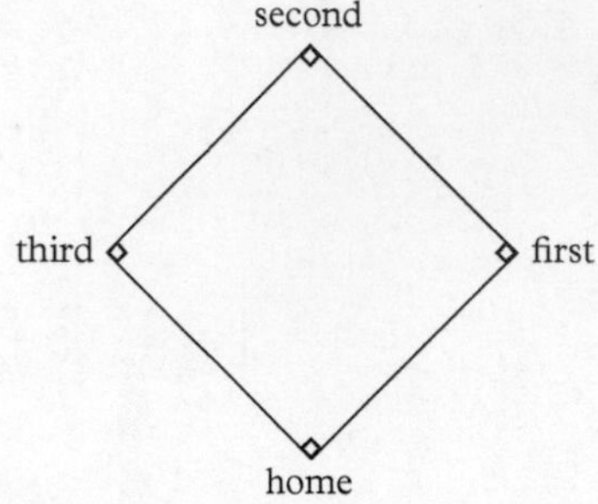

Figure 13 A baseball diamond

17. A book 25 cm tall leans against the side of a bookcase in a dorm room. A bored student pushes the bottom of the book toward the side of the bookcase at a rate of 0.4 cm/s. At what rate is the top of the book moving upward when the top of the book is 9 cm above the shelf?
18. The bases in a baseball diamond (figure 13) are 90 feet apart. A player runs from first base to second base at a rate of 16 ft/s. How fast is the player's distance to third base changing when halfway between first and second?
19. At 12:00 noon, ship A is 40 km north and 2 km east of ship B. Ship A is traveling west at 6 km/h and ship B is traveling east at 10 km/h. How fast is the distance between the ships changing at 1:00 p.m.?
20. A photographer is positioned 6 m from the bottom of a vertical caterpillar's thread (this distance is horizontal). The caterpillar is descending the thread at a rate of 2 cm/min. At what rate is the caterpillar's distance to the photographer changing when the caterpillar is 1.5 m above the level of the photographer?
21. (Love) As a person walks across a bridge at a rate of 5 ft/s, a boat passes directly beneath the person at 10 ft/s. If the bridge is 30 feet above the water, how fast are the person and the boat separating 3 seconds later?
22. (Love) An elevated train on a track 30 feet above the ground crosses the street at a rate of 20 ft/s at the instant that an automobile, approaching at a rate of 30 ft/s, is 40 feet up the street. Find how fast the train and the automobile are separating 2 seconds later.

Not exercise 2, but example 2 from the section narrative.

23. For the submarine in example 2, how fast is the distance between the radar station and the submarine increasing at (a) 1630 hours? (b) 1700 hours? (c) 1800 hours?
24. For the ships in example 3, at what instant in time is the distance between the ships not changing?

This is the same as example 1 except the distance $y = 12$ is changed to $y = \omega$.

25. For the ladder in example 1, suppose the ladder has slid so far that the distance between the ground and the top of the ladder is an infinitesimal amount, ω. How fast is the ladder sliding down

the side of the building at that instant? (Render a real result for your answer.)

26. (GSL) For the ladder in exercise 8, when will the top and bottom of the ladder move at the same rate?

27. (Love) Two trains start from the same point at the same time, one going east at a rate of 40 mph, the other going north at 60 mph. At what rate do they separate?

Related Rates: Non-Pythagorean Relationships

2.9

We continue our study of related rates exercises by exploring more general geometric relationships. The solution method from section 2.8 is still valid.

Expanding sphere

Example 1 *Gas is being pumped into a spherical balloon at a rate of* 5 ft^3/*min. Find the rate at which the radius is changing when the diameter is* 18 *inches.*

Solution As in section 2.8, we use the steps of the related rates solution method.

❶ We begin by drawing a dynamic diagram. ⓐ In figure 1, we draw the objects (balloon) and label the quantities that are not changing (none):

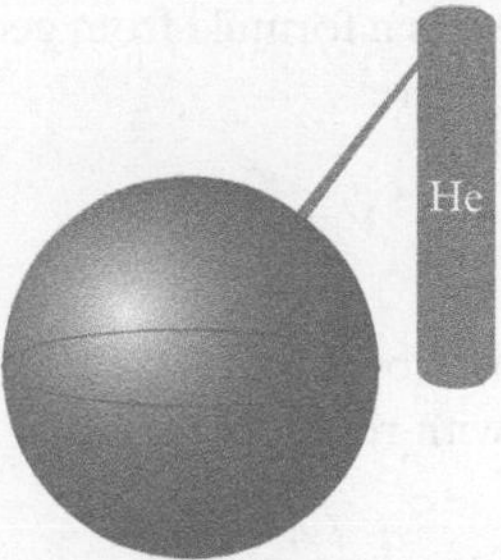

Figure 1 *Gas being pumped into a balloon*

ⓑ Which variables are changing? Sometimes clues come from the rates that are given and requested. The rate that is given is the rate at which gas is being pumped into the balloon, 5 ft^3/min. Gas fills volume, so perhaps the volume is changing. Look also at the units on

Note: "ft" is a unit of distance, "ft^2" is a unit of area, and "ft^3" is a unit of volume.

It is traditional to use the same variable names that are commonly used in geometric formulas. Therefore, the variable V is a good choice to represent a volume and the variable r is a good choice to represent a radius.

Imagining the motion is often much easier than drawing it. The diagram's purpose is to help us visualize the motion, so it is acceptable to leave out difficult-to-draw elements of a diagram if the visualization has been accomplished. But do not use this as an excuse to leave out easy-to-draw elements of the diagram.

that rate: the numerator is cubic feet, which represents a volume. The denominator is minutes, which is a unit of time. Therefore, the rate is change in volume over time, so volume is a variable that is changing. Let's label the volume V. The rate that is requested is the rate at which the radius is changing, so we label the radius r. It is not difficult to draw and label the radius of the balloon in the diagram, but drawing and labeling the volume is more challenging; we omit V from the diagram.

ⓒ We place an arrow in the diagram to indicate how the radius is changing. Adding gas to the balloon causes the balloon to expand, so we draw the arrow outward, indicating an increasing radius. We can also visualize how the balloon's volume is expanding. ⓓ We label the arrow with its rate of change. Although we did not draw an arrow to represent the change in volume, we can write the rate of change $\frac{dV}{dt}$, along with its value, at the side of the diagram. See figure 2.

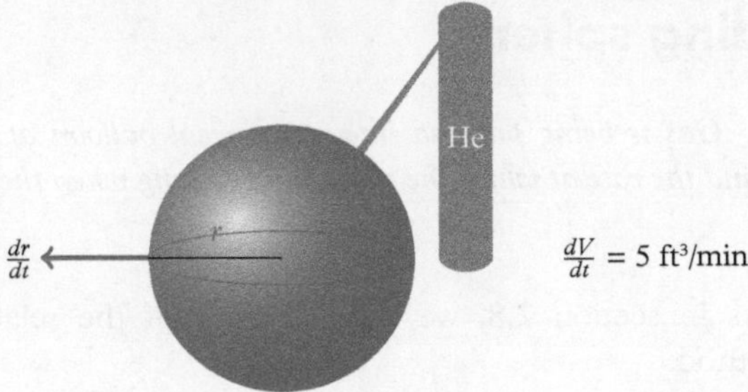

Figure 2 *Gas being pumped into a balloon*

Choosing the correct geometric relationship is key to a related rates exercise. In section 2.8, the geometric relationship was always the Pythagorean theorem, but many other relationships are possible. Well-known or easily found geometric formulas are prime suspects.

❷ Now we need a geometric relationship between the variables of step ❶ⓑ, V and r. There is a formula from geometry that relates the volume and radius of a sphere:

$$V = \frac{4}{3}\pi r^3.$$

❸ Next we differentiate both sides of the equation (the geometric relationship) implicitly, with respect to time t:

$$\frac{d}{dt}(V) = \frac{d}{dt}\left(\frac{4}{3}\pi r^3\right)$$
$$\frac{dV}{dt} = \frac{4}{3}\pi \cdot 3r^2 \cdot \frac{dr}{dt} = 4\pi r^2 \frac{dr}{dt}.$$

❹ We know $\frac{dV}{dt} = 5$ ft³/min:

$$5 = 4\pi r^2 \frac{dr}{dt}.$$

❺ We wish to know $\frac{dr}{dt}$ when the diameter is 18 inches. But, before we rush to use $r = 18$, note that it is the diameter, not the radius, that is 18 inches. Therefore, we need the rate when $r = 9$ inches. But 9 is still not the correct number to use for r, because we are using feet, not inches, in the equation of step ❹. Rewriting r using feet gives $r = \frac{3}{4}$ feet, so we use $r = \frac{3}{4}$:

$$5 = 4\pi\left(\frac{3}{4}\right)^2\frac{dr}{dt}.$$

❻ Solving yields

$$5 = \frac{9}{4}\pi\frac{dr}{dt},$$

$$\frac{dr}{dt} = \frac{5}{\frac{9}{4}\pi} = \frac{20}{9\pi}.$$

❼ The rate at which the radius is changing when the diameter is 18 inches is $\frac{20}{9\pi}$ ft/min ≈ 0.707 ft/min. ■

Including units in the equations beginning in step ❹ can be helpful, because the arithmetic and algebra when units are included should result in the correct units for the final answer:

$$5\,\text{ft}^3/\text{min} = 4\pi\left(\frac{3}{4}\,\text{ft}\right)^2\frac{dr}{dt}$$

$$\frac{dr}{dt} = \frac{5\,\text{ft}^3/\text{min}}{4\pi\frac{9}{16}\,\text{ft}^2} = \frac{20}{9\pi}\,\text{ft/min}.$$

For this reason, many in the sciences prefer that units always be included at each step in equations.

The final units should be units of length (the units of the radius as used in the equations) over units of time.

Instead of using feet in example 1, we can use inches instead. Since $5\,\text{ft}^3/\text{min} = 5 \cdot 12^3\,\text{in}^3/\text{min} = 8640\,\text{in}^3/\text{min}$, we have

$$8640\,\text{in}^3/\text{min} = 4\pi(9\,\text{in})^2\frac{dr}{dt},$$

$$\frac{dr}{dt} = \frac{8640\,\text{in}^3/\text{min}}{324\pi\,\text{in}^2} = \frac{80}{3\pi}\,\text{in/min},$$

which is the same as $\frac{20}{9\pi}$ ft/min.

Because it is much easier to make an error when converting square or cubic units, the preferred method is to keep the units of the rate of change as given (as illustrated in step ❹ of the solution) and convert units of the variable instead (as illustrated in step ❺ of the solution).

CAUTION: MISTAKE TO AVOID
Although 1 ft = 12 in, $1\,\text{ft}^3 \neq 12\,\text{in}^3$. Think of how many cubes with side length 1 inch fit in a cube of side length 1 foot:

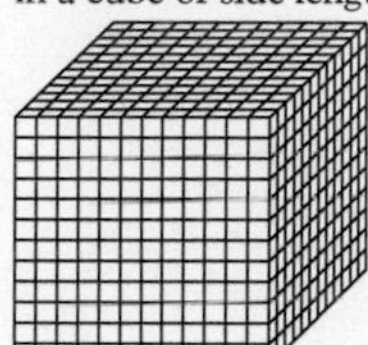

It takes $12 \cdot 12 \cdot 12$ of the small cubes to fill the large cube. In other words, $1\,\text{ft}^3 = (12\,\text{in})^3 = 12^3\,\text{in}^3 = 1728\,\text{in}^3$.

Melting ice

Example 2 *The top of a silo has the shape of a hemisphere of diameter 20 feet. If it is coated uniformly with a layer of ice, and if the thickness of the ice is decreasing at a rate of $\frac{1}{4}$ in/h, how fast is the volume of ice changing when the ice is 2 inches thick? See figure 3.*

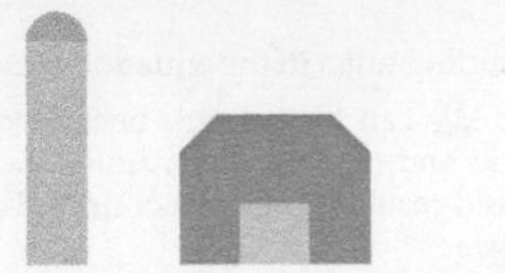

Figure 3 A silo (left) and barn (right). A silo is often used for storing animal feed

Solution ❶ We begin our dynamic diagram by ⓐ drawing the objects (top of silo, ice) and labeling the quantities that are not changing (diameter of top of silo), as in figure 4. A cross-section view is used to help visualize the ice on the outside of the structure.

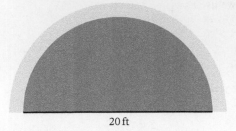

Figure 4 *Ice on the hemispherical top of a silo, cross-sectional view. Thickness of ice not to scale*

The variable t is used for time in related rates problems and should not be used for anything else. The variable r is used for radius and the thickness is not a radius, so it also is not a good choice. Thus, we'll use k.

ⓑ A quantity that is changing is the thickness of the ice. Label the thickness k. The quantity we are asked to find is the rate of change of the volume of the ice; call the volume V. ⓒ The thickness is decreasing, so we can draw arrows inward. ⓓ Finally, we label an arrow with the rate of change of thickness $\frac{dk}{dt}$. Because the thickness is decreasing, $\frac{dk}{dt}$ is negative. See figure 5. Also, we need the radius rather than the diameter, and because the rate is expressed in terms of inches we convert the size of the radius, 10 feet, to inches, 120 inches.

❷ Next we need a geometric formula that relates the variables of step ❶ⓑ, V and k. The formula for the volume of a sphere is $V = \frac{4}{3}\pi r^3$, and the volume of a hemisphere is half that of the sphere:

$$V_{\text{hemisphere}} = \frac{2}{3}\pi r^3.$$

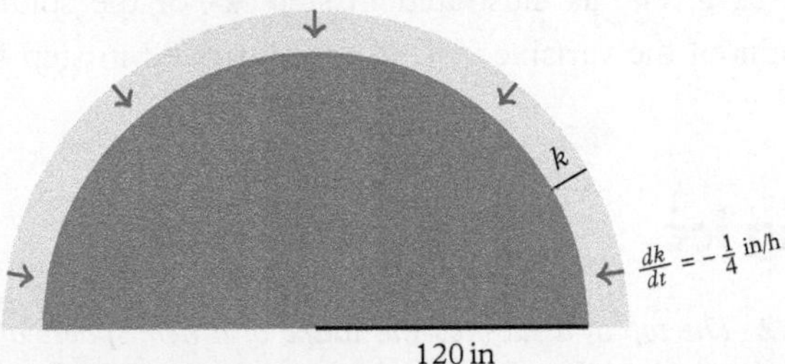

Figure 5 *Ice on the hemispherical top of a silo, cross-sectional view. Thickness of ice not to scale. The ice is melting; the thickness of the ice is decreasing. The units of the radius have been changed (converted from feet to inches) to match the units of the rate of change*

The problem is that we need to relate the variables V and k, not the variables V and r. The formula gives the volume of a hemisphere, and the ice is not a hemisphere. On the other hand, the ice is the difference of two hemispheres. We can find the volume of the ice by taking the volume of the hemisphere that includes the silo top and the ice and then subtracting the volume of the silo top, which is also a hemisphere:

This formula can also be found by searching for the volume of a hemispherical shell.

$$V_{\text{ice}} = V_{\text{silo plus ice}} - V_{\text{silo}}$$
$$V_{\text{ice}} = \frac{2}{3}\pi(120+k)^3 - \frac{2}{3}\pi(120)^3.$$

❸ Next we differentiate both sides with respect to time t:

Note that the right-hand term $\frac{2}{3}\pi(120)^3$ is a constant.

$$\frac{d}{dt}(V) = \frac{d}{dt}\left(\frac{2}{3}\pi(120+k)^3 - \frac{2}{3}\pi(120)^3\right)$$
$$\frac{dV}{dt} = \frac{2}{3}\pi \cdot 3(120+k)^2 \cdot \frac{dk}{dt} - 0.$$

❹ We know that $\frac{dk}{dt} = -\frac{1}{4}$ in/h, and ❺ we want to find $\frac{dV}{dt}$ when $k = 2$ inches:

$$\frac{dV}{dt} = 2\pi(120+2)^2\left(-\frac{1}{4}\right) = -23\,380\ \text{in}^3/\text{h}.$$

Step ❻ is complete.

❼ The volume of the ice is decreasing at a rate of 23 380 in^3/h. ■

23 380 in^3/h = 101.2 gal/h = 383.1 L/h

An alternate solution is to think of the ice as occupying the inside of the silo as well (figure 6). Then labeling the radius of the ice r, we have:

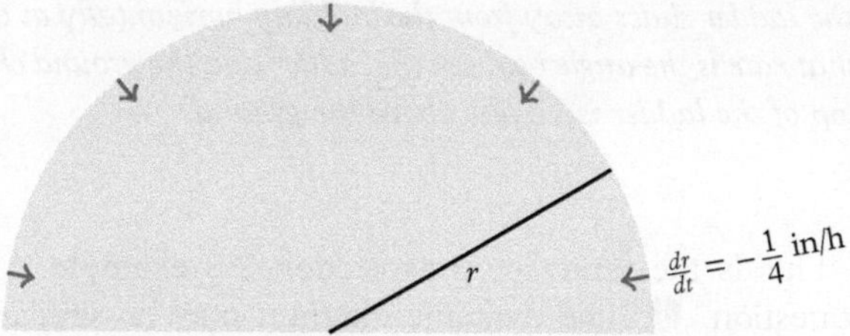

Figure 6 *A hemisphere of ice*

The volume of a hemisphere is the correct geometric formula:

$$V = \frac{2}{3}\pi r^3.$$

Differentiating both sides with respect to time t yields

$$\frac{d}{dt}(V) = \frac{d}{dt}\left(\frac{2}{3}\pi r^3\right)$$

$$\frac{dV}{dt} = \frac{2}{3}\pi \cdot 3r^2 \cdot \frac{dr}{dt} = 2\pi r^2 \frac{dr}{dt}.$$

The given rate is $\frac{dr}{dt} = -\frac{1}{4}$ in/h and we wish to know $\frac{dV}{dt}$ when $r = 122$, giving

$$\frac{dV}{dt} = 2\pi(122)^2\left(-\frac{1}{4}\right) = -23\,380\ \text{in}^3/\text{h}$$

as before.

The alternate solution works because, geometrically, the change in the ice is the same whether the silo is empty or filled with ice, as long as we keep the same rate at which the thickness of the ice is changing.

Sliding ladder, revisited

Example 3 *A ladder 20 feet long leans against a vertical building. If the bottom of the ladder slides away from the building horizontally at a rate of 2 ft/s, at what rate is the angle between the ladder and the ground changing when the top of the ladder is 12 feet above the ground?*

Drawing the diagram with the correct variables is essential for determining the correct geometric relationship.

Solution This is the same setup as section 2.8 example 1, with a different question. ❶ The dynamic diagram may be similar to the one in section 2.8, but because we are asking a different question, we need to reconsider each step. ⓐ We draw the objects (ladder, building, ground) and label the quantities that do not change (length of ladder). ⓑ The distance from the bottom of the ladder to the building is changing and its rate is given; label this distance x. We are asked for the rate at which the angle between the ladder and the ground is changing; label this angle θ. Other quantities may also be changing, but they are irrelevant to this exercise and are not included in the diagram. ⓒ Arrows representing the directions in which x and θ are changing, along with ⓓ labels for those arrows finish the diagram (figure 7):

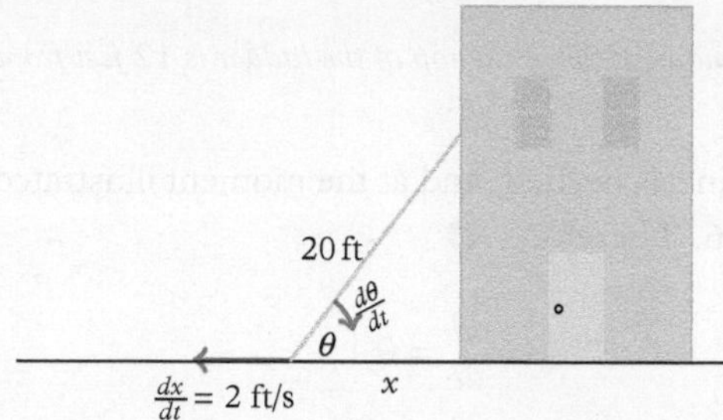

Figure 7 *A 20-foot ladder leaning against a building, with variables x and θ (not y) marked. Marking extraneous variables such as y can lead to confusion about the required geometric relationship*

❷ We need to determine a geometric relationship between the variables of step ❶ⓑ, x and θ. Although we have a right triangle, the geometric relationship we seek is not the Pythagorean theorem; the variables are not two sides of the triangle, but rather one side and an angle. Noticing that x is adjacent to the angle θ and the hypotenuse has length 20 feet, the cosine of the angle gives the appropriate relationship:

$$\cos\theta = \frac{x}{20}.$$

This equation expresses the relationship between the variable with a rate of change that we know and the variable with a rate of change that we wish to find.

❸ Differentiating with respect to time t yields

$$\frac{d}{dt}(\cos\theta) = \frac{d}{dt}\left(\frac{x}{20}\right)$$
$$-\sin\theta \cdot \frac{d\theta}{dt} = \frac{1}{20} \cdot \frac{dx}{dt}.$$

❹ We know $\frac{dx}{dt} = 2$ ft/s:

$$-\sin\theta \cdot \frac{d\theta}{dt} = \frac{1}{20}(2) = 0.1.$$

❺ We wish to know $\frac{d\theta}{dt}$ when the distance between the top of the ladder and the ground is 12 feet. Drawing a static diagram helps us determine the relevant values (figure 8).

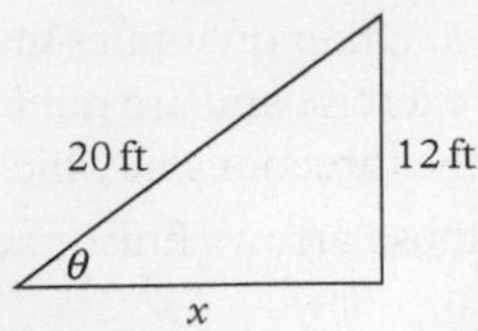

Figure 8 *The moment when the top of the ladder is 12 feet from the ground*

The value of $\sin\theta$ is needed, and at the moment illustrated in figure 8, $\sin\theta = \frac{12}{20} = 0.6$. Therefore, ❻

$$-0.6\frac{d\theta}{dt} = 0.1$$
$$\frac{d\theta}{dt} = \frac{0.1}{-0.6} = -\frac{1}{6}.$$

The units for $\frac{d\theta}{dt}$ are units of angle (radians) over units of time (seconds).

❼ The rate of change of the angle between the ladder and the ground is $-\frac{1}{6}$ rad/s. ■

Streetlight shadows

Example 4 *A light is at the top of a 16-foot pole. A child 5 feet tall walks away from the pole at a rate of 4 ft/s. (a) At what rate is the tip of the child's shadow moving when the child is 18 feet from the pole? (b) At what rate is the length of the child's shadow increasing?*

Solution ❶ We begin drawing a dynamic diagram by ⓐ drawing the objects (pole, light, child, shadow) and labeling the quantities that are not changing (height of pole, height of child). See figure 9.

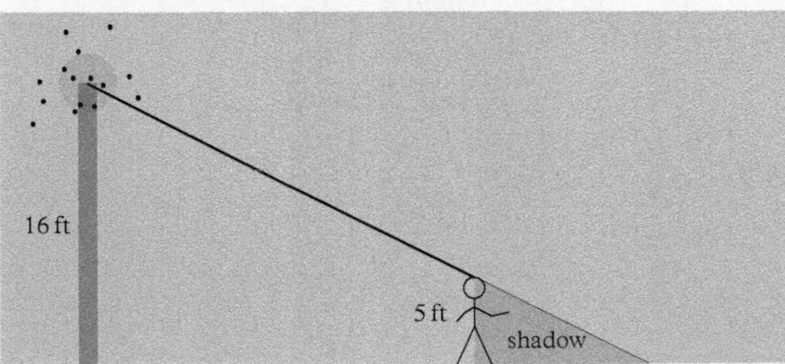

Figure 9 *A light on top of a 16-foot pole and a 5-foot child walking away from the pole, along with summertime insects*

ⓑ Quantities that are changing include the length of the shadow (z) and the distance from the pole to the child (x). ⓒ Both x and z are lengthening, so we draw arrows to the right. ⓓ We also label the arrows with their rates of change, $\frac{dx}{dt}$ (the value of which we are given) and $\frac{dz}{dt}$ (which answers part (b) of the question). See figure 10.

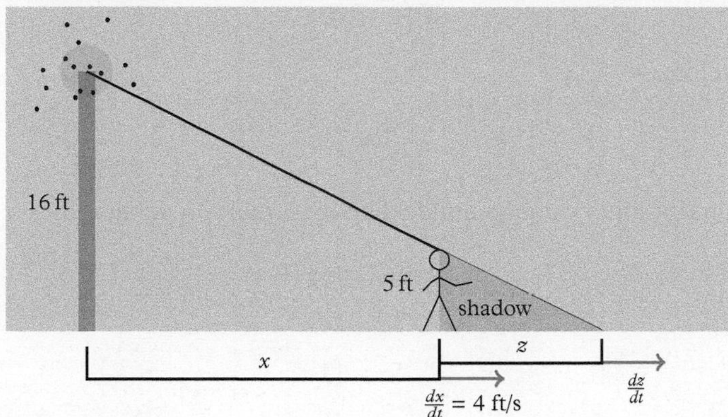

Figure 10 *A light on top of a 16-foot pole and a 5-foot child walking away from the pole, with variables, directions, and rates marked below their spots in the diagram*

❷ Next we need a geometric relationship between the two variables x and z. Even though we have right triangles in the diagram, this is not a Pythagorean relationship because the variables are not two of the three sides of the same triangle. In fact, there is more than one triangle present; using figure 11, the two triangles are $\triangle ABE$ and $\triangle CDE$.

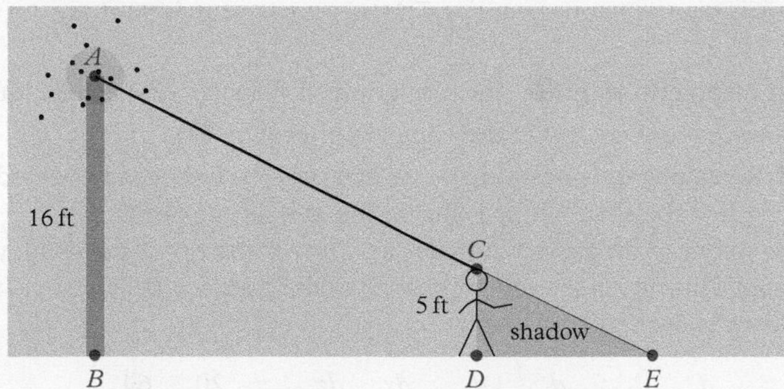

Figure 11 *A light on top of a 16-foot pole and a 5-foot child walking away from the pole, forming two triangles: $\triangle ABE$ and $\triangle CDE$*

The two triangles are similar. Both triangles have right angles (at B and at D), they share the angle at E, and by subtraction (or by the corresponding angles theorem from geometry), the angles at A and C are also congruent. Similar triangles are helpful because the ratios of the sides are the same in similar triangles:

$$\frac{BE}{AB} = \frac{DE}{CD}.$$

Therefore,

$$\frac{x+z}{16} = \frac{z}{5}.$$

This relationship can be simplified by cross-multiplication:

$$\begin{aligned} 5(x+z) &= 16z \\ 5x + 5z &= 16z \\ 5x &= 11z. \end{aligned}$$

❸ Next we differentiate with respect to time t:

$$\begin{aligned} \frac{d}{dt}(5x) &= \frac{d}{dt}(11z) \\ 5\frac{dx}{dt} &= 11\frac{dz}{dt}. \end{aligned}$$

❹ We know that $\frac{dx}{dt} = 4$ ft/s:

$$\begin{aligned} 5 \cdot 4 &= 11\frac{dz}{dt} \\ \frac{20}{11} &= \frac{dz}{dt}, \end{aligned}$$

which completes step ❻ (there is no need for step ❺ because the variables x and z are not in the equation after step ❸).

❼ Rereading the question, we see that part (b) has been answered: the length of the shadow (z) is increasing at a rate of $\frac{20}{11}$ ft/s.

We still need an answer to part (a). How is the tip of the shadow moving? This rate is the rate of change of the distance $BE = x + z$. In other words, we want

$$\frac{d}{dt}(x+z) = \frac{d}{dt}x + \frac{d}{dt}z = \frac{dx}{dt} + \frac{dz}{dt} = 4 + \frac{20}{11} = \frac{64}{11}.$$

The rate at which the tip of the shadow is moving is $\frac{64}{11}$ ft/s. ■

The shadow is lengthening at a constant speed and the tip of the shadow is moving at a constant speed, so the position of the child (18 feet from the pole) is irrelevant.

Rising water

Example 5 *A water tank has the shape of an inverted right circular cone of altitude* 12 *feet and a base radius of* 6 *feet. Given that water is pumped into the tank at a rate of* 10 *gallons per minute, at what rate is the water level rising when the water is* 3 *feet deep?*

1 gallon ≈ 0.1337 ft^3

Solution ❶ We begin our dynamic diagram by ⓐ drawing the objects (tank, water) and labeling the quantities that are not changing (tank radius, tank height). See figure 12.

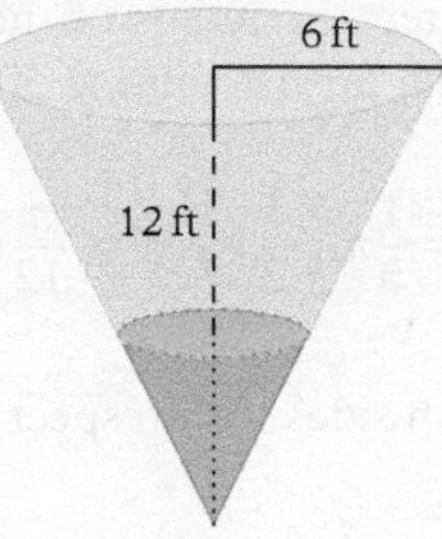

Figure 12 *An inverted conical water tank with an altitude (height) of* 12 *feet and a base radius of* 6 *feet*

ⓑ Water is being pumped into the tank, which affects the volume of water; let's call this variable V. The requested rate of change involves the water level (the distance from the bottom of the tank to the top of the water); let's call this y. ⓒ We need to add an arrow for the change in water level and ⓓ label the arrow with its rate of change. Although we do not write V in the figure, we can write its rate of change at the side of the diagram. The result is in figure 13.

A gallon is a measure of volume.

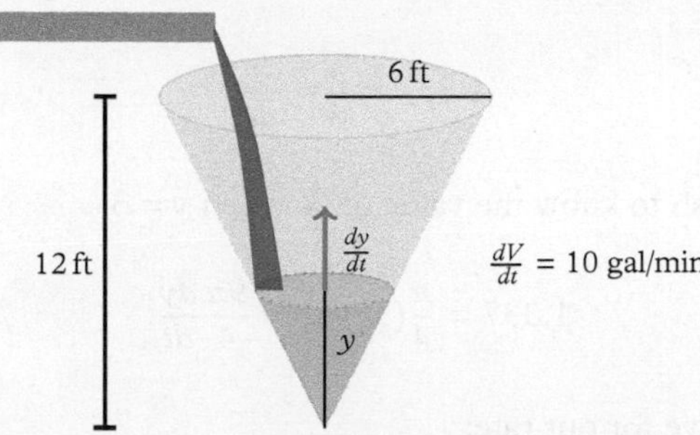

Figure 13 *An inverted conical water tank with an altitude (height) of* 12 *feet, a base radius of* 6 *feet, and a rising water level*

❷ Next we need a geometric relationship between the variables V and y. The volume of a cone is given by

$$V = \frac{1}{3}\pi r^2 h,$$

CAUTION: MISTAKE TO AVOID
Using $r = 6$ and $h = 12$ in the formula for the volume of the cone just gives the volume of the tank (its capacity) and does not help us relate the variables V (volume of the water) and y.

where h is the height of the cone and r is its radius. Because V represents the volume of water, and the water is in the shape of a cone (see figure 13), the formula applies with the height of the cone of water being $h = y$. What about the radius? Notice that the smaller cone (the cone of water) is proportional (similar) to the larger cone (the tank). The radius of the tank is half its height (6 feet vs. 12 feet), so the same is true of the cone of water; the radius of the water is $r = \frac{1}{2}y$. Using these values in the formula for the volume of a cone,

$$V = \frac{1}{3}\pi\left(\frac{1}{2}y\right)^2 y = \frac{\pi}{12}y^3.$$

❸ Differentiating both sides with respect to time t gives

$$\frac{d}{dt}(V) = \frac{d}{dt}\left(\frac{\pi}{12}y^3\right)$$

$$\frac{dV}{dt} = \frac{\pi}{12} \cdot 3y^2 \cdot \frac{dy}{dt} = \frac{\pi}{4}y^2\frac{dy}{dt}.$$

❹ We know that $\frac{dV}{dt} = 10$ gal/min, but we do not use 10 for the value of $\frac{dV}{dt}$. Because we want to use units of feet for the value of y, we need to convert gallons to cubic feet:

$$\frac{dV}{dt} = 10\,\text{gal/min} = 10\,\text{gal/min} \cdot \frac{0.1337\,\text{ft}^3}{1\,\text{gal}} = 1.337\,\text{ft}^3/\text{min}.$$

Our equation then becomes

$$1.337 = \frac{\pi}{4}y^2\frac{dy}{dt}.$$

❺ We wish to know the value of $\frac{dy}{dt}$ when $y = 3$:

$$1.337 = \frac{\pi}{4}(3^2)\frac{dy}{dt} = \frac{9\pi}{4}\frac{dy}{dt}.$$

❻ We solve for our rate:

$$\frac{dy}{dt} = 1.337 \cdot \frac{4}{9\pi} \approx 0.1891.$$

❼ The rate at which the water level is rising when the water is 3 feet deep is 0.1891 ft/min. ■

Examples 1, 2, and 5 used formulas for volume in step ❷. Examples 2 and 5 required a little reasoning with those formulas to finish determining the geometric relationship between the variables we used. Example 3 used a trigonometric relationship. Example 4 used similar triangles. These are just a few of the many types of geometric relationships that can be used when working a related rates exercise.

Exercises 2.9

1. A spherical scoop of ice cream melts in such a way that its radius decreases at a rate of 3 mm/min. At what rate is the volume of the ice cream changing when its diameter is 50 mm?
2. (GSL) Gas is escaping from a spherical balloon at a rate of $1000\,\text{cm}^3/\text{min}$. At the instant when the radius is 10 cm, at what rate is the radius decreasing?
3. A person who is 6 feet tall saunters toward the only working lamppost in an otherwise dark alley at a rate of 3 ft/s. The filament of the bulb is located 15 feet above the level of the street, the street is perfectly horizontal, nothing blocks the light, and so on. At what rate is the length of the person's shadow changing when their distance from the lamppost is 20 feet?

It is customary for many of these issues not to be mentioned; reasonable assumptions may be made when working these exercises.

4. (GSL) An arc light is hung 12 feet directly above a straight horizontal walk on which a child 5 feet tall is walking. How fast is the child's shadow lengthening when the child is walking away from the light at a rate of 168 ft/min?
5. (Love) A light is placed on the ground 30 feet from a building. A person 6 feet tall walks from the light toward the building at a rate of 5 ft/s. Find the rate at which the person's shadow on the wall is shortening when the person is 15 feet from the building.
6. (Love) A person 6 feet tall walks away from a lamppost 12 feet high at a rate of 4.4 ft/s. How fast does the end of the person's shadow move?
7. (Love) Water is flowing into a conical reservoir that is 20 feet deep and 10 feet across at the top at a rate of $15\,\text{ft}^3\,\text{min}$. Find how fast the surface is rising when the water is 8 feet deep.

The conical tank is oriented the same as in example 5.

8. (Love) Water is flowing into a cylindrical tank at a rate of $10\,\text{m}^3/\text{min}$. If the radius of the tank is 3 m, how fast is the surface rising?

9. (Love) A rectangular trough is 10 feet long and 1 foot wide. Find how fast the surface of the water rises if the water flows in at a rate of 2 ft^3/min.

10. (GSL) A horizontal trough 10 feet long has a vertical section in the shape of an isosceles right triangle. If water is poured into it at a rate of 8 ft^3/min, at what rate is the surface of the water rising when the water is 2 feet deep? See figure 14.

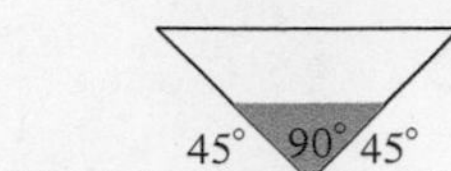

Figure 14 Orientation of the trough for exercise 10. The length of the trough is not pictured

11. (Love) A ladder 10 feet long leans against a wall. If the lower end moves outward at a rate of 1.5 ft/s, find the rate at which the angle between the ladder and the floor is decreasing when the lower end of the ladder is 6 feet from the wall.

Compare to section 2.8 exercise 12.

12. A child flies a kite at a height of 200 feet. The wind is carrying the kite horizontally away from the child at a rate of 16 ft/s. Find the rate of change of the angle between the string and the vertical when the kite is 350 feet away from the child.

13. (GSL) A circular plate of metal expands by heat so that its radius increases at a rate of 0.01 in/s. At what rate is the area of the plate increasing when the radius is 2 inches?

Exercise 14 is typeset with no changes from the original. The abbreviation "cu. ft." stands for cubic feet, or ft^3. The abbreviation lb is for the English unit of weight "pound." In Boyle's law, p is pressure, v is volume, and c is a constant.

14. (GSL) A gas holder contains 1000 cu. ft. of gas at a pressure of 5 lb. per square inch. If the pressure is decreasing at the rate of 0.05 lb. per square inch per hour, find the rate of increase of the volume. (Assume Boyle's law: $pv = c$.)

Compare to section 2.8 exercise 15.

15. A submarine is attempting to surface and is moving vertically upward at a rate of 1.5 m/s. A stationary ship is nearby, 800 m from the point at which the submarine will surface. At what rate is the angle between the vertical and the path from the sub to the ship increasing when the submarine is at a depth of 600 m?

16. As gravel is poured into a pile, the pile of gravel maintains the shape of a cone that is as wide as it is tall. Gravel is poured onto the pile at a rate of 4 m^3/min. At what rate is the height of the pile increasing when the pile is 8 m high?

17. In a lighthouse that is 200 feet from the nearest point P on a straight shoreline, a rotating beacon makes one revolution every 15 seconds. Find the rate at which a ray from the light moves along the shore at a point 400 feet from P.

18. (GSL) A light is 20 feet from a wall and 10 feet above the center of a path that is perpendicular to the wall. A person 6 feet tall is walking on the path toward the wall at a rate of 2 ft/s. When the person is 4 feet from the wall, how fast is the shadow of the person's head moving up the wall?

19. (Love) An arc light hangs at a height of 30 feet above the center of a street 60 feet wide. A person 6 feet tall walks along the sidewalk at a rate of 4 ft/s. How fast is the person's shadow lengthening when the person is 40 feet up the street?

20. (GSL) Water flows from a faucet into a hemispherical basin of diameter 14 inches at a rate of 2 in^3/s. How fast is the water rising when the water is halfway to the top? (The volume of a spherical segment is $\frac{1}{2}\pi r^2 h + \frac{1}{6}\pi h^3$, where h is the altitude of the segment.)

21. (Love) A trapezoidal trough is 12 feet long, 3 feet wide at the top, 2 feet wide at the bottom, and 2 feet deep. If water flows in at a rate of 20 ft^3/min, find how fast the surface is rising when the water is 1 foot deep.

22. (GSL) A horizontal trough 12 feet long has a vertical cross section in the shape of a trapezoid, with the bottom being 3 feet wide and the sides inclined to the vertical at an angle with a sine of $\frac{4}{5}$. At what rate is water being drawn from the trough if the level is falling 0.1 ft/min when the water is 3 feet deep?

Chapter III

Applications of the Derivative

Absolute Extrema

3.1

Consider the function $f(x) = x^2 + 2$ as graphed in figure 1. What

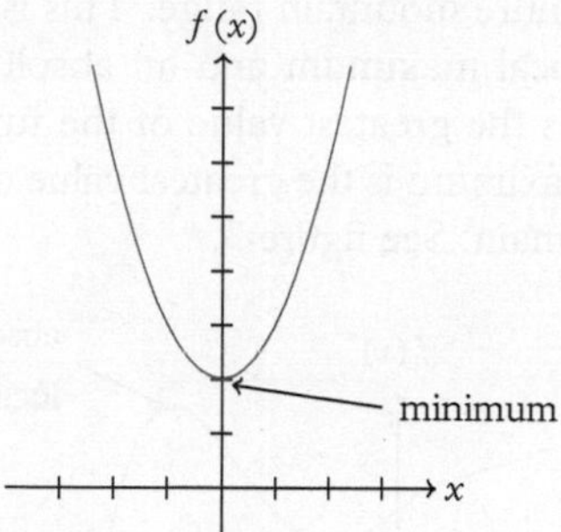

Figure 1 *The graph of $f(x) = x^2 + 2$, which can be generated quickly from the instantly recognized $y = x^2$ shifted upward 2 units. The lowest point on the curve, the* minimum point, *is* (0, 2)

is the smallest value possible for $f(x)$? Recall that values of $f(x)$ are y-coordinates on the graph of the function. A glance at the graph reveals that the answer is two. We call this smallest value the *minimum value* of the function f. We say that the minimum value occurs at the location $x = 0$; we say that the *minimum point* on the curve is (0, 2).

Instead of using the graph, we could also reason that the smallest squared value is zero, and after adding two, the smallest possible value of $f(x) = x^2 + 2$ is two.

In a similar manner, we can look at *maximum values* of functions. For the function $g(x) = -x^2 + 1$ (see figure 2), the maximum value is one, which occurs at $x = 0$. The *maximum point* is (0, 1).

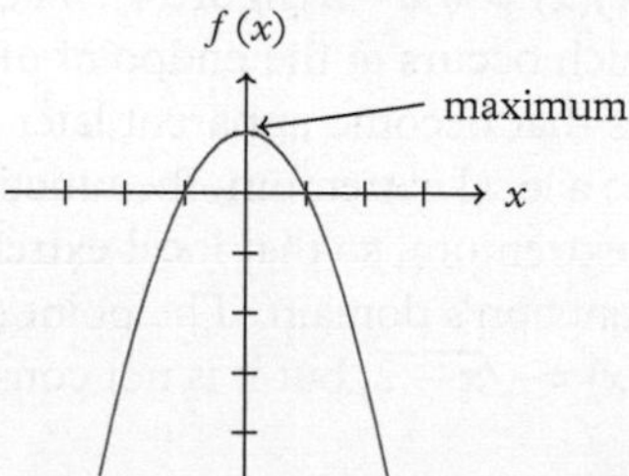

Figure 2 *The graph of $f(x) = -x^2 + 1$, which can be generated quickly from the instantly recognized $y = x^2$ by vertical reflection followed by an upward shift of* 1 *unit. The highest point on the curve, the* maximum point, *is* (0, 1)

Grammar lesson:

singular	plural
minimum	minima
maximum	maxima
extremum	extrema

The word *extremum* is used to refer to either a maximum or a minimum.

Local vs. absolute extrema

Mountain ranges may have several peaks. The mountaintop, or peak, has the maximum elevation of that mountain, but this is just a local view. A more global view is to say the maximum elevation is at the highest peak in the entire mountain range. This is the idea of the difference between a local maximum and an absolute maximum. The absolute maximum is the greatest value of the function anywhere in its domain; a local maximum is the greatest value of the function in its region within the domain. See figure 3.

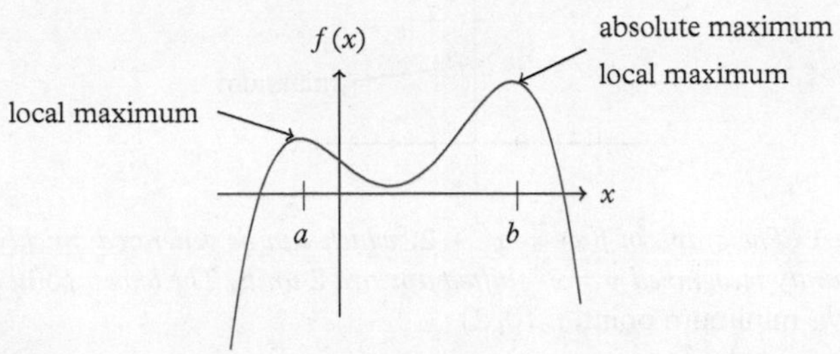

Figure 3 *Local maxima at x = a and x = b, each located visually at the top of a hill. The maximum at x = a is not an absolute maximum because the graph rises higher than that value. The maximum at x = b is the highest the graph ever rises, so it is an absolute maximum*

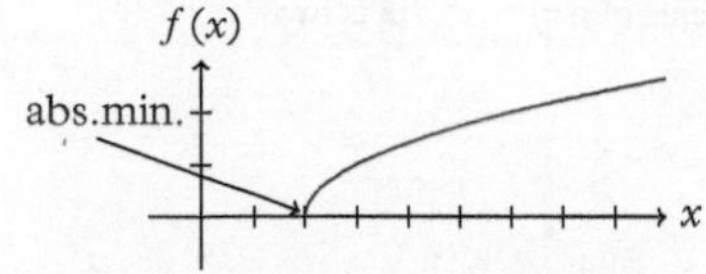

Figure 4 *The graph of $f(x) = \sqrt{x-2}$. The absolute minimum of f is zero, which occurs at $x = 2$. The point $(2, 0)$ is not a local minimum point because it occurs at an endpoint of the function's domain*

The same idea holds for minima. Local minima occur at the bottoms of valleys in the graph, but the absolute minimum point is the lowest point anywhere in the graph.

Absolute extrema can occur at endpoints of a domain. Consider, for instance, the graph of $f(x) = \sqrt{x-2}$ (figure 4). The absolute minimum value of f is zero, which occurs at the endpoint of its domain, $x = 2$. However, for reasons that become apparent later in this section (before theorem 1), to be a local extremum, the function must be defined on both sides of the extremum, so that local extrema cannot occur at an endpoint of the function's domain. The point $(2, 0)$ is an absolute minimum point of $f(x) = \sqrt{x-2}$, but it is not considered to be a local minimum.

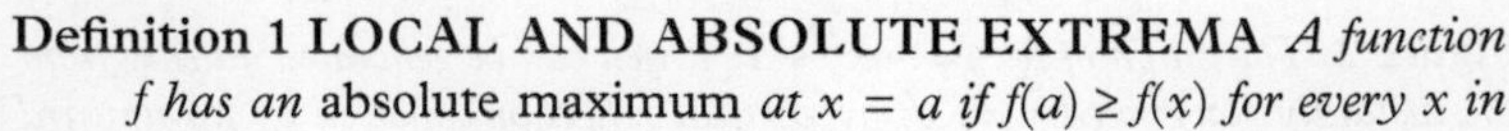

Definition 1 LOCAL AND ABSOLUTE EXTREMA *A function f has an* absolute maximum *at $x = a$ if $f(a) \geq f(x)$ for every x in*

the domain of f. Similarly, f has an **absolute minimum** *at $x = a$ if $f(a) \le f(x)$ for every x in the domain of f.*

A function f has a **local maximum** *at $x = a$ if there is an open interval containing a such that $f(a) \ge f(x)$ for all x in that interval. Similarly, f has a* **local minimum** *at $x = a$ if there is an open interval containing a such that $f(a) \le f(x)$ for all x in that interval. See figure 5.*

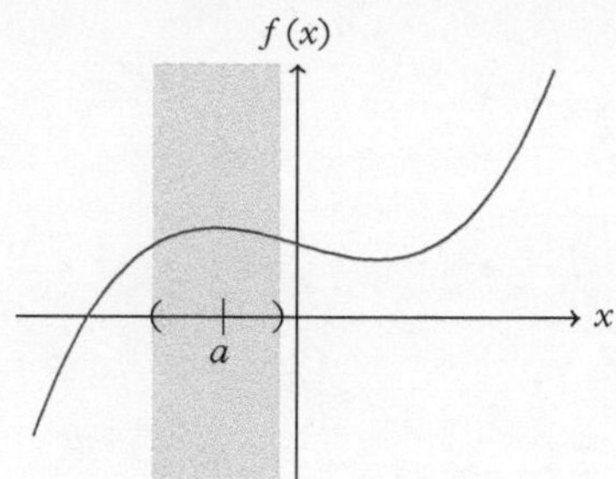

Figure 5 *A function with a local maximum at $x = a$. An open interval containing $x = a$ is pictured. The light-gray region has x-coordinates in this open interval. In this region, the maximum value of the function occurs at $x = a$*

Example 1 *Identify the locations of all local and absolute extrema of the function f using the graph in figure 6.*

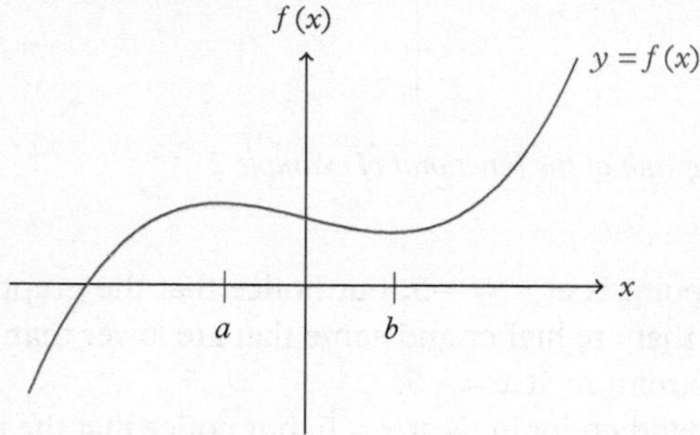

Figure 6 *The graph of the function f of example 1. Because the left- and right-hand ends of the picture are not emphasized with dots, we make the assumption that the graph continues to the right and left indefinitely*

Solution The function f has a local maximum at $x = a$ (top of the hill) and a local minimum at $x = b$ (bottom of the valley). Because part of the graph rises higher than the top of the hill, the maximum at $x = a$ is not an absolute maximum. Because part of the graph falls lower than the bottom of the valley, the minimum at $x = b$ is not an absolute minimum. Assuming the graph continues to rise to the right and fall to the left, there are no absolute maximum or minimum points on the graph.

The solution can be summarized as

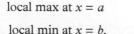

We use *max* as an abbreviation for maximum and *min* as an abbreviation for minimum.

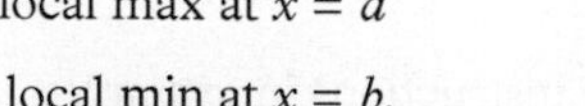

local max at $x = a$

local min at $x = b$. ■

Example 2 *Identify the locations of all local and absolute extrema of the function f using the graph in figure 7.*

Solution No function can have a local extremum at an endpoint, but a function might have an absolute extremum at an endpoint. The

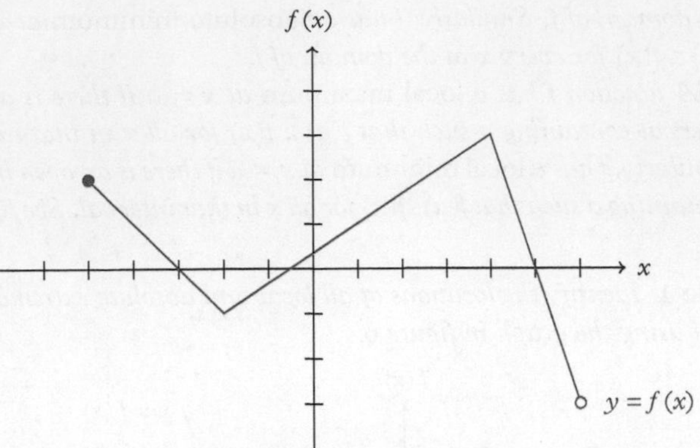

Figure 7 *The graph of the function f of example 2*

left-hand endpoint is at $x = -5$, but notice that the graph has some y-coordinates that are higher and some that are lower than at $x = -5$. There is no extremum at $x = -5$.

The right-hand endpoint is at $x = 6$, but notice that the point is not included in the graph. If it was included, then it would be the absolute minimum point because it is lower than all other points on the graph. But, because the point is not included, it is not a point—much less a minimum point—of the function. There is no extremum at $x = 6$.

There is a bottom of the valley at $x = -2$, so f has a local minimum at $x = -2$. However, this minimum is not an absolute minimum because part of the graph falls below this level.

There is a top of the hill at $x = 4$, so f has a local maximum at $x = 4$. This is also as high as the graph ever rises, so f has an absolute maximum at $x = 4$.

The solution can be summarized as

$$\text{local min at } x = -2$$
$$\text{local and absolute max at } x = 4. \qquad \blacksquare$$

Notice that the instructions in examples 1 and 2 are to identify the locations of the extrema. For this reason we identified x-coordinates and used the word *at*—a word that indicates location. Had the instructions been to find the values of the extrema, the answer to example 2 would be

$$\text{local min } -1 \ \ (\text{at } x = -2)$$
$$\text{local and absolute max } 3 \ \ (\text{at } x = 4).$$

Maximum and minimum values "are" y-coordinates, and they occur "at" x-coordinates.

If the instructions to example 2 asked for the *extreme points* of the function f, the answer would be

$$\text{local minimum point } (-2, -1)$$

$$\text{local and absolute maximum point } (4, 3).$$

When asked for a point, both x- and y-coordinates must be given.

CAUTION: MISTAKE TO AVOID
Learn the meaning of the instructions regarding extrema to ensure that you give the requested information:

request	desired response
locations	"at" x-coordinates
values	y-coordinates
points	(x, y)

Reading Exercise 1 *(a) Does the function graphed in figure 8 have a local minimum at $x = a$? (b) Does the function have an absolute minimum at $x = a$?*

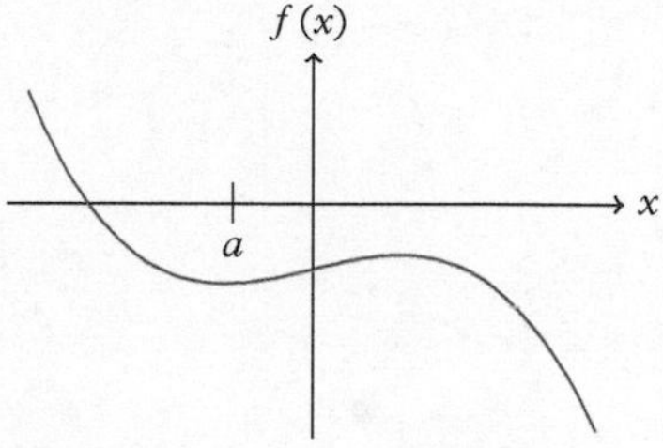

Figure 8 *The graph for reading exercise 1*

Derivatives and local extrema

It turns out that the derivative can help us identify the possible locations of local extrema. Consider the functions graphed in figure 9, with local extrema at $x = a$. In either case, the tangent line at $x = a$ is horizontal, so that $f'(a) = 0$.

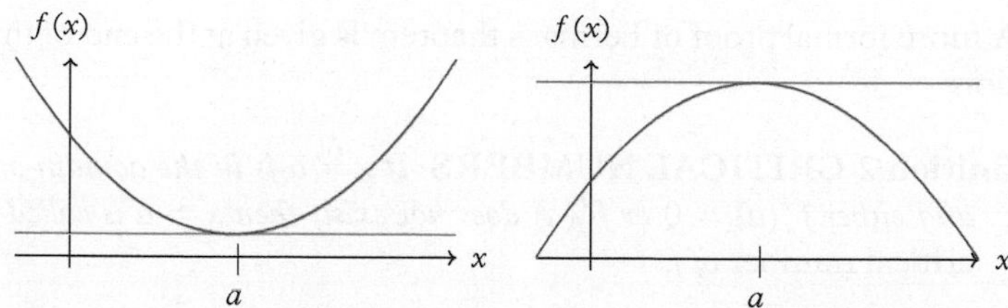

Figure 9 *Functions with local extrema at $x = a$ having horizontal tangent lines at $x = a$*

Another possibility is that the function has a corner at the local extremum point, as in figure 10. In this case, $f'(a)$ does not exist.

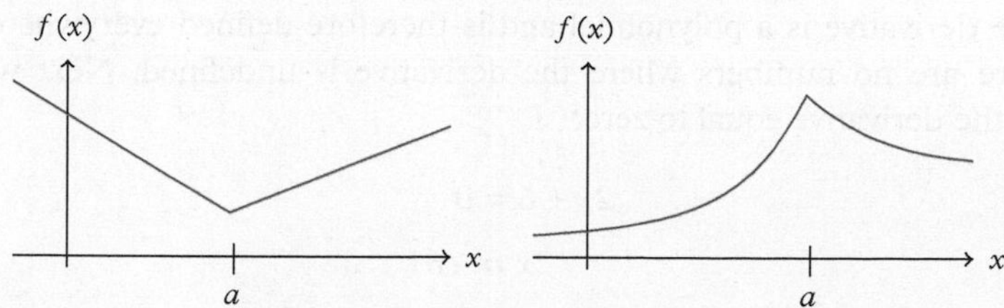

Figure 10 *Functions with local extrema at $x = a$ having corners at $x = a$*

Are there other possibilities? That is, can there be an extremum at $x = a$ such that $f'(a)$ exists but $f'(a) \neq 0$? If so, then $f'(a)$ must be either a negative number or a positive number. The case $f'(a) > 0$ is pictured in figure 11. By local linearity, the function's graph must be

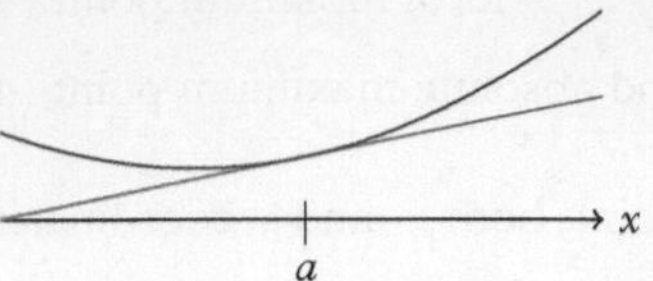

Figure 11 *A function with $f'(a) > 0$ cannot have a local extremum at $x = a$*

indistinguishable from the tangent line if we zoom in far enough. However, to the left of $x = a$, the tangent line is below the point of tangency, so the function must go below the level of the point $(a, f(a))$, whereas to the right of $x = a$, the tangent line is above the point of tangency, so the function must go above the level of the point $(a, f(a))$. There is neither a local minimum nor a local maximum at $x = a$.

Pierre de Fermat, 1601–1665
http://www-history.mcs.st-andrews.ac.uk/Biographies/Fermat.html

Theorem 1 FERMAT'S THEOREM *If f has a local maximum or minimum at $x = a$, and if $f'(a)$ exists, then $f'(a) = 0$.*

Fermat's theorem in calculus is not to be confused with the famous *Fermat's last theorem.* http://www-history.mcs.st-andrews.ac.uk/HistTopics/Fermat's_last_theorem.html

A more formal proof of Fermat's theorem is given at the end of this section.

Definition 2 CRITICAL NUMBERS *If $x = a$ is in the domain of f and either $f'(a) = 0$ or $f'(a)$ does not exist, then $x = a$ is called a* critical number *of f.*

Critical numbers are sometimes called *critical values* (for instance, in GSL) or *critical points.*

Example 3 *Find all critical numbers of the function $f(x) = x^2 + 6x - 4$.*

Solution We need to know where the derivative is zero or undefined. First we calculate the derivative:

$$f'(x) = 2x + 6.$$

Ans. to reading exercise 1
(a) yes, (b) no

The derivative is a polynomial and is therefore defined everywhere; there are no numbers where the derivative is undefined. Next we set the derivative equal to zero:

$$2x + 6 = 0$$
$$2x = -6$$
$$x = -3.$$

The function f has one critical number, $x = -3$. ■

Example 4 *Find all critical numbers of the function* $f(x) = \sqrt[3]{x}$.

Solution We need to know where the derivative is zero or undefined. First we calculate the derivative of $f(x) = \sqrt[3]{x} = x^{1/3}$:

$$f'(x) = \frac{1}{3}x^{-2/3} = \frac{1}{3\sqrt[3]{x^2}}.$$

This derivative features the variable in the denominator and therefore might be undefined at some numbers. Indeed, setting the denominator equal to zero shows that f' is undefined at $x = 0$. Therefore, $x = 0$ is a critical number.

Next we set the derivative equal to zero:

$$\frac{1}{3\sqrt[3]{x^2}} = 0.$$

Multiplying both sides of the equation by $3\sqrt[3]{x^2}$ results in $1 = 0$, and there are no solutions to the equation. The function f has one critical number, $x = 0$. ■

Setting a fraction equal to zero and solving is equivalent to setting the numerator equal to zero and solving:

$$\frac{a}{b} = 0 \text{ if and only if } a = 0.$$

Therefore, if the derivative is a fraction, we find critical values by setting the numerator equal to zero (finding where $f'(x) = 0$) and setting the denominator equal to zero (finding where $f'(x)$ is undefined.)

Reading Exercise 2 *Find all critical numbers of the function* $f(x) = x^2 + x$.

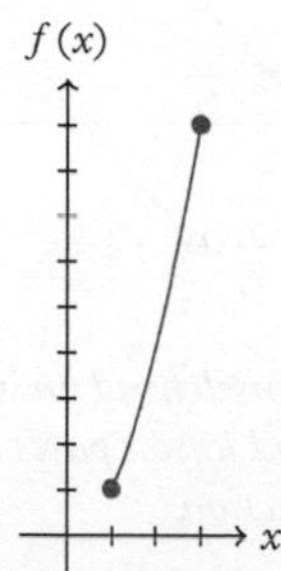

Figure 12 *The graph of* $f(x) = x^2$ *on* $[1, 3]$. *This function has absolute extrema (at the endpoints) where the tangent lines are not horizontal*

Fermat's theorem only applies to local extrema, not to absolute extrema. This is because absolute extrema can occur at endpoints of the domain. For instance, the function $f(x) = x^2$ on $[1, 3]$ has absolute minimum point $(1, 1)$ and absolute maximum point $(3, 9)$ (figure 12). Although the absolute minimum occurs at $x = 1$, since $f'(x) = 2x$, we have $f'(1) = 2 \neq 0$. One may argue whether the (two-sided) derivative technically exists at the endpoint, but the point remains that Fermat's theorem applies only to local extrema, not absolute extrema.

The extreme value theorem

Figure 13 *The graph of $f(x) = x$, which has no extrema. The derivative $f'(x) = 1$ is always defined and not zero, so there are no local extrema. The graph rises to the right and falls to the left without any bounds, so there are no absolute extrema*

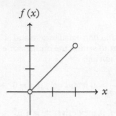

Must functions have extrema? The graph of $f(x) = x$ (figure 13) shows that the answer is no.

Perhaps the reason there are no extrema at all on the graph of $f(x) = x$ is that it goes on forever to the right and to the left. If we restrict ourselves to a bounded interval, must a function have extrema? Let's look at $f(x) = x$ on the bounded interval $0 < x < 2$, graphed in figure 14. There are still no local extrema, nor are there any absolute extrema because the endpoints are missing. There is no largest value because we can always draw closer to the (missing) right-hand endpoint to get a larger value, and no smallest value because we can always draw closer to the (missing) left-hand endpoint to get a smaller value.

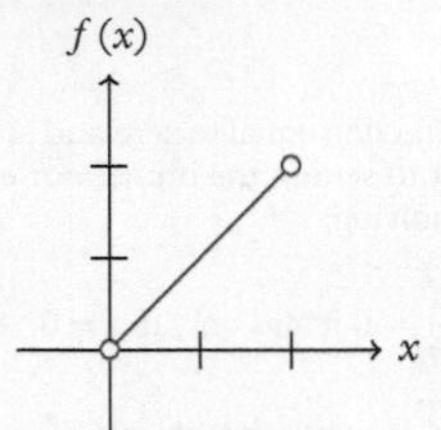

Figure 14 *The graph of $f(x) = x$ on the interval $(0, 2)$, which has no extrema*

In figure 14, the culprit is the open interval. Maybe we should restrict ourselves to closed, bounded intervals. Must a function on a closed, bounded interval have extrema? Again, the answer is no. See figure 15 for a picture of such a function. This may seem like cheating, but discontinuities can cause a function to be barren of extrema. Perhaps continuity needs to be added to our list.

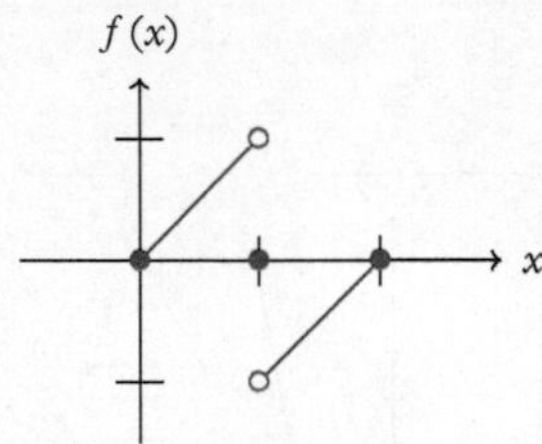

Figure 15 *The graph of a function defined on $[0, 2]$ that has no extrema. The "missing" points at the highest and lowest parts of the graph ensure there is no largest or smallest value of the function*

Last, must a continuous function on a closed, bounded interval have extrema? It is not difficult to come up with such a function that is missing local extrema (try $f(x) = x$ on $[0, 2]$), but these functions must always have absolute extrema (figure 16).

Notice the wording in the extreme value theorem. The maximum value is $f(c)$ (a y-coordinate) at c (an x-coordinate).

Theorem 2 EXTREME VALUE THEOREM *If f is continuous on a closed, bounded interval $[a, b]$, then f attains an absolute maximum value $f(c)$ and an absolute minimum value $f(d)$ at real numbers c and d in $[a, b]$.*

The proof of the extreme value theorem is a topic for a more advanced course.

Ans. to reading exercise 2
$x = -\frac{1}{2}$

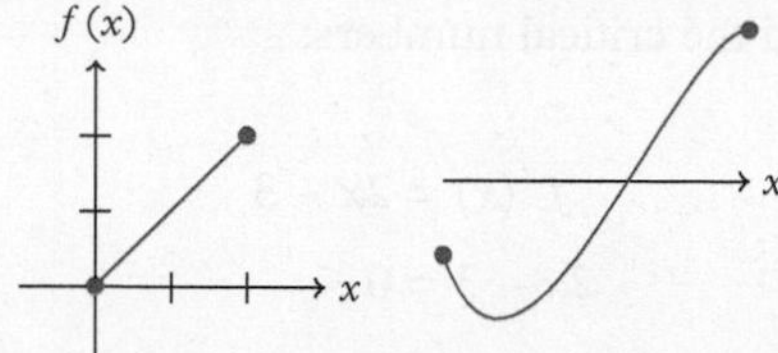

Figure 16 *A continuous function on a closed, bounded interval has endpoints and may not have local extrema (left), but must always have absolute extrema. If neither endpoint is an absolute minimum, then the curve must dip below the level of the left endpoint and come back up to hit the right endpoint in a continuous manner, causing a local and absolute minimum somewhere in between (right)*

Now that we know that a continuous function on a closed, bounded interval must have absolute extrema, how do we find them? Absolute extrema may occur at local extrema. They may also occur at the endpoints of the domain, as in figure 12. Because these are the only places that absolute extrema may occur, the only locations we need to check are the critical numbers and the endpoints. The largest value of f from among these values of x must be the absolute maximum value, and the smallest must be the absolute minimum value.

CLOSED INTERVAL METHOD

To find the absolute extrema of a continuous function on a closed, bounded interval, do the following:

❶ Find the critical numbers.

❷ Evaluate f at the critical numbers and at the endpoints.

❸ The largest of the values from step ❷ is the absolute maximum and the smallest is the absolute minimum.

When finding the critical numbers, we only include numbers in the domain of the function.

Examples of finding absolute extrema

Example 5 *Find the absolute extrema of* $f(x) = x^2 - 3x + 4$ *on* $[-5, 5]$.

Although the natural domain of a polynomial is $(-\infty, \infty)$, there are applications in which the domain must be restricted (usually where certain numbers are not possible or make no sense in the given setting). Here, the domain is restricted to the closed, bounded interval $[-5, 5]$.

Solution This is a continuous function on a closed, bounded interval and the closed interval method applies.

❶ First we find the critical numbers:

$$f'(x) = 2x - 3$$
$$2x - 3 = 0$$
$$2x = 3$$
$$x = \frac{3}{2}.$$

Because $\frac{3}{2}$ is in the interval $[-5, 5]$, it is a critical number.

The derivative is a polynomial and is therefore defined everywhere, so the only critical number is $x = \frac{3}{2}$.

Because the domain of f is restricted to $[-5, 5]$, the domain of the derivative f' is also restricted. This is implied when we say "the derivative is defined everywhere."

❷ We evaluate f at the critical number and at the endpoints:

$$f\left(\frac{3}{2}\right) = \left(\frac{3}{2}\right)^2 - 3\left(\frac{3}{2}\right) + 4 = \frac{7}{4}$$
$$f(-5) = 25 - 3(-5) + 4 = 44$$
$$f(5) = 25 - 15 + 4 = 14.$$

❸ The largest of the values in step ❷ is 44; therefore, f has an absolute maximum of 44 at $x = -5$. The smallest of the values in step ❷ is $\frac{7}{4}$; therefore, f has an absolute minimum of $\frac{7}{4}$ at $x = \frac{3}{2}$. ■

One way of writing the solution is

absolute max 44 at $x = -5$

absolute min $\frac{7}{4}$ at $x = \frac{3}{2}$.

Example 6 *Find the absolute extrema of* $f(x) = x^4 - 8x^2 + 6$ *on* $[-1, 3]$.

Solution This is a continuous function on a closed, bounded interval and the closed interval method applies.

❶ First we find the critical numbers:

$$f'(x) = 4x^3 - 16x$$
$$4x^3 - 16x = 0$$
$$4x(x^2 - 4) = 0$$
$$4x(x + 2)(x - 2) = 0$$
$$x = 0, x = -2, x = 2.$$

Although $x = 0$ and $x = 2$ are in the interval $[-1, 3]$, $x = -2$ is not. Hence, the only critical numbers are $x = 0$ and $x = 2$.

CAUTION: MISTAKE TO AVOID
Don't forget to check to make sure your critical numbers are in the function's interval of definition (the function's domain).

❷ We evaluate f at the critical numbers and at the endpoints:

$$f(0) = 6$$
$$f(2) = 2^4 - 8(2^2) + 6 = -10$$
$$f(-1) = (-1)^4 - 8(-1)^2 + 6 = -1$$
$$f(3) = 3^4 - 8(3^2) + 6 = 15.$$

❸ The largest of the values in step ❷ is 15; therefore, f has absolute max 15 at $x = 3$. The smallest of the values in step ❷ is -10; therefore, f has absolute min -10 at $x = 2$. ■

Example 7 *Find the absolute extrema of* $f(x) = \sqrt{9 - x^2}$.

Solution Because no domain is given explicitly, we must assume the natural domain is to be used. We notice the square root and set what is inside the square root to be at least zero:

$$9 - x^2 \geq 0$$
$$9 \geq x^2$$
$$\sqrt{9} \geq \sqrt{x^2}$$
$$3 \geq |x|$$
$$|x| \leq 3$$
$$-3 \leq x \leq 3.$$

The domain of this function is $[-3, 3]$. Because we have a continuous function on a closed, bounded interval, the closed interval method applies.

❶ To find the critical numbers we find the derivative:

$$f'(x) = \frac{1}{2\sqrt{9 - x^2}} \cdot (-2x) = \frac{-x}{\sqrt{9 - x^2}}.$$

Because the derivative is a fraction, we set both the numerator and the denominator equal to zero. First we solve $f'(x) = 0$ by setting the numerator equal to zero:

CAUTION: MISTAKE TO AVOID
Don't forget there are two types of critical numbers. We need to check not only where $f'(x) = 0$, but also where $f'(x)$ is undefined.

$$-x = 0$$
$$x = 0.$$

Because $x = 0$ is in the interval, it is a critical number.

Next we find where f' is undefined by setting the denominator equal to zero:

$$\sqrt{9-x^2} = 0$$
$$9 - x^2 = 0$$
$$9 = x^2$$
$$x = \pm 3.$$

Both -3 and 3 are in the interval. Our three critical numbers are $x = 0$, $x = 3$, and $x = -3$.

❷ We evaluate f at the critical numbers and at the endpoints:

$$f(0) = \sqrt{9} = 3$$
$$f(3) = \sqrt{9-9} = 0$$
$$f(-3) = \sqrt{9-9} = 0.$$

❸ The largest of the values in step ❷ is 3, so f has an absolute maximum value of 3 at $x = 0$. The smallest of the values in step ❷ is 0, which occurs twice. There is no need to break the tie; the absolute minimum value of f is zero at $x = 3$ and at $x = -3$. ■

There is still only one minimum value of f, which is zero. It occurs at two locations—that is, at two values of x. There are therefore two minimum points: $(-3, 0)$ and $(3, 0)$.

Reading Exercise 3 *Find the absolute extrema of* $f(x) = x^2$ *on* $[-1, 3]$.

Proof of Fermat's theorem

I promised a proof of Fermat's theorem, which says that if f has a local maximum or minimum at $x = a$, and if $f'(a)$ exists, then $f'(a) = 0$.

Proof. Assume that f has a local maximum at $x = a$ and that $f'(a)$ exists. By the definition of a local maximum, there is an open interval (b, c) containing a such that

$$f(a) \geq f(x) \text{ for all real } x \text{ with } b < x < c.$$

By the transfer principle, the same is true for all hyperreal x:

$$f(a) \geq f(x) \text{ for all hyperreal } x \text{ with } b < x < c.$$

Suppose that $f'(a) > 0$. Because f is differentiable at $x = a$, the local linearity theorem applies and

$$f(a+\omega) \approx f(a) + \underbrace{\omega}_{>0} \cdot \underbrace{f'(a)}_{>0} > f(a),$$

which is a contradiction; $f(a+\omega)$ can't be greater than the largest value of the function. Therefore, $f'(a) \not> 0$.

Suppose instead that $f'(a) < 0$. Because f is differentiable at $x = a$, the local linearity theorem applies and

$$f(a-\omega) \approx f(a) + \underbrace{\underbrace{-\omega}_{<0} \cdot \underbrace{f'(a)}_{<0}}_{>0} > f(a),$$

which is also a contradiction; $f(a-\omega)$ can't be greater than the largest value of the function. Therefore, $f'(a) \not< 0$.

Since $f'(a) \not> 0$ and $f'(a) \not< 0$, the only remaining possibility is $f'(a) = 0$, which is what we wanted to prove.

The proof for f having a local minimum at $x = a$ is similar. ■

Exercises 3.1

1–8. Rapid response: is the given statement correct (including correctly worded) about the function f graphed in the following? Answer "correct" or "wrong."

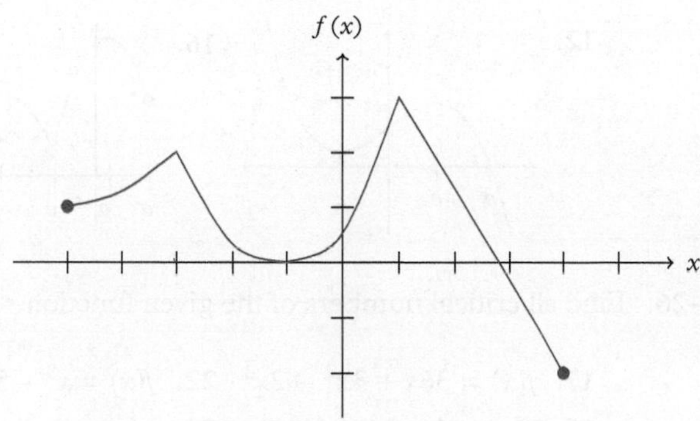

1. f has local maximum point $x = -3$.
2. The absolute maximum value of f is 3.
3. The absolute minimum of f is 4.
4. f has a local minimum at $x = -5$.
5. f has a local minimum at 0.
6. f has a local maximum at $x = 1$.
7. The absolute minimum point of f is $(4, -2)$.
8. f has local minimum 0 at $x = -1$.

Ans. to reading exercise 3
absolute minimum 0 at $x = 0$, absolute maximum 9 at $x = 3$

9–16. Identify the locations of all local and absolute extrema of the function in the graphs that follow.

9.
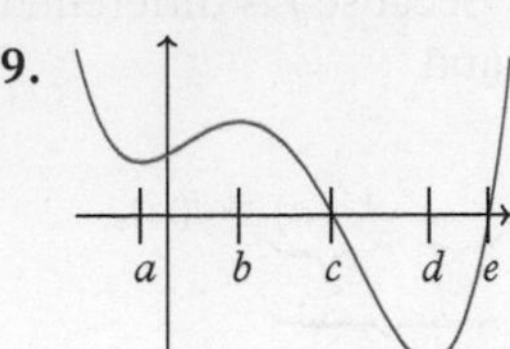

13.
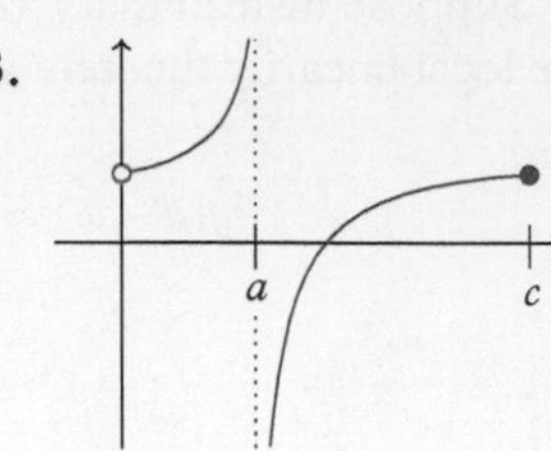

10.
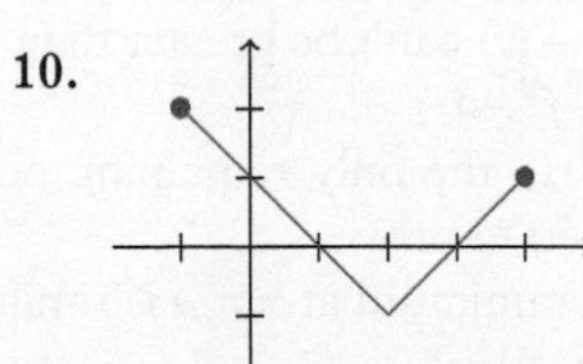

14.
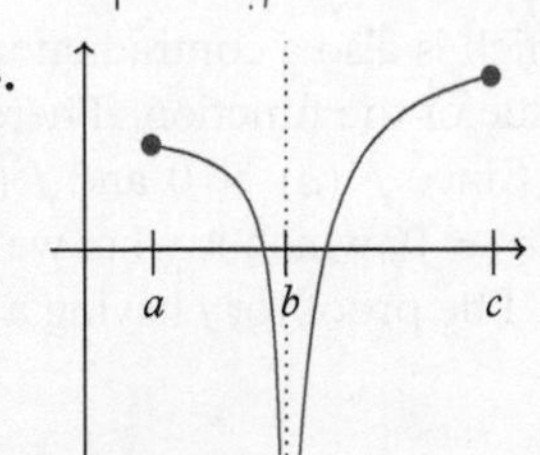

11.
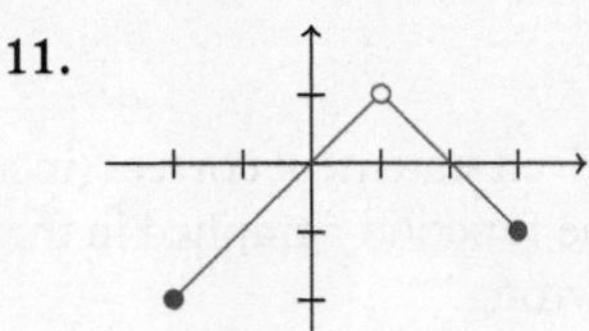

15.
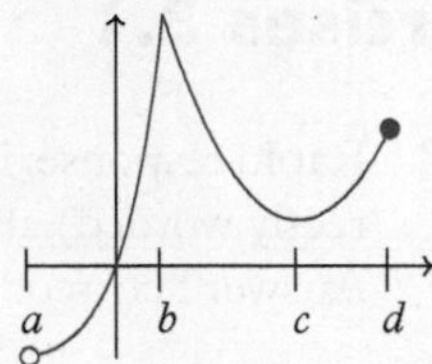

12.
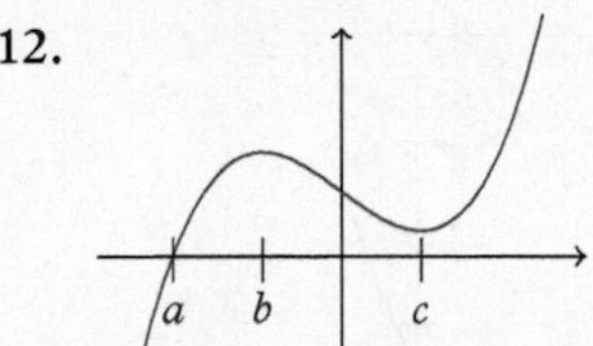

16.
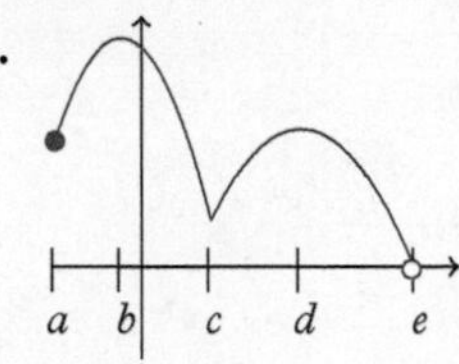

17–26. Find all critical numbers of the given function.

17. $f(x) = 36x + 3x^2 - 2x^3$
18. $f(x) = 4 + 24x^2 + 4x^3$
19. $g(x) = 10x^4 - 20x^3 + 2$
20. $g(x) = 15x^4 - 120x^2 + \pi$
21. $f(t) = 4t^5 - 20t^3 + 17$
22. $f(x) = x^3 - 5x^2 + 12x - 1$
23. $g(y) = 2y^3 + 4y^2 - 3y$
24. $s(t) = 4t^3 - 5t^2 + 8t - 12$
25. $f(x) = \sqrt[3]{x} - x$
26. $h(x) = (x^2 - 3x)^{(-1/3)}$

For exercise 26, the $(-1/3)$ is an exponent.

27–34 The derivative of a function is given. Finish finding the critical numbers.

27. $f'(x) = \dfrac{3x - 6}{\sqrt[3]{x}}$
28. $f'(x) = \dfrac{x + 4}{\sqrt[5]{2x - 7}}$

29. $f'(x) = \dfrac{x^2 - 9}{x^2 + 1}$

30. $g'(x) = x(x^3 - 8)(x^2 - 1)$

31. $g'(x) = x^2\sqrt{x - 4}$

32. $f'(t) = \dfrac{5x - 8}{\sqrt[3]{x^2 + 4x + 10}}$

33. $f'(x) = x \sec x$

34. $f'(x) = \cos x,\ 0 < x < \pi$

35–40. Is the given function's natural domain a closed, bounded interval?

35. $f(x) = \dfrac{1}{x}$

36. $f(x) = \sqrt{x}$

37. $f(x) = \sqrt{1 - x^2}$

38. $g(x) = \sqrt[3]{1 - x^2}$

39. $f(x) = x^5 - 7x^4 + 13x$

40. $f(x) = \sqrt[4]{4x + 5}$

For the remainder of this section, it is implied that the closed interval method should be used when it applies. Graphical solutions are insufficient for these exercises.

41. Find the absolute extrema of $f(x) = x^2$ on $[-1, 2]$.

42. Find the absolute extrema of $f(x) = x^2$ on $[2, 4]$.

43. Find the absolute extrema of $f(x) = x^3$ on $[-1, 2]$.

44. Find the absolute extrema of $f(x) = x^4$ on $[-1, 2]$.

45. Determine the absolute maximum and minimum values of $f(x) = x^3 - 6x + 7$ on $[0, 3]$.

46. Determine the absolute maximum and minimum values of $f(x) = x^3 - 12x + 7$ on $[-4, 1]$.

47. Determine the locations of the absolute maximum and absolute minimum of $f(x) = x^4 - 8x + 1$ on $[-1, 3]$.

48. Determine the locations of the absolute maximum and absolute minimum of $f(x) = x^3 - 4x + 7$ on $[-3, 2]$.

49. Determine the absolute minimum point and the absolute maximum point of $f(x) = x^3 - 9x + 12$ on $[0, 2]$.

50. Determine the absolute minimum point and the absolute maximum point of $f(x) = x^3 - 6x^2 + 5$ on $[-1, 2]$.

51. Find the absolute extrema of $f(x) = x^2 - 4x + 8$ on $[-1, 5]$.

52. Find the absolute extrema of $f(x) = x^3 - 3x + 7$ on $[0, 3]$.

53. Determine the absolute extreme values of $f(x) = 4 + 42x - 7x^2$ on $[0, 10]$.

54. Determine the absolute extreme values of $f(x) = x^2 - 3x + 7$ on $[0, 2]$.

55. Determine the absolute extreme points of $f(x) = \sqrt[3]{4x - 1}$ on $[0, 7]$.

56. Determine the absolute extreme points of $f(x) = \frac{1}{x}$ on $[1, 42]$.

57. Determine the absolute extreme points of $f(x) = \sin x$ on $[0, \frac{3\pi}{4}]$.

58. Determine the absolute extreme points of $f(x) = \cos x$ on $[0, \frac{3\pi}{4}]$.
59. Determine the absolute extrema of $f(x) = \tan x$ on $[-\frac{\pi}{4}, \frac{\pi}{4}]$.
60. Determine the absolute extrema of $f(x) = \sec x$ on $[-\frac{\pi}{4}, \frac{\pi}{4}]$.
61. Find the absolute extrema of $f(x) = 2 - \sqrt{25 - x^2}$.
62. Find the absolute extrema of $f(x) = \sqrt{1 - x^2}$.
63. Find the critical numbers of $f(x) = x - \sqrt{4 - x^2}$.
64. Find the critical numbers of $f(x) = 3x + \sqrt{16 - x^2}$.
65. Find the absolute extrema of $f(x) = x - \sqrt{4 - x^2}$.
66. Find the absolute extrema of $f(x) = 3x + \sqrt{16 - x^2}$.

Mean Value Theorem

3.2

This section contains important theorems that are needed later in this book.

Rolle's theorem

Suppose f is a function on $[a, b]$ such that $f(a) = 0$ and $f(b) = 0$. What does the graph of f look like? Three of the many possibilities are in figure 1. Each graph has at least one horizontal tangent line. In fact,

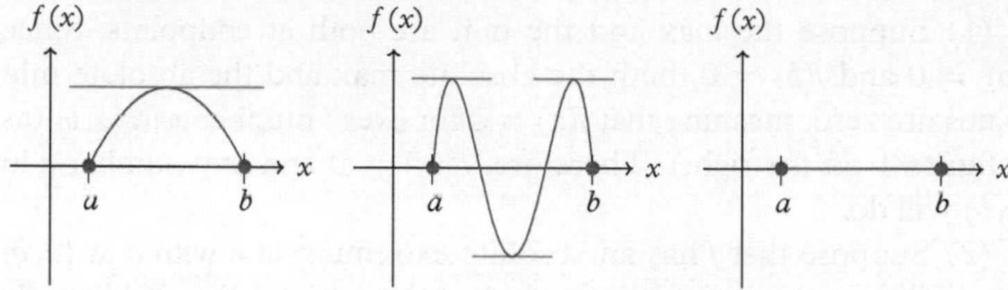

Figure 1 *Functions on* $[a, b]$ *with* $f(a) = f(b) = 0$, *each having one or more horizontal tangent lines*

there are locations c strictly between a and b such that $f'(c) = 0$. There is one such number c in the graph on the left, three in the graph in the middle, and infinitely many in the graph on the right (constant functions have derivative $f'(x) = 0$, so every number c between a and b works).

Must such a function have a horizontal tangent line? No. See figure 2. The graph on the left has a jump discontinuity and the graph in the middle has a removable discontinuity at an endpoint. Therefore,

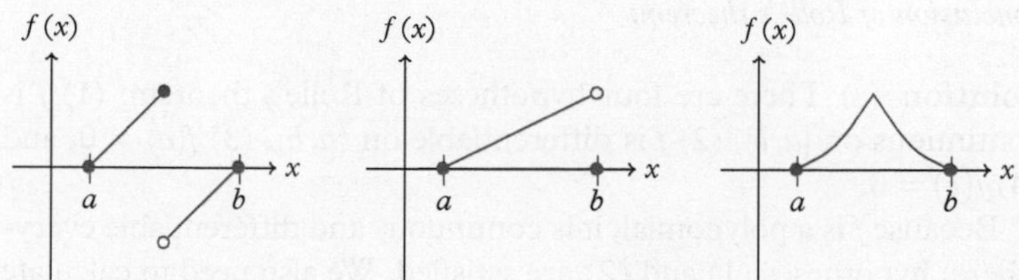

Figure 2 *Functions on* $[a, b]$ *with* $f(a) = f(b) = 0$ *without any horizontal tangent lines*

if we want to guarantee that the function has a horizontal tangent line, we need the function to be continuous on $[a, b]$ (endpoints needed). The graph on the right in figure 2 has a corner and therefore a place of nondifferentiability. Because a corner must have part of the graph to its left and part to its right, a corner cannot occur at an endpoint. Therefore, we need the function to be differentiable on (a, b) (endpoints not needed). Rolle's theorem says that's all we need.

Michel Rolle 1652–1719
http://www-history.mcs.st-andrews.ac.uk/Biographies/Rolle.html

Theorem 3 ROLLE'S THEOREM *If the function f is continuous on $[a, b]$, differentiable on (a, b), $f(a) = 0$, and $f(b) = 0$, then there is at least one number c in (a, b) such that $f'(c) = 0$.*

To avoid confusion, in this context, the intervals $[a, b]$ and (a, b) are usually pronounced "the closed interval a, b" and "the open interval a, b," respectively. Sometimes the words *from* and *to* are also added, as in "the open interval from a to b."

Proof. Since f is continuous on $[a, b]$, f must have both an absolute maximum and an absolute minimum on $[a, b]$ by the extreme value theorem (theorem 2). Then, either (1) both the max and the min are at the endpoints or (2) either the max or the min (or both) occurs at some number c strictly between a and b (that is, in the interval (a, b)).

(1) Suppose the max and the min are both at endpoints. Since $f(a) = 0$ and $f(b) = 0$, both the absolute max and the absolute min values are zero, meaning that $f(x) = 0$ for every number x in $[a, b]$ (as in figure 1 on the right). Therefore, $f'(x) = 0$ and any number c in (a, b) will do.

(2) Suppose that f has an absolute extremum at c with c in (a, b) (possibilities are pictured in figure 1 at the left and the middle). By hypothesis, $f'(c)$ exists. Because this absolute extremum is not at an endpoint, it is also a local extremum. Then, by Fermat's theorem (theorem 1), $f'(c) = 0$.

In either case, there is at least one number c in (a, b) such that $f'(c) = 0$. ■

Reading Exercise 4 *(a) Is $f(x) = x^2$ continuous on $[-4, 7]$ and differentiable on $(-4, 7)$? (b) Is $g(x) = \dfrac{1}{x}$ continuous on $[-4, 7]$ and differentiable on $(-4, 7)$?*

Example 1 *Let $f(x) = -x^2 - 2x + 8$ on $[-4, 2]$. (a) Verify that f satisfies the hypotheses of Rolle's theorem. (b) Find a number c that satisfies the conclusion of Rolle's theorem.*

The *hypotheses* of a theorem are the conditions that must be satisfied for the conclusion to hold. In a statement of the form "if p, then q," the hypotheses are p and the conclusion is q.

Solution (a) There are four hypotheses of Rolle's theorem: (1) f is continuous on $[a, b]$, (2) f is differentiable on (a, b), (3) $f(a) = 0$, and (4) $f(b) = 0$.

Because f is a polynomial, it is continuous and differentiable everywhere; hypotheses (1) and (2) are satisfied. We also need to calculate

$f(a)$ and $f(b)$:

$$f(-4) = -(-4)^2 - 2(-4) + 8 = -16 + 8 + 8 = 0,$$
$$f(2) = -2^2 - 2(2) + 8 = -4 - 4 + 8 = 0.$$

Hypotheses (3) and (4) are also satisfied.

(b) We wish to find a number c in $(-4, 2)$ such that $f'(c) = 0$. First we find $f'(x)$:

$$f'(x) = -2x - 2.$$

Next we set $f'(c) = 0$ and solve:

$$f'(c) = -2c - 2 = 0,$$
$$-2c = 2,$$
$$c = -1.$$

Because $c = -1$ is in the interval $(-4, 2)$, we are finished. ■

CAUTION: MISTAKE TO AVOID
Always check values of c to make sure they are in the interval (a, b), which does not include the endpoints.

Mean value theorem

The mean value theorem is a slanted version of Rolle's theorem. The hypotheses that f is continuous on $[a, b]$ and differentiable on (a, b) are kept, but we remove the requirements that $f(a) = 0$ and $f(b) = 0$. Now our function may or may not have horizontal tangents, as in figure 3, because Rolle's theorem does not apply. But, if we rotate the picture

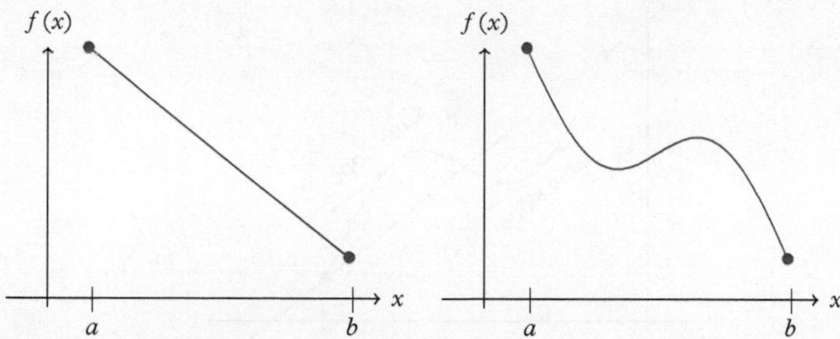

Figure 3 *A function on $[a, b]$ for which (left) there is no horizontal tangent line and (right) there are horizontal tangent lines (tangent lines not pictured)*

(we'll use the one on the right side of figure 3) so that the line from the point $(a, f(a))$ to the point $(b, f(b))$ is "horizontal" (the green line in figure 4), then it looks as if Rolle's theorem must apply and there must be at least one "horizontal" tangent line (orange in figure 4). Next we unrotate the picture. What we have left is the existence of at least one tangent line with the same slope as the line between the endpoints (figure 5). Because the slope of the line between the endpoints is

Ans. to reading exercise 4
(a) yes, polynomial functions are continuous and differentiable everywhere; (b) no, g is not continuous at $x = 0$

$$\frac{f(b) - f(a)}{b - a},$$

then there must be at least one number c in (a, b) such that

In figures 4 and 5, $\frac{f(b) - f(a)}{b - a}$ is the slope of the green line and $f'(c)$ is the slope of the orange lines.

$$f'(c) = \frac{f(b) - f(a)}{b - a}.$$

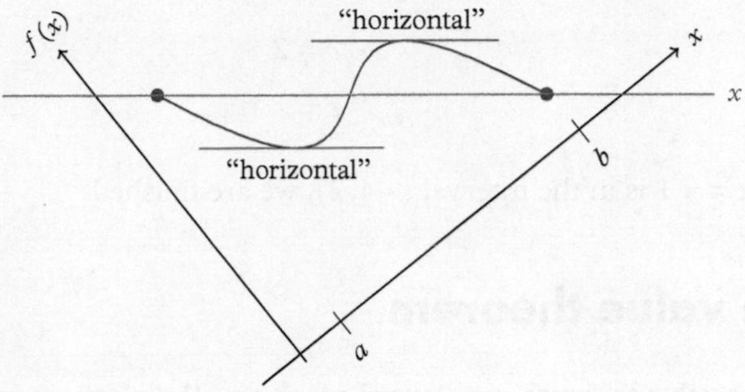

Figure 4 *A rotated picture of a function (blue) so that the line through the endpoints (green) appears horizontal. If the green line is the "new x-axis," then Rolle's theorem seems to apply*

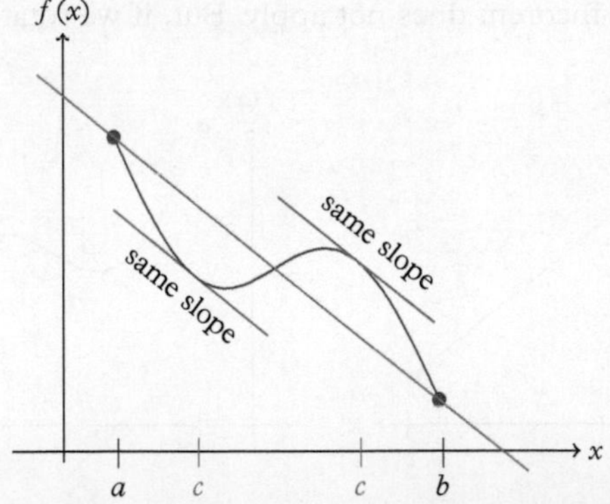

Figure 5 *The unrotated picture; the tangent lines pictured (orange) have the same slope as the line between the endpoints (green). Two possible locations are given for the number c*

Theorem 4 MEAN VALUE THEOREM *If the function f is continuous on $[a, b]$ and differentiable on (a, b), then there is at least one number c in (a, b) such that*

$$f'(c) = \frac{f(b) - f(a)}{b - a}.$$

A more formal proof of the mean value theorem is at the end of this section.

The position function of a moving object is an example of a function that is both continuous and differentiable. Think about it. A jump discontinuity in the position function of a car requires the car to jump instantly from one position to another, which is a physical impossibility. A removable discontinuity requires the car to disappear for an instant and possibly reappear elsewhere at that instant only. An infinite discontinuity requires the car to move completely outside the universe. A corner in the graph (a point of nondifferentiability) requires the velocity to change instantly to a different value and therefore requires an infinite acceleration, which is also an impossibility. These humorous images illustrate the fundamental principle that continuity and differentiability are quite the norm in the physical world. The hypotheses of the mean value theorem are often met.

One exception is a switch that is either on or off, creating a jump discontinuity.

One interpretation of the conclusion of the mean value theorem is that at some point c in the interval, the instantaneous rate of change of the function ($f'(c)$) must be the same as the average rate of change over the entire interval $\left(\frac{f(b) - f(a)}{b - a}\right)$.

Example 2 *A driver travels* 100 *miles in* 2 *hours. Show that, at some point in the trip, the vehicle's speedometer must have read* 50 mi/h.

We also make the assumption that the vehicle's speedometer is working correctly.

Solution The average velocity over the interval is

$$\frac{f(b) - f(a)}{b - a} = \frac{100\,\text{mi}}{2\,\text{h}} = 50\,\text{mi/h}.$$

Recall that if time a is 1:30 and time b is 3:30, and the position at time a is mile marker 82 whereas the position at time b is mile marker 182, the average velocity is found as

$$\frac{182 - 82}{3{:}30 - 1{:}30}.$$

Because the position function f of a vehicle is both continuous and differentiable, the mean value theorem applies and there is at least one time c during the 2-hour interval at which the instantaneous velocity $f'(c)$, which is recorded on the speedometer, is the same as the average velocity 50 mi/h. ■

The conclusion of example 2 is in use when enforcing speed limits by aircraft. Symbols are painted on the pavement at a precise distance (say, $\frac{1}{2}$ mile or 1 mile apart), and a pilot flying over the area times the suspect vehicle as it passes over the symbols. The average velocity is then calculated easily (or looked up in a chart). The conclusion is that the vehicle must have been traveling at that speed at some point during the interval, and therefore a ticket can be issued for driving that speed. The pilot contacts an officer on the ground who can then pursue and ticket the driver.

The claim "I'm just a very efficient driver" and similar defenses are proved wrong by the mean value theorem.

Example 3 *Does the function $f(x) = \sqrt[3]{x}$ on $[0, 5]$ satisfy the hypotheses of the mean value theorem?*

Solution The hypotheses of the mean value theorem are (1) f is continuous on $[a, b]$ and (2) f is differentiable on (a, b).

(1) The function $f(x) = \sqrt[3]{x}$ is a root function and is continuous where it is defined. Because it is defined everywhere, f is continuous everywhere, including the interval $[0, 5]$.

(2) The derivative is

$$f'(x) = \frac{1}{3}x^{-2/3} = \frac{1}{3\sqrt[3]{x^2}},$$

which is undefined at $x = 0$ (for reason of division by zero). The function is therefore not differentiable at $x = 0$. But, it is differentiable everywhere else (because f' is defined everywhere else), so it is differentiable on the open interval $(0, 5)$.

If the interval is instead $[-3, 3]$, then the function is not differentiable on $(-3, 3)$ ($x = 0$ is in this interval) and the hypotheses of the mean value theorem are not satisfied.

The function does satisfy the hypotheses of the mean value theorem. ■

Example 4 *Find a value of c that satisfies the conclusion of the mean value theorem for $f(x) = 1 - x^2$ on $[-1, 2]$.*

Solution Since f is a polynomial, it is differentiable and continuous everywhere, so the hypotheses of the mean value theorem are met. Next we calculate the average rate of change:

$$\frac{f(b) - f(a)}{b - a} = \frac{f(2) - f(-1)}{2 - (-1)} = \frac{-3 - 0}{2 - (-1)} = \frac{-3}{3} = -1.$$

We also need to calculate the derivative, $f'(x) = -2x$. Last, we equate $f'(c)$ with the average rate of change and solve for c:

$$f'(c) = -2c = -1,$$

$$c = \frac{-1}{-2} = \frac{1}{2}.$$

Because $c = \frac{1}{2}$ is in the interval $[-1, 2]$, we are finished. ■

Although the value of c in example 4 is in the exact middle of the interval, this is not always the case. The value of c that satisfies the conclusion of the mean value theorem for the function in example 3 is $\frac{5}{3\sqrt{3}}$, which is close to the left-hand endpoint of the interval.

Reading Exercise 5 *Find a value of c that satisfies the conclusion of the mean value theorem for $f(x) = x^2$ on $[0, 5]$.*

Functions with the same derivative

You may have noticed that many different functions can share the same derivative. For instance, if $f(x) = x^2 + 3$ and $g(x) = x^2 - 1$, their derivatives are $f'(x) = 2x$ and $g'(x) = 2x$. Having the same derivative means their slopes are the same; at any particular value of x, their tangent lines are parallel (see figure 6). Other functions sharing the same derivative include $h(x) = x^2 - 5$ and $j(x) = x^2 + \frac{47}{3}$. In fact, any function of the form

$$f(x) = x^2 + C,$$

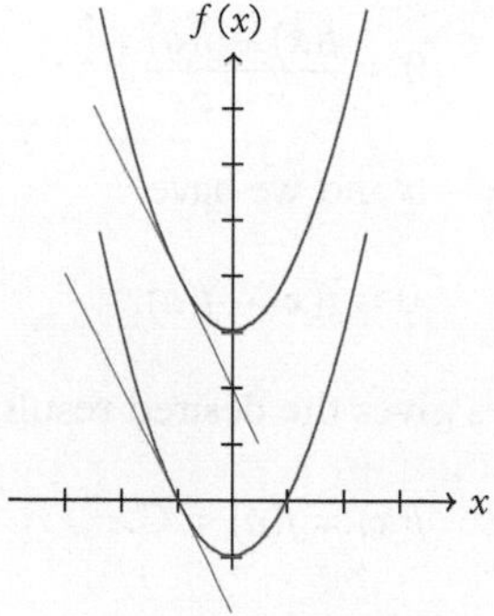

Figure 6 *The graphs of $f(x) = x^2 + 3$ and $g(x) = x^2 - 1$, each with tangent lines at $x = -1$. The tangent lines are parallel*

where C is a constant, has the derivative

$$f'(x) = 2x.$$

A question then arises: are there any others?

To answer this question, we first turn to the related question of which functions have derivative 0. We know the derivative of a constant

Functions with the same derivative have the same slopes. Functions with the same slopes must have the same shape. The functions in figure 6 are parabolas, so any function with the same shape must also be a parabola, opening upward, with the same width, and so on. The only differences include vertical placement from vertical shifts, $+C$.

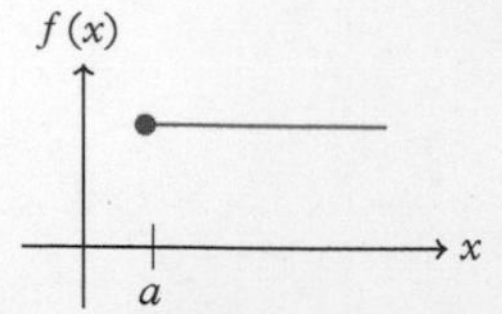

Figure 7 *A function with derivative 0 must be horizontal*

function is zero, but are there any others? Beginning at the point $(a, f(a))$, try to draw a function with derivative $f'(x) = 0$. Because the slope must be zero, the tangent line must always be horizontal and the graph must be level horizontally as well (figure 7). It appears that the only functions with derivative 0 are the constant functions. This crucial fact can be proved using the mean value theorem.

Theorem 5 ZERO-DERIVATIVE THEOREM *If $f'(x) = 0$ for all x in $[a, b]$, then $f(x) = C$ for some real number C.*

Proof. Suppose $f'(x) = 0$ for all x in $[a, b]$. We'll let C be the value of the function at the left endpoint; we set $C = f(a)$. We wish to show that the graph looks like that of figure 7.

Let x be any number in $(a, b]$ (we already know $f(a) = C$). Then, f is differentiable on $[a, x]$ (the derivative is zero) and therefore also continuous on $[a, x]$ (recall that differentiability implies continuity). Therefore, the mean value theorem applies, and there is some c strictly between a and x such that

$$f'(c) = \frac{f(x) - f(a)}{x - a}.$$

But, $f'(c) = 0$; therefore,

$$0 = \frac{f(x) - f(a)}{x - a}.$$

Multiply both sides by $x - a$ and we have

$$0 = f(x) - f(a).$$

Adding $f(a)$ to both sides gives the desired result:

$$f(x) = f(a) = C.$$

The function f is a constant function. ■

Ans. to reading exercise 5
$c = \frac{5}{2}$

Armed with this result, we are now ready to prove that any two functions with the same derivative must differ only by a constant.

Theorem 6 IDENTICAL DERIVATIVES THEOREM *If $f'(x) = g'(x)$ for all x in $[a, b]$, then $f(x) = g(x) + C$ for all x in $[a, b]$, for some real number constant C.*

Proof. Let $h(x) = f(x) - g(x)$. Then,

$$h'(x) = f'(x) - g'(x) = 0,$$

beacause $f'(x)$ and $g'(x)$ are the same. By theorem 5, h is a constant function: $h(x) = C$ for some real number C. Then,

$$h(x) = C,$$
$$f(x) - g(x) = C,$$
$$f(x) = g(x) + C.$$

We conclude that the functions f and g differ only by a constant, as in figure 6. ■

For instance, if $f(x) = x^2 + 3$ and $g(x) = x^2 - 1$, then the constant C is 4:

$$f(x) = g(x) + C$$
$$x^2 + 3 = x^2 - 1 + 4.$$

We were asked a question earlier: although functions of the form $f(x) = x^2 + C$ have derivative $f'(x) = 2x$, are there any other functions with the same derivative? According to theorem 6, the answer is no.

Example 5 *If* $y = x^3$, *then* $y' = 3x^2$. *State every function with the derivative* $3x^2$.

Solution By theorem 6, all such functions must differ by a constant. We know one of them, so they must differ from it by a constant. Functions of the form

$$f(x) = x^3 + C$$

are the only functions with derivative $f'(x) = 3x^2$. ■

Reading Exercise 6 *Recall that* $\frac{d}{dx}x^4 = 4x^3$. *What other functions have derivative* $4x^3$*?*

Proof of the mean value theorem

The proof of the mean value theorem relies on the reasoning given earlier.

Proof. Suppose f is continuous on $[a, b]$ and differentiable on (a, b). To mimic the rotation, we let the line between the endpoints (the green line of figure 4) be represented by the function g. Using the point-slope equation of a line with point $(a, f(a))$ and slope $\frac{f(b) - f(a)}{b - a}$ gives

$$y - f(a) = \frac{f(b) - f(a)}{b - a}(x - a),$$

and adding $f(a)$ to both sides and writing $g(x)$ for y results in

$$g(x) = \frac{f(b) - f(a)}{b - a}(x - a) + f(a).$$

To make this line act as the x-axis (again see figure 4), we subtract $g(x)$ from $f(x)$, making $g(x)$ the zero line. We call this new function h:

$$h(x) = f(x) - g(x).$$

Then,

$$h(a) = f(a) - g(a) = f(a) - \left(\frac{f(b) - f(a)}{b - a}(a - a) + f(a)\right) = 0$$

and

$$h(b) = f(b) - g(b) = f(b) - \left(\frac{f(b) - f(a)}{b - a}(b - a) + f(a)\right)$$

$$= f(b) - (f(b) - f(a) + f(a)) = 0.$$

Because g is a line, it is continuous and differentiable everywhere. Because the difference of continuous functions is continuous and the difference of differentiable functions is differentiable, $h = (f - g)$ is continuous on $[a, b]$ and differentiable on (a, b). The hypotheses of Rolle's theorem are satisfied. Therefore, there is some number c in (a, b) such that $h'(c) = 0$. But,

$$h'(x) = f'(x) - g'(x)$$

and $g'(x)$ is the slope of that line, $\frac{f(b) - f(a)}{b - a}$. Putting these all together,

$$0 = h'(c) = f'(c) - \frac{f(b) - f(a)}{b - a},$$

and rearranging yields

$$\frac{f(b) - f(a)}{b - a} = f'(c),$$

as desired. ■

Exercises 3.2

Ans. to reading exercise 6
functions of the form $f(x) = x^4 + C$ (or, since the word *other* was used, functions of the form $f(x) = x^4 + C$ with $C \neq 0$)

1–8. Does the function satisfy the hypotheses of Rolle's theorem on the given interval?

1. $f(x) = x^2 - 2x$ on $[0, 2]$
2. $f(x) = x^3 - x$ on $[0, 3]$

3. $g(x) = x - \sqrt{x}$ on $[0, 1]$
4. $f(x) = 3x + x^2$ on $[-3, 0]$
5. $f(x) = x - \dfrac{1}{x}$ on $[-1, 1]$
6. $g(x) = \cos x$ on $[-\pi, 0]$
7. $f(x) = \sin x$ on $[0, \pi]$
8. $s(t) = \sin(x^2 - 2x)$ on $[0, 2]$

9–16. Does the function satisfy the hypotheses of the mean value theorem on the given interval?

9. $f(x) = x^2 - 2x$ on $[0, 2]$
10. $f(x) = 5x^{2/3}$ on $[-1, 1]$
11. $f(x) = x - \dfrac{1}{x}$ on $[-1, 1]$
12. $f(x) = x^3 - x$ on $[0, 3]$
13. $g(x) = \cos x$ on $[-\pi, 0]$
14. $g(x) = 5x^{4/3}$ on $[-1, 1]$
15. $s(t) = \sqrt{t}$ on $[0, 2\pi]$
16. $h(x) = 5x^{2/3}$ on $[0, 1]$

17–24. Find all numbers c that satisfy the conclusion of Rolle's theorem.

17. $f(x) = x^3 + 2x^2$ on $[-2, 0]$
18. $g(x) = 5x^2 - 10x$ on $[0, 2]$
19. $f(t) = \cos t$ on $[-\frac{\pi}{2}, \frac{\pi}{2}]$
20. $f(x) = \sin x$ on $[\pi, 2\pi]$
21. $f(x) = 1 - \dfrac{2}{x^2 + 1}$ on $[-1, 1]$
22. $g(x) = x^2 - \sqrt{x}$ on $[0, 1]$
23. $r(z) = \tan(z^2 - z)$ on $[0, 1]$
24. $g(t) = \sin t^2$ on $[0, \sqrt{\pi}]$

25–34. Find all numbers c that satisfy the conclusion of the mean value theorem.

25. $s(t) = t^2 - 7t + 2$ on $[0, 2]$
26. $f(t) = 3t^2 - t$ on $[-1, 5]$
27. $f(x) = 2x^3$ on $[0, 3]$
28. $f(x) = x^3 + 9$ on $[-2, 0]$
29. $f(x) = \dfrac{8}{x}$ on $[1, 4]$

30. $g(x) = x^3 - x$ on $[0, 4]$
31. $g(x) = \sqrt{x}$ on $[0, 9]$
32. $f(x) = \sqrt{x+1}$ on $[0, 3]$
33. $f(t) = t^3 - 1$ on $[-2, 2]$
34. $f(x) = \dfrac{x+2}{x^2+1}$ on $[0, 3]$

35. If $y = 4x^2$, then $y' = 8x$. State every function with derivative $8x$.
36. If $y = 2x^3$, then $y' = 6x^2$. State every function with derivative $6x^2$.
37. If $y = \sin x$, then $y' = \cos x$. State every function with derivative $\cos x$.
38. If $y = \tan x$, then $y' = \sec^2 x$. State every function with derivative $\sec^2 x$.
39. If $y = (5x^2 + 3)^4$, then $y' = 40x(5x^2 + 3)^3$. State every function with derivative $40x(5x^2 + 3)^3$.
40. If $y = \cos(5x - 9)$, then $y' = -5\sin(5x - 9)$. State every function with derivative $-5\sin(5x - 9)$.
41. State every function with derivative $\sec x \tan x$.
42. State every function with derivative $-\sin x$.
43. State every function with derivative $6x^5$.
44. State every function with derivative $5x^4$.
45. State every function with derivative x^5.
46. State every function with derivative 17.
47. A trucker enters a toll road at mile marker 14 at 12:00 noon and exits at mile marker 160 at 2:00 p.m. The speed limit is 70 mph. Does the mean value theorem guarantee that, at some point, the trucker was driving faster than the speed limit?
48. On a fly ball to right field, a baseball player runs from second base to third base, beginning and ending at a velocity of zero. Use a theorem from this section to show that, at some point in between second base and third base, the player's acceleration was zero.
49. (GSL) Given $f(x) = \tan x$, then $f(0) = 0$ and $f(\pi) = 0$. Does Rolle's theorem justify the conclusion that $f'(x)$ vanishes for some value of x between 0 and π? Explain your answer.

The phrase "$f'(x)$ vanishes" means $f'(x) = 0$.

Local Extrema

3.3

In section 3.1 we learned Fermat's theorem, which states that local extrema can occur only at locations in the domain of a function where the derivative is zero or undefined. In this section we further explore the connection between local extrema and derivatives.

Increasing and decreasing functions

Imagine traveling from left to right along the curve pictured in figure 1. You travel uphill before reaching $x = a$, then downhill between $x = a$

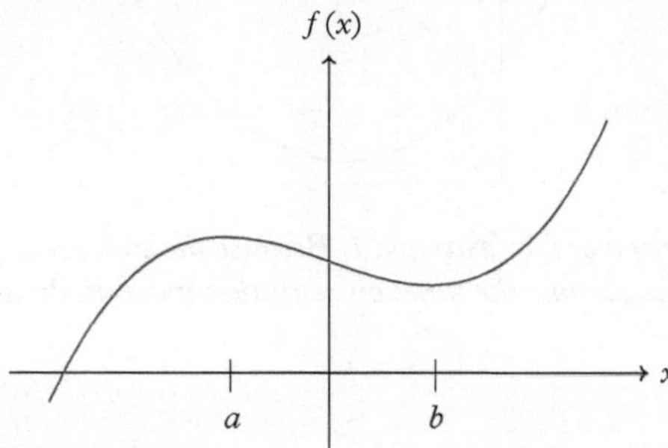

Figure 1 *A function increasing on $(-\infty, a)$, decreasing on (a, b), and increasing on (b, ∞)*

and $x = b$, and then uphill again after $x = b$. When traveling uphill, elevation increases; when traveling downhill, elevation decreases. This is the idea of increasing and decreasing functions. The function of figure 1 is increasing on the interval $(-\infty, a)$, decreasing on the interval (a, b), and increasing on the interval (b, ∞). Of course, traveling the opposite direction reverses increasing and decreasing, so for the sake of consistency, we always use the perspective of traveling along the curve from left to right.

We can also say the function is increasing on $(-\infty, a]$, decreasing on $[a, b]$, and increasing on $[b, \infty)$.

The elevations of the curve are represented by the values of the function. To say that the function is increasing means the values of the function are increasing as we move to the right, and moving to the right means x-coordinates are getting larger. Therefore, we describe an increasing function by saying that if $x_2 > x_1$, then $f(x_2) > f(x_1)$, as in figure 2.

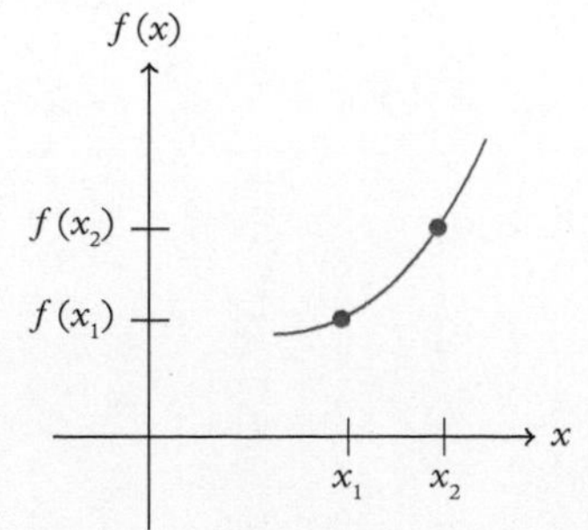

Figure 2 *"A function f is increasing" means that $x_1 < x_2$ implies $f(x_1) < f(x_2)$*

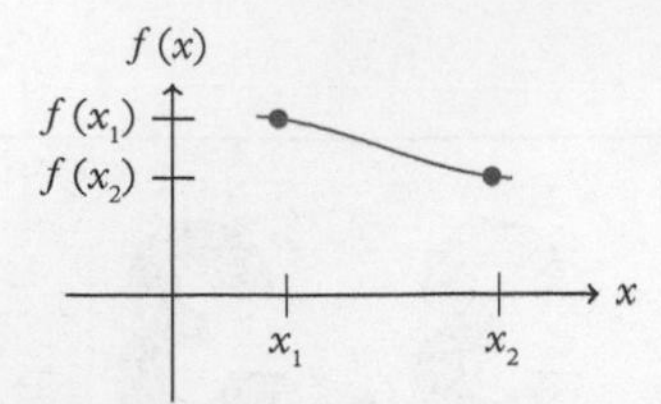

Figure 3 *"A function f is decreasing" means that $x_1 < x_2$ implies $f(x_1) > f(x_2)$*

Similarly, where f is decreasing, $x_2 > x_1$ requires $f(x_2) < f(x_1)$ (figure 3).

Definition 3 INCREASING AND DECREASING *Let f be a function defined throughout an interval I. If for every $x_1 < x_2$ in I we have $f(x_1) < f(x_2)$, then f is* **increasing** *on I. If for every $x_1 < x_2$ in I we have $f(x_1) > f(x_2)$, then f is* **decreasing** *on I.*

Example 1 *For the function f graphed in figure 4, on which intervals is f increasing? On which intervals is f decreasing?*

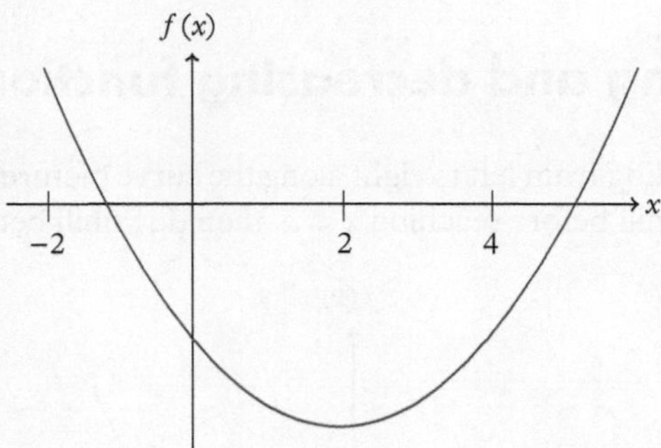

Figure 4 *The graph of f for example 1. Because the endpoints are not emphasized with dots, we assume the function continues indefinitely to the left and to the right*

Solution Inspecting the graph, we see that the function is increasing on the interval $(2, \infty)$ and decreasing on the interval $(-\infty, 2)$. ■

An alternate solution is that f is increasing on $[2, \infty)$ and decreasing on $(-\infty, 2]$. The preference for open intervals is explained in example 2.

Reading Exercise 7 *For the function f graphed in figure 5, on which intervals is f increasing? On which intervals is f decreasing?*

Now consider figure 6, where a few tangent lines are marked. Notice that where f is increasing, the tangent lines have a positive slope and where f is decreasing, the tangent lines have a negative slope. The converse is also true: if $f'(x) > 0$ (positive derivative) throughout an interval, then this indicates the function is increasing on the interval; $f'(x) < 0$ (negative derivative) throughout an interval indicates the function is decreasing on the interval.

Theorem 7 INCREASING AND DECREASING FUNCTIONS

(a) *If $f'(x) > 0$ on an interval I, then f is increasing on the interval I.*

The use of "an interval I" means we could have intervals of any form, such as $[a, b]$, (a, b), (a, ∞), and so on.

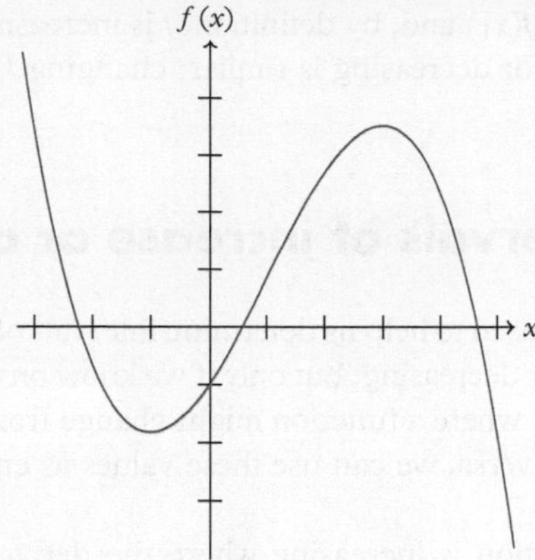

Figure 5 *The graph of f for reading exercise 7*

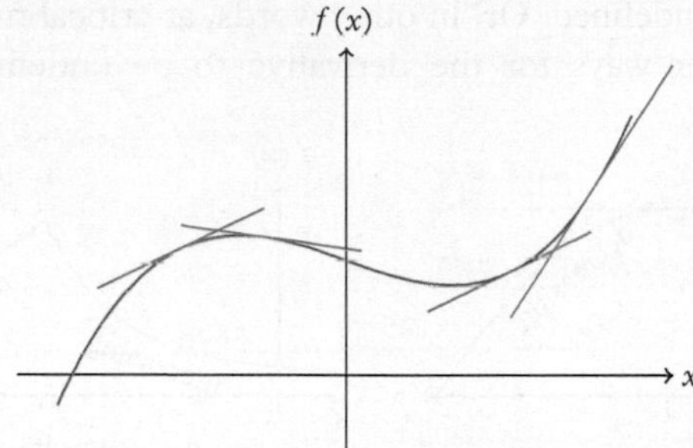

Figure 6 *Where f is increasing, the tangent lines have a positive slope; where f is decreasing, the tangent lines have a negative slope*

(b) If $f'(x) < 0$ on an interval I, then f is decreasing on the interval I.

Proof. (a) Suppose $f'(x) > 0$ on an interval I. Let x_1, x_2 be in the interval with $x_1 < x_2$. Notice that because $f'(x)$ is defined throughout the interval, f is differentiable on $[x_1, x_2]$ (and therefore continuous on $[x_1, x_2]$) and the hypotheses of the mean value theorem are met. Therefore, there is some real number c between x_1 and x_2 such that

Notice that this proof uses the mean value theorem from section 3.2.

$$\frac{f(x_2) - f(x_1)}{x_2 - x_1} = f'(c).$$

However, because c is in our interval I, $f'(c) > 0$. Rearranging the previous equation,

$$f(x_2) - f(x_1) = \underbrace{f'(c)}_{\text{positive}} \underbrace{(x_2 - x_1)}_{\text{positive}} > 0.$$

Therefore, $f(x_2) > f(x_1)$ and, by definition, f is increasing on I.

(b) The proof for decreasing is similar; changing $f'(c)$ to negative yields $f(x_2) < f(x_1)$. ■

Finding intervals of increase or decrease

Theorem 7 can be used to help us determine intervals on which a function is increasing or decreasing, but only if we know on which intervals to look. If we know where a function might change from increasing to decreasing or vice versa, we can use these values as endpoints of our intervals.

Ans. to reading exercise 7
increasing on $(-1, 3)$, decreasing on $(-\infty, -1)$ and $(3, \infty)$; alternate solution: increasing on $[-1, 3]$, decreasing on $(-\infty, -1]$ and $[3, \infty)$

Because a function is increasing where the derivative is positive and decreasing where it is negative, changes may occur where the derivative is neither positive nor negative—that is, where the derivative is zero or undefined. Or, in other words, at critical numbers (see figure 7). Other ways for the derivative to be undefined include

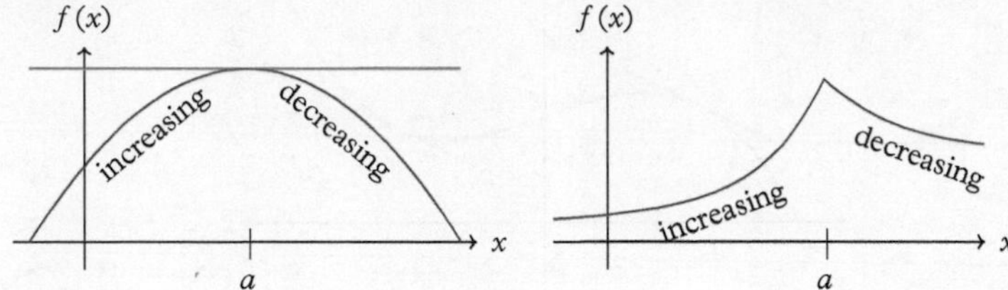

Figure 7 *Functions changing from increasing to decreasing at a critical number a*

where the function itself is not defined or where the function is not continuous, such as at a vertical asymptote or a jump discontinuity (figure 8).

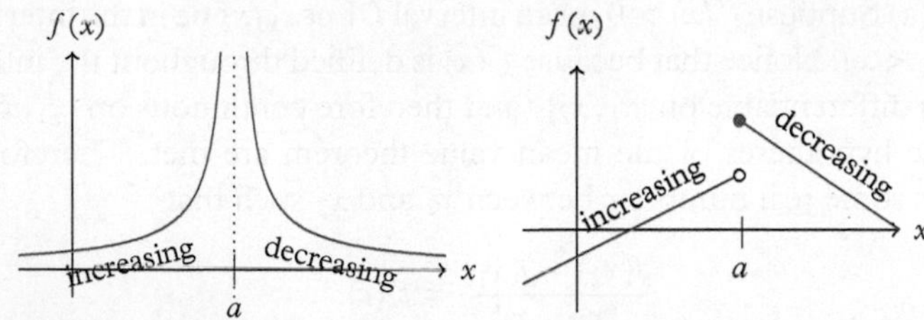

Figure 8 *Functions changing from increasing to decreasing at a discontinuity located at x = a*

We can therefore find intervals of increase and decrease of a function by using the following procedure:

❶ Find the domain of f and any discontinuities of f.

❷ Find the critical numbers of f.

❸ Determine intervals by using the domain and by using the discontinuities and critical numbers as endpoints.

❹ Determine the sign of f' on each interval (positive or negative).

❺ Determine increasing or decreasing from the sign of the derivative.

The next example illustrates one way of organizing the procedure.

Example 2 *On which intervals is $f(x) = x^3 - 3x^2 - 45x + 17\sqrt{5}$ increasing? On which intervals is f decreasing?*

Solution ❶ Because f is a polynomial, its domain is $(-\infty, \infty)$ and there are no discontinuities.

❷ To find the critical numbers, we find the derivative

$$f'(x) = 3x^2 - 6x - 45,$$

set it equal to zero, and solve:

Recall that if factoring is not quickly fruitful, using the quadratic formula is always an option.

$$\begin{aligned} 3x^2 - 6x - 45 &= 0 \\ 3(x^2 - 2x - 15) &= 0 \\ 3(x-5)(x+3) &= 0 \\ x = 5, x &= -3. \end{aligned}$$

❸ Next we determine the intervals using any discontinuities (there are none) and the critical numbers as endpoints. This can be illustrated by drawing a number line, marking the discontinuities and critical numbers, and thinking of using scissors to cut the number line at those points; the resulting pieces are the desired intervals. See figure 9.

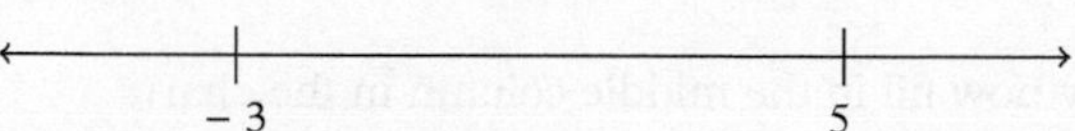

Figure 9 *A number line illustrating the function's domain with all discontinuities and critical numbers marked*

The intervals are $(-\infty, -3)$, $(-3, 5)$, and $(5, \infty)$. It is often helpful to organize the work of steps ❸–❺ in a chart, which so far looks like this:

For our procedure, we wish to have intervals on which the derivative is always positive or always negative. We do not include the endpoints because these are places where the derivative is zero or undefined.

interval	sign of f'	inc/dec
$(-\infty, -3)$		
$(-3, 5)$		
$(5, \infty)$		

❹ There are many ways by which we can determine the sign of the derivative f' on each interval. One way is to use *test points*. Because the derivative is continuous (it is a polynomial), then the only places where it can change sign (from + to − or from − to +) are where f' is zero or undefined, which have been marked as the endpoints of the intervals. Thus, for each of the three intervals, the derivative f' is either positive throughout the entire interval or it is negative throughout the entire interval. If we evaluate the derivative at one point of our choosing (called a *test point*) in the interval and its value is positive, then f' is positive throughout the entire interval. If the value of f' at the test point is negative, then f' is negative throughout the entire interval.

To choose test points, it is helpful to consult the number line that was drawn to determine the intervals. One point in each interval is chosen (see figure 10).

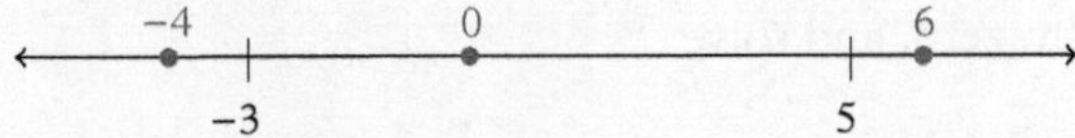

Figure 10 *Test points chosen. We could choose* −3.0784, π, *and* 284 756, *but these values are less convenient*

Next we evaluate the derivative at each of the test points to determine its sign (positive or negative):

$$f'(-4) = 3(-4)^2 - 6(-4) - 45 = 27 > 0 \text{ (positive)}$$
$$f'(0) = -45 < 0 \text{ (negative)}$$
$$f'(6) = 3(6^2) - 6(6) - 45 = 27 > 0 \text{ (positive)}.$$

An alternative is to use the factored form of the derivative,

$$f'(x) = 3(x-5)(x+3),$$

not bothering to determine the exact value at the test point but only considering whether each factor is positive or negative:

$$f'(-4) = + \cdot - \cdot - = +$$
$$f'(0) = + \cdot - \cdot + = -$$
$$f'(6) = + \cdot + \cdot + = +.$$

To save time, this work can be written in the chart.

We may now fill in the middle column in the chart:

interval	sign of f'	inc/dec
$(-\infty, -3)$	+	
$(-3, 5)$	−	
$(5, \infty)$	+	

❺ Determining increasing or decreasing for each interval is accomplished by applying theorem 7. A positive derivative indicates increasing and a negative derivative indicates decreasing:

interval	sign of f'	inc/dec
$(-\infty, -3)$	+	increasing
$(-3, 5)$	−	decreasing
$(5, \infty)$	+	increasing

We conclude that f is increasing on the intervals $(-\infty, -3)$ and $(5, \infty)$, and that f is decreasing on the interval $(-3, 5)$. ■

Although it may seem simpler just to graph the function and proceed as in example 1, not all functions cooperate by having their intervals end at nice, small integers. This procedure ensures we find all the intervals and determine their endpoints correctly.

Example 3 *On which intervals is* $f(x) = \dfrac{1}{x}$ *increasing? On which intervals is f decreasing?*

Solution ❶ The domain of $f(x) = \frac{1}{x}$ must exclude $x = 0$ (division by zero). The domain is $(-\infty, 0) \cup (0, \infty)$ and f has a discontinuity at $x = 0$ (the type of discontinuity is irrelevant to this type of exercise).
❷ The derivative is

$$f'(x) = -x^{-2} = -\frac{1}{x^2}.$$

Because the graph of $f(x) = \frac{1}{x}$ should be on immediate recall, it should be quickly apparent that the answer is that f is decreasing on the intervals $(-\infty, 0)$ and $(0, \infty)$. We use the five steps to illustrate the procedures more clearly.

To find the critical numbers, we set both the numerator and the denominator equal to zero, yielding $1 = 0$ (which has no solutions) and $x^2 = 0$, which has solution $x = 0$. Because this value is not in the domain of the function, f has no critical numbers.

❸ After marking the discontinuities and critical numbers on a number line (figure 11), we see that the intervals are $(-\infty, 0)$ and $(0, \infty)$.

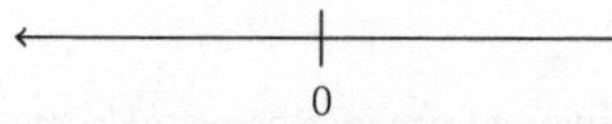

Figure 11 *A number line with all discontinuities and critical numbers marked*

❹ Choosing −1 and 1 as test points (figure 12), we evaluate f' and see that f' is negative on both intervals:

$$f'(-1) = -\frac{1}{(-1)^2} = -1$$

$$f'(1) = -\frac{1}{1^2} = -1.$$

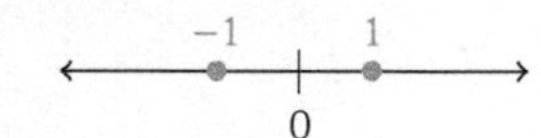

Figure 12 *Test points for example 3*

❺ The completed chart is

interval	sign of f'	inc/dec
$(-\infty, 0)$	−	decreasing
$(0, \infty)$	−	decreasing

and f is decreasing on the intervals $(-\infty, 0)$ and $(0, \infty)$. ■

Implicit in step ❹ is that f' is continuous throughout its domain, so that in each interval either f' is always positive or f' is always negative. This property always holds for the types of functions we consider in calculus.

CAUTION: MISTAKE TO AVOID
In example 3, we do not say that f is decreasing on the interval $(-\infty, \infty)$. To do so violates the definition, for $f(-1) = -1 < f(1) = 1$. Adjacent intervals cannot be combined across a discontinuity.

Reading Exercise 8 *Finish the chart to find the intervals on which f is increasing and the intervals on which f is decreasing:*

interval	sign of f'	inc/dec
$(-\infty, 2)$	$-$	
$(2, 5)$	$+$	
$(5, \infty)$	$-$	

First derivative test

Fermat's theorem states that local extrema may only occur at critical numbers. If you are traveling left to right and arrive at the top of a hill (a local maximum), then you are moving uphill (increasing) on the left side of the hilltop and you then move downhill (decreasing) on the right side of the hilltop (see figure 13). The derivative therefore changes sign from positive to negative as you crest the top of the hill. Similarly, the derivative changes sign from negative to positive at a local minimum (figure 14).

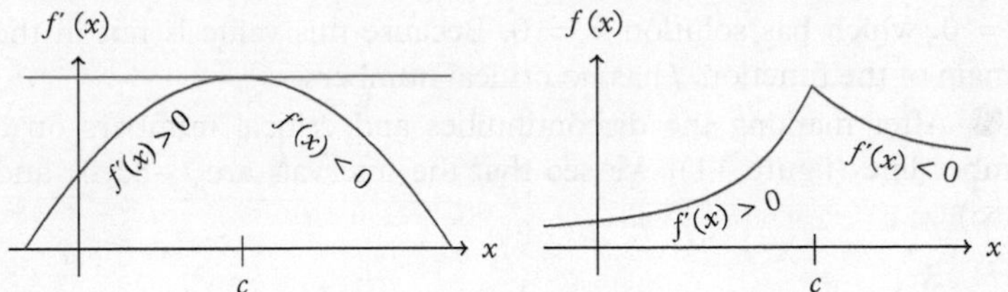

Figure 13 *Functions changing from increasing to decreasing at a critical number c, resulting in a local maximum at x = c. The derivative is positive to the left of x = c and negative to the right of x = c*

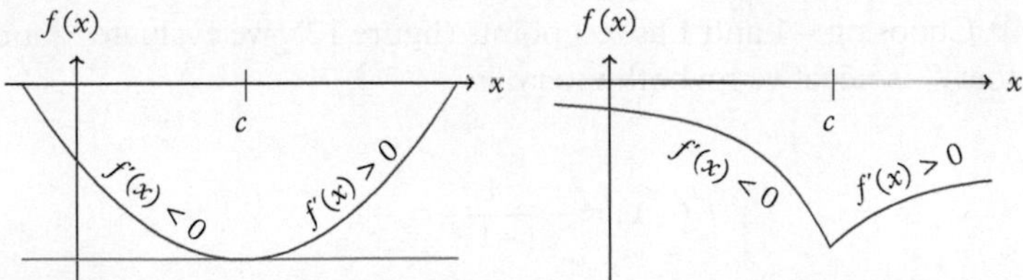

Figure 14 *Functions changing from decreasing to increasing at a critical number c, resulting in a local minimum at x = c. The derivative is negative to the left of x = c and positive to the right of x = c*

If the derivative does not change signs at a critical number, then the function does not have a local maximum or minimum at that critical number (figure 15).

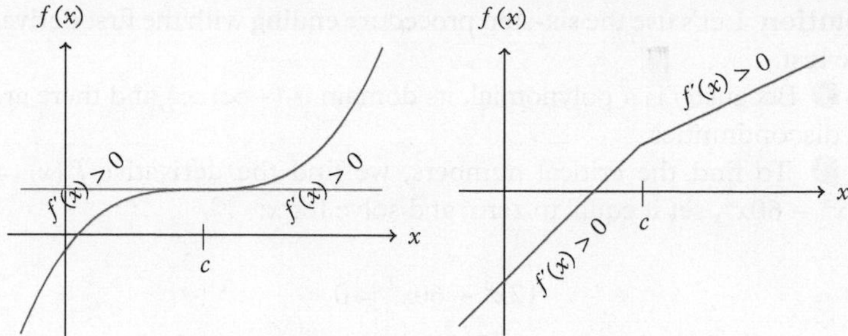

Figure 15 *Functions increasing on both sides of a critical number c lack a local extremum at x = c. The derivative is positive to the left and to the right of x = c*

Theorem 8 FIRST DERIVATIVE TEST *Suppose that $x = c$ is a critical number of a continuous function f.*

(a) *If f' changes sign from positive to negative at c, then f has a local maximum at $x = c$.*

(b) *If f' changes sign from negative to positive at c, then f has a local minimum at $x = c$.*

(c) *If f' does not change signs at c, then f has no local extremum at $x = c$.*

A more formal version of the theorem reads, for part (a), if $f'(x) > 0$ on an interval (a, c) and $f'(x) < 0$ on an interval (c, b), then f has a local maximum at $x = c$.

The five-step procedure for determining intervals of increase and decrease sets us up to use the first derivative test. We simply add a column to the chart and a sixth step to the procedure:

Ans. to reading exercise 8

interval	sign of f'	inc/dec
$(-\infty, 2)$	$-$	decreasing
$(2, 5)$	$+$	increasing
$(5, \infty)$	$-$	decreasing

❻ Use the first derivative test to determine the locations of any local extrema.

The chart for example 2 finishes as

interval	sign of f'	inc/dec	local extrema
$(-\infty, -3)$	$+$	increasing	
			← local max at $x = -3$
$(-3, 5)$	$-$	decreasing	
			← local min at $x = 5$
$(5, \infty)$	$+$	increasing	

Notice the positioning of the lines for local extrema: in between the lines containing the other information.

whereas the chart for example 3 finishes as follows:

interval	sign of f'	inc/dec	local extrema
$(-\infty, 0)$	$-$	decreasing	
			← no extremum
$(0, \infty)$	$-$	decreasing	

In example 3, f is not defined at $x = 0$. Therefore, even if the sign of f' changes, there is no local extremum at $x = 0$ because there is no point there.

Examples of finding local extrema

Example 4 *Find the locations of all local extrema of $f(x) = 3x^4 - 20x^3$.*

Solution Let's use the six-step procedure ending with the first derivative test.

❶ Because f is a polynomial, its domain is $(-\infty, \infty)$ and there are no discontinuities.

❷ To find the critical numbers, we find the derivative $f'(x) = 12x^3 - 60x^2$, set it equal to zero, and solve for x:

$$12x^3 - 60x^2 = 0$$
$$12x^2(x-5) = 0$$
$$x = 0, x = 5.$$

❸ After marking discontinuities and critical numbers on a number line (figure 16), we see that the intervals are $(-\infty, 0)$, $(0, 5)$, and $(5, \infty)$.

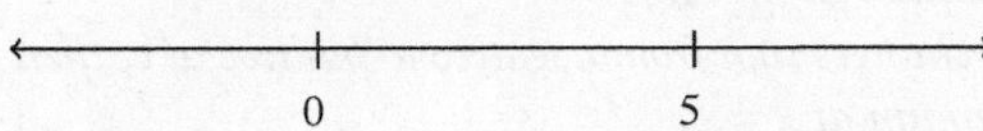

Figure 16 *A number line with all discontinuities and critical numbers marked*

❹ We choose -1, 1, and 6 as test points (figure 17) and evaluate f':

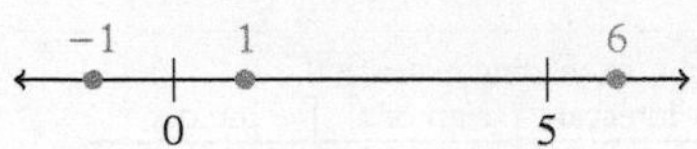

Figure 17 *Convenient test points for example 4*

$$f'(-1) = -72 < 0$$
$$f'(1) = -48 < 0$$
$$f'(6) = 432 > 0.$$

❺–❻ We can complete the chart using theorem 7 and the first derivative test (theorem 8):

interval	sign of f'	inc/dec	local extrema
$(-\infty, 0)$	$-$	decreasing	
			← no extremum
$(0, 5)$	$-$	decreasing	
			← local min at $x = 5$
$(5, \infty)$	$+$	increasing	

We conclude that the only local extremum is a local minimum at $x = 5$. ■

If we were asked for the local extreme values, we calculate

$$f(5) = 3(5^4) - 20(5^3) = -625$$

and conclude that f has a local minimum value of -625 at $x = 5$.
If we were asked for the local extreme points, we conclude that f has a local minimum point of $(5, -625)$.

Reading Exercise 9 *Finish the chart to determine where f has local extrema (assume f is defined at all real numbers):*

interval	*sign of f′*	*inc/dec*	*local extrema*
$(-\infty, -1)$	$+$		
$(-1, 1)$	$-$		
$(1, 5)$	$+$		
$(5, \infty)$	$+$		

Example 5 *Determine all local extreme points on the graph of* $f(x) = \sqrt[3]{x^2}$.

Solution ❶ The domain of f is $(-\infty, \infty)$ and f has no discontinuities.

❷ The derivative is $f'(x) = \frac{2}{3}x^{-1/3} = \frac{2}{3\sqrt[3]{x}}$. To find our critical numbers we set both the numerator and the denominator equal to zero:

Don't forget there are two types of critical numbers: where the derivative is zero and where the derivative is undefined. Setting the denominator of f' equal to zero helps us determine where the derivative is undefined.

$$2 = 0 \qquad\qquad 3\sqrt[3]{x} = 0$$

$$\text{no solutions} \qquad\qquad x = 0.$$

There is one critical number, $x = 0$.

❸–❻ Because cube roots of negative numbers are negative and cube roots of positive numbers are positive, the chart is finished as

interval	sign of f'	inc/dec	local extrema
$(-\infty, 0)$	$-$	decreasing	← local min at $x = 0$
$(0, \infty)$	$+$	increasing	

and we conclude there is a local minimum at $x = 0$. Because we wish to find the extreme points (not just their locations), we also need the y-coordinate of the point:

$$f(0) = \sqrt[3]{0^2} = 0.$$

The local minimum point is $(0, 0)$. ■

Can you picture the graph of the function in example 5 in your mind? If your mental picture looks like a parabola opening upward with vertex at $(0, 0)$, you're wrong. The function does not have a horizontal tangent line at $x = 0$, but rather a corner at $x = 0$, because it fails to be differentiable there. The information in the chart is insufficient to tell us the difference between the two graph shapes in figure 18.

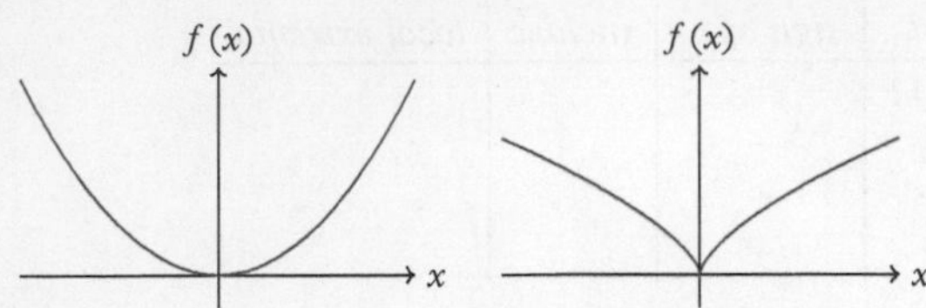

Figure 18 *Two graphs satisfying the information of the following chart:*

interval	f'	inc/dec	local extrema
$(-\infty, 0)$	$-$	dec.	local min at $x = 0$
$(0, \infty)$	$+$	inc.	

Procuring the additional information needed to tell the two situations apart is the topic of the next section. First, however, let's look at one more example.

Example 6 *Determine intervals of increase and decrease as well as the locations of any extrema for* $f(x) = \sqrt{x}$.

This is another function with a graph that should be on instant recall. We should conclude that f is increasing on $[0, \infty)$ and that there are no local extrema, although there is an absolute minimum point of $(0, 0)$. Again, we use the six-step method for the purpose of illustrating the solution procedure.

Solution ❶ The domain of the function is $x \geq 0$, or $[0, \infty)$. Root functions are continuous where defined, so there is no discontinuity.

❷ The derivative is

$$f(x) = \frac{1}{2}x^{-1/2} = \frac{1}{2\sqrt{x}}$$

and the only critical number is $x = 0$.

❸ To determine the intervals, we must take the domain into account. Drawing only the portion of the number line containing the domain, we then mark all discontinuities and critical numbers, as in figure 19. We have only one interval for our chart: $(0, \infty)$.

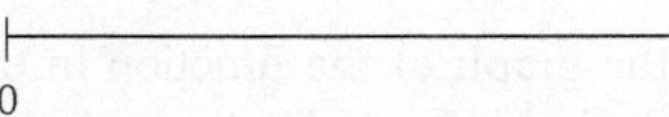

Figure 19 *A number line showing the function's domain with all discontinuities and critical numbers marked*

❹–❻ Choosing 1 as a test point and evaluating f', or simply reasoning that because square roots are always nonnegative, the derivative is always positive on $(0, \infty)$, allows us to finish the chart:

In order for the derivative to change signs, there must be at least two lines in the chart (at least two intervals). Any time the chart has only one line, there cannot be a local extremum.

interval	sign of f'	inc/dec	local extrema
$(0, \infty)$	$+$	increasing	

The function is increasing on $(0, \infty)$ and there are no local extrema on the graph of this function. ■

Reading Exercise 10 *Find the local extrema of $f(x) = x^2$.*

Ans. to reading exercise 9

interval	f'	inc/dec	loc extrema
$(-\infty, -1)$	+	inc	
			loc max at $x = -1$
$(-1, 1)$	−	dec	
			loc min at $x = 1$
$(1, 5)$	+	inc	
			no loc extremum
$(5, \infty)$	+	inc	

Exercises 3.3

1–8. Rapid response: finish the chart. Assume that f is continuous at all real numbers.

1.

interval	sign of f'	inc/dec	local extrema
$(-\infty, -3)$	+		
$(-3, 0)$	−		
$(0, 6)$	−		
$(6, \infty)$	+		

2.

interval	sign of f'	inc/dec	local extrema
$(-\infty, -5)$	−		
$(-5, -3)$	−		
$(-3, 1)$	+		
$(1, \infty)$	−		

3.

interval	sign of f'	inc/dec	local extrema
$(-\infty, 4)$	+		
$(4, 7)$	−		
$(7, \infty)$	+		

4.

interval	sign of f'	inc/dec	local extrema
$(-\infty, 12)$	−		
$(12, 47)$	+		
$(47, \infty)$	−		

5.

interval	sign of f'	inc/dec	local extrema
$(-\infty, 2)$	+		
$(2, \infty)$	+		

6.

interval	sign of f'	inc/dec	local extrema
$(-\infty, \infty)$	+		

7.

interval	sign of f'	inc/dec	local extrema
$(-\infty, 4.3)$	+		
$(4.3, 2\pi)$	−		
$(2\pi, \infty)$	−		

8.

interval	sign of f'	inc/dec	local extrema
$(-\infty, -22)$	−		
$(-22, \sqrt{11})$	+		
$(\sqrt{11}, \infty)$	+		

9–16. On which intervals is the function increasing? On which intervals is the function decreasing?

9.

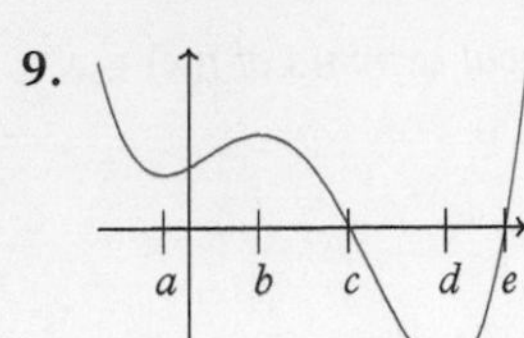

10.

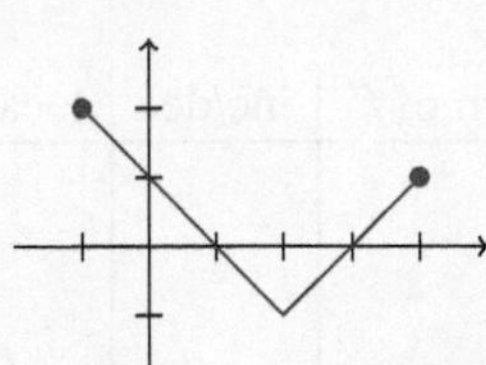

11.

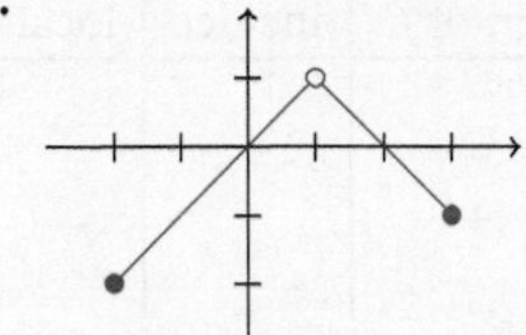

12.

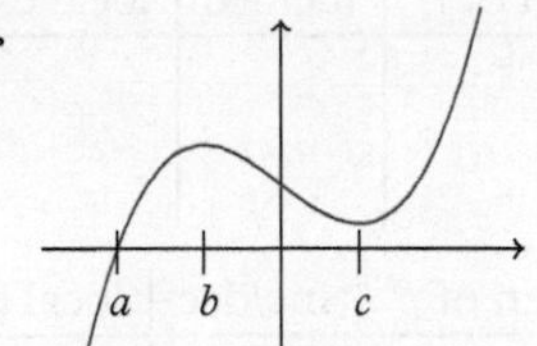

13.

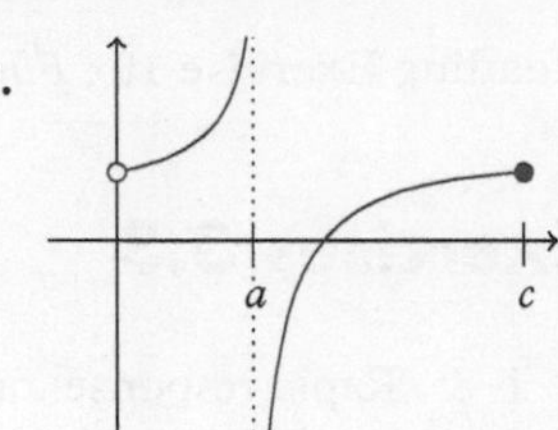

14.

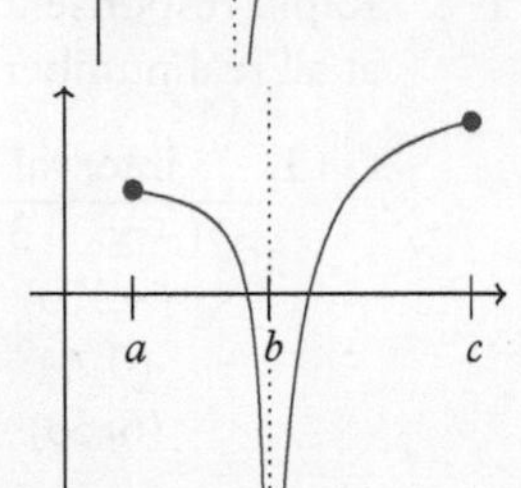

15.

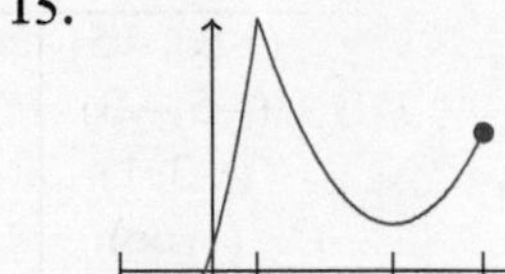

16.

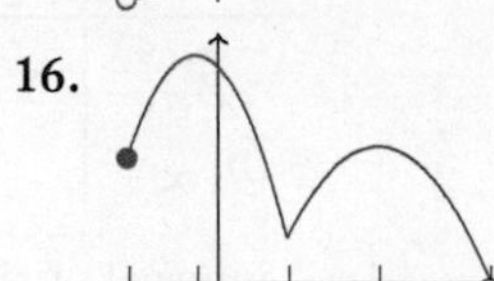

Ans. to reading exercise 10

interval	f'	inc/dec	local extrema
$(-\infty, 0)$	$-$	dec	
			local min at $x = 0$
$(0, \infty)$	$+$	inc	

The function has a local min of 0 at $x = 0$.

17–20. Finish the chart using the additional information provided.

17. The function f is continuous at $x = 4$ but is undefined at $x = 7$.

interval	sign of f'	inc/dec	local extrema
$(0, 4)$	$+$		
$(4, 7)$	$-$		
$(7, \infty)$	$+$		

18. The function f has domain $(0, 4] \cup (7, \infty)$.

interval	sign of f'	inc/dec	local extrema
$(0, 4)$	$+$		
$(7, \infty)$	$-$		

19. The function f is differentiable at $x = 0$ and is continuous but not differentiable at $x = 5$.

interval	sign of f'	inc/dec	local extrema
$(-\infty, 0)$	$+$		
$(0, 5)$	$-$		
$(5, \infty)$	$+$		

20. The function f has an infinite discontinuity at $x = -3$ and a jump discontinuity at $x = 2$.

interval	sign of f'	inc/dec	local extrema
$(-\infty, -3)$	$-$		
$(-3, 2)$	$+$		
$(2, \infty)$	$-$		

21–30. On which intervals is the function increasing? On which intervals is the function decreasing?

It is implied that these exercises should be worked using the five-step method illustrated in this section rather than by guessing from a graphing calculator.

21. $f(x) = \dfrac{x^3}{3} + \dfrac{5x^2}{2} + 4x + 9$

22. $f(x) = \dfrac{x^3}{3} + x^2 - 3x + 7$

23. $h(x) = x^4 - 4x^3 + 7$

24. $g(x) = 15x^4 - 80x^3 - 13$

25. $f(x) = x^3 - 15x - 8$

26. $f(x) = \dfrac{2}{3}x^3 - \dfrac{7}{2}x^2 - 4x + 7$

27. $f(x) = \sin x,\ 0 \le x \le 2\pi$

28. $f(x) = \cos x,\ -\pi \le x \le \pi$

29. $f(x) = \tan x,\ -\pi \le x \le \pi$

30. $s(t) = \sin 2x,\ 0 \le x \le \pi$

31–38. Determine the locations of all local extrema of the given function.

It is implied that these exercises, and those of the remainder of this section, should be worked using the six-step method illustrated in this section rather than by guessing from a graphing calculator.

31. $f(x) = \dfrac{x^3}{3} - \dfrac{x^2}{2} - 6x + 11$

32. $f(x) = \sqrt{3} + 10x + \dfrac{3x^2}{2} - \dfrac{x^3}{3}$

33. $m(x) = 10x^{3/2} - 30x$

34. $s(t) = 4t^5 - 20t^3 + 1$
35. $g(t) = \dfrac{t}{t-3}$
36. $h(x) = \sqrt{3x+1}$
37. $f(x) = \sqrt{9-x^2}$
38. $g(t) = \dfrac{t^2}{t-3}$

39–46. Determine the local extrema of the given function.

39. $g(x) = 15x^4 - 20x^3 + 7$
40. $f(t) = 5t^4 - 40t^3 + 90t^2 + 5$
41. $f(x) = 7 - 3x - 5x^2 - x^3$
42. $f(x) = x + \dfrac{4}{x}$
43. $f(x) = \sqrt{x^2-9}$
44. $s(t) = 2t + \sin t$
45. (GSL) $f(x) = (2+x)^{1/3}(1-x)^{2/3}$
46. (GSL) $f(x) = \dfrac{x+2}{x^2+2x+4}$

47–54. Determine the local extreme points of the given function.

47. $f(t) = 5t^4 + 20t^3 + 12$
48. $g(x) = 2x^3 - 9x^2 - 60x + 4$
49. $k(x) = 15x^4 - 120x^2 + 1$
50. (GSL) $f(x) = 3x^4 - 4x^3 - 12x^2$
51. (GSL) $f(x) = \dfrac{x^2+x+4}{x+1}$
52. (GSL) $f(x) = \dfrac{x^2+x+4}{x^2+2x+4}$
53. (GSL) $f(x) = x^2 + \dfrac{16}{x^2}$
54. (GSL) $f(x) = (2x-6)^{1/3}(x-6)^{2/3}$

Concavity

3.4

The graph of $y = ax^2 + bx + c$ (where $a \neq 0$) is a parabola, and if $a > 0$, the parabola opens upward; if $a < 0$, the parabola opens downward. The nature of these curved shapes, specifically the direction in which the shape is curved, can be explored using derivatives.

Concavity and derivatives

Figure 1 shows several snapshots of a tangent line to the curve $f(x) = x^2 - 3x + 5$. Notice that all the tangent lines lie below the curve. Where this occurs, the curve is called *concave up*.

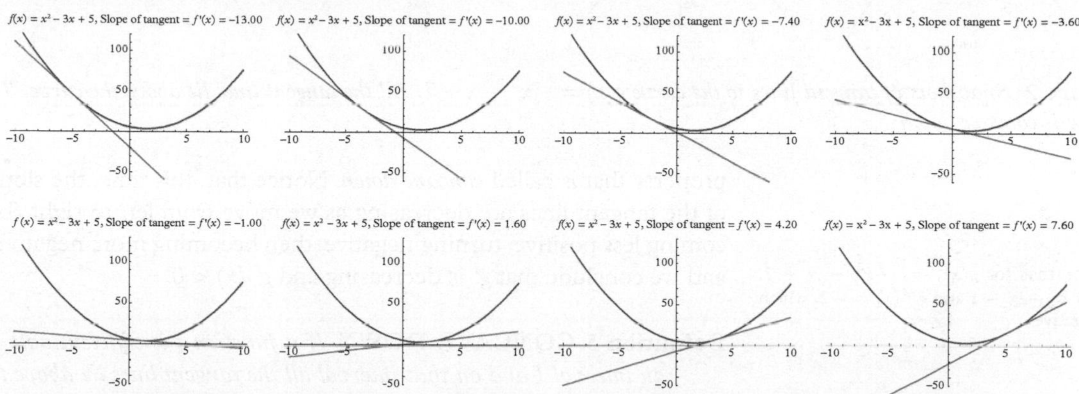

Figure 1 *Snapshots of tangent lines to the curve* $f(x) = x^2 - 3x + 5$. *All the tangent lines lie beneath the curve. The curve is concave up*

Definition 4 CONCAVE UP *If a function f is differentiable on an interval I and on that interval all the tangent lines lie below the curve, the function is* concave up *on I.*

Return to figure 1 and consider the succession of pictures. As the point of tangency moves from left to right, the slopes of the tangent lines are increasing. In the top row, the slope of the tangent line is becoming less negative; in the bottom row, the slope of the tangent

The phrase "the tangent line at $x = a$ lies below the curve" means that for all x in some open interval containing a, except for $x = a$,

$$f(a) + f'(a)(x - a) < f(x).$$

line turns positive and then becomes more positive. The slope of the tangent line is given by the derivative. Therefore, as we move to the right, the quantity $f'(x)$ is increasing. When a function is increasing, its derivative, $f''(x)$ in this case, is positive. We may therefore characterize concave up by equating this notion with the second derivative of the function being positive.

Notice that for $f(x) = x^2 - 3x + 5$, $f'(x) = 2x - 3$ and $f''(x) = 2$, which is positive.

Figure 2 shows several snapshots of a tangent line to the curve $g(x) = -x^2 - x + 7$. Here, the tangent lines lie above the curve, a

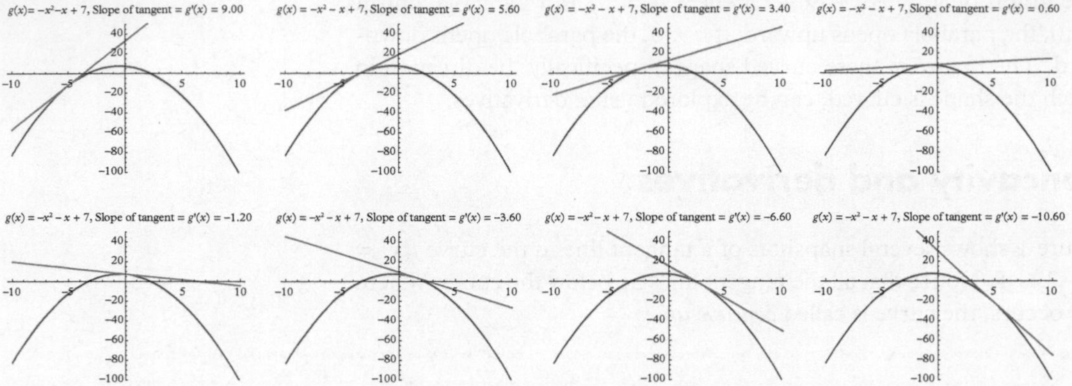

Figure 2 *Snapshots of tangent lines to the curve* $g(x) = -x^2 - x + 7$. *All the tangent lines lie above the curve. The curve is concave down*

property that is called *concave down*. Notice that, this time, the slopes of the tangent lines are decreasing as we move from left to right (becoming less positive, turning negative, then becoming more negative), and we conclude that g' is decreasing and $g''(x) < 0$.

Notice that for $g(x) = -x^2 - x + 7$, $g'(x) = -2x - 1$ and $g''(x) = -2$, which is negative.

Definition 5 CONCAVE DOWN *If a function f is differentiable on an interval I and on that interval all the tangent lines lie above the curve, the function is* concave down *on I.*

Theorem 9 CONCAVITY TEST

(a) If $f''(x) > 0$ *on an interval I, then f is concave up on I.*

(b) If $f''(x) < 0$ *on an interval I, then f is concave down on I.*

If $f''(x) = 0$ on an interval, then the function is not curved up or down, but rather is a straight line on that interval.

The graph of a function may change from an "up shape" to a "down shape" or vice versa; this is called a *change in concavity*. The point at which the change occurs is called an *inflection point*. Figure 3 shows the appearance of a change in concavity at an inflection point. We require the tangent line to exist at an inflection point, so that the tangent line lies above the curve on one side of the point of tangency but below the curve on the other side (see figure 4).

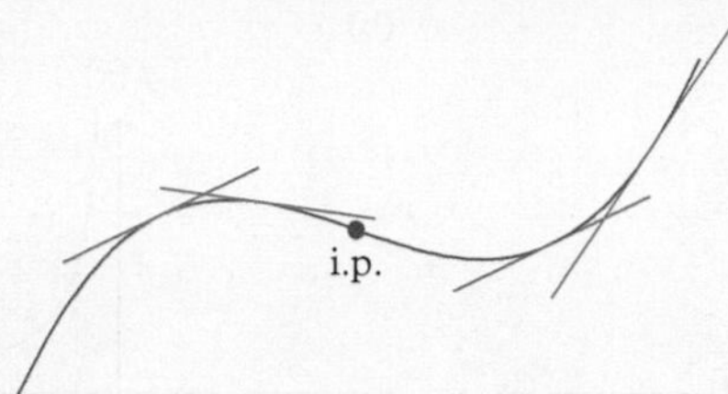

Figure 3 *A function with an inflection point, marked by the blue dot and the abbreviation i.p. To the left of the inflection point, the curve is concave down and tangent lines lie above the curve; to the right of the inflection point, the curve is concave up and the tangent lines lie below the curve*

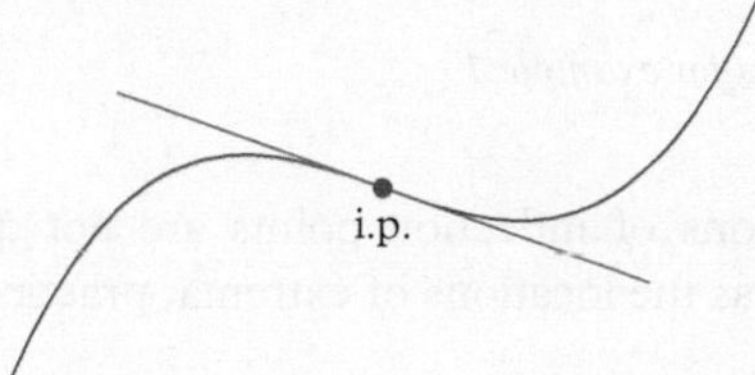

Figure 4 *A function with an inflection point, marked by the blue dot and the abbreviation i.p. The tangent line at the inflection point lies above the curve to the left of the inflection point, but below the curve to its right. The visual effect is an emphasis of the concave-down shape to the left and the concave-up shape to the right of the inflection point*

Definition 6 INFLECTION POINT *An* inflection point *is a point on a curve where the tangent line exists, but for which the curve is concave up on one side of the point and concave down on the other side of the point.*

We do not distinguish between a change from concave up to concave down and a change from concave down to concave up. Either is called a *point of inflection*.

Example 1 *For each function in figure 5, on which intervals is the function concave up? On which intervals is the function concave down? State the locations of any inflection points.*

Solution (a) The function is concave down on $(0, a)$ and concave up on (a, b). There is an inflection point located at $x = a$ because the concavity changes at a point on the graph where the tangent line exists (no corner).

Just as with extrema, the word *at* indicates location, whereas a point requires two coordinates; the inflection point is $(a, 0)$.

(b) The function is concave down on $(-\infty, 0)$ and concave up on $(0, \infty)$. Although the concavity changes at $x = 0$, there is no point at $x = 0$ and therefore no inflection point at $x = 0$. Alternately, because the function is not continuous at $x = 0$, it can have no tangent line at $x = 0$ and therefore no inflection point there. ∎

To be concave down, a function need not have a local maximum; to be concave up a function need not have a local minimum. At issue is whether the tangent lines lie above or below the curve, and this does not require the presence of a horizontal tangent line.

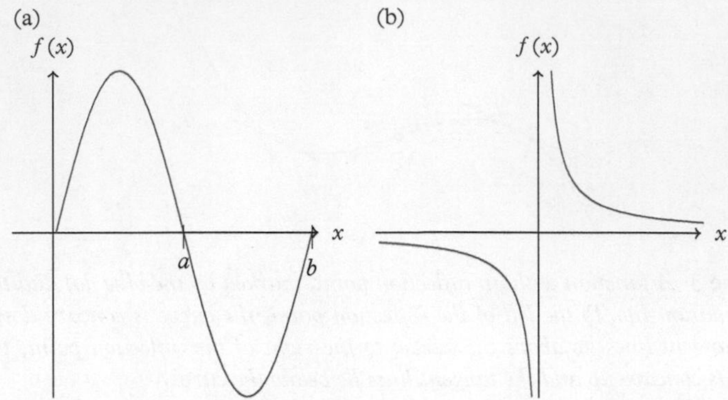

Figure 5 *The graphs for example 1*

Although locations of inflection points are not quite as easy to determine visually as the locations of extrema, practice helps.

Reading Exercise 11 *At which of a, b, or c is the inflection point located? (See figure 6.)*

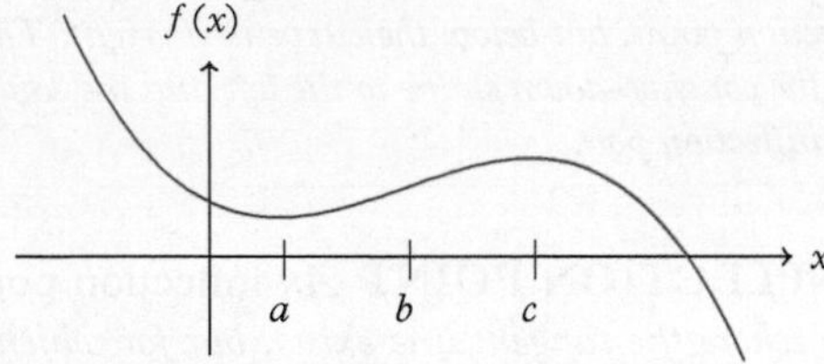

Figure 6 *The function for reading exercise 11*

A change in concavity can occur without the presence of an inflection point. If a function is concave up on one side of a corner and concave down on the other side, then concavity changes but there is no tangent line and therefore no point of inflection at the point of change in concavity:

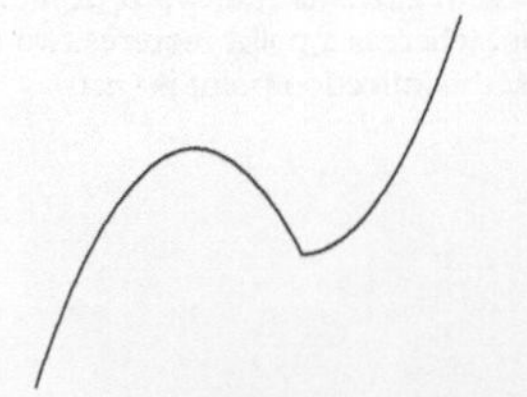

Because the derivative $f'(x)$ is undefined at a corner, so is $f''(x)$. Therefore, it is still true that at a change in concavity, either $f''(x) = 0$ or $f''(x)$ is undefined.

Where might an inflection point occur? Suppose f has an inflection point at $x = a$. Because the tangent line must exist, the function must be differentiable there—that is, $f'(a)$ must exist. Because the tangent line cannot be entirely above or entirely below the curve, $f''(a)$ can be neither positive nor negative. Therefore, either $f''(a) = 0$ or $f''(a)$ fails to exist. This should sound familiar. It is analogous to a critical number, but with the second derivative instead of the first derivative. Theorem 10 plays the role of Fermat's theorem, just as the concavity test plays the role of theorem 7.

Theorem 10 POSSIBLE INFLECTION LOCATIONS *If f has a point of inflection at $x = a$, then either $f''(a) = 0$ or $f''(a)$ does not exist.*

The procedure for finding intervals of concavity is therefore the same as the procedure for finding intervals of increase and decrease, except that we use the second derivative instead of the first. The procedure for identifying locations of points of inflection is likewise like that for identifying local extrema.

Examples of determining concavity and inflection points

The six-step procedure modified to determine concavity and inflection is as follows:

❶ Find the domain of f.

❷ Determine where $f''(x) = 0$ or where $f''(x)$ is undefined.

Values where $f''(x)$ is undefined include locations of discontinuities and corners.

❸ Determine intervals by using the domain and by using the values found in step ❷ as endpoints.

❹ Determine the sign of f'' on each interval (positive or negative).

❺ Determine concavity from the sign of the second derivative (concavity test).

❻ Determine locations of points of inflection using definition 6.

A chart is a convenient way to record the information from steps ❸–❻.

Example 2 *Determine intervals of concavity and locations of inflection points for* $f(x) = \dfrac{x^3}{3} - \dfrac{x^2}{2} - 2x + 5$.

Solution ❶ Because f is a polynomial, its domain is $(-\infty, \infty)$ and it is continuous and there are no corners.

Ans. to reading exercise 11
b

❷ Next we find the second derivative, set it equal to zero, and solve for x:

We also need to know values where f'' is undefined, but there are no such values for a polynomial.

$$f'(x) = x^2 - x - 2$$

$$f''(x) = 2x - 1$$

$$2x - 1 = 0$$

$$2x = 1$$

$$x = \frac{1}{2}.$$

There is one possible inflection location, $x = \frac{1}{2}$.

CAUTION: MISTAKE TO AVOID

Do not call the numbers from step ❷ inflection points (or locations of inflection points). Wait until you know whether concavity changes; determination of inflection points is done in step ❻.

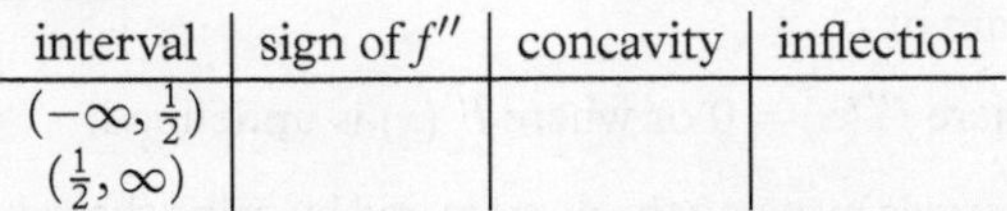

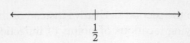

Figure 7 *A number line with all discontinuities, corners, and possible inflection locations marked*

❸ As in the previous section, determining the intervals can be visualized with a number line. We draw the function's domain and mark all the locations determined in step ❷ (figure 7).

The intervals are $(-\infty, \frac{1}{2})$ and $(\frac{1}{2}, \infty)$. We begin our chart with the intervals.

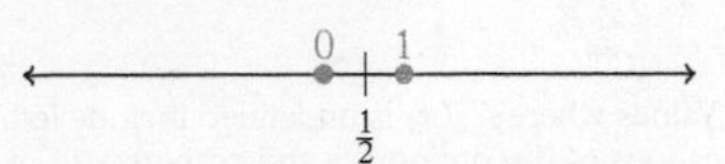

Figure 8 *Convenient test points for example 2*

interval	sign of f''	concavity	inflection
$(-\infty, \frac{1}{2})$			
$(\frac{1}{2}, \infty)$			

❹ Once again we choose a test point in each interval; 0 and 1 are convenient (figure 8).

Evaluating f'' at the test points reveals the sign of f'' in each interval:

Because f'' is continuous where it is defined (as are all the functions we will use), the only places it can change sign are where it is zero or undefined, which are the endpoints of the intervals. Therefore, for each interval, either f'' is positive throughout the interval or f'' is negative throughout the interval. Knowing the sign of f'' at a test point tells us the sign of f'' throughout the interval.

$$f''(0) = -1 < 0$$

$$f''(1) = 1 > 0.$$

interval	sign of f''	concavity	inflection
$(-\infty, \frac{1}{2})$	$-$		
$(\frac{1}{2}, \infty)$	$+$		

❺ Using the concavity test (theorem 9), f is concave down on $(-\infty, \frac{1}{2})$ and concave up on $(\frac{1}{2}, \infty)$.

interval	sign of f''	concavity	inflection
$(-\infty, \frac{1}{2})$	$-$	down	
$(\frac{1}{2}, \infty)$	$+$	up	

The tangent line must exist at an inflection point; but, for a polynomial, the derivative (and therefore the tangent line) always exists.

❻ Because the concavity changes at $x = \frac{1}{2}$, there is an inflection point at $x = \frac{1}{2}$.

Just as with extrema in the increasing/decreasing chart, the lines identifying inflection points are written in between the lines containing the other information.

interval	sign of f''	concavity	inflection
$(-\infty, \frac{1}{2})$	$-$	down	
			$\leftarrow$ i.p. at $x = \frac{1}{2}$
$(\frac{1}{2}, \infty)$	$+$	up	

If we were asked for inflection points (instead of just their locations), we would need to find the y-coordinate of the inflection point. ■

Example 3 *Determine intervals of concavity and find all inflection points for $y = x^{5/3} + 4$.*

Solution ❶ This function involves an odd root, so it is defined everywhere; the domain is $(-\infty, \infty)$.

❷ The derivatives are

$$y' = \frac{5}{3}x^{2/3},$$
$$y'' = \frac{10}{9}x^{-1/3} = \frac{10}{9\sqrt[3]{x}}.$$

Tip: rewriting expressions involving negative exponents as fractions is helpful.

We set both the numerator and the denominator of y'' equal to zero and solve:

$$10 = 0 \qquad\qquad 9\sqrt[3]{x} = 0$$
$$\text{no solutions} \qquad\qquad \sqrt[3]{x} = 0$$
$$x = 0.$$

Setting the numerator equal to zero finds values where $y''(x) = 0$, whereas setting the denominator equal to zero finds values of x where $y''(x)$ is undefined.

There is one possible inflection location, $x = 0$.

❸ The intervals are $(-\infty, 0)$ and $(0, \infty)$.

❹–❺ Recalling that the cube root of a negative number is negative and the cube root of a positive number is positive, then using the concavity test, we can complete these steps.

An alternative to this reasoning is to choose test points such as 1 and -1 and evaluate y'' at the test points.

interval	sign of y''	concavity	inflection
$(-\infty, 0)$	$-$	down	
$(0, \infty)$	$+$	up	

The first part of our task is complete. The function is concave down on $(-\infty, 0)$ and concave up on $(0, \infty)$.

❻ Concavity changes at $x = 0$, but to conclude there is an inflection point at $x = 0$, we need to check that the tangent line exists. Because $y'(0)$ exists, the tangent line exists.

In the event that f is defined at $x = c$ but $f'(c)$ results in division by zero, if concavity changes at $x = c$ there may still be a vertical tangent line at $x = c$ and, by our definition, there is still an inflection point at $x = c$.

interval	sign of y''	concavity	inflection
$(-\infty, 0)$	$-$	down	
			← i.p. at $x = 0$
$(0, \infty)$	$+$	up	

Because we wish to find the inflection points, not just their locations, we calculate

$$y(0) = 0^{5/3} + 4 = 4$$

and conclude that the inflection point is the point $(0, 4)$. ■

Notice that we used each of y, y', and y'' at various times during the solution to example 3. Thoroughly understanding the role of each function is crucial to success without stress.

Example 4 *Determine intervals of concavity and locations of inflection points for* $g(x) = x^4$.

CAUTION: MISTAKE TO AVOID
Saying there is an inflection point at $x = 0$ is incorrect. See step ❻.

Solution ❶ Because g is a polynomial, its domain is $(-\infty, \infty)$.

❷ We calculate $g'(x) = 4x^3$ and $g''(x) = 12x^2$. Setting $g''(x) = 0$ and solving for x yields the possible inflection location $x = 0$.

❸ The intervals are $(-\infty, 0)$ and $(0, \infty)$.

❹–❺ Notice that $g''(x) = 12x^2$ is positive (except at $x = 0$).

interval	sign of g''	concavity	inflection
$(-\infty, 0)$	$+$	up	
$(0, \infty)$	$+$	up	

We conclude that g is concave up on $(-\infty, 0)$ and on $(0, \infty)$.

Because the tangent line at $x = 0$ lies below the curve, we can combine these two intervals and say that g is concave up on $(-\infty, \infty)$. But, caution should be exercised in combining adjacent intervals with the same concavity, for they cannot be combined across a discontinuity or a corner (the definitions require differentiability throughout the interval).

❻ Because concavity does not change at $x = 0$, there is no inflection point. If desired, this can be denoted in the chart as follows:

interval	sign of g''	concavity	inflection
$(-\infty, 0)$	$+$	up	$\leftarrow$ no i.p.
$(0, \infty)$	$+$	up	

■

Reading Exercise 12 *Finish the following chart to determine where f has inflection points (assume f is differentiable at all real numbers):*

interval	*sign of f''*	*concavity*	*inflection points*
$(-\infty, 0)$	$+$		
$(0, 4)$	$-$		
$(4, \infty)$	$-$		

Second derivative test

Suppose a differentiable function f has a horizontal tangent line at $x = c$ and is concave up at $x = c$ (figure 9, left). Then, f must have a local minimum at $x = c$. (The tangent line $y = f(c)$ lies below the curve near

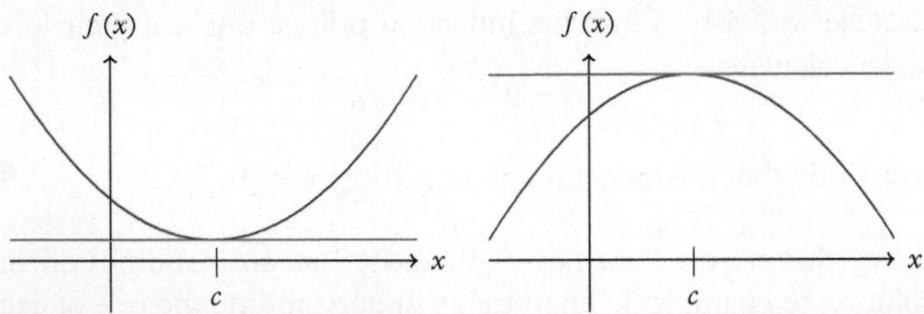

Figure 9 *(left) A function concave up with a horizontal tangent line at $x = c$ has a local minimum at $x = c$; (right) a function concave down with a horizontal tangent line at $x = c$ has a local maximum at $x = c$*

$x = c$, making $f(c)$ the lowest y-coordinate on the curve near $x = c$.) If

f is concave down at $x = c$ instead, then f must have a local maximum there (figure 9, right).

Recalling that a positive second derivative indicates concave up and a negative second derivative indicates concave down yields the next theorem.

CAUTION: MISTAKE TO AVOID
It is easy to switch max/min because we are associating negative with max and positive with min. Mentally visualizing the pictures of figure 9 every time the second derivative test is used can help avoid errors.

Theorem 11 SECOND DERIVATIVE TEST *Suppose f is a function with a second derivative that is continuous on an open interval containing c.*

(a) If $f'(c) = 0$ and $f''(c) > 0$, then f has a local minimum at $x = c$.
(b) If $f'(c) = 0$ and $f''(c) < 0$, then f has a local maximum at $x = c$.

Although it is not always conclusive, the second derivative test is often a quicker method for determining the local extrema of a function. But, because it only refers to critical numbers of the type $f'(x) = 0$, the second derivative test cannot be used to test critical numbers of the type $f'(x)$ undefined. The procedure can be summarized as follows:

CAUTION: MISTAKE TO AVOID
This second derivative test says nothing about what happens when both $f'(c) = 0$ and $f''(c) = 0$. In this case, the test is inconclusive; in fact, three possibilities (local min, local max, and no local extrema) still remain. It is erroneous to conclude there is no local extremum at $x = c$ using this second derivative test.

❶ Determine the critical numbers of f. If there is a critical number of the type $f'(x)$ undefined, stop; the second derivative test cannot be used on such a critical number.

❷ At each critical number c, find $f''(c)$ and use the second derivative test to determine extrema.

Because polynomials are differentiable everywhere, they are good candidates for trying the second derivative test.

Example 5 *Find the local extrema of* $f(x) = \frac{x^3}{3} - \frac{x^2}{2} - 2x + 5$.

Solution ❶ We begin by finding the critical numbers. We calculate $f'(x) = x^2 - x - 2$, then find its zeros:

The **zeros** of a function g are the values of x for which $g(x) = 0$. "Find the zeros of f'" is a shortcut phrase for setting $f'(x) = 0$ and solving for x.

$$x^2 - x - 2 = 0$$
$$(x-2)(x+1) = 0$$
$$x = 2, x = -1.$$

The critical numbers are $x = 2$ and $x = -1$.

❷ We calculate the second derivative

$$f''(x) = 2x - 1,$$

and then evaluate f'' at each critical number:

$$f''(2) = 2(2) - 1 = 3 > 0, \text{ concave up, min}$$

$$f''(-1) = 2(-1) - 1 = -3 < 0, \text{ concave down, max.}$$

CAUTION: MISTAKE TO AVOID
The local minimum and maximum values of f are not 3 and -3. To find the local min and max values, we must calculate $f(2) = \frac{5}{3}$ and $f(-1) = \frac{37}{6}$.

We conclude that f has a local minimum at $x = 2$ and a local maximum at $x = -1$. ■

The second derivative test is quicker than the first derivative test, but the first derivative test applies to a much wider variety of functions. For instance, the second derivative test can be used for section 3.3 example 4, but not for section 3.3 examples 5 and 6. When the second derivative test is inconclusive, such as for $f(x) = x^3$, $g(x) = x^4$, and $h(x) = -x^4$ at $x = 0$, the first derivative test is still needed to give the answers of no extremum, a local minimum, and a local maximum, respectively.

Ans. to reading exercise 12

interval	concavity	infection points
$(-\infty, 0)$	up	
		$\leftarrow$ i.p. at $x = 0$
$(0, 4)$	down	
$(4, \infty)$	down	

Reading Exercise 13 *Given $f'(4) = 0$ and $f''(4) > 0$ (and that the other hypotheses of the second derivative test are met), does f have a local min or a local max at $x = 4$?*

Exercises 3.4

1–4. Rapid response: finish the chart. Assume that f is differentiable at all real numbers.

1.

interval	sign of f''	concavity	inflection points
$(-\infty, -3)$	+		
$(-3, 0)$	$-$		
$(0, 6)$	$-$		
$(6, \infty)$	+		

2.

interval	sign of f''	concavity	inflection points
$(-10, -5)$	$-$		
$(-5, -3)$	$-$		
$(-3, 1)$	+		
$(1, \infty)$	$-$		

3.

interval	sign of f''	concavity	inflection points
$(-\infty, 4)$	+		
$(4, 7)$	$-$		
$(7, 54)$	+		

4.

interval	sign of f''	concavity	inflection points
$(-\infty, 12)$	$-$		
$(12, 47)$	$+$		
$(47, \infty)$	$-$		

5–12. On which intervals is the function concave up? On which intervals is the function concave down? Identify the locations of inflection points.

5.

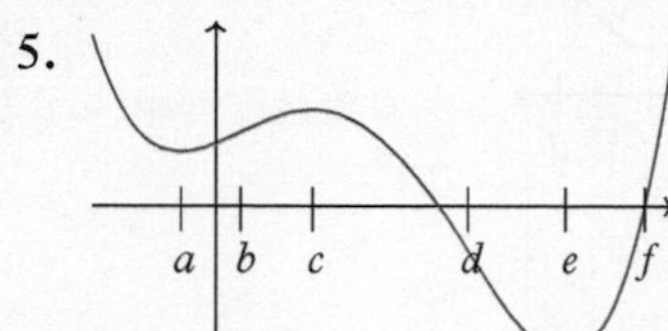

6.

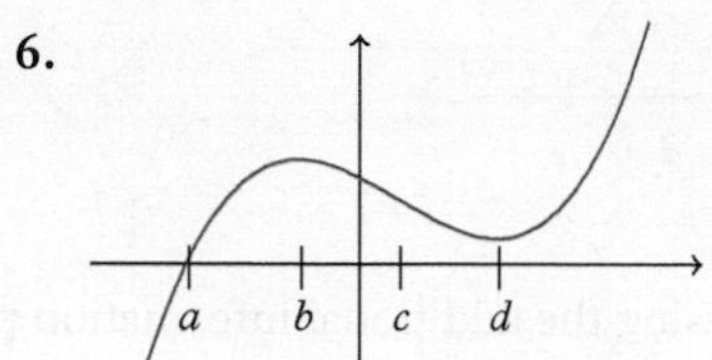

7.

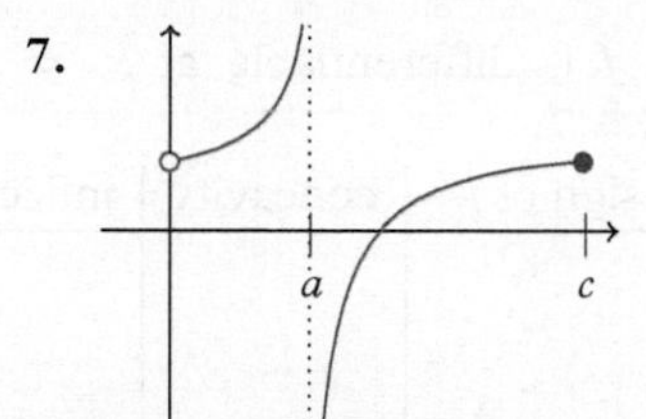

8.

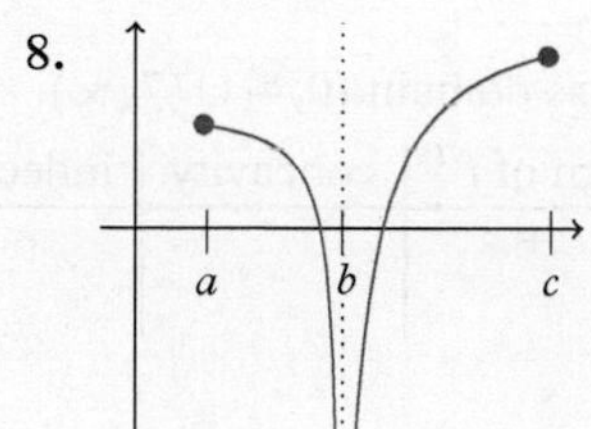

9.

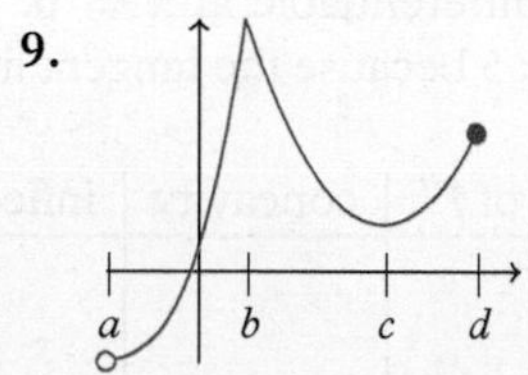

10.

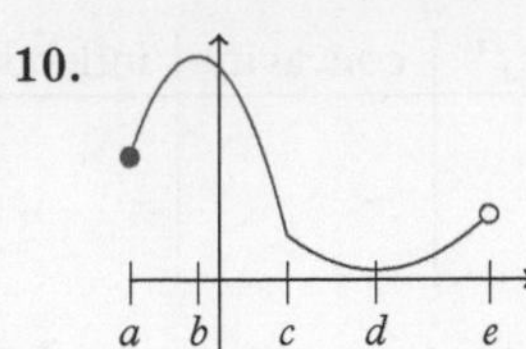

11.

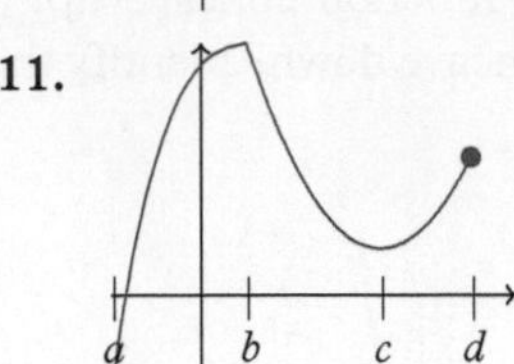

12.

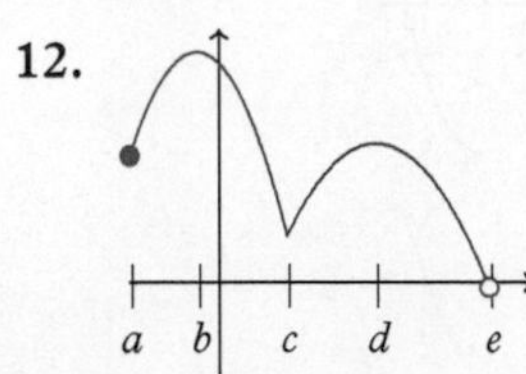

Ans. to reading exercise 13
Yes, f has a local min at $x = 4$.

13–16. Finish the chart using the additional information provided.

13. The function f is differentiable at $x = 4$, but is undefined at $x = 7$.

interval	sign of f''	concavity	inflection points
$(0, 4)$	+		
$(4, 7)$	−		
$(7, \infty)$	+		

14. The function f has domain $(0, 4] \cup (7, \infty)$.

interval	sign of f''	concavity	inflection points
$(0, 4)$	+		
$(7, \infty)$	−		

15. The function f is differentiable at $x = 0$, but is not differentiable at $x = 5$ because the tangent line there is vertical.

interval	sign of f''	concavity	inflection points
$(-\infty, 0)$	+		
$(0, 5)$	−		
$(5, \infty)$	+		

16. The function f has a corner at $x = -3$ and a jump discontinuity at $x = 2$.

interval	sign of f''	concavity	inflection points
$(-\infty, -3)$	$-$		
$(-3, 2)$	$+$		
$(2, \infty)$	$-$		

17–24. Determine intervals of concavity and locations of inflection points for the given function. Use the methods of this section, even for functions with graphs that should be on instant recall.

Q. What does a prison dentist fill?
A. A con cavity.

17. $f(x) = \sqrt{x}$

18. $f(x) = \dfrac{1}{x^2}$

19. $g(x) = \sqrt[3]{x}$

20. $s(t) = \cos t,\ 0 \le t \le 2\pi$

21. $f(t) = \sin t,\ -\pi \le t \le \pi$

22. $f(x) = \tan x$ on $(-\frac{\pi}{2}, \frac{\pi}{2})$

23. $f(x) = \sec x$ on $(-\frac{\pi}{2}, \frac{\pi}{2})$

24. $g(y) = y^{2/7}$

25–34. Determine intervals of concavity and locations of inflection points for the given function.

It is implied that the methods of this section should be used and that guesses from a graphing calculator are not acceptable.

25. $f(x) = 2x^3 - 4x^2 + 9x - 11$

26. $f(t) = -t^3 + 2t^2 - 1$

27. $s(t) = 3t^4 - t^3 + 4t^2$

28. $g(x) = x^4 + x^3 - 9x^2 + 3x + 7$

29. $g(x) = 3x^5 - 5x^4 + 7x$

30. $f(x) = x^5 + 5x^4 + 12x - 9$

31. $h(x) = \sqrt{9 - x^2}$

32. $f(x) = \dfrac{x}{x-5}$

33. $f(x) = x^2 - \dfrac{1}{x}$

34. $s(t) = -\sin x - \dfrac{x^2}{4}$

35–38. Determine the intervals on which the function is concave up, the intervals on which the function is concave down, and the inflection points of the given function.

35. $g(x) = x^5 - 10x^4 + 30x^3 + 5x + \sqrt{3}$

36. $f(x) = 4x + 8x^3 + x^4$

37. $f(x) = 4x^{5/2} - 15x^2$

38. $g(x) = 15\sqrt{x} + x^{5/2}$

39–42. Use the second derivative test on the given critical numbers. (Assume the critical numbers are of type $f'(c) = 0$; answer local max, local min, or inconclusive).

39. $f''(x) = 2x - 5$, critical numbers $x = 0$ and $x = 5$

40. $f''(x) = 2x - 1$, critical numbers $x = -2$ and $x = 3$

41. $f''(x) = -8x^{-3}$, critical numbers $x = \pm\sqrt{2}$
42. $f''(x) = -\sin x$, critical numbers $x = \frac{\pi}{3}$ and $x = \frac{2\pi}{3}$

43–52. Use the second derivative test to find the local extreme points of the given function.

43. $f(x) = x^3 + 3x^2 - 2$
44. $y = x^3 - 3x + 4$
45. $y = 2 + 12x + 3x^2 - 2x^3$
46. $f(x) = 3x - 2x^2 - \dfrac{4x^3}{3}$
47. (GSL) $g(x) = 3x^4 - 4x^3 - 12x^2 + 2$
48. $f(x) = x^4 - 4x^2 + 4$
49. $m(x) = 10x^{3/2} - 30x$
50. (GSL) $g(t) = \dfrac{t^2}{t-3}$
51. (GSL) $f(x) = x^2 + \dfrac{16}{x}$
52. $f(x) = \sin x$, $0 \le x \le 2\pi$

53–55. If $y = f(x)$ represents the national debt of a country, then the deficit (or surplus) represents the rate of change of the debt. Answer the following questions about the graph of $y = f(x)$ in terms of the deficit (or surplus), when appropriate.

53. What is the significance of an inflection point?
54. What is the significance of a local maximum?
55. What is the significance of an x-intercept?

Hint: playing with powers of x can help.

56–58. The second derivative test can be extended to apply to the first nonzero derivative. Make conjectures regarding what we can conclude about the critical number c under the following circumstances.

56. $f'(c) = 0, f''(c) = 0, f'''(c) = 0, f^{(4)}(c) > 0$
57. $f'(c) = 0, f''(c) = 0, f'''(c) = 0, f^{(4)}(c) < 0$
58. $f'(c) = 0, f''(c) = 0, f'''(c) \neq 0$

59. Let $y = f(x)$ represent the national debt. Which derivative of f (first, second, ...) is described by the word “slowly” in the following sentence?

The rate of growth in the deficit is increasing more slowly.

Curve Sketching: Polynomials

3.5

Polynomials have domain $(-\infty, \infty)$ and are continuous and differentiable everywhere. With no discontinuities or corners, the graphs are connected and smooth. The process of finding intervals of increase/decrease, local extrema, intervals of concavity, and inflection points is relatively quick compared to many other types of functions, and a rough sketch of the graph of a polynomial can be drawn by hand using that information alone.

It is my opinion that the modern tendency toward almost complete reliance on the graphing calculator to graph functions stunts visual processing development, removing one key tool for understanding calculus.

Turning information into a sketch

In sections 3.3 and 3.4 we learned how to *collect* the necessary information to complete a graph. What is needed now is the ability to turn this information into a sketch of a graph. A procedure that works well is as follows:

1. Decide on appropriate scales for the x- and y-axes, then draw and mark the axes.
2. Plot the local extreme points and points of inflection. Label local maximum points with *max* above the plotted point, local minimum points with *min* below the point, and inflection points with *i.p.* beside or otherwise near the point. Other points may be added if desired (candidates include x- or y-intercepts).
3. Beginning with the left-most inflection point, draw the curve to the left, using the correct concavity (up or down) and passing through any local extremum appropriately. If the tangent line is horizontal at the point of inflection, take this into account. Repeat from the right-most inflection point, drawing to the right, then repeat between each pair of inflection points.
4. Check the graph by verifying that intervals of increase and decrease are represented correctly.

Although y-intercepts are easy to calculate, x-intercepts of polynomials can be difficult or even impossible to find exactly by hand.

Between inflection points, the curve can be drawn left to right or right to left as desired.

Example 1 *By hand, sketch the graph of the polynomial* $y = x^3 - 3x^2 - 24x + 18$*, given the following information:*

critical numbers $x = -2$, $x = 4$

interval	*sign of* f'	*inc/dec*	*local extrema*
$(-\infty, -2)$	+	*increasing*	
			← *local max at* $x = -2$
$(-2, 4)$	−	*decreasing*	
			← *local min at* $x = 4$
$(4, \infty)$	+	*increasing*	

interval	*sign of* f''	*concavity*	*inflection*
$(-\infty, 1)$	−	*down*	
			← *i.p. at* $x = 1$
$(1, \infty)$	+	*up*	

local maximum point $(-2, 46)$, *local minimum point* $(4, -62)$, *inflection point* $(1, -8)$

Solution ❶ Our graph must include the extreme points and the inflection points. Therefore, the y-axis must stretch below -62 and above 46. Perhaps marking the scale by increments of 20 and going from -80 to 60 works.

The x-axis must stretch to the left of -2 and to the right of 4, but by how much? Nice results can usually be obtained by taking the distance between the two left-most of the extreme and inflection points (in this case the distance between -2 and 1, which is 3) and leaving a little more room than this to the left of the left-most of those points (starting, say, 4 units to the left of -2), and doing the same on the right side (distance between 1 and 4 is 3, end 4 units right of 4). Let's mark our x-axis from -6 to 8, marking by increments of 2. See figure 1.

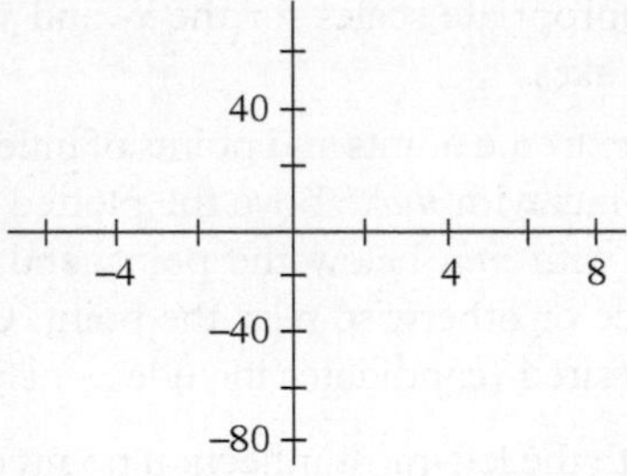

Figure 1 *Scales marked*

Labeling the points is a helpful visual aid.

❷ Next, we plot the extreme points and points of inflection (figure 2).

❸ We are now ready to start drawing the curve. We start at the left-most inflection point, $(1, -8)$, and draw to the left. In other words, we draw the portion of the graph that is for the interval $(-\infty, 1)$. This portion of the graph is concave down. Starting at the inflection point,

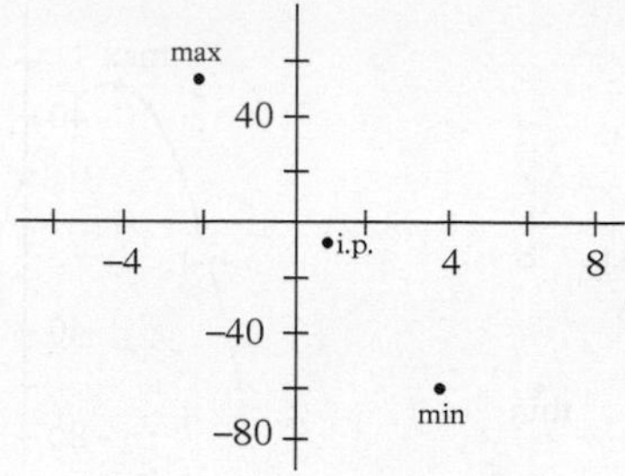

Figure 2 *Extreme points and points of inflection plotted*

we begin drawing toward the local max, but we aim a little high (as if tossing a ball) so that the curve can be drawn concave down; see figure 3 (left). Continuing to the local max and beyond we must keep a concave-down shape (figure 3 (right)).

Tip: drawing the portion of the curve lying to the left of the inflection point all in one motion can help maintain the concave-down shape.

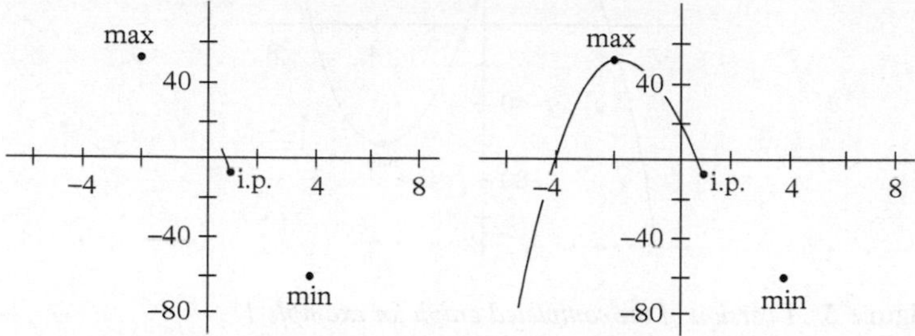

Figure 3 *The portion of the curve to the left of the left-most inflection point is drawn, from right to left. We start toward the local max, aiming high*

Next we draw from the right-most inflection point, $(1, 8)$, to the right. This portion of the curve, over the interval $(1, \infty)$, is concave up. But, because we have already drawn the curve to the other side of the inflection point, we must also be careful to match the slope there in order not to create a corner. We start drawing in the direction mandated by the existing slope (figure 4, left) but then bend the curve to match the required concave-up shape, passing through the local min (figure 4, right).

The graph is complete, although we could consider extending the left and right ends of the graph a little longer so that it is clear that the graph continues in the indicated direction and shape indefinitely (figure 5).

❹ We finish by checking that the intervals of increase and decrease are represented correctly. A check of either figure 4 (right) or figure 5 shows that the function is increasing on $(-\infty, -2)$ and on $(4, \infty)$, and that it is decreasing on $(-2, 4)$, matching the information as given. ■

Notice that we do not refer to increasing and decreasing while actually drawing the graph. Instead, concavity is the driving indicator.

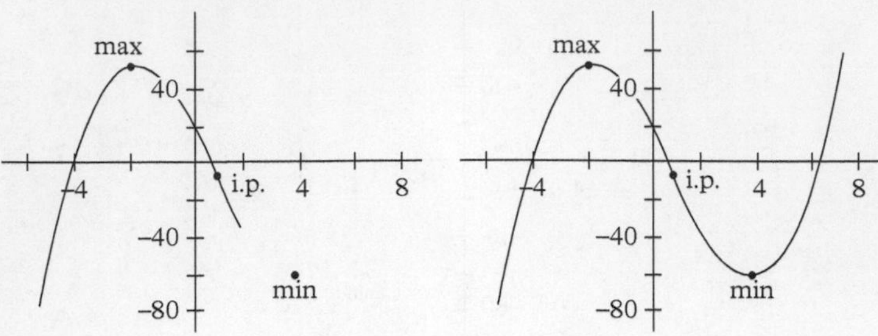

Figure 4 *The portion of the curve to the right of the right-most inflection point is drawn, from left to right. We must avoid creating a corner at the inflection point, so the slope as we begin drawing is already determined*

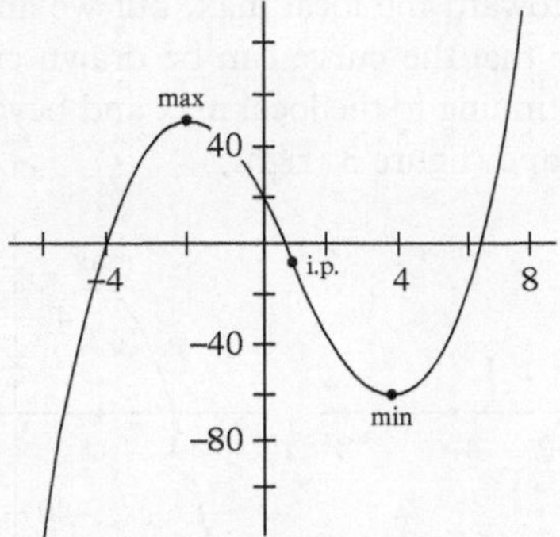

Figure 5 *A version of the completed graph for example 1*

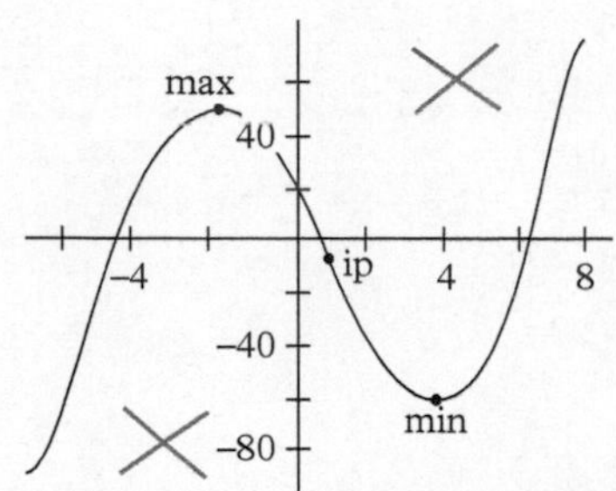

CAUTION: MISTAKE TO AVOID
Failure to maintain the correct concavity at the tips of the graph is a common error. In this drawing for example 1, the graph is (incorrectly) concave up on the far left and (also incorrectly) concave down on the far right.

The polynomial of example 1 has y-intercept 18 and x-intercepts -4.04, 0.703, and 6.34. An examination of figure 5 shows that we got pretty close, even without knowing those intercepts (confession: I had noticed the y-intercept—it's hard not to—and that may have, at least subconsciously, affected the direction in which I began drawing the curve at the beginning of step ❸). Plotting more points certainly increases the accuracy of the graph, but we have a very good by-hand sketch of the curve knowing the coordinates of only three points!

Example 2 *By hand, sketch the graph of the polynomial* $y = x^4 - 12x^3 + 30x^2 - 28x + 11$*, given the following information:*

critical numbers $x = 1$, $x = 7$

interval	*sign of* f'	*inc/dec*	*local extrema*
$(-\infty, 1)$	$-$	*decreasing*	
$(1, 7)$	$-$	*decreasing*	$\leftarrow$ *local min at* $x = 7$
$(7, \infty)$	$+$	*increasing*	

interval	*sign of* f''	*concavity*	*inflection*
$(-\infty, 1)$	$+$	*up*	
			$\leftarrow$ *i.p. at* $x = 1$
$(1, 5)$	$-$	*down*	
			$\leftarrow$ *i.p. at* $x = 5$
$(5, \infty)$	$+$	*up*	

local minimum point $(7, -430)$, *inflection points* $(1, 2)$ *and* $(5, -254)$

Tip: more room is often needed in the y-direction beyond inflection points than extrema.

Solution ❶ Because the y-coordinates of local extrema and inflection points range from -430 to 2, let's mark the y-axis in increments of 100 from -500 to 200. With x-coordinates of extreme and inflection points of 1, 5, and 7, let's mark the x-axis from -4 to 10, by 2s.

The distance between the two left-most coordinates is 4, so moving 5 to the left of 1 yields -4. The distance between the two right-most coordinates is 2, and moving 3 to the right of 7 gives 10.

❷ The axes with extrema and inflection points marked are in figure 6.

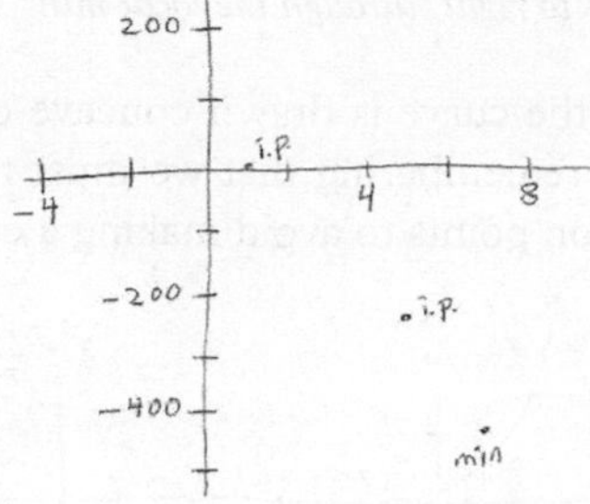

Figure 6 *Extreme points and points of inflection plotted*

❸ We begin the sketch of the curve to the left of the left-most inflection point, $(1, 2)$. The graph is concave up on $(-\infty, 1)$. Notice, however, that $x = 1$ is also a critical number of the polynomial, and therefore the tangent line there is horizontal. We begin drawing from the point $(1, 2)$, trying for a horizontal tangent line there,drawing right to left, concave up (figure 7).

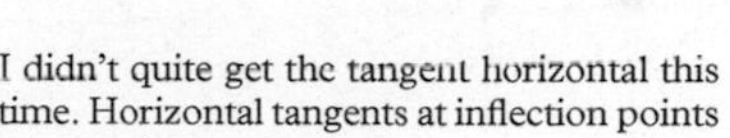
I didn't quite get the tangent horizontal this time. Horizontal tangents at inflection points can be tricky sometimes!

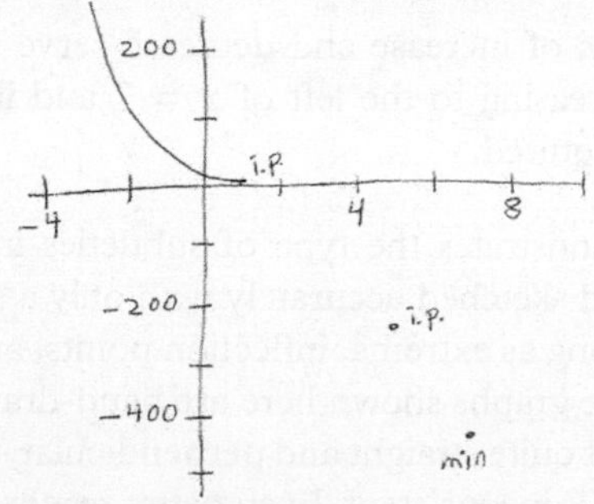

Figure 7 *The portion of the curve to the left of the left-most inflection point is drawn, from right to left. The tangent line at* $x = 1$ *should be horizontal*

Next comes the portion of the curve to the right of the right-most inflection point. This is drawn left to right, bottoming out at the local minimum and then rising, keeping the concave-up shape (figure 8).

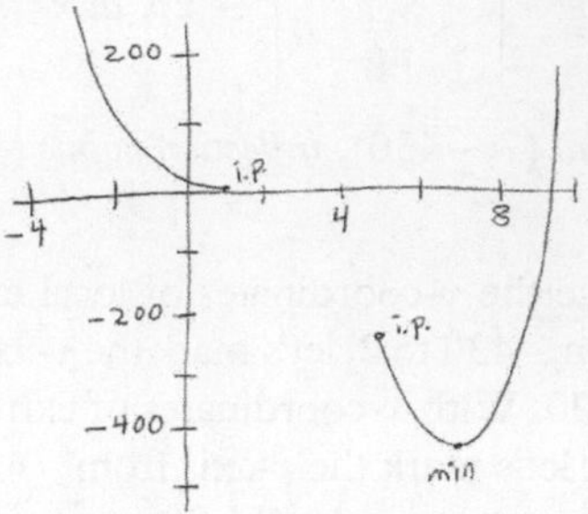

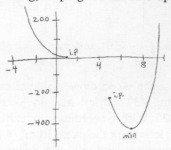

Figure 8 *The portion of the curve to the right of the right-most inflection point is added, drawn from left to right, through the local min*

The final piece of the curve is drawn concave down between the two inflection points, remembering that we must match the existing slopes at these inflection points to avoid making a corner (figure 9).

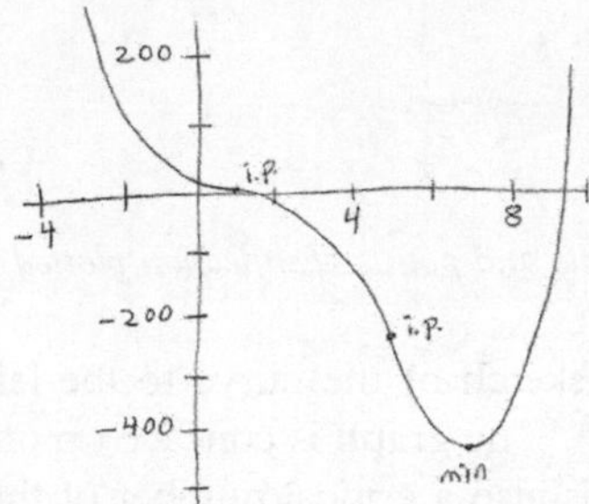

Figure 9 *The middle portion of the curve is added, making sure the slopes match at the inflection points and the shape of this portion is concave down. This completes the sketch of the curve*

❹ The intervals of increase and decrease serve as our accuracy check. We are decreasing to the left of $x = 7$ and increasing to the right of $x = 7$, as required. ■

Example 2 demonstrates the type of subtleties in the graph that can be detected and sketched accurately with only a small number of plotted points, as long as extrema, inflection points, and concavity are known. Because the graphs shown here are hand-drawn on blank paper, the axes are not quite straight and perpendicular, and the distance between tick marks is inconsistent. Even better results can be obtained with the use of graph paper.

GSL: "Coördinate plotting paper should be employed." (Note the spelling from a century ago!)

The next two examples run through the entire process.

Polynomial sketch: cubic

Example 3 *By hand, sketch the graph of the polynomial* $f(x) = x^3 - 3x^2 - 9x + 4$.

A cubic polynomial is a polynomial of degree 3.

Solution Let's start by collecting information about intervals of increase/decrease and local extrema. Some of the steps used in section 3.3 be skipped in the case of a polynomial. We start by calculating the derivative, setting it equal to zero, and solving for x:

$$f'(x) = 3x^2 - 6x - 9$$
$$3x^2 - 6x - 9 = 0$$
$$3(x^2 - 2x - 3) = 0$$
$$3(x-3)(x+1) = 0$$
$$x = 3, x = -1.$$

Of course, the quadratic formula may be used to solve quadratic equations if a factorization is not quickly apparent. Many other tools for solving polynomial equations exist and can be used, such as formulas/methods for solving cubic and quartic equations (usually too long to be practical). Methods for helping the factoring process for higher-degree polynomials include the rational roots theorem, which is sometimes tedious but not difficult to perform by hand, and is usually accompanied by synthetic division. Some polynomial equations cannot be solved using only rational numbers and extraction of roots (square roots, cube roots, and so on). In such cases, seeking approximate solutions is the norm, often with the help of technology.

We then choose test points (figure 10) and evaluate f' using the factored form of f' when convenient:

$$f'(-2) = 3(-5)(-1) > 0$$
$$f'(0) = -9 < 0$$
$$f'(4) = 3(1)(5) > 0.$$

We can fill in the remainder of the chart with ease:

interval	sign of f'	inc/dec	local extrema
$(-\infty, -1)$	+	increasing	
			$\leftarrow$ local max at $x = -1$
$(-1, 3)$	$-$	decreasing	
			$\leftarrow$ local min at $x = 3$
$(3, \infty)$	+	increasing	

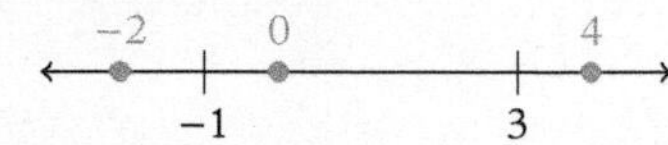

Figure 10 *Convenient test points for the increasing/decreasing chart*

Next we determine intervals of concavity and inflection points. We calculate the second derivative, set it equal to zero, and solve for x:

$$f''(x) = 6x - 6$$
$$6x - 6 = 0$$
$$x = 1.$$

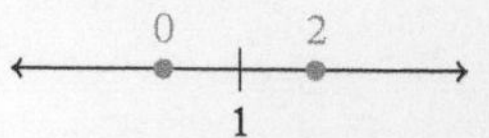

Figure 11 *Convenient test points for the concavity chart*

Again we form the intervals, pick test points (figure 11), and evaluate f'':

$$f''(0) = -6 < 0$$

$$f''(2) = 6 > 0.$$

We finish the chart:

interval	sign of f''	concavity	inflection
$(-\infty, 1)$	$-$	down	$\leftarrow$ i.p. at $x = 1$
$(1, \infty)$	$+$	up	

Because we wish to graph the function, knowing the locations of extrema and inflection points is not enough; we need to know y-coordinates as well. We therefore evaluate our function at -1, 3, and 1:

$$f(-1) = (-1)^3 - 3(-1)^2 - 9(-1) + 4 = 9,$$

$$f(3) = 3^3 - 3(3^2) - 9(3) + 4 = -23,$$

$$f(1) = 1^3 - 3 \cdot 1^2 - 9(1) + 4 = -7.$$

We now have three points to place on the graph: a local maximum $(-1, 9)$, a local minimum $(3, -23)$, and an inflection point $(1, -7)$.

All that remains is to produce the graph itself. We follow the same four steps used in examples 1 and 2.

❶ From our extrema and inflection points, marking the y-axis by 5s from -30 to 20 seems reasonable, as does marking the x-axis by 1s from -4 to 6. After plotting and marking the axes, and ❷ the local extrema and inflection points, we arrive at figure 12.

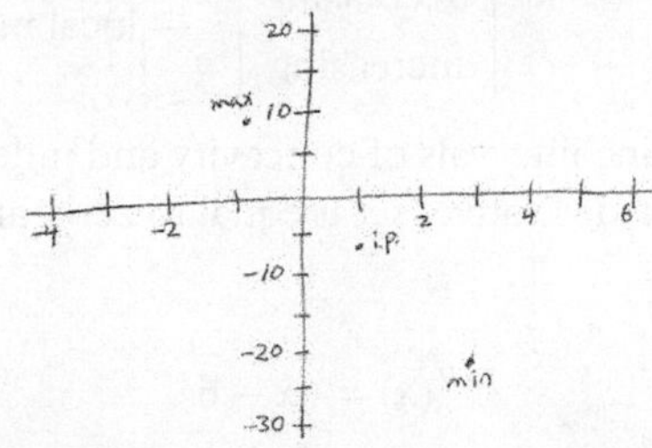

Figure 12 *Extreme points and points of inflection plotted*

❸ Next we draw the portion of the curve to the left of the left-most inflection point. This portion of the curve is concave down (figure 13). We finish the sketch by drawing the portion of the curve to the right of

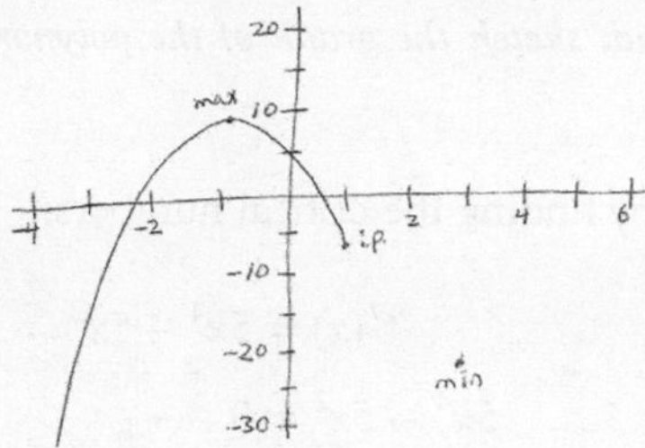

Figure 13 *The portion of the curve to the left of the inflection point is drawn, from right to left, through the local max*

the right-most inflection point (figure 14), using a concave up shape, making sure to match the slope at the inflection point to avoid drawing a corner.

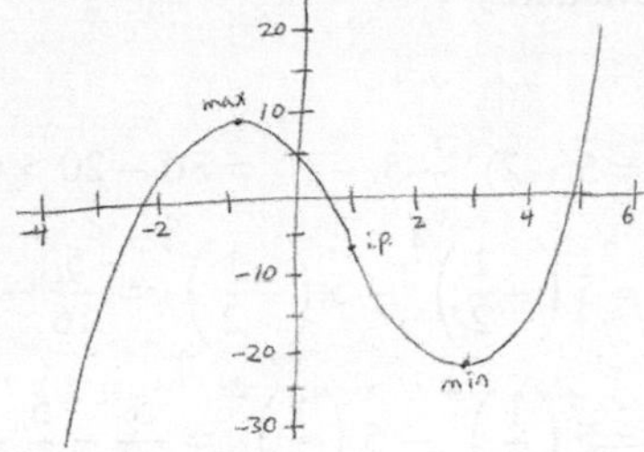

Figure 14 *The portion of the curve to the right of the inflection point is added, drawn from left to right, through the local min*

❹ A check of the increasing/decreasing chart verifies the graph represents this information correctly. ■

The easily calculated y-intercept in example 3 is 4, and including this plotted point increases accuracy somewhat. The less easily calculated x-intercepts are approximately -2.12, 0.399, and 4.73, and again the hand-drawn graph gets pretty close!

Graphs of cubics are fairly easy to classify. Because the derivative of a degree 3 polynomial has degree 2 and the second derivative has degree 1, there is always exactly one inflection point. There are, at most, two critical numbers. If there are two, then the curve has both a local max and a local min. If there are no critical numbers, then the curve has no local extrema. If there is exactly one critical number, then it occurs at the inflection point and the curve has no local extrema.

Polynomial sketch: quintic

Graphs of higher-degree polynomials can contain subtleties of the type exhibited by example 2.

A quintic polynomial has degree 5; a quartic polynomial has degree 4.

Example 4 *By hand, sketch the graph of the polynomial* $f(x) = x^5 - \frac{5}{3}x^3 + 1$.

Solution We start by finding the critical numbers:

$$f'(x) = 5x^4 - 5x^2$$

$$5x^4 - 5x^2 = 0$$

$$5x^2(x^2 - 1) = 0$$

$$5x^2(x-1)(x+1) = 0$$

$$x = 0, x = 1, x = -1.$$

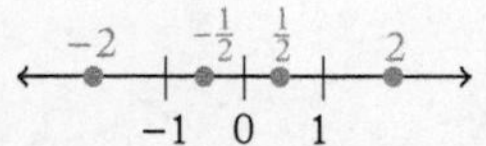

Figure 15 *Convenient test points for the increasing/decreasing chart*

The critical numbers are ± 1 and 0. Choosing test points of ± 2 and $\pm\frac{1}{2}$ (figure 15), we evaluate f',

$$f'(-2) = 5(-2)^4 - 5(-2)^2 = 80 - 20 > 0$$

$$f'\left(-\frac{1}{2}\right) = 5\left(-\frac{1}{2}\right)^4 - 5\left(-\frac{1}{2}\right)^2 = \frac{5}{16} - \frac{5}{4} < 0$$

$$f'\left(\frac{1}{2}\right) = 5\left(\frac{1}{2}\right)^4 - 5\left(\frac{1}{2}\right)^2 = \frac{5}{16} - \frac{5}{4} < 0$$

$$f'(2) = 5(2)^4 - 5(2)^2 = 80 - 20 > 0$$

and then complete the increasing/decreasing chart:

interval	sign of f'	inc/dec	local extrema
$(-\infty, -1)$	+	increasing	
			← local max at $x = -1$
$(-1, 0)$	−	decreasing	
$(0, 1)$	−	decreasing	
			← local min at $x = 1$
$(1, \infty)$	+	increasing	

We then turn our attention to the second derivative:

$$f''(x) = 20x^3 - 10x$$

$$20x^3 - 10x = 0$$

$$10x(2x^2 - 1) = 0$$

$$x = 0,\ 2x^2 = 1$$

$$x = \pm\sqrt{\frac{1}{2}}.$$

The possible inflection locations are $x = 0$, $x = \sqrt{\frac{1}{2}}$, and $x = -\sqrt{\frac{1}{2}}$. Choosing test points ± 1 and ± 0.1 (figure 16), we evaluate f'',

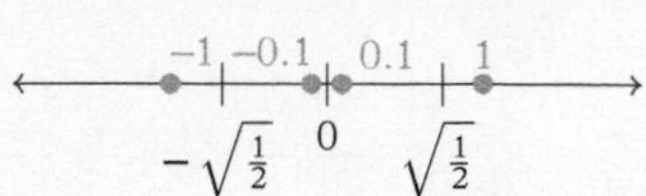

Figure 16 *Convenient test points for the concavity chart*

$$f''(-1) = 20(-1)^3 - 10(-1) = -20 + 10 < 0$$
$$f''(-0.1) = 20(-0.1)^3 - 10(-0.1) = -0.020 + 1 > 0$$
$$f''(0.1) = 20(0.1)^3 - 10(0.1) = 0.020 - 1 < 0$$
$$f''(1) = 20(1)^3 - 10(1) = 20 - 10 > 0$$

and finish the concavity chart:

interval	sign of f''	concavity	inflection
$\left(-\infty, -\frac{1}{\sqrt{2}}\right)$	$-$	down	
			$\leftarrow$ i.p. at $x = -\frac{1}{\sqrt{2}}$
$\left(-\frac{1}{\sqrt{2}}, 0\right)$	$+$	up	
			$\leftarrow$ i.p. at $x = 0$
$\left(0, \frac{1}{\sqrt{2}}\right)$	$-$	down	
			$\leftarrow$ i.p. at $x = \frac{1}{\sqrt{2}}$
$\left(\frac{1}{\sqrt{2}}, \infty\right)$	$+$	up	

Last, we evaluate f at our extrema and inflection points:

$$f(\ 1) = (\ 1)^5 \ \ \frac{5}{3}(\ 1)^3 + 1 = \frac{5}{3}$$
$$f(1) = \cdots = \frac{1}{3}$$
$$f\left(-\frac{1}{\sqrt{2}}\right) = \cdots = \frac{7\sqrt{2}}{24} + 1 \approx 1.412$$
$$f(0) = 1$$
$$f\left(\frac{1}{\sqrt{2}}\right) = \cdots = 1 - \frac{7\sqrt{2}}{24} \approx 0.5875.$$

Exact values here may be nice, but remembering that our task is to graph the function, all we need are approximate values anyway. Using a calculator to get a decimal approximation is definitely acceptable.

We have a local maximum point $(-1, \frac{5}{3})$, a local minimum point $(1, \frac{1}{3})$, and inflection points $(0, 1)$ and approximately $(-0.7071, 1.412)$ and $(0.7071, 0.5875)$.

We now have our information and are ready to sketch the graph.

❶ –❷ All five points we have calculated are close to the origin. We may still mark the axes by 1s, but if we do so the tick marks need to be spread fairly widely apart to separate the various plotted points (figure 17).

❸ We draw the curve to the left of the left-most inflection point, drawing concave down and passing through the local max, then we draw the curve to the right of the right-most inflection point, which is concave up and passes through the local min. See figure 18.

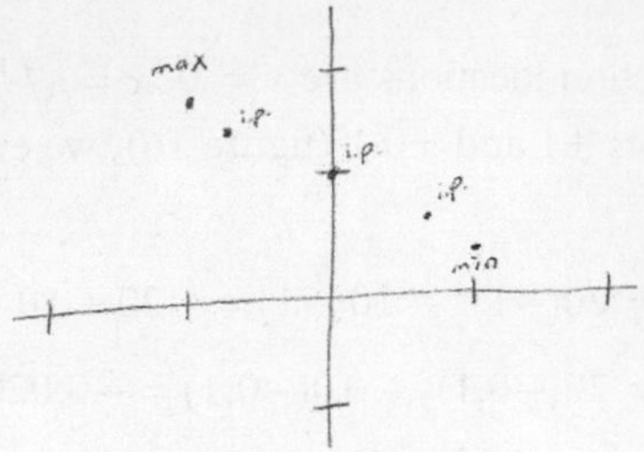

Figure 17 *Extreme points and points of inflection plotted*

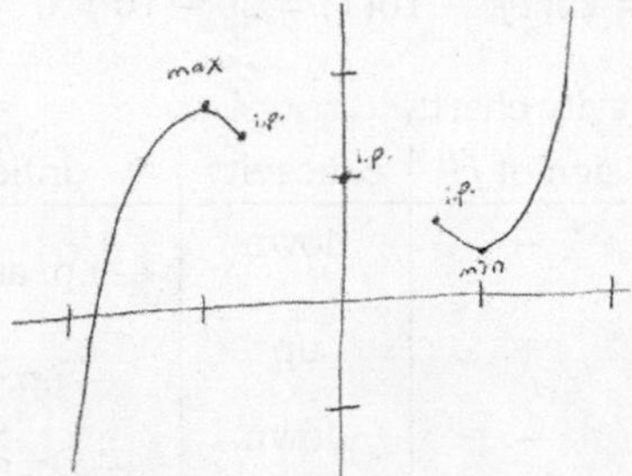

Figure 18 *The portions of the curve to the left of the left-most inflection point and to the right of the right-most inflection point, drawn from the inflection points outward*

Last, we draw the inside portions using the correct concavity on each piece, noting that since $x = 0$ is both a critical number and an inflection point, the tangent line there must be horizontal. The slopes must also match the existing slopes at each inflection point. See figure 19.

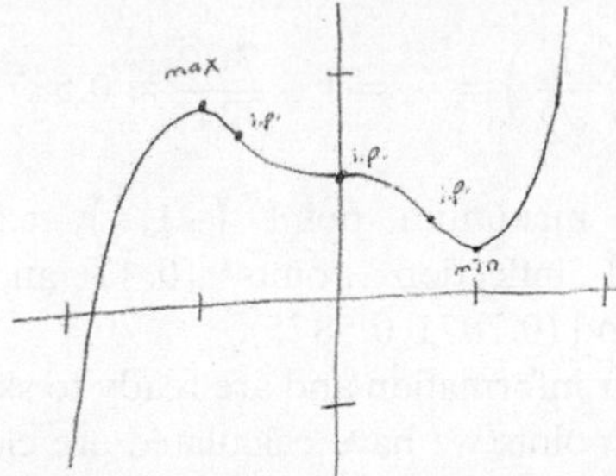

Figure 19 *The inside portions of the curve are added. Existing slopes must be matched, the tangent line at* $x = 0$ *must be horizontal, and the correct concavity must be observed*

❹ We check that the graph shows the function is increasing on $(-\infty, -1)$ and on $(1, \infty)$, and is decreasing elsewhere, and then declare the graph complete. ■

Graphing polynomials using technology

You might ask: Why spend all this time to graph a polynomial by hand when technology can do it for us much more quickly? One answer is that the experience of producing a few such graphs by hand does wonders for one's ability to process visually the results given by technology, and cements the connection between the visual and the algebraic. A second answer is that practicing the process by hand helps provide the skills needed to use technology to find all the features of a function's graph even when they are initially hidden or outside the bounds of naively chosen plot boxes.

Just as learning how to sand a piece of wood does not usually involve jumping straight to the use of a power tool, in mathematics the development of skills, insight, and creativity sometimes involves tedious activities, activities that in practice can be accomplished more easily by the use of more sophisticated tools.

Example 5 *Use a computer algebra system (CAS) to graph the function* $f(x) = x^5 - 181x^4 + 6514x^3 - 4390x^2 + 327x + 273$, *showing all its features (use more than one picture if necessary).*

Solution We still begin by finding the critical numbers and possible inflection locations so that we know on what interval we should graph the function. Using a CAS to solve the equation $f'(x) = 0$ results in the critical numbers $x = 0.040975$, $x = 0.415378$, $x = 35.2089$, and $x = 109.135$ (decimal approximations used; the exact values are not necessary for the purpose of graphing the function). Using the CAS to solve the equation $f''(x) = 0$ results in the possible inflection locations $x = 0.227515$, $x = 22.4592$, and $x = 85.9133$. Using the CAS to produce a plot of the function on an interval containing all these values of x, and playing with that interval to find something that looks good, results in figure 20.

It is possible to graph this polynomial by hand using the procedure of examples 3 and 4 (although solving the polynomial equations takes algebra we have not examined), but it is much more reasonable to use technology to perform the task.

Recall that in this section we only work with polynomials, so there is no need to look for the other type of critical number (where $f'(x)$ does not exist).

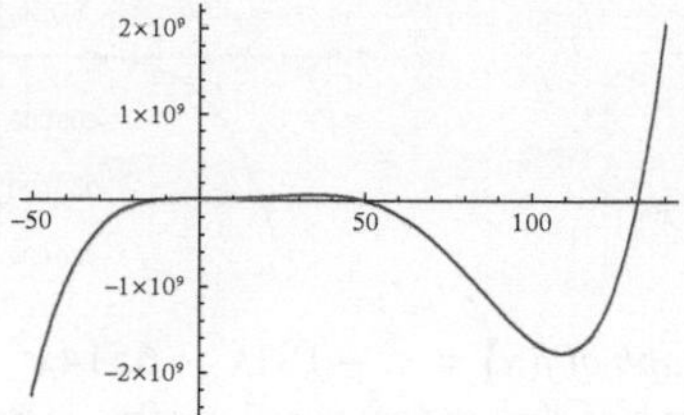

Figure 20 *The graph of* $f(x) = x^5 - 181x^4 + 6514x^3 - 4390x^2 + 327x + 273$ *over an interval containing all the critical numbers and possible inflection locations*

We were asked to show all the features of the graph. Because it is somewhat difficult to tell from figure 20 whether there are extrema at the left-most critical numbers and possible inflection locations, we can produce additional graphs near these locations, as in figure 21.

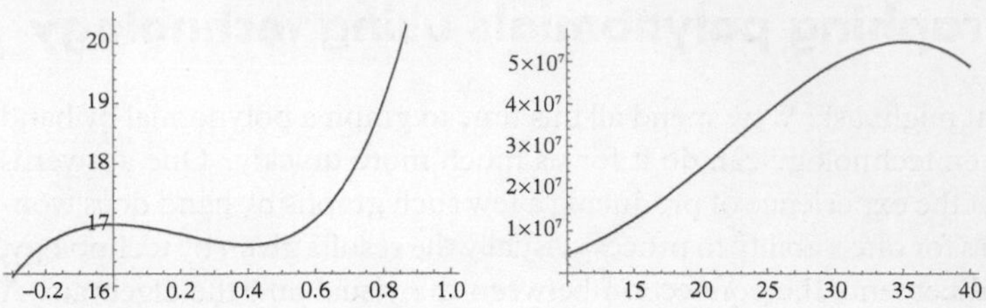

Figure 21 *Graphs of* $f(x) = x^5 - 181x^4 + 6514x^3 - 4390x^2 + 327x + 273$ *showing local extrema and inflection points not easily discernible in figure 20*

All local extreme points and inflection points are now apparent from the three graphs of the function. If desired, intervals of increase/decrease and intervals of concavity can also be recorded using the graphs, critical numbers, and inflection locations. ■

Although we asked the CAS to perform the calculations, we still used calculus in the solution of example 5. Could we instead just skip the calculus and use graphs to find everything? Let's answer this question by comparing the graphs of two polynomials of degree 5: the polynomial of example 5, $f(x) = x^5 - 181x^4 + 6514x^3 - 4390x^2 + 327x + 273$, and the polynomial $g(x) = 6x^5 - 3x^4 + 10x^3 - 6x^2 + 17$.

If we skip calculus and just look at graphs, we need an interval over which to graph. The interval $[-10, 10]$ is a commonly used interval for this purpose. Figure 22 shows the graphs of f and g on this interval.

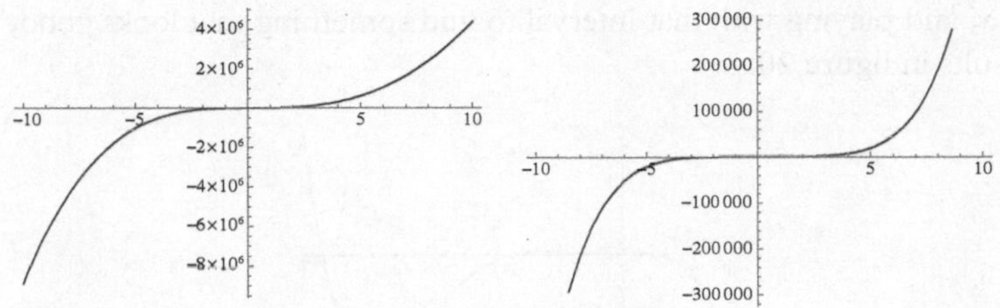

Figure 22 *(left) the graph of* $f(x) = x^5 - 181x^4 + 6514x^3 - 4390x^2 + 327x + 273$ *and (right) the graph of* $g(x) = 6x^5 - 3x^4 + 10x^3 - 6x^2 + 17$, *each on the interval* $[-10, 10]$

From these graphs, there appear to be inflection points and possibly horizontal tangent lines near $x = 0$, so we zoom in near there to get a closer look. For each function, graphing over the interval $[-0.3, 1]$ is instructive (figure 23).

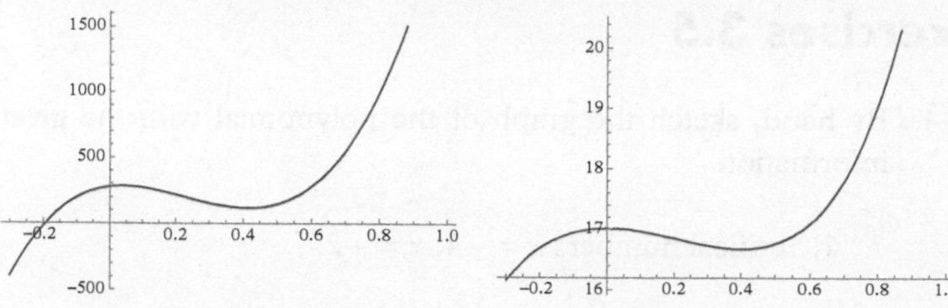

Figure 23 *(left) the graph of* $f(x) = x^5 - 181x^4 + 6514x^3 - 4390x^2 + 327x + 273$ *and (right) the graph of* $g(x) = 6x^5 - 3x^4 + 10x^3 - 6x^2 + 17$, *each on the interval* $[-0.3, 1]$. *Although f has additional extrema not shown in this graph, g does not*

We know from example 5 that the function *f* has features not shown on the graphs in figures 22 and 23. We might reason that because a polynomial of degree 5 has a derivative of degree 4, there can be as many as four local extrema, and because only two are apparent in figure 23, we can simply expand our graphing interval until we find the other two, eventually arriving at figure 20.

The problem with the reasoning in the previous paragraph is that the graph of function *g*, which has basically the same shape as the graph of *f* over the intervals shown in figures 22 and 23, has no additional extrema or inflection points. Hence, search as you may, there are no others to be found. Perhaps you counter with eventually you just give up the search, if no such additional extrema are found within a reasonable distance. The problem is determining a reasonable distance, for it is possible to find quintic polynomials with graphs that look like those in figures 22 and 23 with additional extrema as far away as we wish! There is no way to tell whether the search for additional extrema should continue or whether we have already found them all. The conclusion is clear: *graphing alone is insufficient to ensure finding all extrema and inflection points on the graph of a polynomial function. Calculus is required!*

The time may come—possibly by the time you read this—when mathematically powered search engines perform the calculus automatically and provide all the viewpoints necessary for visualizing all extrema and inflection points of polynomial functions or even other types of functions. But, until technology can reach into and manipulate our brains to provide instant insight and understanding (a scary thought for many reasons), we still benefit from graphing a few polynomials by hand, using calculus.

Exercises 3.5

These exercises are adapted from Love.

1–4. By hand, sketch the graph of the polynomial with the given information.

1. critical numbers $x = -4$, $x = -2$

interval	sign of f'	inc/dec	local extrema
$(-\infty, -4)$	$-$	decreasing	
			$\leftarrow$ local min at $x = -4$
$(-4, -2)$	$+$	increasing	
			$\leftarrow$ local max at $x = -2$
$(-2, \infty)$	$-$	decreasing	

interval	sign of f''	concavity	inflection
$(-\infty, -3)$	$+$	up	
			$\leftarrow$ i.p. at $x = -3$
$(-3, \infty)$	$-$	down	

local minimum point $\left(-4, \frac{28}{3}\right)$, local maximum point $\left(-2, \frac{32}{3}\right)$, inflection point $(-3, 10)$

2. critical numbers $x = -\frac{1}{6}$, $x = \frac{3}{2}$

interval	sign of f'	inc/dec	local extrema
$(-\infty, -\frac{1}{6})$	$+$	increasing	
			$\leftarrow$ local max at $x = -\frac{1}{6}$
$(-\frac{1}{6}, \frac{3}{2})$	$-$	decreasing	
			$\leftarrow$ local min at $x = \frac{3}{2}$
$(\frac{3}{2}, \infty)$	$+$	increasing	

interval	sign of f''	concavity	inflection
$(-\infty, \frac{2}{3})$	$-$	down	
			$\leftarrow$ i.p. at $x = \frac{2}{3}$
$(\frac{2}{3}, \infty)$	$+$	up	

local maximum point $\left(-\frac{1}{6}, \frac{250}{81}\right)$, local minimum point $\left(\frac{3}{2}, 0\right)$, inflection point $\left(\frac{2}{3}, \frac{125}{81}\right)$

3. critical numbers $x = -1$, $x = \frac{5}{4}$

interval	sign of f'	inc/dec	local extrema
$(-\infty, -1)$	$-$	decreasing	
$(-1, \frac{5}{4})$	$-$	decreasing	$\leftarrow$ local min at $x = \frac{5}{4}$
$(\frac{5}{4}, \infty)$	$+$	increasing	

interval	sign of f''	concavity	inflection
$(-\infty, -1)$	$+$	up	$\leftarrow$ i.p. at $x = -1$
$(-1, \frac{1}{2})$	$-$	down	$\leftarrow$ i.p. at $x = \frac{1}{2}$
$(\frac{1}{2}, \infty)$	$+$	up	

local minimum point $\left(\frac{5}{4}, -\frac{2187}{256}\right)$, inflection points $(-1, 0)$ and $\left(\frac{1}{2}, -\frac{81}{16}\right)$

4. critical numbers $x = \pm 1$, $x = \pm\frac{1}{\sqrt{5}}$

interval	sign of f'	inc/dec	local extrema
$(-\infty, -1)$	$+$	increasing	$\leftarrow$ local max at $x = -1$
$(-1, -\frac{1}{\sqrt{5}})$	$-$	decreasing	$\leftarrow$ local min at $x = -\frac{1}{\sqrt{5}}$
$(-\frac{1}{\sqrt{5}}, \frac{1}{\sqrt{5}})$	$+$	increasing	$\leftarrow$ local max at $x = \frac{1}{\sqrt{5}}$
$(\frac{1}{\sqrt{5}}, 1)$	$-$	decreasing	$\leftarrow$ local min at $x = 1$
$(1, \infty)$	$+$	increasing	

interval	sign of f''	concavity	inflection
$(-\infty, -\sqrt{\frac{3}{5}})$	$-$	down	$\leftarrow$ i.p. at $x = -\sqrt{\frac{3}{5}}$
$(-\sqrt{\frac{3}{5}}, 0)$	$+$	up	$\leftarrow$ i.p. at $x = 0$
$(0, \sqrt{\frac{3}{5}})$	$-$	down	$\leftarrow$ i.p. at $x = \sqrt{\frac{3}{5}}$
$(\sqrt{\frac{3}{5}}, \infty)$	$+$	up	

local maximum points $(-1, 0)$ and approximately $(0.447, 0.286)$, local minimum points $(1, 0)$ and approximately $(-0.447, -0.286)$, inflection points $(0, 0)$ and approximately $(-0.775, -0.124)$ and $(0.775, 0.124)$

It is implied that these exercises require completion of the increasing/decreasing and concavity charts as part of the information-gathering process. Adapted from Love.

5–18. By hand, sketch the graph of the given polynomial.

5. $y = \frac{1}{3}x^3 - 2x^2 + 3x - 2$

6. $y = 4 - 3x^2 - x^3$

7. $f(x) = 1 - 6x - 9x^2 - 4x^3$

8. $g(x) = 2 + 3x + \dfrac{x^2}{4} - \dfrac{x^3}{6}$

9. $g(x) = 2x^3 - 3$

10. $f(x) = \frac{2}{3}x^3 - \frac{3}{2}x$

11. $y = \frac{1}{4}x^4 - \frac{1}{2}x^2$

12. $y = x^2 + \frac{1}{3}x^3 - \frac{1}{4}x^4$

13. $f(x) = \frac{1}{12}(x^4 + 6x^2 - 7)$

14. $f(x) = 3x^4 - 4x^3$

15. $g(x) = \frac{1}{5}x^5 - \frac{1}{4}x^4$

16. $y = \frac{1}{4}x^4 - \frac{4}{3}x^3 + \frac{3}{2}x^2$

17. $y = \frac{1}{5}x^5 - \frac{2}{3}x^3 + x + 1$

18. $g(x) = \frac{1}{5}x^5 - \frac{2}{3}x^3 + 2x$

19–26. Use a CAS to graph the given function, showing all its features. Use more than one picture if necessary.

When solving equations, a CAS usually gives complex solutions as well as real solutions to the equation. Complex solutions (solutions of the form $a + bi$) can be ignored. However, the use of numerical methods by a CAS may result in a number that is listed as having a very small imaginary part (such as $4.73287 + 1.256 \times 10^{-12}\,i$), but is actually real. Such numbers should be considered in your graphical search for extrema and inflection points.

19. $h(x) = x^5 + 85x^4 - 7323x^3 + 2730x^2 + 6764x$

20. $s(x) = x^5 - 181x^4 + 6514x^3 - 69\,430x^2 + 193\,200x$

21. $f(x) = 6x^5 - 315x^4 + 4610x^3 - 6300x^2$

22. $f(x) = -3x^5 + 13x^4 - 127x^3 + 99x^2 - 15x + 21$

23. $g(x) = x^6 - 43x^5 - 2379x^3 + 478x^2 + 85\,974x - 2298$

24. $f(x) = x^6 + 196x^5 - 946x^4 - 832x^3 + 6055x^2 + 7464x$

25. $m(x) = x^6 - 76x^5 - 7852x^4 - 60\,050x^3 + 571\,662x^2 - 931\,284x$

26. $k(x) = x^6 + 96x^5 - 4546x^4 - 1132x^3 + 62\,055x^2 + 108\,864x$

27–34. Use a CAS to find (a) extreme points, (b) inflection points, (c) intervals of increase and intervals of decrease, and (d) intervals of concavity for the given function.

These are the same functions as exercises 19–26. Decimal approximations are acceptable and expected.

27. $h(x) = x^5 + 85x^4 - 7323x^3 + 2730x^2 + 6764x$

28. $s(x) = x^5 - 181x^4 + 6514x^3 - 69\,430x^2 + 193\,200x$

29. $f(x) = 6x^5 - 315x^4 + 4610x^3 - 6300x^2$

30. $f(x) = -3x^5 + 13x^4 - 127x^3 + 99x^2 - 15x + 21$

31. $g(x) = x^6 - 43x^5 - 2379x^3 + 478x^2 + 85\,974x - 2298$

32. $f(x) = x^6 + 196x^5 - 946x^4 - 832x^3 + 6055x^2 + 7464x$

33. $m(x) = x^6 - 76x^5 - 7852x^4 - 60\,050x^3 + 571\,662x^2 - 931\,284x$

34. $k(x) = x^6 + 96x^5 - 4546x^4 - 1132x^3 + 62\,055x^2 + 108\,864x$

Limits at Infinity

3.6

In section 1.4 we studied vertical asymptotes using limits. In this section we do the same for horizontal asymptotes.

This section does not use derivatives and could be covered after section 1.5.

Horizontal asymptotes and limits at infinity

Consider the graph of $y = 4 + \frac{1}{x}$. You should be able to form a mental image of this graph fairly quickly because we can start with the graph of $y = \frac{1}{x}$, which should be on instant recall, and shift up 4 units (figure 1).

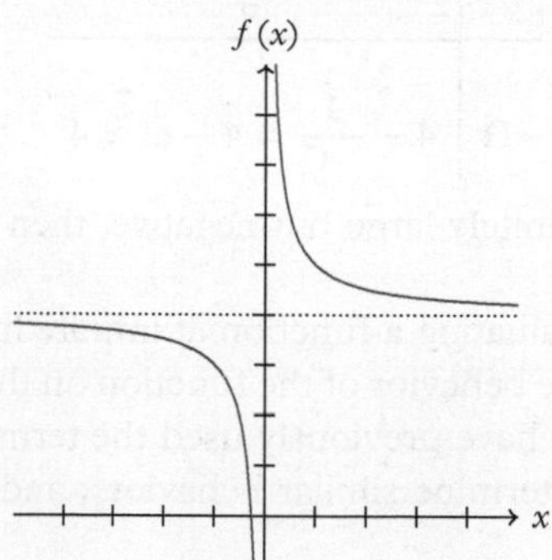

Figure 1 *The instantly recognized* $y = \frac{1}{x}$ *shifted up* 4 *units*

The dotted line, which is not part of the graph of the function, represents a *horizontal asymptote* with equation $y = 4$. Just as with a vertical asymptote, the idea is that the function approaches the asymptote, and the dotted line serves as an aid in sketching the graph of the function.

How do we interpret a horizontal asymptote? The idea for $y = 4 + \frac{1}{x}$ is that as x gets "large," y gets "close" to 4. This idea can be explored numerically by calculating values of y as our values of x get larger:

Notice how the magnitude of the chosen values of x increases.

x	y
10	4.1
100	4.01
1000	4.001

It appears to be the case that as x gets larger, y gets ever closer to 4. This seems evident in this case, but what about more complicated functions? How do we know that something different doesn't happen as x gets even larger? The problem is that *large* and *close* are fairly ambiguous terms. Here is where the hyperreal numbers come to our rescue again. Instead of just using a "large" real number, we use an infinite hyperreal number such as Ω:

Recall that Ω and ω are reciprocals of one another.

x	y
Ω	$4 + \frac{1}{\Omega} = 4 + \omega \approx 4$

Now we see that if x is infinitely large, then y is infinitely close to 4. This gives precise meanings to *large* and *close*!

Notice also that $4+\omega$ is just a little larger than 4, so that on the right side of the graph in figure 1, we are approaching the horizontal asymptote from above.

The same phenomenon happens when we look at the left side of the graph, where the x-coordinates are negative. We still choose an infinite number, but it needs to be negative, such as $-\Omega$:

x	y
$-\Omega$	$4 + \frac{1}{-\Omega} = 4 - \omega \approx 4$

Therefore, if x is infinitely large but negative, then y is still infinitely close to 4.

Notice this time that $4 - \omega$ is a little less than 4, so that on the left side of the graph in figure 1, we are approaching the horizontal asymptote from below.

It appears that evaluating a function at infinite hyperreals gives us information about the behavior of the function on the far right and far left of the graph. We have previously used the term *limit* for calculations that help us determine similar behaviors, and we use this term again here.

Definition 7 LIMITS AT INFINITY

(a) *If $f(\Omega)$ is defined and renders the same real result L for every positive infinite hyperreal Ω, then we write*

$$\lim_{x\to\infty} f(x) = L.$$

When L is a real number (i.e., not ∞ or $-\infty$), then we say that the limit exists and that f has a horizontal asymptote of $y = L$ (on the right).

(b) *If $f(-\Omega)$ is defined and renders the same real result L for every positive infinite hyperreal Ω, then we write*

$$\lim_{x\to-\infty} f(x) = L.$$

When L is a real number (i.e., not ∞ or $-\infty$), then we say that the limit exists and that f has a horizontal asymptote of $y = L$ (on the left).

Notice that if Ω is positive, then $-\Omega$ is negative, so that in part (b) we are evaluating f at a negative infinite hyperreal.

For the purposes of calculation we write

$$\lim_{x\to\infty} f(x) = f(\Omega),$$

$$\lim_{x\to-\infty} f(x) = f(-\Omega),$$

with the understanding that Ω is an arbitrary positive infinite hyperreal and that we must render the same real result for every such Ω or the limit does not exist.

Examples of limits at infinity

The evaluate–approximate–render (EAR) procedure is used for this type of limit as well.

Example 1 *Determine* $\lim\limits_{x\to\infty} \dfrac{2x+8}{x^3+4x+2}$.

Solution We evaluate the expression at $x = \Omega$:

$$\begin{aligned}\lim_{x\to\infty} \frac{2x+8}{x^3+4x+2} &= \frac{2\Omega+8}{\Omega^3+4\Omega+2}\\ &\approx \frac{2\Omega}{\Omega^3} = \frac{2}{\Omega^2} = 2\omega^2\\ &\doteq 0.\end{aligned}$$

Recall that the larger the power of Ω, the higher the level of the number.

Recall that an infinitesimal such as $2\omega^2$ renders the real result 0.

The limit is 0. ■

The graph of the function with the limit that is taken in example 1 has a horizontal asymptote of $y = 0$ (on the right).

Since $2\omega^2 > 0$, the function approaches the horizontal asymptote $y = 0$ on the right side from above.

Example 2 *Evaluate* $\lim\limits_{x\to-\infty} \dfrac{x^2+4}{3x^2+5x-9}$.

Solution We evaluate the expression at $x = -\Omega$:

CAUTION: MISTAKE TO AVOID
Failure to use parentheses with $-\Omega$ often results in sign errors.

$$\lim_{x\to-\infty} \frac{x^2+4}{3x^2+5x-9} = \frac{(-\Omega)^2+4}{3(-\Omega)^2+5(-\Omega)-9} = \frac{\Omega^2+4}{3\Omega^2-5\Omega-9}$$

$$\approx \frac{\Omega^2}{3\Omega^2} = \frac{1}{3}.$$

In the limit calculation of example 2, it is not immediately apparent whether the exact value before approximation is a little larger or a little smaller than $\frac{1}{3}$, so we need additional reasoning to determine whether the function approaches the horizontal asymptote (on the left) from above or from below.

The limit is $\frac{1}{3}$. ■

The graph of the function with the limit that is taken in example 2 has a horizontal asymptote of $y = \frac{1}{3}$ (on the left).

Example 3 *Find the limit:* $\lim\limits_{x\to\infty} \dfrac{5x^3-2x+7}{x^2+7x-1}$.

Solution We evaluate the expression at $x = \Omega$:

$$\lim_{x\to\infty} \frac{5x^3-2x+7}{x^2+7x-1} = \frac{5\Omega^3-2\Omega+7}{\Omega^2+7\Omega-1}$$

$$\approx \frac{5\Omega^3}{\Omega^2} = 5\Omega$$

$$\doteq \infty.$$

Recall that if the result of an approximation is positive infinite, then we render the real result ∞.

The limit is ∞. ■

There do exist functions, some of which we study in chapter 5, that have horizontal asymptotes on one side only. In other words, a function can have a horizontal asymptote on the left side even if it does not have one on the right side.

Because the limit in example 3 is ∞, the limit does not exist and there is no horizontal asymptote on the right side of the graph of the function.

Some algebra textbooks give (without proof) a method for determining the horizontal asymptotes of a rational function by considering the degrees of the polynomials in the numerator and the denominator and, if necessary, the leading coefficients. If you have studied this method, then reconsider examples 1, 2, and 3 and see if you can determine why this method works.

Reading Exercise 14 *Find* $\lim\limits_{x\to\infty} \dfrac{4x^2+7x-9}{2x^2-1}$.

Example 4 *Calculate* $\lim\limits_{x\to\infty} (5x^2 - 4x^3)$.

Solution We evaluate the expression at $x = \Omega$:

$$\lim_{x\to\infty}(5x^2 - 4x^3) = 5\Omega^2 - 4\Omega^3$$
$$\approx -4\Omega^3$$
$$\doteq -\infty.$$

The limit is $-\infty$. ■

The calculation shows there is no horizontal asymptote (on the right) on the graph of the polynomial $y = 5x^2 - 4x^3$. In fact, it is not difficult to see that the same is true for any polynomial.

Although there are no horizontal asymptotes for polynomials, limits at infinity still describe end behavior. For instance, $\lim_{x\to\infty}(5x^2 - 4x^3) = -\infty$ means that on the far right side of the graph, the polynomial's values plunge downward toward $-\infty$.

Polynomials do not have horizontal asymptotes.

Recall that if the result of an approximation is negative infinite, then we render the real result $-\infty$.

For the degree n polynomial

$$a_nx^n + a_{n-1}x^{n-1} + \cdots + a_1x + a_0,$$

finding limits at infinity gives

$$\lim_{x\to\infty}(a_nx^n + a_{n-1}x^{n-1} + \cdots)$$
$$= a_n\Omega^n + a_{n-1}\Omega^{n-1} + \cdots$$
$$\approx a_n\Omega^n,$$

which renders eithers ∞ or $-\infty$, depending on the value of a_n. Similarly,

$$\lim_{x\to-\infty}(a_nx^n + a_{n-1}x^{n-1} + \cdots) \approx a_n(-\Omega)^n,$$

which renders either ∞ or $-\infty$, depending on the values of n and a_n.

Examples of finding asymptotes

Determining whether a function has a horizontal asymptote (or two) is accomplished by checking the limits at infinity. Since $\lim_{x\to\infty} f(x)$ may be different from $\lim_{x\to-\infty} f(x)$, both limits should be checked.

For a rational function, if there is a horizontal asymptote on one side (left or right), then the function has the same horizontal asymptote on the other side. (For those readers who are students, check with your instructor to find out whether both limits must still be checked for rational functions.)

Example 5 *Find all horizontal asymptotes on the graph of* $f(x) = \dfrac{3x - 7}{4x + \sqrt[3]{x^3 + 5x}}$.

Solution We first check the right side by evaluating the limit as $x \to \infty$:

Because the function of example 5 involves a cube root, it is not a rational function.

$$\lim_{x\to\infty}\frac{3x - 7}{4x + \sqrt[3]{x^3 + 5x}} = \frac{3\Omega - 7}{4\Omega + \sqrt[3]{\Omega^3 + 5\Omega}}$$
$$\approx \frac{3\Omega}{4\Omega + \sqrt[3]{\Omega^3}} = \frac{3\Omega}{4\Omega + \Omega} = \frac{3\Omega}{5\Omega} = \frac{3}{5}.$$

CAUTION: MISTAKE TO AVOID
Recall that we may approximate underneath (inside) roots, but this must take place entirely inside the root. In this case, the 5Ω inside the root can be discarded, but the 4Ω outside the root must be kept. After the root is taken, we approximate the (simplified) expression again if necessary. The approximation principle still applies.

The function has a horizontal asymptote on the right, $y = \frac{3}{5}$. We also need to check the left side to see if the result is different. We therefore evaluate the limit as $x \to -\infty$:

$$\lim_{x\to-\infty}\frac{3x-7}{4x+\sqrt[3]{x^3+5x}}=\frac{3(-\Omega)-7}{4(-\Omega)+\sqrt[3]{(-\Omega)^3+5(-\Omega)}}$$
$$=\frac{-3\Omega-7}{-4\Omega+\sqrt[3]{-\Omega^3-5\Omega}}$$
$$\approx\frac{-3\Omega}{-4\Omega+\sqrt[3]{-\Omega^3}}=\frac{-3\Omega}{-4\Omega-\Omega}=\frac{-3\Omega}{-5\Omega}=\frac{3}{5}.$$

There is an alternate procedure that allows both sides to be checked simultaneously: calculate $f(A)$ for an arbitrary infinite hyperreal A, which could be either positive or negative. In this case, we have

$$\lim_{x\to\pm\infty}\frac{3x-7}{4x+\sqrt[3]{x^3+5x}}=\frac{3A-7}{4A+\sqrt[3]{A^3+5A}}$$
$$\approx\frac{3A}{4A+\sqrt[3]{A^3}}=\frac{3A}{4A+A}=\frac{3A}{5A}=\frac{3}{5}.$$

Then, because we rendered the same real result for every such A, both limits (as $x\to\infty$ and as $x\to-\infty$) agree and the function has a horizontal asymptote of $x=\frac{3}{5}$ on both sides. However, if we get stuck (for instance with $\frac{A}{\sqrt{A^2}}$, which can be either 1 or -1), we need to resort to checking both sides separately.

The function has a horizontal asymptote of $y=\frac{3}{5}$ on the left side as well. ∎

Reading Exercise 15 *Find any horizontal asymptotes on the graph of* $y=\dfrac{4x-1}{x+3}$.

Example 6 *Find all asymptotes on the graph of* $y=\dfrac{\sqrt{4x^2+2}}{3x-1}$.

Solution The phrase "all asymptotes" includes both horizontal and vertical asymptotes.

Let's start with finding any vertical asymptotes, which is a review from section 1.4. Recall that vertical asymptotes in an algebraic function can occur only where the denominator is zero. Solving $3x-1=0$ yields $x=\frac{1}{3}$, but this does not yet guarantee that we have a vertical asymptote; the limits must be checked. First we check the limit from the right using $x=\frac{1}{3}+\omega$:

Other possibilities for zero denominators include removable discontinuities and jump discontinuities.

$$\lim_{x\to\frac{1}{3}^+}\frac{\sqrt{4x^2+2}}{3x-1}=\frac{\sqrt{4(\frac{1}{3}+\omega)^2+2}}{3(\frac{1}{3}+\omega)-1}=\frac{\sqrt{4(\frac{1}{9}+\frac{2}{3}\omega+\omega^2)+2}}{1+3\omega-1}$$
$$\approx\frac{\sqrt{\frac{4}{9}+2}}{3\omega}=\frac{\sqrt{2+\frac{4}{9}}}{3}\Omega$$
$$\doteq\infty.$$

There is a vertical asymptote, $x=\frac{1}{3}$.

Although we have already determined the vertical asymptote, if we wish to graph the function we need to check the limit from the left as well, using $x=\frac{1}{3}-\omega$:

Ans. to reading exercise 14
2

$$\lim_{x \to \frac{1}{3}^-} \frac{\sqrt{4x^2+2}}{3x-1} = \frac{\sqrt{4(\frac{1}{3}-\omega)^2+2}}{3(\frac{1}{3}-\omega)-1} = \frac{\sqrt{4(\frac{1}{9}-\frac{2}{3}\omega+\omega^2)+2}}{1-3\omega-1}$$

$$\approx \frac{\sqrt{\frac{4}{9}+2}}{-3\omega} = -\frac{\sqrt{2+\frac{4}{9}}}{3}\Omega$$

$$\doteq -\infty.$$

Determining horizontal asymptotes requires that we find the limits as $x \to \infty$ and as $x \to -\infty$. We start with the right-hand side, evaluating at $x = \Omega$:

$$\lim_{x \to \infty} \frac{\sqrt{4x^2+2}}{3x-1} = \frac{\sqrt{4\Omega^2+2}}{3\Omega-1}$$

$$\approx \frac{\sqrt{4\Omega^2}}{3\Omega} = \frac{2\Omega}{3\Omega} = \frac{2}{3}.$$

Although $\sqrt{x^2} = |x|$, because Ω is positive we can ignore the absolute values:

$$\sqrt{4\Omega^2} = 2\sqrt{\Omega^2} = 2|\Omega| = 2\Omega.$$

The function has the horizontal asymptote $y = \frac{2}{3}$ on the right.

Next we check the left side, evaluating at $x = -\Omega$:

$$\lim_{x \to -\infty} \frac{\sqrt{4x^2+2}}{3x-1} = \frac{\sqrt{4(-\Omega)^2+2}}{3(-\Omega)-1} = \frac{\sqrt{4\Omega^2+2}}{-3\Omega-1}$$

$$\approx \frac{\sqrt{4\Omega^2}}{-3\Omega} = \frac{2\Omega}{-3\Omega} = -\frac{2}{3}.$$

CAUTION: MISTAKE TO AVOID
Recall that

$$\sqrt{(-4)^2} \neq -4.$$

Similarly,

$$\sqrt{(-\Omega)^2} \neq -\Omega.$$

Just as $\sqrt{(-4)^2} = \sqrt{16} = 4$, we square first and then take the square root:

$$\sqrt{(-\Omega)^2} = \sqrt{\Omega^2} = \Omega.$$

The function has the horizontal asymptote $y = -\frac{2}{3}$ on the left.

To summarize, the function has three asymptotes: the vertical asymptote $x = \frac{1}{3}$ and the two horizontal asymptotes $y = \frac{2}{3}$ (on the right) and $y = -\frac{2}{3}$ (on the left). ■

When a function exhibits two different horizontal asymptotes, the dotted lines indicating the asymptotes should not be drawn all the way across the picture.The dotted line for the asymptote on the right is drawn on the right side only and the one for the asymptote on the left is drawn on the left side only, as in figure 2.

The reason that a continuous function (continuous where defined) may not cross a vertical asymptote is that it is not defined at that value of x. The infinite discontinuity prevents us from connecting the pieces on either side of the asymptote, but there is no corresponding restriction on horizontal asymptotes. The y-coordinate of a horizontal

Notice that vertical asymptotes have equations that are vertical lines (form $x = c$) and horizontal asymptotes have equations that are horizontal lines (form $y = c$).

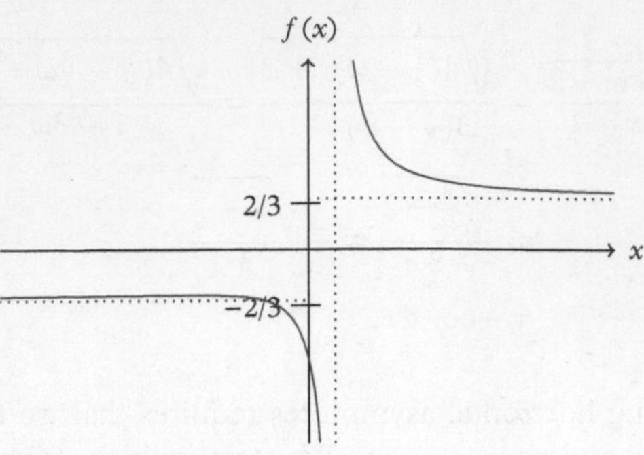

Figure 2 *The graph of the function in example 6, with dotted lines indicating the vertical and horizontal asymptotes. Notice that on the left side of the graph, the graph crosses the horizontal asymptote and approaches that asymptote from above, not from below*

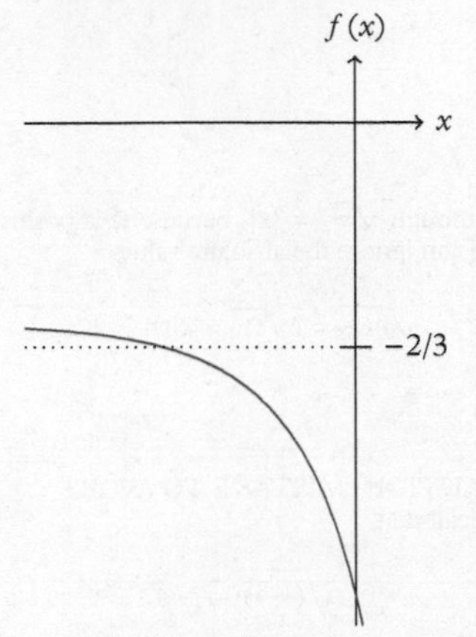

Figure 3 *A portion of the graph of the function in example 6, showing the crossing of the horizontal asymptote. The curve then approaches the asymptote from above*

asymptote does not arise from division by zero or any similar hole in the function's domain. Therefore, a function might cross a horizontal asymptote, as in figure 2. Because the scale makes this crossing difficult to see, a close-up view of the crossing is given in figure 3.

Limits at infinity: trig examples

Example 7 *Evaluate* $\lim\limits_{x\to\infty} \dfrac{\sin x}{x}$.

Solution We evaluate the expression at $x = \Omega$:

Recall that by the transfer principle,

$$-1 \le \sin\Omega \le 1.$$

$$\lim_{x\to\infty} \frac{\sin x}{x} = \frac{\sin\Omega}{\Omega} = \underbrace{\underbrace{(\sin\Omega)}_{\text{real level}} \cdot \underbrace{\omega}_{\text{infinitesimal}}}_{\text{infinitesimal}} \doteq 0.$$

The limit is 0. ■

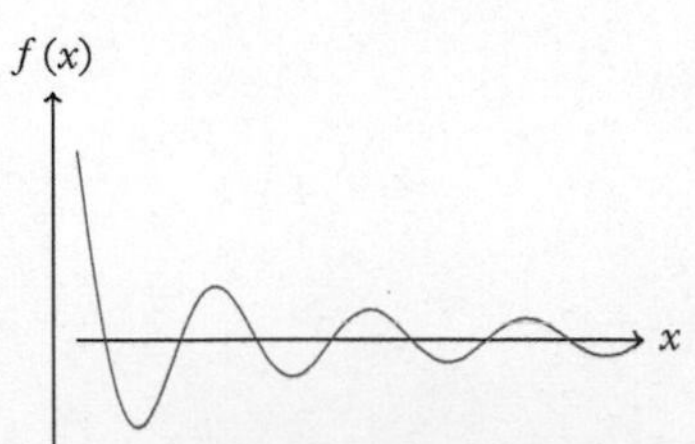

Figure 4 *the graph of* $y = \frac{\sin x}{x}$ *to the right of* $x = 2$*, which crosses its horizontal asymptote* $y = 0$ *in an oscillatory manner. Notice that the heights of the oscillations are smaller for larger values of x, so that the graph of the function still approaches the asymptote*

Example 7 shows that the function $y = \frac{\sin x}{x}$ has a horizontal asymptote on the right of $y = 0$. Because $(\sin\Omega) \cdot \omega$ may be either positive or negative, the function does not approach the asymptote from only above or from only below, but rather oscillates across the horizontal asymptote, crossing it infinitely many times! See figure 4.

It turns out that the squeeze theorem (chapter 1 theorem 5) is also valid for this type of limit (the proof is omitted). This gives an alternate method for determining the limit in example 7. We start by remembering that

$$-1 \le \sin x \le 1.$$

Because we are taking the limit as x goes to ∞, we may assume x is positive. Dividing by x then gives

$$-\frac{1}{x} \le \frac{\sin x}{x} \le \frac{1}{x}.$$

For $\lim_{x\to-\infty}$, we use negative values of x and switch the direction of the inequality, resulting in

$$-\frac{1}{x} \ge \frac{\sin x}{x} \ge \frac{1}{x}.$$

We take the limits of the expressions on the left,

$$\lim_{x\to\infty} -\frac{1}{x} = -\frac{1}{\Omega} = -\omega \doteq 0,$$

and on the right,

$$\lim_{x\to\infty} \frac{1}{x} = \frac{1}{\Omega} = \omega \doteq 0.$$

Because the two limits are the same, the limit of the expression in the middle is "squeezed" to the same value:

$$\lim_{x\to\infty} \frac{\sin x}{x} = 0.$$

Ans. to reading exercise 15
$y = 4$ (on both the right and the left)

Example 8 *Determine* $\lim\limits_{x\to\infty} \dfrac{x \sin x}{4x+5}$.

Solution We evaluate the expression at $x = \Omega$:

$$\lim_{x\to\infty} \frac{x \sin x}{4x+5} = \frac{\Omega \sin \Omega}{4\Omega + 5} \approx \frac{\Omega \sin \Omega}{4\Omega} = \frac{\sin \Omega}{4}.$$

At this point we are stuck. Because $\sin \Omega$ can be any value between -1 and 1, $\frac{\sin \Omega}{4}$ can be any value between $-\frac{1}{4}$ and $\frac{1}{4}$. Because we do not render the same real result for every positive infinite hyperreal Ω, the limit does not exist (DNE). We write

$$\lim_{x\to\infty} \frac{x \sin x}{4x+5} \text{ DNE.}$$ ■

CAUTION: MISTAKE TO AVOID
When we approximate, we throw away terms, not factors. Although $\Omega + \sin \Omega$ has two terms and we may discard the lower-level term, approximating as $\Omega + \sin \Omega \approx \Omega$, the expression $\Omega \cdot \sin \Omega$ has just one term (with two factors) and therefore has no term to be discarded.

Because the limit does not exist, the graph of $y = \dfrac{x \sin x}{4x+5}$ does not have a horizontal asymptote (on the right). See figure 5.

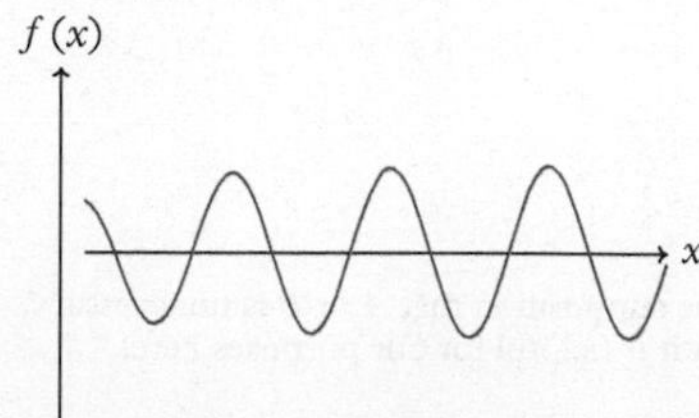

Figure 5 *The graph of* $y = \dfrac{x \sin x}{4x+5}$ *to the right of* $x = 2$*, which does not exhibit a horizontal asymptote. Compare to figure 4*

The approximation principle still applies!

Any time we approximate with hyperreals, in any setting, the approximation principle still applies.

Example 9 *Calculate* $\lim_{x\to\infty}\left(x-\frac{3x^2-6}{3x+4}\right)$.

Solution We begin by evaluating the expression at $x = \Omega$:

$$\lim_{x\to\infty}\left(x-\frac{3x^2-6}{3x+4}\right)=\Omega-\frac{3\Omega^2-6}{3\Omega+4}.$$

Next it seems reasonable to approximate the fraction by approximating the numerator and the denominator, which gives

$$\approx\Omega-\frac{3\Omega^2}{3\Omega}=\Omega-\Omega=0,$$

which violates the approximation principle. We approximated (terms were thrown away) and the result is zero.

The problem here is not the fraction on the right, because it does not approximate as zero. Instead, the problem is the subtraction that yields zero. Therefore, the subtraction must be performed exactly before we approximate.

We must try something different. Some sort of exact arithmetic must be used to change the form of the expression before we approximate. Let's get a common denominator and perform the subtraction:

$$\begin{aligned}\Omega-\frac{3\Omega^2-6}{3\Omega+4}&=\Omega\cdot\frac{3\Omega+4}{3\Omega+4}-\frac{3\Omega^2-6}{3\Omega+4}\\&=\frac{3\Omega^2+4\Omega}{3\Omega+4}-\frac{3\Omega^2-6}{3\Omega+4}=\frac{4\Omega+6}{3\Omega+4}.\end{aligned}$$

By subtracting before approximating, we eliminated the problem. We no longer have a violation of the approximation principle.

Now we can try approximation again:

$$\approx\frac{4\Omega}{3\Omega}=\frac{4}{3}.$$

The limit is $\frac{4}{3}$. ■

Suppose that a continuous function f has the horizontal asymptote $y = k$ on the right and $k \neq 0$. Then, for any positive infinite hyperreal Ω, $f(\Omega) \doteq k$. Now suppose that A and B are two different positive infinite hyperreals with $A \approx B$. We then have

The supposition that $A \approx B$ is unnecessary, but it is helpful for our purposes here.

$$f(A) \doteq k,$$
$$f(B) \doteq k.$$

Because k is a nonzero real number, by the definition of rendering, $f(A) \approx k$ and $f(B) \approx k$. Therefore, they are approximately equal to one another:

$$f(A) \approx f(B).$$

This justifies the following theorem, which is a complement to chapter 1 theorem 16.

Theorem 12 APPROXIMATING INSIDE A CONTINUOUS FUNCTION WITH A HORIZONTAL ASYMPTOTE

(a) *If f is a continuous function with horizontal asymptote $y = k$ on the right, $k \neq 0$, $A \approx B$, and B is a positive infinite hyperreal, then $f(A) \approx f(B)$.*

(b) *If f is a continuous function with horizontal asymptote $y = k$ on the left, $k \neq 0$, $-A \approx -B$, and B is a positive infinite hyperreal, then $f(-A) \approx f(-B)$.*

Theorem 12 is helpful in chapter 5.

Exercises 3.6

1–8. Rapid response: state the equation of the horizontal asymptote, if any, indicated by the limit.

1. $\lim_{x\to\infty} \dfrac{x^2 + 3x - 9}{4x^2 - 7} = \dfrac{1}{4}$
2. $\lim_{x\to-\infty} \dfrac{x^2 - 8x + 6}{4x^3 - 7x^2 + x} = 0$
3. $\lim_{x\to-\infty} f(x) = 14$
4. $\lim_{x\to\infty} g(x) = -\infty$
5. $\lim_{x\to\infty} (x^2 - 7) = \infty$
6. $\lim_{x\to\infty} g(x) = \pi$
7. $\lim_{x\to 3} f(x) = 4$
8. $\lim_{x\to 2^+} g(x) = \infty$

9–14. Rapid response: state whether the limit indicates a horizontal asymptote, a vertical asymptote, or neither. (Assume that a and b are real numbers.)

No one limit can indicate both a vertical and a horizontal asymptote, so "both" is not listed as a possible answer.

9. $\lim_{x\to a} f(x) = \infty$

10. $\lim_{x\to\infty} f(x) = a$

11. $\lim_{x\to b} f(x) = a$

12. $\lim_{x\to b^-} f(x) = \infty$

13. $\lim_{x\to\infty} f(x) = \infty$

14. $\lim_{x\to -\infty} f(x) = -b$

15–18. Find (a) $\lim_{x\to\infty} f(x)$ and (b) $\lim_{x\to -\infty} f(x)$ from the graph of $y = f(x)$.

15.

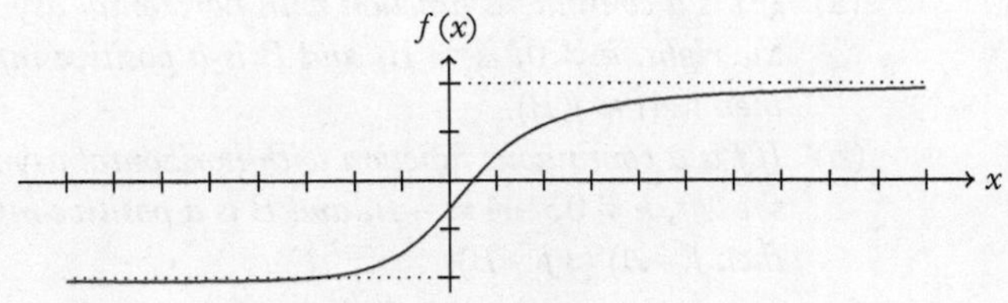

16.

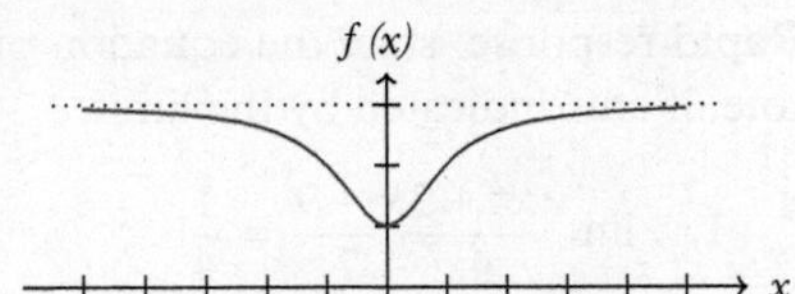

17.

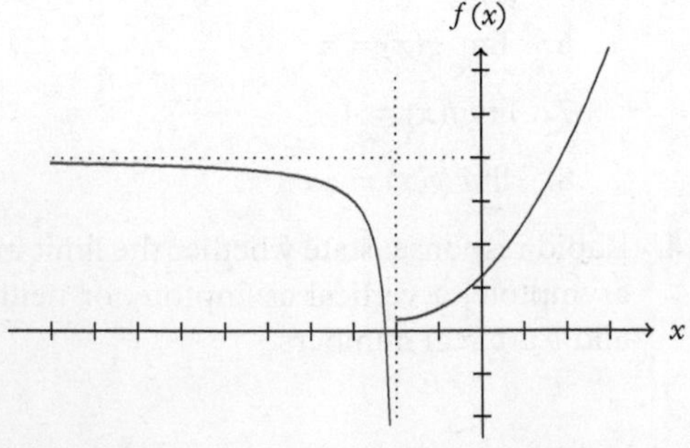

18.

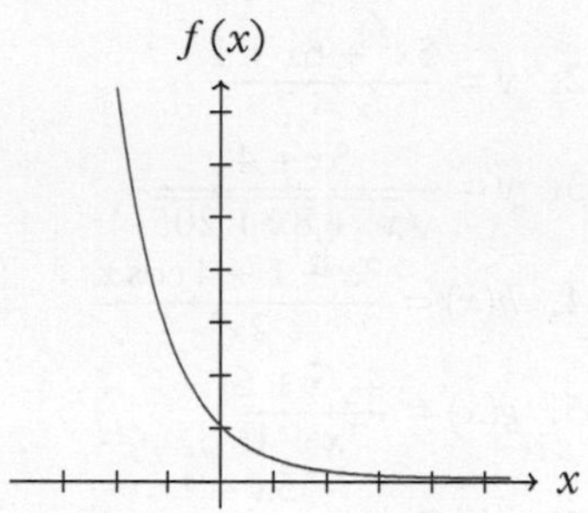

19–40. Calculate the limit.

Remember to watch for violations of the approximation principle.

19. $\lim_{x\to\infty} \frac{x^2 - 6x + 7}{5x^3 + 2x^2 + 4}$

20. $\lim_{x\to\infty} \frac{8x^5 - 7x^3 + x^2}{3x^5 - 9x^2 + 1}$

21. $\lim_{x\to\infty} \frac{5x^3 + 2x^2 + 4}{x^2 - 6x + 7}$

22. $\lim_{x\to\infty} \frac{3x^5 - 9x^2 + 1}{8x^5 - 7x^3 + x^2}$

23. $\lim_{x\to-\infty} \frac{x^2 - 6x + 7}{5x^3 + 2x^2 + 4}$

24. $\lim_{x\to\infty} \frac{3x^2 - 4\sqrt{x} + 1}{4x^2 + x^{1.9}}$

25. $\lim_{x\to\infty} (2x^3 - 3x^2)$

26. $\lim_{x\to-\infty} 4x^5$

27. $\lim_{x\to\infty} \left(4x - 7 + \frac{1}{x}\right)$

28. $\lim_{x\to-\infty} \frac{2x^{27} - x^{11}}{3x - 4}$

29. $\lim_{x\to-\infty} \frac{4x^2 - 3x + 12}{7x^2 - 11x}$

30. $\lim_{x\to-\infty} \frac{2x - 114}{x^2 + 9x}$

31. $\lim_{x\to-\infty} \frac{6x\cos x}{7x^2 + 1}$

32. $\lim_{x\to\infty} \left(4 + \frac{1}{2x}\sin x\right)$

33. $\lim_{x\to\infty} \frac{4x^2 - 7x^3 + 2x}{3x^2 - 11}$

34. $\lim_{x\to-\infty} \left(x^2 + \frac{x^5 - x^4}{1 - x^3}\right)$

35. $\lim_{x\to\infty} \frac{2x^{7/3} + x^2 - 7x}{x^{9/4} - 4x^{16/7} + 5x^{7/3}}$

36. $\lim_{x\to\infty} x^{-1/2}$

37. $\lim_{x\to-\infty} \frac{6x^2\cos x^2}{x^2 + 4}$

38. $\lim_{x\to\infty} \frac{9x^3 - 3x + 12}{x^2 - 4x - 5x^3}$

39. $\lim_{x\to\infty} \left(\frac{3x^4 + 11x + 14}{x^2 + 1} - 3x^2\right)$

40. $\lim_{x\to\infty} \frac{2x\sin x}{3x + 11}$

41–50. Find all horizontal asymptotes on the graph of the given function.

It is implied that limits should be used.

41. $f(x) = \dfrac{4x^3 - 4x^2 + 1}{x^3 - 5}$

42. $y = \dfrac{5x^2 + 6x + 1}{x - 7}$

43. $y = \dfrac{3x + 4}{\sqrt{x^2 + 8x + 205}}$

44. $h(x) = \dfrac{3x + 1 + 4\cos x}{2x^2}$

45. $g(x) = \dfrac{4\sqrt{x} + 6}{7x + 1}$

46. $f(x) = \dfrac{5x - 1}{\sqrt{x^2 + 12x + 927}}$

47. $g(x) = \dfrac{x^3 - 6x + 16}{\sqrt[3]{x^5 - 7x^2 + 197}}$

48. $s(t) = \dfrac{9t^2 - \sqrt{2t^2 + 1}}{4t^2 + \sqrt{7t^4 - 8t^3}}$

49. $f(x) = \dfrac{\sin(2x^3 + \sqrt{x})}{5x}$

50. $y = 3 - \dfrac{\cos x}{x}$

51–56. Find all asymptotes on the graph of the given function.

51. $f(x) = \dfrac{5x - 7}{x - 2}$

52. $f(x) = \dfrac{x - 4}{x^2 - 1}$

53. $f(x) = \dfrac{2x^2 - 5x - 3}{x^2 + x - 12}$

54. $f(x) = \dfrac{1}{x + 1} - \dfrac{1}{x + 4}$

55. $g(x) = \dfrac{x^3 - 7x^2 + 1}{x^2 + 1}$

56. $g(x) = \dfrac{(x + 1)(x + 2)(x + 3)}{(x + 4)(x + 5)}$

57–62. Find the given limit.

57. (GSL) $\lim\limits_{h\to\infty} \dfrac{3h + 2xh^2 + x^2h^3}{4 - 3xh - 2x^3h^3}$

58. $\lim\limits_{x\to\infty} \left(\dfrac{x^2 + 3x}{2x + 1} - \dfrac{2x + 7}{4}\right)$

59. $\lim\limits_{x\to\infty} \tan\left(\dfrac{x^2 + 5}{3x^2 - 1}\right)$

60. $\lim\limits_{x\to\infty} \dfrac{2x + \cos x}{5x - \sin x}$

61. $\lim\limits_{x\to\infty} x \sin\dfrac{3}{x}$

62. $\lim\limits_{x\to\infty} \dfrac{x^3 \sin\frac{1}{x^2}}{2 + 4x}$

63. Let $f(x) = a_n x^n + a_{n-1}x^{n-1} + \cdots + a_1 x + a_0$ be a polynomial of degree n, with n being even. Give conditions on a_n for which (a) $\lim_{x\to\infty} f(x) = \infty$, (b) $\lim_{x\to\infty} f(x) = -\infty$, (c) $\lim_{x\to-\infty} f(x) = \infty$, and (d) $\lim_{x\to-\infty} f(x) = -\infty$.

64. Let $f(x) = a_n x^n + a_{n-1}x^{n-1} + \cdots + a_1 x + a_0$ be a polynomial of degree n, with n being odd. Give conditions on a_n for which (a) $\lim_{x\to\infty} f(x) = \infty$, (b) $\lim_{x\to\infty} f(x) = -\infty$, (c) $\lim_{x\to-\infty} f(x) = \infty$, and (d) $\lim_{x\to-\infty} f(x) = -\infty$.

65. For the submarine of section 2.8 example 2, imagine that the trip could extend infinitely far. How fast is the distance between the radar station and the submarine increasing Ω minutes later?

66. Evaluate $\lim_{n\to\infty}\left(n + \frac{1}{2} - \sqrt{n(n+1)}\right)$. (This is the limiting value of the difference between the arithmetic mean of n and $n+1$ and the geometric mean of n and $n+1$.)

Hint: the answer is not $\frac{1}{2}$. Watch for subtle violations of the approximation principle.

67. The attraction between two atoms is commonly modeled by the simple Lennard-Jones potential,

This exercise is adapted from materials developed by Brian Faulkner.

The answer to this exercise is not provided.

$$U(r) = \frac{A}{r^{12}} - \frac{B}{r^6},$$

where U is the potential energy between two molecules (in electron volts), r is the distance between these two molecules (in nanometers), and A and B are empirical parameters that depend on the substance under consideration. The place that this potential energy is minimized is the place atoms settle in the long term. (a) What units must A have? What units must B have? (b) What happens to the potential energy when the atoms are very far apart? (c) What happens to the potential energy when the atoms are very close together? (d) At what distance of separation do the atoms settle in the long term? (Write this distance r_0 in terms of A and B.)

Curve Sketching: General Functions

3.7

When sketching the graphs of polynomials, we used information on extrema, inflection points, and intervals of concavity. We also used information on intervals of increase and decrease to check the graph. This was enough because polynomials are continuous and differentiable, and have no asymptotes. For nonpolynomials, there are other considerations.

Turning information into a sketch

The information we can collect about a function using calculus has now grown to a rather long list:

- discontinuities, including vertical asymptotes, removable discontinuities, and jump discontinuities
- corners or vertical tangents (continuous but not differentiable)
- intervals of increase/decrease
- extreme points
- intervals of concavity
- inflection points
- horizontal asymptotes

If desired, additional information such as x- or y-intercepts or other individual points can be added. The issue becomes how to take this information and turn it into a sketch. To keep the process organized without having to memorize a too-detailed 20-step procedure, we use the following steps:

❶ Plot extreme points, points of inflection, and other points such as intercepts (when expedient).

❷ Produce graph snippets for any discontinuities found (based on limit information), as practiced in sections 1.4 and 1.5. Draw the dotted lines for any horizontal asymptotes as well (also based on limit information).

If there are infinitely many extrema and inflection points, such as with $y = \sin x$, a representative graph window that illustrates the pattern (if there is one) is chosen and only the portion of the graph in that window is drawn. Similar advice applies to later steps as well.

❸ Identify and plot points where the function is continuous but not differentiable.

❹ Identify the pieces of the curve to be drawn. Pieces begin and end at discontinuities, points from step ❸, and inflection points.

❺ Draw the pieces of the curve one at a time, using the correct concavity, passing through relevant points appropriately (i.e., ensuring extrema are represented correctly), and connecting graph snippets. For the left- and right-most pieces, which should be drawn first, any horizontal asymptotes must also be approached.

For the examples and exercises we examine, step ❸ is needed only occasionally.

Example 1 *Sketch the graph of a function f for which we have found the following information: f has an x-intercept at* $x = -3$, *a local minimum at* $(-1, -5)$, *and an inflection point at* $(0, -3)$; *f is concave up on the intervals* $(-\infty, -2)$ *and* $(-2, 0)$; *and f is concave down on the interval* $(0, \infty)$, $\lim_{x\to\infty} f(x) = -1$, $\lim_{x\to-\infty} f(x) = -1$, *and* $\lim_{x\to-2} f(x) = \infty$.

Solution ❶ We begin by plotting the x-intercept, local min, and inflection point. We leave plenty of extra room on all sides. See figure 1.

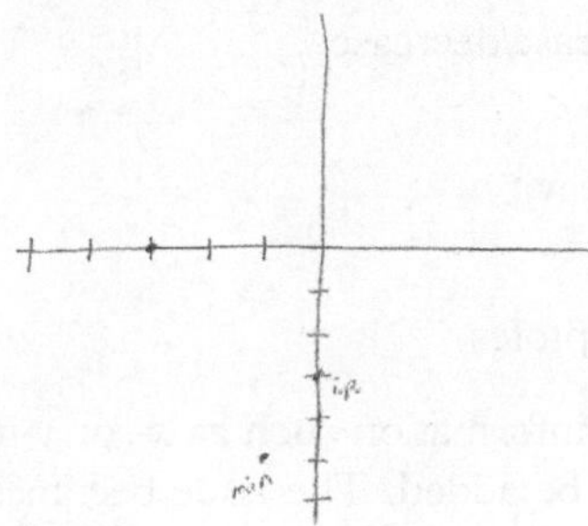

Figure 1 *Points plotted*

❷ Because $\lim_{x\to-2} f(x) = \infty$, there is a vertical asymptote (infinite discontinuity) at $x = -2$. We draw a dotted line at $x = -2$. Because the two-sided limit is ∞, we draw graph snippets next to the asymptote at the top of the picture. Since $\lim_{x\to\infty} f(x) = -1$ and $\lim_{x\to-\infty} f(x) = -1$, there is a horizontal asymptote $y = -1$ on both the right and left sides. We also draw that dotted line. See figure 2.

❸ No continuous nondifferentiable points are given.

❸ From the concavity and inflection point information, f is differentiable everywhere except possibly $x = -2$. Because f has an infinite discontinuity at $x = -2$, f is differentiable everywhere it is continuous.

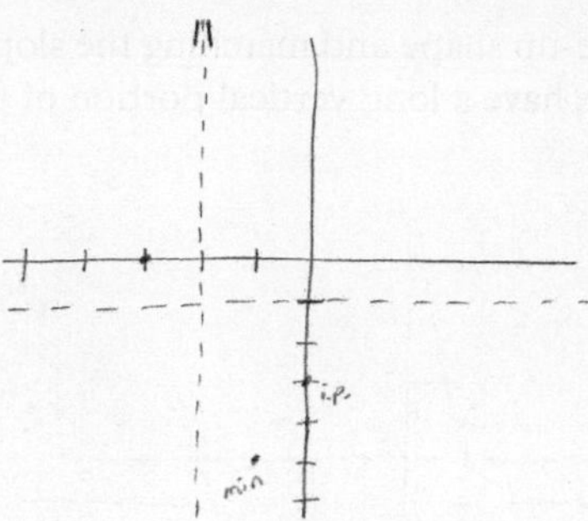

Figure 2 *Asymptotes and graph snippets added. Because we do not yet know whether we will approach the horizontal asymptote from above or below, we do not draw graph snippets for the horizontal asymptote*

❹ There is one discontinuity (at $x = -2$) and one inflection point (at $x = 0$). Our three pieces are to the left of $x = -2$ (the interval $(-\infty, -2)$), to the right of $x = 0$ (the interval $(0, \infty)$), and between the two (the interval $(-2, 0)$).

❺ We draw the outside pieces first. The piece on the left is concave up, must go through the plotted point, and must approach both asymptotes. Beginning at the graph snippet for the vertical asymptote, we draw right to left, downward through the point and toward the horizontal asymptote, keeping the concave-up shape.

The piece on the right is concave down and must approach the horizontal asymptote. Beginning at the inflection point, we start the curve upward toward the horizontal asymptote and then approach it, always keeping the concave-down shape. See figure 3.

Recall that concave up does not require us to hit a minimum and come back up; we just need to keep the tangent lines below the curve. Likewise, concave down does not require us to hit a maximum and come back down; we just need to keep the tangent lines above the curve.

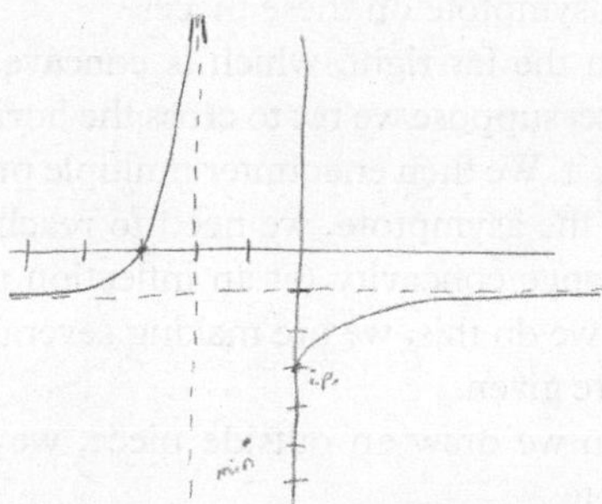

Figure 3 *Portions of graph on left and right drawn*

Next we draw the piece in the middle, which is concave up. We need to connect with the graph snippet, pass through the minimum, and connect to the inflection point. We draw the curve accordingly,

If the slope at the inflection point was not drawn steeply enough for these criteria to be met, then the previous step may need to be redrawn with this in mind.

keeping the concave-up shape and matching the slope at the inflection point (figure 4). We have a long vertical portion of the curve to draw

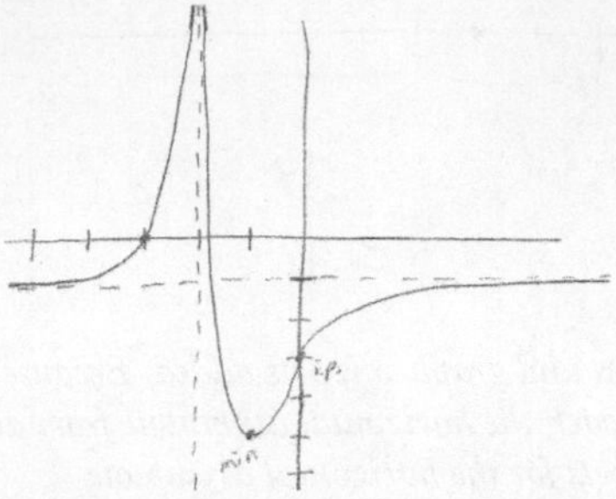

Figure 4 *Middle portion added, completing the graph*

between the graph snippet next to the vertical asymptote and the local min, so it's okay if the graph looks nearly straight vertically there as long as it does not ever have the wrong concavity. Slight imperfections (interpreted carefully) in the sketch resulting from an unsteady hand may be hard to avoid and are generally not of importance. But be careful not to have the graph tails move away from horizontal asymptotes, as cautioned in figure 5. ■

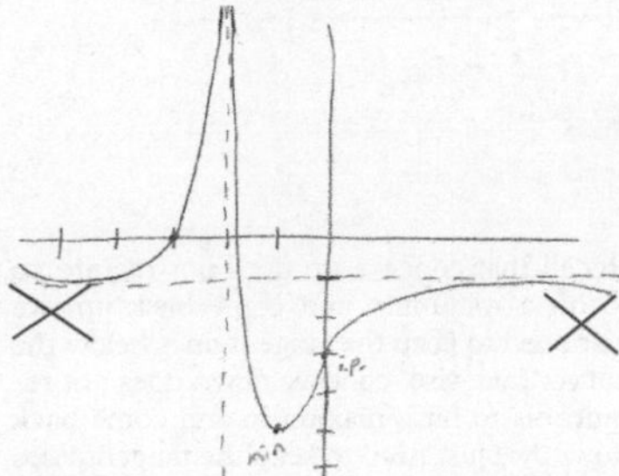

Figure 5 *If the tails of the graph move away from a horizontal asymptote as pictured, then not only are the limits and horizontal asymptote misrepresented, but also additional local extrema are created*

CAUTION: MISTAKE TO AVOID

In section 3.6 it was noted that curves can cross horizontal asymptotes, even infinitely often. When sketching the curve for example 1, we crossed the horizontal asymptote when drawing the middle piece (see figure 4). But, we did not cross the horizontal asymptote when drawing the outside pieces. Why not? How do we know the graph does not cross the asymptote on these pieces?

For the piece on the far right, which is concave down and starts below the asymptote, suppose we try to cross the horizontal asymptote before approaching it. We then encounter multiple problems. To come back down toward the asymptote, we need to reach a local max and we also need to change concavity (at an inflection point) to concave up; see figure 6. If we do this, we are making several violations of the information we were given.

Therefore, when we draw an outside piece, we cannot cross the horizontal asymptote.

DRAWING OUTSIDE PIECES

When drawing outside pieces of a continuous function (to the right of all discontinuities, corners, inflection points, and local

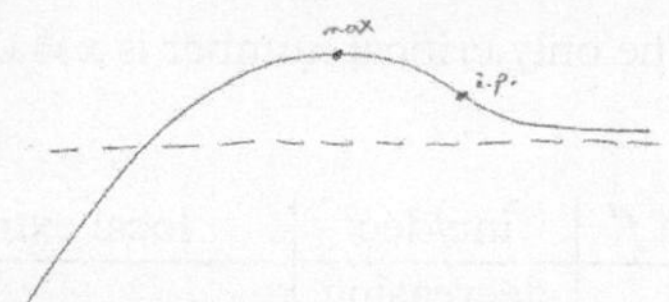

Figure 6 *Drawing a curve crossing a horizontal asymptote to approach from the other side requires an additional extremum and inflection point*

extrema, or to the left of all such), we cannot cross a horizontal asymptote. The asymptote must be approached from the current side (above or below).

Outside pieces do not always exist; see figure 7.

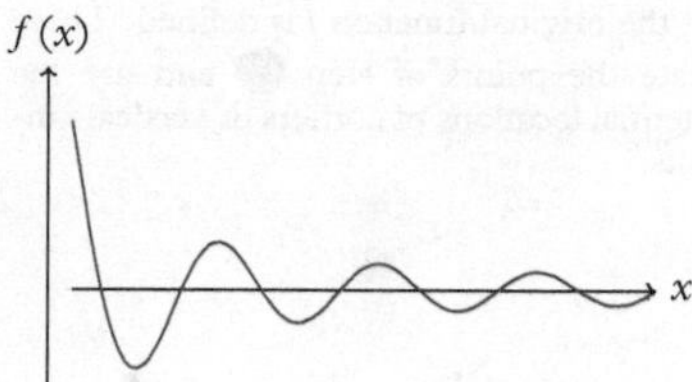

Figure 7 *When the graph of a function crosses its horizontal asymptote in an oscillatory manner, there are no outside pieces, for there are always more local extrema*

Complete curve-sketching examples

Example 1 provided the information needed to sketch the curve. What if we need to gather this information? The complete process is lengthy but highly informative. The next example illustrates the entire process for a relatively simple function.

Example 2 *Sketch the graph of* $f(x) = \sqrt[3]{x^2}$.

Solution For any function, a good place to start is to determine its domain. Because f contains only an odd root and no denominator, it is defined on all real numbers.

Some CASs designed to work in the setting of complex numbers may interpret $x^{2/3}$ differently than $\sqrt[3]{x^2}$. Caution must be used to craft the input carefully to ensure the domain is interpreted the way it is meant.

Our earlier list of information that calculus can provide gives us a list of what to explore. We start with discontinuities. Because root functions are continuous where defined, the function is continuous everywhere; there are no discontinuities.

Next on the list is corners or vertical tangents, which may be identified as part of the process of finding intervals of increase/decrease and extreme points. We defer this item momentarily.

For intervals of increase/decrease, we need to find the derivative and identify critical numbers. Differentiating, we have

$$f'(x) = \frac{2}{3}x^{-1/3} = \frac{2}{3\sqrt[3]{x}}.$$

To find critical numbers, we set both the numerator and the denominator equal to zero and solve; $2 = 0$ has no solutions and $3\sqrt[3]{x} = 0$

To solve, divide both sides by 3 and then cube both sides.

has solution $x = 0$. The only critical number is $x = 0$, and our chart is completed easily:

interval	sign of f'	inc/dec	local extrema
$(-\infty, 0)$	$-$	decreasing	$\leftarrow$ local min at $x = 0$
$(0, \infty)$	$+$	increasing	

For the purpose of graphing, we need the extreme points and not just their locations, so we find the y-coordinate as well:

$$f(0) = \sqrt[3]{0^2} = 0.$$

The local minimum point is $(0, 0)$. Notice that because f' is undefined at $x = 0$, there is no horizontal tangent line at this local min. In other words, the local min might be at a corner in the graph.

Recall that critical numbers include not just where $f'(x) = 0$ and the tangent line is horizontal, but also where f' is not defined but the original function f is defined. These locate the points of step ❸ and are the potential locations of corners or vertical tangents.

We continue by finding intervals of concavity and inflection points. The second derivative is

$$f''(x) = -\frac{2}{9}x^{-4/3} = -\frac{2}{9x^{4/3}} = -\frac{2}{9\sqrt[3]{x^4}}.$$

Possible inflection locations are found by setting both the numerator and the denominator equal to zero and solving; again, $2 = 0$ has no solutions, but $9x^{4/3} = 0$ has the solution $x = 0$. We set up the chart with two intervals.

interval	sign of f''	concavity	inflection
$(-\infty, 0)$	$-$	down	$\leftarrow$ no i.p.
$(0, \infty)$	$-$	down	

We see there are no inflection points.

Last on our list is horizontal asymptotes, which are detected using limits:

$$\lim_{x\to\infty} \sqrt[3]{x^2} = \sqrt[3]{\Omega^2} = \Omega^{2/3} \doteq \infty$$

and

$$\lim_{x\to-\infty} \sqrt[3]{x^2} = \sqrt[3]{(-\Omega)^2} = \sqrt[3]{\Omega^2} = \Omega^{2/3} \doteq \infty.$$

Because these limits are not real numbers, there are no horizontal asymptotes on the graph of the function.

The information-gathering stage is over and we may proceed to graph the function.

❶ There is one point to plot: a local min that might be a corner. We plot the point $(0, 0)$ and below it write both *min* and *corner?* (see figure 8).

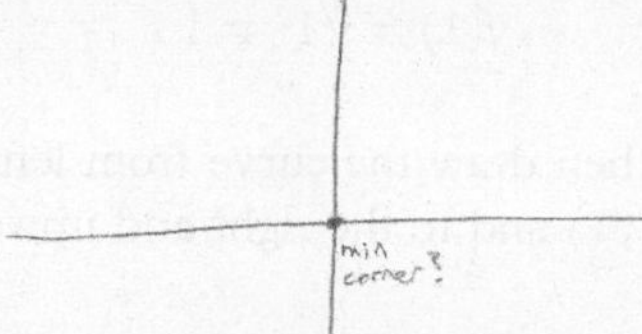

Figure 8 *Point plotted*

❷ There are no discontinuities or horizontal asymptotes.

❸ There is a point at which the function is continuous but not differentiable, identified earlier as the point $(0, 0)$. This point is also the local min and has already been plotted on the graph.

❹ There are no discontinuities and no inflection points, but there is a point from step ❸—namely, $(0, 0)$. Our two pieces are to the left of $x = 0$ and to the right of $x = 0$.

❺ To the left of $x = 0$, the curve is concave down. We have no other points to go through. What other information do we have? The curve is decreasing on this interval, and $(0, 0)$ is a local min. We also know there is no horizontal asymptote; as $x \to -\infty$, our y-coordinates rise to ∞. Although it is not necessary, sometimes plotting an additional point can help. Let's choose something simple, such as $x = -1$:

We also can surmise that because $f'(0)$ is undefined, the curve might have slopes that become infinite at $(0, 0)$.

$$f(-1) = \sqrt[3]{(-1)^2} = \sqrt[3]{1} = 1.$$

We plot the point $(-1, 1)$ and then draw the curve from right to left, from the local min through $(-1, 1)$ and to the left and upward, concave down (figure 9).

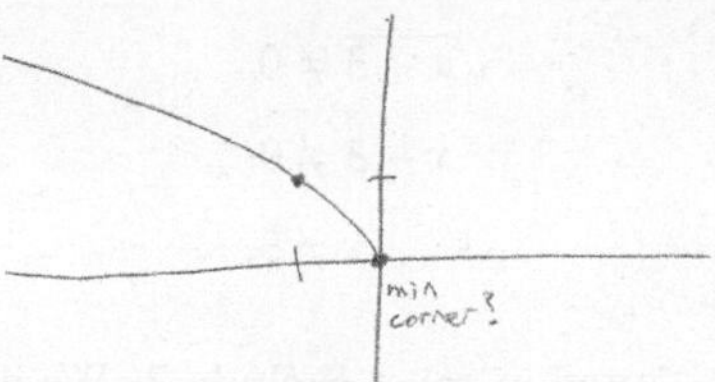

Figure 9 *Left side of the graph drawn from the local min leftward and upward, through the added point, concave down*

To the right of $x = 0$ is similar. The curve is concave down and increasing. According to our limit calculation, as $x \to \infty$ the y-coordinates rise to ∞. Choosing to plot the point with $x = 1$, we find

$$f(1) = \sqrt[3]{1^2} = 1$$

and plot $(1, 1)$. We then draw the curve from left to right, from the local min through $(1, 1)$ and to the right and upward, concave down (figure 10).

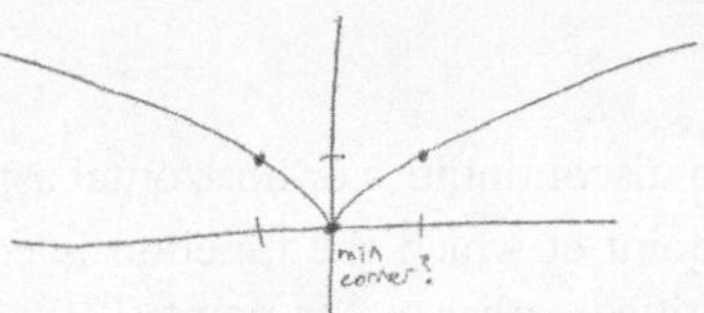

Figure 10 *Right side of the graph added, drawn from the local min rightward and upward, through the added point, concave down*

The graph looks symmetric, and it is. Notice that we do have a corner, not a tangent line, at the local min $(0, 0)$. ■

The final example of this section illustrates a few of the many additional issues that can arise when sketching the graphs of functions. In practice, such graphs are usually explored using technology, but knowing how to work such examples by hand is indispensable to understanding how to work with the technology and understand the results.

Example 3 *Sketch the graph of* $f(x) = \dfrac{x}{\sqrt{x-3}}$.

When reading this example, pay special attention to the implications of the domain, which must be considered several times.

Solution Once again we start by finding the function's domain. We must avoid division by zero:

$$\begin{aligned} \sqrt{x-3} &\neq 0 \\ x-3 &\neq 0 \\ x &\neq 3. \end{aligned}$$

The domain of the function must exclude 3. We also have a square root, which requires the inside expression to be nonnegative:

$$\begin{aligned} x-3 &\geq 0 \\ x &\geq 3. \end{aligned}$$

Combining the information, we see that the domain is $x > 3$, or $(3, \infty)$.

Next we look for discontinuities, including vertical asymptotes. We are dividing by zero when $x = 3$, so there may be a vertical asymptote there. We need to check a limit to see. However, notice that the domain includes only numbers to the right of $x = 3$, not to the left; therefore, we can only take the limit from the right:

$$\lim_{x\to 3^+} \frac{x}{\sqrt{x-3}} = \frac{3+\omega}{\sqrt{3+\omega-3}} = \frac{3+\omega}{\sqrt{\omega}} \approx \frac{3}{\sqrt{\omega}} = 3\sqrt{\Omega} \doteq \infty.$$

If we try the other side, we have

$$\lim_{x\to 3^-} \frac{x}{\sqrt{x-3}} = \frac{3-\omega}{\sqrt{3-\omega-3}} = \frac{3-\omega}{\sqrt{-\omega}},$$

which is undefined (square root of a negative number).

Because the limit is infinite, there is a vertical asymptote at $x = 3$.

For intervals of increase/decrease and extrema, we need to know the derivative, which we calculate using the quotient rule and chain rule:

$$f'(x) = \frac{\sqrt{x-3}\cdot 1 - x\cdot \frac{1}{2\sqrt{x-3}}\cdot 1}{\left(\sqrt{x-3}\right)^2} = \frac{\sqrt{x-3} - \frac{x}{2\sqrt{x-3}}}{x-3}.$$

Further simplification is in order, especially because we eventually need f''. One obvious simplification is to eliminate the compound fraction by multiplying the numerator and the denominator by $2\sqrt{x-3}$:

$$= \frac{\sqrt{x-3} - \frac{x}{2\sqrt{x-3}}}{x-3}\cdot\frac{2\sqrt{x-3}}{2\sqrt{x-3}}$$

$$= \frac{\sqrt{x-3}\cdot 2\sqrt{x-3} - \frac{x}{2\sqrt{x-3}}\cdot 2\sqrt{x-3}}{(x-3)2\sqrt{x-3}}$$

$$= \frac{2(x-3)-x}{2(x-3)^{3/2}} = \frac{x-6}{2(x-3)^{3/2}}.$$

Now we may find our critical numbers by setting both the numerator and the denominator equal to zero. Setting $x - 6 = 0$ yields $x = 6$, whereas $2(x-3)^{3/2} = 0$ gives $x = 3$. Since $x = 3$ is not in the domain of f, the only critical number is $x = 6$.

When setting up the chart, we must remember to include only those values in the domain of the function. We therefore only have two intervals. We complete the chart using the test points illustrated in figure 11:

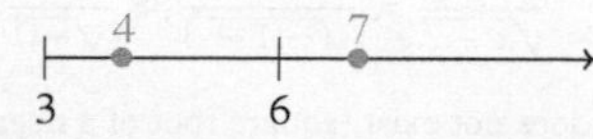

Figure 11 *Intervals and convenient test points for the increasing/decreasing chart. Because the domain is $x > 3$, we do not use an interval to the left of 3*

interval	sign of f'	inc/dec	local extrema
$(3, 6)$	−	decreasing	← local min at $x = 6$
$(6, \infty)$	+	increasing	

We finish finding the local min by determining its y-coordinate:

If we wish, we can rationalize the denominator:

$$\frac{6}{\sqrt{3}} \cdot \frac{\sqrt{3}}{\sqrt{3}} = \frac{6\sqrt{3}}{3} = 2\sqrt{3}.$$

However, all we really need to locate the point on the graph is a decimal approximation.

$$f(6) = \frac{6}{\sqrt{6-3}} = \frac{6}{\sqrt{3}} \approx 3.4.$$

The local minimum point is approximately $(6, 3.4)$.

Intervals of concavity and inflection points require the second derivative:

Factor out the term $(x-3)^{1/2}$ in the numerator.

Use laws of exponents to simplify $\dfrac{(x-3)^{1/2}}{(x-3)^3}$.

$$\begin{aligned} f''(x) &= \frac{2(x-3)^{3/2} \cdot 1 - (x-6) \cdot 3(x-3)^{1/2} \cdot 1}{\left(2(x-3)^{3/2}\right)^2} \\ &= \frac{2(x-3)^{3/2} - 3(x-6)(x-3)^{1/2}}{4(x-3)^3} \\ &= \frac{(x-3)^{1/2}\left(2(x-3)^1 - 3(x-6)\right)}{4(x-3)^3} \\ &= \frac{2x-6-3x+18}{4(x-3)^{5/2}} = \frac{-x+12}{4(x-3)^{5/2}}. \end{aligned}$$

Possible inflection locations are then found by setting the numerator equal to zero (solution, $x = 12$) and the denominator equal to zero (solution, $x = 3$), and because $x = 3$ is not in the domain of f, the only possible inflection location is $x = 12$. Once again, the chart has only two intervals. The test points are illustrated in figure 12.

4 19
3 12

Figure 12 *Intervals and convenient test points for the concavity chart. Because the domain is $x > 3$, we do not use an interval to the left of 3*

interval	sign of f''	concavity	inflection
$(3, 12)$	+	up	$\leftarrow$ i.p. at $x = 12$
$(12, \infty)$	−	down	

We finish finding the point of inflection by finding its y-coordinate:

$$f(12) = \frac{12}{\sqrt{12-3}} = \frac{12}{\sqrt{9}} = 4.$$

The inflection point is $(12, 4)$.

Next we check for horizontal asymptotes. Because f is not defined to the left of $x = 3$, there cannot be a horizontal asymptote on the left side. We check only the right side:

If we try checking the left side, we have

$$\lim_{x\to-\infty} \frac{x}{\sqrt{x-3}} = \frac{-\Omega}{\sqrt{-\Omega-3}} \approx \frac{-\Omega}{\sqrt{-\Omega}},$$

which does not exist (square root of a negative number).

$$\lim_{x\to\infty} \frac{x}{\sqrt{x-3}} = \frac{\Omega}{\sqrt{\Omega-3}} \approx \frac{\Omega}{\sqrt{\Omega}} = \sqrt{\Omega} \doteq \infty.$$

There is no horizontal asymptote on the graph of the function.

The information-gathering stage is over (finally!) and we may proceed to graph the function.

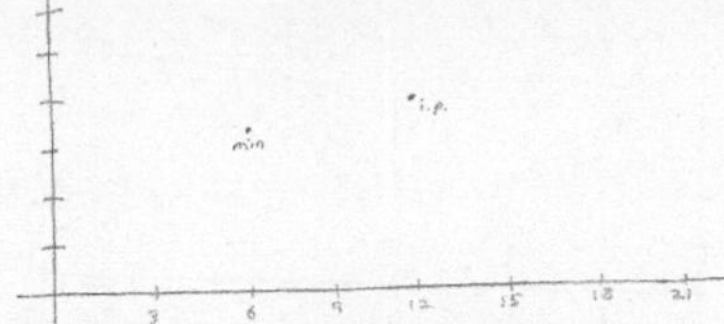

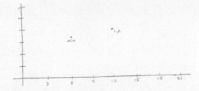

Figure 13 *Points plotted*

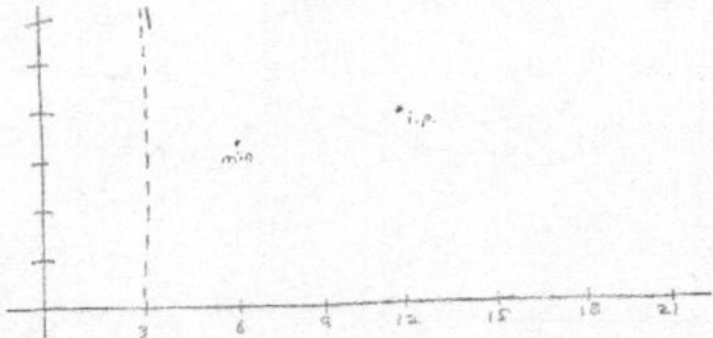

Figure 14 *asymptote and graph snippet added*

❶ We plot the local min and inflection point (figure 13).

❷ We draw the vertical asymptote $x = 3$, with a graph snippet at the top of the picture on the right side of the vertical asymptote (remember, the function is not defined to the left of $x = 3$). See figure 14.

❸ There were no critical values of the type f' undefined.

❹ One piece of the graph is from the vertical asymptote to the inflection point and the other piece is to the right of the inflection point.

❺ Drawing the outside pieces first means drawing the right-side piece first. This piece is concave down and increasing, and rises toward ∞ as $x \to \infty$; there is no horizontal asymptote to approach. Plotting one additional point helps:

$$f(19) = \frac{19}{\sqrt{19-3}} = \frac{19}{4}.$$

We plot the point $(19, 4.75)$, then draw the curve from the inflection point through $(19, 4.75)$ and onward to the right and up, concave down (figure 15).

We finish by drawing the other piece, starting with the inflection point (to match the slope), passing through the min, and connecting to the graph snippet next to the vertical asymptote, keeping the concave-up shape (figure 16).

The graph is now complete. ■

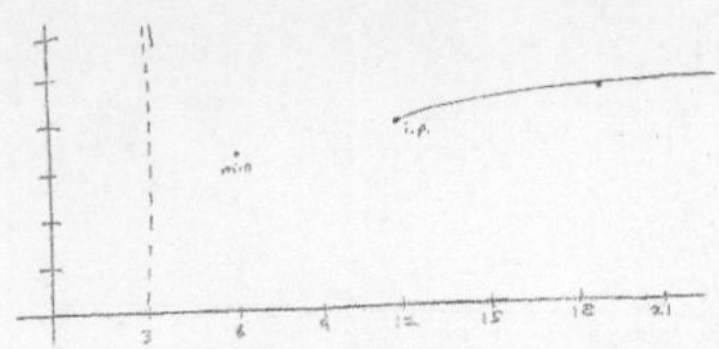

Figure 15 *The right side of the graph is added, including an additional point to help*

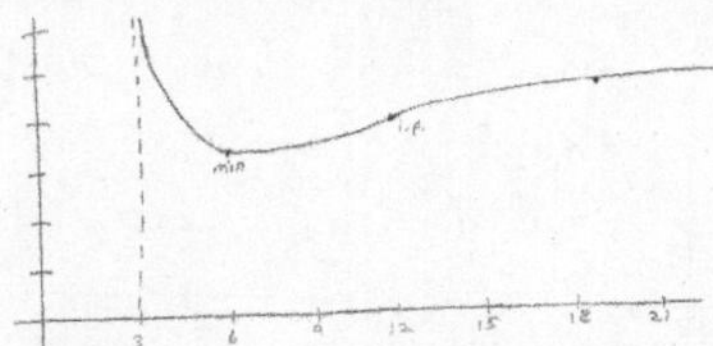

Figure 16 *The finished graph*

The graph in figure 16 may look unusual, but that's the point. Even when sketching an unusual-looking function by hand, calculus can provide all the necessary details!

As in section 3.5, where we graphed polynomials, the use of a CAS can reduce significantly the time and effort required to detect, visualize, and understand all the features of a graph, as long as one knows how to look for them.

Exercises 3.7

1–10. Sketch the graph of a function f with the properties stated.

1. f has a local max at $(0, 4)$, a local min at $(-3, -3)$, and inflection points at $(1, 2)$, $(-2, 0)$, and $(-4, -2)$; $\lim_{x\to\infty} f(x) = 1$ and $\lim_{x\to-\infty} f(x) = -1$; f is concave up on the intervals $(-4, -2)$ and $(1, \infty)$; and f is concave down on the intervals $(-\infty, -4)$ and $(-2, 1)$.
2. f has a local max $(0, 0)$ and inflection points $(-2, -2)$ and $(2, -2)$, f is concave down on the interval $(-2, 2)$, f is concave up on the intervals $(-\infty, -2)$ and $(2, \infty)$, and $\lim_{x\to\infty} f(x) = -5$ and $\lim_{x\to-\infty} f(x) = -5$.
3. f has an x-intercept at $x = 3$, a local max $(-1, 3)$, and an inflection point $(-2, 2)$; f is concave up on the interval $(-\infty, -2)$; f is concave down on the intervals $(-2, 1)$ and $(1, \infty)$; and $\lim_{x\to\infty} f(x) = 1$, $\lim_{x\to-\infty} f(x) = 1$, and $\lim_{x\to 1} f(x) = -\infty$.

4. f has an x-intercept at $x = -2$, a local min $(1, 1)$, and an inflection point $(3, 2)$; $\lim_{x\to -1^-} f(x) = -\infty$, $\lim_{x\to -1^+} f(x) = \infty$, $\lim_{x\to\infty} f(x) = 3$, and $\lim_{x\to -\infty} f(x) = 3$; f is concave down on the intervals $(-\infty, -1)$ and $(3, \infty)$; and f is concave up on the interval $(-1, 3)$.
5. f has an x-intercept at $x = 3$, a local min $(1, 1)$, and an inflection point $(-1, 2)$; f is concave up on the interval $(-1, 2)$; f is concave down on the intervals $(-\infty, -1)$ and $(2, \infty)$; and $\lim_{x\to\infty} f(x) = \infty$, $\lim_{x\to -\infty} f(x) = 3$, $\lim_{x\to 2^-} f(x) = \infty$, and $\lim_{x\to 2^+} f(x) = -\infty$.
6. f has x-intercepts $(-2, 0)$ and $(1, 0)$ but has no local extrema or inflection points; $\lim_{x\to -1^-} f(x) = -\infty$, $\lim_{x\to -1^+} f(x) = \infty$, $\lim_{x\to -\infty} f(x) = 2$, and $\lim_{x\to\infty} f(x) = -1$; f is concave down on the interval $(-\infty, -1)$; and f is concave up on the interval $(-1, \infty)$.
7. f has a local min $(0, 0)$ and inflection points $(-2, 2)$ and $(2, 2)$, f is concave up on the interval $(-2, 2)$, f is concave down on the intervals $(-\infty, -2)$ and $(2, \infty)$, and $\lim_{x\to\infty} f(x) = \lim_{x\to -\infty} f(x) = 4$.
8. f has local minimum points $(0, 0)$ and $(2, 0)$, and an inflection point $(3, 1)$; $\lim_{x\to 1} f(x) = 1$; f is undefined at $x = 1$; $\lim_{x\to\infty} f(x) = 2$ and $\lim_{x\to -\infty} f(x) = \infty$; f is concave up on the intervals $(-\infty, 1)$ and $(1, 3)$; and f is concave down on the interval $(3, \infty)$.
9. f has an x-intercept $(-3, 0)$, a local max $(3, 0)$, a local min $(0, -3)$, and an inflection point $(-1, -2)$; $\lim_{x\to 2^-} f(x) = 1$, $\lim_{x\to 2^+} f(x) = -1$, $\lim_{x\to -\infty} f(x) = 1$, and $\lim_{x\to\infty} f(x) = -\infty$; f is concave up on the interval $(-1, 2)$; and f is concave down on the intervals $(-\infty, -1)$ and $(2, \infty)$.
10. f passes through the points $(-4, 2)$ and $(4, 2)$; f has a local min $(0, 3)$; $\lim_{x\to 2} f(x) = \infty$, $\lim_{x\to -2} f(x) = \infty$, and $\lim_{x\to -\infty} f(x) = \lim_{x\to\infty} f(x) = 0$; and f is concave up on the intervals $(-\infty, -2)$, $(-2, 2)$, and $(2, \infty)$.

11–18. Sketch the graph of the function.

11. $f(x) = \dfrac{1}{x}$
12. $f(x) = \dfrac{1}{x^2}$
13. $f(x) = \sqrt{x}$
14. $f(x) = \cos x,\ 0 \le x \le 2\pi$
15. $f(x) = \sin x,\ 0 \le x \le 2\pi$
16. $f(x) = \sqrt{4 - x}$
17. $f(x) = \dfrac{2}{x - 3}$
18. $f(x) = \sqrt[3]{x}$

Although the graphs of these functions should either be on instant recall or are shifts, reflections, and so on, thereof, the procedures of examples 2 and 3 should be used.

19–30. Sketch the graph of the function.

19. (Love) $f(x) = \dfrac{1}{1+x^2}$

20. (Love) $f(x) = \dfrac{x}{1+x^2}$

21. $f(x) = \dfrac{x-2}{x+1}$

22. $f(x) = \dfrac{x+3}{x-2}$

23. (Love) $f(x) = \dfrac{2x}{x^4+3}$

24. (Love) $f(x) = \dfrac{3x^4+1}{x^4+1}$

25. $f(x) = \sqrt{4-x^2}$

26. $f(x) = \sqrt{1+x^3}$

27. $f(x) = \sqrt[3]{4-x^2}$

28. $f(x) = \sqrt[3]{1+x^2}$

29. (Love) $f(x) = \dfrac{4x+4}{x^2+2x+5}$

30. (Love) $f(x) = \dfrac{1}{x^3-3x}$

This group of exercises is from Love, published before the invention of computers. It is possible to do each exercise by hand. Students who pride themselves on their algebraic prowess may wish to try one without technology. Some are not that much more sophisticated than exercises 19–30, whereas others are much more difficult to complete all the details by hand.

31–40. Use a CAS to graph the given function, showing all its features. Use more than one picture if necessary.

31. $f(x) = \dfrac{3x}{x^2-x+1}$

32. $f(x) = \dfrac{2-x}{3-x^2}$

33. $f(x) = \dfrac{x^2}{1+x^4}$

34. $f(x) = \dfrac{4x^3}{x^4+27}$

35. $f(x) = \dfrac{1}{x^3+3x^2}$

36. $f(x) = \dfrac{2x+1}{3x^2+8x+4}$

37. $f(x) = \dfrac{1}{x(x-1)(x-2)}$

38. $f(x) = \dfrac{-x^3}{2+9x+12x^2}$

39. $f(x) = \dfrac{(x^2-1)^2}{x+2}$

40. $f(x) = \dfrac{(x^2-1)^3}{x}$

Decimal approximations are acceptable and expected.

41–50. Use a CAS to find (a) vertical asymptotes, (b) horizontal asymptotes, (c) extreme points, (d) inflection points, (e) intervals of increase and intervals of decrease, and (f) intervals of concavity for the function graphed in a previous exercise.

41. exercise 31

42. exercise 32

43. exercise 33

44. exercise 34

45. exercise 35

46. exercise 36

47. exercise 37

48. exercise 38

49. exercise 39

50. exercise 40

Optimization

3.8

We have studied local and absolute extrema of functions. There are a very large number of practical settings in which one might wish to find these extrema, such as determining the minimum amount of material for completing a project, the maximum sustainable yield of a renewable resource, or the product price that yields a maximum profit. The process of finding such extrema is called *optimization*, for we are trying to find an "optimum," or best, value of the variable.

These exercises are also called *max-min problems*.

Optimization example: maximum enclosed area

The general solution strategy for an optimization problem is to determine the quantity to be optimized, make that quantity the value of a function, and then find the extreme values of that function. Let's examine the details of this strategy using an example.

Example 1 *A farmer's child has purchased a piglet. The farmer has given the child* 60 *feet of fencing left over from another project. Using the side of the barn as one side of a rectangular pig pen, the child wishes to enclose the largest area possible. What dimensions should be used?*

Solution Perhaps the first step in solving an optimization problem is to recognize that it is an optimization problem. This is accomplished by noticing that the stated task involves the largest, the smallest, the greatest, the best, the maximum, the minimum, or [insert optimum word here]. In this case, *largest* is the word used to indicate an optimum value.

As when working any word problem, we draw a picture, if possible. The pig pen is described as a rectangle, so we draw a rectangle. We are looking at the ground from above (a top view, or aerial view). We depict a barn along one side of the rectangle. See figure 1.

We sometimes encounter an optimization exercise in which no picture is helpful.

Although nothing in this example is changing (this is not a related rates exercise), so our diagram is static (diagrams for related rates exercises are dynamic), it is still helpful to visualize various possibilities. We are told that there is 60 feet of fencing to make the rectangular pig pen.

This visualization can be accomplished mentally and does not have to be drawn on paper.

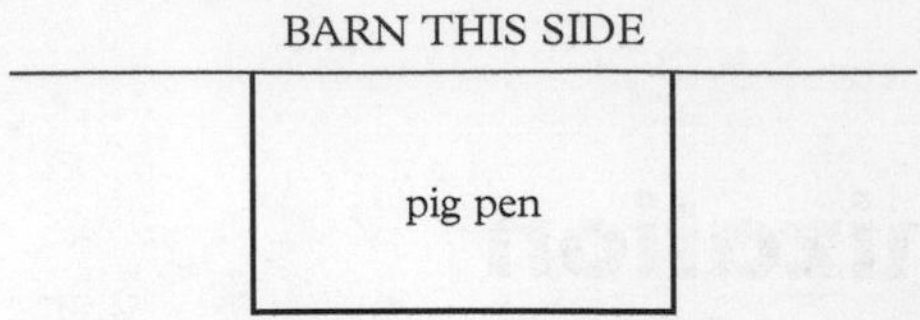

Figure 1 *A fenced-in rectangular pig pen using the barn as one side, viewed from above. The fence occupies three sides of the rectangle*

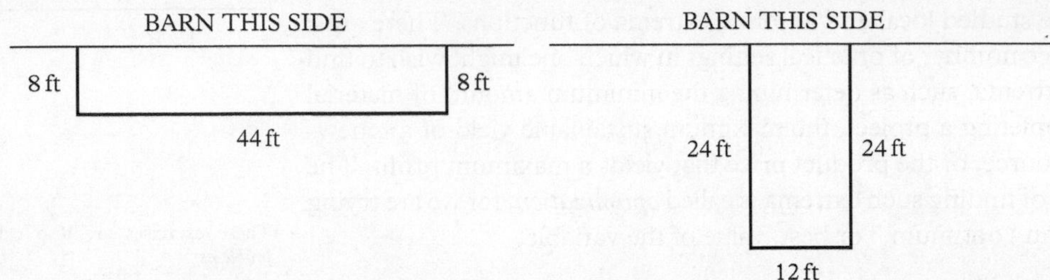

Figure 2 *Two possible configurations of the rectangular pig pen using 60 feet of fencing. The option on the left encloses more area than the option on the right*

We could make the pig pen very wide but not very long (figure 2, left), very long but not very wide (figure 2, right), or something in between.

We cannot draw all possible configurations and check their areas, for there are infinitely many possibilities. For this reason, we introduce one or more variables to help. Let's use ℓ for length and w for width, as in figure 3. Then area, which is the quantity we wish to maximize, is given by

$$A = \ell \cdot w.$$

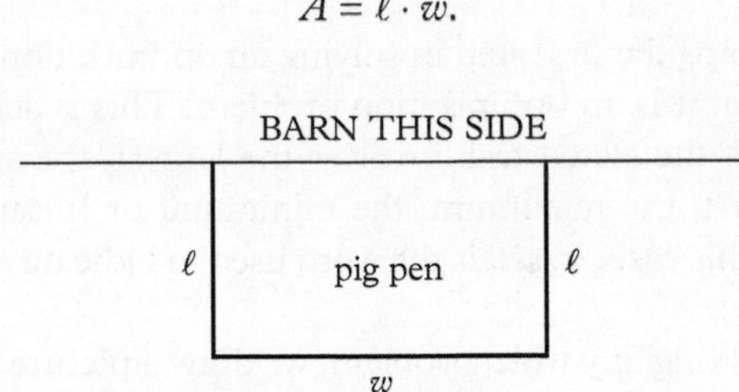

Figure 3 *Variables have been added to the original diagram*

In function notation, we have $A(\ell, w) = \ell \cdot w$, so that area A is a function of two variables, ℓ and w.

The problem with what we just did is that we have two variables, not one. The methods of this chapter help us find extrema of functions of one variable; extrema for functions of two variables is a topic for much later in the study of calculus. We therefore wish to eliminate one of the variables if we can. Thankfully, there is a relationship that can help. Because we have a total of 60 feet of fencing, the three sides add up to 60 feet, or

$$\ell + \ell + w = 60.$$

If we solve for one of the variables, it may be replaced in our diagram and function:

It seems reasonable to assume the maximum area requires that all the fencing be used.

$$2\ell + w = 60$$
$$w = 60 - 2\ell.$$

We replace w with $60 - 2\ell$ in the diagram (figure 4) and rewrite our area function as well:

$$A(\ell) = \ell \cdot (60 - 2\ell).$$

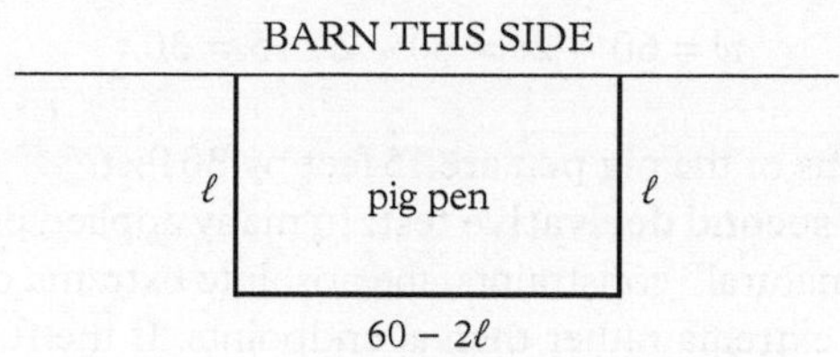

Figure 4 *The variable w has been replaced by* $60 - 2\ell$

Now all that is left is to find the maximum value of the area function. There are three methods for completing this portion of the problem. Although we only need to use one of the methods, all three methods are demonstrated.

This example requests "the largest area," so we want to find the maximum area. We therefore make area the value of our function.

Method 1: closed interval method. We wish to find the absolute maximum value of our function—in this case, the largest possible area. The closed interval method helps us do this, but we need a continuous function (the area function is a polynomial, so it is continuous) on a closed, bounded interval. To find the closed, bounded interval, we look at the possible values of the variable, ℓ. A consultation of our figures helps us see that ℓ cannot be negative—that is, $\ell \geq 0$. Also, ℓ cannot be more than 30 feet, for we have only 60 feet of fencing. Therefore, $0 \leq \ell \leq 30$, and our function is only valid on the interval $[0, 30]$.

To carry out the closed interval method, we first find the critical numbers of our function:

$$A(\ell) = \ell(60 - 2\ell) = 60\ell - 2\ell^2$$
$$A'(\ell) = 60 - 4\ell$$
$$60 - 4\ell = 0$$
$$60 = 4\ell$$
$$15 = \ell.$$

Critical numbers are not always integers, which is why we cannot always find extrema just by trying several possibilities.

We have one critical number: $\ell = 15$. Next we evaluate the function at the endpoints and critical number, and choose the largest:

$$A(0) = 0,$$
$$A(15) = 15(60 - 2 \cdot 15) = 450,$$
$$A(30) = 30(60 - 2 \cdot 30) = 0.$$

The maximum area is 450 ft^2 when $\ell = 15$.

We finish the example by rereading the question to ensure the final form of our answer is what was requested. We are asked for the dimensions (notice the plural!) of the rectangle, which means length and width. We refer back to our formula for w in terms of ℓ:

$$w = 60 - 2\ell = 60 - 2 \cdot 15 = 30.$$

The dimensions of the pig pen are 15 feet by 30 feet.

Method 2: second derivative test. In many applied problems that feature only "natural" constraints, the absolute extrema of a function occur at local extrema rather than at endpoints. If the function to be optimized is continuous and there is only one critical number *and* that critical number is a local extremum of the correct type (a local max if we are trying to maximize the function, or a local min if we are trying to minimize the function), then that local extremum must be the absolute extremum (try drawing some pictures to convince yourself of this fact). Local extrema may be detected by either the first or second derivative test. If the second derivative is not difficult to calculate, then the second derivative test is the shortest method to use.

The advantage of the second derivative test over the closed interval method is that we usually do not need to determine the domain of the function.

To carry out the second derivative test, we first find the critical numbers of the function:

$$A(\ell) = \ell(60 - 2\ell) = 60\ell - 2\ell^2$$
$$A'(\ell) = 60 - 4\ell$$
$$60 - 4\ell = 0$$
$$60 = 4\ell$$
$$15 = \ell.$$

We have one critical number: $\ell = 15$. Next we calculate the second derivative and evaluate it at the critical number:

$$A''(\ell) = -4$$
$$A''(15) = -4 < 0, \text{ max.}$$

The maximum area occurs when $\ell = 15$.

We finish the example by rereading the question to ensure the final form of our answer is what was requested. We are asked for the dimensions (notice the plural!) of the rectangle, which means length and width. We refer back to our formula for w in terms of ℓ:

This paragraph is repeated from method 1 so that each solution method is self-contained, allowing easier reference later.

$$w = 60 - 2\ell = 60 - 2 \cdot 15 = 30.$$

The dimensions of the pig pen are 15 feet by 30 feet.

Method 3: first derivative test. As stated in method 2, if the function to be optimized is continuous and there is only one critical number *and* that critical number is a local extremum of the correct type (a local max if we are trying to maximize the function, or a local min if we are trying to minimize the function), then that local extremum must be the absolute extremum. We may detect the local extremum using the first derivative test.

To carry out the first derivative test, we first find the critical numbers of the function:

In the setting of optimization exercises, the first derivative test nearly always takes more time and effort than the second derivative test and is therefore not recommended. This method is demonstrated only for the sake of completeness.

$$\begin{aligned} A(\ell) &= \ell(60 - 2\ell) = 60\ell - 2\ell^2 \\ A'(\ell) &= 60 - 4\ell \\ 60 - 4\ell &= 0 \\ 60 &= 4\ell \\ 15 &= \ell. \end{aligned}$$

We have one critical number: $\ell = 15$.

Our chart then has two intervals, and using the test points $\ell = 14$ and $\ell = 16$, we can complete the chart:

interval	sign of f'	inc/dec	local extrema
$(-\infty, 15)$	$+$	increasing	$\leftarrow$ local max at $\ell = 15$
$(15, \infty)$	$-$	decreasing	

The intervals stated in the chart are for the function's natural domain. Instead, we may determine the domain of the function using the constraints of the exercise as in method 1, and then use the intervals $[0, 15)$ and $(15, 30]$.

The maximum area occurs when $\ell = 15$.

We finish the example by rereading the question to ensure the final form of our answer is what was requested. We are asked for the dimensions (notice the plural!) of the rectangle, which means length and width. We refer back to our formula for w in terms of ℓ:

This paragraph is repeated from method 1 so that each solution method is self-contained, allowing easier reference later.

$$w = 60 - 2\ell = 60 - 2 \cdot 15 = 30.$$

The dimensions of the pig pen are 15 feet by 30 feet. ■

Although it is not part of our formal solution method, much can be learned by thinking about the geometry, especially the relative dimensions, of solutions to optimization exercises. For instance, in example 1, the amount of "east–west" fencing is the same as the amount of "north–south" fencing. Making such observations at the conclusion of each exercise can help us notice overall patterns and possibly even general truths that can help guide design intuition.

Although it is true that the area function of example 1 is quadratic and its graph is a parabola opening down so that the critical number must be a local (and absolute) max, not all functions to be optimized are analyzed so easily. Therefore **always** check critical numbers to ensure the correct type of extremum.

For the aircraft example, see Biddle, Wayne, "Skeleton Alleged in the Stealth Bomber's Closet," *Science* 50 No. 4905 (May 12, 1989), 650–651.

Notice that in each of the three methods, we did not just assume the critical number yielded a maximum; we checked to make sure we had a max and not a min (or not an extremum at all). This should be done *every time* for every critical number. One famous case involving two critical numbers in which the max and min were switched involved the design of stealth aircraft, a rather expensive item upon which to make such a mistake! It made the design for that particular feature the worst possible design instead of the best possible design.

SOLVING OPTIMIZATION PROBLEMS

❶ Draw a picture (if applicable); visualize different possibilities.

❷ Introduce variables and label the picture appropriately.

❸ Determine the quantity to be optimized and express that quantity as a function of the variable(s) of step ❷.

❹ If necessary, eliminate variables so the function of step ❸ is a function of one variable.

❺ Use either the closed interval method or the second derivative test to detect extrema.

❻ Reread the question to determine the final form of the answer.

Optimization example: maximum volume

Our next classical example involves making a box.

Example 2 *An open-top box is to be made from a sheet of cardboard measuring* 11 *inches by* 17 *inches (legal size) by cutting squares from the corners and folding up the sides. What size squares should be cut from the corners to maximize the volume of the box?*

Solution First we recognize this as an optimization problem because we are asked to "maximize the volume."

❶ We draw a picture of a rectangular sheet of cardboard with its corners cut out (figure 5.) Visualizing different possibilities, we

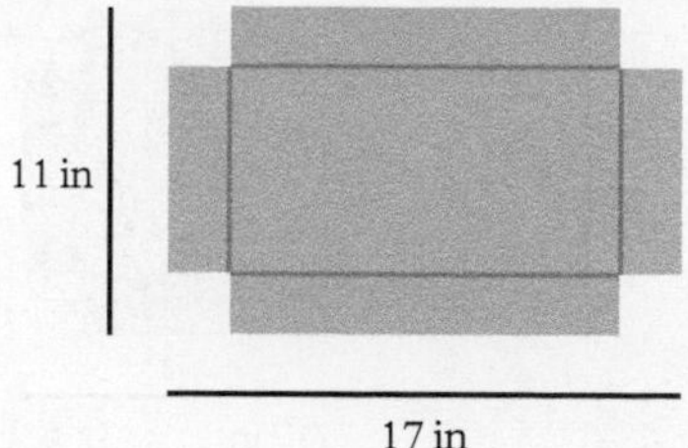

Figure 5 *A 17-inch-by-11-inch piece of cardboard with the corners removed and edges creased (shown as dark-brown lines) for the purpose of folding up the sides of the box. Flaps for gluing are not shown*

realize that cutting out small squares results in a wide, long, short box, whereas cutting out large squares results in a tall box that is less wide and long. Although we do not need to draw these diagrams if we can picture them mentally, they are given in figure 6.

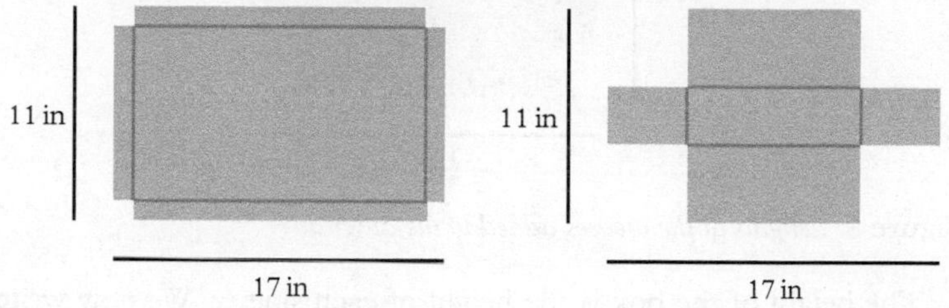

Figure 6 *A 17-inch-by-11-inch piece of cardboard with the corners removed and edges creased for the purpose of folding up the sides of the box. Cutting small squares from the corners (left) gives a box with short sides and a large base, whereas cutting large squares from the corners (right) gives a box with tall sides but a small base*

❷ The fundamental quantity that seems to vary is the size of square that we cut from the corners. A square has the same length as width, so let's call that length and width x. We then label the (original) diagram using the variable x; see figure 7.

❸ The quantity to be optimized (in this case, maximized) is the volume of the box. The volume of a rectangular box is length times width times height:

$$V = \ell \cdot w \cdot h.$$

Consulting the diagram, we see that after the sides are folded up, the base of the box is the rectangle inside the creases. Thus, the length

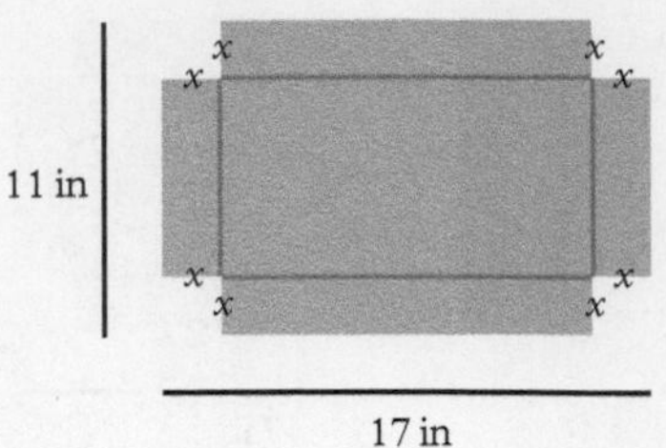

Figure 7 *The variable x is added to the diagram. The squares cut from each corner have length and width x*

and width of the box are the lengths of these creases, 17 inches minus $2x$ inches and 11 inches minus $2x$ inches. It may be helpful to write these dimensions in the diagram as well; see figure 8.

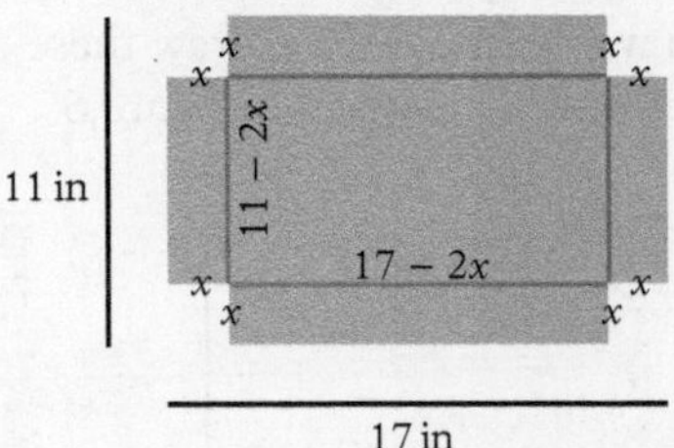

Figure 8 *Lengths of the creases added to the diagram*

The height of the box is the height of each side, x. We now write the volume of the box in terms of the variable x, completing step ❹:

$$V = \ell \cdot w \cdot h = (17 - 2x)(11 - 2x)x = 187x - 56x^2 + 4x^3.$$

We could use the second derivative test instead.

❺ To use the closed interval method, we need to know the domain of the function. Consulting the diagram, we see that x must not be negative, and the largest square that can be cut from every corner has half the side length of 11 inches. The domain is therefore $[0, 5.5]$.

We also need to find the critical numbers:

As always, when a quadratic equation cannot be factored quickly, we turn to the quadratic formula.

$$V'(x) = 187 - 112x + 12x^2$$

$$187 - 112x + 12x^2 = 0$$

$$x = \frac{112 \pm \sqrt{112^2 - 4(187)(12)}}{2(12)}$$

$$= \frac{28 \pm \sqrt{223}}{6}$$

$$x = 2.18, \; x = 7.16.$$

Because $x = 7.16$ is not in the function's domain, the only critical number is $x = 2.18$.

Next we evaluate the function at the endpoints and the critical number:

$$V(0) = 0,$$
$$V(5.5) = 0,$$
$$V(2.18) = 183.$$

The maximum volume is 183 in^3 when $x \approx 2.18$.

❻ Rereading the question, we see that the size of the squares is required. To maximize the volume, we cut squares of side length 2.18 inches from each corner. ■

Author's note: during the summer between high school and college, I worked part-time at a box factory. The process was almost exactly as described in this example, except that flaps were left for gluing. A sheet of cardboard was put into a press, which perforated the sheet where the corners (or other scrap) were to be removed, and a crease was placed in the cardboard where the folds were to be made to create the sides of the box. My job was to remove the corners (remove the scrap)—all day long in a unair-conditioned metal building—during a hot Kansas summer. After the manual scrap removal, a machine did the folding and gluing.

Optimization example: best path

Example 3 *A utility company needs to run a pipeline from its power plant on one side of a straight, 900-m-wide river to a manufacturer on the other side of the river, 3000 m downstream. It costs \$50/m to install the pipeline along the ground, but \$200/m to install the pipeline across (under) the river. At what point should the pipe emerge from the river to minimize the cost of the pipeline?*

Solution First we recognize this as an optimization problem because we are asked to minimize the cost.

❶ We draw a picture of a utility on the bank of a straight river, with a manufacturer on the opposite side of the river but downstream. We also label the width of the river (900 m) and the downstream distance to the manufacturer (3000 m). See figure 9. Visualizing different possibilities, we see the pipeline could go straight to the opposite shore to have the least amount of pipe under water (figure 10, top). The pipeline could also go directly to the manufacturer, remaining under

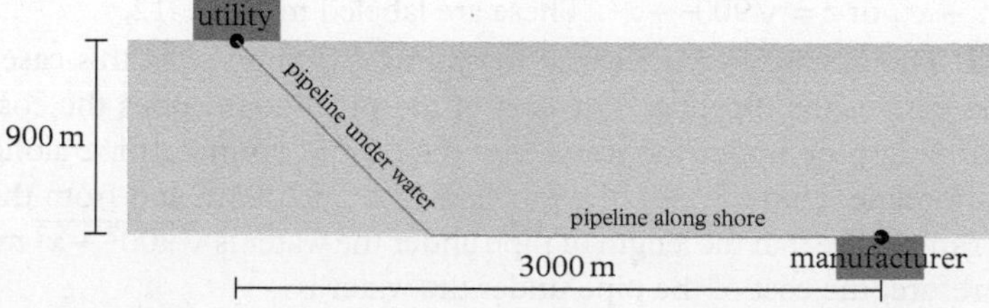

Figure 9 *A utility on one side of a river and a manufacturer on the other side, with a pipeline connecting the two as seen from above (aerial view)*

water the entire route, to have the least total amount of pipe (figure 10, bottom). But we are not asked to minimize the amount of pipe under water or minimize the total length of pipe; instead, we are asked to minimize cost. It seems as if the least cost prompts us to follow a route like that in figure 9.

Minimizing the wrong value could literally be costly! (It could cost more than the minimum cost.)

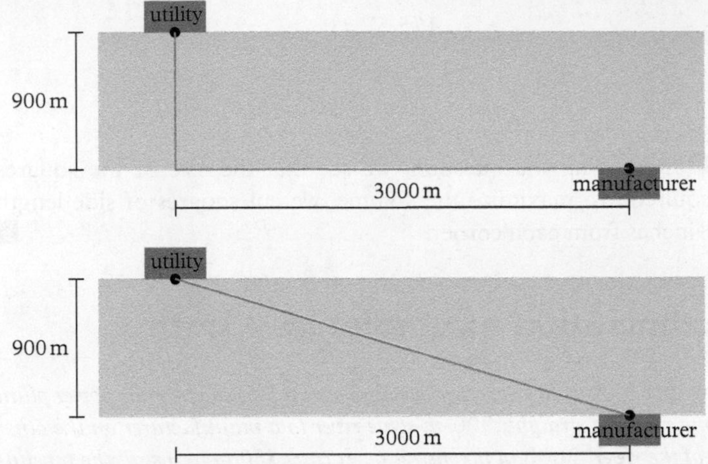

Figure 10 *Two possible routes for the pipeline, neither of which minimizes the total cost of the pipeline. (top) The pipeline goes straight across the river, minimizing the amount of pipe under water. (bottom) The pipeline goes directly to the manufacturer, minimizing the total length of the pipe*

❷ The variable amount in figure 9 appears to be the spot at which the pipe emerges from the river, which is in fact what we are asked for. Let's let x represent the distance downstream from the utility at which the pipe emerges, and label this distance in the diagram; see figure 11. We can then determine and label other lengths as well. The length of the pipe along the shore is $(3000-x)$ m, whereas the "vertical leg" of the right triangle is the width of the river, 900 m. The length of the hypotenuse can be found using the Pythagorean theorem: $c^2 = 900^2 + x^2$, or $c = \sqrt{900^2 + x^2}$. These are labeled in figure 12.

Alternately, we can let x represent the length of pipe under the water or the length of pipe along the ground.

❸ The quantity we are asked to optimize (minimize in this case) is the cost of the pipeline. The cost of the pipeline includes the cost of running pipe under the water and the cost of running pipe along the shoreline. Under water, the pipeline cost is \$200/m, and from the diagram we see that the length of pipe under the water is $\sqrt{900^2 + x^2}$ m. Therefore, the cost of the pipe under the water is

$$200\sqrt{900^2 + x^2}.$$

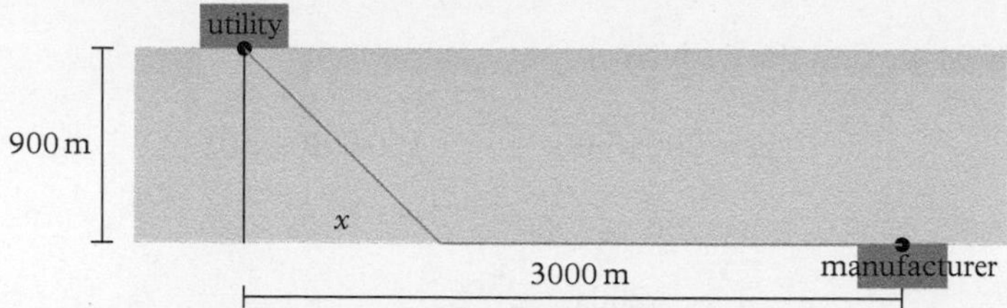

Figure 11 *Our variable x, which represents the distance downshore from the utility at which the pipe emerges from the river, is added to the diagram. Another line is added as well, so that a right triangle can be seen*

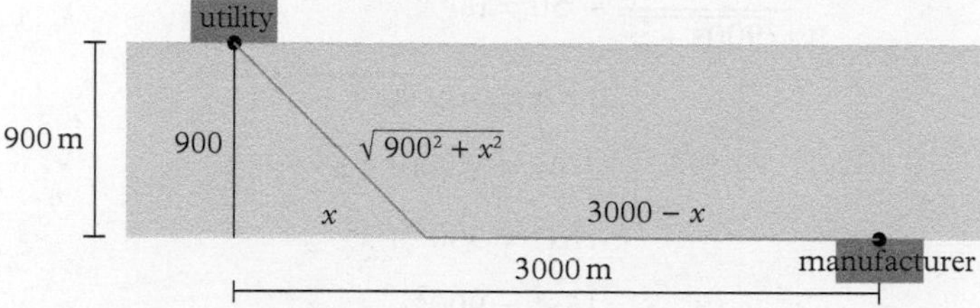

Figure 12 *Other lengths are added to the diagram*

The cost of the pipeline along the ground is $50/m, and from the diagram we see that the length of pipe along the shoreline is $(3000 - x)$ m. Therefore, the cost of the pipe along the ground is

$$50(3000 - x).$$

The total cost of the pipeline is therefore

$$C(x) = 200\sqrt{900^2 + x^2} + 50(3000 - x).$$

❺ To use the closed interval method, we need to know the domain of the function. Inspecting figures 12 and 10, we see that although we could allow x to be a negative number and emerge from the river upstream, this certainly does not produce the lowest cost, so we assume $x \geq 0$. Similarly, the pipe could go underwater past the manufacturer and then along the shore back to the factory; but again, this certainly does not produce the lowest cost, so we assume $x \leq 3000$. Our domain is $[0, 3000]$.

The second derivative test could be used instead.

Next we find the derivative

$$C(x) = 200\sqrt{900^2 + x^2} + 150\,000 - 50x$$

$$C'(x) = 200 \cdot \frac{1}{2}(900^2 + x^2)^{-1/2} \cdot 2x - 50$$

$$= \frac{200x}{\sqrt{900^2 + x^2}} - 50$$

and set it equal to zero to find the critical numbers:

$$\frac{200x}{\sqrt{900^2 + x^2}} - 50 = 0$$

$$200x = 50\sqrt{900^2 + x^2}$$

$$4x = \sqrt{900^2 + x^2}$$

$$16x^2 = 900^2 + x^2$$

$$15x^2 = 900^2$$

$$x^2 = 54\,000$$

$$x = \sqrt{54\,000} \approx 232.4.$$

We also need to determine where $C'(x)$ might be undefined, but the equation $\sqrt{900^2 + x^2} = 0$ has no solutions. There is one critical number: $x \approx 232.4$.

We evaluate the cost function at the critical number and endpoints of the domain:

These cost figures are rounded to four significant digits. Additional precision can be used if desired.

$$C(0) = 200\sqrt{900^2} + 150\,000 = \$330\,000,$$

$$C(3000) = 200\sqrt{900^2 + 3000^2} + 150\,000 - 50(3000) = \$626\,400,$$

$$C(232.4) = 200\sqrt{900^2 + 232.4^2} + 150\,000 - 50(232.4) = \$324\,300.$$

The minimum cost is \$324 300 when $x = 232.4$.

❻ Rereading the question, we see that we wish to know where the pipe should emerge from the river, which is 232.4 m downstream from the power plant. ■

Optimization example: maximum profit

See section 2.7.

Recall that for the revenue, cost, and profit functions from economics, *marginal* means "derivative." Recall also that profit is revenue minus cost:

$$P(x) = R(x) - C(x).$$

If we wish to maximize profit, then we need to find the critical numbers of the profit function—that is, where $P'(x) = 0$:

$$P'(x) = R'(x) - C'(x)$$
$$R'(x) - C'(x) = 0$$
$$R'(x) = C'(x).$$

If marginal revenue is greater than marginal cost, then selling one more unit increases revenue more than cost and results in a greater profit. If marginal revenue is less than marginal cost, then selling one less unit decreases revenue less than it decreases cost, resulting in a greater profit. Therefore, profit is maximized at a point where marginal revenue equals marginal cost. This reasoning is widely known and applied.

In other words, the critical numbers of the profit function occur where marginal revenue equals marginal cost. The traditional solution method for profit maximization problems is to equate marginal revenue and marginal cost. Because we also wish to check to ensure that profit is maximized rather than minimized, we still form the profit function and determine its maximum.

Example 4 *Each month we can sell as many widgets as we can make for \$12 each. The cost, in dollars, of making x widgets is given by*

$$C(x) = 10\,000 + 7x - 0.002x^2 + \frac{1}{3} \cdot 10^{-6}x^3.$$

How many widgets should we make to maximize profit?

Note that $C(0) = 10\,000$, which represents the cost of producing no widgets. These are the *fixed costs*, costs that are incurred regardless of whether any production occurs. The $7x$ represents the cost of labor, materials, and similar items per widget. However, the efficiency of mass production results in a lower per-widget cost when more are produced. The terms $7x - 0.002x^2$ can be factored as $(7 - 0.002x)x$, so the cost per widget is $7 - 0.002x$. Offsetting this efficiency is the inefficiency of large organizations (more layers of management are required). Although such inefficiency may be modeled more accurately with an exponential term (exponential functions are discussed in chapter 5), here we use a cubic term, $\frac{1}{3} \cdot 10^{-6}x^3$, for this purpose.

Solution Because we wish to maximize profit, this is an optimization problem.

❶–❷ Notice that there is nothing geometric about this problem. No picture seems applicable, so we don't draw one. Furthermore, the relevant variable has already been introduced; x is the number of widgets made in 1 month.

❸ The quantity to be maximized is profit. We are given the cost function $C(x)$, and we know that $P(x) = R(x) - C(x)$. Therefore, we need to find the revenue function, R. We are told that we can sell as many widgets as we can make for \$12 each, and the number of widgets made is x. Therefore revenue is $R(x) = 12x$. Thus, the profit function is

$$P(x) = R(x) - C(x) = 12x - (10\,000 + 7x - 0.002x^2 + \frac{1}{3} \cdot 10^{-6}x^3)$$
$$= -10\,000 + 5x + 0.002x^2 - \frac{1}{3} \cdot 10^{-6}x^3.$$

❹ (Because P is a function of just one variable, step 4 is unnecessary).

❺ For polynomial functions, the second derivative test is often the quickest solution method. We first identify critical numbers:

The quadratic formula could be used in lieu of factoring.

$$P'(x) = 5 + 0.004x - 10^{-6}x^2$$
$$0 = 5 + 0.004x - 0.000001x^2$$
$$0 = (0.001x - 5)(-0.001x - 1)$$
$$x = 5000, \ x = -1000.$$

Because we cannot make a negative number of widgets, there is only one critical number: $x = 5000$. To classify this potential extremum, we calculate the second derivative and evaluate it at the critical number:

$$P''(x) = 0.004 - 0.000002x$$
$$P''(5000) = 0.004 - 0.000002(5000)$$
$$= 0.004 - 0.01 = -0.006 < 0, \text{ max.}$$

❻ Rereading the question, we see that we need the number of widgets made. The maximum profit occurs when 5000 widgets are made (and sold). ■

Optimization example: minimum material

Example 5 *A cylindrical can must have volume* 100 cm^3. *What dimensions should be used to minimize the amount of material used?*

Solution We notice the phrase "minimize the amount of material used" and conclude that this is an optimization problem.

❶–❷ We are told the can is cylindrical, so we draw a cylindrical can (figure 13).We are asked for the dimensions to use, which include the can's height and radius, so we visualize various possible shapes, such as tall and thin or short and wide (figure 14).

❸ We wish to minimize the amount of material used to make the can. The material of the can includes the top, bottom, and side of the can. Assuming a uniform thickness of the material, the material used is proportional to the surface area of the can. The formula for the surface area (SA) of a cylinder is

The material of the can is not proportional to the volume because the can is hollow inside, not a solid (filled-in) cylinder of metal.

$$SA = 2\pi rh + 2\pi r^2.$$

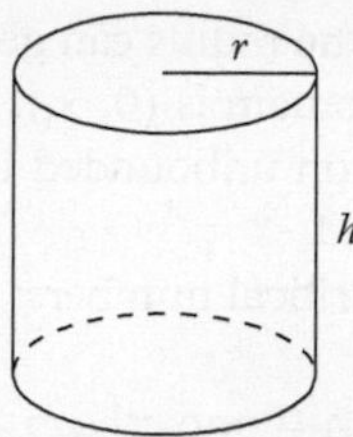

Figure 13 *A cylindrical can with radius and height marked*

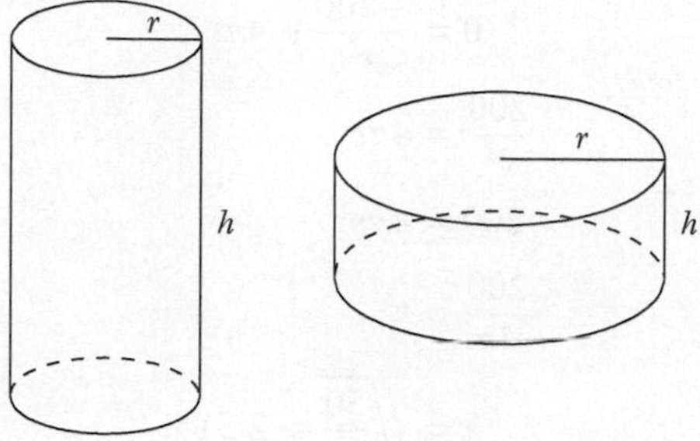

Figure 14 *Cylindrical cans with various shapes*

❹ Because our function to be minimized contains two variables, r and h, we want to eliminate one of them. This requires some other information that gives a relationship between the two variables. Checking the statement of the problem, we see that the volume of the can must be $100\,\text{cm}^3$. Using the formula for the volume of a cylinder, we have

$$100 = V = \pi r^2 h.$$

We can solve this equation for one of the variables. Solving for h gives

$$h = \frac{100}{\pi r^2}.$$

Substituting this expression for h in our formula for surface area gives

$$SA(r) = 2\pi r\left(\frac{100}{\pi r^2}\right) + 2\pi r^2 = \frac{200}{r} + 2\pi r^2.$$

❺ To use the closed interval method, we need to find the domain of the function. We cannot have a negative radius, so $r \geq 0$. We also cannot use $r = 0$ because this causes us to divide by zero in our surface area function. But, a check of figures 13 and 14 reveals that there

If we solve the volume equation for r instead of h, we have $r^2 = \frac{100}{\pi h}$, or

$$r = \sqrt{\frac{100}{\pi h}}.$$

In this form, without further simplification, the resulting derivative requires the quotient rule inside the chain rule inside the product rule. Moral of the story: if solving for one variable results in messy algebra, try solving for the other variable instead.

is no maximum value of r; the radius can get very large as long as the height is very small. The domain is $(0, \infty)$. Because the closed interval method does not apply on unbounded intervals, we must use the second derivative test instead.

We start by finding the critical numbers:

$$\begin{aligned} SA(r) &= 200r^{-1} + 2\pi r^2 \\ SA'(r) &= -200r^{-2} + 4\pi r \\ 0 &= \frac{-200}{r^2} + 4\pi r \\ \frac{200}{r^2} &= 4\pi r \\ 200 &= 4\pi r^3 \\ \frac{200}{4\pi} &= r^3 \\ r &= \sqrt[3]{\frac{50}{\pi}} \approx 2.515. \end{aligned}$$

Because $r = 0$ is not in the domain of the function, $r = 0$ is not a critical number.

There is one critical number: $r \approx 2.515$. Next we calculate the second derivative and evaluate it at the critical number:

$$\begin{aligned} SA''(r) &= 400r^{-3} + 4\pi = \frac{400}{r^3} + 4\pi \\ SA''(2.515) &= \frac{400}{2.515^3} + 4\pi > 0, \text{ min.} \end{aligned}$$

The calculation of $SA''(2.515)$ does not need to be finished because all we need to know is that the number is positive, which could in fact have been observed when calculating $SA''(r)$.

The surface area function has a minimum at $r \approx 2.515$.

❻ Rereading the question, we see that we want the dimensions (plural) of the can. We have the radius, so we need to calculate the height. Using the earlier formula for h,

$$h = \frac{100}{\pi r^2} = \frac{100}{\pi\left(\sqrt[3]{\frac{50}{\pi}}\right)^2} \approx 5.031.$$

The dimensions that minimize the material used for the can are a radius of 2.515 cm and a height of 5.031 cm. ■

Notice that the height of the can appears to be about twice the radius. In fact, the exact expression for h can be simplified to show that it is equal to twice the radius. It is true, in general, that the surface area of a cylinder is minimized when height and diameter are equal.

So why is it that cans on grocery store shelves do not always have equal height and diameter? There are other factors that were not considered here, such as the rims at the top and bottom of the can. Also, marketing comes into play. A tuna can that does not look like a tuna can might be overlooked easily on the shelves and would be at a competitive disadvantage.

Exercises 3.8

It is implied that the procedures of this section should be used to solve each exercise, including checking for type of extremum using the closed interval method or second derivative test.

1. A rural homeowner has found 400 feet of unused wire fencing behind the tractor shed. The goat needs to be fenced in to keep it from eating the garden. The homeowner decides to fence a rectangular area using one side of the tractor shed for one side of the rectangle, and the wire fencing for the other three sides. What dimensions should be used to maximize the area inside the fence?
2. An enclosure is to be made from 160 feet of straight fencing material. Using the sheer rock face of a cliff as one side of a rectangle, the fencing will be used to form the other three sides of the rectangular enclosure. What dimensions should be used to enclose the largest possible area?
3. A dog owner has 90 feet of fencing to make a rectangular pen. The garage will be used as one side of the pen. What dimensions should be used to fence in the largest possible area?
4. For the situation in exercise 3, if a semicircular fence is made instead, would the enclosed area be larger than the best rectangular pen? By how much?
5. A cylindrical can must have a volume of $200\,\text{in}^3$. What dimensions should be used to minimize the amount of material used?
6. A cylindrical can must have a volume of $320\,\text{in}^3$. What dimensions should be used to minimize the amount of material used?
7. An oil refinery is located on the north bank of a straight river that is 2 km wide. A pipeline is to be constructed from the refinery to storage tanks located on the south bank of the river 6 km east of the refinery. The cost of laying the pipe is \$400 000/km over land to a point P on the north bank and \$800 000/km under the river to the tanks. To minimize the cost of the pipeline, where should P be located?
8. Repeat exercise 7 using a cost of only \$600 000/km under the river.

Some of the exercises taken from GSL or Love are modified by giving a specific value to a dimension, rather than an unspecified constant.

9. A farmer with a small number of cattle wants to fence a rectangular field with the field divided in half by a fence parallel to one of the sides of the rectangle. The farmer wishes to enclose a total area of 450 000 ft^2. What dimensions should be used for the rectangle to minimize the amount of fencing used?
10. (Love) A rectangular field of area 28 800 m^2 is to be enclosed and divided into three lots by parallels to one of the sides. What should be the dimensions of the field if the amount of fencing is to be a minimum?
11. An open-top box is to be made from a sheet of cardboard measuring 25 inches by 10 inches by cutting squares from the corners and folding up the sides. What size squares should be cut to maximize the volume of the box?

The box of exercise 12 is an open-top box.

12. (Love) Find the volume of the largest box that can be made by cutting equal squares out of the corners of a piece of cardboard 15 inches by 24 inches and turning up the sides.
13. (GSL) It is desired to make an open-top box of greatest possible volume from a square piece of tin with a side that is 18 inches, by cutting equal squares out of the corners and then folding up the tin to form the sides. What should be the length of a side of the squares cut out?
14. A square-based, open-top box needs to contain a volume of 25 ft^3. A metal alloy forms the sides and base; there is no top to the container. What dimensions should be used for the container to minimize the material used?
15. If we produce x widgets in a month, then we can sell them for $4200 - x$ dollars each. The cost of making x widgets in a month is given by $C(x) = 25\,000 + 600x - x^2 + 0.02x^3$. (a) How many widgets should we make each month to maximize profit? (b) What is that maximum profit?
16. Each week we can sell as many widgets as we can make for \$112 each. The cost (in dollars) of making x widgets in a week is given by $C(x) = 1200 + 46x - 0.014x^2 + 0.00033x^3$. (a) How many widgets should we make each week to maximize profit? (b) What is that maximum profit?
17. If we produce x mini-widgets each month, then we can sell them for $660 - 0.5x$ dollars each. The cost of making x mini-widgets in a month is given by $C(x) = 35\,000 + 540x - 1.5x^2 + 0.004x^3$. (a) When the correct number of mini-widgets is produced to maximize profit, what is the price of a mini-widget? (b) What is the maximum monthly profit?

18. Repeat exercise 17 for a cost function of $C(x) = 44\,000 + 540x - 2.5x^2 + 0.004x^3$ (higher fixed costs, but greater efficiency of mass production—that is, costs per widget that decrease more rapidly with greater production).

19. A college student is staying at a house on a beach that is along a straight shoreline. The student receives a call from a friend who is on an island 6 miles down the shore and $\frac{1}{2}$ mile offshore, challenging the student to run and swim to the island in less than 1 hour. If the student can swim at a pace of 25 min/mi and run at a pace of 8 min/mi, can the student meet the challenge? How far should the student run down the shoreline before beginning to swim toward the island to minimize the time required?

20. An athlete who can swim 6 km/h and run 20 km/h is competing in a race with a starting line on an island's beach 2 km off the straight mainland shore to a finish line 10 km down that shore. (If an athlete crosses the water swimming directly toward the shore, the race involves 2 km of swimming and 10 km of running). How far from the finish line should the athlete emerge from the water to minimize the time required to complete the race?

21. (GSL) What should be the diameter of a tin can holding 1 quart (58 in^3) and requiring the least amount of tin if the can is open at the top?

For exercise 21, assume the can is cylindrical in shape.

22. (GSL) A telephone company finds there is a net profit of $15 per instrument if an exchange has 1000 subscribers or less. If there are more than 1000 subscribers, the profits per instrument decrease 1¢ for each subscriber above that number. How many subscribers would give the maximum net profit?

23. (GSL) Assuming that the strength of a beam with a rectangular cross section varies as the product of the breadth and the square of the depth, what are the dimensions of the strongest beam that can be sawed out of a round log with a diameter of 1 m?

24. (Love) The stiffness of a rectangular beam is proportional to the product of the breadth and the cube of the depth. Find the shape of the stiffest beam that can be cut from a log of diameter 1 m.

25. (Love) A Norman window consists of a rectangle surmounted by a semicircle. If the perimeter of the window is to be 20 feet, what dimensions give the most light?

26. (Love) A person in a rowboat 6 miles from shore desires to reach a point on the shore at a distance of 10 miles from the person's current position. If the individual can walk 4 mph and row 3 mph, where should the person land to reach the destination in the shortest possible time?

27. (GSL) For a certain article, the increase in the number of pounds consumed is proportional to the decrease in the tax on each pound. If the consumption is m pounds when there is no tax and n pounds when the tax is t dollars per pound, find the tax that should be imposed on each pound to bring in the most revenue.

28. (GSL) A radio manufacturer finds that it can sell x instruments per week at p dollars each, where $5x = 375 - 3p$. The cost of production is $(500 + 15x + \frac{1}{5}x^2)$ dollars. Show that the maximum profit is obtained when the production is about 30 instruments per week.

29. (GSL) A steel plant is capable of producing x tons per day of a low-grade steel and y tons per day of a high-grade steel, where $y = \dfrac{40 - 5x}{10 - x}$. If the fixed market price of low-grade steel is half that of high-grade steel, how many tons of low-grade steel should be produced per day for maximum revenue?

30. (GSL) Find the point on the curve $y = \dfrac{x^2}{2}$ that is nearest to the point $(4, 1)$.

Newton's Method

3.9

Not all equations can be solved using exact algebraic methods. There are a variety of methods for approximating solutions to equations to any desired degree of accuracy, including the bisection method presented in section 1.7. A more efficient method of approximating solutions to equations is the topic of this section.

The idea of Newton's method

Suppose we wish to solve an equation of the form $f(x) = 0$. Graphically, this is the same as asking where the curve $y = f(x)$ crosses the x-axis, because we are finding where $y = 0$ on that graph (figure 1).

The equation $x^7 - x^4 + 2x - 5 = 0$ is in the form $f(x) = 0$, with $f(x) = x^7 - x^4 + 2x - 5$.

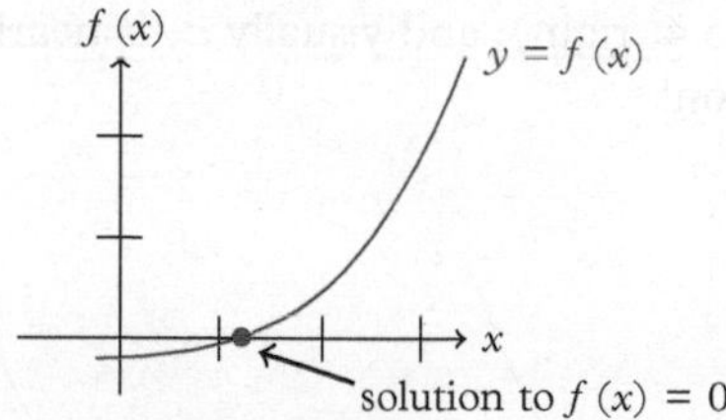

Figure 1 *The solution to $f(x) = 0$ is the x-intercept on the graph of $y = f(x)$*

From the graph we might get a rough estimate of the solution. In figure 1, we see the solution is somewhere between $x = 1$ and $x = 2$, and a guess might be $x = 1.25$. But, whatever our eyes tell us is just a guess; we are likely off at least a little bit. What Newton's method does is to take that initial guess and refine it, producing a better estimate, and then repeat to form a yet better estimate, continuing to repeat until we get as close as we want.

To illustrate the procedure, suppose the initial guess is $x = a_1$, as pictured in figure 2. Consider the tangent line to the curve at $x = a_1$. This tangent line crosses the x-axis at a point between the initial guess a_1 and the solution we are seeking, which gives us a better estimate of the solution. Let's call that number a_2 (figure 3). Notice how much closer to the solution we are at a_2 than we were at a_1. Repeat (figure

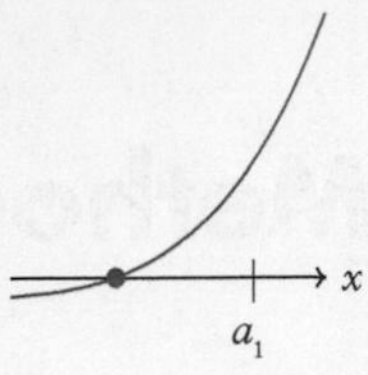

Figure 2 *An initial guess $x = a_1$ to the right of the solution*

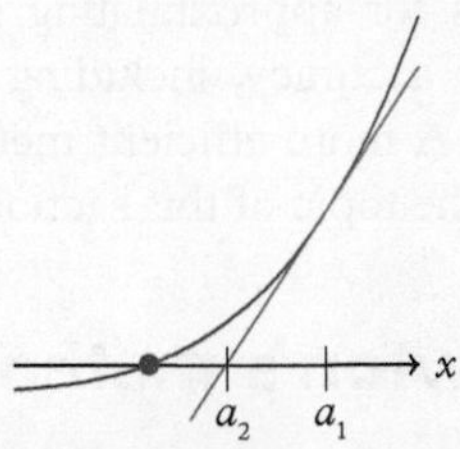

Figure 3 *Following the tangent line (orange) to a better estimate of the solution*

4, left), and call the new point a_3, which is even closer to the solution. Repeat again (figure 4, right), and visually a_4 is nearly indistinguishable from the solution!

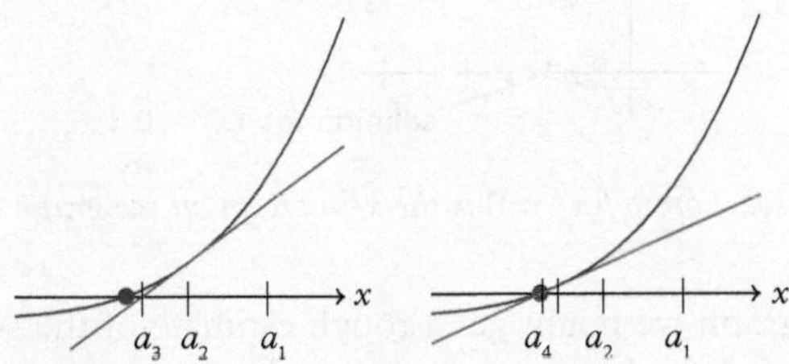

Figure 4 *Following the tangent line (orange) to a better estimate of the solution*

Newton was a great physicist as well as a mathematician and is widely considered to be one of the cofounders of the subject of calculus.

Isaac Newton 1643–1727
http://www-history.mcs.st-andrews.ac.uk/Biographies/Newton.html

This efficient process for estimating solutions to equations is called *Newton's method*. This method is widely used in software packages today.

To implement the method algebraically, consider how we might calculate a_2 from the initial guess a_1 (using figure 3 as a guide). First we need the formula for the tangent line at $x = a_1$, which we know is also called the *linearization of f at a_1*:

$$y = f(a_1) + f'(a_1)(x - a_1).$$

The tangent line crosses the x-axis where its y-coordinate is zero:

$$0 = f(a_1) + f'(a_1)(x - a_1).$$

We need the value of x where this happens, so we solve for x:

$$\begin{aligned} 0 &= f(a_1) + f'(a_1)x - f'(a_1)a_1 \\ f'(a_1)x &= f'(a_1)a_1 - f(a_1) \\ x &= \frac{f'(a_1)a_1 - f(a_1)}{f'(a_1)} = \frac{f'(a_1)a_1}{f'(a_1)} - \frac{f(a_1)}{f'(a_1)} \\ &= a_1 - \frac{f(a_1)}{f'(a_1)}. \end{aligned}$$

This value of x, where the tangent line crosses the x-axis, is what we called a_2. Therefore,

$$a_2 = a_1 - \frac{f(a_1)}{f'(a_1)}.$$

The next step, as in figure 4 (left), is to repeat the process, using a_2 in place of a_1. The exact same algebra as used previously produces the solution

$$x = a_2 - \frac{f(a_2)}{f'(a_2)}$$

for where the tangent line crosses the x-axis. We called this a_3:

$$a_3 = a_2 - \frac{f(a_2)}{f'(a_2)}.$$

We can repeat this process again and again, and each time the algebra is the same. For as far as we wish to go,

$$a_{n+1} = a_n - \frac{f(a_n)}{f'(a_n)}.$$

A process such as this one, where each value is based on the previous value, is called an *iterative process* or an *iterative method*. The formula itself is called a *recursive formula*.

NEWTON'S METHOD

To approximate a solution to the equation $f(x) = 0$, do the following:

During the first few centuries after Newton's method was introduced, the graphs were drawn by hand and the arithmetic was also done by hand (as in paper and pen or pencil). Although tedious, if one needed a solution to four or five significant digits, this was the best method available.

❶ Identify the function $y = f(x)$.

❷ Use technology (a calculator or computer software) to graph the function $y = f(x)$.

❸ From the graph, make an initial guess of the value of an x-coordinate where the curve crosses the x-axis. Call that guess a_1.

What a $f(a_n)$tastic formula!

❹ Use the recursive formula

$$a_{n+1} = a_n - \frac{f(a_n)}{f'(a_n)}$$

repeatedly to find $a_2, a_3, a_4, \ldots$ until the desired accuracy is obtained.

The last two steps may be repeated if there are multiple solutions to the equation.

Newton's method example

This equation is an example of one that cannot be solved using exact algebraic methods because of its mixture of trig and polynomial terms.

Example 1 *Use Newton's method to estimate a solution to the equation* $\sin x = -2x + 3$ *accurate to four decimal places.*

Solution We use the four steps of Newton's method listed previously.

❶ Because Newton's method is for solving an equation of the form $f(x) = 0$, we rewrite our equation in that form,

$$\sin x + 2x - 3 = 0,$$

and assign $f(x) = \sin x + 2x - 3$.

❷ Next we use technology to graph the function (figure 5).

❸ The graph appears to cross the x-axis at about $x = 1$. Therefore, we set $a_1 = 1$.

There is usually a good deal of latitude with regard to which value to choose for a_1. Using $a_1 = 2$ or $a_1 = 0$ still works for this example, taking only one more use of the recursive formula. There is no need for zooming in and choosing a more precise estimate for a_1.

❹ The recursive formula

$$a_{n+1} = a_n - \frac{f(a_n)}{f'(a_n)}$$

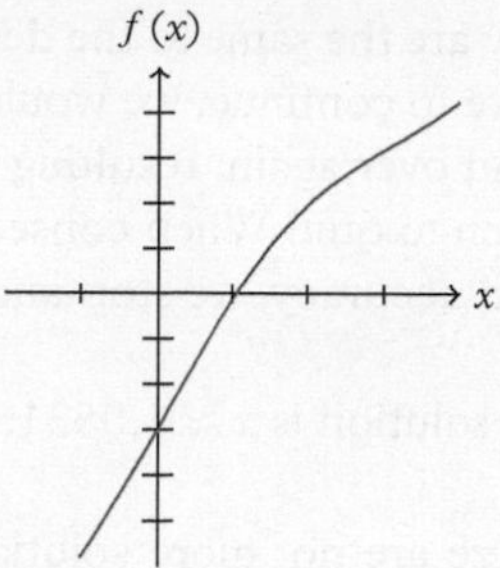

Figure 5 *The graph of* $f(x) = \sin x + 2x - 3$

is used to find $a_2, a_3, \ldots$. The formula involves the derivative, so we calculate f':

$$f'(x) = \cos x + 2.$$

Notice that we are to evaluate each function at a_n:

$$f(a_n) = \sin a_n + 2a_n - 3,$$
$$f'(a_n) = \cos a_n + 2.$$

The recursive formula to be used is

$$a_{n+1} = a_n - \frac{\sin a_n + 2a_n - 3}{\cos a_n + 2}.$$

We already have $a_1 = 1$, so we begin our calculations by finding a_2. Using the recursive formula with $n = 1$ gives

Because we want to find a_2 and the formula gives us a_{n+1}, we use the formula with $n = 1$.

$$a_2 = a_1 - \frac{\sin a_1 + 2a_1 - 3}{\cos a_1 + 2} = 1 - \frac{\sin 1 + 2(1) - 3}{\cos 1 + 2} \approx 1.0624.$$

If you are following along with your calculator and you got $a_2 = 1.3275$, then your calculator is in degree mode. Change to radian mode.

The value of a_2 is written to a precision of four decimal places because that is the desired accuracy specified in the problem statement. We continue by calculating a_3, using the recursive formula with $n = 2$:

$$a_3 = a_2 - \frac{\sin a_2 + 2a_2 - 3}{\cos a_2 + 2} = 1.0624 - \frac{\sin 1.0624 + 2(1.0624) - 3}{\cos 1.0624 + 2} \approx 1.0631.$$

The next iterate is found by using the recursive formula with $n = 3$:

The word *iterate* refers to one of the values found using an iterative process. Here, $a_2, a_3, \ldots$ are iterates.

$$a_4 = a_3 - \frac{\sin a_3 + 2a_3 - 3}{\cos a_3 + 2} = 1.0631 - \frac{\sin 1.0631 + 2(1.0631) - 3}{\cos 1.0631 + 2} \approx 1.0631.$$

Notice that $a_4 = a_3$; they are the same to the desired accuracy of four decimal places. If we were to continue, we would just repeat the exact same calculation over and over again, resulting in 1.0631 every time. That's how we know when to quit! When consecutive iterates give the same value to the desired accuracy, we stop and use that value as the estimate of the solution.

We conclude that the solution is $x \approx 1.0631$. ■

STOPPING CRITERIA FOR NEWTON'S METHOD

Newton's method ends when consecutive iterates give the same value to the desired level of accuracy.

How do we know there are not more solutions to the equation in example 1? After all, we know that sometimes features of graphs can be hidden and show up when graphed on wider scales or when zoomed in to much smaller scales. Could there be more solutions that we do not see in figure 5? This is one of the drawbacks to Newton's method. Often, there is no assurance that we have found all the solutions. Therefore, exact solution methods are preferred when available.

For this particular example, however, we can use other methods to assure ourselves that there are no other solutions. Note that the derivative is $f'(x) = \cos x + 2$. Since

$$-1 \le \cos x \le 1,$$

the value of the derivative must be positive:

$$1 \le \cos x + 2 \le 3.$$

Therefore, the function f is increasing on $(-\infty, \infty)$ and can only cross the x-axis once.

TECHNOLOGY SHORTCUT

If your calculator has a previous answer button (perhaps labeled ANS), then it can be used to calculate quickly successive iterates of a recursive formula. First type the initial value a_1 (1 for example 1) into the calculator and hit ENTER. Then enter the recursive formula, using the ANS button for a_n (for example 1, type `ANS-(sin(ANS)+2ANS-3)/(cos(ANS)+2)`, being careful with parentheses) and hit ENTER repeatedly until successive values are the same to the desired level of accuracy.

Reading Exercise 16 *For solving the equation* $\cos x + x^3 = 0$ *using Newton's method, (a) identify* $f(x)$ *and (b) state the recursive formula to use.*

Newton's method example: two solutions

Example 2 *Use Newton's method to estimate all solutions to the equation* $x^6 + 3x^3 - 7x = 10$*, accurate to six decimal places.*

Polynomial equations of degree 5 and higher often are not solvable using only the algebraic operations of addition, subtraction, multiplication, division, and extraction of roots. Such equations are prime candidates for the use of Newton's method.

Solution ❶ Rewriting the equation with all terms on the same side of the equation,

$$x^6 + 3x^3 - 7x - 10 = 0,$$

we set

$$f(x) = x^6 + 3x^3 - 7x - 10.$$

We could also move all terms to the other side of the equation,

$$0 = 10 - x^6 - 3x^3 + 7x,$$

and use

$$f(x) = -x^6 - 3x^3 + 7x + 10$$

instead. The graph is then reflected across the x-axis, but its x-intercepts are the same.

❷ A graph of f is in figure 6. used, playing with the graphing window to reveal approximate locations of x-intercepts may be necessary.

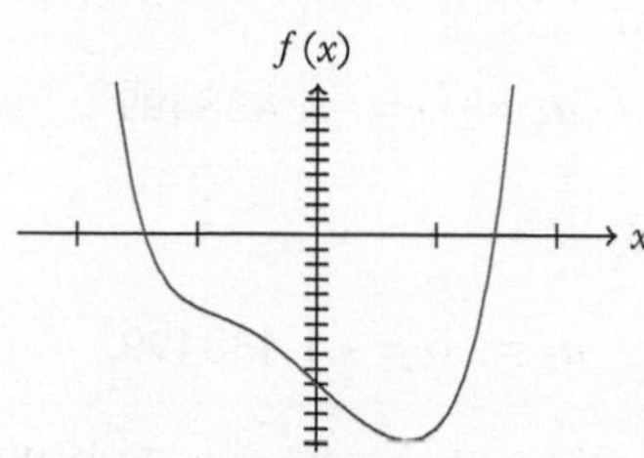

Figure 6 *The graph of* $f(x) = x^6 + 3x^3 - 7x - 10$

❸ There appear to be two solutions, one of which is between $x = -1$ and $x = -2$, and the other of which is between $x = 1$ and $x = 2$. For the solution on the left, we could use $a_1 = -1$ or $a_1 = -2$, but the x-intercept appears closer to $x = -1.5$, so we set $a_1 = -1.5$. For the solution on the right, using $a_1 = 1$ or $a_1 = 2$ works, but setting $a_1 = 1.5$ is quicker.

Just graphing the function is no guarantee that we have identified all solutions. As with example 1, it is possible to use methods from calculus (in this case, finding inflection points and using additional reasoning) to see that there are no additional solutions. In general, this is not always possible.

❹ The recursive formula requires knowledge of f':

$$f'(x) = 6x^5 + 9x^2 - 7.$$

The recursive formula is

$$a_{n+1} = a_n - \frac{a_n^6 + 3a_n^3 - 7a_n - 10}{6a_n^5 + 9a_n^2 - 7}.$$

Using $a_1 = -1.5$, we use the recursive formula to find a_2:

$$\begin{aligned} a_2 &= a_1 - \frac{a_1^6 + 3a_1^3 - 7a_1 - 10}{6a_1^5 + 9a_1^2 - 7} \\ &= -1.5 - \frac{(-1.5)^6 + 3(-1.5)^3 - 7(-1.5) - 10}{6(-1.5)^5 + 9(-1.5)^2 - 7} \\ &\approx -1.445358. \end{aligned}$$

The answer is given to a precision of six decimal places because that is the requested accuracy. By *precision* we mean the resolution to which the number is written; rounding to six decimal places describes the precision. *Accuracy* is the difference between the estimated value and the actual value, which the problem specifies as desired to be within six decimal places.

We repeat to find a_3:

$$a_3 = a_2 - \frac{a_2^6 + 3a_2^3 - 7a_2 - 10}{6a_2^5 + 9a_2^2 - 7}$$

$$= -1.445358 - \frac{(-1.445358)^6 + 3(-1.445358)^3 - 7(-1.445358) - 10}{6(-1.445358)^5 + 9(-1.445358)^2 - 7}$$

$$\approx -1.438593.$$

Ans. to reading exercise 16
(a) $f(x) = \cos x + x^3$, (b) $a_{n+1} = a_n - \dfrac{\cos a_n + a_n^3}{-\sin a_n + 3a_n^2}$

Continuing yields

$$a_4 = \cdots = -1.438499$$

and

$$a_5 = \cdots = -1.438499.$$

Using $a_1 = -1$ requires us to go as far as a_9; using $a_1 = -2$ requires us to go to a_8. If each iterate is typed into the calculator separately, going a few more iterates is added work. If the technology shortcut is used instead, the extra iterates take only an extra second or two.

Because the last two values match, we conclude that a solution to the equation is $x \approx -1.438499$.

The tedium of typing each iterate's formula into a calculator can be avoided by using the technology shortcut given in the margin after example 1. Let's use the shortcut to find the second solution:

ⓐ To use the initial guess $a_1 = 1.5$, type `1.5` into the calculator and hit ENTER:

```
1.5
```

The number 1.5 (the initial guess) should appear as the result of the calculation so that it can be used as the "previous answer" for the first use of the recursive formula.

ⓑ Type the recursive formula using ANS (or whatever command is used for the previous answer) for a_n. Parentheses are required around the numerator and denominator of the fraction:

```
ANS-(ANS^6+3ANS^3-7ANS-10)/(6ANS^5+9ANS^2-7)
```

ⓒ Then hit ENTER repeatedly until the values match to the required number of decimal places:

The values listed are a_2, a_3, and a_4.

```
1.482731137
1.482259487
1.482259145
```

Because the last two values match to six decimal places, we conclude that a solution to the equation is $x \approx 1.482259$. ■

Hitting ENTER one more time gives `1.482259145`, and we have the answer to 10 significant digits!

When does Newton's method work?

Newton's method works in most situations when an appropriate initial guess is chosen. However, a bad guess can doom Newton's method to failure.

To see why, let's first look at situations in which Newton's method works very well. Figure 7 illustrates four situations in which each iterate of Newton's method gives a next iterate closer to the solution: concave up and increasing, starting to the right of the solution; concave down and increasing starting to the left of the solution; concave up and decreasing starting to the left of the solution; and concave down and decreasing starting to the right of the solution.

The proof that under these situations Newton's method always succeeds is beyond the scope of this book.

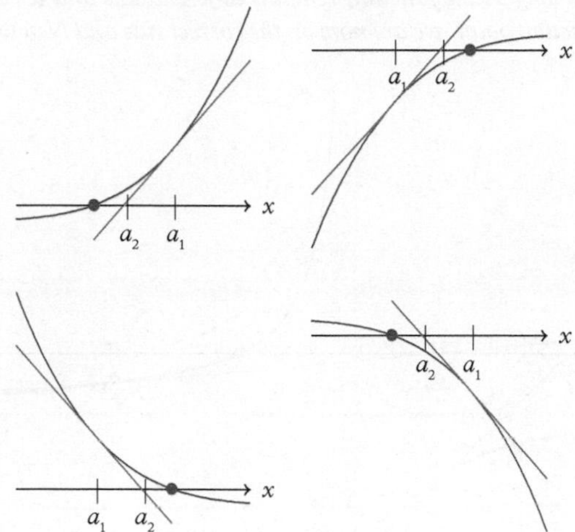

Figure 7 *Four situations in which a_2 is closer to the solution than a_1. Successive iterates continue to get closer to the solution, and Newton's method succeeds*

Notice that for every combination of concavity (up/down) and direction (increase/decrease) on an interval including the solution point, there is a correct side on which an initial guess leads to Newton's method succeeding. But, what if we start on the wrong side of the solution? In this case, the first application of the procedure throws us to the opposite side of the solution, as in figure 8. As long as we don't land too far away, we arrive on the correct side of the solution and Newton's method works just fine from there, continuing to bring us closer to the solution. This is why, as long as our initial guess is reasonably close to the solution, as in examples 1 and 2, Newton's method succeeds no matter from which side of the solution we start.

If we begin with the initial guess a_1 on the wrong side and the tangent line throws us too far away, Newton's method can fail. Figure 9 is an example where the tangent line at a_1 crosses the x-axis on the other side of a vertical asymptote, and then successive applications of the recursive formula only throw us farther and farther to the left, increasingly far from the solution point.

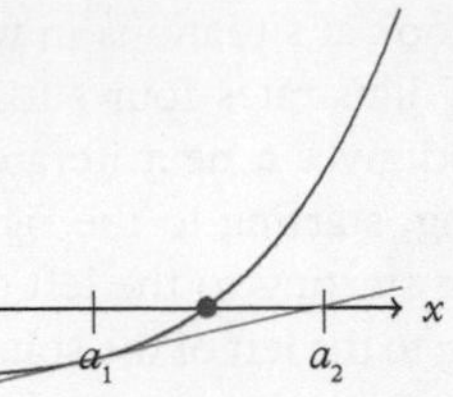

Figure 8 *Starting with a_1 on the wrong side throws a_2 to the other side of the solution. As long as the function remains differentiable and we do not cross beyond an inflection point, we are now on the correct side and Newton's method still succeeds*

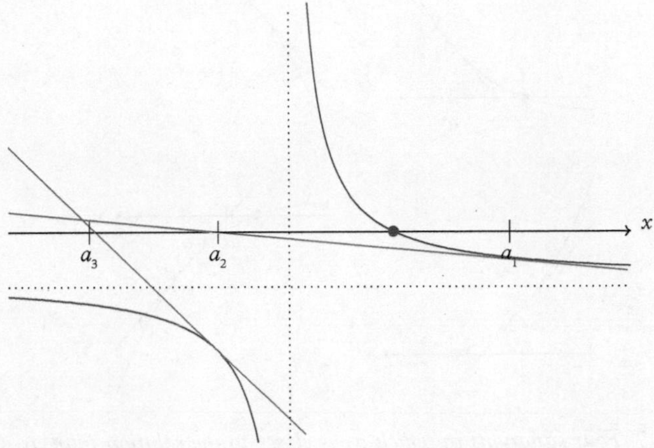

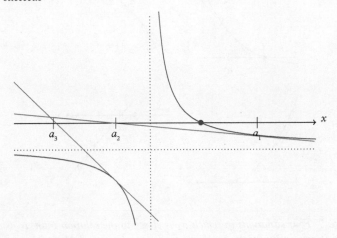

Figure 9 *A poor choice of a_1 dooms Newton's method to failure; $a_2, a_3, \ldots$ move farther left and away from the solution point. Instead, choosing a_1 between the vertical asymptote and the solution leads to success*

Another possibility is that a poor choice of a_1 produces an a_2 that is closer to a different solution than the one intended. In this case, Newton's method may give an answer, but not the solution being sought.

Yet another way in which Newton's method can fail is for the recursive formula to be undefined. Since

$$a_2 = a_1 - \frac{f(a_1)}{f'(a_1)},$$

if $f'(a_1) = 0$, then we are dividing by zero and a_2 is undefined. This corresponds to choosing a_1 where the tangent line is horizontal, and then the tangent line never crosses the x-axis (see figure 10).

There are other ways in which Newton's method can fail, some of which are explored in the exercises.

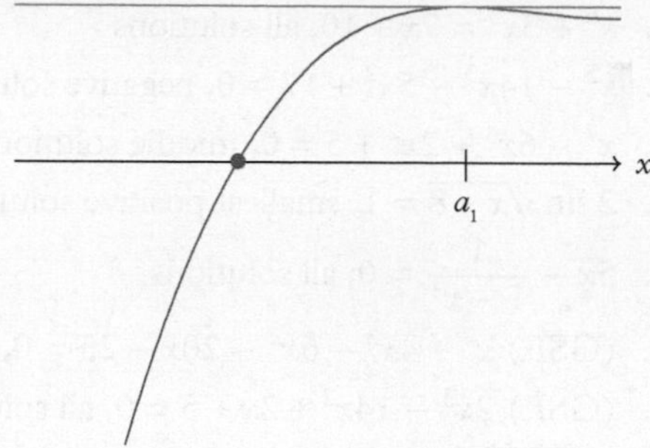

Figure 10 *When the tangent line at a_1 is horizontal it never crosses the x-axis and a_2 is undefined*

Exercises 3.9

1–10. Use Newton's method to approximate the indicated solution to the equation to four decimal places, using the provided initial guess a_1. State the values $a_2, a_3, \ldots$ used to determine the solution.

1. $x^5 - 4x^4 + 2x^3 - 11x + 157 = 0$, $a_1 = -2$
2. $x^5 - 32x^3 + 1000x^2 - 114 = 0$, $a_1 = -12$
3. $x^6 - 14x^3 + 61x^2 = 13$, $a_1 = 1$
4. $x^6 - 14x^3 + 61x^2 = 13$, $a_1 = -1$
5. $30x^4 + 13\,830x^2 + 1125 = 1260x^3 + 12\,600x$, $a_1 = 16$
6. $30x^4 + 13\,830x^2 + 1125 = 1260x^3 + 12\,600x$, $a_1 = 24$
7. $3x^2 = 5\sqrt{2 + \sin x}$, $a_1 = -1$
8. $\dfrac{4x + 15}{x - 12}$, $a_1 = -5$
9. $1 = x^2 \sin x$, $a_1 = 1$
10. $(6x - 11)^{14} - 311 = 0$, $a_1 = 2.1$

For exercise 10, try using each of $a_1 = 2.0$, $a_1 = 1.9$, and $a_1 = 1.8$ to see what happens.

11–22. Use Newton's method to approximate the indicated solutions to the equation to four decimal places. State your initial guess a_1 and the values $a_2, a_3, \ldots$ used to determine the solution.

11. $x^7 - 3x^2 + 5 = 0$, all solutions
12. $x^6 + 2x^4 - 3x - 9 = 0$, negative solution
13. $x^3 - 6x - 2\cos x = 0$, positive solution
14. $x^3 - 5x + 2\sin x = 0$, positive solution
15. $x^2 = 5 + \sin x$, all solutions

16. $x^6 + 3x^3 = 7x + 10$, all solutions
17. $x^5 - 14x^3 - 5x^2 + 17 = 0$, negative solution
18. $x^7 - 6x^4 + 2x^2 + 5 = 0$, middle solution
19. $2\sin\sqrt{x+8} = 1$, smallest positive solution
20. $5x - \dfrac{1}{1-x^3} = 0$, all solutions
21. (GSL) $x^4 + 4x^3 - 6x^2 - 20x - 23 = 0$, all solutions
22. (GSL) $2x^3 - 14x^2 + 2x + 5 = 0$, all solutions

GSL, in an era before electronic calculating devices, only required the solutions to two decimal places. But graphs had to be drawn by hand (such as by plotting points for a rough sketch).

23. Approximate $\sqrt[7]{4911}$ by using Newton's method to approximate a solution to the equation $x^7 = 4911$ to six decimal places.
24. Approximate $\sqrt[5]{2214}$ using Newton's method to five decimal places.

Although we use technology for such approximations now, throughout history, many different procedures for approximating roots of numbers by hand were used.

25. (GSL) A solid wooden sphere of specific gravity S and diameter d sinks in water to a depth h. If $x = \dfrac{h}{d}$, then $2x^3 - 3x^2 + S = 0$. Find x for a maple ball for which $S = 0.786$. Give the answer to three decimal places.
26. (GSL) The inner radius r and outer radius R (in inches) of the hollow steel driving shaft of a steamer transmitting H horse power at a speed of N revolutions per minute satisfy the relation

$$R^4 - r^4 = \frac{33HR}{N\pi^2}.$$

If $H = 2500$, $N = 160$, and $r = 6$, find R. Give the answer to three decimal places.
27. Use Newton's method to approximate the critical numbers of $f(x) = x^5 + 20x^2 - 60x + 17$ to two decimal places.
28. Use Newton's method to approximate the critical numbers of $g(x) = \frac{1}{2}x^2 - x + 1 - \cos x$ to three decimal places.

29–32. For the given function and initial value a_1, follow the Newton's method procedure until it is clear that it fails, then give the reason for the failure. Graph the function and appropriate tangent lines to illustrate.

29. $h(x) = \dfrac{2-x}{1+x}$, $a_1 = 5$
30. $f(x) = \sqrt{x} - \frac{1}{2}$, $a_1 = 4$
31. $g(x) = 4x^4 - x^3 - 8x^2 + 5x + 4$, $a_1 = 1$
32. $k(x) = \sqrt[3]{x}$, $a_1 = 1$

Chapter IV

Integration

Antiderivatives

4.1

Many mathematical operations have inverse operations. Subtraction is the inverse of addition, division is the inverse of multiplication, and extracting roots is the inverse of exponentiation. In algebra, factoring is inverse to multiplying polynomials.

Issues often arise for inverse operations that were not present in the original operation. For instance, we can multiply any two numbers, but we cannot always divide any two numbers (division by zero is undefined). We can square any number, but if we consider only real numbers, then we cannot take the square root of negative numbers. Any pair of polynomials can be multiplied, but not every polynomial can be factored (for instance, $x^2 + 1$ cannot be factored using only real numbers).

Some of these operations can be accomplished using complex numbers.

Differentiation also has an inverse operation: *antidifferentiation*. Just as with other inverse operations, we encounter issues that were not present with differentiation.

Antiderivatives: reversing the differentiation process

Consider the function $y = x^2 + 7x - 4$. Its derivative is $y' = 2x + 7$. Now consider the reverse question: if

$$y' = 2x + 7,$$

what is y? One quick answer is the function with the derivative we just took:

This is just one of the many possible answers to the question.

$$y = x^2 + 7x - 4.$$

This is the idea of antidifferentiation: we start with the derivative and ask what the original function might be, reversing the process of differentiation.

Definition 1 ANTIDERIVATIVE *A function F is called an* antiderivative *of f on an interval I if $F'(x) = f(x)$ for all x in the interval I.*

Notice the phrase *an antiderivative*, not *the antiderivative*. This is because there can be more than one antiderivative for the same function.

Example 1 *For the function $f(x) = 2x + 7$, (a) find an antiderivative of f, then (b) find another antiderivative of f, then (c) find yet another antiderivative of f.*

Definition 1 is for an antiderivative "on an interval I." When no interval is specified, we assume an appropriate interval is implied. For this example, the interval $(-\infty, \infty)$ can be used. For exercises of the form "find an antiderivative," an interval is often not stated explicitly in the exercise or in the solution.

Solution (a) Based on definition 1, a function with a derivative that is $f(x) = 2x + 7$ is

$$F(x) = x^2 + 7x - 4.$$

(b) Another function with a derivative that is $f(x) = 2x + 7$ is

$$F(x) = x^2 + 7x - 3,$$

since

We use the same function name to emphasize the fulfillment of the definition. Otherwise, it is best to use a different name for each of these functions.

$$F'(x) = 2x + 7.$$

(c) Yet another function with a derivative that is $f(x) = 2x + 7$ is

$$F(x) = x^2 + 7x + 428.72,$$

since

$$F'(x) = 2x + 7. \quad \blacksquare$$

It should be clear that we can end the function F with any real number we want! There are infinitely many antiderivatives of $f(x) = 2x + 7$—namely, any function of the form

$$F(x) = x^2 + 7x + C,$$

where C is a real number. All these functions have the same derivative $F'(x) = 2x + 7$, and any of these functions could serve as answers to example 1.

We know there are infinitely many antiderivatives of the function $f(x) = 2x + 7$, but have we found all of them? Could there be even more? Could there be some other function, perhaps some function we have not yet studied, that also has the same derivative? Not according to theorem 6 of section 3.2, which was a consequence of the mean value theorem. In that section we learned that functions with the same derivative have the same slopes, and therefore the same shape, so the functions must be of the same type. We have found all the functions with derivative $2x + 7$.

See the subsection "Functions with the same derivative" in section 3.2 for more details.

Definition 2 MOST GENERAL ANTIDERIVATIVE *The* most general antiderivative *of f is the collection of all possible antiderivatives of f.*

Example 2 *Find the most general antiderivative of $f(x) = 2x + 7$.*

Solution Based on the previous discussion, all antiderivatives of f are of the form $F(x) = x^2 + 7x + C$. The collection of all these functions forms the most general antiderivative of f. ■

Compare the solutions to examples 1 and 2. The many possible solutions to example 1 are individual functions, whereas the only solution to example 2 is a collection (a set) of functions.

Example 3 *Find the most general antiderivative of $f(x) = \cos x$.*

Solution To find the most general antiderivative, we start by finding an antiderivative. That is, we need a function with a derivative that is $f(x) = \cos x$. Can you think of one? We learned derivative formulas for trigonometric functions, and one of these formulas is

$$\frac{d}{dx} \sin x = \cos x.$$

Therefore, $F(x) = \sin x$ is an antiderivative of $f(x) = \cos x$. Other antiderivatives must differ from this one by a constant, so the most general antiderivative is the collection of functions

$$F(x) = \sin x + C. \quad ■$$

Other antiderivatives include $F(x) = \sin x + 12$ and $F(x) = \sin x - \sqrt{77}$.

The expression "$+ C$" takes care of both addition and subtraction of real numbers. This is because the value of C may be any real number. If C is a negative number, such as -11, then "$+ C$" is the same as "$+ (-11)$," which can be rewritten as "-11."

Next we examine a new notation and a new name for the most general antiderivative.

Indefinite integrals

An *indefinite integral* is another name for the most general antiderivative.

Definition 3 INDEFINITE INTEGRAL *Suppose F is an antiderivative of f on the interval I. Then the* indefinite integral *of f on the interval I, written $\int f(x)\,dx$, is the most general antiderivative of f:*

$$\int f(x)\,dx = F(x) + C,$$

where C is a real number.

The notation $\int f(x)\,dx$ can be read "the integral of f of x with respect to x" or "the integral of f of x dx," where the latter letters are pronounced by their names individually, but with a pause before the dx.

An indefinite integral is not one function but a collection of infinitely many functions, one for each real number C. The reasoning behind this name and notation is apparent later in this chapter.

Examples 2 and 3 can be rewritten using the indefinite integral notation.

The function "inside" the integral is called the *integrand*. The integrand for part (a) is $2x+7$ and the integrand for part (b) is $\cos x$.

Example 4 *Evaluate (a)* $\int (2x+7)\,dx$ *and (b)* $\int \cos x\,dx$.

Solution (a) We know the most general antiderivative from example 2 and only need to rewrite that solution using the indefinite integral notation:

$$\int (2x+7)\,dx = x^2 + 7x + C.$$

(b) Similarly, we rewrite the solution to example 3:

$$\int \cos x\,dx = \sin x + C. \qquad \blacksquare$$

Antiderivative rules

(1) refers to the sum and difference rules whereas (2) refers to the constant multiple rule.

We learned to take derivatives of polynomials by learning the formula for the derivative of x^n and then learning that (1) the derivative can be taken term-by-term and (2) constant multiples can be carried along to the derivative. The approach to finding antiderivatives of polynomials is much the same.

We start by seeking a power rule for antiderivatives. Perhaps reversing a derivative of a power helps. Reversing the derivative exercise

Any derivative calculation can be reversed to give an antiderivative result, as illustrated here:

$$\frac{d}{dx}x^4 = 4x^3$$
$$\int 4x^3\,dx = x^4 + C.$$

$$\frac{d}{dx}x^4 = 4x^3$$

gives the antiderivative result

$$\int 4x^3\,dx = x^4 + C.$$

We want a power rule for antidifferentiation, so it would be nice to know $\int x^3\,dx$ instead. To get a function with a derivative that is x^3, dividing the original function by 4 does the trick:

As in the previous margin equation, we reverse the derivative calculation to give an antiderivative result in this manner:

$$\frac{d}{dx}\left(\frac{x^4}{4}\right) = x^3$$
$$\int x^3\,dx = \frac{x^4}{4} + C.$$

$$\frac{d}{dx}\left(\frac{x^4}{4}\right) = \frac{d}{dx}\left(\frac{1}{4}\cdot x^4\right) = \frac{1}{4}\cdot 4x^3 = x^3.$$

Reversing this exercise gives the antidifferentiation result

$$\int x^3\,dx = \frac{x^4}{4} + C.$$

Notice that going from x^3 to $\frac{x^4}{4}$ requires adding 1 to the exponent (changing from 3 to 4) and then dividing by the new exponent (dividing by 4). This process works in general. Because

$$\frac{d}{dx}\left(\frac{x^{n+1}}{n+1}\right) = \frac{1}{n+1}\cdot(n+1)x^n = x^n,$$

To avoid division by zero, this derivative calculation assumes $n \neq -1$.

the antiderivative power rule is

$$\int x^n\,dx = \frac{x^{n+1}}{n+1} + C.$$

ANTIDERIVATIVE POWER RULE

For any real number $n \neq -1$,

$$\int x^n\,dx = \frac{x^{n+1}}{n+1} + C.$$

Recall that the general power rule for derivatives has not yet been proved completely. So far, we have only the proof for rational exponents. As a result, the same is true for the antiderivative power rule. However, just as with derivatives, we continue to use the antiderivative power rule for all exponents (except -1).

Example 5 *Evaluate the integral (a)* $\int x^6\,dx$ *and (b)* $\int x^{-3}\,dx$.

Solution (a) Using the antiderivative power rule with $n = 6$ gives

$$\int x^6\,dx = \frac{x^7}{7} + C.$$

(b) Using the antiderivative power rule with $n = -3$ gives

$$\int x^{-3}\,dx = \frac{x^{-2}}{-2} + C.$$ ■

Since $x^{-3} = \frac{1}{x^3}$ is undefined at $x = 0$, the interval for this antiderivative cannot include zero. For instance, the interval could be $(-\infty, 0)$ or $(0, \infty)$.

CAUTION: MISTAKE TO AVOID
Recall that $-3 + 1 = -2$:

$$\int x^{-3}\,dx \neq \frac{x^{-4}}{-4} + C.$$

Reading Exercise 1 *Evaluate the integral* $\int x^{12}\,dx$.

In the derivative calculation

$$\frac{d}{dx}4x^7 = 4\cdot 7x^6,$$

the constant multiple rule is used to carry the 4 along to the derivative. Going backward for an antiderivative means the same thing happens; constant multiples are carried along to the antiderivative as well.

DERIVATION OF THE ANTIDERIVATE CONSTANT MULTIPLE RULE

To prove the antiderivative constant multiple rule, suppose

$$\int f(x)\,dx = F(x) + C,$$

meaning that F is an antiderivative of f, or $F'(x) = f(x)$. If k is any real number, then

$$\frac{d}{dx}k \cdot F(x) = k \cdot F'(x) = k \cdot f(x).$$

Reversing,

$$\int k{\cdot}f(x)\,dx = k{\cdot}F(x) + C = k{\cdot}\int f(x)\,dx,$$

and the proof is complete.

ANTIDERIVATIVE CONSTANT MULTIPLE RULE

For any real number k,

$$\int k \cdot f(x)\,dx = k \cdot \int f(x)\,dx,$$

assuming the latter integral exists.

Example 6 *Determine* $\int 11x^4\,dx$.

Solution Using the antiderivative power rule and the antiderivative constant multiple rule,

$$\int 11x^4\,dx = 11 \cdot \frac{x^5}{5} + C = \frac{11}{5}x^5 + C. \quad \blacksquare$$

Reading Exercise 2 *Determine* $\int \frac{1}{2}x^4\,dx$.

The sum rule is summarized as "the derivative of a sum is the sum of the derivatives." This gives us the ability to calculate

$$\frac{d}{dx}\left(x^3 + x^7\right) = 3x^2 + 7x^6$$

by taking the derivative of each term. Reversing this calculation gives

$$\int \left(3x^2 + 7x^6\right)\,dx = x^3 + x^7 + C,$$

which is the same as

$$\int 3x^2\,dx + \int 7x^6\,dx.$$

In other words, the antiderivative of a sum is the sum of the antiderivatives.

When finding the sum of two indefinite integrals, we do not have to worry about adding the $+\,C$ twice. One reason is that "any real number" plus "any real number" is still "any real number." A more formal way of seeing this is that the indefinite integral is *an* antiderivative plus a real number; therefore, we can add the specific antiderivatives first and then add the $+\,C$ at the very end.

ANTIDERIVATIVE SUM RULE

$$\int (f(x) + g(x))\,dx = \int f(x)\,dx + \int g(x)\,dx.$$

DERIVATION OF THE ANTIDERIVATE SUM RULE

To prove the antiderivative sum rule, suppose

$$\int f(x)\,dx = F(x) + C$$

Similarly, the antiderivative of a difference is the difference of the antiderivatives.

ANTIDERIVATIVE DIFFERENCE RULE

$$\int (f(x) - g(x))\, dx = \int f(x)\, dx - \int g(x) dx.$$

and

$$\int g(x)\, dx = G(x) + C.$$

Then, $F'(x) = f(x)$ and $G'(x) = g(x)$. By the sum rule for derivatives,

$$\frac{d}{dx}(F(x) + G(x)) = F'(x) + G'(x) = f(x) + g(x).$$

Therefore,

$$\int (f(x) + g(x))\, dx = F(x) + G(x) + C = \int f(x) dx + \int g(x) dx.$$

The proof of the antiderivative difference rule is similar.

Example 7 *Find the value of the indefinite integral for (a)* $\int (x^2 + x^5)\, dx$ *and (b)* $\int (4x^5 - 2x^3)\, dx$.

Solution (a) Using the antiderivative sum rule and the antiderivative power rule,

$$\int (x^2 + x^5)\, dx = \int x^2\, dx + \int x^5\, dx = \frac{x^3}{3} + \frac{x^6}{6} + C.$$

Although there are two integrals, we only need to write the $+ C$ at the end one time.

(b) Using the antiderivative difference rule, the antiderivative constant multiple rule, and the antiderivative power rule,

$$\int (4x^5 - 2x^3)\, dx = \int 4x^5\, dx - \int 2x^3\, dx = 4 \cdot \frac{x^6}{6} - 2 \cdot \frac{x^4}{4} + C = \frac{2}{3}x^6 - \frac{1}{2}x^4 + C. \quad \blacksquare$$

Ans. to reading exercise 1

$\frac{x^{13}}{13} + C$

Reading Exercise 3 *Evaluate* $\int (x^3 - 4x^9)\, dx$.

Note that since $\frac{d}{dx}3x = 3$, reversing gives the indefinite integral $\int 3\, dx = 3x + C$. This works for any constant.

Ans. to reading exercise 2

$\frac{1}{10}x^5 + C$

ANTIDERIVATIVE OF A CONSTANT

For any real number k,

$$\int k\, dx = kx + C.$$

This formula comes from reversing

$$\frac{d}{dx}kx = k.$$

Example 8 *Evaluate* $\int 632\, dx$.

Solution Because 632 is a constant,

CAUTION: MISTAKE TO AVOID
Although the derivative of a constant is 0, the antiderivative of a constant is not 0:

$$\int 632\,dx \neq 0 + C.$$

$$\int 632\,dx = 632x + C. \quad \blacksquare$$

These rules are sufficient to allow us to find antiderivatives for all polynomials (and some nonpolynomials). After a little practice, it is traditional to learn to calculate antiderivatives of polynomials in one step.

Example 9 *Calculate* $\int (2x^5 + 9x^4 - \frac{1}{5}x^3 + 2x + 11.5)\,dx$.

Solution Using the various antiderivative rules,

$$\int (2x^5 + 9x^4 - \frac{1}{5}x^3 + 2x + 11.5)\,dx$$

$$= 2 \cdot \frac{x^6}{6} + 9 \cdot \frac{x^5}{5} - \frac{1}{5} \cdot \frac{x^4}{4} + 2 \cdot \frac{x^2}{2} + 11.5x + C$$

$$= \frac{1}{3}x^6 + \frac{9}{5}x^5 - \frac{1}{20}x^4 + x^2 + 11.5x + C. \quad \blacksquare$$

Reading Exercise 4 *Calculate* $\int (5x^2 - 3x + 2)\,dx$.

Manipulating integrands

We have now reversed several derivative rules and given their antiderivative analogs, including the power rule, sum rule, difference rule, and constant multiple rule. Other derivative rules, such as the product rule, quotient rule, and chain rule, are more difficult to reverse. Although they have their analogs in antidifferentiation, they are not as straightforward to apply as recognizing a product or a quotient and using the associated rule. For this reason, we study them in chapter 7. In the meantime, if we wish to find antiderivatives of products, quotients, or compositions, we must change the form of the expression into something we can antidifferentiate using the rules we have at our disposal.

Ans. to reading exercise 3

$\frac{x^4}{4} - \frac{2}{5}x^{10} + C$

Example 10 *Evaluate* $\int x(x^2 - 3x)\,dx$.

Solution Because we have no product rule for antidifferentiation, we cannot evaluate the antiderivative in its current form. Instead, let's perform the multiplication in the integrand:

When manipulating the integrand (the expression inside the integral), the integral symbol at the beginning and the *dx* at the end do not change; only the expression inside changes to something equivalent.

$$\int x(x^2 - 3x)\,dx = \int (x^3 - 3x^2)\,dx.$$

Now the integrand is in a form we can handle using the available rules; there is no product in the integral on the right. Using the antiderivative difference rule, the antiderivative constant multiple rule, and the antiderivative power rule, we have

$$\int (x^3 - 3x^2)\,dx = \frac{x^4}{4} - 3 \cdot \frac{x^3}{3} + C$$
$$= \frac{x^4}{4} - x^3 + C. \qquad \blacksquare$$

Example 11 *Evaluate* $\int \frac{x^5 - 4x}{2x^3}\,dx.$

The expression $\frac{x^5 - 4x}{2x^3}$ has one term in its denominator—namely, $2x^3$. An expression of the form $\frac{2x^3}{x^5 - 4x}$ has two terms in its denominator, x^5 and $4x$ (recall that terms are separated by a + or −). Performing the division in an expression with two terms in the denominator is more difficult.

Solution Because we have no quotient rule for integration, we cannot evaluate the integral in its current form. We need to manipulate the expression inside the integral (the integrand) into a form for which we can use the available rules. Because there is only one term in the denominator, we may perform the division easily:

$$\int \frac{x^5 - 4x}{2x^3}\,dx = \int \left(\frac{1}{2}x^2 - 2x^{-2}\right) dx.$$

Details of the division:

$$\frac{x^5 - 4x}{2x^3} = \frac{x^5}{2x^3} - \frac{4x}{2x^3}$$
$$= \frac{1}{2}x^{5-3} - 2x^{1-3}$$
$$= \frac{1}{2}x^2 - 2x^{-2}.$$

Now the integrand is in a form we can handle using the available rules; there is no quotient (see margin note) in the integral on the right. Using the antiderivative difference rule, the antiderivative constant multiple rule, and the antiderivative power rule, we have

We are only concerned about quotients of variable expressions. In this context, $\frac{1}{2}$ doesn't count as a quotient because it is a constant.

$$\int \left(\frac{1}{2}x^2 - 2x^{-2}\right) dx = \frac{1}{2} \cdot \frac{x^3}{3} - 2 \cdot \frac{x^{-1}}{-1} + C$$
$$= \frac{1}{6}x^3 + \frac{2}{x} + C. \qquad \blacksquare$$

Example 12 *Evaluate* $\int (2x - 1)^2\,dx.$

Solution Because we have no chain rule for antidifferentiation, we cannot evaluate the integral in its current form. However, this issue can be fixed if we go ahead and square the integrand:

Just as with differentiation, the antiderivative power rule is for "just plain x" to a power. The antiderivative power rule does not apply directly to $\int (2x - 1)^2\,dx$ because we are squaring $2x - 1$, not x.

$$\int (2x - 1)^2\,dx = \int (4x^2 - 4x + 1)\,dx.$$

Now the integrand is in a form we can handle using the available rules; there is no composition in the integral on the right. Using

Ans. to reading exercise 4

$$\frac{5x^3}{3} - \frac{3x^2}{2} + 2x + C$$

the antiderivative difference rule, the antiderivative sum rule, the antiderivative constant multiple rule, the antiderivative power rule, and the rule for the antiderivative of a constant, we have

$$\int (4x^2 - 4x + 1)\,dx = 4 \cdot \frac{x^3}{3} - 4 \cdot \frac{x^2}{2} + 1 \cdot x + C$$
$$= \frac{4}{3}x^3 - 2x^2 + x + C. \qquad \blacksquare$$

In my experience observing integral calculations, more errors occur as a result of applying a nonexistent "rule" than failing to know a rule.

THE RULES RULE

Knowing what "rules" do not exist is just as important as knowing which rules do exist for calculating antiderivatives.

Trig antiderivatives

As previously stated, all derivative formulas can be reversed to provide antiderivative formulas. An example involving trigonometric functions is the derivative formula

$$\frac{d}{dx}\tan x = \sec^2 x,$$

which yields

$$\int \sec^2 x\,dx = \tan x + C.$$

When the derivative formula results in the use of a negative sign, a slight adjustment is helpful. Reversing $\frac{d}{dx}\cos x = -\sin x$ gives $\int(-\sin x)\,dx = \cos x + C$, which is not convenient to use because of the negative sign. If we differentiate negative cosine instead, we get a more convenient form. Reversing

Details:

$\frac{d}{dx}(-\cos x) = -(-\sin x)$
$= \sin x$

$$\frac{d}{dx}(-\cos x) = \sin x$$

yields the antidifferentiation formula

$$\int \sin x\,dx = -\cos x + C.$$

All six trig derivatives can be reversed in this manner.

A FEW TRIG ANTIDERIVATIVES

derivative formula	more convenient form	indefinite integral formula
$\frac{d}{dx}\sin x = \cos x$		$\int \cos x\,dx = \sin x + C$
$\frac{d}{dx}\cos x = -\sin x$	$\frac{d}{dx}(-\cos x) = \sin x$	$\int \sin x\,dx = -\cos x + C$
$\frac{d}{dx}\tan x = \sec^2 x$		$\int \sec^2 x\,dx = \tan x + C$
$\frac{d}{dx}\csc x = -\csc x\cot x$	$\frac{d}{dx}(-\csc x) = \csc x\cot x$	$\int \csc x\cot x\,dx = -\csc x + C$
$\frac{d}{dx}\sec x = \sec x\tan x$		$\int \sec x\tan x\,dx = \sec x + C$
$\frac{d}{dx}\cot x = -\csc^2 x$	$\frac{d}{dx}(-\cot x) = \csc^2 x$	$\int \csc^2 x\,dx = -\cot x + C$

The indefinite integral formulas (antiderivative formulas) in the right-most column should be memorized, because each formula is used repeatedly for the remainder of this book.

Example 13 *Evaluate the indefinite integral* $\int (4\sin x + \sec^2 x)\,dx$.

Solution Using the formulas for the antiderivatives of $\sin x$ and $\sec^2 x$, along with the antiderivative sum rule and the antiderivative constant multiple rule, gives

$$\int (4\sin x + \sec^2 x)\,dx = 4(-\cos x) + \tan x + C$$

$$= -4\cos x + \tan x + C. \qquad \blacksquare$$

CAUTION: MISTAKE TO AVOID
Writing $4 - \cos x + \tan x + C$ is incorrect, because it turns the multiplication $4(-\cos x)$ into a subtraction: $4 - \cos x$.

Reading Exercise 5 *Evaluate the indefinite integral* $\int 6\sec x\tan x\,dx$.

Example 14 *Which of the following integrands are not in a form that can be evaluated with the rules and formulas given so far? (a)* $\int \sin x\cos x\,dx$, *(b)* $\int \csc x\cot x\,dx$, *(c)* $\int \sin 5x\,dx$, *(d)* $\int \tan x\,dx$, *and (e)* $\int \sin^2 x\,dx$.

Each of the integrals of example 14 can be evaluated, but most require techniques that we learn in section 4.6, and in chapters 5 and 7.

Solution (a) Although we can evaluate $\int \sin x\,dx$ and $\int \cos x\,dx$, we do not have a product rule for integration (just as in example 10) and we are stuck. The integral is not in a form that can be antidifferentiated using the current rules and formulas.

(b) Although the integrand is a product, the integral can be evaluated because it is in the list of integration formulas (antiderivative formulas).

(c) Because $5x$ is "not just plain x," the formula for $\int \sin x\,dx$ does not apply directly and we are stuck. The integral is not in a form that can be antidifferentiated using the current rules and formulas.

(d) Notice that $\int \tan x\,dx$ is not in our list of trig antiderivatives. We are stuck. The integral is not in a form that can be antidifferentiated using the current rules and formulas.

Even using a trig identity to write

$$\int \tan x\,dx = \int \frac{\sin x}{\cos x}\,dx$$

is no help because we have no quotient rule for integration (just as in example 11).

(e) A quick search reveals there is no formula in the list for handling $\int \sin^2 x\,dx$. We cannot combine the formula for $\int \sin x\,dx$ with the antiderivative power rule because the antiderivative power rule is for "just plain x" to a power, not for a trig function to a power. We are stuck. The integral is not in a form that can be antidifferentiated using the current rules and formulas. ■

Example 14 demonstrates that we have a long way to go to integrate (find antiderivatives of) the many functions we commonly encounter. Doing so turns out to be too tall of a task. Some antiderivatives are not expressible using only the "elementary" types of functions, and expanding our list of types of functions only increases the problem. Although this presents difficulties not seen with differentiation, we eventually encounter several techniques that can help us expand our antidifferentiation repertoire greatly.

Recall that by using the derivative rules (including the product, quotient, and chain rules), we can find derivatives (where they exist) of all the common functions.

Exercises 4.1

The importance of understanding exercises 1–12 cannot be overemphasized!

1–12. Rapid response: is the integrand in a form for which the antiderivative rules and formulas of this section can be used?

1. $\int \frac{x^2-5x}{4x+12}\,dx$
2. $\int (x^2+3x)(x-5)\,dx$
3. $\int (x^3-4)^5\,dx$
4. $\int (x^2+7x-9)\,dx$
5. $\int x^2 \sin x\,dx$
6. $\int \frac{\cos x}{x^2}\,dx$
7. $\int (3x^5-4x+x^{-5})\,dx$
8. $\int \sec x\,dx$
9. $\int (x^2-12\sec x\tan x)\,dx$
10. $\int \cos(3x-5)\,dx$
11. $\int \sec 4x \tan 4x\,dx$
12. $\int \sqrt{x^2+5}\,dx$

13–32. Evaluate the integral.

Ans. to reading exercise 5
$6\sec x + C$

13. $\int x^8\,dx$

14. $\int x^7\,dx$

15. $\int x^{-4}\,dx$

16. $\int \frac{1}{x^5}\,dx$

17. $\int \sqrt{x}\,dx$

18. $\int \sqrt[3]{x}\,dx$

19. $\int 5x^2\,dx$

20. $\int 7x^{12}\,dx$

21. $\int \frac{6}{x^2}\,dx$

22. $\int 15\,dx$

23. $\int (x^2 + 5x)\,dx$

24. $\int (x^3 - x)\,dx$

25. $\int (x^4 + 7x^3 - x + 6)\,dx$

26. $\int (5x^3 - 3x^2 + 7x - 14)\,dx$

27. $\int (x^2 - \sqrt{11})\,dx$

28. $\int (\sqrt{5}x^2 - \pi)\,dx$

29. $\int (2x^3 - 5x^2 + 1)\,dx$

30. $\int \left(5\sqrt{x} - \pi^2\right)\,dx$

31. $\int \left(5x^2 - 2x + 7 - \sqrt{x} + \frac{1}{x^3}\right)dx$

32. $\int \left(\frac{4}{7}x^2 + \frac{1}{3}x - \frac{1}{5}\right)\,dx$

Why the parentheses when an integrand has more than one term? One reason has historical roots. When the differential was used rather than the derivative, antidifferentials were used instead of antiderivatives. To write a differential correctly, order of operations dictates that we write $(x^3 - x)\,dx$ rather than $x^3 - x\,dx$, because the latter only multiplies the differential dx by the last term and not by both terms.

33–40. Find the requested antiderivative.

33. $\int \sin x\,dx$

34. $\int \cos x\,dx$

35. $\int \sec x \tan x\,dx$

36. $\int \sec^2 x\,dx$

37. $\int (4\cos x - 7\csc^2 x)\,dx$

38. $\int -3\csc x \cot x\,dx$

39. $\int (4 + \cos x)\,dx$

40. $\int (3x^2 - 4\sin x)\,dx$

41–54. Manipulate the integrand into a form for which the available rules and formulas apply, and then finish evaluating the integral.

41. $\int (x-4)(x+1)\,dx$

42. $\int \frac{x^3 - 4x^2 + 1}{x^2}\,dx$

43. $\int \frac{x^2 + 5x}{x}\,dx$

44. $\int (x+1)^3\,dx$

45. $\int (x-1)^2\,dx$

46. $\int 3x^4(x^2 - 7)\,dx$

47. $\int \sqrt{(3x+1)^4}\,dx$

48. $\int (3x - 5x^2(x+6))\,dx$

49. $\int \tan x \cos x\,dx$

50. $\int \frac{1}{\cos^2 x}\,dx$

51. $\int \sec x \frac{\sin x}{\cos x}\,dx$

52. $\int \sqrt{x^4 + 2x^2 + 1}\,dx$

53. $\int \frac{x^2 + 3}{\sqrt{x}}\,dx$

54. $\int 5\sin^2 x \csc x\,dx$

55–78. Evaluate the integral, if possible. If not possible, state "not possible with available rules and formulas."

55. $\int (x^3 - 4x^2 + 7x - 1)\,dx$

56. $\int (7x^2 - 11x + 9)\,dx$

57. $\int (\cos x + 3\sqrt{x})\,dx$

58. $\int -2\sec x \tan x\,dx$

59. $\int \frac{1}{x^7}\,dx$

60. $\int 4\sqrt[3]{x}\,dx$

61. $\int \frac{1}{x}\,dx$

62. $\int \sin 7x\,dx$

63. $\int 6\csc^2 x\,dx$
64. $\int (7x + \sin x)\,dx$
65. $\int x^3(4x - 9)\,dx$
66. $\int 2(x + 3)^2\,dx$
67. $\int \frac{3}{4x^2}\,dx$
68. $\int \sec x\,dx$
69. $\int (5x - 2)^2\,dx$
70. $\int \frac{x^3 - 1}{x^2}\,dx$
71. $\int x\cos x\,dx$
72. $\int \frac{x^2}{x^3 - 1}\,dx$
73. $\int \frac{x^3 - 7x}{4x}\,dx$
74. $\int 4\pi\sqrt{5}\cos x\,dx$
75. $\int (4 + \sqrt{17.2})\,dx$
76. $\int (2x - 1)(3x + 5)\,dx$
77. $\int \left(\frac{1}{\sin^2 x} + \frac{1}{x^2}\right)\,dx$
78. $\int \frac{4x + 5}{\sqrt{x}}\,dx$

79–86. Find an antiderivative of the given function.

79. $f(x) = x^2$
80. $f(x) = 4x + 2$
81. $f(x) = \cos x$
82. $f(x) = \sqrt{x}$
83. $f(x) = \pi^3$
84. $f(x) = x\sqrt{x}$
85. $f(x) = x^3 - 4x^{-3}$
86. $f(x) = \sin 5$

87. Find an antiderivative of $f(x) = ax^2 + bx + c$.
88. Evaluate $\int \sin x\cos x\tan x\cot x\sec x\csc x\,dx$.
89. What does an antiderivative of acceleration represent?
90. Evaluate $\int (y^2 - 5)\,dy$.
91. Find three different antiderivatives of $f(x) = 5x - 8$.
92. Find three different antiderivatives of $f(x) = 1987\sin x$.
93. Write an antiderivative formula based on exercise 61 in section 2.4.

Finite Sums

4.2

During your study of geometry, you encountered several area formulas, such as the area of a rectangle ($A = wh$), the area of a triangle ($A = \frac{1}{2}bh$), and the area of a trapezoid ($A = \frac{b_1 + b_2}{2}h$). A few even involved objects with curved boundaries, such as the area of a circle ($A = \pi r^2$) or a circular sector ($A = \frac{1}{2}r^2\theta$). But, what if we want to know the area of a more general region, such as the area between the x-axis and the curve $y = x^2$ from $x = 1$ to $x = 5$, as illustrated in figure 1?

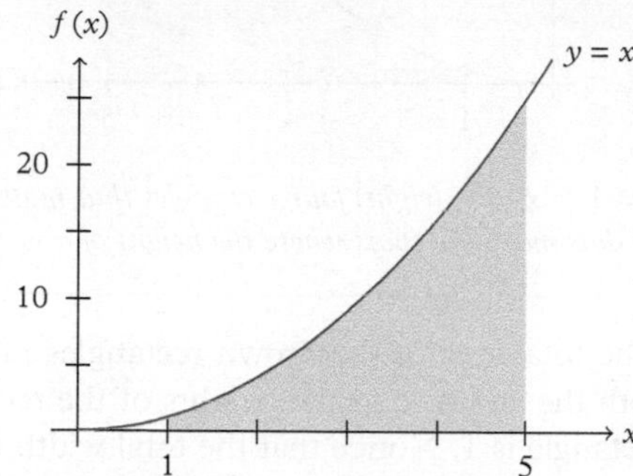

Figure 1 *The graph of the curve $y = x^2$ (blue), with the region underneath the curve down to the x-axis shaded between $x = 1$ and $x = 5$ (green). In general, the areas of regions bounded by curves are not given in geometry*

None of the area formulas you studied in geometry apply to this region. How, then, do we proceed? Although this question is not fully answered until later in this chapter, exploration of the necessary ideas begins now.

Estimating areas using rectangles: left- and right-hand endpoints

The area of the green-shaded region in figure 1 is commonly described as the *area under the curve* $y = x^2$ from $x = 1$ to $x = 5$. When we use this terminology, we need a curve that lies on or above the x-axis, and the region is then between the x-axis and the curve.

Area is measured in terms of *square units*. For instance, a rectangle with width 4 units and height 7 units has area

$$A = wh = 4 \text{ units} \cdot 7 \text{ units} = 28 \text{ units}^2.$$

Notice that "units" is treated like a variable, and when it is multiplied by itself, units · units = units^2. In this text, sometimes units are left out until we write the final answer. For instance,

$$A = wh = 4 \cdot 7 = 28 \text{ units}^2.$$

Although the area formulas from geometry do not apply to this region directly, we can use the formulas to attempt to estimate the area. Perhaps the easiest area formula is the one for a rectangle. To use the rectangle area formula, we need to decide how many rectangles to use and where to place them. If we use $n = 4$ rectangles of width 1 and let the height of each rectangle match the height of the curve at the upper left corner of the rectangle (figure 2, right), we can then estimate the area under the curve by finding the total area of the four rectangles.

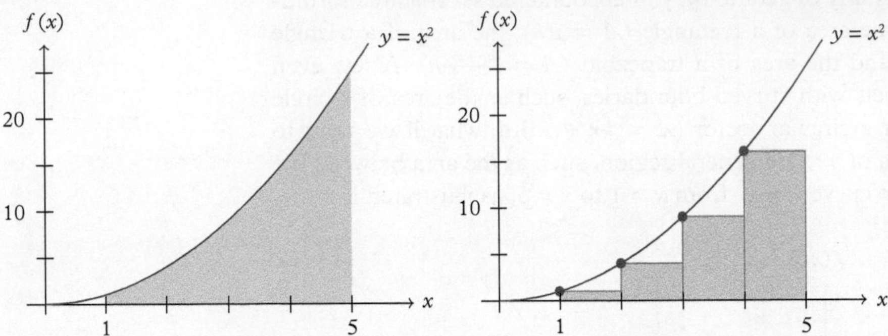

Figure 2 *(left) The region under the curve $y = x^2$ from $x = 1$ to $x = 5$, (right) four rectangles that match the height of the curve at the upper left corner of each rectangle. The red dots mark the spots where the height of a rectangle matches the height of the curve*

To calculate the total area of the brown rectangles in figure 2, we need to know both the widths and the heights of the rectangles. The width of each rectangle is 1. Notice that the total width of the region is $5 - 1 = 4$ and there are four rectangles. In general, if the region goes from $x = a$ to $x = b$ and there are n rectangles, the width of each rectangle is given by

We denote the width of each rectangle by Δx.

$$\text{width} = \Delta x = \frac{b - a}{n}.$$

Here, the width of each rectangle is $\Delta x = \frac{b-a}{n} = \frac{5-1}{4} = 1$.

We can think of each rectangle as being placed on a *subinterval* of the region's interval. In figure 2, left, the green region's interval along the x-axis is the interval from $x = 1$ to $x = 5$, or the interval $[1, 5]$. The brown rectangles in figure 2, right, are placed on the subintervals $[1, 2]$, $[2, 3]$, $[3, 4]$, and $[4, 5]$. The heights of the rectangles then match the heights of the curve at the left-hand endpoint of each subinterval—that is, at $x = 1$, $x = 2$, $x = 3$, and $x = 4$. Because the curve is $y = x^2$, the heights of the rectangles are $y = 1^2 = 1$, $y = 2^2 = 4$, $y = 3^2 = 9$, and $y = 4^2 = 16$ (figure 3).

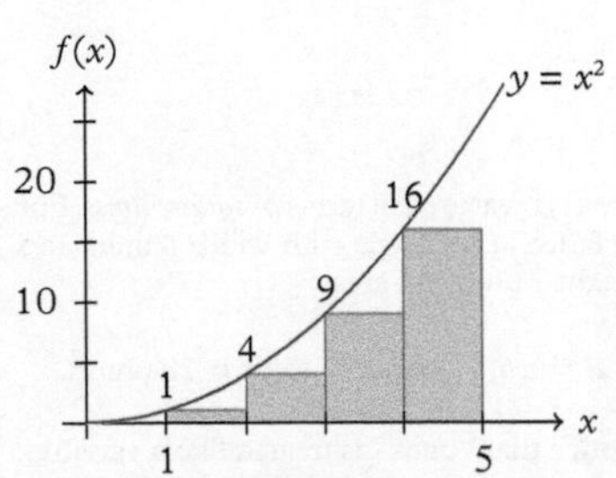

Figure 3 *The heights of the brown rectangles are written near where the rectangle matches the curve*

We are finally ready to calculate the total area of the brown rectangles. Using the formula $A = wh$ for the area of a rectangle, the areas of the rectangles are $1 \cdot 1$, $1 \cdot 4$, $1 \cdot 9$, and $1 \cdot 16$. The total area is

$$1 \cdot 1 + 1 \cdot 4 + 1 \cdot 9 + 1 \cdot 16 = 30 \text{ units}^2.$$

Unfortunately, this isn't really the answer we were originally seeking. The total area of the brown rectangles is not the same as the area of the green region, because some of the green region has been left out

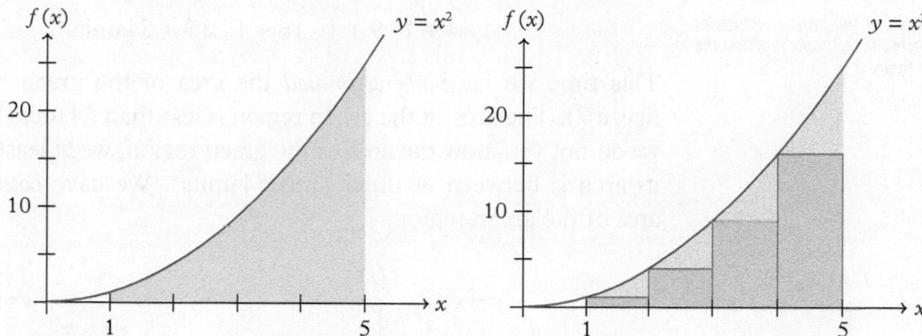

Figure 4 *The area of the green region (left) and the area of the brown rectangles (right) are not the same; the difference between the two is the gray region*

(see figure 4). We have *underestimated* the area of the green region. The area of the green region is larger than 30 units^2.

Now let's try it again, but instead of matching the height of the curve at the upper *left* corner of the rectangle, let's match the height of the curve at the upper *right* corner of the rectangle, as in figure 5.

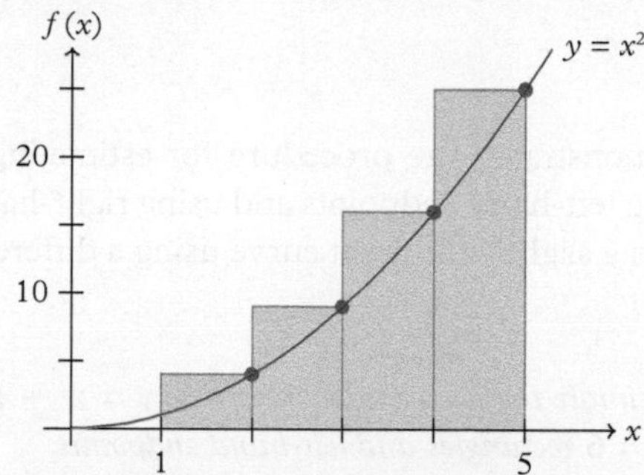

Figure 5 *Four rectangles that match the height of the curve at the upper right corner of each rectangle. The red dots mark the spots where the height of a rectangle matches the height of the curve. Compare to figure 2*

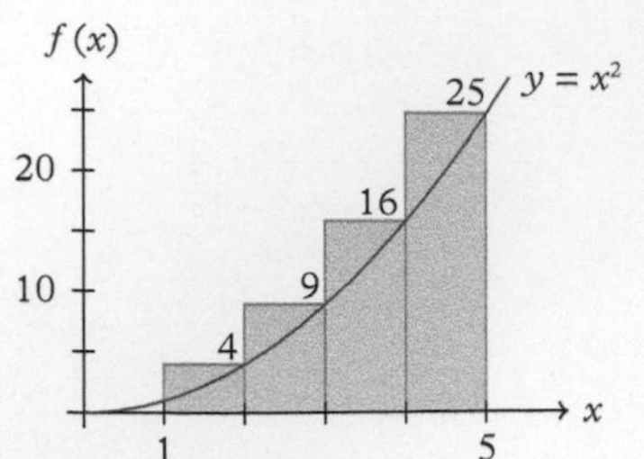

Figure 6 *The heights of the brown rectangles are written near where the rectangle matches the curve. Compare to figure 3*

We are still using four rectangles, so the width of each rectangle remains $\Delta x = \frac{b-a}{n} = \frac{5-1}{4} = 1$. The subintervals on which the rectangles lie are also the same as before: $[1,2]$, $[2,3]$, $[3,4]$, and $[4,5]$. This time, however, the heights of the rectangles match the height of the curve at the right-hand endpoint of each subinterval—that is, at $x = 2$, $x = 3$, $x = 4$, and $x = 5$. Because the curve is $y = x^2$, the heights of the rectangles are $y = 2^2 = 4$, $y = 3^2 = 9$, $y = 4^2 = 16$, and $y = 5^2 = 25$ (figure 6).

The total area of the brown rectangles is

$$1 \cdot 4 + 1 \cdot 9 + 1 \cdot 16 + 1 \cdot 25 = 54\,\text{units}^2.$$

This time we have *overestimated* the area of the green region (see figure 7). The area of the green region is less than 54 units^2. Although we do not yet know the area of the green region, we at least know that its area is between 30 units^2 and 54 units^2. We have *bounds* for the area of the green region.

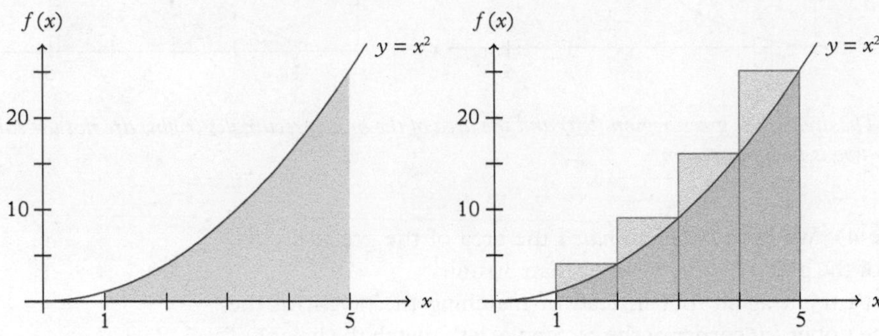

Figure 7 *The area of the green region (left) and the area of the rectangles (right) are not the same; the difference between the two is the gray region. Compare to figure 4*

We have demonstrated the procedure for estimating an area under a curve using left-hand endpoints and using right-hand endpoints. Let's practice on a slightly different curve using a different number of rectangles.

Example 1 *Estimate the area under the curve $y = x^2 + 1$ from $x = -1$ to $x = 2$ using $n = 6$ rectangles and left-hand endpoints.*

Although sketching the region and the approximating rectangles is not requested specifically, doing so helps us develop an understanding of the process as well as acquire the intuition necessary for the next few sections.

Solution ❶ The first task is to sketch the curve and the region whose area we wish to estimate. One method for sketching curves is to plot points, and we eventually need to know the heights of the rectangles, so it makes sense to find a few points on the curve now:

x	$y = x^2 + 1$
-1	2
0	1
1	2
2	5

The curve and region are plotted in figure 8.

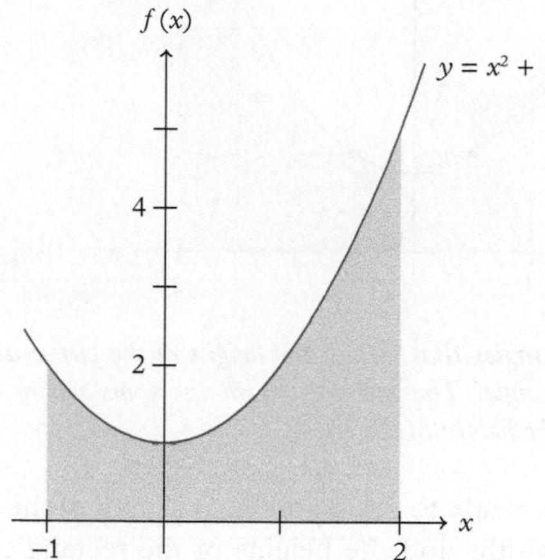

Figure 8 *The graph of the curve $y = x^2 + 1$ (blue), with the region beneath the curve shaded between $x = -1$ and $x = 2$ (green)*

❷ Next we wish to sketch the approximating rectangles. The width of each rectangle is $\Delta x = \frac{b-a}{n} = \frac{2-(-1)}{6} = \frac{1}{2}$. The subintervals are therefore $[-1, -0.5]$, $[-0.5, 0]$, $[0, 0.5]$, $[0.5, 1]$, $[1, 1.5]$, and $[1.5, 2]$. The heights of the rectangles need to match the heights of the curve at the top left corner of each rectangle. The result is in figure 9.

❸ The third step is to determine the heights of the rectangles. They must match the height of the curve $y = x^2 + 1$ at the left-hand endpoints, so we evaluate the curve at $x = -1$, $x = -0.5$, $x = 0$, $x = 0.5$, $x = 1$, and $x = 1.5$. We calculated some of these in step ❶; we just need to fill in the rest.

x	$y = x^2 + 1$
-1	2
-0.5	1.25
0	1
0.5	1.25
1	2
1.5	3.25

If you are having trouble with step ❷, try starting each rectangle by drawing a portion of the left and right sides first, but not all the way up to the curve. For the left-most rectangle, the relevant portion of the diagram looks like this:

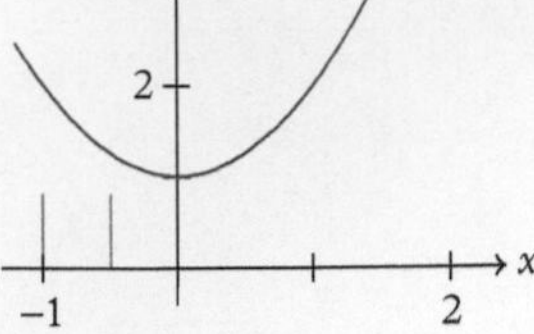

Next, because we are using left-hand endpoints, draw the left side of the rectangle up to the curve:

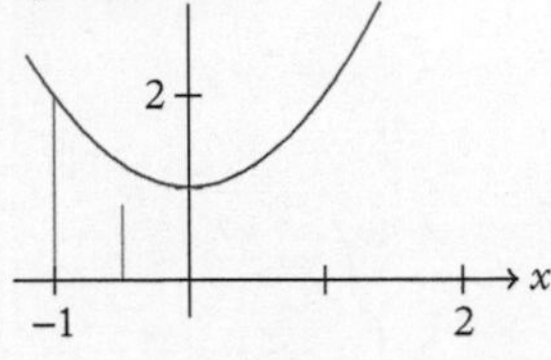

Then, draw the top of the rectangle at the same height as the side you just finished:

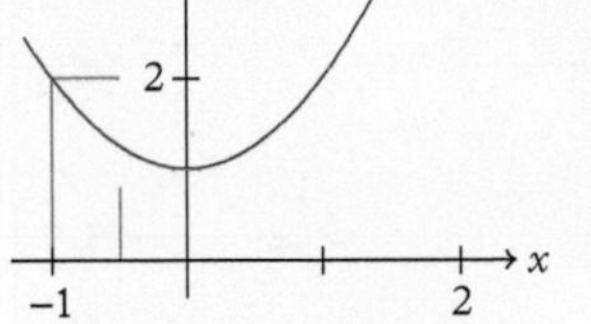

Finish by drawing the other side of the rectangle:

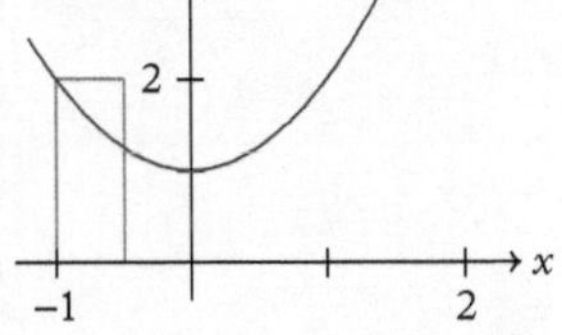

Notice that $x = 2$ is left out of the table here, because we do not have a rectangle with the same height as the curve at $x = 2$ (see figure 9).

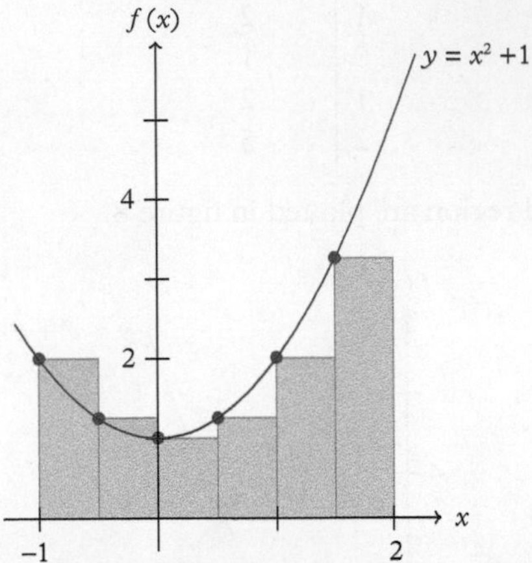

Figure 9 *Six rectangles that match the height of the curve at the upper left corner of each rectangle. The red dots mark the spots where the height of a rectangle matches the height of the curve*

❹ We are now ready to calculate the total area of the rectangles by multiplying the widths and the heights of the rectangles and adding them all up:

$$\frac{1}{2}\cdot 2+\frac{1}{2}\cdot 1.25+\frac{1}{2}\cdot 1+\frac{1}{2}\cdot 1.25+\frac{1}{2}\cdot 2+\frac{1}{2}\cdot 3.25 = 5.375\ \text{units}^2.$$

We estimate the area of the region (the green region of figure 8) to be $5.375\ \text{units}^2$. ■

Notice that in figure 9, two of the rectangles include area that is not in the green region of figure 8, but that there is also a portion of the green region above the other four rectangles that is left out. It is not immediately apparent whether the estimate of $5.375\ \text{units}^2$ is an overestimate or an underestimate.

Reading Exercise 6 *For the region under the curve given in figure 10, draw four rectangles for approximating the area using (a) left-hand endpoints and (b) right-hand endpoints.*

More rectangles

Estimating the area of one the green regions above using only four or six rectangles is not very compelling. Visually, whether we choose

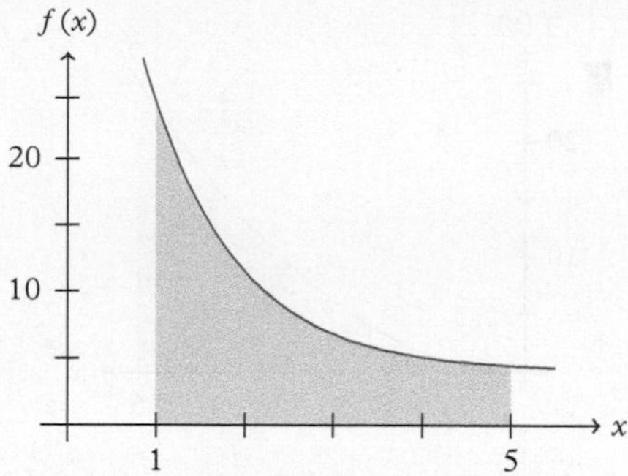

Figure 10 *The graph for reading exercise 6*

the left-hand endpoints or the right-hand endpoints, the areas of the brown rectangles do not seem very close to the area we want to find. Consider, for instance, figure 2 (repeated as figure 11 for convenience). It turns out that the area of the brown rectangles is approximately 27.4% less than the area of the green region we wished to find. This is not an insignificant difference!

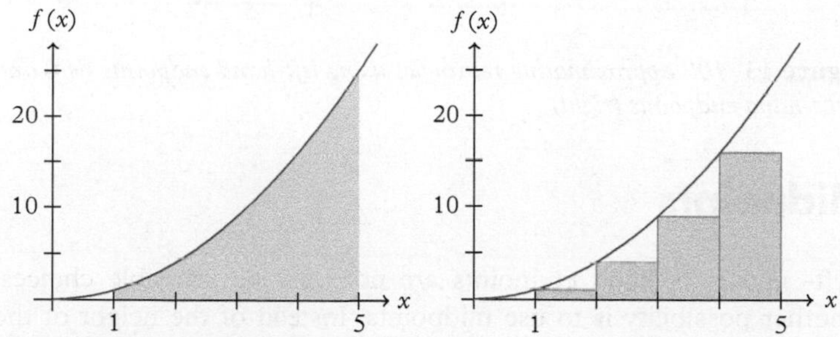

Figure 11 *(left) The region under the curve $y = x^2$ from $x = 1$ to $x = 5$, (right) four approximating rectangles using left-hand endpoints*

Using more rectangles can help. For instance, using $n = 8$ rectangles instead of just four eliminates some of the "missing" area (figure 12).

The use of $n = 100$ rectangles (figure 13) makes it clear that we can get very good estimates of the area under the curve by using enough rectangles.

Perhaps you can see where this is going. Stay tuned for the conclusion in the next section! However, for the remainder of this section we continue to use a finite number of rectangles.

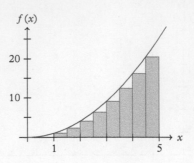

Figure 12 *Eight approximating rectangles using left-hand endpoints. Compared to the use of four rectangles (figure 11, right), the white space between the tops of the rectangles and the curve is much smaller*

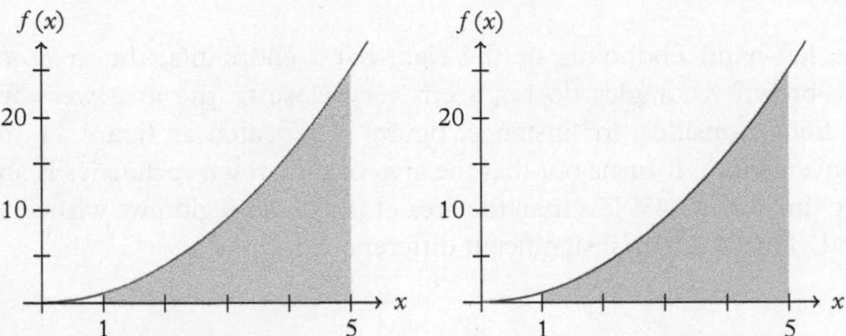

With 100 rectangles and left-hand endpoints, it turns out that the area under the curve is underestimated by about 1.16%. The picture in figure 13 makes the error look even smaller than that, but most of the difference is covered up by the thickness of the curve.

Figure 13 *100 approximating rectangles using left-hand endpoints (left) and right-hand endpoints (right)*

Midpoints

Left- and right-hand endpoints are not the only possible choices. Another possibility is to use midpoints. Instead of the height of the rectangle matching the curve at the top left corner or the top right corner, we make the height of the rectangle match the height of the curve at the midpoint of the top side of the rectangle. Returning again to $y = x^2$ from $x = 1$ to $x = 5$ with $n = 4$ rectangles, the approximating rectangles appear as in figure 14. Notice there is both area left out by the brown rectangles as well as extra area included in the brown rectangles. This happens often when using midpoints, and this is the reason why midpoints often give a much better estimate of the area we are seeking.

Because we are once again using $n = 4$, the width of each rectangle is 1 and the subintervals are $[1, 2]$, $[2, 3]$, $[3, 4]$, and $[4, 5]$, as before. What changes is the heights of the rectangles. Instead of finding the height of the curve at $x = 1, 2, 3$, and 4 as we did for left-hand endpoints, or at $x = 2, 3, 4$, and 5 as we did for right-hand endpoints,

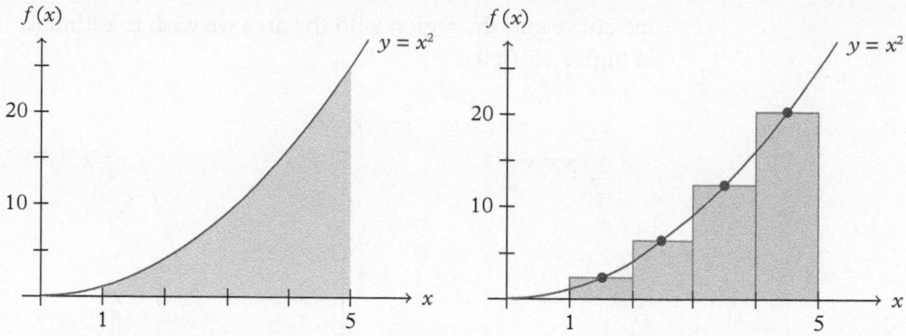

Figure 14 *(left) The region under the curve $y = x^2$ from $x = 1$ to $x = 5$, (right) four rectangles that match the height of the curve at the midpoint of the top side of each rectangle. The red dots mark the spots where the height of a rectangle matches the height of the curve. Compare to figures 2 and 5*

we instead find the heights of the curve at the midpoints of the subintervals—that is, at $x = 1.5$, $x = 2.5$, $x = 3.5$, and $x = 4.5$:

x	$y = x^2$
1.5	2.25
2.5	6.25
3.5	12.25
4.5	20.25

Now that we have the widths and heights of the rectangles, we multiply to find the areas and add the results to get the total area of the brown rectangles:

$$1 \cdot 2.25 + 1 \cdot 6.25 + 1 \cdot 12.25 + 1 \cdot 20.25 = 41 \text{ units}^2.$$

This is still not the exact area of the green region. Is this an underestimate or an overestimate of the area of the green region? It is hard to tell from figure 14, and I won't give away the answer just yet. This is one disadvantage of the midpoint method; it is (normally) not useful for determining bounds for the area. But, it more than makes up for that with increased accuracy. In fact, the midpoint estimate of 41 units^2 obtained by using just $n = 4$ rectangles is closer to the actual area than the estimates obtained by using $n = 100$ rectangles with either left-hand endpoints or right-hand endpoints! Considering figure 13, this is quite remarkable.

Ans. to reading exercise 6

(a)

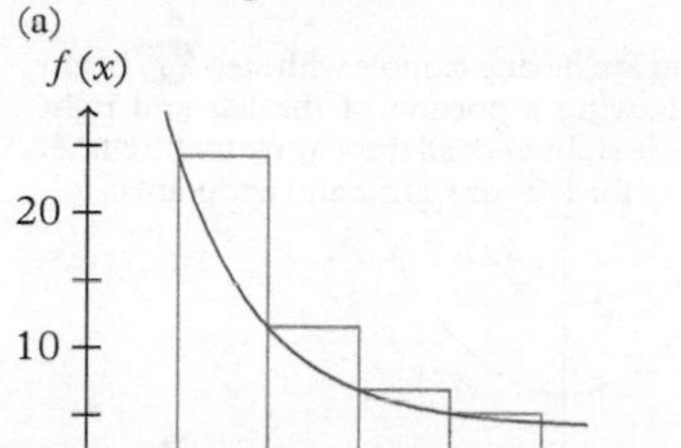

(b)

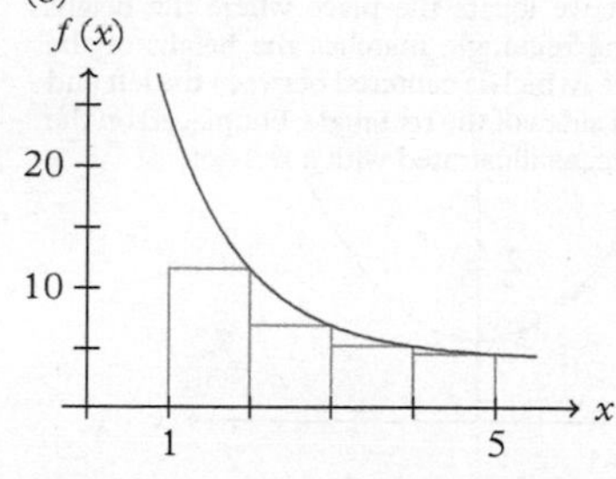

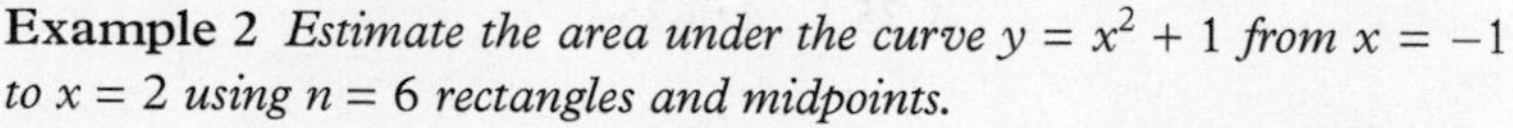

Example 2 *Estimate the area under the curve $y = x^2 + 1$ from $x = -1$ to $x = 2$ using $n = 6$ rectangles and midpoints.*

Solution We may use the same steps as in example 1, replacing left-hand endpoints with midpoints. The result of step ❶, sketching

The importance of greater accuracy for less work is not just for hand computations. Computational efficiency is important in computer algorithms as well.

the curve and the region with the area we wish to estimate, is repeated in figure 15 (left).

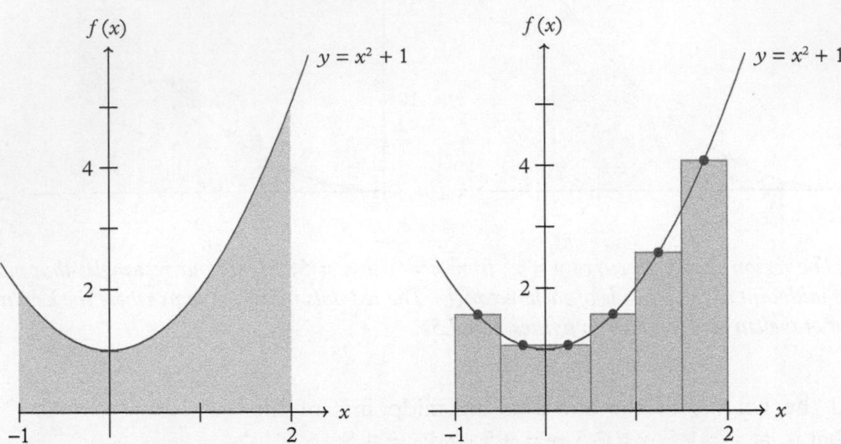

Figure 15 *The graph of the curve $y = x^2 + 1$, with (left) the region beneath the curve shaded between $x = -1$ and $x = 2$ and (right) six rectangles that match the height of the curve at the midpoint of the top side of each rectangle*

If you are having trouble with step ❷, begin by drawing a portion of the left and right sides first, but not all the way up to the curve, just as for left- or right-hand endpoints:

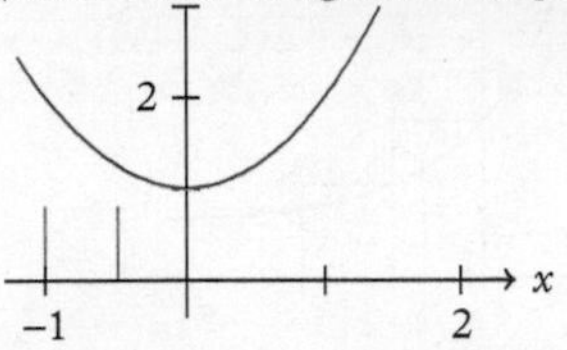

Next we locate the place where the height of the rectangle matches the height of the curve, which is centered between the left and right sides of the rectangle, but placed on the curve, as illustrated with a red dot:

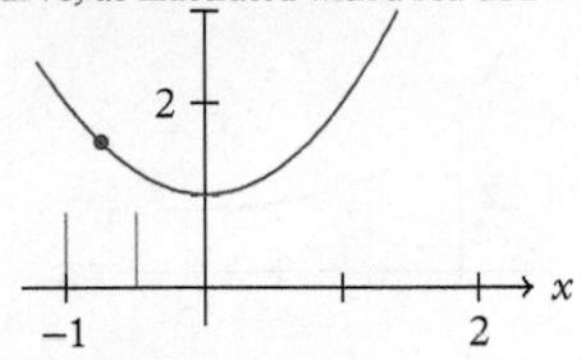

We then draw the top of the rectangle through the red dot:

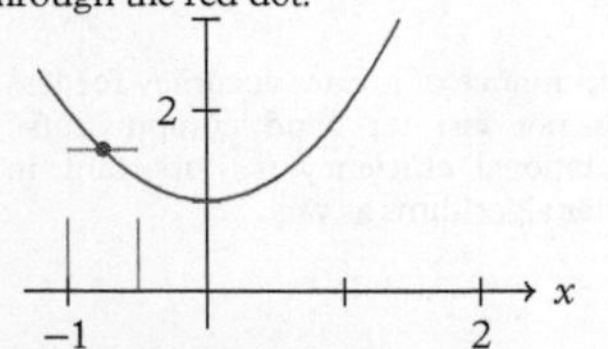

The rectangle can then be completed by finishing its left and right sides (not pictured).

❷ Before sketching the approximating rectangles we need to know the widths of the rectangles. As in example 1, the widths are $\Delta x = \frac{b-a}{n} = \frac{2-(-1)}{6} = \frac{1}{2}$, and the subintervals are $[-1, -0.5]$, $[-0.5, 0]$, $[0, 0.5]$, $[0.5, 1]$, $[1, 1.5]$, and $[1.5, 2]$. The heights of the rectangles need to match the heights of the curve at the midpoint of the top side of each rectangle. The result is in figure 15 (right).

❸ The third step is to determine the heights of the rectangles. They must match the curve $y = x^2 + 1$ at the midpoints of the subintervals. The first subinterval is $[-1, -0.5]$, and its midpoint is $\frac{-1+-0.5}{2} = -0.75$. The other subinterval midpoints are -0.25, 0.25, 0.75, 1.25, and 1.75. These are the values of x at which we evaluate the curve:

x	$y = x^2 + 1$
−0.75	1.5625
−0.25	1.0625
0.25	1.0625
0.75	1.5625
1.25	2.5625
1.75	4.0625

❹ The total area of the approximating rectangles is found by multiplying widths by heights and adding the results:

$$\frac{1}{2} \cdot 1.5625 + \frac{1}{2} \cdot 1.0625 + \frac{1}{2} \cdot 1.0625 + \frac{1}{2} \cdot 1.5625 + \frac{1}{2} \cdot 2.5625 + \frac{1}{2} \cdot 4.0625$$
$$= 5.9375 \text{ units}^2.$$

We estimate the area of the region to be 5.9375 units^2. ■

The error in the estimate (the difference between the estimated area and the not-yet-computed correct area) for example 1 is 10 times as high as the error in the estimate for example 2. Using midpoints is worth the effort!

Reading Exercise 7 *For the region under the curve given in figure 16, draw four rectangles for approximating the area using midpoints.*

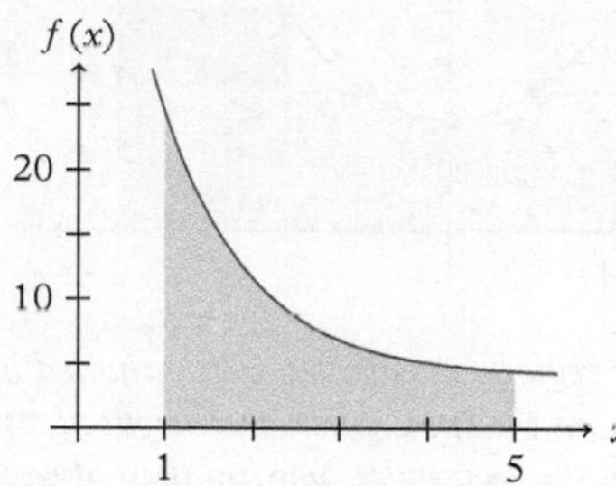

Figure 16 *The graph for reading exercise 7*

Upper and lower estimates

In example 1, the tops of the approximating rectangles were not always below the curve, neither were they always above the curve (figure 9). As a result, we could not tell immediately whether the total area of the rectangles was an underestimate or an overestimate. Left- and right-hand endpoints did not give us bounds for the area under the curve.

One way to rectify this situation and find bounds for the area under the curve is to make the rectangles as tall as possible while not going over the top of the curve (*lower estimate*) and then to make the rectangles as short as possible without the tops of the rectangles going below the curve (*upper estimate*).

Example 3 *Estimate the area under the curve $y = x^2 + 1$ from $x = -1$ to $x = 2$ using $n = 6$ rectangles, finding (a) a lower estimate and (b) an upper estimate.*

Upper estimates are also called *upper sums*. Lower estimates are also called *lower sums*.

Solution (a) We use the same steps as in examples 1 and 2, replacing left-hand endpoints or midpoints with a lower estimate. The result of step ❶, sketching the curve and the region with the area we wish to estimate, is repeated in figure 17, left.

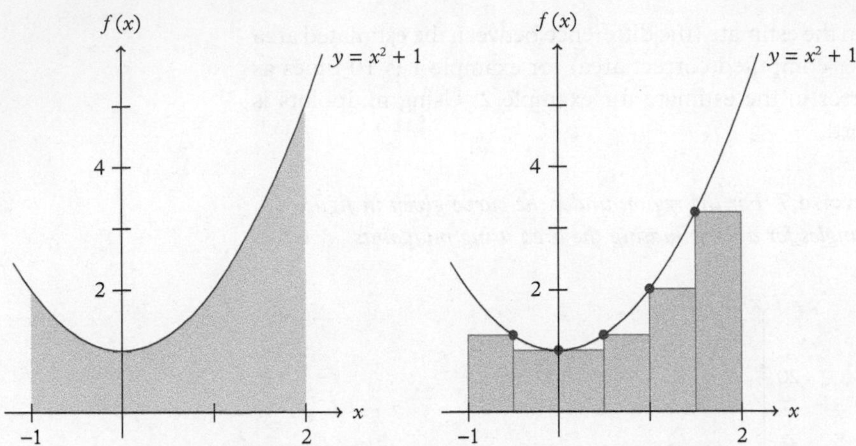

Figure 17 *The graph of the curve $y = x^2 + 1$, with (left) the region beneath the curve shaded between $x = -1$ and $x = 2$ and (right) six rectangles for a lower sum. The red dots mark the spots where the height of a rectangle matches the height of the curve. These dots are not always on the left side of the rectangles, nor are they always on the right side of the rectangles. There are only five dots for the six rectangles because one of the dots (at $x = 0$) serves as the matching spot on the curve for two rectangles. Compare to figures 9 and 15*

❷ Before sketching the approximating rectangles, we need to know the widths of the rectangles. As in examples 1 and 2, the widths are $\Delta x = \frac{b-a}{n} = \frac{2-(-1)}{6} = \frac{1}{2}$, and the subintervals are $[-1, -0.5]$, $[-0.5, 0]$, $[0, 0.5]$, $[0.5, 1]$, $[1, 1.5]$, and $[1.5, 2]$. For a lower estimate, the rectangles need to be as tall as possible while not going over the top of the curve. The result is in figure 17 (right).

The height of each rectangle in figure 18 is the minimum value of the curve on that subinterval.

❸ The third step is to determine the heights of the rectangles. This time, rather than give a table of values, let's mark the heights on the curve (figure 18). Because we are not using left-hand endpoints or right-hand endpoints or midpoints consistently to get the height of the curve, it is easier to see which values to include or not include from the diagram than it is from the table of values. The heights of the rectangles are the heights of the curve at $x = -0.5$, $x = 0$ (for two rectangles), $x = 0.5$, $x = 1$, and $x = 1.5$.

❹ Consulting figure 18, we calculate the total area of the approximating rectangles by multiplying widths by heights and adding the results:

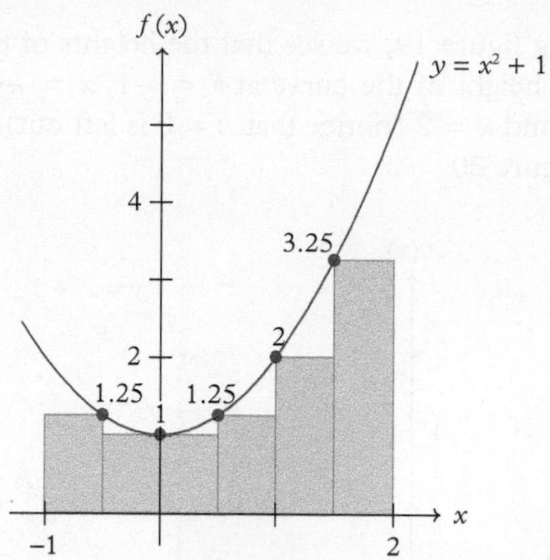

Figure 18 *Heights of the rectangles are labeled. Two rectangles have height* 1

$$\frac{1}{2}\cdot 1.25+\frac{1}{2}\cdot 1+\frac{1}{2}\cdot 1+\frac{1}{2}\cdot 1.25+\frac{1}{2}\cdot 2+\frac{1}{2}\cdot 3.25=4.875\text{ units}^2.$$

Notice that there are two rectangles of height 1, and both must be included in the calculation.

This is a lower estimate (an underestimate). The area of the region under the curve is greater than 4.875 units^2.

(b) ❷ For an upper sum, our rectangles must be as short as possible while not allowing the tops of the rectangles to go below the curve. An alternate description is to think of the curve as the floor and let the tops of the rectangles rest on the floor. The resulting rectangles are drawn in figure 19.

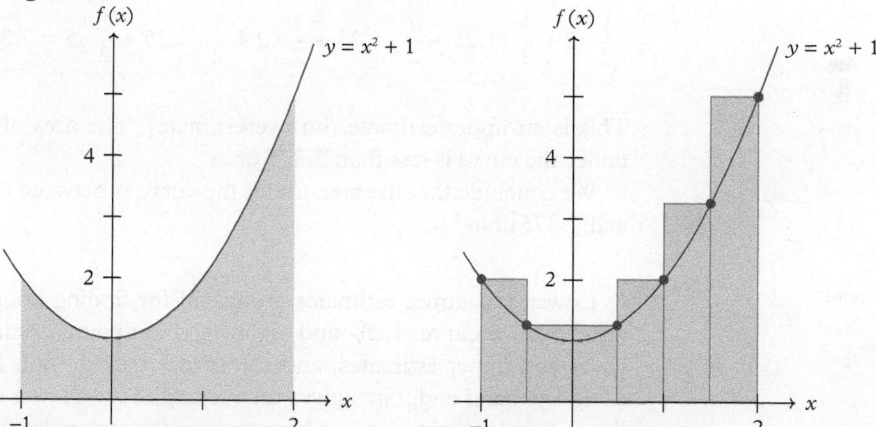

Figure 19 *The graph of the curve* $y = x^2 + 1$, *with (left) the region beneath the curve shaded between* $x = -1$ *and* $x = 2$ *and (right) six rectangles for an upper sum. The red dots mark the spots where the height of a rectangle matches the height of the curve. These dots are not always on the left side of the rectangles, nor are they always on the right side of the rectangles. Notice there is no dot at* $x = 0$. *Compare to figure 17*

❸ Consulting figure 19, we see that the heights of the rectangles must match the height of the curve at $x = -1$, $x = -0.5$, $x = 0.5$, $x = 1$, $x = 1.5$, and $x = 2$ (notice that $x = 0$ is left out). The heights are marked in figure 20.

The height of each rectangle in figure 20 is the maximum value of the curve on that subinterval.

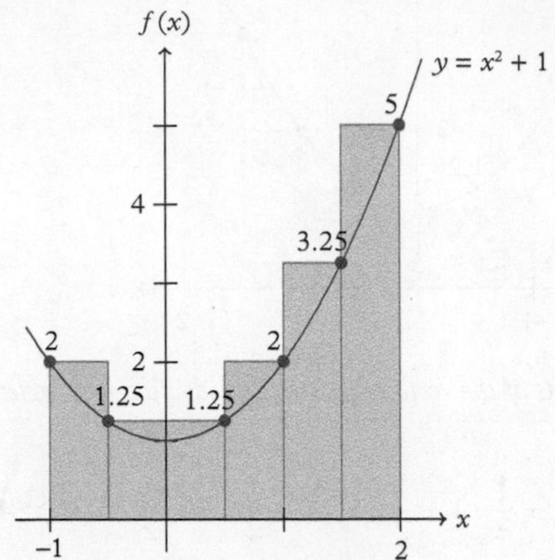

Figure 20 *Heights of the rectangles are labeled*

❹ Consulting figure 20, we calculate the total area of the approximating rectangles by multiplying widths by heights and adding the results:

$$\frac{1}{2} \cdot 2 + \frac{1}{2} \cdot 1.25 + \frac{1}{2} \cdot 1.25 + \frac{1}{2} \cdot 2 + \frac{1}{2} \cdot 3.25 + \frac{1}{2} \cdot 5 = 7.375 \text{ units}^2.$$

This is an upper estimate (an overestimate). The area of the region under the curve is less than 7.375 units^2.

We conclude that the area under the curve is between 4.875 units^2 and 7.375 units^2. ■

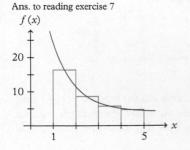

Ans. to reading exercise 7

Lower and upper estimates are useful for finding bounds for the area under a curve. Left- and right-hand endpoints sometimes give lower and upper estimates, and sometimes they do not. In figure 2, using left-hand endpoints gives the rectangles for a lower estimate; in figure 5, using right-hand endpoints gives the rectangles for an upper estimate. However, in figure 9, using left-hand endpoints does not give rectangles for an upper or lower estimate.

Reading Exercise 8 *For the region under the curve given in figure 21, draw four rectangles for finding (a) a lower estimate and (b) an upper estimate of the area under the curve.*

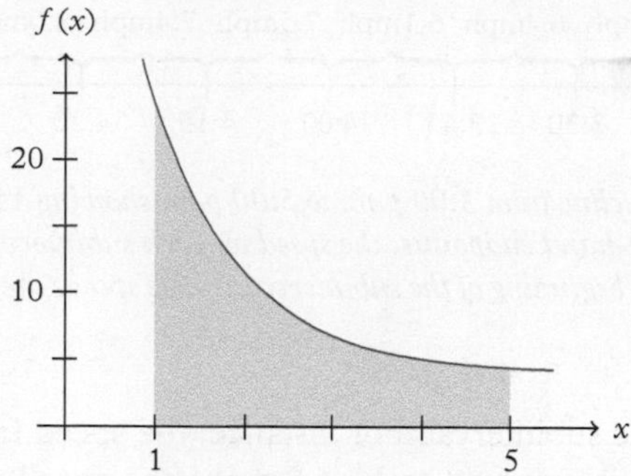

Figure 21 *The graph for reading exercise 8*

Estimating distance traveled

The formula distance = rate · time, or $d = rt$, is very similar to the formula $A = wh$. The quantity on the left is found by multiplying the two quantities on the right. Just as area can be estimated using *finite sums,* distances can be estimated using finite sums as well.

The area estimates derived earlier were found by adding (summing) the areas of rectangles. We used only a finite number of rectangles (we did not use an infinite number of rectangles), hence the term *finite sums.*

Example 4 *A herd of buffalo is migrating. The speed at which the lead animal is traveling is recorded every 15 minutes and is given in the following table. (a) Give an estimate of the distance traveled from 3:00 p.m. to 5:00 p.m. using left-hand endpoints. (b) Give an upper estimate of the distance traveled from 3:00 p.m. to 5:00 p.m.*

time(p.m.)	speed(mph)
3:00	6.8
3:15	7.2
3:30	6.4
3:45	6.1
4:00	7.5
4:15	7.4
4:30	8.0
4:45	6.6
5:00	7.9

Ans. to reading exercise 8

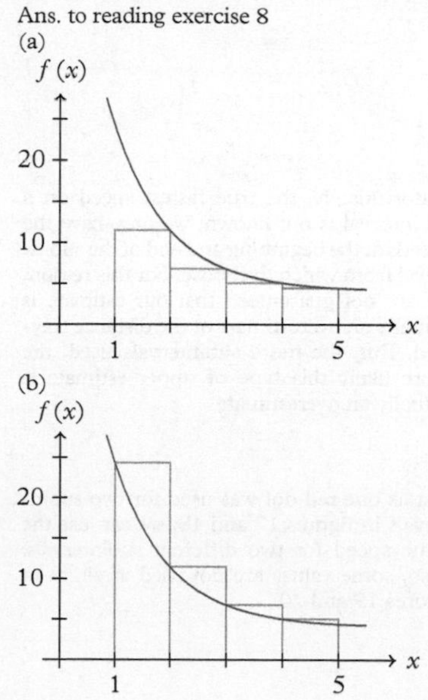

Solution (a) What does it mean to use left-hand endpoints for a problem such as this? A diagram helps. Figure 22 shows a timeline from 3:00 p.m. to 5:00 p.m., with 15-minute subintervals marked. Using left-hand endpoints means we assume the speed on a subinterval is the speed at the left-hand endpoint of the subinterval—that is, at the

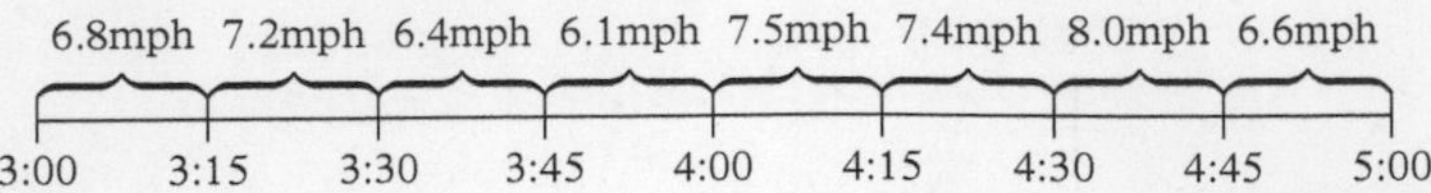

Figure 22 *A timeline from 3:00 p.m. to 5:00 p.m. showing 15-minute subintervals. Using left-hand endpoints, the speed on each subinterval is assumed to be the speed at the beginning of the subinterval. These speeds are indicated above each subinterval*

If right-hand endpoints are used, the speed is assumed to be the speed at the right-hand endpoint of the subinterval—that is, at the end of the subinterval. Then, from 3:00 p.m. to 3:15 p.m., we assume the speed of travel is 7.2 mph (the speed at 3:15 p.m.).

beginning of the subinterval. For instance, the speed from 3:00 p.m. until 3:15 p.m. is assumed to be 6.8 mph (the speed at 3:00 p.m.), whereas the speed from 3:15 p.m. until 3:30 p.m. is assumed to be 7.2 mph (the speed at 3:15 p.m.).

Because $d = rt$, we find the distance traveled during a subinterval by multiplying the rate (speed) by the time (aka the width of the subinterval). The distance traveled from 3:00 p.m. to 3:15 p.m. is

Because units of speed are miles per hour, we use hour as the unit of time: 15 minutes $= \frac{1}{4}$ hour.

$$d = rt = 6.8 \text{ mi/h} \cdot \frac{1}{4} \text{ h} = 1.7 \text{ mi}.$$

The total distance traveled is then estimated by adding the distance traveled over each subinterval:

Notice how similar this calculation is to an area estimate! In fact, if we think of the speeds as the heights of rectangles, and the widths of subintervals as the widths of rectangles, the calculation is identical.

$$\begin{aligned} d &= 6.8 \cdot \frac{1}{4} + 7.2 \cdot \frac{1}{4} + 6.4 \cdot \frac{1}{4} + 6.1 \cdot \frac{1}{4} \\ &\quad + 7.5 \cdot \frac{1}{4} + 7.4 \cdot \frac{1}{4} + 8.0 \cdot \frac{1}{4} + 6.6 \cdot \frac{1}{4} \\ &= 14 \text{ mi}. \end{aligned}$$

The estimate of distance traveled using left-hand endpoints is 14 miles.

(b) Now we repeat the process, but replace left-hand endpoints with an upper estimate. For an upper estimate, on each subinterval we wish to use the fastest speed. We therefore choose the larger of the speeds at the beginning and end of the subinterval and assume this is the speed over the entire subinterval. For instance, for the subinterval 3:00 p.m. until 3:15 p.m., the speeds from which to choose are 6.8 mph (the speed at 3:00 p.m.) and 7.2 mph (the speed at 3:15 p.m.). The larger of the two speeds is 7.2 mph, so we use this speed for the entire subinterval. For the subinterval 3:15 p.m. until 3:30 p.m., the speeds from which to choose are 7.2 mph (the speed at 3:15 p.m.) and 6.4 mph (the speed at 3:30 p.m.). The largest of the two is 7.2 mph, so we use this speed for the entire subinterval. The resulting timeline, with speeds marked for an upper estimate, is in figure 23. The upper estimate of distance traveled is then calculated by multiplying speeds (rates) by lengths of subintervals (times) and adding the results:

Unfortunately, the true fastest speed on a subinterval is not known; we only have the speeds at the beginning and end of the subinterval from which to choose. For this reason, we are not guaranteed that our estimate is actually an overestimate of the distance traveled. But, the more subintervals used, the more likely this type of upper estimate is actually an overestimate.

Just as one red dot was used for two subintervals in figures 17 and 18, we can use the same speed for two different subintervals. Also, some values are not used at all, as in figures 19 and 20.

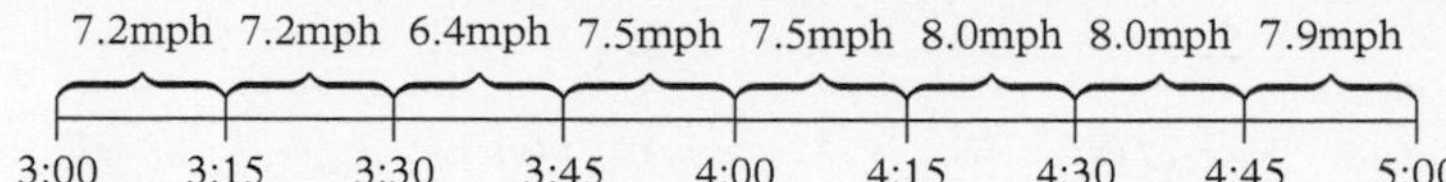

Figure 23 *A timeline from 3:00 p.m. to 5:00 p.m. showing 15-minute subintervals. For an upper estimate, the speed on each subinterval is assumed to be the larger of the speed at the beginning of the subinterval and the speed at the end of the subinterval. These speeds are indicated above each subinterval*

$$\begin{aligned} d &= 7.2 \cdot 0.25 + 7.2 \cdot 0.25 + 6.4 \cdot 0.25 + 7.5 \cdot 0.25 \\ &\quad + 7.5 \cdot 0.25 + 8.0 \cdot 0.25 + 8.0 \cdot 0.25 + 7.9 \cdot 0.25 \\ &= 14.925 \text{ mi.} \end{aligned}$$

Conclusion: it is likely that the distance traveled from 3:00 p.m. until 5:00 p.m. is less than 14.925 mi. ■

Exercises 4.2

1–10. Rapid response: state whether the given rectangles are for left-hand endpoints, right-hand endpoints, midpoints, upper estimate, lower estimate, or none of the above. Name all that apply.

1.

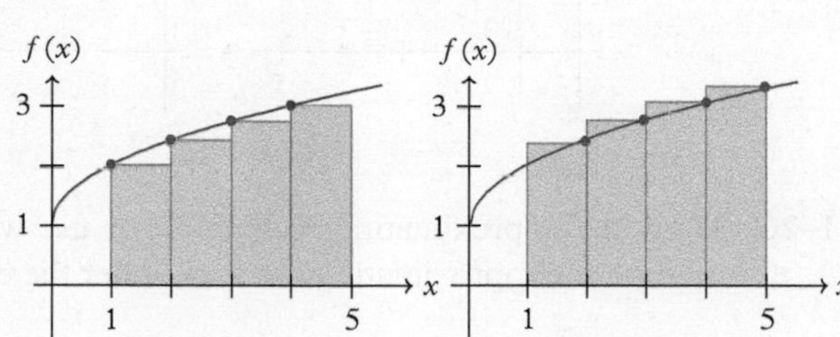

2.

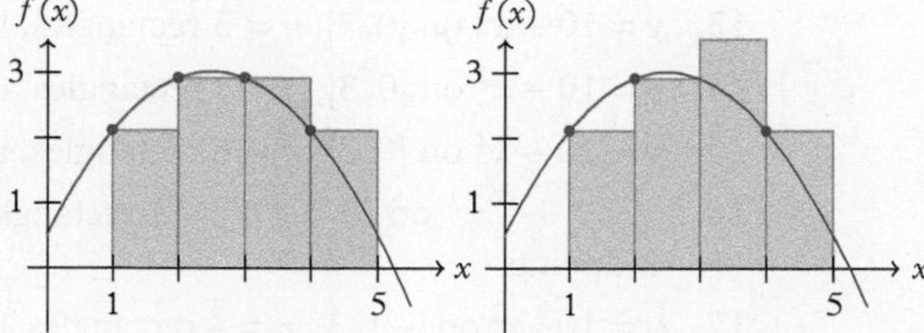

5. **8.**

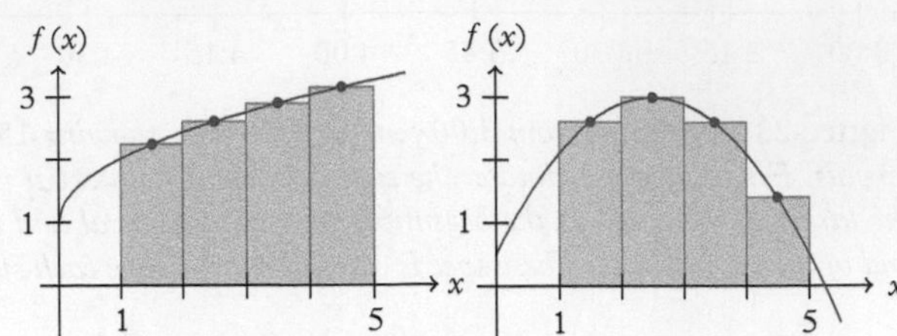

6. **9.**

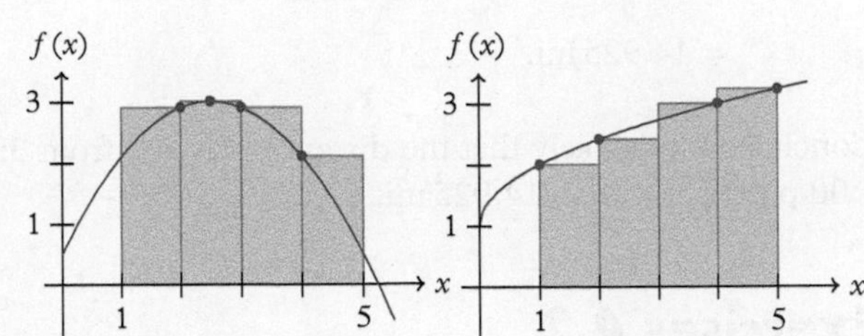

7. **10.**

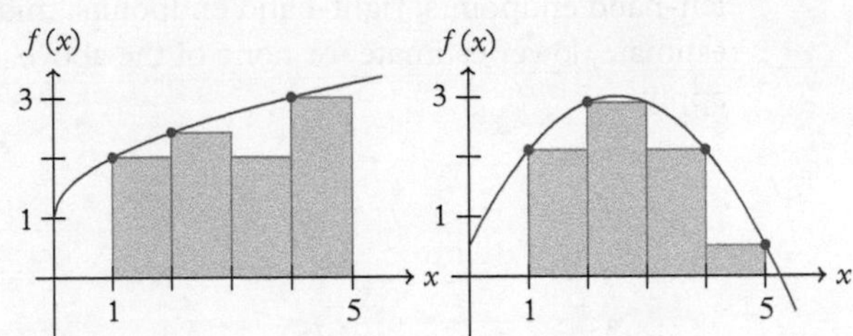

Only the drawing is required for exercises 11–20. You are not required to calculate the approximation.

11–20. Draw the approximating rectangles for use with the stated method for approximating the area under the curve.

11. $y = 10 - x^2$ on $[0, 3]$, $n = 3$ rectangles, left-hand endpoints

12. $y = 10 - x^2$ on $[0, 3]$, $n = 3$ rectangles, midpoints

13. $y = 10 - x^2$ on $[0, 3]$, $n = 3$ rectangles, lower estimate

14. $y = 10 - x^2$ on $[0, 3]$, $n = 3$ rectangles, upper estimate

15. $y = 10 - x^2$ on $[0, 3]$, $n = 6$ rectangles, midpoints

16. $y = 10 - x^2$ on $[0, 3]$, $n = 6$ rectangles, right-hand endpoints

17. $y = 10 - x^2$ on $[-1, 3]$, $n = 4$ rectangles, upper estimate

18. $y = 10 - x^2$ on $[-1, 3]$, $n = 4$ rectangles, lower estimate

19. $y = 4 - x^2$ on $[-2, 2]$, $n = 4$ rectangles, right-hand endpoints

20. $y = \sqrt{x}$ on $[0, 4]$, $n = 4$ rectangles, left-hand endpoints

21–40. Approximate the area under the curve using the given method and number of rectangles.

21. $y = 2 + \sqrt{x}$ from $x = 0$ to $x = 5$, $n = 5$ rectangles, left-hand endpoints

22. $y = 2 + \sqrt{x}$ from $x = 0$ to $x = 5$, $n = 5$ rectangles, right-hand endpoints

23. $y = 2 + \sqrt{x}$ from $x = 0$ to $x = 5$, $n = 5$ rectangles, midpoints

24. $y = 2 + \sqrt{x}$ from $x = 0$ to $x = 5$, $n = 10$ rectangles, left-hand endpoints

25. $y = 2 + \sqrt{x}$ from $x = 0$ to $x = 5$, $n = 10$ rectangles, right-hand endpoints

26. $y = 2 + \sqrt{x}$ from $x = 0$ to $x = 5$, $n = 10$ rectangles, midpoints

27. $y = \cos x$ from $x = -\frac{\pi}{2}$ to $x = \frac{\pi}{2}$, $n = 4$ rectangles, left-hand endpoints

28. $y = \cos x$ from $x = -\frac{\pi}{2}$ to $x = \frac{\pi}{2}$, $n = 4$ rectangles, right-hand endpoints

29. $y = \cos x$ from $x = -\frac{\pi}{2}$ to $x = \frac{\pi}{2}$, $n = 4$ rectangles, upper estimate

30. $y = \cos x$ from $x = -\frac{\pi}{2}$ to $x = \frac{\pi}{2}$, $n = 4$ rectangles, lower estimate

31. $y = \cos x$ from $x = -\frac{\pi}{2}$ to $x = \frac{\pi}{2}$, $n = 6$ rectangles, right-hand endpoints

32. $y = \cos x$ from $x = -\frac{\pi}{2}$ to $x = \frac{\pi}{2}$, $n = 6$ rectangles, left-hand endpoints

33. $y = \cos x$ from $x = -\frac{\pi}{2}$ to $x = \frac{\pi}{2}$, $n = 6$ rectangles, lower estimate

34. $y = \cos x$ from $x = -\frac{\pi}{2}$ to $x = \frac{\pi}{2}$, $n = 6$ rectangles, upper estimate

35. $y = 10 + x^3$ on $[-2, 6]$, $n = 4$ rectangles, midpoints

36. $y = 10 + x^3$ on $[-2, 6]$, $n = 8$ rectangles, lower estimate

37. $y = 10 + x^3$ on $[-2, 6]$, $n = 8$ rectangles, upper estimate

38. $y = \sqrt{1+x^4}$ on $[-2, 2]$, $n = 4$ rectangles, midpoints

39. $y = \sqrt{1+x^4}$ on $[-2, 2]$, $n = 6$ rectangles, lower estimate

40. $y = \sqrt{1+x^4}$ on $[-2, 2]$, $n = 6$ rectangles, upper estimate

41–46. Find bounds for the area under the curve by finding lower and upper estimates using the given number of rectangles.

41. $y = \sqrt{5x}$ on $[3, 7]$, $n = 4$ rectangles

42. $y = \frac{1}{x}$ on $[2, 7]$, $n = 5$ rectangles

43. $y = \frac{1}{1+x^2}$ on $[-2, 2]$, $n = 4$ rectangles

44. $y = \sqrt{9-x^2}$ on $[-3, 3]$, $n = 3$ rectangles

45. $y = \frac{1}{1+x^2}$ on $[-2, 2]$, $n = 8$ rectangles

46. $y = \sqrt{9-x^2}$ on $[-3, 3]$, $n = 6$ rectangles

47–50. The buffalo herd of example 4 is on the move again the next morning. Give an estimate of the distance traveled from 9:00 a.m. until 11:00 a.m. using the speeds in the following table and the indicated method:

time (a.m.)	speed (mph)
9:00	5.7
9:15	6.6
9:30	6.8
9:45	6.0
10:00	6.5
10:15	7.1
10:30	7.3
10:45	6.2
11:00	6.4

47. left-hand endpoints **49.** lower estimate

48. right-hand endpoints **50.** upper estimate

51–54. You are on a sailboat in the Atlantic Ocean, all your GPS devices and phones were swept overboard in a storm, your instrument panel does not work, and it is so cloudy you cannot see the sun, moon, or stars. You resort to dead reckoning to determine your position. You have kept a due-west course (according to your compass), and you determine your speed every half hour using knots in a rope let out into the water.

According to a former student, dead reckoning is still a required subject to obtain a captain's license.

Use the following data to estimate the distance you have traveled from 8:00 a.m. until 12:00 noon, using the specified method:

time (a.m.)	speed (knots)
8:00	6
8:30	8
9:00	7
9:30	9
10:00	11
10:30	12
11:00	11
11:30	9
12:00 noon	8

51. right-hand endpoints **53.** upper estimate

52. left-hand endpoints **54.** lower estimate

The unit of speed for exercises 41–44 is *knot* (abbreviation kn), which is the same as nautical mile per hour. The unit of distance is nautical mile (abbreviation nmi).

55. Estimate the area above the curve $y = x^2 - 4$ between $x = 0$ and $x = 2$ using $n = 4$ rectangles and left-hand endpoints.

56. The curve $y = x^3$ is symmetric about the origin so that the area between the curve and the x-axis (the area under the curve) between $x = 0$ and $x = 1$ appears to be matched by the area between the curve and the x-axis (the area above the curve) between $x = -1$ and $x = 0$. (a) Use $n = 4$ rectangles and midpoints to calculate a finite sum for the function on the interval $[-1, 1]$, keeping the positive and negative values of the curve as the "heights" of the rectangles. (b) Repeat for the interval $[-1, 3]$ and (c) for the interval $[-3, 1]$. (d) What do these finite sums seem to represent?

57. In figure 2, the approximating rectangles for left-hand endpoints also serve as the rectangles for a lower estimate. Draw a curve on an interval and choose a number of rectangles so that the approximating rectangles for midpoints also serve as the rectangles for a lower estimate.

Areas and Sums

4.3

Before we proceed with the next step in finding areas, we need to study some notation.

Summation notation

Can we write $1 + 2 + 3 + \cdots + 100$ without the ellipsis (the $\cdots$)? Yes, using *summation notation*:

Σ is the capital Greek letter *sigma*.

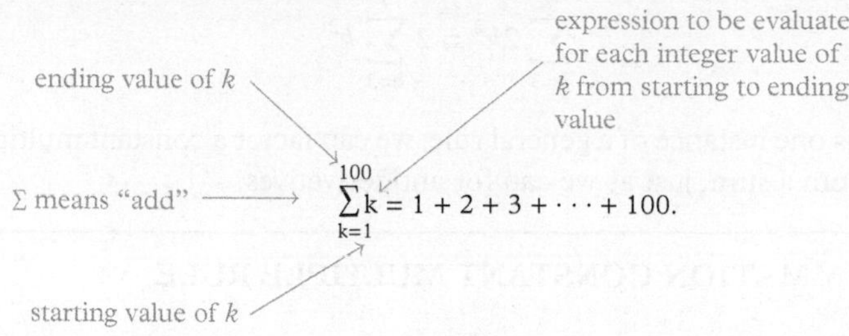

With the notation $1+2+3+\cdots+100$, we are required to infer the pattern from the numbers given. With the notation $\sum_{k=1}^{100} k$, there is nothing to infer; the notation indicates precisely which terms are to be added

Example 1 *Rewrite without summation notation:* $\sum_{k=5}^{8} \sqrt{k}$.

Solution The terms of the sum are the values of the expression $\sqrt{k}$. The values of k start with $k = 5$, end with $k = 8$, and use each integer in between; that is, we use $k = 5, 6, 7$, and 8:

We are only asked to rewrite the sum, not to calculate its value.

$$\sum_{k=5}^{8} \sqrt{k} = \sqrt{5} + \sqrt{6} + \sqrt{7} + \sqrt{8}. \qquad \blacksquare$$

Example 2 *Evaluate* $\sum_{k=3}^{7} k^2$.

Solution The terms to add are the values of k^2, beginning with $k = 3$ and ending with $k = 7$:

To evaluate the sum means to add all the terms and state the resulting value.

$$\sum_{k=3}^{7} k^2 = 3^2 + 4^2 + 5^2 + 6^2 + 7^2 = 135. \qquad \blacksquare$$

Example 3 *Evaluate* $\sum_{k=3}^{7} 2k^2$.

Solution The terms to add are the values of $2k^2$, beginning with $k = 3$ and ending with $k = 7$:

$$\sum_{k=3}^{7} 2k^2 = 2 \cdot 3^2 + 2 \cdot 4^2 + 2 \cdot 5^2 + 2 \cdot 6^2 + 2 \cdot 7^2.$$

Rather than just finish the calculation from here, notice that we can factor out the 2 from each term:

We calculated $3^2 + 4^2 + 5^2 + 6^2 + 7^2 = 135$ in example 2.

$$= 2(3^2 + 4^2 + 5^2 + 6^2 + 7^2) = 2 \cdot 135 = 270. \quad \blacksquare$$

In other words,

$$\sum_{k=3}^{7} 2k^2 = 2\sum_{k=3}^{7} k^2.$$

This is one instance of a general rule: we can factor a constant multiple out from a sum, just as we can for antiderivatives.

In the formula, the a_k is any function of k.

SUMMATION CONSTANT MULTIPLE RULE

For any real number c,

$$\sum_{k=m}^{n} c \cdot a_k = c\sum_{k=m}^{n} a_k.$$

PROOF OF THE SUMMATION CONSTANT MULTIPLE RULE

The proof of the summation constant multiple rule is like the solution of example 3: $\sum_{k=m}^{n} c \cdot a_k = c \cdot a_m + c \cdot a_{m+1} + \cdots + c \cdot a_n = c(a_m + a_{m+1} + \cdots + a_n) = c\sum_{k=m}^{n} a_k$.

The summation constant multiple rule is really just the distributive property from algebra rewritten for use with summation notation. There are sum and difference rules for summations as well. These rules are essentially versions of the associative and commutative properties from algebra.

SUMMATION SUM RULE

$$\sum_{k=m}^{n} (a_k + b_k) = \sum_{k=m}^{n} a_k + \sum_{k=m}^{n} b_k.$$

PROOF OF THE SUMMATION SUM RULE

The proof of the summation sum rule: $\sum_{k=m}^{n} (a_k + b_k) = (a_m + b_m) + (a_{m+1} + b_{m+1}) + \cdots + (a_n + b_n) = (a_m + a_{m+1} + \cdots + a_n) + (b_m + b_{m+1} + \cdots + b_n) = \sum_{k=m}^{n} a_k + \sum_{k=m}^{n} b_k$.

SUMMATION DIFFERENCE RULE

$$\sum_{k=m}^{n}(a_k - b_k) = \sum_{k=m}^{n} a_k - \sum_{k=m}^{n} b_k.$$

PROOF OF THE SUMMATION DIFFERENCE RULE

The proof of the summation difference rule: $\sum_{k=m}^{n}(a_k - b_k) = (a_m - b_m) + (a_{m+1} - b_{m+1}) + \cdots + (a_n - b_n) = (a_m + a_{m+1} + \cdots + a_n) - (b_m + b_{m+1} + \cdots + b_n) = \sum_{k=m}^{n} a_k - \sum_{k=m}^{n} b_k.$

Example 4 *Evaluate* $\sum_{k=1}^{4}\left(3k^2 - k\right)$.

Solution The terms to add are the values of $3k^2 - k$, beginning with $k = 1$ and ending with $k = 4$:

$$\begin{aligned}\sum_{k=1}^{4}\left(3k^2 - k\right) &= (3 \cdot 1^2 - 1) + (3 \cdot 2^2 - 2) + (3 \cdot 3^2 - 3) + (3 \cdot 4^2 - 4)\\ &= 2 + 10 + 24 + 44\\ &= 80.\end{aligned}$$

Alternately, we could have used the summation constant multiple rule and the summation difference rule:

Use of the summation rules is more important in a different form later in this section.

$$\begin{aligned}\sum_{k=1}^{4}\left(3k^2 - k\right) &= 3\sum_{k=1}^{4} k^2 - \sum_{k=1}^{4} k\\ &= 3(1^2 + 2^2 + 3^2 + 4^2) - (1 + 2 + 3 + 4)\\ &= 3(30) - 10 = 80.\end{aligned}$$

■

Reading Exercise 9 *Evaluate* $\sum_{k=2}^{5} 4k$.

Writing summation notation

What if we have a sum written with ellipses and we wish to write it using summation notation? The key is to look for something in each term that is changing.

Example 5 *Write* $\omega + 2\omega + 3\omega + 4\omega + \cdots + 48\omega$ *using summation notation.*

Solution The terms being added are ω, 2ω, 3ω, and so on. Do you see something that is changing from term to term? The changing value is emphasized in bold:

Remember that $1\omega = \omega$, so the pattern still works for that term.

$$1\omega + 2\omega + 3\omega + 4\omega + \cdots + 48\omega.$$

The changing values are consecutive integers beginning with 1 and ending with 48, so we let these be the values of k. The terms are then of the form $k\omega$, which goes after Σ:

$$\omega + 2\omega + 3\omega + 4\omega + \cdots + 48\omega = \sum_{k=1}^{48} k\omega.$$

■

Example 6 *Write* $(5 \cdot 3\omega)^2 + (6 \cdot 3\omega)^2 + (7 \cdot 3\omega)^2 + (8 \cdot 3\omega)^2 + \cdots + (50 \cdot 3\omega)^2$ *using summation notation.*

Solution What is changing from term to term? The 3ω is present in every term and is not changing. The exponent 2 is the same in every term. The changing value is emphasized in bold:

$$(\mathbf{5} \cdot 3\omega)^2 + (\mathbf{6} \cdot 3\omega)^2 + (\mathbf{7} \cdot 3\omega)^2 + (\mathbf{8} \cdot 3\omega)^2 + \cdots + (\mathbf{50} \cdot 3\omega)^2 .$$

The changing values are consecutive integers from 5 to 50, so we represent these values by k. The terms being added are therefore $(k \cdot 3\omega)^2$, and the sum can be written as

$$\sum_{k=5}^{50} (k \cdot 3\omega)^2.$$

■

Ans. to reading exercise 9
56

Example 7 *Write* $(1 + \omega)^3 + (1 + 2\omega)^3 + (1 + 3\omega)^3 + (1 + 4\omega^3) + \cdots + (1 + 70\omega)^3$ *using summation notation.*

Solution This time the 1+, the ω, and the exponent 3 are not changing, but the coefficient on ω is changing, as emphasized in bold:

Once again, the numbers in bold are consecutive integers. There are techniques for handling nonconsecutive integers, such as when the changing numbers (the numbers in bold) increase by two each time, and it is not difficult to do so. For our purposes in this section, however, we only need to work with consecutive integers.

$$(1 + \mathbf{1}\omega)^3 + (1 + \mathbf{2}\omega)^3 + (1 + \mathbf{3}\omega)^3 + (1 + \mathbf{4}\omega^3) + \cdots + (1 + \mathbf{70}\omega)^3.$$

The terms are $(1+k\omega)^3$ from $k = 1$ to $k = 70$, and the sum is written as

$$\sum_{k=1}^{70} (1 + k\omega)^3.$$

■

Reading Exercise 10 *Write* $(4 \cdot 2\omega)^3 + (5 \cdot 2\omega)^3 + (6 \cdot 2\omega)^3 + \cdots + (12 \cdot 2\omega)^3$ *using summation notation.*

Helpful summation formulas

At the beginning of this section we looked at the sum of the first 100 positive integers $\left(\sum_{k=1}^{100} k = 1 + 2 + \cdots + 100\right)$ but we did not calculate that sum. Here is a quick method:

$$\begin{array}{rcccccccc} S &=& 1 &+& 2 &+& \cdots &+& 100 \\ S &=& 100 &+& 99 &+& \cdots &+& 1 \\ \hline 2S &=& 101 &+& 101 &+& \cdots &+& 101 \end{array}$$

$$\underbrace{\qquad\qquad\qquad\qquad}_{100 \text{ times}}$$

Therefore, $2S = 100(101)$ and $S = \dfrac{100(101)}{2}$. This method can be used to prove the following formula. For any positive integer n,

$$\sum_{k=1}^{n} k = \frac{n(n+1)}{2}.$$

Then, the sum of the first 100 integers uses the formula with $n = 100$,

$$\sum_{k=1}^{100} k = \frac{100(100+1)}{2} = 5050$$

and the sum of the first 3000 integers uses the formula with $n = 3000$,

In other words, $1 + 2 + 3 + \cdots + 3000 = 4\,501\,500$.

$$\sum_{k=1}^{3000} k = \frac{3000(3000+1)}{2} = 4\,501\,500.$$

Similar formulas exist for the sums of squares, cubes, and so on.

SUMMATION FORMULAS

For any positive integer n,

$$\sum_{k=1}^{n} 1 = n,$$

$$\sum_{k=1}^{n} k = \frac{n(n+1)}{2},$$

The first formula is easily seen to be true:

$$\sum_{k=1}^{n} 1 = \underbrace{1 + 1 + \cdots + 1}_{n \text{ times}} = n.$$

The other formulas can be proved using mathematical induction, which is often taught in an introduction to proofs course.

$$\sum_{k=1}^{n} k^2 = \frac{n(n+1)(2n+1)}{6}, \text{ and}$$

$$\sum_{k=1}^{n} k^3 = \left(\frac{n(n+1)}{2}\right)^2.$$

In the past, we used Ω to represent a positive infinite hyperreal number, but because the previous summation formulas require an integer n, we must add the restriction that Ω is also an integer. Such numbers are sometimes called *hypernatural numbers*.

To extend the fun, what happens if we use a positive infinite hyperreal integer Ω for n? The transfer principle guarantees the formulas are still valid. We have

Recall that the higher exponents on Ω are on higher levels, and we can discard the lower-level terms when approximating.

$$\sum_{k=1}^{\Omega} 1 = \Omega,$$

$$\sum_{k=1}^{\Omega} k = \frac{\Omega(\Omega+1)}{2} = \frac{\Omega^2+\Omega}{2} \approx \frac{\Omega^2}{2},$$

$$\sum_{k=1}^{\Omega} k^2 = \frac{\Omega(\Omega+1)(2\Omega+1)}{6} = \frac{2\Omega^3+3\Omega^2+\Omega}{6} \approx \frac{2\Omega^3}{6} = \frac{\Omega^3}{3}, \text{ and}$$

$$\sum_{k=1}^{\Omega} k^3 = \left(\frac{\Omega(\Omega+1)}{2}\right)^2 = \cdots \approx \frac{\Omega^4}{4}.$$

In each line of these formulas, compare the exponent on k in the summation on the left to the approximation on the right and you should notice a pattern. The pattern continues for higher exponents. For instance, the summation formula for seventh powers is

Ans. to reading exercise 10

$\sum_{k=4}^{12} (k \cdot 2\omega)^3$

$$\sum_{k=1}^{n} k^7 = \frac{n^2(n+1)^2(3n^4+6n^3-n^2-4n+2)}{24},$$

and because the transfer principle says the formula is valid for positive infinite hyperreal integers as well,

$$\sum_{k=1}^{\Omega} k^7 = \cdots \approx \frac{3\Omega^8}{24} = \frac{\Omega^8}{8}.$$

Notice how much these formulas resemble the antiderivative power rule! We prove the formula in general for larger exponents in section 10.4, but let's state and use the formula now.

SUM OF POWERS APPROXIMATION FORMULA

For any real number $m > -1$ and any positive infinite hyperreal integer Ω,

$$\sum_{k=1}^{\Omega} k^m \approx \frac{\Omega^{m+1}}{m+1}.$$

The sum of powers approximation formula is introduced and proved in the article Dawson, C. Bryan, "A New Extension of the Riemann Integral," *American Mathematical Monthly* 125 No. 2 (February 2018), 130–140, doi: https://doi.org/10.1080/00029890.2018.1401832.

Reading Exercise 11 *Approximate* $\sum_{k=1}^{\Omega} k^4$.

Omega sums

The summation constant multiple rule, summation sum rule, and summation difference rule are also valid for sums to Ω (by the transfer principle). Using these rules along with the sum of powers approximation formula helps us evaluate a variety of sums. Because these sums have ending value Ω, they are called Ω *sums* (*omega sums*).

The term *omega sum* was coined by my colleague Troy Riggs. Omega sums are a type of *hyperfinite sum*, so named because they exhibit many of the properties of finite sums, such as obeying the same sum rules.

OMEGA SUM CONSTANT MULTIPLE RULE

For any positive infinite hyperreal integer Ω and any hyperreal c,

$$\sum_{k=1}^{\Omega} c \cdot a_k = c \sum_{k=1}^{\Omega} a_k.$$

Example 8 *Evaluate* $\sum_{k=1}^{\Omega} 4\omega^2 k$.

Solution Using the omega sum constant multiple rule (first line) and then the sum of powers approximation formula (second line),

$$\sum_{k=1}^{\Omega} 4\omega^2 k = 4\omega^2 \sum_{k=1}^{\Omega} k$$

$$\approx 4\cancel{\omega^2} \cdot \frac{\cancel{\Omega^2}}{2} = 2.$$

■

In an omega sum, it is implied that Ω is a positive infinite hyperreal integer.

The number $4\omega^2$ is a constant, so it can factor out of the sum. The number k is not constant; it is a number that is changing.

Recall that ω and Ω are reciprocals—that is, $\Omega = \frac{1}{\omega}$. Therefore, ω^2 and Ω^2 cancel, since $\omega^2 \cdot \Omega^2 = \omega^2 \cdot \frac{1}{\omega^2} = 1$.

Example 9 *Evaluate* $\sum_{k=1}^{\Omega}(3\omega k)^2 \cdot 7\omega$.

The *summand* is the expression that comes after Σ.

Solution Fist let's simplify the summand:

$$\sum_{k=1}^{\Omega}(3\omega k)^2 \cdot 7\omega = \sum_{k=1}^{\Omega} 9\omega^2 k^2 \cdot 7\omega = \sum_{k=1}^{\Omega} 63\omega^3 k^2.$$

The factors that do not involve k are constant and can be factored out. The factor k^2 is not constant, but rather changes with k.

Now we can use the omega sum constant multiple rule followed by the sum of powers approximation formula:

$$\sum_{k=1}^{\Omega} 63\omega^3 k^2 = 63\omega^3 \sum_{k=1}^{\Omega} k^2$$

As in example 8, whenever the exponents on ω and Ω are the same, they cancel.

$$\approx 63\cancel{\omega^3} \cdot \frac{\cancel{\Omega^3}}{3} = 21. \quad \blacksquare$$

OMEGA SUM SUM RULE

For any positive infinite hyperreal integer Ω,

$$\sum_{k=1}^{\Omega}(a_k + b_k) = \sum_{k=1}^{\Omega} a_k + \sum_{k=1}^{\Omega} b_k.$$

Explanation of name: This is a sum rule for omega sums, so it is the "(omega sum) (sum rule)." Similarly, the difference rule for omega sums is the "(omega sum) (difference rule)."

Ans. to reading exercise 11

$\frac{\Omega^5}{5}$

OMEGA SUM DIFFERENCE RULE

For any positive infinite hyperreal integer Ω,

$$\sum_{k=1}^{\Omega}(a_k - b_k) = \sum_{k=1}^{\Omega} a_k - \sum_{k=1}^{\Omega} b_k.$$

The next example of an omega sum makes use of the sum rule.

Example 10 *Evaluate* $\sum_{k=1}^{\Omega}(2 + 5\omega k)^2 \cdot \omega$.

Solution As in the previous example, we begin by manipulating the summand algebraically:

$$\sum_{k=1}^{\Omega}(2+5\omega k)^2\cdot\omega=\sum_{k=1}^{\Omega}(4+20\omega k+25\omega^2k^2)\omega$$
$$=\sum_{k=1}^{\Omega}(4\omega+20\omega^2k+25\omega^3k^2).$$

To use the sum of powers approximation formula, we need expressions of the form $\sum_{k=1}^{\Omega}k^m$. Completing the operations of squaring and multiplying by ω are necessary steps toward accomplishing this goal.

Now we can use the omega sum sum rule to break this into three separate omega sums:

$$=\sum_{k=1}^{\Omega}4\omega+\sum_{k=1}^{\Omega}20\omega^2k+\sum_{k=1}^{\Omega}25\omega^3k^2.$$

We can then use the omega sum constant multiple rule on each of the omega sums, followed by the sum of powers approximation formula on each of the omega sums:

With antiderivatives, we often skipped writing some steps, such as the sum rule and constant multiple rule. With practice, it is easy to skip some steps here as well and go straight from

$$\sum_{k=1}^{\Omega}(4\omega+20\omega^2k+25\omega^3k^2)$$

to

$$\approx 4\omega\cdot\Omega+20\omega^2\cdot\frac{\Omega^2}{2}+25\omega^3\cdot\frac{\Omega^3}{3}.$$

$$=4\omega\sum_{k=1}^{\Omega}1+20\omega^2\sum_{k=1}^{\Omega}k+25\omega^3\sum_{k=1}^{\Omega}k^2$$
$$\approx 4\omega\cdot\Omega+20\omega^2\cdot\frac{\Omega^2}{2}+25\omega^3\cdot\frac{\Omega^3}{3}$$
$$=4+10+\frac{25}{3}=\frac{67}{3}.$$ ■

The omega sums in examples 8–10 are typical of the sums we encounter when solving area problems.

Reading Exercise 12 *Evaluate* $\sum_{k=1}^{\Omega}(2\omega k)^2\cdot 3\omega$.

Areas revisited

Let's return to the idea of trying to find the area under the curve $y=x^2$ between $x=1$ and $x=5$. Estimating the area using $n=8$ rectangles and right-hand endpoints, the width of each rectangle is $\Delta x=\frac{b-a}{n}=\frac{5-1}{8}=0.5$.

Instead of following the usual procedure, consider the locations of the right-hand endpoints of the subintervals in relation to the widths of the rectangles, as marked in figure 1. The right-hand endpoints are $1,2,3,\ldots,8$ rectangle widths to the right of $x=1$. Because the width is 0.5, the endpoints can be calculated as $1+1\cdot0.5$, $1+2\cdot0.5$, $1+3\cdot0.5$,

..., 1+8·0.5. The heights of the rectangles must match the height of the curve at the right-hand endpoints. The curve is $y = x^2$, so the heights of the curve are $(1 + 1 \cdot 0.5)^2$, $(1 + 2 \cdot 0.5)^2$, $(1 + 3 \cdot 0.5)^2$, ..., $(1 + 8 \cdot 0.5)^2$. The total area of the rectangles is found by multiplying heights by widths and adding the results. However, rather than calculating a numerical answer, let's write the result in summation notation. Noting the pattern in the heights of the rectangles, the total area is

$$\sum_{k=1}^{8} \underbrace{(1 + k \cdot 0.5)^2}_{\text{heights}} \cdot \underbrace{0.5}_{\text{widths}}.$$

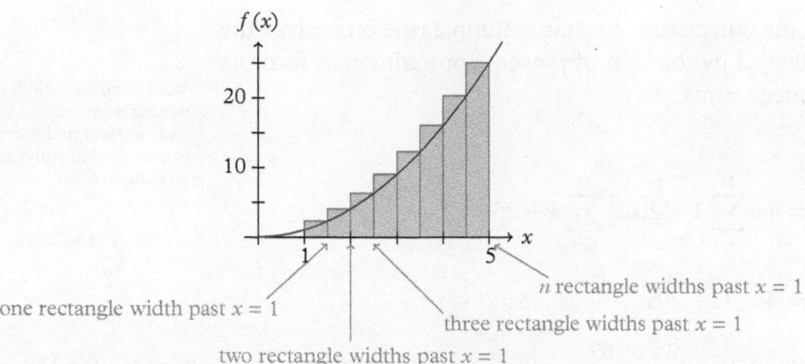

Figure 1 *Eight approximating rectangles using right-hand endpoints. The right-hand endpoints are* 1, 2, 3, ..., 8 *rectangle widths past* $x = 1$

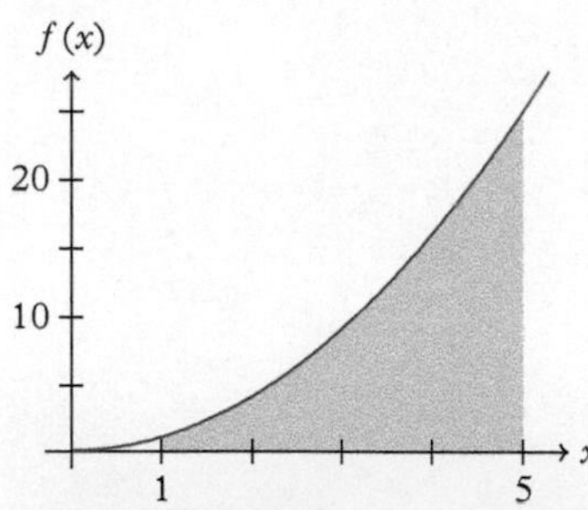

Figure 2 *Infinitely many* (Ω) *approximating rectangles using right-hand endpoints. The individual rectangles can only be seen when zoomed in to an infinitesimal scale. The total area of the rectangles appears to be equal to the area under the curve. Silly remark: drawing this takes a very, very long time*

Calculating the value of that sum gives us an overestimate of the area under the curve. Figure 1 makes it clear that the total area of the rectangles and the area under the curve are different. As discussed in the previous section, using more rectangles makes the estimate better.

Now for the key idea that you may have already figured out: if using more rectangles is better, why not use infinitely many rectangles? This ought to eliminate any "extra" area and give us the exact area under the curve! The shaded rectangles appear as depicted in figure 2.

Using Ω rectangles, where Ω is a positive infinite hyperreal integer, gives a width of

$$\Delta x = \frac{b - a}{n} = \frac{5 - 1}{\Omega} = 4\omega.$$

Because the approximating rectangles have infinitesimal width, there is not enough room for the straight tops of the rectangles to stray more than infinitesimally from the curve itself. Any real difference between the area of the rectangles and the area under the curve is therefore eliminated.

The intuitive arguments given here rely on using a continuous curve.

The right-hand endpoints are $1, 2, 3, \ldots, \Omega$ rectangle widths to the right of $x = 1$, or $1 + 1 \cdot 4\omega, 1 + 2 \cdot 4\omega, 1 + 3 \cdot 4\omega, \ldots, 1 + \Omega \cdot 4\omega$. The curve is $y = x^2$, so the heights of the rectangles are

Compare to the previous calculations involving $n = 8$ rectangles.

$$(1 + 1 \cdot 4\omega)^2, (1 + 2 \cdot 4\omega)^2, (1 + 3 \cdot 4\omega)^2, \ldots, (1 + \Omega \cdot 4\omega)^2.$$

The total area of the rectangles is found by multiplying heights by widths and adding the results. In summation notation, this is

Ans. to reading exercise 12
4

$$\text{area under the curve} = \sum_{k=1}^{\Omega} \underbrace{(1 + k \cdot 4\omega)^2}_{\text{heights}} \cdot \underbrace{4\omega}_{\text{widths}}.$$

This is an omega sum! The area under the curve is therefore

We always do the indicated arithmetic on the "heights" of the rectangles (line 1) and multiply by the widths (line 2) before using the sum rules (line 3) and the sum of powers approximation formula (line 4).

$$\begin{aligned}
\sum_{k=1}^{\Omega}(1 + k \cdot 4\omega)^2 \cdot 4\omega &= \sum_{k=1}^{\Omega}(1 + 8\omega k + 16\omega^2 k^2)4\omega \\
&= \sum_{k=1}^{\Omega}(4\omega + 32\omega^2 k + 64\omega^3 k^2) \\
&= 4\omega \sum_{k=1}^{\Omega} 1 + 32\omega^2 \sum_{k=1}^{\Omega} k + 64\omega^3 \sum_{k=1}^{\Omega} k^2 \\
&\approx 4\cancel{\omega}\cancel{\Omega} + 32\cancel{\omega^2}\frac{\cancel{\Omega^2}}{2} + 64\cancel{\omega^3}\frac{\cancel{\Omega^3}}{3} \\
&= 4 + 16 + \frac{64}{3} = \frac{124}{3} \text{ units}^2.
\end{aligned}$$

This result ($41.\overline{3}$ units^2) is the exact area under the curve. Compare this value to the estimates obtained in the previous section.

Perhaps now you see why we studied writing sums in summation notation, summation rules, the sum of powers approximation formula, and omega sums. They are all needed to help us determine the exact area under the curve!

We didn't calculate the hyperreal number that represents the exact area of the approximating rectangles. Instead, we used approximation (specifically, the sum of powers approximation formula) to determine an appropriate real number that represents the exact area under the curve. If the two numbers differ, they differ by an infinitesimal amount.

Example 11 *Determine the area under the curve* $y = x^3$ *over the interval* $[0, 2]$.

Solution We use $n = \Omega$ rectangles and right-hand endpoints to determine the area.

We could use left-hand endpoints or midpoints or upper estimates or lower estimates instead, and the same answer for the area is obtained. Right-hand endpoints are used because the resulting work is simpler.

❶ Widths: the widths of the rectangles are

$$\Delta x = \frac{b-a}{n} = \frac{2-0}{\Omega} = 2\omega.$$

❷ Endpoints: the right-hand endpoints are always $1, 2, 3, \ldots, \Omega$ rectangle widths from $x = a$. Our endpoints are

$$0 + 1 \cdot 2\omega,\ 0 + 2 \cdot 2\omega,\ 0 + 3 \cdot 2\omega,\ \ldots,\ 0 + \Omega \cdot 2\omega.$$

❸ Heights: the heights of the rectangles must match the height of the curve at the endpoints. Since $y = x^3$, the heights are

$$(1 \cdot 2\omega)^3,\ (2 \cdot 2\omega)^3,\ (3 \cdot 2\omega)^3,\ \ldots,\ (\Omega \cdot 2\omega)^3.$$

❹ Form the omega sum: the area under the curve is found by multiplying heights by widths and adding the results:

$$\text{area under the curve} = \sum_{k=1}^{\Omega} \underbrace{(k \cdot 2\omega)^3}_{\text{heights}} \cdot \underbrace{2\omega}_{\text{widths}}.$$

To evaluate the omega sum, we do the indicated arithmetic on the heights (line 1), multiply by the widths (line 2), use the sum rules (line 3), and use the sum of powers approximation formula (line 4).

❺ Evaluate the omega sum:

$$\begin{aligned}\sum_{k=1}^{\Omega} (k \cdot 2\omega)^3 \cdot 2\omega &= \sum_{k=1}^{\Omega} k^3 \cdot 8\omega^3 \cdot 2\omega \\ &= \sum_{k=1}^{\Omega} 16\omega^4 k^3 \\ &= 16\omega^4 \sum_{k=1}^{\Omega} k^3 \\ &\approx 16\not{\omega^4} \frac{\not{\Omega^4}}{4} \\ &= 4\ \text{units}^2.\end{aligned}$$

The (exact) area under the curve is 4 units^2. ■

Reading Exercise 13 *Find the area under the curve $y = x^4$ from $x = 0$ to $x = 2$.*

Exercises 4.3

1–8. Rewrite without summation notation.

1. $\sum_{k=2}^{5} \frac{1}{k}$
2. $\sum_{k=1}^{3} \sin 2k$
3. $\sum_{k=-2}^{1} k^2$
4. $\sum_{k=5}^{9} \frac{k+1}{k}$
5. $\sum_{k=5}^{12} (k-2)$
6. $\sum_{k=-4}^{-1} k^3$
7. $\sum_{k=0}^{4} \sqrt{5k+1}$
8. $\sum_{k=1}^{6} (2k+1)^2$

9–16. Evaluate the sum.

9. $\sum_{k=2}^{5} (k-1)^2$
10. $\sum_{k=1}^{4} (2k)^2$
11. $\sum_{k=1}^{4} \sin\left(\frac{k\pi}{2}\right)$
12. $\sum_{k=2}^{4} \left(\frac{1}{k} - k\right)$
13. $\sum_{k=3}^{7} (4k + k^2)$
14. $\sum_{k=0}^{4} \cos\left(\frac{k\pi}{4}\right)$
15. $\sum_{k=12}^{13} 5k$
16. $\sum_{k=25}^{25} 6\sqrt{k}$

17–26. Write the expression using summation notation.

17. $5^4 + 6^4 + 7^4 + \cdots + 125^4$
18. $\frac{1}{2} + \frac{1}{3} + \frac{1}{4} + \cdots + \frac{1}{9}$
19. $\sqrt{\omega} + \sqrt{2\omega} + \sqrt{3\omega} + \cdots + \sqrt{14\omega}$
20. $\omega + \omega^2 + \omega^3 + \cdots + \omega^{27}$
21. $(4+5\omega)^3 + (4+6\omega)^3 + (4+7\omega)^3 + \cdots + (4+99\omega)^3$
22. $(6 \cdot 8\omega)^2 + (7 \cdot 8\omega)^2 + (8 \cdot 8\omega)^2 + \cdots + (22 \cdot 8\omega)^2$
23. $\tan(3 + \frac{\pi\omega}{4}) + \tan(3 + \frac{2\pi\omega}{4}) + \tan(3 + \frac{3\pi\omega}{4}) + \cdots + \tan(3 + \frac{9\pi\omega}{4})$
24. $(7+2\omega)^{-4} + (7+3\omega)^{-4} + (7+4\omega)^{-4} + \cdots + (7+1229\omega)^{-4}$

25. $(1 + 2\omega)^2 + (1 + 3\omega)^2 + (1 + 4\omega)^2 + \cdots + (1 + \Omega \cdot \omega)^2$
26. $(1 \cdot 2\omega)^2 \cdot \omega + (2 \cdot 2\omega)^2 \cdot \omega + (3 \cdot 2\omega)^2 \cdot \omega + \cdots + (\Omega \cdot 2\omega)^2 \cdot \omega$

27–34. Evaluate the omega sum.

27. $\sum_{k=1}^{\Omega} (2\omega k)^3 \cdot 5\omega$
28. $\sum_{k=1}^{\Omega} (6\omega k)^2 \cdot 2\omega$
29. $\sum_{k=1}^{\Omega} (2 + \omega k)^2 \cdot \omega$
30. $\sum_{k=1}^{\Omega} (1 + \omega k)^3 \cdot 5\omega$
31. $\sum_{k=1}^{\Omega} \left(7 - \frac{1}{3}\omega k\right) \cdot 2\omega$
32. $\sum_{k=1}^{\Omega} (0.3 - 0.1\omega k)^2 \cdot 20\omega$
33. $\sum_{k=1}^{\Omega} \sqrt{4\omega k} \cdot 3\omega$
34. $\sum_{k=1}^{\Omega} (5\omega k)^{3/2} \cdot \omega$

Ans. to reading exercise 13
$\frac{32}{5}$ units^2

35–44. Determine the area under the curve over the indicated interval.

35. $y = x^2$ from $x = 0$ to $x = 5$
36. $y = 3x^2$ from $x = 0$ to $x = 3$
37. $y = x^2$ on $[2, 5]$
38. $y = x^3$ on $[1, 2]$
39. $y = x^7$ from $x = 0$ to $x = 2$
40. $y = x^4$ from $x = -2$ to $x = 2$
41. $y = x + 4$ on $[1, 5]$
42. $y = 2x - 1$ on $[3, 7]$
43. $y = \sqrt{x}$ from $x = 0$ to $x = 4$
44. $y = \begin{cases} \frac{1}{\sqrt{x}}, & x > 0 \\ 0, & x = 0 \end{cases}$ on $[0, 1]$

Because right-hand endpoints are used, for exercise 44, the value of the function at $x = 0$ is irrelevant; only $y = \frac{1}{\sqrt{x}}$ matters.

45. (a) Evaluate the omega sum $\sum_{k=1}^{\Omega} (-1 + 2\omega k) \cdot 2\omega$ using the method of examples 8–10. (b) Because the answer to part (a) shows that

the approximation principle has been violated (by use of the sum of powers approximation formula, something was thrown away and the result was zero), use the exact formulas $\sum_{k=1}^{\Omega} 1 = \Omega$ and $\sum_{k=1}^{\Omega} k = \dfrac{\Omega^2 + \Omega}{2}$ to evaluate the sum.

46. Evaluate the omega sum $\sum_{k=1}^{\Omega} (-1 + 2\omega k)^2$.

Definite Integral

4.4

Finding the area under a curve is not the only application of the omega sums studied in the previous section. In the finite sums section we saw that distance calculations can be very similar to area calculations. If we use Ω time intervals, the result is an omega sum.

When we studied the derivative, we saw that slopes of tangent lines and velocities both used limits of difference quotients. Because there were multiple applications of the same calculation, we extracted that calculation from its context and called it the derivative, studied it, simplified its calculation through learning various rules, and eventually applied it to other settings. We are in the same situation now with omega sums. It is time to extract that calculation from its context, name it, study it, simplify its calculation through learning various rules, and eventually apply it to other geometric and physical settings.

Definition of the definite integral

Recall the procedure for finding the area under a curve $y = f(x)$ over the interval $[a, b]$. The procedure begins with ❶ finding the width of the intervals. Using $n = \Omega$ rectangles, the width of each rectangle is

$$\Delta x = \frac{b-a}{n} = \frac{b-a}{\Omega} = (b-a)\omega.$$

❷ The right-hand endpoints are are always $1, 2, 3, \ldots, \Omega$ rectangle widths from $x = a$. Calling these endpoints $x_1, x_2, x_3, \ldots, x_\Omega$, we have $x_1 = a + 1 \cdot \Delta x$, $x_2 = a + 2 \cdot \Delta x$, $x_3 = a + 3 \cdot \Delta x$, …, $x_\Omega = a + \Omega \cdot \Delta x$. In other words,

$\Delta x = \frac{b-a}{\Omega} = (b-a)\omega$

$$x_k = a + k\Delta x.$$

❸ The heights of the rectangles match the height of the curve $y = f(x)$ at the endpoints x_k, hence the heights are $f(x_k)$. We then ❹ form the omega sum. The area under the curve is found by multiplying heights by widths and adding the results:

$x_k = a + k\Delta x$

$$\text{area under the curve} = \sum_{k=1}^{\Omega} \underbrace{f(x_k)}_{\text{heights}} \cdot \underbrace{\Delta x}_{\text{widths}}.$$

The procedure finishes with ❺, evaluating the omega sum.

These formulas form the basis of the definition of a definite integral.

When defining the definite integral, most calculus textbooks use what is also called the *Riemann integral.* The definite integral presented here is called the *omega integral* (or Ω integral) and is introduced in the article Dawson, C. Bryan, "A New Extension of the Riemann Integral," *American Mathematical Monthly* 125 No. 2 (February 2018), 130–140,. doi: https://doi.org/10.1080/00029890.2018.1401832.

Definition 4 DEFINITE INTEGRAL *Let f be a function defined on* $[a, b]$. *If* $\sum_{k=1}^{\Omega} f(x_k)\Delta x$ *renders the same real result L for every positive infinite hyperreal integer* Ω *(where* $\Delta x = (b-a)\omega$ *and* $x_k = a + k\Delta x$ *for* $k = 1, 2, 3, \ldots, \Omega$*), then we write*

$$\int_a^b f(x)\,dx = L$$

and call $\int_a^b f(x)\,dx$ *the* definite integral of *f* from *a* to *b*.

When L is a real number (i.e., not ∞ *or* $-\infty$*), then we say that f is* integrable *on* $[a, b]$.

The Riemann integral is based on limits of Riemann sums instead of omega sums.

Georg Friedrich Bernhard Riemann 1826–1866

http://www-history.mcs.st-andrews.ac.uk/Biographies/Riemann.html

For the purposes of calculation we write

$$\int_a^b f(x)\,dx = \sum_{k=1}^{\Omega} f(x_k)\Delta x$$

with the understanding that Ω is an arbitrary positive infinite hyperreal integer and we must render the same real result for every such Ω or the integral does not exist.

Example 1 *Use the definition of the definite integral to evaluate* $\int_0^3 x^2\,dx$.

Solution Comparing $\int_a^b f(x)\,dx$ (notation from the definition) to $\int_0^3 x^2\,dx$, we see that $a = 0$, $b = 3$, and $f(x) = x^2$. We then use the formulas to proceed:

$$\Delta x = (b-a)\omega = (3-0)\omega = 3\omega,$$

$$x_k = a + k\Delta x = 0 + k \cdot 3\omega = k \cdot 3\omega, \text{ and}$$

$$f(x_k) = (k \cdot 3\omega)^2.$$

The formula for the definite integral gives

$$\int_0^3 x^2\,dx = \sum_{k=1}^{\Omega} f(x_k)\,\Delta x = \sum_{k=1}^{\Omega} \underbrace{(k \cdot 3\omega)^2}_{f(x_k)} \cdot \underbrace{3\omega}_{\Delta x}$$

$$= \sum_{k=1}^{\Omega} k^2 \cdot 9\omega^2 \cdot 3\omega$$

$$= \sum_{k=1}^{\Omega} 27\omega^3 k^2$$

$$\approx 27\omega^3 \frac{\Omega^3}{3} = 9.$$

The value of the definite integral is 9. ■

Although we are no longer in the context of areas, the procedure is the same; only the notation is new. The first line is the same as multiplying heights ($f(x_k)$) by widths (Δx) and adding the results. The second line performs the indicated arithmetic on the heights, the third line multiplies by the widths, and the fourth line uses the sum rules and the sum of powers approximation formula.

Notice that "units^2" is not present in the final answer because we extracted the calculation from the context of area. Unless a context is given, no units are appropriate.

Reading Exercise 14 *Use the definition of the definite integral to evaluate* $\int_0^4 x^2\,dx$.

Notation

You have no doubt noticed the similarity between the names and notation for the indefinite integral (most general antiderivative) and the definite integral. There are some differences. For instance, the result of an indefinite integral is a collection of functions, whereas the result of a definite integral is a real number (assuming the integral exists). However, we have already seen that there is a connection between the two types of integral. The sum of powers approximation formula, which was used in the calculation of the definite integral in example 1, is very reminiscent of the antiderivative power rule. The connection is made even stronger in the next section.

The notation $\int_a^b f(x)\,dx$ is read "the integral from a to b of f of x dx" or "the integral from a to b of f of x with respect to x." Other terminology associated with the notation is pointed out here:

The only difference between reading $\int f(x)\,dx$ and $\int_a^b f(x)\,dx$ is the insertion of the phrase "from a to b" in the latter.

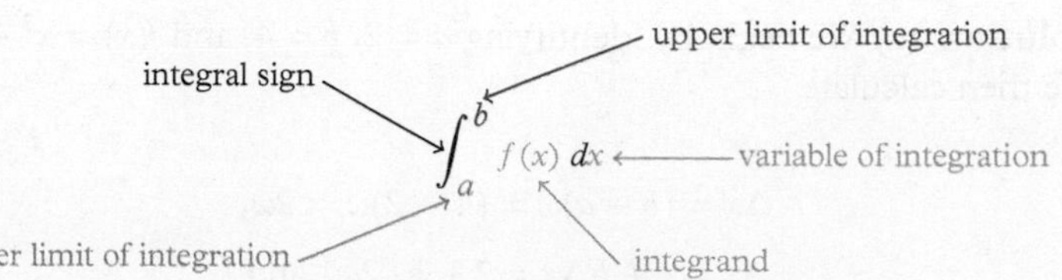

The *limits of integration* are not limits as in chapter 1. Here we are using *limit* synonymously with *edge* or *boundary*.

The integral sign (integral symbol) is an elongated S (for "sum") and is attributed to Leibniz.

The interval $[a, b]$ is also called the *interval of integration*.

The variable of integration is useful for telling the difference between the variable being used and other symbols representing constants. It also lets us know that a and b are values of the variable x, a fact that is important later in this chapter.

Why the differential dx? The notation arose when differentials were dominant instead of the use of the derivative. One would use the differential $x^2\,dx$ rather than the function value x^2.

Eventually, we encounter other uses of the definite integral. Such uses are recognized by the development of an omega sum, so it's helpful to practice rewriting an omega sum as a definite integral.

Example 2 *Express the sum as a definite integral on the given interval:* $\displaystyle\sum_{k=1}^{\Omega}\left(x_k^3-\sqrt{x_k}\right)\Delta x$ *on* $[2,5]$.

Solution Compare the omega sum notation to the definite integral notation and the procedure is clear: we replace the summation symbol with the integral symbol, we insert the lower limit of integration 2 and the upper limit of integration 5, we replace each x_k with the variable x, and we replace the Δx with dx:

$$\int_2^5\left(x^3-\sqrt{x}\right)\,dx. \qquad \blacksquare$$

Ans. to reading exercise 14
$\frac{64}{3}$
Selected details:
$\Delta x=(b-a)\omega=(4-0)\omega=4\omega$
$x_k=a+k\Delta x=0+k\cdot 4\omega=k\cdot 4\omega$
$f(x_k)=(k\cdot 4\omega)^2$
$\int_0^4 x^2\,dx=\sum_{k=1}^{\Omega}f(x_k)\Delta x$
$=\sum_{k=1}^{\Omega}(k\cdot 4\omega)^2\cdot 4\omega=\cdots=\frac{64}{3}$

Reading Exercise 15 *Express the sum as a definite integral on the given interval:* $\displaystyle\sum_{k=1}^{\Omega}\left(x_k^2+\sin x_k\right)\Delta x$ *on* $[1,4]$.

Net area

One interpretation of the definite integral, per the previous section, is the area under a curve, provided the curve lies on or above the x-axis throughout the interval of integration.

Because we are asked to evaluate a definite integral, the phrase "use the definition of an integral" refers to the definition of a definite integral.

Example 3 *(a) Use the definition of an integral to evaluate* $\int_2^4 (x^2-4)\,dx$. *(b) Give an interpretation of the integral as an area.*

Solution (a) We begin by identifying $a=2$, $b=4$, and $f(x)=x^2-4$. We then calculate

$$\Delta x=(b-a)\omega=(4-2)\omega=2\omega,$$

$$x_k=a+k\Delta x=2+k\cdot 2\omega,\ \text{and}$$

$$f(x_k)=(2+k\cdot 2\omega)^2-4.$$

The formula for a definite integral gives

$$\begin{aligned}\int_2^4 \left(x^2 - 4\right) dx &= \sum_{k=1}^{\Omega} f(x_k)\Delta x = \sum_{k=1}^{\Omega} \left((2 + k \cdot 2\omega)^2 - 4\right) \cdot 2\omega \\ &= \sum_{k=1}^{\Omega} \left(4 + 8\omega k + 4\omega^2 k^2 - 4\right) \cdot 2\omega \\ &= \sum_{k=1}^{\Omega} \left(8\omega k + 4\omega^2 k^2\right) 2\omega \\ &= \sum_{k=1}^{\Omega} \left(16\omega^2 k + 8\omega^3 k^2\right) \\ &\approx 16\omega^2 \frac{\Omega^2}{2} + 8\omega^3 \frac{\Omega^3}{3} \\ &= 8 + \frac{8}{3} = \frac{32}{3}.\end{aligned}$$

CAUTION: MISTAKE TO AVOID
Do not write

$$\sum_{k=1}^{\Omega}(2 + k \cdot 2\omega)^2 - 4 \cdot 2\omega,$$

for this does not multiply heights by widths; it only multiplies the width 2ω by the 4 and not the rest of the expression for the heights. Note the correct use of parentheses to the left.

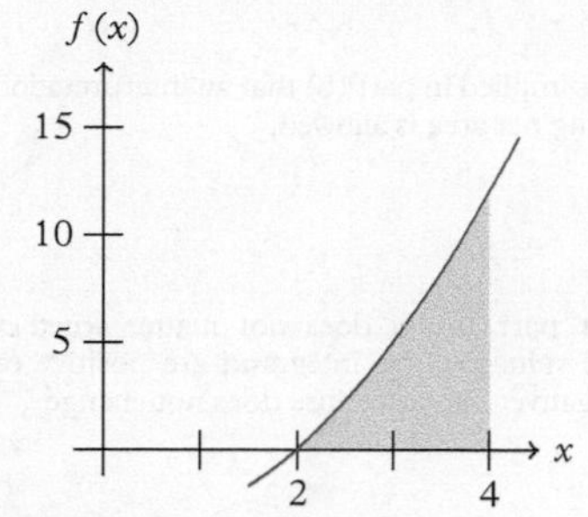

Figure 1 *The region under the curve $y = x^2 - 4$ (blue) between $x = 2$ and $x = 4$ (green)*

(b) Since $f(x) = x^2 - 4$ is on or above the x-axis throughout the interval $[2, 4]$, the area under the curve $y = x^2 - 4$ from $x = 2$ to $x = 4$ is $\frac{32}{3}$ units^2 (figure 1). ■

But what if the curve passes below the x-axis? There is nothing in the definition of a definite integral to prevent us from using such a curve. How do we interpret the result?

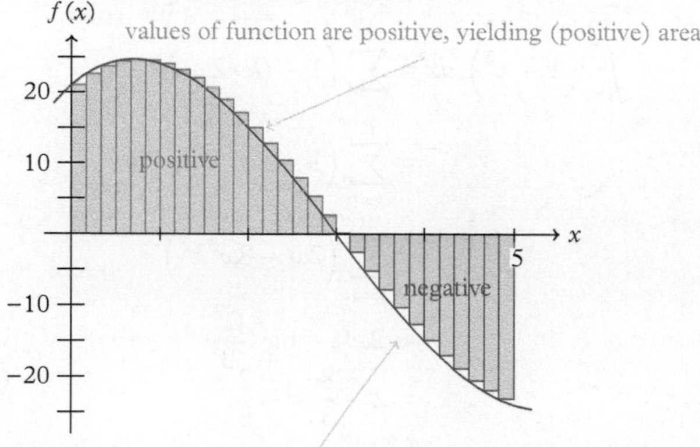

Figure 2 *The region between the curve $y = x^3 - 9x^2 + 11x + 21$ and the x-axis between $x = 0$ and $x = 5$, shown with a finite number of approximating rectangles (shaded). Using an infinite number of rectangles, the definite integral sums the areas of the green rectangles (where the curve lies above the x-axis) and subtracts the areas of the pink rectangles (where the curve lies below the x-axis), resulting in the net area under the curve*

The issue is that where the curve passes below the x-axis, the values of the function are negative (figure 2). Because the "heights" of the rectangles match the values on the curve, these "heights" are negative. Instead of the areas of the rectangles, we are then adding the negatives of the areas of the rectangles. This counts against the total, just as losses count against gains when adding up profits. The omega sum adds positive values where the curve is above the x-axis to negative values where the curve is below the x-axis. The resulting value of the integral is the *net area under the curve*—that is, the area of the green region minus the area of the pink region.

Ans. to reading exercise 15

$\int_1^4 \left(x^2 + \sin x\right)\, dx$

Unlike area, net area is allowed to be negative.

It is implied in part (b) that an interpretation using net area is allowed.

Example 4 *(a) Use the definition of an integral to evaluate* $\int_0^2 \left(1 - x^2\right) dx$*. (b) Give an interpretation of the integral as an area.*

Solution (a) We begin by identifying $a = 0$, $b = 2$, and $f(x) = 1 - x^2$. We then calculate

For part (a), it does not matter whether the values of the integrand are positive or negative; the procedure does not change.

$$\Delta x = (b - a)\omega = (2 - 0)\omega = 2\omega,$$
$$x_k = a + k\Delta x = 0 + k \cdot 2\omega, \text{ and}$$
$$f(x_k) = 1 - (k \cdot 2\omega)^2.$$

The formula for a definite integral gives

$$\begin{aligned}\int_0^2 \left(1 - x^2\right) dx &= \sum_{k=1}^{\Omega} \left(1 - (k \cdot 2\omega)^2\right) \cdot 2\omega \\ &= \sum_{k=1}^{\Omega} \left(1 - 4\omega^2 k^2\right) \cdot 2\omega \\ &= \sum_{k=1}^{\Omega} \left(2\omega - 8\omega^3 k^2\right) \\ &\approx 2\omega\Omega - 8\omega^3 \frac{\Omega^3}{3} \\ &= 2 - \frac{8}{3} = -\frac{2}{3}.\end{aligned}$$

(b) The net area under the curve $y = 1 - x^2$ between $x = 0$ and $x = 2$ is $-\frac{2}{3}$ units^2. ■

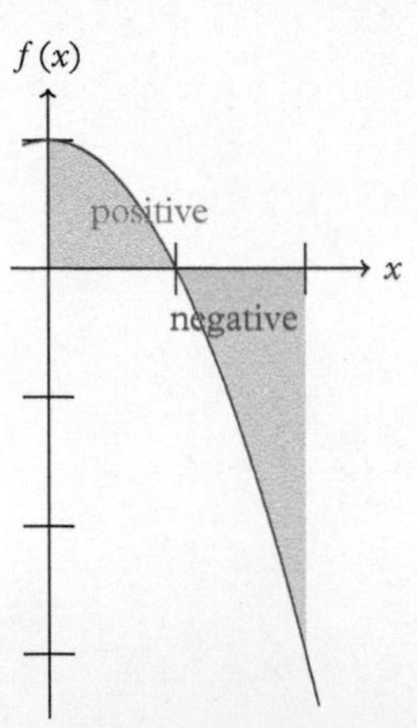

Figure 3 *The region between the curve* $y = 1 - x^2$ *and the x-axis between* $x = 0$ *and* $x = 2$ *(shaded). The definite integral takes the area of the green region (where the curve lies above the x-axis) and subtracts the area of the pink region (where the curve lies below the x-axis), resulting in the net area under the curve*

The negative net area under the curve in example 4 means the area of the pink region (where the curve is below the x-axis) is greater than the area of the green region (where the curve is above the x-axis); see figure 3. When the area of the green region is larger, as in figure 2, the net area is positive.

The net area interpretation is still valid even when the curve does not pass below the x-axis. For instance, an alternative answer to example 3(b) is that the net area under the curve $y = x^2 - 4$ from $x = 2$ to $x = 4$ is $\frac{32}{3}$ units^2.

Properties of definite integrals

In definition 4, the definite integral $\int_a^b f(x)\,dx$ was defined on an interval $[a, b]$, meaning that $a < b$. The possibilities $a = b$ and $a > b$ are defined next.

Definition 5 DEFINITE INTEGRAL, SPECIAL CASES

(a) Let f be a function defined at $x = a$. Then,

$$\int_a^a f(x)\,dx = 0.$$

(b) Let f be a function defined on $[a, b]$. Then,

$$\int_b^a f(x)\,dx = -\int_a^b f(x)\,dx.$$

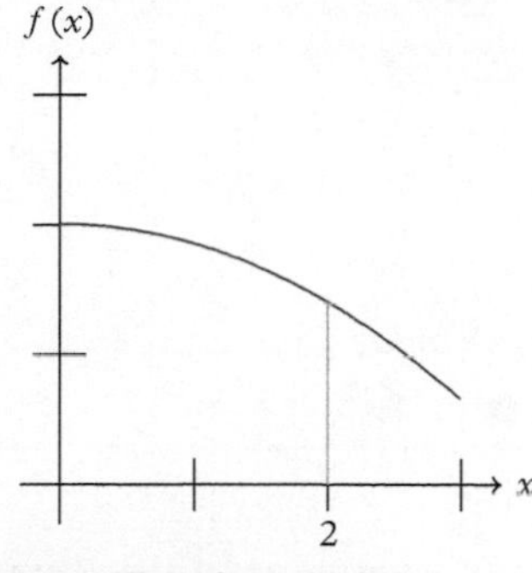

Figure 4 *The region under the curve $y = 2 - 0.15x^2$ between $x = 2$ and $x = 2$ represented as a line segment (green), which has no width*

To explain part (a), think of the area under the curve $y = f(x)$ from $x = a$ to $x = a$, as in figure 4. The result is a line segment, which has no width and therefore no area. The area is zero. (Alternately, using the prior procedure, $\Delta x = (a - a)\omega = 0$, and the result of the definite integral is zero.)

To explain part (b), try the prior procedure. We have $\Delta x = (a - b)\omega$, which is negative. Everything else is the same as before (but with the x_k's in reverse order and one other inconsequential difference), and the negative carries through to the end result.

Although the prior procedure can be carried out for $\int_a^a f(x)\,dx$ or $\int_b^a f(x)\,dx$, definition 4 does not apply to these integrals, hence the need for definition 5.

Example 5 *Evaluate* $\displaystyle\int_4^4 \frac{x^2 + 7x - 9}{\sqrt{x^3 - 8}}\,dx.$

Solution Because the upper and lower limits of integration are the same (both are 4), the integral's value is 0:

$$\int_4^4 \frac{x^2 + 7x - 9}{\sqrt{x^3 - 8}}\,dx = 0. \quad \blacksquare$$

Many of the omega sum rules translate to definite integral rules. For instance,

$$\int_a^b k \cdot f(x)\, dx = \sum_{k=1}^{\Omega} k \cdot f(x_k)\Delta x = k \sum_{k=1}^{\Omega} f(x_k)\Delta x = k \int_a^b f(x)\, dx,$$

proving the definite integral constant multiple rule.

DEFINITE INTEGRAL CONSTANT MULTIPLE RULE

If $\int_a^b f(x)\, dx$ exists, then

$$\int_a^b kf(x)\, dx = k \int_a^b f(x)\, dx.$$

The proofs of the sum and difference rules are similar.

DEFINITE INTEGRAL SUM RULE

If $\int_a^b f(x)\, dx$ and $\int_a^b g(x)\, dx$ both exist, then

$$\int_a^b (f(x) + g(x))\, dx = \int_a^b f(x)\, dx + \int_a^b g(x)\, dx.$$

DEFINITE INTEGRAL DIFFERENCE RULE

If $\int_a^b f(x)\, dx$ and $\int_a^b g(x)\, dx$ both exist, then

$$\int_a^b (f(x) - g(x))\, dx = \int_a^b f(x)\, dx - \int_a^b g(x)\, dx.$$

Example 6 *Given* $\int_0^2 f(x)\, dx = 3$ *and* $\int_0^2 g(x)\, dx = 5$, *find (a)* $\int_2^0 f(x)\, dx$ *and (b)* $\int_0^2 (13f(x) - g(x))\, dx$.

Solution (a) Since $\int_2^0 f(x)\, dx$ has the lower and upper limits of integration reversed from what was given, we use definition 5 part (b):

$$\int_2^0 f(x)\,dx = -\int_0^2 f(x)\,dx = -3.$$

(b) Using the definite integral difference rule followed by the definite integral constant multiple rule,

$$\begin{aligned}\int_0^2 (13f(x) - g(x))\,dx &= \int_0^2 13f(x)\,dx - \int_0^2 g(x)\,dx \\ &= 13\int_0^2 f(x)\,dx - \int_0^2 g(x)\,dx \\ &= 13 \cdot 3 - 5 = 34.\end{aligned}$$ ■

Because the graph of a constant function $f(x) = c$ is a horizontal line, the area under the curve (if c is positive) is the area of a rectangle, as in figure 5. The height of the rectangle is c and the width is $b - a$, so its area is $c(b - a)$. In definite integral notation, $\int_a^b c\,dx = c(b - a)$. A more formal proof can be written using the procedure of example 1.

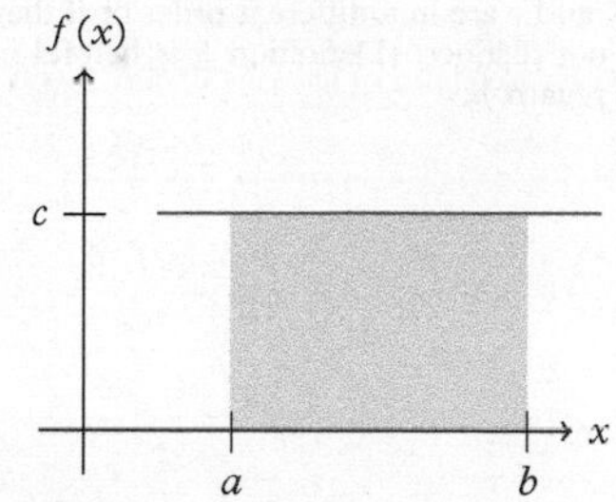

Figure 5 *The region under a constant curve $y = c$ between $x = a$ and $x = b$ (green) has area $c(b - a)$*

DEFINITE INTEGRAL OF A CONSTANT

For any real number c and any interval $[a, b]$,

$$\int_a^b c\,dx = c(b - a).$$

Example 7 *Evaluate* $\int_1^7 12\,dx$.

Solution Because 12 is a constant, we use the rule for the definite integral of a constant:

$$\int_1^7 12\,dx = 12(7 - 1) = 72.$$ ■

Reading Exercise 16 *Evaluate* $\int_{-2}^4 3\,dx$.

Our intuition about area is that, in figure 6, the area of the region on the left plus the area of the region on the right is the total area of the combined region. In definite integral notation, $\int_a^b f(x)\,dx + \int_b^c f(x)\,dx =$

$\int_a^c f(x)\,dx$. The proof of this *additive property* is complicated enough that it is omitted.

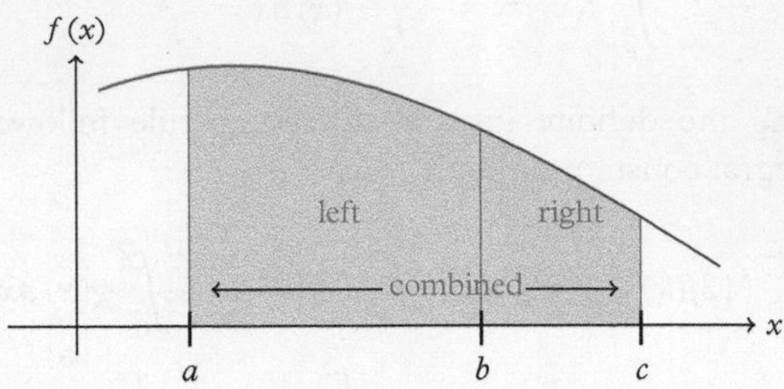

Figure 6 *The area under the curve from $x = a$ to $x = b$ (left region) plus the area under the curve from $x = b$ to $x = c$ (right region) is the area under the curve from $x = a$ to $x = c$ (combined region)*

As long as f is continuous throughout all the intervals, the additive property holds even if a, b, and c are in a different order or if they are not distinct. (Definition 5 is helpful in this regard.)

ADDITIVE PROPERTY

If f is continuous on $[a, c]$ and $a < b < c$, then

$$\int_a^b f(x)\,dx + \int_b^c f(x)\,dx = \int_a^c f(x)\,dx.$$

Example 8 *Given $\int_1^6 f(x)\,dx = 4$ and $\int_6^9 f(x)\,dx = 7$, determine $\int_1^9 f(x)\,dx$.*

The two sides of the additive property formula are written in reverse order for convenience.

Solution Using the additive property with $a = 1$, $b = 6$, and $c = 9$,

$$\int_1^9 f(x)\,dx = \int_1^6 f(x)\,dx + \int_6^9 f(x)\,dx = 4 + 7 = 11. \quad \blacksquare$$

As long as any two of the three integrals in the additive property formula are given, the third can be determined.

Example 9 *Given $\int_0^3 f(x)\,dx = 2$ and $\int_0^7 f(x)\,dx = 5$, determine $\int_3^7 f(x)\,dx$.*

Solution The limits of integration on the integrals include the three numbers 0, 3, and 7. Written in numerical order, they are the values of a, b, and c in the additive property: $a = 0$, $b = 3$, and $c = 7$. The additive property formula then gives

$$\int_0^3 f(x)\,dx + \int_3^7 f(x)\,dx = \int_0^7 f(x)\,dx,$$

$$2 + \int_3^7 f(x)\,dx = 5, \text{ and}$$

$$\int_3^7 f(x)\,dx = 5 - 2 = 3. \quad \blacksquare$$

Reading Exercise 17 *Given* $\int_1^4 f(x)\,dx = 3$ *and* $\int_4^7 f(x)\,dx = -1$, *find* $\int_1^7 f(x)\,dx$.

A few inequalities are helpful as well.

Ans. to reading exercise 16
18

DEFINITE INTEGRAL INEQUALITIES

If $\int_a^b f(x)\,dx$ and $\int_a^b g(x)\,dx$ both exist, then

(a) if $f(x) \geq 0$ on $[a, b]$, then $\int_a^b f(x)\,dx \geq 0$;

(b) if $f(x) \leq g(x)$ on $[a, b]$, then $\int_a^b f(x)\,dx \leq \int_a^b g(x)\,dx$;

(c) if $m \leq f(x) \leq M$ on $[a, b]$, then $m(b-a) \leq \int_a^b f(x)\,dx \leq M(b-a)$.

The proof of (a) involves looking at $\sum_{k=1}^{\Omega} f(x_k)\Delta x$ and noticing that there are no negative numbers in the sum, so the sum cannot be negative. Part (b) can be proved by looking at $\int_a^b (g(x) - f(x))\,dx$ and using part (a) and the definite integral difference rule. Part (c) can be proved from part (b) and by using the formula for the definite integral of a constant.

The proof of the following theorem is beyond the scope of this text.

Theorem 1 INTEGRABILITY OF CONTINUOUS FUNCTIONS *If f is continuous on $[a, b]$ then f is integrable.*

In other words, if f is continuous on $[a, b]$, then the definite integral $\int_a^b f(x)\,dx$ exists; it is a real number. Calculating this real number, however, is another matter altogether.

Exercises 4.4

1–6. Rapid response: is the given expression a definite integral or an indefinite integral?

1. $\int_1^5 (3x^2 - 7)\,dx$
2. $\int \sqrt{x}\,dx$
3. $\int (3x^2 - 7)\,dx$
4. $\int_{-3}^{-1} \sqrt{x^2+1}\,dx$
5. $\int 14\,dx$
6. $\int_{12}^{13} 14\,dx$

7–12. Rapid response: what type of object is the expression? Is it a collection of functions or is it a real number? Assume the integrals exist.

7. $\int_4^7 x^5\,dx$
8. $\int f(x)\,dx$
9. $\int_4^7 f(x)\,dx$
10. $\int \frac{1}{x}\,dx$
11. $\int x^5\,dx$
12. $\int_a^b \sqrt{x}\,dx$

Answer to reading exercise 17
2

13–26. Use the definition to evaluate the definite integral.

13. $\int_0^1 x^9\,dx$
14. $\int_0^7 x^2\,dx$
15. $\int_2^5 -x^2\,dx$
16. $\int_3^4 5x^2\,dx$
17. $\int_{-4}^{0} 3x^2\,dx$
18. $\int_{-1}^{1} (2x^2 - 7)\,dx$
19. $\int_0^3 (x^2 + 5)\,dx$
20. $\int_0^2 (3x^2 - 4)\,dx$
21. $\int_1^5 (10 - x^2)\,dx$
22. $\int_1^4 x^3\,dx$
23. $\int_0^4 x^{1.5}\,dx$
24. $\int_{-2}^{-1} (x^2 - 4x + 11)\,dx$
25. $\int_0^3 t^4\,dt$
26. $\int_1^3 y^2\,dy$

Exercises 25 and 26 change the name of the variable of integration, but otherwise they are identical to integrals with respect to *x*.

27–32. Give an interpretation of the definite integral using area, for the integral from the given exercise.

27. exercise 13
28. exercise 18
29. exercise 15
30. exercise 20
31. exercise 21
32. exercise 14

33–40. Express the sum as a definite integral on the given interval.

33. $\sum_{k=1}^{\Omega} (x_k^5 - 4x_k + 1)\Delta x$ on $[3, 19]$

34. $\displaystyle\sum_{k=1}^{\Omega} \frac{x_k^2+1}{x_k^3}\Delta x$ on $[1,9]$

35. $\displaystyle\sum_{k=1}^{\Omega} (\cos(x_k+\pi)+\sin(4x_k))\Delta x$ on $[\frac{\pi}{2}, \frac{3\pi}{2}]$

36. $\displaystyle\sum_{k=1}^{\Omega} 19\sqrt[3]{1+x_k}\,\Delta x$ on $[-1,1]$

37. $\displaystyle\sum_{k=1}^{\Omega} 4x_k\sqrt{1-x_k^4}\,\Delta x$ on $[0,1]$

38. $\displaystyle\sum_{k=1}^{\Omega} (9-x_k+12x_k\tan x_k)\Delta x$ on $[-1,0]$

39. $\displaystyle\sum_{k=1}^{\Omega} \left((f(x_k))^2+(g(x_k))^2\right)\Delta x$ on $[a,b]$

40. $\displaystyle\sum_{k=1}^{\Omega} \sqrt{1+f'(x_k)^2}\,\Delta x$ on $[a,b]$

41–52. Evaluate the integral given that $\int_1^5 f(x)\,dx = 4$ and $\int_1^5 g(x)\,dx = 10$.

41. $\int_1^5 4f(x)\,dx$
42. $\int_1^5 -g(x)\,dx$
43. $\int_1^5 (f(x)-2g(x))\,dx$
44. $\int_5^1 4f(x)\,dx$
45. $\int_5^1 3g(x)\,dx$
46. $\int_1^5 -3\,dx$
47. $\int_1^5 6\,dx$
48. $\int_1^5 \left(\frac{3}{2}f(x)-7g(x)\right)\,dx$
49. $\int_5^5 6f(x)\,dx$
50. $\int_5^1 2\,dx$
51. $\int_1^5 (\sqrt{11}f(x)-5)\,dx$
52. $\int_1^1 (-3g(x)+15)\,dx$

53. Evaluate $\int_7^7 \left(77x^7+777\sqrt[7]{x}\right)\,dx$.

54. Evaluate $\int_0^0 \tan(x^2+3x\pi)\,dx$.

55. Given $\int_3^7 f(x)\,dx = 12$ and $\int_7^{10} f(x)\,dx = 4$, find $\int_3^{10} f(x)\,dx$.

56. Given $\int_1^5 f(x)\,dx = -5$ and $\int_5^8 f(x)\,dx = -2$, find $\int_1^8 f(x)\,dx$.

57. Given $\int_3^7 g(x)\,dx = 6$ and $\int_5^7 g(x)\,dx = 1$, find $\int_3^5 g(x)\,dx$.

58. Given $\int_{-3}^5 f(x)\,dx = 17$ and $\int_3^5 f(x)\,dx = 10$, find $\int_{-3}^3 f(x)\,dx$.

59. Given $\int_4^2 f(x)\,dx = 6$ and $\int_2^8 f(x)\,dx = 2$, find $\int_4^8 f(x)\,dx$.

60. Given $\int_6^8 3f(x)\,dx = 12$ and $\int_5^8 2f(x)\,dx = 20$, find $\int_6^5 f(x)\,dx$.

61. Prove the definite integral sum rule.

62. Prove the definite integral difference rule.

63. Prove that $\int_a^b c\,dx = c(b-a)$.

64. Draw a picture that illustrates the truth of definite integral inequality (b).

65. Draw a picture that illustrates the truth of definite integral inequality (c).

66. (a) Use the definition of a definite integral to evaluate $\int_0^2 x^2\,dx$. (b) Evaluate $\int x^2\,dx$. (c) Using the function $H(x) = \frac{x^3}{3}$, find $H(2)$ and compare to the answer to part (a). (d) Use the definition of a definite integral to evaluate $\int_0^5 x^2\,dx$. (e) Find $H(5)$ and compare to the answer to part (d). (f) Use the definition of a definite integral to evaluate $\int_2^5 x^2\,dx$. (g) How does the answer to part (f) compare to the answers to parts (c) and (e)? Hint: try using the additivity property.

The answer to part (g) anticipates the amazing theorems of the next section.

67. This exercise has been moved to another section.

68. Use the definition of a definite integral to evaluate the definite integral $\int_0^{\pi/2} \sin x\,dx$. Use Lagrange's trigonometric identity

The technique of this exercise was contributed by Kyle A. James, an undergraduate mathematics and meteorology student at the University of Oklahoma. The exercises that follow use the same technique but different identities.

$$\sum_{k=1}^{n} \sin(k\theta) = \frac{\cos\left(\frac{1}{2}\theta\right) - \cos\left(n\theta + \frac{1}{2}\theta\right)}{2\sin\left(\frac{1}{2}\theta\right)},$$

and then simplify the result using the sine of a sum formula, approximating and simplifying as needed.

69. Use the definition of a definite integral to evaluate the definite integral $\int_0^{\pi/2} \sin x\,dx$. Use the identity

$$\sum_{k=1}^{n} \sin(k\theta) = \frac{\sin\left(\frac{n}{2}\theta\right)\sin\left(\frac{n+1}{2}\theta\right)}{\sin\left(\frac{1}{2}\theta\right)}.$$

70. Use the definition of a definite integral to evaluate the definite integral $\int_0^{\pi/2} \cos x\,dx$. Use Lagrange's trigonometric identity

$$\sum_{k=1}^{n} \cos(k\theta) = -\frac{1}{2} + \frac{\sin\left(n\theta + \frac{1}{2}\theta\right)}{2\sin\left(\frac{1}{2}\theta\right)}.$$

71. Use the definition of a definite integral to evaluate the definite integral $\int_a^b \cos x\, dx$. Use the identity

$$\cos\phi + \cos(\phi+\theta) + \cos(\phi+2\theta) + \cdots + \cos(\phi+n\theta) = \frac{\sin\left(\frac{(n+1)\theta}{2}\right)\cdot\cos\left(\phi+\frac{n\theta}{2}\right)}{\sin\frac{\theta}{2}}.$$

For this identity, see Knapp, Michael P., "Sines and Cosines of Angles in Arithmetic Progression," *Mathematics Magazine* 82 No. 5 (December 2009), 371–372, doi: https://doi.org/10.4169/002557009X478436

Compare to the answer to exercise 66(g).

72. Use the definition of a definite integral to evaluate the definite integral $\int_a^b \sin x\, dx$. Use the identity

$$\sin\phi + \sin(\phi+\theta) + \sin(\phi+2\theta) + \cdots + \sin(\phi+n\theta) = \frac{\sin\left(\frac{(n+1)\theta}{2}\right)\cdot\sin\left(\phi+\frac{n\theta}{2}\right)}{\sin\frac{\theta}{2}}.$$

Fundamental Theorem of Calculus

4.5

The sum of powers approximation formula hints at a relationship between antiderivatives and definite integrals. The fundamental theorem of calculus (FTC) makes this relationship explicit.

Fundamental theorem of calculus, part I

When we studied antiderivatives, we encountered functions with antiderivatives we did not know how to find. Part I of the fundamental theorem of calculus (FTC I) says that for continuous functions, antiderivatives always exist, even if we do not have a name for them. The theorem also says how to build such an antiderivative using the definite integral, which is helpful in chapter 5.

Theorem 2 FUNDAMENTAL THEOREM OF CALCULUS, PART I *If f is continuous on $[a, b]$, then the function*

$$F(x) = \int_a^x f(t)\,dt$$

is differentiable on $[a, b]$ and $F'(x) = f(x)$. In other words, F is an antiderivative of f.

Notice the phrase "an antiderivative," not "the antiderivative." Recall that any function with an antiderivative has infinitely many of them; F is just one of the antiderivatives of f.

Before examining the proof, we need to explore the function F.

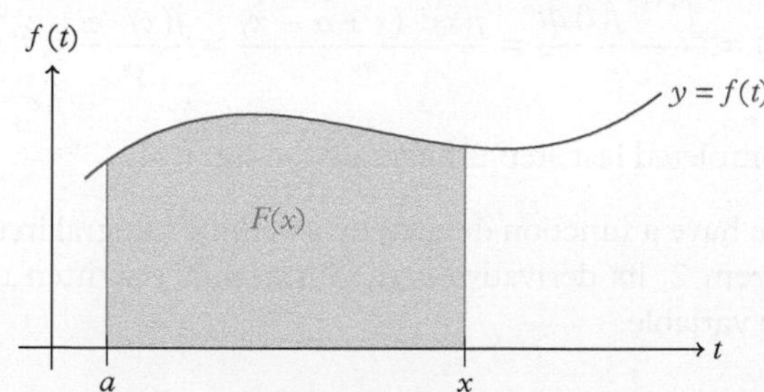

Figure 1 *The value of $F(x)$ is the area under the curve $y = f(t)$ from $t = a$ to $t = x$. Notice that theorem 2 requires f to be continuous*

We start with a different function, f, with a variable that is t rather than x. If the curve $y = f(t)$ is on or above the horizontal axis (the t-axis), then the integral can be interpreted as the area under the curve, as in figure 1. $F(x)$ is then the area under the curve $y = f(t)$ from $t = a$ to $t = x$. Similarly, $F(x + 1)$ is the area under the curve $y = f(t)$ from $t = a$ to $t = x + 1$ (figure 2).

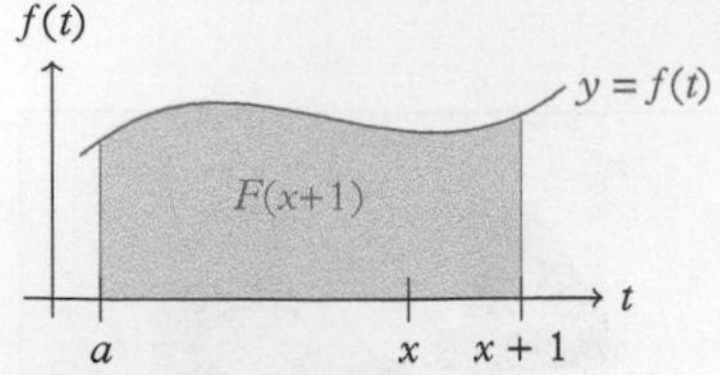

Figure 2 *The value of* $F(x + 1)$ *is the area under the curve* $y = f(t)$ *from* $t = a$ *to* $t = x+1$

The details of how the transfer principle is used in this proof—for instance, the use of hyperreals as limits of integration—are omitted so the intuitive ideas are not obscured by the details.

Proof. As hypothesized, let f be continuous on $[a, b]$. By theorem 1, f is integrable, so the value $F(x)$ exists for each x in $[a, b]$. That is, F is defined on $[a, b]$. We need to show that $F'(x) = f(x)$. Using the definition of a derivative and our definition of the function F,

$$F'(x) = \frac{F(x+\alpha) - F(x)}{\alpha} = \frac{\int_a^{x+\alpha} f(t)\,dt - \int_a^x f(t)\,dt}{\alpha},$$

remembering that α is an arbitrary infinitesimal and can be positive or negative.

By the additive property of definite integrals,

$$\int_a^{x+\alpha} f(t)\,dt - \int_a^x f(t)\,dt = \int_x^{x+\alpha} f(t)\,dt,$$

even if α and f are not as pictured in figure 3. Then,

$$F'(x) = \frac{\int_x^{x+\alpha} f(t)\,dt}{\alpha}.$$

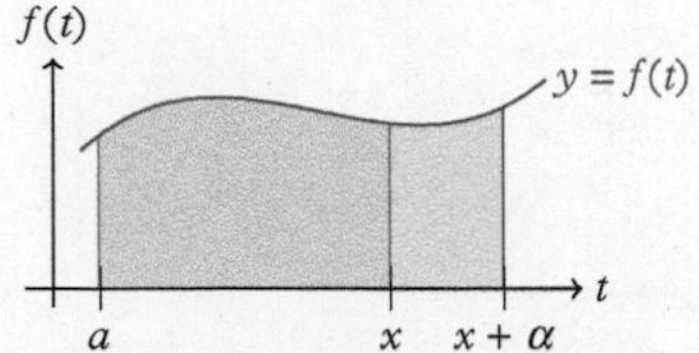

Figure 3 *The relationship between* $F(x)$ *and* $F(x + \alpha)$*, pictured with* α *positive and* f *on or above the horizontal axis. The value of* $F(x)$ *is the area under the curve* $y = f(t)$ *from* $t = a$ *to* $t = x$ *(green), the value of* $F(x + \alpha)$ *is the area under the curve from* $t = a$ *to* $t = x + \alpha$ *(entire shaded region), and the difference between the two is the area under the curve* $y = f(t)$ *from* $t = x$ *to* $t = x + \alpha$ *(pink). In the figure,* α *is made too large so that the pink region can be seen. Because* α *is infinitesimal, the pink region is actually infinitesimally thin and impossible to see at this scale*

Recall that "f is continuous" means that $f(x + \text{infinitesimal}) \doteq f(x)$ for any infinitesimal. Because α is infinitesimal, all the values of the function between the limits of integration are approximately $f(x)$, so the rule for a definite integral of a constant can be used:

$$F'(x) = \frac{\int_x^{x+\alpha} f(t)\,dt}{\alpha} = \frac{f(x)\cdot(x+\alpha-x)}{\alpha} = \frac{f(x)\cdot\alpha}{\alpha} = f(x).$$

The just-completed last step is illustrated in figure 4. ■

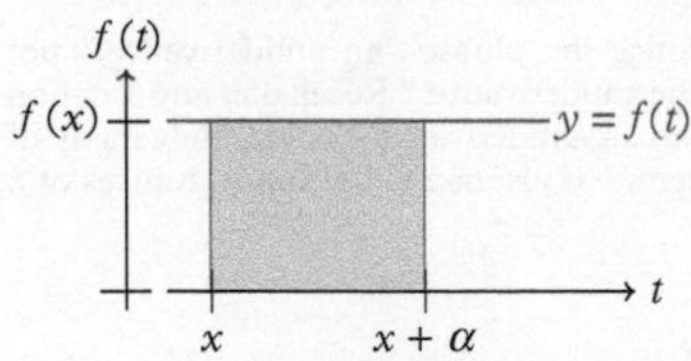

Figure 4 *An infinitesimal scale on the horizontal axis makes* x *and* $x+\alpha$ *distinguishable; a real scale on the vertical axis makes* $f(x)$ *visible. Because* f *is continuous,*

$$f(x + \textit{infinitesimal}) \doteq f(x),$$

and the graph of the function appears flat given these scales. The height of the resulting rectangle is $f(x)$ *and the width is* α*, so the area of the green region is* $f(x)\cdot\alpha$*:*

$$\int_x^{x+\alpha} f(t)\,dt = f(x)\cdot\alpha.$$

When we have a function defined by a definite integral in the manner of theorem 2, its derivative is the integrand, rewritten using the appropriate variable.

Example 1 *Find* $F'(x)$ *for* $F(x) = \int_2^x (t^2 + \cos t)\,dt$.

Solution Using the notation of FTC I, $a = 2$ and $f(t) = t^2 + \cos t$. Because $F'(x) = f(x)$, we simply rewrite f using the variable x instead of t:

$$F'(x) = f(x) = x^2 + \cos x. \qquad \blacksquare$$

Other notations for functions and derivatives can be used as well, as can different variables of integration. However, the function's independent variable (x in example 1) and the variable of integration (t in example 1) must be distinct.

Example 2 *Determine $\frac{dy}{dx}$ for $y = \int_{-3}^{x} \sqrt{z^2 + 5}\, dz$.*

Solution The variable of integration in this example is z instead of the t in FTC I, but this doesn't matter. Our function is called y instead of $F(x)$, so asking for $\frac{dy}{dx}$ is the same as asking for $F'(x)$. We therefore rewrite the integrand with an x instead of the variable z:

$$\frac{dy}{dx} = \sqrt{x^2 + 5}. \qquad \blacksquare$$

Notice that in the formulas of FTC I, we have "just plain x." As with any other derivative formula, if we are given something other than "just plain x," the chain rule is required. The use of the chain rule in this setting is explored in exercise 83.

Reading Exercise 18 *Find $F'(x)$ for $F(x) = \displaystyle\int_1^x (t + 5 - \sqrt{t})\, dt$.*

Fundamental theorem of calculus, part II

Part I of the fundamental theorem of calculus says that given a function f that is continuous on $[a, b]$, and defining

$$F(x) = \int_a^x f(t)\, dt,$$

then F is an antiderivative of f. Using the formula for F at the left endpoint a of our interval,

$$F(a) = \int_a^a f(t)\, dt = 0,$$

because the limits of integration are the same (using definition 5). We can also use the formula at the right endpoint b of our interval:

$$F(b) = \int_a^b f(t)\,dt.$$

This is the net area under the curve $y = f(t)$ from $t = a$ to $t = b$. Can the antiderivative F give us a quicker means to calculate definite integrals? The problem is that there are infinitely many antiderivatives of f, and we don't happen to know ahead of time which one of them is F in particular. But, this problem can be remedied.

Let H be any antiderivative of f (it doesn't matter which one). Then, $H'(x) = f(x)$. We know from chapter 3 theorem 6 that any two antiderivatives of f must differ by a constant, and because F and H are both antiderivatives of f, then

$$H(x) = F(x) + C$$

Here, C is a specific (albeit unspecified) real number.

for some constant C. Now we calculate $H(b) - H(a)$:

$$\begin{aligned} H(b) - H(a) &= (F(b) + C) - (F(a) + C) \\ &= F(b) - F(a) \\ &= \int_a^b f(t)\,dt - 0 \\ &= \int_a^b f(t)\,dt. \end{aligned}$$

Then, *any* antiderivative of our choosing can be used to calculate the value of the definite integral!

Ans. to reading exercise 18
$F'(x) = x + 5 - \sqrt{x}$

Some say that theorem 3 is the most important computational discovery in history, for it was essential to the development of many aspects of modern science and engineering. Definite integrals aren't just for finding areas; they are useful in many other applications, some of which appear in later chapters of this text.

Theorem 3 FUNDAMENTAL THEOREM OF CALCULUS, PART II *If f is continuous on $[a, b]$ and H is any antiderivative of f on $[a, b]$, then*

$$\int_a^b f(t)\,dt = H(b) - H(a).$$

This area was calculated using omega sums in section 4.3.

We now return to an example we worked before: find the area under the curve $y = x^2$ from $x = 1$ to $x = 5$. The requested area is represented by the definite integral $\int_1^5 x^2\,dx$. According to the fundamental theorem of calculus, part II (FTC II), we need an antiderivative of x^2;

accordingly, we let $H(x) = \frac{x^3}{3}$. Then,

$$\int_1^5 x^2\,dx = H(5) - H(1) = \frac{5^3}{3} - \frac{1^3}{3} = \frac{124}{3},$$

and the area under the curve is $\frac{124}{3}$ units^2. This is much quicker than using the definition of a definite integral (using omega sums)!

Any other antiderivative can be used instead. For instance, if $H(x) = \frac{x^3}{3} + 17$, then

$$\int_1^5 x^2\,dx = H(5) - H(1) = \left(\frac{5^3}{3} + 17\right) - \left(\frac{1^3}{3} + 17\right) = \frac{124}{3}$$

Whichever "+ C" we use is added with $H(b)$, but subtracted right back out with $H(a)$.

A notational convention helps shorten the process even further. We use an *evaluation bar* (the vertical bar, |, also called an *evaluation symbol*) to signify evaluating an expression at the top number and at the bottom number and subtracting the results:

$$\left.\frac{x^3}{3}\right|_1^5 = \frac{5^3}{3} - \frac{1^3}{3} = \frac{124}{3}.$$

CAUTION: MISTAKE TO AVOID
For an evaluation bar, the order matters. It is the result of the top number minus the result of the bottom number.

The entire computation is then written as follows:

$$\int_1^5 x^2\,dx = \underbrace{\left.\frac{x^3}{3}\right|_1^5}_{\substack{\text{antiderivative of integrand,}\\ \text{with evaluation symbol}}} = \underbrace{\frac{5^3}{3} - \frac{1^3}{3}}_{\substack{\text{antiderivative evaluated}\\ \text{at top number minus}\\ \text{antiderivative evaluated}\\ \text{at bottom number}}} = \frac{124}{3}.$$

When the antiderivative is taken we drop the integral notation and use the evaluation symbol instead.

Example 3 *Evaluate* $\int_2^3 5x^3\,dx$.

Solution Since $y = 5x^3$ is continuous on $[2, 3]$, we use FTC II:

$$\int_2^3 5x^3\,dx = 5\cdot\left.\frac{x^4}{4}\right|_2^3 = 5\cdot\frac{3^4}{4} - 5\cdot\frac{2^4}{4} = \frac{325}{4}. \quad ■$$

The steps are ❶ find an antiderivative of the integrand and write it alongside the evaluation symbol, ❷ evaluate the antiderivative at the top number and subtract the antiderivative evaluated at the bottom number, and ❸ finish the arithmetic.

When rewriting an expression algebraically, all else stays the same. If we rewrite an integrand, the rest of the integral does not change. If we rewrite the antiderivative before evaluating, we carry along the evaluation bar.

Example 4 *Evaluate* $\int_0^6 \sqrt{x}\,dx$.

Solution Since $y = \sqrt{x}$ is continuous on $[0, 6]$, we use FTC II:

$$\begin{aligned}\int_0^6 \sqrt{x}\,dx &= \int_0^6 x^{1/2}\,dx \\ &= \left.\frac{x^{3/2}}{\frac{3}{2}}\right|_0^6 \\ &= \left.\frac{2}{3}x^{3/2}\right|_0^6 \\ &= \frac{2}{3}\cdot 6^{3/2} - \frac{2}{3}\cdot 0^{3/2} = \frac{2}{3}\cdot 6^{3/2} \\ &= \frac{2}{3}\cdot 6\sqrt{6} = 4\sqrt{6}. \quad \blacksquare\end{aligned}$$

Algebraic manipulations leave the rest of the expression unchanged.
In line 1, the integrand is rewritten, but all else stays the same. In line 2, we found the antiderivative, so we drop the integral notation and use the evaluation bar. In line 3, the expression in front of the evaluation bar is rewritten, and the evaluation bar remains unchanged. In line 4, the expression is evaluated at the top and bottom numbers, and the results are subtracted, so the evaluation bar is dropped.

Reading Exercise 19 *Evaluate* $\int_1^3 x^4\,dx$.

Examples 3 and 4 can be worked using the definition of an integral (using omega sums); FTC II makes the solutions a little shorter. The real advantage of FTC II comes from the ability to calculate easily those definite integrals that are difficult or impossible to handle using the definition.

Example 5 *Evaluate* $\int_0^{2\pi} \sin x\,dx$.

Solution Since $y = \sin x$ is continuous on $[0, 2\pi]$, we use FTC II:

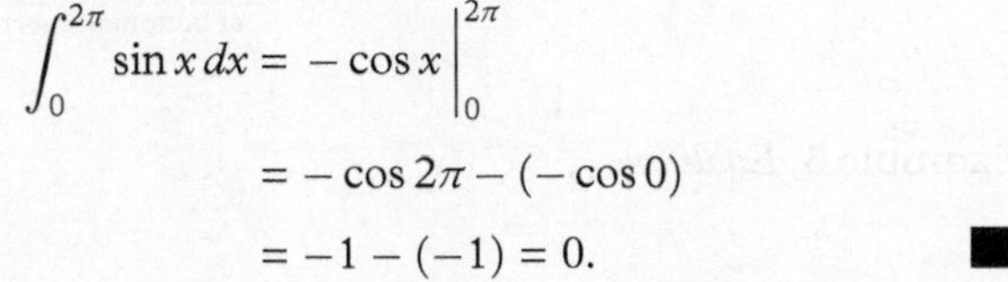

$$\begin{aligned}\int_0^{2\pi} \sin x\,dx &= \left.-\cos x\right|_0^{2\pi} \\ &= -\cos 2\pi - (-\cos 0) \\ &= -1 - (-1) = 0. \quad \blacksquare\end{aligned}$$

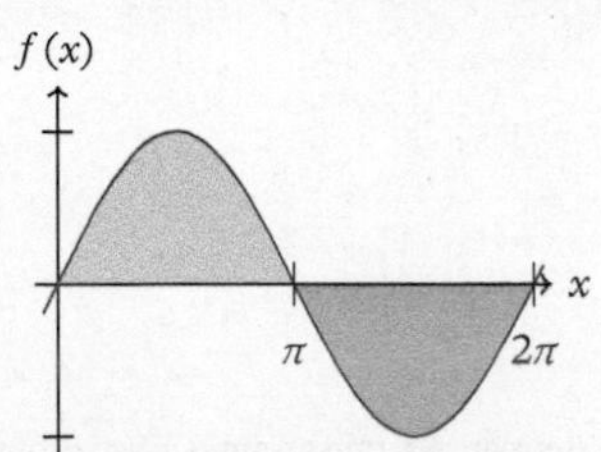

Figure 5 *The region between the curve* $y = \sin x$ *and the x-axis between* $x = 0$ *and* $x = 2\pi$ *(shaded). The definite integral takes the area of the green region (where the curve lies above the x-axis) and subtracts the area of the pink region (where the curve lies below the x-axis), resulting in the net area under the curve. As a result of symmetry, the net area is* 0 units^2

An interpretation of the solution to example 5 is that the net area under the curve $y = \sin x$ from $x = 0$ to $x = 2\pi$ is 0 units^2, as illustrated in figure 5.

Recall that not all integrands are written in a form for which the antiderivative rules and formulas at our disposal can be used. In this case, we can sometimes manipulate the integrand to rewrite it in a form for which we can determine an antiderivative.

Example 6 *Find* $\int_{-2}^5 (1 - x)^2\,dx$.

Solution Since $y = (1 - x)^2$ is continuous on $[-2, 5]$, we use FTC II. However, the antiderivative power rule is for "just plain *x*," and we have $1-x$ to a power instead. Let's rewrite the integrand by performing the squaring operation before taking the antiderivative:

$$\int_{-2}^{5}(1-x)^2\,dx = \int_{-2}^{5}(1-2x+x^2)\,dx$$
$$= \left(x - x^2 + \frac{x^3}{3}\right)\Big|_{-2}^{5}$$
$$= 5 - 25 + \frac{125}{3} - \left(-2 - 4 + \frac{-8}{3}\right)$$
$$= \cdots = \frac{91}{3}. \quad \blacksquare$$

Line 1: the integrand is rewritten to facilitate the correct determination of an antiderivative.

CAUTION: MISTAKE TO AVOID
When subtracting the value of the antiderivative evaluated at the bottom number, be sure to use parentheses if the expression contains more than one term.

Example 7 *Calculate* $\int_{-\frac{\pi}{4}}^{0} \tan x \cos x\, dx.$

Solution Since $y = \tan x \cos x$ is continuous on $[-\pi/4, 0]$, we use FTC II. However, this expression is not in our list of antiderivative formulas, nor is there an antiderivative product rule. Rewriting tangent in terms of sines and cosines can help:

Rewriting trigonometric functions in terms of sines and cosines sometimes leads to helpful simplifications.

$$\int_{-\frac{\pi}{4}}^{0} \tan x \cos x\, dx = \int_{-\frac{\pi}{4}}^{0} \frac{\sin x}{\cos x}\cos x\, dx$$
$$= \int_{-\frac{\pi}{4}}^{0} \sin x\, dx$$
$$= -\cos x\Big|_{-\frac{\pi}{4}}^{0}$$
$$= -\cos 0 - \left(-\cos\left(-\frac{\pi}{4}\right)\right) = -1 + \frac{\sqrt{2}}{2}. \quad \blacksquare$$

Antiderivative problems involving sine and cosine are prone to sign errors; watch your negatives carefully.

Use of parentheses can help distinguish between the negative from the FTC II formula and the negative from the antiderivative.

The name of the variable of integration does not affect whether FTC II can be used.

Example 8 *Find the area under the curve* $y = \frac{1}{t^2}$ *from* $t = 1$ *to* $t = 4$.

Solution Since $y = \frac{1}{t^2}$ is above the x-axis on the interval $[1, 4]$, the area under the curve is found by evaluating the definite integral $\int_1^4 \frac{1}{t^2}\, dt$.

Ans. to reading exercise 19
$\frac{242}{5} = 48.4$

Since $y = \frac{1}{t^2}$ is continuous on $[1, 4]$, we use FTC II:

$$\int_1^4 \frac{1}{t^2}\,dt = \int_1^4 t^{-2}\,dt$$
$$= \left.\frac{t^{-1}}{-1}\right|_1^4$$
$$= \left.-\frac{1}{t}\right|_1^4$$
$$= -\frac{1}{4} - (-1) = \frac{3}{4}.$$

Units are included in the answer to the problem because it is placed in a context that requires them.

The area under the curve is $\frac{3}{4}$ units^2. ■

CAUTION: MISTAKE TO AVOID
Because FTC II requires that the integrand be continuous on the interval of integration, we must check for continuity every time. Failure to do so may result in a wrong answer:

$$\int_{-1}^1 t^{-2}\,dt \neq \left.\frac{t^{-1}}{-1}\right|_{-1}^1 = \cdots = -2.$$

Because the curve (the graph of the integrand) is above the x-axis, the definite integral cannot be negative.

Example 9 *Evaluate* $\int_{-1}^1 \frac{1}{t^2}\,dt$.

Solution Since $y = \frac{1}{t^2}$ is *not* continuous on $[-1, 1]$ (it is undefined at $t = 0$ for reason of division by zero), we cannot use FTC II. *We cannot evaluate this integral at this time.* ■

In section 7.11 we learn how to handle some integrals containing a discontinuity.

Although FTC II is stated for the smaller number as the lower limit of integration and the larger number as the upper limit of integration, because of definition 5, it still works even when the smaller number is on top and the larger number is on the bottom.

Example 10 *Find* $\int_4^1 x^2\,dx$.

Solution Since $y = x^2$ is continuous on $[1, 4]$, we use FTC II:

$$\int_4^1 x^2\,dx = \left.\frac{x^3}{3}\right|_4^1$$
$$= \frac{1^3}{3} - \frac{4^3}{3} = -21.$$
■

Reading Exercise 20 *Evaluate* $\int_0^\pi \cos x\,dx$.

Exercises 4.5

1–10. Rapid response: does FTC II apply to the integral?

1. $\int_3^7 4x^2\,dx$
2. $\int_6^{17} (x^8 + x^3)\,dx$
3. $\int_{-5}^8 \frac{1}{x^2}\,dx$
4. $\int_{-5}^{-2} \frac{1}{x^2}\,dx$
5. $\int_{-3\pi}^{3\pi} \cos x\,dx$
6. $\int_0^7 \frac{1}{x^4}\,dx$
7. $\int_{-3\pi}^{3\pi} \tan x\,dx$
8. $\int_4^0 \sqrt{x}\,dx$
9. $\int_{-4}^{-1} \frac{1}{x-3}\,dx$
10. $\int_0^9 \sec x\,dx$

11–20. Evaluate the definite integral.

11. $\int_1^4 x^3\,dx$
12. $\int_4^9 6x\,dx$
13. $\int_1^2 (6x^2 - 2x + 5)\,dx$
14. $\int_1^3 (x^2 + 7x - 1)\,dx$
15. $\int_1^2 (2x^2 + x - 3)\,dx$
16. $\int_1^{10} (x^2 - x^{-2})\,dx$
17. $\int_0^{\pi/4} \sec^2 x\,dx$
18. $\int_0^{\pi/4} \sec x \tan x\,dx$
19. $\int_1^{-1} x^{4/3}\,dx$
20. $\int_8^3 (x+5)\,dx$

Ans. to reading exercise 20
0

21–30. Find the net area under the curve.

21. $y = x^2 + 1$ from $x = -3$ to $x = 1$
22. $y = x^3$ from $x = 1$ to $x = 2$
23. $y = x^3 - 6x + 3$ from $x = 1$ to $x = 4$
24. $y = 3x^2 + 4x - 2$ from $x = 1$ to $x = 3$
25. $y = 4 - x^2$ from $x = -2$ to $x = 2$
26. $y = \sqrt{x}$ from $x = 4$ to $x = 9$
27. $y = \frac{3}{x^2}$ from $x = -4$ to $x = -2$
28. $y = -2x$ from $x = 2$ to $x = 11$
29. $y = \sin x + \cos x$ from $x = 0$ to $x = \pi$
30. $y = 1 - 2\sin x$ from $x = \frac{\pi}{2}$ to $x = \pi$

31–40. Manipulate the integrand into a form to which the currently available antiderivative formulas and rules apply, then finish evaluating the integral.

31. $\int_0^1 (4x-7)^2\,dx$

32. $\int_1^3 2(4-x)^2\,dx$

33. $\int_1^4 \frac{x^5-7x^3+1}{x^2}\,dx$

34. $\int_0^{\pi/4} \cos^2 x \sec x\,dx$

35. $\int_0^5 \sqrt{4x}\,dx$

36. $\int_{0.5}^{0.8} \frac{10x^3-7x}{x^3}\,dx$

37. $\int_{-0.2}^{0.3} \tan x \cot x\,dx$

38. $\int_5^1 \frac{1}{4x^{-1}}\,dx$

39. $\int_2^1 \frac{x^2-2x-15}{x+3}\,dx$

40. $\int_0^2 (x+1)^3\,dx$

41–70. Evaluate the integral, if possible. If not possible, say why (either "FTC II does not apply" or "the available antiderivative formulas and rules are insufficient to determine an antiderivative").

41. $\int_9^{16} \frac{1}{\sqrt{x}}\,dx$

42. $\int_2^4 8x^3\,dx$

43. $\int_{-3}^7 \frac{2}{x^3}\,dx$

44. $\int_0^4 \sqrt{9x}\,dx$

45. $\int_2^{-1} -4x^3\,dx$

46. $\int_6^0 (1+x^2)\,dx$

47. $\int_{-4}^{-4} \frac{x^2-7}{\sqrt{x+19}}\,dx$

48. $\int_1^2 \frac{1}{x-3}\,dx$

49. $\int_0^{\pi} \sin x\,dx$

50. $\int_3^{-3} \frac{4}{x^4}\,dx$

51. $\int_0^{12} \sqrt{2x+1}\,dx$

52. $\int_0^6 (1+x)^2\,dx$

53. $\int_0^{\pi} \sec x \tan x\,dx$

54. $\int_{\pi/4}^{3\pi/4} \cot x \sin x\,dx$

55. $\int_2^5 \frac{1}{3x^{-2}}\,dx$

56. $\int_0^{\pi/3} \tan x\,dx$

57. $\int_1^4 (1-\sqrt{x})^2\,dx$

58. $\int_{-1}^5 \sqrt{x^2+2x+1}\,dx$

59. $\int_2^{\infty} (2x+5)^2\, dx$

60. $\int_1^2 \frac{-x^2-6}{x^2}\, dx$

61. $\int_{30}^{100} 11\, dx$

62. $\int_{-2\pi}^{0} \frac{1}{2}\cos x\, dx$

63. $\int_{\pi/3}^{2\pi/3} 8\sin x\, dx$

64. $\int_0^{\pi/3} \sec^2 x\, dx$

65. $\int_{\pi/3}^{2\pi/3} \sin 8x\, dx$

66. $\int_1^{-1} \frac{5x^2+10x-1}{5}\, dx$

67. $\int_{-1}^{1} (x+2x)^4\, dx$

68. $\int_{-1}^{5} x^{-2}\, dx$

69. $\int_{-4}^{-2} \sqrt{x^2}\, dx$

70. $\int_0^1 \frac{2}{(x-4)^{-2}}\, dx$

71–80. Differentiate the given function.

71. $y = \int_{-\sqrt{15}}^{x} \frac{7\cos t}{5t^4+8}\, dt$

72. $y = \int_{273}^{x} \frac{7+t^2}{5t^4+\sin t}\, dt$

73. $F(x) = \int_3^x \frac{\sin(t^4+t)}{5t+18}\, dt$

74. $F(x) = \int_{-2}^{x} \sqrt{\frac{t-7}{t+1}}\, dt$

75. $F(t) = \int_0^t (3x+5x^2+x^4)^7\, dx$

76. $F(x) = \int_{-0.002}^{x} (t+3)^{4\sin 5}\, dt$

77. $F(x) = \int_x^4 \tan t^4 \cos 3t\, dt$

78. $F(x) = \int_x^0 \frac{1+\frac{2}{t}}{3-\frac{1}{1+t}}\, dt$

79. $F(x) = \int_0^x \frac{1-t}{\sqrt{t^4+8t^2+22}}\, dt$

80. $G(y) = \int_3^y (4z^2-\sqrt{z})\, dz$

81. What interval $[a, b]$ can be used for the domain of the function $F(x) = \int_1^x \frac{1}{t^2 - 4}\, dt$?

82. What interval $[a, b]$ can be used for the domain of the function $F(x) = \int_1^x \frac{1}{t}\, dt$?

83. The function $F(x) = \int_3^{x^2} \sqrt{3 + \sin t}\, dt$ has an upper limit of integration of x^2 instead of "just plain x." Find $F'(x)$ using the chain rule. Hint: $F(x)$ can be written as the composition of two functions:

$$G(x) = \int_3^x \sqrt{3 + \sin t}\, dt \text{ and } H(x) = x^2.$$

84. Differentiate $F(x) = \int_1^{4x^2+6x} \cos^5(4t^2 + 7)\, dt$.

85. Find $F''(x)$ for $F(x) = \int_3^x \frac{1}{\sqrt{t^2 - 1}}\, dt$.

86. Find $F'''(x)$ for $F(x) = \int_1^x \sin t \cos t\, dt$.

Substitution for Indefinite Integrals

4.6

In section 4.1 we reversed a few derivative rules to give antiderivative rules. One of the rules we have not yet reversed is the chain rule, which is needed for derivatives involving "not just plain x," such as $\frac{d}{dx}\sin(4x-7)$ or $\frac{d}{dx}(x^2-9x+3)^4$. It is the purpose of this section to introduce a technique for finding antiderivatives (an *integration technique*) based on reversing the chain rule. Unfortunately, we will see that, although this helps with some instances of "not just plain x," such as $\int \sin(4x-7)\,dx$, it does not help with others, such as $\int (x^2-9x+3)^4\,dx$.

Because the chain rule is often not remembered as a formula, a review of its formula is in order. One form is

$$\frac{d}{dx}f(g(x)) = f'(g(x))g'(x).$$

For $\frac{d}{dx}\sin(4x-7)$, we designate $f(x) = \sin x$ and $g(x) = 4x-7$. Then, $f'(x) = \cos x$, $g'(x) = 4$, and

$$\begin{aligned}\frac{d}{dx}\sin(4x-7) &= \frac{d}{dx}f(g(x))\\ &= f'(g(x))g'(x)\\ &= \cos(4x-7)\cdot 4.\end{aligned}$$

Substitution: reversing the chain rule

The derivative calculation

$$\frac{d}{dx}\cos(x^2+3) = -\sin(x^2+3)\cdot 2x = -2x\sin(x^2+3)$$

can be reversed to give the antiderivative result

$$\int -2x\sin(x^2+3)\,dx = \cos(x^2+3) + C.$$

However, if we start from scratch given the integral

$$\int -2x\sin(x^2+3)\,dx,$$

how do we proceed to find the antiderivative if we do not already know it? The integrand contains both a product ($-2x$ times $\sin(x^2+3)$) and a composition (sine of "not just plain x"). We need a clear strategy for how to handle such integrands.

To this end, recall the chain rule formula:

$$\frac{d}{dx}f(g(x)) = f'(g(x))\cdot g'(x).$$

Suppose F is an antiderivative of f—that is, $F'(x) = f(x)$. We can rewrite the chain rule using F as

$$\frac{d}{dx}F(g(x)) = f(g(x)) \cdot g'(x).$$

Reversing the derivative calculation yields the antiderivative result

$$\int f(g(x)) \cdot g'(x)\, dx = F(g(x)) + C.$$

Next we make the substitution

Recall that the derivative is $\frac{du}{dx} = g'(x)$, but the differential rearranges this expression to give $du = g'(x)\, dx$.

$$u = g(x)$$
$$du = g'(x)\, dx$$

into the antiderivative formula:

$$\int f(u)\, du = F(u) + C.$$

This integral makes sense given that F is an antiderivative of f. It also points the way to our procedure, for it gives us an easier integral to evaluate. The method of *substitution* takes an integral $\int f(g(x))g'(x)\, dx$ involving "not just plain x" (the expression $g(x)$), ❶ makes the previous substitution to obtain an integral $\int f(u)\, du$ with "just plain u," ❷ finds its (most general) antiderivative $F(u) + C$ using previously available antiderivative rules and formulas, and then ❸ reverses the substitution, giving the antiderivative we are seeking: $F(g(x)) + C$.

Example 1 *Evaluate* $\int 3x^2(x^3 + 7)^4\, dx$.

We also do not have an antiderivative product rule, but the first barrier to tackle is always "not just plain x."

Solution Because we have "not just plain x" to a power, the antiderivative power rule does not apply and we need to try the method of substitution.

The "not just plain x" part is often in parentheses to a power, under a square root, inside a trig function, and so on, and is recognized in the same manner as recognizing when the chain rule is required for differentiation.

❶ The key to making a helpful substitution is to let the "not just plain x" part be u. Here, this is the expression inside the parentheses:

$$\mathbf{u = x^3 + 7}$$

(use of bold and gray explained shortly). Next we must calculate the differential du:

$$du = 3x^2\, dx.$$

The use of bold and gray helps illustrate how the substitution is carried out;

$$\int 3x^2(\mathbf{x^3+7})^4\,dx$$

becomes

$$\int (\mathbf{u})^4\,du.$$

Using $du = 3x^2\,dx$ explains the portion of the substitution in gray, and using $u = x^3 + 7$ explains the portion of the substitution in bold. Everything else stays the same. Notice that the resulting integral is much easier than the one we started with!

A successful substitution switches completely from the old variable (in this case, x) to the new variable (in this case, u).

❷ Next we integrate (find an antiderivative):

$$\int u^4\,du = \frac{u^5}{5} + C.$$

❸ Last, we return to the original variable by using the substitution formula $u = x^3 + 7$:

$$= \frac{(x^3+7)^5}{5} + C.$$

Recall that we can always check our work by taking the derivative:

$$\frac{d}{dx}\frac{1}{5}(x^3+7)^5 = \frac{1}{5}\cdot 5(x^3+7)^4\cdot 3x^2$$
$$= 3x^2(x^3+7)^4,$$

which is the original integrand as required.

We conclude that

$$\int 3x^2(x^3+7)^4\,dx = \frac{(x^3+7)^5}{5} + C. \quad ■$$

Example 2 *Evaluate* $\int -\cos^3 x \sin x\,dx$.

Solution It is often helpful to rewrite powers of trig functions using parentheses to help us interpret the integrand correctly. We want to evaluate

$$\int -(\cos x)^3 \sin x\,dx.$$

Now it is apparent that we have "not just plain x" to a power, so we use substitution and let what is inside the parentheses be u:

We need the chain rule to differentiate $(\cos x)^3$, so we need substitution to antidifferentiate it.

$$\mathbf{u = \cos x}.$$

Next we calculate the differential du:

$$du = -\sin x\,dx.$$

Step ❶ is finished in the first line, step ❷ is in the second line, and step ❸ is in the third line.

Then,

$$\int -(\mathbf{cos\,x})^3 \sin x\,dx = \int (\boldsymbol{u})^3\,du$$
$$= \frac{u^4}{4} + C$$
$$= \frac{\cos^4 x}{4} + C. \quad \blacksquare$$

Reading Exercise 21 *Evaluate* $\int 4x^3(x^4+7)^4\,dx$.

More substitution examples

The next example explores what we can do if we run into a snag when making the substitution.

Example 3 *Calculate* $\int s^3(s^4+7)^{12}\,ds$.

Solution Once again we have an integrand with "not just plain s" to a power, so we try substitution and let u be the "not just plain s" part:

$$\mathbf{u} = \mathbf{s^4+7}.$$

CAUTION: MISTAKE TO AVOID
The expression for du must be calculated from u using the derivative. Do not merely look for whatever is "left over" in the integrand.

Next we calculate the differential du:

$$du = 4s^3\,ds.$$

Now we have a problem. The substitution requires replacing $4s^3\,ds$ with du, but the 4 is missing from the integrand in $\int s^3(s^4+7)^{12}\,ds$. However, we can always multiply any expression by one without changing its value, so we can multiply by $4\cdot\frac{1}{4}$ to obtain

$$\int 4\cdot\frac{1}{4}\cdot s^3(s^4+7)^{12}\,ds.$$

Now everything we need for the substitution is present and we can move forward:

The expression in gray is replaced with du, the expression in bold is replaced with u, and everything not colored gray or bold stays the same.

$$\int 4\cdot\frac{1}{4}\cdot s^3(\mathbf{s^4+7})^{12}\,ds. = \int \frac{1}{4}(\boldsymbol{u})^{12}\,du$$
$$= \frac{1}{4}\cdot\frac{u^{13}}{13} + C$$
$$= \frac{(s^4+7)^{13}}{52} + C. \quad \blacksquare$$

As long as we are off only by a constant factor in our *du*, then the trick of example 3 applies (multiply and divide by the missing constant factor) and the substitution works.

Reading Exercise 22 *Evaluate* $\int x^4(x^5+1)^7\,dx$.

The first three examples featured "not just plain *x*" to a power. Just as with the chain rule, other circumstances requiring substitution include an integrand with a trig function of "not just plain *x*."

We have an antiderivative rule for sine:

$$\int \sin x\,dx = -\cos x + C.$$

This rule is for "just plain *x*," not for any other expression of *x*. It does not apply directly to $\int \sin 4x\,dx$, $\int \sin\sqrt{x}\,dx$, or similar expressions; substitution (and possibly other techniques as well) are required.

Example 4 *Determine* $\displaystyle\int \frac{\sin\left(\frac{1}{x}\right)}{x^2}\,dx$.

Solution The integrand features a trig function of "not just plain *x*," so we use substitution and let *u* be the "not just plain *x*" part:

$$\mathbf{u} = \frac{\mathbf{1}}{\mathbf{x}} = x^{-1}.$$

CAUTION: MISTAKE TO AVOID
We do not have an antiderivative quotient rule. Finding the antiderivative of the numerator and the denominator separately is incorrect.

We then calculate the differential *du*:

$$du = -x^{-2}\,dx = -\frac{1}{x^2}\,dx.$$

Because we need a negative to substitute *du* that is not present in the integrand, we use a variant of the same trick as in example 3:

Ans. to reading exercise 21
$\frac{(x^4+7)^5}{5} + C$

$$\begin{aligned}\int \frac{\sin\left(\frac{1}{x}\right)}{x^2}\,dx &= \int --\frac{\sin\left(\frac{1}{x}\right)}{x^2}\,dx \\ &= \int -\sin \mathbf{u}\,du \\ &= \cos u + C \\ &= \cos\frac{1}{x} + C.\end{aligned}$$ ■

The expression in gray is replaced with *du*, the expression in bold is replaced with *u*, and everything not colored gray or bold stays the same.

"Not just plain *x*" under a square root also calls for the use of substitution.

Example 5 *Evaluate* $\int x\sqrt{x^2-3}\,dx$.

Solution The integrand contains a square root of "not just plain *x*," so we use substitution and let *u* be the "not just plain *x*" part:

$$u = x^2 - 3.$$

Examples 1, 2, 3, and 5 all have integrands containing products, but this is not how we recognize the need for substitution. Substitution is required when "not just plain *x*" is encountered where the relevant antiderivative formula requires "just plain *x*." In other words, substitution is required when an integrand contains a composition of functions.

As always, we calculate the differential du:

$$du = 2x\,dx.$$

CAUTION: MISTAKE TO AVOID
Forgetting the dx when calculating du (or being lazy and skipping it) is the root cause of many errors. For those of you who wish to maximize performance, you must be diligent in using the correct notation.

Because du contains a 2, we need a 2 that is not present in the integrand. We multiply and divide the integrand by 2 and then proceed with the substitution:

Tip: a constant that is not used as part of du can be moved in front of the integral by use of the antiderivative constant multiple rule.

$$\begin{aligned}\int x\sqrt{x^2-3}\,dx &= \frac{1}{2}\int 2x\sqrt{x^2-3}\,dx \\ &= \frac{1}{2}\int \sqrt{u}\,du \\ &= \frac{1}{2}\cdot\frac{u^{3/2}}{3/2}+C = \frac{u^{3/2}}{3}+C \\ &= \frac{(x^2-3)^{3/2}}{3}+C. \qquad \blacksquare\end{aligned}$$

Ans. to reading exercise 22
$\frac{(x^5+1)^8}{40}+C$

Integrands requiring substitution do not always have products or quotients. Remember, we recognize the need for substitution by the presence of a composition, the presence of "not just plain x" when the relevant antiderivative formula contains "just plain x."

Example 6 *Evaluate* $\int \sin\left(3z-\frac{\pi}{2}\right)\,dz$.

Solution We have a trig function of "not just plain z," so we use substitution:

$$u = 3z - \frac{\pi}{2}$$

$$du = 3\,dz.$$

Tip: never refer to the integrand when determining du; du is **calculated**, not **chosen.**

We need a 3 for du, so we multiply and divide the integrand by 3 and proceed with the substitution:

$$\begin{aligned}\int \sin\left(3z-\frac{\pi}{2}\right)\,dz &= \frac{1}{3}\int 3\sin\left(3z-\frac{\pi}{2}\right)\,dz \\ &= \frac{1}{3}\int \sin u\,du \\ &= -\frac{1}{3}\cos u + C \\ &= -\frac{1}{3}\cos\left(3z-\frac{\pi}{2}\right)+C. \qquad \blacksquare\end{aligned}$$

Don't forget to return to the original variable. Leaving the answer in terms of the substitution variable u is incorrect.

Reading Exercise 23 *Evaluate* $\int \cos(4x)\,dx$.

Substitutions sometimes fail

Sometimes we can recognize that substitution appears to be required, yet substitution is not successful. For instance, consider

$$\int \sin x^2\,dx.$$

The integrand is a trig function of "not just plain x," so we try substitution, letting u be the "not just plain x" part:

$$u = x^2$$

$$du = 2x\,dx.$$

In order for the substitution to work, we need to substitute du for $2x\,dx$. The problem is that there is no $2x$ in the integrand. If what was missing was just a constant, such as in examples 3–6, then we could adjust by multiplying and dividing by that constant. Unfortunately, $2x$ is not just a constant; the variable x is also missing, causing the substitution to fail.

Why do we stop there and say that the substitution fails? We could try to go forward using the same trick:

$$\int \sin x^2\,dx = \int 2x \cdot \frac{1}{2x} \cdot \sin x^2\,dx.$$

It is occasionally possible to move forward in this manner and complete a substitution, but we don't explore such examples until chapter 7.

We then substitute u for x^2 (bold) and du for $2x\,dx$ (gray), leaving all else unchanged:

$$\int 2x \cdot \frac{1}{2x} \cdot \sin \mathbf{x^2}\,dx = \int \frac{1}{2x} \cdot \sin \boldsymbol{u}\,du.$$

CAUTION: MISTAKE TO AVOID
When using substitution, never write an integral expression containing both variables. Such an expression is not in a form that can be integrated.

This is an *incomplete substitution* because of the continued presence of the variable x, and the substitution still fails. Substitution requires us to switch completely to a function of the new variable u. In the development of the method of substitution, the switch is made to $\int f(u)\,du$, which no longer contains any reference to the variable x.

If substitution doesn't work, then what must we do to find the requested antiderivative? Unfortunately, there are times when an antiderivative involves a function that is not in our list of "elementary" functions. The solution to $\int \sin x^2\,dx$ involves a function you have never studied.

Chapters 5 and 7 include the study of additional functions and additional integration techniques that can be used when substitution is not appropriate. But, it is not possible to expand our list to enough functions that all antiderivatives can be calculated involving only functions from the list.

To summarize, if the substitution formula for du contains a variable expression that is not present in the integrand, then the substitution fails.

Example 7 *Does the substitution succeed or fail? (a)* $\int x^2 \cos x^3\, dx$, $u = x^3$, *(b)* $\int \sqrt{x^3 + 7x}\, dx$, $u = x^3 + 7x$.

Details of the substitution:

$$\int x^2 \cos x^3\, dx = \frac{1}{3}\int 3x^2 \cos x^3\, dx$$
$$= \frac{1}{3}\int \cos u\, du.$$

Solution (a) Substitution is needed because the integrand has a trig function of "not just plain x," and u is the "not just plain x" part, so the setup is correct. We still need to calculate du for the substitution:

$$du = 3x^2\, dx.$$

We only need to adjust for a missing constant (3), so the substitution succeeds.

(b) Substitution is needed because the integrand contains a square root of "not just plain x," and u is the "not just plain x" part, so the setup is correct. Calculate du:

$$du = (3x^2 + 7)\, dx.$$

Because the variable expression $3x^2 + 7$ is not present in the integrand, the substitution fails. ■

The solution to the integral in part (b) involves a function you have not studied.

Ans. to reading exercise 23
$\frac{1}{4}\sin(4x) + C$

Another way for a substitution to fail is if the integrand contains additional expressions that cannot be converted successfully to expressions in the new variable u.

Example 8 *Does the substitution succeed or fail? (a)* $\int \frac{4x^3 + 2x}{x+1} \cos(x^4 + x^2)\, dx$, $u = x^4 + x^2$, *(b)* $\int 2x(x^2 + 4)\sin(x^2 + 4)\, dx$, $u = x^2 + 4$.

Solution (a) Substitution is needed because the integrand contains a trig function of "not just plain x," and because u is the "not just plain x" part, the setup is correct. Next we calculate du,

$$du = (4x^3 + 2x)\, dx,$$

which is present in the integrand. We proceed with the substitution:

$$\int \frac{4x^3 + 2x}{x+1} \cos(\mathbf{x}^4 + \mathbf{x}^2)\, dx = \int \frac{1}{x+1} \cos(\boldsymbol{u})\, du.$$

If an integral cannot be expressed only in terms of the new variable u, then the substitution fails.

The problem is that there is no match for $x + 1$ involving either u or du, so the substitution fails.

(b) Substitution is needed because the integrand contains a trig function of "not just plain x," and because u is the "not just plain x" part, the setup is correct. Then,

$$du = 2x\, dx$$

and

$$\int 2x(\mathbf{x}^2+4)\sin(\mathbf{x}^2+4)\,dx = \int \boldsymbol{u}\sin\boldsymbol{u}\,du.$$

Notice that although u was chosen to be used for the "not just plain x" part inside the trig function, it can also be used elsewhere in the expression when completing the substitution.

The substitution succeeds. ∎

Even when a substitution succeeds, we are not guaranteed the ability to evaluate the resulting integral. For example 8 part (b), the resulting integral is $\int u \sin u\, du$, which contains a product and is not in a form we can integrate using the available antiderivative rules and formulas. We will be able to evaluate this integral later, in chapter 7, but there are always situations in which the resulting integral is still not one we can evaluate.

Let's end this section with two substitutions that should never be used. The substitution $u = k$ for a constant k (for example, $u = 4$ or $u = 17$) gives $du = 0\,dx$ and is never successful. The substitution $u = x$ gives $du = dx$ and simply changes the name of the variable from x to u without any other changes. This substitution is always successful, but never helpful.

TWO SUBSTITUTIONS TO AVOID

(1) $u = k$ for a constant k
(2) $u = x$

Exercises 4.6

1–10. Rapid response: is substitution needed?

Some of the integrals in exercises 1–10 require integration techniques we have not yet studied. You are not asked to evaluate the integral, just to determine whether substitution is needed.

1. $\int 3x^2\cos(x^3+12)\,dx$

2. $\int \sqrt{4x-9}\,dx$

3. $\int x^2\cos x\,dx$

4. $\int (6x+1)\sqrt{3x^2+x}\,dx$

5. $\int \sec x \tan x\,dx$

6. $\int \frac{x+4}{x-1}\,dx$

7. $\int (4x-7)^8\,dx$

8. $\int x^2 \sin 6\,dx$

9. $\int \frac{1}{\sqrt{6x-11}}\,dx$

10. $\int \tan \pi x\,dx$

11–20. Does the substitution succeed or fail?

11. $\int \sin(7x-92)\,dx,\ u = 7x-92$

12. $\int \sin(7x^2-92)\,dx,\ u = 7x^2-92$

13. $\int \sqrt{6x^3+12}\,dx,\ u = 6x^3+12$

14. $\int \sqrt{6x+12}\,dx,\ u = 6x+12$

15. $\int \sec 5x \tan 5x\,dx,\ u = 5x$

16. $\int 5x \sec 5x \tan 5x\,dx,\ u = 5x$

17. $\int x^3(x^2-17x+1)^{14}\,dx,\ u = x^2-17x+1$

18. $\int \cos x\sqrt{\sin x}\,dx,\ u = \sin x$

19. $\int 2x\sqrt{\sin x^2}\cos x^2\,dx$, $u = \sin x^2$
20. $\int \sqrt{x}\cos(7x+5)\,dx$, $u = 7x+5$

Not all of these integrals require substitution.

21–44. Evaluate the integral.

21. $\int x^3 \sin(x^4+12)\,dx$
22. $\int x\cos x^2\,dx$
23. $\int x^2(x^3+5)^{14}\,dx$
24. $\int (x^2+\cos x)\,dx$
25. $\int \cos(2x+5)\,dx$
26. $\int x^2\sin(x^3+5)\,dx$
27. $\int (2x+5\cos x)\,dx$
28. $\int \sqrt{9x-10}\,dx$
29. $\int \sec 7x\tan 7x\,dx$
30. $\int \sin(3x-4)\,dx$
31. $\int \frac{3}{\sqrt{2x+7}}\,dx$
32. $\int \sec^2\left(\pi x+\frac{\pi}{2}\right)\,dx$
33. $\int 3x\cos(x^2+1)\,dx$
34. $\int \frac{x}{\sqrt{x^2+5}}\,dx$
35. $\int (3x^2+8x)\sqrt{x^3+4x^2}\,dx$
36. $\int (x^2+4)^2\,dx$
37. $\int \sqrt{x^3}\,dx$
38. $\int (3x+7)^{10}\,dx$
39. $\int \frac{-\sin x}{\cos^2 x}\,dx$
40. $\int \frac{4x}{(x^2-1)^3}\,dx$
41. $\int (6x+3)\sin(x^2+x)\,dx$
42. $\int \frac{\sec^2 x}{\tan^2 x}\,dx$
43. $\int \left(x^2-3\sqrt{x}+x^2\sqrt{x^3+4}\right)\,dx$
44. $\int (15x^2+40x)(x^3+4x^2+1)^{22}\,dx$

45–58. Evaluate the integral or state that it cannot be done with the available antiderivative formulas, rules, and techniques.

45. $\int x^2\sin x^3\,dx$
46. $\int x\cos x^2\,dx$
47. $\int \sqrt{y}(y^2-8y+2)\,dy$
48. $\int \frac{\sqrt{z^3-7z}}{3z^2-7}\,dz$
49. $\int \frac{1}{\sqrt{9-x^2}}\,dx$
50. $\int \tan^4 x\sec^2 x\,dx$
51. $\int \csc^2(6z-15)\,dz$
52. $\int \csc(\pi r)\cot(\pi r)\,dr$
53. $\int s^4\sin s^3\,ds$
54. $\int \frac{14}{x-5}\,dx$
55. $\int (x-\sin x)(1-\cos x)\,dx$
56. $\int y^5\sqrt[4]{y^6+1}\,dy$
57. $\int (4x-3)\cos^2(2x^2-3x)\,dx$
58. $\displaystyle\int \frac{(3x^2-14x)\sqrt{x^3-7x^2}}{(x^3-7x^2)^3}\,dx$

59. Sometimes an algebraic manipulation along with a substitution is helpful. Evaluate $\int 2x^3(x^2+3)^{15}\,dx$ by using the substitution $u = x^2+3$ and rewriting $2x^3$ as $2x\cdot x^2$, then replacing the x^2 with $u-3$.

60. Use the technique of exercise 59 to evaluate $\displaystyle\int \frac{x^5}{\sqrt{x^3-9}}\,dx$.

Substitution for Definite Integrals

4.7

The fundamental theorem of calculus, part II allows us to evaluate definite integrals by finding antiderivatives. This crucial step in the process is often the majority of the work; what follows is usually just arithmetic. Because substitution helps us find antiderivatives, it helps us evaluate definite integrals as well.

Substitution with definite integrals: method 1

If a definite integral is complicated enough, then ❶ the antiderivative step can be written as a separate problem, to be followed by ❷ application of the fundamental theorem.

Keep this in mind as we learn additional techniques of integration in chapter 7.

Example 1 *Evaluate* $\int_0^{\pi/4} \sin\left(2z - \frac{\pi}{2}\right)\, dz$.

Solution Recall that we should always check for continuity before attempting to use FTC II, and since $\sin(2z - \frac{\pi}{2})$ is continuous everywhere, we may proceed. ❶ We find an antiderivative first; that is, we calculate $\int \sin\left(2z - \frac{\pi}{2}\right)\, dz$. Because we have sine of "not just plain z," substitution is called for:

$$\begin{aligned} u &= 2z - \frac{\pi}{2} \\ du &= 2\, dz. \end{aligned}$$

Then,

$$\begin{aligned} \int \sin\left(2z - \frac{\pi}{2}\right) dz &= \frac{1}{2} \int 2 \sin\left(2z - \frac{\pi}{2}\right) dz \\ &= \frac{1}{2} \int \sin u\, du \\ &= -\frac{1}{2} \cos u + C \\ &= -\frac{1}{2} \cos\left(2z - \frac{\pi}{2}\right) + C. \end{aligned}$$

Although we only need one antiderivative to use FTC II, when we write an indefinite integral $\int f(x)\, dx$, the answer is the most general antiderivative and therefore $+\, C$ is required.

❷ Next we use FTC II to evaluate the definite integral using an antiderivative calculated in step ❶:

As discussed previously, any one antiderivative will do when applying FTC II; there is no need to use the + C written in the answer to step ❶.

$$\begin{aligned}\int_0^{\pi/4} \sin\left(2z - \frac{\pi}{2}\right)\, dz &= -\frac{1}{2}\cos\left(2z - \frac{\pi}{2}\right)\Bigg|_0^{\pi/4} \\ &= -\frac{1}{2}\cos 0 - \left(-\frac{1}{2}\cos\left(-\frac{\pi}{2}\right)\right) \\ &= \cdots = -\frac{1}{2}.\end{aligned}$$

■

Because method 1 consists solely of combining two previous procedures, it is not difficult to learn.

Reading Exercise 24 *Use method 1 to evaluate* $\int_1^2 \sqrt{5x-1}\, dx$.

Substitution with definite integrals: method 2

Although method 1 is well-suited for evaluating any definite integral for which substitution is needed, an alternate procedure is used more often because of its relative efficiency. The only drawback is that more care must be taken with its notation.

Method 2 combines portions of the two steps of example 1 as follows. We begin by recognizing that substitution is required and write

$$u = 2z - \frac{\pi}{2}$$
$$du = 2\, dz.$$

The promised efficiency is accomplished by switching the limits of integration from values of the original variable (z) to values of the substitution variable (u):

In method 1, these calculations occur during evaluation. See the evaluation bar and the expression that precedes it.

$$z = \frac{\pi}{4}: \quad u = 2\cdot\frac{\pi}{4} - \frac{\pi}{2} = 0$$
$$z = 0: \quad u = 2\cdot 0 - \frac{\pi}{2} = -\frac{\pi}{2}.$$

When we then carry out the substitution, we also switch the limits of integration:

$$\int_0^{\pi/4} \sin\left(2z - \frac{\pi}{2}\right) dz = \frac{1}{2}\int_0^{\pi/4} 2\sin\left(2z - \frac{\pi}{2}\right) dz$$

$$= \frac{1}{2}\int_{-\pi/2}^{0} \sin u \, du.$$

The lower limit stays the lower limit and the upper limit stays the upper limit when switching variables, regardless of which number is smaller or larger.

The calculation then continues in the new variable as its own definite integral without needing to switch back to the original variable, making the calculations shorter:

$$= -\frac{1}{2}\cos u\Big|_{-\pi/2}^{0}$$

$$= -\frac{1}{2}\cos 0 - \left(-\frac{1}{2}\cos\left(-\frac{\pi}{2}\right)\right)$$

$$= \cdots = -\frac{1}{2}.$$

Comparing these calculations to those of example 1 helps us see how all the necessary steps remain, but unnecessary steps are discarded. However, care must be taken always to remember to switch the limits of integration to values of the new variable.

Example 2 *Find* $\displaystyle\int_1^4 \frac{1}{2\sqrt{y}(1+\sqrt{y})^2}\, dy.$

Solution We first note that the integrand is continuous on $[1, 4]$, therefore FTC II applies and we may proceed.

Details of continuity: $\sqrt{y}$ requires $y \geq 0$, and avoiding division by zero requires $y \neq 0$ (notice that the equation $1 + \sqrt{y} = 0$ has no solutions). The domain is $(0, \infty)$, which includes $[1, 4]$.

Because we have "not just plain y" to a power, substitution is needed:

$$u = 1 + \sqrt{y}$$

$$du = \frac{1}{2\sqrt{y}}\, dy.$$

Details of the derivative:

$$\frac{d}{dy}(1 + \sqrt{y}) = \frac{d}{dy}(1 + y^{1/2})$$

$$= \frac{1}{2}y^{-1/2} = \frac{1}{2\sqrt{y}}.$$

Using method 2, we also calculate the u values of the limits of integration:

$$y = 4: \quad u = 1 + \sqrt{4} = 3$$

$$y = 1: \quad u = 1 + \sqrt{1} = 2.$$

Tip: write the calculation for the upper limit on top and the lower limit on bottom to avoid confusion.

We then perform the substitution (including the change of the limits of integration) and calculate the value of the definite integral using the new variable:

In line 1, the substitution is made, including switching from values of the original variable y (the limits of integration 1 and 4) to values of the substitution variable u (the limits of integration 2 and 3). The new definite integral is then calculated easily.

Details of the antiderivative:

$$\int \frac{1}{u^2}\,du = \int u^{-2}\,du = \frac{u^{-1}}{-1} + C = -\frac{1}{u} + C.$$

$$\int_1^4 \frac{1}{2\sqrt{y}(1+\sqrt{y})^2}\,dy = \int_2^3 \frac{1}{u^2}\,du = -\frac{1}{u}\bigg|_2^3 = -\frac{1}{3} - \left(-\frac{1}{2}\right) = \frac{1}{6}. \quad \blacksquare$$

Although the correct answer can be obtained by keeping the original limits of integration but switching back to the original variable before evaluating the definite integral, the resulting work is still incorrect because it says that two unequal expressions are equal. Said colloquially, two wrongs don't make a right, they make two wrongs.

The most common mistake made when using method 2 is to fail to switch the limits of integration when making the substitution. If the original limits of integration are kept, then the resulting definite integral does not yield the correct answer, as illustrated using the integral from example 2:

$$\int_1^4 \frac{1}{u^2}\,du = -\frac{1}{u}\bigg|_1^4 = -\frac{1}{4} - (-1) = \frac{3}{4} \neq \frac{1}{6}.$$

Another common mistake when evaluating definite integrals using substitution is to combine methods 1 and 2 into something that is neither method and gives the wrong answer. To avoid such errors, it is best always to use the same method, either method 1 or method 2. After this section, this text uses method 2 when evaluating definite integrals using substitution.

Example 3 *Evaluate* $\displaystyle\int_0^2 \frac{1}{(7-3x)^2}\,dx.$

The integrand is discontinuous at $x = \frac{7}{3}$, which thankfully is not in our interval of integration.

Solution Because the integrand is continuous on $[0, 2]$, FTC II applies and we may proceed.

Because the integrand involves "not just plain x" to a power, substitution is needed:

$$u = 7 - 3x$$
$$du = -3\,dx.$$

We calculate the new limits of integration:

$$x = 2: \quad u = 7 - 3(2) = 1$$
$$x = 0: \quad u = 7 - 3(0) = 7.$$

Then,

$$\begin{aligned}\int_0^2 \frac{1}{(7-3x)^2}\,dx &= -\frac{1}{3}\int_0^2 -3\frac{1}{(7-3x)^2}\,dx \\ &= -\frac{1}{3}\int_7^1 \frac{1}{u^2}\,du \\ &= -\frac{1}{3}\left(-\frac{1}{u}\right)\bigg|_7^1 = \frac{1}{3u}\bigg|_7^1 \\ &= \frac{1}{3} - \frac{1}{21} = \frac{6}{21} = \frac{2}{7}.\end{aligned}$$ ■

In line 1, we adjust the constants to set up the substitution; the limits of integration do not change because the variable is still x. In line 2, the substitution makes the switch to the variable u and therefore the limits of integration change to values of u.

After substitution, the integrands in examples 2 and 3 are identical—namely, $\frac{1}{u^2}$—even though the original integrands seem quite different at first glance.

Because so many learners of calculus make the mistake of forgetting to change the limits of integration to values of the substitution variable u, the warning given after example 2 is repeated. What happens if we don't switch the limits of integration to values of u in example 3? We have

$$-\frac{1}{3}\int_0^2 \frac{1}{u^2}\,du = \frac{1}{3u}\bigg|_0^2 = \frac{1}{6} - \frac{1}{0},$$

which is undefined!

Tip: if you come across an undefined expression when evaluating a definite integral after a substitution is made, check to see if you forgot to switch the limits of integration. Another possibility is that you forgot to check the original integrand for continuity.

In summary, there are three pieces to completing a substitution with definite integrals: (1) switch to du (in gray in the previous section), switch to u (in bold in the previous section), and (3) switch the limits of integration to values of u.

Reading Exercise 25 *Use method 2 to evaluate* $\int_0^4 2x\sqrt{x^2+9}\,dx$.

Proof of validity of method 2

To prove that switching the limits of integration to values of the substitution variable u works as indicated, return to the setup of the justification for substitution for indefinite integrals in the previous section, where F is an antiderivative of f and

$$\int f(g(x))\cdot g'(x)\,dx = F(g(x)) + C.$$

We made the substitution

$$u = g(x)$$
$$du = g'(x)\,dx$$

Ans. to reading exercise 24
Using substitution with $u = 5x - 1$,

$$\begin{aligned}\int \sqrt{5x-1}\,dx &= \frac{1}{5}\int 5\sqrt{5x-1}\,dx \\ &= \frac{1}{5}\int u^{1/2}\,du = \frac{1}{5}\frac{u^{3/2}}{3/2} + C \\ &= \frac{2}{15}(5x-1)^{3/2} + C.\end{aligned}$$

Then,

$$\begin{aligned}\int_1^2 \sqrt{5x-1}\,dx &= \frac{2}{15}(5x-1)^{3/2}\bigg|_1^2 \\ &= \cdots = \frac{38}{15}.\end{aligned}$$

and saw that

$$\int f(u)\,du = F(u) + C.$$

Now suppose that we wish to evaluate

$$\int_a^b f(g(x)) \cdot g'(x)\,dx.$$

As long as the integrand is continuous on $[a, b]$, FTC II applies and

Here we use the fact that $F(g(x))$ is an antiderivative of $f(g(x))g'(x)$.

$$\int_a^b f(g(x)) \cdot g'(x)\,dx = F(g(x))\Big|_a^b$$
$$= F(g(b)) - F(g(a)).$$

What if, instead, we switch to values of u in the manner of method 2 and then evaluate that definite integral? We have

$$x = b: \quad u = g(b)$$
$$x = a: \quad u = g(a),$$

and then

Here we use the fact that $F(u)$ is an antiderivative of $f(u)$.

$$\int_{g(a)}^{g(b)} f(u)\,du = F(u)\Big|_{g(a)}^{g(b)}$$
$$= F(g(b)) - F(g(a)),$$

which is the correct answer per the previous calculation. Therefore, method 2 always yields the correct answer.

Exercises 4.7

1–10. Evaluate the definite integral using (a) method 1 and (b) method 2.

1. $\int_0^2 6x^2(2x^3 - 1)^4\,dx$
2. $\int_4^5 2x\sqrt{x^2 - 16}\,dx$
3. $\int_0^{\pi/2} 5\cos(5x - \pi)\,dx$
4. $\int_0^1 (r^3 + 1)(r^4 + 4r)^{11}\,dr$
5. $\int_0^1 \sqrt{3y + 1}\,dy$
6. $\int_0^{\pi/2} \sin(2\theta - \pi)\,d\theta$
7. $\int_{-\pi}^{\pi} (\sin x)^7 \cos x\,dx$
8. $\int_1^0 (z^5 - 1)^3 z^4\,dz$
9. $\int_0^2 3t^2\sqrt{t^3 + 1}\,dt$
10. $\int_0^{\pi/4} \sec^2 6x\,dx$

11–40. Evaluate the definite integral, if possible. If not possible, say why (either "FTC II does not apply" or "the available antiderivative formulas, rules, and techniques are insufficient to determine an antiderivative").

Ans. to reading exercise 25
$\frac{196}{3} = 65 + \frac{1}{3}$
Some details: use $u = x^2 + 9$. Then,

$$\int_0^4 2x\sqrt{x^2+9}\,dx = \int_9^{25} \sqrt{u}\,du = \left.\frac{u^{3/2}}{3/2}\right|_9^{25} = \ldots.$$

11. $\int_0^1 2x(x^2+1)^7\,dx$
12. $\int_2^{10} \sqrt{2y+5}\,dy$
13. $\int_0^{\pi/2} \sqrt{\sin x}\cos x\,dx$
14. $\int_2^3 \frac{4}{(2x-7)^2}\,dx$
15. $\int_{-\pi/4}^{0} \frac{\sec x \tan x}{\sec^5 x}\,dx$
16. $\int_0^{\sqrt{\pi}} 4x^2\cos x\,dx$
17. $\int_1^4 (t^2-5)\sqrt{t}\,dt$
18. $\int_0^{\sqrt{\pi}} 4x\cos x^2\,dx$
19. $\int_{-\sqrt{\pi}}^{\sqrt{\pi}} \sin x^2\,dx$
20. $\int_0^5 \frac{3}{(2z-8)^4}\,dz$
21. $\int_0^{\pi} \frac{\cos s}{(2+\sin s)^3}\,ds$
22. $\int_5^6 \frac{3}{(2t-8)^4}\,dt$
23. $\int_1^2 (4x-3)^5\,dx$
24. $\int_0^{\pi/3} (1-\cos x)^3 \sin x\,dx$
25. $\int_0^{2\pi} \sec 4x \tan 4x\,dx$
26. $\int_1^4 \left(\sqrt{x} + \frac{1}{x^3}\right)\,dx$
27. $\int_{\pi/3}^{\pi} (\sin\theta - \cos\theta)\,d\theta$
28. $\int_0^{\pi^2} \frac{\cos\sqrt{r}}{\sqrt{r}}\,dr$
29. $\int_2^3 x(x^2-4)^{2/3}\,dx$
30. $\int_0^{\pi} \sec\frac{y}{4}\tan\frac{y}{4}\,dy$
31. $\int_0^1 \frac{(4+\sqrt{x})^9}{\sqrt{x}}\,dx$
32. $\int_0^{4\pi} -\csc^2 x\,dx$
33. $\int_1^4 \frac{(3+\sqrt{y})^5}{2\sqrt{y}}\,dy$
34. $\int_{-1}^{0} \sqrt[4]{3x+5}\,dx$
35. $\int_0^2 \sqrt{x^4+5x^2+3}\,dx$
36. $\int_0^{\pi/8} (4s+\cos 4s)\,ds$
37. $\int_{-1}^{0} \frac{1}{x^2+4x+4}\,dx$
38. $\int_1^2 x^3\sqrt{x^4+x}\,dx$
39. $\int_0^{\pi^2/4} (\cos\sqrt{y})^4 \cdot \frac{\sin\sqrt{y}}{\sqrt{y}}\,dy$
40. $\int_{\sqrt{\pi-1}}^{\sqrt{2\pi-1}} x\cos(x^4+2x^2+1)\,dx$

41–48. Find the net area under the curve.

41. $y = 5\sqrt{5x+4}$ from $x = 0$ to $x = 9$
42. $y = (2x-1)^7$ from $x = 1$ to $x = 2$
43. $y = \sin(x-\pi)$ from $x = \pi$ to $x = 2\pi$
44. $y = x\cos 67$ from $x = 0$ to $x = 3$
45. $y = (3x-7)\sqrt{\pi+1}$ from $x = 2$ to $x = 3$
46. $y = x^2\sqrt{x^3+2}$ from $x = \sqrt[3]{2}$ to $x = \sqrt[3]{7}$
47. $y = (2x-3)(x^2-3x+1)^5$ from $x = 0$ to $x = 1$
48. $y = \sqrt{\tan x}\sec^2 x$ from $x = 0$ to $x = \frac{\pi}{4}$

Numerical Integration, Part I

4.8

When calculating the value of a definite integral, our first choice is to use the fundamental theorem of calculus, where we begin by finding an antiderivative of the integrand:

$$\int_1^5 x^2\,dx = \left.\frac{x^3}{3}\right|_1^5 = \cdots = \frac{124}{3}.$$

Next try $\displaystyle\int_1^2 \frac{\sin x}{x}\,dx$. There is no quotient rule for finding antiderivatives, and substitution does not help. How do we proceed if we cannot determine an antiderivative?

Exploring the options

An antiderivative for $y = \frac{\sin x}{x}$ cannot be found using the available antiderivative formulas, rules, and techniques. Our first thought might be that we should investigate additional antiderivative formulas, rules, and techniques to determine an antiderivative. There is nothing wrong with this thought, and much effort is expended in chapters 5 and 7 toward this end. But, expanding our formulas, rules, and techniques, and even our available list of types of functions has its limitations. Computer software can give an antiderivative for $y = \frac{\sin x}{x}$ involving a function you have not yet encountered, but it does not give an antiderivative for $y = \sqrt{x^2 - 4x^5 + x^7}$ in a form useful with the FTC.

The functions we have studied so far using calculus are part of the collection of *elementary functions,* which include any function that can be built from polynomial, trigonometric, exponential, logarithmic, and inverse trigonometric functions using the operations of addition, subtraction, multiplication, division, exponentiation (including extraction of roots), and composition. Expanding the list of available functions does not solve our problem; there are always functions with antiderivatives that are of a type not in the list.

Example 1 *Use technology to attempt to evaluate the indefinite integral $\int \frac{\sin x}{x}\,dx$ and state whether the result is an elementary function, a function of some other type, or no result is given.*

By "use technology," I mean to use a computer algebra system (CAS), a website, a mobile app, or similar software-based technology that is capable of performing the task.

Solution Mathematica gives the solution SinIntegral(x). This function is not of a type we have encountered previously (it is not an elementary function), so we answer "a function of some other type." ∎

Mathematica is a CAS.

Example 2 *Use technology to attempt to evaluate the indefinite integral $\int \sqrt{x^2 - 4x^5 + x^7}\, dx$ and state whether the result is an elementary function, a function of some other type, or no result is given.*

Solution Mathematica gives the solution $\int \sqrt{x^2 - 4x^5 + x^7}\, dx$, which is just a repeat of the command it was asked to execute. It did not give an antiderivative, so we answer "no result is given." ■

Example 3 *Use technology to attempt to evaluate the indefinite integral $\int x \sin x\, dx$ and state whether the result is an elementary function, a function of some other type, or no result is given.*

Some CASs give just one antiderivative rather than the most general antiderivative.

Solution Mathematica gives the solution $-x \cos x + \sin x$, which is an elementary function (it is a combination of polynomial and trig functions). ■

The search for antiderivatives was successful in examples 1 and 3, so we use the FTC to evaluate definite integrals for these functions. For instance,

$$\int_0^{\pi} x \sin x\, dx = (-x \cos x + \sin x)\Big|_0^{\pi} = \cdots = \pi$$

If desired, a decimal approximation to a quantity such as SinIntegral(2) − SinIntegral(1) can be obtained using the same technology that gave the antiderivative in the first place.

and

$$\int_1^2 \frac{\sin x}{x}\, dx = \text{SinIntegral}(x)\Big|_1^2 = \text{SinIntegral}(2) - \text{SinIntegral}(1).$$

The search for an antiderivative was not successful in example 2, even though we used technology. What do we do, then, if we wish to evaluate $\int_2^3 \sqrt{x^2 - 4x^5 + x^7}\, dx$?

The FTC is not the only approach we have used in this chapter. The definition of an integral used omega sums, but this is usually more difficult than using the FTC. If we cannot find an antiderivative for a function, then using omega sums is very likely not going to succeed either. Going back even further, we used finite sums (left-hand endpoints, right-hand endpoints, midpoints, and so on) to estimate areas. As long as we can evaluate a function, we can calculate finite sums, so that approach could work.

Trapezoid rule

We can determine this area easily using the FTC, as at the beginning of this section. This example is used to illustrate the principles involved in the trapezoid rule with simple calculations.

In section 4.2 we used finite sums to estimate the area under the curve $y = x^2$ from $x = 1$ to $x = 5$—that is, $\int_1^5 x^2\, dx$. In particular, rectangles were used. The horizontal tops of the bars did not follow the nonhorizontal curve, and for $n = 4$ rectangles neither left-hand endpoints nor right-hand endpoints gave very accurate results, as recalled by figure 1.

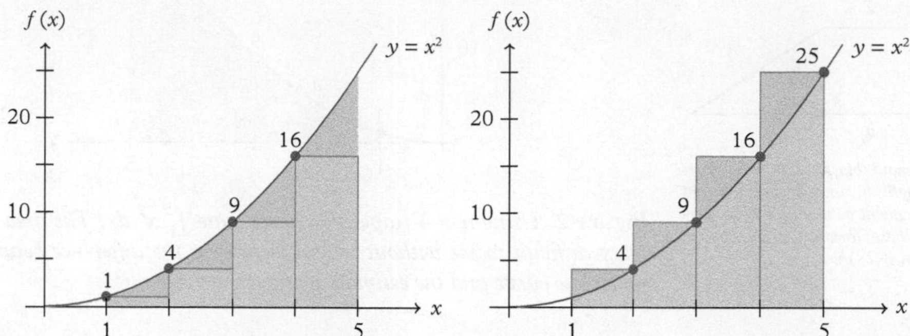

Figure 1 *Using $n = 4$ rectangles to estimate $\int_1^5 x^2\,dx$ using (left) left-hand endpoints and (right) right-hand endpoints. The area of the region that is shaded gray is the difference between the area under the curve and the estimate using rectangles*

The area estimate for left-hand endpoints was

$$1 \cdot 1 + 1 \cdot 4 + 1 \cdot 9 + 1 \cdot 16 = 30\,\text{units}^2$$

whereas the area estimate for right-hand endpoints was

$$1 \cdot 4 + 1 \cdot 9 + 1 \cdot 16 + 1 \cdot 25 = 54\,\text{units}^2.$$

Instead of requiring the tops of the bars to be horizontal, what if we drew a straight line from the point on the curve at the left-hand endpoint to the point on the curve at the right-hand endpoint? The resulting bars are trapezoids instead of rectangles, and the result is shown in figure 2. This clearly gives a much better estimate of the value of the integral than either the left- or right-hand estimate.

Our next task is to determine the total area of the four trapezoids pictured in figure 2. Recall that the area of a trapezoid (see figure 3) is $\frac{1}{2}(b_1 + b_2)h$.

The four approximating trapezoids in figure 2 are oriented differently than in figure 3; the parallel sides are vertical instead of horizontal. Figure 4 illustrates the calculation of the area for one of the trapezoids. The "height" h of the trapezoid is $\Delta x = 1$ (what we were previously calling the "width" for rectangles), whereas the lengths of the bases are the values of the function at the left-hand endpoint and the right-hand endpoint, and have lengths $b_2 = 9$ and $b_1 = 16$.

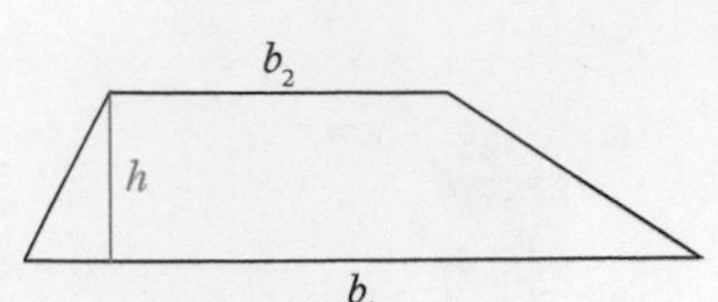

Figure 3 *A trapezoid (black), with bases (parallel sides) of length b_1 and b_2, and height (length of a line segment between the bases perpendicular to both bases (orange)) h. The area of a trapezoid is $\frac{1}{2}(b_1 + b_2)h$*

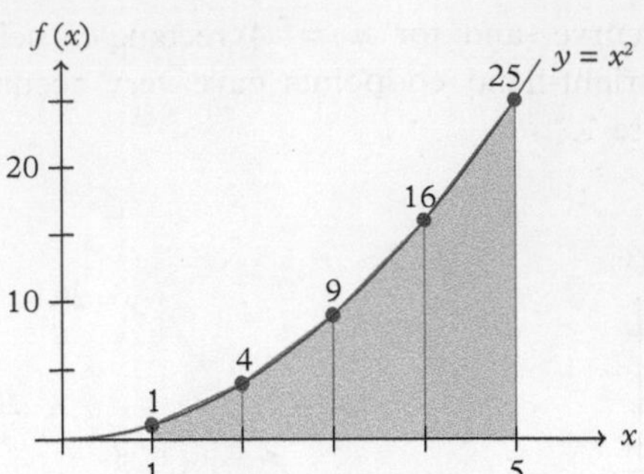

Figure 2 *Using n = 4 trapezoids to estimate $\int_1^5 x^2\,dx$. The area shaded gray (very difficult to see without magnification) is the difference between the area under the curve and the estimate using trapezoids*

Using the formula for the area of a trapezoid, the area of the trapezoid in figure 4 is $\frac{1}{2}(b_1 + b_2) \cdot h = \frac{1}{2}(9 + 16) \cdot 1$.

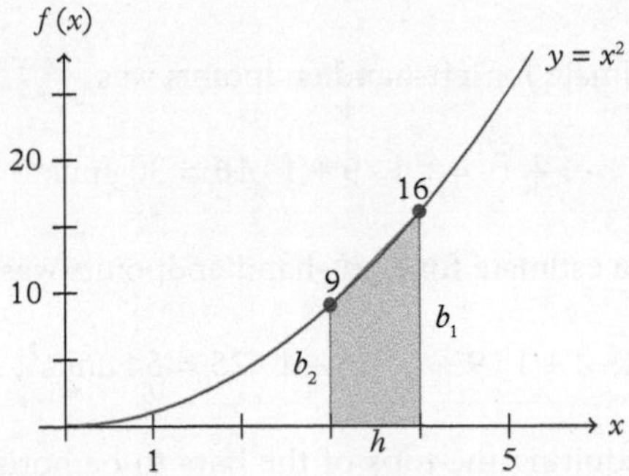

Figure 4 *One of the approximating trapezoids, with h, b_1, and b_2 labeled. Compared to the trapezoid in figure 3, the vertical/horizontal orientation is reversed. Also, one side of the trapezoid is perpendicular to the bases*

The total area of all the trapezoids in figure 2 is

In T_4, the T is for *trapezoid* and the 4 is the number of trapezoids used.

In line 1, the formula for the area of a trapezoid is used for each trapezoid. In line 2, the $\frac{1}{2}$ in front of each term and the 1 at the end of each term from line 1 are factored out.

$$\begin{aligned} T_4 &= \frac{1}{2}(1+4)\cdot 1 + \frac{1}{2}(4+9)\cdot 1 + \frac{1}{2}(9+16)\cdot 1 + \frac{1}{2}(16+25)\cdot 1 \\ &= \frac{1}{2}\cdot 1\cdot(1+4+4+9+9+16+16+25) \\ &= 42 \text{ units}^2. \end{aligned}$$

Notice that the area of 42 units^2 given by four trapezoids is the average of the left- and right-hand estimates (30 and 54) for four rectangles.

Notation break: before stating the trapezoid rule for approximating $\int_a^b f(x)\,dx$, we need to pause to examine some notation. Just as with rectangles, the x-coordinates forming the left- and right-hand endpoints are $x_0, x_1, \ldots, x_n$, with $x_0 = a$ and $x_n = b$. Corresponding to these are the y-coordinates $y_0, y_1, \ldots, y_n$, with $y_k = f(x_k)$. These values are illustrated in figure 5.

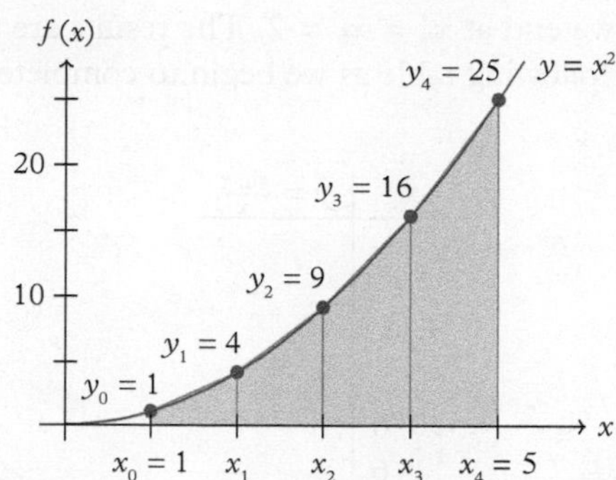

Figure 5 *Notation for use with the trapezoid rule; here, $n = 4$*

TRAPEZOID RULE

To approximate $\int_a^b f(x)\,dx$ using n trapezoids, use

$$T_n = \frac{1}{2}\Delta x(y_0 + 2y_1 + 2y_2 + \cdots + 2y_{n-1} + y_n),$$

where $y_0, y_1, \ldots, y_n$ are the values of f at the endpoints of the n equally spaced subintervals, and $\Delta x = \frac{b-a}{n}$ is the width of each subinterval.

To understand the formula, compare it to the second line of the previous calculation of T_4. The $\frac{1}{2}$ originates from the formula for the area of a trapezoid. The width Δx is 1 in the calculation. Inside the parentheses in the calculation, $y_0 = 1$ and $y_n = y_4 = 25$ each appear only once, whereas $y_1 = 4$, $y_2 = 9$, and $y_3 = 16$ each appear twice.

Trapezoid rule example

Example 4 *Use the trapezoid rule with $n = 6$ to estimate* $\displaystyle\int_1^2 \frac{\sin x}{x}\,dx.$

Solution The trapezoid rule can be completed using the following steps:

❶ Calculate Δx. Using $a = 1$ and $b = 2$ from the limits of integration,

The trapezoid rule can be executed without drawing a picture. If desired, a picture in the style of figure 2 can be drawn.

$$\Delta x = \frac{b-a}{n} = \frac{2-1}{6} = \frac{1}{6}.$$

❷ Calculate the x values $x_0, x_1, \ldots, x_n$ to begin a chart of values of the function. We begin with $x_0 = a = 1$, and then add $\Delta x = \frac{1}{6}$ repeatedly until we end at $x_n = x_6 = 2$. The results are placed in the x column of the following table as we begin to complete our chart of values:

These are the values of $x_0, x_1, \ldots, x_n$. In particular, $x_0 = 1$, $x_1 = 7/6$, $x_2 = 8/6$, ..., $x_6 = 2$.

x	$y = \frac{\sin x}{x}$
1	
7/6	
8/6	
9/6	
10/6	
11/6	
2	

❸ Calculate the y values $y_0, y_1, \ldots, y_n$ to complete the chart of values:

The y values are found just as when plotting points. The function is evaluated at each value of x.

Don't forget to keep your calculator in radian mode when calculating values of trigonometric functions.

It is customary to use the same number of decimal places (alternately, the same number of significant digits) for each y value as well as for the final answer in step ❹.

x	$y = \frac{\sin x}{x}$	
1	0.84147	y_0
7/6	0.78810	y_1
8/6	0.72895	y_2
9/6	0.66500	y_3
10/6	0.59724	y_4
11/6	0.52676	y_5
2	0.45465	y_6

❹ Calculate T_n using the formula. We have

Line 1 is the trapezoid rule formula with $n = 6$; line 2 inputs the calculated values. Notice that the first and last y values are not doubled.

$$\begin{aligned} T_6 &= \frac{1}{2} \cdot \Delta x \cdot (y_0 + 2y_1 + \cdots + 2y_5 + y_6) \\ &= \frac{1}{2} \cdot \frac{1}{6} \cdot \Big(0.84147 + 2(0.78810) + 2(0.72895) + 2(0.66500) \\ &\quad + 2(0.59724) + 2(0.52676) + 0.45465\Big) \\ &= 0.65902. \end{aligned}$$

Because the task is to estimate the value of a definite integral rather than to estimate an area, units are not included in the answer.

We conclude that $\int_1^2 \frac{\sin x}{x}\, dx \approx T_6 = 0.65902.$ ■

Reading Exercise 26 *Use the trapezoid rule with $n = 5$ to estimate $\int_1^2 \frac{\cos x}{x}\, dx$, given the following values:*

x	$\frac{\cos x}{x}$
1	0.54030
1.2	0.30196
1.4	0.12141
1.6	−0.01825
1.8	−0.12622
2	−0.20807

We used charts such as the one in reading exercise 26 when studying finite sums. For left-hand endpoints, our calculation does not use the last value in the list, whereas for right-hand endpoints, our calculation leaves out the first value in the list. The trapezoid rule uses all values in the list.

Earlier, using the result of example 1, we calculated the exact value of the integral in example 4 as

$$\int_1^2 \frac{\sin x}{x}\,dx = \cdots = \text{SinIntegral}(2) - \text{SinIntegral}(1),$$

which to five decimal places is 0.65933. The trapezoid rule gave the correct value to the first three decimal places!

The difference between the approximation given by the trapezoid rule and the actual value of the definite integral is called the *error* in the trapezoid rule approximation, denoted E_T. For example 4 we have

$$E_T = T_6 - \text{actual value} = 0.65902 - 0.65933 = -0.00031.$$

Not bad! A positive value of E_T means the trapezoid rule overestimated the value of the integral; a negative value of E_T indicates an underestimate.

For comparison purposes, the *relative error* is more important. The relative error is calculated as

$$\frac{T_6 - \text{actual value}}{\text{actual value}} = \frac{0.65902 - 0.65933}{0.65933} = -0.00047,$$

which, as a percentage, is -0.047%.

ERROR IN TRAPEZOID RULE APPROXIMATION

$$E_T = T_n - \text{actual value}$$

Using a calculator

The task of approximating the value of an integral using a procedure such as the trapezoid rule is called *numerical integration.* There are many methods of numerical integration, some of which can be learned in a course on numerical analysis. Methods implemented in calculators and in computer software are typically more complicated than the ones taught in this text, but they tend to give more accurate answers more quickly (less computing time) and are therefore preferred.

Many calculators can give a numerical approximation to the value of a definite integral. For the TI-83 or TI-84, the command to use is `fnInt`, found under the MATH menu. The syntax to use is

`fnInt`(integrand, X, lower limit, upper limit).

Here, "lower limit" and "upper limit" refer to the limits of integration.

Example 5 *Use a calculator to estimate* $\displaystyle\int_1^2 \frac{\sin x}{x}\,dx$.

Solution Using a TI-83 or TI-84, we enter the command

```
fnInt(sin(X)/X,X,1,2),
```

which gives the result

```
.6593299064.
```

We assume the result is accurate to the number of digits displayed. This is usually a good assumption. However, if you wish to state the answer to fewer digits than those displayed by the calculator, use rounding.

We conclude that $\displaystyle\int_1^2 \frac{\sin x}{x}\,dx \approx 0.65933.$ ■

Reading Exercise 27 *Use a calculator to estimate* $\displaystyle\int_2^3 \sqrt{x^2 - 4x^5 + x^7}\,dx.$

Ans. to reading exercise 26

$$T_5 = \frac{1}{2}\cdot\frac{1}{5}\cdot\Big(0.54030 + 2(0.30196) + 2(0.12141) + 2(-0.01825) + 2(-0.12622) + (-0.20807)\Big) = 0.08900$$

Error bound for the trapezoid rule

Calculators such as the TI-84 that do not use a built-in CAS use a numerical method to give values for definite integrals even when we can do them easily by hand with the FTC, such as for $\int_1^5 x^2\,dx$.

In example 5, the calculator did not find an antiderivative and use the FTC to give its answer. Instead, it used a numerical method. We used a numerical method (the trapezoid rule) in example 4 and were only accurate to three decimal places. How do we know we can trust the value given by the calculator? How does it know that its answer is accurate to the number of decimal places it gives? The answer lies in knowing a theorem giving a maximum value for the error.

Theorem 4 ERROR BOUND FOR THE TRAPEZOID RULE
When estimating $\int_a^b f(x)\,dx$ *using the trapezoid rule with n trapezoids, if* f'' *is continuous and* $|f''(x)| \le K$ *throughout the interval* $[a, b]$, *then the error* E_T *in the trapezoid rule satisfies*

$$|E_T| \le \frac{K(b-a)^3}{12n^2}.$$

The proof of this theorem can be found in books on numerical analysis.

Our estimate of $\int_1^5 x^2\,dx$ using the trapezoid rule with $n = 4$ was $T_4 = 42$. We know the actual value of that integral from using FTC, but suppose we did not know this value. Then, theorem 15 gives a bound for the error. We calculate this bound in the next example.

A bound for the error tells us not how close we are, but how far away we might be. For instance, if our estimate is $T_4 = 42$ and we know that $|E_T| \le 2$, then we know the value of the integral is somewhere in the interval $[40, 44]$.

Example 6 *Determine the error bound for using the trapezoid rule with* $n = 4$ *to estimate* $\int_1^5 x^2\,dx$.

Solution To use the formula $|E_T| \leq \frac{K(b-a)^3}{12n^2}$, we need the values of n, a, b, and K. We are given $n = 4$, $a = 1$, and $b = 5$, so we only lack K.

❶ Determine a suitable value of K. To determine K, we need the second derivative of the integrand:

$$f(x) = x^2,$$
$$f'(x) = 2x,$$
$$f''(x) = 2.$$

Notice that f'' is continuous everywhere because it is a constant function. This hypothesis should always be checked.

The value of K is supposed to be a number that is at least as large as $|f''(x)|$ throughout the interval $[1, 5]$, so using $K = 2$ suffices.

❷ Calculate the error bound. Using the values we have already determined,

$$|E_T| \leq \frac{2(5-1)^3}{12 \cdot 4^2} = \frac{2}{3}.$$

■

We can conclude from example 6 that $T_4 = 42$ is not more than $\frac{2}{3}$ from the actual value of $\int_1^5 x^2\,dx$. However, what if we wanted to be within 0.001 of the actual value? We cannot change the values of a, b, and K, but we *can* change the value of n. Notice that the larger the value of n, the smaller the error bound given by the formula. In practice, this is how the calculator or computer software knows when to stop. It chooses a value of n that is large enough to make the error bound as small as needed to get the desired accuracy in the result.

Some iterative numerical methods use *stopping criteria* instead. The procedure used for choosing n depends on the numerical method being used.

Example 7 *We wish to use the trapezoid rule to estimate* $\int_1^5 x^2\,dx$ *to within* 0.001. *How large must n be so that the error bound from theorem 4 is less than* 0.001?

Ans. to reading exercise 27
16.2649

Solution We want a value of n so that

$$\frac{K(b-a)^3}{12n^2} \leq 0.001.$$

We have $a = 1$ and $b = 5$. We also have $K = 2$ from the solution to example 6. Now we need

$$\frac{2(5-1)^3}{12n^2} \leq 0.001$$

and we can solve for n:

$$\frac{128}{12n^2} \leq 0.001$$

$$\frac{128}{12} \leq 0.001n^2$$

$$\frac{128}{12(0.001)} \leq n^2$$

$$\sqrt{\frac{128}{12(0.001)}} \leq n$$

$$n \geq 103.28.$$

Because we need $n \geq 103.28$, using $n = 103$ is insufficient.

We can use $n = 104$. ■

Neither example 6 nor example 7 required knowledge of the actual value of the definite integral. Error bounds such as the one given in theorem 4 can be calculated before the trapezoid rule (or other numerical method) is carried out, and then we can be sure the result has the desired accuracy.

Using 104 trapezoids seems like a lot of work for a human. Although 104 trapezoids is not an issue for a computer, numerical methods that offer more accuracy for less work are preferred and can make a noticeable difference in performance in some cases.

In example 6, the calculation of K was simple. An example with a more complicated second derivative is next.

Example 8 *Determine an error bound for using the trapezoid rule with* $n = 6$ *to estimate* $\int_1^2 \frac{\sin x}{x}\,dx$.

Solution ❶ We begin the determination of K by finding the second derivative of the integrand:

$$f(x) = \frac{\sin x}{x},$$

$$f'(x) = \frac{x\cos x - \sin x}{x^2},$$

$$f''(x) = \frac{x^2\left(x(-\sin x) + 1\cdot\cos x - \cos x\right) - (x\cos x - \sin x)2x}{x^4}$$

$$= \cdots = \frac{(2 - x^2)\sin x - 2x\cos x}{x^3}.$$

Notice that f'' has a discontinuity at $x = 0$ (division by zero), but this is not in the interval of integration $[1, 2]$. The hypothesis of f'' being continuous throughout $[a, b]$ is met.

Now we need to determine the largest possible value of

$$|f''(x)| = \left|\frac{(2-x^2)\sin x - 2x\cos x}{x^3}\right|$$

on the interval $[1, 2]$. This will be our value of K. Because this is a continuous function on a closed and bounded interval, we can use the closed interval method of section 3.1. To save time and effort, however, let's look at a graph (figure 6) and pick a sufficient value of K from there.

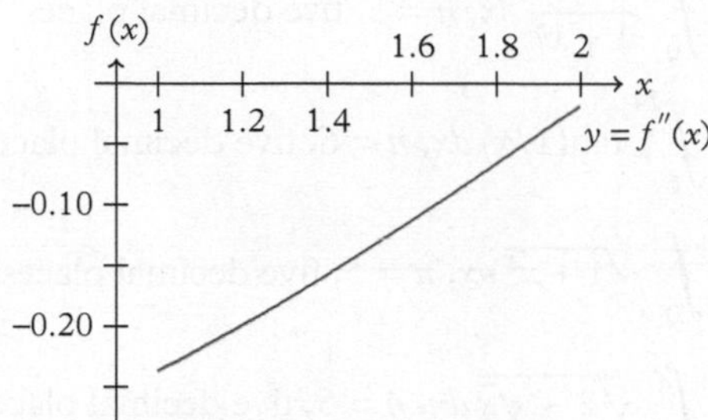

Figure 6 *The graph of $y = f''(x)$ for example 8 over the interval of integration $[1, 2]$. The value chosen for K can be any number greater than the absolute value of every y-coordinate on the graph*

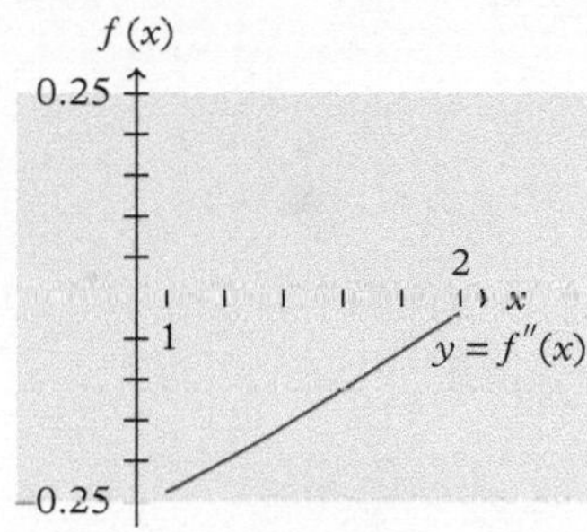

Figure 7 *The graph of $y = f''(x)$ for example 8 over the interval of integration $[1, 2]$, with shading between $y = -0.25$ and $y = 0.25$. Because the graph of $y = f''(x)$ lies completely within the shaded region, we may choose* $K = 0.25$

The y-coordinates appear to range from nearly -0.25 to perhaps -0.02. One way to interpret our task is that we must choose K so that the y-coordinates on the graph all lie between $-K$ and K. Choosing $K = 0.25$ is sufficient; see figure 7.

❷ Next we calculate the error bound. Using $n = 6$, $a = 1$, $b = 2$, and $K = 0.25$ yields

$$|E_T| \le \frac{0.25(2-1)^3}{12 \cdot 6^2} = 0.0005787. \qquad ■$$

In example 4, we used the trapezoid rule with $n = 6$ to estimate the value of $\int_1^2 \frac{\sin x}{x}\,dx$ as $T_6 = 0.65902$. In example 8, we see that the error bound is $|E_T| \le 0.00058$. In other words, the value of the integral must be within 0.00058 of 0.65902, or

We use the same number of decimal places in the error bound as we do for T_n when combining the results:

$$0.65902 - 0.00058 = 0.65844$$
$$0.65902 + 0.00058 = 0.65960.$$

$$0.65844 \le \int_1^2 \frac{\sin x}{x}\,dx \le 0.65960.$$

Indeed, the actual value of the integral (to five decimal places) is 0.65933, which satisfies the inequality.

Exercises 4.8

1–16. Use the trapezoid rule with the given value of n to estimate the value of the integral to the requested number of decimal places.

1. $\int_3^5 \sqrt{x^3 - 4}\, dx$, $n = 8$, three decimal places

2. $\int_3^5 \sqrt{2 + x^3}\, dx$, $n = 6$, three decimal places

3. $\int_0^1 \frac{1}{1 + x^5}\, dx$, $n = 5$, five decimal places

4. $\int_6^{24} \tan(1/x)\, dx$, $n = 6$, five decimal places

5. $\int_0^1 \sqrt{1 + x^4}\, dx$, $n = 5$, five decimal places

6. $\int_0^1 \sqrt{2 - \sqrt{x}}\, dx$, $n = 5$, five decimal places

7. $\int_1^{1.06} \sin(x^3 - x)\, dx$, $n = 6$, seven decimal places

8. $\int_0^2 2^{(x^3 - x)}\, dx$, $n = 8$, three decimal places

9. $\int_{-1}^0 2^{(-x^2)}\, dx$, $n = 4$, five decimal places

10. $\int_{-1}^2 \sin(-x^2)\, dx$, $n = 6$, five decimal places

11. $\int_4^9 \cos \sqrt{x}\, dx$, $n = 5$, four decimal places

12. $\int_0^{\pi} \sqrt{x} \cos x\, dx$, $n = 4$, five decimal places

13. $\int_{1.4}^{2.2} \cos x^2\, dx$, $n = 8$, four decimal places

14. $\int_0^3 x^2 \cos x\, dx$, $n = 10$, four decimal places

15. $\int_{0.09}^{0.21} \frac{1}{x^6 - 7x^{3.4}}\, dx$, $n = 6$, three decimal places

16. $\int_{-6}^{-3} \frac{1}{1+\sin x}\,dx$, $n = 6$, five decimal places

17–32. (a) Use a calculator to estimate the value of the integral. (b) Using the answer from (a) as the actual value of the integral, find the error E_T in using the trapezoid rule from the corresponding previous exercise to the same number of decimal places as used for the trapezoid rule.

Exercise 17 corresponds to exercise 1, exercise 18 corresponds to exercise 2, and so on.

17. $\int_{3}^{5} \sqrt{x^3 - 4}\,dx$

18. $\int_{3}^{5} \sqrt{2 + x^3}\,dx$

19. $\int_{0}^{1} \frac{1}{1+x^5}\,dx$

20. $\int_{6}^{24} \tan(1/x)\,dx$

21. $\int_{0}^{1} \sqrt{1+x^4}\,dx$

22. $\int_{0}^{1} \sqrt{2 - \sqrt{x}}\,dx$

23. $\int_{1}^{1.06} \sin(x^3 - x)\,dx$

24. $\int_{-1}^{0} 2^{(x^3 - x)}\,dx$

25. $\int_{-1}^{0} 2^{(-x^2)}\,dx$

26. $\int_{-1}^{2} \sin(-x^2)\,dx$

27. $\int_{4}^{9} \cos\sqrt{x}\,dx$

28. $\int_{0}^{\pi} \sqrt{x}\cos x\,dx$

29. $\int_{1.4}^{2.2} \cos x^2\,dx$

30. $\int_{0}^{3} x^2 \cos x\,dx$

31. $\int_{0.09}^{0.21} \frac{1}{x^6 - 7x^{3.4}}\,dx$

32. $\int_{-6}^{-3} \frac{1}{1+\sin x}\,dx$

33–42. (a) Using the procedure of example 8, determine an error bound for using the trapezoid rule for the given integral and value of n. (b) Verify that the error E_T (as calculated in a corresponding exercise from exercises 17–32) satisfies that bound. (c) How large must n be so that the error bound is less than 0.00001?

33. $\int_{3}^{5} \sqrt{x^3 - 4}\,dx$, $n = 8$

34. $\int_{3}^{5} \sqrt{2 + x^3}\,dx$, $n = 6$

35. $\int_{0}^{1} \frac{1}{1+x^5}\,dx$, $n = 5$

36. $\int_{6}^{24} \tan(1/x)\,dx$, $n = 6$

37. $\int_{4}^{9} \cos\sqrt{x}\,dx$, $n = 5$

38. $\int_{-1}^{2} \sin(-x^2)\,dx$, $n = 6$

39. $\int_{1.4}^{2.2} \cos x^2 \, dx$, $n = 8$

40. $\int_{0}^{3} x^2 \cos x \, dx$, $n = 10$

41. $\int_{0.09}^{0.21} \frac{1}{x^6 - 7x^{3.4}} \, dx$, $n = 6$

42. $\int_{-6}^{-3} \frac{1}{1 + \sin x} \, dx$, $n = 6$

43–58. Use technology to attempt to evaluate the indefinite integral. State whether the result is an elementary function, a function of some other type, or no result is given.

43. $\int \sqrt{x^3 - 4} \, dx$

44. $\int \sqrt{2 + x^3} \, dx$

45. $\int \sqrt{1 + x^4} \, dx$

46. $\int \sqrt{2 - \sqrt{x}} \, dx$

47. $\int \sin(x^3 - x) \, dx$

48. $\int 2^{(x^3 - x)} \, dx$

49. $\int 2^{(-x^2)} \, dx$

50. $\int \sin(-x^2) \, dx$

51. $\int \cos \sqrt{x} \, dx$

52. $\int \sqrt{x} \cos x \, dx$

53. $\int \cos x^2 \, dx$

54. $\int x^2 \cos x \, dx$

55. $\int \frac{1}{x^6 - 7x^{3.4}} \, dx$

56. $\int \frac{1}{1 + \sin x} \, dx$

57. $\int \frac{1}{x^6 - 7x^3} \, dx$

58. $\int \frac{1}{x + \sin x} \, dx$

59. A herd of buffalo is migrating. The speed at which the lead animal is traveling is recorded every 15 minutes and is given in the following table. (a) Use the trapezoid rule to give an estimate of the distance traveled from 9:00 a.m. until 11:00 a.m. (b) Compare your answer to those of section 4.2 exercises 47–50.

time (a.m.)	speed (mph)
9:00	5.7
9:15	6.6
9:30	6.8
9:45	6.0
10:00	6.5
10:15	7.1
10:30	7.3
10:45	6.2
11:00	6.4

60. You are on a sailboat in the Atlantic Ocean, all your GPS devices and phones were swept overboard in a storm, your instrument panel does not work, and it is so cloudy you cannot see the sun, moon, or stars. You resort to dead reckoning to determine your position. You have kept a due-west course (according to your compass), and you determine your speed every half hour using knots in a rope let out into the water. (a) Use the following data to estimate the distance you have traveled from 8:00 a.m. until 12:00 noon, using the trapezoid rule. (b) Compare your answer to those of section 4.2 exercises 51–54.

time (a.m.)	speed (knots)
8:00	6
8:30	8
9:00	7
9:30	9
10:00	11
10:30	12
11:00	11
11:30	9
12:00 noon	8

61. Show that if f is an integrable function with $\int_a^b f(x)\,dx \neq 0$ and Ω is a positive infinite hyperreal integer, then $T_\Omega \approx \int_a^b f(x)\,dx$.

This exercise shows that $\lim_{n\to\infty} T_n = \int_a^b f(x)\,dx$. As n becomes infinite the error E_T becomes infinitesimal.

62. Show that for any value of n, the trapezoid rule always gives the average (arithmetic mean) of the left- and right-hand estimates.

Numerical Integration, Part II

4.9

The trapezoid rule is not the only method of numerical integration that can be performed easily by hand, nor is it usually the most accurate for a given amount of work. Two additional methods are discussed in this section.

Midpoint rule

In section 4.2 we estimated areas using various methods, including midpoints. The midpoint rule is merely a restatement of the use of midpoints. The next example is both a reminder of what we did in section 4.2 and a vehicle to introduce new notation reminiscent of the trapezoid rule.

Example 1 *Use midpoints with $n = 5$ to estimate $\int_1^2 \frac{\sin x}{x}\,dx$ to five decimal places.*

Solution We organize the solution using the same steps as for the trapezoid rule.

❶ Calculate Δx:

$$\Delta x = \frac{b-a}{n} = \frac{2-1}{5} = \frac{1}{5} = 0.2.$$

❷ Calculate the x values. Using $\Delta x = 0.2$, $a = 1$, and $b = 2$, the endpoints of the five subintervals are $x_0 = 1$, $x_1 = 1.2$, $x_2 = 1.4$, $x_3 = 1.6$, $x_4 = 1.8$, and $x_5 = 2$. But, we do not evaluate the function at the endpoints of the intervals; we use the midpoints instead. These midpoints are $\frac{x_0 + x_1}{2} = 1.1$, $\frac{x_1 + x_2}{2} = 1.3, \ldots,$ $\frac{x_4 + x_5}{2} = 1.9$. We begin the chart of values with these midpoints as our x values:

x	$y = \frac{\sin x}{x}$
1.1	
1.3	
1.5	
1.7	
1.9	

❸ Calculate the heights of the rectangles to complete the chart of values. Because these y-coordinates are from the midpoints rather than the endpoints, we call these heights $h_1, h_2, \ldots, h_n$:

In the notation of the previous section, y_0 is the value of the function at x_0, y_1 is the value of the function at x_1, and so on. Let's not confuse matters by redefining the y_k's.

x	$y = \frac{\sin x}{x}$	
1.1	0.81019	h_1
1.3	0.74120	h_2
1.5	0.66500	h_3
1.7	0.58333	h_4
1.9	0.49805	h_5

❹ Calculate M_5:

Let's use M for midpoint just as we used T for trapezoid, with the subscript giving the value of n.

Recall that we estimate the area using midpoints by adding the areas of rectangles—that is, by summing the widths (Δx) times the heights ($h_1, \ldots, h_5$).

$$\begin{aligned} M_5 &= 0.2(0.81019) + 0.2(0.74120) + 0.2(0.66500) + 0.2(0.58333) \\ &\quad + 0.2(0.49805) \\ &= 0.2(0.81019 + 0.74120 + 0.66500 + 0.58333 + 0.49805) \\ &= 0.65955. \end{aligned}$$

We conclude that $\displaystyle\int_1^2 \frac{\sin x}{x} \approx 0.65955.$ ■

The midpoint rule summarizes the previous steps in the same manner as the trapezoid rule.

MIDPOINT RULE

To approximate $\int_a^b f(x)\,dx$ using midpoints with n rectangles, use

$$M_n = \Delta x(h_1 + h_2 + \cdots + h_n),$$

where $h_1, h_2, \ldots, h_n$ are the values of f at the midpoints of n equally spaced subintervals, and $\Delta x = \frac{b-a}{n}$ is the width of each subinterval.

The formula follows the next-to-last line of the calculation of M_5 in example 1.

The error E_M in using the midpoint rule is calculated in the same manner as for the trapezoid rule: $E_M = M_n -$ actual value.

ERROR IN MIDPOINT RULE APPROXIMATION

$E_M = M_n -$ actual value

Example 2 *Using a calculator's value as the actual value of the integral, determine the error E_M in using the midpoint rule with $n = 5$ to approximate* $\displaystyle\int_1^2 \frac{\sin x}{x}\,dx.$

Solution ❶ Determine $\int_1^2 \frac{\sin x}{x}\,dx$ using a calculator. We did this in section 4.8 example 5:

Most calculators also use a numerical method to estimate the value of the integral, but let's make the assumption that the value given by the calculator is accurate.

$$\int_1^2 \frac{\sin x}{x}\,dx \approx 0.65933.$$

❷ Calculate E_M:

M_5 was calculated in example 1.

$$\begin{aligned} E_M &= M_5 - \text{actual value} \\ &= 0.65955 - 0.65933 \\ &= 0.00022. \end{aligned}$$ ■

Noticc that the error for using the midpoint rule with $n = 5$, 0.00022, is less in absolute value than the error for using the trapezoid rule with $n = 6$, -0.00031. Fewer calculations, more accuracy!

The error in using the trapezoid rule was calculated in the previous section.

Error bound for the midpoint rule

Although the midpoint rule does not always have a smaller error than the trapezoid rule for a given value of n, it does happen often. The error bound gives further evidence of this, as it is half the error bound for the trapezoid rule.

Theorem 5 ERROR BOUND FOR THE MIDPOINT RULE
When estimating $\int_a^b f(x)\,dx$ using the midpoint rule with n rectangles, if f'' is continuous and $|f''(x)| \leq K$ throughout the interval $[a, b]$, then the error E_M in the midpoint rule satisfies

The proof of this theorem can be found in books on numerical analysis.

$$|E_M| \leq \frac{K(b-a)^3}{24n^2}.$$

This error bound is nearly identical to the error bound for the trapezoid rule. The only exception is a 24 in the denominator instead of a 12.

Because their error bound theorems are so similar, calculating an error bound for the midpoint rule is nearly identical to calculating an error bound for the trapezoid rule.

Example 3 *Determine an error bound for using the midpoint rule with $n = 5$ to estimate $\int_1^2 \frac{\sin x}{x}\,dx$.*

Step ❶ is a repeat of the work in section 4.8 example 8, but is included again here for your convenience.

Solution ❶ We begin the determination of K by finding the second derivative of the integrand:

$$f(x) = \frac{\sin x}{x},$$

$$f'(x) = \frac{x\cos x - \sin x}{x^2},$$

$$f''(x) = \frac{x^2\left(x(-\sin x) + 1\cdot\cos x - \cos x\right) - (x\cos x - \sin x)2x}{x^4}$$

$$= \cdots = \frac{(2-x^2)\sin x - 2x\cos x}{x^3}.$$

Notice that f'' has a discontinuity at $x = 0$ (division by zero), but this is not in the interval of integration $[1, 2]$. The hypothesis of f'' being continuous throughout $[a, b]$ is met.

Now we need to determine the largest possible value of

$$|f''(x)| = \left|\frac{(2-x^2)\sin x - 2x\cos x}{x^3}\right|$$

on the interval $[1, 2]$. This will be our value of K. Let's look at a graph of $y = f''(x)$ (figure 1) to make the determination.

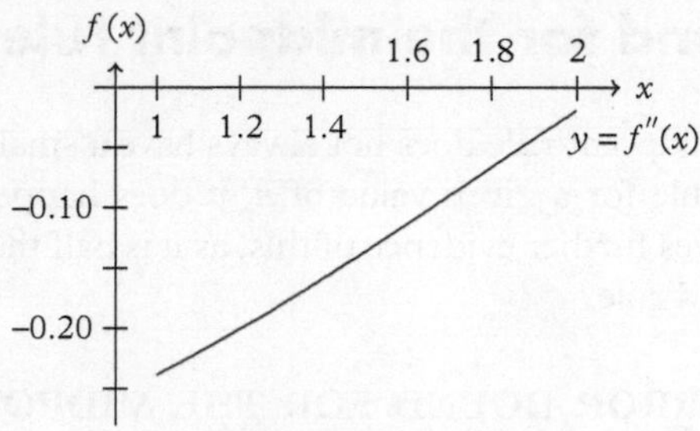

Figure 1 *The graph of $y = f''(x)$ for example 3 over the interval of integration $[1, 2]$. The value chosen for K may be any number greater than the absolute value of every y-coordinate on the graph*

The y-coordinates appear to range from nearly -0.25 to perhaps -0.02, so let's choose $K = -0.25$.

❷ Next we calculate the error bound. Using $n = 5$, $a = 1$, $b = 2$, and $K = 0.25$ yields

The error bound for using the trapezoid rule with $n = 6$ was 0.0005787.

$$|E_M| \le \frac{0.25(2-1)^3}{24\cdot 5^2} = 0.0004167. \qquad ■$$

The error E_M calculated in example 2 was 0.00022, which (as expected) is less than the error bound 0.0004167 of example 3.

Simpson's rule

The trapezoid rule depends on matching a curve's values at both the left and right endpoints of equally spaced subintervals and then following a straight line between these points, as illustrated in figure 2. In the figure, it is apparent that the straight-line tops of the trapezoidal

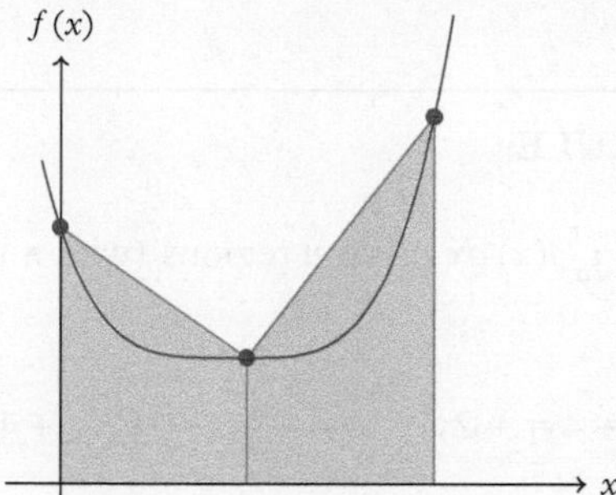

Figure 2 *Using $n = 2$ trapezoids to estimate $\int_a^b f(x)\,dx$. The area of the shaded region overestimates the area under the curve. The straight lines at the top of the trapezoids do not follow the curve very well*

regions do not follow the curve very well, leading to an overestimate of the value of the integral.

Idea: what if we allow the tops of the brown regions to be curved? Parabolas are curved; perhaps we could allow the tops of the regions

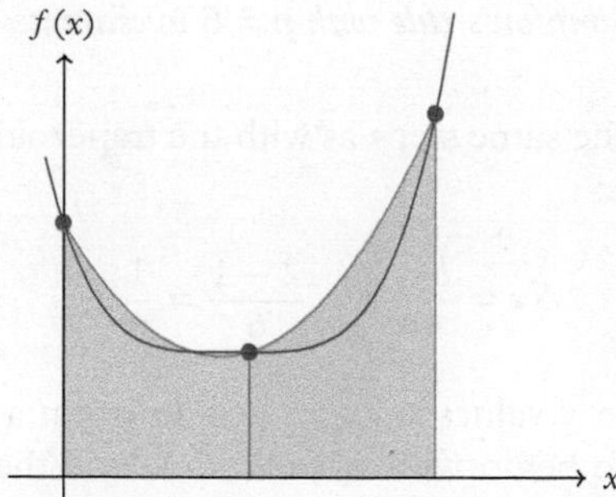

Figure 3 *Using a parabola-topped region to estimate $\int_a^b f(x)\,dx$. The area of the shaded region is closer to the area under the curve than that given by the trapezoids of figure 2*

to be parabolas, as in figure 3. Although we still appear to have an overestimate of the area under the curve, we appear to have a better estimate.

The full development of this method of estimation is not difficult, but it is a little tedious and is therefore saved for the end of this section. However, one of the details should be mentioned here: although it takes two points to determine a line, it takes three points to determine a parabola. Therefore, we use one parabola for every two subintervals (the three points matching the curve in figure 3 yield one parabola), so we must use an even number of subintervals for what is called Simpson's rule.

Simpson's rule is named after Thomas Simpson, who popularized the method, but credit for the rule is due to Isaac Newton. On the other hand, the modern iterative form of Newton's method (see section 3.9) is due to Simpson.

Thomas Simpson 1710–1761
http://www-history.mcs.st-andrews.ac.uk/Biographies/Simpson.html

SIMPSON'S RULE

To approximate $\int_a^b f(x)\,dx$ using n regions (with n being an even number), use

$$S_n = \frac{1}{3}\Delta x(y_0 + 4y_1 + 2y_2 + 4y_3 + \cdots + 2y_{n-2} + 4y_{n-1} + y_n),$$

where $y_0, y_1, \ldots, y_n$ are the values of f at the endpoints of the n equally spaced subintervals, and $\Delta x = \frac{b-a}{n}$ is the width of each subinterval.

Calculations using Simpson's rule follow the same pattern as for the trapezoid rule, just with a slightly different formula. For the "middle" values, factors of 4 and 2 alternate, beginning and ending with 4. Also notice the $\frac{1}{3}$, where as the trapezoid rule has $\frac{1}{2}$.

Steps ❶–❸ are identical to the trapezoid rule and are repeated here from section 4.8 example 4.

Example 4 *Use Simpson's rule with* $n = 6$ *to estimate* $\displaystyle\int_1^2 \frac{\sin x}{x}\,dx$.

Solution We use the same steps as with the trapezoid rule.

❶ Calculate Δx:

$$\Delta x = \frac{b-a}{n} = \frac{2-1}{6} = \frac{1}{6}.$$

❷ Calculate the x values $x_0, x_1, \ldots, x_n$ to begin a chart of values of the function. We begin with $x_0 = a = 1$, and then add $\Delta x = \frac{1}{6}$ repeatedly until we end at $x_n = x_6 = 2$. The results are placed in the x column of the following chart of values:

x	$y = \frac{\sin x}{x}$
1	
7/6	
8/6	
9/6	
10/6	
11/6	
2	

❸ Calculate the y values $y_0, y_1, \ldots, y_n$ to complete the chart of values:

x	$y = \frac{\sin x}{x}$	
1	0.84147	y_0
7/6	0.78810	y_1
8/6	0.72895	y_2
9/6	0.66500	y_3
10/6	0.59724	y_4
11/6	0.52676	y_5
2	0.45465	y_6

❹ Calculate S_n using the formula. We have

$$S_6 = \frac{1}{3} \cdot \Delta x \cdot (y_0 + 4y_1 + 2y_2 + 4y_3 + 2y_4 + 4y_5 + y_6)$$

$$= \frac{1}{3} \cdot \frac{1}{6} \cdot \Big(0.84147 + 4(0.78810) + 2(0.72895) + 4(0.66500) + 2(0.59724) + 4(0.52676) + 0.45465\Big)$$

$$= 0.65933.$$

Line 1 is the Simpson's rule formula with $n = 6$; line 2 inputs the calculated values. Notice the alternating factors of 4 and 2, beginning and ending with 4, for all the middle values. Just as with the trapezoid rule, the first and last y values are not multiplied by an additional factor.

We conclude that $\int_1^2 \frac{\sin x}{x}\, dx \approx S_6 = 0.65933.$ ■

The error E_S in using Simpson's rule is calculated in the same manner as for the trapezoid and midpoint rules: $E_S = S_n -$ actual value.

ERROR IN SIMPSON'S RULE APPROXIMATION

$E_S = S_n -$ actual value

Example 5 *Using a calculator's value as the actual value of the integral, determine the error E_S in using Simpson's rule with $n = 6$ to approximate $\int_1^2 \frac{\sin x}{x}\, dx$.*

Solution ❶ Determine $\int_1^2 \frac{\sin x}{x}\, dx$ using a calculator:

$$\int_1^2 \frac{\sin x}{x}\, dx \approx 0.65933.$$

❷ Calculate E_S:

$$E_S = S_6 - \text{actual value} = 0.65933 - 0.65933 = 0.00000.$$

S_6 was calculated in example 4.

Writing $E_S = 0.00000$ communicates something different than writing $E_S = 0$. The former indicates the number of decimal places to which E_S is zero. The latter can be misinterpreted easily as indicating that the value of E_S is identical to the exact value of the integral.

To five decimal places, the error in using Simpson's rule with $n = 6$ is zero. ■

Error bound for Simpson's rule

The amount of work in using Simpson's rule for $n = 6$ was exactly the same as for using the trapezoid rule with $n = 6$, but Simpson's rule had much greater accuracy. This is also reflected in the error bound.

The proof of this theorem can be found in books on numerical analysis.

Theorem 6 ERROR BOUND FOR SIMPSON'S RULE *When estimating $\int_a^b f(x)\,dx$ using Simpson's rule with n regions, if $f^{(4)}$ is continuous and $|f^{(4)}(x)| \le K$ throughout the interval $[a, b]$, then the error E_S in Simpson's rule satisfies*

$$|E_S| \le \frac{K(b-a)^5}{180n^4}.$$

Example 6 *Determine an error bound for using Simpson's rule with $n = 6$ to estimate* $\displaystyle\int_1^2 \frac{\sin x}{x}\,dx$.

Solution As for both the trapezoid rule and the midpoint rule, we already have all the necessary values for determining the error bound, except for K.

❶ Determine K. We need $f^{(4)}$, the fourth derivative of the integrand. Calculating by hand or technology results in an expression equivalent to the following:

This is the same procedure as determining K for the trapezoid and midpoint rules, but with the fourth derivative instead of the second derivative.

$$f^{(4)}(x) = \frac{\sin x}{x} + \frac{4\cos x}{x^2} - \frac{12\sin x}{x^3} - \frac{24\cos x}{x^4} + \frac{24\sin x}{x^5}.$$

Now we need to determine the largest possible value of

$$|f^{(4)}(x)| = \left|\frac{\sin x}{x} + \frac{4\cos x}{x^2} - \frac{12\sin x}{x^3} - \frac{24\cos x}{x^4} + \frac{24\sin x}{x^5}\right|$$

on the interval $[1, 2]$. This will be our value of K. Let's look at a graph of $y = f^{(4)}$ (figure 4) to make the determination. It appears that $K = 0.14$ should work.

❷ Next we calculate the error bound. Using $n = 6$, $a = 1$, $b = 2$, and $K = 0.14$ yields

$$|E_S| \le \frac{0.14(2-1)^5}{180 \cdot 6^4} = 6.00137 \times 10^{-7} = 0.000000600137. \quad \blacksquare$$

The error bound indicates that S_6 should be accurate to six decimal places (subject to rounding). To capitalize on this accuracy, we need

No matter what accuracy is indicated by the error bound, we should not claim that our trapezoid rule, midpoint rule, or Simpson's rule estimates carry more accuracy than indicated by the actual number of decimal places used in our calculations.

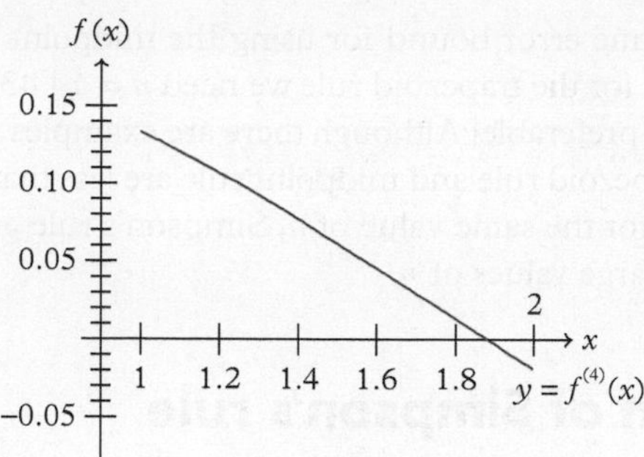

Figure 4 *The graph of* $y = f^{(4)}(x)$ *for example 6 over the interval of integration* $[1, 2]$. *The value chosen for K can be any number greater than the absolute value of every y-coordinate on the graph*

to go back and recalculate all previous values to more than the five decimal places that we used.

The reason the error bounds for Simpson's rule get so small so quickly is the presence of n^4 in the denominator, compared to the n^2 in the denominator of the error bounds for the trapezoid and midpoint rules.

Example 7 *We wish to use Simpson's rule to estimate* $\int_1^2 \frac{\sin x}{x}\,dx$. *How large must n be so that the error bound from theorem 6 is less than* 10^{-10}?

The calculator estimate obtained in section 4.8 example 5 was given to 10 decimal places, hence the use of 10^{-10} in this example.

Solution We want a value of n so that

$$\frac{K(b-a)^5}{180n^4} \le 10^{-10}.$$

Using $a = 1$, $b = 2$, and $K = 0.14$ (from example 6), we solve for n:

$$\frac{0.14(2-1)^5}{180n^4} \le 10^{-10}$$

$$\frac{0.14}{180} \le 10^{-10} n^4$$

$$\frac{0.14}{180 \cdot 10^{-10}} \le n^4$$

$$n \ge \sqrt[4]{\frac{0.14}{180 \cdot 10^{-10}}} = 52.81.$$

We require a number that's even larger than 52.81. Therefore, we can use $n = 54$. ■

Remember that Simpson's rule requires n to be an even number, so we cannot use $n = 53$.

To get the same error bound for using the midpoint rule requires $n = 10\,207$, and for the trapezoid rule we need $n = 14\,434$. Simpson's rule is definitely preferable! Although there are examples for which the errors in the trapezoid rule and midpoint rule are less than the error in Simpson's rule for the same value of n, Simpson's rule generally wins for sufficiently large values of n.

Derivation of Simpson's rule

When developing the formula for the trapezoid rule, we needed the formula for the area of a trapezoid. When developing the formula for the midpoint rule, we used the formula for the area of a rectangle. For Simpson's rule, we need the area of parabola-topped regions. Furthermore, we wish to be able to determine this area from only the heights of the curve at three locations (the endpoints of the two subintervals). In other words, we need to be able to determine the area of the shaded region in figure 5 from the values y_0, y_1, and y_2.

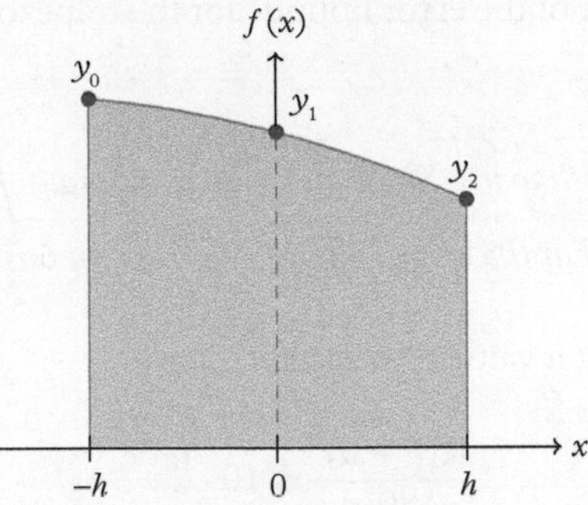

Figure 5 *Two regions for Simpson's rule, topped by one parabola. For convenience, the subintervals are* $[-h, 0]$ *and* $[0, h]$*—meaning,* $\Delta x = h$*. The goal is to determine the shaded area in terms of* y_0*,* y_1*, and* y_2

The equation of a parabola is of the form $y = ax^2 + bx + c$. Between $x = -h$ and $x = h$, the net area under the curve is

$$\int_{-h}^{h} \left(ax^2 + bx + c\right)\, dx = \left(\frac{ax^3}{3} + \frac{bx^2}{2} + cx\right)\Bigg|_{-h}^{h}$$

$$= \frac{a}{3}h^3 + \frac{b}{2}h^2 + ch - \left(-\frac{a}{3}h^3 + \frac{b}{2}h^2 - ch\right)$$

Verifying the equality in line 3 is easier from right to left than from left to right.

$$= \frac{2a}{3}h^3 + 2ch = \frac{h}{3}\left(2ah^2 + 6c\right).$$

We also know that the values of y_0, y_1, and y_2 are the values of $y = ax^2 + bx + c$ at $x = -h$, $x = 0$, and $x = h$, respectively—that is,

$$y_0 = y(-h) = ah^2 - bh + c,$$
$$y_1 = y(0) = c, \text{ and}$$
$$y_2 = y(h) = ah^2 + bh + c.$$

Next, use these values to calculate $y_0 + 4y_1 + y_2$:

$$\begin{aligned} y_0 + 4y_1 + y_2 &= ah^2 - bh + c + 4c + ah^2 + bh + c \\ &= 2ah^2 + 6c. \end{aligned}$$

Comparing this calculation to the net area under the curve calculated earlier, we see that

$$\text{net area under the curve} = \frac{h}{3}\left(2ah^2 + 6c\right) = \frac{h}{3}(y_0 + 4y_1 + y_2).$$

Next we note that in these calculations, the width of each subinterval is $\Delta x = h$, so we rewrite the area formula as

$$\text{net area under the curve} = \frac{\Delta x}{3}(y_0 + 4y_1 + y_2),$$

finishing the task of expressing the net area under the curve in terms of the y values y_0, y_1, and y_2.

The final step is to add the areas of all the regions, as in figure 6.

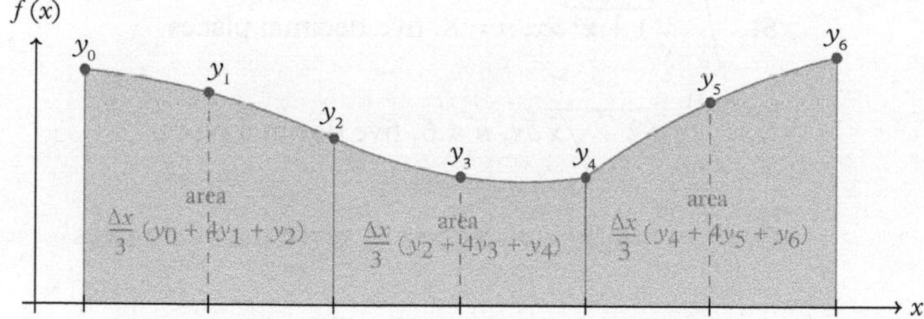

Figure 6 *Shaded regions for Simpson's rule for n = 6, with three parabolas. The net area under each parabola is noted*

For the case $n = 6$ as pictured in figure 6, the total area is

$$\begin{aligned} &\frac{\Delta x}{3}(y_0 + 4y_1 + y_2) + \frac{\Delta x}{3}(y_2 + 4y_3 + y_4) + \frac{\Delta x}{3}(y_4 + 4y_5 + y_6) \\ &= \frac{\Delta x}{3}(y_0 + 4y_1 + 2y_2 + 4y_3 + 2y_4 + 4y_5 + y_6). \end{aligned}$$

Following this pattern, the formula for any even value of n should be clear.

Exercises 4.9

1–16. (a) Use Simpson's rule with the given value of n to estimate the value of the integral to the requested number of decimal places. (b) Using a calculator's value as the actual value of the integral, determine the error E_S in the estimate of part (a). (c) If the same value of n was used for a trapezoid rule estimate in a section 4.8 exercise, compare the error E_S for using Simpson's rule with the error E_T for using the trapezoid rule.

Calculator values for these integrals and values of E_T for the trapezoid rule were part of exercises 17–32 of section 4.8.

1. $\int_3^5 \sqrt{x^3 - 4}\, dx$, $n = 8$, three decimal places

2. $\int_3^5 \sqrt{2 + x^3}\, dx$, $n = 6$, three decimal places

3. $\int_0^1 \frac{1}{1 + x^5}\, dx$, $n = 4$, five decimal places

4. $\int_6^{24} \tan(1/x)\, dx$, $n = 6$, five decimal places

5. $\int_0^1 \sqrt{1 + x^4}\, dx$, $n = 8$, five decimal places

6. $\int_0^1 \sqrt{2 - \sqrt{x}}\, dx$, $n = 6$, five decimal places

7. $\int_1^{1.06} \sin(x^3 - x)\, dx$, $n = 6$, seven decimal places

8. $\int_0^2 2^{(x^3 - x)}\, dx$, $n = 8$, three decimal places

9. $\int_{-1}^0 2^{(-x^2)}\, dx$, $n = 4$, five decimal places

10. $\int_{-1}^2 \sin(-x^2)\, dx$, $n = 6$, five decimal places

11. $\int_4^9 \cos \sqrt{x}\, dx$, $n = 10$, four decimal places

12. $\int_0^{\pi} \sqrt{x}\cos x\,dx$, $n = 4$, five decimal places

13. $\int_{1.4}^{2.2} \cos x^2\,dx$, $n = 8$, four decimal places

14. $\int_0^3 x^2 \cos x\,dx$, $n = 10$, four decimal places

15. $\int_{0.09}^{0.21} \frac{1}{x^6 - 7x^{3.4}}\,dx$, $n = 6$, three decimal places

16. $\int_{-6}^{-3} \frac{1}{1 + \sin x}\,dx$, $n = 6$, five decimal places

17–26. (a) Determine an error bound using Simpson's rule for the given integral and value of n. (b) How large must n be so that the error bound is less than 0.00001? (c) Compare the answer to (b) to the required value of n for the trapezoid rule from section 4.8 exercises 33(c)–42(c).

17. $\int_3^5 \sqrt{x^3 - 4}\,dx$, $n = 8$

18. $\int_3^5 \sqrt{2 + x^3}\,dx$, $n = 6$

19. $\int_0^1 \frac{1}{1 + x^5}\,dx$, $n = 4$

20. $\int_6^{24} \tan(1/x)\,dx$, $n = 6$

21. $\int_4^9 \cos\sqrt{x}\,dx$, $n = 10$

22. $\int_{-1}^{2} \sin(-x^2)\,dx$, $n = 6$

23. $\displaystyle\int_{1.4}^{2.2} \cos x^2\, dx,$

$n = 8$

24. $\displaystyle\int_{0}^{3} x^2 \cos x\, dx,$
$n = 10$

25. $\displaystyle\int_{0.09}^{0.21} \frac{1}{x^6 - 7x^{3.4}}\, dx,$

$n = 6$

26. $\displaystyle\int_{-6}^{-3} \frac{1}{1 + \sin x}\, dx,$
$n = 6$

Calculator values for these integrals and values of E_T for the trapezoid rule were part of exercises 17–32 of section 4.8.

27–42. (a) Use the midpoint rule with the given value of n to estimate the value of the integral to the requested number of decimal places. (b) Using a calculator's value as the actual value of the integral, determine the error E_M in the estimate of part (a). (c) Compare the error E_M for using the midpoint rule with the error E_T for using the trapezoid rule.

27. $\displaystyle\int_{3}^{5} \sqrt{x^3 - 4}\, dx$, $n = 8$, three decimal places

28. $\displaystyle\int_{3}^{5} \sqrt{2 + x^3}\, dx$, $n = 6$, three decimal places

29. $\displaystyle\int_{0}^{1} \frac{1}{1 + x^5}\, dx$, $n = 5$, five decimal places

30. $\displaystyle\int_{6}^{24} \tan(1/x)\, dx$, $n = 6$, five decimal places

31. $\displaystyle\int_{0}^{1} \sqrt{1 + x^4}\, dx$, $n = 5$, five decimal places

32. $\displaystyle\int_{0}^{1} \sqrt{2 - \sqrt{x}}\, dx$, $n = 5$, five decimal places

33. $\displaystyle\int_{1}^{1.06} \sin(x^3 - x)\, dx$, $n = 6$, seven decimal places

34. $\displaystyle\int_{0}^{2} 2^{(x^3 - x)}\, dx$, $n = 8$, three decimal places

35. $\int_{-1}^{0} 2^{(-x^2)}\, dx$, $n = 4$, five decimal places

36. $\int_{-1}^{2} \sin(-x^2)\, dx$, $n = 6$, five decimal places

37. $\int_{4}^{9} \cos\sqrt{x}\, dx$, $n = 5$, four decimal places

38. $\int_{0}^{\pi} \sqrt{x}\cos x\, dx$, $n = 4$, five decimal places

39. $\int_{1.4}^{2.2} \cos x^2\, dx$, $n = 8$, four decimal places

40. $\int_{0}^{3} x^2 \cos x\, dx$, $n = 10$, four decimal places

41. $\int_{0.09}^{0.21} \frac{1}{x^6 - 7x^{3.4}}\, dx$, $n = 6$, three decimal places

42. $\int_{-6}^{-3} \frac{1}{1 + \sin x}\, dx$, $n = 6$, five decimal places

43. A herd of buffalo is migrating. The speed at which the lead animal is traveling is recorded every 15 minutes and is given in the following table. (a) Use Simpson's rule to give an estimate of the distance traveled from 9:00 a.m. until 11:00 a.m. (b) Compare your answer to that of section 4.8 exercise 59. Which do you think is a better estimate? Why?

time (a.m.)	speed (mph)
9:00	5.7
9:15	6.6
9:30	6.8
9:45	6.0
10:00	6.5
10:15	7.1
10:30	7.3
10:45	6.2
11:00	6.4

44. You are on a sailboat in the Atlantic Ocean, all your GPS devices and phones were swept overboard in a storm, your instrument panel does not work, and it is so cloudy you cannot see the sun, moon, or stars. You resort to dead reckoning to determine your position. You have kept a due-west course (according to your compass), and you determine your speed every half hour using knots in a rope let out into the water. (a) Use the following data to estimate the distance you have traveled from 8:00 a.m. until 12:00 noon, using Simpson's rule. (b) Compare your answer to that of section 4.8 exercise 60. Which do you think is a better estimate? Why?

time (a.m.)	speed (knots)
8:00	6
8:30	8
9:00	7
9:30	9
10:00	11
10:30	12
11:00	11
11:30	9
12:00 noon	8

45. Exercises 1–16 may give the impression that using Simpson's rule always results in a smaller error than using the trapezoid rule, but this is not always the case, as this exercise aims to show. (a) Use the trapezoid rule with $n = 6$ to estimate $\int_{-3}^{3} 2^{(-x^2)}\,dx$ (use four decimal places for all answers). (b) Use Simpson's rule with $n = 6$ to estimate $\int_{-3}^{3} 2^{(-x^2)}\,dx$. (c) Using a calculator's value as the actual value of the integral, calculate E_T and E_S. Which has the smaller error, the trapezoid rule or Simpson's rule?

46. Exercises 27–42 may give the impression that the error in the midpoint rule is always about half that of the trapezoid rule and opposite in sign, but this is not always the case, as this exercise aims to show. (a) Use the trapezoid rule with $n = 4$ to estimate $\int_0^1 \cos(16\pi x)\,dx$. (b) Use the midpoint rule with $n = 4$ to estimate $\int_0^1 \cos(16\pi x)\,dx$. (c) Evaluate $\int_0^1 \cos(16\pi x)\,dx$. (d) Calculate E_T and E_M and compare.

This exercise is based on Dawson, Bryan, "Proposed Problem 99," *Math Horizons* V No. 4 (April 1998), 32, doi: https://doi.org/10.1080/10724117.1998.11975062, and Woltermann, Michael, "Solution to Problem 99," *Math Horizons VI* No. 2 (November 1998), 34, doi: https://doi.org/10.1080/10724117.1998.11975085

47. (a) Use Simpson's rule with $n = 4$ to estimate $\int_1^5 x^3\,dx$. (b) Evaluate the integral and calculate E_S. (c) Use Simpson's rule with $n = 4$ to estimate $\int_0^1 (ax^3 + bx^2 + cx + d)\,dx$. (d) Evaluate the integral and calculate E_S. (e) What is the accuracy of Simpson's rule for a cubic polynomial? Explain by calculating the error bound.

48. Show that if f is an integrable function with $\int_a^b f(x)\,dx \neq 0$ and Ω is a positive infinite hyperreal integer, then $S_\Omega \approx \int_a^b f(x)\,dx$.

This exercise shows that $\lim_{n\to\infty} S_n = \int_a^b f(x)\,dx$. As n becomes infinite, the error E_S becomes infinitesimal. Remark: this exercise is more difficult than the corresponding exercise for the trapezoid rule.

Initial Value Problems and Net Change

4.10

Integration—both definite and indefinite—has applications beyond determining areas. Two broad application categories are initial value problems (IVPs) and net change. Many additional applications are covered in later chapters.

Initial value problems

In the most basic variation of an IVP, we are given information about a function's derivative along with an *initial condition*, which amounts to one point on the graph of the function. For instance, suppose we wish to determine a function f for which $f'(x) = x^2 + 5$ and $f(1) = 4$. Because f is an antiderivative of f', we use an indefinite integral to help:

$$f(x) = \int (x^2 + 5)\, dx = \frac{x^3}{3} + 5x + C.$$

The issue is that the indefinite integral yields not one function, but infinitely many functions, some of which are graphed in figure 1.

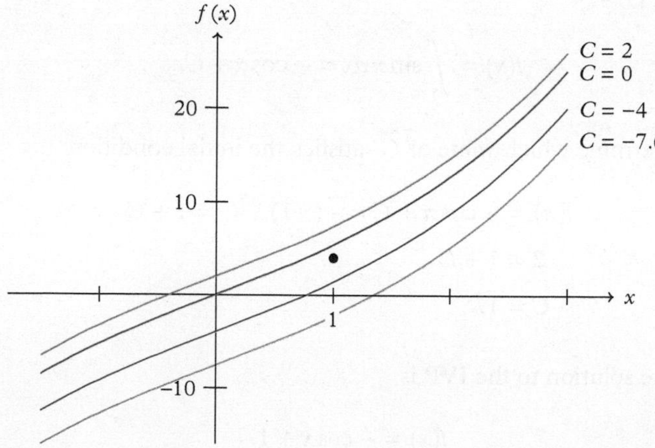

Figure 1 *The graph of* $f(x) = \frac{x^3}{3} + 5x + C$ *for* $C = 0$ *(blue),* $C = 2$ *(green),* $C = -4$ *(red), and* $C = -7.6$ *(brown), none of which pass through the point* $(1, 4)$ *(black dot). The infinitely many additional choices of* C *are not pictured*

The various functions depicted in figure 1 all have the same shape; their graphs are shifted vertically from one another, depending on the value of C. Since $f(1) = 4$, the function f must pass through the point $(1, 4)$. Returning to the figure, the idea is to determine which vertical shift (which value of C) allows the function to pass through the desired point. This can be accomplished by evaluating $f(1)$ as we have it so far, setting the result equal to 4, and solving for C:

In line 1, $f(1)$ is evaluated using the expression for $f(x)$ obtained from the antiderivative calculation. In line 2, the expression is set equal to 4. In line 3, the resulting equation is solved for C.

$$\begin{aligned} f(1) &= \frac{1^3}{3} + 5(1) + C = \frac{16}{3} + C \\ 4 &= \frac{16}{3} + C \\ C &= -\frac{4}{3}. \end{aligned}$$

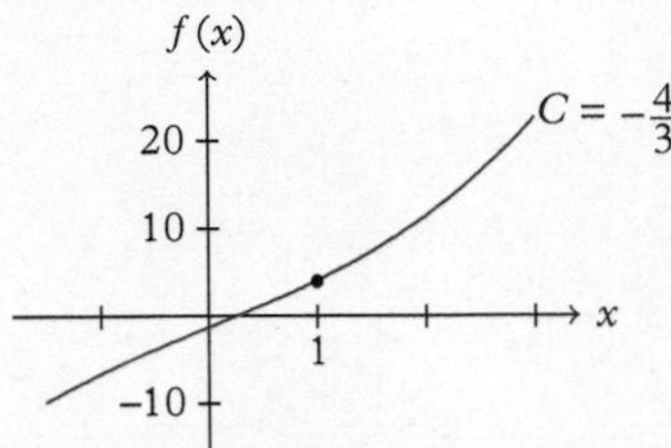

Figure 2 *The graph of* $f(x) = \frac{x^3}{3} + 5x - \frac{4}{3}$ *(blue), which passes through the point* $(1, 4)$ *(black dot). This function has the desired properties—namely, that* $f'(x) = x^2 + 5$ *and* $f(1) = 4$

The function f must correspond to $C = -\frac{4}{3}$, and therefore our function must be

$$f(x) = \frac{x^3}{3} + 5x - \frac{4}{3}.$$

Figure 2 shows the function passing through the point $(1, 4)$ as desired. We call this function the *solution to the IVP*.

The steps to solving this basic form of an IVP are ❶ antidifferentiate the given derivative, ❷ determine which value of C satisfies the initial condition, and ❸ write the final answer.

Example 1 *Solve the IVP* $f'(x) = \sin x$, $f(\pi) = 2$.

Solution ❶ Antidifferentiate:

$$f(x) = \int \sin x \, dx = -\cos x + C.$$

❷ Determine which value of C satisfies the initial condition:

Line 1 plugs the x-coordinate of the initial condition into the expression obtained in step ❶ line 2 sets the expression obtained in line 1 equal to the y-coordinate of the initial condition and line 3 solves the expression in line 2 for C.

$$\begin{aligned} f(\pi) &= -\cos \pi + C = -(-1) + C = 1 + C \\ 2 &= 1 + C \\ C &= 1. \end{aligned}$$

❸ The solution to the IVP is

In step ❸ we rewrite the function f found in step ❶ using the value of C found in step ❷.

$$f(x) = -\cos x + 1.$$ ■

Reading Exercise 28 *Find* $f(x)$ *if* $f'(x) = x^2$ *and* $f(1) = 0$.

IVP: gravity

Acceleration resulting from gravity near the surface of the earth is approximately 9.8 m/s^2, or 32 ft/s^2, toward the center of the earth. Ignoring air resistance and other factors, gravity gives us derivative information for falling objects. Specifically, we know the value of the second derivative of the position function.

There are some mountains that are massive enough to affect gravity measurably, in the sense that objects do not fall directly toward the center of the Earth, but rather slightly offline toward the mountain.

In chapter 2 we used derivatives to find velocity and acceleration from position. Here we solve IVPs to find a position function from acceleration and initial conditions.

Recall that if s is the position function, $s' = v$ is velocity and $s'' = v' = a$ is acceleration.

Example 2 *A projectile is fired vertically upward from a height of* 4 *feet above the ground with initial velocity (muzzle velocity)* 25 ft/s. *When does the projectile hit the ground? Ignore air resistance.*

The term *initial velocity* refers to the velocity at time $t = 0$—that is, at the time the projectile is fired. This is the source of the term *initial condition.*

Solution The projectile hits the ground when its height above the ground (its position) is zero. We therefore need to determine the projectile's position function.

Because units listed in the problem are feet and not meters, we use 32 ft/s^2 for the acceleration resulting from gravity. Then,

$$s''(t) = v'(t) = a(t) = -32.$$

Using height above the ground for the position function, zero is ground level, positive coordinates are up, and negative coordinates are down. Because the direction of acceleration resulting from gravity is toward the earth (downward), acceleration is negative. Because the projectile is fired upward, initial velocity is positive.

Using $t = 0$ as the time when the projectile is fired, we have the initial condition for velocity

$$v(0) = 25.$$

We therefore have the setup of an IVP for finding the velocity function (we know its derivative and an initial condition).

❶ Antidifferentiate:

$$v(t) = \int -32\,dt = -32t + C.$$

❷ Determine which value of C satisfies the initial condition:

$$\begin{aligned} v(0) &= -32(0) + C = C \\ 25 &= C. \end{aligned}$$

Using the initial condition $v(0) = 25$, line 1 evaluates the antiderivative at $t = 0$, line 2 sets the result equal to 25, and line 3 (solving for C) is unnecessary.

❸ The velocity function is

$$v(t) = -32t + 25.$$

Recall that it is the position function we wish to know. Now we know its derivative (the velocity function) and we have an initial condition. The projectile's position when it was fired was 4 feet above the ground—that is,

$$s(0) = 4.$$

Notice the familiar $-16t^2$ that was used in previous examples and exercises for falling objects using English units. The source of this repeatedly used term, as well as for $-4.9t^2$ when using SI units, is acceleration resulting from gravity.

We can solve this new IVP to find the position function.

❶ Antidifferentiate:

$$s(t) = \int (-32t + 25)\, dt = -16t^2 + 25t + C.$$

❷ Determine which value of C satisfies the initial condition:

CAUTION: MISTAKE TO AVOID
It is **not** always the case that when the initial condition is $f(0) = k$ that $C = k$.

$$s(0) = -16(0)^2 + 25(0) + C = C$$
$$4 = C.$$

❸ The position function is

$$s(t) = -16t^2 + 25t + 4.$$

Now we can determine when the projectile hits the ground. We wish to know the time ("when") that the position ($s(t)$) is zero (ground level). We therefore set the position function to zero and solve for time t:

$$-16t^2 + 25t + 4 = 0.$$

Because the equation is quadratic and cannot be factored easily, we use the quadratic formula:

Ans. to reading exercise 28
$f(x) = \frac{x^3}{3} - \frac{1}{3}$

$$t = \frac{-25 \pm \sqrt{25^2 - 4(-16)(4)}}{2(-16)},$$

giving

$$t = -0.1463, t = 1.7088.$$

Because negative values of time t correspond to before the projectile is fired, we eliminate $t = -0.1463$ and conclude the projectile hits the ground at $t = 1.7088$ seconds after the projectile is fired. ■

Units for time are seconds, as given in the acceleration resulting from gravity and the initial velocity.

Notice that in example 2 we solved not just one, but two, IVPs when going from acceleration to position. As in previous exercises, we had to differentiate twice to go from position to acceleration. When we differentiated twice, we called the result the *second derivative;* when we solve two IVPs with initial conditions at the same value of t, we call the overall problem a *second-order IVP.*

The second-order IVP solved in example 2 can be rewritten as $s''(t) = -32, s'(0) = 25, s(0) = 4$.

Net change

One interpretation of the definite integral is the net area under the curve, which is not the same thing as the total area between the curve and the x-axis. Likewise, in the setting of distance traveled instead of area, there is a difference between net change in position and total distance traveled.

Consider a vehicle with an internal combustion engine. The total distance traveled since its last oil change is important. The net change in position is essentially zero if the vehicle is returned to its owner's driveway every evening, but this is irrelevant. The oil needs to be changed after a certain total distance traveled. On the other hand, if we are traveling from home to the airport, the net change in position is important and the total distance traveled is irrelevant. What matters is that we arrive at the airport, which is a fixed distance from the starting location, regardless of whether we had to turn around to return home to get forgotten luggage or had to take an unexpected detour to bypass a flooded road.

An internal combustion engine typically runs on gasoline or diesel fuel.

Net change in position and total distance traveled are different concepts, each of which has applications.

Let $s(t)$ be the position of a particle on a number line. The velocity is its derivative: $v(t) = s'(t)$. Reversing, position is an antiderivative of velocity and, by FTC II,

$$\int_a^b v(t)\,dt = s(t)\Big|_a^b = s(b) - s(a),$$

which is the difference between the position at time $t = b$ and the position at time $t = a$, the *net change in position* from $t = a$ to $t = b$. The definite integral of velocity is the net change in position.

Generically, the definite integral of a rate of change (such as the velocity) is the net change.

NET CHANGE INTERPRETATION OF THE DEFINITE INTEGRAL

The definite integral of a rate of change F' is the net change in F:

$$\int_a^b F'(t)\,dt = F(b) - F(a).$$

The time period over which the net change occurs is from $t = a$ to $t = b$.

This is a restatement of FTC II and therefore requires F' to be continuous. It is sometimes referred to as the *net change theorem*.

Example 3 *The rate of growth of a retirement account in dollars per year is $r'(t)$. What does $\int_{2025}^{2040} r'(t)\,dt$ represent?*

Solution We are given the rate of change (rate of growth) of what quantity? A retirement account. The definite integral of the rate of change is the net change, so the integral represents the net change in the retirement account from 2025 to 2040. ■

More specifically, $r(t)$ is the value of the retirement account, in dollars, with units of time t given as years. The integral represents the net change in value of the retirement account, measured in dollars, from year 2025 to year 2040.

If the retirement account of example 3 is invested in stocks and bonds, the value of the account changes constantly, up and down on a minute-by-minute basis. The net change does not take the total distance traveled of the value of the account up and down and up and down, because this is not relevant. Instead, it simply gives the difference between the account value in 2040 and the account value in 2025 (presumably at a specified time in those years).

Reading Exercise 29 *The rate of change of temperature of a casserole in an oven is $T'(t)$, given in degrees Fahrenheit per minute. What does $\int_0^{90} T'(t)\,dt$ represent?*

Displacement vs. total distance traveled

Back to the trip to the airport. If we never turned around and just drove straight to the airport, the total distance traveled and the net change in position (also called *displacement*) are identical. It is only when turning around and backtracking or taking a detour that the two quantities differ.

For the purpose of intuition, assume the road is straight.

Suppose a particle is traveling on a number line. In order for displacement to be, in absolute value, less than the total distance traveled, the particle has to change directions from forward to backward or from backward to forward. These changes in direction are only possible when the particle stops—that is, when the velocity is zero. Then, between such stops, the distance traveled is the same as the (absolute value of the) net change in position and can therefore be calculated using the definite integral of velocity.

Traveling on a number line (one dimension) is important here. When traveling in a plane (two dimensions), the particle can change directions without ever stopping.

Example 4 *A particle moves on a number line so that its velocity is $v(t) = (4t - 10)$ m/s. (a) Find the particle's displacement over the time period $0 \le t \le 3$. (b) Find the total distance traveled during that time period.*

The displacement of the particle is the difference between its starting and ending positions (locations). By contrast, borrowing a term from a vehicle, the total distance traveled by the particle is the difference between the starting and ending numbers on its odometer.

Solution (a) We are asked for the net change in position (displacement) of the particle from $t = 0$ to $t = 3$. By the net change interpretation of the definite integral,

$$s(3) - s(0) = \int_0^3 v(t)\,dt = \int_0^3 (4t - 10)\,dt$$

$$= (2t^2 - 10t)\Big|_0^3 = \cdots = -12\,\text{m}.$$

Units for a definite integral $\int_a^b f(x)\,dx$ are units of the integrand $f(x)$ times units of the variable x. For this integral, units are the units of velocity times units of the variable t (time):

$$\frac{\text{meters}}{\text{seconds}} \cdot \text{seconds} = \text{meters}.$$

The particle's displacement on $[0, 3]$ is -12 m, meaning the particle's position at time $t = 3$ is 12 m to the left of its position at time $t = 0$.

(b) To find the total distance traveled, we must determine when the particle changed direction. Stopping happens when the velocity is zero, so we set the velocity equal to zero and solve for time t:

$$v(t) = 4t - 10 = 0$$

$$4t = 10$$

$$t = \frac{5}{2}.$$

This is identical to finding the critical numbers of the position function. We set its derivative, $v(t)$, equal to zero and solve the resulting equation. Determining the intervals over which we must find displacement is the same process as determining the intervals for the charts used in determining intervals of increase/decrease.

This gives us two intervals on which to find the displacement, $\left(0, \frac{5}{2}\right)$ and $\left(\frac{5}{2}, 3\right)$:

$$\int_0^{5/2} (4t - 10)\,dt = (2t^2 - 10t)\Big|_0^{5/2} = \cdots = -12.5\,\text{m},$$

$$\int_{5/2}^3 (4t - 10)\,dt = (2t^2 - 10t)\Big|_{5/2}^3 = \cdots = 0.5\,\text{m}.$$

Note that the net change in position, or displacement, is $-12.5\,\text{m} + 0.5\,\text{m} = -12\,\text{m}$. The net change pays attention to the signs ($\pm$) of the quantities, whereas the total distance traveled adds the absolute values of those quantities. In other words, the net change takes into account direction to arrive at change in position whereas the total distance traveled does not care about direction at all.

Therefore, from $t = 0$ to $t = \frac{5}{2}$, the particle moved 12.5 m backward (to the left), whereas from $t = \frac{5}{2}$ to $t = 3$, the particle moved forward (to the right) 0.5 m. The total distance traveled is therefore

$$12.5\,\text{m} + 0.5\,\text{m} = 13\,\text{m}.$$

■

There is a relationship between IVPs, which deal with the indefinite integral, and net change, which deals with the definite integral. To illustrate, an alternate approach to working part (a) of example 4 is first to determine the position function $s(t)$ and then find $s(3) - s(0)$. But, without an initial condition, we do not have an IVP and we cannot find $s(t)$ exactly. The beauty of the net change approach used in the solution to example 4 is that any antiderivative of velocity can be used; we do not need to determine $s(t)$ specifically.

Ans. to reading exercise 29
the net change in temperature of the casserole in the first 90 minutes (i.e., between time $t = 0$ and time $t = 90$)

Reading Exercise 30 *A particle moves on a number line with velocity* $v(t) = 5t$ *ft/s. What is the displacement of the particle over the time period* $2 \le t \le 4$?

Exercises 4.10

1–14. Solve the IVP.

1. $f'(x) = x^3, f(-1) = 2$
2. $f'(x) = 3x^2 - 5, f(0) = 11$
3. $f'(x) = \sin(2x), f(0) = 5$
4. $f'(x) = \sec x \tan x\,, f(0) = 0$
5. $\frac{dy}{dx} = 2x - 3, y(1) = 5$
6. $\frac{dy}{dx} = \sqrt{x}, y(4) = 1$
7. $\frac{dy}{dx} = \dfrac{2x^2 - 4x}{x}, y(1) = 5$
8. $\frac{dy}{dx} = x^{-2}(3 + x^3), y(10) = 14$
9. $\frac{dy}{dx} = \dfrac{2x^4 - x - 2}{x^3}, y(1) = 4$
10. $\frac{dy}{dx} = \dfrac{5x - 1}{\sqrt{x}}, y(1) = 1$
11. $f'(x) = x\sec^2 x^2, f(\sqrt{\pi}) = 0$
12. $f'(x) = \dfrac{\cos\sqrt{x}}{\sqrt{x}}, f(\pi^2) = 2$
13. $f''(x) = x + 2, f'(3) = 1, f(3) = 0$
14. $f''(x) = 3x^2, f'(1) = 1, f(1) = 2$

interpretation exercises

15–45. 15. The rate of change of the volume of water in a lake, in gallons per hour, is given by $W'(t)$. What does $\int_0^{24} W'(t)\,dt$ represent?

16. The rate of growth in an animal population, measured in animals per year, is given by $P'(t)$. What does $\int_1^7 P'(t)\,dt$ represent?

17. Give an interpretation of $\int_{-10}^{-5} y'(t)\,dt$, where $y'(t)$ is the rate of change per week in the percentage of likely voters in support of Proposition 1 ($t = -2$ means 2 weeks before the election).

18. Give an interpretation of $\int_0^{120} g'(t)\,dt$, where $g'(t)$ is the rate of change in the level of an enzyme t minutes after administering a drug.

particle motion IVPs

19. A particle moves along a number line with velocity $v(t) = 3t^2 + 5$. The position of the particle at time $t = 1$ is $s(1) = 4$. Find the position function.

20. A particle moves along a number line with velocity $v(t) = 3t + 1$. The position of the particle at time $t = 2$ is $s(2) = 15$. Find the position function.

21. A particle moves along a number line with velocity $v(t) = t - 7$. The position of the particle at time $t = 3$ is $s(3) = \frac{1}{2}$. Find the position function.

22. A particle moves along a number line with velocity $v(t) = t^2 - 1$. The position of the particle at time $t = 1$ is $s(1) = 2$. Find the position function.

23. A particle moves along a number line with acceleration $a(t) = 3 + 2t$. The velocity of the particle at time $t = 1$ is $v(1) = 6$. Find the velocity function.

Ans. to reading exercise 30
30 feet

24. A particle moves along a number line with acceleration $a(t) = 4\cos t$. The velocity of the particle at time $t = 0$ is $v(0) = 6$. Find the velocity function.

vertical projectile motion IVPs

Ignore air resistance for all of these exercises.

25. A projectile is fired vertically upward from a platform 3 m above the ground with initial velocity 10 m/s. When does the projectile hit the ground?

26. A projectile is fired vertically upward from ground level with initial velocity 100 m/s. When does the projectile hit the ground?

27. A model rocket is fired vertically upward from the ground with initial velocity 40 ft/s. When does the rocket reach its maximum height?

28. A model rocket is fired vertically upward from the ground with initial velocity 30 ft/s. How high does the rocket go before beginning to fall?

29. Repeat exercise 25 for a projectile on Mars, where the acceleration resulting from gravity is $3.71\ \mathrm{m/s^2}$.

30. Repeat exercise 26 for a projectile on Eris, where the acceleration resulting from gravity is $0.83\ \mathrm{m/s^2}$.

displacement, other net change, and total distance traveled exercises

31. A particle moves on a number line so that its velocity is $v(t) = (7-2t)$ m/s. (a) Find the particle's displacement over the time period $0 \le t \le 5$. (b) Find the total distance traveled during this time period.

32. A particle moves on a number line so that its velocity is $v(t) = (4-t^2)$ m/s. (a) Find the particle's displacement over the time period $1 \le t \le 3$. (b) Find the total distance traveled during this time period.

33. A particle moves on a number line so that its velocity is $v(t) = 5\cos 2t$ m/s. (a) Find the particle's displacement over the time period $0 \le t \le \pi$. (b) Find the total distance traveled during this time period.

34. A particle moves on a number line so that its velocity is $v(t) = t^2$ m/s. (a) Find the particle's displacement over the time period $-1 \le t \le 3$. (b) Find the total distance traveled during this time period.

35. A particle moves on a number line so that its acceleration is $v(t) = (t+1)$ m/s. Find the net change in velocity over the time period $0 \le t \le 5$.

36. A quantity $r(t)$ changes over time (where $t \ge 1$) at the rate of $r'(t) = \frac{1}{\sqrt{t}}$. What is the quantity's net change from $t = 4$ to $t = 25$?

miscellaneous exercises

37. (GSL) At every point of a certain curve, $y'' = \frac{12}{x^3}$. Find the equation of the curve if it passes through the point $(1, 0)$ and is tangent to the line $6x+y = 6$ at that point.

38. (GSL) At every point of a certain curve, $y'' = x$. Find the equation of the curve if it passes through the point $(3, 0)$ and has the slope $\frac{7}{2}$ at that point.

39. (GSL) Given $\frac{dy}{dx} = x\sqrt{100 - x^2}$ and $y = 0$ when $x = 0$, find the value of y when $x = 8$.

40. (GSL) Given $\frac{dy}{dx} = 2x + 1$ and $y = 7$ when $x = 1$, find the value of y when $x = 3$.

The *inclination* of a curve at a point $(x, f(x))$ is the angle made between the tangent line and the horizontal, as measured in standard position.

41. (GSL) Find the equation of the curve with second derivative $y'' = \frac{3}{\sqrt{x+3}}$ and that passes through the point $(1, 1)$ with an inclination of $45°$.

42. Find the equation of the curve with second derivative $y'' = (2x + 1)^3$ and that is tangent to the x-axis at the origin.

43. A particle moves on a number line so that its velocity is $v(t) = \sin t$ m/s. (a) Find the particle's displacement over the time period $0 \le t \le 200\pi$. (b) Find the total distance traveled during this time period.

44. A particle moves on a number line so that its acceleration is $v(t) = (t - 3)$ m/s, with velocity 4 m/s at time $t = 0$. Find the total distance traveled during the time interval $[1, 7]$.

45. It was noted in section 4.1 that an indefinite integral is valid on an interval I throughout which the antiderivative relationship must hold. In particular, the functions must be defined throughout the interval. For IVPs such as those in exercises 1 and 3, for which the given derivative is defined for all real numbers, the interval can be taken to be $(-\infty, \infty)$. Find an appropriate interval for which the solution to the IVP is valid for (a) exercise 9 and for (b) exercise 11.

Chapter V

Transcendental Functions

Logarithms, Part I

5.1

We know there are many functions for which the available antiderivative formulas, rules, and techniques are not sufficient to find an antiderivative. This chapter explores new functions that expand our ability to integrate. First up is the antiderivative for $f(x) = \frac{1}{x}$.

Introducing the natural logarithmic function

The antiderivative power rule states that

$$\int x^n\,dx = \frac{x^{n+1}}{n+1} + C$$

for $n \neq -1$. Then,

$$\int \frac{1}{x}\,dx = \int x^{-1}\,dx$$

cannot be worked using the antiderivative power rule. Why not? Because we would be dividing by zero. We would have $\frac{x^0}{0}$, which is undefined.

The fundamental theorem of calculus, part I (FTC I) says that if a function f is continuous on $[a, b]$ then

$$F(x) = \int_a^x f(t)\,dt$$

is an antiderivative of f. Then, setting $f(t) = \frac{1}{t}$ allows us to build an antiderivative F through the use of the definite integral! The new function is called the *natural logarithmic function*, or *natural log* for short.

Definition 1 NATURAL LOGARITHMIC FUNCTION *For* $x > 0$,

$$\ln x = \int_1^x \frac{1}{t}\,dt.$$

If you have studied this function in an algebra course using a different definition, for the purposes of this section, pretend this is a new function about which you know nothing.

The expression "$\ln x$" is read "natural log x" or "the natural logarithm of x." It can also be read using the names of the letters: "el - en ex."

FTC I uses a closed interval $[a, b]$, yet here we use an open interval $(0, \infty)$ for the domain of $y = \ln x$. To see why this is justified, think of defining $\ln x$ at $x = \frac{1}{4}$. We use the interval $[a, b] = [0.1, 30]$ and then apply FTC I. For an even smaller value of x, move the interval. For defining $\ln x$ at $x = 0.02$, use the interval $[0.01, 30]$. The same trick allows $\ln x$ to be defined for arbitrarily large values of x as well.

The restriction that $x > 0$ is necessary to ensure continuity on the interval of integration. Since $f(t) = \frac{1}{t}$ has a discontinuity at $t = 0$ and the lower limit of integration is positive, then the upper limit of integration must also be positive.

Now by FTC I we have the antiderivative we seek: $\frac{d}{dx} \ln x = \frac{1}{x}$.

DERIVATIVE OF THE NATURAL LOGARITHM

$$\frac{d}{dx} \ln x = \frac{1}{x}$$

Example 1 *Find* $\frac{d}{dx} x \ln x$.

Notationally, "ln" is like "sin" or other trig function names. Recognizing the need for the product rule, the chain rule, and so on for the natural log is the same as for trig functions.

Solution Using the product rule,

$$\frac{d}{dx} x \ln x = x \cdot \frac{1}{x} + \ln x \cdot 1 = 1 + \ln x. \quad \blacksquare$$

Combined with the chain rule and other derivative rules, the formula for the derivative of the natural log allows us to differentiate all functions involving natural logarithms.

DERIVATIVE OF NATURAL LOG (CHAIN RULE FORM)

$$\frac{d}{dx} \ln u = \frac{1}{u} \frac{du}{dx}$$

Example 2 *Find* $f'(x)$ *for* $f(x) = 2 \ln(4x - 7)$.

The constant multiple rule is also used here.

Solution Because we see the natural log of "not just plain x," the chain rule form of the derivative of the natural log is needed:

$$\begin{aligned} f'(x) &= 2 \cdot \frac{d}{dx} \ln(4x - 7) = 2 \cdot \frac{1}{4x - 7} \cdot \frac{d}{dx}(4x - 7) \\ &= 2 \cdot \frac{1}{4x - 7} \cdot 4 = \frac{8}{4x - 7}. \quad \blacksquare \end{aligned}$$

Example 3 *Find* y' *for* $y = \ln \sqrt{x}$.

There is an alternate solution to this problem in example 6.

Solution Again we have the natural log of "not just plain x," so we use the chain rule form of the derivative of the natural log:

$$y' = \frac{1}{\sqrt{x}} \cdot \frac{d}{dx}\sqrt{x} = \frac{1}{\sqrt{x}} \cdot \frac{1}{2\sqrt{x}} = \frac{1}{2x}. \quad \blacksquare$$

Reading Exercise 1 *Find y' for $y = \ln(4x)$.*

Graph of y = ln x

Now that we have defined a new function, we need to gain familiarity with it. When you needed to gain familiarity with trig functions such as $y = \sin x$, this task was likely accomplished in three parts: (1) evaluate the function by hand and with a calculator, (2) graph the function, and (3) perform algebraic manipulations with the function, such as developing and using trigonometric identities. These same three steps can be used to become comfortable with any function.

With only a couple of notable exceptions, which are pointed out along the way, logarithms are evaluated here using a calculator. The ln button on a calculator is used in the same fashion as the sin and cos buttons. Try it. For ln(12.7) you should get 2.5416.

Why the order of the initials "ln" instead of "nl" for natural logarithm? In Latin and many other languages, adjectives come after nouns; the Latin term is *logarithmus naturalis*.

Reading Exercise 2 *Use a calculator to find (a)* $\ln 427.3$ *and (b)* $\ln 10\,589\,000$.

The second task in gaining familiarity is to graph the function. We can explore the graph of $y = \ln x$ using the methods developed in chapter 3 for sketching a curve.

The definition of $y = \ln x$ requires $x > 0$. This gives us the domain. In interval notation, the domain is $(0, \infty)$.

Any function defined through FTC I is differentiable (the derivative of $y = \ln x$ is $y = \frac{1}{x}$); by the theorem that says differentiability implies continuity, our function is continuous throughout its domain.

The function has an x-intercept at $x = 1$. This is seen by applying the definition for $x = 1$:

$$\ln 1 = \int_1^1 \frac{1}{t}\, dt = 0,$$

because the limits of integration are the same. This is one value of the function that should be learned (memorized):

ln 1 = 0.

$$\ln 1 = 0.$$

We therefore know a point on the graph: $(1, 0)$.

Because $x = 0$ is not in the domain of the function, there is no y-intercept.

Next on the list is to determine intervals of increase/decrease and local extrema for $y = \ln x$. We find critical numbers by setting the derivative equal to zero,

$$\frac{1}{x} = 0,$$

The other type of critical number to look for is where the derivative is undefined. The only value for which the derivative is undefined is at $x = 0$, but this is not in the domain of the function and therefore is not a critical number.

which has no solutions, so there are no critical numbers and therefore no local extrema.

With no critical numbers, there is only one interval to use for finding intervals of increase/decrease—namely, the domain $(0, \infty)$. Because $\frac{1}{x}$ is positive whenever x is positive, our function $y = \ln x$ is increasing on $(0, \infty)$:

interval	sign of f'	inc/dec	local extrema
$(0, \infty)$	+	increasing	none

Ans. to reading exercise 1
$y' = \frac{1}{4x} \cdot 4 = \frac{1}{x}$

Next comes concavity and inflection points, for which we need the second derivative. Since $y' = \frac{1}{x} = x^{-1}$,

$$y'' = -x^{-2} = -\frac{1}{x^2}.$$

Setting $y'' = 0$,

$$-\frac{1}{x^2} = 0,$$

The other type of potential inflection point to look for is where the second derivative is undefined, but this value ($x = 0$) is not in the domain of the function.

which yields no solutions. There are therefore no points of inflection, and there is only one interval for our chart:

interval	sign of f''	concavity	inflection
$(0, \infty)$	−	down	none

So far, we have an function that is increasing and concave down throughout its domain $(0, \infty)$, and it passes through the point $(1, 0)$. This is enough to get a rough sketch; see figure 1.

Figure 1 brings up two questions: first, is there a horizontal asymptote or does it continue to rise (albeit slowly) as does $y = \sqrt{x}$? Second, is there a vertical asymptote at $x = 0$? It turns out that there is a vertical asymptote but no horizontal asymptote, although we must explore some algebraic manipulations before we can make these conclusions.

Ans. to reading exercise 2
(a) 6.057, (b) 16.175

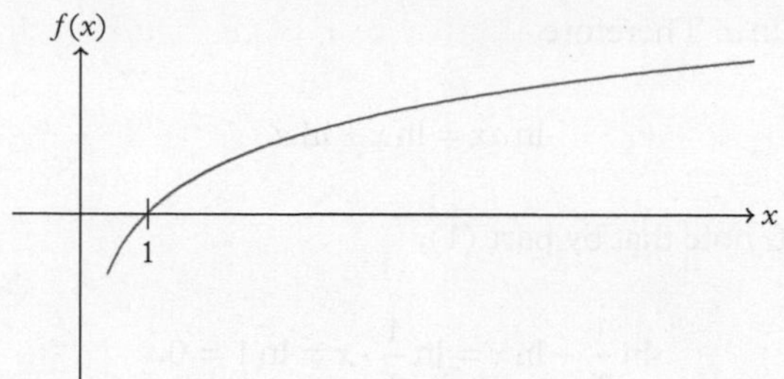

Figure 1 *A first sketch of the graph of $y = \ln x$, which is increasing and concave down on $(0, \infty)$, and passes through the point $(1, 0)$. There are no local extrema or points of inflection on the graph of the function. Because the function is continuous throughout its domain, it can be drawn in one piece, and because the function is differentiable throughout its domain, there are no corners on the graph*

Laws of logarithms

Three important rules for manipulating logarithmic expressions are called *laws of logarithms*. These laws are not the same as for manipulating trig functions, and other functions have yet different laws for algebraic manipulations. Care must be taken not just to assume that functions can be manipulated any way one wants; any algebraic manipulations for a function must be proved for that function.

You may have studied laws of logarithms before, but any proof given in an algebra class using a different definition of logarithm is not valid here. Proofs based on definition 1 are required.

Theorem 1 LAWS OF LOGARITHMS *For any positive real numbers a and x and rational number n,*

(1) $\ln ax = \ln a + \ln x$,

(2) $\ln \frac{a}{x} = \ln a - \ln x$, *and*

(3) $\ln x^n = n \ln x$.

Part (3) of the theorem is true for any real number n, but the proof here is valid only for rational n.

Proof. (1) Begin by calculating two derivatives:

$$\frac{d}{dx} \ln x = \frac{1}{x}$$

$$\frac{d}{dx} \ln ax = \frac{1}{ax} \cdot a = \frac{1}{x}.$$

Here we are taking the derivative with respect to x, so a is a constant.

The two functions $y = \ln x$ and $y = \ln ax$ have the same derivative and must therefore differ by a constant:

Notice that the last portion of the proof of part (1) resembles solving an initial value problem (IVP).

$$\ln ax = \ln x + C.$$

This is true for all x, including $x = 1$:

$$\ln a \cdot 1 = \ln 1 + C = 0 + C = C,$$

Although we have been treating x as a variable and a as a constant, there is no such distinction necessary in this final statement; it works for any positive number (any value of the variable) x and any positive number a.

hence $C = \ln a$. Therefore,

$$\ln ax = \ln x + \ln a.$$

(2) First, note that by part (1),

$$\ln \frac{1}{x} + \ln x = \ln \frac{1}{x} \cdot x = \ln 1 = 0,$$

which can be rearranged to conclude that

$$\ln \frac{1}{x} = -\ln x.$$

Once more applying part (1),

$$\ln \left(\frac{a}{x}\right) = \ln \left(a \cdot \frac{1}{x}\right) = \ln a + \ln \frac{1}{x} = \ln a - \ln x.$$

So far we have only proved the power rule for rational exponents. Therefore (so far), this proof of part (3) is also valid only for rational exponents.

(3) As in part (1), we start with two derivative calculations:

$$\frac{d}{dx} \ln x^n = \frac{1}{x^n} \cdot nx^{n-1} = n \cdot \frac{x^{n-1}}{x^n} = n \cdot \frac{1}{x}$$

$$\frac{d}{dx} n \ln x = n \cdot \frac{1}{x}.$$

Because the two functions have the same derivative, they must differ by a constant:

$$\ln x^n = n \ln x + C.$$

Again using $x = 1$ and solving for C, we have

$$\begin{aligned} \ln 1^n &= n \ln 1 + C \\ \ln 1 &= n \ln 1 + C \\ 0 &= n \cdot 0 + C \\ 0 &= C. \end{aligned}$$

Therefore,

$$\ln x^n = n \ln x. \qquad \blacksquare$$

Using the laws of logarithms

The original importance of logarithms was their ability to simplify calculations. The laws of logarithms take arithmetic operations down a notch; the log of a product is the sum of the logs, and addition is easier than multiplication. Likewise, the log of a quotient is the difference of the logs, and subtraction is easier than division. Logs even take exponentiation down to multiplication. By reducing complexity, these algebraic manipulations can be used for simplifications in calculus procedures as well. First, however, we need to review the algebraic use of laws of logarithms.

The laws of logarithms can be used from left to right, resulting in an expansion of the expression, or they can be used from right to left to combine expressions. Let's look at one example of each. If you studied logarithms in an algebra class, then these examples should be review.

Example 4 *Express* $2\ln x + \ln 3 - \ln y$ *as a single logarithm.*

Solution This task is accomplished by moving from the right side of each law of logarithms to the left side:

$$\begin{aligned} 2\ln x + \ln 3 - \ln y &= \ln x^2 + \ln 3 - \ln y \\ &= \ln 3x^2 - \ln y \\ &= \ln \frac{3x^2}{y}. \end{aligned}$$ ■

Line 1 uses law (3), line 2 uses law (1), and line 3 uses law (2).

In the solution to example 4, law (3) was used first, before laws (1) and (2). This order is important, because

$$2\ln x + \ln 3 \neq 2(\ln x + \ln 3) = 2(\ln 3x).$$

CAUTION: MISTAKE TO AVOID
When combining logarithms, always use law (3) first, if applicable. Otherwise, you might violate the order of operations and create an expression not equal to the original expression.

Using law (1) first to get the right-hand side of the previous expression is essentially an order-of-operations error, or equivalent to an error in writing parentheses.

Example 5 *Expand the expression* $\ln\left(\dfrac{(3x+4)^4(x^2+4)}{\cos x}\right)$.

Solution The order of operations dictates how an expression is interpreted and therefore affects the order of use of the laws of logarithms. But, instead of working from inside outward, as in the order of operations, we work from outside inward to expand the expression:

Line 1 uses law (2), line 2 uses law (1), and line 3 uses law (3).

Why this order? Law (3) cannot be used in line 1 or in line 2 because the preceding expression was not yet of the form $\ln(\text{stuff})^4$.

$$\begin{aligned}\ln\left(\frac{(3x+4)^4(x^2+4)}{\cos x}\right) &= \ln\left((3x+4)^4(x^2+4)\right) - \ln\cos x \\ &= \ln(3x+4)^4 + \ln(x^2+4) - \ln\cos x \\ &= 4\ln(3x+4) + \ln(x^2+4) - \ln\cos x. \quad \blacksquare\end{aligned}$$

CAUTION: MISTAKE TO AVOID

$\ln(a+b) \neq \ln a + \ln b$

Notice where the expansion stops in example 5. Because there is no law of logarithms for handling the log of a sum using the standard operations, we cannot expand $\ln(3x+4)$ or $\ln(x^2+4)$.

Reading Exercise 3 *Expand* $\ln\left(\frac{(x+5)^7(x+1)}{x^3}\right)$.

Differentiating logs, simplified

Because logarithms take arithmetic operations down a notch, expanding an expression using the laws of logarithms can lead to simpler differentiation.

Example 6 *(example 3 revisited) Differentiate* $y = \ln\sqrt{x}$.

Solution First rewrite the expression using laws of logarithms:

$$y = \ln\sqrt{x} = \ln x^{1/2} = \frac{1}{2}\ln x.$$

Then,

$$y' = \frac{1}{2}\cdot\frac{1}{x} = \frac{1}{2x}. \quad \blacksquare$$

The solution in example 3 uses the chain rule, but the solution in example 6 only uses the constant multiple rule. The next example demonstrates how to avoid the product and quotient rules.

Example 7 *Differentiate* $y = \ln\left(\frac{(x+4)^5(x-2)}{x^2}\right)$.

Solution Rather than use the chain rule inside the product rule inside the quotient rule inside the chain rule, expand the expression using the laws of logarithms first:

$$\begin{aligned} y &= \ln\left(\frac{(x+4)^5(x-2)}{x^2}\right) \\ &= \ln\left((x+4)^5(x-2)\right) - \ln x^2 \\ &= \ln(x+4)^5 + \ln(x-2) - \ln x^2 \\ &= 5\ln(x+4) + \ln(x-2) - 2\ln x. \end{aligned}$$

Line 2 uses law (2), line 3 uses law (1), and line 4 uses law (3). Because we do not have a law for log of a sum or a difference, the expansion is finished after line 4.

Now the derivative is much simpler to compute:

$$y' = 5 \cdot \frac{1}{x+4} \cdot 1 + \frac{1}{x-2} \cdot 1 - 2 \cdot \frac{1}{x} = \frac{5}{x+4} + \frac{1}{x-2} - \frac{2}{x}. \quad ■$$

Although the chain rule is still needed, the product and quotient rules are avoided.

Try computing the derivative for example 7 without using the laws of logarithms first. Not only is the derivative much more complicated to take, the resulting expression needs much simplification to arrive at the nice expression at the end of the previous solution.

Reading Exercise 4 *Differentiate* $y = \ln \sqrt[3]{x}$.

Asymptotes on $y = \ln x$

The laws of logarithms can help us determine whether there are asymptotes on the graph of $y = \ln x$.

Ans. to reading exercise 3
$7\ln(x+5) + \ln(x+1) - 3\ln x$

For the horizontal asymptote, we need to calculate

$$\lim_{x\to\infty} \ln x = \ln \Omega,$$

and thus we need to know something about $\ln \Omega$. Because $y = \ln x$ is an increasing function and $\Omega > 10^n$ for any natural number n,

$$\begin{aligned} \ln \Omega &> \ln 10^n \\ &= n \ln 10 \\ &> n. \end{aligned}$$

In line 1, we use $y = \ln x$ is increasing, line 2 uses law (3), and line 3 uses $\ln 10 = 2.3026 > 1$.

Therefore, $\ln \Omega$ is larger than any real number. It is positive infinite and renders ∞. The limit calculation, in full, becomes

$$\lim_{x\to\infty} \ln x = \ln \Omega \doteq \infty.$$

There is no horizontal asymptote on the graph of $y = \ln x$.

We can only check the limit from the right because the domain of $y = \ln x$ is $x > 0$.

Line 2 uses law of logarithms (3) and line 3 uses the earlier fact that $\ln \Omega \doteq \infty$.

We also need to check for a vertical asymptote at $x = 0$:

$$\begin{aligned}\lim_{x\to 0^+} \ln x &= \ln(0+\omega) = \ln \omega = \ln \frac{1}{\Omega} = \ln \Omega^{-1} \\ &= -\ln \Omega \\ &\doteq -\infty.\end{aligned}$$

There is a vertical asymptote at $x = 0$ on the graph of $y = \ln x$.

The graph of $y = \ln x$ is revisited in figure 2. Although the graph

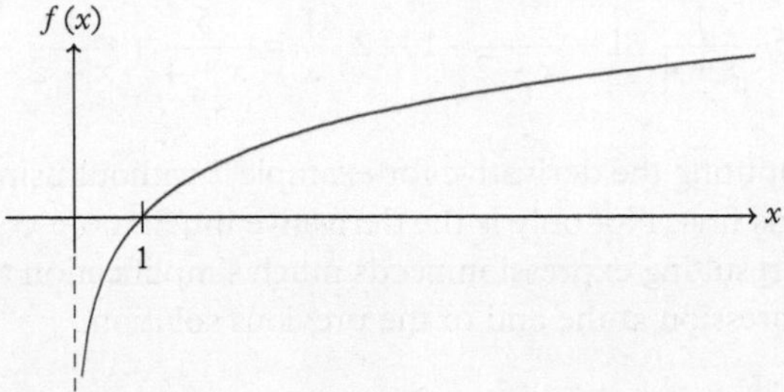

Figure 2 *The graph of $y = \ln x$, which is continuous, differentiable, increasing and concave down on $(0, \infty)$, and passes through the point $(1, 0)$. There is a vertical asymptote at $x = 0$, but there is no horizontal asymptote*

rises to ∞, it goes there very, very slowly. A few points on the graph are in the following chart:

x	$y = \ln x$
427	6.057
10 589 000	16.175
3 617 824 005	22.009

In fact, at the scale of the printed picture in figure 2, to get a y-coordinate a half meter higher we need to move along the x-axis so far that we leave our galaxy!

Limits with logs

Back to earth. The calculations before figure 2 show how to handle certain logarithmic expressions:

LOGARITHMS OF HYPERREALS

For any positive infinitesimal ω and any positive infinite hyperreal Ω,

$$\ln(\omega) \doteq -\infty$$
$$\ln(\Omega) \doteq \infty.$$

The log of a positive infinitesimal renders $-\infty$, whereas the log of a positive infinite hyperreal renders $+\infty$.

These facts are quite useful for determining limits. But, first we need to practice rendering.

Ans. to reading exercise 4
$y' = \frac{1}{3x}$

Example 8 *Render a real result for the expression, given (as usual) that ω is a positive infinitesimal: (a)* $\ln(3\omega - \omega^4)$ *(b)* $\ln(5\sqrt{\Omega} - \Omega + 0.7\Omega^2)$.

Solution (a) Because the log of a positive infinitesimal renders $-\infty$ and $3\omega - \omega^4 \approx 3\omega$ is a positive infinitesimal,

$$\ln(3\omega - \omega^4) \doteq -\infty.$$

(b) Because the log of a positive infinite hyperreal renders ∞ and $5\sqrt{\Omega} - \Omega + 0.7\Omega^2 \approx 0.7\Omega^2$ is a positive infinite hyperreal,

$$\ln(5\sqrt{\Omega} - \Omega + 0.7\Omega^2) \doteq \infty.$$ ■

Example 9 *Find* $\lim_{x \to 1^+} \ln(x^2 - 1)$.

Solution As always, our first thought is to try to evaluate using continuity:

Just as with other functions we have studied, $y = \ln x$ is continuous throughout its domain.

$$\lim_{x \to 1^+} \ln(x^2 - 1) = \ln(1^2 - 1) = \ln 0,$$

which is undefined, so evaluating using continuity fails.

Therefore, to find the limit, we evaluate the expression at $x = 1 + \omega$ for an arbitrary positive infinitesimal ω:

This is a one-sided limit, so we use ω.

$$\begin{aligned}\lim_{x \to 1^+} \ln(x^2 - 1) &= \ln\left((1 + \omega)^2 - 1\right) \\ &= \ln(1 + 2\omega + \omega^2 - 1) = \ln(2\omega + \omega^2) \\ &\doteq -\infty.\end{aligned}$$ ■

Since $2\omega + \omega^2 \approx 2\omega$ is a positive infinitesimal, $\ln(2\omega + \omega^2)$ renders $-\infty$.

Before rendering in the solution to example 9, could we have written $\ln(2\omega + \omega^2) \approx \ln(2\omega)$? Because the value inside the logarithm is infinitesimal, chapter 1 theorem 16 (approximating inside a continuous function) does not apply. However, a proof similar to chapter 2 theorem 6 (approximation inside root functions) can be used to show that we may approximate inside logarithms even when that inside quantity is infinitesimal or infinite. The details of the proof are exercises.

If the quantity inside the logarithm is approximately a positive real number other than one, then chapter 1 theorem 16 applies. The only remaining cases are when the quantity inside the logarithm is a positive infinitesimal or is positive infinite.

Theorem 2 APPROXIMATION INSIDE LOGARITHMS
If $x \approx y$, $y > 0$, *and* $y \neq 1$, *then* $\ln x \approx \ln y$.

Example 10 *Find* $\lim_{x \to 1} \ln\left(\frac{3x}{(x-1)^2}\right)$.

Evaluating using continuity fails for reason of division by zero. This is a two-sided limit, so we use α.

Solution To find the limit, we evaluate the expression at $1 + \alpha$ for an arbitrary infinitesimal α:

$$\begin{aligned}\lim_{x \to 1} \ln\left(\frac{3x}{(x-1)^2}\right) &= \ln\left(\frac{3(1+\alpha)}{(1+\alpha-1)^2}\right)\\ &= \ln\left(\frac{3+3\alpha}{\alpha^2}\right)\\ &\approx \ln\left(\frac{3}{\alpha^2}\right) = \ln\left(3A^2\right)\\ &\doteq \infty.\end{aligned}$$ ■

Because $3A^2$ is a positive infinite hyperreal, $\ln\left(3A^2\right)$ renders ∞.

Example 11 *Find* $\lim_{x\to 4} \ln \sqrt{x-1}$.

Solution Evaluating using continuity yields

$$\lim_{x \to 4} \ln \sqrt{x-1} = \ln \sqrt{4-1} = \ln \sqrt{3}.$$

If desired, the answer can be simplified using laws of logarithms to $\frac{1}{2} \ln 3$, or a calculator can be used to determine a numerical approximation to the value: 0.5493. ■

Exercises 5.1

1–10. Rapid response: does the calculation use the laws of logarithms correctly? Answer "correct" or "wrong."

1. $\ln x + \ln(x+1) = \ln x(x+1)$
2. $\ln(x+1) = \ln x + \ln 1$
3. $\ln(3x-4) = \ln 3x - \ln 4$
4. $\ln 3x - \ln 4 = \ln \frac{3x}{4}$
5. $\ln 3x - \ln 4 = \frac{\ln 3x}{\ln 4}$
6. $6 \ln 4x = \ln(4x)^6$
7. $6 \ln 4x = \ln 4x^6$
8. $\ln(x+3)(x-5)^2 = 2\ln(x+3)(x-5)$

9. $(\ln x)^3 = 3\ln x$

10. $\ln(x^3) = 3\ln x$

11–18. Rewrite the expression using only one logarithm.

11. $4\ln x - 56\ln(x-11)$

12. $7\ln(3x+1) + \ln(x^2+5x)$

13. $13\ln x - 4\ln(x+1) + 2\ln(x-1)$

14. $\frac{1}{2}\ln x^3 + \ln 5$

15. $\ln x + \ln x^2 + \ln x^3 + \ln x^4$

16. $\ln 8 - \ln 4 + \ln 3$

17. $2\ln(x^2+2x+1) - 3\ln(x+1)$

18. $1 - (\ln 3x + \ln 5x)$

19–26. Expand the expression using the laws of logarithms.

19. $\ln\left(\frac{x-4}{(x+6)^5\sin x}\right)$

20. $\ln\left(\frac{x^3(x-2)^2}{\sqrt{x+1}}\right)$

21. $\ln\left(\frac{2x^5}{x^2-9}\right)$

22. $\ln\left(\left(\frac{(9x+4)^3}{x\sqrt{x+1}}\right)^5\right)$

23. $\ln\,(x+\sqrt{x})\,(1+x)$

24. $\ln\left(\frac{(x^2+3)^2}{5x^3}\right)$

25. $\ln\left(\frac{\frac{x}{x+1}+2}{3x+5}\right)$

26. $\ln\,(3\sec^2 x)$

27–46. Differentiate the function.

27. $y = \frac{x}{\ln x}$

28. $f(x) = \ln\sqrt{x^3+4}$

29. $f(x) = (1+\ln x)^2$

30. $y = \ln x\cos x$

31. $y = \ln\left(\frac{x+4}{x^2-x}\right)^3$

32. $y = \ln\cos x$

33. $y = t^2 - \ln 4$

34. $y = \ln\left(\frac{x^3}{4x+1}\right)$

35. $f(x) = \ln\left(\frac{(x-1)^2}{3x+7}\right)$

36. $f(x) = \ln\left(\frac{x+3}{4x-1}\right)^3$

37. $y = x^2\ln(5x^2)$

38. $y = \ln(x\ln x)$

39. $y = \ln\sin\theta$

40. $g(z) = z^4 - (\ln 6)z^2$

41. $y = \ln\left((x-3)^2(2x+1)^{14}\right)$

42. $g(z) = z^4 - \ln(6z^2)$

43. $y = \ln\left(\sqrt{\frac{x-3}{4x+1}}\right)$

44. $y = \sqrt{\ln\left(\frac{x-3}{4x+1}\right)}$

45. $f(x) = \ln x^{3x}$

46. $f(t) = \ln(2 + \sin^2\sqrt{4t+1})$

47–58. Find the limit.

47. $\lim_{x \to 2^+} \ln(4x - 8)$

48. $\lim_{x \to \infty} \ln \sqrt{x}$

49. $\lim_{x \to \infty} \ln(4x - 8)$

50. $\lim_{x \to 5^+} \dfrac{\ln(x - 5)}{x}$

51. $\lim_{x \to 3} \ln(3x - 8)$

52. $\lim_{x \to 0} \ln |x|$

53. $\lim_{x \to 4} \ln \dfrac{x + 3}{(x - 4)^2}$

54. $\lim_{x \to \infty} \left(\ln \dfrac{1}{x} + \ln 3x^4 \right)$

55. $\lim_{x \to \infty} \left(\ln x - \ln(4x^2 + 1) \right)$

56. $\lim_{x \to 1} \dfrac{x - \ln x}{x + 1}$

57. $\lim_{x \to 4^-} \ln(2 - \sqrt{x})$

58. $\lim_{x \to 0^-} \ln \dfrac{\frac{1}{x} - \frac{1}{2}}{x - 2}$

59–62. Find the equation of the tangent line to the curve at the given point.

59. $f(x) = \ln(x^2)$ at $(1, 0)$

60. $f(x) = x \ln x$ at $(1, 0)$

61. $y = x + 3\ln(x - 1)$ at $(5, 5 + 3\ln 4)$

62. $y = \sqrt{\ln x}$ at $(2, \sqrt{\ln 2})$

63–66. Use shifting and reflection (from the graph of $y = \ln x$) to sketch the graph of the function.

63. $y = \ln(x - 3)$

64. $y = 1 + \ln x$

65. $y = -\ln x$

66. $y = -\ln(x + 2)$

Here we are using B as the capital Greek letter "beta," so we can represent its reciprocal by lowercase beta, β.

Here we are using K as the capital Greek letter "kappa," so we can represent its reciprocal by lowercase kappa, κ.

67. Let ω be a positive infinitesimal. Because $\ln \omega \doteq -\infty$, there must be some positive infinite hyperreal B such that $\ln \omega = -B$. Use this fact to evaluate $\lim_{x \to 0^+} \frac{1}{\ln x}$.

68. Let Ω be a positive infinite hyperreal. Because $\ln \Omega \doteq \infty$, there must be some positive infinite hyperreal K such that $\ln \Omega = K$. Use this fact to evaluate $\lim_{x \to \infty} \frac{1}{\ln x}$.

69. Prove theorem 2 for the case y is positive infinite, following the general pattern of the proof of chapter 2 theorem 6. Hints: use laws of logarithms to simplify $\frac{\ln x - \ln y}{\ln y}$. What are the sizes of the numbers in the numerator and the denominator of that expression?

70. Repeat exercise 69 for the case y is a positive infinitesimal.

Logarithms, Part II

5.2

The previous section began the story of the antiderivative of $y = \frac{1}{x}$. We built an antiderivative for $y = \frac{1}{x}$ where $x > 0$ using the definite integral, but what about $y = \frac{1}{x}$ where $x < 0$? This question is answered in this section.

Domains of logarithmic functions

The definition

$$\ln x = \int_1^x \frac{1}{t}\,dt, x > 0$$

from the previous section made it clear that $\ln x$ is defined only for positive values of the input x. Similar to the restriction that we can only take square roots of nonnegative numbers, we may only take logs of positive numbers.

These statements are valid in the context of the real numbers and are not necessarily valid in other contexts.

Example 1 *Find the domain of* $y = \ln(4x - 7)$.

Solution Because we may only take logs of positive numbers, the expression inside the logarithm must be positive:

$$4x - 7 > 0.$$

We solve this inequality for x to find the values of x that work:

$$4x > 7$$

$$x > \frac{7}{4}.$$

In order for the function to be defined, we must have $x > \frac{7}{4}$, so this forms our domain. The answer is $x > \frac{7}{4}$ or, in interval notation, $(\frac{7}{4}, \infty)$. ■

Example 2 *Let* $f(x) = \ln(-x)$. *Find the domain of* f.

Solution Once again, what is inside the logarithm must be positive:

$$-x > 0.$$

Another way to think of this result is as follows: $f(5) = \ln(-5)$, which is undefined, so 5 is not in the domain of f. But, $f(-3) = \ln(-(-3)) = \ln 3$, which is defined, so -3 is in the domain of f.

We then solve the inequality for x by multiplying both sides by -1, which reverses the direction of the inequality:

$$-1(-x) < -1(0)$$
$$x < 0.$$

The domain of f is $(-\infty, 0)$. ■

The graph of $y = \ln(-x)$ is the reflection across the y-axis of the graph of $y = \ln x$ (see figure 1).

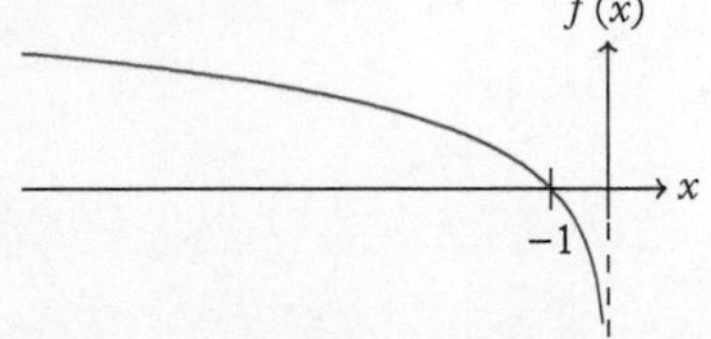

Figure 1 *The graph of* $y = \ln(-x)$ *is the reflection of the graph of* $y = \ln x$ *across the y-axis*

Reading Exercise 5 *Find the domain of* $f(x) = \ln(5x - 2)$.

An antiderivative of $y = \frac{1}{x}$, $x \neq 0$

Example 3 *Find* $f'(x)$ *for* $f(x) = \ln(-x)$.

Solution Using the chain rule,

$$f'(x) = \frac{1}{-x} \cdot (-1) = \frac{1}{x}.$$ ■

Because the domain of $f(x) = \ln(-x)$ is $x < 0$ (example 2), example 3 shows that

$$\frac{d}{dx} \ln(-x) = \frac{1}{x}, x < 0.$$

Coupling this with our result from the previous section,

$$\frac{d}{dx} \ln x = \frac{1}{x}, x > 0,$$

gives two different results for an antiderivative of $y = \frac{1}{x}$, depending on whether x is positive or negative. Can these results be combined? Yes, because

$$|x| = \begin{cases} -x, & \text{if } x \leq 0 \\ x, & \text{if } x > 0. \end{cases}$$

Then, as illustrated in figure 2,

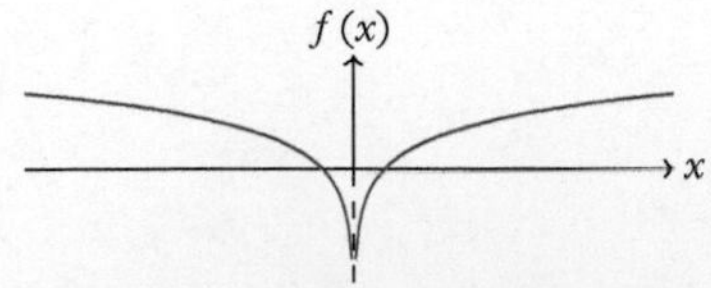

Figure 2 *The graph of* $y = \ln|x|$, *which has a vertical asymptote at* $x = 0$

$$\ln|x| = \begin{cases} \ln(-x), & \text{if } x < 0 \\ \ln x, & \text{if } x > 0. \end{cases}$$

Using the two previous derivative results,

$$\frac{d}{dx}\ln|x| = \frac{1}{x}, \quad x \neq 0.$$

Reversing this derivative result gives the antiderivative result we are seeking.

$\frac{d}{dx}\ln|x| = \frac{1}{x}, \quad x \neq 0$

ANTIDERIVATIVE OF *1/x*

$$\int \frac{1}{x}\,dx = \ln|x| + C$$

The formula can be restated using u to make it clear how it can be used with substitution:

$$\int \frac{1}{u}\,du = \ln|u| + C.$$

There is still no quotient rule for antidifferentiation, but this formula can help us find antiderivatives for some quotients.

Example 4 *Evaluate* $\int \frac{1}{2x-3}\,dx.$

Solution We now have an antiderivative formula for $1/x$, but this is 1 over "not just plain x," so the formula does not apply directly. Perhaps substitution will help; as always, the "not just plain x" part is our u:

$$u = 2x - 3$$

$$du = 2\,dx.$$

The need for substitution here is harder to recognize than when "not just plain x" is taken to a power or is inside trig functions or a square root. Practice is essential, especially because we eventually have more formulas and techniques for integrating rational functions.

Adjusting for the constant 2, we make the substitution:

$$\int \frac{1}{2x-3}\,dx = \frac{1}{2}\int \frac{1}{2x-3}\cdot 2\,dx = \frac{1}{2}\int \frac{1}{u}\,du.$$

We then finish by using the formula and returning to the original variable x:

$$= \frac{1}{2}\ln|u| + C$$

$$= \frac{1}{2}\ln|2x-3| + C. \qquad \blacksquare$$

The solution can be checked by differentiation:

$$\frac{d}{dx}\left(\frac{1}{2}\ln|2x-3| + C\right)$$

$$= \frac{1}{2}\cdot\frac{1}{2x-3}\cdot 2 + 0$$

$$= \frac{1}{2x-3},$$

which is the original integrand, as required.

In the formula $\int \frac{1}{u}\,du$, the u is in the denominator whereas du is in the numerator. If you recognize that an integrand is a fraction where the derivative of the denominator is in the numerator (allowing for adjustment by a constant), then a substitution of u = denominator is helpful.

In some sources, $\int \frac{1}{u}\,du$ is occasionally written $\int \frac{du}{u}$.

Example 5 *Find* $\displaystyle\int \frac{2x}{x^2+1}\,dx.$

Practice looking for this pattern. For rational functions, a numerator with degree one less than the denominator might indicate using substitution.

Solution Notice that the denominator is $x^2 + 1$ and the numerator is its derivative, $2x$. We therefore try substitution, letting u be the denominator:

$$u = x^2 + 1$$
$$du = 2x\,dx.$$

Ans. to reading exercise 5
$x > \frac{2}{5}$

Making the substitution and continuing from there,

$$\int \frac{2x}{x^2+1}\,dx = \int \frac{1}{u}\,du = \ln|u| + C = \ln|x^2+1| + C. \quad \blacksquare$$

Tip: when in doubt, leave the absolute values. Both $\ln|x^2+1| + C$ and $\ln(x^2+1) + C$ are correct solutions to example 5.

Because the quantity x^2+1 is always positive, the absolute values can be dropped in the solution to example 5. In other words, the solution can be written as $\ln(x^2 + 1) + C$.

Example 6 *Find* $\displaystyle\int \frac{\sec^2 x}{4+5\tan x}\,dx.$

Solution The derivative of $\tan x$, which is in the denominator, is $\sec^2 x$, which is in the numerator. We therefore let the denominator be u:

CAUTION: MISTAKE TO AVOID
Remember that du must be calculated from u; it is **not** just what's left over in the integral.

$$u = 4 + 5\tan x$$
$$du = 5\sec^2 x\,dx.$$

We adjust for the constant, make the substitution, antidifferentiate, and go back to the original variable:

In line 1, an adjustment is made for the needed constant 5. In line 2, the substitution is made. In line 3, the most general antiderivative is found. In line 4, we return to the original variable.

$$\begin{aligned}\int \frac{\sec^2 x}{4+5\tan x}\,dx &= \frac{1}{5}\int \frac{5\sec^2 x}{4+5\tan x}\,dx \\ &= \frac{1}{5}\int \frac{1}{u}\,du \\ &= \frac{1}{5}\ln|u| + C \\ &= \frac{1}{5}\ln|4+5\tan x| + C. \quad \blacksquare\end{aligned}$$

Unlike the solution to example 5, the absolute values in the solution to example 6 are required because what's inside, $4 + 5\tan x$, is not always positive.

Reading Exercise 6 *Find* $\int \frac{3x^2}{x^3+5}\,dx.$

Example 7 *Evaluate* $\int_2^4 \frac{1}{x-6}\,dx.$

Solution Because the integrand is continuous on $[2, 4]$, the fundamental theorem of calculus, part II (FTC II) applies and we may continue.

The integrand is not continuous at $x = 6$, but this value is not in the interval of integration.

Any time the numerator is a constant and the denominator is linear (of the form $ax + b$), the substitution u = denominator works:

$$u = x - 6$$
$$du = dx.$$

Because this is a definite integral, we also calculate the new limits of integration:

$$x = 4: \quad u = 4 - 6 = -2$$
$$x = 2: \quad u = 2 - 6 = -4.$$

Next we make the substitution and continue:

$$\begin{aligned}\int_2^4 \frac{1}{x-6}\,dx &= \int_{-4}^{-2} \frac{1}{u}\,du\\ &= \ln|u|\Big|_{-4}^{-2}\\ &= \ln 2 - \ln 4\\ &= \ln\frac{2}{4} = \ln\frac{1}{2} = \ln 2^{-1} = -\ln 2.\end{aligned}$$ ■

In line 1, the substitution is carried out, including the switch to the new limits of integration. In line 2, an antiderivative is found. In line 3, the antiderivative is evaluated. In line 4, laws of logarithms are used to simplify the solution. Line 4 is not necessary to answer the question, but such simplifications can be helpful.

Notice what happens in the solution to example 7 if we forget the absolute values. We have $\ln u\Big|_{-4}^{-2} = \ln(-2) - \ln(-4)$, which is undefined! The absolute values are absolutely required.

CAUTION: MISTAKE TO AVOID
Forgetting the absolute values in $\int \frac{1}{u}\,du = \ln|u| + C$ can result in logs of negative numbers.

Example 8 *Find* $\int \frac{1}{\sqrt{y}(\sqrt{y}+3)}\,dy.$

Solution We have a fraction, so let's try continuing the recent theme by letting u be the denominator:

$$u = \sqrt{y}(\sqrt{y}+3).$$

Simplifying first to $u = y + 3\sqrt{y}$ makes finding du a little easier.

In this form, we need the product rule for calculating du:

$$du = \left(\sqrt{y}\cdot\frac{1}{2\sqrt{y}} + (\sqrt{y}+3)\frac{1}{2\sqrt{y}}\right)dy = \cdots = \left(1+\frac{3}{2\sqrt{y}}\right)dy.$$

If one substitution fails, there might still be a different substitution that works.

There is no $1+\frac{3}{2\sqrt{y}}$ in the integrand, so the substitution fails.

Does this mean the available antiderivative formulas, rules, and techniques are insufficient to determine an antiderivative? No, it means that the substitution we tried failed. It turns out that a different substitution works. Instead of letting the entire denominator be u, let's use just the part in parentheses:

Ans. to reading exercise 6
$\ln|x^3+5|+C$ (the absolute values are necessary). Details:
$u = x^3+5$, $du = 3x^2\,dx$

$$\int \frac{3x^2}{x^3+5}\,dx = \int \frac{1}{u}\,du = \ln|u|+C$$
$$= \ln|x^3+5|+C$$

$$u = \sqrt{y}+3$$
$$du = \frac{1}{2\sqrt{y}}\,dy.$$

Because u is only what is in the parentheses, the remaining $\sqrt{y}$ in the denominator is available for du and we need only adjust for the constant:

Line 1 adjusts for the constant $\frac{1}{2}$; line 2 makes the substitution. In line 3, we antidifferentiate. Line 4 returns to the original variable. In line 5, absolute values are not necessary because $\sqrt{y}+3$ is always positive. Line 5 is optional.

$$\begin{aligned}\int \frac{1}{\sqrt{y}(\sqrt{y}+3)}\,dy &= 2\int \frac{1}{2\sqrt{y}(\sqrt{y}+3)}\,dy\\ &= 2\int \frac{1}{u}\,du\\ &= 2\ln|u|+C\\ &= 2\ln|\sqrt{y}+3|+C\\ &= 2\ln(\sqrt{y}+3)+C.\end{aligned}$$ ■

The moral of the story: sometimes integration is a process of trying different approaches until one works.

On the other hand, sometimes the available antiderivative formulas, rules, and techniques are insufficient to find an antiderivative, so a balance must be struck between continuing to try different ideas and knowing when to quit.

Antiderivatives of tangent and cotangent

Reversing the derivative rules for trig functions did not reveal how to antidifferentiate tangent and cotangent. Let's rectify that omission now.

To evaluate

$$\int \tan x\,dx,$$

we rewrite tangent in terms of sine and cosine:

$$= \int \frac{\sin x}{\cos x}\, dx.$$

Do you see a potentially helpful substitution? Let u be the denominator:

$$u = \cos x$$
$$du = -\sin x\, dx.$$

Adjust for the constant, make the substitution, and continue:

$$\begin{aligned}\int \tan x\, dx &= \int \frac{\sin x}{\cos x}\, dx = -\int \frac{-\sin x}{\cos x}\, dx \\ &= -\int \frac{1}{u}\, du \\ &= -\ln|u| + C \\ &= -\ln|\cos x| + C.\end{aligned}$$

Line 1 rewrites and adjusts for the constant, line 2 substitutes, line 3 antidifferentiates, and line 4 returns to the original variable.

Traditionally, the answer is written in a different form, using the laws of logarithms, algebra, and a trig identity:

$$\begin{aligned}\int \tan x\, dx &= -\ln|\cos x| + C \\ &= \ln|\cos x|^{-1} + C \\ &= \ln\left|\frac{1}{\cos x}\right| + C \\ &= \ln|\sec x| + C.\end{aligned}$$

This form may be easier to remember because of the connection already made between tangent and secant in derivative and antiderivative formulas.

ANTIDERIVATIVE OF TANGENT

$$\int \tan x\, dx = \ln|\sec x| + C$$

The antiderivative formula

$$\int \tan x\, dx = -\ln|\cos x| + C$$

is also correct, but is not traditionally used.

The formula for $\int \cot x\, dx$ is derived in a similar manner, rewriting using sine and cosine:

$$\int \cot x \, dx = \int \frac{\cos x}{\sin x} \, dx.$$

Substituting, we use

$$u = \sin x$$
$$du = \cos x \, dx.$$

Then,

$$\begin{aligned}\int \cot x \, dx &= \int \frac{\cos x}{\sin x} \, dx \\ &= \int \frac{1}{u} \, du \\ &= \ln |u| + C \\ &= \ln |\sin x| + C \\ &= -(-\ln |\sin x|) + C \\ &= -\ln |\sin x|^{-1} + C \\ &= -\ln \left| \frac{1}{\sin x} \right| + C \\ &= -\ln |\csc x| + C.\end{aligned}$$

Line 1 rewrites the integrand, line 2 substitutes, line 3 antidifferentiates, and line 4 returns to the original variable. Then, to obtain the traditional form, use algebra (lines 5–7) and a trig identity (line 8).

ANTIDERIVATIVE OF COTANGENT

$$\int \cot x \, dx = -\ln |\csc x| + C$$

The antiderivative formula

$$\int \cot x \, dx = \ln |\sin x| + C$$

is also correct, but is not traditionally used.

Example 9 *Evaluate* $\int 2x \tan(x^2) \, dx$.

Solution We see tangent of "not just plain x" and therefore use substitution:

$$u = x^2$$
$$du = 2x \, dx.$$

Then,

$$\int 2x\tan(x^2)\,dx = \int \tan u\,du$$
$$= \ln|\sec u| + C$$
$$= \ln|\sec x^2| + C. \qquad \blacksquare$$

Although $\sec^2 x$ is always positive, $\sec x^2$ is not always positive; the absolute values are required.

Reading Exercise 7 *Evaluate* $\int 3\cot x\,dx$.

Logarithmic differentiation

Logarithmic differentiation is a way of using laws of logarithms and implicit differentiation to avoid complicated combinations of the product, quotient, and chain rules. The technique is illustrated in the following example.

Example 10 *Find y' given* $y = \dfrac{4x^2 - 1}{(2x+3)^{4/3}(x+1)}$.

Solution We could use the quotient, product, and chain rules to find the derivative using previous techniques. Or, we could ❶ take the natural logarithm of both sides of the expression and expand using laws of logarithms, ❷ differentiate both sides implicitly with respect to x, and ❸ solve for y'.

These three steps are collectively called *logarithmic differentiation*.

❶ Take the natural log of both sides and expand:

$$\ln y = \ln \frac{4x^2 - 1}{(2x+3)^{4/3}(x+1)}$$
$$\ln y = \ln(4x^2 + 1) - \ln\left((2x+3)^{4/3}(x+1)\right)$$
$$\ln y = \ln(4x^2 + 1) - \left(\ln(2x+3)^{4/3} + \ln(x+1)\right)$$
$$\ln y = \ln(4x^2 + 1) - \frac{4}{3}\ln(2x+3) - \ln(x+1).$$

Line 1 takes the ln of both sides, line 2 expands using law of logarithms (2), line 3 uses law (1), and line 4 uses law (3) and algebra.

❷ Next we differentiate both sides implicitly with respect to x. The right-hand side has only the variable x, so this is the same as taking its derivative with the usual procedures. The left-hand side has an expression involving y, so we take the derivative with respect to y and multiply by y'. This gives

$$\frac{1}{y}\cdot y' = \frac{1}{4x^2+1}\cdot 8x - \frac{4}{3}\cdot\frac{1}{2x+3}\cdot 2 - \frac{1}{x+1}\cdot 1.$$

For logarithmic differentiation, the left-hand side is always $\ln y$, so in step ❷ its derivative is always $\frac{1}{y}\cdot y'$.

❸ To solve for y', we only need to multiply both sides of the equation by y:

Step ❸ always consists of multiplying both sides by y and then replacing y with its value in terms of x.

$$y' = y\left(\frac{8x}{4x^2+1} - \frac{8}{3(2x+3)} - \frac{1}{x+1}\right).$$

Replacing y with its value in terms of x (that is, using the original equation) completes the process:

This form of the solution is nicer than the one obtained by using the quotient, product, and chain rules.

$$y' = \frac{4x^2-1}{(2x+3)^{4/3}(x+1)}\left(\frac{8x}{4x^2+1} - \frac{8}{3(2x+3)} - \frac{1}{x+1}\right). \blacksquare$$

Because steps ❷ and ❸ are always the same, logarithmic differentiation is easier to learn than it looks. A fair question, though, is how much does logarithmic differentiation really help if we can already take the derivatives anyway? The nicer form for the solution is a benefit, but later in this chapter there are functions for which logarithmic differentiation is very helpful. Practicing the technique now pays off later.

Ans. to reading exercise 7
$-3\ln|\csc x| + C$

Exercises 5.2

1–12. Rapid response: what substitution can you try?

1. $\displaystyle\int \frac{2}{(5x-7)^3}\,dx$
2. $\displaystyle\int x^2 \sin x^3\,dx$
3. $\displaystyle\int \frac{2}{(5x-7)^2}\,dx$
4. $\displaystyle\int \frac{\sin\sqrt{x}}{\sqrt{x}}\,dx$
5. $\displaystyle\int \frac{2}{5x-7}\,dx$
6. $\displaystyle\int \frac{\sin(\ln x)}{x}\,dx$
7. $\displaystyle\int 2x(x^2+4)^5\,dx$
8. $\displaystyle\int \frac{x^4}{x^5+3}\,dx$
9. $\displaystyle\int \cos x(\sin x)^5\,dx$
10. $\displaystyle\int \frac{1+\sec x\tan x}{x+\sec x}\,dx$
11. $\displaystyle\int \frac{(\ln x)^5}{x}\,dx$
12. $\displaystyle\int \frac{2x-3}{x^2-3x+4}\,dx$

13–20. Find the domain of the function.

13. $f(x) = \ln(5x-3)$
14. $y = \ln(7-2x)$
15. $y = \ln(x^2+5)$
16. $g(x) = \sqrt{2x}\ln(1-x)$
17. $g(x) = \frac{1}{\ln(x+4)}$
18. $y = \ln\left((x^3(2-x)^4\right)$
19. $f(t) = \ln\frac{t-2}{7-t}$
20. $f(x) = \frac{1}{\ln(x^2+1)}$

21–50. Evaluate the integral, if possible. If not possible, say why (either "FTC II does not apply" or "the available antiderivative formulas, rules, and techniques are insufficient to determine an antiderivative").

21. $\displaystyle\int \frac{2}{x+1}\,dx$

22. $\displaystyle\int \frac{4}{6x-3}\,dx$

23. $\displaystyle\int \frac{-3}{5x+2}\,dx$

24. $\displaystyle\int_{-1}^{7} \frac{1}{x-8}\,dx$

25. $\displaystyle\int_{3}^{6} \frac{2}{7-x}\,dx$

26. $\displaystyle\int_{2}^{3} \frac{10}{2x-5}\,dx$

27. $\displaystyle\int_{0}^{4} \frac{4}{x-3}\,dx$

28. $\displaystyle\int \frac{3}{4-x}\,dx$

29. $\displaystyle\int \tan(9x-\pi)\,dx$

30. $\displaystyle\int \frac{3}{(4-x)^2}\,dx$

31. $\displaystyle\int \frac{2}{x^2+4}\,dx$

32. $\displaystyle\int \frac{3x}{4-x^2}\,dx$

33. $\displaystyle\int \frac{2x}{x^2+4}\,dx$

34. $\displaystyle\int \cot(5x+1)\,dx$

35. $\displaystyle\int_{0}^{1} \frac{2x^3+x}{x^4+x^2+5}\,dx$

36. $\displaystyle\int \frac{\sin(\ln x)}{x}\,dx$

37. $\displaystyle\int \tan\left(\frac{x+4}{3}\right)dx$

38. $\displaystyle\int_{0}^{4} \frac{1}{(x-5)^2}\,dx$

39. $\displaystyle\int_{1}^{2} \frac{(\ln x)^2}{x}\,dx$

40. $\displaystyle\int \frac{10x^4+56x^3-28}{x^5+7x^4-14x}\,dx$

41. $\displaystyle\int \frac{\cot\left(\frac{1}{x}\right)}{x^2}\,dx$

42. $\displaystyle\int x^2 \cot x^3\,dx$

43. $\displaystyle\int \frac{\ln\sqrt{x}}{x}\,dx$

44. $\displaystyle\int_{1}^{3} \frac{x+1}{x^2}\,dx$

45. $\displaystyle\int \frac{x^3-7x^2+4}{x}\,dx$

46. $\displaystyle\int 2\tan 14x\,dx$

47. $\displaystyle\int \frac{1}{x\sqrt{\ln x}}\,dx$

48. $\displaystyle\int \frac{\cos x}{(\sin x)(\ln \sin x)}\,dx$

49. $\displaystyle\int \frac{\cos(\ln x^5)}{x}\,dx$

50. $\displaystyle\int \frac{3x^3-7x}{x^4-7x^2+x}\,dx$

51–56. Use logarithmic differentiation to find the derivative of the function.

51. $y = \dfrac{(3x+5)^4(1+x^2)^3}{(x-2)^7}$

52. $y = \dfrac{(x^2+x+2)^3}{(2x+9)^4(x-1)^5}$

53. $y = (14x-3)^2(12x+5)^7(1-2x)^3$

54. $y = \dfrac{x^4(3x+x^2)^4}{(9x+2)^7}$

55. $y = \left(\dfrac{x+\sin x}{(3x+2)(x-\sqrt{x})^5}\right)^2$

56. $y = \left(\left((2+x)^2(9-x)^4\right)^3(4x+1)^5\right)^2$

57–60. Solve the IVP.

57. $y' = \frac{1}{x}$, $y(2) = 5$

58. $y' = \tan x$, $y(\frac{\pi}{4}) = 0$

59. $y' = \cot(2x-\pi)$, $y(\frac{\pi}{4}) = 1$

60. $y' = \frac{6}{3x+1}$, $y(0) = 2$

61. Use the definition of integral and the approximation formula $\displaystyle\sum_{k=1}^{\Omega}\frac{1}{k} \approx \ln\Omega$ to evaluate $\displaystyle\int_0^1 \frac{1}{x}\,dx$.

Because FTC II is not being used, the discontinuity at $x = 0$ is not an issue.

62. Repeat exercise 61 for $\displaystyle\int_0^3 \frac{2x+1}{x}\,dx$.

63. Recall that $\displaystyle\int \frac{1}{x}\,dx = \ln|x| + C$ and $\displaystyle\int \frac{1}{y}\,dy = \ln|y| + C$. Evaluate $\displaystyle\int \frac{1}{\text{cabin}}\,d\text{cabin}$.

64. Use the techniques of section 3.7 to sketch the graph of $f(x) = \ln\left(\dfrac{x+1}{x^2-7}\right)$.

Inverse Functions

5.3

In an algebra course you may have studied the idea of an *inverse function*. This section begins with a review of that algebra before turning to the calculus of inverse functions.

Inverse functions: review

One conceptual representation of a function is as a machine that takes inputs and produces outputs. We can think of the line to the left of the box in figure 1 as a conveyor belt moving in the direction of the arrow; place a real number x (an input) on the conveyor belt, it moves into the machine, and what comes out on the other end is a real number y (an output).

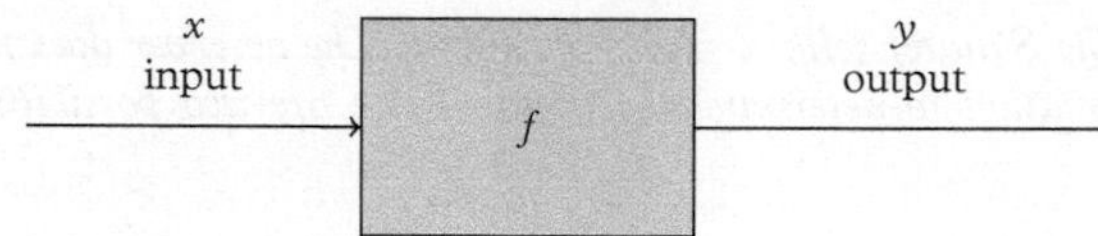

Figure 1 *Machine diagram of a function. Inputs are often represented as x-coordinates and outputs are often represented as y-coordinates*

As an example, consider the function $f(x) = 4x$, which we could call *The Quadrupler*. If we place 3 on the conveyor belt as an input, then the output is 12, as in figure 2.

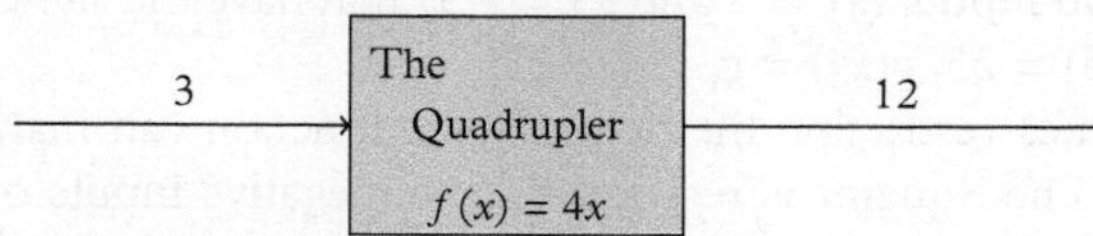

Figure 2 *The Quadrupler with* $x = 3$ *as input*

Now picture yourself standing at the end of the conveyor belt looking at the output of The Quadrupler (figure 3). Can you determine what the input is? If the output is 16, what is the input? This is the idea of the inverse function; we start with the original function's output and determine its input. The process is reversed.

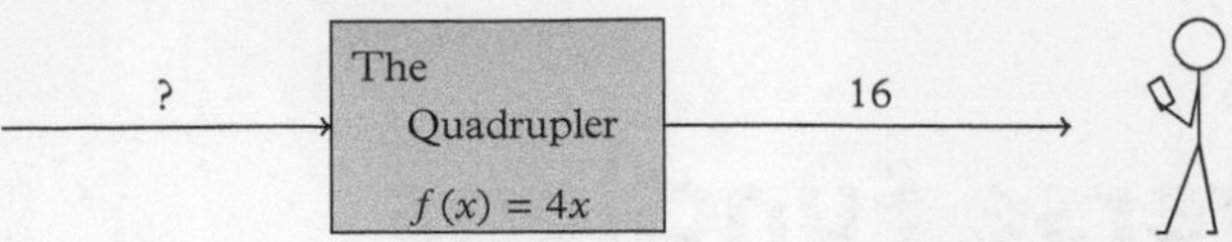

Figure 3 *The Quadrupler with $x = 16$ as output. The observer's task is to determine the input*

In figure 3, the answer is not difficult to obtain; the input must be 4. However, not all functions are as nice as The Quadrupler. If a function gives the same output for two different inputs, then determining which input is used can be impossible. For the function $g(x) = x^2$ with the output 25, the observer does not know whether the input is 5 or -5 (see figure 4). But as long as no two inputs give the same output, we can complete the task. Such a function is called a *one-to-one function.*

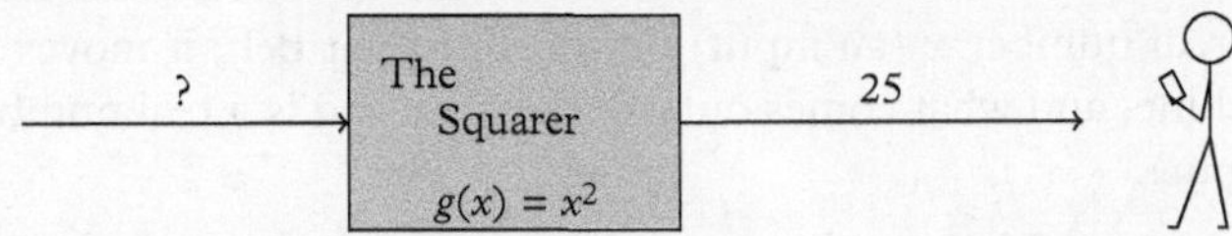

Figure 4 *The Squarer with $x = 25$ as output. The observer does not have enough information to determine the input. There are two possibilities: -5 and 5*

This definition can be summarized as saying that different inputs ($x_1 \neq x_2$) give different outputs ($f(x_1) \neq f(x_2)$).

Definition 2 ONE-TO-ONE FUNCTION *A function f is* one-to-one *(one input to one output) on an interval I if, whenever $x_1, x_2 \in I$ and $x_1 \neq x_2$, then we have $f(x_1) \neq f(x_2)$.*

Here, $x_1 \neq x_2$ but $g(x_1) = g(x_2)$, so definition 2 is not satisfied.

The function $f(x) = 4x$ is one-to-one on the interval $(-\infty, \infty)$. The function $g(x) = x^2$ is not one-to-one on the interval $(-\infty, \infty)$ because there are two inputs ($x_1 = 5$ and $x_2 = -5$) that have the same output ($g(x_1) = g(5) = 25$, $g(x_2) = g(-5) = 25$).

Sometimes restricting the domain of a function can make a difference. If The Squarer is restricted to nonnegative inputs only and the output is 25, then we know that the input is 5 (figure 5). For this modified function we can determine the input from the output, and we say that $g(x) = x^2$ is one-to-one on the interval $[0, \infty)$.

Whether a function is one-to-one can also be understood from the function's graph through the *horizontal line test*. Because outputs are represented by y-coordinates and inputs are represented by x-coordinates, to be one-to-one requires that no horizontal line intersect

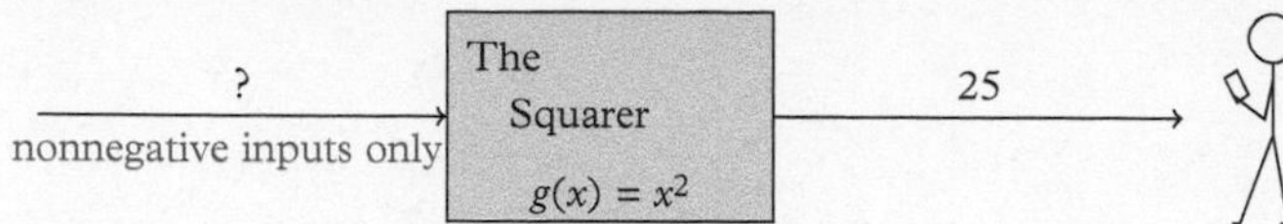

Figure 5 *The Squarer with domain restricted to* $[0, \infty)$*, with* $x = 25$ *as output. Because the inputs cannot be negative, the input cannot be* -5*. The observer is able to determine the input is 5*

the graph at more than one point. Otherwise, there would be two or more inputs (x-coordinates) giving the same output (y-coordinate).

> **HORIZONTAL LINE TEST**
>
> Given the graph of a function, if there is a horizontal line that intersects the graph at more than one point, then the function is not one-to-one. If there is no such line, then the function is one-to-one.

Example 1 *For each graph in figure 6, use the horizontal line test to determine whether the function represented by the graph is one-to-one.*

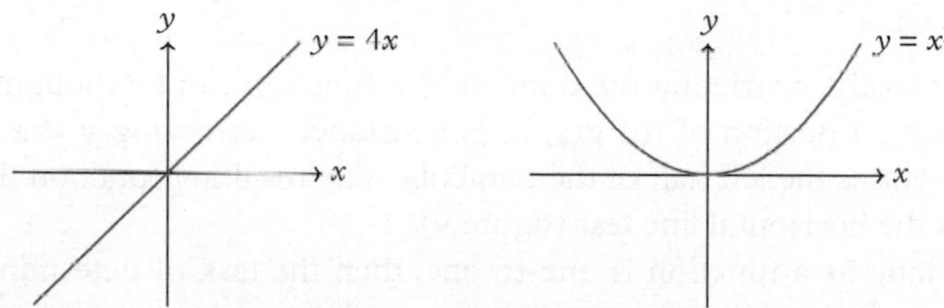

Figure 6 *Graphs for example 1*

Solution (a) Any horizontal line (gray in figure 7) intersects the graph at, at most, one point. The graph passes the horizontal line test, so the function represented by the graph is one-to-one.

(b) There is a horizontal line that intersects the graph at more than one point (see figure 8), so the graph fails the horizontal line test. The function represented by the graph is not one-to-one. (The function can be described as two-to-one—two inputs to one output—but we only care whether the function is one-to-one.) ■

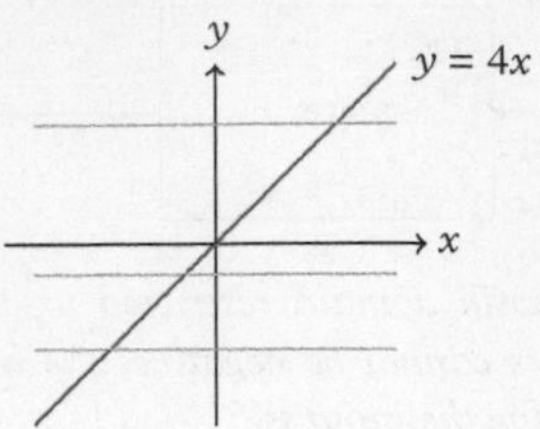

Figure 7 *Each horizontal line (gray) intersects the graph of example 1(a) at, at most, one point. The horizontal line test is satisfied and the function is one-to-one*

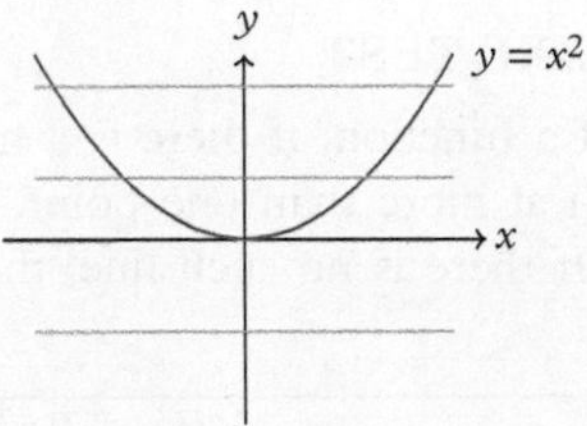

Figure 8 *Some of the horizontal lines (gray) intersect the graph of example 1(b) at more than one point. The graph fails the horizontal line test and the function is not one-to-one. The fact that other horizontal lines intersect the graph at just one point or do not intersect the graph at all is irrelevant*

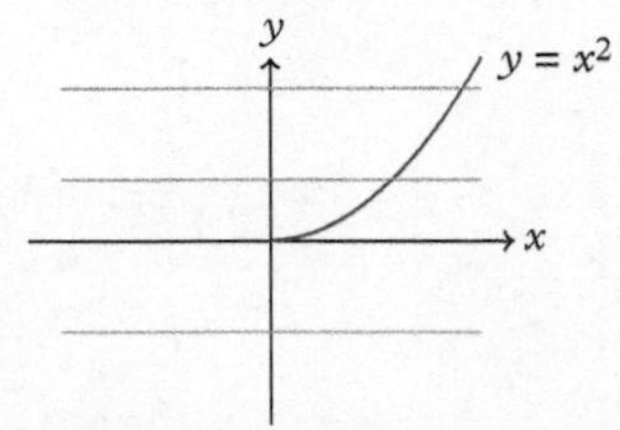

Figure 9 *The graph of $y = x^2$ restricted to the domain $[0, \infty)$. Each horizontal line (gray) intersects the graph at, at most, one point. The horizontal line test is satisfied and the function is one-to-one*

Pictorially, restricting the domain of a function can be thought of as erasing a portion of the graph. For instance, restricting $y = x^2$ to $[0, \infty)$ erases the left half of the parabola. The resulting function then passes the horizontal line test (figure 9).

As long as a function is one-to-one, then the task of determining the input from a given output, as in figures 3–5, can be accomplished. In other words, one-to-one functions have inverses. The inverse function looks at the output from f and determines the input from f. This process is a function in its own right, one that reverses the original function.

Definition 3 INVERSE FUNCTION *Let f be a one-to-one function with domain A and range B. The* inverse function f^{-1} *then has domain B and range A and is defined by*

$$f^{-1}(b) = a \text{ if and only if } f(a) = b.$$

Recall that the domain is the set of possible inputs to a function and the range is the set of all outputs from the function. Because the inverse function f^{-1} swaps inputs and outputs, it swaps the domain set and the range set.

An inverse function switches the inputs and outputs of the original function:

$$f^{-1}(y) = x \text{ if and only if } f(x) = y.$$

input output input output

CAUTION: MISTAKE TO AVOID
An inverse function is not a reciprocal:

$$f^{-1}(x) \neq \frac{1}{f(x)}.$$

The reason for using notation that makes an inverse function look like a reciprocal is understood more easily in an upper-level course.

Reading Exercise 8 *For each graph in figure 10, is the function one-to-one?*

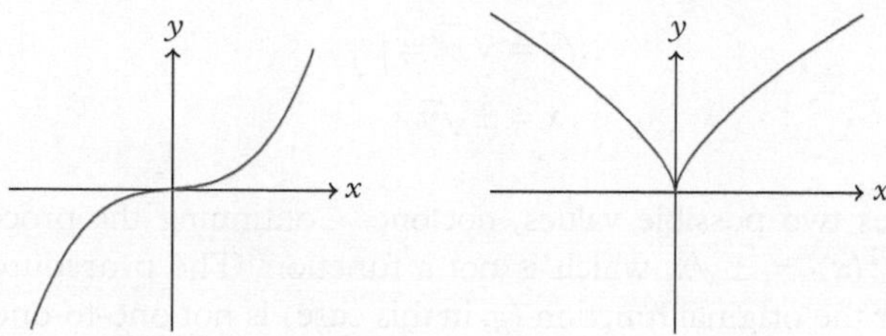

Figure 10 *The graphs for reading exercise 8*

Finding inverses

Sometimes, inverse functions can be determined algebraically.

Example 2 *Find $f^{-1}(x)$ for $f(x) = 4x$.*

Solution We wish to determine the input x from a given output y.

The function $f(x) = 4x$ is pictured in figures 2, 3, and 6.

❶ Begin by writing y in place of $f(x)$ (if necessary):

$$y = 4x.$$

❷ Solve the equation for the variable x:

$$y = 4x$$

$$\frac{1}{4}y = x.$$

Step ❷ determines the input to the function f (the value of x) in terms of the output from the function f (the value of y) and usually comprises the bulk of the work.

The two sides of the equation can be re-ordered as $x = \frac{1}{4}y$ to emphasize the idea of the input x into function f being found from the output y from function f.

❸ Write the inverse function. As in the sentence after definition 3, we replace x with $f^{-1}(y)$:

$$f^{-1}(y) = \frac{1}{4}y.$$

Addition and subtraction are inverse operations; multiplication and division are inverse operations. The inverse of the function $f(x) = 4x$, which multiplies numbers by 4, is the function $f^{-1}(x) = \frac{x}{4}$, which divides numbers by 4.

Because f^{-1} is a function and it is traditional to use x as the input variable in a function, it is traditional to rewrite the inverse function as

$$f^{-1}(x) = \frac{1}{4}x. \qquad \blacksquare$$

What happens if we try this same procedure for a function that is not one-to-one? For $g(x) = x^2$, we have

$$y = x^2$$
$$\sqrt{y} = \sqrt{x^2} = |x|$$
$$x = \pm\sqrt{y}.$$

A function must have just one output for each input. Writing $g^{-1}(x) = \pm\sqrt{x}$ gives two outputs ($\sqrt{x}$ and $-\sqrt{x}$) for each input, so it is not a function.

This gives two possible values, not one. Continuing the procedure yields $g^{-1}(x) = \pm\sqrt{x}$, which is not a function. The procedure fails whenever the original function (g, in this case) is not one-to-one.

By contrast, the procedure should work for the restricted domain function $g(x) = x^2$ on $[0, \infty)$, because this function is one-to-one.

Example 3 *Find* $g^{-1}(x)$ *for* $g(x) = x^2$ *on* $[0, \infty)$.

Solution ❶ Write y in place of $g(x)$:

$$y = x^2.$$

❷ Solve the equation for the variable x:

$$x = \sqrt{y}.$$

Ans. to reading exercise 8
(a) yes, (b) no

Obtaining $x = \sqrt{y}$ instead of $x = \pm\sqrt{y}$ is a result of the restricted domain $[0, \infty)$. Because x cannot be negative, $x = -\sqrt{y}$ is not a possibility.

Exponentiation and taking roots can be considered inverse operations. The inverse of the function $g(x) = x^2$ on $[0, \infty)$, which squares positive numbers, is the function $g^{-1}(x)$, which takes square roots.

❸ Write the inverse function:

$$f^{-1}(x) = \sqrt{x}. \qquad \blacksquare$$

You may assume that $f(x) = x^3$ is one-to-one.

Reading Exercise 9 *Find* $f^{-1}(x)$ *for* $f(x) = x^3$.

As emphasized already, the inverse function reverses the inputs and outputs of the original function. For the function $f(x) = 4x$ of example 2, $f(3) = 12$ whereas $f^{-1}(12) = 3$ (figure 11). Performing the function followed by the inverse function gets us back to where we started: $f^{-1}(f(3)) = f^{-1}(12) = 3$. In general,

$$f^{-1}(f(x)) = x.$$

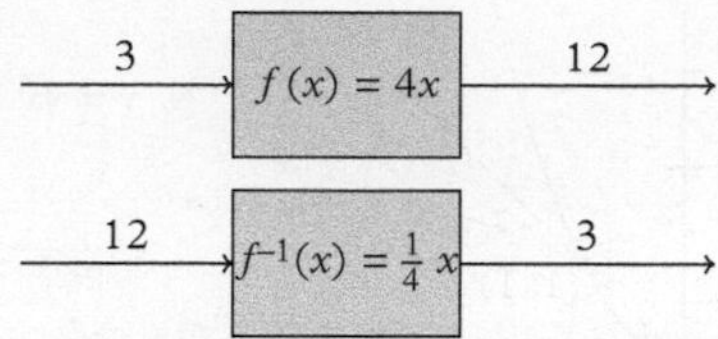

Figure 11 *The inverse function f^{-1} reverses the inputs and outputs from f. If $f(3) = 12$, then $f^{-1}(12) = 3$*

COMPOSITION OF A FUNCTION WITH ITS INVERSE

If f is a one-to-one function and x is a real number in the domain of f, then

$$f^{-1}(f(x)) = x.$$

Graphs of inverse functions

Because f^{-1} reverses the inputs and outputs of f, the x- and y-coordinates are reversed on the graphs, as illustrated in figure 12.

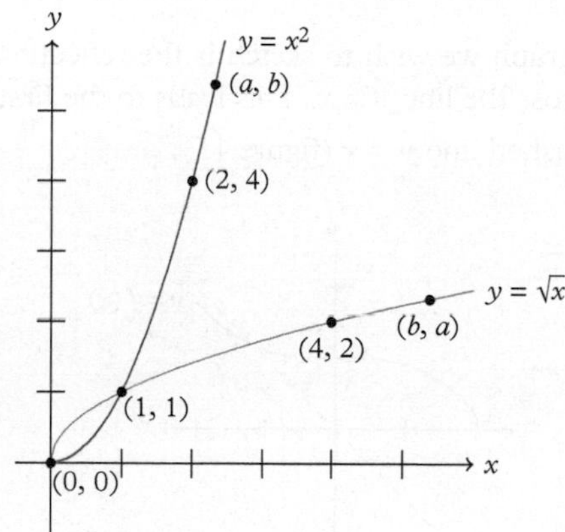

Figure 12 *The function $g(x) = x^2$ on $[0, \infty)$ (blue) and its inverse $g^{-1}(x) = \sqrt{x}$ (green). For each point (a, b) on the graph of g, the point (b, a) is on the graph of g^{-1}*

If $(2, 4)$ is on the graph of f, then $(4, 2)$ is on the graph of f^{-1}. If $(3, 9)$ is on the graph of f, then $(9, 3)$ is on the graph of f^{-1}. However, if the x- and y-coordinates are both the same, reversing them results in the same point. For the functions g and g^{-1} of figure 12, the point $(1, 1)$

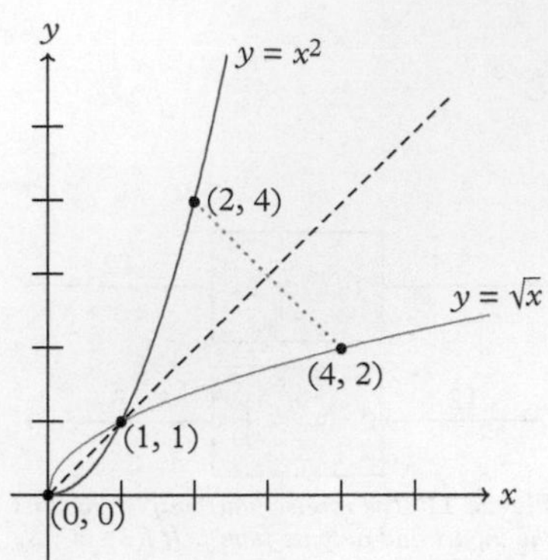

Figure 13 *The graph of g^{-1} (green) is the reflection of the graph of g (blue) across the line $y = x$ (black dashed line). The point $(4, 2)$ is the mirror image of the point $(2, 4)$ across this line, as illustrated by the brown dotted segment*

is on the graph of g, and reversing coordinates means that $(1, 1)$ is also on the graph of g^{-1}. The same is true for $(0, 0)$. These two points are on the line $y = x$, which plays a special role in relating the graphs of a function and its inverse (figure 13). The graph of a function f and its inverse f^{-1} are reflections across the line $y = x$.

Example 4 *The graph of a function $y = f(x)$ is given in figure 14. Sketch the graph of $y = f^{-1}(x)$.*

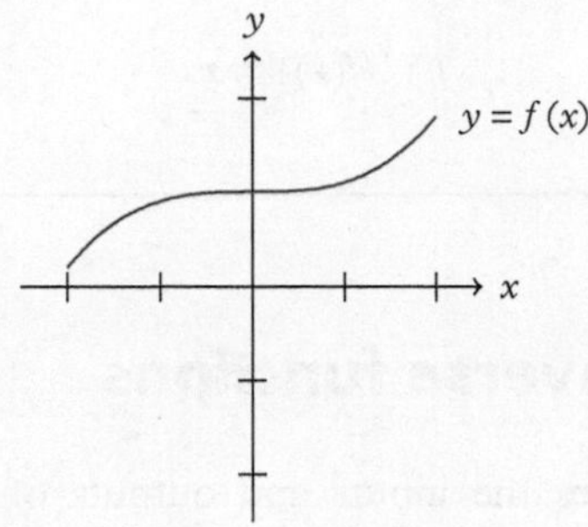

Figure 14 *The graph of $y = f(x)$ for example 4*

Ans. to reading exercise 9
$f^{-1}(x) = \sqrt[3]{x}$

Solution The graph we wish to sketch is the reflection of the graph that is given, across the line $y = x$. This leads to the first step.

❶ Draw a dashed line $y = x$ (figure 15).

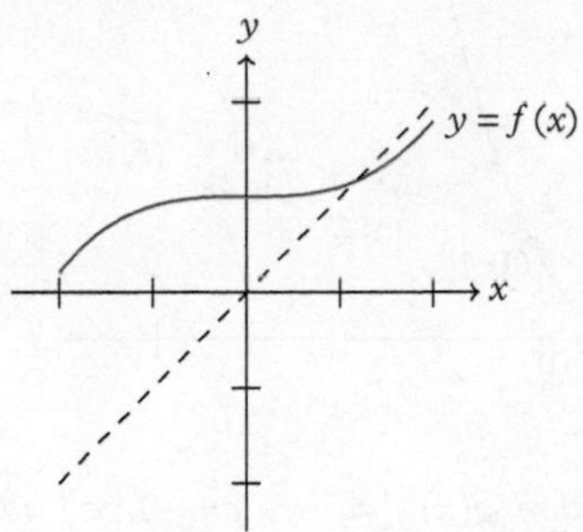

Figure 15 *The graph of $y = f(x)$ (blue), with line $y = x$ (black dashed line) added*

We may also reverse coordinates to get points on the graph of f^{-1}, suggesting the second step.

❷ Choose a small number of points along the graph of f, reverse coordinates to get points on the graph of f^{-1}, and plot those points (figure 16).

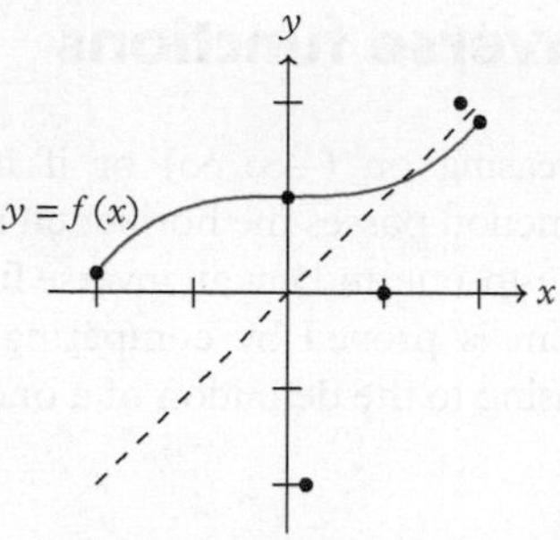

Figure 16 *Points are chosen along the graph of $y = f(x)$, and corresponding points on the graph of $y = f^{-1}(x)$ are plotted. When a point is very far from the dashed line $y = x$, reversing coordinates is more accurate than trying to guess at the location of the mirror image across the line. For instance, the point $(0, 1)$ becomes the point $(1, 0)$. If a point is close to the dashed line $y = x$ (as at the upper right corner), then producing the mirror image across the line might be easier*

❸ Draw the curve through the plotted points, keeping in mind that (1) the graph of f^{-1} should be the mirror image of the graph of f across the line $y = x$ and (2) where the graph of f crosses the line $y = x$, the graph of f^{-1} also crosses the line (figure 17).

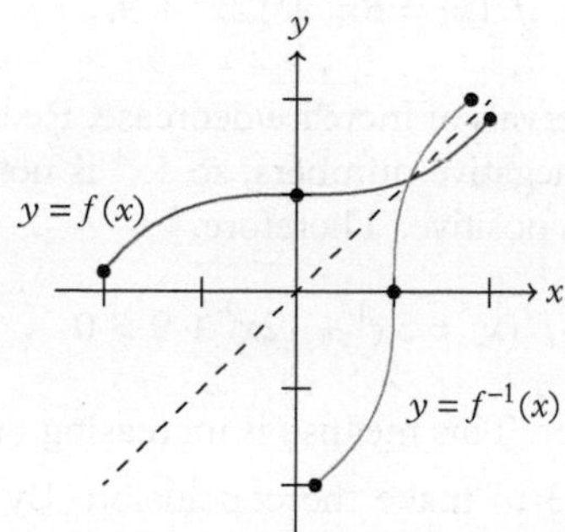

Figure 17 *The graph of $y = f^{-1}(x)$ (green) is added to the sketch*

If you wish to check that your answer is reasonable, turn your paper 45° counterclockwise to align the middle of your vision with the dashed line $y = x$ and check the symmetry of the two graphs. You might even consider performing step ❸ with your paper rotated. ■

Reading Exercise 10 *The point* $(4, 7)$ *is on the graph of the one-to-one function f. Name a point on the graph of* f^{-1}.

Calculus of inverse functions

If a function is increasing on $(-\infty, \infty)$ or if it is decreasing on $(-\infty, \infty)$, then the function passes the horizontal line test, and therefore the function is one-to-one and has an inverse function (figure 18). The following theorem is proved by comparing the definitions of increasing and decreasing to the definition of a one-to-one function.

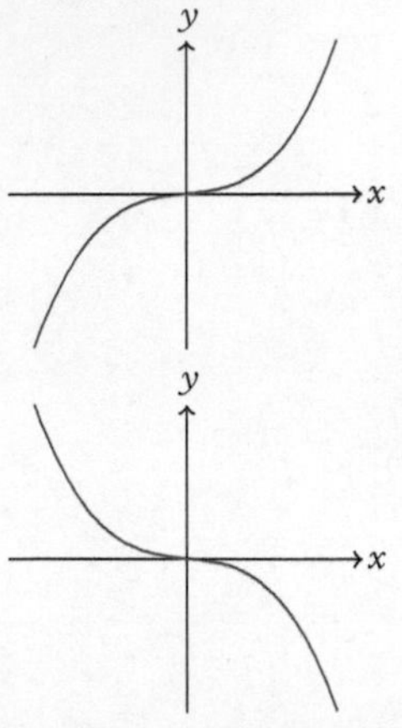

Figure 18 *(Top) a function that is increasing on* $(-\infty, \infty)$ *and (bottom) a function is decreasing on* $(-\infty, \infty)$*. Each satisfies the horizontal line test, meaning that the function is one-to-one and therefore has an inverse*

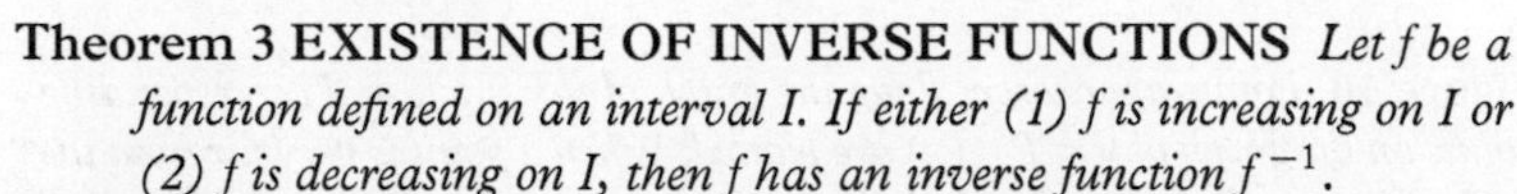

Theorem 3 EXISTENCE OF INVERSE FUNCTIONS *Let f be a function defined on an interval I. If either (1) f is increasing on I or (2) f is decreasing on I, then f has an inverse function* f^{-1}.

We know how to use the derivative to determine whether a function is increasing or decreasing. Putting this together with theorem 3 allows us to show that a function must have an inverse.

Example 5 *Let* $f(x) = x^5 + 4x^3 + 9x - 4$*. Show that* f^{-1} *exists.*

Solution We use derivative information to determine where the function is increasing or decreasing and then apply theorem 3.

❶ Find the derivative:

$$f'(x) = 5x^4 + 12x^2 + 9.$$

To show that f^{-1} exists means the same thing as showing that f has an inverse.

Recall that *nonnegative* means zero or positive.

❷ Determine intervals of increase/decrease. Recall that even powers always yield nonnegative numbers, so $5x^4$ is nonnegative, $12x^2$ is nonnegative, and 9 is positive. Therefore,

$$f'(x) = 5x^4 + 12x^2 + 9 > 0$$

An alternative is to follow the procedure of section 3.3. Trying to solve $f'(x) = 0$ results in no solutions and therefore just one interval for the chart: $(-\infty, \infty)$.

for every real number. This means f is increasing on $(-\infty, \infty)$.

❸ Use theorem 3 to make the conclusion. By theorem 3, f^{-1} exists. ■

Look again at figures 13 and 17. Recall that the graph of f^{-1} is the reflection of the graph of f across the line $y = x$. If the graph of f can be drawn in one piece without picking the pencil up off the paper, shouldn't the same be true of its mirror image? This suggests that if a one-to-one function is continuous, then its inverse function is also continuous.

Theorem 4 CONTINUITY OF INVERSE FUNCTIONS *If f is a one-to-one continuous function defined on an interval, then its inverse function f^{-1} is also continuous.*

Further justification of this theorem is at the end of this section, but a full proof is saved for an upper-level course.

Next, let's turn our attention to the derivative of an inverse function. In example 2 we learned that if $f(x) = 4x$, then $f^{-1}(x) = \frac{1}{4}x$. The slopes of these lines, 4 and $\frac{1}{4}$, are reciprocals of one another. Is this always true? Interpreted appropriately, it is.

To illustrate, we return to the functions of example 3: $g(x) = x^2$ on $[0, \infty)$ and $g^{-1}(x) = \sqrt{x}$. The point $(3, 9)$ is on the graph of g, and reversing coordinates gives the point $(9, 3)$ on the graph of g^{-1}. Calculate the slope of the tangent line to the curve $y = g(x)$ at $x = 3$:

$$g(x) = x^2$$
$$g'(x) = 2x$$
$$g'(3) = 2(3) = 6.$$

Calculate the slope of the tangent line to the curve $y = g^{-1}(x)$ at $x = 9$:

$$g^{-1}(x) = \sqrt{x}$$
$$(g^{-1})'(x) = \frac{1}{2\sqrt{x}}$$
$$(g^{-1})'(9) = \frac{1}{2\sqrt{9}} = \frac{1}{6}.$$

Ans. to reading exercise 10
$(7, 4)$

The two slopes we calculated are reciprocals of one another! Why? Compare the graphs of the one-to-one function $y = f(x)$ and its inverse $y = f^{-1}(x)$ in figure 19. The point $(k, f(k))$ is on the graph of f, so point $(f(k), k)$ is on the graph of f^{-1}. Reversing coordinates does nothing to change the geometry in the picture. Reversing the coordinates of the tangent line to $y = f(x)$ at $x = k$ results in the tangent line to the curve $y = f^{-1}(x)$ at $x = f(k)$. Now consider the segments labeled "f_{run}" and "f_{rise}." The slope of a line is the "rise" over the "run," as illustrated along the tangent line to $y = f(x)$ at $x = k$. But, reversing coordinates makes the horizontal segment f_{run} in the picture of f a vertical segment in the picture of f^{-1}, and it makes the vertical segment f_{rise} in the picture of f a horizontal segment in the picture of f^{-1}. Although the slope of the tangent line to $y = f(x)$ at $x = k$ is

$$f'(k) = \frac{f_{\text{rise}}}{f_{\text{run}}},$$

Geometric shapes are *invariant* under rotations, reflections, and translations. Rotate a rectangle by 60° and it is still a rectangle; reflect it in a mirror and it is still a rectangle. A tangent line is still a tangent line in the mirror

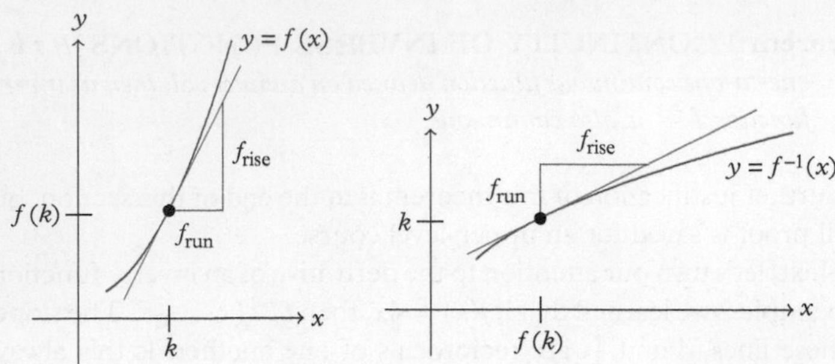

Figure 19 *(left) the graph of $y = f(x)$ (blue) along with its tangent line (orange) at $x = k$ and (right) the graph of $y = f^{-1}(x)$ (blue) and its tangent line (orange) at $x = f(k)$. The x- and y-coordinates of the objects in the graph on the right are swapped with those of the objects in the graph on the left, resulting in the vertical/horizontal positions of f_{rise} and f_{run} also being swapped.*

the slope of the tangent line to $y = f^{-1}(x)$ at $x = f(k)$ is

$$(f^{-1})'(f(k)) = \frac{f_{\text{run}}}{f_{\text{rise}}}.$$

The two slopes are reciprocals of one another.

Theorem 5 DERIVATIVE RULE FOR INVERSE FUNCTIONS
Let f be a one-to-one differentiable function on an interval I. If k is in the interval I and $f'(k) \neq 0$, then

$$(f^{-1})'(f(k)) = \frac{1}{f'(k)}.$$

Alternate version of the formula: If $f(a) = b$, then

$$(f^{-1})'(b) = \frac{1}{f'(a)}.$$

Proof. Using α as an arbitrary infinitesimal, since $f'(k) \neq 0$, the number $\alpha f'(k)$ is also infinitesimal. Because $f'(k)$ exists, by the local linearity theorem,

$$f(k + \alpha) \approx f(k) + \alpha f'(k).$$

We also need to know that f^{-1} is continuous so that we can approximate "inside" this function, but f is differentiable and therefore continuous (chapter 2 theorem 1), so we can then apply the continuity of inverse functions theorem.

Using $\alpha f'(k)$ as our arbitrary infinitesimal in calculating the derivative of f^{-1}, we have

Starting with some arbitrary infinitesimal, we can divide it by $f'(k)$ and call the result α. Our original arbitrary infinitesimal is then the same as $\alpha \cdot f'(k)$.

$$
\begin{aligned}
(f^{-1})'(f(k)) &= \frac{f^{-1}(f(k) + \alpha f'(k)) - f^{-1}(f(k))}{\alpha f'(k)} \\
&\approx \frac{f^{-1}(f(k+\alpha)) - k}{\alpha f'(k)} \\
&= \frac{k + \alpha - k}{\alpha f'(k)} = \frac{\alpha}{\alpha f'(k)} = \frac{1}{f'(k)}. \quad \blacksquare
\end{aligned}
$$

Line 1 uses the definition of a derivative with function f^{-1}, the real-number input $f(k)$, and infinitesimal $\alpha f'(k)$. Line 2 uses the result of the local linearity theorem to approximate inside the continuous function f^{-1} (see chapter 1 theorem 16), as well as the composition of a function with its inverse (formula from this section). Line 3 again uses the composition of a function with its inverse and then simplifies.

Informally, we use theorem 5 by using the alternate version of the formula

$$(f^{-1})'(b) = \frac{1}{f'(a)},$$

where $f(a) = b$. We need to know that the function is one-to-one, but as long as we can make the calculations in the formula, the other conditions of the theorem are met for the types of functions we study in calculus.

Example 6 *Let $g(x) = x^2$ on $[0, \infty)$. Use theorem 5 to find the value of $(g^{-1})'(9)$. Hint: $g(3) = 9$.*

Solution We wish to use the formula

$$(g^{-1})'(b) = \frac{1}{g'(a)}$$

to find $(g^{-1})'(9)$. Then, $b = 9$. The formula requires $g(a) = b$; so, by the hint, $a = 3$. The formula requires $g'(a)$, and we calculate

$$
\begin{aligned}
g'(x) &= 2x \\
g'(3) &= 2(3) = 6.
\end{aligned}
$$

Then,

$$(g^{-1})'(9) = \frac{1}{g'(3)} = \frac{1}{6}. \quad \blacksquare$$

The formula $(g^{-1})'(b) = \frac{1}{g'(a)}$ is used with $b = 9$, $a = 3$.

We calculated this same value $(g^{-1})'(9) = \frac{1}{6}$ before, first by determining g^{-1}, then calculating its derivative $(g^{-1})'$, and then evaluating that derivative at 9. The advantage of theorem 5 is that it applies even if we are not able to determine the inverse function.

Without the hint, could you have found the input to f that gives the output -4?

Example 7 *Let* $f(x) = x^5 + 4x^3 + 9x - 4$. *Find* $(f^{-1})'(-4)$. *Hint:* $f(0) = -4$.

Solution Checking that f is one-to-one was accomplished in example 5.

Using $b = -4$ and $a = 0$, we calculate

$$f'(x) = 5x^4 + 12x^2 + 9$$

$$f'(0) = 9.$$

Then,

$$(f^{-1})'(-4) = \frac{1}{f'(0)} = \frac{1}{9}. \quad \blacksquare$$

Calculating the inverse function $y = f^{-1}(x)$ for $f(x) = x^5+4x^3+9x-4$ in the manner of examples 2 and 3 is not possible. Yet we were still able to calculate a value of the derivative of that unknown function!

Reading Exercise 11 *The function* $f(x) = x^5+2x^3+2x+5$ *is one-to-one. Find* $(f^{-1})'(5)$. *Hint:* $f(0) = 5$.

Near-proof of the continuity of inverse functions theorem

Version 2 of the definition of continuity is used here.

Intuitively, continuity means that if $x = k+\alpha$ is infinitely close to k, then $f(x) = f(k+\alpha)$ is infinitely close to $f(k)$.

We can come close to a proof of the continuity of inverse functions theorem as follows. Suppose $y = f(x)$ is continuous at $x = k$. Then, for any infinitesimal α, $f(k + \alpha) \doteq f(k)$. Another way to write this is that there is an infinitesimal β such that

$$f(k + \alpha) = f(k) + \beta.$$

Then,

Line 1 uses substitution from the previous formula, line 2 uses the composition of a function with its inverse, line 3 uses the fact that α is an infinitesimal, and line 4 uses the composition of a function with its inverse.

$$\begin{aligned} f^{-1}(f(k) + \beta) &= f^{-1}(f(k+\alpha)) \\ &= k + \alpha \\ &\doteq k \\ &= f^{-1}(f(k)). \end{aligned}$$

This is almost what we need to show that f^{-1} is continuous at $f(k)$. The issue is that we need to know that $f^{-1}(f(k) + \beta) \doteq f^{-1}(f(k))$ for every infinitesimal β, but so far we have only shown that it happens for infinitesimals β that arise in the manner just mentioned. This missing step takes additional work that is not included here.

Exercises 5.3

1–6. Rapid response: is the function represented by the graph a one-to-one function?

1.

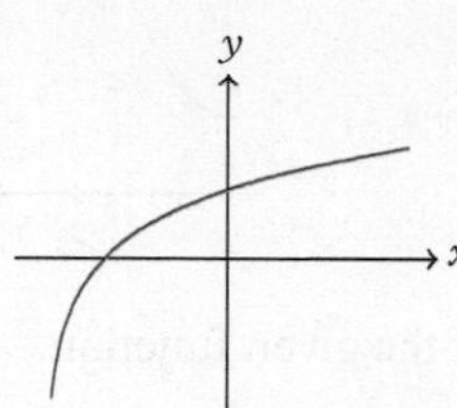

4.

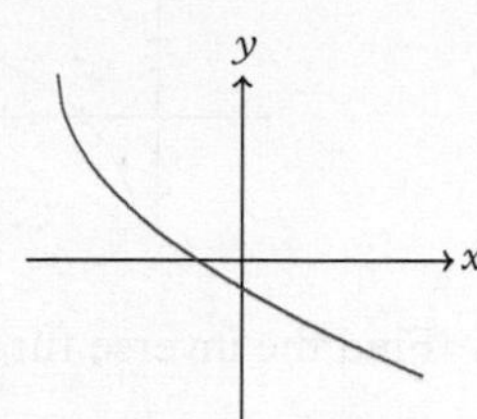

2.

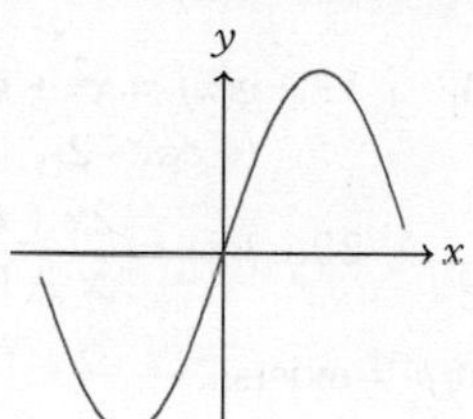

5.

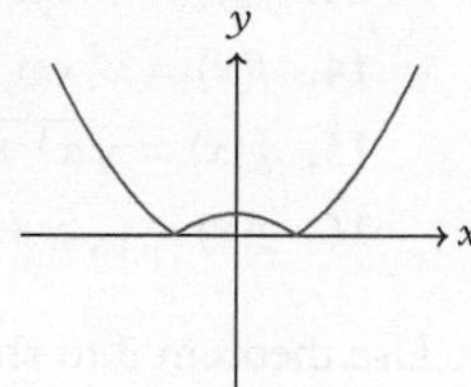

3.

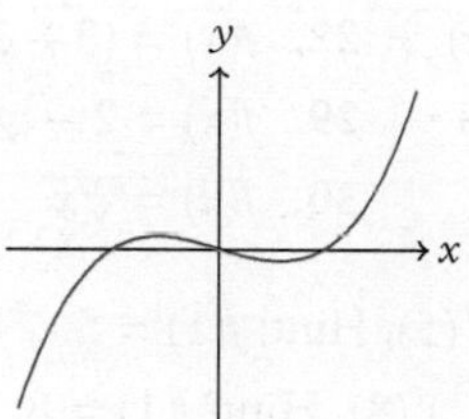

6.

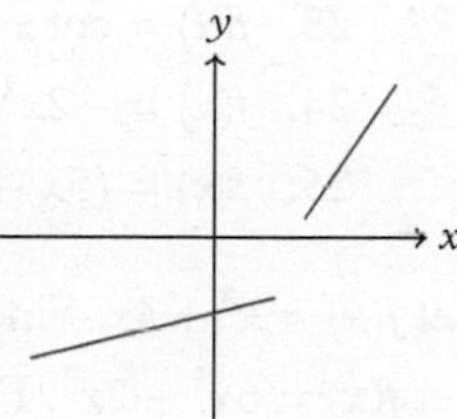

7–10. The graph of a function $y = f(x)$ is given. Sketch the graph of $y = f^{-1}(x)$.

7.

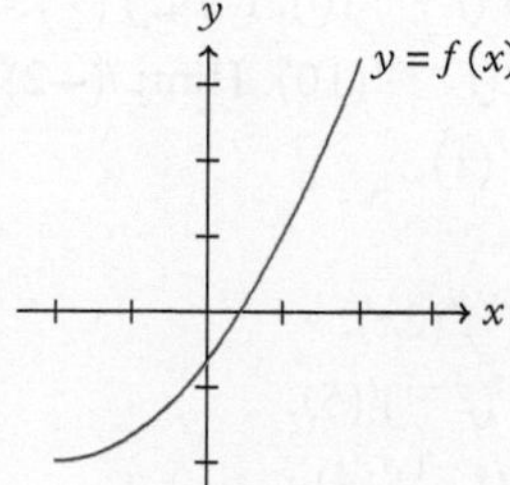

8.

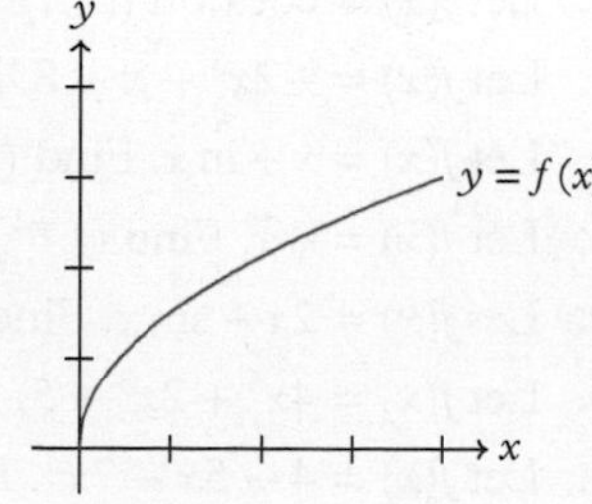

9.

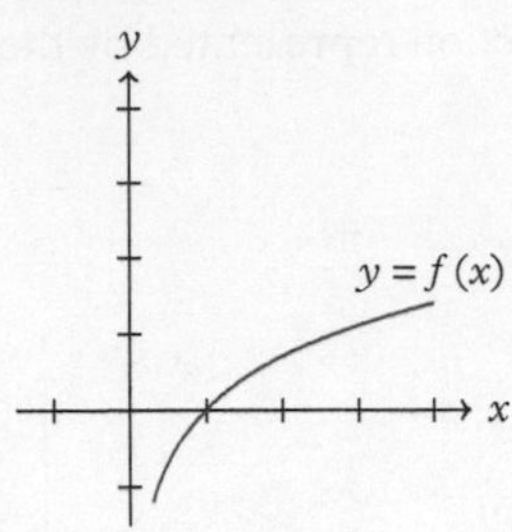

10.

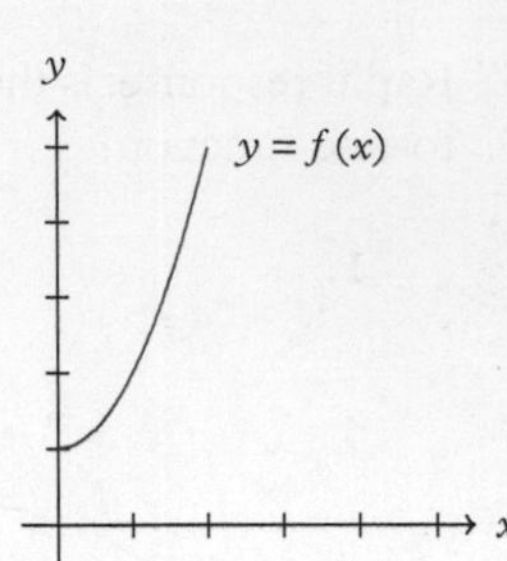

11–20. Find the inverse function of the given function.

11. $f(x) = 7x - 5$

12. $g(x) = \frac{1}{3}x + 9$

13. $f(x) = x^4$ on $[0, \infty)$

14. $f(x) = x^4$ on $(-\infty, 0]$

15. $g(x) = \sqrt{x^3 + 5}$

16. $f(x) = (\sqrt{x} + 4)^2$

17. $k(x) = \dfrac{1}{3 - x}$

18. $h(x) = \dfrac{x + 2}{x}$

19. $g(x) = x^2 + 4x + 4$ on $(-\infty, -2]$

20. $f(x) = \dfrac{2x + 6}{x - 1}$

Ans. to reading exercise 11
$(f^{-1})'(5) = \dfrac{1}{f'(0)} = \dfrac{1}{2}$

21–30. Use theorem 3 to show that f^{-1} exists.

21. $f(x) = x^3 + 4x$

22. $f(x) = 6x^5 + 2x^3$

23. $f(x) = \cot x$ on $(0, \pi)$

24. $f(x) = -2x^3 - x - 8$

25. $f(x) = (5x - 2)^3$

26. $f(x) = x \ln x$ on $(1, \infty)$

27. $f(x) = x + \ln x$

28. $f(x) = (3 - 2x)^3$

29. $f(x) = 2 - \sqrt{x}$

30. $f(x) = \sqrt[3]{x}$

The functions in exercises 31–38 were shown to be one-to-one in previous exercises.

31. Let $f(x) = x^3 + 4x$. Find $(f^{-1})'(5)$. Hint: $f(1) = 5$.

32. Let $f(x) = 6x^5 + 2x^3$. Find $(f^{-1})'(8)$. Hint: $f(1) = 8$.

33. Let $f(x) = (5x - 2)^3$. Find $(f^{-1})'(-8)$. Hint: $f(0) = -8$.

34. Let $f(x) = (3 - 2x)^3$. Find $(f^{-1})'(125)$. Hint: $f(-1) = 125$.

35. Let $f(x) = \cot x$ on $(0, \pi)$. Find $(f^{-1})'(0)$. Hint: $f\left(\frac{\pi}{2}\right) = 0$.

36. Let $f(x) = -2x^3 - x - 8$. Find $(f^{-1})'(10)$. Hint: $f(-2) = 10$.

37. Let $f(x) = x + \ln x$. Find $(f^{-1})'(1)$.

38. Let $f(x) = \sqrt[3]{x}$. Find $(f^{-1})'(2)$.

For exercises 39–42, don't forget to verify that the function is one-to-one so that the inverse exists.

39. Let $f(x) = 2x + \sin x$. Find $(f^{-1})'(2\pi)$.

40. Let $f(x) = 4x^5 + 2x^3 + 5$. Find $(f^{-1})'(5)$.

41. Let $f(x) = 4 - 5x - 7x^3$. Find $(f^{-1})'(4)$.

42. Let $f(x) = -5x + 1 + \cos x$. Find $(f^{-1})'(2)$.

43. The function $f(x) = \sin x$ is not one-to-one on the interval $(-\infty, \infty)$. Name an interval over which the function is one-to-one.

44. The function $g(x) = x^3 - 3x$ is not one-to-one on the interval $(-\infty, \infty)$. Name an interval over which the function is one-to-one.

45. Consider $y = \tan x$ on $[-\frac{\pi}{2}, \frac{\pi}{2}]$. (a) Use the derivative to verify that the function is one-to-one. (b) Sketch the graph of the function on a piece of paper. Be sure to include the x- and y-axes and something (such as scale markings) to indicate which direction is positive and which is negative. Mark any asymptotes clearly. (c) Hold the paper in front of a mirror and rotate it so that the x- and y-coordinates are switched, in such a way that positive and negative coordinates look correct in the mirror (positive horizontal coordinates to the right, negative to the left; positive vertical coordinates up, negative vertical coordinates down). The result should be the graph of the inverse function. Sketch this graph on a separate piece of paper.

46. Repeat exercise 45 for $y = x^3$.

Exponentials

5.4

In an algebra course, it is typical to study exponential functions before studying logarithmic functions. Exponential functions are easy to describe because you learned to calculate with exponents long before taking algebra; then, logarithmic functions are defined using inverse functions. Laws of logarithms are derived from laws of exponents.

The issue that arises with this approach is how to define exponentiation with irrational exponents. Exponentiation with natural-number exponents is repeated multiplication: $5^3 = 5 \cdot 5 \cdot 5$. Exponentiation with negative exponents is defined using reciprocals: $5^{-3} = \frac{1}{5^3}$. Exponentiation with fractional exponents is defined using roots: $5^{1/3} = \sqrt[3]{5}$ and $5^{7/3} = \sqrt[3]{5^7}$. But what is 5^{π}? What is $5^{\sqrt{3}}$? Typical approaches to describing such numbers in an algebra class involve what amounts to an intuitive notion of limit. One way or another, calculus is needed to define exponentiation with irrational numbers.

An approach that solves this problem is first to use the definite integral to define natural logs, as was done earlier in this chapter, and then use the opposite approach taken in algebra: define the natural exponential function from the natural logarithm.

The natural exponential function

Consider the graph of $y = \ln x$ (figure 1). Because the y-coordinates on the graph rise to ∞ (no horizontal asymptote) and fall to $-\infty$ (along the vertical asymptote), it appears that each real-number y-coordinate appears once on the graph of the function. The fact that no y-coordinate is skipped is a result of the continuity of the function (through the use of the intermediate value theorem), and the fact that no y-coordinate is hit twice is because the function is increasing, so it never turns back downward to repeat a y-coordinate.

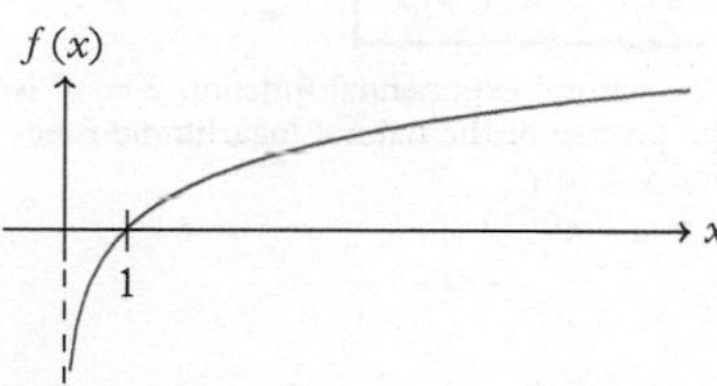

Figure 1 *The graph of $y = \ln x$, which is continuous, differentiable, increasing and concave down on $(0, \infty)$, and passes through the point $(1, 0)$. There is a vertical asymptote at $x = 0$, but there is no horizontal asymptote*

The narrative of this section is written in a manner that allows section 5.3 to be skipped. Marginal notes include comments meant for those who have read section 5.3.

The natural logarithmic function $y = \ln x$ is one-to-one and therefore has an inverse because it is increasing throughout its domain.

> Every real number is a y-coordinate on the graph of $y = \ln x$ exactly once.

We use this fact to define the natural exponential function. We begin by defining the number e, as illustrated in figure 2.

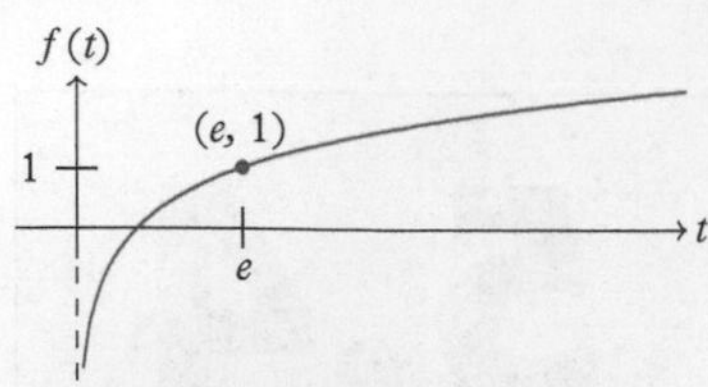

Figure 2 *The graph of* $y = \ln t$, *with horizontal coordinate e resulting in a vertical coordinate of* 1

The use of the letter e for this number is due to Leonhard Euler, who also introduced the mathematical use of the familiar symbols π, i, Σ, and $f(x)$.

Leonhard Euler, 1707–1783
http://www-history.mcs.st-andrews.ac.uk/Biographies/Euler.html

Definition 4 THE NUMBER KNOWN AS *e* *The number e is the number with the natural log* 1; *that is, e is the number for which* $\ln e = 1$.

It turns out that e is an irrational number with the value

$$e = 2.71828182845904523536\ldots.$$

Personal story: while taking calculus as a college freshman, I was told that it was a neat party trick to be able to recite e to 15 decimal places. Believe it or not, I actually tried it, but the results were not what I had hoped for.

Using law of logarithms (3) and definition 4, $\ln e^3 = 3\ln e = 3\cdot 1 = 3$ and $\ln e^{4/5} = \frac{4}{5}\ln e = \frac{4}{5}\cdot 1 = \frac{4}{5}$. In fact, for every rational number x, $\ln e^x = x$. The continuity of the natural logarithmic function leads us to surmise that perhaps this should be true of every real number x, not just rational numbers (figure 3). This "should" leads us to a way of defining e^x for irrational values of x.

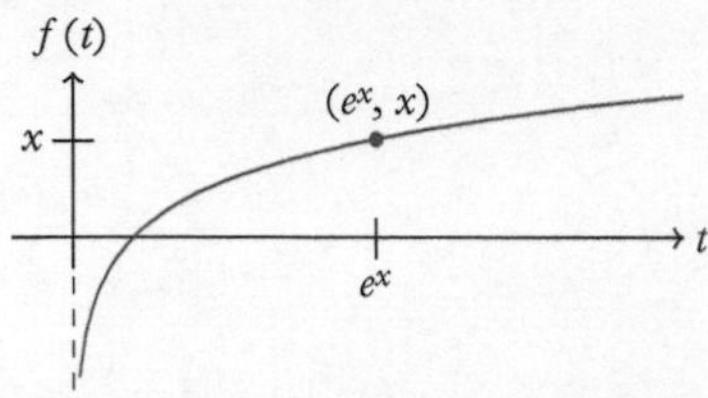

Figure 3 *the graph of* $y = \ln t$, *with horizontal coordinate* e^x *resulting in a vertical coordinate of* x

Definition 5 NATURAL EXPONENTIAL FUNCTION *For any real number x,* e^x *is the number for which* $\ln e^x = x$.

Pause and appreciate what was just accomplished. We have given meaning to numbers such as e^π and $e^{\sqrt{21}}$. Defining irrational exponents on numbers other than e is accomplished in the next section.

$\ln e = 1 \quad \ln e^x = x$

The natural exponential function $y = e^x$ is the inverse of the natural logarithmic function $y = \ln x$.

Because definition 5 is for any real number x, the domain of $y = e^x$ is $(-\infty, \infty)$.

Calculus of the natural exponential function

For $y = e^x$, what is y'? We desire a derivative formula for this function. Because the natural exponential function comes from the natural logarithmic function, perhaps using natural logs can help. Take the natural log of both sides of the equation $y = e^x$ to obtain

This equation also uses definition 5.

$$\ln y = \ln e^x = x.$$

Now we have restated the explicitly defined function $y = e^x$ as an implicitly defined function $\ln y = x$. We proceed to find y' by differentiating implicitly:

$$\ln y = x$$
$$\frac{1}{y} \cdot y' = 1$$
$$y' = y$$
$$y' = e^x.$$

Line 2 differentiates implicitly with respect to x, line 3 solves for y', and line 4 replaces y with its expression in x.

The natural exponential function is its own derivative!

DERIVATIVE OF THE NATURAL EXPONENTIAL FUNCTION

$$\frac{d}{dx}e^x = e^x$$

For $y = e^x$, we have $y' = e^x$. Then, $y'' = e^x$ as well; continuing, $y''' = e^x, \ldots, y^{(4297)} = e^x, \ldots$. All derivatives of e^x are e^x.

Example 1 *Find* y' *for* $y = xe^x$.

Solution We recognize the product $x \cdot e^x$ and use the product rule:

The presence of x in two places in the same term (that is, not separated by a + or −) often indicates the need for the product rule.

$$\frac{d}{dx}x \cdot e^x = x\frac{d}{dx}(e^x) + e^x \cdot \frac{d}{dx}(x)$$
$$= x \cdot e^x + e^x \cdot 1 = (x+1)e^x. \quad ■$$

If the exponent is not "just plain x," the chain rule is needed.

CHAIN RULE VERSION OF THE DERIVATIVE OF THE NATURAL EXPONENTIAL FUNCTION

$$\frac{d}{dx}e^u = e^u\frac{du}{dx}$$

Example 2 *Find* y' *for* $y = e^{4x^2+3x-8}$.

Solution The formula for $\frac{d}{dx}e^x = e^x$ is for an exponent of "just plain x." Because our function has an exponent that is not "just plain x," the chain rule is required:

$$\frac{d}{dx}e^{4x^2+3x-8} = e^{4x^2+3x-8} \cdot \frac{d}{dx}(4x^2 + 3x - 8)$$
$$= e^{4x^2+3x-8} \cdot (8x + 3). \quad ■$$

To differentiate a function of the form $y = e^u$ with respect to x, we repeat the exponential function and multiply by the derivative of the exponent.

Reading Exercise 12 *Find* $f'(x)$ *for* $f(x) = e^{5x}$.

As always, we can reverse the derivative formula to reveal an antiderivative formula:

ANTIDERIVATIVE OF THE NATURAL EXPONENTIAL FUNCTION

$$\int e^x \, dx = e^x + C$$

Example 3 *Evaluate* $\int e^{3x} \, dx$.

Solution Because the antiderivative formula requires "just plain x" and our exponent is $3x$, substitution is needed:

Any time the exponent in an exponential function is anything other than just plain x, try substitution with u as the exponent.

$$u = 3x$$
$$du = 3 \, dx.$$

Then,

In line 1, adjust for the required constant 3; in line 2, substitute; in line 3, antidifferentiate; and in line 4, return to the original variable.

$$\begin{aligned} \int e^{3x} \, dx &= \frac{1}{3} \int 3e^{3x} \, dx \\ &= \frac{1}{3} \int e^u \, du \\ &= \frac{1}{3} e^u + C \\ &= \frac{1}{3} e^{3x} + C. \end{aligned}$$ ■

Integrals similar to that of example 3 come up often enough that it is convenient to memorize a formula for $\int e^{ax} \, dx$. The derivation replaces 3 with a and follows the previous solution.

ANTIDERIVATIVE OF e^{ax}

$$\int e^{ax} \, dx = \frac{1}{a} e^{ax} + C$$

Reading Exercise 13 *Evaluate* $\int e^{5x} \, dx$.

Example 4 *Calculate* $\int_0^1 x e^{x^2} \, dx$.

Since $y = e^x$ is differentiable throughout its domain $(-\infty, \infty)$, it is continuous as well (because differentiability implies continuity). Therefore, FTC II applies and we may proceed.

Solution Recognizing that the exponent is not "just plain x," we use substitution, setting u as the exponent:

$$u = x^2$$
$$du = 2x \, dx.$$

We calculate the new limits of integration:

$$x = 1: \quad u = 1$$
$$x = 0: \quad u = 0.$$

We adjust for the missing constant, substitute, and integrate:

$$\begin{aligned}\int_0^1 xe^{x^2}\,dx &= \frac{1}{2}\int_0^1 2xe^{x^2}\,dx \\ &= \frac{1}{2}\int_0^1 e^u\,du \\ &= \frac{1}{2}e^u\Big|_0^1 \\ &= \frac{1}{2}e^1 - \frac{1}{2}e^0 = \frac{1}{2}e - \frac{1}{2} = \frac{1}{2}(e-1).\end{aligned}$$ ■

Recall that we do not need to return to the original variable when making a substitution in a definite integral.

Ans. to reading exercise 12
$f'(x) = 5e^{5x}$

Graph of $y = e^x$

As noted earlier, if $y = e^x$, then $\ln y = x$. In other words, the relationship $y = e^x$ is the same as the relationship $x = \ln y$. Comparing this relationship to $y = \ln x$, we see that the x- and y-coordinates are reversed. So, instead of the point $(1, 0)$ on the graph of $y = \ln x$, we have the point $(0, 1)$ on the graph of $y = e^x$ (the graph of $x = \ln y$); instead of the point $(e, 1)$ on the graph of $y = \ln x$, we have the point $(1, e)$ on the graph of $y = e^x$. Reversing all the coordinates gives the graph in figure 4.

If you studied the "exponential form" and the "logarithmic form" of expressions in algebra, then rewriting $y = e^x$ as $\ln y = x$ should be familiar to you.

$e^0 = 1$	$e^1 = e$

Look at one of the graphs in figure 4 and then rotate this page 90° counterclockwise. You should see a version of the other graph with left and right switched.

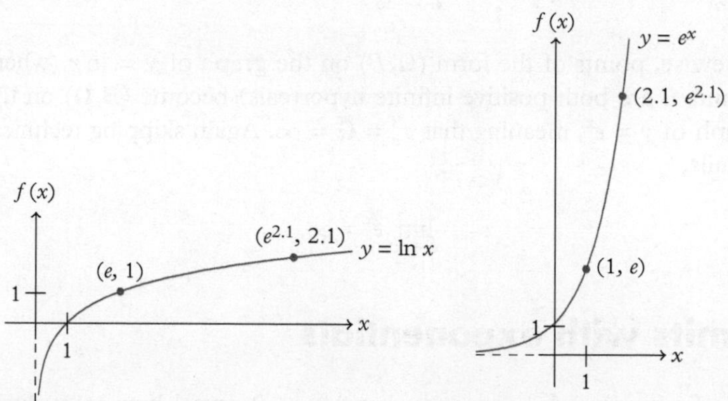

Figure 4 *(left) The graph of* $y = \ln x$*, including points* $(1, 0)$*,* $(e, 1)$*, and* $(e^{2.1}, 2.1)$*, and vertical asymptote* $x = 0$*; (right) the graph of* $y = e^x$*, including the points* $(0, 1)$*,* $(1, e)$*, and* $(2.1, e^{2.1})$*, and horizontal asymptote* $y = 0$

Because $y = \ln x$ and $y = e^x$ are inverses of one another, the graph of $y = e^x$ can be found from the graph of $y = \ln x$ by reflecting the latter across the line $y = x$.

Because all the x-coordinates on the graph of $y = \ln x$ are positive, all the y-coordinates on the graph of $y = e^x$ must also be positive. In other words, $e^x > 0$.

For any real number x,

$$e^x > 0.$$

Ans. to reading exercise 13
$\frac{1}{5}e^{5x} + C$

Using the techniques of chapter 3 to explore the graph of $y = e^x$, we can verify what we see in figure 4. Since

$$y' = e^x > 0$$

throughout the domain of the function, $y = e^x$ is increasing on $(-\infty, \infty)$. Since $e^x = 0$ has no solutions (because $e^x > 0$), there are no critical numbers and therefore no extrema. Since

$$y'' = e^x > 0,$$

the function is concave up on $(-\infty, \infty)$ and there are no inflection points.

Reversing the x- and y-coordinates also transforms the asymptote $x = 0$ on the graph of $y = \ln x$ to the asymptote $y = 0$ on the graph of $y = e^x$. This can be seen in the limits as well. Points of the form $(\omega, -B)$ on the graph of $y = \ln x$ (where ω is a positive infinitesimal and B is a positive infinite hyperreal) become $(-B, \omega)$ on the graph of $y = e^x$, meaning that $e^{-B} = \omega \doteq 0$. Skipping a few technical details, we see that

$$\lim_{x\to-\infty} e^x = 0.$$

$\lim_{x\to-\infty} e^x = 0$

$\lim_{x\to\infty} e^x = \infty$

Likewise, points of the form (Ω, B) on the graph of $y = \ln x$ (where Ω and B are both positive infinite hyperreals) become (B, Ω) on the graph of $y = e^x$, meaning that $e^B = \Omega \doteq \infty$. Again skipping technical details,

$$\lim_{x\to\infty} e^x = \infty.$$

Limits with exponentials

The limits $\lim_{x\to\infty} e^x = \infty$ and $\lim_{x\to-\infty} e^x = 0$ reveal how to evaluate exponentials of infinite hyperreals.

EXPONENTIALS OF INFINITE HYPERREALS

For any positive infinite hyperreal Ω,

$$e^{\Omega} \doteq \infty$$
$$e^{-\Omega} \doteq 0.$$

These facts are useful when evaluating limits.

Example 5 *Evaluate (a)* $\lim_{x\to 0^+} e^{4/x}$ *and (b)* $\lim_{x\to 0^-} e^{4/x}$.

Solution (a) Using $x = 0 + \omega$ for an arbitrary positive infinitesimal ω,

Because the exponent 4Ω is a positive infinite hyperreal, the expression renders ∞.

$$\lim_{x\to 0^+} e^{4/x} = e^{4/\omega} = e^{4\Omega} \doteq \infty.$$

(b) Using $x = 0 - \omega$ for an arbitrary positive infinitesimal ω,

Because the exponent -4Ω is a negative infinite hyperreal, the expression renders zero.

$$\lim_{x\to 0^-} e^{4/x} = e^{4/(-\omega)} = e^{-4\Omega} \doteq 0.$$

■

Laws of exponents

You are familiar with the laws of exponents, having used them in arithmetic and algebra. You may have also seen proofs of the laws of exponents, proofs that were valid only for rational exponents because only rational exponents were defined. If we wish to use the laws of exponents for irrational exponents as well, then we must prove these laws still apply when using definition 5.

Theorem 6 LAWS OF EXPONENTS *For any real numbers m and n,*
(1) $e^m e^n = e^{m+n}$ *and*
(2) $\frac{e^m}{e^n} = e^{m-n}$.

These laws of exponents are stated for the natural exponential function only. More general laws of exponents are stated and proved in the next section.

Proof. (1) The strategy for this proof is to show that the natural log of the left-hand side is equal to the natural log of the right-hand side and then draw the desired conclusion.

Using law of logarithms (1),

$$\ln\left(e^m \cdot e^n\right) = \ln e^m + \ln e^n.$$

Using definition 5 yields

$$= m + n.$$

We can now use definition 5 again, this time reversing the order and using $m + n$ for the x in the definition to obtain

$$= \ln e^{m+n}.$$

To summarize, we now have $\ln(e^m e^n) = \ln\left(e^{m+n}\right)$. We also know that the natural logarithmic function is one-to-one (there cannot be two different inputs into the natural logarithmic function that give the same output; so, if $\ln a = \ln b$, then $a = b$). Therefore, what is inside our two logarithms must be the same number—that is,

If $\ln a = \ln b$, then $a = b$.

$$e^m e^n = e^{m+n}.$$

(2) Using law of logarithms (2) this time, along with definition 5, gives

$$\ln \frac{e^m}{e^n} = \ln e^m - \ln e^n$$

$$= m - n$$

$$= \ln e^{m-n}.$$

Therefore, $\frac{e^m}{e^n} = e^{m-n}$. ■

There is a third law of logarithms,

$$\left(e^m\right)^n = e^{mn}.$$

Although this law is also true for all real numbers m and n, this proof has to wait because we still have not yet defined irrational exponents on numbers other than e. But, with this law, we can have an irrational exponent n on the real number e^m.

An additional property of exponents is that $e^{\ln x} = x$. This can be seen by using definition 5 (replacing x with $\ln x$) to obtain

$$\ln\left(e^{\ln x}\right) = \ln x.$$

Then, as before, what is inside the natural logs must be equal:

$$e^{\ln x} = x.$$

> For any positive real number x,
>
> $$e^{\ln x} = x.$$

Compare this formula (and the earlier formula $\ln e^x = x$) to the formula for the composition of a function with its inverse from section 5.3.

Algebra with logs and exponentials

The properties of logs and exponents can be used to simplify expressions and to solve equations involving exponents.

The algebraic manipulations shown in these examples should be review for those of you who studied exponentials and logarithms in algebra.

Example 6 *Simplify* $\ln e^{\tan x}$.

Solution Using definition 5,

$$\ln e^{\tan x} = \tan x. \quad \blacksquare$$

Example 7 *Simplify* $e^{3 \ln 7}$.

Solution The law $e^{\ln x} = x$ does not apply directly to the expression $e^{3 \ln 7}$. The expression is not of the form $e^{\ln \text{stuff}}$. But, this is easy to rectify using law of logarithms (3) first:

$$e^{3 \ln 7} = e^{\ln 7^3}.$$

Now the expression is in the form $e^{\ln \text{stuff}}$ and the law applies. The full calculation from start to finish is

$$e^{3 \ln 7} = e^{\ln 7^3} = 7^3 = 343. \quad \blacksquare$$

Example 8 *Solve for* x*:* $e^{2-4x} = 1$.

Solution Any time a variable expression is in an exponent, it can be brought down using law of logarithms (3) or definition 5. The strategy is to take the natural log of both sides of the equation:

Exponentiating both sides of an equation is also useful sometimes.

$$\ln e^{2-4x} = \ln 1$$

$$2 - 4x = 0$$

$$x = \frac{1}{2}. \quad \blacksquare$$

Line 1 takes the logs of both sides of the original equation; line 2 uses $\ln e^x = x$ (left side) and $\ln 1 = 0$ (right side) and line 3 finishes solving the equation for x.

Reading Exercise 14 *Solve for* x*:* $e^{x+1} = 1$.

Simplifying an expression is sometimes helpful in other types of problems.

Example 9 *Evaluate* $\int e^{2\ln x}\,dx$.

We begin by using the same strategy used in examples 3 and 4.

Solution Recognizing that the exponent is not "just plain x," we try substitution, setting u as the exponent:

$$u = 2\ln x$$
$$du = 2\cdot\frac{1}{x}\,dx = \frac{2}{x}\,dx.$$

Unfortunately, there is no $\frac{1}{x}$ for substituting du, and the substitution fails.

But, all is not lost! Simplifying the integrand in the manner of example 7 helps:

Line 1 uses law of logs (3), line 2 uses $e^{\ln x} = x$, and line 3 finishes by using the antiderivative power rule.

$$\int e^{2\ln x}\,dx = \int e^{\ln x^2}\,dx$$
$$= \int x^2\,dx$$
$$= \frac{x^3}{3} + C. \qquad \blacksquare$$

Even though the expression $\frac{x^3}{3}$ is defined on $(-\infty,\infty)$, the solution is valid only on $(0,\infty)$ because the integrand is only defined for positive values of x.

A caution about notation

Because $f(x) = e^x$ is a continuous function, by chapter 1 theorem 16, we may approximate in the exponent of this exponential function ("inside" the exponential function) as long as the resulting exponent is real. For instance,

$$e^{3+4\alpha^2} \approx e^3.$$

However, this theorem does not say what happens if the exponent, is infinite. According to the laws of exponents, $e^{\Omega+3} = e^{\Omega}\cdot e^3$, which is not approximately equal to e^{Ω}:

CAUTION: MISTAKE TO AVOID
Approximating in an exponent is incorrect if the resulting exponent is infinite.

$$e^{\Omega+3} \not\approx e^{\Omega}.$$

Although the two quantities $e^{\Omega+3}$ and e^{Ω} are not approximately equal, they do render the same real result: ∞. Thus, rendering may sometimes be used in place of approximation to avoid errors.

Example 10 *Evaluate* $\lim_{x\to\infty} e^{(x^4+3)/(x^2-2x)}$.

Solution Since

$$\frac{\Omega^4+3}{\Omega^2-2\Omega} \approx \frac{\Omega^4}{\Omega^2} = \Omega^2,$$

this quantity is positive infinite. Recalling that a positive infinite power of e renders ∞, we have

$$\lim_{x\to\infty} e^{(x^4+3)/(x^2-2x)} = e^{(\Omega^4+3)/(\Omega^2-2\Omega)} \doteq \infty. \quad \blacksquare$$

As tempting as it is to write $\lim_{x\to\infty} e^{(x^4+3)/(x^2-2x)} = e^{(\Omega^4+3)/(\Omega^2-2\Omega)} \approx e^{\Omega^2} \doteq \infty$, especially because the correct answer is obtained, this approximation is not correct. The featured solution uses correct notation to reach the correct answer.

Exercises 5.4

1–8. Rapid response: without using a calculator, state the value of each expression.

1. $\ln e^4$
2. $e^{\ln 6}$
3. e^0
4. $\ln e$
5. $e^{2\ln 5}$
6. $\sin(\ln 1)$
7. $\ln \frac{1}{e}$
8. $e^{-\ln 2}$

9–16. Simplify the expression.

9. $\ln e^{4x-7}$
10. $e^{\ln(x\sqrt{x-1})}$
11. $e^{\ln x - \ln 4}$
12. $\ln\left(e^x e^3\right)$
13. $e^{4-x} + e^{\ln(4+x)}$
14. $e^{7\ln(x-1)}$
15. $\ln\left(\frac{e^{14x}}{e^{\sin x}}\right)$
16. $\ln 4e^{x-7}$

17–22. Solve the equation.

17. $e^{1-x^2} = 2$
18. $e^{2+x} = e^{1-x}$
19. $e^{2\ln x} = 4$
20. $\ln x^2 = 4$
21. $\ln(2-x) = 1$
22. $e^{3x} = -4$

23–32. Evaluate the limit.

Ans. to reading exercise 14
$x = -1$

23. $\lim_{x\to -1^+} e^{x/(x+1)}$
24. $\lim_{x\to\infty} e^{(x^2-7)/(x+2)}$
25. $\lim_{x\to 0} e^{x/(x+1)}$
26. $\lim_{x\to 2} e^{(x+3)/((x-2)^2)}$
27. $\lim_{x\to -\infty} \sin e^x$
28. $\lim_{x\to 5} e^{x+\ln(x-4)}$
29. $\lim_{x\to -\infty} \left(\frac{3}{x} - e^x\right)$
30. $\lim_{x\to\infty} \ln\left(\frac{x+1}{x+2} + e^x\right)$
31. $\lim_{x\to\infty} \frac{e^{3x+1}}{2+e^{2x}}$
32. $\lim_{x\to 0^+} \frac{e^x}{x}$

33–46. Differentiate the function.

33. $f(x) = x^2 + e^4$
34. $f(x) = e^{4x}$
35. $y = \frac{e^x}{4x+3}$
36. $f(t) = e^t \sin t$
37. $g(x) = 5e^{4x+3}$
38. $y = e^{6\ln x}$
39. $f(x) = e^{x\sin x}$
40. $f(x) = \cos\left(e^{4x^2}\right)$
41. $y = xe^{x^2}$
42. $g(y) = e^{\sin 4y}$
43. $g(t) = \frac{t^2 e^{\sqrt{5t-8}}}{4(3t^2-5)^6}$
44. $y = \ln\left(3 + e^{7x}\right)$
45. $f(x) = x^2 \ln\left(xe^{6/x}\right)$
46. $f(x) = \sqrt{3x^2 \sin^3(2\pi x e^{7x})}$

47–62. Evaluate the integral.

All of these integrals can be evaluated successfully.

47. $\int 4e^{5x}\,dx$
48. $\int e^{6x}\,dx$
49. $\int e^{x/2}\,dx$
50. $\int \frac{e^{-4x}}{3}\,dx$
51. $\int \frac{e^{\sqrt{x}}}{2\sqrt{x}}\,dx$
52. $\int xe^{x^2}\,dx$
53. $\int_0^1 e^{4x-1}\,dx$
54. $\int_1^4 e^{\ln\sqrt{x}}\,dx$
55. $\int e^{\tan x}\sec^2 x\,dx$
56. $\int e^x \sec e^x \tan e^x\,dx$
57. $\int \ln e^{\sin x}\,dx$
58. $\int e^{\sec x}\sec x\tan x\,dx$
59. $\int \frac{e^{\sin x}}{\sec x}\,dx$
60. $\int (2x+3)e^{3x}e^{x^2}\,dx$
61. $\int e^x\sqrt{2+e^x}\,dx$
62. $\int \frac{3e^{3x}+1}{e^{3x}+x}\,dx$

63–66. Find the equation of the tangent line to the function at the given value of x.

63. $y = e^{3x}$ at $x = 0$

64. $y = e^{2x}$ at $x = \ln 5$

65. $y = \left(e^{x+1} + 1\right)^2$ at $x = -1$

66. $y = \dfrac{1}{1 + e^x}$ at $x = 0$

67–70. Solve the IVP.

67. $y' = x + e^x$, $y(0) = 3$

68. $y' = \pi e^{\sqrt{2}x}$, $y(\sqrt{2}) = 0$

69. $y' = e^{5 \ln x}$, $y(1) = \frac{7}{6}$

70. $y' = \dfrac{\pi \cos\left(\pi e^{-t}\right)}{e^t}$, $y(0) = 2$

71. Find y'' and y''' for $y = xe^x$ and conjecture the value of $y^{(42)}$.

72. Find all local extrema on the graph of $y = x \ln x$.

73. For a physics exercise, you and four of your friends are asked to use the formula that relates the charge on a discharging capacitor to how long it has been discharging. You missed class and they all remember different formulas:

Exercise 73 is adapted from materials developed by Brian Faulkner.

(i) $Q(t) = \frac{Q_o RC}{t}$
(ii) $Q(t) = Q_0 \left(1 - \frac{t}{RC}\right)$
(iii) $Q(t) = Q_0 e^{-\frac{t}{RC}}$
(iv) $Q(t) = e^{-\frac{t}{RC}}$

where $Q(t)$ is the charge on the capacitor at time t (in coulombs), Q_0 is the initial charge on the capacitor (in coulombs), R is the resistance in the discharging circuit (in ohms), C is the capacitance of the capacitor (in farads), and t is the time since the switch has been opened (in seconds). No outside aid (books, notes, internet) is allowed, and you must decide which formula to use. You decide that any reasonable formula has to have all of the following properties: (a) the charge on the capacitor starts out at a value of Q_0, just as the switch is thrown; (b) the charge on the capacitor should always be decreasing and (c) a very long time after the switch is thrown, the capacitor fully discharges. Use the three criteria to determine which formula(s) must be incorrect (state each criterion that is violated), and which formula(s) meet all the criteria and should be considered.

74. Explore $\lim_{x\to\infty} \frac{e^{x+5}}{e^x}$ as follows: (a) Simplify $\frac{e^{x+5}}{e^x}$ using laws of exponents. Use the result to calculate the limit. (b) Calculate the limit using $\frac{e^{\Omega+5}}{e^{\Omega}}$ and approximate incorrectly by approximating the exponent. Is the result the same as part (a)? (c) Draw a conclusion from your answers to parts (a) and (b).

General Exponentials

5.5

In the previous section we defined $e^{\sqrt{3}}$, but $5^{\sqrt{3}}$ remains undefined. We tackle this next.

General exponential functions

What is $5^{\sqrt{3}}$? Can we express this quantity in terms of something that is defined already? If so, it could form the basis for a new definition. The keys are to use the formula $e^{\ln x} = x$ and what we know about laws of exponents.

First, using $e^{\ln x} = x$, we know that $5 = e^{\ln 5}$. We should then get

$$5^{\sqrt{3}} = (e^{\ln 5})^{\sqrt{3}}.$$

Remember that we have not yet defined $5^{\sqrt{3}}$, and until we do so, this equation is exploratory and not fact.

In the previous section we did not justify the well-known third law of exponents $(e^m)^n = e^{mn}$. But, assuming we can, then the next step in the previous calculation is

$$5^{\sqrt{3}} = (e^{\ln 5})^{\sqrt{3}} = e^{\sqrt{3}\ln 5}.$$

Notice that the resulting expression is defined, for it is e to a real-number exponent! This can form the basis for our definition, where we write a for 5 and x for $\sqrt{3}$ so that it applies in general.

Definition 6 EXPONENTIAL FUNCTION WITH BASE a *For any* $a > 0$ *and any real number* x,

$$a^x = e^{x \ln a}.$$

We require $a > 0$ so that $\ln a$ is defined. Some exponents on negative numbers, such as $(-1)^{1/2} = \sqrt{-1}$, remain undefined because we are not using complex numbers.

Example 1 *Rewrite using definition 6: (a)* 3^π, *(b)* $\sqrt{2}^{\sqrt{2}}$, *and (c)* 4^2.

Solution (a) Using $a^x = e^{x\ln a}$ yields

$$3^\pi = e^{\pi \ln 3}.$$

Recall that, so far, we have only proved $n \ln x = \ln x^n$ for rational numbers n, so it is not yet appropriate to write $3^\pi = e^{\pi \ln 3} = e^{\ln 3^\pi}$.

(b) Using $a^x = e^{x \ln a}$ yields

$$\sqrt{2}^{\sqrt{2}} = e^{\sqrt{2} \ln \sqrt{2}}.$$

The quantity in part (c) was defined previously. Whenever a new definition applies to something defined previously, it must be consistent. Using law of logarithms (3), $e^{2 \ln 4} = e^{\ln 4^2} = 4^2$, with the last step justified by the formula $e^{\ln x} = x$. Therefore, the new definition does appear to be consistent with previous knowledge.

(c) Using $a^x = e^{x \ln a}$ yields

$$4^2 = e^{2 \ln 4}.$$ ■

Reading Exercise 15 *Rewrite the expression* $6^{\sqrt{5}}$ *using definition 6.*

Calculus of general exponentials

Let $f(x) = a^x$ for $a > 0$. Then,

$$f(x) = a^x = e^{x \ln a}$$

by definition 6, and we can take its derivative using previously known rules and formulas:

In line 1, the derivative rule for the natural exponential is used, along with the chain rule. Line 2 uses definition 6.

$$f'(x) = e^{x \ln a} \cdot \ln a$$
$$= a^x \ln a.$$

DERIVATIVE OF A GENERAL EXPONENTIAL

For $a > 0$,

$$\frac{d}{dx} a^x = a^x \ln a.$$

Note that if $a = e$, the formula gives $\frac{d}{dx} e^x = e^x \ln e = e^x \cdot 1$.

Finding derivatives involving general exponential functions is similar to finding derivatives for natural exponentials, with the exception that we must multiply by $\ln a$.

Example 2 *Find* y' *for (a)* $y = 4^x$ *and (b)* $y = 4^{2x-7}$.

Solution (a) Using the derivative rule for a general exponential function with $a = 4$,

The chain rule is not required because we have, exactly, the form $\frac{d}{dx} a^x$, using $a = 4$.

$$y' = 4^x \ln 4.$$

(b) Because the exponent is "not just plain x," the chain rule is required:

$$y' = \left(4^{2x-7}\ln 4\right)\cdot 2$$
$$= 2\cdot 4^{2x-7}\ln 4. \qquad \blacksquare$$

The ln 4 comes from the derivative rule for the general exponential function, whereas the 2 comes from the chain rule. We multiply by the derivative of the exponent, just as we did for the natural exponential function.

Reading Exercise 16 *Find y' for $y = 3^{5x+7}$.*

As always, the derivative formula can be reversed to give an antiderivative formula:

$$\int a^x \ln a\,dx = a^x + C.$$

Details:

$$\frac{d}{dx}a^x = a^x \ln a$$
$$\int a^x \ln a\; dx = a^x + C.$$

Noting that $\ln a$ is a constant, we rewrite using the antiderivative constant multiple rule:

$$\ln a \int a^x\,dx = a^x + C.$$

Dividing both sides by $\ln a$ gives the antiderivative formula

$$\int a^x\,dx = \frac{1}{\ln a}\cdot a^x + C.$$

Although $\int a^x\,dx = \frac{a^x + C}{\ln a} = \frac{a^x}{\ln a} + \frac{C}{\ln a}$ is correct, the quantity $\frac{C}{\ln a}$ is a constant and therefore it is customary to rewrite the expression as given.

ANTIDERIVATIVE OF A GENERAL EXPONENTIAL

For $a > 0$ with $a \neq 1$,

$$\int a^x\,dx = \frac{1}{\ln a}\cdot a^x + C.$$

Since $\ln 1 = 0$, we must exclude $a = 1$ to avoid division by zero.

Finding antiderivatives involving general exponential functions is similar to finding antiderivatives for natural exponentials, with the exception that we must divide by $\ln a$.

Ans. to reading exercise 15
$6^{\sqrt{5}} = e^{\sqrt{5}\ln 6}$

Example 3 *Evaluate (a) $\int 3^x\,dx$ and (b) $\int x\cdot 3^{x^2-7}\,dx$.*

Solution (a) Using the formula for the antiderivative of a general exponential function with $a = 3$,

Substitution is not required because the exponent is "just plain x." The form of the expression matches the antiderivative formula exactly.

$$\int 3^x\,dx = \frac{1}{\ln 3} \cdot 3^x + C.$$

(b) Just as with the natural exponential function, if the exponent is not "just plain x," then substitution is needed:

$$u = x^2 - 7$$
$$du = 2x\,dx.$$

Then,

In line 1, adjust for the required constant 2; in line 2, substitute; in line 3, antidifferentiate using the rule for a general exponential with $a = 3$; in line 4, return to the original variable. If desired, 2 ln 3 can be rewritten as ln 9.

$$\begin{aligned}\int x \cdot 3^{x^2-7}\,dx &= \frac{1}{2}\int 2x \cdot 3^{x^2-7}\,dx \\ &= \frac{1}{2}\int 3^u\,du \\ &= \frac{1}{2} \cdot \frac{1}{\ln 3} \cdot 3^u + C \\ &= \frac{1}{2\ln 3} \cdot 3^{x^2-7} + C.\end{aligned}$$ ■

Graph of $y = a^x$

If $a > 0$, then $a^x > 0$.

Since $a^x = e^{x\ln a}$, the graph of $y = a^x$ is similar to that of $y = e^x$. For instance, because e to a power is positive, so is a to a power (for $a > 0$), so the graph lies above the x-axis. However, whether $\ln a$ is a positive number or a negative number makes a difference. Because $\ln a$ is positive if $a > 1$ and negative if $a < 1$, we have two cases, as graphed in figure 1.

Beginning with the graph on the left in figure 1, if $a > 1$, then $\ln a > 0$. The derivative $y' = a^x \ln a > 0$, and a positive derivative means the function $y = a^x$ is increasing (on $(-\infty, \infty)$). Similarly, $y'' = a^x \ln a{\cdot}\ln a = a^x(\ln a)^2 > 0$, and the function is concave up on $(-\infty, \infty)$. Checking for horizontal asymptotes, we have

$$\lim_{x\to\infty} a^x = a^\Omega = e^{\Omega\ln a} \doteq \infty,$$

because the exponent $\Omega \ln a$ is positive infinite. There is no horizontal asymptote on the right. But,

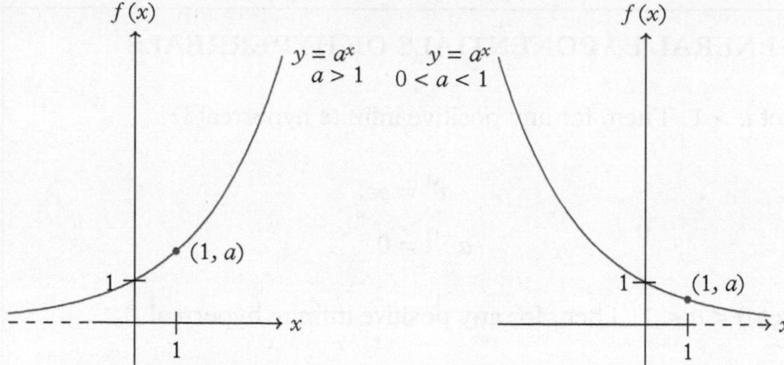

Figure 1 *(left) The graph of the increasing function $y = a^x$ with $a > 1$; (right) the graph of the decreasing function $y = a^x$ with $0 < a < 1$. Both graphs include the points $(0, 1)$ and $(1, a)$, with the height of the latter dependent on whether $a > 1$ or $a < 1$. Also, $y = 0$ is a horizontal asymptote on both graphs, but on the left for $a > 1$ and on the right for $0 < a < 1$*

Ans. to reading exercise 16
$y' = 3^{5x+7} \cdot \ln 3 \cdot 5$

$$\lim_{x \to -\infty} a^x = a^{-\Omega} = e^{-\Omega \ln a} \doteq 0,$$

because the exponent $-\Omega \ln a$ is negative infinite; $y = 0$ is a horizontal asymptote on the left.

For the graph on the right, we have $0 < a < 1$. Then, $\ln a < 0$ and many properties are reversed. We then have $y' = a^x \ln a < 0$, so the function is decreasing. We still have $y'' = a^x(\ln a)^2 > 0$, and the function is concave up. A check of the limits shows that, this time, the horizontal asymptote is on the right side. These limits (and the previous limits) are summarized in the following:

If $a > 1$, then

$$\lim_{x \to \infty} a^x = \infty \text{ and } \lim_{x \to -\infty} a^x = 0.$$

If $0 < a < 1$, then

$$\lim_{x \to \infty} a^x = 0 \text{ and } \lim_{x \to -\infty} a^x = \infty.$$

Limits with general exponentials

As we have seen, evaluating limits involving general exponentials depends on whether $a > 1$ or $0 < a < 1$.

GENERAL EXPONENTIALS OF HYPERREALS

Let $a > 1$. Then, for any positive infinite hyperreal Ω,

$$a^{\Omega} \doteq \infty,$$
$$a^{-\Omega} \doteq 0.$$

Since $e > 1$, the formulas for e^{Ω} and $e^{-\Omega}$ agree with the formulas for $a > 1$.

Let $0 < a < 1$. Then, for any positive infinite hyperreal Ω,

$$a^{\Omega} \doteq 0,$$
$$a^{-\Omega} \doteq \infty.$$

These formulas can be inferred from the associated limits given previously. They can also be proved using definition 6.

As with the natural exponential, these facts are useful for evaluating limits.

Example 4 *Evaluate (a)* $\lim_{x\to\infty} 0.6^{4x}$ *and (b)* $\lim_{x\to 3^+} 4^{1/(x-3)}$.

Solution (a) Using an arbitrary positive infinite hyperreal Ω,

$$\lim_{x\to\infty} 0.6^{4x} = 0.6^{4\Omega} \doteq 0.$$

With $0 < a < 1$ and a positive infinite exponent, the expression renders zero.

(b) Evaluating using continuity fails (division by zero), so we use $x = 3 + \omega$ for an arbitrary positive infinitesimal ω:

$$\lim_{x\to 3^+} 4^{1/(x-3)} = 4^{1/(3+\omega-3)} = 4^{1/\omega} = 4^{\Omega} \doteq \infty. \quad \blacksquare$$

With $a > 1$ and a positive infinite exponent, the expression renders ∞.

Laws of exponents for general exponentials

When studying the natural exponential function $y = e^x$, we proved only two of the three familiar laws of exponents. Next we prove all three laws of exponents for general exponentials.

Theorem 7 LAWS OF EXPONENTS *Let $a > 0$. Then, for any real numbers m and n,*

(1) $a^m a^n = a^{m+n}$,
(2) $\frac{a^m}{a^n} = a^{m-n}$, *and*
(3) $(a^m)^n = a^{mn}$.

Proof. The tools at our disposal are the definition ($a^x = e^{x \ln a}$) and the two laws of exponents for the natural exponential that we have already proved ($e^m e^n = e^{m+n}$ and $e^m / e^n = e^{m-n}$). It turns out that this is all we need.

(1) For any positive number a and any real numbers m and n,

$$\begin{aligned} a^m a^n &= e^{m \ln a} e^{n \ln a} \\ &= e^{m \ln a + n \ln a} \\ &= e^{(m+n) \ln a} \\ &= a^{m+n}. \end{aligned}$$

Line 1 uses the definition; line 2 uses law of exponents (1) for natural exponentials, which has already been proved; line 3 uses factoring in the exponent; and line 4 uses the definition to go back to an exponential with base a.

The proof of law (2) is similar and is left for you as an exercise.

We are now ready to prove law (3), which has not yet been proved for natural exponentials either. We begin by using the definition for the quantity inside the parentheses:

$$(a^m)^n = (e^{m \ln a})^n.$$

Now whatever number it happens to be, the quantity in parentheses is being raised to a power. We can use the definition again, this time with the quantity in the parentheses playing the role of the base a, yielding

$$e^{n \ln e^{m \ln a}}.$$

Then, in the exponent, we can use the property $\ln e^x = x$ to simplify, resulting in

$$e^{nm \ln a}.$$

Last, we use the definition (backward) one more time to obtain

$$a^{nm},$$

completing the proof. ■

The entire derivation, written without all the explanation, is

$$\begin{aligned} (a^m)^n &= (e^{m \ln a})^n \\ &= e^{n \ln e^{m \ln a}} \\ &= e^{nm \ln a} \\ &= a^{nm} \\ &= a^{mn}. \end{aligned}$$

We can now perform algebra such as

$$(3^{\sqrt{2}})^{\sqrt{2}} = 3^{\sqrt{2} \cdot \sqrt{2}} = 3^2 = 9,$$

with the confidence that all the familiar laws of exponents have been proved for any positive real number base and all real-number exponents.

Finally, because law (3) works for any $a > 0$, it works for e. In other words, $(e^m)^n = e^{mn}$ for any real numbers m and n.

Comparing derivative rules

We now have more than one derivative rule involving functions with exponents, including the power rule $\frac{d}{dx}x^n = nx^{n-1}$ and the rule for the derivative of a general exponential $\frac{d}{dx}a^x = a^x \ln a$. How can we tell which rule to use?

Example 5 *Find y' for*

$$\text{(a) } y = x^2, \text{ (b) } y = 2^x, \text{ (c) } y = 2^2, \text{ and (d) } y = x^x.$$

Solution (a) For $y = x^2$, the variable is in the base of the expression and the exponent is a constant. This fits the form of the power rule, and we have

Using the derivative rule for a general exponential gives an incorrect answer for part (a).

$$y' = 2x.$$

(b) For $y = 2^x$, the variable is in the exponent and the base is a constant. This fits the form of the rule for the derivative of a general exponential, and we have

Using the power rule gives an incorrect answer for part (b).

$$y' = 2^x \ln 2.$$

(c) For $y = 2^2$, both the base and the exponent are constants, so neither the power rule nor the rule for a general exponential apply. Instead, $y = 2^2 = 4$ is a constant function, and applying the rule for a constant function yields

Using only the power rule or the rule for a general exponential gives an incorrect answer for part (c). The same is true for part (d).

$$y' = 0.$$

(d) For $y = x^x$, both the base and the exponent contain the variable x, so neither the power rule nor the rule for a general exponential apply. The power rule requires a constant in the exponent, whereas the rule for a general exponential requires a constant in the base. There are two options for proceeding.

Option 1: use definition 6 to rewrite y and then use the derivative rule for natural exponentials. This gives

$$y = x^x = e^{x \ln x}$$

and

$$\begin{aligned} y' &= e^{x\ln x}\left(x \cdot \frac{1}{x} + 1 \cdot \ln x\right) \\ &= e^{x\ln x}(1 + \ln x) \\ &= x^x(1 + \ln x). \end{aligned}$$

Line 1 uses $\frac{d}{dx}e^x = e^x$, but because the exponent is not "just plain x," we need to multiply by the derivative of the exponent. The exponent is a product, so the product rule is used. Line 2 simplifies inside the parentheses. Line 3 uses definition 6 (backward) to rewrite $e^{x\ln x}$ in the original form x^x.

Option 2: use logarithmic differentiation. Beginning with $y = x^x$, we ❶ take the natural log of both sides and apply the third law of logarithms to obtain

$$\ln y = \ln(x^x) = x \ln x.$$

The third law of logarithms has only been proved for rational exponents, but we prove it for any real-number exponent later in this section. Its use in logarithmic differentiation is justified.

❷ Differentiating both sides implicitly with respect to x gives

$$\frac{1}{y}y' = x \cdot \frac{1}{x} + 1 \cdot \ln x = 1 + \ln x.$$

❸ Solving for y' gives

$$\begin{aligned} y' &= y(1 + \ln x) \\ &= x^x(1 + \ln x), \end{aligned}$$

Don't forget to replace y with its original expression x^x so that the final answer is written explicitly as a function of x.

which is the same answer we obtained using option 1. ■

The lesson to remember from part (d) of exercise 5 is that if the variable appears in both the base and the exponent of the expression, neither the power rule nor the rule for general exponentials can be used.

Reading Exercise 17 *Find y' for $y = x^{2x}$.*

Power rule, general case

The power rule was one of the first derivative rules that we encountered, but it has been proved in pieces—first for positive integer exponents, then for negative integer exponents (using the quotient rule), and then for rational exponents (using implicit differentiation). We are now ready for the final piece, a proof that includes irrational exponents.

> **GENERAL POWER RULE**
>
> Let $f(x) = x^n$ for a real number n. Then, $f'(x) = nx^{n-1}$.

If n is irrational, the domain of the function $f(x) = x^n$ is $x \geq 0$—that is, $[0, \infty)$.

Proof. For $f(x) = x^n$ where $x > 0$,

Line 1 uses definition 6, line 2 uses the derivative rule for e^x along with the chain rule (and other rules), line 3 uses definition 6 again (backward), and line 4 uses laws of exponents.

$$\begin{aligned} \frac{d}{dx}x^n &= \frac{d}{dx}e^{n\ln x} \\ &= e^{n\ln x}\cdot n\cdot\frac{1}{x} \\ &= x^n\cdot n\cdot\frac{1}{x} \\ &= nx^{-1}x^n = nx^{n-1}. \end{aligned}$$

Combined with previous cases of the power rule, the proof is now complete. ■

Example 6 *Find y' for $y = 4x^\pi$.*

Because π is irrational, previous cases of the power rule are not sufficient to justify this calculation.

Solution Using the general power rule and the constant multiple rule,

$$y' = 4\pi x^{\pi-1}.$$ ■

Exercises 5.5

1–10. Rewrite using definition 6.

1. $14^{\sqrt{3}}$
2. 8^{11}
3. 6^e
4. $(\ln 3)^\pi$
5. 4^{3x}
6. 101^0
7. 9^1
8. 4^{27x^2}
9. $(2+\sin x)^{x-2}$
10. $(5x^2)^{3x-1}$

11–36. Differentiate the given function.

11. $y = x^{\sqrt{2}}$
12. $y = 16^x$
13. $y = \sqrt{2}^x$
14. $y = x2^x$
15. $f(x) = 17^{4x+8}$
16. $y = 3x^{-\sqrt{7}}$
17. $f(t) = \sqrt{9+2^t}$
18. $y = \dfrac{6^{t-7}}{6^{2t+5}}$
19. $y = \ln(3+4^x)$
20. $f(x) = \cos(4-2^x)$
21. $g(t) = 5^t\,5^{3t-7}$
22. $g(y) = 6^{4y-8}$
23. $f(x) = \dfrac{3x-4}{2^x}$
24. $y = (x+5^x)^7$

Ans. to reading exercise 17
$x^{2x}(2+2\ln x)$
Details, option 1:

$$\begin{aligned} y &= x^{2x} = e^{2x\ln x} \\ y' &= e^{2x\ln x}(2x\cdot\frac{1}{x} + \ln x\cdot 2) \\ &= x^{2x}(2+2\ln x) \end{aligned}$$

Details, option 2:

$$\begin{aligned} \ln y &= \ln x^{2x} = 2x\ln x \\ \frac{1}{y}y' &= 2x\cdot\frac{1}{x} + \ln x\cdot 2 = 2+2\ln x \\ y' &= y(2+2\ln x) = x^{2x}(2+2\ln x) \end{aligned}$$

25. $y = (4x-7)^5\, 5^{4x-7}$

26. $s(t) = 3^{\sqrt{t}}$

27. $f(t) = t4^t \sin t$

28. $g(x) = x^{x\ln 2}$

29. $y = x^{2x-1}$

30. $f(x) = \dfrac{2^x}{x\ln 2}$

31. $g(y) = (3+\sqrt{5})^{1-\sin y}$

32. $h(k) = (k^3+3k)(3^k+k)^3$

33. $h(x) = 4x^{4x}$

34. $f(t) = (\tan t)^t$

35. $s(t) = (3t-1)^t$

36. $y = (x^3)^x$

37–46. Evaluate the integral.

All of these integrals can be evaluated successfully.

37. $\displaystyle\int 6^x\,dx$

38. $\displaystyle\int_0^1 2^x\,dx$

39. $\displaystyle\int_0^{\pi} 2^{\sin x}\cos x\,dx$

40. $\displaystyle\int \left(\frac{3}{4}\right)^{4x} dx$

41. $\displaystyle\int \frac{5^{\ln x}}{x}\,dx$

42. $\displaystyle\int 5^3\sec^2(4x+1)\,dx$

43. $\displaystyle\int (6-0.4^x)^3\, 0.4^x\,dx$

44. $\displaystyle\int \frac{2^{\sqrt{x}}}{\sqrt{x}}\,dx$

45. $\displaystyle\int \frac{5^{3x}}{5^{x-4}}\,dx$

46. $\displaystyle\int 4^x\sqrt{5+2^{2x}}\,dx$

47–58. Evaluate the limit.

47. $\displaystyle\lim_{x\to 3}\frac{2^x-1}{x+3}$

48. $\displaystyle\lim_{x\to 0^-} 0.22^{1/x}$

49. $\displaystyle\lim_{x\to\infty} 2.3^{4x+5}$

50. $\displaystyle\lim_{x\to 1^+}\frac{3^x-x}{x+4}$

51. $\displaystyle\lim_{x\to\infty} 0.7^{4x+5}$

52. $\displaystyle\lim_{x\to 3} 4^{1/(x-3)^2}$

53. $\displaystyle\lim_{x\to 0^-} 2^{1/x}$

54. $\displaystyle\lim_{x\to-\infty}\left(\frac{1}{x}+3^x\right)$

55. $\displaystyle\lim_{x\to 1^+}\left(\frac{13}{15}\right)^{(x-2)/(x-1)}$

56. $\displaystyle\lim_{x\to 1^+}\frac{3^{2-x}}{x-1}$

57. $\displaystyle\lim_{x\to 2}\frac{4^x}{x-2}$

58. $\displaystyle\lim_{x\to\infty}\frac{2^{x+1}}{3^{x+1}}$

59. Find the area under the curve $y = 3^x$ between $x = 0$ and $x = 1$.

60. Find the differential dy for $y = (x+1)3^{x+2}$.

61. Find the equation of the tangent line to the curve $y = 4 + 2^x$ at the point $(2, 8)$.

62. Solve the IVP $y' = 3^{4x-1}$, $y(0) = 4$.

63. Show that $y = a^x$ is concave up for any positive $a \neq 1$.

64. Prove law of exponents (2).

65. (a) Use definition 6 to determine the value of 1^x for any real number x. Simplify. (b) Evaluate the derivative of $y = 1^x$ using the formula $\frac{d}{dx}a^x = a^x \ln a$ and simplify. (c) Evaluate the derivative of $y = 1^x$ using the result of part (a) and a derivative formula from a previous chapter. (d) Compare the results of (b) and (c). Is the formula $\frac{d}{dx}a^x = a^x \ln a$ valid when $a = 1$?

66. Use definition 6 to derive the formulas $\lim_{x\to-\infty} a^x = \infty$ and $\lim_{x\to\infty} a^x = 0$ for $0 < a < 1$.

General Logarithms

5.6

We finish our development of logarithmic and exponential functions by using their inverse relationship once again.

General logarithmic functions

The graphs of the general exponential functions in section 5.5 figure 1 pass the horizontal line test. In other words, whether $0 < a < 1$ or $a > 1$, $f(x) = a^x$ is a one-to-one function and has an inverse function. Recalling that $y = e^x$ and $y = \ln x$ are inverses of one another, the inverse of a general exponential is written as a general logarithm.

If $a = 1$, $f(x) = a^x = 1^x = 1$ is a constant function and fails the horizontal line test.

Definition 7 LOGARITHMIC FUNCTION WITH BASE *a*

For any positive number $a \neq 1$, $g(x) = \log_a x$ is the inverse function of $f(x) = a^x$. In other words,

$$y = \log_a x \text{ if and only if } a^y = x.$$

Recall that the inverse function reverses x- and y-coordinates. So instead of $y = a^x$, the coordinates are reversed to $x = a^y$ for the inverse function. The general logarithmic notation merely rewrites this expression.

The expressions $y = \log_a x$ and $a^y = x$ are called the *logarithmic form* and *exponential form* of the expression, respectively.

Example 1 *(a) Rewrite $y = \log_3 7$ in exponential form and (b) rewrite $4^{2x} = 17$ in logarithmic form.*

Solution (a) Using definition 7, the exponential form of $y = \log_3 7$ is

$$3^y = 7.$$

(b) Using definition 7 in the other direction, the logarithmic form of $4^{2x} = 17$ is

$$\log_4 17 = 2x. \qquad \blacksquare$$

Logarithms, although not in their modern form, were introduced by Napier.
John Napier 1550–1617
http://www-history.mcs.st-andrews.ac.uk/Biographies/Napier.html

Rewriting an exponential equation in logarithmic form can help solve the equation. For instance, after rewriting the equation $4^{2x} = 17$ from example 1(b) as $\log_4 17 = 2x$, we merely need to divide both sides by 2 to find the solution $x = \frac{1}{2}\log_4 17$.

Reading Exercise 18 *(a) Rewrite* $y = \log_4 12$ *in exponential form and (b) rewrite* $5^x = 9$ *in logarithmic form.*

$\log_e x = \ln x$

Applying definition 7 when $a = e$ says that $y = \log_e x$ if and only if $e^y = x$. We also know that $e^y = x$ is the same relationship as $y = \ln x$. Therefore, $\log_e x = \ln x$.

Because it is an inverse function, the graph of $y = \log_a x$ can be found from the graph of $y = a^x$ by switching the x- and y-coordinates. For $a > 1$, the graph of $y = \log_a x$ is increasing and looks similar to the graph of $y = e^x$; see figure 1. The horizontal asymptote on the graph of $y = a^x$, when reflected across the mirror $y = x$, becomes

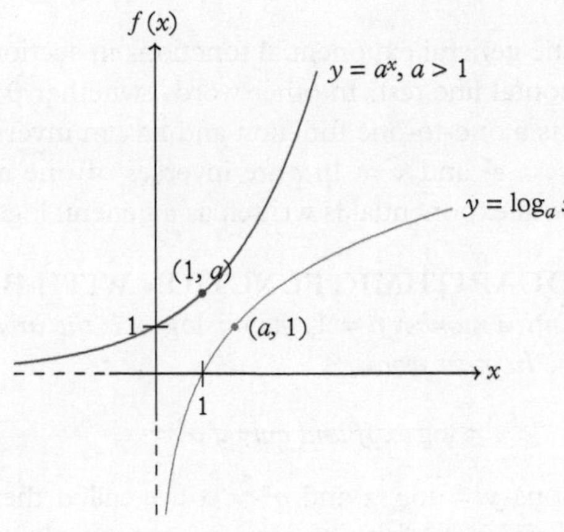

Figure 1 *The graph of the function* $y = a^x$ *with* $a > 1$ *(blue) along with the graph of its inverse function* $y = \log_a x$ *(green). The point* $(1, a)$ *of* $y = a^x$ *corresponds to the point* $(a, 1)$ *of* $y = \log_a x$*; the horizontal asymptote of* $y = a^x$ *corresponds to the vertical asymptote of* $y = \log_a x$*. Both functions are increasing*

the vertical asymptote on the graph of $y = \log_a x$. We therefore have the limit

$$\lim_{x\to 0^+} \log_a x = -\infty.$$

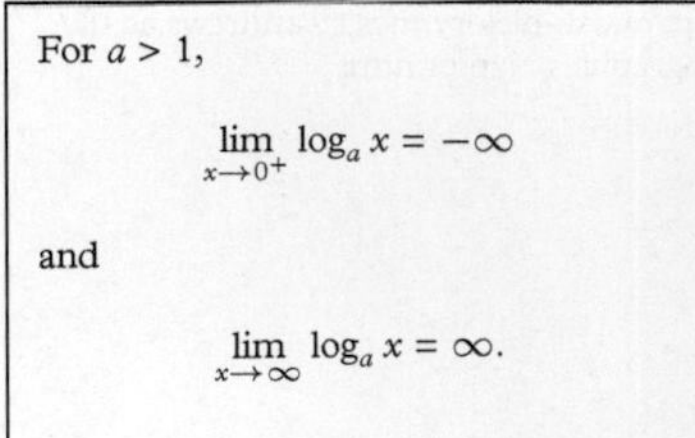

For $a > 1$,

$$\lim_{x\to 0^+} \log_a x = -\infty$$

and

$$\lim_{x\to\infty} \log_a x = \infty.$$

From $\lim_{x\to\infty} a^x = \infty$, we see that for $y = a^x$, positive infinite x-coordinates give positive infinite y-coordinates. Swapping coordinates shows the same is true for $y = \log_a x$—that is,

$$\lim_{x\to\infty} \log_a x = \infty.$$

Just as with the natural logarithmic function, the domain of $y = \log_a x$ is $[0, \infty)$.

For $0 < a < 1$, both functions are decreasing and we have the limits

$$\lim_{x \to 0^+} \log_a x = \infty$$

and

$$\lim_{x \to \infty} \log_a x = -\infty.$$

See figure 2.

For $0 < a < 1$,

$$\lim_{x \to 0^+} \log_a x = \infty$$

and

$$\lim_{x \to \infty} \log_a x = -\infty.$$

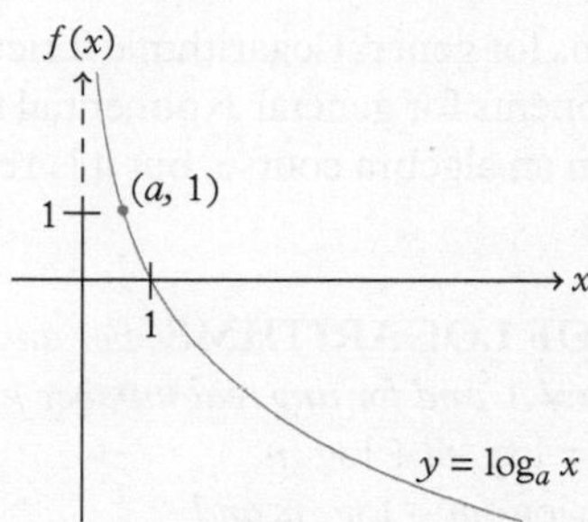

Figure 2 *The graph of the function* $y = \log_a x$ *with* $0 < a < 1$. *The function is decreasing and the vertical asymptote is approached upward. There is no horizontal asymptote*

These limits can also be expressed in terms of hyperreals.

GENERAL LOGARITHMS OF HYPERREALS

For any real number $a > 1$, any positive infinitesimal ω, and any positive infinite hyperreal Ω,

$$\log_a \omega \doteq -\infty$$

and

$$\log_a \Omega \doteq \infty.$$

For any real number a with $0 < a < 1$, any positive infinitesimal ω, and any positive infinite hyperreal Ω,

$$\log_a \omega \doteq \infty$$

and

$$\log_a \Omega \doteq -\infty.$$

Ans. to reading exercise 18
(a) $4^y = 12$, (b) $x = \log_5 9$

Evaluating limits with general logarithms works the same as evaluating limits with natural logarithms, although we must pay attention to whether $a > 1$ or $0 < a < 1$ when rendering a real result. We look at examples of this later in this section.

Laws of logarithms for general logarithmic functions

The laws of logarithms for general logarithmic functions can be proved from the laws of exponents for general exponential functions. You may have seen this done in an algebra course, but it is repeated here for the sake of completeness.

Theorem 8 LAWS OF LOGARITHMS *For any positive numbers a, n, and m with $a \neq 1$ and for any real number k,*
(1) $\log_a mn = \log_a m + \log_a n$,
(2) $\log_a \frac{m}{n} = \log_a m - \log_a n$, *and*
(3) $\log_a m^k = k \log_a m$.

Proof. Each law of logs involves one or both of $\log_a m$ and $\log_a n$. Let's call these numbers p and q, respectively:

$$p = \log_a m,$$
$$q = \log_a n.$$

These expressions can be rewritten in exponential form as

$$m = a^p,$$
$$n = a^q.$$

Using law of exponents (1),

$$mn = a^p a^q = a^{p+q}.$$

The expression we are rewriting is $mn = a^{p+q}$.

Now we can rewrite this expression in logarithmic form as

$$\log_a mn = p + q.$$

By substitution,

$$\log_a mn = \log_a m + \log_a n$$

and law of logarithms (1) holds true.

Continuing using these same facts and using law of exponents (2),

$$\frac{m}{n} = \frac{a^p}{a^q} = a^{p-q}.$$

Rewriting in logarithmic form and using substitution yields

The expression we are rewriting is $\frac{m}{n} = a^{p-q}$.

$$\log_a \frac{m}{n} = p - q = \log_a m - \log_a n,$$

proving law of logarithms (2).

Last, by law of exponents (3),

$$m^k = (a^p)^k = a^{pk}.$$

Rewriting in logarithmic form and using substitution gives

The expression we are rewriting is $m^k = a^{pk}$.

$$\log_a m^k = pk = k \cdot \log_a m,$$

yielding law of logarithms (3). ■

For natural logarithms, our previous proof of law (3) was valid only when the exponent k was a rational number. But, $\ln x = \log_e x$, so our proof of law (3) here holds for natural logarithms as well. For any real-number exponent k and positive number x,

$$\ln x^k = k \ln x.$$

Change of base property

The next theorem expresses general logarithms in terms of natural logarithms.

Theorem 9 CHANGE OF BASE PROPERTY *For any positive number $a \neq 1$ and any positive number x,*

$$\log_a x = \frac{\ln x}{\ln a}.$$

Proof. By definition 7, if $y = \log_a x$, then $a^y = x$. Take the natural logarithm of both sides and then solve for y:

$$\ln a^y = \ln x$$

$$y \ln a = \ln x$$

$$y = \frac{\ln x}{\ln a}.$$

By substitution (because each expression is equal to y),

$$\log_a x = \frac{\ln x}{\ln a}.$$

■

The use of the natural logarithm in the proof was just for convenience. We could have used $\log_{10}$ (or any other base) instead, and the proof works the same.

CHANGE OF BASE PROPERTY EXPANDED

For any positive numbers $a \neq 1$, $b \neq 1$, and x,

$$\log_a x = \frac{\ln x}{\ln a} = \frac{\log_{10} x}{\log_{10} a} = \frac{\log_b x}{\log_b a}.$$

Many calculators have two logarithm buttons, one for the natural logarithm labeled "LN" and one for the *common logarithm*, the logarithm base 10, labeled "LOG." The change of base property makes it possible to use either button to find the value of any general logarithm.

Example 2 *Use a calculator to find* $\log_4 7$.

Solution Using the change of base property,

$$\log_4 7 = \frac{\ln 7}{\ln 4}.$$

The numerator and denominator can each be calculated using the LN button, resulting in

$$\log_4 7 = \frac{\ln 7}{\ln 4} = 1.4037.$$

Refer to your calculator's manual if there is any question regarding which logarithm is performed by a particular button. Some calculators or software applications use LOG to refer to the natural logarithm instead of the common logarithm.

An alternate solution using the LOG button is

$$\log_4 7 = \frac{\log_{10} 7}{\log_{10} 4} = 1.4037.$$

■

Example 3 *Solve* $4^x = 7$.

Solution When we wish to solve for a variable that is in an exponent, taking logarithms of both sides and using law of logarithms (3) is often helpful:

$$\ln 4^x = \ln 7$$
$$x \ln 4 = \ln 7$$
$$x = \frac{\ln 7}{\ln 4} = 1.4037.$$

Use of a calculator can verify that (at least approximately) $4^{1.4037} = 7$. ■

Using the common logarithm, $\log_{10}$, instead of the natural log is just as convenient if that button is available on your calculator.

Compare the calculations in examples 2 and 3. The logarithmic form of the equation $4^x = 7$ is $x = \log_4 7$. This illustrates a way of thinking about general logarithms: the number $\log_4 7$ is the exponent on 4 that gives you 7.

The number $\log_a c$ is the exponent on a that gives you c.

Reading Exercise 19 *Use a calculator to find* $\log_8 15$.

Calculus with general logarithmic functions

The change of base formula allows us to find derivatives and integrals involving general logarithms by using our previous rules for the natural logarithm instead.

Example 4 *Find* $\frac{dy}{dx}$ *for* $y = \log_2 x$.

Solution We first rewrite y using the change of base formula:

$$y = \log_2 x = \frac{\ln x}{\ln 2} = \frac{1}{\ln 2} \cdot \ln x.$$

Because $\frac{1}{\ln 2}$ is a constant, we use the constant multiple rule and the derivative rule for the natural logarithm to obtain

$$y' = \frac{1}{\ln 2} \cdot \frac{1}{x} = \frac{1}{x \ln 2}.$$ ■

Replacing 2 with a in example 4 yields a proof of the derivative rule for general logarithmic functions.

Because $y = \log_a x$ is differentiable throughout its domain, it is also continuous throughout its domain.

DERIVATIVE OF GENERAL LOGARITHMIC FUNCTIONS

For any positive $a \neq 1$,

$$\frac{d}{dx} \log_a x = \frac{1}{x \ln a}.$$

DERIVATIVE OF GENERAL LOGARITHMIC FUNCTIONS (CHAIN RULE FORM)

For any positive $a \neq 1$,

$$\frac{d}{dx} \log_a u = \frac{1}{u \ln a} \cdot \frac{du}{dx}.$$

As illustrated in example 4, use of the derivative rule for general logarithmic functions can be avoided by using the change of base property instead. Nevertheless, we use the rule in the next example.

Example 5 *Differentiate* $f(x) = \log_7(\sin 5x)$.

Solution Using the chain rule form of the derivative rule for general logarithmic functions with $a = 7$ and $u = \sin 5x$ gives

$$\begin{aligned} f'(x) &= \frac{1}{(\sin 5x) \ln 7} \cdot \cos 5x \cdot 5 \\ &= \frac{5}{\ln 7} \cot 5x. \end{aligned}$$

■

Reading Exercise 20 *Differentiate* $y = \log_4(x^2 + 5x + 7)$.

Reversing the derivative rule

$$\frac{d}{dx} \log_a x = \frac{1}{x \ln a}$$

results in the antiderivative rule

$$\int \frac{1}{x \ln a}\, dx = \log_a x + C.$$

In the same manner as with the natural logarithm, the rule can be expanded beyond positive values of x by using absolute values, resulting in

$$\int \frac{1}{x \ln a}\, dx = \log_a |x| + C.$$

Ans. to reading exercise 19
1.3023

However, this does not add any new power to our antiderivative arsenal, because the antiderivative constant multiple rule can be used just as easily;

$$\int \frac{1}{x \ln a}\, dx = \frac{1}{\ln a} \int \frac{1}{x}\, dx = \frac{1}{\ln a} \cdot \ln|x| + C.$$

The change of base property shows that our two antiderivative results are equivalent.

Example 6 *Evaluate* $\int \frac{\log_{10} x}{x}\, dx.$

Solution To handle the general logarithm, we rewrite it in terms of natural logarithms using the change of base formula:

$$\int \frac{\log_{10} x}{x}\, dx = \int \frac{\frac{\ln x}{\ln 10}}{x}\, dx$$
$$= \frac{1}{\ln 10} \int \frac{\ln x}{x}\, dx.$$

This integral should be familiar to you. Although we do not yet have an antiderivative formula for $\ln x$, and there is no antiderivative quotient rule, we use substitution:

$$u = \ln x$$
$$du = \frac{1}{x}\, dx.$$

Then,

$$\frac{1}{\ln 10} \int \frac{\ln x}{x}\, dx = \frac{1}{\ln 10} \int u\, du$$
$$= \frac{1}{\ln 10} \frac{u^2}{2} + C$$
$$= \frac{1}{2 \ln 10} (\ln x)^2 + C.$$

If desired, the solution can be rewritten in terms of $\log_{10} x$ by using the change of base formula. ■

An alternative solution is to start directly with the substitution:

$$u = \log_{10} x$$
$$du = \frac{1}{x \ln 10}\, dx.$$

We then adjust for the needed constant, make the substitution, and continue:

$$\int \frac{\log_{10} x}{x}\, dx = \ln 10 \int \frac{\log_{10} x}{x \ln 10}\, dx$$
$$= \ln 10 \int u\, du$$
$$= (\ln 10) \frac{u^2}{2} + C$$
$$= \frac{\ln 10}{2} (\log_{10} x)^2 + C.$$

CAUTION: MISTAKE TO AVOID
Rewriting $(\ln x)^2$ as $2 \ln x$ is an incorrect attempt to use law of logarithms (3):

$$(\ln x)^2 \neq 2 \ln x.$$

Limits involving general exponentials can be calculated using the earlier formulas for general logarithms of hyperreals.

Example 7 *Evaluate* $\lim_{x \to 2^-} \log_4 \frac{2 - x}{x}.$

Solution Using $x = 2 - \omega$ for an arbitrary positive infinitesimal ω and approximating inside the logarithm gives

Continuity of the general logarithmic function forms most of the justification for approximating inside the logarithm. The remaining technical details in justifying this step are saved for an upper-level course.

$$\lim_{x\to 2^-} \log_4 \frac{2-x}{x} = \log_4 \frac{2-(2-\omega)}{2-\omega} = \log_4 \frac{\omega}{2-\omega}$$
$$\approx \log_4 \frac{\omega}{2}$$
$$\doteq -\infty.$$

In the last step, we rendered $-\infty$ because the expression inside the logarithm is infinitesimal and $a = 4 > 1$. ■

Example 8 *Calculate* $\lim_{x\to\infty} \log_{4/5} \frac{2}{x}$.

Solution Replacing x with an arbitrary positive infinite hyperreal Ω yields

$$\lim_{x\to\infty} \log_{4/5} \frac{2}{x} = \log_{4/5} \frac{2}{\Omega}$$
$$= \log_{4/5} 2\omega$$
$$\doteq \infty.$$

In the last step, we rendered ∞ because the expression inside the logarithm is infinitesimal and $a = 4/5 < 1$. ■

An alternate solution is to use the change of base property at some point during the calculation to switch to the natural logarithm. Making the switch at the beginning gives

$$\lim_{x\to\infty} \log_{4/5} \frac{2}{x} = \lim_{x\to\infty} \frac{\ln \frac{2}{x}}{\ln 4/5}$$
$$= \frac{\ln \frac{2}{\Omega}}{\ln 0.8}$$
$$= \frac{\ln 2\omega}{\ln 0.8}.$$

Because 2ω is infinitesimal, $\ln 2\omega \doteq -\infty$. Therefore, the number $\ln 2\omega$ must be negative infinite, or $-B$ for some positive infinite hyperreal B. The expression then becomes

$$= \frac{-B}{\ln 0.8},$$

and because $\ln(0.8)$ is also a negative number, the expression is positive infinite and renders ∞:

$$\frac{-B}{\ln 0.8} \doteq \infty.$$

This alternate solution method is more difficult to perform correctly and to write correctly than the original solution to example 8.

Alternate definition of e

Traditionally, many high school algebra textbooks introduce the number e informally through the use of a compound interest formula and the as-yet-undiscussed concept of limit. Theorem 10 shows that the resulting value is the same as our number e.

Theorem 10 THE NUMBER KNOWN AS e **(ALTERNATE VERSION)** *An alternate definition of the number e is given by*

$$e = \lim_{x\to 0} (1+x)^{1/x}.$$

Proof. Let $f(x) = \ln x$. Then, $f'(x) = \frac{1}{x}$, and $f'(1) = 1$. The definition of a derivative must give us the same result. For any infinitesimal α,

$$\begin{aligned} 1 = f'(1) &= \frac{f(1+\alpha) - f(1)}{\alpha} \\ &= \frac{\ln(1+\alpha) - \ln 1}{\alpha} \\ &= \frac{1}{\alpha}\ln(1+\alpha) \\ &= \ln(1+\alpha)^{1/\alpha}. \end{aligned}$$

We also know that

$$\ln e = 1$$

and that $y = \ln x$ is a continuous one-to-one function. Therefore, the real number rendered by $(1+\alpha)^{1/\alpha}$ must be e. Then,

$$\lim_{x \to 0}(1+x)^{1/x} = (1+\alpha)^{1/\alpha} \doteq e. \qquad \blacksquare$$

Ans. to reading exercise 20
$y = \frac{1}{(x^2 + 5x + 7)\ln 4}(2x + 5)$.

Technically, the expression $\frac{f(1+\alpha) - f(1)}{\alpha}$ just renders the real result $f'(1)$, so what we have is $\ln(1+\alpha)^{1/\alpha} \doteq 1$.

By the transfer property, the laws of logarithms work with hyperreal numbers as well.

The expressions inside the natural logarithmic function with a result of 1 (or that renders 1), both $(1+\alpha)^{1/\alpha}$ and e, must be the same real number (or at least render the same real number).

Calculating exponentials by hand

We end this section with a topic that has nothing to do with calculus but everything to do with my transition into an old geezer, starting sentences with the phrase, "When I was your age...."

In elementary school you learned how to add, subtract, multiply, and divide numbers without calculators, using only paper and pencil. You also learned to exponentiate with integer exponents, such as 12^3. But, did you learn how to find values such as $4890^{15.7}$ without using a calculator? For most students, the answer is no.

The procedure uses a table of common logarithms, scientific notation, laws of logarithms, and multiplication. Let's look at this method using the following example.

Actually, when I was a teenager, we did have electronic calculators that performed these calculations, but the algebra textbook used in my high school still taught how to do it by hand.

Example 9 *Without using a calculator, calculate (give a decimal approximation for)* $4890^{15.7}$.

Solution Let's call the number we wish to calculate x. The strategy is to calculate the common logarithm of x (using tables of values instead

The use of tables of common logarithms to simplify calculations is largely a result of the work of Henry Briggs.
Henry Briggs 1561–1630
http://www-history.mcs.st-andrews.ac.uk/Biographies/Briggs.html

Recall that the common logarithm is the logarithm base 10.

of a calculating device) and then to determine the value of x from its common logarithm. Let's begin:

$$x = 4890^{15.7}$$

$$\log_{10} x = \log_{10} 4890^{15.7}$$

$$= 15.7 \log_{10} 4890.$$

With the use of law of logarithms (3), we can already see the advantage of using logarithms. Instead of the exponentiation $4890^{15.7}$, we have the multiplication $15.7 \cdot \log_{10} 4890$, and multiplication is an operation we can perform by hand. But why use base 10 logarithms? Because every number can be written easily in scientific notation (a number at least as large as 1 but less than 10, times a power of 10):

$$15.7 \log_{10} 4890 = 15.7 \log_{10}(4.890 \cdot 10^3).$$

Laws of logarithms can then be used to rewrite the number so that we only need to know the logarithm of a number between 1 and 10:

$$15.7 \log_{10}(4.890 \cdot 10^3) = 15.7(\log_{10} 4.890 + \log_{10} 10^3)$$

$$= 15.7(\log_{10} 4.89 + 3).$$

The exponent on 10 that gives you 10^3 is 3. This is one of the steps for which $\log_{10}$ is convenient.

It is easy to overlook the importance of this step. Limiting the range of numbers for which we need to be able to find logarithms makes a finite-length table of logarithms sufficient!

The next step is to use a table of common logarithms (see table 1) to obtain the value 0.6893 for $\log_{10} 4.89$. We then have

Table 1 gives logarithms of two decimal-place numbers. So, if necessary, we round any value to the nearest hundredth before using the table.

$$15.7(0.6893 + 3) = 15.7(3.6893),$$

which can be calculated by hand:

$$\begin{array}{r} 3.6893 \\ \times \quad 15.7 \\ \hline 258251 \\ 184465 \\ 36893 \\ \hline 57.92201 \end{array}$$

Thus,

$$\log_{10} x = 57.9220,$$

Table 1 only gives four-decimal-place logarithms. So, when necessary, we round to four decimal places before finding the *antilogarithm*.

and it remains to determine x by reversing the previous steps. Using the table "backward," we find the number with the common logarithm that yields the digits to the right of the decimal point, 9220 (called the *mantissa*), which is 8.36. The digits to the left of the decimal point (called the *characteristic*) give the exponent on 10, and our number is $x = 4890^{15.7} = 8.36 \cdot 10^{57}$. ■

Before the advent of calculating devices, books were printed containing tables of logarithms to many more decimal places than given in table 1. How are the numbers in the tables calculated? This is a story for another day.

Table 1 *Table of Common Logarithms*

	0.00	0.01	0.02	0.03	0.04	0.05	0.06	0.07	0.08	0.09
1.0	0.0000	0.0043	0.0086	0.0128	0.0170	0.0212	0.0253	0.0294	0.0334	0.0374
1.1	0.0414	0.0453	0.0492	0.0531	0.0569	0.0607	0.0645	0.0682	0.0719	0.0755
1.2	0.0792	0.0828	0.0864	0.0899	0.0934	0.0969	0.1004	0.1038	0.1072	0.1106
1.3	0.1139	0.1173	0.1206	0.1239	0.1271	0.1303	0.1335	0.1367	0.1399	0.1430
1.4	0.1461	0.1492	0.1523	0.1553	0.1584	0.1614	0.1644	0.1673	0.1703	0.1732
1.5	0.1761	0.1790	0.1818	0.1847	0.1875	0.1903	0.1931	0.1959	0.1987	0.2014
1.6	0.2041	0.2068	0.2095	0.2122	0.2148	0.2175	0.2201	0.2227	0.2253	0.2279
1.7	0.2304	0.2330	0.2355	0.2380	0.2405	0.2430	0.2455	0.2480	0.2504	0.2529
1.8	0.2553	0.2577	0.2601	0.2625	0.2648	0.2672	0.2695	0.2718	0.2742	0.2765
1.9	0.2788	0.2810	0.2833	0.2856	0.2878	0.2900	0.2923	0.2945	0.2967	0.2989
2.0	0.3010	0.3032	0.3054	0.3075	0.3096	0.3118	0.3139	0.3160	0.3181	0.3201
2.1	0.3222	0.3243	0.3263	0.3284	0.3304	0.3324	0.3345	0.3365	0.3385	0.3404
2.2	0.3424	0.3444	0.3464	0.3483	0.3502	0.3522	0.3541	0.3560	0.3579	0.3598
2.3	0.3617	0.3636	0.3655	0.3674	0.3692	0.3711	0.3729	0.3747	0.3766	0.3784
2.4	0.3802	0.3820	0.3838	0.3856	0.3874	0.3892	0.3909	0.3927	0.3945	0.3962
2.5	0.3979	0.3997	0.4014	0.4031	0.4048	0.4065	0.4082	0.4099	0.4116	0.4133
2.6	0.4150	0.4166	0.4183	0.4200	0.4216	0.4232	0.4249	0.4265	0.4281	0.4298
2.7	0.4314	0.4330	0.4346	0.4362	0.4378	0.4393	0.4409	0.4425	0.4440	0.4456
2.8	0.4472	0.4487	0.4502	0.4518	0.4533	0.4548	0.4564	0.4579	0.4594	0.4609
2.9	0.4624	0.4639	0.4654	0.4669	0.4683	0.4698	0.4713	0.4728	0.4742	0.4757

continued

Table 1 *Continued*

	0.00	0.01	0.02	0.03	0.04	0.05	0.06	0.07	0.08	0.09
3.0	0.4771	0.4786	0.4800	0.4814	0.4829	0.4843	0.4857	0.4871	0.4886	0.4900
3.1	0.4914	0.4928	0.4942	0.4955	0.4969	0.4983	0.4997	0.5011	0.5024	0.5038
3.2	0.5051	0.5065	0.5079	0.5092	0.5105	0.5119	0.5132	0.5145	0.5159	0.5172
3.3	0.5185	0.5198	0.5211	0.5224	0.5237	0.5250	0.5263	0.5276	0.5289	0.5302
3.4	0.5315	0.5328	0.5340	0.5353	0.5366	0.5378	0.5391	0.5403	0.5416	0.5428
3.5	0.5441	0.5453	0.5465	0.5478	0.5490	0.5502	0.5514	0.5527	0.5539	0.5551
3.6	0.5563	0.5575	0.5587	0.5599	0.5611	0.5623	0.5635	0.5647	0.5658	0.5670
3.7	0.5682	0.5694	0.5705	0.5717	0.5729	0.5740	0.5752	0.5763	0.5775	0.5786
3.8	0.5798	0.5809	0.5821	0.5832	0.5843	0.5855	0.5866	0.5877	0.5888	0.5899
3.9	0.5911	0.5922	0.5933	0.5944	0.5955	0.5966	0.5977	0.5988	0.5999	0.6010
4.0	0.6021	0.6031	0.6042	0.6053	0.6064	0.6075	0.6085	0.6096	0.6107	0.6117
4.1	0.6128	0.6138	0.6149	0.6160	0.6170	0.6180	0.6191	0.6201	0.6212	0.6222
4.2	0.6232	0.6243	0.6253	0.6263	0.6274	0.6284	0.6294	0.6304	0.6314	0.6325
4.3	0.6335	0.6345	0.6355	0.6365	0.6375	0.6385	0.6395	0.6405	0.6415	0.6425
4.4	0.6435	0.6444	0.6454	0.6464	0.6474	0.6484	0.6493	0.6503	0.6513	0.6522
4.5	0.6532	0.6542	0.6551	0.6561	0.6571	0.6580	0.6590	0.6599	0.6609	0.6618
4.6	0.6628	0.6637	0.6646	0.6656	0.6665	0.6675	0.6684	0.6693	0.6702	0.6712
4.7	0.6721	0.6730	0.6739	0.6749	0.6758	0.6767	0.6776	0.6785	0.6794	0.6803
4.8	0.6812	0.6821	0.6830	0.6839	0.6848	0.6857	0.6866	0.6875	0.6884	0.6893
4.9	0.6902	0.6911	0.6920	0.6928	0.6937	0.6946	0.6955	0.6964	0.6972	0.6981
5.0	0.6990	0.6998	0.7007	0.7016	0.7024	0.7033	0.7042	0.7050	0.7059	0.7067
5.1	0.7076	0.7084	0.7093	0.7101	0.7110	0.7118	0.7126	0.7135	0.7143	0.7152
5.2	0.7160	0.7168	0.7177	0.7185	0.7193	0.7202	0.7210	0.7218	0.7226	0.7235
5.3	0.7243	0.7251	0.7259	0.7267	0.7275	0.7284	0.7292	0.7300	0.7308	0.7316
5.4	0.7324	0.7332	0.7340	0.7348	0.7356	0.7364	0.7372	0.7380	0.7388	0.7396
5.5	0.7404	0.7412	0.7419	0.7427	0.7435	0.7443	0.7451	0.7459	0.7466	0.7474
5.6	0.7482	0.7490	0.7497	0.7505	0.7513	0.7520	0.7528	0.7536	0.7543	0.7551

continued

Table 1 *Continued*

	0.00	0.01	0.02	0.03	0.04	0.05	0.06	0.07	0.08	0.09
5.7	0.7559	0.7566	0.7574	0.7582	0.7589	0.7597	0.7604	0.7612	0.7619	0.7627
5.8	0.7634	0.7642	0.7649	0.7657	0.7664	0.7672	0.7679	0.7686	0.7694	0.7701
5.9	0.7709	0.7716	0.7723	0.7731	0.7738	0.7745	0.7752	0.7760	0.7767	0.7774
6.0	0.7782	0.7789	0.7796	0.7803	0.7810	0.7818	0.7825	0.7832	0.7839	0.7846
6.1	0.7853	0.7860	0.7868	0.7875	0.7882	0.7889	0.7896	0.7903	0.7910	0.7917
6.2	0.7924	0.7931	0.7938	0.7945	0.7952	0.7959	0.7966	0.7973	0.7980	0.7987
6.3	0.7993	0.8000	0.8007	0.8014	0.8021	0.8028	0.8035	0.8041	0.8048	0.8055
6.4	0.8062	0.8069	0.8075	0.8082	0.8089	0.8096	0.8102	0.8109	0.8116	0.8122
6.5	0.8129	0.8136	0.8142	0.8149	0.8156	0.8162	0.8169	0.8176	0.8182	0.8189
6.6	0.8195	0.8202	0.8209	0.8215	0.8222	0.8228	0.8235	0.8241	0.8248	0.8254
6.7	0.8261	0.8267	0.8274	0.8280	0.8287	0.8293	0.8299	0.8306	0.8312	0.8319
6.8	0.8325	0.8331	0.8338	0.8344	0.8351	0.8357	0.8363	0.8370	0.8376	0.8382
6.9	0.8388	0.8395	0.8401	0.8407	0.8414	0.8420	0.8426	0.8432	0.8439	0.8445
7.0	0.8451	0.8457	0.8463	0.8470	0.8476	0.8482	0.8488	0.8494	0.8500	0.8506
7.1	0.8513	0.8519	0.8525	0.8531	0.8537	0.8543	0.8549	0.8555	0.8561	0.8567
7.2	0.8573	0.8579	0.8585	0.8591	0.8597	0.8603	0.8609	0.8615	0.8621	0.8627
7.3	0.8633	0.8639	0.8645	0.8651	0.8657	0.8663	0.8669	0.8675	0.8681	0.8686
7.4	0.8692	0.8698	0.8704	0.8710	0.8716	0.8722	0.8727	0.8733	0.8739	0.8745
7.5	0.8751	0.8756	0.8762	0.8768	0.8774	0.8779	0.8785	0.8791	0.8797	0.8802
7.6	0.8808	0.8814	0.8820	0.8825	0.8831	0.8837	0.8842	0.8848	0.8854	0.8859
7.7	0.8865	0.8871	0.8876	0.8882	0.8887	0.8893	0.8899	0.8904	0.8910	0.8915
7.8	0.8921	0.8927	0.8932	0.8938	0.8943	0.8949	0.8954	0.8960	0.8965	0.8971
7.9	0.8976	0.8982	0.8987	0.8993	0.8998	0.9004	0.9009	0.9015	0.9020	0.9025
8.0	0.9031	0.9036	0.9042	0.9047	0.9053	0.9058	0.9063	0.9069	0.9074	0.9079
8.1	0.9085	0.9090	0.9096	0.9101	0.9106	0.9112	0.9117	0.9122	0.9128	0.9133
8.2	0.9138	0.9143	0.9149	0.9154	0.9159	0.9165	0.9170	0.9175	0.9180	0.9186

continued

Table 1 *Continued*

	0.00	0.01	0.02	0.03	0.04	0.05	0.06	0.07	0.08	0.09
8.3	0.9191	0.9196	0.9201	0.9206	0.9212	0.9217	0.9222	0.9227	0.9232	0.9238
8.4	0.9243	0.9248	0.9253	0.9258	0.9263	0.9269	0.9274	0.9279	0.9284	0.9289
8.5	0.9294	0.9299	0.9304	0.9309	0.9315	0.9320	0.9325	0.9330	0.9335	0.9340
8.6	0.9345	0.9350	0.9355	0.9360	0.9365	0.9370	0.9375	0.9380	0.9385	0.9390
8.7	0.9395	0.9400	0.9405	0.9410	0.9415	0.9420	0.9425	0.9430	0.9435	0.9440
8.8	0.9445	0.9450	0.9455	0.9460	0.9465	0.9469	0.9474	0.9479	0.9484	0.9489
8.9	0.9494	0.9499	0.9504	0.9509	0.9513	0.9518	0.9523	0.9528	0.9533	0.9538
9.0	0.9542	0.9547	0.9552	0.9557	0.9562	0.9566	0.9571	0.9576	0.9581	0.9586
9.1	0.9590	0.9595	0.9600	0.9605	0.9609	0.9614	0.9619	0.9624	0.9628	0.9633
9.2	0.9638	0.9643	0.9647	0.9652	0.9657	0.9661	0.9666	0.9671	0.9675	0.9680
9.3	0.9685	0.9689	0.9694	0.9699	0.9703	0.9708	0.9713	0.9717	0.9722	0.9727
9.4	0.9731	0.9736	0.9741	0.9745	0.9750	0.9754	0.9759	0.9763	0.9768	0.9773
9.5	0.9777	0.9782	0.9786	0.9791	0.9795	0.9800	0.9805	0.9809	0.9814	0.9818
9.6	0.9823	0.9827	0.9832	0.9836	0.9841	0.9845	0.9850	0.9854	0.9859	0.9863
9.7	0.9868	0.9872	0.9877	0.9881	0.9886	0.9890	0.9894	0.9899	0.9903	0.9908
9.8	0.9912	0.9917	0.9921	0.9926	0.9930	0.9934	0.9939	0.9943	0.9948	0.9952
9.9	0.9956	0.9961	0.9965	0.9969	0.9974	0.9978	0.9983	0.9987	0.9991	0.9996

To find $\log_{10} 4.89$, use row 4.8 and column 0.09. The number at the intersection of that row and column is the desired value: 0.6893.
To find the number with the common logarithm of 0.9220, choose the closest number inside the table: 0.9222. Add the corresponding row and column headings, 8.3 and 0.06, to get the desired number: 8.36.

Exercises 5.6

1–8. Rewrite in exponential form.

1. $y = \log_9 5$
2. $y = \log_{11} 0.003$
3. $y = \log_3 \left(4x^2 + 1\right)$
4. $y = x \log_{12} 5$
5. $5y = \log_{0.5} \left(\sqrt{5}x\right)$
6. $s = \log_a 5t^3$
7. $8r^2 = 3 \log_{10} \left(s - s^2\right)$
8. $y + y^2 = \log_4(x + 3)$

9–16. Rewrite in logarithmic form.

9. $2^{x-4} = 11$
10. $5^{x^2+2} = 7$
11. $y = 3^x$
12. $8 = 4^v 4^3$
13. $5^{2x+1} = x$
14. $7y = 2^{x+4}$
15. $s = 3 \cdot 7^t$
16. $2 + 5^x = 13$

17–24. Use a calculator to find the requested value.

17. $\log_{12} 49.6$
18. $\log_9 15$
19. $\log_{0.2} 2847$
20. $\log_{2.37} 984.11$
21. $\log_4 \left(1.946 \cdot 10^{18}\right)$
22. $\log_2(\log_3 5)$
23. $3 \log_7 \sin 14.48\pi$
24. $(\log_5 2)^6$

25–36. Find the derivative of the function.

25. $y = \log_2 \left(x^3 - 7x + 1\right)$
26. $x(t) = \log_5(t^2 + t)$
27. $f(x) = 4x^2 \log_5(x+3)^{10}$
28. $y = \sqrt{x + 1 + \log_2 x}$
29. $y = \dfrac{\log_7(2x+3)}{x^2}$
30. $y = \log_2 (x \sin x)$
31. $y = \left(x + \log_2(6x^2)\right)^4$
32. $y = \log_{10} \left(\frac{1}{(8x+3)^{17}}\right)$
33. $f(x) = 3^x \log_3(9x+2)$
34. $f(x) = x + x^2 \log_8 x$
35. $y = \sin^2 \sqrt{4^x \log_2 5x}$
36. $y = \log_{1/2} \left(\log_3(x^5 \tan x)\right)$

37–44. Evaluate the integral.

37. $\displaystyle\int \frac{1}{x \ln 3}\, dx$
38. $\displaystyle\int \log_3 12\, dx$
39. $\displaystyle\int \frac{3^x \ln 3 + 5}{3^x + 5x}\, dx$
40. $\displaystyle\int \frac{3 \log_2 x}{x \ln 2}\, dx$
41. $\displaystyle\int \frac{2x \log_2(x^2+1)}{x^2+1}\, dx$
42. $\displaystyle\int \frac{1}{x\sqrt{\log_3 x}}\, dx$
43. $\displaystyle\int \frac{\sin(\log_2 x)}{x}\, dx$
44. $\displaystyle\int x^{-1} \log_3 x^2\, dx$

45–56. Evaluate the limit.

45. $\lim_{x\to 1^+} \log_{1/3}(x^2 - 1)$

46. $\lim_{x\to -1^+} \log_3(1 + x)$

47. $\lim_{x\to \infty} \log_2 \dfrac{-3}{1 - x^2}$

48. $\lim_{x\to -1^+} \log_3(2 + x)$

49. $\lim_{x\to 4^+} \log_{0.6}(x - 4)^{-1}$

50. $\lim_{x\to \infty} \log_{0.86} \dfrac{x^2 + 3x - 1}{x^3 - 4x}$

51. $\lim_{x\to -\infty} \log_4(1 - x)$

52. $\lim_{x\to 0} \log_{3.86} \dfrac{1}{x^2}$

53. $\lim_{x\to 3} \log_{7421} \dfrac{5x - 9}{2x}$

54. $\lim_{x\to -\infty} \log_{1/4} \sqrt{x^4 - 3x}$

55. $\lim_{x\to \infty} \log_{\pi/4} \dfrac{x^3 + 4x - 7}{4x^2 + 5x - 1}$

56. $\lim_{x\to 2\pi^+} \log_4 \sin x$

57. Use implicit differentiation to find $\frac{dy}{dx}$ for $x^2 \log_2 y + 3x = y^3$.

58. Find $\frac{d^2}{dx^2}\left(\log_2 x^3\right)$.

59. Show that the graph of $y = \log_a x$, where $a > 1$, is concave down on $(0, \infty)$.

60. Show that the graph of $y = \log_a x$, where $0 < a < 1$, is concave up on $(0, \infty)$.

61–66. Use table 1 to calculate the requested value by hand.

61. $6.27^{1.04}$

62. $1.93^{2.1}$

63. $376^{10.2}$

64. $112^{6.3}$

65. $0.0722^{3.2}$

66. $72\ 400^{-2.16}$

Exponential Growth and Decay

5.7

One of the reasons that exponential functions are studied is for their many applications. A few of these applications are explored here, including some you may have seen when studying algebra.

Exponential change

Many quantities change at a rate proportional to their size. If you have \$1000 in an account earning 4% interest annually, then your account balance grows by \$40 this year; but if you have \$10 000 in that account, your balance grows by \$400. The amount of interest earned, the rate of growth of money in the account, is proportional to the amount of money in the account. If a herd of 100 goats grows by 50 goats through reproduction, then we expect a herd of 1000 goats grows by 500. The rate of growth of the number of goats is proportional to the number of goats in the herd.

Another name for the number of goats in the herd is the *population* of goats.

The key to a mathematical interpretation of the previous paragraph is "*rate of change*." We know that a rate of change is a derivative. To say that a population P changes (over time) at a rate proportional to its size is to say that the derivative of P with respect to time t, $\frac{dP}{dt}$, is proportional to P:

$$\frac{dP}{dt} = kP,$$

where k is a proportionality constant.

The equation $\frac{dP}{dt} = kP$ is an example of a *differential equation*—that is, an equation that involves a derivative. Although differential equations can become quite complicated to solve, this one is not difficult if we treat the quantities dP and dt as differentials. Let's try isolating everything with the variable P on the left side of the equation and everything involving the variable t on the right side of the equation:

$$\frac{dP}{dt} = kP$$
$$dP = kP\,dt$$
$$\frac{1}{P}\,dP = k\,dt.$$

We succeeded in separating the two variables P and t, one to each side of the equation. (Where a constant such as k goes does not matter; it can be moved to either side.) A differential equation in which this can be done is called *separable.*

During the 18th and 19th centuries, the differential was used instead of the derivative, hence the name *differential equation*. Indefinite integrals were antidifferentials instead of antiderivatives, and the step of finding antidifferentials of each side of the equation was rather natural. When derivatives are emphasized, additional justification is needed for this procedure. This justification is found in textbooks on differential equations.

The next step in solving a separable differential equation is to place an integral symbol in front of each side of the equation and integrate. We then have

$$\int \frac{1}{P}\,dP = \int k\,dt$$
$$\ln|P| + C_1 = kt + C_2.$$

The constants of integration for the two integrals do not have to be the same, hence the use of C_1 and C_2. But, subtracting C_1 from both sides yields

This illustrates why we can go straight from $\int \frac{1}{P}\,dP = \int k\,dt$ to $\ln|P| = kt + C$, including the constant of integration only on one side when solving a separable differential equation.

$$\ln|P| = kt + C_2 - C_1 = kt + C,$$

where we may replace the constant $C_2 - C_1$ with the constant C.

Our goal is to solve for P, which currently resides inside a logarithm. Exponentiating each side and using the formula $e^{\ln x} = x$ helps:

Line 1 exponentiates both sides of $\ln|P| = kt + C$, line 2 uses $e^{\ln x} = x$, and line 3 uses law of exponents (1) and the fact that $|x| = \pm x$.

$$e^{\ln|P|} = e^{kt+C}$$
$$|P| = e^{kt+C}$$
$$P = \pm e^{kt}e^{C}.$$

Because C is a constant, so is e^C and also $\pm e^C$. Replacing the latter with D brings us to

Here, D is a constant. Traditionally, it is acceptable to write C in place of D, even though it is not the "same" constant as C in the previous line.

$$P = De^{kt}$$

and we have solved the differential equation.

Next, suppose we also know an initial condition, such as the population at time 0:

The constant P_0 is read "P naught."

$$P(0) = P_0.$$

We then apply this initial condition to solve for the constant D:

$$P_0 = P(0) = De^{k \cdot 0} = D \cdot 1 = D.$$

Substitution of P_0 for D yields

$$P = P_0 e^{kt}.$$

The descriptive phrase *exponential change* arises from the presence of the natural exponential function in the formula.

LAW OF EXPONENTIAL CHANGE

If the rate of change of a quantity P is proportional to P, then

$$P = P_0 e^{kt},$$

where P is the population at time t, P_0 is the population at time $t = 0$ (the initial population), and k is a proportionality constant, also known as a *growth constant* when $k > 0$ or a *decay constant* when $k < 0$.

Exponential change is often called *exponential growth* when $k > 0$ and *exponential decay* when $k < 0$.

When applied to interest that is compounded continuously, the formula is sometimes written $A = A_0 e^{kt}$, where k is the rate of interest to be compounded continuously, A is the amount of money in the account at time t, and A_0 is the *principal*, the initial amount of money in the account.

Exponential change example: growth

Example 1 *A population of bacteria increases at a rate proportional to its size. Suppose the population increases by* 50% *in* 40 *minutes. How long does it take the population to double its initial size?*

Solution Because the population increases at a rate proportional to its size, the law of exponential change applies and our formula is

$$P = P_0 e^{kt},$$

where P is the population at time t, P_0 is the population at time $t = 0$, and k is the growth constant (the population is growing).

CAUTION: MISTAKE TO AVOID
Although it is tempting to reason that if the population increases 50% in 40 minutes that it should increase 100% in 80 minutes, this is wrong. During the next 40 minutes, the population does increase by 50%, but it's 50% of a larger population. The total increase from the initial amount in 80 minutes is more than 100%. The growth is compounded.

❶ Determine k. When the growth constant k is unknown, the first step in answering an exponential change question is to determine the value of k. Look for information describing or inferring how fast the population is growing. In this case, we are told the population increases by 50% in 40 minutes. Whatever the initial population P_0 is, the population after 40 minutes, $P(40)$, is 50% greater, or $1.50P_0$. We put this information into the exponential change formula and solve for k:

We may use any unit of time that is convenient, but we must be consistent throughout the problem.

$$1.5P_0 = P_0 e^{k \cdot 40}$$

$$1.5 = e^{40k}.$$

Line 1 uses $P = P_0 e^{kt}$ with $P = 1.5P_0$ when $t = 40$; line 2 divides both sides by P_0.

Notice that the quantity P_0 no longer appears in our equation, and the only remaining variable is k. Because k is in an exponent, we take the natural logarithm of both sides and continue to solve for k:

$$\ln 1.5 = \ln e^{40k}$$

$$\ln 1.5 = 40k$$

$$\frac{\ln 1.5}{40} = k$$

$$k = 0.01014.$$

The algebra needed to find k in an exponential change problem often follows this same pattern.

When using a calculator to determine a decimal approximation, be sure to use at least four significant digits. Recall that we begin counting significant digits at the leftmost nonzero digit. The number 0.01014 has four significant digits.

Now that we have determined k, the exponential change formula is written

$$P = P_0 e^{0.01014t}.$$

❷ Determine the requested quantity. After determining k, we need to re-read the question to determine which quantity is sought. We wish to know how long (a value of time t) it takes for the population to double its initial size—that is, when $P = 2P_0$. Therefore, we substitute $2P_0$ for P in our exponential change formula and solve for time t:

$$2P_0 = P_0 e^{0.01014t}$$

$$2 = e^{0.01014t}$$

$$\ln 2 = \ln e^{0.01014t} = 0.01014t$$

$$t = \frac{\ln 2}{0.01014}$$

$$= 68.36 \text{ min.}$$

In line 1, use $P = P_0 e^{0.01014t}$ and replace P with $2P_0$. The remaining algebra is much the same as in step ❶. In line 2, divide both sides by P_0. In line 3, take the natural logarithm of both sides because the variable is in the exponent, then simplify.

It takes 68.36 minutes for the population of bacteria to double its initial size. ■

Reading Exercise 21 *Solve the equation for k:* $1.25P_0 = P_0 e^{20k}$.

Suppose in example 1 that the initial population was just 10 bacteria. Our function is then

$$P(t) = 10e^{0.01014t}.$$

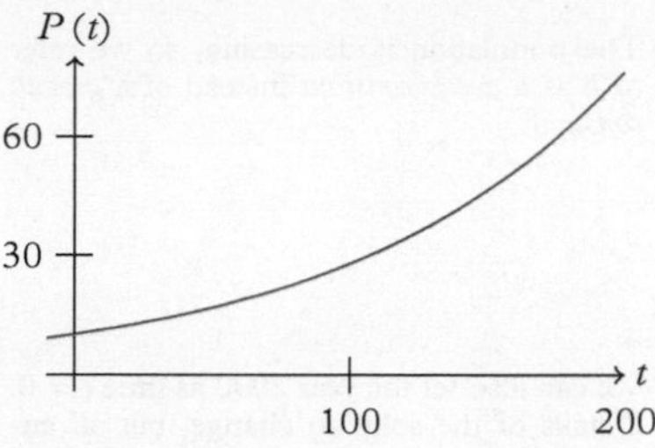

Figure 1 The graph of $P(t) = 10e^{0.01014t}$ on $[0, 200]$

A graph of the population of bacteria for the first 200 minutes is in figure 1. The graph is increasing because the first derivative is positive; $P'(t) = 0.1014e^{0.01014t} > 0$. In fact, all the derivatives are positive. Not only is the function increasing, the rate at which it is increasing is increasing, and that rate is increasing, and so on. The graph in figure 1 may look relatively tame, but a graph of the first 24 hours (figure 2) illustrates just how quickly an exponential function can grow.

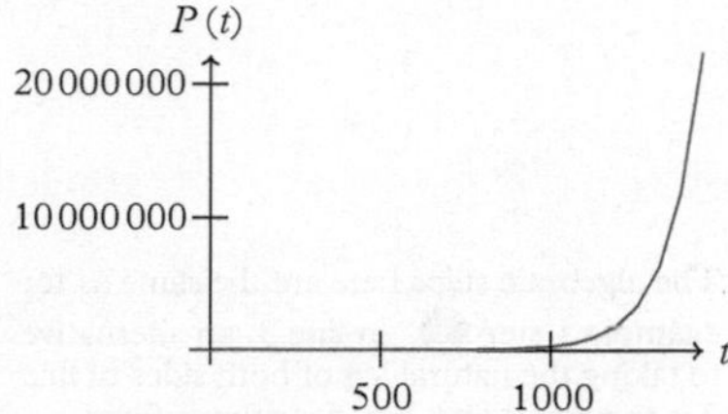

Figure 2 The graph of $P(t) = 10e^{0.01014t}$ on $[0, 1440]$. Note that 1440 minutes is 24 hours

The rate of growth, the rate of the rate of growth, and so on continue to grow more and more quickly, so that eventually the number seems to explode beyond comprehension. In fact, after 8 days, the number of bacteria is more than the number of atoms in the earth! The phrase "*grows exponentially*" is in common use (and misuse) in everyday language to describe this phenomenon. Although the exponential growth model works very well for some applications up to a certain point, unrestrained exponential growth is usually not possible. One model that puts a cap on the size of a population is studied in section 9.8.

Exponential change example: decay

A population or amount can also decrease (decay) exponentially. I graduated high school in Parsons, Kansas, a city with a population that has been decreasing for many decades.

Example 2 *Assume that the rate of change of the population of Parsons, Kansas is proportional to the population. Census data are as follows:*

year	*population*
1980	12 898
2000	11 514

(a) "Predict" the population in 2010. *(b) Find the half-life of Parsons, Kansas.*

Solution Because we are to assume that the population changes at a rate proportional to the population, the law of exponential change applies and our formula is

$$P = P_0 e^{kt},$$

The population is decreasing, so we refer to k as a *decay constant* instead of a *growth constant*.

where P is the population at time t, P_0 is the population at time $t = 0$, and k is the decay constant. We set time $t = 0$ as we wish. Choosing 1980 as time $t = 0$, we use $P_0 = 12\,898$. The model is

$$P = 12\,898e^{kt}.$$

We can also set the year 2000 as time $t = 0$. Details of the solution change, but all answers are the same. It makes no sense to set year 0 as time $t = 0$ when our data is so far removed from that time.

❶ Determine k. The decay constant is not given directly, so we look for information that describes the change in population. The change in population from 1980 to 2000 gives us this information. The 1980 population is P_0, and this information is already in our model. The year 2000 corresponds to time $t = 20$, so we use $P = 11\,514$ when $t = 20$ to find k:

The algebraic steps here are the same as for example 1 step ❶. In line 3, an alternative to taking the natural log of both sides of line 2 is to rewrite line 2 in logarithmic form.

$$\begin{aligned} 11\,514 &= 12\,898e^{k\cdot 20} \\ \frac{11\,514}{12\,898} &= e^{20k} \\ \ln\frac{11\,514}{12\,898} &= \ln e^{20k} = 20k \\ k &= \frac{\ln\left(\frac{11\,514}{12\,898}\right)}{20} = -0.005675. \end{aligned}$$

The negative value of k is expected and corresponds to a decreasing population. The model is now

$$P = 12\,898e^{-0.005675t}.$$

(a) ❷ Determine the requested quantity. The population in the year 2010 is the value of P when $t = 30$:

Because 1980 corresponds to $t = 0$, 2010 corresponds to $t = 2010 - 1980 = 30$.

$$P = 12\,898e^{-0.005675(30)} = 10\,879.$$

We "predict" the population in 2010 will be 10 879.

(b) ❷ Determine the requested quantity. The *half-life* of a quantity is the time it takes until half the initial population is left. We therefore wish to find the value of t for which $P = \frac{1}{2}P_0 = 6449$:

The term *half-life* is commonly used in the setting of radioactive decay. The amount of a radioactive element remaining at time t changes at a rate proportional to the amount, so it follows the law of exponential change. Disclaimer: I am not attempting to imply that the town in which I was raised is radioactive. It was, in fact, a nice place to grow up.

$$\begin{aligned} 6449 &= 12\,898e^{-0.005675t} \\ \frac{1}{2} &= e^{-0.005675t} \\ \ln 0.5 &= \ln e^{-0.005675t} = -0.005675t \\ t &= \frac{\ln 0.5}{-0.005675} = 122 \text{ years.} \end{aligned}$$ ■

Ans. to reading exercise 21
$k = \frac{\ln 1.25}{20} = 0.01116$

The solution $P = 10\,879$ in part (a) of example 2 can be compared to the census value for that year: 10 500. The prediction is not bad, but it is also not perfect. Real life is complicated enough that any mathematical model is likely not completely accurate. Often, the culprit is an assumption that is not correct. For example 2, we assumed that the rate of change of the population is proportional to the population, with that proportion staying constant (k is a constant). In reality, the growth rate of a city varies over time, because circumstances change. Although the population of Parsons, Kansas, has declined nearly every census since 1920, the one exception came in 1950, after an ammunition plant was built for World War II. The 122-year half-life calculated in part (b) might seem reasonable for now, but conditions will likely change over the course of more than a century. Mathematical models are still useful for planning purposes, but the further removed a prediction is from the scope of its data, the less reliable it is.

A quote attributed to George Box is "all models are wrong but some are useful." Box, George E.P. "Robustness in the strategy of scientific model building," 201–236. In Launer, R.L., and Wilkinson, G.N. (eds.). *Robustness in Statistics*. New York, NY: Academic Press, 1979, doi: 10.1016/B978-0-12-438150-6.50018-2

George Edward Pelham Box 1919–2013 https://mathshistory.st-andrews.ac.uk/Biographies/Box/

The growth rate of a population of bacteria declines when the available resources are depleted. The interest rate on an account might vary according to economic conditions. Some experiments seem to indicate that the rate of decay of a radioactive element can vary at extreme temperatures. But, as long as we operate within conditions in which the model's assumptions are fairly accurate, the model should be useful.

Continuously compounded interest example

An assumption in the development of the law of exponential change is that the function is differentiable and hence continuous. Technically, the population of Parsons, Kansas (example 2), is not a continuous function. There were never exactly 11 367.02 residents of the city. The same is true for the bacteria population of example 1. However, the exponential model is still useful for making predictions. The same is true for continuously compounded interest. Although the interest is not technically added continually to the account balance, the rate of growth of an account can still be described in these terms.

Example 3 *If* \$1000 *is invested in a retirement account that earns* 6% *interest compounded continuously, how much is that investment worth in* 50 *years?*

Although the returns on investments in stocks fluctuate wildly, the total inflation-adjusted annual rate of return from the broad stock market from 1950 to 2009 was approximately 7%, and the exponential growth model is reasonable to use for the purposes of long-term planning.

Solution Continuously compounded interest follows the law of exponential change. The initial amount $A_0 = 1000$ (the principal) is given. We use $A = A_0e^{kt} = 1000e^{kt}$.

❶ Determine k. This time, the rate of change k is given as 6%, or $k = 0.06$. The model is therefore

$$A = 1000e^{0.06t}.$$

❷ Determine the requested quantity. We wish to know the amount in the account after 50 years, so we wish to know the value of A when $t = 50$:

Unless otherwise specified, the time unit used for t in the continuously compounded interest formula is a year.

$$A = 1000e^{0.06(50)} = \$20\,085.54.$$ ■

An alternate derivation of the continuously compounded interest formula is given at the end of this section.

Newton's law of cooling

Newton's law of cooling states that the rate of change of an object's temperature is proportional to the difference between the object's temperature and the temperature of the surrounding medium. Labeling variables, we let T be the temperature of the object at time t; T_m, the temperature of the surrounding medium; T_0, the initial temperature of the object (the temperature at time $t = 0$); and k, the proportionality constant. Newton's law of cooling can then be written as the differential equation

The surrounding medium is air, water, oil, or whatever surrounds the object.

$$\frac{dT}{dt} = k(T - T_m).$$

To understand the differential equation, reread the first sentence and interpret each item according to its variables: "the rate of change of an object's temperature" ($\frac{dT}{dt}$) "is" (=) "proportional to" ($k\cdot$) "the difference between" (−) "the object's temperature" (T) "and the temperature of the surrounding medium" (T_m).

This differential equation is also separable, and—following the pattern used for the law of exponential change—we can solve the equation and then apply the initial condition $T(0) = T_0$ to determine a useful formula (a mathematical model) for the temperature T.

The variables in the differential equation are T and t; k and T_m are constants. We therefore wish to have any mention of T on one side of the equation and any mention of t on the other side; we do not care which side k or T_m is on. Then,

The constant k depends on factors such as the material used in the object, the shape and size of the object, and the substance in the surrounding medium. Although the temperature T_m of the surrounding medium can change, we assume that it does not.

$$\frac{dT}{dt} = k(T - T_m)$$

$$dT = k(T - T_m)\,dt$$

$$\frac{1}{T - T_m}\,dT = k\,dt.$$

Next we integrate:

$$\int \frac{1}{T - T_m}\,dT = \int k\,dt.$$

The integrand on the left is not quite $\frac{1}{T}$, so we use substitution:

$$u = T - T_m$$
$$du = dT.$$

We then make the substitution and complete the integration for the integral on the left side:

$$\int \frac{1}{u}\,du = \ln|u| = \ln|T - T_m| + C.$$

Returning to the differential equation, integration of both sides yields

$$\ln|T - T_m| = kt + C.$$

The variable for which we wish to solve, T, is inside a logarithm, so we exponentiate both sides of the equation and continue to solve:

$$e^{\ln|T-T_m|} = e^{kt+C}$$
$$|T - T_m| = e^{kt}e^{C}$$
$$T - T_m = \pm e^{C}e^{kt}$$
$$T - T_m = De^{kt}$$
$$T = T_m + De^{kt}.$$

Line 2 uses the formula $e^{\ln x} = x$ and law of exponents (1), line 3 uses $|x| = \pm x$, and line 4 relabels the constant $\pm e^{C}$ as the constant D.

Now that we have solved the differential equation, we apply the initial condition $T(0) = T_0$ to determine the value of the constant D. Using $T = T_0$ when $t = 0$ gives

$$T_0 = T_m + De^{k(0)}$$
$$T_0 = T_m + D$$
$$D = T_0 - T_m.$$

Substitution puts the finishing touch on our formula:

$$T = T_m + (T_0 - T_m)e^{kt}.$$

An alternate, and perhaps better, name is *Newton's law of temperature change,* because the law applies both to objects getting cooler and objects getting warmer.

NEWTON'S LAW OF COOLING

Let T be the temperature of an object at time t, let T_m be the constant temperature of the surrounding medium, and let T_0 be the initial temperature of the object. Then,

$$T = T_m + (T_0 - T_m)e^{kt},$$

where k is a proportionality constant.

When solving problems involving the law of exponential change, we did not re-solve the differential equation each time; we used the formula. The same is true when solving problems involving Newton's law of cooling.

Newton's law of cooling example

Example 4 *A glass of milk is at 35°F when poured. Ten minutes later, its temperature is 42°F. If room temperature is 70°F, how long does it take the milk to warm to 60°F?*

Solution Newton's law of cooling applies. We have an object (a glass of milk) sitting in a surrounding medium (air). Let T be the temperature of the glass of milk. We are given the initial temperature $T_0 = 35°F$ and the temperature of the surrounding medium $T_m = 70°F$. Using the Newton's law of cooling formula as our model, we have

$$T = T_m + (T_0 - T_m)e^{kt} = 70 + (35 - 70)e^{kt} = 70 - 35e^{kt}.$$

The steps to solve this example are the same as for exponential change problems.

The proportionality constant k can be called a *change constant*. It controls the rate at which the temperature changes.

❶ Determine k. If we are not given the value of the proportionality constant k, we use information that describes or infers the rate at which the temperature is changing. We are told that after 10 minutes, the temperature has changed to 42°F, which means that when $t = 10$ (this fixes our unit of time as a minute), $T = 42$:

$$42 = 70 - 35e^{k(10)}.$$

The variable for which we wish to solve, k, is in an exponent, indicating that we need to take logarithms of each side of the equation. But, if we do so now, the right-hand side is

$$\ln\left(70 - 35e^{10k}\right),$$

and this does not simplify easily. We do not have a law of logarithms for the logarithm of a difference. Instead, we first isolate the exponential e^{10k} on one side of the equation by itself. Only after that do we take logarithms of both sides:

$$\begin{aligned} 42 &= 70 - 35e^{10k} \\ -28 &= -35e^{10k} \\ \frac{-28}{-35} &= e^{10k} \\ \ln\left(\frac{4}{5}\right) &= \ln\left(e^{10k}\right) = 10k \\ k &= \frac{\ln 0.8}{10} = -0.022314. \end{aligned}$$

As with exponential change problems, at least four significant digits should be used when rounding off a calculator's answer. Five significant digits are used here.

The model is now

$$T = 70 - 35e^{-0.022314t}.$$

❷ Determine the requested quantity. We reread the question and see that we wish to know how long (how much time) it takes for $T = 60°F$. We therefore replace T with 60 and solve for t, using the same algebraic steps as in step ❶ :

$$\begin{aligned} 60 &= 70 - 35e^{-0.022314t} \\ -10 &= -35e^{-0.022314t} \\ \frac{-10}{-35} &= e^{-0.022314t} \\ \ln\left(\frac{2}{7}\right) &= \ln e^{-0.022314t} = -0.022314t \\ t &= \frac{\ln\left(\frac{2}{7}\right)}{-0.022314} = 56.14 \text{ min}. \end{aligned}$$

In line 3, the right-hand side is $e^{-0.022314t}$. At this point (and not before), we have succeeded in isolating the exponential on one side of the equation by itself. We are now ready to take logarithms of both sides of the equation in line 4.

It takes 56.14 minutes from the time the glass of milk is poured for it to warm to $60°F$. ■

Reading Exercise 22 *Solve the equation for k:* $400 = 100 + (800 - 100)e^{4k}$.

Alternate derivation of the continuously compounded interest formula

A formula for compound interest is usually developed in high school algebra textbooks. Denote the principal (the initial investment) by A_0, let the (annual) rate of interest be given by r, and let n be the number of times compounded per year ($n = 12$ if interest is compounded monthly, $n = 4$ for quarterly compounding, and so on). The amount A in the account after t years is given by

$$A = A_0 \left(1 + \frac{r}{n}\right)^{nt}.$$

Many banks use $n = 360$ instead of $n = 365$ for the number of days in a year. This "standardizes" a month as 30 days or a quarter as 90 days.

The formula is valid for any natural number n. If we let $n = 365$, then we are compounding the interest daily. To compound the interest every minute, we use $n = 525\,600$. Continuous compounding means that instead of days or minutes or even milliseconds, the compounding periods are infinitesimally short. Using $n = \Omega$ for an arbitrary positive infinite hyperreal integer Ω gives

$$\begin{aligned} A &= A_0 \left(1 + \frac{r}{\Omega}\right)^{\Omega t} \\ &= A_0 \left[\left(1 + \frac{r}{\Omega}\right)^{\Omega / r}\right]^{rt} \\ &= A_0 e^{rt}. \end{aligned}$$

From the proof of theorem 10, $(1+\alpha)^{1/\alpha} \doteq e$ for any infinitesimal α. Using this fact with $\alpha = \frac{r}{\Omega}$, the quantity inside the square brackets is e.

Exercises 5.7

For problems involving bacteria, assume the population grows at a rate proportional to its size.

1. A population of bacteria increases by 40% in 1 hour. How long does it take to double its initial size?
2. A research assistant in a lab needs a culture of 35 000 bacteria. A culture is started with 20 bacteria. After 4 hours, the culture contains 136 bacteria. How long does it take to get the desired number of bacteria?
3. An account initially containing \$30 000 earns 4% interest compounded continuously. (a) What is the balance after 10 years? (b) How long does it take for the balance to reach \$100 000?
4. An account initially containing \$4200 earns 3.5% interest compounded continuously. (a) What is the balance after 4 years? (b) How long does it take the balance to triple?
5. At the time this is written, the Federal Reserve Board of the United States has a target inflation rate of 2%. If inflation is a constant 2%

compounded continuously, how long does it take the general price level to double?

6. Suppose that the prices of existing homes in a particular city follow the law of exponential change. If a home purchased for \$120 000 is worth \$172 000 10 years later, what is the rate (compounded continuously) at which the prices are changing?

Ans. to reading exercise 22
$k = \frac{\ln(3/7)}{4} = -0.2118$

7. Census data for a city are given in the following chart. Assuming the population changes at a rate proportional to its size, predict the population in the year 2030.

year	population
1970	104 272
2010	126 910

8. Census data for a city are given in the following chart. Assuming the population changes at a rate proportional to its size, predict the population in the year 2040.

year	population
1970	381 562
2010	347 683

9. A brick is heated to a temperature of 650°*F*. It is hung by a wire outside where the temperature is 40°*F*. Ten minutes later the temperature of the brick is 530°*F*. How long does it take for the brick to cool to 200°*F*?

10. You hear the ice cream truck and immediately run to it to purchase an ice cream sandwich, which has a temperature of 5°*F*. A neighbor begins an urgent conversation and distracts you; you place the ice cream sandwich on the porch rail and forget to eat it. The outside temperature is 95°*F*. Five minutes later, the temperature of the ice cream sandwich is 15°*F*. How long does it take for the ice cream sandwich to warm to 32°*F*?

11. A person pours a cup of coffee. When poured, the coffee has a temperature of 180°*F*. The coffee drinker takes the cup outside immediately to the front porch, where the temperature is 55°*F*. Ten minutes later, the coffee has cooled to 150°*F*. What is the temperature of the coffee after 26 minutes?

Interpret "after 26 minutes" as being 26 minutes from the time the cup of coffee is poured.

12. A person takes a hot dog from the refrigerator and puts it in a pot of boiling water. The hot dog has a starting temperature of 5°*C* and the boiling water is at 100°*C*. Five minutes later, the hot dog is at 35°*C*. What is the temperature of the hot dog after 12 minutes?

Interpret "after 12 minutes" as being 12 minutes from the time the hot dog is placed into the pot of boiling water.

13. A cup of 100°*F* coffee is left on a front porch at 9:00 a.m. At 10:00 a.m., the coffee has cooled to 60°*F*. What is the temperature of the coffee at 11:00 a.m. if the temperature on the front porch holds steady at 30°*F*?

14. At midnight at a crime scene, a large pot of chili at 212°*F* appears to have been cooking for a long time. A fresh bowl of chili with a temperature of 140°*F*, next to a clean spoon and napkin, and an unopened sleeve of crackers, is sitting on the counter. Thirty minutes later, the bowl of chili on the counter has cooled to 100°*F*. The thermostat in the kitchen is set at 66°*F*. How long before midnight was the bowl of chili taken from the pot?

This means that the rate of change of the amount of raw sugar is proportional to the amount of raw sugar.

15. (GSL) In the inversion of raw sugar, the time-rate of change varies as the amount of raw sugar remaining. If, after 10 hours, 1000 pounds of raw sugar has been reduced to 800 pounds, how much raw sugar remains after 24 hours?

In GSL, instead of "according to the law of exponential change," it reads "according to the compound-interest law."

16. (GSL) Atmospheric pressure p at points above the earth's surface as a function of the altitude h above sea level changes according to the law of exponential change. Assuming $p = 15$ pounds per square inch (psi) when $h = 0$, and $p = 10$ psi when $h = 10\,000$ feet, find p (a) when $h = 5\,000$ feet and; (b) when $h = 15\,000$ feet.

17. The half-life of Pu-241 (an isotope of plutonium) is 14.4 years. If the initial amount of Pu-241 is 0.0425 mg, how much remains after 100 years?

The half-life of caffeine varies from one individual to another and can be affected by factors such as heavy cigarette smoking or pregnancy.

18. The half-life of caffeine in an individual's bloodstream is 6 hours. If that individual drinks a large soda containing 60 mg caffeine at 3:00 p.m., how much caffeine remains in the bloodstream at 11:00 p.m.?

19. The initial amount of a radioactive isotope is 50 mg. After 90 days, the remaining amount of that isotope is 44.6 mg. Determine the half-life of the isotope.

20. The initial amount of a radioactive isotope is 4.50 mg. After 3.2 years, the remaining amount of that isotope is 2.66 mg. Determine the half-life of the isotope.

21. (GSL) A boat moving in still water is subject to a retardation proportional to its velocity at any instant. At the instant power was lost, the boat was moving with a velocity of 5 mph. Now, the boat continues to drift in still water, and 1 minute later, the velocity is 3 mph. When does the boat's velocity drop below 1 mph?

22. The *rule of 72* states that if an account earns $m\%$ interest, then the balance doubles after approximately $72/m$ years. In other words that is, doubling takes 24 years at 3% interest or 9 years at 8% interest. Are these reasonable approximations? What if the interest rate is 24%? Compare to exercise 5.

The rule of 72 is convenient for quick mental approximations.

Inverse Trigonometric Functions

5.8

In section 5.3 we took a function that did not pass the horizontal line test, $y = x^2$ on $(-\infty, \infty)$, and restricted its domain so that it did pass the horizontal line test. For the resulting function, $y = x^2$ on $[0, \infty)$, we then found its inverse function, $y = \sqrt{x}$. In this section we do the same for the trigonometric functions. We restrict the domains so that the functions are one-to-one and have inverses, then we define (and explore) the inverse trigonometric functions.

Inverse sine

We begin with the sine function. The function $y = \sin x$ is not one-to-one; it fails the horizontal line test (see figure 1) and does not have an inverse function. We can fix this problem by restricting the domain of

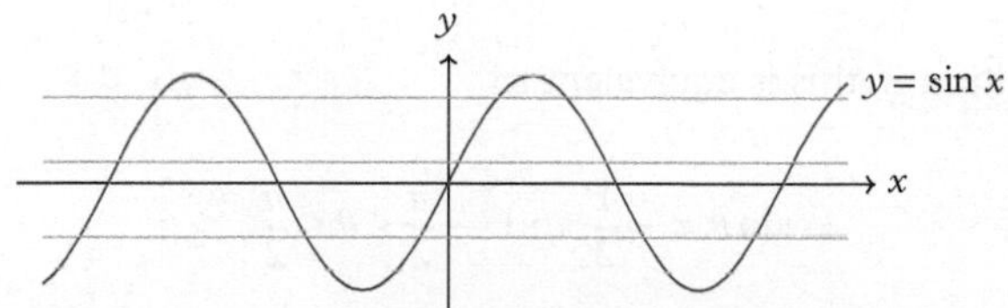

Figure 1 *The graph of $y = \sin x$, which fails the horizontal line test because there are horizontal lines (gray) that intersect the graph in more than one point*

the function. The values of $y = \sin x$ range from -1 to 1 and we need to choose a domain interval over which all the values are attained. The interval $\left[-\frac{\pi}{2}, \frac{\pi}{2}\right]$ will do; see figure 2.

Other choices that work include $\left[\frac{\pi}{2}, \frac{3\pi}{2}\right]$ and $\left[-\frac{11\pi}{2}, -\frac{9\pi}{2}\right]$, but restricting the domain to $\left[-\frac{\pi}{2}, \frac{\pi}{2}\right]$ is the most convenient.

Now we have a function, $y = \sin x$ on $\left[-\frac{\pi}{2}, \frac{\pi}{2}\right]$, that is one-to-one and therefore has an inverse function. We name this inverse function the *inverse sine* function, written symbolically as $\sin^{-1}$. It is defined by reversing the x- and y-coordinates from the sine function.

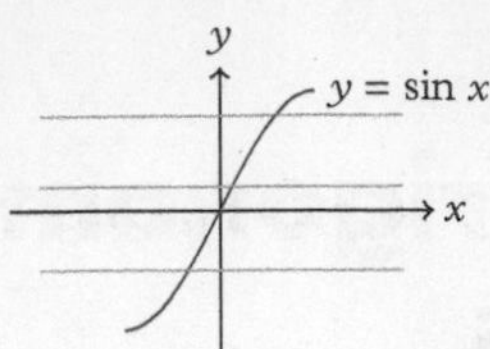

Figure 2 *The graph of* $y = \sin x$ *on* $\left[-\frac{\pi}{2}, \frac{\pi}{2}\right]$, *which passes the horizontal line test because any horizontal line (gray) intersects the graph at, at most, one point*

CAUTION: MISTAKE TO AVOID
The -1 in $\sin^{-1} x$ is not an exponent:

$$\sin^{-1} x \neq (\sin x)^{-1}$$
$$\sin^{-1} x \neq \frac{1}{\sin x}.$$

Definition 8 INVERSE SINE *The inverse sine function is defined by*

$$y = \sin^{-1} x \text{ if and only if } \sin y = x \text{ and } -\frac{\pi}{2} \leq y \leq \frac{\pi}{2}.$$

An alternate name for the inverse sine function is *arcsine:*

$$\arcsin x = \sin^{-1} x.$$

The definition can be used to help find values of the inverse sine function.

If you studied inverse trigonometric functions in a trigonometry or precalculus course, then this example is a review.

Example 1 *Without using a calculator, find* $\sin^{-1}\left(-\frac{1}{2}\right)$.

Solution Let

$$\theta = \sin^{-1}\left(-\frac{1}{2}\right).$$

We could use y instead of θ to match the symbols used in definition 8. The symbol θ is commonly used for angles, and we may think of $\sin^{-1} x$ as the angle whose sine is x.

By definition 8, this is equivalent to

$$\sin \theta = -\frac{1}{2} \text{ and } -\frac{\pi}{2} \leq \theta \leq \frac{\pi}{2}.$$

In other words, we wish to know the angle θ in quadrant I or IV whose sine is $-\frac{1}{2}$. It is not necessary to think in terms of angles, but it is convenient. In fact, drawing a diagram is helpful. Because $\sin \theta$ is negative, the angle must be in the fourth quadrant, so we draw a fourth-quadrant angle and label the sides to represent $\sin \theta = -\frac{1}{2}$ (figure 3). The length of the third side of the triangle is found using the Pythagorean theorem;

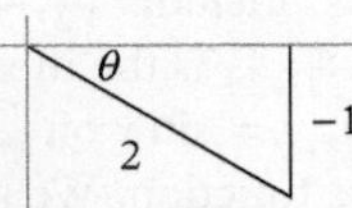

Figure 3 *A triangle with angle* θ *in standard position for which* $\sin \theta = -\frac{1}{2}$

$x^2 + (-1)^2 = 2^2$, so $x^2 = 4 - 1 = 3$ and $x = \sqrt{3}$. This value can now be placed in the diagram (figure 4).

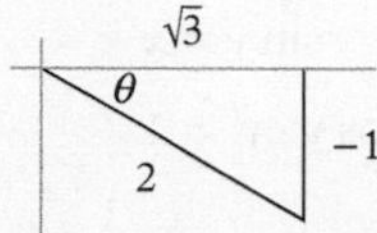

Figure 4 *The length of the third side of the triangle is added to the diagram*

The 30°–60°–90° and 45°–45°–90° special triangles should be recognizable for use in these exercises.

You should recognize this triangle as one with an angle θ that is 30°, or $\frac{\pi}{6}$ radians. Because it is in the fourth quadrant, the angle is negative, and $\theta = -\frac{\pi}{6}$. Therefore,

$$\sin^{-1}\left(-\frac{1}{2}\right) = -\frac{\pi}{6}. \qquad \blacksquare$$

The graph of the inverse sine function can be produced from the graph of $y = \sin x$ on $\left[-\frac{\pi}{2}, \frac{\pi}{2}\right]$ in figure 2 by swapping x- and y-coordinates; see figure 5.

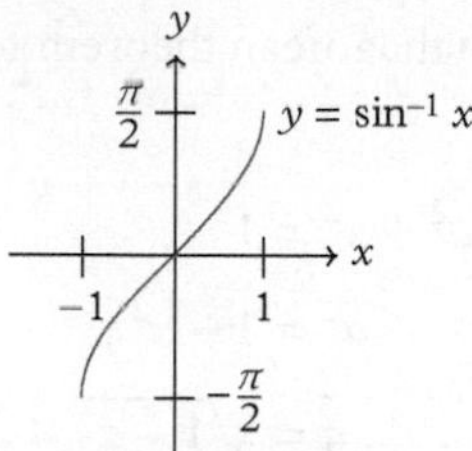

Figure 5 *The graph of* $y = \sin^{-1} x$. *The domain of the function is* $[-1, 1]$

Features of the graph such as concavity or the presence of vertical tangent lines can be inferred from the corresponding features on the graph of $y = \sin x$, but they can also be verified through the use of derivatives. Our next task is to differentiate $y = \sin^{-1} x$.

Derivative of $y = \sin^{-1} x$

Let

$$y = \sin^{-1} x.$$

Then, by definition 8,

$$\sin y = x \text{ and } \frac{-\pi}{2} \leq y \leq \frac{\pi}{2}.$$

We wish to find $\frac{dy}{dx}$. Using the previous equation, we can do so using implicit differentiation! We have

Line 2 differentiates both sides implicitly with respect to x; line 3 solves for y'.

$$\sin y = x$$
$$\cos y \cdot y' = 1$$
$$y' = \frac{1}{\cos y}.$$

Because we want to know $\frac{dy}{dx}$ explicitly in terms of just the variable x, we need to try to rewrite $\cos y$. What we know about y is that $\sin y = x$. Following the pattern of example 2, we draw a triangle to illustrate what we know (figure 6).

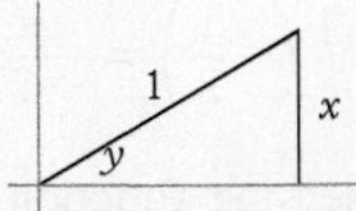

Figure 6 *A triangle with angle y in standard position for which* $\sin y = x = \frac{x}{1}$

Now we can use the Pythagorean theorem to determine the length a of the third side:

Because y must be a first- or fourth-quadrant angle, the value of a must be positive; we use the positive square root in line 3.

$$a^2 + x^2 = 1^2$$
$$a^2 = 1 - x^2$$
$$a = \sqrt{1 - x^2}.$$

We place this value in the diagram, as illustrated in figure 7. Then,

The same result for $\cos y$ is obtained if we draw the diagram with a fourth-quadrant angle y.

$$\cos y = \frac{\sqrt{1 - x^2}}{1} = \sqrt{1 - x^2}$$

and substitution yields

$$y' = \frac{1}{\sqrt{1 - x^2}}.$$

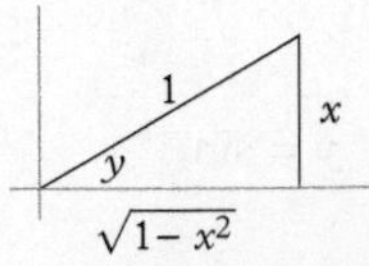

Figure 7 *The length of the horizontal side is added to the diagram*

DERIVATIVE OF INVERSE SINE

$$\frac{d}{dx}\sin^{-1}x = \frac{1}{\sqrt{1-x^2}}$$

DERIVATIVE OF INVERSE SINE (CHAIN RULE FORM)

$$\frac{d}{dx}\sin^{-1}u = \frac{1}{\sqrt{1-u^2}}\cdot\frac{du}{dx}$$

Example 2 *Find $\frac{dy}{dx}$ given $y = \sin^{-1}(4x+2)$.*

Solution Because we have the inverse sine of "not just plain x," we use the chain rule form of the derivative rule for inverse sine with $u = 4x + 2$:

$$\frac{dy}{dx} = \frac{1}{\sqrt{1-(4x+2)^2}}\cdot 4 = \frac{4}{\sqrt{1-(4x+2)^2}}. \quad \blacksquare$$

Reading Exercise 23 *Differentiate $y = \sin^{-1}(x^2 + 4x)$.*

Inverse cosine

The development of inverse cosine follows the same pattern as inverse sine. The function $y = \cos x$ on $(-\infty, \infty)$ is not one-to-one (figure 8).

The same interval used for $y = \sin x$ does not work here. Instead, let's choose $[0, \pi]$. The graph of $y = \cos x$ on $[0, \pi]$, which is one-to-one and therefore has an inverse, is shown in figure 9.

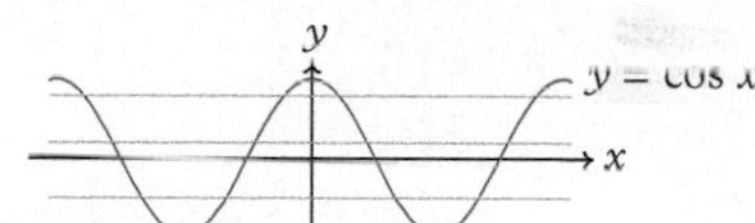

Figure 8 *The graph of $y = \cos x$, which fails the horizontal line test because there are horizontal lines (gray) that intersect the graph at more than one point*

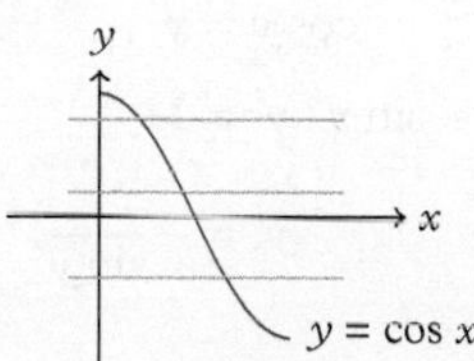

Figure 9 *The graph of $y = \cos x$ on $[0, \pi]$, which passes the horizontal line test because any horizontal line (gray) intersects the graph at, at most, one point. Also, every possible y-coordinate from the graph of $y = \cos x$ is attained*

Definition 9 INVERSE COSINE *The inverse cosine function is defined by*

$$y = \cos^{-1}x \text{ if and only if } \cos y = x \text{ and } 0 \le y \le \pi.$$

An alternate name for the inverse cosine function is *arccosine:*

$$\arccos x = \cos^{-1}x.$$

Finding values of the inverse cosine function is similar to the procedure in example 2, except we use the first and second quadrants.

Switching x- and y-coordinates on the graph of $y = \cos x$ on $[0, \pi]$ (figure 9) results in the graph of $y = \cos^{-1} x$ (figure 10).

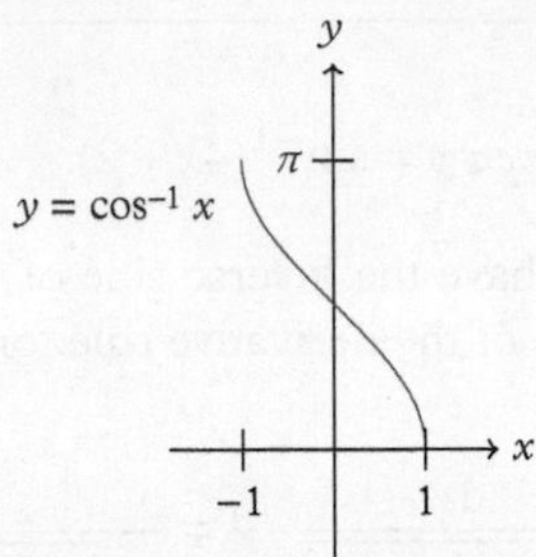

Figure 10 *The graph of $y = \cos^{-1} x$. The domain of the function is $[-1, 1]$. Swapping x- and y-coordinates results in the point $(0, 1)$ on the graph of $y = \cos x$ becoming the point $(1, 0)$ on the graph of $y = \cos^{-1} x$, the point $(\pi, -1)$ on the graph of $y = \cos x$ becoming the point $(-1, \pi)$ on the graph of $y = \cos^{-1} x$, and so on*

The derivative of $y = \cos^{-1} x$ is found using implicit differentiation:

$$\begin{aligned} y &= \cos^{-1} x \\ \cos y &= x \\ -\sin y \cdot y' &= 1 \\ y' &= -\frac{1}{\sin y}. \end{aligned}$$

Ans. to reading exercise 23
$y' = \frac{2x + 4}{\sqrt{1 - (x^2 + 4x)^2}}$

To express $\sin y$ in terms of x, we draw a diagram representing $\cos y = x$ (figure 11). Now we can use the Pythagorean theorem to determine the length b of the third side:

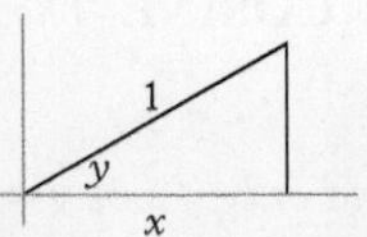

Figure 11 *A triangle with angle y in standard position for which* $\cos y = x$

$$x^2 + b^2 = 1^2$$
$$b^2 = 1 - x^2$$
$$b = \sqrt{1 - x^2}.$$

Because y must be a first- or second-quadrant angle, the value of b must be positive. We use the positive square root in line 3.

We place this value in the diagram, as illustrated in figure 12. Then,

$$\sin y = \frac{\sqrt{1 - x^2}}{1} = \sqrt{1 - x^2}$$

The same result for $\sin y$ is obtained if we draw the diagram with a second-quadrant angle y.

and substitution yields

$$y' = -\frac{1}{\sqrt{1 - x^2}}.$$

A shorter, but less visual, alternative to finding $\sin y$ is to use the Pythagorean identity $\sin^2 y + \cos^2 y = 1$, substitute x for $\cos y$ to obtain $\sin^2 y + x^2 = 1$, then solve for $\sin y$, noting that sine is nonnegative in the first and second quadrants.

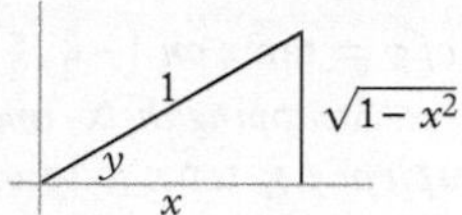

Figure 12 *A triangle with angle y in standard position for which* $\cos y = x$

DERIVATIVE OF INVERSE COSINE

$$\frac{d}{dx}\cos^{-1} x = -\frac{1}{\sqrt{1 - x^2}}$$

DERIVATIVE OF INVERSE COSINE (CHAIN RULE FORM)

$$\frac{d}{dx}\cos^{-1} u = -\frac{1}{\sqrt{1 - u^2}} \cdot \frac{du}{dx}$$

Example 3 *Find* y' *for* $y = x\cos^{-1} x$.

Solution Using the product rule and the rule for the derivative of inverse cosine,

Recognizing the need for the product rule involving an inverse trig function is the same skill as recognizing the need for the product rule involving a trig function.

$$y' = x \cdot \frac{-1}{\sqrt{1 - x^2}} + \cos^{-1} x \cdot 1$$
$$= \frac{-x}{\sqrt{1 - x^2}} + \cos^{-1} x.$$ ■

Reading Exercise 24 *Find* y' *for* $y = \dfrac{\cos^{-1} x}{x}$.

Inverse tangent

Restricting the domain of $y = \tan x$ to $\left(-\frac{\pi}{2}, \frac{\pi}{2}\right)$ results in a one-to-one function with an inverse function that is called $y = \tan^{-1} x$. Their graphs are in figure 13.

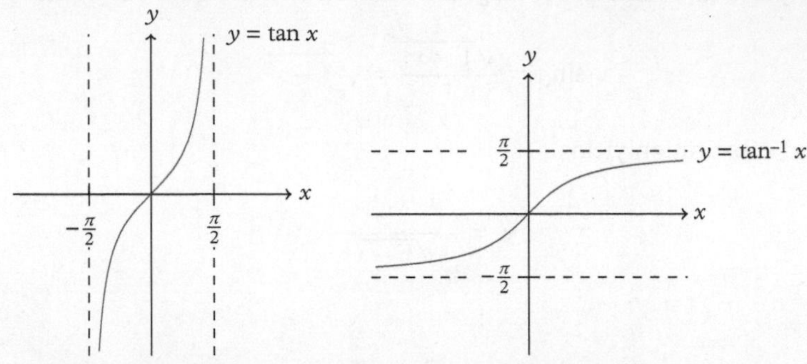

Figure 13 *(left) The graph of $y = \tan x$ on $\left(-\frac{\pi}{2}, \frac{\pi}{2}\right)$ and (right) the graph of its inverse function $y = \tan^{-1} x$. Swapping the x- and y-coordinates causes the vertical asymptotes on the graph of $y = \tan x$ to become horizontal asymptotes on the graph of $y = \tan^{-1} x$. The domain of $y = \tan^{-1} x$ is $(-\infty, \infty)$*

Definition 10 INVERSE TANGENT *The inverse tangent function is defined by*

$$y = \tan^{-1} x \text{ if and only if } \tan y = x \text{ and } -\frac{\pi}{2} < y < \frac{\pi}{2}.$$

An alternate name for the inverse tangent function is *arctangent:*

$$\arctan x = \tan^{-1} x.$$

The derivative of $y = \tan^{-1} x$ is found using implicit differentiation in the same manner as for inverse sine and inverse cosine, and is left to you as an exercise.

DERIVATIVE OF INVERSE TANGENT

$$\frac{d}{dx} \tan^{-1} x = \frac{1}{x^2 + 1}$$

DERIVATIVE OF INVERSE TANGENT (CHAIN RULE FORM)

$$\frac{d}{dx} \tan^{-1} u = \frac{1}{u^2 + 1} \cdot \frac{du}{dx}$$

Example 4 *Find $f'(x)$ for $f(x) = \ln \tan^{-1} e^{3x}$.*

The implied parentheses are as follows:

$$\ln\left(\tan^{-1}\left(e^{3x}\right)\right).$$

Note that law of logarithms (3) does not apply because the -1 is not an exponent.

Solution Using the chain rule multiple times, we have

$$\begin{aligned} f'(x) &= \frac{1}{\tan^{-1} e^{3x}} \cdot \frac{1}{(e^{3x})^2 + 1} \cdot e^{3x} \cdot 3 \\ &= \frac{3e^{3x}}{(e^{6x} + 1) \tan^{-1} e^{3x}}. \end{aligned}$$

■

Reading Exercise 25 *Differentiate* $y = \tan^{-1}(3x)$.

Limits with inverse tangent

From the graph of inverse tangent (figure 13), we see that the function has two horizontal asymptotes. They give rise to the limits

$$\lim_{x\to\infty} \tan^{-1} x = \frac{\pi}{2}$$

and

$$\lim_{x\to-\infty} \tan^{-1} x = -\frac{\pi}{2}.$$

Ans. to reading exercise 24

$$y' = \frac{x \cdot \frac{-1}{\sqrt{1-x^2}} - \cos^{-1} x \cdot 1}{x^2} = \frac{\frac{-x}{\sqrt{1-x^2}} - \cos^{-1} x}{x^2}$$

$$\lim_{x\to\infty} \tan^{-1} x = \frac{\pi}{2}$$

$$\lim_{x\to-\infty} \tan^{-1} x = -\frac{\pi}{2}$$

These limits also tell us how to evaluate the inverse tangent at infinite hyperreals.

INVERSE TANGENT OF INFINITE HYPERREALS

For any positive infinite hyperreal Ω,

$$\tan^{-1} \Omega \doteq \frac{\pi}{2}$$

and

$$\tan^{-1}(-\Omega) \doteq -\frac{\pi}{2}.$$

The same is true for approximation:

$$\tan^{-1} \Omega \approx \frac{\pi}{2}$$

and

$$\tan^{-1}(-\Omega) \approx -\frac{\pi}{2}.$$

Example 5 *Find* $\lim_{x\to-4^-} \tan^{-1}\left(\frac{1}{2x+8}\right)$.

Solution Evaluating using continuity fails, so we use $x = -4 - \omega$ for an arbitrary positive infinitesimal ω:

$$\begin{aligned}\lim_{x\to-4^-} \tan^{-1}\left(\frac{1}{2x+8}\right) &= \tan^{-1}\left(\frac{1}{2(-4-\omega)+8}\right)\\ &= \tan^{-1} \frac{1}{-2\omega}\\ &= \tan^{-1}\left(-\frac{1}{2}\Omega\right)\\ &\doteq -\frac{\pi}{2}.\end{aligned}$$

■

Because the expression inside the inverse tangent in line 3 is negative infinite, we render $-\frac{\pi}{2}$.

Because the inverse tangent function has nonzero horizontal asymptotes on both the right and the left, chapter 3 theorem 12 (approximating inside a continuous function with a horizontal asymptote) applies. We may approximate inside the inverse tangent function even when the quantity inside the function is infinite.

As always, the approximation principle still applies.

Example 6 *Find* $\lim_{x \to -\infty} \tan^{-1}\left(\frac{4x^2+1}{3-x}\right)$.

Solution For any positive infinite hyperreal Ω,

$$\begin{aligned}\lim_{x \to -\infty} \tan^{-1}\left(\frac{4x^2+1}{3-x}\right) &= \tan^{-1}\left(\frac{4(-\Omega)^2+1}{3-(-\Omega)}\right) = \tan^{-1}\left(\frac{4\Omega^2+1}{3+\Omega}\right) \\ &\approx \tan^{-1}\left(\frac{4\Omega^2}{\Omega}\right) = \tan^{-1}(4\Omega) \\ &\doteq \frac{\pi}{2}. \end{aligned}$$ ■

Line 2 uses chapter 3 theorem 12 to approximate inside the inverse tangent function. For line 3, because the expression inside the inverse tangent is positive infinite, we render $\frac{\pi}{2}$.

Inverse cotangent, secant, and cosecant

There is not universal agreement on which domain restriction should be used for developing the inverse cotangent, inverse secant, and inverse cosecant functions. For example, some sources restrict $y = \cot x$ to $(0, \pi)$ (as is done here in the exercises), whereas others use $\left(-\frac{\pi}{2}, 0\right) \cup \left(0, \frac{\pi}{2}\right)$. The resulting functions may be given the same name, but they are not identical; graphs, derivatives, limits, and more can differ. Therefore, caution should be taken when determining which inverse cotangent function (or inverse secant or inverse cosecant) is being used.

Because of this ambiguity, only the inverse sine, cosine, and tangent are used in the remainder of this book.

Ans. to reading exercise 25
$y' = \frac{1}{(3x)^2+1} \cdot 3 = \frac{3}{1+9x^2}$

Integral formulas: inverse sine and cosine

New derivative formulas give rise to new antiderivative formulas.

ANTIDERIVATIVE GIVING INVERSE SINE

$$\int \frac{1}{\sqrt{1-x^2}}\,dx = \sin^{-1} x + C$$

Details:

$$\frac{d}{dx}\sin^{-1} x = \frac{1}{\sqrt{1-x^2}}$$

$$\int \frac{1}{\sqrt{1-x^2}}\,dx = \sin^{-1} x + C.$$

Example 7 *Evaluate* $\int \frac{5}{\sqrt{1-x^2}}\,dx.$

Solution Using the antiderivative constant multiple rule and the antiderivative giving inverse sine formula,

$$\int \frac{5}{\sqrt{1-x^2}}\,dx = 5\int \frac{1}{\sqrt{1-x^2}}\,dx = 5\sin^{-1}x + C. \quad \blacksquare$$

Notice that even though we have the square root of "not just plain x," substitution is not needed. The reason is that we are not using the antiderivative power rule, but rather the exact version of the antiderivative giving the inverse sine. We have $\frac{1}{\sqrt{1-x^2}}$ with "just plain x."

Reversing the derivative formula for inverse cosine gives the antiderivative formula

$$\int \frac{-1}{\sqrt{1-x^2}}\,dx = \cos^{-1}x + C.$$

But, this formula does not add anything to our antiderivative arsenal. The same function can be integrated using the method of example 7.

Example 8 *Evaluate* $\int \frac{-1}{\sqrt{1-x^2}}\,dx.$

Solution Using the antiderivative constant multiple rule and the antiderivative formula giving the inverse sine,

$$\int \frac{-1}{\sqrt{1-x^2}}\,dx = -\int \frac{1}{\sqrt{1-x^2}}\,dx = -\sin^{-1}x + C. \quad \blacksquare$$

We now have two different antiderivatives for the same function. Does this mean that $-\sin^{-1}x$ and $\cos^{-1}x$ are equal? No! Any two antiderivatives of a function must differ by a constant, but that constant need not be zero. In this case,

$$\cos^{-1}x = -\sin^{-1}x + \frac{\pi}{2}.$$

This identity can be proved using the definitions of inverse sine and cosine, and a cofunction identity from trigonometry.

Reading Exercise 26 *Find* $\int \frac{3}{\sqrt{1-x^2}}\,dx.$

Example 9 *Find* $\int \frac{x}{\sqrt{1-x^4}}\,dx.$

Solution Because the denominator is not exactly $\sqrt{1-x^2}$, the antiderivative giving the inverse sine formula does not apply directly.

The substitution

$$u = 1 - x^4$$
$$du = -4x^3\,dx$$

does not succeed because there is no x^3 in the integrand.

But, $x^4 = (x^2)^2$, and we rewrite the original integral as

$$\int \frac{x}{\sqrt{1-x^4}}\,dx = \int \frac{x}{\sqrt{1-(x^2)^2}}\,dx.$$

The "not just plain x" idea applies! Try substitution:

$$u = x^2$$
$$du = 2x\,dx.$$

We have x in the numerator of the integrand, so we need to adjust for the constant 2. Performing the substitution and then using the antiderivative giving the inverse sine,

Line 1 adjusts for the needed constant, line 2 performs the substitution, line 3 uses the antiderivative giving the inverse sine formula, and line 4 returns to the original variable.

$$\begin{aligned}\int \frac{x}{\sqrt{1-(x^2)^2}}\,dx &= \frac{1}{2}\int \frac{2x}{\sqrt{1-(x^2)^2}}\,dx \\ &= \frac{1}{2}\int \frac{1}{\sqrt{1-u^2}}\,du \\ &= \frac{1}{2}\sin^{-1} u + C \\ &= \frac{1}{2}\sin^{-1} x^2 + C.\end{aligned}$$

■

Example 10 *Evaluate* $\int \frac{x}{\sqrt{1-x^2}}\,dx.$

Solution Rewriting the integrand as

$$\int \frac{x}{\sqrt{1-x^2}}\,dx = \int x \cdot \frac{1}{\sqrt{1-x^2}}\,dx$$

is not helpful. Although we can integrate each of the expressions x and $\frac{1}{\sqrt{1-x^2}}$, we have no antiderivative product rule.

As we progress in our study of integration, we will sometimes find that the first strategy we think of does not work. In this case (after checking for errors), do not try to make a round peg fit into a square hole. Try a different strategy.

Switching strategies, it is still the case that the denominator is the square root of "not just plain x." Let's try substitution:

$$u = 1 - x^2$$
$$du = -2x\,dx.$$

Because x is in the numerator of the integrand, we need only adjust for the constant and the substitution succeeds:

$$\begin{aligned}\int \frac{x}{\sqrt{1-x^2}}\,dx &= -\frac{1}{2}\int \frac{-2x}{\sqrt{1-x^2}}\,dx \\ &= -\frac{1}{2}\int \frac{1}{\sqrt{u}}\,du = -\frac{1}{2}\int u^{-1/2}\,du \\ &= -\frac{1}{2}\frac{u^{1/2}}{1/2} + C = -u^{1/2} + C \\ &= -\sqrt{1-x^2} + C.\end{aligned}$$ ■

Line 1 adjusts for the needed constant, line 2 performs the substitution, line 3 uses the antiderivative power rule, and line 4 returns to the original variable.

Ans. to reading exercise 26
$3\sin^{-1}x + C$

A careful study of the difference between examples 9 and 10 can be very instructive. The two integrands are very similar but require different solution strategies. Do not be concerned if you are unable to immediately "see" the right strategy every time.

Example 11 *Find* $\int \frac{(\sin^{-1}x)^2}{\sqrt{1-x^2}}\,dx.$

Solution Noticing that we have "not just plain x" to a power, we try substitution:

$$u = \sin^{-1}x$$

$$du = \frac{1}{\sqrt{1-x^2}}\,dx.$$

You might also have "seen" the function $\sin^{-1}x$ and its derivative $\frac{1}{\sqrt{1-x^2}}$, suggesting substitution.

Proceeding with the substitution,

Compare this example to example 6 in section 5.6.

$$\begin{aligned}\int \frac{(\sin^{-1}x)^2}{\sqrt{1-x^2}}\,dx &= \int u^2\,du \\ &= \frac{u^3}{3} + C \\ &= \frac{1}{3}\left(\sin^{-1}x\right)^3 + C.\end{aligned}$$ ■

Because the -1 is not an exponent, the solution should **not** be written as $\frac{1}{3}\sin^{-3}x + C$.

Integral formulas: inverse tangent

Next we reverse the derivative formula for inverse tangent.

ANTIDERIVATIVE GIVING INVERSE TANGENT

$$\int \frac{1}{x^2+1}\,dx = \tan^{-1}x + C$$

Details:

$$\frac{d}{dx}\tan^{-1}x = \frac{1}{x^2+1}$$

$$\int \frac{1}{x^2+1}\,dx = \tan^{-1}x + C.$$

What if the denominator is $x^2 + 25$ instead of $x^2 + 1$? The next formula is helpful.

The version 2 formula can be proved by differentiating the right-hand side (treating a as a constant) to obtain the integrand.

ANTIDERIVATIVE GIVING INVERSE TANGENT, VERSION 2

For any real number a,

$$\int \frac{1}{x^2 + a^2}\, dx = \frac{1}{a} \tan^{-1}\left(\frac{x}{a}\right) + C.$$

Example 12 *Integrate* $\int \frac{1}{x^2 + 25}\, dx.$

We could also use $a = -5$, but it is usually more convenient to choose the positive value for a.

Solution Using version 2 of the antiderivative giving the inverse tangent formula with $a = 5$, we have

$$\int \frac{1}{x^2 + 25}\, dx = \frac{1}{5} \tan^{-1}\left(\frac{x}{5}\right) + C.$$ ■

Reading Exercise 27 *Find* $\int \frac{1}{4 + x^2}\, dx.$

Next, matters are complicated by placing a coefficient on x^2.

Example 13 *Evaluate* $\int \frac{1}{16 + 9x^2}\, dx.$

Solution Because of the 9, the integrand is not of the form $\frac{1}{x^2 + a^2}$. In other words, we have $9x^2$ instead of x^2, a "not just plain x" situation. So, we try substitution:

$$u = 3x \text{ (thus } u^2 = 9x^2)$$
$$du = 3\, dx.$$

We need to adjust for the constant 3, substitute, and integrate:

Line 1 adjusts for the constant; line 2 substitutes; line 3 uses version 2 of the antiderivative giving the inverse tangent formula, with $a = 4$; and line 4 returns to the original variable.

$$\begin{aligned}\int \frac{1}{16 + 9x^2}\, dx &= \frac{1}{3} \int \frac{1}{16 + (3x)^2}\, 3\, dx \\ &= \frac{1}{3} \int \frac{1}{16 + u^2}\, du \\ &= \frac{1}{3} \cdot \frac{1}{4} \tan^{-1}\left(\frac{u}{4}\right) + C \\ &= \frac{1}{12} \tan^{-1}\left(\frac{3x}{4}\right) + C.\end{aligned}$$ ■

There is an alternate solution to example 13 that does not use substitution. Instead, we factor a 9 out of the denominator to put the integrand in the form $\frac{1}{x^2+a^2}$:

$$\begin{aligned}\int \frac{1}{16+9x^2}\,dx &= \int \frac{1}{9\left(\frac{16}{9}+x^2\right)}\,dx \\ &= \frac{1}{9}\int \frac{1}{\frac{16}{9}+x^2}\,dx \\ &= \frac{1}{9}\cdot\frac{1}{\frac{4}{3}}\tan^{-1}\left(\frac{x}{\frac{4}{3}}\right)+C,\end{aligned}$$

which is equivalent algebraically to the original solution.

Line 3 uses version 2 of the antiderivative giving the inverse tangent formula with $a = \frac{4}{3}$.

Review: trig composed with inverse trig

We close this section by reviewing a type of exercise that can be found in many books on trigonometry.

Example 14 *Calculate* $\cos\left(\arctan\left(\frac{1}{5}\right)\right)$.

Solution Thinking of $\arctan\frac{1}{5}$ as an angle with a tangent that is $\frac{1}{5}$, we let $\theta = \arctan\frac{1}{5}$ and try to find $\cos\theta$. As with example 1, a diagram is helpful. We draw a triangle with angle θ for which $\tan\theta = \frac{1}{5}$ (figure 14). Next we use the Pythagorean theorem to find the third side; $c^2 = 5^2 + 1^2 = 26$, so $c = \sqrt{26}$. Placing this value in the diagram results in figure 15. Then,

Since $\tan^{-1}\frac{1}{5} = \theta$, by definition 10 $\tan\theta = \frac{1}{5}$, with θ in quadrant I or quadrant IV.

$$\cos\left(\arctan\left(\frac{1}{5}\right)\right) = \cos\theta = \frac{5}{\sqrt{26}}.$$

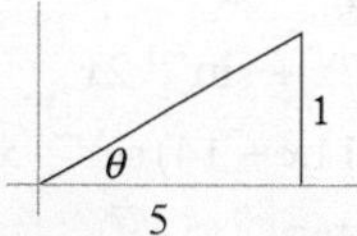

Figure 14 *Triangle with angle θ in standard position for which* $\tan\theta = \frac{1}{5}$. *The angle is drawn in quadrant I because the tangent is positive. We draw it in quadrant IV if the tangent is negative*

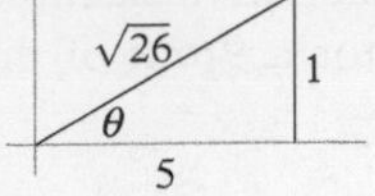

$\sqrt{26}$ 1 θ 5

Figure 15 *A triangle with angle θ in standard position for which* $\tan\theta = \frac{1}{5}$

Ans. to reading exercise 27
$\frac{1}{2}\tan^{-1}\frac{x}{2} + C$

■

Exercises 5.8

1–6. Rapid response: rewrite the expression using a trigonometric function instead of an inverse trigonometric function.

1. $y = \tan^{-1} 3$
2. $\theta = \cos^{-1}\frac{1}{2}$
3. $0.4 = \cos^{-1} x$
4. $t + 4 = \sin^{-1} 3u$
5. $y = \sin^{-1}(x - 2)$
6. $\tan^{-1}(0.3x^2) = 7$

7–12. Rapid response: rewrite the expression using an inverse trigonometric function instead of a trigonometric function.

7. $y = \tan 3$
8. $\theta = \cos\frac{1}{2}$
9. $0.4 = \cos x$
10. $t + 4 = \sin 3u$
11. $y = \sin(x - 2)$
12. $\tan(0.3x^2) = 7$

13–20. Evaluate the expression without using a calculator.

13. $\cos^{-1}\frac{1}{2}$
14. $\sin^{-1}\frac{1}{\sqrt{2}}$
15. $\tan^{-1} -\frac{1}{\sqrt{3}}$
16. $\cos^{-1}\frac{\sqrt{3}}{2}$
17. $\sin\left(\arccos\frac{3}{5}\right)$
18. $\sin(\arctan 7)$
19. $\sec\left(\sin^{-1}\frac{1}{7}\right)$
20. $\tan\left(\cos^{-1}\left(-\frac{1}{4}\right)\right)$

21–40. Differentiate the function.

21. $y = 5\tan^{-1} 3x$
22. $y = \sin 2x + \sin^{-1} 2x$
23. $f(x) = (11x + 14)\sin^{-1} x$
24. $f(x) = x\tan^{-1}(5\sqrt{7})$
25. $f(t) = \dfrac{t^2}{\sin^{-1} 5t}$
26. $y = \cos^{-1}(x^2 - 3x)$
27. $g(t) = \left(\cos^{-1}(0.00014\pi e)\right)$
28. $s(t) = t^2\tan^{-1} t^2$

29. $y = \cos\left((4x+6)^5\right)$

30. $y = \cos^{-1}\left((4x+6)^5\right)$

31. $f(x) = \sin^{-1}\left(7x^3+1\right)$

32. $y = \dfrac{\sin^{-1} x}{\sqrt{1-x^2}}$

33. $s(x) = \cos^{-1}\left(\dfrac{x^2}{x^3+5}\right)$

34. $y = \tan^{-1} e^{4x^2-3}$

35. $y = \sin^{-1} e^{3x}$

36. $g(x) = \cos^{-1} \sqrt{x+1}$

37. $y = \tan^{-1} \ln\left(2+\cos x^2\right)$

38. $y = \sqrt{2+\sin^{-1}(7x^3+1)}$

39. $y = 4^{\sin^{-1} x} + \left(\sin^{-1} x\right)^4$

40. $f(x) = \ln\left(\dfrac{\sin^{-1}(3x)}{4x-9}\right)$

41–60. Evaluate the integral.

41. $\displaystyle\int \frac{12}{36+x^2}\,dx$

42. $\displaystyle\int \frac{\sqrt{3}}{16+x^2}\,dx$

43. $\displaystyle\int \frac{4}{\sqrt{1-x^2}}\,dx$

44. $\displaystyle\int \frac{7x}{x^2+1}\,dx$

45. $\displaystyle\int \frac{4x}{\sqrt{1-x^2}}\,dx$

46. $\displaystyle\int \frac{7}{x^2+1}\,dx$

47. $\displaystyle\int \frac{1}{36+4x^2}\,dx$

48. $\displaystyle\int \frac{\cos^{-1} x}{\sqrt{1-x^2}}\,dx$

49. $\displaystyle\int \frac{5}{\sqrt{1-(1+5x)^2}}\,dx$

50. $\displaystyle\int \frac{1}{9+9x^2}\,dx$

51. $\displaystyle\int \frac{x^2}{\sqrt{1-x^6}}\,dx$

52. $\displaystyle\int \frac{1}{-3\sqrt{1-x^2}}\,dx$

53. $\displaystyle\int \frac{1}{x\sqrt{1-(\ln x)^2}}\,dx$

54. $\displaystyle\int \frac{2x+1}{\sqrt{1-(x^2+x)^2}}\,dx$

55. $\displaystyle\int \frac{\sqrt{\tan^{-1} x}}{1+x^2}\,dx$

56. $\displaystyle\int \frac{8x^3}{8+x^8}\,dx$

57. $\displaystyle\int \frac{\sin^2(5x)+\cos^2(5x)}{7+x^2}\,dx$

58. $\displaystyle\int \frac{\cos 3x}{\sqrt{1-\sin^2 3x}}\,dx$

59. $\displaystyle\int \frac{1}{\sqrt{4-4x^2}}\,dx$

60. $\displaystyle\int \frac{e^x}{1+e^{2x}}\,dx$

61–70. Evaluate the limit.

61. $\lim_{x\to5^+} \tan^{-1}\left(\frac{1}{x-5}\right)$

62. $\lim_{x\to5^-} \tan^{-1}\left(\frac{1}{x-5}\right)$

63. $\lim_{x\to\infty} \tan^{-1}\left(\frac{2-5x^2}{x+3}\right)$

64. $\lim_{x\to\infty} \tan^{-1}\left(\frac{x^2}{x-5}\right)$

65. $\lim_{x\to1} \sqrt{x^2+\tan^{-1}(x^2-1)}$

66. $\lim_{x\to4} \tan^{-1}\left(\frac{x-3}{(x-4)^2}\right)$

67. $\lim_{x\to-\infty} \tan^{-1}\left(\frac{\sqrt{x^4+1}}{x-1}\right)$

68. $\lim_{x\to4} \frac{\tan^{-1}(x-3)}{x-3}$

69. $\lim_{x\to\infty} 4\tan^{-1}\left(\frac{x^3+4}{2x+1}\right)$

70. $\lim_{x\to0^-} x\tan^{-1}\left(\frac{2}{x}\right)$

71. Is $y = \sin^{-1} x$ differentiable at $x = 1$? Give both an algebraic reason and a graphical reason for your answer.

72. Use the methods of section 3.4 to determine intervals of concavity and inflection points for $y = \sin^{-1} x$.

73. Use the methods of section 3.4 to determine intervals of concavity and inflection points for $y = \tan^{-1} x$.

74. Use implicit differentiation to derive the formula $\frac{d}{dx}\tan^{-1} x = \frac{1}{x^2+1}$.

75. Prove version 2 of the antiderivative giving the inverse tangent formula.

76. Evaluate $\int \frac{1}{\sqrt{a^2-x^2}}\,dx$, where a is a real number. Write a “version 2” of the antiderivative giving the inverse sine formula.

77–82. Restricting the cotangent function to the interval $(0, \pi)$, explore the inverse cotangent function.

77. Write the definition of $\cot^{-1} x$.

78. Reverse x- and y-coordinates from the graph of $y = \cot x$ on $(0, \pi)$ to graph $y = \cot^{-1} x$.

79. Use the graph to determine $\lim_{x\to\infty} \cot^{-1} x$ and $\lim_{x\to-\infty} \cot^{-1} x$.

80. Use implicit differentiation to find the derivative of $y = \cot^{-1} x$.

81. Write the chain rule form of the differentiation formula.

82. Reverse the differentiation formula to produce an antiderivative formula. Does this help us integrate any additional functions?

Hyperbolic and Inverse Hyperbolic Functions

5.9

We continue our survey of additional functions with functions that are useful in engineering and that help expand our catalog of antiderivatives.

Hyperbolic functions

We begin with the definition of the hyperbolic cosine function.

Definition 11 HYPERBOLIC COSINE *For any real number x,*

$$\cosh x = \frac{e^x + e^{-x}}{2}.$$

The "h" in cosh is for hyperbolic, and the "cos" is for cosine.

The hyperbolic cosine, often pronounced by its abbreviation "cosh," is the average of the exponential functions $y = e^x$ and $y = e^{-x}$, which can help us visualize the graph of hyperbolic cosine (figure 1).

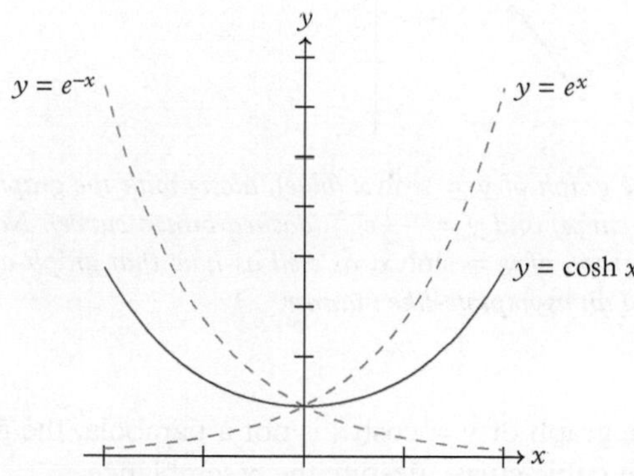

Figure 1 *The graph of* $y = \cosh x$ *(blue), along with the graphs of* $y = e^x$ *(dashed green curve) and* $y = e^{-x}$ *(dashed brown curve). Notice the point* $(0, 1)$ *on the graph*

$\cosh 0 = 1$

Although the graph of hyperbolic cosine might remind you of a parabola, it is not a parabola. Instead, the shape is called a *catenary*.

The familiar shape of utility wires hanging between two poles or jump ropes dangling from two hands is that of a catenary. Another famous catenary is the Gateway Arch in St. Louis, Missouri.

The hyperbolic sine function is similar, but with subtraction instead of addition in the definition.

Definition 12 HYPERBOLIC SINE *For any real number x,*

$$\sinh x = \frac{e^x - e^{-x}}{2}.$$

The "h" in sinh is for hyperbolic, and the "sin" is for sine.

Hyperbolic sine is often pronounced by adding a "c" to its abbreviation, as in "sinch" (sounds like the word *cinch*). Among the ways of understanding the graph of $y = \sinh x$ (figure 2) is to add the two functions $y = \frac{1}{2}e^x$ and $y = -\frac{1}{2}e^{-x}$. When one of these two functions approaches zero, the graph of $y = \sinh x$ approaches the other function in an asymptote-like manner.

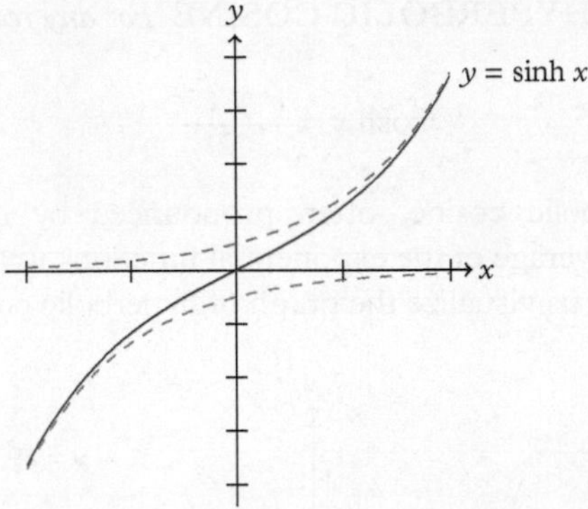

Figure 2 *The graph of* $y = \sinh x$ *(blue), along with the graphs of* $y = \frac{1}{2}e^x$ *(dashed green curve) and* $y = -\frac{1}{2}e^{-x}$ *(dashed brown curve). Notice the point* $(0, 0)$ *on the graph of* $y = \sinh x$*, as well as how that graph approaches the other graphs in an asymptote-like manner*

Just as the graph of $y = \cosh x$ is not a parabola, the graph of $y = \sinh x$ is not a cubic curve, despite the resemblance.

The other four hyperbolic functions are hyperbolic tangent, hyperbolic cotangent, hyperbolic secant, and hyperbolic cosecant. Following the ratio and reciprocal identities in trigonometry gives us a way of defining these four functions, two of which are graphed in figure 3.

Definition 13 FOUR MORE HYPERBOLIC FUNCTIONS *For each real number x for which the expression is defined,*

$$\tanh x = \frac{\sinh x}{\cosh x} \qquad \operatorname{sech} x = \frac{1}{\cosh x}.$$
$$\coth x = \frac{\cosh x}{\sinh x} \qquad \operatorname{csch} x = \frac{1}{\sinh x}.$$

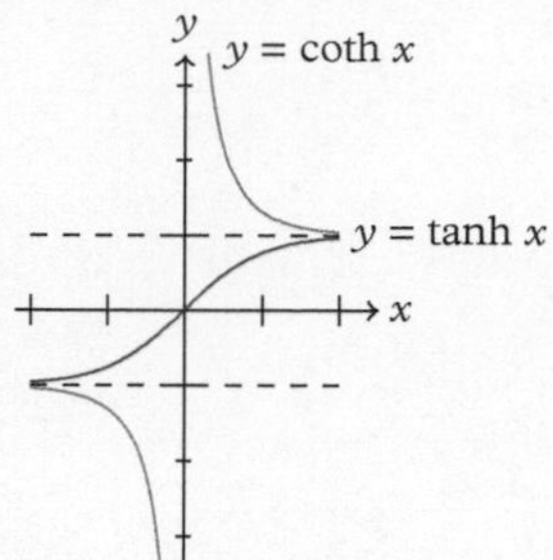

Figure 3 *The graphs of* $y = \tanh x$ *(blue) and* $y = \coth x$ *(green), with horizontal asymptotes* $y = 1$ *and* $y = -1$. *There is a vertical asymptote* $x = 0$ *on the graph of* $y = \coth x$. *The hyperbolic tangent curve is not an inverse tangent curve, despite the resemblance*

Some of the similarities between the trigonometric functions and the hyperbolic functions are apparent already, and others become apparent later. But why "hyperbolic?"

The trigonometric functions are sometimes called *circular functions* because of their relationship to the unit circle $x^2 + y^2 = 1$. Points of the form $(\cos\theta, \sin\theta)$ lie on the unit circle. One explanation of this is that with $x = \cos\theta$ and $y = \sin\theta$, the trigonometric identity $\cos^2\theta + \sin^2\theta = 1$ shows that the equation $x^2 + y^2 = 1$ is satisfied (figure 4, left).

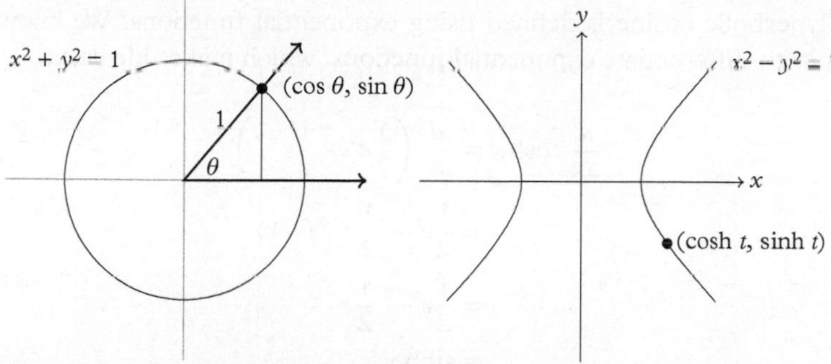

Figure 4 *(left) The coordinates of a point on the unit circle are* $(\cos\theta, \sin\theta)$*; (right) the coordinates of a point on the unit hyperbola are* $(\cosh t, \sinh t)$*, although t is not interpreted as an angle. This is the origin of the names circular functions and hyperbolic functions*

Similarly, points of the form $(\cosh t, \sinh t)$ lie on the *unit hyperbola* $x^2 - y^2 = 1$ (figure 4, right), which can be seen from the identity $\cosh^2 t - \sinh^2 t = 1$. When $x = \cosh t$ and $y = \sinh t$, the identity shows the equation $x^2 - y^2 = 1$ is satisfied.

Just as with the trigonometric functions, the notation $\cosh^2 t$ means the same as $(\cosh t)^2$.

Example 1 *Derive the hyperbolic identity* $\cosh^2 t - \sinh^2 t = 1$.

There are many trigonometric identities; likewise, there are many *hyperbolic identities*. Some can be proved using the definitions and algebra. Some are easier using previously proven identities.

Solution We use the definitions of hyperbolic cosine and sine as well as some algebra:

$$\begin{aligned}
\cosh^2 t - \sinh^2 t &= \left(\frac{e^t + e^{-t}}{2}\right)^2 - \left(\frac{e^t - e^{-t}}{2}\right)^2 \\
&= \frac{1}{4}\left(e^t + e^{-t}\right)^2 - \frac{1}{4}\left(e^t - e^{-t}\right)^2 \\
&= \frac{1}{4}\left(e^{2t} + 2e^t e^{-t} + e^{-2t}\right) - \frac{1}{4}\left(e^{2t} - 2e^t e^{-t} + e^{-2t}\right) \\
&= \frac{1}{4}\left(e^{2t} + 2 + e^{-2t}\right) - \frac{1}{4}\left(e^{2t} - 2 + e^{-2t}\right) \\
&= \frac{1}{4}e^{2t} + \frac{1}{2} + \frac{1}{4}e^{-2t} - \frac{1}{4}e^{2t} + \frac{1}{2} - \frac{1}{4}e^{-2t} \\
&= 1.
\end{aligned}$$

■

Derivatives of hyperbolic functions

Hyperbolic cosine is defined using exponential functions. We know how to differentiate exponential functions, which makes life easy:

Line 1 uses the definition of $\cosh x$; line 2 differentiates the exponential functions, using the chain rule as necessary; and line 4 uses the definition of $\sinh x$.

$$\begin{aligned}
\frac{d}{dx}\cosh x &= \frac{d}{dx}\left(\frac{1}{2}e^x + \frac{1}{2}e^{-x}\right) \\
&= \frac{1}{2}e^x + \frac{1}{2}e^{-x}(-1) \\
&= \frac{1}{2}e^x - \frac{1}{2}e^{-x} \\
&= \sinh x.
\end{aligned}$$

Surprised? Don't be. The similarities to the trigonometric functions are myriad and are continually helpful. In chapter 2 we found the derivative of secant by using the derivative of cosine and the quotient rule. We can do the same for hyperbolic secant:

Line 1 uses the definition of sech x, line 2 uses the quotient rule, line 3 simplifies and rearranges, and line 4 uses the definitions of sech x and $\tanh x$.

$$\begin{aligned}
\frac{d}{dx}\operatorname{sech} x &= \frac{d}{dx}\frac{1}{\cosh x} \\
&= \frac{\cosh x \cdot 0 - 1 \cdot \sinh x}{(\cosh x)^2} \\
&= \frac{-1}{\cosh x} \cdot \frac{\sinh x}{\cosh x} \\
&= -\operatorname{sech} x \, \tanh x.
\end{aligned}$$

The derivatives of the other hyperbolic functions are left to you as exercises.

DERIVATIVES OF HYPERBOLIC FUNCTIONS

$$\frac{d}{dx}\cosh x = \sinh x \qquad \frac{d}{dx}\operatorname{sech} x = -\operatorname{sech} x \tanh x$$

$$\frac{d}{dx}\sinh x = \cosh x \qquad \frac{d}{dx}\operatorname{csch} x = -\operatorname{csch} x \coth x$$

$$\frac{d}{dx}\tanh x = \operatorname{sech}^2 x \qquad \frac{d}{dx}\coth x = -\operatorname{csch}^2 x$$

Notice the similarities to the formulas for trig derivatives, with the only differences being that none of the formulas on the left have a negative sign whereas all the formulas on the right do. Armed with this knowledge you may already have the formulas memorized!

Calculating derivatives of hyperbolic functions is nearly identical to calculating derivatives of trigonometric functions.

Example 2 *Differentiate* $y = x^2 \sinh(4x + 3)$.

Solution This one's a cinch (sorry; I couldn't resist) using the derivative rule for $\sinh x$, the product rule, and the chain rule:

$$\begin{aligned} y' &= x^2 \cosh(4x+3) \cdot 4 + 2x \cdot \sinh(4x+3) \\ &= 4x^2 \cosh(4x+3) + 2x \sinh(4x+3). \end{aligned}$$ ■

Reading Exercise 28 *Find the derivative of* $y = \sinh(4x)$.

Integrals of hyperbolic functions

Reversing the derivative formulas results in antiderivative formulas.

INTEGRALS OF HYPERBOLIC FUNCTIONS

$$\int \sinh x\,dx = \cosh x + C \qquad \int \operatorname{sech} x \tanh x\,dx = -\operatorname{sech} x + C$$

$$\int \cosh x\,dx = \sinh x + C \qquad \int \operatorname{csch} x \coth x\,dx = -\operatorname{csch} x + C$$

$$\int \operatorname{sech}^2 x\,dx = \tanh x + C \qquad \int \operatorname{csch}^2 x\,dx = -\coth x + C$$

Calculating integrals of hyperbolic functions is nearly identical to calculating integrals of trigonometric functions.

Notice that ln 2 is a constant.

Example 3 *Evaluate* $\int 4\cosh(3x - \ln 2)\,dx$.

Solution Because we have hyperbolic cosine of "not just plain x," we use substitution:

$$u = 3x - \ln 2$$
$$du = 3\,dx.$$

Continuing,

Line 1 adjusts for the required constant 3, line 2 makes the substitution, line 3 uses the antiderivative formula for cosh, and line 4 returns to the original variable.

$$\begin{aligned}\int 4\cosh(3x - \ln 2)\,dx &= \frac{1}{3}\int 4\cosh(3x - \ln 2)\,3\,dx \\ &= \frac{1}{3}\int 4\cosh u\,du = \frac{4}{3}\int \cosh u\,du \\ &= \frac{4}{3}\sinh u + C \\ &= \frac{4}{3}\sinh(3x - \ln 2) + C.\end{aligned}$$

■

Reading Exercise 29 *Find* $\int \sinh(4x + 1)\,dx$.

Example 4 *Find* y' *for* $y = \ln(\cosh x)$.

Since $\cosh x > 0$ on $(-\infty, \infty)$, the domain of $\ln\cosh x$ is $(-\infty, \infty)$.

Solution Using the chain rule and the derivative formulas for the natural logarithmic function and hyperbolic cosine,

$$y' = \frac{1}{\cosh x}\cdot\sinh x = \tanh x.$$

■

Example 4 shows how to integrate the hyperbolic tangent function. Integrating hyperbolic cotangent is similar, although separate calculations are needed for $x > 0$ and for $x < 0$.

MORE INTEGRALS OF HYPERBOLIC FUNCTIONS

$$\int \tanh x\,dx = \ln(\cosh x) + C \qquad \int \coth x\,dx = \ln|\sinh x| + C$$

Inverse hyperbolic functions

Four of the hyperbolic functions are one-to-one; the ones that are not may be restricted to $[0, \infty)$ to be one-to-one. The inverse hyperbolic functions, one of which is graphed in figure 5, are defined using the pattern of definition 3.

Ans. to reading exercise 28
$y' = 4\cosh(4x)$

Definition 14 INVERSE HYPERBOLIC FUNCTIONS *The inverse hyperbolic functions are defined by*

$y = \sinh^{-1} x$ *if and only if* $\sinh y = x$,

$y = \tanh^{-1} x$ *if and only if* $\tanh y = x$,

$y = \coth^{-1} x$ *if and only if* $\coth y = x$,

$y = \operatorname{csch}^{-1} x$ *if and only if* $\operatorname{csch} y = x$,

$y = \cosh^{-1} x$ *if and only if* $\cosh y = x$ *and* $y \geq 0$, *and*

$y = \operatorname{sech}^{-1} x$ *if and only if* $\operatorname{sech} y = x$ *and* $y \geq 0$.

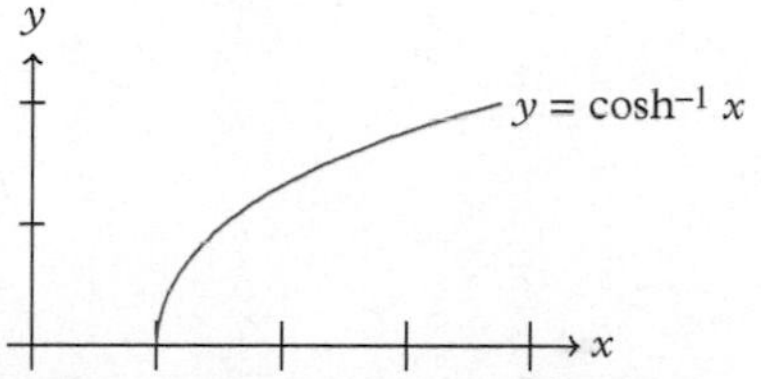

Figure 5 *The graph of* $y = \cosh^{-1} x$, *which is not a square root curve, despite the resemblance*

Because $\cosh x = \frac{e^x + e^{-x}}{2}$ and the inverse of $y = e^x$ is $y = \ln x$, it is possible to express $\cosh^{-1} x$ using logarithms as

$$\cosh^{-1} x = \ln\left(x + \sqrt{x^2 - 1}\right).$$

Similar formulas hold for the other inverse hyperbolic functions.

Derivatives of inverse hyperbolic functions

Another strategy is to differentiate the logarithmic formulas for the inverse hyperbolic functions. For instance,

$$\begin{aligned}\frac{d}{dx}\cosh^{-1}x &= \frac{d}{dx}\ln\left(x+\sqrt{x^2-1}\right)\\ &= \frac{1}{x+\sqrt{x^2-1}}\cdot\left(1+\frac{1}{2\sqrt{x^2-1}}\cdot 2x\right)\\ &= \frac{1+\frac{x}{\sqrt{x^2-1}}}{x+\sqrt{x^2-1}}\\ &= \frac{\sqrt{x^2-1}+x}{(x+\sqrt{x^2-1})\sqrt{x^2-1}}\\ &= \frac{1}{\sqrt{x^2-1}}.\end{aligned}$$

One strategy for differentiating inverse hyperbolic functions is the same as for differentiating inverse trigonometric functions: we use implicit differentiation and then find a way of expressing the result in terms of the variable x. For instance, if $y = \sinh^{-1} x$, then

$$\begin{aligned}\sinh y &= x\\ \cosh y\cdot\frac{dy}{dx} &= 1\\ \frac{dy}{dx} &= \frac{1}{\cosh y}.\end{aligned}$$

Although we cannot draw triangles to interpret $\cosh y$ geometrically, we do have an identity at our disposal. Using this identity and $\sinh y = x$ from the previous equation,

Line 1 states the hyperbolic identity from example 1, applied to the input y; line 2 substitutes x for $\sinh y$.

$$\begin{aligned}\cosh^2 y - \sinh^2 y &= 1\\ \cosh^2 y - x^2 &= 1\\ \cosh^2 y &= 1 + x^2\\ \cosh y &= \sqrt{1+x^2}.\end{aligned}$$

The positive square root is used because cosh is never negative. Using substitution, we conclude

$$\frac{dy}{dx} = \frac{1}{\sqrt{1+x^2}}.$$

This and other derivatives of inverse hyperbolic functions are summarized next.

Ans. to reading exercise 29
$\frac{1}{4}\cosh(4x+1)+C$

DERIVATIVES OF INVERSE HYPERBOLIC FUNCTIONS

$$\frac{d}{dx}\sinh^{-1}x = \frac{1}{\sqrt{1+x^2}} \qquad \frac{d}{dx}\operatorname{csch}^{-1}x = \frac{-1}{|x|\sqrt{1+x^2}}$$

$$\frac{d}{dx}\cosh^{-1}x = \frac{1}{\sqrt{x^2-1}} \qquad \frac{d}{dx}\operatorname{sech}^{-1}x = \frac{-1}{x\sqrt{1-x^2}}$$

$$\frac{d}{dx}\tanh^{-1}x = \frac{1}{1-x^2},\ |x|<1 \qquad \frac{d}{dx}\coth^{-1}x = \frac{1}{1-x^2},\ |x|>1$$

To see why $\tanh^{-1}$ and $\coth^{-1}$ have non overlapping domains, see figure 3.

Example 5 *Find y' for $y = \operatorname{csch}^{-1} 2^\theta$.*

Solution Using the derivative formulas for the inverse hyperbolic cosecant and general exponentials, as well as the chain rule, yields

$$y' = \frac{-1}{|2^\theta|\sqrt{1+(2^\theta)^2}} \cdot 2^\theta \ln 2 = \frac{-\ln 2}{\sqrt{1+2^{2\theta}}}.$$

The cancellation shown here works because 2^θ is always positive: $|2^\theta| = 2^\theta$. ■

Reading Exercise 30 *Differentiate $g(x) = 4\sinh^{-1} x$.*

Integrals with inverse hyperbolic functions

The two derivative formulas for inverse hyperbolic functions that are most useful to reverse and to use as antiderivative formulas are the ones for inverse hyperbolic sine and cosine.

ANTIDERIVATIVES GIVING INVERSE HYPERBOLICS

$$\int \frac{1}{\sqrt{1+x^2}}\,dx = \sinh^{-1} x + C$$

$$\int \frac{1}{\sqrt{x^2-1}}\,dx = \cosh^{-1} x + C$$

ANTIDERIVATIVES GIVING INVERSE HYPERBOLICS, VERSION 2

For $a > 0$,

$$\int \frac{1}{\sqrt{a^2+x^2}}\,dx = \sinh^{-1}\left(\frac{x}{a}\right) + C;$$

for $0 < a < x$,

$$\int \frac{1}{\sqrt{x^2-a^2}}\,dx = \cosh^{-1}\left(\frac{x}{a}\right) + C.$$

These antiderivative formulas are similar to integrals involving inverse trigonometric functions. The issues in recognizing and using the formulas are therefore the same.

Example 6 *Evaluate* $\displaystyle\int_0^{1/3} \frac{6}{\sqrt{1+9x^2}}\,dx.$

Solution The integrand is similar to $\frac{1}{\sqrt{1+x^2}}$, but with $9x^2$ instead of x^2. We therefore try substitution:

$$u = 3x \text{ (then } u^2 = 9x^2)$$

$$du = 3\,dx.$$

One method of determining the substitution to make is to write $u^2 = 9x^2$ first, then take square roots and try $u = 3x$.

If your first thought is to try the substitution $u = 1 + 9x^2$, you calculate $du = 18x\,dx$, and without an x in the numerator, you conclude the substitution fails.

Because we have a definite integral, we also need to change the limits of integration from values of x to values of u:

$$x = \frac{1}{3}: \quad u = 3\left(\frac{1}{3}\right) = 1$$

$$x = 0: \quad u = 3(0) = 0.$$

Then,

Line 1 adjusts for the constant 3 appearing in du; line 2 makes the substitution, including changing the limits of integration to values of u; line 3 uses the antiderivative giving the inverse hyperbolic sine formula; and line 5 uses the fact that $\sinh^{-1} 0 = 0$ (true, since $\sinh 0 = 0$).

$$\begin{aligned} \int_0^{1/3} \frac{6}{\sqrt{1+9x^2}}\,dx &= \int_0^{1/3} \frac{2 \cdot 3}{\sqrt{1+9x^2}}\,dx \\ &= \int_0^1 \frac{2}{\sqrt{1+u^2}}\,du \\ &= 2\sinh^{-1} u\Big|_0^1 \\ &= 2\sinh^{-1} 1 - 2\sinh^{-1} 0 \\ &= 2\sinh^{-1} 1 = 1.76275. \end{aligned}$$ ■

$\sinh^{-1} 0 = 0$

The TI-83/TI-84 calculators have hyperbolic and inverse hyperbolic functions listed in the "catalog" of functions. If you are using a calculator without the inverse hyperbolic functions, you may calculate values using the equivalent natural logarithmic formula instead.

If the integrand is $\frac{6}{\sqrt{1-9x^2}}$ instead, the formula to use is the antiderivative giving inverse sine. If the integrand is $\frac{6}{\sqrt{9x^2-1}}$, we use the antiderivative giving inverse hyperbolic cosine. These three formulas are similar, and the form of the expression determines which is to be used.

Example 7 *Evaluate* $\displaystyle\int_{e^2}^{e^3} \frac{1}{x\sqrt{(\ln x)^2 - 1}}\,dx.$

Solution We have "not just plain x" squared, which suggests we might try substitution:

Noticing that we have the square root of "not just plain x" suggests the substitution $u = (\ln x)^2 - 1$. Then, $du = 2\ln x \cdot \frac{1}{x}\,dx$, and, because there is no $\ln x$ in the numerator of the integrand, the substitution fails.

$$u = \ln x$$

$$du = \frac{1}{x}\,dx.$$

With the x in the denominator of the integrand, we are in luck; it appears the substitution will succeed. Next we calculate the new limits of integration:

Ans. to reading exercise 30

$$g'(x) = \frac{4}{\sqrt{1+x^2}}$$

$$x = e^3: \quad u = \ln e^3 = 3$$

$$x = e^2: \quad u = \ln e^2 = 2.$$

Then,

$$\int_{e^2}^{e^3} \frac{1}{x\sqrt{(\ln x)^2 - 1}}\,dx = \int_2^3 \frac{1}{\sqrt{u^2 - 1}}\,du$$

$$= \cosh^{-1} u \Big|_2^3$$

$$= \cosh^{-1} 3 - \cosh^{-1} 2 \approx 0.4458.$$ ■

Line 1 performs the substitution; line 2 uses the antiderivative giving the inverse hyperbolic cosine formula.

Another fact that can be useful is that since $\cosh 0 = 1$, $\cosh^{-1} 1 = 0$.

$\cosh^{-1} 1 = 0$

Reading Exercise 31 *Integrate* $\displaystyle\int \frac{1}{\sqrt{9 + x^2}}\,dx.$

Exercises 5.9

1–6. Rapid response: rewrite the expression using a hyperbolic function instead of an inverse hyperbolic function.

1. $y = \tanh^{-1} 0.3$
2. $t = \cosh^{-1} \frac{3}{2}$
3. $1.4 = \cosh^{-1} x$
4. $t + 4 = \sinh^{-1} 3u$
5. $y = \sinh^{-1}(x - 2)$
6. $\tanh^{-1}(0.3x^2) = \frac{1}{7}$

7–12. Rapid response: rewrite the expression using an inverse hyperbolic function instead of a hyperbolic function.

7. $y = \tanh 0.3$
8. $t = \cosh \frac{3}{2}$
9. $1.4 = \cosh x$
10. $t + 4 = \sinh 3u$
11. $y = \sinh(x - 2)$
12. $\tanh(0.3x^2) = \frac{1}{7}$

13–28. Differentiate the function.

13. $f(x) = 6\tanh(4x - 11)$
14. $g(x) = \sinh\left(x^2 - 6x + 4\right)$
15. $f(x) = 6\tanh^{-1}(4x - 11)$
16. $g(x) = \sinh^{-1}\left(x^2 - 6x + 4\right)$
17. $y = \dfrac{2^x}{\sinh x}$
18. $y = \sin^{-1} \sinh x$
19. $y = xe^{\cosh x}$
20. $f(x) = \ln\left(1 + \cosh\sqrt{x - 7}\right)$
21. $f(t) = \left(\sinh^{-1}(t^2 + 4)\right)^5$
22. $y = \cosh\left(\sin\left(e^{3x-4}\right)\right)$

23. $y = \cosh^{-1} \sqrt{x}$

24. $y = x \tanh^{-1}(2x+1)$

25. $x = t \operatorname{sech} t - t^2 \operatorname{csch} t$

26. $s(t) = \operatorname{csch}\left(\frac{2t+7}{t^2-1}\right)$

27. $y = x^{\tanh x}$

28. $y = \operatorname{sech}^{-1} \sqrt{x^3 - 8}$

29–66. Evaluate the integral or state that it cannot be done with the available antiderivative formulas, rules, and techniques.

Not all integrals in these exercises use formulas from this section.

29. $\displaystyle\int \frac{1}{\sqrt{1+(2x)^2}}\,dx$

30. $\displaystyle\int \frac{2x}{\sqrt{1-x^4}}\,dx$

31. $\displaystyle\int \frac{1}{\sqrt{1-(2x)^2}}\,dx$

32. $\displaystyle\int \frac{2x}{\sqrt{1+x^4}}\,dx$

33. $\displaystyle\int \frac{1}{\sqrt{(2x)^2-1}}\,dx$

34. $\displaystyle\int \frac{2x}{1+x^4}\,dx$

35. $\displaystyle\int \frac{1}{1+(2x)^2}\,dx$

36. $\displaystyle\int \frac{2x}{\sqrt{x^4-1}}\,dx$

37. $\displaystyle\int \frac{1}{\sqrt{1+2x}}\,dx$

38. $\displaystyle\int \frac{2x^3}{\sqrt{1-x^4}}\,dx$

39. $\displaystyle\int \frac{x}{\sqrt{1+(2x)^2}}\,dx$

40. $\displaystyle\int \frac{2x^2}{\sqrt{1-x^4}}\,dx$

41. $\displaystyle\int \frac{x+1}{\sqrt{1+(2x)^2}}\,dx$

42. $\displaystyle\int \frac{2}{\sqrt{1-x}}\,dx$

43. $\displaystyle\int \frac{x^2}{\sqrt{1+(2x)^2}}\,dx$

44. $\displaystyle\int \frac{2x^3+2x}{\sqrt{1-x^4}}\,dx$

45. $\displaystyle\int \left(\cosh x - \operatorname{sech}^2 x\right)\,dx$

46. $\displaystyle\int_0^1 (\sinh x + \operatorname{sech} x \tanh x)\,dx$

47. $\displaystyle\int \frac{5}{\sqrt{1+(2x+3)^2}}\,dx$

48. $\displaystyle\int \frac{\cosh x}{1+\sinh^2 x}\,dx$

49. $\displaystyle\int \frac{\cos x}{\sqrt{1+\sin^2 x}}\,dx$

50. $\displaystyle\int \frac{e^3}{\sqrt{16+x^2}}\,dx$

51. $\displaystyle\int \tanh(6x-8)\,dx$

52. $\displaystyle\int \frac{\cosh^{-1} x}{\sqrt{4x^2-4}}\,dx$

53. $\displaystyle\int_3^4 \frac{7}{3\sqrt{x^2-8}}\,dx$

54. $\displaystyle\int \frac{2}{\sqrt{(4x+1)^2-1}}\,dx$

55. $\displaystyle\int e^{\coth x} \operatorname{csch}^2 x\,dx$

56. $\displaystyle\int \frac{6}{(\cosh^{-1} 3x)\sqrt{9x^2-1}}\,dx$

57. $\displaystyle\int_{\sqrt{5}}^3 x\cosh(x^2-5)\,dx$

58. $\displaystyle\int_e^{e^2} \frac{\operatorname{csch}(\ln x)\coth(\ln x)}{x}\,dx$

59. $\int \frac{1}{\sqrt{9x^2 - 36}}\,dx$

60. $\int x \sinh x^2\,dx$

61. $\int \operatorname{csch} x\,dx$

62. $\int x^2 \sinh x^2\,dx$

63. $\int \frac{2^x}{\sqrt{4^x - 1}}\,dx$

64. $\int \frac{1}{x^{1/3}\sqrt{4 + x^{4/3}}}\,dx$

65. $\int \sinh(\ln x)\,dx$

66. $\int \frac{\cosh^2 x - \sinh^2 x}{\sqrt{1 - 4x^2}}\,dx$

67–74. Derive the hyperbolic identity.

Ans. to reading exercise 31
$\sinh^{-1}(\frac{x}{3}) + C$ (Hint: use a "version 2" formula with $a = 3$)

67. $\sinh 2x = 2 \sinh x \cosh x$

68. $\cosh^2 x = \frac{\cosh 2x + 1}{2}$

69. $\sinh^2 x = \frac{\cosh 2x - 1}{2}$

70. $\cosh^2 x + \sinh^2 x = \cosh 2x$

71. $1 - \tanh^2 x = \operatorname{sech}^2 x$

72. $1 + \operatorname{csch}^2 x = \coth^2 x$

73. $\sinh(x + y) = \sinh x \cosh y + \cosh x \sinh y$

74. $\cosh(x + y) = \cosh x \cosh y + \sinh x \sinh y$

75–80. Prove the derivative formula.

Hyperbolic identities may be helpful for these derivations.

75. $\frac{d}{dx} \sinh x = \cosh x$

76. $\frac{d}{dx} \operatorname{csch} x = -\operatorname{csch} x \coth x$

77. $\frac{d}{dx} \coth x = -\operatorname{csch}^2 x$

78. $\frac{d}{dx} \tanh x = \operatorname{sech}^2 x$

79. $\frac{d}{dx} \tanh^{-1} x = \frac{1}{1 - x^2}$

80. $\frac{d}{dx} \coth^{-1} x = \frac{1}{1 - x^2}$

81. Find the equation of the tangent line to $y = \sinh x$ at the point $(0, 0)$.

82. Find the equation of the tangent line to $y = \cosh^{-1} x$ at the point $(1, \cosh^{-1} 1)$.

83. Find the differential dy for $y = \frac{\sinh^{-1} x}{x}$.

84. Find the differential dy for $y = \tanh^{-1} \frac{1}{x}$.

85. Use the methods of section 3.4 to determine intervals of concavity and inflection points for $y = \sinh x$.

86. Use the methods of section 3.4 to determine intervals of concavity and inflection points for $y = \tanh x$.

87. Use figure 3 to determine the values of $\lim_{x\to-\infty} \tanh x$ and $\lim_{x\to\infty} \tanh x$. Write rules for rendering $\tanh \Omega$ and $\tanh(-\Omega)$ for an arbitrary positive infinite hyperreal Ω.

88. Use figure 1 to determine the values of $\lim_{x\to-\infty} \cosh x$ and $\lim_{x\to\infty} \cosh x$. Write rules for rendering $\cosh \Omega$ and $\cosh(-\Omega)$ for an arbitrary positive infinite hyperreal Ω.

89. (GSL) Given $\tanh x = \frac{4}{5}$, find the values of the other hyperbolic functions.

90. Prove the version 2 formula $\int \frac{1}{\sqrt{a^2+x^2}}\, dx = \sinh^{-1}\left(\frac{x}{a}\right) + C$ by differentiating the right-hand side to obtain the integrand.

91. Prove the version 2 formula $\int \frac{1}{\sqrt{x^2-a^2}}\, dx = \cosh^{-1}\left(\frac{x}{a}\right) + C$ by differentiating the right-hand side to obtain the integrand.

92. (compiled from GSL) The Gudermannian function, which is helpful in the Mercator projection in cartography, is defined by

$$\text{gd } x = \tan^{-1}(\sinh x).$$

Prove that

$$\frac{d}{dx}\text{gd } x = \text{sech } x.$$

This exercise is adapted from materials developed by Brian Faulkner.

93. The rate at which an ocean wave travels can be modeled by

$$c(h) = \sqrt{\frac{g\lambda}{2\pi}\tanh\left(\frac{2\pi h}{\lambda}\right)},$$

where c is the velocity of the wave in meters per second, λ is the wavelength of the wave in meters, h is the depth of the water in meters, and $g = 9.8\text{m/s}^2$ is the acceleration resulting from gravity at the earth's surface. (a) Verify that the units on each side of the equation are the same. (b) Find a simplified deep-water formula by finding $\lim_{h\to\infty} c(h)$. (c) Oceans are not infinitely deep. What does "deep" mean for using the formula you found in part (b)? Deep compared to what? (d) Find the equation of the tangent line to $f(x) = \tanh x$ at the point $(0, 0)$. (e) Use the result of part (d) to find a simplified shallow-water formula. (f) Using $\lambda = 8.5$ m, at what depth is the formula from part (e) no longer accurate to within 2%?

Comparing Rates of Growth

5.10

Some functions grow faster than others. In many applications, knowing how rates of growth compare is an important component of decision making. In this section we explore the growth of several types of functions.

Faster growth means higher-level numbers

Consider the growth of the functions $y = x$ and $y = x^2$, graphed in figure 1. Beginning at $x = 0$, as the x-coordinate increases, at first the graph of $y = x$ is higher than the graph of $y = x^2$. This changes at

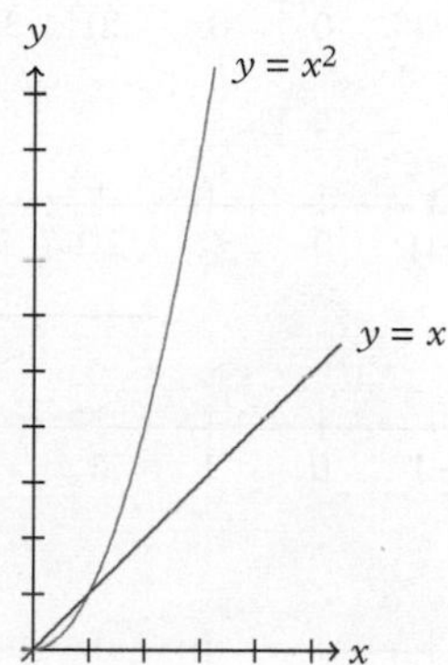

Figure 1 *The graph of $y = x$ (blue) and $y = x^2$ (green). The graph of $y = x^2$ is growing much faster than the graph of $y = x$*

$x = 1$, where the graph of $y = x^2$ catches up. Afterward, the graph of $y = x^2$ gets higher much more quickly; its values (y-coordinates) are growing faster. In fact, at a little past $x = 3$, the graph of $y = x^2$ surpasses the top of the figure, whereas the graph of $y = x$ lags well behind.

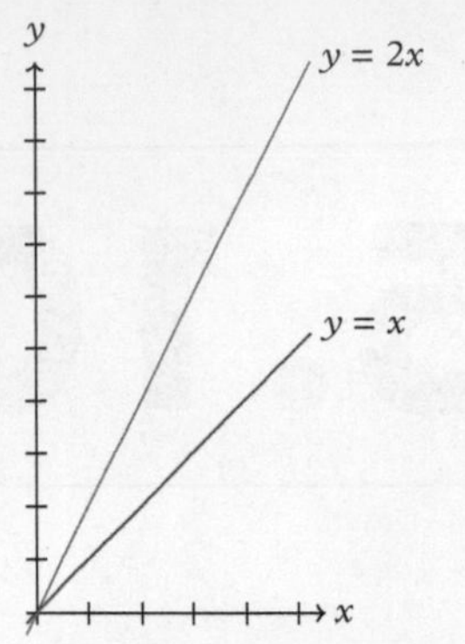

Figure 2 *The graph of $y = x$ (blue) and $y = 2x$ (green). For every $x > 0$, the graph of $y = 2x$ is twice as high as the graph of $y = x$.*

Numbers are compared by their ratio. Comparing 25 to 5, calculating $\frac{25}{5} = 5$, we note that 25 is five times as large as 5. Comparing 2Ω to Ω, we calculate $\frac{2\Omega}{\Omega} = 2$ and conclude that 2Ω is twice as large as Ω. Comparing Ω^2 to Ω, we calculate $\frac{\Omega^2}{\Omega} = \Omega$ and conclude that Ω^2 is infinitely larger than Ω.

The difference in growth between $y = x$ and $y = x^2$ is not just one of amount, but of type. The comparison between $y = x$ and $y = 2x$ (figure 2) is one of amount; both functions grow linearly (each graph is a line), but $y = 2x$ is always twice as high as $y = x$. Comparing $y = x^2$ to $y = x$, however, the multiple also keeps growing. Although $y = x^2$ is twice as high as $y = x$ when $x = 2$ (y-coordinate 4 vs. y-coordinate 2), $y = x^2$ is five times as high as $y = x$ when $x = 5$ (25 vs. 5), and a thousand times as high at $x = 1000$ (1 000 000 vs. merely 1000). The growth of $y = x^2$ is quadratic, which is a different type of growth than linear.

These comparisons can be stretched out infinitely far. When $x = \Omega$, the graph of $y = x^2$ is infinitely many times as high as the graph of $y = x$ (Ω^2 vs. the much smaller Ω). The growth is so different that at Ω, the

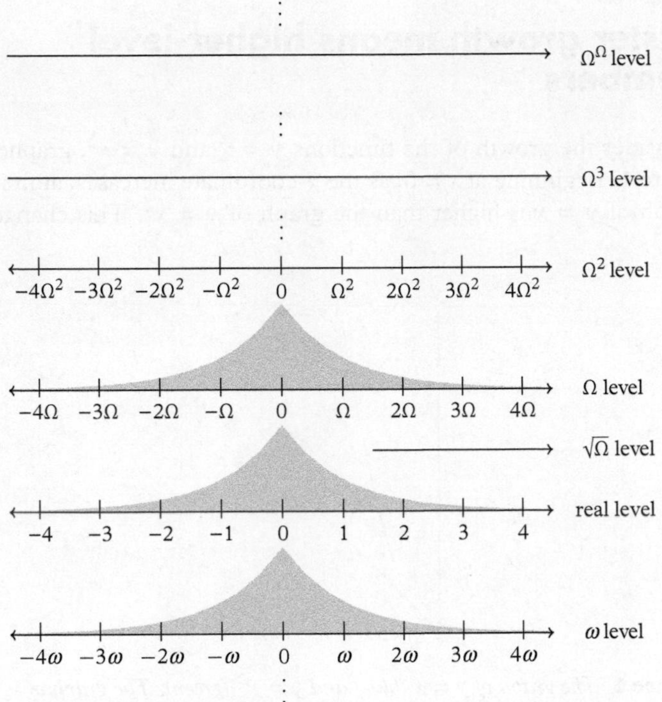

Figure 3 *A few of the infinitesimal and infinite levels of hyperreal numbers. Note that Ω and 2Ω appear on the same number line—namely, the Ω level. The number Ω^2 is not on the same number line as Ω; it is on a higher level*

y-coordinates are on different levels! Not so when comparing $y = 2x$ to $y = x$. At $x = \Omega$, the graph of $y = 2x$ is still just twice as high as the graph of $y = x$ (2Ω vs. Ω), and their values are on the same level.

Figure 3 is a repeat of a figure from chapter 1 that describes some of the different levels of hyperreals.

We are now ready to define what is meant when we say that one function grows faster than another.

Definition 15 FASTER GROWTH *Let $y = f(x)$ and $y = g(x)$ be two functions that are positive on (b, ∞). Then, f* **grows faster than** *g if $f(\Omega)$ is on a higher level than $g(\Omega)$ for an arbitrary positive infinite hyperreal Ω.*

This is equivalent to saying that f grows faster than g if $\lim_{x\to\infty} \frac{f(x)}{g(x)} = \infty$.

If $f(\Omega)$ and $g(\Omega)$ are on the same level, then we say that f and g *have the same type of growth.*

Equivalently, f and g have the same type of growth if $\lim_{x\to\infty} \frac{f(x)}{g(x)} = L$ for some nonzero real number L.

Example 1 *Which function grows faster than the other? (a) $f(x) = 3x^2 - 7$ vs. $g(x) = 6x^3 - 7x$ and (b) $p(x) = \sqrt{16x^6 - 3}$ vs. $q(x) = x^3$.*

Solution To determine which function has faster growth, we evaluate each function at an arbitrary positive infinite hyperreal Ω and check the levels of the results. The highest level indicates the fastest growth.

(a) We check f first:

$$f(\Omega) = 3\Omega^2 - 7 \approx 3\Omega^2,$$

which is on the Ω^2 level. Next we check g:

Recall that we may approximate when checking the level of a number, and that the name of the level is the name of the number line on which it resides. Any real-number multiple is not part of the name of the level.

$$g(\Omega) = 6\Omega^3 - 7\Omega \approx 6\Omega^3,$$

which is on the Ω^3 level. Because the Ω^3 level is the higher of the two levels, g grows faster than f.

We can also say that f grows slower than g.

(b) We check p first:

$$p(\Omega) = \sqrt{16\Omega^6 - 3} \approx \sqrt{16\Omega^6} = 4\Omega^3,$$

which is on the Ω^3 level. Next we check q:

$$q(\Omega) = \Omega^3,$$

which is also on the Ω^3 level. The functions p and q have the same type of growth. ■

Reading Exercise 32 *Which function grows faster than the other: $f(x) = 2975x^4 - x^2 + 2$ or $g(x) = 0.0003x^5 + 1$?*

By now you may have surmised an important conclusion: for polynomials, the higher the degree, the faster the growth. Cubic growth (degree 3 polynomial) is faster than quadratic growth (degree 2 polynomial), which is faster than linear growth, and so on.

> **POLYNOMIAL GROWTH RATES**
>
> A polynomial function p grows faster than a polynomial function q if and only if $p(x)$ has a higher degree than $q(x)$.

Faster growth means eventually higher values

Example 2 *Which function grows faster than the other:* $f(x) = 0.02x + 1$ *or* $g(x) = \sqrt{5x+1}$?

Solution We evaluate each function at an arbitrary positive infinite hyperreal Ω and check the level of the results. Checking f first,

$$f(\Omega) = 0.02\Omega + 1 \approx 0.02\Omega,$$

which is on the Ω level. Also,

$$g(\Omega) = \sqrt{5\Omega + 1} \approx \sqrt{5\Omega} = \sqrt{5}\sqrt{\Omega},$$

which is on the $\sqrt{\Omega}$ level (the $\Omega^{1/2}$ level). The higher of the two levels is Ω, so f grows faster. ■

Although in example 2 f grows faster and the two functions "start" tied at a height of 1 when $x = 0$, we can't conclude that the graph of f is always higher than the graph of g (figure 4). "Growing faster" tells us properties of the functions as $x \to \infty$, not necessarily as $0 \le x \le 10.5$. Because f eventually is infinitely many times as high as g (this is our definition of faster growth), f must eventually surpass g, then reach double the height of g, triple the height of g, and so on, but "eventually" might not occur until large values of x. Graph the two functions for sufficiently large values of x and we see f dominating g (figure 5).

The graphs in figures 4 and 5 demonstrate why decisions depend not just upon which grows faster, but also upon the relevant interval of x values. If f and g represent the amount in an investment account after x months, then you will not be alive when f surpasses g, so you might not care that f grows faster in the (very) long run.

Growth in investment accounts tends to be of a different type than the growth exhibited by these two functions. Nevertheless, this application demonstrates the idea that the timescale may override the importance of the type of growth.

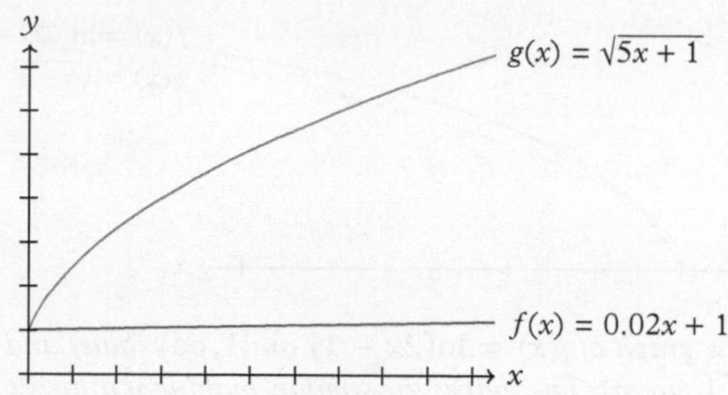

Figure 4 *The graph of $f(x) = 0.02x + 1$ (blue), which grows faster than the graph of $g(x) = \sqrt{5x+1}$ (green). Although g is higher for these values of x, f eventually surpasses g.*

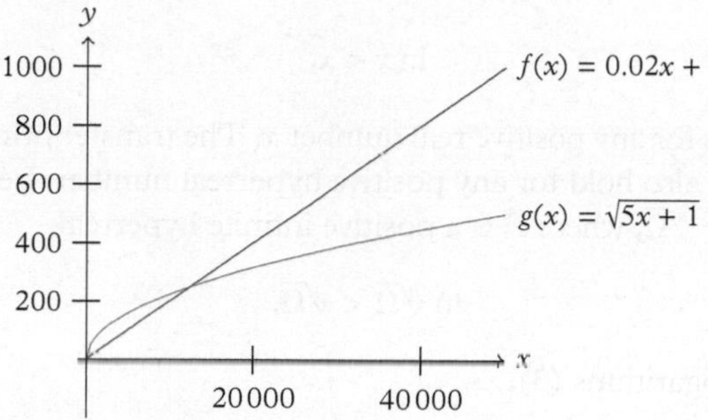

Figure 5 *The graph of $f(x) = 0.02x + 1$ (blue), which grows faster than the graph of $g(x) = \sqrt{5x+1}$ (green). This viewing window includes the points where f surpasses g and then reaches a height twice that of g.*

Logarithmic growth rates

Take another look at figure 1. Looking at the shape of the curves, it should make sense to you that quadratic growth is faster than linear growth. Similarly, the shape of the curves in figure 5 indicates that square root growth is slower than linear growth. But, what if, visually, it is difficult to tell the difference in the shapes? Although the shape of a logarithmic curve and a square root curve are different, it may be difficult to tell at first glance if one grows faster than the other (figure 6). To be able to answer questions about logarithmic growth, we need to know where $\ln \Omega$ fits in the scheme of different levels of hyperreal numbers. We tackle this next.

Ans. to reading exercise 32
g grows faster than f

Because $y = \ln x$ is concave down throughout its domain, the graph lies below its tangent line at $x = 1$, which has the equation $y = x - 1$ (figure 7). Therefore,

$$\ln x \leq x - 1.$$

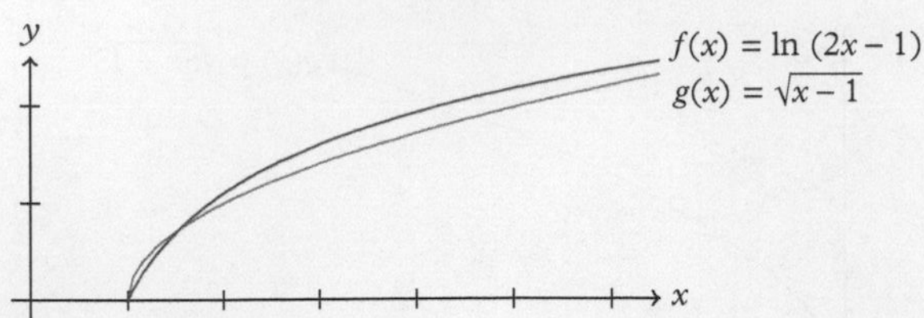

Figure 6 *The graph of $f(x) = \ln(2x-1)$ on $[1, \infty)$ (blue) and the graph of $g(x) = \sqrt{x-1}$ (green). The shapes are similar, making it difficult to tell which, if either, grows faster in the long run.*

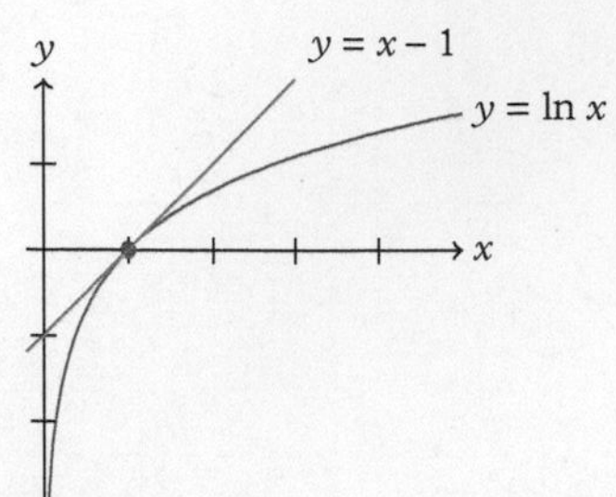

Figure 7 *The curve $y = \ln x$ (blue), which lies below the tangent line $y = x - 1$ (orange) at $x = 1$*

Since $x - 1 < x$, we also have

$$\ln x < x,$$

which holds for any positive real number x. The transfer principle says that it must also hold for any positive hyperreal number. Let's choose the number $\sqrt[n]{\Omega}$, where Ω is a positive infinite hyperreal:

$$\ln \sqrt[n]{\Omega} < \sqrt[n]{\Omega}.$$

By law of logarithms (3),

$$\ln \sqrt[n]{\Omega} = \frac{1}{n} \ln \Omega,$$

which is on the same level as $\ln \Omega$. Now we have

$$\frac{1}{n} \ln \Omega < \sqrt[n]{\Omega},$$

so that $\ln \Omega$ is on the same or a lower level than $\sqrt[n]{\Omega}$. Because this is true for any n (where n is a (real) natural number), $\ln \Omega$ must be on a lower level than any root of Ω, as depicted in figure 8.

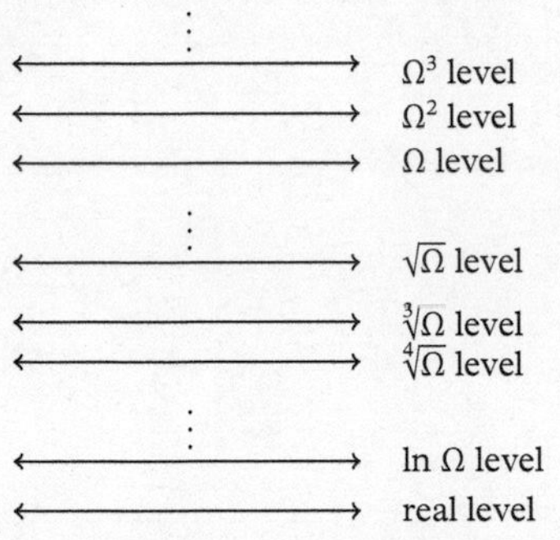

Figure 8 *Various levels of numbers, including $\ln \Omega$. The scales on the number lines and the shading have been removed in this figure. There are more levels between the ones pictured*

Example 3 *Which function grows faster than the other: $f(x) = \ln(2x-1)$ or $g(x) = \sqrt{x-1}$?*

Solution We evaluate each function at an arbitrary positive infinite hyperreal Ω and check the level of the results. Checking f first,

$$\begin{aligned} f(\Omega) &= \ln(2\Omega - 1) \\ &\approx \ln(2\Omega) \\ &= \ln 2 + \ln \Omega \\ &\approx \ln \Omega, \end{aligned}$$

Line 2 approximates using theorem 2; line 3 uses law of logarithms (1); and line 4 approximates again, using the fact that $\ln 2$ is a real number but $\ln \Omega$ is infinite.

which is on the $\ln\Omega$ level. Also,

$$g(\Omega) = \sqrt{\Omega - 1} \approx \sqrt{\Omega},$$

which is on the $\sqrt{\Omega}$ level. The higher of the two levels is $\sqrt{\Omega}$, so g grows faster. ■

Check figure 6 again. We see that g was above f at first, then fell below; but, because g grows faster, it must eventually rise above f again, for good.

LOGARITHMIC GROWTH

Logarithmic functions grow slower than root functions.

Despite the similarities in shape seen in figure 6, logarithmic growth is of a different, slower type than root function growth.

Many values are on the $\ln\Omega$ level. Using law of logarithms (3),

$$\ln\Omega^k = k\ln\Omega,$$

which is on the $\ln\Omega$ level for any real number k. Using the change of base property,

$$\log_a \Omega = \frac{\ln\Omega}{\ln a} = \frac{1}{\ln a}\cdot\ln\Omega,$$

which is on the $\ln\Omega$ level for any positive $a \neq 1$. To get an even lower level (but still infinite), we have to nest the natural logarithms; $\ln(\ln\Omega)$ is on a lower level than $\ln\Omega$, and $\ln(\ln(\ln\Omega))$ is on an even lower level. The situation is summarized in figure 9.

We know that $\ln\Omega$ is on a lower level than Ω for any positive infinite hyperreal Ω. Apply this fact to the positive infinite hyperreal $\ln\Omega$, and we conclude that $\ln(\ln\Omega)$ is on a lower level than $\ln\Omega$.

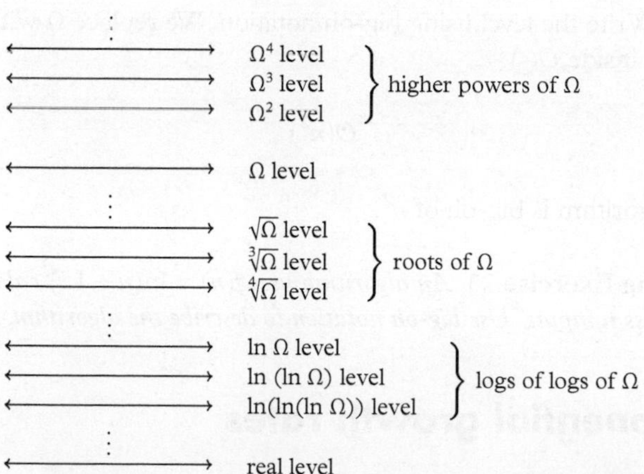

Figure 9 *Groups of levels of numbers are shown. Each wide arrow represents a number line at the stated level. There are more levels between the ones pictured*

Computer science: big-oh notation

Modern computing on large data sets with billions or trillions (or even more) of items has revolutionized society, and is the force behind many tasks that affect our daily lives. An *algorithm* (a step-by-step method for accomplishing a task) usually takes longer to perform if the number of input items n is larger. The number of calculations to be performed is often a function of the number of input items n, and the rate of growth of that function helps us understand the efficiency of the algorithm and therefore helps us choose between competing algorithms.

A notation commonly used to describe the rate of growth of an algorithm is *big-oh notation.* If an algorithm with n input items can be completed in $f(n)$ calculations and $f(\Omega)$ is on the $\ln\Omega$ level, then we say the algorithm is $O(\ln n)$, read "big-oh of natural log n." If $f(\Omega)$ is on the Ω^2 level, we say that the algorithm is $O(n^2)$. *Logarithm-time* algorithms are generally preferred to *polynomial-time* algorithms because the number of required calculations grows much more slowly as the number of inputs increases.

It is customary to write $O(\log n)$ instead of $O(\ln n)$. Also, big-oh notation means that the algorithm's growth is, at most, as fast as that type, and can possibly be a lower rate; think $\leq$.

Example 4 *An algorithm uses* $f(n) = \sqrt{12n^4 + 3n^2 + 28}$ *calculations to process n inputs. Use big-oh notation to describe the algorithm.*

Solution ❶ Determine the level of $f(\Omega)$. Using an arbitrary positive infinite hyperreal Ω,

$$f(\Omega) = \sqrt{12\Omega^4 + 3\Omega^2 + 28} \approx \sqrt{12\Omega^4} = \sqrt{12}\Omega^2,$$

which is on the Ω^2 level.

❷ Write the level using big-oh notation. We replace Ω with n and place it inside $O(\ )$:

$$O(n^2).$$

The algorithm is big-oh of n^2. ■

Reading Exercise 33 *An algorithm uses* $f(n) = \ln(n + 15)$ *calculations to process n inputs. Use big-oh notation to describe the algorithm.*

Exponential growth rates

Exponential growth was discussed in section 5.7. How does exponential growth compare to quadratic growth or other polynomial growth?

We can answer this question using our knowledge of logarithmic growth.

Let $a > 1$ be a real number. We know that $\log_a \Omega$ is on the same level as $\ln \Omega$, and therefore on a lower level than Ω. Then, for any real number k,

We are assuming that Ω is a positive infinite hyperreal.

$$k \log_a \Omega < \Omega.$$

The number $k \log_a \Omega$ is on the $\ln \Omega$ number line.

Using law of logarithms (3),

$$\log_a \Omega^k < \Omega.$$

Since $a > 1$, the function $y = a^x$ is increasing. Therefore, exponentiating both sides of the previous equation gives

$$a^{\log_a \Omega^k} < a^{\Omega},$$

and simplifying the left-hand side gives

$$\Omega^k < a^{\Omega}.$$

Then, for any base $a > 1$ and for any real-number exponent k, the number a^{Ω} is on at least as high a level as Ω^k. Because this works for every k, the number a^{Ω} must be on a higher level than every Ω^k.

EXPONENTIAL GROWTH

Exponential growth is faster than polynomial growth. For any $a > 1$, $y = a^x$ grows faster than any polynomial function $y = g(x)$.

Example 5 *Which function grows faster than the other:* $f(x) = 0.04e^x$ *or* $g(x) = 11x^{2877} + x^{119}$*?*

Solution We evaluate each function at an arbitrary positive infinite hyperreal Ω and check the level of the results. Checking f first,

$$f(\Omega) = 0.04e^{\Omega},$$

which is on the e^{Ω} level. Also,

$$g(\Omega) = 11\Omega^{2877} + \Omega^{199} \approx 11\Omega^{2877},$$

which is on the Ω^{2877} level. The higher of the two levels is e^{Ω}, so f grows faster. ■

How do two different exponentials compare? The larger the base, the higher the level. Since

Recall $\left(\frac{4}{3}\right)^{\Omega}$ renders infinity since $\frac{4}{3} > 1$, but $\left(\frac{3}{4}\right)^{\Omega} \doteq 0$ since $\frac{3}{4} < 1$.

$$\frac{4^{\Omega}}{3^{\Omega}} = \left(\frac{4}{3}\right)^{\Omega} \doteq \infty,$$

4^{Ω} must be on a higher level than 3^{Ω}.

EXPONENTIAL GROWTH: COMPARING BASES

The larger the base, the faster the exponential growth. If $1 < a < b$, then $y = b^x$ grows faster than $y = a^x$.

An again-expanded version of the diagram of different levels of hyperreals is presented in figure 10.

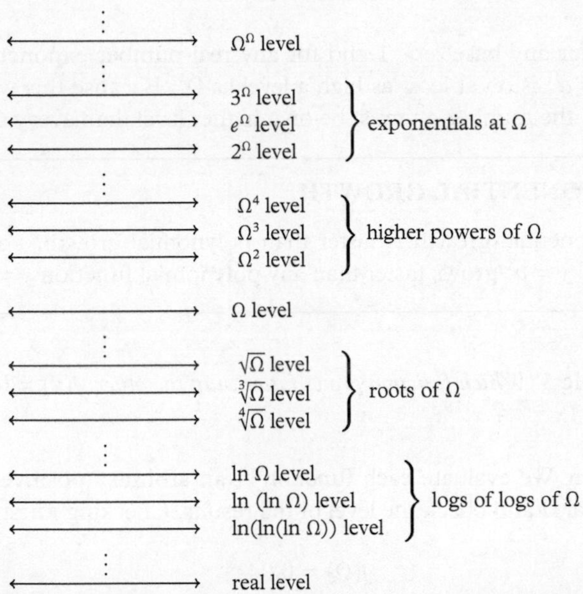

Figure 10 *Groups of levels of numbers are shown. Each wide arrow represents a number line at the stated level. There are more levels between the ones pictured*

Ans. to reading exercise 33
$O(\log n)$ (alternately $O(\ln n)$)

Example 6 *Which function grows faster than the other:* $f(x) = 2^x + x^{27}$ *or* $g(x) = 4e^x + x^3$*?*

Solution We evaluate each function at an arbitrary positive infinite hyperreal Ω and check the level of the results. Checking f first,

$$f(\Omega) = 2^{\Omega} + \Omega^{27}$$
$$\approx 2^{\Omega},$$

When approximating in line 2, we throw away the lower-level term Ω^{27} and keep the higher-level term 2^{Ω}.

which is on the 2^{Ω} level. Next,

$$g(\Omega) = 4e^{\Omega} + \Omega^3$$
$$\approx 4e^{\Omega},$$

When approximating in line 2, we throw away the lower-level term Ω^3 and keep the higher-level term $4e^{\Omega}$.

which is on the e^{Ω} level. Although both levels are exponentials at Ω, the larger the base, the higher the level. Therefore, e^{Ω} is the higher level, and g grows faster than f. ■

Reading Exercise 34 *Which function grows faster than the other: $f(x) = 4e^x + x^2$ or $g(x) = 3^x + 1$?*

How does $2^{3\Omega}$ or $e^{0.03\Omega}$ compare to other exponential-at-Ω levels? The multipliers 3 and 0.03 disguise the base of the exponential and can therefore be misleading regarding the level and rate of growth. For instance, using law of exponents (3), we see

$$2^{3\Omega} = (2^3)^{\Omega} = 8^{\Omega}$$

is on the 8^{Ω} level, whereas

$$e^{0.03\Omega} = (e^{0.03})^{\Omega}$$

is on the $(e^{0.03})^{\Omega}$ level, noting that $e^{0.03}$ is around 1.03045.

Because $1.03045 < e^{0.03}$, it is technically incorrect to say that $e^{0.03\Omega}$ is on the 1.03045^{Ω} level; the former is on a higher level than the latter.

Example 7 *Which function grows faster than the other: $f(x) = \sqrt{7} \cdot 3^x + 492x^{11}$ or $g(x) = 2^{2x} + \sqrt{x}$?*

Solution We evaluate each function at an arbitrary positive infinite hyperreal Ω and check the level of the results. Checking f first,

$$f(\Omega) = \sqrt{7} \cdot 3^{\Omega} + 492\Omega^{11}$$
$$\approx \sqrt{7} \cdot 3^{\Omega},$$

When approximating in line 2, we throw away the lower-level term $492\Omega^{11}$ and keep the higher-level term $\sqrt{7} \cdot 3^{\Omega}$.

which is on the 3^Ω level. Next,

Before approximating, we use law of exponents (3) to make it clear on which level each number exists, even if we already know which term is to be thrown away.

$$\begin{aligned} g(\Omega) &= 2^{2\Omega} + \sqrt{\Omega} \\ &= (2^2)^\Omega + \sqrt{\Omega} \\ &= 4^\Omega + \sqrt{\Omega} \\ &\approx 4^\Omega, \end{aligned}$$

which is on the 4^Ω level. The higher of the levels is 4^Ω, so g grows faster than f. ■

These examples also give insight into the power of larger interest rates.

Example 8 *Which function grows faster than the other:* $f(x) = 200e^{0.04x}$ *or* $g(x) = 82\,900e^{0.03x}$*?*

Solution We evaluate each function at an arbitrary positive infinite hyperreal Ω and check the level of the results. Checking f first,

Because there is no lower-level term to throw away, there is no need for an approximation step.

$$\begin{aligned} f(\Omega) &= 200e^{0.04\Omega} \\ &= 200(e^{0.04})^\Omega, \end{aligned}$$

which is on the $(e^{0.04})^\Omega$ level. Also,

$$\begin{aligned} g(\Omega) &= 82\,900e^{0.03\Omega} \\ &= 82\,900(e^{0.03})^\Omega, \end{aligned}$$

Details: $e^{0.04} = 1.04081$ (approximately), whereas $e^{0.03} = 1.03045$ (approximately).

which is on the $(e^{0.03})^\Omega$ level. Since $e^{0.04} > e^{0.03}$, the higher of the two levels is $(e^{0.04})^\Omega$ and f grows faster than g. ■

According to example 8, an account with principal \$200 earning 4% interest compounded continuously grows faster than an account with principal \$82 900 earning 3% interest compounded continuously. This is not a surprise; 4% is a faster rate of growth than 3%, and the principal is irrelevant to this determination. What might come as a surprise is that it grows faster with the meaning of this section. The amount in the account with the higher interest rate is eventually larger than the amount in the account with the smaller interest rate, then twice the amount, then 10 times the amount, and so on, regardless of the difference in principal between the two accounts.

The time frame is still relevant, though, to any decision making. It takes more than 600 years for the \$200-principal account with 4% interest to overtake the \$82 900-principal account with 3% interest. In practice, other factors such as risk may also be relevant.

EXPONENTIAL GROWTH: COMPARING RATES

When two functions grow exponentially with growth constants k_1 and k_2, the function with the larger growth constant grows faster. That is, if A and B are positive numbers and $k_1 > k_2$, then $y = Ae^{k_1 t}$ grows faster than $y = Be^{k_2 t}$.

The fun never ends: even more levels ...

Describing the different levels, and even the different types of levels, of hyperreals is a never-ending task. Levels between the ones in figure 10, above the ones in figure 10, and below them abound. Sometimes it is simple to discover where a number fits in the level scheme; at other times, it takes a little creativity to figure this out. But, learning the relative placement of the levels in figure 10 is well worth the effort.

Ans. to reading exercise 34
g grows faster than f

Exercises 5.10

1–6. Rapid response: which type of growth is faster?

1. linear or logarithmic
2. exponential or quadratic
3. logarithmic or exponential
4. polynomial or exponential
5. square root or linear
6. square root or logarithmic

7–14. Rapid response: which number is on a higher level?

7. e^{Ω} or Ω^3
8. $\sqrt{\Omega}$ or $\ln \Omega$
9. Ω or $\ln \Omega$
10. 3^{Ω} or 5^{Ω}
11. $3^{2\Omega}$ or $2^{3\Omega}$
12. 4^{Ω} or $2^{2\Omega}$
13. $\ln(\Omega)$ or $\ln\left((\Omega)^3\right)$
14. $\ln \Omega$ or $\ln(\ln \Omega)$

15–20. Rapid response: algorithms are described in big-oh notation. Which is preferred for a sufficiently large number of inputs n?

15. $O(n^2)$ or $O(n^3)$
16. $O(e^n)$ or $O(2^n)$
17. $O(n)$ or $O(\log n)$
18. $O(2^n)$ or $O(n^2)$

19. $O(n)$ or $O(\sqrt{n})$

20. $O(\log\log n)$ or $O(\log\log\log n)$

It is implied that the type of work shown in the examples of this section is required to complete the exercises.

21–36. Which function grows faster than the other?

21. $f(x) = x^2 + 7x - 9$, $g(x) = x^4 + x^2 + 1$

22. $f(x) = 0.02x^3 - 17x^2$, $g(x) = \sqrt{x+3}$

23. $f(x) = \sqrt{\dfrac{x^4 + 3x - 11}{x}}$, $g(x) = \dfrac{x^2+5}{\sqrt{x}}$

24. $f(x) = 3^x$, $g(x) = (x^{20} + 4x)^2$

25. $f(x) = 2^x + 87x^4$, $g(x) = 5^x + 12$

26. $f(x) = 2^x + 5^x$, $g(x) = 3^x + 4^x$

27. $f(x) = 4^{x+3}$, $g(x) = 5^{x+267}$

28. $f(x) = 2^{4x+1}$, $g(x) = 4^{2x+5}$

29. $f(x) = \sqrt[3]{8x+2}$, $g(x) = \ln 15x$

30. $f(x) = \log_5(7x + 18)$, $g(x) = \ln(x^3 + 5)$

31. $f(x) = \log_{17}(6x)$, $g(x) = \log_{11}(6x)$

32. $f(x) = (2^x + x)^2$, $g(x) = \sqrt{3^x + x^2}$

33. $f(x) = 10\,000e^{0.052x}$, $g(x) = 400e^{0.066x}$

34. $f(x) = \log_4(x^3 - x)$, $g(x) = \sqrt{x^{1.4} + x}$

35. $f(x) = \ln\sqrt{x}$, $g(x) = \sqrt{\ln x}$

36. $f(x) = x^x$, $g(x) = 57^{3x}$

37–44. An algorithm uses $f(n)$ calculations to process n inputs. Use big-oh notation to describe the algorithm.

37. $f(n) = 5n^2 + 6n + 11$

38. $f(n) = \sqrt{n^4 + 5n^3}$

39. $f(n) = 3 \cdot 2^n + 7n$

40. $f(n) = 2(3^n + 4) + n$

41. $f(n) = \ln(n^3 + 4n)$

42. $f(n) = 6n + \sqrt{n+1}$

43. $f(n) = \sqrt{n^2 + 6n - 1}$

44. $f(n) = \ln(7n + n^4)$

For example, for the number Ω^7, the answer is "between the Ω^4 level and the 2^Ω level." Some of these exercises are straightforward, but others require more creativity.

45–54. Using the levels depicted explicitly in figure 10, describe the location of the level of the given number.

45. 5^Ω

46. $\Omega^{1.7}$

47. $\Omega^{0.2}$

48. $\ln(\ln(\ln(\ln\Omega)))$

49. $3^{\ln\Omega}$

50. $\Omega^{\Omega+1}$

51. $\ln(\Omega^\Omega)$

52. $\Omega^2\ln\Omega$

53. $\Omega!$

54. $(2\Omega)!$

55. Consider the functions of figure 6—namely, $f(x) = \ln(2x - 1)$ and $g(x) = \sqrt{x - 1}$. Use technology to produce a graph similar to figure 5, showing g rising well above f.

The functions in figure 6 "start" at $x = 1$, so graph on an interval of the form $[1, b]$.

56. The function $f(t) = 400e^{0.05t}$ grows faster than the function $g(t) = 2500e^{0.03t}$. (a) Use technology to produce a graph similar to figure 5, showing f rising well above g. (b) Give an interpretation of financial accounts earning interest compounded continuously.

57. Which function grows faster: $f(x) = x \cdot 3^x$ or $g(x) = 3.2^x$?

Limits with Transcendental Functions: L'Hospital's Rule, Part I

5.11

We begin by reviewing some limit calculations and the idea of an indeterminate form.

Limits and indeterminate forms

We have learned that our first thought when evaluating a limit (at a real number, not at infinity) is to try evaluating using continuity. For instance,

$$\lim_{x\to 1}\frac{x+2}{x-4}=\frac{1+2}{1-4}=\frac{3}{-3}=-1$$

The form of this expression is $\frac{\text{non zero real number}}{\text{non zero real number}}$, which is a real number.

and

$$\lim_{x\to 1}\frac{x-1}{x-4}=\frac{0}{-3}=0.$$

The form of this expression is $\frac{0}{\text{non zero real number}}$, which is zero.

If, however, the denominator is zero, evaluating using continuity fails and we use infinitesimals. Trying

$$\lim_{x\to 1^+}\frac{x-4}{x-1}=\frac{-3}{0}$$

The form of this expression is $\frac{\text{non zero real number}}{0}$, which is always infinite.

gives an undefined expression, so we use infinitesimals instead. The limit is one-sided, and using $x=1+\omega$ gives

Recall that for two-sided limits, we use α; for one-sided limits, we use ω.

$$\lim_{x\to 1^+}\frac{x-4}{x-1}=\frac{1+\omega-4}{1+\omega-1}=\frac{-3+\omega}{\omega}\approx\frac{-3}{\omega}=-3\Omega\doteq-\infty.$$

Evaluating using continuity gives us the answer or nearly gives us the answer in each of the three previous limits. Even with $\frac{-3}{0}$, we can tell the answer is infinite. But, if we have $\frac{0}{0}$, we can't tell anything about the limit just from this information. No possible answer can be eliminated. It is an *indeterminate form*.

Trying to evaluate using continuity results in

$$\lim_{x\to1}\frac{x^2-1}{x-1}=\frac{1^2-1}{1-1}=\frac{0}{0}.$$

An example of a limit of the form $\frac{0}{0}$ with a value that is a real number is

$$\lim_{x\to1}\frac{x^2-1}{x-1}=\frac{(1+\alpha)^2-1}{1+\alpha-1}=\frac{1+2\alpha+\alpha^2-1}{\alpha}=\frac{2\alpha+\alpha^2}{\alpha}\approx\frac{2\alpha}{\alpha}=2.$$

Trying to evaluate using continuity results in

$$\lim_{x\to0^+}\frac{\sin x}{x^2}=\frac{\sin 0}{0^2}=\frac{0}{0}.$$

Although evaluating using continuity fails, it still provides information. We see the limit has the indeterminate form $\frac{0}{0}$.

An example of a limit of the form $\frac{0}{0}$ with a value that is infinite is

$$\lim_{x\to0^+}\frac{\sin x}{x^2}=\frac{\sin\omega}{\omega^2}\approx\frac{\omega}{\omega^2}=\frac{1}{\omega}=\Omega\doteq\infty.$$

Another indeterminate form is $\frac{\pm\infty}{\pm\infty}$. This information alone does not give us any clue regarding the value of the limit. It could be a real number, as with

$$\lim_{x\to\infty}\frac{x^3+3x+7}{2x^3+1}=\frac{\Omega^3+3\Omega+7}{2\Omega^3+1}\approx\frac{\Omega^3}{2\Omega^3}=\frac{1}{2},$$

The expressions $\frac{\Omega^3+3\Omega+7}{2\Omega^3+1}$ and $\frac{\Omega}{\sqrt{\Omega-1}}$ have a positive infinite number in both the numerator and the denominator, and are said to be of the form $\frac{\infty}{\infty}$.

or something infinite, as with

$$\lim_{x\to\infty}\frac{x}{\sqrt{x-1}}=\frac{\Omega}{\sqrt{\Omega-1}}\approx\frac{\Omega}{\sqrt{\Omega}}=\sqrt{\Omega}\doteq\infty.$$

Recall that EAR stands for evaluate, approximate, render.

The EAR procedure was sufficient to evaluate the previous five limits involving indeterminate forms, but what about $\lim_{x\to0}\frac{5^x-1}{x}$? Using continuity results in

$$\lim_{x\to0}\frac{5^x-1}{x}=\frac{5^0-1}{0}=\frac{0}{0},$$

$\frac{0}{0}$ is an indeterminate form.

which is indeterminate. Using $x=0+\alpha=\alpha$ begins with

$$\lim_{x\to0}\frac{5^x-1}{x}=\frac{5^\alpha-1}{\alpha},$$

CAUTION: MISTAKE TO AVOID
Although $\alpha\doteq0$, it cannot be approximated as zero; that is, $\alpha\not\approx0$. For this limit, trying it anyway violates the approximation principle since $5^0-1=1-1=0$.

but how do we approximate $5^\alpha-1$? Our current knowledge is insufficient to carry out the EAR procedure on this limit (although we discuss how to do this in chapter 10).

Similarly,

$$\lim_{x\to\infty}\frac{\ln x}{x}=\frac{\ln\Omega}{\Omega},$$

$\frac{\infty}{\infty}$ is an indeterminate form.

which is of the indeterminate form $\frac{\infty}{\infty}$. Knowing that $\ln\Omega$ is on a lower level than Ω means the quotient is infinitesimal, so it renders zero:

$$\lim_{x\to\infty}\frac{\ln x}{x}=\frac{\ln\Omega}{\Omega}\doteq0.$$

However, there are many, many different levels of hyperreals, and what if we cannot determine whether the numerator or the denominator is on the higher level?

A new rule that applies to indeterminate forms, including those that involve transcendental functions, can help.

L'Hospital's Rule

L'Hospital's rule helps us evaluate limits involving the indeterminate forms $\frac{0}{0}$ and $\frac{\pm\infty}{\pm\infty}$ through the use of derivatives.

An alternate spelling is L'Hôpital.

Theorem 11 L'HOSPITAL'S RULE *Suppose that both (1) f and g are differentiable and (2) $g'(x) \neq 0$ everywhere on an open interval containing b, except possibly at $x = b$. Suppose further that $\lim_{x\to b} \frac{f(x)}{g(x)}$ is of the indeterminate form $\frac{0}{0}$ or $\frac{\pm\infty}{\pm\infty}$. Then,*

$$\lim_{x\to b} \frac{f(x)}{g(x)} = \lim_{x\to b} \frac{f'(x)}{g'(x)}$$

provided this limit exists. The same applies to one-sided limits and limits at infinity.

L'Hospital's rule appeared in the very first calculus textbook, by the Marquis de L'Hospital, although it was actually discovered by Johann Bernoulli.

Guillaume François Antoine Marquis de L'Hospital 1661–1704
http://www-history.mcs.st-andrews.ac.uk/Biographies/De_LHopital.html

Johann Bernoulli 1667–1748
http://www-history.mcs.st-andrews.ac.uk/Biographics/Bernoulli_Johann.html

Proof. The full proof is beyond the scope of this book, but let's prove one special case, where $f(b) = g(b) = 0$, both f and g are differentiable at $x = b$, and $g'(b) \neq 0$. In this case, we have

If $f(b) = g(b) = 0$, then attempting to evaluate using continuity shows that $\lim_{x\to b} \frac{f(x)}{g(x)}$ has the indeterminate form $\frac{0}{0}$.

$$\begin{aligned} \lim_{x\to b} \frac{f(x)}{g(x)} &= \frac{f(b+\alpha)}{g(b+\alpha)} \\ &\approx \frac{f(b) + \alpha f'(b)}{g(b) + \alpha g'(b)} \\ &= \frac{\alpha f'(b)}{\alpha g'(b)} = \frac{f'(b)}{g'(b)} \\ &= \lim_{x\to b} \frac{f'(x)}{g'(x)}. \end{aligned}$$

Line 1 evaluates the limit using $x = b + \alpha$ for an arbitrary infinitesimal α, line 2 uses the local linearity theorem, line 3 uses the supposition that $f(b) = g(b) = 0$, and line 4 uses evaluation of the limit using continuity (backward).

The proof of the other, more difficult cases can be found in textbooks for upper-level courses in analysis. ■

This proof was brought to my attention by my colleague Troy Riggs.

The technical conditions in the hypotheses of the theorem are nearly always met as long as the calculation

L'Hospital's rule allows us to replace $f(x)$ with $f'(x)$ and replace $g(x)$ with $g'(x)$, which might lead to an easier limit to calculate.

$$\lim_{x\to b}\frac{f(x)}{g(x)} = \lim_{x\to b}\frac{f'(x)}{g'(x)}$$

can be carried out. Therefore, when encountering an indeterminate expression $\frac{0}{0}$ or $\frac{\pm\infty}{\pm\infty}$, we often immediately apply l'Hospital's rule by taking the derivatives of the numerator and the denominator within the limit expression.

Example 1 *Evaluate* $\lim\limits_{x\to 0}\dfrac{5^x - 1}{x}$.

Solution As always, the first step is to try evaluating the limit using continuity:

$$\lim_{x\to 0}\frac{5^x - 1}{x} = \frac{5^0 - 1}{0} = \frac{0}{0},$$

and evaluating using continuity fails (division by zero). But, this informs us that we have the indeterminate form $\frac{0}{0}$, hence we can apply l'Hospital's rule by taking the derivative of both the numerator and the denominator:

CAUTION: MISTAKE TO AVOID
Do not use the quotient rule. L'Hospital's rule says to take the derivatives of the numerator and denominator separately.

$$\lim_{x\to 0}\frac{5^x - 1}{x} = \lim_{x\to 0}\frac{\frac{d}{dx}(5^x - 1)}{\frac{d}{dx}(x)} = \lim_{x\to 0}\frac{5^x \ln 5}{1}.$$

Notice that we still have a limit to take after applying l'Hospital's rule. Using continuity on this limit gives

The entire calculation, from beginning to end, is written as

$$\lim_{x\to 0}\frac{5^x - 1}{x} = \lim_{x\to 0}\frac{5^x \ln 5}{1} = 5^0 \ln 5 = \ln 5.$$

$$= 5^0 \ln 5 = \ln 5.$$

The limit's value is ln 5. ■

As demonstrated by example 1, l'Hospital's rule expands our ability to calculate limits. The rule also works on limits that we can already calculate using the EAR procedure.

Example 2 *Evaluate* $\lim\limits_{x\to 1}\dfrac{x^2 - 1}{x - 1}$.

Solution Evaluating using continuity gives

$$\lim_{x\to 1}\frac{x^2 - 1}{x - 1} = \frac{0}{0}.$$

Because we have the indeterminate form $\frac{0}{0}$, we may use l'Hospital's rule:

In line 1, to apply l'Hospital's rule, inside the limit expression, we replace the numerator $x^2 - 1$ with its derivative $2x$ and the denominator $x - 1$ with its derivative 1. In line 2, we evaluate the resulting limit using continuity.

$$\lim_{x\to 1} \frac{x^2-1}{x-1} = \lim_{x\to 1} \frac{\frac{d}{dx}(x^2-1)}{\frac{d}{dx}(x-1)} = \lim_{x\to 1} \frac{2x}{1}$$
$$= \frac{2\cdot 1}{1} = 2,$$

which is the same solution we obtained using infinitesimals earlier in this section. ■

As in example 2, l'Hospital's rule can be used whenever it applies. It is not necessary (unless instructed otherwise) to use it only as a last resort.

Reading Exercise 35 *Find* $\lim_{x\to 1} \frac{\ln x}{x-1}$.

Theorem 11 states that l'Hospital's rule applies to limits at infinity as well.

Example 3 *Evaluate* $\lim_{x\to\infty} \frac{\ln x}{x}$.

Solution Using $x = \Omega$ for an arbitrary positive infinite hyperreal Ω gives

$$\lim_{x\to\infty} \frac{\ln x}{x} = \frac{\ln \Omega}{\Omega}.$$

The number in the numerator, $\ln \Omega$, renders ∞, as does the number in the denominator, Ω; the expression is of the form $\frac{\infty}{\infty}$. Therefore, we can go back to the beginning and use l'Hospital's rule:

To apply l'Hospital's rule, inside the limit expression, we replace the numerator $\ln x$ with its derivative $\frac{1}{x}$, and we replace the denominator x with its derivative 1. We then evaluate the resulting limit.

$$\lim_{x\to\infty} \frac{\ln x}{x} = \lim_{x\to\infty} \frac{\frac{d}{dx}\ln x}{\frac{d}{dx}x} = \lim_{x\to\infty} \frac{\frac{1}{x}}{1}$$
$$= \frac{1}{\Omega} = \omega \doteq 0.$$

■

After using l'Hospital's rule to obtain a different, hopefully easier limit, the new limit can be calculated using continuity, as was done in examples 1 and 2, or it can be evaluated using infinitesimals/hyperreals, as was done in example 3. We can even use l'Hospital's rule again.

Example 4 *Calculate* $\lim\limits_{x\to\infty} \dfrac{e^x}{-x^2}$.

Solution As with any limit at (positive) infinity, we begin by using $x = \Omega$ for an arbitrary positive infinite hyperreal Ω:

$$\lim_{x\to\infty} \frac{e^x}{-x^2} = \frac{e^\Omega}{-\Omega^2}.$$

We can finish evaluating this limit without using l'Hospital's rule. Because e^Ω is on a higher level than Ω^2, the expression is infinite. Because it is also negative, it renders $-\infty$:

$$\lim_{x\to\infty} \frac{e^x}{-x^2} = \frac{e^\Omega}{-\Omega^2} \doteq -\infty.$$

Because the numerator e^Ω is (positive) infinite and the denominator $-\Omega^2$ is (negative) infinite, the expression is of the indeterminate form $\frac{\infty}{-\infty}$. If we wish, we can use l'Hospital's rule:

Inside the limit expression, we replaced the numerator e^x with its derivative, and we replaced the denominator $-x^2$ with its derivative.

$$\lim_{x\to\infty} \frac{e^x}{-x^2} = \lim_{x\to\infty} \frac{e^x}{-2x}.$$

This new limit is also of the form $\frac{\infty}{-\infty}$, so we use l'Hospital's rule again:

$$= \lim_{x\to\infty} \frac{e^x}{-2}.$$

Now this limit is finished easily using $x = \Omega$:

This time, the limit is not of the form $\frac{\pm\infty}{\pm\infty}$, so l'Hospital's rule does not apply to this limit.

$$= \frac{e^\Omega}{-2} = -\frac{1}{2}e^\Omega \doteq -\infty. \quad \blacksquare$$

The entire calculation of example 4 is written compactly as

$$\lim_{x\to\infty} \frac{e^x}{-x^2} = \lim_{x\to\infty} \frac{e^x}{-2x} = \lim_{x\to\infty} \frac{e^x}{-2} = \frac{e^\Omega}{-2} \doteq -\infty.$$

When applying l'Hospital's rule, we keep the limit notation; when evaluating the limit using hyperreals or using continuity, we drop the limit notation.

The pattern in this calculation using l'Hospital's rule is not difficult to see, and l'Hospital's rule may be applied as many times as needed.

Example 5 *Evaluate* $\lim\limits_{x\to\infty} \dfrac{e^x}{x^{247} + 69x^{112} - 14x^{84} + 13}$.

Solution Recognizing that the expression is of the form $\frac{\pm\infty}{\pm\infty}$, we apply l'Hospital's rule 247 times:

$$\lim_{x\to\infty} \frac{e^x}{x^{247} + 69x^{112} - 14x^{84} + 13} = \cdots$$
$$= \lim_{x\to\infty} \frac{e^x}{\text{positive constant}} \doteq \infty. \quad \blacksquare$$

An alternate calculation not using l'Hospital's rule is

$$\lim_{x\to\infty} \frac{e^x}{x^{247} + 69x^{112} - 14x^{84} + 13}$$
$$= \frac{e^\Omega}{\Omega^{247} + 69\Omega^{112} - 14\Omega^{84} + 13}$$
$$\approx \frac{e^\Omega}{\Omega^{247}}$$
$$\doteq \infty,$$

because the numerator is on a higher level than the denominator (the e^Ω level vs. the Ω^{247} level).

Reading Exercise 36 *Find* $\lim_{x\to\infty} \frac{x^2}{e^x}$.

Example 5 illustrates an alternate way of showing that e^Ω is on a higher level than Ω^n for any natural number n. Since

$$\lim_{x\to\infty} \frac{e^x}{\text{polynomial}(x)} = \pm\infty$$

for any polynomial, the function $y = e^x$ grows faster than any polynomial function. In a similar manner, example 3 shows that $\ln\Omega$ is on a lower level than Ω.

USING LIMITS TO COMPARE LEVELS

If $f(\Omega)$ and $g(\Omega)$ are infinite and

$$\lim_{x\to\infty} \frac{f(x)}{g(x)} = 0,$$

then $f(\Omega)$ is on a lower level than $g(\Omega)$. If $f(\Omega)$ and $g(\Omega)$ are infinite and

$$\lim_{x\to\infty} \frac{f(x)}{g(x)} = \pm\infty,$$

then $f(\Omega)$ is on a higher level than $g(\Omega)$.

L'Hospital's rule is for indeterminate forms only

Theorem 11 only says that the l'Hospital's rule procedure applies when the limit has an indeterminate form $\frac{0}{0}$ or $\frac{\pm\infty}{\pm\infty}$. If the limit is not of one of those forms, l'Hospital's rule does not apply.

Ans. to reading exercise 35
1

Example 6 *Calculate* $\lim_{x\to\pi} \frac{\tan x}{x}$.

Solution Evaluating using continuity gives

$$\lim_{x\to\pi} \frac{\tan x}{x} = \frac{\tan\pi}{\pi} = \frac{0}{\pi} = 0.$$

The form of this expression is $\frac{0}{\text{non zero real number}}$, which is not indeterminate. L'Hospital's rule does not apply.

Evaluating using continuity succeeded, therefore l'Hospital's rule is not needed. ■

What if we try the l'Hospital's rule procedure on the limit of example 6, even though it doesn't apply? We get the wrong answer:

CAUTION: MISTAKE TO AVOID
L'Hospital's rule may give the wrong answer if the limit is not one of the indeterminate forms $\frac{0}{0}$ or $\frac{\pm\infty}{\pm\infty}$.

$$\lim_{x\to\pi} \frac{\tan x}{x} = \lim_{x\to\pi} \frac{\sec^2 x}{1} = (\sec\pi)^2 = (-1)^2 = 1. \quad \text{(It should be zero.)}$$

L'Hospital's rule might not be easier

Sometimes, l'Hospital's rule is not the easiest approach. Consider this limit from GSL:

$$\lim_{x\to 0^+} \frac{\ln\sin 2x}{\ln\sin x}.$$

It turns out that this limit is of the form $\frac{-\infty}{-\infty}$, so we can try l'Hospital's rule:

$$\lim_{x\to 0^+} \frac{\ln \sin 2x}{\ln \sin x} = \lim_{x\to 0^+} \frac{\frac{1}{\sin 2x}\cdot \cos 2x \cdot 2}{\frac{1}{\sin x}\cdot \cos x} = \lim_{x\to 0^+} \frac{2\cot 2x}{\cot x}.$$

It can be verified (and should be before using l'Hospital's rule again) that this limit is also of the form $\frac{\pm\infty}{\pm\infty}$. Another application of l'Hospital's rule gives

$$= \lim_{x\to 0^+} \frac{-4\csc^2 2x}{-\csc^2 x},$$

which once again is of the form $\frac{\pm\infty}{\pm\infty}$. Trying l'Hospital's rule again and again only gives expressions that evaluate to $\csc 0$ and $\cot 0$, which are undefined (infinite). So, let's back up to the very first application of l'Hospital's rule and use different simplifying algebra:

$$\lim_{x\to 0^+} \frac{\ln \sin 2x}{\ln \sin x} = \lim_{x\to 0^+} \frac{\frac{1}{\sin 2x}\cdot \cos 2x \cdot 2}{\frac{1}{\sin x}\cdot \cos x} = \lim_{x\to 0^+} \frac{2\sin x \cos 2x}{\sin 2x \cos x}.$$

This time we have the indeterminate form $\frac{0}{0}$, and using l'Hospital's rule for this expression is helpful:

$$= \lim_{x\to 0^+} \frac{2\sin x\cdot(-\sin 2x)\cdot 2 + \cos 2x\cdot 2\cos x}{\sin 2x\cdot(-\sin x) + \cos x\cdot \cos 2x\cdot 2} = \frac{0+2}{0+2} = 1.$$

Alternately, from the limit

$$\lim_{x\to 0^+} \frac{2\cot 2x}{\cot x},$$

the cotangent double-angle identity can be used, followed by additional simplification and evaluating using continuity:

$$\begin{aligned}\lim_{x\to 0^+} \frac{2\cot 2x}{\cot x} &= \lim_{x\to 0^+} \frac{2\cdot \frac{\cot^2 x - 1}{2\cot x}}{\cot x} \\ &= \lim_{x\to 0^+} \frac{\cot^2 x - 1}{\cot^2 x} = \lim_{x\to 0^+}\left(1 - \frac{1}{\cot^2 x}\right) \\ &= \lim_{x\to 0^+}(1 - \tan^2 x) = 1 - 0 = 1.\end{aligned}$$

Although all these calculations demonstrate the usefulness of trying different approaches, for this limit it is simpler to use the EAR procedure (use infinitesimals):

$$\lim_{x\to 0^+} \frac{\ln \sin 2x}{\ln \sin x} = \frac{\ln \sin 2\omega}{\ln \sin \omega} \approx \frac{\ln 2\omega}{\ln \omega} = \frac{\ln \omega + \ln 2}{\ln \omega} \approx \frac{\ln \omega}{\ln \omega} = 1.$$

Line 1 uses $x = 0 + \omega$ for a positive infinitesimal ω; for line 2, recall that $\sin \alpha \approx \alpha$ for any infinitesimal α; line 3 uses law of logarithms (1); and line 4 approximates by keeping the infinite term $\ln \omega$ (which renders $-\infty$) and by throwing away the real-number term $\ln 2$.

It is often not possible to tell by looking at a limit problem which approach is the easiest or quickest, so do not be surprised if you need to back up and try a different approach.

Ans. to reading exercise 36
0

Exercises 5.11

1–8. Rapid response: is the limit of an indeterminate form $\frac{0}{0}$ or $\frac{\pm\infty}{\pm\infty}$?

1. $\lim_{x\to 2} \frac{x^2-4}{x-2}$
2. $\lim_{x\to 1^+} \frac{\sqrt{x}-1}{\ln x}$
3. $\lim_{x\to -\infty} \frac{x^2-4}{x-2}$
4. $\lim_{x\to \infty} \frac{\sqrt{x}-1}{\ln x}$
5. $\lim_{x\to 0} \frac{x^2-4}{x-2}$
6. $\lim_{x\to 3} \frac{\sqrt{x}-1}{\ln x}$
7. $\lim_{x\to 0^+} \frac{\ln x}{\frac{1}{x}}$
8. $\lim_{x\to \infty} \frac{e^{-x}}{x}$

9–38. Evaluate the limit.

These limits are not necessarily of an indeterminate form. Although some of these limits do not require l'Hospital's rule, you should use l'Hospital's rule when possible for the purpose of practice (unless instructed otherwise).

9. $\lim_{x\to 0} \frac{1-\cos x}{4x}$
10. $\lim_{x\to 0} \frac{1-\cosh x}{x}$
11. $\lim_{x\to 2} \frac{x^2-4}{x^3-3x-2}$
12. $\lim_{x\to 2} \frac{2^x-4}{x-2}$
13. $\lim_{x\to -1} \frac{x+1}{x^2+1}$
14. $\lim_{x\to \infty} \frac{2^x-4}{x-2}$
15. $\lim_{x\to \infty} \frac{x+1}{x^2+1}$
16. $\lim_{x\to 0} \frac{\sinh x}{\sin x}$
17. $\lim_{x\to 0} \frac{x^2+1-\cosh x}{5x^2}$
18. $\lim_{x\to 0} \frac{e^x-1}{\cos x}$
19. $\lim_{x\to \infty} \frac{x^2}{e^{2x}}$
20. $\lim_{x\to 0} \frac{x-\sinh x}{3x^2}$

21. $\lim_{x\to 0} \frac{5^x - 3^x}{x}$

22. $\lim_{x\to -\infty} \frac{3^{-x}}{x^2}$

23. $\lim_{x\to 0} \frac{e^x - 1}{\sin x}$

24. $\lim_{x\to 0} \frac{\tan^{-1} x}{x}$

25. $\lim_{x\to -\infty} \frac{3x^2}{e^{2x}}$

26. $\lim_{x\to \infty} \frac{2x + \sqrt{x}}{\ln 2^x}$

27. $\lim_{x\to 0} \frac{\sin^{-1} x}{x}$

28. $\lim_{x\to 0} \frac{\sinh^{-1} x}{x^2}$

29. $\lim_{x\to \infty} \frac{\tanh^{-1} \frac{1}{x}}{e^{-x}}$

30. $\lim_{x\to 1^+} \frac{\ln x - 1}{\sqrt{x}}$

31. (GSL) $\lim_{x\to \infty} \frac{x + \ln x}{x \ln x}$

32. (GSL) $\lim_{x\to 3} \frac{\sqrt{3x} - \sqrt{12 - x}}{2x - 3\sqrt{19 - 5x}}$

33. (GSL) $\lim_{x\to 0^+} \frac{\cot x}{\ln x}$

34. (GSL) $\lim_{x\to 0} \frac{\cot 2x}{\cot 3x}$

35. (GSL) $\lim_{x\to \pi/2} \frac{\ln \sin x}{(\pi - 2x)^2}$

36. (GSL) $\lim_{x\to 0} \frac{e^x + \sin x - 1}{\ln(1 + x)}$

37. (GSL) $\lim_{x\to 2} \frac{\sqrt{16x - x^4} - 2\sqrt[3]{4x}}{2 - \sqrt[4]{2x^3}}$

38. (GSL) $\lim_{x\to 0^+} \frac{\ln x}{\csc x}$

39. Find all vertical asymptotes, if any, on the graph of $y = \frac{2^x - x - 1}{x^2 - x}$.

40. Find all vertical asymptotes, if any, on the graph of $y = \frac{\tan^{-1} x}{x^3 - x^2}$.

41. Find all horizontal asymptotes, if any, on the graph of $y = \dfrac{2^x + x + 1}{x^2 + x}$.

42. Find all horizontal asymptotes, if any, on the graph of $y = \dfrac{e^x + x^2}{e^x}$.

43. Show that $\ln \Omega$ is on a lower level than $\sqrt[10]{\Omega}$ by using l'Hospital's rule to calculate $\lim_{x \to \infty} \dfrac{\ln x}{\sqrt[10]{x}}$.

44. Show that $\cosh \Omega$ is on a higher level than Ω by using l'Hospital's rule to calculate $\lim_{x \to \infty} \dfrac{\cosh x}{x}$.

45. The very first calculus textbook was l'Hospital's *Analyse des infiniment petits, pour l'intelligence des lignes courbes,* published in 1696. The first example of the use of l'Hospital's rule (rewritten in modern notation) was to find

$$\lim_{x \to a} \frac{\sqrt{2a^3x - x^4} - a\sqrt[3]{a^2x}}{a - \sqrt[4]{ax^3}},$$

where a is a constant. Show that this limit is of the form $\frac{0}{0}$. Use l'Hospital's rule once, then evaluate using continuity and simplify to obtain the value $\frac{16a}{9}$.

An English translation of the title is *Analysis of the Infinitely Small for the Understanding of Curved Lines.*

46. A l'Hospital's rule problem by the 18th-century mathematician Leonhard Euler was to evaluate the limit

$$\lim_{x \to 1} \frac{x^x - x}{1 - x + \ln x}.$$

Please evaluate this limit.

L'Hospital's Rule, Part II: More Indeterminate Forms

5.12

L'Hospital's rule can be carried out for limits with the indeterminate forms $\frac{0}{0}$ or $\frac{\pm\infty}{\pm\infty}$. For limits with other indeterminate forms, the strategy is to rewrite the limit into a form to which l'Hospital's rule applies.

Indeterminate form $0 \cdot \infty$

Informally, the indeterminate form $0 \cdot \infty$ involves two competing ideas:

$$0 \cdot \text{anything} = 0 \text{ and}$$
$$\text{anything} \cdot \infty = \infty.$$

More precise versions of these ideas are that

$$0 \cdot k = 0$$

for any hyperreal k and

$$r \cdot \Omega \doteq \infty$$

for any positive real number r and any positive infinite hyperreal Ω.

But, in a limit in which we move infinitesimally away from zero and the infinite quantity is an infinite hyperreal, any result is possible for their product. Examples include

$$4\omega^2 \cdot \Omega^3 = 4\Omega \doteq \infty,$$
$$4\omega^3 \cdot \Omega^2 = 4\omega \doteq 0, \text{ and}$$
$$4\omega^2 \cdot \Omega^2 = 4.$$

The infinitesimals render zero and the positive infinite hyperreals render ∞, so these expressions are representative of limits of the indeterminate form $0 \cdot \infty$.

Because expressions of the form $0 \cdot \infty$ can yield anything, the form $0 \cdot \infty$ is indeterminate.

$0 \cdot \infty$ is an indeterminate form.

To apply l'Hospital's rule, we must have an expression of the form $\frac{0}{0}$ or $\frac{\pm\infty}{\pm\infty}$. These hypotheses of theorem 11 still apply even when we run across other indeterminate forms. But, we can usually rewrite an expression of the form $0 \cdot \infty$ into an expression to which l'Hospital's rule applies.

Example 1 *Evaluate* $\lim_{x\to-\infty} e^x x^2$.

Solution ❶ Try evaluating using continuity (for limits at real numbers) or using EAR (for limits at infinity). The limit is to $-\infty$, so we begin by using $x = -\Omega$ for an arbitrary positive infinite hyperreal Ω:

$$\lim_{x\to-\infty} e^x x^2 = e^{-\Omega}(-\Omega)^2 = e^{-\Omega}\Omega^2.$$

Since $e^{-\Omega} \doteq 0$ and $\Omega^2 \doteq \infty$, the expression is of the indeterminate form $0 \cdot \infty$.

❷ If the expression is of the indeterminate form $0 \cdot \infty$, we rewrite the expression so that l'Hospital's rule applies. Rewriting e^x as $\frac{1}{e^{-x}}$ does the trick:

This limit can be evaluated without the use of l'Hospital's rule, using the same algebraic trick as needed to apply the rule:

$$\lim_{x\to-\infty} e^x x^2 = e^{-\Omega}\Omega^2 = \frac{\Omega^2}{e^{\Omega}} \doteq 0,$$

with the last step justified by noting that Ω^2 is on a lower level than e^{Ω} and therefore the fraction is infinitesimal.

$$\lim_{x\to-\infty} e^x x^2 = \lim_{x\to-\infty} \frac{x^2}{e^{-x}}.$$

Now let's try evaluating the limit again:

$$\lim_{x\to-\infty} \frac{x^2}{e^{-x}} = \frac{(-\Omega)^2}{e^{-(-\Omega)}} = \frac{\Omega^2}{e^{\Omega}},$$

and we have the form $\frac{\infty}{\infty}$.

❸ We use l'Hospital's rule to evaluate the limit and it yields

$$\lim_{x\to-\infty} \frac{x^2}{e^{-x}} = \lim_{x\to-\infty} \frac{\frac{d}{dx}x^2}{\frac{d}{dx}e^{-x}} = \lim_{x\to-\infty} \frac{2x}{-e^{-x}},$$

which is of the form $\frac{-\infty}{-\infty}$. One more application of l'Hospital's rule yields

Because e^{Ω} is infinite, $\frac{2}{e^{\Omega}}$ is infinitesimal and renders zero.

$$= \lim_{x\to-\infty} \frac{\frac{d}{dx}(2x)}{\frac{d}{dx}(-e^{-x})} = \lim_{x\to-\infty} \frac{2}{e^{-x}} = \frac{2}{e^{\Omega}} \doteq 0$$

and we have succeeded in evaluating the limit. ■

Although the calculation of the solution to example 1 can be shortened to

$$\lim_{x\to-\infty} e^x x^2 = \lim_{x\to-\infty} \frac{x^2}{e^{-x}} = \lim_{x\to-\infty} \frac{2x}{-e^{-x}} = \lim_{x\to-\infty} \frac{2}{e^{-x}} \doteq 0,$$

care should be taken at each step to ensure that l'Hospital's rule is used only when the limit is in an appropriate indeterminate form, and it is easy to mistake the form of such an expression.

What if, instead of rewriting e^x as $\frac{1}{e^{-x}}$, we choose to rewrite x^2 as $\frac{1}{x^{-2}}$? Then, instead of the form $\frac{\infty}{\infty}$, the resulting expression is of the form $\frac{0}{0}$:

$$\lim_{x\to-\infty} e^x x^2 = \lim_{x\to-\infty} \frac{e^x}{x^{-2}} = \frac{e^{-\Omega}}{(-\Omega)^{-2}} = \frac{e^{-\Omega}}{\omega^2},$$

Recall that $\Omega^{-2} = \frac{1}{\Omega^2} = \omega^2$.

and, as advertised, both the numerator and the denominator render zero. However, this time, repeated applications of l'Hospital's rule continue to yield the form $\frac{0}{0}$ and are not helpful for evaluating the limit:

$$\lim_{x\to-\infty} e^x x^2 = \lim_{x\to-\infty} \frac{e^x}{x^{-2}} = \lim_{x\to-\infty} \frac{e^x}{-2x^{-3}} = \lim_{x\to-\infty} \frac{e^x}{6x^{-4}} = \cdots.$$

There are usually two options for rewriting the indeterminate form $0 \cdot \infty$ to make use of l'Hospital's rule. If one option is not fruitful, try the other.

If rewriting a limit of the form $0 \cdot \infty$ to the form $\frac{0}{0}$ is not helpful, try rewriting to the form $\frac{\pm\infty}{\pm\infty}$, and vice versa.

Reading Exercise 37 *Find* $\lim_{x\to\infty} xe^{-x}$.

Indeterminate form $\infty - \infty$

If we stop midstream to consider the form of the limit, we can see that

$$\lim_{x\to\infty} (2x - x) = 2\Omega - \Omega$$

is of the form $\infty - \infty$. This form is indeterminate because any result is possible. Examples include

$\infty - \infty$ is an indeterminate form.

$$\lim_{x\to\infty} (2x - x) = 2\Omega - \Omega = \Omega \doteq \infty,$$

$$\lim_{x\to\infty} (x - 2x) = \Omega - 2\Omega = -\Omega \doteq -\infty, \text{ and}$$

$$\lim_{x\to\infty} ((x + 3) - x) = (\Omega + 3) - \Omega = 3.$$

These limits are simple to evaluate using the EAR procedure, but for more complicated limits of the form $\infty - \infty$, a strategy that is often fruitful is to simplify and/or rewrite the expression to obtain a form to which l'Hospital's rule applies, and then continue as before.

Example 2 *Evaluate* $\lim_{x\to 0}(\csc x - \cot x)$.

Solution ❶ Try evaluating using continuity (for limits at real numbers) or using EAR (for limits at infinity). Because the limit is to zero, we try evaluating using continuity:

$$\lim_{x\to 0}(\csc x - \cot x) = \csc 0 - \cot 0,$$

which is undefined; both $\csc 0$ and $\cot 0$ are undefined. You may recall that the graphs of both cosecant and cotangent have vertical asymptotes at $x = 0$, indicating the possibility of the form $\infty - \infty$.

From the right, the form is $\infty - \infty$; from the left, the form is $-\infty - (-\infty)$, which amounts to the same thing.

❷ If the expression is of the indeterminate form $\infty - \infty$, simplify and/or rewrite the expression so that l'Hospital's rule applies. Using the familiar ratio and reciprocal identities, we have

Sometimes, simplifying and/or rewriting helps the EAR procedure succeed, or even allows evaluating using continuity to succeed.

$$\lim_{x\to 0}(\csc x - \cot x) = \lim_{x\to 0}\left(\frac{1}{\sin x} - \frac{\cos x}{\sin x}\right) = \lim_{x\to 0}\frac{1-\cos x}{\sin x}.$$

Trying to evaluate using continuity shows that the limit is now of the form $\frac{0}{0}$ and we can use l'Hospital's rule.

Rewriting tan, cot, sec, and csc in terms of sin and cos should now be a familiar strategy.

❸ Use l'Hospital's rule to evaluate the limit. L'Hospital's rule yields

$$\lim_{x\to 0}\frac{1-\cos x}{\sin x} = \lim_{x\to 0}\frac{\frac{d}{dx}(1-\cos x)}{\frac{d}{dx}\sin x} = \lim_{x\to 0}\frac{\sin x}{\cos x}$$

$$= \frac{\sin 0}{\cos 0} = \frac{0}{1} = 0.$$

We conclude that $\lim_{x\to 0}(\csc x - \cot x) = 0$. ■

Trying the EAR procedure for the limit of example 2 gives

$$\lim_{x\to 0}(\csc x - \cot x) = \csc \alpha - \cot \alpha$$

$$= \frac{1}{\sin \alpha} - \frac{\cos \alpha}{\sin \alpha}.$$

Using $\sin \alpha \approx \alpha$ and $\cos \alpha \approx 1$ gives

$$\approx \frac{1}{\alpha} - \frac{1}{\alpha} = 0,$$

which violates the approximation principle because we approximated and obtained zero. The answer is correct only by coincidence. Fixing this difficulty without using l'Hospital's rule is possible, but requires knowledge from chapter 10.

Notice that in step ❸ of the solution to example 2 we used a common denominator to transform the difference $\frac{1}{\sin x} - \frac{\cos x}{\sin x}$ into the quotient $\frac{1-\cos x}{\sin x}$. Such algebra is common to many exercises involving the indeterminate form $\infty - \infty$.

Reading Exercise 38 *Find* $\lim_{x\to \pi/2}(\sec x - \tan x)$.

Indeterminate forms 0^0, ∞^0, and 1^∞

For the form 0^0, we have competing familiar ideas:

$$\text{anything}^0 = 1 \text{ and}$$
$$0^{\text{anything}} = 0.$$

Which is it? Once again, we have an indeterminate form.

Powers can be tricky, but one strategy we have used previously is to take logarithms and make use of law of logarithms (3) to turn the exponentiation into a product, such as with logarithmic differentiation. To use this strategy, we need a theorem.

Ans. to reading exercise 37
0

0^0 is an indeterminate form.

Theorem 12 A STRATEGY FOR LIMITS OF POWERS

If

$$\lim_{x\to b} \ln f(x) = L$$

for a real number L, then

$$\lim_{x\to b} f(x) = e^L.$$

The same is true for one-sided limits and limits at infinity.

An analogous statement for L being ∞ or $-\infty$ is also true.

Proof. Suppose $\lim_{x\to b} \ln f(x) = L$. Then, for any infinitesimal α,

$$\ln f(b+\alpha) \doteq L.$$

Since $y = e^x$ is a continuous function,

$$e^{\ln f(b+\alpha)} \doteq e^L.$$

But,

$$e^{\ln f(b+\alpha)} = f(b+\alpha),$$

and by substitution

$$f(b+\alpha) \doteq e^L.$$

Details for the curious: any number that renders L must be of the form $L + \beta$, where β is either infinitesimal or zero. Then, using $\ln f(b+\alpha) = L + \beta$,

$$f(b+\alpha) = e^{\ln f(b+\alpha)} = e^{L+\beta} \doteq e^L,$$

with the last step accomplished using the definition of continuity.

Because this formula holds for any infinitesimal α, then by definition of limit we have

$$\lim_{x\to b} f(x) = e^L.$$

The proofs for one-sided limits and limits at infinity are similar. ■

According to theorem 12, if we can find $\lim_{x\to b} \ln f(x)$ (perhaps with the help of l'Hospital's rule), then we can find $\lim_{x\to b} f(x)$.

Example 3 *Evaluate* $\lim_{x\to 0^+} x^x$.

CAUTION: MISTAKE TO AVOID
Failure to recognize that 0^0 is indeterminate could result in the wrong answer. Assuming either $0^0 = 0$ or $0^0 = 1$ is incorrect.

Solution Trying to evaluate using continuity yields

$$\lim_{x\to 0^+} x^x = 0^0,$$

which is indeterminate. We therefore follow the strategy for limits of powers.

Following the strategy of theorem 12, we calculate $\lim_{x\to b} \ln f(x)$ to find the value of L.

❶ Take the log of the expression and simplify. We begin this step by inserting ln and then using law of logarithms (3):

$$\lim_{x\to 0^+} \ln x^x = \lim_{x\to 0^+} x \ln x.$$

Ans. to reading exercise 38
0

❷ Evaluate the resulting limit. Evaluating using continuity still fails because $\ln 0$ is undefined. Using infinitesimals (with $x = 0 + \omega$ because the limit is one-sided) gives

$$= \omega \ln \omega,$$

Note that the infinitesimal ω renders zero and $\ln \omega$ renders $-\infty$.

which is of the indeterminate form $0 \cdot (-\infty)$. How do we handle this indeterminate form? See example 1. Following the steps of that example, we rewrite the limit to get the form $\frac{0}{0}$ or $\frac{\pm\infty}{\pm\infty}$:

After rewriting the limit, we check the form to see if we can use l'Hospital's rule. Checking the form may require starting the EAR procedure.

$$\lim_{x\to 0^+} x \ln x = \lim_{x\to 0^+} \frac{\ln x}{\frac{1}{x}}$$

$$= \frac{\ln \omega}{\frac{1}{\omega}} = \frac{\ln \omega}{\Omega},$$

which is of the form $\frac{-\infty}{\infty}$. Now we can use l'Hospital's rule and finish evaluating the limit:

$$\lim_{x\to0^+}\frac{\ln x}{\frac{1}{x}} = \lim_{x\to0^+}\frac{\frac{d}{dx}\ln x}{\frac{d}{dx}\frac{1}{x}} = \lim_{x\to0^+}\frac{\frac{1}{x}}{-\frac{1}{x^2}}$$

$$= \lim_{x\to0^+}\frac{-x^2}{x} = \lim_{x\to0^+}(-x)$$

$$= 0.$$

Line 1 uses l'Hospital's rule, line 2 simplifies algebraically, and line 3 uses continuity to evaluate the limit.

We have now calculated the value of L.

❸ Use theorem 12 to calculate the value of the original limit. By step ❷, $L = 0$. Therefore,

$$\lim_{x\to0^+} x^x = e^0 = 1. \blacksquare$$

We began using theorem 12 by calculating $\lim_{x\to b}\ln f(x)$, with a value that is called L. We finish using theorem 12 by writing the original limit $\lim_{x\to b} f(x) = e^L$, using the value of L we just calculated.

Reading Exercise 39 *Find* $\lim_{x\to0^+} x^{2x}$.

With a little cleverness, the limit of example 3 can be evaluated using the EAR procedure:

$$\lim_{x\to0^+} x^x = \omega^\omega$$

$$= e^{\omega\ln\omega}$$

$$= e^{\frac{1}{\Omega}\ln\Omega^{-1}}$$

$$= e^{-\frac{\ln\Omega}{\Omega}}$$

$$= e^{\text{infinitesimal}}$$

$$\doteq e^0 = 1.$$

Line 1 uses $x = 0 + \omega$ to start the EAR procedure, line 2 uses definition 6, line 4 uses law of exponents (3), line 5 uses the fact that $\ln\Omega$ is on a lower level than Ω, and line 6 uses the continuity of $y = e^x$.

Changing to another indeterminate form, why is ∞^0 indeterminate? After all, for any positive infinite hyperreal Ω,

$$\Omega^0 = 1.$$

However, if the exponent is not exactly zero but is infinitesimal instead, as in most limit calculations, we can obtain other answers:

$$\left(e^\Omega\right)^\omega = e^1 = e;$$

but, because $e^\Omega \doteq \infty$ and $\omega \doteq 0$, the expression on the left is of the form ∞^0. Because we can obtain different answers with this form, the form ∞^0 is indeterminate.

∞^0 is an indeterminate form.

Why not ask for the two-sided limit? If we allowed $x < 0$, then $\cot x < 0$, and $\ln(\cot x)$ is not defined.

Because the indeterminate form ∞^0 is a power expression, the strategy uses theorem 12, just as in example 3.

Instead of recalling the graph, we could begin the EAR procedure to check the form:

$$\lim_{x\to 0^+} (\cot x)^{\sin x} = (\cot \omega)^{\sin \omega}.$$

Since

$$\cot \omega = \frac{\cos \omega}{\sin \omega} \approx \frac{1}{\omega} = \Omega \doteq \infty$$

and

$$\sin \omega \approx \omega \doteq 0,$$

the form of the expression is ∞^0.

Following the strategy of theorem 12, we calculate $\lim_{x\to b} \ln f(x)$ to find the value of L.

Example 4 *Calculate* $\lim_{x\to 0^+} (\cot x)^{\sin x}$.

Solution Trying to evaluate using continuity yields

$$\lim_{x\to 0^+} (\cot x)^{\sin x} = (\cot 0)^{\sin 0},$$

but $\cot 0$ is undefined. Recalling that there is a vertical asymptote on the graph of $y = \cot x$ at $x = 0$ and that $\sin 0 = 0$, we see that the form of the expression is ∞^0, which is indeterminate. We therefore follow the strategy for limits of powers.

❶ Take the log of the expression and simplify. We have

$$\lim_{x\to 0^+} \ln\left((\cot x)^{\sin x}\right) = \lim_{x\to 0^+} (\sin x)\ln(\cot x).$$

❷ Evaluate the resulting limit. As before, evaluating using continuity fails because $\cot 0$ is undefined. Beginning the EAR procedure, we use $x = 0 + \omega$ and obtain

$$\lim_{x\to 0^+} (\sin x)\ln(\cot x) = (\sin \omega)\ln(\cot \omega).$$

Because $\cot \omega$ is positive infinite, so is its natural log; we also know that $\sin \omega \approx \omega \doteq 0$. We therefore have the indeterminate form $0 \cdot \infty$. Handling this form requires rewriting the limit to get the form $\frac{0}{0}$ or $\frac{\pm\infty}{\pm\infty}$. Rather than take the reciprocal of a logarithm, let's take the reciprocal of the trig function:

Ans. to reading exercise 39
1

$$\lim_{x\to 0^+} (\sin x)\ln(\cot x) = \lim_{x\to 0^+} \frac{\ln(\cot x)}{\csc x},$$

which is of the form $\frac{\infty}{\infty}$ (we already know the numerator is positive infinite, and we see quickly that the denominator is positive infinite as well). We are finally set up to use l'Hospital's rule and evaluate the limit:

Line 1 carries out l'Hospital's rule; line 2 simplifies, switches to sines and cosines, and simplifies again; and line 3 evaluates using continuity. It is sometimes easiest to simplify the expression before evaluating the limit.

We have now calculated the value of L.

$$\begin{aligned}\lim_{x\to 0^+} \frac{\ln(\cot x)}{\csc x} &= \lim_{x\to 0^+} \frac{\frac{d}{dx}\ln(\cot x)}{\frac{d}{dx}\csc x} = \lim_{x\to 0^+} \frac{\frac{1}{\cot x}(-\csc^2 x)}{-\csc x \cot x} \\ &= \lim_{x\to 0^+} \frac{\csc x}{\cot^2 x} = \lim_{x\to 0^+} \frac{\frac{1}{\sin x}}{\frac{\cos^2 x}{\sin^2 x}} = \lim_{x\to 0^+} \frac{\sin x}{\cos^2 x} \\ &= \frac{0}{1^2} = 0.\end{aligned}$$

❸ Use theorem 12 to calculate the value of the original limit. By step ❷, $L = 0$. Therefore,

$$\lim_{x\to 0^+} (\cot x)^{\sin x} = e^0 = 1.$$

■

The last indeterminate form to examine is 1^∞. Although

$$1^\Omega = 1$$

for any positive infinite hyperreal Ω, moving slightly away from one gives different answers:

An alternate method of seeing that, 1^∞ is indeterminate is that, informally, using definition 6 gives $1^\infty = e^{\infty \ln 1} = e^{\infty \cdot 0}$, which has an indeterminate form in the exponent.

$$0.99^\Omega \doteq 0$$

(think a^Ω with $a < 1$) and

$$1.01^\Omega \doteq \infty$$

(think a^Ω with $a > 1$). Although 0.99 and 1.01 are not infinitesimally close to one (as happens in a limit), this still illustrates the truth that 1^∞ is an indeterminate form. Because 1^∞ is a power, evaluating limits with this indeterminate form follows the pattern of examples 3 and 4. An example of this indeterminate form is given in example 6.

1^∞ is an indeterminate form.

Although 0^0, ∞^0, and 1^∞ are indeterminate forms, not all exponential expressions are indeterminate.

Example 5 *Evaluate* $\lim_{x\to 0^+} x^{x+1}$.

Solution Evaluating using continuity yields

$$\lim_{x\to 0^+} x^{x+1} = 0^1 = 0.$$

The form 0^1 is not indeterminate; its value is zero.

Sometimes evaluating using continuity works. ■

Another alternate definition of e

As stated in section 5.6, many high school algebra textbooks introduce the number e informally through the use of a compound interest formula and the as-yet-undiscussed concept of limit. A second version of this limit is in the next example.

Example 6 *Evaluate* $\lim_{x\to\infty} \left(1 + \frac{1}{x}\right)^x$.

Solution Trying the EAR procedure, we use $x = \Omega$ for an arbitrary positive infinite hyperreal Ω, obtaining

An expression in the proof of theorem 10 shows we can evaluate $(1+\omega)^{\Omega}$ directly as e, but this is not the point of this example.

$$\lim_{x\to\infty}\left(1+\tfrac{1}{x}\right)^x = \left(1+\tfrac{1}{\Omega}\right)^{\Omega} = (1+\omega)^{\Omega},$$

which is of the indeterminate form 1^{∞} (since $1+\omega \approx 1$ and $\Omega \doteq \infty$). We therefore follow the strategy for limits of powers.

❶ Take the log of the expression and simplify. We have

$$\lim_{x\to\infty}\ln\left(\left(1+\tfrac{1}{x}\right)^x\right) = \lim_{x\to\infty} x\ln\left(1+\tfrac{1}{x}\right).$$

❷ Evaluate the resulting limit. We begin with

$$\begin{aligned}\lim_{x\to\infty} x\ln\left(1+\tfrac{1}{x}\right) &= \Omega\cdot\ln\left(1+\tfrac{1}{\Omega}\right) = \Omega\cdot\ln(1+\omega)\\ &\approx \Omega\cdot\ln 1 = \Omega\cdot 0 = 0,\end{aligned}$$

which violates the approximation principle. Although the result of this calculation is not valid, we can certainly recognize that the limit is of the indeterminate form $\infty\cdot 0$. This prompts us to rearrange the form of the limit to take advantage of l'Hospital's rule:

As in other examples, we avoid moving the log and instead move the other part of the expression to the denominator.

$$\lim_{x\to\infty} x\ln\left(1+\tfrac{1}{x}\right) = \lim_{x\to\infty}\frac{\ln\left(1+\frac{1}{x}\right)}{\frac{1}{x}},$$

which is of the form $\frac{0}{0}$. Now we may try l'Hospital's rule:

Line 1 carries out l'Hospital's rule; line 2 results from canceling the factors of $\frac{-1}{x^2}$ from the numerator and the denominator; and in lines 3 and 4, the limit is evaluated using the EAR procedure.

$$\begin{aligned}\lim_{x\to\infty}\frac{\ln\left(1+\frac{1}{x}\right)}{\frac{1}{x}} &= \lim_{x\to\infty}\frac{\frac{d}{dx}\ln\left(1+\frac{1}{x}\right)}{\frac{d}{dx}\frac{1}{x}} = \lim_{x\to\infty}\frac{\frac{1}{1+\frac{1}{x}}\cdot\left(0+\frac{-1}{x^2}\right)}{-\frac{1}{x^2}}\\ &= \lim_{x\to\infty}\frac{1}{1+\frac{1}{x}}\\ &= \frac{1}{1+\frac{1}{\Omega}} = \frac{1}{1+\omega}\\ &\approx \frac{1}{1} = 1.\end{aligned}$$

❸ Use theorem 12 to calculate the value of the original limit. By step ❷, $L = 1$. Therefore,

$$\lim_{x\to\infty}\left(1+\tfrac{1}{x}\right)^x = e^1 = e.$$

■

Example 6 serves as the proof of the following theorem.

Theorem 13 A LIMIT EXPRESSION YIELDING e *An alternate definition of the number e is given by*

$$e = \lim_{x\to\infty} \left(1 + \tfrac{1}{x}\right)^x.$$

Exercises 5.12

1–12. Rapid response: is the given form indeterminate?

1. $\frac{1}{0}$
2. $\frac{0}{0}$
3. $\frac{0}{1}$
4. $\frac{\infty}{-\infty}$
5. $0 \cdot \infty$
6. $\frac{\infty}{0}$
7. $\infty + \infty$
8. $\infty - \infty$
9. 0^0
10. 1^∞
11. ∞^0
12. ∞^1

13–44. Evaluate the limit.

Some of these exercises do not involve indeterminate forms. Some do not require the use of l'Hospital's rule.

13. $\lim_{x\to\infty} \left(\ln x - \ln(4x^2 + 1)\right)$
14. $\lim_{x\to\infty} \frac{1}{x} \ln x$
15. $\lim_{x\to 0^+} \frac{1}{x} \ln x$
16. (GSL) $\lim_{x\to\frac{\pi}{2}} \left(x \tan x - \frac{\pi}{2} \sec x\right)$
17. $\lim_{x\to 0^+} \left(\ln x - \ln \frac{x^2}{x+5}\right)$
18. $\lim_{x\to 0^+} \frac{1}{x} - \ln x$
19. (GSL) $\lim_{x\to 0} \frac{\pi}{x} \tan \frac{\pi x}{2}$
20. $\lim_{x\to 0} (\sin x)^{e^x}$
21. (GSL) $\lim_{x\to 1} x^{1/(1-x)}$
22. $\lim_{x\to\infty} \left(\ln(3x + 1) - \ln \sqrt{x^2 + 1}\right)$
23. $\lim_{x\to 0^+} x^{\sqrt{x}}$
24. (GSL) $\lim_{x\to\infty} x \sin \frac{4}{x}$

25. $\lim_{x\to\infty}\left(x-\sqrt{x^2+1}\right)$

26. $\lim_{x\to 0}\frac{\tan x}{x^2+2x}$

27. (GSL) $\lim_{x\to 0^+} x\ln(\sin x)$

28. (GSL) $\lim_{x\to 1}\left(\frac{1}{\ln x}-\frac{x}{\ln x}\right)$

29. $\lim_{x\to 0}(\cos x)^{\sin x}$

30. (GSL) $\lim_{x\to\infty}\left(\frac{2}{x}+1\right)^x$

31. (GSL) $\lim_{x\to 0}\left(e^x+x\right)^{1/x}$

32. (GSL) $\lim_{\theta\to\frac{\pi}{4}}(1-\tan\theta)\sec 2\theta$

33. $\lim_{x\to 0}\frac{4x+x^{5/3}}{\tan x}$

34. $\lim_{x\to 0^+} x^{6x}$

35. $\lim_{x\to\infty}(\ln x)^{1/x}$

36. (GSL) $\lim_{x\to\infty}\left(\cos\frac{2}{x}\right)^x$

37. (GSL) $\lim_{x\to\frac{\pi}{2}}(\pi-2x)\tan x$

38. (GSL) $\lim_{x\to\infty}\left(\cos\frac{2}{x}\right)^{x^2}$

39. $\lim_{x\to 0}(5x+1)^{\cot x}$

40. (GSL) $\lim_{x\to\infty}\left(\cos\frac{2}{x}\right)^{x^3}$

41. (GSL) $\lim_{x\to 2}\frac{x^2-4}{x^2}\tan\frac{\pi x}{4}$

42. (GSL) $\lim_{x\to 1}\left(\frac{2}{x^2-1}-\frac{1}{x-1}\right)$

43. (GSL) $\lim_{x\to 0}\left(\frac{\pi}{4x}-\frac{\pi}{2x(e^{\pi x}+1)}\right)$

44. (GSL) $\lim_{x\to 0}\left(\frac{1}{x}\right)^{\sin x}$

45. What type of discontinuity does the function $f(x) = \frac{1}{\ln(1+x)} - \frac{1}{x}$ have at $x = 0$?

46. The graph of $y = \tan x$ has a vertical asymptote at $x = \frac{\pi}{2}$. Does $g(x) = (\sin x)^{\tan x}$ also have a vertical asymptote at $x = \frac{\pi}{2}$?

47. Find any horizontal asymptotes on the graph of $y = \left(1 + \frac{4}{x}\right)^x$.

Functions without End

5.13

Take out a fresh sheet of paper and pick up a pencil. Draw x- and y-axes. Now draw a continuous curve that does not violate the vertical line test. You have just drawn the graph of a continuous function.

What is your function's "formula?" This is a hard question! Can you write out a formula for the function with the graph you drew, using only the types of functions we have studied so far, allowing yourself any combination of these functions that you can dream up? Perhaps. If you drew a straight line you could. If you intentionally drew some other curve, such as a sine curve or a parabola, it is possible, but what if you simply drew something—a curve without any specific intention, just some generic, unknown curve? Then chances are that writing out a formula that matches the curve exactly (and is not just an approximation) is impossible. There are many other types of functions that we have not yet studied, even many that have important, practical applications in physics, engineering, probability and statistics, and other settings.

Feel free to ignore any imperfections inherent to the drawing process.

The tools of calculus are helpful in understanding functions, even new types of functions when the need arises. In this chapter-ending section we practice such explorations. The idea of the following examples is not to learn a specific new function, but rather to learn how to explore new functions when we encounter them.

Sine integral: definition and derivative

This chapter began by introducing the natural logarithmic function. It was first noted that, at the time, we had not yet found an antiderivative for $y = \frac{1}{x}$. We then built one using FTC I. We can build a new type of function in this manner beginning with any continuous function—in particular, when the function's antiderivative is not expressible (using addition, subtraction, multiplication, division, exponentiation including root-taking, and composition) in terms of other known function types. This is not the only way to build a new type of function, but it is the way we build new functions in this section.

Many noncontinuous functions may also be used.

Recall that functions expressible using these operations beginning with polynomial, trig, inverse trig, logarithmic, and exponential functions are called *elementary functions.*

It can be shown that the antiderivative of $\frac{\sin x}{x}$ is not an elementary function.

Consider $f(x) = \dfrac{\sin x}{x}$. We do not know an antiderivative for f; the available antiderivative formulas, rules, and techniques are not sufficient to determine an antiderivative.

How do we know to ask this particular question? Exploring a function is not a rote procedure, with five steps to complete and we're done. There is a large list of questions we have learned to ask about functions, questions that just might yield insight. Which questions are fruitful vary from one function to another.

The function f is not continuous at $x = 0$. What type of discontinuity does it have there? We can take a limit to find out:

$$\lim_{x \to 0} \frac{\sin x}{x} = \frac{\sin \alpha}{\alpha} \approx \frac{\alpha}{\alpha} = 1.$$

Because the limit is a real number, f has a removable discontinuity at $x = 0$. So why not remove it? The function

A name used for this function is *sine cardinal*, giving rise to the abbreviation sinc.

$$g(x) = \operatorname{sinc} x = \begin{cases} \dfrac{\sin x}{x}, & x \neq 0 \\ 1, & x = 0 \end{cases}$$

is continuous on $(-\infty, \infty)$.

We are now ready to make a definition of a new function.

Example 1 *Use FTC I to define a new function that is an antiderivative of* $g(x) = \begin{cases} \dfrac{\sin x}{x}, & x \neq 0 \\ 1, & x = 0 \end{cases}$. *Call the new function* sine integral, *with abbreviation* Si.

Solution Because g is a continuous function, we may define a new function as

$$\operatorname{Si} x = \int_0^x g(t)\, dt = \int_0^x \begin{cases} \dfrac{\sin t}{t}, & t \neq 0 \\ 1, & t = 0 \end{cases} dt.$$

Choosing a lower limit of integration different from zero results in a different antiderivative, one that differs from this function by a constant.

The lower limit of integration can be any number we want; we chose zero. ■

Example 2 *Use FTC I to calculate the derivative of* $y = \operatorname{Si} x$.

Solution Using FTC I gives

$$\frac{d}{dx} \operatorname{Si} x = g(x) = \operatorname{sinc} x = \begin{cases} \dfrac{\sin x}{x}, & x \neq 0 \\ 1, & x = 0 \end{cases}.$$ ■

For simplicity, let's use the formula

$$\frac{d}{dx} \operatorname{Si} x = \frac{\sin x}{x},$$

which is valid for all $x \neq 0$.

Example 3 *Write the chain rule form of the derivative of the sine integral function.*

Solution The chain rule form of the derivative is

$$\frac{d}{dx}\operatorname{Si} u = \frac{\sin u}{u} \cdot \frac{du}{dx}. \quad \blacksquare$$

Example 4 *Differentiate* $f(x) = x^2 \operatorname{Si}(4x - 3)$.

Solution Using the product rule along with the chain rule form of the derivative of sine integral gives

$$f'(x) = x^2 \cdot \frac{\sin(4x-3)}{4x-3} \cdot 4 + 2x \cdot \operatorname{Si}(4x-3). \quad \blacksquare$$

Sine integral: exploration using the derivative

Now that we have the derivative of the function, we can use the techniques of chapter 3 to shed light on features of the graph of $y = \operatorname{Si} x$.

Example 5 *Determine intervals of increase and decrease, and the locations of any local extrema on the graph of* $y = \operatorname{Si} x$.

Solution Using

$$y' = \frac{\sin x}{x},$$

we set the derivative equal to zero to find the critical numbers:

$$\begin{aligned} \frac{\sin x}{x} &= 0 \\ \sin x &= 0 \\ x &= n\pi, \; n = 0, \pm 1, \pm 2, \pm 3, \ldots. \end{aligned}$$

Line 1 sets the derivative equal to zero, line 2 multiplies both sides of the equation by the denominator to clear the fraction (or simply think of setting the numerator equal to zero), and line 3 gives the solution to the trigonometric equation.

Next we use the critical numbers $n\pi$ to form our chart. (Notes on the critical numbers: when solving $\frac{\sin x}{x} = 0$, the solution $x = 0$ is extraneous because $\frac{\sin x}{x}$ is undefined at $x = 0$. We might then look for the second type of critical number—namely, where the derivative is undefined. Because the expression $\frac{\sin x}{x}$ is undefined at $x = 0$, we might conclude that $x = 0$ is also a critical number. But, there is another complication. The derivative is actually sinc x, and $\operatorname{sinc} 0 = 1 \neq 0$, so $x = 0$ is not a critical number after all. However, using an extra critical number in our chart does not keep us from making correct conclusions about locations of local extrema.)

Because there are infinitely many critical numbers, we only use a few lines for the chart to get an idea of the pattern. The following chart has been completed using some critical numbers:

interval	sign of f'	inc/dec	local extrema
⋮			
			← local min at $x = -3\pi$
$(-3\pi, -2\pi)$	+	increasing	
			← local max at $x = -2\pi$
$(-2\pi, -\pi)$	−	decreasing	
			← local min at $x = -\pi$
$(-\pi, 0)$	+	increasing	
$(0, \pi)$	+	increasing	
			← local max at $x = \pi$
$(\pi, 2\pi)$	−	decreasing	
			← local min at $x = 2\pi$
$(2\pi, 3\pi)$	+	increasing	
			← local max at $x = 3\pi$
⋮			

There is no local extremum at $x = 0$, where the sign of f' does not change. (Returning to the explanation about critical numbers, $x = 0$ isn't a critical number, so there is no local extremum there.)

For the sign of f', we want the sign of $\frac{\sin x}{x}$. When $x > 0$, the sign of $\frac{\sin x}{x}$ is the same as the sign of $\sin x$, the graph of which you should recall easily. Therefore, f' is positive on $(0, \pi)$, negative on $(\pi, 2\pi)$, positive on $(2\pi, 3\pi)$, and so on, alternating the sign on each interval. When $x < 0$, the negative denominator means the sign of $\frac{\sin x}{x}$ is the opposite of the sign of $\sin x$; therefore, f' is positive on $(-\pi, 0)$, negative on $(-2\pi, -\pi)$, positive on $(-3\pi, -2\pi)$, and so on, alternating the sign on each interval.

We conclude there are infinitely many local extrema. ■

Infinitely many local extrema should not be too surprising, for there are infinitely many local extrema on the graph of $y = \sin x$.

Example 6 *Determine intervals of concavity and locations of inflection points on the graph of* $y = \operatorname{Si} x$.

Solution Possible locations of inflection points are determined by solving the equation $y'' = 0$. We first calculate the second derivative using the quotient rule:

$$y'' = \frac{x \cdot \cos x - \sin x \cdot 1}{x^2}.$$

We then try to solve:

$$\frac{x \cos x - \sin x}{x^2} = 0$$

$$x \cos x - \sin x = 0.$$

Stuck? There's a good reason why. Solutions to equations mixed with polynomials and trig functions usually cannot be determined exactly. We must pause to look for approximate solutions to the equation. ■

When exploring a new type of function (or any function, for that matter) you might run into roadblocks. What do we try when we cannot solve an equation exactly? One tool that can help is Newton's method.

Example 7 *Use Newton's method to approximate the value of the smallest positive solution to $x\cos x - \sin x = 0$.*

Solution ❶ The equation we wish to solve is

$$x\cos x - \sin x = 0,$$

which is in the correct form. For the Newton's method procedure, we assign $f(x) = x\cos x - \sin x$.

The numbered steps here are the ones used for the discussion of Newton's method in section 3.9.

❷ Next we use technology to graph the function f (figure 1).

❸ The graph appears to cross the x-axis at $x = 0$ (which is not positive) and at about $x = 4.5$ and $x = 7.8$. We want the location of the smallest positive solution, so we use an initial guess of $a_1 = 4.5$.

❹ The recursive formula is

$$a_{n+1} = a_n - \frac{f(a_n)}{f'(a_n)}.$$

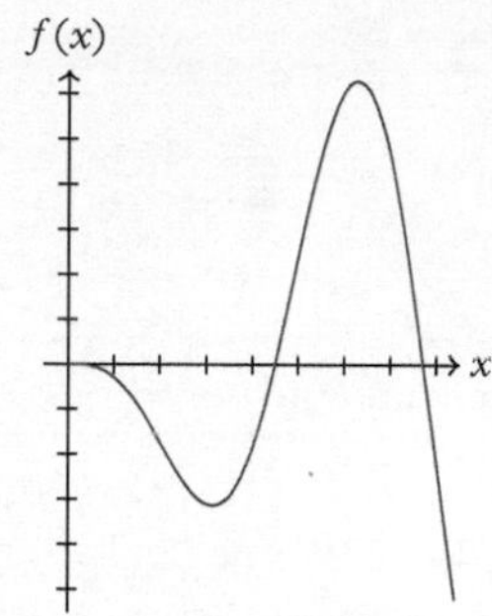

Figure 1 *The graph of $f(x) = x\cos x - \sin x$, for use with Newton's method. This is not the graph of the sine integral function*

We need the derivative:

$$f'(x) = x(-\sin x) + 1 \cdot \cos x - \cos x = -x\sin x.$$

The recursive formula to be used is

$$a_{n+1} = a_n - \frac{a_n \cos a_n - \sin a_n}{-a_n \sin a_n}.$$

Calculating successive values of a_n gives

$$a_1 = 4.5$$
$$a_2 = 4.49342$$
$$a_3 = 4.49341$$
$$a_4 = 4.49341.$$

Since $a_3 = a_4$ to five decimal places, we conclude that the smallest positive solution to the equation is $x \approx 4.49341$. ■

Using Newton's method to produce other solutions to the equation gives solutions $x = \pm 4.49341$, $x = \pm 7.72525$, $x = \pm 10.9041$, $x = \pm 14.0662$, ... Now we can return to finding intervals of concavity and locations of inflection points.

Example 8 *(example 6 continued) Determine intervals of concavity and locations of inflection points on the graph of* $y = \text{Si } x$.

Solution Using the possible inflection points $x = 0$, $x = \pm 4.49341$, and $x = \pm 7.72525$ to form intervals for our chart to get an idea of the pattern, and using test points or a graph to determine the sign of f'' on each interval, results in the following chart:

interval	sign of y''	concavity	inflection point (i.p.)
⋮			
			← i.p. at $x = -7.72525$
$(-7.72525, -4.49341)$	$-$	down	
			← i.p. at $x = -4.49341$
$(-4.49341, 0)$	$+$	up	
			← i.p. at $x = 0$
$(0, 4.49341)$	$-$	down	
			← i.p. at $x = 4.49341$
$(4.49341, 7.72525)$	$+$	up	
			← i.p. at $x = 7.72525$
⋮			

There are infinitely many inflection points on the graph of $y = \text{Si } x$. ■

The graph of sine integral

Now that we know the locations of local extrema and inflection points on the graph of $y = \text{Si } x$, as well as intervals of increase/decrease and concavity, we can try to graph the function. To do so requires plotting these inflection points and local extreme points, which requires not just their locations (their x-coordinates), but also their y-coordinates. We therefore wish to evaluate the function $y = \text{Si } x$ at each of the x-coordinates.

Let's start with the inflection point located at $x = 0$. Its y-coordinate is

$$y(0) = \text{Si } 0 = \int_0^0 \begin{cases} \dfrac{\sin t}{t}, & t \neq 0 \\ 1, & t = 0 \end{cases} dt = 0,$$

because the lower and upper limits of integration are identical.

So far so good. But what about the y-coordinate for the local max located at $x = \pi$? This is

$$y(\pi) = \mathrm{Si}\,\pi = \int_0^{\pi} \begin{cases} \dfrac{\sin t}{t}, & t \neq 0 \\ 1, & t = 0 \end{cases} dt.$$

How do we evaluate this integral? We have an antiderivative, so we can try the usual method (the FTC II)? This gives

$$\mathrm{Si}\,\pi = \int_0^{\pi} \begin{cases} \dfrac{\sin t}{t}, & t \neq 0 \\ 1, & t = 0 \end{cases} dt = \mathrm{Si}\,x\Big|_0^{\pi} = \mathrm{Si}\,\pi - \mathrm{Si}\,0 = \mathrm{Si}\,\pi - 0 = \mathrm{Si}\,\pi,$$

which gets us nowhere! This effort was doomed to fail from the start. Can you figure out why?

How do we evaluate an integral when FTC II does not help? We use numerical integration!

Example 9 *Use Simpson's rule with* $n = 6$ *to estimate the value of*

$$\mathrm{Si}\,\pi = \int_0^{\pi} \begin{cases} \dfrac{\sin t}{t}, & t \neq 0 \\ 1, & t = 0 \end{cases} dt.$$

Solution ❶ Calculate Δx:

The numbered steps here are the same as the ones used when discussing Simpson's rule in section 4.9.

$$\Delta x = \frac{b-a}{n} = \frac{\pi - 0}{6} = \frac{\pi}{6}.$$

❷ Calculate the x-values $x_0, x_1, \ldots, x_n$ to begin a chart of values of the function. We use $x_0 = a = 0$ and add $\Delta x = \frac{\pi}{6}$ repeatedly until we end at $x_n = x_6 = \pi$:

x	$y = \frac{\sin x}{x}$ $(x \neq 0)$, $y = 1$ $(x = 0)$
0	
$\pi/6$	
$2\pi/6$	
$3\pi/6$	
$4\pi/6$	
$5\pi/6$	
π	

The x-coordinate fractions were purposely left unreduced.

❸ Calculate the y-values $y_0, y_1, \ldots, y_n$ to complete the chart of values:

The integrand is a piecewise-defined function, so we must use the appropriate rule for each value of x. For $x = 0$, we use the rule $y = 1$. For all other values of x, we use the rule $y = \frac{\sin x}{x}$.

x	$y = \frac{\sin x}{x}\ (x \neq 0), y = 1\ (x = 0)$	
0	1	y_0
$\pi/6$	0.95493	y_1
$2\pi/6$	0.82699	y_2
$3\pi/6$	0.63662	y_3
$4\pi/6$	0.41350	y_4
$5\pi/6$	0.19099	y_5
π	0	y_6

❸ Calculate S_n using the formula. We have

Using a calculator's numerical integration capability to approximate the integral gives $\text{Si}\ \pi \approx 1.85194$. Using this as the actual value, the error E_S in using Simpson's rule with $n = 6$ to approximate the integral is 0.00005.

$$\begin{aligned} S_6 &= \frac{1}{3} \cdot \Delta x \cdot (y_0 + 4y_1 + 2y_2 + 4y_3 + 2y_4 + 4y_5 + y_6) \\ &= \frac{1}{3} \cdot \frac{\pi}{6} \cdot \Big(1 + 4(0.95493) + 2(0.82699) + 4(0.63662) \\ &\qquad + 2(0.41350) + 4(0.19099) + 0\Big) \\ &= 1.85199. \end{aligned}$$

We conclude that $\text{Si}\ \pi \approx 1.85199$. ■

Now we know that a local maximum point on the graph of $y = \text{Si}\ x$ is approximately $(\pi, 1.85199)$.

Continuing in this manner to construct the graph of $y = \text{Si}\ x$ by hand is tedious. We can use technology to help.

Example 10 *Use a computer algebra system (CAS) to graph* $y = \text{Si}\ x$ *on* $[-20, 20]$.

Solution The graph produced by Mathematica is in figure 2. Notice the locations of local extrema (including the local max point $(\pi, 1.85199)$), locations of inflection points, and other information that we previously deduced as they appear on the graph. ■

Inspecting the graph can lead us to suspect other facts. For instance, it appears the graph has an absolute maximum at $x = \pi$ (where we know we have a local maximum) and an absolute minimum at $x = -\pi$.

We also might suspect there are horizontal asymptotes.

Example 11 *Use a CAS to determine whether there are any horizontal asymptotes on the graph of* $y = \text{Si}\ x$.

Solution Using a CAS to calculate the required limits, we get

$$\lim_{x \to \infty} \text{Si}\ x = \frac{\pi}{2}$$

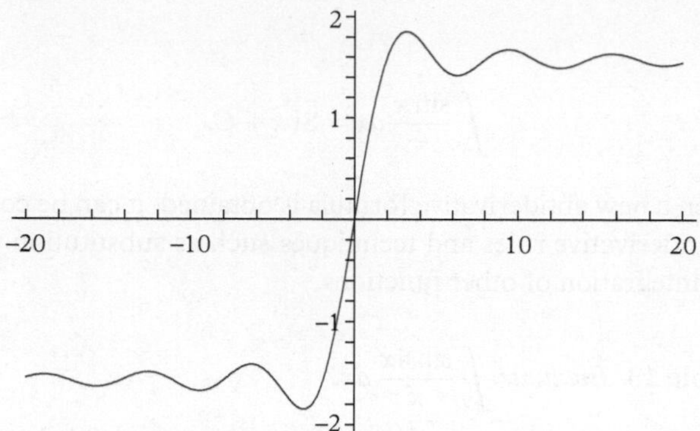

Figure 2 *The graph of $y = \text{Si } x$ on $[-20, 20]$ as produced by Mathematica*

and

$$\lim_{x\to-\infty} \text{Si } x = -\frac{\pi}{2}.$$

There are horizontal asymptotes of $y = \frac{\pi}{2}$ (on the right) and $y = -\frac{\pi}{2}$ (on the left). ■

So why not skip all this "unnecessary" work and just use the CAS to get the graph in the first place? This isn't an unreasonable question. But, even using a CAS, if we wish to know the locations of the local extrema we see in the graph, we still must know to solve $y' = 0$. The same holds for $y'' = 0$ for locations of inflection points, which still requires Newton's method. Even if the CAS does some of this work automatically, there is still value in knowing what (at least what type) of thing the software is doing behind the scenes, such as performing some numerical method to approximate values of Si x to produce the graph. Understanding the principles of calculus is often a necessary part of explorations using a CAS.

Sine integral: antiderivatives

Example 12 *Reverse the derivative formula for sine integral to obtain an antiderivative formula.*

Solution Reversing

$$\frac{d}{dx}\text{Si } x = \frac{\sin x}{x}$$

gives

$$\int \frac{\sin x}{x}\,dx = \operatorname{Si} x + C. \qquad \blacksquare$$

When a new antiderivative formula is obtained, it can be combined with antiderivative rules and techniques such as substitution to allow for the integration of other functions.

Example 13 *Evaluate* $\displaystyle\int \frac{\sin 4x}{x}\,dx.$

Solution This looks like $\frac{\sin x}{x}$, but with $4x$ inside the trig function instead of "just plain x," so substitution is needed:

$$u = 4x$$
$$du = 4\,dx.$$

We then make the substitution, evaluate the resulting integral, and return to the original variable:

Line 1 multiplies and divides by the required constant 4; notice the placement of the 4 in the denominator alongside the x instead of placing it in front of the integral as $\frac{1}{4}$. Line 2 makes the substitution. Note that we need $\frac{\sin u}{u}$, so we replace $4x$ in the denominator with u. Line 3 uses the sine integral antiderivative formula; line 4 returns to the original variable.

$$\begin{aligned}\int \frac{\sin 4x}{x}\,dx &= \int \frac{\sin 4x}{4x}\cdot 4\,dx\\ &= \int \frac{\sin u}{u}\,du\\ &= \operatorname{Si} u + C\\ &= \operatorname{Si} 4x + C.\end{aligned} \qquad \blacksquare$$

Pay attention to the manner in which the constant was adjusted for when making the substitution in example 13.

Let's look at one more example.

Example 14 *Evaluate* $\displaystyle\int \frac{(\operatorname{Si} x)^3 \sin x}{6x}\,dx.$

Solution Noticing that we have "not just plain x" to a power, we use substitution:

$$u = \operatorname{Si} x$$
$$du = \frac{\sin x}{x}\,dx.$$

With no need to adjust for a constant, we make the substitution, evaluate the resulting integral, and go back to the original variable:

$$\begin{aligned}\int \frac{(\text{Si } x)^3 \sin x}{6x}\,dx &= \int \frac{u^3}{6}\,du \\ &= \frac{u^4}{24} + C \\ &= \frac{(\text{Si } x)^4}{24} + C.\end{aligned}$$

■

The purpose of the examples in this section is not just to learn the function Si x. Rather, it is to learn how to explore a new function. The exercises in this section have this same purpose—to practice exploring new functions.

Exercises 5.13

1–28. Explore the Fresnel S and Fresnel C functions.

1. Use FTC I to define a new function that is an antiderivative of $g(x) = \sin\left(\frac{\pi}{2}x^2\right)$. Call the new function *Fresnel S*, and denote it $S(x)$.
2. Use FTC I to define a new function that is an antiderivative of $h(x) = \cos\left(\frac{\pi}{2}x^2\right)$. Call the new function *Fresnel C*, and denote it $C(x)$.
3. Use FTC I to calculate the derivative of $y = S(x)$.
4. Use FTC I to calculate the derivative of $y = C(x)$.
5. Write the chain rule form of the derivative of the Fresnel S function.
6. Write the chain rule form of the derivative of the Fresnel C function.
7. Differentiate $f(x) = \dfrac{S(x^2+2)}{4x}$.
8. Differentiate $f(x) = \sqrt{4 + C(5x+1)}$.
9. Determine intervals of increase and decrease, and the locations of any local extrema on the graph of $y = S(x)$.
10. Determine intervals of increase and decrease, and the locations of any local extrema on the graph of $y = C(x)$.
11. Determine intervals of concavity and locations of inflection points on the graph of $y = S(x)$.

12. Determine intervals of concavity and locations of inflection points on the graph of $y = C(x)$.
13. There is an inflection point on the graph of $y = S(x)$ at $x = 1$. Use Simpson's rule with $n = 6$ to estimate the value of $S(1)$, which is the y-coordinate of that inflection point.
14. There is an inflection point on the graph of $y = C(x)$ at $x = 2$. Use Simpson's rule with $n = 8$ to estimate the value of $C(2)$, which is the y-coordinate of that inflection point.
15. Use a CAS to graph $y = S(x)$ on $[-5, 5]$.
16. Use a CAS to graph $y = C(x)$ on $[-5, 5]$.
17. Use a CAS to determine whether there are any horizontal asymptotes on the graph of $y = S(x)$.
18. Use a CAS to determine whether there are any horizontal asymptotes on the graph of $y = C(x)$.
19. Use the results of exercise 17 to write a statement about S evaluated at infinite hyperreals.
20. Use the results of exercise 18 to write a statement about C evaluated at infinite hyperreals.
21. (a) Use the graph of $y = S(x)$ to decide whether there appears to be an absolute maximum or an absolute minimum value of S. (b) If so, where do they occur? (c) What are these values?
22. (a) Use the graph of $y = C(x)$ to decide whether there appears to be an absolute maximum or an absolute minimum value of C. (b) If so, where do they occur? (c) What are these values?
23. Compare the spacing of the local extrema on the graph of $y = S(x)$ to the spacing of the local extrema on the graph of $y = \operatorname{Si} x$. What do you notice?
24. What is the relationship between the local extrema of $y = S(x)$ and the inflection points of $y = C(x)$?
25. Differentiate $y = x \cdot S(x) + \frac{1}{\pi} \cos\left(\frac{\pi}{2}x^2\right)$. Write a corresponding antiderivative formula.
26. Differentiate $y = x \cdot C(x) - \frac{1}{\pi} \sin\left(\frac{\pi}{2}x^2\right)$. Write a corresponding antiderivative formula.
27. Evaluate $\int \sin x^2 \, dx$.

28. Evaluate $\int \cos x^2\, dx$.

29–40. Explore the erf function, as defined in exercise 29.

29. Use FTC I to define a new function that is an antiderivative of $g(x) = \frac{2}{\sqrt{\pi}} e^{-x^2}$. Call the new function the *error function*, with the abbreviation *erf*.

30. Use FTC I to calculate the derivative of $y = \operatorname{erf} x$.

31. Write the chain rule form of the derivative of the error function.

32. Differentiate $y = e^{-x^2} \operatorname{erf} x$.

33. Determine intervals of increase and decrease, and the locations of any local extrema on the graph of $y = \operatorname{erf} x$.

34. Determine intervals of concavity and locations of inflection points on the graph of $y = \operatorname{erf} x$.

35. Use a CAS to determine whether there are any horizontal asymptotes on the graph of $y = \operatorname{erf} x$.

36. Using the results of exercises 33–35 sketch by hand the graph of $y = \operatorname{erf} x$. Then, use a CAS to graph $y = \operatorname{erf} x$ on $[-5, 5]$ and compare.

37. Use the trapezoid rule with $n = 8$ to estimate the value of erf 2.

38. Use the results of exercise 35 to write a statement about erf evaluated at infinite hyperreals.

39. (a) Use the results of previous exercises to determine whether there is an absolute maximum or an absolute minimum value of erf. (b) If so, where do they occur? (c) What are these values?

40. Evaluate $\int (\sin x) e^{-\cos^2 x}\, dx$.

41. Evaluate $\int \frac{2 \sin x^2}{x}\, dx$.

42. Evaluate $\int_1^2 \frac{\sin x}{x}\, dx$, giving an exact value in terms of the function Si, then estimate using Simpson's rule with $n = 4$.

43. Use the results of example 11 to write a statement about Si evaluated at infinite hyperreals.

44. Verify that $y = \operatorname{Si} x$ has an absolute maximum at $x = \pi$, as follows. Use the graph of $y = \frac{\sin x}{x}$ to state why each "hump" on the graph of $y = \frac{\sin x}{x}$ is shorter than the previous one. Then, use the net area under that curve to explain why the curve $y = \operatorname{Si} x$ cannot "bounce back" as high as before.

Chapter VI

Applications of Integration

Area between Curves

6.1

Integration has many applications. Some of these applications, such as net area under a curve, are geometric, whereas others are found in the fields of physics, engineering, economics, and more. In this chapter we begin studying some of these applications. Up first is the area of a region between two curves.

Area between curves: definition

Consider the region between the curves $f(x) = x + 2$ and $g(x) = x^2$, from $x = -1$ to $x = 2$ as pictured in figure 1. What is the area of this

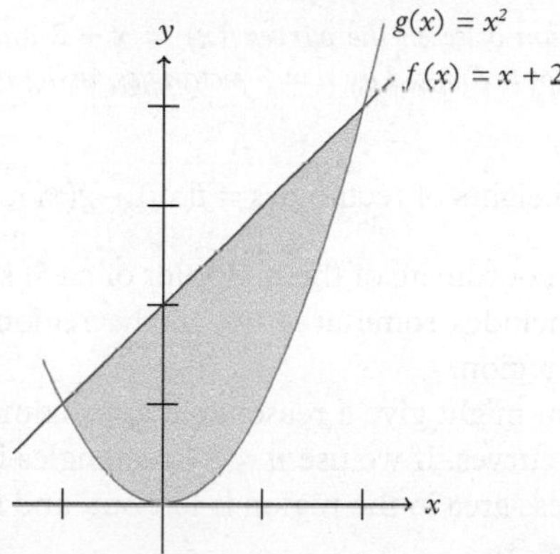

Figure 1 *The region (light green) between the curves $f(x) = x + 2$ (blue) and $g(x) = x^2$ (green), from $x = -1$ to $x = 2$*

region? It is not the area under a curve, so it is not quite the same as what we have done before. But, it is very similar, so it is reasonable to try a similar approach to determine the area.

The steps used here to develop a formula for the area between two curves are illustrative of the general approach we shall use for developing other formulas in this chapter.

The area between the curves can be approximated using a finite sum. Using $n = 6$ rectangles and midpoints, we let the width of each rectangle be

$$\text{width of rectangles} = \Delta x = \frac{b-a}{n} = \frac{2-(-1)}{6} = \frac{1}{2},$$

By using Δx, we are "slicing" along the x-axis.

and we let the height of each rectangle be the difference between the y-coordinate of the top curve and the y-coordinate of the bottom curve, each evaluated at the midpoint of the interval (figure 2), represented symbolically by

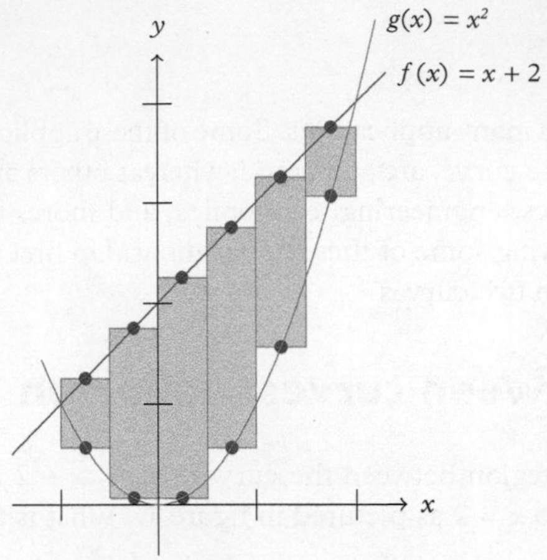

Figure 2 *The region between the curves* $f(x) = x + 2$ *and* $g(x) = x^2$ *from* $x = -1$ *to* $x = 2$*, approximated by* $n = 6$ *rectangles, using midpoints*

$$\text{heights of rectangles} = f(x_k) - g(x_k),$$

We can calculate the values of the midpoints $x_1, x_2, \ldots, x_6$; calculate the heights of the rectangles and so on; and finish this estimate, but doing so is not necessary for the task at hand.

where x_k is the x-coordinate of the midpoint of each subinterval. This approximation includes some area not in the region and leaves out some area in the region.

This finite sum might give a reasonable approximation to the area between the two curves. If we use $n = 24$ rectangles instead, we get a better estimate; less area in the region is left out, and less "extra" area is included (figure 3).

The orientation of these rectangles can be described as "vertical" (see figure 3). With an infinitesimal width but a real-number-level height, the rectangles are much longer vertically than horizontally.

If using more rectangles helps, what if we use infinitely many rectangles? Then the width of each rectangle is infinitesimal and the area in the region that is left out, as well as the extra area that is included, is also infinitesimal. After rendering a real result, we ought to get the exact area!

Let Ω be an arbitrary positive infinite hyperreal integer. Using Ω rectangles and right-hand endpoints, the width of each rectangle is $\Delta x = \frac{b-a}{\Omega} = (b-a)\omega$, and the endpoints are $x_k = a + k\Delta x$ (sound familiar?). As mentioned, the heights of the rectangles are $f(x_k) - g(x_k)$. The total area of the rectangles is the sum of the heights times the widths, or

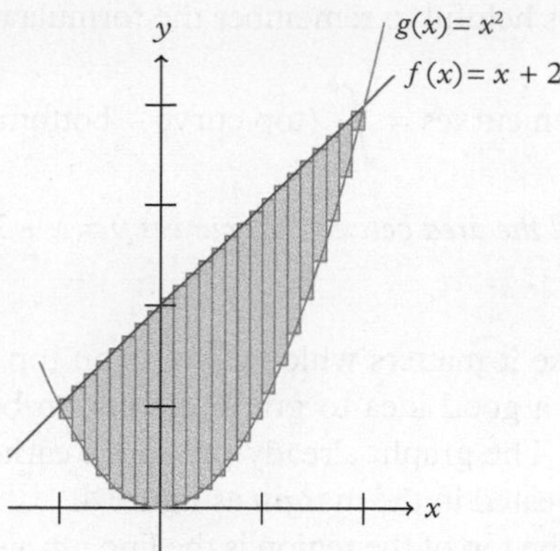

Figure 3 *The region between the curves $f(x) = x + 2$ and $g(x) = x^2$ from $x = -1$ to $x = 2$, approximated by $n = 24$ rectangles, using midpoints. This appears to give a better estimate than that using $n = 6$ rectangles*

$$\text{area} = \sum_{k=1}^{\Omega} (f(x_k) - g(x_k))\,\Delta x.$$

This is an omega sum!

In section 4.4 we learned how to express an omega sum as a definite integral. The process is to replace the summation symbol with an integral symbol, replace Δx with dx, remove the subscript (change each x_k to x), and insert the limits of integration. Doing so with our area formula gives

The omega sum for finding the specific area between the curves depicted in figure 1 is evaluated at the end of this section.

$$\text{area} = \int_a^b (f(x) - g(x))\,dx,$$

where $y = f(x)$ is the curve on top and $y = g(x)$ is the curve on the bottom. Then, all the power of the fundamental theorem of calculus, part II (FTC II), can be used to find these areas!

Definition 1 AREA BETWEEN CURVES *Let $y = f(x)$ and $y = g(x)$ be two curves for which $f(x) \geq g(x)$ throughout the interval $[a, b]$. The* area between the curves *from $x = a$ to $x = b$ is then defined as*

$$\textit{area between curves} = \int_a^b (f(x) - g(x))\,dx.$$

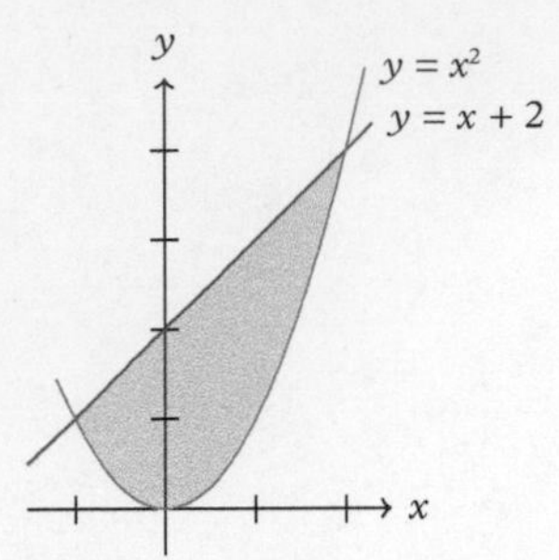

Figure 4 *The region (light green) between the curves $y = x + 2$ (blue) and $y = x^2$ (green) from $x = -1$ to $x = 2$. The top curve is $y = x + 2$; the bottom curve is $y = x^2$*

It is sometimes helpful to remember the formula as

$$\text{area between curves} = \int_a^b (\text{top curve} - \text{bottom curve})\, dx.$$

Example 1 *Find the area between the curves $y = x + 2$ and $y = x^2$ from $x = -1$ to $x = 2$.*

Solution Because it matters which curve is on top and which is on the bottom, it is a good idea to graph the region before attempting any calculations. The graph, already produced earlier in this section as figure 1, is repeated in the margin as figure 4.

The curve at the top of the region is the line $y = x+2$ and the curve at the bottom of the region is the parabola $y = x^2$. Then,

Line 1 uses the formula in definition 1, with the top curve as f and the bottom curve as g; lines 2 and 3 use FTC II to evaluate the integral.

Because the context is area, we include appropriate units of area. When no specific unit of measure is indicated, we write *units*^{2}.

$$\begin{aligned}\text{area between curves} &= \int_{-1}^{2} \left(x + 2 - x^2\right) dx \\ &= \left(\frac{x^2}{2} + 2x - \frac{x^3}{3}\right)\Bigg|_{-1}^{2} \\ &= \cdots = \frac{9}{2}\ \text{units}^2.\end{aligned}$$

Notice that the lower limit of integration is the left-most x-coordinate of the region, -1, whereas the upper limit of integration is the right-most x-coordinate of the region, 2.

We conclude that the area between the two curves is $\frac{9}{2}$ units^2. ■

The two curves do not have to meet on the left at $x = a$ or on the right at $x = b$.

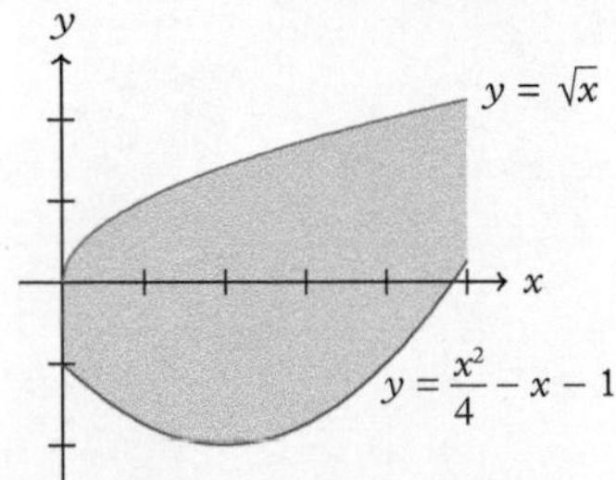

Figure 5 *The region (light green) between the curves $y = \frac{x^2}{4} - x - 1$ (blue) and $y = \sqrt{x}$ (green), from $x = 0$ to $x = 5$*

Example 2 *Find the area between the curves $y = \frac{x^2}{4} - x - 1$ and $y = \sqrt{x}$ from $x = 0$ to $x = 5$.*

Solution We begin by graphing each curve to determine which is on top and which is on bottom (figure 5). We conclude that $y = \sqrt{x}$ is the top curve and $y = \frac{x^2}{4} - x - 1$ is the bottom curve.

Now we use definition 1 to calculate the area:

CAUTION: MISTAKE TO AVOID
In line 1, notice the parentheses around the expression for the bottom curve, $\frac{x^2}{4} - x - 1$. Failure to use parentheses leads to sign errors and therefore an incorrect area if the bottom curve's expression has more than one term.

$$\begin{aligned}\text{area between curves} &= \int_0^5 \left(\sqrt{x} - \left(\frac{x^2}{4} - x - 1\right)\right) dx \\ &= \int_0^5 \left(\sqrt{x} - \frac{x^2}{4} + x + 1\right) dx \\ &= \left(\frac{x^{3/2}}{3/2} - \frac{x^3}{12} + \frac{x^2}{2} + x\right)\Bigg|_0^5\end{aligned}$$

$$= \frac{2 \cdot 5\sqrt{5}}{3} - \frac{125}{12} + \frac{25}{2} + 5 - (0 - 0 + 0 + 0)$$
$$= \cdots = \frac{85 + 40\sqrt{5}}{12} = 14.5369.$$

We conclude that the area between the two curves is approximately 14.5369 units^2. ■

Reading Exercise 1 *Find the area between the curves $y = x^2$ and $y = x$ from $x = 0$ to $x = 1$ (figure 6).*

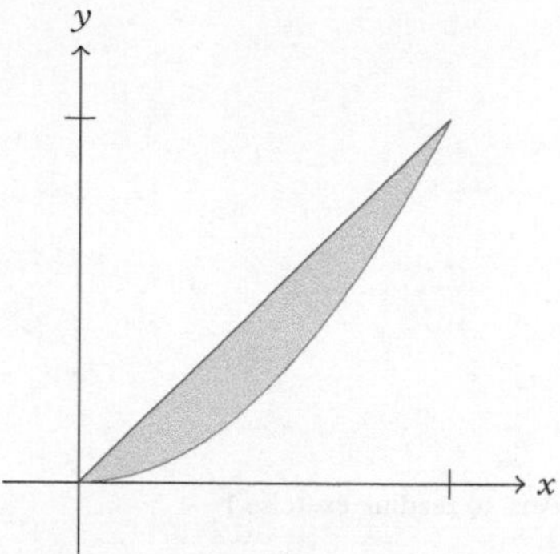

Figure 6 *The region (light green) between the curves $y = x$ (blue) and $y = x^2$ (green) from $x = 0$ to $x = 1$, for reading exercise. 1*

Area enclosed by two curves

An alternate description of the region in example 1 (figure 4) is "the region *bounded by* the two curves $y = x^2$ and $y = x + 2$." It can also be described as the region *enclosed by* the two curves. Such a description implies that the region extends from where the curves meet on the left, $x = -1$, to where they meet on the right, $x = 2$. There are just two boundaries of the region: the two curves.

By way of contrast, the region in example 2 (figure 5) cannot be described as bounded just by the two curves because the two curves do not meet at $x = 0$ or at $x = 5$. There are four boundaries, including the two curves and the two vertical lines $x = 0$ and $x = 5$. To describe this region by its boundaries, we must mention all four: "the region bounded by $y = \frac{x^2}{4} - x - 1$, $y = \sqrt{x}$, $x = 0$, and $x = 5$."

When a region is described as enclosed by (or bounded by) two curves, to find its area we must determine the upper and lower limits of integration by determining where (at which x-coordinates) the boundaries intersect.

Example 3 *Find the area of the region bounded by the curves $y = 7 - 2x^2$ and $y = x^2 + 4$.*

Solution First we must determine the region with the area we wish to find. Because the region is to be bounded (enclosed) by the two curves, we need to know where the curves intersect. To find where the curves intersect is to find where their y-coordinates match, so we set these equal to one another and solve for x:

$$7 - 2x^2 = x^2 + 4$$
$$3 = 3x^2$$
$$1 = x^2$$
$$x = \pm 1.$$

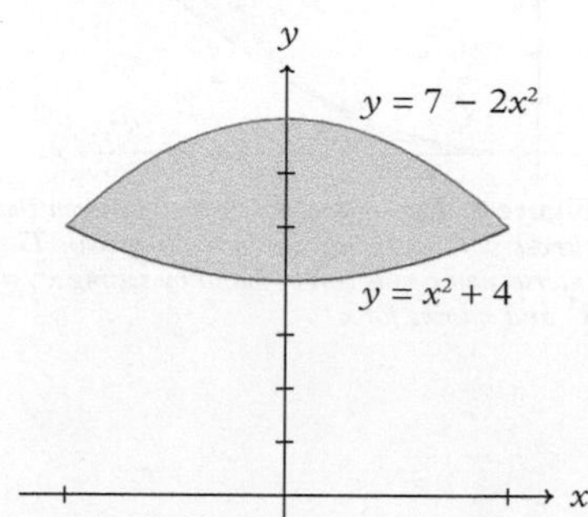

Figure 7 *The region (light green) between the curves $y = x^2 + 4$ (blue) and $y = 7 - 2x^2$ (green)*

Now we may produce a graph of the region easily (figure 7).

The top curve is $y = 7 - 2x^2$ and the bottom curve is $y = x^2 + 4$. We are now ready to use the formula to find the area of the region, integrating from $x = -1$ to $x = 1$:

$$\begin{aligned}\text{area} &= \int_{-1}^{1} \Big(\underbrace{\left(7 - 2x^2\right)}_{\text{top}} - \underbrace{\left(x^2 + 4\right)}_{\text{bottom}} \Big)\, dx \\ &= \int_{-1}^{1} \left(-3x^2 + 3\right) dx \\ &= \left(-x^3 + 3x\right)\Big|_{-1}^{1} \\ &= (-1^3 + 3) - (-(-1)^3 + (-3)) = \cdots = 4.\end{aligned}$$

We conclude that the area of the region is 4 units^2. ■

Ans. to reading exercise 1

$\frac{1}{6}$

Reading Exercise 2 *Where do the curves $y = x^2 + 3$ and $y = x + 5$ intersect?*

Area between curves: more complicated regions

The formula in definition 1 requires the same curve to be on top throughout the entire interval. What if the curves switch places?

Example 4 *Find the area between the curves $y = x^2$ and $y = x^3$ from $x = 0$ to $x = 2$.*

Solution We begin by graphing the two curves from $x = 0$ to $x = 2$ (figure 8).

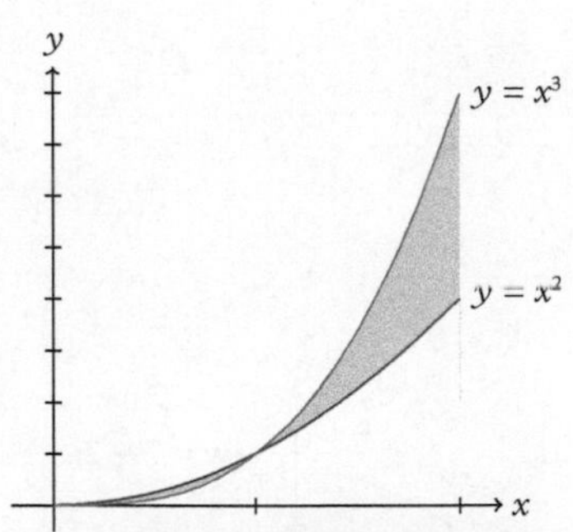

Figure 8 *The region (light green) between the curves $y = x^2$ (blue) and $y = x^3$ (green). The intersection points can be found by setting $x^2 = x^3$ and solving for x*

Which curve is on top changes at $x = 1$, where the two curves intersect. In the region from $x = 0$ to $x = 1$, the curve $y = x^2$ is on top; in the region from $x = 1$ to $x = 2$, the curve $y = x^3$ is on top. Reasoning that the total area between the curves ought to be the sum of the areas of the two regions, we calculate them separately. The area of the region from $x = 0$ to $x = 1$ is

$$\begin{aligned}\text{area} &= \int_{0}^{1} (\underset{\text{top}}{x^2} - \underset{\text{bot}}{x^3})\, dx \\ &= \left(\frac{x^3}{3} - \frac{x^4}{4}\right)\Big|_{0}^{1} \\ &= \frac{1}{3} - \frac{1}{4} = \frac{1}{12}.\end{aligned}$$

The area of the region from $x = 1$ to $x = 2$ is

$$\text{area} = \int_1^2 (\underset{\text{top}}{x^3} - \underset{\text{bot}}{x^2})\, dx$$

$$= \left(\frac{x^4}{4} - \frac{x^3}{3}\right)\Bigg|_1^2$$

$$= \frac{16}{4} - \frac{8}{3} - \left(\frac{1}{4} - \frac{1}{3}\right) = \frac{17}{12}.$$

The total area of the combined region between the curves is

$$\int_0^1 \left(x^2 - x^3\right) dx + \int_1^2 \left(x^3 - x^2\right) dx = \frac{1}{12} + \frac{17}{12} = \frac{18}{12} = 1.5.$$

We conclude that the area of the region between the curves from $x = 0$ to $x = 2$ is 1.5 units^2. ■

CAUTION: MISTAKE TO AVOID
Ignoring a change in which curve is on top results in an incorrect answer:

$$\int_0^2 \left(x^3 - x^2\right) dx$$

$$= \left(\frac{x^4}{4} - \frac{x^3}{3}\right)\Bigg|_0^2$$

$$= \frac{16}{4} - \frac{8}{3} = \frac{4}{3} \neq 1.5.$$

Reading Exercise 3 *Which curve is on top for which interval? Use $y = x^3$ and $y = x$, from $x = -1$ to $x = 1$.*

A similar issue can arise for a region bounded by more than two curves.

Example 5 *Find the area of the region bounded by the curves $y = \sqrt{x}$, $y = 1$, and $y = 4 - \frac{1}{2}x$.*

Solution Sometimes graphing the curves by plotting points can quickly reveal the points of intersection. The three curves enclose exactly one region, as shown in figure 9. The region has three boundaries, and which curve is on top changes at $x = 4$.

The points of intersection can also be found by equating the y-coordinates of the curves two at a time—that is, by solving the equations $\sqrt{x} = 1$, $\sqrt{x} = 4 - \frac{1}{2}x$, and $1 = 4 - \frac{1}{2}x$.

It is sometimes helpful to draw typical *slices* to help determine how to set up an integral. When developing the formula in definition 1, the rectangles we used had infinitesimal widths; we were "slicing along the x-axis." These rectangles that are infinitesimally thin, with heights that are on the real-number level, are also called *vertical elements*. Placing these vertical elements in our drawing helps us visualize the interval for which $y = \sqrt{x}$ is on top (from $x = 1$ to $x = 4$) and the interval for which $y = 4 - \frac{1}{2}x$ is on top (from $x = 4$ to $x = 6$). One vertical element for each case is drawn in figure 9. Just as in example 4, we need two integrals to find the area.

Ans. to reading exercise 2
$x = -1, x = 2$

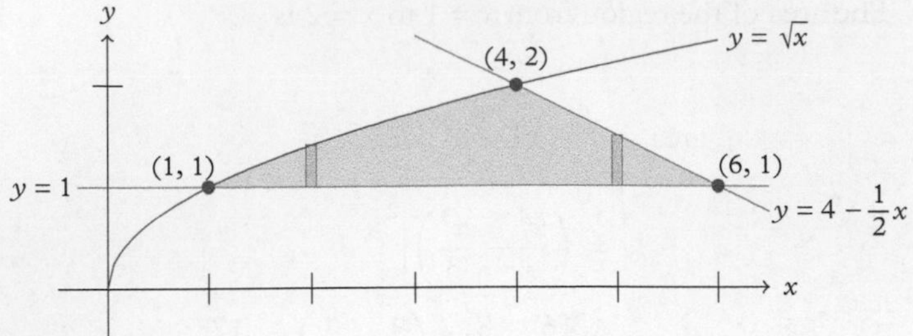

Because we cannot draw something with infinitesimal width and still see it, a drawn vertical element must look as if it is a rectangle for a finite sum—that is, a rectangle that looks like it belongs in figure 2 or figure 3, but we still think of it as having infinitesimal width.

Figure 9 *The region (light green) bounded by the curves $y = \sqrt{x}$ (blue), $y = 4 - \frac{1}{2}x$ (green), and $y = 1$ (brown), with points of intersection marked (red dots). Typical vertical elements (light-brown rectangles), also called slices, are drawn to help illustrate which curve is on the top and which is on the bottom for different portions of the region*

The area of the portion of the region from $x = 1$ to $x = 4$, where $y = \sqrt{x}$ is on the top and $y = 1$ is on the bottom, is

$$\int_1^4 \Big(\underbrace{\sqrt{x}}_{\text{top}} - \underbrace{1}_{\text{bot}}\Big)\, dx = \left(\frac{x^{3/2}}{3/2} - x\right)\bigg|_1^4$$
$$= \frac{2}{3}\cdot 4\sqrt{4} - 4 - \left(\frac{2}{3} - 1\right)$$
$$= \cdots = \frac{5}{3}.$$

When geometric formulas can be used to find the area of a region, they should match what integration gives us; otherwise, something is wrong! This portion of the region is a triangle (all three of its sides are straight lines) with width 2 and height 1, so its area is

$$\text{area} = \frac{1}{2}ab = \frac{1}{2}(2)(1) = 1\text{unit}^2.$$

The area of the portion of the region from $x = 4$ to $x = 6$, where $y = 4 - \frac{1}{2}x$ is on the top and $y = 1$ is on the bottom, is

$$\int_4^6 \Big(\underbrace{4 - \frac{1}{2}x}_{\text{top}} - \underbrace{1}_{\text{bot}}\Big)\, dx = \int_4^6 \left(3 - \frac{1}{2}x\right) dx$$
$$= \left(3x - \frac{1}{4}x^2\right)\bigg|_4^6$$
$$= \cdots = 1.$$

The total area of the combined region bounded by the curves is therefore

$$\int_1^4 (\sqrt{x} - 1)\, dx + \int_4^6 \left(4 - \frac{1}{2}x - 1\right) dx = \frac{5}{3} + 1 = \frac{8}{3}.$$

We conclude that the area bounded by the curves is $\frac{8}{3}$ units^2. ■

Ans. to reading exercise 3
$y = x^3$ is on top for $[-1, 0]$, whereas $y = x$ is on top for $[0, 1]$.

Area between curves, sideways

It is possible to switch the roles of x and y when finding the area between curves. Instead of using vertical elements, we use horizontal elements; instead of integrating with respect to the variable x, we integrate with respect to the variable y. To visualize how this works, look again at figures 1–3, but with your head turned sideways (90° to the left, so that your left ear points downward and your right ear points upward). The resulting rectangles are wide (widths = right curve − left curve), but have infinitesimal height. Again, the roles of x and y are simply switched, hence the following formula.

When integrating with respect to y, we are slicing along the y-axis (instead of the x-axis).

AREA BETWEEN CURVES (HORIZONTAL ELEMENTS)

Let $x = f(y)$ and $x = g(y)$ be two curves for which $f(y) \geq g(y)$ throughout the interval $a \leq y \leq b$. The area between the curves from $y = a$ to $y = b$ is then given by

$$\text{area between curves} = \int_a^b (f(y) - g(y))\, dy.$$

Compare this statement to definition 1. The only difference is that x and y are swapped.

It is sometimes helpful to remember this formula as

$$\text{area between curves} = \int_{y=a}^{y=b} (\text{right curve} - \text{left curve})\, dy.$$

Example 6 *Find the area of the region enclosed by* $x - y^2 = 0$ *and* $x + 2y^2 = 3$.

Solution As before, we first need to use a graph to determine the region with the area we wish to find. To graph the curve $x - y^2 = 0$, we can try solving the equation for y, but the presence of y^2 means this curve is not written where y is a function of x. Instead, we can solve for x, thinking of x as a function of y:

$$x = y^2.$$

Recall that this curve is simple to graph; it is a parabola with a vertex at the origin, opening to the right. In a similar manner, the curve $x + 2y^2 = 3$ can be written easily where x is a function of y:

$$x = 3 - 2y^2.$$

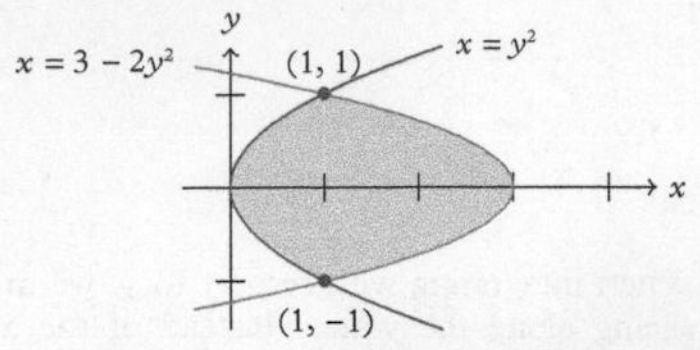

Figure 10 *The region (light green) enclosed by the curves $x - y^2 = 0$ (blue) and $x + 2y^2 = 3$ (green)*

This function is also a parabola, this time opening to the left. If points of intersection are not apparent from plotting points, they can be found by equating the x-coordinates (that is, setting $y^2 = 3-2y^2$) and solving for y. See figure 10.

Next we need to set up the integration. Placing vertical elements in the diagram (figure 11, left), we see that two integrals are needed (one on each side of the intersection points, where the top and bottom curves change). We also need to solve the curves for y in terms of x; but, because we have already written the curves so that x is a function of y, we are set to use horizontal elements instead of vertical elements, integrating with respect to y instead of with respect to x. Placing a horizontal element in the diagram (figure 11, right), we see that only one integral is needed; there is no change in which curve is on the right and which is on the left. Let's make that choice.

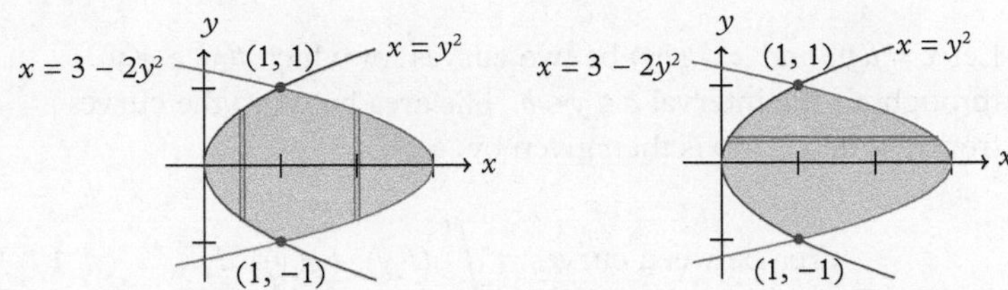

Figure 11 *(left) Vertical elements are placed in the diagram. Because which curve is on the top and which curve is on the bottom changes, two vertical elements are shown, and two integrals are required. (right) A horizontal element is placed in the diagram. Because which curve is on the right and which curve is on the left does not change, only one horizontal element is shown, and only one integral is required*

We therefore use $\int_{y=a}^{y=b}$ (right curve − left curve) dy, where the right curve is $x = 3 - 2y^2$, the left curve is $x = y^2$, and we integrate from $y = -1$ to $y = 1$ (from the lowest y-coordinate of the region to the largest):

$$\begin{aligned}\text{area} &= \int_{-1}^{1} \Big(\underbrace{(3-2y^2)}_{\text{right}} - \underbrace{y^2}_{\text{left}}\Big)\,dy \\ &= \int_{-1}^{1} \left(3 - 3y^2\right)\,dy \\ &= \left(3y - y^3\right)\Big|_{-1}^{1} \\ &= 3 - 1 - (-3 - (-1)) = 4.\end{aligned}$$

We conclude that the area of the region is 4 units^2. ■

If we choose to use vertical elements in example 6, we need to solve the equations for y. Solving $x - y^2 = 0$ yields $y = \pm\sqrt{x}$, and solving $x + 2y^2 = 3$ yields $y = \pm\sqrt{\frac{3-x}{2}}$.

The details are omitted for this alternate solution.

For the region between $x = 0$ and $x = 1$, the curve $y = \sqrt{x}$ is on the top whereas the curve $y = -\sqrt{x}$ is on the bottom. The area is

$$\int_0^1 \left(\sqrt{x} - (-\sqrt{x})\right)\, dx = \cdots = \frac{4}{3}.$$

For the region between $x = 1$ and $x = 3$, the curve $y = \sqrt{\frac{3-x}{2}}$ is on the top whereas the curve $y = -\sqrt{\frac{3-x}{2}}$ is on the bottom. The area is

$$\int_1^3 \left(\sqrt{\frac{3-x}{2}} - \left(-\sqrt{\frac{3-x}{2}}\right)\right) dx = \cdots = \frac{8}{3},$$

and notice that the integral requires the use of substitution. The total area is the sum of these two integrals, or 4 units^2, which of course is the same answer as the earlier solution.

It should be clear that using vertical elements is more work than using horizontal elements for example 6. When working the exercises, choose wisely!

Omega sums and areas between curves

At the beginning of this section we used geometric reasoning to form an omega sum for an area between curves. We then rewrote the omega sum as an integral, and we have used that integral formula ever since. But, it is sometimes possible to complete the calculation of an area between curves using only an omega sum. Let's demonstrate this by finding the area sought in example 1.

Using an omega sum to find the area of a region between two curves is the same idea as using the definition of definite integral instead of FTC II.

Example 7 *Use an omega sum to calculate the area between the curves $y = x + 2$ and $y = x^2$ from $x = -1$ to $x = 2$.*

Solution The idea is to use infinitely many rectangles so that each has infinitesimal width. We therefore use Ω rectangles (for an arbitrary positive infinite hyperreal integer Ω).

Using the formulas developed at the beginning of this section,

$$\Delta x = (b - a)\omega = \frac{2 - (-1)}{\Omega} = 3\omega$$

and

$$x_k = a + k\Delta x = -1 + k \cdot 3\omega.$$

As in example 1, the top curve is $f(x) = x+2$ and the bottom curve is $g(x) = x^2$; see figure 4.

The area of the region is the sum of the heights $(f(x_k) - g(x_k))$ times the widths (Δx):

$$\begin{aligned}
\text{area} &= \sum_{k=1}^{\Omega} \Bigg(\underbrace{\underbrace{(-1+3\omega k)+2}_{f(x_k)} - \underbrace{(-1+3\omega k)^2}_{g(x_k)}}_{\text{heights}} \Bigg) \cdot \underbrace{\underbrace{3\omega}_{\Delta x}}_{\text{widths}} \\
&= \sum_{k=1}^{\Omega} \left(-1 + 3\omega k + 2 - (1 - 6\omega k + 9\omega^2 k^2)\right) 3\omega \\
&= \sum_{k=1}^{\Omega} \left(9\omega k - 9\omega^2 k^2\right) 3\omega \\
&= \sum_{k=1}^{\Omega} \left(27\omega^2 k - 27\omega^3 k^2\right) \\
&\approx 27\omega^2 \frac{\Omega^2}{2} - 27\omega^3 \frac{\Omega^3}{3} \\
&= \frac{27}{2} - \frac{27}{3} = \frac{27}{6} = \frac{9}{2}.
\end{aligned}$$

As expected, we conclude that the area between the curves is $\frac{9}{2}$ units^2. ■

Why units^2?

There are some units of area that are not written as length-units squared, such as acres and hectares.

Lengths are measured in units such as centimeters (cm) or feet (ft). Areas are measured in square units, such as cm^2 or ft^2. For instance, a rectangle measuring 5 cm by 2 cm has area

$$5\,\text{cm} \cdot 2\,\text{cm} = 10\,\text{cm}^2.$$

But, when using an integral to find area, it might look as if the units are not squared. If the units of example 1 are centimeters, then the integral might seem to be $\int_{-1}^{2} \left((x+2)\,\text{cm} - x^2\,\text{cm}\right) dx = \int_{-1}^{2} \left(x + 2 - x^2\right) \text{cm}\, dx$. Where does cm^2 come from?

This question is answered more easily by considering the omega sum from which the integral comes. Using work from example 7, the heights of the rectangles are $\left((-1+3\omega k)+2-(-1+3\omega k)^2\right)$ cm, and the widths are 3ω cm. The area is

$$\sum_{k=1}^{\Omega} \underbrace{\left((-1+3\omega k)+2-(-1+3\omega k)^2\right) \text{cm}}_{\text{heights}} \cdot \underbrace{3\omega \text{ cm}}_{\text{widths}}$$

$$= \sum_{k=1}^{\Omega} \left((-1+3\omega k)+2-(-1+3\omega k)^2\right) \cdot 3\omega \text{ cm}^2,$$

so the units are cm^2.

Why is this not apparent in the integral? Because the widths, which are 3ω cm in the omega sum, are the Δx, which becomes dx in the integral. In other words, units are "hidden" in the dx. In fact, in a definite integral, the dx has the same units as the variable x. The earlier integral that included units is written better as

$$\int_{-1}^{2} \left(x+2-x^2\right) \text{cm}\, dx \,\text{cm},$$

making it clear that the resulting units are cm^2.

Including the units of dx is essential to determining units for applications of integration. For instance, consider the application to position, velocity, and acceleration. Suppose a particle has acceleration $a(t) = (3\sqrt{t}+5)$ cm/s^2. The velocity can then be found by integrating acceleration. The dt in the integral has the same units as the variable t, which is seconds (in this setting, the variable t is time.) Then,

This type of reasoning is sometimes called *dimensional analysis*.

$$v(t) = \int (3\sqrt{t}+5) \text{ cm/s}^2 \, dt \,\text{s},$$

and the units are $\frac{\text{cm}}{\text{s}^2} \cdot \text{s} = \frac{\text{cm}}{\text{s}}$, as expected.

Exercises 6.1

1–10. Rapid response: to find the area of the shaded region, which looks easier: integrating with respect to x or integrating with respect to y?

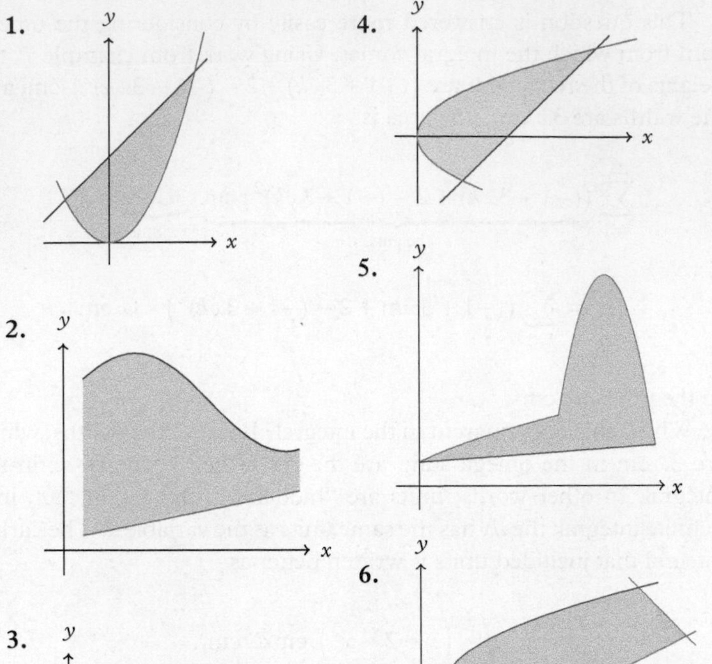

Reality check: because area is positive, if your answer is negative you have made an error, either in setting up the integration or in evaluating the integral(s).

7–14. Find the area of the shaded region.

7.

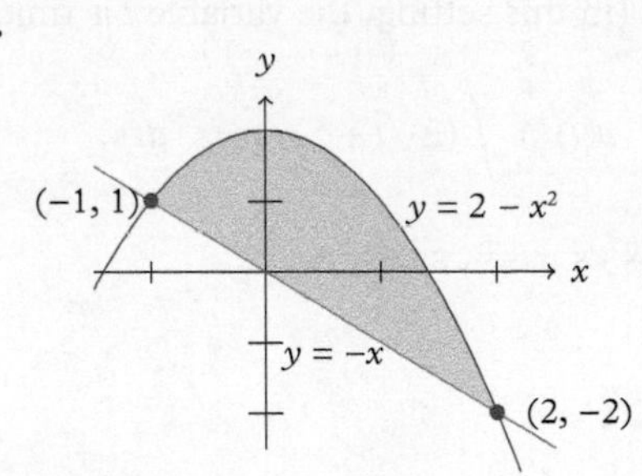

8.

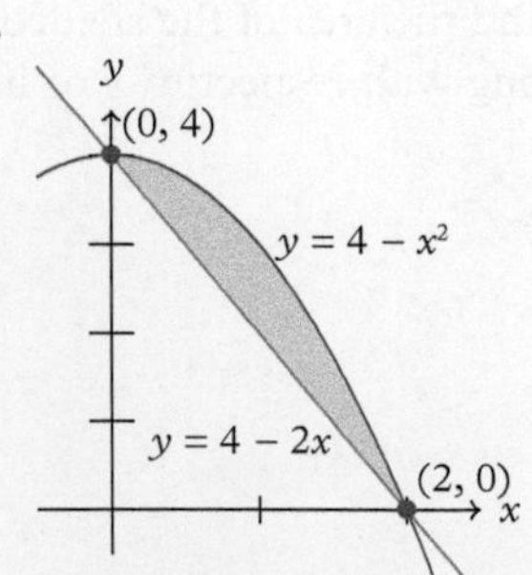

9.

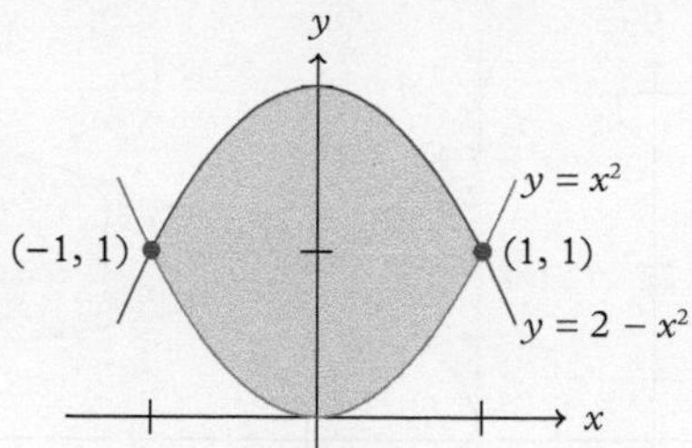

10.

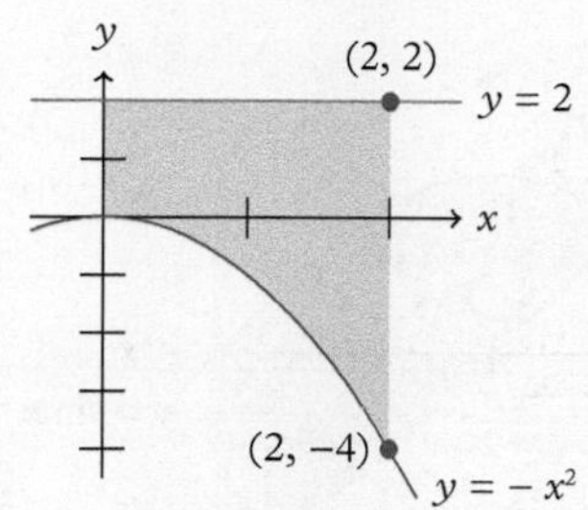

11.

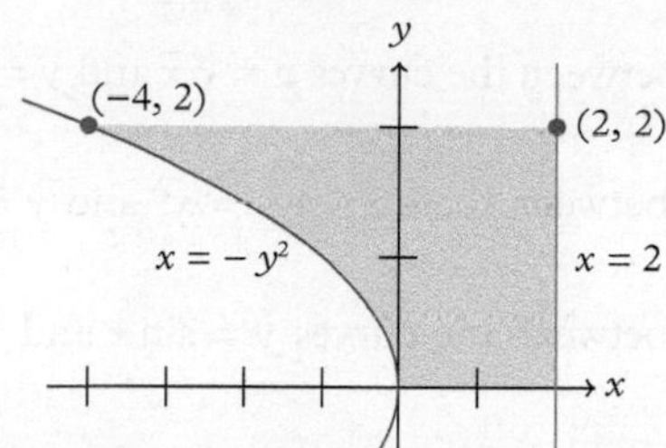

12.

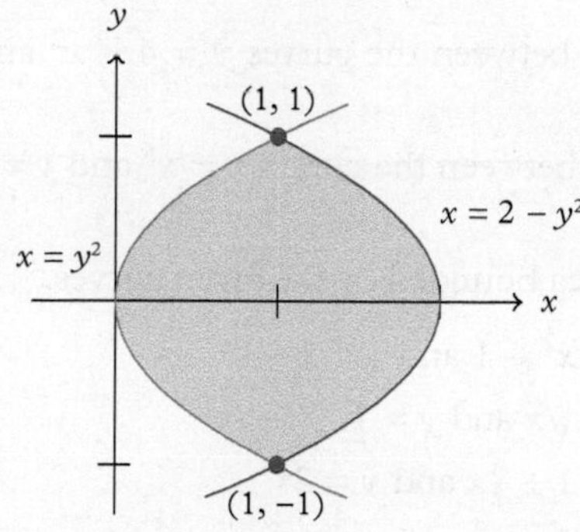

13.

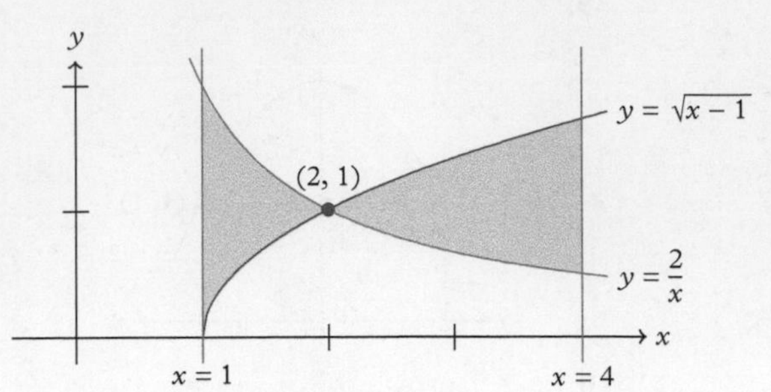

14.

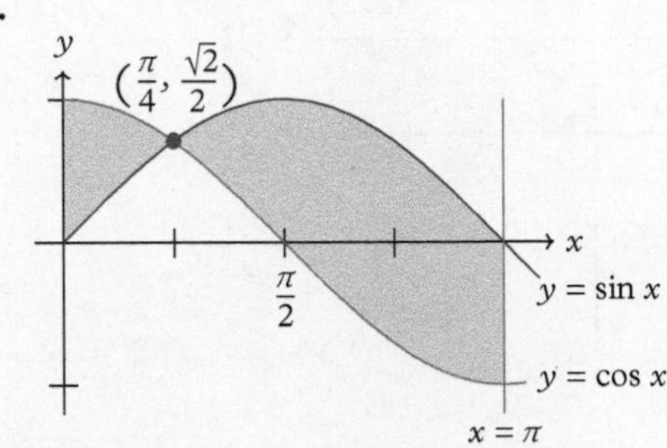

15. Find the area between the curves $y = \sqrt[3]{x}$ and $y = x$ from $x = -1$ to $x = 1$.

16. Find the area between the curves $y = x^2$ and $y = 1$ from $x = 0$ to $x = 1$.

17. Find the area between the curves $y = \sin x$ and $y = x$ from $x = -\pi$ to $x = \pi$.

18. Find the area between the curves $y = e^x$ and $y = e^{-x}$ from $x = -2$ to $x = 1$.

19. Find the area between the curves $y = 4 - x^2$ and $y = 3x$ from $x = 0$ to $x = 3$.

20. Find the area between the curves $y = x^5$ and $y = x$ from $x = -1$ to $x = 1$.

21–34. Find the area bounded by the given curves.

21. $y = x^2 - 1$ and $y = x + 1$

22. $y = \sqrt{x}$ and $y = \frac{x}{2}$

23. $y = 1 + \frac{3}{2}x$ and $y = 2^x$

24. $y = 5 - x^2$ and $y = 1$

25. (GSL) $y = 6x - x^2$ and $y = x$

26. $y = x^3$ and $y = x$

27. (GSL) $y^2 = 4x$ and $2x - y = 4$

28. (GSL) $y^2 = 2x$ and $x - y = 4$
29. $x = y^2$ and $y = x - 2$
30. $y = x^2$ and $y = 2 - |x|$
31. $y = 3x$, $y = 4 - \sqrt{x}$, and $y = \frac{1}{2}x$
32. $y = 2x - x^2$, $y = 2x$, and $y = 4 - 2x$
33. $y = \sqrt{x}$, $y = \dfrac{1}{x^2}$, and $y = \frac{1}{8}x$ from $x = 0$ to $x = 2$
34. $x = (y - 2)^2 + 1$, $y = \frac{1}{2}x$, and $x = 9 - y$
35. (GSL) Find the area bounded by the parabola $y = 6 + 4x - x^2$ and the chord joining $(-2, -6)$ and $(4, 6)$.

For "chord" think "line segment."

36. (GSL) Find the area bounded by the curve $x^2 y = x^2 - 1$ and the lines $y = 1$, $x = 1$, and $x = 4$.
37. (GSL) Find the area bounded by the curve $y = x^3 - 9x^2 + 24x - 7$, the y-axis, and the line $y = 29$.
38. (GSL) Find the area bounded by the curves $y = x(1 \pm \sqrt{x})$ and the line $x = 4$.
39. Use an omega sum to calculate the area of the region in exercise 7.
40. Use an omega sum to calculate the area of the region in exercise 8.

Volumes, Part I

6.2

One geometric application of integration is the calculation of areas. Another is the calculation of volumes.

Volume of a solid of revolution

Consider a region under a curve $y = f(x)$ on an interval $[a, b]$ such as the one in figure 1. What happens if we choose to rotate this region

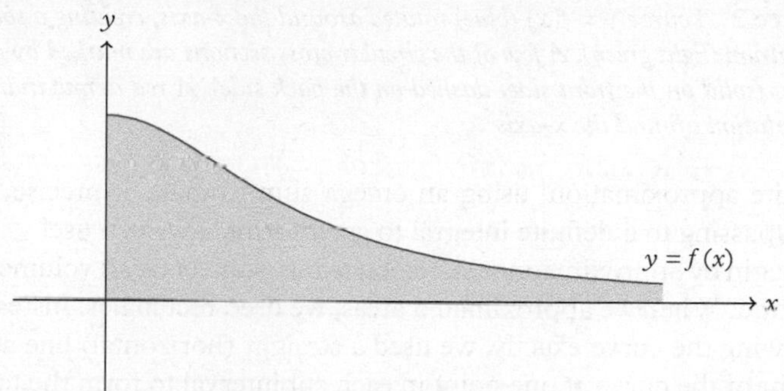

Figure 1 *A curve $y = f(x)$ (blue) and the region between the curve and the x-axis (light green)*

about the x-axis? The region rotates through space, creating a three-dimensional solid figure called a *solid of revolution* (figure 2).

A solid of revolution has circular cross sections and is the typical type of object created on a lathe. Among the many objects you may have seen that were turned on a lathe are baseball bats, candlesticks, and table legs. A lathe spins a block of wood (or metal or other material) along an axis of rotation (the x-axis in figure 2), and a cutting tool is (eventually) placed along the curve to shape the object as desired.

A person who uses a lathe to create wooden objects is called a *wood turner*. A skilled wood turner can make beautiful, artistic objects.

How might we find the volume of the object in figure 2? In your study of geometry you may have learned formulas for the volume of a sphere, a rectangular prism, a cone, a cylinder, or other similar objects, but this short list does not include the shape in figure 2. Does this situation sound familiar? Let's follow the strategy of thinking through

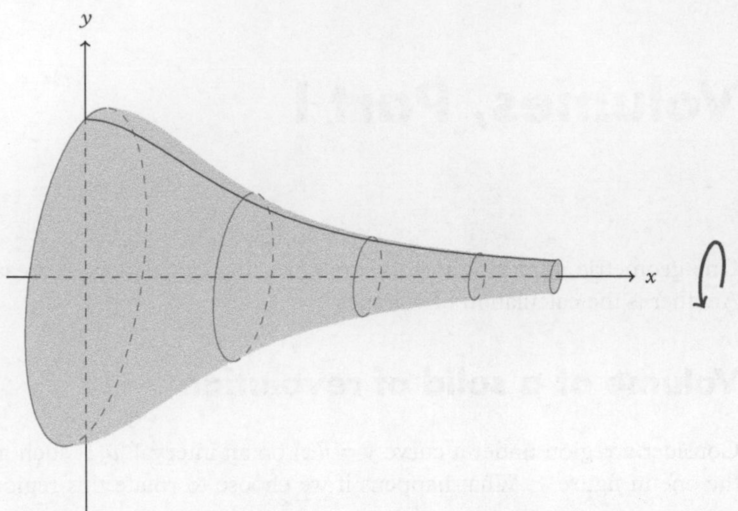

Figure 2 *A curve $y = f(x)$ (blue) rotated around the x-axis, creating a solid of revolution (light green). A few of the circular cross sections are marked by green circles (solid on the front side, dashed on the back side). A red arrow indicates the rotation around the x-axis*

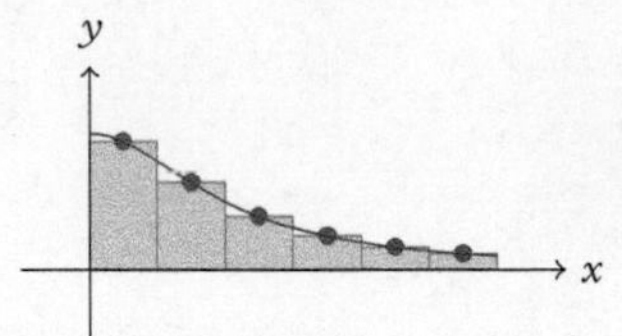

Figure 3 *An illustration of approximating the area under a curve using $n = 6$ rectangles. Notice that the tops of the rectangles are horizontal lines that match the curve at a point in the subinterval*

The formula for the volume of a cylinder is

$$V = \pi r^2 h,$$

where r is the radius of the cylinder and h is its height.

a finite approximation, using an omega sum to make it precise, and then passing to a definite integral to get a formula we can use!

Begin by approximating the region using n objects with volumes we can find. When we approximated areas, we used rectangles. Instead of following the curve exactly, we used a straight (horizontal) line at the height of the curve at one point in each subinterval to form the top of the rectangle (figure 3); let's try the same idea here. Rotating such a line about the x-axis gives a thin cylinder, reminiscent of a coin. Figure 4 shows the result using $n = 6$ subintervals. Because we have a formula for the volume of a cylinder, we can complete the approximation by finding the volume of each cylinder and then add these volumes to find the total volume. But, a look at figure 4 shows we are leaving out some volume in the solid of revolution and including some volume that is not in the solid of revolution, so we won't have the exact volume we are seeking. Using more cylinders makes for a better estimate; $n = 15$ cylinders is pictured in figure 5. If more cylinders make for a better estimate, why not use infinitely many? This should give us the exact volume of the solid of revolution. To this end, let Ω be an arbitrary positive infinite hyperreal integer and slice along the x-axis, letting

$$\Delta x = \frac{b-a}{\Omega} = (b-a)\omega$$

be the width of each subinterval. Then,

$$x_k = a + k\Delta x$$

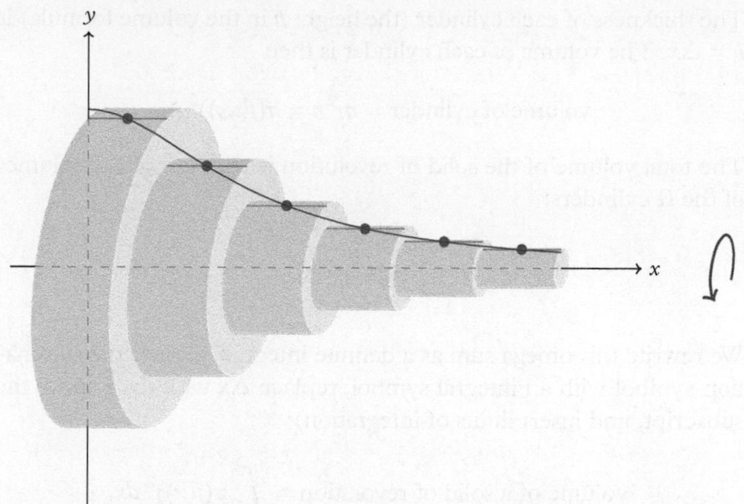

Figure 4 *A solid of revolution is approximated using n = 6 cylinders. The solid is formed by rotating y = f(x) (blue) about the x-axis. The radius of each cylinder pictured here matches the height of the curve at the midpoint of the subinterval; the resulting horizontal line (brown) is then rotated about the x-axis to form an approximating cylinder*

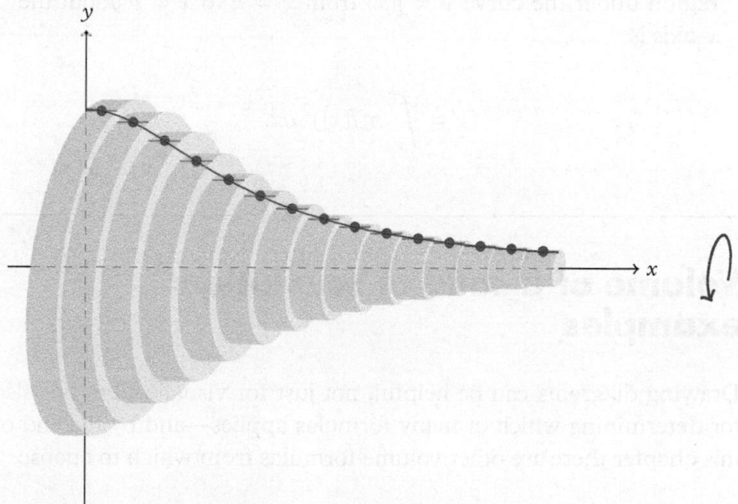

Figure 5 *The solid of revolution approximated using n = 15 cylinders*

provides the endpoints of each subinterval (let's use right-hand endpoints, as usual). The radius of each cylinder (see figure 6) is the height of the curve at the right-hand endpoint:

$$\text{radius of cylinder} = r = f(x_k).$$

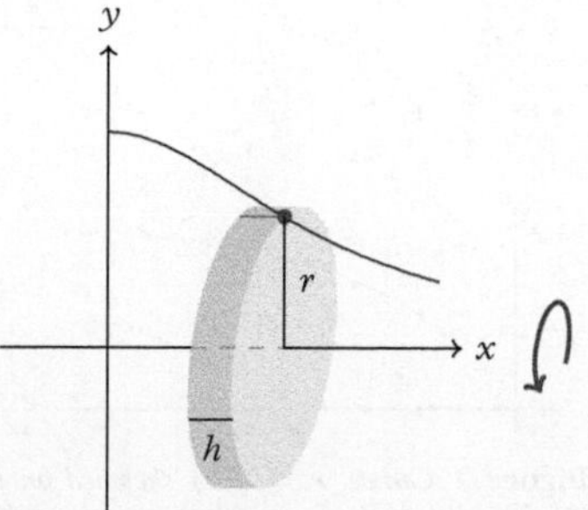

Figure 6 *One of the approximating cylinders, with radius r and thickness h marked*

If units of length are centimeters, then the radius and thickness are each measured in centimeters, and the volume of each cylinder is

$$\pi \cdot (f(k_k)\text{ cm})^2 \cdot \Delta x\text{ cm} = \pi(f(x_k))^2 \Delta x\text{ cm}^3.$$

The thickness of each cylinder (the height h in the volume formula) is $h = \Delta x$. The volume of each cylinder is then

$$\text{volume of cylinder} = \pi r^2 h = \pi(f(x_k))^2 \Delta x.$$

The total volume of the solid of revolution is the sum of the volumes of the Ω cylinders:

$$\sum_{k=1}^{\Omega} \pi(f(x_k))^2 \Delta x.$$

We rewrite this omega sum as a definite integral (replace the summation symbol with an integral symbol, replace Δx with dx, remove the subscript, and insert limits of integration):

$$\text{volume of a solid of revolution} = \int_a^b \pi(f(x))^2\, dx.$$

Technically speaking, we need to define volume before making use of formulas to find volume. One approach is to use this formula as a definition.

VOLUME OF A SOLID OF REVOLUTION (ABOUT THE *X*-AXIS)

The volume of a solid of revolution generated by revolving the region under the curve $y = f(x)$ from $x = a$ to $x = b$ about the x-axis is

$$V = \int_a^b \pi(f(x))^2\, dx.$$

Use this formula when the radius of a circular cross section extends from the x-axis to the curve.

Volume of a solid of revolution: examples

Drawing diagrams can be helpful, not just for visualization, but also for determining which of many formulas applies—and by the end of this chapter there are other volume formulas from which to choose.

Example 1 *Find the volume of the solid of revolution generated by revolving the region under the curve $y = \sqrt{x}$ from $x = 1$ to $x = 4$ about the x-axis.*

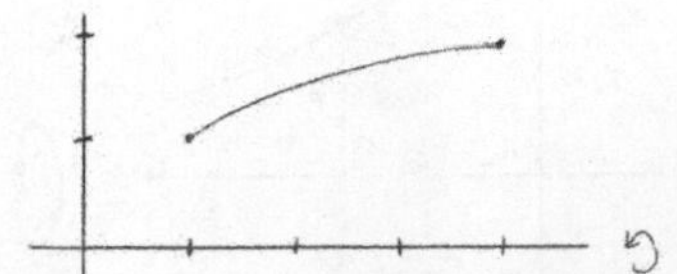

Figure 7 *Curve $y = f(x)$ sketched on the specified interval, along with the rotation arrow marking the axis of rotation*

Solution Rather than jump straight to the formula, let's begin by sketching the solid with the volume we wish to find. First we sketch the curve $y = \sqrt{x}$ on the given interval, $[1, 4]$, then we signify the axis of rotation by adding a rotation arrow around the x-axis (figure 7).

Next we sketch the reflection of the curve below the x-axis. That is, we sketch $y = -\sqrt{x}$ from $x = 1$ to $x = 4$. This gives the bottom edge of the solid of revolution (figure 8).

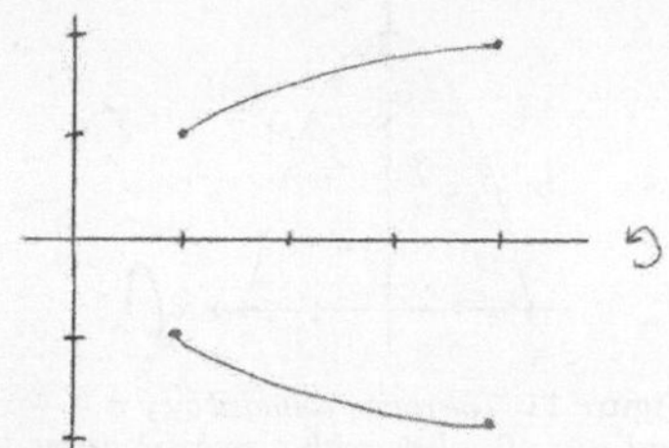

Figure 8 *The bottom edge of the solid of revolution, $y = -f(x)$ on the specified interval, is added*

Look again at figure 2. The illusion of three dimensions is given on paper by drawing a few of the circular cross sections. For all but the right-most cross section, the portion of the cross sections that appears on the front side is drawn with a solid line, whereas the portion on the back side is drawn with a dashed line. The perspective in this figure is such that we see all of the right-most cross section, so all of it is drawn using a solid line. Notice that the circular cross sections are drawn more like ellipses (they are taller than they are wide), but at the same ratio, so they all give the illusion of being circles. Cross sections are added in figure 9. At this point, you should be able to visualize the solid.

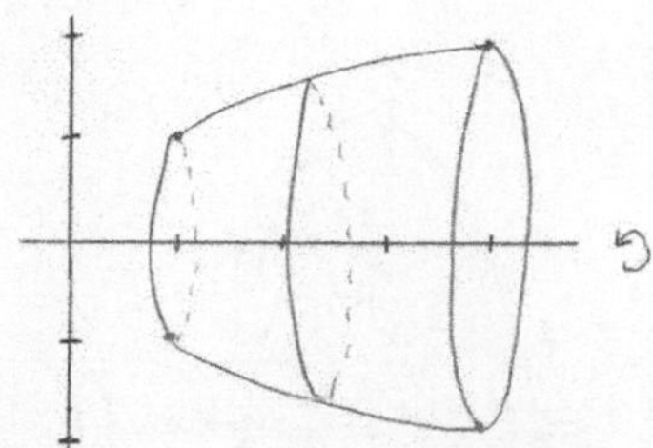

Figure 9 *Circular cross sections are added, with the curve $y = f(x)$ at the top of each cross section and the curve $y = -f(x)$ at the bottom*

Last, it is helpful to draw the radius of one of the cross sections to help determine which formula to use. The radius is drawn from the axis of rotation (the x-axis in this example) to the perimeter of the circle (the curve $y = \sqrt{x}$ in this example). The completed sketch is in figure 10.

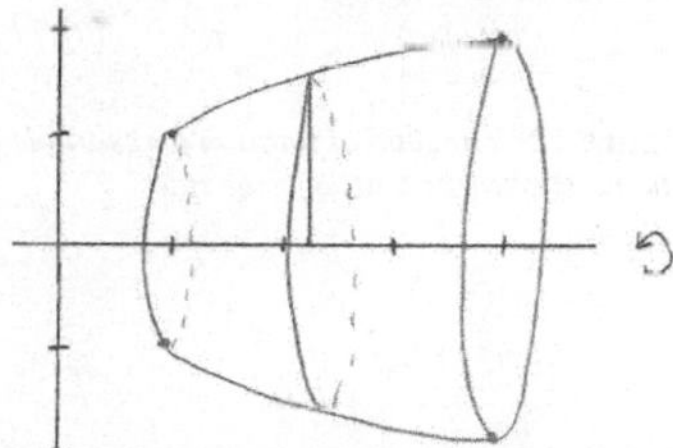

Figure 10 *A radius of a cross section is added to help to determine which formula to use*

The formula for the volume of a solid of revolution about the x-axis applies. Therefore,

$$\begin{aligned} V &= \int_1^4 \pi\left(\sqrt{x}\right)^2\,dx \\ &= \int_1^4 \pi x\,dx \\ &= \frac{x^2}{2}\pi\Big|_1^4 \\ &= 8\pi - \frac{1}{2}\pi = \frac{15}{2}\pi \approx 23.562. \end{aligned}$$

We conclude that the volume of the solid is $\frac{15}{2}\pi$ units^3. ■

Because this is a volume, we use cubic units, as explained earlier.

The region to be rotated can be described in different ways. As long as the region is the region between a curve and the x-axis between $x = a$ and $x = b$, the previous formula applies.

Example 2 *Find the volume of the solid of revolution generated by revolving the region bounded by $y = 4 - x^2$ and $y = 0$ about the x-axis.*

Solution To determine the region to be rotated about the x-axis, we need to know where the two curves intersect. Equating their y-coordinates gives

$$4 - x^2 = 0$$
$$4 = x^2$$
$$x = \pm 2.$$

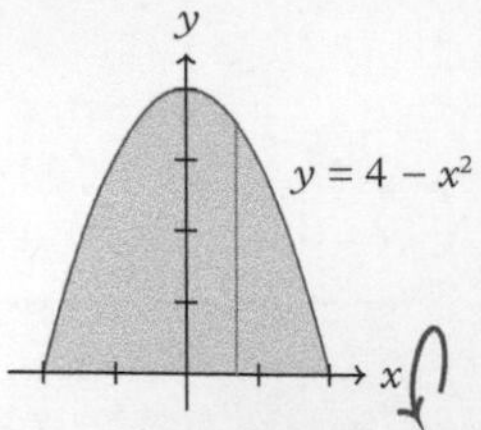

Figure 11 *The region bounded by $y = 4 - x^2$ and $y = 0$, along with a rotation arrow. A radius of a cross section has also been added*

Sketching the two curves from $x = -2$ to $x = 2$ reveals the region to rotate (figure 11).

Next we draw the solid of revolution; see example 1 for tips on sketching the object by hand. The solid of revolution, along with typical cross sections, is depicted in figure 12. A segment representing the radius of a cross section is added to figure 11 because it is easier to visualize there than it is in figure 12.

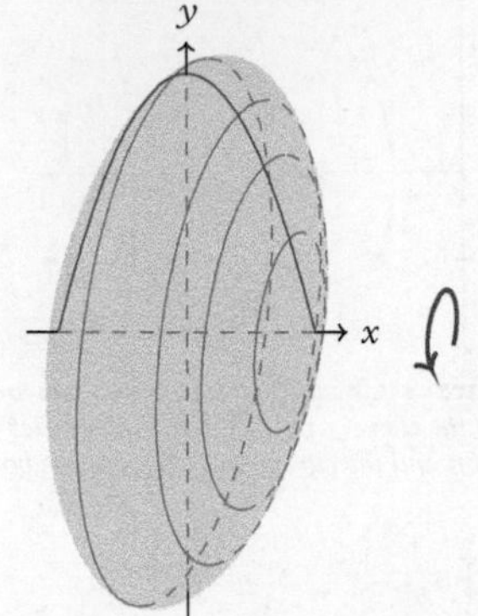

Figure 12 *The solid of rotation for example 2. The perspective is a little to the side*

The formula for the volume of a solid of revolution about the x-axis applies. Therefore,

$$\begin{aligned} V &= \int_{-2}^{2} \pi(4 - x^2)^2\, dx \\ &= \int_{-2}^{2} \pi\left(16 - 8x^2 + x^4\right) dx \\ &= \pi\left(16x - \frac{8}{3}x^3 + \frac{1}{5}x^5\right)\Bigg|_{-2}^{2} \\ &= \cdots = \frac{512\pi}{15} \approx 107.23. \end{aligned}$$

We conclude that the volume is $\frac{512\pi}{15}$ units^3. ■

Reading Exercise 4 *A solid is generated by revolving $y = x^2$ about the x-axis from $x = 0$ to $x = 2$. Find its volume.*

Rotating about the y-axis

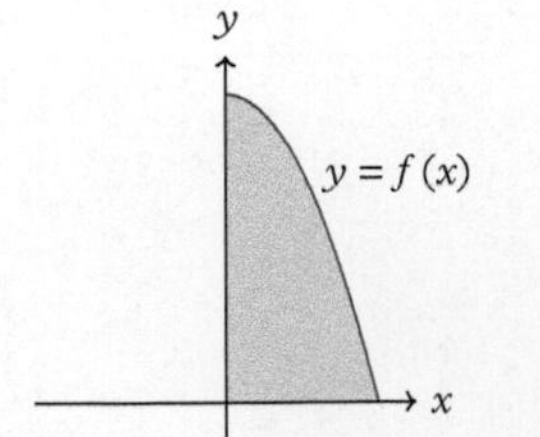

Figure 13 *The region under the curve $y = f(x)$*

Consider the region under the curve $y = f(x)$ as shown in figure 13. We can rotate this region about the x-axis to get a solid of revolution and then find its volume. What if, instead, we rotate the region about the y-axis? What does the solid of revolution look like, and how do we find its volume?

The result of rotating the region about the y-axis is given in figure 14. The solid has a different shape than if we rotate the region about the x-axis, so it might have a different volume as well. However, the approach we use is still the same. We approximate the volume by using cylinders. Notice in figure 14 that the circular cross

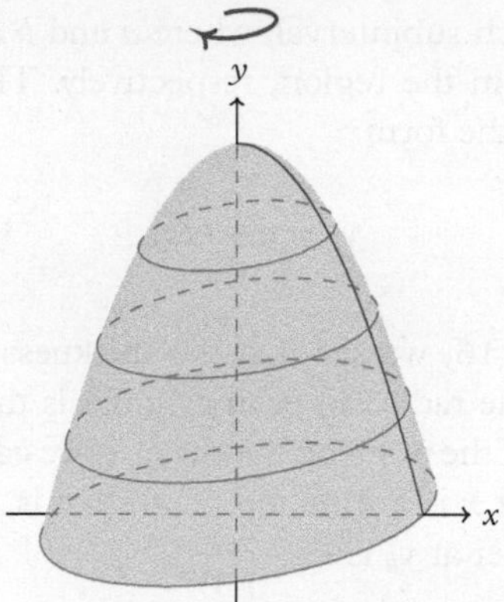

Figure 14 *A curve (blue) rotated around the y-axis, creating a solid of revolution (light green). A few of the circular cross sections are marked by green circles (solid on the front side, dashed on the back side). A red arrow indicates the rotation around the y-axis*

sections are now oriented horizontally; our approximating cylinders must have this same orientation. Using a finite sum to approximate the volume is pictured in figure 15.

As before, we can get a better approximation by using more cylinders, and we can get the exact volume by using infinitely many cylinders. Once again, we let Ω be an arbitrary positive infinite hyperreal integer and use Ω cylinders, but this time we are slicing our cylinders thinly along the y-axis, not the x-axis. We therefore use

$$\Delta y = \frac{b-a}{\Omega} = (b-a)\omega$$

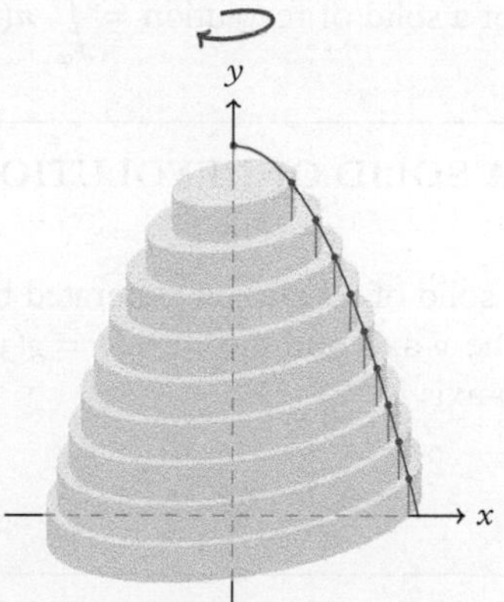

Figure 15 *The solid of revolution approximated using $n = 10$ cylinders. The top cylinder has radius 0 and is not visible*

for the width of each subinterval, where a and b are the smallest and largest values of y in the region, respectively. The endpoints of the subintervals are of the form

$$y_k = a + k\Delta y.$$

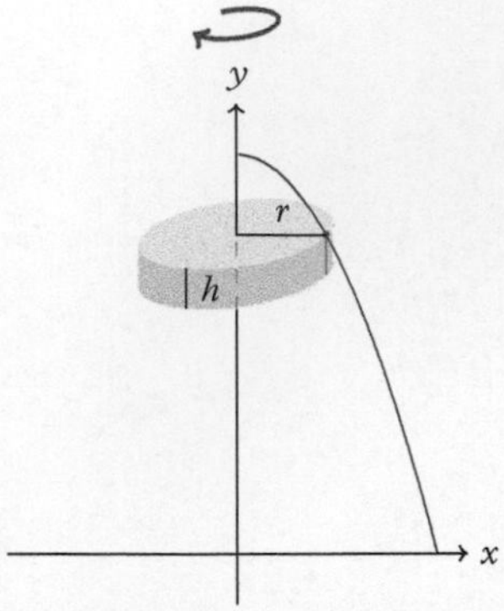

Figure 16 *One of the approximating cylinders, with radius r and thickness h marked*

Referencing figure 16, we see that the thickness of each cylinder is $h = \Delta y$, whereas the radius of each cylinder is the x-coordinate (not the y-coordinate) of the point on the curve. If we can write the equation of the curve so that x is a function of y—that is, $x = g(y)$—then the radius of the cylinder at y_k is

$$\text{radius of cylinder} = r = g(y_k).$$

This gives us all we need to find the volume of each cylinder:

$$\text{volume of cylinder} = \pi r^2 h = \pi(g(y_k))^2 \Delta y.$$

The total volume of the solid of revolution is the sum of the volumes of the Ω cylinders,

$$\sum_{k=1}^{\Omega} \pi(g(y_k))^2 \Delta y.$$

Rewriting this omega sum as a definite integral (replacing the summation symbol with an integral sign, replacing Δy with dy, removing the subscript, and inserting the limits of integration) gives the desired formula:

Ans. to reading exercise 4
$\int_0^2 \pi(x^2)^2\,dx = \cdots = \frac{32}{5}\pi \text{ units}^3 \approx$ 20.1 units^3

$$\text{volume of a solid of revolution} = \int_a^b \pi(g(y))^2\,dy.$$

Use this formula when the radius of a circular cross section is from the y-axis to the curve.

VOLUME OF A SOLID OF REVOLUTION (ABOUT THE Y-AXIS)

The volume of a solid of revolution generated by revolving the region between the y-axis and the curve $x = g(y)$ from $y = a$ to $y = b$ about the y-axis is

$$V = \int_a^b \pi(g(y))^2\,dy.$$

Notice that the two formulas for the volume of a solid of revolution, one when rotating about the x-axis and one when rotating about the y-axis, are really the same formula. The only differences is that the roles of x and y have been swapped. This should make sense intuitively, because turning your head 90° to visualize swapping the x- and y-axes certainly doesn't change the volume of the object. This gives rise to a general principle that we exploit often when writing geometric formulas.

Technically, swapping the x- and y-axes requires a reflection; a rotation does not complete the job.

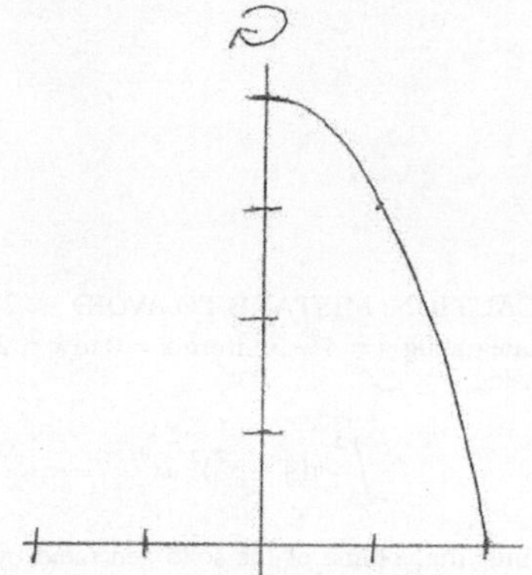

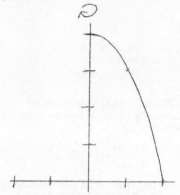

Figure 17 *Curve sketched on the specified interval, along with a rotation arrow marking the axis of rotation*

SWAPPING X AND Y

When the orientation of a geometric application changes from x to y, then the roles of x and y are swapped in the formula: dx becomes dy, functions of x are replaced with functions of y, and limits of integration as values of x are replaced with values of y. Where applicable, top and bottom are replaced by right and left, axes of rotation are swapped, and so on.

Example 3 *A solid is generated by revolving the region under the curve* $y = 4 - x^2$ *from* $x = 0$ *to* $x = 2$*, about the y-axis. Find its volume.*

Solution As we did in example 1, we begin by sketching the solid with the volume we wish to find. First we sketch the curve $y = 4 - x^2$ from $x = 0$ to $x = 2$, then we signify the axis of rotation by adding a rotation arrow around the y-axis (figure 17).

Next we sketch the reflection of the curve across the y-axis, which is on the left side of the y-axis. Note that because the curve hits the axis of rotation, its reflection also goes through that same point. See figure 18.

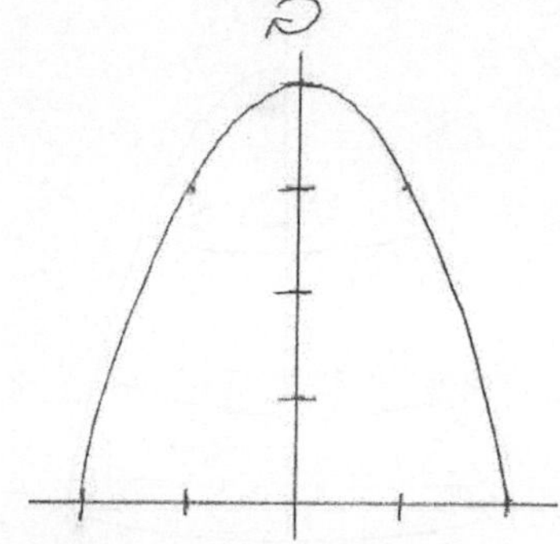

Figure 18 *The left edge of the solid is added*

Next we need to draw a few circular cross sections. Because we are rotating about the y-axis, these cross sections are horizontal. Using a perspective similar to those in figures 14 and 15, we draw the front of the cross sections lower than the back of the cross sections, and we draw them solid on the front side and dashed on the back side. As in example 1, what we draw is more like an ellipse than a circle, but it serves to give the illusion of depth to the drawing and allows us to interpret the drawing as having circular cross sections (figure 19). If there was a top circular cross section (in this example, the solid comes to a point), it would be drawn solid in front and back because it would be completely visible in the chosen perspective.

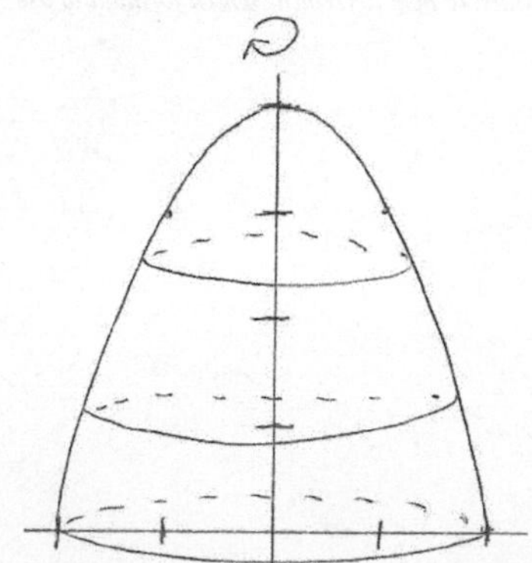

Figure 19 *Circular cross sections are added*

The final step in the drawing is to include the radius of a cross section. The radius is always drawn from the axis of rotation (the y-axis in this example) to the perimeter of the cross-sectional circle. The completed sketch is in figure 20.

Because we are rotating about the y-axis, we use the formula

$$V = \int_a^b \pi(g(y))^2\, dy.$$

This formula uses $x = g(y)$ for the curve. But, the equation of the curve is given to us as $y = 4 - x^2$, which is in the wrong form. We need to rearrange the equation of the curve to express the variable x as a function of y:

CAUTION: MISTAKE TO AVOID
Integrating $y = 4 - x^2$ from $x = 0$ to $x = 2$ using

$$\int_0^2 \pi(4 - x^2)^2\, dx$$

finds the volume of the solid generated by revolving about the x-axis, not the y-axis.

$$y = 4 - x^2$$
$$x^2 = 4 - y$$
$$x = \pm\sqrt{4 - y}.$$

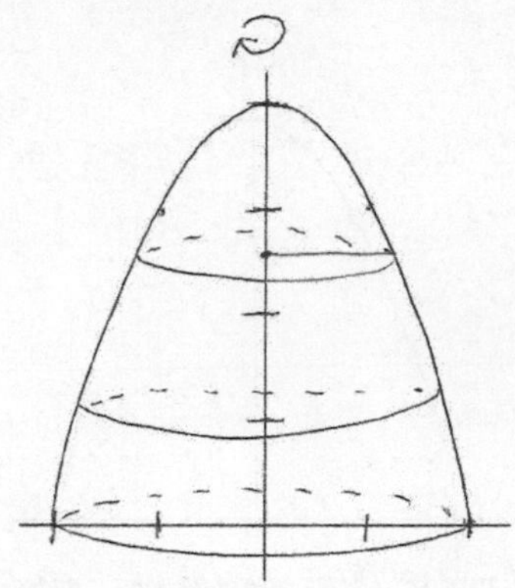

Figure 20 *The radius of a cross section is added to help determine which formula to use*

Looking again at figure 17, notice that the curve has positive x-coordinates, so we use the positive square root, $g(y) = \sqrt{4 - y}$, in the volume formula.

Also note that the limits of integration in the formula are values of y, not x. The region we are rotating has y-coordinates from $y = 0$ to $y = 4$, so these serve as the limits of integration. We are finally ready to set up our integral and find the volume:

$$\begin{aligned} V &= \int_0^4 \pi(\sqrt{4 - y})^2\, dy \\ &= \int_0^4 \pi(4 - y)\, dy \\ &= \pi\left(4y - \frac{y^2}{2}\right)\Big|_0^4 \\ &= \cdots = 8\pi\,\text{units}^3 \approx 25.1\,\text{units}^3. \end{aligned}$$

We conclude that the volume of the solid is $8\pi\,\text{units}^3$. ■

Reading Exercise 5 *A solid is generated by revolving $y = x^2$ about the y-axis, from $y = 0$ to $y = 4$. Find its volume.*

The axis of rotation indicates which formula to use. In examples 1 and 2, we rotated about the x-axis and used the formula $V = \int_a^b \pi(f(x))^2\, dx$; in example 3, we rotated about the y-axis and used the formula $V = \int_a^b \pi(g(y))^2\,dy$. Another way to tell which formula to use is based on the orientation of the radius of a cross section. In figure 10, the radius is a vertical segment. Vertical quantities are labeled y, so the radius is of the form $y = f(x)$ and we use the formula $V = \int_a^b \pi(f(x))^2\,dx$. In figure 20, the radius is a horizontal segment. Horizontal quantities are labeled x, so the radius is of the form $x = g(y)$ and we use the formula $V = \int_a^b \pi(g(y))^2\,dy$.

Volumes by cross-sectional area

In each of the formulas $V = \int_a^b \pi(f(x))^2\,dx$ and $V = \int_a^b \pi(g(y))^2\,dy$, the integrand is of the form

$$\pi \cdot \text{radius}^2.$$

The cross sections of the solids in these formulas are circles, and a look back at the development of the formulas shows that the function being integrated is the area of a cross section. This holds true even if the cross sections are not circular; volumes can be found by integrating the cross-sectional area.

VOLUMES BY CROSS-SECTIONAL AREA

A solid lying between $x = a$ and $x = b$ with x cross sections of area $A(x)$ has volume

$$\text{volume} = \int_a^b A(x)\,dx.$$

A solid lying between $y = a$ and $y = b$ with y cross sections of area $A(y)$ has volume

$$\text{volume} = \int_a^b A(y)\,dy.$$

Example 4 *A box slide (figure 21) in an outdoor play ground has vertical cross sections in the shape of a square with side length 80 cm. A measuring*

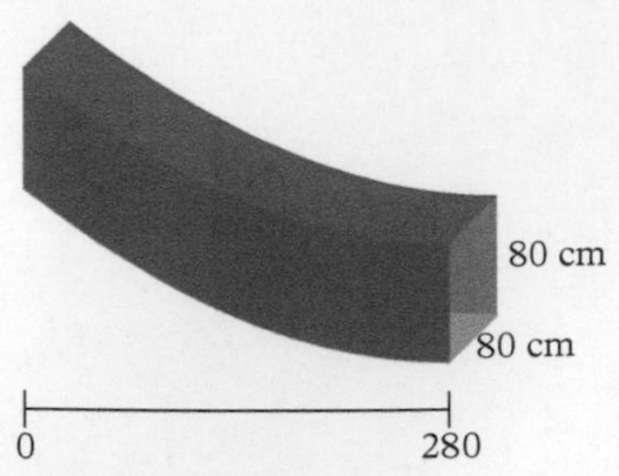

Figure 21 *The slide of example 4, with square cross sections.*

tape placed along the (horizontal) ground shows that the slide lies above a length of 280 cm—that is, you may consider it as lying from $x = 0$ to $x = 280$ on the x-axis. What is the volume enclosed by the slide?

Solution Knowing that the cross sections of the slide are squares, we can find the area of a cross section easily:

$$A = 80\text{ cm} \cdot 80\text{ cm} = 80^2\text{ cm}^2.$$

Then, using the formula for volume by cross-sectional area (with $a = 0$, $b = 280$) yields

$$\begin{aligned} \text{volume} &= \int_a^b A(x)\,dx \\ &= \int_0^{280} 80^2\,dx = 80^2x\Big|_0^{280} = 80^2 \cdot 280 - 0 \\ &= 1\,792\,000\text{ cm}^3. \end{aligned}$$

■

CAVALIERI'S PRINCIPLE

If the cross-sectional area functions $A_1(x)$ and $A_2(x)$ of two solids are equal, then the two solids have the same volume.

Bonaventura Francesco Cavalieri 1598–1647
http://www-history.mcs.st-andrews.ac.uk/Biographies/Cavalieri.html

It seems that the only difference between the slide of example 4 and a rectangular box of side lengths 80 cm, 80 cm, and 280 cm is that the slide is "curved." The volume of such a box is $V = lwh = 80 \cdot 80 \cdot 280\text{ cm}^3$, which exactly matches the end of the calculation of the volume enclosed by the slide. This is an illustration of *Cavalieri's principle*, which states that any two objects with identical cross sections have the same volume. An examination of the formulas for volumes by cross-sectional area should convince you of the compatibility of these ideas.

The calculation of the cross-sectional area in example 4 was relatively simple; the cross sections were all squares of the same size. Cross sections can vary, however, as illustrated in the final example of this section.

Figure 22 *Tree trunk with wedge removed. As pictured, the lower cut is horizontal whereas the upper cut is angled downward toward the middle of the tree*

Example 5 *Find the volume of a wedge of wood cut from a tree of diameter 14 inches, given that the width of the wedge is 7 inches and its maximum thickness is 3.5 inches.*

Solution A tree trunk with a wedge of wood removed (perhaps as part of the process for cutting down a dead tree) is pictured in figure 22. The wedge (figure 23) is not a solid of revolution, so those formulas do not apply. But, if we can find a formula for the area of a cross section, we can integrate this function and find the volume.

Figure 23 *The wedge that has been removed from the tree*

Figure 24 *Three of the many possible ways of slicing the wedge, resulting in different cross-sectional shapes*

There are often many different ways to slice an object, and the resulting slices may have different shapes (figure 24). If we slice the wedge horizontally (parallel to the lower cut) as on the left in figure 24, the cross sections are portions of a circular region, but not semicircles. If we slice the wedge vertically and parallel to the tree's diameter, as in the middle in figure 24, the cross sections are rectangles. If we slice the wedge vertically but perpendicular to the tree's diameter, as on the right in figure 24, the cross sections are triangles. The formulas for the area of a rectangle and a triangle are familiar, so slicing in one of these ways appears to be a good strategy. Let's choose the rectangles (figure 24, middle).

Ans. to reading exercise 5
$\int_0^4 \pi(\sqrt{y})^2\,dx = \cdots = 8\pi \text{ units}^3 \approx 25.1 \text{ units}^3$.

To find a function $A(x)$ or $A(y)$ for the area of a cross section, we need to place the wedge in a coordinate system. Because the diameter of the tree is given, the implication is that the tree trunk is cylindrical. In other words, a horizontal slice of the tree is a circle. Equations of circles centered at the origin are simple, so we place the center of the circle at the origin. The radius of the tree is 7 inches, so the equation of the outside edge of the tree is

$$x^2 + y^2 = 7^2 = 49.$$

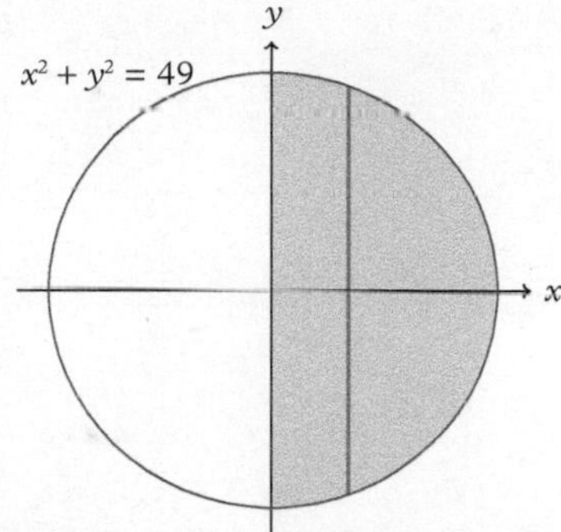

Figure 25 *A top view of the wedge of wood (light brown) in a coordinate system, with the equation of the outer edge of the tree. A vertical slice parallel to the tree's diameter is indicated by the darker brown line*

Because the wedge has width 7 inches, the two cuts producing the wedge meet at a diameter of the tree. A top view of the wedge can be placed on the right half of the circle. See figure 25.

Because we are slicing parallel to the straight edge of the wedge (again, reference the middle diagram in figure 24), we are slicing along x in figure 25. The rectangular cross section extends from the top curve to the bottom curve. This is a length in the y-direction; therefore, we need to solve the equation of the curve for y:

$$x^2 + y^2 = 49$$
$$y^2 = 49 - x^2$$
$$y = \pm\sqrt{49 - x^2}.$$

Figure 26 *A side view of the wedge. This view looks like a triangle; the given dimensions are width 7 inches and height (maximum thickness of the wedge) 3.5 inches*

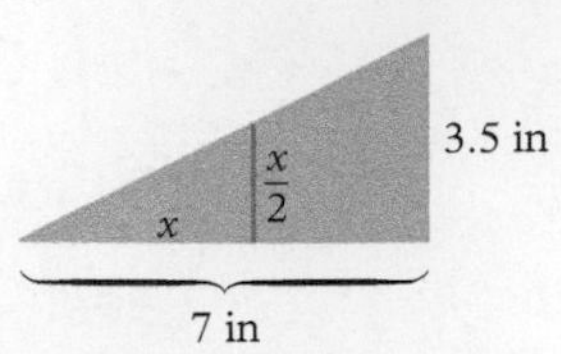

Figure 27 *A triangle with height half its width. At each cross section (darker brown segment), the height is still half the corresponding width*

The distance is top – bottom, or

$$\sqrt{49-x^2} - (-\sqrt{49-x^2}) = 2\sqrt{49-x^2},$$

giving the width of the rectangular cross section.

The next challenge is to find the height of each rectangular cross section. A straight-on side view of the wedge with its slice is helpful (figure 26). The height of the triangle in figure 26 is half its length. The same, therefore, is true of the cross section; the height of the rectangular cross section is half of its x-coordinate there (see figure 27), or $\frac{x}{2}$. The area of the cross section is therefore

$$A(x) = (2\sqrt{49-x^2})\left(\frac{x}{2}\right) = x\sqrt{49-x^2}.$$

Referencing figure 25, we see that the x-coordinates of the wedge extend from $x = 0$ to $x = 7$, so these are the limits of integration. The volume we are seeking is

$$V = \int_a^b A(x)\,dx = \int_0^7 x\sqrt{49-x^2}\,dx.$$

Using substitution, we let $u = 49 - x^2$ and calculate $du = -2x\,dx$. Then,

$$V = \int_0^7 x\sqrt{49-x^2}\,dx = -\frac{1}{2}\int_{49}^{0} u^{1/2}\,du = \frac{1}{2}\int_0^{49} u^{1/2}\,du$$
$$= \left.\frac{u^{3/2}}{3}\right|_0^{49} = \cdots = \frac{343}{3}\text{ in}^3.$$

The volume of the wedge of wood is $\frac{343}{3}$ in^3. ■

Often, the majority of the work is in visualizing the cross sections and setting up the integral. In the solution to example 5, we made a choice regarding which way we slice the solid. This choice then dictated the remainder of the procedure, including the integral we used to find the volume. What if we had made a different choice? The area of a cross section might be different, and therefore the integral might be different. But, the volume of the solid does not change.

To illustrate the fact that the same volume can be obtained by slicing in a different manner, consider the triangular cross sections on the right side of figure 24. Following the same procedure as for the rectangular cross sections, we place the wedge in a coordinate system, as in figure 25, but this time the cross sections are perpendicular to the straight edge of the wedge (figure 28).

The base of the triangular cross section stretches from the y-axis to the curve, and its length is a distance in the x-direction. We therefore solve the equation of the curve for the variable x:

$$x^2 + y^2 = 49$$
$$x^2 = 49 - y^2$$
$$x = \pm\sqrt{49 - y^2}.$$

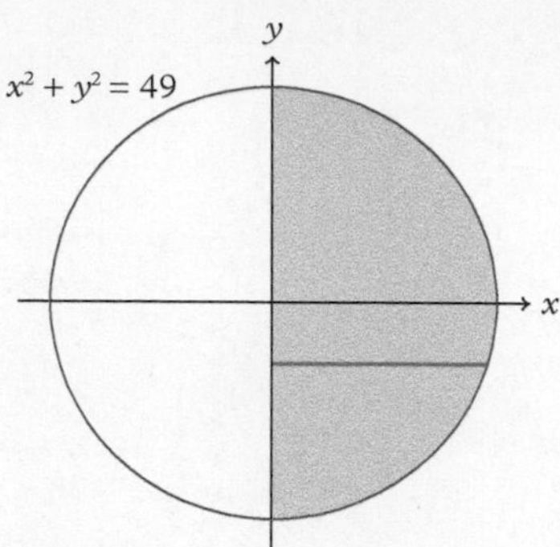

Figure 28 *A top view of the wedge of wood (light brown) in a coordinate system, with the equation of the outer edge of the tree. A vertical slice perpendicular to the tree's diameter is indicated by the darker brown line*

The desired distance is right – left, or

$$\sqrt{49 - y^2} - 0 = \sqrt{49 - y^2}.$$

Figure 29 *A side view of the wedge. In contrast to figure 26, where the rectangular cross section is viewed on its edge as a line segment, here the triangular cross section (darker brown) is visible as a triangle*

Now we need to know the height of a triangular cross section. Again, a side view is helpful; see figure 29. Just as before, the height of the triangle is half its width, making the height of the triangular cross section

$$\frac{1}{2}\sqrt{49 - y^2}.$$

Using the formula $A = \frac{1}{2}bh$ for the area of a triangle gives the area of a cross section as

$$A(y) = \frac{1}{2} \cdot \sqrt{49 - y^2} \cdot \frac{1}{2}\sqrt{49 - y^2} = \frac{1}{4}(49 - y^2).$$

Consulting figure 28, we see that the y-coordinates of the wedge extend from $y = -7$ to $y = 7$, providing the limits of integration. The volume of the wedge is then

$$V = \int_a^b A(y)\, dy = \int_{-7}^{7} \frac{1}{4}(49 - y^2)\, dy$$
$$= \frac{49}{4}y - \frac{1}{12}y^3 \Big|_{-7}^{7} = \cdots = \frac{343}{3} \text{ in}^3,$$

which is (of course) the same result we already obtained.

Notice that this alternate solution produced an integral that did not require substitution, whereas the integral in the first solution to example 5 did require substitution. This is a general principle worth noting: if there is a choice for setting up an integral with respect to x or y, one integral might be easier to evaluate than the other.

Exercises 6.2

1–10. Find the volume of the solid of revolution generated by revolving the region under the given curve about the x-axis.

1. $y = x^2$ from $x = 1$ to $x = 3$
2. $y = x^3$ from $x = 0$ to $x = 2$
3. $y = x^2 + 1$ from $x = -1$ to $x = 1$
4. $y = 2 - x^2$ from $x = -1$ to $x = 1$
5. $y = \dfrac{1}{x}$ from $x = 2$ to $x = 5$
6. $y = \dfrac{1}{\sqrt{x}}$ from $x = 1$ to $x = e$
7. $y = e^x$ from $x = 0$ to $x = 1$
8. $y = 2^x$ from $x = -1$ to $x = 0$
9. $y = \sqrt{\sin 3x}$ from $x = 0$ to $x = \frac{1}{6}\pi$
10. $y = x\sqrt{\cos x^3}$, from $x = 0$ to $x = \sqrt[3]{\pi/2}$

11–18. Find the volume of the solid of revolution generated by revolving the region bounded by the curves about the x-axis.

Some of these exercises may require more than one integral.

11. $y = x$, $x = 9$, and $y = 0$
12. $y = 0$, $y = \sqrt{x}$, and $x = 9$
13. $y = 9 - x^2$ and $y = 0$
14. $y = -x^2 + 7x - 10$ and $y = 0$
15. $y = x^2 - 4$ and $y = 0$
16. $y = x^4 - 1$ and $y = 0$
17. $y = 0$, $y = 2x + 4$, and $y = 4 - x$
18. $y = 0$, $y = x$, and $y = 2 - x$

19–26. Find the volume of the solid of revolution generated by revolving the given region about the y-axis.

19. the portion of the region between the y-axis and the curve $x = y^2$ that lies between $y = 0$ and $y = 3$
20. the portion of the region between the y-axis and the curve $x = 2 - \sqrt{y}$ that lies between $y = 0$ and $y = 1$
21. the region bounded by $x = 0$, $y = 4$, and $y = x^2$

22. the region bounded by $x = 0$, $y = 16$, and $y = \sqrt{x}$
23. the region under the curve $y = \sqrt{9 - \frac{9}{4}x^2}$ from $x = 0$ to $x = 2$
24. the region under the curve $y = (x - 2)^2 - 1$ from $x = 0$ to $x = 1$
25. the region bounded by $xy = 1$, $y = 1$, $y = 4$, and $x = 0$
26. the region bounded by $x = 0$, $y = 1$, and $y = x^3$

27–34. A solid of revolution is generated by revolving the given region about the given axis. (a) Set up the integral for determining the volume of the solid. (b) Use Simpson's rule with $n = 6$ to estimate the volume of the solid. (c) Use a calculator or computer algebra system (CAS) to find the volume of the solid.

Although we eventually study techniques for determining antiderivatives for some of the integrands in these exercises (in chapter 7, to be precise), others have antiderivatives that are not elementary functions.

27. the region under the curve $y = \ln x$ from $x = 1$ to $x = 4$, about the x-axis
28. the region under the curve $y = x \cos x^2$ from $x = 0$ to $x = 0.3$, about the x-axis
29. the region bounded by $y = e^x \sin x$, $y = 0$, and $x = 3$, about the x-axis
30. the region bounded by $y = x + \sin x$, $y = 0$, and $x = 6$, about the x-axis
31. the region under the curve $y = \cosh x$ from $x = 0$ to $x = 2$, about the x-axis
32. the region under the curve $y = 2 + \sinh x$ from $x = -1$ to $x = 2$, about the x-axis
33. the region bounded by $y = \sin x$, $x = 0$, and $y = 1$, about the y-axis
34. the region bounded by $x = 0$, $y = 4$, and $y = e^x$, about the y-axis

35. A tube slide has circular vertical cross sections of radius 30 cm. The slide lies above a horizontal length of 200 cm. Use integration to find the volume enclosed by the slide.
36. An enclosed slide has trapezoidal vertical cross sections, each with a lower base of 70 cm, an upper base of 60 cm, and a height of 45 cm. The slide lies above a horizontal length of 300 cm. Use integration to find the volume enclosed by the slide.

For exercises 35 and 36, do not use Cavalieri's principle, even though it applies. Use integration formulas instead.

Reminder: exercises marked "(GSL)" are taken from or slightly modified from the classic texts Granville, W.A., Smith, P.F., and Longley, W.R. Elements of Differential and Integral Calculus, rev. ed. Boston, MA: Ginn and Company, 1941; or Granville, W.A. Elements of the Differential and Integral Calculus. Boston, MA: Ginn and Company, 1904. Similarly, exercises marked "(Love)" are from Love, C.E. Differential and Integral Calculus. New York, NY: MacMillan, 1926.

37. Give the volume of the slide in exercise 35 in terms of cubic meters by (a) converting the answer from cubic centimeters to cubic meters and by (b) converting the lengths given in the problem from centimeters to meters and then recalculating the integral.

38. Give the volume of the slide in exercise 36 in terms of cubic meters by (a) converting the answer from cubic centimeters to cubic meters and by (b) converting the lengths given in the problem from centimeters to meters and then recalculating the integral.

39. A solid lies between $x = 0$ and $x = 10$ with cross sections perpendicular to the x-axis in the shape of semicircles of radius x. Find the volume of the solid.

40. A solid lies between $x = 0$ and $x = 10$ with cross sections perpendicular to the x-axis in the shape of squares of side length $\sqrt{x}$. Find the volume of the solid.

41. (Love) A carpenter chisels a square hole of side 2 inches through a round post of radius 2 inches, with the axis of the hole intersecting that of the post at right angles. Find the volume of the piece of wood cut out. (Use numerical integration if necessary.)

42. (GSL) The base of a solid is a circle of radius r. All sections perpendicular to a fixed diameter of the base are squares. Find the volume of the solid.

43. A solid is generated by revolving the region under the curve $y = x - 1$ from $x = 1$ to $x = 3$ about the x-axis. (a) Find the volume of the solid using integration. (b) Find the volume of the solid using the formula for the volume of a cone. (c) Compare the answers to parts (a) and (b).

44. A solid is generated by revolving the region under the curve $y = 2x + 1$ from $x = 1$ to $x = 3$ about the x-axis. (a) Find the volume of the solid using integration. (b) Find the volume of the solid using the formula for the volume of a frustum of a cone. (c) Compare the answers to parts (a) and (b).

45. The region between the x-axis and the semicircle $y = \sqrt{r^2 - x^2}$ is rotated about the x-axis. (a) Find the volume of the resulting solid. (b) What is the shape of that solid of revolution?

46. The formula $V = \int_a^b A(x)\,dx$ for the volume of a solid lying between $x = a$ and $x = b$ with x cross sections of area $A(x)$

can be justified using omega sums. Let Ω be an arbitrary positive infinite hyperreal integer and slice along the x-axis. Determine (a) Δx and (b) x_k. (c) Estimate the volume of a slice using right-hand endpoints. (d) Add the volumes of the slices together to find the total volume. (e) Rewrite the omega sum as a definite integral.

47. The wedge in example 5 was cut from a tree with a diameter of 14 inches, had a width of 7 inches, and had a maximum thickness of 3.5 inches. Suppose the upper cut was made a half inch too low, so that the cut did not go all the way to the tree's diameter, instead resulting in a wedge of width 6 inches and a maximum thickness of 3 inches. Find (or estimate) the volume of the wedge, using numerical integration if necessary.

I have such a wedge of wood, cut from a dead tree.

Volumes, Part II

6.3

Our study of volumes continues by considering rotations about lines other than the coordinate axes as well as rotations of regions between curves.

Solids of revolution: rotations about $y=k$ or $x=k$

In the previous section we found the volume of a solid of revolution generated by rotating the region between a curve and the x-axis about the x-axis, as in figure 1, but what happens if we use a different horizontal line as the axis of rotation? Horizontal lines have equations of the form $y = k$. If we rotate the region between the curve and the line $y = k$ about that same line $y = k$ (figure 2), we still get a solid of revolution, just not the same one as when rotating about the x-axis.

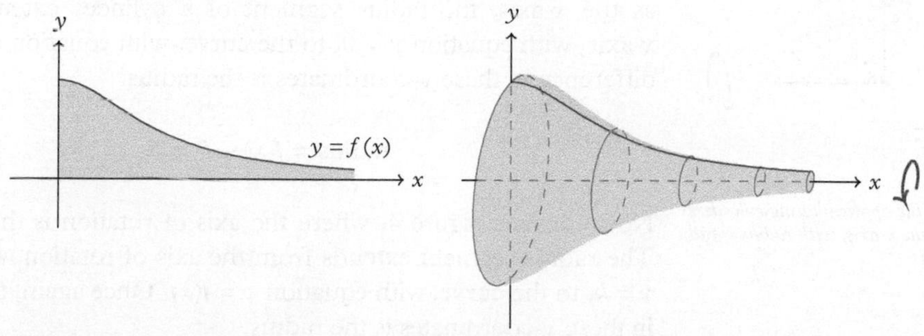

Figure 1 *(left) The region between a curve y = f(x) and the x-axis; (right) that region rotated around the x-axis, creating a solid of revolution*

Although the same curve is pictured in figures 1 and 2, the axis of rotation, which serves as the center of each circular cross section, is in a different place in figure 2 than when we developed our previous volume formulas. Therefore, the radius of the cross section is different, and the formula for finding the volume is necessarily different as well.

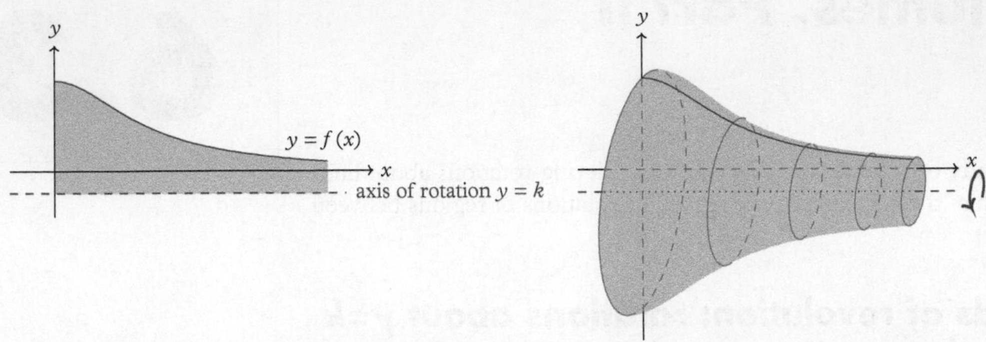

Figure 2 *(left) The region between a curve $y = f(x)$ and the line $y = k$; (right) that region rotated around the axis of rotation $y = k$*

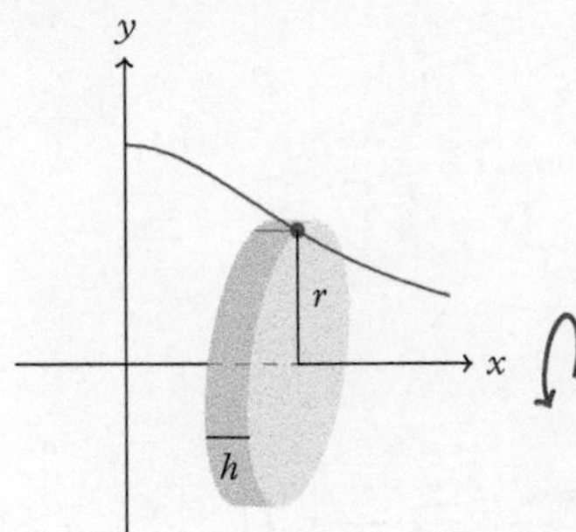

Figure 3 *One of the approximating cylinders for rotating about the x-axis, with radius r and thickness h marked*

To see more clearly how the radius of a cross section differs from before, recall how we determined the radius in section 6.2. The relevant diagram is repeated here as figure 3. With the axis of rotation as the x-axis, the radius segment of a cylinder extends from the x-axis, with equation $y = 0$, to the curve, with equation $y = f(x)$. The difference in these y-coordinates is the radius:

$$\text{radius} = f(x) - 0 = f(x).$$

Now consider figure 4, where the axis of rotation is the line $y = k$. The radius segment extends from the axis of rotation, with equation $y = k$, to the curve, with equation $y = f(x)$. Once again, the difference in these y-coordinates is the radius:

$$\text{radius} = f(x) - k.$$

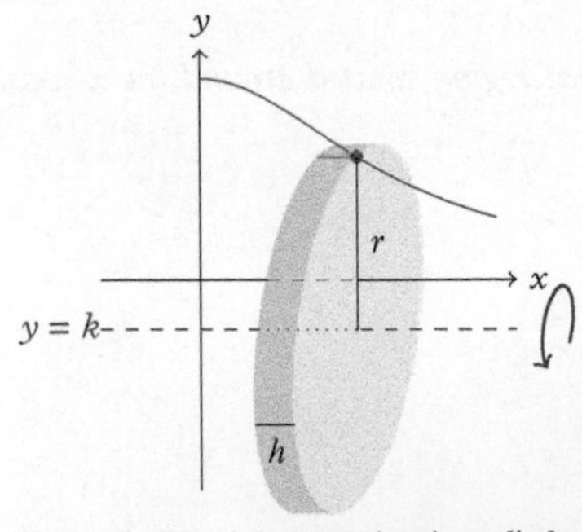

Figure 4 *One of the approximating cylinders for rotating about the line $y = k$, with radius r and thickness h marked; the radius segment extends from $y = k$ to $y = f(x)$*

Because the cross sections are still circular, the cross-sectional area is now

$$A(x) = \pi(\text{radius})^2 = \pi(f(x) - k)^2.$$

Therefore, the volume of the solid is

$$V = \int_a^b A(x)\, dx = \int_a^b \pi(f(x) - k)^2\, dx.$$

VOLUME OF A SOLID OF REVOLUTION (ABOUT $y = k$)

The volume of a solid of revolution generated by revolving the region between the curve $y = f(x)$ and the horizontal line $y = k$ from $x = a$ to $x = b$ about the line $y = k$ is

$$V = \int_a^b \pi(f(x) - k)^2 \, dx.$$

Use this formula when the radius of a circular cross section is from the horizontal line $y = k$ to the curve.

To rotate about the line $x = k$, we merely need to swap the roles of x and y, as described in section 6.2. The resulting formula is next.

VOLUME OF A SOLID OF REVOLUTION (ABOUT $x = k$)

The volume of a solid of revolution generated by revolving the region between the curve $x = g(y)$ and the vertical line $x = k$ from $y = a$ to $y = b$ about the line $x = k$ is

$$V = \int_a^b \pi(g(y) - k)^2 \, dy.$$

Use this formula when the radius of a circular cross section is from the vertical line $x = k$ to the curve.

Example 1 *A solid is generated by revolving $y = 4 - x^2$ about the line $y = -1$ from $x = 0$ to $x = 2$. Find its volume.*

Solution As in example 1 in section 6.2, we begin by sketching the solid with the volume we wish to find. First we sketch the curve $y = 4 - x^2$ on the given interval, $[0, 2]$, then we draw the axis of rotation $y = -1$ and add a rotation arrow (figure 5).

Next we sketch the reflection of the curve across the axis of rotation. One method is to pick a point on the curve (such as $(0, 4)$), determine its distance to the axis of rotation ($4 - (-1) = 5$ units from the axis of rotation), then plot a point at that distance on the other side of the axis of rotation (5 units below $y = -1$, or the point $(0, -6)$) using the same x-coordinate. We repeat this for a few other points. (The point $(1, 3)$ on the curve is 4 units above the axis of rotation, so we plot $(1, -5)$,

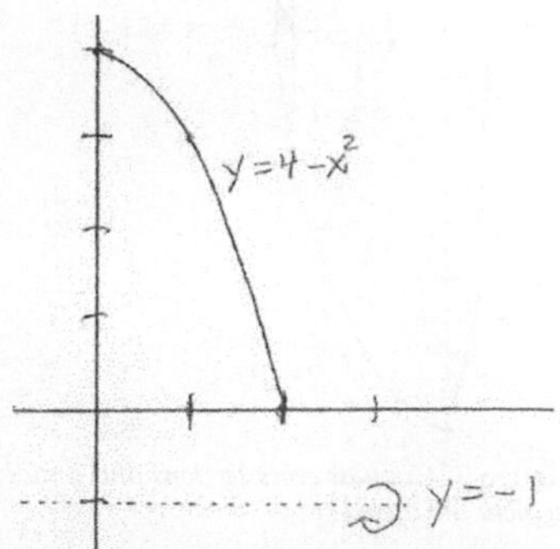

Figure 5 *Curve $y = 4 - x^2$ sketched on the specified interval $[0, 2]$, along with the axis of rotation $y = -1$ and a rotation arrow marking the axis of rotation*

which is 4 units below the axis of rotation; similarly we plot $(2, -2)$.) See figure 6.

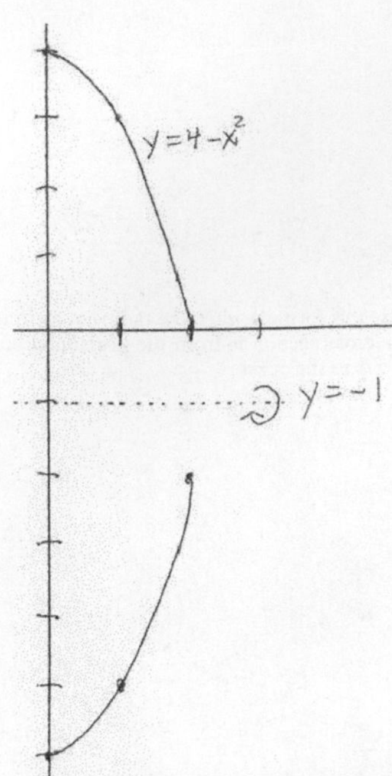

Figure 6 *The reflection of the curve across the axis of rotation is added*

We finish the sketch of the solid by adding a few circular cross sections as well as a segment representing the radius of a typical cross section. Notice that the radius segment starts at the axis of rotation and goes to the curve; see figure 7.

We now have four formulas for finding the volume of a solid of revolution, one each for rotating about the x-axis, for rotating about the y-axis, for rotating about the horizontal line $y = k$, and for rotating about the vertical line $x = k$. Here we are rotating about the horizontal line $y = -1$. Using the formula for rotating about the line $y = k$ with $k = -1$, $a = 0$, $b = 2$, and $f(x) = 4 - x^2$ gives

$$V = \int_a^b \pi(f(x) - k)^2\, dx = \int_0^2 \pi\left(4 - x^2 - (-1)\right)^2 dx$$

$$= \pi \int_0^2 \left(25 - 10x^2 + x^4\right) dx$$

$$= \pi \left(25x - 10\frac{x^3}{3} + \frac{x^5}{5}\right)\Bigg|_0^2$$

$$= \cdots = \frac{446\pi}{15}\text{ units}^3 \approx 93.4\text{ units}^3.$$

We conclude that the volume of the solid is $\frac{446\pi}{15}$ units^3. ■

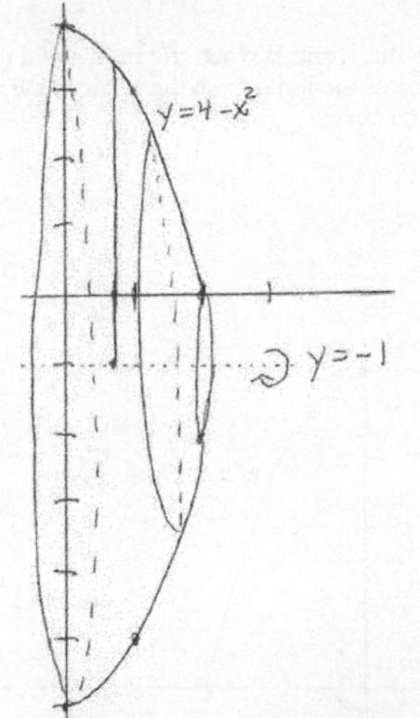

Figure 7 *Circular cross sections and a radius segment are added*

Example 2 *A solid is generated by revolving $y = \sqrt{x}$ about the line $x = 3$ from $x = 4$ to $x = 9$. Find the volume of the solid.*

Solution A sketch of the solid, using the same steps as in example 1, is in figure 8. As always, the circular cross sections are drawn perpendicularly to the axis of rotation; because the axis of rotation is vertical, the cross sections are oriented horizontally.

Because the axis of rotation is $x = 3$, we use the formula for rotating about the line $x = k$ with $k = 3$:

$$V = \int_a^b \pi\left(g(y) - k\right)^2 dy.$$

This formula uses $x = g(y)$. Instead, we are given the curve in the form $y = f(x)$—that is, $y = \sqrt{x}$. We therefore need to solve the equation for x:

$$y = \sqrt{x},$$
$$y^2 = x.$$

Any of the volume formulas may require rearranging an equation or similar efforts. Always check the formula's components for compatibility.

We use $g(y) = y^2$. Also notice that the formula is for the region from $y = a$ to $y = b$. Instead, the region is described from $x = 4$ to $x = 9$. A peek at figure 8 reveals that we need the y-coordinates corresponding to $x = 4$ and $x = 9$. Since $y = \sqrt{x}$, the region extends from $y = 2$ to $y = 3$, as in the diagram. We are finally ready to use the formula to calculate the volume:

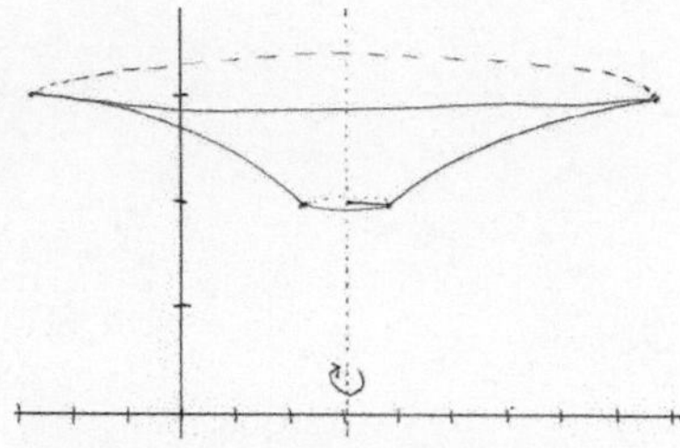

Figure 8 *The solid generated by the curve $y = \sqrt{x}$ on $[4, 9]$ with an axis of rotation of $x = 3$, including two circular cross sections and a radius segment. Notice that different scales are used on the x- and y-axes, so care must be taken when interpreting the sketch*

$$V = \int_a^b \pi\,(g(y) - k)^2\,dy = \int_2^3 \pi\left(y^2 - 3\right)^2\,dy$$
$$= \pi \int_2^3 \left(y^4 - 6y^2 + 9\right)\,dy$$
$$= \pi\left(\frac{y^5}{5} - 2y^3 + 9y\right)\Big|_2^3$$
$$= \cdots = \frac{66\pi}{5}\ \text{units}^3 \approx 41.469\ \text{units}^3.$$

We conclude that the volume is $\frac{66\pi}{5}$ units^3. ■

Reading Exercise 6 *A solid is generated by revolving $y = x^2$ about the line $y = -1$ from $x = 0$ to $x = 1$. Find its volume.*

Washer-shaped cross sections

An object turned on a lathe does not have to be solid all the way to the axis of rotation. Some of the "inside" can be carved out to make a bowl, a vase, or some other hollow object. The effect is the same as rotating the region between two curves about an axis of rotation.

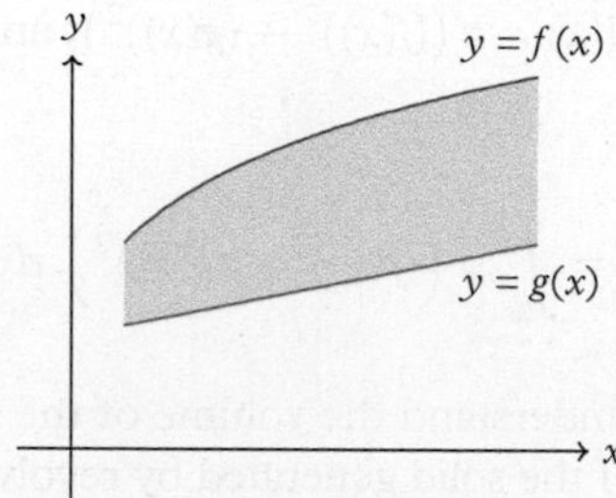

Figure 9 *A region (light green) between two curves (blue)*

Consider the region between two curves $y = f(x)$ and $y = g(x)$, as pictured in figure 9. If this region is rotated about the x-axis, then the top curve ($y = f(x)$ in figure 9) rotates around the outside of the resulting solid, whereas the inner curve ($y = g(x)$) rotates around the inside of the solid (figure 10). A cross section of the solid is not circular.

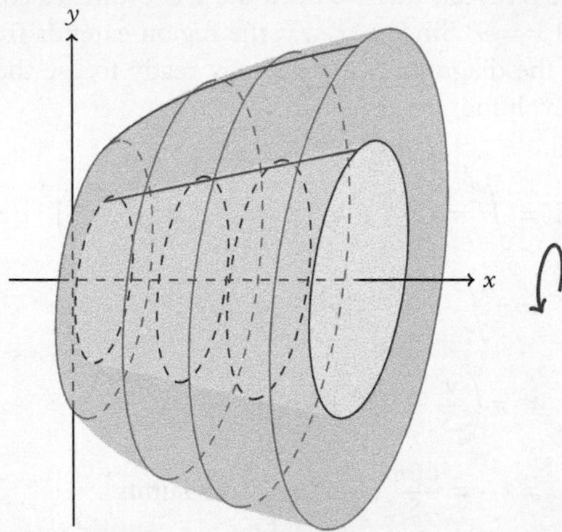

Figure 10 *A solid generated by revolving a region between two curves about the x-axis. A cross section of the solid is an annulus (think "washer" or "flattened donut")*

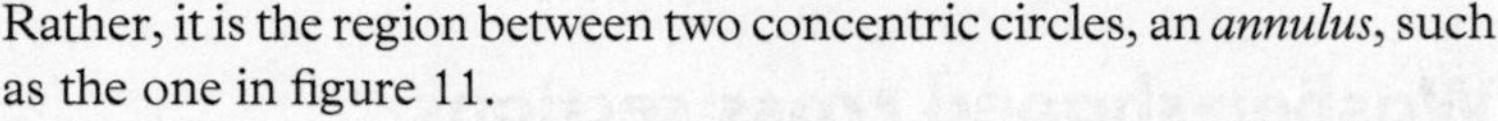

Rather, it is the region between two concentric circles, an *annulus*, such as the one in figure 11.

How do we find the volume of the solid? If we can find the area $A(x)$ of a cross section, we can use $V = \int_a^b A(x)\,dx$. Each cross section is an annulus, and the formula for the area of an annulus is $A = \pi(R^2 - r^2)$ (see figure 11). The outer radius is $R = f(x)$ whereas the inner radius is $r = g(x)$ (each radius is the distance from the axis of rotation, where $y = 0$, to the curve, where $y = f(x)$ or $y = g(x)$). Therefore, the area of the cross section is $A(x) = \pi\left((f(x))^2 - (g(x))^2\right)$, and the volume of the solid is

$$V = \int_a^b \pi\left((f(x))^2 - (g(x))^2\right)\,dx.$$

Another way to understand the volume of the solid in figure 10 is to take the volume of the solid generated by revolving the outer curve $y = f(x)$ around the x-axis, which is $V = \int_a^b \pi(f(x))^2\,dx$, and then

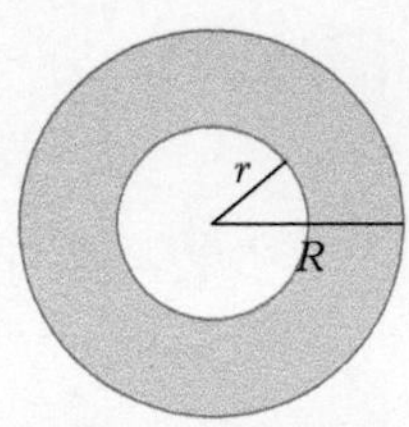

Figure 11 *An annulus (light green), which is the region between two circles with the same center. Its area is the area of the outer circle minus the area of the inner circle:*

$$A = \pi R^2 - \pi r^2 = \pi(R^2 - r^2).$$

Ans. to reading exercise 6

$\int_0^1 \pi\left(x^2 - (-1)\right)^2\,dx = \cdots =$

$\frac{28}{15}\pi$ units^3 ≈ 5.86 units^3

subtract the volume of the "hole" generated by revolving the inner curve $y = g(x)$ around the x-axis, which is $V = \int_a^b \pi(g(x))^2\,dx$, to get

$$\begin{aligned} V &= \int_a^b \pi(f(x))^2\,dx - \int_a^b \pi(g(x))^2\,dx \\ &= \int_a^b \left(\pi(f(x))^2 - \pi(g(x))^2\right) dx \\ &= \int_a^b \pi\left((f(x))^2 - (g(x))^2\right) dx. \end{aligned}$$

See figure 12.

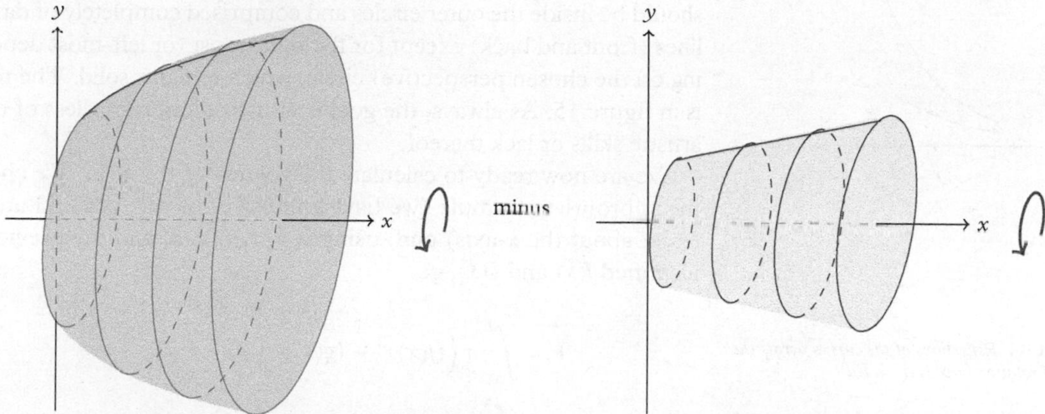

Figure 12 *The volume of the solid on the left minus the volume of the solid on the right is the volume of the solid in figure 10*

> **VOLUME OF A SOLID OF REVOLUTION WITH ANNULAR CROSS SECTIONS (ROTATION ABOUT THE X-AXIS)**
>
> The volume of a solid of revolution generated by revolving the region between the curves $y = f(x)$ and $y = g(x)$ from $x = a$ to $x = b$ about the x-axis, where $f(x) \geq g(x) \geq 0$ or $f(x) \leq g(x) \leq 0$, is
>
> $$V = \int_a^b \pi\left((f(x))^2 - (g(x))^2\right) dx.$$

The word *annular* is the adjectival form of the noun *annulus*.

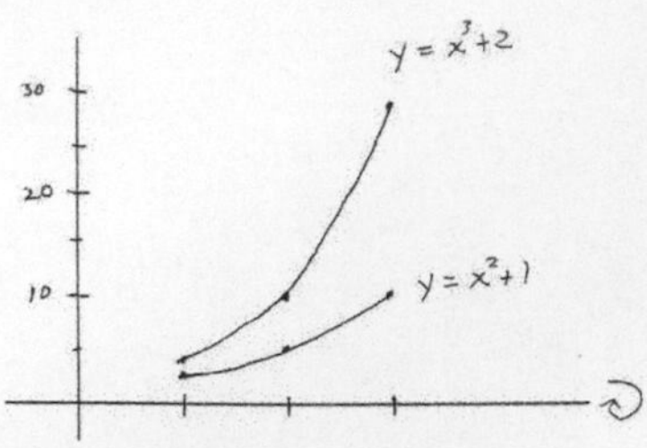

Figure 13 *Curves bounding the region to be rotated*

Example 3 *Find the volume of the solid generated by revolving the region between the curves $y = x^2 + 1$ and $y = x^3 + 2$ between $x = 1$ and $x = 3$ about the x-axis.*

Solution The steps involved in drawing the solid are the same as for other solids of revolution. First, we sketch the curves and/or region to be rotated and indicate the axis of rotation (figure 13).

At this point, it should be clear that the outer curve is $f(x) = x^3 + 2$ and the inner curve is $g(x) = x^2 + 1$. Next we sketch the reflections of both curves across the axis of rotation (the x-axis), as in figure 14.

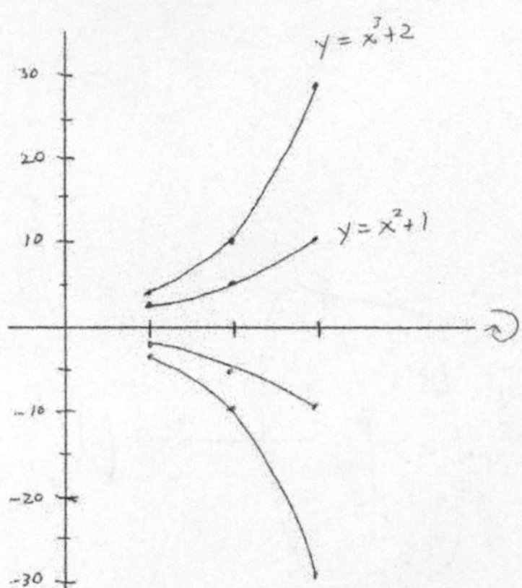

Figure 14 *Reflections of the curves across the axis of rotation have been added*

To draw the cross sections, we begin with the outside circles, the ones that connect the outside curve with its reflection. (Recall that what is drawn on the paper is like an ellipse, and hidden portions are drawn using dashed lines.) We then draw the inner circles, which should be inside the outer circles and comprised completely of dashed lines (front and back) except for the right-most (or left-most depending on the chosen perspective) circle, which remains solid. The result is in figure 15. As always, the goal is visualization, regardless of one's artistic skills or lack thereof.

We are now ready to calculate the volume of the solid. We choose the appropriate formula (we have annular cross sections and are rotating about the x-axis) and, using $a = 1$, $b = 3$, and the previously identified $f(x)$ and $g(x)$, get

$$\begin{aligned} V &= \int_a^b \pi\left((f(x))^2 - (g(x))^2\right) dx \\ &= \int_1^3 \pi\left((x^3+2)^2 - (x^2+1)^2\right) dx \\ &= \pi \int_1^3 \left(x^6 + 4x^3 + 4 - (x^4 + 2x^2 + 1)\right) dx \\ &= \pi \int_1^3 \left(x^6 - x^4 + 4x^3 - 2x^2 + 3\right) dx \\ &= \pi\left(\frac{x^7}{7} - \frac{x^5}{5} + x^4 - \frac{2x^3}{3} + 3x\right)\Bigg|_1^3 \\ &= \cdots = \frac{34\,918\pi}{105} \text{ units}^3 \approx 1044.7 \text{ units}^3. \end{aligned}$$

We conclude that the volume is $\frac{34\,918\pi}{105}$ units^3. ■

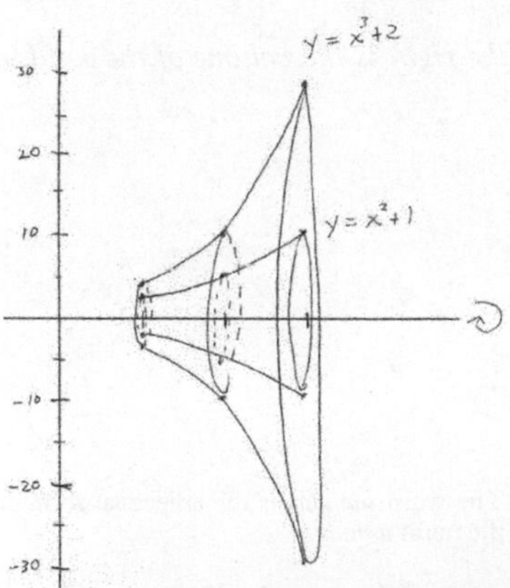

Figure 15 Annular cross sections are added. Compare to figure 10

Reading Exercise 7 *Find the volume of the solid generated by revolving the region between the curves $y = x$ and $y = x^2$ between $x = 0$ and $x = 1$, about the x-axis.*

Exercises 6.3

1–10. A solid is generated by revolving the given curve about the line. Find the volume of the solid.

1. $y = 3x$ from $x = 1$ to $x = 5$, about $y = 1$
2. $y = 2x + 1$ from $x = 0$ to $x = 2$, about $y = -1$
3. $y = 3x$ from $x = 1$ to $x = 5$, about $x = 1$
4. $y = 2x + 1$ from $x = 0$ to $x = 2$, about $x = -1$
5. $y = 3 + \sqrt{\sin x}$ from $x = 0$ to $x = \pi$, about $y = 3$
6. $x = \sqrt{y + 4}$ from $y = 0$ to $y = 5$, about $x = -3$
7. $x = y^3$ from $y = 1$ to $y = 2$, about $x = 10$
8. $y = x^2 + 4$ from $x = 1$ to $x = 3$, about $y = 1$
9. $xy = 1$ from $y = 1$ to $y = 2$, about $y = -2$
10. $x = \ln y$ from $y = 1$ to $y = e$, about $y = 5$

11–22. A solid is generated by revolving the region between the two curves about the given axis. Find the volume of the solid.

Some of these exercises might require the answer to exercise 35. You might want to complete exercise 35 before you tackle exercises 11–22.

11. between $y = 2 - x$ and $y = x^2$ from $x = 0$ to $x = 1$, about the x-axis
12. between $y = x$ and $y = \sqrt{x}$ from $x = 1$ to $x = 4$, about the x-axis
13. between $y = x + 1$ and $y = x^3$ from $x = 0$ to $x = 1$, about the x-axis
14. between $y = \sqrt{4 + \cosh x}$ and $y = 2$ from $x = 0$ to $x = 3$, about the x-axis
15. between $y = 1 + \sec x$ and $y = \sqrt{2 \sec x}$ from $x = 0$ to $x = \frac{\pi}{4}$, about the x-axis
16. between $y = x + 1$ and $y = 2x^2$ from $x = 0$ to $x = 1$, about the x-axis
17. between $y = x^3$ and $y = x$ from $x = -1$ to $x = 0$, about the x-axis
18. between $y = x^2 - 5$ and $y = -1$ from $x = 0$ to $x = 2$, about the x-axis
19. between $x = 2$ and $x = y^3$ from $y = 0$ to $y = 1$, about the y-axis
20. between $x = 4 - y$ and $x = \sqrt{y} + 4$ from $y = 0$ to $y = 4$, about the y-axis

21. between $y = \dfrac{1}{\sqrt{x^2+1}}$ and $y = \dfrac{x}{2}$ from $x = 0$ to $x = 1$, about the x-axis
22. between $y = \dfrac{1}{\sqrt[4]{1-x^2}}$ and $y = 1$ from $x = -\frac{1}{2}$ to $x = \frac{1}{2}$, about the x-axis

23–30. A solid is generated by revolving the region bounded by the curves about a line. Find the volume of the solid.

23. bounded by $y = 1$, $y = e^x$, and $x = 1$, about $y = 1$
24. bounded by $y = x^2$ and $y = 4$, about $y = 4$
25. (GSL) bounded by $y = 3$ and $y = 4x - x^2$, about $y = 3$
26. bounded by $x = 2\sqrt{y}$, $y = 9$, and $x = 0$, about the x-axis
27. bounded by $x = y^2$ and $y = x$, about the x-axis
28. bounded by $y = \sqrt{x}$, $x = 1$, and $y = -x$, about $x = 1$
29. bounded by $y = x^3$ and $y = 4x$, for nonnegative x only, about the y-axis
30. bounded by $y = x^2$ and $y = 4x - 3$, about the y-axis

Ans. to reading exercise 7

$\int_0^1 \pi\left((x)^2 - (x^2)^2\right)\,dx = \cdots = \frac{2}{15}\pi \text{ units}^3 \approx 0.419 \text{ units}^3$

31–34. For the given solid, (a) set up the integral for determining the volume of the solid, (b) use the trapezoid rule with $n = 5$ to estimate the volume of the solid, and (c) use a calculator or CAS to find the volume of the solid.

31. bounded by $y = 1$, $y = e^x$, and $x = 1$, about $x = 1$
32. the region between $y = \ln x$ and $y = 3 - x$ from $x = 1$ to $x = 2$, about the x-axis
33. the region bounded by $y = \sin^{-1} x$, $x = 1$, and $y = 0$, about the y-axis
34. the curve $y = \sec x$ between $x = 0$ and $x = \frac{\pi}{4}$, about $y = -1$

35. Rewrite the formula "volume of a solid of revolution with annular cross sections (rotation about the x-axis)" for revolving about the y-axis instead.

36. The formula $V = \int_a^b \pi\left((f(x))^2 - (g(x))^2\right)\,dx$ for the volume of a solid of revolution generated by revolving the region between the curves $y = f(x)$ and $y = g(x)$ from $x = a$ and $x = b$, about the x-axis, can be justified using omega sums. Assume that $f(x) \geq g(x) \geq 0$. Let Ω be an arbitrary positive infinite hyperreal integer and slice along the

x-axis. Determine (a) Δx and (b) x_k. (c) Estimate the volume of a slice using right-hand endpoints, where each slice is in the shape of a washer (an annulus with thickness). (d) Add the volumes of the slices together to find the total volume. (e) Rewrite the omega sum as a definite integral.

Shell Method for Volumes

6.4

In this section we study yet another method for determining the volume of a solid of revolution.

Slicing parallel to the axis of rotation

Consider the region beneath the curve $y = 1 - x^2 + 2x$ between $x = 1$ and $x = 2$, as pictured in figure 1. Rotate the region about the y-axis

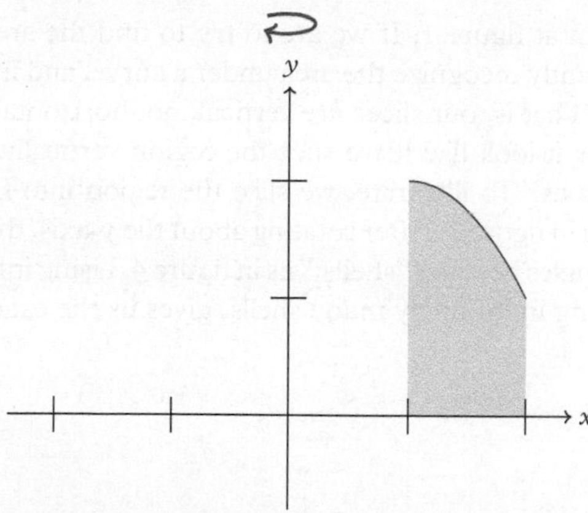

Figure 1 *The region (light green) beneath the curve $y = 1 - x^2 + 2x$ (blue), between $x = 1$ and $x = 2$, to be rotated about the y-axis. Notice that the region does not adjoin the axis of rotation*

and the result is a solid with a "hole" in it, as pictured in figure 2. Horizontal cross sections of the solid are annular ("washers"), so we already have formulas for finding the volume of the solid. This requires two integrals, one for the top half of the solid where the curve $y = 1 - x^2 + 2x$ is on the right, and one for the bottom half of the solid where $x = 2$ is on the right. For the top-half integral, we need to solve $y = 1 - x^2 + 2x$ for the variable x, which is doable for this function but can be impossible for more complicated functions.

Recall that the methods of the previous two sections required us to slice perpendicularly to the axis of rotation, so we need to use horizontal cross sections to use those formulas, resulting in integrating with respect to the variable y.

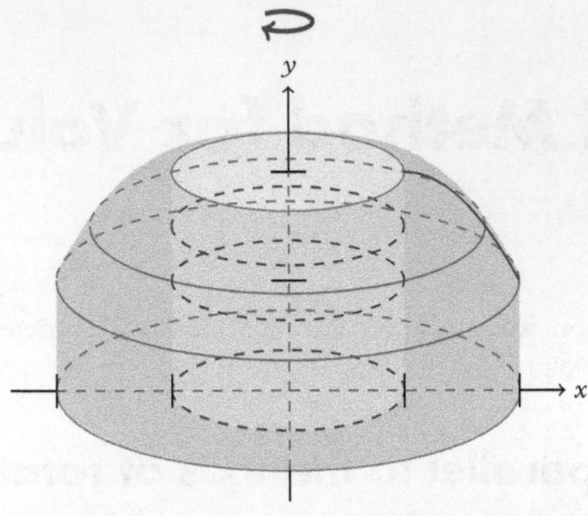

Figure 2 *The solid generated by revolving about the y-axis the region beneath the curve* $y = 1 - x^2 + 2x$ *between* $x = 1$ *and* $x = 2$. *Notice the "hole" in the object*

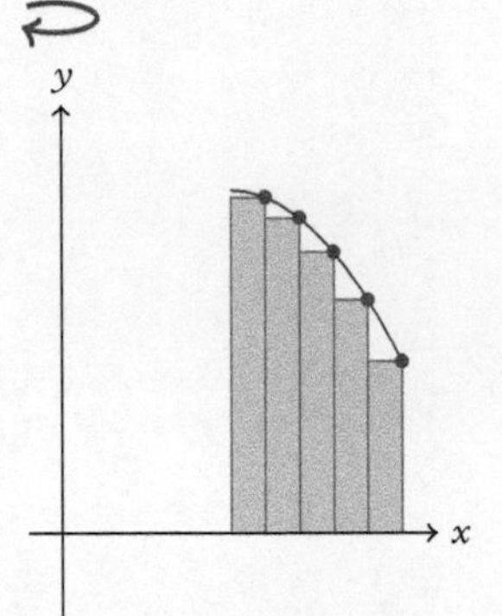

Figure 3 *The region under the curve* $y = 1 - x^2 + 2x$ *between* $x = 1$ *and* $x = 2$, *approximated by five rectangles, using right-hand endpoints*

Look again at figure 1. If we are to try to find the area of this region, we instantly recognize the area under a curve, and integrate with respect to x. That is, our slices are vertical, not horizontal.

What does it look like if we slice the region vertically, then rotate about the y-axis? To illustrate, we slice the region into finitely many rectangles, as in figure 3. After rotating about the y-axis, these approximating rectangles become "shells," as in figure 4. Using infinitely many slices, resulting in infinitely many shells, gives us the exact volume of the solid.

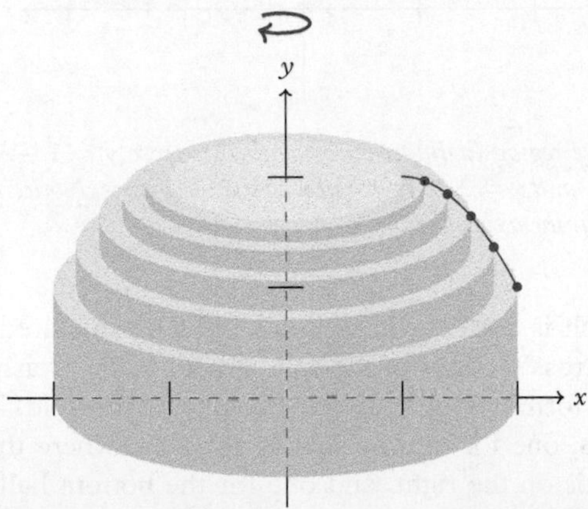

Figure 4 *The solid of figure 2, approximated using five shells, one for each of the rectangles in figure 3. Notice that we are slicing parallel to the axis of rotation*

As usual, let Ω be an arbitrary positive infinite hyperreal integer and slice along the x-axis, letting $\Delta x = \frac{b-a}{\Omega} = (b-a)\omega$ be the width of each subinterval. In figure 5, we see that the width of a subinterval is the thickness of a shell:

Compare the shell in figure 5 to an approximating rectangle from figure 3.

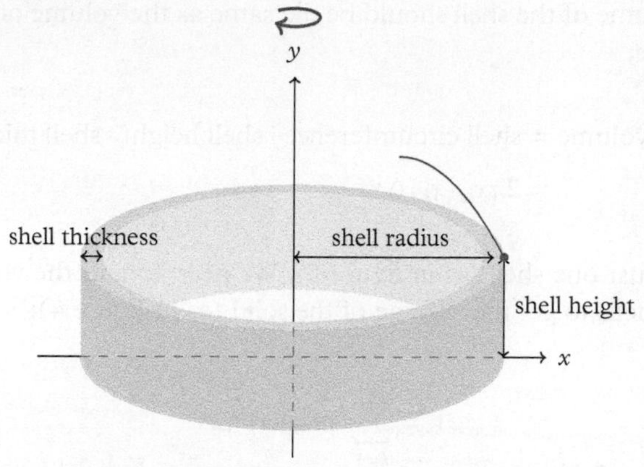

Figure 5 *One approximating shell, with the thickness, radius, and height labeled*

$$\text{shell thickness} = \Delta x.$$

If we look again at figure 5, we see that the radius of each shell is the distance from the axis of rotation (the y-axis) to the shell, which is the x-coordinate where the shell is, giving

Recall that $x_k = a + k\Delta x$ is the right-hand endpoint of each subinterval.

$$\text{shell radius} = x_k.$$

The shell height is the height of the curve at x_k—that is,

$$\text{shell height} = f(x_k).$$

Next we need to know the formula for the volume of a shell. Imagine cutting the shell parallel to the axis of rotation, then stretching it out lengthwise like a box. The result is shown in figure 6.

If we actually stretch out the shell into a box, it deforms slightly, because the inner circumference of the shell is a little less than the outer circumference of the shell. But, if the shell's thickness is infinitesimal, the two circumferences are approximately equal and any deformation can be ignored.

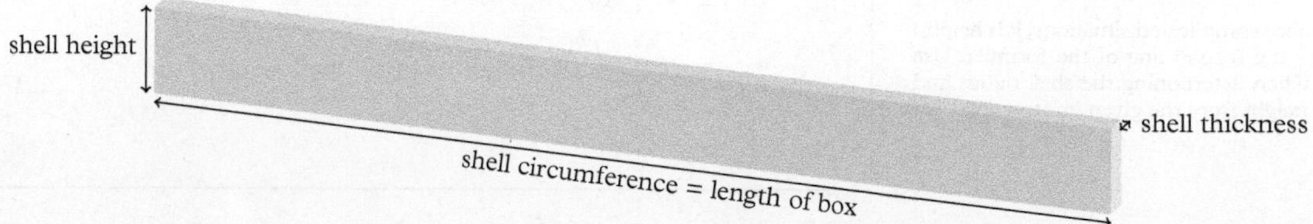

Figure 6 *The result of cutting the shell parallel to the axis of rotation and then flattening it out to form a box. (The result is not quite to scale; it should be a little longer.)*

The circumference of a circle is $2\pi r$; therefore,

$$\text{shell circumference} = 2\pi \cdot \text{shell radius} = 2\pi x_k.$$

The volume of the shell should be the same as the volume of the box; therefore,

$$\begin{aligned}\text{shell volume} &= \text{shell circumference} \cdot \text{shell height} \cdot \text{shell thickness} \\ &= 2\pi x_k f(x_k) \Delta x.\end{aligned}$$

This is just one shell (as in figure 5). We need to add the volumes of all the shells to get the volume of the solid (as in figure 4):

$$V = \sum_{k=1}^{\Omega} 2\pi x_k f(x_k) \Delta x.$$

Rewriting the omega sum as a definite integral (replacing the summation symbol with an integral symbol, replacing Δx with dx, removing the subscripts, and inserting limits of integration) gives the desired volume formula:

$$\text{volume} = \int_a^b 2\pi x \cdot f(x)\, dx.$$

SHELL METHOD FOR VOLUME OF A SOLID OF REVOLUTION (ABOUT THE *Y*-AXIS)

The volume of a solid of revolution generated by revolving the region under the curve $y = f(x)$ from $x = a$ to $x = b$ about the y-axis is

$$\begin{aligned}V &= \int_a^b 2\pi x \cdot f(x)\, dx \\ &= \int_a^b 2\pi(\text{shell radius})(\text{shell height})\, dx.\end{aligned}$$

Use this formula when rotating a region under a curve about the *y*-axis, slicing parallel to the *y*-axis.

For more complicated situations, it is helpful to use the second line of the formula. Usc care when determining the shell radius and shell height from the given information.

For the solid in figure 2, we have $f(x) = 1 - x^2 + 2x$, $a = 1$, and $b = 2$. The volume is calculated as

$$\begin{aligned} V &= \int_a^b 2\pi x \cdot f(x)\,dx \\ &= \int_1^2 2\pi x\left(1 - x^2 + 2x\right)\,dx \\ &= \int_1^2 2\pi\left(x - x^3 + 2x^2\right)\,dx \\ &= 2\pi\left(\frac{x^2}{2} - \frac{x^4}{4} + \frac{2x^3}{3}\right)\Bigg|_1^2 \\ &= \cdots = \frac{29\pi}{6}\,\text{units}^3 \approx 15.184\,\text{units}^3. \end{aligned}$$

Shell method examples: rotating about the y-axis

Example 1 *Find the volume of the solid generated by revolving the region bounded by the x-axis, $x = 0$, $x = 1$, and $y = x^2 + 1$, about the y-axis.*

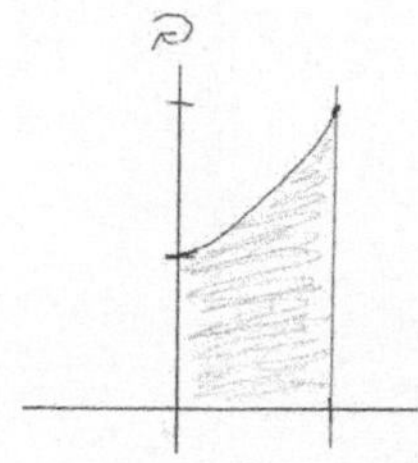

Figure 7 *Curve $y = x^2 + 1$ sketched on the specified interval $[0, 1]$, along with a rotation arrow marking the axis of rotation*

Solution We begin our diagram (figure 7) by sketching the region to be rotated along with a rotation arrow. We now have two choices for how to slice: perpendicular to the axis of rotation, as in the previous two sections, or parallel to the axis of rotation. Which should we use?

If we slice perpendicular to the axis of rotation (figure 8), we see there are two portions of the solid to consider. In the lower half, the slice extends from the axis of rotation to the line $x = 1$, indicating use of the formula for circular cross sections (disks) in section 6.2. In the upper half, the slice extends from the curve to the line $x = 1$, indicating use of the formula for annular cross sections (washers) in section 6.3. Each of the resulting integrals is with respect to y. To continue, we need to solve the equation $y = x^2 + 1$ for the variable x.

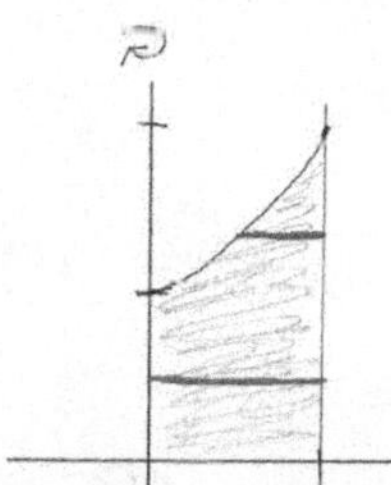

Figure 8 *The region to be rotated, along with slices perpendicular to the axis of rotation. Notice that a slice in the upper half goes between different lines/curves than a slice in the lower half*

If, instead, we slice parallel to the axis of rotation (figure 9), we see that the region to be rotated is the region under the curve $y = x^2 + 1$ from $x = 0$ to $x = 1$, which sets up correctly for the shell method; only one integral is needed, and the integral is with respect to x. There is no need to rewrite the equation $y = x^2 + 1$. This second method appears to be the easiest option.

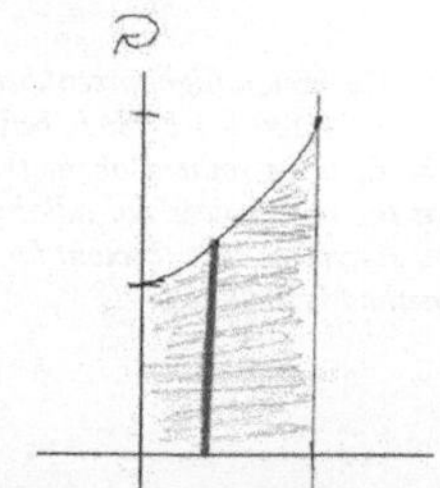

Figure 9 *The region to be rotated, along with a slice parallel to the axis of rotation. Notice that every such slice goes from the x-axis to the curve, so only one integral is needed*

Choosing the shell method, we need to determine the shell radius and the shell height. Because shells can be difficult to draw, let's illustrate them by marking the diagram without sketching the solid or the shells (figure 10). The shell radius always extends from the axis of

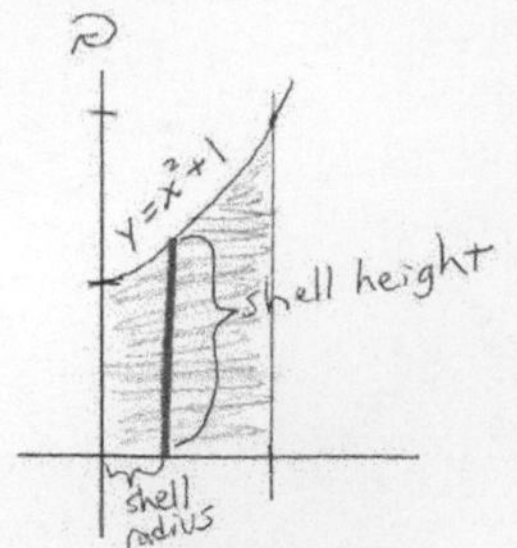

Figure 10 *The region to be rotated, with the axis of rotation, shell radius, and shell height marked for use with the shell method*

rotation to the slice; this is the x-coordinate of the shell, so we use x for the shell radius. The shell height is the height of the slice, which extends from the x-axis to the curve, and is therefore the equation of the curve; we use $f(x) = x^2 + 1$ for the shell height. Also, using $a = 0$ and $b = 1$,

$$
\begin{aligned}
V &= \int_a^b 2\pi x \cdot f(x)\, dx \\
&= \int_0^1 2\pi x \left(x^2 + 1\right) dx = \int_0^1 2\pi \left(x^3 + x\right) dx \\
&= 2\pi \left(\frac{x^4}{4} + \frac{x^2}{2}\right)\Bigg|_0^1 \\
&= \cdots = \frac{3}{2}\pi\,\text{units}^3 \approx 4.712\,\text{units}^3.
\end{aligned}
$$

We conclude that the volume of the solid is $\frac{3}{2}\pi\,\text{units}^3$. ■

As with example 1, the next example reads like an exercise that could have been included in the previous section. Part of the task for this material is to learn to discern from a diagram whether to use disks/washers by slicing perpendicular to the axis of rotation or to use shells by slicing parallel to the axis of rotation.

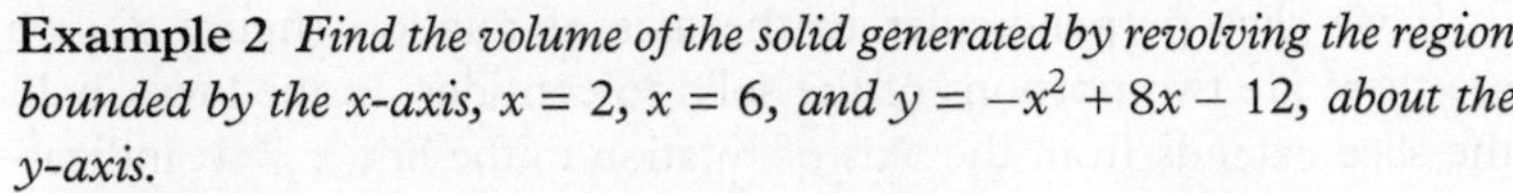

Example 2 *Find the volume of the solid generated by revolving the region bounded by the x-axis, $x = 2$, $x = 6$, and $y = -x^2 + 8x - 12$, about the y-axis.*

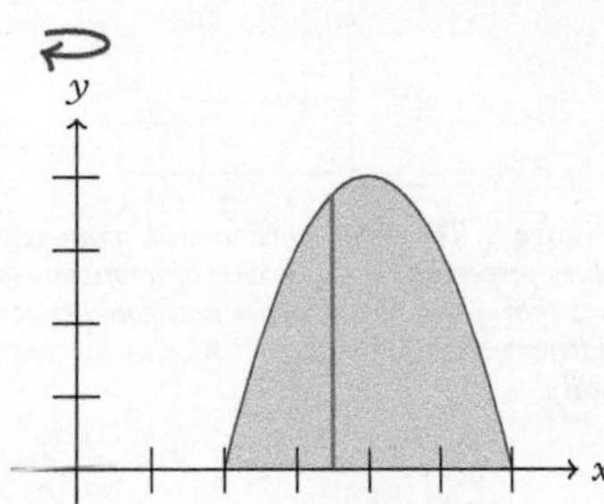

Figure 11 *The region (light green) beneath the curve $y = -x^2 + 8x - 12$ (blue), between $x = 2$ and $x = 6$, to be rotated about the y-axis. Notice that the region does not adjoin the axis of rotation. A typical slice (brown) for use with the shell method is included*

Solution We begin as always by graphing the region to be rotated (figure 11). When plotting the curve $y = -x^2 + 8x - 12$ between $x = 2$ and $x = 6$, notice that $x = 2$ and $x = 6$ are the x-intercepts of the curve. Include the rotation arrow. Although we might not sketch the solid, we should be able to visualize the shape of the solid. In this case, it's somewhat reminiscent of a bundt cake.

Notice that the region is the region under a curve and that we are rotating about the y-axis. This is exactly the setup for using the shell method for finding the volume. The slices are parallel to the axis of rotation (vertical, as in figure 11). After placing a vertical slice in the diagram, we can determine that the shell radius is x and the shell height is $f(x) = -x^2 + 8x - 12$. Then,

$$\begin{aligned} V &= \int_a^b 2\pi x \cdot f(x)\,dx \\ &= \int_2^6 2\pi x\left(-x^2+8x-12\right)\,dx \\ &= 2\pi\int_2^6 \left(-x^3+8x^2-12x\right)\,dx \\ &= 2\pi\left(-\frac{x^4}{4}+\frac{8x^3}{3}-6x^2\right)\Bigg|_2^6 \\ &= \cdots = \frac{256\pi}{3}\text{ units}^3 \approx 268\text{ units}^3. \end{aligned}$$

An alternative is to use a horizontal slice (perpendicular to the axis of rotation), which extends from the left side of the curve to the right side of the curve. Cross sections are washers and we need to solve the equation of the curve for the variable x, using the quadratic equation to obtain $x = 4 \pm \sqrt{4-y}$. We then need to determine which of these options is the right side of the curve and which is the left; use the formula for washers; simplify the integrand by squaring binomials and subtracting; then integrate an expression involving square roots, which requires substitution to obtain the answer $\frac{256\pi}{3}$ units^3. In other words, using the shell method is simpler for this solid.

We conclude that the volume of the solid is $\frac{256\pi}{3}$ units^3. ■

Reading Exercise 8 *Set up the integral (no need to evaluate the integral) for using the shell method to find the volume of the solid generated by revolving the region under the curve $y = \sqrt{x}$ between $x = 1$ and $x = 5$ about the y-axis; see figure 12.*

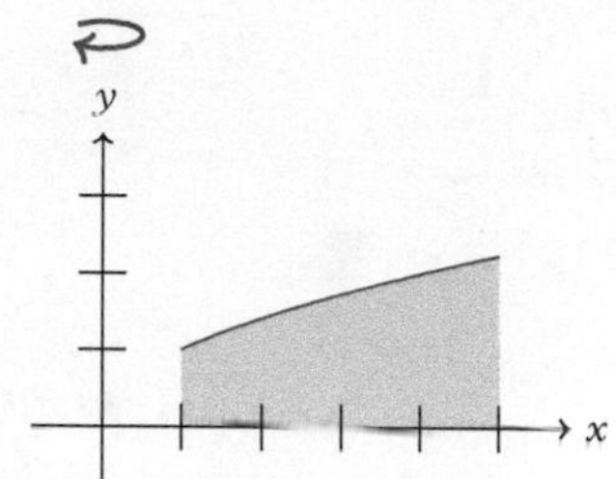

Figure 12 *The figure for reading exercise 8*

Shell method: rotating regions between curves

If the region to be rotated is not the region under a curve, but rather the region between two curves, care must be taken to determine the correct shell height.

Example 3 *Use the shell method to determine the volume of the solid generated by revolving the region bounded by the curves $y = \sqrt{x}$, $y = x$, $x = 1$, and $x = 4$, about the y-axis.*

Solution A sketch of the region bounded by the curves (figure 13) shows that it is not the region *under* a curve, but rather the region *between* two curves. We draw a typical vertical slice (parallel to the axis of rotation) and notice that it extends from the curve $y = \sqrt{x}$ to the curve $y = x$. The shell height is therefore the upper curve minus the lower curve:

$$\text{shell height} = x - \sqrt{x}.$$

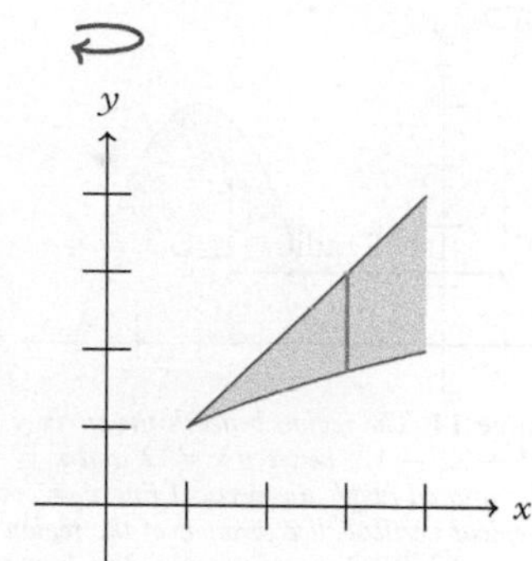

Figure 13 *The region between the curves $y = \sqrt{x}$ and $y = x$, between $x = 1$ and $x = 4$, along with a rotation arrow and a typical vertical slice (brown) of the region. Using the shell method, a slice that is parallel to the axis of rotation represents the shell height*

The shell radius, which is the distance from the axis of rotation to the location of the shell, is still x. Then, using the second line in the shell method formula,

$$\begin{aligned}
V &= \int_a^b 2\pi(\text{shell radius})(\text{shell height})\,dx \\
&= \int_1^4 2\pi x\left(x - \sqrt{x}\right)dx \\
&= \int_1^4 2\pi\left(x^2 - x^{3/2}\right)dx \\
&= 2\pi\left(\frac{x^3}{3} - \frac{x^{5/2}}{5/2}\right)\Bigg|_1^4 \\
&= \cdots = \frac{258\pi}{15}\,\text{units}^3 = \frac{86\pi}{5}\,\text{units}^3 \approx 54.035\,\text{units}^3.
\end{aligned}$$

We conclude that the volume of the solid is $\frac{86\pi}{5}$ units^3. ■

Shell method: rotating about x=k

Rotating about the vertical line $x = k$ instead of the y-axis changes the shell radius.

Example 4 *Use the shell method to find the volume of the solid generated by revolving the region bounded by the x-axis, $x = 2$, $x = 6$, and $y = -x^2 + 8x - 12$, about the line $x = -1$.*

Solution We begin as always by sketching the region to be rotated, which is the same as in example 2. Because the axis of rotation (the line $x = -1$) is not one of the coordinate axes, we draw that axis of rotation as well as the rotation arrow. To use the shell method, we add a typical slice parallel to the axis of rotation. See figure 14.

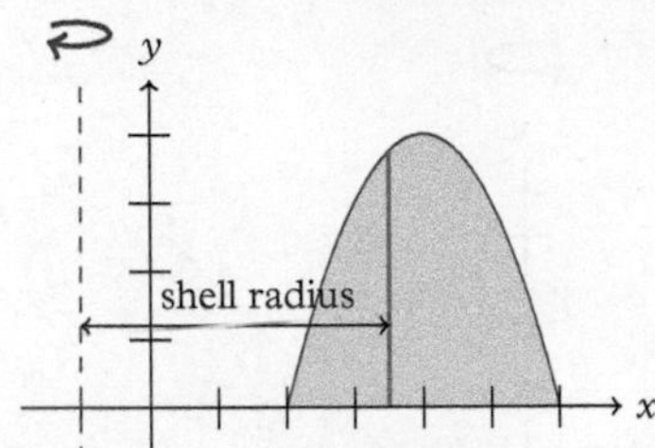

Figure 14 *The region beneath the curve $y = -x^2 + 8x - 12$, between $x = 2$ and $x = 6$, to be rotated about the vertical line $x = -1$. A typical vertical slice (brown) of the region is shown, as is the shell radius, extending from the axis of rotation to the slice. Compare to example 2 and figure 11*

What changes from our earlier examples? The shell height is the same as in example 2; the region to be rotated is the region under a curve, and therefore the shell height is $y = f(x) = -x^2 + 8x - 12$. The shell radius, however, is different from example 2. The shell radius extends from the axis of rotation to the slice—that is, from the line $x = -1$ to the x-coordinate of the slice. The shell radius is the difference of the two:

$$\text{shell radius} = \text{right} - \text{left} = x - (-1) = x + 1.$$

Ans. to reading exercise 8

$\int_1^5 2\pi x\sqrt{x}\,dx = \int_1^5 2\pi x^{3/2}\,dx$ (shell radius $= x$, shell height $= \sqrt{x}$)

The volume is then

$$V = \int_a^b 2\pi(\text{shell radius})(\text{shell height})\,dx$$

$$= \int_2^6 2\pi(x+1)\left(-x^2+8x-12\right)\,dx$$

$$= \int_2^6 2\pi\left(-x^3+8x^2-12x-x^2+8x-12\right)\,dx$$

$$= \int_2^6 2\pi\left(-x^3+7x^2-4x-12\right)\,dx$$

$$= 2\pi\left(\frac{-x^4}{4}+\frac{7x^3}{3}-2x^2-12x\right)\Big|_2^6$$

$$= \cdots = \frac{320\pi}{3}\text{ units}^3 \approx 335\text{ units}^3.$$

We conclude that the volume of the solid is $\frac{320\pi}{3}$ units^3. ■

Notice that because the radius is larger than in example 2, the volume is larger.

Shell method: rotating about the x-axis

Recall that we can swap the orientation of a geometric application of integration by swapping the roles of x and y. Doing so for our previous work with the shell method gives the following formula.

> **SHELL METHOD FOR VOLUME OF A SOLID OF REVOLUTION (ABOUT THE *X*-AXIS)**
>
> The volume of a solid of revolution generated by revolving the region between the y-axis and the curve $x = g(y)$ from $y = a$ to $y = b$ about the x-axis is
>
> $$V = \int_a^b 2\pi y \cdot g(y)\,dy$$
>
> $$= \int_a^b 2\pi(\text{shell radius})(\text{shell height})\,dy.$$

Use this formula when rotating a region between the y-axis and a curve about the x-axis, slicing parallel to the x-axis.

In the second line of the formula, the shell radius extends from the axis of rotation to the slice and is therefore a vertical distance. The shell height is measured along a slice parallel to the axis of rotation and is therefore a horizontal distance.

Using the shell method when the axis of rotation is horizontal is, in theory, not more difficult than using the shell method when the axis of rotation is vertical; the geometric analysis is identical. Practically, however, it can sometimes be challenging to ensure the correct variables are being used in each step. The key to success is to learn the geometric relationships, such as "the shell height is measured along a slice parallel to the axis of rotation," rather than only memorizing whether x or y should be used in the formula.

Example 5 *Find the volume of the solid generated by revolving the region bounded by $y = \sqrt{x}$, $x = -y$, and $y = 2$, about the x-axis.*

Solution We begin by sketching the region. Although we can slice the region vertically (perpendicular to the axis of rotation) and use the formula for annular cross sections (washers), with one integral for the left side of the solid and one integral for the right side, let's use the shell method instead. We slice horizontally (parallel to the axis of rotation), as shown in figure 15.

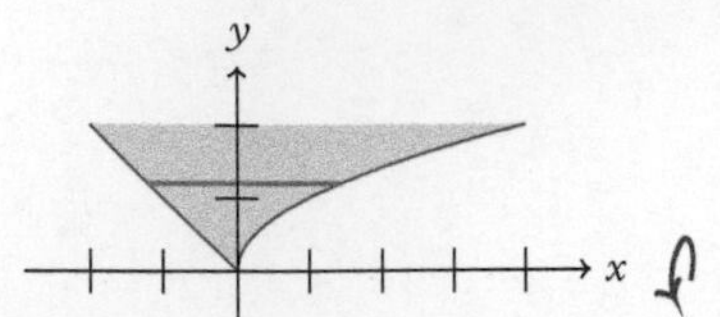

Figure 15 *The region bounded by $x = -y$, $y = \sqrt{x}$, and $y = 2$, to be rotated about the x-axis. A typical slice for the shell method, parallel to the axis of rotation, is shown (brown)*

Because we are rotating about the x-axis, the shell method formula integrates with respect to y. To use

$$V = \int_a^b 2\pi y \cdot g(y)\, dy,$$

we need to have the region between the y-axis and the curve $x = g(y)$. Looking at figure 15, we see that we do not have this type of region. Instead, the region is between two curves, as in example 3 and figure 13. Therefore, we need to use the more general second line of the formula:

$$V = \int_a^b 2\pi(\text{shell radius})(\text{shell height})\, dy.$$

Instead of using top minus bottom, as in example 3, we use right minus left. Each of these curves must be expressed as $x = g(y)$—that is, x as a function of y. The curve on the left is in that form already: $x = -y$. The curve on the right is not, so we must solve that equation for x:

$$y = \sqrt{x}$$
$$y^2 = x.$$

We use $x = y^2$ for the curve on the right. Then,

$$\text{shell height} = \text{right} - \text{left} = y^2 - (-y) = y^2 + y.$$

The shell radius, which extends from the axis of rotation (x-axis) to the slice, is a vertical distance and is the y-coordinate of the slice. Thus,

$$\text{shell radius} = y,$$

as in the first line of the formula for the shell method rotating about the x-axis. Last, to find a and b, we need the values of y over which the region lies, which is from $y = 0$ to $y = 2$. We use $a = 0$ and $b = 2$. Then,

$$\begin{aligned} V &= \int_a^b 2\pi(\text{shell radius})(\text{shell height})\, dy \\ &= \int_0^2 2\pi y \left(y^2 + y\right)\, dy \\ &= 2\pi \int_0^2 \left(y^3 + y^2\right)\, dy \\ &= 2\pi \left(\frac{y^4}{4} + \frac{y^3}{3}\right)\bigg|_0^2 \\ &= \cdots = \frac{40\pi}{3}\ \text{units}^3 \approx 41.9\,\text{units}^3. \end{aligned}$$

We conclude that the volume of the solid is $\frac{40\pi}{3}$ units^3. ■

Reading Exercise 9 *Set up the integral (no need to evaluate the integral) for using the shell method to find the volume of the solid generated by revolving the region bounded by the curves $y = x^3$, $y = 8$, and $x = 0$, about the x-axis; see figure 16.*

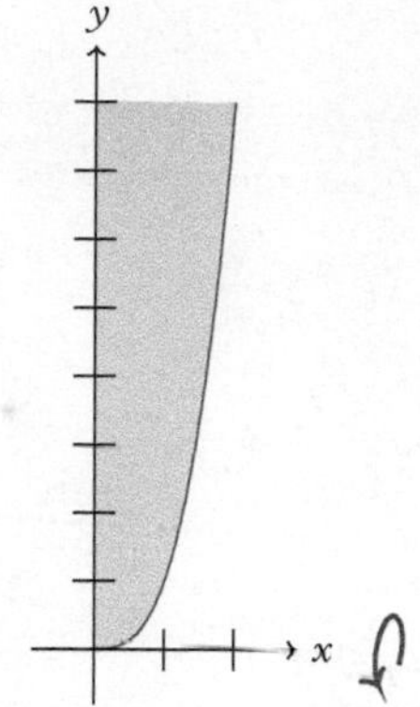

Figure 16 *The figure for reading exercise 9*

Volume: summary of methods

As is the case for many applications of integration, often the most difficult part of the task is to set up the integral correctly. With disks, washers, and shells all used for determining volumes, a choice of method must sometimes be made.

When drawing the diagram, there is a choice of orientation for a slice. This orientation, in turn, affects the method to be used.

These guides are meant to be used as helpful reminders in addition to, not in place of, the steps demonstrated in the examples of sections 6.2–6.4.

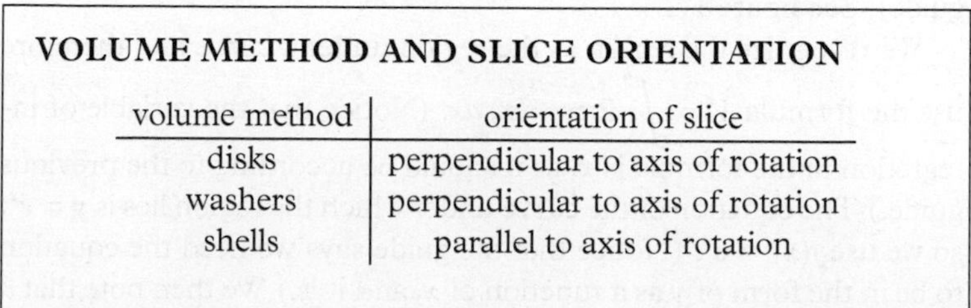

VOLUME METHOD AND SLICE ORIENTATION

volume method	orientation of slice
disks	perpendicular to axis of rotation
washers	perpendicular to axis of rotation
shells	parallel to axis of rotation

The orientation of a slice indicates the variable of integration. The variable of integration determines the form we must use for equations of curves, with y as a function of x or x as a function of y. We may need to solve an equation for x or for y to place it in the required form.

SLICE ORIENTATION, INTEGRAL VARIABLE, AND FORM OF EQUATIONS

orientation of slice	variable of integration	form of equations
vertical	$\int\, dx$	$y = f(x)$
horizontal	$\int\, dy$	$x = g(y)$

When slices lie perpendicular to the axis of rotation, we still must determine whether the formula for disks or the formula for washers should be used.

SLICE LOCATION AND METHOD

location of slice	volume method
always touches axis of rotation	disks
does not touch axis of rotation	washers

After setting up the integral, we may or may not be able to determine an antiderivative using the available formulas, rules, and techniques. If we cannot determine an antiderivative, we can use numerical integration, such as the trapezoid rule, Simpson's rule, a calculator, or a CAS.

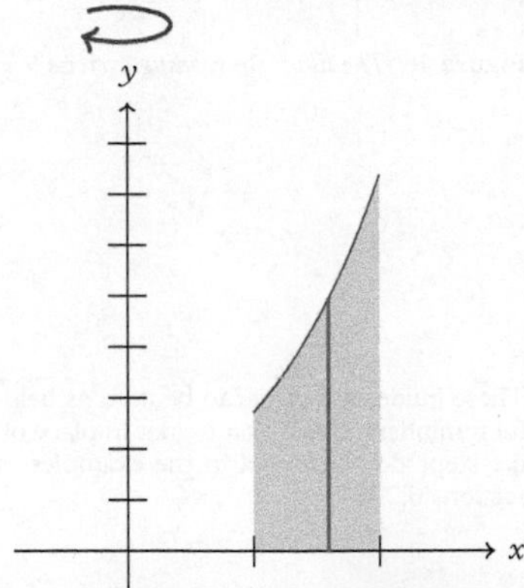

Figure 17 *The region bounded by $y = e^x$, $x = 1$, $x = 2$, and $y = 0$, along with a rotation arrow and a typical slice, for the shell method*

Example 6 *Use the shell method to determine the volume of the solid generated by revolving the region bounded by $x = 1$, $x = 2$, $y = 0$, and $y = e^x$, about the y-axis. Use a calculator to evaluate the integral if necessary.*

Solution We sketch the region and add a rotation arrow. For the shell method, we add a typical slice parallel to the axis of rotation (per the guide). See figure 17.

We recognize the region as the region under a curve and therefore use the formula $V = \int_a^b 2\pi x \cdot f(x)\, dx$. (Notice that the variable of integration in the formula is x, as it should be according to the previous guide.) The equation of the curve under which the region lies is $y = e^x$, so we use $f(x) = e^x$. (Notice that the guide says we need the equation to be in the form of y as a function of x, and it is.) We then note that a and b must be values of x, and (consulting the diagram) we conclude that $a = 1$ and $b = 2$. Then,

$$V = \int_a^b 2\pi x \cdot f(x)\,dx$$

$$= \int_1^2 2\pi x e^x\,dx = 2\pi \int_1^2 x e^x\,dx.$$

Unfortunately, the available formulas, rules, and techniques are not sufficient to find an antiderivative of xe^x; the integrand is a product that cannot be simplified, we do not have an antiderivative product rule, and substitution does not help. We therefore resort to a calculator to evaluate the integral (as instructed), determining the volume of the solid to be 46.4268 units^3. ■

We expand our list of techniques of integration in chapter 7. At that time, we will be able to find an antiderivative for this integrand.

Exercises 6.4

Ans. to reading exercise 9
$\int_0^8 2\pi y \sqrt[3]{y}\,dy = \int_0^8 2\pi y^{4/3}\,dy$ (shell radius $= y$, shell height $= \sqrt[3]{y}$)

1–8. Rapid response: a region to be rotated about an axis is given, along with a rotation arrow and a typical slice. (a) Which method is being used to determine the volume of the solid? Choose from among disks, washers, and shells. (b) The resulting integral(s) for determining the volume of the solid are with respect to which variable? Choose from either x or y. (c) How many integrals are needed to find the volume of the solid?

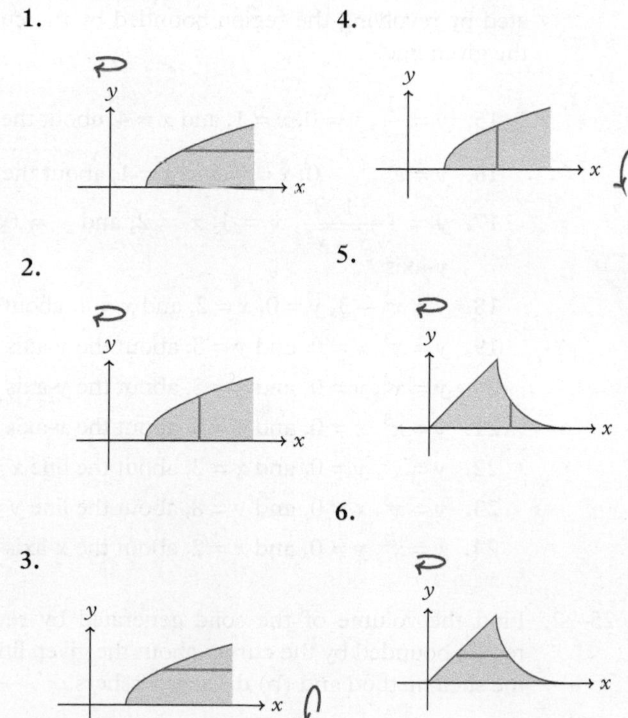

7.

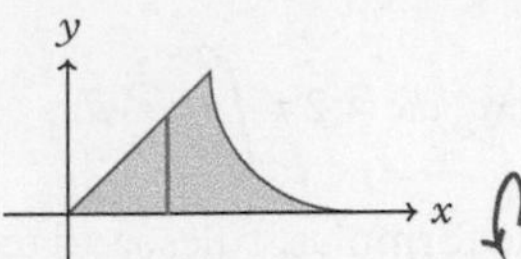

8. 

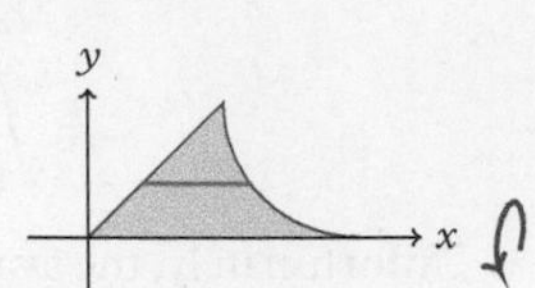

9–14. A solid is generated by revolving the region bounded by the curves about the given line. (a) Set up the integral for using the shell method to find the volume of the solid. (b) Use a calculator to determine the value of the integral.

9. $y = \sin x$ for $0 \le x \le \frac{\pi}{2}$, $y = 0$, and $x = \frac{\pi}{2}$, about the y-axis

10. $y = \ln x^2$, $x = 1$, $x = \sqrt{2}$, and $y = 0$, about the y-axis

11. $y = \sin x$ for $0 \le x \le \frac{\pi}{2}$, $y = 0$, and $x = \frac{\pi}{2}$, about the x-axis

12. $y = \ln x^2$, $x = 1$, $x = \sqrt{2}$, and $y = 0$, about the x-axis

13. $y = \sqrt{1 + x^2 + 15x^4}$, $x = 2$, $x = 17.7$, and $y = -1$, about the y-axis

14. $y = x$ and $y = \tan x$ for $0 \le x \le \frac{\pi}{4}$, about the y-axis

For these and higher-numbered exercises, evaluate the integral without resorting to numerical integration (i.e., the calculator).

15–24. Use the shell method to find the volume of the solid generated by revolving the region bounded by the curves about the given line.

15. $y = \dfrac{1}{x^2}$, $y = 0$, $x = 1$, and $x = 4$, about the y-axis

16. $y = e^{x^2}$, $y = 0$, $x = 0$, and $x = 1$, about the y-axis

17. $y = \dfrac{1}{\sqrt{5 - x^2}}$, $x = 1$, $x = 2$, and $y = 0$, about the y-axis

18. $y = x^2 - 3$, $y = 0$, $x = 2$, and $x = 4$, about the y-axis

19. $y = x^3$, $x = 0$, and $y = 8$, about the y-axis

20. $y = x^2$, $y = 0$, and $x = 2$, about the y-axis

21. $y = x^3$, $x = 0$, and $y = 8$, about the x-axis

22. $y = x^2$, $y = 0$, and $x = 2$, about the line $x = 3$

23. $y = x^3$, $x = 0$, and $y = 8$, about the line $y = -1$

24. $y = x^2$, $y = 0$, and $x = 2$, about the x-axis

25–32. Find the volume of the solid generated by revolving the region bounded by the curves about the given line using (a) the shell method and (b) disks or washers.

25. $y = 2\sqrt{x}$, $x = 9$, and $y = 0$, about the y-axis

26. $y = x^3$ and $y = x$, about the x-axis

27. $x = y^2$ and $y = x$, about the x-axis

28. $y = 4x^{3/2}$, $x = 4$, and $y = 0$, about the y-axis

29. $y = x^2$ for $0 \leq x \leq 1$, $y = 2 - x$, and $x = 0$, about the x-axis

30. $y = x^3$, $x = 2$, and $y = 0$, about the x-axis

31. $y = x^2$, $x = 1$, $x = 3$, and $y = 1$, about the y-axis

32. $x = \sqrt{y}$, $y = 4$, and $x = 0$, about the y-axis

33–42. Find the volume of the solid generated by revolving the region bounded by the curves about the given line using any appropriate method.

33. $y = \sqrt{4 + x^2}$, $x = 0$, $x = 1$, and $y = 0$, about the x-axis

34. $y = x^2 + 1$, $x = 0$, $x = 2$, and $y = 0$, about the x-axis

35. $y = \sqrt{4 + x^2}$, $x = 0$, $x = 1$, and $y = 0$, about the y-axis

36. $y = x^2 + 1$, $x = 0$, $x = 2$, and $y = 0$, about the y-axis

37. $y = \frac{-1}{x}$, $x = 1$, $x = 3$, and $y = 0$, about the y-axis

38. $y = x^2 - 4x + 3$ and $y = 0$, about the y-axis

39. $x = \frac{7}{y + y^3}$, $x = 0$, $y = 1$, and $y = 3$, about the x-axis

40. $y = 4 - x^2$ for $2 \leq x \leq 3$, $y = 0$, and $x = 3$, about the y-axis

41. $y = 4 - x^2$ and $y = 0$, about the x-axis

42. $y = x^3$, $y = 1$, and $x = 0$, about the x-axis

6.5

Work, Part I

As its name suggests, homework can be work. Although various types of tasks require effort, there is a specific meaning of the word *work* in physics. In this section we begin to use calculus to explore this concept.

Force

Imagine a 10-kg child sitting in a chair. Now imagine instead that a 20-kg child is sitting in that chair. With either child sitting in the chair, there is a force on the chair—a downward force—that is not present when the chair is empty (assuming, that is, that we are on earth and not in space).

There are subtleties when discussing space and gravity that are ignored here.

Our experience indicates that the force on the chair from the 20-kg child is greater than the force on the chair from the 10-kg child. But, the mass of the child is not the entire story regarding the force on the chair; otherwise, the force would be the same on earth and in space. What else is involved?

Earth's gravity creates an acceleration toward the center of the planet. This is why we observe objects falling toward the ground. It is the acceleration resulting from gravity that plays a role in the force on the chair. Think about it this way: if someone removes the chair on which you are sitting, what happens? You start moving toward the ground. You were not moving, then you are; this is a change in velocity. Because acceleration is the rate of change of velocity with respect to time, it is an acceleration that you experience. This acceleration toward the ground is the result of gravity. The force "felt" by the chair is influenced by gravity attempting to accelerate the child (or you) toward the ground.

To summarize, the greater the mass of the child, the greater the force on the chair, and acceleration also plays a role in the force on the chair. This conclusion is formalized in Newton's second law of motion, which states force equals mass times acceleration:

$$F = ma.$$

NEWTON'S SECOND LAW OF MOTION
Force equals mass times acceleration:

$$F = ma.$$

Common units of force are the pound, abbreviated lb in the English system, and the newton, abbreviated N in the SI system. Consistent with $F = ma$, the relationship between SI units for force, mass, and acceleration is

$$1\,\text{N} = 1\,\text{kg} \cdot 1\,\text{m/s}^2.$$

The abbreviation lb for pound comes from the Latin word *libra,* meaning "scale" or "balance" (as in the measuring instrument). The libra was an ancient Roman measure corresponding to a little under three quarters of a pound.

Notice that a pound is not a unit of mass, but rather a unit of force. Meanwhile, a kilogram is a unit of mass. Although we often "convert" pounds to kilograms, such a conversion is only valid for a specific location's acceleration resulting from gravity, such as near sea level on the earth or similarly on the surface of the moon. In fact, "weight," which is a force, is measured properly in units of pounds but not kilograms. Mass is measured properly in units of kilograms but not pounds. The commonly used term *weightlessness* refers correctly to the force on the object, not its mass. After all, it is not the amount of substance in the object that changes in a weightless environment. Meanwhile, instead of saying that a dieter is "losing weight," it is more accurate to say that the dieter is "losing mass."

Have you heard of the astronaut diet? Move to the moon and lose 120 pounds. The acceleration resulting from gravity is less on the moon than on the earth, so the force is less. Although you do not lose kilograms (mass) when you move to the moon, you lose pounds (force).

Work with constant force

Now imagine lifting the 10-kg child to shoulder height, or lifting the 20-kg child to shoulder height. Which is more work? Lifting the child involves counteracting the force trying to pull the child to the ground. The gravitational force on the 20-kg child is greater than the gravitational force on the 10-kg child. The greater the force, the greater the work required.

Now imagine lifting the 10-kg child for a distance of 1 m and for a distance of 2 m. Which is more work? It also makes sense that the greater the distance moved, the greater the work required.

It turns out that work can be defined as the product of force and distance:

$$W = F \cdot d.$$

Work equals force times distance:

$$W = F \cdot d.$$

Typical units of work are the foot-pound (abbreviated ft · lb) in the English system and the joule (abbreviated J, same as a newton-meter; $1\,\text{J} = 1\,\text{N} \cdot \text{m}$) in the SI system.

$1\,\text{ft} \cdot \text{lb} \approx 1.356\,\text{J}$

To determine the amount of work performed, we multiply the force by the distance.

Example 1 *A weightlifter lifts a 200-pound weight 7 feet. How much work is performed?*

In this text, we only lift objects and not lower them, so the work performed is positive.

Solution The weight 200 pounds is a force. Seven feet is a distance. The weight is constant during the 7-foot lift. Then,

If the force (in this case, the weight of 200 pounds) is not changing as the object moves, then it is constant.

$$\text{work} = \text{force} \cdot \text{distance}$$

$$W = 200\,\text{lb} \cdot 7\,\text{ft} = 1400\,\text{ft} \cdot \text{lb}.$$

We conclude that the work performed is $1400\,\text{ft} \cdot \text{lb}$. ■

Again, we read ft · lb as "foot-pounds."

Example 2 *A weightlifter lifts a 90-kg mass 2.1 m. How much work is performed?*

Solution The mass 90 kg is *not* a force. To find work, we need force, but we also know that force is mass times acceleration. Assuming the weightlifter is on the earth (a reasonable assumption), the acceleration resulting from gravity is $9.8\,\text{m/s}^2$. The force resulting from gravity is then

A more precise value for the acceleration resulting from gravity on earth is $9.807\,\text{m/s}^2$. We use the value $9.8\,\text{m/s}^2$ for calculations in this book because it is relatively easy to remember.

$$F = ma$$

$$F = 90\,\text{kg} \cdot 9.8\,\text{m/s}^2 = 882\,\text{N}.$$

In the second line, we not only multiply the numerals, $90 \cdot 9.8 = 882$, but also the units: $\text{kg} \cdot \text{m/s}^2 = \text{N}$.

(Note: we say that the weight of the object being lifted is 882 N, whereas its mass is 90 kg.) Now we are ready to calculate work:

$$W = Fd$$

$$W = 882\,\text{N} \cdot 2.1\,\text{m} = 1852.2\,\text{J}.$$

In the second line, we multipy the units: $\text{N} \cdot \text{m} = \text{J}$.

We conclude that the work performed is 1852.2 J. ■

Reading Exercise 10 *A pet owner lifts a 50-pound bag of dog food 3 feet. How much work is performed?*

Work with variable force

In examples 1 and 2, the weight of the object being lifted did not change while the object moved; the force was constant. How do we handle a variable force?

Suppose we are hauling a liquid construction compound to the roof of a building using a bucket, rope, and pulley, as in figure 1. The weight of the bucket remains the same throughout the lift, as does the weight of the liquid construction compound, as long as the bucket is not leaking. However, observe that the amount of rope between the pulley and the bucket changes, and therefore the weight of the rope that is being lifted changes! The method of examples 1 and 2 can be used to determine the work performed in lifting the bucket and construction compound to the roof, but it cannot be used to determine the work performed when lifting the rope.

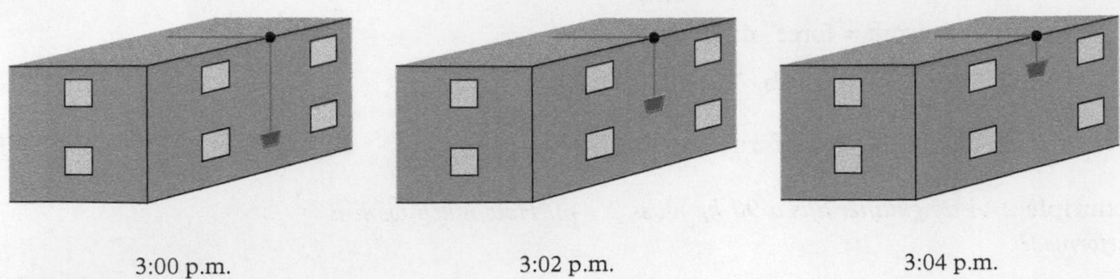

Figure 1 *A bucket is hauled to the roof of a building. As time progresses, the amount of rope hanging over the side of the building diminishes*

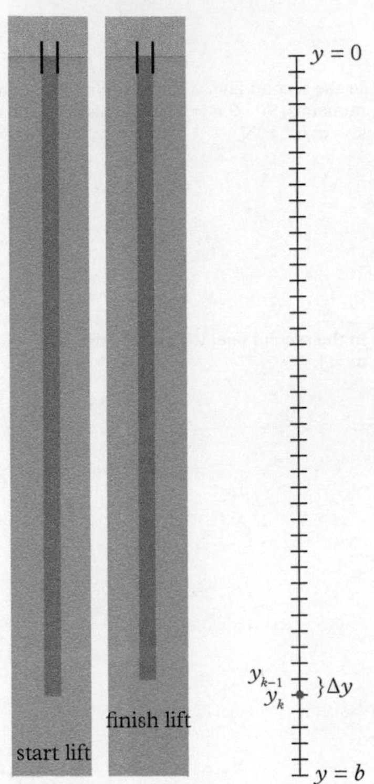

Figure 2 *A view of the rope hanging over the side of the building to y_k (left) and then after being lifted to y_{k-1} (center), along with a vertical number line (right). The rope is lifted a distance of Δy. Using the "right-hand endpoint" (the largest coordinate in the subinterval) to approximate (red dot), the work performed during this short lift is $f(y_k) \cdot \Delta y$, if f represents the force (the weight of the rope)*

Ans. to reading exercise 10
150 ft · lb

Looking at figure 1, we see that the amount of rope changes significantly over time; the weight of the rope is much less at 3:04 p.m. than it is at 3:00 p.m. Imagine instead that the distance lifted is small, as represented in figure 2 as Δy. In this case, the amount of rope (and therefore the weight of rope) does not change much during this short lift.

In fact, if we lift just an infinitesimal distance, the amount of rope does not seem to change at all. We should be able to find the work performed by multiplying the weight of the rope (the force) by the distance lifted. If we add up the work required to lift each infinitesimal distance, all along the length of the rope, then we get the total work required. Let's formalize this idea next.

Let Ω be an arbitrary positive infinite hyperreal integer. Chop the interval up into Ω pieces and let Δy represent the (infinitesimal) length of each of the resulting subintervals. Let $f(y)$ be the weight of the rope (the force) when the rope extends to a length of y. The work performed by lifting the rope from a length of y_k to a length of y_{k-1} (a distance of Δy) is then

$$\begin{aligned} W &= F \cdot d \\ &= f(y_k)\Delta y. \end{aligned}$$

This is the work for lifting the rope just one subinterval. We need to lift it all Ω subintervals, so the total work performed is

$$\text{total work} = \sum_{k=1}^{\Omega} f(y_k)\Delta y.$$

This is an omega sum! We can rewrite it as a definite integral by replacing the summation with an integral symbol, replacing Δy with dy, dropping the subscript, and inserting limits of integration:

$$W = \int_a^b f(y)\, dy.$$

Whatever the context, if we wish to find the amount of work performed by a variable force $f(y)$, the analysis is the same, and the total work required is $\int_a^b f(y)\,dy$.

WORK, VARIABLE FORCE

The work performed from $y = a$ to $y = b$, where the force is $f(y)$, is

$$W = \int_a^b f(y)\,dy.$$

Using the variable x instead of y is virtually identical. The work performed from $x = a$ to $x = b$, where the force is $f(x)$, is $W = \int_a^b f(x)\,dx$.

Example 3 *You haul a liquid construction compound to the roof of your building using a bucket, rope, and pulley. How much work is done, assuming the bucket weighs 2 pounds; the rope,* $\frac{1}{10}$ *lb/ft; the liquid, 20 pounds; and it is lifted 24 feet?*

Solution The weight of the bucket does not change while being lifted, so the work performed on the bucket is

$$\begin{aligned} W &= F \cdot d \\ &= 2\,\text{lb} \cdot 24\,\text{ft} = 48\,\text{ft} \cdot \text{lb}. \end{aligned}$$

Likewise, the weight of the liquid does not change as it is lifted, so the work performed on the liquid is

$$W = F \cdot d = 20\,\text{lb} \cdot 24\,\text{ft} = 480\,\text{ft} \cdot \text{lb}.$$

As the rope is lifted to the roof, the length of the rope diminishes, so the amount of rope being lifted changes and is not constant. To use

$$W = \int_a^b f(y)\,dy,$$

we need to have an expression for the force $f(y)$, which is the weight of the rope at y. We therefore need two items: (1) an understanding of the weight of the rope and (2) coordinates for y that correspond to the problem.

For item (1), the weight of the rope per foot is given as $\frac{1}{10}$ lb/ft. In other words, 20 feet of rope weighs

The amount the rope weighs per unit length is called its *linear weight density*.

$$20\,\text{ft} \cdot \frac{1}{10}\,\text{lb/ft} = 2\,\text{lb}.$$

Notice $\text{ft} \cdot \frac{\text{lb}}{\text{ft}} = \frac{\cancel{\text{ft}} \cdot \text{lb}}{\cancel{\text{ft}}} = \text{lb}$.

Then, y feet of rope weighs

$$y\,\text{ft} \cdot \frac{1}{10}\,\text{lb/ft} = \frac{1}{10}y\,\text{lb}.$$

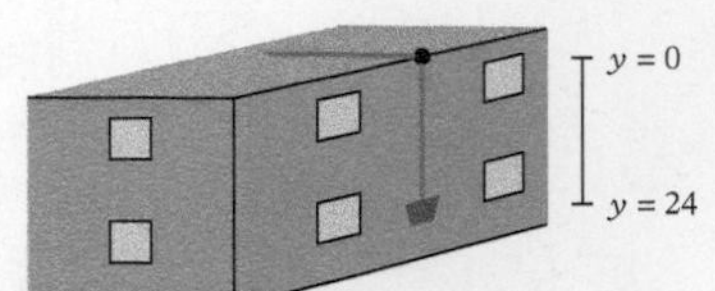

Figure 3 *A coordinate axis for y is placed alongside the rope. We choose to let the value of y be the length of the rope*

For item (2), we need to place a number line—coordinates—for y into the diagram (figure 3). It is convenient to let y represent the length of rope when the bucket is at that position. At the bottom, $y = 24$; at the top, $y = 0$. The weight of the rope at a coordinate of y is then

$$f(y) = \frac{1}{10}y,$$

measured in pounds. Because the rope extends from $y = 0$ to $y = 24$, we use $a = 0$, $b = 24$. (The subinterval length $\Delta y = (b - a)\omega$ must be a positive quantity because it represents a distance. We integrate from the smaller y-coordinate to the larger y-coordinate so that $b - a > 0$.) Then,

$$\begin{aligned} W &= \int_a^b f(y)\,dy \\ &= \int_0^{24} \frac{1}{10}y\,dy \\ &= \frac{1}{20}y^2\Big|_0^{24} \\ &= \frac{576}{20} - 0 = 28.8\,\text{ft} \cdot \text{lb}. \end{aligned}$$

Since $\int_a^b f(y)\,dy = -\int_b^a f(y)\,dy$, the only effect of placing the limits of integration in the wrong order is changing the sign on the result and obtaining a negative amount of work performed. Knowing that the amount of work performed in a lift must be positive helps identify this error.

The units for this integral are units of work. Because we used pounds for force and feet for distance, our units are foot-pounds. Alternately, the units can be seen by including them in the integral:

$$W = \int_0^{24} \frac{1}{10}y\,\text{lb}\,dy\,\text{ft} = \cdots = 28.8\,\text{ft} \cdot \text{lb}.$$

To find the total work performed, we add the work performed on the bucket, construction compound, and rope to get

$$\begin{aligned} \text{total work} &= 48\,\text{ft} \cdot \text{lb} + 480\,\text{ft} \cdot \text{lb} + 28.8\,\text{ft} \cdot \text{lb} \\ &= 556.8\,\text{ft} \cdot \text{lb}. \end{aligned}$$

Using multiplication for constant forces and integration for the variable force, the total work is determined to be 556.8 foot-pounds. ■

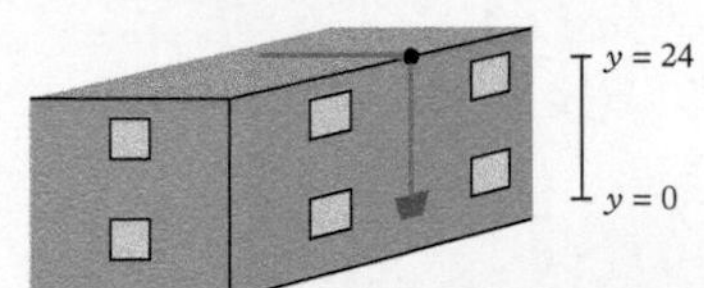

Figure 4 *A coordinate axis for y is placed alongside the rope. The function f representing the weight of the rope must be correct for our chosen coordinate system*

In figure 3, the vertical scale has zero at the top and positive y-coordinates below it—that is, the positive direction is downward. This is the opposite of our normal orientation, where positive is upward on a vertical axis. What if, as in figure 4, we choose to put $y = 24$ at the top and $y = 0$ at the bottom? Then the weight of the rope $f(y)$ is not $\frac{1}{10}y$. At the top, when $y = 24$, the weight of the rope (in pounds) is $f(24) = 0$, whereas at the bottom, when $y = 0$, the weight of the rope is $f(0) = 2.4$. At the coordinate y, the length of rope is $(24 - y)$ feet, and therefore we use $f(y) = \frac{1}{10}(24 - y)$.

The work performed in lifting the rope is then

$$\begin{aligned} W &= \int_a^b f(y)\,dy \\ &= \int_0^{24} \frac{1}{10}(24 - y)\,dy = \int_0^{24} (2.4 - 0.1y)\,dy \\ &= \left(2.4y - 0.05y^2\right)\Big|_0^{24} \\ &= 2.4(24) - 0.05(24^2) - 0 = 28.8\,\text{ft}\cdot\text{lb}, \end{aligned}$$

We still integrate from $y = 0$ to $y = 24$ so that the distance Δy is positive.

just as in the original solution.

Although marking the coordinate axis as in figure 4 and using the function $f(y) = \frac{1}{10}y$ (which matches the coordinate axis in figure 3) may result in the correct value, the method used is still incorrect.

Reading Exercise 11 *Fifty feet of heavy cable weighing 3 lb/ft is dangling from the roof of a tall building. How much work is performed in lifting the cable to the roof?*

Work: springs

Hooke's law states the force needed to stretch or compress a spring.

Robert Hooke 1635–1703
http://www-history.mcs.st-andrews.ac.uk/Biographies/Hooke.html

HOOKE'S LAW

The force needed to stretch or compress a spring x units from its resting position (and hold it there) is proportional to x. That is,

$$F = kx,$$

where $k > 0$ is called the *spring constant* or *force constant* for the spring. When stretching the spring, $x > 0$; when compressing the spring, $x < 0$.

In physics, the focus is often on the force the spring exerts on whatever is attached to it. Here we use the force exerted on the spring instead (same magnitude, different sign).

A spring can be stretched or compressed by pulling or pressing on the spring. In figure 5 (left), a spring is at its natural length, or resting position. Suppose we take our finger and pull on the spring to stretch it. If you have ever done this, you may have noticed that the farther you stretch the spring, the harder it is to pull it. This is Hooke's law; the farther the spring is stretched, the greater the force required to stretch it. The same is true when compressing the spring. The farther you

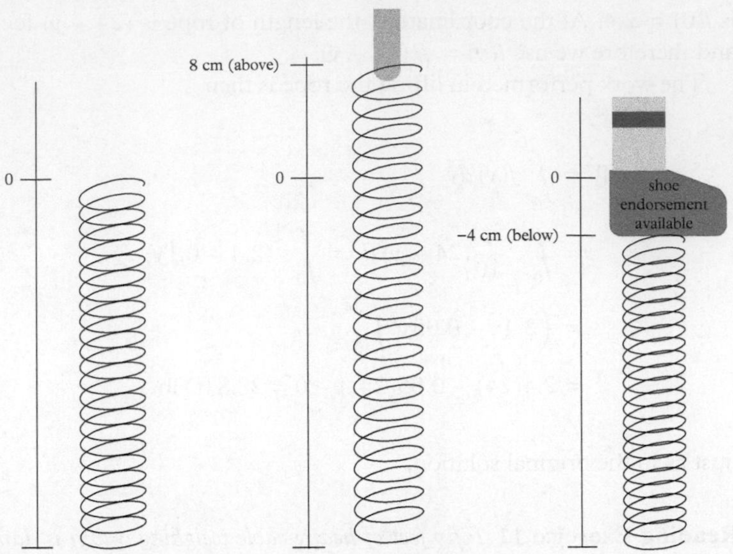

Figure 5 *(left) A spring at its natural length—in the resting position; (center) the spring stretched* 8 cm *using a finger, to* $x = 8$*; (right) the spring compressed 4 cm using a foot, to* $x = -4$

compress the spring by stepping on it, the greater the force required to compress it.

The force required to stretch or compress a spring varies from one spring to another. A child's toy spring is stretched more easily than a car spring. This means the spring constant k for a car spring is greater than the spring constant for a toy spring.

Although it is possible to pull or press with greater force than kx, to avoid complications let's assume the force exerted on the spring is always exactly kx.

Because the force required to stretch or compress a spring is not constant, it is a variable force. Therefore, the work performed by stretching or compressing the spring is found by $W = \int_a^b F(x)\,dx$, where F is the force function. But, Hooke's law gives us this function! Using $F(x) = kx$ and $x = 0$ for the resting position, per Hooke's law, gives

Ans. to reading exercise 11
$\int_0^{50} 3y\,dy = 3750 \text{ ft} \cdot \text{lb}$

$$W = \int_a^b kx\,dx.$$

This equation is for the work done by the object on the spring, which is always positive in this text.

The work done by the spring on the object is negative in this text.

Example 4 *Find the work required to compress a spring from its natural length of 18 inches to a length of 12 inches if the force constant is $k = 200$ lb/ft.*

Solution It helps to draw a quick diagram and place a coordinate axis in the picture. Figure 6 shows such a figure, with the spring at rest on the left and the spring compressed on the right.

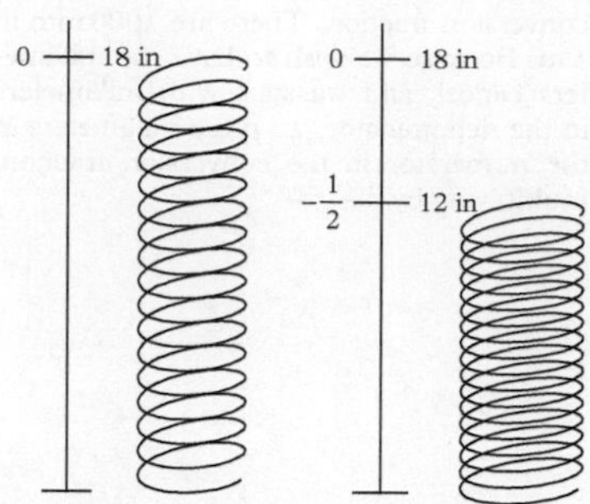

Figure 6 *(left) The spring of example 4 in its resting position; (right) the same spring compressed to a shorter length. The values marked on the left on a coordinate axis are the values of the variable x, in feet; the values marked on the right are the length of the spring in inches*

The spring constant $k = 200$ lb/ft is given. What else do we need to know? The values of a and b are needed. Before choosing values for a and b, we need to know which units to use for the variable x, which is the distance from the resting position. Because the force constant is given in terms of pounds per foot, let's use feet as the unit for the variable x.

We are compressing the spring from its natural length of 18 inches to a length of 12 inches. When using this setup for Hooke's law, we always use $x = 0$ for the natural length, so $a = 0$. We are compressing to 6 inches below the resting position. But, because our units for x are in feet, this gives $x = -\frac{1}{2}$ (recall 1 foot = 12 inches). We therefore use $b = -\frac{1}{2}$.

CAUTION: MISTAKE TO AVOID
Make sure that all units match! Check for this *every time.*

The work performed in compressing the spring is therefore

$$W = \int_a^b kx\,dx$$

$$= \int_0^{-1/2} 200x\,dx$$

$$= 100x^2\Big|_0^{-1/2}$$

$$= 100\left(\frac{1}{4}\right) - 0 = 25\text{ ft}\cdot\text{lb}.$$

The units for this integral are foot-pounds. To see this, write the units in the integral:

$$\int_0^{-1/2} 200\text{ lb/ft}\cdot x\text{ ft}\,dx\text{ ft}.$$

Multiplying these units gives

$$\frac{\text{lb}}{\not{\text{ft}}}\cdot\not{\text{ft}}\cdot\text{ft} = \text{ft}\cdot\text{lb}.$$

We conclude that the work required to compress the spring to a length of 12 inches is 25 ft · lb. ■

Reading Exercise 12 *How much work is required to stretch a spring from its natural length of 3 feet to a length of 4 feet given that its force constant is 50 lb/ft?*

Example 5 *A car spring has a natural length (resting position) of 35 cm. Its spring rate (spring constant) is $k = 26$ N/mm. How many joules of work is performed in stretching the spring from a length of 35.5 cm to a length of 36 cm?*

In this context, the spring constant is often referred to as the *spring rate.*

Solution Sometimes it is helpful to perform all necessary unit conversions first. Because we are asked to provide the answer in joules,

We need to change from millimeters to meters, so these are the units to use in our conversion fraction. There are 1000 mm in 1 m. Because we wish to have the millimeters cancel, and we start with millimeters in the denominator, we place millimeters in the numerator in the conversion fraction, multiplying by $\frac{1000\text{ mm}}{1\text{ m}}$.

and a joule is a newton-meter, we need to change all measurements of centimeters and millimeters to meters. We have

$$k = 26\frac{\text{N}}{\text{mm}} \cdot \frac{1000\text{ mm}}{1\text{ m}} = \frac{26\,000\text{ N}}{\text{m}}.$$

The natural length of the spring is

$$35\text{ cm} \cdot \frac{1\text{ m}}{100\text{ cm}} = 0.35\text{ m}.$$

We then wish to know the work performed when stretching the spring from a length of 0.355 m to a length of 0.36 m.

Now we are ready to sketch a diagram of the springs, and we can include the appropriate units in the diagram (which, by the way, is a reason to do all the unit conversions before drawing the diagram). The beginning length of the spring, which is not its natural length, is depicted on the left in figure 7. The natural length of 0.35 m corresponds to $x = 0$, so the length of 0.355 m corresponds to $x = 0.005$. The ending length of the spring, depicted on the right in figure 7, is 0.36 m, which corresponds to $x = 0.010$.

Figure 7 *(left) The spring of example 5 at a length of 0.355 m, which is longer than its natural length; (right) the same spring stretched to a greater length. The values marked on the left on a coordinate axis are the values of the variable x, whereas the values marked on the right are the lengths of the spring. The natural length is also marked in each picture. The axes are not to scale*

Having gathered all the necessary information, we calculate the work performed:

$$\begin{aligned} W &= \int_a^b kx\,dx \\ &= \int_{0.005}^{0.010} 26\,000x\,dx \\ &= 13\,000x^2 \Big|_{0.005}^{0.010} \\ &= 1.3 - 0.325 = 0.975\text{ N}\cdot\text{m}. \end{aligned}$$

We conclude that the work performed in stretching the spring from a length of 35.5 cm to a length of 36 cm is 0.975 J. ■

Because we used units of newtons per meter for k and units of meters for values of x, we can be confident that the units of the answer are

$$\text{N/m}\cdot\text{m}\cdot\text{m} = \text{N}\cdot\text{m} = \text{J}.$$

The spring constant is not always given. Sometimes it must be calculated from the given information.

Example 6 *The force required to stretch a spring from its natural length of 18 inches to a length of 20 inches is 100 pounds. How much work does it take to stretch the spring from a length of 21 inches to a length of 24 inches?*

Solution Because we are using English units, we choose to answer in units of foot-pounds. This time we need to convert all measurements from inches to feet as we encounter them.

Ans. to reading exercise 12

$$\int_0^1 50x\,dx = 25\text{ ft}\cdot\text{lb}$$

In addition, this time we must calculate the spring constant k. We are told that a force of 100 pounds stretches the spring 2 inches, which is equal to $\frac{1}{6}$ feet. Using Hooke's law,

$$\begin{aligned} F &= kx \\ 100\,\text{lb} &= k \cdot \frac{1}{6}\,\text{ft} \\ 600\,\text{lb/ft} &= k. \end{aligned}$$

Next we draw a diagram of the beginning and ending positions for which we wish to calculate the work, using the natural length of 18 inches = 1.5 feet to correspond to $x = 0$; see figure 8.

Using $k = 600$, $a = 0.25$, and $b = 0.5$, we have

$$\begin{aligned} W &= \int_a^b kx\,dx \\ &= \int_{0.25}^{0.5} 600x\,dx \\ &= 300x^2 \Big|_{0.25}^{0.5} \\ &= \cdots = 56.25\,\text{ft} \cdot \text{lb}. \end{aligned}$$

We conclude that the work performed in stretching the spring from a length of 21 inches to a length of 24 inches is 56.25 ft · lb. ■

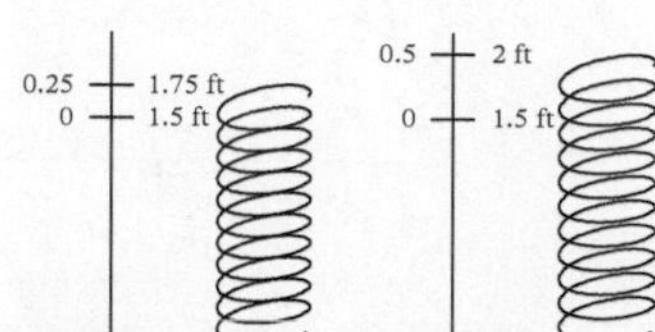

Figure 8 *(left) The spring of example 6 at a length of 21 inches = 1.75 feet; (right) the same spring stretched to a length of 24 inches = 2 feet. The natural length of 18 inches = 1.5 feet corresponds to x = 0. The values marked on the left on each coordinate axis are the values of the variable x, whereas the values marked on the right are the lengths of the spring*

Exercises 6.5

1. A store employee lifts a box weighing 45 pounds a distance of 4 feet to place it on a shelf. How much work is performed on the box?
2. A soldier lifts a 180-pound injured colleague 4 feet. How much work is performed on the injured colleague?
3. A drone lifts a package 4 m. The combined mass of the drone and package is 13 kg. Determine the work performed.
4. A forklift lifts a pallet of supplies a distance of 3.2 m. The total mass being lifted is 1100 kg. Determine the work performed.
5. A group of rescuers uses a rope weighing 0.5 lb/ft to lift a victim from a sinkhole. The victim weights 200 pounds and must be lifted vertically 40 feet. Assume that at the beginning of the lift there is 40 feet of rope to lift and at the end there is 0 feet of rope to lift. How much work is performed?

6. A 20-foot-long chain dangles from the side of a ship. The chain weighs 5 lb/ft. How much work is performed in lifting the chain back to the deck from which it dangles?
7. A rope of length 15 feet hangs from a treehouse. The rope weighs 2 lb/ft. How much work is performed lifting the entire rope back into the treehouse?
8. A 500-pound anchor lies at the bottom of the sea floor 60 feet below its upward position. It is attached by a chain that weighs 4 lb/ft. How much work is done in lifting the anchor to its upward position?
9. While the water in a pool is being pumped out, a neighbor steals the pool using four helicopters with strong cables attached to the corners of the pool to lift the pool 20 feet in the air, before moving it to the neighbor's yard (pretend this is possible). Find the work performed on the pool if the pool is lifted at a constant rate, water is pumped out at a constant rate, the pool and water weigh 60 000 pounds when the helicopters started lifting, and weigh 40 000 pounds by the time the pool is 20 feet in the air.
10. A bucket of water is hauled to the roof of a building using a rope and pulley. The bucket weights 1 pound and the rope weighs 1.5 lb/ft. Unfortunately, the bucket has a leak. When the lift begins, the water in the bucket weighs 22 pounds. At the end of the lift, when the last of the rope is pulled in, only half of the water remains in the bucket. The length of the lift is 33 feet. Find the work performed in lifting the bucket, rope, and water.
11. A person with mass 80 kg is lifted 27 m into a hovering helicopter with the use of a cable and winch. The linear mass density of the cable is 0.72 kg/m. How much work is performed on the person and the cable?
12. A person with mass 80 kg is lifted 12 m into a hovering helicopter with the use of a cable and winch. The weight of the cable per meter is 11.88 N/m. How much work is performed on the person and the cable?
13. A spring with force constant 10 lb/ft is stretched from its natural length of 60 inches to a length of 66 inches. How much work is performed?
14. How much work is required to stretch a spring from its natural length of 20 inches to a length of 24 inches, given that its force constant is 180 lb/ft?
15. A force of 200 pounds stretches a spring from its natural length of 6 feet to a length of 10 feet. How much work is performed?

16. A force of magnitude 6 pounds compresses a spring from its natural length of 1.6 feet to a length of 1.4 feet. How much work is performed?
17. A force of magnitude 10 pounds compresses a spring 3 inches from its natural length. How much work is necessary to stretch this spring 6 inches from its natural length?
18. A force of 30 pounds stretches a spring from its natural length of 6 feet to a length of 8 feet. How much work is required to compress the spring from its natural length to a length of 5 feet?
19. A spring with force constant 0.7 N/m is compressed from its natural length of 11 cm to a length of 9 cm. How much work is performed?
20. A vehicle spring with a spring rate of $k = 14$ N/mm and a natural length of 30 cm is compressed from a length of 29 cm to a length of 27 cm. How much work is performed?
21. A vehicle spring with a spring rate of $k = 44$ N/mm and a natural length of 61 cm is compressed from a length of 55 cm to a length of 49 cm. How much work is performed?
22. A spring with a force constant of 0.04 N/m is stretched from its natural length of 12.8 mm to a length of 15.5 mm. How much work is performed?
23. An object is moved in the presence of a resistive force of size $F(x) = x^2$ N from $x = 0$ m to $x = 5$ m. What is the work performed on the object?

The N in the force equation is not a variable, but rather the unit of measure: newton.

24. An object is moved in the presence of a resistive force of size $F(x) = \frac{1}{x^2}$ N from $x = 2$ m to $x = 5$ m. What is the work performed on the object?
25. A sail is lifted using ropes and pulleys from a non deployed position to a fully deployed one. The sail is rectangular: 14 m tall by 6 m wide. The sail is made from a 0.35-kg/m^2 cloth. How much work is performed when lifting the sail?

The mass of the cloth per unit of area is called the *areal mass density* of the cloth.

26. (GSL) A well is 100 feet deep. A bucket, weighing 3 pounds, has a volume of 2 ft^3. The bucket is filled with water at the bottom of the well and is then raised at a constant rate of 5 ft/s to the top. Neglecting the weight of the rope, what work is done when raising the bucket if it is discovered that the water is leaking out at a constant rate of 0.01 ft^3/s? (A cubic foot of water weighs 62.4 pounds.)

27–28. (GSL) *Work done by an expanding gas.* If a gas in a cylinder expands against a piston head from volume v_0 ft^3 to v_1 ft^3, the external work done in foot-pounds is

$$\text{work} = \int_{v_0}^{v_1} p\,dv,$$

(GSL) Isothermal expansion occurs when the temperature remains constant. Then, $n = 1$ and the pressure–volume relation is $pv = p_o v_0 = p_1 v_1$.

where p is pressure in pounds per square foot and pv^n is a constant, with the exponent n being a constant.

27. A quantity of gas with an initial volume of 16 ft^3 and a pressure of 60 lb/in^2 expands until the pressure is 30 lb/in^2. Determine the final volume and the work done by the gas if the law is $pv = C$.
28. Repeat exercise 27 if the law is $pv^n = C$, assuming $n = 1.2$.

Work, Part II

6.6

We continue our study of work, this time as it applies to pumping fluids.

Work: pumping fluids

Several formulas in this chapter have been derived by slicing and forming omega sums, but we have not yet extended this pleasure to examples and exercises. It is time to rectify this situation!

Important information to gather in a fluid-pumping problem includes (1) the shape and dimensions of the object occupied by the fluid, (2) the depth of the fluid, (3) the location of the pump (height to which the fluid is pumped), and (4) the density of the fluid.

Example 1 *Find the work required to pump all the water out of a basement, given that the basement has a rectangular floor plan of dimensions* 10 *feet by* 20 *feet, the walls are vertical, the water is* 2 *feet deep in the basement, and the water is to be pumped through a window that is* 6 *feet above the floor of the basement. The weight density of the water is* $62.4\,\mathrm{lb/ft^3}$.

The *weight density* of a fluid, also called the *volumetric weight density* is the weight of the fluid per unit of volume.

Solution As always, a diagram helps. In fact, it's helpful to draw two diagrams: one that is a floor plan of the basement, indicating its shape and dimensions, and one that is a side view of the basement, indicating the depth of the water and the height to which the fluid is to be pumped.

The basement floor plan is shown in figure 1. Because the walls of the basement are vertical, any horizontal cross section of the basement has the same area as the floor:

$$\text{cross-sectional area} = 20\,\mathrm{ft}\cdot 10\,\mathrm{ft} = 200\,\mathrm{ft}^2.$$

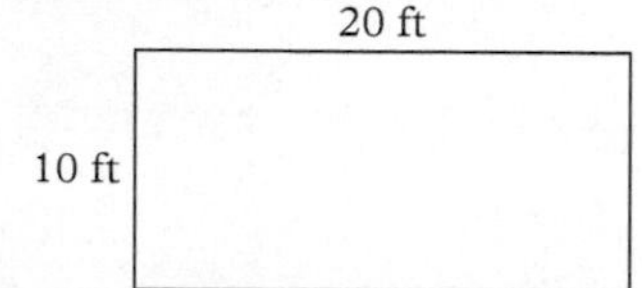

Figure 1 *Basement floor plan. Only horizontal dimensions are placed in this figure*

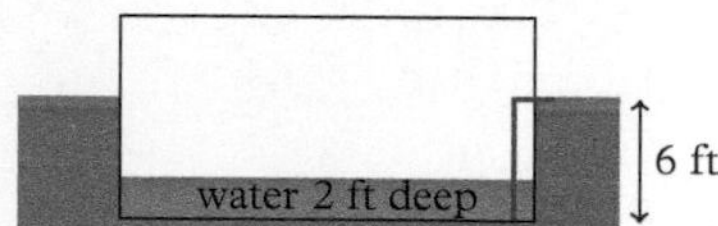

Figure 2 *Basement side view. Only vertical dimensions are placed in this picture*

The side view of the basement is in figure 2. Here we begin to see the issue we have in trying to determine work, which is force times distance; not all of the water is pumped the same distance. The water at the top must only be lifted 4 feet, whereas the water at the bottom must be lifted the full 6 feet.

The next task is to set up a (vertical) number line to give coordinates for our variable. This can be done in more than one way. Let's make $y = 0$ correspond to the bottom of the basement. Using units of feet for distance, $y = 2$ is at the top of the water and $y = 6$ is the height to which we are pumping the water. See figure 3.

Figure 3 *A coordinate axis is added to the basement side view*

The trick in any fluid-pumping problem is to slice horizontally. Why? Because on an infinitesimally thin slice, the distance to be lifted does not change; there is not enough thickness in the slice for the position of the water to be different at the top of the slice than at the bottom of the slice (up to an approximation, anyway). As usual, we'll use Ω slices, and let $\Delta y = (b - a)\omega$ and $y_k = a + k\Delta y$.

We do not need to use the formulas for Δy or y_k.

Figure 4 illustrates the slice at a vertical coordinate y_k, with thickness Δy, to be pumped to $y = 6$. The distance the slice must be moved

Technically, the "slice at a vertical coordinate y_k" actually lies between y_{k-1} and y_k, and has thickness Δy. Using right-hand endpoints means we use y_k in our formulas.

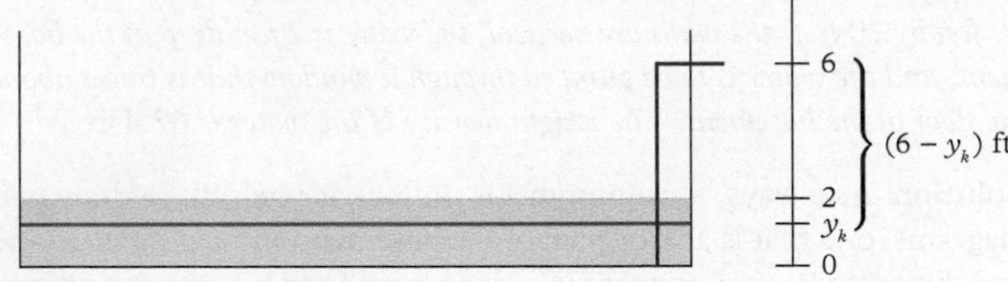

Figure 4 *One slice of water (darker blue), from the water in the basement (lighter blue). The slice extends throughout the basement and has three dimensions: width* 20 *feet and length* 10 *feet as in the basement floor plan, and height* Δy *ft, which is the vertical thickness of the slice*

is therefore

$$\text{slice distance moved} = (6 - y_k)\text{ ft.}$$

Work is force times distance; we still need the force, which is the weight of the water in the slice. The weight of the water is the weight density of the water (units: pounds/cubic feet) times the volume of the water (units: cubic feet). The volume of the water is the volume of the slice, which is a rectangular box with horizontal dimensions 20 feet by 10 feet and vertical dimension Δy feet, which gives

slice volume = cross-sectional area · Δy

$$\text{slice volume} = 20\,\text{ft} \cdot 10\,\text{ft} \cdot \Delta y\,\text{ft}$$
$$= 200\Delta y\,\text{ft}^3,$$

which is the previously calculated cross-sectional area times the slice thickness Δy. The weight of a slice is therefore

$$\text{slice weight} = \text{slice volume} \cdot \text{weight density of liquid}$$
$$= 200\Delta y\,\text{ft}^3 \cdot 62.4\,\text{lb}/\text{ft}^3$$
$$= 12\,480\Delta y\,\text{lb}.$$

Last, the work moving this one slice is

$$\text{slice work} = \text{slice weight} \cdot \text{slice distance}$$
$$= 12\,480\Delta y\,\text{lb} \cdot (6 - y_k)\,\text{ft}$$
$$= (74\,880 - 12\,480 y_k)\Delta y\,\text{ft} \cdot \text{lb}.$$

However, this is just one slice. To get the work performed on all the slices, we must add them all up. Assuming there are Ω slices (so the slice thickness is infinitesimal), the total work is

$$\sum_{k=1}^{\Omega} (74\,880 - 12\,480 y_k)\Delta y \text{ ft} \cdot \text{lb}.$$

This is an omega sum! To write the sum as an integral, we need to know a and b, which is the extent of the y-coordinates where the slices (the water) lie. Consulting figure 3, we see the water extends from $y = 0$ to $y = 2$, and we use $a = 0$ and $b = 2$.

We then write the sum as an integral and evaluate:

$$\text{work} = \int_0^2 (74\,880 - 12\,480 y)\ dy$$
$$= \left(74\,880 y - 6\,240 y^2\right)\Big|_0^2$$
$$= \cdots = 124\,800\,\text{ft} \cdot \text{lb}.$$

We conclude that the work needed to pump the water out of the basement is $124\,800\,\text{ft} \cdot \text{lb}$. ■

> slice weight
> = slice volume · weight density of liquid

> slice work = slice weight · slice distance

Including units in all calculations up through the omega sum, and performing the correct algebra on those units, is helpful to avoid errors.

A unit of power, 1 hp is approximately equal to the amount of power that one horse can produce during sustained activity.

The answer of 124 800 ft · lb seems like a lot of work, but it's not as much as you might think. One-horsepower (1 hp)-output pumps are readily available at home improvement and hardware stores. Since

$$1\text{ hp} = 550\text{ ft}\cdot\text{lb/s},$$

such a pump could produce the necessary work in

$$\frac{124\,800\text{ ft}\cdot\text{lb}}{550\text{ ft}\cdot\text{lb/s}} = 226.9\text{ s},$$

For this reason, some basement floors are designed to slope toward a drain basin containing a sump pump.

which is a little less than 4 minutes. One should not expect a dry basement in this short of time, however, because of complicating factors when the water level gets fairly low, such as the horizontal spread of the water or the water level dropping below the size of the input tube.

The steps in the solution to example 1 can be summarized as follows:

❶ Draw a cross-sectional diagram.

❷ Draw a side-view diagram.

❸ Introduce a vertical number line (coordinate system) to the side-view diagram.

❹ Place a typical slice in the side-view diagram, at coordinate y_k.

❺ Calculate the slice distance moved.

❻ Calculate the slice volume.

❼ Calculate the slice weight.

❽ Calculate the slice work.

❾ Add the slices to get an omega sum.

❿ Rewrite as an integral and evaluate.

Let's use these steps in the next example.

An "upright" cylinder implies circular horizontal cross sections and vertical walls.

Example 2 *An upright cylindrical tank with height* 5 *m and radius* 3 *m is half full of a chemical with weight density* 8700 N/m^3. *How much work needs to be performed to pump the chemical out of the tank (to the level of the top of the tank)?*

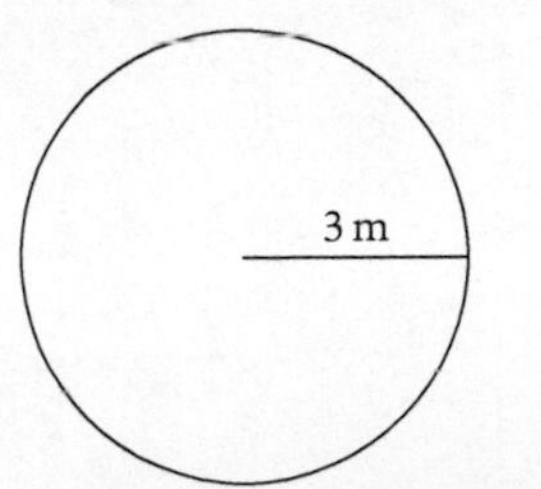

Figure 5 *Top view (horizontal cross section) of tank, including horizontal dimensions*

Solution ❶ Draw a cross-sectional diagram. This is a view from the top, which shows a circle with a diameter of 3 m. Only horizontal dimensions are included. See figure 5.

❷ Draw a side-view diagram. Only vertical dimensions are included. The depth of the chemical in the tank and the height the liquid is to be pumped are included. See figure 6.

❸ Introduce a vertical number line to the side-view diagram. We choose to let $y = 0$ be at the bottom of the tank. The units are meters. The line is present in figure 6.

❹ Place a typical slice in the side-view diagram, at coordinate y_k. See figure 6.

❺ Calculate the slice distance moved. We wish to move the slice from $y = y_k$ to $y = 5$, thus

$$\text{slice distance moved} = (5 - y_k)\,\text{m}.$$

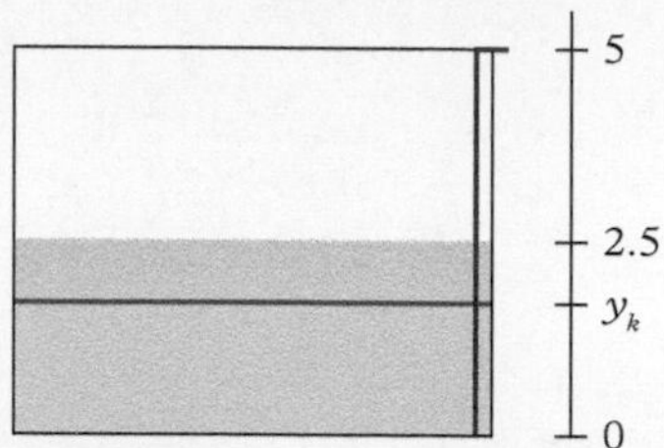

Figure 6 ❷ *Side view of the half-full tank,* ❸ *including vertical coordinates.* ❹ *A typical slice (darker blue) is shown at* y_k

❻ Calculate the slice volume. Cross sections are circular (figure 5), and the area of a circle is $A = \pi r^2$, so

$$\begin{aligned}\text{slice volume} &= \text{cross-sectional area} \cdot \Delta y\\ &= \pi(3\,\text{m})^2 \cdot \Delta y\,\text{m}\\ &= 9\pi\Delta y\,\text{m}^3.\end{aligned}$$

❼ Calculate the slice weight. The chemical's specific weight is $8700\,\text{N/m}^3$; therefore,

Weight density is also called *specific weight*.

$$\begin{aligned}\text{slice weight} &= \text{slice volume} \cdot \text{weight density of liquid}\\ &= 9\pi\Delta y\,\text{m}^3 \cdot 8700\,\text{N/m}^3 = 78\,300\pi\Delta y\,\text{N}.\end{aligned}$$

❽ Calculate the slice work:

$$\begin{aligned}\text{slice work} &= \text{slice weight} \cdot \text{slice distance}\\ &= 78\,300\pi\Delta y\,\text{N} \cdot (5 - y_k)\,\text{m}\\ &= \pi(391\,500 - 78\,300 y_k)\Delta y\,\text{J}.\end{aligned}$$

❾ Add the slices to get the omega sum. The total work performed is

$$\sum_{k=1}^{\Omega} \pi(391\,500 - 78\,300 y_k)\Delta y\,\text{J}.$$

This omega sum is in a form conducive to rewriting it as an integral. For fun, we could evaluate the omega sum directly by using $\Delta y = (b-a)\omega = 2.5\omega$ and $y_k = a + k\Delta x = 2.5\omega k$, then proceed as in section 4.3.

❿ Rewrite as an integral and evaluate. As illustrated in figure 6, the chemical in the tank is from $y = 0$ to $y = 2.5$ (the tank is half full), so we may use $a = 0$ and $b = 2.5$. To rewrite the integral, we replace the summation by an integral symbol, replace Δy with dy, remove the subscript, and insert limits of integration. We continue by evaluating the integral:

$$\int_0^{2.5} \pi(391\,500 - 78\,300y)\,dy$$
$$= \pi(391\,500y - 39\,150y^2)\Big|_0^{2.5}$$
$$= \cdots = 2\,306\,125\,\text{J}.$$

The work required to pump the chemical out of the tank is 2 306 125 J. ■

Tanks can come in various shapes, including some with cross-sectional areas that are not constant, such as spherical tanks or conical tanks. This just adds to the fun (or challenge, or ...) of step ❻. Enjoy!

Exercises 6.6

1. A vat in the shape of a rectangular box is 20 feet long, 8 feet wide, and 5 feet high. A chemical with weight density 57 lb/ft^3 fills the vat to a level of 4 feet (i.e., filled to within 1 foot of the top). How much work is required to pump the chemical out of the vat (to the level of the top of the vat)?
2. An in-ground swimming pool is in the shape of a cylinder with a diameter of 16 feet. The pool is 6 feet deep (from the bottom of the pool to the deck above it), but is only filled with water to a depth of 5 feet. What is the work required to pump all the water out of the pool (to the level of the deck)? Water weighs 62.4 lb/ft^3.
3. A cistern in the shape of a cylinder (flat circular bottom) has a diameter of 10 feet and is 12 feet high. The water in the cistern is 8 feet deep (i.e., filled to within 4 feet of the top). How much work is required to pump the water to the level of the top of the cistern to empty it? The weight density of water is 62.4 lb/ft^3.
4. A vat of oil is in the shape of a rectangular box with a width of 5 feet, a length of 20 feet, and a height of 2 feet. The vat is full of oil with a weight density of 56 lb/ft^3. How much work is done when pumping the oil over the top of the vat?
5. Diesel fuel is stored underground in an upright cylindrical tank (horizontal cross sections are circular) with a radius of 1.5 m and a height of 3 m. The tank is full and must be pumped out to a height of 2 m above the top of the tank. The weight density of the fuel is 8134 N/m^3. How much work is required?
6. A cylindrical cup with a radius of 4 cm and a height of 15 cm contains a diet soft drink with a weight density of 9770 N/m^3. The

For some popular brands of soft drinks, the regular drink has a greater density than water whereas the diet drink has a lower density than water.

cup is filled to 1 cm below the top. How much work is required to pump all the soda to a height of 5 cm above the top of the cup (such as through a straw)?

7. Water (weight density, $9800\,\text{N/m}^3$) has filled a cubical drainage basin measuring 2 m on each side. How much work is required to pump out half the water (to the level of the top of the basin)?

8. A trench 1.7 m deep, 0.8 m wide, and 15.8 m long is filled with muddy water (weight density, $16\,600\,\text{N/m}^3$) that must be pumped out (to the height of the top of the trench) so that the project can continue. How much work is required?

9. A conical tank with a height of 6 m, a radius at the top of 2 m, and a radius at the bottom of 0 m is filled to a depth of 4 m with a chemical with a weight density of $10\,000\,\text{N/m}^3$. How much work is required to pump the liquid out of the tank (to the height of the top of the tank)?

10. A saltwater aquarium is in the shape of a sphere with a radius of 3 feet. The aquarium is half full of water. How much work must be done to pump all the water through a hole in the top of the sphere? Assume the weight density of saltwater is $64\,\text{lb/ft}^3$.

11. (GSL) A hemispherical cistern with a radius of 3 feet is full of water. Two workers, A and B, are to pump it out, each doing half the work. What is the depth d of the water when the first worker is finished?

The top of the cistern is a circle with a radius of 3 feet. The pumping is to the level of the top of the cistern. Assume the water weighs $62.4\,\text{lb/ft}^3$.

Average Value of a Function

6.7

We have studied geometric applications of integration as well as applications from physics. Our next application of integration is related to statistics.

Computing the average value of a function

Do you recall how an *average* is computed? To find the average value of a list of numbers, we add the numbers in the list and divide by however many there are. For instance, given the list 5, 7, and 15, the average value is

In statistics, the term used for this concept of average is the *arithmetic mean*.

$$\frac{5+7+15}{3} = 9.$$

Given the list 2, 3, 6, 9, and 11, the average is

$$\frac{2+3+6+9+11}{5} = 6.2.$$

The formula for the average of n numbers is written as

$$\text{average} = \frac{x_1 + x_2 + \cdots + x_n}{n}.$$

The symbol used for the mean of a population is the Greek letter μ; the symbol used for the mean of a (random) sample is $\bar{x}$.

What if there are infinitely many numbers to average? This situation occurs when finding the average value of a function. Consider $f(x) = x^2$ on the interval $[-1, 2]$, as pictured in figure 1. There are infinitely many different y-coordinates on the graph of the function. How do we add them all up?

We can approximate (estimate) the average by choosing some of the y-coordinates on the graph and averaging these numbers. One approach is to use n subintervals and choose the value at the right-hand endpoint of each subinterval. (Sound familiar?) Using $n = 6$ and

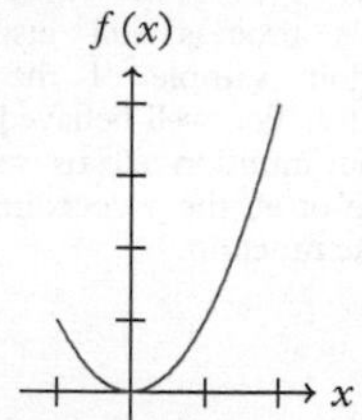

Figure 1 *The graph of* $f(x) = x^2$ *on* $[1, 2]$. *The y-coordinates include every real number between* 0 *and* 4

$n = 12$ is pictured in figure 2. The averages are computed as

The numbers we wish to average are the y-coordinates on the graph, and the y-coordinate on the graph at $x = x_k$ is $f(x_k)$, the value of the function at x_k.

$$\frac{f(x_1) + f(x_2) + \cdots + f(x_6)}{6} = \frac{(-0.5)^2 + 0^2 + (0.5)^2 + 1^2 + 1.5^2 + 2^2}{6} = \frac{7.75}{6} \approx 1.292$$

Each of these averages is called a *finite approximation to the average value of the function*, just as we computed finite approximations to areas under a curve earlier using omega sums to calculate the "exact" value, which served as our definition of area.

and

$$\frac{f(x_1) + f(x_2) + \cdots + f(x_{12})}{12} = \frac{(-0.75)^2 + (-0.5)^2 + (-0.25)^2 + \cdots + 2^2}{12} = \frac{13.625}{12} \approx 1.135.$$

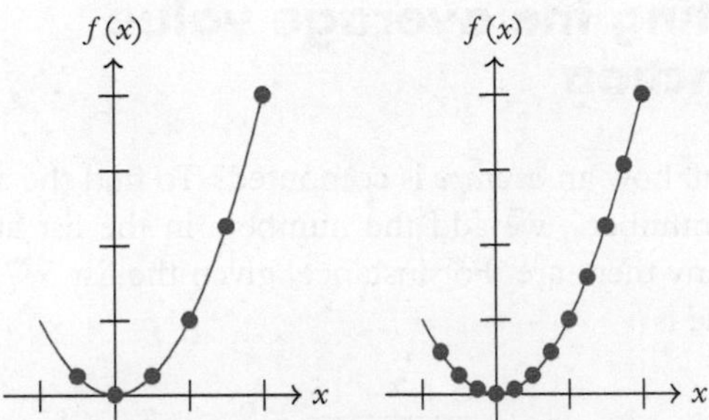

Figure 2 *(left) Six equally spaced points (red dots) on the graph of $f(x) = x^2$ on $[1, 2]$ and (right) 12 equally spaced points, each using right-hand endpoints of the subintervals. Each is a nonrandom sample of the y-coordinates on the graph of f*

Instead of using $n = 6$ or $n = 12$ points on the graph, why not use infinitely many points? This should give us complete coverage of the graph of the function instead of just spotty coverage (in this case, the red dots would appear to cover the graph completely), and thus give us the true average value of the function instead of just a finite approximation.

Technically, even with infinitely many points, this process still just produces a nonrandom sample of the function's y-coordinates. For well-behaved functions, however, our intuition tells us we should get the average of all the y-coordinates on the graph of the function.

Following the usual routine, we use Ω subintervals (where Ω is a positive infinite hyperreal integer) and let $\Delta x = (b - a)\omega$ and $x_k = a + k\Delta x$. We then add the resulting values of the function and divide by Ω to get

$$f_{\text{avg}} = \frac{f(x_1) + f(x_2) + \cdots + f(x_\Omega)}{\Omega} = \frac{\sum_{k=1}^{\Omega} f(x_k)}{\Omega} = \sum_{k=1}^{\Omega} f(x_k) \cdot \omega.$$

This is almost an omega sum—if only we had Δx instead of ω. But, since $\Delta x = (b-a)\omega$, which can be rearranged to get $\omega = \frac{1}{b-a}\Delta x$, we can continue:

To be an omega sum that can be rewritten as an integral, the expression must contain a Δx that can be replaced by a dx.

$$\begin{aligned} f_{\text{avg}} &= \sum_{k=1}^{\Omega} f(x_k)\cdot\omega \\ &= \sum_{k=1}^{\Omega} f(x_k)\cdot\frac{1}{b-a}\Delta x \\ &= \frac{1}{b-a}\sum_{k=1}^{\Omega} f(x_k)\Delta x. \end{aligned}$$

This is an omega sum! The average value of the function is therefore

$$f_{\text{avg}} = \frac{1}{b-a}\int_a^b f(x)\,dx.$$

In this line, the omega sum is rewritten as a definite integral by replacing the summation with an integral symbol, replacing Δx with dx, dropping the subscript, and inserting limits of integration.

Definition 2 AVERAGE VALUE OF A FUNCTION *The average value (average y-coordinate) of the function f on the interval* $[a,b]$ *is*

$$f_{\text{avg}} = \frac{1}{b-a}\int_a^b f(x)\,dx,$$

provided the integral exists.

Although this definition works for any integrable function f, it is customary to restrict it to "nice" functions, such as those that are continuous, where the intuition used to derive the formula applies.

Example 1 *Find the average value of the function* $f(x) = x^2$ *on the interval* $[-1,2]$.

Solution Using $a=-1$ and $b=2$,

$$\begin{aligned} f_{\text{avg}} &= \frac{1}{b-a}\int_a^b f(x)\,dx \\ &= \frac{1}{2-(-1)}\int_{-1}^{2} x^2\,dx = \frac{1}{3}\cdot\frac{x^3}{3}\bigg|_{-1}^{2} = \frac{8}{9}-\frac{-1}{9} = 1. \end{aligned}$$

The average value of the function is 1. ■

In an applied setting, the units to use are the same as the units in the function f. Without a context, we do not state units for the average value of the function.

Reading Exercise 13 *Find the average value of* $f(x) = x^2$ *on* $[0,2]$.

Any function that can be integrated can be averaged.

Example 2 *Find the average value of* $f(x) = x^2\cos x^3$ *on* $[0,\pi]$.

Solution The average value is

$$f_{\text{avg}} = \frac{1}{\pi - 0}\int_0^{\pi} x^2 \cos x^3\, dx.$$

Recognizing cosine of "not just plain x," we use substitution:

$$u = x^3$$
$$du = 3x^2\, dx.$$

Next we calculate the new limits of integration:

$$x = \pi:\quad u = \pi^3$$
$$x = 0:\quad u = 0$$

Then,

Line 1 is from a previous calculation; in line 2, we adjust for the constant 3 needed for substitution; and in line 3, we make the substitution, changing from values of x to values of u.

$$\begin{aligned} f_{\text{avg}} &= \frac{1}{\pi}\int_0^{\pi} x^2 \cos x^3\, dx \\ &= \frac{1}{\pi}\cdot\frac{1}{3}\int_0^{\pi} 3x^2 \cos x^3\, dx \\ &= \frac{1}{3\pi}\int_0^{\pi^3} \cos u\, du \\ &= \frac{1}{3\pi}\sin u\Big|_0^{\pi^3} \\ &= \cdots = \frac{1}{3\pi}\sin \pi^3 \approx -0.04226. \end{aligned}$$

The average value of the function is $\frac{1}{3\pi}\sin \pi^3$. ■

Average value: geometric interpretation

The formula for the average value of a function has a geometric interpretation. Rearrange the formula to solve for the integral:

$$f_{\text{avg}} = \frac{1}{b-a}\int_a^b f(x)\, dx$$

$$(b-a)f_{\text{avg}} = \int_a^b f(x)\, dx.$$

The right-hand side of the latter equation is the familiar (net) area under the curve, whereas the left-hand side is the area of a rectangle with width $(b-a)$ and height f_{avg}. These two areas should be the same! In figure 3, the two regions are shown simultaneously.

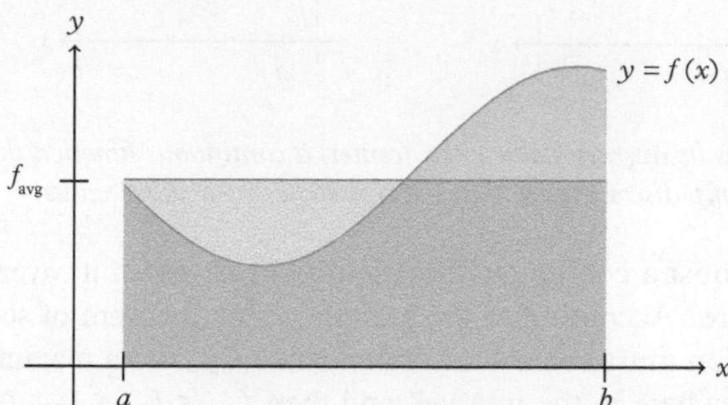

Figure 3 *The graph of $y = f(x)$ (blue) on $[a, b]$ and a horizontal line at height $y = f_{avg}$ (brown) on the same interval. The region under the curve $y = f(x)$ is shaded light blue whereas the region under the brown line is shaded tan. The region under both is a combined color*

The area under the curve $y = f(x)$ is seen to be the same as the area of the rectangle with height f_{avg}. Alternately, the tan-only region that lies in the rectangle but not under the curve has the same area as the light-blue-only region that lies under the curve but not in the rectangle.

Yet another way to see the area comparison is to imagine figure 3 as a thin cross section of sand in a glass box, such as in an ant farm. Begin with the sand showing a valley and a hill, following the curve $y = f(x)$. Now shake the box back and forth horizontally to let the sand settle as a flat line; the resulting level of the top of the sand is at f_{avg}.

Ans. to reading exercise 13
$$f_{avg} = \frac{1}{2-0}\int_0^2 x^2\,dx = \frac{4}{3}$$

Mean value theorem for integrals

If you live in a city where the average number of children per household is 1.372, then no one is "average." There is no household with exactly 1.372 children. It is not automatic that an average value actually occurs, but under certain conditions it will.

Given a function f on an interval $[a, b]$, must its average value occur somewhere? If the function is continuous, the answer is yes. Take another look at figure 3, in which the curve pictured is continuous. Does the curve cross the line $y = f_{avg}$? Figure 4 shows this can occur at one value or at more than one value; but, if the function is not continuous, it might not happen at all.

"More than one" includes infinitely many.

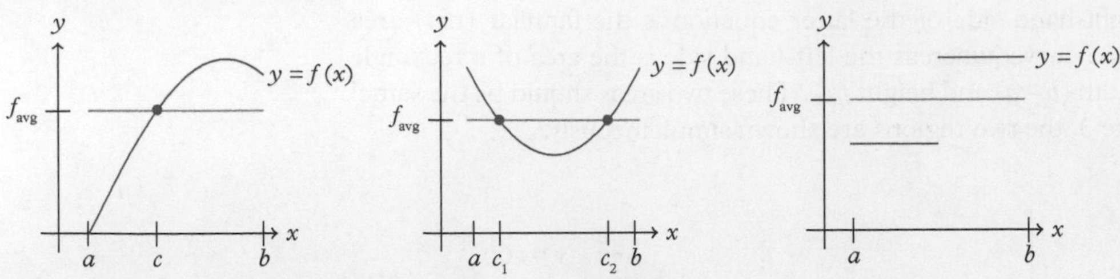

Figure 4 *(left) A continuous function that matches its average value once, (center) a continuous function that matches its average value twice, (right) a function with a jump discontinuity that never matches its average value*

Why must a continuous function on $[a, b]$ match its average value somewhere? According to the extreme value theorem of section 3.1, the function must attain a minimum value $f_{\min}$ and a maximum value $f_{\max}$ somewhere in the interval, and then $f_{\min} \le f_{\text{avg}} \le f_{\max}$ (the graph must contain points both below and above the average line). The intermediate value theorem of section 1.7 says the function must attain every number in between $f_{\min}$ and $f_{\max}$, so it must match the value f_{avg} somewhere.

Exception to parenthetical remark: if f is a constant function, it matches its average value at every point in the interval.

Theorem 1 MEAN VALUE THEOREM FOR INTEGRALS *If f is continuous on $[a, b]$, then there is at least one number c in $[a, b]$ such that*

$$f(c) = f_{\text{avg}} = \frac{1}{b-a}\int_a^b f(x)\,dx.$$

Example 3 *Find a number c in $[-1, 2]$ such that $f(c) = f_{\text{avg}}$ for $f(x) = x^3$.*

Solution First we calculate f_{avg}:

$$f_{\text{avg}} = \frac{1}{2-(-1)}\int_{-1}^{2} x^3\,dx = \frac{1}{3}\frac{x^4}{4}\bigg|_{-1}^{2} = \cdots = \frac{5}{4}.$$

Next, we solve the equation $f(c) = f_{\text{avg}}$:

$$f(c) = f_{\text{avg}}$$
$$c^3 = \frac{5}{4}$$
$$c = \sqrt[3]{\frac{5}{4}} \approx 1.077.$$

We conclude that $f\left(\sqrt[3]{\frac{5}{4}}\right) = f_{\text{avg}}$. ■

Reading Exercise 14 *Find a number c in the interval $[0, 2]$ such that $f(c) = f_{\text{avg}}$ for $f(x) = x^2$.*

Exercises 6.7

1–16. Find the average value of the function on the interval.

1. $f(x) = \sqrt{x}$ on $[0, 4]$
2. $f(x) = \cosh x$ on $[-2, 0]$
3. $f(x) = \frac{1}{x}$ on $[1, 4]$
4. $f(x) = e^x$ on $[0, 2]$
5. $f(x) = \sin x$ on $[0, \frac{\pi}{2}]$
6. $f(x) = \frac{1}{x}$ on $[-6, -1]$
7. $f(x) = x^2 - 1$ on $[1, 3]$
8. $f(x) = \dfrac{1}{1 + x^2}$ on $[-1, 1]$
9. $f(x) = \dfrac{(\sin^{-1} x)^4}{\sqrt{1 - x^2}}$ on $[-\frac{1}{2}, 0]$
10. $f(x) = \dfrac{x}{x^2 + 1}$ on $[2, 5]$
11. $f(x) = \dfrac{x^2}{\sqrt{x^3 + 1}}$ on $[-1, 2]$
12. $f(x) = x^2 \sinh x^3$ on $[0, 2]$
13. $f(x) = x3^{x^2}$ on $[0, 2]$
14. $f(x) = e^{3x} \sin e^{3x}$ on $[\frac{1}{2}, \frac{3}{4}]$
15. $f(x) = \dfrac{\tan(\ln x)}{2x}$ on $[1, e^{\pi/4}]$
16. $f(x) = \dfrac{\sqrt{\tan^{-1} 3x}}{1 + 9x^2}$ on $[0, \frac{1}{3}]$

17. (a) Estimate f_{avg} visually on the following graph (on $[0, 3]$), then (b) calculate f_{avg} using integration and compare to your answer to part (a). How close is your visual estimate?

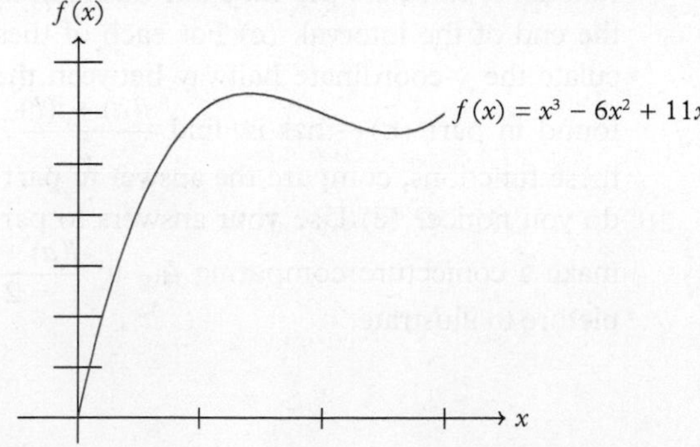

18. (a) Estimate f_{avg} visually on the following graph (on $[1, 3]$), then (b) calculate f_{avg} using integration and compare to your answer to part (a). How close is your visual estimate?

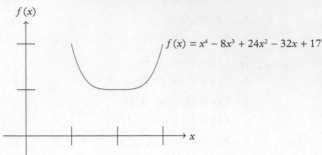

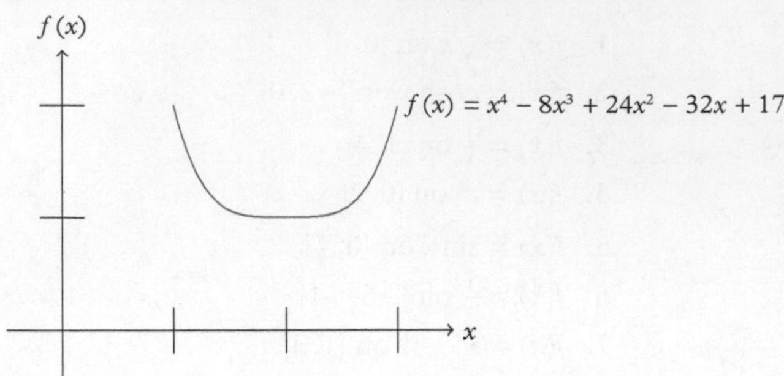

19–26. Find every number c in the interval such that $f(c) = f_{avg}$.

19. $f(x) = x^2$ on $[1, 4]$

20. $f(x) = \sqrt{x}$ on $[1, 4]$

21. $f(x) = e^x$ on $[0, 3]$

22. $f(x) = \cos x$ on $[0, \frac{\pi}{2}]$

23. $f(x) = x^3 - 1$ on $[0, 1]$

24. $f(x) = 2x^2 - 1$ on $[-1, 1]$

25. $f(x) = x^2$ on $[-3, 3]$

26. $f(x) = \frac{1}{x}$ on $[1, e]$

Ans. to reading exercise 14
$c = \frac{2}{\sqrt{3}}$

27. (a) Are the functions in exercises 1 and 5 concave up or concave down on the given interval? (b) For each of these functions, calculate the function values at the beginning and the end of the interval. (c) For each of these functions, calculate the y-coordinate halfway between the function values found in part (b)—that is, find $\frac{f(a) + f(b)}{2}$. (d) For each of these functions, compare the answer to part (c) to f_{avg}. What do you notice? (e) Use your answers to parts (a) and (d) to make a conjecture comparing f_{avg} to $\frac{f(a) + f(b)}{2}$ and draw a picture to illustrate.

28. (a) Are the functions in exercises 3 and 7 concave up or concave down on the given interval? (b) For each of these functions, calculate the function values at the beginning and the end of the interval. (c) For each of these functions, calculate the y-coordinate halfway between the function values found in part (b)—that is, find $\frac{f(a) + f(b)}{2}$. (d) For each of these functions, compare the answer to part (c) to f_{avg}. What do you notice? (e) Use your answers to parts (a) and (d) to make a conjecture comparing f_{avg} to $\frac{f(a) + f(b)}{2}$ and draw a picture to illustrate.

29. Some say that "no one is average." Can this be true in light of the mean value theorem for integrals? Use figure 2, figure 3, or figure 4 in your explanation.

30. Let $s(t)$ be the position function of a particle moving on a number line. Using the formula in section 1.8, the average velocity of the particle from $t = a$ to $t = b$ is $\dfrac{s(b) - s(a)}{b - a}$. Use definition 2 to calculate the average value of the velocity function on the interval $[a, b]$.

31. (GSL) A point is taken at random on a straight line of length a. Prove (a) that the average area of the rectangle with sides that are the two segments is $\frac{1}{6}a^2$ and (b) that the average value of the sum of the squares on the two segments is $\frac{2}{3}a^2$.

The modern terminology is *a line segment of length* a.

Chapter VII

Techniques of Integration

Algebra for Integration

7.1

We have antiderivative formulas for a wide variety of functions. We also have some antiderivative rules, such as the antiderivative sum and difference rules, and the antiderivative constant multiple rule. We do not have an antiderivative product rule or quotient rule. The chain rule was reversed to gain an integration (antiderivative) technique: substitution. These formulas, rules, and techniques alone are not sufficient for integrating even some fairly simple functions. For instance, we do not yet know how to evaluate $\int \ln x \, dx$ or $\int x \sin x \, dx$. This chapter expands our list of integration techniques. After working through this chapter, there will still be functions we cannot antidifferentiate, but the list will be smaller.

Let's begin by studying some algebraic manipulations that put an integrand into a form we can handle.

Review: long division of polynomials

Long division of polynomials is similar to long division of integers, including terminology. As a reminder, if we perform $4119 \div 12$ by long division,

$$\begin{array}{r} 343 \\ 12\overline{)4119} \\ \underline{36} \\ 51 \\ \underline{48} \\ 39 \\ \underline{36} \\ 3 \end{array}$$

the *divisor* is 12, the *dividend* is 4119, the *quotient* is 343, and the *remainder* is 3. When long division of integers terminates, the remainder is less than the divisor.

The stopping criteria is what distinguishes long division of integers from long division resulting in as many decimal places as we wish.

You also may remember rewriting the result as

$$\frac{4119}{12} = 343 + \frac{3}{12},$$

which follows the form

$$\frac{\text{dividend}}{\text{divisor}} = \text{quotient} + \frac{\text{remainder}}{\text{divisor}}.$$

We call a fraction with a numerator larger than or equal to the denominator, such as $\frac{4119}{12}$, an *improper fraction*, and we call a fraction in which the numerator is smaller than the denominator a *proper fraction.* Looking at the result, the idea of long division of integers is to turn an improper fraction into an integer plus a proper fraction.

The same process works for long division of polynomials, but in place of the decimal digits we place powers of the variable. For instance, to divide $2x^3 - 4x^2 + 3x - 1$ by $x + 3$, we begin by writing the dividend and divisor in the long division symbol just as we do for dividing two integers. Each is written in descending order of exponents:

Always include a place for terms of each exponent from the degree of the polynomial down to a constant. If the dividend is $6x^3 - 7x + 5$, you may write the divided in the long division symbol as $6x^3 + 0x^2 - 7x + 5$.

$$x+3\,\overline{)\;2x^3 \quad -4x^2 \quad +3x \quad -1}.$$

We then ask ourselves: How many of the largest power term of the divisor goes into the largest power term of the dividend? That is, how many x go into $2x^3$. The answer is $2x^2$, which we place in the quotient:

One way to get the $2x^2$ is to divide the $2x^3$ by x,

$$\frac{2x^3}{x} = 2x^2.$$

$$\begin{array}{r|llll} & & 2x^2 & & \\ \hline x+3 & 2x^3 & -4x^2 & +3x & -1. \end{array}$$

Next we multiply $2x^2$ by the divisor $x + 3$ and subtract, carrying down the next term from the dividend:

$$\begin{array}{r|llll} & & 2x^2 & & \\ \hline x+3 & 2x^3 & -4x^2 & +3x & -1 \\ & -(2x^3 & +6x^2) & & \\ \cline{2-3} & & -10x^2 & +3x. & \end{array}$$

We repeat this as many times as needed until what remains has a smaller degree than the divisor. This criterion is not met yet, so we repeat, dividing $-10x^2$ by x to get $-10x$, which is placed in the quotient:

$$\begin{array}{r|llll} & & 2x^2 & -10x & \\ \hline x+3 & 2x^3 & -4x^2 & +3x & -1 \\ & -(2x^3 & +6x^2) & & \\ \cline{2-3} & & -10x^2 & +3x. & \end{array}$$

We then multiply $-10x$ by the divisor and subtract, again carrying down the next term:

$$
\begin{array}{r|llll}
 & 2x^2 & -10x & & \\ \cline{2-5}
x+3 & 2x^3 & -4x^2 & +3x & -1 \\
 & -(2x^3 & +6x^2) & & \\ \cline{2-3}
 & & -10x^2 & +3x & \\
 & & -(-10x^2 & -30x) & \\ \cline{3-4}
 & & & 33x & -1.
\end{array}
$$

The degree of what remains and the degree of the divisor are the same, so we continue:

We do not stop until the degree of the remainder is less than the degree of the divisor.

$$
\begin{array}{r|lllll}
 & 2x^2 & -10x & +33 & \\ \cline{2-5}
x+3 & 2x^3 & -4x^2 & +3x & -1 \\
 & -(2x^3 & +6x^2) & & \\ \cline{2-3}
 & & -10x^2 & +3x & \\
 & & -(-10x^2 & -30x) & \\ \cline{3-4}
 & & & 33x & -1 \\
 & & & -(33x & +99) \\ \cline{4-5}
 & & & & -100
\end{array} .
$$

Now the degree of what remains is less than the degree of the divisor, so we stop. The remainder is -100 and the quotient is $2x^2 - 10x + 33$. Rewriting using the form

$$\frac{\text{dividend}}{\text{divisor}} = \text{quotient} + \frac{\text{remainder}}{\text{divisor}}$$

yields

$$\frac{2x^3 - 4x^2 + 3x - 1}{x+3} = 2x^2 - 10x + 33 + \frac{-100}{x+3}.$$

Imagine integrating the expression on the left. That looks intimidating. Imagine integrating the expression on the right. This is just a polynomial plus a fraction which can be integrated easily using the natural log.

Integrating using long division

A fraction of polynomials is improper if the numerator has a degree at least as large as the denominator. Long division allows us to rewrite an improper fraction as a polynomial plus a proper fraction, which is very helpful for integration.

Example 1 *Evaluate* $\displaystyle\int \frac{x^2 + 3x - 4}{x+5}\,dx.$

STRATEGY

If the integrand is an improper fraction, use long division to rewrite it as a simpler integral.

Solution The integrand is a rational function with the degree of the numerator larger than the degree of the denominator. We can therefore simplify the integrand using long division:

The steps used to perform long division of polynomials are described in detail in the previous subsection.

$$\begin{array}{r|rrr} & x & -2 & \\ \hline x+5 & x^2 & +3x & -4 \\ & -(x^2 & +5x) & \\ \hline & & -2x & -4 \\ & & -(-2x & -10) \\ \hline & & & 6. \end{array}$$

Here we rewrite using

$$\frac{\text{dividend}}{\text{divisor}} = \text{quotient} + \frac{\text{remainder}}{\text{divisor}}.$$

Using the result of the long division process,

$$\frac{x^2+3x-4}{x+5} = x-2+\frac{6}{x+5}.$$

Replacing the integrand with the expression on the right results in

More details of integrating the fraction $\frac{6}{x+5}$ are given in example 2.

$$\int \frac{x^2+3x-4}{x+5}\,dx = \int \left(x-2+\frac{6}{x+5}\right)dx$$
$$= \frac{x^2}{2} - 2x + 6\ln|x+5| + C. \quad \blacksquare$$

Reading Exercise 1 *Evaluate* $\int \frac{x^2+4}{x-1}\,dx.$

Substitution: $u = x + k$

In example 1, the use of substitution for integrating $\frac{6}{x+5}$ was not demonstrated. These details are shown next.

Example 2 *Evaluate* $\int \frac{6}{x+5}\,dx.$

Solution Using the substitution

$$u = x+5$$
$$du = 1\,dx,$$

we have

$$\int \frac{6}{x+5}\,dx = \int \frac{6}{u}\,du$$
$$= 6\ln|u| + C$$
$$= 6\ln|x+5| + C,$$

as used in example 1. $\blacksquare$

In a substitution where $du = 1\,dx$, there is no portion of the original expression that disappears into du during the substitution, nor is there a constant for which to adjust. The integral can be completed "naively" without the use of substitution. In other words,

Check these answers using differentiation. The chain rule requires multiplication by the derivative of what is inside, but this is just one in these examples.

$$\int \frac{2}{x-17}\,dx = 2\ln|x-17| + C$$

and

$$\int \sin(x+4)\,dx = -\cos(x+4) + C.$$

But be careful! If the coefficient of x is anything other than one, du is not just dx and integrating naively gives an incorrect answer.

CAUTION: MISTAKE TO AVOID
Use this strategy only when "not just plain x" is of the form $x + k$. In other words, if $du \neq 1\,dx$, substitution is still needed.

Evaluating $\int \frac{1}{ax+b}\,dx$

The next example has a subtle but very important change from example 2.

Example 3 *Evaluate* $\int \frac{6}{2x+5}\,dx.$

Solution Using the substitution

CAUTION: MISTAKE TO AVOID
The answer is **not** $6\ln|2x+5| + C$.

$$u = 2x+5$$
$$du = 2\,dx,$$

we have

$$\begin{aligned}\int \frac{6}{2x+5}\,dx &= \frac{1}{2}\int \frac{6}{2x+5}\cdot 2\,dx \\ &= \frac{1}{2}\int \frac{6}{u}\,du \\ &= \frac{1}{2}\cdot 6\ln|u| + C \\ &= 3\ln|2x+5| + C.\end{aligned}$$

Line 1, adjustment made for needed constant 2; line 2, substitution made (including $2\,dx$ being replaced with du).

The answer can be checked using differentiation:

$$\frac{d}{dx}3\ln|2x+5| = 3\cdot\frac{1}{2x+5}\cdot 2 = \frac{6}{2x+5},$$

as expected. ■

Ans. to reading exercise 1

$\frac{x^2}{2} + x + 5\ln|x-1| + C$

It is important to understand the difference between examples 2 and 3. The form of the denominator ($x + 5$ vs. $2x + 5$) is the key, regardless of whether the coefficient of x is one or something else. The integral

$$\int \frac{4}{x-9}\,dx = 4\ln|x-9| + C$$

is quick, whereas

CAUTION: MISTAKE TO AVOID
Those of you who do this substitution mentally have a much greater error rate than those of you who write out the substitution.

$$\int \frac{4}{5x+9}\,dx = \cdots = \frac{4}{5}\ln|5x+9| + C$$

requires substitution.

Reading Exercise 2 *Integrate* $\int \frac{4}{3x-1}\,dx$.

Review: completing the square

The key to remembering how completing the square works, and why, is to remember how to square a binomial, specifically in the form

$$(x+a)^2 = x^2 + 2ax + a^2,$$

examples of which include

$$(x+1)^2 = x^2 + 2x + 1$$
$$(x+2)^2 = x^2 + 4x + 4$$
$$(x+3)^2 = x^2 + 6x + 9$$
$$(x+4)^2 = x^2 + 8x + 16.$$

Suppose we start with the expression

$$x^2 + 6x.$$

What needs to be added to make it a perfect square? The answer, of course, is 9. But, how do we come up with 9 without referring to the previous list? Compare the formula to the expression:

Adding 9 makes for a perfect square:

$$x^2 + 6x + 9 = (x+3)^2.$$

$$(x+a)^2 = x^2 + 2ax + a^2$$
$$x^2 + 6x.$$

We have $2a = 6$, or $a = 3$. The number we then need to add is a^2, which is 9.

Completing the square helps us rewrite an expression in the form $(x+a)^2 + k$, which is helpful for integration.

Example 4 *Complete the square on* $x^2 + 5x + 12$.

To "complete the square" means to rewrite the expression in the form $(x+a)^2+k$, where a and k are numbers.

Solution ❶ We begin by moving the constant to the right, leaving space for another number:

$$x^2 + 5x \qquad + 12.$$

❷ Find half the coefficient of x; call it a:

$$a = \tfrac{5}{2}.$$

❸ Add and subtract a^2 in the original expression:

Because we are adding and subtracting the same number, the value of the expression is not changed.

$$x^2 + 5x + \tfrac{25}{4} + 12 - \tfrac{25}{4}.$$

❹ Factor the first three terms into the perfect square $(x+a)^2$:

$$= (x + \tfrac{5}{2})^2 + 12 - \tfrac{25}{4}.$$

❺ Combine the remaining constants.

$$= (x + \tfrac{5}{2})^2 + \tfrac{23}{4}.$$

The result of completing the square is $(x + \frac{5}{2})^2 + \frac{23}{4}$. ■

If the coefficient of x^2 is not one, factor out that coefficient from the x^2 and x terms, and complete the square inside the parentheses.

Example 5 *Complete the square on* $4x^2 + 32x + 9$.

Solution ❶ We begin by moving the constant to the right, leaving space for another number. ❶ ⓐ Then, factor out the coefficient of x^2, but not from the constant term:

Ans. to reading exercise 2
$\frac{4}{3}\ln|3x-1| + C$

$$4x^2 + 32x \qquad + 9$$

$$4(x^2 + 8x \qquad) + 9.$$

❷ Find half the coefficient of x; call it a:

$$a = 4.$$

❸ Add a^2 inside the parentheses, then balance the equation by subtracting the appropriate quantity at the end:

By adding 16 inside the parentheses, we really added $4 \cdot 16 = 64$ to the expression. We therefore also subtract 64 from the expression so that its value remains unchanged.

$$4(x^2 + 8x + 16) + 9 - 64.$$

❹ Factor the quantity inside the parentheses into the perfect square $(x + a)^2$:

$$= 4(x + 4)^2 + 9 - 64.$$

❺ Combine the remaining constants:

$$= 4(x + 4)^2 - 55.$$

■

The result of completing the square is $4(x + 4)^2 - 55$.

Integration using completing the square

Before using completing the square to help with integration, let's review some integrals with denominators that are quadratic polynomials.

Recall that a quadratic polynomial is a polynomial of degree 2. An example is $4 + x^2$.

Example 6 *Evaluate* $\displaystyle\int \frac{3}{4 + x^2}\, dx.$

Solution Using version 2 of the antiderivative giving the inverse tangent formula with $a = 2$,

$$\int \frac{3}{4 + x^2}\, dx = 3 \int \frac{1}{2^2 + x^2}\, dx = 3 \cdot \frac{1}{2} \tan^{-1}\left(\frac{x}{2}\right) + C. \quad ■$$

The next example uses the same formula but with substitution.

Example 7 *Evaluate* $\displaystyle\int \frac{1}{1 + (3x - 4)^2}\, dx.$

Solution Noticing "not just plan x" to a power, we use substitution:

$$u = 3x - 4$$
$$du = 3\, dx.$$

Adjusting for the constant 3, we proceed to integrate:

Line 1 adjusts for the required constant, line 2 completes the substitution, line 3 uses the antiderivative giving the inverse tangent formula, and line 4 returns to the original variable.

$$\begin{aligned}\int \frac{1}{1+(3x-4)^2}\,dx &= \frac{1}{3}\int \frac{1}{1+(3x-4)^2}\cdot 3\,dx\\ &= \frac{1}{3}\int \frac{1}{1+u^2}\,du\\ &= \frac{1}{3}\tan^{-1} u + C\\ &= \frac{1}{3}\tan^{-1}(3x-4) + C. \quad \blacksquare\end{aligned}$$

The denominator in example 7 can be expanded as

$$1+(3x-4)^2 = 1+9x^2-24x+16 = 9x^2-24x+17.$$

We may then integrate

$$\int \frac{1}{9x^2-24x+17}\,dx$$

from the equivalent form

$$\int \frac{1}{1+(3x-4)^2}\,dx.$$

This is how completing the square helps! If we start with a quadratic denominator containing an x term, such as $9x^2 - 24x + 17$, by completing the square we may be able to write it in a form conducive to the method of example 7.

Example 8 *Evaluate* $\displaystyle\int \frac{1}{x^2+6x+14}\,dx.$

Solution Recognizing a quadratic denominator that includes an x term (middle term), we begin by completing the square; $a = \frac{6}{2} = 3$, $a^2 = 9$, and we have

$$\begin{aligned}&x^2+6x \qquad +14\\ &= x^2+6x+9+14-9\\ &= (x+3)^2+5.\end{aligned}$$

We can now rewrite the integrand in a form like that of example 7:

$$\int \frac{1}{x^2+6x+14}\,dx = \int \frac{1}{5+(x+3)^2}\,dx.$$

Since $du = dx$, substitution can be safely avoided. In such cases, the substitution steps may be shown anyway.

Proceeding as in example 7, we recognize the need for substitution:

$$u = x + 3$$
$$du = 1\,dx.$$

We finish evaluating the integral using substitution and version 2 of the antiderivative giving the inverse tangent formula with $a = \sqrt{5}$:

$$= \int \frac{1}{5 + u^2}\,du$$
$$= \frac{1}{\sqrt{5}} \tan^{-1}\left(\frac{u}{\sqrt{5}}\right) + C$$
$$= \frac{1}{\sqrt{5}} \tan^{-1}\left(\frac{x+3}{\sqrt{5}}\right) + C.$$

Completing the square helps us complete the integration. ■

STRATEGY

If the denominator of an integrand contains a quadratic expression that includes the middle term (the x term), it might help to complete the square.

Reading Exercise 3 *Integrate* $\displaystyle\int \frac{1}{x^2 + 4x + 5}\,dx.$

The same strategy is available when the quadratic is under a square root.

Example 9 *Evaluate* $\displaystyle\int \frac{7}{\sqrt{x^2 + 10x + 9}}\,dx.$

Solution Recognizing a quadratic that includes an x term, we begin by completing the square; $a = \frac{10}{2} = 5$, $a^2 = 25$, and we have

$$x^2 + 10x \qquad + 9$$
$$= x^2 + 10x + 25 + 9 - 25$$
$$= (x+5)^2 - 16.$$

We rewrite the integrand using the result:

$$\int \frac{7}{\sqrt{x^2 + 10x + 9}}\,dx = \int \frac{7}{\sqrt{(x+5)^2 - 16}}\,dx.$$

Recognizing "not just plain x" to a power, we use substitution:

$$u = x + 5$$
$$du = dx.$$

Substitution takes us to

$$= \int \frac{7}{\sqrt{u^2 - 16}}\, du.$$

Recognizing the antiderivative giving the inverse hyperbolic cosine formula, we use $a = 4$ and obtain

$$= 7\cosh^{-1}\left(\frac{u}{4}\right) + C$$

$$= 7\cosh^{-1}\left(\frac{x+5}{4}\right) + C. \qquad \blacksquare$$

The integral in example 9 is relatively straightforward if the usefulness of completing the square is recognized at first. After that, recognizing substitution is not difficult, and then the integrand is ready for an antiderivative formula. Much of this chapter is similar in that recognizing algebraic forms—and knowing what to do to handle these forms—is important.

Exercises 7.1

1–4. Rewrite as a polynomial plus a proper fraction.

1. $\dfrac{x^2 + 9x - 18}{x - 6}$
2. $\dfrac{x^3 - 7x + 4}{x + 5}$
3. $\dfrac{x^3 - 2x + 4}{x^2 + 3x}$
4. $\dfrac{4x^3 + 8x - 15}{2x + 1}$

5–8. Complete the square.

5. $x^2 - 8x + 11$
6. $x^2 + 9x + 21$
7. $5x^2 + 2x - 4$
8. $3x^2 - 6x + 1$

Ans. to reading exercise 3
$\tan^{-1}(x + 2) + C$

9–30. Evaluate the integral.

9. $\displaystyle\int \frac{x^2 - 5x + 7}{x - 2}\, dx$
10. $\displaystyle\int \frac{3x^2 + 4x}{x + 1}\, dx$
11. $\displaystyle\int \frac{21x^2 + 2x + 5}{3x - 1}\, dx$
12. $\displaystyle\int \frac{2x - 9}{x - 4}\, dx$
13. $\displaystyle\int \frac{2x^2 + 25}{x^2 + 9}\, dx$
14. $\displaystyle\int \frac{1}{1 + (4x + 9)^2}\, dx$

15. $\displaystyle\int \frac{x}{x^2+9}\,dx$

16. $\displaystyle\int \frac{x^2+3}{x^2+1}\,dx$

17. $\displaystyle\int \frac{2x-3}{x+\frac{1}{2}}\,dx$

18. $\displaystyle\int \frac{1}{x^2+10x+26}\,dx$

19. $\displaystyle\int \frac{1}{x^2+2x+2}\,dx$

20. $\displaystyle\int \frac{4}{x^2-6x+18}\,dx$

21. $\displaystyle\int \frac{\sqrt{7}}{x^2-6x+16}\,dx$

22. $\displaystyle\int \frac{3x^3+5}{2x+1}\,dx$

23. $\displaystyle\int \frac{1}{\sqrt{x^2+8x+17}}\,dx$

24. $\displaystyle\int \frac{1}{\sqrt{x^2+4x+5}}\,dx$

25. $\displaystyle\int \frac{1}{\sqrt{10x-24-x^2}}\,dx$

26. $\displaystyle\int \frac{4}{\sqrt{x^2+4x+3}}\,dx$

27. $\displaystyle\int \frac{3x^3-24x^2+51x+5}{x^2-8x+17}\,dx$

28. $\displaystyle\int \frac{1}{3x^2+12x+28}\,dx$

29. $\displaystyle\int \frac{2}{\sqrt{9x^2+30x-56}}\,dx$

30. $\displaystyle\int \frac{x^3-x^2+5x+12}{x^2-2x+7}\,dx$

Integration by Parts

7.2

The chain rule for derivatives was reversed to obtain the technique of integration known as substitution. In this section we reverse the product rule.

Integration by parts: the formula

The product rule states that if f and g are differentiable at x, then

$$\frac{d}{dx}(f(x) \cdot g(x)) = f(x)g'(x) + g(x)f'(x).$$

Knowing that every derivative formula can be reversed to give an integral formula, we get

$$\int \left(f(x)g'(x) + g(x)f'(x)\right)\, dx = f(x) \cdot g(x) + C.$$

We rewrite

$$\frac{d}{dx}\text{left side} = \text{right side}$$

as

$$\int \text{right side}\, dx = \text{left side} + C.$$

Using the antiderivative sum rule on the left-hand side gives

$$\int f(x)g'(x)\, dx + \int g(x)f'(x)\, dx = f(x)g(x) + C.$$

This equation can be rearranged by subtracting $\int g(x)f'(x)\, dx$ from both sides:

$$\int f(x)g'(x)\, dx = f(x)g(x) - \int g(x)f'(x)\, dx.$$

Because the right side now contains an integral, the $+ C$ can be considered part of the answer to this integral. It is not necessary to write "$+ C$" at the end of the equation.

What did this accomplish? If $\int f(x)g'(x)\, dx$ is a difficult integral but $\int g(x)f'(x)\, dx$ is an easier integral, then the method helps. We call this technique of integration *integration by parts.*

> **INTEGRATION BY PARTS**
>
> If the functions are integrable,
>
> $$\int f(x)g'(x)\, dx = f(x)g(x) - \int g(x)f'(x)\, dx.$$

> The integration-by-parts formula can be remembered as
>
> $$\int fg'\, dx = fg - \int gf'\, dx.$$

The integration-by-parts formula can be written in a differential format as

$$\int u\, dv = uv - \int v\, du.$$

Integration by parts: basic examples

Example 1 *Use integration by parts to evaluate* $\int x \sin x \, dx$.

Solution The integration-by-parts formula applies to an integral of the form $\int f(x)g'(x)\,dx$. The integrand must be a product of two functions, f and g'. Our integrand is the product of x and $\sin x$. Try using

$$f(x) = x \qquad\qquad g'(x) = \sin x.$$

The right-hand side of the integration-by-parts formula uses $f'(x)$ and $g(x)$, so they must be calculated. We find $f'(x)$ by taking the derivative of $f(x)$, and we find $g(x)$ by finding an antiderivative of $g'(x)$. We write them below our f and g':

Any antiderivative of $g'(x)$ suffices. We do not need all of them, so there is no need to write "$+ C$" at the end of $g(x)$.

$$\begin{aligned} f(x) &= x & g'(x) &= \sin x \\ f'(x) &= 1 & g(x) &= -\cos x. \end{aligned}$$

Applying the integration-by-parts formula,

Line 1 is the integration-by-parts formula; line 2 applies this formula with the functions f, g', f', and g; line 3 simplifies algebraically; and line 4 completes the integration and gives the final answer, which includes $+ C$.

$$\begin{aligned} \int f(x)g'(x)\,dx &= f(x)g(x) - \int g(x)f'(x)\,dx \\ \int x\sin x\,dx &= x(-\cos x) - \int (-\cos x)\cdot 1\,dx \\ &= -x\cos x + \int \cos x\,dx \\ &= -x\cos x + \sin x + C. \end{aligned}$$ ∎

In example 1, we began with an integral that contained a product, but by using integration by parts, we reached an integral that did not contain a product, $\int \cos x\,dx$, which we could integrate easily. Although the right-hand side's integral is $\int g(x)f'(x)\,dx$, which is a product, since $f'(x) = 1$, the "product" melts away into a single factor!

Going from $f(x) = x$ to $f'(x) = 1$ makes for a simpler integral. Meanwhile, changing from $g'(x)$ to $g(x)$ merely changes from $\sin x$ to $-\cos x$, which is not more complicated. This is the key: let f be a factor that has a simpler derivative, and let g be the rest of the expression, which hopefully does not have a more complicated antiderivative.

Which types of functions have less-complicated derivatives? Some examples are polynomials, logarithmic functions, and inverse trig functions:

polynomial	$f(x) = x^2$	$f'(x) = 2x$
logarithmic	$f(x) = \ln x$	$f'(x) = \frac{1}{x}$
inverse trig	$f(x) = \tan^{-1} x$	$f'(x) = \frac{1}{1+x^2}$.

For polynomials the degree gets smaller; for logarithmic and inverse trig, the derivative is algebraic. These functions are good candidates for f in integration by parts.

On the other hand, exponential, trig, and hyperbolic functions do not have less-complicated derivatives:

exponential	$f(x) = e^x$	$f'(x) = e^x$
trig	$f(x) = \sin x$	$f'(x) = \cos x$
hyperbolic	$f(x) = \cosh x$	$f'(x) = \sinh x$

The derivatives of these functions are the same type of function. These functions are not good candidates for f in integration by parts.

STRATEGY

If the integrand is a product of a polynomial function with an exponential, trig, or hyperbolic function, try integration by parts with the polynomial factor as f.

Reading Exercise 4 *Evaluate* $\int x \cos x \, dx$.

Example 2 *Evaluate* $\int x^2 \cos x \, dx$.

Solution We recognize a product of a polynomial (x^2) with a trig function ($\cos x$). We therefore try integration by parts with the polynomial factor as f:

$$f(x) = x^2.$$

We let $g'(x)$ be the rest of the integrand:

The product $f(x)g'(x)$ must be equal to the integrand.

$$g'(x) = \cos x.$$

Next we differentiate f and integrate g':

Because we have g' and want g, we are integrating:

$$g(x) = \int \cos x \, dx = \sin x + C.$$

(Because g may be any antiderivative of g', the $+\ C$ is not necessary and is omitted when writing it in the problem solution.)

$$f(x) = x^2 \qquad g'(x) = \cos x$$
$$f'(x) = 2x \qquad g(x) = \sin x.$$

Applying the integration-by-parts formula,

$$\int f(x)g'(x)\,dx = f(x)g(x) - \int g(x)f'(x)\,dx$$
$$\int x^2 \cos x \, dx = x^2(\sin x) - \int (\sin x)(2x)\,dx$$
$$= x^2 \sin x - 2\int x \sin x \, dx.$$

Unfortunately, what happened in example 1 doesn't happen here. The new integral is still a product. But, not all is lost! The new integral is the integral we just evaluated in example 1, so we know how to evaluate it. We just need to use integration by parts again.

Compare the use of the word *part* here to the name of the technique, *integration by parts.*

Repeating the work from example 1, we let f be the polynomial part and g' the trig part, then calculate f' and g:

$$f(x) = x \qquad g'(x) = \sin x$$
$$f'(x) = 1 \qquad g(x) = -\cos x.$$

Using the integration-by-parts formula for $\int x \sin x \, dx$, we continue the original calculation as

Line 1 is repeated from the previous integral calculation; line 2 uses the integration by parts formula, $\int fg' \, dx = fg - \int gf' \, dx$, on the integral $\int x \sin x \, dx$; line 3 uses algebra to simplify; and line 4 completes the integration.

$$\begin{aligned}\int x^2 \cos x \, dx &= x^2 \sin x - 2 \int x \sin x \, dx \\ &= x^2 \sin x - 2 \left(x(-\cos x) - \int (-\cos x) 1 \, dx \right) \\ &= x^2 \sin x + 2x \cos x - 2 \int \cos x \, dx \\ &= x^2 \sin x + 2x \cos x - 2 \sin x + C,\end{aligned}$$

which is the antiderivative we were asked to find. ■

Tabular integration

Evaluating $\int x \sin x \, dx$ requires integration by parts once. Evaluating $\int x^2 \cos x \, dx$ requires integration by parts twice. Care to guess how many applications of integration by parts are required to evaluate $\int x^3 \sin x \, dx$? Each application of parts with the polynomial part as f reduces the degree of the polynomial term by one, so $\int x^3 \sin x \, dx$ requires three applications of parts.

A careful study of example 2 reveals there is some redundancy in writing all the details of parts twice. A more efficient way of writing multiple applications of integration by parts is called *tabular integration*, which is shown in the following example.

Example 3 *Evaluate* $\int x^5 \sin x \, dx$.

Solution In each application of parts, the polynomial part is f, and we take its derivative. In each application of parts, the trig part is g', and we integrate it. This can be arranged in two columns—one where we take derivatives and the other, antiderivatives:

Ans. to reading exercise 4
$x \sin x + \cos x + C$

part 1, derivatives		part 2, integrals
x^5	+	$\sin x$
$5x^4$	−	$-\cos x$
$20x^3$	+	$-\sin x$
$60x^2$	−	$\cos x$
$120x$	+	$\sin x$
120	−	$-\cos x$
0		$-\sin x.$

The derivative column begins with the polynomial part of the integrand, x^5. In each row, a derivative is taken, and we continue until we reach zero. The integral column begins with the trig (or exponential or hyperbolic) part of the integrand, $\sin x$. In each row, an antiderivative is taken (be careful of the $\pm$ signs on the antiderivatives!) until we have the same number of rows as for the derivatives. Next we draw slanted lines from an entry in the left-hand column to an entry in the right-hand column one row lower. The table is finished by writing alternating signs, + and −, on those lines, beginning with +.

We are now ready to write the final answer. The first term is the product of the factors linked by the first slanted line, $x^5(-\cos x)$, with + in front; the second term is the product of the factors linked by the second slanted line, $5x^4(-\sin x)$, with − in front; and we continue through all the slanted lines to get

The products of the factors linked by the slanted lines are from the $f(x)g(x)$ in the integration-by-parts formula. Compare to how parts are written in example 1. The alternating + and − signs written on the slanted lines are caused by the subtraction (the −) of the integral on the right side of the integration-by-parts formula. Other comparisons of this tabulating method to multiple applications of parts are left to you.

$$\int x^5 \sin x\,dx = +x^5(-\cos x) - 5x^4(-\sin x) + 20x^3(\cos x) - 60x^2(\sin x) + 120x(-\cos x) - 120(-\sin x) + C$$

$$= -x^5\cos x + 5x^4\sin x + 20x^3\cos x - 60x^2\sin x - 120x\cos x + 120\sin x + C.$$

Once the procedure is learned, it is much faster than writing out the full details of parts several times. ■

Some of you may even prefer to use tabular integration on the integral in example 2.

Substitution vs. parts

When calculating a derivative, every product requires the product rule. For integrals, we do not use parts every time we see a product.

For instance,

$$\int \sec x \tan x \, dx$$

contains a product, but there is an antiderivative formula for this:

$$\int \sec x \tan x \, dx = \sec x + C.$$

Next we look at another instance of integrating a product for which integration by parts is not the method to use.

Example 4 *Evaluate* $\int x \sin x^2 \, dx$.

Solution (First attempt.) Recognizing the product of a polynomial and a trig function, we try integration by parts as in examples 1–3. We let the polynomial part be f and the trig part be g', then calculate f' and g:

$$f(x) = x \qquad g'(x) = \sin x^2$$

$$f'(x) = 1 \qquad g(x) = ??$$

When finding g from g', all the usual rules of antidifferentiation apply, such as the need for substitution if there is a trig function of "not just plain x."

Finding $g(x)$ is a problem. We need

$$g(x) = \int \sin x^2 \, dx,$$

which requires substitution because we have a trig function of "not just plain x." Then,

$$u = x^2$$

$$du = 2x \, dx,$$

and there is no x in the integral to use for our du. We are stuck, and this attempt at integration does not work. ■

Look again at the integrand of example 4. There is the x we need for substitution! In fact, we should recognize from the beginning that substitution is needed and try that first.

STRATEGY

Substitution trumps parts. If substitution is needed, it should be tried first.

Solution (Second attempt.) Recognizing a trig function of "not just plain x," we use substitution:

$$u = x^2$$

$$du = 2x \, dx.$$

We then adjust for the constant, make the substitution, and integrate:

$$\begin{aligned}\int x \sin x^2\, dx &= \frac{1}{2}\int 2x \sin x^2\, dx \\ &= \frac{1}{2}\int \sin u\, du \\ &= \frac{1}{2}(-\cos u) + C \\ &= -\frac{1}{2}\cos x^2 + C,\end{aligned}$$

and we integrate successfully without needing to use parts. ■

Integration by parts: exponential times trig

As stated earlier, polynomial, logarithmic, and inverse trig parts are good candidates for f when integrating by parts, and exponential, trig, and hyperbolic parts are not. What if we have a product of exponential and trig functions? The answer is to perform parts twice and then solve for the integral.

Example 5 *Evaluate* $\int e^x \sin x\, dx$.

Solution Recognizing a product, we try integration by parts. Let e^x be f and $\sin x$ be g'. We then calculate f' and g:

$$\begin{aligned} f(x) &= e^x & g'(x) &= \sin x \\ f'(x) &= e^x & g(x) &= -\cos x. \end{aligned}$$

Next we use the integration by parts formula:

$$\begin{aligned}\int e^x \sin x\, dx &= e^x(-\cos x) - \int(-\cos x)e^x\, dx \\ &= -e^x \cos x + \int e^x \cos x\, dx.\end{aligned}$$

A lot of good that did! The new integral is still exponential times trig. Let's try parts again anyway, still using the exponential part as f and the trig part as g':

$$\begin{aligned} f(x) &= e^x & g'(x) &= \cos x \\ f'(x) &= e^x & g(x) &= \sin x. \end{aligned}$$

Switching the roles of the parts—that is, using $f(x) = \cos x$ and $g'(x) = e^x$—just undoes the first application of parts. Keep the same roles for the exponential and trig parts.

Using the integration-by-parts formula again, we continue the earlier calculation and get

Line 1 repeats the result of the previous integral calculation; line 2 replaces $\int e^x \cos x\,dx$ using parts.

$$\begin{aligned}\int e^x \sin x\,dx &= -e^x \cos x + \int e^x \cos x\,dx \\ &= -e^x \cos x + e^x \sin x - \int e^x \sin x\,dx.\end{aligned}$$

Aren't we back where we started? Not quite. Look again at the previous equation. We have the same integral on both sides of the equation. If we treat this integral like a variable, we can solve for it and have our answer:

Line 1 repeats the previous equation, line 2 adds $\int e^x \sin x\,dx$ to both sides, and line 3 divides both sides by 2 and adds the $+C$. (Technically, the $+C$ should also be written at the end of line 2, but it is omitted here to help emphasize the algebra being done on that line.)

$$\begin{aligned}\int e^x \sin x\,dx &= -e^x \cos x + e^x \sin x - \int e^x \sin x\,dx \\ 2\int e^x \sin x\,dx &= -e^x \cos x + e^x \sin x \\ \int e^x \sin x\,dx &= \frac{-e^x \cos x + e^x \sin x}{2} + C.\end{aligned}$$ ■

An alternative solution to example 5 is to integrate by parts twice with the trig as f and exponential as g', both times. The strategy is the same: integrate by parts twice and solve for the integral. The only practical advantage to choosing one part for f over the other is when the chain rule or substitution must be used.

STRATEGY

If the integrand is exponential times trig (or hyperbolic), use parts twice, then solve for the integral. To be successful, the roles cannot be switched; for instance, use exponential as f both times.

Integration by parts: logs, inverse trig, and inverse hyperbolic

Integration by parts can be helpful even when there is no visible product. Many integrals involving logarithms can be integrated by parts with the logarithm as f. The same is true for inverse trig and inverse hyperbolic functions.

STRATEGY

If the integrand contains a logarithmic function, integration by parts with the logarithmic part as f might help. The same is true for inverse trig and inverse hyperbolic functions.

Example 6 *Evaluate* $\int \ln x\,dx$.

Solution The integrand is not presented as a product, but we can still consider it to be $\ln x \cdot 1$, which is in the form of a product. Let's try parts as follows:

$$\begin{aligned} f(x) &= \ln x & g'(x) &= 1 \\ f'(x) &= \frac{1}{x} & g(x) &= x.\end{aligned}$$

Then,

$$\int f(x)g'(x)\,dx = f(x)g(x) - \int g(x)f'(x)\,dx$$

$$\int \ln x \cdot 1\,dx = (\ln x)x - \int x \cdot \frac{1}{x}\,dx$$

$$= x\ln x - \int 1\,dx$$

$$= x\ln x - x + C.$$

We conclude that $\int \ln x\,dx = x\ln x - x + C$, which serves as a new antiderivative formula. ■

ANTIDERIVATIVE OF NATURAL LOG

$$\int \ln x\,dx = x\ln x - x + C$$

Combining substitution and parts

Integration by parts and substitution can be used together on the same integral.

Example 7 *Evaluate* $\int e^{\sqrt{x}}\,dx.$

Solution Recognizing an exponent of "not just plain x," we try substitution, letting u be that exponent:

$$u = \sqrt{x}$$

$$du = \frac{1}{2\sqrt{x}}\,dx.$$

We need $\frac{1}{2\sqrt{x}}$ for the du, and it is not present in the integrand. Following the same strategy as adjusting for a constant, we try multiplying and dividing by this expression, but we need to be careful. The multiplication and division must take place completely inside the integral:

$$\int e^{\sqrt{x}}\,dx = \int \frac{2\sqrt{x}}{2\sqrt{x}} \cdot e^{\sqrt{x}}\,dx$$

$$= \int 2u \cdot e^{u}\,du.$$

Although this substitution can be completed, it is often the case that multiplying and dividing by a variable expression does not result in a successful substitution. As mentioned, caution is necessary.

CAUTION: MISTAKE TO AVOID
Never pull a variable expression from inside an integral to the outside:

$$\int e^{\sqrt{x}}\,dx \neq 2\sqrt{x}\int \frac{1}{2\sqrt{x}}e^{\sqrt{x}}\,dx.$$

Only constants can be pulled out front, as justified by the antiderivative constant multiple rule.

Above left: line 2 carries out the substitution. The $2\sqrt{x}$ in the denominator disappears into the du. The $2\sqrt{x}$ in the numerator becomes $2u$.

Now we recognize a product of a polynomial ($2u$) and an exponential (e^u), which indicate the use of integration by parts. We let the polynomial part be f and the exponential part be g', then calculate f' and g:

Notice that we write the functions as $f(u)$ and $g'(u)$, not $f(x)$ and $g'(x)$, because the variable in use is u.

$$f(u) = 2u \qquad g'(u) = e^u$$
$$f'(u) = 2 \qquad g(u) = e^u.$$

Next we use the integration-by-parts formula;

$$\int 2ue^u \, du = 2ue^u - \int e^u \cdot 2 \, du$$
$$= 2ue^u - 2e^u + C.$$

At this point we are not yet finished. The original variable is x, so we must undo the substitution by replacing u with $\sqrt{x}$. Our final answer is

$$\int e^{\sqrt{x}} \, dx = 2\sqrt{x}\, e^{\sqrt{x}} - 2e^{\sqrt{x}} + C.$$

The answer can be simplified further if desired. ■

Integration by parts: definite integrals

In my opinion, the easiest method for evaluating a definite integral using integration by parts is first to evaluate the indefinite integral (find the antiderivative) as a separate problem and then evaluate the definite integral. In other words, don't include the limits of integration until the antidifferentiation is complete.

Example 8 *Evaluate* $\displaystyle\int_0^1 xe^x \, dx.$

Solution ❶ Evaluate the indefinite integral. We rewrite the integral without the limits of integration and seek an antiderivative:

$$\int xe^x \, dx.$$

The integrand is the product of a polynomial (x) and an exponential (e^x), and therefore we integrate by parts with the polynomial part as f:

$$f(x) = x \qquad g'(x) = e^x$$
$$f'(x) = 1 \qquad g(x) = e^x.$$

Next we use the integration-by-parts formula:

$$\begin{aligned}\int xe^x\,dx &= xe^x - \int e^x \cdot 1\,dx\\ &= xe^x - e^x + C.\end{aligned}$$

❷ Return to the indefinite integral. Returning to the original problem now that we have an antiderivative,

$$\begin{aligned}\int_0^1 xe^x\,dx &= \left(xe^x - e^x\right)\Big|_0^1\\ &= e - e - (0 - 1) = 1.\end{aligned}$$ ■

The integral in example 8 is not that complicated, so it is not too difficult to keep notation and calculation straight while writing the calculation directly as a definite integral throughout the problem. But, as illustrated by some of the earlier examples, some integrals can become rather involved, and in these instances, keeping notation and calculation for a definite integral correct for the entire duration of the evaluation can be more trouble than it is worth. The approach in example 8 is highly recommended.

Exercises 7.2

1–10. Rapid response: is integration by parts the right approach? If so, what are f and g'?

1. $\int x\cos x\,dx$

2. $\int x\cos x^2\,dx$

3. $\int x^2\cos x\,dx$

4. $\int xe^x\,dx$

5. $\int xe^{x^2}\,dx$

6. $\int x^2e^x\,dx$

7. $\int e^x\sin x\,dx$

8. $\int x^3\sinh x\,dx$

9. $\int \sin^{-1} x\,dx$

10. $\int x\ln x\,dx$

11–54. Integrate.

11. $\int x \cos x \, dx$

12. $\int xe^x \, dx$

13. $\int x^2 \cos x \, dx$

14. $\int x^2 e^x \, dx$

15. $\int x \cos x^2 \, dx$

16. $\int xe^{x^2} \, dx$

17. $\int x^4 \cos x \, dx$

18. $\int x^7 e^x \, dx$

19. $\int 2x \sinh x \, dx$

20. $\int 4x \sec^2 x \, dx$

21. $\int (4x^2 - 3x + 1)e^x \, dx$

22. $\int x \cos 4x \, dx$

23. $\int xe^{2x} \, dx$

24. $\int (3x - 5) \sin x \, dx$

25. $\int \sin^{-1} x \, dx$

26. $\int x^3 \sin x^4 \, dx$

27. $\int_0^4 xe^x \, dx$

28. $\int x \ln x \, dx$

29. $\int x^2 \ln x \, dx$

30. $\int (x^4 + 2x) \cos x \, dx$

31. $\int e^{2x} \sin x \, dx$

32. $\int \cos^{-1} x \, dx$

33. $\int_0^\pi 2x^4 \sin x \, dx$

34. $\int e^x \sinh x \, dx$

35. $\int 3^x \cosh x \, dx$

36. $\int e^{4x} \sin 3x \, dx$

37. $\int 2^{\sin x} \cos x \, dx$

38. $\int_1^3 x^2 \cosh x \, dx$

39. $\int e^x \sin e^x \, dx$

40. $\int x^5 \cosh x^3 \, dx$

41. $\int 3x^4 e^x \, dx$

42. $\int e^x \cos 4x \, dx$

43. $\int \cosh \sqrt{x} \, dx$

44. $\int_0^1 x^3 \sin x \, dx$

45. $\int x^3 e^{x^2} \, dx$

46. $\int 2^x \cos 2^x \, dx$

47. $\int \ln x^7 \, dx$

48. $\int \sin \sqrt{x} \, dx$

49. $\int x \cos^{-1} x^2 \, dx$

50. $\int x^2 \tan^{-1} x^3 \, dx$

51. $\int e^x \cos 5x\, dx$

52. $\int x^7 \ln x\, dx$

53. $\int 3x^2 \tan^{-1} x\, dx$

54. $\int (\sin x)(\cos x)e^{\sin x}\, dx$

55. Find the volume of the solid of revolution generated by revolving the region under the curve $y = \sin x + 2$ from $x = 0$ to $x = \pi$, about the y-axis.

56. Find the volume of the solid of revolution generated by revolving the region under the curve $y = \cos x + \sin x$ from $x = 0$ to $x = \frac{\pi}{2}$, about the y-axis.

57. (GSL) Find the volume of the solid of revolution generated by revolving the region bounded by $y = xe^x$, $y = 0$, and $x = 1$, about the x-axis.

58. Find the volume of the solid of revolution generated by revolving the region bounded by $y = e^x$, $y = e$, and $x = 0$, about the y-axis.

59–64. Integrate.

59. $\int (2x^5 - 3x^2 + 7)(e^{2x} - x)\, dx$

60. $\int x^2(\sin x + 2\cos x)\, dx$

61. (GSL) $\int \frac{\ln(x+1)}{\sqrt{x+1}}\, dx$

62. (GSL) $\int \frac{x^2}{e^x}\, dx$

63. (Love) $\int x\sqrt{1+x}\, dx$

64. (Love) $\int x^3\sqrt{9-x^2}\, dx$

Trigonometric Integrals

7.3

In this section we expand our ability to evaluate integrals involving the trigonometric functions.

$\int \sec x\,dx$

There are still two of the basic six trigonometric functions for which we have not yet found an antiderivative. One is secant. The trick in evaluating $\int \sec x\,dx$ is to multiply the numerator and the denominator by $\sec x + \tan x$:

$$\int \sec x\,dx = \int \sec x \cdot \frac{\sec x + \tan x}{\sec x + \tan x}\,dx$$

$$= \int \frac{\sec^2 x + \sec x \tan x}{\sec x + \tan x}\,dx.$$

Now compare the denominator of the integrand to its numerator. The numerator is the derivative of the denominator! With this clue, we use substitution, allowing the denominator to be u:

$$u = \sec x + \tan x$$

$$du = (\sec x \tan x + \sec^2 x)\,dx.$$

Then, completing the substitution and finishing the integration yields

$$= \int \frac{1}{u}\,du$$

$$= \ln|u| + C$$

$$= \ln|\sec x + \tan x| + C.$$

The formula $\int \sec x\,dx = \ln|\sec x + \tan x| + C$ may be used like any other antiderivative formula.

ANTIDERIVATIVE OF SECANT

$$\int \sec x\,dx = \ln|\sec x + \tan x| + C$$

Example 1 *Evaluate* $\int \sec 3x\,dx$.

Solution Recognizing a trig function of "not just plain x," we use substitution:

$$u = 3x$$
$$du = 3\,dx.$$

We then adjust for the constant and use the antiderivative of the secant formula:

Line 1 adjusts for the constant, line 2 carries out the substitution, line 3 uses the antiderivative of the secant formula, and line 4 returns to the original variable.

$$\begin{aligned}\int \sec 3x\,dx &= \frac{1}{3}\int 3\sec 3x\,dx\\ &= \frac{1}{3}\int \sec u\,du\\ &= \frac{1}{3}\ln|\sec u + \tan u| + C\\ &= \frac{1}{3}\ln|\sec 3x + \tan 3x| + C.\end{aligned}$$

■

Reading Exercise 5 *Evaluate* $\int (\sec x + \tan x)\,dx$.

The derivation of an antiderivative for cosecant is similar to that of secant and is left as an exercise. This completes the task of finding antiderivatives for the six basic trigonometric functions, bringing our list of trig antiderivative formulas to the following:

LIST OF TRIGONOMETRIC ANTIDERIVATIVE FORMULAS

$\int \sin x\,dx = -\cos x + C$	$\int \csc x\,dx = -\ln|\csc x + \cot x| + C$
$\int \cos x\,dx = \sin x + C$	$\int \sec x = \ln|\sec x + \tan x| + C$
$\int \tan x\,dx = \ln|\sec x| + C$	$\int \cot x\,dx = -\ln|\csc x| + C$
$\int \sec^2 x\,dx = \tan x + C$	$\int \csc^2 x\,dx = -\cot x + C$
$\int \sec x\tan x\,dx = \sec x + C$	$\int \csc x\cot x\,dx = -\csc x + C$

$\int (\cos^n x)(\sin^m x)\,dx$, n or m odd

An example of an integral of this form that's not new helps motivate the strategy.

This is of the form $\int (\cos^n x)(\sin^m x)\,dx$ with $n = 1$, $m = 2$; notice that n is odd.

Example 2 *Evaluate* $\int \sin^2 x \cos x\,dx$.

Solution Rewriting $\sin^2 x$ as $(\sin x)^2$,

$$\int (\sin x)^2 \cos x\,dx,$$

helps us notice that we have "not just plain x" to a power, indicating the need for substitution:

$$u = \sin x$$
$$du = \cos x\,dx.$$

Then,

$$\int (\sin x)^2 \cos x\,dx = \int u^2\,du$$
$$= \frac{u^3}{3} + C = \frac{\sin^3 x}{3} + C. \quad \blacksquare$$

What makes the substitution work in example 2 is that there is a $\cos x$ for the du. This strategy can work in general. If there is an odd power of either $\sin x$ or $\cos x$, then one of them can be saved for the du and substitution works.

Although this substitution can be carried out for even values of n and m, the remaining portions of the strategy do not work.

Before turning to the next example, a reminder of a trigonometric identity is in order. The Pythagorean identity for sine and cosine can be rearranged in multiple ways:

$$\sin^2\theta + \cos^2\theta = 1$$
$$\sin^2\theta = 1 - \cos^2\theta$$
$$\cos^2\theta = 1 - \sin^2\theta.$$

Then, any even power of sine or cosine can be rewritten in terms of the other function; for instance,

$$\cos^8 x = (\cos^2 x)^4 = (1 - \sin^2 x)^4$$

Ans. to reading exercise 5
$\ln|\sec x + \tan x| + \ln|\sec x| + C$ (This expression can be rewritten using a law of logarithms to

$$\ln|\sec^2 x + \sec x \tan x| + C.$$

Compare this expression to derivative formulas. Interesting, isn't it?)

and

$$\sin^6 x = (\sin^2 x)^3 = (1 - \cos^2 x)^3.$$

Example 3 *Evaluate* $\int \sin^2 x \cos^3 x \, dx.$

Implied parentheses are as follows:

$$\int \sin^2 x \cos^3 x \, dx = \int (\sin^2 x)(\cos^3 x) \, dx.$$

Solution Noticing the odd power of cosine, let's "save" one cosine to use with *du* as follows:

$$\int \sin^2 x \cos^3 x \, dx = \int \sin^2 x \cos^2 x \cos x \, dx.$$

The idea now is to use the substitution,

$$u = \sin x$$
$$du = \cos x \, dx,$$

to complete the integral. We have the *du*; this is the $\cos x \, dx$ at the end of the integral. The $\sin^2 x$ is u^2. But, what about the remaining $\cos^2 x$? This is where the Pythagorean identity comes in. First we rewrite $\cos^2 x$ as $1 - \sin^2 x$, then we can complete the substitution:

Line 1 is repeated from the beginning of the solution; line 2 rewrites using the Pythagorean identity; line 3 completes the substitution, leaving just a polynomial in *u*, which is integrated easily; line 4 uses algebra; line 5 finds an antiderivative; and line 6 returns to the original variable.

$$\begin{aligned} \int \sin^2 x \cos^3 x \, dx &= \int \sin^2 x \cos^2 x \cos x \, dx \\ &= \int \sin^2 x (1 - \sin^2 x) \cos x \, dx \\ &= \int u^2 (1 - u^2) \, du \\ &= \int (u^2 - u^4) \, du \\ &= \frac{u^3}{3} - \frac{u^5}{5} + C \\ &= \frac{\sin^3 x}{3} - \frac{\sin^5 x}{5} + C. \end{aligned}$$ ■

STRATEGY

For odd powers of $\cos x$—that is, $(\cos x)^{2k+1}$, rewrite as

$$\begin{aligned} (\cos x)^{2k+1} &= (\cos x)^{2k} \cos x \\ &= (\cos^2 x)^k \cos x \\ &= (1 - \sin^2 x)^k \cos x, \end{aligned}$$

then use substitution with $u = \sin x$.

The strategy of example 3 works with odd powers of sine as well.

Example 4 *Evaluate* $\int \sin^5 x \cos^2 x \, dx.$

Solution Following the strategy for odd powers of sine, we begin with

$$\int \sin^5 x \cos^2 x\, dx = \int \sin^4 x \cos^2 x \sin x\, dx$$
$$= \int (\sin^2 x)^2 \cos^2 x \sin x\, dx$$
$$= \int (1 - \cos^2 x)^2 \cos^2 x \sin x\, dx.$$

Continuing the strategy, we use substitution with $u = \cos x$:

$$u = \cos x$$
$$du = -\sin x\, dx.$$

Adjusting for the constant ($-\sin x$ is the same as $-1 \cdot \sin x$), the integral calculation continues as

$$= -\int (1 - \cos^2 x)^2 \cos^2 x(-\sin x)\, dx$$
$$= -\int (1 - u^2)^2 u^2\, du$$
$$= -\int (1 - 2u^2 + u^4)u^2\, du$$
$$= -\int (u^2 - 2u^4 + u^6)\, du$$
$$= -\frac{u^3}{3} + \frac{2u^5}{5} - \frac{u^7}{7} + C$$
$$= -\frac{\cos^3 x}{3} + \frac{2\cos^5 x}{5} - \frac{\cos^7 x}{7} + C.$$

There is no need to try to simplify the answer. ■

STRATEGY

For odd powers of $\sin x$—that is, $(\sin x)^{2k+1}$, rewrite as

$$(\sin x)^{2k+1} = (\sin x)^{2k} \sin x$$
$$= (\sin^2 x)^k \sin x$$
$$= (1 - \cos^2 x)^k \sin x,$$

then use substitution with $u = \cos x$.

Line 1 adjusts for the constant by the addition of the negative signs in front of the integral and with the $\sin x$, line 2 completes the substitution, lines 3 and 4 perform algebra to place the polynomial in u in a form that is easy to integrate, line 5 antidifferentiates, and line 6 returns to the original variable.

If the exponents on $\sin x$ and $\cos x$ are both odd, then either strategy may be used.

Reading Exercise 6 *Evaluate* $\int \sin^4 x \cos^3 x\, dx$.

Review: $\int \sin kx\, dx$ and $\int \cos kx\, dx$

Some integrals involving simple substitutions are so common that it may be worth memorizing a formula to avoid repeating the substitution.

Example 5 *Evaluate* $\int \sin 3x\, dx$.

Solution Recognizing a trig function of "not just plain x," we use substitution:

$$u = 3x$$
$$du = 3\, dx.$$

Then,

$$\begin{aligned}\int \sin 3x\, dx &= \frac{1}{3}\int 3 \sin 3x\, dx\\ &= \frac{1}{3}\int \sin u\, du\\ &= \frac{1}{3}(-\cos u) + C\\ &= -\frac{1}{3}\cos 3x + C.\end{aligned}$$

■

Replacing every occurrence of the number 3 with a generic constant k in example 5 results in the following antiderivative formula. A similar result is derived easily for other trig functions as well. It is worth the effort to memorize these as you would any other antiderivative formula.

TRIG ANTIDERIVATIVES OF *KX*

$$\int \sin kx\, dx = -\frac{1}{k}\cos kx + C$$
$$\int \cos kx\, dx = \frac{1}{k}\sin kx + C$$

Example 6 *Evaluate* $\int \cos 7x\, dx$.

Solution Using the previous antiderivative formula with $k = 7$,

$$\int \cos 7x\, dx = \frac{1}{7}\sin 7x + C.$$

■

$\int (\cos^n x)(\sin^m x)\,dx$, *n* and *m* even

Just as with *n* or *m* odd, there are trigonometric identities that make the task of integration possible. Here, they are variations of the half-angle formulas for sine and cosine:

$$\sin^2\theta = \frac{1}{2}(1 - \cos 2\theta)$$

$$\cos^2\theta = \frac{1}{2}(1 + \cos 2\theta).$$

These identities can help take even powers of sine and cosine and rewrite them in terms of cosines of *kx* instead, without the exponents. This gives us a strategy for integrating even powers of sines and cosines (and their products).

Example 7 *Evaluate* $\int \sin^4 x\,dx$.

The same strategy works for $\int \sin^4 x \cos^2 x\,dx$. The algebra, of course, has different details, but the strategy is the same.

Solution First we rewrite $\sin^4 x$ as $(\sin^2 x)^2$ to facilitate the use of the identity. Then, applying the half-angle identity for sine with $\theta = x$ and using algebra gives

$$\begin{aligned}\int \sin^4 x\,dx &= \int (\sin^2 x)^2\,dx\\ &= \int \left(\frac{1}{2}(1-\cos 2x)\right)^2 dx\\ &= \int \frac{1}{4}(1-\cos 2x)^2\,dx\\ &= \int \frac{1}{4}(1 - 2\cos 2x + \cos^2 2x)\,dx\\ &= \int \left(\frac{1}{4} - \frac{1}{2}\cos 2x + \frac{1}{4}\cos^2 2x\right) dx.\end{aligned}$$

Line 1 prepares for using the half-angle identity, line 2 applies that identity with $\theta = x$, and lines 3 and 4 use algebra to expand the resulting square.

The reason for applying algebra to square the expression in the integral (lines 3 and 4 in the calculation) is that otherwise the expression is not in a form for which the available rules, formulas, and techniques apply. But, have we really accomplished our goal? There is still an even power of cosine in our integral: $\cos^2 2x$. Notice, however, that it is an exponent of 2, whereas the original integrand contained an exponent of 4. This is an improvement. Perhaps if we apply the half-angle formula for cosine (with $\theta = 2x$) we can make even more progress. Let's try:

Ans. to reading exercise 6
$\frac{\sin^5 x}{5} - \frac{\sin^7 x}{7} + C$

$$= \int \left(\frac{1}{4} - \frac{1}{2}\cos 2x + \frac{1}{4}\left(\frac{1}{2}(1+\cos 4x)\right)\right) dx$$
$$= \int \left(\frac{1}{4} - \frac{1}{2}\cos 2x + \frac{1}{8} + \frac{1}{8}\cos 4x \right) dx.$$

This time the resulting expression does not contain any powers of sine or cosine, just expressions of the form $\cos kx$, which we can integrate easily! The strategy works. Continuing the integration, we have

$$= \frac{1}{4}x - \frac{1}{2}\cdot\frac{1}{2}\sin 2x + \frac{1}{8}x + \frac{1}{8}\cdot\frac{1}{4}\sin 4x + C$$
$$= \frac{3}{8}x - \frac{1}{4}\sin 2x + \frac{1}{32}\sin 4x + C,$$

and we are finished. ■

STRATEGY

For even powers of both sine and cosine, use the half-angle identities and expand algebraically, repeating as needed, to rewrite the integrand so that only terms of the form $\cos kx$, with no exponents, remain.

This strategy works only when there are no odd exponents.

CAUTION: MISTAKE TO AVOID
Do not mix these strategies! They are separate cases. To integrate

$$\int (\cos^n x)(\sin^m x)\,dx,$$

if either n or m is odd, use the odd strategy; only use the even strategy when both exponents are even.

Integrating powers of tangent and secant

Although a little less straightforward than odd powers of sine and cosine, the same ideas apply to powers of tangent and secant. This time, the available substitutions are

$$u = \tan x$$
$$du = \sec^2 x\,dx$$

and

$$u = \sec x$$
$$du = \sec x \tan x\,dx.$$

Therefore, we either need to "save" $\sec^2 x$ or $\sec x \tan x$ for the du. Meanwhile, the helpful identities are variations of the Pythagorean identity for tangent and secant:

$$\tan^2 x = \sec^2 x - 1$$
$$\sec^2 x = \tan^2 x + 1.$$

Some trial and error is inevitable with powers of tangent and secant, so if one substitution is not successful, try the other.

STRATEGY

To integrate powers of tangent and secant, use substitution with either $u = \sec x$ or $u = \tan x$. Helpful identities are

$$\tan^2 x = \sec^2 x - 1$$
$$\sec^2 x = \tan^2 x + 1.$$

Example 8 *Evaluate* $\int \sec^6 x\,dx$.

Solution The integrand is a power of secant. We therefore try to save a $\sec^2 x$ or a $\sec x \tan x$ for our *du*, and with no tangent present it appears that it is easier to save a $\sec^2 x$:

$$\int \sec^6 x\,dx = \int \sec^4 x \sec^2 x\,dx.$$

The substitution that matches our choice is

$$u = \tan x$$

$$du = \sec^2 x\,dx.$$

We saved the $\sec^2 x$ for the *du*, but for the substitution to be successful, everything else must be placed in terms of $\tan x$. We therefore need to rewrite $\sec^4 x$ in terms of $\tan x$. This is where the identity $\sec^2 x = \tan^2 x + 1$ comes in:

$$\int \sec^4 x \sec^2 x\,dx = \int (\sec^2 x)^2 \cdot \sec^2 x\,dx$$

$$= \int (\tan^2 x + 1)^2 \sec^2 x\,dx.$$

Notice that when applying the identity in line 2, we only applied it inside the parentheses. Remember, we are saving the latter $\sec^2 x$ for the *du*, so we leave it alone.

Now we have the proper setup for the substitution. The *du* is present, and everything else is in terms of *u*—that is, $\tan x$. Making the substitution and continuing to evaluate the integral, we have

$$= \int (u^2 + 1)^2\,du$$

$$= \int \left(u^4 + 2u^2 + 1\right)\,du$$

$$= \frac{u^5}{5} + \frac{2u^3}{3} + u + C$$

$$= \frac{\tan^5 x}{5} + \frac{2\tan^3 x}{3} + \tan x + C. \qquad \blacksquare$$

Example 9 *Evaluate* $\int \tan^3 x\,dx$.

Solution The integrand is a power of tangent. We therefore try to save a $\sec^2 x$ or a $\sec x \tan x$ for our *du*. Both of these choices require the presence of secant, which is not present in the integrand in its current form. To get it, we can use the identity $\tan^2 x = \sec^2 x - 1$. We therefore separate a factor of $\tan^2 x$ for this purpose and use the identity:

$$\begin{aligned}\int \tan^3 x\,dx &= \int \tan x \tan^2 x\,dx \\ &= \int \tan x(\sec^2 x - 1)\,dx \\ &= \int (\tan x \sec^2 x - \tan x)\,dx.\end{aligned}$$

Now we are left with a different problem. If we wish to use the substitution

$$u = \tan x$$

$$du = \sec^2 x\,dx,$$

the $\sec^2 x$ must be a factor of the entire integrand, and it is not. The fix is to use the antiderivative difference rule to separate the terms into two integrals, one of which is set up for substitution and the other of which doesn't need it:

$$= \int \tan x \sec^2 x\,dx - \int \tan x\,dx.$$

Let's work with $\int \tan x \sec^2 x\,dx$ first. Using the chosen substitution, we have

$$\begin{aligned}\int \tan x \sec^2 x\,dx &= \int u\,du \\ &= \frac{u^2}{2} + C \\ &= \frac{\tan^2 x}{2} + C.\end{aligned}$$

The antiderivative formula for tangent is

$$\int \tan x\,dx = \ln|\sec x| + C.$$

Combining our results, the solution is

$$\int \tan^3 x\,dx = \frac{\tan^2 x}{2} - \ln|\sec x| + C.$$ ■

Examples 8 and 9 should be studied not for the exact steps used, but rather for the thought processes that led to successful integration. Because trial and error is necessary, fruitful ideas for the "trial" are essential.

Have you noticed that although several of the integrals in this section have involved products, we haven't yet used integration by parts? This is about to change. Although many integrals of powers of tangent and secant can be integrated using substitutions and identities, some require a different approach.

Example 10 *Evaluate* $\int \sec^3 x\,dx$.

Solution The trick to this integral is to try integration by parts as follows:

$$f(x) = \sec x \qquad g'(x) = \sec^2 x$$
$$f'(x) = \sec x \tan x \qquad g(x) = \tan x.$$

Using the integration-by-parts formula,

$$\begin{aligned}\int \sec^3 x\,dx &= \int \sec x \sec^2 x\,dx \\ &= \sec x \tan x - \int \tan x \cdot \sec x \tan x\,dx \\ &= \sec x \tan x - \int \sec x \tan^2 x\,dx.\end{aligned}$$

Next we rewrite using a Pythagorean identity, and then use the antiderivative difference rule, and the antiderivative formula for secant:

$$\begin{aligned}&= \sec x \tan x - \int \sec x(\sec^2 x - 1)\,dx \\ &= \sec x \tan x - \int (\sec^3 x - \sec x)\,dx \\ &= \sec x \tan x + \int \sec x\,dx - \int \sec^3 x\,dx \\ &= \sec x \tan x + \ln|\sec x + \tan x| - \int \sec^3 x\,dx.\end{aligned}$$

Now we are back to the original integral, $\int \sec^3 x\,dx$. But, we've had this happen before with integration-by-parts problems; we simply solve for the integral! To this end,

$$\int \sec^3 x\,dx = \sec x \tan x + \ln|\sec x + \tan x| - \int \sec^3 x\,dx$$

$$2\int \sec^3 x\,dx = \sec x \tan x + \ln|\sec x + \tan x| + C$$

$$\int \sec^3 x\,dx = \frac{1}{2}(\sec x \tan x + \ln|\sec x + \tan x|) + C. \blacksquare$$

ANTIDERIVATIVE OF SECANT CUBED

$$\int \sec^3 x\,dx$$
$$= \frac{1}{2}(\sec x \tan x + \ln|\sec x + \tan x|) + C$$

The integral of example 10 is difficult enough, and it occurs often enough in applied problems, that it is worth noting the result rather than having to rederive it every time.

$\int$ sin *mx* cos *nx* *dx*

Read carefully. We are not returning to an earlier topic. Here, the m and n are not exponents; the integrals are of the form

$$\int (\sin(mx))(\cos(nx))\,dx.$$

Once again, a trigonometric identity comes to our rescue. This time, it is a product-to-sum identity:

$$\sin mx \cos nx = \tfrac{1}{2}(\sin(m-n)x + \sin(m+n)x)\,.$$

Example 11 *Evaluate* $\int \sin(2x)\cos(7x)\,dx$.

Solution Using the product-to-sum identity with $m = 2$ and $n = 7$ as well as the antiderivative formula for $\sin kx$, we have

Line 1 uses a product-to-sum identity, line 2 simplifies, line 3 uses the antiderivative formula for $\sin kx$, and line 4 simplifies.

$$\int \sin(2x)\cos(7x)\,dx = \int \frac{1}{2}(\sin(2-7)x + \sin(2+7)x)\,dx$$
$$= \frac{1}{2}\int (\sin(-5x) + \sin(9x))\,dx$$
$$= \frac{1}{2}\left(-\frac{1}{-5}\cos(-5x) + \left(-\frac{1}{9}\cos(9x)\right)\right) + C$$
$$= \frac{1}{10}\cos(-5x) - \frac{1}{18}\cos(9x) + C. \blacksquare$$

For integrals of the form

$$\int \sin mx \sin nx\,dx$$

or

$$\int \cos mx \cos nx \, dx,$$

use the same strategy as that shown in example 11, making use of the identity

$$\sin mx \sin nx = \tfrac{1}{2}\left(\cos(m-n)x - \cos(m+n)x\right)$$

or

$$\cos mx \cos nx = \tfrac{1}{2}\left(\cos(m-n)x + \cos(m+n)x\right),$$

as appropriate.

STRATEGY

For integrands of the form $\int \sin mx \cos nx \, dx$, $\int \sin mx \sin nx \, dx$, or $\int \cos mx \cos nx \, dx$, use a product-to-sum identity.

Reading Exercise 7 *Evaluate* $\int \sin 5x \cos 3x \, dx$.

Exercises 7.3

1–30. Evaluate the integral.

1. $\int \sin^3 x \cos^2 x \, dx$
2. $\int \cos^2 x \sin x \, dx$
3. $\int \sin 3x \cos 2x \, dx$
4. $\int \cos^3 x \sin^3 x \, dx$
5. $\int \sin^6 x \cos x \, dx$
6. $\int \sin^2 2x \cos^3 2x \, dx$
7. $\int \sin^5 x \cos^3 x \, dx$
8. $\int \sin^3 x \cos^4 x \, dx$
9. $\int \cos^2 x \, dx$
10. $\int \sin 3x \cos 4x \, dx$
11. $\int \cos^4 x \, dx$
12. $\int \sin^2 x \, dx$
13. $\int x^2 (\sin^8 x^3)(\cos^3 x^3) \, dx$
14. $\int \cos^2 x \sin^2 x \, dx$
15. $\int \sin 4x \sin 7x \, dx$
16. $\int \left(\cos^2 x + \sin^2 x\right) dx$
17. $\int \sec^4 x \, dx$
18. $\int \cos 5x \cos 2x \, dx$
19. $\int \tan^2 x \, dx$
20. $\int \tan^3 x \sec x \, dx$

21. $\displaystyle\int \tan^3 x \sec^3 x\,dx$

22. $\displaystyle\int \tan^3 x \sec^4 x\,dx$

23. $\displaystyle\int \sec(4x+9)\,dx$

24. $\displaystyle\int x^2 \sec x^3\,dx$

25. $\displaystyle\int \csc x\,dx$

26. $\displaystyle\int \csc^3 x\,dx$

27. $\displaystyle\int x \sin 2x \cos 5x\,dx$

28. $\displaystyle\int x \sin^2 x\,dx$

29. $\displaystyle\int \sin^2(3x-4)\cos^2(3x-4)\,dx$

30. $\displaystyle\int \frac{\sec^4(\ln x)\tan(\ln x)}{x}\,dx$

Ans. to reading exercise 7
$-\frac{1}{4}\cos 2x - \frac{1}{16}\cos 8x + C$

31. Find the volume of the solid of revolution generated by revolving the region under the curve $y = \sin x + 2$ from $x = 0$ to $x = \pi$, about the x-axis.

32. Find the volume of the solid of revolution generated by revolving the region under the curve $y = \cos x + \sin x$ from $x = 0$ to $x = \frac{\pi}{2}$, about the x-axis.

33–36. (Love) Evaluate the integral.

33. $\displaystyle\int \frac{\sin^3 x}{\cos^2 x}\,dx$

34. $\displaystyle\int \frac{\cos^3 x}{\sin^4 x}\,dx$

35. $\displaystyle\int \cos^7 \theta\,d\theta$

36. $\displaystyle\int \sin^6 \theta\,d\theta$

37–44. (GSL) Evaluate the integral.

37. $\displaystyle\int \frac{\sin^5 y}{\sqrt{\cos y}}\,dy$

38. $\displaystyle\int \frac{\cos^5 t}{\sqrt[3]{\sin t}}\,dt$

39. $\displaystyle\int \csc^4 \frac{x}{4}\,dx$

40. $\displaystyle\int \cot^3 2x \csc 2x\,dx$

41. $\displaystyle\int \frac{\sec^4 t}{\tan^3 t}\,dt$

42. $\displaystyle\int \frac{\sec^4 x}{\sqrt{\tan x}}\,dx$

43. $\displaystyle\int \left(\sin^2 x + \cos x\right)^2\,dx$

44. $\displaystyle\int \left(\sqrt{\cos\theta} - 2\sin\theta\right)^2\,d\theta$

45. Show that $\displaystyle\int \cos kx\,dx = \frac{1}{k}\sin kx + C$.

(GSL) This average value is frequently used in the theory of alternating currents.

46. (GSL) Find the average value of $\sin^2 x$ between $x = 0$ and $x = \pi$.

Trigonometric Substitution

7.4

Many years ago, you learned to calculate the area of a circle from the formula $A = \pi r^2$. For instance, the circle $x^2 + y^2 = 4$, pictured in figure 1, has radius $r = 2$ and therefore area $A = \pi(2^2) = 4\pi$ units^2. Likely, no justification was given for the area formula when you first learned it. Can the formula be justified using integration? This question is soon answered using a technique called *trigonometric substitution.*

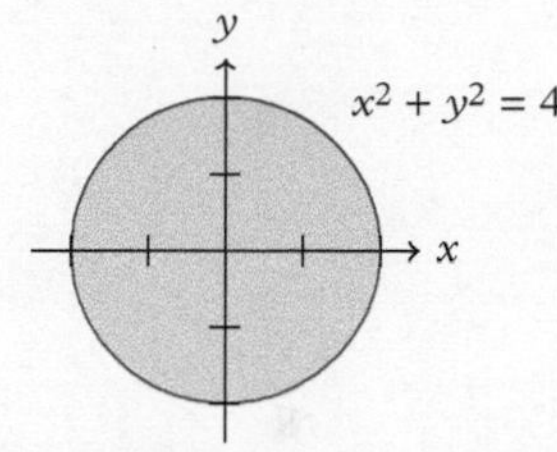

Figure 1 *The circle $x^2 + y^2 = 4$ (blue) with the region inside shaded light green. The area of the region inside the circle is 4π units^2*

A motivating example

The area of the circle in figure 1 should be twice the area of the semicircle in figure 2, which is the area under a curve. Which curve? We can solve for y to answer this question:

$$x^2 + y^2 = 4$$

$$y^2 = 4 - x^2$$

$$y = \pm\sqrt{4 - x^2}.$$

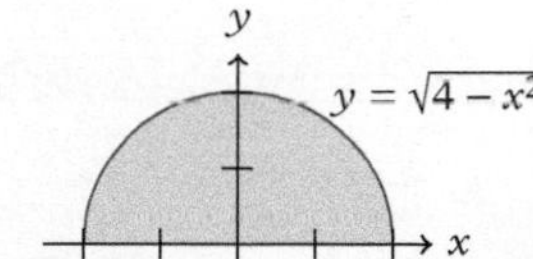

Figure 2 *The semicircle $y = \sqrt{4 - x^2}$ (blue) with the region under the curve shaded light green. The area of the region is half that of the circle*

Because we want the top half of the circle, which lies on and above the x-axis, we choose $y = \sqrt{4 - x^2}$. Again, we wish to find the area under this curve. Therefore, we wish to find $\int_{-2}^{2} \sqrt{4 - x^2}\, dx$, but this is just the area of the semicircle, and the area of the entire circle in figure 1 is twice that amount, or

$$\text{area of the circle} = 2\int_{-2}^{2} \sqrt{4 - x^2}\, dx.$$

How do we evaluate this integral? If the square root appears in the denominator instead, we recognize the antiderivative giving the inverse sine. However, it's in the numerator, so this doesn't help us. We have no antiderivative formulas that apply. Noticing the square root of "not just plain x," we might try substitution, but $u = 4 - x^2$ means $du = -2x\,dx$, and there is no x outside the square root to use with du. The integral does not appear to be set up to use integration by parts. We seem to be out of options.

The trick to solving this problem does not involve a combination of previous techniques, but it *does* borrow ideas from several of them. For instance, a technique that is similar to substitution is used. The integral looks similar to an inverse sine integral, and inverse sine makes an appearance. Just as with trigonometric integrals, trig identities are helpful. The entire process is called *trigonometric substitution* and is explained as we work through the example.

We begin by making the substitution

$$\begin{aligned} x &= 2\sin\theta \\ dx &= 2\cos\theta\, d\theta. \end{aligned}$$

Notice that this is not a regular substitution. Instead of declaring u as a function of x, we are declaring x as a function of the variable θ. It is more of a replacement than our previous technique of substitution. The rules are somewhat similar, in that we calculate the differential dx from the formula for x, and we must determine the new limits of integration as values of the variable θ, but the criteria for choosing the trig substitution is different from choosing a regular substitution and is explained in the next subsection.

It is here that many of you with the bad habit of not writing the dx or the du at the end of an integral will regret this inattention to proper notation. Omitting the portion of the expression which replaces dx is a common error when using trig substitution.

To carry out the trig substitution, we replace x with $2\sin\theta$ and we replace dx with $2\cos\theta\, d\theta$. This gives

$$2\int_{-2}^{2} \sqrt{4 - x^2}\, dx = 2\int_{?}^{?} \sqrt{4 - (2\sin\theta)^2} \cdot 2\cos\theta\, d\theta.$$

How do we find the new limits of integration, which must be values of the variable θ? We need to find the value of θ that corresponds to a given value of x. It helps to solve the substitution formula for the variable θ:

This solution assumes θ is an angle between $-\frac{\pi}{2}$ and $\frac{\pi}{2}$. This is important later.

$$\begin{aligned} x &= 2\sin\theta \\ \frac{x}{2} &= \sin\theta \\ \theta &= \sin^{-1}\frac{x}{2}. \end{aligned}$$

(There's the presence of the inverse sine!) Now we can calculate the new limits of integration:

$$x = 2: \quad \theta = \sin^{-1}\frac{2}{2} = \sin^{-1} 1 = \frac{\pi}{2}$$

$$x = -2: \quad \theta = \sin^{-1}\frac{-2}{2} = \sin^{-1}(-1) = -\frac{\pi}{2}.$$

The new integral is therefore

$$2\int_{-\pi/2}^{\pi/2} \sqrt{4-(2\sin\theta)^2}\cdot 2\cos\theta\, d\theta.$$

This looks worse than when we started! But, simplification is in order. First we square the term inside the parentheses and then we factor out the resulting common factor and extract the square root:

$$\begin{aligned}2\int_{-\pi/2}^{\pi/2} \sqrt{4-(2\sin\theta)^2}\cdot 2\cos\theta\, d\theta &= 2\int_{-\pi/2}^{\pi/2} \sqrt{4-4\sin^2\theta}\cdot 2\cos\theta\, d\theta\\ &= 2\int_{-\pi/2}^{\pi/2} \sqrt{4(1-\sin^2\theta)}\cdot 2\cos\theta\, d\theta.\\ &= 2\int_{-\pi/2}^{\pi/2} 2\sqrt{1-\sin^2\theta}\cdot 2\cos\theta\, d\theta.\end{aligned}$$

Now a Pythagorean identity can be used. Recalling that $\cos^2\theta = 1 - \sin^2\theta$, we have

$$= 2\int_{-\pi/2}^{\pi/2} 2\sqrt{\cos^2\theta}\cdot 2\cos\theta\, d\theta.$$

Eliminating the square root is in our sights! But first we should recall the algebraic fact that

$$\sqrt{x^2} = |x|,$$

which means that

$$\sqrt{\cos^2\theta} = |\cos\theta|.$$

We always set up a trig substitution so this issue can be ignored, in the manner explained here. In practice, we can go straight from $\sqrt{\cos^2\theta}$ to $\cos\theta$.

We wish that the absolute values weren't there—and our wish is granted! This is because we have already (implicitly) made the assumption that $-\frac{\pi}{2} \le \theta \le \frac{\pi}{2}$ when we solved the substitution formula for θ. This means θ is a first- or fourth-quadrant angle, where the cosine is always nonnegative. Thus, $|\cos\theta| = \cos\theta$, and the integral simplifies to

$$= 2\int_{-\pi/2}^{\pi/2} 2\cos\theta \cdot 2\cos\theta \, d\theta$$

$$= 8\int_{-\pi/2}^{\pi/2} \cos^2\theta \, d\theta.$$

Finally we have an integral we know how to evaluate. Because we see an even power of cosine, we use a half-angle identity and complete the integration from there:

$$8\int_{-\pi/2}^{\pi/2} \cos^2\theta \, d\theta = 8\int_{-\pi/2}^{\pi/2} \frac{1}{2}(1+\cos 2\theta)\, d\theta$$

$$= 4\int_{-\pi/2}^{\pi/2} (1+\cos 2\theta)\, d\theta$$

$$= \left(4\theta + 4\cdot\frac{1}{2}\sin 2\theta\right)\Bigg|_{-\pi/2}^{\pi/2}$$

$$= 2\pi + 0 - (-2\pi + 0) = 4\pi \,\text{units}^2,$$

Replacing 4 with r^2 at the beginning of this area calculation results in $A = \pi r^2$, verifying the formula.

as promised!

This was pages of work and explanation for one integral, but don't be intimidated. The steps are the same with any trigonometric substitution. When you know the correct substitution to use (where it applies) and the trig identity that pairs with it, the square root (or $\frac{3}{2}$ power, and so on) can always be eliminated and the integral completed.

Table of trigonometric substitutions

When do we use a trigonometric substitution? Which substitution do we use? Which identity is needed to help eliminate a square root? What must be true about θ so we don't have to worry about absolute values and we can use inverse trig functions to find values of θ? These questions are all answered in table 1.

Table 1 *Table of trigonometric substitutions*

expression	substitution	identity	interval
$a^2 - x^2$ or $\sqrt{a^2 - x^2}$	$x = a\sin\theta$	$1 - \sin^2\theta = \cos^2\theta$	$-\frac{\pi}{2} \le \theta \le \frac{\pi}{2}$
$a^2 + x^2$ or $\sqrt{a^2 + x^2}$	$x = a\tan\theta$	$1 + \tan^2\theta = \sec^2\theta$	$-\frac{\pi}{2} < \theta < \frac{\pi}{2}$
$x^2 - a^2$ or $\sqrt{x^2 - a^2}$	$x = a\sec\theta$	$\sec^2\theta - 1 = \tan^2\theta$	$0 \le \theta < \frac{\pi}{2}$ or $\pi \le \theta < \frac{3\pi}{2}$

When we evaluated

$$2\int_{-2}^{2}\sqrt{4-x^2}\,dx,$$

the expression was of the form $\sqrt{a^2-x^2}$ with $a=2$, and we used the substitution $x=2\sin\theta$. We then used the identity $1-\sin^2\theta=\cos^2\theta$ to simplify the integral and eliminate the square root. The interval $-\frac{\pi}{2}\le\theta\le\frac{\pi}{2}$ was used twice, once in solving the substitution formula for θ using inverse sine, and once when reducing $\sqrt{\cos^2\theta}$ to $\cos\theta$. The entire strategy of evaluating the integral is laid out in a single line in the table of trigonometric substitutions.

STRATEGY

If the integrand is of a form given in the table of trigonometric substitutions, use that row of the table to determine the substitution to use. After making the substitution, use the identity to eliminate the square root or otherwise simplify the expression so the integral can be evaluated. Use the interval as necessary to determine values of θ (definite integral) or to return to the original variable (indefinite integral).

Example 1 *Evaluate* $\displaystyle\int\frac{1}{(1+x^2)^2}\,dx.$

Solution The integrand almost looks like we could use the antiderivative giving the inverse tangent formula, but there is an extra $(\,)^2$ that prevents its use. Instead, we recognize that there is an expression inside the parentheses of the form a^2+x^2, with $a=1$. This indicates the use of a trigonometric substitution.

❶ Make the substitution. Using the second column row of the table of trigonometric substitutions, we make the substitution

$$x=\tan\theta$$
$$dx=\sec^2\theta\,d\theta.$$

Just as we calculated du with regular substitution, after we determine x, we must calculate the differential dx by taking a derivative.

Making the substitution in the integral yields

$$\int\frac{1}{(1+x^2)^2}\,dx=\frac{1}{(1+\tan^2\theta)^2}\cdot\sec^2\theta\,d\theta.$$

Both x and dx are replaced during the substitution.

❷ Use the identity. After we make the substitution, we begin to look for a way to simplify the integral using the identity. Here, the identity to use is

$$1+\tan^2\theta=\sec^2\theta,$$

and seeing $1 + \tan^2\theta$ inside the parentheses, we replace it with $\sec^2\theta$ and simplify:

Line 1 uses the identity and line 2 simplifies.

Tip: in a trig substitution, the correct direction to use the identity is to replace the left-hand side of the identity by the right-hand side, not the other way around. We replace $1 + \tan^2\theta$ with $\sec^2\theta$. We do not replace $\sec^2\theta$ with $1 + \tan^2\theta$.

$$= \int \frac{1}{(\sec^2\theta)^2} \cdot \sec^2\theta \, d\theta$$

$$= \int \frac{\sec^2\theta}{\sec^4\theta} \, d\theta = \int \frac{1}{\sec^2\theta} \, d\theta.$$

❸ Integrate. After using the identity and simplifying, the next task is to evaluate the integral. How do we evaluate $\int \frac{1}{\sec^2\theta} \, d\theta$? Sometimes another trig identity helps. In this case, since $\frac{1}{\sec\theta} = \cos\theta$, the integral simplifies to

$$= \int \cos^2\theta \, d\theta,$$

which we have seen before. We use a half-angle identity and finish evaluating the integral:

$$= \int \frac{1}{2}(1 + \cos 2\theta) \, d\theta$$

$$= \frac{1}{2}\theta + \frac{1}{2}\frac{\sin 2\theta}{2} + C$$

$$= \frac{1}{2}\theta + \frac{1}{4}\sin 2\theta + C.$$

❹ Return to the original variable (indefinite integrals only). This task can be completed by using the trig substitution formula, $x = \tan\theta$. Solving this equation for θ gives

$$\theta = \tan^{-1} x.$$

Replacing θ with $\tan^{-1} x$ gives the solution:

$$= \frac{1}{2}\tan^{-1} x + \frac{1}{4}\sin\left(2\tan^{-1} x\right) + C.$$

Although this expression is a solution to the integral, it is traditional to write the solution in a simplified form. To do so requires another trig identity and a diagram. These two substeps are described next.

❹ ⓐ If necessary, use the double-angle identity for sine:

$$\sin 2\theta = 2\sin\theta\cos\theta.$$

If $\sin 2\theta$ appears in the solution, rewrite using the double-angle identity

$$\sin 2\theta = 2\sin\theta\cos\theta.$$

Returning to the solution in the variable θ, we have

$$\frac{1}{2}\theta + \frac{1}{4}\sin 2\theta + C = \frac{1}{2}\theta + \frac{1}{4}(2\sin\theta\cos\theta)$$
$$= \frac{1}{2}\theta + \frac{1}{2}\sin\theta\cos\theta + C.$$

❹ ⓑ Draw a triangle and label the sides. Using the substitution formula and treating θ like an angle, we draw a triangle in the first quadrant and label the angle θ, the opposite side x, and the adjacent side 1, so that $\tan\theta = \frac{x}{1} = x$. The third side can then be found using the Pythagorean theorem:

$$a^2 + b^2 = c^2$$
$$1^2 + x^2 = c^2$$
$$\sqrt{1 + x^2} = c.$$

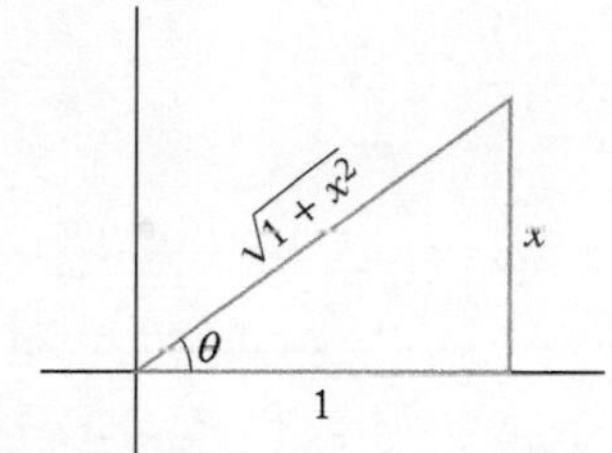

Figure 3 *A diagram of the relationship between x and θ. The triangle may always be drawn in the first quadrant*

The completed diagram is in figure 3.

The diagram allows us to determine $\sin\theta$ and $\cos\theta$ using the usual opposite over hypotenuse and adjacent over hypotenuse, yielding the simplified solution

$$\int \frac{1}{(1+x^2)^2}\,dx = \frac{1}{2}\theta + \frac{1}{2}\sin\theta\cos\theta + C$$
$$= \frac{1}{2}\tan^{-1}x + \frac{1}{2}\cdot\frac{x}{\sqrt{1+x^2}}\cdot\frac{1}{\sqrt{1+x^2}} + C$$
$$= \frac{1}{2}\tan^{-1}x + \frac{x}{2(1+x^2)} + C. \quad \blacksquare$$

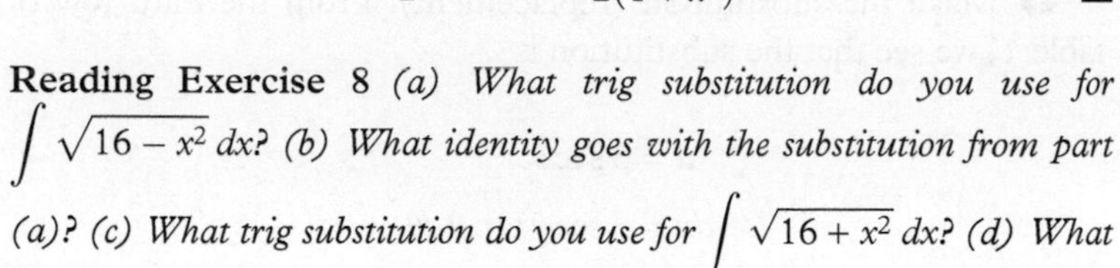

Reading Exercise 8 *(a) What trig substitution do you use for* $\int \sqrt{16 - x^2}\,dx$*? (b) What identity goes with the substitution from part (a)? (c) What trig substitution do you use for* $\int \sqrt{16 + x^2}\,dx$*? (d) What identity goes with the substitution from part (c)?*

Multiple-technique example

Sometimes more than one technique of integration is needed to evaluate an integral.

Example 2 *Evaluate* $\displaystyle\int \frac{e^t}{(e^{2t} - 9)^{3/2}}\,dt.$

Solution Recognizing "not just plain x" to a power, we try the substitution

$$u = e^{2t} - 9$$

$$du = e^{2t} \cdot 2\, dt.$$

The problem is that we need e^{2t} for du, and we only have e^t in the numerator. This substitution fails.

What else can we try? Taking a clue from the previous substitution attempt, what if we tried substituting $u = e^t$ instead? That is, try

$$u = e^t$$

$$du = e^t\, dt.$$

Now we have what we need for du, but how do we handle e^{2t}? The answer comes from using laws of exponents:

$$e^{2t} = \left(e^t\right)^2 = u^2.$$

Substitute:

$$\int \frac{e^t}{(e^{2t} - 9)^{3/2}}\, dt = \int \frac{1}{(u^2 - 9)^{3/2}}\, du.$$

Next we recognize an expression of the form $u^2 - a^2$ with $a = 3$, indicating the use of trig substitution.

❶ Make the substitution (replacement). From the third row of table 1, we see that the substitution is

Because our variable is u, we use u in place of x throughout the trig substitution process.

$$u = 3 \sec \theta$$

$$du = 3 \sec \theta \tan \theta\, d\theta.$$

Making the substitution gives

Both u and du are replaced.

$$\int \frac{1}{(u^2 - 9)^{3/2}}\, du = \int \frac{1}{(9 \sec^2 \theta - 9)^{3/2}} \cdot 3 \sec \theta \tan \theta\, d\theta.$$

❷ Use the identity. According to table 1, the identity to use is

$$\sec^2 \theta - 1 = \tan^2 \theta.$$

Factoring a 9 inside the parentheses allows us to use the identity:

Ans. to reading exercise 8
(a) $x = 4\sin\theta$, (b) $1 - \sin^2\theta = \cos^2\theta$, (c) $x = 4\tan\theta$, and (d) $1 + \tan^2\theta = \sec^2\theta$

$$= \int \frac{1}{(9(\sec^2\theta - 1))^{3/2}} \cdot 3\sec\theta\tan\theta\,d\theta$$
$$= \int \frac{1}{(9\tan^2\theta)^{3/2}} \cdot 3\sec\theta\tan\theta\,d\theta.$$

Now we simplify:

$$= \int \frac{1}{27\tan^3\theta} \cdot 3\sec\theta\tan\theta\,d\theta$$
$$= \frac{1}{9}\int \frac{\sec\theta}{\tan^2\theta}\,d\theta.$$

Line 1 calculates

$$9^{3/2} = (9^{1/2})^3 = 3^3 = 27$$

and

$$(\tan^2\theta)^{\frac{3}{2}} = \tan^{2\cdot\frac{3}{2}}\theta = \tan^3\theta.$$

❸ Integrate. We now see an integral with powers of tangent and secant. However, trying to save out a $\sec\theta\tan\theta$ for use with substitution undoes the trig substitution, so this is not helpful. There is no $\sec^2\theta$ to use in a du, and trying to get one by changing $\tan^2\theta$ to $\sec^2\theta - 1$ undoes the use of the identity in step ❷.

When all else fails in a trig integral, try changing all trig functions to sines and cosines and simplifying:

Tip: don't undo the trig substitution or the identity. Going backward is not helpful. This time, rewriting in terms of $\sin\theta$ and $\cos\theta$ is helpful.

$$\frac{1}{9}\int \frac{\sec\theta}{\tan^2\theta}\,d\theta = \frac{1}{9}\int \frac{\frac{1}{\cos\theta}}{\frac{\sin^2\theta}{\cos^2\theta}}\,d\theta$$
$$= \frac{1}{9}\int \frac{1}{\cos\theta} \cdot \frac{\cos^2\theta}{\sin^2\theta}\,d\theta$$
$$= \frac{1}{9}\int \frac{\cos\theta}{\sin^2\theta}\,d\theta.$$

Perhaps now a strategy has appeared to you. This integrand has powers of sine and cosine, with an odd power of cosine. This suggests using the substitution

The variable u is already in use in this problem for a different substitution. Do not reuse a variable; choose a different letter (here, we choose w).

$$w = \sin\theta$$
$$dw = \cos\theta\,d\theta.$$

We then have

Line 1 makes the substitution, line 2 rewrites for use with the power rule, line 3 finds an antiderivative (Finally! Let there be rejoicing in the streets!), and line 4 simplifies.

$$\begin{aligned}\frac{1}{9}\int \frac{\cos\theta}{\sin^2\theta}\,d\theta &= \frac{1}{9}\int \frac{1}{w^2}\,dw \\ &= \frac{1}{9}\int w^{-2}\,dw \\ &= \frac{1}{9}\cdot\frac{w^{-1}}{-1}+C \\ &= -\frac{1}{9w}+C.\end{aligned}$$

❹ Return to the original variable. We now need to retrace our steps through all three substitutions to return to the variable t. Undoing the last substitution is simple enough:

$$= -\frac{1}{9\sin\theta}+C.$$

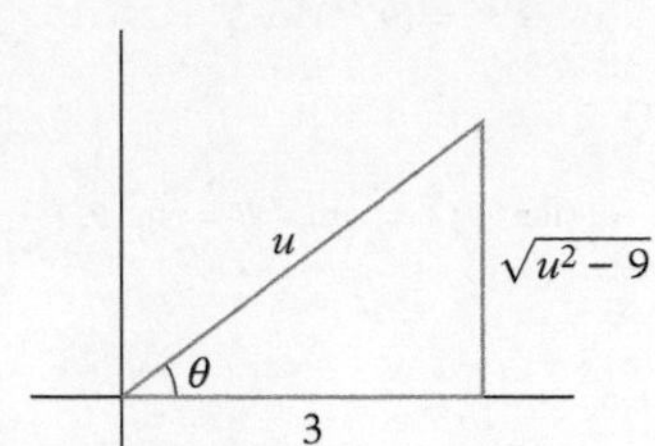

Figure 4 *A diagram of the relationship between u and θ*

Undoing the trig substitution is next and, as in example 1, this requires the use of a diagram (figure 4).

Since $u = 3\sec\theta$, solving for the trig function results in

$$\sec\theta = \frac{u}{3}.$$

Labeling the hypotenuse u and the adjacent side 3 makes $\sec\theta = \frac{u}{3}$. The Pythagorean theorem gives the other leg a value of $\sqrt{u^2-9}$. Then, $\sin\theta$ is the opposite over the hypotenuse, bringing us to

$$\begin{aligned}-\frac{1}{9\sin\theta}+C &= -\frac{1}{9\cdot\frac{\sqrt{u^2-9}}{u}}+C \\ &= -\frac{u}{9\sqrt{u^2-9}}+C.\end{aligned}$$

Undoing the first substitution finishes the solution:

$$= -\frac{e^t}{9\sqrt{e^{2t}-9}}+C. \qquad \blacksquare$$

The three examples referred to are the area calculation at the beginning of the section and examples 1 and 2.

The three examples of this section were chosen to illustrate collectively all the varieties of trig substitution and the difficulties that can arise. Do not be discouraged. Not all exercises are as long as the ones in this narrative.

Exercises 7.4

1–8. (a) Which trigonometric substitution might be useful? (b) Which identity goes with that trig substitution?

1. $\displaystyle\int \frac{\sqrt{x^2+9}}{x^4}\,dx$
2. $\displaystyle\int \sqrt{x^2-25}\,dx$
3. $\displaystyle\int \frac{1}{(x^2-4)^2}\,dx$
4. $\displaystyle\int \frac{x^2}{\sqrt{x^2+25}}\,dx$
5. $\displaystyle\int \frac{\sqrt{x^2-7}}{x}\,dx$
6. $\displaystyle\int \frac{x}{(7-x^2)^{3/2}}\,dx$
7. $\displaystyle\int \frac{x^2}{(16-x^2)^{3/2}}\,dx$
8. $\displaystyle\int \frac{\sqrt{6-x^2}}{x^2}\,dx$

9–38. Evaluate the integral.

9. (GSL) $\displaystyle\int \frac{1}{(x^2+2)^{3/2}}\,dx$
10. (GSL) $\displaystyle\int \frac{x^2}{\sqrt{x^2-6}}\,dx$
11. (GSL) $\displaystyle\int \frac{1}{(5-x^2)^{3/2}}\,dx$
12. (GSL) $\displaystyle\int \frac{t^2}{\sqrt{4-t^2}}\,dt$
13. $\displaystyle\int_0^3 \frac{x^2}{\sqrt{9-x^2}}\,dx$
14. (GSL) $\displaystyle\int \frac{u^2}{(9-u^2)^{3/2}}\,du$
15. $\displaystyle\int \frac{1}{\sqrt{6-x^2}}\,dx$
16. $\displaystyle\int \frac{1}{x^2+9}\,dx$
17. (GSL) $\displaystyle\int \frac{x^2}{(x^2+8)^{3/2}}\,dx$
18. $\displaystyle\int_0^2 \sqrt{16-x^2}\,dx$
19. (GSL) $\displaystyle\int \frac{1}{x\sqrt{x^2+4}}\,dx$
20. $\displaystyle\int \frac{x}{\sqrt{x^2+4}}\,dx$
21. (GSL) $\displaystyle\int \frac{1}{y^2\sqrt{y^2-7}}\,dy$
22. (GSL) $\displaystyle\int \frac{1}{x\sqrt{25-x^2}}\,dx$
23. $\displaystyle\int \frac{x}{(x^2-25)^{3/2}}\,dx$
24. (GSL) $\displaystyle\int \frac{1}{x^2\sqrt{5-x^2}}\,dx$
25. (GSL) $\displaystyle\int \frac{1}{x^3\sqrt{x^2-9}}\,dx$
26. $\displaystyle\int_{-3}^0 \frac{2}{\sqrt{x^2+9}}\,dx$
27. $\displaystyle\int_{-3}^3 \frac{2}{(9+x^2)^{3/2}}\,dx$
28. (GSL) $\displaystyle\int \frac{\sqrt{16-t^2}}{t^2}\,dt$

29. (GSL) $\int \frac{1}{(x^2-3)^{3/2}}\,dx$

30. (GSL) $\int \frac{\sqrt{y^2-9}}{y}\,dy$

31. (GSL) $\int \frac{1}{x^4\sqrt{x^2-5}}\,dx$

32. (GSL) $\int \frac{\sqrt{x^2+9}}{x^6}\,dx$

33. $\int \frac{x^2}{(x^2+16)^2}\,dx$

34. (GSL) $\int \frac{x^2}{\sqrt{x^2+5}}\,dx$

35. (GSL) $\int \frac{\sqrt{100-u^2}}{u}\,du$

36. $\int \frac{8}{x(x^2-4)^2}\,dx$

37. (GSL) $\int \frac{1}{x^3\sqrt{4-x^2}}\,dx$

38. (GSL) $\int \frac{1}{x^3\sqrt{x^2+1}}\,dx$

39. Evaluate $\int \frac{1}{\sqrt{a^2-x^2}}\,dx$ using trigonometric substitution and compare the result to the antiderivative giving the inverse sine formula.

40. Evaluate $\int \frac{1}{a^2+x^2}\,dx$ using trigonometric substitution and compare the result to the antiderivative giving the inverse tangent formula, version 2.

41. (modified from GSL) Find the volume of the solid of revolution generated by revolving the region inside the circle $x^2+(y-5)^2=9$, about the x-axis.

GSL calls this donut-shaped solid a *ring solid, anchor ring,* or *torus.*

42. (modified from GSL) Repeat exercise 41 by replacing 9 with a^2 and 5 with b to derive the formula $V=2\pi^2a^2b$.

43. Find the average y-coordinate of the semicircle $y=\sqrt{100-x^2}$.

Partial Fractions, Part I

7.5

In this section we expand our ability to integrate rational functions.

Two examples

Example 1 *Evaluate* $\int \left(\frac{2}{x-5} - \frac{3}{2x+7} \right) dx.$

Solution Applying the antiderivative difference rule, we need to evaluate

$$\int \frac{2}{x-5}\, dx - \int \frac{3}{2x+7}\, dx.$$

For the integral on the left, we recognize an antiderivative giving a natural log, and because the coefficient on x is one, we may safely evaluate without writing the details of substitution:

$$\int \frac{2}{x-5}\, dx = 2 \ln|x-5| + C.$$

For the integral on the right, we use the substitution

$$\begin{aligned} u &= 2x+7 \\ du &= 2\, dx, \end{aligned}$$

resulting in

This type of substitution should be familiar enough that, after this example, the result of such an integral is stated without referring explicitly to substitution or writing its details.

$$\begin{aligned} \int \frac{3}{2x+7}\, dx &= \frac{3}{2} \int \frac{1}{2x+7} \cdot 2\, dx \\ &= \frac{3}{2} \int \frac{1}{u}\, du \\ &= \frac{3}{2} \ln|u| + C \\ &= \frac{3}{2} \ln|2x+7| + C. \end{aligned}$$

Combining these results gives

$$\int \left(\frac{2}{x-5} - \frac{3}{2x+7}\right) dx = 2\ln|x-5| - \frac{3}{2}\ln|2x+7| + C,$$

completing the task. ■

The two fractions in the integrand of example 1 can be subtracted by finding a common denominator:

$$\begin{aligned}\frac{2}{x-5} - \frac{3}{2x+7} &= \frac{2(2x+7)}{(x-5)(2x+7)} - \frac{3(x-5)}{(2x+7)(x-5)} \\ &= \frac{4x+14-(3x-15)}{(x-5)(2x+7)} \\ &= \frac{x+29}{2x^2-3x-35}.\end{aligned}$$

This result brings us to the next example.

Example 2 *Evaluate* $\displaystyle\int \frac{x+29}{2x^2-3x-35}\,dx.$

Solution Using the algebra of the previous paragraph in reverse,

$$\int \frac{x+29}{2x^2-3x-35}\,dx = \cdots = \int \left(\frac{2}{x-5} - \frac{3}{2x+7}\right) dx,$$

the solution to which we calculated in example 1 as

$$2\ln|x-5| - \frac{3}{2}\ln|2x+7| + C.$$

■

Easy! Or is it? Take a look again at the calculation before example 2 and try reversing the last step, going from

$$\frac{x+29}{2x^2-3x-35}$$

to the previous expression

$$\frac{4x+14-(3x-15)}{(x-5)(2x+7)}.$$

The denominator is simple enough; it's just factoring. But, how do we know to rewrite $x+29$ as $4x+14-(3x-15)$? There lies the problem. The answer is to use the algebraic technique known as *partial fractions*.

The method of partial fractions

Partial fractions is an algebraic technique for "un-adding" fractions. To add (or subtract) fractions, we begin by finding a common denominator and continue as demonstrated in the previous section. To reverse this process is the goal of partial fractions. Not specific to integration only, the same algebra can be used in any context in which we wish to decompose a fraction.

Let's examine the method of partial fractions using the integrand of example 2:

$$\frac{x+29}{2x^2-3x-35}.$$

❶ Factor the denominator. We have

$$2x^2-3x-35=(2x+7)(x-5).$$

❷ Write the form of the decomposition. As long as the denominator factors into a product of nonrepeated linear factors, then the decomposition should contain each of these linear factors in a denominator, with a constant in the numerator, summed as follows:

In the next section, we look at other cases, including when the factorization of the denominator contains a repeated linear factor such as $(2x+1)^3$ or an irreducible quadratic factor such as (x^2+4).

$$\frac{A}{2x+7}+\frac{B}{x-5}.$$

We now wish to determine the values of the constants A and B.

❸ Equate the form of the decomposition with the original fraction and solve for the unknown coefficients. We wish to solve

$$\frac{A}{2x+7}+\frac{B}{x-5}=\frac{x+29}{2x^2-3x-35}$$

for A and B. To do so, we begin by clearing the fractions, multiplying both sides of the equation by the denominator of the original fraction:

In line 1, we multiply by the same quantity on both sides. It is just more convenient to write the nonfactored denominator by the original fraction and the factored denominator on the other side.

$$(2x+7)(x-5)\cdot\left(\frac{A}{2x+7}+\frac{B}{x-5}\right)=\frac{x+29}{2x^2-3x-35}\cdot(2x^2-3x-35)$$

$$A(x-5)+B(2x+7)=x+29.$$

There are two methods for proceeding, and each should be practiced.

Method 1. Expand the side of the equation containing the unknown constants (A and B) and collect like terms, then equate coefficients on powers of x from the two sides of the equation. Solve the resulting simultaneous equations.

We expand and collect like terms:

$$Ax - 5A + 2Bx + 7B = x + 29$$
$$(A + 2B)x - 5A + 7B = x + 29.$$

Next we equate coefficients. The coefficient on x should be the same on both sides of the equation; therefore, $A+2B = 1$. The constant term should be the same on both sides of the equation; therefore, $-5A + 7B = 29$. We now have two equations with two unknowns (A and B):

$$A + 2B = 1$$
$$-5A + 7B = 29.$$

Any of the various methods for solving simultaneous linear equations may be used. This time we choose the "addition method." Multiplying both sides of the first equation by 5 and adding eliminates A,

$$\begin{array}{rcrcr} 5A & + & 10B & = & 5 \\ -5A & + & 7B & = & 29 \\ \hline & & 17B & = & 34, \end{array}$$

and we have $B = 2$. Returning to one of the equations (the first is convenient), we can solve for A:

$$A + 2B = 1$$
$$A + 2(2) = 1$$
$$A = -3.$$

Method 2 is a variation of what is sometimes called the *cover-up method* (or the *Heaviside cover-up method*).

For method 2, we begin differently than method 1. Hence, we are returning to the form of the equation given before method 1.

Method 2. Use carefully chosen values of x to determine the coefficients. Because the equation

$$A(x - 5) + B(2x + 7) = x + 29$$

must be true for each value of x, it must be true for $x = 5$. Substituting 5 for x yields

$$A(0) + B(17) = 34$$
$$B = 2.$$

This is quicker than in method 1. Why? The chosen value $x = 5$ makes the term $A(x - 5)$ zero, eliminating one of the variables so that we can solve for the other variable quickly. With this in mind, we choose $x = -\frac{7}{2}$ next, and substitute it into the equation:

Solving $2x + 7 = 0$ gives $x = -\frac{7}{2}$.

$$A\left(-\tfrac{7}{2} - 5\right) + B(0) = -\tfrac{7}{2} + 29$$
$$A\left(-\tfrac{17}{2}\right) = \tfrac{51}{2}$$
$$A = -3.$$

❹ Write the partial fraction decomposition. Using either method 1 or method 2, we arrive at $A = -3$, $B = 2$. Then,

$$\frac{x + 29}{2x^2 - 3x - 35} = \frac{A}{2x + 7} + \frac{B}{x - 5} = \frac{-3}{2x + 7} + \frac{2}{x - 5},$$

as desired. We can then integrate as we did in example 2.

Note: The method of partial fractions described here works for proper fractions only.

The method of partial fractions can transform the integral

$$\int \frac{x - 7}{3x^2 + 13x + 4}\,dx$$

into the simpler integral given in reading exercise 9.

Reading Exercise 9 *Evaluate* $\displaystyle\int \left(\frac{1}{x + 4} - \frac{2}{3x + 1}\right) dx.$

Partial fractions: complete examples

Example 3 *Use partial fractions to evaluate* $\displaystyle\int \frac{x + 4}{x^2 - 2x - 15}\,dx.$

Solution First we rewrite the integrand using the method of partial fractions.

❶ Factor the denominator. We have

$$x^2 - 2x - 15 = (x - 5)(x + 3).$$

❷ Write the form of the decomposition. The denominator factored into two nonrepeated linear factors, so the form of the decomposition is

$$\frac{A}{x - 5} + \frac{B}{x + 3}.$$

❸ Equate the form of the decomposition with the original fraction and solve for the unknown coefficients. Equating and simplifying (multiplying both sides by the denominator of the original fraction) results in

$$\frac{x+4}{x^2-2x-15} = \frac{A}{x-5} + \frac{B}{x+3},$$

$$x+4 = A(x+3) + B(x-5).$$

Both methods should be practiced, even though method 2 is quickest for the problems in this section.

For nonrepeated linear factors, method 2 is usually the quickest. Choosing $x = -3$ and substituting in the previous equation gives

$$1 = 0 + B(-8)$$
$$-\tfrac{1}{8} = B.$$

The value $x = -3$ is chosen because it makes $(x + 3)$ zero; the value $x = 5$ is chosen because it makes $(x - 5)$ zero.

Choosing $x = 5$ gives

$$9 = A(8) + 0$$
$$\tfrac{9}{8} = A.$$

Partial fractions is an algebraic method. It is applied in the evaluation of the integral. The conclusion of partial fractions is shown separately to emphasize that fact, but it is not incorrect to go straight to the integral.

❹ Write the partial fraction decomposition. We conclude that

$$\frac{x+4}{x^2-2x-15} = \frac{\frac{9}{8}}{x-5} + \frac{-\frac{1}{8}}{x+3}.$$

Now that the method of partial fractions is concluded, we rewrite the integral and proceed with its evaluation:

$$\begin{aligned}\int \frac{x+4}{x^2-2x-15}\,dx &= \int \left(\frac{\frac{9}{8}}{x-5} + \frac{-\frac{1}{8}}{x+3}\right) dx \\ &= \frac{9}{8}\int \frac{1}{x-5}\,dx - \frac{1}{8}\int \frac{1}{x+3}\,dx \\ &= \frac{9}{8}\ln|x-5| - \frac{1}{8}\ln|x+3| + C. \quad ■\end{aligned}$$

The next example is similar to example 3, but it uses method 1 instead.

Ans. to reading exercise 9
$\ln|x+4| - \frac{2}{3}\ln|3x+1| + C$

Example 4 *Use partial fractions to evaluate* $\displaystyle\int \frac{12}{x^2+2x-35}\,dx.$

Solution ❶ Factor the denominator. We have

$$x^2 + 2x - 35 = (x-5)(x+7).$$

❷ Write the form of the decomposition. The denominator factored as two nonrepeated linear factors, so the form of the decomposition is

$$\frac{A}{x-5}+\frac{B}{x+7}.$$

❸ Equate the form of the decomposition with the original fraction and solve for the unknown coefficients. Equating and simplifying results in

$$\frac{12}{x^2+2x-35}=\frac{A}{x-5}+\frac{B}{x+7}$$
$$12=A(x+7)+B(x-5).$$

Line 1 equates the original fraction with the form of the decomposition; line 2 simplifies by multiplying both sides of the equation by $(x-5)(x+7)$.

Following method 1, we expand the right side of the equation further and collect like terms:

$$12=Ax+7A+Bx-5B$$
$$12=(A+B)x+7A-5B.$$

When equating coefficients, we see that x does not appear on the left side of the equation, but it does on the right side. But, noting that $12=0\cdot x+12$, means we can still equate the coefficients on x, obtaining $0=A+B$. Equating the constants gives $12=7A-5B$. Our simultaneous equations are

Even though method 2 is quicker for the exercises in this section, both methods should be practiced.

$$A+B=0$$
$$7A-5B=12.$$

One option to solve simultaneous equations is the "substitution method." We solve an equation for one of the variables, then insert that expression into the other equation. To this end, solving the first equation for B gives

The substitution method for solving simultaneous linear equations is not to be confused with the integration technique called substitution.

$$B=-A,$$

and then replacing B with $-A$ in the other equation gives

$$7A-5(-A)=12$$
$$12A=12$$
$$A=1.$$

Then,

$$B = -A = -1.$$

Having finished partial fractions, we move straight to the integral to be evaluated:

Line 1 uses the result of partial fractions to rewrite the integrand; line 2 finishes the integration.

$$\int \frac{12}{x^2 + 2x - 35}\,dx = \int \left(\frac{1}{x-5} + \frac{-1}{x+7} \right) dx$$
$$= \ln|x-5| - \ln|x+7| + C. \qquad \blacksquare$$

In section 7.1 we integrated rational functions, including some that looked similar to example 4. With a quadratic in the denominator and a middle term, why not just try completing the square? With $a = \frac{2}{2} = 1$ and $a^2 = 1$, we have

$$\begin{aligned} &x^2 + 2x \qquad - 35 \\ =&x^2 + 2x + 1 - 35 - 1 \\ =&(x+1)^2 - 36. \end{aligned}$$

The integral becomes

$$\int \frac{12}{x^2 + 2x - 35}\,dx = \int \frac{12}{(x+1)^2 - 36}\,dx.$$

Using the substitution

$$u = x + 1$$
$$du = dx$$

brings us to

$$\int \frac{12}{(x+1)^2 - 36}\,dx = 12 \int \frac{1}{u^2 - 36}\,du.$$

Now comes the problem. In section 7.1, after completing the square, we finished integrating using the inverse tangent, which requires the form

We used inverse sine and inverse cosh as well, but they required the presence of a square root.

$$\int \frac{1}{u^2 + a^2}\,dx,$$

not

$$\int \frac{1}{u^2 - a^2}\,dx.$$

We have not presented a formula applicable to this integral, hence the need for partial fractions.

So how do we know when to use partial fractions and when to use completing the square? The answer is found by factoring the denominator. If the denominator can be factored into two linear factors (without using complex numbers), we use partial fractions. Otherwise, we use completing the square if necessary.

We can complete the integration with a combination of $\coth^{-1}$ and $\tanh^{-1}$, but the result is not as simple or as satisfying as the solution presented in example 4.

Reading Exercise 10 *For* $\int \frac{x-2}{x^2+5x+6}\,dx$*, should we try partial fractions or completing the square?*

STRATEGY

If the integrand is a (proper) rational function with a quadratic denominator, try factoring the denominator. If the denominator factors into linear factors (real, not complex), use partial fractions. Otherwise, complete the square (if necessary).

Partial fractions: improper fractions

If the integrand is an improper fraction, first use long division to rewrite the integrand as a polynomial plus a proper fraction. Partial fractions can then be completed on what remains, if necessary.

Example 5 *Evaluate* $\int \frac{x^3+1}{x^2-3x+2}\,dx.$

Solution The integrand is an improper fraction (degree 3 over degree 2). We begin by using long division to rewrite the integrand.

$$\begin{array}{r|rrrr} & & & x & +3 \\ \hline x^2-3x+2 & x^3 & & & +1 \\ & -(x^3 & -3x^2 & +2x) & \\ \hline & & 3x^2 & -2x & +1 \\ & & -(3x^2 & -9x & +6\ ) \\ \hline & & & 7x & -5 \end{array}$$

As a result of long division, the integral can be rewritten as

$$\int \frac{x^3+1}{x^2-3x+2}\,dx = \int \left(x+3+\frac{7x-5}{x^2-3x+2}\right)dx.$$

We still need to integrate $\frac{7x-5}{x^2-3x+2}$, which is a (proper) rational function with a quadratic denominator. Following the strategy, we try to factor the denominator:

$$x^2-3x+2 = (x-1)(x-2).$$

Because the denominator factored into linear factors, we try partial fractions. Step ❶, factoring the denominator, is complete.

❷ Write the form of the decomposition:

$$\frac{A}{x-1} + \frac{B}{x-2}.$$

❸ Equate and simplify:

$$\frac{7x-5}{x^2-3x+2} = \frac{A}{x-1} + \frac{B}{x-2}$$
$$7x - 5 = A(x-2) + B(x-1).$$

Choosing $x = 1$ gives

$$2 = A(-1) + 0$$
$$-2 = A$$

and choosing $x = 2$ gives

$$9 = 0 + B(1)$$
$$9 = B.$$

We rewrite the integral again, this time using the results of the method of partial fractions:

$$\int \left(x + 3 + \frac{7x-5}{x^2-3x+2} \right) dx = \int \left(x + 3 + \frac{-2}{x-1} + \frac{9}{x-2} \right) dx.$$

Ans. to reading exercise 10
partial fractions, because the denominator factors as $x^2 + 5x + 6 = (x+2)(x+3)$

Now every term in the integrand is in a form for which we can find an antiderivative, and the solution is

$$\frac{x^2}{2} + 3x - 2\ln|x-1| + 9\ln|x-2| + C. \qquad \blacksquare$$

Partial fractions example: three linear factors

As long as the denominator of a rational function factors into non-repeated linear factors, the methods of this section apply, no matter how large the degree of the denominator.

Example 6 *Evaluate* $\displaystyle\int \frac{3x^2 + 7x - 2}{x^3 - x}\, dx.$

Solution The integrand is a (proper) rational function. Because the degree of the denominator is larger than two, we suspect that partial fractions may need to be used.

❶ Factor the denominator:

$$x^3 - x = x(x^2 - 1) = x(x - 1)(x + 1).$$

❷ Write the form of the decomposition. The denominator factored into three nonrepeated linear factors, so the form of the decomposition is

$$\frac{A}{x} + \frac{B}{x-1} + \frac{C}{x+1}.$$

❸ Equate the form of the decomposition with the original fraction and solve for the unknown coefficients. Equating gives

$$\frac{A}{x} + \frac{B}{x-1} + \frac{C}{x+1} = \frac{3x^2 + 7x - 2}{x^3 - x}.$$

To clear the fractions, we multiply both sides of the equation by the denominator, the factored form of which is $x(x - 1)(x + 1)$:

Line 2 distributes the quantity $x(x-1)(x+1)$ through the parentheses containing the three fractions. Cancellations are shown, with results in the last line.

$$x(x-1)(x+1) \cdot \left(\frac{A}{x} + \frac{B}{x-1} + \frac{C}{x+1}\right) = \frac{3x^2 + 7x - 2}{x^3 - x} \cdot x(x-1)(x+1)$$

$$\cancel{x}(x-1)(x+1) \cdot \frac{A}{\cancel{x}} + x\cancel{(x-1)}(x+1) \cdot \frac{B}{\cancel{x-1}} + x(x-1)\cancel{(x+1)} \cdot \frac{C}{\cancel{x+1}}$$
$$= 3x^2 + 7x - 2$$

$$(x-1)(x+1)A + x(x+1)B + x(x-1)C = 3x^2 + 7x - 2.$$

To use method 2 to solve for A, B, and C, choose $x = 0$, $x = 1$, and $x = -1$. Using $x = 0$ in the equation gives

$$(-1)(1)A + 0 + 0 = -2$$
$$A = 2.$$

Using $x = 1$ in the equation gives

$$0 + 1(2)B + 0 = 8$$
$$B = 4.$$

Using $x = -1$ in the equation gives

$$0 + 0 + (-1)(-2)C = -6$$
$$C = -3.$$

The result of the method of partial fractions allows us to evaluate the integral:

$$\int \frac{3x^2 + 7x - 2}{x^3 - x}\, dx = \int \left(\frac{2}{x} + \frac{4}{x-1} + \frac{-3}{x+1} \right) dx$$
$$= 2 \ln|x| + 4 \ln|x - 1| - 3 \ln|x + 1| + C. \quad \blacksquare$$

Multiple-technique example

Partial fractions may be combined with other methods as necessary.

Example 7 *Evaluate* $\displaystyle\int \frac{3 \cos t}{\sin^2 t + 5 \sin t + 6}\, dt.$

Solution Noticing $\sin t$ in the denominator and $\cos t$ in the numerator, we suspect that substitution is helpful. We try

$$u = \sin t$$
$$du = \cos t\, dt.$$

Making the substitution,

$$\int \frac{3 \cos t}{\sin^2 t + 5 \sin t + 6}\, dt = \int \frac{3}{u^2 + 5u + 6}\, du.$$

Next we recognize a rational function with a quadratic denominator, which we can factor:

$$u^2 + 5u + 6 = (u + 3)(u + 2).$$

Proceeding with partial fractions, we equate the fraction with the form of its partial fraction decomposition, then simplify:

$$\frac{3}{u^2 + 5u + 6} = \frac{A}{u+3} + \frac{B}{u+2}$$
$$3 = A(u + 2) + B(u + 3).$$

Choosing $u = -2$ gives

$$3 = 0 + B(1)$$
$$B = 3.$$

Choosing $u = -3$ gives

$$3 = (-1)A + 0$$
$$A = -3.$$

Then,

$$\begin{aligned}\int \frac{3}{u^2 + 5u + 6}\,du &= \int \left(\frac{-3}{u+3} + \frac{3}{u+2}\right) du \\ &= -3\ln|u+3| + 3\ln|u+2| + C \\ &= -3\ln|3 + \sin t| + 3\ln|2 + \sin t| + C.\end{aligned}$$

Line 1 rewrites the integral using the result of partial fractions, line 2 integrates, and line 3 returns to the original variable, switching the order of the terms inside the absolute values for the sake of clarity.

The absolute values may be removed if desired because the quantities inside are always positive. ■

Exercises 7.5

1–6. Should we try partial fractions or completing the square?

1. $\displaystyle\int \frac{2}{x^2 + 3x + 2}\,dx$
2. $\displaystyle\int \frac{3}{x^2 - x - 12}\,dx$
3. $\displaystyle\int \frac{2}{x^2 + 4x + 7}\,dx$
4. $\displaystyle\int \frac{4}{x^2 - 7}\,dx$
5. $\displaystyle\int \frac{2}{x^2 - 4}\,dx$
6. $\displaystyle\int \frac{1}{x^2 + 2x + 5}\,dx$

7–12. Write the form of the partial fraction decomposition.

7. $\displaystyle\int \frac{2x - 1}{x^2 + 3x - 4}\,dx$
8. $\displaystyle\int \frac{x - 214}{2x^2 - x - 1}\,dx$
9. $\displaystyle\int \frac{x^2 + 1}{x^3 + 6x^2 + 5x}\,dx$
10. $\displaystyle\int \frac{2}{x^3 - 25x}\,dx$
11. $\displaystyle\int \frac{4}{x(x-1)(x-4)(x+7)}\,dx$
12. $\displaystyle\int \frac{3}{(x^2 - 4)(x + 5)}\,dx$

13–42. Evaluate the integral.

13. $\displaystyle\int \frac{4}{(x+3)(x+5)}\,dx$

14. $\displaystyle\int \frac{x+1}{(x-2)(x+9)}\,dx$

15. $\displaystyle\int \frac{2x-1}{(3x-1)(4x+1)}\,dx$

16. $\displaystyle\int \frac{-4}{(2x+3)(5x-1)}\,dx$

17. $\displaystyle\int \frac{7x+14}{x^2+3x-10}\,dx$

18. $\displaystyle\int \frac{x+10}{x^2+5x+4}\,dx$

19. $\displaystyle\int \frac{-2x-10}{x^2+6x+8}\,dx$

20. $\displaystyle\int \frac{6x-17}{x^2-9x+14}\,dx$

21. $\displaystyle\int \frac{x}{x^2-4x+3}\,dx$

22. $\displaystyle\int \frac{3}{x^2+6x+8}\,dx$

23. $\displaystyle\int \frac{x^2}{x^2-4x+3}\,dx$

24. $\displaystyle\int \frac{3}{x^3+6x^2+8x}\,dx$

25. $\displaystyle\int \frac{x^2+3}{x^3-4x}\,dx$

26. $\displaystyle\int \frac{x^3}{x^2+6x+8}\,dx$

27. $\displaystyle\int \frac{2x}{x^2-4}\,dx$

28. $\displaystyle\int \frac{29x+3}{35x^2+8x-3}\,dx$

29. $\displaystyle\int \frac{5x+1}{x^2-100}\,dx$

30. $\displaystyle\int \frac{3}{x^2-49}\,dx$

31. $\displaystyle\int \frac{2x-17}{2x^2-x-15}\,dx$

32. $\displaystyle\int \frac{3x^2}{x^3-1}\,dx$

33. $\displaystyle\int \frac{2}{x^2+4x+5}\,dx$

34. $\displaystyle\int \frac{-7}{x^2+2x+2}\,dx$

35. $\displaystyle\int \frac{6x-2\sqrt{7}}{x^2-7}\,dx$

36. (GSL) $\displaystyle\int \frac{3x^2+11x+2}{(x+3)(x^2-1)}\,dx$

37. (GSL) $\displaystyle\int \frac{3z+7}{(z+1)(z+2)(z+3)}\,dz$

38. $\displaystyle\int \frac{2\sinh x}{\cosh^2 x-9}\,dx$

39. $\displaystyle\int \frac{4e^x}{e^{2x}+8e^x+7}\,dx$

40. $\displaystyle\int \frac{4x-3\sqrt{13}}{x^2-\sqrt{13}x}\,dx$

41. $\displaystyle\int \frac{\ln x}{x((\ln x)^2-4)}\,dx$

42. $\displaystyle\int \frac{2}{\sqrt{x}\,(x-3\sqrt{x}+2)}\,dx$

43. (GSL) Evaluate $\displaystyle\int \frac{2t^4+3t^3-20t-28}{(t^2-4)(2t-1)}\,dt.$

44. (GSL) Evaluate $\displaystyle\int \frac{y^4-3y^3}{(y^2-1)(y-2)}\,dy.$

45. (Love) Evaluate $\int \frac{2x^3 + x - 1}{x^3 + x^2 - 4x - 4}\, dx.$

46. (Love) Evaluate $\int \frac{x^4}{x^3 + 2x^2 - x - 2}\, dx.$

47. (Love) Evaluate $\int \frac{e^{3x}}{1 - e^{2x}}\, dx.$

48. (Love) Evaluate $\int \frac{x}{x^4 + 4x^2 + 3}\, dx.$

49. (Love) Derive an alternate formula for $\int \csc\theta\, d\theta$ by noticing that

$$\csc\theta = \frac{1}{\sin\theta} = \frac{\sin\theta}{\sin^2\theta} = \frac{\sin\theta}{1 - \cos^2\theta},$$

then integrating the latter expression using substitution and partial fractions.

50. (Love) Derive an alternate formula for $\int \sec\theta$ in the manner of exercise 49.

Partial Fractions, Part II

7.6

We examine the remaining cases for partial fractions in this section.

Partial fraction forms

So far, the only form of the partial fraction decomposition we have encountered is when the denominator factors into distinct (nonrepeated) linear factors such as $(x-7)(x+4)(x-3)$. However, repeated linear factors such as $(x-2)^3$ and irreducible quadratic factors such as (x^2+9) are also possible. Knowing how to deal with each type of factor allows us to use partial fractions for any rational expression with a factored denominator.

There are multiple ways to decompose a fraction if the denominator contains repeated linear factors or irreducible quadratic factors. The form we use is the one that is most helpful for integration.

FORM OF THE PARTIAL FRACTION DECOMPOSITION

For each factor in the denominator of a rational expression, the term or terms of the partial fraction decomposition are as follows:

(a) Nonrepeated linear factor $(ax+b)$: use

$$\frac{A}{ax+b}.$$

(b) Repeated linear factor $(ax+b)^n$: use

$$\frac{A}{ax+b}+\frac{B}{(ax+b)^2}+\cdots+\frac{C}{(ax+b)^n}.$$

(c) Nonrepeated irreducible quadratic factor ax^2+bx+c: use

$$\frac{Ax+B}{ax^2+bx+c}.$$

(d) Repeated irreducible quadratic factor $(ax^2+bx+c)^n$: use

$$\frac{Ax+B}{ax^2+bx+c}+\frac{Cx+D}{(ax^2+bx+c)^2}+\cdots+\frac{Ex+F}{(ax^2+bx+c)^n}.$$

An *irreducible quadratic* factor is a quadratic that cannot be further factored into linear factors (without using complex numbers).

Example 1 *Write the form of the partial fraction decomposition for* $\frac{x^2 - 3x + 8}{(x-2)^3(x^2+9)}$.

Solution For the repeated linear factor $(x-2)^3$, according to case (b) we need a constant over each power of this linear factor. We should include one term that is a constant over the factor with exponent 1, another term that is a constant over the factor squared, continuing until we reach the exponent n, which in this case is 3. For the irreducible quadratic factor $(x^2 + 9)$, according to case (c), we need to include a term with that denominator and a linear expression in the numerator. The form of the partial fraction decomposition is

Although a difference of squares can be factored, a sum of squares cannot (in the real numbers). Therefore, $x^2 + 9$ is an irreducible quadratic.

$$\frac{A}{x-2} + \frac{B}{(x-2)^2} + \frac{C}{(x-2)^3} + \frac{Dx+E}{x^2+9}. \qquad \blacksquare$$

How do we know if a quadratic expression can be factored further or if it is irreducible? Having a difficult time finding a factorization is not the same as knowing that no such factorization exists. The quadratic formula can be used to make the determination. If $ax^2 + bx + c = 0$ has no real solutions, then $ax^2 + bx + c$ is irreducible; if the equation has one or more real solutions, then $ax^2 + bx + c$ can be factored into two linear factors (not necessarily distinct).

An alternative to using the quadratic formula is to complete the square. If the result is of the form $(x \pm a)^2 - k$, then the expression can be factored; if the result is of the form $(x \pm a)^2 + k$, then the expression is irreducible (both times assuming $k > 0$).

Example 2 *Factor the expression or state that it is irreducible for (a)* $x^2 + 8x + \frac{63}{4}$ *and (b)* $x^2 + 8x + \frac{67}{4}$.

Solution (a) If factoring is not quickly successful, try using the quadratic formula to solve $x^2 + 8x + \frac{63}{4} = 0$:

$$\begin{aligned} x &= \frac{-8 \pm \sqrt{64 - 4(1)(\frac{63}{4})}}{2(1)} \\ &= \frac{-8 \pm \sqrt{64 - 63}}{2} \\ &= \frac{-8 \pm 1}{2} \\ &= -\frac{7}{2}, -\frac{9}{2}. \end{aligned}$$

Because the equation has real solutions, the expression is not irreducible. It factors as

In algebra, the *factor theorem* states that for a polynomial $f(x)$, if $f(c) = 0$, then $x - c$ is a factor of the polynomial. Therefore, $x - (-\frac{7}{2})$ and $x - (-\frac{9}{2})$ are factors of the polynomial.

$$x^2 + 8x + \frac{63}{4} = \left(x + \frac{7}{2}\right)\left(x + \frac{9}{2}\right).$$

(b) If factoring is not quickly successful, try using the quadratic formula to solve $x^2 + 8x + \frac{67}{4} = 0$:

$$\begin{aligned} x &= \frac{-8 \pm \sqrt{64 - 4(1)(\frac{67}{4})}}{2(1)} \\ &= \frac{-8 \pm \sqrt{64 - 67}}{2} \\ &= \frac{-8 \pm \sqrt{-3}}{2}. \end{aligned}$$

As we can see, the equation has no real solutions. The expression is an irreducible quadratic. ■

Example 3 *Write the form of the partial fraction decomposition for* $\dfrac{x^3 - 7}{(x^2 + 6x + 9)(x^2 - x - 6)(x^2 + 4x + 7)}$.

Solution For every quadratic expression in the denominator, we try to factor further. We have

$$x^2 + 6x + 9 = (x + 3)^2$$

and

$$x^2 - x - 6 = (x - 3)(x + 2).$$

Because we can't find a factorization of $x^2 + 4x + 7$ quickly, we try the quadratic formula:

$$\begin{aligned} x &= \frac{-4 \pm \sqrt{16 - 4(1)(7)}}{2(1)} \\ &= \frac{-4 \pm \sqrt{-12}}{2}, \end{aligned}$$

and the expression is an irreducible quadratic. Using cases (a), (b), and (c) of the directions for the form of the partial fraction decomposition, we have

$$\frac{A}{x+3} + \frac{B}{(x+3)^2} + \frac{C}{x-3} + \frac{D}{x+2} + \frac{Ex+F}{x^2+4x+7}. \quad ■$$

We use a constant over each nonrepeated linear term, a constant over each power of the repeated linear term $(x+3)^2$, and a linear expression over the irreducible quadratic term $x^2 + 4x + 7$. The order of the terms, and the name of the constants A, B, and so on, is not important.

Reading Exercise 11 *Write the form of the partial fraction decomposition for* $\dfrac{x^2 + x - 9}{(x+2)(x+5)^2(x^2+1)}$.

Partial fractions example: repeated linear factor

The steps in the method of partial fractions remain the same. The only modifications are the additional cases in step ❷ (writing the form of the partial fraction decomposition) and evaluating the resulting integrals.

Example 4 *Evaluate* $\displaystyle\int \frac{3x^2+1}{(x+1)(x+4)^2}\,dx.$

The degree of the numerator is 2, whereas the degree of the denominator is 3. To count the degree of the denominator, we count $(x+4)^2$ as degree 2 and $(x+1)$ as degree 1, then add the degrees of the factors.

Solution Recognizing a (proper) rational function for which other methods (substitution, formulas) do not seem applicable, the factored denominator indicates the use of partial fractions. Step ❶, factoring the denominator, has already been completed.

❷ Write the form of the decomposition. Using case (a) for the nonrepeated linear factor $(x+1)$ and case (b) for the repeated linear factor $(x+4)^2$, the form of the partial fraction decomposition is

$$\frac{A}{x+1}+\frac{B}{x+4}+\frac{C}{(x+4)^2}.$$

❸ Equate the form of the decomposition with the original fraction and solve for the unknown coefficients. Equating yields

$$\frac{A}{x+1}+\frac{B}{x+4}+\frac{C}{(x+4)^2}=\frac{3x^2+1}{(x+1)(x+4)^2}.$$

Next we multiply both sides of the equation by the denominator $(x+1)(x+4)^2$ and simplify:

Line 2 distributes $(x+1)(x+4)^2$ to the three terms inside the parentheses and indicates which factors cancel.

$$(x+1)(x+4)^2\left(\frac{A}{x+1}+\frac{B}{x+4}+\frac{C}{(x+4)^2}\right)=\frac{3x^2+1}{(x+1)(x+4)^2}\cdot(x+1)(x+4)^2$$

$$\cancel{(x+1)}(x+4)^2\frac{A}{\cancel{x+1}}+(x+1)(x+4)^{\cancel{2}}\frac{B}{\cancel{x+4}}+(x+1)\cancel{(x+4)^2}\frac{C}{\cancel{(x+4)^2}}=3x^2+1$$

$$A(x+4)^2+B(x+1)(x+4)+C(x+1)=3x^2+1.$$

Following method 2, we choose values of x that make a factor zero; choosing $x=-4$ gives

$$0+0+C(-3)=3(16)+1$$

$$C=-\frac{49}{3}$$

and choosing $x=-1$ gives

$$A(9)+0+0=4$$

$$A=\frac{4}{9}.$$

Have you noticed the difference from before, when we only dealt with non-repeated linear factors? There are three coefficients to solve for—A, B, and C—but only two distinct factors for which to choose convenient values of x. We found the values of A and C quickly, but not B.

One way to proceed is to abandon method 2 and go back to method 1, and often this is the best approach. However, this time, there is a quicker way. Recalling that the previous equation is true for every value of x, we can choose some other convenient value of x, such as $x = 0$, and substitute the values of A and C already found. Then,

$$\begin{aligned} A(x+4)^2 + B(x+1)(x+4) + C(x+1) &= 3x^2 + 1 \\ A(4^2) + B(1)(4) + C(1) &= 1 \\ \left(\frac{4}{9}\right)(16) + 4B + -\frac{49}{3} &= 1 \\ B &= \frac{23}{9}. \end{aligned}$$

Line 1 repeats the equation we are using, line 2 substitutes zero for x, line 3 substitutes the already calculated values of A and C, and line 4 completes the calculation of the value of B.

❹ Write the partial fraction decomposition. We have

$$\frac{3x^2+1}{(x+1)(x+4)^2} = \frac{\frac{4}{9}}{x+1} + \frac{\frac{23}{9}}{x+4} + \frac{-\frac{49}{3}}{(x+4)^2}.$$

All that is left is to rewrite and evaluate the integral. We wish to evaluate

$$\int \left(\frac{\frac{4}{9}}{x+1} + \frac{\frac{23}{9}}{x+4} + \frac{-\frac{49}{3}}{(x+4)^2} \right) dx.$$

The first two terms are integrated easily; the third requires substitution, so we work with it first. Using

$$\begin{aligned} u &= x + 4 \\ du &= dx, \end{aligned}$$

we have

$$\begin{aligned} \int \frac{-\frac{49}{3}}{(x+4)^2}\, dx &= -\frac{49}{3} \int \frac{1}{u^2}\, du = -\frac{49}{3} \int u^{-2}\, du \\ &= -\frac{49}{3} \frac{u^{-1}}{-1} + C = \frac{49}{3u} + C \\ &= \frac{49}{3(x+4)} + C. \end{aligned}$$

CAUTION: MISTAKE TO AVOID

The third term is not of the form $\int \frac{1}{u}\, du$, and therefore its antiderivative does not involve a natural log. The work shown here is typical of the antiderivative step after partial fractions involving case (b), repeated linear factors.

Therefore, the original integral can be evaluated as

Ans. to reading exercise 11
$\frac{A}{x+2}+\frac{B}{x+5}+\frac{C}{(x+5)^2}+\frac{Dx+E}{x^2+1}$

$$\int \frac{3x^2+1}{(x+1)(x+4)^2}\,dx = \frac{4}{9}\ln|x+1| + \frac{23}{9}\ln|x+4| + \frac{49}{3(x+4)} + C. \quad \blacksquare$$

Reading Exercise 12 *Using partial fractions yieldes the integral* $\int\left(\frac{2}{x-1}+\frac{3}{x+2}+\frac{-5}{(x+2)^2}\right)dx$. *Finish the integration.*

Partial fractions examples: irreducible quadratic factor

Example 5 *Evaluate* $\int \frac{2x^2+x+13}{(x-5)(x^2+9)}\,dx.$

Solution Recognizing a (proper) rational function for which other methods (substitution, formulas) do not seem applicable, the factored denominator indicates the use of partial fractions. Step ❶ has been completed.

❷ Write the form of the decomposition. Since x^2+9 is a sum of squares, it is irreducible. Using cases (a) and (c), the form of the partial fraction decomposition is

$$\frac{A}{x-5}+\frac{Bx+C}{x^2+9}.$$

❸ Equate the form of the decomposition with the original fraction and solve for the unknown coefficients. Equating and simplifying gives

$$\frac{A}{x-5}+\frac{Bx+C}{x^2+9}=\frac{2x^2+x+13}{(x-5)(x^2+9)}$$

$$A(x^2+9)+(Bx+C)(x-5)=2x^2+x+13.$$

Method 1 is generally the best option when an irreducible quadratic is involved. Therefore, we expand the left-hand side and collect like terms:

$$Ax^2+9A+Bx^2-5Bx+Cx-5C=2x^2+x+13$$

$$(A+B)x^2+(-5B+C)x+9A-5C=2x^2+x+13.$$

Equating coefficients on the x^2, x, and constant terms between the two sides of the equation gives the simultaneous equations

$$A + B = 2$$
$$-5B + C = 1$$
$$9A - 5C = 13.$$

Using the substitution method to solve the simultaneous equations, we rearrange the first and second equations to

This algebraic strategy is often the quickest option for solving the simultaneous linear equations resulting from partial fractions with a linear and irreducible quadratic factor.

$$A = 2 - B$$
$$C = 1 + 5B$$

and substitute into the third equation to solve for B:

$$9(2 - B) - 5(1 + 5B) = 13$$
$$B = 0.$$

Using the previously rearranged equations then yields $A = 2$ and $C = 1$.

Skipping step ❹, we move directly to rewriting the integral and evaluating:

$$\int \frac{2x^2 + x + 13}{(x-5)(x^2+9)}\,dx = \int \left(\frac{2}{x-5} + \frac{1}{x^2+9}\right) dx$$
$$= 2\ln|x-5| + \frac{1}{3}\tan^{-1}\frac{x}{3} + C. \quad ■$$

In example 5, the irreducible quadratic led to the use of the antiderivative giving the inverse tangent formula. There are other possibilities. For instance, if we have a term of the form

$$\frac{Ax + B}{x^2 + 4}$$

and find $A = 2$, $B = 0$, we need to evaluate

$$\int \frac{2x}{x^2+4}\,dx,$$

for which we use substitution with $u = x^2 + 4$, $du = 2x\,dx$.

Ans. to reading exercise 12
$2\ln|x-1| + 3\ln|x+2| + \frac{5}{x+2} + C$

Reading Exercise 13 *Evaluate* $\displaystyle\int \frac{1/3}{x+1} + \frac{2x}{x^2+1} + \frac{3}{x^2+1}\,dx.$

There are scenarios in which substitution, completing the square, and the antiderivative giving the inverse tangent formula are all needed. Let's practice finishing such an integral in the next example.

Example 6 *Partial fractions has resulted in the need to evaluate the integral* $\int \frac{x-4}{x^2+10x+41}\,dx$. *Complete the integration.*

Solution Because the method of partial fractions has been completed, the denominator $x^2 + 10x + 41$ must be irreducible (it is). If the numerator includes just a constant and no x term, we use completing the square and the antiderivative giving the inverse tangent formula. However, the presence of x in the numerator indicates a different strategy.

STRATEGY

When partial fractions has resulted in an integral of the form linear over quadratic, try substitution first, then—on what remains—try completing the square if necessary.

Let's first try substitution, with u as the denominator:

$$u = x^2 + 10x + 41$$
$$du = (2x + 10)\,dx.$$

Because du contains $2x$ and the numerator of the integral just contains x, we need to adjust for the constant:

$$\begin{aligned}\int \frac{x-4}{x^2+10x+41}\,dx &= \frac{1}{2}\int \frac{2(x-4)}{x^2+10x+41}\,dx\\ &= \frac{1}{2}\int \frac{2x-8}{x^2+10x+41}\,dx.\end{aligned}$$

Now the numerator is $2x-8$, but for du we need $2x+10$. It seems that substitution fails. Or does it? Try this:

Line 1 adds and subtracts the constant 10 needed for substitution; line 2 simplifies, resulting in the linear term needed for substitution plus another constant; line 3 rewrites the integrand as two fractions, one of which is set up for substitution; and line 4 uses the antiderivative sum rule.

$$\begin{aligned}\frac{1}{2}\int \frac{2x-8}{x^2+10x+41}\,dx &= \frac{1}{2}\int \frac{2x+10-8-10}{x^2+10x+41}\,dx\\ &= \frac{1}{2}\int \frac{2x+10-18}{x^2+10x+41}\,dx\\ &= \frac{1}{2}\int \left(\frac{2x+10}{x^2+10x+41} + \frac{-18}{x^2+10x+41}\right)dx\\ &= \frac{1}{2}\int \frac{2x+10}{x^2+10x+41}\,dx + \frac{1}{2}\int \frac{-18}{x^2+10x+41}\,dx.\end{aligned}$$

Although we now have two integrals, one of them is set up for substitution and the other has just a constant in the numerator and can therefore be handled by completing the square and using the antiderivative giving the inverse tangent formula.

The reason for doing substitution first is now clear. We do not know which constant will be in the numerator of the integral for which we use inverse tangent until after the substitution is set up. Also, calculating du is easier before completing the square on the denominator.

Using substitution on the integral on the left,

$$\begin{aligned}\frac{1}{2}\int \frac{2x+10}{x^2+10x+41}\,dx &= \frac{1}{2}\int \frac{1}{u}\,du\\ &= \frac{1}{2}\ln|u| + C\\ &= \frac{1}{2}\ln\left|x^2+10x+41\right| + C.\end{aligned}$$

Line 1 makes the substitution $u = x^2 + 10x + 41$, $du = (2x + 10)\,dx$; line 2 is the antiderivative step; and line 3 goes back to the original variable.

For the remaining integral, we first complete the square on the denominator. With $a = \frac{10}{2} = 5$, we add and subtract $a^2 = 25$ to get

$$\begin{aligned}&x^2 + 10x \qquad + 41\\ &= x^2 + 10x + 25 + 41 - 25\\ &= (x+5)^2 + 16.\end{aligned}$$

Rewriting the integrand and using the antiderivative giving the inverse tangent formula then gives

$$\begin{aligned}\frac{1}{2}\int \frac{-18}{x^2+10x+41}\,dx &= -9\int \frac{1}{(x+5)^2+16}\,dx\\ &= -9\cdot\frac{1}{4}\tan^{-1}\frac{x+5}{4} + C.\end{aligned}$$

We may now state the final result as

$$\begin{aligned}\int \frac{x-4}{x^2+10x+41}\,dx =& \frac{1}{2}\ln\left|x^2+10x+41\right|\\ &- \frac{9}{4}\tan^{-1}\frac{x+5}{4} + C.\end{aligned}$$

■

Now let's look at an example in which partial fractions results in an integral of the type practiced in example 6.

Example 7 *Evaluate* $\displaystyle\int \frac{2x-1}{(x+3)(x^2+3x+3)}\,dx.$

Solution Recognizing a (proper) rational function for which other methods (substitution, formulas) do not seem applicable, the factored denominator indicates the use of partial fractions.

Ans. to reading exercise 13

$\frac{1}{3}\ln|x+1| + \ln|x^2+1| + 3\tan^{-1}x + C$

❶ Factor the denominator. This step is not complete until we are sure the denominator does not factor further. Specifically, we need to know whether $x^2 + 3x + 3$ can be factored or whether it is irreducible.

Because all that matters is whether the quantity under the square root is negative, we can just check the *discriminant*

$$\sqrt{b^2 - 4ac} = \sqrt{9 - 4(1)(3)} = \sqrt{-3},$$

which is not real. The quadratic is irreducible. (If the discriminant is real we still need to factor, and the full quadratic formula helps.)

We check by using the quadratic formula to solve $x^2 + 3x + 3 = 0$:

$$\begin{aligned} x &= \frac{-3 \pm \sqrt{9 - 4(1)(3)}}{2(1)} \\ &= \frac{-3 \pm \sqrt{-3}}{2}, \end{aligned}$$

and the quadratic is irreducible.

❷ Write the form of the decomposition. Using cases (a) and (c), the form of the partial fraction decomposition is

$$\frac{A}{x+3} + \frac{Bx + C}{x^2 + 3x + 3}.$$

❸ Equate the form of the decomposition with the original fraction and solve for the unknown coefficients. Equating and simplifying gives

$$\frac{A}{x+3} + \frac{Bx + C}{x^2 + 3x + 3} = \frac{2x - 1}{(x+3)(x^2 + 3x + 3)}$$

$$A(x^2 + 3x + 3) + (Bx + C)(x + 3) = 2x - 1.$$

Following method 1 because an irreducible quadratic is involved, we expand the left-hand side and collect like terms;

$$Ax^2 + 3Ax + 3A + Bx^2 + 3Bx + Cx + 3C = 2x - 1$$

$$(A + B)x^2 + (3A + 3B + C)x + 3A + 3C = 2x - 1.$$

Equating coefficients on the x^2, x, and constant terms between the two sides of the equation gives the simultaneous equations

$$\begin{aligned} A + B &= 0 \\ 3A + 3B + C &= 2 \\ 3A + 3C &= -1. \end{aligned}$$

Using the substitution method to solve the simultaneous equations, we rearrange the first and third to

$$\begin{aligned} B &= -A \\ C &= -\frac{1}{3} - A \end{aligned}$$

and substitute into the second equation to solve for A:

$$\begin{aligned} 3A + 3(-A) + \left(-\frac{1}{3} - A\right) &= 2 \\ A &= -\frac{7}{3}. \end{aligned}$$

Using the previously rearranged equations then yields

$$B = -\left(-\frac{7}{3}\right) = \frac{7}{3}$$
$$C = -\frac{1}{3} - \left(-\frac{7}{3}\right) = 2.$$

Skipping step ❹, we move directly to the integral:

$$\int \frac{2x - 1}{(x+3)(x^2+3x+3)}\,dx = \int \left(\frac{-\frac{7}{3}}{x+3} + \frac{\frac{7}{3}x + 2}{x^2+3x+3}\right) dx$$
$$= \int \frac{-\frac{7}{3}}{x+3}\,dx + \int \frac{\frac{7}{3}x+2}{x^2+3x+3}\,dx.$$

The integral on the left evaluates as

$$\int \frac{-\frac{7}{3}}{x+3}\,dx = -\frac{7}{3}\ln|x+3| + C.$$

The integral on the right is of the form practiced in example 6; the numerator is linear and the denominator is irreducible quadratic. Therefore, we begin its evaluation by trying substitution:

$$u = x^2 + 3x + 3$$
$$du = (2x+3)\,dx.$$

Adjusting for the constant 2 might be easier if we factor out the current coefficient of $\frac{7}{3}$ first:

$$\int \frac{\frac{7}{3}x+2}{x^2+3x+3}\,dx = \int \frac{\frac{7}{3}\left(x+\frac{6}{7}\right)}{x^2+3x+3}\,dx$$
$$= \frac{7}{3}\int \frac{x+\frac{6}{7}}{x^2+3x+3}\,dx$$
$$= \frac{1}{2}\cdot\frac{7}{3}\int \frac{2\left(x+\frac{6}{7}\right)}{x^2+3x+3}\,dx$$
$$= \frac{7}{6}\int \frac{2x+\frac{12}{7}}{x^2+3x+3}\,dx.$$

Line 1 factors out $\frac{7}{3}$ from the numerator (notice that $\frac{2}{\frac{7}{3}} = 2\cdot\frac{3}{7} = \frac{6}{7}$), line 2 uses the antiderivative constant multiple rule, line 3 adjusts for the needed constant 2, and line 4 distributes 2 through the parentheses.

For substitution, we need a numerator of $2x+3$, so we add and subtract 3 in the numerator and rewrite as two separate integrals:

$$\begin{aligned}
\frac{7}{6}\int \frac{2x+\frac{12}{7}}{x^2+3x+3}\,dx &= \frac{7}{6}\int \frac{2x+3+\frac{12}{7}-3}{x^2+3x+3}\,dx \\
&= \frac{7}{6}\int \frac{2x+3-\frac{9}{7}}{x^2+3x+3}\,dx \\
&= \frac{7}{6}\int \frac{2x+3}{x^2+3x+3}\,dx + \frac{7}{6}\int \frac{-\frac{9}{7}}{x^2+3x+3}\,dx.
\end{aligned}$$

The integral on the left may be completed using the substitution:

$$\begin{aligned}
\frac{7}{6}\int \frac{2x+3}{x^2+3x+3}\,dx &= \frac{7}{6}\int \frac{1}{u}\,du \\
&= \frac{7}{6}\ln|u| + C \\
&= \frac{7}{6}\ln|x^2+3x+3| + C.
\end{aligned}$$

The remaining integral has a constant in the numerator and an irreducible quadratic in the denominator, so we complete the square and use the antiderivative giving the inverse tangent formula. To complete the square, we compute $a = \frac{3}{2}$ and then add and subtract $a^2 = \frac{9}{4}$, yielding

$$\begin{aligned}
&x^2 + 3x \quad + 3 \\
&= x^2 + 3x + \frac{9}{4} + 3 - \frac{9}{4} \\
&= \left(x + \frac{3}{2}\right)^2 + \frac{3}{4}.
\end{aligned}$$

We then use the result to rewrite the integral and continue with the antiderivative giving the inverse tangent formula:

$$\begin{aligned}
\frac{7}{6}\int \frac{-\frac{9}{7}}{x^2+3x+3}\,dx &= -\frac{3}{2}\int \frac{1}{(x+\frac{3}{2})^2+\frac{3}{4}}\,dx \\
&= -\frac{3}{2}\cdot\frac{1}{\frac{\sqrt{3}}{2}}\tan^{-1}\frac{x+\frac{3}{2}}{\frac{\sqrt{3}}{2}} + C \\
&= -\sqrt{3}\tan^{-1}\frac{2x+3}{\sqrt{3}} + C.
\end{aligned}$$

Line 2 uses the antiderivative giving the inverse tangent formula with $a^2 = \frac{3}{4}$, or $a = \frac{\sqrt{3}}{2}$; line 3 simplifies the compound fractions.

The final task is to collect all the pieces of the solution:

$$\begin{aligned}
&\int \frac{2x-1}{(x+3)(x^2+3x+3)}\,dx \\
&= -\frac{7}{3}\ln|x+3| + \frac{7}{6}\ln|x^2+3x+3| \\
&\quad - \sqrt{3}\tan^{-1}\frac{2x+3}{\sqrt{3}} + C.
\end{aligned}$$

■

An additional type of integral that can result from a partial fraction decomposition is from case (d), an example of which is

$$\int \frac{5}{(x^2+4)^2}\,dx.$$

Trigonometric substitution is required. This integral is similar to the one in example 1 of section 7.4. If the numerator is linear instead, such as

$$\int \frac{3x+5}{(x^2+4x+19)^2},$$

the strategy is similar to that of examples 6 and 7. Use substitution first, and then the remaining integral (if any) should have a constant numerator, for which the strategy is to complete the square (if necessary), followed by trigonometric substitution.

Exercises 7.6

1–6. Write the form of the partial fraction decomposition.

1. $\int \frac{x^2-3}{(x-4)^3(x+1)}\,dx$

2. $\int \frac{x^4-17}{x^2(x+3)^3}\,dx$

3. $\int \frac{x^3+4x-7}{(x^2+5)(x^2-4)}\,dx$

4. $\int \frac{2}{(x^2-1)(x+1)}\,dx$

5. $\int \frac{1}{(x^2+2x+19)^2(x-1)(x+3)^2}\,dx$

6. $\int \frac{x-1}{(x^2+3)^2(x^2+4x+5)}\,dx$

7–28. Evaluate the integral.

7. $\int \frac{x-1}{(x+1)(x+3)^2}\,dx$

8. $\int \frac{x^2}{(x-2)(x+1)^2}\,dx$

9. $\int \frac{x^2-3x+1}{(x+4)(x^2+1)}\,dx$

10. $\int \frac{2x+5}{(x+1)(x^2+16)}\,dx$

11. $\int \frac{-2x^2+x-3}{x^3-x^2}\,dx$

12. (GSL) $\int_0^4 \frac{9x^2}{(2x+1)(x+2)^2}\,dx$

13. (GSL) $\int_0^5 \frac{x^2-3}{(x+2)(x+1)^2}\,dx$

14. $\int \frac{6x^2+x-6}{x^3+3x^2}\,dx$

15. $\int \frac{3x^2 + 2x + 12}{x^3 + 4x}\,dx$

16. $\int \frac{5x^2 - 4x + 4}{x^3 - 2x^2 + x}\,dx$

17. $\int \frac{3x^2 + 16}{x^4 + 4x^2}\,dx$

18. $\int \frac{2x^2 - x + 14}{x^3 + 7x}\,dx$

19. $\int \frac{2x + 7}{x^2 + 4}\,dx$

20. $\int \frac{4x^2 + 7}{x^4 + 7x^2}\,dx$

21. $\int \frac{9x^2 + 4x - 4}{x^3 + x^2 - 2x}\,dx$

22. $\int \frac{4x - 6}{x^2 + 9}\,dx$

23. $\int \frac{x - 1}{x^2 + 4x + 5}\,dx$

24. $\int \frac{x + 5}{x^2 - 6x + 13}\,dx$

25. $\int \frac{1}{x^4 - 1}\,dx$

26. $\int \frac{4x^2 - 18}{x^3 + 5x^2 + 6x}\,dx$

27. $\int \frac{3x^2 - 5x - 4}{x^3 + 4x^2 + 4x}\,dx$

28. $\int \frac{12}{x^4 - 16}\,dx$

29. (Love) Evaluate $\int \frac{1}{(1 + e^x)^2}\,dx.$

30. (Love) Evaluate $\int \frac{1}{\cos\theta \sin^2\theta}\,d\theta.$

31. (Love) Evaluate $\int \frac{\tan^{-1} x}{x^3}\,dx.$

32. (Love) Evaluate $\int \frac{\tan^{-1} x}{x^2(1 + x^2)}\,dx.$

33. (Love) Derive an alternate formula for $\int \sec^3\theta\,d\theta$ by noticing that

$$\sec^3\theta = \frac{1}{\cos^3\theta} = \frac{\cos\theta}{\cos^4\theta} = \frac{\cos\theta}{(1 - \sin^2\theta)^2},$$

then integrating the latter expression using substitution and partial fractions.

34. (Love) Derive an alternate formula for $\int \csc^3\theta\,d\theta$ in the manner of exercise 33.

Any calculus student who can complete exercise 35 by hand correctly on the first try is in possession of very rare skill. Caution: attempting this exercise can be very time-consuming.

35. Evaluate $\int \frac{x^6 + 2x^5 - 11x^4 - 65x^3 - 66x^2 + 104x + 323}{(x - 1)(x^2 + 4x + 7)^2}\,dx.$

Other Techniques of Integration

7.7

Our list of integration techniques now includes substitution, integration by parts, trigonometric substitution, and partial fractions. Algebraic manipulations of integrands have proved helpful, such as long division when faced with an improper rational function or the use of trigonometric identities. These are the most common techniques of integration, but there remain many additional techniques of integration, more than can be covered in a calculus textbook.

In this section there is a small sampling of techniques that were once covered in standard calculus textbooks, but they no longer commonly appear. (Perhaps you wish that some techniques we have already covered would join this list.) The techniques presented here, as well as many of the examples, all appear in GSL.

Rationalizing fractional powers

If the integrand contains fractional powers of "just plain x," then it can be transformed into a rational function of a different variable by the substitution

$$x = z^n,$$

where n is the least common denominator of the fractional exponents appearing in the integrand. This type of substitution is a replacement, similar to trigonometric substitution.

Among the many calculus textbooks in which this technique appeared is the textbook from which I learned calculus, *Calculus with Analytic Geometry*, 2nd edition, by Earl W. Swokowski. Published by Prindle, Weber & Schmidt, Boston, 1979.

Example 1 *Evaluate* $\displaystyle\int \frac{x^{\frac{1}{2}}}{1 + x^{\frac{3}{4}}}\,dx.$

Solution The integrand contains fractional powers of "just plain x." Following the instructions, we determine the least common denominator of the fractional exponents, $\frac{1}{2}$ and $\frac{3}{4}$. The least common denominator is 4. We then make the substitution

$$x = z^4$$

$$dx = 4z^3\,dz.$$

Because we need to replace $x^{\frac{1}{2}}$ and $x^{\frac{3}{4}}$, we calculate these quantities in terms of z:

In each line, laws of exponents are used. The exponents are $4 \cdot \frac{1}{2} = 2$ and $4 \cdot \frac{3}{4} = 3$.

$$x^{\frac{1}{2}} = \left(z^4\right)^{\frac{1}{2}} = z^2$$
$$x^{\frac{3}{4}} = \left(z^4\right)^{\frac{3}{4}} = z^3.$$

Then,

Line 1 replaces $x^{\frac{1}{2}}$ with z^2, replaces $x^{\frac{3}{4}}$ with z^3, and replaces dx with $4z^3\,dz$. Line 2 simplifies and uses the antiderivative constant multiple rule.

$$\begin{aligned}\int \frac{x^{\frac{1}{2}}}{1+x^{\frac{3}{4}}}\,dx &= \int \frac{z^2}{1+z^3}\cdot 4z^3\,dz\\ &= 4\int \frac{z^5}{1+z^3}\,dz.\end{aligned}$$

Because the result of this type of substitution is a rational function, it is called a *rationalizing substitution*.

As promised, the integrand is now a rational function of the variable z. We now proceed using previous techniques. First we notice the improper fraction and divide:

$$\begin{array}{r} z^2 \\ z^3+1\,\big)\,\overline{z^5} \\ \underline{-(z^5+z^2)} \\ -z^2 \end{array}$$

Then,

Line 1 rewrites using the result of long division; line 2 uses the antiderivative difference rule and finds the antiderivative of z^2.

$$\begin{aligned}4\int \frac{z^5}{1+z^3}\,dz &= 4\int \left(z^2 - \frac{z^2}{z^3+1}\right)dz\\ &= 4\cdot\frac{z^3}{3} - 4\int \frac{z^2}{z^3+1}\,dz.\end{aligned}$$

The remaining integral can be evaluated using the substitution

$$u = z^3 + 1$$
$$du = 3z^2\,dz.$$

Then,

Line 1 adjusts for the constant, line 2 makes the substitution, and line 4 returns to the variable z.

$$\begin{aligned}4\int \frac{z^2}{z^3+1}\,dz &= 4\cdot\frac{1}{3}\int \frac{3z^2}{z^3+1}\,dz\\ &= \frac{4}{3}\int \frac{1}{u}\,du\\ &= \frac{4}{3}\ln|u| + C\\ &= \frac{4}{3}\ln\left|z^3+1\right| + C.\end{aligned}$$

Putting the results together brings us to

$$\frac{4}{3}z^3 - \frac{4}{3}\ln\left|z^3 + 1\right| + C.$$

We still need to return to the original variable. We use the previously calculated $z^3 = x^{\frac{3}{4}}$, and the final solution is

$$\int \frac{x^{\frac{1}{2}}}{1 + x^{\frac{3}{4}}}\,dx = \frac{4}{3}x^{\frac{3}{4}} - \frac{4}{3}\ln\left|x^{\frac{3}{4}} + 1\right| + C.$$

■

An alternative is to solve the substitution formula for z:

$$x = z^4$$

$$z = x^{\frac{1}{4}}.$$

We then replace each instance of z with $x^{\frac{1}{4}}$.

In the event that the integrand contains fractional powers of $a + bx$ only, we follow the same idea, but with the substitution

$$a + bx = z^n,$$

where n is the least common denominator of the fractional exponents of $a + bx$.

This can also be accomplished by first making a substitution of the form $u = a + bx$ and then following the method of example 1, but the procedure presented here is a little more compact.

Example 2 *Evaluate* $\displaystyle\int \frac{1}{(5x+7)^{\frac{3}{2}} + (5x+7)^{\frac{1}{2}}}\,dx.$

Solution The integrand contains fractional powers of $5x + 7$. The least common denominator of the fractional exponents $\frac{3}{2}$ and $\frac{1}{2}$ is 2. We then make the substitution

$$5x + 7 = z^2.$$

To determine the replacement for dx, we first solve for x:

$$\begin{aligned} 5x + 7 &= z^2 \\ 5x &= z^2 - 7 \\ x &= \tfrac{1}{5}z^2 - \tfrac{7}{5} \\ dx &= \tfrac{2}{5}z\,dz. \end{aligned}$$

We also need to determine replacements for $(5x+7)^{\frac{3}{2}}$ and $(5x+7)^{\frac{1}{2}}$:

$$(5x+7)^{\frac{3}{2}} = \left(z^2\right)^{\frac{3}{2}} = z^3$$

$$(5x+7)^{\frac{1}{2}} = \left(z^2\right)^{\frac{1}{2}} = z.$$

Making the substitution allows us to finish the integral:

Line 1 replaces $(5x+7)^{\frac{3}{2}}$ with z^3, replaces $(5x+7)^{\frac{1}{2}}$ with z, and replaces dx with $\frac{2}{5}z\,dz$; line 2 simplifies; line 3 is the antiderivative step; and line 4 returns to the original variable using the previously calculated $z = (5x+7)^{\frac{1}{2}}$.

$$\begin{aligned}\int \frac{1}{(5x+7)^{\frac{3}{2}}+(5x+7)^{\frac{1}{2}}}\,dx &= \int \frac{1}{z^3+z}\cdot\frac{2}{5}z\,dz\\ &= \frac{2}{5}\int \frac{1}{z^2+1}\,dx\\ &= \frac{2}{5}\tan^{-1} z + C\\ &= \frac{2}{5}\tan^{-1}\sqrt{5x+7}+C. \qquad \blacksquare\end{aligned}$$

Although partial fractions did not come up in examples 1 or 2, it should not be difficult to imagine the need for partial fractions arising when making a rationalizing substitution.

Rational trigonometric integrands: the substitution $\tan\frac{u}{2} = z$

A type of rationalizing substitution can be used for some expressions involving trigonometric functions. If the integrand is a composition of the form rational function of a trig function, the substitution

$$\tan\frac{u}{2} = z$$

can transform the integrand into a rational expression. The trick is knowing what to do with $\sin u$ and $\cos u$ in such a substitution. Using the half-angle identities $\sin^2\theta = \frac{1}{2}(1-\cos 2\theta)$ and $\cos^2\theta = \frac{1}{2}(1+\cos 2\theta)$ with $\theta = \frac{u}{2}$,

$$\tan^2\frac{u}{2} = \frac{\sin^2\frac{u}{2}}{\cos^2\frac{u}{2}} = \frac{\frac{1}{2}(1-\cos u)}{\frac{1}{2}(1+\cos u)} = \frac{1-\cos u}{1+\cos u}.$$

Applying the substitution formula, we get

$$z^2 = \frac{1-\cos u}{1+\cos u}.$$

Next we solve this equation for $\cos u$:

$$\begin{aligned}z^2(1+\cos u) &= 1-\cos u\\ z^2 + z^2\cos u &= 1-\cos u\\ (z^2+1)\cos u &= 1-z^2\\ \cos u &= \frac{1-z^2}{1+z^2}.\end{aligned}$$

Calculating $\sin^2 u$ is next on the list:

$$\begin{aligned}\sin^2 u &= 1 - \cos^2 u \\ &= 1 - \left(\frac{1-z^2}{1+z^2}\right)^2 \\ &= \frac{(1+z^2)^2}{(1+z^2)^2} - \frac{(1-z^2)^2}{(1+z^2)^2} \\ &= \frac{(1+z^2)^2 - (1-z^2)^2}{(1+z^2)^2} \\ &= \frac{1+2z^2+z^4-(1-2z^2+z^4)}{(1+z^2)^2} \\ &= \frac{4z^2}{(1+z^2)^2}.\end{aligned}$$

Line 1 is a Pythagorean identity, line 2 uses the previously calculated $\cos u = \frac{1-z^2}{1+z^2}$; line 3 rewrites the expression to get a common denominator, line 4 performs the subtraction with that common denominator, line 5 expands the numerator, and line 6 simplifies the numerator.

Taking the square roots,

$$\sin u = \frac{2z}{1+z^2}.$$

The reason for $2z$ and not $\pm 2z$ or $-2z$ is determined by ensuring the signs $(+)$ of $\sin u$ and $\tan u$ align correctly in the various quadrants.

Last, we need du:

$$\begin{aligned}\tan\frac{u}{2} &= z \\ \frac{u}{2} &= \tan^{-1} z \\ u &= 2\tan^{-1} z \\ du &= \frac{2}{1+z^2}\,dz.\end{aligned}$$

Line 1 is the substitution formula, line 2 rewrites using the definition of inverse tangent, line 3 finishes solving for u, and line 4 calculates du based on the formula for u.

Although we have written the intended substitution as $\tan\frac{u}{2} = z$, when making the substitution, we use the equivalent facts derived therefrom—namely,

$$\cos u = \frac{1-z^2}{1+z^2}, \qquad \sin u = \frac{2z}{1+z^2}, \qquad du = \frac{2}{1+z^2}\,dz.$$

For slightly differing derivations of these formulas, see GSL or the aforementioned text of Swokowski.

Example 3 *Evaluate* $\displaystyle\int \frac{1}{2+\cos x}\,dx.$

Solution The integrand contains a trig function, but not a product of powers of trig functions or a similar integrand, which we encountered

Let $f(x) = \frac{1}{2+x}$, which is a rational function, and $g(x) = \cos x$, which is a trig function. Then, the composition

$$f \circ g(x) = f(\cos x) = \frac{1}{2 + \cos x}$$

is a rational function of a trig function.

earlier. Instead, the integrand is a composition of the form rational function of a trig function. With $u = x$ in the just-derived formulas, we apply the substitution

$$\tan \frac{x}{2} = z,$$

along with

$$\cos x = \frac{1 - z^2}{1 + z^2}, \qquad \sin x = \frac{2z}{1 + z^2}, \qquad dx = \frac{2}{1 + z^2}\, dz.$$

Then,

Line 1 uses the substitution formulas, lines 2 and 3 simplify, line 4 uses the antiderivative giving the inverse tangent formula, and line 5 returns to the original variable using the substitution formula $z = \tan \frac{x}{2}$.

$$\begin{aligned}
\int \frac{1}{2 + \cos x}\, dx &= \int \frac{1}{2 + \frac{1 - z^2}{1 + z^2}} \cdot \frac{2}{1 + z^2}\, dz \\
&= \int \frac{2}{2(1 + z^2) + 1 - z^2}\, dz = \int \frac{2}{2 + 2z^2 + 1 - z^2}\, dz \\
&= 2 \int \frac{1}{z^2 + 3}\, dz \\
&= \frac{2}{\sqrt{3}} \tan^{-1} \frac{z}{\sqrt{3}} + C \\
&= \frac{2}{\sqrt{3}} \tan^{-1} \left(\frac{\tan \frac{x}{2}}{\sqrt{3}} \right) + C.
\end{aligned}$$

■

Although the solution to example 3 contains a trig function inside an inverse trig function, the expression cannot be simplified further because of the division by $\sqrt{3}$ inside the inverse trig function.

The next example combines substitution with a trig rationalizing substitution to demonstrate the usefulness of writing the latter formulas with the variable u.

Example 4 *Evaluate* $\displaystyle\int \frac{1}{5 + 4 \sin 2x}\, dx.$

Solution Even before noticing that the integrand is a composition of the form "rational function of a trig function," we should notice that we have a trig function of "not just plain x," indicating the need for (regular) substitution. Making the substitution

$$u = 2x$$
$$du = 2\, dx$$

gives

$$\int \frac{1}{5+4\sin 2x}\,dx = \frac{1}{2}\int \frac{1}{5+4\sin 2x}\cdot 2\,dx = \frac{1}{2}\int \frac{1}{5+4\sin u}\,du.$$

Now we are ready for the rationalizing substitution

$$\tan\frac{u}{2} = z,$$

along with

$$\cos u = \frac{1-z^2}{1+z^2}, \qquad \sin u = \frac{2z}{1+z^2}, \qquad du = \frac{2}{1+z^2}\,dz.$$

We have

$$\begin{aligned}\frac{1}{2}\int \frac{1}{5+4\sin u}\,du &= \frac{1}{2}\int \frac{1}{5+4\cdot\frac{2z}{1+z^2}}\cdot\frac{2}{1+z^2}\,dz \\ &= \frac{1}{2}\int \frac{2}{5(1+z^2)+8z}\,dz = \int \frac{1}{5+5z^2+8z}\,dz.\end{aligned}$$

Having obtained a rational integrand, we use previous methods. The denominator is a quadratic, so we try factoring. Finding no quick factorization, we check to determine whether the denominator is an irreducible quadratic. Because the discriminant of the quadratic formula gives

$$\sqrt{b^2-4ac} = \sqrt{8^2-4(5)(5)} = \sqrt{-36},$$

the denominator is irreducible. Seeing the middle term $8z$, we try completing the square:

$$\begin{aligned}&5z^2+8z \quad +5 \\ &= 5\left(z^2+\frac{8}{5}z\right)+5 \\ &= 5\left(z^2+\frac{8}{5}z+\left(\frac{4}{5}\right)^2\right)+5-\frac{16}{5} \\ &= 5\left(z+\frac{4}{5}\right)^2+\frac{9}{5}.\end{aligned}$$

Line 2 factors the coefficient 5 from the terms involving z, but not the constant term. Line 3 uses $2a = \frac{8}{5}$, resulting in $a^2 = \left(\frac{4}{5}\right)^2$, and because we added $5\cdot\left(\frac{4}{5}\right)^2 = 5\cdot\frac{16}{25} = \frac{16}{5}$, we must subtract that quantity as well. Line 4 factors and simplifies.

Using the result to rewrite the integral, we then use the antiderivative giving the inverse tangent formula and work our way back to the original variable:

Line 2 factors 5 from the denominator of the integrand; line 3 uses the antiderivative giving the inverse tangent formula with $a = \frac{3}{5}$; line 4 simplifies, including

$$\frac{z+\frac{4}{5}}{\frac{3}{5}} = \left(z+\frac{4}{5}\right)\cdot\frac{5}{3} = \frac{5\left(z+\frac{4}{5}\right)}{3} = \frac{5z+4}{3};$$

line 5 returns to the variable u using $z = \tan\frac{u}{2}$; and line 6 returns to the original variable x using $u = 2x$.

$$\begin{aligned}
\int \frac{1}{5z^2+8z+5}\,dz &= \int \frac{1}{5\left(z+\frac{4}{5}\right)^2+\frac{9}{5}}\,dz \\
&= \frac{1}{5}\int \frac{1}{\left(z+\frac{4}{5}\right)^2+\frac{9}{25}}\,dz \\
&= \frac{1}{5}\cdot\frac{1}{\frac{3}{5}}\tan^{-1}\frac{z+\frac{4}{5}}{\frac{3}{5}} + C \\
&= \frac{1}{3}\tan^{-1}\frac{5z+4}{3} + C \\
&= \frac{1}{3}\tan^{-1}\left(\frac{4+5\tan\frac{u}{2}}{3}\right) + C \\
&= \frac{1}{3}\tan^{-1}\left(\frac{4+5\tan x}{3}\right) + C. \qquad \blacksquare
\end{aligned}$$

The reciprocal substitution

Another substitution that is sometimes useful is the *reciprocal substitution*

$$x = \frac{1}{z}$$

$$dx = -\frac{1}{z^2}\,dz,$$

along with the following associated algebraic manipulations:

$$z = \sqrt{z^2} \text{ when } z > 0 \text{ and}$$

$$-z = \sqrt{z^2} \text{ when } z < 0.$$

In contrast to many other techniques of integration, there is no easily written rule that states when this substitution is helpful. Nevertheless, we look at its use in the next example.

Example 5 *Evaluate* $\displaystyle\int_1^3 \frac{\sqrt{9-x^2}}{x^4}\,dx.$

This integral can be evaluated using the trigonometric substitution $x = 3\sin\theta$, resulting in $\frac{1}{9}\int \frac{\cos^2\theta}{\sin^4\theta}\,d\theta$, which is rewritten using trig identities as $\frac{1}{9}\int \cot^2\theta\csc^2\theta\,d\theta$, which in turn is evaluated using the substitution $u = \cot\theta$. The featured solution shown here is shorter.

Solution We choose to try the reciprocal substitution

$$x = \frac{1}{z}$$

$$dx = -\frac{1}{z^2}\,dz.$$

We also need to change the limits of integration from values of x to values of z, which are just the reciprocals 1 and $\frac{1}{3}$. This gives

Line 1 makes the substitution, including changing the limits of integration to values of the variable z; line 2 simplifies.

$$\int_1^3 \frac{\sqrt{9-x^2}}{x^4}\,dx = \int_1^{\frac{1}{3}} \frac{\sqrt{9-\frac{1}{z^2}}}{\frac{1}{z^4}} \cdot \frac{-1}{z^2}\,dz$$

$$= \int_1^{\frac{1}{3}} \sqrt{9-\frac{1}{z^2}} \cdot \frac{z^4}{1} \cdot \frac{-1}{z^2}\,dz = -\int_1^{\frac{1}{3}} z^2\sqrt{9-\frac{1}{z^2}}\,dz.$$

The fraction inside the square root can be eliminated by making use of $z = \sqrt{z^2}$, which is valid because the values of z are positive (if we use negative values of z, then we need to use $-z = \sqrt{z^2}$). Using one z from z^2 for this purpose, we continue:

The values of z are positive because we are integrating from $z = 1$ to $z = \frac{1}{3}$.

Line 1 rewrites z^2 as $z \cdot z$ to facilitate line 2, which uses $z = \sqrt{z^2}$. Line 3 simplifies.

$$-\int_1^{\frac{1}{3}} z^2\sqrt{9-\frac{1}{z^2}}\,dz = -\int_1^{\frac{1}{3}} z \cdot z\sqrt{9-\frac{1}{z^2}}\,dz$$

$$= -\int_1^{\frac{1}{3}} z\sqrt{z^2\left(9-\frac{1}{z^2}\right)}\,dz$$

$$= -\int_1^{\frac{1}{3}} z\sqrt{9z^2-1}\,dz.$$

This integral is set up nicely for the substitution

In the original integrand, the substitution for the "not just plain x" under the square root ($u = 9 - x^2$) does not work. For the new integral, this type of substitution does work. This is what the reciprocal substitution accomplished for us.

$$u = 9z^2 - 1$$

$$du = 18z\,dz.$$

We need to calculate the new limits of integration:

$$z = \frac{1}{3}: \qquad u = 9\left(\frac{1}{9}\right) - 1 = 0$$

$$z = 1: \qquad u = 9 - 1 = 8.$$

Line 1 adjusts for the constant; line 2 makes the substitution, including switching the limits of integration to values of the variable u; and line 3 contains the antiderivative step.

Continuing,

$$-\int_1^{\frac{1}{3}} z\sqrt{9z^2-1}\,dz = -\frac{1}{18}\int_1^{\frac{1}{3}} 18z\sqrt{9z^2-1}\,dz$$

$$= -\frac{1}{18}\int_8^0 u^{\frac{1}{2}}\,du$$

$$= -\frac{1}{18}\frac{u^{\frac{3}{2}}}{\frac{3}{2}}\Bigg|_8^0$$

$$= 0 - \left(-\frac{1}{18}\frac{8^{\frac{3}{2}}}{\frac{3}{2}}\right) = \frac{8^{\frac{3}{2}}}{27}$$

$$= \frac{16\sqrt{2}}{27}. \qquad \blacksquare$$

For indefinite integrals, the cases $z > 0$ and $z < 0$ must be handled separately, making the reciprocal substitution a less-convenient choice than with definite integrals.

Although this section added a few integration techniques to our storehouse of knowledge, there remain many others that are not mentioned in this book. Modern calculus courses and textbooks also contain material not in older calculus textbooks. Inclusion is a matter of choice of material and determining which topics are generally considered to be of the greatest benefit. Such choices change as needs in other disciplines change, and as technology changes. If you peek at a calculus textbook 50 years from now, it may look quite different from the one you are reading today.

Exercises 7.7

1–32. Evaluate the integral.

1. $\displaystyle\int \frac{x}{(3+5x)^{\frac{3}{2}}}\,dx$

2. (GSL) $\displaystyle\int \frac{x^{\frac{3}{2}} - x^{\frac{1}{3}}}{6x^{\frac{1}{4}}}\,dx$

3. (GSL) $\displaystyle\int \frac{1}{x - x^{\frac{4}{3}}}\,dx$

4. $\displaystyle\int \frac{1}{1+\sqrt[3]{x-5}}\,dx$

5. (GSL) $\displaystyle\int \frac{x^2}{(4x+1)^{\frac{5}{2}}}\,dx$

6. $\displaystyle\int x(7+x)^{\frac{1}{3}}\,dx$

7. (GSL) $\displaystyle\int \frac{1}{1+\sin\theta+\cos\theta}\,d\theta$

8. (GSL) $\displaystyle\int \frac{1}{13-5\cos x}\,dx$

9. (GSL) $\displaystyle\int \frac{1}{5+4\cos x}\,dx$

10. (GSL) $\displaystyle\int \frac{1}{4+5\cos x}\,dx$

11. (GSL) $\displaystyle\int \frac{\sqrt{x+1}+1}{\sqrt{x+1}-1}\,dx$

12. (GSL) $\displaystyle\int \frac{1}{\sin x+\tan x}\,dx$

13. (GSL) $\displaystyle\int_0^{\frac{1}{2}} \frac{1}{(2t)^{\frac{1}{2}}\left(9+(2t)^{\frac{1}{3}}\right)}\,dt$

14. (GSL) $\displaystyle\int \frac{1}{(x-2)^{\frac{1}{2}}-(x-2)^{\frac{3}{4}}}\,dx$

15. (GSL) $\displaystyle\int \frac{1}{\cot\theta+\csc\theta}\,d\theta$

16. (GSL) $\displaystyle\int \frac{x}{(2x+3)^{\frac{4}{3}}}\,dx$

17. $\displaystyle\int \frac{x^2}{3+\cos x^3}\,dx$

18. (GSL) $\displaystyle\int_0^1 \frac{x^{\frac{3}{2}}}{x+1}\,dx$

19. (GSL) $\displaystyle\int_0^{\frac{\pi}{2}} \frac{1}{2+\sin x}\,dx$

20. $\displaystyle\int \frac{1}{2\sin 6x-\cos 6x+3}\,dx$

21. (GSL) $\displaystyle\int \frac{t+5}{(t+4)\sqrt{t+2}}\,dt$

22. (GSL) $\displaystyle\int_0^{\frac{\pi}{2}} \frac{1}{12+13\cos x}\,dx$

23. (GSL) $\int_1^{64} \frac{1}{2\sqrt{t}+\sqrt[3]{t}}\,dt$

24. (GSL) $\int_0^3 \frac{1}{(x+2)\sqrt{x+1}}\,dx$

25. (GSL) $\int \frac{1}{x+2\sqrt{x}+5}\,dx$

26. (GSL) $\int \frac{1}{x^{\frac{5}{8}}-x^{\frac{1}{8}}}\,dx$

27. (GSL) $\int \frac{5x+9}{(x-9)x^{\frac{3}{2}}}\,dx$

28. (GSL) $\int \frac{1}{4\sec x+5}\,dx$

29. (GSL) $\int \frac{\cos\theta}{5-3\cos\theta}\,d\theta$

30. (GSL) $\int \frac{\sqrt{x}}{x^3+2x^2-3x}\,dx$

31. (GSL) $\int \frac{1}{1+2\sin x}\,dx$

32. (GSL) $\int \frac{\sin\theta}{5+4\sin\theta}\,d\theta$

33–36. Use the substitution $x = \frac{1}{z}$ to evaluate the integral.

33. $\int_{\sqrt{2}}^{2} \frac{1}{x\sqrt{x^2-1}}\,dx$

34. $\int_{\frac{1}{2}}^{1} \frac{1}{x\sqrt{x^2+1}}\,dx$

35. $\int_{-\frac{1}{\sqrt{3}}}^{-\frac{1}{\sqrt{8}}} \frac{1}{x^2\sqrt{x^2+1}}\,dx$

36. $\int_{-\sqrt{\frac{28}{3}}}^{-\sqrt{\frac{63}{8}}} \frac{1}{x^2\sqrt{x^2-7}}\,dx$

37–38. Use the substitution $u = e^x$ to evaluate the integral.

37. $\int \frac{1}{e^x-1}\,dx$

38. (GSL) $\int_0^1 \frac{1}{e^x+e^{-x}}\,dx$

39. In section 7.2 we reversed the product rule to develop the integration-by-parts formula, which had the form

$$\int f(x)g'(x)\,dx = \text{expression} - \int \text{expression}.$$

Reverse the quotient rule to develop an integration formula of the form

$$\int \frac{f'(x)}{g(x)}\,dx = \text{expression} + \int \text{expression}.$$

40. Use the integration formula developed in exercise exercise 39 to evaluate the integral $\displaystyle\int \frac{\sin x}{e^x}\,dx$.

This integral can also be evaluated using integration by parts.

Strategy for Integration

7.8

An integral in the context of others like it is one thing. An integral in the wild is quite another. It is easy to become complacent about knowing how to approach an integral when we spend one day working only integration-by-parts integrals and another day working only trigonometric substitution integrals. When presented with an integral out of that daily context, how do we know which strategy to apply? How do we know the order in which to check strategies for usefulness? This is our task in this section.

Order for trial and error

Sometimes we look at an integral and immediately recognize what to try. If that attempt works, good! But, if it does not work, or if we do not immediately recognize an appropriate strategy, we need an order for evaluating various strategies—a checklist—that can help us determine the best strategy for approaching the integral. One such checklist follows:

Strategy for Integration Checklist

❶ Is there an antiderivative formula that directly applies (without substitution)? If so, try it.

❷ Do you recognize "not just plain x" to a power, under a square root, in an exponent, inside a trig function, inside a logarithm, inside a hyperbolic function, or inside …? If so, try a substitution.

❸ Does the integrand contain a transcendental function (such as $\ln x$ or $\tan^{-1} x$) and its derivative in a product? If so, try substitution with u as the transcendental function.

❹ Does the integrand contain a log, inverse trig, or inverse hyperbolic function? If so, try integrating by parts with f as the log, inverse trig, or inverse hyperbolic function.

❺ Does the integrand contain a product of a polynomial with trig, exponential, or hyperbolic functions? If so, try integrating by parts with f as the polynomial part. If the polynomial is degree 2 or greater, consider using tabular integration.

6. Is the integrand a product of exponential, trig, and/or hyperbolic functions? If so, try integrating by parts twice and solving for the integral.
7. Is the integrand a product (or quotient) of powers of sine and cosine? If so, follow the strategy for n or m odd, or the strategy for both n and m even, as appropriate.
8. Is the integrand a product (or quotient) of powers of tangent and secant? If so, try the associated strategy.
9. Is the integrand of the form $\sin mx \sin nx$ or a similar product? If so, try a product-to-sum identity.
10. Does the integrand contain trigonometric functions but the previous strategies do not apply? If so, try switching to all sines and cosines and simplifying, or try a trig identity.
11. Is the integrand an improper rational function? If so, use long division to rewrite as a polynomial plus a proper fraction.
12. Is the integrand a proper rational function for which the denominator can be factored? If so, try partial fractions.
13. Does the integrand contain an expression of the form a^2+x^2 to a power or—including if it is under a square root or to a fractional power—an expression of the form $a^2 - x^2$ or $x^2 - a^2$? If so, try trigonometric substitution.
14. Does the integrand contain a quadratic with a middle term mx? If so, try completing the square.
15. Try something else. For instance, try simplifying the integrand or rewriting it in a different form. Try a creative substitution, or try manipulating the integrand to make a substitution work (but remember that variables cannot be moved outside an integral).

Sometimes more than one technique must be used. After applying one strategy, another trip through the checklist might be necessary to integrate what remains. Also, if the integrand is a sum or difference of terms, the checklist can be applied separately to each term.

Applying the strategy

In the examples that follow, we don't try to finish an integration; the details of the various techniques of integration have already been covered. Instead, we focus on identifying an appropriate strategy.

For many of these integrals, we might recognize an appropriate approach immediately. If we do not, the solution to each example describes how to move through the checklist to determine an integration strategy. Occasionally the result of a strategy is given so the next step can be examined.

Example 1 *Evaluate* $\displaystyle\int \frac{\sin^3 x}{\cos x}\,dx$.

Solution ❶ No antiderivative formula applies directly.

❷ There is "not just plain x" to a power in the integrand. Trying $u = \sin x$, $du = \cos x\,dx$ requires a $\cos x$ in the numerator, which is not present. The substitution fails, at least for the moment.

Part of a successful strategy is knowing what not to do. Because there is no antiderivative power rule or antiderivative quotient rule, we cannot antidifferentiate the numerator and the denominator separately.

❸–❻ These options do not apply.

❼ The integrand contains a quotient of powers of sine and cosine, with n and m odd. Save $\sin x$ for du and try the substitution $u = \cos x$. This strategy succeeds. ■

When we say "this strategy succeeds," we mean that using the option in the checklist and carrying out the details correctly result in the antiderivative. We are not saying the evaluation of the integral has been completed.

Example 2 *Evaluate* $\displaystyle\int \frac{e^{\tan^{-1} x}}{1+x^2}\,dx$.

Solution ❶ No antiderivative formula applies directly.

❷ There is "not just plain x" in the exponent. Try $u = \tan^{-1} x$. This strategy succeeds. ■

Example 3 *Evaluate* $\displaystyle\int x^4 \ln x\,dx$.

Solution ❶ No antiderivative formula applies directly.

❷ There is no "not just plain x" for which substitution is required.

❸ Although $\ln x$ is present, its derivative $\frac{1}{x}$ is not present.

❹ Since $\ln x$ is in the integrand, we try integrating by parts with $f(x) = \ln x$. This strategy succeeds. ■

Example 4 *Evaluate* $\displaystyle\int \frac{x-1}{x^2-4x-5}\,dx$.

Solution ❶ No antiderivative formula applies directly.

❷ There is no "not just plain x" for which substitution is required.

❸– ❿ These options do not apply.

⓫ The integrand is a rational function, but it is proper.

⓬ The integrand is a proper rational function with a denominator that can be factored: $x^2 - 4x - 5 = (x-5)(x+1)$. Try partial fractions. This strategy succeeds. ■

Example 5 *Evaluate* $\int x \sin x^2\, dx$.

Solution ❶ No antiderivative formula applies directly.

❷ There is "not just plain x" inside a trig function. Try the substitution $u = x^2$, $du = 2x\, dx$. This strategy succeeds. ■

Example 6 *Evaluate* $\int x^2 \sin x\, dx$.

Compare the integrals in examples 5 and 6. Notice that checking for substitution comes before checking for integration by parts.

Solution ❶ No antiderivative formula applies directly.

❷ There is no "not just plain x" for which substitution is required.

❸–❹ These options do not apply.

❺ The integrand is the product of a polynomial (x^2) with a trig function ($\sin x$), so we try integrating by parts with $f(x) = x^2$. (Tabular integration is a good choice because the degree of the polynomial is 2.) This strategy succeeds. ■

Example 7 *Evaluate* $\int e^{x+e^x}\, dx$.

Solution ❶ No antiderivative formula applies directly.

❷ There is "not just plain x" in the exponent, so we try substitution with $u = x + e^x$, $du = (1 + e^x)\, dx$. Since du needs $1 + e^x$, which is not present in the integrand, this substitution fails.

❸– ⓮ These options do not apply.

⓯ Try something else. Perhaps we can try to simplify or rewrite the integrand in a different form. Using laws of exponents, we get

$$\int e^{x+e^x}\, dx = \int e^x e^{e^x}\, dx.$$

Start over. After rewriting the integrand, we go back through the strategy to see if anything applies.

❶ No antiderivative formula applies directly.

❷ There is "not just plain x" in the exponent of e^{e^x}, so we try substitution with $u = e^x$, $du = e^x\, dx$. This time the expression in du is present in the integrand. This strategy succeeds. ■

Example 8 *Evaluate* $\int \frac{x^2}{\sqrt{4 - x^2}}\, dx$.

Solution ❶ No antiderivative formula applies directly.

❷ There is "not just plain x" under the square root, so we try substitution with $u = 4 - x^2$, $du = -2x\,dx$. Since du requires an x and x^2 is present instead, this substitution fails.

Remember that du must always be calculated by taking the derivative of u. Making du whatever is "left over" is a great strategy for obtaining the wrong answer.

❸–❿ These options do not apply.

⓫ – ⓬ These options also do not apply. Because of the square root, the integrand is not a rational function.

The integrand is algebraic, but not rational.

⓭ The integrand contains an expression of the form $a^2 - x^2$ under a square root. Try trigonometric substitution. This strategy succeeds. ■

Example 9 *Evaluate* $\displaystyle\int \frac{x}{\sqrt{4-x^2}}\,dx.$

Solution ❶ No antiderivative formula applies directly.

❷ There is "not just plain x" under the square root, so we try substitution with $u = 4-x^2$, $du = -2x\,dx$. Unlike the previous example, this integrand contains the x needed for du. This strategy succeeds. ■

Example 10 *Evaluate* $\displaystyle\int \ln\sqrt{7}\,dx.$

Solution ❶ Because the integrand is a constant, we use the antiderivative of a constant rule. This strategy succeeds. ■

The solution is $(\ln\sqrt{7})x + C$.

Example 11 *Evaluate* $\displaystyle\int \ln\sqrt{x}\,dx.$

Solution ❶ No antiderivative formula applies directly.

❷ There is "not just plain x" inside a logarithm, so we try substitution with $u = \sqrt{x}$, $du = \frac{1}{2\sqrt{x}}\,dx$. Not seeing the $\sqrt{x}$ in a denominator to use with du, this substitution does not succeed.

❸ Although there is a natural log in the integrand, its derivative is not present in the integrand.

❹ The integrand contains a natural log, so we try integration by parts with $f(x) = \ln\sqrt{x}$. This strategy succeeds. ■

An alternative is to use laws of logarithms to simplify the integrand first. In this approach, integration by parts is still needed, with $f(x) = \ln x$.

Example 12 *Evaluate* $\displaystyle\int \sin 5x \cot 5x\,dx.$

Solution ❶ No antiderivative formula applies directly.

❷ There is "not just plain x" inside a trig function, so we try the substitution $u = 5x$, $du = 5\,dx$. This substitution can be carried out, but it does not solve the main problem. The integrand still contains $\sin u \cot u$.

❸–❺ These options do not apply.

Technically, integration by parts can be used on this integral, but the use of step ❿ is still required.

❻ This option also does not apply. The integrand is a product of two trig functions, which is not what is meant by option ❻.

❼–❽ These options do not apply.

❾ This option does not apply directly. The available product-to-sum identities are for working with products of sines and cosines.

❿ The integrand contains trigonometric functions and the previous strategies do not apply. We therefore try switching to all sines and cosines and simplifying. Following this strategy leads to

$$\int \cos 5x\, dx.$$

Having applied this option, we start over with the new integral:

❶ There is an antiderivative formula for $\cos kx$. This strategy succeeds. ■

Example 13 *Evaluate* $\int e^x \cos 5x\, dx.$

Solution ❶ No antiderivative formula applies directly.

❷ There is "not just plain x" inside a trig function, so we try the substitution $u = 5x$, $du = 5\, dx$. Carrying out this substitution results in

$$\frac{1}{5}\int e^{u/5} \cos u\, du,$$

and we still have "not just plain x" in an exponent. Substitution does not help.

❸–❺ These options do not apply.

❻ This integrand is a product of an exponential function and a trig function, so we integrate by parts twice and solve for the integral. This strategy succeeds. ■

Remember to use the trig part as f both times or use the exponential part as f both times; once each gets us nowhere.

Example 14 *Evaluate* $\int \frac{5}{9+x^2}\, dx.$

Trigonometric substitution also succeeds for this integral, but it is much more work. This is why antiderivative formulas are option ❶ in the checklist, whereas trig substitution is farther down the list, at option ⓭.

Solution ❶ The antiderivative giving the inverse tangent formula applies. This strategy succeeds. ■

Example 15 *Evaluate* $\int x^5 e^{x^3}\, dx.$

Solution ❶ No antiderivative formula applies directly.

❷ There is "not just plain x" in an exponent, so we try the substitution $u = x^3$, $du = 3x^2\,dx$. The result is the integral

Rewriting $x^5 = x^2x^3$ helps.

$$\frac{1}{3}\int ue^u\,du.$$

Having applied this option, we start over with the new integral.

❶ No antiderivative formula applies directly.

❷ There is no "not just plain u" for which substitution is required.

❸–❹ These options do not apply.

❺ The integrand contains the product of a polynomial (u) and an exponential (e^u), so we try integrating by parts, with $f(u) = u$, $g'(u) = e^u$. This strategy succeeds. ■

Example 16 *Evaluate* $\displaystyle\int \frac{x^2+1}{x^2+3x+2}\,dx.$

Solution ❶ No antiderivative formula applies directly.

❷ There is no "not just plain x" for which substitution is required.

❸– ❿ These options do not apply.

⓫ The integrand is an improper rational function, so we use long division. The result is

$$\int \left(1 + \frac{-3x-1}{x^2+3x+2}\right) dx.$$

Having applied this option, we start over with the new integral. The antiderivative sum rule applies and, although integrating the constant 1 is casy, we still need to integrate the other term. We apply the checklist to $\displaystyle\int \frac{-3x-1}{x^2+3x+2}\,dx.$

❶ No antiderivative formula applies directly.

❷ There is no "not just plain x" for which substitution is required.

❸– ⓫ These options do not apply.

⓬ The integrand is a proper rational function with a denominator that factors as $(x+1)(x+2)$. We apply partial fractions. This strategy succeeds. ■

Example 17 *Evaluate* $\displaystyle\int \cos 2x \cos 6x\,dx.$

Solution ❶ No antiderivative formula applies directly.

❷ There is ‘"not just plain x" inside a trig function. The problem is that there are two different such occurrences. Using either $u = 2x$,

$du = 2\,dx$ or $u = 6x$, $du = 6\,dx$ still leaves a trig function of "not just plain x," so substitution does not help.

❸–❺ These options do not apply.

Actually, the technique in option ❻ works, but it takes more effort than using option ❾.

❻ This option also does not apply. The integrand is a product of two trig functions, which is not what is meant by option ❻.

❼–❽ These options do not apply.

❾ The integrand is of the form $\cos mx \cos nx$, so we try a product-to-sum identity. This strategy succeeds. ■

Example 18 *Evaluate* $\displaystyle\int \frac{x}{x^4 + x^2 + 1}\,dx.$

Solution ❶ No antiderivative formula applies directly.

❷ There is no "not just plain x" for which substitution is required.

❸– ❿ These options do not apply.

⓫ The integrand is a proper rational function, so this step does not apply.

⓬ The integrand is a proper rational function. We try to factor the denominator, but this doesn't succeed.

Any polynomial of degree greater than 2 has a factorization into linear and quadratic factors. Actually finding these factors is sometimes very difficult or even impossible.

⓭ This option does not apply.

⓮ This option does not apply because the denominator has degree greater than 2.

⓯ Try something else. In this case a creative substitution helps. Noticing that the powers of x in the denominator are all even, we try the substitution $u = x^2$, $du = 2x\,dx$. The result is the integral

An expression that can be rewritten as a quadratic using a substitution is called *quadratic in form.*

$$\frac{1}{2}\int \frac{1}{u^2 + u + 1}\,du.$$

Having applied this option, we start over with the new integral.

❶ No antiderivative formula applies directly.

❷ There is no "not just plain u" for which substitution is required.

❸– ❿ These options do not apply.

⓫ The integrand is a proper rational function, so this option does not apply.

⓬ The integrand is a proper rational function. We try to factor the denominator, but determine that it is an irreducible quadratic.

⓭ This option does not apply.

⓮ The denominator is a quadratic with a middle term (a term of degree 1), so we try completing the square. This strategy succeeds. ■

Example 19 *Evaluate* $\int \tan^3 x \sec x\,dx$.

Solution ❶ No antiderivative formula applies directly.

❷ There is no "not just plain x" for which substitution is required.

❸–❺ These options do not apply.

❻ This option also does not apply. The integrand is a product of two trig functions, which is not what is meant by option ❻.

❼ This option does not apply.

❽ The integrand is a product of powers of tangent and secant. Trying the associated strategy succeeds. ■

Example 20 *Evaluate* $\int \sin\sqrt{x}\,dx$.

Solution ❶ No antiderivative formula applies directly.

❷ There is "not just plain x" inside a trig function, so we try the substitution $u = \sqrt{x}$, $du = \frac{1}{2\sqrt{x}}\,dx$. Not seeing the $\sqrt{x}$ in a denominator to use with du, this substitution does not succeed.

❸–❾ These options do not apply.

❿ The integrand contains trig functions and the previous strategies do not apply. Switching to all sines and cosines does not apply (we already have only a sine function), and no identity seems to apply.

⓫ – ⓮ These options do not apply.

⓯ Try something else. Nothing seems to apply but the aforementioned substitution, so perhaps we can manipulate the integrand to make the substitution work:

$$\int \sin\sqrt{x}\,dx = \int \frac{2\sqrt{x}}{2\sqrt{x}} \sin\sqrt{x}\,dx = \int 2u \sin u\,du.$$

Having applied this option, we start over with the new integral.

❶ No antiderivative formula applies directly.

❷ There is no "not just plain u" for which substitution is required.

❸–❹ These options do not apply.

❺ The integrand is the product of a polynomial ($2u$) with a trig function ($\sin u$), so we integrate by parts with $f(u) = 2u$, $g'(u) = \sin u$. This strategy succeeds. ■

Example 21 *Evaluate* $\int \frac{\sin^{-1} x}{\sqrt{1 - x^2}}\,dx$.

Solution ❶ No antiderivative formula applies directly.

❷ There is "not just plain x" under a square root, so we try $u = 1 - x^2$, $du = -2x\,dx$. The x required by the du is not present, so the substitution fails.

❸ The integrand contains a transcendental function, $\sin^{-1} x$, and its derivative, $\frac{1}{\sqrt{1-x^2}}$, in a product. We therefore try substitution with $u = \sin^{-1} x$. This strategy succeeds. ■

Example 22 *Evaluate* $\int \sin^{-1} x\,dx$.

Solution ❶ No antiderivative formula applies directly.

❷ There is no "not just plain x" for which substitution is required.

❸ There is a transcendental function, but its derivative is not present.

❹ The integrand contains an inverse trig function. We try integrating by parts with $f(x) = \sin^{-1} x$ (and $g'(x) = 1$). This strategy succeeds. ■

Sorting through every item in the checklist, one at a time, for every integral, is the stuff that computer algorithms are made of. It is not necessary for a human to work through the checklist when we see immediately the proper technique to use. Skill in recognizing form is gained from practice, until it seems that choosing an integration technique is more an intuitive art than a long, drawn-out process. Nevertheless, the checklist is a useful tool when we are confronted with an integral that defies immediate recognition, or when the technique we think will work does not.

As you practice, pay attention to which parts of the checklist you actually use. These are the ones that are your top priority to learn.

Exercises 7.8

Unlike the examples presented in this section, the evaluation of these integrals should be completed.

1–50. Evaluate the integral.

1. $\int \frac{\tan\sqrt{x}}{\sqrt{x}}\,dx$

2. $\int \frac{x^2 - 3x + 5}{2x + 3}\,dx$

3. $\int \sin 3x \sin 5x\,dx$

4. $\int \frac{2}{x^2 + 4x + 5}\,dx$

5. $\int \sin 3x \sec^2 3x\,dx$

6. $\int \frac{2}{x^2 - x - 6}\,dx$

7. $\int \sin^3 x \cos^5 x\, dx$

8. $\int \dfrac{2x^2}{x^2 - x - 6}\, dx$

9. $\int x^2 \sin 3x\, dx$

10. $\int \dfrac{x - 5}{x^2 + 6x + 9}\, dx$

11. $\int \sin^{-1} 3x\, dx$

12. $\int \dfrac{7}{\sqrt{9 - x^2}}\, dx$

13. $\int \dfrac{5}{\sqrt{4 - x}}\, dx$

14. $\int \dfrac{7x}{\sqrt{9 - x^2}}\, dx$

15. $\int \dfrac{5}{\sqrt{4 - x^2}}\, dx$

16. $\int \dfrac{7x^2}{\sqrt{9 - x^2}}\, dx$

17. $\int_0^1 \dfrac{5x}{\sqrt{4 - x^2}}\, dx$

18. $\int \dfrac{\sqrt{x^2 - 25}}{x^2}\, dx$

19. $\int \dfrac{5}{4 - x^2}\, dx$

20. $\int \sec(6x + 7)\, dx$

21. $\int \dfrac{5}{(4 - x^2)^{3/2}}\, dx$

22. $\int x \sec^2 x\, dx$

23. $\int \dfrac{\tan^{-1} x}{1 + x^2}\, dx$

24. $\int 5x^3 e^{(x^4 - 1)}\, dx$

25. $\int e^x \cosh 5x\, dx$

26. $\int 5x^3 e^x\, dx$

27. $\int \tan^3 x \sec^4 x\, dx$

28. $\int e^{3x} \cos x\, dx$

29. $\int \dfrac{x^2}{x^2 - 6x + 8}\, dx$

30. $\int \sin^3 x \cos^4 x\, dx$

31. $\int \dfrac{x}{x^2 - 6x + 9}\, dx$

32. $\int \sin^4 x \cos^2 x\, dx$

33. $\int \dfrac{1}{x^2 - 6x + 10}\, dx$

34. $\int \cos^3 x\, dx$

35. $\int x^3 \sin x^2\, dx$

36. $\int \sec^4 x \tan x\, dx$

37. $\int_0^{1/2} x e^{2x-1}\, dx$

38. $\int_1^2 \dfrac{\ln x}{x}\, dx$

39. $\int \cos x \ln \sin x\, dx$

40. $\int \dfrac{3}{x - 2}\, dx$

41. $\int \dfrac{\sinh x + e^x}{\cosh x + e^x}\, dx$

42. $\int_0^3 \dfrac{5}{x^2 + 9}\, dx$

43. $\int \cos^4 x\, dx$

44. $\int \tan^{-1} x\, dx$

45. $\int \frac{3}{x^3 + 25x}\, dx$

46. $\int x \ln x\, dx$

47. $\int \frac{x^3}{(x^2 - 4)^{3/2}}\, dx$

48. $\int \sin 4x \cos 2x\, dx$

49. $\int \frac{x}{x^4 + 1}\, dx$

50. $\int \frac{x^2}{\sqrt{1 - x^6}}\, dx$

51. (GSL) Evaluate $\int \left(e^{2x} - 2x\right)^2 dx$.

52. (GSL) Evaluate $\int \frac{1}{e^x - 4e^{-x}}\, dx$.

53. (GSL) Evaluate $\int \frac{\sin\theta}{(1 - \cos\theta)^3}\, d\theta$.

54. (Love) Evaluate $\int \frac{x^2 - 2}{1 + 6x - x^3}\, dx$.

55. (GSL) Evaluate $\int \ln(1 - \sqrt{x})\, dx$.

56. (GSL) Evaluate $\int \sqrt{\frac{\sin^{-1} x}{1 - x^2}}\, dx$.

Tables of Integrals and Use of Technology

7.9

After practicing various techniques of integration, you may now have an appreciation for the large difference between differentiation, in which the derivative of any elementary function can be found using formulas and a few relatively simple rules, and integration, in which a large collection of sophisticated techniques are insufficient to determine many of the antiderivatives we seek. It is natural, therefore, to seek the help of technology.

The fact that a process and its inverse process can have very different levels of difficulty, such as antidifferentiation having a greater level of difficulty than differentiation, is not always a disadvantage. Internet security and the commerce and convenience that it enables is not possible without similar mathematical situations. One public key cryptosystem in common use is based on the simplicity of multiplying prime numbers compared to the much greater difficulty of factoring the resulting number.

Using technology

For decades, from room-size computers to early microcomputers to laptops to smartphones and beyond, computer algebra systems (CASs) have been available to assist in mathematical computations. We don't even have to know a software package's picky syntax anymore, because these software packages have become more user friendly, and search engine capabilities have made the task so easy, that many antiderivatives can be found merely by asking your favorite assistant verbally. However, there are still advantages to using a CAS in typed form, and the relative ease of written precision compared to using verbal language may make those advantages persist.

How long will it take for this passage to seem dated? Next week? Next century? As I write this, it is impossible to know.

Another fact that may persist is that the algorithms implemented in software to compute antiderivatives are not always the same algorithms taught in calculus classes. For this reason, those of you using a CAS to check an answer may be confused by the result. One strategy for handling the discrepancy is based on the fact that any two antiderivatives of a function must differ by a constant. Take the answer given by the CAS, subtract your antiderivative, and determine whether the answer is a constant.

Example 1 *A student has computed the value of* $\int \sin^3 x \cos^5 x \, dx$ *as*

$$\frac{\sin^4 x}{4} - \frac{\sin^6 x}{3} + \frac{\sin^8 x}{8} + C.$$

The student asks a CAS to compute the answer and obtains

$$-\frac{3}{128}\cos 2x - \frac{1}{256}\cos 4x + \frac{1}{384}\cos 6x + \frac{1}{1024}\cos 8x + C.$$

Are the two answers equivalent?

CAUTION: MISTAKE TO AVOID
CASs do not always simplify results; you might have to ask them to do so. Even then, the answer may not be fully simplified. If the result is a constant, you may safely conclude the two results are equivalent; if the result does not look like a constant, you cannot necessarily conclude they are not equivalent.

Solution Ask the CAS to compute the difference between the two answers—that is, calculate

$$\begin{aligned} &-\frac{3}{128}\cos 2x - \frac{1}{256}\cos 4x + \frac{1}{384}\cos 6x + \frac{1}{1024}\cos 8x \\ &-\left(\frac{\sin^4 x}{4} - \frac{\sin^6 x}{3} + \frac{\sin^8 x}{8}\right) \\ &= -\frac{73}{3072}, \end{aligned}$$

which is a constant. The two answers are equivalent. ■

In example 1 the student's answer was correct. What does it look like if the student's answer is incorrect?

Example 2 *A student has computed the value of* $\int e^x \cosh 5x\,dx$ *as*

$$\frac{1}{24}e^x \cosh 5x - \frac{5}{24}e^x \sinh 5x + C.$$

The student asks a CAS to compute the answer and obtains

$$-\frac{1}{8}e^{-4x} + \frac{1}{12}e^{6x} + C.$$

Are the two answers equivalent?

An alternative is to compute the derivative of the student expression and subtract the integrand; the difference should be zero. The remainder of the process is the same.

Solution Ask the CAS to compute the difference between the two answers—that is, calculate

$$\begin{aligned} &-\frac{1}{8}e^{-4x} + \frac{1}{12}e^{6x} - \left(-\frac{1}{24}e^x \cosh 5x - \frac{5}{24}e^x \sinh 5x\right) \\ &= \frac{5}{24}e^{-4x}\left(-1 + e^{10x}\right). \end{aligned}$$

The result of $\frac{5}{24}e^{-4x}\left(-1 + e^{10x}\right)$ certainly appears not to be a constant, and the CAS does not simplify the expression further. One tactic to help verify that the result is not a constant is to graph it; see figure 1.

The graph of a constant function is a horizontal line, but this graph is definitely not a horizontal line, and we conclude that the difference between the two expressions is not a constant. The two expressions are therefore not equivalent. (It turns out that there is a sign error in the student expression.) ■

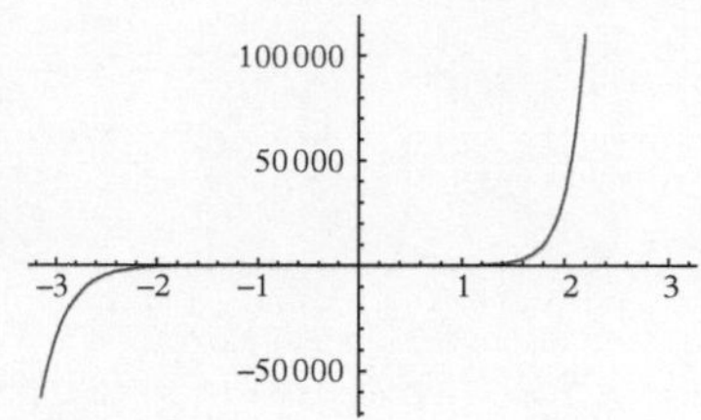

Figure 1 *The graph of* $\frac{5}{24}e^{-4x}\left(-1+e^{10x}\right)$*, produced by a CAS*

Important note: if the graph of the difference between the two expressions is depicted as random fluctuations of extremely small magnitude (sometimes called *noise*), then the two expressions are most likely equal (or equivalent).

Tables of integrals

What did people do before technology was readily available? Just as you may have compiled a list of antiderivative formulas to help you organize your material, during the development of calculus, much larger lists of antiderivative formulas were compiled to make finding antiderivatives more efficient. Entire books were published containing nothing but antiderivative formulas! Such a list is called a *table of integrals*. Tables of integrals have appeared in calculus texts for centuries, and although their need has passed, the tradition has not. Table 1, a table of integrals (which looks more like a list) appears on the next few pages, organized by type of function.

Using a formula from a table of integrals is no different than using any other antiderivative formula. Formulas in the table are written using the variable u; all other letters are assumed to be constants. The constants may be either positive or negative, unless specifically indicated otherwise. The majority of the formulas in the table of integrals are copied from GSL.

Example 3 *Use a formula from the table of integrals (see Table 1) to evaluate* $\int \sqrt{9+x^2}\,dx$*.*

Solution Because the integrand contains $9+x^2$ we look in the section of the table with the heading "Forms Containing a^2+u^2." There we find formula #89:

$$\int \sqrt{a^2+u^2}\,du = \frac{u}{2}\sqrt{a^2+u^2}+\frac{a^2}{2}\ln\left(u+\sqrt{a^2+u^2}\right)+C.$$

Using the formula with x in place of u and with $a=3$, we have

Using $u = x$ in a formula does not require substitution.

$$\int \sqrt{9+x^2}\,dx = \frac{x}{2}\sqrt{9+x^2}+\frac{9}{2}\ln\left(x+\sqrt{9+x^2}\right)+C.$$ ■

As evidenced by example 3, using a table of integrals is much easier than trigonometric substitution. Many other techniques of integration can also be avoided using a table of integrals, including some we do not study in this textbook. It may not be as powerful or as quick as a CAS, but in the precomputer age, a table of integrals served the same purpose.

Table 1 *A partial table of integrals, by type*

Trigonometric Forms

1. $\int \sin u \, du = -\cos u + C$
2. $\int \cos u \, du = \sin u + C$
3. $\int \tan u \, du = \ln|\sec u| + C$
4. $\int \csc u \, du = -\ln|\csc u + \cot u| + C$
5. $\int \sec u \, du = \ln|\sec u + \tan u| + C$
6. $\int \cot u \, du = -\ln|\csc u| + C$
7. $\int \sec^2 u \, du = \tan u + C$
8. $\int \sec u \tan u \, du = \sec u + C$
9. $\int \csc^2 u \, du = -\cot u + C$
10. $\int \csc u \cot u \, du = -\csc u + C$
11. $\int \sin^2 u \, du = \frac{1}{2}u - \frac{1}{4}\sin 2u + C$
12. $\int \cos^2 u \, du = \frac{1}{2}u + \frac{1}{4}\sin 2u + C$
13. $\int \sec^3 u \, du = \frac{1}{2}(\sec u \tan u + \ln|\sec u + \tan u|) + C$
14. $\int \csc^3 u \, du = -\frac{1}{2}(\csc u \cot u - \ln|\csc u - \cot u|) + C$
15. $\int \tan^2 u \, du = -u + \tan u + C$
16. $\int \cot^2 u \, du = -u - \cot u + C$
17. $\int u \sin u \, du = \sin u - u \cos u + C$
18. $\int u \cos u \, du = \cos u + u \sin u + C$
19. $\int e^{au} \sin nu \, du = \dfrac{e^{au}(a \sin nu - n \cos nu)}{a^2 + n^2} + C$
20. $\int e^{au} \cos nu \, du = \dfrac{e^{au}(n \sin nu + a \cos nu)}{a^2 + n^2} + C$

Inverse Trigonometric Forms

21. $\int \sin^{-1} u \, du = u \sin^{-1} u + \sqrt{1 - u^2} + C$
22. $\int \cos^{-1} u \, du = u \cos^{-1} u - \sqrt{1 - u^2} + C$
23. $\int \tan^{-1} u \, du = u \tan^{-1} u - \ln\sqrt{1 + u^2} + C$
24. $\int \cot^{-1} u \, du = u \cot^{-1} u + \ln\sqrt{1 + u^2} + C$
25. $\int \sec^{-1} u \, du = u \sec^{-1} u - \cosh^{-1} u + C$

Table 1 *Continued*

26. $\int \csc^{-1} u\,du = u\csc^{-1} u + \cosh^{-1} u + C$

27. $\int \left(\sin^{-1} u\right)^2 du = -2u + 2\sqrt{1-u^2}\sin^{-1} u + u\left(\sin^{-1} u\right)^2 + C$

28. $\int \left(\cos^{-1} u\right)^2 du = -2u - 2\sqrt{1-u^2}\cos^{-1} u + u\left(\cos^{-1} u\right)^2 + C$

29. $\int u\sin^{-1} u\,du = \frac{1}{4}u\sqrt{1-u^2} + \frac{1}{4}(2u^2-1)\sin^{-1} u + C$

30. $\int u\cos^{-1} u\,du = -\frac{1}{4}u\sqrt{1-u^2} + \frac{1}{2}u^2\cos^{-1} u + \frac{1}{4}\sin^{-1} u + C$

31. $\int u\tan^{-1} u\,du = -\frac{1}{2}u + \frac{1}{2}\tan^{-1} u + \frac{1}{2}u^2\tan^{-1} u + C$

Exponential and Logarithmic Forms

32. $\int e^{au}\,du = \frac{e^{au}}{a} + C$

33. $\int b^{au}\,du = \frac{b^{au}}{a\ln b} + C$

34. $\int ue^{au}\,du = \frac{au-1}{a^2}e^{au} + C$

35. $\int \ln u\,du = u\ln u - u + C$

36. $\int u^n \ln u\,du = u^{n+1}\left(\frac{\ln u}{n+1} - \frac{1}{(n+1)^2}\right) + C$

37. $\int \frac{1}{u\ln u}\,du = \ln|\ln u| + C$

Hyperbolic Forms

38. $\int \sinh u\,du = \cosh u + C$

39. $\int \cosh u\,du = \sinh u + C$

40. $\int \tanh u\,du = \ln\cosh u + C$

41. $\int \operatorname{csch} u\,du = \ln\left|\tanh\frac{1}{2}u\right| + C$

42. $\int \operatorname{sech} u\,du = \tan^{-1}|\sinh u| + C$

43. $\int \coth u\,du = \ln|\sinh u| + C$

44. $\int \operatorname{sech}^2 u\,du = \tanh u + C$

45. $\int \operatorname{sech} u\tanh u\,du = -\operatorname{sech} u + C$

46. $\int \operatorname{csch}^2 u\,du = -\coth u + C$

47. $\int \operatorname{csch} u\coth u\,du = -\operatorname{csch} u + C$

48. $\int \sinh^2 u\,du = -\frac{1}{2}u + \frac{1}{4}\sinh 2u + C$

49. $\int \cosh^2 u\,du = \frac{1}{2}u + \frac{1}{4}\sinh 2u + C$

50. $\int \tanh^2 u\,du = u - \tanh u + C$

51. $\int \coth^2 u\,du = u - \coth u + C$

52. $\int u\sinh u\,du = -\sinh u + u\cosh u + C$

53. $\int u\cosh u\,du = -\cosh u + u\sinh u + C$

54. $\int e^{au}\sinh nu\,du = \frac{e^{au}(a\sinh nu - n\cosh nu)}{a^2-n^2} + C$

55. $\int e^{au}\cosh nu\,du = \frac{e^{au}(a\cosh nu - n\sinh nu)}{a^2-n^2} + C$

continued

Table 1 *Continued*

Inverse Hyperbolic Forms

56. $\int \sinh^{-1} u\, du = u \sinh^{-1} u - \sqrt{1+u^2} + C$

57. $\int \cosh^{-1} u\, du = u \cosh^{-1} u - \sqrt{u^2-1} + C$

58. $\int \tanh^{-1} u\, du = u \tanh^{-1} u + \frac{1}{2} \ln\left|1-u^2\right| + C$

59. $\int \coth^{-1} u\, du = u \coth^{-1} u + \frac{1}{2} \ln\left|1-u^2\right| + C$

60. $\int \operatorname{sech}^{-1} u\, du = u \operatorname{sech}^{-1} u + \sin^{-1} u + C$

61. $\int \operatorname{csch}^{-1} u\, du = u \operatorname{csch}^{-1} u + \left|\sinh^{-1} u\right| + C$

62. $\int \left(\sinh^{-1} u\right)^2 du = 2u - 2\sqrt{1+u^2} \sinh^{-1} u + u\left(\sinh^{-1} u\right)^2 + C$

63. $\int \left(\cosh^{-1} u\right)^2 du = 2u - 2\sqrt{u^2-1} \cosh^{-1} u + u\left(\cosh^{-1} u\right)^2 + C$

64. $\int u \sinh^{-1} u\, du = -\frac{1}{4} u\sqrt{1+u^2} + \frac{1}{4} \sinh^{-1} u + \frac{1}{2} u^2 \sinh^{-1} u + C$

65. $\int u \cosh^{-1} u\, du = -\frac{1}{4} u\sqrt{u^2-1} + \frac{1}{2} u^2 \cosh^{-1} u - \frac{1}{4} \ln\left|u + \sqrt{u^2-1}\right| + C$

66. $\int u \tanh^{-1} u\, du = \frac{1}{2} u + \frac{1}{2} u^2 \tanh^{-1} u + \frac{1}{4} \ln|1-u| - \frac{1}{4} \ln|1+u| + C$

Elementary Rational Forms

67. $\int \frac{1}{u}\, du = \ln|u| + C$

68. $\int \frac{1}{1+u^2}\, du = \tan^{-1} u + C$

69. $\int \frac{1}{\sqrt{1-u^2}}\, du = \sin^{-1} u + C$

70. $\int \frac{1}{\sqrt{1+u^2}}\, du = \sinh^{-1} u + C$

71. $\int \frac{1}{\sqrt{u^2-1}}\, du = \cosh^{-1} u + C$

Rational Forms Containing $a + bu$

72. $\int (a+bu)^n\, du = \frac{(a+bu)^{n+1}}{b(n+1)} + \boldsymbol{C}\ (\boldsymbol{n} \neq -1)$

73. $\int \frac{1}{a+bu}\, du = \frac{1}{b} \ln|a+bu| + C$

74. $\int \frac{u}{a+bu}\, du = \frac{1}{b^2}\left(a + bu - a\ln|a+bu|\right) + C$

75. $\int \frac{u^2}{a+bu}\, du = \frac{1}{b^3}\left(\frac{1}{2}(a+bu)^2 - 2a(a+bu) + a^2 \ln|a+bu|\right) + C$

76. $\int \frac{u}{(a+bu)^2}\, du = \frac{1}{b^2}\left(\frac{a}{a+bu} + \ln|a+bu|\right) + C$

Table 2 *Continued*

77. $$\int \frac{u^2}{(a+bu)^2}\,du = \frac{1}{b^3}\left(a + bu - \frac{a^2}{a+bu} - 2a\ln|a+bu|\right) + C$$

78. $$\int \frac{u}{(a+bu)^3}\,du = \frac{1}{b^2}\left(-\frac{1}{a+bu} + \frac{a}{2(a+bu)^2}\right) + C$$

79. $$\int \frac{1}{u(a+bu)}\,du = -\frac{1}{a}\ln\left|\frac{a+bu}{u}\right| + C$$

80. $$\int \frac{1}{u^2(a+bu)}\,du = -\frac{1}{au} + \frac{b}{a^2}\ln\left|\frac{a+bu}{u}\right| + C$$

81. $$\int \frac{1}{u(a+bu)^2}\,du = \frac{1}{a(a+bu)} - \frac{1}{a^2}\ln\left|\frac{a+bu}{u}\right| + C$$

Forms Containing $\sqrt{a+bu}$

82. $$\int u\sqrt{a+bu}\,du = -\frac{2(2a-3bu)(a+bu)^{3/2}}{15b^2} + C$$

83. $$\int u^2\sqrt{a+bu}\,du = \frac{2(8a^2 - 12abu + 15b^2u^2)(a+bu)^{3/2}}{105b^3} + C$$

84. $$\int \frac{u}{\sqrt{a+bu}}\,du = -\frac{2(2a-bu)\sqrt{a+bu}}{3b^2} + C$$

85. $$\int \frac{u^2}{\sqrt{a+bu}}\,du = \frac{2(8a^2 - 4abu + 3b^2u^2)\sqrt{a+bu}}{15b^3} + C$$

86. $$\int \frac{1}{u\sqrt{a+bu}}\,du = \frac{1}{\sqrt{a}}\ln\left|\frac{\sqrt{a+bu}-\sqrt{a}}{\sqrt{a+bu}+\sqrt{a}}\right| + C \quad (\text{for } a > 0)$$

87. $$\int \frac{1}{u\sqrt{a+bu}}\,du = \frac{2}{\sqrt{-a}}\tan^{-1}\sqrt{\frac{a+bu}{-a}} + C \quad (\text{for } a < 0)$$

Forms Containing $a^2 + u^2$ ($a > 0$)

88. $$\int \frac{1}{a^2 + b^2u^2}\,du = \frac{1}{ab}\tan^{-1}\frac{bu}{a} + C$$

89. $$\int \sqrt{a^2+u^2}\,du = \frac{u}{2}\sqrt{a^2+u^2} + \frac{a^2}{2}\ln\left(u + \sqrt{a^2+u^2}\right) + C$$

90. $$\int u^2\sqrt{a^2+u^2}\,du = \frac{u}{8}(a^2 + 2u^2)\sqrt{a^2+u^2} - \frac{a^4}{8}\ln\left(u + \sqrt{a^2+u^2}\right) + C$$

91. $$\int \frac{1}{\sqrt{a^2+u^2}}\,du = \ln\left(u + \sqrt{a^2+u^2}\right) + C$$

92. $$\int \frac{1}{(a^2+u^2)^{3/2}}\,du = \frac{u}{a^2\sqrt{a^2+u^2}} + C$$

93. $$\int \frac{u^2}{\sqrt{a^2+u^2}}\,du = \frac{u}{2}\sqrt{a^2+u^2} - \frac{a^2}{2}\ln\left(u + \sqrt{a^2+u^2}\right) + C$$

continued

Table 1 *Continued*

94. $\displaystyle\int \frac{u^2}{(a^2+u^2)^{3/2}} = -\frac{u}{\sqrt{a^2+u^2}} + \ln\left(u+\sqrt{a^2+u^2}\right) + C$

95. $\displaystyle\int \frac{1}{u\sqrt{a^2+u^2}}\,du = -\frac{1}{a}\ln\left|\frac{a+\sqrt{a^2+u^2}}{u}\right| + C$

96. $\displaystyle\int \frac{1}{u^2\sqrt{a^2+u^2}}\,du = -\frac{\sqrt{a^2+u^2}}{a^2u} + C$

97. $\displaystyle\int \frac{1}{u^3\sqrt{a^2+u^2}}\,du = -\frac{\sqrt{a^2+u^2}}{2a^2u^2} + \frac{1}{2a^3}\ln\left|\frac{a+\sqrt{a^2+u^2}}{u}\right| + C$

98. $\displaystyle\int \frac{\sqrt{a^2+u^2}}{u}\,du = \sqrt{a^2+u^2} - a\ln\left|\frac{a+\sqrt{a^2+u^2}}{u}\right| + C$

99. $\displaystyle\int \frac{\sqrt{a^2+u^2}}{u^2}\,du = -\frac{\sqrt{a^2+u^2}}{u} + \ln\left(u+\sqrt{a^2+u^2}\right) + C$

Forms Containing $u^2 - a^2$ $(a > 0)$

100. $\displaystyle\int \sqrt{u^2-a^2}\,du = \frac{u}{2}\sqrt{u^2-a^2} - \frac{a^2}{2}\ln\left|u+\sqrt{u^2-a^2}\right| + C$

101. $\displaystyle\int \frac{1}{\sqrt{u^2-a^2}}\,du = \ln\left|u+\sqrt{u^2-a^2}\right| + C$

102. $\displaystyle\int \frac{1}{(u^2-a^2)^{3/2}}\,du = \frac{u}{-a^2\sqrt{u^2-a^2}} + C$

103. $\displaystyle\int \frac{u^2}{\sqrt{u^2-a^2}}\,du = \frac{u}{2}\sqrt{u^2-a^2} + \frac{a^2}{2}\ln\left|u+\sqrt{u^2-a^2}\right| + C$

104. $\displaystyle\int \frac{u^2}{(u^2-a^2)^{3/2}}\,du = -\frac{u}{\sqrt{u^2-a^2}} + \ln\left|u+\sqrt{u^2-a^2}\right| + C$

105. $\displaystyle\int \frac{1}{u\sqrt{u^2-a^2}}\,du = \frac{1}{a}\sec^{-1}\frac{u}{a} + C$

106. $\displaystyle\int \frac{1}{u^2\sqrt{u^2-a^2}}\,du = \frac{\sqrt{u^2-a^2}}{a^2u} + C$

107. $\displaystyle\int \frac{1}{u^3\sqrt{u^2-a^2}}\,du = \frac{\sqrt{u^2-a^2}}{2a^2u^2} + \frac{1}{2a^3}\sec^{-1}\frac{|u|}{a} + C$

108. $\displaystyle\int \frac{\sqrt{u^2-a^2}}{u}\,du = \sqrt{u^2-a^2} - a\sec^{-1}\frac{|u|}{a} + C$

109. $\displaystyle\int \frac{\sqrt{u^2-a^2}}{u^2}\,du = -\frac{\sqrt{u^2-a^2}}{u} + \ln\left|u+\sqrt{u^2-a^2}\right| + C$

Table 1 *Continued*

Forms Containing $a^2 - u^2$ $(a > 0)$

110. $\displaystyle\int \frac{1}{\sqrt{a^2-u^2}}\,du = \sin^{-1}\frac{u}{a} + C$

111. $\displaystyle\int \sqrt{a^2-u^2}\,du = \frac{u}{2}\sqrt{a^2-u^2} + \frac{a^2}{2}\sin^{-1}\frac{u}{a} + C$

112. $\displaystyle\int \frac{1}{(a^2-u^2)^{3/2}}\,du = \frac{u}{a^2\sqrt{a^2-u^2}} + C$

113. $\displaystyle\int \frac{u^2}{\sqrt{a^2-u^2}}\,du = -\frac{u}{2}\sqrt{a^2-u^2} + \frac{a^2}{2}\sin^{-1}\frac{u}{a} + C$

114. $\displaystyle\int \frac{u^2}{(a^2-u^2)^{3/2}}\,du = \frac{u}{\sqrt{a^2-u^2}} - \sin^{-1}\frac{u}{a} + C$

115. $\displaystyle\int \frac{1}{u\sqrt{a^2-u^2}}\,du = -\frac{1}{a}\ln\left|\frac{a+\sqrt{a^2-u^2}}{u}\right| + C$

116. $\displaystyle\int \frac{1}{u^2\sqrt{a^2-u^2}}\,du = -\frac{\sqrt{a^2-u^2}}{a^2u} + C$

117. $\displaystyle\int \frac{1}{u^3\sqrt{a^2-u^2}}\,du = -\frac{\sqrt{a^2-u^2}}{2a^2u^2} - \frac{1}{2a^3}\cosh^{-1}\frac{a}{u} + C$

118. $\displaystyle\int \frac{\sqrt{a^2-u^2}}{u}\,du = \sqrt{a^2-u^2} - a\cosh^{-1}\frac{a}{u} + C$

119. $\displaystyle\int \frac{\sqrt{a^2-u^2}}{u^2}\,du = -\frac{\sqrt{a^2-u^2}}{u} - \sin^{-1}\frac{u}{a} + C$

Forms Containing $a + bu \pm cu^2$ $(c > 0)$

120. $\displaystyle\int \frac{1}{a+bu+cu^2}\,du = \frac{2}{\sqrt{4ac-b^2}}\tan^{-1}\left(\frac{2cu+b}{\sqrt{4ac-b^2}}\right) + C \text{ (for } b^2 < 4ac)$

121. $\displaystyle\int \frac{1}{a+bu+cu^2}\,du = \frac{1}{\sqrt{b^2-4ac}}\ln\left|\frac{2cu+b-\sqrt{b^2-4ac}}{2cu+b+\sqrt{b^2-4ac}}\right| + C \text{ (for } b^2 > 4ac)$

122. $\displaystyle\int \frac{1}{a+bu-cu^2}\,du = \frac{1}{\sqrt{b^2+4ac}}\ln\left|\frac{\sqrt{b^2+4ac}+2cu-b}{\sqrt{b^2+4ac}-2cu+b}\right| + C$

123. $\displaystyle\int \sqrt{a+bu+cu^2}\,du = \frac{2cu+b}{4c}\sqrt{a+bu+cu^2} - \frac{b^2-4ac}{8c\sqrt{c}}\ln\left|2cu+b+2\sqrt{c}\sqrt{a+bu+cu^2}\right| + C$

124. $\displaystyle\int \sqrt{a+bu-cu^2}\,du = \frac{2cu-b}{4c}\sqrt{a+bu-cu^2} + \frac{b^2+4ac}{8c\sqrt{c}}\sin^{-1}\left(\frac{2cu-b}{\sqrt{b^2+4ac}}\right) + C$

continued

Table 1 *Continued*

Trigonometric Reduction Formulas

125. $\displaystyle\int \sin^n u\,du = -\frac{\sin^{n-1} u \cos u}{n} + \frac{n-1}{n}\int \sin^{n-2} u\,du$

126. $\displaystyle\int \cos^n u\,du = \frac{\cos^{n-1} u \sin u}{n} + \frac{n-1}{n}\int \cos^{n-2} u\,du$

127. $\displaystyle\int \tan^n u\,du = \frac{\tan^{n-1} u}{n-1} - \int \tan^{n-2} u\,du$

128. $\displaystyle\int \cot^n u\,du = -\frac{\cot^{n-1} u}{n-1} - \int \cot^{n-2} u\,du$

129. $\displaystyle\int \sec^n u\,du = \frac{\sin u}{(n-1)\cos^{n-1} u} + \frac{n-2}{n-1}\int \sec^{n-2} u\,du$

130. $\displaystyle\int \csc^n u\,du = -\frac{\cos u}{(n-1)\sin^{n-1} u} + \frac{n-2}{n-1}\int \csc^{n-2} u\,du$

131. $\displaystyle\int \cos^m u \sin^n u\,du = \frac{\cos^{m-1} u \sin^{n+1} u}{m+n} + \frac{m-1}{m+n}\int \cos^{m-2} u \sin^n u\,\boldsymbol{du}$

132. $\displaystyle\int \cos^m u \sin^n u\,du = -\frac{\sin^{n-1} u \cos^{m+1} u}{m+n} + \frac{n-1}{m+n}\int \cos^m u \sin^{n-2} u\,\boldsymbol{du}$

133. $\displaystyle\int \frac{\cos^m u}{\sin^n u}\,du = -\frac{\cos^{m+1} u}{(n-1)\sin^{n-1} u} - \frac{m-n+2}{n-1}\int \frac{\cos^m u}{\sin^{n-2} u}\,du$

134. $\displaystyle\int \frac{\cos^m u}{\sin^n u}\,du = \frac{\cos^{m-1} u}{(m-n)\sin^{n-1} u} + \frac{m-1}{m-n}\int \frac{\cos^{m-2} u}{\sin^n u}\,du$

135. $\displaystyle\int \frac{\sin^n u}{\cos^m u}\,du = \frac{\sin^{n+1} u}{(m-1)\cos^{m-1} u} - \frac{n-m+2}{m-1}\int \frac{\sin^n u}{\cos^{m-2} u}\,du$

136. $\displaystyle\int \frac{\sin^n u}{\cos^m u}\,du = -\frac{\sin^{n-1} u}{(n-m)\cos^{m-1} u} + \frac{n-1}{n-m}\int \frac{\sin^{n-2} u}{\cos^m u}\,du$

137. $\displaystyle\int e^{au} \cos^n u\,du = \frac{e^{au}\cos^{n-1} u(a\cos u + n\sin u)}{a^2+n^2} + \frac{n(n-1)}{a^2+n^2}\int e^{au}\cos^{n-2} u\,du$

138. $\displaystyle\int e^{au} \sin^n u\,du = \frac{e^{au}\sin^{n-1} u(a\sin u - n\cos u)}{a^2+n^2} + \frac{n(n-1)}{a^2+n^2}\int e^{au}\sin^{n-2} u\,du$

139. $\displaystyle\int u^m \cos au\,du = \frac{u^{m-1}}{a^2}(au\sin au + m\cos au) - \frac{m(m-1)}{a^2}\int u^{m-2}\cos au\,du$

140. $\displaystyle\int u^m \sin au\,du = \frac{u^{m-1}}{a^2}(m\sin au - au\cos au) - \frac{m(m-1)}{a^2}\int u^{m-2}\sin au\,du$

141. $\displaystyle\int \frac{1}{\cos^m u \sin^n u}\,du = \frac{1}{(m-1)\sin^{n-1} u\cos^{m-1} u} + \frac{m+n-2}{m-1}\int \frac{1}{\cos^{m-2} u\sin^n u}\,du$

142. $\displaystyle\int \frac{1}{\cos^m u \sin^n u}\,du = -\frac{1}{(n-1)\sin^{n-1} u\cos^{m-1} u} + \frac{m+n-2}{n-1}\int \frac{1}{\cos^m u\sin^{n-2} u}\,du$

Table 1 *Continued*

Miscellaneous Reduction Formulas
143. $\int u^n e^{au}\, du = \frac{u^n e^{au}}{a} - \frac{n}{a}\int u^{n-1} e^{au}\, du$
144. $\int u^m (\ln u)^n\, du = \frac{u^{m+1}}{m+1}(\ln u)^n - \frac{n}{m+1}\int u^m (\ln u)^{n-1}\, du$
145. $\int e^{au} \ln u\, du = \frac{e^{au}\ln u}{a} - \frac{1}{a}\int \frac{e^{au}}{u}\, du$
146. $\int u^m \sqrt{a+bu}\, du = \frac{2u^m (a+bu)^{3/2}}{b(2m+3)} - \frac{2am}{b(2m+3)}\int u^{m-1}\sqrt{a+bu}\, du$
147. $\int \frac{u^m}{\sqrt{a+bu}}\, du = \frac{2u^m\sqrt{a+bu}}{b(2m+1)} - \frac{2am}{b(2m+1)}\int \frac{u^{m-1}}{\sqrt{a+bu}}\, du$
148. $\int \frac{\sqrt{a+bu}}{u^m}\, du = -\frac{(a+bu)^{3/2}}{a(m-1)u^{m-1}} - \frac{b(2m-5)}{2a(m-1)}\int \frac{\sqrt{a+bu}}{u^{m-1}}\, du$
149. $\int (u^2+a^2)^{n/2}\, du = \frac{u(u^2+a^2)^{n/2}}{n+1} + \frac{na^2}{n+1}\int (u^2+a^2)^{(n/2)-1}\, du \; (n \neq -1)$
150. $\int (a^2-u^2)^{n/2}\, du = \frac{u(a^2-u^2)^{n/2}}{n+1} + \frac{a^2 n}{n+1}\int (a^2-u^2)^{(n/2)-1}\, du \; (n \neq -1)$
151. $\int (u^2-a^2)^{n/2}\, du = \frac{u(u^2-a^2)^{n/2}}{n+1} - \frac{na^2}{n+1}\int (u^2-a^2)^{(n/2)-1}\, du \; (n \neq -1)$

Example 4 *Use a table of integrals (see Table 1) to evaluate* $\int x^2\sqrt{7+3x}\, dx$.

Solution Looking at the table under the heading "Forms Containing $\sqrt{a+bu}$," we find formula #83:

$$\int u^2\sqrt{a+bu}\, du = \frac{2(8a^2 - 12abu + 15b^2u^2)(a+bu)^{3/2}}{105b^3} + C.$$

This fits our integrand, with $a = 7$ and $b = 3$. Therefore,

$$\begin{aligned}\int x^2\sqrt{7+3x}\, dx &= \frac{2(8(49) - 12(7)(3)x + 15(9)x^2)(7+3x)^{3/2}}{105(27)} + C\\ &= \frac{(784 - 504x + 270x^2)(7+3x)^{3/2}}{2835} + C. \quad \blacksquare\end{aligned}$$

Sometimes substitution is needed in conjunction with the table of integrals.

Example 5 *Use a table of integrals to evaluate* $\int x \sinh^{-1} 4x\, dx$.

Solution Searching under "Inverse Hyperbolic Forms," we find formula #64:

$$\int u \sinh^{-1} u\, du = -\tfrac{1}{4}u\sqrt{1+u^2} + \tfrac{1}{4}\sinh^{-1} u + \tfrac{1}{2}u^2 \sinh^{-1} u + C.$$

Alas, the formula is not for $\int u \sinh^{-1} au\, du$; if it was, we could use the formula as is and use $a = 4$. Instead, to fit the formula we have, we must use $u = 4x$—that is, we must make the substitution

$$u = 4x$$
$$du = 4\, dx.$$

Adjusting for the constant, making the substitution, and using formula #64 gives

Line 1 adjusts for the constant; line 2 makes the substitution, including replacing x with $\frac{u}{4}$ (hint: solve $u = 4x$ for x); line 3 simplifies; line 4 applies formula #64; line 5 goes back to the original variable; and line 6 simplifies.

$$\begin{aligned}
\int x \sinh^{-1} 4x\, dx &= \frac{1}{4}\int x \sinh^{-1} 4x \cdot 4\, dx \\
&= \frac{1}{4}\int \frac{u}{4} \sinh^{-1} u\, du \\
&= \frac{1}{16}\int u \sinh^{-1} u\, du \\
&= \frac{1}{16}\left(-\frac{1}{4}u\sqrt{1+u^2} + \frac{1}{4}\sinh^{-1} u + \frac{1}{2}u^2 \sinh^{-1} u\right) + C \\
&= \frac{1}{16}\left(-\frac{1}{4}(4x)\sqrt{1+16x^2} + \frac{1}{4}\sinh^{-1} 4x + \frac{1}{2}(16x^2)\sinh^{-1} 4x\right) + C \\
&= -\frac{1}{16}x\sqrt{1+16x^2} + \frac{1}{64}\sinh^{-1} 4x + \frac{1}{2}x^2 \sinh^{-1} 4x + C.
\end{aligned}$$ ■

Substitution was not necessary in examples 3 and 4 because we could use "just plain x" in place of u. In example 5, we needed substitution because $u = 4x$.

The next example demonstrates the use of a reduction formula.

Example 6 *Use a table of integrals to evaluate* $\int \tan^5 x\, dx$.

Solution Looking under "Trigonometric Forms," we do not find a formula for $\int \tan^5 x\, dx$. However, there is also a section titled

"Trigonometric Reduction Formulas," and there we find formula #127:

$$\int \tan^n u\,du = \frac{\tan^{n-1} u}{n-1} - \int \tan^{n-2} u\,du.$$

Applying the formula with $n = 5$ gives

$$\int \tan^5 x\,dx = \frac{\tan^4 x}{4} - \int \tan^3 x\,dx.$$

Although we still do not find a formula for $\int \tan^3 x\,dx$ in table 1, we did reduce the degree of the integrand (hence the name *reduction formula*). Applying formula #127 again, this time with $n = 3$, brings more progress:

$$= \frac{\tan^4 x}{4} - \left(\frac{\tan^2 x}{2} - \int \tan x\,dx\right)$$

$$= \frac{\tan^4 x}{4} - \frac{\tan^2 x}{2} + \int \tan x\,dx.$$

Now all that remains is an integral with a formula that is in the list, as #3. Applying formula #3 finishes the job and we conclude that

$$\int \tan^5 x\,dx = \frac{\tan^4 x}{4} - \frac{\tan^2 x}{2} + \ln|\sec x| + C. \quad \blacksquare$$

Exercises 7.9

1–30. (a) Use a table of integrals to evaluate the integral. (b) Use technology to evaluate the integral. (c) If the answers to parts (a) and (b) do not match, use technology to establish the equivalence of the two answers.

Because different technologies may give different antiderivatives, the answers to parts (b) and (c) are not provided at the back of the book.

1. $\int x^{17} \ln x\,dx$
2. $\int \frac{1}{\sqrt{25+x^2}}\,dx$
3. $\int \frac{x}{(4+9x)^2}\,dx$
4. $\int \frac{1}{x^2\sqrt{x^2-8}}\,dx$
5. $\int \frac{x^2}{(4-9x)^2}\,dx$
6. $\int 4e^{-2x} \cos 3x\,dx$
7. $\int 9x \sinh 3x\,dx$
8. $\int \frac{x}{(3+7x)^3}\,dx$

9. $\int e^{5x} \sin 7x \, dx$

10. $\int \sqrt{3 + 10x + x^2} \, dx$

11. $\int \frac{42}{x^3 \sqrt{x^2 - 9}} \, dx$

12. $\int \left(\sin^{-1} 5x\right)^2 \, dx$

13. $\int \frac{1}{2 + 5x - 3x^2} \, dx$

14. $\int \frac{4}{5x^2 \sqrt{100 - x^2}} \, dx$

15. $\int \frac{5x^2}{\sqrt{3 + 4x}} \, dx$

16. $\int 5x \tanh^{-1} x \, dx$

17. $\int \frac{5x^2}{\sqrt{3 + 4x^2}} \, dx$

18. $\int 74xe^{12x} \, dx$

19. $\int \sec^5 x \, dx$

20. $\int \sin^4 x \, dx$

21. $\int \frac{x^3}{\sqrt{1 + 3x}} \, dx$

22. $\int \frac{11}{\sqrt{1 + x^2}} \, dx$

23. $\int x^4 (\ln x)^3 \, dx$

24. $\int \cos^4 x \sin^2 x \, dx$

25. $\int 5x^4 \sin^{-1} x^5 \, dx$

26. $\int x^4 \sqrt{4 + 5x} \, dx$

27. $\int \frac{5x^4}{x^5 (3 + 7x^5)} \, dx$

28. $\int x^3 \coth^2 x^4 \, dx$

29. $\int \frac{6}{9 + 16x^2} \, dx$

30. $\int \frac{\cos x}{\sin x (3 + 8 \sin x)} \, dx$

31. Using the method of section 7.7 example 2, derive a table-of-integrals-style formula for $\int \frac{u}{(a + bu)^{\frac{3}{2}}} \, du$.

32. Use trigonometric substitution, trig identities, substitution, and integration by parts to evaluate the integral in formula #104 of the table of integrals. Then, use algebra (including rationalizing the denominator) and laws of logarithms to finish deriving the formula given in the table of integrals.

Type I Improper Integrals

7.10

All of our definite integrals and all of our applications of integration have always involved intervals of the form $[a, b]$, where a and b are real numbers. This is about to change.

Improper integrals, type I: integrating to infinity

Consider the region under the curve $f(x) = \frac{1}{x^2}$ to the right of $x = 1$, which is pictured in part in figure 1. What is the area of the region? The region is certainly infinitely long; it extends over the interval $[1, \infty)$. Is the area of the region also infinite?

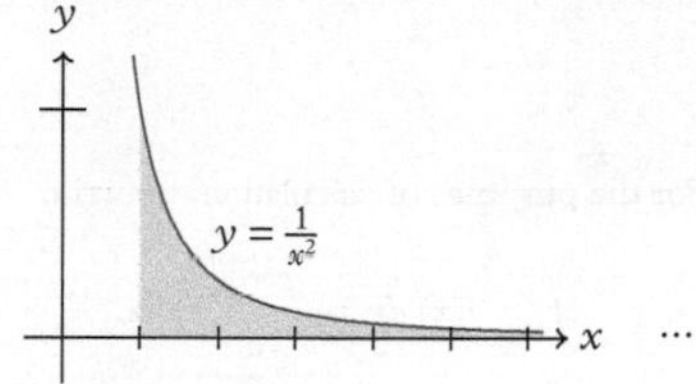

Figure 1 *A portion of the region under the curve $y = \frac{1}{x^2}$ to the right of $x = 1$. This margin is too small to contain the entire region*

The usual procedure for calculating the area under a curve is to integrate over the appropriate interval. Doesn't that make this area

$$\int_1^{\infty} \frac{1}{x^2}\, dx?$$

The problem is that we need to integrate to infinity. The fundamental theorem of calculus, part II (FTC II), which we have been using to evaluate definite integrals, does not apply directly to such an interval. Therefore, it is called an *improper integral of type I*.

There is another type of improper integral, hence the labeling type I and type II.

What do we need to do to evaluate this integral? We apply FTC II anyway, even though it doesn't apply directly. This is where the transfer principle comes to the rescue again. The integrand is continuous on $[1, \infty)$, and therefore the FTC applies on any interval of the form $[1, r]$ for a real number r larger than one. By the transfer principle, it therefore applies to intervals of the form $[1, \Omega]$ for any hyperreal larger than one. Then, for any positive infinite hyperreal Ω,

$$\begin{aligned}\int_1^{\Omega} \frac{1}{x^2}\, dx &= -\frac{1}{x}\bigg|_1^{\Omega} \\ &= -\frac{1}{\Omega} - (-1) = -\omega + 1 \\ &\approx 1.\end{aligned}$$

Because Ω is arbitrary (that is, because we obtain the same result for any positive infinite hyperreal Ω), it appears that the area of the region is 1 unit^2. Even though the region is infinitely long, the area of the region is finite!

The reasoning of the previous paragraphs leads us to the following definition.

Definition 1 IMPROPER INTEGRALS OF TYPE I *(1) Let f be continuous on the interval (a, ∞). If $\int_a^{\Omega} f(x)\,dx$ renders the same real result L for every positive infinite hyperreal Ω, then we write*

$$\int_a^{\infty} f(x)\,dx = L.$$

For the purposes of calculation, we write

$$\int_a^{\infty} f(x)\,dx = \int_a^{\Omega} f(x)\,dx.$$

(2) Let f be continuous on the interval $(-\infty, b)$. If $\int_{-\Omega}^{b} f(x)\,dx$ renders the same real result L for every positive infinite hyperreal Ω, then we write

$$\int_{-\infty}^{b} f(x)\,dx = L.$$

For the purposes of calculation, we write

$$\int_{-\infty}^{b} f(x)\,dx = \int_{-\Omega}^{b} f(x)\,dx.$$

If either (1) or (2) is satisfied, when L is a real number (i.e., not ∞ or $-\infty$), then we say the integral converges. *If such an integral does not converge, then we say that it* diverges.

Applying the definition to the integral $\int_1^{\infty} \frac{1}{x^2}\,dx$, from our earlier calculation we conclude

$$\int_1^{\infty} \frac{1}{x^2}\,dx = 1$$

and say that the integral converges.

Technically, we also need to expand the definition of area to include improper integrals. We do so implicitly by this remark rather than make it explicit.

Another region that is infinitely long but increasingly thin is conquered next.

Example 1 *Determine the area of the region under the curve $y = \frac{1}{x}$ to the right of $x = 1$.*

Solution The region is pictured in part in figure 2. We wish to calculate

$$\int_1^{\infty} \frac{1}{x}\,dx.$$

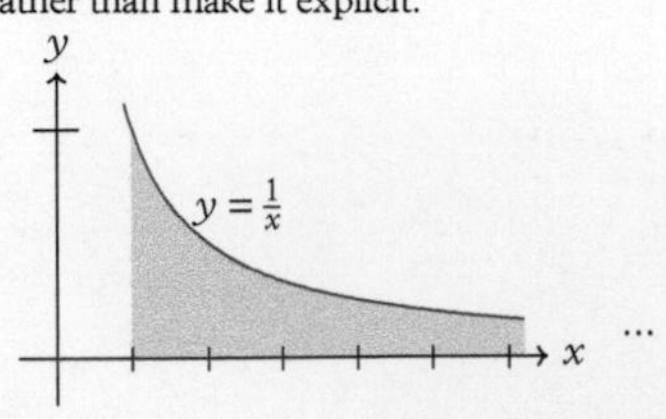

Figure 2 *A portion of the region under the curve $y = \frac{1}{x}$ to the right of $x = 1$. Compare this region to the region in figure 1; this curve is higher, so the area should be greater*

For any positive infinite hyperreal Ω,

$$\begin{aligned}\int_1^{\infty} \frac{1}{x}\,dx &= \int_1^{\Omega} \frac{1}{x}\,dx \\ &= \ln x\Big|_1^{\Omega} \\ &= \ln\Omega - \ln 1 = \ln\Omega \\ &\doteq \infty.\end{aligned}$$

Line 1 applies definition 1 and line 2 uses FTC II. Recalling the graph of $y = \ln x$ helps with line 4.

The area under the curve $y = \frac{1}{x}$ to the right of $x = 1$ is infinite. ■

Because the value of the integral is ∞, the integral diverges.

The curves in figures 1 and 2 are similar. Each function is positive, decreasing, and concave up on $[1, \infty)$, and each function has the horizontal asymptote $y = 0$. Yet, one has a finite area to the right of $x = 1$ whereas the other has an infinite area to the right of $x = 1$.

Example 2 *Evaluate* $\displaystyle\int_{-\infty}^{2} \frac{1}{x-3}\,dx.$

The integrand is not continuous at $x = 3$, but this number is not in the interval $(-\infty, 2)$. The conditions to apply definition 1 are met.

Solution For any positive infinite hyperreal Ω,

$$\begin{aligned}\int_{-\infty}^{2} \frac{1}{x-3}\,dx &= \int_{-\Omega}^{2} \frac{1}{x-3}\,dx \\ &= \ln|x-3|\Big|_{-\Omega}^{2} \\ &= \ln|-1| - \ln|-\Omega-3| \\ &\approx -\ln\Omega \\ &\doteq -\infty.\end{aligned}$$

Line 1 applies definition 1 and line 2 is the antiderivative step. In line 4, notice that $\ln|-1| = \ln 1 = 0$, that approximating inside the natural log is allowed, and that $|-\Omega| = \Omega$. Line 5 uses $\ln\Omega \doteq \infty$.

■

Because the value of the integral is $-\infty$, the interval diverges.

Reading Exercise 14 *Evaluate* $\displaystyle\int_1^{\infty} \frac{1}{x^3}\,dx.$

Substitution with type I improper integrals

The next example demonstrates the use of substitution in an improper integral.

Technical note: the procedure described before definition 1 relies on the transfer principle to make use of FTC II. Because it just requires an antiderivative, any valid antiderivative technique is allowed.

Example 3 *Evaluate* $\displaystyle\int_3^{\infty} xe^{-x^2}\,dx.$

Solution Recognizing an exponent of "not just plain x," we see the need for substitution. This time let's choose to determine an antiderivative first—that is, to calculate $\int xe^{-x^2}\,dx$. Using the substitution

By taking this approach, we do not have to keep track of the limits of integration while determining the antiderivative.

$$u = -x^2$$
$$du = -2x\,dx,$$

we have

Line 1 adjusts for the constant; line 2 makes the substitution; line 3 is the antiderivative step, and $+C$ is included because it is written as an indefinite integral; and line 4 returns to the original variable.

$$\begin{aligned}\int xe^{-x^2}\,dx &= -\frac{1}{2}\int -2xe^{-x^2}\,dx\\ &= -\frac{1}{2}\int e^u\,du\\ &= -\frac{1}{2}e^u + C\\ &= -\frac{1}{2}e^{-x^2} + C.\end{aligned}$$

Using this antiderivative result and definition 1,

Line 1 uses definition 1; line 2 is the antiderivative step, using the antiderivative calculated previously; line 3 uses FTC II; and line 4 uses the fact that e to a negative infinite power renders zero, which you might remember from the horizontal asymptote on the left side of the graph of $y = e^x$.

$$\begin{aligned}\int_3^{\infty} xe^{-x^2}\,dx &= \int_3^{\Omega} xe^{-x^2}\,dx\\ &= -\frac{1}{2}e^{-x^2}\Big|_3^{\Omega}\\ &= -\frac{1}{2}e^{-\Omega^2} - \left(-\frac{1}{2}e^{-9}\right)\\ &\doteq 0 + \frac{1}{2}e^{-9} = \frac{1}{2e^9}.\end{aligned}$$

■

An alternate way to write the solution to example 3 is to keep track of the limits of integration carefully when making the substitution. With

$$u = -x^2$$
$$du = -2x\,dx,$$

we calculate the new limits of integration as

CAUTION: MISTAKE TO AVOID
The new limits of integration should be calculated after using definition 1. We use the specific hyperreal Ω, which is a number, not the generic symbol ∞, which is not a number.

$$x = \Omega{:}\quad u = -\Omega^2$$
$$x = 3{:}\quad u = -9.$$

Then,

$$
\begin{aligned}
\int_3^{\infty} xe^{-x^2}\,dx &= \int_3^{\Omega} xe^{-x^2}\,dx \\
&= -\frac{1}{2}\int_3^{\Omega} -2xe^{-x^2}\,dx \\
&= -\frac{1}{2}\int_{-9}^{-\Omega^2} e^u\,du \\
&= -\frac{1}{2}e^u\Big|_{-9}^{-\Omega^2} \\
&= -\frac{1}{2}e^{-\Omega^2} - \left(-\frac{1}{2}e^{-9}\right) \\
&\doteq 0 + \frac{1}{2}e^{-9} = \frac{1}{2e^9}.
\end{aligned}
$$

Line 1 uses definition 1; line 2 adjusts for the constant; line 3 makes the substitution, including switching to the new limits of integration; line 4 is the antiderivative step; and lines 5 and 6 are identical to the last portion of the featured solution to example 3.

Ans. to reading exercise 14
$\frac{1}{2}$

Integrating from $-\infty$ to ∞

When a function is continuous on $(-\infty, \infty)$, can we integrate to infinity on both sides? The answer is yes, but we must be careful. The traditional procedure is to split the integral into two integrals, each of which must converge for the sum to converge.

Definition 2 IMPROPER INTEGRALS FROM $-\infty$ TO ∞ *Let f be continuous on $(-\infty, \infty)$ and let c be a real number. If both $\int_{-\infty}^{c} f(x)\,dx$ and $\int_{c}^{\infty} f(x)\,dx$ converge, then*

$$\int_{-\infty}^{\infty} f(x)\,dx = \int_{-\infty}^{c} f(x)\,dx + \int_{c}^{\infty} f(x)\,dx$$

and we say that the integral converges. Otherwise, the integral diverges.

In the event that $\int_{-\infty}^{\infty} f(x)\,dx$ converges, we obtain the same answer using any real number c. We can therefore choose whatever splitting point is most convenient.

Example 4 *Evaluate* $\int_{-\infty}^{\infty} \frac{1}{1+x^2}\,dx.$

Solution Let's begin by choosing zero as a convenient place to split the integrals:

Definition 2 is applied here.

$$\int_{-\infty}^{\infty} \frac{1}{1+x^2}\,dx = \int_{-\infty}^{0} \frac{1}{1+x^2}\,dx + \int_{0}^{\infty} \frac{1}{1+x^2}\,dx.$$

For the first integral,

Line 1 uses definition 1, line 2 is the antiderivative step, line 3 uses $\tan^{-1} 0 = 0$ and line 4 uses $\tan^{-1}(-\Omega) \doteq -\frac{\pi}{2}$.

$$\begin{aligned}\int_{-\infty}^{0} \frac{1}{1+x^2}\,dx &= \int_{-\Omega}^{0} \frac{1}{1+x^2}\,dx \\ &= \tan^{-1} x\Big|_{-\Omega}^{0} \\ &= \tan^{-1} 0 - \tan^{-1}(-\Omega) = -\tan^{-1}(-\Omega) \\ &\doteq -\left(-\frac{\pi}{2}\right) = \frac{\pi}{2}.\end{aligned}$$

For the integral on the right,

Line 1 uses definition 1, line 2 is the antiderivative step, line 3 uses $\tan^{-1} 0 = 0$ and line 4 uses $\tan^{-1} \Omega \doteq \frac{\pi}{2}$.

$$\begin{aligned}\int_{0}^{\infty} \frac{1}{1+x^2}\,dx &= \int_{0}^{\Omega} \frac{1}{1+x^2}\,dx \\ &= \tan^{-1} x\Big|_{0}^{\Omega} \\ &= \tan^{-1} \Omega - \tan^{-1} 0 = \tan^{-1} \Omega \\ &\doteq \frac{\pi}{2}.\end{aligned}$$

Adding the results,

This step is also from definition 2.

$$\int_{-\infty}^{\infty} \frac{1}{1+x^2}\,dx = \int_{-\infty}^{0} \frac{1}{1+x^2}\,dx + \int_{0}^{\infty} \frac{1}{1+x^2}\,dx = \frac{\pi}{2} + \frac{\pi}{2} = \pi. \quad ■$$

Why not just do the integral all at once? After all,

If $\int_{-\infty}^{\infty} f(x)\,dx$ converges, then $\int_{-\Omega}^{\Omega} f(x)\,dx$ always gives the correct answer. But, see the next margin comment.

$$\begin{aligned}\int_{-\Omega}^{\Omega} \frac{1}{1+x^2}\,dx &= \tan^{-1} x\Big|_{-\Omega}^{\Omega} = \tan^{-1} \Omega - \left(\tan^{-1}(-\Omega)\right) \\ &\doteq \frac{\pi}{2} - \left(-\frac{\pi}{2}\right) = \pi,\end{aligned}$$

which is the value we calculated in example 4. The problem is that it doesn't always work so nicely.

Consider

$$\int_{-\infty}^{\infty} x\,dx.$$

Our calculus definitions have always required that the same result be obtained no matter which hyperreals are used. What complicates matters here is that there are two places to use infinite hyperreals, and nothing says they must be the same. For instance, we could try

$$\int_{-\Omega}^{\Omega} x\,dx = \left.\frac{x^2}{2}\right|_{-\Omega}^{\Omega} = \frac{\Omega^2}{2} - \frac{(-\Omega)^2}{2} = 0$$

or

$$\int_{-\Omega}^{\Omega^2} x\,dx = \left.\frac{x^2}{2}\right|_{-\Omega}^{\Omega^2} = \frac{\left(\Omega^2\right)^2}{2} - \frac{(-\Omega)^2}{2} \approx \frac{\Omega^4}{2} \doteq \infty$$

or

$$\int_{-3\Omega^5}^{\Omega} x\,dx = \left.\frac{x^2}{2}\right|_{-3\Omega^5}^{\Omega} = \frac{\Omega^2}{2} - \frac{\left(-3\Omega^5\right)^2}{2} \approx -\frac{9\Omega^{10}}{2} \doteq -\infty.$$

These three results show that we do not render the same real result for every choice of infinite hyperreals, despite getting a value of zero for $\int_{-\Omega}^{\Omega} x\,dx$. We can only conclude that $\int_{-\infty}^{\infty} x\,dx$ does not exist—that is, the integral diverges.

If $\int_{-\infty}^{\infty} f(x)\,dx$ diverges, then the value given by $\int_{-\Omega}^{\Omega} f(x)\,dx$ is not the only value that can be calculated. Therefore, only calculating $\int_{-\Omega}^{\Omega} f(x)\,dx$ is not sufficient.

Reading Exercise 15 *Write* $\int_{-\infty}^{\infty} \frac{2}{x^2+4}\,dx$ *as a sum of two improper integrals (don't integrate, just do the first step of writing it as a sum of two integrals).*

Divergent but not to infinity

We do not require an infinite value to say that an integral diverges. The other possibility is that the calculation does not render the same real result for every positive infinite hyperreal Ω.

Example 5 *Evaluate* $\int_{0}^{\infty} \cos x\,dx$.

Solution For any positive infinite hyperreal Ω,

$$\int_{0}^{\Omega} \cos x\,dx = \left.\sin x\right|_{0}^{\Omega} = \sin\Omega - \sin 0 = \sin\Omega.$$

The sine wave varies from -1 to 1 with period 2π in the real numbers, and by the transfer principle it does the same in the hyperreal numbers. Every value in this interval is a possible value of $\sin \Omega$.

Unfortunately, the value of $\sin \Omega$ can be any real number from -1 to 1:

$$-1 \leq \sin \Omega \leq 1.$$

Because we do not render the same real result for every positive infinite hyperreal Ω, the integral value does not exist; the integral diverges. ■

p-Test for integrals

We have seen that $\int_1^\infty \frac{1}{x^2}\,dx$ converges whereas $\int_1^\infty \frac{1}{x}\,dx$ diverges. Theorem 1 summarizes results for similar functions.

Theorem 1 *p*-TEST FOR INTEGRALS *If $p > 1$, $\int_1^\infty \frac{1}{x^p}\,dx$ converges with value $\frac{1}{p-1}$. If $p \leq 1$, the integral diverges.*

The proof of the p-test is given at the end of this section. First let's look at its use with a few examples.

Example 6 *Does the interval converge or diverge? (a) $\int_1^\infty \frac{1}{x^2}\,dx$, (b) $\int_1^\infty \frac{1}{x}\,dx$, (c) $\int_1^\infty \frac{1}{x^{14}}\,dx$, (d) $\int_1^\infty \frac{1}{\sqrt{x}}\,dx$.*

Solution (a) By the p-test for integrals with $p = 2 > 1$, $\int_1^\infty \frac{1}{x^2}\,dx$ converges.

(b) By the p-test for integrals with $p = 1$, $\int_1^\infty \frac{1}{x}\,dx$ diverges.

(c) By the p-test for integrals with $p = 14 > 1$, $\int_1^\infty \frac{1}{x^{14}}\,dx$ converges.

(d) By the p-test for integrals with $p = \frac{1}{2} < 1$, $\int_1^\infty \frac{1}{\sqrt{x}}\,dx$ diverges. ■

The lower limit of integration does not matter, as long as it is positive. Consider the region under the curve $y = \frac{1}{x^p}$ between $x = 1$ and $x = 3$, and to the right of $x = 3$, as illustrated in figure 3. The infinitely long region B might have a finite area or it might have an infinite area. However, region A is a finite region and has a finite area, regardless of

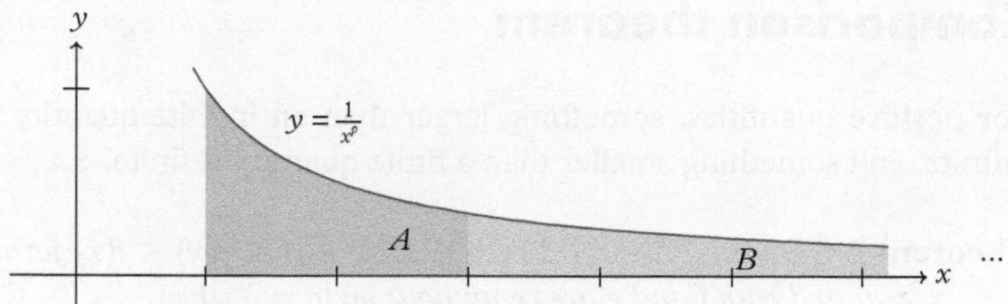

Figure 3 *The region A under the curve* $y = \dfrac{1}{x^p}$ *between* $x = 1$ *and* $x = 3$ *(darker green), and a portion of the region B under the curve to the right of* $x = 3$ *(light green)*

the value of p. Therefore if $\displaystyle\int_1^\infty \frac{1}{x^p}\,dx$ is infinite, so is $\displaystyle\int_3^\infty \frac{1}{x^p}\,dx$, and if $\displaystyle\int_1^\infty \frac{1}{x^p}\,dx$ is finite, so is $\displaystyle\int_3^\infty \frac{1}{x^p}\,dx$. In summary, the p-test can be applied to $\displaystyle\int_a^\infty \frac{1}{x^p}\,dx$ for any $a > 0$.

Example 7 *Does* $\displaystyle\int_{12}^\infty \frac{1}{x^{14}}\,dx$ *converge or diverge?*

Solution The lower limit of integration, 12, does not matter; the p-test still applies. Because $p = 14 > 1$, the integral converges. ■

Although the p-test applies to this integral for determining convergence or divergence of the integral, the p-test only gives the value of a convergent integral if the lower limit of integration is one.

Multiplying by a constant also does not affect the convergence or divergence of an integral. Multiply a finite area by 328 and it stays finite; multiply an infinite area by $\frac{1}{2918}$ and it is still infinite.

Example 8 *Does* $\displaystyle\int_{27}^\infty \frac{3}{\sqrt{x}}\,dx$ *converge or diverge?*

Solution Neither the lower limit of integration 27 nor the numerator 3 matter; the p-test still applies. Because $p = \frac{1}{2} < 1$, the integral diverges. ■

Reading Exercise 16 *Determine convergence or divergence of the integral using the p-test: (a)* $\displaystyle\int_1^\infty \frac{1}{x^4}\,dx$, *(b)* $\displaystyle\int_{12}^\infty \frac{1}{\sqrt[3]{x}}\,dx$.

Ans. to reading exercise 15 $\displaystyle\int_{-\infty}^0 \frac{2}{x^2+4}\,dx + \int_0^\infty \frac{2}{x^2+4}\,dx$ (Replacing zero with a different real number is also correct, as long as the same number is used in both places.)

Comparison theorem

For positive quantities, something larger than an infinite quantity is infinite, and something smaller than a finite quantity is finite.

Theorem 2 COMPARISON THEOREM *If* $0 \le g(x) \le f(x)$ *for all* $x \ge a$, *and both* f *and* g *are continuous on* $[a, \infty)$, *then*

(1) *if* $\int_a^\infty f(x)\,dx$ *converges, so does* $\int_a^\infty g(x)\,dx$, *and*

(2) *if* $\int_a^\infty g(x)\,dx$ *diverges, then so does* $\int_a^\infty f(x)\,dx$.

The comparison theorem is illustrated in figure 4. Suppose both $y = f(x)$ and $y = g(x)$ meet the conditions of the comparison theorem, so that f lies above g. If the region under the curve $y = f(x)$ has a finite area, then g must also, for it is smaller. If the region under the curve $y = g(x)$ has an infinite area, then f must also, for it is larger.

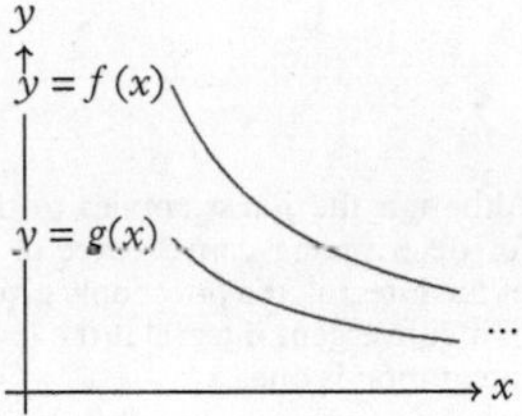

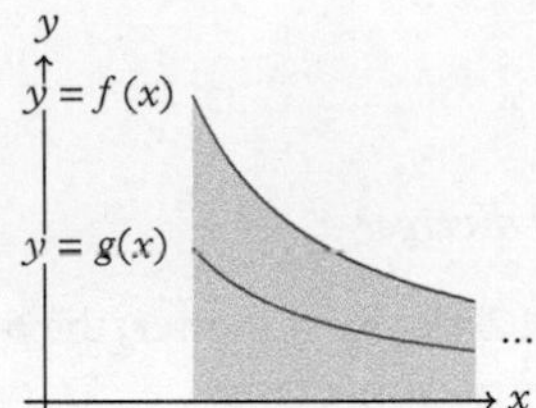

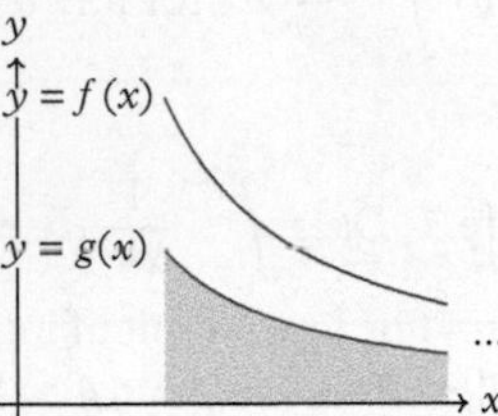

Figure 4 *(left) Two curves with* $0 \le g(x) \le f(x)$. *(middle) The two curves with the region under the curve* $y = f(x)$ *shaded. If the area shaded is finite, then so is the area under the curve* $y = g(x)$, *because it is a smaller quantity. (right) The two curves with the region under the curve* $y = g(x)$ *shaded. If the area shaded is infinite, then so is the area under the curve* $y = f(x)$, *because it is a larger quantity*

Before working an example, we need to recall how to compare fractions. Assuming the numerators are the same, the larger the denominator, the smaller the fraction:

$$\frac{3}{17} < \frac{3}{5}.$$

Example 9 *Determine the convergence or divergence of*

$$\int_1^\infty \frac{1}{x^4 + x + 3}\,dx.$$

Solution The strategy here is to compare the integral to one for which convergence or divergence is known. For instance, we know that

$$\int_1^\infty \frac{1}{x^4}\,dx$$

converges by the p-test for integrals, because $p = 4 > 1$.

In light of the p-test for integrals, comparisons to $\int_a^{\infty} \frac{1}{x^p}\,dx$ are convenient.

The numerators of the two integrands are the same; we need to compare the denominators. Because x is positive (we are only concerned about the interval $(1, \infty)$), $x + 3 > 0$. Thus, $x^4 + x + 3 > x^4$. Therefore,

The larger denominator makes for the smaller fraction.

$$\frac{1}{x^4 + x + 3} < \frac{1}{x^4}.$$

By the comparison theorem,

$$\int_1^{\infty} \frac{1}{x^4 + x + 3}\,dx$$

converges (it is smaller than a convergent integral). ■

Sometimes more than one comparison can be made, but not all are helpful. For the integral in example 9, notice that

$$\frac{1}{x^4 + x + 3} < \frac{1}{x},$$

since $x^4 + x + 3 > x$ (for positive x). But, $\int_1^{\infty} \frac{1}{x}\,dx$ diverges (p-test, $p = 1$), and being less than something infinite tells us nothing. Neither part (1) nor part (2) of the comparison theorem have been satisfied. We can draw no conclusion from this comparison.

CAUTION: MISTAKE TO AVOID
The comparison theorem draws a conclusion for only two cases out of four possible cases. Smaller than infinite could be either infinite or finite; larger than finite could be either infinite or finite. In these two cases, no conclusion can be drawn.

Example 10 *Determine the convergence or divergence of* $\int_3^{\infty} \frac{\ln x}{x}\,dx$.

Solution This time it is convenient to keep the same denominator and let the numerator change. We make the comparison

For this example it would not be too difficult to use substitution and evaluate the improper integral as we did in example 3. We choose to illustrate the use of the comparison test instead.

$$\frac{\ln x}{x} > \frac{1}{x}.$$

(The comparison works since $\ln x > 1$ when $x > 3$). But, $\int_3^{\infty} \frac{1}{x}\,dx$ diverges by the p-test (because $p = 1$). By the comparison theorem,

$$\int_3^{\infty} \frac{\ln x}{x}\,dx$$

diverges (it is larger than one that diverges). ■

Proof of the *p*-test for integrals

Ans. to reading exercise 16
(a) $p = 4 > 1$, the integral converges;
(b) $p = \frac{1}{3} < 1$, the integral diverges.

We wish to prove that $\int_1^\infty \frac{1}{x^p}\,dx$ converges if $p > 1$ and diverges if $p \le 1$.

Proof. For $p \neq 1$,

$$\int_1^\infty \frac{1}{x^p}\,dx = \int_1^\Omega \frac{1}{x^p}\,dx = \int_1^\Omega x^{-p}\,dx$$

$$= \left.\frac{x^{-p+1}}{-p+1}\right|_1^\Omega$$

$$= \frac{\Omega^{-p+1}}{-p+1} - \frac{1}{-p+1}.$$

For the case $p < 1$, the quantity $-p+1$ is positive. Therefore, Ω^{-p+1} is infinite, and we may continue as

$$\approx \frac{\Omega^{-p+1}}{-p+1} \doteq \infty,$$

and the integral diverges.

For the case $p > 1$, the quantity $-p+1$ is negative; $-(-p+1) = p-1$ is positive. Therefore,

$$\Omega^{-p+1} = \frac{1}{\Omega^{-(-p+1)}} = \frac{1}{\Omega^{p-1}} = \omega^{p-1}$$

is infinitesimal. We may therefore continue the integral calculation as

$$\approx -\frac{1}{-p+1} = \frac{1}{p-1},$$

which is finite, and the integral converges.

The case $p = 1$ was example 1. ■

Exercises 7.10

1–22. Evaluate the integral.

1. $\int_2^\infty \frac{7}{x^3}\,dx$

2. $\int_4^\infty \frac{3}{x^2}\,dx$

3. $\int_2^\infty \frac{1}{\sqrt{x}}\,dx$

4. $\int_0^\infty \frac{1}{16+x^2}\,dx$

5. $\int_{-\infty}^\infty \frac{1}{4x^2+4}\,dx$

6. $\int_{-\infty}^5 \frac{1}{x-7}\,dx$

7. $\int_0^\infty \frac{-5}{3x+1}\,dx$

8. $\int_{-\infty}^0 e^{3x}\,dx$

9. $\int_{-\infty}^3 \frac{1}{x^2+9}\,dx$

10. $\int_{-\infty}^\infty \frac{1}{1+9x^2}\,dx$

11. $\int_{-\infty}^{-5} \frac{1}{x^2-9}\,dx$

12. $\int_{-\infty}^\infty \frac{2x}{1+x^2}\,dx$

13. $\int_0^\infty 3x^2e^{-x^3}\,dx$

14. $\int_2^\infty \left(\frac{4}{x-1}+\frac{3}{x+2}\right)dx$

15. $\int_{-\infty}^\infty \frac{3}{x^2-10x+29}\,dx$

16. $\int_5^\infty \left(\frac{1}{x+4}-\frac{1}{x-3}\right)dx$

17. $\int_{-\infty}^\infty \frac{2x}{\sqrt{x^2+4}}\,dx$

18. $\int_0^\infty \sin 3x\,dx$

19. $\int_\pi^\infty \frac{\cos x}{2+\sin x}\,dx$

20. $\int_0^\infty \frac{2x+1}{x^2+7x+10}\,dx$

21. $\int_1^\infty \frac{\sin\frac{1}{x}}{x^2}\,dx$

22. $\int_0^\infty \operatorname{sech} x \tanh x\,dx$

23–34. Use theorem 1 or theorem 2 to determine whether the integral converges or diverges, or state that neither theorem applies.

23. $\int_1^\infty \frac{7}{x^3}\,dx$

24. $\int_{\sqrt{5}}^\infty \frac{2}{x^2}\,dx$

25. $\int_{71}^\infty \frac{2}{\sqrt[3]{x^4}}\,dx$

26. $\int_{\sqrt{5}}^\infty \frac{2}{x^2+3x}\,dx$

27. $\int_4^\infty \frac{1}{\sqrt[3]{x}}\,dx$

28. $\int_{\sqrt{5}}^\infty \frac{4}{x}\,dx$

29. $\int_1^\infty \frac{7}{x^3+x^2}\,dx$

30. $\int_{\sqrt{5}}^\infty \frac{2}{\sqrt{x}}\,dx$

31. $\displaystyle\int_{1000}^{\infty} \frac{1}{\sqrt[3]{x}-8}\,dx$

32. $\displaystyle\int_{\sqrt{5}}^{\infty} \frac{2}{\sqrt{x+1}}\,dx$

33. $\displaystyle\int_{1000}^{\infty} \frac{1}{\sqrt[3]{x}+8}\,dx$

34. $\displaystyle\int_{\sqrt{5}}^{\infty} \frac{2}{\sqrt{x-1}}\,dx$

35. Determine the area under the curve $y = \dfrac{2}{x^2}$ to the right of $x = 3$.

36. Determine the area under the curve $y = \dfrac{1}{x^{3/2}}$ to the right of $x = 4$.

37. The region under the curve $y = \dfrac{2}{x^2}$ to the right of $x = 3$ is rotated about the x-axis. Find the volume of the resulting solid of revolution.

38. The region under the curve $y = \dfrac{1}{x^{3/2}}$ to the right of $x = 4$ is rotated about the x-axis. Find the volume of the resulting solid of revolution.

39. Find the average value of $f(x) = \frac{1}{x}$ on $[1, \infty)$.

40. Find the average value of $f(x) = \ln x$ on $[1, \infty)$.

41. (GSL) Evaluate $\displaystyle\int_{1}^{\infty} \frac{1}{(1+x)^{3/2}}\,dx$.

42. (GSL) Evaluate $\displaystyle\int_{1}^{\infty} \frac{1}{x\sqrt{2x^2-1}}\,dx$.

43. (Love) Evaluate $\displaystyle\int_{-\infty}^{\infty} \frac{1}{(9+x^2)^{3/2}}\,dx$.

44. (Love) Evaluate $\displaystyle\int_{1}^{\infty} \frac{1}{x(x^2+1)}\,dx$.

45. If $\displaystyle\int_{-\infty}^{\infty} f(x)\,dx$ diverges but $\displaystyle\int_{-\Omega}^{\Omega} f(x)\,dx$ renders the same real result L for every positive infinite hyperreal Ω, then we call L the *Cauchy principal value* of the integral. (a) Calculate the Cauchy principal value of $\displaystyle\int_{-\infty}^{\infty} \sqrt[3]{x}\,dx$. (b) Calculate the Cauchy principal value of $\displaystyle\int_{-\infty}^{\infty} x\,dx$.

46. (a) Compute $\int_{-\infty}^{0} x\,dx$. (b) Compute $\int_{0}^{\infty} x\,dx$. (c) Add the results from (a) and (b). Is this result an indeterminate form? Use the answer to discuss $\int_{-\infty}^{\infty} x\,dx$.

47. An alternate definition for $\int_{-\infty}^{\infty} f(x)\,dx$ is as follows: Let f be continuous on the interval $(-\infty, \infty)$. If $\int_{-B}^{\Omega} f(x)\,dx$ renders the same real result for every positive infinite hyperreal Ω and every positive infinite hyperreal B, then we write $\int_{-\infty}^{\infty} f(x)dx = L$. Note that for purposes of computation, Ω could be on a higher level than B, or B could be on a higher level than Ω, or the two could be on the same level. All must be considered, if applicable, when approximating. (a) Use this definition to compute $\int_{-\infty}^{\infty} \frac{1}{1+x^2}\,dx$. (b) Use this definition to compute $\int_{-\infty}^{\infty} x\,dx$.

48. When using the classical definition of $\int_{a}^{\infty} f(x)\,dx$, the computation begins with

$$\int_{a}^{\infty} f(x)\,dx = \lim_{k\to\infty}\left(\int_{a}^{k} f(x)\,dx\right).$$

Reconcile this computation with definition 1.

Type II Improper Integrals

7.11

Sometimes a region that is infinitely long horizontally can have a finite area. The same is true of a region that is infinitely long vertically.

Improper integrals, type II: handling discontinuities

Consider the region under the curve $f(x) = \frac{1}{x^2}$ between $x = 0$ and $x = 1$, partially pictured in figure 1. The function f has an infinite discontinuity at $x = 0$, a fact that we can verify by taking a limit:

$$\lim_{x\to 0^+} \frac{1}{x^2} = \frac{1}{(0+\omega)^2} = \frac{1}{\omega^2} = \Omega^2 \doteq \infty.$$

What is the area of the region? Is it infinite?

The usual method of calculating the area under the curve is to evaluate the integral

$$\int_0^1 \frac{1}{x^2}\, dx.$$

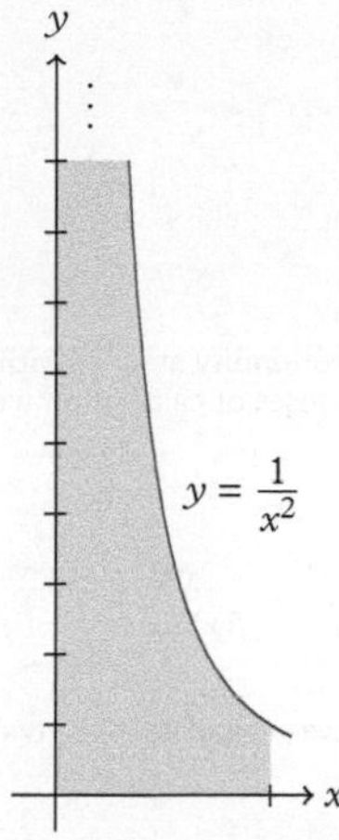

Figure 1 *A portion of the region under the curve $y = \frac{1}{x^2}$ between $x = 0$ and $x = 1$. The region extends infinitely upward along the vertical asymptote $x = 0$. Turn this page 90° counterclockwise and the region appears similar to those explored by improper integrals of type I*

Once again, we are confronted with an integral for which FTC II does not apply directly. FTC II requires that the integrand be continuous on $[a, b]$, which for this integral is $[0, 1]$, but we have a discontinuity at $x = 0$. However, if we move over just slightly to the right of the vertical asymptote, from there to $x = 1$, the function is continuous. By "just slightly," we could mean "infinitesimally." For any positive infinitesimal ω, the function is continuous on $[\omega, 1]$, and FTC II applies (invoking the transfer principle again, because ω is not a real number). Now we evaluate that integral:

$$\int_\omega^1 \frac{1}{x^2}\, dx = -\frac{1}{x}\bigg|_\omega^1 = -1 - \left(-\frac{1}{\omega}\right) = -1 + \Omega \approx \Omega \doteq \infty.$$

The area is infinite, and we say the integral diverges. Such an integral is called an *improper integral of type II* and, as usual, we require the same real result be rendered for every positive infinitesimal ω.

Definition 3 IMPROPER INTEGRALS OF TYPE II

With a discontinuity at $x = a$, for the purposes of calculation we write

$$\int_a^b f(x)\,dx = \int_{a+\omega}^b f(x)\,dx.$$

(1) *Let f be continuous on $(a, b]$ but not at $x = a$. If $\int_{a+\omega}^b f(x)\,dx$ renders the same real result L for every positive infinitesimal ω, then we write*

$$\int_a^b f(x)\,dx = L.$$

With a discontinuity at $x = b$, for the purposes of calculation we write

$$\int_a^b f(x)\,dx = \int_a^{b-\omega} f(x)\,dx.$$

(2) *Let f be continuous on $[a, b)$ but not at $x = b$. If $\int_a^{b-\omega} f(x)\,dx$ renders the same real result L for every positive infinitesimal ω, then we write*

$$\int_a^b f(x)\,dx = L.$$

If either (1) or (2) is satisfied when L is a real number (i.e., not ∞ or $-\infty$), we say the integral converges. *If such an integral does not converge, we say it* diverges.

With a discontinuity at $x = c$ with $a < c < b$, for the purposes of calculation we write

$$\int_a^b f(x)\,dx = \int_a^{c-\omega} f(x)\,dx + \int_{c+\omega}^b f(x)\,dx,$$

although we calculate the two integrals separately.

(3) *Let f be continuous on $[a, c)$ and $(c, b]$ but not at $x = c$. Then,*

$$\int_a^b f(x)\,dx = \int_a^c f(x)\,dx + \int_c^b f(x)\,dx,$$

provided these two integrals both converge.

The manner in which the integral must be rewritten can be understood visually. In each picture in figure 2, a function has a discontinuity depicted by a vertical asymptote. Because FTC II can only be applied on an interval where the function is continuous, the location of the discontinuity within the interval $[a, b]$ dictates where we can integrate.

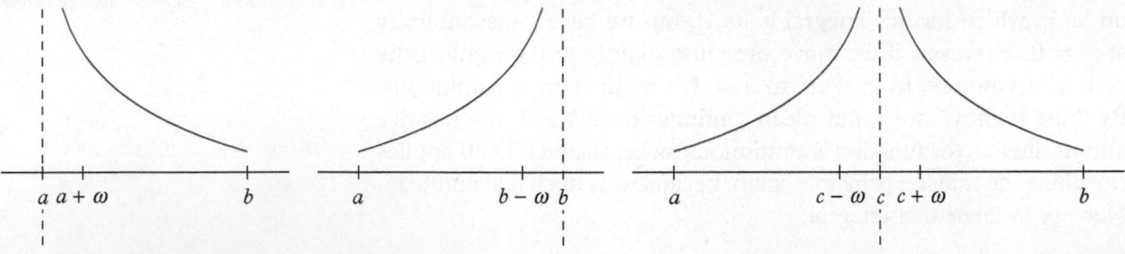

Figure 2 *(left) A function with a discontinuity at $x = a$, continuous on $[a+\omega, b]$; (middle) a function with a discontinuity at $x = b$, continuous on $[a, b - \omega]$; (right) a function with a discontinuity at $x = c$, continuous on $[a, c - \omega]$ and on $[c + \omega, b]$. In all three views, ω is a positive infinitesimal. The pictures are not to scale so that the intervals for rewriting the integral may be visualized*

For instance, with a discontinuity at $x = a$, we must move infinitesimally to the right of the discontinuity to avoid it, so we set the lower limit of integration to $a + \omega$. With a discontinuity at $x = b$, we must move infinitesimally to the left to avoid it, so we set the upper limit of integration to $b - \omega$. With a discontinuity in the middle, we must use two integrals, one ending just shy of the discontinuity and the other beginning just beyond it.

Example 1 *Do the first step. That is, rewrite the integral (if necessary) and do not integrate:* $\int_2^5 \frac{1}{2-x}\,dx.$

Solution Examining the integrand, we see that it has a discontinuity at $x = 2$. The interval is $[2, 5]$. Because the discontinuity is at the left endpoint, it matches the picture on the left in figure 2 and we rewrite the integral as

Setting $2 - x = 0$ and solving for x identifies the discontinuity.

$$\int_2^5 \frac{1}{2-x}\,dx = \int_{2+\omega}^5 \frac{1}{2-x}\,dx. \quad \blacksquare$$

Example 2 *Do the first step. That is, rewrite the integral (if necessary) and do not integrate:* $\int_2^5 \frac{1}{x^2-25}\,dx.$

Solution This time the integrand has two discontinuities, one at $x = 5$ and one at $x = -5$. The discontinuity at $x = -5$ is not in the interval $[2, 5]$, so it is irrelevant. The discontinuity at $x = 5$ is at the right endpoint of the interval, so it matches the picture in the middle of figure 2. We rewrite the integral as

$$\int_2^5 \frac{1}{x^2-25}\,dx = \int_2^{5-\omega} \frac{1}{x^2-25}\,dx. \quad \blacksquare$$

Example 3 *Do the first step. That is, rewrite the integral (if necessary) and do not integrate:* $\int_2^5 \frac{1}{(x-4)^2}\,dx.$

Solution This integrand has a discontinuity at $x = 4$ which is inside the interval $[2, 5]$, matching the picture on the right in figure 2. Accordingly, we rewrite the integral in two pieces,

$$\int_2^5 \frac{1}{(x-4)^2}\,dx = \int_2^{4-\omega} \frac{1}{(x-4)^2}\,dx + \int_{4+\omega}^5 \frac{1}{(x-4)^2}\,dx. \quad \blacksquare$$

Reading Exercise 17 *Do the first step. That is, rewrite the integral (if necessary) and do not integrate:* $\int_1^4 \frac{2}{x-2}\,dx.$

Type II improper integrals: examples

Example 4 *Evaluate* $\int_{-7}^{-2} \frac{1}{x+2}\,dx.$

CAUTION: MISTAKE TO AVOID
Before applying FTC II, the integrand must always be checked for discontinuities.

Solution Checking the integrand for discontinuities, we identify a discontinuity at $x = -2$. The discontinuity is located at the right endpoint of the interval $[-7, -2]$, meaning we have an improper integral of type II. Integrating from $x = -7$ to $x = -2 - \omega$,

Line 1 rewrites using definition 3 (see also figure 2, middle), line 2 is the antiderivative step, line 3 uses FTC II, and line 4 simplifies. For line 5, recall that the natural log of a positive infinitesimal renders $-\infty$.

$$\begin{aligned}
\int_{-7}^{-2} \frac{1}{x+2}\,dx &= \int_{-7}^{-2-\omega} \frac{1}{x+2}\,dx \\
&= \ln|x+2|\Big|_{-7}^{-2-\omega} \\
&= \ln|-2-\omega+2| - \ln|-7+2| \\
&= \ln\omega - \ln 5 \\
&\doteq -\infty.
\end{aligned}$$

This integral diverges. ■

Example 5 *Evaluate* $\int_0^5 \frac{1}{x+3}\,dx.$

Solution Checking the integrand for discontinuities, we identify a discontinuity at $x = -3$. This discontinuity is not in the interval $[0, 5]$, so the integral is not improper. We may proceed as usual:

$$\int_0^5 \frac{1}{x+3}\,dx = \ln|x+3|\Big|_0^5 = \ln 8 - \ln 3 = \ln\frac{8}{3}. \quad \blacksquare$$

Substitution and other integration techniques can be used for improper integrals.

Example 6 *Evaluate* $\int_1^2 \frac{4}{\sqrt{x-1}}\,dx.$

Solution Noticing both a square root and a denominator, we check the domain of the integrand. We need both $\sqrt{x-1} \neq 0$ and $x - 1 \geq 0$. The domain is therefore $x > 1$, or $(1, \infty)$. The function has a discontinuity at $x = 1$, which is at the left endpoint of the interval $[1, 2]$. We begin by rewriting the integral as in example 1:

In addition to identifying any discontinuities, we check the domain to ensure the function is defined (and continuous) on the rest of the interval.

$$\int_1^2 \frac{4}{\sqrt{x-1}}\, dx = \int_{1+\omega}^2 \frac{4}{\sqrt{x-1}}\, dx.$$

The antiderivative step is next. Noticing "not just plain x" inside the square root, we use substitution:

$$u = x - 1$$
$$du = dx.$$

Before making the substitution we also calculate the new limits of integration:

$$x = 2: \quad u = 2 - 1 = 1$$
$$x = 1 + \omega: \quad u = 1 + \omega - 1 = \omega.$$

Making the substitution and continuing to evaluate gives

$$\int_{1+\omega}^2 \frac{4}{\sqrt{x-1}}\, dx = \int_\omega^1 \frac{4}{\sqrt{u}}\, du = \left.\frac{4u^{1/2}}{\frac{1}{2}}\right|_\omega^1 = \left.8\sqrt{u}\right|_\omega^1 = 8 - 8\sqrt{\omega} \approx 8.$$

Ans. to reading exercise 17
$\int_1^{2-\omega} \frac{2}{x-2}\, dx + \int_{2+\omega}^4 \frac{2}{x-2}\, dx$

This integral converges. ■

At the beginning of this section, we evaluated an improper integral of type II with integrand $\frac{1}{x^2}$ that diverged. In example 6, we evaluated an improper integral of type II with integrand $\frac{4}{\sqrt{u}}$ that converged. Note that this is the opposite of our conclusions about improper integrals of type I from the p-test. The p-test for integrals applies only to improper integrals of type I, not of type II.

CAUTION: MISTAKE TO AVOID
The p-test for integrals does not apply to type II improper integrals.

Example 7 *Evaluate* $\int_0^1 \frac{2}{3-4x}\, dx.$

Solution We begin by locating the discontinuity: $x = \frac{3}{4}$. Because $\frac{3}{4}$ is inside the interval $[0, 1]$, the integral is improper and we rewrite as

$$\int_0^1 \frac{2}{3-4x}\, dx = \int_0^{\frac{3}{4}-\omega} \frac{2}{3-4x}\, dx + \int_{\frac{3}{4}+\omega}^1 \frac{2}{3-4x}\, dx.$$

CAUTION: MISTAKE TO AVOID
Whenever two integrals are required to calculate an improper integral, the two integrals must be evaluated separately. Both must converge for the improper integral to converge.

Recall that when an improper integral of type I requires two integrals, we evaluate each integral separately. If either integral diverges, then the entire improper integral diverges and we are finished. The same is true for improper integrals of type II, and for the same reasons. If we combine the arithmetic, we may cancel infinite terms incorrectly and obtain an incorrect answer.

Choosing $\int_0^{\frac{3}{4}-\omega} \frac{2}{3-4x}\, dx$ first, we notice that the denominator contains $-4x$ and substitution is required:

$$u = 3 - 4x$$
$$du = -4\, dx.$$

The new limits of integration are

$$x = \tfrac{3}{4} - \omega: \quad u = 3 - 4(\tfrac{3}{4} - \omega) = 4\omega$$
$$x = 0: \quad u = 3.$$

Making the substitution and continuing to evaluate,

Line 1 adjusts for the constant; line 2 makes the substitution, including the new limits of integration; line 3 is the antiderivative step; line 4 uses FTC II; and line 5 uses the fact that the natural log of a positive infinitesimal renders $-\infty$.

$$\begin{aligned}
\int_0^{\frac{3}{4}-\omega} \frac{2}{3-4x}\, dx &= -\frac{1}{4}\int_0^{\frac{3}{4}-\omega} \frac{2}{3-4x}(-4)\, dx \\
&= -\frac{1}{4}\int_3^{4\omega} \frac{2}{u}\, du \\
&= -\tfrac{1}{2} \ln|u| \Big|_3^{4\omega} \\
&= -\tfrac{1}{2}\ln 4\omega - (-\tfrac{1}{2}\ln 3) \\
&\doteq -(-\infty) = \infty.
\end{aligned}$$

Technically, we cannot do arithmetic on the symbol ∞, but it is convenient to keep track of $\pm$ signs in the manner demonstrated in line 5.

Because this integral diverges, we are finished. There is no need to evaluate the second integral. The integral $\int_0^1 \frac{2}{3-4x}\, dx$ diverges. ■

Reading Exercise 18 *Evaluate* $\int_1^2 \frac{4}{x-1}\, dx.$

Divergent but not to infinity

As with improper integrals of type I, we do not require an infinite result for an improper integral of type II to diverge. If the calculation does not render the same real result for every positive infinitesimal ω, the integral diverges.

Example 8 *Evaluate* $\displaystyle\int_0^1 \frac{\cos\frac{1}{x}}{-x^2}\,dx.$

Solution Checking the integrand for discontinuities, we identify a discontinuity at $x = 0$. The discontinuity is at the left end of the interval $[0, 1]$, so we rewrite the integral as

Notice that checking for discontinuities is the first thing to do, even before making any needed substitutions.

$$\int_0^1 \frac{\cos\frac{1}{x}}{-x^2}\,dx = \int_\omega^1 \frac{\cos\frac{1}{x}}{-x^2}\,dx.$$

Next we notice there is a trig function of "not just plain x," indicating the need for substitution:

$$u = \frac{1}{x}$$
$$du = -\frac{1}{x^2}\,dx.$$

The new limits of integration are

$$x = 1: \quad u = 1$$
$$x = \omega: \quad u = \frac{1}{\omega} = \Omega.$$

Making the substitution and continuing,

$$\begin{aligned}\int_\omega^1 \frac{\cos\frac{1}{x}}{-x^2}\,dx &= \int_\Omega^1 \cos u\,du\\ &= \sin u\Big|_\Omega^1\\ &= \sin 1 - \sin\Omega.\end{aligned}$$

Because $\sin\Omega$ can be any number in the interval $[-1, 1]$, we do not render the same real result for every positive infinitesimal ω. The integral diverges. ■

For a more complete explanation, see the margin comment for section 7.10 example 5.

Did you notice in the solution to example 8 that, after substitution, we obtained the integral

$$\int_\Omega^1 \cos u\,du?$$

What is notable about this integral is that it involves Ω, as if we were evaluating an improper integral of type I. It looks like it comes from the integral

Ans. to reading exercise 18
∞ (diverges)

$$\int_{\infty}^{1} \cos u \, du.$$

The substitution $u = \frac{1}{x}$ can turn an improper integral of one type into an improper integral of the other type! As long as the improper integral is rewritten before applying substitution, the calculation can continue without noticing the change in type.

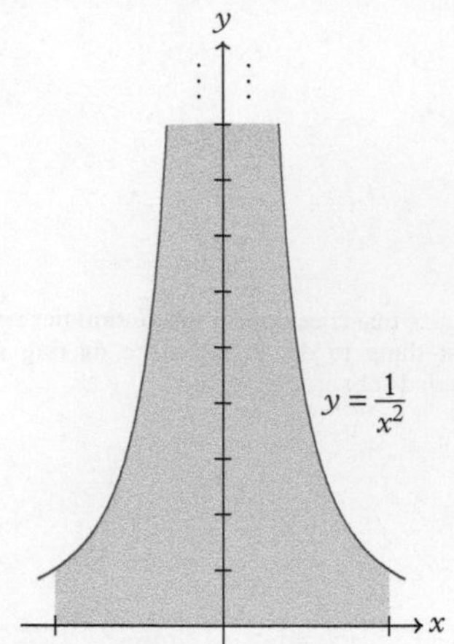

Figure 3 *A portion of the region under the curve $y = \frac{1}{x^2}$ between $x = -1$ and $x = 1$. The region extends infinitely upward along the vertical asymptote $x = 0$*

Type II improper integrals: why?

Suppose we wish to determine the area of the region under the curve $y = \frac{1}{x^2}$ between $x = -1$ and $x = 1$. This region, pictured in part in figure 3, has a positive amount of (net) area because the curve lies entirely above the x-axis.

As usual, we wish to determine the value of

$$\int_{-1}^{1} \frac{1}{x^2} \, dx.$$

This is an improper integral of type II because the integrand is discontinuous at $x = 0$, which is in the interval $[-1, 1]$. What happens if we ignore this fact? We can still calculate a value:

CAUTION: MISTAKE TO AVOID
I purposely made an incorrect calculation (right) to illustrate the consequences.

$$\int_{-1}^{1} \frac{1}{x^2} \, dx = -\frac{1}{x}\bigg|_{-1}^{1} = -\frac{1}{1} - \left(-\frac{1}{-1}\right) = -2.$$

Whatever the area of the region is, it is definitely not negative! Ignoring the discontinuity gives an obviously incorrect answer. Although this might catch your attention in this example, an incorrect answer might not always be this obvious. It is important to get into the habit of always checking an integrand for discontinuities.

CAUTION: MISTAKE TO AVOID
Before applying FTC II, the integrand must always be checked for discontinuities.

The correct solution is to rewrite as

$$\int_{-1}^{1} \frac{1}{x^2} \, dx = \int_{-1}^{-\omega} \frac{1}{x^2} \, dx + \int_{\omega}^{1} \frac{1}{x^2} \, dx.$$

The integral on the right has already been calculated at the beginning of this section. It diverges, so the original integral diverges. The area is infinite.

Reading Exercise 19 *What is wrong with this calculation?*

$$\int_1^4 \frac{2}{x-2}\,dx = 2\ln|x-2|\Big|_1^4$$
$$= 2\ln|4-2| - 2\ln|1-2|$$
$$= 2\ln 2 - 2\ln 1$$
$$= 2\ln 2.$$

Exercises 7.11

1–12. Do the first step. That is, rewrite the integral (if necessary) and do not integrate.

1. $\displaystyle\int_0^6 \frac{1}{x-6}\,dx$
2. $\displaystyle\int_1^3 \frac{1}{x-3}\,dx$
3. $\displaystyle\int_0^7 \frac{1}{x+7}\,dx$
4. $\displaystyle\int_1^3 \frac{1}{x-4}\,dx$
5. $\displaystyle\int_{-2}^1 \frac{1}{x^2-4}\,dx$
6. $\displaystyle\int_1^3 \frac{1}{x-2}\,dx$
7. $\displaystyle\int_0^{10} \frac{1}{\sqrt[3]{x}-2}\,dx$
8. $\displaystyle\int_{-2}^4 \frac{x}{1-\sqrt{x^2+1}}\,dx$
9. $\displaystyle\int_0^{\pi/2} \tan x\,dx$
10. $\displaystyle\int_0^{\pi/2} \csc x\,dx$
11. $\displaystyle\int_1^4 \frac{1}{(x-3)(x+2)}\,dx$
12. $\displaystyle\int_0^1 \ln(1-x)\,dx$

13–32. Evaluate the integral.

13. $\displaystyle\int_5^9 \frac{1}{\sqrt{x-5}}\,dx$
14. $\displaystyle\int_2^5 \frac{1}{x-5}\,dx$
15. $\displaystyle\int_1^3 \frac{1}{x-3}\,dx$
16. $\displaystyle\int_2^5 \frac{2x}{\sqrt{x^2-4}}\,dx$
17. $\displaystyle\int_{-1}^2 \frac{1}{x^4}\,dx$
18. $\displaystyle\int_2^5 \frac{1}{x-1}\,dx$
19. $\displaystyle\int_0^1 \frac{1}{x-2}\,dx$
20. $\displaystyle\int_1^{10} (x-1)^{-1/2}\,dx$

21. $\int_0^1 \frac{1}{\sqrt{x}}\,dx$

22. $\int_{-2}^1 \frac{2}{3x}\,dx$

23. $\int_0^9 \frac{1}{x^{3/2}}\,dx$

24. $\int_0^1 \frac{1}{x^3}\,dx$

25. $\int_0^1 \frac{e^{1/x}}{x^2}\,dx$

26. $\int_{-1}^1 \frac{1}{x^2}\,dx$

27. $\int_{-1}^8 \frac{1}{\sqrt[3]{x}}\,dx$

28. $\int_{-1}^0 \frac{e^{1/x}}{x^2}\,dx$

29. $\int_2^5 \frac{1}{x^2-3x+2}\,dx$

30. $\int_2^5 \frac{1}{\sqrt[3]{x-4}}\,dx$

31. $\int_0^1 \frac{\sin(\ln x)}{x}\,dx$

32. $\int_0^{\pi/2} \sec x\,dx$

33. (GSL) Evaluate $\int_0^6 \frac{2x}{(x^2-4)^{2/3}}\,dx$.

34. (GSL) Evaluate $\int_1^5 \frac{x}{\sqrt{5-x}}\,dx$.

35. (GSL) Evaluate $\int_1^2 \frac{1}{x^2\sqrt{4-x^2}}\,dx$.

Ans. to reading exercise 19
The integrand is not continuous at $x = 2$, which is inside the interval $[1, 4]$. Therefore, FTC II does not apply directly. In fact, the answer $2\ln 2$ is wrong.

36. (GSL) Evaluate $\int_0^3 \frac{x^2}{\sqrt{9-x^2}}\,dx$.

37. If an integrand has discontinuities at both ends of the interval, split the integral at a point inside the interval into two integrals using the additive property. If each of the improper integrals converges, the sum converges. Practice this technique by writing

$$\int_{-1}^1 \frac{1}{\sqrt{1-x^2}}\,dx = \int_{-1}^0 \frac{1}{\sqrt{1-x^2}}\,dx + \int_0^1 \frac{1}{\sqrt{1-x^2}}\,dx$$

and finish evaluating the integral.

38. Use the procedure of exercise 37 to evaluate $\int_{-2}^2 \frac{2x}{x^2-4}\,dx$.

39. The procedures given in this section work not just with infinite discontinuities, but with jump and removable discontinuities as well. (a) Verify that $f(x) = \frac{x^2-x}{|x-1|}$ has a jump discontinuity at $x = 1$. (b) Evaluate $\int_0^2 \frac{x^2-x}{|x-1|}\,dx$. Hint: for each integral, rewrite $|x-1|$ as $x-1$ or $-(x-1)$ as appropriate.

The type II improper integral procedures also apply to oscillatory discontinuities, but the antiderivatives needed to carry out the procedures tend to be nonelementary functions.

40. (a) Verify that $f(x) = \dfrac{x^2 - 4}{x - 2}$ has a removable discontinuity at $x = 2$. (b) Evaluate $\displaystyle\int_1^3 \frac{x^2 - 4}{x - 2}\, dx$.

41. Some improper integrals can be calculated using an omega sum. If the discontinuity is at the left endpoint of the interval, then because the omega sum uses right-hand endpoints, there is no division-by-zero error. Also, the sum of powers approximation formula $\displaystyle\sum_{k=1}^{\Omega} k^m \approx \frac{\Omega^{m+1}}{m + 1}$ holds as long as $m > -1$. Use an omega sum to calculate the value of $\displaystyle\int_0^1 \frac{1}{\sqrt{x}}\, dx$.

42. Let f be continuous on $[a, b)$ but not at $x = b$. The classical definition of an improper integral of type II for this case is

$$\int_a^b f(x)\, dx = \lim_{k \to b^-} \left(\int_a^k f(x)\, dx \right).$$

Reconcile this definition with definition 3.

Chapter VIII

Alternate Representations: Parametric and Polar Curves

Parametric Equations

8.1

We have studied motion on a number line. If a particle's position on a number line at time t is given by $s(t)$, then its velocity at time t is $v(t) = s'(t)$, and its acceleration at time t is $a(t) = v'(t) = s''(t)$. Regardless of whether our number line is positioned horizontally or, as in the case of a falling object, vertically, a particle moving on a number line is moving in only one dimension. What if, instead, the particle is moving in two dimensions?

Describing motion in two dimensions

Consider an insect moving along a tabletop, as in figure 1. The insect might not move in a straight line, so motion along a number line might not be sufficient to describe the insect's path. If we place an xy-coordinate system on the tabletop, the insect's path might not trace out a function, so writing y as a function of x is also not sufficient to describe the motion.

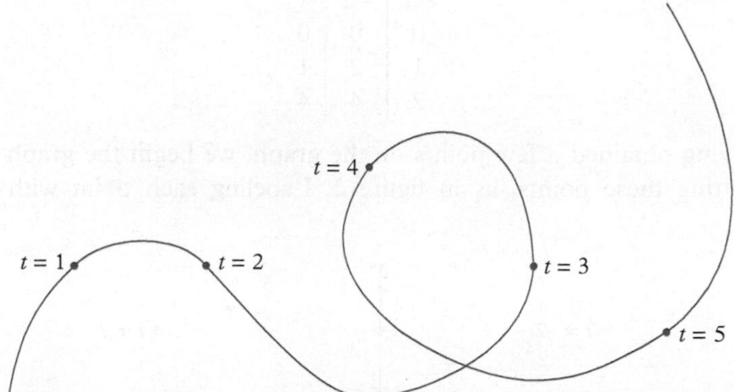

Figure 1 *The path of an insect crawling on a table; the view is from above. Some points are marked and labeled with the time at which the insect was located at that point*

The idea of *parametric equations* is to describe the location of a point in the plane at time t by describing both its x-coordinate and its y-coordinate at time t. Because the insect of figure 1 can only be at one

Ignore the fact that the insect is larger than a single point.

point at any particular time t, then the x-coordinate of the point must be a function of t, and the y-coordinate of the point must be a function of t. In other words, we use two functions, one for each coordinate, to describe the motion.

An example of a curve described by parametric equations is

$$x = 2t,$$
$$y = t^2.$$

We call t a *parameter*. We often think of t as representing time, although in a particular application the parameter t might represent something different. The points on the curve are of the form $(x, y) = (2t, t^2)$.

What does the graph of this curve look like? We can begin to answer this question by plotting points, just as we did when we first learned to graph functions. For parametric equations, we choose values of the parameter t and then calculate the corresponding values of the variables x and y. For instance, when $t = -1$, $x = 2(-1) = -2$ and $y = (-1)^2 = 1$. Then, $(-2, 1)$ is the point on the graph corresponding to $t = -1$. A table containing a few values of t and the corresponding points on the graph follows.

Even if t represents time, negative values are allowed. If t is measured in hours and $t = 0$ corresponds to 6 a.m., then $t = -1$ is 1 hour earlier, which is 5 a.m. that same day.

Choosing values of t is done using the same criteria as choosing values of x when plotting the graph of a function. Here, because one of the coordinate functions is $y = t^2$, we choose both negative and positive values of t.

t	x	y
-2	-4	4
-1	-2	1
0	0	0
1	2	1
2	4	4

Having obtained a few points on the graph, we begin the graph by plotting these points, as in figure 2. Labeling each point with its

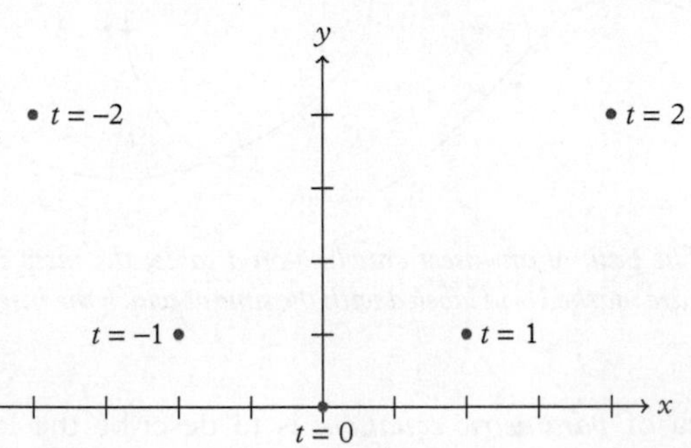

Figure 2 *Points on the curve* $x = 2t$, $y = t^2$

"time stamp" allows us to see the order in which the points should be connected to form the curve.

As long as each of the coordinate functions is continuous on the interval of values chosen for t, it is legitimate to connect the points as you do for a continuous function. However, we do not automatically connect the plotted points from left to right; instead, to draw the curve we follow the order of the parameter t. (In this case the order is left-to-right, but in the next example it will not be.)

CAUTION: MISTAKE TO AVOID
When graphing continuous parametric equations, the correct order to connect plotted points might not be from left to right. Use the order of the values of the parameter t.

The drawn curve is depicted in figure 3. Notice that arrows have

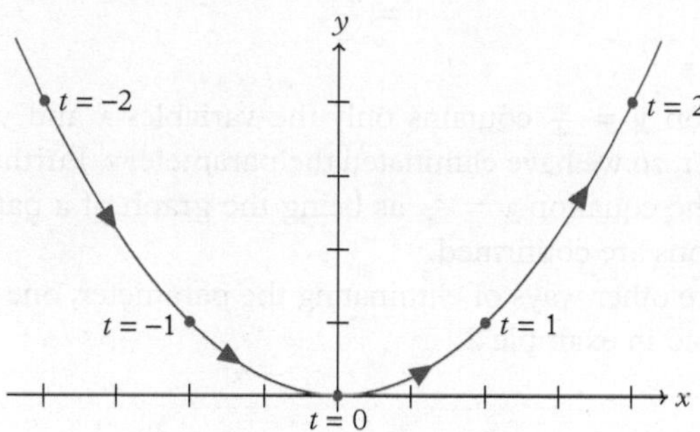

Figure 3 *The parametrically defined curve $x = 2t$, $y = t^2$, including time stamps (values of t corresponding to a few of the points) and direction arrows. The arrows and dots are not part of the curve but are added to increase visual clarity*

been added to indicate the direction the curve is traversed.

Although some parametric curves take paths more closely resembling that of figure 1, the curve in figure 3 looks like a parabola. Is it a parabola? We tackle this question next.

Eliminating the parameter

The curve in figure 3 appears to be a parabola. We know what the equation of a parabola looks like in terms of the two variables x and y, but the curve in figure 3 is given parametrically as $x = 2t$, $y = t^2$, which uses three variables (x, y, and t). If we can rewrite the equations in such a way that t is no longer present—a process called *eliminating the parameter*—then perhaps we can recognize the equation and use previous knowledge to identify the curve.

One method of eliminating the parameter is to solve one of the coordinate functions for the variable t and then replace t with the resulting expression in the other coordinate function. Choosing the easiest option, we solve $x = 2t$ for t:

The other option is to solve $y = t^2$ for t, obtaining $t = \pm\sqrt{y}$. Substituting $\pm\sqrt{y}$ for t in the other coordinate function then gives $x = 2t = 2(\pm\sqrt{y}) = \pm 2\sqrt{y}$. Since $x = \pm 2\sqrt{y}$ does not contain the parameter t, we have eliminated the parameter successfully. Although this equation is equivalent to $y = \frac{x^2}{4}$, the equation $x = \pm 2\sqrt{y}$ is not quite as easily recognized as a parabola.

$$x = 2t$$
$$\frac{x}{2} = t.$$

Next we substitute $\frac{x}{2}$ for t in the other coordinate function:

$$y = t^2$$
$$y = \left(\frac{x}{2}\right)^2$$
$$y = \frac{x^2}{4}.$$

The equation $y = \frac{x^2}{4}$ contains only the variables x and y, and does not contain t, so we have eliminated the parameter t. Furthermore, we recognize the equation $y = \frac{x^2}{4}$ as being the graph of a parabola, and our suspicions are confirmed.

There are other ways of eliminating the parameter, one of which is demonstrated in example 2.

Parametric equations: example

In the next example, the domain of the variable t is specified.

Example 1 *Graph the parametrically defined curve* $x = 2\sin t$, $y = \cos t$, $0 \le t \le \pi$.

In chapter 4, when we wanted n subintervals for a finite sum, we set $\Delta x = \frac{b-a}{n}$ and called it the *width* of a subinterval. We then calculated $a, a + \Delta x, a + 2\Delta x, \ldots$ until we reached b. The step size plays the same role as Δx in choosing values of t.

Solution The domain of the variable t is $0 \le t \le \pi$, so we should at least plot the points corresponding to $t = 0$ and $t = \pi$, the endpoints of the domain. A *step size* of $t = \frac{\pi}{4}$ results in the table of values below.

t	x	y
0	0	1
$\frac{\pi}{4}$	$\sqrt{2}$	$\frac{1}{\sqrt{2}}$
$\frac{\pi}{2}$	2	0
$\frac{3\pi}{4}$	$\sqrt{2}$	$-\frac{1}{\sqrt{2}}$
π	0	-1

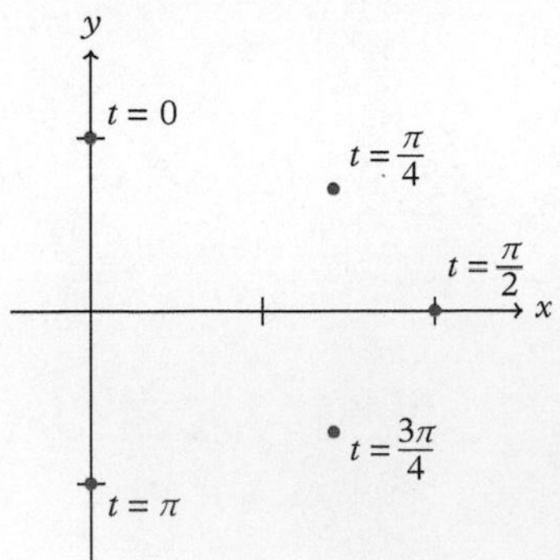

Figure 4 *The points given in the table are plotted along with labels indicating the value of the parameter t at each point*

Next we plot points and label them by values of t, as in figure 4.

The coordinate functions $x = 2\sin t$ and $y = \cos t$ are continuous for all values of t, so we may draw the parametric curve in a continuous fashion. Following the order of the values of t from $t = 0$ to $t = \pi$, we draw the curve and place direction arrows as in figure 5.

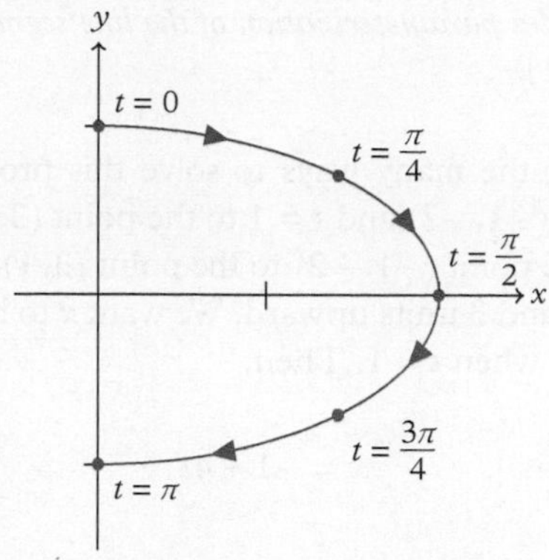

Figure 5 *The completed curve $x = 2\sin t$, $y = \cos t$, including direction arrows* ■

The graph of a parametric curve does not have to pass the vertical line test. The vertical line test applies to y as a function of x

Reading Exercise 1 *Graph $x = 2\sin t$, $y = 3\cos t$, $0 \le t \le 2\pi$.*

When parametric equations involve trigonometric functions, sometimes an identity can be helpful when eliminating the parameter. For instance, if the equations can be solved for $\sin t$ and $\cos t$, then the identity $\sin^2\theta + \cos^2\theta = 1$ can be used.

Example 2 *Eliminate the parameter for $x = 2\sin t$, $y = 3\cos t$.*

Solution We solve the first equation for $\sin t$ and the second for $\cos t$, resulting in

$$\frac{x}{2} = \sin t,$$
$$\frac{y}{3} = \cos t.$$

Substituting these expressions into the identity $\sin^2 t + \cos^2 t = 1$ yields

$$\left(\frac{x}{2}\right)^2 + \left(\frac{y}{3}\right)^2 = 1$$
$$\frac{x^2}{4} + \frac{y^2}{9} = 1,$$

which is the equation of an ellipse. ■

Notice that example 2 uses the parametric equations from reading exercise 1. Check the answer to the reading exercise to verify that the graph appears to be an ellipse. Similarly, the graph in figure 5 produced for example 1 is a portion of an ellipse; there, the domain $0 \le t \le \pi$ traverses only half of the ellipse.

Parameterization of a line segment

When we have parametric equations, we can produce a graph (as in example 1) or we can try eliminating the parameter (as in example 2), both of which help identify the curve. Reversing this idea, what if we start with an identified curve and try to find a parameterization?

Example 3 *Find a parameterization of the line segment with endpoints* $(-1, -2)$ *and* $(3, 1)$.

Solution Among the many ways to solve this problem is to assign $t = 0$ to the point $(-1, -2)$ and $t = 1$ to the point $(3, 1)$, as in figure 6. To move from the point $(-1, -2)$ to the point $(3, 1)$, we must move 4 units to the right and 3 units upward. We want x to be -1 when $t = 0$, and 4 units larger when $t = 1$. Then,

Taking the point corresponding to $t = 0$ as (x_0, y_0) and the point corresponding to $t = 1$ as (x_1, y_1), we can calculate the horizontal "run" by finding $x_1 - x_0$ and the vertical "rise" by finding $y_1 - y_0$.

$$x = -1 + 4t$$

works. Similarly, we can use

$$y = -2 + 3t,$$

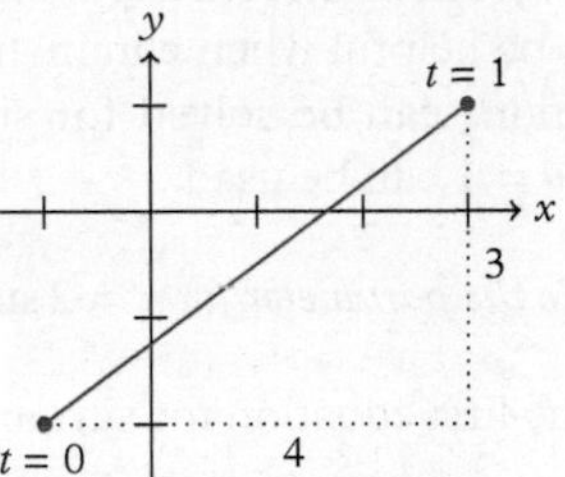

Figure 6 *The line segment with endpoints* $(-1, -2)$ *and* $(3, 1)$ *(blue), with dotted lines indicating the horizontal "run" of* 4 units *and the vertical "rise" of* 3 units *between the endpoints*

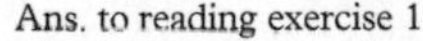
Ans. to reading exercise 1

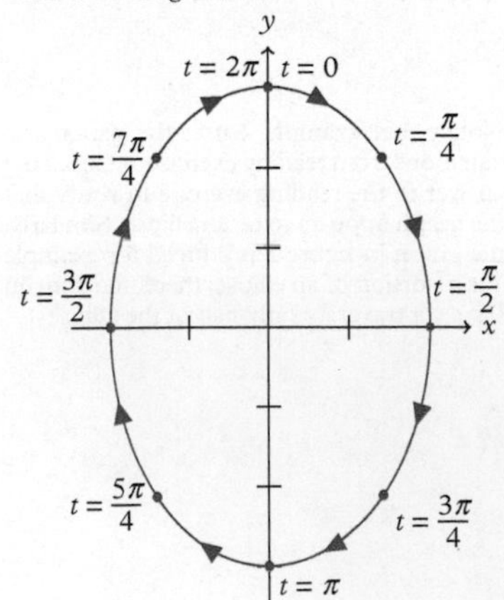

so that when $t = 0$, y is -2; when $t = 1$, y is 3 units larger. The parameterization of the line segment is stated as $x = -1 + 4t$, $y = -2 + 3t$, $0 \le t \le 1$. ■

In the solution to example 3, we verified that the parametric equations give the correct points when $t = 0$ and when $t = 1$, but we did not verify that the curve is a (portion of a) line. This can be accomplished by eliminating the parameter. Solving the first equation for t gives

$$x = -1 + 4t$$

$$\frac{x+1}{4} = t.$$

Replacing t in the second equation gives

$$y = -2 + 3t$$

$$y = -2 + 3\left(\frac{x+1}{4}\right)$$

$$y = \frac{3}{4}x - \frac{5}{4},$$

which is the equation of a line.

The procedure in example 3 can be followed for an arbitrary line segment to produce the following formulas.

PARAMETERIZATION OF A LINE SEGMENT WITH ENDPOINTS (h, k) AND $(h + a, k + b)$

A parameterization of the line segment with endpoints (h, k) and $(h + a, k + b)$ is

$$x = h + at,$$

$$y = k + bt,$$

$$0 \le t \le 1.$$

There are infinitely many different parameterizations of any given line segment. These formulas give one option.

The numbers a and b do not have to be positive; they can be any real number. The same is true of h and k.

Let's look at the use of these formulas in the following example.

Example 4 *Find a parameterization of the line segment between the points* $(-3, 5)$ *and* $(4, 1)$.

Solution It does not matter which point is chosen for $t = 0$ and which is chosen for $t = 1$. If we choose $(4, 1)$ for $t = 0$, we have $h = 4$, $k = 1$. To find the values of a and b, we subtract coordinates:

As before, we label the point corresponding to $t = 0$ as (x_0, y_0) and the point corresponding to $t = 1$ as (x_1, y_1).

$$a = x_1 - x_0 = -3 - 4 = -7,$$

$$b = y_1 - y_0 = 5 - 1 = 4.$$

A parameterization of the line is then

$$x = 4 - 7t$$

$$y = 1 + 4t$$

$$0 \le t \le 1. \qquad \blacksquare$$

Reading Exercise 2 *Find a parameterization of the line segment between the points* $(2, 5)$ *and* $(8, 4)$.

Parameterization of a circle

In trigonometry, the coordinates x and y of a point on the unit circle are given by $x = \cos\theta$ and $y = \sin\theta$, where θ is the angle in standard position with terminal side passing through (x, y) (figure 7). As the angle θ increases from $\theta = 0$ to $\theta = 2\pi$, the point (x, y) travels around the circle exactly once. All we have to do to get a parameterization of

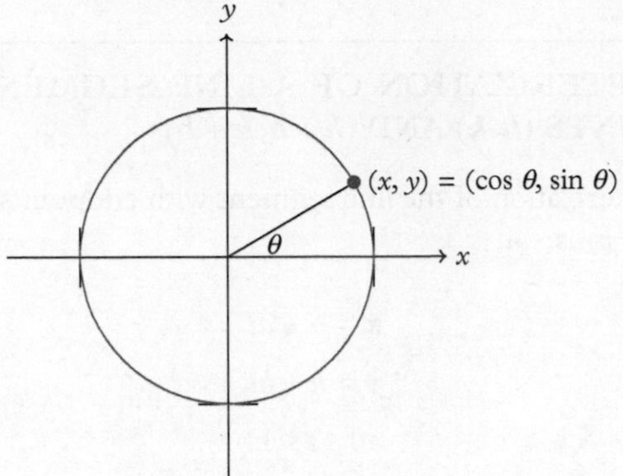

Figure 7 *The unit circle*

Or, we could just say that θ is the parameter we are using.

the circle is to replace the angle θ by the parameter t:

$$x = \cos t$$

$$y = \sin t.$$

As t varies throughout the real numbers, the point (x, y) continues traveling around the circle. To traverse the circle exactly once, we may choose any interval for t of length 2π, such as $0 \le t \le 2\pi$ or $-3\pi \le t \le -\pi$ or $\sqrt{2} + 5\pi \le t \le \sqrt{2} + 7\pi$.

The unit circle has radius 1. If we wish to parameterize a circle with radius r centered at the origin, we merely need to multiply the x and y coordinates by r:

$$x = r\cos t$$

$$y = r\sin t.$$

To shift the circle's center to the point (h, k) as in figure 8, we need to

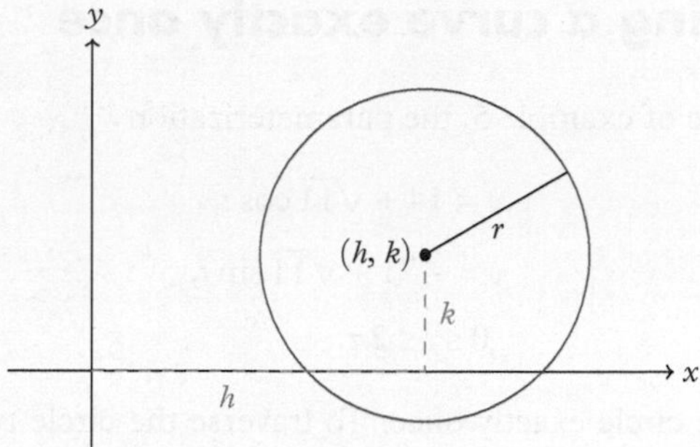

Figure 8 *A circle of radius r and center (h, k)*

add h and k to the x and y coordinates, respectively:

$$x = h + r\cos t$$

$$y = k + r\sin t.$$

> **PARAMETERIZATION OF A CIRCLE WITH CENTER (h, k) AND RADIUS r**
>
> A parameterization of the circle with center (h, k) and radius r that traverses the circle one time between $t = 0$ and $t = 2\pi$ is
>
> $$x = h + r\cos t,$$
>
> $$y = k + r\sin t.$$

There are infinitely many different parameterizations of any given circle. These formulas give one option. Other options include swapping $\sin t$ and $\cos t$; the effect is to start at a different point and travel clockwise instead of counterclockwise. Replacing t with $2t$ changes the "speed" at which we travel the circle, as does replacing t with t^3. The options are endless (literally).

Example 5 *Find a parameterization of the circle with center* $(14, -71)$ *and radius* $\sqrt{11}$.

Solution Using the parameterization of a circle formulas with $h = 14$, $k = -71$, and $r = \sqrt{11}$, we have

$$x = 14 + \sqrt{11}\cos t$$

$$y = -71 + \sqrt{11}\sin t. \quad \blacksquare$$

Ans. to reading exercise 2
If $(2, 5)$ is chosen for $t = 0$: $x = 2 + 6t$, $y = 5 - t$, $0 \le t \le 1$. If $(8, 4)$ is chosen as $t = 0$: $x = 8 - 6t$, $y = 4 + t$, $0 \le t \le 1$.

Reading Exercise 3 *Find a parameterization of the circle with center* $(1, 3)$ *and radius 8.*

Traversing a curve exactly once

For the circle of example 5, the parameterization

$$x = 14 + \sqrt{11}\cos t,$$
$$y = -71 + \sqrt{11}\sin t,$$
$$0 \le t \le 2\pi$$

traverses the circle exactly once. To traverse the circle twice, we can use $0 \le t \le 4\pi$ instead. Another parameterization that traverses the circle exactly once is

$$x = 14 + \sqrt{11}\cos 2t,$$
$$y = -71 + \sqrt{11}\sin 2t,$$
$$0 \le t \le \pi.$$

Although finding values of the parameter t that traverse a curve exactly once is relatively simple for a circle, it can be more difficult for other curves. One way to explore a parameterized curve is through graphing the curve, as we did in example 1.

Example 6 *Find an interval of values of the parameter t that traverses the parametric curve* $x = 3\cos t + \cos 1.25t$, $y = 3\sin t - \sin 1.5t$ *exactly once.*

Solution Because $\cos t$ and $\sin t$ have period 2π, we might suspect that using $0 \le t \le 2\pi$ causes the curve to begin to repeat. The graph of $x = 3\cos t + \cos 1.25t$, $y = 3\sin t - \sin 1.5t$, $0 \le t \le 2\pi$ is in figure 9.

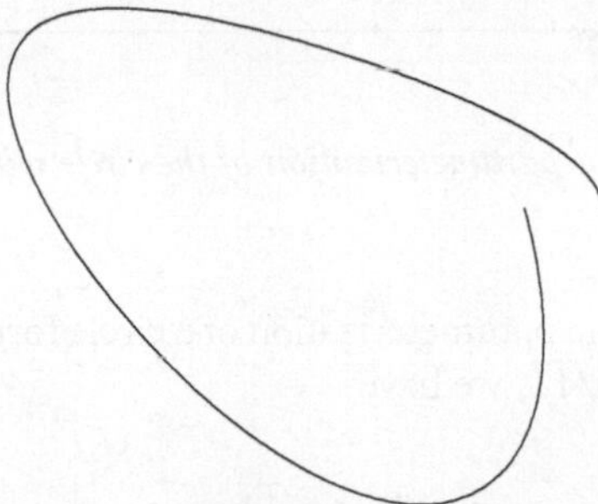

Figure 9 *The graph of the parametric curve* $x = 3\cos t + \cos 1.25t$, $y = 3\sin t - \sin 1.5t$, *for* $0 \le t \le 2\pi$. *The point corresponding to* $t = 0$ *is the right-most endpoint (loose end); the curve travels counterclockwise and the left-most endpoint corresponds to* $t = 2\pi$

It missed! Instead of looping back to the beginning, the curve comes close but does not close up. The points for $t = 0$ and for $t = 2\pi$ are not the same.

The problem is that not all the arguments of the trig functions are t; one is $1.25t$ and another is $1.5t$. Continuing to explore the curve graphically, we expand the graph to include $0 \le t \le 4\pi$ (figure 10).

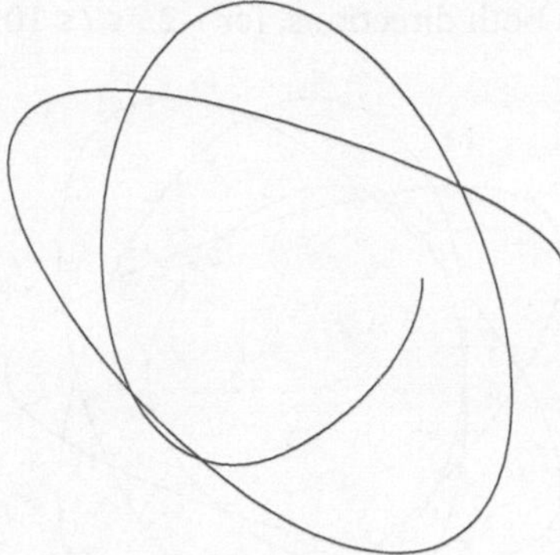

Figure 10 *The graph of the parametric curve* $x = 3\cos t + \cos 1.25t$, $y = 3\sin t - \sin 1.5t$, *for* $0 \le t \le 4\pi$. *The point corresponding to* $t = 0$ *is the right-most endpoint (loose end); the curve travels counterclockwise and the left-most endpoint corresponds to* $t = 4\pi$

As you may have already concluded, producing intricate graphs such as these by hand can be quite challenging and time-consuming. The advantage of using technology is clear.

It still misses! The curve does not close up. The nature of exploration is that it is exploratory (of course!), so we might as well keep trying. The graphs for $0 \le t \le 6\pi$ and $0 \le t \le 8\pi$ are in figure 11.

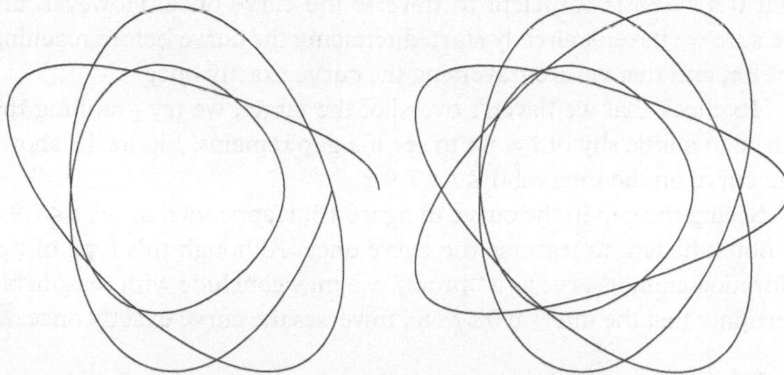

Figure 11 *The graph of the parametric curve* $x = 3\cos t + \cos 1.25t$, $y = 3\sin t - \sin 1.5t$, *(left) for* $0 \le t \le 6\pi$ *and (right) for* $0 \le t \le 8\pi$

Success! Although we still did not return to the starting point at $t = 6\pi$, we did at $t = 8\pi$. But is the task finished?

Ans. to reading exercise 3
Options include $x = 1 + 8\cos t$, $y = 3 + 8\sin t$.

Just because we returned to the starting point does not mean we must begin repeating the curve. Look again at the graph on the left in figure 11 and compare it to figure 9. The curve at $t = 6\pi$ returned to the same point as for $t = 2\pi$. Instead of repeating after $t = 6\pi$, it continues tracing new parts of the curve. We cannot be sure we are repeating a previously graphed portion of the curve unless we try it for an even more expansive interval of values of t. Figure 12 shows the curve expanded in both directions, for $-2\pi \le t \le 10\pi$.

Checking only positive values of t, or failing to expand in both directions, might miss a portion of a parametric curve.

Figure 12 *The graph of the parametric curve* $x = 3\cos t + \cos 1.25t$, $y = 3\sin t - \sin 1.5t$, *for* $-2\pi \le t \le 10\pi$. *The graph appears to be identical to the graph on the right in figure 11*

This time, we obtain the exact same graph as the right side of figure 11. We do appear to be retracing previously graphed portions of the curve. We may safely conclude (although not with 100% surety) that $0 \le t \le 8\pi$ is sufficient to traverse the curve once. However, are we sure we haven't already started retracing the curve before reaching $t = 8\pi$, and that we are traversing the curve exactly once?

To check that we haven't overshot the target, we try graphing the curve to a little shy of $t = 8\pi$ to see if a gap remains. Figure 13 shows the curve on the interval $0 \le t \le 7.9\pi$.

Seeing the gap in the curve in figure 13, it appears that $0 \le t \le 7.9\pi$ is not sufficient to traverse the curve once. Although this type of exploration cannot serve as a "proof," we may conclude with reasonable certainty that the interval $0 \le t \le 8\pi$ traverses the curve exactly once. ■

The solution method in example 6 is an example of *exploratory mathematics*. Accompanied by an appropriate amount of reasoning, exploratory mathematics can give a degree of surety as to the truth of a statement. In cases when a formal proof is difficult or impossible, a tentative conclusion is still useful. Or perhaps we might want to try a little exploration before investing the time needed to produce a

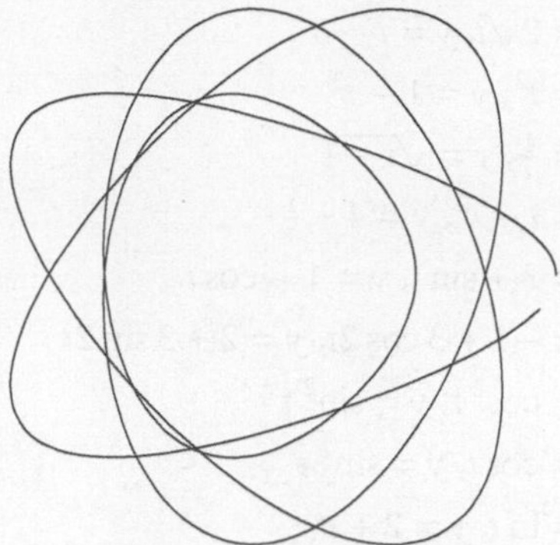

Figure 13 *The graph of the parametric curve $x = 3\cos t + \cos 1.25t$, $y = 3\sin t - \sin 1.5t$, for $0 \le t \le 7.9\pi$. The graph is not identical to the graph on the right in figure 11. There is a small gap in this picture of the curve*

formal proof. Explorations never completely replace proofs, but they have their place in anyone's set of mathematical tools.

Exercises 8.1

1–14. Graph the parametrically defined curve (without using graphing technology).

1. $x = 2\sqrt{t}, y = t - 3$

2. $x = t^2, y = 1 - t^3, -2 \le t \le 2$

3. $x = \cos t, y = \sin 2t, 0 \le t \le 2\pi$

4. $x = t - t^2, y = 4 + \frac{1}{t}, 1 \le t \le 4$

5. $x = \frac{1}{t}, y = \sqrt{t-1}, 1 \le t \le 5$

6. $x = t\cos t, y = \sin t, t \ge 0$

7. $x = 3 + \sin t, y = 1 - \cos t$

8. $x = -1 + 3\cos 2t, y = 2 + 3\sin 2t, 0 \le t \le \frac{3\pi}{4}$

9. $x = 3(t - \sin t), y = 3(1 - \cos t)$

10. $x = \cos t^2, y = \sin t^2, 0 \le t \le \sqrt{2\pi}$

11. $x = \cos^2 t, y = \sin^2 t$

12. $x = \cos t, y = \sin^2 t$

13. $x = \ln t, y = 2 + t^2$

14. $x = 4\cos^3 t, y = 4\sin^3 t$

15–28. Eliminate the parameter.

Some of these curves have been graphed in previous exercises. Do you think it would have been helpful to work these exercises before graphing these curves?

15. $x = 2\sqrt{t}, y = t - 3$

16. $x = t^2, y = 1 - t^3$

17. $x = \frac{1}{t}, y = \sqrt{t-1}$

18. $x = t - t^2, y = 4 + \frac{1}{t}$

19. $x = 3 + \sin t, y = 1 - \cos t$

20. $x = -1 + 3\cos 2t, y = 2 + 3\sin 2t$

21. $x = \cos^2 t, y = \sin^2 t$

22. $x = \cos t, y = \sin^2 t$

23. $x = \ln t, y = 2 + t^2$

24. $x = 4\cos^3 t, y = 4\sin^3 t$

25. $x = \cosh^2 3t, y = 2 - \sinh^2 3t$

26. $x = \tan\frac{4}{t}, y = 2 - \sec^2\frac{4}{t}$

27. $x = 4\sin t^3, y = 5\cos^2 t^3$

28. $x = \ln t, y = \ln 5t^2$

29–36. Find a parameterization of the line segment with the given endpoints.

29. $(4,7), (8,9)$

30. $(-2,5), (3,6)$

31. $(-2,5), (3,1)$

32. $(0,-4), (5,0)$

33. $(2,7), (5,7)$

34. $(-3,7), (2,\frac{1}{2})$

35. $(9,22), (-3,6)$

36. $(1,5), (1,9)$

37–44. Find a parameterization of the circle with the following description.

37. center $(2,7)$, radius 6

38. center $(-1,5)$, radius 11

39. center $(1,0)$, radius 3

40. center $(\ln 4, 5)$, radius $\sqrt{5}$

41. center $(1,5)$ through the point $(3,4)$

42. center $(-3,0)$ through the point $(1,8)$

43. endpoints of diameter $(-2,9)$ and $(6,11)$

44. center $(2,7)$ tangent to $y = 3$

Some of these exercises can be worked quickly without the use of technology. Some of the curves have been explored in previous exercises. Not all are closed curves.

45–56. Find an interval of values of the parameter t that traverses the parametric curve exactly once.

45. $x = 3 + \sin t, y = 1 - \cos t$

46. $x = -1 + 3\cos 2t, y = 2 + 3\sin 2t$

47. $x = \sin 2t, y = \cos^2 t$
48. $x = 4\cos^3 t, y = 4\sin^3 t$
49. $x = \cos^2 t, y = \sin^2 t$
50. $x = \cos t, y = \sin^2 t$
51. $x = \sin t, y = \cos 2.5t$
52. $x = \dfrac{1}{2 + \cos t}, y = \dfrac{1}{\sin^2 4t - 3}$
53. $x = \cosh^2 3t, y = 2 - \sinh^2 3t$
54. $x = \tan t, y = \sec t$, portion of the curve above the x-axis only
55. $x = e^{\sin t^2}, y = \cos t^2$
56. $x = \ln(1.1 + \cos t), y = \ln(1.1 + \sin 1.5t)$

57. The parametric curve of exercises 11, 21, and 49 is a line segment. (a) Give a similar parameterization of the line segment with endpoints $(3, 0)$ and $(0, 3)$. (b) Give a similar parameterization of the line segment with endpoints $(-3, 0)$ and $(0, 3)$. (c) Give a similar parameterization of a segment of the line $y = 4x - 5$ and give the endpoints of the segment.
58. In example 6 it was determined that the parametric curve $x = 3\cos t + \cos 1.25t, y = 3\sin t - \sin 1.5t$ is traversed exactly once on the interval $0 \le t \le 8\pi$. (a) Show that each of the functions $\cos t$, $\cos 1.25t$, $\sin t$, and $\sin 1.5t$ have the same value at $t + 8\pi$ as at t. (b) Does this happen for $t + 6\pi$? Does it happen for $t + k\pi$ with $k < 8$?
59. (a) Graph $y = x^2$ for $-3 \le x \le 3$. (b) Graph the parametric curve $x = t, y = t^2, -3 \le t \le 3$. (c) Find a parameterization for the curve given by $y = \sqrt{x + 4}$. (d) Find a parameterization for the curve given by $y = \dfrac{1}{1 + x^2}$. (e) Can any function $y = f(x)$ be parameterized?

Tangents to Parametric Curves

8.2

In this section we continue our study of parametric curves by exploring tangent lines.

Parametric curves: tangent lines

Consider the graph of a parametric curve such as the one in figure 1.

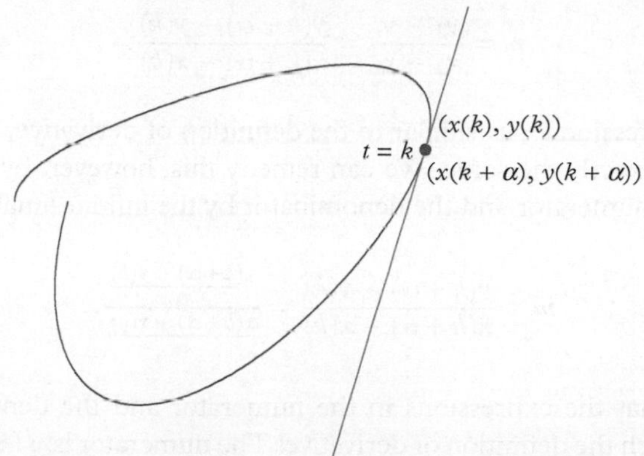

Figure 1 *The graph of a parametric curve (blue) with coordinate functions $x(t)$ and $y(t)$, along with its tangent line at $t = k$ (orange). When plotting the point corresponding to $t = k$, we evaluate the coordinate functions to find its x and y coordinates, resulting in $(x(k), y(k))$. The same procedure applied to $t = k + \alpha$ yields the point $(x(k + \alpha), y(k + \alpha))$. For an infinitesimal α, the two points should be infinitely close (assuming continuity), and therefore they are depicted as being at the same location. The direction in which the parametric curve is traversed is irrelevant, hence no direction arrows are included*

A tangent line to a parametric curve follows the same idea as a tangent line to the graph of a function. The slope of the line is still the rise over the run, the slope formula

$$m = \frac{y_2 - y_1}{x_2 - x_1}$$

still applies, and we still need two points (x_1, y_1) and (x_2, y_2) to apply the slope formula. Using $x(t)$ and $y(t)$ as the two coordinate functions for the parametric curve, and supposing we desire the tangent line at $t = k$, one of the two points is

$$(x(k), y(k)).$$

As we did with tangent lines to function curves, we can scoot over just a smidgen, an infinitesimal amount, to find a second point that lies geometrically at the same place as the first point. Using α as an arbitrary infinitesimal, the second point is

To guarantee that $(x(k+\alpha), y(k+\alpha))$ lies at the same place as $(x(k), y(k))$, the coordinate functions must be continuous.

$$(x(k+\alpha), y(k+\alpha)).$$

Applying the slope formula to our two points, using $(x(k), y(k))$ as (x_1, y_1) and $(x(k+\alpha), y(k+\alpha))$ as (x_2, y_2), yields

$$m = \frac{y_2 - y_1}{x_2 - x_1} = \frac{y(k+\alpha) - y(k)}{x(k+\alpha) - x(k)}.$$

This expression looks similar to the definition of derivative, although it is not exactly the same. We can remedy this, however, by dividing both the numerator and the denominator by the infinitesimal α:

$$m = \frac{y(k+\alpha) - y(k)}{x(k+\alpha) - x(k)} = \frac{\frac{y(k+\alpha) - y(k)}{\alpha}}{\frac{x(k+\alpha) - x(k)}{\alpha}}.$$

Notice that the expressions in the numerator and the denominator both match the definition of derivative! The numerator is $y'(k)$ and the denominator is $x'(k)$, where both derivatives are taken with respect to the variable t. Therefore, the slope of the tangent line at $t = k$ is

The definition of derivative also requires that we render the same real result for every infinitesimal α.

$$m = \frac{y'(k)}{x'(k)},$$

The derivative is with respect to the variable t because both k and $k + \alpha$ are values of t.

assuming this quantity exists.

SLOPE OF THE TANGENT LINE TO A PARAMETRIC CURVE

Let $x(t)$, $y(t)$ be a parametrically defined curve. If both x and y are differentiable at $t = k$ and $x'(k) \neq 0$, then the slope of the

The restriction $x'(k) \neq 0$ is for the purpose of avoiding division by zero. Because differentiability implies continuity, the coordinate functions $x(t)$ and $y(t)$ are continuous as well.

tangent line to the curve at $t = k$ is given by

$$m = \frac{y'(k)}{x'(k)},$$

where both derivatives are taken with respect to the variable t.

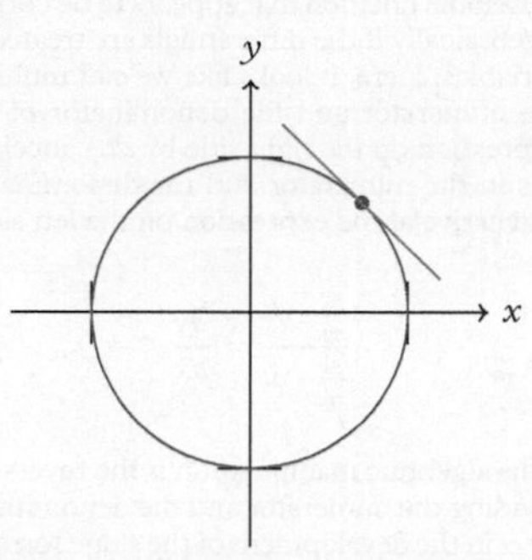

Figure 2 *The parametric curve (blue) of example 1, which is a circle centered at* $(0,0)$ *with radius 1, along with its tangent line (orange) at* $t = \frac{\pi}{4}$

Example 1 *Find the slope of the tangent line to the curve* $x = \cos t$, $y = \sin t$ *at* $t = \frac{\pi}{4}$.

Solution First we calculate x' and y':

$$\frac{dx}{dt} = x' = -\sin t,$$

$$\frac{dy}{dt} = y' = \cos t.$$

Next we use the formula for the slope of the tangent line to a parametric curve (with $k = \frac{\pi}{4}$):

$$m = \frac{y'\left(\frac{\pi}{4}\right)}{x'\left(\frac{\pi}{4}\right)} = \frac{\cos\frac{\pi}{4}}{-\sin\frac{\pi}{4}} = \frac{\frac{1}{\sqrt{2}}}{-\frac{1}{\sqrt{2}}} = -1.$$

CAUTION: MISTAKE TO AVOID
Notice that x' is in the denominator, whereas y' is in the numerator. A common mistake is switching the two derivatives.

We conclude that the slope of the tangent line is -1, which is corroborated nicely by figure 2. ■

Notice the brief use of Leibniz notation in the solution to example 1, making clear that the derivatives are taken with respect to the variable t. Leibniz notation can be used further. We can replace $m = \frac{y'\left(\frac{\pi}{4}\right)}{x'\left(\frac{\pi}{4}\right)}$ with

$$m = \frac{\left.\frac{dy}{dt}\right|_{t=\frac{\pi}{4}}}{\left.\frac{dx}{dt}\right|_{t=\frac{\pi}{4}}} = \left.\frac{\frac{dy}{dt}}{\frac{dx}{dt}}\right|_{t=\frac{\pi}{4}}.$$

Because the slope m is given by the derivative $\frac{dy}{dx}$, it should be clear that for parametric curves,

$$\frac{dy}{dx} = \frac{\frac{dy}{dt}}{\frac{dx}{dt}}.$$

This is another instance of a formula written in Leibniz notation that appears to be correct algebraically if the differentials are treated as variables. Here, it looks like we can multiply the numerator and the denominator of the expression on the right side by dt, cancel the dt's in the numerator and the denominator, and arrive at the expression on the left side:

$$\frac{\frac{dy}{\not{dt}} \cdot \not{dt}}{\frac{dx}{\not{dt}} \cdot \not{dt}} = \frac{dy}{dx}.$$

This algebraic manipulation is the reverse of dividing the numerator and the denominator by α in the development of the slope formula for parametric curves.

DERIVATIVE $\frac{dy}{dx}$ FOR A PARAMETRIC CURVE

Let $x(t)$, $y(t)$ be a parametrically defined curve. Then,

$$\frac{dy}{dx} = \frac{\frac{dy}{dt}}{\frac{dx}{dt}}$$

wherever the latter expression exists.

Example 2 *Determine $\frac{dy}{dx}$ for the parametric curve $x = \ln t$, $y = 2 + t^2$.*

Solution Using the derivative formula for parametric curves,

$$\frac{dy}{dx} = \frac{\frac{dy}{dt}}{\frac{dx}{dt}} = \frac{2t}{\frac{1}{t}} = 2t^2.$$

■

Finding equations of tangent lines follows the usual pattern. Find the slope, find the point, and use the point-slope form of the equation of a line.

Example 3 *Find the equation of the tangent line to the curve $x = \frac{1}{t}$, $y = \sqrt{t-1}$ at $t = 2$.*

❶ Find the slope (calculate the derivative, evaluate the derivative).

Solution First we notice that the curve is defined parametrically. We therefore begin by calculating the derivative $\frac{dy}{dx}$ using the derivative formula for parametric curves:

$$\frac{dy}{dx} = \frac{\frac{dy}{dt}}{\frac{dx}{dt}} = \frac{\frac{1}{2\sqrt{t-1}}}{-\frac{1}{t^2}} = -\frac{t^2}{2\sqrt{t-1}}.$$

Next we evaluate the derivative at $t = 2$ to find the slope of the tangent line:

$$m = \left.\frac{dy}{dx}\right|_{t=2} = -\frac{4}{2\sqrt{1}} = -2.$$

❷ Find the point.

We also need to determine the point of tangency:

$$(x(2), y(2)) = \left(\frac{1}{2}, 1\right).$$

We then finish by using the point-slope form of the equation of a line:

❸ Use the point-slope form of the equation of a line.

$$y - y_1 = m(x - x_1)$$

$$y - 1 = -2\left(x - \frac{1}{2}\right)$$

$$y = -2x + 2.$$

We conclude that the equation of the tangent line to the curve at $t = 2$ is $y = -2x + 2$. ■

Reading Exercise 4 *Find the slope of the tangent line to the curve $x = t^2$, $y = \cos t$ when $t = \pi/2$.*

Multiple tangent lines at a single point

Because a parametric curve can cross itself, there can be more than one tangent line at the same point. Let's look at this with an example.

Example 4 *Show that the curve $x = \sin \pi t$, $y = t^2 - 1$ has two tangents at the point $(0, 0)$ and find their equations.*

Solution First we notice that the curve is defined parametrically. We know how to find slopes of tangent lines to parametric curves at values of t, but no value of t is given. A point is given instead. We therefore need to determine what value(s) of t yield the point $(0, 0)$. To do this, we need the x-coordinate equation to yield the desired x-coordinate and we need the y-coordinate equation to yield the desired y-coordinate—that is,

When we plotted points to graph a parametric curve, we began with values of t and found the values of x and y. The setup was

t	x	y
0		
1		
2		

Here, we know x and y, but want to find values of t that give them. The setup is

t	x	y
?	0	0

$$\sin \pi t = 0,$$

$$t^2 - 1 = 0,$$

simultaneously.

The easiest of the two equations to solve is $t^2 - 1 = 0$:

$$t^2 = 1$$

$$t = \pm 1.$$

Solving $y = t^2 - 1 = 0$ for t brings us to this situation:

t	x	y
−1	?	0
1	?	0

There are two values of t for which the y-coordinate is zero.

Having solved one of the two equations, we check each of the resulting values of t to determine which of them satisfies the other equation:

Now we need to check the x-coordinates to make sure that they are also zero.

$$t = -1: \quad x = \sin(\pi(-1)) = \sin(-\pi) = 0$$

$$t = 1: \quad x = \sin(\pi(1)) = \sin\pi = 0.$$

We arrive at

t	x	y
-1	0	0
1	0	0

and have the values of t we seek.

We now know that two different values of t yield the point $(0, 0)$.

We are also asked to find the equations of the tangent lines, which may be accomplished by using the process of example 3 for each value of t.

The derivative formula for parametric curves gives

$$\frac{dy}{dx} = \frac{\frac{dy}{dt}}{\frac{dx}{dt}} = \frac{2t}{\pi\cos\pi t}.$$

The slope of the tangent line for $t = -1$ is

$$m = \left.\frac{dy}{dx}\right|_{t=-1} = \frac{2(-1)}{\pi\cos(-\pi)} = \frac{-2}{\pi(-1)} = \frac{2}{\pi}.$$

Ans. to reading exercise 4
$-1/\pi$

The point of tangency is $(0, 0)$, and the equation of the tangent line (for $t = -1$) is

$$y - 0 = \frac{2}{\pi}(x - 0)$$

$$y = \frac{2}{\pi}x.$$

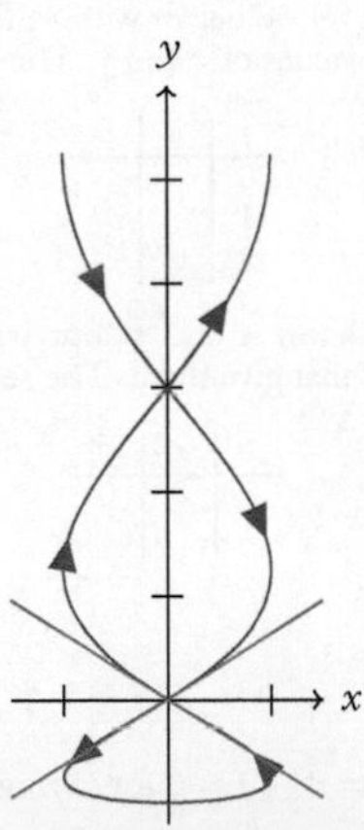

Figure 3 *The parametrically defined curve* $x = \sin\pi t$, $y = t^2 - 1$ *of example 4. The curve passes through the point* $(0, 0)$ *twice, traveling in different directions and resulting in two tangent lines (orange)*

Repeating the process for $t = 1$, the slope of the tangent line is

$$m = \left.\frac{dy}{dx}\right|_{t=1} = \frac{2(1)}{\pi\cos\pi} = \frac{2}{\pi(-1)} = -\frac{2}{\pi},$$

and the equation of the tangent line is

$$y - 0 = -\frac{2}{\pi}(x - 0)$$

$$y = -\frac{2}{\pi}x.$$

The two tangent lines have different equations, so the curve has two tangents at the point $(0, 0)$, as we were asked to show. The tangents are pictured in figure 3. ■

Reading Exercise 5 *Find the equation of the tangent line to the curve* $x = t^2$, $y = 4t$ *at the point* $(9, 12)$.

Second derivatives with parameterized curves

Return to the parametric curve of example 2, which was $x = \ln t$, $y = 2 + t^2$. We calculated $\frac{dy}{dx} = 2t^2$. How might we calculate the second derivative $\frac{d^2y}{dx^2}$?

Before answering this question, let's consider again the process of finding $\frac{dy}{dx}$. We begin with $y = 2 + t^2$, which is y expressed in terms of the variable t. To find the derivative of y with respect to the variable x, we use the derivative formula for parametric curves, which is rewritten slightly as

The variable y is emphasized in bold to highlight its position in the formulas.

$$\frac{d}{dx}(\mathbf{y}) = \frac{\frac{d}{dt}(\mathbf{y})}{\frac{dx}{dt}}.$$

To calculate the second derivative, we start with $\frac{dy}{dx} = 2t^2$, which is $\frac{dy}{dx}$ expressed in terms of the variable t. We wish to find

$$\frac{d^2y}{dx^2} = \frac{d}{dx}\left(\frac{dy}{dx}\right),$$

which is the derivative of $\frac{dy}{dx}$ with respect to the variable x. The setup is exactly the same as for the first derivative, but with $\frac{dy}{dx}$ in place of y. Therefore,

$$\frac{d}{dx}\left(\mathbf{\frac{dy}{dx}}\right) = \frac{\frac{d}{dt}\left(\mathbf{\frac{dy}{dx}}\right)}{\frac{dx}{dt}}.$$

Using the same emphasis, the bold $\frac{dy}{dx}$ occupies the same position held previously by the bold y, leaving all else the same.

In other words, it is the same procedure as taking the first derivative, but we use $\frac{dy}{dx}$ instead of y:

Compare this calculation to that of example 2.

$$\frac{d}{dx}\left(\frac{dy}{dx}\right) = \frac{\frac{d}{dt}\left(\frac{dy}{dx}\right)}{\frac{dx}{dt}} = \frac{4t}{\frac{1}{t}} = 4t^2.$$

SECOND DERIVATIVE $\frac{d^2y}{dx^2}$ FOR A PARAMETRIC CURVE

Let $x(t)$, $y(t)$ be a parametrically defined curve. Then,

$$\frac{d^2y}{dx^2} = \frac{\frac{d}{dt}\left(\frac{dy}{dx}\right)}{\frac{dx}{dt}}$$

wherever the latter expression exists.

As with the first derivative formula, the result is a function of the variable t.

Example 5 *Find* $\frac{d^2y}{dx^2}$ *for* $x = \cos t$, $y = \sin t$.

Solution Before we can find the second derivative, we must find the first derivative. Using the derivative formula for parametric curves,

The advantage of simplifying the expression is that we do not need to use the quotient rule when finding the second derivative.

$$\frac{dy}{dx} = \frac{\frac{dy}{dt}}{\frac{dx}{dt}} = \frac{\cos t}{-\sin t} = -\cot t.$$

Next we use the second derivative formula for parametric curves:

$$\frac{d^2y}{dx^2} = \frac{\frac{d}{dt}\left(\frac{dy}{dx}\right)}{\frac{dx}{dt}} = \frac{-(-\csc^2 t)}{-\sin t} = -\frac{\csc^2 t}{\sin t}.$$

If we wish, we simplify the result to $\frac{d^2y}{dx^2} = -\csc^3 t$. ■

Ans. to reading exercise 5
$y = \frac{2}{3}x + 6$
Details:
When are we at $(9, 12)$?
$t^2 = 9$ gives $t = \pm 3$. When $t = 3$, $y = 12$, so this works. When $t = -3$, $y = -12$, so this does not work. We are at $(9, 12)$ only when $t = 3$:

$$\frac{dy}{dx} = \frac{4}{2t} = \frac{2}{t}$$

$$\left.\frac{dy}{dx}\right|_{t=3} = \frac{2}{3}$$

$$y - 12 = \frac{2}{3}(x - 9).$$

Reading Exercise 6 *Find* $\frac{d^2y}{dx^2}$ *for* $x = t^2$, $y = t^3$.

Exercises 8.2

The names of some of these curves are given. See the *Famous Curves Index* for more information. https://www-history.mcs.st-andrews.ac.uk/Curves/Curves.html

1–16. Find (a) the slope of the tangent line and (b) the equation of the tangent line to the parametric curve at the indicated value of t.

1. $x = t^2 - 3$, $y = t^4 + t^2 - 5$, at $t = -1$
2. $x = 3t^2 - 5t$, $y = t^3 + t + 4$, at $t = 1$
3. $x = \frac{t^2 + 1}{t^2 - 1}$, $y = (4t + 1)^3$, at $t = 2$
4. $x = \sqrt{6t - 1}$, $y = (t + 3)(t - 7)^4$, at $t = 3$
5. $x = 3 + \sin t$, $y = -1 + \cos t$, at $t = \frac{\pi}{4}$
6. $x = 1 + 3\cos t$, $y = 2 + 3\sin t$, at $t = \frac{\pi}{3}$
7. $x = 2^t + t$, $y = t^2 - 9$, at $t = 0$
8. $x = 5t^3 - 6t$, $y = e^{t^2}$, at $t = 0$
9. $x = 2\cos t + \cos 2t$, $y = 2\sin t - \sin 2t$, at $t = \frac{\pi}{2}$ (tricuspoid)
10. $x = 2\cos t - \cos 2t$, $y = 2\sin t - \sin 2t$, at $t = -\frac{\pi}{2}$ (cardioid)
11. $x = \sin^{-1} t$, $y = 3t^2 + 1$, at $t = \frac{1}{2}$
12. $x = t\sin t$, $y = t^2\cos t$, at $t = \pi$

13. $x = \dfrac{3t}{1+t^3}$, $y = \dfrac{3t^2}{1+t^3}$, at $t = 1$ (folium of Descartes)

14. $x = 4\cos^3 t$, $y = 4\sin^3 t$, at $t = \frac{3\pi}{4}$ (astroid)

15. $x = \operatorname{sech} t$, $y = t - \tanh t$, at $t = \ln 2$ (tractrix)

16. $x = 3t$, $y = \dfrac{3}{1+t^2}$, at $t = -\frac{1}{2}$ (witch of Agnesi)

Agnesi was the author of a famous 18th-century book on differential calculus.
Maria Gaëtana Agnesi, 1718–1799
https://www-history.mcs.st-andrews.ac.uk/Biographies/Agnesi.html

17–22. Find $\dfrac{dy}{dx}$.

17. $x = 3t^5 - 8t^2 + 17$, $y = t^{-3}$

18. $x = 4\sqrt{t^3}$, $y = \dfrac{t^2 - 4t}{t^2}$

19. $x = \sin^2 t$, $y = \cos^2 2t$

20. $x = \cosh^2(3t+1)$, $y = 3te^t$

21. $x = (t+4)(t+5)$, $y = t(t+6)^3$

22. $x = \ln\left((t+4)^2 e^t\right)$, $y = \ln\left(\dfrac{t^2}{t^3 - 4}\right)$

23–30. Find $\dfrac{d^2y}{dx^2}$.

23. $x = \sin t$, $y = t^2$

24. $x = \ln t$, $y = t^3 - 4t$

25. $x = \cosh 3t$, $y = \sinh 3t$

26. $x = (4-t)^2$, $y = t^{-1}$

27. $x = 4t$, $y = \cosh t^2$

28. $x = \tan^{-1} t$, $y = t^2 + 3t$

29. $x = 3t + \frac{1}{2}t^2$, $y = \dfrac{2t}{3+t}$

30. $x = e^{2t}$, $y = \dfrac{t}{e^{3t}}$

31–38. Find the equations of all tangent lines to the curve at the indicated point.

31. $x = t^2 + 2t$, $y = t^3 - 19t$, at $(15, -30)$

32. $x = t^2 - 4t$, $y = 2t^3 - 26t$, at $(5, 24)$

33. $x = t^2 + 2t$, $y = t^3 - 19t$, at $(8, -30)$

34. $x = t^2 - 4t$, $y = 2t^3 - 26t$, at $(-3, -24)$

35. $x = t^3 + t^2 - 6t + 1$, $y = \sqrt{t^2 + 3t + 4}$, at $(1, 2)$

36. $x = \sin 3t \cos t$, $y = \sin 3t \sin t$, at $(0, 0)$

37. $x = \sin 3t$, $y = \cos 2t$, at $\left(0, -\frac{1}{2}\right)$

38. $x = \cos t$, $y = \sin 2t$, at $(0, 0)$

Ans. to reading exercise 6
$\frac{d^2y}{dx^2} = \frac{\frac{3}{2}}{2t} = \frac{3}{4t}$

39–42. (GSL) Find every point on the curve at which there is a horizontal tangent line.

39. $x = t^2 - 2t, y = t^3 - 12t$

40. $x = 3 - 4\sin t, y = 4 + 3\cos y$

41. $x = \sin 2t, y = \sin t$

42. $x = \cos^4 t, y = \sin^4 t$

The solution to each of these exercises consists of one picture, which includes both the curve and its tangent line.

43–46. Graph the parametric curve along with the tangent line to the curve at the indicated value of t.

43. $x = 5 + 2\cos t, y = 3 + 2\sin t$ at $t = \frac{\pi}{2}$

44. $x = \sin 3t, y = \cos 3t$ at $t = -\frac{\pi}{4}$

45. $x = \sqrt{t}, y = t - 4$ at $t = 1$

46. $x = t^2 - 1, y = t^2 + t$ at $t = 1$

47–50. Use technology to graph the parametric curve along with the tangent line to the curve at the indicated value of t.

The solution to each of these exercises consists of one picture, which includes both the curve and its tangent line.

47. $x = 2.5\cos t + \cos 2.25t$, $y = 2.5\sin t - \sin 2.5t$ at $t = \frac{21\pi}{4}$

48. $x = 2.5\cos t + \cos 2.25t$, $y = 2.5\sin t - \sin 2.25t$ at $t = \frac{21\pi}{4}$

49. $x = 5\cos t - 3\cos\frac{5}{3}t$, $y = 5\sin t - 3\sin\frac{5}{3}t$ at $t = \frac{23\pi}{4}$ (epicycloid)

50. $x = 7\cos t - 7\cos\frac{7}{5}t$, $y = 7\sin t - 7\sin\frac{7}{5}t$ at $t = \frac{23\pi}{4}$ (epitrochoid)

A wide variety of graphs of curves similar to the ones requested here using technology can be produced by hand using the drawing toy Spirograph, which I greatly enjoyed using as a child.

51. (a) Use the pattern from the development of the second derivative formula to write a formula for $\frac{d^3y}{dx^3}$ for a parametric curve.
(b) Find $\frac{d^3y}{dx^3}$ for $x = \cos t, y = \sin t$.

52. (a) Using the result of part (a) of exercise 51, find $\frac{d^3y}{dx^3}$ for $x = \ln t$, $y = 2 + t^2$. (b) Eliminate the parameter and write y as a function of x. (c) Use the result of (b) to calculate y'''. (d) Show that the answers to parts (a) and (c) are equivalent.

53. Without eliminating the parameter, determine the interval(s) of the variable t for which the graph of the parametric curve $x = 2^t$, $y = 1 - t^3$ is concave up and the interval(s) for which it is concave down.

54. Without eliminating the parameter, determine the interval(s) of the variable t for which the graph of the parametric curve $x = \cos t$, $y = \sin t$, $0 \le t \le 2\pi$ is concave up and the interval(s) for which it is concave down.

55. (a) Use the formula $m = \dfrac{y(k+\alpha) - y(k)}{x(k+\alpha) - x(k)}$ to calculate the slope of the tangent line to the curve $x = t^3 - 3t^2 + 3t + 4$, $y = \frac{1}{t}$ at $t = 1$. (Use $k = 1$ in the formula.) (b) Use the result of part (a) to describe the tangent line. (c) Give an equation of the tangent line.

The formula $m = \dfrac{y(k+\alpha) - y(k)}{x(k+\alpha) - x(k)}$, where α represents an arbitrary infinitesimal, was an intermediate step in the development of the formula for the slope of the tangent line to a parametric curve.

56. (a) Use the formula $m = \dfrac{y(k+\alpha) - y(k)}{x(k+\alpha) - x(k)}$ to calculate the slope of the tangent line to the curve $x = t - t^2$, $y = 4 + \frac{1}{t}$ at $t = \frac{1}{2}$. (Use $k = \frac{1}{2}$ in the formula.) (b) Give an equation of the tangent line. (c) Use technology to graph the curve. Then, looking at the graph, determine the signs (positive or negative) of the slopes of the tangent lines on either side of the point corresponding to $t = \frac{1}{2}$. (d) Repeat part (c) using the curve (and the value of t) of exercise 55. (e) Based on the results of parts (c) and (d), assuming the tangent line to a parametric curve at $t = k$ is vertical and there is no corner in the graph at $t = k$, explain the relationship between (i) properties of the graph and (ii) being able to render ∞ or $-\infty$ in the calculation in part (a) or being "stuck."

57. (a) For the parametric curve $x = t^3 - 3t^2 + 3t + 4$, $y = \frac{1}{t}$, find $\left.\dfrac{dx}{dt}\right|_{t=1}$ and $\left.\dfrac{dy}{dt}\right|_{t=1}$. (b) For the parametric curve $x = 3(t - \sin t)$, $y = 3(1 - \cos t)$, find $\left.\dfrac{dx}{dt}\right|_{t=2\pi}$ and $\left.\dfrac{dy}{dt}\right|_{t=2\pi}$. (c) The graph of the parametric curve of part (b) exhibits a corner (a cusp) at $t = 2\pi$ (see section 8.1, exercise 9). The graph of the parametric curve of part (a) does not have a corner at $t = 1$ (see exercise 56, part (d)). The condition $\left.\dfrac{dx}{dt}\right|_{t=k} = 0$ is enough to deny the existence of $\frac{dy}{dx}$ at $t = k$. Is the condition enough to guarantee a corner at $t = k$? (d) Write a conjecture about a condition that must happen in order for a corner to be present at $t = k$.

Polar Coordinates

8.3

There is more than one way to describe the location of a point. One method uses the familiar Cartesian coordinate system that you first encountered in an algebra course, or perhaps even earlier. In this section we study a different method: the polar coordinate system.

Polar coordinates

Cartesian coordinates (figure 1) are akin to giving directions in a city laid out in a grid. To travel from the origin $(0, 0)$ to the point $(3, 1)$, we would say to travel three blocks east and one block north. Because there are obstacles (buildings) in the way of traveling on a direct line to the place we wish to go, giving east–west and north–south directions in this environment makes sense.

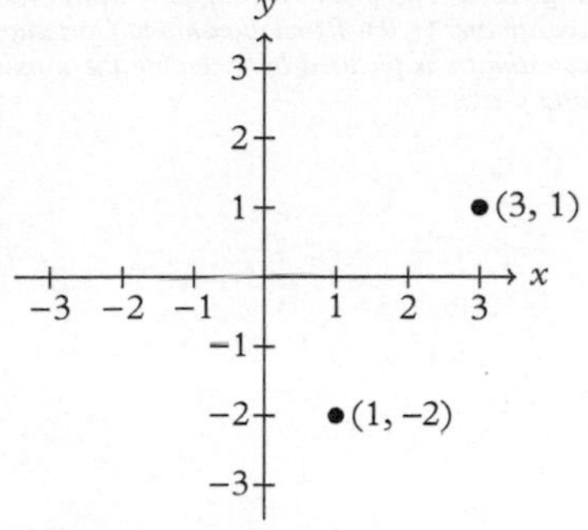

Figure 1 *The Cartesian coordinate system, with points* $(3, 1)$ *and* $(1, -2)$ *plotted*

Suppose, instead, that we are in the middle of an open field, and we wish to state the location of an object relative to where we are standing. Rather than giving east–west and north–south distances, we point out the correct direction—any direction we wish, because there are no obstacles in the way—and then give the distance to the object. "Turn 30 degrees to your left and walk 100 paces" suffices to give the object's location. Specifying an angle and a distance is the idea of polar coordinates.

In the polar coordinate system (see figure 2), we describe the direction as an angle θ given in standard position, measured in radians.

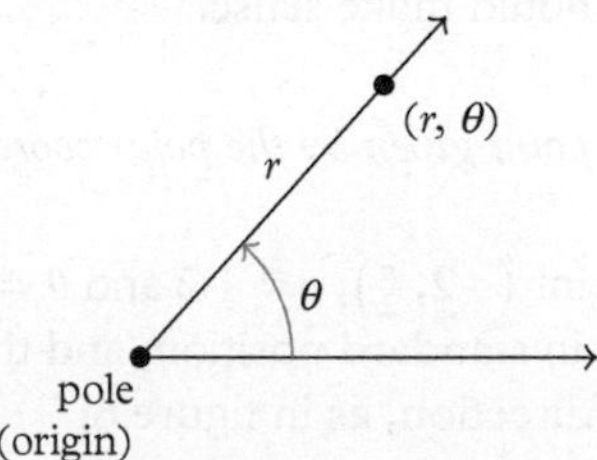

Figure 2 *The polar coordinate system, with coordinates* (r, θ)*. Beginning at the origin, or pole, the angle* θ*, given in standard position, describes the direction to the point, and r describes the distance to the point*

As always, positive angles describe counterclockwise rotation whereas negative angles describe clockwise rotation. The "distance" is described using a real number r, and it includes negative values of r (more on this in a moment). We still use ordered pair notation for the coordinates, using the order (r, θ). The term *origin* refers to the point $(0, 0)$, although it also goes by the name *pole* in polar coordinates. We may also refer to the initial side and terminal side of the angle in the same manner as in trigonometry. The relationship to the x- and y-axes is pictured in figure 3.

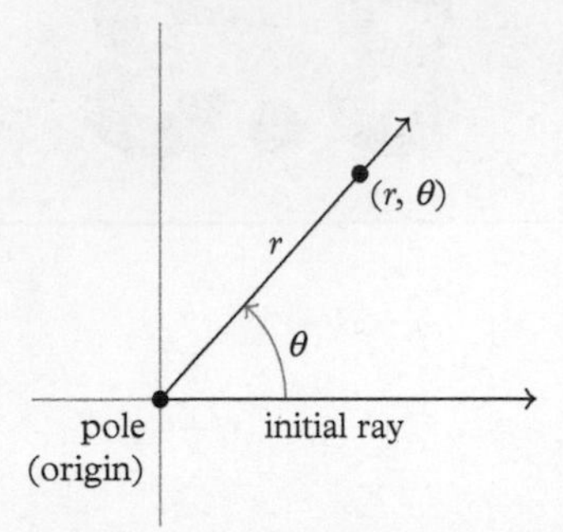

Figure 3 *The polar coordinate system, with coordinates (r, θ). Its relationship to Cartesian coordinates is pictured by including the x-axis and y-axis*

Example 1 *Plot the point given by the polar coordinates* $\left(1, \frac{\pi}{4}\right)$.

Solution The coordinates are in the order (r, θ); hence, $r = 1$ and $\theta = \frac{\pi}{4}$. We draw an angle measuring $\frac{\pi}{4}$ radians in standard position,

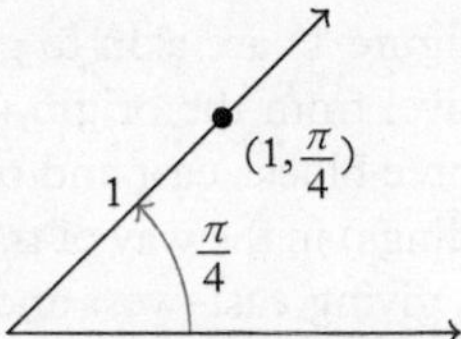

Figure 4 *The point with polar coordinates* $\left(1, \frac{\pi}{4}\right)$

and then locate our point 1 unit from the origin along the terminal side of the angle. See figure 4 for the solution. ■

In example 1, the distance of 1 unit is measured along the terminal side of the angle; it is not an x-coordinate. One way of thinking about how to locate the point is to begin with a horizontal number line, then rotate that number line by the angle $\theta = \frac{\pi}{4}$. The point $\left(1, \frac{\pi}{4}\right)$ is located at $r = 1$ on the rotated number line, as shown in figure 5.

By thinking of a rotated number line, the placement of points with negative values of r should make sense.

Example 2 *Plot the point given by the polar coordinates* $\left(-2, \frac{\pi}{2}\right)$.

Solution For the point $\left(-2, \frac{\pi}{2}\right)$, $r = -2$ and $\theta = \frac{\pi}{2}$. We draw an angle measuring $\frac{\pi}{2}$ radians in standard position, and then locate our point 2 units in the negative direction, as in figure 6. ■

The rotated-number-line picture of the location of the point in example 2 is given in figure 7.

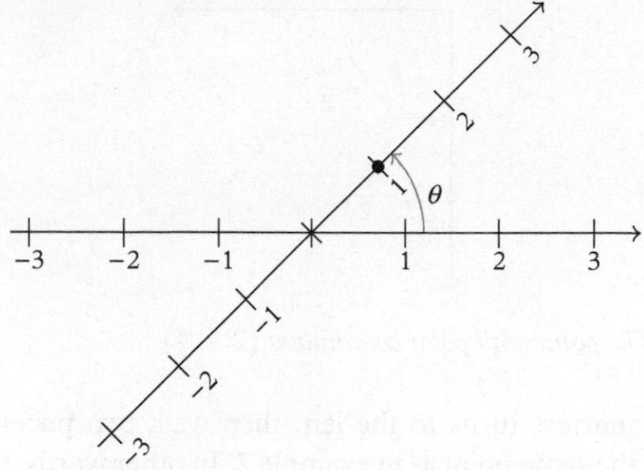

Figure 5 *A horizontal number line rotated by the angle $\theta = \frac{\pi}{4}$, with the point $\left(1, \frac{\pi}{4}\right)$ marked. The pole (origin) is the center of the rotation*

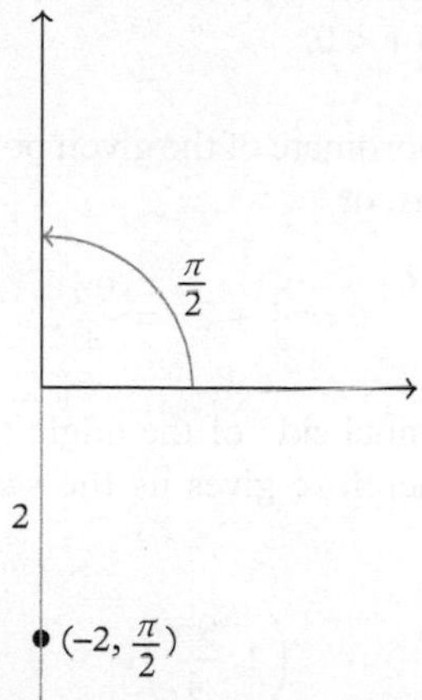

Figure 6 *The point with polar coordinates $\left(-2, \frac{\pi}{2}\right)$*

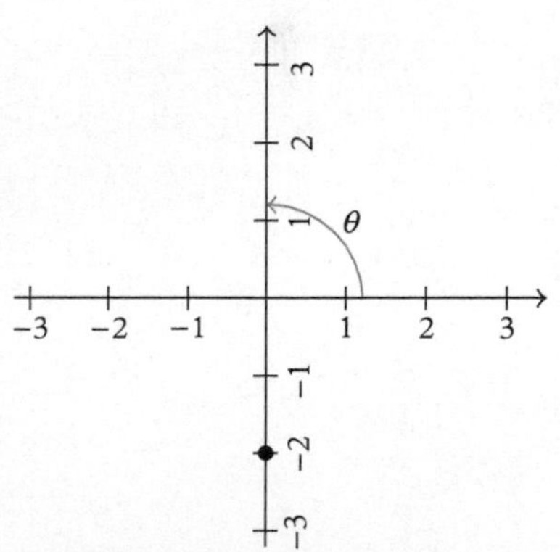

Figure 7 *A horizontal number line rotated by the angle $\theta = \frac{\pi}{2}$, with the point $\left(-2, \frac{\pi}{2}\right)$ marked*

Returning to the idea of giving directions in an open field, the polar coordinates of example 2 (figure 6) is equivalent to giving the directions "face north, then walk two paces backward." The immediate question in the mind of the person receiving the directions is likely to be, "Why didn't you just tell me to face south and walk two paces forward?" (figure 8). This illustrates an interesting feature of polar coordinates: the coordinates of a point are not unique! With rectangular (Cartesian) coordinates, there is one and only one way to write the coordinates of a given point. With polar coordinates, there are many ways (infinitely many, in fact) to write the coordinates of any given point. We could tell a person standing at the origin facing east to rotate two

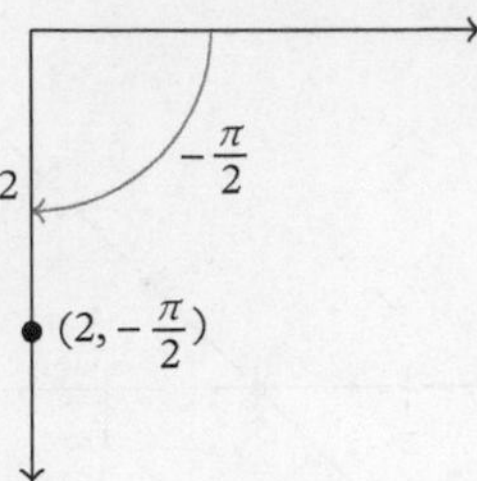

Figure 8 *The point with polar coordinates* $\left(2, -\frac{\pi}{2}\right)$

Asking the person to rotate $10\frac{3}{4}$ turns is more entertaining.

and three-quarters turns to the left, then walk two paces, and they would find the same point as in example 2. In other words, the ordered pairs $\left(-2, \frac{\pi}{2}\right)$, $\left(2, -\frac{\pi}{2}\right)$, and $\left(2, \frac{11\pi}{2}\right)$ all represent the same point in the polar coordinate system.

Example 3 *Give another set of polar coordinates for the point* $\left(1, \frac{\pi}{4}\right)$*, for which (a)* $r > 0$ *and (b)* $r < 0$.

Solution (a) The θ-coordinate of the given point is $\theta = \frac{\pi}{4}$. By rotating an additional 2π radians, or

$$\theta = \frac{\pi}{4} + 2\pi = \frac{9\pi}{4},$$

we have the same terminal side of the angle (see figure 9). Keeping the same value of r therefore gives us the same point. The desired coordinates are

$$\left(1, \frac{9\pi}{4}\right).$$

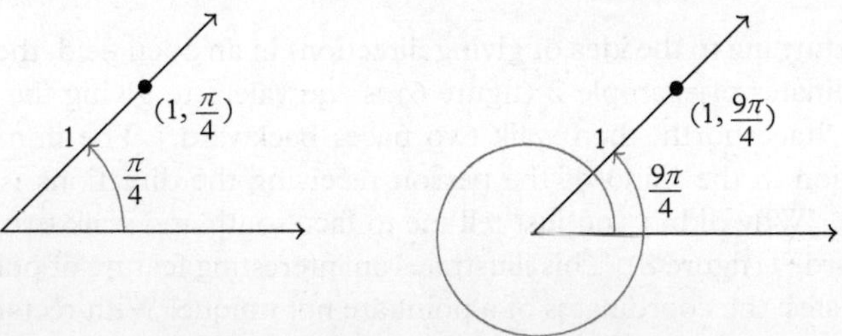

Figure 9 *(left) The point with polar coordinates* $\left(1, \frac{\pi}{4}\right)$*; (right) the point with polar coordinates* $\left(1, \frac{9\pi}{4}\right)$

Other possible solutions include adding any other multiple of 2π radians to θ, a few of which are $\left(1, \frac{17\pi}{4}\right)$, $\left(1, \frac{-7\pi}{4}\right)$, and $\left(1, \frac{4001\pi}{4}\right)$.

(b) To get the same point using a negative r-coordinate, we need to change the sign of r. This can be accomplished by an additional half-rotation, so that the number line is rotated halfway around, switching the locations of positive and negative coordinates. See figure 10. The

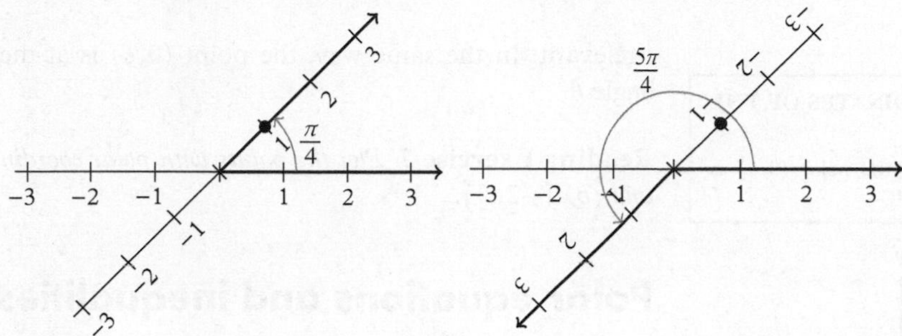

Figure 10 *(left) A horizontal number line rotated by the angle* $\theta = \frac{\pi}{4}$, *with the point* $\left(1, \frac{\pi}{4}\right)$ *marked; (right) a horizontal number line rotated by the angle* $\theta = \frac{5\pi}{4}$, *with the point* $\left(-1, \frac{5\pi}{4}\right)$ *marked*

new θ-coordinate is found by adding π radians (a half-rotation),

$$\theta = \frac{\pi}{4} + \pi = \frac{5\pi}{4}.$$

The desired coordinates are

$$\left(-1, \frac{5\pi}{4}\right).$$

Other possible solutions are found by adding a multiple of 2π to the θ-coordinate of the existing solution, such as $\left(-1, \frac{13\pi}{4}\right)$, $\left(-1, \frac{205\pi}{4}\right)$, and $\left(-1, \frac{-11\pi}{4}\right)$. ■

Example 4 *Plot the point given by the polar coordinates* $(0, 3)$.

Solution We are given $r = 0$ and $\theta = 3$. Because 3 is a little less than π, the angle is a little less than a half rotation. With $r = 0$, the point is located 0 unit from the origin along that line, so the point is at the origin. ■

Back in the open field again, if we tell someone to face northwest and then look down because they are already standing at the desired location, they may wonder why we specified the direction at all. It is

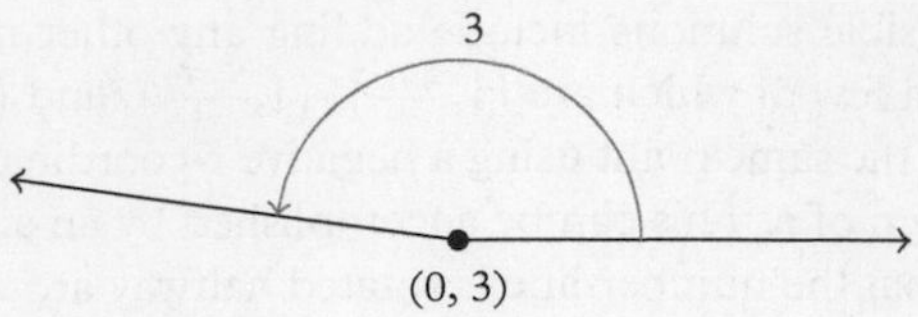

Figure 11 *The point with polar coordinates* $(0, 3)$

irrelevant. In the same way, the point $(0, \theta)$ is at the origin for any angle θ.

> **POLAR COORDINATES OF THE ORIGIN**
>
> For any angle θ, the point $(0, \theta)$ is located at the origin.

Reading Exercise 7 *Plot the points with polar coordinates (a)* $\left(1, \frac{3\pi}{2}\right)$ *and (b)* $\left(-2, \frac{\pi}{4}\right)$.

Polar equations and inequalities

In Cartesian coordinates, the graph of an equation is the set of points that satisfies the relationship given by the equation. For instance, the graph of $y = x^2 - 1$ (figure 12) is the set of all points of the form (x, y) for which $y = x^2 - 1$. In set notation, the graph is the set of points

$$\left\{(x, y) \mid y = x^2 - 1\right\}.$$

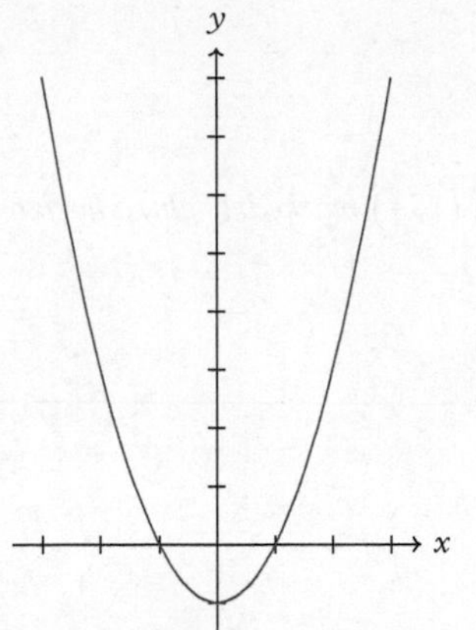

Figure 12 *The graph of* $y = x^2 - 1$. *Points on the graph satisfy the equation. For instance, the point* $(-2, 3)$ *is on the graph because, for* $x = -2, y = 3, 3 = (-2)^2 - 1$

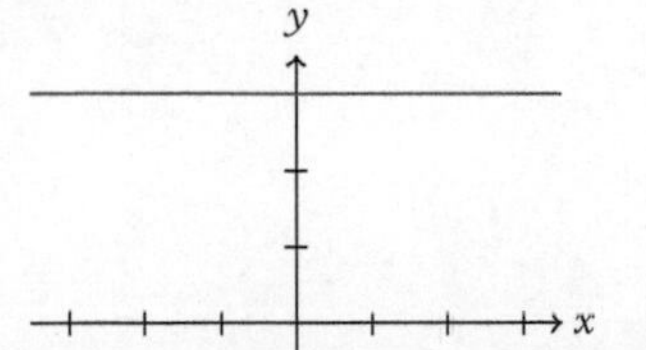

Figure 13 *The graph of* $y = 3$, *consisting of all points with y-coordinate 3*

Plotting points, playing the connect-the-dot game for continuous functions, determining properties such as concavity, and recognizing equation types such as those giving rise to parabolas or lines are all helpful strategies for drawing a graph by hand.

An equation is not required to use both variables. For instance, the horizontal line $y = 3$ (figure 13) consists of every point with a y-coordinate of 3; the x-coordinate is any real number.

Graphs of polar equations follow the same ideas, using the variables r and θ. The graph consists of all points of the form (r, θ) that satisfy the equation.

Example 5 *Graph the polar equation* $r = 3$.

Solution The graph consists of all points with r-coordinate 3. We therefore want all points for which the distance from the pole is 3. This is the description of a circle centered at the origin with radius 3, which is easy to draw. The solution is in figure 14. ■

Another way to think of the solution to example 5 is to use the idea of a rotating number line, which we saw in figure 5. Because the angle can be anything, we follow the location of the 3 on the number line as it makes a full rotation (figure 15).

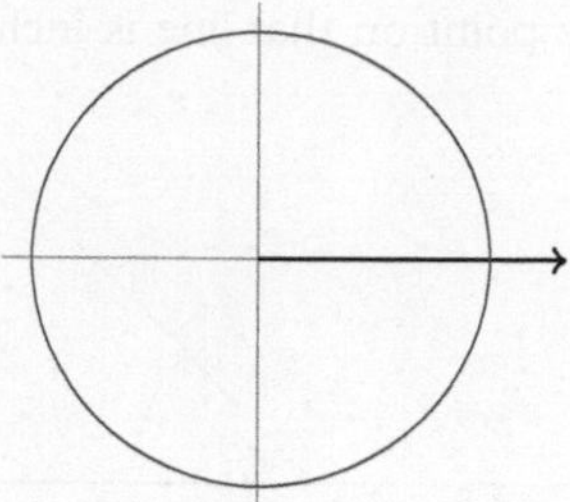

Figure 14 *The graph of r = 3 (blue), which is a circle centered at the origin with radius 3. The initial ray is drawn in black. For the purposes of comparison, the x- and y-axes are also included (gray)*

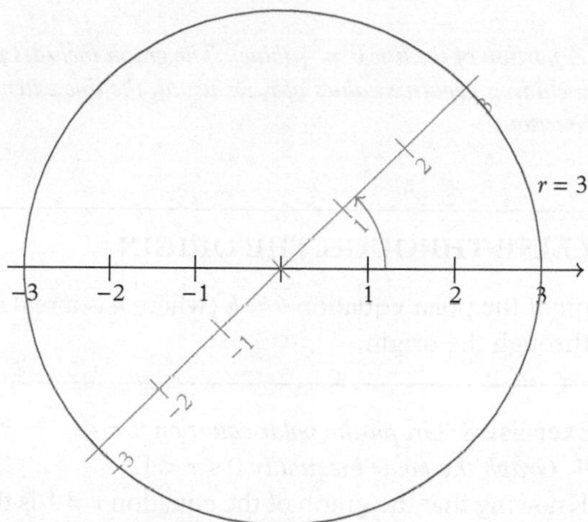

Figure 15 *The circle r = 3 (blue), traced out by the number 3 on the number line as it makes a full rotation. The rotating number line (gray) is illustrated partway through the rotation*

Now use figure 15 to follow the number -3 on the number line for a full rotation. The same circle of radius 3 is traced out! The graph of the equation $r = -3$ is the same as the graph of $r = 3$.

> **POLAR CIRCLE CENTERED AT THE ORIGIN**
>
> The graph of the polar equation $r = k$ (for $k \neq 0$) is a circle of radius $|k|$ centered at the origin.

Example 6 *Graph the polar equation* $\theta = \frac{\pi}{4}$.

Solution The graph consists of all points with θ-coordinate $\frac{\pi}{4}$. Thinking of the rotated number line, because r may be any number (positive,

Ans. to reading exercise 7

(a)

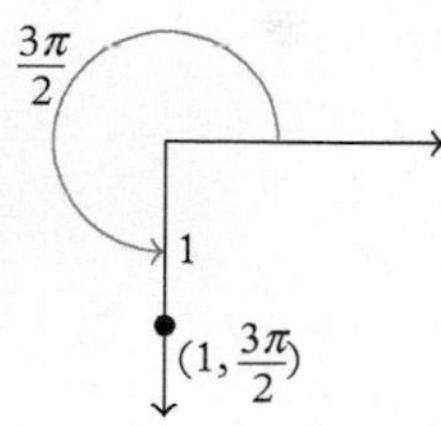

(b)

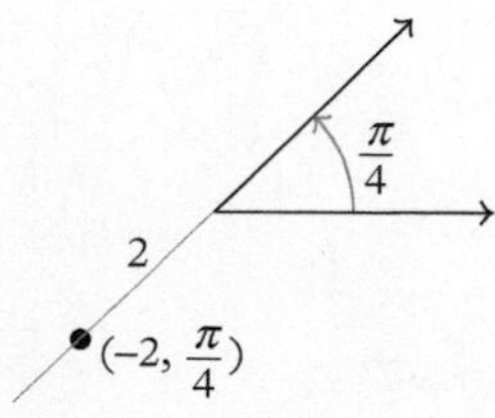

(**Note:** the scales on the two pictures are not the same.)

negative, zero), every point on that line is included. The result is in figure 16. ■

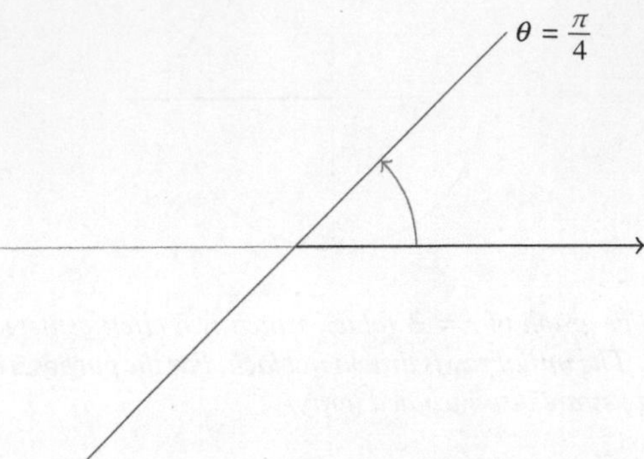

Figure 16 *A portion of the line $\theta = \frac{\pi}{4}$ (blue). The graph includes points for all values of r, including negative values of r. As usual, the line extends infinitely far in both directions*

POLAR LINE THROUGH THE ORIGIN

The graph of the polar equation $\theta = k$ (where k is a real number) is a line through the origin.

Reading Exercise 8 *Graph the polar equation $\theta = \frac{\pi}{2}$.*

Example 7 *Graph the polar inequality $0 \le r < 1$.*

Solution Knowing that the graph of the equation $r = 1$ is the set of all points 1 unit from the origin, points satisfying the inequality $0 \le r < 1$ are all points less than 1 unit from the origin—that is, the interior of the circle $r = 1$. The solution is in figure 17. The circle $r = 1$ is dashed to indicate that it is not included in the region. ■

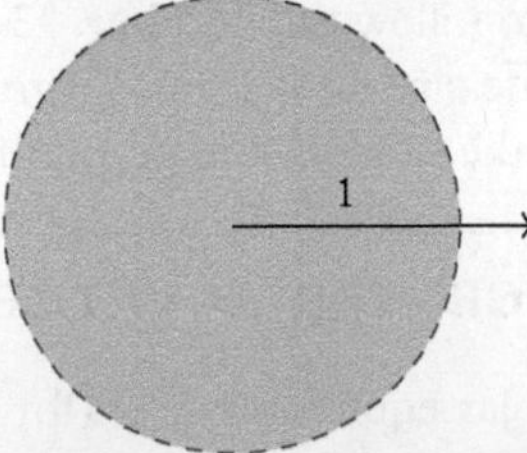

Figure 17 *The graph of $0 \le r < 1$ (blue region), which is the interior of the circle centered at the origin with radius 1. The initial ray is drawn in black. The boundary of the circle is dashed to indicate that it is not included*

Just as with Cartesian coordinates, to indicate that a line or curve is included in a region, it is drawn solid; to indicate that it is not included in the region, we use a dashed line. For instance, the region $0 \le r \le 1$ has the boundary drawn solid as in figure 18.

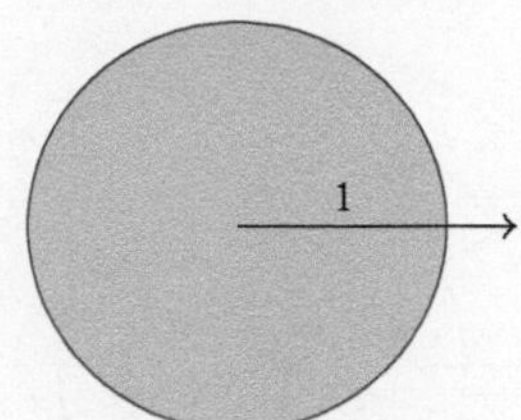

Figure 18 *The graph of $0 \le r \le 1$ (blue region), which is the interior of the circle centered at the origin with radius 1. The boundary of the circle is solid to indicate that it is included; compare this figure to figure 17*

Example 8 *Graph the polar inequality $\frac{\pi}{4} < \theta \le \frac{\pi}{2}$.*

Solution We begin by noting that the line $\theta = \frac{\pi}{4}$ should be dashed because it is not included in the region, whereas the line $\theta = \frac{\pi}{2}$ should be solid because it is included in the region. We also want all points with θ-coordinates between $\frac{\pi}{4}$ and $\frac{\pi}{2}$ to be shaded; see figure 19. (Think of

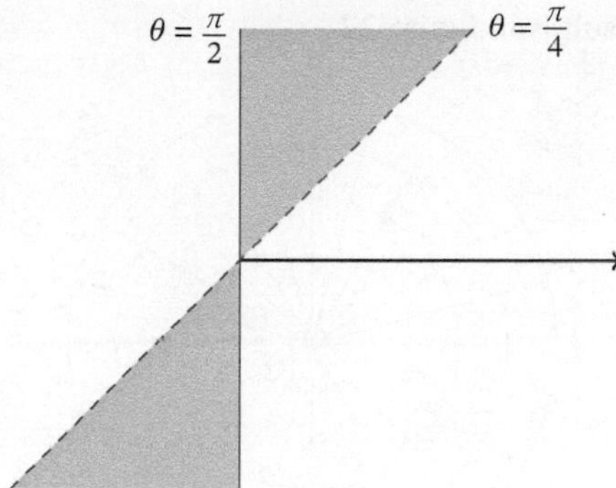

Figure 19 *The region $\frac{\pi}{4} < \theta \le \frac{\pi}{2}$ (blue). The line $\theta = \frac{\pi}{4}$ is dashed to indicate it is not included, whereas the line $\theta = \frac{\pi}{2}$ is solid to indicate it is included. Note that both positive and negative values of r are included. The lines and the region extend infinitely far in both directions. The initial ray is included for reference*

the rotating number line, and where the number line is located as it rotates from $\theta = \frac{\pi}{4}$ to $\theta = \frac{\pi}{2}$; this is the region to be shaded.) ■

Polar graph paper

You may be familiar with graph paper for the Cartesian coordinate system. Traditionally, the paper has light-blue lines laid out on evenly spaced, constant values of the x- and y-coordinates, so points are located easily. As a reminder, plotting the points with Cartesian coordinates $(3, 1)$ and $(1, -2)$ using graph paper is illustrated in figure 20.

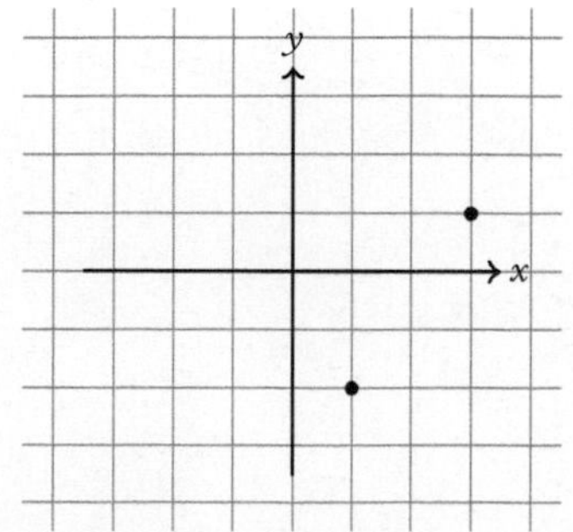

Figure 20 *Graph paper for Cartesian coordinates, with points $(3, 1)$ and $(1, -2)$ plotted*

Printable files containing polar graph paper are also easy to find. The paper has markings laid out on evenly spaced, constant values of the r- and θ-coordinates, so that points are located easily. The polar

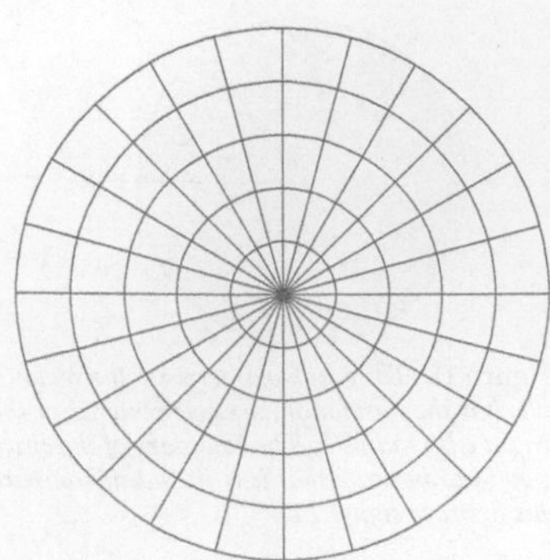

Figure 21 *Graph paper for polar coordinates. The concentric circles are for r-coordinates of 1, 2, 3, ... The radial lines are for angles (θ-coordinates) in multiples of $\frac{\pi}{12}$ radians*

graph paper in figure 21 has θ-coordinates marked in increments of $\frac{\pi}{12}$ radians.

Example 9 *Use polar graph paper to plot the points (a)* $\left(3, \frac{2\pi}{3}\right)$ *and (b)* $\left(-4, \frac{11\pi}{4}\right)$.

Solution (a) For the point $\left(3, \frac{2\pi}{3}\right)$, we first locate $\theta = \frac{2\pi}{3}$. Because the grid lines are laid out in increments of $\frac{\pi}{12}$, we count

$$\frac{2\pi}{3} \cdot \frac{12}{\pi} = \frac{24}{3} = 8$$

grid lines counterclockwise from the initial ray. Then, along that grid line, we locate $r = 3$, which is on the third circle (counting from the origin). The result is in figure 22.

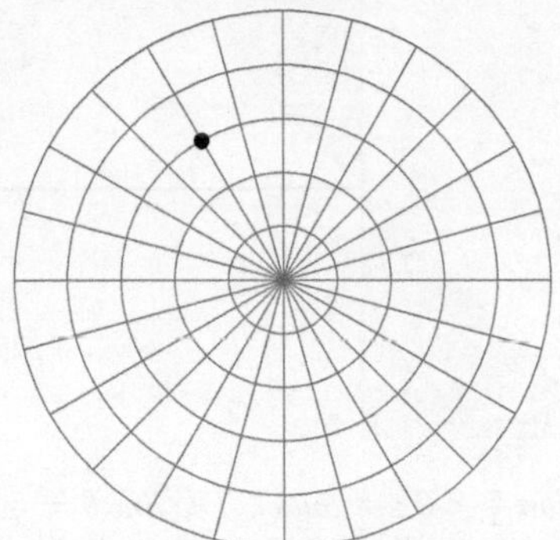

Figure 22 *The point with polar coordinates* $\left(3, \frac{2\pi}{3}\right)$ *plotted on polar graph paper*

Adding or subtracting a multiple of 2π radians to obtain $0 \le \theta < 2\pi$ is often convenient.

(b) For the point $\left(-4, \frac{11\pi}{4}\right)$, we first locate $\theta = \frac{11\pi}{4}$. Noticing that θ is given as an angle larger than 2π, we subtract 2π and rewrite the point as

$$\left(-4, \frac{3\pi}{4}\right).$$

Next we count

$$\frac{3\pi}{4} \cdot \frac{12}{\pi} = \frac{36}{4} = 9$$

grid lines counterclockwise from the initial ray. Then, along that grid line, we locate $r = -4$, which is on the fourth circle but in the opposite direction from the ninth grid line. The result is in figure 23.

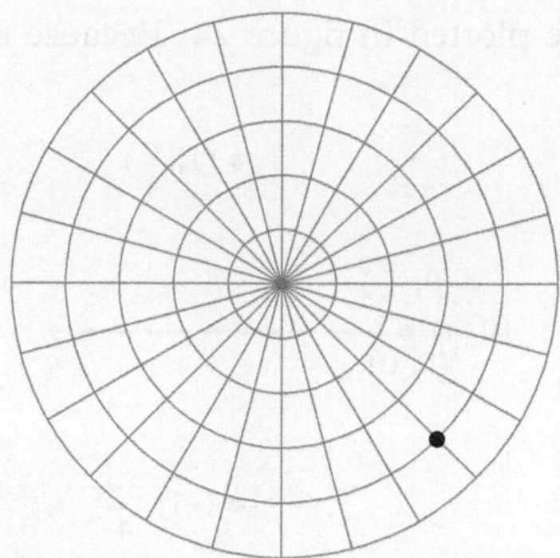

Figure 23 *The point with polar coordinates* $\left(-4, \frac{11\pi}{4}\right)$*, plotted on polar graph paper* ■

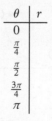

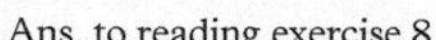

Ans. to reading exercise 8

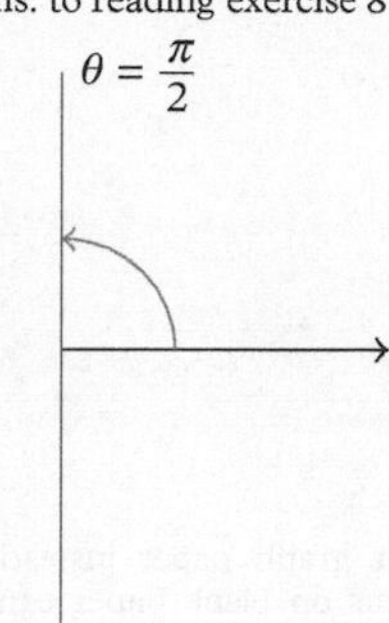

Graphing polar curves

When you first learned to graph functions in Cartesian coordinates, you plotted points and connected the dots. The same procedure applies to polar functions, where the variable r is written as a function of the variable θ.

For a polar function $r = f(\theta)$, the independent variable is θ and the dependent variable is r.

Example 10 *Graph the polar curve* $r = \sin 2\theta$.

Solution We begin by plotting points. We need to choose values of the independent variable θ. Because the function involves a trig function of 2θ, it seems reasonable to choose values of θ ranging from $\theta = 0$ to $\theta = \pi$.

θ	r
0	
$\frac{\pi}{4}$	
$\frac{\pi}{2}$	
$\frac{3\pi}{4}$	
π	

Next we determine the values of r by using the equation $r = \sin 2\theta$.

θ	r
0	0
$\frac{\pi}{4}$	1
$\frac{\pi}{2}$	0
$\frac{3\pi}{4}$	-1
π	0

It is unfortunate that the traditional notation for Cartesian coordinates places the independent variable on the left, (x, y), whereas the traditional notation for polar coordinates places the independent variable on the right, (r, θ). Here, we tabulate the values by always placing the independent variable on the left in the table. Therefore, the order must be switched when writing the corresponding points to be plotted. In this case, the points are $(0, 0)$, $\left(1, \frac{\pi}{4}\right)$, $\left(0, \frac{\pi}{2}\right)$, $\left(-1, \frac{3\pi}{4}\right)$, and $(0, \pi)$.

The five points are plotted in figure 24. Because some of the points

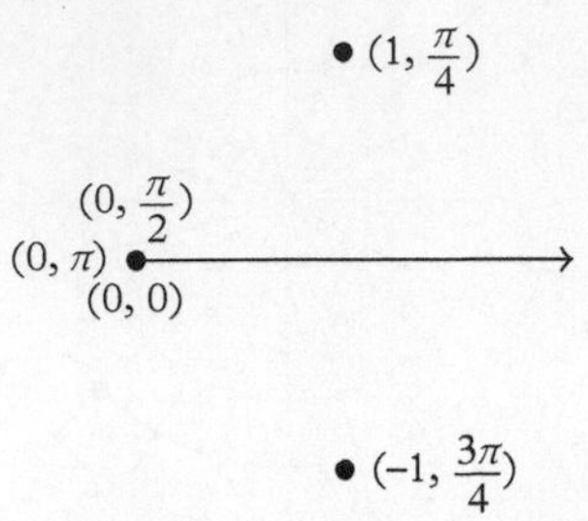

$(1, \frac{\pi}{4})$

$(0, \frac{\pi}{2})$

$(0, \pi)$

$(0, 0)$

$(-1, \frac{3\pi}{4})$

Figure 24 *Points on the graph of* $r = \sin 2\theta$. *The initial ray is included for reference*

coincide, there are only three distinct points with which to work. How do we draw the curve from this information? We don't. More information is needed. There are two options. One is to fill in the gaps by choosing other values of θ, obtaining more points to plot (this is what most graphing technology does). The second option is to use additional knowledge of the trig functions to fill in the gaps.

Using polar graph paper instead of eyeballing points on blank paper can be very helpful with the first option. See the comments after the solution to this example.

Following the second option, note that as θ goes from $\theta = 0$ to $\theta = \frac{\pi}{4}$, the values of $r = \sin 2\theta$ increase from 0 to 1. Consider the rotating number line as it rotates from $\theta = 0$ to $\theta = \frac{\pi}{4}$. The value of r increases from 0 to 1. Physically, you can imagine allowing your pencil to slide from 0 to 1 while the ruler is rotating. The resulting portion of the graph is in figure 25.

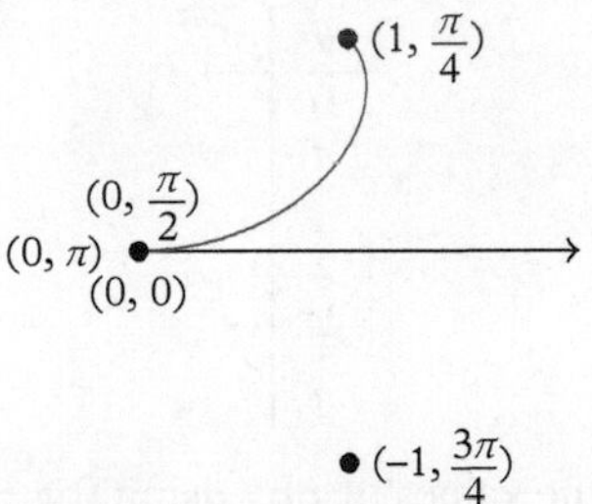

Figure 25 *The portion of the graph of* $r = \sin 2\theta$ *for which* $0 \le \theta \le \frac{\pi}{4}$. *The initial ray is included for reference*

Continuing, the values of r decrease from 1 to 0 as θ goes from $\frac{\pi}{4}$ to $\frac{\pi}{2}$, so the curve returns back to the origin. From $\theta = \frac{\pi}{2}$ to $\theta = \frac{3\pi}{4}$, the values of r turn negative, going from 0 to -1. As the number line is rotating through these values of θ, positive values of r are to the upper left and negative values of r are to the lower right. Then, in the next segment, the values of r return to the origin. The result is in figure 26.

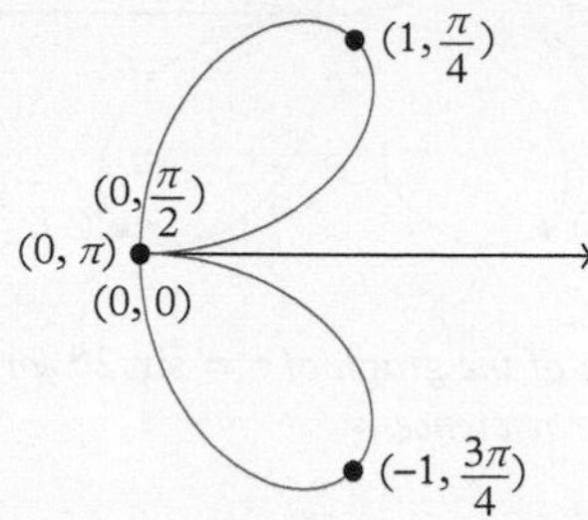

Figure 26 *The portion of the graph of $r = \sin 2\theta$ for which $0 \leq \theta \leq \pi$. The initial ray is included for reference*

Figure 26 looks good, but are we finished? Do we begin retracing the curve if we expand our interval of values of θ? Or do we trace out additional parts of the curve that have not yet been drawn? One method to find out is to expand our interval and see what happens. A continuation of the earlier table to $\theta = 2\pi$ is as follows.

In the same manner that graphing technology can be used to determine an interval of values of the independent variable t to trace a parametric curve exactly once, graphing technology can be used to determine an interval of values of the independent variable θ to trace a polar curve exactly once.

θ	r
$\frac{5\pi}{4}$	1
$\frac{3\pi}{2}$	0
$\frac{7\pi}{4}$	1
2π	0

These points are added in figure 27. Because some of the points are

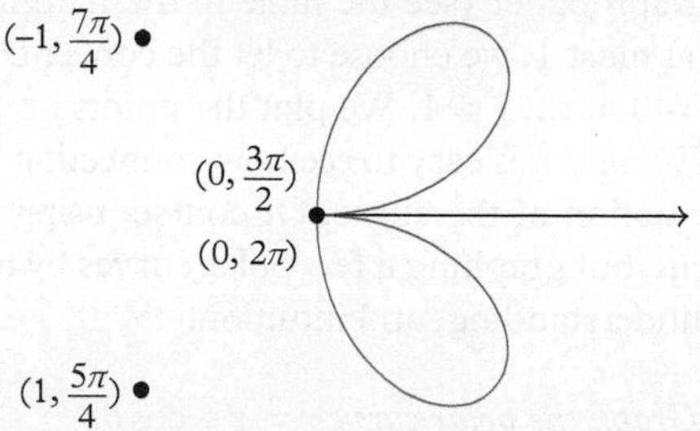

Figure 27 *More points added to the sketch*

new, it is obvious there is more to the curve than what is in figure 26. Using the "second option" method we sketch the portion of the curve from $\theta = \pi$ to $\theta = 2\pi$, which is added in figure 28. Continuing to use

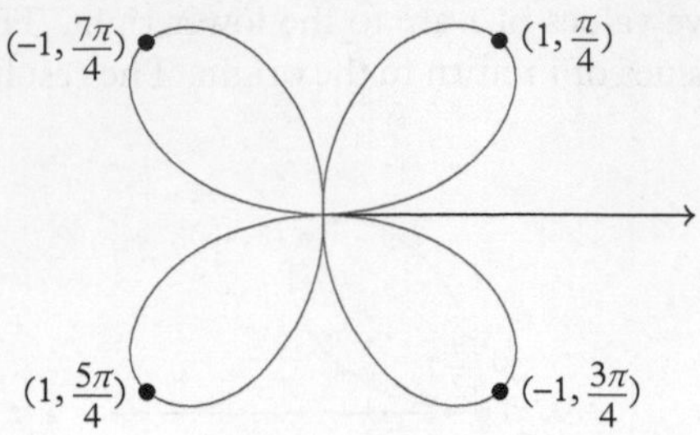

Figure 28 *The portion of the graph of* $r = \sin 2\theta$ *for which* $0 \le \theta \le 2\pi$*. The initial ray is included for reference*

the "second option" method, and picturing the rotating number line and the corresponding values of r, we see that we begin to retrace the curve after $\theta = 2\pi$. Figure 28 shows the complete solution. The name of this curve is the *four-leaved rose.* ■

CAUTION: MISTAKE TO AVOID
Polar curves involving trigonometric functions may or may not begin to retrace the curve after one period of the trig function. An interval for tracing the curve exactly once may be shorter or longer than the period of the trigonometric function.

Why was $\theta = 0$ to $\theta = \pi$ insufficient to trace out the entire curve $r = \sin 2\theta$ in example 10? Although the values of r begin to repeat after $\theta = \pi$ (because the period of $\sin 2\theta$ is π), the rotating number line is not in the same position. After rotating by an angle of π, positive and negative are switched; therefore, even though we are repeating the values of r, the points are not in the same place. But, by going from $\theta = 0$ to $\theta = 2\pi$, the rotating number line returns to its original position, and the values of r begin to repeat yet again.

θ	r
0	0
$\frac{\pi}{12}$	$\frac{1}{2}$
$\frac{2\pi}{12}$	$\frac{\sqrt{3}}{2}$
$\frac{3\pi}{12}$	1
$\frac{4\pi}{12}$	$\frac{\sqrt{3}}{2}$
$\frac{5\pi}{12}$	$\frac{1}{2}$
$\frac{6\pi}{12}$	0

Using polar graph paper for more precise location of plotted points can sometimes be a good alternative to using the "second method" featured in the solution to example 10. For the upper-right "leaf" of the rose, we first calculate points by increments of $\frac{\pi}{12}$ to match the design of the graph paper (see the table in the margin). Because the values of r are at most 1, we choose to let the concentric circles represent $r = 0.2$, $r = 0.4$, …, $r = 1$. We plot the points on the polar graph paper (figure 29), and it is easy to see how connecting these dots gives a good representation of the curve. Of course, using technology can be more efficient, but graphing a few polar curves by hand is good for building skill, understanding, and intuition.

Example 11 *Graph the polar curve* $r = 1 + \cos\theta$*.*

After plotting points and sketching the curve, don't forget to check to see if the chosen values of θ are sufficient.

Solution The period of the trigonometric function is 2π, so let's try values of θ ranging from $\theta = 0$ to $\theta = 2\pi$:

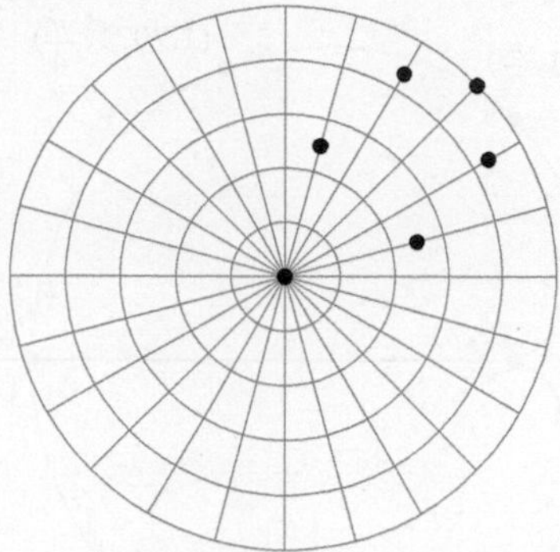

Figure 29 *Points on the first leaf of the four-leaved rose* $r = \sin 2\theta$

θ	r
0	2
$\frac{\pi}{4}$	$1 + \frac{\sqrt{2}}{2}$
$\frac{\pi}{2}$	1
$\frac{3\pi}{4}$	$1 - \frac{\sqrt{2}}{2}$
π	0
$\frac{5\pi}{4}$	$1 - \frac{\sqrt{2}}{2}$
$\frac{3\pi}{2}$	1
$\frac{7\pi}{4}$	$1 + \frac{\sqrt{2}}{2}$
2π	2

Noting that $1 + \frac{\sqrt{2}}{2} \approx 1.7071$ and $1 - \frac{\sqrt{2}}{2} \approx 0.2929$, we plot the points. All values of r are nonnegative, so it is not difficult to connect the points in the proper order. The final result is in figure 30. All that remains is to determine whether the graph is complete. One method is to plot additional points. Doing so confirms that we begin retracing the curve. An alternative is to reason that because the period of the trigonometric function and the period of the rotating number line are both 2π, then we must repeat points after an interval of length 2π. ■

View figure 30 by turning the page a quarter-rotation clockwise (or by turning your head a quarter-rotation counterclockwise) and it is not hard to see why the curve in example 11 is called a *cardioid* (heart-shaped graph).

The word *cardioid* has the same root source as the word *cardiac*.

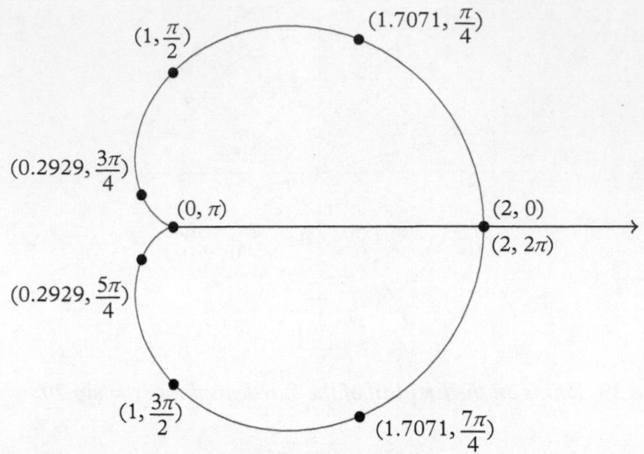

Figure 30 *The portion of the graph of* $r = 1 + \cos\theta$ *for which* $0 \le \theta \le 2\pi$. *The initial ray is included for reference*

Exercises 8.3

1–12. Plot the point for the given polar coordinates.

1. $(3, \frac{\pi}{3})$

2. $(1, \frac{\pi}{2})$

3. $(-1, \frac{7\pi}{4})$

4. $(-5, \frac{3\pi}{4})$

5. $(1.5, \frac{-3\pi}{2})$

6. $(0, \frac{3\pi}{4})$

7. $(0, \frac{3\pi}{2})$

8. $(2, -\frac{4\pi}{3})$

9. $(\pi, 0)$

10. $(-3, \frac{17\pi}{2})$

11. $(2, \frac{22\pi}{3})$

12. $(\frac{\pi}{2}, 1)$

13–20. Give another set of polar coordinates for the given point, for which (a) $r > 0$ and (b) $r < 0$.

13. $(17, \frac{\pi}{3})$

14. $(45, \frac{\pi}{4})$

15. $(4, -\frac{5\pi}{6})$

16. $(-5, -\frac{\pi}{2})$

17. $(-3, 4\pi)$

18. $(2, \frac{13\pi}{4})$

19. $(1.4, \frac{2\pi}{7})$

20. $(\pi, 3\pi)$

21–36. Graph the polar equation or inequality.

21. $r = 4$

22. $r = 2$

23. $\theta = \frac{2\pi}{3}$

24. $\theta = \frac{11\pi}{4}$

25. $r = -2$

26. $r = -3$

27. $\theta = -\frac{\pi}{2}$

28. $1 \le r \le 3$

29. $r \ge 3$

30. $\pi \le \theta < \frac{4\pi}{3}$

31. $0 \le r < 4$

32. $\pi \le \theta < \frac{4\pi}{3}, r \le 0$

33. $0 < \theta < \frac{3\pi}{4}$

34. $\pi \le \theta < \frac{4\pi}{3}, r > 2$

35. $0 < \theta < \frac{3\pi}{4}, r \ge 0$

36. $r = 0$

37–46. Graph the polar curve.

37. $r = 3\sin\theta$

38. $r = 2\cos\theta$

39. $r = 3 - 2\sin\theta$

40. $r = 1 - 2\cos\theta$ (limaçon)

41. $r = \sin 3\theta$

42. $r = \cos 2\theta$

43. $r = \sin\frac{1}{3}\theta$

44. $r = \cos^2\theta$

45. $r = \dfrac{2}{\theta}, \pi \le \theta \le 6\pi$ (hyperbolic spiral)

46. $r = \dfrac{\theta}{2}, 0 \le \theta \le 6\pi$ (spiral of Archimedes)

47–54. Find an interval of values of the variable θ for which the polar curve is traversed exactly once.

Some of these exercises are difficult without the use of technology.

47. $r = 6\cos\theta$

48. $r = \sin\frac{1}{4}\theta$

49. $r = \sqrt{\cos 2\theta}$ (lemniscate of Bernoulli)

50. $r = 4\cos\theta\sin^2\theta$ (double folium)

51. $r = 1 + 2\sin\frac{1}{2}\theta$ (Freeth's nephroid)

52. $r = \sqrt{\dfrac{25 - 24\tan^2\theta}{1 - \tan^2\theta}}$ (Devil's curve)

53. $r = \cos\theta\sin\frac{1}{2}\theta$

54. $r = \cos\theta + \sin 2.5\theta$

Tangents to Polar Curves

8.4

Before discussing tangents to polar curves, we must discuss converting between types of coordinates.

Polar–rectangular conversions

Consider a point represented by the rectangular coordinates (x, y) and the polar coordinates (r, θ), as in figure 1. By dropping a perpendicular

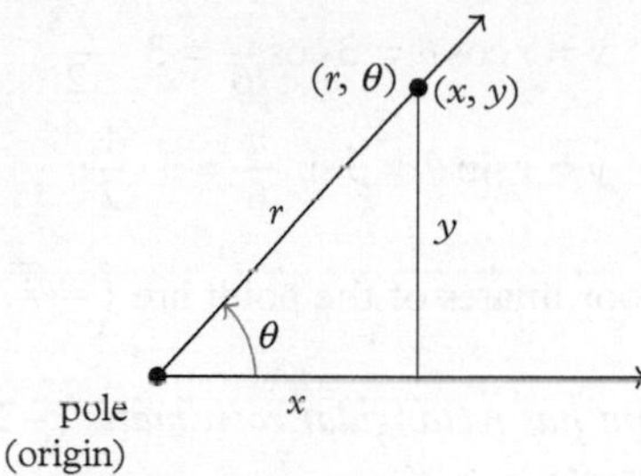

Figure 1 *A point with polar coordinates* (r, θ) *and rectangular coordinates* (x, y)*. The lengths r, x, and y, and the angle* θ *are also labeled*

from the point to the x-axis, a right triangle is formed that illustrates the relationship between the two types of coordinates. Notice that $\sin\theta = \frac{y}{r}$. Rearranging, we have

$$y = r\sin\theta.$$

Similarly, from the diagram we see that $\cos\theta = \frac{x}{r}$; therefore,

$$x = r\cos\theta.$$

Two more relationships in the diagram are helpful. First, from the Pythagorean theorem,

$$x^2 + y^2 = r^2.$$

Second,

$$\tan\theta = \frac{y}{x}.$$

POLAR AND RECTANGULAR COORDINATE CONVERSION FORMULAS

$$y = r\sin\theta \qquad x = r\cos\theta$$

$$x^2 + y^2 = r^2 \qquad \tan\theta = \frac{y}{x}$$

The use of the word *the* to refer to rectangular coordinates is correct because rectangular coordinates are unique.

Example 1 *A point has polar coordinates* $\left(3, \frac{\pi}{6}\right)$. *Find the rectangular coordinates of the point.*

Solution We are given that $r = 3$ and $\theta = \frac{\pi}{6}$ for the point. Because we have r and θ but wish to know x and y, the formulas $y = r\sin\theta$ and $x = r\cos\theta$ are the ones to use:

$$x = r\cos\theta = 3\cos\frac{\pi}{6} = 3\cdot\frac{\sqrt{3}}{2},$$

$$y = r\sin\theta = 3\sin\frac{\pi}{6} = 3\cdot\frac{1}{2}.$$

The rectangular coordinates of the point are $\left(\frac{3\sqrt{3}}{2}, \frac{3}{2}\right)$. ■

Using the word *the* in front of "polar coordinates" is incorrect, because polar coordinates are not unique.

Example 2 *A point has rectangular coordinates* $(-2\sqrt{3}, 2)$. *Find polar coordinates of the point.*

Solution We are given that $x = -2\sqrt{3}$ and $y = 2$. Because we have x and y but wish to determine an appropriate r and θ, the formulas $r^2 = x^2 + y^2$ and $\tan\theta = \frac{y}{x}$ are the ones to use. First we find r:

Recall that there are two solutions to $r^2 = 16$: $r = 4$ and $r = -4$. This fact is important shortly.

$$r^2 = x^2 + y^2 = (-2\sqrt{3})^2 + 2^2 = 12 + 4 = 16,$$
$$r = 4.$$

To find θ, we begin with

$$\tan\theta = \frac{y}{x} = \frac{2}{-2\sqrt{3}} = -\frac{1}{\sqrt{3}}.$$

One solution to this trigonometric equation is

This is not the only solution to the equation $\tan\theta = -\frac{1}{\sqrt{3}}$.

$$\theta = \tan^{-1}\left(-\frac{1}{\sqrt{3}}\right) = -\frac{\pi}{6}.$$

Unfortunately, the coordinates we have found, $\left(4, -\frac{\pi}{6}\right)$, are not correct! Why? Our point has rectangular coordinates $(-2\sqrt{3}, 2)$, so it is in the second quadrant, but the point with polar coordinates $\left(4, -\frac{\pi}{6}\right)$ is in the fourth quadrant (see figure 2). The issue is that the inverse tangent function always returns a first- or fourth-quadrant angle, when our point may be in any quadrant.

One way to proceed is to solve the trigonometric equation $\tan\theta = -\frac{1}{\sqrt{3}}$ subject to the constraint that θ is a second-quadrant angle, but a quicker solution is available. Look again at figure 2 and notice that the two points are the same distance from the origin but in opposite directions. Instead of correcting the value of θ, we may instead correct the value of r. The coordinates

$$\left(-4, -\frac{\pi}{6}\right)$$

work. ■

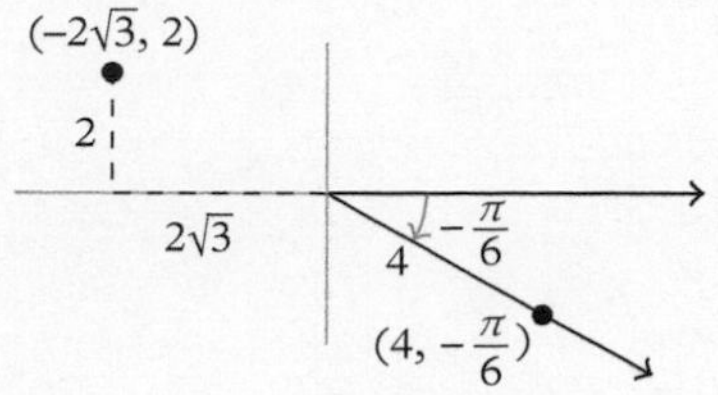

Figure 2 *The point with rectangular coordinates $(-2\sqrt{3}, 2)$ and the point with polar coordinates $(4, -\frac{\pi}{6})$ are not identical*

As seen in example 2, the equations $r^2 = x^2 + y^2$ and $\tan\theta = \frac{y}{x}$ each have multiple solutions. A choice of solutions that matches a given point can be summarized as follows.

INVERSE TANGENT METHOD FOR CONVERTING FROM RECTANGULAR TO POLAR COORDINATES

For a first- or fourth-quadrant point, use the formulas

$$r = \sqrt{x^2 + y^2}, \qquad \theta = \tan^{-1}\frac{y}{x}.$$

For a second- or third-quadrant point, use the formulas

$$r = -\sqrt{x^2 + y^2}, \qquad \theta = \tan^{-1}\frac{y}{x}.$$

For a point on the x-axis, use $r = x$ and $\theta = 0$. For a point on the y-axis, use $r = y$ and $\theta = \frac{\pi}{2}$. Other solutions are also possible.

Reading Exercise 9 *A point has the rectangular coordinates $(-3, 3)$. Find polar coordinates of the point.*

Examples 1 and 2 illustrated converting points from one coordinate system to the other. Next we look at converting equations.

Example 3 *Convert $r\cos\theta = 0$ to Cartesian coordinates.*

Recall that *Cartesian coordinates* is another name for *rectangular coordinates*.

Solution Using the conversion formula $x = r\cos\theta$, we may simply replace $r\cos\theta$ with x to arrive at the Cartesian equation

$$x = 0.$$ ■

Not all such exercises are as simple as example 3, but the idea is the same. We use the equations at our disposal to rewrite expressions in terms of r and θ to expressions in terms of x and y.

Example 4 *Rewrite the polar equation* $r^2 \sin 2\theta = 2$ *as a Cartesian equation.*

Solution None of the conversion formulas at our disposal involve an angle other than θ. Therefore, it is beneficial to rewrite $\sin 2\theta$ in terms of the angle θ, if possible. The trig identity $\sin 2\theta = 2\sin\theta\cos\theta$ does the trick:

Line 1 is the original equation; line 2 uses the trig identity $\sin 2\theta = 2\sin\theta\cos\theta$.

$$r^2 \sin 2\theta = 2$$

$$r^2 \cdot 2\sin\theta\cos\theta = 2.$$

Noticing the r^2, we might be tempted to use the conversion formula $r^2 = x^2 + y^2$. But, then we'd be out of luck with $\sin\theta$ and $\cos\theta$. The strategy is to look for ways to eliminate θ first, then deal with r if necessary. The available conversion formulas involving $\sin\theta$ and $\cos\theta$ also require an r. Let's rearrange our equation accordingly:

STRATEGY
Try to eliminate θ first.

$$r\sin\theta \cdot r\cos\theta \cdot 2 = 2.$$

Now the equation is in a form where the conversion formulas are easily applied:

Line 1 replaces $r\sin\theta$ with y and replaces $r\cos\theta$ with x; line 2 simplifies.

$$y \cdot x \cdot 2 = 2$$

$$xy = 1.$$

Having eliminated r and θ, the equation $xy = 1$ is the desired Cartesian equation. ■

Although rewriting polar equations as Cartesian equations sometimes requires ingenuity, rewriting from Cartesian form to polar form is quite straightforward using the conversion formulas $x = r\cos\theta$ and $y = r\sin\theta$.

Example 5 *Rewrite* $x^2 - y^2 = 1$ *as a polar equation.*

Solution Replacing x with $r\cos\theta$ and y with $r\sin\theta$ gives

$$(r\cos\theta)^2 - (r\sin\theta)^2 = 1.$$

This solution is sufficient. If we wish, we can simplify the expression. Other correct answers include

$$r^2\cos^2\theta - r^2\sin^2\theta = 1$$

and

$$r^2\left(\cos^2\theta - \sin^2\theta\right) = 1.$$ ■

Ans. to reading exercise 9 using the inverse tangent method, $\left(-\sqrt{18}, -\frac{\pi}{4}\right) = \left(-3\sqrt{2}, -\frac{\pi}{4}\right)$; alternate solutions include $\left(3\sqrt{2}, \frac{3\pi}{4}\right)$

Reading Exercise 10 *Rewrite $y = x^2 + 1$ as a polar equation.*

Tangents to polar curves

In the previous section we graphed polar curves of the form $r = f(\theta)$, where the variable r is a function of the variable θ. How can we find a tangent to such a curve, as in figure 3? If we allow ourselves to write the

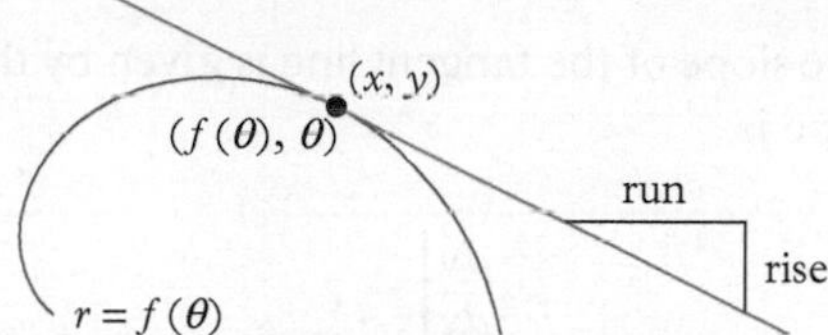

Figure 3 *A polar curve $r = f(\theta)$ (blue) along with its tangent line (orange) at $(f(\theta), \theta)$. The slope of the tangent line is still the rise over the run*

equation of the tangent line in rectangular coordinates, then knowledge of the slope of the tangent line and the rectangular coordinates (x, y) of the point of tangency is sufficient. The slope of the tangent line is still the rise over the run, or the change in y divided by the change in x. This is still the derivative $\frac{dy}{dx}$. The task, then, is to determine that derivative.

Relating the curve's polar equation to the rectangular coordinate quantities x and y is accomplished using the coordinate conversion formulas $x = r\cos\theta$ and $y = r\sin\theta$. Combined with $r = f(\theta)$, the formulas become

$$y = r\sin\theta = f(\theta)\sin\theta,$$

$$x = r\cos\theta = f(\theta)\cos\theta.$$

Now we have the two variables x and y in terms of the variable θ. Or is it the parameter θ? The previous equations are parametric equations for x and y! Using the derivative formula for a parametric curve,

$$\frac{dy}{dx} = \frac{\frac{dy}{d\theta}}{\frac{dx}{d\theta}}.$$

The quantities $\frac{dy}{d\theta}$ and $\frac{dx}{d\theta}$ can be calculated using the product rule:

$$\frac{dy}{dx} = \frac{\frac{dy}{d\theta}}{\frac{dx}{d\theta}} = \frac{f'(\theta)\sin\theta + f(\theta)\cos\theta}{f'(\theta)\cos\theta - f(\theta)\sin\theta}.$$

DERIVATIVE $\frac{dy}{dx}$ FOR A POLAR CURVE

For the polar curve $r = f(\theta)$,

$$\frac{dy}{dx} = \frac{f'(\theta)\sin\theta + f(\theta)\cos\theta}{f'(\theta)\cos\theta - f(\theta)\sin\theta}.$$

There are two different derivatives mentioned in the formula. One is $\frac{dy}{dx}$. The other is $f'(\theta)$, by which we mean $\frac{d}{d\theta}f(\theta)$.

As always, the slope of the tangent line is given by the derivative $\frac{dy}{dx}$. At $\theta = k$ the slope is

$$\left.\frac{dy}{dx}\right|_{\theta=k}.$$

Example 6 *Find the slope of the tangent line to the four-leaved rose* $r = \sin 2\theta$ *when* $\theta = \frac{\pi}{4}$.

Solution We wish to find the slope of the tangent line to a polar curve $r = f(\theta)$, where

$$f(\theta) = \sin 2\theta.$$

To use the derivative formula we also need $f'(\theta)$, which is

$$f'(\theta) = 2\cos 2\theta.$$

Using the formula for the derivative $\frac{dy}{dx}$ of a polar curve,

$$\begin{aligned}\frac{dy}{dx} &= \frac{f'(\theta)\sin\theta + f(\theta)\cos\theta}{f'(\theta)\cos\theta - f(\theta)\sin\theta} \\ &= \frac{2\cos 2\theta \cdot \sin\theta + \sin 2\theta \cdot \cos\theta}{2\cos 2\theta \cdot \cos\theta - \sin 2\theta \cdot \sin\theta}.\end{aligned}$$

We wish to know the slope of the tangent line at $\theta = \frac{\pi}{4}$, so we evaluate $\frac{dy}{dx}$ at $\theta = \frac{\pi}{4}$:

The $\frac{\pi}{2}$ in the expression comes from the angle $2\theta = 2 \cdot \frac{\pi}{4}$.

$$\left.\frac{dy}{dx}\right|_{\theta=\frac{\pi}{4}} = \frac{2\cos\frac{\pi}{2}\sin\frac{\pi}{4} + \sin\frac{\pi}{2}\cos\frac{\pi}{4}}{2\cos\frac{\pi}{2}\cos\frac{\pi}{4} - \sin\frac{\pi}{2}\sin\frac{\pi}{4}}$$

$$= \frac{0 \cdot \frac{\sqrt{2}}{2} + 1 \cdot \frac{\sqrt{2}}{2}}{0 \cdot \frac{\sqrt{2}}{2} - 1 \cdot \frac{\sqrt{2}}{2}} = -1.$$

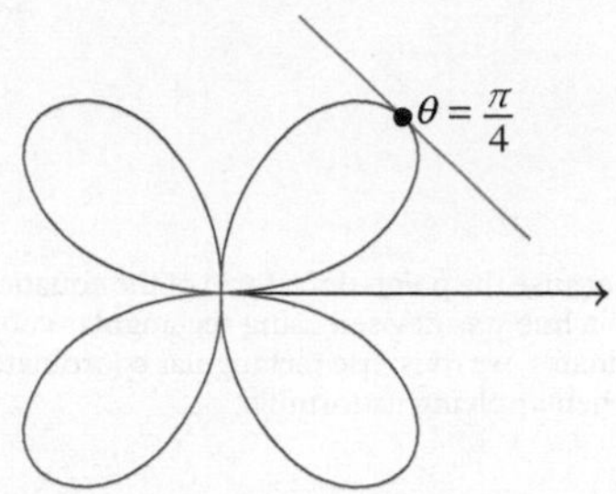

Figure 4 *The four-leaved rose $r = \sin 2\theta$ (blue), along with its tangent line at $\theta = \frac{\pi}{4}$ (orange)*

The slope of the tangent line at $\theta = \frac{\pi}{4}$ is -1. The curve, which was drawn in section 8.3 example 10, and its tangent line are pictured in figure 4. ■

Reading Exercise 11 *Find the slope of the tangent line to the curve $r = \sin 4\theta$ when $\theta = \dfrac{\pi}{4}$.*

Ans. to reading exercise 10
$r \sin\theta = r^2\cos^2\theta + 1$

Example 7 *Find an equation of the tangent line to the curve $r = 1 + \cos\theta$ where $\theta = \frac{\pi}{2}$.*

Why "an" equation of "the" tangent line rather than "the" equation? Although there is just one tangent line, we have more than one coordinate system in which to express an equation of the line. Let's use the rectangular coordinate system in our solution.

Solution We begin by finding the slope of the tangent line using

$$f(\theta) = 1 + \cos\theta$$

and

$$f'(\theta) = -\sin\theta.$$

Using the formula for the derivative $\frac{dy}{dx}$ of a polar curve,

$$\frac{dy}{dx} = \frac{f'(\theta)\sin\theta + f(\theta)\cos\theta}{f'(\theta)\cos\theta - f(\theta)\sin\theta}$$

$$= \frac{-\sin\theta\sin\theta + (1+\cos\theta)\cos\theta}{-\sin\theta\cos\theta - (1+\cos\theta)\sin\theta}.$$

The slope of the tangent line at $\theta = \frac{\pi}{2}$ is then

$$\left.\frac{dy}{dx}\right|_{\theta=\frac{\pi}{2}} = \frac{-\sin\frac{\pi}{2}\sin\frac{\pi}{2} + (1+\cos\frac{\pi}{2})\cos\frac{\pi}{2}}{-\sin\frac{\pi}{2}\cos\frac{\pi}{2} - (1+\cos\frac{\pi}{2})\sin\frac{\pi}{2}}$$

$$= \frac{-1 \cdot 1 + 1 \cdot 0}{-1 \cdot 0 - 1 \cdot 1} = \frac{-1}{-1} = 1.$$

The polar coordinates of the point of tangency are easily determined:

$$(f(\theta), \theta) = (1 + \cos\tfrac{\pi}{2}, \tfrac{\pi}{2}) = (1, \tfrac{\pi}{2}).$$

Notice the order $(f(\theta), \theta)$, because polar coordinates are written in the form (r, θ).

But, instead of polar coordinates, we need to know the rectangular coordinates of the point of tangency, which are

$$x = r\cos\theta = 1 \cdot \cos\tfrac{\pi}{2} = 0,$$
$$y = r\sin\theta = 1 \cdot \sin\tfrac{\pi}{2} = 1.$$

Because the point-slope form of the equation of a line was devised using rectangular coordinates, we must use rectangular coordinates when applying its formula.

Using the point-slope form of the equation of a line with point $(0, 1)$ and slope 1, we have

$$y - 1 = 1(x - 0)$$
$$y = x + 1.$$

The equation of the tangent line in rectangular coordinates is $y = x + 1$. ■

If desired, the solution to example 7 can be rewritten as a polar equation using the method of example 5 to obtain

$$r\sin\theta = r\cos\theta + 1$$

or, equivalently,

$$r = \frac{1}{\sin\theta - \cos\theta}.$$

More polar circles

In the previous section we noted that the graph of the polar equation $r = k$ is a circle centered at the origin with radius $|k|$. Next we study polar equations of a few more (but not all) circles.

Consider the polar equation

$$r = 2a\cos\theta$$

Ans. to reading exercise 11 $\dfrac{4\cos\pi\sin\frac{\pi}{4} + \sin\pi\cos\frac{\pi}{4}}{4\cos\pi\cos\frac{\pi}{4} - \sin\pi\sin\frac{\pi}{4}} = 1$

for a (nonzero) real number a. To show that the equation represents a circle, we need to rewrite it as a rectangular equation. Applying the conversion formula $x = r\cos\theta$, which can be rearranged as $\cos\theta = \frac{x}{r}$, gives

$$r = 2a\frac{x}{r}.$$

This equation can be rearranged as

$$r^2 = 2ax.$$

Applying the conversion formula $r^2 = x^2 + y^2$ gives us the rectangular equation

$$x^2 + y^2 = 2ax.$$

To see whether this is the equation of a circle, we complete the square on the variable x:

$$\begin{aligned} x^2 + y^2 &= 2ax \\ x^2 - 2ax \qquad + y^2 &= 0 \\ x^2 - 2ax + a^2 + y^2 &= a^2 \\ (x-a)^2 + y^2 &= a^2. \end{aligned}$$

The coefficient on x in line 2 is $-2a$. To complete the square, we take half that coefficient ($-a$) and square it ($(-a)^2$) to get a^2, which we add to both sides of the equation in line 3. Factoring (line 4) finishes the procedure.

The latter equation is the equation of a circle centered at $(a, 0)$ with radius $|a|$; such a circle passes through the origin (see figure 5, left).

Recall that the rectangular coordinates $(a, 0)$ and the polar coordinates $(a, 0)$ represent the same point.

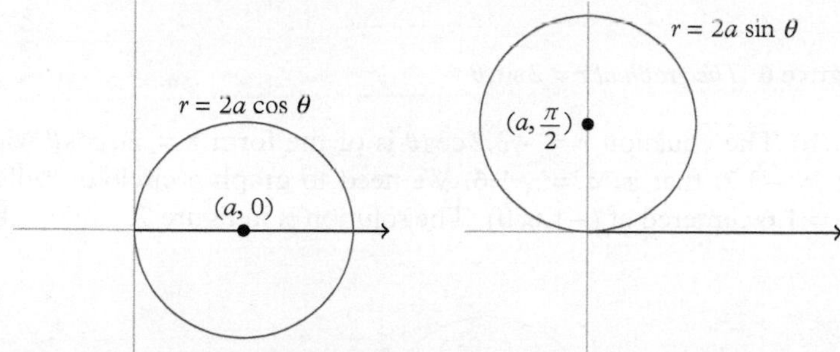

Figure 5 *(left) The graph of $r = 2a\cos\theta$, which is a circle centered at $(a, 0)$ with radius $|a|$, and (right) the graph of $r = 2a\sin\theta$, which is a circle centered at $(a, \frac{\pi}{2})$ with radius $|a|$. In each view, the initial ray is drawn in black. For the purposes of comparison, the x- and y-axes are also included (gray). The pictures show $a > 0$, although a could also be negative*

A similar derivation shows that $r = 2a\sin\theta$ is a circle of radius $|a|$ through the origin centered at the point with polar coordinates $(a, \frac{\pi}{2})$ (figure 5, right).

You might recall from your study of geometry that the tangent line to a point on a circle is perpendicular to the radius segment at that point. Therefore, in figure 5, the circle $r = 2a\cos\theta$ is tangent to the y-axis at the origin whereas the circle $r = 2a\sin\theta$ is tangent to the x-axis at the origin.

SOME POLAR CIRCLES PASSING THROUGH THE ORIGIN

The graph of the polar equation $r = 2a\cos\theta$ is a circle of radius $|a|$ centered at $(a, 0)$. The graph of the polar equation $r = 2a\sin\theta$

is a circle of radius $|a|$ centered at $(a, \frac{\pi}{2})$. The origin $(0, 0)$ is a point on these circles.

Example 8 *Graph the circles given by the polar equations (a)* $r = 2\sin\theta$ *and (b)* $r = -3.2\cos\theta$.

Solution (a) The equation $r = 2\sin\theta$ is of the form $r = 2a\sin\theta$ with $a = 1$. We need to graph a circle of radius 1 centered at $(1, \frac{\pi}{2})$. The solution is in figure 6.

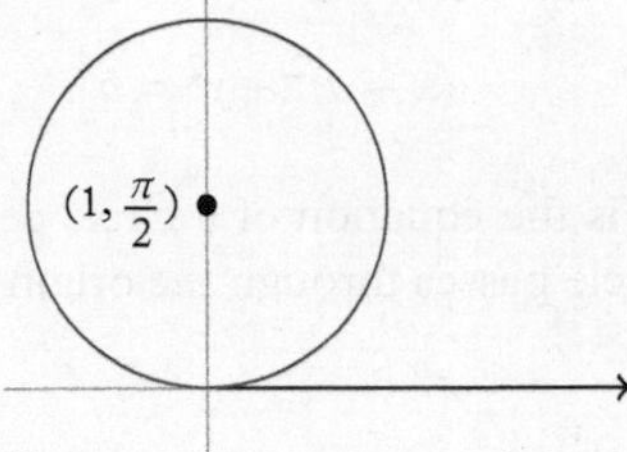

Figure 6 *The graph of* $r = 2\sin\theta$

(b) The equation $r = -3.2\cos\theta$ is of the form $r = 2a\cos\theta$ with $2a = -3.2$; that is, $a = -1.6$. We need to graph a circle of radius $|a| = 1.6$ centered at $(-1.6, 0)$. The solution is in figure 7. ■

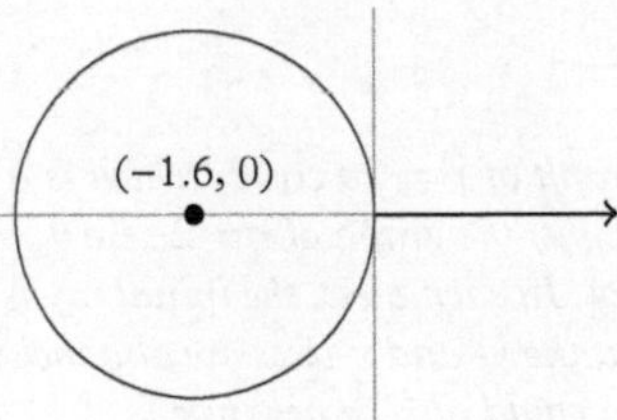

Figure 7 *The graph of* $r = -3.2\cos\theta$

Right, left, above, and *below* are in reference to the origin.

It is helpful to associate cosine with x, where $a > 0$ means the center is to the right and $a < 0$ means the center is to the left. Similarly, it is helpful to associate sine with y, where $a > 0$ means the center is above and $a < 0$ means the center is below.

Exercises 8.4

1–10. Polar coordinates of a point are given. Find the rectangular coordinates of the point.

1. $(3, \pi)$

2. $(2, \frac{3\pi}{2})$

3. $(-4, \frac{\pi}{4})$

4. $(0, \frac{2\pi}{3})$

5. $(0, \frac{3\pi}{7})$

6. $(-1, \frac{\pi}{4})$

7. $(2, -\frac{22\pi}{3})$

8. $(2.5, \frac{22}{7})$

9. $(1.2, 7.3)$

10. $(-6, -\frac{3\pi}{4})$

11–20. The rectangular coordinates of a point are given. Find polar coordinates of the point.

11. $(-4, 0)$

12. $(0, 5)$

13. $(-2, -2\sqrt{3})$

14. $(-3, -3)$

15. $(6, 6)$

16. $(1, 4)$

17. $(\sqrt{75}, -5)$

18. $(-3, \sqrt{27})$

19. $(-2, 5)$

20. $(2, \frac{\pi}{2})$

21–26. Rewrite the rectangular equation as a polar equation.

21. $x^2 + y^2 = 5$

22. $y = 3x - 2$

23. $2x + 7y = 1$

24. $xy^2 + 2y = x^2$

25. $\frac{x}{y} = 3$

26. $x = y^2$

27–32. Rewrite the polar equation as a rectangular equation.

27. $r \sin\theta + r \cos\theta = 11$

28. $r^2 = 3 - r \cos\theta$

29. $r^{-1} = 2\cos\theta + 4\sin\theta$

30. $r = 3 \sec\theta$

31. $\tan\theta = 1$

32. $r = 3\sin\theta$

33–38. Graph the polar circle quickly.

33. $r = 4\cos\theta$

34. $r = 6\sin\theta$

35. $r = \sin\theta$

36. $r = -4\cos\theta$

37. $r = -3\sin\theta$

38. $r = 3$

39–42. Find the derivative $\frac{dy}{dx}$ of the polar curve.

39. $r = \tan\theta$

40. $r = \sec\theta$

41. $r = \theta^3$

42. $r = 4$

43–48. Find the slope of the tangent line to the polar curve at the given value of θ.

43. $r = \tan\theta$ at $\theta = \frac{\pi}{4}$

44. $r = 1 - 2\sin\theta$ at $\theta = \pi$

45. $r = \frac{2}{\theta}$ at $\theta = 2\pi$

46. $r = \cos 4\theta$ at $\theta = \pi$

47. (GSL) $r = e^{\theta}$ at $\theta = 0$

48. $r = 4$ at $\theta = \frac{4\pi}{3}$

49–56. (a) Find an equation of the tangent line to the polar curve at the given value of θ. (b) Graph the curve along with the tangent line from part (a).

Graphs of these curves were exercises in section 8.3.

49. $r = 3\sin\theta$ at $\theta = \frac{\pi}{3}$

50. $r = 2\cos\theta$ at $\theta = \frac{\pi}{4}$

51. $r = 3 - 2\sin\theta$ at $\theta = \pi$

52. $r = 1 - 2\cos\theta$, at $\theta = \frac{\pi}{2}$

53. $r = \sin 3\theta$, at $\theta = \frac{11\pi}{6}$

54. $r = \cos 2\theta$, at $\theta = \frac{3\pi}{4}$

55. $r = \sin\frac{1}{3}\theta$, at $\theta = \frac{3\pi}{2}$

56. $r = \cos^2\theta$ at $\theta = \frac{\pi}{3}$

57. Find all the tangent lines to the curve $r = \sin 3\theta$ at the point $(0, 0)$. (This curve was graphed in the previous section.)

58. Find all the tangent lines to the curve $r = \sin\frac{1}{3}\theta$ at the point with rectangular coordinates $(0, \frac{1}{2})$. (This curve was graphed in the previous section.)

59. Beginning with the general form of the equation of a line $ax+by+c = 0$, (a) rewrite the equation as a polar equation and rearrange it to the form $r = f(\theta)$ to find the polar form of the equation of a line. (b) Determine a formula for the slope of the line in part (a).

60. Using the formula for the derivative $\frac{dy}{dx}$ for a polar curve and the formula $x = f(\theta)\cos\theta$, apply the formula for the second derivative for a parametric curve to derive the formula for the second derivative $\frac{d^2y}{dx^2}$ for a polar curve.

Use θ in place of t.

61. Using the results of exercise 59, determine the slope of the line $r = \dfrac{5}{4\cos\theta - 7\sin\theta}$.

62. (a) Use the formula derived in exercise 60 to find $\frac{d^2y}{dx^2}$ for the polar curve $r = \sin 2\theta$. (b) Use the result of part (a) to determine whether the curve is concave up or concave down at $\theta = \frac{\pi}{4}$.

Conic Sections

8.5

There are two approaches to the study of geometry. The ancient Greeks worked with *synthetic geometry*, which contains no coordinates (figure 1). No equations were used and no algebra was required. The alternative is *analytic geometry*, which does use coordinates; objects can be described algebraically using equations.

Consider the definition of circle (which you may have learned when studying geometry), which contains no reference to coordinates or equations.

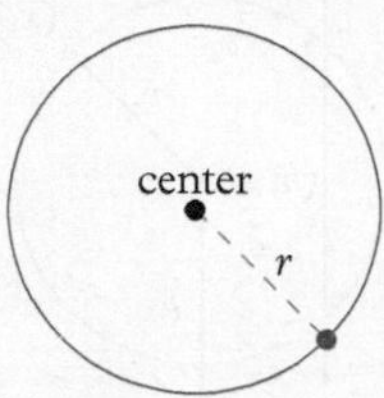

Figure 1 *A synthetic diagram of a circle, with no coordinates or equations given. The circle (blue) is shown along with a segment (brown, dashed) from the center of the circle to a point on the circle, with length r equal to the radius of the circle. The center (black) of the circle is not a point on the circle. A circle can be drawn with a compass*

Definition 1 CIRCLE *A* circle *is the set of all points equidistant from a fixed point, called the* center. *The constant distance is called the* radius.

During your study of algebra, you most likely encountered equations of circles. A circle centered at the origin with radius r has the equation $x^2 + y^2 = r^2$. Why? Consider the circle in figure 2. Any point on the circle with coordinates (x, y) must be r units from the origin $(0, 0)$, and because the x- and y-coordinate directions are perpendicular (they meet at a right angle), the Pythagorean theorem applies to the right triangle shown in the diagram, yielding the desired equation.

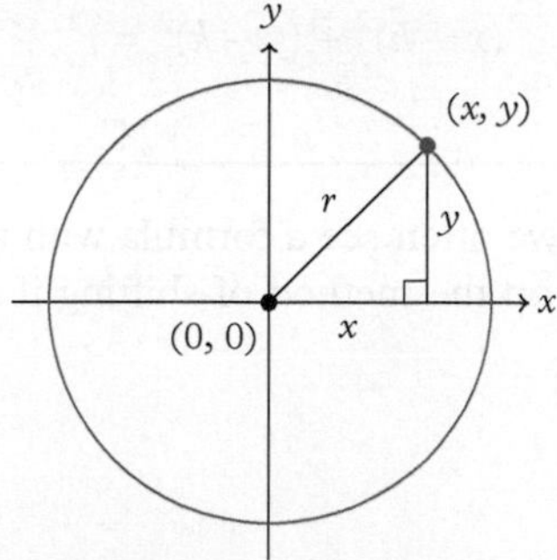

Figure 2 *An analytic diagram of a circle. A point with coordinates (x, y) is located on a circle (blue) with radius r. The right triangle pictured has side lengths x and y, and hypotenuse length r. The relationship $x^2 + y^2 = r^2$ holds for any point (x, y) on the circle*

This process of using a synthetic definition for a geometric curve and using algebra to develop an analytic description is the theme of this section.

Review: circles

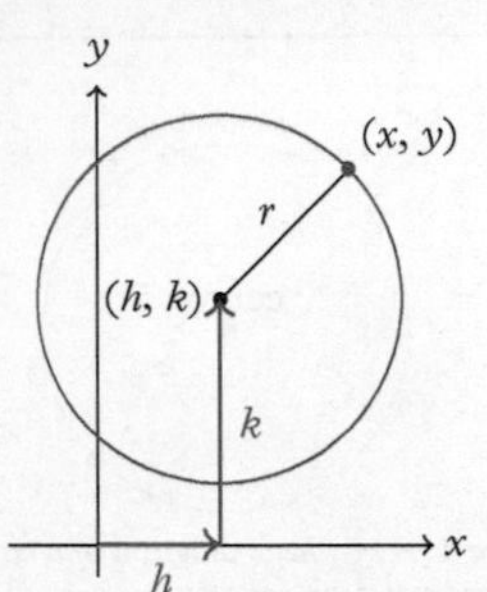

Figure 3 *A point (x, y) on a circle (blue) with radius r and center (h, k). The distance between (x, y) and (h, k) must be r for any point (x, y) on the circle*

What if we wish to know the equation of a circle with some other center, such as (h, k)? The quick answer is to begin with the equation $x^2 + y^2 = r^2$ of a circle with center $(0, 0)$ and radius r and then shift h units horizontally and k units vertically, as in figure 3. Similar to the shifting described in section 0.5, we replace x with $x - h$ and replace y with $y - k$ to arrive at

$$(x - h)^2 + (y - k)^2 = r^2.$$

Alternately, we could use the distance formula to calculate the distance between the point (x, y) on the circle and the center (h, k), then equate the result to the known distance r. Squaring both sides yields the formula

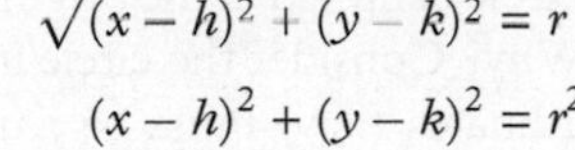

$$\sqrt{(x - h)^2 + (y - k)^2} = r$$

$$(x - h)^2 + (y - k)^2 = r^2.$$

The distance formula is derived from the Pythagorean theorem, so these two derivations are essentially the same.

> **EQUATION OF A CIRCLE**
>
> The equation of a circle centered at (h, k) with radius r is
>
> $$(x - h)^2 + (y - k)^2 = r^2.$$

In what follows, we often see a formula with a center or vertex at $(0, 0)$ and then rely on the method of shifting if that point is located elsewhere.

Parabolas

You know that the graph of a quadratic function $f(x) = ax^2 + bx + c$ is a parabola. Parabolas were studied for about two millennia before the advent of analytic geometry using a synthetic definition such as the one that follows.

Definition 2 PARABOLA *A* parabola *is the set of all points equidistant from a fixed point (called the* focus*) and a fixed line (called the* directrix*).*

It is more consistent with ancient Greek geometry to state the definition in terms of segments having the same length instead of equal distances. The difference is subtle, but the Greeks did not consider the length of a segment to be a number in the same way that we think of distance as a number. The modern terminology of equal distances is convenient for deriving equations.

Notice that the style of the definition is similar to that of a circle, with no coordinates or equations given. A synthetic picture of a parabola is given in figure 4.

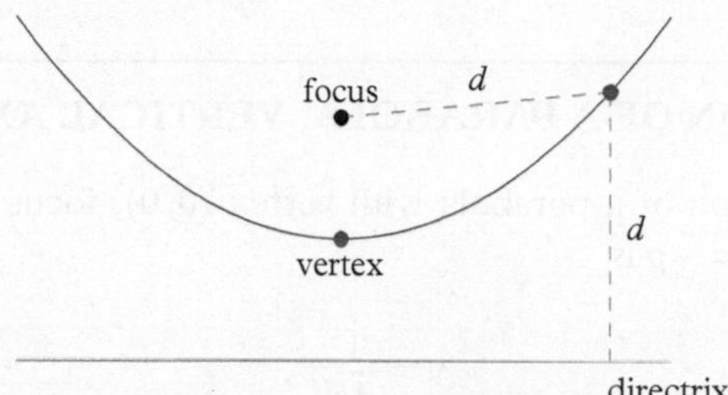

Figure 4 *A synthetic diagram of a parabola (blue), along with its directrix (green line), vertex, and focus. The distance from a point on the parabola to the focus is the same as the distance from the point to the directrix, measured along a line perpendicular to the directrix. The line through the focus and vertex (not pictured) is called the* axis *of the parabola. The axis is perpendicular to the directrix*

To find a formula for an equation of a parabola, we must introduce coordinates to the diagram, as in figure 5. Placing the vertex at the point $(0, 0)$ and focus p units above the vertex at the point $(0, p)$, the directrix must be p units below the vertex and have the equation $y = -p$. The distance from the point (x, y) to the point $(0, p)$, as calculated

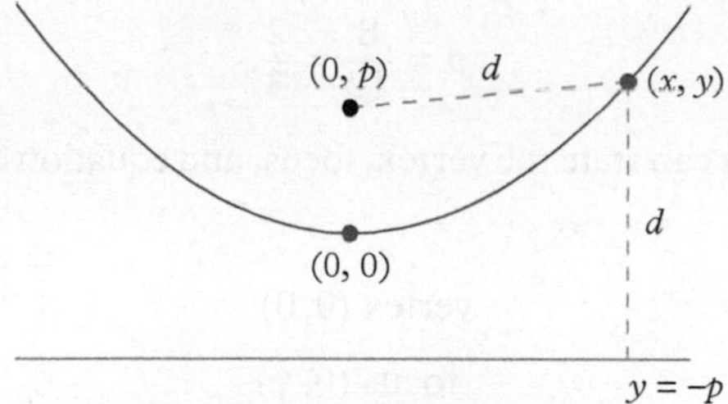

Figure 5 *An analytic diagram of a parabola (blue), along with its directrix* $y = -p$ *(green line), vertex* $(0, 0)$*, and focus* $(0, p)$*; a positive value of* p *is shown. The distance from a point* (x, y) *on the parabola to the focus is the same as the distance from the point to the directrix, measured along a line perpendicular to the directrix. Equating these distances yields an equation for the parabola*

by the distance formula, must then be the same as the vertical distance between the point (x, y) and the line $y = -p$, which is the difference in their y-coordinates. Therefore,

The left side of line 1 is the distance formula applied to points (x, y) and $(0, p)$, whereas the right side is the difference in the y-coordinates of the point (y) and the line ($-p$). Line 2 squares both sides, line 3 expands, line 4 simplifies, and line 5 rewrites in terms of the variable y.

$$\begin{aligned}\sqrt{(x-0)^2+(y-p)^2} &= y-(-p)\\ x^2+(y-p)^2 &= (y+p)^2\\ x^2+y^2-2yp+p^2 &= y^2+2yp+p^2\\ x^2 &= 4yp\\ y &= \frac{x^2}{4p}.\end{aligned}$$

EQUATION OF A PARABOLA, VERTICAL AXIS

The equation of a parabola with vertex $(0, 0)$, focus $(0, p)$, and directrix $y = -p$ is

$$y = \frac{1}{4p}x^2.$$

You may recall that $y = ax^2$ is the equation of a parabola through the origin, opening upward if $a > 0$ and opening downward if $a < 0$. The equation given here is consistent with that earlier experience. Expressing the coefficient on x^2 as $\frac{1}{4p}$, instead of merely a, aids in the identification of the focus and directrix.

Example 1 *Find the vertex, focus, and directrix of the parabola* $y = \frac{3x^2}{8}$.

Solution Notice that the equation of the parabola $y = \frac{3}{8}x^2$ matches the form of the equation of a parabola with a vertical axis as given in the boxed equation, $y = \frac{1}{4p}x^2$. We simply need to equate $\frac{3}{8}$ with $\frac{1}{4p}$ to find the value of p:

$$\begin{aligned}\frac{3}{8} &= \frac{1}{4p}\\ 8 &= 12p\\ p &= \frac{8}{12} = \frac{2}{3}.\end{aligned}$$

Knowing p, we can state the vertex, focus, and equation of the directrix as

We use vertex $(0, 0)$, focus $(0, p)$, and directrix $y = -p$ for $p = \frac{2}{3}$.

vertex $(0, 0)$

focus $(0, \frac{2}{3})$

directrix $y = -\frac{2}{3}$. ■

Reading Exercise 12 *Find the vertex, focus, and directrix of the parabola* $y = \frac{1}{8}x^2$.

Any parabola of the form $(y - k) = \frac{1}{4p}(x - h)^2$ is similar to $y = \frac{1}{4p}x^2$, but is shifted. Instead of learning more general formulas for such cases, let's look at the use of shifting in the next example.

Example 2 *Find the vertex, focus, and directrix of the parabola* $12(y-7)=(x+1)^2$.

Solution The parabola $12(y-7)=(x+1)^2$ is similar to the parabola $12y=x^2$, but is shifted up 7 units to account for the $y-7$ and is shifted left 1 unit to account for the $(x+1)$, which is the same as $(x-(-1))$. First we find the vertex, focus, and directrix of the parabola $12y=x^2$, then we shift to determine the vertex, focus, and directrix of the original parabola.

Recall that for $x-h$, we shift right h units. Also, shifting right -1 unit is the same as shifting left 1 unit. For $y-k$, we shift up k units.

To find the value of p for $12y=x^2$, we first rearrange to the form $y=\frac{1}{4p}x^2$. Solving for y gives

$$y=\frac{1}{12}x^2.$$

Then, $\frac{1}{12}=\frac{1}{4p}$; hence,

$$4p=12$$
$$p=3.$$

For the parabola $12y=x^2$, we then have

$$\text{vertex } (0,0)$$
$$\text{focus } (0,3)$$
$$\text{directrix } y=-3.$$

We use vertex $(0,0)$, focus $(0,p)$, and directrix $y=-p$ for $p=3$.

All that remains is to shift left 1 unit and up 7 units to get

$$\text{vertex } (-1,7)$$
$$\text{focus } (-1,10)$$
$$\text{directrix } y=4.$$

The results are illustrated in figure 6. ■

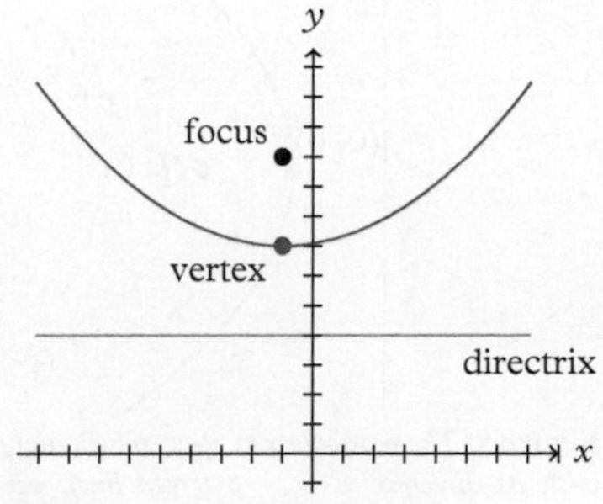

Figure 6 *The parabola of example 2*

You may be used to encountering the equation of a parabola in the form $y=ax^2+bx+c$. For instance, the parabola of example 2 can be rewritten as

$$y=\frac{1}{12}x^2+\frac{1}{6}x+\frac{85}{12}.$$

To find the vertex, focus, and directrix of such a parabola, our strategy is, first, to rewrite the equation in the form used in example 2, which often requires completing the square:

$$y = \frac{1}{12}x^2 + \frac{1}{6}x + \frac{85}{12}$$
$$y = \frac{1}{12}(x^2 + 2x \qquad) + \frac{85}{12}$$
$$y = \frac{1}{12}(x^2 + 2x + 1) + \frac{85}{12} - \frac{1}{12}$$
$$y = \frac{1}{12}(x + 1)^2 + \frac{84}{12}.$$

The next step is to isolate the quantity of the form $(x - h)^2$ on the right-hand side:

$$y - 7 = \frac{1}{12}(x + 1)^2$$
$$12(y - 7) = (x + 1)^2.$$

We then proceed as in example 2 to determine the vertex, focus, and directrix.

In the equation of a parabola with a vertical axis, if p is negative, the parabola opens downward.

Ans. to reading exercise 12
vertex $(0, 0)$, focus $(0, 2)$, directrix $y = -2$

Example 3 *A parabola has vertex* $(0, 0)$ *and focus* $(0, -1)$. *Find the equation of the directrix and the equation of the parabola.*

Solution Because the parabola has vertex $(0, 0)$ and focus $(0, -1)$, it fits the criteria for a parabola with a vertical axis with $p = -1$. Therefore, the equation of the directrix is $y = -(-1)$ or $y = 1$, and the equation of the parabola is $y = -\frac{1}{4}x^2$. See figure 7. ■

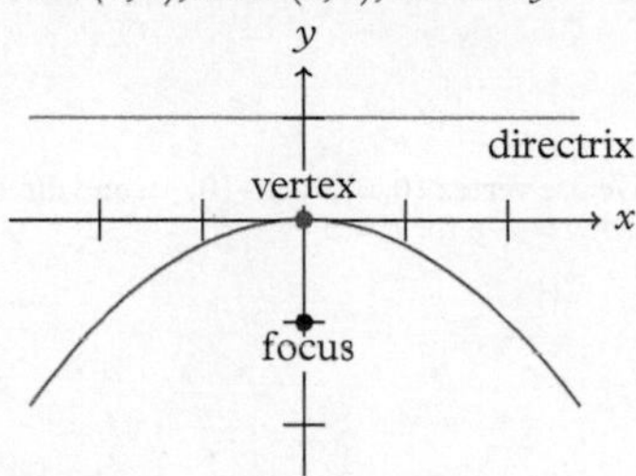

Figure 7 *The parabola of example 3*

Changing the orientation of the parabola to horizontal requires switching the roles of x and y. The derivation is very similar to the one for a vertical axis and you will explore it in the exercises at the end of this section.

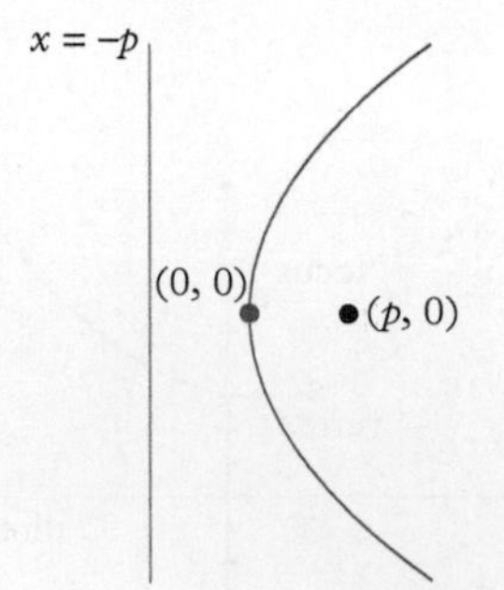

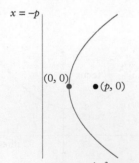

Figure 8 *The parabola* $x = \frac{1}{4p}y^2$ *(blue), along with its directrix* $x = -p$ *(green line), vertex* $(0, 0)$, *and focus* $(p, 0)$. *A positive value of* p *is shown. The axis of the parabola (the line containing the vertex and focus) is horizontal*

EQUATION OF A PARABOLA, HORIZONTAL AXIS

The equation of a parabola with vertex $(0, 0)$, focus $(p, 0)$, and directrix $x = -p$ is

$$x = \frac{1}{4p}y^2.$$

Figure 8 illustrates a parabola with a horizontal axis. For $p > 0$, the parabola opens to the right; for $p < 0$, the parabola opens to the left. Working exercises for a horizontal axis is basically the same as for a vertical axis.

Ellipses

Next we turn to the study of ellipses. An ellipse has not just one focus, but rather two foci.

The plural of *focus* is *foci*.

Definition 3 ELLIPSE *An* ellipse *is the set of points for which the sum of the distances to two fixed points (foci) is constant.*

Just as with a circle and a parabola, the synthetic definition of ellipse does not reference coordinates or equations. A synthetic picture of an ellipse is in figure 9.

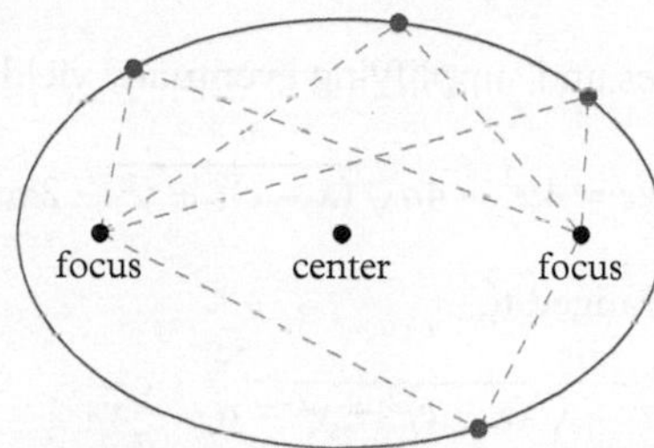

Figure 9 *A synthetic diagram of an ellipse (blue), along with its two foci. Each complete path (dashed brown line) from a focus to a point on the ellipse to the other focus has the same length as any other such path. The line through the two foci (not pictured) is sometimes called the* focal axis *of the ellipse. The* center *of the ellipse is on the focal axis, halfway between the two foci*

To draw an ellipse, place a thumb tack at each focus and tie the ends of a string on each thumb tack. Use a pencil to stretch the string tight. The tightened string should then look like one of the dashed brown paths in figure 9, and the pencil should be located at a point on the ellipse. Because the length of the string doesn't change, wherever the pencil is located is a point on the ellipse. Keep drawing for a full rotation and the ellipse is complete.

To find an equation for an ellipse, we must introduce coordinates, as in figure 10. Place the foci at $(-c, 0)$ and $(c, 0)$, and let the constant length of the path be $2a$. Then, for any point (x, y) on the ellipse, the distance from (x, y) to $(-c, 0)$ plus the distance from (x, y) to $(c, 0)$ is $2a$. Using the distance formula, we have

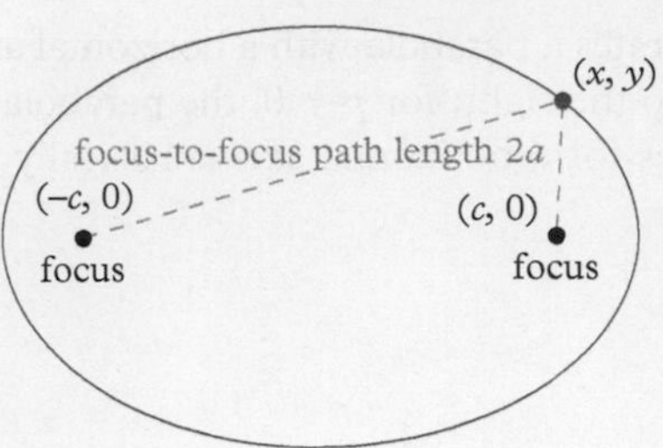

Figure 10 *An analytic diagram of an ellipse (blue), along with its two foci at* $(\pm c, 0)$. *The length of a path (dashed brown line) from a focus to a point on the ellipse to the other focus is 2a. See also figure 11*

$$\sqrt{(x-(-c))^2+(y-0)^2}+\sqrt{(x-c)^2+(y-0)^2}=2a.$$

Rearranging gives

$$\sqrt{(x+c)^2+y^2}=2a-\sqrt{(x-c)^2+y^2}.$$

Supplying the algebraic details of this derivation is left to you as exercise 70 at the end of this section.

Squaring both sides and simplifying eventually yields

$$2xc=4a^2-4a\sqrt{(x-c)^2+y^2}-2xc,$$

which can be rearranged to

$$\sqrt{(x-c)^2+y^2}=a-\frac{c}{a}x.$$

Squaring both sides again, simplifying, rearranging, and factoring yields

$$\left(1-\frac{c^2}{a^2}\right)x^2+y^2=a^2-c^2,$$

which can be rearranged to

$$\frac{x^2}{a^2}+\frac{y^2}{a^2-c^2}=1.$$

Now, define b^2 by

$$b^2=a^2-c^2,$$

If $a^2=b^2$, then the equation $\frac{x^2}{a^2}+\frac{y^2}{a^2}=1$ can be rearranged to $x^2+y^2=a^2$, which is a circle centered at the origin with radius a. We also have $c=0$, meaning that both foci of the ellipse are at the center of the circle.

and the equation reaches its final form:

$$\frac{x^2}{a^2}+\frac{y^2}{b^2}=1.$$

Solving $b^2 = a^2 - c^2$ for c gives $c = \sqrt{a^2 - b^2}$. Because the points $(\pm a, 0)$ and $(0, \pm b)$ satisfy the equation, they are points on the ellipse.

Because we use $c = \sqrt{a^2 - b^2}$ for a horizontal focal axis, we must have $a^2 > b^2$.

EQUATION OF AN ELLIPSE, HORIZONTAL FOCAL AXIS

The equation of an ellipse with foci $(\pm c, 0)$ and path length $2a$ is

$$\frac{x^2}{a^2} + \frac{y^2}{b^2} = 1,$$

with $c = \sqrt{a^2 - b^2}$.

An ellipse has two vertices, not four. The points $(0, \pm b)$ are not vertices of the ellipse.

Furthermore, we call the points $(\pm a, 0)$ the *vertices* of the ellipse. The *major axis* is the line segment with endpoints $(-a, 0)$ and $(a, 0)$, and it contains the foci. The *minor axis* is the line segment with endpoints $(0, -b)$ and $(0, b)$. Since $a^2 > b^2$, the major axis is longer than the minor axis. See figure 11.

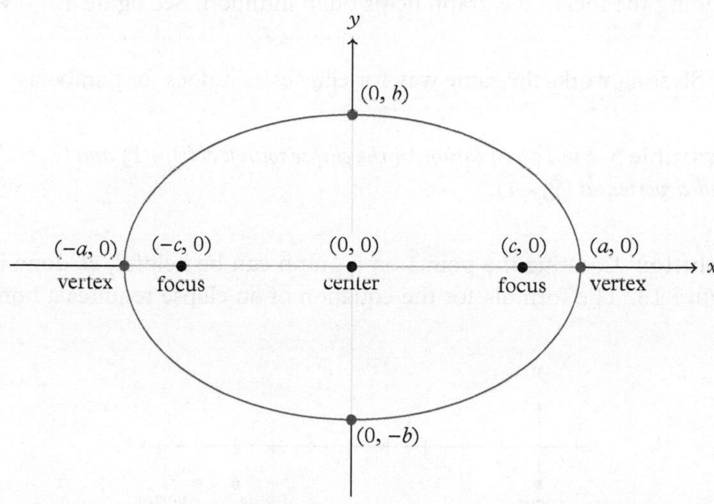

Figure 11 *An analytic diagram of an ellipse (blue), along with its two foci at* $(\pm c, 0)$*, center at* $(0, 0)$*, and vertices at* $(\pm a, 0)$*. The major axis (green line segment) and minor axis (yellow line segment) are also highlighted. Compare to figure 15*

Example 4 *Find the vertices and foci for the ellipse* $\frac{x^2}{36} + \frac{y^2}{8} = 1$ *and sketch the graph.*

Which denominator is larger indicates the orientation of the ellipse (compare to example 6).

Solution The equation $\frac{x^2}{36} + \frac{y^2}{8} = 1$ fits the form $\frac{x^2}{a^2} + \frac{y^2}{b^2}$ where $a^2 > b^2$, so the graph is an ellipse with a horizontal focal axis. We have

Because the length of the path, $2a$, must be positive, a must be positive. We make the same choice for b.

$$a^2 = 36, \quad a = 6$$

$$b^2 = 8, \quad b = \sqrt{8}$$

$$c = \sqrt{a^2 - b^2} = \sqrt{28} = 2\sqrt{7}.$$

The foci are $(\pm c, 0)$ and the vertices are $(\pm a, 0)$, and we state the desired points as

$$\text{foci } (2\sqrt{7}, 0) \text{ and } (-2\sqrt{7}, 0),$$

$$\text{vertices } (6, 0) \text{ and } (-6, 0).$$

To sketch the graph, place points at the vertices and at $(0, \pm b) = (0, \pm\sqrt{8})$, and then sketch the ellipse with these four points as a guide. Adding the foci to the graph helps build intuition. See figure 12. ■

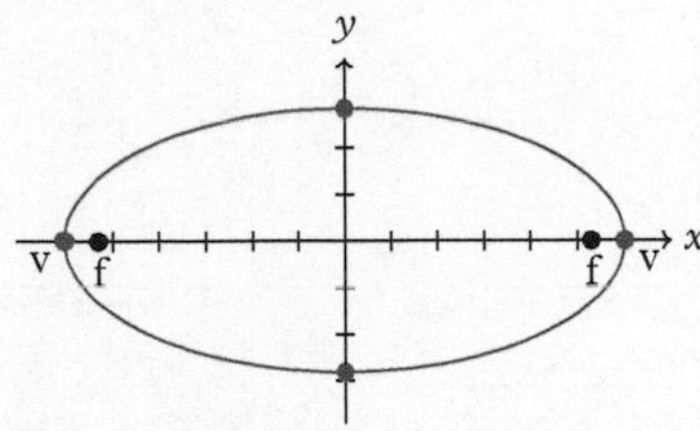

Figure 12 *The ellipse of example 4. The endpoints of the major axis (vertices) and endpoints of the minor axis are indicated in blue; they can be used as a guide for sketching the ellipse. The vertices and foci are labeled v and f, respectively*

Shifting works the same way for ellipses as it does for parabolas.

Example 5 *Find an equation for the ellipse with foci* $(0, -1)$ *and* $(8, -1)$, *and a vertex at* $(9, -1)$.

Solution Locating the points on a graph can be helpful, as done in figure 13. The formula for the equation of an ellipse requires a hori-

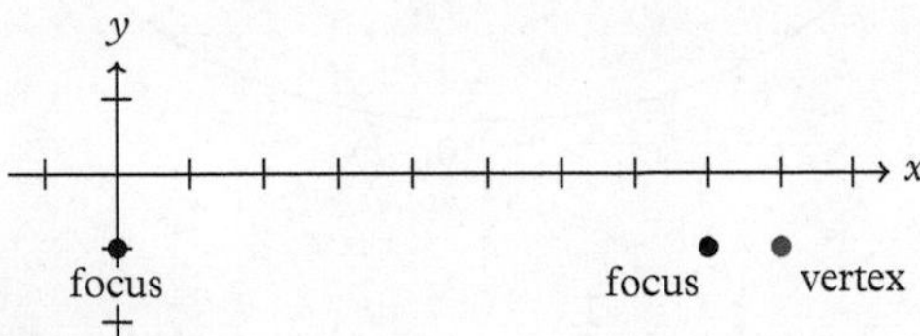

Figure 13 *The given points are plotted and labeled*

zontal focal axis, which we have, as verified in figure 13. The formula also requires the center of the ellipse to be located at the origin. The center is halfway between the foci. Averaging coordinates gives the center at the point $(4, -1)$. Therefore, the formula does not apply, but perhaps shifting can help.

Shifting 4 units to the left and 1 unit up places the center at the origin, the foci at $(\pm 4, 0)$, and the known vertex at $(5, 0)$, as in figure 14. Now the formula for the equation of an ellipse applies. Using the

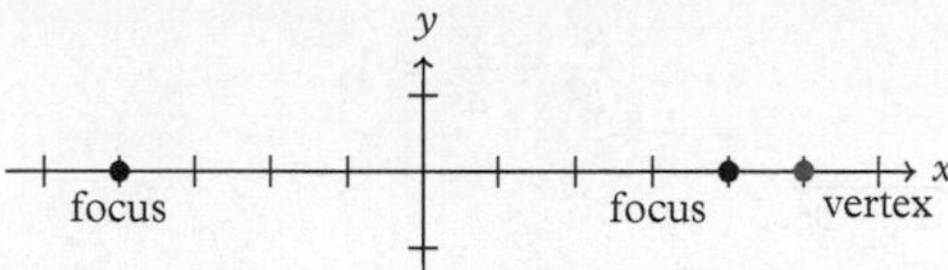

Figure 14 *The given points are shifted to place the center at the origin*

coordinates of the foci we have $c = 4$, and the coordinates of the known vertex tells us that $a = 5$. We may then calculate b^2 as

$$b^2 = a^2 - c^2 = 25 - 16 = 9.$$

The equation of the shifted ellipse is $\frac{x^2}{a^2} + \frac{y^2}{b^2} = 1$, giving

$$\frac{x^2}{25} + \frac{y^2}{9} = 1.$$

We finish by shifting back to the original ellipse, which requires us to shift right 4 units and down 1 unit. Replacing x with $x - 4$ and y with $y - (-1)$ results in

$$\frac{(x-4)^2}{25} + \frac{(y+1)^2}{9} = 1,$$

which is the equation we seek. ■

Reading Exercise 13 *Find the equation of the ellipse with foci* $(12, 0)$ *and* $(-12, 0)$, *and vertices* $(-13, 0)$ *and* $(13, 0)$.

To change the orientation of the ellipse to vertical, we place the foci at $(0, \pm c)$ so that the focal axis is vertical. Keeping the same constant path length $2a$, the derivation of a formula goes much the same as before, including setting $b^2 = a^2 - c^2$ so that $c = \sqrt{a^2 - b^2}$ and having $a^2 > b^2$. The resulting formula looks very similar, but this time the larger quantity is in the denominator of the y^2 term.

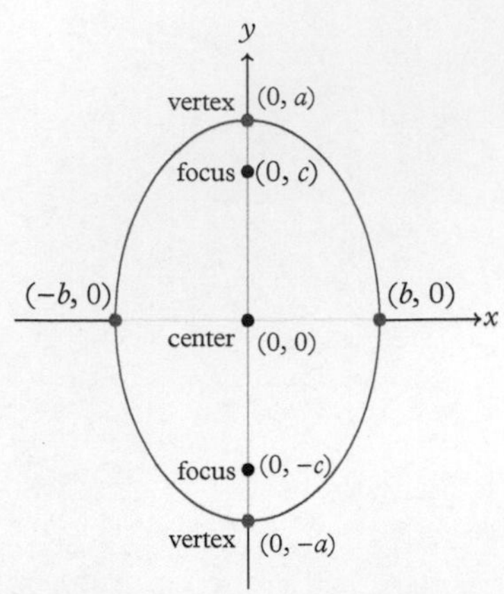

Figure 15 *An analytic diagram of an ellipse (blue) with a vertical focal axis, along with its two foci at* $(0, \pm c)$*, center at* $(0, 0)$*, and vertices at* $(0, \pm a)$*. The major axis (green line segment) and minor axis (yellow line segment) are also highlighted. Compare to figure 11*

Compare to example 4. The larger denominator indicates the orientation of the ellipse. If the denominator of the x^2 term is larger, the focal axis is horizontal. If the denominator of the y^2 term is larger, the focal axis is vertical.

The values of a, b, and c are identical to those of example 4. The larger of the two denominators is a^2, the smaller is b^2.

EQUATION OF AN ELLIPSE, VERTICAL FOCAL AXIS

The equation of an ellipse with foci $(0, \pm c)$ and path length $2a$ is

$$\frac{x^2}{b^2} + \frac{y^2}{a^2} = 1,$$

with $c = \sqrt{a^2 - b^2}$.

The vertices of the ellipse are located at $(0, \pm a)$, which still serve as the endpoints of the major axis. The minor axis is the line segment with endpoints $(-b, 0)$ and $(b, 0)$. Since $a^2 > b^2$, the major axis is still longer than the minor axis. See figure 15.

Example 6 *Find the vertices and foci for the ellipse* $\frac{x^2}{8} + \frac{y^2}{36} = 1$ *and sketch the graph.*

Solution The equation $\frac{x^2}{8} + \frac{y^2}{36} = 1$ fits the form $\frac{x^2}{b^2} + \frac{y^2}{a^2}$ where $a^2 > b^2$, so the graph is an ellipse with a vertical focal axis. We have

$$a^2 = 36, \quad a = 6$$
$$b^2 = 8, \quad b = \sqrt{8}$$
$$c = \sqrt{a^2 - b^2} = \sqrt{28} = 2\sqrt{7}.$$

The foci are $(0, \pm c)$ and the vertices are $(0, \pm a)$, and we state the desired points as

$$\text{foci } (0, 2\sqrt{7}) \text{ and } (0, -2\sqrt{7}),$$
$$\text{vertices } (0, 6) \text{ and } (0, -6).$$

To sketch the graph, place points at the vertices and at $(\pm b, 0) = (\pm\sqrt{8}, 0)$ and then sketch the ellipse with these four points as a guide. See figure 16. ■

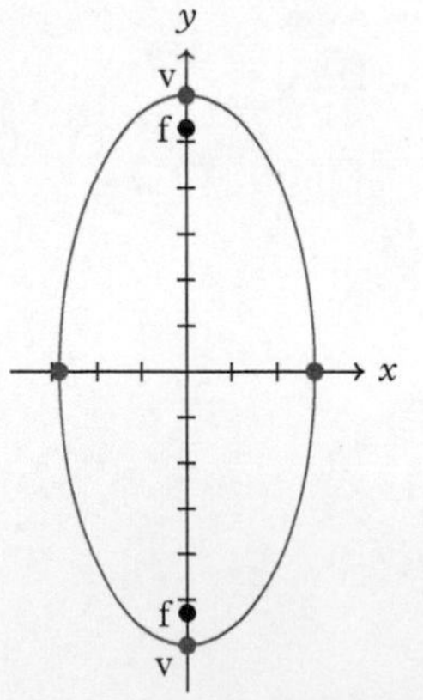

Figure 16 *The ellipse of example 6. The endpoints of the major axis (vertices) and endpoints of the minor axis are indicated by blue dots. They can be used as a guide for sketching the ellipse. The vertices and foci are labeled v and f, respectively*

The equations in examples 4 and 6 appear, at first, to have the same form. The graphs of the two ellipses can be produced identically, using the denominator of the x^2 term to locate the points at the left and right extremities of the ellipse, and using the denominator of the y^2 term to locate the upper and lower extremities of the ellipse, then using these four points as a guide for sketching. However, which of these four points are the vertices differs based on the relative sizes of the denominators. The longer axis is the major axis with endpoints that are the vertices, whereas the endpoints of the minor axis (the shorter axis) are not vertices. The foci always appear on the major axis, never the minor axis.

Ans. to reading exercise 13
$$\frac{x^2}{169} + \frac{y^2}{25} = 1$$

Hyperbolas

Hyperbolas are defined in a manner similar to ellipses, but using the difference instead of the sum.

Definition 4 HYPERBOLA *A* hyperbola *is the set of points for which the difference of the distances to two fixed points (foci) is constant.*

Once again we have a definition without reference to coordinates or equations. A synthetic picture of a hyperbola is in figure 17.

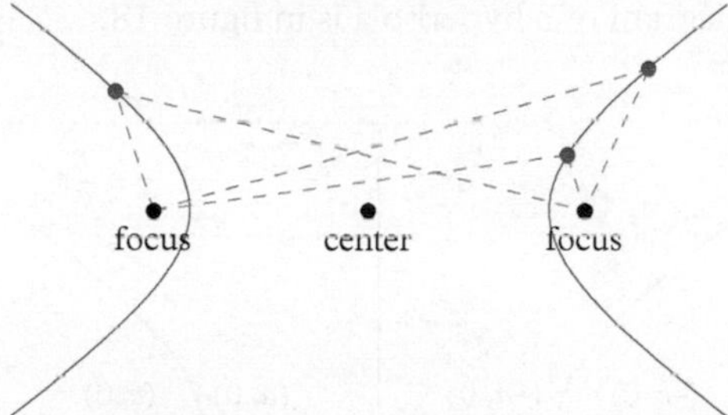

Figure 17 *A synthetic diagram of a hyperbola (blue), along with its two foci. The difference between the lengths of the longer (dashed brown) segment and the shorter (dashed green) segment from a point on the hyperbola to the two foci is the same for each point on the hyperbola. The line through the two foci (not pictured) is sometimes called the* focal axis *of the hyperbola. The* center *of the hyperbola is on the focal axis, halfway between the two foci*

A hyperbola has two *branches*, one associated with each focus. The intersection of the focal axis with a branch of the hyperbola is a *vertex* of the hyperbola. To determine an equation for a hyperbola, we use the coordinates $(\pm c, 0)$ for the foci, which then places the center at $(0, 0)$ and makes the focal axis horizontal. From any point (x, y) on the hyperbola, we set the common difference in distances to the foci as $2a$. The derivation is very much the same as for an ellipse, with the exception that since $c^2 > a^2$, we set $b^2 = c^2 - a^2$, making $c = \sqrt{a^2 + b^2}$.

Unlike an ellipse, there is no requirement about the relative sizes of a^2 and b^2. Either can be larger, and they can be the same.

The resulting equation of the hyperbola is $\frac{x^2}{a^2} - \frac{y^2}{b^2} = 1$.

EQUATION OF A HYPERBOLA, HORIZONTAL FOCAL AXIS

The equation of a hyperbola with foci $(\pm c, 0)$ and common difference $2a$ is

$$\frac{x^2}{a^2} - \frac{y^2}{b^2} = 1,$$

with $c = \sqrt{a^2 + b^2}$.

The coordinates of the vertices are $(\pm a, 0)$, similar to an ellipse. For a hyperbola, however, the vertices are between the foci instead of the other way around. Each branch of the hyperbola approaches asymptotes with equations $y = \frac{b}{a}x$ and $y = -\frac{b}{a}x$, which are lines through the center of the hyperbola. The asymptotes can be used as graphing aids. An analytic diagram of a hyperbola is in figure 18.

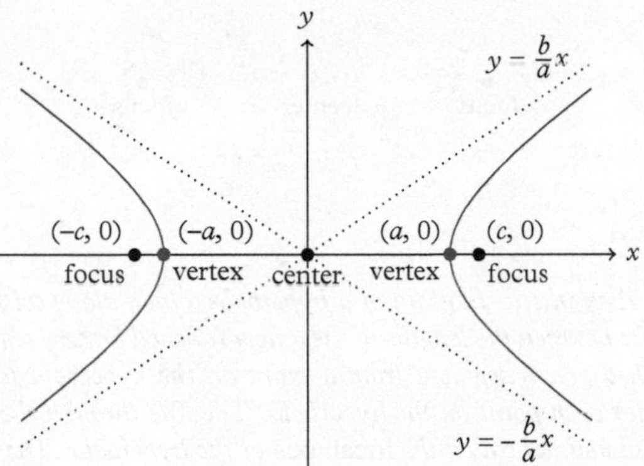

Figure 18 *An analytic diagram of a hyperbola (blue) with a horizontal focal axis. The foci are marked with black dots and the vertices with blue dots. The asymptotes are indicated by dotted lines*

The orientation of a hyperbola is decided by which variable's coefficient is positive. For $\frac{x^2}{a^2} - \frac{y^2}{b^2} = 1$, the coefficient on x^2 is positive and the focal axis is horizontal. For $\frac{y^2}{a^2} - \frac{x^2}{b^2} = 1$, the coefficient on y^2 is positive and the focal axis is vertical. Again, the derivation is similar to the horizontal case and is also left to you as an exercise.

> **EQUATION OF A HYPERBOLA, VERTICAL FOCAL AXIS**
>
> The equation of a hyperbola with foci $(0, \pm c)$ and common difference $2a$ is
>
> $$\frac{y^2}{a^2} - \frac{x^2}{b^2} = 1,$$
>
> with $c = \sqrt{a^2 + b^2}$.

For a vertical focal axis, the hyperbola's vertices are located at $(0, \pm a)$, and the equations of the asymptotes are $y = \pm\frac{a}{b}x$. See figure 19.

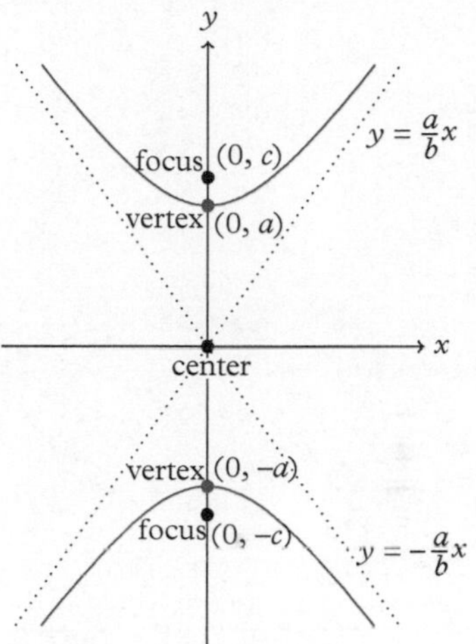

Figure 19 *An analytic diagram of a hyperbola (blue) with a vertical focal axis. The foci are marked with black dots and the vertices with blue dots. The asymptotes are indicated by dotted lines*

Example 7 *Find the foci and vertices of the hyperbola* $\frac{y^2}{9} - \frac{x^2}{4} = 1$, *and sketch the graph of the hyperbola.*

Solution The equation has the form $\frac{y^2}{a^2} - \frac{x^2}{b^2} = 1$, so the graph is a hyperbola with a vertical focal axis. We have $a = 3$, $b = 2$, and $c = \sqrt{9+4} = \sqrt{13}$. The foci $(0, \pm c)$ and vertices $(0, \pm a)$ are

$$\text{foci } (0, \sqrt{13}) \text{ and } (0, -\sqrt{13}),$$

$$\text{vertices } (0, 3) \text{ and } (0, -3).$$

The asymptotes are $y = \pm\frac{a}{b}x = \pm\frac{3}{2}x$. To sketch the graph, we plot the vertices, foci, and asymptotes, and use the asymptotes as a guide to draw the two branches of the hyperbola (figure 20). ■

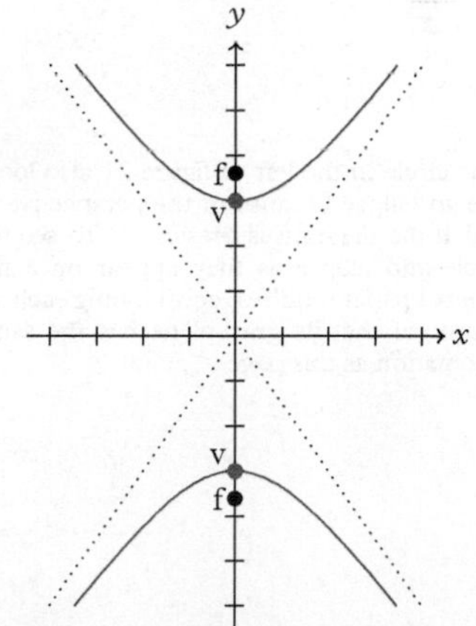

Figure 20 *The hyperbola of example 7. The foci and vertices are labeled f and v, respectively*

A quick method for drawing the asymptotes in figure 20 is to locate the four corners of a rectangular box—for this example, $(\pm 2, \pm 3)$—and draw the asymptotes through opposite corners. The values 2 and 3 come from the denominators of x^2 and y^2, respectively, and this works regardless of the orientation of the hyperbola. The vertices are located at the midpoints of two of the sides of the box, but be careful to determine the orientation of the hyperbola correctly to determine which two sides.

Reading Exercise 14 *Find the foci of the hyperbola* $\frac{x^2}{9} - \frac{y^2}{16} = 1$.

Why the name conic sections?

Why are circles, parabolas, ellipses, and hyperbolas called *conic sections?* The answer is that each can be described as a planar cut, or section, of a cone.

A *right circular cone* has circular cross sections when sliced perpendicular to the cone's axis. The axis of the cone is a line through the center of each of the circular cross sections. The left side of figure 21 shows a right circular cone with a vertical axis, along with a horizontal plane intersecting the cone. The intersection of the cone and plane is a circle.

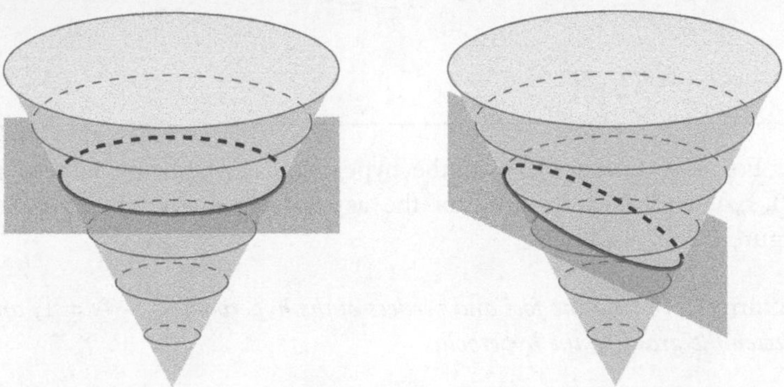

Figure 21 *(left) The intersection of a right circular cone (green) and a plane perpendicular to the cone's axis (gray) is a circle (blue). (right) A plane cutting through the cone, but not perpendicular to the cone's axis, intersects the cone in an ellipse*

Next we tilt the plane so that it is no longer horizontal and cut through the cone, as shown in the right side of figure 21. The resulting intersection is an ellipse. However, if we tilt the cutting plane far enough, a finite-length cut no longer reaches completely through the cone and the result is no longer an ellipse.

Figure 22 shows two views of the same plane cutting a cone. As seen in the view on the left, the plane is parallel to the "left side" of the cone. Therefore, the cut never reaches the left side, so it does not complete an ellipse; one end is left open, as in a parabola. Walk to the right around the structure and you soon arrive at the perspective of figure 22 (right), where the intersection is seen more clearly to be a parabola.

The circle in the left of figure 21 also looks like an ellipse because of the perspective in which the diagram is presented. To see the circle and ellipse as they appear on a flat sheet of paper requires repositioning each diagram so that its gray plane has the same orientation as this page.

Using a *double-napped cone*, which consists of a right circular cone and its reflection, we can obtain a hyperbola by tilting the plane to cut through both nappes of the cone (figure 23). This differs from a parabola, where the plane is positioned such that it only cuts through one nappe of the cone.

Each of the assertions in this subsection can be proved algebraically, but it requires knowledge of three-dimensional coordinate geometry.

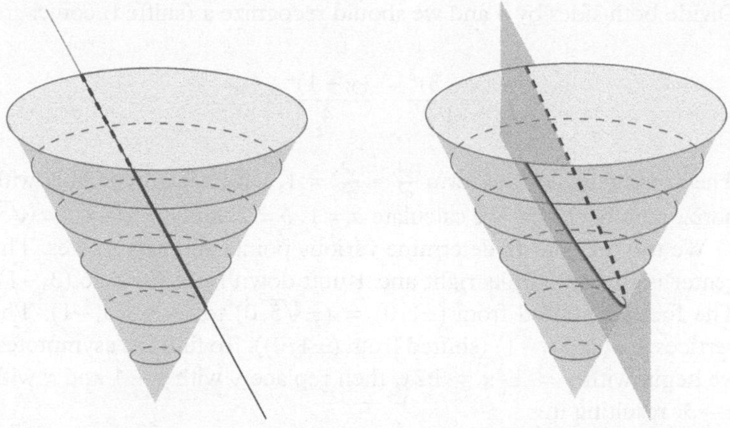

Figure 22 *Two views of the intersection of a right circular cone (green) and a plane (gray). The plane is parallel to the "left side" of the cone. (left) When viewed with the plane edge-on, the plane appears to be a line that is parallel to the left side of the cone. (right) A rotated view offers a different perspective, one in which the intersection, a parabola, is seen more clearly*

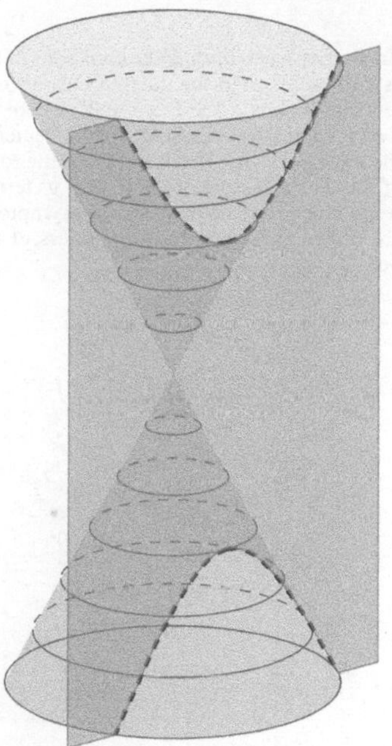

Figure 23 *A double-napped cone (green) with a plane (gray) cutting through both nappes. The intersection (blue) is a hyperbola*

Ans. to reading exercise 14
$(-5, 0)$ and $(5, 0)$

A more general example

The term *conic* or *conic section* is used to refer to any of the intersections of a cone and a plane. To identify a conic is to say whether it is a circle, an ellipse, and so on.

Example 8 *Find each focus of the conic* $4x^2 - y^2 - 24x - 2y + 31 = 0$*, identify the conic, and sketch its graph.*

Solution To identify the conic we need to rearrange the equation so that it is in one of the forms we recognize for a circle, ellipse, parabola, or hyperbola, or at least so that it can be shifted to be placed in one of these forms. With shifting in mind, we begin by completing the square in both the x and y variables:

$$\begin{aligned} 4x^2 - 24x \qquad - y^2 - 2y \qquad &= -31 \\ 4(x^2 - 6x \qquad) - (y^2 + 2y \qquad) &= -31 \\ 4(x^2 - 6x + 9) - (y^2 + 2y + 1) &= -31 + 4(9) - 1 \\ 4(x - 3)^2 - (y + 1)^2 &= 4. \end{aligned}$$

As long as the conic is not degenerate (see the exercises at the end of the section) and is in a reasonably simplified or factored form, the conic can be identified prior to the type of algebraic manipulation demonstrated in example 8. Returning to the original expression $4x^2 - y^2 - 24x - 2y + 31 = 0$, because both of the variables x and y are squared, the conic is not a parabola. Because x^2 and y^2 do not have the same coefficient, the conic is not a circle. Because the coefficients on x^2 and y^2 have different signs, the conic is a hyperbola (the same signs means an ellipse).

Divide both sides by 4 and we should recognize a (shifted) conic:

$$\frac{(x-3)^2}{1} - \frac{(y+1)^2}{4} = 1.$$

The conic matches the form $\frac{x^2}{a^2} - \frac{y^2}{b^2} = 1$, which is a hyperbola with horizontal focal axis. We calculate $a = 1$, $b = 2$, and $c = \sqrt{1+4} = \sqrt{5}$.

We use the shift to determine various points and asymptotes. The center is shifted 3 units right and 1 unit down from $(0, 0)$ to $(3, -1)$. The foci are shifted from $(\pm c, 0) = (\pm\sqrt{5}, 0)$ to $(3 \pm \sqrt{5}, -1)$. The vertices are $(3 \pm 1, -1)$ (shifted from $(\pm 1, 0)$). To find the asymptotes, we begin with $y = \pm\frac{b}{a}x = \pm 2x$, then replace y with $y + 1$ and x with $x - 3$, resulting in

An alternative to finding the equations of the asymptotes is to locate the four corners of a rectangular box, $(3 \pm 1, -1 \pm 2)$, where the 3 and -1 are the coordinates of the center of the hyperbola, and the 1 and 2 are the roots of the denominators of the x^2 and y^2 terms, respectively. As stated earlier, the asymptotes are drawn through opposite corners of the box.

$$\begin{aligned}(y+1) &= 2(x-3)\\ y &= 2x - 7\end{aligned}$$

and

$$\begin{aligned}(y+1) &= -2(x-3)\\ y &= -2x + 5.\end{aligned}$$

The graph (see figure 24) of the hyperbola is sketched by plotting the vertices and asymptotes, and by using the asymptotes as a guide to draw the curve.

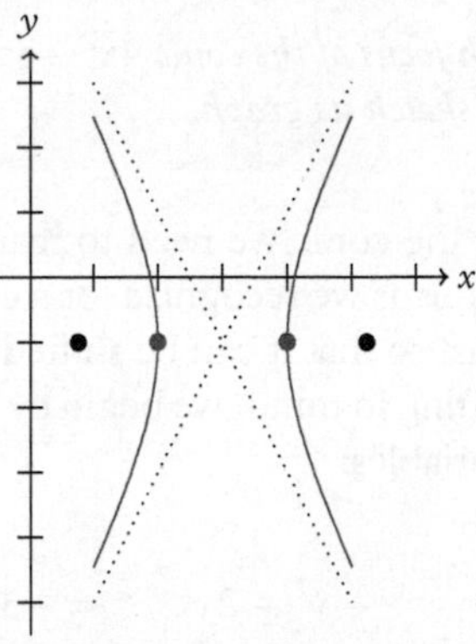

Figure 24 *The hyperbola (blue) of example 8. The foci are marked with black dots and the vertices with blue dots. The asymptotes are indicated by dotted lines*

■

Quadratic equations of the form $Ax^2 + Bxy + Cy^2 + Dx + Ey + F = 0$ containing a nonzero xy term are also conic sections. The axis or focal axis of such a conic is neither vertical nor horizontal. Identification of properties of the conic such as foci or vertices may require a rotation of the coordinate system. The algebra of rotations is not difficult, just tedious. It is not demonstrated here.

The conic is a parabola if $B^2 - 4AC = 0$, an ellipse if $B^2 - 4AC < 0$, and a hyperbola if $B^2 - 4AC > 0$.

Reflective properties of conics

Conic sections have remarkable reflective properties with many important applications.

Suppose the surface of a parabola is made of reflective material and that a signal to be reflected approaches the parabola parallel to its axis, as in the left side of figure 25. The signal reflects off the surface of

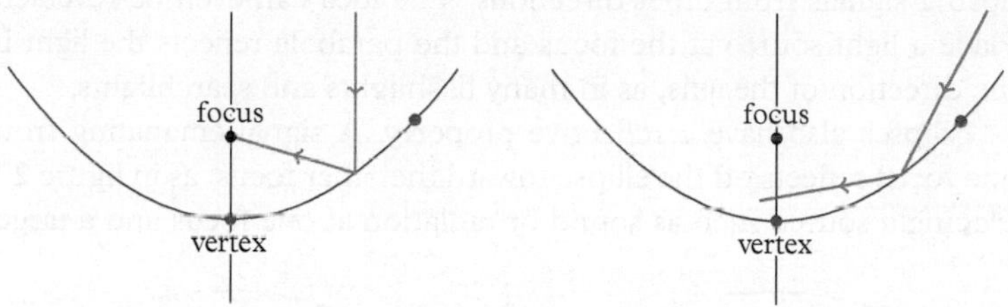

Figure 25 *(left) A signal (brown) parallel to the axis of the parabola reflects toward the focus. (right) A signal (green) not parallel to the axis of the parabola does not reflect toward the focus*

the parabola directly toward the focus of the parabola. However, if the signal is not parallel to the axis of the parabola, as shown in the right side of figure 25, the signal does not reflect toward the focus.

A parabolic microphone is designed to take advantage of this reflective property, with a reflective surface on the parabola and a microphone located at the focus. Consider a sound source, for instance on a playing field during a sporting event, and a parabolic mic located out of the field of play, some distance away. The sound travels in all directions away from the source, and there may be many other sounds in the environment. However, if the axis of the parabola is pointed toward the desired sound source, then the microphone "hears" a wide swath of the sound coming from the desired source (figure 26), but it does not hear the sounds coming from other sources because they do not reflect toward the microphone at the focus. Although the visual effect in figure 26 is striking, it is even more so in three dimensions. Rotate the parabola around its axis and you have an entire surface that

The microphone may pick up a very thin slice of other sounds coming directly at the microphone, but it pales in comparison to the volume reflected from the desired source.

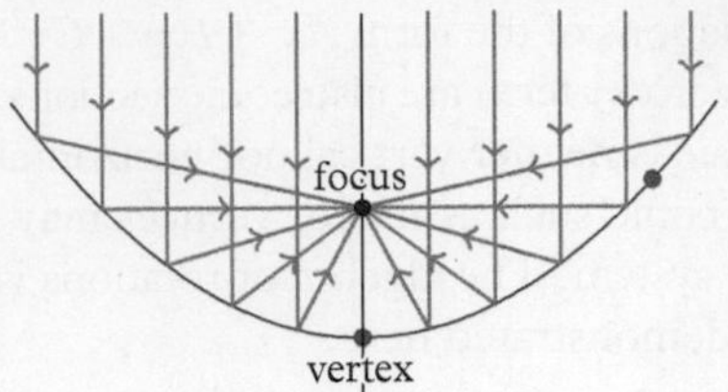

Figure 26 *Signals from the direction of the parabola's axis all reflect toward the focus, so that a microphone located at the focus picks up a very strong signal from that direction. Only a comparatively very weak signal from other directions is picked up*

reflects toward the focus. This design is used not only in parabolic microphones, but also in satellite dishes and other similar settings where the desire is to capture a signal from one particular direction while ignoring signals from other directions. The idea can even be reversed. Place a light source at the focus and the parabola reflects the light in the direction of the axis, as in many flashlights and searchlights.

Ellipses also have a reflective property. A signal emanating from one focus reflects off the ellipse toward the other focus, as in figure 27. Placing a source such as sound or radiation at one focus and a target

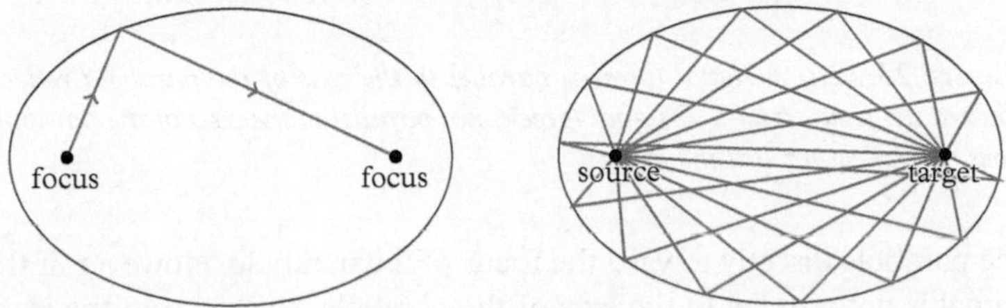

Figure 27 *A signal from one focus of an ellipse reflects off the ellipse toward the other focus, allowing energy from a source at one focus to be transferred to a target at the other focus*

at the other focus then allows energy from the source to be transferred to the target with minimal effect other than at the target. This is, of course, accomplished using a three-dimensional version, which may be formed by revolving the ellipse around the focal axis. Medical applications include the use of sound (the source at one focus) to blast a kidney stone (the target placed at the other focus) into small pieces so that it can pass through the ureter, a procedure called *lithotripsy*.

Other applications of conics, including hyperbolas, abound in such areas as telescope design and 20th-century navigation systems. An additional application of ellipses is described in the next section.

Exercises 8.5

1–14. Rapid response: which type of conic section is given by the equation? Give the best answer from among circle, parabola, ellipse, or hyperbola.

1. $x^2 + y^2 = 4$
2. $\frac{x^2}{7} + \frac{y^2}{12} = 1$
3. $x = 4y^2$
4. $\frac{x^2}{7} - \frac{y^2}{12} = 1$
5. $\frac{x^2}{5} + \frac{y^2}{9} = 1$
6. $\frac{x^2}{7} + \frac{y}{12} = 1$
7. $\frac{(x-5)^2}{1} + \frac{(y-1)^2}{9} = 1$
8. $\frac{x^2}{5} + \frac{y^2}{5} = 1$
9. $\frac{y^2}{5} - \frac{x^2}{9} = 1$
10. $x^2 + 6y^2 = 36$
11. $6x^2+2y^2-5x+9y = 11$
12. $6x^2-2y^2-5x+9y = 11$
13. $6x^2 - 5x + 9y = 11$
14. $6x^2+6y^2-5x+9y = 11$

15–18. Find the vertex, focus, and directrix of the parabola.

15. $y = \frac{1}{3}x^2$
16. $x = \frac{y^2}{12}$
17. $y^2 = 5x$
18. $x^2 = \frac{y}{12}$

19–26. Find the equation of the parabola.

19. focus $(0, 4)$, directrix $y = -4$
20. focus $(0, -3)$, directrix $y = 3$
21. vertex $(0, 4)$, directrix $y = -4$
22. focus $(-3, 0)$, directrix $x = 3$
23. focus $(4, 0)$, vertex $(7, 0)$
24. vertex $(0, -3)$, directrix $y = 3$
25. focus $(2, 6)$, directrix $y = 4$
26. focus $(2, 6)$, directrix $x = 4$

27–30. (a) Find the vertices and foci for the ellipse, and (b) sketch the graph of the ellipse.

27. $\frac{x^2}{25} + \frac{y^2}{16} = 1$

28. $\frac{x^2}{16} + \frac{y^2}{25} = 1$

29. $\frac{(x-3)^2}{4} + \frac{(y-1)^2}{81} = 1$

30. $\frac{(x+2)^2}{16} + \frac{(y-5)^2}{9} = 1$

If only one focus is listed, this does not mean the ellipse has just one focus; it means that listing the second focus is unnecessary for determining the equation of the ellipse. The same is true if only one vertex is listed.

31–36. Find an equation of the ellipse.

31. foci $(\pm 7, 0)$, vertex $(8, 0)$

32. foci $(\pm 7, 0)$, vertex $(-10, 0)$

33. focus $(0, 4)$, vertices $(0, \pm 10)$

34. focus $(0, -3)$, vertex $(0, -5)$, center $(0, 0)$

35. foci $(2, 8)$ and $(2, 0)$, vertex $(2, -1)$

36. vertices $(0, 0)$ and $(9, 0)$, focus $(7, 0)$

37–40. (a) Find the vertices, foci, and asymptotes of the hyperbola; and (b) sketch the graph of the hyperbola.

37. $\frac{x^2}{4} - \frac{y^2}{7} = 1$

38. $\frac{y^2}{0.25} - \frac{x^2}{0.04} = 1$

39. $\frac{y^2}{4} - \frac{(x-2)^2}{4} = 1$

40. $x^2 - y^2 = 1$

41–46. Find an equation of the hyperbola.

41. foci $(\pm 3, 0)$, vertex $(2, 0)$
42. vertices $(\pm 3, 0)$, focus $(-5, 0)$
43. vertex $(4, 0)$, asymptotes $y = \pm \frac{5}{4}x$
44. vertex $(0, -3)$, focus $(0, -5)$, center $(0, 0)$
45. vertices $(0, 2)$ and $(0, 10)$, focus $(0, 11)$
46. foci $(3, -5)$ and $(29, -5)$, vertex $(4, -5)$

47–66. If the given equation is for a circle, state the center and radius; for a parabola, state the focus, vertex, and directrix; for an ellipse, state the foci and vertices; for a hyperbola, state the foci, vertices, and equations of asymptotes. If the equation is not one of these conic sections, state "none of the above."

47. $y = 5x^2$

48. $6y = x^2$

49. $4x^2 + 11y^2 = 1$

50. $4x^2 + 11y^2 = 44$

51. $5(x-1) + 6(y+2) = 1$

52. $\frac{x^2}{9} - y^2 - 1 = 0$

53. $\frac{y^2}{8} - \frac{x^2}{2} = 2$

54. $x + 1 = (y-5)^2$

55. $y - 4 = \frac{1}{8}(x-3)^2$

56. $\frac{(x+2)^2}{4} + \frac{(y-6)^2}{4} = 16$

57. $\frac{(x-3)^2}{16} + \frac{(y+8)^2}{16} = 1$

58. $\frac{(x+2)^2}{4} + \frac{(y-6)^2}{9} = 16$

59. $\frac{(x-3)^2}{16} + \frac{(y+8)^2}{9} = 1$

60. $x^2 - 2y^3 - 2x + 4 = 0$

61. $\frac{(x-3)^2}{16} - \frac{(y+8)^2}{16} = 1$

62. $x^2 - 2y^2 - 2x + 4 = 0$

63. $x + 2y^2 - 6y + 5 = 0$

64. $x^2 - 2y - 2x + 4 = 0$

65. $\frac{1}{2}x^2 + 5y^2 - 3x + 20y - \frac{1}{2} = 0$

66. $x^2 + 9y^2 + 4x - 18y - 3 = 0$

67. Find the equation of the tangent line to the ellipse $\frac{x^2}{8} + \frac{y^2}{50} = 1$ at the point $(2, -5)$.

68. Find the equation of the tangent line to the hyperbola $\frac{x^2}{5} - \frac{y^2}{10} = 1$ at the point $(\sqrt{7}, -2)$.

69. Use the distance formula and the definition of parabola to derive the equation of a parabola with vertex $(0, 0)$, focus $(p, 0)$, and directrix $x = -p$.

70. Supply the missing details in the derivation of the equation of an ellipse with a horizontal focal axis.

71. The intersection of a double-napped cone and a plane can result in a circle, an ellipse, a hyperbola, or a parabola. Other possibilities, called *degenerate conic sections*, exist. Using figures 21, 22, and 23 as a guide for your imagination, describe three possibilities not given in these figures.

72. Use the distance formula and the definition of hyperbola to derive the equation of a hyperbola with foci $(\pm c, 0)$ and common difference $2a$, using $b^2 = c^2 - a^2$.

73. The area of a circle with radius 1, such as $x^2 + y^2 = 1$, is π. Use the following steps to show that the area of the ellipse $x^2+4y^2 = 1$ is $\frac{\pi}{2}$. (a) Solve the equation of the circle $x^2+y^2 = 1$ for y. (b) Using Ω subintervals, write an omega sum for the area of the circle. Do not evaluate the omega sum. (c) Solve the equation of the ellipse $x^2 + 4y^2 = 1$ for y. (d) Using Ω subintervals, write an omega sum for the area of the ellipse. Do not evaluate the omega sum. (e) Knowing that the value of the omega sum in part (b) must be π, determine the value of the omega sum in part (d) to find the area of the ellipse.

Cavalieri used slicing arguments similar to this one to find areas of geometric objects. Instead of using what we call *omega sums* to add the areas of infinitely many rectangles of infinitesimal width, Cavalieri reasoned in the same manner using *indivisibles*, which were a precursor to the current understanding of an infinitesimal.

74. Following the steps in exercise 73, compare the area of the ellipse $\frac{x^2}{a^2} + \frac{y^2}{b^2} = 1$ to the area of the circle $x^2 + y^2 = a^2$ to show that the area of the ellipse is πab.

Conic Sections in Polar Coordinates

8.6

In this section we continue our study of conic sections through the use of polar coordinates.

Eccentricity

You may have noticed that in an ellipse that is somewhat circular, the foci are located close to the center of the ellipse, whereas in an ellipse that looks elongated (long and flat), the foci are located close to the vertices. The placement of the foci relative to the vertices is called the *eccentricity* of the conic section.

Because distances are nonnegative, the eccentricity of a conic cannot be negative.

Definition 5 ECCENTRICITY *The eccentricity e of an ellipse or hyperbola is given by*

$$e = \frac{\textit{distance between foci}}{\textit{distance between vertices}} = \frac{c}{a}.$$

The eccentricity of a parabola is defined to be $e = 1$.

For an ellipse, the distance between the foci is less than the distance between the vertices (figure 1), so the eccentricity of an ellipse is less than 1. Recalling that $c = \sqrt{a^2 - b^2}$, the eccentricity of an ellipse is rewritten as

$$\text{ellipse eccentricity} = \frac{\sqrt{a^2 - b^2}}{a}.$$

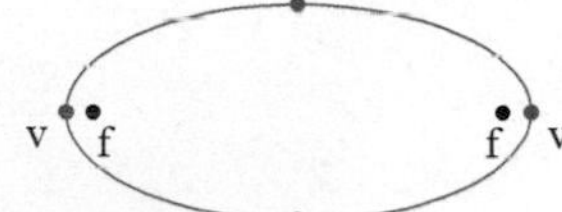

Figure 1 *The foci of an ellipse are closer together than the vertices. Vertices and foci are depicted as v and f, respectively*

A circle is an ellipse with foci that are both at the center; so, in a circle, the distance between the foci is 0, making the eccentricity 0. Another way to see this is that a circle is an ellipse with $a^2 = b^2$, making $c = 0$ and $e = 0$.

In a hyperbola, the distance between the foci is greater than the distance between the vertices (figure 2), so the eccentricity of a hyperbola is greater than 1. Recalling that $c = \sqrt{a^2 + b^2}$, the eccentricity of a hyperbola is rewritten as

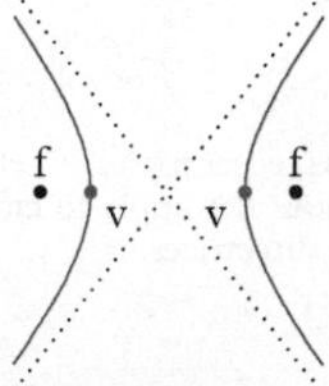

Figure 2 *The foci of a hyperbola are father apart than the vertices. Vertices and foci are depicted as v and f, respectively*

$$\text{hyperbola eccentricity} = \frac{\sqrt{a^2 + b^2}}{a}.$$

A parabola has only one focus and one vertex, so the distances between the foci and between the vertices are not defined. A reason for defining $e = 1$ for a parabola is described in the next subsection, after theorem 1.

Figure 3 shows the classification of conics by eccentricity.

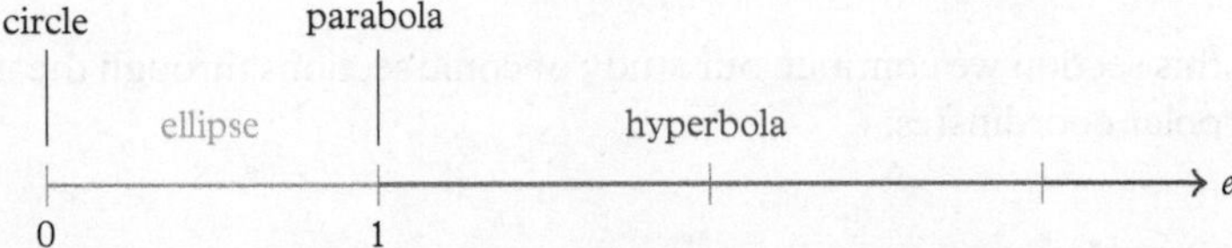

Figure 3 *Type of conic as classified by eccentricity. For a circle,* $e = 0$*. For an ellipse,* $0 < e < 1$*. For a parabola,* $e = 1$*. For a hyperbola,* $e > 1$

The closer the eccentricity of an ellipse is to zero, the more circular the ellipse appears. The closer the eccentricity of an ellipse is to one, the more elongated the ellipse appears.

The orbits of planets, comets, and other objects about the sun are elliptical, with the center of the sun at a focus (not the center!) of the ellipse. The earth's orbit is nearly circular, with an eccentricity of 0.0167. Halley's comet, by contrast, has a very elongated orbit, with an eccentricity of 0.967.

Because of the influence of objects other than the sun, such as other planets, the orbit of an object around the sun is not a perfect ellipse. However, for most orbiting objects, an ellipse provides a very good model of the orbit. The eccentricity of an orbit can also vary over time.

Reading Exercise 15 *Identify the conic by its eccentricity: (a)* $e = 1.37$*, (b)* $e = 0.37$*, (c)* $e = 0$*, and (d)* $e = 1$*.*

Directrices

A parabola has a directrix. Ellipses and hyperbolas also have directrices.

The word *directrices* is the plural form of the word *directrix*.

Definition 6 DIRECTRIX *A* **directrix** *of an ellipse or a hyperbola is a line perpendicular to the focal axis located* $\pm\frac{a}{e}$ *units from the conic's center, where a and e are as described previously.*

A circle has eccentricity $e = 0$, so the definition does not apply to circles. Circles do not have directrices.

The property of parabolas that the distance from a point to the focus is the same as the distance from the point to the directrix is generalized for ellipses and hyperbolas using eccentricity. The proof is outlined in the exercises.

Theorem 1 FOCUS–DIRECTRIX–ECCENTRICITY RELATIONSHIP *For any point P on a parabola, ellipse, or hyperbola, the distance from P to a focus is the eccentricity times the distance from P to the associated directrix.*

For theorem 1 to be true for a parabola, the eccentricity must be one. This is not the only reason, but it is one of the reasons why we define $e = 1$ for a parabola.

Whether the conic is an ellipse, hyperbola, or parabola, a vertex is always between a focus and a directrix, with the conic curving around the focus away from the directrix, as shown in figure 4. For an ellipse, a vertex is closer to a focus than a directrix; for a hyperbola, a vertex is closer to a directrix than a focus.

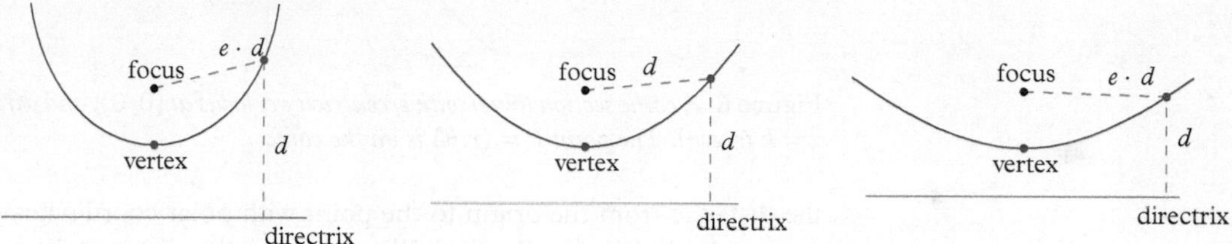

Figure 4 *Three conics with different eccentricities, each with the same distance between the focus and the vertex. The ellipse (left) has $e < 1$, so points (including the vertex) are closer to the focus than the directrix. The hyperbola (right) has $e > 1$, so points are farther from the focus than the directrix. The parabola (center) has $e = 1$, so points are equidistant from the focus and the directrix*

With an ellipse, the directrices are on the outside; with a hyperbola, the directrices are between the branches. Figure 5 illustrates a complete ellipse with both its foci and its directrices, as well as a hyperbola with its foci and directrices.

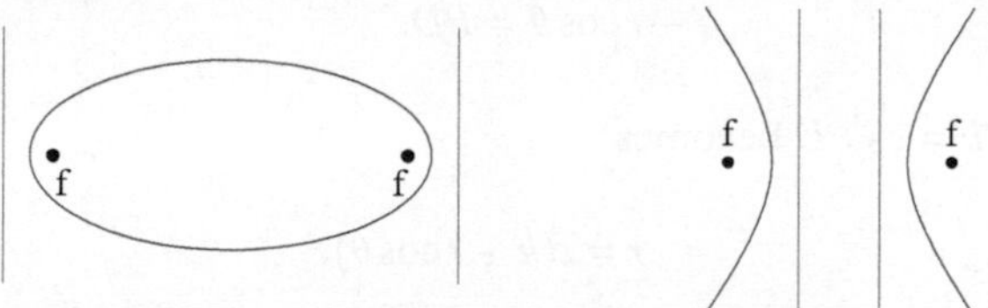

Figure 5 *(left) An ellipse and (right) a hyperbola, each with its two foci and associated directrices. The directrices (green) are not part of the conic (blue)*

Polar equation of a conic section, e>0

Theorem 1 helps us derive a polar equation for a conic section. Suppose our conic section has a focus located at the point $(0, 0)$ and a

Ans. to reading exercise 15
(a) hyperbola, (b) ellipse, (c) circle, (d) parabola

directrix with equation $x = k$ (then the conic cannot be a circle; we must have $e > 0$). Let the point P be a point on the conic with polar coordinates (r, θ), as in figure 6. By theorem 1, $PF = e \cdot PD$. Because

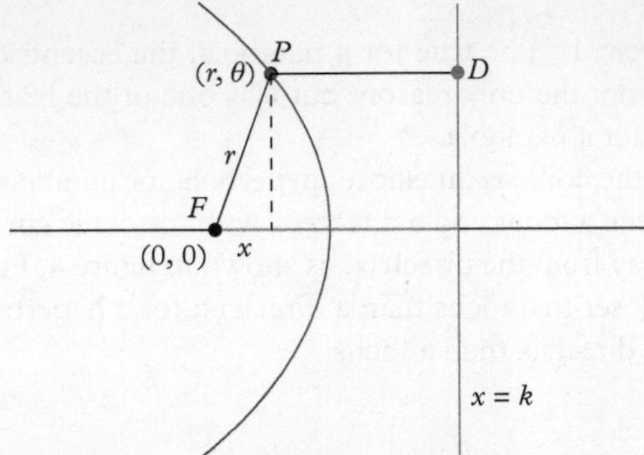

Figure 6 *A conic section (blue) with eccentricity e, focus at* $(0, 0)$, *and directrix* $x = k$ *(green). The point* $P = (r, \theta)$ *is on the conic*

the distance from the origin to the point with polar coordinates (r, θ) is r, we have $PF = r$. To find PD, note that the distance from F to the directrix (length of the horizontal gray segment) is the same as the length x plus the length PD. The directrix is k units from the origin, so the horizontal gray segment has length k, and we have

$$k = x + PD.$$

We also know that $x = r\cos\theta$; hence,

$$k = r\cos\theta + PD$$

$$k - r\cos\theta = PD.$$

Then, $PF = e \cdot PD$ becomes

$$r = e(k - r\cos\theta),$$

and solving for r gives the desired equation:

$$r = ek - er\cos\theta$$

$$r + er\cos\theta = ek$$

$$r(1 + e\cos\theta) = ek$$

$$r = \frac{ek}{1 + e\cos\theta}.$$

This equation must hold for every point P on the conic.

POLAR EQUATION OF A CONIC

A polar equation of an ellipse, parabola, or hyperbola with eccentricity e, focus at the origin, and directrix $x = k$ is

$$r = \frac{ek}{1 + e\cos\theta}.$$

Equations of conics given in the previous section had the vertex of a parabola or the center of an ellipse or hyperbola at the origin, rather than the focus at the origin. The placement of this conic is not the same as for those equations.

The vertices of the conic are located where $\theta = 0$ and $\theta = \pi$.

For a parabola, r is undefined at one of the values of θ, resulting in only one vertex.

If desired, the equation of the directrix can be rewritten using polar coordinates as well. Substituting $x = r\cos\theta$ into $x = k$ results in $r = k\sec\theta$ for the polar equation of the directrix.

Example 1 *Find the eccentricity, identify the conic, give an equation of a directrix, and determine the vertices for*

$$r = \frac{3}{2 + \cos\theta}.$$

Solution A polar equation is given, but not quite in the form $r = \frac{ek}{1 + e\cos\theta}$. In particular, the denominator needs to have the form $1 + e\cos\theta$, but we currently have $2 + \cos\theta$. Dividing the numerator and the denominator by 2 puts the denominator in the required form:

CAUTION: MISTAKE TO AVOID
The denominator of the polar equation **must** be in the form $1 + e\cos\theta$ (or a form corresponding to an alternate orientation) before determining e or k.

$$r = \frac{3}{2 + \cos\theta} \cdot \frac{\frac{1}{2}}{\frac{1}{2}} = \frac{\frac{3}{2}}{1 + \frac{1}{2}\cos\theta}.$$

Comparing to the form $r = \frac{ek}{1 + e\cos\theta}$ reveals that $e = \frac{1}{2}$; thus, the conic is an ellipse. We still need an equation of a directrix, which means we need to find k. Comparing numerators,

Compare the denominators to determine e.

$$ek = \tfrac{3}{2}$$
$$\tfrac{1}{2}k = \tfrac{3}{2}$$
$$k = 3.$$

Therefore, a directrix has the equation $x = 3$.

The directrix found here is the one associated with the focus $(0, 0)$. The other directrix can be found by sketching the ellipse (using only the information gathered here) and applying symmetry.

To find the vertices, we evaluate the polar equation at $\theta = 0$ and at $\theta = \pi$:

$$r = \frac{3}{2 + \cos 0} = 1$$
$$r = \frac{3}{2 + \cos\pi} = 3.$$

The polar coordinates of the vertices are $(1, 0)$ and $(3, \pi)$.

We conclude that the conic is an ellipse with $e = \frac{1}{2}$, vertices $(1, 0)$ and $(3, \pi)$, and a directrix with equation $x = 3$. ■

The polar form of the directrix is also given easily instead. Since $k = 3$, the polar equation of the directrix is $r = 3 \sec \theta$.

Reading Exercise 16 *What is the eccentricity of the conic* $r = \dfrac{3}{1 + 1.5 \cos \theta}$?

When we developed the equation $r = \frac{ek}{1 + e \cos \theta}$, the conic had a horizontal axis and directrix on the right. Changing the orientation of the conic results in a denominator of $1 - e \cos \theta$, $1 + e \sin \theta$, or $1 - e \sin \theta$, as in figure 7.

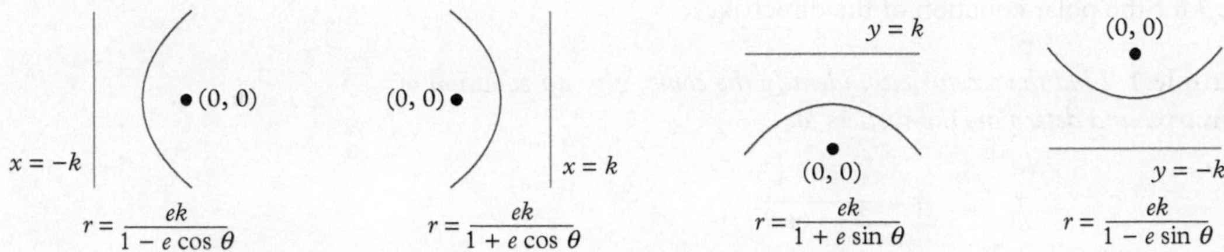

Figure 7 *Four orientations of a conic section with resulting polar equations*

For a vertical axis, the vertices are found by evaluating the equation at $\theta = \frac{\pi}{2}$ and $\theta = \frac{3\pi}{2}$.

Example 2 *Find the eccentricity, identify the conic, give an equation of a directrix, and determine the vertices for*

$$r = \frac{10}{5 - 6 \sin \theta}.$$

Solution The equation most closely fits the form $r = \frac{ek}{1 - e \sin \theta}$, so the conic's axis is vertical. Comparing denominators, we see the need to divide the numerator and the denominator by 5:

$$r = \frac{10}{5 - 6 \sin \theta} \cdot \frac{\frac{1}{5}}{\frac{1}{5}} = \frac{2}{1 - \frac{6}{5} \sin \theta}.$$

We read the eccentricity from the latter expression as $e = \frac{6}{5}$, so the conic is a hyperbola.

We do not use $ek = 10$ when comparing numerators. We must use the expression with $1 \pm \cdots$ in the denominator.

Next we find k by comparing numerators:

$$ek = 2$$
$$\frac{6}{5}k = 2$$
$$k = \frac{5}{3}.$$

For this form of the equation, a directrix is $y = -k$, or $y = -\frac{5}{3}$.

Because the orientation of the conic is vertical, we determine the vertices by evaluating the equation at $\theta = \frac{\pi}{2}$ and $\theta = \frac{3\pi}{2}$:

$$r = \frac{10}{5 - 6\sin\frac{\pi}{2}} = -10$$

$$r = \frac{10}{5 - 6\sin\frac{3\pi}{2}} = \frac{10}{11}.$$

Remembering the equation of the directrix that corresponds to the polar form of the conic is not as difficult as it looks. Merely associate $\cos\theta$ with x and $\sin\theta$ with y, then borrow the + or − preceding the trig term.

We conclude that the conic is a hyperbola with eccentricity $e = \frac{6}{5}$, a directrix at $y = -\frac{5}{3}$, and vertices having polar coordinates $(-10, \frac{\pi}{2})$ and $(\frac{10}{11}, \frac{3\pi}{2})$. ■

A quick sketch would show that the vertex that is between the focus $(0, 0)$ and its directrix $y = -\frac{5}{3}$ is $(\frac{10}{11}, \frac{3\pi}{2})$.

Example 3 *Find a polar equation for a conic with focus* $(0, 0)$, *directrix* $x = -2$, *and eccentricity* 17.

Solution With directrix $x = -k$ we use the formula $r = \frac{ek}{1 - e\cos\theta}$. With $k = 2$ and $e = 17$, the desired equation is

$$r = \frac{34}{1 - 17\cos\theta}.$$ ■

Ans. to reading exercise 16
$e = 1.5$

Alternate polar form for ellipses

In the previous section, we located the vertices and foci of an ellipse with a center at the origin at $(\pm a, 0)$ and $(\pm c, 0)$ (using Cartesian coordinates) as labeled in figure 8.

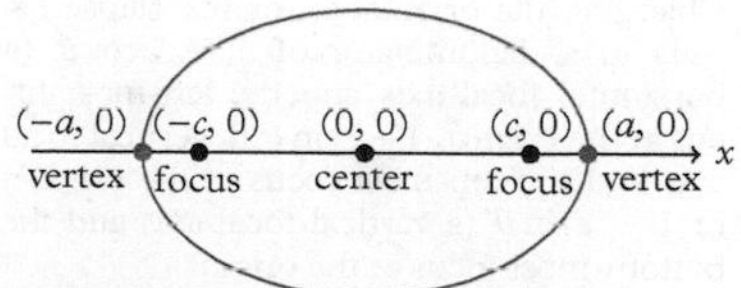

Figure 8 *An ellipse labeled with Cartesian coordinates*

To match the positioning of the conic used to develop the polar equation, we need to shift figure 8's ellipse c units to the left so that the right-most focus is at the origin. See figure 9. The polar form of the equation of the conic then applies, so that $r = \frac{ek}{1 + e\cos\theta}$. The point corresponding to $\theta = 0$ is the right-most vertex, and for this point

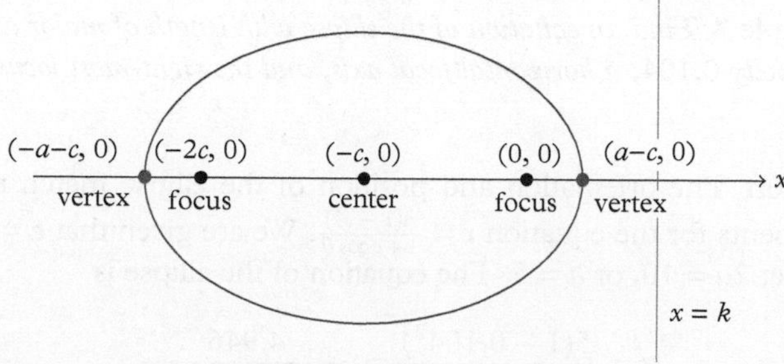

Figure 9 *An ellipse with a horizontal focal axis and the right-most focus at the origin labeled with Cartesian coordinates. The directrix associated with the focus* $(0, 0)$ *is shown, but the other directrix is not shown*

$$\frac{ek}{1 + e\cos 0} = \frac{ek}{1 + e}.$$

However, this vertex is $a - c$ units from the origin; hence,

$$r = \frac{ek}{1 + e} = a - c.$$

Rearranging the definition of eccentricity $e = \frac{c}{a}$ gives $c = ae$. Substituting $c = ae$ into the previous formula and solving for ek, we have

$$\frac{ek}{1 + e} = a - ae$$

$$\frac{ek}{1 + e} = a(1 - e)$$

$$ek = a(1 - e)(1 + e) = a(1 - e^2).$$

Substituting the latter quantity for ek in the polar form of a conic results in

$$r = \frac{a(1 - e^2)}{1 + e\cos\theta}.$$

ALTERNATE FORM OF THE POLAR EQUATION OF AN ELLIPSE

The equation of an ellipse with eccentricity e, length of major axis $2a$, a horizontal focal axis, and the right-most focus at the origin is

$$r = \frac{a(1 - e^2)}{1 + e\cos\theta}.$$

Changing the orientation of the ellipse results in a denominator of $1 - e\cos\theta$ (a horizontal focal axis and the left-most focus at the origin), $1 + e\sin\theta$ (a vertical focal axis and the top-most focus at the origin), or $1 - e\sin\theta$ (a vertical focal axis and the bottom-most focus at the origin).

Example 4 *Find an equation of the ellipse with length of major axis* 10, *eccentricity* 0.104, *a horizontal focal axis, and the right-most focus at the origin.*

Solution The orientation and position of the ellipse match the requirements for the equation $r = \frac{a(1 - e^2)}{1 + e\cos\theta}$. We are given that $e = 0.104$ and that $2a = 10$, or $a = 5$. The equation of the ellipse is

$$r = \frac{5(1 - 0.104^2)}{1 + 0.104\cos\theta} = \frac{4.946}{1 + 0.104\cos\theta}.$$

■

Reading Exercise 17 *Find an equation of the ellipse with length of major axis* 14*, eccentricity* 0.04*, a horizontal focal axis, and the right-most focus at the origin.*

It was mentioned that the orbit of an object around the sun is (approximately) an ellipse, with the center of the sun at one focus. If we position a coordinate system to match figure 9 with the sun at the focus $(0, 0)$, then the vertex closest to the sun is $a - c$ units from the sun whereas the vertex farthest from the sun is $a + c$ units from the sun. The closest position to the sun is called the *perihelion* and the farthest is called the *aphelion*.

The words *perihelion* and *aphelion* have a Greek origin, from "peri" (near), "apo" (away from), and "helios" (sun). For an object orbiting the earth, we instead use the terms *perigee* and *apogee*.

The alternate form of the polar equation of an ellipse uses the values a and e. Since $c = ae$, calculating the distances at perihelion $(a - c)$ and aphelion $(a + c)$ is not difficult.

Example 5 *The semimajor axis of earth's orbit is approximately* 92 956 000 *miles (that is,* $a = 92\,956\,000$*). Its eccentricity is approximately* 0.0167*. (a) Find an equation representing the earth's orbit. (b) Find the earth's distance to the sun at perihelion and at aphelion.*

Unless otherwise noted, all distances and eccentricities relating to orbiting objects should be understood to be approximate (perhaps rounded) and reflect the current estimate as of the time of writing.

Solution (a) We are given $a = 92\,956\,000$ and $e = 0.0167$. We can use the alternate polar form of the equation of an ellipse:

$$r = \frac{a(1 - e^2)}{1 + e\cos\theta} = \frac{92\,956\,000(1 - 0.0167^2)}{1 + 0.0167\cos\theta} = \frac{92\,930\,000}{1 + 0.0167\cos\theta}.$$

(b) We calculate c first:

$$c = ae = 92\,956\,000 \cdot 0.0167 = 1\,552\,000.$$

Then, the requested distances are

The actual aphelion and perihelion distances vary from year to year.

$$\text{perihelion: } a - c = 92\,956\,000 - 1\,552\,000 = 91\,404\,000 \text{ mi}$$

$$\text{aphelion: } a + c = 92\,956\,000 + 1\,552\,000 = 94\,508\,000 \text{ mi.} \quad \blacksquare$$

The earth's perihelion, its closest approach to the sun, occurs around January 4. Aphelion occurs around July 5. This serves to moderate temperatures in the northern hemisphere, because the earth is farther from the sun in the early summer and closer to the sun in the early winter. The effect is opposite in the southern hemisphere, however, causing greater temperature variation. The equation of the earth's orbit found in example 5 is graphed in figure 10.

The average dates of perihelion and aphelion change slowly, so that thousands of years from now, the effect could be reversed, with the southern hemisphere enjoying the greater moderation.

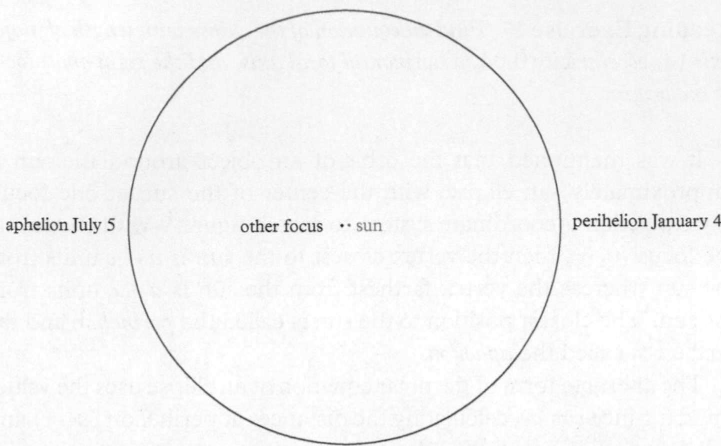

Ans. to reading exercise 17

$$r = \frac{7(1 - 0.04^2)}{1 + 0.04\cos\theta} = \frac{6.9888}{1 + 0.04\cos\theta}$$

Figure 10 *The elliptic but nearly circular orbit of the earth (blue), the sun (orange), and the other focus of the ellipse. The sun is to scale. Because the earth's diameter is less than one-hundredth that of the sun, the earth is not seen if it is plotted to scale. The perihelion and aphelion positions are at the vertices of the ellipse*

Exercises 8.6

1–8. Rapid response: identify the conic by its eccentricity.

1. $e = 1$
2. $e = 42$
3. $e = 0.004$
4. $e = 0$
5. $r = \dfrac{14}{1 - 3\cos\theta}$
6. $r = \dfrac{14}{1 + 3\cos\theta}$
7. $r = \dfrac{14}{6 + 3\cos\theta}$
8. $r = \dfrac{14}{3 + 3\cos\theta}$

9–18. (a) Find the eccentricity, (b) identify the conic, (c) give an equation of a directrix, and (d) determine the vertices.

9. $r = \dfrac{6}{1 + 2\cos\theta}$
10. $r = \dfrac{3}{1 - \sin\theta}$
11. $r = \dfrac{7}{4 - 3\cos\theta}$
12. $r = \dfrac{1}{5 + 8\cos\theta}$
13. $r = \dfrac{4}{\frac{1}{2} + \sin\theta}$
14. $r = \dfrac{4}{3 - 9\sin\theta}$

15. $r = \dfrac{57}{2\sqrt{3} + \pi\cos\theta}$ **17.** $r = \dfrac{12}{4 - 4\sin\theta}$

16. $r = \dfrac{3}{5 - \frac{1}{2}\cos\theta}$ **18.** $r = \dfrac{\csc\theta}{\cot\theta + \frac{1}{3}\csc\theta}$

19–28. Find a polar equation for a conic with a focus at $(0, 0)$ and additional information as given.

When a directrix is given, it is the directrix associated with the focus $(0, 0)$.

19. directrix $x = 10$, eccentricity 0.02

20. directrix $x = 4$, eccentricity 3

21. directrix $y = -2$, eccentricity 14

22. directrix $y = 3$, eccentricity 0.3

23. parabola, directrix $x = -3$

24. circle, point $(3, 4)$

25. directrix $y = 1$, eccentricity $\frac{1}{3}$

26. directrix $x = -11$, eccentricity 1

27. directrix $y = -\frac{11}{5}$, distance between vertices 12, distance between foci 10

28. directrix $y = -\frac{11}{6}$, distance between foci 12, distance between vertices 10

29–34. Find a polar equation of an ellipse with the given properties.

29. length of major axis 48, eccentricity $\frac{1}{2}$, a horizontal focal axis, the right-most focus at the origin

30. length of major axis 6, the eccentricity 0.2, a vertical focal axis, the top-most focus $(0, 0)$

31. semimajor axis length 10, eccentricity 0.99, a vertical focal axis, the top-most focus at the origin

32. a horizontal focal axis, the left-most focus at the origin, eccentricity $\frac{1}{100}$, length of major axis 9700

33. focus $(0, 0)$, vertices $(2, 0)$ and $(-10, 0)$

34. eccentricity 0.4, vertex $(1, 0)$, right-most focus $(0, 0)$

35. The planet Awasis, which orbits the star Nikawiy, was discovered in 2010 and given its current name in 2019. The semimajor axis of its orbit has length 1.32 AU and its eccentricity

An *astronomical unit* (AU) is approximately the length of the semimajor axis of the earth, about 92 956 000 miles. It is officially defined as 149 597 870 700 m.

is 0.255. (a) Find an equation of the orbit of Awasis. (b) Determine Awasis' furthest distance from Nikawiy.

36. The dwarf planet Eris, the discovery of which was announced in 2003, has about the same mass as Pluto but, on average, is nearly twice as far from the sun. Its semimajor axis is 67.67 AU and its eccentricity is 0.44177. (a) Find an equation of the orbit of Eris. (b) Find its distance to the sun at perihelion and aphelion.

37. The dwarf planet Ceres, which—like Pluto—was once considered to be a planet, was discovered in 1801. Its semimajor axis is 2.769 AU and its eccentricity is 0.076. (a) Find an equation of the orbit of Ceres. (b) Find its distance to the sun at perihelion and aphelion.

38. The discovery of planet CoRoT-9 b, orbiting the star CoRoT-9, was announced in 2010. The planet's semimajor axis is 0.407 AU and its eccentricity is 0.11. (a) Find an equation of the orbit of the planet. (b) Determine the planet's closest distance to the star it orbits.

39. Find the eccentricity of the ellipse $\dfrac{x^2}{30} + \dfrac{y^2}{14} = 1$.

40. Find the eccentricity of the hyperbola $\dfrac{x^2}{11} - \dfrac{y^2}{14} = 1$.

41. Find the equation of the tangent line to the ellipse $r = \dfrac{30}{10 + \cos\theta}$ at $\theta = \frac{\pi}{2}$.

42. Find the equation of the tangent line to the conic $r = \dfrac{2}{1 + 2\cos\theta}$ at $\theta = -\frac{\pi}{4}$.

43. (a) Rewrite the polar equation of a conic $r = \frac{ek}{1 + e\cos\theta}$ for a parabola, recalling that $e = 1$. (b) Verify that the equation is undefined at $\theta = \pi$, so that only one vertex exists. (c) Determine the location of the vertex of the parabola and verify that it is equidistant from the focus and the directrix.

44. Derive the formula $r = \frac{ek}{1 - e\sin\theta}$ for a conic with eccentricity e, focus $(0, 0)$, and associated directrix $y = -k$.

45. Prove theorem 1 for an ellipse with center $(0, 0)$, focus $(c, 0)$, vertex $(a, 0)$, and directrix $x = \frac{a}{e}$ (figure 11) using the following steps: (a) Let (x, y) be a point on the ellipse $\dfrac{x^2}{a^2} + \dfrac{y^2}{b^2} = 1$. Solve this equation for y^2. (b) Write an expression for the distance between a point (x, y) on the ellipse and the focus $(c, 0)$. Replace y^2 using your answer to part (a). (c) Expand the quantity $\left(\frac{c}{a}x - a\right)^2$. (d) Show that the quantity under the square root in

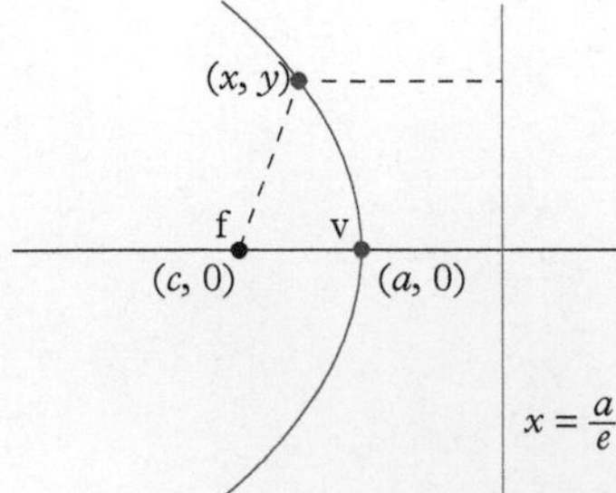

Figure 11 *An ellipse with center* $(0, 0)$ *(not shown), focus* $(c, 0)$*, vertex* $(a, 0)$*, and directrix* $x = \frac{a}{e}$*. A focus (f) and vertex (v) are shown*

your answer to part (b) is equivalent to your answer in part (c), using $c^2 = a^2 - b^2$ as necessary. (e) Use part (d) to rewrite the distance from the point to the focus in a simplified form. Use $e = \frac{c}{a}$ to simplify even further. (f) Write an expression for the distance between (x, y) and the directrix $x = \frac{a}{e}$. (g) Form the ratio of your answer to part (e) over your answer to part (f), which is the ratio of the quantity distance from the point to the focus to the quantity distance from the point to the directrix. Simplify this ratio to obtain e, which is the conclusion of theorem 1.

46. Prove theorem 1 for a hyperbola with center $(0, 0)$, focus $(c, 0)$, vertex $(a, 0)$, and directrix $x = \frac{a}{e}$.

Chapter IX

Additional Applications of Integration

Arc Length

9.1

In this chapter we return to our study of applications of integration. We begin by studying the length of a path.

Arc length: how long is a curved path?

Consider walking along the path in figure 1 from the point $(a, f(a))$ to the point $(b, f(b))$. How long is the walk, not as the crow flies—which is the shortest path distance along a straight line and is found using the distance formula $d = \sqrt{(x_2 - x_1)^2 + (y_2 - y_1)^2}$—but as the length of the curve $y = f(x)$ between the two points?

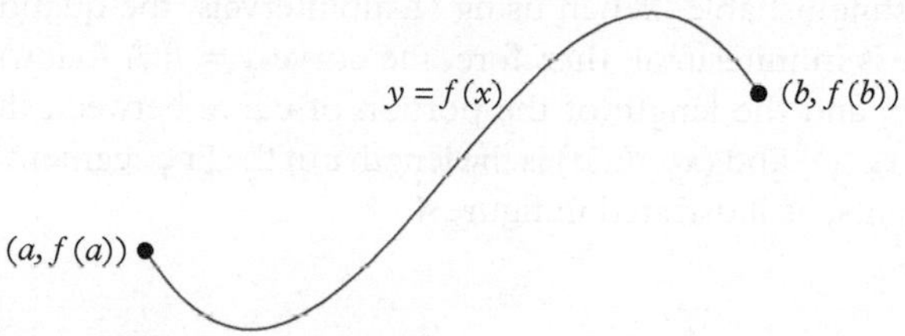

Figure 1 *A curve $y = f(x)$ from $x = a$ to $x = b$*

Once again we find ourselves in the position of wanting to deal with a curved object, and only having tools from geometry for dealing with straight objects. Our strategy has been to break up the interval $[a, b]$ into smaller pieces and approximate using straight objects. In figure 2, the interval is broken into five subintervals, with straight-line segments between points on the curve. Adding the lengths of the five line segments, which can be computed using the distance formula, gives an approximation to the length of the curve.

It seems reasonable to assume that using $n = 10$ subintervals (10 line segments, as in figure 3) gives a better approximation, and $n = 100$ is even better, so why not use $n = \Omega$ subintervals? Using infinitely many subintervals seems as though it should give us the true length of the curve. But, there is one issue to account for. In figure 2, it is apparent that every approximating line segment is always shorter than the curved path it is meant to approximate. How do we know that

Many computer algebra systems (CASs) graph functions by plotting a large number of points and "connecting the dots" with straight line segments. The computer-generated graphs in this book are produced in this manner. We are using the same basic idea to approximate the curve's length.

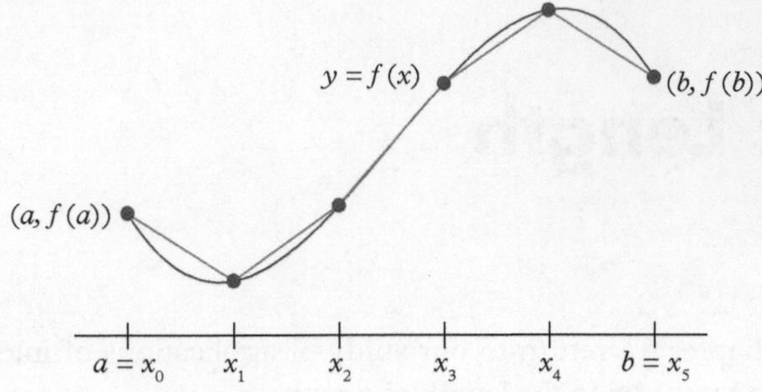

Figure 2 *A curve $y = f(x)$ (blue) from $x = a$ to $x = b$. The interval $[a, b]$ is partitioned into five subintervals. The length of the curve is approximated by the sum of the lengths of the five line segments (brown)*

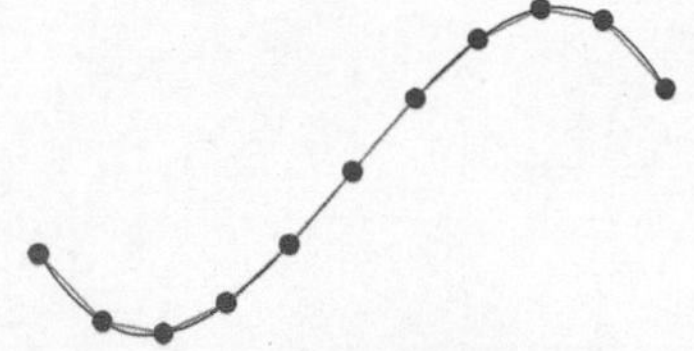

Figure 3 *A curve $y = f(x)$ (blue) from $x = a$ to $x = b$. The interval $[a, b]$ is partitioned into 10 subintervals. The length of the curve is approximated by the sum of the lengths of the 10 line segments (brown). These line segments follow the curve more closely than in figure 2, so this approximation should be better*

infinitely many subintervals fixes this issue? The answer comes from considering local linearity.

The concept of local linearity studied in chapter 3 states that, on an infinitesimal scale, a differentiable function and its tangent line are indistinguishable. When using Ω subintervals, the quantity $\Delta x = (b - a)\omega$ is infinitesimal; therefore, the curve $y = f(x)$ follows its tangent line, and the length of the portion of curve between the points $(x_{k-1}, f(x_{k-1}))$ and $(x_k, f(x_k))$ is the length c of the line segment between these points, as illustrated in figure 4.

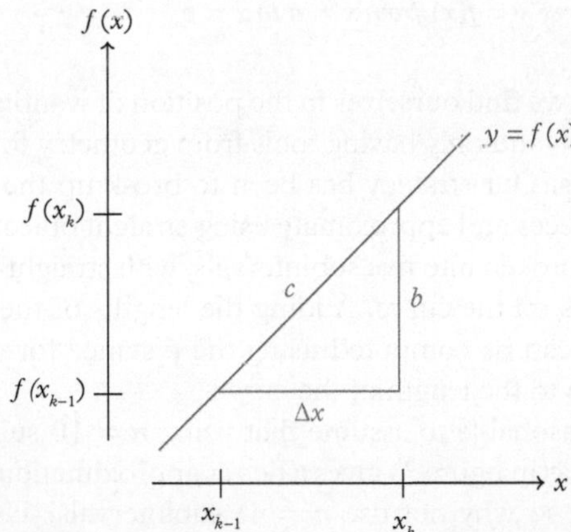

Figure 4 *Infinitesimally close to x_k, the curve $y = f(x)$ and its tangent line at $x = x_k$ (orange) are indistinguishable. The length of the portion of the curve between $x = x_{k-1}$ and $x = x_k$ is the length of this segment of its tangent line*

Using right-hand endpoints as usual, the slope of the tangent line in figure 4 is

$$f'(x_k) = \frac{b}{\Delta x}.$$

Then,

$$b = f'(x_k)\Delta x.$$

Instead of labeling the quantities in figure 4 Δx, b, and c, they are traditionally labeled dx, dy, and ds instead. The Pythagorean relationship is then restated as

$$(ds)^2 = (dx)^2 + (dy)^2.$$

By the Pythagorean theorem,

$$(\Delta x)^2 + b^2 = c^2,$$

and we solve for the length c:

$$\begin{aligned} c &= \sqrt{(\Delta x)^2 + b^2} = \sqrt{(\Delta x)^2 + (f'(x_k)\Delta x)^2} \\ &= \sqrt{(\Delta x)^2 + (f'(x_k))^2(\Delta x)^2} = \sqrt{(\Delta x)^2(1 + (f'(x_k))^2)} \\ &= \sqrt{1 + (f'(x_k))^2} \cdot \Delta x. \end{aligned}$$

In line 1, the positive square root is used because the length c is positive, and b is replaced by $f'(x_k)\Delta x$ from an earlier equation. In line 2, $(\Delta x)^2$ is factored out. In line 3, the square root of $(\Delta x)^2$ is extracted.

Having found the length of the curve on one subinterval, we add the lengths over all the subintervals to obtain the total length of the curve:

$$\text{length of curve} = \sum_{k=1}^{\Omega} \sqrt{1 + (f'(x_k))^2} \cdot \Delta x.$$

Analysis of the derivation of the formula reveals that the unit of measure of the length of the curve is the same unit of measure as the variables x and y. The analysis is left to you to do as an exercise.

This is an omega sum! Following the usual procedure for rewriting as a definite integral (replacing the summation symbol with the integral symbol, dropping the subscripts, replacing Δx with dx, and inserting the limits of integration), we have

$$\text{length of curve} = \int_a^b \sqrt{1 + (f'(x))^2}\, dx.$$

ARC LENGTH

If f' is continuous on $[a, b]$, then the length of the curve $y = f(x)$ on $[a, b]$ is

$$L = \int_a^b \sqrt{1 + (f'(x))^2}\, dx.$$

We add the condition that f' be continuous to guarantee integrability.

Technically speaking, this formula should be presented as the definition of arc length. For the remainder of this chapter all such formulas are presented as formulas. The issue of whether they should be written as definitions is ignored.

Example 1 *Find the length of the curve $y = x^{3/2}$ from $x = 0$ to $x = 9$.*

Before using the arc length formula to find a solution, we explore the function's graph (figure 5) to determine a reasonable answer. Because the curve rises from a y-coordinate of 0 to a y-coordinate

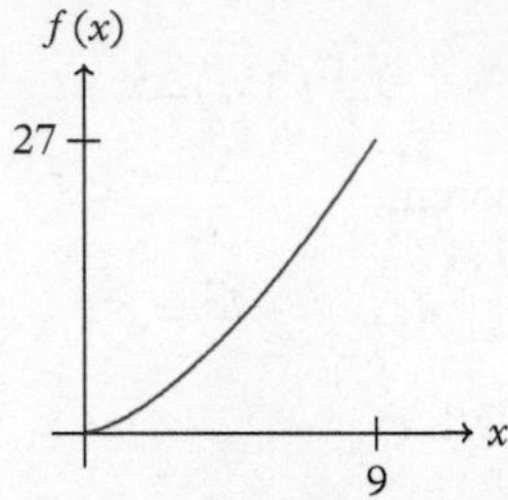

Figure 5 *The graph of $y = x^{3/2}$ between $x = 0$ and $x = 9$*

of 27, the length of the curve must be at least 27 units. In fact, the length of the curve must be at least as large as the distance between the endpoints $(0, 0)$ and $(9, 27)$, which is

$$\sqrt{(9-0)^2 + (27-0)^2} = \sqrt{810} = 28.46.$$

Solution Because the arc length formula uses $f'(x)$, we first calculate this derivative:

$$y = f(x) = x^{3/2}$$
$$f'(x) = \frac{3}{2}x^{1/2}.$$

The length of the curve is then

$$L = \int_a^b \sqrt{1 + (f'(x))^2}\, dx$$
$$= \int_0^9 \sqrt{1 + (\tfrac{3}{2}x^{1/2})^2}\, dx$$
$$= \int_0^9 \sqrt{1 + \tfrac{9}{4}x}\, dx.$$

If the instructions had been to set up the integral for finding the length of the curve but not to evaluate the integral, we would stop here.

Recognizing a square root of "not just plain x," we try substitution:

$$u = 1 + \tfrac{9}{4}x$$
$$du = \tfrac{9}{4}\, dx.$$

We also calculate the new limits of integration:

$$x = 9: \quad u = 1 + \tfrac{9}{4}(9) = \tfrac{85}{4}$$
$$x = 0: \quad u = 1 + \tfrac{9}{4}(0) = 1.$$

Adjusting for the constant, making the substitution, and completing the integration yields

$$\begin{aligned} L &= \tfrac{4}{9}\int_0^9 \sqrt{1 + \tfrac{9}{4}x} \cdot \tfrac{9}{4}\,dx \\ &= \tfrac{4}{9}\int_1^{85/4} \sqrt{u}\,du \\ &= \tfrac{4}{9} \cdot \frac{u^{3/2}}{\frac{3}{2}}\Bigg|_1^{85/4} = \tfrac{8}{27}u^{3/2}\Bigg|_1^{85/4} \\ &= \tfrac{8}{27}\left(\left(\tfrac{85}{4}\right)^{3/2} - 1\right) \approx 28.73 \text{ units}, \end{aligned}$$

which, as expected, is larger than 28.46 units. ■

The relative difference between the straight-line distance 28.46 units and the curved-path distance 28.73 units is just less than 1%. It is apparent from figure 5 that the curve does not stray far from the straight-line path, so just a little "slack" is all that is needed.

Reading Exercise 1 *Set up the integral for finding the length of the curve $y = x^3$ from $x = 1$ to $x = 4$. Do not integrate.*

Arc length example: extracting the square root

Example 2 *Find the length of the curve $x = \frac{y^4}{8} + \frac{1}{4y^2}$, $1 \le y \le 2$.*

Solution Notice that the roles of x and y are switched. Swapping x and y, the length of the curve $x = f(y)$ for $a \le y \le b$ is $L = \int_a^b \sqrt{1 + (f'(y))^2}\,dy$.

Because the arc length formula uses $f'(y)$, we first calculate this derivative:

$$f(y) = \tfrac{1}{8}y^4 + \tfrac{1}{4}y^{-2}$$
$$f'(y) = \tfrac{1}{2}y^3 - \tfrac{1}{2}y^{-3}.$$

The length of the curve is then

Line 1 uses the arc length formula for $x = f(y)$, line 2 squares the quantity in the parentheses, line 3 simplifies the next-to-last term under the square root, and line 4 performs the arithmetic $1 - \frac{1}{2}$.

$$
\begin{aligned}
L &= \int_1^2 \sqrt{1 + \left(\tfrac{1}{2}y^3 - \tfrac{1}{2}y^{-3}\right)^2}\, dy \\
&= \int_1^2 \sqrt{1 + \tfrac{1}{4}y^6 - 2 \cdot \tfrac{1}{2}y^3 \cdot \tfrac{1}{2}y^{-3} + \tfrac{1}{4}y^{-6}}\, dy \\
&= \int_1^2 \sqrt{1 + \tfrac{1}{4}y^6 - \tfrac{1}{2} + \tfrac{1}{4}y^{-6}}\, dy \\
&= \int_1^2 \sqrt{\tfrac{1}{4}y^6 + \tfrac{1}{2} + \tfrac{1}{4}y^{-6}}\, dy.
\end{aligned}
$$

Now something really nice happens. The quantity under the square root is a perfect square! Look at the quantity from line 1 of the previous calculation inside the parentheses and the result in line 3 of squaring that quantity. Instead of $-\frac{1}{2}$, in line 4 we now have $+\frac{1}{2}$, and the only change necessary to refactor the expression is to use a + instead of a −. Factoring this perfect square allows us to extract the square root:

In line 2 we use $\left(\frac{1}{2}y^3 + \frac{1}{2}y^{-3}\right)$ and not $-\left(\frac{1}{2}y^3 + \frac{1}{2}y^{-3}\right)$ because for $1 \le y \le 2$, the former is positive.

$$
\begin{aligned}
&= \int_1^2 \sqrt{\left(\tfrac{1}{2}y^3 + \tfrac{1}{2}y^{-3}\right)^2}\, dy \\
&= \int_1^2 \left(\tfrac{1}{2}y^3 + \tfrac{1}{2}y^{-3}\right) dy.
\end{aligned}
$$

Now that the square root has been extracted, the integral is in a form that we can integrate easily:

$$
\begin{aligned}
&= \left(\frac{1}{2}\frac{y^4}{4} + \frac{1}{2}\frac{y^{-2}}{-2}\right)\Bigg|_1^2 = \left(\frac{1}{8}y^4 - \frac{1}{4y^2}\right)\Bigg|_1^2 \\
&= \cdots = \frac{33}{16} \text{ units.}
\end{aligned}
$$

The length of the curve is $\frac{33}{16}$ units. ■

Manipulating the integrand to extract the square root is accomplished only in a small number of carefully constructed scenarios. In addition to the previous scenario, where $1 - \frac{1}{2} = \frac{1}{2}$ was a key simplification, identities such as $1 + \tan^2\theta = \sec^2\theta$ and $1 + \sinh^2 x = \cosh^2 x$ can be used to extract a square root in arc length problems.

CAUTION: MISTAKE TO AVOID
Many readers make the mistake of assuming that an integrand must be factorable into a perfect square and use "creative algebra" (i.e., make algebraic errors) to force it to be so. It is reasonable to look for ways to extract the square root, but knowingly making an algebraic error to force the issue is never a good strategy.

As illustrated in example 1, sometimes substitution works. Trig substitution is another technique that is helpful with arc length exercises. However, the number of exercises that can be worked by hand is more limited than with some of the other applications of integration we have studied. For this reason, numerical integration—either by hand using a technique such as Simpson's rule or by using technology such as a calculator or CAS—is often necessary.

Ans. to reading exercise 1

$$L = \int_1^4 \sqrt{1 + (3x^2)^2}\,dx$$
$$= \int_1^4 \sqrt{1 + 9x^4}\,dx$$

Reading Exercise 2 *Simplify the expression* $\sqrt{1 + (\frac{1}{2}y^5 - \frac{1}{2}y^{-5})^2}$.

The arc length function

Consider again the curve of figure 5. Instead of finding the length of the curve from $x = 0$ to $x = 9$, what if we want to know the length of the curve from $x = 0$ to an arbitrary point along the curve—say, at $x = t$? The answer is represented in the form of a function, the *arc length function.*

> **ARC LENGTH FUNCTION**
>
> If f' is continuous on an interval and a and t are numbers in that interval, then the arc length function $y = s(t)$ giving the length of the curve $y = f(x)$ from $x = a$ to $x = t$ is
>
> $$s(t) = \int_a^t \sqrt{1 + (f'(x))^2}\,dx.$$

Example 3 *Find the arc length function that gives the length of the curve* $y = x^{3/2}$ *from* $x = 0$ *to* $x = t$.

The solution to this example is very much the same as for example 1, but with the number 9 replaced with the variable t.

Solution Because the formula for finding the arc length function uses $f'(x)$, we first calculate this derivative:

$$y = f(x) = x^{3/2}$$
$$f'(x) = \frac{3}{2}x^{1/2}.$$

The arc length function is then

$$s(t) = \int_a^t \sqrt{1 + (f'(x))^2}\,dx$$
$$= \int_0^t \sqrt{1 + (\tfrac{3}{2}x^{1/2})^2}\,dx$$
$$= \int_0^t \sqrt{1 + \tfrac{9}{4}x}\,dx.$$

Recognizing a square root of "not just plain x," we try substitution:

$$u = 1 + \tfrac{9}{4}x$$
$$du = \tfrac{9}{4}\,dx.$$

We also calculate the new limits of integration:

$$x = t: \quad u = 1 + \tfrac{9}{4}(t)$$
$$x = 0: \quad u = 1 + \tfrac{9}{4}(0) = 1.$$

Ans. to reading exercise 2
$\frac{1}{2}y^5 + \frac{1}{2}y^{-5}$
Details:

$$\sqrt{1 + (\frac{1}{2}y^5 - \frac{1}{2}y^{-5})^2}$$
$$= \sqrt{1 + \frac{1}{4}y^{10} - \frac{1}{2} + \frac{1}{4}y^{-10}}$$
$$= \sqrt{\frac{1}{4}y^{10} + \frac{1}{2} + \frac{1}{4}y^{-10}}$$
$$= \sqrt{(\frac{1}{2}y^5 + \frac{1}{2}y^{-5})^2}$$
$$= \frac{1}{2}y^5 + \frac{1}{2}y^{-5}$$

Adjusting for the constant and making the substitution yields

$$s(t) = \tfrac{4}{9}\int_0^t \sqrt{1 + \tfrac{9}{4}x} \cdot \tfrac{9}{4}\,dx$$
$$= \tfrac{4}{9}\int_1^{1+\frac{9}{4}t} \sqrt{u}\,du.$$

To complete the integration we may treat t as if it is a constant because it is not the variable of integration (it is neither x nor u):

$$= \tfrac{4}{9} \cdot \frac{u^{3/2}}{\frac{3}{2}}\Bigg|_1^{1+\frac{9}{4}t} = \tfrac{8}{27}u^{3/2}\Bigg|_1^{1+\frac{9}{4}t}$$
$$= \tfrac{8}{27}\left(\left(1 + \tfrac{9}{4}t\right)^{3/2} - 1\right).$$

We conclude that the arc length function is given by

$$s(t) = \tfrac{8}{27}\left(\left(1 + \tfrac{9}{4}t\right)^{3/2} - 1\right). \qquad ■$$

The value of t has to be in the domain of the function $y = x^{3/2}$, so we must have $t \geq 0$.

We now know that the length of the curve $y = x^{3/2}$ from $x = 0$ to $x = t$ is given by $s(t) = \frac{8}{27}\left(\left(1 + \frac{9}{4}t\right)^{3/2} - 1\right)$ for any value of t. For instance, the solution to example 1 is $s(9) = \frac{8}{27}\left(\left(1 + \frac{9}{4}(9)\right)^{3/2} - 1\right) = \frac{8}{27}\left(\left(\frac{85}{4}\right)^{3/2} - 1\right)$. More than merely evaluating the length for various values of t, the arc length function also gives us a convenient method for solving other types of problems.

Example 4 *A wire of length* 5 m *is to be placed along the curve* $y = x^{3/2}$ *with one end at the point* $(0, 0)$. *Determine the point on the curve where the other end of the wire is located.*

Solution Rephrasing, we wish to find the point on the curve $y = x^{3/2}$ for which the length of the curve from $x = 0$ to $x = t$ is 5 m, as shown in figure 6. This is the point for which the arc length function's value is 5. In other words, we wish to solve the equation $s(t) = 5$.

Notice that this is the same curve as in example 3, and we have already calculated the arc length function to be

$$s(t) = \tfrac{8}{27}\left(\left(1 + \tfrac{9}{4}t\right)^{3/2} - 1\right).$$

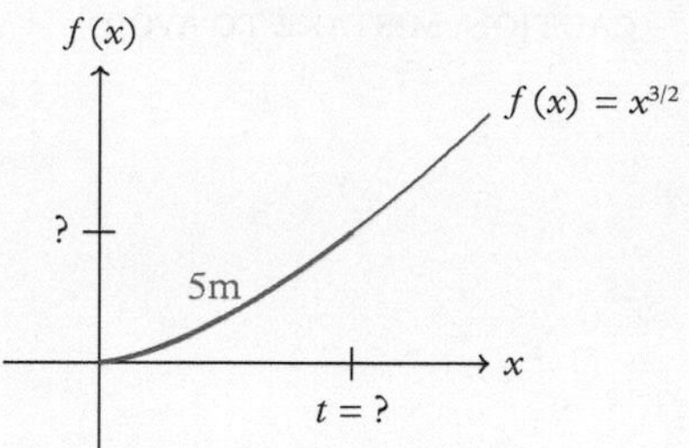

Figure 6 *The graph of $y = x^{3/2}$, on which lies a wire (red) of length 5 m with one end at the point $(0, 0)$*

Setting the arc length equal to 5 and solving for t, we have

$$\frac{8}{27}\left(\left(1+\frac{9}{4}t\right)^{3/2} - 1\right) = 5$$

$$\left(1+\frac{9}{4}t\right)^{3/2} - 1 = 5\cdot\frac{27}{8} = \frac{135}{8}$$

$$\left(1+\frac{9}{4}t\right)^{3/2} = \frac{135}{8} + 1 = \frac{143}{8}$$

$$1+\frac{9}{4}t = \left(\frac{143}{8}\right)^{2/3}$$

$$\frac{9}{4}t = \left(\frac{143}{8}\right)^{2/3} - 1$$

$$t = \frac{4}{9}\left(\left(\frac{143}{8}\right)^{2/3} - 1\right) = 2.594.$$

The x-coordinate of the end of the wire is (approximately) 2.594. We also want the y-coordinate, which is

$$y = 2.594^{3/2} = 4.178.$$

The end of the wire is at the point $(2.594, 4.178)$, as illustrated in figure 7. ■

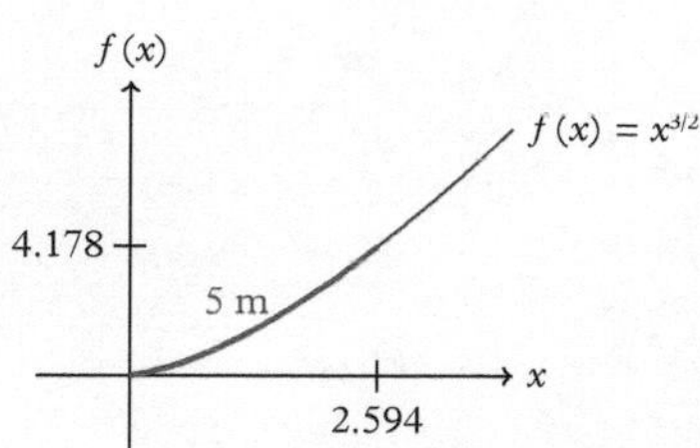

Figure 7 *The graph of $y = x^{3/2}$, on which lies a wire (red) of length 5 m with one end at the point $(0, 0)$. The coordinates of the other end of the wire are found by setting the arc length function equal to 5 m and solving the equation*

Arc length: a strategy that doesn't work

At the beginning of this section we developed the formula for arc length using straight line segments between points on the curve, as in figures 2 and 3. The line segments were not all horizontal. By way of contrast, when developing the formula for area under a curve and the formula for volume using the disk method, we approximated the curve with horizontal line segments. Why didn't we use horizontal segments this time?

CAUTION: MISTAKE TO AVOID

Figure 8 illustrates why. If all the line segments are horizontal, then the lengths of the segments always adds up to the total width of the interval, $b - a$. These segments only capture the horizontal width of the curve, completely ignoring any vertical movement in the curve.

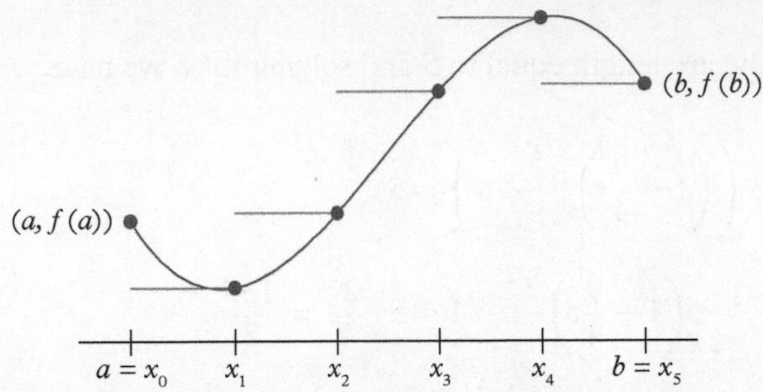

Figure 8 *A curve $y = f(x)$ (blue) from $x = a$ to $x = b$. The interval $[a, b]$ is partitioned into five subintervals. Horizontal line segments (brown) do not give an approximation to the length of the curve, for the sum of the lengths of horizontal segments is always equal to the width of the interval and does not represent the length of the curve*

It is important to remember that although partitioning an interval into infinitely many pieces and forming an omega sum that represents a geometric or physical situation are powerful tools for developing useful formulas, the details of each situation are unique and must be considered carefully. Otherwise, we could end up with a formula that gives us a different quantity than the one we seek.

Exercises 9.1

1–6. (a) Set up the integral for finding the length of the curve on the given interval. (b) Use a calculator to determine the value of the integral.

1. $y = 2x^5 - 3x^2 + 1,\ 0 \le x \le 3$
2. $y = x^3,\ -1 \le x \le 1$
3. $y = \tan^{-1} x$ between $x = 0$ and $x = 1$
4. $y = x + \sin x$ between $x = 0$ and $x = \pi$
5. $y = e^x$ on $[0, \frac{1}{2}\ln 3]$
6. $y = \ln x$ on $[\sqrt{3}, \sqrt{8}]$

It is implied that numerical integration should not be used on these problems; an antiderivative must be calculated.

7–18. Find the length of the curve on the given interval.

7. $y = x^{3/2}$ from $x = 1$ to $x = 4$
8. $y = \dfrac{2x^{3/2}}{3}$, $0 \le x \le 1$
9. $y = \cosh x$, $0 \le x \le 3$
10. $y = \ln(\cos x)$, $0 \le x \le \frac{\pi}{3}$
11. (GSL) $y = \dfrac{x^3}{6} + \dfrac{1}{2x}$ from $x = 1$ to $x = 3$
12. (GSL) $y = \frac{1}{4}x^2 - \frac{1}{2}\ln x$ from $x = 1$ to $x = 2$
13. (GSL) $y = \ln\sec x$ from the point $(0, 0)$ to the point $(\frac{\pi}{3}, \ln 2)$
14. $y = \ln(\cos x)$, $\pi \le x \le \frac{5\pi}{4}$
15. $y = -3x^2$ from the point $(1, -3)$ to the point $(4, -48)$
16. $y = 1 - x^{2/3}$ from the point $(1, 0)$ to the point $(8, -3)$
17. $y = \dfrac{x^4}{8} + \dfrac{1}{4x^2}$, $-2 \le x \le -1$
18. (GSL) $y = \ln(1 - x^2)$ from $x = 0$ to $x = \frac{1}{2}$

19–22. Find the arc length function that gives the length of the curve from $x = 1$ to $x = t$.

For some of these exercises, the variable t must be restricted to $t > 0$.

19. $y = \dfrac{1}{4}x^2 - \dfrac{1}{2}\ln x$
20. $y = \dfrac{x^3}{6} + \dfrac{1}{2x}$
21. $y = x^3 + \dfrac{1}{12x}$
22. $y = \dfrac{1}{5}\cosh 5x$

23. A wire of length 5 m is to be placed along the curve $y = \frac{1}{4}x^2 - \frac{1}{2}\ln x$ with one end at the point $(1, \frac{1}{4})$. Using Newton's method to solve the necessary equation, determine the point on the curve where the other end of the wire is located.
24. A wire of length 5 m is to be placed along the curve $y = \frac{1}{5}\cosh 5x$ with one end at the point $(1, \frac{1}{5}\cosh 5)$. Determine the point on the curve where the other end of the wire is located.
25. A utility wire hangs between a pole (at $x = -5$) and a house (at $x = 4$) that are 9 m apart according to the equation $y = 10\cosh(.1x) - 6$ on the interval $[-5, 4]$, where y represents the height (in meters) above the flat, horizontal ground. (a) What is the minimum clearance between the wire and the ground? (b) What is the difference in height of the wire at the pole and at the house? (c) How long is the wire between the pole and the house?

26. A utility wire is to be strung between poles 30 m apart across a stretch of flat, horizontal ground. To account for wind, movement of utility poles, and other factors, a minimum of 1 m slack between poles is desired; that is, 31 m of wire is to be strung between poles. The minimum clearance of the wire above the ground should be 8 m. The equation of a catenary, the shape of the wire between the poles, is $y = b + a\cosh\frac{x}{a}$ on $[-15, 15]$. Using the minimum values, (a) determine the equation of the catenary that matches the minimum values (that is, determine appropriate values for a and b) and (b) determine how high on the poles the wire must be attached. Use technology when necessary.

27. Find the length of the arc of the circle $x^2 + y^2 = 9$ between the points $\left(-\frac{3}{2}, \frac{3\sqrt{3}}{2}\right)$ and $\left(\frac{3}{2}, \frac{3\sqrt{3}}{2}\right)$.

28. (GSL) Find the length of the arch of the parabola $y = 4x - x^2$, which lies above the x-axis.

29. (GSL) Find the length between $x = a$ and $x = b$ of the curve $y = \ln\dfrac{e^x + 1}{e^x - 1}$.

30. (Love) Find the length of the curve $y = \frac{4}{5}x^{5/4}$ between the origin and the point with $x = 4$.

31. (a) Use the arc length formula to determine the length of the curve $y = ax + b$ between $x = 0$ and $x = 1$. (b) Because the "curve" is actually a straight line, use the distance formula to determine the distance between the endpoints of the segment and verify that the result is the same as in part (a).

32. (GSL) Find the length of the arc of the parabola $y^2 = 2ax$ from the vertex to one extremity of the latus rectum.

Treat a as a constant. A *latus rectum* of a conic section is the chord (line segment between points on the curve) through a focus parallel to the associated directrix.

33. Find the arc length of exercise 6 without using technology as follows: inside the square root, add by getting a common denominator, extract a square root, make the substitution $u = x^2 + 1$, make the replacement substitution $u = v^2$, and use long division and partial fractions to determine the solution.

34. Find the arc length of exercise 5 without using technology by making the substitution $u = 1 + e^{2x}$, after which the integral should be identical to an intermediate step in exercise 33.

35. Suppose the unit for measuring length in the x- and y-directions is centimeters. (a) What is the unit of measure for the variables b and Δx in figure 4? (b) What is the unit of measure for the quantity $f'(x_k) = \frac{b}{\Delta x}$? (c) What is the unit of measure for the quantity $c = \sqrt{1 + (f'(x_k))^2} \cdot \Delta x$?

(d) What is the unit of measure for the length of the curve $\sum_{k=1}^{\Omega} \sqrt{1 + (f'(x_k))^2} \cdot \Delta x$?

36. In a roundabout manner, it is possible to calculate the length of the curve in example 1 using only omega sums. (a) For the curve $f(x) = x^{3/2}$ on the interval $[0, 9]$, calculate Δx and x_k as usual. (b) Use the formula $L = \sum_{k=1}^{\Omega} \sqrt{1 + (f'(x_k))^2} \cdot \Delta x$ to form the necessary omega sum, but do not evaluate the sum. (c) Calculate the omega sum for finding the area under the curve $f(x) = \frac{4}{9}\sqrt{x}$ on the interval $[1, \frac{85}{4}]$. (d) Compare the omega sums in parts (b) and (c).

Now the (equivalent) task is to calculate the area under the curve in part (c). Do so as follows: (e) Use an omega sum to calculate the area under the curve $f(x) = \frac{4}{9}\sqrt{x}$ on the interval $[0, \frac{85}{4}]$. (f) Use an omega sum to calculate the area under the curve $f(x) = \frac{4}{9}\sqrt{x}$ on the interval $[0, 1]$. (g) Subtract the result in part (f) from the result in part (e) to find the area requested in part (c), which must be the same as the length of the curve in part (a).

It is possible to start with the omega sum from part (b) and reverse-engineer a function and interval (those of part (c)) that create the omega sum.

Areas and Lengths in Polar Coordinates

9.2

Let's now turn our attention to the geometry of polar curves. First we tackle area.

Review: area of a circular sector

A *sector* of a circle is a region in the circle's interior between two radii, such as the shaded regions in figure 1; think "pizza slices."

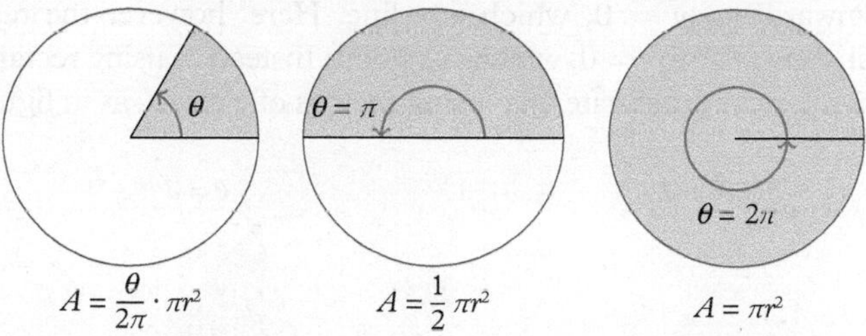

$A = \frac{\theta}{2\pi} \cdot \pi r^2$ $\quad A = \frac{1}{2}\pi r^2$ $\quad A = \pi r^2$

Figure 1 *Sectors (shaded light green) of circles (blue). A sector is a region between two radii (black) of a circle*

The area of a circle is known to be $A = \pi r^2$, where r is the radius of the circle. The sector in the middle of figure 1 covers half the circle's interior, so its area is $A = \frac{1}{2}\pi r^2$. This proportionality argument works in general. The proportion of a full rotation represented by the angle is the same as the proportion of the circle's area represented by the sector. Because the proportion of a full rotation is $\frac{\theta}{2\pi}$, the area of a sector is

$$A = \frac{\theta}{2\pi} \cdot \pi r^2 = \frac{1}{2} r^2 \theta.$$

AREA OF A SECTOR OF A CIRCLE
The area of a sector with angle θ and radius r is

$$A = \frac{1}{2} r^2 \theta.$$

It is implied that the angle θ satisfies $0 \le \theta \le 2\pi$ so that the region does not overlap itself.

Areas in polar coordinates

Suppose we are given a polar curve $r = f(\theta)$ and wish to know the area of a region enclosed by the curve or a portion of the curve, such as shaded in figure 2. The region under a curve $y = f(x)$ stretches

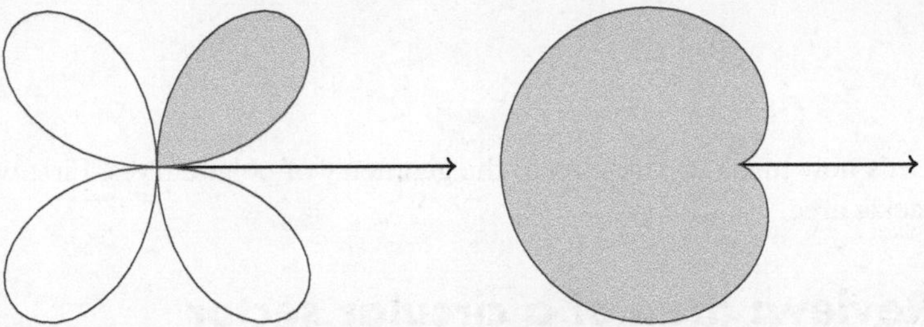

Figure 2 *(left) The four-leaved rose* $r = \sin 2\theta$, *with the region inside one leaf shaded, and (right) the cardioid* $r = 1 - \cos\theta$, *with the interior shaded*

"downward" to $y = 0$, which is a line. Here, however the regions stretch "inward" to $r = 0$, which is a point. Instead of using rectangles, we therefore approximate area using sectors of a circle, as in figure 3.

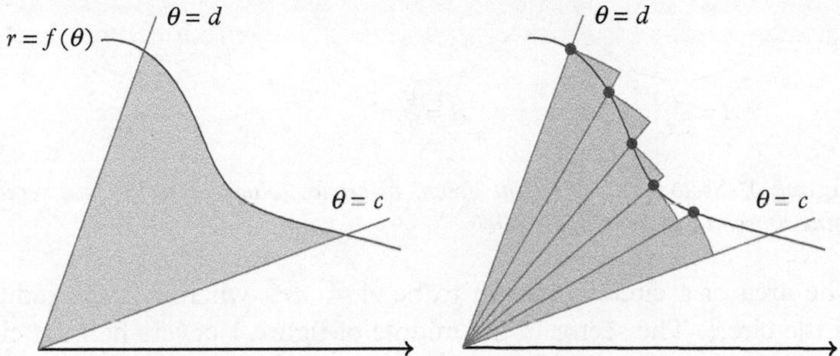

Figure 3 *(left) The region "inside" the polar curve* $r = f(\theta)$ *between* $\theta = c$ *and* $\theta = d$; *(right) the same region approximated using* $n = 5$ *sectors of circles. Notice that the outside edges of the sectors are curved, not straight, because they follow an arc of a circle*

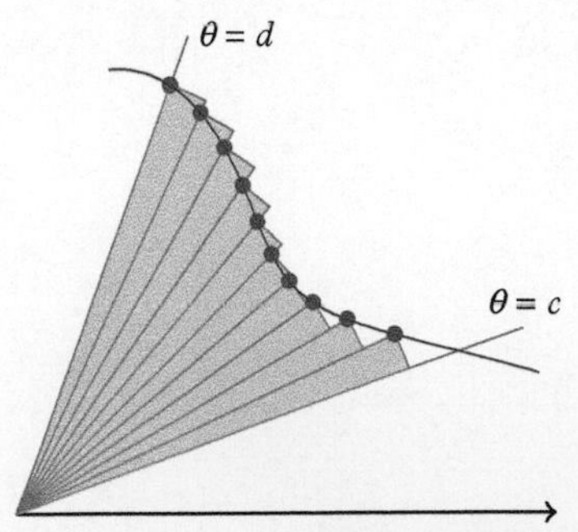

Figure 4 *The region inside the polar curve* $r = f(\theta)$ *between* $\theta = c$ *and* $\theta = d$ *approximated using* $n = 10$ *sectors of circles. Compare to figure 3*

Figure 3 uses $n = 5$ sectors to approximate the area inside the curve $r = f(\theta)$ between $\theta = c$ and $\theta = d$. Using $n = 10$ sectors, as in figure 4, appears to reduce the "extra" area and "missing" area, which should make for a better approximation. As you may have anticipated, the idea is to use infinitely many sectors to get the exact area inside the curve.

Using $n = \Omega$ sectors, we slice the variable θ and have

$$\Delta\theta = \frac{d-c}{\Omega} = (d-c)\omega.$$

The "right-hand endpoints" (the equivalent in this context is the farther counterclockwise endpoints) is found with

$$\theta_k = c + k \cdot \Delta\theta.$$

The radius of a sector matches the value of the polar curve at the right-hand endpoint:

$$r_{\text{sector}} = f(\theta_k).$$

The area of a sector is

$$A_{\text{sector}} = \frac{1}{2}r^2\theta = \frac{1}{2}\,(f(\theta_k))^2\,\Delta\theta,$$

because the angle in a sector is $\Delta\theta$. The total area inside the curve is found by adding the areas of all the sectors:

Recall that a unit of measure of an angle, such as a radian, is "dimensionless" and therefore does not appear in the dimensional analysis for this quantity. The area is measured by square units.

$$A_{\text{total}} = \sum_{k=1}^{\Omega} \frac{1}{2}\,(f(\theta_k))^2\,\Delta\theta.$$

Surprise! (Or, by now, not so surprisingly.) We have an omega sum. Rewriting as a definite integral (replacing the summation symbol with the integral symbol, dropping the subscript, replacing $\Delta\theta$ with $d\theta$, and inserting the limits of integration) gives the desired form of the formula:

$$A = \int_c^d \frac{1}{2}(f(\theta))^2\,d\theta.$$

POLAR AREA

The area inside a polar curve $r = f(\theta)$ between $\theta = c$ and $\theta = d$ is given by

$$A = \int_c^d \frac{1}{2}(f(\theta))^2\,d\theta.$$

It is implied that the values of d and c do not allow the region to overlap itself.

Example 1 *Find the area inside one loop of the four-leaved rose* $r = \sin 2\theta$.

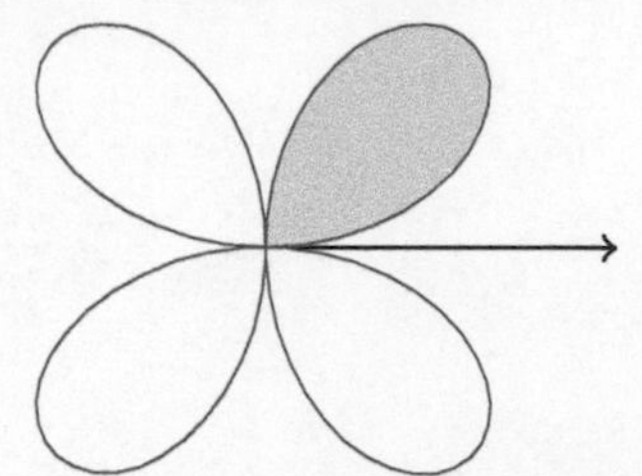

Figure 5 *The four-leaved rose $r = \sin 2\theta$ of example 1, with one loop (one leaf) shaded*

Plotting points by hand or with the help of technology is a common method of determining the values of θ.

Another choice is to use $c = 2\pi$ and $d = \frac{5\pi}{2}$. There are infinitely many other appropriate choices as well. All appropriate choices result in the same answer.

Solution ❶ Graph the curve and determine the region. This is a curve we have graphed before. Because the four leaves appear to be the same size, it does not matter which leaf we choose. The region with the area we want to find is shaded in figure 5.

❷ If necessary, determine the values of θ. From section 8.3 example 10, we know that $0 \leq \theta \leq \frac{\pi}{2}$ produces the loop around the shaded region. We therefore use $c = 0$ and $d = \frac{\pi}{2}$.

❸ Use the formula to compute the area. Applying the formula with $f(\theta) = \sin 2\theta$, $c = 0$, and $d = \frac{\pi}{2}$, we have

$$A = \int_c^d \frac{1}{2}(f(\theta))^2\, d\theta = \int_0^{\frac{\pi}{2}} \frac{1}{2}(\sin 2\theta)^2\, d\theta = \frac{1}{2}\int_0^{\frac{\pi}{2}} \sin^2 2\theta\, d\theta.$$

Recalling our techniques of integration, we use the identity $\sin^2 x = \frac{1}{2}(1 - \cos 2x)$ to continue:

$$= \frac{1}{2}\int_0^{\frac{\pi}{2}} \frac{1}{2}(1 - \cos 4\theta)\, d\theta = \frac{1}{4}\int_0^{\frac{\pi}{2}} (1 - \cos 4\theta)\, d\theta$$

$$= \frac{1}{4}\left(\theta - \frac{\sin 4\theta}{4}\right)\Bigg|_0^{\frac{\pi}{2}}$$

$$= \frac{1}{4}\left(\frac{\pi}{2} - \frac{\sin 2\pi}{4}\right) - \frac{1}{4}\left(0 - \frac{\sin 0}{4}\right) = \frac{\pi}{8}\text{ units}^2.$$

■

Reading Exercise 3 *Find the area inside the polar curve* $r = \sqrt{\cos\theta}$ *from* $\theta = 0$ *to* $\theta = \dfrac{\pi}{2}$.

Lengths in polar coordinates

To derive a formula for the length of a polar curve $r = f(\theta)$ from $\theta = c$ to $\theta = d$, we don't need to start from scratch. We already know relationships between Cartesian and polar coordinates, such as

$$x = r\cos\theta = f(\theta)\cos\theta$$

and

$$\frac{dy}{dx} = \frac{f'(\theta)\sin\theta + f(\theta)\cos\theta}{f'(\theta)\cos\theta - f(\theta)\sin\theta}.$$

We also know the arc length formula in Cartesian coordinates, which is rewritten as

For the arc length formula, we rewrite the derivative using Leibniz notation.

$$L = \int_a^b \sqrt{1 + \left(\frac{dy}{dx}\right)^2}\, dx.$$

Making the replacement substitution

$$x = f(\theta)\cos\theta$$
$$dx = (f'(\theta)\cos\theta - f(\theta)\sin\theta)\, d\theta$$

gives

$$L = \int_a^b \sqrt{1 + \left(\frac{dy}{dx}\right)^2}\, dx$$
$$= \int_c^d \sqrt{1 + \left(\frac{f'(\theta)\sin\theta + f(\theta)\cos\theta}{f'(\theta)\cos\theta - f(\theta)\sin\theta}\right)^2} \cdot (f'(\theta)\cos\theta - f(\theta)\sin\theta)\, d\theta.$$

In addition to using the replacement substitution formula to replace dx and the derivative for a polar curve formula to replace $\frac{dy}{dx}$, we change the limits of integration to values of the variable θ.

This is not as scary as it seems. Notice that the quantity outside the square root is the same as the denominator inside the square root. This allows us to simplify the expression:

The potential sign error made by replacing a quantity with a sign that is unknown with the square root of that quantity squared (which is always positive) is mitigated by the placement of c and d where $c < d$. Because the integrand is nonnegative, the result is a nonnegative length, as required.

$$= \int_c^d \sqrt{\left(1 + \left(\frac{f'(\theta)\sin\theta + f(\theta)\cos\theta}{f'(\theta)\cos\theta - f(\theta)\sin\theta}\right)^2\right)(f'(\theta)\cos\theta - f(\theta)\sin\theta)^2}\, d\theta$$
$$= \int_c^d \sqrt{(f'(\theta)\cos\theta - f(\theta)\sin\theta)^2 + (f'(\theta)\sin\theta + f(\theta)\cos\theta)^2}\, d\theta.$$

The quantity inside the square root can be simplified further. Working with just this quantity, we square each of the expressions, simplify, factor, and use a Pythagorean identity:

The first square from line 1 is expanded in line 2, whereas the second square from line 1 is expanded in line 3. The resulting terms may not be written in the order you expect. In line 4, the quantities $(f(\theta))^2$ and $(f'(\theta))^2$ are factored out. In line 5, the identity $\sin^2\theta + \cos^2\theta = 1$ is applied.

$$\begin{aligned}
&(f'(\theta)\cos\theta - f(\theta)\sin\theta)^2 + (f'(\theta)\sin\theta + f(\theta)\cos\theta)^2 \\
&= (f(\theta))^2\sin^2\theta + \cos^2\theta(f'(\theta))^2 - \cancel{2f(\theta)\sin\theta\cos\theta f'(\theta)} \\
&\quad + (f(\theta))^2\cos^2\theta + \sin^2\theta(f'(\theta))^2 + \cancel{2f(\theta)\sin\theta\cos\theta f'(\theta)} \\
&= (f(\theta))^2(\sin^2\theta + \cos^2\theta) + (f'(\theta))^2(\cos^2\theta + \sin^2\theta) \\
&= (f(\theta))^2 + (f'(\theta))^2.
\end{aligned}$$

This may seem like a lot of messy algebra, but it is still easier than developing the formula from scratch by forming an omega sum.

Replacing the quantity from the integral with this last equivalent, we arrive at our simplified formula:

$$L = \int_c^d \sqrt{(f(\theta))^2 + (f'(\theta))^2}\, d\theta.$$

The condition that f' be continuous is added to guarantee integrability.

POLAR ARC LENGTH

If f' is continuous on $[c, d]$, then the length of the polar curve $r = f(\theta)$ on $[c, d]$ is

$$L = \int_c^d \sqrt{(f(\theta))^2 + (f'(\theta))^2}\, d\theta.$$

Example 2 *(a) Set up the integral for finding the length of the polar curve* $r = \ln\theta$, $1 \le \theta \le \frac{\pi}{2}$. *(b) Use technology to evaluate the integral.*

Solution (a) The polar arc length formula uses $r = f(\theta)$; therefore, we assign

$$f(\theta) = \ln\theta.$$

We also need $f'(\theta)$, which is

$$f'(\theta) = \frac{1}{\theta}.$$

Using the polar arc length formula,

$$L = \int_1^{\frac{\pi}{2}} \sqrt{(\ln\theta)^2 + \left(\frac{1}{\theta}\right)^2}\, d\theta.$$

Ans. to reading exercise 3

$$\int_0^{\pi/2} \frac{1}{2}\left(\sqrt{\cos\theta}\right)^2 d\theta = \int_0^{\pi/2} \frac{1}{2}\cos\theta\, d\theta = \cdots = \frac{1}{2}$$

(b) Using technology, we find

$$\int_1^{\frac{\pi}{2}} \sqrt{(\ln\theta)^2 + \left(\frac{1}{\theta}\right)^2}\, d\theta = 0.48037.$$

The length of the curve is 0.48037 units (approximately). ■

Reading Exercise 4 *Set up the integral to find the length of the polar curve* $r = 1 + 2\cos\theta$ *from* $\theta = 0$ *to* $\theta = \pi$. *Do not integrate.*

In example 2 the interval of values of θ was given. If such an interval is not provided, then we must add a preliminary step to the solution method.

Example 3 *Find the length of the polar curve $r = 3\sin\theta$.*

Solution ❶ If necessary, determine the interval of values of θ. Because an interval of values of θ is not specified, we must determine an appropriate interval that traverses the curve exactly once, which is a task we practiced in chapter 8. The easiest way to do this is to create a graph. To complete this step by hand, we notice that $r = 3\sin\theta$ is of the form $r = 2a\sin\theta$ with $a = \frac{3}{2}$, so the graph is a circle through the origin of radius $\frac{3}{2}$ and center $(\frac{3}{2}, \frac{\pi}{2})$. But, to determine appropriate values for θ, plotting points is still needed. Given that the argument of the trig function is θ, it seems reasonable to try $0 \le \theta \le 2\pi$:

This step is sometimes more time-consuming than setting up the integral and performing the integration. Use of technology can help shorten the process.

Although we can quickly determine the length of the curve as the circumference of a circle of radius $\frac{3}{2}$, which is $2\pi \cdot \frac{3}{2} = 3\pi$ units, we want to practice using the polar arc length formula.

θ	$r = 3\sin\theta$
0	0
$\frac{\pi}{2}$	3
π	0
$\frac{3\pi}{2}$	-3
2π	0

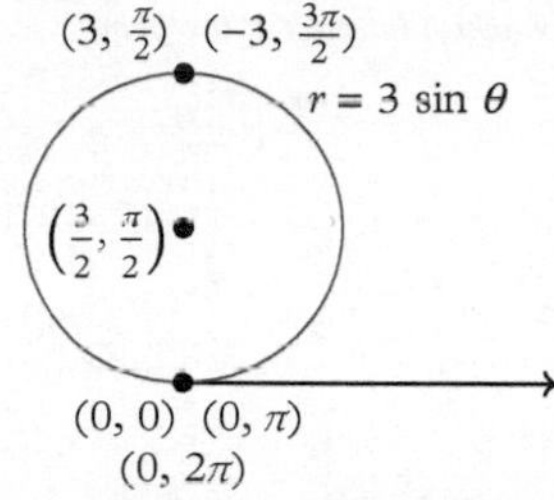

Figure 6 *The curve $r = 3\sin\theta$ of example 3*

Plotting those points on the circle gives the graph in figure 6.

Tracing around the circle through the plotted points, we see that from $\theta = 0$ to $\theta = 2\pi$ we actually traverse the circle twice! We do not use $0 \le \theta \le 2\pi$; we use $0 \le \theta \le \pi$.

❷ Set up the integral. We begin by identifying $f(\theta)$ and $f'(\theta)$:

$$f(\theta) = 3\sin\theta$$
$$f'(\theta) = 3\cos\theta.$$

Using the polar arc length formula with the limits of integration calculated in step ❶, and f, f' as just calculated,

$$L = \int_0^\pi \sqrt{(3\sin\theta)^2 + (3\cos\theta)^2}\, d\theta.$$

❸ Evaluate the integral. Simplifying the integrand using the Pythagorean identity makes the integral relatively simple:

We always evaluate the integral by finding an antiderivative and using the fundamental theorem of calculus, part II unless instructed specifically to use technology.

$$\begin{aligned}\int_0^\pi \sqrt{(3\sin\theta)^2 + (3\cos\theta)^2}\, d\theta &= \int_0^\pi \sqrt{9\sin^2\theta + 9\cos^2\theta}\, d\theta \\ &= \int_0^\pi 3\sqrt{\sin^2\theta + \cos^2\theta}\, d\theta \\ &= \int_0^\pi 3\, d\theta = 3\theta\Big|_0^\pi = 3\pi - 3\cdot 0 \\ &= 3\pi \text{ units.}\end{aligned}$$

Ans. to reading exercise 4
$\int_0^\pi \sqrt{(1 + 2\cos\theta)^2 + (-2\sin\theta)^2}\, d\theta$

The length of the curve is 3π units. ■

We could have made other choices in step ❶ of example 3. From figure 6, it is apparent that $\frac{\pi}{2} \le \theta \le \frac{3\pi}{2}$ and $\pi \le \theta \le 2\pi$ are among the many other possibilities. Using a different interval of θ changes the limits of integration in step ❷, but does not change the final answer as long as the curve is traversed exactly once on the chosen interval.

Intersections of polar curves

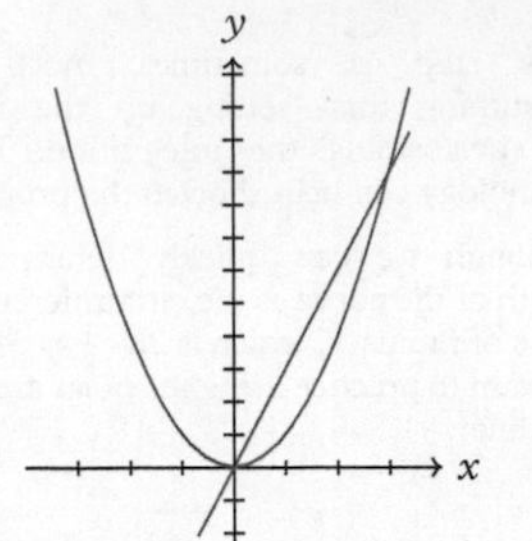

Figure 7 *The Cartesian curves $y = x^2$ and $y = 3x$, which intersect at two points*

Recall that in Cartesian coordinates we find intersections of curves by setting the curves' expressions equal to one another and solving for x. For instance, if we wish to find the intersection points of the parabola $y = x^2$ and the line $y = 3x$ (see figure 7), we solve the following equation:

$$x^2 = 3x$$

$$x^2 - 3x = 0$$

$$x(x - 3) = 0$$

$$x = 0, 3.$$

Either curve's expression can then be evaluated at these x-coordinates to find the y-coordinates of the intersection points.

Unfortunately, the same procedure is not always adequate for intersections of polar curves. To illustrate why, let's try to find all the points of intersection of the polar curves $r = 1 + \cos\theta$ and $r = 1 - \cos\theta$.

According to the previous procedure, we set the curves' expressions equal to one another and solve for θ:

Recall that often the first step in solving a trigonometric equation is to solve for a trig function of the variable. In this case, we first solve for $\cos\theta$.

$$1 + \cos\theta = 1 - \cos\theta$$

$$\cos\theta = -\cos\theta$$

$$2\cos\theta = 0$$

$$\cos\theta = 0.$$

CAUTION: MISTAKE TO AVOID
Continuing the calculation by writing $\theta = \cos^{-1} 0 = \frac{\pi}{2}$ is inadequate because inverse cosine only returns values of θ between 0 and π (inclusive), neglecting other solutions.

One method for solving $\cos\theta = 0$ is to recall the graph of $y = \cos x$ and remember where it crosses the x-axis. The solutions are

$$\theta = \frac{\pi}{2} + 2\pi k$$

and

$$\theta = -\frac{\pi}{2} + 2\pi k,$$

The solutions may also be expressed as $\theta = \frac{\pi}{2} + \pi k$, where k is any integer.

where k is any integer.

We can then find the r-coordinates of the points of intersection by evaluating either curve's expression at the values of θ that solve the equation. A few such points are calculated here using the first set of solutions with $k = 0, 1, 2$:

θ	$1 + \cos\theta$
$\frac{\pi}{2}$	1
$\frac{5\pi}{2}$	1
$\frac{9\pi}{2}$	1

θ	$1 - \cos\theta$
$\frac{\pi}{2}$	1
$\frac{5\pi}{2}$	1
$\frac{9\pi}{2}$	1

Of course it does not matter which curve is selected. The matching tables of values are the result of solving the equation $1 + \cos\theta = 1 - \cos\theta$. What you may or may not suspect is that, rather than three different points, these are all the same point! The coordinates $(1, \frac{\pi}{2})$, $(1, \frac{5\pi}{2})$ and $(1, \frac{9\pi}{2})$ are different polar representations of one point.

Using the second set of solutions gives a second distinct point: $(1, \frac{-\pi}{2}) = (1, \frac{3\pi}{2}) = \cdots$.

This procedure has netted two distinct points of intersection. There's only one problem: the two curves intersect at three distinct points, not just two! The procedure for solving an equation that works so well for Cartesian curves (when the equation can be solved) misses one of the intersection points for these polar curves. Why? And how do we find the third point of intersection?

It turns out that our best choice for detecting intersection points of polar curves is to graph the curves and look for intersections.

Example 4 *Find all points of intersection of the two cardioids $r = 1 + \cos\theta$ and $r = 1 - \cos\theta$.*

Solution We begin by plotting points (table 1). The resulting graphs are in figure 8.

Three points of intersection are visible. One point of intersection

Table 1 *Tables of plotted points for the curves of example 4*

θ	$1 + \cos\theta$	θ	$1 - \cos\theta$
0	2	0	0
$\frac{\pi}{2}$	1	$\frac{\pi}{2}$	1
π	0	π	2
$\frac{3\pi}{2}$	1	$\frac{3\pi}{2}$	1
2π	2	2π	0

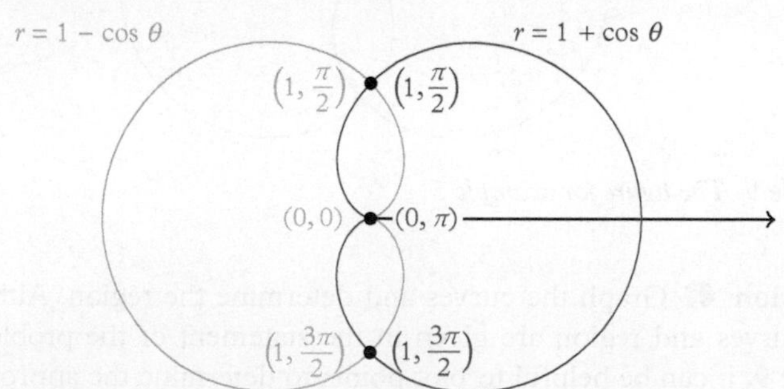

Figure 8 *The polar curves $r = 1 + \cos\theta$ (blue) and $r = 1 - \cos\theta$ (green). There appear to be three points of intersection*

is the point $(1, \frac{\pi}{2})$. Another is the point $(1, \frac{3\pi}{2})$. The third point of intersection is at the origin. ■

In example 4, the two curves reach the point $(1, \frac{\pi}{2})$ at the same value of θ—namely, $\frac{\pi}{2}$. This is why setting the two curves' expressions equal to one another and solving for θ (which we did before example 4) identified this point successfully. The same is true for the point $(1, \frac{3\pi}{2})$, which also appears in the list of points to plot for each curve. However, notice that the origin is not reached by the two curves at the same value of θ: $r = 1 + \cos\theta$ is at the origin at $\theta = \ldots, -\pi, \pi, 3\pi, \ldots$ whereas $r = 1 - \cos\theta$ is at the origin at $\theta = \ldots, -2\pi, 0, 2\pi, \ldots$. This is why solving the equation $1 - \cos\theta = 1 + \cos\theta$ for θ does not identify this point. There is no single value of θ that gives an r-coordinate of zero for both curves.

The same phenomenon can happen at any point. For instance, one curve could give $(2, \frac{\pi}{4})$ whereas the other gives $(-2, \frac{5\pi}{4})$, which are different sets of polar coordinates for the same point. Solving equations is helpful for identifying some points of intersection of two polar curves, but graphing the curves is generally a better procedure for identifying all intersection points.

Additional examples

Example 5 *Find the area of the shaded region in figure 9.*

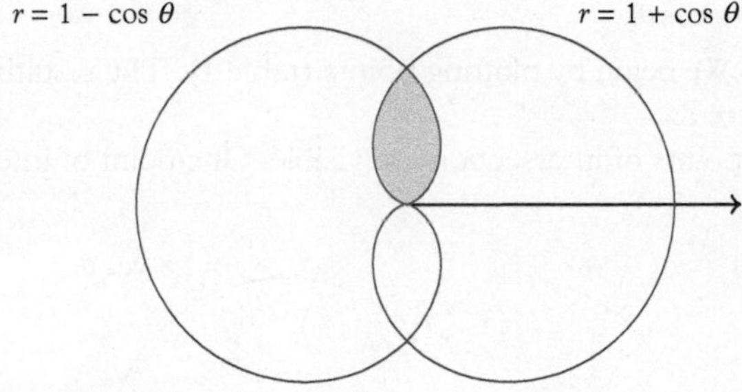

Figure 9 *The figure for example 5*

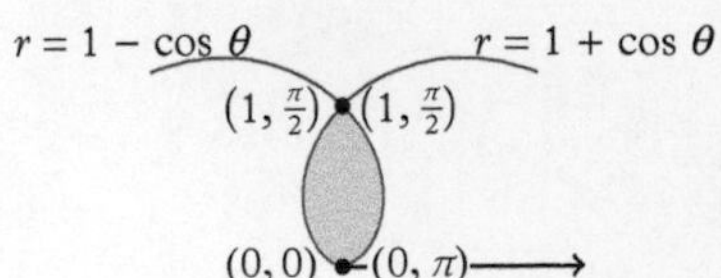

Figure 10 *Points of intersection identified*

Solution ❶ Graph the curves and determine the region. Although the curves and region are given in the statement of the problem in figure 9, it can be helpful to plot points to determine the appropriate points of intersection. This was done in example 4; see figure 10.

It is also helpful to notice that the region is not bounded by just one curve, but two. The right half of the region is "inside" the curve $r = 1 - \cos\theta$. (Radial lines from the origin to that curve sweep out the right half of the region.) The left half of the region is inside the curve $r = 1 + \cos\theta$.

❷ If necessary, determine the values of θ. Using the coordinates at the points of intersection for the curve $r = 1 - \cos\theta$, the values of θ for the right half of the region are $0 \le \theta \le \frac{\pi}{2}$. Using the coordinates at the points of intersection for the curve $r = 1 + \cos\theta$, the values of θ for the left half of the region are $\frac{\pi}{2} \le \theta \le \pi$.

The steps used here are the same as those used in example 1.

❸ Use the formula to compute the area. For the right half of the region, we apply the formula with $f(\theta) = 1 - \cos\theta$, $c = 0$, and $d = \frac{\pi}{2}$, and evaluate the integral:

$$
\begin{aligned}
A &= \int_c^d \frac{1}{2}(f(\theta))^2\,d\theta = \int_0^{\frac{\pi}{2}} \frac{1}{2}(1-\cos\theta)^2\,d\theta \\
&= \int_0^{\frac{\pi}{2}} \frac{1}{2}(1 - 2\cos\theta + \cos^2\theta)\,d\theta \\
&= \int_0^{\frac{\pi}{2}} \frac{1}{2}\left(1 - 2\cos\theta + \frac{1}{2}(1+\cos 2\theta)\right)\,d\theta \\
&= \int_0^{\frac{\pi}{2}} \left(\frac{3}{4} - \cos\theta + \frac{1}{4}\cos 2\theta\right)\,d\theta \\
&= \left(\frac{3}{4}\theta - \sin\theta + \frac{1}{8}\sin 2\theta\right)\Bigg|_0^{\frac{\pi}{2}} \\
&= \frac{3}{4}\cdot\frac{\pi}{2} - \sin\frac{\pi}{2} + \frac{1}{8}\sin\pi - (0 - 0 + 0) \\
&= \left(\frac{3}{8}\pi - 1\right)\ \text{units}^2.
\end{aligned}
$$

Line 3 uses the identity $\cos^2\theta = \frac{1}{2}(1+\cos 2\theta)$; line 5 uses the formula $\int \cos kx\,dx = \frac{\sin kx}{k} + C$.

Similarly, the area of the left half of the region is

$$
\begin{aligned}
A &= \int_c^d \frac{1}{2}(f(\theta))^2\,d\theta = \int_{\frac{\pi}{2}}^{\pi} \frac{1}{2}(1+\cos\theta)^2\,d\theta \\
&= \cdots = \int_{\frac{\pi}{2}}^{\pi} \left(\frac{3}{4} + \cos\theta + \frac{1}{4}\cos 2\theta\right)\,d\theta \\
&= \left(\frac{3}{4}\theta + \sin\theta + \frac{1}{8}\sin 2\theta\right)\Bigg|_{\frac{\pi}{2}}^{\pi} \\
&= \cdots = \left(\frac{3}{8}\pi - 1\right)\ \text{units}^2.
\end{aligned}
$$

The symmetry of the region indicates that the area of the left half should be the same as the area of the right half. However, evaluating the two halves separately is a good quality control procedure. If the two calculations produce the same result, then it is less likely that an error has been made.

Adding the areas of the two portions of the region results in a total area of $\left(\frac{3}{4}\pi - 2\right)$ units^2. ■

Example 6 *Find the perimeter of the shaded region in figure 11.*

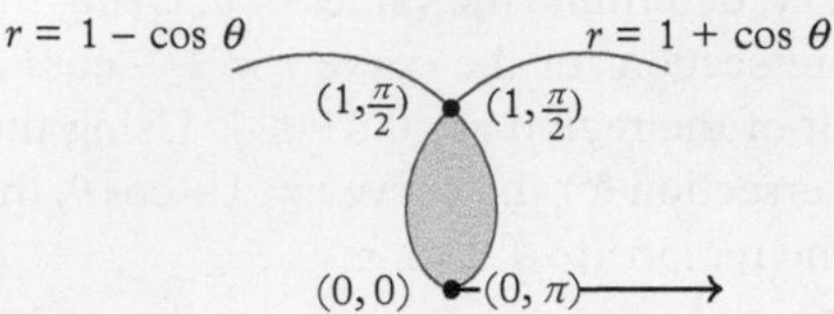

Figure 11 *The figure for example 6*

Solution This is the same region as the one in example 5. There we saw that the curve bounding the right half of the region is $r = 1 - \cos\theta$ for $0 \le \theta \le \frac{\pi}{2}$. With $f(\theta) = 1 - \cos\theta$ and $f'(\theta) = \sin\theta$, the polar arc length formula gives

Line 3 uses the identity $\cos^2\theta + \sin^2\theta = 1$ and simplifies, then factors out $2 = 4 \cdot \frac{1}{2}$. Line 4 uses the identity $\frac{1}{2}(1 - \cos 2x) = \sin^2 x$ with $x = \frac{1}{2}\theta$ (we have often used this same identity in the other direction). Line 5 takes the square root, being careful to make sure the resulting trig function is positive throughout the interval of integration. Line 6 uses the antiderivative formula $\displaystyle\int \sin kx = -\frac{\cos kx}{k} + C$.

$$\begin{aligned}
L &= \int_0^{\frac{\pi}{2}} \sqrt{(1 - \cos\theta)^2 + \sin^2\theta\, d\theta} \\
&= \int_0^{\frac{\pi}{2}} \sqrt{1 - 2\cos\theta + \cos^2\theta + \sin^2\theta\, d\theta} \\
&= \int_0^{\frac{\pi}{2}} \sqrt{2 - 2\cos\theta}\, d\theta = \int_0^{\frac{\pi}{2}} \sqrt{4\left(\frac{1}{2}(1 - \cos\theta)\right)}\, d\theta \\
&= \int_0^{\frac{\pi}{2}} \sqrt{4\sin^2\frac{1}{2}\theta\, d\theta} \\
&= \int_0^{\frac{\pi}{2}} 2\sin\frac{1}{2}\theta\, d\theta \\
&= \left.\frac{-2\cos\frac{1}{2}\theta}{\frac{1}{2}}\right|_0^{\frac{\pi}{2}} = -4\cos\frac{1}{2}\theta\Big|_0^{\frac{\pi}{2}} \\
&= -4\cos\frac{\pi}{4} - (-4\cos 0) = (4 - 2\sqrt{2}) \text{ units.}
\end{aligned}$$

Similarly, the curve bounding the left half of the region is $f(\theta) = 1 + \cos\theta$ for $\frac{\pi}{2} \le \theta \le \pi$, and the polar arc length formula gives

$$\begin{aligned}
L &= \int_{\frac{\pi}{2}}^{\pi} \sqrt{(1 + \cos\theta)^2 + (-\sin\theta)^2}\, d\theta \\
&= \cdots = \int_{\frac{\pi}{2}}^{\pi} \sqrt{4\cos^2\frac{1}{2}\theta\, d\theta} \\
&= \cdots = (4 - 2\sqrt{2}) \text{ units.}
\end{aligned}$$

Adding the two lengths, the perimeter of the region is $(8 - 4\sqrt{2})$ units. ■

Exercises 9.2

1–4. (a) Use the polar area formula to find the area inside the circle. (b) Use a formula from geometry to find the area of the circle and verify that the answer is identical to part (a).

1. $r = 4$ **3.** $r = 4 \sin \theta$

2. $r = -2$ **4.** $r = -2 \cos \theta$

5. Find the area inside the polar curve $r = 3 - 2 \sin \theta$.

6. Find the area inside all leaves of the rose $r = \sin 5\theta$.

7. Find the area inside one leaf of the rose $r = 2 \sin 3\theta$.

8. Find the area enclosed by the limaçon $r = 1 - 2 \cos \theta$.

9. Find the area inside the polar curve $r = \sqrt{\sin 2\theta}$, $0 \le \theta \le \frac{\pi}{2}$.

10. Find the area inside the portion of the spiral $r = \theta$, $0 \le \theta \le \pi$.

11–14. (a) Use the polar arc length formula to find the length of the circle. (b) Use a formula from geometry to find the circumference of the circle and verify that the answer is identical to part (a).

11. $r = -3$ **13.** $r = 4 \cos \theta$

12. $r = 5$ **14.** $r = 3 \sin \theta$

15–16. (a) Use the polar arc length formula to find the length of the line segment. (b) Determine Cartesian coordinates for the endpoints of the segment, use the distance formula from geometry to find the length of the line segment, and verify that the answer is identical to part (a).

15. $r = \sec \theta$, $0 \le \theta \le \frac{\pi}{3}$ **16.** $r = \csc \theta$, $\frac{\pi}{4} \le \theta \le \frac{\pi}{2}$

17. Find the length of the polar curve $r = e^{\theta}$, $0 \le \theta \le 1$.

18. Find the length of the polar curve $r = 3^{2\theta}$, $0 \le \theta \le 1$.

19. (GSL) Find the length of the spiral of Archimedes $r = 3\theta$ from the origin to the end of the first revolution.

20. (GSL) Find the length of the curve $r = \sec^2 \frac{\theta}{2}$ from $\theta = 0$ to $\theta = \frac{\pi}{2}$.

21–26. Using technology for assistance, (a) graph the polar curve, (b) set up the integral for finding the indicated area, and (c) evaluate the integral.

In some of the GSL exercises, the region is changed from the original problem. The resulting integrals were to be evaluated without the use of technology. If you enjoy a challenge, you might try one of the more difficult ones by hand. Also, GSL used the lowercase Greek letter ρ (rho) instead of r for polar coordinates and equations.

21. $r = 2 + \sec\theta$, area inside the loop by the origin (conchoid)

22. (GSL) $r = \frac{1}{2} + \cos 2\theta$, area inside the right-most loop

23. (GSL) $r = \frac{1}{2} + \cos 2\theta$, area inside the upper small loop

24. (GSL) $r = 2 + \sin 3\theta$, area enclosed by the curve

25. (GSL) $r = \cos 3\theta - \cos\theta$, area enclosed by one loop

26. (GSL) $r = \cos 3\theta - 2\cos\theta$, area enclosed by the curve

27–32. Using technology for assistance, (a) graph the polar curve, (b) set up the integral for finding the indicated length, and (c) evaluate the integral.

For the trisectrix of Maclaurin, the value of r at $\theta = 0$ is undefined, but the limit exists. When finding the length of the curve, assume that the removable discontinuity in the curve has been removed—that is, that the point that appears to be there in the graph is a point on the curve.

27. $r = \cos\theta(4\sin^2\theta - 1)$, length of the curve (trifolium)

28. $r = \dfrac{\sin 3\theta}{\sin 2\theta}$, length of the loop (trisectrix of Maclaurin)

29. $r = \cos^3 \frac{\theta}{3}$, length of the inner loop (Cayley's sextic)

30. $r = 1+2\sin\frac{\theta}{2}$, length of the curve (Freeth's nephroid)

31. $r = \cos 3\theta - \cos\theta$, length of the curve

32. $r = \cos 3\theta - 2\cos\theta$, length of the curve

33–38. (GSL) Calculate the area that the curves in each of the following pairs have in common.

The region with the desired area is the region that happens to lie inside both curves.

33. $r = 3\cos\theta$, $r = 1 + \cos\theta$

34. $r = 1 + \cos\theta$, $r = 1$

35. $r = 1 - \cos\theta$, $r = \sin\theta$

36. $r = \sqrt{2\cos 2\theta}$, $r = 1$

37. $r = \sqrt{6}\cos\theta$, $r = 3\sqrt{\cos 2\theta}$

38. $r = \sqrt{\cos 2\theta}$, $r = \sqrt{\sin 2\theta}$

39–42. Calculate the length of the indicated portion of a curve.

39. the portion of the circle $r = \frac{\sqrt{3}}{3}\cos\theta$ that lies inside the right-most leaf of the four-leaved rose $r = \cos 2\theta$

40. the portion of the cardioid $r = 1+\cos\theta$ that lies inside the circle $r = 1$

41. the portion of the cardioid $r = 1 - \cos\theta$ that lies outside the circle $r = \sin\theta$

42. the portion of the circle $r = 1$ that lies outside the lemniscate of Bernoulli $r = \sqrt{2\cos 2\theta}$

Jacob Bernoulli 1655–1705
https://www-history.mcs.st-andrews.ac.uk/Biographies/Bernoulli_Jacob.html

43. (GSL) Show that the area bounded by any two radii vectors of the hyperbolic spiral $r = \dfrac{a}{\theta}$ is proportional to the difference between the lengths of these radii.

44. Form an omega sum for calculating the area inside the circle $r = 3$ and then evaluate the omega sum. (Use Ω sectors and so on.)

Surface Area

9.3

If we rotate a region in the xy-plane about the x-axis or the y-axis, we obtain a solid of revolution. Possibilities for the region include the region under the curve $y = f(x)$ between $x = a$ and $x = b$, and the region bounded between certain curves. However, what if we only rotate a curve, not a region? Instead of creating a solid of revolution, we create a hollow *surface of revolution.*

Rotating only the curve means *not* rotating what's underneath. We just get the "outside" of the object, not its "inside."

Think of a surface of revolution as being the "outside" surface of a solid of revolution. The amount of paint or stain required to cover the surface of the object is proportional to its surface area. Alternately, you can think of a surface of revolution as being like a vase or a lampshade with circular cross sections. A real-life vase or lampshade has a certain thickness to its material, but the material used to make the object is proportional to the surface area.

Area of a surface of revolution, horizontal axis

Consider a curve $y = f(x)$ between $x = a$ and $x = b$. Rotate the curve about the x-axis as in figure 1. How do we find the area of the resulting surface? Proceeding as usual, we subdivide the interval $[a, b]$ into

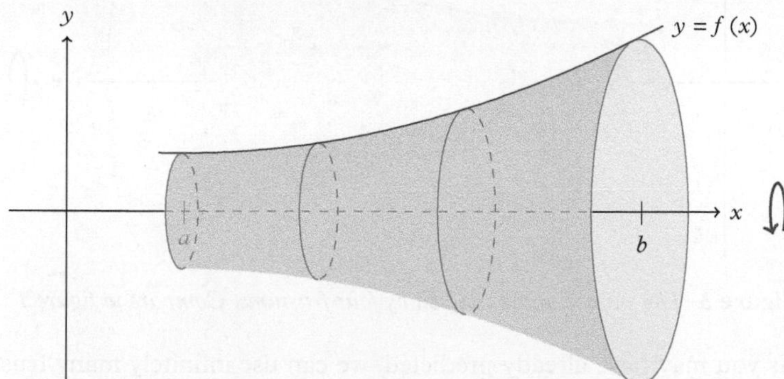

Figure 1 *A surface of revolution formed by rotating the curve $y = f(x)$ between $x = a$ and $x = b$ about the x-axis. The object is hollow*

smaller pieces and approximate. Instead of following the curve, we approximate each piece of the curve by a straight line segment, as we did for arc length. Rotating that line segment about the x-axis results in a surface such as the one shown in figure 2.

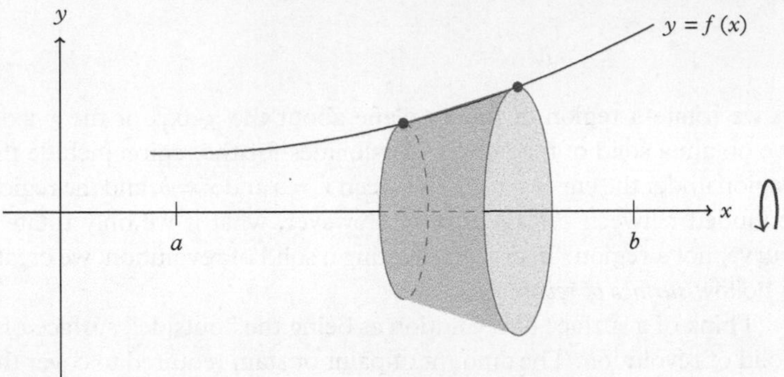

Figure 2 *An approximating frustum of a cone (brown). If you look sideways at the picture (roll your head left toward your left shoulder), you can see that the object looks like a lampshade*

Because the frustum of figure 2 is oriented with a horizontal axis, the "top" that is cut off is not above the remaining object, but to the left of it.

The name of this shape is a *frustum of a right circular cone*, which is a type of slice of a cone (think of a cone with the "top"—where the vertex is—cut off). If we approximate the entire surface using frustums, we get a good approximation of the area of the surface, as illustrated in figure 3. The more frustums we use, the better the approximation.

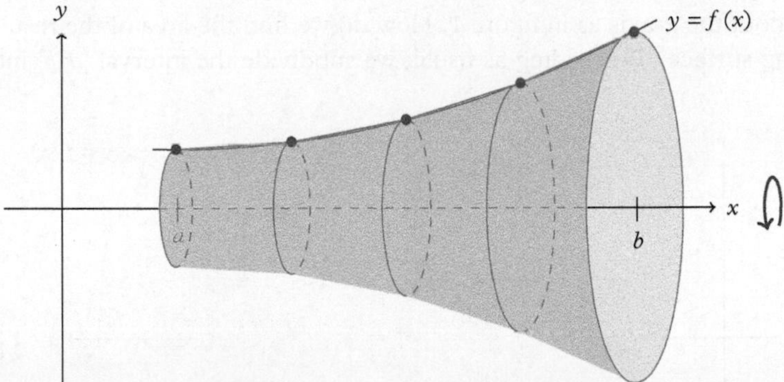

Figure 3 *The surface approximated by four frustums. Compare to figure 1*

As you may have already predicted, we can use infinitely many frustums to calculate the exact area of the surface. If we add the surface areas of all these frustums, we need to know the frustum surface area formula from geometry, which is

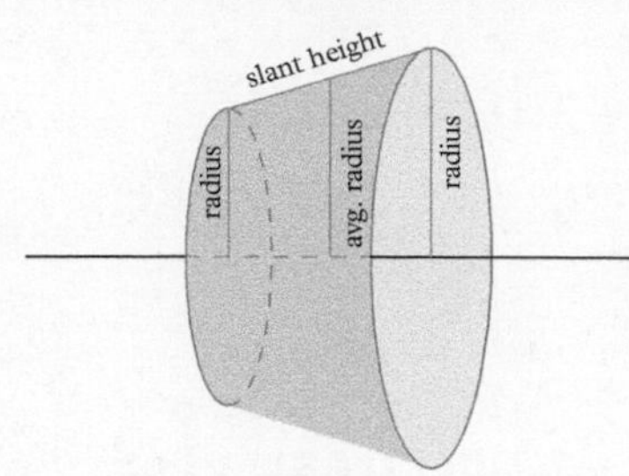

Figure 4 *The surface area of a frustum (think lampshade, so that only the lateral surface is considered) is SA = 2π · average radius · slant height*

$$\text{surface area } (SA) = 2\pi \cdot \text{average radius} \cdot \text{slant height},$$

as in figure 4.

When using Ω subintervals, what is the surface area for each slice (each frustum)? Referring to figures 2 and 4 as a guide, we see that the average radius is the height of the curve, or $f(x_k)$. The slant height is a quantity we calculated when studying arc length, when we derived the formula $\sqrt{1 + (f'(x_k))^2}\,\Delta x$. Thus, the surface area for a slice is

$$SA_{\text{slice}} = 2\pi \cdot \text{avg. radius} \cdot \text{slant height} = 2\pi \cdot f(x_k) \cdot \sqrt{1 + (f'(x_k))^2}\,\Delta x.$$

Identifying $f(x_k)$ with the average radius has a technicality that is explored in exercise 39.

The total surface area is then found by adding the surface areas of the slices, resulting in the omega sum

$$SA_{\text{total}} = \sum_{k=1}^{\Omega} 2\pi \cdot f(x_k) \cdot \sqrt{1 + (f'(x_k))^2}\,\Delta x.$$

Rewriting as a definite integral (dropping the subscript, replacing Δx with dx, replacing the summation symbol with the integral symbol, and inserting the limits of integration) gives us the desired formula:

$$SA = \int_a^b 2\pi f(x)\sqrt{1 + (f'(x))^2}\,dx.$$

As with the arc length formula, adding the condition that f' is continuous on $[a, b]$ guarantees integrability. The requirement is also related to the technicality of identifying $f(x_k)$ with the average radius, which is explored in exercise 39.

AREA OF A SURFACE OF REVOLUTION, HORIZONTAL AXIS OF ROTATION

If f' is continuous on $[a, b]$, then the area of the surface obtained by rotating $y = f(x)$ about the x-axis is

$$SA = \int_a^b 2\pi f(x)\sqrt{1 + (f'(x))^2}\,dx.$$

Example 1 *The curve $y = x^3$ between $x = 0$ and $x = 1$ is rotated about the x-axis. Find the surface area.*

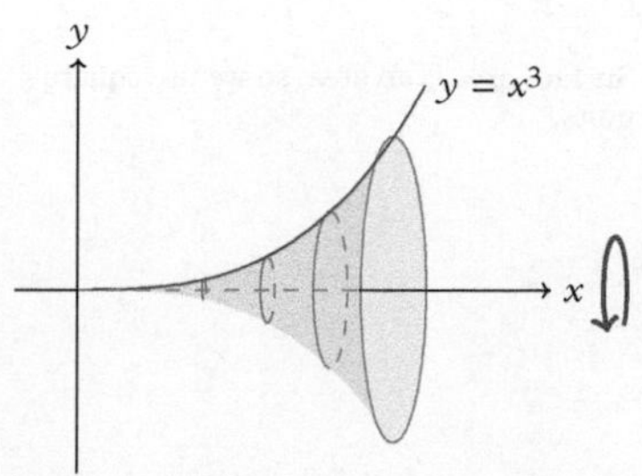

Figure 5 *The surface of revolution of example 1*

Solution As always, it is good practice to sketch the object of interest, which in this case is a surface of revolution (figure 5). (Sketching a surface of revolution is basically the same as sketching a solid of revolution.) The formula for the area of a surface of revolution with a horizontal axis applies, with $f(x) = x^3$ and $f'(x) = 3x^2$. Then,

$$\begin{aligned} SA &= \int_a^b 2\pi f(x)\sqrt{1+(f'(x))^2}\,dx \\ &= \int_0^1 2\pi x^3\sqrt{1+(3x^2)^2}\,dx \\ &= \int_0^1 2\pi x^3\sqrt{1+9x^4}\,dx. \end{aligned}$$

We recognize the need for substitution:

$$\begin{aligned} u &= 1+9x^4 \\ du &= 36x^3\,dx. \end{aligned}$$

Next we calculate the new limits of integration:

$$\begin{aligned} x = 1:&\quad u = 1+9\cdot 1^4 = 10 \\ x = 0:&\quad u = 1+9\cdot 0^4 = 1. \end{aligned}$$

We make the substitution and evaluate the resulting integral:

Line 1 adjusts for the constant, line 2 makes the substitution, line 3 evaluates the integral, and line 4 gives a decimal approximation of the area.

$$\begin{aligned} &= \frac{1}{18}\int_0^1 36\pi x^3\sqrt{1+9x^4}\,dx \\ &= \frac{\pi}{18}\int_1^{10}\sqrt{u}\,du \\ &= \frac{\pi}{18}\cdot\frac{u^{3/2}}{3/2}\bigg|_1^{10} = \frac{\pi}{27}\left(10^{3/2}-1\right)\text{ units}^2 \\ &= 3.56312\,\text{units}^2. \end{aligned}$$

Surface area is an area, so we use square units.

The area of the surface is 3.56312 units^2. ■

As with arc length, the surface area formula often results in integrals that must be evaluated numerically.

Example 2 *The curve $y = x\sin x$ from $x = 0$ to $x = \pi$ is rotated about the x-axis. (a) Set up the integral for determining the area of the resulting surface. (b) Use a calculator to determine the value of the integral.*

Solution The surface of revolution is pictured in figure 6.

(a) We use the formula for the area of a surface of revolution with a horizontal axis with $f(x) = x \sin x$ and $f'(x) = x \cos x + \sin x$:

$$SA = \int_a^b 2\pi f(x)\sqrt{1 + (f'(x))^2}\,dx$$
$$= \int_0^\pi 2\pi x \sin x\sqrt{1 + (x\cos x + \sin x)^2}\,dx.$$

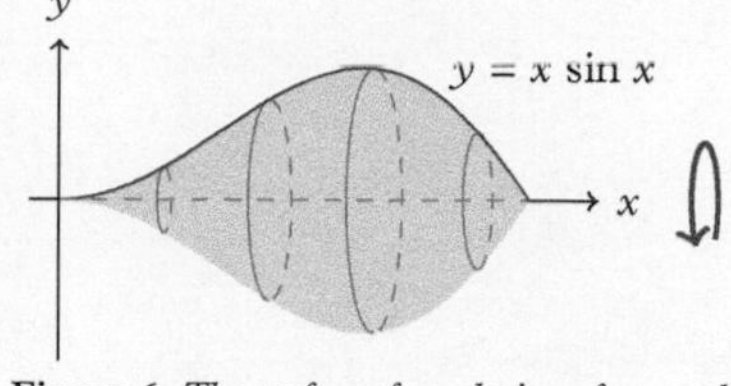

Figure 6 *The surface of revolution of example 2. The object is hollow*

(b) Using numerical integration on a calculator gives 29.9649 units^2. ■

Reading Exercise 5 *The curve $y = x^4$ between $x = 2$ and $x = 7$ is rotated about the x-axis. Set up the integral for finding the area of the resulting surface. Do not integrate.*

Area of a surface of revolution, vertical axis

Next we rotate the curve $y = f(x)$ between $x = a$ and $x = b$ about the y-axis (instead of the x-axis) to create a surface of revolution (figure 7). What changes? We can still approximate the surface using frustums,

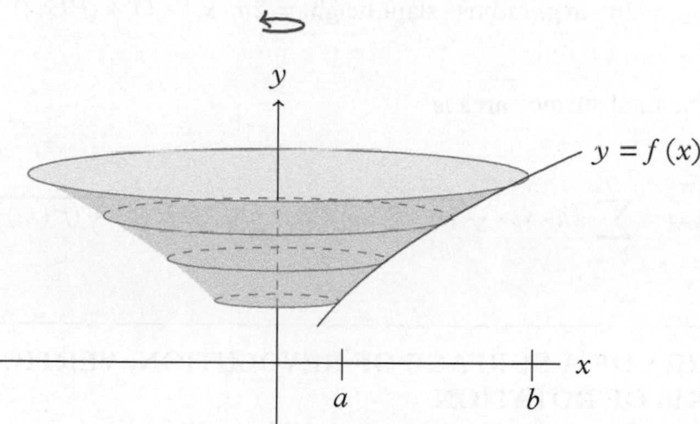

Figure 7 *A surface of revolution (green) created by revolving $y = f(x)$ between $x = a$ and $x = b$ about the y-axis. The object is hollow. Compare to figure 1*

but because the axis of the frustum is the axis of rotation, the orientation of the frustums changes (figure 8). We still slice x, subdividing the interval from $x = a$ to $x = b$ into Ω pieces. The slant height of the frustum in figure 8 is still the length of the segment between approximating

We use $\Delta x = \frac{b-a}{\Omega} = (b - a)\omega$, as always.

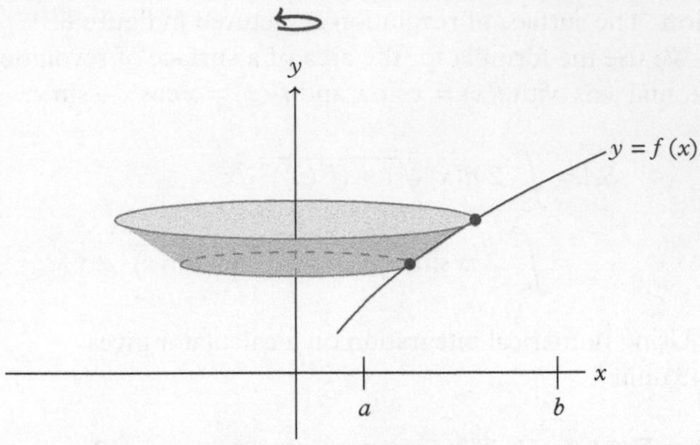

Figure 8 *An approximating frustum (brown). The axis of the frustum is the same as the axis of rotation. Compare to figure 2*

points (the red dots in figure 8), so the formula $\sqrt{1+(f'(x_k))^2}\,\Delta x$ applies. However, now the average radius of the frustum is no longer a vertical distance, but rather a horizontal distance. So instead of using $f(x_k)$ for the average radius, we use x_k. The surface area for the frustum is then

Identifying x_k with the average radius has a technicality that is explored in exercise 40.

$$SA_{\text{slice}} = 2\pi \cdot \text{avg. radius} \cdot \text{slant height} = 2\pi \cdot x_k \cdot \sqrt{1+(f'(x_k))^2}\,\Delta x,$$

and the total surface area is

$$SA_{\text{total}} = \sum_{k=1}^{\Omega} 2\pi \cdot x_k \cdot \sqrt{1+(f'(x_k))^2}\,\Delta x = \int_a^b 2\pi x\sqrt{1+(f'(x))^2}\,dx.$$

Adding the condition that f' is continuous on $[a, b]$ guarantees integrability. The requirement is also related to the technicality of identifying x_k with the average radius, which is explored in exercise 40.

AREA OF A SURFACE OF REVOLUTION, VERTICAL AXIS OF ROTATION

If f' is continuous on $[a, b]$, then the area of the surface obtained by rotating $y = f(x)$ between $x = a$ and $x = b$ about the y-axis is

$$SA = \int_a^b 2\pi x\sqrt{1+(f'(x))^2}\,dx.$$

Example 3 *The curve $y = 3 + \cosh x$ from $x = 0$ to $x = 1$ is rotated about the y-axis. Find the area of the resulting surface.*

Solution The surface is graphed in figure 9.

There are now two surface area formulas from which to choose, one for rotating about the x-axis and one for rotating about the y-axis. Because our curve is rotated about the y-axis, we use

$$SA = \int_a^b 2\pi x\sqrt{1 + (f'(x))^2}\, dx.$$

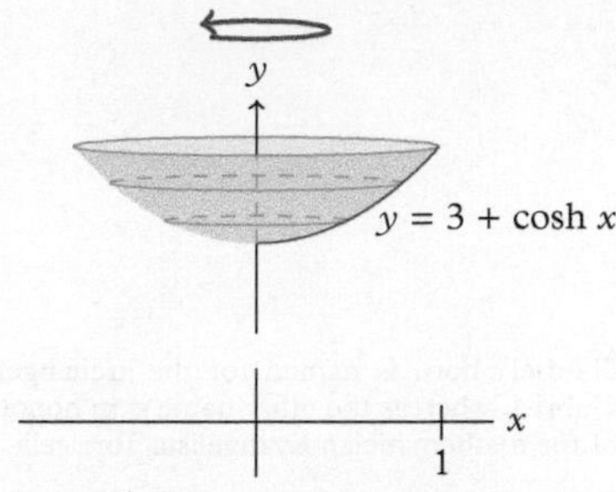

Figure 9 *The surface of example 3*

Noting that $y' = \sinh x$, we have

$$\begin{aligned} SA &= \int_0^1 2\pi x\sqrt{1 + \sinh^2 x}\, dx \\ &= \int_0^1 2\pi x\sqrt{\cosh^2 x}\, dx \\ &= 2\pi \int_0^1 x\cosh x\, dx. \end{aligned}$$

Line 2 uses a hyperbolic identity. For line 3, note that $\cosh x$ is positive and therefore $\sqrt{\cosh^2 x} = \cosh x$.

The integrand is a product of a polynomial (x) and a hyperbolic function ($\cosh x$), indicating the use of integration by parts. To integrate $\int x\cosh x\, dx$, we set f to be the polynomial part and g' to be the hyperbolic part:

$$\begin{aligned} f(x) &= x & g'(x) &= \cosh x \\ f'(x) &= 1 & g(x) &= \sinh x. \end{aligned}$$

Ans. to reading exercise 5

$$\begin{aligned} SA &= \int_2^7 2\pi x^4\sqrt{1 + (4x^3)^2}\, dx \\ &= \int_2^7 2\pi x^4\sqrt{1 + 16x^6}\, dx \end{aligned}$$

Then,

$$\begin{aligned} \int x\cosh x\, dx &= x\sinh x - \int \sinh x\, dx \\ &= x\sinh x - \cosh x + C. \end{aligned}$$

Returning to our original integral,

$$\begin{aligned} SA &= 2\pi \int_0^1 x\cosh x\, dx \\ &= 2\pi(x\sinh x - \cosh x)\Big|_0^1 \\ &= 2\pi(\sinh 1 - \cosh 1) - 2\pi(0 - 1) \\ &= 2\pi(\sinh 1 - \cosh 1 + 1)\ \text{units}^2. \end{aligned}$$

The surface area is approximately 3.972 units^2. ■

Reading Exercise 6 *The curve $y = x^4$ between $x = 2$ and $x = 7$ is rotated about the y-axis. Set up the integral for finding the area of the resulting surface. Do not integrate.*

Improper surface area

Gabriel's horn is named for the archangel Gabriel, whereas the other name is in honor of the mathematician Evangelista Torricelli.

Evangelista Torricelli 1608–1647

http://www-history.mcs.st-andrews.ac.uk/Biographies/Torricelli.html

Imagine an infinitely long surface generated by revolving the curve $y = \frac{1}{x}$ for $x \geq 1$ about the x-axis. The surface resembles an infinitely long trumpet, often called *Gabriel's horn* or *Torricelli's trumpet*, with the opening of the bell on the left and the mouthpiece infinitely far to the right. What is the surface area of this object? Although our formula is for a closed, bounded interval $[a, b]$, we can try the formula on $[1, \infty)$. With $f(x) = \frac{1}{x}$ and $f'(x) = -\frac{1}{x^2}$, we get

$$SA = \int_a^b 2\pi f(x)\sqrt{1 + (f'(x))^2}\, dx$$

$$= \int_1^\infty 2\pi \frac{1}{x}\sqrt{1 + \left(-\frac{1}{x^2}\right)^2}\, dx,$$

In general, applying a formula meant for a bounded interval $[a, b]$ to an infinite interval must be done with care. The idea is to proceed as we did for improper integrals, considering what happens on the interval $[1, \Omega]$ for an arbitrary positive infinite hyperreal Ω. If the formula applies for each interval $[1, \Omega]$ and the same result is obtained for every such Ω, then we say the result holds for the interval $[1, \infty)$. But, in some applications, this procedure does not give satisfactory results.

which is an improper integral of type I. Using the positive infinite hyperreal number Ω gives

$$= 2\pi \int_1^\Omega \frac{1}{x}\sqrt{1 + \frac{1}{x^4}}\, dx.$$

It is difficult to find an antiderivative for this integrand, but the comparison theorem from chapter 7 (theorem 2) helps. Noting that $\frac{1}{x^4}$ is always positive on our interval, we have

It is not necessary to use the comparison theorem in the exercises at the end of this section.

$$\frac{1}{x}\sqrt{1 + \frac{1}{x^4}} > \frac{1}{x}\sqrt{1 + 0} = \frac{1}{x}.$$

Replacing the original integrand with the smaller integrand $\frac{1}{x}$ gives

$$2\pi \int_1^\Omega \frac{1}{x}\, dx = 2\pi \ln x \Big|_1^\Omega$$

$$= 2\pi \ln \Omega$$

$$\doteq \infty.$$

Because our integral is larger than one that diverges to infinity, the surface area of Gabriel's horn is infinite.

Perhaps this result is not surprising. We expect that an infinitely long object has an infinite surface area. But a surprise is in the works when we calculate the volume inside Gabriel's horn:

$$
\begin{aligned}
V &= \int_a^b \pi (f(x))^2 \, dx = \int_1^\infty \pi \left(\frac{1}{x}\right)^2 \, dx = \int_1^\Omega \pi x^{-2} \, dx \\
&= \left.\frac{-\pi}{x}\right|_1^\Omega \\
&= \frac{-\pi}{\Omega} - \frac{-\pi}{1} = -\pi\omega + \pi \\
&\approx \pi \, \text{units}^3.
\end{aligned}
$$

The surface area is infinite, but the volume is finite! This paradox is often described using paint. Gabriel's horn can only hold a finite amount of paint, but it takes an infinite amount of paint to cover the surface! Like any paradox, there is a satisfactory resolution. Rather than spoil it now, a resolution is given in the answers to the exercises at the end of this book.

Ans. to reading exercise 6

$$
\begin{aligned}
SA &= \int_2^7 2\pi x \sqrt{1 + (4x^3)^2} \, dx \\
&= \int_2^7 2\pi x \sqrt{1 + 16x^6} \, dx
\end{aligned}
$$

A few more hints

Simplifying an integrand is often helpful. If an integrand involves a product of square roots, then recalling that $\sqrt{a} \cdot \sqrt{b} = \sqrt{a \cdot b}$ is helpful.

Example 4 *Simplify* $\sqrt{100 - x^2}\sqrt{1 + \frac{x^2}{100 - x^2}}$.

Solution Using $\sqrt{a} \cdot \sqrt{b} = \sqrt{a \cdot b}$, we have

$$
\begin{aligned}
\sqrt{100 - x^2}\sqrt{1 + \frac{x^2}{100 - x^2}} &= \sqrt{(100 - x^2)\left(1 + \frac{x^2}{100 - x^2}\right)} \\
&= \sqrt{(100 - x^2) \cdot 1 + \cancel{(100 - x^2)} \cdot \frac{x^2}{\cancel{100 - x^2}}} \\
&= \sqrt{100 - x^2 + x^2} \\
&= \sqrt{100}.
\end{aligned}
$$

■

In the latter square root of line 1, instead of seeing an expression of the form $(a - b)(c + d)$ and thinking "foil," instead look at the expression as having the form $k(c + d)$, which is expanded in the next line as $kc + kd$ to facilitate the cancellation that makes the expression much simpler.

Sometimes a creative substitution is helpful. For the integral

$$\int x\sqrt{1 + 25x^4} \, dx,$$

the substitution $u = 1+25x^4$ fails, since $du = 100x^3\,dx$ and the x^3 is not present. However, the substitution $u = 5x^2$ works. For this substitution, $du = 10x\,dx$ and we only need to adjust for the constant. We also note that $u^2 = 25x^4$, so the quantity under the square root becomes $1 + u^2$, yielding

$$\int x\sqrt{1+25x^4}\,dx = \frac{1}{10}\int 10x\sqrt{1+25x^4}\,dx$$
$$= \frac{1}{10}\int \sqrt{1+u^2}\,du.$$

The integral can then be completed by hand using trig substitution and integration by parts, or by using a table of integrals.

When simplifying a square root of a perfect square, we must consider the sign of the quantity over the interval of integration. For instance,

$$\int_3^7 \sqrt{(2x-5)^2}\,dx = \int_3^7 (2x-5)\,dx$$

since $2x - 5 > 0$ on the interval $[3, 7]$, but

$$\int_0^1 \sqrt{(2x-5)^2}\,dx = \int_0^1 -(2x-5)\,dx$$

since $2x - 5 < 0$ on the interval $[0, 1]$.

Last, instead of extracting a quantity from a square root, sometimes it is helpful to put a quantity inside a square root.

Example 5 *Simplify the integrand of* $\int_2^5 x\sqrt{1+\frac{4}{x^2}}\,dx.$

Had the limits of integration been $x = -4$ to $x = -2$, then we would have used $-x = \sqrt{x^2}$.

Solution Because x is positive over the interval $[2, 5]$, we know that

$$x = \sqrt{x^2}.$$

We may then simplify the integrand as

$$x\sqrt{1+\frac{4}{x^2}} = \sqrt{x^2}\sqrt{1+\frac{4}{x^2}} = \sqrt{x^2 + \cancel{x^2}\cdot\frac{4}{\cancel{x^2}}} = \sqrt{x^2+4}.$$

■

A mnemonic device

One way of remembering surface area formulas is through the use of the arc length function $s(x) = \int_a^x \sqrt{1 + (f'(t))^2}\, dt$. By the fundamental theorem of calculus, part I, $\frac{ds}{dx} = \sqrt{1 + (f'(x))^2} = \sqrt{1 + \left(\frac{dy}{dx}\right)^2}$. The differential ds is then

For the sake of convenience, the roles of the variables t and x written here are reversed from the formula given in section 9.1. A careful examination of the statement of the arc length function should convince you that this is not an issue.

$$ds = \sqrt{1 + \left(\frac{dy}{dx}\right)^2}\, dx.$$

With $y = f(x)$, the surface area formula for rotation about the x-axis is rewritten as

Limits of integration are omitted for the shortcut formulas. Appropriate values must be inserted when setting up the integral to be computed.

$$SA_{x\text{-axis}} = \int 2\pi y\, ds,$$

whereas the formula for rotation about the y-axis is rewritten as

$$SA_{y\text{-axis}} = \int 2\pi x\, ds.$$

Although these formulas may be easier to memorize, for computational purposes there is no advantage; the variable of integration is still x (or y, if the roles of x and y are reversed), not s, and ds must be rewritten in terms of the variable of integration.

These shortcut formulas are also useful in the next section.

Exercises 9.3

1–10. A curve on an interval is rotated about an axis. (a) Set up the integral for determining the area of the resulting surface. (b) Use a calculator to determine the value of the integral to four significant digits.

1. $y = \sin^{-1} x$ between $x = 0$ and $x = \frac{1}{2}$, about the x-axis
2. $y = \ln x$ between $x = 1$ and $x = 4$, about the x-axis
3. $y = -\tan x$ between $x = -\frac{\pi}{4}$ and $x = 0$, about the x-axis
4. $y = x \cosh x$ between $x = 1$ and $x = 4$, about the x-axis

5. $y = e^{x^2+1}$ from $x = 0$ to $x = 2$, about the y-axis
6. $y = \sin x$ from $x = 0$ to $x = \frac{\pi}{4}$, about the y-axis
7. $y = x^4$, $0 \le x \le 1$, about the x-axis
8. $y = x^4$, $0 \le x \le 1$, about the y-axis
9. $f(x) = \frac{1}{x}$, $x = 1$ to $x = 4$, about the y-axis
10. $f(x) = \sin^2 x$, $x = \frac{\pi}{2}$ to $x = \pi$, about the x-axis

11–22. Find the area of the surface of revolution generated by the given curve, interval, and axis of rotation.

It is implied that numerical integration should not be used for these problems; an antiderivative must be calculated.

11. $y = 2x$, $x = 0$ to $x = 2$, about the y-axis
12. $y = 2x$, $x = 0$ to $x = 2$, about the x-axis
13. $y = \sqrt{4 - x}$, $x = 0$ to $x = 4$, about the x-axis
14. (GSL) $y = \sqrt{4 - x^2}$, $x = 1$ to $x = 2$, about the x-axis
15. (GSL) $y = 3\sqrt{x}$, $x = 0$ to $x = 4$, about the x-axis
16. (GSL) $y = \frac{1}{9}x^3$, $x = 0$ to $x = 2$, about the x-axis
17. (GSL) $y = \sqrt{24 - 4x}$, $x = 3$ to $x = 6$, about the x-axis
18. $y = x^2$, $x = 0$ to $x = 3$, about the y-axis
19. $y = \frac{1}{2}\ln x - \frac{1}{4}x^2$, $x = 1$ to $x = 2$, about the y-axis
20. (Love) $y = \frac{1}{6}x^3 + \frac{1}{2x}$, $x = 1$ to $x = 3$, about the y-axis
21. (Love) $y = \frac{1}{6}x^3 + \frac{1}{2x}$, $x = 1$ to $x = 3$, about the x-axis
22. $y = \cosh x$, $x = 0$ to $x = 1$, about the x-axis

23–28. Find the area of the surface of revolution generated by the given curve, interval, and axis of rotation. Use a table of integrals as necessary.

23. (Love) $y = \sin x$, $0 \le x \le \pi$, about the x-axis
24. (GSL) $y = \frac{1}{6}x^2$, $0 \le x \le 4$, about the x-axis
25. $y = e^{2x}$, $x = -\ln\sqrt{2}$ to $x = 0$, about the x-axis
26. $y = \ln x$, $x = 1$ to $x = 4$, about the y-axis
27. (GSL) $y = x^3$, $x = 0$ to $x = 2$, about the y-axis
28. $y = e^x$, $x = 0$ to $x = 1$, about the x-axis

29–30. A surface of revolution is generated by revolving the curve about an axis. Use an improper integral to find the area of the surface.

29. (GSL) $y = e^{-x}$ for $x \ge 0$, about the x-axis
30. (Love) $y = \ln x$ for $x = 0$ to $x = 1$, about the y-axis

31–36. If we consider surface area to be defined by the formulas given in this section, then we can use these formulas to derive surface area formulas for geometric objects. For each object, an equation of a curve is given that can be rotated about an axis to generate the object. Determine the surface area. All letters other than x and y represent constants.

It may be necessary to solve the given equation for y. If rotating about the x-axis, using positive square roots (if applicable) is sufficient.

31. sphere; use $x^2 + y^2 = r^2$, about the x-axis

32. torus; use $x^2 + (y - b)^2 = r^2$, where $b < r$, about the x-axis

33. cone; use $y = mx$ from $x = 0$ to $x = h$, about the x-axis (lateral surface area)

34. (Love) zone cut from a sphere by two parallel planes at a distance h apart; use $x^2 + y^2 = r^2$ from $x = a$ to $x = a + h$, about the x-axis

35. cylinder; use $y = r$ from $x = 0$ to $x = h$, about the x-axis (lateral surface area)

36. frustum; use $y = mx + b$ from $x = 0$ to $x = a$, about the x-axis (lateral surface area)

37. (Love) Find the area of the surface formed by revolving the four-cusped hypocycloid $x^{2/3} + y^{2/3} = a^{2/3}$ about the x-axis.

Hint: use symmetry, integrating from $x = 0$ to $x =$ (something) and doubling that value.

38. (GSL) Find the area of the surface formed by revolving the curve $y = \frac{x}{2}\sqrt{x^2 - 1} + \frac{1}{2}\ln(x - \sqrt{x^2 - 1})$, from $x = 2$ to $x = 5$, about the y-axis. (Hint: simplify y'.)

39. In the narrative when deriving the formula for the area of a surface of revolution with a horizontal axis of rotation, we stated that the average radius of a frustum is $f(x_k)$. Using right-hand endpoints, this is actually the radius on the right side of the frustum pictured in figure 4. The radius on the left side of the frustum is $f(x_{k-1})$. (a) Write an expression for the average radius of the frustum. (b) Noting that $f(x_{k-1}) = f(x_k - \Delta x)$ and that Δx is an infinitesimal, rewrite this expression using the local linearity formula. (c) Using the result of part (b), show that the average radius of the frustum is $f(x_k) - \frac{1}{2}\Delta x f'(x_k)$. (d) Form the correct omega sum for the surface area. (e) By distributing through the parentheses, rewrite the sum from part (d) as a difference of two sums, one of which is the omega sum given in the narrative.

(*Now the objective is to show that the second sum is infinitesimal.*) (f) The second sum has an "extra" Δx inside. Factor it out of the sum. (g) Rewrite the expression as a definite integral minus Δx times a definite integral.

(If f' is continuous, then the integrand of the second integral of part (g) is a product and composition of continuous functions and is therefore continuous (chapter 1 theorem 9). Continuous functions are integrable (chapter 4 theorem 1). Therefore, the value of the definite integral is a real number.) (h) Using the fact that Δx is infinitesimal and that the second integral is a real number, approximate the quantity. All that remains is the surface area formula.

40. Repeat exercise 39 for rotation about the y-axis.

41. When an object is painted, the paint usually has (approximately) a uniform thickness. Use this idea to resolve the Gabriel's horn paradox.

42. Notice that the derivative of the volume of a sphere (with respect to the variable r) is the surface area. Explain.

Lengths and Surface Areas with Parametric Curves

9.4

We now turn our attention to determining arc length and surface area when the curve is given in parametric form.

Lengths of parametric curves

To derive a formula for the length of a parametric curve $x = f(t)$, $y = g(t)$ from $t = c$ to $t = d$, we don't need to start from scratch. We already know that

$$\frac{dy}{dx} = \frac{\frac{dy}{dt}}{\frac{dx}{dt}} = \frac{g'(t)}{f'(t)}.$$

We also know the arc length formula in Cartesian coordinates, which is rewritten as

$$L = \int_a^b \sqrt{1 + \left(\frac{dy}{dx}\right)^2}\, dx.$$

For the arc length formula, the derivative is rewritten using Leibniz notation.

Making the replacement substitution

$$x = f(t) \qquad y = g(t)$$
$$dx = f'(t)\, dt \qquad dy = g'(t)\, dt$$

The replacements $y = g(t)$ and $dy = g'(t)\, dt$ are not used in this derivation. However, technically, they are part of the substitution.

gives

$$\begin{aligned} L &= \int_a^b \sqrt{1 + \left(\frac{dy}{dx}\right)^2}\, dx \\ &= \int_c^d \sqrt{1 + \frac{(g'(t))^2}{(f'(t))^2}} \cdot f'(t)\, dt \end{aligned}$$

Line 2 makes the substitution of $\frac{g'(t)}{f'(t)}$ for $\frac{dy}{dx}$ and $f'(t)\, dt$ for dx; line 3 gets a common denominator. In line 4, the potential sign error made by replacing $f'(t)$, with a sign that is unknown, with $\sqrt{(f'(t))^2}$, which is always positive, is mitigated by the placement of c and d where $c < d$. Because the integrand is nonnegative, the result is a nonnegative length, as required. Line 5 simplifies and line 6 rewrites using Leibniz notation.

$$= \int_c^d \sqrt{\frac{(f'(t))^2 + (g'(t))^2}{(f'(t))^2}} \cdot f'(t)\, dt$$

$$= \int_c^d \sqrt{\frac{(f'(t))^2 + (g'(t))^2}{(f'(t))^2} \cdot (f'(t))^2}\, dt$$

$$= \int_c^d \sqrt{(f'(t))^2 + (g'(t))^2}\, dt$$

$$= \int_c^d \sqrt{\left(\frac{dx}{dt}\right)^2 + \left(\frac{dy}{dt}\right)^2}\, dt.$$

PARAMETRIC ARC LENGTH FORMULA

If $\frac{dx}{dt}$ and $\frac{dy}{dt}$ are continuous on $[c, d]$, then the length of the parametric curve $x = f(t)$, $y = g(t)$ from $t = c$ to $t = d$ is

$$L = \int_c^d \sqrt{\left(\frac{dx}{dt}\right)^2 + \left(\frac{dy}{dt}\right)^2}\, dt,$$

provided the curve is traversed only once.

If you walk two laps along a path, you have walked twice the length of the path. The formula finds the length walked between $t = c$ and $t = d$. To find the length of the path, we need to ensure that from $t = c$ to $t = d$, we only walk once around the path.

Example 1 *(a) Set up the integral for finding the length of the parametric curve $x = 4\cos t$, $y = 3\sin t$. (b) Use a calculator to determine the value of the integral.*

The parametric curve $x = a\cos t$, $y = b\sin t$ is an ellipse, as is $x = a\sin t$, $y = b\cos t$. Either is traversed exactly once between $t = 0$ and $t = 2\pi$, but the direction and starting point differ.

Solution (a) We calculate $\frac{dx}{dt} = -4\sin t$ and $\frac{dy}{dt} = 3\cos t$. Because $\frac{dx}{dt}$ and $\frac{dy}{dt}$ are continuous, the parametric arc length formula may be used. The curve is traversed exactly once between $t = 0$ and $t = 2\pi$. (This can be verified by plotting points or by graphing the curve.) Then,

$$L = \int_c^d \sqrt{\left(\frac{dx}{dt}\right)^2 + \left(\frac{dy}{dt}\right)^2}\, dt$$

$$= \int_0^{2\pi} \sqrt{(-4\sin t)^2 + (3\cos t)^2}\, dt$$

$$= \int_0^{2\pi} \sqrt{16\sin^2 t + 9\cos^2 t}\, dt.$$

(b) Using a calculator to evaluate the integral gives a length of 22.1035 units. ■

The integrand can be simplified somewhat using $\sin^2\theta + \cos^2\theta = 1$, giving $\int_0^{2\pi} \sqrt{16\sin^2 t + 9\cos^2 t}\, dt = \int_0^{2\pi} \sqrt{7\sin^2 t + 9\sin^2 t + 9\cos^2 t}\, dt = \int_0^{2\pi} \sqrt{9 + 7\sin^2 t}\, dt$. However, this does not help us determine an antiderivative.

Reading Exercise 7 *Set up the integral for finding the length of the curve $x = \sin t$, $y = 3 + e^t$ from $t = 0$ to $t = 12\pi$. Do not integrate.*

Example 2 *Find the length of the curve* $x = \frac{1}{9}(6t+9)^{\frac{3}{2}}$, $y = \frac{1}{2}t^2$ *from* $t = -1$ *to* $t = 2$.

Solution The curve is given parametrically. We calculate

$$\frac{dx}{dt} = \frac{1}{9} \cdot \frac{3}{2}(6t+9)^{\frac{1}{2}} \cdot 6 = \sqrt{6t+9}$$

and

$$\frac{dy}{dt} = t.$$

Because $\frac{dx}{dt}$ and $\frac{dy}{dt}$ are continuous on $[-1, 2]$, the parametric arc length formula applies. Then,

When an interval of values for t is specified, it is implied that the curve is traversed only once on that interval. If desired, this fact can be verified using the procedures described in chapter 8.

$$\begin{aligned} L &= \int_c^d \sqrt{\left(\frac{dx}{dt}\right)^2 + \left(\frac{dy}{dt}\right)^2}\, dt \\ &= \int_{-1}^{2} \sqrt{(\sqrt{6t+9})^2 + t^2}\, dt \\ &= \int_{-1}^{2} \sqrt{6t+9+t^2}\, dt \\ &= \int_{-1}^{2} \sqrt{(t+3)^2}\, dt \\ &= \int_{-1}^{2} (t+3)\, dt \\ &= \left(\tfrac{1}{2}t^2 + 3t\right)\Big|_{-1}^{2} \\ &= 8 - \left(-\tfrac{5}{2}\right) = \tfrac{21}{2} \text{ units.} \end{aligned}$$

The quantity $\sqrt{(t+3)^2}$ can be simplified to $t+3$ because $t+3$ is nonnegative on the interval $[-1, 2]$. When extracting a square root of a perfect square, the sign of the simplified quantity should always be checked.

The length of the curve is $\frac{21}{2}$ units. ■

In example 2 the quantity inside the square root was a perfect square and therefore the integrand could be simplified. In general, the same integration strategies apply for parametric arc length exercises as for Cartesian arc length exercises.

Surface areas using parametric curves

Suppose the parametric curve $x = f(t)$, $y = g(t)$ from $t = c$ to $t = d$ is rotated about the x-axis. How do we find the area of the resulting surface? Once again, we begin with the Cartesian formula, which for surface area is

$$SA = \int_a^b 2\pi f(x)\sqrt{1 + (f'(x))^2}\, dx,$$

where $f(x)$ represents the y-coordinate of the curve (from the average radius of an approximating frustum) and $\sqrt{1 + (f'(x))^2}\, dx$ is from the arc length (the slant height of the frustum). Having already derived the parametric arc length formula, we know how this part turns out. The y-coordinate of the parametric curve is given by $y = g(t)$. Making these changes results in

Ans. to reading exercise 7

$$L = \int_0^{12\pi} \sqrt{\cos^2 t + e^{2t}}\, dt$$

$$SA_{\text{parametric, } x\text{-axis}} = \int_c^d 2\pi y\sqrt{\left(\frac{dx}{dt}\right)^2 + \left(\frac{dy}{dt}\right)^2}\, dt.$$

Notice that this integral is with respect to the variable t. As a result, each of the quantities y, $\frac{dx}{dt}$, and $\frac{dy}{dt}$ must be expressed in terms of the variable t to complete the integration.

Just as with arc length, we require that $\frac{dx}{dt}$ and $\frac{dy}{dt}$ are both continuous, and that the curve is traversed only once.

PARAMETRIC SURFACE AREA FORMULA, HORIZONTAL AXIS OF ROTATION

If $\frac{dx}{dt}$ and $\frac{dy}{dt}$ are continuous on $[c, d]$, then the area of the surface obtained by rotating the parametric curve $x = f(t)$, $y = g(t)$ from $t = c$ to $t = d$ about the x-axis is

$$SA = \int_c^d 2\pi y\sqrt{\left(\frac{dx}{dt}\right)^2 + \left(\frac{dy}{dt}\right)^2}\, dt,$$

provided the curve is traversed only once.

This surface area formula is also remembered as

$$SA_{x\text{-axis}} = \int 2\pi y\, ds,$$

where

$$ds = \sqrt{\left(\frac{dx}{dt}\right)^2 + \left(\frac{dy}{dt}\right)^2}\, dt$$

is the arc length differential.

A similar analysis applies to adapting the Cartesian surface area formula for rotating about the y-axis for use with a parametric curve.

PARAMETRIC SURFACE AREA FORMULA, VERTICAL AXIS OF ROTATION

If $\frac{dx}{dt}$ and $\frac{dy}{dt}$ are continuous on $[c, d]$, then the area of the surface obtained by rotating the parametric curve $x = f(t)$, $y = g(t)$ from $t = c$ to $t = d$ about the y-axis is

The formula

$$SA_{y\text{-axis}} = \int 2\pi x\, ds$$

applies here as well.

$$SA = \int_c^d 2\pi x \sqrt{\left(\frac{dx}{dt}\right)^2 + \left(\frac{dy}{dt}\right)^2}\, dt,$$

provided the curve is traversed only once.

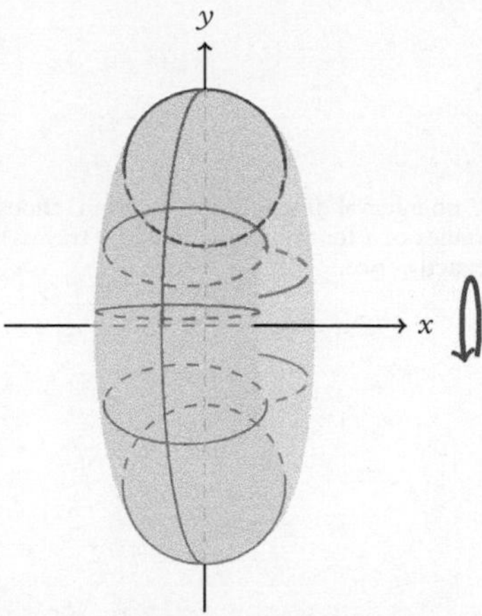

Figure 1 *The surface of example 3 (A green donut? I'll pass.)*

Example 3 *Rotate the curve $x = \sin t$, $y = 2 + \cos t$ for $0 \le t \le 2\pi$ about the x-axis and find the surface area.*

Solution The parametric curve $x = \sin t$, $y = 2 + \cos t$ is a circle of radius 1 centered at $(0, 2)$. Rotate the curve about the x-axis and the result resembles a donut (see figure 1).

To find the surface area, we begin by calculating the derivatives

$$\frac{dx}{dt} = \cos t, \qquad \frac{dy}{dt} = -\sin t.$$

Because $\frac{dx}{dt}$ and $\frac{dy}{dt}$ are continuous, we may use a parametric surface area formula. Because the rotation is about the x-axis, we have

$$\begin{aligned} SA &= \int_c^d 2\pi y \sqrt{\left(\frac{dx}{dt}\right)^2 + \left(\frac{dy}{dt}\right)^2}\, dt \\ &= \int_0^{2\pi} 2\pi(2 + \cos t)\sqrt{(\cos t)^2 + (-\sin t)^2}\, dt \\ &= \int_0^{2\pi} 2\pi(2 + \cos t)\sqrt{1}\, dt \\ &= 2\pi(2t + \sin t)\Big|_0^{2\pi} \\ &= 2\pi(4\pi + \sin 2\pi) - 2\pi(0 + \sin 0) \\ &= 8\pi^2 \text{ units}^2. \end{aligned}$$

In line 2, y is replaced with $2 + \cos t$, $\frac{dx}{dt}$ with $\cos t$, and $\frac{dy}{dt}$ with $-\sin t$. Line 3 uses $\cos^2 t + \sin^2 t = 1$, line 4 finds the antiderivative, line 5 evaluates, and line 6 simplifies.

The area of the surface, which should be proportional to the amount of glazing to put on the donut, is $8\pi^2$ units^2. ■

The name of the geometric shape (the surface) in this example is *torus*.

Reading Exercise 8 *The curve $x = 1 + 4t$, $y = \ln t$ between $t = 1$ and $t = 7$ is rotated about the x-axis. Set up the integral for finding the area of the resulting surface. Do not integrate.*

Exercises 9.4

1–8. (a) Set up the integral for determining the length of the parametric curve. (b) Use a calculator to determine the value of the integral to four significant digits.

If no interval of values for t is given, choose values of t for which the curve is traversed exactly once.

1. $x = \sin^{-1} t$, $y = e^{3t}$, $-\frac{1}{2} \le t \le \frac{1}{2}$
2. $x = \sqrt[3]{t+7}$, $y = \sin 3t$, $t = 0$ to $t = 2$
3. $x = \dfrac{4t-1}{5t^2+11}$, $y = t^2 \cosh t$, $t = 1$ to $t = 3$
4. $x = (3t + \sqrt{t})^4$, $y = te^t$, $t = 1$ to $t = 4$
5. $x = 3\cos t$, $y = 5\sin t$
6. $x = \sin 2t$, $y = \cos 6t$
7. $x = (3t+1)2^t$, $y = \sin(3t+1)$, $-2 \le t \le 2$
8. $x = \sin 2t$, $y = 2\cos 4t$ (a game of catch)

9–16. A parametric curve is rotated about an axis. (a) Set up the integral for determining the area of the resulting surface. (b) Use a calculator to determine the value of the integral to four significant digits.

9. $x = 4 - t^2$, $y = \cosh t$, $t = 0$ to $t = 2$, about the x-axis
10. $x = t^2 + 7$, $y = \ln t$, $t = 1$ to $t = 5$, about the y-axis
11. $x = 4 - t^2$, $y = \cosh t$, $t = 0$ to $t = 2$, about the y-axis
12. $x = t^2 + 7$, $y = \ln t$, $t = 1$ to $t = 5$, about the x-axis
13. $x = \sqrt{t+9}$, $y = \sqrt{t^3+1}$, $0 \le t \le 2$, about the x-axis
14. $x = t^3 + 5t$, $y = t^4 + t^2 + 1$, $t = -2$ to $t = 4$, about the x-axis
15. $x = \sqrt{t+9}$, $y = \sqrt{t^3+1}$, $0 \le t \le 2$, about the y-axis
16. $x = t^3 + 5t$, $y = t^4 + t^2 + 1$, $t = 0$ to $t = 4$, about the y-axis

17–26. Find the length of the parametric curve.

17. (Love) $x = 5t^2$, $y = 5t^3$, $t = 0$ to $t = \sqrt{5}$
18. $x = 3 + 5\cos t$, $y = -1 + 5\sin t$
19. $x = \sin^2 t$, $y = \cos^2 t$, $0 \le t \le \frac{\pi}{2}$
20. $x = 2t^6 - 1$, $y = 3t^4 + 5$, $0 \le t \le 2$
21. (Love) $x = \sin^3 t$, $y = \cos^3 t$, $t = \frac{\pi}{2}$ to $t = \pi$ (one arch of a four-cusped hypocycloid)

22. (GSL) $x = \cos t + t\sin t$, $y = \sin t - t\cos t$, $t = 0$ to $t = \sqrt{2}$

23. (GSL) $x = e^t \sin t$, $y = e^t \cos t$, $t = 0$ to $t = \frac{\pi}{2}$

24. $x = \cos t - \sin t$, $y = 1 + \cos t + \sin t$, $t = 0$ to $t = \pi$

25. (Love) $x = t - \sin t$, $y = 1 - \cos t$, $0 \le t \le 2\pi$ (one arch of a cycloid)

26. $x = 2\sqrt{1 + t^2}$, $y = \tan^{-1} t$, $-1 \le t \le 1$

Ans. to reading exercise 8

$$SA = \int_1^7 2\pi \ln t \sqrt{16 + \frac{1}{t^2}}\, dt$$

27–36. A parametric curve is rotated about an axis to generate a surface of revolution. Determine the area of the surface.

27. $x = 4$, $y = \ln t$, $1 \le t \le e^2$, about the x-axis

28. $x = 3 + 2t$, $y = 1 - 5t$, $0 \le t \le 1$, about the y-axis

29. (Love) $x = \tan^2 t$, $y = 2\tan t$, $t = 0$ to $t = \frac{\pi}{4}$, about the x-axis

30. $x = \frac{1}{9}(6t+9)^{\frac{3}{2}}$, $y = \frac{1}{2}t^2 + 18$, from $t = 0$ to $t = 1$, about the x-axis

31. (Love) $x = \cos^3 t$, $y = \sin^3 t$, from $t = 0$ to $t = \pi$, about the x-axis

Hint for exercise 31: use symmetry, integrating from $t = 0$ to $t =$ (something) and doubling that value. Compare to section 9.3 exercise 37.

32. $x = \cos^3 t$, $y = \sin^3 t$, from $t = 0$ to $t = \frac{1}{2}\pi$, about the y-axis

33. (GSL) $x = t - \sin t$, $y = 1 - \cos t$, $t = 0$ to $t = 2\pi$, about the x-axis

34. $x = \cos t - \sin t$, $y = 1 + \cos t + \sin t$, $t = 0$ to $t = \pi$, about the x-axis

35. (GSL) $x = e^t \sin t$, $y = e^t \cos t$, $0 \le t \le \frac{\pi}{2}$, about the y-axis

36. (GSL) $x = e^t \sin t$, $y = e^t \cos t$, $0 \le t \le \frac{\pi}{2}$, about the x-axis

37–38. Find the length of the parametric curve. Use a table of integrals as necessary.

37. $x = \tan^2 t$, $y = 2\tan t$, $t = -\frac{\pi}{4}$ to $t = \frac{\pi}{4}$

38. $x = 2\cot t$, $y = \csc^2 t$, $t = \frac{1}{2}\pi$ to $t = \frac{3}{4}\pi$

39–40. For each object, a parametric equation of a curve is given that can be rotated about an axis to generate the object. Determine the surface area. All letters other than x and y represent constants.

Compare to exercises 31 and 32 of section 9.3.

39. sphere; use $x = r\cos t$, $y = r\sin t$, $0 \le t \le \pi$, about the x-axis

40. torus; use $x = a\sin t$, $y = b + a\cos t$, $0 \le t \le 2\pi$, about the x-axis

41. Use the parametric arc length formula to derive the formula for the circumference of the circle $x = r\cos t$, $y = r\sin t$.

42. (GSL) Find the area of the surface generated by rotating $x = 2\cos t - \cos 2t$, $y = 2\sin t - \sin 2t$, $t = 0$ to $t = \pi$, about the x-axis. (This challenging problem can be completed by hand without using a table of integrals.)

43. Evaluate the integral of exercise 19 by eliminating the parameter and using the Cartesian arc length formula.

44. Evaluate the integral of exercise 8 by hand by eliminating the parameter and using the Cartesian arc length formula. A table of integrals may also come in handy.

Hydrostatic Pressure and Force

9.5

The deeper you dive, the greater pressure you feel from the water surrounding you. Storage tanks must be built strong enough to hold the fluid they are meant to contain. If the walls of the tank cannot withstand the force, they may fail. In this section we study the calculus of fluid pressure and force.

Fluid pressure

Fluid pressure, also called *hydrostatic pressure*, is proportional to depth. When you are swimming under water in a pool, it is as if the weight of all the fluid above you is pressing on you equally from all directions. The deeper you are, the more water there is above you, so you feel more pressure. Weight, of course, involves gravity. The density of the fluid also plays a role; there is more pressure at the same depth in salt water than there is in fresh water.

Put your swimming pool on the moon (inside an enclosure of some type, of course) and you will not feel as much pressure as you would on earth at the same depth.

The two factors of gravity and density of the fluid can be combined to form the *weight density* of a fluid:

The density of a fluid can also vary according to its temperature, but the effect is small enough to ignore in this discussion.

$$\text{weight density} = (\text{density of the fluid}) \cdot (\text{acceleration from gravity}).$$

For instance, the density of water is $1000\,\text{kg/m}^3$. Using $9.8\,\text{m/s}^2$ as the acceleration resulting from gravity, we have

$$\begin{aligned}\text{weight density of water} &= 1000\,\text{kg/m}^3 \cdot 9.8\,\text{m/s}^2 = 9800\,\text{kg}\cdot\text{m}/(\text{m}^3\cdot\text{s}^2)\\ &= 9800\ \text{N/m}^3.\end{aligned}$$

In the English system, the weight density of water is $62.4\,\text{lb/ft}^3$. Weight densities of several fluids are given in table 1.

Table 1 *Weight densities of fluids*

fluid	weight density SI (N/m^3)	weight density Eng. (lb/ft^3)
water	9800	62.4
salt water	10 030	63.9
diesel	8580	54.6
kerosene	7900	50.0
olive oil	8950	57.0
ethyl alcohol	7740	49.3
milk	10 130	64.5

Eng. = English system; SI = International System of Units.

Having combined density with gravity allows us to write pressure as the product of weight density and depth:

$$\text{pressure} = \text{weight density} \cdot \text{depth}$$

or, symbolically,

$$P = \delta \cdot d,$$

where P is pressure, δ is the weight density of the fluid, and d is the depth at which we are determining the pressure. Typical units in the International System of Units (SI) system are therefore

$$(\mathrm{N/m^3}) \cdot \mathrm{m} = \mathrm{N/m^2} = \mathrm{Pa},$$

where Pa is the symbol for the unit of measure *pascal*, named in honor of Blaise Pascal. The form of the unit of pressure is weight (newton) per area (square meter). Similarly, typical units for pressure in the English system are pounds per square foot or pounds per square inch, which often goes by the abbreviation psi (pounds per square inch).

Blaise Pascal 1623–1662
http://www-history.mcs.st-andrews.ac.uk/Biographies/Pascal.html

Example 1 *A plastic drinking cup is* 11.4 *cm tall. If the cup is full of water, what is the pressure at the bottom of the cup?*

Notice that what is not given, and what we do not need to know, is the shape of the cup or its horizontal dimensions. We only need to know the vertical dimension—that is, the depth of the water in the cup. Nor do we need to know the volume or weight of water in the cup, just the weight density of the water.

Solution To use the formula $P = \delta \cdot d$, we need to know the weight density of water (the fluid in the cup), which according to table 1 is $\delta = 9800\,\mathrm{N/m^3}$. We are given the depth of the water (from the top surface of the fluid to the location where we want to know the pressure) as $d = 11.4$ cm. Noticing the differing units centimeters and meters, we need to do a quick conversion of units: $d = 0.114$ m. Then,

Details: $d = 11.4\,\cancel{\mathrm{cm}} \cdot \frac{1\,\mathrm{m}}{100\,\cancel{\mathrm{cm}}} = \frac{11.4}{100}$ m.

$$P = \delta \cdot d = 9800\,\mathrm{N/m^3} \cdot 0.114\,\mathrm{m} = 1117\,\mathrm{N/m^2} = 1117\,\mathrm{Pa}.$$

The pressure on the bottom of the cup is 1117 Pa. ■

As a point of reference, 1117 Pa = 0.1620 psi.

Reading Exercise 9 *Find the fluid pressure at the bottom of a fresh water aquarium if the depth of the water is* 18 *inches.*

Fluid force

The form of the units for pressure gives a clue regarding how we can calculate fluid force from fluid pressure. Because pressure is measured in units of force divided by area, it must be the case that force is pressure times area:

Fluid force is also known as *hydrostatic force*.

$$\text{fluid force} = \text{pressure} \cdot \text{area}$$

or, symbolically,

$$F = P \cdot A = \delta \cdot d \cdot A.$$

As always, units of force include newtons (SI) and pounds (English).

Example 2 *A plastic drinking cup is* 11.4 *cm tall with a circular bottom of diameter* 6.0 *cm. If the cup is full of water, what is the force of the water on the bottom of the cup?*

Solution All along the bottom of the cup, the pressure is the same because it is at a constant depth. We calculated this pressure in example 2, with the result $P = 1117\,\mathrm{N/m^2}$. The area of the bottom of the cup is not difficult to find using the formula for the area of a circle: $A = \pi r^2$. The diameter of the cup is 6 cm, so the radius is

This time we are given the horizontal dimensions of the bottom of the cup, but we are still not given other information about the cup's size or shape. For instance, we do not know whether the cup is wider at the top than at the bottom. We do not have enough information to determine the volume of water in the cup or the weight of the water in the cup.

$$r = 3\,\mathrm{cm} = 0.03\,\mathrm{m}.$$

The area of the bottom of the cup is then

$$A = \pi r^2 = \pi \cdot (0.03\,\mathrm{m})^2 = 0.0009\pi\,\mathrm{m}^2 = 0.002827\,\mathrm{m}^2.$$

The force on the bottom of the cup is therefore

$$F = P \cdot A = 1117\,\mathrm{N/m^2} \cdot 0.002827\,\mathrm{m}^2 = 3.158\,\mathrm{N}. \qquad \blacksquare$$

CAUTION: MISTAKE TO AVOID
If the units are not kept consistent—for instance, if we have square meters in the denominator but square centimeters in the numerator—the units do not cancel as shown here.

Note that we did not calculate the weight of water in the cup. The fluid force on the bottom of the cup is not necessarily identical to the weight of water in the cup.

Fluid force on a vertical plate

Consider a tank of water as in figure 1. At the top of the water, the depth and pressure are both zero; but, as we move deeper into the water, the depth and the pressure increase. In example 2 we found the fluid force on the bottom of a cup by multiplying pressure and area, which worked because the pressure did not vary along the bottom of the cup. However, what if we want to find the force exerted by the fluid on the side of the tank as pictured in figure 1? Along the side of the tank, the depth and the pressure vary. We cannot simply multiply the pressure by the area of the side of the tank, because the pressure varies; pressure is not the same all along the side of the tank.

Ans. to reading exercise 9
pressure $= 62.4\,\mathrm{lb/ft^3} \cdot 1.5\,\mathrm{ft} = 93.6\,\mathrm{lb/ft^2}$

We have faced a similar scenario before! When finding the amount of work performed in pumping fluid from a tank, the distance the fluid needed to be moved varied. We solved the problem by using infinitesimally thin slices (see figure 2), oriented such that the vertical dimension of the slice is infinitesimal. We can take the same approach

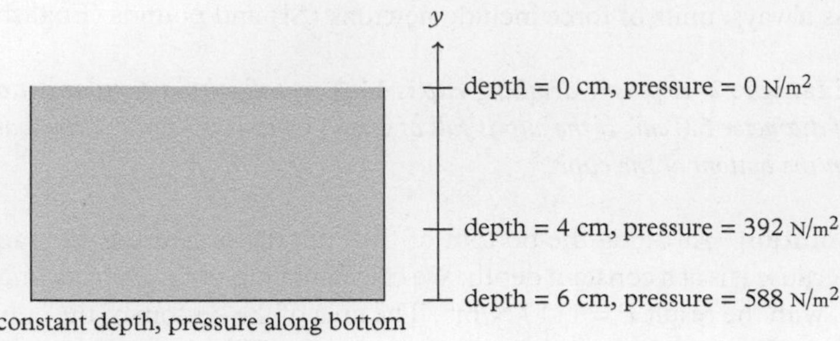

Figure 1 *A tank containing water. The fluid pressure of the water anywhere on the bottom of the tank is 588 N/m^2, but the pressure along the side of the tank varies according to depth*

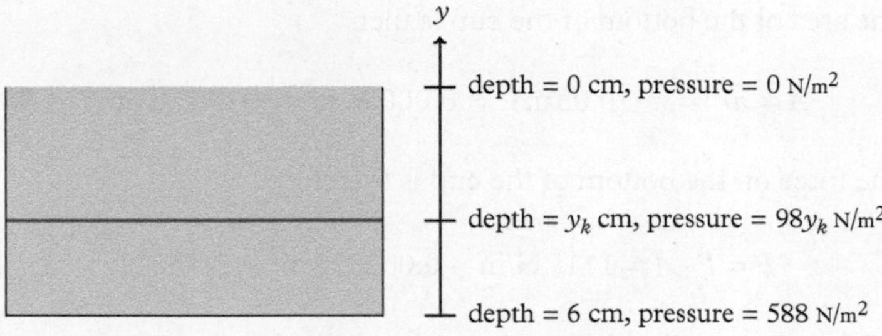

Figure 2 *A tank of water (light blue) and an infinitesimally thin horizontal slice (darker blue). The vertical dimension of the slice is* Δy

here, reasoning that on an infinitesimally thin slice, the depth only varies infinitesimally and the pressure is (approximately) constant on this slice.

The steps involved in finding the force that a fluid exerts on a vertical object (such as a tank wall) are illustrated in the next example.

Just as with pumping fluid from a tank, each situation may involve a different shape and therefore we approach each such problem individually, slicing and forming an omega sum and rewriting as an integral.

Example 3 *A truck hauls milk in a horizontal right circular cylindrical tank with a diameter of 6 feet. How much force does the milk exert on each end of the tank when the tank is half full?*

Solution ❶ Draw a diagram. The end of the tank is drawn in figure 3, including its shape and the location of the fluid. (For clarification of "right circular cylindrical," see figure 4.)

❷ Place a vertical number line (or perhaps a coordinate system) in the diagram. Because the equation of a circle is simplest when the origin is placed at the center of the circle, we make that choice. The vertical number line is marked accordingly.

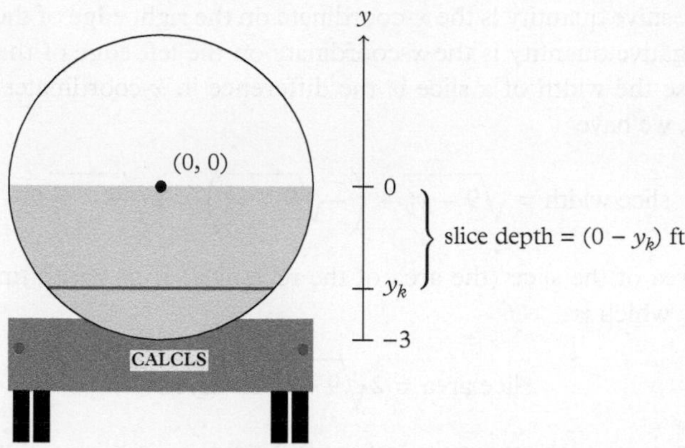

Figure 3 *An end view (rear view) of the milk truck of example 3, with the coordinate system introduced. The tank is half full of milk (light blue). A typical slice (dark blue) is drawn at a y-coordinate of y_k. The depth of a slice is the difference between the y-coordinate at the top of the fluid (0) and the y-coordinate representing the slice, y_k*

Figure 4 *The tank containing milk. "Right circular cylindrical" means a cylinder with cross sections that are circular and for which the axis through the center of these circles is perpendicular ("right") to the plane of each of these circles*

❸ Place a typical slice in the diagram, at y-coordinate y_k.

❹ Calculate the slice pressure. The weight density of milk is 64.5 lb/ft^3 and the depth of the slice (see figure 3) is $(0 - y_k)$ ft $= -y_k$ ft. Therefore,

$$\text{slice pressure} = 64.5\,\text{lb/ft}^3 \cdot (-y_k)\,\text{ft} = -64.5 y_k\,\text{lb/ft}^2.$$

❺ Calculate the slice area. Although the left and right edges of the slice illustrated in figure 3 are not vertical (they follow the curve of the circle), because the slice is infinitesimally thin, we can approximate the area of the slice by using a rectangle. The height of a slice is always Δy (see figure 5). The width of a slice is the distance between the right- and left-hand endpoints of the slice, which is the difference in their x-coordinates. This amount differs based on the location of the slice. In figure 3, it is clear that the slices are wider toward the top of the fluid and thinner toward the bottom. However, we have a relationship between x and y, given by the equation of the circle:

$$x^2 + y^2 = 3^2 = 9.$$

We can solve this equation for x to find the x-coordinates on the circle for a given y-coordinate:

$$x = \pm\sqrt{9 - y^2}.$$

Δy

width = right − left

Figure 5 *An approximating slice. Its area is its width times its height. Because we always slice using the same orientation, the height of a slice is always Δy*

Calculating the width of a slice is sometimes (but not always) the most challenging task of working a fluid force problem.

If we set up our coordinate system differently, then the equation of the circle is different. The choice of coordinate system has consequences!

The positive quantity is the x-coordinate on the right edge of the slice; the negative quantity is the x-coordinate on the left edge of the slice. Because the width of a slice is the difference in x-coordinates when $y = y_k$, we have

$$\text{slice width} = \sqrt{9 - y_k^2} - \left(-\sqrt{9 - y_k^2}\right) = 2\sqrt{9 - y_k^2}\,\text{ft}.$$

The area of the slice (the area of the rectangle) is its width times its height, which is

$$\text{slice area} = 2\sqrt{9 - y_k^2}\,\text{ft} \cdot \Delta y\,\text{ft}.$$

Carrying the units all the way through step ❻ is a good method for ensuring you have not skipped a step or performed it incorrectly. If your units are something other than units of force at the end of this step, then you have erred.

❻ Calculate the slice force. Since $F = P \cdot A$, the force on a slice is

$$\text{slice force} = -64.5y_k\,\text{lb/ft}^2 \cdot 2\sqrt{9 - y_k^2}\,\Delta y\,\text{ft}^2 = -129y_k\sqrt{9 - y_k^2}\,\Delta y\,\text{lb}.$$

❼ Add the slices to get an omega sum. The total force on the side of the tank is

We always use Ω slices, where Ω is a positive infinite hyperreal integer.

$$\text{total force} = \sum_{k=1}^{\Omega} -129y_k\sqrt{9 - y_k^2}\,\Delta y\,.$$

❽ Rewrite as an integral and evaluate. The milk in the tank extends from $y = -3$ to $y = 0$ (see figure 3). As always, we rewrite the omega sum as a definite integral by replacing the summation symbol with the integral symbol, dropping the subscripts, replacing Δy with dy, and inserting the limits of integration:

$$\text{force} = \int_{-3}^{0} -129y\sqrt{9 - y^2}\,dy.$$

Making the substitution

$$u = 9 - y^2$$
$$du = -2y\,dy$$

results in

Line 1 rewrites 129 as $64.5 \cdot 2$ and then makes the substitution, including replacing values of y with values of u; line 2 is the antiderivative step; and line 3 finishes the evaluation.

$$\begin{aligned} \int_{-3}^{0} 64.5(-2y)\sqrt{9 - y^2}\,dy &= \int_{0}^{9} 64.5u^{1/2}\,du \\ &= 64.5\frac{u^{3/2}}{3/2}\bigg|_0^9 \\ &= \cdots = 1161\,\text{lb}. \end{aligned}$$

The force of the milk on each end of the tank is 1161 lb. ■

If the top of the milk had been at any level other than the middle of the tank, then the resulting integral would have required trig substitution.

Reading Exercise 10 *Set up the integral for finding the fluid force given that the slice pressure is* $62.4(3 - y_k)\,\text{lb}/\text{ft}^2$, *the slice area is* $4\,\Delta y\,\text{ft}^2$, *and the water is from* $y = 0$ *to* $y = 3$.

Let's look at one more example.

Example 4 *A gutter is full of rainwater. What is the fluid force on the end of the gutter if the dimensions are as shown in figure 6?*

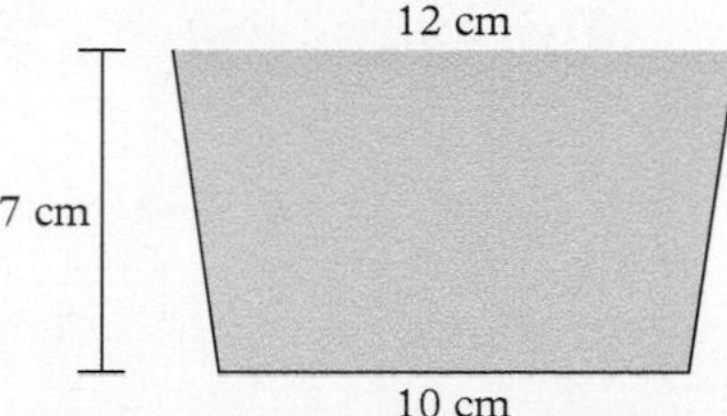

Figure 6 *The gutter full of rainwater for example 4*

Solution ❶ Draw a diagram. This step is already provided in figure 6, although we need to relabel lengths using meters.

❷ Place a vertical number line (or perhaps a coordinate system) in the diagram. This time we mark the bottom of the gutter as the y-coordinate 0 and the top as the y-coordinate 0.07, as measured in meters.

❸ Place a typical slice in the diagram, at y-coordinate y_k. See figure 7 for the results of steps ❷ and ❸.

Tip: if English units are used, always label the diagram in terms of feet. If SI units are used, always label the diagram in terms of meters. Although other choices are possible, this recommendation is made for the purpose of minimizing unit conversion errors.

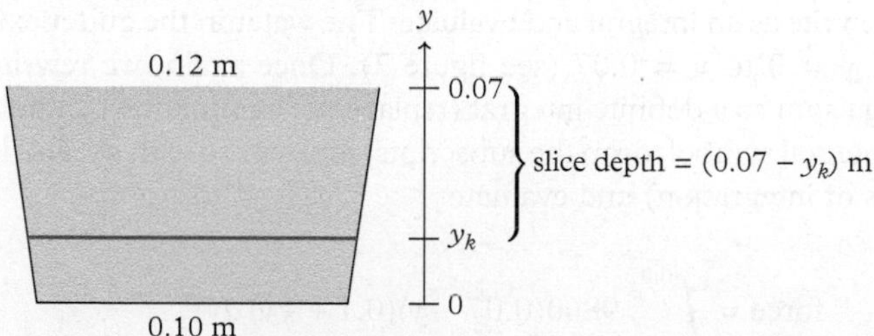

Figure 7 *A vertical scale and a typical slice are added to the diagram. Other choices can be made for the vertical scale, although this changes the details of the solution (but it does not change the final result). For instance, we could label the top of the gutter* $y = 0$ *and the bottom* $y = -0.07$, *but then the depth of the slice would be* $(0 - y_k)$ m

❹ Calculate the slice pressure. The weight density of water is $9800\,\text{N}/\text{m}^3$ and the depth of the slice (see figure 7) is $(0.07 - y_k)$ m. Therefore

$$\text{slice pressure} = 9800\,\text{N}/\text{m}^3 \cdot (0.07 - y_k)\,\text{m} = 9800(0.07 - y_k)\,\text{N}/\text{m}^2.$$

❺ Calculate the slice area. As always, the slice height is Δy, and this time it is measured in meters. To calculate the slice width, notice that the width is 0.1 m at $y = 0$. The gutter gets wider at a rate of 0.02 m of width for every 0.07 m of height, or a rate of $\frac{0.02}{0.07} = \frac{2}{7}$. (This analysis works only because the sides of the gutter are straight lines, so that the gain in width is linear in relation to the y-coordinate.) Therefore,

$$\text{slice width} = \left(0.1 + \tfrac{2}{7}y_k\right)\text{ m}$$

Check: if $y_k = 0$, then width $= \left(0.1 + \frac{2}{7}(0)\right)$ m $= 0.1$ m. If $y_k = 0.07$, then width $= \left(0.1 + \frac{2}{7}(0.07)\right)$ m $= 0.12$ m.

and

$$\text{slice area} = \left(0.1 + \tfrac{2}{7}y_k\right)\text{ m}\cdot\Delta y\,\text{m}.$$

❻ Calculate the slice force. Since $F = P \cdot A$, the force on a slice is

$$\begin{aligned}\text{slice force} &= 9800\,(0.07 - y_k)\text{ N/m}^2 \cdot \left(0.1 + \tfrac{2}{7}y_k\right)\Delta y\,\text{m}^2\\ &= 9800(0.07 - y_k)\left(0.1 + \tfrac{2}{7}y_k\right)\Delta y\,\text{N}.\end{aligned}$$

❼ Add the slices to get an omega sum. The total force on the end of the gutter is

$$\text{total force} = \sum_{k=1}^{\Omega} 9800(0.07 - y_k)\left(0.1 + \tfrac{2}{7}y_k\right)\Delta y.$$

❽ Rewrite as an integral and evaluate. The water in the gutter extends from $y = 0$ to $y = 0.07$ (see figure 7). Once again, we rewrite the omega sum as a definite integral (replace the summation symbol with the integral symbol, drop the subscripts, replace Δy with dy, and insert limits of integration) and evaluate:

Ans. to reading exercise 10

$$\int_0^3 62.4(3 - y)\cdot 4\,dy$$

$$\begin{aligned}\text{force} &= \int_0^{0.07} 9800(0.07 - y)(0.1 + \tfrac{2}{7}y)\,dy\\ &= 9800\int_0^{0.07}\left(0.007 - 0.08y - \tfrac{2}{7}y^2\right)dy\\ &= 9800\left(0.007y - 0.04y^2 - \tfrac{2}{21}y^3\right)\Big|_0^{0.07}\\ &= 9800(0.00049 - 0.000196 - 0.0000327) - 0\\ &= 2.561\text{ N}.\end{aligned}$$

If the vertical scale is chosen to have $y = 0$ at the top of the gutter and $y = -0.07$ at the bottom, then the integral is

$$\int_{-0.07}^{0} 9800(-y)(0.12 + \tfrac{2}{7}y)\,dy.$$

The value of the integral is still 2.561 N.

The force of the rainwater on the end of the gutter is 2.561 N. ■

Reading Exercise 11 *Complete step ❺ (find the area of the slice) as depicted in figure 8.*

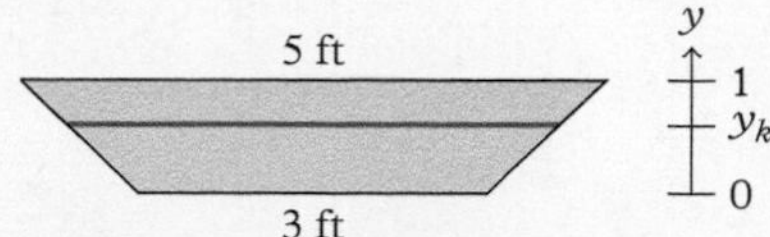

Figure 8 *The figure for reading exercise 11*

Compare the solutions to examples 3 and 4 and you should find that the main difference is in the calculation of the slice width. A good diagram, including all necessary coordinates and dimensions, is one of the keys to successfully calculating the slice width.

Exercises 9.5

1–6. Rapid response: determine an expression for the depth of the slice depicted in the diagram. All scales are measured in meters.

1. 4.

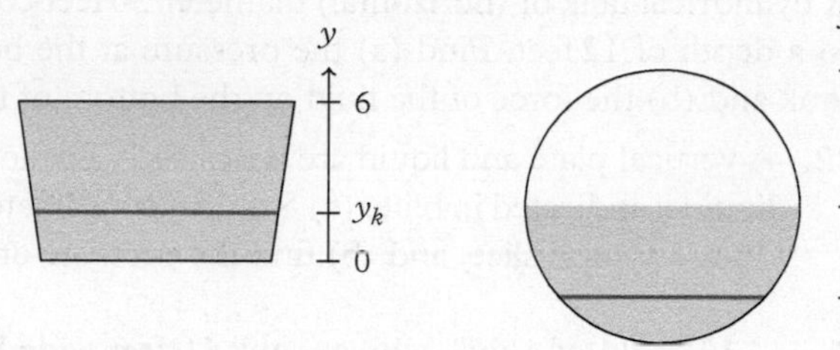

2. 5.

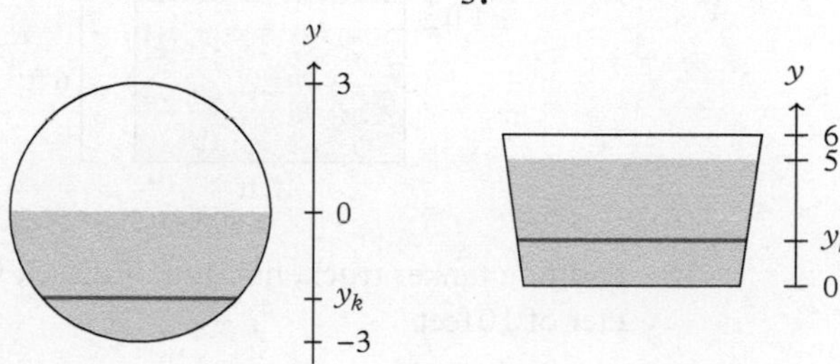

3. 6.

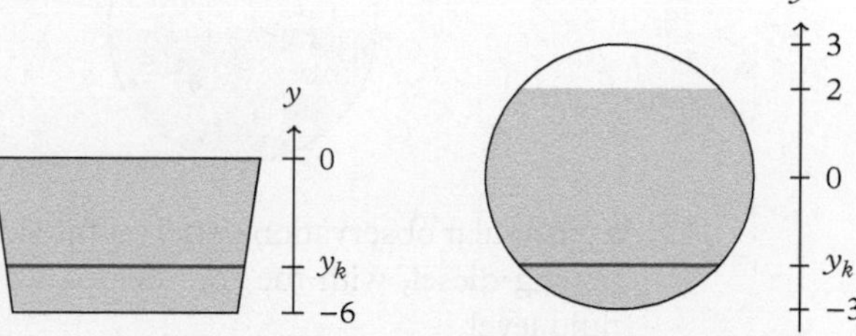

7–12. Using the diagram from the indicated exercise and the given fluid, determine the hydrostatic pressure at the slice. Use SI units.

7. exercise 1, saltwater
8. exercise 2, olive oil
9. exercise 3, kerosene
10. exercise 4, diesel
11. exercise 5, ethyl alcohol
12. exercise 6, water

13. A swimming pool is 20 feet wide and 40 feet long, and the fresh water is 5 feet deep. (a) What is the pressure at the bottom of the pool? (b) What is the force of the water on the bottom of the pool?

14. A freshwater aquarium is 3 m long and 1 m wide, with water 0.7 m deep. (a) What is the pressure at the bottom of the aquarium? (b) What is the force of the water on the bottom of the aquarium?

15. An ocean-faring submarine has a circular hatch of diameter 1.8 m. The hatch is oriented horizontally, and it is at a depth of 600 m. Find (a) the pressure at that depth and (b) the force of the fluid on the hatch.

Ans. to reading exercise 11
$(3 + 2y_k)\Delta y\ \text{ft}^2$

16. A cylindrical tank of (horizontal) diameter 30 feet contains diesel to a depth of 12 feet. Find (a) the pressure at the bottom of the tank and (b) the force of the fluid on the bottom of the tank.

17–22. A vertical plate and liquid are described. The location of the liquid is indicated in blue. (a) Set up a coordinate system and draw a typical slice, and (b) find the pressure on that slice.

17. end of a milk storage tank 10 feet wide by 6 feet tall, filled to within 1 foot of the top

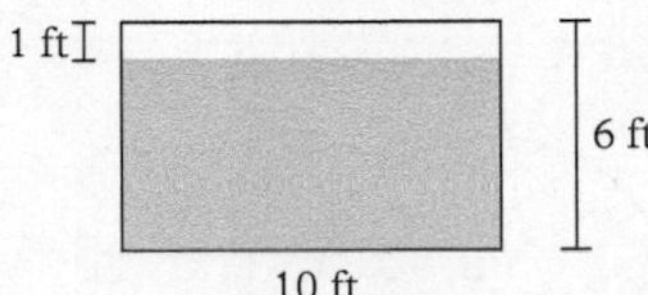

18. end of a tanker truck, half full of diesel, with a diameter of 10 feet

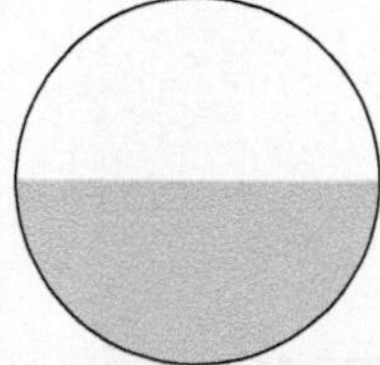

19. a triangular observation panel on the side of a vat containing diesel, with the top located 20 cm below the fluid level

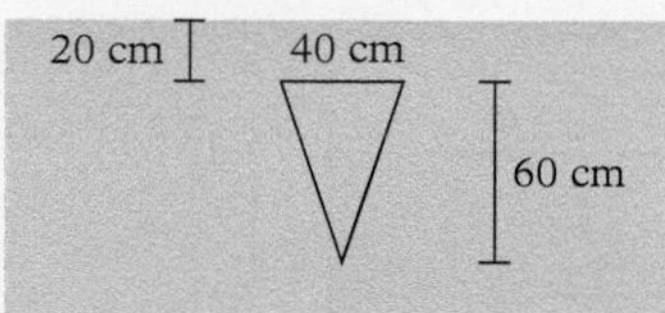

20. a vertical plate submerged in saltwater, with the top of the plate 1 m below the surface

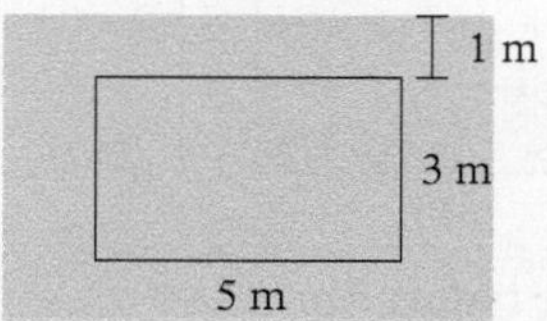

The vertical plate is rectangular; its outline is in the diagram.

21. a rectangular release gate in a dam. The dimensions of the gate are 15 m wide and 2.5 m tall. The top of the gate is 5 m below the top of dam, and the water level is at the top of the release gate.

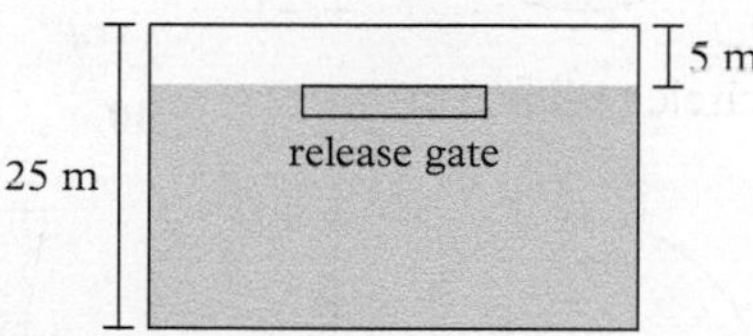

22. a circular valve gate in a vat of olive oil. The gate diameter is 2 feet and the bottom of the valve gate is 1 foot above the bottom of the vat.

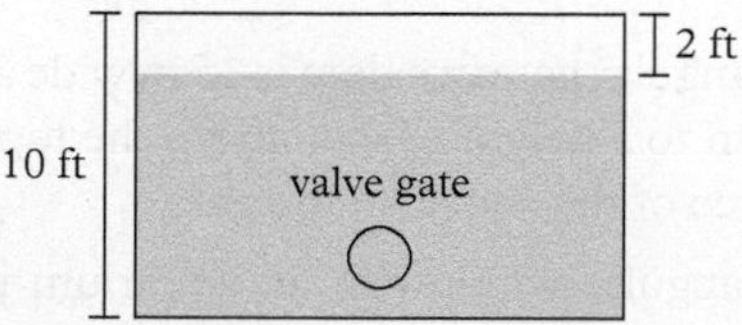

23–30. Using the coordinate system in the diagram and other information if provided, determine the area of the slice located at y-coordinate y_k.

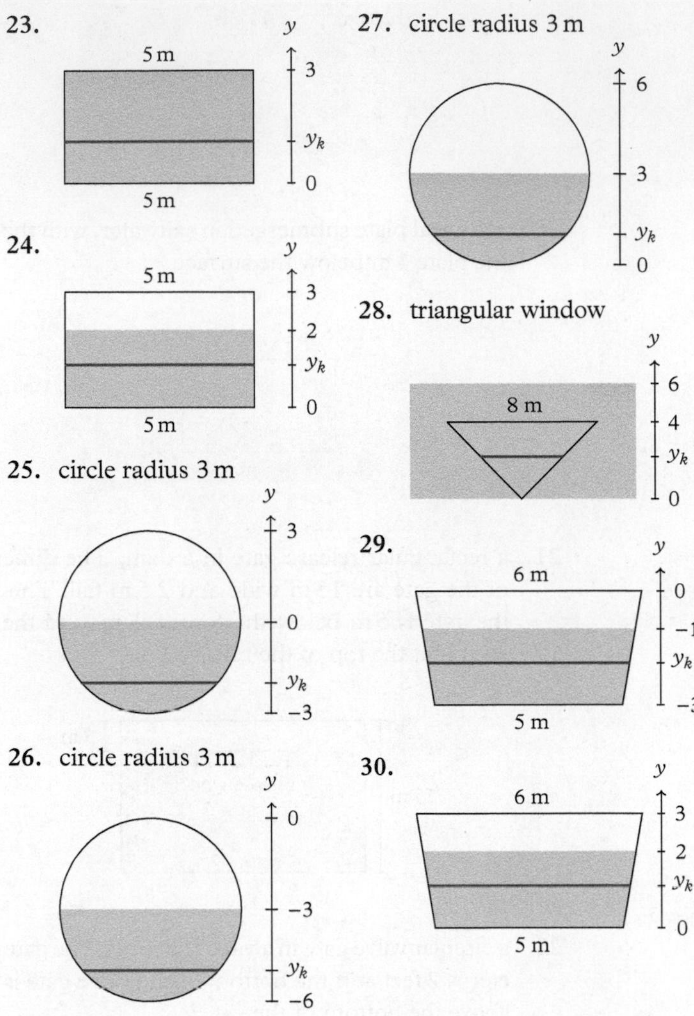

31. A rectangular gate in a dam is 15 m wide and 2.5 m tall. Water fills the dam to a height of 4 m above the top of the gate. Determine the force of the water on the gate.

32. A rectangular window in an aquarium is 5 feet wide and 3 feet tall. The aquarium is filled with water to a height of 1 foot above the top of the window. Determine the hydrostatic force on the window.

33. A circular gate in a dam has a diameter of 4 m. Water fills the dam to the level of the middle of the circular gate. Determine the hydrostatic force on the gate.

34. A circular window in a polar bear exhibit has a diameter of 3 m. Water fills the exhibit to the middle of the window. Determine the force of the water on the window.

35. (GSL) Each end of a horizontal oil tank is an ellipse with a horizontal axis that is 12 feet long and a vertical axis that is 6 feet long. Calculate the fluid force on one end when the tank is half full of oil weighing 60 lb/ft^3.

36. The end of a vat of kerosene is shaped like the parabola $y = \frac{1}{3}x^2$ from $x = -3$ to $x = 3$, measured in meters. The vat is filled to a height of $\frac{4}{3}$ m. Find the hydrostatic force on the end of the vat.

37. The end of a container of olive oil is a trapezoid with the dimensions shown in the diagram in the margin. The container is filled to a height of 25 cm. Determine the fluid force on the end of the container.

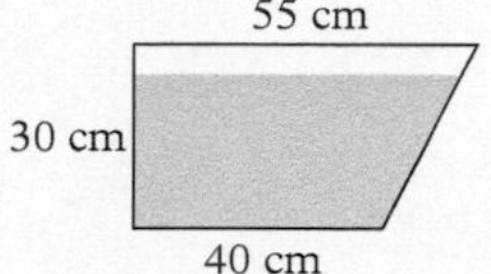

38. (Love) What force must be withstood by a trapezoidal dam 100 feet long at the top, 80 feet long at the bottom, and 20 feet deep? (Assume the fluid behind the dam is water.)

39. The end of a container of diesel has the shape of a rectangle with a semicircle added to each end, as shown in the following diagram. The container is half full. Determine the fluid force on the end of the container.

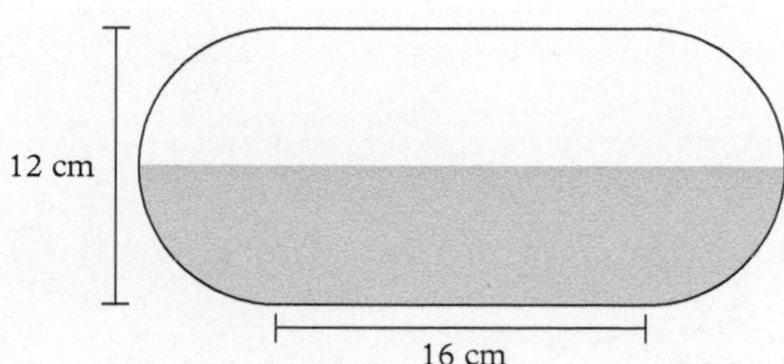

40. Repeat exercise 39 with all the dimensions doubled. Because this quadruples the area of the end of the container, is the fluid force also quadrupled?

41. Calculate the hydrostatic force on the end of the milk truck of example 3 if the tank is full of milk.

42. Calculate the force of water on the window of the polar bear exhibit of exercise 34 if the water is filled to the top of the window.

43. (GSL) A vertical cylindrical tank of diameter 30 feet and height 50 feet is full of water. Find the fluid force on the curved surface.

44. (GSL) The vertical end of a water trough is an isosceles triangle 5 feet across the top and 5 feet deep. Calculate the force on the end when the trough is full of water.

45. Rewrite the omega sum of step ❼ in example 4 by substituting appropriate expressions for Δy and y_k. Then, evaluate the sum.

Centers of Mass

9.6

If you ever tried to balance a pencil or some other object on a finger, then you have already tried to find an object's *center of mass*, or balancing point. You may have first attempted to find the balancing point of the object visually (aka graphically), and then tweaked your answer based upon which direction the object started to fall—an experimental approach. Perhaps you made repeated adjustments (an iterative process) until you found the perfect balancing point. In this section we investigate centers of mass using calculus.

Point-mass systems

If two children with the same mass stand carefully on the ends of a seesaw (teeter-totter) they can make it balance, as in figure 1. However,

As a child, I often stood on a seesaw, trying to balance with friends, but such actions are not recommended for the sake of safety.

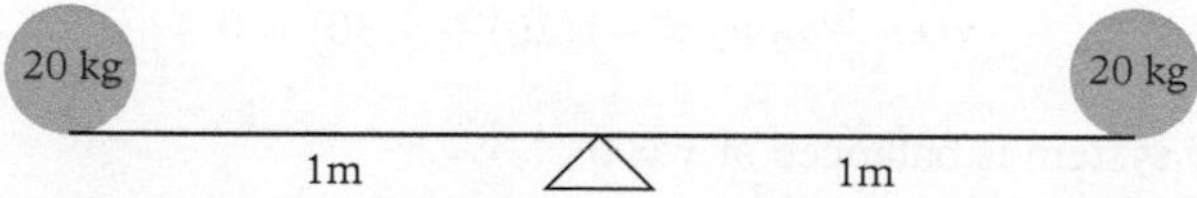

Figure 1 *Two well-rounded children on a seesaw. The children have the same mass and are the same distance from the fulcrum, so the system is balanced*

if one of the children is larger than the other and the children both stand on the ends of the seesaw, then the end with the larger child falls to the ground, while the end with the smaller child is lifted in the air (figure 2). What can the two children of different masses do

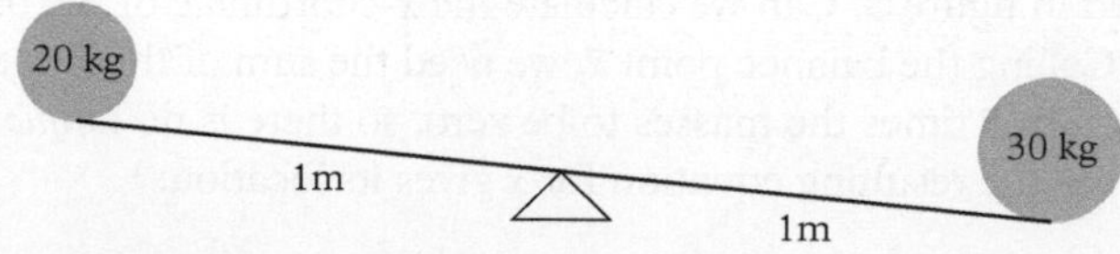

Figure 2 *An unbalanced system*

to make the seesaw balance? The larger child can stand closer to the fulcrum, as in figure 3. If the product of the mass and the distance is

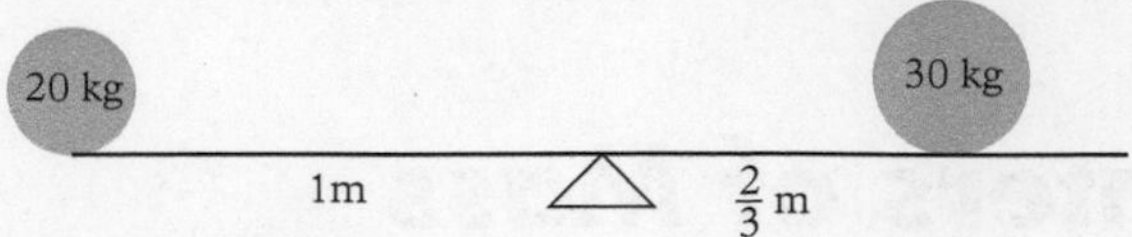

Figure 3 *A balanced system*

the same on both sides of the fulcrum, the system balances. In figure 3, the mass on the left side is $m_1 = 20\,\text{kg}$ at a distance of $d_1 = 1\,\text{m}$; the mass on the right side is $m_2 = 30\,\text{kg}$ at a distance of $d_2 = \frac{2}{3}\,\text{m}$. Then, $d_1 m_1 = 20\,\text{kg} \cdot \text{m} = d_2 m_2$ and the system is in balance.

Balancing on a seesaw is easier if the seesaw is a plank of wood from one end of the seesaw to the other. A seesaw consisting of a metal rod with a seat on each end is not conducive to standing closer to the fulcrum. In any case, I don't recommend trying this balancing act. Please follow all playground safety regulations.

The previous calculation seems to be predicated on the idea that each mass is situated at one spot on the seesaw (at one distance from the fulcrum). If you have ever tried balancing on a seesaw, recall how leaning forward or backward just a bit changes the balance. The horizontal position on the seesaw where your center of mass is located is what matters (vertical position does not matter). We consider each object to be a *point mass*.

Now place the point masses in a coordinate system, as in figure 4. We place mass m_1 at x-coordinate x_1 and mass m_2 at x-coordinate x_2; the system balances at $x = 0$ if $x_1 m_1 + x_2 m_2 = 0$. For instance, in figure 4 we have

$$x_1 m_1 + x_2 m_2 = -1(20) + \tfrac{2}{3}(30) = 0$$

and the system is balanced at $x = 0$.

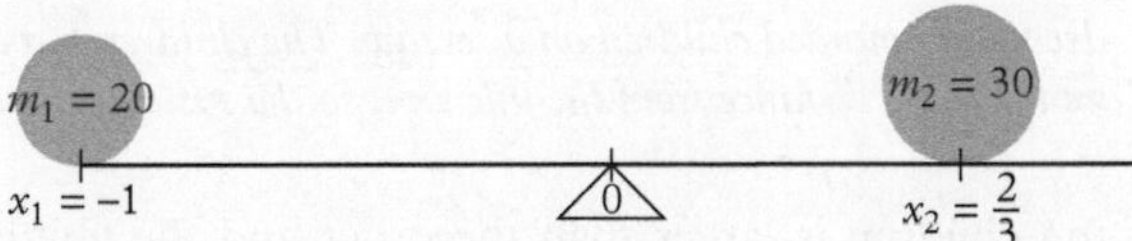

Figure 4 *Two point masses in a coordinate system, with a fulcrum at* $x = 0$

The symbol $\overline{x}$ is read "x bar."

Now try a more complicated system of several point masses, as illustrated in figure 5. Can we calculate the x-coordinate of the balance point? Calling the balance point $\overline{x}$, we need the sum of the signed distances from $\overline{x}$ times the masses to be zero, so there is no *torque* about $\overline{x}$. Solving the resulting equation for $\overline{x}$ gives its location:

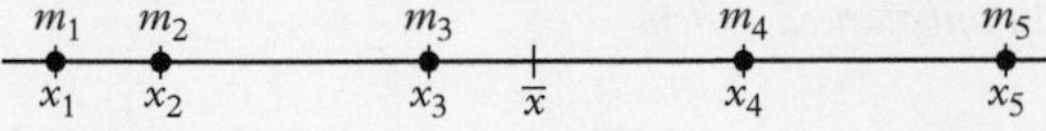

Figure 5 *Several point masses in a coordinate system, with balance point* $\overline{x}$

$$\sum_{k=1}^{n}(x_k - \overline{x})m_k = 0$$

$$\sum_{k=1}^{n}(x_k m_k - \overline{x}m_k) = 0$$

$$\sum_{k=1}^{n} x_k m_k - \sum_{k=1}^{n} \overline{x}m_k = 0$$

$$\sum_{k=1}^{n} x_k m_k - \overline{x}\sum_{k=1}^{n} m_k = 0$$

$$\sum_{k=1}^{n} x_k m_k = \overline{x}\sum_{k=1}^{n} m_k$$

$$\text{center of mass} = \overline{x} = \frac{\sum_{k=1}^{n} x_k m_k}{\sum_{k=1}^{n} m_k}.$$

CENTER OF MASS FOR A POINT-MASS SYSTEM

For a system of n objects of mass m_k located at x-coordinate x_k for $k = 1, \ldots, n$, the center of mass is located at the x-coordinate $\overline{x}$ where

$$\overline{x} = \frac{\sum_{k=1}^{n} x_k m_k}{\sum_{k=1}^{n} m_k}.$$

The quantity $\sum_{k=1}^{n} x_k m_k$ is called the *total moment about the origin* or the *moment of the system about the origin*, and the quantity $\sum_{k=1}^{n} m_k$ is called the *total mass* or the *system mass*.

The formula for the center of mass is identical to the formula for a *weighted average* in statistics. In fact, this connection to a center of mass is the source of the adjective *weighted* in the term *weighted average*.

The total moment about the origin is the magnitude of the tendency for the system to pivot about a fulcrum placed at the origin. To be balanced at $\overline{x}$, we want the moment about $\overline{x}$ to be zero—that is, $\sum_{k=1}^{n}(x_k - \overline{x})m_k = 0$.

Example 1 *Find the center of mass for the system given in figure 6.*

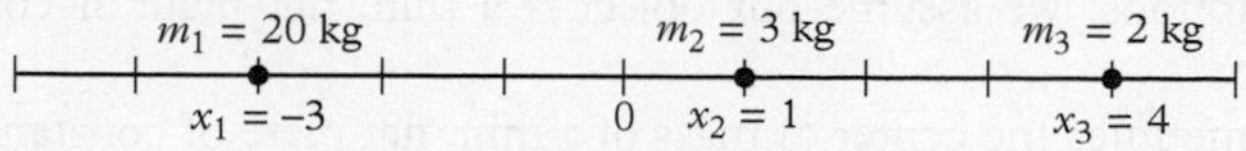

Figure 6 *The point-mass system of example 1. The number line is measured in meters*

Solution Using the formula for the center of mass for a point-mass system, we have

$$\begin{aligned}\overline{x} &= \frac{\sum_{k=1}^{n} x_k m_k}{\sum_{k=1}^{n} m_k}\\ &= \frac{-3(20) + 1(3) + 4(2)}{20 + 3 + 2}\\ &= \frac{-49}{25} = -1.96.\end{aligned}$$

The center of mass is at the x-coordinate -1.96. ■

The balancing point is so far left because the mass m_1 is much larger than the other masses.

Note that in example 1, the moment of the system about the origin is $-49\,\text{kg} \cdot \text{m}$ and the system mass is $25\,\text{kg}$.

Reading Exercise 12 *Find the center of mass for the system given in figure 7.*

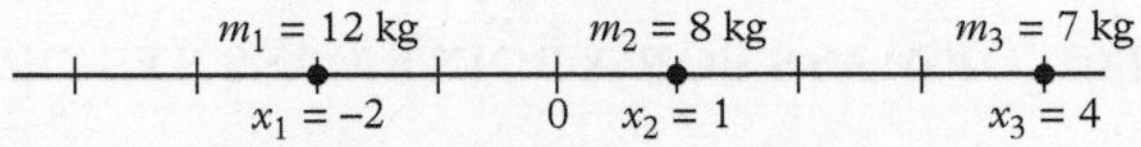

$m_1 = 12$ kg $m_2 = 8$ kg $m_3 = 7$ kg

$x_1 = -2$ 0 $x_2 = 1$ $x_3 = 4$

Figure 7 *The point-mass system of reading exercise 12. The number line is measured in meters*

Center of mass: thin flat plates of constant density

Figure 8 *A cutout yard decoration*

Perhaps you have seen a yard decoration cut from a sheet of plywood and painted, such as the snowman depicted in figure 8. If laid flat and placed on a cart to carry to storage, we need the horizontal center of mass to lie inside the boundaries of the cart. To find this balance point, we need to know two coordinates, not just one. Both the x- and y-coordinates of the center of mass are required.

If the plywood for the snowman cutout is thicker in some places than in others, this complicates the task of finding the center of mass. However, the thickness of plywood is relatively close to being a constant, and its composition is also fairly consistent. Let's make these assumptions: we assume our object is a thin, flat plate of constant density.

Sometimes the center of mass of a thin, flat plate of constant density is obvious. Consider a thin flat plate in the shape of a circle. Isn't the center of mass located at the center of the circle? Similarly, for a

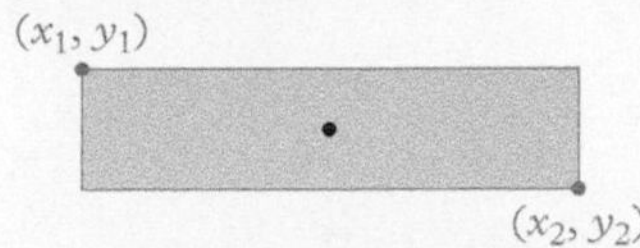

Figure 9 *A thin, flat plate of constant density in the shape of a rectangle. The center of mass is marked by a black dot. Geometrically, it is at the intersection of the diagonals of the rectangle*

rectangle, the center of mass is at the "center" of the rectangle, halfway between the left and right sides, and halfway between the top and bottom. The center of mass for the rectangle in figure 9 is located at the point

$$\left(\frac{x_1 + x_2}{2}, \frac{y_1 + y_2}{2}\right).$$

But what about the center of mass of a thin, flat plate in the shape of the region in figure 10? This is not so obvious. The y-coordinates extend from $y = 0$ to $y = 4$; however, with more of the plate below $y = 2$ than above it, the center of mass must be below $y = 2$. To determine the location of the center of mass we turn to calculus and once more begin by slicing the figure and approximating with regions with a center of mass that we do know how to find: rectangles.

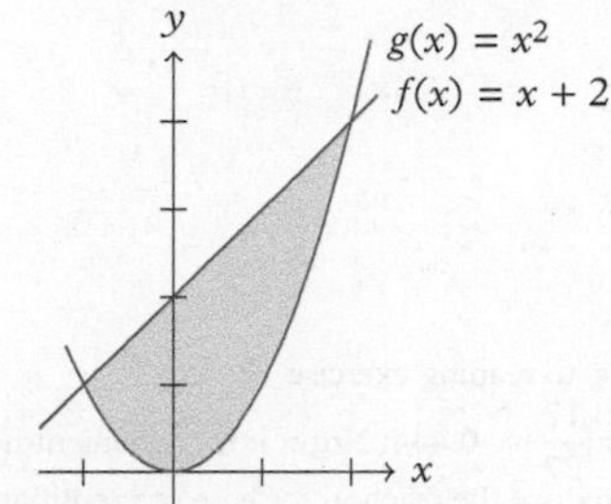

Figure 10 *The region between the curves $f(x) = x + 2$ and $g(x) = x^2$, from $x = -1$ to $x = 2$*

Slicing along the x-axis, we can approximate the region by rectangles in the manner of figure 11. Next, imagine each rectangle to be a

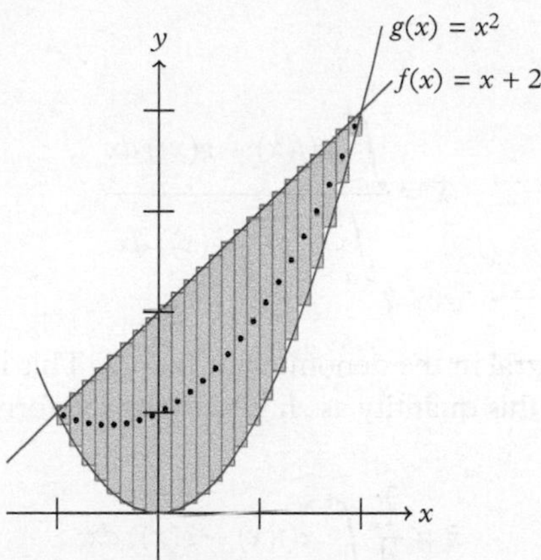

Figure 11 *The region between the curves $f(x) = x + 2$ and $g(x) = x^2$ approximated by rectangles. The center of mass of each rectangle is noted by a black dot*

child standing along a seesaw. We can approximate the x-coordinate of the center of mass for this region by using the formula for the center of mass for a point-mass system! All we need to know are appropriate values of x_k and m_k to use in this formula. For x_k, we use the right-hand endpoints, as always. Although we don't know the mass m_k of a slice, because the plate has constant density, the mass is proportional to the area of the rectangle. The area of a slice is the width Δx times the height, which is $f(x_k) - g(x_k)$, the difference in y-coordinates between the top and bottom curves. Therefore, we use

Technically, because the center of mass is the midpoint of the interval, we should use the midpoints instead of the right-hand endpoints. This detail is explored in exercise 26, as is the use of a proportionality constant in calculating the mass.

$$m_k = (f(x_k) - g(x_k))\Delta x.$$

Using $n = \Omega$ slices and applying the formula for the center of mass for a point-mass system, we have

$$\overline{x} = \frac{\sum_{k=1}^{n} x_k m_k}{\sum_{k=1}^{n} m_k} = \frac{\sum_{k=1}^{\Omega} x_k (f(x_k) - g(x_k))\Delta x}{\sum_{k=1}^{\Omega} (f(x_k) - g(x_k))\Delta x}.$$

Ans. to reading exercise 12

$\overline{x} = \dfrac{12}{27} \approx 0.444$. **Note:** it is convenient not to reduce the fraction for ease of identifying the moment of the system about the origin and the system mass.

Double the fun! That's two omega sums, one in the numerator and one in the denominator. Rewrite each omega sum as a definite integral (replace the summation symbol with the integral symbol, drop the subscripts, replace Δx with dx, and insert limits of integration) and we have

$$\overline{x} = \frac{\int_a^b x(f(x) - g(x))\,dx}{\int_a^b (f(x) - g(x))\,dx}.$$

We saw the integral in the denominator before. This is the area of the region. Writing this quantity as A, we rewrite the formula as

$$\overline{x} = \frac{1}{A}\int_a^b x(f(x) - g(x))\,dx.$$

We still need the y-coordinate of the center of mass of the region. Using the same approach and still slicing along the x-axis, the value of y_k to use is the y-coordinate of the center of mass of a slice. Because this is the average of the y-coordinates at the top and the bottom of the slice, we have

$$y_k = \frac{f(x_k) + g(x_k)}{2}.$$

Because the slices have not changed, we still use $m_k = (f(x_k) - g(x_k))\Delta x$. Applying the formula for the center of mass for a point-mass system, we then have

$$\overline{y} = \frac{\sum_{k=1}^{n} y_k m_k}{\sum_{k=1}^{n} m_k} = \frac{\sum_{k=1}^{\Omega} \frac{1}{2}(f(x_k) + g(x_k))(f(x_k) - g(x_k))\Delta x}{\sum_{k=1}^{\Omega} (f(x_k) - g(x_k))\Delta x}.$$

Once again rewriting as an integral and noticing that the denominator is the area of the region,

$$\overline{y} = \frac{\int_a^b \frac{1}{2}(f(x) + g(x))(f(x) - g(x))\,dx}{\int_a^b (f(x) - g(x))\,dx} = \frac{1}{A}\int_a^b \frac{1}{2}\left((f(x))^2 - (g(x))^2\right) dx.$$

CENTER OF MASS FOR A THIN, FLAT PLATE OF CONSTANT DENSITY

The center of mass of a thin, flat plate of constant density located between $y = f(x)$ (top curve) and $y = g(x)$ (bottom curve) from $x = a$ to $x = b$ is the point $(\overline{x}, \overline{y})$ where

$$\overline{x} = \frac{1}{A}\int_a^b x(f(x) - g(x))\,dx,$$

$$\overline{y} = \frac{1}{A}\int_a^b \frac{1}{2}\left((f(x))^2 - (g(x))^2\right) dx.$$

For a thin, flat plate of constant density, the center of mass $(\overline{x}, \overline{y})$ is the same point as what is called in geometry the *centroid* of the region.

Example 2 *Find the center of mass of a thin, flat plate of constant density covering the region enclosed by the parabolas* $y = x^2 - 3$ *and* $y = -2x^2$.

Solution ❶ Sketch the region. We begin by identifying where the curves intersect:

Sketching the region, including finding points of intersection, is identical to the process used when finding the area between the curves.

$$x^2 - 3 = -2x^2$$
$$3x^2 = 3$$
$$x = \pm 1$$

The region is sketched in figure 12.

❷ Find the area between the curves. We identify the top curve as $f(x) = -2x^2$ and the bottom curve as $g(x) = x^2 - 3$. Then,

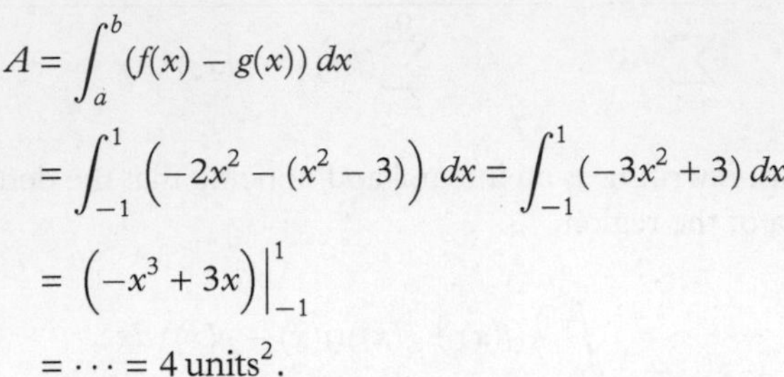
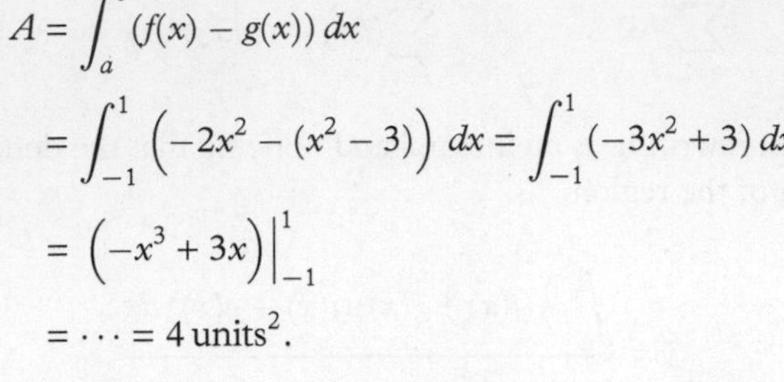

$$A = \int_a^b (f(x) - g(x))\,dx$$

$$= \int_{-1}^{1} \left(-2x^2 - (x^2 - 3)\right) dx = \int_{-1}^{1} (-3x^2 + 3)\,dx$$

$$= \left(-x^3 + 3x\right)\Big|_{-1}^{1}$$

$$= \cdots = 4\,\text{units}^2.$$

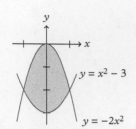

Figure 12 *The region enclosed by the curves* $y = -2x^2$ *and* $y = x^2 - 3$

❸ Determine $\overline{x}$. Using the same top and bottom curves, and the same limits of integration as for finding the area between the curves, along with the value of A found in step ❷, we use the formula for the center of mass for a thin, flat plate of constant density:

$$\overline{x} = \frac{1}{A}\int_a^b x(f(x) - g(x))\,dx$$

$$= \frac{1}{4}\int_{-1}^{1} x\left(-2x^2 - (x^2 - 3)\right) dx = \frac{1}{4}\int_{-1}^{1} \left(-3x^3 + 3x\right) dx$$

$$= \frac{1}{4}\left(\frac{-3x^4}{4} + \frac{3x^2}{2}\right)\Bigg|_{-1}^{1}$$

$$= \cdots = 0.$$

A quick look at the region's symmetry about the y-axis should convince you that $\overline{x}$ must be zero.

❹ Determine $\overline{y}$. Continuing the use of the formula for the center of mass for a thin, flat plate of constant density,

$$\overline{y} = \frac{1}{A}\int_a^b \frac{1}{2}\left((f(x))^2 - (g(x))^2\right) dx$$

$$= \frac{1}{4}\int_{-1}^{1} \frac{1}{2}\left((-2x^2)^2 - (x^2 - 3)^2\right) dx$$

$$= \frac{1}{8}\int_{-1}^{1} \left(4x^4 - (x^4 - 6x^2 + 9)\right) dx = \frac{1}{8}\int_{-1}^{1} \left(3x^4 + 6x^2 - 9\right) dx$$

$$= \frac{1}{8}\left(\frac{3x^5}{5} + \frac{6x^3}{3} - 9x\right)\Bigg|_{-1}^{1}$$

$$= \cdots = -\frac{8}{5} = -1.6.$$

❺ Write the results from steps ❸ and ❹ as a point:

$$\text{center of mass} = (0, -1.6).$$

To check the reasonableness of the answer, we plot the center of mass in the graph of the region and see if it appears to be correct. See figure 13. ■

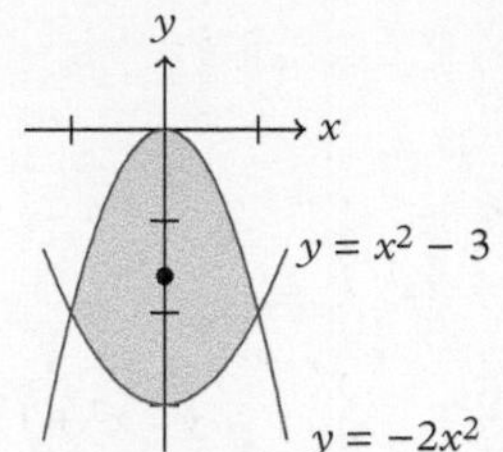

Figure 13 *The centroid (center of mass) is added to the region as a check on the reasonableness of the answer*

Reading Exercise 13 *Set up the integrals for finding the values of* $\overline{x}$ *and* $\overline{y}$ *for the region in figure 14, given that the area between the curves is* $\frac{5}{12}$*. Do not integrate.*

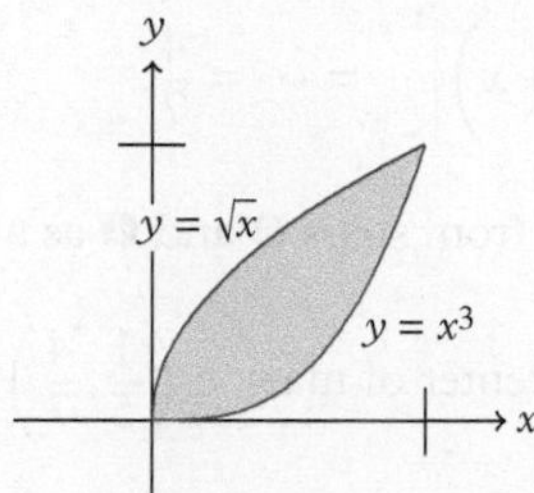

Figure 14 *The region for reading exercise 13*

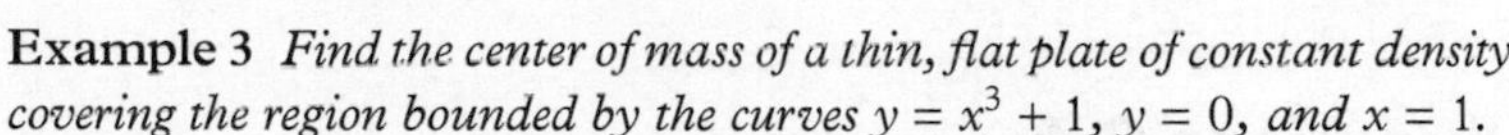

Example 3 *Find the center of mass of a thin, flat plate of constant density covering the region bounded by the curves* $y = x^3 + 1$, $y = 0$, *and* $x = 1$.

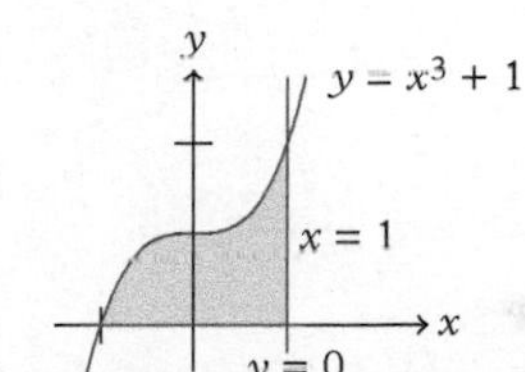

Figure 15 *The region enclosed by the curves* $y = x^3 + 1$, $y = 0$, *and* $x = 1$*. Inspection of the region tells us to expect the center of mass to be to the right of* $x = 0$ *and perhaps above* $y = 0.5$

Solution ❶ Sketch the region. The result is in figure 15.

❷ Find the area between the curves. We identify the top curve as $f(x) = x^3 + 1$ and the bottom curve as $y = 0$. Then,

$$A = \int_a^b (f(x) - g(x))\,dx = \int_{-1}^{1} \left(x^3 + 1\right) dx$$

$$= \left(\frac{x^4}{4} + x\right)\Bigg|_{-1}^{1}$$

$$= \cdots = 2 \text{ units}^2.$$

❸ Determine $\overline{x}$. Using the same top and bottom curves, and the same limits of integration as for step ❷, we use the formula for the center of mass for a thin, flat plate of constant density:

$$\overline{x} = \frac{1}{A}\int_a^b x(f(x) - g(x))\,dx$$

$$= \frac{1}{2}\int_{-1}^{1} x\left(x^3 + 1 - 0\right) dx = \frac{1}{2}\int_{-1}^{1}\left(x^4 + x\right) dx$$

$$= \frac{1}{2}\left(\frac{x^5}{5} + \frac{x^2}{2}\right)\Bigg|_{-1}^{1} = \cdots = \frac{1}{5}.$$

When the curves are polynomials, as in examples 2 and 3, using exact arithmetic usually involves finding common denominators and is tedious (and therefore is perhaps more prone to error), but it is not difficult.

❹ Determine $\overline{y}$. Continuing the use of the formula for the center of mass for a thin, flat plate of constant density,

$$\begin{aligned}\overline{y} &= \frac{1}{A}\int_a^b \frac{1}{2}\left((f(x))^2 - (g(x))^2\right)\,dx\\ &= \frac{1}{2}\int_{-1}^{1} \frac{1}{2}\left((x^3+1)^2 - 0^2\right)\,dx = \frac{1}{4}\int_{-1}^{1}\left(x^6 + 2x^3 + 1\right)\,dx\\ &= \frac{1}{4}\left(\frac{x^7}{7} + \frac{x^4}{2} + x\right)\Bigg|_{-1}^{1} = \cdots = \frac{4}{7}.\end{aligned}$$

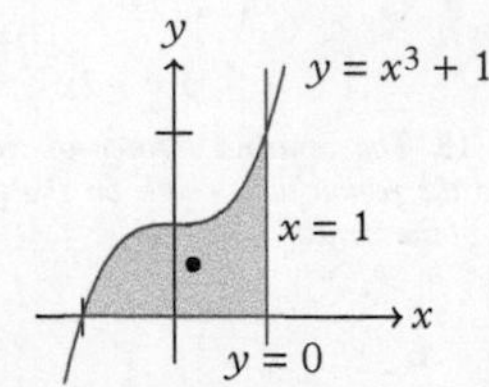

Figure 16 *The centroid found in example 3 is added to the region as a check for the reasonableness of the answer*

❺ Write the results from steps ❸ and ❹ as a point:

$$\text{center of mass} = \left(\frac{1}{5}, \frac{4}{7}\right).$$ ■

In figure 16, the center of mass is added to the graph of the region from example 3.

Exercises 9.6

1–4. (a) Find the center of mass for the point-mass system. (b) Identify the moment of the system about the origin. (c) Identify the system mass.

1. Units: feet, pounds

$m_1 = 3$ at $x_1 = -4$; $m_2 = 2$ at $x_2 = -1$; $m_3 = 5$ at $x_3 = 3$ (origin marked 0)

2. Units: kilograms, meters

$m_1 = 3$ at $x_1 = -5$; $m_2 = 6$ at $x_2 = -2$; $m_3 = 2$ at $x_3 = 0$; $m_4 = 50$ at $x_4 = 5$

3. Units: grams, millimeters

$m_1 = 43$ at $x_1 = -36$; $m_2 = 85$ at $x_2 = -5$; $m_3 = 266$ at $x_3 = 14$; $m_4 = 37$ at $x_4 = 41$

4. Units: kilograms, centimeters

$m_1 = 2$ at $x_1 = -5$; $m_2 = 3$ at $x_2 = 0$; $m_3 = 4$ at $x_3 = 12$

Ans. to reading exercise 13

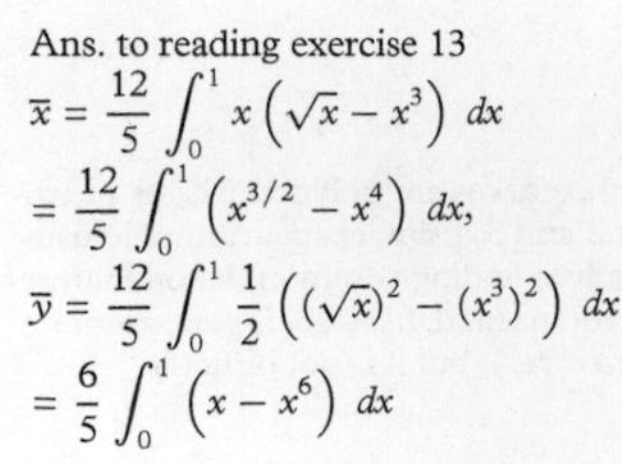

$$\overline{x} = \frac{12}{5}\int_0^1 x\left(\sqrt{x} - x^3\right)\,dx = \frac{12}{5}\int_0^1\left(x^{3/2} - x^4\right)\,dx,$$

$$\overline{y} = \frac{12}{5}\int_0^1 \frac{1}{2}\left((\sqrt{x})^2 - (x^3)^2\right)\,dx = \frac{6}{5}\int_0^1\left(x - x^6\right)\,dx$$

5–8. (a) Set up the integrals for finding the values of $\overline{x}$ and $\overline{y}$ for the shaded region, given the area A between the two curves. (b) Use technology to evaluate the integrals.

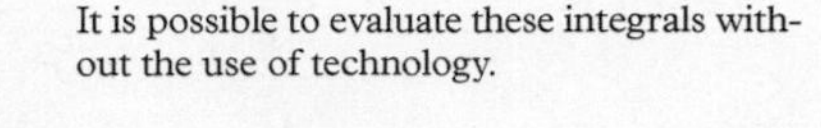
It is possible to evaluate these integrals without the use of technology.

5. $A = -\frac{8}{15} + \frac{\pi}{4}$

$y = \sin^2 x$

$y = \sin^5 x$

$\frac{\pi}{2}$

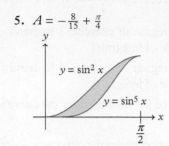

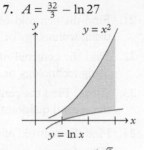

6. $A = \frac{e-2}{2e}$

$y = e^x$

$y = x + 1$

7. $A = \frac{32}{3} - \ln 27$

$y = x^2$

$y = \ln x$

8. $A = \sinh 2 - \frac{4\sqrt{2}}{3}$

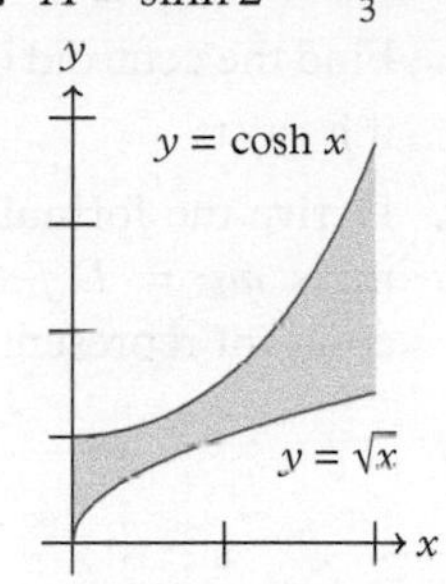

9–10. Sketch the region. Estimate visually and place a point in the region where you think the center of mass is located.

9. region between $y = \cos x$ and $y = x^2 - 1$, from $x = -1$ to $x = 1$

10. region bounded by $y = x$, $y = -x^3$, and $x = 1$

11–20. Determine the center of mass of a thin, flat plate of constant density covering the region described.

11. region under the curve $y = x^3$ from $x = 0$ to $x = 2$

12. region under the curve $y = 4 - x^2$ from $x = -2$ to $x = 2$

13. region bounded by $y = x^2$ and $y = x + 2$

14. region bounded by $y = x$ and $y = x^2$

15. region bounded by $y = 4x - x^2$ and $y = x$

16. region between the curves $y = x$ and $y = \sqrt{x}$ from $x = 1$ to $x = 4$

17. region under the curve $y = \sin x$ from $x = 0$ to $x = \pi$

18. region bounded by $y = \sqrt{4 - x^2}$ and $y = 0$

19. (GSL) portion of the region inside the ellipse $\frac{x^2}{36} + \frac{y^2}{9} = 1$ that lies in the first quadrant

20. region under the curve $y = \frac{1}{1 + x^2}$ from $x = -1$ to $x = 1$

21. Find the centroid of the region of exercise 7 by hand (without using technology or a table of integrals).

22. Find the centroid of the region of exercise 6 by hand (without using technology or a table of integrals).

23. (Love) Find the centroid of the region under the curve $y = e^x$ in the second quadrant.

24. Find the centroid of the region under the curve $y = \frac{1}{x^3}$ on $[1, \infty)$, if it exists.

25. Find the centroid of the region under the curve $y = \frac{1}{x^2}$ on $[1, \infty)$, if it exists.

Hint: follow the pattern of parts (d)–(h) of exercise 39 in section 9.3.

26. Derive the formula for $\overline{x}$ by using the midpoints $x_k - \frac{1}{2}\Delta x$ and mass $m_k = D(f(x_k) - g(x_k))\Delta x$, where D is a proportionality constant representing the plate's density and thickness.

9.7 Applications to Economics

Many different topics in mathematics have applications in economics. We have seen applications of the derivative such as marginal cost and marginal revenue. In this section we explore some of the economics applications of integration.

Introduction: supply and demand

You may already be familiar with the ideas of *supply* and *demand*. The lower the price of an item, the more of it consumers are willing to purchase. With quantity as the horizontal axis and price as the vertical axis, the *demand curve d* has a negative slope because a lower price is required to entice consumers to purchase a larger quantity (figure 1). On the supply side, the opposite is true; the higher the price of an item, the more of it producers are willing to provide. The *supply curve s* has a positive slope. Where demand and supply match, at the intersection

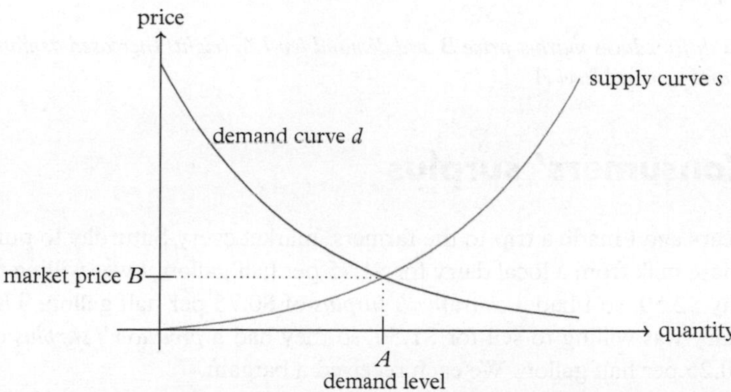

Figure 1 *A demand curve (blue) and a supply curve (green). The point of intersection of the two curves is (A, B), where A is the demand level and B is the market price*

of the two curves, we call the vertical coordinate the *market price* and label it B, and we call the horizontal coordinate the *demand level* and label it A. If the price was set below B, then demand is larger than

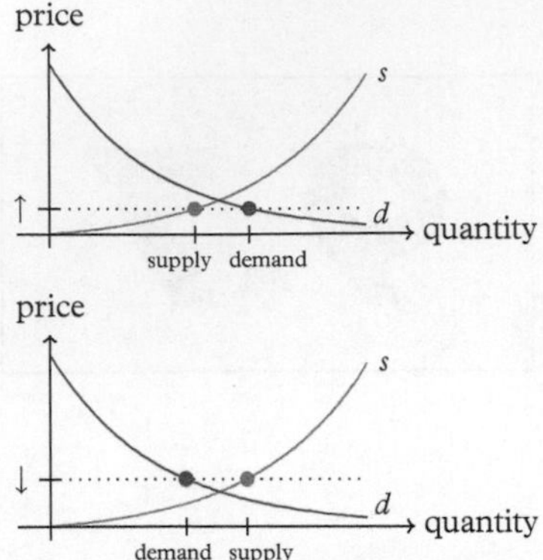

Figure 2 *(top) If the price level is set too low, demand is greater than supply and there is upward pressure on the price. (bottom) If the price level is set too high, supply is greater than demand and there is downward pressure on the price*

supply (figure 2, top) and there is upward pressure on the price. If the price is set above B, then supply is greater than demand and there is downward pressure on the price (figure 2, bottom).

Changes in the supply curve or the demand curve affect market price. Sometimes this can be the result of a consumer fad. Late during my elementary school years, "everyone" started chewing sunflower seeds. This shifted the demand curve to the right (figure 3, right), so that market price increased and producers provided more. The next summer when we visited family in the Texas panhandle, there were far more fields of sunflowers than we ever noticed before. Nearly a half century later, the coronavirus pandemic resulted in many stay-at-home orders and greatly reduced travel, shifting the demand curve for crude oil to the left and reducing the market price (figure 3, left)—and the OPEC cartel and other countries reached a deal to reduce production.

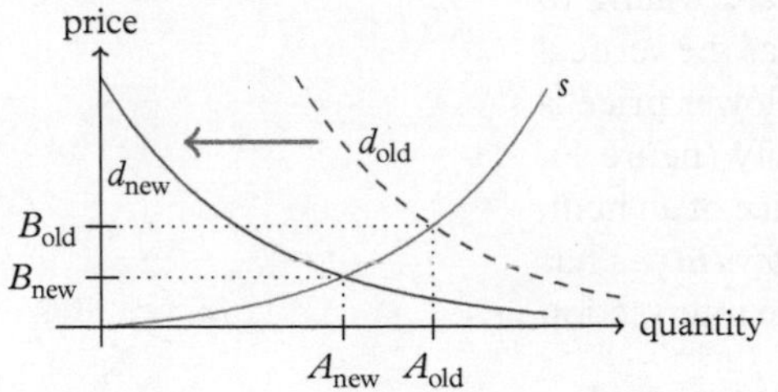

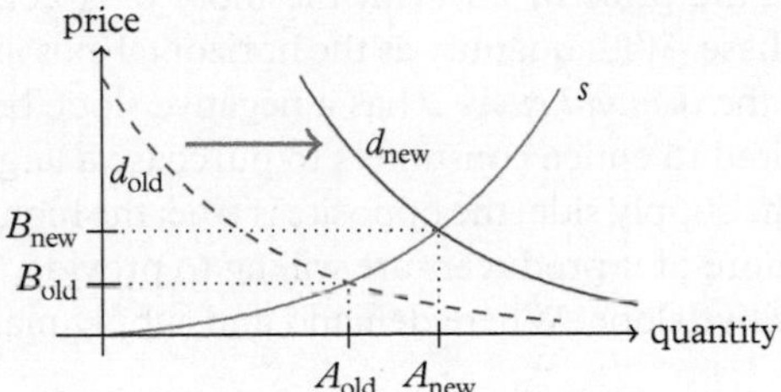

Figure 3 *(left) Reduced demand (leftward shift) reduces market price B and demand level A. (right) Increased demand (rightward shift) increases market price B and demand level A*

Consumers' surplus

Years ago I made a trip to the farmers' market every Saturday to purchase milk from a local dairy for \$1.75 per half gallon. I was willing to pay \$2.50, so I had a *consumer's surplus* of \$0.75 per half gallon. The dairy was willing to sell for \$1.50, so they had a *producer's surplus* of \$0.25 per half gallon. We each received a bargain.

The dairy's willingness to sell for \$1.50 is speculation on my part.

The marketplace is comprised of not just one consumer and one producer, but many consumers and many producers. How do we measure the total amount of consumers' surplus? The key to answering this question lies in understanding some of the details of the supply-and-demand graph.

Consider figure 4. The points toward the top left of the demand curve mean that only a few consumers (a small x-coordinate on the graph) are willing to pay a high price (a large y-coordinate on the graph) for the item. As we move down and to the right along the curve, the price gets lower and more consumers are willing to make the purchase. However, these consumers are not all paying different prices; everyone who makes a purchase is paying the market price B. Hence an individual consumer's surplus is the difference between the price the individual is willing to pay, which is a y-coordinate on the demand curve, and the market price, B, as illustrated by the brown slice in figure 4. An individual consumer's surplus comprises a small

Also, some consumers may be willing to purchase a larger quantity of the item; this is equivalent to more consumers being willing to make a purchase.

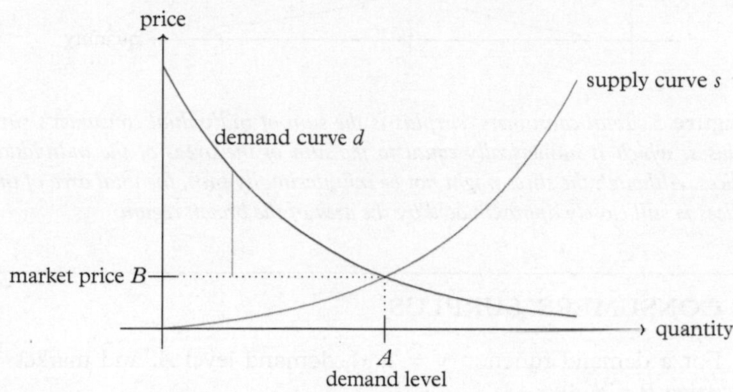

Figure 4 *An individual consumer's surplus (brown slice) is placed in the supply and demand diagram. The width of a slice is 1 (one consumer, assuming a quantity purchased of 1 unit), so the area of the slice is numerically equal to that consumer's surplus*

portion—not quite an infinitesimally thin slice, but akin to it—of the total consumers' surplus. The consumers making the purchase are those at and to the left of A, because they are willing to pay at least B for the item. Add together all these individual surpluses and the result is the area of the region between the demand curve $y = d(x)$ and the market price line $y = B$, from $x = 0$ to $x = A$, as in figure 5. Therefore,

$$\text{consumers' surplus} = \int_0^A (d(x) - B)\, dx.$$

Notice that the plural possessive is used in the statement of the formula.

Because an individual consumer's surplus is measured in monetary units, so is the total consumers' surplus. Last, notice that the market price B is the y-coordinate on both the demand curve and the supply curve at the demand level A; that is,

$$B = d(A) = s(A).$$

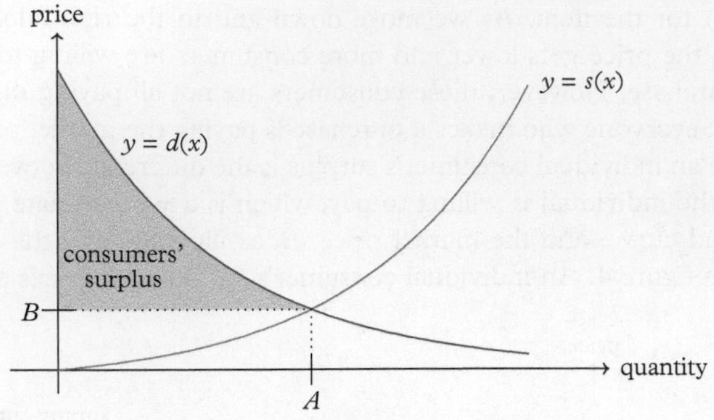

Figure 5 *Total consumers' surplus is the sum of individual consumer's surpluses, which is numerically equal to the sum of the areas of the individual slices. Although the slices might not be infinitesimally thin, the total area of the slices is still closely approximated by the area of the brown region*

The comments in the description of figure 5 address the issue of continuity. Often, quantity is not truly a continuous variable, and the demand and supply functions are, in reality, comprised of discrete chunks and are not continuous. However, in a marketplace consisting of thousands (or millions) of consumers, one consumer is so small by comparison that we can model the marketplace using continuous functions and apply infinitesimal reasoning out of convenience without sacrificing accuracy.

CONSUMERS' SURPLUS

For a demand function $y = d(x)$, demand level A, and market price $B = d(A)$,

$$\text{consumers' surplus} = \int_0^A (d(x) - B)\, dx.$$

Example 1 *For the demand function* $d(x) = 200 - \frac{1}{2}x$ *and demand level* $x = 300$, *find the consumers' surplus.*

Solution We are given $A = 300$. We also need B, which may be calculated as

$$B = d(A) = d(300) = 200 - \frac{1}{2}(300) = 50.$$

Using the consumers' surplus formula,

$$\begin{aligned}\text{consumers' surplus} &= \int_0^A (d(x) - B)\, dx\\ &= \int_0^{300} \left(200 - \frac{1}{2}x - 50\right) dx\\ &= \left(150x - \frac{x^2}{4}\right)\Big|_0^{300}\\ &= \cdots = \$22\,500.\end{aligned}$$

If the market price B is given instead of the demand level A, we set $B = d(x)$ and solve for x. The solution to the equation is A. If neither A nor B are given, but $s(x)$ is given, we set $d(x) = s(x)$ and solve for x. The solution is the value of A. Then, we find B as in example 1.

The total consumers' surplus is \$22 500. ■

Reading Exercise 14 *For the demand function $d(x) = 300 - x$ and demand level $x = 200$, find the consumers' surplus.*

Example 2 *For the demand function $d(x) = 27 \cdot 1.003^{-0.02x}$ and the market price \$12, find the consumers' surplus.*

Solution We are given $B = 12$, but not A. To find A, we solve the equation $B = d(A)$ for A:

We use the same algebraic steps as for determining the growth constant k in an exponential change problem (see section 5.7).

$$\begin{aligned} B &= d(A) \\ 12 &= 27 \cdot 1.003^{-0.02A} \\ \frac{12}{27} &= 1.003^{-0.02A} \\ \ln\frac{12}{27} &= \ln 1.003^{-0.02A} = -0.02A \ln 1.003 \\ A &= \frac{\ln\frac{12}{27}}{-0.02 \ln 1.003} = 13\,536. \end{aligned}$$

We are now ready to use the formula for consumers' surplus:

$$\begin{aligned} \text{consumers' surplus} &= \int_0^A (d(x) - B)\, dx \\ &= \int_0^{13\,536} \left(27 \cdot 1.003^{-0.02x} - 12\right) dx \\ &= \left(27 \cdot \frac{1.003^{-0.02x}}{-0.02 \ln 1.003} - 12x\right)\Bigg|_0^{13\,536} \\ &= \cdots = \$87\,946. \end{aligned}$$

Line 3 uses the antiderivative formula $\int b^{au}\, du = \frac{b^{au}}{a \ln b} + C$, which is formula #33 in the table of integrals in section 7.9, table 1. This is a convenient formula to know.

The total consumers' surplus is \$87 946. ■

Producers' surplus

As indicated earlier, an individual producer's surplus is the difference between the market price B and the price at which the producer is willing to sell, which is a y-coordinate on the supply curve. The total producers' surplus is then the area between the market price line $y = B$ and the supply curve $y = s(x)$, from $x = 0$ to $x = A$, as in figure 6.

The producers willing to sell at price B are represented in the supply curve as being at or to the left of A.

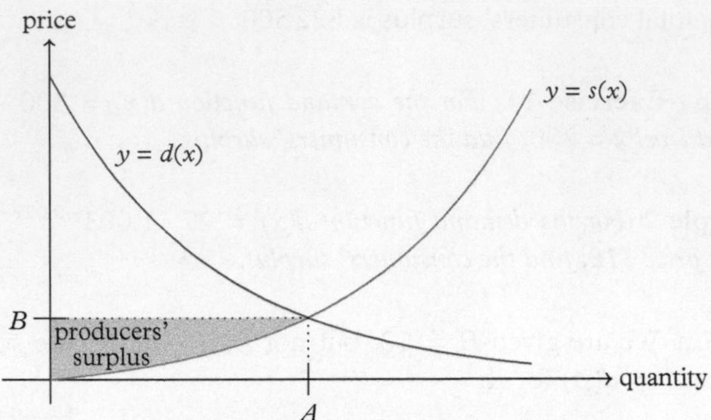

Figure 6 *Total producers' surplus is the sum of individual producer's surpluses, resulting in the area of the brown region. Compare to figure 5*

> **PRODUCERS' SURPLUS**
>
> For a supply function $y = s(x)$, demand level A, and market price $B = s(A)$,
>
> $$\text{producers' surplus} = \int_0^A (B - s(x))\, dx.$$

Ans. to reading exercise 14
\$20 000
Details:
$B = d(200) = 100$
consumers' surplus $= \int_0^{200} (300 - x - 100)\, dx = \cdots = \$20{,}000$

The same options present themselves here as for consumers' surplus problems: either A or B can be provided, as could neither if both d and s are given instead.

Example 3 *For the supply function* $s(x) = 0.4x$ *and the demand level* $x = 200$*, find the producers' surplus.*

Solution We are given $A = 200$. We also need B, which may be calculated as

$$B = s(A) = s(200) = 0.4(200) = 80.$$

Using the producers' surplus formula,

$$\begin{aligned}\text{producers' surplus} &= \int_0^A (B - s(x))\, dx \\ &= \int_0^{200} (80 - 0.4x)\, dx \\ &= 80x - 0.2x^2 \Big|_0^{200} \\ &= \cdots = \$8000.\end{aligned}$$

The total producers' surplus is \$8000. ■

Of course, both consumers' and producers' surpluses can be found given the demand and supply curves.

Example 4 *Find consumers' surplus and producers' surplus when* $d(x) = 120 - 0.16x$ *and* $s(x) = 0.08x$.

Solution Because neither the market price nor the demand level are given, both must be calculated. We begin by noting that, at the demand level, $d(A) = s(A)$. We solve this equation to find A:

$$120 - 0.16A = 0.08A$$
$$120 = 0.24A$$
$$A = \frac{120}{0.24} = 500.$$

The demand level is $A = 500$. Now, either d or s can be used to find the market price B:

$$B = s(500) = 0.08(500) = 40.$$

Then,

$$\begin{aligned}\text{consumers' surplus} &= \int_0^A (d(x) - B)\, dx \\ &= \int_0^{500} (120 - 0.16x - 40)\, dx \\ &= \left(80x - 0.08x^2\right)\Big|_0^{500} \\ &= \cdots = \$20\,000\end{aligned}$$

and

$$\begin{aligned}\text{producers' surplus} &= \int_0^A (B - s(x))\, dx \\ &= \int_0^{500} (40 - 0.08x)\, dx \\ &= \left(40x - 0.04x^2\right)\Big|_0^{500} \\ &= \cdots = \$10\,000.\end{aligned}$$

The consumers' surplus is \$20 000 and the producers' surplus is half that amount, or \$10 000. ■

Gini coefficient of income distribution

Suppose in a particular country that the 20% of the population with the lowest incomes earn 5% of the country's total income, that the next-lowest 20% of income earners earn 8% of the country's total income, the middle 20% earn 12% of the total income, the next-highest 20% earn 25% of the total income, and the highest-earning 20% of the population earn 50% of the country's total income. This information can be rephrased as the cumulative income earned by the lowest 20%, 40%, 60%, 80%, and 100% of the population, and is shown in table 1.

Note that the percentages earned by each group must be larger than the previous group, because they are earning higher incomes.

Table 1 *Income quintiles for a fictional country*

population proportion	income proportion
20%	5%
40%	13%
60%	25%
80%	50%
100%	100%

The values in the right-hand column of table 1 are called *income quintiles*, consistent with the use of the term *quintiles* in statistics.

The income quintiles are plotted in figure 7. These points are part of the *Lorenz curve*, introduced in 1905 by economist Max Lorenz. Proportion is used in the diagram rather than percentage. The points

Lorenz, M. O., "Methods of Measuring the Concentration of Wealth," *Publications of the American Statistical Association* 9 No. 70 (June 1905), 209–219, doi: 10.2307/2276207.

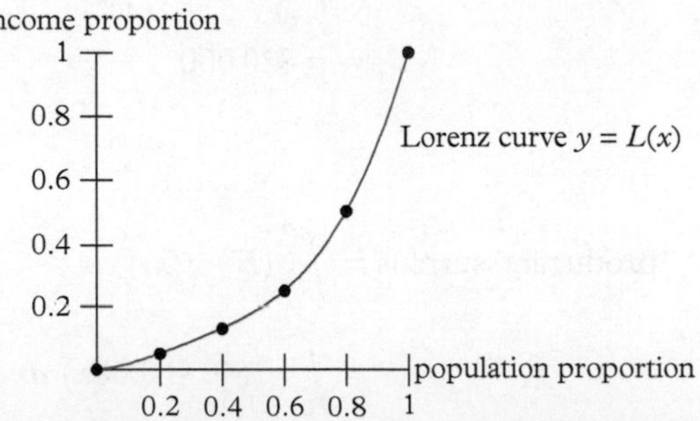

Figure 7 *The Lorenz curve (blue) with income quantiles (black points) plotted. Because proportion of income increases as we move to the right, the slopes are increasing, and therefore the Lorenz curve is concave up. The horizontal and vertical scales are marked in terms of proportion (= percentage ÷100) rather than percentage*

(0, 0) and (1, 1) must always be on the Lorenz curve, because 0% of the population earns 0% of the income and 100% of the population earns 100% of the income.

If there is perfect equality of income so that everyone earns the same amount, then any $k\%$ of the population earns $k\%$ of the income, and the Lorenz curve is $y = x$. If there is perfect inequality of income when one person gets all the income and no one else has any income, then the curve resembles $y = 0$ and then is (nearly) vertical to reach (1, 1). Neither of these extremes (figure 8) are possible. The Lorenz curve for a country lies between them.

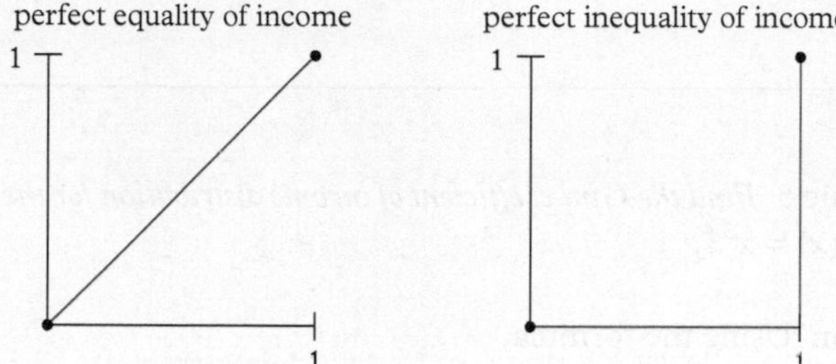

Figure 8 *Lorenz curves for (left) perfect equality and (right) perfect inequality of income*

In 1912, Corrado Gini developed a fairly simple way to measure income inequality: measure the area of the region that lies between perfect equality (the line $y = x$) and the Lorenz curve (figure 9). For perfect equality of income, the two curves are identical and the area is 0. For perfect inequality, the region is a triangle with base and height 1,

Corrado Gini 1884–1965
http://www-history.mcs.st-andrews.ac.uk/Biographies/Gini.html

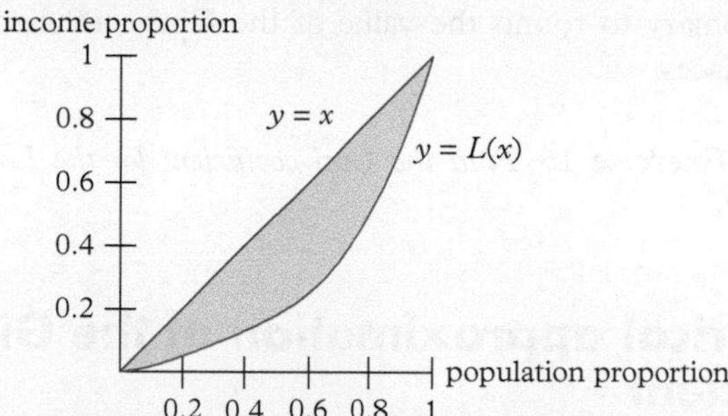

Figure 9 *The Gini coefficient of income distribution is equal to twice the area of the region (green) between $y = x$ and the Lorenz curve $y = L(x)$. For perfect equality of income, the Gini coefficient is 0; for perfect inequality of income, the Gini coefficient is* 1

and the area is $\frac{1}{2}$. Double the area and the possible values range from 0 (perfect equality of income) to 1 (perfect inequality of income); the larger the value, the more inequality there is.

The Gini coefficient of income distribution is also known as the *Gini index*. Some sources prefer to scale the index from 0 to 100, in which case the formula is $200 \int_0^1 (x - L(x))\, dx$.

GINI COEFFICIENT OF INCOME DISTRIBUTION

For a Lorenz curve $y = L(x)$, the Gini coefficient of income distribution is

$$\text{Gini coefficient} = 2 \int_0^1 (x - L(x))\, dx.$$

Example 5 *Find the Gini coefficient of income distribution for the Lorenz curve* $L(x) = x^{3.4}$.

Notice that $L(0) = 0$ and $L(1) = 1$, as required.

Solution Using the formula,

$$\begin{aligned} \text{Gini coefficient} &= 2 \int_0^1 (x - L(x))\, dx \\ &= 2 \int_0^1 \left(x - x^{3.4}\right) dx \\ &= 2 \left(\frac{x^2}{2} - \frac{x^{4.4}}{4.4}\right)\Bigg|_0^1 \\ &= \cdots = 0.545. \end{aligned}$$

It is customary to round the value of the Gini coefficient to three decimal places. ■

Reading Exercise 15 *Find the Gini coefficient for the Lorenz curve* $L(x) = x^{2.7}$.

Numerical approximation of the Gini coefficient

Although the equation of a country's Lorenz curve is usually not available, income quintiles are readily available (for instance, at wolframalpha.com or worldbank.org). However, income quintiles are values of the Lorenz curve—meaning, we have a few values of $L(x)$. The quintiles from table 1 are rewritten as values of the Lorenz function in table 2. Table 2 resembles the information we needed to perform

Table 2 *The income quintiles of table 1 rewritten as function values*

x	$L(x)$
0	0
0.2	0.05
0.4	0.13
0.6	0.25
0.8	0.50
1	1

numerical integration in chapter 4 using, for example, the trapezoid rule or Simpson's rule. Because the quintiles are for $n = 5$, we cannot use Simpson's rule (recall that we need an even value of n to apply Simpson's rule). Although we can use the trapezoid rule, it turns out there is a better option. A numerical integration rule has been designed specifically for estimating the Gini coefficient.

NUMERICAL APPROXIMATION OF THE GINI COEFFICIENT

A numerical approximation of the Gini coefficient, where $L(x)$ represents the Lorenz curve, is

$$\text{Gini coefficient} = \frac{25}{144}\Big(3(0.2 - L(0.2)) + 2(0.4 - L(0.4)) + 2(0.6 - L(0.6)) + 3(0.8 - L(0.8))\Big).$$

Gerber, Leon, "A Quintile Rule for the Gini Coefficient," *Mathematics Magazine* 80 No. 2 (2007), 133–135, doi: http://doi.org/10.1080/0025570X.2007.11953468

Example 6 *Approximate the Gini coefficient for a country with the income distribution given in the following table.*

population proportion	income proportion
20%	5%
40%	13%
60%	25%
80%	50%
100%	100%

Solution These are the same income quintiles given in table 1 and rewritten as values of the Lorenz function in table 2. Using the numerical approximation formula, we have

$$\begin{aligned}\text{Gini coefficient} &= \tfrac{25}{144}\Big(3(0.2 - L(0.2)) + 2(0.4 - L(0.4))\\ &\quad + 2(0.6 - L(0.6)) + 3(0.8 - L(0.8))\Big)\\ &= \tfrac{25}{144}\Big(3(0.2 - 0.05) + 2(0.4 - 0.13)\\ &\quad + 2(0.6 - 0.25) + 3(0.8 - 0.5)\Big)\\ &= \cdots = 0.450.\end{aligned}$$

■

Ans. to reading exercise 15

$\text{Gini coefficient} = 2\int_0^1 (x - x^{2.7})\,dx$

$= \cdots = 0.459$

There are various opinions regarding which Gini index is optimal. Settling such a question is beyond the scope of this book.

Exercises 9.7

1. For a certain product, the demand function is $d(x) = 3250 - 0.2x$. For a demand level of $x = 200$, find the market price.
2. For a certain product, the demand function is $d(x) = 400 - 3x$. For a demand level of $x = 500$, find the market price.
3. For a certain product, the demand function is $d(x) = 14 - 0.02x$. If the market price is \$10, what is the demand level?
4. For a certain product, the demand function is $d(x) = 50 - 0.1x$. If the market price is \$47, what is the demand level?
5. For a certain product, the demand function is $d(x) = 420 \cdot 1.02^{-0.0042x}$. If the market price is \$250, what is the demand level?
6. For a certain product, the demand function is $d(x) = 30 \cdot 1.008^{-0.0022x}$ and the supply function is $s(x) = 10 \cdot 1.07^{0.003x}$. Determine the demand level A.

7–12. Use the given information to determine the consumers' surplus.

7. demand function $d(x) = 400 - 0.2x$, demand level $x = 100$
8. demand function $d(x) = 1000 - 2x$, market price \$800

9. demand function $d(x) = 82 \cdot 1.024^{-0.0015x}$, market price \$20
10. demand function $d(x) = 10 + \frac{100}{x+10}$, demand level $x = 40$
11. demand function $d(x) = \frac{30\,000}{x+50}$, market price \$2.46
12. demand function $d(x) = 320 \cdot 1.001^{-0.0015x}$, market price \$62

13–18. Use the given information to determine the producers' surplus.

13. supply function $s(x) = 2 + 0.01x$, demand level $x = 430$
14. supply function $s(x) = 10+0.004x$, market price \$20
15. supply function $s(x) = 13 + 0.000\,031x^2$, market price \$20
16. supply function $s(x) = 0.000\,000\,27x^3$, demand level $x = 800$
17. supply function $s(x) = 3.17e^{0.0003x}$, demand level $x = 5990$
18. supply function $s(x) = 22.19e^{0.000\,147x}$, market price \$40

19–24. Use the demand function $d(x)$ and supply function $s(x)$ to determine (a) the consumers' surplus and (b) the producers' surplus.

19. $d(x) = 1000 - 0.05x$, $s(x) = 200 + 0.03x$
20. $d(x) = 42 - 0.01x$, $s(x) = 0.004x$
21. $d(x) = 67e^{-0.0026x}$, $s(x) = 11e^{0.0046x}$
22. $d(x) = 8e^{-0.011x}$, $s(x) = 2e^{0.003x}$
23. $d(x) = 112 \cdot 1.4^{-0.000\,274x}$, $s(x) = 14 \cdot 2.1^{0.000\,035\,1x}$
24. $d(x) = 295 \cdot 1.03^{-0.0041x}$, $s(x) = 67 \cdot 1.19^{0.000\,383x}$

25–34. Find the Gini coefficient of income distribution for the given Lorenz curve.

25. $L(x) = x^{1.9}$
26. $L(x) = x^4$
27. $L(x) = \frac{3}{4}x + \frac{1}{4}x^3$
28. $L(x) = \frac{1}{3}x + \frac{2}{3}x^2$
29. $L(x) = \frac{5^x - 1}{4}$
30. $L(x) = 2^x - 1$

31. $L(x) = \dfrac{e^x - 1}{e - 1}$

32. $L(x) = \frac{1}{2} \cdot 3^x - \frac{1}{2}$

33. $L(x) = x^p, p > 1$

34. $L(x) = \dfrac{a^x - 1}{a - 1}, a > 1$

35–38. Use the numerical formula to approximate the Gini coefficient for a country with the given income distribution.

35.

population proportion	income proportion
20%	8%
40%	20%
60%	35%
80%	60%
100%	100%

36.

population proportion	income proportion
20%	12%
40%	28%
60%	50%
80%	74%
100%	100%

37.

population proportion	income proportion
20%	18%
40%	37%
60%	57%
80%	78%
100%	100%

38.

population proportion	income proportion
20%	2%
40%	6%
60%	14%
80%	30%
100%	100%

39. (a) Using the Lorenz curve $L(x) = \dfrac{e^x - 1}{e - 1}$, determine the values of $L(0.2)$, $L(0.4)$, $L(0.6)$, and $L(0.8)$. (b) Use the numerical approximation formula and the results from part (a) to calculate the Gini coefficient. (c) Compare to your answer to exercise 31.

40. (a) Using the Lorenz curve $L(x) = \frac{1}{3}x + \frac{2}{3}x^2$, determine the values of $L(0.2)$, $L(0.4)$, $L(0.6)$, and $L(0.8)$. (b) Use the numerical approximation formula and the results from part (a) to calculate the Gini coefficient. (c) Compare to your answer to exercise 28.

41. (a) Find the income quintiles for the country of your birth (alternately, for the country in which your location of birth is currently situated) and use the numerical approximation formula to estimate the Gini coefficient. (b) Compare your result to the official estimate of the Gini coefficient for the country.

Official estimates of the Gini coefficient may use more data points than just the quintiles, and therefore may vary from the numerical estimate found in part (a).

42. A certain product has demand function $d(x) = 52e^{-0.000\,24x}$ and supply function $s(x) = 10 + 0.000\,095x^2$. (a) Use Newton's method to determine the demand level A. (b) Find the producers' surplus.

43. Determine the effect on demand level A and market price B if the supply curve shifts to the right.

44. Determine the effect on demand level A and market price B if the supply curve shifts to the left.

45. Using the trapezoid rule to estimate the Gini coefficient numerically always produces an underestimate. Why?

Logistic Growth

9.8

In this section we return to the study of growth. Although the development of formulas requires the use of integration, the application of the formulas does not.

Unrestrained growth

In section 5.7 we developed the law of exponential change under the assumption that the rate of change of a population P is proportional to its size—that is, $\frac{dP}{dt} = kP$. With P_0 representing the initial population (the population at time $t = 0$), we derived the relationship

$$P = P_0 e^{kt},$$

where k is a constant representing the rate of growth (if $k > 0$) or decay (if $k < 0$).

If $k > 0$, the exponential growth of this model is unbounded because

$$\lim_{t \to \infty} e^{kt} = e^{k\Omega} \doteq \infty.$$

In some settings, this is not an issue. For instance, there is no natural ceiling on prices under inflation. The amount of currency needed to purchase an item could keep doubling without end. This is *unrestrained growth.*

In practice, under hyperinflation, the number of zeroes on the currency eventually gets large enough that the currency's value is reset. But, there is still no natural ceiling on the price level.

Some populations, though, cannot grow unrestrained without bound. The number of deer in a hundred-acre wood cannot reach one trillion, for there is not enough room for them to fit, let alone enough resources for food and shelter. Such a population may grow rapidly for a time, but eventually the population levels off near the *carrying capacity* of the environment. The law of exponential change seems to fit when the population is growing rapidly, but it is not a good model for the population in the long term. We need a different model.

Restrained growth

The *logistic model* of population growth assumes that the rate of change of a population is proportional not only to the size of the population, but also to the amount of room it has left to grow,

This equation is called the *logistic differential equation.*

$$\frac{dP}{dt} = kP\left(1 - \frac{P}{M}\right),$$

where P is the population at time t, M is the carrying capacity of the environment (the maximum sustainable population), and $k > 0$ is a growth constant. When the population P is relatively small compared to the carrying capacity M, then $1 - \frac{P}{M}$ is close to one and the rate of change is close to kP, resembling exponential growth. When the population is nearly as large as the carrying capacity M, then $1 - \frac{P}{M}$ is close to zero and the rate of change is instead very small.

The logistic differential equation is solved in a manner similar to the derivation of the law of exponential change. We begin by separating the variables, treating dP and dt as separate quantities, and isolating the variables P and t on opposite sides of the equation:

$$\frac{dP}{dt} = kP\left(1 - \frac{P}{M}\right)$$

$$dP = kP\left(1 - \frac{P}{M}\right)\, dt$$

$$\frac{1}{P\left(1 - \frac{P}{M}\right)}\, dP = k\, dt$$

$$\frac{M}{P(M-P)}\, dP = k\, dt.$$

In line 4 we simplify the compound fraction by multiplying the numerator and the denominator by M.

Next we place an integral symbol in front of each side of the equation. As stated in section 5.7, this technique, called *separation of variables*, is justified in a course on differential equations:

$$\int \frac{M}{P(M-P)}\, dP = \int k\, dt.$$

To evaluate the integral on the left side, we need to use partial fractions. (The need for partial fractions is recognized more easily if we replace the variable of integration P by our usual variable x, and replace the constant M with some number—say, 7—which gives $\int \frac{7}{x(7-x)}\, dx$.) The denominator is already factored, and the form of the partial fraction decomposition is

$$\frac{A}{P} + \frac{B}{M-P}.$$

Equating the form with the original fraction gives

$$\frac{A}{P} + \frac{B}{M-P} = \frac{M}{P(M-P)}$$
$$A(M-P) + BP = M.$$

Choosing $P = 0$, we have $AM = M$ and therefore $A = 1$. Choosing $P = M$, we have $BM = M$ and therefore $B = 1$. We can now rewrite the integral and continue:

$$\int \left(\frac{1}{P} + \frac{1}{M-P}\right) dP = \int k\,dt$$
$$\ln|P| - \ln|M-P| = kt + C$$
$$\ln\left|\frac{P}{M-P}\right| = kt + C$$
$$\left|\frac{P}{M-P}\right| = e^{kt+C} = e^C e^{kt}$$
$$\frac{P}{M-P} = \pm e^C e^{kt} = De^{kt}.$$

In line 1 on the left side, P is the variable and M is a constant; in line 1 on the right side, t is the variable and k is a constant. In line 2 on the left side, the negative sign comes from the substitution $u = M - P$, $du = -1 \cdot dP$. In line 5, the constant $\pm e^C$ is replaced with the constant D.

At this point let's pause the derivation and determine the value of the constant D. If the population is P_0 at time $t = 0$, then using the latter equation we have

$$\frac{P_0}{M-P_0} = De^0 = D.$$

Here, P is replaced with P_0 and t is replaced with zero.

With this in mind, we continue with the previous calculation:

$$\frac{P}{M-P} = De^{kt}$$
$$P = (M-P)De^{kt} = MDe^{kt} - PDe^{kt}$$
$$P + PDe^{kt} = MDe^{kt}$$
$$P(1 + De^{kt}) = MDe^{kt}$$
$$P = \frac{MDe^{kt}}{1 + De^{kt}}.$$

The expression may be simplified further by multiplying the numerator and the denominator by $\frac{1}{De^{kt}}$ to get

$$P = \frac{M}{\frac{1}{D}e^{-kt} + 1}.$$

We also replace the constant $\frac{1}{D}$ by the constant A, finally arriving at

$$P = \frac{M}{1 + Ae^{-kt}},$$

where

The formula for A is the reciprocal of the earlier formula for D.

$$A = \frac{M - P_0}{P_0}.$$

LOGISTIC GROWTH

If the rate of change of a quantity P is jointly proportional to P and $\left(1 - \frac{P}{M}\right)$, where M is the carrying capacity, then

$$P = \frac{M}{1 + Ae^{-kt}},$$

where $A = \frac{M - P_0}{P_0}$, P_0 is the value of P when $t = 0$, and $k > 0$ is a growth constant.

Comparison of exponential and logistic models

We have said that the logistic model closely follows the law of exponential change at first, and that as the population nears the carrying capacity, the population changes very slowly. Let's look at the following example to see this situation in action.

Example 1 *A rodent infestation begins with an initial population of* 24 *in an environment with a carrying capacity of* 2950. *After* 40 *days, the population has risen to* 59. *(a) Assuming the population follows the law of exponential change, what is the population after* 100 *days? (b) Assuming the population follows the logistic model, what is the population after* 100 *days? (c) Assuming the population follows the law of exponential* *change, what is the population after* 1000 *days? (d) Assuming the population follows the logistic model, what is the population after* 1000 *days?*

Solution We are given the initial population $P_0 = 24$ and the carrying capacity $M = 2950$. For parts (a) and (c), we follow the solution method for the exponential growth model from section 5.7.

(a) ❶ Determine k. We use the additional information given, which is that $P = 59$ when $t = 40$:

$$\begin{aligned} P &= P_0 e^{kt} \\ 59 &= 24e^{k(40)} \\ \tfrac{59}{24} &= e^{40k} \\ \ln \tfrac{59}{24} &= 40k \\ k &= \frac{\ln \frac{59}{24}}{40} = 0.02249. \end{aligned}$$

❷ Determine the requested quantity. We wish to know the population after 100 days:

$$P = P_0 e^{kt} = 24e^{0.02249 \cdot 100} = 227.$$

(b) Following the same steps with similar algebra, we apply the formula for logistic growth. But, first we add a preliminary step.

⓿ Calculate A:

$$A = \frac{M - P_0}{P_0} = \frac{2950 - 24}{24} = 121.92.$$

❶ Determine k. We use the additional information given, which is that $P = 59$ when $t = 40$:

$$\begin{aligned} P &= \frac{M}{1 + Ae^{-kt}} \\ 59 &= \frac{2950}{1 + 121.92e^{-k(40)}} \\ 59 + 59 \cdot 121.92e^{-40k} &= 2950 \\ 7193e^{-40k} &= 2891 \\ e^{-40k} &= \frac{2891}{7193} \\ -40k &= \ln \frac{2891}{7193} \\ k &= \frac{\ln \frac{2891}{7193}}{-40} = 0.02279. \end{aligned}$$

Similar to part (a), the procedure is to isolate the exponential e^{-40k} on one side of the equation (accomplished in line 5), take the natural log of both sides (line 6), and then solve for k (line 7).

❷ Determine the requested quantity. We wish to know the population after 100 days:

$$P = \frac{M}{1 + Ae^{-kt}} = \frac{2950}{1 + 121.92e^{-0.02279 \cdot 100}} = 219.$$

(c) Using the law of exponential change, we wish to know the population after 1000 days:

$$P = P_0 e^{kt} = 24e^{0.02249 \cdot 1000} \approx 140\,400\,000\,000.$$

(d) Using the logistic growth model, we wish to know the population after 1000 days:

$$P = \frac{M}{1 + Ae^{-kt}} = \frac{2950}{1 + 121.92e^{-0.02279 \cdot 1000}} = 2950.$$ ■

Notice that the growth constants are also close, with $k = 0.02249$ for the exponential model and $k = 0.02279$ for the logistic model.

Let's pause to consider the results of example 1. The answers to parts (a) and (b) are very similar, with the exponential and logistic growth models giving 227 and 219 rodents after 100 days. However, the answers to parts (c) and (d) are not close at all. The exponential model predicts more than 140 billion rodents after 1000 days, whereas the logistic model predicts that the rodent population is at the carrying capacity of 2950, which is a much more realistic answer. The population functions of the two models are graphed in figure 1. As

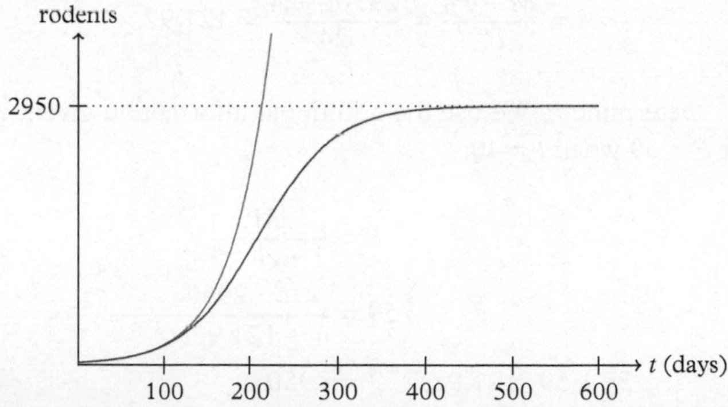

Figure 1 *Population functions for example 1 produced by the law of exponential change (green) and the logistic model (blue). The horizontal line $y = 2950$ represents the carrying capacity M*

promised, the logistic model closely follows the exponential model at first, but then the rate of increase slows and eventually drops to practically zero (no increase). The exponential model produces a curve that is always concave up, whereas the logistic model produces a curve with an inflection point and a horizontal asymptote.

Logistic growth example: declining population

The law of exponential change applies not only to increasing populations, but also to decreasing quantities as well, such as the amount of a radioactive isotope in a rock. Can the logistic growth model also handle declining populations? The answer is yes, if the initial population is greater than the carrying capacity.

One scenario in which a population can be greater than the carrying capacity is if the carrying capacity changes. Whether slow or sudden, changes to an environment, such as a reduction in available resources or the introduction of a competing species, can reduce an environment's carrying capacity. The logistic model then predicts a decrease in the population, which approaches the carrying capacity asymptotically from above.

Example 2 *A sudden loss of available food causes the carrying capacity of the rodents of example 1 to be reduced to* 1000. *If the population of rodents was* 2950 *at the time of the change, how long does it take the population to drop to* 1200 *rodents? Assume the value of the growth constant k has not changed.*

Solution At the time of the change (with which we associate $t = 0$), the population is $P_0 = 2950$. We are also given $M = 1000$ and that k has not changed from example 1; thus, $k = 0.02279$.

⓿ Calculate A:

$$A = \frac{M - P_0}{P_0} = \frac{1000 - 2950}{2950} = -0.66102.$$

❶ Determine k. The value of k is provided.

❷ Determine the requested quantity. We wish to know the value of t when $P = 1200$:

The algebraic procedure for solving for t is identical to the procedure for solving for k. We isolate $e^{-0.02279t}$ on one side of the equation by itself (accomplished in line 6), take logs of both sides (line 7), and solve for t (line 8).

$$P = \frac{M}{1 + Ae^{-kt}}$$

$$1200 = \frac{1000}{1 - 0.66102e^{-0.02279t}}$$

$$1200(1 - 0.66102e^{-0.02279t}) = 1000$$

$$1200 - 793.22e^{-0.02279t} = 1000$$

$$-793.22e^{-0.02279t} = -200$$

$$e^{-0.02279t} = \frac{200}{793.22}$$

$$-0.02279t = \ln \frac{200}{793.22}$$

$$t = \frac{\ln \frac{200}{793.22}}{-0.02279} = 60.46 \text{ days.}$$

The rodent population shrinks to 1200 in approximately 60 days. ■

The rodent population function $P = \frac{1000}{1 - 0.66102e^{-0.02279t}}$ of example 2 is graphed in figure 2. As promised, the curve approaches the carrying capacity of 1000 asymptotically from above.

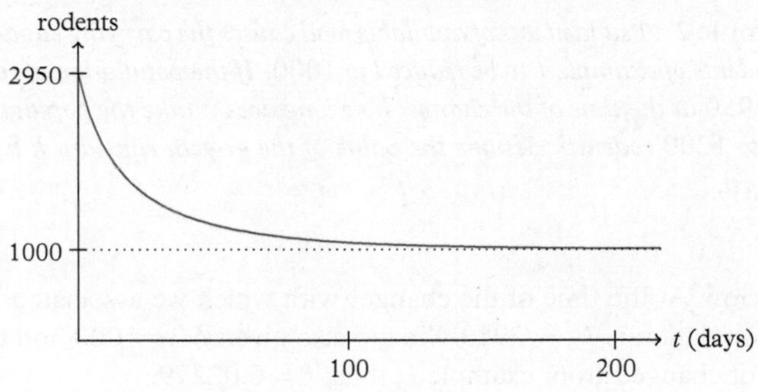

Figure 2 *The graph of the rodent population function (blue) of example 2*

The graphs of the logistic curves in figures 1 and 2 (blue in both figures) are typical of the shapes of logistic curves. When the population P is less than the carrying capacity M, we have $\frac{P}{M} < 1$; thus, $1 - \frac{P}{M} > 0$ and $\frac{dP}{dt} = kP\left(1 - \frac{P}{M}\right) > 0$. The positive derivative indicates an increasing function, as exhibited in figure 1. However, when the population P is larger than the carrying capacity M, we have $\frac{P}{M} > 1$; hence, $1 - \frac{P}{M} < 0$ and $\frac{dP}{dt} = kP\left(1 - \frac{P}{M}\right) < 0$. The negative derivative indicates a decreasing function, as exhibited in figure 2.

Recall that $\frac{dP}{dt} = kP\left(1 - \frac{P}{M}\right)$ is the logistic differential equation that we used to derive the logistic growth formulas.

Have you heard ...?

The spread of a rumor can be modeled using logistic growth. If the rumor starts with one person telling two people, and those individuals each tell two more, and then they tell two more, the spread resembles exponential growth. Eventually, some of those spreading the rumor find that their listeners have heard the rumor before, so the spread slows down. And when nearly everyone has heard it, the spread becomes slower because there are very few new people to tell (it's old news!).

Example 3 *An organization has 650 employees in an office building. At 9:00 a.m., 13 employees hear a rumor. By 10:30 a.m., the rumor is known by 47 employees. At what time will 90% of the employees have heard the rumor? Assume logistic growth.*

Solution We are given that $P_0 = 13$ (meaning $t = 0$ corresponds to 9:00 a.m.) and $M = 650$.

The carrying capacity, or maximum number of employees who have heard the rumor, is the total number of employees.

⓪ Calculate A:

$$A = \frac{M - P_0}{P_0} = \frac{650 - 13}{13} = 49.$$

❶ Determine k. We use the additional information given, that $P = 47$ when $t = 1.5$ (meaning, t is measured in hours):

Alternatively, we can measure t in minutes and use $P = 47$ when $t = 90$.

$$\begin{aligned} P &= \frac{M}{1 + Ae^{-kt}} \\ 47 &= \frac{650}{1 + 49e^{-k(1.5)}} \\ 47 + 2303e^{-1.5k} &= 650 \\ 2303e^{-1.5k} &= 603 \\ e^{-1.5k} &= \frac{603}{2303} \\ -1.5k &= \ln\frac{603}{2303} \\ k &= \frac{\ln\frac{603}{2303}}{-1.5} = 0.8934. \end{aligned}$$

❷ Determine the requested quantity. We wish to know t when 90% of the employees have heard the rumor—that is, when $P = 0.90 \cdot 650 = 585$:

$$\begin{aligned} P &= \frac{M}{1+Ae^{-kt}} \\ 585 &= \frac{650}{1+49e^{-0.8934t}} \\ 585 + 28\,665e^{-0.8934t} &= 650 \\ 28\,665e^{-0.8934t} &= 65 \\ e^{-0.8934t} &= \frac{65}{28\,665} \\ -0.8934t &= \ln \frac{65}{28\,665} \\ t &= \frac{\ln \frac{65}{28\,665}}{-0.8934} = 6.816. \end{aligned}$$

The time when 90% of the employees have heard the rumor is 6 hours and $0.816 \times 60 = 49$ minutes after 9:00 a.m., or 3:49 p.m. ■

The manner of spread of a rumor is that each person can be "infected" by the rumor only once; hearing the rumor a second time has no effect. Assuming no memory loss, when someone is infected, that person remains infected. There are a finite number of individuals that can become infected (the carrying capacity). This is the perfect setup for the logistic growth model.

If a virus can be described by the characteristics of the previous paragraph, then it can also be modeled well by logistic growth. Other models used for the spread of a virus include susceptible–infected–recovered (SIR) models.

Even if a virus does not meet these characteristics, a logistic growth model is a better fit than an exponential growth model.

The study of population dynamics doesn't end here: predator–prey models, considerations for harvesting, or for immigration and emigration—the list goes on. There is no shortage of interesting ideas to explore.

Exercises 9.8

1–8. Rapid response: which is a better choice to model the given quantity: an exponential model, a logistic model, or neither?

1. the number of people on a social media platform who have viewed a cat video
2. the general price level in a well-controlled economy with 2% inflation

3. the population of a rat species on an island after accidental colonization from a ship
4. the number of individuals in a community who have been exposed to a virus
5. the balance in a money-market account with a fixed interest rate and no deposits or withdrawals
6. the cumulative number of times that you have brushed your teeth in your lifetime
7. the temperature of a cup of coffee left on the table
8. the population of an invasive fish species in a waterway

9–22. Assume logistic growth to answer each question.

9. A bacteria is introduced into an environment with a carrying capacity of 1 000 000. If the initial population is 1200 and after 4 hours the population has risen to 6900, what is the population after 12 hours?
10. A bacteria is introduced into an environment with a carrying capacity of 4 000 000. If the initial population is 1200 and after 4 hours the population has risen to 6900, what is the population after 12 hours?
11. A weed is introduced into an undisturbed meadow. The initial population is 14 plants. After 6 years, the population has reached 37 plants. Assuming a carrying capacity of 687 plants, how long does it take for the number of plants to reach 500?
12. A weed is introduced into an undisturbed meadow. The initial population is 14 plants. After 6 years, the population has reached 37 plants. Assuming a carrying capacity of 2190 plants, how long does it take for the number of plants to reach 500?
13. A store has a customer base of 12 400 individuals. The store introduces a new loyalty program and 4700 customers sign up immediately. After 3 months, the loyalty program has enrolled 8200 customers. How long is it until the program's participation reaches 10 000?
14. The number of people on a social media platform interested in watching a cat video is 3 400 000. Suppose that the initial number of viewers is 32 and that after 14 days the number of viewers has reached 1191. How long does it take the number of viewers to reach 1 000 000?

We are counting the number of distinct viewers, not the total number of views.

15. A rumor is spreading through an organization of 1280 employees. At 9:00 a.m., 25 people have heard the rumor. At 9:30 a.m., 46 have heard the rumor. At noon, the organization makes an announcement with the same content as the rumor. Did more employees hear the news from the rumor mill or from the official announcement?

16. A conspiracy theory is spreading through a network of 8100 like-minded persons. Forty days before an election, 184 members of the network have heard the theory. Thirty days before the election, this number has risen to 388. Will the majority have heard the theory by the time of the election?

17. A mammal species has colonized an environment with a carrying capacity of 37 400 with an initial population of 42 animals. After 4 years, the population has grown to 156. What is the population after 10 years?

18. An invasive fish species has been introduced to a reservoir with a carrying capacity of 280 000. Assume $P_0 = 125$ and that after 11 months the population has grown to 8104. What is the population after 36 months?

19. A sudden loss of resources has reduced the carrying capacity for the mammals of exercise 17 to 9200. Assume that the growth constant k has not changed and that the population reached 37 400 at the time of the change. How long does it take the population to drop to 20 000?

20. The water level in the reservoir of exercise 18 has been purposefully reduced and the carrying capacity for the invasive fish species has changed to 115 000. Assume that the growth constant k has not changed and that the population reached 280 000 at the time of the change. How long does it take the population to drop to 150 000?

21. A virus is capable of spreading at a growth rate of 5% per day (that is, $k = 0.05$). In a community of 1000 susceptible individuals, the virus has initially infected 10 of them. (a) How long does it take for the number of infected individuals to double to 20? (b) When the number infected has reached 300, how much longer does it take to double to 600?

22. A virus is capable of spreading at a growth rate of 8% per day (that is, $k = 0.08$). In a community of 120 000 susceptible individuals, the virus has initially infected 40 of them. (a) How long does it take for the number of infected individuals to increase by an order of magnitude to 400? (b) How much longer does it take to increase another order of magnitude to 4000? (c) And how much longer to reach 40 000? (e) Why is there no part (d)?

23. (a) Answer the question of exercise 11 using an exponential model. (b) Compare the answer to part (a) to the answer from exercise 11.

24. (a) Answer the question of exercise 14 using an exponential model. (b) Compare the answer to part (a) to the answer from exercise 14.

25. Determine the time t at which the logistic population curve of example 1 reaches an inflection point.

26. The algebraic steps in examples 1(b) and 3 for determining the growth constant k are identical. Repeat these steps to determine a formula for k in terms of M, A, P_1, and t_1, where we are given the additional information that the population is P_1 at time t_1.

27. By setting $M = \Omega$, show that if the carrying capacity is infinite, the logistic model is identical to the exponential model.

28. In example 1, the value of k for the logistic model is slightly larger than the value of k for the exponential model. Why?

Chapter X

Sequences and Series

Sequences

10.1

This chapter begins with a discussion of sequences.

What is a sequence?

Informally, a *sequence* is an infinitely long list of numbers, such as

$$1, 3, 5, 7, 9, \ldots,$$

The ellipses (the ...) indicate that the sequence continues forever without end. The comma after the ellipses is punctuation for the sentence, and it is not part of the notation for the sequence. The sequence by itself, with no punctuation (no period at the end of this sentence), is written as

$$1, 3, 5, 7, 9, \ldots$$

the sequence of odd numbers. The individual numbers are called *terms* of the sequence. The first term of the previous sequence is 1, the second term is 3, the third term is 5, and so on. If we label the nth term of the sequence a_n, then we can write a formula for the nth term:

$$a_n = 2n - 1.$$

This is an example of an *arithmetic sequence*, a sequence in which the difference between successive terms is a constant (in this case, 2).

Additional examples of sequences include

$$1, \frac{1}{2}, \frac{1}{3}, \frac{1}{4}, \frac{1}{5}, \ldots,$$

for which the pattern seems to be

$$b_n = \frac{1}{n},$$

and

$$2, 3, 5, 7, 11, 13, 17, \ldots,$$

the sequence of prime numbers, for which there is no known algebraic formula for the nth term.

Have you encountered a test question of the type: What is the next term in the sequence 1, 3, 5, 7, 9, ...? The intended answer to the question is the next odd number, 11. The idea is to assess your ability to recognize a pattern, which is a useful skill. But which pattern? It turns out that for any real number you may wish to use as the sixth term of the sequence, there is a polynomial formula of degree 5 that matches the first five numbers and the number you selected. For instance, if you want 12 to be the next term in the sequence, you could use the formula $a_n = \frac{1}{120}n^5 - \frac{1}{8}n^4 + \frac{17}{24}n^3 - \frac{15}{8}n^2 + \frac{257}{60}n - 2$. The first six terms of this sequence are 1, 3, 5, 7, 9, and 12.

Returning to the sequence $a_n = 2n - 1$, notice that we are using the formula in the same way we evaluate a function. We choose a value for n and evaluate the expression $2n - 1$ for that value of n. For instance, the fifth term of the sequence is $a_5 = 2(5) - 1 = 9$. However, rather than choose any real number, we are restricting the possible choices of n to the natural numbers **N**—that is, positive integers. These observations bring us to the formal definition of a sequence.

Definition 1 SEQUENCE *A* sequence *is a function with domain* **N.**

Although we can write our sequence using function notation as $f(n) = 2n - 1$ and evaluate the third term as $f(3) = 2(3) - 1 = 5$, we use the notation $a_n = 2n - 1$ to emphasize the restriction of the domain to the positive integers.

Example 1 *Write the first four terms of the sequence* $a_n = (-2)^n$.

Notice that a_1 is the first term of the sequence, a_4 is the fourth term of the sequence, and a_n is the *n*th term of the sequence. We use the notation (a_n) to refer to the sequence as a whole, although we also refer to a sequence by the formula that generates it, such as "the sequence given by $a_n = (-2)^n$."

Solution Replacing *n* with 1, 2, 3, and 4, we have

$$a_1 = (-2)^1 = -2$$

$$a_2 = (-2)^2 = 4$$

$$a_3 = (-2)^3 = -8$$

$$a_4 = (-2)^4 = 16.$$

This is an example of a *geometric sequence.* Instead of always adding the same number to obtain the next term, as in an arithmetic sequence, we always multiply by the same number to get the next term. In this case, we are multiplying by -2.

The first four terms of the sequence are

$$-2, 4, -8, 16.$$

(Note that those are the first four terms. To write the sequence in list form, we add the ellipses: $-2, 4, -8, 16, \ldots$. The fourth dot is the period at the end of the sentence.) ■

Reading Exercise 1 *Write the first four terms of the sequence* $a_n = \dfrac{n}{n+1}$.

Recursively defined sequences

Another way to specify a sequence is to provide one or more initial terms of the sequence and then state a rule for determining additional terms. For instance, we can define the sequence (a_n) using

$$a_1 = 3, a_n = a_{n-1} + 10 \text{ for } n \geq 2.$$

The initial term is $a_1 = 3$, whereas $a_n = a_{n-1} + 10$ is called a *recursive formula.* With $n = 2$ in the recursive formula, we have

$$\begin{aligned} a_2 &= a_{2-1} + 10 \\ &= a_1 + 10 \\ &= 3 + 10 \\ &= 13. \end{aligned}$$

In line 1, we replace the n in the recursive formula with 2; in line 2, we do the arithmetic in the subscript; in line 3, we replace a_1 with its value 3; and in line 4, we complete the arithmetic.

We can use the formula repeatedly to get as many terms as we wish:

$$a_3 = a_2 + 10 = 13 + 10 = 23,$$

$$a_4 = a_3 + 10 = 23 + 10 = 33,$$

and so on.

Because this sequence is

$$3, 13, 23, 33, 43, \ldots,$$

the pattern is apparent and we can easily determine a formula for the nth term:

$$a_n = 10(n-1) + 3 = 10n - 7.$$

Sometimes, however, determining a formula for the nth term is more difficult.

Example 2 *Determine the first 10 terms of the recursively defined sequence* $a_1 = 1$, $a_2 = 1$, $a_n = a_{n-1} + a_{n-2}$ *for* $n \geq 3$.

Solution We use the recursive formula beginning with $n = 3$:

Here, the initial terms are $a_1 = 1$ and $a_2 = 1$, and the recursive formula is $a_n = a_{n-1} + a_{n-2}$.

$$a_3 = a_{3-1} + a_{3-2} = a_2 + a_1 = 1 + 1 = 2,$$

$$a_4 = a_{4-1} + a_{4-2} = a_3 + a_2 = 2 + 1 = 3,$$

$$a_5 = a_{5-1} + a_{5-2} = a_4 + a_3 = 3 + 2 = 5.$$

However, it is quicker to notice the pattern given by the recursive formula: that each term a_n is the sum of the previous two terms in the sequence, a_{n-1} and a_{n-2}. Then,

$$a_6 = 5 + 3 = 8,$$

$$a_7 = 8 + 5 = 13.$$

The first 10 terms of the sequence are

$$1, 1, 2, 3, 5, 8, 13, 21, 34, \text{ and } 55.$$

This sequence is known as the *Fibonacci sequence*. ■

Leonardo of Pisa, also known as Fibonacci, is credited with introducing Hindu-Arabic numerals to Europe and advocating for their use instead of Roman numerals.

Leonardo Pisano Fibonacci, ca. 1170–1250 http://www-history.mcs.st-andrews.ac.uk /Biographies/Fibonacci.html

The Newton's method formula $a_{n+1} = a_n - \dfrac{f(a_n)}{f'(a_n)}$ for $n \geq 1$, along with an initial guess a_1, is another example of defining a sequence using recursion.

Plotting sequences

There are two common methods for visualizing a sequence. Let's look at an example.

Ans. to reading exercise 1
$\frac{1}{2}, \frac{2}{3}, \frac{3}{4}, \frac{4}{5}$

Example 3 *Plot the sequence* $a_n = \dfrac{n}{n+1}$.

Solution Before plotting the sequence, we calculate its first few terms:

This is the same sequence as in reading exercise 1.

$$\frac{1}{2}, \frac{2}{3}, \frac{3}{4}, \frac{4}{5}, \ldots .$$

The first method for plotting a sequence produces a one-dimensional picture. We plot the first few terms of the sequence on a number line, labeling the terms $a_1, a_2, \ldots$. See figure 1.

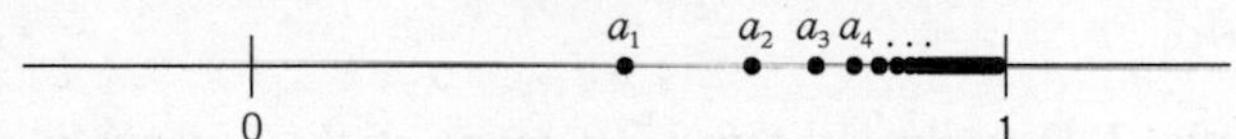

Figure 1 *The first 100 terms of the sequence* $a_n = \frac{n}{n+1}$ *plotted on a number line. Each term is represented by a dot on the number line. The dots become indistinguishable from one another as they get closer to* 1

The second method for plotting a sequence produces a two-dimensional picture. Thinking of the sequence as a function (table 1), we plot points as we would for any other function. We don't connect the dots because this is not a continuous function on an interval of real numbers. See figure 2.

Table 1 *Table of points for the function* $a_n = \frac{n}{n+1}$.

This table of values is like those we use to graph functions. The left-hand column is the horizontal coordinate (think "*x*-coordinate"), and the right-hand column is the vertical coordinate (think "*y*-coordinate").

n	$a_n = \frac{n}{n+1}$
1	$\frac{1}{2} = 0.500$
2	$\frac{2}{3} = 0.667$
3	$\frac{3}{4} = 0.750$
4	$\frac{4}{5} = 0.800$
$\vdots$	$\vdots$

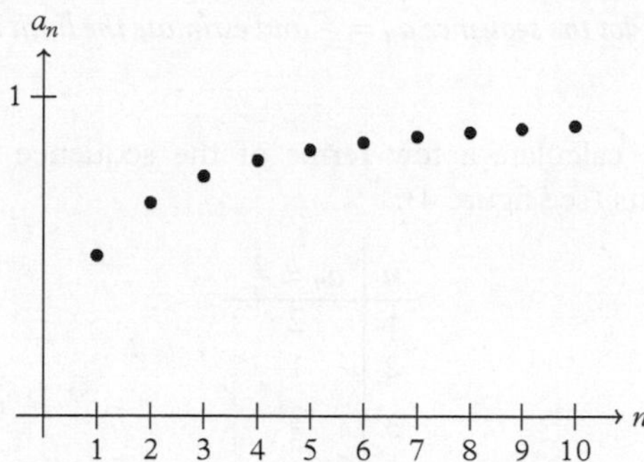

Figure 2 *The first 10 terms of the sequence $a_n = \frac{n}{n+1}$ plotted in two dimensions. The horizontal axis represents the number (index, subscript) n of the term, and the vertical coordinate represents the value a_n of the term*

Although it is good to practice both methods of plotting sequences, we usually use the second method and produce a two-dimensional picture. ■

Most graphing calculators have a sequence mode and can produce the two-dimensional pictures used in this section.

Reading Exercise 2 *Plot the sequence $a_n = \frac{1}{n}$.*

In figure 2, the picture of the sequence resembles the graph of a function $y = f(x)$ with a horizontal asymptote of $y = 1$. For such a function, we write $\lim_{x\to\infty} f(x) = 1$. In a similar manner, we say the limit of the sequence is one and write

$$\lim_{n\to\infty} a_n = 1.$$

Figure 3 is a repeat of figure 2, with the addition of a dotted horizontal line representing the sequence limit. Although we do not call this a horizontal asymptote, the graphical interpretation of the limit is very much the same.

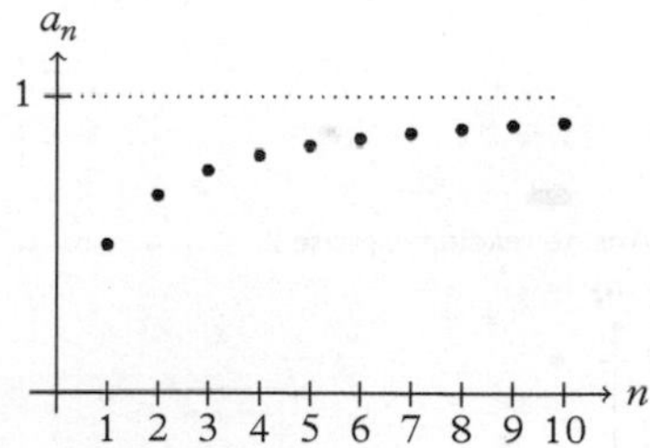

Figure 3 *A dotted line representing the limit of the sequence is inserted*

The one-dimensional picture in figure 1 also depicts the limit of the sequence, but in a different manner. As we move farther in the sequence, the dots representing the terms get closer and closer to 1 on the number line. If we continue indefinitely, the position of the dots appears indistinguishable from 1.

We look at sequence limits formally in the next section and calculate them algebraically, but for now we study them graphically.

Estimating sequence limits graphically

As stated previously, estimating the limit of a sequence graphically is the same skill as estimating the limit of a function graphically at infinity.

Example 4 *Plot the sequence $a_n = \frac{2}{n}$ and estimate the limit of the sequence graphically.*

Solution We calculate a few terms of the sequence and plot the resulting points (see figure 4):

n	$a_n = \frac{2}{n}$
1	2
2	1
3	$\frac{2}{3}$
4	$\frac{1}{2}$
⋮	⋮

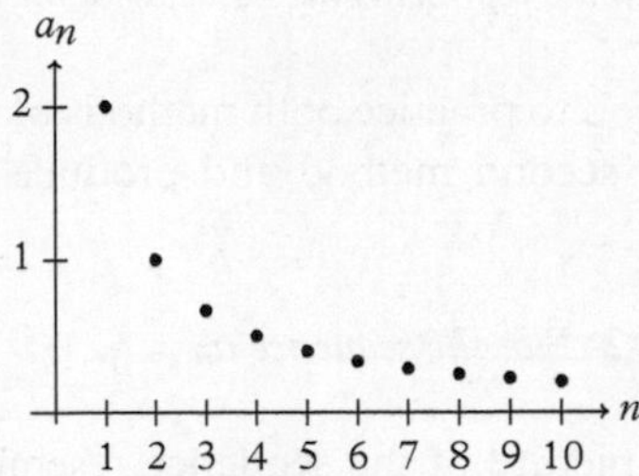

Figure 4 *The sequence for example 4*

It appears that the limit of the sequence is zero—that is,

$$\lim_{n\to\infty} a_n = \lim_{n\to\infty} \frac{2}{n} = 0.$$

■

Ans. to reading exercise 2

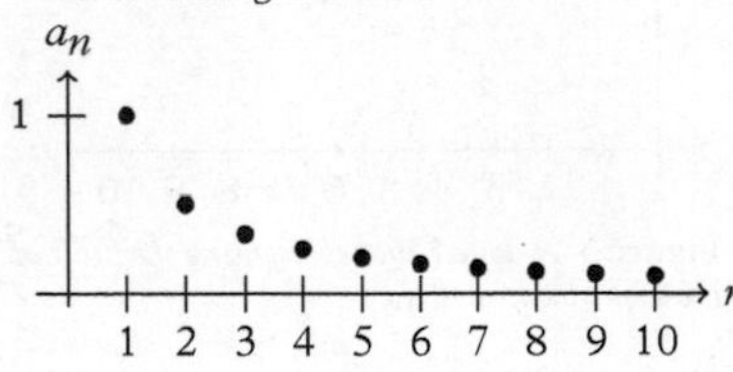

When the limit of a sequence is a real number, as in example 4, we say the sequence *converges*. Otherwise, we say the sequence *diverges*.

Example 5 *Plot the sequence $a_n = n$ and estimate the limit of the sequence graphically.*

Solution The sequence is plotted in figure 5. Because the values of the function appear to rise without bound, we conclude that the limit of the sequence is ∞—that is,

$$\lim_{n\to\infty} a_n = \infty.$$

We can write $\lim_{n\to\infty} n = \infty$ instead.

This sequence diverges. ■

In figure 4, each term of the sequence is smaller than the previous term. We call such a sequence *decreasing*. In figure 5, the opposite is true. When each term is larger than the previous term, we call the sequence *increasing*.

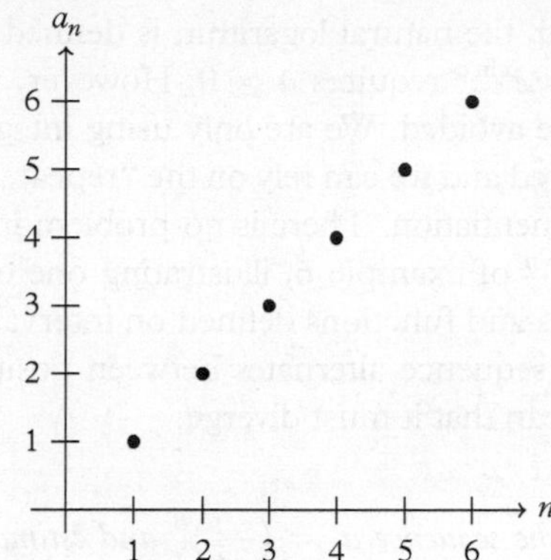

Figure 5 *The sequence for example 5*

Definition 2 MONOTONE SEQUENCE *A sequence* (a_n) *is* decreasing *if* $a_n < a_{n-1}$ *for every* $n > 1$. *The sequence is* increasing *if* $a_n > a_{n-1}$ *for every* $n > 1$. *If a sequence is increasing or if it is decreasing, it is called* monotone.

Contrary to our work with functions, we do not say that a sequence is increasing on some interval and decreasing on some other interval. We only say that a sequence is increasing if its value increases from one term to the next, every time, throughout the entire sequence. The same is true for decreasing; it must be true throughout the entire sequence.

Example 6 *Plot the sequence* $a_n = (-1)^n$ *and estimate the limit of the sequence graphically.*

Solution Since $(-1)^n = 1$ when n is even and $(-1)^n = -1$ when n is odd, the sequence alternates, as in figure 6.

Because the sequence values do not approach one number, the limit does not exist (DNE)—that is,

Alternatively, we could write $\lim_{n\to\infty} a_n$ DNE.

$$\lim_{n\to\infty} (-1)^n \text{ DNE}.$$

The sequence diverges. ■

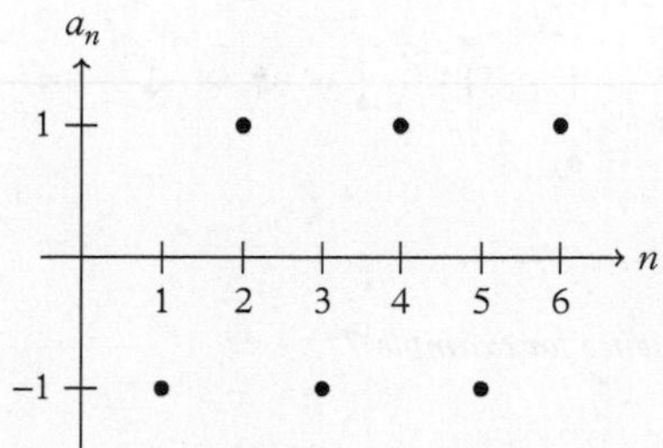

Figure 6 *The sequence for example 6*

The sequence in example 6 is neither increasing nor decreasing, and therefore it is not monotone.

In chapter 5 we defined exponential functions $f(x) = a^x$ only for positive values of a. The reason for this can be seen in two ways. First, even-numbered roots of negative numbers such as $(-1)^{\frac{1}{2}} = \sqrt{-1}$ are undefined. Second, the natural logarithm is defined only for positive numbers, so $a^x = e^{x \ln a}$ requires $a > 0$. However, with a sequence, these problems are avoided. We are only using integer exponents, so no roots are involved and we can rely on the "repeated multiplication" definition of exponentiation. There is no problem in defining the sequence $a_n = (-1)^n$ of example 6, illustrating one of the differences between sequences and functions defined on intervals.

Just because a sequence alternates between positive and negative terms does not mean that it must diverge.

Example 7 *Plot the sequence* $a_n = \left(-\frac{1}{2}\right)^n$ *and estimate the limit of the sequence graphically.*

Solution Once again we calculate a few values of the sequence and plot the resulting points (figure 7):

n	$a_n = \left(-\frac{1}{2}\right)^n$
1	$-\frac{1}{2}$
2	$\frac{1}{4}$
3	$-\frac{1}{8}$
4	$\frac{1}{16}$
5	$-\frac{1}{32}$

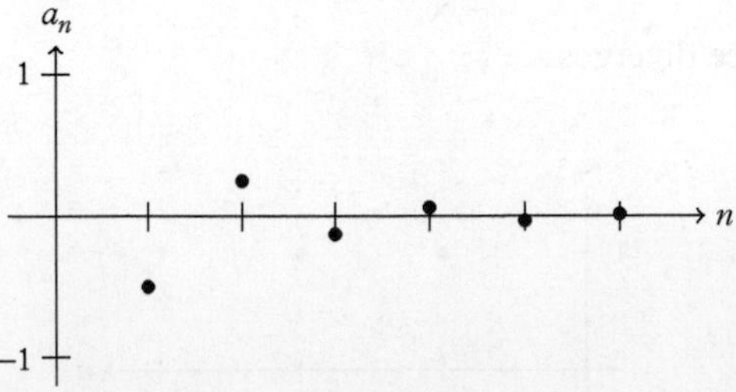

Figure 7 *The sequence for example 7*

Although the terms of the sequence are alternating positive and negative, they still appear to be getting closer (arbitrarily close) to zero. We conclude that the limit of the sequence is zero—that is,

This sequence is not monotone (it is not an increasing sequence, nor is it a decreasing sequence).

$$\lim_{n\to\infty} a_n = \lim_{n\to\infty} \left(-\tfrac{1}{2}\right)^n = 0.$$

This sequence converges. ■

Reading Exercise 3 *Estimate graphically the limit of the sequence (a_n) pictured in figure 8.*

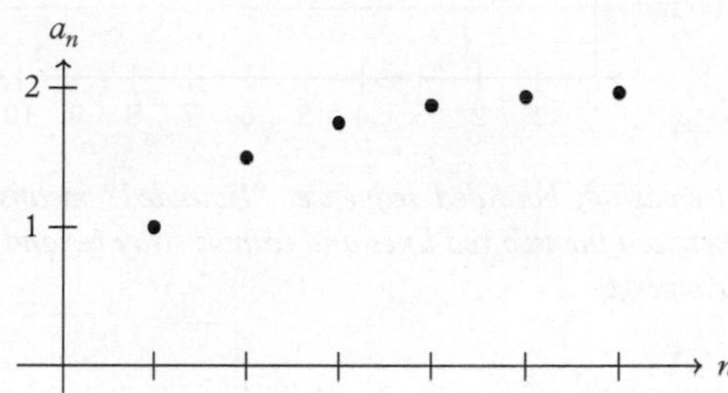

Figure 8 *The sequence (a_n) for reading exercise 3*

Monotone convergence theorem

Examples 5 and 6 illustrate two different ways in which a sequence can diverge. In example 5, the sequence $a_n = n$ is unbounded and the limit is ∞. In example 6, the sequence $a_n = (-1)^n$ is bounded, but it bounces back and forth between two values, so that the limit still does not exist.

What if a sequence is bounded (to avoid the type of divergence in example 5) and monotone (to avoid bouncing back and forth between two values as in example 6)? Is this enough to guarantee that the sequence converges?

The phrase *monotone decreasing* has the same meaning as *decreasing*.

Figure 9 illustrates a monotone decreasing sequence that is bounded. The bounds are shown by red lines; the values of the sequence must remain between the two bounds. With infinitely many remaining terms that can only go lower, and a finite amount of room in which to work, the sequence can't avoid settling down toward one particular limit. Such a sequence must converge. The limit might not be exactly where the red line representing a lower bound has been drawn (figure 10), but the limit exists nonetheless.

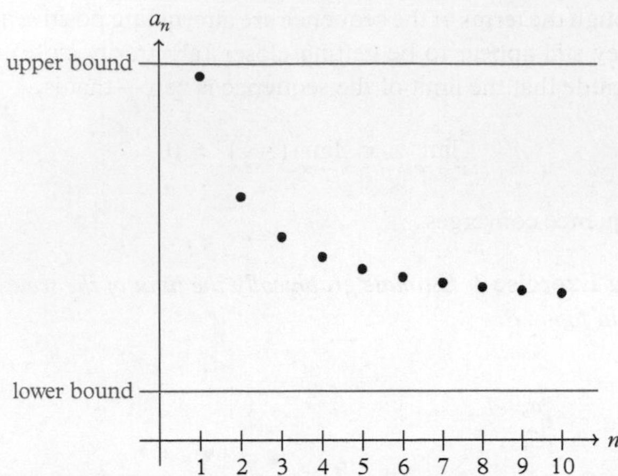

The bounds in figure 9 are not necessarily placed at the optimum locations. For instance, since the sequence is decreasing it can't possibly go above the value of the first term a_1; the upper bound could have been drawn lower than it is.

Figure 9 *A decreasing, bounded sequence. "Bounded" means the sequence must lie on or between the two red lines and cannot stray beyond them. Such a sequence must converge*

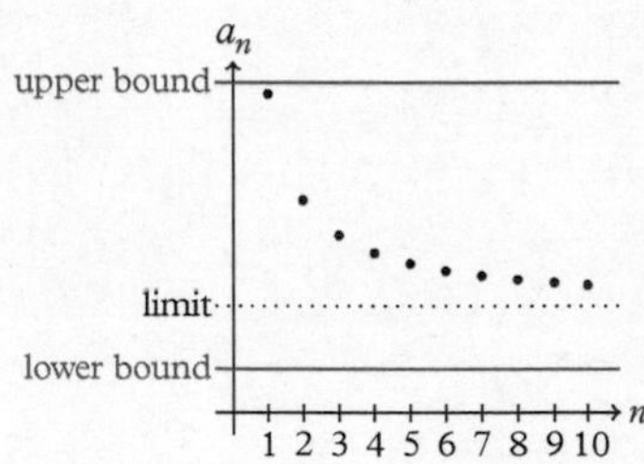

Figure 10 *A dotted line is added to illustrate the location of the limit of the sequence. The red line illustrating a lower bound could have been placed as high as the dotted line*

The same must be true of a bounded, monotone increasing sequence (figure 11). Whether increasing or decreasing, any bounded, monotone sequence must converge. The formal proof of the theorem is usually presented in an upper-level course on analysis.

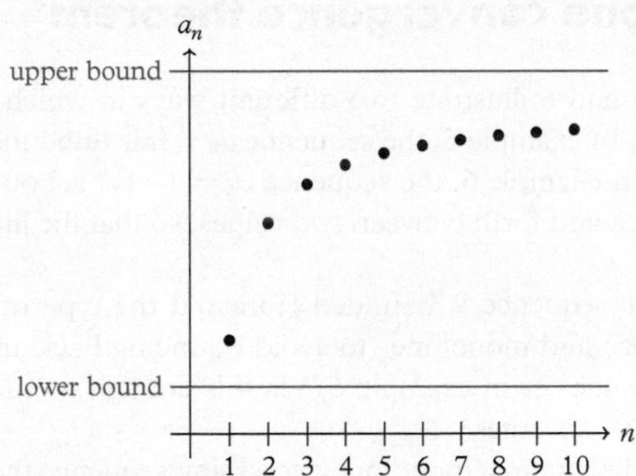

Figure 11 *An increasing, bounded sequence. Such a sequence must converge*

Ans. to reading exercise 3
$\lim_{n\to\infty} a_n = 2$

Theorem 1 MONOTONE CONVERGENCE THEOREM *Every bounded, monotone sequence converges.*

Reading Exercise 4 *Does the sequence pictured in figure 2 appear to be monotone? Does it appear to be bounded?*

Exercises 10.1

1–6. Rapid response: is this a sequence?

1. 2, 4, 8, 16, 32, 64, 128
2. 2, 4, 8, 16, 32, 64, 128, …
3. a_n
4. (a_n)
5. $a_1 = 1,\ a_n = \dfrac{3 + a_{n-1}}{4}$
6. $\lim_{n\to\infty} \dfrac{3}{n}$

7–18. Give the first six terms of the sequence.

7. $a_n = \dfrac{1}{2n-1}$
8. $a_n = \dfrac{2}{n+3}$
9. $a_n = \dfrac{(-1)^n}{2n-1}$
10. $a_n = (-2)^{n+1}$
11. $a_n = \dfrac{\sqrt{2}\cos\left(\frac{(2n-1)\pi}{4}\right)}{2n-1}$
12. $a_n = \dfrac{\sin\left(\frac{(2n+1)\pi}{4}\right)}{n}$
13. $a_n = \dfrac{2n!}{(2n)!}$
14. $a_n = \dfrac{2^n}{n!}$

Reminder: $7! = 7 \cdot 6 \cdot 5 \cdot 4 \cdot 3 \cdot 2 \cdot 1$

15. $a_1 = 1,\ a_n = \dfrac{a_{n-1}}{n}$
16. $a_1 = 3,\ a_n = \dfrac{1}{a_{n-1}}$
17. $a_1 = 2,\ a_2 = 3,\ a_n = 2a_{n-1} - a_{n-2}$
18. $a_1 = 2,\ a_2 = 3,\ a_n = 2a_{n-1} + a_{n-2}$

19–26. Give a formula for the nth term of the sequence.

19. $\dfrac{-2}{5}, \dfrac{-2}{6}, \dfrac{-2}{7}, \dfrac{-2}{8}, \ldots$
20. $\dfrac{1}{3}, \dfrac{1}{4}, \dfrac{1}{5}, \dfrac{1}{6}, \ldots$

21. $\frac{2}{5}, \frac{-2}{6}, \frac{2}{7}, \frac{-2}{8}, \ldots$

22. $\frac{1}{3}, -\frac{1}{4}, \frac{1}{5}, -\frac{1}{6}, \ldots$

23. $\frac{1}{3}, \frac{1}{9}, \frac{1}{27}, \frac{1}{81}, \ldots$

24. $\frac{1}{4}, \frac{3}{7}, \frac{5}{10}, \frac{7}{13}, \ldots$

25. $\frac{1}{3}, \frac{1}{5}, \frac{1}{7}, \frac{1}{9}, \ldots$

26. $e^{-2}, e^{-4}, e^{-6}, e^{-8}, \ldots$

27–36. (a) Plot the sequence. (b) Does the sequence appear to converge or diverge? (c) Estimate the limit of the sequence graphically. (d) Is the sequence monotone? (e) Is the sequence bounded?

27. $a_n = \frac{n+3}{2n}$

28. $a_n = \left(\frac{1}{2}\right)^n$

29. $a_n = \frac{n^2+3}{2n}$

30. $a_n = \frac{(-1)^n n}{n+1}$

31. $a_n = \frac{(-1)^n 2n}{n^2+3}$

32. $a_n = \frac{(-1)^n}{n+1}$

33. $a_n = \frac{(-1)^{n+1}(n+3)}{2n}$

34. $a_n = \frac{3^n}{n!}$

35. $a_n = \frac{5n^2}{2^n}$

36. $a_n = \frac{e^n}{n^5}$

37–40. Use technology to plot the sequence. Include at least 10 terms.

37. $a_n = \sin\left(n^2\right)$

38. $a_n = \frac{4n-27}{n-6.5}$

39. $a_n = \dfrac{(n^2+1)2^n}{n!}$

40. $a_n = \ln\left(\left(-\frac{3}{5}\right)^n + 5\right)$

41–46. The On-Line Encyclopedia of Integer Sequences (OEIS, located at https://oeis.org/) is useful for identifying where a sequence of integers may have come from. For instance, if you enter the numbers 1, 1, 2, 3, 5, 8, 13, 21 in the search bar at this site, the top result is sequence A000045, the Fibonacci numbers, with a very large number of comments, references, and links.

Use the OEIS database (a) to identify the sequence and, if available, (b) give a formula (possibly a recursive formula) for the sequence or (c) state a description of the sequence.

41. 2, 1, 3, 4, 7, 11, 18, 29, …

42. 1, 1, 2, 4, 9, 21, 51, 127, 323, …

43. 6, 28, 496, 8 128, 33 550 336, …

44. 0, 1, 2, 5, 12, 29, 70, 169, 408, …

45. 70, 836, 4 030, 5 830, 7 192, …

46. 1, 2, 21, 22, 3, 31, 32, 321, …

Ans. to reading exercise 4
Yes, the sequence appears to be monotone (increasing) and bounded.

47. Look up "Lagrange interpolating polynomial" and apply the formula for the points $(1,1)$, $(2,3)$, $(3,5)$, $(4,7)$, $(5,9)$, and $(6,12)$. Simplify. Using n as the variable, you should be able to obtain the formula located in a marginal note toward the beginning of this section. The first six terms of the resulting sequence are 1, 3, 5, 7, 9, and 12.

48. Repeat exercise 47 to find a formula for a sequence with the first six terms 1, 3, 5, 7, 9, and 15.

Sequence Limits

10.2

In the previous section we estimated the limit of a sequence graphically. Graphical estimation has its limitations, as evidenced by the difficulty in discerning whether the sequence in figure 1 converges or diverges. Do the terms of the sequence get arbitrarily close to zero? Or do they

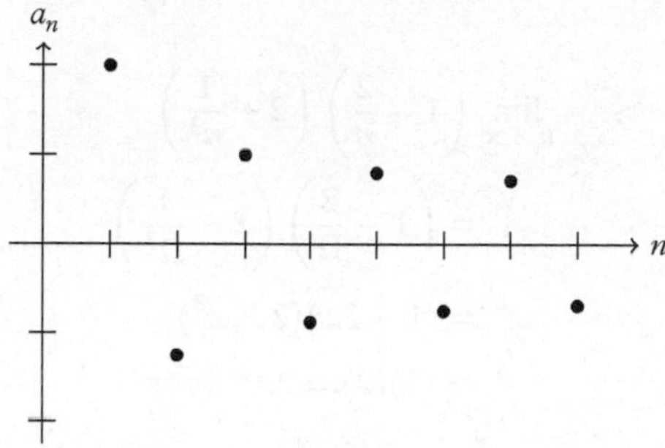

Figure 1 *A sequence. Does the sequence converge or diverge?*

stay away from zero while continuing to alternate positive and negative, so that the limit does not exist? We need a more precise method for determining the limit of a sequence.

Sequence limit definition

We have already noted the similarity between estimating the limit of a sequence graphically and estimating the limit of a function at (positive) infinity graphically. The definitions of the two kinds of limit are likewise similar. The main difference is that, for a sequence limit, the positive infinite hyperreal number Ω must also be an integer.

Definition 3 SEQUENCE LIMIT *Let (a_n) be a sequence. If a_Ω renders the same real result L for every positive infinite hyperreal integer Ω, then we write*

$$\lim_{n \to \infty} a_n = L,$$

and we call L the **limit** *of the sequence. If L is a real number (that is, not ∞ or $-\infty$), then we say the sequence (a_n)* **converges**. *Otherwise, we say (a_n)* **diverges**.

If we want to know what ultimately happens to the terms of a sequence—not at the 1000th term or the 1 000 000th term, but infinitely far out—we look at the "capital omegath" term of the sequence.

For the purpose of calculations, we write

$$\lim_{n\to\infty} a_n = a_\Omega.$$

In other words, we replace the n with Ω.

Let's look at the use of the definition in example 1.

Example 1 *Find* $\lim_{n\to\infty} a_n$ *for the sequence* $a_n = \left(1-\frac{2}{n}\right)\left(2+\frac{1}{n^3}\right)$.

Solution We begin by replacing n with a positive infinite hyperreal integer Ω. All the usual rules of working with hyperreal numbers apply. Then,

$$\lim_{n\to\infty}\left(1-\frac{2}{n}\right)\left(2+\frac{1}{n^3}\right) = \left(1-\frac{2}{\Omega}\right)\left(2+\frac{1}{\Omega^3}\right) = (1-2\omega)(2+\omega^3) \approx (1)(2) = 2.$$

Because the value of the limit is a real number, we say the sequence converges. ■

To match the definition of sequence limit more closely, we can rephrase the calculation as a computation of the capital omegath term:

$$a_\Omega = \left(1-\frac{2}{\Omega}\right)\left(2+\frac{1}{\Omega^3}\right) = (1-2\omega)(2+\omega^3) = 2-4\omega+\omega^3-2\omega^4 \approx 2.$$

Because we render the same real result, 2, for every positive infinite hyperreal integer Ω, we write

$$\lim_{n\to\infty} a_n = 2.$$

As expected, the calculation of sequence limits can be very similar to the calculation of limits of functions at infinity.

Example 2 *Find* $\lim_{n\to\infty} \frac{2n^2+4n-6}{3n^2+5}$.

Solution Using an arbitrary positive infinite hyperreal integer Ω,

$$\lim_{n\to\infty} \frac{2n^2+4n-6}{3n^2+5} = \frac{2\Omega^2+4\Omega-6}{3\Omega^2+5} \approx \frac{2\Omega^2}{3\Omega^2} = \frac{2}{3}.$$

The limit of the sequence is $\frac{2}{3}$. ■

Change the n to an x, and this looks just like a function limit at infinity. Unless otherwise noted, when we write such an expression using the letter n, we are implying that the expression is the nth term of a sequence.

Reading Exercise 5 *Evaluate* $\lim_{n\to\infty} \frac{n^2+3}{n^3-7}$.

Sequence limits: handling $(-1)^n$

As noted in the previous section, although we defined $f(x) = a^x$ only for positive values of a (for any real number x), the expression $(-1)^n$

is defined for any integer n and may be used in a sequence definition. How do we handle the limit of such a sequence? The key is to recall that $(-1)^n$ is either 1 or -1, depending on whether n is even or odd.

Example 3 *Evaluate* $\lim_{n\to\infty} \dfrac{n+(-1)^n}{3n+1}$.

Solution We begin by replacing n with a positive infinite hyperreal integer Ω:

$$\lim_{n\to\infty} \frac{n+(-1)^n}{3n+1} = \frac{\Omega+(-1)^\Omega}{3\Omega+1}.$$

Because $(-1)^\Omega$ is either 1 or -1, it is on the real-number level. Then, becasue the numerator is on the Ω level, we discard the $(-1)^\Omega$ when we approximate:

This is another instance of the transfer principle at work. Because $(-1)^n$ is either 1 or -1 when the exponent is a real-number integer, the same is true when the exponent is a hyperreal-number integer.

$$\approx \frac{\Omega}{3\Omega} = \frac{1}{3}.$$

The limit of the sequence $a_n = \frac{n+(-1)^n}{3n+1}$ is $\frac{1}{3}$. ∎

In the next example, the sequence is tweaked by changing + to ·.

Example 4 *Evaluate* $\lim_{n\to\infty} \dfrac{n\cdot(-1)^n}{3n+1}$.

Solution We begin by replacing n with a positive infinite hyperreal integer Ω:

In contrast to example 3, the numerator here consists of just one term (there is no addition or subtraction of terms), so there are no lower-level terms to discard. Therefore, in line 2, we leave the numerator alone when approximating. In line 3, we simplify by canceling a factor of Ω in the numerator and the denominator.

$$\begin{aligned}\lim_{n\to\infty} \frac{n(-1)^n}{3n+1} &= \frac{\Omega(-1)^\Omega}{3\Omega+1}\\ &\approx \frac{\cancel{\Omega}(-1)^\Omega}{3\cancel{\Omega}}\\ &= \frac{(-1)^\Omega}{3}.\end{aligned}$$

Now what? The result of the calculation depends on whether Ω is even or odd. If Ω is even, $(-1)^\Omega = 1$ and we have

Just as with real-number integers, there are even hyperreal-number integers and odd hyperreal-number integers.

$$= \frac{1}{3}.$$

If Ω is odd, $(-1)^\Omega = -1$ and we have

$$= \frac{-1}{3}.$$

Because we do not render the same real result for every positive infinite hyperreal integer Ω (sometimes we get $\frac{1}{3}$ and sometimes we get $-\frac{1}{3}$), the limit does not exist. This sequence diverges. ■

We can write the conclusion as $\lim_{n\to\infty} \frac{n\cdot(-1)^n}{3n+1}$ DNE.

Sequence limits: approximating inside continuous functions

Ans. to reading exercise 5

0

Details: $\lim_{n\to\infty} \frac{n^2+3}{n^3-7} = \frac{\Omega^2+3}{\Omega^3-7} \approx \frac{\Omega^2}{\Omega^3} = \frac{1}{\Omega} = \omega \doteq 0$

Conditions under which we may approximate inside a continuous function are given chapter 1 theorem 16, chapter 2 theorem 6 (approximating inside root functions), chapter 3 theorem 12 (approximating inside a continuous function with a horizontal asymptote), and chapter 5 theorem 2 (approximation inside logarithms). These theorems are sometimes needed when calculating sequence limits.

Example 5 *Evaluate* $\lim_{n\to\infty} \sqrt{\frac{2n^2+4}{n^2}}$.

Solution For any positive infinite hyperreal integer Ω,

$$\lim_{n\to\infty} \sqrt{\frac{2n^2+4}{n^2}} = \sqrt{\frac{2\Omega^2+4}{\Omega^2}}$$
$$\approx \sqrt{\frac{2\Omega^2}{\Omega^2}}$$
$$= \sqrt{2}.$$ ■

In line 2, we approximate inside the square root, which is allowed according to chapter 2 theorem 6.

What if the conditions of none of the theorems are met? The next example illustrates an alternative if the approximation principle is violated.

Example 6 *Evaluate* $\lim_{n\to\infty} \ln\left(1-\frac{1}{n}\right)$.

Solution For any positive infinite hyperreal integer Ω,

$$\lim_{n\to\infty} \ln\left(1-\frac{1}{n}\right) = \ln\left(1-\frac{1}{\Omega}\right)$$
$$= \ln(1-\omega)$$
$$\approx \ln 1$$
$$= 0.$$

In line 3 we attempt to approximate inside the natural log, which is allowed as long as we do not violate the approximation principle, per chapter 5 theorem 2.

Because the approximation yields zero, we have violated the approximation principle. However, because no additional arithmetic is being performed afterward, the definition of continuity allows us to render a real result instead of using approximation:

$$\begin{aligned}\lim_{n\to\infty} \ln\left(1-\frac{1}{n}\right) &= \ln\left(1-\frac{1}{\Omega}\right)\\ &= \ln(1-\omega)\\ &\doteq \ln 1\\ &= 0.\end{aligned}$$

Line 3 relies directly on version 2 of the definition of continuity (chapter 1 definition 9). Since $y = \ln x$ is continuous at $x = 1$, $\ln(1+\alpha)$ must render the real result ln 1 for any infinitesimal α.

Although the result of the calculation is the same, the use of approximation is incorrect. The notation used in the second calculation is correct, and the limit of the sequence is zero. ■

Reading Exercise 6 *Evaluate* $\lim_{n\to\infty} \cos\left(\pi - \frac{1}{n}\right)$.

The squeeze theorem for sequences

The idea of the squeeze theorem still applies in the context of sequences. If $\lim_{n\to\infty} a_n = \lim_{n\to\infty} c_n$ and $a_n \le b_n \le c_n$ for all n, then the sequence (b_n) must have the same limit. See figure 2.

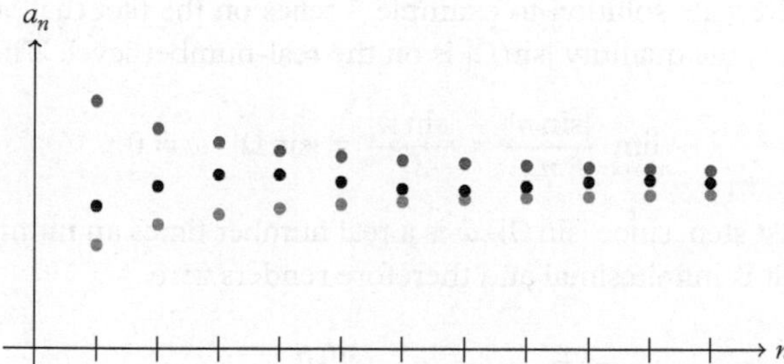

Figure 2 *A sequence (b_n) (black) "squeezed" between the sequences (a_n) (orange) and (c_n) (red). If the limits of the orange and red sequences are the same, then the limit of the black sequence must also have that same value*

Theorem 2 SQUEEZE THEOREM FOR SEQUENCES *If $a_n \le b_n \le c_n$ for every positive integer n and $\lim_{n\to\infty} a_n = \lim_{n\to\infty} c_n = L$, then $\lim_{n\to\infty} b_n = L$.*

The conclusion of the theorem holds as long as the "'squeeze" eventually happens—that is, as long as there is some point after which the sequence (b_n) always lies between the other two.

Example 7 *Evaluate* $\lim_{n\to\infty} \frac{|\sin n|}{n}$.

The strategy here is to work with a quantity for which we know bounds, which in this case is the numerator. Everything else, which in this case is the denominator, can be kept the same in all portions of the inequality.

Solution Since $-1 \le \sin n \le 1$, if we take the absolute value then $0 \le |\sin n| \le 1$. Dividing by n,

$$\frac{0}{n} \le \frac{|\sin n|}{n} \le \frac{1}{n}.$$

This inequality sets up the use of the squeeze theorem for sequences, with $a_n = \frac{0}{n} = 0$, $b_n = \frac{|\sin n|}{n}$, and $c_n = \frac{1}{n}$. Next, we calculate the limits of the sequences on the left and right sides:

This shows that $\lim_{n\to\infty} a_n = 0$.

$$\lim_{n\to\infty} \frac{0}{n} = \lim_{n\to\infty} 0 = 0$$

and

This shows that $\lim_{n\to\infty} c_n = 0$.

$$\lim_{n\to\infty} \frac{1}{n} = \frac{1}{\Omega} = \omega \doteq 0.$$

Since $\lim_{n\to\infty} a_n = 0$ and $\lim_{n\to\infty} c_n = 0$, the squeeze theorem for sequences says that $\lim_{n\to\infty} b_n = 0$.

Because these two limits are equal, the conditions for applying the squeeze theorem for sequences are met and we conclude that

$$\lim_{n\to\infty} \frac{|\sin n|}{n} = 0. \quad \blacksquare$$

An alternate solution to example 7 relies on the fact that since $0 \le |\sin \Omega| \le 1$, the quantity $|\sin \Omega|$ is on the real-number level. Then,

$$\lim_{n\to\infty} \frac{|\sin n|}{n} = \frac{|\sin \Omega|}{\Omega} = |\sin \Omega| \cdot \omega \doteq 0.$$

In the last step, since $|\sin \Omega| \cdot \omega$ is a real number times an infinitesimal, the result is infinitesimal and therefore renders zero.

Ans. to reading exercise 6

-1

Details: $\lim_{n\to\infty} \cos\left(\pi - \frac{1}{n}\right) = \cos\left(\pi - \frac{1}{\Omega}\right) = \cos(\pi - \omega) \approx \cos \pi = -1$. The rendering symbol can be used in place of the approximation symbol; either one is correct.

Reading Exercise 7 *Evaluate* $\lim_{n\to\infty} \frac{\sin n}{n}$.

Sequence limits: level analysis

Returning to the alternate solution to example 7, consider the intermediate step $\frac{|\sin \Omega|}{\Omega}$. As already observed, the numerator is between -1 and 1, and therefore is on the real-number level. The denominator is on the Ω level. Therefore, the numerator is on a lower level than the denominator:

$$\frac{|\sin\Omega| \leftarrow \text{lower level (real level)}}{\Omega \quad \leftarrow \text{higher level } (\Omega \text{ level})} \doteq 0.$$

If the numerator is on a lower level than the denominator, the fraction is infinitesimal and renders zero. If the numerator is on a higher level than the denominator, the fraction is infinite and renders ∞ if positive or $-\infty$ if negative. This type of reasoning is called *level analysis*.

Example 8 *Evaluate* $\lim_{n\to\infty} \frac{n}{e^n - 15}$.

Solution For any positive infinite hyperreal integer Ω,

$$\begin{aligned}\lim_{n\to\infty} \frac{n}{e^n - 15} &= \frac{\Omega}{e^\Omega - 15} \\ &\approx \frac{\Omega \leftarrow \text{lower level}}{e^\Omega \leftarrow \text{higher level}} \\ &\doteq 0.\end{aligned}$$

The sequence converges with limit zero. ■

Example 9 *Evaluate* $\lim_{n\to\infty} \frac{-\sqrt{n+1}}{\ln n^2}$.

Solution For any positive infinite hyperreal integer Ω,

$$\begin{aligned}\lim_{n\to\infty} \frac{-\sqrt{n+1}}{\ln n^2} &= \frac{-\sqrt{\Omega+1}}{\ln \Omega^2} \\ &\approx \frac{-\sqrt{\Omega}}{\ln \Omega^2} \\ &= \frac{-\sqrt{\Omega} \leftarrow \text{higher level } (\Omega^{\frac{1}{2}} \text{ level})}{2\ln\Omega \leftarrow \text{lower level } (\ln\Omega \text{ level})} \\ &\doteq -\infty.\end{aligned}$$

Line 3 uses the laws of logarithms to simplify, making the level of the number more apparent.

The sequence diverges. ■

A partial listing of levels of hyperreal numbers is given in section 5.10 figure 10. See also section 1.1.

Before looking at the next example, we need to know where the $\Omega!$ level fits. First, notice that

The capital omega factorial level applies under the condition that Ω is an integer.

The numerator and the denominator in line 1 each contain Ω factors, allowing us to rewrite the expression as shown in line 2.

$$\frac{2^\Omega}{\Omega!} = \frac{2\cdot 2\cdot 2\cdots 2}{\Omega(\Omega-1)(\Omega-2)\cdots 2\cdot 1}$$
$$= \frac{2}{\Omega}\cdot\frac{2}{\Omega-1}\cdots\frac{2}{2}\cdot\frac{2}{1},$$

which is infinitesimal. Therefore $\Omega!$ is on a higher level than 2^Ω. The same is true for all the exponentials at Ω.

We can compare $\Omega!$ to Ω^Ω in the same manner:

$$\frac{\Omega^\Omega}{\Omega!} = \frac{\Omega\cdot\Omega\cdots\Omega}{\Omega(\Omega-1)(\Omega-2)\cdots 2\cdot 1} = \frac{\Omega}{\Omega}\cdot\frac{\Omega}{\Omega-1}\cdots\frac{\Omega}{2}\cdot\frac{\Omega}{1},$$

which is infinite. Therefore, $\Omega!$ is on a lower level than Ω^Ω. A partial picture of levels showing the placement of $\Omega!$ is in figure 3.

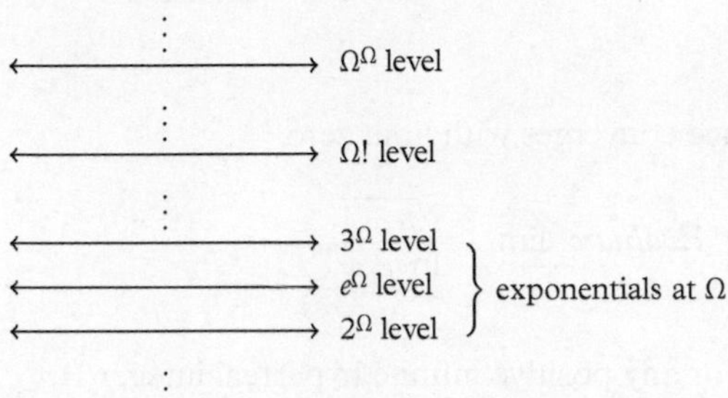

Figure 3 *A partial listing of levels of numbers showing the placement of $\Omega!$. Each wide arrow represents a number line at the stated level. There are more levels between the ones pictured*

Ans. to reading exercise 7

0

Details: $\lim_{n\to\infty}\frac{\sin n}{n} = \frac{\sin\Omega}{\Omega} \doteq 0$, because $\sin\Omega$ is a real number between -1 and 1, and the denominator is infinite. Alternately, use the squeeze theorem for sequences.

Example 10 *Evaluate* $\lim_{n\to\infty}\frac{n!+2^n}{5^n+n^5}$.

Solution For any positive infinite hyperreal integer Ω,

On the right side of line 1, in the numerator, $\Omega!$ is the highest-level term so the 2^Ω can be discarded when approximating the numerator in line 2.

$$\lim_{n\to\infty}\frac{n!+2^n}{5^n+n^5} = \frac{\Omega!+2^\Omega}{5^\Omega+\Omega^5}$$
$$\approx \frac{\Omega!}{5^\Omega}\begin{array}{l}\leftarrow \text{higher level}\\ \leftarrow \text{lower level}\end{array}$$
$$\doteq \infty.$$

The sequence diverges. ■

The similar function rule

Review: suppose we want to check the function $f(x) = \dfrac{x}{e^x - 15}$ for horizontal asymptotes. For the right side, we check using the limit

$$\lim_{x\to\infty} \frac{x}{e^x - 15} = \frac{\Omega}{e^\Omega - 15}.$$

This calculation can be continued:

$$\approx \frac{\Omega \leftarrow \text{lower level}}{e^\Omega \leftarrow \text{higher level}} \doteq 0.$$

Pausing to notice that the form is $\frac{\infty}{\infty}$, we could instead apply l'Hospital's rule:

$$\begin{aligned}\lim_{x\to\infty} \frac{x}{e^x - 15} &= \lim_{x\to\infty} \frac{1}{e^x}\\ &= \frac{1}{e^\Omega}\\ &\doteq 0.\end{aligned}$$

L'Hospital's rule is applied in line 1; the expression in line 2 is infinitesimal and renders zero.

Compare this function limit to the sequence limit of example 8. Because it seems to be the same limit, could we have used l'Hospital's rule to complete example 8? The following theorem answers this question in the affirmative.

Theorem 3 SIMILAR FUNCTION RULE *Suppose f is a function such that* $f(n) = a_n$ *for every natural number n. If* $\lim_{x\to\infty} f(x) = L$, *then we must also have* $\lim_{n\to\infty} a_n = L$.

Any technique for finding the limit of a function can be used to find the limit of a sequence, provided the similar function rule applies. This includes the use of l'Hospital's rule.

Example 11 *Use l'Hospital's rule to evaluate* $\lim_{n\to\infty} \dfrac{\ln n}{n}$.

Solution For any positive infinite hyperreal integer Ω,

$$\lim_{n\to\infty} \frac{\ln n}{n} = \frac{\ln \Omega}{\Omega}.$$

Recognizing the form $\frac{\infty}{\infty}$, we use l'Hospital's rule:

$$\begin{aligned}\lim_{n\to\infty} \frac{\ln n}{n} &= \lim_{n\to\infty} \frac{\frac{1}{n}}{1}\\ &= \lim_{n\to\infty} \frac{1}{n}\\ &= \frac{1}{\Omega} = \omega \doteq 0.\end{aligned}$$

■

Technically, we are relying on the similar function rule here, with the function $f(x) = \frac{\ln x}{x}$. Rather than write this out formally, we continue using the variable n when applying l'Hospital's rule.

The similar function rule also allows us to use the strategy for limits of powers (chapter 5 theorem 12).

Example 12 *Evaluate* $\lim_{n\to\infty} \sqrt[n]{n}$.

Solution Trying

$$\lim_{n\to\infty} \sqrt[n]{n} = \lim_{n\to\infty} n^{\frac{1}{n}} = \sqrt[\Omega]{\Omega} = \Omega^{\frac{1}{\Omega}} = \Omega^{\omega}$$

does not help if we do not know how to render a real result for any of these expressions. Following the similar function rule, we instead evaluate

$$\lim_{x\to\infty} x^{\frac{1}{x}}.$$

As you may recall, the strategy for limits of powers consists of three steps.

❶ Take the log of the expression and simplify. Simplifying involves the use of the third law of logarithms:

$$\lim_{x\to\infty} \ln\left(x^{\frac{1}{x}}\right) = \lim_{x\to\infty}\left(\frac{1}{x}\ln x\right).$$

Notice it is not said that Ω is an integer. When using the similar function rule, we are calculating a function limit.

❷ Evaluate the resulting limit. For any positive infinite hyperreal Ω,

$$\lim_{x\to\infty}\left(\frac{1}{x}\ln x\right) = \frac{1}{\Omega}\ln\Omega = \frac{\ln\Omega \leftarrow \text{lower level}}{\Omega \leftarrow \text{higher level}} \doteq 0.$$

❸ Use chapter 5 theorem 12 to to calculate the value of the original limit. Because we took the natural log of the expression, we reverse this change using the natural exponential:

$$\lim_{x\to\infty} x^{\frac{1}{x}} = e^{0 \leftarrow \text{answer from step ❷}} = 1.$$

Then, by the similar function rule, the limit of our sequence is also 1. ■

In example 12 we learned that

$$\lim_{x\to\infty} x^{\frac{1}{x}} = 1.$$

However, for any positive infinite hyperreal Ω,

$$\lim_{x\to\infty} x^{\frac{1}{x}} = \sqrt[\Omega]{\Omega} = \Omega^{\frac{1}{\Omega}} = \Omega^{\omega}.$$

Because these expressions must render the value of the limit,

$$\sqrt[\Omega]{\Omega} = \Omega^{\frac{1}{\Omega}} = \Omega^{\omega} \doteq 1.$$

CONVENIENT RENDERING FACTS

$$\sqrt[\Omega]{\Omega} = \Omega^{\frac{1}{\Omega}} = \Omega^{\omega} \doteq 1$$

List of common sequence limits

It helps to learn a few of the more common sequence limits so that we don't have to derive these limits every time.

LIST OF COMMON SEQUENCE LIMITS

1. $\lim\limits_{n\to\infty} \dfrac{\ln n}{n} = 0$
2. $\lim\limits_{n\to\infty} a^{\frac{1}{n}} = 1$, for $a > 0$
3. $\lim\limits_{n\to\infty} a^n = 0$, for $|a| < 1$
4. $\lim\limits_{n\to\infty} \sqrt[n]{n} = 1$
5. $\lim\limits_{n\to\infty} \left(1 + \dfrac{a}{n}\right)^n = e^a$
6. $\lim\limits_{n\to\infty} \dfrac{a^n}{n!} = 0$

Some of the limits in the list have already been derived. Others are exercises for you to tackle on your own.

Example 13 *Evaluate the limit: (a)* $\lim\limits_{n\to\infty} \sqrt[n]{3}$, *(b)* $\lim\limits_{n\to\infty} \left(\dfrac{1}{4}\right)^n$, *(c)* $\lim\limits_{n\to\infty} \left(1 + \dfrac{3}{n}\right)^n$, *and (d)* $\lim\limits_{n\to\infty} \dfrac{3^n}{n!}$.

Solution (a) Using common limit (2) with $a = 3$,

$$\lim_{n\to\infty} \sqrt[n]{3} = \lim_{n\to\infty} 3^{\frac{1}{n}} = 1.$$

(b) Using common limit (3) with $a = \frac{1}{4}$,

$$\lim_{n\to\infty} \left(\frac{1}{4}\right)^n = 0.$$

(c) Using common limit (5) with $a = 3$,

$$\lim_{n\to\infty} \left(1 + \frac{3}{n}\right)^n = e^3.$$

(d) Using common limit (6) with $a = 3$,

$$\lim_{n\to\infty} \frac{3^n}{n!} = 0.$$

■

Exercises 10.2

1–30. Evaluate the limit.

1. $\lim_{n\to\infty} \frac{4n-5}{n^2-11}$
2. $\lim_{n\to\infty} \frac{n^2-7n+1}{n^2+4}$
3. $\lim_{n\to\infty} \frac{4n-5}{\sqrt{n}-11}$
4. $\lim_{n\to\infty} \frac{n^2-7n+1}{n+4}$
5. $\lim_{n\to\infty} \tan^{-1} \frac{4^n}{n^4}$
6. $\lim_{n\to\infty} \frac{n^2-7n+1}{n^3+4}$
7. $\lim_{n\to\infty} \frac{n^3-14n}{2-4n^3}\left(2-\frac{4}{n^3}\right)$
8. $\lim_{n\to\infty} \frac{(-1)^n(n^2-7n+1)}{n^2+4}$
9. $\lim_{n\to\infty} (-1)^n \frac{6n+5}{n-4}$
10. $\lim_{n\to\infty} \frac{n^2-7n+(-1)^n}{n^2+4}$
11. $\lim_{n\to\infty} (-1)^n \frac{6n+5}{n^2-4}$
12. $\lim_{n\to\infty} \frac{3n^2+17}{\sqrt{2n^4+4n}}$
13. $\lim_{n\to\infty} (-2)^n \frac{6n+5}{n^2-4}$
14. $\lim_{n\to\infty} \left(\frac{4}{7}\right)^n$
15. $\lim_{n\to\infty} \cos\left(\frac{\pi n}{3^n}\right)$
16. $\lim_{n\to\infty} \frac{n!}{4^n}$
17. $\lim_{n\to\infty} \cos(\pi n)$
18. $\lim_{n\to\infty} \cos\left(\frac{1}{n}\right)$
19. $\lim_{n\to\infty} \left(\frac{7}{4}\right)^{1/n}$
20. $\lim_{n\to\infty} \left(\frac{4}{7}\right)^{1/n}$
21. $\lim_{n\to\infty} \left(\frac{7}{4}\right)^n$
22. $\lim_{n\to\infty} e^{3n/(n+2)}$
23. $\lim_{n\to\infty} \frac{2^n+(-1)^n}{5^n}$
24. $\lim_{n\to\infty} \frac{7^n+n}{4^n}$
25. $\lim_{n\to\infty} \frac{2^n}{n^5}$
26. $\lim_{n\to\infty} \ln\left(\frac{1}{n}\right)$
27. $\lim_{n\to\infty} \left(1+\frac{0.3}{n}\right)^n$
28. $\lim_{n\to\infty} \frac{n^3}{e^n}$
29. $\lim_{n\to\infty} \frac{n!+2^n}{n^n+n!}$
30. $\lim_{x\to\infty} \frac{7^n}{3^{2n}}$

31–40. Evaluate the limit of the sequence.

31. $a_n = e^{(7n-5)/((1/4)^n+8n)}$

32. $a_n = \dfrac{\left(\frac{7}{4}\right)^n}{n!}$

33. $a_n = \left(1 + \dfrac{0.3}{n}\right)^{2n}$

34. $a_n = \left(1 + \dfrac{1/2}{n}\right)^n$

35. $a_n = \dfrac{2^n}{n^{5!}}$

36. $a_n = \left(1 - \dfrac{1/2}{n}\right)^n$

37. $a_n = \dfrac{2^n}{5n!}$

38. $a_n = \left(1 + \dfrac{4}{n}\right)^n$

39. $a_n = \sqrt[n]{4n}$

40. $a_n = \left(4 + \dfrac{4}{n}\right)^n$

41. Suppose $\sqrt[n]{\frac{1}{2}} \leq b_n \leq \sqrt[n]{n}$ for every positive integer n. Find $\lim_{n\to\infty} b_n$.

42. Suppose $5 - \dfrac{e^n}{4^n + 1} \leq b_n \leq 3 + \dfrac{6n}{3n - 1}$ for every positive integer n. Find $\lim_{n\to\infty} b_n$.

43. Derive common limit (2).

44. Derive common limit (3).

45. Derive common limit (5).

46. Derive common limit (6).

Infinite Series

10.3

We have seen that a sequence is an infinitely long list of numbers. For instance, the sequence $a_n = \frac{1}{2^n}$ can be written as

$$\frac{1}{2}, \frac{1}{4}, \frac{1}{8}, \frac{1}{16}, \frac{1}{32}, \dots.$$

By contrast, an *infinite series* can be described informally as an infinitely long arithmetic problem, the sum of all the terms of a sequence:

The word *infinite* is often omitted from the term. If I refer to a *series,* I mean an *infinite series.*

$$\frac{1}{2} + \frac{1}{4} + \frac{1}{8} + \frac{1}{16} + \frac{1}{32} + \cdots.$$

Using summation notation, this series is written as

$$\sum_{n=1}^{\infty} \frac{1}{2^n}.$$

Recall that the symbol ∞ does not represent a specific hyperreal number.

Example 1 *Is it a sequence or an infinite series? (a)* $1+\frac{1}{2}+\frac{1}{3}+\frac{1}{4}+\frac{1}{5}+\cdots$; *(b)* $1, \frac{1}{2}, \frac{1}{3}, \frac{1}{4}, \frac{1}{5}, \dots$.

Solution (a) The expression $1 + \frac{1}{2} + \frac{1}{3} + \frac{1}{4} + \frac{1}{5} + \cdots$ is an infinitely long sum, so it is an infinite series. (b) Because $1, \frac{1}{2}, \frac{1}{3}, \frac{1}{4}, \frac{1}{5}, \dots$ is an infinitely long list of numbers, it is a sequence. ■

For a sequence, we are interested in its limit: $\lim_{n\to\infty} a_n$. For an infinite series, we are instead interested in its sum: $\sum_{n=1}^{\infty} a_n$. Convergence or divergence of a sequence is based on whether the limit of the sequence exists (is a real number); convergence or divergence of an infinite series is based on whether the sum exists.

Distinguishing between the limit of a sequence and the sum of an infinite series is very important throughout the remainder of this chapter.

Reading Exercise 8 *Sequence or series? (a)* $1 + \frac{1}{4} + \frac{1}{9} + \frac{1}{16} + \cdots$; *(b)* $1, \frac{1}{4}, \frac{1}{9}, \frac{1}{16}, \dots$.

Infinite series: defining the sum

Suppose we wish to evaluate the infinite series

$$\sum_{n=1}^{\infty} \frac{1}{2^n} = \frac{1}{2} + \frac{1}{4} + \frac{1}{8} + \frac{1}{16} + \cdots .$$

In other words, we wish to find the value of this sum. Adding finitely many numbers is never a problem, and we start there. However, adding infinitely many? This is more of a challenge.

Let's define a *partial sum* S_k as the sum of the first k terms of the infinite series:

$$S_k = \sum_{n=1}^{k} \frac{1}{2^n}.$$

The first four partial sums are

$$S_1 = \frac{1}{2}$$
$$S_2 = \frac{1}{2} + \frac{1}{4} = \frac{3}{4}$$
$$S_3 = \frac{1}{2} + \frac{1}{4} + \frac{1}{8} = \frac{7}{8}$$
$$S_4 = \frac{1}{2} + \frac{1}{4} + \frac{1}{8} + \frac{1}{16} = \frac{15}{16}.$$

The first partial sum S_1 is the first term of the series. The second partial sum S_2 is the sum of the first two terms of the series. The third partial sum S_3 is the sum of the first three terms of the series.

Do you see the pattern? The denominators are the powers of 2, whereas the numerators are one less than the denominators:

$$S_k = \frac{2^k - 1}{2^k}.$$

Notice there are now two different sequences in this discussion. One is the sequence (a_n) of the terms being added,

$$\frac{1}{2}, \frac{1}{4}, \frac{1}{8}, \frac{1}{16}, \ldots ;$$

the other is the sequence (S_k) of partial sums,

$$\frac{1}{2}, \frac{3}{4}, \frac{7}{8}, \frac{15}{16}, \ldots .$$

The sum of the first 100 terms of the series is

$$S_{100} = \frac{2^{100} - 1}{2^{100}}$$
$$= 0.99999999999999999999999999999992111390948\ldots,$$

which is very close to 1.

As we add more terms of the series, it seems the sum gets closer to 1. This is reminiscent of the idea of a limit. One way to think of the sum of an infinite series is that it is the limit of the sequence (S_k) of partial sums.

Technically, the infinite series is a sum, but the phrases "sum of the series" and "sum of the infinite series" are used to refer to the sum.

Alternately, we can still think of the sum of the series as the sum of "all" infinitely many terms. To this end, add the first Ω terms of the series (where Ω is a positive infinite hyperreal integer) and we have

Using the formula for S_k with a positive infinite hyperreal integer instead of a real-number integer is an application of the transfer principle.

$$S_\Omega = \frac{2^\Omega - 1}{2^\Omega} \approx \frac{2^\Omega}{2^\Omega} = 1.$$

Because we render the same real result for any such Ω, we call this result the sum of the series and write

$$\sum_{n=1}^{\infty} \frac{1}{2^n} = 1.$$

These ideas form the basis of our definition of the sum of an infinite series.

Ans. to reading exercise 8
(a) series; (b) sequence

Definition 4 SERIES CONVERGENCE AND DIVERGENCE

For an infinite series $\sum_{n=1}^{\infty} a_n$*, the kth* partial sum *is*

$$S_k = \sum_{n=1}^{k} a_n.$$

If S_Ω renders the same real result L for every positive infinite hyperreal integer Ω, then we call L the sum *of the series and write*

$$\sum_{n=1}^{\infty} a_n = L.$$

This definition is equivalent to saying that the sum of a series is the limit of its partial sums—that is,

$$\sum_{n=1}^{\infty} a_n = \lim_{k\to\infty} S_k.$$

If L is a real number (that is, not ∞ or $-\infty$), then we say the series converges. *Otherwise, we say the series* diverges.

For the purposes of calculation, we write

$$\sum_{n=1}^{\infty} a_n = S_\Omega$$

or

$$\sum_{n=1}^{\infty} a_n = \lim_{k\to\infty} S_k.$$

Even relatively simple-looking series can be difficult to evaluate, and as a result we will spend much of our effort in this chapter determining whether a series converges or diverges without necessarily determining the value of a convergent series. For now, however, let's turn our attention to a few instances in which the sum can be determined.

Geometric series

A geometric series is recognized when there is a number r for which each term of the series is r times the previous term. Recursively, for a geometric series $a_{n+1} = r \cdot a_n$.

Definition 5 GEOMETRIC SERIES *A* geometric series *is a series of the form*

$$a + ar + ar^2 + ar^3 + \cdots = \sum_{n=1}^{\infty} ar^{n-1}.$$

We call r the common ratio *of the geometric series.*

Example 2 *Is the series geometric? If so, find the first term a and common ratio r. (a)* $\frac{3}{5} + \frac{3}{25} + \frac{3}{125} + \frac{3}{625} + \cdots$*; (b)* $\frac{3}{5} + \frac{3}{10} + \frac{3}{15} + \frac{3}{20} + \cdots$.

Reminder: $\frac{\frac{3}{25}}{\frac{3}{5}} = \frac{3}{25} \cdot \frac{5}{3} = \frac{1}{5}$.

Solution (a) The first term of the series is $a = \frac{3}{5}$. To find the common ratio, r, we divide successive terms:

$$\frac{\frac{3}{25}}{\frac{3}{5}} = \frac{1}{5} \qquad \frac{\frac{3}{125}}{\frac{3}{25}} = \frac{1}{5} \qquad \frac{\frac{3}{625}}{\frac{3}{125}} = \frac{1}{5} \cdots.$$

Because the ratio between successive terms is the same each time, the series is geometric, with the common ratio $r = \frac{1}{5}$.

(b) The first term of the series is $a = \frac{3}{5}$. To find the common ratio r, we divide successive terms:

$$\frac{\frac{3}{10}}{\frac{3}{5}} = \frac{1}{2} \qquad \frac{\frac{3}{15}}{\frac{3}{10}} = \frac{2}{3} \qquad \frac{\frac{3}{20}}{\frac{3}{15}} = \frac{3}{4} \cdots.$$

Because the ratio between successive terms is not the same each time, the series is not geometric. ■

In summation notation, the series of example 2(a) is

$$\sum_{n=1}^{\infty} \frac{3}{5^n} = \sum_{n=1}^{\infty} \frac{3}{5}\left(\frac{1}{5}\right)^{n-1}.$$

The form of the summand is a constant times an exponential function of n, which is the same form given in definition 5. By contrast, the series of example 2(b) is

$$\sum_{n=1}^{\infty} \frac{3}{5n},$$

which has a polynomial denominator rather than an exponential. Such a series is not geometric.

Our only tool for finding the sum of a series is to investigate the partial sums. To this end, for a geometric series $\sum_{n=1}^{\infty} ar^{n-1}$, the sum of the first k terms is

$$S_k = a + ar + ar^2 + \cdots + ar^{k-1}.$$

Notice that the second term has an exponent of 1 on r, the third term has an exponent of 2 on r, and so on. The exponent for the kth term is $k - 1$.

Multiplying both sides of this equation by r gives

$$rS_k = ar + ar^2 + ar^3 + \cdots + ar^k.$$

Next we align our two equations and subtract:

$$\begin{aligned} rS_k &= \qquad ar + ar^2 + ar^3 + \cdots + ar^{k-1} + ar^k \\ -S_k &= -(a + ar + ar^2 + ar^3 + \cdots + ar^{k-1}) \\ \hline (r-1)S_k &= -a \qquad\qquad\qquad\qquad\qquad\qquad + ar^k. \end{aligned}$$

Now we solve for S_k:

$$\begin{aligned} (r-1)S_k &= ar^k - a \\ (r-1)S_k &= a(r^k - 1) \\ S_k &= \frac{a(r^k - 1)}{r - 1}. \end{aligned}$$

The sum of the first Ω terms for a positive infinite hyperreal integer Ω is

$$S_\Omega = \frac{a(r^\Omega - 1)}{r - 1}.$$

As long as $|r| < 1$, we have $r^\Omega \doteq 0$ (see common sequence limit (3)). Then,

Common sequence limit (3) says that $\lim_{n\to\infty} r^n = 0$ for $|r| < 1$. Therefore, r^Ω must render zero.

$$S_\Omega \doteq \frac{a(0-1)}{r-1} = \frac{-a}{r-1} = \frac{a}{1-r}.$$

SUM OF A GEOMETRIC SERIES

If $|r| < 1$,

$$\sum_{n=1}^{\infty} ar^{n-1} = \frac{a}{1-r}.$$

If $|r| \geq 1$, the series diverges.

When we know the first term a and the common ratio r of a geometric series, then the only remaining question before applying the formula is whether $|r| < 1$.

Example 3 *Determine the sum or state that the series diverges:* $\sum_{n=1}^{\infty} 5 \cdot 0.4^{n-1} = 5 + 2 + 0.8 + 0.32 + \cdots$.

Solution The series $\sum_{n=1}^{\infty} 5 \cdot 0.4^{n-1}$ is in the form of a geometric series with $a = 5$ and $r = 0.4$. Since $|r| = 0.4 < 1$, the formula applies and

$$\sum_{n=1}^{\infty} 5 \cdot 0.4^{n-1} = \frac{a}{1-r} = \frac{5}{1-0.4} = \frac{5}{0.6} = \frac{25}{3} = 8.\overline{3}.$$ ■

Example 4 *Determine the sum or state that the series diverges:* $\sum_{n=1}^{\infty} \frac{1}{(-2)^n}$.

Solution We begin by writing out the first few terms of the series:

$$\sum_{n=1}^{\infty} \frac{1}{(-2)^n} = \frac{1}{-2} + \frac{1}{4} + \frac{1}{-8} + \frac{1}{16} + \cdots$$
$$= -\frac{1}{2} + \frac{1}{4} - \frac{1}{8} + \frac{1}{16} - \cdots.$$

For the common ratio r, notice that each term of the series is the previous term multiplied by $-\frac{1}{2}$. Alternately, compute a few successive ratios in the manner of example 2.

This is a geometric series, with the first term $a = -\frac{1}{2}$ and the common ratio $r = -\frac{1}{2}$. Since $|r| = \frac{1}{2} < 1$, the formula applies and

$$\sum_{n=1}^{\infty} \frac{1}{(-2)^n} = \frac{a}{1-r} = \frac{-\frac{1}{2}}{1-(-\frac{1}{2})} = -\frac{1}{3}.$$ ■

Example 5 *Determine the sum or state that the series diverges:* $\sum_{n=1}^{\infty} 5 \cdot 3^n$.

Solution The first few terms of the series are

$$\sum_{n=1}^{\infty} 5 \cdot 3^n = 15 + 45 + 135 + 405 + \cdots .$$

This is a geometric series, with the first term $a = 15$ and the common ratio $r = 3$. Since $|r| = 3 \geq 1$, the series diverges. ■

Notice the exponential form of the summands in each of examples 3, 4, and 5. Each is a geometric series.

Repeated decimals can be explored using geometric series.

Example 6 *Determine the sum or state that the series diverges:* $0.3 + 0.03 + 0.003 + 0.0003 + \cdots = 0.3333\cdots = 0.\overline{3}$.

Solution This is a geometric series, with the first term $a = 0.3$ and the common ratio $r = 0.1$. Since $|r| = 0.1 < 1$, the series converges and

$$0.\overline{3} = \frac{a}{1-r} = \frac{0.3}{1-0.1} = \frac{0.3}{0.9} = \frac{1}{3}.$$ ■

Reading Exercise 9 *Determine the sum or state that the series diverges:* $\sum_{n=1}^{\infty} 2\left(\frac{1}{4}\right)^{n-1}$.

Telescoping series

Partial fractions can help evaluate some series.

Example 7 *Evaluate* $\sum_{n=1}^{\infty} \frac{2}{n(n+1)}$.

Solution The first few terms of this series are

$$\sum_{n=1}^{\infty} \frac{2}{n(n+1)} = \frac{2}{2} + \frac{2}{6} + \frac{2}{12} + \frac{2}{20} + \cdots$$
$$= 1 + \frac{1}{3} + \frac{1}{6} + \frac{1}{10} + \cdots .$$

Alternately, notice that the summand's denominator is a polynomial instead of an exponential, meaning the series is not geometric.

Computing the first two ratios between successive terms is enough to conclude that the series is not geometric.

The trick for this series is to use partial fractions to rewrite the summand. The variable is n instead of x, but the algebraic procedure is the same. The denominator is already factored, so we are ready for the form of the partial fraction decomposition:

$$\frac{2}{n(n+1)} = \frac{A}{n} + \frac{B}{n+1}.$$

Clear the fractions:

$$n(n+1) \cdot \frac{2}{n(n+1)} = \left(\frac{A}{n} + \frac{B}{n+1}\right) \cdot n(n+1)$$
$$2 = A(n+1) + Bn.$$

Choosing $n = -1$, we have

$$2 = A(0) + B(-1)$$
$$B = -2.$$

Choosing $n = 0$, we have

$$2 = A(1) + B(0)$$
$$A = 2.$$

Our series is then rewritten as

$$\sum_{n=1}^{\infty} \frac{2}{n(n+1)} = \sum_{n=1}^{\infty} \left(\frac{2}{n} - \frac{2}{n+1}\right).$$

Now let's try writing out the first few terms of the rewritten series and something quite nice appears to happen:

CAUTION: MISTAKE TO AVOID
Cancellation of terms in an infinite series can often be misleading. The use of partial sums is required if errors are to be avoided.

$$\left(\frac{2}{1} - \cancel{\frac{2}{2}}\right) + \left(\cancel{\frac{2}{2}} - \cancel{\frac{2}{3}}\right) + \left(\cancel{\frac{2}{3}} - \cancel{\frac{2}{4}}\right) + \cdots = 2.$$

Unfortunately, writing cancellation in this manner can sometimes lead to incorrect conclusions, so it is much better to work with partial sums. The sum of the first k terms is

$$S_k = \left(\frac{2}{1} - \frac{2}{2}\right) + \left(\frac{2}{2} - \frac{2}{3}\right) + \left(\frac{2}{3} - \frac{2}{4}\right) + \dots$$
$$+ \left(\frac{2}{k-1} - \frac{2}{k}\right) + \left(\frac{2}{k} - \frac{2}{k+1}\right)$$
$$= 2 - \frac{2}{k+1},$$

where the right-side term in one set of parentheses cancels the left-side term in the next set of parentheses, leaving the very first and very last terms uncancelled.

Have you ever watched a pirate movie where a sailor has a pocket telescope? The scope is made of several sections, each of which slides into the next section that is slightly larger. Extend it to its full length, and the eyepiece is at the small end with the lens at the larger end. Fold it up, and only the largest section is visible. Compare this to the cancellation here and you can see why this is called a *telescoping series*.

Now that we have computed and simplified the partial sum S_k, we are ready to compute the sum of the series. Letting Ω be a positive infinite hyperreal integer,

$$\sum_{n=1}^{\infty} \frac{2}{n(n+1)} = S_\Omega = 2 - \frac{2}{\Omega+1} \approx 2 - \frac{2}{\Omega} = 2 - 2\omega \approx 2.$$

We may now safely conclude that the series converges with sum 2. ■

Ans. to reading exercise 9
$\frac{8}{3}$; a geometric series with $a = 2$, $r = \frac{1}{4}$

Why go to the trouble of using partial sums when we already knew the sum was 2 before we calculated S_k? The answer is this cancellation can be deceiving. For instance, it seems that

$$\sum_{n=1}^{\infty} (-1)^n = (-1) + 1 + (-1) + 1 + (-1) + 1 - \dots = 0.$$

CAUTION: MISTAKE TO AVOID
This calculation is incorrect.

But, this is incorrect. This series diverges! The key is to look at even-numbered partial sums and odd-numbered partial sums. When k is even,

$$S_k = (-1) + 1 + (-1) + 1 - \dots + (-1) + 1 = 0.$$

But, when k is odd,

$$S_k = (-1) + 1 + (-1) + 1 - \dots + (-1) + 1 + (-1) = -1.$$

Then, depending on whether the positive infinite hyperreal integer Ω is even or odd, we have either $S_\Omega = 0$ or $S_\Omega = -1$. Because we do not render the same real result for every positive infinite hyperreal integer Ω, the series diverges.

Reading Exercise 10 *Evaluate* $\sum_{n=1}^{\infty} \left(\frac{3}{n} - \frac{3}{n+1}\right)$.

Test for divergence

What happens to the terms of a convergent series? Consider the series $\sum_{n=1}^{\infty} \frac{2}{n(n+1)}$ of example 7. Instead of looking at the sum of the series, look at the limit of the sequence $\left(\frac{2}{n(n+1)}\right)$, which is

$$\lim_{n\to\infty} \frac{2}{n(n+1)} = \frac{2}{\Omega(\Omega+1)} = \frac{2}{\Omega^2+\Omega} \approx \frac{2}{\Omega^2} = 2\omega^2 \doteq 0.$$

It is worthwhile to pause and note the difference between the sum of the series and the limit of the sequence:

The limit on the right was just calculated; the sum on the left was calculated in example 7.

$$\sum_{n=1}^{\infty} \frac{2}{n(n+1)} = 2, \qquad \lim_{n\to\infty} \frac{2}{n(n+1)} = 0.$$

These are not the same quantities, and it is important to understand the difference between them and to know what each notation means.

The fact that the limit of the sequence containing the terms of the convergent series is zero is not unusual. Exploring this idea further, the following are the first few terms of the convergent series from examples 3, 4, and 6:

In line 1, the sum of the first four terms of the series is $5+2+0.8+0.32 = 8.12$. There are still infinitely many terms left and the sum is only supposed to be $8.\overline{3}$, so the remaining terms must be small. The sum of the first 10 terms of the series is 8.33246, which is already very close to the sum of all infinitely many terms. For a series to converge, the individual terms must get smaller and smaller, with a limit of zero.

$$\sum_{n=1}^{\infty} 5\cdot 0.4^{n-1} = 5+2+0.8+0.32+\cdots = 8.\overline{3}$$

$$\sum_{n=1}^{\infty} \frac{1}{(-2)^n} = -\frac{1}{2}+\frac{1}{4}-\frac{1}{8}+\frac{1}{16}-\cdots = -\frac{1}{3}$$

$$\sum_{n=1}^{\infty} 0.3\left(\frac{1}{10}\right)^{n-1} = 0.3+0.03+0.003+0.0003+\cdots = 0.\overline{3} = \frac{1}{3}.$$

In each case, the terms of the series seem to be approaching zero. This is always the case.

Theorem 4 TERMS OF A CONVERGENT SERIES *If* $\sum_{n=1}^{\infty} a_n$ *converges, then* $\lim_{n\to\infty} a_n = 0$.

Proof. Suppose that $\sum_{n=1}^{\infty} a_n$ has sum L. Then, by definition, for any positive infinite hyperreal integer Ω, the capital omegath partial sum renders the real result L:

$$S_\Omega \doteq L.$$

Consider $\Omega - 1$. Subtracting one from an integer still leaves an integer; subtracting one from an infinite number still leaves an infinite number. Then, $\Omega - 1$ is also a positive infinite hyperreal integer, so

$$S_{\Omega-1} \doteq L$$

as well.

Next, note that

$$S_\Omega = S_{\Omega-1} + a_\Omega.$$

This may help you understand the equation to the left:

$$S_4 = (a_1 + a_2 + a_3) + a_4$$
$$= S_3 + a_4.$$

Replace 4 with Ω and you have the desired equation.

Rearranging,

$$a_\Omega = S_\Omega - S_{\Omega-1} \doteq L - L = 0.$$

Because this is true for any positive infinite hyperreal integer Ω,

$$\lim_{n\to\infty} a_n = a_\Omega \doteq 0. \quad ■$$

Theorem 4 gives us a means to conclude that a series diverges. If the limit of the terms does not exist, or if the limit is not zero, then the series diverges.

TEST FOR DIVERGENCE

If $\lim_{n\to\infty} a_n \neq 0$ or the limit does not exist, then $\sum_{n=1}^{\infty} a_n$ diverges.

The test for divergence does not say anything about what happens if $\lim_{n\to\infty} a_n = 0$. In this case, the test is inconclusive.

Starting now, the test for divergence should be the first task to run through mentally when determining whether a series converges or diverges. If the limit of the terms of the series is not zero, then the series diverges.

Example 8 *Determine the sum or state that the series diverges:* $\sum_{n=1}^{\infty} \frac{3n+2}{5n}$.

Solution We try the test for divergence by calculating the limit of the terms of the series. For any positive infinite hyperreal integer Ω,

$$\lim_{n\to\infty} \frac{3n+2}{5n} = \frac{3\Omega+2}{5\Omega} \approx \frac{3\Omega}{5\Omega} = \frac{3}{5} \neq 0.$$

By the test for divergence, the series diverges. ■

Ans. to reading exercise 10
3

CAUTION: MISTAKE TO AVOID
We did not calculate the sum of the series. The sum of this series is infinite, not $\frac{3}{5}$.

Example 9 *Determine the sum or state that the series diverges:* $\sum_{n=1}^{\infty} n = 1 + 2 + 3 + 4 + \cdots$.

Solution We try the test for divergence by calculating the limit of the terms of the series. For any positive infinite hyperreal integer Ω,

$$\lim_{n\to\infty} n = \Omega \doteq \infty \neq 0.$$

By the test for divergence, this series diverges. ■

Example 10 *Determine the sum or state that the series diverges:* $\sum_{n=1}^{\infty} \frac{1}{n} = 1 + \frac{1}{2} + \frac{1}{3} + \frac{1}{4} + \frac{1}{5} + \cdots$.

Solution We try the test for divergence by calculating the limit of the terms of the series. For any positive infinite hyperreal integer Ω,

$$\lim_{n\to\infty} \frac{1}{n} = \frac{1}{\Omega} = \omega \doteq 0.$$

CAUTION: MISTAKE TO AVOID
The test for divergence never concludes that a series converges. If the limit is zero, the test is inconclusive. Other means must be used to determine the convergence or divergence of the series.

Because the limit of the terms of the sequence is zero, the test for divergence is inconclusive. (We revisit this example later in this section and prove that the series diverges.) ■

Example 11 *Determine the sum or state that the series diverges:* $\sum_{n=1}^{\infty} (-1)^n = -1 + 1 - 1 + 1 - 1 + 1 - \cdots$.

Solution We try the test for divergence by calculating the limit of the terms of the series. For any positive infinite hyperreal integer Ω,

$$\lim_{n\to\infty} (-1)^n = (-1)^{\Omega},$$

See also the discussion after example 7.

which is 1 if Ω is even or -1 if Ω is odd. Because the expression does not render the same real result for every Ω, the limit does not exist. By the test for divergence, the series diverges. ■

Reading Exercise 11 *Does* $\sum_{n=1}^{\infty} \frac{n}{n+1}$ *diverge?*

Harmonic series

The series $\sum_{n=1}^{\infty} \frac{1}{n} = 1 + \frac{1}{2} + \frac{1}{3} + \frac{1}{4} + \cdots$ is called the *harmonic series.* In example 10 it was noted that the test for divergence is inconclusive. The series is not geometric, nor does it appear to be telescoping. We need a different idea.

There is a connection to music. The ratios of lengths of strings producing certain harmonies are terms of the harmonic series.

One relatively simple way to show that the harmonic series diverges is to group the terms and show that the sum is larger than a sum that is infinite:

$$1 + \frac{1}{2} + \underbrace{\frac{1}{3} + \frac{1}{4}}_{> \frac{1}{4} + \frac{1}{4}} + \underbrace{\frac{1}{5} + \frac{1}{6} + \frac{1}{7} + \frac{1}{8}}_{> \frac{1}{8} + \frac{1}{8} + \frac{1}{8} + \frac{1}{8}} + \underbrace{\frac{1}{9} + \cdots + \frac{1}{16}}_{> \frac{1}{16} + \cdots + \frac{1}{16}} + \underbrace{\frac{1}{17} + \cdots + \frac{1}{32}}_{> \frac{1}{32} + \cdots + \frac{1}{32}} + \cdots$$

$$> 1 + \frac{1}{2} + \frac{2}{4} + \frac{4}{8} + \frac{8}{16} + \frac{16}{32} + \cdots$$

$$= 1 + \frac{1}{2} + \frac{1}{2} + \frac{1}{2} + \frac{1}{2} + \frac{1}{2} + \cdots = \infty.$$

Although the sum of the harmonic series is infinite, it diverges very slowly. The sum of the first million terms of the series is approximately 14.3927, and the sum of the first billion terms is only about 21.3005.

This also helps explain why we say that the test for divergence is inconclusive if $\lim_{n\to\infty} a_n = 0$. The limit of the terms of the series from example 7 is zero and the series converges. However, the limit of the terms of the harmonic series is also zero, yet the harmonic series diverges. Therefore, knowing $\lim_{n\to\infty} a_n = 0$ is not in and of itself enough evidence to determine whether the series $\sum_{n=1}^{\infty} a_n$ converges or diverges.

Series rules

The summation sum rule, summation difference rule, and summation constant multiple rule of section 4.3 can be extended to infinite series using the transfer principle.

SERIES RULES

Suppose that $\sum_{n=1}^{\infty} a_n = A$; $\sum_{n=1}^{\infty} b_n = B$; and that A, B, and c are real numbers. Then,

PROOF OF THE SERIES CONSTANT MULITPLE RULE

Suppose $\sum_{n=1}^{\infty} a_n = A$. Let Ω be a positive infinite hyperreal integer. The capital omegath partial sum of this series then renders A—that is, $S_\Omega \doteq A$. By the summation constant multiple rule of section 4.3, for any real positive integer k,

$$\sum_{n=1}^{k} ca_n = c\sum_{n=1}^{k} a_n.$$

Then, by the transfer principle,

$$\sum_{n=1}^{\Omega} ca_n = c\sum_{n=1}^{\Omega} a_n = cS_\Omega \doteq cA.$$

Because this holds for any such Ω, by definition 4,

$$\sum_{n=1}^{\infty} ca_n = cA.$$

$$\sum_{n=1}^{\infty}(a_n + b_n) = A + B = \sum_{n=1}^{\infty} a_n + \sum_{n=1}^{\infty} b_n,$$

$$\sum_{n=1}^{\infty}(a_n - b_n) = A - B = \sum_{n=1}^{\infty} a_n - \sum_{n=1}^{\infty} b_n, \text{ and}$$

$$\sum_{n=1}^{\infty} ca_n = cA = c\sum_{n=1}^{\infty} a_n.$$

Example 12 *Evaluate* $\displaystyle\sum_{n=1}^{\infty}\left(\frac{1}{2^n} + \frac{2}{n(n+1)}\right)$.

The series $\sum_{n=1}^{\infty}\left(\frac{1}{2^n} + \frac{2}{n(n+1)}\right)$ is neither telescoping nor geometric. However, the result of using the series sum rule is the sum of a geometric series and a telescoping series, which allows us to evaluate the combined series.

Solution In example 7 we concluded that

$$\sum_{n=1}^{\infty}\frac{2}{n(n+1)} = 2.$$

The series $\displaystyle\sum_{n=1}^{\infty}\frac{1}{2^n} = \frac{1}{2} + \frac{1}{4} + \frac{1}{8} + \cdots$ is geometric, with the first term $a = \frac{1}{2}$ and the common ratio $r = \frac{1}{2}$. Hence,

$$\sum_{n=1}^{\infty}\frac{1}{2^n} = \frac{a}{1-r} = \frac{\frac{1}{2}}{1-\frac{1}{2}} = 1.$$

Then, by the series sum rule (with $A = 1$ and $B = 2$),

$$\sum_{n=1}^{\infty}\left(\frac{1}{2^n} + \frac{2}{n(n+1)}\right) = \sum_{n=1}^{\infty}\frac{1}{2^n} + \sum_{n=1}^{\infty}\frac{2}{n(n+1)}$$
$$= 1 + 2 = 3. \quad \blacksquare$$

The series rules show that the sum or difference of two convergent series is convergent. Another consequence of the series rules is that the sum or difference of a convergent series and a divergent series diverges.

Use care when evaluating sums or differences of two divergent series. They may either converge or diverge. The ideas of indeterminate forms apply.

Example 13 *Evaluate* $\displaystyle\sum_{n=1}^{\infty}\left(\frac{1}{2^n} + \frac{1}{n}\right)$.

Ans. to reading exercise 11
Yes, it diverges by the test for divergence: $\displaystyle\lim_{n\to\infty}\frac{n}{n+1} = 1 \neq 0.$

Solution We learned in example 12 that $\displaystyle\sum_{n=1}^{\infty}\frac{1}{2^n} = 1$. We also know that the harmonic series $\displaystyle\sum_{n=1}^{\infty}\frac{1}{n}$ diverges to ∞. Because the sum of a convergent series and a divergent series diverges, $\sum_{n=1}^{\infty}\left(\frac{1}{2^n}+\frac{1}{n}\right)$ diverges. ■

More specifically, $\sum_{n=1}^{\infty}\left(\frac{1}{2^n}+\frac{1}{n}\right)=\infty$.

Starting points

All the series in examples 1 through 13 begin at $n = 1$. It is not necessary to start there.

Example 14 *Evaluate* $\sum_{n=5}^{\infty}\frac{4}{3^n}$.

Solution The series terms are formed from a constant and an exponential, so we suspect the series is geometric. The first few terms are

$$\sum_{n=5}^{\infty}\frac{4}{3^n}=\frac{4}{243}+\frac{4}{729}+\frac{4}{2187}+\frac{4}{6561}+\cdots.$$

The series is geometric, with the first term $a=\frac{4}{243}$ and the common ratio $r=\frac{1}{3}$. The sum of the series is

$$\sum_{n=5}^{\infty}\frac{4}{3^n}=\frac{a}{1-r}=\frac{\frac{4}{243}}{1-\frac{1}{3}}=\frac{2}{81}.$$ ■

Intuitively, inserting or removing a finite number of terms from a series changes its value only by a finite amount, meaning that the convergence or divergence of the resulting series is not affected. This is true; it is a consequence of the series sum and difference rules.

Example 15 *Evaluate* $\sum_{n=23}^{\infty}\frac{1}{n}$.

Solution This is the harmonic series with a few terms removed at the beginning. Because the harmonic series diverges, this series also diverges. ■

Removing infinitely many terms from a series may or may not affect its convergence or divergence. The reasoning of example 15 does not apply to the series $\sum_{n=1}^{\infty}\frac{1}{2n}=\frac{1}{2}+\frac{1}{4}+\frac{1}{6}+\frac{1}{8}+\cdots$, which is the harmonic series with every other term removed. However, the series constant multiple rule does apply, and the series does in fact diverge.

Exercises 10.3

1–6. Rapid response: what is it?

1. $2, 4, 8, 16, 32, 64, 128, \ldots$
2. $2 + 4 + 8 + 16 + 32 + 64 + 128 + \cdots$
3. a_n
4. (a_n)
5. $\displaystyle\sum_{n=1}^{\infty} a_n$
6. $\displaystyle\lim_{n\to\infty} a_n$

7–14. (a) Is the series geometric? If so, (b) state the first term a and the common ratio r, and (c) state whether the series converges or diverges.

7. $\dfrac{e}{2} + \dfrac{e}{4} + \dfrac{e}{6} + \dfrac{e}{8} + \cdots$
8. $\dfrac{e}{2} + \dfrac{e}{6} + \dfrac{e}{18} + \dfrac{e}{54} + \cdots$
9. $\displaystyle\sum_{n=1}^{\infty} \frac{4}{3^n}$
10. $\displaystyle\sum_{n=1}^{\infty} 3\left(\frac{5}{3}\right)^{n-1}$
11. $\displaystyle\sum_{n=6}^{\infty} \frac{3^n}{4}$
12. $\displaystyle\sum_{n=1}^{\infty} n\left(\frac{5}{3}\right)^{n-1}$
13. $\displaystyle\sum_{n=1}^{\infty} \frac{4}{3+n}$
14. $\displaystyle\sum_{n=3}^{\infty} 4^n$

15–36. Determine the sum or state that the series diverges.

15. $\displaystyle\sum_{n=12}^{\infty} \frac{1}{n}$
16. $\displaystyle\sum_{n=1}^{\infty} 0.6^n$
17. $\displaystyle\sum_{n=1}^{\infty} 21\left(\frac{4}{9}\right)^{n-1}$
18. $\displaystyle\sum_{n=2}^{\infty} 6^n$
19. $\displaystyle\sum_{n=1}^{\infty} \sin n$
20. $\displaystyle\sum_{n=1}^{\infty} \left(2^{1/n} - 2^{1/(n+1)}\right)$
21. $\displaystyle\sum_{n=4}^{\infty} \frac{0.24^n}{1.55}$
22. $\displaystyle\sum_{n=1}^{\infty} 2^{1/n}$

23. $\sum_{n=1}^{\infty}\left(\tan^{-1} n - \tan^{-1}(n+1)\right)$

24. $\sum_{n=1}^{\infty}\frac{6}{n}$

25. $\sum_{n=1}^{\infty}\frac{1}{(n+5)(n+6)}$

26. $\sum_{n=12}^{\infty}\frac{6}{n}$

27. $\sum_{n=1}^{\infty}\frac{3^n+1}{5^n}$

28. $\sum_{n=1}^{\infty}\ln\left(\frac{n}{n+1}\right)$

29. $\sum_{n=1}^{\infty}2\cdot\frac{3^n+1}{5^n}$

30. $\sum_{n=1}^{\infty}\ln\left(\frac{2n}{n+1}\right)$

31. $\sum_{n=1}^{\infty}\frac{6^n+2}{5^n}$

32. $\sum_{n=0}^{\infty}\frac{2^n\ 3^n}{4^n}$

33. $\sum_{n=1}^{\infty}\frac{3n+4n^3}{n^3+3n^2+2n}$

34. $\sum_{n=1}^{\infty}\frac{6}{n^2+2n}$

35. $\sum_{n=1}^{\infty}\frac{3n+4}{n^3+3n^2+2n}$

36. $\sum_{n=1}^{\infty}\frac{6n^2}{n^2+2n}$

37. Write the number $0.\overline{9}$ as a geometric series in the style of example 6 and then find the sum of that series.

38. Express $0.\overline{49}$ as a geometric series and find the sum of that series.

39. Express the number $3.7\overline{15}$ as a fraction as follows: rewrite the number as $3.7 + 0.0\overline{15}$, evaluate a geometric series for $0.0\overline{15}$, and add $\frac{37}{10}$.

40. Express the number $0.0\overline{345}$ as a geometric series and find the sum of that series.

41. Express the number $141.\overline{41}$ as a geometric series and find the sum of that series.

42. (GSL) Determine the sum or state that the series diverges: $\sum_{n=1}^{\infty}\frac{2n+1}{(n+1)(n+2)(n+3)}$.

43. (GSL) Determine the sum or state that the series diverges: $\sum_{n=1}^{\infty}\frac{n}{(n+1)(n+2)}$.

44. Prove the series sum rule.

45. Show that a geometric series with first term $a \neq 0$ and common ratio $r = 1$ diverges.

46. Suppose you wish to walk a path of length 1 km. Before making it to the end of the path, you must make it to the halfway point; that is, you must walk $\frac{1}{2}$ km. You must then walk half of the remaining distance, or an additional $\frac{1}{4}$ km; and again half of the remaining distance, or an additional $\frac{1}{8}$ km; and then an additional $\frac{1}{16}$ km; and so on forever. Because you have infinitely many tasks to perform before reaching the end of the path, it is impossible to reach the end of the path—or so says Zeno. This is a variant of one of Zeno's paradoxes, designed to argue that motion is merely an illusion and is not actually possible. Discuss Zeno's reasoning in relation to the topics of this section.

Zeno of Elea, ca. 490–425 BC
http://www-history.mcs.st-andrews.ac.uk/Biographies/Zeno_of_Elea.html

47. If you drop a ball so that it bounces vertically off the ground, the ball does not bounce back to its original height. Each successive bounce is to a shorter height. As the bounces get shorter in height, they also get shorter in time, until eventually the bounces come so close together that they cannot be distinguished from one another, followed by the ball coming to rest. Why does the ball not bounce forever? Geometric series give the answer. For this exercise, ignore such factors as air resistance and spin. (a) Suppose a ball bounces from the ground (position $s = 0$) with an initial upward velocity of v_0 m/s . Using an acceleration resulting from gravity of $-9.8\,\text{m/s}^2$, solve the initial value problem to determine the position function $s(t)$ of the ball. (b) At what time does the ball return to the ground? That is, how long is it between bounces? (c) Upon impact with the ground, the ball loses kinetic energy. Part of this energy is translated into sound (the sound of a bounce). The *coefficient of restitution* can be used to describe the relationship between the initial velocity of the ball on one bounce and the initial velocity of the ball on its next bounce. Suppose the initial velocity of our ball begins at 5 m/s and the coefficient of restitution is 0.95, meaning that the initial upward velocity of the ball when it leaves the ground next is $5 \cdot 0.95$ m/s, followed by $5 \cdot 0.95^2$ m/s, and so on. Assuming the ball continues to bounce in this same manner infinitely many times, write a geometric series for the total time elapsed as it bounces. (d) Evaluate the geometric series.

48. For the ball of exercise 47, determine the ball's total distance traveled while bouncing.

49. Suppose $\sum_{n=1}^{\infty} a_n$ diverges and that $c \neq 0$ is a real number. Must $\sum_{n=1}^{\infty} ca_n$ diverge? (In other words, must a constant multiple of a divergent series diverge?)

50. Because the sum of the harmonic series is infinite, it must be the case that $\sum_{k=1}^{\Omega} \frac{1}{k}$ is infinite for every positive infinite hyperreal integer Ω. Use this fact and the definition of definite integral to evaluate the (improper) integral $\int_0^1 \frac{1}{x}\,dx$.

This exercise is adapted from an article by Ghrist, Michelle, "The IRS Uses Geometric Series?" *MAA Focus* **39** No. 4 (August/September 2019), 14–15.

51. A church holds a raffle for an all-terrain vehicle with a fair market value of \$4000. The church announces that it will also pay income tax on the prize, which is subject to withholding at a rate of 24% of the value of the prize. But, this adds 24% of \$4000 to the value of the prize, which is also taxable. Paying the withholding tax on this amount adds even more to the prize. (a) Use a geometric series to determine the total amount of withholding the church must pay. (b) What percentage of the value of the prize should be withheld?

Specifically, this applies to a noncash prize.

Integral Test

10.4

In section 4.2 we used finite sums to estimate the area under a curve. These ideas can be extended to a relationship between infinite sums (infinite series) and the area under a curve on an infinitely long interval—that is, a type I improper integral.

The integral test: convergence or divergence of a series

In a manner reminiscent of the similar function rule, consider an infinite series $\sum_{n=1}^{\infty} a_n$ and a function f for which $f(n) = a_n$ for every natural number n. Suppose further that the function f is positive, decreasing, and continuous on the interval $[1, \infty)$, as in figure 1. The height of the curve at $x = 1$ is a_1, the height of the curve at $x = 2$ is a_2, and so on, as represented by red dots in the figure.

For $\sum_{n=1}^{\infty} \frac{n+1}{2^n}$, we can use $f(x) = \frac{x+1}{2^x}$; but, for $\sum_{n=1}^{\infty} \frac{n+1}{n!}$, we have not studied a function of the form $x!$ that is defined for every real number $x > 1$. So, in general, such a function f might not always exist. When a similar function f does exist, it might not be positive or decreasing or continuous.

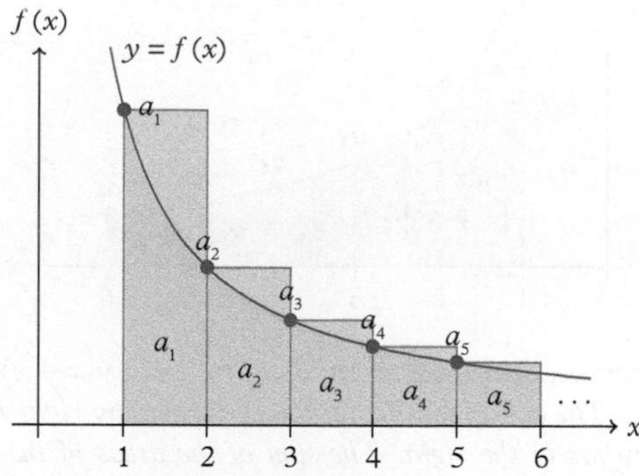

Figure 1 *A positive, decreasing, continuous function $y = f(x)$ on $[1, \infty)$, for which $f(n) = a_n$. The curve and the rectangles formed by left-hand endpoints extend infinitely far to the right. The sum of the areas of the rectangles is greater than the area under the curve; the excess area is indicated in gray*

In a manner reminiscent of estimating the area under a curve using left-hand endpoints, figure 1 shows rectangles of width 1 beginning at $x = 1$, except this time there are infinitely many such rectangles.

Because we have a decreasing function, left-hand endpoints create an overestimate of the area under the curve; the "extra" area is the gray area in figure 1. Because the width of each rectangle is 1, the area of each rectangle is the same as its height (areas are labeled inside the rectangles, whereas the heights are labeled at the red dots). The sum of the areas of the rectangles, which is the sum of the a_n, is then greater than the area under the curve:

$$\sum_{n=1}^{\infty} a_n > \int_1^{\infty} f(x)\, dx.$$

If the area under the curve happens to be infinite (if the type I improper integral diverges), then the sum must also be infinite. The series must also diverge.

By contrast, using right-hand endpoints produces an underestimate, as in figure 2. If we do not include the rectangle for a_1, we see that

$$\sum_{n=2}^{\infty} a_n < \int_1^{\infty} f(x)\, dx.$$

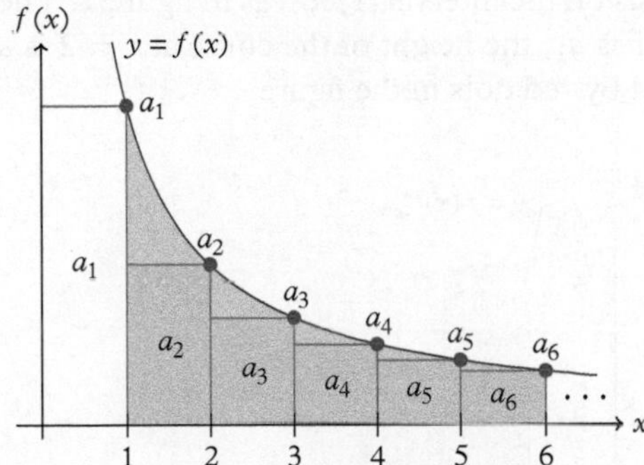

Figure 2 *A positive, decreasing, continuous function $y = f(x)$ on $[1, \infty)$, for which $f(n) = a_n$. The curve and the rectangles formed by right-hand endpoints extend infinitely far to the right. The sum of the areas of the rectangles (not including the rectangle of area a_1) is less than the area under the curve; the missing area is indicated in gray*

If the integral converges, then the area under the curve is finite. Because the sum of the areas of the rectangles (not including a_1) is less than the area under the curve, then this quantity is also finite. Adding the value of a_1 isn't enough to change this; therefore, the series also converges.

These facts are summarized by the integral test.

INTEGRAL TEST

If $\sum_{n=1}^{\infty} a_n$ is an infinite series; if the function f is positive, decreasing, and continuous on $[1, \infty)$; and if $f(n) = a_n$ for every natural number n, then either

(1) both $\int_1^{\infty} f(x)\,dx$ and $\sum_{n=1}^{\infty} a_n$ converge or

(2) both diverge.

As difficult as integration can be, finding the exact value of the sum of a series can sometimes be even more difficult. When it applies, the integral test gives us an option at least to be able to tell whether the sum is finite.

Integral test examples

To use the integral test on a series $\sum_{n=1}^{\infty} a_n$, we must identify a function f and verify that it is (1) positive, (2) decreasing, and (3) continuous on the interval $[1, \infty)$. We may then calculate the type I improper integral to determine convergence or divergence of the series.

Example 1 *Use the integral test to determine whether* $\sum_{n=1}^{\infty} \frac{3}{n+2}$ *converges or diverges.*

Solution To apply the integral test, we begin by rewriting the summand as a function:

$$f(x) = \frac{3}{x+2}.$$

Next we must verify three properties of the function on the interval $[1, \infty)$.

(1) Positive. The numerator is positive, and since $x \geq 1$, so is the denominator. The function is positive on $[1, \infty)$.

(2) Decreasing. The numerator is constant, so the fraction is decreasing as long as the denominator is increasing. Since $y = x + 2$ is a line with a positive slope, it is increasing. Therefore, the function f is decreasing on $[1, \infty)$.

An alternative is to calculate the derivative and verify that it is negative on the interval $[1, \infty)$. To this end,

$$f'(x) = \frac{(x+2)\cdot 0 - 3\cdot 1}{(x+2)^2} = \frac{-3}{(x+2)^2} < 0.$$

(3) Continuous. The function is not continuous at $x = -2$ (division by zero), but this number is not in the interval $[1, \infty)$. The function is continuous on $[1, \infty)$.

Now that we have verified the hypotheses of the test are met, we calculate the integral:

$$\begin{aligned}\int_1^{\infty} \frac{3}{x+2}\,dx &= \int_1^{\Omega} \frac{3}{x+2}\,dx \\ &= 3\ln|x+2|\Big|_1^{\Omega} \\ &= 3\ln(\Omega+2) - 3\ln 3 \\ &\approx 3\ln\Omega \\ &\doteq \infty.\end{aligned}$$

When the integral test applies, we are seeking to determine whether the integral converges or diverges. We then apply the same conclusion to the series.

The integral diverges, and therefore the series diverges. ■

Reading Exercise 12 *The hypotheses of the integral test are met for* $\sum_{n=1}^{\infty} \frac{1}{n+4}$. *Complete the test.*

Notice that the summand is undefined at $n = 1$.

If the series begins at some number other than $n = 1$, then the interval for the integral test begins at that value.

Example 2 *Apply the integral test to* $\sum_{n=2}^{\infty} \frac{1}{n \ln n}$.

Solution To apply the integral test, we begin by rewriting the summand as a function:

$$f(x) = \frac{1}{x \ln x}.$$

Next we must verify three properties of the function on the interval $[2, \infty)$.

(1) Positive. The numerator is positive, and since $x \geq 2$, so is the denominator. The function is positive on $[2, \infty)$.

Alternately,

$$f'(x) = \frac{-(x \cdot \frac{1}{x} + \ln x)}{(x \ln x)^2} < 0$$

because $x \geq 2$, and the function is decreasing.

(2) Decreasing. The numerator is constant. In the denominator, both $y = x$ and $y = \ln x$ are increasing functions, so the denominator is increasing and the fraction (the function) is decreasing.

(3) Continuous. The domain of the function requires both $x > 0$ (so that the logarithm is defined) and $x \neq 1$ (to avoid dividing by

zero because $\ln 1 = 0$). Neither requirement poses a problem on our interval; the function is continuous on $[2, \infty)$.

Now that we have verified the hypotheses of the test are met, we calculate the integral using the interval $[2, \infty)$:

$$\int_2^{\infty} \frac{1}{x \ln x}\, dx = \int_2^{\Omega} \frac{1}{x \ln x}\, dx.$$

Using the substitution

$$u = \ln x$$
$$du = \frac{1}{x}\, dx$$

gives

$$\begin{aligned} &= \int_{\ln 2}^{\ln \Omega} \frac{1}{u}\, du \\ &= \ln |u| \Big|_{\ln 2}^{\ln \Omega} \\ &= \ln \ln \Omega - \ln \ln 2 \\ &\approx \ln \ln \Omega \\ &\doteq \infty. \end{aligned}$$

The series diverges. ■

p-series

Example 3 *Use the integral test to determine whether* $\sum_{n=1}^{\infty} \frac{1}{n^3}$ *converges or diverges.*

Solution First we set $f(x) = \frac{1}{x^3} = x^{-3}$.

(1) Positive. On $[1, \infty)$, the function is positive.

(2) Decreasing. Since $f'(x) = -3x^{-2} = \frac{-3}{x^2} < 0$ on $[1, \infty)$, f is decreasing on $[1, \infty)$.

(3) Continuous. Although f is not continuous at $x = 0$, it is continuous on the interval $[1, \infty)$.

All that remains is to determine the convergence or divergence of the integral $\int_1^{\infty} \frac{1}{x^3}\, dx$. By the p-test for integrals from section 7.10, theorem 1, since $p = 3 > 1$, the integral converges. Therefore $\sum_{n=1}^{\infty} \frac{1}{n^3}$ converges. ■

Ans. to reading exercise 12
$\int_1^{\infty} \frac{1}{x+4}\,dx = \cdots = \infty$, so the series diverges.

The p-test for integrals applies to integrals of the form $\int_1^{\infty} \frac{1}{x^p}\,dx$. As long as $p > 0$, the hypotheses of the integral test are met for $\sum_{n=1}^{\infty} \frac{1}{n^p}$ in exactly the same manner as in example 3. We therefore conclude that the series converges for $p > 1$ and diverges for $0 < p \leq 1$. If $p \leq 0$, then the test for divergence applies and the series diverges. We call such a series a *p-series*.

Theorem 5 p-SERIES *The series*

$$\sum_{n=1}^{\infty} \frac{1}{n^p}$$

converges if $p > 1$ and diverges if $p \leq 1$.

Example 4 *Does the series converge or diverge? (a)* $\sum_{n=1}^{\infty} \frac{1}{n^2}$, *(b)* $\sum_{n=1}^{\infty} \frac{1}{n}$, *(c)* $\sum_{n=3}^{\infty} \frac{15}{n^4}$.

Solution (a) The series $\sum_{n=1}^{\infty} \frac{1}{n^2}$ is a p-series with $p = 2 > 1$; the series converges.

We already know that the series in example 4(b), which is the harmonic series, diverges.

(b) The series $\sum_{n=1}^{\infty} \frac{1}{n}$ is a p-series with $p = 1 \leq 1$; the series diverges.

(c) Convergence or divergence of a series is not affected by the starting point ($n = 3$ instead of $n = 1$) or by a constant multiple (by the series constant multiple rule). Therefore, we treat $\sum_{n=3}^{\infty} \frac{15}{n^4}$ the same as $\sum_{n=1}^{\infty} \frac{1}{n^4}$, which is a p-series with $p = 4 > 1$, and the series converges. ■

Although the two series $\sum_{n=3}^{\infty} \frac{15}{n^4}$ and $\sum_{n=1}^{\infty} \frac{1}{n^4}$ both converge, they have different sums. They can be treated the same for purposes of convergence and divergence only.

Reading Exercise 13 *Does* $\sum_{n=1}^{\infty} \frac{1}{\sqrt{n}}$ *converge or diverge?*

Estimating sums using the integral test

In example 3 we reasoned that because $\int_1^\infty \frac{1}{x^3}\,dx$ converges, $\sum_{n=1}^\infty \frac{1}{n^3}$ also converges. However, we did not compute the sum of the series. This is because the value of the integral and the sum of the series are not the same. In fact, although

$$\int_1^\infty \frac{1}{x^3}\,dx = -\frac{1}{2x^2}\bigg|_1^\Omega = -\frac{1}{2\Omega^2} - \left(-\frac{1}{2}\right) \approx \frac{1}{2},$$

the sum of the series is obviously larger:

$$\sum_{n=1}^\infty \frac{1}{n^3} = 1 + \frac{1}{8} + \frac{1}{27} + \cdots > 1.$$

This relationship is consistent with figure 1, from which we concluded

$$\sum_{n=1}^\infty a_n > \int_1^\infty f(x)\,dx.$$

However, figure 2 gives an upper bound for the sum of the series. Although we did not include a_1 when we noticed that

$$\sum_{n=2}^\infty a_n < \int_1^\infty f(x)\,dx,$$

including a_1 results in

$$\sum_{n=1}^\infty a_n < a_1 + \int_1^\infty f(x)\,dx.$$

Combining the results yields a method for estimating the value of the series.

ESTIMATING A SUM USING THE INTEGRAL TEST

If the integral test has concluded that the series $\sum_{n=1}^\infty a_n$ converges, then

$$\int_1^\infty f(x)\,dx < \sum_{n=1}^\infty a_n < a_1 + \int_1^\infty f(x)\,dx.$$

For $\sum_{n=1}^{\infty} \frac{1}{n^3}$, the first term is $a_1 = 1$. Then,

$$\int_1^{\infty} \frac{1}{x^3}\,dx < \sum_{n=1}^{\infty} \frac{1}{n^3} < 1 + \int_1^{\infty} \frac{1}{x^3}\,dx$$

$$\frac{1}{2} < \sum_{n=1}^{\infty} \frac{1}{n^3} < 1 + \frac{1}{2} = \frac{3}{2}.$$

We may not know the sum of the series, but at least we know something about it. And summing a few terms of the series can give a better lower bound as well. Advanced methods show that

Because all the terms of the series are positive, the sum of the series must be greater than the sum of however many terms we wish to add. The sequence of partial sums looks similar to figure 11 in section 10.1.

$$\sum_{n=1}^{\infty} \frac{1}{n^3} = 1.20205690315959\ldots,$$

which is indeed between 0.5 and 1.5.

Example 5 *The integral test has shown that* $\sum_{n=1}^{\infty} \left(\frac{3}{3n+1} - \frac{1}{n+5} \right)$ *converges. Estimate the sum using the integral test.*

This series is neither geometric nor telescoping. Determining the exact value of the sum of the series requires advanced methods.

Solution First we calculate the value of the integral, which wasn't given:

$$\begin{aligned}\int_1^{\infty} \left(\frac{3}{3x+1} - \frac{1}{x+5} \right) dx &= \int_1^{\Omega} \left(\frac{3}{3x+1} - \frac{1}{x+5} \right) dx \\ &= \left(\ln|3x+1| - \ln|x+5|\right)\Big|_1^{\Omega} \\ &= \ln \frac{3x+1}{x+5}\Big|_1^{\Omega} \\ &= \ln \frac{3\Omega+1}{\Omega+5} - \ln \frac{4}{6} \\ &\approx \ln 3 - \ln \frac{2}{3}.\end{aligned}$$

Next we calculate the first term of the series, $\frac{7}{12}$. Then, using the formula for estimating a sum using the integral test,

$$\ln 3 - \ln \frac{2}{3} < \sum_{n=1}^{\infty} \left(\frac{3}{3n+1} - \frac{1}{n+5} \right) < \frac{7}{12} + \ln 3 - \ln \frac{2}{3}$$

$$1.504 < \sum_{n=1}^{\infty} \left(\frac{3}{3n+1} - \frac{1}{n+5} \right) < 2.087. \qquad \blacksquare$$

The exact value of the sum of the series, which can be calculated by a computer algebra system (CAS), is

$$\frac{-43 + 10\pi\sqrt{3} + 90\ln 3}{60},$$

or approximately 1.83815, which is indeed between 1.504 and 2.087.

Ans. to reading exercise 13
diverge (p-series with $p = \frac{1}{2} \leq 1$)

Reading Exercise 14 *We used the integral test to conclude that* $\sum_{n=1}^{\infty} a_n$ *converges. Given that* $a_1 = 0.4$ *and* $\int_1^{\infty} f(x)\,dx = 1.5$, *estimate the sum of the series.*

Proof of the sum of powers approximation formula

Consider $\sum_{n=1}^{\infty} n^2 = 1 + 4 + 9 + 16 + \cdots$. Although $f(x) = x^2$ has the properties that $f(n) = n^2$ for every natural number n, f is positive on $[1, \infty)$, and f is continuous on $[1, \infty)$, the integral test still does not apply because f is increasing on $[1, \infty)$, not decreasing. We can try the test for divergence (which is the first thing we should try anyway, even before the integral test). In doing so, we have

$$\lim_{n\to\infty} n^2 = \Omega^2 \doteq \infty \neq 0,$$

and the series diverges.

Alternately, $\sum_{n=1}^{\infty} n^2 = \sum_{n=1}^{\infty} \frac{1}{n^{-2}}$ is a p-series with $p = -2 \leq 1$, and therefore the series diverges.

If this series looks vaguely familiar, there is a reason for it. We used the capital omegath partial sum of this series, $\sum_{n=1}^{\Omega} n^2$, when evaluating omega sums (although we used the variable k instead of n). As you may recall, a key step in evaluating omega sums is the use of the sum of powers approximation formula that, when applied in this case, is $\sum_{k=1}^{\Omega} k^2 \approx \frac{\Omega^3}{3}$.

In section 4.3 we only proved the sum of powers approximation formula for a few exponents. The proof presented in this section is valid for all positive exponents.

The proof of the sum of powers approximation formula is similar to the proof of the integral test. Figure 3 is similar to figures 1 and 2, but with an increasing function instead of a decreasing function.

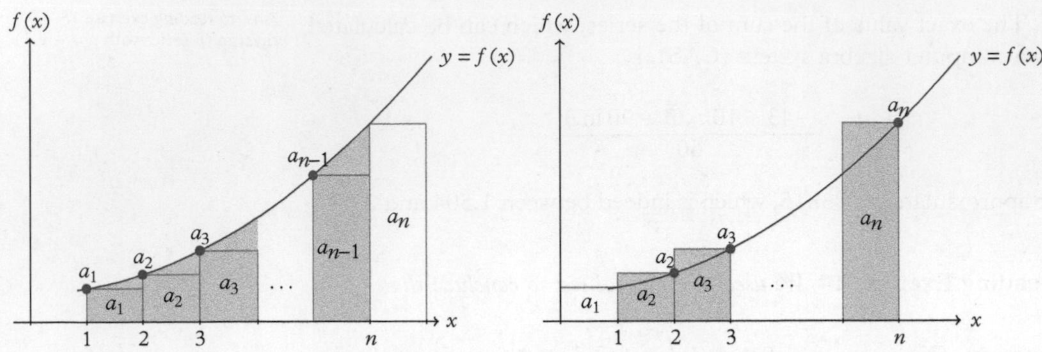

Figure 3 *(left) When left-hand endpoints are used to estimate the area under an increasing curve from x = 1 to x = n, the result is an underestimate; (right) when right-hand endpoints are used to estimate the area under an increasing curve from x = 1 to x = n, the result is an overestimate. The areas of the color-filled rectangles comprise the estimates of the area under the curve; the white-filled rectangles are not included in the estimates*

Proof. Suppose $y = f(x)$ is a positive, continuous, increasing function on $[1, \infty)$ such that $f(k) = a_k$ for every natural number k. Let n be a (real) natural number. We see from the left side of figure 3 that

$$\sum_{k=1}^{n-1} a_k < \int_1^n f(x)\, dx.$$

Adding a_n to both sides,

$$\sum_{k=1}^{n} a_k < a_n + \int_1^n f(x)\, dx.$$

And from the right side of figure 3,

$$\int_1^n f(x)\, dx < \sum_{k=2}^{n} a_k.$$

Adding a_1 to both sides,

$$a_1 + \int_1^n f(x)\, dx < \sum_{k=1}^{n} a_k.$$

Ans. to reading exercise 14

$1.5 < \sum_{n=1}^{\infty} a_n < 1.9$

Combining inequalities, for any natural number n,

$$a_1 + \int_1^n f(x)\,dx < \sum_{k=1}^{n} a_k < a_n + \int_1^n f(x)\,dx.$$

By the transfer principle, this inequality must also be true if we replace n with a positive infinite hyperreal integer Ω:

$$a_1 + \int_1^\Omega f(x)\,dx < \sum_{k=1}^{\Omega} a_k < a_\Omega + \int_1^\Omega f(x)\,dx.$$

If $m > 0$ then $f(x) = x^m$ is positive, continuous, and increasing on $[1, \infty)$, and the previous formula applies, with $a_k = f(k) = k^m$. Then, $a_1 = 1^m = 1$, $a_\Omega = \Omega^m$, and

$$1 + \int_1^\Omega x^m\,dx < \sum_{k=1}^{\Omega} k^m < \Omega^m + \int_1^\Omega x^m\,dx.$$

Next we evaluate the integral:

$$\int_1^\Omega x^m\,dx = \left.\frac{x^{m+1}}{m+1}\right|_1^\Omega = \frac{\Omega^{m+1}}{m+1} - \frac{1}{m+1} \approx \frac{\Omega^{m+1}}{m+1}.$$

The approximation is valid as long as $m + 1 > 0$, so that Ω^{m+1} is infinite.

Replacing the integral with its value in the earlier inequality gives

$$1 + \frac{\Omega^{m+1}}{m+1} < \sum_{k=1}^{\Omega} k^m < \Omega^m + \frac{\Omega^{m+1}}{m+1}.$$

However, both thc left and right sides of the inequality are approximately $\dfrac{\Omega^{m+1}}{m+1}$. Therefore, the quantity between them must approximate to the same value, yielding the sum of powers approximation formula

$$\sum_{k=1}^{\Omega} k^m \approx \frac{\Omega^{m+1}}{m+1}.$$

This proves the formula for any positive real exponent m. The proof can be modified to show that the formula is true for all real $m > -1$. ■

Exercises 10.4

1–8. Rapid response: does the series converge or diverge?

1. $\sum_{n=1}^{\infty} \frac{1}{n^2}$

2. $\sum_{n=1}^{\infty} \frac{1}{n^{1.01}}$

3. $\sum_{n=1}^{\infty} \frac{1}{n^{0.99}}$

4. $\sum_{n=1}^{\infty} \frac{1}{n}$

5. $\sum_{n=1}^{\infty} \frac{5}{n^2}$

6. $\sum_{n=1}^{\infty} \frac{-1}{n^2}$

7. $\sum_{n=1}^{\infty} \left(\frac{3}{n^2} + \frac{4}{n^5}\right)$

8. $\sum_{n=1275}^{\infty} \frac{\sqrt{n}}{n^3}$

9–16. Are the hypotheses of the integral test met for the series and function?

9. $\sum_{n=1}^{\infty} \frac{1}{n^2}, f(x) = \frac{1}{x^5}$

10. $\sum_{n=1}^{\infty} n^3, f(x) = x^3$

11. $\sum_{n=1}^{\infty} \frac{1}{(-n)^3}$, $f(x) = \frac{1}{(-x)^3}$

12. $\sum_{n=1}^{\infty} \frac{(-1)^n}{n^3}$, $f(x) = \frac{(-1)^x}{x^3}$

13. $\sum_{n=1}^{\infty} \frac{1}{n - \frac{8}{3}}, f(x) = \frac{1}{x - \frac{8}{3}}$

14. $\sum_{n=1}^{\infty} e^{-n}, f(x) = e^{-x}$

15. $\sum_{n=1}^{\infty} \frac{1}{\sqrt{n} + 1}, f(x) = \frac{1}{\sqrt{x} + 1}$

16. $\sum_{n=1}^{\infty} \sin\left(\frac{1}{n}\right), f(x) = \sin \frac{1}{x}$

See the solution to example 5 for the desired format of an answer to part (b).

17–28. (a) Use the integral test to determine whether the series converges or diverges. (b) If the series converges, estimate the sum.

17. $\sum_{n=1}^{\infty} \frac{1}{5n}$

18. $\sum_{n=1}^{\infty} \frac{1}{n+5}$

19. $\sum_{n=1}^{\infty} n^{-4}$

20. $\sum_{n=1}^{\infty} \frac{1}{1+n^2}$

21. $\sum_{n=1}^{\infty} \frac{6n+16}{n^2+4n+3}$

22. $\sum_{n=1}^{\infty} n^{-3/2}$

23. $\sum_{n=1}^{\infty} 2^{-n}$

24. $\sum_{n=7}^{\infty} \frac{3+2n}{n-5}$

25. $\sum_{n=3}^{\infty} \frac{1}{2n-5}$

26. $\sum_{n=1}^{\infty} \frac{3+6n}{n^2+5n+4}$

27. $\sum_{n=1}^{\infty} \frac{n}{1+n^2}$

28. $\sum_{n=2}^{\infty} \frac{(\tan^{-1} n)^2}{1+n^2}$

29–46. If possible, determine whether the series converges or diverges; otherwise, state "convergence or divergence cannot be determined using available techniques."

You are not required to use the integral test for these exercises. You may use a different test where applicable.

29. $\sum_{n=1}^{\infty} \frac{3+n}{4+n}$

30. $\sum_{n=1}^{\infty} \left(\frac{\pi}{6}\right)^n$

31. $\sum_{n=1}^{\infty} \frac{3}{4+n}$

32. $\sum_{n=1}^{\infty} \frac{\pi}{n^6}$

33. $\sum_{n=1}^{\infty} \frac{3}{4^n}$

34. $\sum_{n=1}^{\infty} (1+n)^{-\frac{4}{3}}$

35. $\sum_{n=1}^{\infty} \frac{3}{n^4}$

36. $\sum_{n=1}^{\infty} \frac{e^n}{n}$

37. $\sum_{n=1}^{\infty} \frac{3}{n!}$

38. $\sum_{n=1}^{\infty} \frac{e^n}{n!}$

39. $\sum_{n=1}^{\infty} \frac{1}{\sqrt{2n+1}}$

40. $\sum_{n=1}^{\infty} \frac{1}{n^2+5n+4}$

41. $\sum_{n=1}^{\infty} \frac{n}{2^{(n^2)}}$

42. $\sum_{n=1}^{\infty} \frac{\sin n}{n^2}$

43. $\sum_{n=1}^{\infty} ne^{-n}$

44. $\sum_{n=3}^{\infty} \frac{n}{\sqrt{n^2-7}}$

45. $\sum_{n=1}^{\infty} \frac{n^2+4n+2}{n^4+2n^3+n^2}$

46. $\sum_{n=1}^{\infty} \frac{n+1}{n^2+3n}$

47. Use the test for divergence to show that the p-series $\sum_{n=1}^{\infty} \frac{1}{n^p}$ diverges if $p < 0$.

An example of a divergent type I integral that is not infinite is $\int_0^\infty \sin x\,dx$.

48. A type I improper integral can diverge without being infinite. In the narrative of this section, when we derived the integral test, we assumed that if the type I improper integral is divergent then it must be infinite. Why? Give an intuitive explanation.

49. A more useful estimate of the sum of a series that is convergent by the integral test can be derived as follows: let S_k be the kth partial sum of $\sum_{n=1}^{\infty} a_n$, let L be the sum of the series, and let f be the function from the integral test. (a) Write $L - S_k$ in summation notation. (b) Draw a picture similar to figure 1, showing rectangles with the area that is the answer to part (a). Your diagram should use left-hand endpoints to match the curve $y = f(x)$. From the diagram, determine an improper integral with a value that is less than the answer to part (a). (c) Draw a picture similar to figure 2, showing rectangles with the area that is the answer to part (a). Your diagram should use right-hand endpoints to match the curve $y = f(x)$. From the diagram, determine an improper integral with a value that is greater than the answer to part (a). (d) Write a compound inequality containing the results of parts (b) and (c), with $L - S_k$ as the middle quantity. This inequality shows how close the kth partial sum S_k must be to the sum L of the series. (e) Use algebra to manipulate the compound inequality of part (d) so that it features L as the middle quantity. This inequality provides an estimate of the sum of the series.

50. Use the results of exercise 49 to estimate $\sum_{n=1}^{\infty} \frac{1}{n^2}$ as follows. Give all answers to five decimal places (six significant digits). (a) Use a CAS to determine S_{100}, the sum of the first 100 terms of the series. (b) Determine $\int_{100}^{\infty} \frac{1}{x^2}\,dx$ and $\int_{101}^{\infty} \frac{1}{x^2}\,dx$. (c) Apply the results of parts (a) and (b) to the inequality of part (e) of exercise 49. (d) Leonhard Euler rather famously showed in 1735 that $\sum_{n=1}^{\infty} \frac{1}{n^2} = \frac{\pi^2}{6}$. Compare this result to the answer to part (c).

Comparison Tests

10.5

Determining the convergence or divergence of the series $\sum_{n=1}^{\infty} \frac{n}{1+n^2}$ was an exercise in the previous section, one that was carried out using the integral test. Suppose instead that we try to reason the answer by comparing the series to $\sum_{n=1}^{\infty} \frac{n}{n^2} = \sum_{n=1}^{\infty} \frac{1}{n}$, which is the harmonic series. In the long run, the terms of the two series are very close, so perhaps the two series have the same answer. And because the harmonic series diverges, perhaps $\sum_{n=1}^{\infty} \frac{n}{1+n^2}$ does as well. This is, indeed, the correct answer as given by the integral test.

"In the long run" evokes the image of a limit at infinity. Try calculating both $\lim_{n\to\infty} \frac{n}{1+n^2}$ and $\lim_{n\to\infty} \frac{n}{n^2}$ and you will see the similarity.

In this section we develop methods for carrying out such comparisons. The goal is to avoid more difficult techniques such as integration.

The comparison test

Suppose the series $\sum_{n=1}^{\infty} c_n$ consists of positive terms and the series converges. What do we see if we graph the sequence of partial sums? Each partial sum must be larger than the partial sums preceding it, because we keep adding positive numbers to the running total (the partial sum). The sequence of partial sums is an increasing sequence that converges to the sum of the series, as shown in figure 1.

Now suppose we have a second series, $\sum_{n=1}^{\infty} a_n$, which also consists of positive terms, but for which each term a_n is smaller than the corresponding term c_n of the first series. In other words,

This development also works for $0 \le a_n \le c_n$.

$$0 < a_n < c_n$$

for each natural number n. Then, as we form the partial sums of the series $\sum_{n=1}^{\infty} a_n$, we are adding smaller numbers than for the first series,

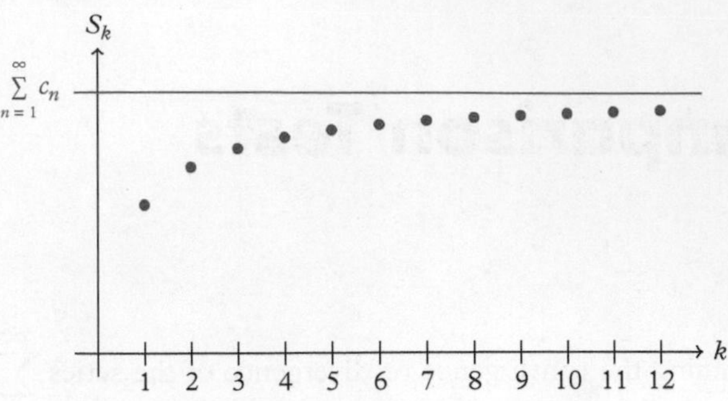

Figure 1 *The sequence of partial sums S_k of the convergent, positive-termed series $\sum_{n=1}^{\infty} c_n$*

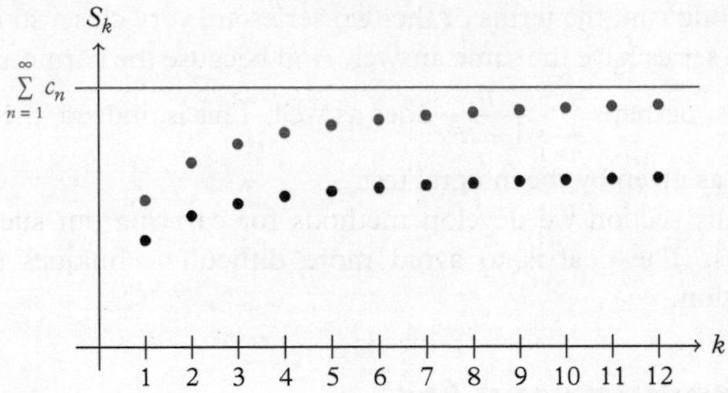

Figure 2 *The sequence of partial sums (red) of the convergent, positive-termed series $\sum_{n=1}^{\infty} c_n$, along with the sequence of partial sums (black) of the series $\sum_{n=1}^{\infty} a_n$, where $0 < a_n < c_n$ for each natural number n*

and therefore the partial sums of this series stay below the partial sums of the first series, as in figure 2.

Remove the partial sums of the first series (figure 3) and the graph of the partial sums of this second series looks remarkably like figure 11 in section 10.1. The sequence of partial sums of the series $\sum_{n=1}^{\infty} a_n$ is monotone increasing, and it is bounded by the sum of the series $\sum_{n=1}^{\infty} c_n$. By the monotone convergence theorem, this sequence of partial sums must converge, and therefore the series $\sum_{n=1}^{\infty} a_n$ converges.

The advantage of this observation is that if it is easy to see that the series $\sum_{n=1}^{\infty} c_n$ converges (such as a p-series with $p > 1$ or a geometric

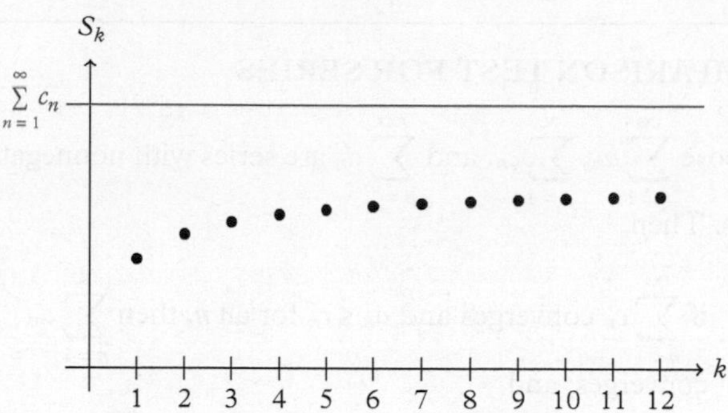

Figure 3 *The sequence of partial sums of the series* $\sum_{n=1}^{\infty} a_n$, *where* $0 < a_n < c_n$, *is a bounded, increasing sequence. Compare to figure 11 section 10.1*

series with $|r| < 1$), then we know that the series $\sum_{n=1}^{\infty} a_n$ must also converge.

In a similar manner, suppose $\sum_{n=1}^{\infty} d_n$ has positive terms and it diverges. If $\sum_{n=1}^{\infty} a_n$ is a larger series—that is, if $a_n \geq d_n$ for every n—then the larger series must also be infinite; see figure 4.

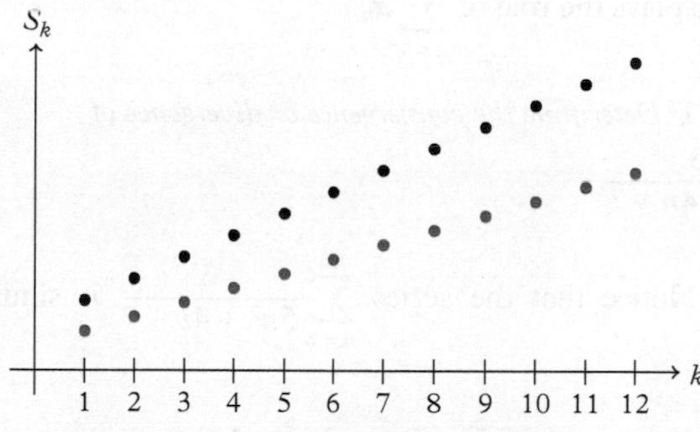

Figure 4 *The sequence of partial sums (red) of the positive-termed, divergent series* $\sum_{n=1}^{\infty} d_n$, *along with the sequence of partial sums (black) of the larger series* $\sum_{n=1}^{\infty} a_n$. *The larger series must also diverge*

This test, which is sometimes called the *direct comparison test,* can be remembered as saying that for positive-termed series, "less than convergent is convergent" and "greater than divergent is divergent."

COMPARISON TEST FOR SERIES

Suppose $\sum_{n=1}^{\infty} a_n$, $\sum_{n=1}^{\infty} c_n$, and $\sum_{n=1}^{\infty} d_n$ are series with nonnegative terms. Then,

(a) if $\sum_{n=1}^{\infty} c_n$ converges and $a_n \le c_n$ for all n, then $\sum_{n=1}^{\infty} a_n$ converges; and

(b) if $\sum_{n=1}^{\infty} d_n$ diverges and $a_n \ge d_n$ for all n, then $\sum_{n=1}^{\infty} a_n$ diverges.

Comparison test examples

Three observations help us use the comparison test. First, recall that although $10 > 2$, $\frac{1}{10} < \frac{1}{2}$. The larger the denominator, the smaller the fraction. Second, it often helps to compare to a series for which the terms are similar "in the long run"—that is, the terms are approximately the same when $n = \Omega$. Third, for the purposes of the comparison test, the series for which we wish to know convergence or divergence plays the role of $\sum_{n=1}^{\infty} a_n$.

Example 1 *Determine the convergence or divergence of*

$$\sum_{n=1}^{\infty} \frac{3}{5n^2 + 4n + 7}.$$

Solution Notice that the series $\sum_{n=1}^{\infty} \frac{3}{5n^2 + 4n + 7}$ is similar to the series

$$\sum_{n=1}^{\infty} \frac{3}{5n^2} = \frac{3}{5} \sum_{n=1}^{\infty} \frac{1}{n^2},$$

which is a convergent p-series ($p = 2 > 1$). The terms of both series are positive, so the comparison test may be applicable. All that remains is to determine how the terms of the two series compare. Since $n > 0$,

How was $\sum_{n=1}^{\infty} \frac{3}{5n^2}$ chosen? If we replace n with Ω and approximate, the terms are $\frac{3}{5\Omega^2 + 4\Omega + 7} \approx \frac{3}{5\Omega^2}$. Sometimes (but not always) a good choice for the comparison series is made by pretending that n is a positive infinite hyperreal and discarding lower-level terms.

$$5n^2 + 4n + 7 > 5n^2.$$

Therefore,

$$3 \cdot \frac{1}{5n^2 + 4n + 7} < 3 \cdot \frac{1}{5n^2}.$$

The series we are testing, $\sum_{n=1}^{\infty} \frac{3}{5n^2 + 4n + 7}$, has smaller terms than the convergent p-series $\sum_{n=1}^{\infty} \frac{3}{5n^2}$, and by the comparison test part (a) (think "smaller than convergent is convergent"), the series $\sum_{n=1}^{\infty} \frac{3}{5n^2 + 4n + 7}$ also converges. ■

If our series was $\sum_{n=1}^{\infty} \frac{3}{5n^2 - 4n - 7}$ instead, then we have

$$5n^2 - 4n - 7 < 5n^2$$

and

$$3 \cdot \frac{1}{5n^2 - 4n - 7} > 3 \cdot \frac{1}{5n^2},$$

When taking reciprocals, the direction of the inequality changes. Multiplying both sides by the positive number 3 does not affect the direction of the inequality.

meaning that the series $\sum_{n=1}^{\infty} \frac{3}{5n^2 - 4n - 7}$ has larger terms (not smaller) than the convergent p-series $\sum_{n=1}^{\infty} \frac{3}{5n^2}$. But, then the hypotheses of the comparison test are not satisfied, and the comparison test does not yield a conclusion regarding whether the series converges or diverges.

Example 2 *Determine the convergence or divergence of* $\sum_{n=3}^{\infty} \frac{\ln n}{n}$.

Solution Although the integral test can be used, let's use the comparison test instead. First, note that since $n > 3$,

$$\ln n > 1.$$

For this series, pretending that n is a positive infinite hyperreal and discarding lower-level terms does not help us find a comparison series because there are no lower-level terms to discard.

Dividing both sides of the inequality by the positive number n does not change the direction of the inequality.

Therefore,

$$\frac{\ln n}{n} > \frac{1}{n}.$$

Hence, the terms of the series $\sum_{n=3}^{\infty} \frac{\ln n}{n}$ are each larger than the terms of the harmonic series $\sum_{n=3}^{\infty} \frac{1}{n}$, which we know diverges. Then, by the comparison test part (b) (think "greater than divergent is divergent"), the series $\sum_{n=3}^{\infty} \frac{\ln n}{n}$ also diverges. ■

Starting the harmonic series at $n = 3$ does not affect its convergence or divergence, as discussed in section 10.3 example 15.

In examples 1 and 2, the comparison series was a p-series. Geometric series also work well as comparison series.

Example 3 *Determine the convergence or divergence of* $\sum_{n=1}^{\infty} \frac{2^n}{3^n + 2}$.

Solution The series $\sum_{n=1}^{\infty} \frac{2^n}{3^n + 2}$ is similar to the series

$$\sum_{n=1}^{\infty} \frac{2^n}{3^n} = \sum_{n=1}^{\infty} \left(\frac{2}{3}\right)^n,$$

which is a convergent geometric series ($r = \frac{2}{3}$ and $\left|\frac{2}{3}\right| < 1$). The terms of both series are positive, so the comparison test is applicable. All that remains is to determine how the terms of the two series compare. Since

$$3^n + 2 > 3^n,$$

we have

$$2^n \cdot \frac{1}{3^n + 2} < 2^n \cdot \frac{1}{3^n}.$$

The direction of the inequality changes when we take reciprocals.

The series we are testing, $\sum_{n=1}^{\infty} \frac{2^n}{3^n + 2}$, has smaller terms that the convergent geometric series $\sum_{n=1}^{\infty} \frac{2^n}{3^n}$, and by the comparison test part (a), the series $\sum_{n=1}^{\infty} \frac{2^n}{3^n + 2}$ also converges. ■

Think "smaller than convergent is convergent." Checking this relationship every time can help you avoid common errors.

The next example explores yet another scenario and helps motivate the next test.

Example 4 *Determine the convergence or divergence of* $\sum_{n=2}^{\infty} \frac{2}{n^2-1}$.

Solution The series $\sum_{n=2}^{\infty} \frac{2}{n^2-1}$ is similar to the series

$$\sum_{n=2}^{\infty} \frac{2}{n^2} = 2 \cdot \sum_{n=2}^{\infty} \frac{1}{n^2},$$

which is a convergent p-series ($p = 2 > 1$; starting the series at $n = 2$ is irrelevant). The terms of both series are positive, so the comparison test may be applicable. All that remains is to determine how the terms of the two series compare. Since

$$n^2 - 1 < n^2,$$

we have

$$2 \cdot \frac{1}{n^2-1} > 2 \cdot \frac{1}{n^2}.$$

The series we are testing, $\sum_{n=2}^{\infty} \frac{2}{n^2-1}$, has larger terms than the convergent p-series $\sum_{n=2}^{\infty} \frac{2}{n^2}$. But, "larger than convergent" is inconclusive; the comparison test does not apply. We must use some other means to determine the convergence or divergence of the series. ■

Notice that the summand is undefined for $n = 1$.

The direction of the inequality changes when we take reciprocals.

The integral test can be used to determine the convergence or divergence of this series, but we soon have a more convenient test to use. Stay tuned.

Reading Exercise 15 *Determine the convergence or divergence of* $\sum_{n=1}^{\infty} \frac{2}{5n^2+2n+5}$.

The level comparison test

Consider the harmonic series $\sum_{n=1}^{\infty} \frac{1}{n}$, which diverges, and the p-series $\sum_{n=1}^{\infty} \frac{1}{n^2}$, which converges. Although the terms of both series tend toward zero, the terms of $\sum_{n=1}^{\infty} \frac{1}{n^2}$ go toward zero much more quickly than the terms of $\sum_{n=1}^{\infty} \frac{1}{n}$, as in figure 5. The laggardly pace at which the terms of the harmonic series go toward zero give it time to accumulate to an infinite quantity slowly, whereas the much faster pace at which

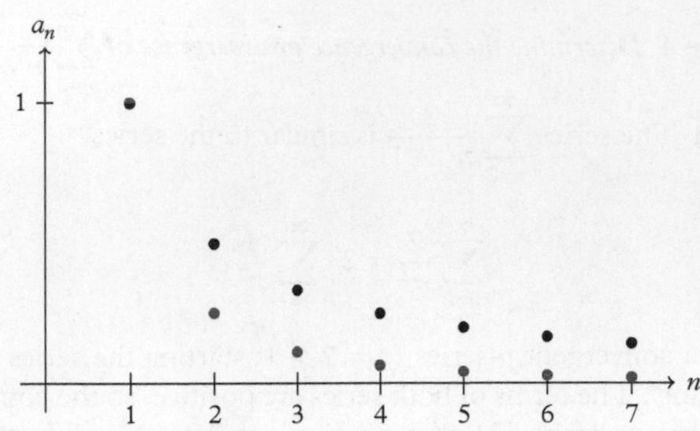

Figure 5 *The terms (not the partial sums) of the series* $\sum_{n=1}^{\infty} \frac{1}{n}$ *(black) and the terms of the series* $\sum_{n=1}^{\infty} \frac{1}{n^2}$ *(red). The terms of the two series begin at the same value, but as n grows, the red dots become much closer to zero than the black dots*

the terms of $\sum_{n=1}^{\infty} \frac{1}{n^2}$ go toward zero do not allow for much accumulation. Intuitively, knowing how fast the terms go toward zero ought to tell us whether a positive-termed series converges or diverges—if the terms go toward zero at all. If they do not go toward zero, then the test for divergence applies.

Toward this end, consider the denominators of these two series. Saying that $\frac{1}{n^2}$ goes toward zero much more quickly than $\frac{1}{n}$ does is akin to saying that n^2 grows much more quickly than n. When we discussed relative rates of growth, such comparisons were made using an arbitrary positive infinite hyperreal Ω. Because Ω^2 is on a higher level than Ω (written $\Omega^2 \gg \Omega$), n^2 grows faster than n, and hence $\frac{1}{n^2}$ goes toward zero more quickly than $\frac{1}{n}$.

> **NOTATION FOR HIGHER AND LOWER LEVELS**
>
> If a and b are two hyperreal numbers, the notation $a \gg b$ means that a is on a higher level than b, and the notation $a \ll b$ means that a is on a lower level than b.

To formalize this idea, let's measure the rate at which the terms of a series $\sum_{n=1}^{\infty} a_n$ go toward zero by the quantity $\frac{1}{a_\Omega}$, the reciprocal of the omegath term of the series, where Ω is an arbitrary positive infinite hyperreal integer. For instance, for the harmonic series $\sum_{n=1}^{\infty} \frac{1}{n}$,

$$\frac{1}{a_\Omega} = \frac{1}{\frac{1}{\Omega}} = \Omega$$

and for the series $\sum_{n=1}^{\infty} \frac{1}{n^2}$,

$$\frac{1}{a_\Omega} = \frac{1}{\frac{1}{\Omega^2}} = \Omega^2.$$

The fact that $\Omega^2 \gg \Omega$ captures the idea that the terms of the series $\sum_{n=1}^{\infty} \frac{1}{n^2}$ are going toward zero faster than the terms of the harmonic series.

Ans. to reading exercise 15
converges; use the comparison test to compare to $\sum_{n=1}^{\infty} \frac{2}{5n^2} = \frac{2}{5}\sum_{n=1}^{\infty} \frac{1}{n^2}$, a p-series with $p = 2 > 1$.

Now suppose that $\sum_{n=1}^{\infty} c_n$ is a convergent, positive-termed series and $\sum_{n=1}^{\infty} a_n$ is another positive-termed series for which

$$\frac{1}{a_\Omega} \gg \frac{1}{c_\Omega}$$

for every positive infinite hyperreal integer Ω. The terms of the series $\sum_{n=1}^{\infty} a_n$ are then going toward zero more quickly than the terms of the convergent series $\sum_{n=1}^{\infty} c_n$ (figure 6). Since $\frac{1}{a_\Omega} \gg \frac{1}{c_\Omega}$ for every positive

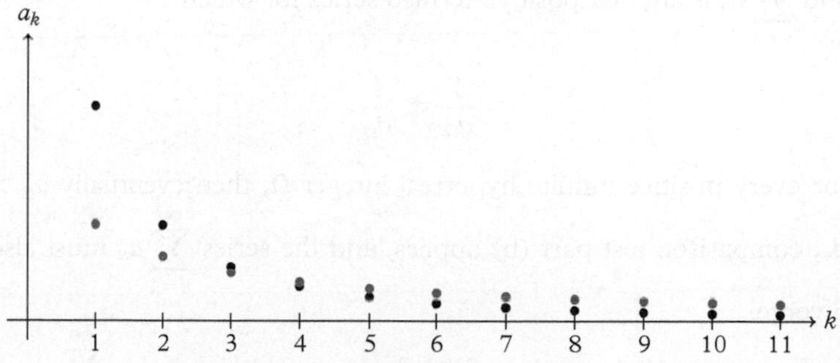

Figure 6 *The terms of a series $\sum_{n=1}^{\infty} a_n$ (black) that go toward 0 more quickly than the terms of a comparison series $\sum_{n=1}^{\infty} c_n$ (red). Assuming $\frac{1}{a_\Omega} \gg \frac{1}{c_\Omega}$ then even if they begin above, eventually the terms of the series $\sum_{n=1}^{\infty} a_n$ must fall below the terms of the comparison series and stay there*

infinite hyperreal integer Ω, as a consequence of the transfer principle, it must be the case that for some natural number N, $\frac{1}{a_n} > \frac{1}{c_n}$ for every $n \geq N$. By taking reciprocals, $a_n < c_n$ for every $n \geq N$ (in figure 6 we

The full details of this use of the transfer principle require application of a theorem that is beyond the scope of this book.

could use $N = 4$). Then, by the comparison test part (a), we conclude the series $\sum_{n=1}^{\infty} a_n$ must also converge.

The series $\sum_{n=1}^{\infty} c_n$ can be any positive-termed convergent series, including p-series $\sum_{n=1}^{\infty} \frac{1}{n^p}$ for which $p > 1$. For such a series we have

$$\frac{1}{c_\Omega} = \frac{1}{\frac{1}{\Omega^p}} = \Omega^p.$$

Hence, as long as

$$\frac{1}{a_\Omega} \gg \Omega^p$$

The comparison must hold for every positive infinite hyperreal integer Ω.

for $p > 1$, then we know the series $\sum_{n=1}^{\infty} a_n$ converges!

In a similar manner, if $\sum_{n=1}^{\infty} d_n$ is a positive-termed divergent series and $\sum_{n=1}^{\infty} a_n$ is another positive-termed series for which

$$\frac{1}{a_\Omega} \ll \frac{1}{d_\Omega}$$

This time, the terms of $\sum_{n=1}^{\infty} a_n$ are going toward zero more slowly than the terms of the divergent series $\sum_{n=1}^{\infty} d_n$.

for every positive infinite hyperreal integer Ω, then eventually $a_n > d_n$, comparison test part (b) applies, and the series $\sum_{n=1}^{\infty} a_n$ must also diverge.

In section 10.4 example 2 we learned that the series $\sum_{n=2}^{\infty} d_n = \sum_{n=2}^{\infty} \frac{1}{n \ln n}$ diverges. For that series we have

Note that because $\ln \Omega$ is infinite, the quantity $\Omega \ln \Omega$ is on a higher level than Ω.

$$\frac{1}{d_\Omega} = \frac{1}{\frac{1}{\Omega \ln \Omega}} = \Omega \ln \Omega.$$

Hence, as long as

$$\frac{1}{a_\Omega} \ll \Omega \ln \Omega,$$

The comparison must hold for every positive infinite hyperreal integer Ω.

the series $\sum_{n=1}^{\infty} a_n$ diverges.

LEVEL COMPARISON TEST FOR SERIES

Suppose $\sum_{n=1}^{\infty} a_n$ is a series with nonnegative terms. Then,

(a) if $\frac{1}{a_\Omega}$ is in the *convergence zone* for every positive infinite hyperreal integer Ω, $\sum_{n=1}^{\infty} a_n$ converges; and

(b) if $\frac{1}{a_\Omega}$ is in the *divergence zone* for every positive infinite hyperreal integer Ω, $\sum_{n=1}^{\infty} a_n$ diverges.

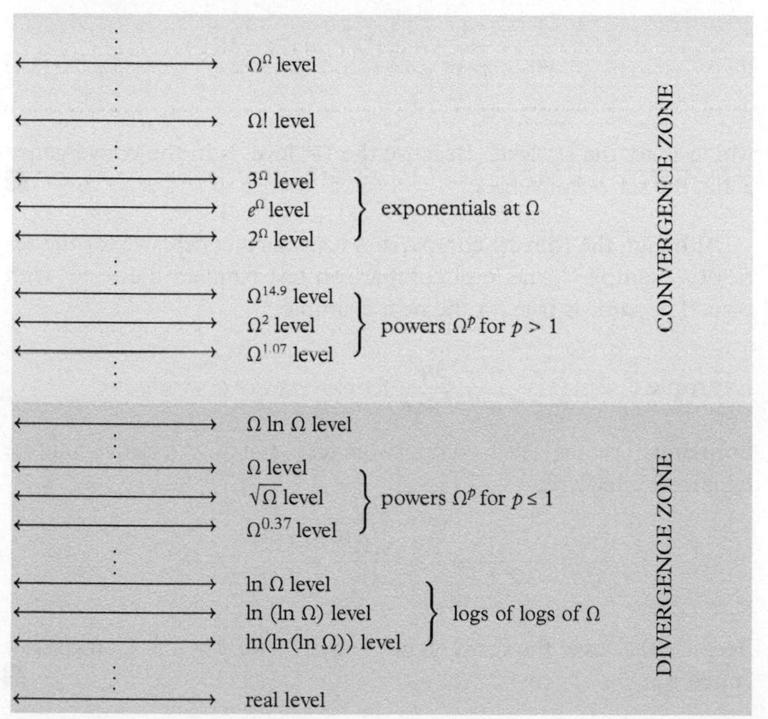

Figure 7 *Convergence zone (light green) and divergence zone (pink). A series* $\sum_{n=1}^{\infty} a_n$ *converges if* $\frac{1}{a_\Omega}$ *is always in the convergence zone; it diverges if* $\frac{1}{a_\Omega}$ *is always in the divergence zone*

If $\frac{1}{a_\Omega}$ is at or below the real level, then the test for divergence also applies.

The convergence zone includes Ω^p for $p > 1$ and higher levels such as exponentials at Ω (figure 7). The divergence zone includes $\Omega \ln \Omega$, Ω^p for $p \le 1$, and lower levels. We examine levels near the boundary between the convergence and divergence zones in a bit more detail later in this section.

Level comparison test examples

Let's begin by reprising example 4.

Example 5 *Determine the convergence or divergence of* $\displaystyle\sum_{n=2}^{\infty} \frac{2}{n^2 - 1}$.

Solution To use the level comparison test, we calculate $\frac{1}{a_\Omega}$ and determine whether this quantity is in the convergence zone or the divergence zone (if we can tell). For any positive infinite hyperreal integer Ω,

Recall that when determining the level of a hyperreal number, we approximate first.

$$\frac{1}{a_\Omega} = \frac{1}{\frac{2}{\Omega^2-1}} = \frac{1}{2}(\Omega^2 - 1) \approx \frac{1}{2}\Omega^2,$$

which is on the Ω^2 level. Because the Ω^2 level is in the convergence zone, the series converges. ■

Although the (direct) comparison test was inconclusive for the series of example 4, the level comparison test handled the series with ease. The same is true for the next example.

Example 6 *Test* $\displaystyle\sum_{n=4}^{\infty} \frac{n^2 - 3n}{\sqrt[3]{n^{10} - 4n^2}}$ *for convergence or divergence.*

Solution Try the level comparison test. For any positive infinite hyperreal integer Ω,

$$\frac{1}{a_\Omega} = \frac{1}{\frac{\Omega^2 - 3\Omega}{\sqrt[3]{\Omega^{10} - 4\Omega^2}}} \approx \frac{\sqrt[3]{\Omega^{10}}}{\Omega^2} = \frac{\Omega^{10/3}}{\Omega^2} = \Omega^{4/3}.$$

Because $\Omega^{4/3}$ is in the convergence zone (Ω^p for $p = \frac{4}{3} > 1$), the series converges. ■

The series $\sum_{n=1}^{\infty} \frac{1}{5n-3} = \frac{1}{2} + \frac{1}{7} + \frac{1}{12} + \cdots$ contains some but not all of the terms of the harmonic series. But, does it leave out enough of the harmonic series' terms to converge?

Example 7 *Determine the convergence or divergence of* $\sum_{n=1}^{\infty} \frac{1}{5n-3}$.

Solution Try the level comparison test. For any positive infinite hyperreal integer Ω,

$$\frac{1}{a_\Omega} = \frac{1}{\frac{1}{5\Omega - 3}} = 5\Omega - 3 \approx 5\Omega,$$

which is on the Ω level. Because the Ω level is in the divergence zone, the series diverges. ■

An alternative is to use the (direct) comparison test, comparing to the series $\sum_{n=1}^{\infty} \frac{1}{5n}$.

If $\frac{1}{a_\Omega} \doteq 0$, then the quantity is infinitesimal and is below the real-number level. Therefore, it is in the divergence zone.

Example 8 *Determine the convergence or divergence of* $\sum_{n=1}^{\infty} \frac{1+3^n}{1+2^n}$.

Solution Try the level comparison test. For any positive infinite hyperreal integer Ω,

$$\frac{1}{a_\Omega} = \frac{1}{\frac{1+3^\Omega}{1+2^\Omega}} = \frac{1+2^\Omega}{1+3^\Omega} \approx \frac{2^\Omega}{3^\Omega} = \left(\frac{2}{3}\right)^\Omega \doteq 0.$$

Recall from section 5.5, on general exponentials, that for $0 < a < 1$, $a^\Omega \doteq 0$.

The quantity is infinitesimal and therefore is in the divergence zone. We conclude the series diverges. ■

An alternate solution is to apply the test for divergence. Since

$$\lim_{n\to\infty} \frac{1+3^n}{1+2^n} = \frac{1+3^\Omega}{1+2^\Omega} \approx \frac{3^\Omega}{2^\Omega} = \left(\frac{3}{2}\right)^\Omega \doteq \infty \neq 0,$$

the series diverges. Note the similarity in calculations between the level comparison test and the test for divergence.

Sometimes a series is amenable to both the comparison test and the level comparison test. When this is the case, as in example 7, the level comparison test is usually more straightforward to apply.

Example 9 *Determine the convergence or divergence of* $\sum_{n=1}^{\infty} \frac{1}{n^n}$.

Solution Try the level comparison test. For any positive infinite hyperreal integer Ω,

$$\frac{1}{a_\Omega} = \frac{1}{\frac{1}{\Omega^\Omega}} = \Omega^\Omega,$$

An alternate solution is to use the (direct) comparison test, comparing to either the convergent p-series $\sum_{n=1}^{\infty} \frac{1}{n^2}$ or the convergent geometric series $\sum_{n=1}^{\infty} \frac{1}{2^n}$, each for $n \geq 2$.

which is in the convergence zone. The series converges. ■

Reading Exercise 16 *Determine the convergence or divergence of* $\displaystyle\sum_{n=1}^{\infty} \frac{2}{5n^2 - 3}$.

When sine or cosine are involved it helps to recall that $-1 \le \sin\theta \le 1$, and the same is true for cosine.

By the transfer principle, $-1 \le \sin\theta \le 1$ is true not just for real θ, but for any hyperreal number as well.

Example 10 *Test the series* $\displaystyle\sum_{n=1}^{\infty} (4 + \cos n)n^{-3}$ *for convergence or divergence.*

Solution Try the level comparison test. For any positive infinite hyperreal integer Ω,

$$\frac{1}{a_\Omega} = \frac{1}{(4 + \cos\Omega)\Omega^{-3}} = \frac{\Omega^3}{4 + \cos\Omega},$$

which is on the Ω^3 level because the denominator is a real number (between 3 and 5). The series converges. ■

Locating a level in the correct zone

Next we reprise example 2, which we worked earlier using the comparison test.

Example 11 *Determine the convergence or divergence of* $\displaystyle\sum_{n=3}^{\infty} \frac{\ln n}{n}$.

Solution Try the level comparison test. For any positive infinite hyperreal integer Ω,

$$\frac{1}{a_\Omega} = \frac{1}{\frac{\ln\Omega}{\Omega}} = \frac{\Omega}{\ln\Omega}.$$

A more compact way of writing the solution is

$$\frac{1}{a_\Omega} = \frac{1}{\frac{\ln\Omega}{\Omega}} = \frac{\Omega}{\ln\Omega} \ll \Omega,$$

and the series diverges by the level comparison test.

Although $\frac{\Omega}{\ln\Omega}$ is not listed in the diagram of figure 7, because $\ln\Omega$ is infinite, we know that $\frac{\Omega}{\ln\Omega}$ is on a lower level than Ω. Because Ω is in the divergence zone, so is $\frac{\Omega}{\ln\Omega}$, and the series diverges. ■

Regardless of whether the solution featured in example 11 is preferred over the solution featured in example 2, the lesson to be learned is that we sometimes must be creative in determining the placement of a level in the convergence or divergence zone. Knowing the relative placement of a level is often enough.

SUFFICIENCY OF RELATIVE PLACEMENT

If $\frac{1}{a_\Omega} > B$ and B is in the convergence zone, then so is $\frac{1}{a_\Omega}$.
If $\frac{1}{a_\Omega} < B$ and B is in the divergence zone, then so is $\frac{1}{a_\Omega}$.

The justification for these statements is found by examining the development of the level comparison test.

Example 12 *Test* $\sum_{n=1}^{\infty} \frac{1 \cdot 4 \cdot 7 \cdots (3n-2)}{13 \cdot 23 \cdot 33 \cdots (10n+3)}$ *for convergence or divergence.*

Ans. to reading exercise 16 converges (use the level comparison test; the comparison test is inconclusive)

Solution Try the level comparison test. For any positive infinite hyperreal integer Ω,

$$\begin{aligned}\frac{1}{a_\Omega} &= \frac{13 \cdot 23 \cdot 33 \cdots (10\Omega + 3)}{1 \cdot 4 \cdot 7 \cdots (3\Omega - 2)} \\ &= \frac{13}{1} \cdot \frac{23}{4} \cdot \frac{33}{7} \cdots \frac{10\Omega + 3}{3\Omega - 2} \\ &> 3 \cdot 3 \cdot 3 \cdots 3 \\ &= 3^\Omega.\end{aligned}$$

Each fraction in line 2 is larger than 3, including the last one, because $3(3\Omega - 2) = 9\Omega - 6 < 10\Omega + 3$. Because there are Ω fractions in line 2, then there are Ω 3s in line 3, the product of which is 3^Ω.

Because 3^Ω is in the convergence zone and $\frac{1}{a_\Omega} > 3^\Omega$, by the sufficiency of relative placement, $\frac{1}{a_\Omega}$ is also in the convergence zone. The series converges. ■

Refining the boundary

Because there are infinitely many levels between any two given levels, it is difficult to define the boundary precisely between the convergence zone and the divergence zone. Some additional detail of levels in each zone is given in figure 8. The groups of levels added here are known

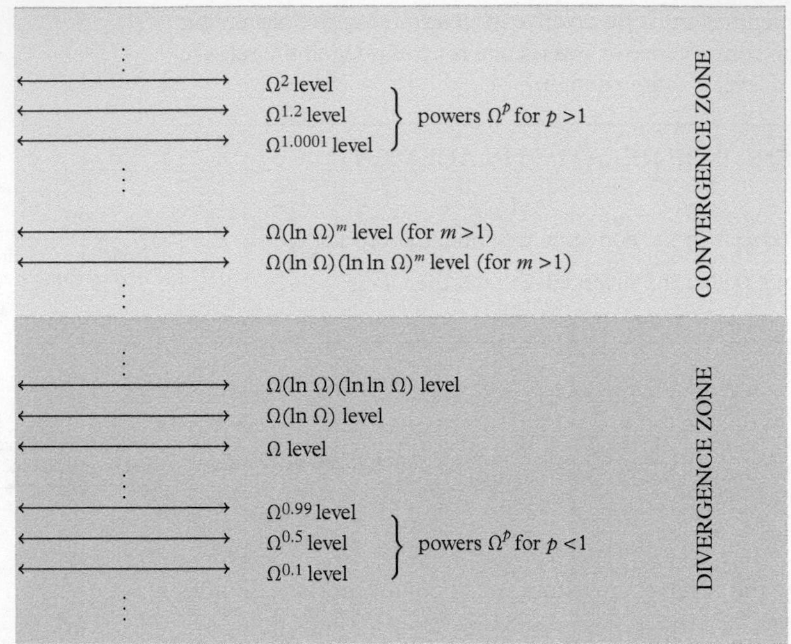

Figure 8 *More detail near the boundary of the convergence zone (light green) and divergence zone (pink). There are still levels between the ones listed in this diagram*

The Abel Prize is awarded annually by the King of Norway and is considered by some to be the equivalent of a Nobel Prize (there is no Nobel Prize in mathematics). It is named in honor of the Norwegian mathematician Niels Abel.

Niels Henrik Abel, 1802–1829
https://www-history.mcs.st-andrews.ac.uk/Biographies/Abel.html

as the Abel-Dini Scale. Just as Ω^1 is in the divergence zone, but any higher power of Ω (such as $\Omega^{1.0000001}$) is in the convergence zone, the divergence zone contains $\Omega(\ln\Omega)^1$. But, increasing this exponent even a small amount (for instance, to $\Omega(\ln\Omega)^{1.003}$) pushes the quantity into the convergence zone. The integral test can be used to provide proofs.

Exercises 10.5

The assumption here is that the answer should apply for every natural number n.

1–8. Rapid response: fill in the space between expressions with $<$ or $>$, or state that neither applies.

1. $n^2+7 \qquad n^2$
2. $\dfrac{1}{n^2+7} \qquad \dfrac{1}{n^2}$
3. $n^2-7 \qquad n^2$
4. $\dfrac{1}{n^2-7} \qquad \dfrac{1}{n^2}$
5. $\dfrac{1}{n+\sqrt{n}+7} \qquad \dfrac{1}{n}$
6. $\dfrac{n+3}{n^2+7} \qquad \dfrac{n}{n^2}$
7. $\dfrac{1}{n+\sqrt{n}-7} \qquad \dfrac{1}{n}$
8. $\dfrac{n-3}{n^2+7} \qquad \dfrac{n}{n^2}$

9–12. Rapid response: (a) to what other series might you compare the given series? (b) Does the comparison series converge or diverge?

9. $\sum_{n=1}^{\infty} \frac{5}{n^2+3}$

10. $\sum_{n=8}^{\infty} \frac{1}{2^n - 157.1}$

11. $\sum_{n=3}^{\infty} \frac{n+5}{n^2-8}$

12. $\sum_{n=4}^{\infty} \frac{6n^3+12n}{\sqrt{n}+n-5}$

13–26. Use the comparison test to determine convergence or divergence of the series, if possible. If not possible, state "the comparison test is inconclusive."

Do not use the level comparison test on these exercises.

13. $\sum_{n=1}^{\infty} \frac{1}{n^3+4}$

14. $\sum_{n=1}^{\infty} \frac{1}{n^2+7}$

15. $\sum_{n=1}^{\infty} \frac{1}{n^3+4n}$

16. $\sum_{n=16}^{\infty} \frac{1}{\sqrt{n}-\frac{7}{2}}$

17. $\sum_{n=3}^{\infty} \frac{1}{n^3-4n}$

18. $\sum_{n=2}^{\infty} \frac{1}{n-\sqrt{n}}$

19. $\sum_{n=1}^{\infty} \frac{n+3}{n^2}$

20. $\sum_{n=4}^{\infty} \frac{n-3}{n^2+7}$

21. $\sum_{n=1}^{\infty} \frac{3-\cos n}{\sqrt{n}}$

22. $\sum_{n=2}^{\infty} \frac{4^n}{3^n-5}$

23. $\sum_{n=1}^{\infty} \frac{3^n}{4^n+5}$

24. $\sum_{n=1}^{\infty} \frac{|\sin n|}{n^3}$

25. $\sum_{n=1}^{\infty} \frac{\left(4.73+\sin\frac{1}{n+1}\right)^n}{2^n}$

26. $\sum_{n=1}^{\infty} \left(2+\frac{3}{n}\right)\left(\frac{1}{2}\right)^n$

27–36. Use the level comparison test to determine the convergence or divergence of the series.

27. $\sum_{n=1}^{\infty} \frac{1}{n^3+4}$

28. $\sum_{n=1}^{\infty} \frac{1}{n^2+7}$

29. $\sum_{n=3}^{\infty} \frac{1}{n^3-4n}$

30. $\sum_{n=4}^{\infty} \frac{n-3}{n^2+7}$

31. $\sum_{n=1}^{\infty} \frac{n+3}{n^2}$

32. $\sum_{n=2}^{\infty} \frac{4^n}{3^n-5}$

33. $\displaystyle\sum_{n=1}^{\infty} \frac{3-\cos n}{\sqrt{n}}$

34. $\displaystyle\sum_{n=1}^{\infty} \frac{|\sin n|}{n^3}$

35. $\displaystyle\sum_{n=1}^{\infty} \frac{3^n}{4^n+5}$

36. $\displaystyle\sum_{n=1}^{\infty} \left(2+\frac{3}{n}\right)\left(\frac{1}{2}\right)^n$

37–62. Determine the convergence or divergence of the series.

37. $\displaystyle\sum_{n=1}^{\infty} \frac{n^2+7}{4n^3-5n+8}$

38. $\displaystyle\sum_{n=8}^{\infty} \frac{1}{2^n-157.1}$

39. $\displaystyle\sum_{n=1}^{\infty} \frac{n^2+8}{\sqrt{n^5+7n^3-1}}$

40. $\displaystyle\sum_{n=2}^{\infty} \frac{n^2+2n}{\sqrt{n}-1}$

41. $\displaystyle\sum_{n=1}^{\infty} \frac{2^n+6n}{5^n+n^5}$

42. $\displaystyle\sum_{n=1}^{\infty} \frac{(n-1)(n+1)}{(n+3)(n+5)(n+7)}$

43. $\displaystyle\sum_{n=1}^{\infty} \frac{(n+3)(n^4+3n)}{(n^2+5)^3}$

44. $\displaystyle\sum_{n=1}^{\infty} e^{-n}2^n$

45. $\displaystyle\sum_{n=1}^{\infty} \frac{\sin\left(\frac{1}{n}\right)}{n}$

46. $\displaystyle\sum_{n=4}^{\infty} \frac{6n^3+12n}{\sqrt{n}+n-5}$

47. $\displaystyle\sum_{n=2}^{\infty} \frac{1}{n^{3/2}\ln n}$

48. $\displaystyle\sum_{n=1}^{\infty} \frac{\sin\left(\frac{1}{\sqrt{n}}\right)}{\sqrt{n}}$

49. $\displaystyle\sum_{n=1}^{\infty} \frac{n!}{2^n}$

50. $\displaystyle\sum_{n=1}^{\infty} \left(\frac{n+1}{n^2+1}\right)^n$

51. $\displaystyle\sum_{n=0}^{\infty} \frac{n+5^n}{2^n}$

52. $\displaystyle\sum_{n=1}^{\infty} \frac{3^n+5}{2^{2n}+3n}$

53. $\displaystyle\sum_{n=1}^{\infty} \frac{n!}{(n+2)!}$

54. $\displaystyle\sum_{n=1}^{\infty} \frac{3^n}{n!\cdot 5^n}$

55. $\displaystyle\sum_{n=1}^{\infty} \left(\frac{(3n+1)^2 n^2}{(2n+5)^4}\right)^n$

56. $\displaystyle\sum_{n=1}^{\infty} \frac{1}{n\ln\left(\frac{n^2+5}{n+3}\right)}$

57. $\displaystyle\sum_{n=1}^{\infty} \frac{\ln(n^2+6)}{n^3+1}$

58. $\displaystyle\sum_{n=1}^{\infty} \left(3^{\frac{n+1}{2n+3}}-1\right)^n$

59. $\displaystyle\sum_{n=2}^{\infty} \frac{(\sqrt{n}+6)^2}{n^2\ln n}$

60. $\displaystyle\sum_{n=1}^{\infty} \frac{3^n}{n!}$

61. $\displaystyle\sum_{n=0}^{\infty} \frac{2\cdot 5\cdot 8\cdots(3n+2)}{5\cdot 12\cdot 19\cdots(7n+5)}$

62. $\displaystyle\sum_{n=1}^{\infty} \frac{1\cdot 3\cdot 5\cdots(2n-1)}{2\cdot 5\cdot 8\cdot(3n-1)}$

63. (a) Use the integral test to show that $\sum_{n=3}^{\infty} \frac{1}{n(\ln n)(\ln \ln n)}$ diverges. (b) Compute $\frac{1}{a_\Omega}$ and conclude that the result is in the divergence zone. Compare to figure 8.

64. (a) Use the integral test to show that $\sum_{n=3}^{\infty} \frac{1}{n(\ln n)(\ln \ln n)^{1.02}}$ converges. (b) Compute $\frac{1}{a_\Omega}$ and conclude that the result is in the convergence zone. Compare to figure 8.

65. Try the level comparison test for $\sum_{n=1}^{\infty} \frac{(1.4 + \sin n)^n}{2^n}$. All the terms are positive. Why is the test inconclusive?

66. Reading exercise 10 was to "evaluate $\sum_{n=1}^{\infty} \left(\frac{3}{n} - \frac{3}{n+1}\right)$." (a) Use the level comparison test to determine the convergence or divergence of the series. (b) Is the answer to part (a) sufficient for completing the reading exercise as stated?

Alternating Series

10.6

Consider the series

$$1 - \frac{1}{3} + \frac{1}{9} - \frac{1}{27} + \frac{1}{81} - \frac{1}{243} + \cdots,$$

which is geometric with the first term $a_1 = 1$ and the common ratio $r = -\frac{1}{3}$. Because $\left|-\frac{1}{3}\right| < 1$, the series converges, and we may easily calculate its sum to be $\frac{3}{4}$. However, what if a series is not geometric and the terms of the series alternate? The integral test, the comparison test, and the level comparison test do not apply directly because the terms are not all positive. We need another test.

Alternating series: definition

We can rewrite the geometric series with the first term $a_1 = 1$ and the common ratio $r = -\frac{1}{3}$ using summation notation as

$$\sum_{n=1}^{\infty} ar^{n-1} = \sum_{n=1}^{\infty} 1 \cdot \left(-\frac{1}{3}\right)^{n-1} = \sum_{n=1}^{\infty} \left(-\frac{1}{3}\right)^{n-1}.$$

Because we can also rewrite $-\frac{1}{3}$ as $-1 \cdot \frac{1}{3}$ and we know from algebra that $(ab)^n = a^n b^n$, we can rewrite the expression as

$$\sum_{n=1}^{\infty} \left(-\frac{1}{3}\right)^{n-1} = \sum_{n=1}^{\infty} (-1)^{n-1} \left(\frac{1}{3}\right)^{n-1}.$$

Each term of the series is now written as the product of the positive number $\left(\frac{1}{3}\right)^{n-1}$ and the number $(-1)^{n-1}$, which is 1 if n is odd and -1 if n is even. We can switch to odd terms negative and even terms positive by using $(-1)^n$ instead. By this method, we can recognize an alternating series as one with terms that are the product of an alternating sign factor and a positive factor.

Note that $(-1)^{n+1}$ and $(-1)^{n-1}$ are equivalent.

Definition 6 ALTERNATING SERIES *An* alternating series *is a series of the form* $\sum_{n=1}^{\infty}(-1)^n b_n$ *or* $\sum_{n=1}^{\infty}(-1)^{n+1} b_n$*, where* $b_n > 0$ *for all* n.

With the series beginning at $n = 1$, we use $(-1)^n$ for a negative first term and $(-1)^{n+1}$ for a positive first term.

Example 1 *Is* $\sum_{n=1}^{\infty}(-1)^n \cdot \frac{1}{n}$ *an alternating series?*

Solution Yes, because $\frac{1}{n} > 0$ for all n. ■

The series $\sum_{n=1}^{\infty}(-1)^{n+1} \cdot \frac{1}{n} = 1 - \frac{1}{2} + \frac{1}{3} - \frac{1}{4} + \cdots$ is also alternating. It is called the *alternating harmonic series.*

Example 2 *Is* $\sum_{n=1}^{\infty}(-1)^n \sin n$ *an alternating series?*

Solution Because $\sin n$ is not always positive, the answer appears to be no. We can check by writing out a few terms of the series, using numerical approximations:

The two consecutive negative terms mean the series is not alternating.

$$\sum_{n=1}^{\infty}(-1)^n \sin n = -0.8415 + 0.9093 - 0.1411 - 0.7568 + 0.9589 \cdots.$$

The terms change sign (and do so infinitely many times), but they do not alternate signs every other term, so this series is not an alternating series. ■

Reading Exercise 17 *Are these an alternating series? (a)* $\sum_{n=1}^{\infty}(-1)^{n+1}\frac{1}{2^n}$*; (b)* $\sum_{n=1}^{\infty}\sin\left((2n+1)\frac{\pi}{4}\right)$.

The alternating series test

When we write an alternating series in this form, with $a_n = (-1)^{n+1}b_n$, then the b's are the absolute values of the a's; that is, $b_n = |a_n|$.

Stated more technically, (b_n) is a decreasing sequence of positive numbers.

Consider an alternating series $\sum_{n=1}^{\infty} a_n = \sum_{n=1}^{\infty}(-1)^{n+1}b_n$, where $b_n > 0$ for all n. Add the condition that $b_{n+1} \leq b_n$ for each n—that is, the b's are decreasing. How can we visualize the sum? Start at zero and add the first term a_1, which is positive; the result is the first partial sum S_1. This is visualized by a long green arrow in figure 1. Add the next term a_2, which is a negative number, and the partial sum goes down (red arrow in figure 1). Add a_3, which is a positive number, and the partial sum goes back up (short green arrow). Notice that the length of an arrow is the value of b_n, and the lengths are decreasing. The direction of the arrow is given by the sign of $(-1)^{n+1}$, which alternates. If the lengths of the arrows are decreasing and they are alternating directions, then each succeeding red arrow doesn't go as far left as the preceding one,

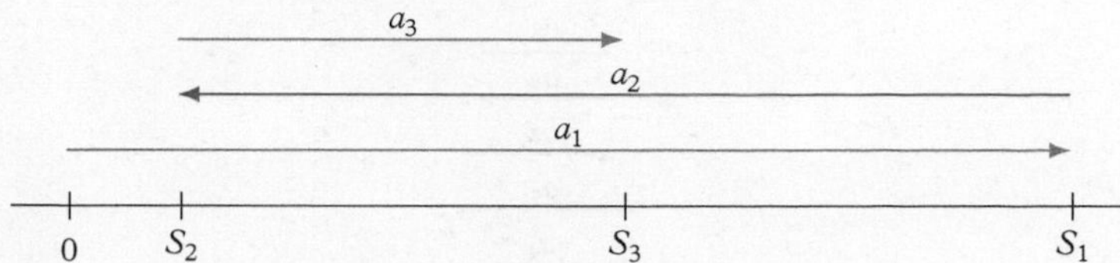

Figure 1 *The first three partial sums of an alternating series* $\sum_{n=1}^{\infty} a_n$ *with positive first term. The green arrows represent positive terms; the red arrows, negative terms. The partial sums* S_1*,* S_2*, and* S_3 *are aligned with the ends of the arrow tips*

and each succeeding green arrow doesn't go as far right as the previous one. The result might look like figure 2, in which the partial sums seem to converge to a sum S.

Since $b_n \geq b_{n+1}$, $|a_n| \geq |a_{n+1}|$ (the length of each arrow is less than or equal to the length of the previous arrow). Therefore, if a_n is positive and a_{n+1} is negative, then $a_n + a_{n+1} > 0$. Adding these two terms increases the partial sum, and adding two more does the same. The even-numbered partial sums (at the tips of the red arrows) are getting larger. Analogously, the odd-numbered partial sums (at the tips of the green arrows) are getting smaller.

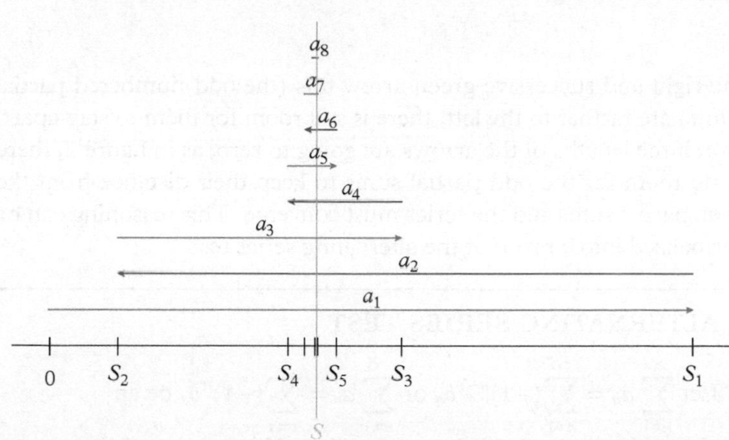

Figure 2 *The first eight partial sums of an alternating series* $\sum_{n=1}^{\infty} a_n$ *with a positive first term. The green arrows represent positive terms; the red arrows, negative terms. The partial sums appear to converge to* S*, represented by the gray line*

What could prevent an alternating series from converging? We already know that some alternating series diverge. For instance, in section 10.3 example 11 we used the test for divergence to conclude that $1 - 1 + 1 - 1 + 1 - 1 + \cdots$ diverges. If the terms of the series do not go to zero—that is, if the lengths of the arrows in the diagram do not go to zero—then the series diverges. In figure 3, even though successive red arrow tips (the even-numbered partial sums) are farther to

Ans. to reading exercise 17
(a) yes; (b) no, the series is $\frac{\sqrt{2}}{2} - \frac{\sqrt{2}}{2} - \frac{\sqrt{2}}{2} + \frac{\sqrt{2}}{2} + \frac{\sqrt{2}}{2} - \cdots$ and does not change sign from each term to the next every time.

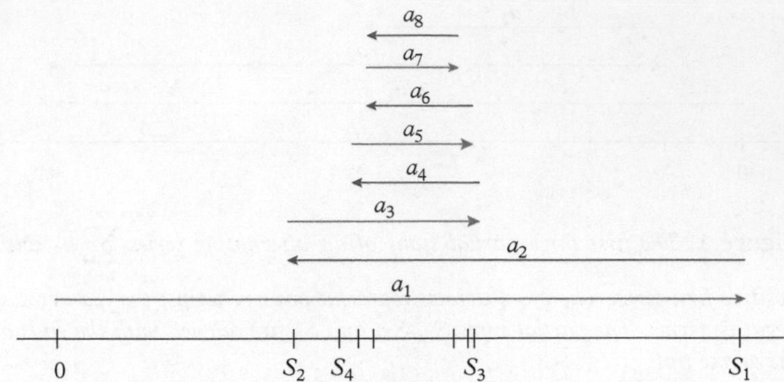

Figure 3 *The first eight partial sums of an alternating series $\sum_{n=1}^{\infty} a_n$ with a positive first term. The green arrows represent positive terms; the red arrows, negative terms. Because the lengths of the arrows are not going to zero, the partial sums do not converge*

Clarification: the tip of the arrow labeled a_6 is farther to the right than the tip of the arrow labeled a_4, and the tip of the a_8 arrow is even farther to the right, and so on.

the right and successive green arrow tips (the odd-numbered partial sums) are farther to the left, there is still room for them to stay apart. But, if the lengths of the arrows are going to zero, as in figure 2, there is no room for the odd partial sums to keep their distance from the even partial sums and the series must converge. This reasoning can be formalized into a proof of the alternating series test.

ALTERNATING SERIES TEST

Let $\sum_{n=1}^{\infty} a_n = \sum_{n=1}^{\infty}(-1)^{n+1} b_n$ or $\sum_{n=1}^{\infty} a_n = \sum_{n=1}^{\infty}(-1)^n b_n$ be an alternating series (with $b_n > 0$ for all n). If

(1) $b_{n+1} \leq b_n$ for all n and

(2) $\lim_{n\to\infty} b_n = 0$,

then the series $\sum_{n=1}^{\infty} a_n$ converges.

Condition (1) can be restated as saying that (b_n) is a decreasing sequence. If condition (2) is not met, then the series diverges by the test for divergence.

Example 3 *Determine the convergence or divergence of the alternating harmonic series, $\sum_{n=1}^{\infty}(-1)^{n+1} \cdot \frac{1}{n}$.*

Solution Because the terms are not all positive, we cannot use the integral test, comparison test, or level comparison test. The series is not geometric, telescoping, or a p-series. However, the series *is* alternating, and therefore we can try the alternating series test. If the two conditions are met, then the series converges.

⓪ Identify b_n. For this series, $b_n = \frac{1}{n}$.

Recall that we require $b_n > 0$ for all n. This requirement is met.

❶ Is $b_{n+1} \le b_n$ for all n? Since $b_{n+1} = \frac{1}{n+1} < \frac{1}{n} = b_n$, yes.

❷ Is $\lim_{n\to\infty} b_n = 0$? We have

$$\lim_{n\to\infty} b_n = \lim_{n\to\infty} \frac{1}{n} = \frac{1}{\Omega} = \omega \doteq 0.$$

Because the two conditions are met, the series $\sum_{n=1}^{\infty}(-1)^{n+1}\cdot\frac{1}{n}$ converges by the alternating series test. ■

Example 4 *Test* $\sum_{n=1}^{\infty}(-1)^n \cdot \frac{n}{n+1}$ *for convergence or divergence.*

Solution We recognize that the series is alternating, so we try the alternating series test.

⓪ Identify b_n. For this series, $b_n = \frac{n}{n+1}$.

Note that $b_n > 0$ for all n.

❶ Is $b_{n+1} \le b_n$ for all n? The first few values of b_n are $\frac{1}{2}, \frac{2}{3}, \frac{3}{4}, \frac{4}{5}, \ldots$, which is not a decreasing sequence; this condition is not met.

Alternate version of step ❶: write out (b_n) as $1, \frac{1}{2}, \frac{1}{3}, \frac{1}{4}, \frac{1}{5}, \ldots$, which is obviously decreasing.

Because one of the conditions is not met, the alternating series test does not apply. We suspect the series diverges, but our statement of the alternating series test does not list divergence as a possible conclusion. Let's try the test for divergence:

$$\lim_{n\to\infty} \frac{n}{n+1} = \frac{\Omega}{\Omega+1} \approx \frac{\Omega}{\Omega} = 1 \neq 0.$$

Whenever condition (1) of the alternating series test is not met, try the test for divergence. The test for divergence may be applied to (b_n).

The series $\sum_{n=1}^{\infty}(-1)^n \cdot \frac{n}{n+1}$ diverges by the test for divergence. ■

What does the green/red arrow diagram look like for the series of example 4? Because (b_n) is an increasing sequence, the lengths of the arrows get longer, so the partial sums are growing farther apart; see figure 4.

Example 5 *Test* $\sum_{n=2}^{\infty}\frac{(-1)^{n-1}}{\ln n}$ *for convergence or divergence.*

Solution We recognize that the series is alternating, so we try the alternating series test.

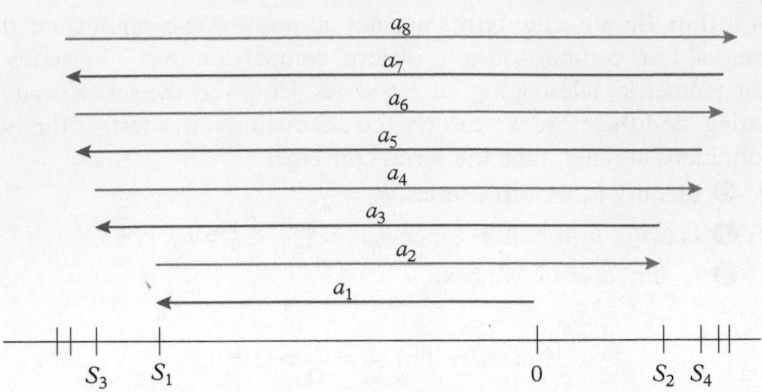

Figure 4 *The first eight partial sums of the alternating series* $\sum_{n=1}^{\infty}(-1)^n \cdot \frac{n}{n+1}$, *which has a negative first term. The green arrows represent positive terms; the red arrows, negative terms. The lengths of the arrows are increasing, causing the odd-numbered partial sums to grow more negative and the even-numbered partial sums to grow more positive. Because the partial sums do not converge, the series diverges*

The phrase "for all n" applies to the values of n in the summation, which are the natural numbers beginning with 2.

⓿ Identify b_n. For this series, $b_n = \frac{1}{\ln n}$. Since $n \geq 2$, $\ln n > 0$ and therefore $b_n > 0$ for all n.

❶ Is $b_{n+1} \leq b_n$ for all n? Since $y = \ln x$ is an increasing function, $\ln(n+1) > \ln n$ and therefore $\frac{1}{\ln(n+1)} < \frac{1}{\ln n}$. The condition is met.

❷ Is $\lim_{n\to\infty} b_n = 0$? We have

Because $\ln \Omega$ is infinite, the expression $\frac{1}{\ln \Omega}$ is infinitesimal and renders zero.

$$\lim_{n\to\infty} b_n = \lim_{n\to\infty} \frac{1}{\ln n} = \frac{1}{\ln \Omega} \doteq 0.$$

Because both conditions are met, the series $\sum_{n=2}^{\infty} \frac{(-1)^{n-1}}{\ln n}$ converges by the alternating series test. ■

Reading Exercise 18 *Do we need the alternating series test for* $\sum_{n=1}^{\infty} \frac{3n+1}{n+1}$*?*

Estimating sums of alternating series

Take another look at figure 2. The sum of the series, S, is to the right of all the red arrow tips and to the left of all the green arrow tips. A portion of this diagram is recreated in figure 5. Look at the distance between

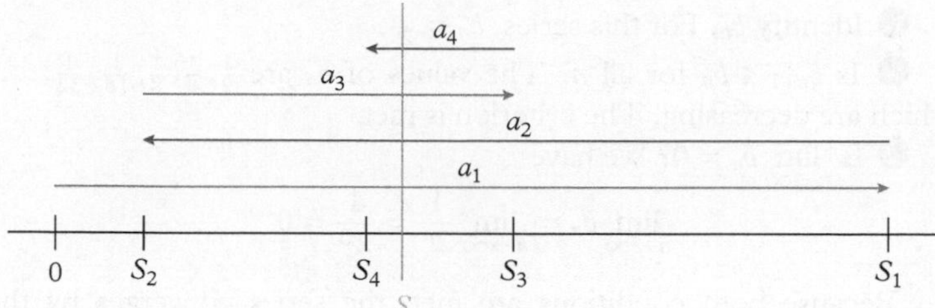

Figure 5 *The first four partial sums of an alternating series* $\sum_{n=1}^{\infty} a_n$ *with a positive first term. The difference between the sum* S *of the series and a partial sum* S_n *is less than the length of the next arrow,* $|a_{n+1}|$

S and the first partial sum S_1 on the number line. That distance is less than the length of the red arrow representing a_2. Why? Because the red arrow must extend beyond (to the left of) S. For the same reason, the distance between S and S_3 is less than the length of the red arrow representing a_4. Similarly, the distance between S and S_2 is less than the length of the green arrow representing a_3, because the green arrow must extend beyond (to the right of) S. These facts can be summarized by the equation

$$|S - S_n| \leq |a_{n+1}| = b_{n+1}, \text{ for all } n.$$

The distance between S and S_n is given by $|S - S_n|$; the length of the next arrow is $|a_{n+1}|$.

This argument, when combined with the justification of the alternating series test, forms the essence of the proof of the following theorem.

Theorem 6 ESTIMATING THE SUM OF AN ALTERNATING SERIES *If the alternating series* $\sum_{n=1}^{\infty}(-1)^n b_n$ *or* $\sum_{n=1}^{\infty}(-1)^{n+1} b_n$ *converges to the sum* S*, then*

$$|S - S_n| \leq b_{n+1} \textit{ for any } n.$$

In the case $b_{n+1} < b_n$ for all n (rather than allowing $\leq$), $|S - S_n| < b_{n+1}$ for any n.

Using theorem 6 requires that the alternating series converge, which can be checked by the alternating series test. We then add terms of the series until we reach a term a_{n+1} with an absolute value that is less than the desired precision for the estimate. The partial sum S_n (or any partial sum thereafter) is used as an estimate of the sum.

If $b_{n+1} = |a_{n+1}|$ is less than the desired precision, then so is the even smaller value $|S - S_n|$, which is the distance between the actual sum of all the terms, S, and the partial sum we use as an estimate, S_n.

Example 6 *Use theorem 6 to estimate* $\sum_{n=1}^{\infty}(-1)^{n+1} \cdot \frac{1}{2^n}$ *to within* 0.001.

Solution We recognize that the series is alternating and proceed to the alternating series test to determine whether the series converges.

Alternately, $b_{n+1} = \frac{1}{2^{n+1}} < \frac{1}{2^n} = b_n$.

Because 2^Ω is infinite, $\frac{1}{2^\Omega}$ is infinitesimal and renders zero.

Ans. to reading exercise 18
no; the terms of the series do not alternate signs, so the alternating series test does not apply. (The terms also do not go to zero, and therefore the test for divergence applies.)

⓿ Identify b_n. For this series, $b_n = \frac{1}{2^n}$.

❶ Is $b_{n+1} \le b_n$ for all n? The values of b_n are $\frac{1}{2}, \frac{1}{4}, \frac{1}{8}, \frac{1}{16}, \frac{1}{32}, \ldots,$ which are decreasing. The criterion is met.

❷ Is $\lim_{n\to\infty} b_n = 0$? We have

$$\lim_{n\to\infty} b_n = \lim_{n\to\infty} \frac{1}{2^n} = \frac{1}{2^\Omega} \doteq 0.$$

Because both conditions are met, the series converges by the alternating series test.

❸ Add terms of the series until the next term is equal to or smaller than (in absolute value) the desired precision of the estimate. The first few terms of the series are

$$\frac{1}{2} - \frac{1}{4} + \frac{1}{8} - \frac{1}{16} + \frac{1}{32} - \frac{1}{64} + \frac{1}{128} - \frac{1}{256} + \frac{1}{512} - \frac{1}{1024} + \cdots.$$

Since $\frac{1}{1024} < \frac{1}{1000} = 0.001$, we stop before including this term. Therefore, our estimate of $\sum_{n=1}^{\infty} (-1)^{n+1} \cdot \frac{1}{2^n}$ is

$$\begin{aligned} S_9 &= \frac{1}{2} - \frac{1}{4} + \frac{1}{8} - \frac{1}{16} + \frac{1}{32} - \frac{1}{64} + \frac{1}{128} - \frac{1}{256} + \frac{1}{512} \\ &= 0.333984. \end{aligned}$$

■

The series in example 6 is geometric, with the first term $\frac{1}{2}$ and the common ratio $r = -\frac{1}{2}$. The sum of the series is therefore

$$\frac{a}{1-r} = \frac{\frac{1}{2}}{1 - (-\frac{1}{2})} = \frac{1}{3}.$$

Notice that

$$|S - S_9| = |\tfrac{1}{3} - 0.333984| = 0.000651,$$

which is indeed less than 0.001, as promised.

Example 7 *Estimate the sum of the alternating harmonic series* $\sum_{n=1}^{\infty} \frac{(-1)^{n+1}}{n}$ *to within* 0.2.

Solution In example 3 we determined that the alternating harmonic series converges, so we are ready for step ❸.

❸ Add terms of the series until the next term is equal to or smaller than (in absolute value) the desired precision of the estimate. The first few terms of the series are

$$1 - \frac{1}{2} + \frac{1}{3} - \frac{1}{4} + \frac{1}{5} - \dots,$$

and because $\frac{1}{5} = 0.2$, we stop before including this term. Our estimate is

$$S_4 = 1 - \frac{1}{2} + \frac{1}{3} - \frac{1}{4} = 0.58333.$$ ■

Given the instructions to this example, including additional terms unnecessarily does no harm. In fact, the inclusion of additional terms improves the estimate. However, the concept of theorem 6 is best learned by including the minimal possible number of terms.

We look at determining the sum of the alternating harmonic series later in this chapter.

It is often possible to determine ahead of time how many terms must be used to produce the desired estimate. For instance, if we wanted to modify example 7 to estimate the sum of the alternating harmonic series to within 0.001, we need to add terms of the series until $\frac{1}{n}$ reaches 0.001. Since $\frac{1}{1000} = 0.001$, adding the first 999 terms of the series is sufficient. This is quite tedious by hand, but with technology, the estimate 0.693647 is obtained easily.

Example 8 *How many terms of the convergent alternating series* $\sum_{n=1}^{\infty} \frac{(-1)^n}{n^2 + 3n + 5}$ *must be added to estimate the sum of the series to within 0.002?*

Solution We need to determine when $\frac{1}{n^2 + 3n + 5} \leq 0.002$. We begin by solving the equation $\frac{1}{n^2 + 3n + 5} = 0.002$:

$$\frac{1}{n^2 + 3n + 5} = 0.002$$

$$\frac{1}{0.002} = n^2 + 3n + 5$$

$$0 = n^2 + 3n - 495.$$

We resort to the quadratic formula:

$$n = \frac{-3 \pm \sqrt{3^2 - 4(1)(-495)}}{2(1)} = \frac{-3 \pm \sqrt{1989}}{2}$$

$$n = 20.799, n = -23.799.$$

The negative solution is discarded. Using $n = 20.799$, we conclude that the next term's absolute value, b_{21}, is smaller than 0.002, and therefore adding the first 20 terms of the series is sufficient to produce the desired estimate. ■

A quick calculation shows that $b_{20} = \frac{1}{20^2 + 3 \cdot 20 + 5} = 0.00215$ and $b_{21} = \frac{1}{21^2 + 3 \cdot 21 + 5} = 0.00196$. Summing the first 19 terms of the series is not enough to satisfy theorem 6, but summing the first 20 terms is enough.

Reading Exercise 19 *How many terms of the convergent alternating series* $\sum_{n=1}^{\infty} \frac{(-1)^n}{n^2}$ *need to be added to estimate the sum of the series to within 0.001?*

Another alternating series test example

Example 9 *Determine the convergence or divergence of*

$$\sum_{n=1}^{\infty} \frac{(-1)^{n+1}(n+3)}{n^2+4n}.$$

Solution We recognize that the series is alternating, so we try the alternating series test.

⓿ Identify b_n. For this series, $b_n = \frac{n+3}{n^2+4n}$.

❶ Is $b_{n+1} \leq b_n$ for all n? The first few values are $\frac{4}{5}, \frac{5}{12}, \frac{6}{21}, \frac{7}{32}, \ldots$. Because both the numerators and the denominators are increasing, it is not immediately clear that the values are decreasing. However, we have another tool at our disposal: the derivative! If the derivative of the function $f(x) = \frac{x+3}{x^2+4x}$ is negative (or merely nonpositive) for $x \geq 1$, then $b_{n+1} \leq b_n$ for all n. To this end we calculate the derivative using the quotient rule:

Because the first finitely many terms of the series cannot affect the convergence or divergence of the series, an eventually negative derivative (negative for $x > k$ for some k) is sufficient.

$$f'(x) = \frac{(x^2+4x)(1) - (x+3)(2x+4)}{(x^2+4x)^2}$$

$$= \frac{x^2+4x-2x^2-10x-12}{(x^2+4x)^2}$$

$$= \frac{-x^2-6x-12}{(x^2+4x)^2},$$

which is clearly negative for $x > 0$ (negative numerator, positive denominator). The condition is satisfied.

❷ Is $\lim_{n\to\infty} b_n = 0$? We have

$$\lim_{n\to\infty} b_n = \lim_{n\to\infty} \frac{n+3}{n^2+4n} = \frac{\Omega+3}{\Omega^2+4\Omega} \approx \frac{\Omega}{\Omega^2} = \frac{1}{\Omega} = \omega \doteq 0.$$

Because both conditions are met, the series $\sum_{n=1}^{\infty} \frac{(-1)^{n+1}(n+3)}{n^2+4n}$ converges by the alternating series test. ■

Exercises 10.6

1–8. Rapid response: is the series alternating?

1. $\sum_{n=1}^{\infty}(-1)^{n+1}\frac{n}{e^n}$

2. $\sum_{n=1}^{\infty}\frac{2^n}{3^n+1}$

3. $\sum_{n=1}^{\infty}\frac{(-2)^n}{3^n+1}$

4. $\sum_{n=1}^{\infty}\frac{(-1)^n}{-n-3}$

5. $\sum_{n=1}^{\infty}\cos\left((2n+1)\frac{\pi}{4}\right)\frac{n}{e^n}$

6. $\sum_{n=1}^{\infty}\cos\left((4n+1)\frac{\pi}{4}\right)\frac{n}{e^n}$

7. $\sum_{n=1}^{\infty}(-1)^{2n}\frac{n}{e^n}$

8. $\sum_{n=1}^{\infty}(-1)^{3n}\frac{n}{e^n}$

9–32. Test the series for convergence or divergence.

9. $\sum_{n=1}^{\infty}(-1)^n\frac{1}{5n-1}$

10. $\sum_{n=1}^{\infty}(-1)^{n+1}\frac{1}{3n+5}$

11. $\sum_{n=1}^{\infty}\frac{(-1)^{n+1}}{n^3}$

12. $\sum_{n=1}^{\infty}\frac{(-1)^n}{\sqrt{n}}$

13. $\sum_{n=1}^{\infty}\frac{(-1)^n}{\ln(n^3+1)}$

14. $\sum_{n=1}^{\infty}(-1)^{n+1}(1+\frac{1}{n})$

15. $\sum_{n=1}^{\infty}(-1)^{n+1}\frac{1}{-2n-5}$

16. $\sum_{n=2}^{\infty}(-1)^n\frac{3}{n\ln n}$

17. $\sum_{n=1}^{\infty}\frac{(-1)^n}{\cosh n}$

18. $\sum_{n=1}^{\infty}\frac{(-2)^n}{-1+3^n}$

19. $\sum_{n=1}^{\infty}(-1)^n\frac{2^n}{2^n+1}$

20. $\sum_{n=1}^{\infty}(-1)^n\ln\left(1+\frac{1}{n}\right)$

21. $\sum_{n=1}^{\infty}(-1)^{n-1}\frac{1}{n^{0.87}}$

22. $\sum_{n=1}^{\infty}\frac{(-1)^{n+2}(n+3)}{n+1}$

23. $\sum_{n=1}^{\infty}\frac{(-1)^n}{2-\frac{1}{n^2}}$

24. $\sum_{n=1}^{\infty}\frac{(-1)^n 2^n}{3^n}$

25. $\sum_{n=2}^{\infty}\frac{(-1)^n\sqrt{n}}{n-1}$

26. $\sum_{n=2}^{\infty}(-1)^n\frac{n}{e^n}$

Ans. to reading exercise 19
31 (note that $\sqrt{1000} = 31.62$)

27. $\displaystyle\sum_{n=1}^{\infty}(-2)^n \frac{1}{n+2}$

28. $\displaystyle\sum_{n=1}^{\infty}\cos(\pi n)\frac{2}{8n+1}$

29. $\displaystyle\sum_{n=1}^{\infty}\sin\left((2n+1)\frac{\pi}{2}\right)\frac{4}{n^2+n}$

30. $\displaystyle\sum_{n=1}^{\infty}(-1)^n \frac{n^2+4}{3n^2+1}$

31. $\displaystyle\sum_{n=1}^{\infty}\frac{n+6}{(-1)^n(n^3+2)}$

32. $\displaystyle\sum_{n=1}^{\infty}(-1)^{n+1}\frac{2n-1}{n^2+2n+4}$

33–38. How many terms of the series do we need to sum to estimate the series to within 0.001?

33. $\displaystyle\sum_{n=1}^{\infty}\frac{(-1)^n}{3n}$

34. $\displaystyle\sum_{n=1}^{\infty}\frac{(-1)^n}{7n}$

35. $\displaystyle\sum_{n=1}^{\infty}\frac{(-1)^n}{3^n}$

36. $\displaystyle\sum_{n=1}^{\infty}\frac{(-1)^n}{\sqrt{n}}$

37. $\displaystyle\sum_{n=1}^{\infty}\frac{(-1)^n}{5n^2+20}$

38. $\displaystyle\sum_{n=1}^{\infty}\frac{(-1)^n}{10^n+1}$

39. (a) Use theorem 6 to estimate $\displaystyle\sum_{n=1}^{\infty}\frac{(-1)^n}{3^n}$ to within 0.001. (b) Use your knowledge of geometric series to find the sum of the series. (c) Calculate the difference in your answers to (a) and (b) and verify that the difference is less than 0.001.

40. Estimate $\displaystyle\sum_{n=1}^{\infty}\frac{(-1)^n}{10^n+1}$ to within 0.001.

41. Estimate $\displaystyle\sum_{n=1}^{\infty}\frac{(-1)^{n+1}}{n^5}$ to within 0.0004.

42. Estimate $\displaystyle\sum_{n=1}^{\infty}\frac{(-1)^{n+1}}{n^6}$ to within 0.0002.

43. Consider the series $\displaystyle\sum_{n=2}^{\infty}(-1)^{n+1}\frac{4}{n^3-1}$. (a) How many terms of the series must be added to estimate the sum to within 0.000001? (b) Use technology to add the required number of terms to obtain the estimate.

44. Consider the alternating harmonic series $\displaystyle\sum_{n=1}^{\infty}(-1)^{n+1}\frac{1}{n}$. (a) How many terms of the series must be added to estimate the sum to within 0.0000014? (b) Use technology to add the required number of terms to obtain the estimate.

45. Draw an arrow diagram of the type given in figures 1–5 in which the alternating series converges to a sum S, but has values of n for which $|S - S_n| = b_{n+1}$.

46. Find a series that meets the conditions of the alternating series test and for which there are values of n satisfying $|S - S_n| = b_{n+1}$. Either (a) give a formula for the series or (b) write out enough terms of the series that the pattern is apparent.

Ratio and Root Tests

10.7

In this section we examine two more tests of series: the ratio and root tests. First, however, we need to refine the notion of convergence further.

Absolute and conditional convergence

We know that the harmonic series $\sum_{n=1}^{\infty} \frac{1}{n}$ diverges; the sum of its terms is infinite. But, by alternating the signs of its terms, we can gain convergence; $\sum_{n=1}^{\infty} \frac{(-1)^{n+1}}{n}$ converges by the alternating series test. The movement to the right (the positive terms) and the movement to the left (the negative terms), each of which totals an infinite distance, is balanced enough that the sum is finite.

By contrast, take a series such as $\sum_{n=1}^{\infty} \frac{1}{n^2}$, which is a p-series with $p = 2 > 1$ and therefore converges. The sum of its terms is finite. If we then alternate the signs of its terms to form $\sum_{n=1}^{\infty} \frac{(-1)^{n+1}}{n^2}$, which also converges by the alternating series test, convergence seems manifest from the outset. The total movement to the right and the total movement to the left are each finite and no balancing act is necessary.

Each of these alternating series converges, but somehow the convergence doesn't feel the same. The convergence of the alternating series $\sum_{n=1}^{\infty} \frac{(-1)^{n+1}}{n^2}$ seems more fundamental, whereas the convergence of the alternating series $\sum_{n=1}^{\infty} \frac{(-1)^{n+1}}{n}$ seems much more fragile. These two situations are given names in the following definition.

Definition 7 ABSOLUTE AND CONDITIONAL CONVERGENCE *Let* $\sum_{n=1}^{\infty} a_n$ *be a series. If* $\sum_{n=1}^{\infty} |a_n|$ *converges,* *then we say that* $\sum_{n=1}^{\infty} a_n$ *is* **absolutely convergent**. *If* $\sum_{n=1}^{\infty} |a_n|$ *diverges but* $\sum_{n=1}^{\infty} a_n$ *converges, then we say that* $\sum_{n=1}^{\infty} a_n$ *is* **conditionally convergent**.

The series $\sum_{n=1}^{\infty} \frac{(-1)^{n+1}}{n^2}$ is absolutely convergent; the alternating harmonic series $\sum_{n=1}^{\infty} \frac{(-1)^{n+1}}{n}$ is conditionally convergent.

Perhaps an easier way to understand definition 7 is with the *decision tree* in figure 1.

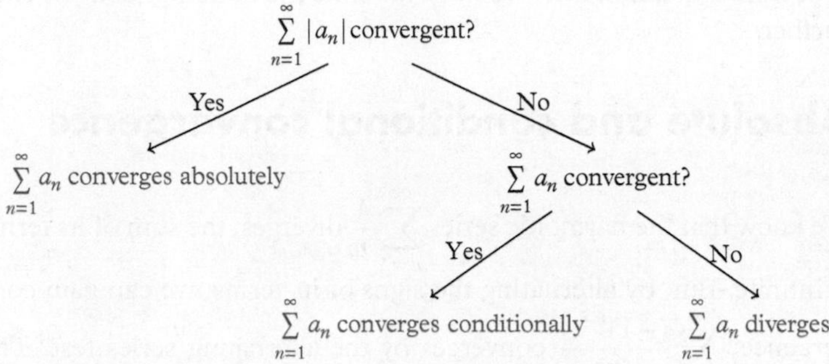

Figure 1 *A decision tree for choosing among absolute convergence, conditional convergence, and divergence. Conclusions are in red*

The decision tree of figure 1 provides the steps we will use when working examples.

The fact that an absolutely convergent series is convergent is proved at the end of this section.

Absolute and conditional convergence: examples

We now have not just two but three options to choose from when testing a series.

In the previous section we were asked whether an alternating series converges or diverges. Now there are two types of convergence from which to choose. The alternating series test concludes that a series converges, but it does not give the type of convergence.

Example 1 *Does* $\sum_{n=1}^{\infty} \frac{(-2)^{n+1}}{n+5^n}$ *converge absolutely, converge conditionally, or diverge?*

Solution Following the decision tree of figure 1, we first test for absolute convergence.

❶ Does $\sum_{n=1}^{\infty} |a_n|$ converge? We test the series $\sum_{n=1}^{\infty} \frac{2^{n+1}}{n+5^n}$. Applying the level comparison test, for any positive infinite hyperreal integer Ω,

Because the terms of $\sum_{n=1}^{\infty} |a_n|$ are not negative, the level comparison test can always be tried.

$$\frac{1}{a_\Omega} = \frac{1}{\frac{2^{\Omega+1}}{\Omega+5^\Omega}} = \frac{\Omega + 5^\Omega}{2^{\Omega+1}} \approx \frac{5^\Omega}{2^{\Omega+1}} = \frac{5^\Omega}{2 \cdot 2^\Omega} = \frac{1}{2}\left(\frac{5}{2}\right)^\Omega,$$

which is on the $\left(\frac{5}{2}\right)^{\Omega}$ level and therefore in the convergence zone. Because $\sum_{n=1}^{\infty} |a_n|$ converges, the series $\sum_{n=1}^{\infty} \frac{(-2)^{n+1}}{n+5^n}$ converges absolutely. ■

The portion of the decision tree used for example 1 is as follows:

$\sum_{n=1}^{\infty} |a_n|$ convergent?

Yes

$\sum_{n=1}^{\infty} a_n$ converges absolutely

A different path through the decision tree is illustrated next.

Example 2 *Does* $\sum_{n=2}^{\infty} (-1)^{n+1} \frac{\ln n}{n}$ *converge absolutely, converge conditionally, or diverge?*

Solution ❶ Does $\sum_{n=2}^{\infty} |a_n|$ converge? We test the series $\sum_{n=2}^{\infty} \frac{\ln n}{n}$, trying the level comparison test. For any positive infinite hyperreal integer Ω,

Compare to section 10.5 example 11.

$$\frac{1}{a_\Omega} = \frac{1}{\frac{\ln \Omega}{\Omega}} = \frac{\Omega}{\ln \Omega} \ll \Omega,$$

which is in the divergence zone. The series does not converge absolutely. Following the decision tree, we see the need to ask another question.

Because $\sum_{n=2}^{\infty} \frac{\ln n}{n}$ diverges, the series $\sum_{n=2}^{\infty} (-1)^{n+1} \frac{\ln n}{n}$ does not converge absolutely. But, per the decision tree of figure 1, the series might still converge conditionally.

❷ Does $\sum_{n=2}^{\infty} a_n$ converge? We test the series $\sum_{n=2}^{\infty} (-1)^{n+1} \frac{\ln n}{n}$. Because the series is alternating, we may try the alternating series test.

Identify b_n. For this series, $b_n = \frac{\ln n}{n}$.

Is $b_{n+1} \leq b_n$ for all n? The first few values are $\frac{\ln 2}{2}, \frac{\ln 3}{3}, \frac{\ln 4}{4} \ldots$, which is not immediately obviously decreasing because both the numerator and the denominator are increasing. Let's try the derivative, using $f(x) = \frac{\ln x}{x}$:

To avoid confusion, the steps of the alternating series test are not numbered in this example.

$$f'(x) = \frac{x \cdot \frac{1}{x} - \ln x \cdot 1}{x^2} = \frac{1 - \ln x}{x^2},$$

which is negative as long as $\ln x > 1$. The condition is met (eventually, at least).

Is $\lim_{n \to \infty} b_n = 0$? We have

$$\lim_{n \to \infty} \frac{\ln n}{n} = \frac{\ln \Omega}{\Omega} \doteq 0.$$

Because $\ln \Omega$ is on a lower level than Ω, the fraction $\frac{\ln \Omega}{\Omega}$ is infinitesimal and renders zero.

This condition is also met and, by the alternating series test, the series $\sum_{n=2}^{\infty} (-1)^{n+1} \frac{\ln n}{n}$ converges.

To state "the series $\sum_{n=2}^{\infty} (-1)^{n+1} \frac{\ln n}{n}$ converges" as the final answer to example 2 is inadequate because the instructions request that the type of convergence (absolute or conditional) be given.

We conclude that the series $\sum_{n=2}^{\infty}(-1)^{n+1}\frac{\ln n}{n}$ converges conditionally. ■

The portion of the decision tree used for example 2 is as follows:

$\sum_{n=1}^{\infty}|a_n|$ convergent?

No

$\sum_{n=1}^{\infty} a_n$ convergent?

Yes

$\sum_{n=1}^{\infty} a_n$ converges conditionally

Reading Exercise 20 *Does* $\sum_{n=1}^{\infty}\frac{(-1)^n}{n^3}$ *converge absolutely, converge conditionally, or diverge?*

Ratio test

The level comparison test gives one way of determining how fast the terms of a series go to zero. An indirect way of measuring this speed is to determine the ratio of each term to the previous term. A geometric series has a common ratio between terms, but in other series the ratios vary. Knowing that a geometric series converges if the common ratio r (in absolute value) < 1 and diverges if $r > 1$ tells how the ratios must work to produce convergence.

RATIO TEST

Let $\sum_{n=1}^{\infty} a_n$ be a series.

(a) If $\lim_{n\to\infty}\left|\frac{a_{n+1}}{a_n}\right| = L < 1$, then $\sum_{n=1}^{\infty} a_n$ converges absolutely.

(b) If $\lim_{n\to\infty}\left|\frac{a_{n+1}}{a_n}\right| = L > 1$ (or if the limit is ∞), then $\sum_{n=1}^{\infty} a_n$ diverges.

The ratio test is inconclusive if $\lim_{n\to\infty}\left|\frac{a_{n+1}}{a_n}\right| = 1$. The series could converge absolutely, converge conditionally, or diverge.

The ratio test can be proved by comparison to a geometric series, as alluded to earlier. Suppose $\lim_{n\to\infty}\left|\frac{a_{n+1}}{a_n}\right| = L < 1$. Then, eventually (beyond $n = N$), all the ratios $\left|\frac{a_{n+1}}{a_n}\right|$ must be less than some number r, where $0 \le L < r < 1$ (one possibility is illustrated in figure 2). Then,

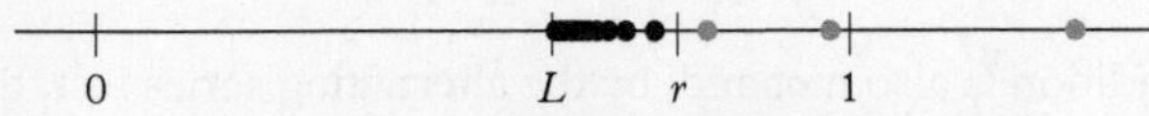

Figure 2 *An example of ratios* $\left|\frac{a_{n+1}}{a_n}\right|$ *with limit* L*, where* $0 < L < 1$*, represented by dots. The first few ratios are gray dots; those beyond* N *are black. Eventually (where* $n \ge N$*), the ratios are all less than some number* r *with* $L < r < 1$

each $|a_{n+1}| < r|a_n|$, from which we can deduce that if $n > N$,

$$|a_n| < |a_N| \cdot r^{n-N}.$$

Since

$$\sum_{n=N}^{\infty} |a_N| r^{n-N}$$

is a geometric series with the first term $|a_N|$ and the common ratio r with $0 < r < 1$, it converges. By the comparison test, $\sum_{n=N}^{\infty} |a_n|$ also converges, and because the first finitely many terms of a series do not affect convergence or divergence, so does $\sum_{n=1}^{\infty} |a_n|$. We conclude that the original series $\sum_{n=1}^{\infty} a_n$ converges absolutely. The proof of part (b) of the ratio test is similar.

To illustrate $|a_n| < |a_N| \cdot r^{n-N}$, suppose $N = 12$ and $n = 15$. Using $|a_{n+1}| < r|a_n|$,

$$|a_{15}| < r|a_{14}| < r \cdot r|a_{13}| < r \cdot r \cdot r|a_{12}|,$$

which is rewritten as $|a_{15}| < r^3|a_{12}| = |a_N| r^{15-12}$.

Ratio test examples

The ratio test requires the calculation of $\lim_{n\to\infty} \left|\frac{a_{n+1}}{a_n}\right|$. It is often convenient to rewrite this limit as

$$\lim_{n\to\infty} \left| a_{n+1} \cdot \frac{1}{a_n} \right|.$$

Ans. to reading exercise 20

converges absolutely (the series $\sum_{n=1}^{\infty} \frac{1}{n^3}$ is a convergent p-series)

Example 3 *Does* $\sum_{n=1}^{\infty} \frac{n(-2)^n}{3^{n+1}}$ *converge absolutely, converge conditionally, or diverge?*

Solution Let's try the ratio test:

$$\begin{aligned}
\lim_{n\to\infty} \left| a_{n+1} \cdot \frac{1}{a_n} \right| &= \lim_{n\to\infty} \left| \frac{(n+1)(-2)^{n+1}}{3^{(n+1)+1}} \cdot \frac{3^{n+1}}{n(-2)^n} \right| \\
&= \lim_{n\to\infty} \left| \frac{n+1}{n} \cdot \frac{(-2)^{n+1}}{(-2)^n} \cdot \frac{3^{n+1}}{3^{n+2}} \right| \\
&= \lim_{n\to\infty} \left| \frac{n+1}{n} \cdot (-2) \cdot \frac{1}{3} \right| \\
&= \frac{\Omega + 1}{\Omega} \cdot \frac{2}{3} \\
&\approx \frac{2\Omega}{3\Omega} = \frac{2}{3} < 1.
\end{aligned}$$

Because the limit is less than one, the series converges absolutely. ■

In line 1, the quantity $\frac{1}{a_n}$ is the reciprocal of a_n, so the reciprocal of $\frac{n(-2)^n}{3^{n+1}}$ is used. Line 2 reorganizes the quantities in line 1. This step is not necessary, but it can be helpful. Laws of exponents are used to obtain line 3. The limit is taken in line 4; in the previous lines (before n is replaced with Ω), the notation $\lim_{n\to\infty}$ must be kept. (Whether algebraic simplification is performed before taking the limit, as shown here, or after taking the limit is a matter of preference.) Also in line 4, note that $|-2| = 2$.

The conclusion of the ratio test is based on the value of the limit. If the limit is less than one, the series converges absolutely; if the limit is greater than one, the series diverges.

Example 4 *Determine whether the series* $\sum_{n=1}^{\infty} \frac{2^n + 7}{5^n - 3}$ *converges absolutely, converges conditionally, or diverges.*

Solution Let's try the ratio test:

On the right side of line 1, the absolute values may be omitted because the terms of the series are all positive. In line 2, the limit is taken (by replacing n with Ω) before performing any algebraic simplification (this is a matter of preference). On the right side of line 3, the expression is reorganized. In line 4, laws of exponents are used to simplify the expression.

$$\begin{aligned}
\lim_{n\to\infty} \left| a_{n+1} \cdot \frac{1}{a_n} \right| &= \lim_{n\to\infty} \frac{2^{n+1} + 7}{5^{n+1} - 3} \cdot \frac{5^n - 3}{2^n + 7} \\
&= \frac{2^{\Omega+1} + 7}{5^{\Omega+1} - 3} \cdot \frac{5^{\Omega} - 3}{2^{\Omega} + 7} \\
&\approx \frac{2^{\Omega+1}}{5^{\Omega+1}} \cdot \frac{5^{\Omega}}{2^{\Omega}} = \frac{2^{\Omega+1}}{2^{\Omega}} \cdot \frac{5^{\Omega}}{5^{\Omega+1}} \\
&= 2 \cdot \frac{1}{5} < 1.
\end{aligned}$$

Because the limit is less than one, the series converges absolutely. ■

For a positive-termed series such as the one in example 4, the series $\sum_{n=1}^{\infty} |a_n|$ is the same as the series $\sum_{n=1}^{\infty} a_n$. The terms *converge* and *converge absolutely* are therefore synonymous for such a series. If we follow the decision tree of figure 1, we see there are only two possible results (not three) for a positive-termed series.

Example 5 *Does* $\sum_{n=1}^{\infty} \frac{n!}{e^{2n}}$ *converge absolutely, converge conditionally, or diverge?*

Solution Let's try the ratio test:

In line 2, the quantity $(n + 1)!$ is rewritten as $(n + 1) \cdot n!$. In line 3, the $n!$ factors cancel and laws of exponents are used to simplify $\frac{e^{2n}}{e^{2n+2}}$ (notice that the result is a denominator of e^2, not e). Line 4 can be rewritten as

$$\frac{\Omega + 1}{e^2} \approx \frac{\Omega}{e^2} > 1$$

to avoid using the nonnumeric symbol ∞ in the computation. Line 4 can also be rewritten as

$$\frac{\Omega + 1}{e^2} > 1;$$

all we need to know is how the limit compares to one.

$$\begin{aligned}
\lim_{n\to\infty} \left| a_{n+1} \cdot \frac{1}{a_n} \right| &= \lim_{n\to\infty} \left| \frac{(n+1)!}{e^{2(n+1)}} \cdot \frac{e^{2n}}{n!} \right| \\
&= \lim_{n\to\infty} \frac{(n+1)n!}{n!} \cdot \frac{e^{2n}}{e^{2n+2}} \\
&= \lim_{n\to\infty} \frac{n+1}{e^2} \\
&= \frac{\Omega + 1}{e^2} \approx \frac{\Omega}{e^2} \doteq \infty > 1.
\end{aligned}$$

Because the limit is greater than one, the series diverges. ■

Reading Exercise 21 *Simplify* $\left|\frac{(n+1)!(-3)^{n+1}}{4^{n+3}} \cdot \frac{4^{n+2}}{n!(-3)^n}\right|$. *Assume that* $n \geq 1$ *is an integer.*

If the ratio test limit is one, then the test is inconclusive and a different test must be used.

Example 6 *Determine the convergence or divergence of* $\sum_{n=1}^{\infty} \frac{n^2}{(n+1)^4}$.

Solution Let's try the ratio test:

$$\lim_{n\to\infty}\left|a_{n+1} \cdot \frac{1}{a_n}\right| = \lim_{n\to\infty} \frac{(n+1)^2}{(n+2)^4} \cdot \frac{(n+1)^4}{n^2}$$

$$= \frac{(\Omega+1)^2(\Omega+1)^4}{(\Omega+2)^4\Omega^2}$$

$$\approx \frac{\Omega^2\Omega^4}{\Omega^4\Omega^2} = 1.$$

Because the limit is one, the ratio test is inconclusive. We must try something else.

The terms of the series are all positive, so we try the level comparison test. For any positive infinite hyperreal integer Ω,

$$\frac{1}{a_\Omega} = \frac{1}{\frac{\Omega^2}{(\Omega+1)^4}} = \frac{(\Omega+1)^4}{\Omega^2} \approx \frac{\Omega^4}{\Omega^2} = \Omega^2,$$

which is in the convergence zone. The series converges. ■

Yet another approach is to try the (direct) comparison test. Note that

$$\frac{n^2}{(n+1)^4} < \frac{n^2}{n^4} = \frac{1}{n^2}.$$

Since $\sum_{n=1}^{\infty} \frac{1}{n^2}$ is a convergent p-series, the series $\sum_{n=1}^{\infty} \frac{n^2}{(n+1)^4}$ converges by the comparison test.

When does the ratio test succeed in yielding a conclusion (as in examples 3–5), and when is it inconclusive (as in example 6)? The answer lies in the algebraic form of the terms of the series. When the terms of the series include exponentials or factorials, the ratio test usually succeeds, with algebraic simplifications similar to those in examples 3–5. When the terms of the series are rational functions of n (polynomial over polynomial), as in example 6, the ratio test is inconclusive and should be avoided.

One might ask what the harm is in trying the ratio test where it will be inconclusive, because we then just try something else, as in example 6. The problem is that an algebraic error might change the limit to something other than one, yielding an erroneous conclusion. Trying something that cannot succeed but could fail is a recipe for regret!

Rational functions of n are not the only expressions for which the ratio test is inconclusive. Consider trying to apply the ratio test to the series of example 2: $\sum_{n=2}^{\infty}(-1)^{n+1}\frac{\ln n}{n}$. We have

$$\lim_{n\to\infty}\left|a_{n+1} \cdot \frac{1}{a_n}\right| = \lim_{n\to\infty}\left|\frac{(-1)^{n+2}\ln(n+1)}{n+1} \cdot \frac{n}{(-1)^{n+1}\ln n}\right|$$

$$= \frac{1 \cdot \ln(\Omega+1)}{\Omega+1} \cdot \frac{\Omega}{1 \cdot \ln\Omega}$$

$$\approx \frac{\ln\Omega}{\Omega} \cdot \frac{\Omega}{\ln\Omega} = 1,$$

Notice that because powers of -1 are either 1 or -1, their absolute values are 1.

and the ratio test is inconclusive. All that work for nothing! And this is not just an expression. Because we eliminated no possible outcomes, we then need to proceed as we did in example 2, without skipping any steps.

Ratio test vs. level comparison test

For example 4, the featured solution used the ratio test. An alternate solution is to use the level comparison test. The terms of the series $\sum_{n=1}^{\infty} \frac{2^n+7}{5^n-3}$ are all positive. For any positive infinite hyperreal integer Ω,

$$\frac{1}{a_\Omega} = \frac{1}{\frac{2^\Omega+7}{5^\Omega-3}} = \frac{5^\Omega-3}{2^\Omega+7} \approx \frac{5^\Omega}{2^\Omega} = \left(\frac{5}{2}\right)^\Omega,$$

which is in the convergence zone. The series converges absolutely. The computation for the level comparison test is shorter and a little easier than the computation for the ratio test, but neither is very difficult.

Similarly, for example 3 the featured solution used the ratio test, but the level comparison test can also be used. The series is alternating, so we proceed as in examples 1 and 2, checking for absolute convergence. We test the series $\sum_{n=1}^{\infty} \frac{n \cdot 2^n}{3^{n+1}}$ using the level comparison test. For any positive infinite hyperreal integer Ω,

$$\frac{1}{a_\Omega} = \frac{1}{\frac{\Omega \cdot 2^\Omega}{3^{\Omega+1}}} = \frac{3^{\Omega+1}}{\Omega \cdot 2^\Omega} = 3 \cdot \frac{3^\Omega}{\Omega \cdot 2^\Omega} = 3 \cdot \frac{\left(\frac{3}{2}\right)^\Omega}{\Omega},$$

which is on the $\frac{\left(\frac{3}{2}\right)^\Omega}{\Omega}$ level. But is this level in the convergence zone? Because $\Omega \ll 1.1^\Omega$,

$$\frac{\left(\frac{3}{2}\right)^\Omega}{\Omega} \gg \frac{\left(\frac{3}{2}\right)^\Omega}{1.1^\Omega} = \left(\frac{1.5}{1.1}\right)^\Omega,$$

which is in the convergence zone. We conclude the original series converges absolutely, matching the conclusion from the ratio test.

Which solution is easier? The ratio test required some messy algebra, but the level comparison test required some nontrivial reasoning to determine whether the result is in the convergence zone. This is one example where the ratio test might be preferred.

This reasoning can be used to show, in general, that if the numerator of an expression is in the "exponentials at Ω" category of levels in the convergence zone and the denominator is in the "powers Ω^p" category of levels (see section 10.5 figure 7), then the entire expression is in the convergence zone. A similar technique can be used for comparing some of the other categories of levels, but caution is required near the boundary of the convergence and divergence zones.

Root test

There is one more test to add to our toolbox.

Ans. to reading exercise 21
$\frac{3(n+1)}{4}$

> **ROOT TEST**
>
> Let $\sum_{n=1}^{\infty} a_n$ be a series.
>
> (a) If $\lim_{n\to\infty} \sqrt[n]{|a_n|} = L < 1$, then $\sum_{n=1}^{\infty} a_n$ converges absolutely.
>
> (b) If $\lim_{n\to\infty} \sqrt[n]{|a_n|} = L > 1$ (or if the limit is ∞), then $\sum_{n=1}^{\infty} a_n$ diverges.

The root test is inconclusive if $\lim_{n\to\infty} \sqrt[n]{|a_n|} = 1$. The series could converge absolutely, converge conditionally, or diverge.

The proof of the root test is similar to that of the ratio test. If $\lim_{n\to\infty} \sqrt[n]{|a_n|} = L < 1$, then eventually (beyond $n = N$) all the values of $\sqrt[n]{|a_n|}$ must be less than some number r, where $0 < r < 1$. Then, $\left(\sqrt[n]{|a_n|}\right)^n < r^n$; that is, $|a_n| < r^n$. Because the geometric series $\sum_{n=1}^{\infty} r^n$ is geometric with $0 < r < 1$, the series converges, and therefore so does $\sum_{n=1}^{\infty} |a_n|$ by the comparison test. The series $\sum_{n=1}^{\infty} a_n$ converges absolutely. The proof of part (b) is similar.

The comparison is valid only for $n \geq N$, but finitely many terms do not affect the convergence or divergence of a series.

Example 7 *Does* $\sum_{n=1}^{\infty}\left(-\frac{2n+1}{n+3}\right)^n$ *converge absolutely, converge conditionally, or diverge?*

Solution Let's try the root test:

$$\begin{aligned}
\lim_{n\to\infty} \sqrt[n]{|a_n|} &= \lim_{n\to\infty} \sqrt[n]{\left|\left(-\frac{2n+1}{n+3}\right)^n\right|} \\
&= \lim_{n\to\infty} \sqrt[n]{\left|\left(-\frac{2n+1}{n+3}\right)\right|^n} \\
&= \lim_{n\to\infty} \left|-\frac{2n+1}{n+3}\right| \\
&= \lim_{n\to\infty} \frac{2n+1}{n+3} \\
&= \frac{2\Omega+1}{\Omega+3} \approx \frac{2\Omega}{\Omega} = 2 > 1.
\end{aligned}$$

Line 2 uses the algebraic fact that $|c^n| = |c|^n$; line 3 uses the algebraic fact that $\sqrt[n]{|c|^n} = |c|$. Just as for the ratio test, it is a matter of preference whether algebraic simplification is performed before or after replacing n with Ω. However, before this replacement, the notation $\lim_{n\to\infty}$ is required.

The conclusions of the root test are identical to those of the ratio test. If the limit is less than one, the series converges absolutely. If the limit is greater than one, the series diverges. If the limit equals one, the test is inconclusive.

Because the limit is greater than one, the series diverges by the root test. ■

The series of example 7 seems tailor-made to fit the root test, but other tests can be used as well. The test for divergence works, as it always does when the ratio test or root test concludes divergence. The computation for the level comparison test is similar to the computation for the test for divergence, so it works. Although the ratio test computation is a little bit longer, it is not much more difficult. In the next section we look at choosing among the various options.

In example 7 the algebraic form $\sqrt[n]{|c|^n}$ made algebraic simplification before taking the limit convenient. In the next example, which is a repeat of example 4, the sums in the numerator and the denominator make approximation attractive, so we take the limit before simplification.

The observation that the form of the numerator and the denominator are sums, making approximation attractive, also applies to the ratio test, as in the solution to example 4. It is not necessary to follow such conventions, but it can make the computations proceed more smoothly.

Example 8 *Determine whether the series* $\sum_{n=1}^{\infty} \frac{2^n+7}{5^n-3}$ *converges absolutely, converges conditionally, or diverges.*

Solution Let's try the root test:

In line 2, as with the ratio test, the limit can be taken (by replacing n with Ω) before performing any algebraic simplification, when desired. In line 3, approximation inside roots is allowable.

$$\begin{aligned}\lim_{n\to\infty} \sqrt[n]{|a_n|} &= \lim_{n\to\infty} \sqrt[n]{\left|\frac{2^n+7}{5^n-3}\right|} \\ &= \sqrt[\Omega]{\frac{2^\Omega+7}{5^\Omega-3}} \\ &\approx \sqrt[\Omega]{\frac{2^\Omega}{5^\Omega}} = \frac{\sqrt[\Omega]{2^\Omega}}{\sqrt[\Omega]{5^\Omega}} \\ &= \frac{2}{5} < 1.\end{aligned}$$

Because the limit is less than one, the series converges absolutely. ■

There are two additional facts that are sometimes helpful when applying the root test. The first is adapted from section 10.2 example 12:

$$\sqrt[\Omega]{\Omega} = \Omega^{\frac{1}{\Omega}} = \Omega^{\omega} \approx 1.$$

Similarly, if c is any positive real number,

$$\sqrt[\Omega]{c} \approx 1.$$

We have $\sqrt[\Omega]{c} = c^{\frac{1}{\Omega}} = c^{\omega}$, which renders $c^0 = 1$. Because the result is a real number, approximation can be used instead of rendering.

To illustrate, let's reexamine example 3.

Example 9 *Does* $\sum_{n=1}^{\infty} \frac{n(-2)^n}{3^{n+1}}$ *converge absolutely, converge conditionally, or diverge?*

Solution This time, let's try the root test:

$$\begin{aligned}
\lim_{n\to\infty} \sqrt[n]{|a_n|} &= \lim_{n\to\infty} \sqrt[n]{\left|\frac{n(-2)^n}{3^{n+1}}\right|} \\
&= \lim_{n\to\infty} \sqrt[n]{\frac{n \cdot |-2|^n}{3 \cdot 3^n}} \\
&= \lim_{n\to\infty} \frac{\sqrt[n]{n}\sqrt[n]{|-2|^n}}{\sqrt[n]{3}\sqrt[n]{3^n}} \\
&= \lim_{n\to\infty} \frac{\sqrt[n]{n} \cdot |-2|}{\sqrt[n]{3} \cdot 3} \\
&= \frac{\sqrt[\Omega]{\Omega} \cdot 2}{\sqrt[\Omega]{3} \cdot 3} \\
&\approx \frac{1 \cdot 2}{1 \cdot 3} = \frac{2}{3} < 1.
\end{aligned}$$

In line 2, using the laws of exponents to rewrite 3^{n+1} as $3 \cdot 3^n$ is preparation for using $\sqrt[n]{|c|^n} = |c|$ in line 4. Line 6 uses two approximation facts— namely, $\sqrt[\Omega]{\Omega} \approx 1$ and $\sqrt[\Omega]{c} \approx 1$ for any positive real number c.

Because the limit is less than one, the series converges absolutely by the root test. ■

The solution presented in example 9 may or may not be perceived as simpler than the solution to example 3, but it does provide an alternative that some may prefer.

Notice from the repeated examples that the values of the limits for the ratio test and the root test are identical ($\frac{2}{5}$ in examples 4 and 8, and $\frac{2}{3}$ in examples 3 and 9). The decision criteria are also identical. This is not a mere coincidence. Both tests are, in some sense, a generalization of the criteria for geometric series, with proofs that rely on geometric series in an identical manner.

Reading Exercise 22 *Use the root test to determine the convergence or divergence of* $\sum_{n=1}^{\infty} \left(\frac{2n+1}{n+2}\right)^n$.

Rearrangements

You may recall the commutative property of addition, which states that we may add a (finite) list of numbers in any order we please. For instance, to find the sum of the numbers 4, 8, 11, −3, −5, and 10, we may perform the arithmetic in that order and get

$$4 + 8 + 11 - 3 - 5 + 10 = 25$$

or we may rearrange the sum as

$$8 - 3 - 5 + 10 + 11 + 4 = 25,$$

which is quicker to perform mentally.

But, what if we wish to add infinitely many numbers, as we do in an infinite sum? Can we still reorder the numbers any way we please and get the same answer? If the series is absolutely convergent, the answer is yes. However, if the series is conditionally convergent, it turns out that by reordering the terms, we get any answer we want!

Let's examine this idea using the alternating harmonic series

$$1 - \frac{1}{2} + \frac{1}{3} - \frac{1}{4} + \frac{1}{5} - \frac{1}{6} + \cdots,$$

which is conditionally convergent. First, recall that the harmonic series

$$1 + \frac{1}{2} + \frac{1}{3} + \frac{1}{4} + \frac{1}{5} + \frac{1}{6} + \cdots = \infty$$

The analogous fact is true of any conditionally convergent series $\sum_{n=1}^{\infty} a_n$; the series $\sum_{n=1}^{\infty} |a_n| = \infty$ diverges.

diverges. Then, the sum of the alternating harmonic series' positive terms

$$1 + \frac{1}{3} + \frac{1}{5} + \frac{1}{7} + \frac{1}{9} + \cdots = \infty$$

and the sum of its negative terms

$$-\frac{1}{2} - \frac{1}{4} - \frac{1}{6} - \frac{1}{8} - \frac{1}{10} - \cdots = -\infty$$

must both be infinite. Why? If neither quantity is infinite, then the entire sum is finite and the series converges absolutely, which it doesn't. If one is infinite and the other is not, then the entire sum is infinite and the series diverges, which it doesn't.

Now suppose we want to reorder the terms to get a sum of 100. We start by adding enough positive terms (in order of the list) so that the sum exceeds 100. More precisely, we add terms until the sum first passes 100, and then we stop:

Computing N may be impractical, but such a value does exist.

$$1 + \frac{1}{3} + \frac{1}{5} + \cdots + \frac{1}{N}.$$

(Note that because the sum of the positive terms is infinite and we have only used up enough to get to more than 100, the sum of the remaining positive terms is still infinite.) Next we include negative terms (in order of the list) until the sum first drops back to less than 100:

$$1 + \frac{1}{3} + \frac{1}{5} + \cdots + \frac{1}{N} - \frac{1}{2} - \frac{1}{4} - \cdots - \frac{1}{M}.$$

(The negative terms we are using have a finite sum, so there is still an infinite sum remaining.)

Interesting tidbit: if we follow this procedure with the alternating harmonic series and a desired sum of 100, we only need to include one negative term each time, but we need many positive terms each time.

Now we repeat, adding more positive terms until the sum again passes 100,

$$1 + \frac{1}{3} + \frac{1}{5} + \cdots + \frac{1}{N} - \frac{1}{2} - \frac{1}{4} - \cdots - \frac{1}{M} + \frac{1}{N+2} + \cdots + \frac{1}{K},$$

then add negative terms to drop the sum back to less then 100,

$$1 + \cdots + \frac{1}{N} - \frac{1}{2} - \cdots - \frac{1}{M} + \frac{1}{N+2} + \cdots + \frac{1}{K} - \frac{1}{M+2} - \cdots - \frac{1}{H},$$

then more positive terms until the sum once again exceeds 100, and so on. We never run out of terms to include because we only use up a finite sum each time. And because we are including terms in the order of each list (positive term list and negative term list), the amount by which we can deviate from 100 (either more or less) continues to shrink because this amount is less than the absolute value of the last term included. The limit of the partial sums must be the desired final sum of 100.

Ans. to reading exercise 22

series diverges (because $\lim_{n\to\infty} \sqrt[n]{|a_n|} = 2 > 1$)

Any sum can be obtained using this procedure, and it works for any conditionally convergent series. The term *conditional* convergence is descriptive in the sense that the sum is conditioned (is dependent) on the order of the terms.

A final detail

Our work thus far assumes that an absolutely convergent series converges, but this fact is not automatic from the definition. It must be proved.

Theorem 7 ABSOLUTE CONVERGENCE IMPLIES CONVERGENCE *If* $\sum_{n=1}^{\infty} a_n$ *is an absolutely convergent series, then the series converges.*

Proof. Because every number is either the same as its absolute value or the same as the negative of its absolute value, for each n,

$$-|a_n| \leq a_n \leq |a_n|.$$

Adding $|a_n|$ to all parts of the compound inequality gives

$$0 \leq a_n + |a_n| \leq 2|a_n|.$$

Because $\sum_{n=1}^{\infty} a_n$ is absolutely convergent, by definition $\sum_{n=1}^{\infty} |a_n|$ converges. Then, by the series constant multiple rule, so does $\sum_{n=1}^{\infty} 2|a_n|$. The terms of the series $\sum_{n=1}^{\infty} (a_n + |a_n|)$ are then nonnegative and are not more than the terms of the comparison series $\sum_{n=1}^{\infty} 2|a_n|$. So, by the comparison test, the series $\sum_{n=1}^{\infty} (a_n + |a_n|)$ converges. Then, by the series difference rule, the original series

$$\sum_{n=1}^{\infty} a_n = \sum_{n=1}^{\infty} (a_n + |a_n| - |a_n|) = \sum_{n=1}^{\infty} (a_n + |a_n|) - \sum_{n=1}^{\infty} |a_n|$$

is the difference of two convergent series and therefore converges. ■

Exercises 10.7

1–8. Rapid response: state whether the ratio test or the root test is being used and interpret the result of the test, choosing from the following options: the series $\sum_{n=1}^{\infty} a_n$ converges absolutely, converges conditionally, diverges, or the test is inconclusive.

1. $\lim_{n\to\infty}\left|\frac{a_{n+1}}{a_n}\right| = \frac{4}{3}$
2. $\lim_{n\to\infty}\left|\frac{a_{n+1}}{a_n}\right| = 0.73$
3. $\lim_{n\to\infty}\left|\frac{a_{n+1}}{a_n}\right| = \infty$
4. $\lim_{n\to\infty}\left|\frac{a_{n+1}}{a_n}\right| = 1$
5. $\lim_{n\to\infty}\sqrt[n]{|a_n|} = 0.73$
6. $\lim_{n\to\infty}\sqrt[n]{|a_n|} = \frac{4}{3}$
7. $\lim_{n\to\infty}\sqrt[n]{|a_n|} = 1$
8. Which of the given options could not possibly be the correct interpretation to one of these tests?

9–12. Set up the limit for using the ratio test on the given series. Do not simplify. Do not evaluate the limit.

9. $\sum_{n=1}^{\infty}\frac{n\cdot 2^n}{(n+1)^2}$
10. $\sum_{n=1}^{\infty}\frac{4^n}{2+n!}$
11. $\sum_{n=1}^{\infty}\frac{n!}{3^{2n}}$
12. $\sum_{n=1}^{\infty}\frac{4^{n+1}}{n\cdot 3^n}$

13–20. Simplify the given expression, assuming that $n \geq 1$ is an integer.

13. $\left|\frac{3^{n+1}}{n+1}\cdot\frac{n}{3^n}\right|$
14. $\left|\frac{(-2)^{n+1}}{n+1}\cdot\frac{n}{(-2)^n}\right|$
15. $\left|\frac{(n+1)!}{(n+2)^2}\cdot\frac{(n+1)^2}{n!}\right|$
16. $\left|\frac{4^{2(n+1)}}{(n+1)!}\cdot\frac{n!}{4^{2n}}\right|$
17. $\sqrt[n]{\left|\frac{2^{n+1}}{3^n}\right|}$
18. $\left|\frac{(n+1)^2}{(3(n+1))!}\cdot\frac{(3n)!}{n^2}\right|$
19. $\left|\frac{(n+2)\ln(n+1)}{2^n}\cdot\frac{2^{n-1}}{(n+1)\ln n}\right|$
20. $\sqrt[n]{\left|\frac{(-3)^n}{4^{n+1}}\right|}$

21–30. Perform the ratio test on the given series.

21. $\sum_{n=1}^{\infty}\frac{3^n+n}{5^n}$
22. $\sum_{n=1}^{\infty}\frac{n}{2^n}$
23. $\sum_{n=1}^{\infty}\frac{3^n\cdot n}{5^n}$
24. $\sum_{n=1}^{\infty}\frac{n!}{2^n}$
25. $\sum_{n=1}^{\infty}\frac{(-3)^{n+1}}{n!}$
26. $\sum_{n=1}^{\infty}\frac{2^n+4^n}{3^n}$

27. $\sum_{n=1}^{\infty} \frac{n^2 3^{2n}}{4^{n+1}}$

28. $\sum_{n=1}^{\infty} \frac{n3^{2n}}{(-9)^n}$

29. $\sum_{n=1}^{\infty} \frac{2^{2n}+1}{n4^n}$

30. $\sum_{n=1}^{\infty} \frac{n!(n+3)!}{(2n)!}$

31–40. Perform the root test on the given series.

31. $\sum_{n=1}^{\infty} \left(-\frac{2n+5}{6n+7}\right)^n$

32. $\sum_{n=1}^{\infty} \left(\frac{n+5}{10n+1}\right)^n$

33. $\sum_{n=1}^{\infty} \left(\frac{2n+5}{6n+7}\right)^{2n}$

34. $\sum_{n=12}^{\infty} \left(\frac{n+5}{10-n}\right)^n$

35. $\sum_{n=1}^{\infty} \left(\frac{2n+5}{\ln n}\right)^n$

36. $\sum_{n=1}^{\infty} \left(\frac{2+\sin n}{n}\right)^n$

37. $\sum_{n=1}^{\infty} \frac{3^n+n}{5^n}$

38. $\sum_{n=1}^{\infty} \frac{6^n+5}{(-5)^n+6}$

39. $\sum_{n=1}^{\infty} \frac{(-4)^n+2}{3^n-2}$

40. $\sum_{n=1}^{\infty} \frac{n3^{2n}}{(-9)^n}$

41–52. Does the series converge absolutely, converge conditionally, or diverge?

These exercises are repeated from section 10.6, where the instructions asked to test for convergence or divergence. This time there are three options from which to choose.

41. $\sum_{n=1}^{\infty} (-1)^n \frac{1}{5n-1}$

42. $\sum_{n=1}^{\infty} (-1)^{n+1} \frac{1}{3n+5}$

43. $\sum_{n=1}^{\infty} \frac{(-1)^{n+1}}{n^3}$

44. $\sum_{n=1}^{\infty} \frac{(-1)^n}{\sqrt{n}}$

45. $\sum_{n=1}^{\infty} \frac{(-1)^n}{\ln(n^3+1)}$

46. $\sum_{n=1}^{\infty} (-1)^{n+1}(1+\frac{1}{n})$

47. $\sum_{n=1}^{\infty} (-1)^n \frac{2^n}{2^n+1}$

48. $\sum_{n=2}^{\infty} (-1)^n \frac{3}{n\ln n}$

49. $\sum_{n=2}^{\infty} \frac{(-1)^n\sqrt{n}}{n-1}$

50. $\sum_{n=1}^{\infty} \frac{(-1)^n 2^n}{3^n}$

51. $\sum_{n=1}^{\infty} (-2)^n \frac{1}{n+2}$

52. $\sum_{n=1}^{\infty} \cos(\pi n) \frac{2}{8n+1}$

53–72. Does the series converge absolutely, converge conditionally, or diverge?

53. $\displaystyle\sum_{n=1}^{\infty}\left(\frac{11n}{16n+255}\right)^n$

54. $\displaystyle\sum_{n=1}^{\infty}\frac{7^{n+1}}{9^n}$

55. $\displaystyle\sum_{n=1}^{\infty}\frac{(-1)^n 3.02^{n+1}}{5.7^n}$

56. $\displaystyle\sum_{n=1}^{\infty}\left(\frac{4n+7}{5n-3}\right)^n$

57. $\displaystyle\sum_{n=1}^{\infty}\sqrt{2}\left(-\frac{3}{4}\right)^n$

58. $\displaystyle\sum_{n=1}^{\infty}\frac{(-1)^{n+1}n^2}{3^n}$

59. $\displaystyle\sum_{n=1}^{\infty}\frac{n^3(-4)^{n+2}}{n!}$

60. $\displaystyle\sum_{n=1}^{\infty}\frac{(-1)^{n+1}5^{n-1}}{(n+1)^2 4^{n+2}}$

61. $\displaystyle\sum_{n=1}^{\infty}\frac{(\frac{1}{2})^n+3n}{n+2}$

62. $\displaystyle\sum_{n=1}^{\infty}\frac{n^3+4n+5}{6n^4+7}$

63. $\displaystyle\sum_{n=1}^{\infty}\frac{5^n+4}{(-8)^n+1}$

64. $\displaystyle\sum_{n=1}^{\infty}\frac{4^{n+1}}{n3^{2n}}$

65. $\displaystyle\sum_{n=1}^{\infty}\frac{n!}{3n+6}$

66. $\displaystyle\sum_{n=1}^{\infty}\frac{(-3)^n}{n3^{n+1}}$

67. $\displaystyle\sum_{n=1}^{\infty}\frac{2^{2n}+1}{n(-4)^n}$

68. $\displaystyle\sum_{n=1}^{\infty}\frac{3^n}{n!}$

69. $\displaystyle\sum_{n=1}^{\infty}\frac{(4n)!}{2^{3n}}$

70. $\displaystyle\sum_{n=1}^{\infty}\frac{4^n}{2+n!}$

71. $\displaystyle\sum_{n=1}^{\infty}\frac{1}{\cosh n}$

72. $\displaystyle\sum_{n=1}^{\infty}\frac{3^n+4}{2^n+n^2}$

73. Does the series converge absolutely, converge conditionally, or diverge? $\displaystyle\sum_{n=1}^{\infty}a_n$, where $a_1=84.6$, $a_{n+1}=\frac{8n^2+1}{n+7}\cdot a_n$ for $n\geq 1$.

74. Does the series converge absolutely, converge conditionally, or diverge? $\displaystyle\sum_{n=1}^{\infty}a_n$, where $a_1=12$, $a_{n+1}=\frac{2n+5}{7n-2}\cdot a_n$ for $n\geq 1$.

75. Does the series converge absolutely, converge conditionally, or diverge? $\displaystyle\sum_{n=1}^{\infty}a_n$, where $a_1=5$, $a_{n+1}=\frac{\sqrt{n^2+6}}{5n+2}\cdot a_n$ for $n\geq 1$.

76. Does the series converge absolutely, converge conditionally, or diverge? $\displaystyle\sum_{n=1}^{\infty}a_n$, where $a_1=-3$, $a_{n+1}=\frac{2n+3}{n+1}\cdot a_n$ for $n\geq 1$.

77. (GSL) Test the series

$$\frac{1}{1}+\frac{1\cdot 3}{1\cdot 4}+\frac{1\cdot 3\cdot 5}{1\cdot 4\cdot 7}+\cdots+\frac{1\cdot 3\cdot 5\cdots(2n-1)}{1\cdot 4\cdot 7\cdots(3n-2)}+\cdots.$$

78. (GSL) Test the series

$$\frac{1}{1}+\frac{1\cdot 2}{1\cdot 3}+\frac{1\cdot 2\cdot 3}{1\cdot 3\cdot 5}+\frac{1\cdot 2\cdot 3\cdot 4}{1\cdot 3\cdot 5\cdot 7}+\cdots.$$

79. Follow the procedure of the "rearrangements" subsection with the alternating harmonic series and the desired sum 2. (a) State the terms of the series that must be added to surpass a sum of 2. Give a decimal approximation to this sum to four decimal places. (b) State the negative terms that must be added to drop the sum back to less than 2. Give a decimal approximation of the current partial sum to four decimal places. (c) State the positive terms that must be added to bring the partial sum back up to more than 2. Give a decimal approximation of the current partial sum to four decimal places. (d) Repeat (b), (c), and then (b) once more.

80. Follow the procedure of the "rearrangements" subsection with the alternating harmonic series and the desired sum -0.6. Because the desired sum is negative, begin by using negative terms (values from the negative term list). (a) State the negative terms that must be added to bring the partial sum to less than -0.6. Give a decimal approximation to this sum to four decimal places. (b) State the positive terms that must be added to bring the partial sum up to greater than -0.6. Give a decimal approximation of the current partial sum to four decimal places. (c) State the negative terms that must be added to bring the partial sum back down to less than -0.6. Give a decimal approximation of the current partial sum to four decimal places. (d) Repeat (b), (c), and then (b) once more.

In other words, we start with $-\frac{1}{2}$.

Strategy for Testing Series

10.8

When should we try the ratio test? The level comparison test? Is the integral test ever the method of choice? We now have many different tests for determining the convergence (absolute or conditional) or divergence of a series. And when more than one method might work, sometimes one test is easier than the others. Determining how to proceed is the task for this section.

Strategy checklist

Sometimes we can look at a series and recognize immediately what we need to try. If this attempt works, good! But, if it does not work, if the test is inconclusive or if we get bogged down in the details, we need an order for evaluating various strategies—a checklist—that helps us make a determination regarding the best strategy for approaching the series. One such checklist follows.

Instructions

1. If the terms of the series are positive, use the checklist. For positive-termed series, *converges* and *converges absolutely* are synonymous.
2. If the series is alternating, check for absolute convergence (or divergence) using the checklist. If the result is that the series converges absolutely, the task is complete. If the result of using the ratio test, root test, test for divergence, or geometric series test is that the series diverges, the task is complete. But, if the result of the level comparison test, the comparison test, the p-series test, or the integral test on the series with terms $|a_n|$ is that the series $\sum_{n=1}^{\infty} |a_n|$ diverges, then the alternating series does not converge absolutely. Use the alternating series test to check for conditional convergence.

If the terms of a series are not all positive but the series is not alternating, check for absolute convergence in the same manner as for an alternating series. However, the alternating series test does not directly apply if the series is not alternating.

Strategy for Testing Series Checklist

❶ Is the series geometric? (If so, it converges absolutely for $|r| < 1$ and diverges for $|r| \geq 1$.)

❷ Is it a p-series?

❸ Are the terms of the form $(f(n))^n$? Try the root test or the level comparison test.

❹ Are factorials present? Try the ratio test (usually the easiest method) or the level comparison test.

❺ Are the terms rational functions of n? Try the level comparison test.

❻ Are exponentials present in products or quotients with polynomials? Try the ratio test. If inconclusive, try the level comparison test.

❼ Are exponentials present but ❻ does not apply? Try the ratio test, root test, or level comparison test, any of which could turn out to be the easiest method.

❽ Did you forget to check if the terms go to zero? Try the test for divergence. If inconclusive (if the terms go to zero), try the level comparison test.

In step ❽, because the computation for the level comparison test is the reciprocal of the computation for the test for divergence, it takes little additional effort to apply. As a result, some may prefer to try this step first.

❾ Are the terms defined recursively? Try the ratio test.

❿ Do the terms contain trig or inverse trig functions? Try the level comparison test or the comparison test, or check for a disguised alternating series.

⓫ Last resorts: ⓐ try the integral test, ⓑ check for a telescoping series (partial fractions may help), ⓒ try using a CAS, or ⓓ try more advanced sources (books/internet).

Steps ⓫ ⓒ and ⓫ ⓓ may not be allowed on an exam.

Remember that the checklist is meant to be your servant, not your master. If you suddenly recognize something and wish to try it, go ahead! For instance, you do not need to wait until step ❽ before trying the test for divergence. But, if you have no ideas, the checklist is helpful. With practice, you may find yourself recognizing what to try more quickly and are less likely to need to resort to the checklist.

Applying the strategy

In the examples that follow, we don't carry out a test. The details of the various tests for series have already been covered. Instead, we focus on identifying an appropriate strategy.

For many of these series, an appropriate approach might be immediately recognizable. However, in case it isn't, the solution to each example discusses moving through the checklist to determine a strategy for testing the series. Occasionally, the result of a strategy is given so the next step can be discussed.

Example 1 *Test the series* $\sum_{n=1}^{\infty} \frac{2n+7}{5n-3}$.

Solution ❶ The series is not geometric.

❷ This is not a p-series.

❸ The terms are not of the form $(f(n))^n$.

❹ Factorials are not present.

❺ The terms are a rational function of n (polynomial over polynomial). Try the level comparison test. The strategy succeeds. ■

By saying "the strategy succeeds," we mean that carrying out the designated tests correctly (without making an error) results in a correct determination of the status of the series (absolutely convergent, conditionally convergent, or divergent).

If you recognized immediately that the terms of the series in example 1 do not go to zero and thus would try the test for divergence, you judged well. This strategy also works and is comparable in difficulty to the level comparison test. Comments about alternate strategies appear in the margin.

Example 2 *Test the series* $\sum_{n=1}^{\infty} \left(\frac{2n+7}{5n-3}\right)^n$.

Solution ❶ The series is not geometric.

❷ This is not a p-series.

❸ The terms are of the form $(f(n))^n$. Try the root test. The strategy succeeds. ■

The level comparison test succeeds and is similar in difficulty. The ratio test succeeds but is more difficult.

Example 3 *Test the series* $\sum_{n=1}^{\infty} \frac{2n+7}{5n^2-3}$.

Solution ❶ The series is not geometric.

❷ This is not a p-series.

❸ The terms are not of the form $(f(n))^n$.

❹ Factorials are not present.

❺ The terms are a rational function of n (polynomial over polynomial). Try the level comparison test. The strategy succeeds. ■

The ratio and root tests are both inconclusive, as is the test for divergence. The comparison test succeeds (compare to $\sum_{n=1}^{\infty} \frac{2n}{5n^2} = \frac{2}{5}\sum_{n=1}^{\infty} \frac{1}{n}$), but it requires more thought.

Compare the solutions to examples 1 and 3. The featured solutions are identical. Although the test for divergence works for example 1, it does not work for example 3. The checklist is designed to provide a relatively efficient method for making decisions regarding which test to use in the absence of other ideas. It does not always mean that using a different strategy, such as using the test for divergence in example 1, is inadvisable.

Example 4 *Test the series* $\sum_{n=1}^{\infty}(-1)^n \frac{2n+7}{5n^2-3}$.

Solution This series is alternating, so we first check for absolute convergence, testing the series $\sum_{n=1}^{\infty} \frac{2n+7}{5n^2-3}$.

❶ The series is not geometric.

❷ This is not a p-series.

❸ The terms are not of the form $(f(n))^n$.

❹ Factorials are not present.

❺ The terms are a rational function of n (polynomial over polynomial). Try the level comparison test. The result is diverge.

The ratio test, root test, and test for divergence are all inconclusive. The comparison test can be used in place of the level comparison test.

Back to the instructions for using the checklist. Because the level comparison test concludes that $\sum_{n=1}^{\infty} \frac{2n+7}{5n^2-3}$ diverges, we know the alternating series $\sum_{n=1}^{\infty}(-1)^n \frac{2n+7}{5n^2-3}$ does not converge absolutely. We therefore perform the alternating series test to check for conditional convergence. The strategy succeeds. ■

Example 5 *Test the series* $\sum_{n=1}^{\infty} \frac{2^n+7}{5^n-3}$.

Solution ❶ The series is not geometric.

❷ This is not a p-series.

❸ The terms are not of the form $(f(n))^n$.

❹ Factorials are not present.

❺ The terms are not a rational function of n.

❻ There are exponentials, but not in products or quotients with polynomials.

❼ The terms include exponentials. We try the ratio test, the root test, or the level comparison test. Each test succeeds. The ratio test takes a little longer to perform than the others. ■

Example 6 *Test the series* $\sum_{n=1}^{\infty} \frac{2^n \cdot 7}{5^n \cdot 3}$.

Solution ❶ The series is geometric with $r = \frac{2}{5}$. The strategy succeeds. ■

The ratio test, root test, and level comparison test also succeed and are still fairly simple, although the geometric test is simplest.

Example 7 *Test the series* $\sum_{n=1}^{\infty} \frac{2^n \cdot (n^7+1)}{5^n \cdot (n^3+1)}$.

Solution ❶ The series is not geometric.

❷ This is not a p-series.

❸ The terms are not of the form $(f(n))^n$.

❹ Factorials are not present.

❺ The terms are not a rational function of n.

❻ Exponentials are present in products with polynomials. Try the ratio test. The strategy succeeds. ■

The root test is comparable in length, but takes a little more thought. The level comparison test is shorter, but takes more thought.

Example 8 *Test the series* $\sum_{n=1}^{\infty} \frac{n^2 \cdot 7}{n^5 \cdot 3}$.

Solution ❶ The series is not geometric.

❷ This is a p-series:

$$\sum_{n=1}^{\infty} \frac{n^2 \cdot 7}{n^5 \cdot 3} = \frac{7}{3}\sum_{n=1}^{\infty} \frac{n^2}{n^5} = \frac{7}{3}\sum_{n=1}^{\infty} \frac{1}{n^3}.$$

The strategy succeeds. ■

The level comparison test succeeds with a similar amount of work. The ratio and root tests are inconclusive.

Example 9 *Test the series* $\sum_{n=1}^{\infty} \frac{(\cos n^2) \cdot 7}{n^5 \cdot 3}$.

Solution The terms are not all positive, so we first check for absolute convergence, testing the series $\sum_{n=1}^{\infty} \frac{|\cos n^2| \cdot 7}{n^5 \cdot 3}$.

❶ The series is not geometric.

❷ This is not a p-series.

❸ The terms are not of the form $(f(n))^n$.

❹ Factorials are not present.

❺ The terms are not a rational function of n.

❻ Exponentials are not present.

❼ Exponentials are not present.

❽ Try the test for divergence. The terms go to zero. Try the level comparison test. The result is converge.

Back to the instructions: the series converges absolutely and the task is complete. ■

Both the test for divergence and the level comparison test use the fact that $|\cos \Omega^2|$ is a real number. The comparison test can also be used with comparison series $\sum_{n=1}^{\infty} \frac{7}{3n^5}$

Example 10 *Test the series* $\sum_{n=1}^{\infty} \frac{2n+7}{5^n - 3}$.

Solution ❶ The series is not geometric.

❷ This is not a p-series.
❸ The terms are not of the form $(f(n))^n$.
❹ Factorials are not present.
❺ The terms are not a rational function of n.
❻ An exponential is present in a quotient with a polynomial. Try the ratio test. The strategy succeeds. ■

The root test is comparable in length, but takes a little more thought. The level comparison test is shorter, but takes more thought.

Example 11 *Test the series* $\sum_{n=1}^{\infty}(\ln(2n+7)-\ln(5n-3))$.

Because the terms are negative for $n \geq 4$, by the constant multiple rule (with $c = -1$), we may apply strategies for positive terms.

Solution Start by making the "obvious" simplification using the laws of logarithms:

$$\sum_{n=1}^{\infty}(\ln(2n+7)-\ln(5n-3)) = \sum_{n=1}^{\infty}\ln\frac{2n+7}{5n-3}.$$

❶ The series is not geometric.
❷ This is not a p-series.
❸ The terms are not of the form $(f(n))^n$.
❹ Factorials are not present.
❺ The terms are not a rational function of n.
❻ Exponentials are not present.
❼ Exponentials are not present.
❽ Try the test for divergence. This strategy succeeds. ■

The level comparison test also succeeds. The ratio and root tests are inconclusive.

Example 12 *Test the series* $\sum_{n=1}^{\infty}(\ln(2n+7)-\ln(2n+3))$.

Solution Start by making the "obvious" simplification using the laws of logarithms:

Because $\frac{2n+7}{2n+3} > 1$, the terms of this series are all positive.

$$\sum_{n=1}^{\infty}(\ln(2n+7)-\ln(2n+3)) = \sum_{n=1}^{\infty}\ln\frac{2n+7}{2n+3}.$$

❶ The series is not geometric.
❷ This is not a p-series.
❸ The terms are not of the form $(f(n))^n$.
❹ Factorials are not present.
❺ The terms are not a rational function of n.
❻ Exponentials are not present.
❼ Exponentials are not present.

The ratio and root tests are inconclusive.

❽ Try the test for divergence. The test for divergence is inconclusive. Try the level comparison test. The level of $\frac{1}{a_n}$ is difficult to compute.

❾ The terms are not defined recursively.

❿ The terms do not contain trig or inverse trig functions.

⓫ ⓐ Try the integral test; the integral is difficult to evaluate.

⓫ ⓑ Is the series telescoping? As originally written (without the "obvious" simplification) the series is telescoping! This strategy succeeds. ∎

Example 13 *Test the series* $\sum_{n=2}^{\infty} \frac{1}{n + \ln n}$.

Solution ❶ The series is not geometric.

❷ This is not a p-series.

❸ The terms are not of the form $(f(n))^n$.

❹ Factorials are not present.

❺ The terms are not a rational function of n.

❻ Exponentials are not present.

❼ Exponentials are not present.

❽ Try the test for divergence (even though it looks obvious). The terms go to zero. Try the level comparison test. This strategy succeeds. ∎

The ratio and root tests are inconclusive, as is the comparison test if comparing to either $\sum_{n=2}^{\infty} \frac{1}{\ln n}$ or $\sum_{n=2}^{\infty} \frac{1}{n}$.

Example 14 *Test the series* $\sum_{n=2}^{\infty} \frac{1}{n \ln n}$.

Solution ❶ The series is not geometric.

❷ This is not a p-series.

❸ The terms are not of the form $(f(n))^n$.

❹ Factorials are not present.

❺ The terms are not a rational function of n.

❻ Exponentials are not present.

❼ Exponentials are not present.

❽ Try the test for divergence; the terms go to zero. Try the level comparison test. This strategy succeeds. ∎

The ratio and root tests are inconclusive, as is the comparison test if comparing to either $\sum_{n=2}^{\infty} \frac{1}{\ln n}$ or $\sum_{n=2}^{\infty} \frac{1}{n}$.

It was by use of the integral test that we found that $\Omega \ln \Omega$ is in the divergence zone. Using the level comparison test instead of the integral test for example 14 is in the same spirit as recognizing the convergence or divergence of a p-series, which was also developed using the integral test.

Example 15 *Test the series* $\sum_{n=3}^{\infty} \frac{(-4)^n (n-2)}{n!}$.

Solution This series is alternating, so we first check for absolute convergence, testing the series $\sum_{n=3}^{\infty} \frac{4^n (n-2)}{n!}$.

❶ The series is not geometric.

❷ This is not a p-series.

❸ The terms are not of the form $(f(n))^n$.

❹ Factorials are present. Try the ratio test. The result is converges absolutely.

The level comparison test succeeds, but is significantly less straightforward.

Back to the instructions for using the checklist. Because the result is converges absolutely, the task is complete. The strategy succeeds. ■

Example 16 *Test the series* $\sum_{n=3}^{\infty} \frac{(-1)^n}{\sqrt{n-2}}$.

Solution This series is alternating, so we first check for absolute convergence, testing the series $\sum_{n=3}^{\infty} \frac{1}{\sqrt{n-2}}$.

❶ The series is not geometric.

❷ This is not written in the form of a p-series.

This is actually a p-series in disguise. Write out the terms and then rewrite (re-index) in summation notation. The p-series criteria can then be used.

❸ The terms are not of the form $(f(n))^n$.

❹ Factorials are not present.

❺ The terms are not a rational function of n.

The terms may not be a rational function of n, but they are an algebraic function of n, so we might suspect that trying the level comparison test is reasonable.

❻ Exponentials are not present.

❼ Exponentials are not present.

❽ Try the test for divergence. The terms go to zero. Try the level comparison test. The result is diverge.

The ratio and root tests are inconclusive. The comparison test succeeds with comparison series $\sum_{n=3}^{\infty} \frac{1}{\sqrt{n}}$.

Back to the instructions for using the checklist. Because the level comparison test concluded that $\sum_{n=3}^{\infty} \frac{1}{\sqrt{n-2}}$ diverges, we know that the alternating series $\sum_{n=3}^{\infty} \frac{(-1)^n}{\sqrt{n-2}}$ does not converge absolutely. We therefore perform the alternating series test to check for conditional convergence. The strategy succeeds. ■

Example 17 *Test the series* $\sum_{n=1}^{\infty} \frac{n^2+5n}{\sqrt{n+2}}$.

Solution ❶ The series is not geometric.

❷ This is not a p-series.

❸ The terms are not of the form $(f(n))^n$.

❹ Factorials are not present.

❺ The terms are not a rational function of n.

The terms may not be a rational function of n, but they are an algebraic function of n, so we might suspect that trying the level comparison test is reasonable. This strategy succeeds.
The ratio and root tests are inconclusive.

❻ Exponentials are not present.

❼ Exponentials are not present.

❽ Try the test for divergence. This strategy succeeds. ■

Example 18 *Test the series* $\sum_{n=1}^{\infty} \frac{(3-\sin n)^n}{5^n}$.

Solution ❶ The series is not geometric.

❷ This is not a p-series.

❸ The terms are of the form $(f(n))^n$. Try the root test. The root test limit does not exist (the test is inconclusive). Try the level comparison test. The result is not immediately apparent.

❹ Factorials are not present.

❺ The terms are not a rational function of n.

❻ Exponentials are not present.

❼ Exponentials are not present.

❽ Try the test for divergence. The terms go to zero. We already tried the level comparison test.

❾ The terms are not defined recursively.

❿ The terms contain a trig function. Try the comparison test, using the fact that $-1 \le \sin n \le 1$. With comparison series $\sum_{n=1}^{\infty} \frac{4^n}{5^n}$, this strategy succeeds. ■

Using the proof of the root test, it is possible to conclude that the series converges, even though the statement of the root test is inadequate to make this conclusion. It is also possible to recognize that for every value of Ω, $\frac{1}{a_\Omega}$ is in the convergence zone. However, one of the purposes of the checklist is to determine what to do if an attempted test is too difficult or confusing to complete.

As stated earlier, it is not necessary to work through every step of the checklist before choosing a strategy.

Example 19 *Test the series* $\sum_{n=1}^{\infty} a_n$, *where* $a_1 = 19$, $a_{n+1} = \frac{6n^2 + 1}{(2n+1)^2} a_n$ *for* $n \ge 1$.

Solution ❾ The terms are defined recursively. Try the ratio test. This strategy succeeds. ■

Example 20 *Test the series* $\sum_{n=1}^{\infty} \frac{\cos n\pi}{5n^2}$.

Solution The terms are not all positive, so we first check for absolute convergence, testing the series $\sum_{n=1}^{\infty} \frac{|\cos n\pi|}{5n^2}$.

❿ The terms contain a trig function. Any of the three recommended options succeeds. (1) The level comparison test requires recognizing that $\cos(\Omega\pi)$ is a real number; the result is converge. (2) The comparison test with the comparison series $\sum_{n=1}^{\infty} \frac{1}{5n^2}$ has the result of converge. (3) Checking for a disguised alternating series, since

$\cos n\pi = (-1)^n$, we are testing $\sum_{n=1}^{\infty} \frac{|\cos n\pi|}{5n^2} = \sum_{n=1}^{\infty} \frac{1}{5n^2}$, which is a convergent p-series.

Back to the instructions. Because the series converges absolutely by any of the three methods, the task is complete. ∎

As was the case with techniques of integration, skill in recognizing form is gained from practice, until it seems that choosing a series test is more an intuitive art than a long, drawn-out process. Nevertheless, the checklist is a useful tool when confronted with a series that defies immediate recognition or when the technique you thought would work does not.

As you practice, pay attention to which steps from the checklist you need to reference. These steps are your top priority to learn.

Exercises 10.8

1–42. Does the series converge absolutely, converge conditionally, or diverge?

Unlike the examples presented in this section, the chosen tests should be completed.

1. $\sum_{n=1}^{\infty} \frac{3^n}{n!}$
2. $\sum_{n=1}^{\infty} \frac{3^n + 5}{2^n}$
3. $\sum_{n=1}^{\infty} (-1)^{n+1} \frac{2}{\sqrt{n}}$
4. $\sum_{n=1}^{\infty} \frac{4 \cdot 5^n}{7^{n+1}}$
5. $\sum_{n=1}^{\infty} \frac{(-4)^n}{3^n + 6}$
6. $\sum_{n=1}^{\infty} \frac{3}{\sqrt[3]{n}}$
7. $\sum_{n=1}^{\infty} 4^{-n}(3^n + 6)$
8. $\sum_{n=1}^{\infty} \frac{(-1)^n 3}{\sqrt[3]{n+1}}$
9. $\sum_{n=1}^{\infty} \frac{4 \cdot 6^{n+2}}{5^{n+1}}$
10. $\sum_{n=1}^{\infty} \frac{(-2)^n}{n!}$
11. $\sum_{n=1}^{\infty} \frac{-1}{\sqrt{n}}$
12. $\sum_{n=1}^{\infty} \frac{n3^n}{2^{2n+1}}$
13. $\sum_{n=1}^{\infty} \frac{\cos\left(n\frac{\pi}{12}\right)}{n^4}$
14. $\sum_{n=1}^{\infty} \frac{2 + \sin n}{n}$
15. $\sum_{n=1}^{\infty} \left(\frac{2 + \cos n}{n}\right)^n$
16. $\sum_{n=1}^{\infty} \frac{3n + 5}{n^3 + n + 4}$

When an answer to an odd-numbered exercise states "converge absolutely" for a positive-termed series, the answer "converge" is equivalent. Alternate ways of working exercises are not always indicated in the answers.

17. $\displaystyle\sum_{n=1}^{\infty}(-1)^n\frac{n^5 - 7n^3 + 1}{(n^4 + n)^2}$

18. $\displaystyle\sum_{n=1}^{\infty}\left(\frac{1}{\sqrt{2n+5}} - \frac{1}{\sqrt{2n+1}}\right)$

19. $\displaystyle\sum_{n=8}^{\infty}\left(\frac{\sqrt{n}+4}{n-7}\right)^2$

20. $\displaystyle\sum_{n=1}^{\infty}\left(2^{\frac{n+1}{2n-3}} - 1\right)^n$

21. $\displaystyle\sum_{n=1}^{\infty}\frac{(n+3)^{12}e^n 4^{n+1}}{(-10)^{n+2}}$

22. $\displaystyle\sum_{n=1}^{\infty}\frac{e^n n!}{(2n+1)!}$

23. $\displaystyle\sum_{n=1}^{\infty}\frac{(-3)^n}{n(-5)^{n+1}}$

24. $\displaystyle\sum_{n=1}^{\infty}\left(\frac{1}{n^2} + \frac{1}{2^n}\right)$

25. $\displaystyle\sum_{n=1}^{\infty}(\ln(4n+3) - \ln(4n-1))$

26. $\displaystyle\sum_{n=1}^{\infty}\frac{3\cdot 5\cdot 7\cdot 9\cdots(2n+1)}{7\cdot 12\cdot 17\cdot 22\cdots(5n+2)}$

27. $\displaystyle\sum_{n=1}^{\infty}\frac{4^n n!}{(2n-1)!}$

28. $\displaystyle\sum_{n=1}^{\infty}\sin\left(\frac{1}{n^2}\right)$

29. $\displaystyle\sum_{n=1}^{\infty}\frac{1}{n^2\ln n}$

30. $\displaystyle\sum_{n=2}^{\infty}\frac{1}{n(\ln n)^2}$

31. $\displaystyle\sum_{n=1}^{\infty}\sqrt{\frac{2^n+1}{5^n-7}}$

32. $\displaystyle\sum_{n=1}^{\infty}\left(\frac{\sin n}{2n}\right)^n$

33. $\sum_{n=1}^{\infty} \frac{2 \cdot 4 \cdot 6 \cdots (2n)}{1 \cdot 4 \cdot 7 \cdots (3n-2)}$

34. $\sum_{n=1}^{\infty} \frac{\sin\left(n\pi + \frac{\pi}{4}\right)}{n+5}$

35. $\sum_{n=1}^{\infty} \left(\frac{11 - 7\cos n}{25}\right)^n$

36. $\sum_{n=1}^{\infty} \frac{n!}{n^n}$

37. $\sum_{n=2}^{\infty} \frac{(-1)^n}{n\sqrt{\ln n}}$

38. $\sum_{n=1}^{\infty} \frac{n^7 + 3^n}{n^4 + 1.6^n}$

39. $\sum_{n=1}^{\infty} \frac{\left(\tan^{-1} n\right)^n}{\left(2 + \frac{1}{n}\right)^{n+1}}$

40. $\sum_{n=1}^{\infty} \frac{n^7 + n^3}{n^4 + 1.6^n}$

41. $\sum_{n=1}^{\infty} a_n$, where $a_1 = -5$, $a_{n+1} = \frac{(n+3)}{\frac{5}{2} - 3n} a_n$

42. $\sum_{n=1}^{\infty} a_n$, where $a_1 = 3$, $a_{n+1} = \frac{4n+1}{3n^2 - 2n + 7} a_n$

Power Series

10.9

How would you describe a function of the form

$$f(x) = 1 + \frac{1}{2}x + \frac{1}{4}x^2 + \frac{1}{8}x^3 + \cdots ?$$

With no last term, the function appears to be a polynomial with infinite degree. These functions are called *power series.*

Power series: definition

We begin by defining a power series centered at zero.

Definition 8 POWER SERIES, CENTER ZERO *A* power series centered at zero *is a series of the form*

$$\sum_{n=0}^{\infty} c_n x^n = c_0 + c_1 x + c_2 x^2 + c_3 x^3 + \cdots,$$

where the coefficient c_n is a real number for each n.

The phrase "about zero" is sometimes used in place of "centered at zero."

Notice the difference in the placement of n on the coefficients c_n and the powers of the variable x^n. The subscript (bold) is used to tell one constant from another, whereas the superscript (gray) is an exponent on the variable x.

An example of a power series centered at zero is

$$\sum_{n=0}^{\infty} x^n = 1 + x + x^2 + x^3 + \cdots.$$

Notice that the index of a power series often starts at $n = 0$.

Each value of the variable x gives a series of real numbers. For $x = \frac{1}{2}$, the series is

$$\sum_{n=0}^{\infty} x^n = \sum_{n=0}^{\infty} \left(\frac{1}{2}\right)^n = 1 + \frac{1}{2} + \frac{1}{4} + \frac{1}{8} + \cdots,$$

which is a geometric series with $|r| = \frac{1}{2} < 1$, and therefore it converges. For $x = 4$, the series is

$$\sum_{n=0}^{\infty} x^n = \sum_{n=0}^{\infty} 4^n = 1 + 4 + 16 + 64 + \cdots,$$

which is a geometric series with $|r| = 4 > 1$, and therefore it diverges. The series can also be alternating. For $x = -\frac{2}{3}$, the series is

$$\sum_{n=0}^{\infty} x^n = \sum_{n=0}^{\infty} \left(-\frac{2}{3}\right)^n = 1 - \frac{2}{3} + \frac{4}{9} - \frac{8}{27} + \frac{16}{81} - \cdots,$$

which is a geometric series with $|r| = \frac{2}{3} < 1$, and therefore it converges.

By the definition of the term *polynomial*, a polynomial has a finite degree and does not include a power series.

In contrast to a polynomial, which always has domain $(-\infty, \infty) = \mathbf{R}$, the domain of a power series does not always include all real numbers. The domain of a function must only include values for which the expression is defined (is a real number), so the domain only includes values of the variable x for which the series converges. For each value of the variable x, the series $\sum_{n=0}^{\infty} x^n$ is a geometric series with $r = x$, which converges when $|r| = |x| < 1$ and diverges when $|r| = |x| \geq 1$. The domain of this power series is therefore $(-1, 1)$.

The center of a power series does not have to be zero.

The phrase "about a" is sometimes used in place of "centered at a."

Notice that for a power series, the exponents on the variable are never negative.

Definition 9 POWER SERIES, CENTER a *A* power series centered at a *is a series of the form*

$$\sum_{n=0}^{\infty} c_n(x-a)^n = c_0 + c_1(x-a) + c_2(x-a)^2 + c_3(x-a)^3 + \cdots,$$

where the coefficient c_n is a real number for each n.

Notice that the index of this power series cannot start at $n = 0$ because we would be dividing by zero.

Example 1 *For the power series* $\sum_{n=1}^{\infty} \frac{(x-2)^n}{n^2}$, *(a) write the series in expanded form; (b) identify the center of the power series; and (c) identify the first four coefficients c_0, c_1, c_2, and c_3.*

Solution (a) The series in expanded form is

$$\sum_{n=1}^{\infty} \frac{(x-2)^n}{n^2} = (x-2) + \frac{1}{4}(x-2)^2 + \frac{1}{9}(x-2)^3 + \cdots.$$

The center of the power series with terms $c_n(x-a)^n$ is a.

The index of the coefficient is the same as the exponent on $(x-a)$. The coefficient on $(x-2)^3$ is $\frac{1}{9}$; therefore, $c_3 = \frac{1}{9}$. The coefficient c_0 is the constant term, of which there is none here; therefore, $c_0 = 0$.

(b) The center of the power series is 2.

(c) The first four coefficients are $c_0 = 0$, $c_1 = 1$, $c_2 = \frac{1}{4}$, and $c_3 = \frac{1}{9}$. ■

Reading Exercise 23 *Determine the center of the power series (a)* $\sum_{n=1}^{\infty} -\frac{1}{4n}(x-3)^n$ *and (b)* $\sum_{n=0}^{\infty} \frac{n^n x^n}{n!}$.

Examples: convergence of power series

For a particular value of x, a power series is a series of real numbers. The same tests of convergence or divergence that apply to series of real numbers therefore apply to power series.

Example 2 *For what values of x does the power series* $\sum_{n=0}^{\infty} \frac{x^n}{n!}$ *converge?*

Because $0! = 1$, this power series is allowed to start at $n = 0$.

Solution The terms of the series include a factorial, indicating the use of the ratio test:

$$\begin{aligned}
\lim_{n\to\infty} \left| a_{n+1} \cdot \frac{1}{a_n} \right| &= \lim_{n\to\infty} \left| \frac{x^{n+1}}{(n+1)!} \cdot \frac{n!}{x^n} \right| \\
&= \lim_{n\to\infty} \left| \frac{x^{n+1}}{x^n} \cdot \frac{n!}{(n+1)n!} \right| \\
&= \lim_{n\to\infty} \left| \frac{x}{1} \cdot \frac{1}{n+1} \right| = \lim_{n\to\infty} \frac{|x|}{n+1} \\
&= \frac{|x|}{\Omega + 1} \approx \frac{|x|}{\Omega} = |x| \cdot \omega \doteq 0.
\end{aligned}$$

In each line we treat x the same as any specific real number. This includes simplification, approximation, and rendering. In line 3, because n is not negative, we omit the absolute values for $n + 1$. However, because x can be any real number and therefore might be negative, we must keep the absolute value for $|x|$. In line 4, because $|x|$ is a real number, $|x| \cdot \omega$ is infinitesimal and therefore renders zero.

Because the limit is less than one, the series converges absolutely. This is true for any x, and therefore the power series converges for every real number x. ■

The domain of the power series is $(-\infty, \infty)$.

When deciding which test to use for a power series, remember that the variable of interest is n, not x; x is treated like a real-number constant. Just as 5^n is considered an exponential, so is $(x - 2)^n$.

Example 3 *For what values of x does the power series* $\sum_{n=1}^{\infty} \frac{(x-2)^n}{n^2}$ *converge?*

This is the same power series used in example 1.

Solution The form of the terms is an exponential divided by a polynomial, indicating use of the ratio test:

$$\begin{aligned}
\lim_{n\to\infty} \left| a_{n+1} \cdot \frac{1}{a_n} \right| &= \lim_{n\to\infty} \left| \frac{(x-2)^{n+1}}{(n+1)^2} \cdot \frac{n^2}{(x-2)^n} \right| \\
&= \lim_{n\to\infty} \left| \frac{x-2}{1} \right| \cdot \frac{n^2}{(n+1)^2} \\
&= |x-2| \cdot \frac{\Omega^2}{(\Omega+1)^2} \approx |x-2| \frac{\Omega^2}{\Omega^2} = |x-2|.
\end{aligned}$$

In line 2, because the limit has the variable n going to infinity, we replace n with Ω in line 3. The symbol x is treated like a real number for the purposes of the limit.

Remember that x is treated like a real-number constant, so the limit has been found; we just need to interpret the result. The series converges absolutely when the limit is less than one—that is, when $|x - 2| < 1$—and diverges when $|x - 2| > 1$. The test is inconclusive when the limit equals one—that is, when $|x - 2| = 1$. Recalling that the quantity $|x - 2|$ is the distance between x and 2 on the number line, the results can be pictured as in figure 1. The power series converges when x is within

The inequality $|x - 2| < 1$ can also be solved algebraically;

$$-1 < x - 2 < 1$$

$$1 < x < 3.$$

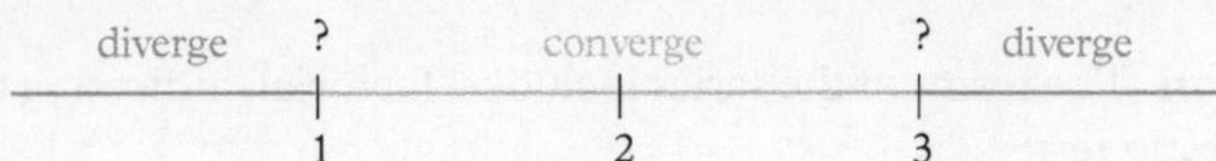

Figure 1 *Values of x (located on the number line) for which the power series* $\sum_{n=1}^{\infty} \frac{(x-2)^n}{n^2}$ *converges (green) and diverges (pink). The converge/diverge status for* $x = 1$ *and* $x = 3$ *has not been determined by the ratio test*

1 unit of 2 and diverges when x is more than 1 unit from 2. The ratio test is inconclusive for $x = 1$ and $x = 3$. These "endpoints" need to be tested individually.

It is easiest to test the right endpoint first. The series to test when $x = 3$ is

$$\sum_{n=1}^{\infty} \frac{(3-2)^n}{n^2} = \sum_{n=1}^{\infty} \frac{1}{n^2},$$

which converges (p-series, $p = 2 > 1$). The series to test when $x = 1$ is

It is possible for the power series to converge at none, one, or both of the endpoints.

$$\sum_{n=1}^{\infty} \frac{(1-2)^n}{n^2} = \sum_{n=1}^{\infty} \frac{(-1)^n}{n^2},$$

which is absolutely convergent (we just tested $\sum_{n=1}^{\infty} \left|\frac{(-1)^n}{n^2}\right| = \sum_{n=1}^{\infty} \frac{1}{n^2}$, which illustrates why it is easiest to check the right endpoint first.)

Ans. to reading exercise 23
(a) 3, (b) 0

We conclude that the series converges for $1 \le x \le 3$. ■

More descriptive terminology is given later in this section, after which we examine additional examples similar to example 3.

Reading Exercise 24 *For what values of x does* $\sum_{n=0}^{\infty} \frac{2x^n}{n!}$ *converge?*

Power series as functions

A function can be defined in terms of a power series. Let's look at this by formalizing an earlier discussion.

Example 4 *Let* $f(x) = \sum_{n=0}^{\infty} x^n$. *Find (a)* $f\left(-\frac{1}{5}\right)$ *and (b)* $f(0.8744)$.

Solution (a) We wish to determine the sum of the series

$$f\left(-\frac{1}{5}\right) = \sum_{n=0}^{\infty}\left(-\frac{1}{5}\right)^n.$$

The series is geometric, with the first term $a = \left(-\frac{1}{5}\right)^0 = 1$ and the common ratio $r = -\frac{1}{5}$. The sum is

$$f\left(-\frac{1}{5}\right) = \frac{a}{1-r} = \frac{1}{1-\left(-\frac{1}{5}\right)} = \frac{1}{\frac{6}{5}} = \frac{5}{6}.$$

(b) We wish to determine the sum of the series

$$f(0.8744) = \sum_{n=0}^{\infty}(0.8744)^n.$$

The series is geometric, with the first term $a = (0.8744)^0 = 1$ and the common ratio $r = 0.8744$. The sum is

$$f(0.8744) = \frac{a}{1-r} = \frac{1}{1-0.8744} = \frac{1}{0.1256} = 7.9618. \quad \blacksquare$$

Because a power series is not always geometric, the task is often more difficult than illustrated in example 4.

What if there was a part (c) to example 4: find $f(0)$? The series is

$$\sum_{n=0}^{\infty} 0^n = 0^0 + 0^1 + 0^2 + 0^3 + \cdots = ? + 0 + 0 + 0 + \cdots.$$

The issue is that 0^0 is an indeterminate form. However, when we expanded the power series earlier, we wrote

$$f(x) = \sum_{n=0}^{\infty} x^n = 1 + x + x^2 + x^3 + \cdots,$$

which means that

$$f(0) = 1 + 0 + 0^2 + 0^3 + \cdots = 1.$$

Pairing these two approaches, we interpret the indeterminate form 0^0 to be 1 in this context of power series.

POWER SERIES, EXPONENT 0

For any power series centered at a,

$$(x-a)^0 = 1$$

for any real number x, even when $x = a$.

What if there was a part (d) to example 4: find $f(x)$? Following the pattern of parts (a) and (b), we wish to determine the sum of the series

$$f(x) = \sum_{n=0}^{\infty} x^n,$$

Recall that the formula for the sum of a geometric series is only valid when the series converges.

which is geometric, with the first term $a = x^0 = 1$ and the common ratio $r = x$. As long as $|r| = |x| < 1$,

$$f(x) = \frac{a}{1-r} = \frac{1}{1-x}.$$

This explicit algebraic formula allows us to graph the function (figure 2), but we must be careful. Instead of graphing $y = \frac{1}{1-x}$ throughout its natural domain $x \neq 1$, we restrict to the domain $(-1, 1)$ because the series only converges for $|x| < 1$.

Ans. to reading exercise 24
the power series converges for every real number x

Determining an explicit algebraic formula for a power series is usually very difficult, but occasionally it is possible.

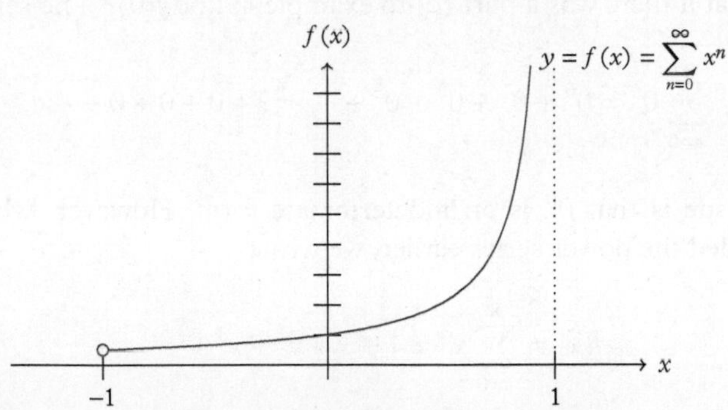

Figure 2 *The graph of the function* $f(x) = \sum_{n=0}^{\infty} x^n$. *The function's domain is* $(-1, 1)$, *and on this domain it matches the graph of* $y = \frac{1}{1-x}$. *On and to the right of the vertical asymptote, the power series diverges with a positive infinite sum. On and to the left of* $x = -1$ *the power series diverges in an infinitely oscillating manner*

Radius and interval of convergence

Take another look at figure 1. The values of x for which the power series of example 3 converge form an interval, $[1, 3]$. The midpoint of the interval, 2, is the same number that is the center of the power series. If we draw a circle centered at 2 with radius 1, the series converges for the values on the number line inside the circle, and diverges for the values on the number line outside the circle. For this reason, we say the *radius of convergence* for this power series is one.

The precise circle to be drawn for a power series is developed more fully in a complex variables course.

The result of example 2 illustrates another possibility—that the power series converges for every real number x. In this case, the *interval of convergence* is $(-\infty, \infty)$ and the radius of convergence is ∞.

It turns out there is only one more possibility: that a power series converges only at its center.

Theorem 8 INTERVAL OF CONVERGENCE CLASSIFICATIONS *For a power series* $\sum_{n=0}^{\infty} c_n(x-a)^n$*, there are only three possibilities:*

(1) the series converges for every real number x,

(2) there is a positive real number R such that the series converges for $|x-a| < R$ and diverges for $|x-a| > R$, or

(3) the series converges only when $x = a$.

For option (b), what happens when $|x - a| = R$ is not specified by the theorem. These "endpoints" must be checked individually for convergence or divergence. See example 6.

We call the number R the *radius of convergence*. For option (a), we say $R = \infty$; for option (c), we say $R = 0$.

The details of a formal proof of theorem 8 are usually saved for a more advanced course, but a partial justification is available using the level comparison test. Suppose a power series $\sum_{n=0}^{\infty} c_n(x-a)^n$ converges at a number $x = k$ (see figure 3). Now choose a number p that is closer to the center a than k, so that $|p-a| < |k-a|$. Taking reciprocals gives

Recall that the distance between p and a is $|p - a|$.

$$\frac{1}{|p-a|} > \frac{1}{|k-a|}.$$

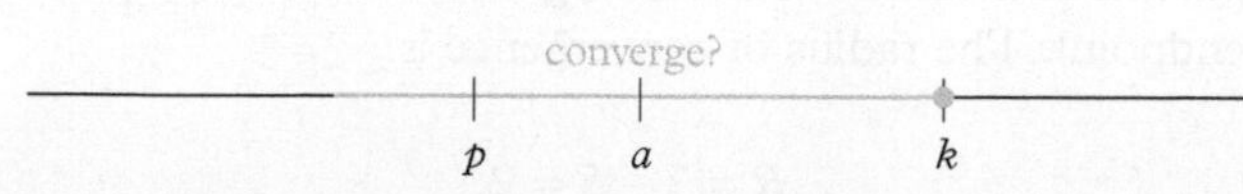

Figure 3 *A number line indicating the location of the center a of a power series and a number k at which the power series converges. If p is closer to the center than k (that is, p is in the green region), then the power series must also converge at p*

The computation for the level comparison test tells us how fast the terms of a series are going toward zero. The computation for $\sum_{n=0}^{\infty} |c_n(k-a)^n|$ results in

The level comparison test is for series with nonnegative terms, so we must use absolute values.

$$\frac{1}{|a_\Omega|} = \frac{1}{|c_\Omega||k-a|^\Omega}.$$

Meanwhile, the computation for $\sum_{n=0}^{\infty} |c_n(p-a)^n|$ gives

$$\frac{1}{|a_\Omega|} = \frac{1}{|c_\Omega||p-a|^\Omega} > \frac{1}{|c_\Omega||k-a|^\Omega},$$

and we conclude the terms of the series $\sum_{n=0}^{\infty} c_n(p-a)^n$ are going toward zero faster than the terms of the series $\sum_{n=0}^{\infty} c_n(k-a)^n$. Because the power series converges at $x = k$, it stands to reason that the power series must also converge at $x = p$. Because this is true for any number p in the green interval in figure 3, we conclude there can be no "gaps" in convergence, and the interval of convergence is symmetric about the center of the power series.

The closer a number is to the center of a power series, the faster the terms of the series go toward zero.

Example 5 *The interval of convergence for a power series is* $[3, 7)$. *(a) What is the center of the power series? (b) What is the radius of convergence?*

Solution (a) The center of the power series is also the midpoint of the interval of convergence. Therefore, the center is

$$\frac{3+7}{2} = 5.$$

(b) The radius of convergence is the distance from the center to the endpoints of the interval of convergence, which is computed using either endpoint. The radius of convergence is

$$R = 7 - 5 = 2. \qquad \blacksquare$$

Reading Exercise 25 *(a) The interval of convergence is* $(5, 9)$. *What is the radius of convergence? (b) The interval of convergence is* $(-\infty, \infty)$. *What is the radius of convergence?*

Examples: radius and interval of convergence

Example 6 *Find the radius and interval of convergence for* $\sum_{n=0}^{\infty} \frac{n^2 x^n}{10^n}$.

Solution Using the ratio test,

$$\begin{aligned}
\lim_{n\to\infty}\left|a_{n+1}\cdot\frac{1}{a_n}\right| &= \lim_{n\to\infty}\left|\frac{(n+1)^2x^{n+1}}{10^{n+1}}\cdot\frac{10^n}{n^2x^n}\right| \\
&= \lim_{n\to\infty}\left|\frac{(n+1)^2}{n^2}\cdot\frac{x^{n+1}}{x^n}\cdot\frac{10^n}{10^{n+1}}\right| \\
&= \lim_{n\to\infty}\frac{(n+1)^2}{n^2}\cdot\left|\frac{x}{1}\right|\cdot\frac{1}{10} \\
&= \frac{(\Omega+1)^2}{\Omega^2}\cdot\frac{|x|}{10}\approx\frac{\Omega^2}{\Omega^2}\cdot\frac{|x|}{10}=\frac{|x|}{10}.
\end{aligned}$$

In line 3, because n is always nonnegative, we omit the absolute value symbols on $\frac{(n+1)^2}{n^2}$. The same is true for $\frac{1}{10}$. However, x can be any real number, including a negative real number, so we keep the absolute values on $\left|\frac{x}{1}\right| = |x|$.

The series converges if the limit is less than one:

$$\begin{aligned}
\frac{|x|}{10} &< 1 \\
|x| &< 10 \\
-10 < x &< 10.
\end{aligned}$$

By theorem 8, this computation is sufficient for determining the endpoints of the interval of convergence.

We still need to check the endpoints for convergence or divergence. For $x = 10$, the series is

Recall that the ratio test is inconclusive when the limit equals one, which happens at the endpoints of the interval of convergence.

$$\sum_{n=0}^{\infty}\frac{n^2 10^n}{10^n} = \sum_{n=0}^{\infty} n^2,$$

which clearly diverges by the test for divergence (the terms are not going toward zero). For $x = -10$, the series is

$$\sum_{n=0}^{\infty}\frac{n^2(-10)^n}{10^n} = \sum_{n=0}^{\infty}(-1)^n n^2,$$

Details:

$$\frac{(-10)^n}{10^n} = \frac{(-1)^n(10)^n}{10^n} = (-1)^n.$$

which also diverges by the test for divergence.

We conclude that the interval of convergence is $(-10, 10)$ and the radius of convergence is $R = 10$. ■

Remember to use parentheses when an endpoint is not included (if the series diverges at that endpoint) and use brackets when an endpoint is included (if the series converges at that endpoint).

Although the ratio test is usually the method of choice when determining the radius and interval of convergence, occasionally the root test is more convenient.

Example 7 *Find the radius and interval of convergence for* $\sum_{n=2}^{\infty} \frac{x^n}{(\ln n)^n}$.

Solution Because the terms have the form $(f(n))^n$, we apply the root test:

$$\lim_{n\to\infty} \sqrt[n]{|a_n|} = \lim_{n\to\infty} \sqrt[n]{\left|\frac{x^n}{(\ln n)^n}\right|}$$

$$= \lim_{n\to\infty} \frac{|x|}{\ln n} = \frac{|x|}{\ln \Omega} \doteq 0.$$

In line 2, because $|x|$ is a real number and $\ln \Omega$ is infinite, the quantity $\frac{|x|}{\ln \Omega}$ is infinitesimal and renders zero.

Because the limit is always less than one, the power series converges for every real number x.

Regardless of the value of x—whether it is 14.7, −347.62, 1 299 876, or anything else—the root test limit is zero and the series converges.

We conclude that the interval of convergence is $(-\infty, \infty)$ and the radius of convergence is $R = \infty$. ■

Ans. to reading exercise 25
(a) $R = 2$, (b) $R = \infty$

Reading Exercise 26 *Find the radius and interval of convergence for* $\sum_{n=0}^{\infty} \frac{x^n}{10^n}$.

Functions as power series

We saw earlier that the power series $f(x) = \sum_{n=0}^{\infty} x^n$ is the same as the function $f(x) = \frac{1}{1-x}$ on the interval $(-1, 1)$. This determination was made using the formula for the sum of a geometric series. However, what if we want to go backward, starting with the function and finding a power series that matches it?

Starting with the function $f(x) = \frac{1}{1-x}$, we note that the form is similar to the formula for the sum of a geometric series. Matching the function to the formula,

$$\frac{1}{1-x} = \frac{a}{1-r},$$

we designate the first term of the series to be $a = 1$ and the common ratio to be $r = x$. The power series

This power series starts with $n = 1$ instead of $n = 0$, but the difference is only cosmetic; the series is identical to $\sum_{n=0}^{\infty} x^n$.

$$f(x) = \sum_{n=1}^{\infty} ar^{n-1} = \sum_{n=1}^{\infty} 1 \cdot x^{n-1} = 1 + x + x^2 + \cdots$$

then matches the function as long as

The formula for the sum of a geometric series is only valid when $|r| < 1$.

$$|r| = |x| < 1,$$

which is the interval of convergence of the power series, $(-1, 1)$.

This technique can be used for many algebraic functions.

Example 8 *Find a power series representation for* $f(x) = \dfrac{x}{1-x^2}$ *and state the interval of convergence.*

Solution Matching the function to the formula for the sum of a geometric series,

$$\frac{x}{1-x^2} = \frac{a}{1-r},$$

we designate the first term of the series to be $a = x$ and the common ratio to be $r = x^2$. The power series

$$f(x) = \sum_{n=1}^{\infty} ar^{n-1} = \sum_{n=1}^{\infty} x \cdot (x^2)^{n-1} = \sum_{n=1}^{\infty} x \cdot x^{2n-2} = \sum_{n=1}^{\infty} x^{2n-1}$$

then matches the function as long as

$$|r| = |x^2| < 1.$$

We solve this inequality to find the interval of convergence:

$$\begin{aligned} |x^2| &< 1 \\ -\sqrt{1} < x &< \sqrt{1} \\ -1 < x &< 1. \end{aligned}$$

In line 2, the solution to the absolute value inequality $|x^n| < k$ is

$$-\sqrt[n]{k} < x < \sqrt[n]{k}.$$

We conclude that a power series representation for $f(x) = \dfrac{x}{1-x^2}$ is $\displaystyle\sum_{n=1}^{\infty} x^{2n-1}$, with an interval of convergence $(-1, 1)$. ■

Reading Exercise 27 *Find a power series representation for* $f(x) = \dfrac{1}{1-x^3}$ *on* $(-1, 1)$.

Sometimes, a little algebraic manipulation is required to match the function to $\frac{a}{1-r}$.

Example 9 *Find a power series representation for* $f(x) = \dfrac{x^3}{1+4x}$ *and state the interval of convergence.*

Ans. to reading exercise 26
interval of convergence $(-10, 10)$, radius of convergence $R = 10$ (The root test, the ratio test, or even the geometric series criteria can be used to make the determination. The geometric series criteria are quickest because the endpoints do not need to be tested.)

Solution Rewriting the function as

$$\frac{x^3}{1+4x} = \frac{x^3}{1-(-4x)}$$

allows us to match the formula for the sum of a geometric series. With

$$\frac{x^3}{1-(-4x)} = \frac{a}{1-r},$$

we designate the first term of the series to be $a = x^3$ and the common ratio to be $r = -4x$. The power series

$$\begin{aligned} f(x) &= \sum_{n=1}^{\infty} ar^{n-1} = \sum_{n=1}^{\infty} x^3 \cdot (-4x)^{n-1} \\ &= \sum_{n=1}^{\infty} x^3(-4)^{n-1}x^{n-1} = \sum_{n=1}^{\infty} (-4)^{n-1}x^{n+2} \end{aligned}$$

then matches the function as long as

$$|r| = |-4x| < 1.$$

Solving the inequality for x,

$$\begin{aligned} |-4x| &< 1 \\ 4|x| &< 1 \\ |x| &< \frac{1}{4} \\ -\frac{1}{4} < x &< \frac{1}{4}. \end{aligned}$$

We conclude that a power series representation for $f(x) = \dfrac{x^3}{1+4x}$ is $\sum_{n=1}^{\infty}(-4)^{n-1}x^{n+2}$, with an interval of convergence $\left(-\frac{1}{4}, \frac{1}{4}\right)$. ■

Although the technique of examples 8 and 9 works for algebraic functions that can be manipulated to match $\frac{a}{1-r}$, the technique does not help us find a power series representation for a nonalgebraic function such as $f(x) = \sin x$ or $f(x) = \ln x$. Another method is needed, and we study it in the next section.

Exercises 10.9

1–7. Rapid response: is it a power series?

1. $\sum_{n=0}^{\infty} nx^n$

2. $\sum_{n=247}^{\infty} nx^n$

3. $\sum_{n=-1}^{\infty} nx^n$

4. $\sum_{n=0}^{\infty} \frac{n}{x^n}$

5. $\sum_{n=0}^{\infty} nx^{2n}$

6. $\sum_{n=0}^{\infty} n(x-3)^n$

7. $\sum_{n=0}^{\infty} n(2x+1)^n$

8–11. Rapid response: what is the center of the power series?

8. $\sum_{n=0}^{\infty} 3n(x-2)^n$

9. $\sum_{n=0}^{\infty} \frac{3}{n!}(x+2)^n$

10. $\sum_{n=0}^{\infty} \frac{x^n}{3}$

11. $\sum_{n=0}^{\infty} (3x-1)^n$

Ans. to reading exercise 27
$\sum_{n=1}^{\infty} (x^3)^{n-1} = \sum_{n=1}^{\infty} x^{3n-3}$

12–15. Rapid response: what is the center of the power series? The interval of convergence is given.

12. $(-3, 3)$

13. $[0, 3)$

14. $[10, 20]$

15. $(-\infty, \infty)$

16–18. Rapid response: what is the radius of convergence of the series?

16. The series converges for $-1 < x < 7$ and diverges elsewhere.

17. The series converges for $x = 5$ and diverges elsewhere.

18. The series converges for all x.

19. For $f(x) = \sum_{n=0}^{\infty} \left(\frac{2}{5}\right)^n (x+1)^n$, find (a) $f(0)$, (b) $f(-1)$, and (c) $f(-3)$.

20. For $f(x) = \sum_{n=0}^{\infty} \frac{(x-2)^n}{10^n}$, find (a) $f(3)$, (b) $f(0)$, and (c) $f(2)$.

21–40. Determine the radius and interval of convergence of the power series.

21. $\sum_{n=0}^{\infty} \frac{x^n}{5^n}$

22. $\sum_{n=0}^{\infty} \frac{(-1)^n x^n}{3^n n!}$

23. $\sum_{n=1}^{\infty} \frac{x^n}{n^5}$

24. $\sum_{n=1}^{\infty} \frac{x^n}{n^2}$

25. $\sum_{n=1}^{\infty} \frac{(x-4)^n}{\sqrt{n}}$

26. $\sum_{n=1}^{\infty} \frac{(x-3)^n}{n}$

27. $\sum_{n=0}^{\infty} \frac{n!(x-1)^n}{3^{n+1}}$

28. $\sum_{n=1}^{\infty} \frac{(x-7)^n}{n4^n}$

29. $\sum_{n=0}^{\infty} \left(\frac{n}{n^3+1}\right)^n x^n$

30. $\sum_{n=0}^{\infty} \sqrt{n}(3x+2)^n$

31. $\sum_{n=0}^{\infty} \frac{n^2(x-3)^n}{6^n}$

32. $\sum_{n=1}^{\infty} \frac{(x+1)^n}{n(n+1)}$

33. $\sum_{n=0}^{\infty} \frac{(2x+1)^n}{n+2}$

34. $\sum_{n=0}^{\infty} \frac{x^{2n}}{3^n}$

35. $\sum_{n=0}^{\infty} \frac{(2-x)^n}{(n+1)!}$

36. $\sum_{n=1}^{\infty} \frac{(x-2)^n}{n^n}$

37. $\sum_{n=0}^{\infty} 3^n(x+4)^n(n-\frac{7}{2})$

38. $\sum_{n=1}^{\infty} \left(\frac{\sin n}{n}\right)^n (3x+7)^n$

39. $\sum_{n=0}^{\infty} \frac{n}{(n+1)4^n} x^{2n+1}$

40. $\sum_{n=0}^{\infty} \left(\frac{(2n^2+1)x}{5n+2}\right)^n$

41–48. Find a power series representation for the function and state the interval of convergence.

41. $f(x) = \frac{4x}{1-x^5}$

42. $f(x) = \frac{72}{1+10x^2}$

43. $f(x) = \frac{1}{1+6x^2}$

44. $f(x) = 2x(1-x^3)^{-1}$

45. $f(x) = \frac{(x+1)^4}{1-(x+1)^2}$

46. $f(x) = \frac{5}{1-(x+3)^4}$

47. $f(x) = \frac{3x}{6+x}$

48. $f(x) = \frac{x^2}{x^7-5}$

49. (GSL, modified) Determine the radius and interval of convergence of the power series

$$x + \frac{2}{1} \cdot \frac{x^3}{3} + \frac{2 \cdot 4}{1 \cdot 3} \cdot \frac{x^5}{5} + \frac{2 \cdot 4 \cdot 6}{1 \cdot 3 \cdot 5} \cdot \frac{x^7}{7} + \cdots.$$

50. (GSL) Determine the radius and interval of convergence of the power series

$$1 + \frac{x^2}{2 \cdot 2^2} + \frac{1 \cdot 3x^4}{2 \cdot 4 \cdot 2^4} + \frac{1 \cdot 3 \cdot 5x^6}{2 \cdot 4 \cdot 6 \cdot 2^6} + \cdots.$$

51. (Love) Determine the radius and interval of convergence of the power series

$$\frac{x^3}{3^2} + \frac{2}{3} \cdot \frac{x^5}{5^2} + \frac{2 \cdot 4}{3 \cdot 5} \cdot \frac{x^7}{7^2} + \cdots.$$

52. (Love) Determine the radius and interval of convergence of the power series

$$1 + \frac{x^2}{2} + \frac{3}{2} \cdot \frac{x^4}{4} + \frac{3 \cdot 5}{2 \cdot 4} \cdot \frac{x^6}{6} + \frac{3 \cdot 5 \cdot 7}{2 \cdot 4 \cdot 6} \cdot \frac{x^8}{8} + \cdots.$$

Taylor and Maclaurin Series

10.10

How do we find the value of ln 4.73? Most likely, we pick up a calculator, enter the command, and read the answer. Before calculating devices, however, there were tables of values, such as the table of logarithms in section 5.6. But, how were the values in the tables calculated? And how does an electronic calculator or a CAS calculate the answer it shows on the screen?

The operations of addition, subtraction, multiplication, and division can be performed by hand with algorithms that are taught in elementary school. Electronic circuits can be designed to perform these same calculations. Values of polynomials can be calculated using these algorithms (or circuits) because polynomials use only the arithmetic operations.

Recall that for positive integer exponents, exponentiation is repeated multiplication.

Trigonometric functions, logarithms, and other transcendental functions are another matter altogether. Historically, many different methods of approximating values of these functions were used, and the production of tables of values was a highly important and extremely time-consuming scientific enterprise. The development of power series representations of functions revolutionized table-making, allowing for greater precision while simultaneously reducing the amount of labor required.

But, you may object, how can adding infinitely many values (after all, it is an infinite series) possibly save labor? A partial answer is that we do not need to use all the terms of the power series. It turns out that a finite number of terms of a power series, which is a polynomial and therefore is calculable by hand or by circuit, is enough to produce the desired accuracy.

Computational efficiency is important not just for human labor, but in computer science as well. If these topics interest you, consider taking a course in numerical analysis.

Finding coefficients for a power series

Suppose we wish to find, for a function f, a power series representation centered at zero:

$$f(x) = \sum_{n=0}^{\infty} c_n x^n = c_0 + c_1 x + c_2 x^2 + c_3 x^3 + c_4 x^4 + c_5 x^5 + \cdots .$$

What we need to know is the value of each coefficient c_n. How can we find these coefficients? It turns out they can be calculated by evaluating the function and its derivatives at the center of the power series.

First we evaluate the function at zero:

$$f(0) = c_0 + c_1 \cdot 0 + c_2 \cdot 0^2 + c_3 \cdot 0^3 + c_4 \cdot 0^4 + c_5 \cdot 0^5 + \cdots = c_0.$$

Next we find the derivative $f'(x)$ and evaluate it at zero:

This act of taking the derivative term-by-term is discussed later in this section.

$$f'(x) = 1c_1 + 2c_2 x + 3c_3 x^2 + 4c_4 x^3 + 5c_5 x^4 + \cdots ,$$

$$f'(0) = c_1 + 2c_2 \cdot 0 + 3c_3 \cdot 0^2 + 4c_4 \cdot 0^3 + 5c_5 \cdot 0^4 + \cdots = c_1.$$

For the purpose of exposing a pattern, when we calculate the second derivative, we do not perform all of the arithmetic:

$$f''(x) = 1 \cdot 2c_2 + 2 \cdot 3c_3 x + 3 \cdot 4c_4 x^2 + 4 \cdot 5c_5 x^3 \cdots ,$$

$$f''(0) = 1 \cdot 2c_2 + 0 + 0 + 0 + \cdots = 1 \cdot 2c_2.$$

The next one is

$$f'''(x) = 1 \cdot 2 \cdot 3c_3 + 2 \cdot 3 \cdot 4c_4 x + 3 \cdot 4 \cdot 5c_5 x^2 + \cdots ,$$

$$f'''(0) = 1 \cdot 2 \cdot 3c_3 + 0 + 0 + \cdots = 1 \cdot 2 \cdot 3c_3,$$

and the next is

$$f^{(4)}(x) = 1 \cdot 2 \cdot 3 \cdot 4c_4 + 2 \cdot 3 \cdot 4 \cdot 5c_5 x + \cdots ,$$

$$f^{(4)}(0) = 1 \cdot 2 \cdot 3 \cdot 4c_4 + 0 + 0 + \cdots = 1 \cdot 2 \cdot 3 \cdot 4c_4.$$

The pattern is clear. For any n, we have

$$f^{(n)}(0) = n!c_n,$$

and solving for the coefficients gives

Since $0! = 1$, the pattern works for $f(0) = 0!c_0$ if we consider the zeroth derivative to be the original function f.

$$c_n = \frac{f^{(n)}(0)}{n!}.$$

Therefore, if a function f has a power series representation centered at zero, the power series representation is

$$f(x) = \sum_{n=0}^{\infty} \frac{f^{(n)}(0)}{n!} x^n,$$

which is called the *Maclaurin series* for the function.

This power series representation, which is a special case of a Taylor series, was popularized in Maclaurin's 1742 calculus textbook *Treatise of Fluxions*, published by T. W. and T. Ruddimans, Edinburgh.
Colin Maclaurin, 1698–1746
http://www-history.mcs.st-andrews.ac.uk/Biographies/Maclaurin.html

Reading Exercise 28 *For a Maclaurin series, (a) what is the formula for c_2? (b) For c_3?*

A similar calculation for a power series centered at a yields the following result, which goes by the name *Taylor series.*

Theorem 9 TAYLOR SERIES *If a function f has a power series representation $f(x) = \sum_{n=0}^{\infty} c_n(x-a)^n$ on the interval of convergence $|x - a| < R$, then the coefficients are given by*

$$c_n = \frac{f^{(n)}(a)}{n!}.$$

Here the radius of convergence R may be either a positive real number or ∞.

A Maclaurin series is a Taylor series centered at zero.

Brook Taylor, 1685–1731
http://www-history.mcs.st-andrews.ac.uk/Biographies/Taylor.html

Maclaurin series examples

Let's start with a relatively simple example.

Example 1 *Find the Maclaurin series for $f(x) = e^x$ and determine its interval of convergence.*

Solution We need to determine the coefficients

$$c_n = \frac{f^{(n)}(0)}{n!}.$$

We need to find the derivatives, evaluate each of them at zero, and divide by $n!$. It is helpful to organize the work in a table (table 1), including enough rows that a pattern can be recognized, if possible.

Now that we have determined the value of c_n, we write out the Maclaurin series:

Theorem 9 has the condition "if a function f has a power series representation." When calculating a Maclaurin or Taylor series we make this assumption. A discussion of when the condition is met is included at the end of this section.

It is not always practical to determine the exact form of the derivative for the nth row. In this case, the row may be omitted.

Table 1 *Calculation of coefficients for example 1*

n	derivatives	derivatives evaluated at the center 0	coefficients
0	$f(x) = e^x$	$f(0) = e^0 = 1$	$c_0 = \frac{f(0)}{0!} = 1$
1	$f'(x) = e^x$	$f'(0) = e^0 = 1$	$c_1 = \frac{f'(0)}{1!} = 1$
2	$f''(x) = e^x$	$f''(0) = e^0 = 1$	$c_2 = \frac{f''(0)}{2!} = \frac{1}{2}$
3	$f'''(x) = e^x$	$f'''(0) = e^0 = 1$	$c_3 = \frac{f'''(0)}{3!} = \frac{1}{6}$
n	$f^{(n)}(x) = e^x$	$f^{(n)}(0) = e^0 = 1$	$c_n = \frac{f^{(n)}(0)}{n!} = \frac{1}{n!}$

Writing the power series in summation notation is helpful for determining the interval of convergence. Writing the power series in expanded form is helpful for interpreting the series. When using the expanded form, the series should end in ellipses (the three dots). Inclusion of the *general term* $\frac{1}{n!}x^n$ requires ellipses on both sides.

$$e^x = \sum_{n=0}^{\infty} c_n x^n = \sum_{n=0}^{\infty} \frac{1}{n!}x^n$$

$$= 1 + x + \frac{1}{2}x^2 + \frac{1}{6}x^3 + \frac{1}{24}x^4 + \cdots + \frac{1}{n!}x^n + \cdots .$$

The first task finished, we turn our attention to finding the interval of convergence of the series. A factorial is present, so let's try the ratio test:

$$\lim_{n\to\infty} \left| a_{n+1} \cdot \frac{1}{a_n} \right| = \lim_{n\to\infty} \left| \frac{x^{n+1}}{(n+1)!} \cdot \frac{n!}{x^n} \right|$$

$$= \lim_{n\to\infty} \frac{|x|}{n+1} = \frac{|x|}{\Omega + 1} \approx \frac{|x|}{\Omega} = |x|\omega \doteq 0.$$

Convergence of the power series is not enough to prove that it converges to the desired value, e^x. This issue is revisited at the end of this section.

Because the limit is always less than one, the power series converges for every real number x. The interval of convergence is $(-\infty, \infty)$. ■

Sometimes recognizing a pattern is easier than writing a formula for the general term (nth term).

Example 2 *Determine the Maclaurin series for* $f(x) = \sin x$.

Solution We find the derivatives, evaluate each of them at zero, and divide by $n!$, including enough rows in the table (table 2) to recognize a pattern. The pattern is apparent because the derivatives repeat in a cycle of length 4. The even-numbered coefficients are all zero, whereas the odd-numbered coefficients alternate in sign. The Maclaurin series, in expanded form, is

The expanded form is often sufficient.

$$\sin x = 0 + x + 0x^2 - \frac{1}{3!}x^3 + 0x^4 + \frac{1}{5!}x^5 + 0x^6 - \frac{1}{7!}x^7 + 0x^8 + \frac{1}{9!}x^9 - \cdots$$

$$= x - \frac{1}{3!}x^3 + \frac{1}{5!}x^5 - \frac{1}{7!}x^7 + \frac{1}{9!}x^9 - \cdots .$$

Table 2 *Calculation of coefficients for example 2*

n	derivatives	derivatives evaluated at the center 0	coefficients
0	$f(x) = \sin x$	$f(0) = \sin 0 = 0$	$c_0 = \frac{f(0)}{0!} = 0$
1	$f'(x) = \cos x$	$f'(0) = \cos 0 = 1$	$c_1 = \frac{f'(0)}{1!} = 1$
2	$f''(x) = -\sin x$	$f''(0) = -\sin 0 = 0$	$c_2 = \frac{f''(0)}{2!} = 0$
3	$f'''(x) = -\cos x$	$f'''(0) = -\cos 0 = -1$	$c_3 = \frac{f'''(0)}{3!} = -\frac{1}{3!}$
4	$f^{(4)}(x) = \sin x$	$f^{(4)}(0) = \sin 0 = 0$	$c_4 = \frac{f^{(4)}(0)}{4!} = 0$

Ans. to reading exercise 28
(a) $\frac{f''(0)}{2!}$, (b) $\frac{f'''(0)}{3!}$

If writing the power series in summation notation is desired, it can be accomplished as

$$\sin x = \sum_{k=0}^{\infty} (-1)^k \frac{1}{(2k+1)!} x^{2k+1}.$$

■

It is helpful to reserve the use of n as the index of the summation for the usual case when the index matches the exponent on the variable. Notice that $2k+1$ is always odd, the factorial matches the exponent, and the signs of the odd-exponent terms alternate.

Reading Exercise 29 *Find the interval of convergence for the Maclaurin series for* $\sin x$.

Finding a Maclaurin series for $f(x) = x^2 \sin x$ by the method of examples 1 and 2 seems more complicated because the derivatives require the use of the product rule. Why can't we just take the Maclaurin series for $\sin x$ and multiply by x^2? The calculation is not difficult,

$$x^2 \sin x = x^2\left(x - \frac{1}{3!}x^3 + \frac{1}{5!}x^5 - \frac{1}{7!}x^7 + \cdots\right)$$
$$= x^3 - \frac{1}{3!}x^5 + \frac{1}{5!}x^7 - \frac{1}{7!}x^9 + \cdots,$$

The first three derivatives are

$$f'(x) = x^2 \cos x + 2x \sin x$$
$$f''(x) = -x^2 \sin x + 4x \cos x + 2 \sin x$$
$$f'''(x) = -x^2 \cos x - 6x \sin x + 6 \cos x.$$

Determining a pattern is difficult.

but is it justified? It turns out that the purely algebraic operations we wish to perform on power series, such as using the distributive property to multiply the series by x^2, all work, at least in the interior of the interval of convergence. (At the endpoints of the interval of convergence, a power series might converge conditionally, which is when we know that weird things can happen, such as being able to reorder the terms of a conditionally convergent series to get any sum we wish.) The radius of convergence does not change.

The justification for most of these operations uses the transfer principle as well as the definition of convergence of a series.

Example 3 *Determine the Maclaurin series for* $f(x) = 4x^3 \sin x$. *What is its interval of convergence?*

We can also use summation notation:

$$4x^3 \sin x = 4x^3 \sum_{k=0}^{\infty}(-1)^k \frac{1}{(2k+1)!}x^{2k+1}$$
$$= \sum_{k=0}^{\infty}(-1)^k \frac{4}{(2k+1)!}x^{2k+4}.$$

Solution Because we already know the Maclaurin series for $\sin x$, we multiply this series by $4x^3$ to obtain

$$\begin{aligned} 4x^3 \sin x &= 4x^3\left(x - \frac{1}{3!}x^3 + \frac{1}{5!}x^5 - \frac{1}{7!}x^7 + \cdots\right) \\ &= 4x^4 - \frac{4}{3!}x^6 + \frac{4}{5!}x^8 - \frac{4}{7!}x^{10} + \cdots. \end{aligned}$$

The radius of convergence is the same as for the Maclaurin series for $\sin x$, $R = \infty$. Therefore, the interval of convergence for this Maclaurin series is $(-\infty, \infty)$. ■

Reading Exercise 30 *Find the Maclaurin series for $f(x) = x^2 e^x$.*

Taylor series example

When might we want to center a power series somewhere other than at zero? One answer is when the function we wish to represent is not defined at $x = 0$.

Example 4 *Find the Taylor series centered at* 1 *for $f(x) = \ln x$ and find its interval of convergence.*

Solution We need to determine the coefficients

$$c_n = \frac{f^{(n)}(1)}{n!}.$$

As we did for a Maclaurin series, we find the derivatives, evaluate each of them at the center 1, and divide by $n!$, including enough rows in the table to recognize a pattern (table 3).

Table 3 *Calculation of coefficients for example 4*

n	derivatives	derivatives evaluated at the center 1	coefficients
0	$f(x) = \ln x$	$f(1) = \ln 1 = 0$	$c_0 = \frac{f(1)}{0!} = 0$
1	$f'(x) = \frac{1}{x} = x^{-1}$	$f'(1) = 1^{-1} = 1$	$c_1 = \frac{f'(1)}{1!} = 1$
2	$f''(x) = (-1)x^{-2}$	$f''(1) = (-1)1^{-2} = -1$	$c_2 = \frac{f''(1)}{2!} = -\frac{1}{2}$
3	$f'''(x) = (-2)(-1)x^{-3}$	$f'''(1) = (-2)(-1)1^{-3} = 2$	$c_3 = \frac{f'''(1)}{3!} = \frac{1}{3}$
4	$f^{(4)}(x) = (-3)(-2)(-1)x^{-4}$	$f^{(4)}(1) = -3!$	$c_4 = \frac{f^{(4)}(1)}{4!} = -\frac{1}{4}$
5	$f^{(5)}(x) = (-4)(-3)(-2)(-1)x^{-5}$	$f^{(5)}(1) = 4!$	$c_4 = \frac{f^{(5)}(1)}{5!} = \frac{1}{5}$

For a Taylor series each term is of the form $c_n(x-a)^n$. For this example, the center is $a = 1$. The Taylor series is

$$\ln x = 0 + 1(x-1) - \frac{1}{2}(x-1)^2 + \frac{1}{3}(x-1)^3 - \frac{1}{4}(x-1)^4 + \frac{1}{5}(x-1)^5 - \cdots = \sum_{n=1}^{\infty}(-1)^{n+1}\cdot\frac{1}{n}(x-1)^n.$$

Notice that the summation starts at $n = 1$ rather than $n = 0$.

We use the ratio test to determine the interval of convergence:

$$\lim_{n\to\infty}\left|a_{n+1}\cdot\frac{1}{a_n}\right| = \lim_{n\to\infty}\left|\frac{(x-1)^{n+1}}{n+1}\frac{n}{(x-1)^n}\right| = \lim_{n\to\infty}\frac{n}{n+1}|x-1| = \frac{\Omega}{\Omega+1}|x-1| \approx \frac{\Omega}{\Omega}|x-1| = |x-1|.$$

Line 1: Multiplying or dividing by a power of -1 is omitted because the absolute value of the result is always 1.

The power series converges if the limit is less than one:

$$|x-1| < 1$$
$$-1 < x-1 < 1$$
$$0 < x < 2.$$

Ans. to reading exercise 29
$(-\infty, \infty)$

We still need to check the endpoints. For $x = 2$, the power series is

$$\sum_{n=1}^{\infty}(-1)^{n+1}\cdot\frac{1}{n}(2-1)^n = \sum_{n=1}^{\infty}(-1)^{n+1}\frac{1}{n}.$$

The alternating harmonic series converges and therefore $x = 2$ is in the interval of convergence. For $x = 0$, the power series is

$$\sum_{n=1}^{\infty}(-1)^{n+1}\cdot\frac{1}{n}(0-1)^n = \sum_{n=1}^{\infty}(-1)^{2n+1}\cdot\frac{1}{n} = \sum_{n=1}^{\infty}-\frac{1}{n},$$

Notice that $2n+1$ is always odd, so $(-1)^{2n+1} = -1$.

which is the negative of the harmonic series. The harmonic series diverges, and therefore $x = 0$ is not in the interval of convergence. The interval of convergence is $(0, 2]$. ■

Ans. to reading exercise 30
$x^2e^x = x^2 + x^3 + \frac{1}{2}x^4 + \frac{1}{6}x^5 + \cdots = \sum_{n=0}^{\infty}\frac{1}{n!}x^{n+2}$

Hidden near the end of the solution to example 4 is a remarkable fact: the sum of the alternating harmonic series is ln 2. It depends, of course, on the power series not only converging, but converging to the desired value, $\ln x$. As stated earlier, we return to this issue at the end of this section.

Taylor polynomials

In section 2.3 we studied tangent lines to functions and noted the concept of local linearity. If we zoom in far enough on the graph of a function, the function and its tangent line become indistinguishable. We can use the tangent line as a reasonable approximation to the function as long as we do not stray too far from the point of tangency.

A line is a polynomial of degree 1, and as such a tangent line is a degree 1 polynomial approximation to a function. Could a polynomial of higher degree offer an even better approximation? For instance, does a "tangent parabola" follow the function's graph more closely, and farther from the point of tangency, than a tangent line? If so, how can we determine these tangent polynomials?

Enter Taylor series.

In expanded form, the Taylor series representation of a function centered at $x = a$ is

$$\begin{aligned} f(x) &= c_0 + c_1(x-a) + c_2(x-a)^2 + c_3(x-a)^3 + \cdots \\ &= f(a) + f'(a)(x-a) + \frac{f''(a)}{2!}(x-a)^2 + \frac{f'''(a)}{3!}(x-a)^3 + \cdots. \end{aligned}$$

The first two terms of the Taylor series expansion, $f(a) + f'(x)(x-a)$, is the linearization of the function at a, identical to the tangent line. In a similar fashion, the first three terms

$$f(a) + f'(a)(x-a) + \frac{f''(a)}{2!}(x-a)^2$$

is a parabola (degree 2 polynomial) that not only matches the slope of the function at a, like the tangent line, but the concavity as well. Using more terms of the Taylor series allows us to match even higher derivatives, and the shape of the polynomial follows the shape of the curve even more closely. This is essentially why a Taylor series converges to the desired function within its interval of convergence. A more formal discussion is provided at the end of this section.

We come back to this less formal discussion after a definition break.

Definition 10 TAYLOR POLYNOMIALS *Let f be a function and let a be a number in the domain of f such that $f^{(n)}(a)$ exists. The* Taylor polynomial of order n centered at *a is the polynomial*

$$T_n(x) = f(a) + f'(a)(x-a) + \frac{f''(a)}{2!}(x-a)^2 + \cdots + \frac{f^{(n)}(a)}{n!}(x-a)^n.$$

The Taylor polynomial of order n consists of the terms of the Taylor series of degree n or less.

Example 5 *Determine the Taylor polynomials of orders 3, 5, and 8 for $f(x) = \sin x$ centered at zero.*

Solution We already know the Taylor series for $f(x) = \sin x$ centered at zero (aka the Maclaurin series):

$$\sin x = x - \frac{1}{3!}x^3 + \frac{1}{5!}x^5 - \frac{1}{7!}x^7 + \frac{1}{9!}x^9 - \cdots.$$

The Taylor polynomial of order 3 consists of the terms of degree 3 or less:

$$T_3(x) = x - \frac{1}{6}x^3.$$

The Taylor polynomial of order 5 consists of the terms of degree 5 or less:

$$T_5(x) = x - \frac{1}{6}x^3 + \frac{1}{120}x^5.$$

The Taylor polynomial of order 8 consists of the terms of degree 8 or less:

$$T_8(x) = x - \frac{1}{6}x^3 + \frac{1}{120}x^5 - \frac{1}{5040}x^7. \qquad \blacksquare$$

Notice that the Taylor polynomial of order 8 has degree 7 because there is no degree 8 term in the Taylor (Maclaurin) series. The Taylor polynomials of order 7 and 8 are identical.

The tangent line to $y = \sin x$ at $x = 0$ is $y = x$. The function and its tangent line are graphed in figure 1. The tangent line is the graph of the Taylor polynomial of order 1. If we extended the line to the left and to the right, it is quickly very far from the graph of the curve.

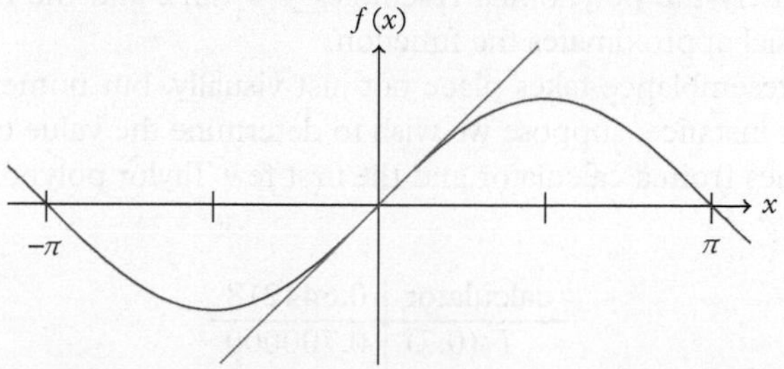

Figure 1 *The curve $y = \sin x$ (blue) with its tangent line $T_1(x) = x$ (orange) at $x = 0$*

The tangent cubic $T_3(x) = x - \frac{1}{6}x^3$, the Taylor polynomial of order 3, is graphed in figure 2. Notice how much more closely the polynomial follows the graph of $y = \sin x$ than the tangent line does.

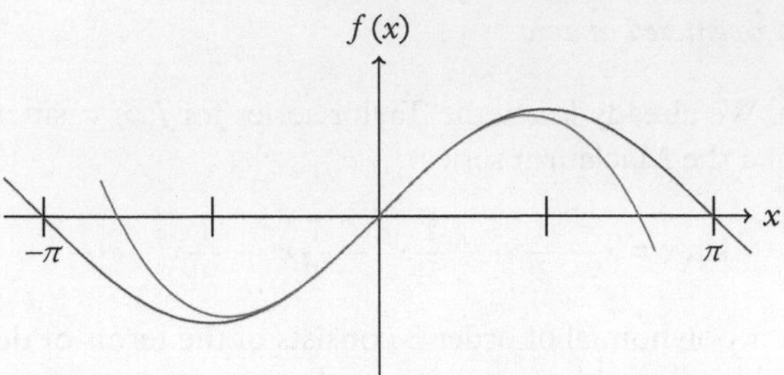

Figure 2 *The curve* $y = \sin x$ *(blue) with its tangent cubic* $T_3(x) = x - \frac{1}{6}x^3$ *(orange) at* $x = 0$*. Compare to figure 1*

The graphs of $T_5(x)$ and $T_7(x)$ follow the curve even more closely (figure 3). The more terms of the Taylor series that are included, the

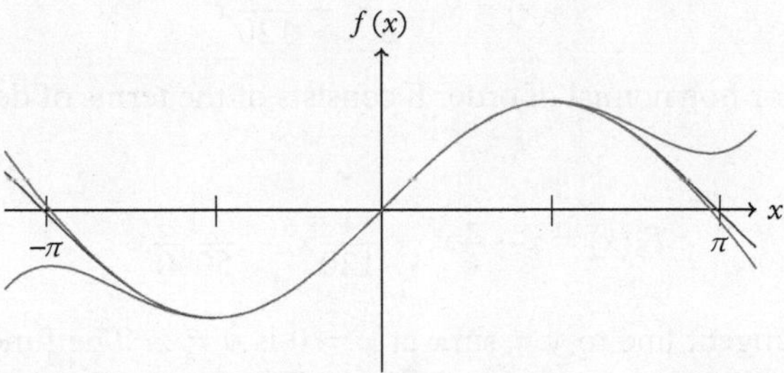

Figure 3 *The curve* $y = \sin x$ *(blue) with its Taylor polynomials* $T_5(x)$ *(orange) and* $T_7(x)$ *(green) at* $x = 0$*. Compare to figures 1 and 2*

more closely the polynomial resembles $y = \sin x$ and the better the polynomial approximates the function.

Any Taylor polynomial, no matter how closely the polynomial resembles the targeted function near the center (the point of tangency), strays far from the function if we move far enough from the center.

The resemblance takes place not just visually, but numerically as well. For instance, suppose we wish to determine the value of $\sin 0.7$. The values from a calculator and the first few Taylor polynomials are as follows:

calculator	0.644218
$T_1(0.7)$	0.700000
$T_3(0.7)$	0.642833
$T_5(0.7)$	0.644234
$T_7(0.7)$	0.644218

The Taylor polynomial of order 7 agrees with the calculator to six decimal places for sin 0.7; the Taylor polynomial of order 11 agrees with all 10 digits displayed on my calculator. Because polynomial values can be computed by hand or by an electronic circuit, a Taylor polynomial (with high enough order in a specified interval of values, and so on) makes a good substitute for a trig function, logarithmic function, exponential function, or other transcendental function.

Example 6 *(a) Use the result of example 4 to determine the Taylor polynomial of order 6 centered at 1 for* $f(x) = \ln x$. *(b) Use that Taylor polynomial to approximate* ln 1.277 *to six decimal places and compare to the value from a calculator. (c) Why is it inappropriate to use that Taylor polynomial to approximate* ln 2.89*?*

Solution (a) Using the terms of the Taylor series of degree 6 or less gives

$$T_6(x) = (x-1)-\frac{1}{2}(x-1)^2+\frac{1}{3}(x-1)^3-\frac{1}{4}(x-1)^4+\frac{1}{5}(x-1)^5-\frac{1}{6}(x-1)^6.$$

(b) The value from a calculator is $\ln 1.277 = 0.244514$. The value of the Taylor polynomial is

$$T_6(1.277) = (1.277-1)-\frac{1}{2}(1.277-1)^2+\cdots-\frac{1}{6}(1.277-1)^6 = 0.244499.$$

The two values match to four decimal places (rounded) but not beyond that.

(c) The number 2.89 is outside the interval of convergence for the Taylor series, which is $(0, 2]$. ■

Limits using series

Taylor series are useful for approximating real values of functions. They are also useful for approximations in the hyperreals. For instance, since

$$\sin x = x - \frac{1}{3!}x^3 + \frac{1}{5!}x^5 - \frac{1}{7!}x^7 + \frac{1}{9!}x^9 - \cdots$$

for every real number x, the same must be true for any infinitesimal α:

$$\sin \alpha = \alpha - \frac{1}{3!}\alpha^3 + \frac{1}{5!}\alpha^5 - \frac{1}{7!}\alpha^7 + \frac{1}{9!}\alpha^9 - \cdots.$$

This series leads to an entire collection of approximation formulas:

The approximation formula in line 1 is not new. The approximation formulas in lines 2 and 3 may be used in lieu of the formula in line 1 whenever it is convenient to do so.

$$\sin\alpha \approx \alpha,$$

$$\sin\alpha \approx \alpha - \frac{1}{6}\alpha^3,$$

$$\sin\alpha \approx \alpha - \frac{1}{6}\alpha^3 + \frac{1}{120}\alpha^5,$$

and more, keeping as many terms of the series as we wish and discarding lower-level terms. These formulas are useful for avoiding violations of the approximation principle.

Example 7 *Evaluate* $\lim_{x\to 0} \dfrac{\sin x - x}{x^3}$.

Solution Using $x = 0+\alpha = \alpha$ for an arbitrary infinitesimal α, we have

$$\lim_{x\to 0} \frac{\sin x - x}{x^3} = \frac{\sin\alpha - \alpha}{\alpha^3}.$$

If we use the approximation formula $\sin\alpha \approx \alpha$, as we have often done before, we get

$$\approx \frac{\alpha - \alpha}{\alpha^3} = \frac{0}{\alpha^3},$$

which is a violation of the approximation principle. Because the form of the expression is $\frac{0}{0}$, we could use l'Hospital's rule, but we could also use a different approximation from the list preceding this example—namely, $\sin\alpha \approx \alpha - \frac{1}{6}\alpha^3$:

It takes three applications of l'Hospital's rule to evaluate this limit.

$$\lim_{x\to 0} \frac{\sin x - x}{x^3} = \frac{\sin\alpha - \alpha}{\alpha^3}$$

$$\approx \frac{\alpha - \frac{1}{6}\alpha^3 - \alpha}{\alpha^3} = \frac{-\frac{1}{6}\alpha^3}{\alpha^3} = -\frac{1}{6}. \blacksquare$$

Example 7 illustrates one of many ways in which knowing a Maclaurin series for a transcendental function is helpful. A list of particularly helpful Maclaurin series is in table 4.

Example 8 *Use the Maclaurin series for* $\ln(1 + x)$ *to write three approximation formulas for* $\ln(1 + \alpha)$.

Solution With $x = \alpha$, the Maclaurin series for $\ln(1 + x)$ is

$$\ln(1+\alpha) = \alpha - \frac{\alpha^2}{2} + \frac{\alpha^3}{3} - \frac{\alpha^4}{4} + \cdots.$$

By discarding lower-level terms, three approximation formulas are

$$\ln(1+\alpha) \approx \alpha,$$

$$\ln(1+\alpha) \approx \alpha - \frac{\alpha^2}{2}, \text{ and}$$

$$\ln(1+\alpha) \approx \alpha - \frac{\alpha^2}{2} + \frac{\alpha^3}{3}. \qquad \blacksquare$$

Any of the approximation formulas in example 8 can be used as needed.

Additional applications of Taylor and Maclaurin series are explored in the exercises. Next, let's turn our attention to more shortcuts for finding Taylor and Maclaurin series.

Table 4 *Table of Maclaurin series*

Series	Radius
$\frac{1}{1-x} = \sum_{n=0}^{\infty} x^n = 1 + x + x^2 + x^3 + \cdots$	$R = 1$
$e^x = \sum_{n=0}^{\infty} \frac{x^n}{n!} = 1 + x + \frac{1}{2!}x^2 + \frac{1}{3!}x^3 + \cdots$	$R = \infty$
$\sin x = \sum_{k=0}^{\infty} (-1)^k \frac{1}{(2k+1)!} x^{2k+1} = x - \frac{1}{3!}x^3 + \frac{1}{5!}x^5 - \frac{1}{7!}x^7 + \cdots$	$R = \infty$
$\cos x = \sum_{k=0}^{\infty} (-1)^k \frac{1}{(2k)!} x^{2k} = 1 - \frac{1}{2!}x^2 + \frac{1}{4!}x^4 - \frac{1}{6!}x^6 + \cdots$	$R = \infty$
$\sin^{-1} x = \sum_{k=0}^{\infty} \frac{(2k)!}{4^k (k!)^2 (2k+1)} x^{2k+1} = x + \frac{1}{6}x^3 + \frac{3}{40}x^5 + \cdots$	$R = 1$
$\cos^{-1} x = \frac{\pi}{2} - \sin^{-1} x = \frac{\pi}{2} - x - \frac{1}{6}x^3 - \frac{3}{40}x^5 - \cdots$	$R = 1$
$\tan^{-1} x = \sum_{k=0}^{\infty} (-1)^k \frac{1}{(2k+1)} x^{2k+1} = x - \frac{1}{3}x^3 + \frac{1}{5}x^5 - \frac{1}{7}x^7 + \cdots$	$R = 1$
$\ln(1+x) = \sum_{n=1}^{\infty} (-1)^{n+1} \frac{x^n}{n} = x - \frac{1}{2}x^2 + \frac{1}{3}x^3 - \frac{1}{4}x^4 + \cdots$	$R = 1$
$\sinh x = \sum_{k=0}^{\infty} \frac{1}{(2k+1)!} x^{2k+1} = x + \frac{1}{3!}x^3 + \frac{1}{5!}x^5 + \frac{1}{7!}x^7 + \cdots$	$R = \infty$
$\cosh x = \sum_{k=0}^{\infty} \frac{1}{(2k)!} x^{2k} = 1 + \frac{1}{2!}x^2 + \frac{1}{4!}x^4 + \frac{1}{6!}x^6 + \cdots$	$R = \infty$
$(1+x)^p = 1 + px + \frac{p(p-1)}{2!}x^2 + \frac{p(p-1)(p-2)}{3!}x^3 + \cdots$	$R = 1$

Derivatives and integrals of power series

The derivative sum and difference rules state that we can take the derivative of a sum or difference term-by-term. For instance,

$$\begin{aligned}\frac{d}{dx}\left(x - \frac{1}{3!}x^3 + \frac{1}{5!}x^5\right) &= \frac{d}{dx}(x) - \frac{d}{dx}\left(\frac{1}{3!}x^3\right) + \frac{d}{dx}\left(\frac{1}{5!}x^5\right)\\ &= 1 - \frac{1}{3!}\cdot 3x^2 + \frac{1}{5!}\cdot 5x^4\\ &= 1 - \frac{1}{2!}x^2 + \frac{1}{4!}x^4.\end{aligned}$$

Line 3 detail: $\frac{1}{3!}\cdot 3x^2 = \frac{1}{\cancel{3}\cdot 2!}\cdot \cancel{3}x^2$.

These derivative rules were proved for adding or subtracting two terms; by repeated use, they apply for finitely many terms. Do they also apply for adding or subtracting infinitely many terms, as in a power series? The answer is yes, but a proof is reserved for a more advanced course.

Theorem 10 DERIVATIVES AND INTEGRALS OF POWER SERIES *If a power series representation of a function $f(x) = \sum_{n=0}^{\infty} c_n(x-a)^n$ has radius of convergence $R > 0$, then*

$$f'(x) = \sum_{n=0}^{\infty} c_n \cdot n(x-a)^{n-1}$$

with radius of convergence R, and

$$\int f(x)\,dx = \sum_{n=0}^{\infty} c_n \cdot \frac{(x-a)^{n+1}}{n+1} + C$$

with radius of convergence R.

Specifying that the radius of convergence is the same does not imply that the interval of convergence is the same, because what happens at the endpoints is not stated.

One way to summarize theorem 10 is that we can differentiate and integrate term-by-term.

The Maclaurin series for $f(x) = \cos x$ can be determined in the same manner as example 2, but term-by-term differentiation gives us a quicker way.

Example 9 *Derive the Maclaurin series for $f(x) = \cos x$ by differentiating the series for* $\sin x$.

Solution Since $\cos x = \frac{d}{dx} \sin x$, differentiating the Maclaurin series for $\sin x$ gives

Because the instructions are to derive the series, merely copying the answer from table 4 is insufficient.

$$\begin{aligned}\cos x &= \frac{d}{dx}\left(x - \frac{1}{3!}x^3 + \frac{1}{5!}x^5 - \frac{1}{7!}x^7 + \cdots\right)\\ &= 1 - \frac{1}{3!} \cdot 3x^2 + \frac{1}{5!}5x^4 - \frac{1}{7!}7x^6 + \cdots\\ &= 1 - \frac{1}{2!}x^2 + \frac{1}{4!}x^4 - \frac{1}{6!}x^6 + \cdots\\ &= \sum_{k=0}^{\infty}(-1)^k \frac{1}{(2k)!}x^{2k}.\end{aligned}$$

In line 2, the derivative is carried out term-by-term. In line 4, note that $2k$ is always even, the factorial matches the exponent, and the factor $(-1)^k$ alternates for the even-exponent terms.

■

An alternate solution to example 9 is to differentiate the Maclaurin series for $\sin x$ in summation notation rather than in expanded form:

$$\begin{aligned}\cos x &= \frac{d}{dx}\left(\sum_{k-0}^{\infty}(-1)^k \frac{1}{(2k+1)!}x^{2k+1}\right)\\ &= \sum_{k=0}^{\infty}\frac{d}{dx}(-1)^k \frac{1}{(2k+1)!}x^{2k+1}\\ &= \sum_{k=0}^{\infty}(-1)^k \frac{1}{(2k+1)!} \cdot (2k+1)x^{2k}\\ &= \sum_{k=0}^{\infty}(-1)^k \frac{1}{(2k)!}x^{2k}.\end{aligned}$$

Line 2 is an application of theorem 10; For line 3, recall that both $(-1)^k$ and $\frac{1}{(2k+1)!}$ are constants with respect to the variable x.

Multiplication and division of power series

Yet another shortcut that is not proved here is that the terms of two power series can be multiplied in the same manner that polynomials can be multiplied. Finding a general pattern for the nth term may be difficult, but finding the first few terms of the series only requires algebra.

Example 10 *Find the first four nonzero terms in the Maclaurin series for* $e^x \sin x$.

Solution Rather than following the procedure of examples 1 and 2, let's take the Maclaurin series from these examples,

$$e^x = 1 + x + \frac{1}{2}x^2 + \frac{1}{6}x^3 + \frac{1}{24}x^4 + \cdots$$

and

$$\sin x = x - \frac{1}{6}x^3 + \frac{1}{120}x^5 - \cdots,$$

and multiply them:

In lines 3–6, each term of the series in line 1 is multiplied by the series of line 2. In lines 7–9, the multiplication from lines 3–5 is carried out, making sure that every term with degree 5 or less is computed. Line 10 collects like terms, one of which is zero.

$$\begin{aligned}(\sin x)(e^x) &= \left(x - \frac{1}{6}x^3 + \frac{1}{120}x^5 - \cdots\right)\cdot \\ &\quad \left(1 + x + \frac{1}{2}x^2 + \frac{1}{6}x^3 + \frac{1}{24}x^4 + \cdots\right) \\ &= x\left(1 + x + \frac{1}{2}x^2 + \frac{1}{6}x^3 + \frac{1}{24}x^4 + \cdots\right) \\ &\quad - \frac{1}{6}x^3\left(1 + x + \frac{1}{2}x^2 + \frac{1}{6}x^3 + \frac{1}{24}x^4 + \cdots\right) \\ &\quad + \frac{1}{120}x^5\left(1 + x + \frac{1}{2}x^2 + \frac{1}{6}x^3 + \frac{1}{24}x^4 + \cdots\right) \\ &\quad - \cdots \\ &= x + x^2 + \frac{1}{2}x^3 + \frac{1}{6}x^4 + \frac{1}{24}x^5 + \cdots \\ &\quad - \frac{1}{6}x^3 - \frac{1}{6}x^4 - \frac{1}{12}x^5 + \cdots \\ &\quad + \frac{1}{120}x^5 + \cdots \\ &= x + x^2 + \frac{2}{6}x^3 + 0x^4 - \frac{4}{120}x^5 + \cdots.\end{aligned}$$

We conclude that the Maclaurin series is

$$e^x \sin x = x + x^2 + \frac{1}{3}x^3 - \frac{1}{30}x^5 + \cdots. \quad \blacksquare$$

The first four nonzero terms of the Maclaurin series are listed, followed by ellipses.

Long division of power series also works, but it must be set up and carried out in the opposite of the usual order, from lower-degree terms to higher-degree terms.

Taylor's inequality

If a function has a power series representation centered at a, then theorem 9 says what the coefficients must be. But, how do we know that a function has a power series representation in the first place? And how closely do the Taylor polynomials approximate the function?

The two questions are related. The difference between the function $f(x)$ and a Taylor polynomial $T_n(x)$ is

$$R_n(x) = f(x) - T_n(x),$$

the *remainder* of the Taylor series. If the Taylor polynomials approximate the function more closely as n increases, then the remainders must be getting smaller. As long as

$$\lim_{n\to\infty} R_n(x) = 0$$

for each x in the interval of convergence, then

$$\lim_{n\to\infty} (f(x) - T_n(x)) = 0$$

and

$$f(x) = \lim_{n\to\infty} T_n(x).$$

The power series therefore converges to the value of the function, as desired. Thus, we only need to know how small the remainders are. One way of expressing this information is through Taylor's theorem, a higher-order analogy of the mean value theorem. For our purposes, however, a simpler expression is sufficient.

For a particular value of x, the value of $T_n(x)$ is the nth partial sum of the series. If $f(x) = \lim_{n\to\infty} T_n(x)$, then the partial sums are converging to the function value.

Theorem 11 TAYLOR'S INEQUALITY *If* $\left|f^{(n+1)}(x)\right| \le M$ *for* $|x-a| \le r$, *then the remainder* $R_n(x)$ *of the Taylor series satisfies*

$$|R_n(x)| \le \frac{M}{(n+1)!}|x-a|^{n+1}$$

for $|x-a| \le r$.

Proof. The proof relies on the fundamental theorem of calculus and the definite integral inequality (c) from section 4.4. The larger the value of n, the more times these theorems are used. Otherwise, each case is similar to the following. Let $n = 1$ and suppose $|f''(x)| \le M$ for $|x-a| \le r$. For any x such that $a \le x \le a+r$,

$$\int_a^x f''(t)\,dt = f'(x) - f'(a)$$

by the fundamental theorem, and

$$\int_a^x (-M)\,dt \le \int_a^x f''(t)\,dt \le \int_a^x M\,dt$$

$$-M(x-a) \le \int_a^x f''(t)\,dt \le M(x-a)$$

by the definite integral inequality (c) and the definite integral of a constant. Therefore,

$$-M(x-a) \le f'(x) - f'(a) \le M(x-a)$$

$$f'(a) - M(x-a) \le f'(x) \le f'(a) + M(x-a)$$

for any x with $a \le x \le a + r$, and we have turned a bound on $f''(x)$ (namely, $|f''(x)| \le M$) into a bound on $f'(x)$ (namely, $f'(a) - M(x-a) \le f'(x) \le f'(a) + M(x-a)$). Now we repeat the process. For any x such that $a \le x \le a + r$,

$$\int_a^x f'(t)\,dt = f(x) - f(a)$$

by the fundamental theorem, and

$$\int_a^x \left(f'(a) - M(t-a)\right)\,dt \le \int_a^x f'(t)\,dt \le \int_a^x \left(f'(a) + M(t-a)\right)\,dt$$

$$\left(f'(a)t - M\frac{(t-a)^2}{2}\right)\Bigg|_a^x \le \int_a^x f'(t)\,dt \le \left(f'(a)t + M\frac{(t-a)^2}{2}\right)\Bigg|_a^x$$

$$f'(a)(x-a) - M\frac{(x-a)^2}{2} \le \int_a^x f'(t)\,dt \le f'(a)(x-a) + M\frac{(x-a)^2}{2}$$

by the definite integral inequality (c) and the fundamental theorem of calculus. Thus,

$$f'(a)(x-a) - M\frac{(x-a)^2}{2} \le f(x) - f(a) \le f'(a)(x-a) + M\frac{(x-a)^2}{2}$$

$$f(a) + f'(a)(x-a) - M\frac{(x-a)^2}{2} \le f(x) \le f(a) + f'(a)(x-a) + M\frac{(x-a)^2}{2}$$

$$T_1(x) - M\frac{(x-a)^2}{2} \le f(x) \le T_1(x) + M\frac{(x-a)^2}{2}$$

$$-M\frac{(x-a)^2}{2} \le R_1(x) \le M\frac{(x-a)^2}{2}$$

$$|R_1(x)| \le M\frac{(x-a)^2}{2},$$

which is the desired conclusion for $n = 1$ and $x > a$. As stated previously, the remaining cases are similar. ■

Earlier in this section we saw that for $f(x) = \sin x$, the value of $T_7(0.7)$ (Taylor polynomial of order 7 centered at zero) agreed with the value of $\sin 0.7$ to six decimal places. Taylor's inequality can be used to give a bound on the error in the approximation ahead of time. We can know how close we must be to the actual value without knowing what the actual value is.

Example 11 *For* $f(x) = \sin x$, *use Taylor's inequality to determine a bound for* $|R_7(0.7)|$, *the difference between* $\sin(0.7)$ *and* $T_7(0.7)$.

Solution The derivatives of $\sin x$ cycle through $\sin x$, $\cos x$, $-\sin x$, and $-\cos x$. Therefore,

$$\left|f^{(n+1)}(x)\right| \leq 1 = M$$

for all x and for any n, meaning that we can use $M = 1$ in Taylor's inequality. With $a = 0$, $n = 7$, and $x = 0.7$, we have

$$|R_7(0.7)| \leq \frac{M}{(n+1)!}|x-a|^{n+1} = \frac{1}{8!}|0.7|^8 = 0.000\,001\,430.$$ ■

We know in advance that we have close to a six-decimal place accuracy. The use of error bounds is how a calculator or CAS is sure that the value it gives is accurate to the number of decimal places shown.

Exercises 10.10

1–6. Rapid response: is the expression a Maclaurin series? If so, state the coefficient c_n.

1. $x + x^2 + x^3 + x^4 + \cdots$
2. $\sum_{n=0}^{\infty} \frac{n}{2^n} x^n$
3. $\sum_{n=0}^{\infty} 2^n n^x$
4. $\sum_{n=0}^{\infty} n \left(\frac{x}{3}\right)^n$
5. $1 + 2(x-2) + 3(x-2)^2 + 4(x-2)^3 + \cdots$
6. $\sum_{n=0}^{\infty} \frac{(x+5)^n}{n!}$

7–10. Rapid response: is the expression a Taylor series? If so, state the center.

7. $1 + 2(x-2) + 3(x-2)^2 + 4(x-2)^3 + \cdots$

8. $\sum_{n=0}^{\infty} \frac{n}{2^n} x^n$

9. $\sum_{n=0}^{\infty} \frac{(x+5)^n}{n!}$

10. $\sum_{n=0}^{\infty} \frac{(x-n)^n}{n!}$

11–16. (a) For the given function, verify the power series formula in the table of Maclaurin series (table 4) using the method of examples 1 and 2. (b) Verify the radius of convergence using the ratio test.

11. $f(x) = \sinh x$

12. $f(x) = \cosh x$

13. $f(x) = \frac{1}{1-x}$

14. $f(x) = \ln(1+x)$

15. $f(x) = (1+x)^p$

16. $f(x) = \tan^{-1} x$

17–20. Find the Maclaurin series for the function by differentiating a Maclaurin series from table 4, as in example 9.

17. $f(x) = \cosh x$

18. $f(x) = \sinh x$

19. $f(x) = \frac{1}{1+x}$

20. $f(x) = \frac{1}{1+x^2}$

21–24. Use the method of example 10 to find the first four nonzero terms of the Maclaurin series.

21. $f(x) = e^x \cos x$

22. $f(x) = e^x \sin^{-1} x$

23. $f(x) = \sin x \cos x$

24. $f(x) = \sinh x \cosh x$

25–32. Find the Maclaurin series for the function using any appropriate method.

25. $f(x) = x^3 \tan^{-1} x$

26. $f(x) = \cosh 2x$

27. $f(x) = \sin x^2$

28. $f(x) = 3x^2 \cosh x$

29. $f(x) = e^{x+5}$

30. $f(x) = \cosh(x-2)$

31. $f(x) = \sin(x+5)$

32. $f(x) = e^x \cosh x$

33–40. (a) Find the Taylor series, (b) find its radius of convergence, and (c) find its interval of convergence.

33. $f(x) = \frac{1}{x}$, center 2

34. $f(x) = \cos x$, center $-\pi$

35. $f(x) = \frac{1}{x}$, center 3

36. $f(x) = \dfrac{1}{x^2}$, center -3

37. $f(x) = \sin x$, center $\frac{\pi}{2}$

38. $f(x) = \sqrt{x}$, center 1

39. $f(x) = \ln x$, center 10

40. $f(x) = \sqrt{x}$, center 9

41–46. Determine the Taylor polynomial.

41. $f(x) = \ln(1 + x)$, center 0, order 4

42. $f(x) = \tan^{-1} x$, center 0, order 5

43. $f(x) = \frac{1}{x}$, center 2, order 3

44. $f(x) = \sqrt{x}$, center 1, order 4

45. $f(x) = \sin x$, center $\frac{\pi}{2}$, order 5

46. $f(x) = \cos x$, center $-\pi$, order 7

The Taylor series for each of these functions is either given in the table of Maclaurin series (table 4) or is requested in a previous exercise.

Clarification: exercise 41 requests the Taylor polynomial of order 4 for $f(x) = \ln(1 + x)$ centered at zero.

47–54. Use a Maclaurin series to write three approximation formulas for the given quantity, assuming α is an infinitesimal.

47. $\cos\alpha$

48. $\sinh\alpha$

49. e^{α}

50. $\sqrt[3]{1+\alpha}$

51. $\sqrt{1+\alpha}$

52. $\cos^{-1}\alpha$

53. $\sin^{-1}\alpha$

54. $\tan^{-1}\alpha$

55–70. Evaluate the limit using approximation formulas.

Do not use l'Hospital's rule for these exercises.

55. $\displaystyle\lim_{x\to 0} \frac{1-\cos x}{4x^2}$

56. $\displaystyle\lim_{x\to 0} \frac{1-\cosh x}{x}$

57. $\displaystyle\lim_{x\to 0} \frac{x^2+1-\cosh x}{5x^2}$

58. $\displaystyle\lim_{x\to 0} \frac{e^x-1}{x\cos x}$

59. $\displaystyle\lim_{x\to 0} \frac{\frac{1}{2}x^2+1-\cosh x}{5x^4}$

60. $\displaystyle\lim_{x\to 0} \frac{\tan^{-1} x}{3x}$

61. $\displaystyle\lim_{x\to 0} \frac{e^x-1}{\sin x}$

62. $\displaystyle\lim_{x\to 0} \frac{e^x+\sin x-1}{\ln(1+x)}$

63. $\displaystyle\lim_{x\to 0^+} \frac{\sin^{-1} x}{x^2}$

64. $\displaystyle\lim_{x\to 0} \left(\frac{1}{\ln(1+x)} - \frac{1}{x}\right)$

65. $\displaystyle\lim_{x\to 1} \left(\frac{1}{\ln x} - \frac{x}{\ln x}\right)$

66. $\displaystyle\lim_{x\to 0} \frac{\sqrt[3]{1+x}-1}{x}$

67. $\lim_{x\to 0} \frac{\sqrt{1+x}-1}{x}$

68. $\lim_{x\to\infty} 3x\sin^{-1}\frac{1}{x}$

69. $\lim_{x\to\infty} \left(x\ln\left(1+\frac{1}{x}\right)-1\right)$

70. $\lim_{x\to 0} \frac{x-\frac{1}{3}x^3-\tan^{-1}x}{x^5}$

71. (a) Determine the Taylor polynomial of order 4, center 0 for $f(x) = e^x$. (b) Use that Taylor polynomial to approximate $e^{0.2}$ to six decimal places. (c) Use a calculator to determine the value of $e^{0.2}$ to six decimal places. (d) Compare the results of parts (b) and (c).

72. (a) Determine the Taylor polynomial of order 7, center 0 for $f(x) = \sinh x$. (b) Use that Taylor polynomial to approximate $\sinh 0.3$ to 10 decimal places. (c) Use a calculator to determine the value of $\sinh 0.3$ to 10 decimal places. (d) Compare the results of parts (b) and (c).

73. (a) Determine the Taylor polynomial of order 4, center 0 for $f(x) = e^x$. (b) Use that Taylor polynomial to approximate $e^{3.6}$ to six significant digits. (c) Use a calculator to determine the value of $e^{3.6}$ to six significant digits. (d) Compare the results of parts (b) and (c). (e) Why is the comparison so different from calculating $e^{0.2}$?

74. (a) Determine the Taylor polynomial of order 7, center 0 for $f(x) = \sinh x$. (b) Use that Taylor polynomial to approximate $\sinh 2.7$ to three decimal places. (d) Compare the results of parts (b) and (c). (e) Why is the comparison so different from calculating $\sinh 0.3$?

75. (a) For $f(x) = \sin x$, use Taylor's inequality to determine a bound for $|R_5(0.1)|$. (b) To how many decimal places must $T_5(0.1)$ agree with $\sin 0.1$? (c) By hand, calculate the value of $\sin 0.1$ to the number of decimal places indicated in part (b).

76. (a) For $f(x) = e^x$, use Taylor's inequality to determine a bound for $|R_6(1)|$. (When determining the value of M, use the easily verifiable fact that $e^x < 3$ for $|x| \leq 1$.) (b) To how many decimal places must $T_6(1)$ agree with e^1? (c) By hand, calculate the value of e to the number of decimal places indicated in part (b).

77. Find the first seven nonzero terms of the Maclaurin series for $f(x) = (x^3 - 1)\sin x$.

78. Find the first six nonzero terms of the Maclaurin series for $(x-3)\ln(1+x)$.

79. Use the level comparison test to determine the convergence or divergence of $\sum_{n=1}^{\infty} \ln\left(1+\frac{1}{n}\right)$.

80. Use the method of examples 1 and 2 to find the Maclaurin series for $f(x) = x^3 + 2x - 7$. What can be said about the Maclaurin series for a polynomial?

81. (a) Write the omega sum for evaluating $\int_1^2 \frac{1}{x}\,dx$ using the definition of definite integral. (b) Match the expression $\frac{\omega}{1+k\omega}$ to the formula for the sum of a geometric series. Write the expression as a geometric series in expanded form. (c) Replace the summand from part (a) with that geometric series in expanded form. (d) Evaluate the omega sum. The result should be an infinite series. (e) Evaluate the infinite series of part (d). It may help to recognize it as the right-hand side of a formula in the table of Maclaurin series (table 4) for a particular value of x.

Exercises 81 and 82, and a variant of their solutions, was provided to me by Josiah Hays, a homeschooled high school student dual-enrolled in first-semester calculus.

82. (a) Repeat exercise 81 for $\int_1^{1.1} \frac{1}{x}\,dx$. (b) Does the same procedure work for $\int_1^{2.1} \frac{1}{x}\,dx$? If not, why not?

83. Use the definition of definite integral and the Maclaurin series for $\sin x$ to evaluate $\int_0^{\pi} \sin x\,dx$.

84. Use the definition of definite integral and the Maclaurin series for $\cos x$ to evaluate $\int_0^{\pi} \cos x\,dx$.

85. (a) Are $1 + \alpha$ and $1 + \alpha^2$ approximately equal? (b) Use the Maclaurin series for $\ln(1 + x)$ to approximate $\ln(1 + \alpha)$ and $\ln(1 + \alpha^2)$. Are these two quantities approximately equal? (c) Compare the results of parts (a) and (b) to chapter 5 theorem 2.

Index

Answers to Odd-numbered Exercises

Section 0.1

1 correct

3 wrong (Wrong order; the smaller number comes first.)

5 wrong (Wrong order or wrong symbol; $-\infty$ can go on the left and ∞ can only go on the right.)

7 correct

9 wrong (The bracket on the right is turned outward instead of inward.)

11 compound inequality

13 separate inequalities

15 compound inequality

17 (a) $(-\infty, \frac{13}{9})$, (b)

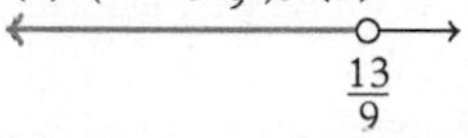

19 (a) $(-\infty, \frac{3}{2}]$, (b)

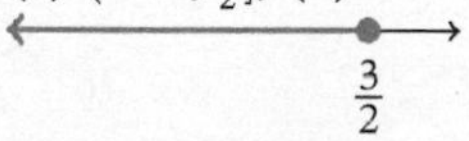

21 (a) $(-\frac{5}{7}, 1)$, (b)

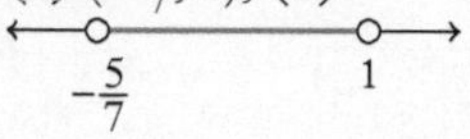

23 (a) $[\frac{9}{2}, \infty)$, (b)

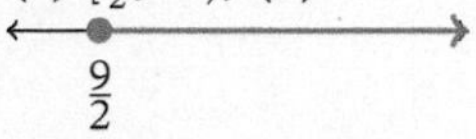

25 (a) $(-\infty, -\frac{11}{7})$, (b)

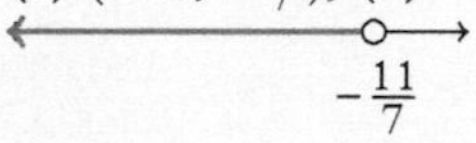

27 $x - |x|$ (0 is incorrect)

29 $12|x|$

31 $-\frac{1}{2} < x < 4$; alternately, $(-\frac{1}{2}, 4)$

33 $x < -\frac{7}{2}$ or $x > \frac{1}{2}$; alternately, $(-\infty, -\frac{7}{2}) \cup (\frac{1}{2}, \infty)$

35 $\frac{8}{27} \leq x \leq \frac{16}{27}$; alternately, $[\frac{8}{27}, \frac{16}{27}]$

37 no solutions (absolute values are not negative, so they cannot be less than -2)

39 $x \leq -\frac{1}{2}$ or $x \geq 2$; alternately, $(-\infty, -\frac{1}{2}] \cup [2, \infty)$

41 12

43 3

45 5

47 $|x - y|$; alternately, $|y - x|$

49 $|(x + 4) - (y - 3)| = |x - y + 7|$; alternately, $|(y - 3) - (x + 4)| = |y - x - 7|$

Section 0.2

1 slope 4, y-intercept -7

3 slope -91, y-intercept 225

5 slope 17, y-intercept 0

7 slope 5, y-intercept -1

9 slope $\frac{2}{3}$, y-intercept 12

11 slope 2, y-intercept 6

13

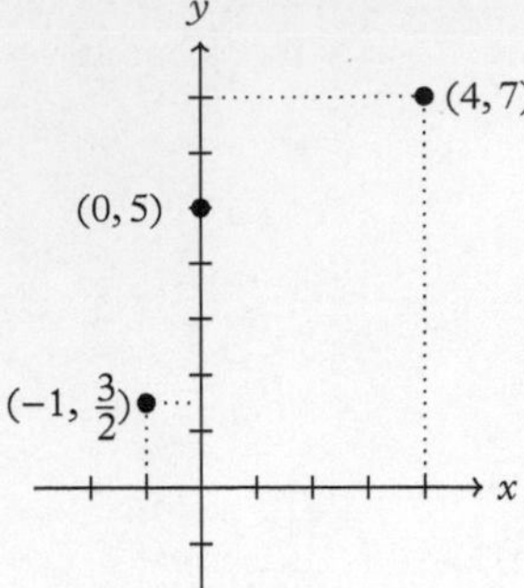

15

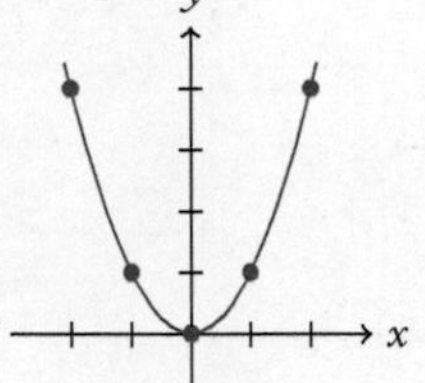

17

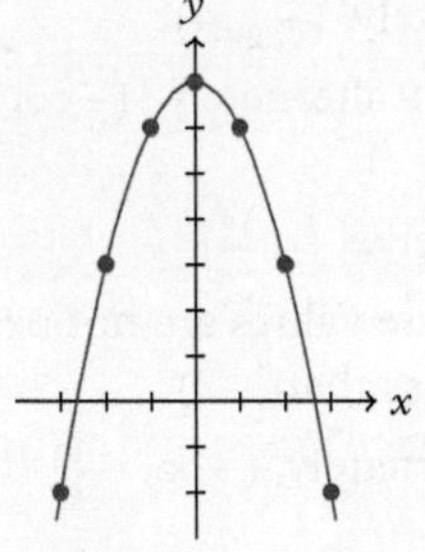

19

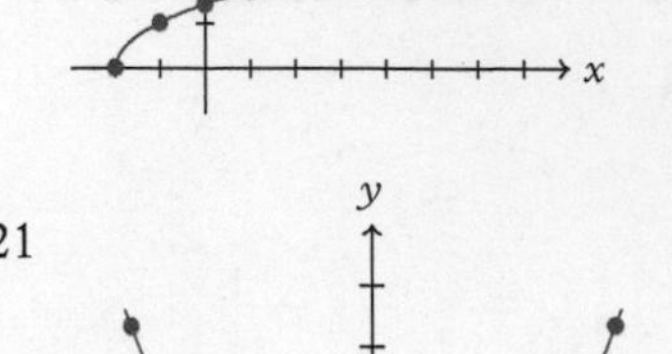

21

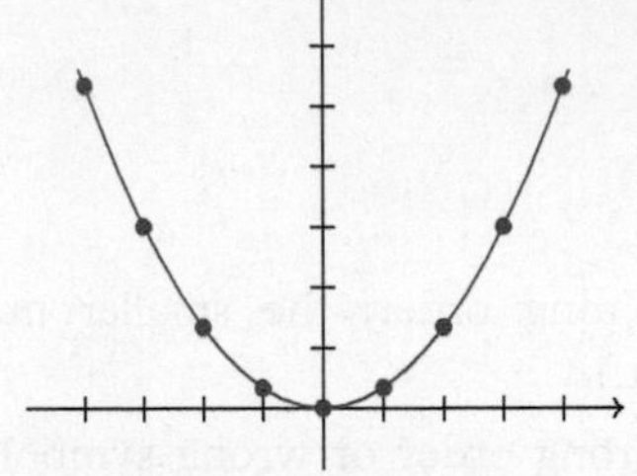

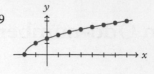

23

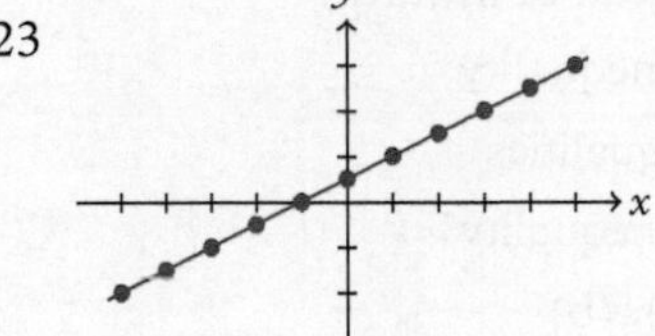

Answers 25–31 The dotted lines are to illustrate the required solution method, but they are not required in the answer itself.

25

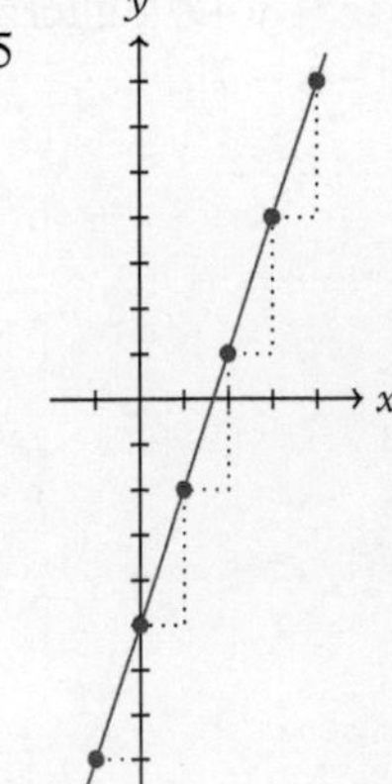

27

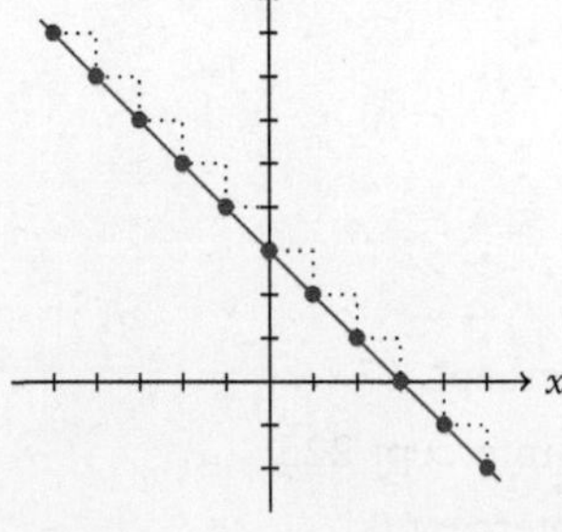

29 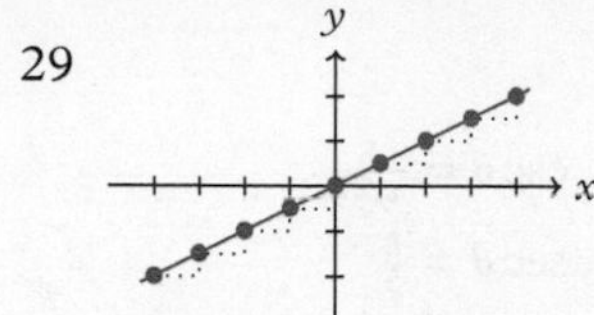

31 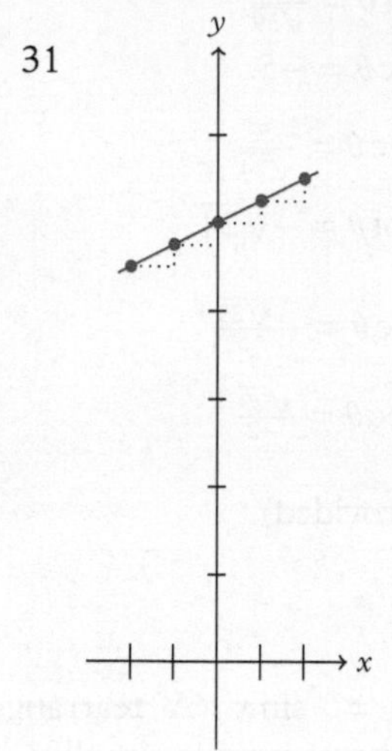

33 x-intercepts at $x = \pm 3$, y-intercept at $y = -9$

35 x-intercept at $x = -\frac{7}{5}$, y-intercept at $y = 7$

37 x-intercept at $x = -2$, y-intercept at $y = \sqrt{2}$

39 $\sqrt{8} = 2\sqrt{2}$

41 $\sqrt{41}$

43 5

45 $\sqrt{(-1-c)^2+4} = \sqrt{(1+c)^2+4} = \sqrt{5+2c+c^2}$

47 17

49 -1

51 $-\frac{4}{5}$

53 undefined

55 0

57 $\frac{9}{8}$

59 $y = 5x + 4$

61 $y = -2x + 7$

63 $y = -\frac{3}{2}x + 4$

65 $y = -3$

67 $x = 8$ (both $y = 8$ and $y = 12$ are incorrect)

69 $y = -2x + 7$

71 $y = 4x - 28$

73 $y = x - 2$

Section 0.3

1 $\frac{\pi}{2}$

3 $-\frac{\pi}{2}$

5 $-\frac{3\pi}{4}$

7 2π

9 $-\frac{\pi}{4}$

11 0

13 0

15 0

17 1

19 -1

21 $\frac{3\pi}{4}$

23 $-\frac{2\pi}{3}$

25 $\frac{7\pi}{6}$

27 210°

29 495°

31 $-300°$

33

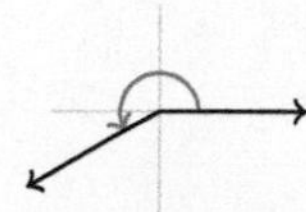

35

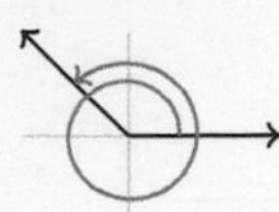

37

39

41 $\sin\theta = \frac{7}{\sqrt{58}}$ $\csc\theta = \frac{\sqrt{58}}{7}$
$\cos\theta = \frac{3}{\sqrt{58}}$ $\sec\theta = \frac{\sqrt{58}}{3}$
$\tan\theta = \frac{7}{3}$ $\cot\theta = \frac{3}{7}$

43 $\sin\theta = \frac{5}{\sqrt{34}}$ $\csc\theta = \frac{\sqrt{34}}{5}$
$\cos\theta = \frac{-3}{\sqrt{34}}$ $\sec\theta = \frac{\sqrt{34}}{-3}$
$\tan\theta = \frac{5}{-3}$ $\cot\theta = \frac{-3}{5}$

45 $\sin\theta = 0.8 = \frac{4}{5}$, $\csc\theta = 1.25 = \frac{5}{4}$, $\cos\theta = 0.6 = \frac{3}{5}$, $\sec\theta = \frac{1}{0.6} = \frac{5}{3} = 1.667$, $\tan\theta = \frac{0.8}{0.6} = \frac{4}{3} = 1.333$, $\cot\theta = 0.75 = \frac{3}{4}$

47 $\sin\theta = 1$, $\csc\theta = 1$, $\cos\theta = 0$, $\sec\theta =$ undefined, $\tan\theta =$ undefined, $\cot\theta = 0$

49 $\sin\theta = -\frac{5}{7}$ $\csc\theta = -\frac{7}{5}$
$\cos\theta = \frac{\sqrt{24}}{7}$ $\sec\theta = \frac{7}{\sqrt{24}}$
$\tan\theta = \frac{-5}{\sqrt{24}}$ $\cot\theta = \frac{\sqrt{24}}{-5}$

51 $\frac{\sqrt{3}}{2}$

53 $-\frac{1}{\sqrt{3}}$

55 $\sqrt{2}$

57 -1

59 -2

61 $\sin\theta = \frac{\sqrt{39}}{8}$ $\csc\theta = \frac{8}{\sqrt{39}}$
(provided) $\sec\theta = \frac{8}{5}$
$\tan\theta = \frac{\sqrt{39}}{5}$ $\cot\theta = \frac{5}{\sqrt{39}}$

63 (provided) $\csc\theta = -5$
$\cos\theta = \frac{\sqrt{24}}{5}$ $\sec\theta = \frac{5}{\sqrt{24}}$
$\tan\theta = \frac{-1}{\sqrt{24}}$ $\cot\theta = -\sqrt{24}$

65 $\sin\theta = \frac{-5}{\sqrt{29}}$ $\csc\theta = -\frac{\sqrt{29}}{5}$
$\cos\theta = \frac{-2}{\sqrt{29}}$ $\sec\theta = \frac{\sqrt{29}}{-2}$
$\tan\theta = \frac{5}{2}$ (provided)

67 $\frac{1}{\cos\theta}$

69 0

71 $\frac{1-\cos^2 x}{\sin x} = \frac{\sin^2 x}{\sin x} = \sin x$ (A rearranged Pythagorean identity was used to simplify.)

73 $\frac{1}{\sin\theta} + 1$

75 (a)

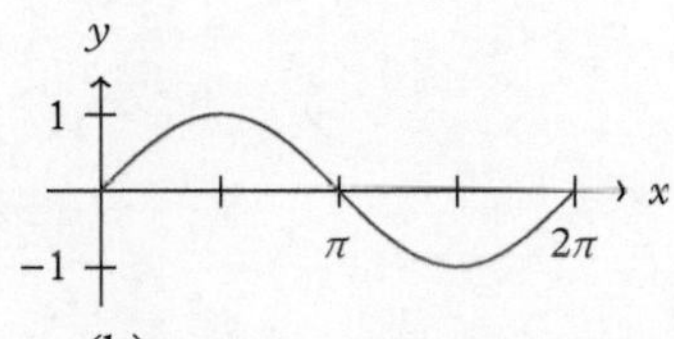

(b)

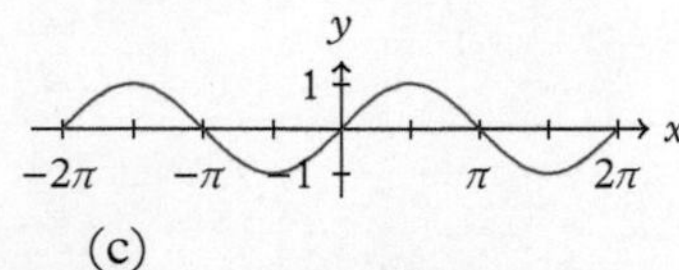

(c)

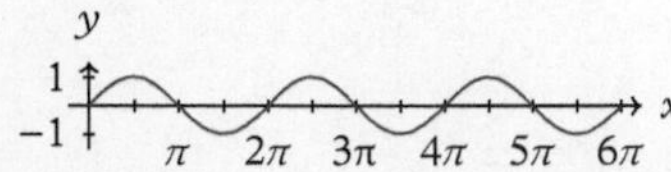

(d)

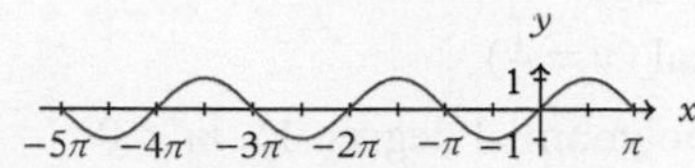

77 (a)

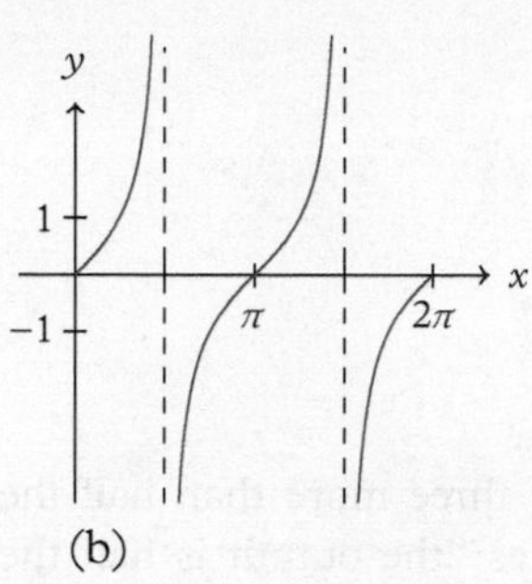

(b)

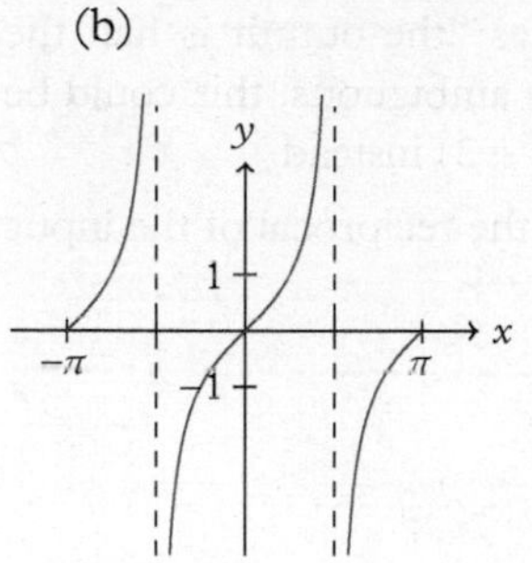

(c)

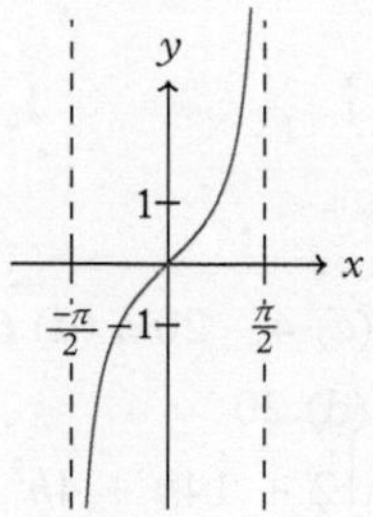

(d)

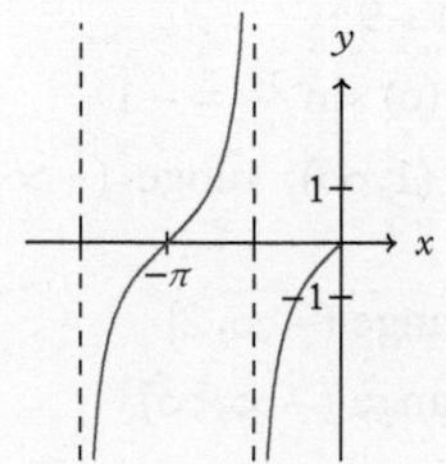

79 (a)

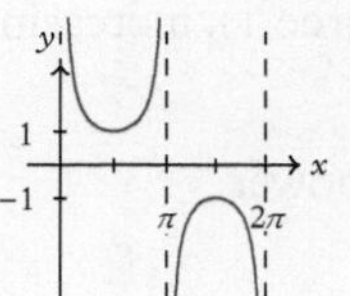

(b)

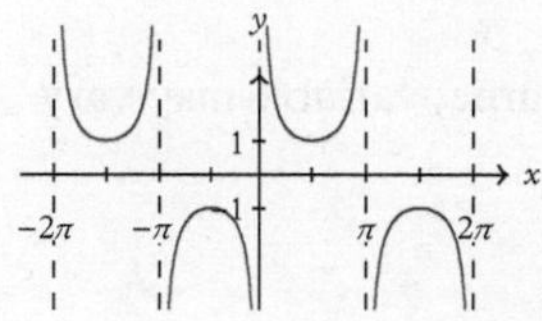

(c)

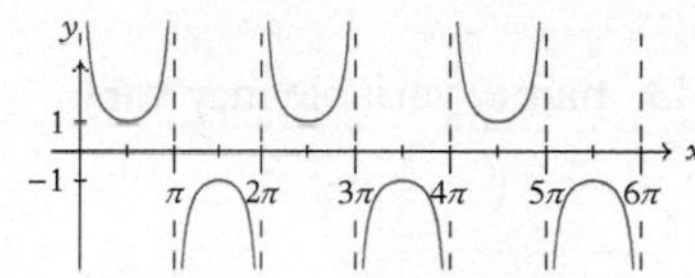

(d)

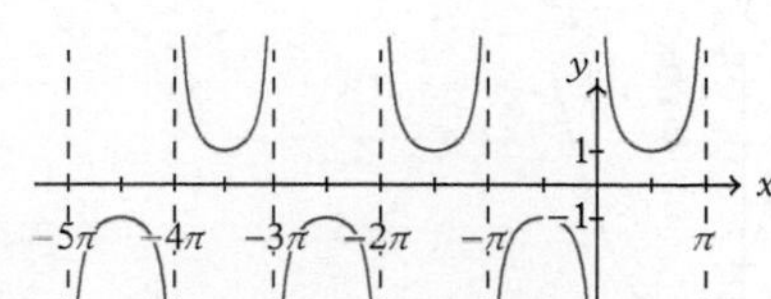

81 $x = k\pi$ for any integer k

83 no solutions

Section 0.4

1 linear (polynomial degree 1), increasing

3 root (power)

5 polynomial degree 7, power

7 reciprocal (power)

9 parabola (polynomial degree 2), opens up

11 root (even root)

13 parabola (polynomial degree 2), $a > 0$

15 reciprocal ($y = \frac{1}{x}$)

17 linear (polynomial degree 1), $m < 0$

19 no

21 yes

23 no

25 yes

Answers 27–31 name, variable may vary

27 $f(x) = 3x$

29 $f(x) = x - 5$

31 $f(x) = x + x^3$

33 the output is five times the input

35 the output is three more than half the input (Reordering as "the output is half the input plus three" is ambiguous; this could be interpreted as $\frac{1}{2}(x + 3)$ instead.)

37 the output is the reciprocal of the input

Answers 39–43 name, variable may vary

39 $f(x) = 2x$

41 $f(x) = x^3$

43 $f(x) = 2x^2$

45

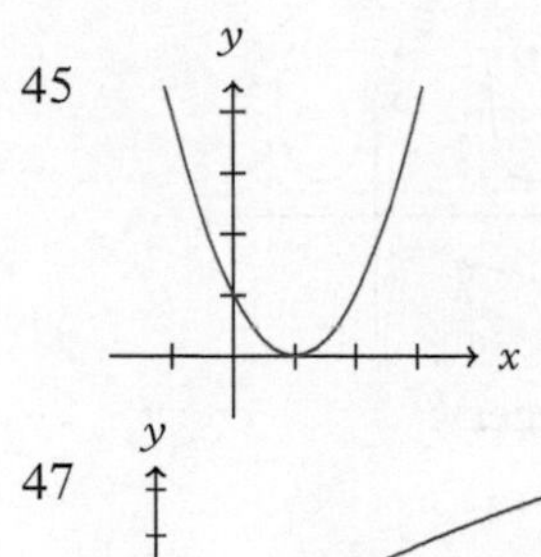

47

49

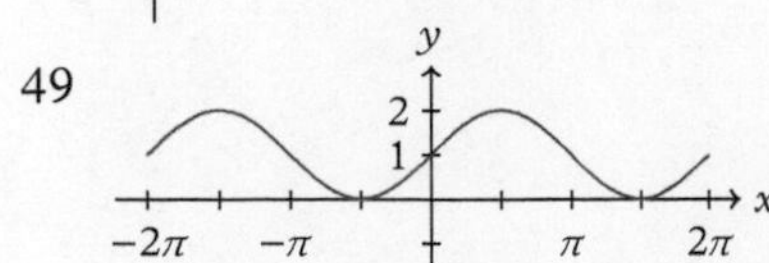

51

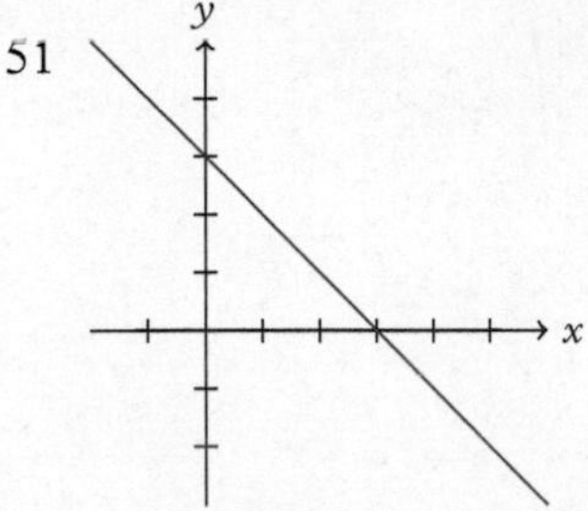

53 (a) −6, (b) $-11 + 5\theta$, (c) $4 - 20x$, (d) $6 - 5x$

55 (a) 17, (b) 17, (c) 17, (d) 20

57 (a) 30, (b) $y + y^2$, (c) $12 + 14h + 4h^2$, (d) $x + 7x^2$

59 (a) −1, (b) −25, (c) −9

61 (a) $\sin \pi = 0$, (b) 0, (c) $\sin \frac{3\pi}{2} = -1$

63 domain $(-\infty, 1) \cup (1, \infty)$, range $(-\infty, 0) \cup (0, \infty)$

65 domain $(-\infty, \infty)$, range $(-\infty, 2]$

67 domain $(-\infty, \infty)$, range $(-\infty, \infty)$

69 $(-\infty, 4]$

71 $(-\infty, 2) \cup (2, 4) \cup (4, \infty)$

73 $[-5, 2) \cup (2, \infty)$

75 $[-3, 3]$

Answers 77–83 if the tick marks on the axes go by anything other than by ones, the scale should be marked accordingly. It is not necessary to label the coordinates of the endpoints, but this is done here for your benefit.

77

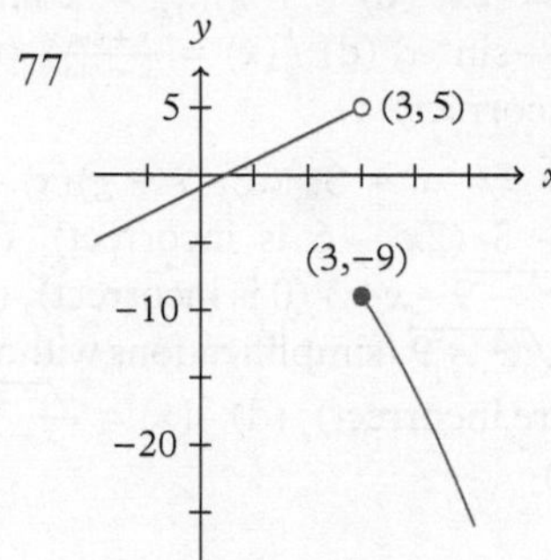

79

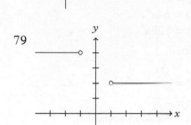

81

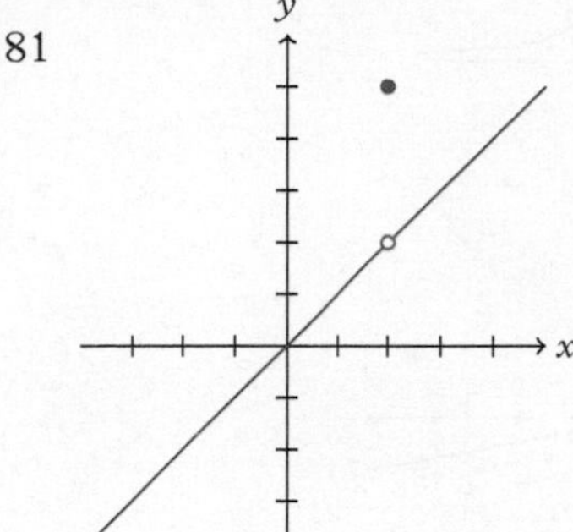

83

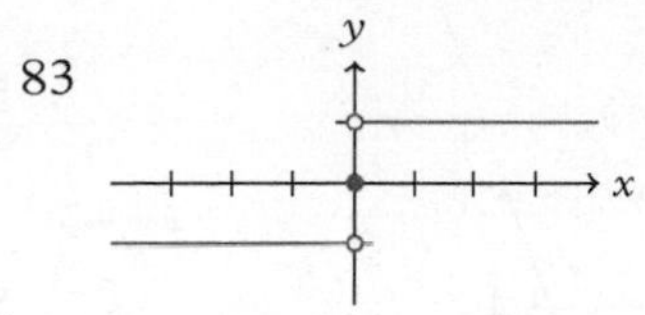

85 $f(x) = \begin{cases} 2, & x \le 0 \\ 1, & x > 0 \end{cases}$

Section 0.5

1 shift

3 shift

5 reflection, stretching

7 reflection, stretching

9 stretching

11 $y = \sqrt{x}$

13 $y = |x|$

15 $y = \sin x$

17 $y = x^2$

19 $y = \frac{1}{x^2}$

21 right

23 up

25 down

27 left

29 up

31

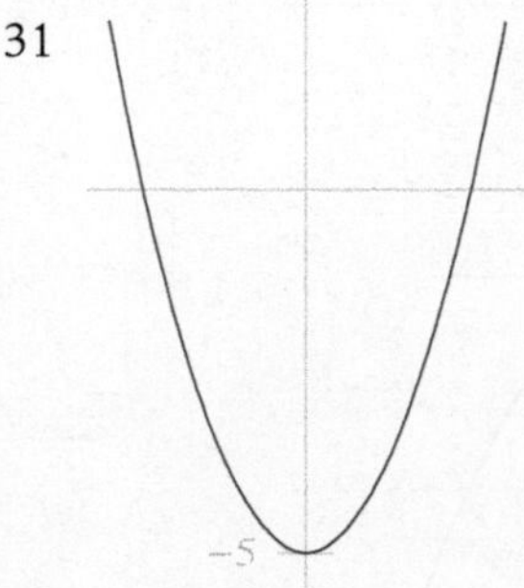

33 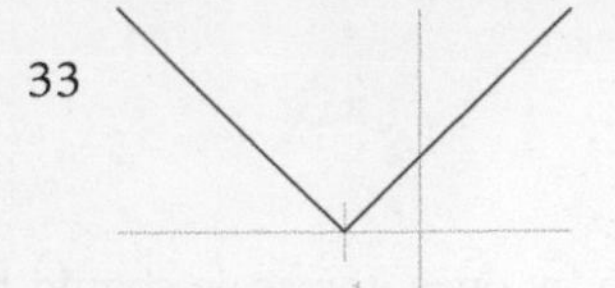

35 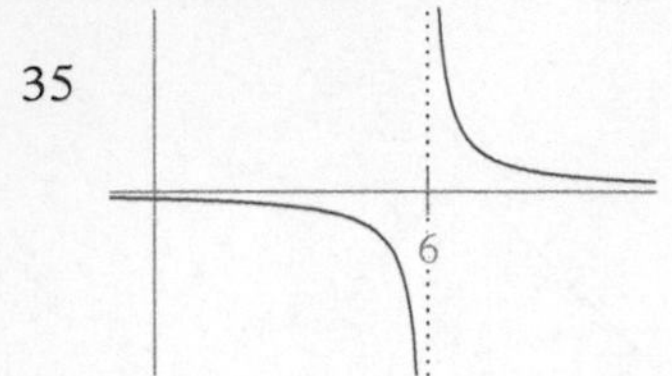

37 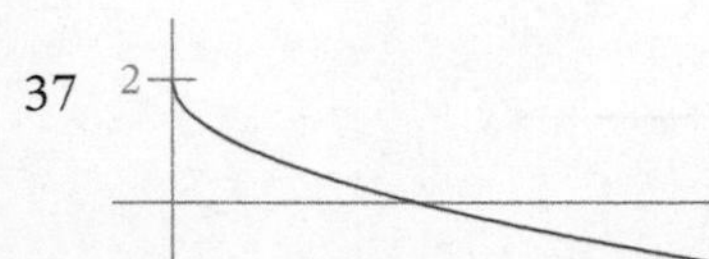

39 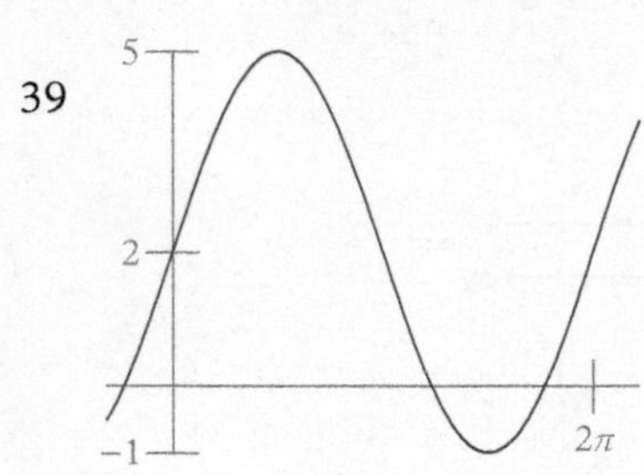

41 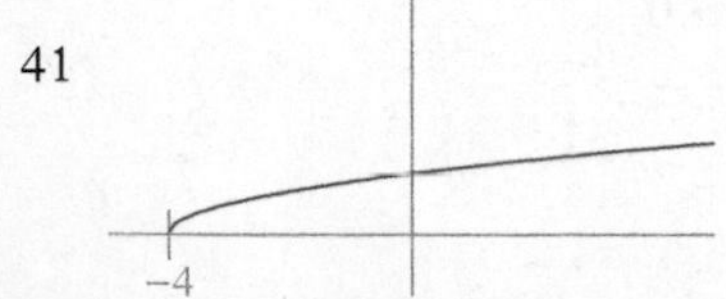

43 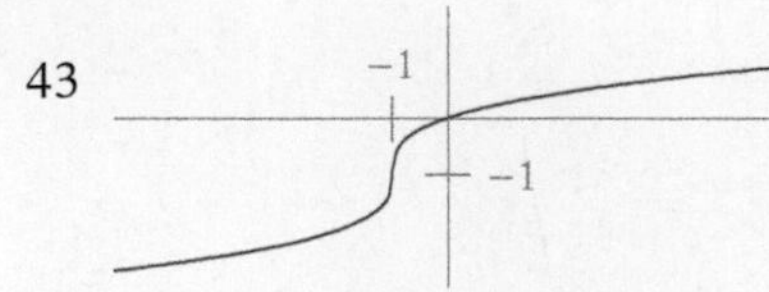

45 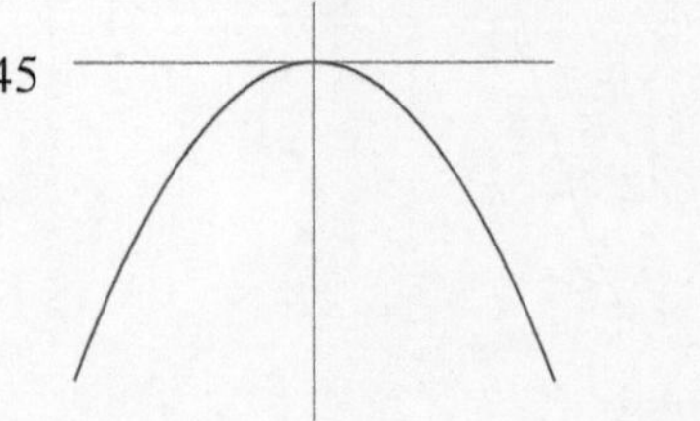

47 (a) $(f+g)(x) = 4x+3$, (b) $(f-g)(x) = 2x+7$, (c) $fg(x) = 3x^2 - x - 10$, (d) $\frac{f}{g}(x) = \frac{3x+5}{x-2}$

49 (a) $(f+g)(x) = \frac{1}{x+1} + x^2 + x$, (b) $(f-g)(x) = \frac{1}{x+1} - x^2 - x$, (c) $fg(x) = \frac{x^2+x}{x+1} = x \ (x \neq -1)$, (d) $\frac{f}{g}(x) = \frac{1}{(x+1)(x^2+x)} = \frac{1}{x(x+1)^2}$

51 (a) $(f+g)(x) = 2x$, (b) $(f-g)(x) = 2\sin x$, (c) $fg(x) = x^2 - \sin^2 x$, (d) $\frac{f}{g}(x) = \frac{x+\sin x}{x-\sin x}$ (no cancellation is correct)

53 Note: $\sqrt{x^2-9} \neq x-3$. (a) $(f+g)(x) = \sqrt{x^2-9} + x - 3$ ($2x-6$ is incorrect), (b) $(f-g)(x) = \sqrt{x^2-9} - x + 3$ (0 is incorrect), (c) $fg(x) = (x-3)\sqrt{x^2-9}$ (simplifications without a square root are incorrect), (d) $\frac{f}{g}(x) = \frac{\sqrt{x^2-9}}{x-3}$ (1 is incorrect)

55 (a) $g(f(0)) = 0$,
(b) $f \circ g(x) = 64x^6$,
(c) $g \circ f(x) = 128x^6$

57 (a) undefined,
(b) $f \circ g(x) = \frac{1}{x^2}$,
(c) $g \circ f(x) = \frac{1}{x^2}$

59 (a) $g(f(0)) = -7$,
(b) $f \circ g(x) = 5x - 35 + \sqrt{x-7}$,
(c) $g \circ f(x) = 5x + \sqrt{x} - 7$

61 (a) $g(f(0)) = 145$,
(b) $f \circ g(x) = |x^2+13| = x^2+13$,
(c) $g \circ f(x) = |x+12|^2 + 1 = (x+12)^2 + 1 = x^2 + 24x + 145$

63 (a) undefined,
(b) $f \circ g(x) = \frac{27}{x^4}$,
(c) $g \circ f(x) = \frac{3}{x^4}$

65 (a) $f(f(x)) = 9x+4$,
(b) $g \circ g(x) = x^4$,
(c) $g \circ f \circ g(x) = 9x^4 + 6x^2 + 1$

67 $y = f(g(x)) = f \circ g(x)$

69 $y = f(g(h(x))) = f \circ g \circ h(x)$

71 $y = g(f(g(x))) = g \circ f \circ g(x)$

73 $y = f(g(x))$, where $f(x) = x^4$, $g(x) = 11x+5$

75 $y = f(g(h(x)))$, where $f(x) = x^2$, $g(x) = \sin x$, $h(x) = x - 3$

77 $y = f(g(h(x)))$, where $f(x) = \frac{1}{x}$, $g(x) = \sqrt{x}$, $h(x) = 19x + 1$

Section 0.6 *for squaring a binomial*

1 wrong

3 correct

5 correct

7 wrong

9 correct

11 wrong

13 wrong

15 wrong

17 wrong

19 correct

21 wrong

23 wrong

25 correct

27 wrong

29 correct

31 $x^2 + 4x + 4$

33 $25x^2 + 30x + 9$

35 $x^2 + 2 + \frac{1}{x^2}$

37 $9 + 6\sqrt{x} + x$

39 $\sin^2\theta + 2\sin\theta\cos\theta + \cos^2\theta$

41 $8x^3 - 36x^2 + 54x - 27$

Section 0.6 *square root of a sum*

1 wrong

3 correct

5 wrong

7 wrong

9 correct

11 wrong

13 correct

15 wrong

17 wrong

19 wrong

21 correct

23 wrong

25 no simplification possible

27 no simplification possible

29 $|x + 18|$ ($x + 18$ is incorrect)

31 no simplification possible

33 1 (Hint: Pythagorean identity)

35 $4x\sqrt{x}$

Section 0.6 *cancellation*

1 correct

3 wrong ($4x^2$ is not a factor of the numerator nor of the denominator.)

5 correct

7 wrong (x is not a factor of the denominator.)

9 wrong ($x + 5$ is not a factor of either the numerator or the denominator.)

11 wrong (The cancellation is correct, but the $7x$ should remain in the denominator.)

13 correct

15 correct

17 wrong (The x is not a factor of $(2x - 7)$ and the x^2 is not a factor of $(4x^2 + 6)$.)

19 wrong (The 9 is not a factor of either the numerator or the denominator.)

21 wrong ($\frac{1}{2x}$ was canceled in the numerator, whereas $2x$ was canceled in the denominator. These factors do not match)

23 wrong ($\sqrt{x}$ was canceled in the numerator, whereas x was canceled in the denominator. These factors do not match)

25 wrong (2 is not a factor of the denominator.)

27 correct, but poorly written; $\frac{\sqrt{x^2+7x}}{\sqrt{x}(x-4)} = \frac{\sqrt{x(x+7)}}{\sqrt{x}(x-4)} = \frac{\cancel{\sqrt{x}}\sqrt{x+7}}{\cancel{\sqrt{x}}(x-4)} = \frac{\sqrt{x+7}}{x-4}$ is better

29 wrong (x is not a factor of the denominator.)

31 $\frac{3x+4}{x-6}$

33 $\frac{x+3}{3x+1}$

35 no simplification possible

37 $\frac{1}{x-3}$

39 $\frac{\sin x}{x+\cos x}$

41 no simplification possible

43 $\frac{2x+5}{x^2}$

45 no simplification possible

Section 0.6 *polynomial equations*

1 $x = -2, x = -5$

3 $x = 4, x = -1$

5 $x = \sqrt{7}, x = -\sqrt{7}$

7 $x = -2$

9 no solutions

11 $x = 7, x = 4$

13 $x = -\frac{1}{2}, x = 3$

15 $x = 0, x = -\frac{8}{7}$

17 $x = \frac{7+\sqrt{33}}{2}, x = \frac{7-\sqrt{33}}{2}$

19 $x = -\frac{3}{4}, x = -\frac{5}{2}$

21 $x = \frac{-9+\sqrt{101}}{10}, x = \frac{-9-\sqrt{101}}{10}$

23 $x = 2$ ($x = -2$ is incorrect)

25 $x = 0, x = 2, x = -2$ (If you began by dividing both sides of the equation by x^3, then you missed the solution $x = 0$. *Never* divide both sides of an equation by a variable expression or you will miss solutions!)

27 $x = 0, x = -5$

29 $x = \sqrt{2}, x = -\sqrt{2}, x = \sqrt{3}, x = -\sqrt{3}$ (This equation is *quadratic in form* and may be factored as $(x^2-3)(x^2-2) = 0$.)

31 $x = 0, x = 3, x = 4$

33 $x = 0, x = \frac{-3+\sqrt{29}}{10}, x = \frac{-3-\sqrt{29}}{10}$

Section 0.6 *inside vs. outside: being careful with function arguments*

1 correct

3 wrong (10 inside the square root is not the same as 10 outside the square root.)

5 wrong ($(7x)^{-1} \neq 7x^{-1} = 7 \cdot \frac{1}{x} = \frac{7}{x}$)

7 correct

9 correct (The applicable identity, $|ab| = |a| \cdot |b|$, was used.)

11 wrong (The x inside one trig function cannot cancel with the x inside another trig function.)

13 wrong (The $4x$ inside one trig function cannot cancel with the $4x$ inside another trig function.)

15 wrong (The $4x$ inside the square root cannot cancel with the $4x$ inside the trig function.)

17 correct (The cancellation took place entirely inside the trig function.)

19 wrong (The $x+2$ inside the trig function cannot cancel the $x+2$ outside the trig function.)

21 $\sin \frac{x}{3}$

23 no simplification possible ($\frac{\sin x}{3}$ is incorrect.)

25 $\frac{1}{\sqrt{x}}$

27 no simplification possible ($\cos x^5$ and $x^5 \cos$ are both incorrect.)

29 $6|x|$ ($-6|x|$ is incorrect)

31 no simplification possible ($\sin 2x$ is incorrect.)

Section 0.6 *trig notation*

1 wrong (The cosine has no argument.)

3 correct

5 wrong (The sine has no argument.)

7 wrong (The correct identity is $\sin 2x = 2\sin x \cos x$.)

9 wrong (The sine has no argument.)

11 wrong (The cosine has no argument.)

13 wrong (The sine has no argument.)

15 wrong (The sine on the right side has no argument.)

17 wrong (Names cannot be canceled; tan is a name, not a number.)

19 wrong (Names cannot be canceled; cos is a name, not a number. Also, the numerator is not $(\cos 3)x$.)

21 correct (The quantity $\cos x$ is a number, so it can be canceled.)

Section 0.6 *parentheses*

1 correct

3 wrong; $(2x+5)^3$

5 wrong; $5x-(2x+5) = 3x-5$

7 wrong; $\frac{2}{3}(2x+5) = \frac{4}{3}x + \frac{10}{3}$

9 wrong; $2(-7x)+5 = -14x+5$

11 wrong; $2\tan x + 5 = 5 + 2\tan x$ (When the term is not inside the trig function, such as the $+5$, it is traditional to write it out front to avoid possible confusion.)

13 $\frac{1}{4x^2-12}$ ($1/4x^2-12$ is incorrect; use parentheses.)

15 $24x^2-72$ ($24x^2-12$ is incorrect; use parentheses.)

17 $2-(4x^2-12) = 14-4x^2$ ($2-4x^2-12 = -10-4x^2$ is incorrect.)

19 $1+2(x+1)+(x+1)^2+\sin(x+1) = x^2+4x+4+\sin(x+1)$

21 $1+2(x^2+5)+(x^2+5)^2+\sin(x^2+5)$

23 $1+6x+9x^2+\sin 3x$

25 $9(-4x+2) = -36x+18$ ($9-4x+2$ is incorrect.)

27 $9(-x^2) = -9x^2$ ($9-x^2$ is incorrect.)

29 $9(-\tan x) = -9\tan x$ ($9-\tan x$ is incorrect.)

31 $9(2x-7) = 18x-63$ ($18x-7$ is incorrect.)

Section 0.6 *when not to distribute*

1 correct

3 wrong ($20x - 10$ is correct.)

5 correct

7 correct

9 wrong

11 wrong ($2 \sin 5x$ is correct.)

13 correct

15 wrong

17 wrong ($\csc(2x + 2)$ is correct.)

19 wrong

21 $15x + 3$

23 $7x \sin x$

25 $7x - 7 \sin x$

27 Keep $\sin(7 + x)$ or use identity to get $\sin 7 \cos x + \cos 7 \sin x$.

29 $3(x + 1)(x + 2)$ or $3x^2 + 9x + 6$

31 $\frac{4x}{5}$

Section 1.1

1 infinitesimal

3 infinitesimal

5 neither

7 neither

9 infinite

11 infinite

13 $14 - 8\omega$

15 $-\frac{1}{4}\omega + \frac{\omega^2}{4}$

17 $9\Omega + 55$

19 $16 - 8\omega + \omega^2$

21 $8 + 12\omega + 6\omega^2 + \omega^3$

23 $3\Omega - 1$

25 $4\Omega - 4$

27 Ω^4

29 Ω^2

31 $\Omega^2 - 2\sqrt{2} + 2\omega^2$

33 $289\,475\,638\,239\,457\,856\sqrt{\omega}$

35 ω

37 $12\Omega^3$

39 $\omega^{3/2}$

41 Ω^9

43 Ω

45 real

47 true

49 45ω cm (The units are required.)

51 4ω cm (The units are required.)

Section 1.2

1 -0.00293%

3 -4.63%

5 $4 + \omega$

7 $3\Omega - 5$

9 $\sqrt{\omega} - \omega$

11 14

13 9Ω

15 5

17 $\frac{4}{5}$

19 $2\omega^{3/2}$

21 ω

23 10

25 $-\frac{3}{2}$

27 4ω

29 $\Omega^{3/2}$

31 $-\frac{1}{28}\sqrt{\Omega}$

33

real level
11 12 13 14
ω^2 level below 13
$13 - 2\omega^2$ $13 - \omega^2$ 13 $13 + \omega^2$
ω^3 level below $13 - \omega^2$
$13 - \omega^2$ $13 - \omega^2 + \omega^3$ $13 - \omega^2 + 2\omega^3$

Section 1.3

1 0

3 0

5 0

7 ∞

9 A, stuck

11 $\frac{1}{5}$

13 0

15 undefined

17 undefined

19 $-\infty$

21 1

23 2

25 $f(3 + \omega) \doteq 0, f(3 - \omega) \doteq 0$

27 $f(\Omega) \doteq \infty, f(-\Omega) \doteq -\infty$

29 $f(\Omega) \approx 1, f(-\Omega) \approx -1$

31 2

Section 1.4

1

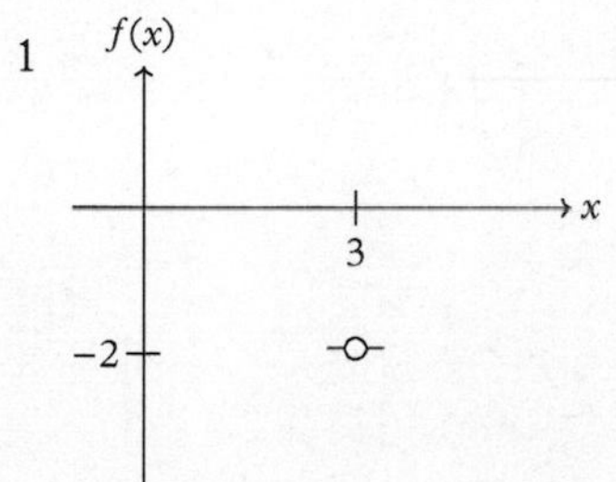

3

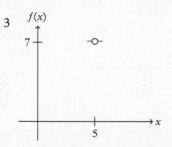

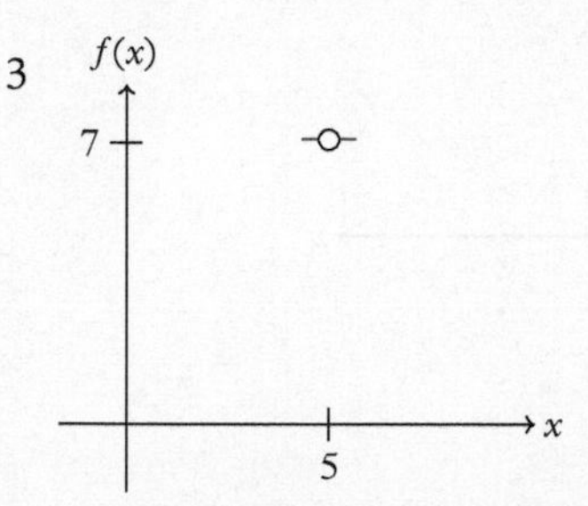

5

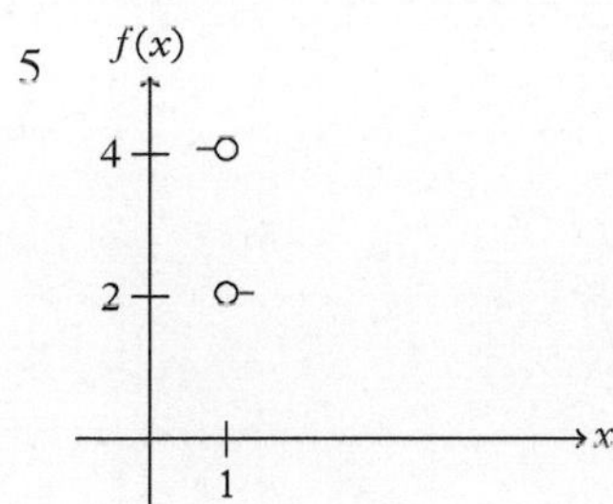

7

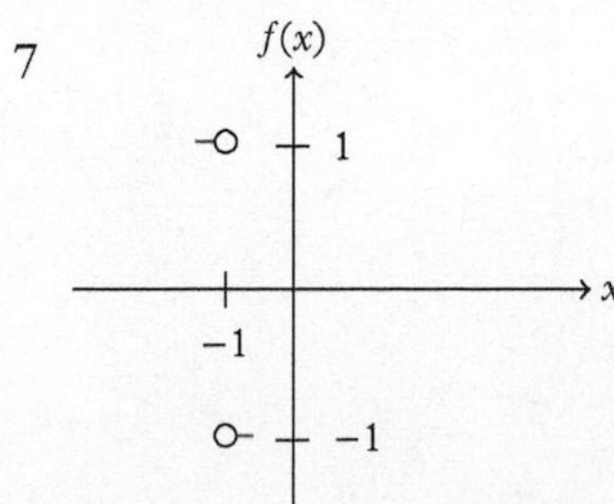

9 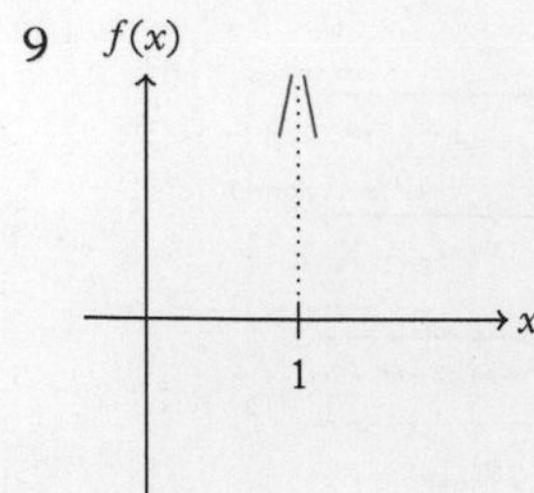

11 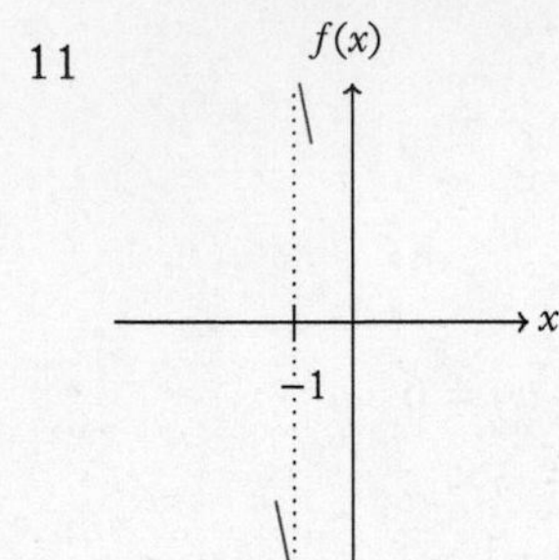

13 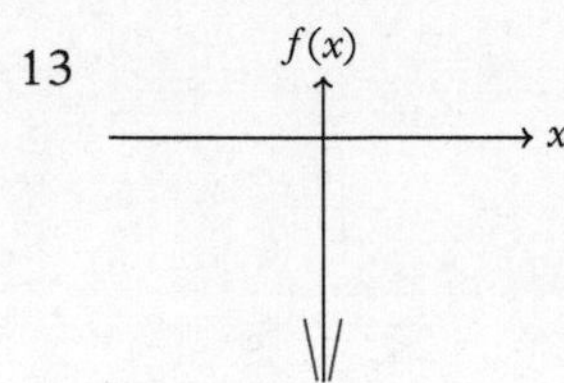

15 $\frac{5}{2}$

17 $-\infty$

19 4

21 $-\infty$

23 $\frac{-3}{2}$

25 $\frac{1}{6}$

27 8

29 12

31 $-\frac{1}{4}$

33 0

35 $\frac{1}{3}$

37 ∞

39 ∞

41 $\lim_{x\to 0} \frac{4}{x}$ DNE, $\lim_{x\to 0^+} \frac{4}{x} = \infty$, $\lim_{x\to 0^-} \frac{4}{x} = -\infty$

43 $\lim_{x\to 3} \frac{x^2-9}{|x-3|}$ DNE, $\lim_{x\to 3^+} \frac{x^2-9}{|x-3|} = 6$, $\lim_{x\to 3^-} \frac{x^2-9}{|x-3|} = -6$

45 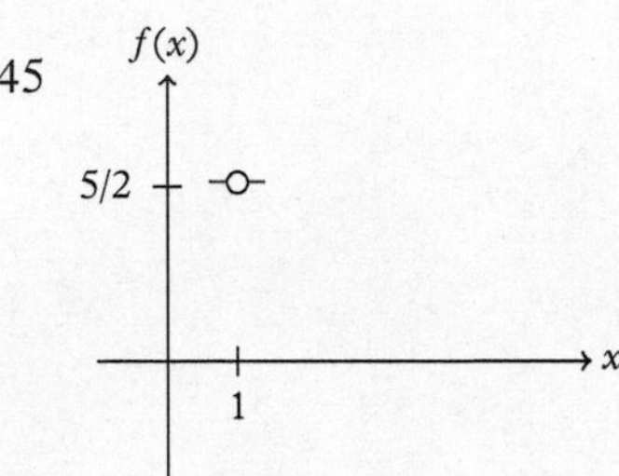

47 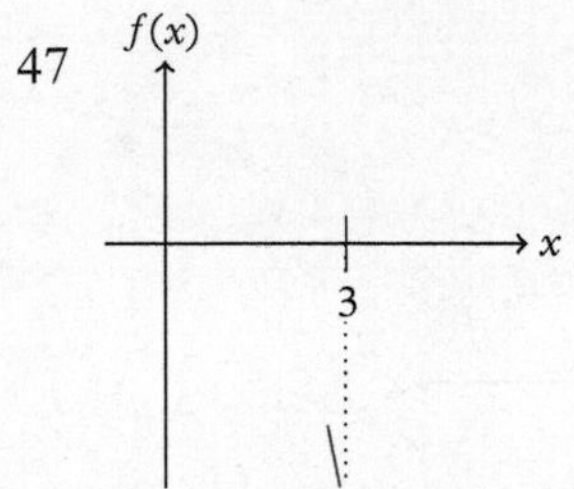

49 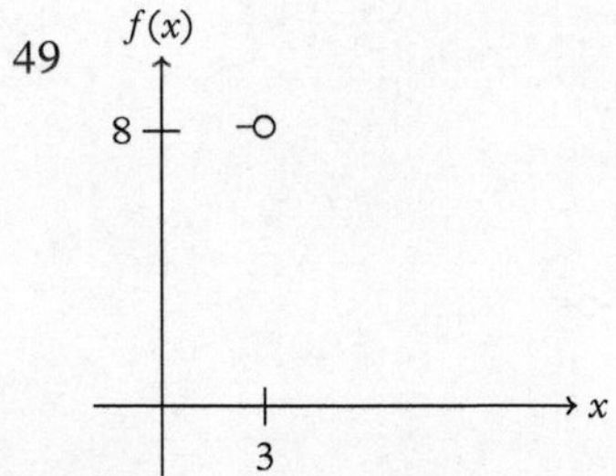

51

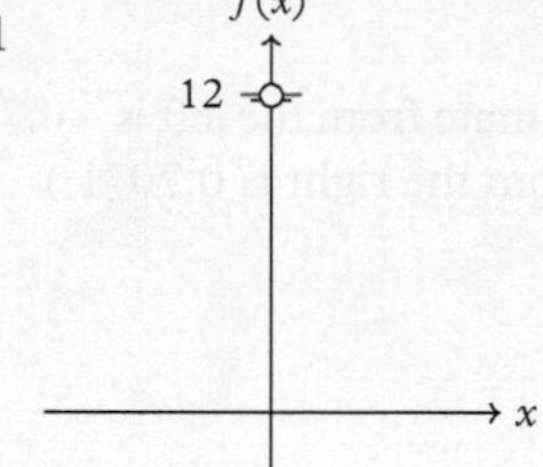

53

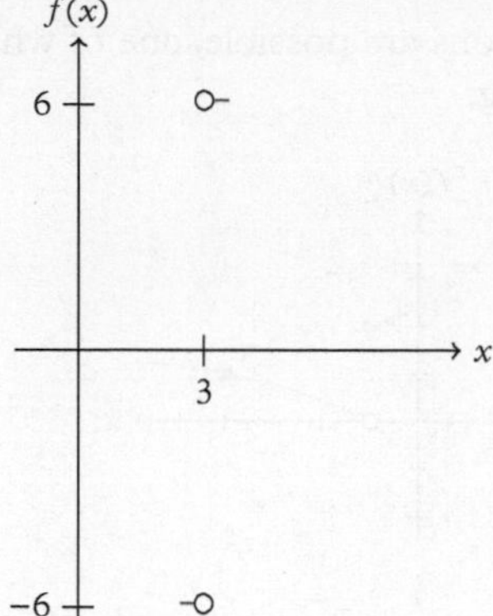

Section 1.5

1 17

3 10

5 12

7 DNE

9 3

11 1

13 2

15 $-\infty$

17 DNE

19 1

21 1

23 DNE

25 -1

27 3

29 $-\infty$

31 DNE

33 0

35 2

37 1

39 Many solutions are possible, one of which is the following:

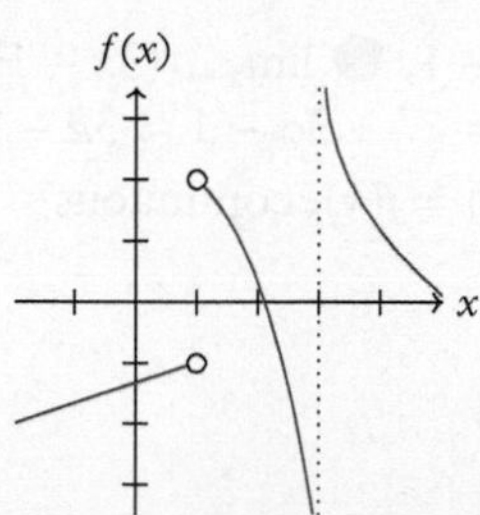

41 Many solutions are possible, one of which is the following:

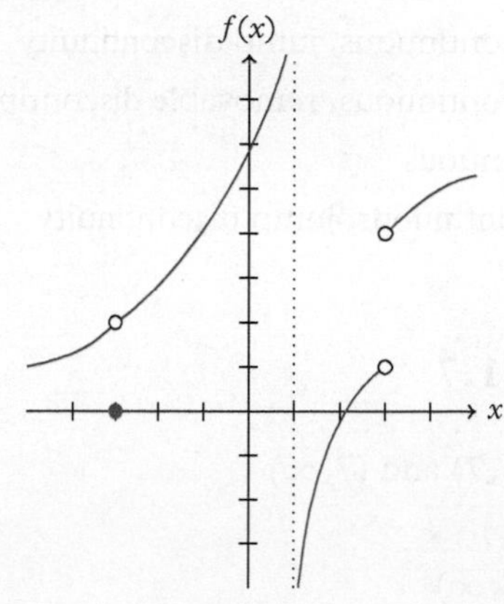

43 Many solutions are possible, one of which is the following:

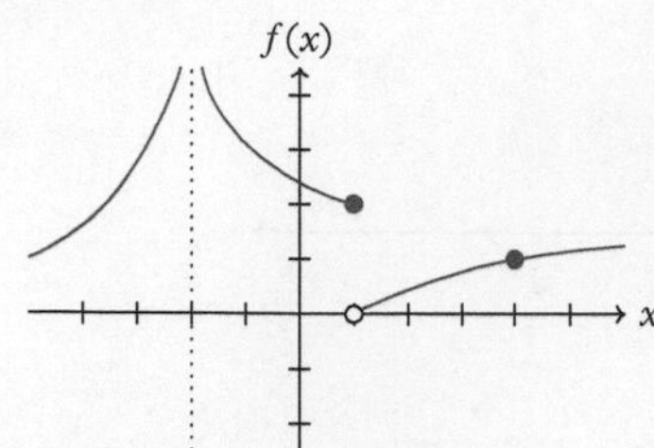

45 -1

47 DNE (The estimate from the left is -0.7071; the estimate from the right is 0.7071.)

49 0

Section 1.6

1 ❶ $f(1) = -6$, ❷ $\lim_{x\to 1}(x-7) = 1+\alpha-7 = -6+\alpha \approx -6$, ❸ $\lim_{x\to 1} f(x) = f(1)$, continuous

3 ❶ $f(0) = -1$, ❷ $\lim_{x\to 0}(x^3-1) = \alpha^3 - 1 \approx -1$, ❸ $\lim_{x\to 0} f(x) = f(0)$, continuous

5 ❶ $f(2) = 6$, ❷ $\lim_{x\to 2}(x^2+x) = (2+\alpha)^2 + (2+\alpha) = 4+4\alpha+\alpha^2+2+\alpha \approx 6$, ❸ $\lim_{x\to 2} f(x) = f(2)$, continuous

7 ❶ $f(1) = -\frac{3}{2}$, ❷ $\lim_{x\to 1}\frac{x+2}{x-3} = \frac{1+\alpha+2}{1+\alpha-3} = \frac{3+\alpha}{-2+\alpha} \approx -\frac{3}{2}$, ❸ $\lim_{x\to 1} f(x) = f(1)$, continuous

9 not continuous, infinite discontinuity

11 not continuous, jump discontinuity

13 not continuous, removable discontinuity

15 continuous

17 discontinuous, jump discontinuity

19 discontinuous, infinite discontinuity

21 not continuous, jump discontinuity

23 continuous

25 not continuous, removable discontinuity

27 continuous

29 not continuous, removable discontinuity

31 not continuous, jump discontinuity

33 not continuous, infinite discontinuity

35 not continuous, infinite discontinuity

37 not continuous, infinite discontinuity

39 not continuous, removable discontinuity

41 not continuous, infinite discontinuity

43 continuous

45 not continuous, jump discontinuity

47 not continuous, removable discontinuity

49 ❶ $f(k) = 3k-1$, ❷ $\lim_{x\to k}(3x-1) = 3(k+\alpha)-1 = 3k+3\alpha-1 \approx 3k-1$, ❸ $\lim_{x\to k} f(x) = f(k)$, continuous

Section 1.7

1 $(-\infty, 7)$ and $(7, \infty)$

3 $[\frac{2}{3}, \infty)$

5 $(-\infty, \infty)$

7 $(-\infty, -1)$, $(-1, 6)$, and $(6, \infty)$

9 $[-2, 2]$

11 $[-1, 0)$ and $(0, \infty)$

13 $(-\infty, \infty)$

15 $\frac{5}{7}$

17 $\frac{1}{17}$

19 ∞

21 -3

23 -1

25 $\frac{1}{\pi^2}$

27 1

29 ❶ $f(x) = \cos(x + \sin x)$, ❷ $f(0) = 1$, $f(1) = -0.267$, ❸ By the intermediate value theorem (IVT), there is a solution between $x = 0$ and $x = 1$.

31 ❶ $f(x) = \sqrt{x+1} - 3\sin x$, ❷ $f(2) = -0.996$, $f(3) = 1.577$, ❸ By the IVT, there is a solution between $x = 2$ and $x = 3$.

33 $x = 0.83\ldots$ (other solutions exist)

35 $x = 2.47\ldots$ (other solutions exist)

37 -1

Section 1.8

1 yes

3 no

5 no

7 equation of tangent line: $y = 2x - 3$
graph:

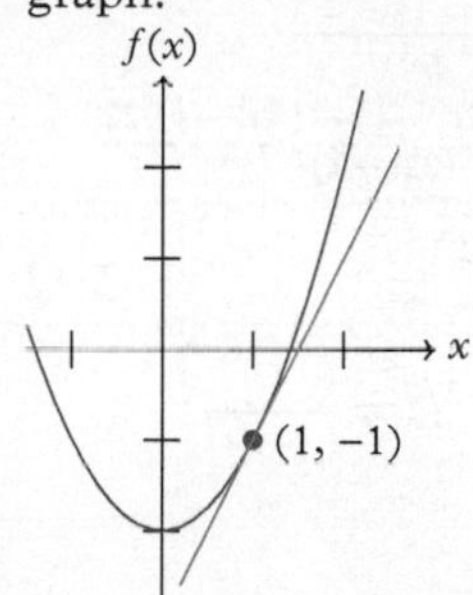

9 -4

11 $y = -4x - 2$

13 $y = \frac{3}{2}x - \frac{1}{2}$

15 $y = -\frac{2}{3}x + 4$

17 $y = 3x + 9$

19 $y = \frac{1}{2}x + \frac{1}{2}$

21 8

23 $-\frac{7}{6}$

25 7 m/s

27 $\frac{1}{4}$ cm/h

29 14 ft/s

31 $\frac{1}{2\sqrt{3}}$ cm/h

33 6 ft/s

35 $\frac{1}{3}$

37 $\frac{1}{4}$

39 average rate of change of weight, $-\frac{3}{8}$ kg/d; average weight loss, $\frac{3}{8}$ kg/d

41 \$0.03/d

43 $-\frac{5}{4}$ min/d

45 $-\frac{1}{192}$ min/d

Section 2.1

1 number

3 function

5 neither

7 not differentiable at $x = -2$ (corner), $x = 0$ (corner), $x = 2$ (corner)

9 not differentiable at $x = -1.5$ (infinite discontinuity), $x = 3$ (vertical tangent)

11 2

13 -3

15 0

17 -1

19 $f'(x) = 2x$

21 $f'(x) = 3$

23 $f'(x) = 2x + 3$

25 The derivative of the sum of two functions is the sum of the derivatives of those two functions.

27 ∞

29 $f'(4)$ DNE

31 0

33 $\frac{dy}{dx} = -\frac{6x}{(x^2+2)^2}$

35 $\frac{d\rho}{d\theta} = \frac{2}{(\theta+2)^2}$

37 $y' = 4x^3$

Section 2.2

1 power rule (positive exponents)

3 power rule (negative exponents) *or* quotient rule

5 power rule and constant multiple rule

7 product rule, sum rule, difference rule, power rule, constant multiple rule, constant rule

9 power rule (general)

11 $y' = 27x^{26}$

13 $f'(x) = 12$

15 $y' = -5x^{-6} = \frac{-5}{x^6}$

17 $s'(t) = 3 - \frac{1}{3}t^{-2/3} - 4t^{-5} = 3 - \frac{1}{3\sqrt[3]{t^2}} - \frac{4}{t^5}$

19 $h'(x) = (x^3 + 2x + 8)(20x^4 - 4x) + (4x^5 - 2x^2 + 38)(3x^2 + 2)$

21 $f'(x) = (12x^3 + 7x - 1)(3x^2 - 9) + (x^3 - 9x - 5)(24x + 7)$

23 $g'(x) = \frac{(4x-9)2x - (x^2-7)4}{(4x+9)^2} = \frac{4x^2+18x+28}{(4x+9)^2}$

25 $f'(t) = 5t^4 - 21t^2 + 4$

27 $\frac{dy}{dx} = \frac{(x^3+5x)(3) - (3x-9)(3x^2+5)}{(x^3+5x)^2} = \frac{-6x^3+27x^2+45}{(x^3+5x)^2}$

29 $y' = 5 + 7x^{-2} = 5 + \frac{7}{x^2}$

31 $\frac{df}{dx} = \frac{5}{2}x^{3/2}$

33 $f'(x) = \frac{(5x^3+3x)4 - (1+4x)(15x^2+3)}{(5x^3+3x)^2} = \frac{-40x^3 - 15x^2 - 3}{(5x^3+3x)^2}$

35 $y' = \frac{(22t-1)(2t^2-\frac{1}{4}) - (\frac{2}{3}t^3 - \frac{1}{4}t)22}{(22t-1)^2} = \frac{\frac{88}{3}t^3 - 2t^2 + \frac{1}{4}}{(22t-1)^2}$

37 $y'' = -4$

39 $f''(x) = -\frac{1}{4}x^{-3/2} = -\frac{1}{4\sqrt{x^3}}$

41 $\frac{d^2y}{dx^2} = \frac{3}{2}x$

43 $60x^2$

45 $f^{(4)}(t) = 840t^3$

47 $\frac{39}{5}x^{\frac{8}{5}} + \frac{7}{3}x^{-\frac{4}{3}} + \frac{24}{7}x^{-\frac{4}{7}}$

51 $x'(f) = 2f + 4$

Section 2.3

1 derivative

3 differential

5 derivative

7 differential

9 $y = 6x - 18$

11 $L(x) = 12 + 6(x - 5)$ (Do not simplify. The answer should be exactly as written, even though it is the same as the tangent line in exercise 9.)

13 $y = -\frac{5}{16}x - \frac{1}{4}$

15 $y = 10x + 29$

17 $y = -3t + 7$ (or $s = -3t + 7$)

19 $y = \frac{1}{3}x + \frac{2}{3}$

21 $y = -7x - 13$

23 $L(x) = -28 + 12(x - (-2))$ (It is acceptable to simplify $(x - (-2))$ to $(x + 2)$; otherwise, your answer should be exactly as written.)

25 $L(x) = \frac{4}{9} + \frac{-4}{27}(x - 2)$; alternately, $L(x) = \frac{4}{9} - \frac{4}{27}(x - 2)$ (Writing $\frac{-12}{81}$ instead of $\frac{-4}{27}$ is also equivalent.)

27 $L(z) = 2 + \frac{-7}{2}(z - 1)$ (Notice the use of the variable z instead of x, because z is the independent variable in the given function. Also, writing $L(z) = 2 - \frac{7}{2}(z - 1)$ is equivalent.)

29 $dy = 2x\,dx$

31 $dy = -44t^{-3}\,dt = -\frac{44}{t^3}\,dt$

33 $dy = (8x - 1)\,dx$

35 $dy = \frac{(\sqrt{x}+1)\frac{1}{2\sqrt{x}} - (\sqrt{x}-1)\frac{1}{2\sqrt{x}}}{(\sqrt{x}+1)^2}\,dx$
$= \frac{1}{\sqrt{x}(\sqrt{x}+1)^2}\,dx$

37 $dy = \left(3 + \frac{1}{(s+1)^2}\right)\,ds$

39 $x = 3, x = -2$

41 $x = 0, x = 1$

43 $x = 5, x = -2$

45 $x = \frac{-6 \pm \sqrt{30}}{6}$

47 no horizontal tangents

49 $x = -5 \pm \sqrt{21}$

51

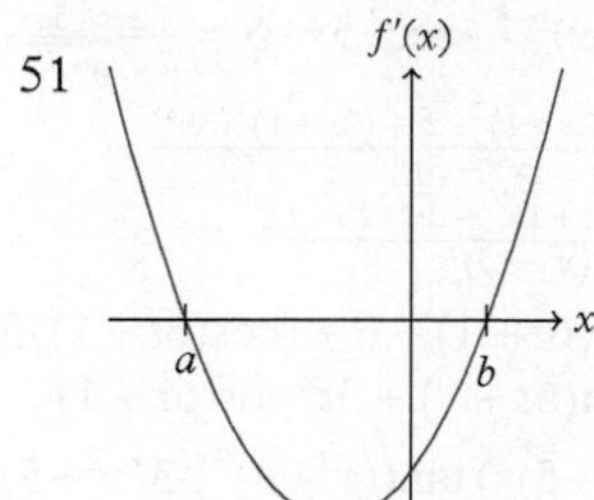

53

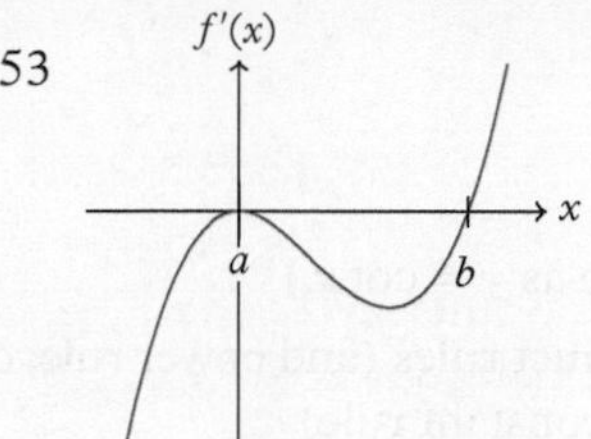

55

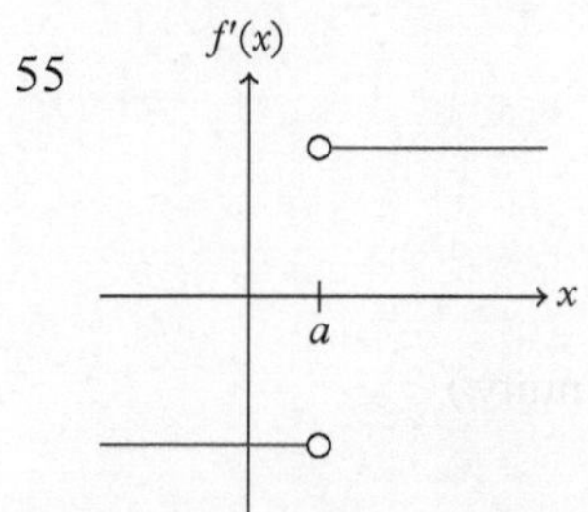

57 $L(x) = 4 + \frac{1}{8}(x - 16)$, $L(15.7) = 3.9625$ (Compare this answer to the value generated by a calculator. It's impressively close.)

59 (a) $y - \frac{1}{a} = -\frac{1}{a^2}(x - a)$; alternately, $y = -\frac{1}{a^2}x + \frac{2}{a}$; (b) x-intercept $2a$ (point $(2a, 0)$), y-intercept $\frac{2}{a}$ (point $(0, \frac{2}{a})$); (c) answers vary (One possibility, noticing that the point of tangency is $(a, \frac{1}{a})$, is that because a is halfway between 0 and $2a$, and $\frac{1}{a}$ is halfway between 0 and $\frac{2}{a}$, the conclusion follows. Another possibility is to calculate the midpoint of the x- and y-intercepts and see that it is the point of tangency.)

Section 2.4

1 product rule

3 none (Rewrite as $y = \cot x$.)

5 quotient, product rules (and power rule, difference rule, constant rule)

7 product rule

9 $\frac{4}{9}$

11 $\frac{5}{7}$

13 ∞

15 $\frac{7}{2}$

17 -1 (Use continuity.)

19 $\frac{25}{36}$

21 $x^2 \sec x \tan x + 2x \sec x$

23 $\frac{(\sec x)(12x^2) - (4x^3 - 7)\sec x \tan x}{\sec^2 x} = \frac{12x^2 - (4x^3 - 7)\tan x}{\sec x}$

25 $5x^2 \cos x + 10x \sin x$

27 $\frac{(\csc y)6 - 6y(-\csc y \cot y)}{\csc^2 y} = \frac{6 + 6y \cot y}{\csc y}$

29 $\frac{(x^3 + 9x)(x \sec x \tan x + \sec x) - (x \sec x)(3x^2 + 9)}{(x^3 + 9x)^2} = \frac{-2x^3 \sec x + (x^4 + 9x^2)\sec x \tan x}{(x^3 + 9x)^2}$

31 $-(t^2 - 4t)\csc^2 t + (2t - 4)\cot t$ (Hint: $g(t) = (t^2 - 4t)\cot t$)

33 $\sin x \sec^2 x + \tan x \cos x = \sin x \sec^2 x + \sin x = \sin x(\sec^2 x + 1)$

35 $y'' = (-\csc x)(-\csc^2 x) + (\cot x)(-(-\csc x \cot x)) = \csc^3 x + \csc x \cot^2 x$

37 $f^{(4)}(x) = \sin x$

39 $f^{(4000)}(x) = \sin x$

41 $y''' = -t(-\sin t) + (\cos t)(-1) - 2\cos t = t \sin t - 3\cos t$

43 $h''(\theta) = (\sin\theta)(-\cos\theta) + (-\sin\theta)\cos\theta + (\cos\theta)(-\sin\theta) + (\cos\theta)(-\sin\theta) = -4\sin\theta\cos\theta$

45 $y = -x + \frac{\pi}{2}$

47 $y = 1$

49 $y = -\pi x + \pi^2$

51 at $x = \frac{\pi}{2} + k\pi$, for any integer k

53 -5 mm/s

55 $dy = (4\theta\cos\theta + 4\sin\theta)\,d\theta$ ($\frac{dy}{d\theta} = 4\theta\cos\theta + 4\sin\theta$ is incorrect.)

57 $L(x) = 0 + 2(x - 0)$ ($y = 2x$ is an incorrect format.)

61 $\frac{d}{dx}\,\text{vers}\, x = \sin x$

Section 2.5

1 product rule

3 chain rule

5 chain rule

7 chain rule as written, product rule if rewritten as $\frac{d}{dx}(\sin x)(\sin x)$

9 no

11 yes

13 as written, yes; if rewritten as $\frac{d}{dx}(36x^2 + 12x + 1)$, no

15 yes (since $\cos^3 x = (\cos x)^3$)

17 yes

19 $y' = -4\sin(4x - 3)$

21 $f'(x) = 28(7x + 5)^3$

23 $g'(x) = \frac{1}{2}(\tan 5x)^{-\frac{1}{2}} \cdot \sec^2 5x \cdot 5 = \frac{5\sec^2 5x}{2\sqrt{\tan 5x}}$

25 $y' = \frac{(x^3 - 5)\cdot 3(2x+1)^2 \cdot 2 - (2x+1)^3(3x^2)}{(x^3 - 5)^2} = \frac{6(x^3 - 5)(2x+1)^2 - 3x^2(2x+1)^3}{(x^3 - 5)^2}$

27 $f'(t) = t^3(-\sin(6t+1)) \cdot 6 + (\cos(6t+1))3t^2 = -6t^3\sin(6t+1) + 3t^2\cos(6t+1)$

29 $g'(x) = \sec((x^2+5)^3)\tan((x^2+5)^3) \cdot 3(x^2+5)^2 \cdot 2x = 6x(x^2+5)^2\sec((x^2+5)^3)\tan((x^2+5)^3)$

31 $h'(x) = (-\csc\sqrt{x^3+x}\cot\sqrt{x^3+x}) \cdot \frac{1}{2}(x^3 + x)^{-\frac{1}{2}}(3x^2 + 1) = \frac{-(3x^2+1)\csc\sqrt{x^3+x}\cot\sqrt{x^3+x}}{2\sqrt{x^3+x}}$

33 $s' = \theta(\cos(2\theta - 3)) \cdot 2 + (\sin(2\theta - 3)) \cdot 1 = 2\theta\cos(2\theta - 3) + \sin(2\theta - 3)$

35 $y' = 4(\tan(x^2+1))^3 \cdot (\sec^2(x^2+1)) \cdot 2x = 8x\sec^2(x^2+1)\tan^3(x^2+1)$

37 $f'(x) = 7(5+\cos x^2)^6(-\sin x^2)(2x) = -14x(5+\cos x^2)^6 \sin x^2$

39 $y' = 2\cos(6-3x)\cdot(-\sin(6-3x))\cdot(-3) = 6\cos(6-3x)\sin(6-3x)$

41 $k'(x) = \frac{(2x+7)\cdot 3(4x^2+1)^2\cdot 8x - (4x^2+1)^3\cdot 2}{(2x+7)^2} = \frac{24x(2x+7)(4x^2+1)^2 - 2(4x^2+1)^3}{(2x+7)^2}$

43 $y' = -\sin\tan\theta\cdot\sec^2\theta = -(\sec^2\theta)(\sin\tan\theta)$

45 $f'(x) = \frac{1}{2}(\cos(x^4+\frac{1}{x}))^{-\frac{1}{2}}(-\sin(x^4+\frac{1}{x}))\cdot(4x^3-x^{-2}) = \frac{-(4x^3-\frac{1}{x^2})\sin(x^4+\frac{1}{x})}{2\sqrt{\cos(x^4+\frac{1}{x})}}$

47 $\frac{dx}{dt} = t^2\cdot\sec(5t-8)^4\tan(5t-8)^4\cdot 4(5t-8)^3\cdot 5+\sec(5t-8)^4\cdot 2t = 20t^2(5t-8)^3\sec(5t-8)^4\tan(5t-8)^4 + 2t\sec(5t-8)^4 = 2t(\sec(5t-8)^4)(10t(5t-8)^3\tan(5t-8)^4+1)$

49 $y'' = 2\sec x\cdot\sec x\tan x = 2\sec^2 x\tan x$

51 $-25\sin 5x$

53 $y = \frac{3}{2}x+\frac{3}{2}$

55 $y = 24x-15$

57 $dy = \frac{2}{\sqrt{4x+1}}\,dx$

59 $dy = (-7\sin 7x)\,dx$

61 $L(x) = -\frac{1}{\sqrt{2}} + \frac{1}{\sqrt{2}}(x-0)$

63 $x = 2$

65 0.0439 h/d = 2.64 min/d

Section 2.6

1 explicitly

3 implicitly

5 explicitly (x is a function of y.)

7 explicitly

9 implicitly

11 $y' = \frac{14x}{15y^2+1}$

13 $y' = \frac{7-4y}{4x-3y^2}$

15 $y' = \frac{7y}{2y^3+7x}$; alternately, $y' = \frac{7}{3y^2+27}$

17 $y' = \frac{-4x}{12y^2+1}$

19 $y' = \frac{5-y^2}{1+2xy}$

21 $\frac{dy}{dx} = \frac{-5x^4}{6(2y-7)^2-1}$

23 $\frac{dy}{dx} = \frac{-3y-4}{3x-2y}$

25 $\frac{dy}{dx} = \frac{1-2x}{-\sin y}$

27 $\frac{dy}{dx} = \frac{2x}{\frac{1}{2\sqrt{y-5}}-2y} = \frac{4x\sqrt{y-5}}{1-4y\sqrt{y-5}}$

29 $\frac{dy}{dx} = \frac{y\cos x-\sin y}{x\cos y-\sin x}$

31 $y'' = \frac{-y+x\left(\frac{-x}{y}\right)}{y^2} = -\frac{y^2+x^2}{y^3}$

33 $y'' = \frac{2(1+\cos y)+2x\sin y\cdot\frac{2x}{1+\cos y}}{(1+\cos y)^2} = \frac{2(1+\cos y)^2+4x^2\sin y}{(1+\cos y)^3}$

35 $y'' = \frac{(4x-5)\left(-48x-4\left(\frac{-24x^2-4y}{4x-5}\right)\right)-4(-24x^2-4y)}{(4x-5)^2} = \frac{-48x(4x-5)+64(x-1)(6x^2+y)}{(4x-5)^2}$

37 $y'' = \frac{-32\left(\frac{1}{32y-17}\right)}{(32y-17)^2} = \frac{-32}{(32y-17)^3}$

39 $\frac{dx}{dt} = \frac{3t^2-3}{2x-1}$ (Hint: $\frac{d}{dt}x \neq 1$)

41 slope $= y'\big|_{(2,1)} = \frac{2(2)-1^3}{3(2)(1^2)+1} = \frac{3}{7}$

43 $y = \frac{5}{4}x+\frac{13}{4}$ (slope $= y'\big|_{(-1,2)} = \frac{2(-1)-4(2)}{4(-1)-2(2)} = \frac{5}{4}$)

45 $L(x) = \sqrt{7} - \frac{3}{\sqrt{7}}(x-3)$

47 $dy = \frac{1-3x^2}{-10y+4}\,dx$ (The answer $\frac{dy}{dx} = \frac{1-3x^2}{-10y+4}$ is in the wrong form and is therefore incorrect.)

Section 2.7

1 position $s(2) = 4$, velocity $v(2) = 10$, acceleration $a(2) = 8$

3 position $s(4) = 224$, velocity $v(4) = -8$, acceleration $a(4) = -32$

5 position $x(2) = 32$, velocity $v(2) = 0$, acceleration $a(2) = -16$

7 position $s(2) = \frac{2}{3}$, velocity $v(2) = \frac{1}{9}$, acceleration $a(2) = -\frac{2}{27}$

9 position $s(2) = 0$, acceleration $a(2) = 32$

11 (a) negative direction (left), (b) $a(2) = s''(2) = 12\,\text{m/s}^2$, (c) at $t = \sqrt{2}$ ($t = -\sqrt{2}$ is not a solution because it was given that $t \geq 0$.), (d) 6 m/s (−6 m/s is incorrect because we want speed, not velocity.)

13 (a) 36 ft/s, (b) $-32\,\text{ft/s}^2$, (c) at $t = \frac{25}{8} = 3.125$ (Hint: the maximum height is when the projectile changes direction from upward to downward), (d) 161.25 ft

15 (a) 7 m/s, (b) at $t = 2$ ($t = -4$ is not valid because we are told that $t \geq 0$.), (c) $\frac{136}{3}$ m (Hint: find the distance traveled from $t = 0$ to $t = 2$ (when it changed directions), the distance traveled from $t = 2$ to $t = 5$, and add the distances.)

17 (a) 14 m/s, (b) $11\,\text{m/s}^2$, (c) $2\,\text{m/s}^3$, (d) at $t = 2$ and at $t = -3$ (Notice $t \geq -10$, so $t = -3$ is valid.)

19 (a) $a(t) = s''(t) = -32\,\text{ft/s}^2$,
(b) $v(1) = s'(1)$
$= -32\,\text{ft/s}$,
(c) $\sqrt{\frac{19}{8}}$ s $= 1.5411$ s after being dropped,
(d) 49.3 ft/s (33.6 mph)

21 (a) $a(2.1) = -32\,\text{ft/s}^2$, (b) $t = 1$ (1 second after being thrown), (c) $t = 3$ ($t = -1$ is not correct because that is before the bottle was thrown.), (d) 64 ft/s (−64 ft/s is incorrect because we want speed, not velocity.)

23 (a) $P'(20\,000) = -\$8/\text{widget}$; (b) no, because profit would decrease

25 (a) $R'(14\,000) = \$6.994/\text{unit}$, (b) At this level of sales, Calculus R Us' revenue will increase by \$6.994 for each additional consulting unit sold.

Section 2.8

1 $\frac{1280}{40\sqrt{2}}$ mi/h ≈ 22.6 mph

3 $250\sqrt{3}$ mph ≈ 433 mph (If your answer is $\frac{1000}{\sqrt{5}}$ mph ≈ 447 mph, then you used the quantity 2 miles for the wrong variable.)

5 $\frac{-126}{\sqrt{157}}$ mph ≈ -10.06 mph

7 $\frac{9}{8}$ ft/s

9 $6\sqrt{5}$ mph ≈ 13.4 mph

11 $\frac{20}{13}$ mph

13 10 ft/min

15 $\frac{150}{\sqrt{650\,000}}$ m/s $= 0.186$ m/s

17 $\frac{0.8\sqrt{544}}{18}$ cm/s $= 1.037$ cm/s

19 $\frac{224}{\sqrt{1796}}$ km/h $= 5.29$ km/h (Hint: think of ship A as stationary and ship B as moving 16 km/h to the east.)

21 $\frac{25}{3}$ ft/s

23 (a) 98.06 m/min, (b) 99.50 m/min, (c) 99.88 m/min

25 ∞ (Because this is clearly impossible, either the bottom of the ladder cannot continue moving at a constant rate or the top of the ladder must lose contact with the side of the building before it hits the ground.)

27 $\frac{520}{\sqrt{52}}$ mph $= 72.1$ mph

Section 2.9

1 $-7500\pi\,\text{mm}^3/\text{min} = -23\,562\,\text{mm}^3/\text{min}$

3 -2 ft/s (The shadow is getting shorter.)

5 The shadow is shortening at a rate of 4 ft/s (The rate of change of the length of the shadow is -4 ft/s.)

7 $\frac{15}{4\pi}$ ft/min = 1.19 ft/min

9 0.2 ft/min

11 The angle is decreasing at a rate of $\frac{3}{16}$ rad/s (The rate of change of the angle is $-\frac{3}{16}$ rad/s.)

13 $0.04\pi\,\text{in}^2/\text{s} = 0.1257\,\text{in}^2/\text{s}$

15 0.0012 rad/s

17 418.9 ft/s

19 0.8 ft/s

21 $\frac{2}{3}$ ft/min

Section 3.1

1 wrong (Points have x- and y-coordinates. The local maximum point is $(-3, 2)$.)

3 wrong (The absolute minimum of f is not the number 4, but it is *at* $x = 4$.)

5 wrong ("at" signifies location, and there is no local min at $x = 0$.)

7 correct

9 local min at $x = a$, local max at $x = b$, local and absolute min at $x = d$

11 absolute min at $x = -2$ (There is no point at $x = 1$, so there is no max there. Local extrema cannot occur at endpoints.)

13 no extrema

15 Local and absolute max at $x = b$, local min at $x = c$ (There is no point at $x = a$ and hence no absolute min there.)

17 $x = 3, x = -2$

19 $x = 0, x = \frac{3}{2}$

21 $t = 0, t = \pm\sqrt{3}$

23 $y = \frac{-8 \pm \sqrt{136}}{12} = \frac{-4 \pm \sqrt{34}}{6}$

25 $x = 0, x = \pm\left(\frac{1}{3}\right)^{3/2} = \pm\frac{1}{3\sqrt{3}}$

27 $x = 0, x = 2$

29 $x = \pm 3$ ($x^2 + 1 = 0$ has no solutions.)

31 $x = 4$ ($x = 0$ is incorrect because it is not in the domain of g'.)

33 $x = 0$ ($\sec x = 0$ has no solutions.)

35 no, domain $(-\infty, 0) \cup (0, \infty)$ is neither closed nor bounded.

37 yes, domain $[-1, 1]$

39 no, domain $(-\infty, \infty)$ is closed but not bounded.

41 absolute min 0 (at $x = 0$), absolute max 4 (at $x = 2$)

43 absolute min -1 (at $x = -1$), absolute max 8 (at $x = 2$)

45 absolute min $\sqrt{2}^3 - 6\sqrt{2} + 7 \approx 1.34$, absolute max 16

47 absolute min at $x = \sqrt[3]{2}$, absolute max at $x = 3$

49 absolute min point $(\sqrt{3}, 12 - 6\sqrt{3})$, absolute max point $(0, 12)$

51 absolute min 4 (at $x = 2$), absolute max 13 (at $x = -1$ and at $x = 5$)

53 absolute min -276, absolute max 67

55 absolute min point $(0, -1)$, absolute max point $(7, 3)$

57 absolute min point $(0, 0)$, absolute max point $(\frac{\pi}{2}, 1)$

59 absolute min -1 (at $x = -\frac{\pi}{4}$), absolute max 1 (at $x = \frac{\pi}{4}$)

61 absolute min -3 (at $x = 0$), absolute max 2 (at $x = 5$ and at $x = -5$)

63 $x = -\sqrt{2}$ ($x = \sqrt{2}$ is not a critical number. If you squared both sides of the equation and got two solutions, this one is extraneous.)

65 absolute min $-2\sqrt{2}$ (at $x = -\sqrt{2}$), absolute max 2 (at $x = 2$)

Section 3.2

1 yes

3 yes (g is not differentiable at $x = 0$, but this is an endpoint of the interval and g is still differentiable on $(0, 1)$.)

5 no (f is not continuous at $x = 0$.)

7 yes

9 yes

11 no (f is not continuous at $x = 0$.)

13 yes

15 yes (s is not differentiable at $t = 0$, but this is an endpoint of the interval and s is still differentiable on $(0, 2\pi)$.)

17 $c = -\frac{4}{3}$ ($c = 0$ is incorrect. It is an endpoint of the interval, but c must be strictly between the endpoints.)

19 $c = 0$ (Other solutions to $-\sin c = 0$ are not in the interval.)

21 $c = 0$

23 $c = \frac{1}{2}$ (Hints: don't forget the chain rule, and $\sec\theta = 0$ has no solutions.)

25 $c = 1$

27 $c = \sqrt{3}$ ($c = -\sqrt{3}$ is not in the interval.)

29 $c = 2$ ($c = -2$ is not in the interval.)

31 $c = \frac{9}{4}$

33 $c = \frac{2}{\sqrt{3}}, c = -\frac{2}{\sqrt{3}}$

35 $f(x) = 4x^2 + C$ (The name of the function is not important for exercises 35–46)

37 $f(x) = \sin x + C$

39 $f(x) = (5x^2 + 3)^4 + C$

41 $f(x) = \sec x + C$

43 $f(x) = x^6 + C$

45 $f(x) = \frac{x^6}{6} + C$

47 yes, because the average speed over the interval was 73 mph

49 no, Rolle's theorem does not apply because f is not continuous at $x = \frac{\pi}{2}$

Section 3.3

1–8 The "sign of f'" column is not given in the answers in order to save space.

1

interval	inc/dec	local extrema
$(-\infty, -3)$	inc	
		local max at $x = -3$
$(-3, 0)$	dec	
$(0, 6)$	dec	
		local min at $x = 6$
$(6, \infty)$	inc	

There is no local extremum at $x = 0$ because f' does not change signs.

3

interval	inc/dec	local extrema
$(-\infty, 4)$	inc	
		local max at $x = 4$
$(4, 7)$	dec	
		local min at $x = 7$
$(7, \infty)$	inc	

5

interval	inc/dec	local extrema
$(-\infty, 2)$	inc	
$(2, \infty)$	inc	

There is no local extremum at $x = 2$ because f' does not change signs.

7

interval	inc/dec	local extrema
$(-\infty, 4.3)$	inc	local max at $x = 4.3$
$(4.3, 2\pi)$	dec	
$(2\pi, \infty)$	dec	

There is no local extremum at $x = 2\pi$ because f' does not change signs.

9 increasing on (a, b), (d, ∞); decreasing on $(-\infty, a)$, (b, d). Closed intervals may be used if desired.

11 increasing on $(-2, 1)$, decreasing on $(1, 3)$. Alternate answer: increasing on $[-2, 1)$, decreasing on $(1, 3]$. The intervals should not include the endpoint 1 because the function is not defined there.

13 increasing on $(0, a)$, (a, c). Alternate answer: increasing on $(0, a)$, $(a, c]$. Neither 0 nor a should be included in the intervals because the function is not defined there. ("Increasing on $(0, c)$" is incorrect.)

15 increasing on (a, b), (c, d); decreasing on (b, c). The endpoints b, c, d may be included in the intervals, but a should not be included.

17

interval	inc/dec	local extrema
$(0, 4)$	inc	local max at $x = 4$
$(4, 7)$	dec	
$(7, \infty)$	inc	

There is no local extremum at $x = 7$ because the function is undefined there.

19

interval	inc/dec	local extrema
$(-\infty, 0)$	inc	local max at $x = 0$
$(0, 5)$	dec	local min at $x = 5$
$(5, \infty)$	inc	

There is a local extremum at $x = 5$. Continuous but nondifferentiable indicates a corner in the graph, and extrema may occur at corners.

21 increasing on $(-\infty, -4)$ and $(-1, \infty)$, decreasing on $(-4, -1)$

23 increasing on $(3, \infty)$, decreasing on $(-\infty, 0)$ and $(0, 3)$

25 increasing on $(-\infty, -\sqrt{5})$ and $(\sqrt{5}, \infty)$, decreasing on $(-\sqrt{5}, \sqrt{5})$

27 increasing on $(0, \frac{\pi}{2})$ and $(\frac{3\pi}{2}, 2\pi)$, decreasing on $(\frac{\pi}{2}, \frac{3\pi}{2})$

29 increasing on $(-\pi, -\frac{\pi}{2})$, $(-\frac{\pi}{2}, \frac{\pi}{2})$, and $(\frac{\pi}{2}, \pi)$. The intervals cannot be combined because of discontinuities at $-\frac{\pi}{2}$ and $\frac{\pi}{2}$.

31 local max at $x = -2$, local min at $x = 3$

33 local min at $x = 4$

35 no local extrema

37 local max at $x = 0$

39 local min 2 at $x = 1$

41 local min -2 at $x = -3$, local max $\frac{202}{27} \approx 7.48148$ at $x = -\frac{1}{3}$

43 no local extrema

45 local min 0 at $x = 1$, local max $\sqrt[3]{4}$ at $x = -1$

47 local min point $(-3, -123)$

49 local min points $(-2, -239)$ and $(2, -239)$, local max point $(0, 1)$

51 local max point $(-3, -5)$, local min point $(1, 3)$

53 local min points $(2, 8)$ and $(-2, 8)$

Section 3.4

1

interval	concavity	inflection points
$(-\infty,-3)$	up	
		i.p. at $x=-3$
$(-3,0)$	down	
$(0,6)$	down	
		i.p. at $x=6$
$(6,\infty)$	up	

3

interval	concavity	inflection points
$(-\infty,4)$	up	
		i.p. at $x=4$
$(4,7)$	down	
		i.p. at $x=7$
$(7,54)$	up	

5 concave up on $(-\infty,b)$ and (d,∞), concave down on (b,d), inflection points at $x=b$ and $x=d$

7 concave up on $(0,a)$, concave down on (a,c), no inflection points

9 concave up on (a,b) and (b,d), no inflection points ("Concave up on (a,d)" is incorrect.)

11 concave up on (b,d), concave down on (a,b), no inflection points (There is a corner and therefore no tangent line at $x=b$.)

13

interval	concavity	inflection points
$(0,4)$	up	
		i.p. at $x=4$
$(4,7)$	down	
$(7,\infty)$	up	

(There is no inflection point at $x=7$ because there is no point at $x=7$.)

15

interval	concavity	inflection points
$(-\infty,0)$	up	
		i.p. at $x=0$
$(0,5)$	down	
		i.p. at $x=5$
$(5,\infty)$	up	

(The tangent line exists at $x=5$, so there is an inflection point there even though the function is not differentiable there.)

17 concave down on $(0,\infty)$

19 concave up on $(-\infty,0)$, concave down on $(0,\infty)$, inflection point at $x=0$

21 concave up on $(-\pi,0)$, concave down on $(0,\pi)$, inflection point at $x=0$

23 concave up on $(-\frac{\pi}{2},\frac{\pi}{2})$

25 concave up on $(\frac{2}{3},\infty)$, concave down on $(-\infty,\frac{2}{3})$, inflection point at $x=\frac{2}{3}$

27 concave up on $(-\infty,\infty)$

29 concave up on $(1,\infty)$, concave down on $(-\infty,0)$ and $(0,1)$, inflection point at $x=1$

31 concave down on $(-3,3)$

33 concave up on $(-\infty,0)$ and $(1,\infty)$, concave down on $(0,1)$, inflection point at $x=1$ (There is no point, and therefore no inflection point, at $x=0$.)

35 concave up on $(0,3)$ and $(3,\infty)$, concave down on $(-\infty,0)$, inflection point $(0,\sqrt{3})$

37 concave up on $(4,\infty)$, concave down on $(0,4)$, inflection point $(4,-112)$

39 local max at $x=0$, local min at $x=5$

41 local max at $x=\sqrt{2}$, local min at $x=-\sqrt{2}$

43 local max point $(-2,2)$, local min point $(0,-2)$

45 local max point $(2,22)$, local min point $(-1,-5)$

47 local max point $(0,2)$, local min points $(-1,-3)$ and $(2,-30)$

49 local min point $(4,-40)$

51 local min point $(2,12)$

53 An inflection point is where the deficit changes from increasing to decreasing or from decreasing to increasing.

55 A miracle occurs! (An x-intercept is where the national debt is zero—that is, when it's all been paid off.)

57 f has a local max at $x=c$.

59 $f^{(4)}$ (I recall hearing a story (possibly an urban legend) about a U.S. politician crafting such a sentence to refer to the seventh derivative.)

Section 3.5

1

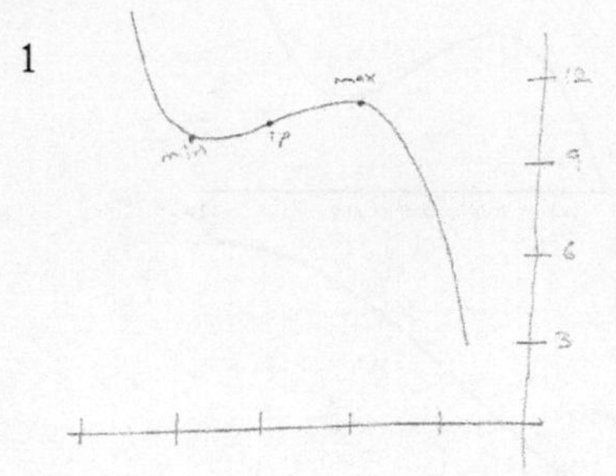

3

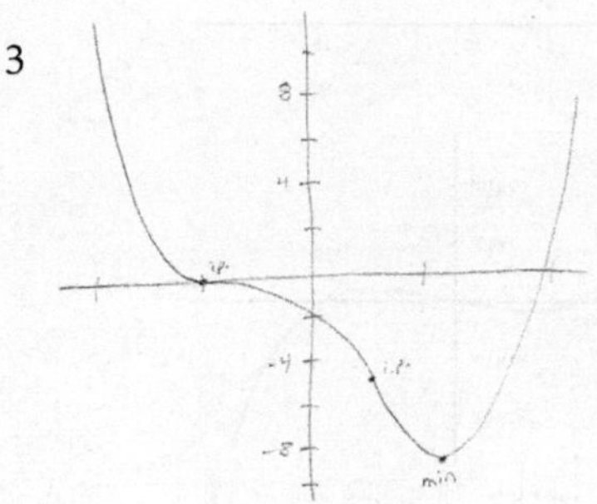

5

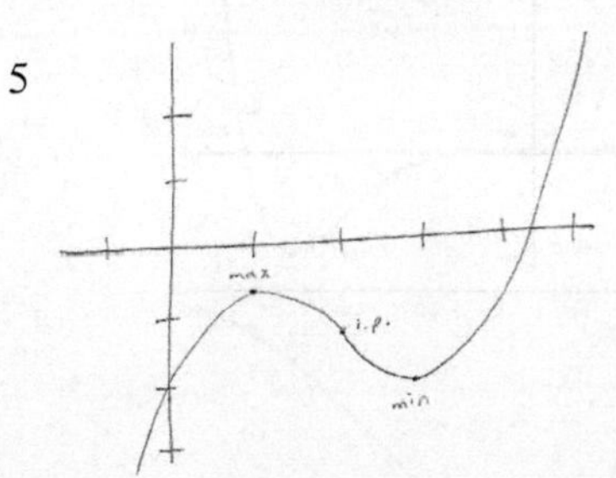

7

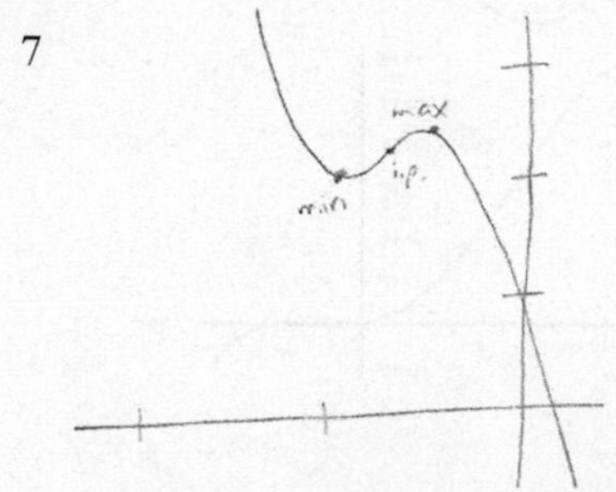

9

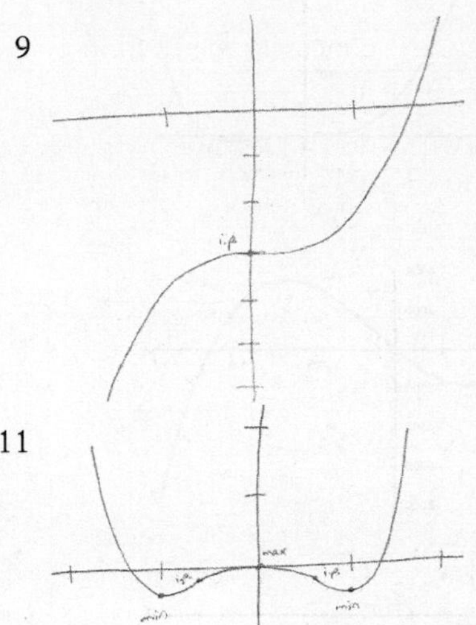

11

13

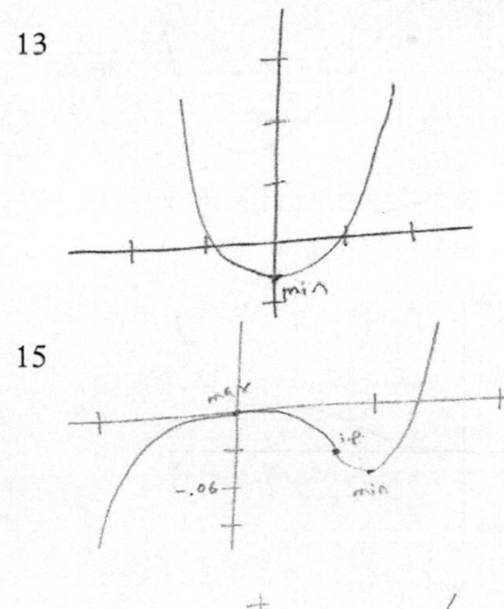

15

17

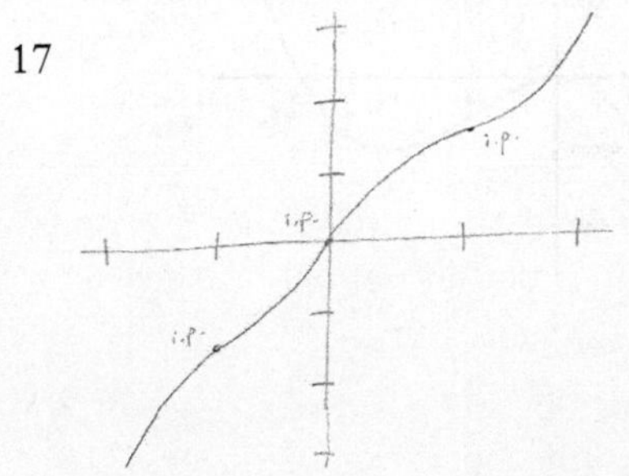

19 

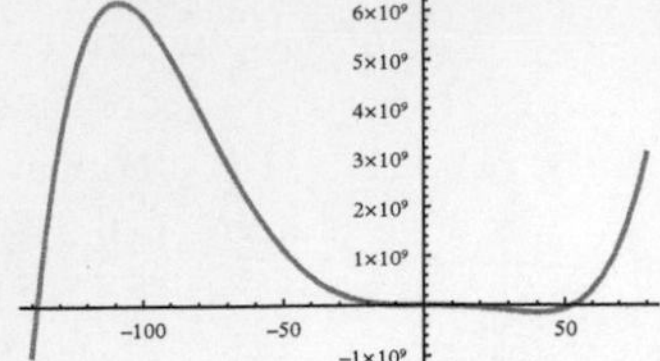
6×10⁹
5×10⁹
4×10⁹
3×10⁹
2×10⁹
1×10⁹
−1×10⁹
−100
−50
50

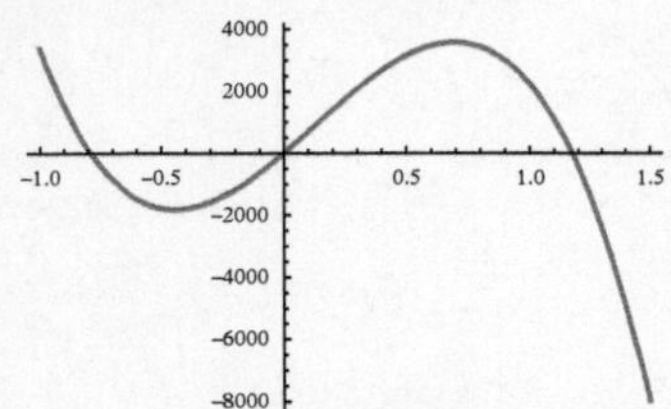
4000
2000
−2000
−4000
−6000
−8000
−1.0
−0.5
0.5
1.0
1.5

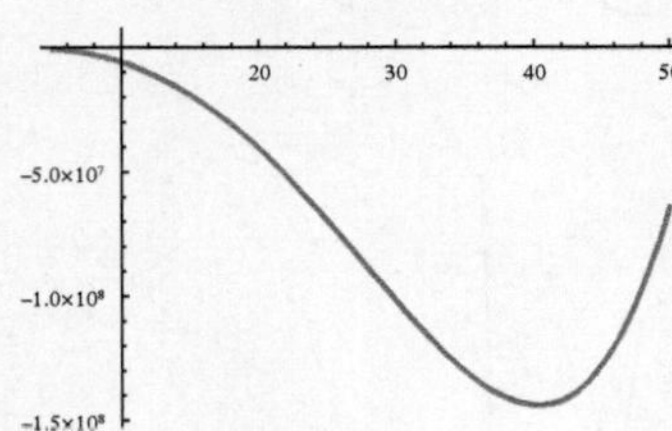
20
30
40
50
−5.0×10⁷
−1.0×10⁸
−1.5×10⁸

21

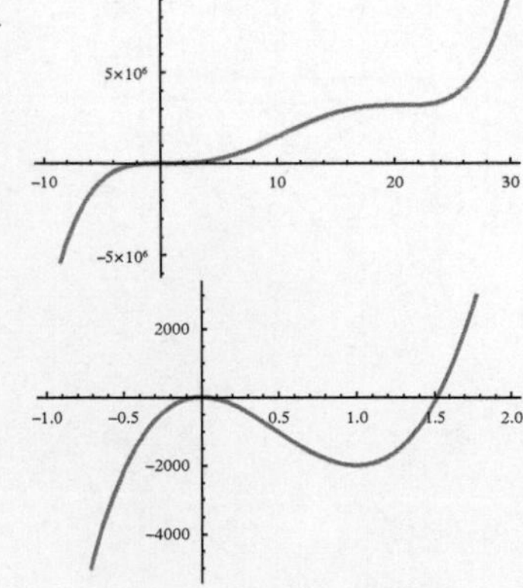
1×10⁷
5×10⁶
−5×10⁶
−10
10
20
30
2000
−2000
−4000
−1.0
−0.5
0.5
1.0
1.5
2.0

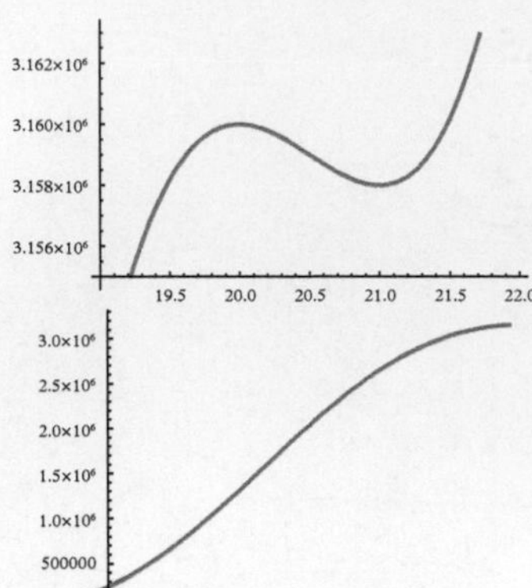
3.162×10⁶
3.160×10⁶
3.158×10⁶
3.156×10⁶
19.5
20.0
20.5
21.0
21.5
22.0
3.0×10⁶
2.5×10⁶
2.0×10⁶
1.5×10⁶
1.0×10⁶
500000
10
15

23

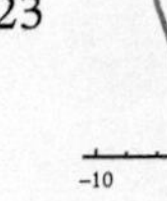

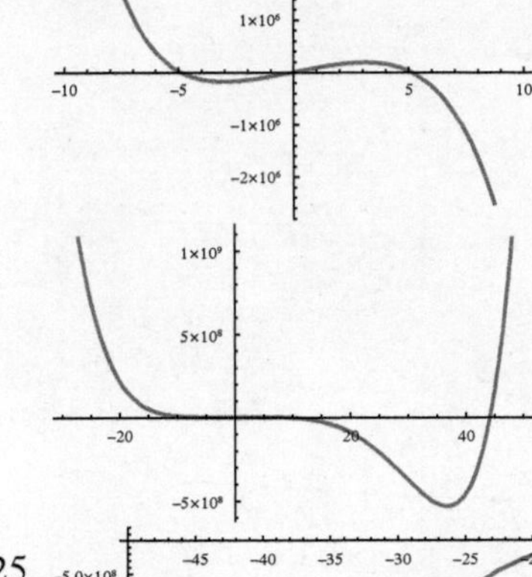
2×10⁶
1×10⁶
−1×10⁶
−2×10⁶
−10
−5
5
10
1×10⁹
5×10⁸
−5×10⁸
−20
20
40

25

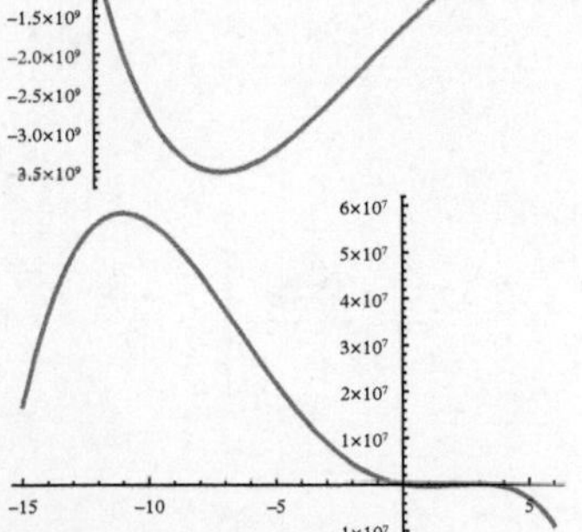
−45
−40
−35
−30
−25
−20
−5.0×10⁸
−1.0×10⁹
−1.5×10⁹
−2.0×10⁹
−2.5×10⁹
−3.0×10⁹
−3.5×10⁹
6×10⁷
5×10⁷
4×10⁷
3×10⁷
2×10⁷
1×10⁷
−1×10⁷
−15
−10
−5
5

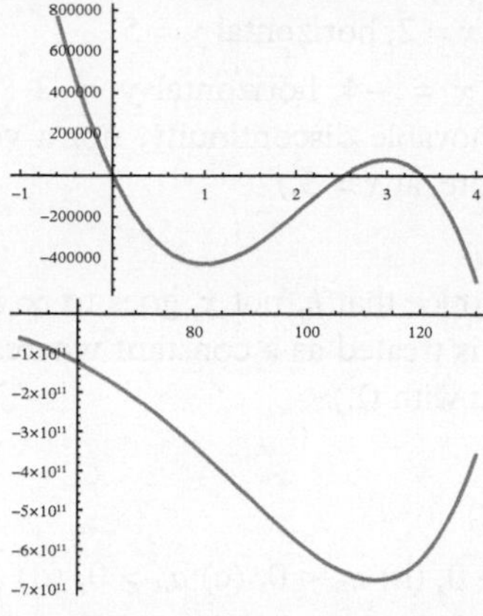

27 (a) local maximum points $(-108.564, 6.12824 \times 10^9)$ and $(0.697535, 3581.36)$, local minimum points $(-0.443177, -1820.79)$ and $(40.3093, -1.4409 \times 10^8)$; (b) inflection points $(-78.8911, 3.84871 \times 10^9)$, $(0.124627, 871.228)$, and $(27.7665, -8.74442 \times 10^7)$; (c) increasing on $(-\infty, -108.564)$, $(-0.443177, 0.697535)$, and $(40.3093, \infty)$; decreasing on $(-108.564, -0.443177)$ and $(0.697535, 40.3093)$; (d) concave up on $(-78.8911, 0.124627)$ and $(27.7665, \infty)$, concave down on $(-\infty, -78.8911)$ and $(0.124627, 27.7665)$

29 (a) local max points $(0, 0)$ and $(20, 3\,160\,000)$, local min points $(1, -1999)$ and $(21, 3\,158\,001)$; (b) inflection points $(0.487508, -980.781)$, $(10.5, 1\,579\,000.5)$, and $(20.5125, 3\,158\,981.8)$; (c) increasing on $(-\infty, 0)$, $(1, 20)$, and $(21, \infty)$; decreasing on $(0, 1)$ and $(20, 21)$; (d) concave up on $(0.487508, 10.5)$ and $(20.5125, \infty)$, concave down on $(-\infty, 0.487508)$ and $(10.5, 20.5125)$

31 (a) local max point $(3.13316, 186\,556)$, local min points $(-3, -180\,507)$ and $(36.7051, -5.3324 \times 10^8)$; (b) inflection points $(0.0669569, 3459.98)$ and $(29.2226, -3.50052 \times 10^8)$; (c) increasing on $(-3, 3.13316)$ and $(36.7051, \infty)$, decreasing on $(-\infty, -3)$ and $(3.13316, 36.7051)$; (d) concave up on $(-\infty, 0.0669569)$ and $(29.2226, \infty)$, concave down on $(0.0669569, 29.2226)$

33 (a) local max points $(-11.0083, 5.83922 \times 10^7)$ and $(3, 76\,005)$; local min points $(-41.8852, -3.51533 \times 10^9)$, $(0.999843, -427\,599)$, and $(112.227, -6.78419 \times 10^{11})$; (b) inflection points $(-32.8147, -2.19617 \times 10^9)$, $(-6.4612, 3.3324 \times 10^7)$, $(2.04499, -167\,310)$, and $(87.8976, -4.42714 \times 10^{11})$; (c) increasing on $(-41.8852, -11.0083)$, $(0.999843, 3)$, and $(112.227, \infty)$; decreasing on $(-\infty, -41.8852)$, $(-11.0083, 0.999843)$, and $(3, 112.227)$; (d) concave up on $(-\infty, -32.8147)$, $(-6.4612, 2.04499)$, and $(87.8976, \infty)$, concave down on $(-32.8147, -6.4612)$ and $(2.04499, 87.8976)$

Section 3.6

1 $y = \frac{1}{4}$

3 $y = 14$

5 no horizontal asymptote

7 no horizontal asymptote. The limit is not as $x \to \infty$ or as $x \to -\infty$.

9 vertical

11 neither

13 neither

15 (a) 2, (b) -2

17 (a) ∞, (b) 4

19 0

21 ∞

23 0

25 ∞

27 ∞ (Note: we can call the line $y = 4x - 7$ a *slant asymptote* on the graph of the function.)

29 $\frac{4}{7}$

31 0

33 $-\infty$

35 $\frac{2}{5}$

37 DNE

39 -3

41 $y = 4$ (on both sides)

43 $y = 3$ (on the right) and $y = -3$ (on the left)

45 $y = 0$ (both sides)

47 none

49 $y = 0$ on the right side only (The domain of the function is $(0, \infty)$.)

51 vertical $x = 2$, horizontal $y = 5$

53 vertical $x = -4$, horizontal $y = 2$ (There is a removable discontinuity, not a vertical asymptote, at $x = 3$.)

55 none

57 $-\frac{1}{2x}$ (Notice that h, not x, goes to ∞ in this limit; x is treated as a constant whereas h is replaced with Ω.)

59 $\tan \frac{1}{3}$

61 3

63 (a) $a_n > 0$, (b) $a_n < 0$, (c) $a_n > 0$, (d) $a_n < 0$

65 $\approx$100 m/min

Section 3.7

1

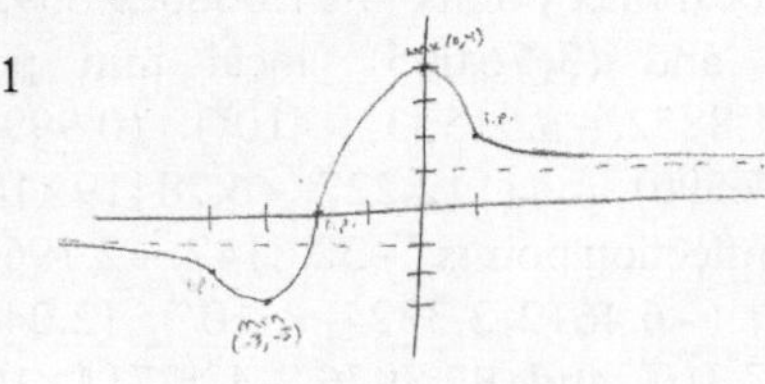

3

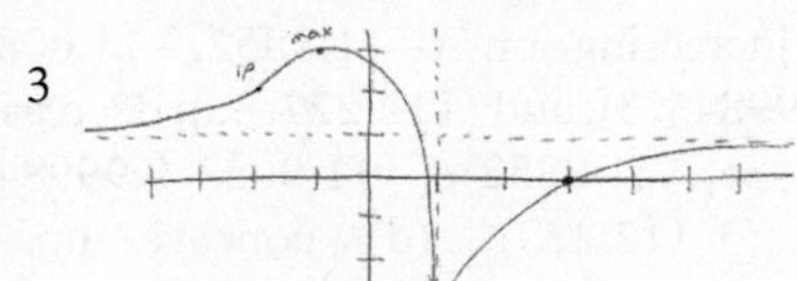

5

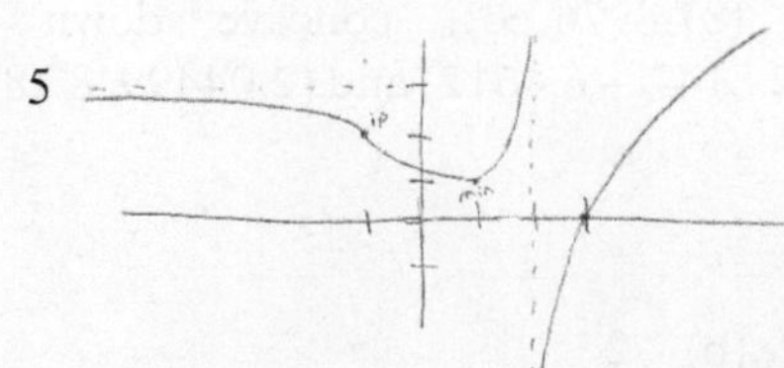

7

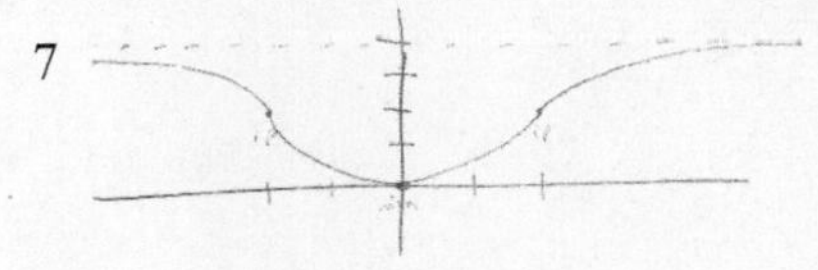

9

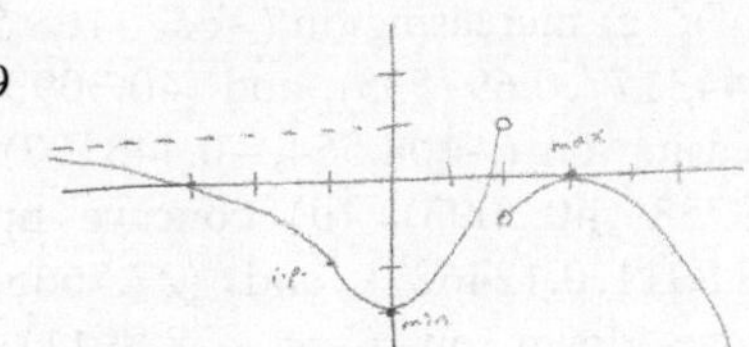

11

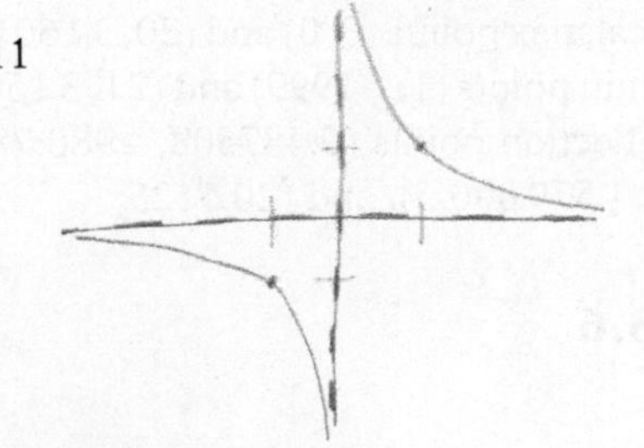

13

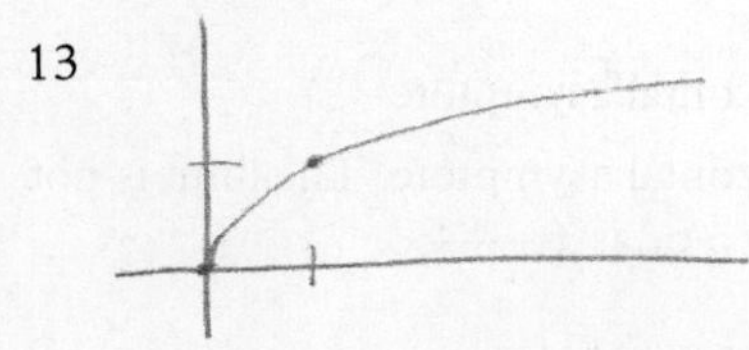

15

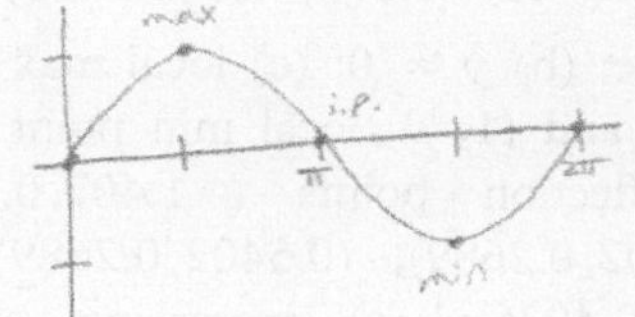

17

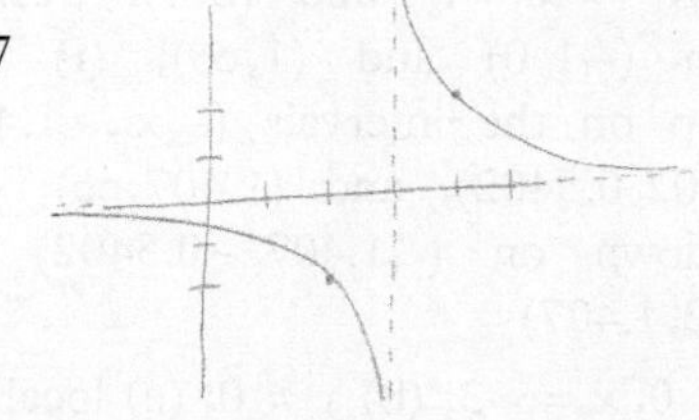

19

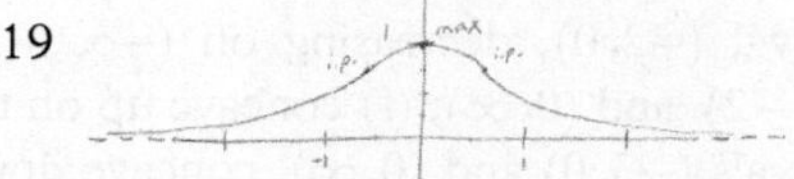

21

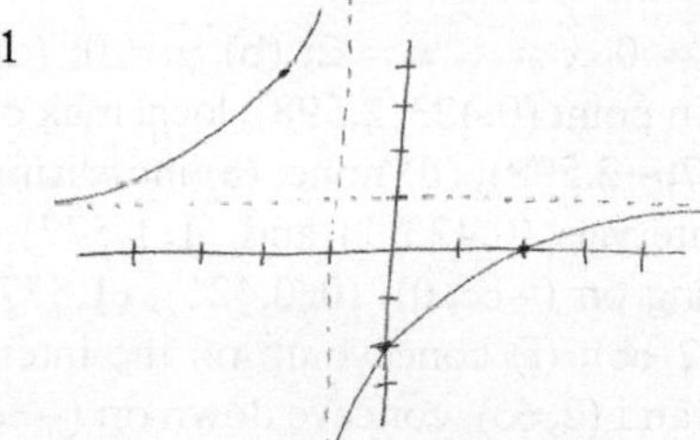

23

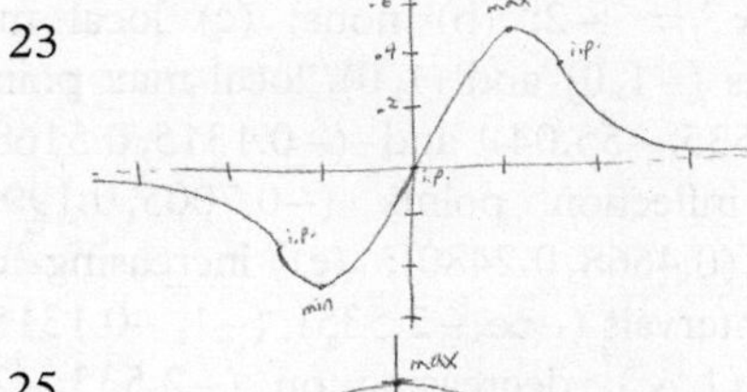

25

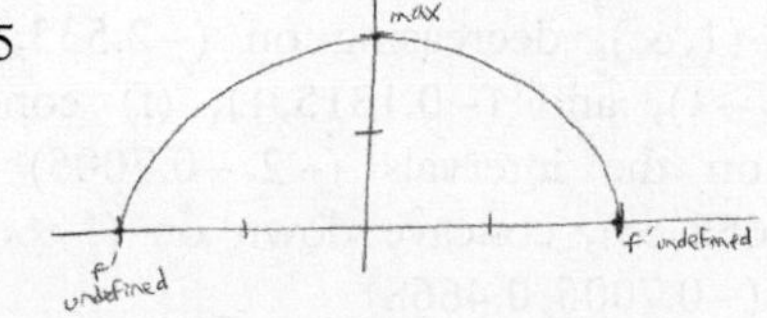

27

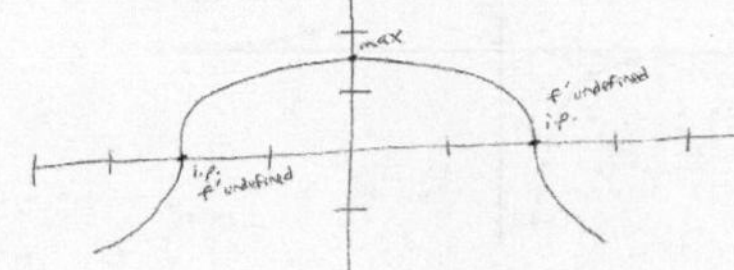

29

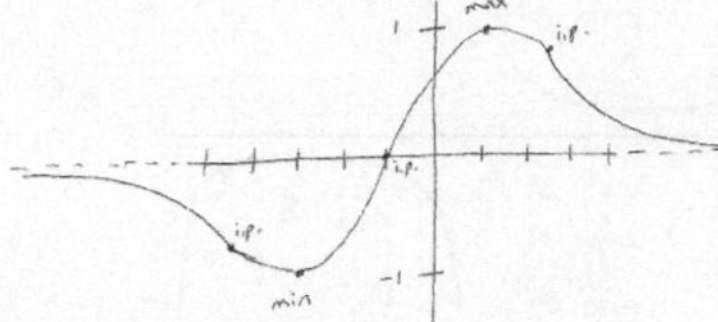

31

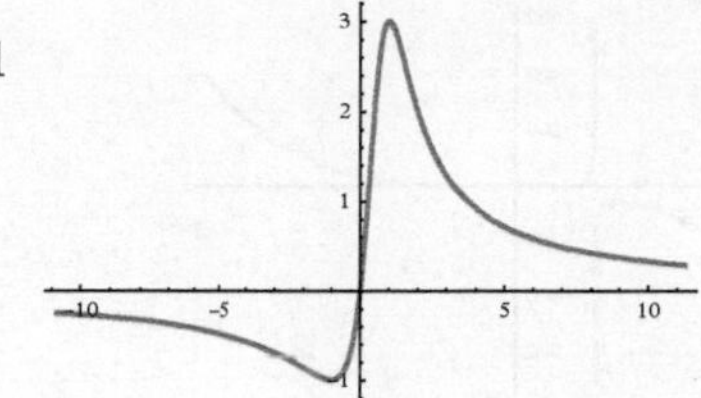

33

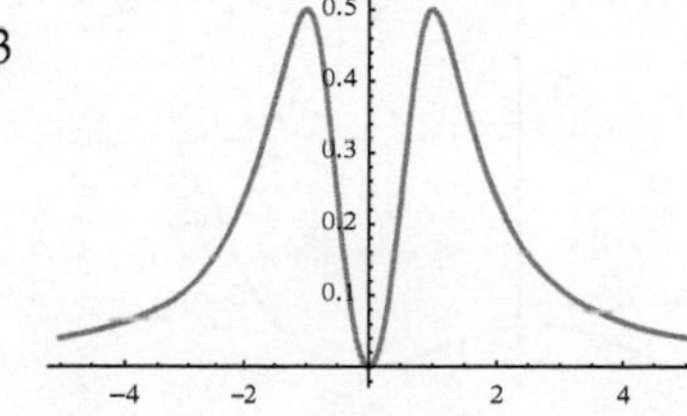

35

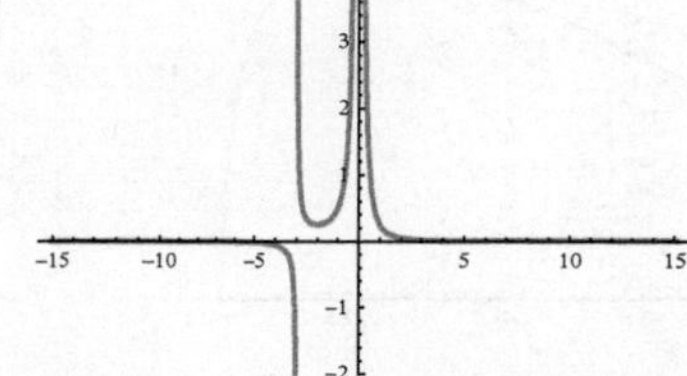

37

39

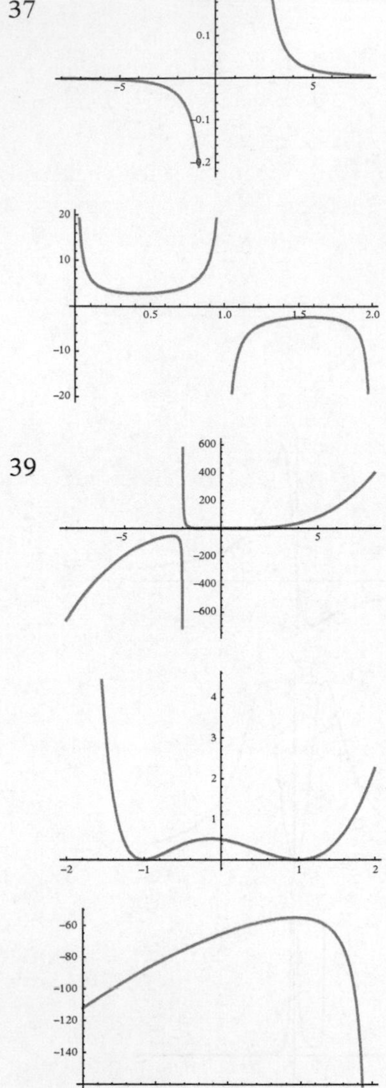

41 (a) none; (b) $y = 0$; (c) local max point $(1,3)$, local min point $(-1,-1)$; (d) inflection points $(1.532, 2.532)$, $(0.3473, 1.3473)$, and $(-1.879, -0.879)$; (e) increasing on the interval $(-1,1)$, decreasing on $(-\infty,-1)$ and $(1,\infty)$; (f) concave up on the intervals $(-1.879, 0.3473)$ and $(1.532,\infty)$, concave down on $(-\infty,-1.879)$ and $(0.3473, 1.532)$

43 (a) none; (b) $y = 0$; (c) local max points $(-1,\frac{1}{2})$ and $(1,\frac{1}{2})$, local min point $(0,0)$; (d) inflection points $(-1.407, 0.4026)$, $(-0.5402, 0.2689)$, $(0.5402, 0.2689)$, and $(1.407, 0.4026)$; (e) increasing on the intervals $(-\infty,-1)$ and $(0,1)$, decreasing on $(-1,0)$ and $(1,\infty)$; (f) concave up on the intervals $(-\infty,-1.407)$, $(-0.5402, 0.5402)$, and $(1.407,\infty)$, concave down on $(-1.407,-0.5402)$ and $(0.5402, 1.407)$

45 (a) $x = 0$, $x = -3$; (b) $y = 0$; (c) local min point $(-2,\frac{1}{4})$; (d) none; (e) increasing on the interval $(-2,0)$, decreasing on $(-\infty,-3)$, $(-3,-2)$, and $(0,\infty)$; (f) concave up on the intervals $(-3,0)$ and $(0,\infty)$, concave down on $(-\infty,-3)$

47 (a) $x = 0$, $x = 1$, $x = 2$; (b) $y = 0$; (c) local min point $(0.423, 2.598)$, local max point $(1.577,-2.598)$; (d) none; (e) increasing on the intervals $(0.423,1)$ and $(1,1.577)$, decreasing on $(-\infty,0)$, $(0,0.423)$, $(1.577,2)$, and $(2,\infty)$; (f) concave up on the intervals $(0,1)$ and $(2,\infty)$, concave down on $(-\infty,0)$ and $(1,2)$

49 (a) $x = -2$; (b) none; (c) local min points $(-1,0)$ and $(1,0)$, local max points $(-2.535,-55.04)$ and $(-0.1315, 0.5168)$; (d) inflection points $(-0.7005, 0.1996)$ and $(0.4668, 0.2480)$; (e) increasing on the intervals $(-\infty,-2.535)$, $(-1,-0.1315)$, and $(1,\infty)$; decreasing on $(-2.535,-2)$, $(-2,-1)$, and $(-0.1315,1)$; (f) concave up on the intervals $(-2,-0.7005)$ and $(0.4668,\infty)$, concave down on $(-\infty,-2)$ and $(-0.7005, 0.4668)$

Section 3.8

1 100 feet by 200 feet

3 22.5 feet by 45 feet

5 radius $r = \sqrt[3]{\frac{100}{\pi}}$ in $\approx$ 3.169 in, height $h = \frac{200}{\pi\left(\sqrt[3]{\frac{100}{\pi}}\right)^2}$ in $\approx$ 6.338 in

7 4.85 km east of the refinery

9 $\sqrt{300\,000}$ feet $\approx$ 548 feet by $\frac{450\,000}{\sqrt{300\,000}}$ feet $\approx$ 822 feet (the divider is parallel to the shorter sides)

11 squares of side length 2.201 inches

13 3 inches

15 (a) 244.9 widgets each month; (b) \$562 877.50

17 (a) \$553.25; (b) −\$2725 (Because the maximum profit is negative, we are losing money and perhaps we should quit making mini-widgets!) These values correspond to $x = 213.5$.

19 Yes, the student can meet the challenge by running 5.831 miles before jumping into the water and swimming the rest of the way, taking a total of 59.84 minutes.

21 $\sqrt[3]{\frac{464}{\pi}}$ in $\approx$ 5.29 in

23 $\frac{1}{\sqrt{3}}$ m by $\sqrt{\frac{2}{3}}$ m

25 width $\approx$ 5.6 feet, height of rectangle $\approx$ 2.8 feet (height of entire window $\approx$ 5.6 feet)

27 $\frac{-mt}{2(n-m)}$ \$/lb

29 $(10 - \sqrt{20})$ tons $\approx$ 5.53 tons

Section 3.9

1 $x = -2.2463$ ($a_2 = -2.2638$, $a_3 = -2.2482$, $a_4 = -2.2465$, $a_5 = a_6 = -2.2463$)

3 $x = 0.4897$ ($a_2 = 0.5930$, $a_3 = 0.4969$, $a_4 = 0.4898$, $a_5 = a_6 = 0.4897$) If your answer is $x = 0$ or if you could not find a solution, then you may be using the wrong function. Use $f(x) = x^6 - 14x^3 + 61x^2 - 13$ (not $\cdots + 13$).

5 $x = 20.1096$ ($a_2 = 19.1412$, $a_3 = 19.8106$, $a_4 = 20.0510$, $a_5 = 20.1061$, $a_6 = a_7 = 20.1096$)

7 Comment: check your a_2 carefully even if you have the correct answer; $x = -1.3024$ ($a_2 = -1.3283$, $a_3 = -1.3025$, $a_4 = a_5 = -1.3024$) If your $a_2 = -1.2862$, then you forgot to use the chain rule. Nevertheless, you should still have the correct final answer, illustrating the robustness of Newton's method. If your $a_2 = -1.5196$ and your final answer is $x = -1.5301$, your calculator is in degree mode; change it to radian mode.

9 $x = 1.0682$ ($a_2 = 1.0713$, $a_3 = a_4 = 1.0682$) If your answer is $x = -9.4360$ or $x = 1.0682$, you have a sign (±) error in your derivative. If your $a_2 = 1.9495$ and your values continue to rise slowly, then your calculator is in degree mode; change it to radian mode. If your $a_2 = 1.4913$ and your values continue to rise slowly, then you forgot to use the product rule.

11–22 Initial guesses and subsequent iterates vary. One or more typical initial guesses are given for each exercise. Your value of a_1 does not need to match one of the values given.

11 $x = -1.0676$ is the only solution. (For $a_1 = -1$, $a_2 = -1.0769$, $a_3 = -1.0678$, $a_4 = a_5 = -1.0676$. Although $a_1 = 1$ works, it is a poor choice and takes several more iterations to produce the solution.)

13 $x = 2.3261$ (For $a_1 = 2$: $a_2 = 2.4052$, $a_3 = 2.3292$, $a_4 = a_5 = 2.3261$; for $a_1 = 3$: $a_2 = 2.4841$, $a_3 = 2.3380$, $a_4 = a_5 = 2.3261$; for $a_1 = 2.5$: $a_2 = 2.3403$, $a_3 = 2.3262$, $a_4 = a_5 = 2.3261$; for $a_1 = 2.3$: $a_2 = 2.3264$, $a_3 = a_4 = 2.3261$. The answer $x = 2.6015$ is obtained by using degree mode instead of radian mode and is incorrect.)

15 $x = -2.0252$ and $x = 2.3847$ (For $a_1 = -2$: $a_2 = -2.0253$, $a_3 = a_4 = -2.0252$; for $a_1 = 2$: $a_2 = 2.4323$, $a_3 = 2.3852$, $a_4 = a_5 = 2.3847$; for $a_1 = 2.5$: $a_2 = 2.3877$, $a_3 = a_4 = 2.3847$.)

17 $x = -3.6022$ (The other two solutions to the equation are positive. For $a_1 = -4$: $a_2 = -3.7052$, $a_3 = -3.6113$, $a_4 = -3.6023$, $a_5 = a_6 = -3.6022$; for $a_1 = -3$: $a_2 = -4.8772$, $a_3 = -4.2186$, $a_4 = -3.8121$, $a_5 = -3.6361$, $a_6 = -3.6032$, $a_7 = a_8 = -3.6022$; for $a_1 = -3.5$: $a_2 = -3.6137$, $a_3 = -3.6023$, $a_4 = a_5 = -3.6022$.)

19 $x = 38.3323$ (For $a_1 = 40$: $a_2 = 38.2449$, $a_3 = 38.3321$, $a_4 = a_5 = 38.3323$; for $a_1 = 38$: $a_2 = 38.3295$, $a_3 = a_4 = 38.3323$.)

21 $x = -4.5956$ and $x = 2.5956$ (Notice the symmetry in the solutions, as well as in the Newton's method iterates! For $a_1 = -5$: $a_2 = -4.6750$, $a_3 = -4.5995$, $a_4 = a_5 = -4.5956$; for $a_1 = -4$: $a_2 = -5.0833$, $a_3 = -4.7052$, $a_4 = -4.6028$, $a_5 = a_6 = -4.5956$; for $a_1 = -4.5$: $a_2 = -4.6021$, $a_3 = a_4 = -4.5956$; for $a_1 = 3$: $a_2 = 2.6750$, $a_3 = 2.5995$, $a_4 = a_5 = 2.5956$; for $a_1 = 2$: $a_2 = 3.0833$, $a_3 = 2.7052$, $a_4 = 2.6028$, $a_5 = a_6 = 2.5956$; for $a_1 = 2.5$: $a_2 = 2.6021$, $a_3 = a_4 = 2.5956$.)

23 $\sqrt[7]{4911} \approx 3.367519$ (For $a_1 = 3$: $a_2 = 3.533804$, $a_3 = 3.389236$, $a_4 = 3.367932$, $a_5 = a_6 = 3.367519$.)

25 $x \approx 0.702$

27 $x \approx -2.36$, $x \approx 1.22$

29 Newton's method fails because $a_2 = -1$ falls on a vertical asymptote and is not in the domain of the function.

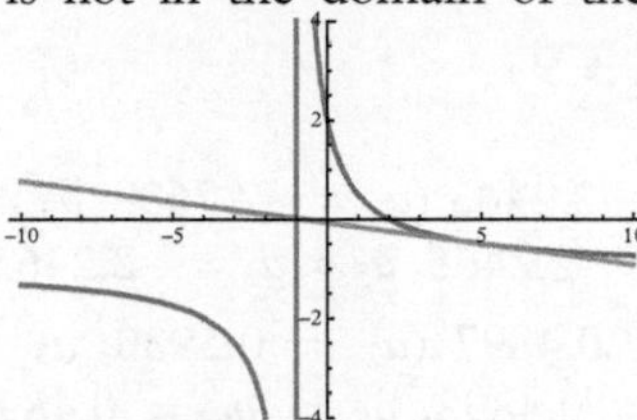

31 Newton's method fails because the iterates bounce back and forth between two values, 1 and -1 ($a_2 = a_4 = a_6 = \cdots = -1$, $a_3 = a_5 = a_7 = \cdots = 1$).

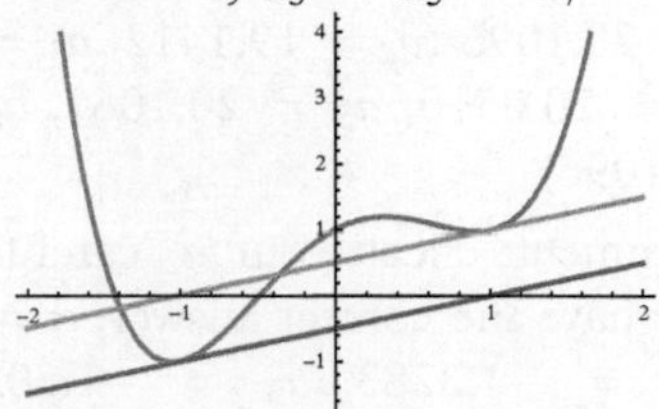

Section 4.1

1 no (quotient)

3 no (composition; "not just plain x" to a power)

5 no (product)

7 yes (The antiderivative difference rule, antiderivative sum rule, antiderivative constant multiple rule, and antiderivative power rule can be used.)

9 yes ($\int \sec x \tan x\, dx$ is in the list of trig antiderivatives.)

11 no (composition; the $4x$ is "not just plain x")

13 $\frac{x^9}{9} + C$ (Note that the $+ C$ is required. The answer is incorrect if the $+ C$ is missing! The same is true for all these exercises.)

15 $\frac{x^{-3}}{-3} + C$ (Recall that adding 1 to -4 results in $-4 + 1 = -3$, not -5.)

17 $\frac{x^{3/2}}{3/2} + C = \frac{2}{3}x^{3/2} + C = \frac{2}{3}x\sqrt{x} + C$ (Recall that $\sqrt{x} = x^{1/2}$, so the antiderivative power rule can be used.)

19 $\frac{5x^3}{3} + C$

21 $\frac{6x^{-1}}{-1} + C = \frac{-6}{x} + C$ (Recall that $\frac{6}{x^2} = 6x^{-2}$, so the antiderivative constant multiple rule and antiderivative power rule can be used.)

23 $\frac{x^3}{3} + 5\frac{x^2}{2} + C$

25 $\frac{x^5}{5} + \frac{7}{4}x^4 - \frac{1}{2}x^2 + 6x + C$

27 $\frac{x^3}{3} - \sqrt{11}x + C$ (Recall that $\sqrt{11}$ is a constant.)

29 $\frac{1}{2}x^4 - \frac{5}{3}x^3 + x + C$

31 $\frac{5x^3}{3} - x^2 + 7x - \frac{2}{3}x^{3/2} - \frac{1}{2x^2} + C$

33 $-\cos x + C$

35 $\sec x + C$

37 $4 \sin x + 7 \cot x + C$

39 $4x + \sin x + C$

41 $\int (x - 4)(x + 1)\, dx = \int (x^2 - 3x - 4)\, dx = \frac{x^3}{3} - \frac{3x^2}{2} - 4x + C$

43 $\int \frac{x^2 + 5x}{x}\, dx = \int (x + 5)\, dx = \frac{x^2}{2} + 5x + C$ (Comment: because the original integrand is undefined at $x = 0$, the interval over which this antiderivative is valid cannot contain $x = 0$.)

45 $\int (x - 1)^2\, dx = \int (x^2 - 2x + 1)\, dx = \frac{x^3}{3} - x^2 + x + C$

47 $\int \sqrt{(3x + 1)^4}\, dx = \int (3x + 1)^2\, dx = \int (9x^2 + 6x + 1)\, dx = 3x^3 + 3x^2 + x + C$

49 $\int \tan x \cos x\, dx = \int \frac{\sin x}{\cancel{\cos x}} \cancel{\cos x}\, dx = \int \sin x\, dx = -\cos x + C$

51 $\int \sec x \frac{\sin x}{\cos x}\, dx = \int \sec x \tan x\, dx = \sec x + C$

53 $\int \frac{x^2 + 3}{\sqrt{x}}\, dx = \int \left(\frac{x^2}{\sqrt{x}} + \frac{3}{\sqrt{x}}\right) dx = \int (x^{3/2} + 3x^{-1/2})\, dx = \frac{2}{5}x^{5/2} + 6x^{1/2} + C$

55 $\frac{x^4}{4} - \frac{4x^3}{3} + \frac{7x^2}{2} - x + C$

57 $\sin x + 2x^{3/2} + C$

59 $\int \frac{1}{x^7}\, dx = \int x^{-7}\, dx = -\frac{1}{6x^6} + C$

61 not possible with available rules and formulas ($\int \frac{1}{x}\, dx = \int x^{-1}\, dx$, but the antiderivative power rule is not valid for an exponent of -1.)

63 $-6 \cot x + C$

65 $\int x^3(4x - 9)\, dx = \int (4x^4 - 9x^3)\, dx = \frac{4x^5}{5} - \frac{9x^4}{4} + C$

67 $\frac{-3}{4x} + C$

69 $\int (5x - 2)^2\, dx = \int (25x^2 - 20x + 4)\, dx = \frac{25x^3}{3} - 10x^2 + 4x + C$

71 not possible with available rules and formulas

73 $\int \frac{x^3 - 7x}{4x}\, dx = \int \left(\frac{1}{4}x^2 - \frac{7}{4}\right) dx = \frac{1}{12}x^3 - \frac{7}{4}x + C$

75 $4x + \sqrt{17.2}x + C$ (The integrand is a constant.)

77 $\int \left(\frac{1}{\sin^2 x} + \frac{1}{x^2}\right) dx = \int (\csc^2 x + x^{-2})\, dx = -\cot x - \frac{1}{x} + C$

79 answers vary. Possible answers include $F(x) = \frac{x^3}{3}$ and $F(x) = \frac{x^3}{3} + 13.8$. Because the question asks for *an* antideriative, the response should be one function and not a collection of functions. The answer $F(x) = \frac{x^3}{3} + C$ is therefore inappropriate.

81 answers vary. Possible answers include $F(x) = \sin x$ and $F(x) = \sin x + 1$. Any answer of the form $F(x) = \sin x + \text{constant}$ is acceptable, but writing $F(x) = \sin x + C$ is not, per note in answer 79.

83 see notes on answers to 79 and 81. One possible answer is $F(x) = \pi^3 x$. Note that π^3 is a constant.

85 see notes in answers 79 and 81. One possible answer is $F(x) = \frac{1}{4}x^4 + 2x^{-2}$.

87 see notes in answers 79 and 81. One possible answer is $\frac{ax^3}{3} + \frac{bx^2}{2} + cx$

89 velocity

91 see notes in answers 79 and 81. One possible answer is the three functions $F_1(x) = \frac{5x^2}{2} - 8x$, $F_2(x) = \frac{5x^2}{2} - 8x + 3$, and $F_3(x) = \frac{5x^2}{2} - 8x - 17$.

93 $\int \sin x\, dx = \text{vers}\, x + C$ (This does not conflict with the usual formula $\int \sin x\, dx = -\cos x + C$, because the two antiderivatives differ by a constant: $\text{vers}\, x = -\cos x + 1$.)

Section 4.2

1 left-hand endpoints, lower estimate

3 right-hand endpoints, upper estimate

5 midpoints

7 none of the above

9 none of the above

11

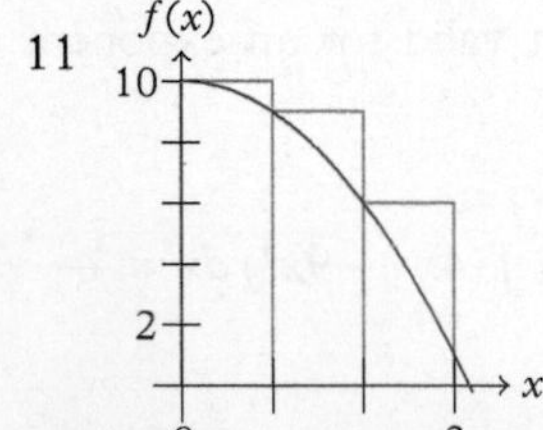

13

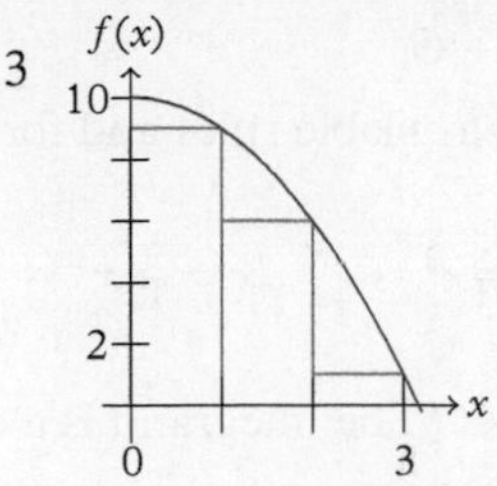

15

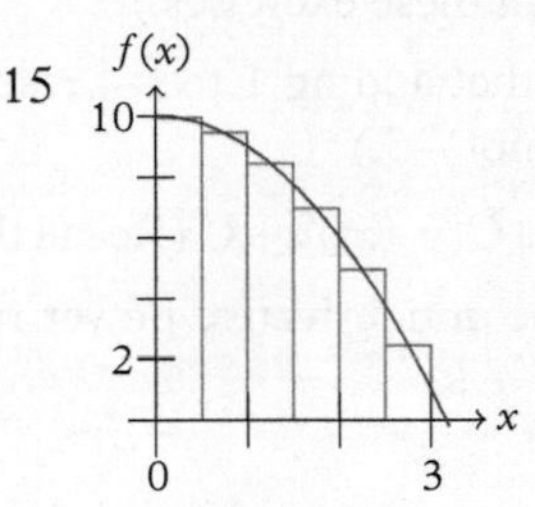

17

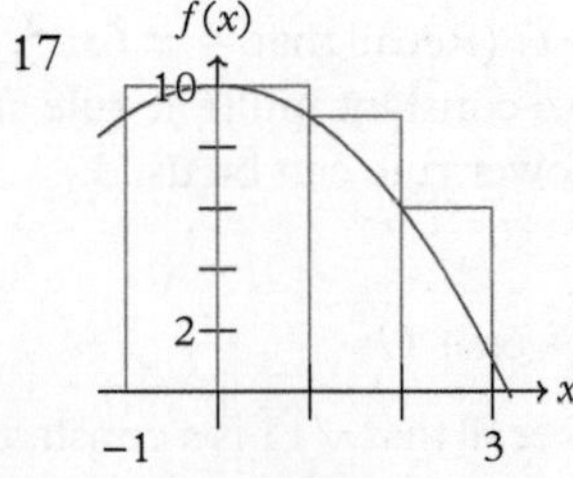

19

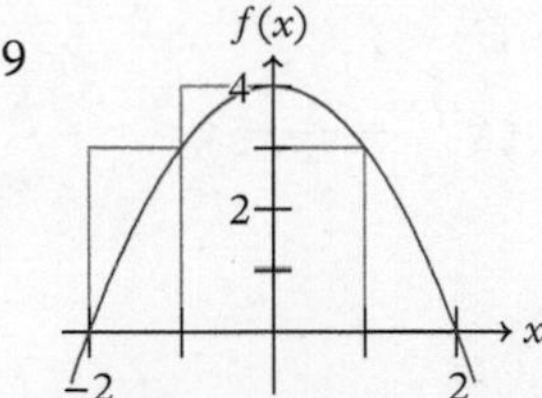

(The rectangle over the subinterval $[1, 2]$ has height 0 and hence is not visible.)

21 16.146 units^2

23 17.505 units^2

25 17.944 units^2

27 $\frac{\pi}{4}(1+\sqrt{2})$ units^2= 1.896 units^2

29 $\frac{\pi}{4}(2+\sqrt{2})$ units^2= 2.682 units^2

31 $\frac{\pi}{6}(2+\sqrt{3})$ units^2= 1.954 units^2

33 $\frac{\pi}{6}(1+\sqrt{3})$ units^2= 1.430 units^2

35 384 units^2

37 520 units^2

39 5.512 units^2

41 The area under the curve is between 18.822 units^2 and 20.865 units^2.

43 The area under the curve is between $\frac{7}{5}$ units^2 = 1.4 units^2 and 3 units^2.

45 The area under the curve is between $\frac{47}{26}$ units^2 = 1.808 units^2 and $\frac{339}{130}$ units^2 = 2.608 units^2.

47 $\frac{1}{4}\cdot 5.7+\frac{1}{4}\cdot 6.6+\frac{1}{4}\cdot 6.8$
$+\frac{1}{4}\cdot 6.0+\frac{1}{4}\cdot 6.5+\frac{1}{4}\cdot 7.1$
$+\frac{1}{4}\cdot 7.3+\frac{1}{4}\cdot 6.2$
= 13.05 mi (If your answer is 52.2 miles, you forgot to multiply by the "width" of each time interval. If your answer is 783 miles, you forgot to convert minutes to hours. If your answer is 14.65 miles, you mistakenly included a ninth 15-minute time period with speed 6.4 mph.)

49 $\frac{1}{4}(5.7+6.6+6.0+6.0+6.5+7.1+6.2+6.2)$ = 12.575 mi

51 $\frac{1}{2}\cdot 8+\frac{1}{2}\cdot 7+\frac{1}{2}\cdot 9$
$+\frac{1}{2}\cdot 11+\frac{1}{2}\cdot 12+\frac{1}{2}\cdot 11$
$+\frac{1}{2}\cdot 9+\frac{1}{2}\cdot 8$
= 37.5 nautical miles (nmi) (If your answer is 75 nmi, you forgot to multiply by the "width" of each time interval. If your answer is 2250 nmi, you forgot to convert minutes to hours. If your answer is 40.5 nmi, you mistakenly included a ninth 30-minute time period with speed 6 knots.)

53 $\frac{1}{2}(8+8+9+11+12+12+11+9) = 40$ nmi

55 −6.25 units^2 (or 6.25 units^2). If left negative, it signifies the area lies beneath the x-axis instead of above the x-axis. Under the mantra that area is always nonnegative, the answer is 6.25 units^2.)

57 answers vary. One possibility is $y = 2+\cos x$ on $[0, 8\pi]$ using $n = 4$ rectangles:

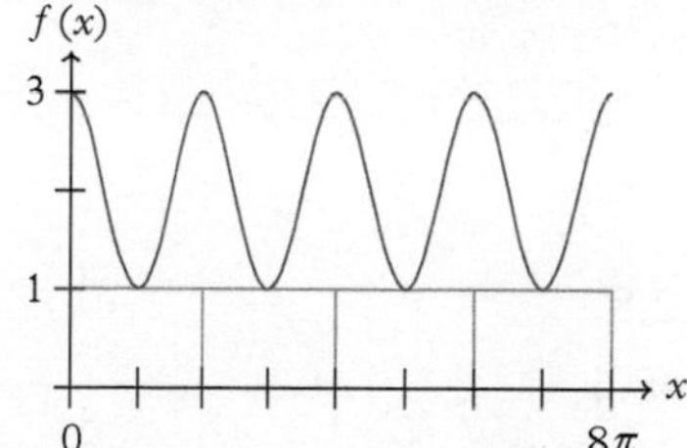

Section 4.3

1 $\frac{1}{2}+\frac{1}{3}+\frac{1}{4}+\frac{1}{5}$

3 $(-2)^2+(-1)^2+0^2+1^2$

5 $3+4+5+6+7+8+9+10$

7 $\sqrt{1}+\sqrt{6}+\sqrt{11}+\sqrt{16}+\sqrt{21}$

9 30

11 0

13 235

15 125

17 $\sum_{k=5}^{125} k^4$

19 $\sum_{k=1}^{14} \sqrt{k\omega}$

21 $\sum_{k=5}^{99} (4+k\omega)^3$

23 $\sum_{k=1}^{9} \tan\left(3 + \frac{k\pi\omega}{4}\right)$

25 $\sum_{k=2}^{\Omega} (1 + k\omega)^2$

27 10

29 $\frac{19}{3}$

31 $\frac{41}{3}$

33 4

35 $\frac{125}{3}$ units^2

37 39 units^2

39 32 units^2

41 28 units^2

43 $\frac{16}{3}$ units^2

45 (a) 0, (b) $2\omega \doteq 0$

Section 4.4

1 definite

3 indefinite

5 indefinite

7 real number

9 real number

11 collection of functions

13 $\frac{1}{10}$

15 −39

17 64

19 24

21 $-\frac{4}{3}$

23 $\frac{64}{5} = 12.8$

25 $\frac{243}{5} = 48.6$

27 The area under the curve $y = x^9$ from $x = 0$ to $x = 1$ is $\frac{1}{10}$ units^2. (An alternate answer: the net area ….)

29 The net area under the curve $y = -x^2$ from $x = 2$ to $x = 5$ is -39^2. (Because this curve lies below the x-axis on the specified interval, an alternate answer is: the area between the curve and the x-axis from $x = 2$ to $x = 5$ is 39 units^2.)

31 The net area under the curve $y = 10 - x^2$ from $x = 1$ to $x = 5$ is $-\frac{4}{3}$ units^2. (Caution: because this curve lies partially above the x-axis and partially below the x-axis over the specified interval, the term *net area* is required. Answers such as "the area under the curve … is $-\frac{4}{3}$ units^2" or "the area between the curve and the x-axis is $-\frac{4}{3}$ units^2" are incorrect.)

33 $\int_3^{19} \left(x^5 - 4x + 1\right)\, dx$ (All portions of the answer, including the dx, are required.)

35 $\int_{\pi/2}^{3\pi/2} (\cos(x + \pi) + \sin(4x))\, dx$

37 $\int_0^1 4x\sqrt{1 - x^4}\, dx$

39 $\int_a^b \left((f(x))^2 + (g(x))^2\right)\, dx$

41 16

43 −16

45 −30

47 24

49 0

51 $4\sqrt{11} - 20$

53 0

55 16

57 5

59 8

69 1

71 $\sin b - \sin a$

Section 4.5

1 yes

3 no; the integrand is not continuous at $x = 0$, which is in the interval $[-5, 8]$.

5 yes

7 no; the integrand is not continuous at several places in the interval, including $x = \pm\frac{\pi}{2}$.

9 yes; although the integrand is not continuous at $x = 3$, this number is not in the interval $[-4, -1]$.

11 $\frac{255}{4}$

13 16

15 $\frac{19}{6}$

17 1

19 $-\frac{6}{7}$

21 $\frac{40}{3}$ units^2 = 13.333 units^2

23 $\frac{111}{4}$ units^2 = 27.75 units^2

25 $\frac{32}{3}$ units^2 = 10.667 units^2

27 $\frac{3}{4}$ units^2

29 2 units^2

31 $\int_0^1 (16x^2 - 56x + 49)\, dx = \cdots = \frac{79}{3}$

33 $\int_1^4 (x^3 - 7x + x^{-2})\, dx = \cdots = 12$

35 $\int_0^5 2\sqrt{x}\, dx = \int_0^5 2x^{1/2}\, dx = \cdots = \frac{20}{3}\sqrt{5}$

37 $\int_{-0.2}^{0.3} 1\, dx = \cdots = 0.5$

39 $\int_2^1 (x - 5)\, dx = \cdots = \frac{7}{2}$

41 2

43 the fundamental theorem of calculus, part II (FTC II) does not apply. (The integrand is not continuous at $x = 0$.)

45 15

47 0 (Upper and lower limits of integration are identical.)

49 2

51 the available antiderivative formulas and rules are insufficient to determine an antiderivative (square root of "not just plain x," not simplifiable).

53 FTC II does not apply. (The integrand is not continuous at $x = \frac{\pi}{2}$; neither secant nor tangent are defined at $\frac{\pi}{2}$.)

55 13 (Rewrite as $\int_2^5 \frac{1}{3}x^2\, dx$.)

57 $\frac{7}{6}$

59 FTC II does not apply. (The limits of integration must be real numbers.)

61 770

63 8

65 the available antiderivative formulas and rules are insufficient to determine an antiderivative (sine of "not just plain x").

67 $\frac{162}{5} = 32.4$

69 6 (Note that $\sqrt{x^2} = |x|$, and because the interval of integration contains only negative values of x, $|x| = -x$ on that interval. Use $\int_{-4}^{-2} -x\, dx$.)

71 $\dfrac{dy}{dt} = \dfrac{7\cos x}{5x^4 + 8}$

73 $F'(x) = \dfrac{\sin(x^4 + x)}{5x + 18}$

75 $F'(t) = (3t + 5t^2 + t^4)^7$ (Notice that the roles of t and x are reversed.)

77 $F'(x) = -\tan x^4 \cos 3x$ (Rewrite as $F(x) = -\int_4^x \tan t^4 \cos 3t\, dt$.)

79 $F'(x) = \dfrac{1 - x}{\sqrt{x^4 + 8x^2 + 22}}$

81 $(-2, 2)$ (Notice the integrand is undefined at $t = \pm 2$ (division by zero), but the integrand must be continuous throughout the interval. Alternate answer: any $[a, b]$ with $-2 < a \le 1 \le b < 2$ with $a \ne b$.)

83 $F'(x) = 2x\sqrt{3 + \sin x^2}$ (Because $F(x) = G(H(x))$, by chain rule, $F'(x) = G'(H(x))H'(x)$).

85 $F''(x) = \dfrac{-x}{(x^2 - 1)^{3/2}}$

Section 4.6

1 yes

3 no

5 no

7 yes (technically no, if you are willing to find the eighth power of $4x - 7$)

9 yes

11 succeed

13 fail

15 succeed

17 fail

19 succeed

21 $-\frac{1}{4}\cos(x^4 + 12) + C$

23 $\frac{(x^3+5)^{15}}{45} + C$

25 $\frac{1}{2}\sin(2x + 5) + C$

27 $x^2 + 5\sin x + C$ (This integral does not require substitution.)

29 $\frac{1}{7}\sec 7x + C$

31 $3\sqrt{2x + 7} + C$

33 $\frac{3}{2}\sin(x^2 + 1) + C$

35 $\frac{2}{3}(x^3 + 4x^2)^{3/2} + C = \frac{2}{3}(x^3 + 4x^2)\sqrt{x^3 + 4x^2} + C$

37 $\frac{2}{5}x^{5/2} + C$ (The substitution $u = x^3$ is unsuccessful, but also unnecessary. Rewrite the integrand as $x^{3/2}$ and use the antiderivative power rule.)

39 $-\frac{1}{\cos x} + C$ (alternate solution not requiring substitution: rewrite as $\int -\tan x \sec x\, dx = -\sec x + C$.)

41 $-3\cos(x^2 + x) + C$

43 $\frac{x^3}{3} - 2x^{3/2} + \frac{2}{9}(x^3 + 4)^{3/2} + C$

45 $-\frac{1}{3}\cos x^3 + C$

47 $\frac{2}{7}y^{7/2} - \frac{16}{5}y^{5/2} + \frac{4}{3}y^{3/2} + C$ (Substitution is not needed.)

49 cannot be done with the available antiderivative formulas, rules, and techniques (Substitution fails. There is a formula for this integral in section 5.8.)

51 $-\frac{1}{6}\cot(6z - 15) + C$

53 cannot be done with the available antiderivative formulas, rules, and techniques (Substitution fails. This integral requires a function that you have not studied.)

55 $\frac{1}{2}(x - \sin x)^2 + C$ (Use substitution with $u = x - \sin x$.)

57 cannot be done with the available antiderivative formulas, rules, and techniques (Substitution succeeds, but a technique from section 7.3 is required to complete the integral.)

59 $\frac{1}{17}(x^2 + 3)^{17} - \frac{3}{16}(x^2 + 3)^{16} + C$

Section 4.7

1 $\frac{759\,376}{5} = 151\,875.2$; (a) $u = 2x^3 - 1$, $du = 6x^2\,dx$, $\int 6x^2(2x^3 - 1)^4\,dx = \int u^4\,du = \frac{u^5}{5} + C = \frac{(2x^3-1)^5}{5} + C$, $\int_0^2 6x^2(2x^3 - 1)^4\,dx = \left.\frac{(2x^3-1)^5}{5}\right|_0^2 = \cdots = \frac{759\,376}{5}$; (b) $u = 2x^3 - 1$, $du = 6x^2\,dx$, $x = 2$: $u = 15$, $x = 0$: $u = -1$, $\int_0^2 6x^2(2x^3 - 1)^4\,dx = \int_{-1}^{15} u^4\,du = \left.\frac{u^5}{5}\right|_{-1}^{15} = \cdots = \frac{759\,376}{5}$.

3 -1; (a) $u = 5x - \pi$, $du = 5\,dx$, $\int 5\cos(5x - \pi)\,dx = \int \cos u\,du = \sin u + C = \sin(5x - \pi) + C$, $\int_0^{\pi/2} 5\cos(5x - \pi)\,dx = \sin(5x - \pi)|_0^{\pi/2} = \cdots = -1$; (b) $u = 5x - \pi$, $du = 5\,dx$, $x = \pi/2$: $u = 3\pi/2$, $x = 0$: $u = -\pi$, $\int_0^{\pi/2} 5\cos(5x - \pi)\,dx = \int_{-\pi}^{3\pi/2} \cos u\,du = \sin u|_{-\pi}^{3\pi/2} = \cdots = -1$.

5 $\frac{14}{9}$; (a) $u = 3y+1$, $du = 3\,dy$, $\int \sqrt{3y+1}\,dy = \frac{1}{3}\int 3\sqrt{3y+1}\,dy = \frac{1}{3}\int \sqrt{u}\,du = \frac{1}{3}\frac{u^{3/2}}{3/2} + C = \frac{2}{9}(3y+1)^{3/2} + C$, $\int_0^1 \sqrt{3y+1}\,dy = \frac{2}{9}(3y+1)^{3/2}\Big|_0^1 = \cdots = \frac{14}{9}$; (b) $u = 3y+1$, $du = 3\,dy$, $y = 1$: $u = 4$, $y = 0$: $u = 1$, $\int_0^1 \sqrt{3y+1}\,dy = \frac{1}{3}\int_0^1 3\sqrt{3y+1}\,dy = \frac{1}{3}\int_1^4 \sqrt{u}\,du = \frac{1}{3}\frac{u^{3/2}}{3/2}\Big|_1^4 = \cdots = \frac{14}{9}$.

7 0; (a) $u = \sin x$, $du = \cos x\,dx$, $\int \sin^7 x \cos x\,dx = \int u^7\,du = \frac{u^8}{8} + C = \frac{\sin^8 x}{8} + C$, $\int_{-\pi}^{\pi} (\sin x)^7 \cos x\,dx = \frac{(\sin x)^8}{8}\Big|_{-\pi}^{\pi} = \cdots = 0$; (b) $u = \sin x$, $du = \cos x\,dx$, $x = \pi$: $u = 0$, $x = -\pi$: $u = 0$; $\int_{-\pi}^{\pi} (\sin x)^7 \cos x\,dx = \int_0^0 u^7\,du = 0$ (The limits of integration are the same!)

9 $\frac{52}{3}$; (a) $u = t^3 + 1$, $du = 3t^2\,dt$, $\int 3t^2\sqrt{t^3+1}\,dt = \int \sqrt{u}\,du = \frac{2}{3}u^{3/2} + C = \frac{2}{3}(t^3+1)^{3/2} + C$, $\int_0^2 3t^2\sqrt{t^3+1}\,dt = \frac{2}{3}(t^3+1)^{3/2}\Big|_0^2 = \cdots = \frac{52}{3}$; (b) $u = t^3 + 1$, $du = 3t^2\,dt$, $t = 2$: $u = 9$, $t = 0$: $u = 1$, $\int_0^2 3t^2\sqrt{t^3+1}\,dt = \int_1^9 \sqrt{u}\,du = \frac{2}{3}u^{3/2}\Big|_1^9 \cdots = \frac{52}{3}$.

11 $\frac{255}{8}$ (substitution, $u = x^2 + 1$. If you answered $\frac{1}{8}$, you forgot to switch the limits of integration to values of u.)

13 $\frac{2}{3}$ (substitution, $u = \sin x$. If you answered $\frac{\pi^{3/2}}{3\sqrt{2}} = 1.31247$, you forgot to switch the limits of integration to values of u.)

15 $-\frac{3}{16}$ (substitution, $u = \sec x$. If you encountered division by zero, you forgot to switch the limits of integration to values of u.)

17 $\frac{272}{21}$ (Substitution not needed.)

19 the available antiderivative formulas, rules, and techniques are insufficient to determine an antiderivative

21 0 (substitution, $u = 2 + \sin s$. If you encountered division by zero, you forgot to switch the limits of integration to values of u.)

23 651 (substitution, $u = 4x - 3$. If you answered $\frac{21}{8}$, you forgot to switch the limits of integration to values of u.)

25 FTC II does not apply (not continuous at several values in $[0, 2\pi]$, including $x = \frac{\pi}{8}$).

27 $\frac{3+\sqrt{3}}{2} = 2.36603$ (Substitution is not needed.)

29 $\frac{3}{10}5^{5/3} = \frac{3}{2}5^{2/3} = 4.38603$ (substitution, $u = x^2 - 4$. If you answered 0.919635, you forgot to switch the limits of integration to values of u.)

31 FTC II does not apply (not continuous at $x = 0$). Ignoring the discontinuity and using substitution with $u = 4 + \sqrt{x}$ yields $\frac{8\,717\,049}{5} = 1\,743\,409.8$, which is the correct answer even though FTC II does not apply. We study such integrals in section 7.11, but ignoring the discontinuity is still incorrect.)

33 $\frac{11\,529}{6} = \frac{3843}{2} = 1921.5$ (substitution, $u = 3 + \sqrt{y}$. If you answered $\frac{1365}{2} = 682.5$, you forgot to switch the limits of integration to values of u.)

35 the available antiderivative formulas, rules, and techniques are insufficient to determine an antiderivative.

37 $\frac{1}{2}$ (Rewrite as $\int_{-1}^{0} \frac{1}{(x+2)^2}\,dx$; substitution, $u = x + 2$. If you encountered division by zero, you forgot to switch the limits of integration to values of u.)

39 $\frac{2}{5}$ (substitution, $u = \cos\sqrt{y}$. If you answered $-\frac{\pi^{10}}{2560} = -36.5813$, you forgot to switch the limits of integration to values of u.)

41 $\frac{670}{3}$ units^2 = 223.33 units^2

43 2 units^2

45 $\frac{\sqrt{\pi+1}}{2}$ units^2 (Substitution is not needed; $\sqrt{\pi+1}$ is a constant)

47 0 units^2

Section 4.8

1 $T_8 = 15.597$

3 $T_5 = 0.88412$

5 $T_5 = 1.09415$ (If y_0, y_1, and so on are found to more than five decimal places, then the end result rounds to $T_5 = 1.09414$.)

7 $T_6 = 0.0038168$

9 $T_4 = 0.80641$

11 $T_5 = -3.9212$

13 $T_8 = -0.5397$

15 $T_6 = -17.364$

17 (a) 15.594, (b) $E_T = 0.003$

19 (a) 0.88831, (b) $E_T = -0.00419$

21 (a) 1.08943, (b) $E_T = 0.00471$

23 (a) 0.0038139, (b) $E_T = 0.0000029$

25 (a) 0.81003, (b) $E_T = -0.00362$

27 (a) −3.9382, (b) $E_T = 0.0170$

29 (a) −0.5455, (b) $E_T = 0.0058$

31 (a) −16.745, (b) $E_T = -0.619$

33 answers vary depending on the choice of K. For $K = 0.32$: (a) $|E_T| \le 0.00333$; (b) $0.003 \le 0.00333$, verified; (c) 147. The smallest allowable K is $K = 0.312178$: (a) $|E_T| \le 0.00325$; (b) $0.003 \le 0.00325$, verified; (c) 145.

35 answers vary depending on the choice of K. For $K = 3.5$: (a) $|E_T| \le 0.01167$; (b) $0.00419 \le 0.01167$, verified; (c) 171. The smallest allowable K is $K = 3.28105$: (a) $|E_T| \le 0.01094$; (b) $0.00419 \le 0.01094$, verified; (c) 166.

37 answers vary depending on the choice of K. For $K = 0.055$: (a) $|E_T| \le 0.0229$; (b) $0.0170 \le 0.0229$, verified; (c) 240. The smallest allowable K is $K = 0.0544247$: (a) $|E_T| \le 0.0227$; (b) $0.0170 \le 0.0227$, verified; (c) 239.

39 answers vary depending on the choice of K. For $K = 14$: (a) $|E_T| \le 0.0093$; (b) $0.0058 \le 0.0093$, verified; (c) 245. The smallest allowable K is $K = 13.8269$: (a) $|E_T| \le 0.0092$; (b) $0.0058 \le 0.0092$, verified; (c) 243.

41 answers vary depending on the choice of K. For $K = 1\,000\,000$: (a) $|E_T| \le 4$; (b) $0.619 \le 4$, verified; (c) 3795. The smallest allowable K is $K = 948\,277$: (a) $|E_T| \le 3.793$; (b) $0.619 \le 3.793$, verified; (c) 3696.

43–58 Answers are given for using Mathematica 11.1. Using a different computer algebra system or other technology might give different results, as might using a later version of Mathematica.

43 function of some other type

45 function of some other type

47 no result is given.

49 function of some other type

51 elementary function

53 function of some other type

55 no result is given.

57 elementary function (Logarithmic functions and inverse trigonometric functions are elementary functions that are studied in chapter 5.)

59 (a) $$T_8 = \frac{1}{2} \cdot \frac{1}{4}\Big(5.7 + 2(6.6) + 2(6.8) + \cdots + 2(6.2) + 6.4\Big) = \frac{1}{8}(105.1) = 13.1375 \text{ mi}$$

(b) the trapezoid rule estimate is halfway between the left- and right-hand estimates, as well as halfway between the lower and upper estimates.

61 $$T_\Omega = \frac{1}{2}\Delta x\Big(f(x_0) + 2f(x_1) + \cdots + 2f(x_{\Omega-1}) + f(x_\Omega)\Big) = \frac{1}{2}\Delta x\Big(f(x_0) - f(x_\Omega)\Big)$$

$$+\sum_{k=1}^{\Omega} 2f(x_k)\Big)$$
$$= \sum_{k=1}^{\Omega} f(x_k)\Delta x$$
$$+\frac{f(x_0)-f(x_\Omega)}{2}\Delta x$$
$$\approx \sum_{k=1}^{\Omega} f(x_k)\Delta x$$
$$= \int_a^b f(x)\,dx$$

Section 4.9

1 (a) $S_8 = 15.594$, (b) $E_S = 0.000$, (c) $E_S < E_T$

3 (a) $S_4 = 0.88903$, (b) $E_S = 0.00072$, (c) not applicable (NA)

5 (a) $S_8 = 1.08943$, (b) $E_S = 0.00000$, (c) NA

7 (a) $S_6 = 0.0038139$, (b) $E_S = 0.0000000$, (c) $E_S < E_T$

9 (a) $S_4 = 0.81006$, (b) $E_S = 0.00003$, (c) $E_S < E_T$

11 (a) $S_{10} = -3.9382$, (b) $E_S = 0.0000$, (c) NA

13 (a) $S_8 = -0.5456$, (b) $E_S = -0.0001$, (c) $E_S < E_T$

15 (a) $S_6 = -16.785$, (b) $E_S = -0.040$, (c) $E_S < E_T$

17 answers vary depending on the choice of K. (a) For $K = 0.6$: $|E_S| \le 0.000\,026\,04$, (b) $n = 12$, (c) much smaller than the 145 required for the trapezoid rule. The smallest allowable K is $K = 0.572527$: (a) $|E_S| \le 0.000\,024\,85$, (b) $n = 12$, (c) much smaller than the 145 required for the trapezoid rule.

19 answers vary depending on the choice of K. For $K = 150$: (a) $|E_S| \le 0.00326$, (b) $n = 18$, (c) much smaller than the 166 required for the trapezoid rule. The smallest allowable K is $K = 125.93$: (a) $|E_S| \le 0.00273$, (b) $n = 18$, (c) much smaller than the 166 required for the trapezoid rule.

21 answers vary depending on the choice of K. For $K = 0.000\,48$: (a) $|E_S| \le 0.000000833$, (b) $n = 6$, (c) much smaller than the 239 required for the trapezoid rule. The smallest allowable K is $K = 0.000474391$: (a) $|E_S| \le 0.000\,000\,824$, (b) $n = 6$, (c) much smaller than the 239 required for the trapezoid rule.

23 answers vary depending on the choice of K. For $K = 320$: (a) $|E_S| \le 0.00014$, (b) $n = 16$, (c) much smaller than the 243 required for the trapezoid rule. The smallest allowable K is $K = 309.083$: (a) $|E_S| \le 0.00014$, (b) $n = 16$, (c) much smaller than the 243 required for the trapezoid rule.

25 answers vary depending on the choice of K. The smallest allowable K is $K = 4.04591 \times 10^9$: (a) $|E_S| \le 0.432$, (b) $n = 88$, (c) much smaller than the 3696 required for the trapezoid rule.

27 (a) $M_8 = 15.593$, (b) $E_M = -0.001$, (c) $|E_M| < |E_T|$

29 (a) $M_5 = 0.89042$, (b) $E_M = 0.00211$, (c) $|E_M| < |E_T|$

31 (a) $M_5 = 1.08707$, (b) $E_M = -0.00236$, (c) $|E_M| < |E_T|$

33 (a) $M_6 = 0.0038124$, (b) $E_M = -0.000001\,5$, (c) $|E_M| < |E_T|$

35 (a) $M_4 = 0.81184$, (b) $E_M = 0.00181$, (c) $|E_M| < |E_T|$

37 (a) $M_5 = -3.9467$, (b) $E_M = -0.0085$, (c) $|E_M| < |E_T|$

39 (a) $M_8 = -0.5485$, (b) $E_M = -0.0030$, (c) $|E_M| < |E_T|$

41 (a) $M_6 = -16.440$, (b) $E_M = 0.305$, (c) $|E_M| < |E_T|$

43 (a) $S_8 = \frac{1}{3} \cdot \frac{1}{4}\big(5.7 + 4(6.6) + 2(6.8) + \cdots + 4(6.2) + 6.4\big) = \frac{1}{12}(156.9) = 13.075$ mi

(b) $S_8 < T_8$; otherwise, answers vary

45 (a) $T_6 = 2.1270$; (b) $S_6 = 2.1680$; (c) $E_T = -0.0011$, $E_S = 0.0399$; the trapezoid rule has the smaller error. (Note: although the error in Simpson's rule is more than 300 times the error in the trapezoid rule for $n = 6$, by $n = 10$, Simpson's rule has the smaller error.)

47 (a) $S_4 = 156$; (b) $\int_1^5 x^3\,dx = 156$, $E_S = 0$; (c) $S_4 = \frac{a}{4} + \frac{b}{3} + \frac{c}{2} + d$; (d) $\int_0^1 (ax^3 + bx^2 + cx + d)\,dx = \frac{a}{4} + \frac{b}{3} + \frac{c}{2} + d$, $E_S = 0$; (e) for a cubic polynomial E_S is always zero. Because $\frac{d^4}{dx^4}(ax^3 + bx^2 + cx + d) = 0$, we can use $K = 0$ in the error-bound formula, giving $|E_S| \le 0$.

Section 4.10

1 $f(x) = \frac{x^4+7}{4}$

3 $f(x) = -\frac{1}{2}\cos 2x + \frac{11}{2}$

5 $y = x^2 - 3x + 7$

7 $y = x^2 - 4x + 8$

9 $y = x^2 + \frac{1}{x} + \frac{1}{x^2} + 1$

11 $f(x) = \frac{1}{2}\tan x^2$

13 $f(x) = \frac{1}{6}x^3 + x^2 - \frac{19}{2}x + 15$

15 the net change in volume of water in the lake from $t = 0$ to $t = 24$ (day 1)

17 the net change in the percentage of likely voters in support of Proposition 1 from 10 weeks before the election to 5 weeks before the election

19 $s(t) = t^3 + 5t - 2$ (Note: $s(t) = t^3 + 5x - 2$ is wrong; the variable is t, not x.)

21 $s(t) = \frac{1}{2}t^2 - 7t + 17$

23 $v(t) = 3t + t^2 + 2$

25 2.3063 seconds after launch

27 $\frac{5}{4}$ seconds after launch (Hint: max height occurs when the rocket stops moving upward and begins moving downward—that is, when velocity is zero.)

29 5.6758 seconds after launch

31 (a) 10 m, (b) $\frac{58}{4}$ m $= \frac{29}{2}$ m $= 14.5$ m

33 (a) 0 m, (b) 10 m

35 $\frac{35}{2}$ m $= 17.5$ m

37 $y = \frac{6}{x} - 6$

39 $\frac{784}{3}$

41 $y = 4(x + 3)^{3/2} - 11x - 20$ (Hint: a 45° inclination means the slope is 1.)

43 (a) 0 m, (b) 400 m

45 (a) $(0, \infty)$ (only discontinuity at $x = 0$; cannot choose $(-\infty, 0)$ because the interval must include $x = 1$), (b) $\left(\sqrt{\frac{\pi}{2}}, \sqrt{\frac{3\pi}{2}}\right)$ (nearest discontinuities on either side of $x = \sqrt{\pi}$)

Section 5.1

1 correct

3 wrong

5 wrong

7 wrong

9 wrong

11 $\ln \frac{x^4}{(x-11)^{56}}$

13 $\ln \frac{x^{13}(x-1)^2}{(x+1)^4}$

15 $\ln x^{10}$; alternate but less desirable answer, $10 \ln x$

17 $\ln \frac{(x^2+2x+1)^2}{(x+1)^3} = \ln(x+1)$

19 $\ln(x-4) - 5\ln(x+6) - \ln \sin x$

21 $\ln 2 + 5\ln x - \ln(x^2 - 9)$ (Note: $5\ln 2x - \ln(x^2 - 9)$ is incorrect.); alternate answer, $\ln 2 + 5\ln x - \ln(x+3) - \ln(x-3)$

23 $\ln(x + \sqrt{x}) + \ln(1+x)$

25 $\ln\left(\frac{x}{x+1} + 2\right) - \ln(3x+5)$ (Note: $\ln x - \ln(x+1) + \ln 2 - \ln(3x+5)$ is incorrect.)

27 $y' = \frac{\ln x - x \cdot \frac{1}{x}}{(\ln x)^2} = \frac{\ln x - 1}{(\ln x)^2}$ (Note: $\frac{\ln x - 1}{2\ln x}$ is incorrect.)

29 $f'(x) = 2(1 + \ln x) \cdot \frac{1}{x} = \frac{2}{x}(1 + \ln x)$

31 $y' = \frac{3}{x+4} - \frac{6x-3}{x^2-x}$; alternate answer, $\frac{1}{\left(\frac{x+4}{x^2-x}\right)^3} \cdot 3\left(\frac{x+4}{x^2-x}\right)^2 \cdot \frac{(x^2-x)-(x+4)(2x-1)}{(x^2-x)^2}$ (The first answer is obtained by expanding the expression before differentiating, whereas the alternate answer is for not using the laws of logarithms.)

33 $y' = 2t$ (Because $\ln 4$ is a constant, it has derivative 0.)

35 $f'(x) = \frac{2}{x-1} - \frac{3}{3x+7}$; alternate answer, $\frac{1}{\left(\frac{(x-1)^2}{3x+7}\right)} \cdot \frac{(3x-7)\cdot 2(x-1) - (x-1)^2 \cdot 3}{(3x+7)^2}$ (the first answer is obtained by expanding the expression before differentiating, whereas the alternate answer is for not using the laws of logarithms.)

37 $y' = x^2 \cdot \frac{1}{5x^2} \cdot 10x + (\ln 5x^2)(2x) = 2x + 2x \ln 5x^2$

39 $y' = \frac{1}{\sin\theta} \cdot \cos\theta = \cot\theta$

41 $y' = \frac{2}{x-3} + \frac{28}{2x+1}$

43 $y' = \frac{1}{2} \cdot \frac{1}{x-3} - \frac{1}{2} \cdot \frac{1}{4x+1} \cdot 4 = \frac{1}{2(x-3)} - \frac{2}{4x+1}$

45 $f'(x) = 3x \cdot \frac{1}{x} + \ln x \cdot 3 = 3 + 3\ln x$

47 $-\infty$

49 ∞

51 0 (Evaluate using continuity.)

53 ∞

55 $-\infty$

57 $-\infty$ (hint: use the conjugate.)

59 $y = 2x - 2$

61 $y = \frac{7}{4}x - \frac{15}{4} + 3\ln 4$

63

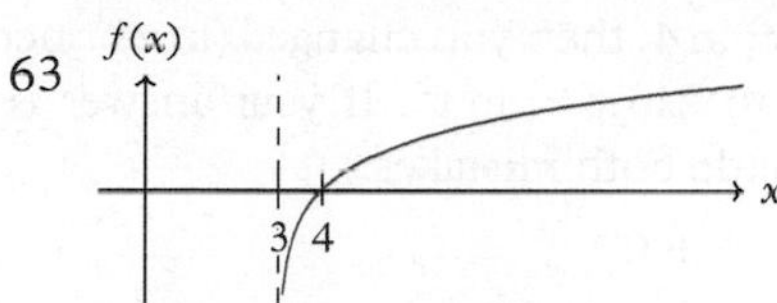

65

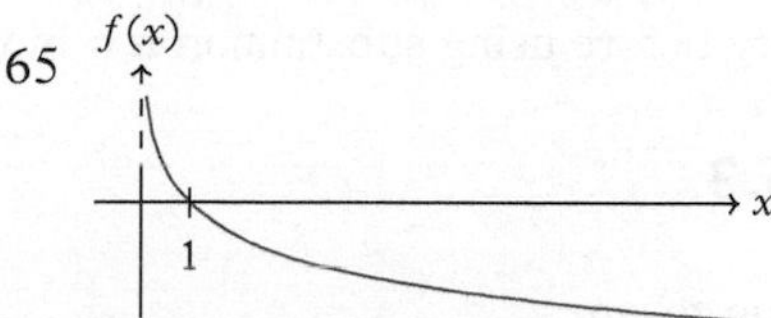

67 0; details: $\lim_{x\to 0^+} \frac{1}{\ln x} = \frac{1}{\ln \omega} = \frac{1}{-B} = -\beta \doteq 0$

Section 5.2

1 $u = 5x - 7$

3 $u = 5x - 7$

5 $u = 5x - 7$

7 $u = x^2 + 4$

9 $u = \sin x$

11 $u = \ln x$

13 $x > \frac{3}{5}$; alternate answer, $(\frac{3}{5}, \infty)$

15 $(-\infty, \infty)$ (Note: $x^2 + 5$ is always positive.)

17 $(-4, -3) \cup (-3, \infty)$ (Don't forget about division by zero and that $\ln 1 = 0$.)

19 $(2, 7)$; alternate answer, $2 < t < 7$ (Hint: rewrite using laws of logarithms.)

21 $2\ln|x+1| + C$

23 $-\frac{3}{5}\ln|5x+2|+C$

25 $2\ln 4$; alternate answers, $4\ln 2$ and $\ln 16$ (If your answer is $-2\ln 6 + 2\ln 3$ or, equivalently, $\ln\frac{9}{36} = \ln\frac{1}{4} = -\ln 4$, then you forgot to switch to values of u.)

27 FTC II does not apply. ($-4\ln 3 = \ln\frac{1}{81}$ is incorrect.)

29 $\frac{1}{9}\ln|\sec(9x-\pi)|+C$

31 the available antiderivative formulas, rules, and techniques are insufficient to determine an antiderivative.

33 $\ln|x^2+4|+C = \ln(x^2+4)+C$

35 $\frac{1}{2}(\ln 7 - \ln 5) = \frac{1}{2}\ln\frac{7}{5} = \ln\sqrt{\frac{7}{5}}$ (If you encountered $\ln 0$, then you forgot to switch to values of u.)

37 $3\ln\left|\sec\frac{x+4}{3}\right|+C$

39 $\frac{1}{3}(\ln 2)^3$ (If your answer is $\frac{7}{3}$, you forgot to change to values of u. If your answer is $2\ln 2 = \ln 4$, then you changed $(\ln x)^2$ incorrectly to $2\ln x = \ln x^2$. If your answer is 2, you made both mistakes.)

41 $\ln\left|\csc\frac{1}{x}\right|+C$

43 $\frac{(\ln x)^2}{4}+C$ (hint: use laws of logarithms to simplify before using substitution $u=\ln x$.)

45 $\frac{x^3}{3}-\frac{7x^2}{2}+4\ln|x|+C$ (Rewrite integrand as $x^2-7x+\frac{4}{x}$.)

47 $2\sqrt{\ln x}+C$

49 $\frac{1}{5}\sin(5\ln x)+C = \frac{1}{5}\sin(\ln x^5)+C$ (hint: use $u=5\ln x$.)

51 $y' = \left(\frac{(3x+5)^4(1+x^2)^3}{(x-2)^7}\right)\cdot\left(\frac{12}{3x+5}+\frac{6x}{1+x^2}-\frac{7}{x-2}\right)$

53 $y' = \left((14x-3)^2(12x+5)^7(1-2x)^3\right)\cdot\left(\frac{28}{14x-3}+\frac{84}{12x+5}-\frac{6}{1-2x}\right)$

55 $y' = \left(\frac{x+\sin x}{(3x+2)(x-\sqrt{x})^5}\right)^2\cdot\left(\frac{2(1+\cos x)}{x+\sin x}-\frac{6}{3x+2}-\frac{10\left(1-\frac{1}{2\sqrt{x}}\right)}{x-\sqrt{x}}\right)$

57 $y = 5-\ln 2+\ln x = 5+\ln\frac{x}{2}$ (Note: because the interval of solution must contain $x=2$, the solution is valid on $(0,\infty)$ and $|x|=x$ on that interval, so the absolute values are not necessary.)

59 $-\frac{1}{2}\ln|\csc(2x-\pi)|+1$

61 ∞

63 houseboat (If you take your natural log cabin and "add it to the sea" ($+C$) or a lake or a river or any other body of water, then it becomes a houseboat.)

Section 5.3

1 yes, one-to-one

3 no, not one-to-one

5 no, not one-to-one

7

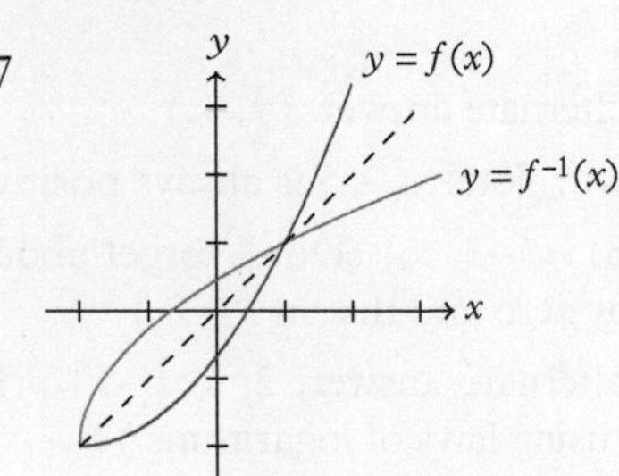

9

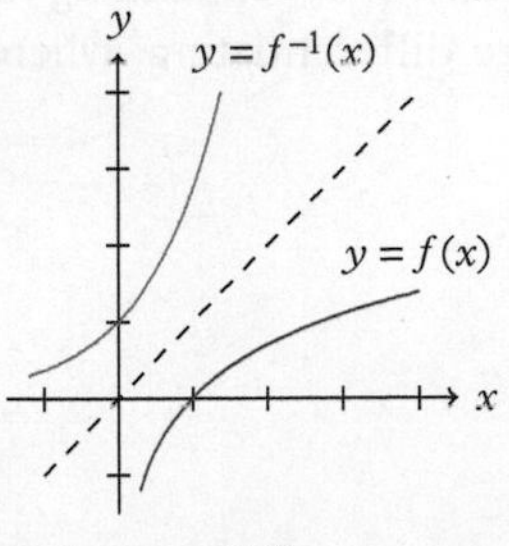

y
$y=f^{-1}(x)$
$y=f(x)$
x

11 $f^{-1}(x) = \frac{x+5}{7}$

13 $f^{-1}(x) = \sqrt[4]{x}$

15 $g^{-1}(x) = \sqrt[3]{x^2-5}$ (Note: because the outputs to the function g are nonnegative, the

domain of g^{-1} is $[0, \infty)$, even though the expression for g^{-1} is defined for all x.)

17 $k^{-1}(x) = 3 - \frac{1}{x}$

19 $g^{-1}(x) = -\sqrt{x} - 2$ (Note: the choice of domain matters for the function g. Notice $g(x) = (x+2)^2$, so we let $y = (x+2)^2$. Then, $\sqrt{y} = \sqrt{(x+2)^2} = |x+2|$. Since $x \le -2$, we have $x + 2 \le 0$. Therefore, $|x+2| = -(x+2)$. Thus, $\sqrt{y} = -x - 2$, yielding $x = -2 - \sqrt{y}$.)

21 $f'(x) = 3x^2 + 4 > 0$, so f is increasing on $(-\infty, \infty)$ and, by theorem 3, f^{-1} exists.

23 $f'(x) = -\csc^2 x \le 0$, so f is decreasing on $(0, \pi)$ and, by theorem 3, f^{-1} exists.

25 $f'(x) = 3(5x-2)^2(5) = 15(5x-2)^2 \ge 0$, so f is increasing on $(-\infty, \infty)$ and, by theorem 3, f^{-1} exists.

27 $f'(x) = 1 + \frac{1}{x} > 0$ (because $x > 0$), so f is increasing on $(0, \infty)$ and, by theorem 3, f^{-1} exists. (Notice that the domain of f is $(0, \infty)$.)

29 $f'(x) = -\frac{1}{2\sqrt{x}} < 0$, so f is decreasing on $(0, \infty)$ and, by theorem 3, f^{-1} exists. (Notice that the domain of f is $[0, \infty)$. The function is decreasing on $[0, \infty)$, but using the derivative to show f is decreasing can only show decreasing on $(0, \infty)$ because f' is not defined at zero.)

31 $(f^{-1})'(5) = \frac{1}{f'(1)} = \frac{1}{7}$

33 $(f^{-1})'(-8) = \frac{1}{f'(0)} = \frac{1}{60}$

35 $(f^{-1})'(0) = \frac{1}{f'(\frac{\pi}{2})} = -1$

37 notice $f(1) = 1$. Then, $(f^{-1})'(1) = \frac{1}{f'(1)} = \frac{1}{2}$.

39 $f'(x) = 2 + \cos x > 0$, so f is increasing on $(-\infty, \infty)$ and, by theorem 3, f^{-1} exists. Notice $f(\pi) = 2\pi$. Then, $(f^{-1})'(2\pi) = \frac{1}{f'(\pi)} = 1$.

41 $f'(x) = -5 - 21x^2 < 0$, so f is decreasing on $(-\infty, \infty)$ and, by theorem 3, f^{-1} exists. Notice $f(0) = 4$. Then, $(f^{-1})'(4) = \frac{1}{f'(0)} = -\frac{1}{5}$.

43 answers may vary. Possibilities include $(-\frac{\pi}{2}, \frac{\pi}{2})$, $(\frac{\pi}{2}, \frac{3\pi}{2})$, and $(-\frac{3\pi}{2}, -\frac{\pi}{2})$.

Section 5.4

1 4

3 1

5 25

7 −1

9 $4x - 7$

11 $\frac{x}{4}$

13 $e^{4-x} + 4 + x$

15 $14x - \sin x$

17 $x = \pm\sqrt{1 - \ln 2}$

19 $x = 2$

21 $x = 2 - e$

23 0

25 1 (Don't forget about evaluating using continuity!)

27 0

29 0

31 ∞

33 $f'(x) = 2x$ (e^4 is a constant.)

35 $y' = \frac{(4x+3)e^x - e^x \cdot 4}{(4x+3)^2} = \frac{(4x-1)e^x}{(4x+3)^2}$

37 $g'(x) = 20e^{4x+3}$

39 $f'(x) = (x\cos x + \sin x)e^{x\sin x}$

41 $y' = x \cdot e^{x^2} \cdot 2x + e^{x^2} \cdot 1 = (2x^2 + 1)e^{x^2}$

43 $g'(t) = [4(3t^2 - 5)^6 \cdot (t^2 \cdot e^{\sqrt{5t-8}} \cdot \frac{1}{2}(5t - 8)^{-1/2} \cdot 5 + e^{\sqrt{5t-8}} \cdot 2t) - t^2 e^{\sqrt{5t-8}} \cdot 24(3t^2 - 5)^5 \cdot 6t]/(4(3t^2 - 5)^6)^2 = \cdots = \frac{\frac{30t^4 - 50t^2}{\sqrt{5t-8}} - 120t^3 - 40t}{16(3t^2 - 5)^7} e^{\sqrt{5t-8}}$

45 $f'(x) = 6 + x + 2x\ln x$ (Hint: first rewrite as $f(x) = x^2(\ln x + \ln e^{6/x}) = \cdots = x^2 \ln x + 6x$);

alternate answer, $f'(x) = x^2 \cdot \frac{1}{xe^{6/x}} \cdot (x \cdot e^{6/x} \cdot -6x^{-2} + e^{6/x} \cdot 1) + (\ln(xe^{6/x}))2x = \cdots = -6 + x + 2x\ln(xe^{6/x})$

47 $\frac{4}{5}e^{5x} + C$

49 $2e^{x/2} + C$

51 $e^{\sqrt{x}} + C$ (substitution, $u = \sqrt{x}$)

53 $\frac{1}{4}(e^3 - e^{-1}) = 4.92941$

55 $e^{\tan x} + C$ (substitution, $u = \tan x$)

57 $-\cos x + C$ (Rewrite as $\int \sin x\,dx$.)

59 $e^{\sin x} + C$ (substitution, $u = \sin x$. Recall $\cos x = \frac{1}{\sec x}$.)

61 $\frac{2}{3}(2 + e^x)^{3/2} + C$ (substitution, $u = 2 + e^x$)

63 $y = 3x + 1$

65 $y = 4x + 8$ (from $y - 4 = 4(x - (-1))$)

67 $y = \frac{x^2}{2} + e^x + 2$

69 $y = \frac{x^6}{6} + 1$ (Hint: simplify y' first.)

71 $y'' = (x + 2)e^x$, $y''' = (x + 3)e^x$, $y^{(42)} = (x + 42)e^x$

73 formula (i) violates criteria (a); formula (ii) violates criteria (c); formula (iii) meets all the criteria and should be considered; formula (iv) violates criteria (a). Choose formula (iii).

Section 5.5

1 $e^{\sqrt{3}\ln 14}$

3 $e^{e\ln 6}$

5 $e^{3x\ln 4}$

7 $e^{1\ln 9}$ $(= e^{\ln 9} = 9)$

9 $e^{(x-2)\ln(2+\sin x)}$

11 $y' = \sqrt{2}x^{\sqrt{2}-1}$

13 $y' = \sqrt{2}^x \ln\sqrt{2} = \sqrt{2}^x \cdot \frac{1}{2}\ln 2 = 2^{\frac{1}{2}x-1}\ln 2$

15 $f'(x) = 17^{4x+8}\,4\ln 17$

17 $f'(t) = \frac{1}{2\sqrt{9+2^t}} \cdot 2^t \ln 2 = \frac{2^{t-1}\ln 2}{\sqrt{9+2^t}}$

19 $y' = \frac{1}{3+4^x} \cdot 4^x \ln 4$

21 $g'(t) = 5^{4t-7}\,4\ln 5$ (Hint: rewrite $g(t) = 5^{4t-7}$.); alternate answer using product rule, $g'(t) = 5^t t^{3t-7}\,3\ln 5 + (5^t\ln 5)5^{3t-7}$

23 $f'(x) = \frac{2^x(3) - (3x-4)2^x\ln 2}{2^{2x}} = \frac{3 - (3x-4)\ln 2}{2^x}$

25 $y' = (4x-7)^5\,5^{4x-7}\,4\ln 5 + 5(4x-7)^4 \cdot 4 \cdot 5^{4x-7}$

27 $f'(t) = 4^t \sin t + t(4^t \ln 4)\sin t + t4^t \cos t$

29 $y' = x^{2x-1}\left(\frac{2x-1}{x} + 2\ln x\right)$ (Notes: $y' = (2x-1)x^{2x-2}$ is incorrect because the power rule does not apply (variable in exponent); $y' = x^{2x-1} \cdot 2\ln x$ is incorrect because the rule for general exponentials does not apply (variable in base). Instead, use logarithmic differentiation or rewrite using definition 6.)

31 $g'(y) = (3 + \sqrt{5})^{1-\sin y}(-\cos y)\ln(3 + \sqrt{5})$

33 $y' = 4x^{4x}(4 + 4\ln x) = 16x^{4x}(1 + \ln x)$ (Notes: if you answered $y' = 4x^{4x}(4 + 4\ln 4x)$, which is incorrect, you may have misinterpreted $4x^{4x}$ as $(4x)^{4x}$. Also, $y' = 16x \cdot x^{4x-1} = 16x^{4x}$ is incorrect because the power rule does not apply (variable in exponent); $y' = 4x^{4x} \cdot 4\ln x$ is incorrect because the rule for general exponentials does not apply (variable in base). Instead, use logarithmic differentiation or rewrite using definition 6.)

35 $s'(t) = (3t-1)^t\left(\frac{3t}{3t-1} + \ln(3t-1)\right)$ (Notes: $s'(t) = t(3t-1)^{t-1} \cdot 3$ is incorrect because the power rule does not apply (variable in exponent); $s'(t) = (3t-1)^t \ln(3t-1)$ is incorrect because the rule for general exponentials does not apply (variable in base). Instead, use logarithmic differentiation or rewrite using definition 6.)

37 $\frac{6^x}{\ln 6} + C$

39 0

41 $\frac{5^{\ln x}}{\ln 5} + C$

43 $\frac{-1}{4\ln 0.4}(6 - 0.4^x)^4 + C$

45 $\frac{5^{2x+4}}{2\ln 5} + C$

47 $\frac{7}{6}$ (Don't forget to try evaluating using continuity!)

49 ∞

51 0

53 0

55 ∞

57 DNE (∞ is incorrect because $16A$ could be either positive or negative.)

59 $\int_0^1 3^x\,dx = \frac{2}{\ln 3}$ units^2

61 $y = (4\ln 2)(x - 2) + 8$; alternately, $y = x\ln 16 + 8 - \ln 256$

63 $y' = a^x \ln a$, $y'' = a^x(\ln a)^2$. Because $a^x > 0$ and $(\ln a)^2 > 0$, $y'' > 0$ on $(-\infty, \infty)$ and $y = a^x$ is concave up on $(-\infty, \infty)$.

65 (a) $1^x = e^{x\ln 1} = e^0 = 1$, (b) $\frac{d}{dx}1^x = 1^x \ln 1 = 1^x \cdot 0 = 0$, (c) $\frac{d}{dx}1 = 0$ (derivative of a constant), (d) the results of (b) and (c) are identical. Yes, the formula is valid.

Section 5.6

1 $9^y = 5$

3 $3^y = 4x^2 + 1$

5 $0.5^{5y} = \sqrt{5x}$

7 $10^{(8/3)r^2} = s - s^2$; alternate answer, $10^{8r^2} = (s - s^2)^3$

9 $\log_2 11 = x - 4$

11 $\log_3 y = x$

13 $\log_5 x = 2x + 1$

15 $\log_7 \frac{s}{3} = t$

17 1.5711

19 −4.9421

21 30.3776

23 −0.003045

25 $y' = \frac{3x^2 - 7}{(x^3 - 7x + 1)\ln 2}$

27 $f'(x) = \frac{40x^2}{(x+3)\ln 5} + 80x\log_5(x+3)$

29 $y' = \frac{\frac{2x}{(2x+3)\ln 7} - 2\log_7(2x+3)}{x^3}$

31 $y' = 4\left(x + \log_2(6x^2)\right)^3\left(1 + \frac{2}{x\ln 2}\right)$; alternate answer, $y' = 4\left(x + \log_2(6x^2)\right)^3\left(1 + \frac{12x}{6x^2\ln 2}\right)$

33 $f'(x) = 3^x \cdot \frac{1}{(9x+2)\ln 3} \cdot 9 + (3^x \ln 3)\log_3(9x+2)$

35 $y' = 2\sin\sqrt{4^x \log_2 5x} \cdot \cos\sqrt{4^x \log_2 5x} \cdot \frac{1}{2\sqrt{4^x\log_2 5x}} \cdot \left(4^x \frac{1}{5x\ln 2} \cdot 5 + (4^x \ln 4)\log_2 5x\right)$

37 $\frac{1}{\ln 3}\ln|x| + C$; alternate answer, $\log_3|x| + C$

39 $\ln|3^x + 5x| + C$ (Hint: let $u = 3^x + 5x$.)

41 $\frac{\ln 2}{2}\left(\log_2(x^2+1)\right)^2 + C$ (Hint: let $u = \log_2(x^2 + 1)$.)

43 $-(\ln 2)\cos(\log_2 x) + C$

45 ∞ (Notice that $a = 1/3 < 1$.)

47 $-\infty$ (Notice that $a = 2 > 1$.)

49 $-\infty$ (Notice that $a = 0.6 < 1$.)

51 ∞ (Notice that $a = 4 > 1$.)

53 0 (Use continuity!)

55 $-\infty$ (Notice that $a = \pi/4 < 1$.)

57 $\frac{dy}{dx} = \frac{-2x\log_2 y - 3}{\frac{x^2}{y\ln 2} - 3y^2}$

59 $y' = \frac{1}{x\ln a} = \frac{1}{\ln a}x^{-1}$, $y'' = -\frac{1}{\ln a}x^{-2} = -\frac{1}{x^2\ln a}$. Because $a > 1$, $\ln a > 0$ and $y'' = -\frac{1}{x^2\ln a} < 0$. Thus, the graph of $y = \log_a x$ is concave down throughout its domain $(0, \infty)$.

61 6.75

63 $1.85 \cdot 10^{26}$

65 0.000222 (Hint: $-3.6528 = -4 + 0.3472$)

Section 5.7

1 2.06 hours (or an additional 1.06 hours)

3 (a) \$44 754.74, (b) $\frac{\ln\left(\frac{10}{3}\right)}{0.04} = 30.1$ years

5 34.66 years

7 140 010 (rounding k to 0.004912 gives 140 011)

9 61.1 minutes from the start (or an additional 51.1 minutes)

11 116.2 °F ($k = -0.02744$)

13 42.86 °F

15 586 lb

17 0.000345 mg ($k = -0.048135$)

19 545.8 days ($k = -0.001270$)

21 after 3.15 minutes ($k = -0.5108$) (Hint: velocity follows the law of exponential change. There is no need to take a derivative or an antiderivative.)

Section 5.8

1 $\tan y = 3$

3 $\cos 0.4 = x$

5 $\sin y = x - 2$

7 $\tan^{-1} y = 3$

9 $\cos^{-1} 0.4 = x$

11 $\sin^{-1} y = x - 2$

13 $\frac{\pi}{3}$

15 $-\frac{\pi}{6}$

17 $\frac{4}{5}$

19 $\frac{7}{\sqrt{48}}$

21 $y' = \frac{15}{1+9x^2}$

23 $f'(x) = \frac{11x+14}{\sqrt{1-x^2}} + 11\sin^{-1} x$

25 $f'(t) = \dfrac{2t\sin^{-1} 5t - \frac{5t^2}{\sqrt{1-25t^2}}}{(\sin^{-1} 5t)^2}$

27 $g'(t) = 0$ (derivative of a constant function)

29 $y' = \left(-\sin((4x+6)^5)\right) \cdot 5(4x+6)^4 \cdot 4$
$= -20(4x+6)^4 \sin(4x+6)^5$

31 $f'(x) = \dfrac{21x^2}{\sqrt{1-(7x^3+1)^2}}$

33 $s'(x) = \dfrac{-1}{\sqrt{1-\left(\frac{x^2}{x^3+5}\right)^2}} \cdot \dfrac{(x^3+5)2x - x^2(3x^2)}{(x^3+5)^2}$
$= \dfrac{x^4 - 10}{(x^3+5)^2\sqrt{1-\frac{x^4}{(x^3+5)^2}}}$

35 $y' = \dfrac{3e^{3x}}{\sqrt{1-e^{6x}}}$

37 $y' = \dfrac{1}{1+(\ln(2+\cos x^2))^2} \cdot \dfrac{1}{2+\cos x^2}\left(-\sin x^2\right)2x$
$= \dfrac{-2x\sin x^2}{(2+\cos x^2)(1+(\ln(2+\cos x^2))^2)}$

39 $y' = \dfrac{4^{\sin^{-1} x}\ln 4}{\sqrt{1-x^2}} + \dfrac{4\left(\sin^{-1} x\right)^3}{\sqrt{1-x^2}}$

41 $12 \cdot \frac{1}{6}\tan^{-1}\frac{x}{6} + C = 2\tan^{-1}\frac{x}{6} + C$

43 $4\sin^{-1} x + C$

45 $-4\sqrt{1-x^2} + C$

47 $\frac{1}{12}\tan^{-1}\frac{x}{3} + C$

49 $\sin^{-1}(1+5x) + C$

51 $\frac{1}{3}\sin^{-1} x^3 + C$

53 $\sin^{-1}\ln x + C$

55 $\frac{2}{3}\left(\tan^{-1} x\right)^{3/2} + C$

57 $\frac{1}{\sqrt{7}}\tan^{-1}\frac{x}{\sqrt{7}} + C$ (Hint: simplify the numerator.)

59 $\frac{1}{2}\sin^{-1} x + C$ (Hint: extract $\sqrt{4} = 2$ from the denominator.)

61 $\frac{\pi}{2}$

63 $-\frac{\pi}{2}$

65 1 (Use continuity.)

67 $-\frac{\pi}{2}$

69 2π (Hint: write as an approximation instead of rendering.)

71 no; algebraic reason: the derivative $y' = \frac{1}{\sqrt{1-x^2}}$ is undefined at $x = 1$ (division by

zero); graphical reason: the tangent line is vertical where $x = 1$.

73 $y'' = \frac{-2x}{(x^2+1)^2}$,

interval	sign y''	concavity
$(-\infty, 0)$	$+$	up
$(0, \infty)$	$-$	down

There is an inflection point at $x = 0$.

75 $\frac{d}{dx}\left(\frac{1}{a}\tan^{-1}\frac{x}{a} + C\right) = \frac{1}{a}\cdot\frac{1}{1+\left(\frac{x}{a}\right)^2}\cdot\frac{1}{a} + 0 = \frac{1}{a^2\left(1+\left(\frac{x}{a}\right)^2\right)} = \frac{1}{a^2+x^2}$

77 the inverse cotangent function is defined by $y = \cot^{-1}x$ if and only if $\cot y = x$ and $0 < y < \pi$.

79 $\lim_{x\to\infty}\cot^{-1}x = 0$, $\lim_{x\to-\infty}\cot^{-1}x = \pi$

81 $\frac{d}{dx}\cot^{-1}u = \frac{-1}{1+u^2}\cdot\frac{du}{dx}$

Section 5.9

1 $\tanh y = 0.3$

3 $\cosh 1.4 = x$

5 $\sinh y = x - 2$

7 $\tanh^{-1} y = 0.3$

9 $\cosh^{-1} 1.4 = x$

11 $\sinh^{-1} y = x - 2$

13 $f'(x) = 24\,\text{sech}^2(4x - 11)$

15 $f'(x) = \frac{24}{1-(4x-11)^2} = \frac{24}{-16x^2+88x-120}$

17 $y' = \frac{2^x(\ln 2)\sinh x - 2^x\cosh x}{\sinh^2 x}$

19 $y' = xe^{\cosh x}\sinh x + e^{\cosh x} = e^{\cosh x}(x\sinh x + 1)$

21 $f'(t) = 5\left(\sinh^{-1}(t^2+4)\right)^4\cdot\frac{1}{\sqrt{1+(t^2+4)^2}}\cdot 2t = \frac{10t\left(\sinh^{-1}(t^2+4)\right)^4}{\sqrt{1+(t^2+4)^2}}$

23 $y' = \frac{1}{\sqrt{(\sqrt{x})^2-1}}\cdot\frac{1}{2\sqrt{x}} = \frac{1}{2\sqrt{x}\sqrt{x-1}} = \frac{1}{2\sqrt{x^2-x}}$

25 $x' = -t\,\text{sech}\,t\tanh t + \text{sech}\,t + t^2\,\text{csch}\,t\coth t - 2t\,\text{csch}\,t$

27 $y' = x^{\tanh x}\left(\frac{\tanh x}{x} + (\ln x)\,\text{sech}^2 x\right)$ (Hint: rewrite as $y = e^{\tanh x\ln x}$ or use logarithmic differentiation.)

29 $\frac{1}{2}\sinh^{-1} 2x + C$

31 $\frac{1}{2}\sin^{-1} 2x + C$

33 $\frac{1}{2}\cosh^{-1} 2x + C$

35 $\frac{1}{2}\tan^{-1} 2x + C$

37 $\sqrt{1+2x} + C$ (Hint: $u = 1 + 2x$)

39 $\frac{\sqrt{1+(2x)^2}}{4} + C$ (Hint: $u = 1 + 4x^2$)

41 $\frac{\sqrt{1+(2x)^2}}{4} + \frac{1}{2}\sinh^{-1} 2x + C$ (Hint: rewrite as $\int\frac{x}{\sqrt{1+(2x)^2}}\,dx + \int\frac{1}{\sqrt{1+(2x)^2}}\,dx$.)

43 cannot be done with the available antiderivative formulas, rules, and techniques

45 $\sinh x - \tanh x + C$

47 $\frac{5}{2}\sinh^{-1}(2x+3) + C$

49 $\sinh^{-1}\sin x + C$

51 $\frac{1}{6}\ln(\cosh(6x-8)) + C$

53 $\frac{7}{3}\cosh^{-1}\frac{4}{\sqrt{8}} - \frac{7}{3}\cosh^{-1}\frac{3}{\sqrt{8}} = 1.2479$

55 $-e^{\coth x} + C$

57 $\frac{1}{2}\sinh 4 = 13.64$

59 $\frac{1}{3}\cosh^{-1}\frac{x}{2} + C$ (Hint: factor 9 from square root and use a version 2 formula with $a = 2$.)

61 cannot be done with the available antiderivative formulas, rules, and techniques

63 $\frac{1}{\ln 2}\cosh^{-1} 2^x + C$ (Hint: rewrite 4^x as $(2^x)^2$.)

65 $\frac{x^2}{4} - \frac{1}{2}\ln|x| + C$ (Hint: rewrite using the definition of sinh, then simplify.)

67 $2\sinh x\cosh x = 2\cdot\frac{e^x-e^{-x}}{2}\cdot\frac{e^x+e^{-x}}{2} = \frac{e^{2x}-e^{-2x}}{2} = \sinh 2x$

69 $\sinh^2 x = \left(\frac{e^x-e^{-x}}{2}\right)^2 = \frac{e^{2x}-2+e^{-2x}}{4} = \frac{e^{2x}+e^{-2x}}{2\cdot 2} - \frac{2}{4} = \frac{\cosh 2x}{2} - \frac{1}{2} = \frac{\cosh 2x - 1}{2}$

71 $\cosh^2 x - \sinh^2 x = 1$, divide by $\cosh^2 x$, which gives $\frac{\cosh^2 x}{\cosh^2 x} - \frac{\sinh^2 x}{\cosh^2 x} = \frac{1}{\cosh^2 x}$; therefore, $1 - \tanh^2 x = \operatorname{sech}^2 x$

73 $\sinh x \cosh y + \cosh x \sinh y$
$= \left(\frac{e^x - e^{-x}}{2}\right)\left(\frac{e^y + e^{-y}}{2}\right) + \left(\frac{e^x + e^{-x}}{2}\right)\left(\frac{e^y - e^{-y}}{2}\right)$
$= \frac{1}{4}\left(e^x e^y + e^x e^{-y} - e^{-x} e^y\right.$
$\left. - e^{-x} e^{-y} + e^x e^y - e^x e^{-y} + e^{-x} e^y - e^{-x} e^{-y}\right)$
$= \frac{1}{4}\left(2e^x e^y - 2e^{-x} e^{-y}\right)$
$= \frac{1}{2}\left(e^{x+y} - e^{-(x+y)}\right) = \sinh(x+y)$

75 $\frac{d}{dx}\sinh x = \frac{d}{dx}\left(\frac{1}{2}e^x - \frac{1}{2}e^{-x}\right) = \frac{1}{2}e^x - \frac{1}{2}e^{-x}(-1) = \frac{1}{2}e^x + \frac{1}{2}e^{-x} = \cosh x$

77 $\frac{d}{dx}\coth x = \frac{d}{dx}\frac{\cosh x}{\sinh x} = \frac{\sinh x \cdot \sinh x - \cosh x \cdot \cosh x}{\sinh^2 x} = 1 - \coth^2 x = -\operatorname{csch}^2 x$

79
$$y = \tanh^{-1} x$$
$$\tanh y = x$$
$$\operatorname{sech}^2 y \frac{dy}{dx} = 1$$
$$\frac{dy}{dx} = \frac{1}{\operatorname{sech}^2 y}$$
$$= \frac{1}{1 - \tanh^2 y}$$
$$= \frac{1}{1 - x^2}$$

81 $y = x$

83 $dy = \frac{\frac{x}{\sqrt{1+x^2}} - \sinh^{-1} x}{x^2}\, dx$

85 $y' = \cosh x$, $y'' = \sinh x$

interval	sign y''	concavity
$(-\infty, 0)$	$-$	down
$(0, \infty)$	$+$	up

There is an inflection point at $x = 0$.

87 $\lim_{x\to-\infty} \tanh x = -1$, $\lim_{x\to\infty} \tanh x = 1$. For any positive infinite hyperreal Ω, $\tanh \Omega \doteq 1$ and $\tanh(-\Omega) \doteq -1$.

89 $\operatorname{sech} x = \frac{3}{5}$, $\cosh x = \frac{5}{3}$, $\sinh x = \frac{4}{3}$, $\operatorname{csch} x = \frac{3}{4}$, $\coth x = \frac{5}{4}$ (Hint: use hyperbolic identities.)

91 $\frac{d}{dx}\cosh^{-1}\left(\frac{x}{a}\right) = \frac{1}{\sqrt{\left(\frac{x}{a}\right)^2 - 1}} \cdot \frac{1}{a} = \frac{1}{\sqrt{\left(\frac{x^2}{a^2} - 1\right)a^2}} = \frac{1}{\sqrt{x^2 - a^2}}$

93 (a) Hint: the units cancel inside the hyperbolic tangent function, and constants such as 2π need not be considered; (b) $c_{\text{deep}} = \sqrt{\frac{g\lambda}{2\pi}}$; (c) the depth h is large compared to the wavelength λ (in practice, $h \geq \frac{1}{2}\lambda$ may be good enough); (d) $y = x$; (e) $c_{\text{shallow}} = \sqrt{gh}$ (Hint: when the depth is small, following the tangent line from part (d) should be approximately correct. Replace $\tanh x$ with x, where x is ....); (f) at approximately 0.473 m

Section 5.10

1 linear

3 exponential

5 linear

7 e^{Ω}

9 Ω

11 $3^{2\Omega}$

13 same level

15 $O(n^2)$

17 $O(\log n)$

19 $O(\sqrt{n})$

21 g grows faster

23 f and g have the same type of growth.

25 g grows faster.

27 g grows faster.

29 f grows faster.

31 f and g have the same type of growth.

33 g grows faster.

35 g grows faster. (Hint: $\sqrt{\Omega}$ is on a lower level than Ω for any positive infinite hyperreal Ω.)

37 $O(n^2)$

39 $O(2^n)$

41 $O(\log n)$

43 $O(n)$

45 between the 3^Ω level and the Ω^Ω level

47 between the $\ln \Omega$ level and the $\sqrt[4]{\Omega}$ level

49 between the Ω level and the Ω^2 level (Hint: use definition 6 and laws of exponents (3) to rewrite as a power of Ω.)

51 between the Ω level and the Ω^2 level

53 between the 3^Ω level and the Ω^Ω level (Hint: write out Ω^Ω, $\Omega!$, and 3^Ω as products of Ω factors.)

55

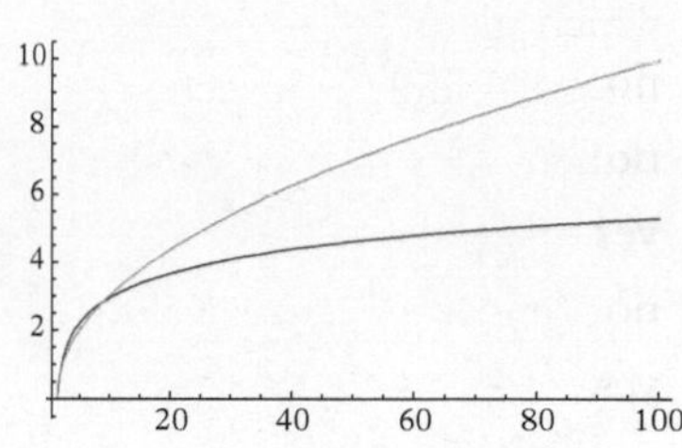

57 g grows faster (hint: rewrite 3.2 as $\frac{3.2}{3} \cdot 3$)

Section 5.11

1 yes, $\frac{0}{0}$

3 yes, $\frac{\infty}{-\infty}$

5 no (The limit can be evaluated using continuity.)

7 yes, $\frac{-\infty}{\infty}$ ($\ln \omega \doteq -\infty$, $\frac{1}{\omega} = \Omega \doteq \infty$)

9 0

11 $\frac{4}{9}$

13 0 (not an indeterminate form; l'Hospital's rule does not apply)

15 0

17 $\frac{1}{10}$

19 0

21 $\ln 5 - \ln 3$

23 1

25 ∞ (not an indeterminate form; l'Hospital's rule does not apply)

27 1

29 ∞ (Hint: rewrite $\frac{\frac{-1}{x^2+1}}{-e^{-x}}$ as $\frac{e^x}{x^2+1}$.)

31 0

33 $-\infty$ (Hint: after the first use of l'Hospital's rule, simplify to $\lim_{x \to 0^+} \frac{x}{-\sin^2 x}$)

35 $-\frac{1}{8}$ (Hint: after the first use of l'Hospital's rule, simplify numerator to $\cot x$.)

37 $\frac{32}{9}$

39 none, but there are two removable discontinuities ($\lim_{x \to 0} \frac{2^x - x - 1}{x^2 - x} = 1 - \ln 2$, $\lim_{x \to 1} \frac{2^x - x - 1}{x^2 - x} = 2\ln 2 - 1$)

41 $y = 0$, on the left

43 use l'Hospital's rule: $\lim_{x \to \infty} \frac{\ln x}{\sqrt[10]{x}} = \lim_{x \to \infty} \frac{\frac{1}{x}}{\frac{1}{10}x^{-9/10}}$; simplify: $= \lim_{x \to \infty} \frac{10x^{9/10}}{x} = \lim_{x \to \infty} \frac{10}{x^{1/10}}$; use EAR (evaluate, approximate, render): $= \frac{10}{\Omega^{1/10}} = 10\omega^{1/10} \doteq 0$. Therefore, $\ln \Omega$ is on a lower level than $\Omega^{1/10}$.

Section 5.12

1 no

3 no

5 yes

7 no

9 yes

11 yes

13 $-\infty$ (Hint: use law of logarithms (2).)

15 $-\infty$ (Notice that $\infty \cdot (-\infty)$ is not an indeterminate form.)

17 ∞

19 $\frac{\pi^2}{2}$ (Hint: rewrite with x in the denominator.)

21 $\frac{1}{e}$

23 1

25 0 (Hint: multiply the numerator and the denominator by the conjugate of the expression.)

27 0 (Hint: after using l'Hospital's rule the first time, rewrite as $\lim_{x\to 0} -\frac{x^2 \cos x}{\sin x}$.)

29 1 (Use continuity.)

31 e^2

33 4 (form $\frac{0}{0}$, as practiced in section 5.11)

35 1

37 2 (Hint: move $\tan x$ to the denominator as $\cot x$.)

39 e^5 (Hint: after performing step ❶, move $\cot x$ to the denominator as $\tan x$.)

41 $-\frac{4}{\pi}$ (Hint: write tangent as sine over cosine.)

43 $\frac{\pi^2}{8}$ (Hint: get a common denominator.)

45 removable ($\lim_{x\to 0} f(x) = \frac{1}{2}$; Hint: after getting a common denominator and using l'Hospital's rule, clear the compound fraction by multiplying the numerator and the denominator by $1 + x$, then use l'Hospital's rule again)

47 $y = e^4$ (on both sides)

Section 5.13

1 $S(x) = \int_0^x \sin\left(\frac{\pi}{2}t^2\right)\, dt$ (The lower limit of integration may be chosen to be any desired real number. If you use a number different from zero, your function will differ from this one by a constant, affecting the answers to exercises 13, 15, 17, 19, and 21(c).)

3 $S'(x) = \sin\left(\frac{\pi}{2}x^2\right)$

5 $\frac{d}{dx}S(u) = \sin\left(\frac{\pi}{2}u^2\right) \cdot \frac{du}{dx}$

7 $f'(x) = \frac{8x^2 \sin\left(\frac{\pi}{2}(x^2+2)^2\right) - 4S(x^2+2)}{16x^2}$

9 Note: the "sign of S'" column is omitted to save space.

interval	inc/dec	local extrema
⋮		
		local min at $x = -\sqrt{6}$
$(-\sqrt{6}, -2)$	inc	
		local max at $x = -2$
$(-2, -\sqrt{2})$	dec	
		local min at $x = -\sqrt{2}$
$(-\sqrt{2}, 0)$	inc	
$(0, \sqrt{2})$	inc	
		local max at $x = \sqrt{2}$
$(\sqrt{2}, 2)$	dec	
		local min at $x = 2$
$(2, \sqrt{6})$	inc	
		local max at $x = \sqrt{6}$
⋮		

(Hint: to solve $\sin\left(\frac{\pi}{2}x^2\right) = 0$, set the argument inside sin equal to $n\pi$ for an integer n.)

11 Note: the "sign of S''" column is omitted to save space (width).

interval	concavity	inflection points
⋮		
		i.p. at $x = -\sqrt{5}$
$(-\sqrt{5}, -\sqrt{3})$	down	
		i.p. at $x = -\sqrt{3}$
$(-\sqrt{3}, -1)$	up	
		i.p. at $x = -1$
$(-1, 0)$	down	
		i.p. at $x = 0$
$(0, 1)$	up	
		i.p. at $x = 1$
$(1, \sqrt{3})$	down	
		i.p. at $x = \sqrt{3}$
$(\sqrt{3}, \sqrt{5})$	up	
		i.p. at $x = \sqrt{5}$
⋮		

(Hint: to solve $\cos\left(\frac{\pi}{2}x^2\right) = 0$, set the argument inside cos equal to $\frac{\pi}{2} + n\pi$ for an integer n.)

13

$$S_6 = \frac{1}{3} \cdot \frac{1}{6}(0 + 4(0.04362) + 2(0.17365) + 4(0.38268) + 2(0.64279) + 4(0.88701) + 1) = 0.43812$$

(See the note in the answer to exercise 1.)

15

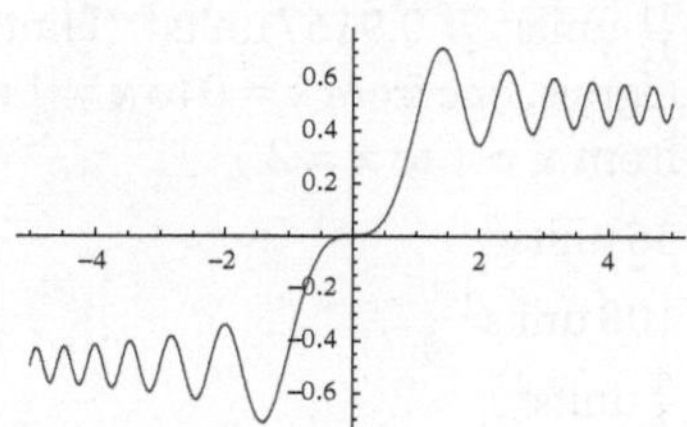

(See the note in the answer to exercise 1.)

17 horizontal asymptotes $y = \frac{1}{2}$ (on the right) and $y = -\frac{1}{2}$ (on the left) (See the note in the answer to exercise 1.)

19 For any positive infinite hyperreal Ω, $S(\Omega) \doteq \frac{1}{2}$ and $S(-\Omega) \doteq -\frac{1}{2}$. (See the note in the answer to exercise 1.)

21 (a) yes, both; (b) absolute max at $x = \sqrt{2}$, absolute min at $x = -\sqrt{2}$; (c) using the TI–84's numerical integration command `fnInt`, $S(\sqrt{2}) = 0.71397$ and $S(-\sqrt{2}) = -0.71397$. Other numerical integration methods of solution are acceptable and may yield slightly different results (See the note in the answer to exercise 1.)

23 the local extrema on the graph of $y = S(x)$ occur closer together as x moves farther from $x = 0$, whereas the local extrema on the graph of $y = \operatorname{Si} x$ are evenly spaced.

25 $y' = x \cdot \sin\left(\frac{\pi}{2}x^2\right) + 1 \cdot S(x) + \frac{1}{\pi}\left(-\sin\left(\frac{\pi}{2}x^2\right)\right) \cdot \frac{\pi}{2} \cdot 2x = S(x)$; $\int S(x)\,dx = xS(x) + \frac{1}{\pi}\cos\left(\frac{\pi}{2}x^2\right) + C$ (If you wish to avoid confusing the constant of integration with the Fresnel C function, you can write $+\,D$ instead of $+\,C$ for the constant of integration.)

27 $\sqrt{\frac{\pi}{2}} \cdot S\left(\sqrt{\frac{2}{\pi}}\,x\right) + C$ (Hint: because you need $\sin\left(\frac{\pi}{2}u^2\right)$, try setting $\frac{\pi}{2}u^2 = x^2$ and solving for u to determine the substitution to make.)

29 $\operatorname{erf} x = \int_0^x \frac{2}{\sqrt{\pi}} e^{-t^2}\,dt$ (The lower limit of integration can be chosen to be any desired real number. If you use a number different from zero, your function will differ from this one by a constant, affecting the answers to exercises 35–38.)

31 $\frac{d}{dx}\operatorname{erf} u = \frac{2}{\sqrt{\pi}} e^{-u^2} \cdot \frac{du}{dx}$

33

interval	sign of erf′	inc/dec
$(-\infty, \infty)$	+	inc

There are no local extrema.

35 horizontal asymptotes $y = 1$ (on the right), $y = -1$ (on the left) (See the note in the answer to exercise 29.)

37

$$T_8 = \frac{1}{2} \cdot \frac{1}{4}(1.12838 + 2(1.06001) + 2(0.87878) + 2(0.64293) + 2(0.41511) + 2(0.23652) + 2(0.11893) + 2(0.05277) + 0.02067) = 0.99489$$

(See the note in the answer to exercise 29.)

39 (a) no, neither exists (the function is increasing on $(-\infty, \infty)$), (b) NA, (c) NA

41 $\mathrm{Si}\, x^2 + C$ (Hint: rewrite as $\int \frac{2x \sin x^2}{x^2}\, dx$.)

43 For any positive infinite hyperreal Ω, $\mathrm{Si}\, \Omega \doteq \frac{\pi}{2}$ and $\mathrm{Si}\,(-\Omega) \doteq -\frac{\pi}{2}$.

Section 6.1

1 x

3 y (Integrating with respect to y requires only one integral, whereas integrating with respect to x requires two integrals.)

5 x (Integrating with respect to x requires two integrals, whereas integrating with respect to y requires three integrals.)

7 $\frac{9}{2}$ units^2

9 $\frac{8}{3}$ units^2

11 $\frac{20}{3}$ units^2

13 $(-\frac{4}{3} + 2\sqrt{3})$ units^2 = 2.13077 units^2 (Hint: use two integrals, one from $x = 1$ to $x = 2$ and the other from $x = 2$ to $x = 4$.)

15 $\frac{1}{2}$ units^2

17 $(\pi^2 - 4)$ units^2

19 $\frac{89}{6}$ units^2 = 14.83 units^2

21 $\frac{9}{2}$ units^2

23 $(5 - \frac{3}{\ln 2})$ units^2 = 0.671915 units^2

25 $\frac{125}{6}$ units^2 = 20.833 units^2

27 9 units^2

29 $\frac{9}{2}$ units^2 (Hint: integrate with respect to y.)

31 $\frac{29}{6}$ units^2 = 4.833 units^2 (Hint: use two integrals, one from $x = 0$ to $x = 1$ and the other from $x = 1$ to $x = 4$.)

33 $\frac{11}{12}$ units^2 = 0.9167 units^2 (Hint: use two integrals, one from $x = 0$ to $x = 1$ and the other from $x = 1$ to $x = 2$.)

35 36 units^2

37 108 units^2

39 $\frac{9}{2}$ units^2

Section 6.2

1 $$\int_1^3 \pi \left(x^2\right)^2 dx = \cdots = \frac{242\pi}{5} \text{ units}^3 = 152.05 \text{ units}^3$$

3 $$\int_{-1}^1 \pi \left(x^2 + 1\right)^2 dx = \cdots = \frac{56\pi}{15} \text{ units}^3 = 11.729 \text{ units}^3$$

5 $$\int_2^5 \pi \frac{1}{x^2}\, dx = \cdots = \frac{3\pi}{10} \text{ units}^3 = 0.94248 \text{ units}^3$$

7 $$\int_0^1 \pi e^{2x}\, dx = \cdots = \frac{\pi}{2}\left(e^2 - 1\right) \text{ units}^3 = 10.036 \text{ units}^3$$

9 $\int_0^{\pi/6} \pi \sin 3x\, dx = \cdots = \frac{\pi}{3}$ units^3
$= 1.0472$ units^3

11 $\int_0^9 \pi x^2\, dx = \cdots = 243\pi$ units^3
$= 763.41$ units^3

13 $\int_{-3}^3 \pi\left(9 - x^2\right)^2 dx = \cdots = \frac{1296\pi}{5}$ units^3
$= 814.30$ units^3

15 $\int_{-2}^2 \pi\left(x^2 - 4\right)^2 dx = \cdots = \frac{512\pi}{15}$ units^3
$= 107.23$ units^3

17 $\int_{-2}^0 \pi(2x+4)^2\, dx + \int_0^4 \pi(4-x)^2\, dx$
$= \cdots = 32\pi$ units^3 $= 100.53$ units^3

19 $\int_0^3 \pi y^4\, dy = \cdots = \frac{243\pi}{5}$ units^3 $= 152.68$ units^3

21 $\int_0^4 \pi(\sqrt{y})^2\, dy = \cdots = 8\pi$ units^3
$= 25.133$ units^3

23 $\int_0^3 \pi\left(4 - \frac{4}{9}y^2\right) dy = \cdots = 8\pi$ units^3
$= 25.133$ units^3

25 $\int_1^4 \frac{\pi}{y^2}\, dy = \cdots = \frac{3\pi}{4}$ units^3 $= 2.3562$ units^3

27 (a) $\int_1^4 \pi(\ln x)^2\, dx$, (b) $S_6 = 8.16303$ units^3, (c) 8.15838 units^3

29 (a) $\int_0^3 \pi e^{2x} \sin^2 x\, dx$, (b) $S_6 = 211.456$ units^3, (c) 208.610 units^3 (If your answer is 1.2536 units^3, your calculator is in degree mode. Switch to radian mode.)

31 (a) $\int_0^2 \pi \cosh^2 x\, dx$, (b) $S_6 = 24.5974$ units^3, (c) 24.5750 units^3

33 (a) $\int_0^1 \pi\left(\sin^{-1} y\right)^2 dy$ ("simplifying" to $\int_0^1 \pi\left(\sin^{-2} y\right) dy$ is incorrect.), (b) $S_6 = 1.5453$ units^3, (c) 1.4684 units^3 (If your answer is 4820.4 units^3, your calculator is in degree mode. Switch to radian mode.)

35 $V = \int_0^{200} \pi(30^2)\, dx = \cdots = 180\,000\pi\,\text{cm}^3$
$= 565\,487\,\text{cm}^3$

37 (a) $V = 180\,000\pi\,\text{cm}^3 \cdot \frac{1\,\text{m}^3}{1\,000\,000\,\text{cm}^3} = 0.18\pi\,\text{m}^3$
$= 0.565\,\text{m}^3$, (b) $V = \int_0^2 \pi(0.3^2)\, dx =$
$\cdots = 0.18\pi\,\text{m}^3 = 0.565\,\text{m}^3$

39 $V = \int_0^{10} \frac{\pi x^2}{2}\, dx = \cdots = \frac{1000\pi}{6}$ units^3
$= 523.60$ units^3

41 $V = \int_1^1 4\sqrt{4 - x^2}\, dx = 15.3\,\text{in}^3$

43 (a) $V = \int_1^3 \pi(x-1)^2\, dx = \cdots = \frac{8\pi}{3}$ units^3; (b) $V = \frac{\pi r^2 h}{3}$, with $r = 2$, $h = 2$, so $V = \frac{8\pi}{3}$ units^3; (c) the answers to (a) and (b) are identical, as expected.

45 (a) $V = \int_{-r}^r \pi\left(\sqrt{r^2 - x^2}\right)^2 dx = \cdots$
$= \frac{4}{3}\pi r^3$ units^3, (b) a sphere of radius r

47 $V = \int_1^7 2\sqrt{49 - x^2}\left(\frac{x}{2} - \frac{1}{2}\right) dx$; an antiderivative cannot be found using the available antiderivative formulas, rules, and techniques. Simpson's rule with $n = 6$ gives $V = 77.456\,\text{in}^3$; Simpson's rule with $n = 20$ gives $V = 79.040\,\text{in}^3$. A better numerical approximation is $V = 79.343\,\text{in}^3$.

Section 6.3

1 $\int_1^5 \pi(3x-1)^2\,dx = \cdots = 304\pi\,\text{units}^3$
$= 955.04\,\text{units}^3$

3 $\int_3^{15} \pi\left(\frac{y}{3}-1\right)^2 dy = \cdots = 64\pi\,\text{units}^3$
$= 201.06\,\text{units}^3$

5 $\int_0^{\pi} \pi\left(3+\sqrt{\sin x}-3\right)^2 dx = \cdots$
$= 2\pi\,\text{units}^3 = 6.2832\,\text{units}^3$

7 $\int_1^2 \pi\left(y^3-10\right)^2 dy = \cdots = \frac{302\pi}{7}\,\text{units}^3$
$= 135.54\,\text{units}^3$

9 $\int_{1/2}^1 \pi\left(\frac{1}{x}-(-2)\right)^2 dx = \cdots$
$= (3+\ln 16)\pi\,\text{units}^3 = 18.135\,\text{units}^3$

11 $\int_0^1 \pi\left((2-x)^2-(x^2)^2\right) dx = \cdots$
$= \frac{32\pi}{15}\,\text{units}^3 = 6.7021\,\text{units}^3$

13 $\int_0^1 \pi\left((x+1)^2-(x^3)^2\right) dx = \cdots$
$= \frac{46\pi}{21}\,\text{units}^3 = 6.8816\,\text{units}^3$

15 $\int_0^{\pi/4} \pi\left((1+\sec x)^2-(\sqrt{2}\sec x)^2\right) dx$
$= \cdots = \left(\pi+\frac{\pi^2}{4}\right)\,\text{units}^3 = 5.6090\,\text{units}^3$

17 $\int_{-1}^0 \pi\left(x^2-(x^3)^2\right) dx = \cdots = \frac{4\pi}{21}\,\text{units}^3$
$= 0.59840\,\text{units}^3$

19 $\int_0^1 \pi\left(8-y^3\right) dy = \cdots = \frac{31\pi}{4}\,\text{units}^3$
$= 24.347\,\text{units}^3$

21 $\int_0^1 \pi\left(\left(\frac{1}{\sqrt{x^2+1}}\right)^2-\left(\frac{1}{2}x\right)^2\right) dx = \cdots$
$= \left(\frac{\pi^2}{4}-\frac{\pi}{12}\right)\,\text{units}^3 = 2.2056\,\text{units}^3$

23 $\int_0^1 \pi\left(e^x-1\right)^2 dx = \cdots$
$= \pi\left(\frac{e^2}{2}-2e+\frac{5}{2}\right)\,\text{units}^3$
$= 2.3812\,\text{units}^3$

25 $\int_1^3 \pi\left(4x-x^2-3\right)^2 dx = \cdots = \frac{16\pi}{15}\,\text{units}^3$
$= 3.3510\,\text{units}^3$

27 $\int_0^1 \pi\left((\sqrt{x})^2-x^2\right) dx = \cdots = \frac{\pi}{6}\,\text{units}^3$
$= 0.52360\,\text{units}^3$

29 $\int_0^8 \pi\left((\sqrt[3]{y})^2-(\tfrac{1}{4}y)^2\right) dy = \cdots$
$= \frac{128\pi}{15}\,\text{units}^3 = 26.808\,\text{units}^3$

31 (a) $\int_1^e \pi(\ln y-1)^2\,dy$, (b) T_5
$= 1.43277\,\text{units}^3$ (Hint: $\Delta x = \frac{e-1}{5}$),
(c) $1.37151\,\text{units}^3$

33 (a) $\int_0^{\pi/2} \pi\left(1^2-\sin^2 y\right) dy$ (If your answer is $\int_0^{\pi/2} \pi(\sin y)^2\,dy$, then you are using the wrong region; regraph the lines and curve to find the correct bounded region. Washers are required for the correct region. Answers to (b) and (c) are the same by coincidence.). (b) $T_5 = 2.46740\,\text{units}^3$. (c) $2.46740\,\text{units}^3$ (If your answer is $4.93357\,\text{units}^3$, your calculator is in degree mode. Change it to radian mode.)

35 the volume of a solid of revolution generated by revolving the region between the curves $x = f(y)$ and $x = g(y)$ from $y = a$ to $y = b$ about the y-axis, where $f(y) \geq g(y) \geq 0$ or $f(y) \leq g(y) \leq 0$, is $V = \int_a^b \pi\left((f(y))^2-(g(y))^2\right) dy$.

Section 6.4

1 (a) washers, (b) y, (c) onc

3 (a) shells, (b) y, (c) one

5 (a) shells, (b) x, (c) two

7 (a) disks, (b) x, (c) two

9 (a) $\int_0^{\pi/2} 2\pi x \sin x \, dx$, (b) 6.2832 units^3

11 (a) $\int_0^1 2\pi y \left(\frac{\pi}{2} - \sin^{-1} y\right) dy$, (b) 2.4674 units^3

13 (a) $\int_2^{17.7} 2\pi x \left(\sqrt{1 + x^2 + 15x^4} - (-1)\right) dx$, (b) 598 117 units^3

15 $\int_1^4 2\pi x \cdot \frac{1}{x^2} \, dx = 2\pi \ln 4$ units^3 ≈ 8.7103 units^3

17 $\int_1^2 2\pi x \frac{1}{\sqrt{5 - x^2}} \, dx = 2\pi$ units^3 ≈ 6.2832 units^3

19 $\int_0^2 2\pi x \left(8 - x^3\right) dx = \frac{96\pi}{5}$ units^3 ≈ 60.319 units^3

21 $\int_0^8 2\pi y \sqrt[3]{y} \, dy = \frac{6\pi}{7} \cdot 8^{7/3}$ units^3 $= \frac{768\pi}{7}$ units^3 ≈ 344.68 units^3

23 $\int_0^8 2\pi (y - (-1)) \sqrt[3]{y} \, dy$ $= \left(\frac{6\pi}{7} \cdot 8^{7/3} + \frac{3\pi}{2} \cdot 8^{4/3}\right)$ units^3 $= \frac{936\pi}{7}$ units^3 ≈ 420.08 units^3

25 (a) $\int_0^9 2\pi x \cdot 2\sqrt{x} \, dx = \frac{1944\pi}{5}$ units^3 ≈ 1221.45 units^3, (b) $\int_0^6 \pi \left(9^2 - \left(\frac{y^2}{4}\right)^2\right) dy$ $= \frac{31\,104\pi}{80}$ units^3 $= \frac{1944\pi}{5}$ units^3 $\approx$ 1221.45 units^3

27 (a) $\int_0^1 2\pi y \left(y - y^2\right) dy = \frac{\pi}{6}$ units^3 ≈ 0.5236 units^3, (b) $\int_0^1 \pi \left((\sqrt{x})^2 - (x)^2\right) dx$ $= \frac{\pi}{6}$ units^3 ≈ 0.5236 units^3

29 (a) $\int_0^1 2\pi y(\sqrt{y}) \, dy + \int_1^2 2\pi y(2 - y) \, dy$ $= \frac{32\pi}{15}$ units^3 ≈ 6.7021 units^3, (b) $\int_0^1 \pi \left((2 - x)^2 - (x^2)^2\right) dx = \frac{32\pi}{15}$ units^3 ≈ 6.7021 units^3

31 (a) $\int_1^3 2\pi x \left(x^2 - 1\right) dx = 32\pi$ units^3 ≈ 100.53 units^3, (b) $\int_1^9 \pi \left(3^2 - (\sqrt{y})^2\right) dy$ $= 32\pi$ units^3 ≈ 100.53 units^3

33 (easiest method: disks) $\int_0^1 \pi \left(\sqrt{4 + x^2}\right)^2 dx$ $= \frac{13\pi}{3}$ units^3 ≈ 13.614 units^3

35 (easiest method: shells) $\int_0^1 2\pi x \left(\sqrt{4 + x^2}\right) dx$ $= \frac{2\pi}{3} \left(5^{3/2} - 4^{3/2}\right)$ units^3 ≈ 6.6609 units^3

37 (easiest method: shells) $\int_0^1 2\pi x \left(0 - \left(-\frac{1}{x}\right)\right) dx = 2\pi$ units^3 ≈ 6.2832 units^3

39 (easiest method: shells) $\int_1^3 2\pi y \left(\frac{7}{y + y^3}\right) dy$ $= 14\pi \left(\tan^{-1} 3 - \tan^{-1} 1\right)$ units^3 $= \left(14\pi \tan^{-1} 3 - \frac{7\pi^2}{2}\right)$ units^3 ≈ 20.392 units^3

41 (easiest method: disks) $\int_{-2}^2 \pi \left(4 - x^2\right)^2 dx =$ $\frac{512\pi}{15}$ units^3 ≈ 107.23 units^3

Section 6.5

1 180 ft · lb

3 509.6 J

5 8400 ft · lb

7 225 ft · lb

9 1 000 000 ft · lb

11 23 740 J (person, 21 168 J; cable, 2 572 J)

13 $\frac{5}{4}$ ft · lb

15 400 ft · lb (Hint: if you used $\int_0^4 200x\,dx$, you forgot to calculate the force constant k.)

17 5 ft · lb

19 1.4 J

21 237.6 J

23 $\frac{125}{3}$ J

25 2017 J

27 95 821 ft · lb (GSL gives the answer to three-decimal-place precision: 95 800 ft·lb.)

Section 6.6

1 109 440 ft·lb

3 $99\,840\pi$ ft · lb = 313 657 ft · lb

5 $192\,166\pi$ J = 603 707 J

7 19 600 J

9 $\frac{640\,000\pi}{9}$ J = 223 402 J

11 1.376 feet

Section 6.7

1 $\frac{4}{3}$

3 $\frac{1}{3}\ln 4$

5 $\frac{2}{\pi}$

7 $\frac{10}{3}$

9 $\frac{2}{5}\left(\frac{\pi}{6}\right)^5 = \frac{\pi^5}{19\,440}$

11 $\frac{2}{3}$

13 $\frac{20}{\ln 3}$

15 $\frac{\ln\sqrt{2}}{2(e^{\pi/4}-1)} \approx 0.1452$

17 (a) answers vary, (b) $\frac{21}{4}$

19 $c = \sqrt{7}$ (The answer $c = -\sqrt{7}$ is not correct because it is not in the interval.)

21 $c = \ln\left(\frac{1}{3}(e^3-1)\right) \approx 1.85$

23 $c = \sqrt[3]{\frac{1}{4}}$

25 $c = \pm\sqrt{3}$ (Both answers must be included.)

27 (a) concave down; (b) for f of exercise 1, $f(0) = 0, f(4) = 2$; for f of exercise 3, $f(0) = 0$, $f(\frac{\pi}{2}) = 1$; (c) for f of exercise 1, 1; for f of exercise 3, $\frac{1}{2}$; (d) for f of exercise 1, $f_{\text{avg}} = \frac{4}{3} > 1$; for f of exercise 3, $f_{\text{avg}} = \frac{2}{\pi} > \frac{1}{2}$; (e) when f is concave down on $[a, b]$, $f_{\text{avg}} > \frac{f(a)+f(b)}{2}$

29 The mean value theorem for integrals requires a continuous function. A continuous function has infinitely many values, but there are only finitely many people. Therefore the mean value theorem for integrals does not apply, and it may be true that "no one is average." (Use of figures 2, 3, or 4 may vary.)

Section 7.1

1 $x + 15 + \frac{72}{x-6}$

3 $x - 3 + \frac{7x+4}{x^2+3x}$

5 $(x-4)^2 - 5$

7 $5(x+\frac{1}{5})^2 - \frac{21}{5}$

9 $\frac{x^2}{2} - 3x + \ln|x-2| + C$

11 $\frac{7x^2}{2} + 3x + \frac{8}{3}\ln|3x-1| + C$

13 $2x + \frac{7}{3}\tan^{-1}\frac{x}{3} + C$

15 $\frac{1}{2}\ln|x^2+9| + C$

17 $2x - 4\ln|x+\frac{1}{2}| + C$

19 $\tan^{-1}(x+1) + C$

21 $\tan^{-1}\frac{x-3}{\sqrt{7}} + C$

23 $\sinh^{-1}(x+4) + C$

25 $\sin^{-1}(x-5) + C$

27 $\frac{3x^2}{2} + 5\tan^{-1}(x-4) + C$

29 $\frac{2}{3}\cosh^{-1}\left(\frac{x+\frac{5}{3}}{3}\right) + C = \frac{2}{3}\cosh^{-1}\left(\frac{3x+5}{9}\right) + C$

Section 7.2

1 yes, $f(x) = x$, $g'(x) = \cos x$

3 yes, $f(x) = x^2$, $g'(x) = \cos x$

5 no (substitution)

7 yes; either is fine, just don't switch roles for second application of parts.

9 yes, $f(x) = \sin^{-1} x$, $g'(x) = 1$

11 $x\sin x + \cos x + C$

13 $x^2\sin x + 2x\cos x - 2\sin x + C$

15 $\frac{1}{2}\sin x^2 + C$

17 $x^4\sin x + 4x^3\cos x - 12x^2\sin x - 24x\cos x + 24\sin x + C = (x^4 - 12x^2 + 24)\sin x + (4x^3 - 24x)\cos x + C$

19 $2x\cosh x - 2\sinh x + C$

21 $(4x^2 - 3x + 1)e^x - (8x-3)e^x + 8e^x + C = (4x^2 - 11x + 12)e^x + C$

23 $\frac{1}{2}xe^{2x} - \frac{1}{4}e^{2x} + C = (\frac{1}{2}x - \frac{1}{4})e^{2x} + C$

25 $x\sin^{-1} x + \sqrt{1-x^2} + C$

27 $3e^4 + 1 \approx 164.8$ (Incorrect answers include $xe^x - e^x + C$ and $3e^4 + 1 + C$.)

29 $\frac{x^3}{3}\ln x - \frac{1}{9}x^3 + C$

31 $\frac{-e^{2x}\cos x + 2e^{2x}\sin x}{5} + C$

33 $2\pi^4 - 24\pi^2 + 96 \approx 53.95$

35 $\frac{3^x\sinh x - (\ln 3)3^x\cosh x}{1-(\ln 3)^2} + C$

37 $\frac{2^{\sin x}}{\ln 2} + C$

39 $-\cos e^x + C$

41 $e^x(3x^4 - 12x^3 + 36x^2 - 72x + 72) + C$

43 $2\sqrt{x}\sinh\sqrt{x} - 2\cosh\sqrt{x} + C$

45 $\frac{1}{2}x^2e^{x^2} - \frac{1}{2}e^{x^2} + C$

47 $7x\ln x - 7x + C$

49 $\frac{x^2}{2}\cos^{-1}x^2 - \frac{1}{2}\sqrt{1-x^4} + C$

51 $\frac{e^x\cos 5x + 5e^x\sin 5x}{26} + C$

53 $x^3\tan^{-1}x - \frac{x^2}{2} + \frac{1}{2}\ln(x^2+1) + C$

55 $(2\pi^3 + 2\pi^2)$ units^3

57 $(\frac{\pi}{4}e^2 - \frac{\pi}{4})$ units^3

59 $(2x^5 - 3x^2 + 7)(\frac{1}{2}e^{2x} - \frac{x^2}{2}) - (10x^4 - 6x)(\frac{1}{4}e^{2x} - \frac{x^3}{6}) + (40x^3 - 6)(\frac{1}{8}e^{2x} - \frac{x^4}{24}) - 120x^2(\frac{1}{16}e^{2x} - \frac{x^5}{120}) + 240x(\frac{1}{32}e^{2x} - \frac{x^6}{720}) - 240(\frac{1}{64}e^{2x} - \frac{x^7}{5040}) + C$; alternate answer, $(2x^5 - 3x^2 + 7)\frac{1}{2}e^{2x} - (10x^4 - 6x)\frac{1}{4}e^{2x} + (40x^3 - 6)\frac{1}{8}e^{2x} - 120x^2\frac{1}{16}e^{2x} + 240x\frac{1}{32}e^{2x} - 240\frac{1}{64}e^{2x} - \frac{2x^7}{7} + \frac{3x^4}{4} - \frac{7x^2}{2} + C$

61 $2\sqrt{x+1}\ln(x+1) - 4\sqrt{x+1} + C$

63 $\frac{2}{3}x(1+x)^{3/2} - \frac{4}{15}(1+x)^{5/2} + C$

Section 7.3

1 $\frac{\cos^5 x}{5} - \frac{\cos^3 x}{3} + C$

3 $-\frac{1}{2}\cos x - \frac{1}{10}\cos 5x + C$

5 $\frac{\sin^7 x}{7} + C$

7 $-\frac{1}{4}\cos^4 x + \frac{1}{3}\cos^6 x - \frac{1}{8}\cos^8 x + C$; alternate answer, $\frac{1}{6}\sin^6 x - \frac{1}{8}\sin^8 x + C$

9 $\frac{1}{2}x + \frac{1}{4}\sin 2x + C$

11 $\frac{3}{8}x + \frac{1}{4}\sin 2x + \frac{1}{32}\sin 4x + C$

13 $\frac{1}{27}\sin^9 x^3 - \frac{1}{33}\sin^{11} x^3 + C$

15 $-\frac{1}{6}\sin(-3x) - \frac{1}{22}\sin(11x) + C = \frac{1}{6}\sin(3x) - \frac{1}{22}\sin(11x) + C$

17 $\frac{1}{3}\tan^3 x + \tan x + C$

19 $-x + \tan x + C$

21 $\frac{1}{5}\sec^5 x - \frac{1}{3}\sec^3 x + C$

23 $\frac{1}{4}\ln|\sec(4x+9) + \tan(4x+9)| + C$

25 $-\ln|\csc x + \cot x| + C$

27 $\frac{1}{6}x\cos(-3x) - \frac{1}{14}x\cos(7x) + \frac{1}{18}\sin(-3x) + \frac{1}{98}\sin(7x) + C$

29 $\frac{1}{8}x - \frac{1}{96}\sin(12x - 16) + C$

31 $\left(8\pi + \frac{9}{2}\pi^2\right)$ units^3

33 $\sec x + \cos x + C$

35 $\sin\theta - \sin^3\theta + \frac{3}{5}\sin^5\theta - \frac{1}{7}\sin^7\theta + C$

37 $-2\sqrt{\cos y} + \frac{4}{5}(\cos y)^{5/2} - \frac{2}{9}(\cos y)^{9/2} + C$

39 $-\frac{4}{3}\cot^3\frac{x}{4} - 4\cot\frac{x}{4} + C$

41 $-\frac{1}{2\tan^2 t} + \ln|\tan t| + C$

43 $\frac{7}{8}x + \frac{2}{3}\sin^3 x + \frac{1}{32}\sin 4x + C$

45 using $u = kx$, $du = k\,dx$, $\displaystyle\int \cos kx\,dx = \frac{1}{k}\int \cos u\,du = \frac{1}{k}\sin u + C = \frac{1}{k}\sin kx + C$; alternate answer, $\frac{d}{dx}\frac{1}{k}\sin kx = \frac{1}{k}\cos kx \cdot k = \cos kx$

Section 7.4

1 (a) $x = 3\tan\theta$, (b) $1 + \tan^2\theta = \sec^2\theta$

3 (a) $x = 2\sec\theta$, (b) $\sec^2\theta - 1 = \tan^2\theta$

5 (a) $x = \sqrt{7}\sec\theta$, (b) $\sec^2\theta - 1 = \tan^2\theta$

7 (a) $x = 4\sin\theta$, (b) $1 - \sin^2\theta = \cos^2\theta$

9 $\frac{x}{2\sqrt{x^2+2}} + C$

11 $\frac{x}{5\sqrt{5-x^2}} + C$

13 $\frac{9\pi}{4}$

15 $\sin^{-1}\frac{x}{\sqrt{6}} + C$

17 $\ln\left|\frac{\sqrt{x^2+8}}{\sqrt{8}} + \frac{x}{\sqrt{8}}\right| - \frac{x}{\sqrt{x^2+8}} + C = \ln\left|\sqrt{x^2+8} + x\right| - \frac{x}{\sqrt{x^2+8}} + C$

19 $-\frac{1}{2}\ln\left|\frac{\sqrt{x^2+4}}{x} + \frac{2}{x}\right| + C = \frac{1}{2}\ln\left|\frac{x}{2+\sqrt{x^2+4}}\right| + C$

21 $\frac{\sqrt{y^2-7}}{7y} + C$

23 $-\frac{1}{\sqrt{x^2-25}} + C$

25 $\frac{1}{54}\sec^{-1}\frac{x}{3} + \frac{\sqrt{x^2-9}}{18x^2} + C$

27 $\frac{2\sqrt{2}}{9}$

29 $-\frac{x}{3\sqrt{x^2-3}} + C$

31 $\frac{\sqrt{x^2-5}}{25x} - \frac{(x^2-5)^{3/2}}{75x^3} + C$

33 $\frac{1}{8}\tan^{-1}\frac{x}{4} - \frac{x}{2(x^2+16)} + C$

35 $-10\ln\left|\frac{10+\sqrt{100-u^2}}{u}\right| + \sqrt{100-u^2} + C$

37 $\frac{1}{16}\left(-\frac{2\sqrt{4-x^2}}{x^2} + \ln\left|\frac{2}{x} - \frac{\sqrt{4-x^2}}{x}\right|\right) + C$

39 (partial solution) use $x = a\sin\theta$, $dx = a\cos\theta\,d\theta$, simplify, $= \cdots = \displaystyle\int 1\,d\theta = \theta + C = \sin^{-1}\frac{x}{a} + C$, which is version 2 of the antiderivative giving inverse sine formula.

41 $90\pi^2$ units^3

43 $\frac{5}{2}\pi \approx 7.854$

Section 7.5

1 partial fractions

3 completing the square

5 partial fractions

7 $\frac{A}{x+4} + \frac{B}{x-1}$ (Swapping A and B is allowed.)

9 $\frac{A}{x} + \frac{B}{x+5} + \frac{C}{x+1}$ (Swapping A, B, and C is allowed.)

11 $\frac{A}{x} + \frac{B}{x-1} + \frac{C}{x-4} + \frac{D}{x+7}$ (Swapping A, B, C, and D is allowed.)

13 $2\ln|x+3| - 2\ln|x+5| + C$

15 $\frac{6}{28}\ln|4x+1| - \frac{1}{21}\ln|3x-1| + C$

17 $3\ln|x+5| + 4\ln|x-2| + C$

19 $\ln|x+4| - 3\ln|x+2| + C$

21 $\frac{3}{2}\ln|x-3| - \frac{1}{2}\ln|x-1| + C$

23 $x + \frac{9}{2}\ln|x-3| - \frac{1}{2}\ln|x-1| + C$

25 $-\frac{3}{4}\ln|x| + \frac{7}{8}\ln|x-2| + \frac{7}{8}\ln|x+2| + C$

27 $\ln|x^2-4| + C = \ln|x+2| + \ln|x-2| + C$ (Substitution is sufficient.)

29 $\frac{51}{20}\ln|x-10| + \frac{49}{20}\ln|x+10| + C$

31 $2\ln|2x+5| - \ln|x-3| + C$

33 $2\tan^{-1}(x+2) + C$ (Partial fractions does not apply. Use completing the square instead.)

35 $2\ln|x-\sqrt{7}| + 4\ln|x+\sqrt{7}| + C$

37 $2\ln|z+1| - \ln|z+2| - \ln|z+3| + C$

39 $-\frac{2}{3}\ln|e^x+7| + \frac{2}{3}\ln|e^x+1| + C$

41 $\frac{1}{2}\ln|2+\ln x| + \frac{1}{2}\ln|-2+\ln x| + C = \frac{1}{2}\ln|(\ln x)^2 - 4| + C$

43 $\frac{1}{2}t^2 + 2t - \ln|t-2| + \ln|t+2| + 5\ln|2t-1| + C$

45 $2x + \frac{17}{12}\ln|x-2| - \frac{19}{4}\ln|x+2| + \frac{4}{3}\ln|x+1| + C$

47 $-e^x - \frac{1}{2}\ln|1-e^x| + \frac{1}{2}\ln|1+e^x| + C$

49 $\int \csc\theta\, d\theta = \frac{1}{2}\ln|1-\cos\theta| - \frac{1}{2}\ln|1+\cos\theta| + C = \frac{1}{2}\ln\left|\frac{1-\cos\theta}{1+\cos\theta}\right| + C$

Section 7.6

1 $\frac{A}{x-4} + \frac{B}{(x-4)^2} + \frac{C}{(x-4)^3} + \frac{D}{x+1}$

3 $\frac{Ax+B}{x^2+5} + \frac{C}{x-2} + \frac{D}{x+2}$

5 $\frac{Ax+B}{x^2+2x+19} + \frac{Cx+D}{(x^2+2x+19)^2} + \frac{E}{x-1} + \frac{F}{x+3} + \frac{G}{(x+3)^2}$

7 $-\frac{1}{2}\ln|x+1| + \frac{1}{2}\ln|x+3| - \frac{2}{x+3} + C$

9 $\frac{29}{17}\ln|x+4| - \frac{6}{17}\ln\left(x^2+1\right) - \frac{3}{17}\tan^{-1}x + C$

11 $2\ln x - \frac{3}{x} - 4\ln|x-1| + C$

13 $-\frac{5}{3} + \ln\frac{7}{2} \approx -0.4139$

15 $3\ln|x| + \tan^{-1}\frac{x}{2} + C$

17 $-\frac{4}{x} - \frac{1}{2}\tan^{-1}\frac{x}{2} + C$

19 $\ln(x^2+4) + \frac{7}{2}\tan^{-1}\frac{x}{2} + C$

21 $2\ln|x| + 3\ln|x-1| + 4\ln|x+2| + C$

23 $\frac{1}{2}\ln(x^2+4x+5) - 3\tan^{-1}(x+2) + C$

25 $\frac{1}{4}\ln|x-1| - \frac{1}{4}\ln|x+1| - \frac{1}{2}\tan^{-1}x + C$

27 $-\ln|x| - 2\ln|x+2| - \frac{3}{x+2} + C$

29 $x - \ln(1+e^x) + \frac{1}{1+e^x} + C$

31 $-\frac{1}{2x^2}\tan^{-1}x - \frac{1}{2x} - \frac{1}{2}\tan^{-1}x + C$

33 $\frac{1}{4}\ln|1+\sin\theta| - \frac{1}{4}\ln|1-\sin\theta| + \frac{1}{4-4\sin\theta} - \frac{1}{4+4\sin\theta} + C = \frac{\sin\theta}{2\cos^2\theta} + \frac{1}{4}\ln\left|\frac{1+\sin\theta}{1-\sin\theta}\right| + C$

35 $\frac{1}{2}x^2 - 5x + 2\ln|x-1| - \frac{13\sqrt{3}}{18}\tan^{-1}\frac{x+2}{\sqrt{3}} - \frac{\frac{5}{6}x+\frac{8}{3}}{x^2+4x+7} + C$

Section 7.7

1 $\frac{10x+12}{25\sqrt{3+5x}} + C$

3 $3\ln\left|x^{\frac{1}{3}}\right| - 3\ln\left|1-x^{\frac{1}{3}}\right| + C = 3\ln\left|\frac{x^{\frac{1}{3}}}{1-x^{\frac{1}{3}}}\right| + C$

5 $\frac{\sqrt{4x+1}}{32} + \frac{1}{16\sqrt{4x+1}} - \frac{1}{96(4x+1)^{\frac{3}{2}}} + C = \frac{6x^2+6x+1}{12(4x+1)^{\frac{3}{2}}} + C$

7 $\ln\left|\tan\frac{\theta}{2} + 1\right| + C$

9 $\frac{2}{3}\tan^{-1}\left(\frac{1}{3}\tan\frac{x}{2}\right) + C$

11 $x + 4\sqrt{x+1} + 4\ln\left|\sqrt{x+1} - 1\right| + C$

13 $3 - 9\tan^{-1}\frac{1}{3}$

15 $\ln\left|1 + \tan^2\frac{\theta}{2}\right| + C = 2\ln\left|\sec\frac{\theta}{2}\right| + C$

17 $\frac{1}{3\sqrt{2}}\tan^{-1}\left(\frac{1}{\sqrt{2}}\tan\frac{x^3}{2}\right) + C$

19 $\frac{\pi}{3\sqrt{3}}$

21 $2\sqrt{t+2} + \sqrt{2}\tan^{-1}\sqrt{\frac{t+2}{2}} + C$

23 $\frac{11}{2} + \frac{3}{8}\ln\frac{3}{5} \approx 5.3084$

25 $\ln\left|x + 2\sqrt{x} + 5\right| - \tan^{-1}\left(\frac{1}{2}(\sqrt{x}+1)\right) + C$

27 $\frac{2}{\sqrt{x}} + 2\ln\left|\sqrt{x} - 3\right| - 2\ln\left|\sqrt{x} + 3\right| + C$

29 $-\frac{\theta}{3} + \frac{5}{6}\tan^{-1}\left(2\tan\frac{\theta}{2}\right) + C$

31 $\frac{1}{\sqrt{3}}\ln\left|\frac{2-\sqrt{3}+\tan\frac{x}{2}}{2+\sqrt{3}+\tan\frac{x}{2}}\right| + C$

33 $\frac{\pi}{12}$

35 1

37 $-x + \ln\left|e^x - 1\right| + C$

39 $\int \frac{f'(x)}{g(x)}\,dx = \frac{f(x)}{g(x)} + \int \frac{f(x)g'(x)}{(g(x))^2}\,dx$

Section 7.8

1 $2\ln\left|\sec\sqrt{x}\right| + C = -2\ln\left|\cos\sqrt{x}\right| + C$

3 $-\frac{1}{4}\sin(-2x) - \frac{1}{16}\sin 8x + C = \frac{1}{4}\sin 2x - \frac{1}{16}\sin 8x + C$

5 $\frac{1}{3\cos 3x} + C = \frac{1}{3}\sec 3x + C$

7 $\frac{1}{4}\sin^4 x - \frac{1}{3}\sin^6 x + \frac{1}{8}\sin^8 x + C$; alternate answer, $\frac{1}{8}\cos^8 x - \frac{1}{6}\cos^6 x + C$; alternate answer, $-\frac{1}{6}\cos^6 x\sin^2 x - \frac{1}{24}\cos^8 x + C$

9 $-\frac{1}{3}x^2\cos 3x + \frac{2}{9}x\sin 3x + \frac{2}{27}\cos 3x + C$

11 $x\sin^{-1}3x + \frac{1}{3}\sqrt{1-9x^2} + C$

13 $-10\sqrt{4-x} + C$

15 $5\sin^{-1}\frac{x}{2} + C$

17 $10 - 5\sqrt{3} \approx 1.3397$

19 $-\frac{5}{4}\ln|2-x| + \frac{5}{4}\ln|2+x| + C = \frac{5}{4}\ln\left|\frac{2+x}{2-x}\right| + C$

21 $\frac{5x}{4\sqrt{4-x^2}} + C$

23 $\frac{1}{2}\left(\tan^{-1}x\right)^2 + C$

25 $-\frac{1}{24}e^x\cosh 5x + \frac{5}{24}e^x\sinh 5x + C$; alternate answer, $\frac{1}{12}e^{6x} - \frac{1}{8}e^{-4x} + C$

27 $\frac{1}{6}\tan^6 x + \frac{1}{4}\tan^4 x + C$; alternate answer, $\frac{1}{6}\sec^6 x - \frac{1}{4}\sec^4 x + C$

29 $x - 2\ln|x-2| + 8\ln|x-4| + C$

31 $\ln|x-3| - \frac{3}{x-3} + C$

33 $\tan^{-1}(x-3) + C$

35 $-\frac{1}{2}x^2\cos x^2 + \frac{1}{2}\sin x^2 + C$

37 $\frac{1}{4e} \approx 0.09197$

39 $\sin x\ln\sin x - \sin x + C$

41 $\ln|\cosh x + e^x| + C = \ln(\cosh x + e^x) + C$

43 $\frac{3}{8}x + \frac{1}{4}\sin 2x + \frac{1}{32}\sin 4x + C$

45 $\frac{3}{25}\ln|x| - \frac{3}{50}\ln(x^2+25) + C = \frac{3}{50}\ln\frac{x^2}{x^2+25} + C$

47 $\frac{-4}{\sqrt{x^2-4}} + \sqrt{x^2-4} + C$

49 $\frac{1}{2}\tan^{-1}x^2 + C$

51 $\frac{4}{3}x^3 + \frac{1}{4}e^{4x} - 2xe^{2x} + e^{2x} + C$

53 $\frac{-1}{2(1-\cos\theta)^2} + C$

55 $\left(-2(1-\sqrt{x}) + (1-\sqrt{x})^2\right)\ln(1-\sqrt{x}) + 2(1-\sqrt{x}) - \frac{(1-\sqrt{x})^2}{2} + C = -\frac{1}{2}x - \sqrt{x} + (x-1)\ln(1-\sqrt{x}) + C$

Section 7.9

1 use formula #36: $x^{18}\left(\frac{\ln x}{18} - \frac{1}{324}\right) + C$

3 use formula #76: $\frac{1}{81}\left(\frac{4}{4+9x} + \ln|4+9x|\right) + C$

5 use formula #77: $-\frac{1}{729}\left(4 - 9x - \frac{16}{4-9x} - 8\ln|4-9x|\right) + C$

7 use formula #52: $-\sinh 3x + 3x\cosh 3x + C$ (The substitution $u = 3x$ is required.)

9 use formula #19: $\frac{e^{5x}(5\sin 7x - 7\cos 7x)}{74} + C$

11 use formula #107: $42\left(\frac{\sqrt{x^2-9}}{18x^2} + \frac{1}{54}\sec^{-1}\frac{|x|}{3}\right) + C$

13 use formula #122: $\frac{1}{7}\ln\left|\frac{6x+2}{-6x+12}\right| + C$

15 use formula #85: $\frac{5(144 - 96x + 96x^2)\sqrt{3+4x}}{960} + C = \frac{(3 - 2x + 2x^2)\sqrt{3+4x}}{4} + C$

17 use formula #93: $\frac{5}{8}\left(x\sqrt{3+4x^2} - \frac{3}{2}\ln\left|2x + \sqrt{3+4x^2}\right|\right) + C$ (The substitution $u = 2x$ is required.)

19 use formula #129 followed by formula #13, or use formula #129 twice followed by formula #5: $\frac{\sin x}{4\cos^4 x} + \frac{3}{8}(\sec x\tan x + \ln|\sec x + \tan x|) + C$

21 use formula #147 followed by formula #85: $\frac{2x^3\sqrt{1+3x}}{21} - \frac{4}{2835}\left(8 - 12x + 27x^2\right)\sqrt{1+3x} + C$

23 use formula #144 twice followed by formula #36: $\frac{x^5}{5}(\ln x)^3 - \frac{3x^5}{25}(\ln x)^2 + \frac{6x^5}{125}\ln x - \frac{6}{625}x^5 + C$

25 use formula #21: $x^5\sin^{-1}x^5 + \sqrt{1-x^{10}} + C$ (The substitution $u = x^5$ is required.)

27 use formula #79: $-\frac{1}{3}\ln\left|\frac{3+7x^5}{x^5}\right| + C$ (The substitution $u = x^5$ is required.)

29 use formula #88: $\frac{1}{2}\tan^{-1}\frac{4x}{3} + C$

31 $\frac{2(2a+bu)}{b^2\sqrt{a+bu}} + C$

Section 7.10

1 $\frac{7}{8}$

3 ∞

5 $\frac{\pi}{4}$

7 $-\infty$

9 $\frac{\pi}{4}$

11 $\frac{1}{6}\ln 4$

13 1

15 $\frac{3\pi}{2}$

17 diverges (Zero is incorrect.)

19 diverges

21 $1 - \cos 1$

23 converges (p-test, $p = 3 > 1$)

25 converges (p-test, $p = \frac{4}{3} > 1$)

27 diverges (p-test, $p = \frac{1}{3} < 1$)

29 converges (comparison test, $\frac{7}{x^3+x^2} < \frac{7}{x^3}$)

31 diverges (comparison test, $\frac{1}{\sqrt[3]{x}-8} > \frac{1}{\sqrt[3]{x}}$)

33 neither theorem applies (The comparison $\frac{1}{\sqrt[3]{x}+8} < \frac{1}{\sqrt[3]{x}}$ is not helpful.)

35 $\frac{2}{3}$ units^2

37 $\frac{4\pi}{81}$ units$^3 \approx 0.1551$ units^3

39 0 (However, because all the function's values are positive, this means all the function's values are above average! Is this possible? For a resolution of this paradox, see Dawson, C. Bryan, "The Lake Wobegon Paradox," *The Mathematical Intelligencer* (**42** No. 4 December 2020), 30–32, doi: https://doi.org/10.1007/s00283-020-09985-x.)

41 $\sqrt{2}$

43 $\frac{2}{9}$

45 (a) 0, (b) 0

47 (a) $\int_{-\infty}^{\infty} \frac{1}{1+x^2}\,dx = \int_{-B}^{\Omega} \frac{1}{1+x^2} = \tan^{-1} x\Big|_{-B}^{\Omega} = \tan^{-1}\Omega - \left(\tan^{-1}(-B)\right) = \frac{\pi}{2} - \left(-\frac{\pi}{2}\right) = \pi$, (b) $\int_{-\infty}^{\infty} x\,dx = \int_{-B}^{\Omega} x\,dx = \frac{x^2}{2}\Big|_{-B}^{\Omega} = \frac{\Omega^2}{2} - \frac{B^2}{2}$. If Ω is on a higher level than B, $\frac{\Omega^2}{2} - \frac{B^2}{2} \approx \frac{\Omega^2}{2} \doteq \infty$. If B is on a higher level than Ω, $\frac{\Omega^2}{2} - \frac{B^2}{2} \approx -\frac{B^2}{2} \doteq -\infty$. Because we do not render the same real result for every choice of Ω and B, the integral diverges. (As soon as different results are obtained, there is no need to consider other cases.)

Section 7.11

1 $\int_0^{6-\omega} \frac{1}{x-6}\,dx$

3 not necessary, no discontinuities in interval

5 $\int_{-2+\omega}^{1} \frac{1}{x^2-4}\,dx$

7 $\int_0^{8-\omega} \frac{1}{\sqrt[3]{x}-2}\,dx + \int_{8+\omega}^{10} \frac{1}{\sqrt[3]{x}-2}\,dx$

9 $\int_0^{\frac{\pi}{2}-\omega} \tan x\,dx$

11 $\int_1^{3-\omega} \frac{1}{(x-3)(x+2)}\,dx + \int_{3+\omega}^{4} \frac{1}{(x-3)(x+2)}\,dx$

13 4

15 $-\infty$

17 diverges (alternate answer, ∞)

19 $-\ln 2$

21 2

23 ∞

25 ∞

27 $\frac{9}{2}$

29 ∞

31 diverges (but not to infinity)

33 $9\sqrt[3]{4}$

35 $\frac{\sqrt{3}}{4}$

37 π (Details: $\int_{-1}^{0} \frac{1}{\sqrt{1-x^2}}\,dx = \int_{-1+\omega}^{0} \frac{1}{\sqrt{1-x^2}}\,dx = \cdots = \frac{\pi}{2}$; $\int_0^1 \frac{1}{\sqrt{1-x^2}}\,dx = \int_0^{1-\omega} \frac{1}{\sqrt{1-x^2}}\,dx = \cdots = \frac{\pi}{2}$; because both converge, $\int_{-1}^{1} \frac{1}{\sqrt{1-x^2}}\,dx = \frac{\pi}{2} + \frac{\pi}{2} = \pi$. Although ignoring the discontinuities gives the correct answer in this particular exercise, the work is still incorrect.)

39 (a) $\lim_{x\to1^-} \frac{x^2-x}{|x-1|} = \cdots = -1$, $\lim_{x\to1^+} \frac{x^2-x}{|x-1|} = \cdots = 1$ (Because the one-sided limits are different real numbers, f has a jump discontinuity at $x = 1$), (b) 1

41 $\Delta x = \omega$, $x_k = k\omega$, $\sum_{k=1}^{\Omega} \frac{1}{\sqrt{k\omega}} \cdot \omega = \sqrt{\omega}\sum_{k=1}^{\Omega} \frac{1}{\sqrt{k}} \approx \sqrt{\omega} \cdot \frac{\Omega^{1/2}}{1/2} = 2$

Section 8.1

1

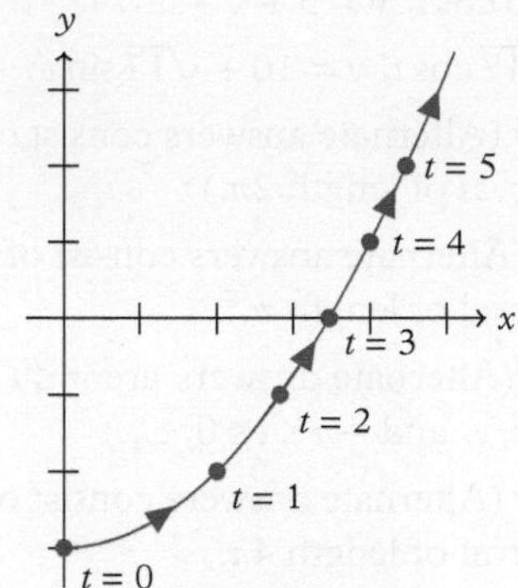

(The domain is $t \geq 0$, and the curve continues indefinitely to the right.)

3

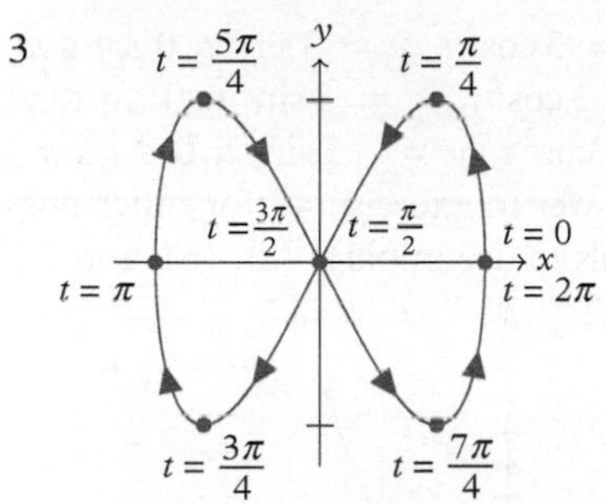

5

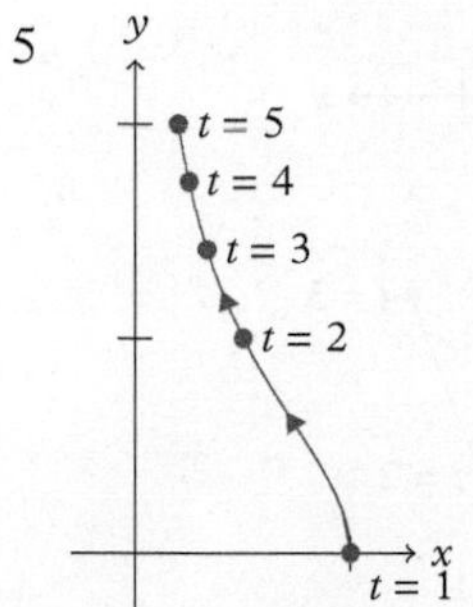

7

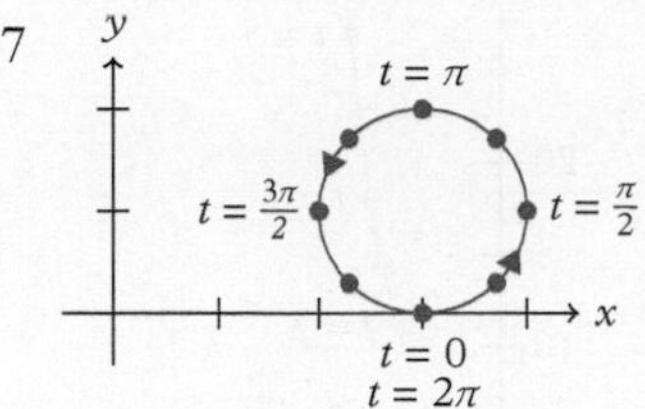

(This curve is a circle centered at $(3, 1)$. Any interval of values for t of length 2π is sufficient to traverse the circle exactly once.)

9

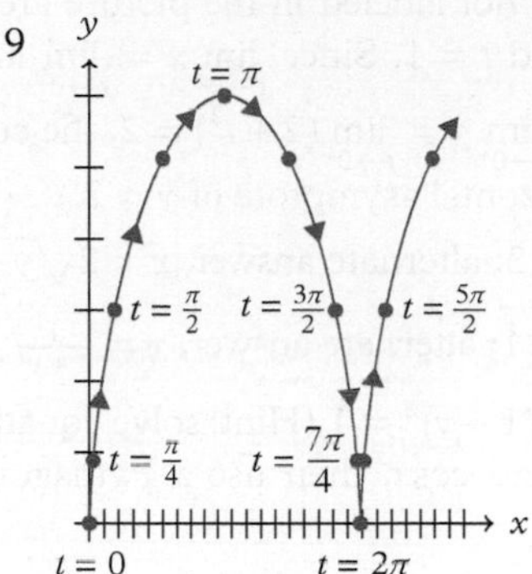

(This graph has "humps" to the left and to the right, continuing indefinitely, moving from left to right. The aspect ratio was chosen to make the graph fit in this column.)

11

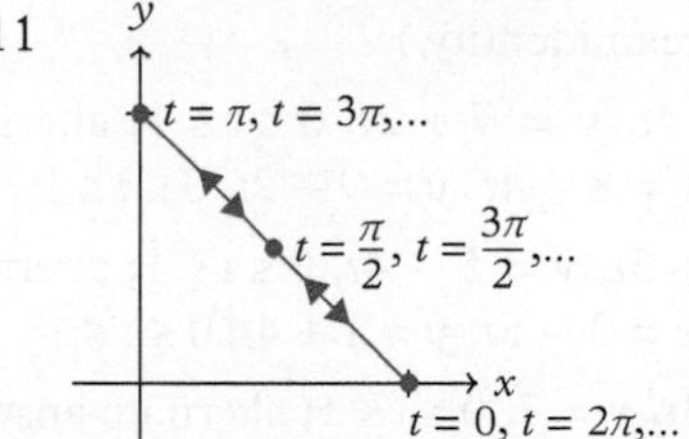

(This graph bounces back and forth between $(0, 1)$ and $(1, 0)$, for both negative and positive values of t.)

13 (The dots not labeled in the picture are for $t = 0.5$ and $t = 1$. Since $\lim_{t\to 0^+} x = \lim_{t\to 0^+} \ln t = -\infty$ and $\lim_{t\to 0^+} y = \lim_{t\to 0^+} (2 + t^2) = 2$, the curve has a horizontal asymptote of $y = 2$.)

15 $y = \frac{1}{4}x^2 - 3$; alternate answer, $x = 2\sqrt{y+3}$

17 $y = \sqrt{\frac{1}{x} - 1}$; alternate answer, $x = \frac{1}{1+y^2}$

19 $(x-3)^2 + (1-y)^2 = 1$ (Hint: solve equations for $\sin t$ and $\cos t$, then use a Pythagorean identity.)

21 $y = 1 - x$; alternate answer, $x + y = 1$

23 $y = 2 + e^{2x}$; alternate answer, $x = \ln\sqrt{y-2}$

25 $y = 3 - x$; alternate answer, $x - (2 - y) = 1$

27 $\left(\frac{x}{4}\right)^2 + \frac{y}{5} = 1$ (Hint: rearrange equations to find $\sin^2 t^3$ and $\cos^2 t^3$, then use a Pythagorean identity.)

29 $x = 4 + 4t$, $y = 7 + 2t$, $0 \le t \le 1$; alternate answer, $x = 8 - 4t$, $y = 9 - 2t$, $0 \le t \le 1$

31 $x = -2 + 5t$, $y = 5 - 4t$, $0 \le t \le 1$; alternate answer, $x = 3 - 5t$, $y = 1 + 4t$, $0 \le t \le 1$

33 $x = 2 + 3t$, $y = 7$, $0 \le t \le 1$; alternate answer, $x = 5 - 3t$, $y = 7$, $0 \le t \le 1$

35 $x = 9 - 12t$, $y = 22 - 16t$, $0 \le t \le 1$; alternate answer, $x = -3 + 12t$, $y = 6 + 16t$, $0 \le t \le 1$

37 $x = 2 + 6\cos t$, $y = 7 + 6\sin t$

39 $x = 1 + 3\cos t$, $y = 3\sin t$

41 $x = 1 + \sqrt{5}\cos t$, $y = 5 + \sqrt{5}\sin t$

43 $x = 2 + \sqrt{17}\cos t$, $y = 10 + \sqrt{17}\sin t$

45 $0 \le t \le 2\pi$ (Alternate answers consist of any other interval of length 2π.)

47 $0 \le t \le \pi$ (Alternate answers consist of any other interval of length π.)

49 $0 \le t \le \pi$ (Alternate answers are $\pi \le t \le 2\pi$, $2\pi \le t \le 3\pi$, …and $-\pi \le t \le 0$, ….)

51 $0 \le t \le 4\pi$ (Alternate answers consist of any other interval of length 4π.)

53 $0 \le t \le \infty$; alternate answer, $-\infty \le t \le 0$

55 $0 \le t \le \sqrt{2\pi}$ (Alternate answers include $\sqrt{\pi} \le t \le \sqrt{3\pi}$, or any other interval of the form $\sqrt{a} \le t \le \sqrt{a + 2\pi}$.)

57 (a) $x = 3\cos^2 t$, $y = 3\sin^2 t$, $0 \le t \le \pi$; (b) $x = -3\cos^2 t$, $y = 3\sin^2 t$, $0 \le t \le \pi$; (c) $x = \frac{5}{4}\cos^2 t$, $y = -5\sin^2 t$, $0 \le t \le \pi$ (See the answer to exercise 49 for other possible intervals to use in parts (a), (b), and (c),)

59 (a)

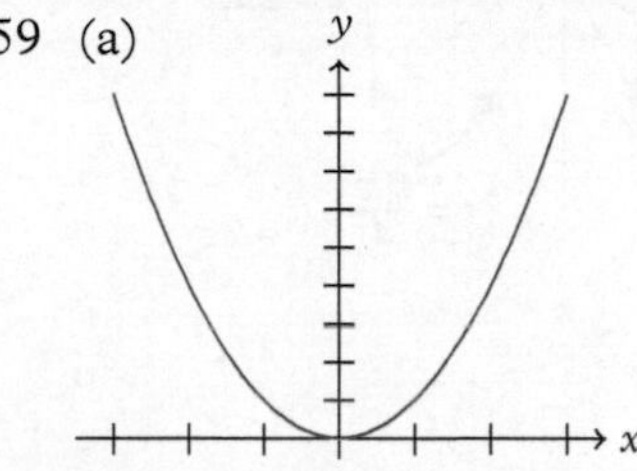

(b)

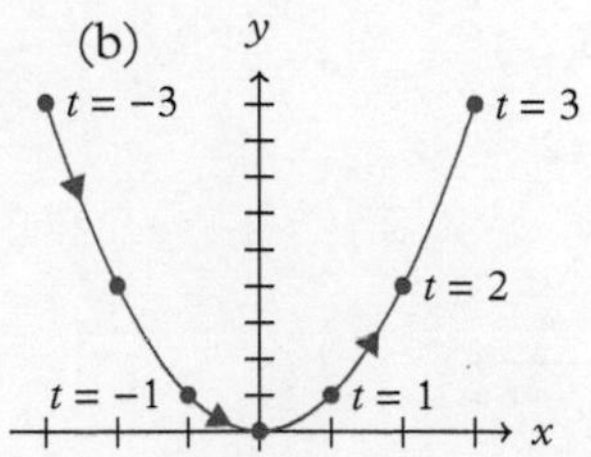

(c) $x = t$, $y = \sqrt{t+4}$; (d) $x = t$, $y = \frac{1}{1+t^2}$; (e) yes, using $x = t$, $y = f(t)$

Section 8.2

1 (a) 3, (b) $y = 3x + 3$

3 (a) $-\frac{2187}{2}$, (b) $y = -\frac{2187}{2}x + \frac{5103}{2}$

5 (a) -1, (b) $y = -x + 2 + \sqrt{2}$

7 (a) 0, (b) $y = -9$

9 (a) -1, (b) $y = -x + 1$

11 (a) $\frac{3\sqrt{3}}{2}$; (b) $y - \frac{7}{4} = \frac{3\sqrt{3}}{2}\left(x - \frac{\pi}{6}\right)$, simplifies to $y = \frac{3\sqrt{3}}{2}x + \frac{21 - 3\sqrt{3}\pi}{12}$

13 (a) -1, (b) $y = -x + 3$

15 (a) $-\frac{3}{4}$, (b) $y = -\frac{3}{4}x + \ln 2$

17 $\frac{dy}{dx} = \frac{-3t^{-4}}{15t^4 - 16t} = \frac{-3}{15t^8 - 16t^5}$

19 $\frac{dy}{dx} = \frac{-4\cos 2t \sin 2t}{2\sin t \cos t} = -4\cos 2t$

21 $\frac{dy}{dx} = \frac{t \cdot 3(t+6)^2 + (t+6)^3}{2t+9} = \frac{(4t+6)(t+6)^2}{2t+9} = \frac{4t^3 + 54t^2 + 216t + 216}{2t+9}$

23 $\frac{d^2y}{dx^2} = \frac{2\cos t + 2t\sin t}{\cos^3 t}$

25 $\frac{d^2y}{dx^2} = \frac{-3\operatorname{csch}^2 3t}{3\sinh 3t} = -\operatorname{csch}^3 3t$

27 $\frac{d^2y}{dx^2} = \frac{\frac{1}{2}t \cdot 2t\cosh t^2 + \frac{1}{2}\sinh t^2}{4} = \frac{1}{4}t^2\cosh t^2 + \frac{1}{8}\sinh t^2$

29 $\frac{d^2y}{dx^2} = \frac{-18}{(3+t)^5}$

31 $y = -7x + 75$ (at $t = -5$) and $y = x - 45$ (at $t = 3$) (Both tangent lines are required.)

33 $y = -\frac{7}{6}x - \frac{62}{3}$ (at $t = 2$) (Note: $t = -4$ gives the wrong point, (8, 12).)

35 $y = -\frac{1}{8}x + \frac{17}{8}$ (at $t = 0$) and $y = -\frac{1}{20}x + \frac{41}{20}$ (at $t = -3$) (Both tangent lines are required.)

37 $y = \frac{\sqrt{3}}{3}x - \frac{1}{2}$ (at $t = \frac{\pi}{3}$ and at other values of t) and $y = -\frac{\sqrt{3}}{3}x - \frac{1}{2}$ (at $t = -\frac{\pi}{3}$ and at other values of t) (Both tangent lines are required.)

39 (8, 16) and (0, −16) (Hint: set $\frac{dy}{dt} = 0$.)

41 (0, −1) and (0, 1)

43

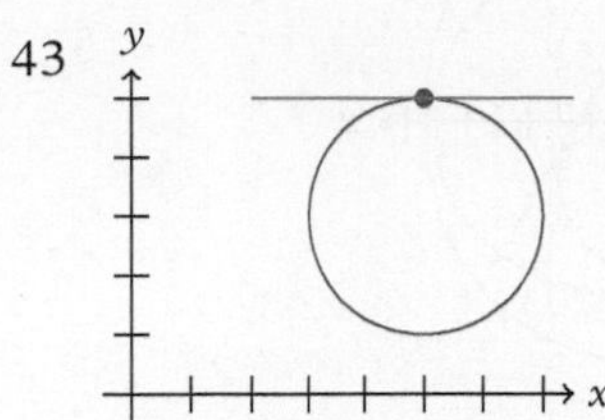

(equation of tangent line, $y = 5$)

45

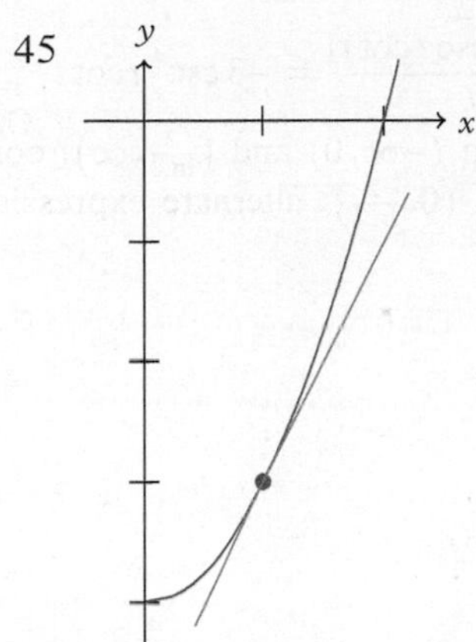

(equation of tangent line, $y = 2x - 5$)

47

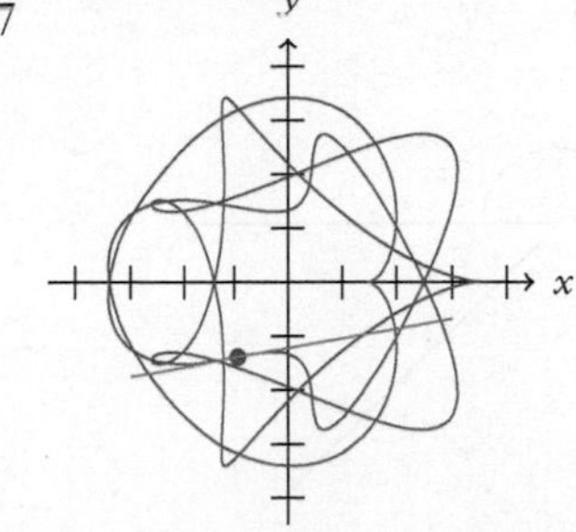

49

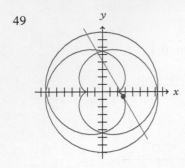

51 (a) $\dfrac{d^3y}{dx^3} = \dfrac{\frac{d}{dt}\left(\frac{d^2y}{dx^2}\right)}{\frac{dx}{dt}}$, (b) $\dfrac{d^3y}{dx^3} = \dfrac{-3(\csc^2 t)(-\csc t \cot t)}{-\sin t} = -3\csc^4 t \cot t$

53 concave up on $(-\infty, 0)$ and $\left(\frac{2}{\ln 2}, \infty\right)$, concave down on $\left(0, \frac{2}{\ln 2}\right)$; alternate expression of answer, concave up on $t \le 0$ and $t \ge \frac{2}{\ln 2}$, concave down on $0 \le t \le \frac{2}{\ln 2}$ (Hint: set $\frac{d^2y}{dx^2} = \frac{-6t \cdot 2^t \ln 2 + 3t^2 \cdot 2^t (\ln 2)^2}{(2^t \ln 2)^3} = 0$, solve for t, ...)

55 (a) $\frac{y(1+\alpha) - y(1)}{x(1+\alpha) - x(1)} = \cdots = -A^2 \doteq -\infty$; (b) the tangent line is vertical, (c) $x = 5$

57 (a) $\left.\frac{dx}{dt}\right|_{t=1} = 0$, $\left.\frac{dy}{dt}\right|_{t=1} = -1$; (b) $\left.\frac{dx}{dt}\right|_{t=2\pi} = 0$, $\left.\frac{dy}{dt}\right|_{t=2\pi} = 0$; (c) no; (d) in order for a corner to be present, we must have both $\frac{dx}{dt} = 0$ and $\frac{dy}{dt} = 0$. (Note: although these conditions must happen for a corner to be present, they do not guarantee the presence of a corner. They are *necessary conditions*, but not *sufficient conditions*.

Section 8.3

1

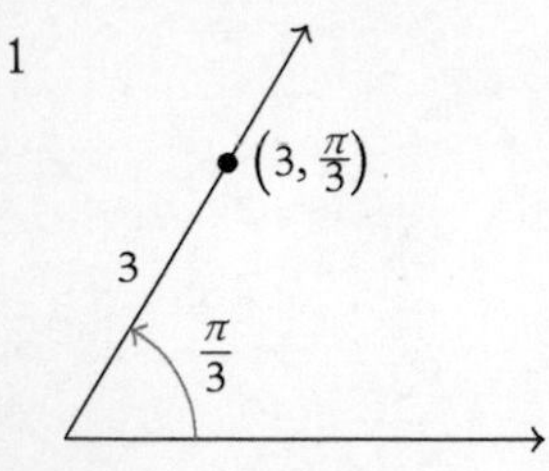

3

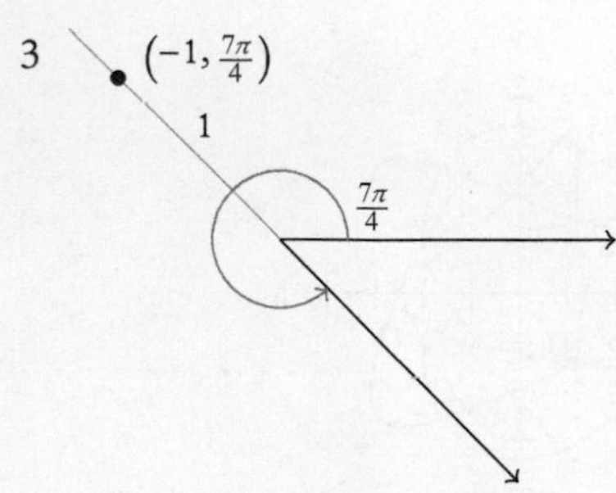

5

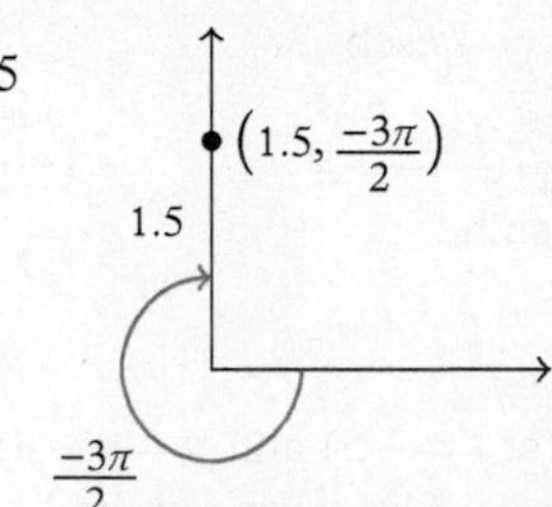

7

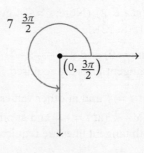

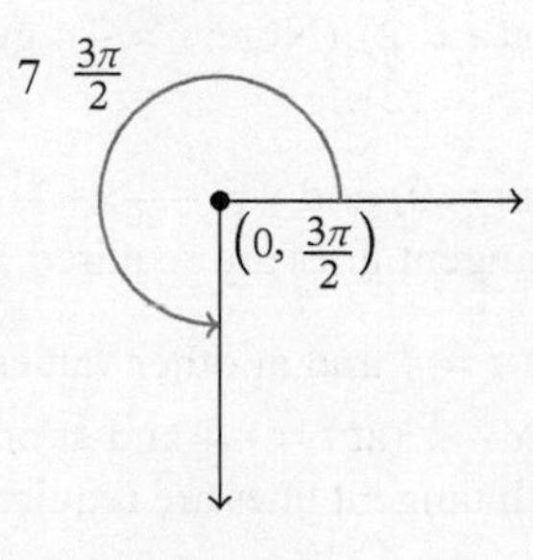

9

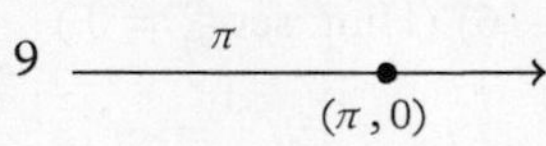

(Note: $r = \pi$, $\theta = 0$)

11 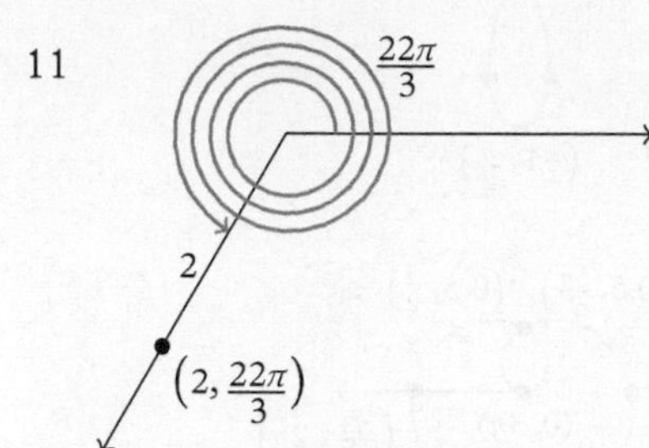

13 (a) $\left(17, \frac{7\pi}{3}\right)$; alternate answers include $\left(17, \frac{13\pi}{3}\right)$, $\left(17, \frac{-5\pi}{3}\right)$, and many others; (b) $\left(-17, \frac{-2\pi}{3}\right)$; alternate answers include $\left(-17, \frac{-8\pi}{3}\right)$, $\left(-17, \frac{4\pi}{3}\right)$, and many others

15 (a) $\left(4, \frac{7\pi}{6}\right)$; alternate answers include $\left(4, \frac{19\pi}{6}\right)$, $\left(4, \frac{-17\pi}{6}\right)$, and many others; (b) $\left(-4, \frac{\pi}{6}\right)$; alternate answers include $\left(-4, \frac{13\pi}{6}\right)$, $\left(-4, \frac{-11\pi}{6}\right)$, and many others

17 (a) $(3, \pi)$; alternate answers include $(3, 3\pi)$, $(3, -\pi)$, and many others; (b) $(-3, 2\pi)$; alternate answers include $(-3, 0)$, $(-3, 6\pi)$, and many others

19 (a) $\left(1.4, \frac{16\pi}{7}\right)$; alternate answers include $\left(1.4, \frac{30\pi}{7}\right)$, $\left(1.4, \frac{-12\pi}{7}\right)$, and many others; (b) $\left(-1.4, \frac{9\pi}{7}\right)$; alternate answers include $\left(-1.4, \frac{23\pi}{7}\right)$, $\left(-1.4, \frac{-5\pi}{7}\right)$, and many others

21 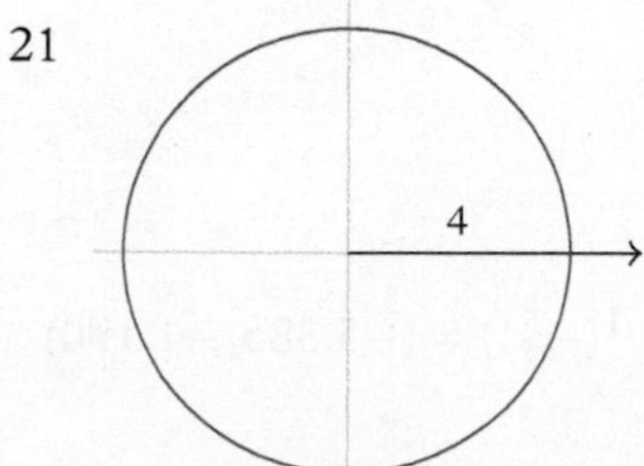

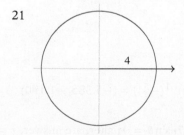

23 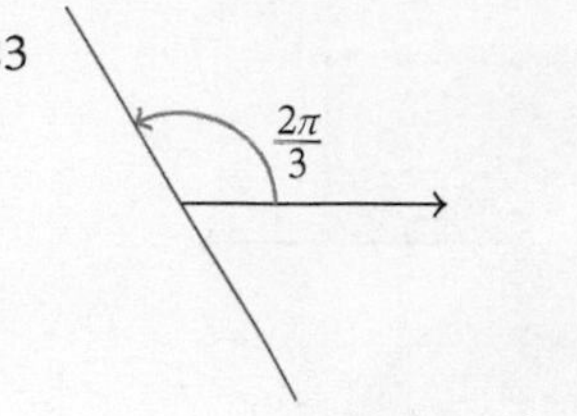

25 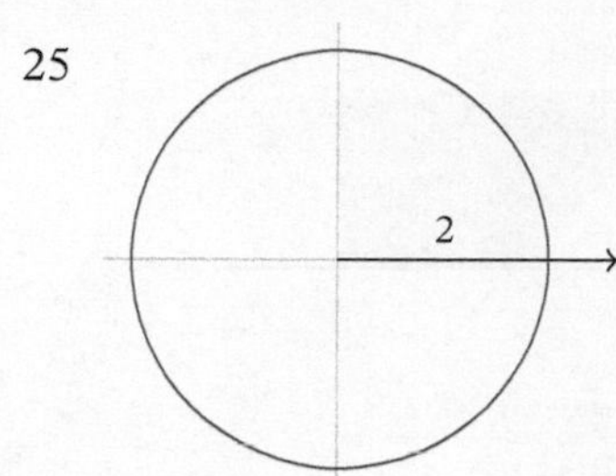

27 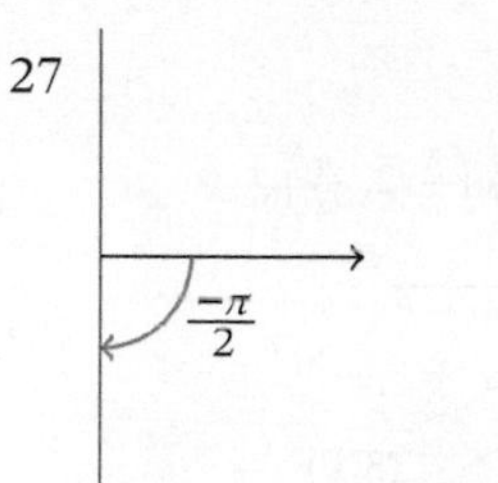

29 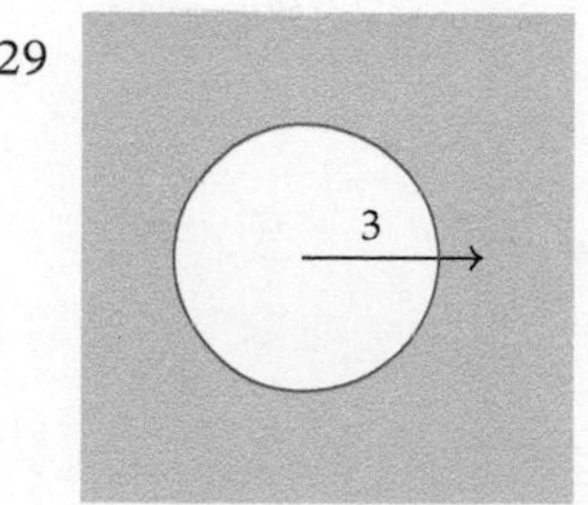

31 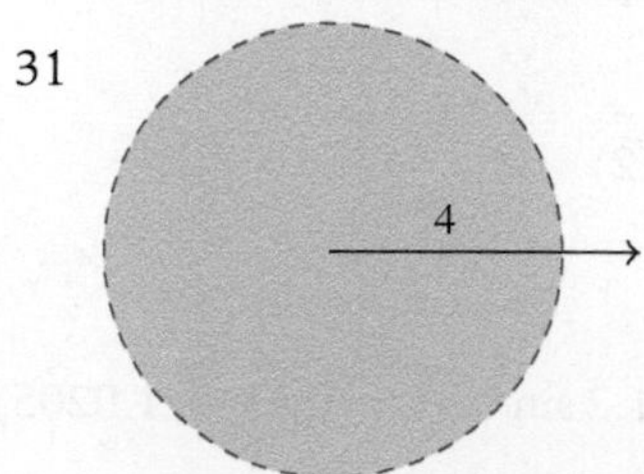

33 $\theta = \frac{3\pi}{4}$

35 $\theta = \frac{3\pi}{4}$

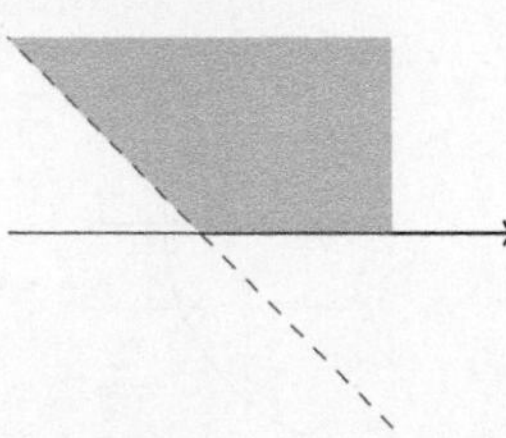

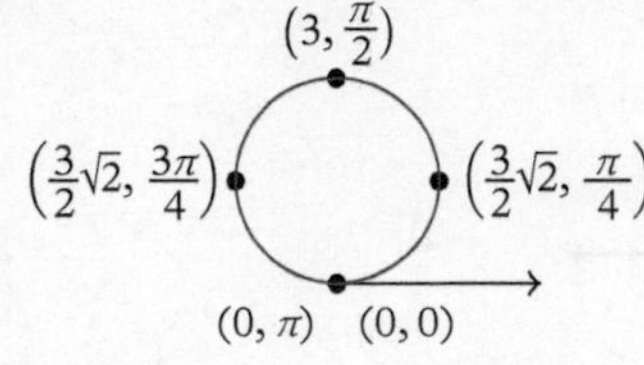

39

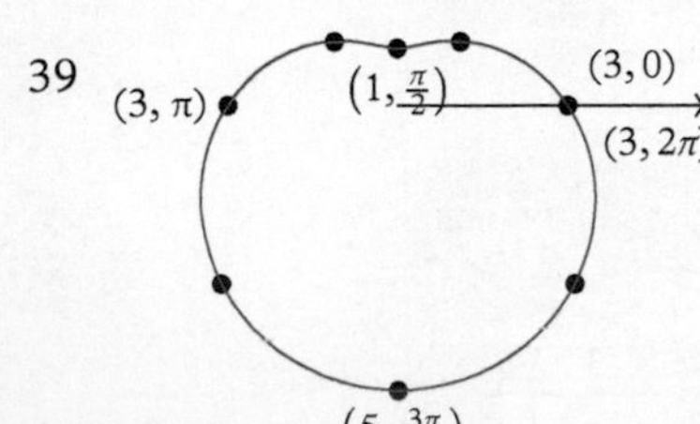

41

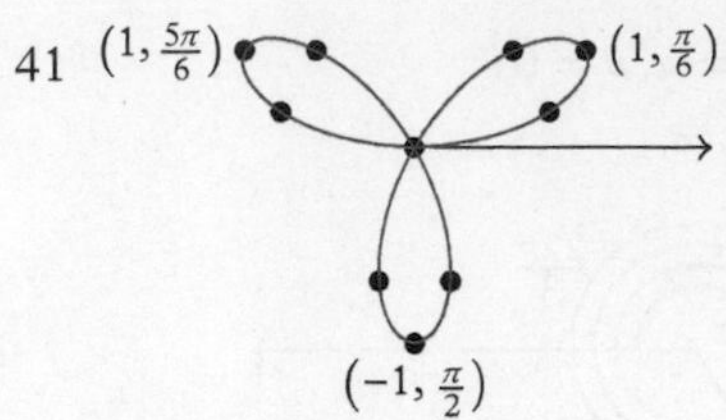

43

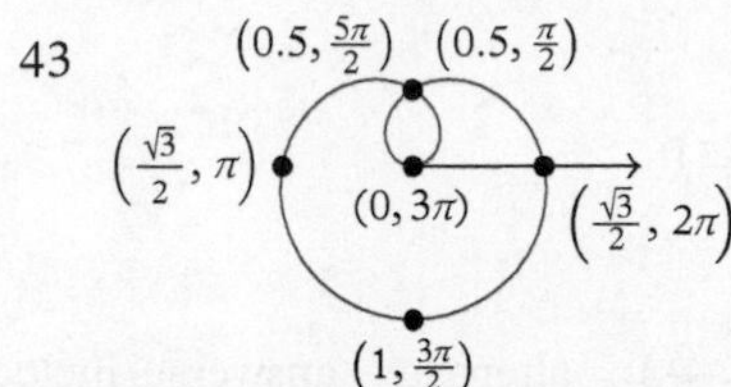

45

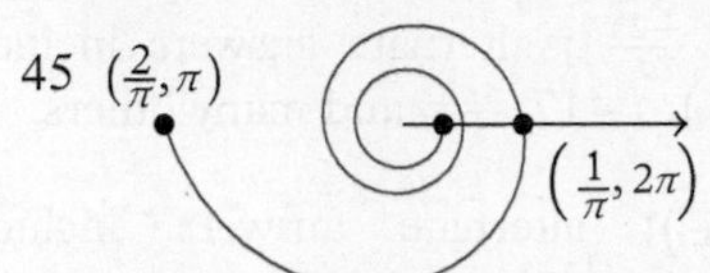

The unlabeled point at the "inside" end is $(\frac{1}{3\pi}, 6\pi)$.

47 $0 \le \theta \le \pi$; alternate answers consist of any other interval of length π

49 $0 \le \theta \le 2\pi$; alternate answers consist of any other interval of length 2π

51 $0 \le \theta \le 4\pi$; alternate answers consist of any other interval of length 4π

53 $0 \le \theta \le 4\pi$; alternate answers consist of any other interval of length 4π

Section 8.4

1 $(-3, 0)$

3 $(-2\sqrt{2}, -2\sqrt{2})$

5 $(0, 0)$

7 $(-1, \sqrt{3})$

9 $(1.2\cos 7.3, 1.2\sin 7.3) = (0.6313, 1.0205)$

11 $(-4, 0)$; alternate answer, $(4, \pi)$

13 $(-4, \frac{\pi}{3})$; alternate answer, $(4, \frac{4\pi}{3})$

15 $(6\sqrt{2}, \frac{\pi}{4})$

17 $(10, -\frac{\pi}{6})$

19 $(-\sqrt{29}, \tan^{-1}(-\frac{5}{2})) = (-5.385, -1.190)$

21 $r^2 = 5$

23 $2r\cos\theta + 7r\sin\theta = 1$; alternate answer, $r = \frac{1}{2\cos\theta + 7\sin\theta}$

25 $\cot\theta = 3$

27 $y + x = 11$; alternate answer, $y = 11 - x$

29 $1 = 2x + 4y$; alternate answer, $y = \frac{1}{4} - \frac{1}{2}x$

31 $\frac{y}{x} = 1$; alternate answer, $y = x$

33

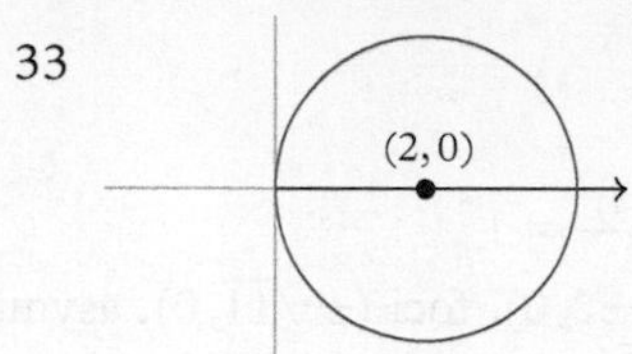

35

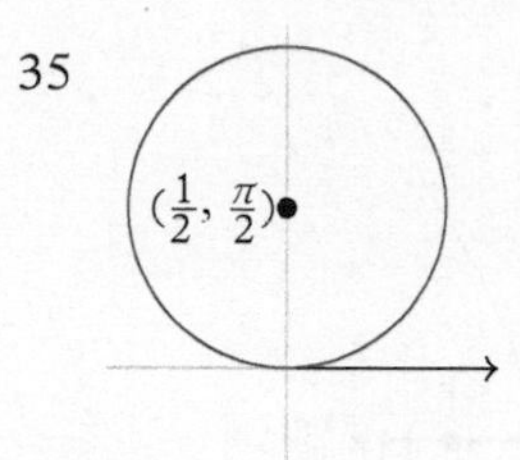

37

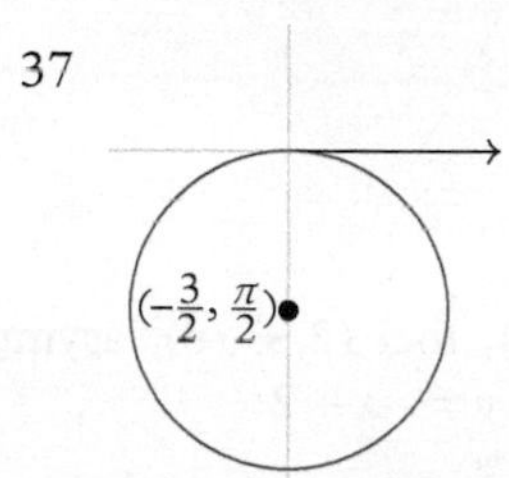

39 $\frac{dy}{dx} = \frac{\sec^2\theta\sin\theta + \tan\theta\cos\theta}{\sec^2\theta\cos\theta - \tan\theta\sin\theta} = \frac{\sec^2\theta\sin\theta + \sin\theta}{\sec\theta - \tan\theta\sin\theta} = \cdots = \tan\theta(\sec^2\theta + 1)$

41 $\frac{dy}{dx} = \frac{3\theta^2\sin\theta + \theta^3\cos\theta}{3\theta^2\cos\theta - \theta^3\sin\theta} = \frac{3\sin\theta + \theta\cos\theta}{3\cos\theta - \theta\sin\theta}$

43 $\left.\frac{dy}{dx}\right|_{\theta=\frac{\pi}{4}} = 3$

45 $\left.\frac{dy}{dx}\right|_{\theta=2\pi} = -2\pi$

47 $\left.\frac{dy}{dx}\right|_{\theta=0} = 1$

49 (a) $y = -\sqrt{3}x + \frac{9}{2}$ (previous step, $y - \frac{9}{4} = -\sqrt{3}(x - \frac{3}{4}\sqrt{3})$),

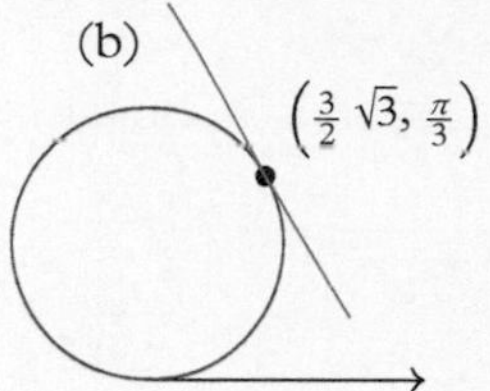

51 (a) $y = \frac{3}{2}x + \frac{9}{2}$ (previous step, $y - 0 = \frac{3}{2}(x - (-3))$),

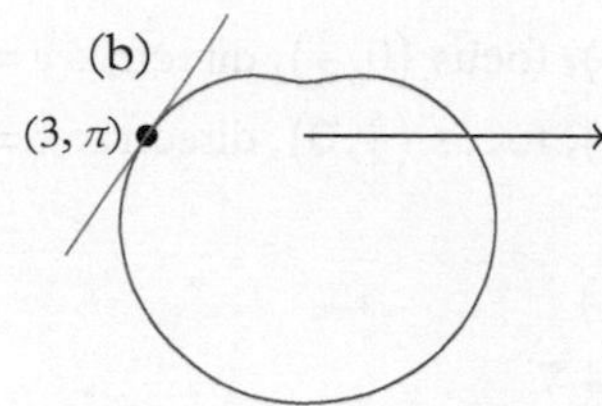

53 (a) $y = \sqrt{3}x + 2$ (previous step, $y - \frac{1}{2} = \sqrt{3}(x - (-\frac{\sqrt{3}}{2}))$),

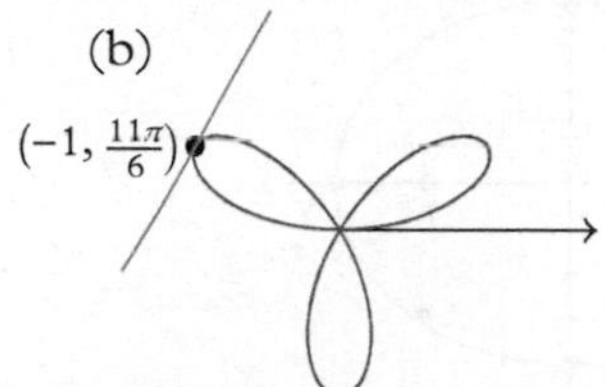

55 (a) $y = -1$ (previous step, $y - (-1) = 0(x - 0)$),

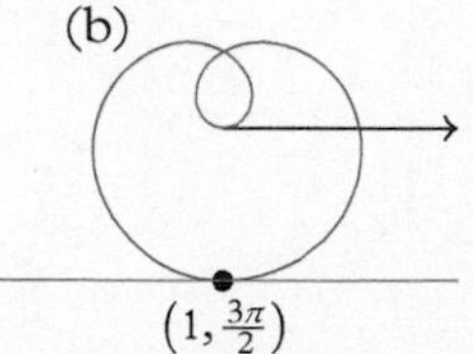

57 $y = 0$ (at $\theta = 0$), $y = \sqrt{3}x$ (at $\theta = \frac{\pi}{3}$), and $y = -\sqrt{3}x$ (at $\theta = \frac{2\pi}{3}$). All three tangent lines are required, although they can be calculated at different values of θ than indicated here.

59 (a) $r = \frac{-c}{a\cos\theta + b\sin\theta}$, (b) slope $= -\frac{a}{b}$

61 $\frac{4}{7}$

Section 8.5

1 circle

3 parabola

5 ellipse

7 ellipse

9 hyperbola

11 ellipse

13 parabola

15 vertex $(0, 0)$, focus $\left(0, \frac{3}{4}\right)$, directrix $y = -\frac{3}{4}$

17 vertex $(0, 0)$, focus $\left(\frac{5}{4}, 0\right)$, directrix $x = -\frac{5}{4}$

19 $y = \frac{1}{16}x^2$

21 $y = \frac{1}{32}x^2 + 4$

23 $x = -\frac{1}{12}y^2 + 7$

25 $y = \frac{1}{4}(x - 2)^2 + 5$

27 (a) vertices $(\pm 5, 0)$, foci $(\pm 3, 0)$

(b)

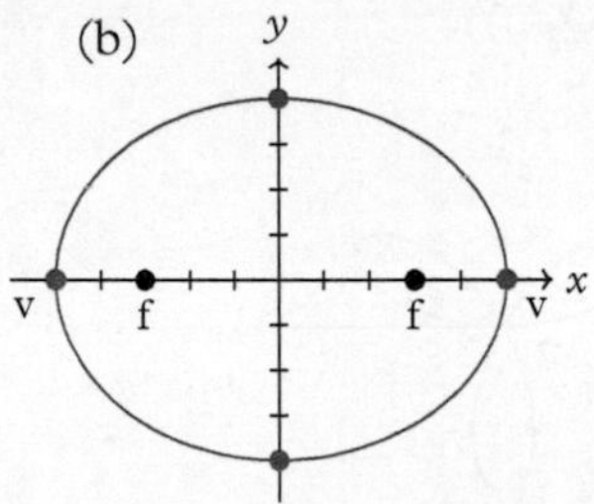

29 (a) vertices $(3, -8)$ and $(3, 10)$, foci $(3, 1 \pm \sqrt{77})$

(b)

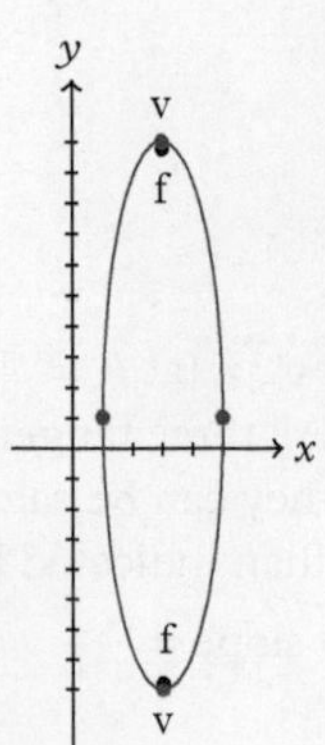

31 $\frac{x^2}{64} + \frac{y^2}{15} = 1$

33 $\frac{x^2}{84} + \frac{y^2}{100} = 1$

35 $\frac{(x-2)^2}{9} + \frac{(y-4)^2}{25} = 1$

37 (a) vertices $(\pm 2, 0)$, foci $(\pm\sqrt{11}, 0)$, asymptotes $y = \pm\frac{\sqrt{7}}{2}x$

(b)

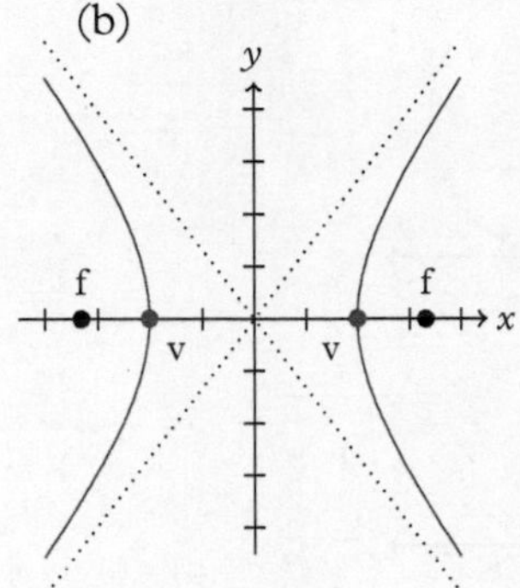

39 (a) vertices $(2, \pm 2)$, foci $(2, \pm\sqrt{8})$, asymptotes $y = x - 2$ and $y = -x + 2$

(b)

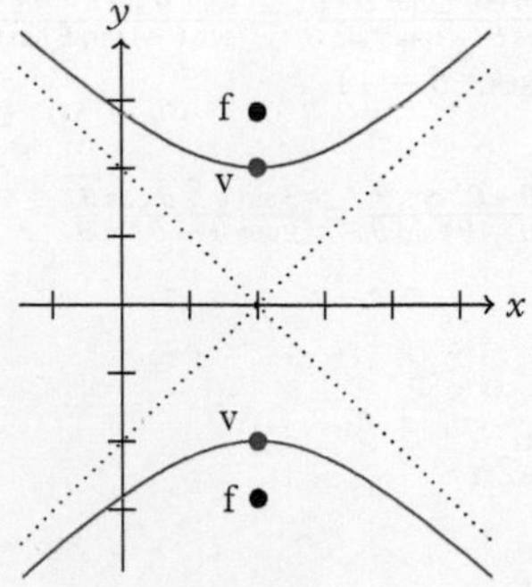

41 $\frac{x^2}{4} - \frac{y^2}{5} = 1$

43 $\frac{x^2}{16} - \frac{y^2}{25} = 1$

45 $\frac{(y-6)^2}{16} - \frac{x^2}{9} = 1$

47 (parabola) focus $\left(0, \frac{1}{20}\right)$, vertex $(0,0)$, directrix $y = -\frac{1}{20}$

49 (ellipse) foci $\left(\pm\sqrt{\frac{7}{44}}, 0\right)$, vertices $\left(\pm\frac{1}{2}, 0\right)$

51 none of the above (It is a line.)

53 (hyperbola) foci $(0, \pm\sqrt{20})$, vertices $(0, \pm 4)$, asymptotes $y = \pm 2x$

55 (parabola) focus $(3, 6)$, vertex $(3, 4)$, directrix $y = 2$

57 (circle) center $(3, -8)$, radius 4

59 (ellipse) foci $(3 \pm \sqrt{7}, -8)$, vertices $(-1, -8)$ and $(7, -8)$

61 (hyperbola) foci $(3 \pm \sqrt{32}, -8)$, vertices $(-1, -8)$ and $(7, -8)$, asymptotes $y = x - 11$ and $y = -x - 5$

63 (parabola) focus $\left(-\frac{5}{8}, \frac{3}{2}\right)$, vertex $\left(-\frac{1}{2}, \frac{3}{2}\right)$, directrix $x = -\frac{3}{8}$

65 (ellipse) foci $(3 \pm \sqrt{45}, -2)$, vertices $(3 \pm \sqrt{50}, -2)$

67 $y - (-5) = \frac{5}{2}(x - 2)$; alternate answer, $y = \frac{5}{2}x - 10$

69

$$\sqrt{(x-p)^2 + (y-0)^2} = x - (-p)$$
$$(x-p)^2 + y^2 = (x+p)^2$$
$$\cancel{x^2} - 2xp + \cancel{p^2} + y^2 = \cancel{x^2} + 2xp + \cancel{p^2}$$
$$y^2 = 4xp$$
$$x = \frac{1}{4p}y^2$$

71 (a) a point, (b) a line, (c) a pair of intersecting lines (For one point, try a plane passing through the vertex of the cones. For one line, try a plane slanted the same as for a parabola but tangent to the cone. For a pair of intersecting lines, try a plane containing the axis of the double-napped cone.)

73 (a) $y = \pm\sqrt{1 - x^2}$, (b) $\sum_{k=1}^{\Omega} 2\sqrt{1 - (-1 + 2\omega k)^2} \cdot 2\omega$, (c) $y = \pm\frac{1}{2}\sqrt{1 - x^2}$, (d) $\sum_{k=1}^{\Omega} \sqrt{1 - (-1 + 2\omega k)^2} \cdot 2\omega$, (e) $\sum_{k=1}^{\Omega} \sqrt{1 - (-1 + 2\omega k)^2} \cdot 2\omega = \frac{1}{2}\sum_{k=1}^{\Omega} 2\sqrt{1 - (-1 + 2\omega k)^2} \cdot 2\omega = \frac{1}{2}\pi$

Section 8.6

1 parabola

3 ellipse

5 hyperbola

7 ellipse

9 (a) $e = 2$, (b) hyperbola, (c) $x = 3$, (d) $(2, 0)$ and $(-6, \pi)$

11 (a) $e = \frac{3}{4}$, (b) ellipse, (c) $x = -\frac{7}{3}$, (d) $(7, 0)$ and $(1, \pi)$

13 (a) $e = 2$, (b) hyperbola, (c) $y = 4$, (d) $\left(\frac{8}{3}, \frac{\pi}{2}\right)$ and $\left(-8, \frac{3\pi}{2}\right)$

15 (a) $e = \frac{\pi}{2\sqrt{3}} = 0.9069$, (b) ellipse, (c) $x = \frac{57}{\pi}$, (d) $\left(\frac{57}{2\sqrt{3} + \pi}, 0\right) = (8.269, 0)$ and $\left(\frac{57}{2\sqrt{3} - \pi}, \pi\right) = (176.7, \pi)$

17 (a) $e = 1$, (b) parabola, (c) $y = -3$, (d) $\left(\frac{3}{2}, \frac{3\pi}{2}\right)$

19 $r = \frac{0.2}{1 + 0.02\cos\theta}$

21 $r = \frac{28}{1 - 14\sin\theta}$

23 $r = \frac{3}{1 - \cos\theta}$

25 $r = \frac{\frac{1}{3}}{1 + \frac{1}{3}\sin\theta} = \frac{1}{3 + \sin\theta}$

27 $r = \frac{\frac{11}{6}}{1 - \frac{5}{6}\sin\theta} = \frac{11}{6 - 5\sin\theta}$

29 $r = \frac{18}{1 + \frac{1}{2}\cos\theta} = \frac{36}{2 + \cos\theta}$

31 $r = \frac{10(1 - 0.99^2)}{1 + 0.99\sin\theta} = \frac{0.199}{1 + 0.99\sin\theta}$

33 $r = \frac{\frac{10}{3}}{1 + \frac{2}{3}\cos\theta} = \frac{10}{3 + 2\cos\theta}$

35 (a) $r = \frac{1.32(1 - 0.255^2)}{1 + 0.255\cos\theta} = \frac{1.234}{1 + 0.255\cos\theta}$, (b) 1.6566 astronomical units (AU)

37 (a) $r = \frac{2.769(1-0.076^2)}{1+0.076\cos\theta} = \frac{2.753}{1+0.076\cos\theta}$; (b) perihelion 2.559 AU, aphelion 2.979 AU

39 $e = \frac{4}{\sqrt{30}}$

41 $y = -\frac{1}{10}x + 3$

43 (a) $r = \frac{k}{1+\cos\theta}$; (b) $\frac{k}{1+\cos\pi} = \frac{k}{0}$, undefined; (c) vertex $\left(\frac{k}{2}, 0\right)$, which is halfway between the focus $(0,0)$ and the directrix $x = k$

45 (a) $y^2 = b^2 - \frac{b^2}{a^2}x^2$, (b) $\sqrt{(x-c)^2 + b^2 - \frac{b^2}{a^2}x^2}$, (c) $\frac{c^2}{a^2}x^2 - 2cx + a^2$, (d) $(x-c)^2 + b^2 - \frac{b^2}{a^2}x^2 = x^2 - 2xc + c^2 + b^2 - \frac{b^2}{a^2}x^2 = x^2\left(1 - \frac{b^2}{a^2}\right) - 2xc + a^2 - b^2 + b^2 = x^2\left(\frac{a^2-b^2}{a^2}\right) - 2xc + a^2 = \frac{c^2}{a^2}x^2 - 2cx + a^2$, (e) $\sqrt{\left(\frac{c}{a}x - a\right)^2} = a - \frac{c}{a}x = a - ex$ (note that $\frac{c}{a}x < x \le a$, so that $a - \frac{c}{a}x > 0$), (f) $\frac{a}{e} - x$, (g) $\frac{a-ex}{\frac{a}{e}-x} = \frac{e(\frac{a}{e}-x)}{\frac{a}{e}-x} = e$

Section 9.1

1 (a) $L = \int_0^3 \sqrt{1 + (10x^4 - 6x)^2}\, dx$, (b) 461.95 units

3 (a) $L = \int_0^1 \sqrt{1 + \left(\frac{1}{1+x^2}\right)^2}\, dx$, (b) 1.2780 units

5 (a) $L = \int_0^{\frac{1}{2}\ln 3} \sqrt{1 + e^{2x}}\, dx$, (b) 0.91785 units

7 $\frac{8}{27}\left(10^{3/2} - \left(\frac{13}{4}\right)^{3/2}\right)$ units

9 sinh 3 units

11 $\frac{14}{3}$ units

13 $\ln(2 + \sqrt{3})$ units

15 $\left(2\sqrt{577} + \frac{1}{12}\ln(24 + \sqrt{577}) - \frac{1}{2}\sqrt{37} - \frac{1}{12}\ln(6 + \sqrt{37})\right)$ units (The table of integrals may be helpful.)

17 $\frac{33}{16}$ units (Hint: since $x < 0$, $\sqrt{(\frac{1}{2}x^3 + \frac{1}{2}x^{-3})^2} = -(\frac{1}{2}x^3 + \frac{1}{2}x^{-3})$.)

19 $s(t) = \frac{1}{4}(t^2 - 1) + \frac{1}{2}\ln t$

21 $s(t) = t^3 - \frac{1}{12t} - \frac{11}{12}$

23 (4.2549, 3.8019)

25 (a) 4 m, (b) $(10\cosh(-0.5) - 10\cosh(0.4))$ m = 0.4655 m, (c) $(10\sinh(0.4) - 10\sinh(-0.5))$ m = 9.318 m

27 π units

29 $\left(a - b + \ln\left|\frac{e^{2b}-1}{e^{2a}-1}\right|\right)$ units (Hints: $L = \int_a^b \sqrt{1 + \left(\frac{-2e^x}{e^{2x}-1}\right)^2}\, dx = \cdots = \int_a^b \sqrt{\frac{(e^{2x}+1)^2}{(e^{2x}-1)^2}}\, dx = \cdots$. Use $u = e^x$ and partial fractions.)

31 (a) $L = \int_0^1 \sqrt{1 + a^2}\, dx = \cdots = \sqrt{1 + a^2}$ units, (b) distance $= \sqrt{(1-0)^2 + (a+b-b)^2} = \sqrt{1+a^2}$ units

33 $(1 + \frac{1}{2}\ln\frac{3}{2})$ units

35 (a) centimeters, (b) none ($\frac{b}{\Delta x}$ is dimensionless), (c) centimeters, (d) centimeters

Section 9.2

1 (a) $A = \int_0^{2\pi} \frac{1}{2}(4)^2\, d\theta = \cdots = 16\pi$ units^2, (b) $A = \pi(4)^2 = 16\pi$ units^2

3 (a) $A = \int_0^{\pi} \frac{1}{2}(4\sin\theta)^2\, d\theta = \cdots = 4\pi$ units^2, (b) $A = \pi(2)^2 = 4\pi$ units^2

5 11π units^2

7 $\frac{\pi}{3}$ units^2

9 $\frac{1}{2}$ units^2

11 (a) $L = \int_0^{2\pi} \sqrt{(-3)^2 + 0^2}\, d\theta = \cdots = 6\pi$ units, (b) $L = 2\pi(3) = 6\pi$ units

13 (a) $L = \int_0^{\pi} \sqrt{(4\cos\theta)^2 + (-4\sin\theta)^2}\, d\theta$ $= \cdots = 4\pi$ units,
(b) $L = 2\pi(2) = 4\pi$ units

15 (a) $L = \int_0^{\pi/3} \sqrt{(\sec\theta)^2 + (\sec\theta\tan\theta)^2}\, d\theta$ $= \cdots = \sqrt{3}$ units;
(b) $(1, 0)$ and $(1, \sqrt{3})$,
$d = \sqrt{(1-1)^2 + (\sqrt{3}-0)^2} = \sqrt{3}$ units

17 $\sqrt{2}(e-1)$ units

19 $\left(3\pi\sqrt{1+4\pi^2} + \frac{3}{2}\ln(2\pi + \sqrt{1+4\pi^2})\right)$ units

21 (a)

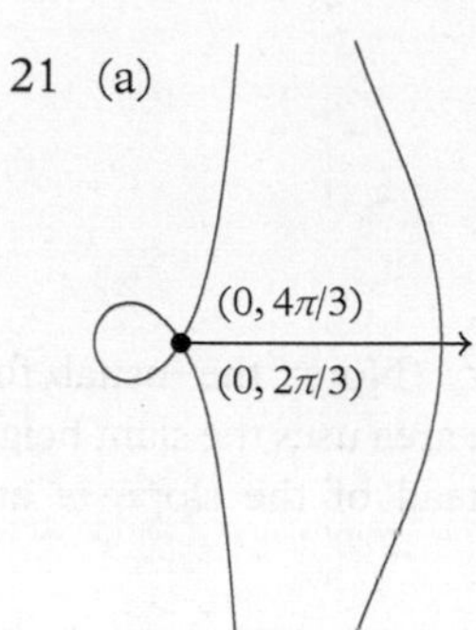

(b) $A = \int_{2\pi/3}^{4\pi/3} \frac{1}{2}(2+\sec\theta)^2\, d\theta$,
(c) $\left(\sqrt{3} + \frac{4\pi}{3} + 2\ln\frac{2-\sqrt{3}}{2+\sqrt{3}}\right)$ units^2 $= 0.65301$ units^2

23 (a)

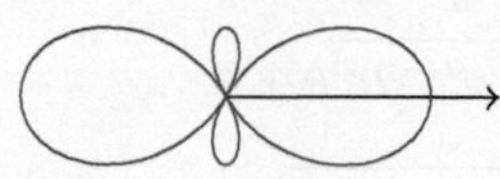

(b) $A = \int_{-2\pi/3}^{-\pi/3} \frac{1}{2}(\frac{1}{2} + \cos 2\theta)^2\, d\theta$; alternate answer, $A = \int_{4\pi/3}^{5\pi/3} \frac{1}{2}(\frac{1}{2} + \cos 2\theta)^2\, d\theta$
(c) $\left(\frac{\pi}{8} - \frac{3\sqrt{3}}{16}\right)$ units$^2 = 0.06794$ units^2

25 (a)

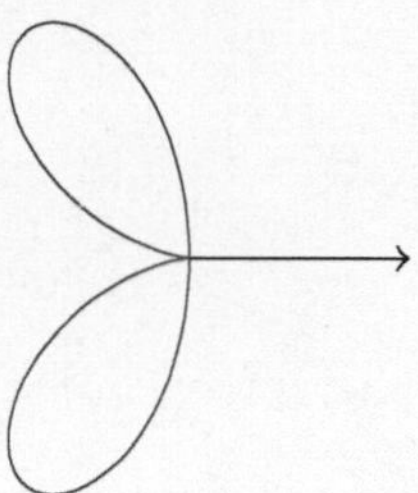

(b) $A = \int_0^{\pi/2} \frac{1}{2}(\cos 3\theta - \cos\theta)^2\, d\theta$, (c) $\frac{\pi}{4}$ units$^2 = 0.7854$ units^2

27 (a)

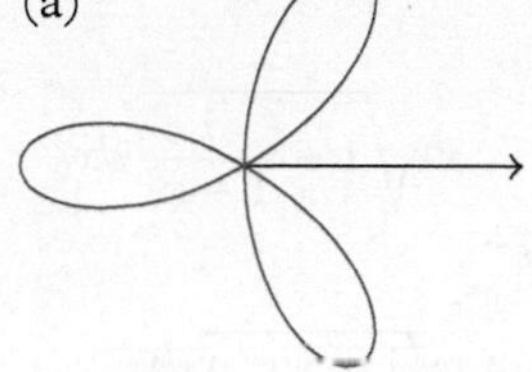

(b) $L = \int_{\pi/6}^{\pi/2} \sqrt{(\cos\theta(4\sin^2\theta - 1))^2 + (8\sin\theta\cos^2\theta - 4\sin^3\theta + \sin\theta)^2}\, d\theta$, (c) 2.227 units

29 (a)

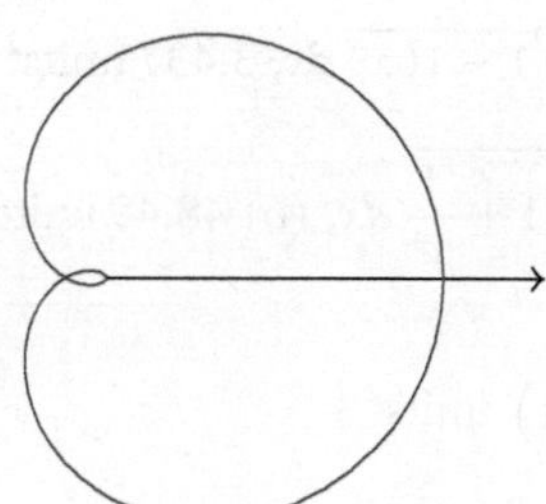

(b) $L = \int_{\pi}^{2\pi} \sqrt{\cos^6\frac{\theta}{3} + \sin^2\frac{\theta}{3}\cos^4\frac{\theta}{3}}\, d\theta$,
(c) $\left(\frac{\pi}{2} - \frac{3\sqrt{3}}{4}\right)$ units $= 0.2718$ units

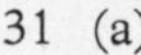

31 (a)

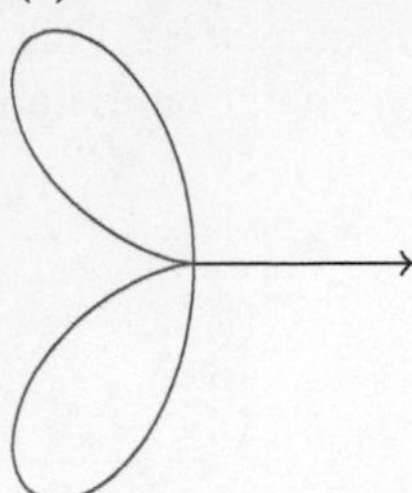

(b) $L = \int_0^{\pi} \sqrt{(\cos 3\theta - \cos\theta)^2 + (-3\sin 3\theta + \sin\theta)^2}\, d\theta$, (c) 7.156 units

33 $\frac{5\pi}{4}$ units^2

35 $\left(\frac{\pi}{2} - 1\right)$ units^2

37 $\left(\frac{9}{2} + \frac{\pi}{2} - \frac{3\sqrt{3}}{2}\right)$ units^2

39 $\frac{\sqrt{3}}{9}\pi$ units

41 $(4 + 2\sqrt{2})$ units

43 for $\theta = b$ to $\theta = c$, the lengths of radii vectors are $r = \frac{a}{b}$ and $r = \frac{a}{c}$. Then, $A = \int_b^c \frac{1}{2}\left(\frac{a}{\theta}\right)^2 d\theta = \cdots = \frac{-a^2}{2c} - \frac{-a^2}{2b} = \frac{a}{2}\left(\frac{a}{b} - \frac{a}{c}\right)$, which is proportional (by $\frac{a}{2}$) to the difference of the lengths.

Section 9.3

1 (a) $\displaystyle\int_0^{\frac{1}{2}} 2\pi(\sin^{-1} x)\sqrt{1 + \frac{1}{1 - x^2}}\, dx$, (b) 1.178 units^2

3 (a) $\displaystyle\int_{-\frac{\pi}{4}}^{0} -2\pi \tan x \sqrt{1 + \sec^4 x}\, dx$, (b) 3.839 units^2

5 (a) $\displaystyle\int_0^2 2\pi x\sqrt{1 + 4x^2 e^{2x^2+2}}\, dx$, (b) 1584 units^2

7 (a) $\displaystyle\int_0^1 2\pi x^4 \sqrt{1 + 16x^6}\, dx$, 3.437 units^2

9 (a) $\displaystyle\int_1^4 2\pi x\sqrt{1 + \frac{1}{x^4}}\, dx$, (b) 48.49 units^2

11 $4\pi\sqrt{5}$ units^2

13 $\frac{\pi}{6}\left(17\sqrt{17} - 1\right)$ units^2

15 49π units^2

17 $\frac{56\pi}{3}$ units^2

19 $\frac{10}{3}\pi$ units^2

21 $\frac{208}{9}\pi$ units^2

23 $\pi\left(2\sqrt{2} + \ln\frac{1+\sqrt{2}}{-1+\sqrt{2}}\right)$ units^2

25 $\frac{\pi}{2}\left(\sqrt{5} + \frac{1}{2}\ln(2 + \sqrt{5})\right) - \frac{\pi}{2}\left(\frac{1}{2}\sqrt{2} + \frac{1}{2}\ln(1 + \sqrt{2})\right)$ units^2

27 $\frac{\pi}{3}\left(6\sqrt{145} + \frac{1}{2}\ln(12 + \sqrt{145})\right)$ units^2

29 $\pi\left(\sqrt{2} + \ln(1 + \sqrt{2})\right)$ units^2

31 $4\pi r^2$ units^2

33 $\pi m h^2\sqrt{1 + m^2}$ units^2 (Note: the usual formula for the surface area uses the slant height ℓ and radius r instead of the slope m and height h.)

35 $2\pi r h$ units^2

37 $\frac{12}{5}\pi a^2$ units^2

39 (a) $\frac{1}{2}(f(x_{k-1}) + f(x_k))$, (b) $f(x_k) - \Delta x f'(x_k)$, (c) $\frac{1}{2}(f(x_{k-1}) + f(x_k)) = \frac{1}{2}(f(x_k) - \Delta x f'(x_k) + f(x_k)) = f(x_k) - \frac{1}{2}\Delta x f'(x_k)$, (d) $SA = \sum_{k=1}^{\Omega} 2\pi\left(f(x_k) - \frac{1}{2}\Delta x f'(x_k)\right) \cdot \sqrt{1 + (f'(x_k))^2}\,\Delta x$, (e) $= \sum_{k=1}^{\Omega} 2\pi f(x_k)\sqrt{1 + (f'(x_k))^2}\,\Delta x - \sum_{k=1}^{\Omega} \pi\Delta x f'(x_k)\sqrt{1 + (f'(x_k))^2}\,\Delta x$,

(f) $= \sum_{k=1}^{\Omega} 2\pi f(x_k)\sqrt{1 + (f'(x_k))^2}\,\Delta x - \Delta x \sum_{k=1}^{\Omega} \pi f'(x_k)\sqrt{1 + (f'(x_k))^2}\,\Delta x$,

(g) $= \int_a^b 2\pi f(x)\sqrt{1 + (f'(x))^2}\, dx - \Delta x \int_a^b \pi f'(x)\sqrt{1 + (f'(x))^2}\, dx$,

(h) $\approx \int_a^b 2\pi f(x)\sqrt{1 + (f'(x))^2}\, dx$

41 The volume of paint inside the infinitely long object is finite only because the diameter of a cross section (slicing perpendicular to the x-axis) gets arbitrarily small as x gets large. When we say it takes an infinite amount of

paint to cover an infinite surface area, we are thinking of a constant thickness of the paint, which doesn't get thinner as the diameter of a cross section shrinks. At $x = \Omega$, the diameter of a cross section is infinitesimal, but the paint on the outside still has the same real-number thickness as at $x = 1$. This is why there is so much more paint on the outside.

Section 9.4

1 (a) $L = \int_{-\frac{1}{2}}^{\frac{1}{2}} \sqrt{\frac{1}{1-t^2} + 9e^{6t}}\, dt$, (b) 4.485 units

3 (a) $L = \int_1^3 \left(\frac{(-20t^2+10t+44)^2}{(5t^2+11)^4} + (t^2 \sinh t + 2t\cosh t)^2 \right)^{\frac{1}{2}} dt$, (b) 89.07 units

5 (a) $L = \int_0^{2\pi} \sqrt{9\sin^2 t + 25\cos^2 t}\, dt$, (b) 25.53 units

7 (a) $L = \int_{-2}^{2} \left(((3t+1)2^t \ln 2 + 3 \cdot 2^t)^2 + 9\cos^2(3t+1) \right)^{\frac{1}{2}} dt$, (b) 32.20 units

9 (a) $SA = \int_0^2 2\pi \cosh t \sqrt{4t^2 + \sinh^2 t}\, dt$, (b) 70.15 units^2

11 (a) $SA = \int_0^2 2\pi(4 - t^2)\sqrt{4t^2 + \sinh^2 t}\, dt$, (b) 59.35 units^2

13 (a) $SA = \int_0^2 2\pi\sqrt{t^3+1}\sqrt{\frac{1}{4t+36} + \frac{9t^4}{4t^3+4}}\, dt$, (b) 25.62 units^2

15 (a) $SA = \int_0^2 2\pi\sqrt{t+9}\sqrt{\frac{1}{4t+36} + \frac{9t^4}{4t^3+4}}\, dt = \int_0^2 2\pi\sqrt{\frac{1}{4} + \frac{9t^4(t+9)}{4t^3+4}}\, dt$, (b) 41.78 units^2

17 $\frac{1675}{27}$ units

19 $\sqrt{2}$ units

21 $\frac{3}{2}$ units (Note: for these values of t, $\sqrt{9\sin^2 t\cos^2 t} = -3\sin t\cos t$.)

23 $\left(\sqrt{2}e^{\pi/2} - \sqrt{2}\right)$ units

25 8 units

27 4π units^2

29 $\left(\frac{16\sqrt{2}\pi}{3} - \frac{8\pi}{3}\right)$ units^2

31 $\frac{12}{5}\pi$ units^2

33 $\frac{64}{3}\pi$ units^2

35 $\frac{2\sqrt{2}\pi}{5}(2e^{\pi} + 1)$ units^2

37 $\left(2\sqrt{2} + \ln\frac{\sqrt{2}+1}{\sqrt{2}-1}\right)$ units

39 $4\pi r^2$ units^2

41 $2\pi r$ units

43 Cartesian equation $y = 1 - x$, $L = \int_0^1 \sqrt{1 + (-1)^2}\, dx = \sqrt{2}$ units

Section 9.5

1 $(6 - y_k)$ m

3 $(0 - y_k)$ m $= -y_k$ m

5 $(5 - y_k)$ m

7 $10\,030(6 - y_k)$ N/m^2

9 $-7900y_k$ N/m^2

11 $7740(5 - y_k)$ N/m^2

13 (a) 312 lb/ft^2, (b) 249 600 lb

15 (a) $6\,018\,000\,\text{N/m}^2$, (b) $4\,874\,500\pi\,\text{N} \approx 15\,313\,900\,\text{N}$ (Reminder: the diameter, not the radius, was given.)

17 Answers vary. The most reasonable answers, in no particular order, are as follows:

(a)

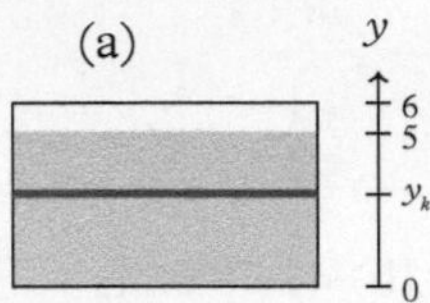

(b) $64.5(5 - y_k)\,\text{lb/ft}^2$;

(a)

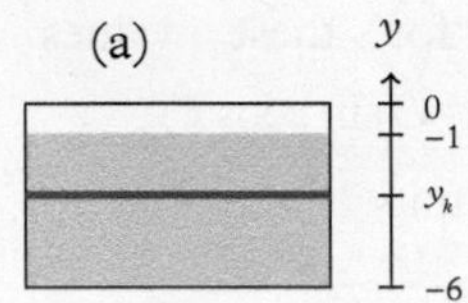

(b) $64.5(-1 - y_k)\,\text{lb/ft}^2$;

(a)

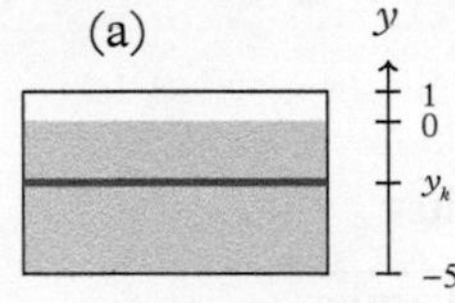

(b) $-64.5y_k\,\text{lb/ft}^2$

19 Answers vary. The most reasonable answers, in no particular order, are as follows:

(a)

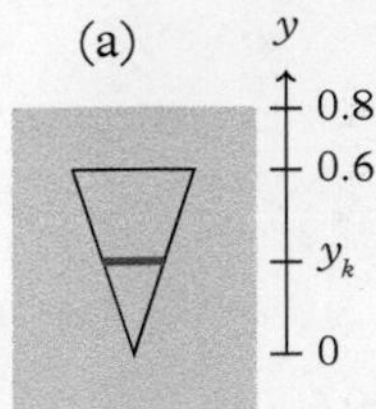

(b) $8580(0.8 - y_k)\,\text{N/m}^2$;

(a)

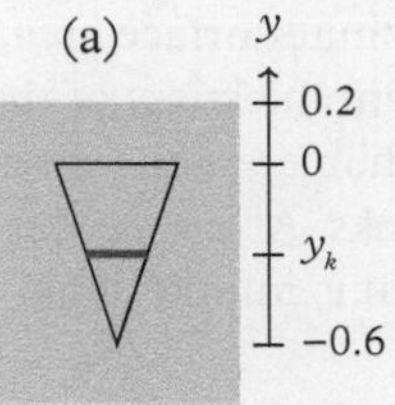

(b) $8580(0.2 - y_k)\,\text{N/m}^2$;

(a)

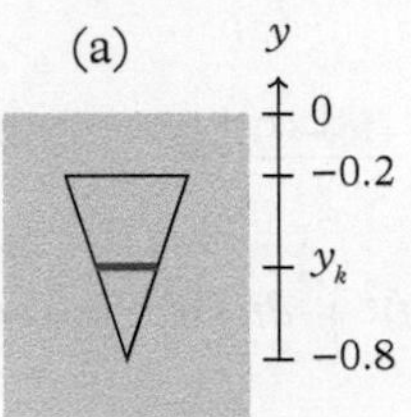

(b) $-8580y_k\,\text{N/m}^2$

21 Answers vary. The most reasonable answers, in no particular order, are as follows:

(a)

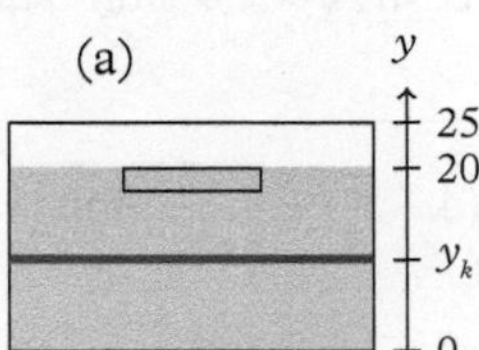

(b) $9800(20 - y_k)\,\text{N/m}^2$;

(a)

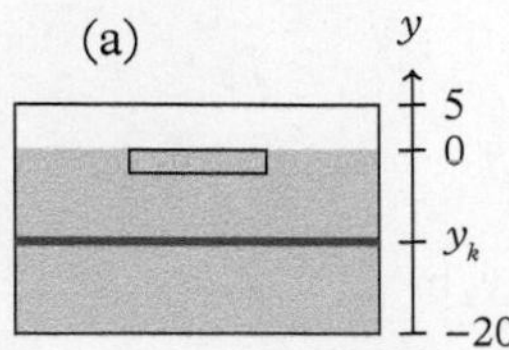

(b) $-9800y_k\,\text{N/m}^2$;

(a)

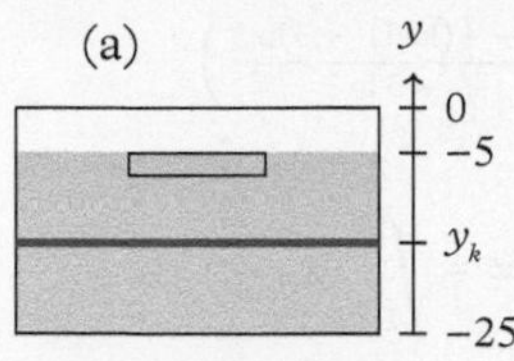

(b) $9800(-5 - y_k)\,\mathrm{N/m^2}$

23 $5\,\Delta y\,\mathrm{m}^2$

25 $2\sqrt{9 - y_k^2}\,\Delta y\,\mathrm{m}^2$

27 $2\sqrt{9 - (y_k - 3)^2}\,\Delta y\,\mathrm{m}^2$

29 $(6 + \frac{1}{3}y_k)\,\Delta y\,\mathrm{m}^2$

31 1 929 375 N (Note that with the top of the water at $y = 0$, the integral is $\int_{-6.5}^{-4} -9800y \cdot 15\,dy$; with the top of the water at $y = 4$, the integral is $\int_{-2.5}^{0} 9800(4 - y) \cdot 15\,dy$.)

33 52 267 N (Note that with the origin in the center of the circle, the integral is $\int_{-2}^{0} -9800y \cdot 2\sqrt{4 - y^2}\,dy$.)

35 2160 lb (Note that with the origin in the center of the ellipse, the integral is $\int_{-3}^{0} -120y \cdot \sqrt{36 - 4y^2}\,dy$.)

37 123.53 N (Possible integrals include $\int_{0}^{0.25} 8950(0.25 - y)(0.4 + \frac{1}{2}y)\,dy$ and $\int_{-0.25}^{0} -8950y(0.525 + \frac{1}{2}y)\,dy$.)

39 3.707 N (Possible integrals include $\int_{-0.06}^{0} -8580y(0.16 + 2\sqrt{0.0036 - y^2})\,dy$.)

41 5471 lb; (Possible integrals include $\int_{-3}^{3} 2\sqrt{9 - y^2}(3 - y)64.5\,dy$.)

43 7 351 300 lb (Possible integrals include $\int_{-50}^{0} -62.4y \cdot 30\pi\,dy$.)

45 $\sum_{k=1}^{\Omega} 9800(0.07 - 0.07\omega k)(0.1 + \frac{2}{7}(0.07\omega k)) \cdot 0.07\omega = \cdots = 2.561\,\mathrm{N}$

Section 9.6

1 (a) $\bar{x} = \frac{1}{10}$, (b) 1 ft · lb, (c) 10 lb

3 (a) $\bar{x} = \frac{3268}{431} \approx 7.58$, (b) 3268 g · mm (= 3.268 kg · mm = 3.268 g · m = 0.003268 kg · m)

5 (a) $\bar{x} = \frac{1}{-\frac{8}{15} + \frac{\pi}{4}} \int_0^{\pi/2} x\left(\sin^2 x - \sin^5 x\right) dx$, $\bar{y} = \frac{1}{-\frac{8}{15} + \frac{\pi}{4}} \int_0^{\pi/2} \frac{1}{2}\left(\sin^4 x - \sin^{10} x\right) dx$; (b) $\bar{x} = 0.8118$, $\bar{y} = 0.4017$

7 (a) $\bar{x} = \frac{1}{\frac{32}{3} - \ln 27} \int_1^3 x(x^2 - \ln x)\,dx$, $\bar{y} = \frac{1}{\frac{32}{3} - \ln 27} \int_1^3 \frac{1}{2}\left(x^4 - (\ln x)^2\right) dx$; (b) $\bar{x} = 2.314$, $\bar{y} = 3.213$

9

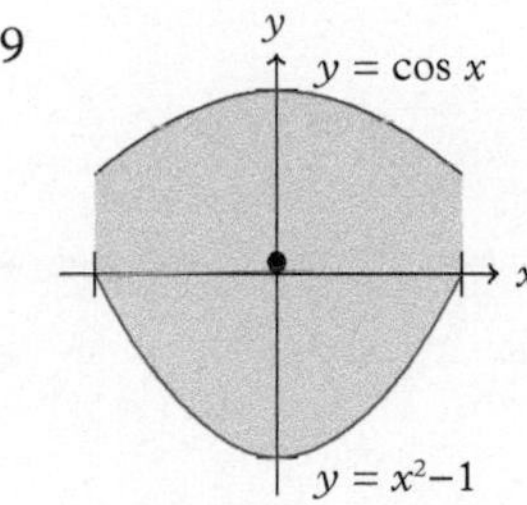

The location of the dot may vary (the actual location is shown here), but the dot should be on the y-axis (based on symmetry) and a little above the x-axis (based on more of the region lying above the x-axis than below).

11 $\left(\frac{8}{5}, \frac{16}{7}\right)$
13 $\left(\frac{1}{2}, \frac{8}{5}\right)$
15 $\left(\frac{3}{2}, \frac{12}{5}\right)$
17 $\left(\frac{\pi}{2}, \frac{\pi}{8}\right)$
19 $\left(\frac{8}{\pi}, \frac{4}{\pi}\right)$
21 $\left(\frac{22 - \frac{9}{2}\ln 3}{\frac{32}{3} - 3\ln 3}, \frac{\frac{111}{5} - \frac{3}{2}(\ln 3)^2 + 3\ln 3}{\frac{32}{3} - 3\ln 3}\right)$
23 $\left(-1, \frac{1}{4}\right)$
25 DNE (since $\overline{x} = \frac{1}{1}\int_1^\infty x\left(\frac{1}{x^2}\right)\,dx = \cdots \doteq \infty$)

Section 9.7

1 $B = \$3210$
3 $A = 200$
5 $A = 6237.7$
7 \$1000
9 \$949 556
11 \$135 026
13 \$924.50
15 \$2217.56
17 \$61 362
19 (a) \$2 500 000, (b) \$1 500 000
21 (a) \$3594, (b) \$3562
23 (a) \$585 509, (b) \$76 958
25 0.310
27 0.125
29 0.257
31 0.164
33 $1 - \frac{2}{p+1}$
35 0.323
37 0.042
39 (a) $L(0.2) = 0.12885$, $L(0.4) = 0.28623$, $L(0.6) = 0.47845$, $L(0.8) = 0.71324$; (b) 0.164; (c) the answers are the same to three (actually more) decimal places.
41 answers vary
43 the demand level A increases and the market price B decreases.
45 Because the Lorenz curve is concave up, using straight line segments for the trapezoid rule cuts off some of the area between $y = x$ and the Lorenz curve.

Section 9.8

1 logistic
3 logistic
5 exponential
7 neither (Newton's law of cooling)
9 188 549
11 29 years (at $t = 28.95$)
13 5 months (at $t = 4.96$)
15 more heard from the official announcement (590 had heard the rumor by noon)
17 1094
19 1 year ($t = 1.015$)
21 (a) 14 days ($t = 14.07$), (b) 25 more days ($100.01 - 74.96$)
23 (a) 22 years ($t = 22.07$), (b) the exponential model predicts the population of weeds reaches 500 plants 7 years sooner than the logistic model does.
25 at $t = 211$
27 $P = \frac{\Omega}{1 + \frac{\Omega - P_0}{P_0}e^{-kt}} \approx \frac{\Omega}{\frac{\Omega}{P_0}e^{-kt}} = \Omega \cdot \frac{P_0}{\Omega} \cdot e^{kt} = P_0 e^{kt}$

Section 10.1

1 no (only finitely many terms)

3 no (a_n is just one term (the nth term) of a sequence.)

5 yes

7 $1, \frac{1}{3}, \frac{1}{5}, \frac{1}{7}, \frac{1}{9}, \frac{1}{11}$

9 $-1, \frac{1}{3}, -\frac{1}{5}, \frac{1}{7}, -\frac{1}{9}, \frac{1}{11}$

11 $1, -\frac{1}{3}, -\frac{1}{5}, \frac{1}{7}, \frac{1}{9}, -\frac{1}{11}$ (Starting with the second term, the sequence alternates two negative terms, two positive terms, ..., and so on.)

13 $1, \frac{1}{6}, \frac{1}{60}, \frac{1}{840}, \frac{1}{15\,120}, \frac{1}{332\,640}$

15 $1, \frac{1}{2}, \frac{1}{6}, \frac{1}{24}, \frac{1}{120}, \frac{1}{720}$

17 2, 3, 4, 5, 6, 7

19 $a_n = \frac{-2}{n+4}$

21 $a_n = \frac{(-1)^n(-2)}{n+4}$; alternate answer, $a_n = \frac{(-1)^{n+1} \cdot 2}{n+4}$

23 $a_n = \frac{1}{3^n}$

25 $a_n = \frac{1}{2n+1}$

27–36 *Some limits may be difficult to determine with a small number of plotted terms. Even convergence or divergence may be difficult to determine. This issue is remedied with the formal definition of limit in section 10.2.*

27 (a)

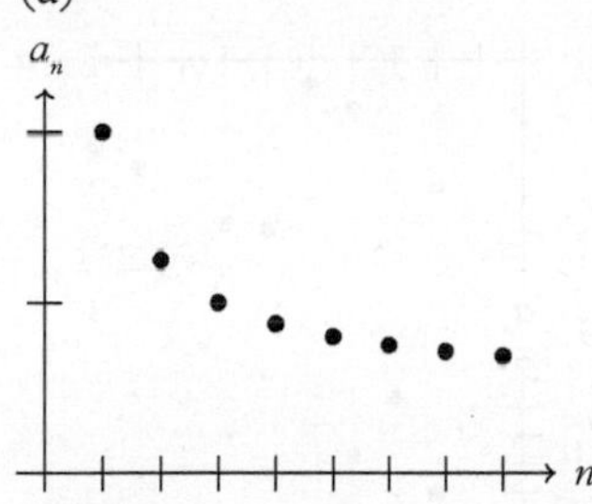

(b) this sequence converges,
(c) $\lim_{n\to\infty} a_n = \frac{1}{2}$,
(d) yes (decreasing), (e) yes

29 (a)

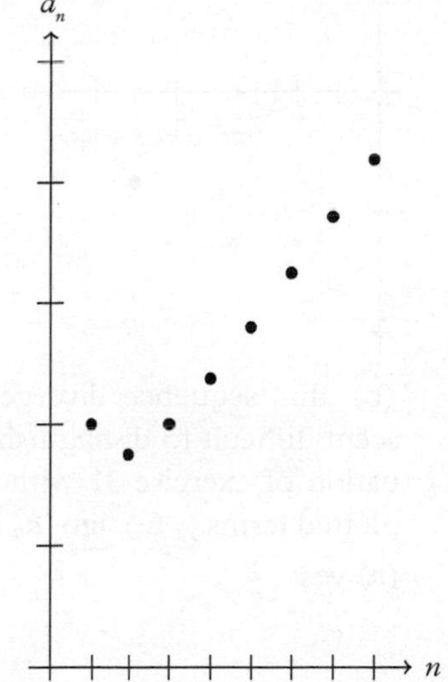

(b) this sequence diverges,
(c) $\lim_{n\to\infty} a_n = \infty$, (d) no, (e) no

31 (a)

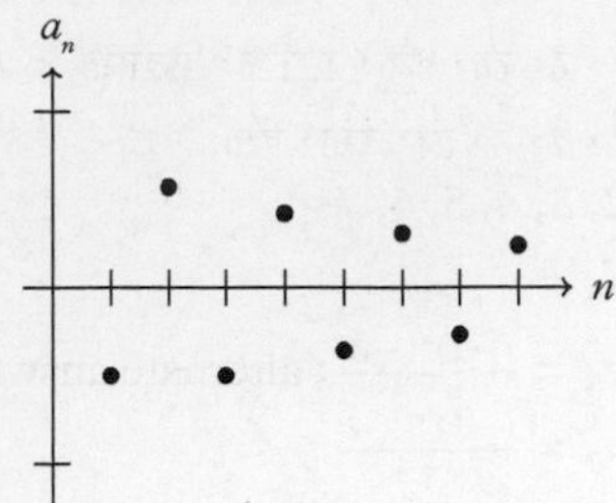

(b) this sequence converges, (c) $\lim_{n\to\infty} a_n = 0$, (d) no, (e) yes

33 (a)

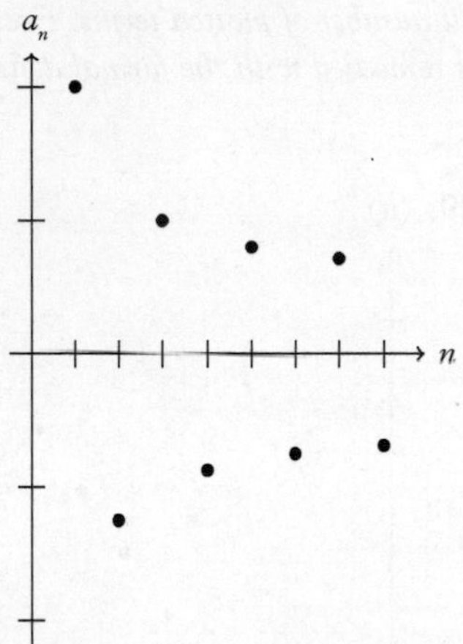

(b) this sequence diverges (This may seem difficult to distinguish from the situation of exercise 31 without additional plotted terms.), (c) $\lim_{n\to\infty} a_n$ DNE, (d) no, (e) yes

35 (a)

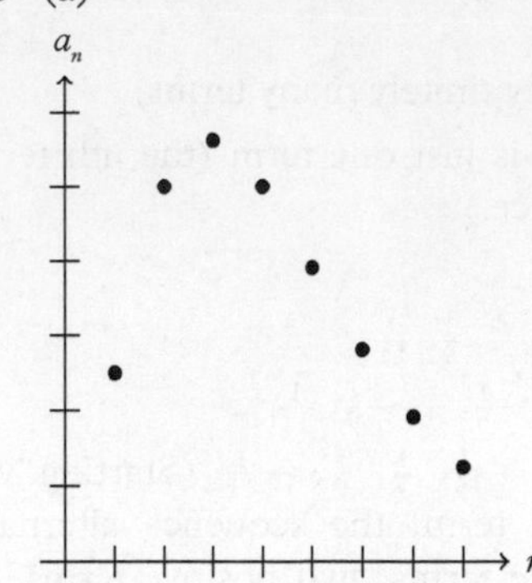

(b) this sequence converges, (c) $\lim_{n\to\infty} a_n = 0$, (d) no, (e) yes

37

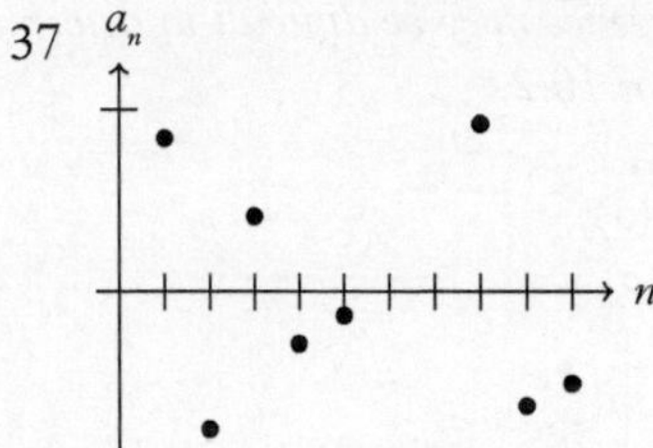

39

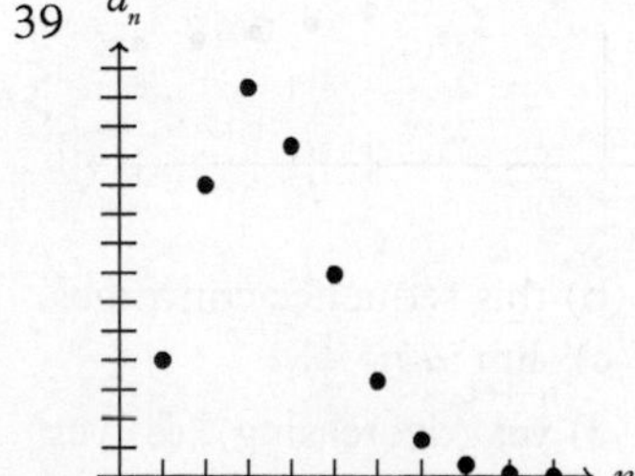

41 (a) Lucas numbers; (b) $L(n) = L(n-1) + L(n-2)$, $L(0) = 2$, $L(1) = 1$; alternate answer, $a_1 = 2$, $a_2 = 1$, $a_n = a_{n-1} + a_{n-2}$ for $n \geq 3$; (c) NA (or answers may vary)

43 (a) perfect numbers, (b) NA, (c) a perfect number is equal to the sum of its proper divisors.

45 (a) weird numbers, (b) NA, (c) weird numbers are abundant but not pseudo-perfect (these are number theory terms).

47 $a_n = \frac{1}{120}n^5 - \frac{1}{8}n^4 + \frac{17}{24}n^3 - \frac{15}{8}n^2 + \frac{257}{60}n - 2$

Section 10.2

1 0

3 ∞

5 $\frac{\pi}{2}$

7 $-\frac{1}{2}$

9 DNE

11 0

13 DNE (infinite but alternating sign; Hint: $(-2)^{\Omega} = (-1)^{\Omega} \cdot 2^{\Omega}$)

15 1

17 DNE

19 1

21 ∞

23 0

25 ∞

27 $e^{0.3}$

29 0

31 $e^{7/8}$

33 $e^{0.6}$

35 ∞

37 0

39 1

41 1

43 $\lim_{n\to\infty} a^{1/n} = a^{1/\Omega} = a^{\omega} \doteq a^0 = 1$ (The rendering step is valid because $f(x) = a^x$ is a continuous function for $a > 0$.)

45 Hints: using the similar function rule, calculate $\lim_{x\to\infty} \left(1 + \frac{a}{x}\right)^x$. Use the strategy for powers. Rewrite $\lim_{x\to\infty} x \ln\left(1 + \frac{a}{x}\right)$ to obtain the indeterminate form $\frac{0}{0}$ and use l'Hospital's rule.

Section 10.3

1 sequence

3 term of a sequence or series

5 series; alternate answer, sum of a series.

7 (a) no, (b)NA, (c) NA

9 (a) yes; (b) $a = \frac{4}{3}$, $r = \frac{1}{3}$; (c) converges

11 (a) yes; (b) $a = \frac{729}{4}$, $r = 3$; (c) diverges

13 (a) no, (b)NA, (c) NA

15 diverges (harmonic, only finitely many terms missing)

17 $\frac{189}{5} = 37.8$ (geometric)

19 diverges (test for divergence)

21 0.002816

23 $-\frac{\pi}{4}$ (If your answer is $\tan^{-1} 1 = \frac{\pi}{4}$, you might have neglected to use partial sums.)

25 $\frac{1}{6}$ (telescoping)

27 $\frac{7}{4}$ (Hint: series sum rule)

29 $2 \cdot \frac{7}{4} = \frac{7}{2}$

31 diverges

33 diverges

35 $\frac{5}{2}$

37 $0.\overline{9} = \sum_{n=1}^{\infty} 9\left(\frac{1}{10}\right)^n = 1$ (Related note: even in the hyperreal numbers, point nine repeating is equal to one)

39 $\frac{3678}{990} = \frac{613}{165}$

41 $141.\overline{41} = 141.4\overline{14} = \sum_{n=1}^{\infty} 140(0.01)^{n-1} = \frac{140}{0.99} = \frac{14\,000}{99}$

43 diverges (Hint: similar to telescoping, but the result differs from the harmonic series by finitely many terms)

45 outline of proof: series is $a + a + a + a + \cdots$; partial sum $S_k = ka$; then, $S_\Omega = \Omega a \doteq \infty$ (for $a > 0$) or $S_\Omega = \Omega a \doteq -\infty$ (for $a < 0$); series diverges

47 (a) $s(t) = -4.9t^2 + v_0 t$, (b) $t = \frac{v_0}{4.9}$ s, (c) total time $\sum_{n=1}^{\infty} \frac{5}{4.9}(0.95)^{n-1}$ s, (d) 20.408 seconds

49 yes, $\sum_{n=1}^{\infty} ca_n$ diverges. (If $\sum_{n=1}^{\infty} ca_n = L$ for a real number L, then by the series constant multiple rule, $\sum_{n=1}^{\infty} a_n = \sum_{n=1}^{\infty} \frac{1}{c} ca_n = \frac{1}{c} \sum_{n=1}^{\infty} ca_n = \frac{1}{c} L$ converges, which is a contradiction.)

51 (a) \$1263.16, (b) 31.5789% (The U.S. Internal Revenue Service rounds this amount to 31.58% in their published regulations at the time of this writing.)

Section 10.4

1 converge (p-series, $p > 1$)

3 diverge (p-series, $p \leq 1$)

5 converge (series constant multiple rule, p-series)

7 converge (series sum rule, series constant multiple rule, p-series)

9 no ($f(n) \neq a_n$)

11 no (negative, increasing on $[1, \infty)$; but $\sum_{n=1}^{\infty} \frac{1}{(-n)^3} = -\sum_{n=1}^{\infty} \frac{1}{n^3}$, and the hypotheses are met for $\sum_{n=1}^{\infty} \frac{1}{n^3}$)

13 no (not continuous (or positive or decreasing) on $[1, \infty)$)

15 yes

17 (a) diverges, (b) NA

19 (a) converges, (b) $\frac{1}{3} < \sum_{n=1}^{\infty} n^{-4} < \frac{4}{3}$

21 (a) diverges, (b) NA

23 (a) converges; (b) $\frac{1}{2\ln 2} < \sum_{n=1}^{\infty} 2^{-n} < \frac{1}{2} + \frac{1}{2\ln 2}$; alternate answer, $0.7213 < \sum_{n=1}^{\infty} 2^{-n} < 1.2213$

25 (a) diverges, (b) NA

27 (a) diverges, (b) NA

29 diverges (test for divergence; the integral test does not apply)

31 diverges (If you concluded the series converges by the test for divergence, your reasoning is incorrect. The test for divergence is inconclusive when the limit is zero.)

33 converges (geometric) (If you concluded the series converges by the test for divergence, your reasoning is incorrect. The test for divergence is inconclusive when the limit is zero.)

35 converges (p-series with $p = 4 > 1$, series constant multiple rule)

37 convergence or divergence cannot be determined using available techniques (The test for divergence is inconclusive, not geometric, not telescoping, not p-series, and the integral test is not available because of $n!$)

39 diverges

41 converges (Hint: substitution)

43 converges (Hint: parts)

45 converges (Hint: partial fractions)

47 $\sum_{n=1}^{\infty} \frac{1}{n^p} = \sum_{n=1}^{\infty} n^{-p}$, $\lim_{n\to\infty} n^{-p} = \Omega^{-p}$, and because $-p$ is positive, this quantity renders $\infty \neq 0$; diverges

49 (a) $L - S_k = \sum_{n=k+1}^{\infty} a_n$, (b) $\int_{k+1}^{\infty} f(x)\,dx$, (c) $\int_{k}^{\infty} f(x)\,dx$, (d) $\int_{k+1}^{\infty} f(x)\,dx < L - S_k < \int_{k}^{\infty} f(x)\,dx$; (e) $S_k + \int_{k+1}^{\infty} f(x)\,dx < L < s_k + \int_{k}^{\infty} f(x)\,dx$

Section 10.5

1 $>$

3 $<$

5 $<$

7 neither applies (although $<$ applies for $n > 49$)

9 (a) $\sum_{n=1}^{\infty} \frac{5}{n^2} = 5 \cdot \sum_{n=1}^{\infty} \frac{1}{n^2}$, (b) converge

11 (a) $\sum_{n=1}^{\infty} \frac{1}{n}$, (b) diverge

13 converges (Compare to $\sum_{n=1}^{\infty} \frac{1}{n^3}$.)

15 converges (Compare to $\sum_{n=1}^{\infty} \frac{1}{n^3}$.)

17 the comparison test is inconclusive (when comparing to the series $\sum_{n=1}^{\infty} \frac{1}{n^3}$).

19 diverges (Compare to $\sum_{n=1}^{\infty} \frac{1}{n}$.)

21 diverges (Compare to $\sum_{n=1}^{\infty} \frac{1}{\sqrt{n}}$.)

23 converges (Compare to $\sum_{n=1}^{\infty} \left(\frac{3}{4}\right)^n$.)

25 diverges (Compare to $\sum_{n=1}^{\infty} \left(\frac{3.73}{2}\right)^n$.)

27 converges ($\frac{1}{a_\Omega} \approx \Omega^3$)

29 converges ($\frac{1}{a_\Omega} \approx \Omega^3$)

31 diverges ($\frac{1}{a_\Omega} \approx \Omega$)

33 diverges ($\frac{1}{a_\Omega}$ is on the $\sqrt{\Omega}$ level.)

35 converges ($\frac{1}{a_\Omega} \approx \left(\frac{4}{3}\right)^\Omega$)

37 diverges

39 diverges

41 converges

43 diverges

45 converges

47 converges

49 diverges

51 diverges (Note that $\left(\frac{2}{5}\right)^\Omega \doteq 0$.)

53 converges

55 converges ($\frac{1}{a_\Omega} \approx \left(\frac{16}{9}\right)^\Omega$)

57 converges

59 diverges

61 converges

63 (a) diverges (Hint: use $u = \ln\ln x$); (b) $\frac{1}{a_\Omega} = \Omega(\ln\Omega)(\ln\ln\Omega)$, which must be in the divergence zone because the series diverges. This quantity is shown in the divergence zone in figure 8.

65 the test is inconclusive. If $\sin\Omega > 0.6$, then $1.4 + \sin\Omega > 2$ and $\frac{1}{a_\Omega} = \left(\frac{2}{1.4+\sin\Omega}\right)^\Omega \doteq 0$ is in the divergence zone. However, if $\sin\Omega < 0.6$, then $1.4 + \sin\Omega < 2$ and $\frac{1}{a_\Omega} = \left(\frac{2}{1.4+\sin\Omega}\right)^\Omega$ is in the convergence zone (take note of whether the quantity in the parentheses is greater or less than one). Because $\frac{1}{a_\Omega}$ is not always in the same zone, the test is inconclusive.

Section 10.6

1 yes

3 yes

5 no (The pattern has two negative terms followed by two positive terms, two negative terms, two positive terms, and so on.)

7 no ($(-1)^{2n} = 1$ for all n)

9 converges

11 converges

13 converges (For step ❶, recall that $y = \ln x$ is an increasing function.)

15 converges ($b_n = \frac{1}{2n+5}$)

17 converges

19 diverges (test for divergence)

21 converges

23 diverges (test for divergence)

25 converges

27 diverges (test for divergence)

29 converges (Note that $\sin\left((2n+1)\frac{\pi}{2}\right) = (-1)^n$.)

31 converges

33 333 terms

35 6 terms

37 13 terms ($5 \cdot 14^2 + 20 = 1000$, can stop before that term)

39 (a) $S_6 = -0.249657$, (b) $-\frac{1}{4}$, (c) $|S - S_6| = 0.000343 < 0.001$

41 $S_4 = 0.9718887$

43 (a) sum through $n = 158$ (Because the series starts at $n = 2$, we are only summing 157 terms.), (b) using S_n to mean the sum from the first term through the $n = 158$ term, $S_{158} = -0.460356720$.

45 answer will vary

Section 10.7

1 ratio test, diverges

3 ratio test, diverges

5 root test, converges

7 root test, inconclusive

9 $\lim_{n\to\infty} \left| \frac{(n+1)\cdot 2^{n+1}}{(n+2)^2} \cdot \frac{(n+1)^2}{n \cdot 2^n} \right|$

11 $\lim_{n\to\infty} \left| \frac{(n+1)!}{3^{2n+2}} \cdot \frac{3^{2n}}{n!} \right|$

13 $\frac{3n}{n+1}$

15 $\frac{(n+1)^3}{(n+2)^2}$

17 $\frac{\sqrt[n]{2}\cdot 2}{3}$

19 $\frac{(n+2)\ln(n+1)}{2(n+1)\ln n}$ (Note: the laws of logarithms cannot simplify $\frac{\ln(n+1)}{\ln n}$.)

21 converges absolutely (ratio test limit $\frac{3}{5}$)

23 converges absolutely (ratio test limit $\frac{3}{5}$)

25 converges absolutely (ratio test limit 0)

27 diverges (ratio test limit $\frac{9}{4}$)

29 inconclusive (ratio test limit 1)

31 converges absolutely (root test limit $\frac{1}{3}$)

33 converges absolutely (root test limit $\frac{1}{9}$)

35 diverges (root test limit ∞)

37 converges absolutely (root test limit $\frac{3}{5}$)

39 diverges (root test limit $\frac{4}{3}$)

41 converges conditionally (Does not converge absolutely by the level comparison test, but converges by the alternating series test. The ratio and root tests are inconclusive.)

43 converges absolutely ($|a_n| = \frac{1}{n^3}$ is a convergent p-series. The level comparison test also works, with $\frac{1}{|a_\Omega|} = \Omega^3$. The ratio and root tests are inconclusive.)

45 converges conditionally (Does not converge absolutely by the level comparison test, but

converges by the alternating series test. The ratio and root tests are inconclusive.)

47 diverges (by test for divergence. The ratio and root tests are inconclusive. The level comparison test shows not absolutely convergent, but does not apply to the original alternating series.)

49 converges conditionally (Does not converge absolutely by the level comparison test, but converges by the alternating series test. The ratio and root tests are inconclusive.)

51 diverges (ratio and root test limits 2; test for divergence also works.)

53 converges absolutely (root test limit $\frac{11}{16}$)

55 converges absolutely (ratio and root test limits $\frac{3.02}{5.7}$, level comparison $\frac{1}{a_\Omega} = \frac{1}{3.02}\left(\frac{5.7}{3.02}\right)^\Omega$)

57 converges absolutely (geometric $|r| = \frac{3}{4}$, ratio and root test limits $\frac{3}{4}$, level comparison $\frac{1}{a_\Omega} = \frac{1}{\sqrt{2}}\left(\frac{4}{3}\right)^\Omega$)

59 converges absolutely (ratio test limit 0)

61 diverges (The ratio and root tests are inconclusive. Notice that $\left(\frac{1}{2}\right)^\Omega \doteq 0$; therefore, $\left(\frac{1}{2}\right)^\Omega + 3\Omega \approx 3\Omega$ (test for divergence limit 3, level comparison $\frac{1}{a_\Omega} = \frac{1}{3}$).)

63 converges absolutely (ratio and root test limits $\frac{5}{8}$, level comparison $\frac{1}{a_\Omega} = \left(\frac{8}{5}\right)^\Omega$)

65 diverges (ratio test limit ∞, test for divergence limit ∞, level comparison $\frac{1}{a_\Omega} \doteq 0$)

67 converges conditionally (The ratio and root tests are inconclusive. Check for absolute convergence by using the level comparison test ($\frac{1}{a_\Omega} = \Omega$) or by using the comparison test (hint: $\frac{2^{2n}+1}{4^n} = \frac{4^n+1}{4^n} > 1$). The alternating series test conditions are met.)

69 diverges (ratio test limit ∞, test for divergence limit ∞, level comparison $\frac{1}{a_\Omega} \doteq 0$)

71 converges absolutely (ratio and root test limits $\frac{1}{e}$, level comparison $\frac{1}{a_\Omega} = \frac{e^\Omega}{2}$)

73 diverges (ratio test limit ∞)

75 converges absolutely (ratio test limit $\frac{1}{5}$)

77 converges (ratio test limit $\frac{2}{3}$)

79 (a) $1+\frac{1}{3}+\frac{1}{5}+\cdots+\frac{1}{15}$, 2.0218; (b) $-\frac{1}{2}$, 1.5218; (c) $+\frac{1}{17}+\frac{1}{19}+\cdots+\frac{1}{41}$, 2.0041; (d) $-\frac{1}{4}$, 1.7541; $+\frac{1}{43}+\frac{1}{45}+\cdots+\frac{1}{69}$, 2.0094; $-\frac{1}{6}$, 1.8428

Section 10.8

1 converge absolutely (ratio test limit 0)

3 converge conditionally (doesn't converge absolutely: p-series with $p = \frac{1}{2}$; converge: alternating series test)

5 diverge (ratio and root test limits $\frac{4}{3}$; test for divergence limit infinite; level comparison test can only conclude doesn't converge absolutely.)

7 converge absolutely (ratio and root test limits $\frac{3}{4}$, level comparison $\frac{1}{a_\Omega} = \left(\frac{4}{3}\right)^\Omega$)

9 diverge (geometric with $r = \frac{6}{5}$, test for divergence limit ∞, ratio and root test limits $\frac{6}{5}$)

11 diverge (p-series with $p = \frac{1}{2}$; the series is not alternating)

13 converge absolutely (level comparison $\frac{1}{a_\Omega} = \frac{\Omega^4}{\cos(\Omega\frac{\pi}{12})}$ on Ω^4 level, comparison test with comparison series $\sum_{n=1}^{\infty}\frac{1}{n^4}$)

15 converge absolutely (root test limit 0)

17 converge absolutely (level comparison $\frac{1}{a_\Omega} = \Omega^3$)

19 diverge (level comparison $\frac{1}{a_\Omega} = \Omega$)

21 diverge (ratio test limit $\frac{4e}{10}$, test for divergence limit infinite)

23 converge absolutely (ratio test limit $\frac{3}{5}$)

25 diverge (telescoping, $S_n = \ln(4n+3) - \ln 3$)

27 converge absolutely (ratio test limit 0)

29 converge absolutely (level comparison $\frac{1}{a_\Omega} = \Omega^2 \ln \Omega \gg \Omega^2$, comparison test with comparison series $\sum_{n=1}^{\infty} \frac{1}{n^2}$)

31 converge absolutely (ratio and root test limits $\sqrt{\frac{2}{5}}$, level comparison $\frac{1}{a_\Omega} = \left(\sqrt{\frac{5}{2}}\right)^\Omega$)

33 converge absolutely (ratio test limit $\frac{2}{3}$)

35 converge absolutely (root test limit DNE, inconclusive; level comparison $\frac{1}{a_\Omega} = \left(\frac{25}{11-7\cos\Omega}\right)^\Omega \gg \left(\frac{25}{18}\right)^\Omega$; comparison test with comparison series $\sum_{n=1}^{\infty} \left(\frac{18}{25}\right)^n$)

37 converge conditionally (doesn't converge absolutely: level comparison $\frac{1}{a_\Omega} = \Omega\sqrt{\ln\Omega} \ll \Omega \ln \Omega$ or integral test integral diverges; converge: alternating series test)

39 converge absolutely (root test limit $\frac{\pi}{4}$)

41 converge absolutely (ratio test limit $\frac{1}{3}$)

Section 10.9

1 yes

3 no (negative exponents not allowed)

5 yes

7 yes (Hint: rewrite $(2x+1)^n$ as $2^n(x+\frac{1}{2})^n$.)

9 -2

11 $\frac{1}{3}$

13 $\frac{3}{2}$

15 the center cannot be determined from the given information

17 $R = 0$

19 (a) $\frac{5}{3}$, (b) 1, (c) $\frac{5}{9}$

21 interval of convergence $(-5, 5)$, radius of convergence $R = 5$

23 interval of convergence $[-1, 1]$, radius of convergence $R = 1$

25 interval of convergence $[3, 5)$, radius of convergence $R = 1$

27 the series converges for $x = 1$ only, radius of convergence $R = 0$

29 interval of convergence $(-\infty, \infty)$, radius of convergence $R = \infty$

31 interval of convergence $(-3, 9)$, radius of convergence $R = 6$

33 interval of convergence $[-1, 0)$, radius of convergence $R = \frac{1}{2}$

35 interval of convergence $(-\infty, \infty)$, radius of convergence $R = \infty$

37 interval of convergence $\left(-\frac{13}{3}, -\frac{11}{3}\right)$, radius of convergence $R = \frac{1}{3}$

39 interval of convergence $(-2, 2)$, radius of convergence $R = 2$

41 $\sum_{n=1}^{\infty} 4x^{5n-4}$, interval of convergence $(-1, 1)$

43 $\sum_{n=1}^{\infty} (-6x^2)^{n-1}$, interval of convergence $\left(-\frac{1}{\sqrt{6}}, \frac{1}{\sqrt{6}}\right)$

45 $\sum_{n=1}^{\infty} (x+1)^{2n+2}$, interval of convergence $(-2, 0)$

47 $\sum_{n=1}^{\infty} \frac{1}{2}x\left(-\frac{1}{6}x\right)^{n-1} = \sum_{n=1}^{\infty} \frac{1}{2}\left(-\frac{1}{6}\right)^{n-1} x^n$, interval of convergence $(-6, 6)$

49 interval of convergence $(-1, 1)$, radius of convergence $R = 1$ (Hint: for endpoints, try the level comparison test.)

51 interval of convergence $[-1, 1]$, radius of convergence $R = 1$ (Hint: for endpoints, try the level comparison test.)

Section 10.10

1 yes, $c_n = 1$

3 no (n^x instead of x^n)

5 no (centered at 2 instead of 0)

7 yes, $a = 2$

9 yes, $a = -5$

11 (a) $\sinh x = x + \frac{1}{3!}x^3 + \frac{1}{5!}x^5 + \cdots = \sum_{k=0}^{\infty} \frac{1}{(2k+1)!}x^{2k+1}$, (b) $R = \infty$ (Note: derivatives cycle between $\sinh x$ and $\cosh x$.)

13 (a) $\frac{1}{1-x} = 1 + x + x^2 + x^3 + \cdots = \sum_{n=0}^{\infty} x^n$, (b) $R = 1$ (Note: $f^{(n)}(x) = n!(1-x)^{-(n+1)}$)

15 (a) $(1+x)^p = 1 + px + \frac{p(p-1)}{2!}x^2 + \frac{p(p-1)(p-2)}{3!}x^3 + \cdots$, (b) $R = 1$ (Note: $f'''(x) = (p-2)(p-1)p(1+x)^{p-3}$)

17 $\cosh x = \frac{d}{dx}\sinh x = \sum_{k=0}^{\infty} \frac{1}{(2k)!}x^{2k}$

19 $\frac{1}{1+x} = \frac{d}{dx}\ln(1+x) = \sum_{n=1}^{\infty}(-1)^{n+1}x^{n-1} = 1 - x + x^2 - x^3 + \cdots$

21 $e^x \cos x = 1 + x - \frac{1}{3}x^3 - \frac{1}{6}x^4 + \cdots$

23 $\sin x \cos x = x - \frac{2}{3}x^3 + \frac{2}{15}x^5 - \frac{4}{315}x^7 + \cdots$

25 $x^3 \tan^{-1} x = \sum_{k=0}^{\infty}(-1)^k \frac{1}{2k+1}x^{2k+4} = x^4 - \frac{1}{3}x^6 + \frac{1}{5}x^8 - \frac{1}{7}x^{10} + \cdots$

27 $\sin x^2 = \sum_{k=0}^{\infty}(-1)^k \frac{1}{(2k+1)!}x^{4k+2} = x^2 - \frac{1}{3!}x^6 + \frac{1}{5!}x^{10} - \frac{1}{7!}x^{14} + \cdots$

29 $e^{x+5} = e^5 e^x = \sum_{n=0}^{\infty} \frac{e^5 x^n}{n!} = e^5 + e^5 x + \frac{e^5}{2!}x^2 + \cdots$

31 $\sin(x+5) = \sin 5 + (\cos 5)x - \frac{\sin 5}{2!}x^2 - \frac{\cos 5}{3!}x^3 + \frac{\sin 5}{4!}x^4 + \cdots$

33 (a) $\frac{1}{x} = \sum_{n=0}^{\infty}(-1)^n 2^{-(n+1)}(x-2)^n = \frac{1}{2} - \frac{1}{4}(x-2) + \frac{1}{8}(x-2)^2 - \frac{1}{16}(x-2)^3 + \cdots$, (b) $R = 2$, (c) $(0, 4)$

35 (a) $\frac{1}{x} = \sum_{n=0}^{\infty}(-1)^n 3^{-(n+1)}(x-3)^n = \frac{1}{3} - \frac{1}{9}(x-3) + \frac{1}{27}(x-3)^2 - \frac{1}{81}(x-3)^3 + \cdots$, (b) $R = 3$, (c) $(0, 6)$

37 (a) $\sin x = 1 - \frac{1}{2!}(x - \frac{\pi}{2})^2 + \frac{1}{4!}(x - \frac{\pi}{2})^4 - \cdots = \sum_{k=0}^{\infty}(-1)^n \frac{1}{(2k)!}(x - \frac{\pi}{2})^{2k}$, (b) $R = \infty$, (c) $(-\infty, \infty)$

39 (a) $\ln x = \ln 10 + \frac{1}{10}(x - 10) - \frac{1}{2 \cdot 10^2}(x - 10)^2 + \frac{1}{3 \cdot 10^3}(x - 10)^3 - \cdots = \ln 10 + \sum_{n=1}^{\infty} \frac{(x-10)^n}{n \cdot 10^n}$, (b) $R = 10$, (c) $(0, 20)$ (Note: the series converges when $x = 0$, but not to the value of the function because $\ln 0$ is undefined.)

41 $T_4(x) = x - \frac{1}{2}x^2 + \frac{1}{3}x^3 - \frac{1}{4}x^4$

43 $T_3(x) = \frac{1}{2} - \frac{1}{4}(x-2) + \frac{1}{8}(x-2)^2 - \frac{1}{16}(x-2)^3$

45 $T_5(x) = 1 - \frac{1}{2!}(x - \frac{\pi}{2})^2 + \frac{1}{4!}(x - \frac{\pi}{2})^4$

47 $\cos\alpha \approx 1$, $\cos\alpha \approx 1 - \frac{1}{2}\alpha^2$, $\cos\alpha \approx 1 - \frac{1}{2}\alpha^2 + \frac{1}{24}\alpha^4$

49 $e^\alpha \approx 1$, $e^\alpha \approx 1 + \alpha$, $e^\alpha \approx 1 + \alpha + \frac{1}{2}\alpha^2$

51 $\sqrt{1+\alpha} = (1+\alpha)^{\frac{1}{2}} \approx 1$, $\sqrt{1+\alpha} \approx 1 + \frac{1}{2}\alpha$, $\sqrt{1+\alpha} \approx 1 + \frac{1}{2}\alpha - \frac{1}{8}\alpha^2$

53 $\sin^{-1}\alpha \approx \alpha$, $\sin^{-1}\alpha \approx \alpha + \frac{1}{6}\alpha^3$, $\sin^{-1}\alpha \approx \alpha + \frac{1}{6}\alpha^3 + \frac{3}{40}\alpha^5$

55 $\frac{1}{8}$

57 $\frac{1}{10}$

59 $-\frac{1}{120}$

61 1

63 ∞

65 -1

67 $\frac{1}{2}$

69 0

71 (a) $T_4(x) = 1 + x + \frac{1}{2}x^2 + \frac{1}{6}x^3 + \frac{1}{24}x^4$, (b) $T_4(0.2) = 1.221400$, (c) $e^{0.2} = 1.221403$, (d) $T_4(0.2)$ agrees with $e^{0.2}$ to five decimal places

73 (a) $T_4(x) = 1 + x + \frac{1}{2}x^2 + \frac{1}{6}x^3 + \frac{1}{24}x^4$; (b) $T_4(3.6) = 25.8544$; (c) $e^{3.6} = 36.5982$; (d) the two numbers do not agree even to one significant digit; (e) because 3.6 is much farther from the center than 0.2, the approximation is not nearly as good

75 (a) $|R_5(0.1)| \le 0.000000001389$, (b) eight decimal places, (c) $\sin 0.1 = 0.09983342$

77 $(x^3 - 1)\sin x = -x + \frac{1}{6}x^3 + x^4 - \frac{1}{120}x^5 - \frac{1}{6}x^6 + \frac{1}{5040}x^7 + \frac{1}{120}x^8 - \cdots$

79 diverges ($\frac{1}{a_\Omega} \approx \Omega$)

81 (a) $\sum_{k=1}^{\Omega} \frac{1}{1+k\omega} \cdot \omega$, (b) $\frac{\omega}{1+k\omega} = \omega - k\omega^2 + k^2\omega^3 - k^3\omega^4 + \cdots$, (c) $\sum_{k=1}^{\Omega} \left(\omega - k\omega^2 + k^2\omega^3 - k^3\omega^4 + \cdots\right)$, (d) $1 - \frac{1}{2} + \frac{1}{3} - \frac{1}{4} + \cdots$, (e) $\ln 2$

83 $\sum_{k=1}^{\Omega}\left(\pi\omega k - \frac{(\pi\omega k)^3}{3!} + \frac{(\pi\omega k)^5}{5!} - \cdots\right)\pi\omega = \cdots \approx \frac{\pi^2}{2!} - \frac{\pi^4}{4!} + \frac{\pi^6}{6!} - \cdots = 1 - \cos\pi = 2.$

85 (a) yes (each ≈ 1); (b) $\ln(1+\alpha) \approx \alpha$, $\ln(1+\alpha^2) \approx \alpha^2$, no; (c) this is why there must be an exception for $y = 1$ in chapter 5 theorem 2.